HANDBOOK OF ENVIRONMENTAL DATA AND ECOLOGICAL PARAMETERS

Editor-in-Chief

S. E. JØRGENSEN

International Society for Ecological Modelling

Distributed throughout the world for the

INTERNATIONAL SOCIETY FOR ECOLOGICAL MODELLING

by

PERGAMON PRESS

OXFORD · NEW YORK · TORONTO · SYDNEY · PARIS · FRANKFURT

ENVIRONMENTAL SCIENCES AND APPLICATIONS

Other titles in the series:

Volume 1	UNITED NATIONS	Desertification: Its Causes and Consequences
Volume 2	FUKUSHIMA, Y.	Science for Better Environment
Volume 3	UNITED NATIONS INSTITUTE FOR TRAINING AND RESEARCH	Alternative Strategies for Desert Development and Management
Volume 4	BISWAS, A. K.	A Strategy for the Ozone Layer
Volume 5	HUSAR, R. B., *et al.*	Sulfur in the Atmosphere
Volume 7	JØRGENSEN, S. E.	State-of-the-Art of Ecological Modelling

ENVIRONMENTAL SCIENCES AND APPLICATIONS

Series Editors: ASIT K. BISWAS
MARGARET R. BISWAS

Volume 6

HANDBOOK OF ENVIRONMENTAL DATA AND ECOLOGICAL PARAMETERS

U.K.	Pergamon Press Ltd., Headington Hill Hall, Oxford OX3 0BW, England
U.S.A.	Pergamon Press Inc., Maxwell House, Fairview Park, Elmsford, New York 10523, U.S.A.
CANADA	Pergamon of Canada, Suite 104, 150 Consumers Road, Willowdale, Ontario M2J 1P9, Canada
AUSTRALIA	Pergamon Press (Aust.) Pty. Ltd., P.O. Box 544, Potts Point, N.S.W. 2011, Australia
FRANCE	Pergamon Press SARL, 24 rue des Ecoles, 75240 Paris, Cedex 05, France
FEDERAL REPUBLIC OF GERMANY	Pergamon Press GmbH, 6242 Kronberg-Taunus, Pferdstrasse 1, Federal Republic of Germany

First edition 1979

British Library Cataloguing in Publication Data

Handbook of environmental data and ecological parameters.
(Environmental sciences and applications; vol. 6)
1. Ecology - Tables.
I. Jørgensen, S. E. II. International Society for Ecological Modelling.
III. Series
574.5'021'2 QH 541 78-41207

ISBN 0-08-023436-4

List of contents, Handbook of environmental Data and Ecological Parameters.

Introduction

When a chemist wants to make calculations on chemical processes, it is possible to find values of constants in a handbook of chemistry - such as solubility, free energy, enthalpy of a great number of compounds are available An engineer can also find the data necessary for construction of a bridge, a road or a machine in a handbook, which simply can be taken from the bookshelf. But if one is dealing with environmental problems, such as preparation of environmental assessments or building ecological models, there has not been any handbook avaible, which gives information on the relevant biological-ecological parameters. We propose to fill this gab by 'Handbook of Environmental Data and Ecological Parameters'.

There is a long way to go before biology is an exact science in line with chemistry and physics, but during the last two decades it has become increasingly common to express biological and ecological scientific results in mathematical or quantitative terms. Currently biological papers does not merely contain observations, but also relations between sets of observations. Furthermore, it is essential to develop such relations if environmental problems are to be solved succesfully. Impact statements can not be prepared without knowing uptake rates, excretion rates and so forth of different organisms. Such parameters however, can only be found at present by extensive litterature search or from review papers. 'Handbook of Environmental Data and Ecological Parameters' is intended to provide ready access to such data easily and quickly.

The handbook will not contain information of local interest such as concentration of pollutants in a specific ecosystem. Firstly, because it would make the handbook too wieldy and secondly, a handbook should contain only data of general interest in the form of rate coefficients, uptake rates, excretion rates, decomposition rates, equilibrium constants, concentration factors, growth rates, half saturation constants, respiration rates, mortality, background concentrations, thermodynamic quantities of biological materials, diffusion coefficients etc.

Although toxicological data is of importance for solution of environmental problems (e.g. how much of a compound can be discharged into the environment without damaning the balance of the ecosystem.

The amount of toxicological data included in this handbook is limited. Only toxicological data of specific importance to environmental problems are included. The are two reasons for this: 1) again the book would simply be too large and wieldy, 2) in Europe and in the USA databases with toxicological parameters exist or are being prepared.

Nevertheless, the influence of heavy metals, pesticides and other compounds on the biological processes of growth, birth, respiration etc is to some extent coverd by the handbook. The toxicological values are all gathered from publications in environmental science; there has been any specific attempt to review the toxicological litterature.

The information provided in the handbook is the result of reviewing numerous international scientific journals as well as books in several languages. It is not attempted to assess how well the data are documented. References to all data are listed in case the user wishes to evaluate the quality of the necessary, evaluate the quality of the data data from the original papers. In selecting litterature to this handbook special notice has been paid to papers published after 1970.

As the field of biology, ecology and environmental science is growing rapidly, it is intended to update the handbook every 2 - 4 years. We hope that the book will be sufficiently well received, to justify an extension of its scope at later editions.

We should also like to receive any comments, criticis mand advice from users of the book. This would enable us to make an improved second edition Especially we will appreciate to get titles of reports or books, which contain informations not included in this edition. We do not claim that the present edition is complete, but that it contains a lot of useful information and has a good coverage within the field of environmental data and ecological parameters. However, we have the ambition to do the next edition more complete and ask the readers to assist us in achieving that. Material that the readers find relevant to the scope of the book can be send to ISEM. (International Society of Ecological Modelling. Langkaer Vaenge. DK-3500 Vaerløse. Denmark).

Copenhagen February 1979

The editorial board

User's Guide

The handbook has seven parts:

A. Compostition and Ecological Parameters of Living Organisms
B. The Ecosphere and Chemical Compounds.
C. Toxicological Effects.
D. Chemical Compounds and the Living Organism.
E. Equations for Biological Processes.
F. Equations for Chemical-Physical Processes.
G. Processes in the Environment.

As seen from the titles of the 7 parts an overlap exists. Therefore it is recommende to use the index rather than to search in one of the seven parts or in one of the 30 chapters.

In part A the emphasis is on the living organisms in the natural environment, and it is possible in this part to find concentrations of natural occuring compounds and elements in the organism. The titles of the chapters in part A is not following a systematic order, but they are based rather upon the practical applicability of the data.

Algae, zooplankton, macrophytes etc. are not systematically homogenous; e.g. is zooplankton not included in the chapter 'invertebrates' and fish are not included in the chapter 'vertebrates'.

Part B is focusing on the Ecosphere including the element cycles and general properties of chemical compounds in the ecosphere, as well pollutants as the natural chemical compounds. Chapter B1 is concerned with the biological half life time of several chemical compounds. In certain cases when the determinations are based upon isotope measurements, and the biological half life time might cover the effective half life time (including the radioactive decay), but this will always be indicated. Chapter B4 - pesticides - includes data about their adsorption equilibrium, uptake and excretion rate of pesticides, although such data to a minor extent also can be found in chapter D5 and G1, if the emphasis has been on the process more than on the compound.

Part C covers the biological effects of chemical compounds and the physical environment. This part of the handbook is touching an item of great importance for environmental problems. However this chapter is far from beeing complete. The handbook is not based on review of the toxicological litterature, but only on environmental and ecological journals and books. All though the handbook contains LD50 and LC50 values, only the most important compounds from an environmental point of view is listed, while the data about the influence of chemical compounds on the ecological parameters are more complete. If this part of the handbook should include a more comprehensive review on the toxicological data, it would easily be an unbalanced handbook of 5000 pages or more. For those who are interested in more detailed information about the toxicology of chemical compounds, the american or european databases on these data should be applied.

Part D contains information of chemical compounds related to species. Such data as background concentrations, concentrations of biological or abiological components in more or less polluted environments, concentration factors for plants, animals, soil or sediment, intake- and excretion rates can be found here.

Part E and F gives information about equations for biological, chemical and physical processes in the environment. Equations which easily can be found in textbooks are not included, but only equations used in ecological models or empirical equations relating properties of biological or abiological components in the environment such as biomass as function of time and temperature for fish, body weight and body length, rate of sulphide oxidation etc.

The last part G is devoted to the dynamics of environment - it means rates of processes. The part contains, however, also euqilibrium data, because a knowledge to the equilibrium often is the point of departure for a dynamical description.

Most tables in the handbook are in the same computerized form. The head of the table gives a two dimensional information about the content of the table e.g. ALGAE: MAX GROWTH RATE. The tables have 4 columns. The first column informs about "the item" e.g. the species, compound etc. The second column will give a value, the third one the conditions under which the measurement was carried out e.g. the temperature, pH, the weight of the item, the number of indepentent determinations etc. The last column gives a number related to the list of references, which gives the user the possibilities to get more informations about the experiments in the original or the review reference.

An expamle:

TABLE A466

ZOOPLANKTON : PRODUCTION			
ITEM:	VALUE:	CONDITIONS:	REF.:
*ASPLANCHNA PRIODONTA	0.0014 G DRY WEIGHT/(DAY*M**2)	PREDATOR MAY-OCT, BIOMASS TURNOVER =2.9 D, NAROCH L., USSR	1731
*ASPLANCHNA PRIODONTA	0.0061 G DRY WEIGHT/(DAY*M**2)	PREDATOR MAY-OCT, BIOMASS TURNOVER =2.5 D, MYASTRO L., USSR	1731
*ASPLANCHNA PRIODONTA	0.0105 G DRY WEIGHT/(DAY*M**2)	PREDATOR MAY-OCT, BIOMASS TURNOVER =3.2 D, BATORIN L., USSR	1731

Part C is organized in a slighly different way. Here consists the head of the table of a chemical compound or element and the biological effect related to the values in the table. e.g. CD: MORTALITY. The item is in this case a species e.g. Mytilus edulis. The value could e.g. be 50 % corresponding to 50 % mortality and in the conditions column is indicated by which concentration, temperatures, pH etc. the 50 % mortality is observed. An example from part C.

TABLE C94

CU : DECREASE IN GROWTH RATE			
ITEM:	VALUE:	CONDITIONS:	REF.:
*ASELLUS MERIDIANUS	0.017 MM/DAY	CONTROL=0.044MM/DAY, CU =0.1 MG/L 10 JUV. ,RIVER GANNAL,U.K.	178
*ASELLUS MERIDIANUS	0.017 MM/DAY	CONTROL=0.033MM/DAY, CU =0.1 MG/L 10 JUV. ,RIVER HAYLE,U.K.	178
*ASELLUS MERIDIANUS	0.028 MM/DAY	CONTROL=0.033MM/DAY, CU =0.1 MG/L 10 JUV. ,RIVER HAYLE,U.K.	178
*ASELLUS MERIDIANUS	0.005 MM/DAY	CONTROL= 0.028MM/DAY, CU =0.1 MG/L 10 JUV. ,RIVER GANNAL,U.K.	178
*ASELLUS MERIDIANUS	0.002 MM/DAY	CONTROL=0.014MM/DAY, CU =0.1 MG/L 10 JUV. ,BREDWELL BROOK,U.K.	178

All computerized tables are in alphabetic order in each of the seven parts, and the items follow in alphabetic order in each table.
Some data could not fit into this scheme, and are presented as tables containing less or more than four columns or in form of diagrams.
The practical applicabality of a handbook requires that most possible information is contained in a limited space. This involves that some general lines have been followed:

1). If the medium is not mentioned under condition, it should be obvious from the context. Water for an algae, soil for an earthworm, air for atmospheric emission etc.
2). The chemical symbols are generelly applied, N for nitrogen etc.
3). The greek letters are spelled , alpha, beta etc.
4). A chemical formula is written on one line e.g. H2O.
5). The general accepted abbreviation for amino acid are applied, LEU for leucine etc.
6). All titles of tables are listed in the front of the handbook.
7). The temperature is understood as degree Kelvin, if not otherwise mentioned.
8). The application of units is not homogenous, but is in most cases in accordance with the reference.
9). Several abbreviations are used (see the list page VLII). For some words two or even more abbreviations migth have been applied, as the abbreviations often are chosen ot fit the space available.
10). Footnotes are marked as a number in a bracket (F007) and refer to a table placed before the subject index.
11). As mentioned above the heads of the tables are two dimensional e.g. "Cd: mortality". the index contains all tables heads in reverse order (e.g. mortality; Cd.). It contains also all species, compounds, elements, effects and reactions etc. listed in the item collumns. The index refers consequently to a table number.
12). The titles of the seven parts and 30 chapters are listed in front of the book.

LIST OF CONTENT

PART A

A1 ALGAE
ASH/BIOMASS RATIO
A2 ALGAE
ASSIMILATION
A3 ALGAE
ATP-C/BIOMASS RATIO
A4 ALGAE
ATP/(CARBON+NITROGEN) RATIO
A5 ALGAE
ATP/BIOMASS RATIO
A6 ALGAE
ATP/C RATIO
A7 ALGAE
ATP/CHLOROPHYLL A RATIO
A8 ALGAE
ATP/N RATIO
A9 ALGAE
ATP/P RATIO
A10 ALGAE
B/BIOMASS RATIO
A11 ALGAE
BIOMASS
A12 ALGAE
BIOTIN/BIOMASS RATIO
A13 ALGAE
C/BIOMASS RATIO
A14 ALGAE
C/BIOTIN RATIO
A15 ALGAE
C/CHLOROPHYLL A RATIO
A16 ALGAE
C/N RATIO
A17 ALGAE
C/N/P RATIO
A18 ALGAE
C/P RATIO
A19 ALGAE
C/SI RATIO
A20 ALGAE
C/VITAMIN B 1 RATIO
A21 ALGAE
C/VITAMIN B 12 RATIO
A22 ALGAE
CA/BIOMASS RATIO
A23 ALGAE
CALORIC CONTENT
A24 ALGAE
CARBOHYDRATE/BIOMASS RATIO
A25 ALGAE
CARBON (INTRACELLULAR)
A26 ALGAE
CARBON FIXATION/NITROGEN FIXATION RATIO
A27 ALGAE
CAROTENOIDS/CHLOROPHYLL A RATIO
A28 ALGAE
CELL DIAMETER
A29 ALGAE
CELL VOLUME
A30 ALGAE
CHLOROPHYLL A/BIOMASS RATIO
A31 ALGAE
CU/BIOMASS RATIO
A32 ALGAE
C2H2 REDUCTION RATE
A33 ALGAE
C2H2 REDUCTION/N2 FIXATION RATIO
A34 ALGAE
DARK UPTAKE
A35 ALGAE
DRY/WET WEIGHT RATIO
A36 ALGAE
EXCRETION OF DISSOLVED ORGANIC CARBON
A37 ALGAE
EXCRETION OF PHOSPHORUS
A38 ALGAE
FAT/BIOMASS RATIO
A39 ALGAE
FE/BIOMASS RATIO
A40 ALGAE
GROWTH RATE
A41 ALGAE
H/BIOMASS RATIO
A42 ALGAE
HALF SATURATION CONSTANT FOR AMMONIUM GROWTH
A43 ALGAE
HALF SATURATION CONSTANT FOR AMMONIUM UPTAKE
A44 ALGAE
HALF SATURATION CONSTANT FOR LIGHT
A45 ALGAE
HALF SATURATION CONSTANT FOR NITRATE GROWTH
A46 ALGAE
HALF SATURATION CONSTANT FOR NITRATE UPTAKE
A47 ALGAE
HALF SATURATION CONSTANT FOR NITRITE UPTAKE
A48 ALGAE
HALF SATURATION CONSTANT FOR NITROGEN GROWTH
A49 ALGAE
HALF SATURATION CONSTANT FOR NITROGEN UPTAKE
A50 ALGAE
HALF SATURATION CONSTANT FOR PHOSPHORUS GR etc
A51 ALGAE
HALF SATURATION CONSTANT FOR PHOSPHORUS UP etc
A52 ALGAE
HALF SATURATION CONSTANT FOR SILICIUM UPTAKE
A53 ALGAE
HALF SATURATION CONSTANT FOR TOTAL NITROGE etcO
A54 ALGAE
HALF SATURATION CONSTANT FOR TOTAL PHOSPHO etcG
A55 ALGAE
K/BIOMASS RATIO
A56 ALGAE
LIPID/BIOMASS RATIO
A57 ALGAE
MAX GROWTH RATE
A58 ALGAE
MAX PHOTOSYNTHETIC EFFICIENCY
A59 ALGAE
MAX RATE OF AMMONIUM UPTAKE
A60 ALGAE
MAX RATE OF NITRATE UPTAKE
A61 ALGAE
MAX RATE OF PHOSPHORUS UPTAKE
A62 ALGAE
MAX RATE OF PHOTOSYNTHESIS
A63 ALGAE
MAX RATE OF SILICIUM UPTAKE
A64 ALGAE
MAX YIELD
A65 ALGAE
MAX YIELD COEFFICIENT
A66 ALGAE
MG/BIOMASS RATIO
A67 ALGAE
MIN CELL N
A68 ALGAE
MIN CELL P
A69 ALGAE
MIN CELL SI
A70 ALGAE
MINIMAL PHOSPHORUS REQUIREMENTS
A71 ALGAE
MN/BIOMASS RATIO
A72 ALGAE
MORTALITY
A73 ALGAE
N/BIOMASS
A74 ALGAE
N/BIOMASS RATIO
A75 ALGAE
N/CHLOROPHYLL A RATIO
A76 ALGAE
N/P RATIO
A77 ALGAE
N/SI RATIO
A78 ALGAE
NA/BIOMASS RATIO
A79 ALGAE
NITROGEN (INTRACELLULAR)
A80 ALGAE
NITROGEN FIXATION
A81 ALGAE
NUTRIENT UTILIZATION CONSTANT
A82 ALGAE
O/BIOMASS RATIO
A83 ALGAE
ORGANIC CONTENT
A84 ALGAE
OXYGEN DEMAND
A85 ALGAE
P/BIOMASS RATIO

B

D

F

I

A261 INVERTEBRATES
MAX STANDING CROP
A262 INVERTEBRATES
MORTALITY
A263 INVERTEBRATES
N/BIOMASS RATIO
A264 INVERTEBRATES
NA/BIOMASS RATIO
A265 INVERTEBRATES
NUMBERS OF EGGS
A266 INVERTEBRATES
O/BIOMASS RATIO
A267 INVERTEBRATES
OVIPOSITION RATE
A268 INVERTEBRATES
P/BIOMASS RATIO
A269 INVERTEBRATES
PRODUCTION/BIOMASS RATIO
A270 INVERTEBRATES
RESPIRATION RATE
A271 INVERTEBRATES
RESPIRATORY COEFFICIENT
A272 INVERTEBRATES
S/BIOMASS RATIO
A273 INVERTEBRATES
SETTLING RATE
A274 INVERTEBRATES
THRESHOLD DENSITY
A275 INVERTEBRATES
TURNOVER TIME OF IODINE
A276 INVERTEBRATES
WET WEIGHT
A277 INVERTEBRATES
WET/DRY WEIGHT RATIO
A278 INVERTEBRATES
PRODUCTION OF EGG VERSUS OXYGEN CONCENTRAT etc
A279 INVERTEBRATES
PRODUCTION OF EMBRYO VERSUS TEMPERATURE
A280 INVERTEBRATES
GROWTH RATE VERSUS FEEDING RATE
A281 INVERTEBRATES
OXYGEN CONSUMPTION VERSUS OXYGEN CONCENTRA etc

K

A282 KORMOPHYTES, MACROPHYTES
ASH/BIOMASS RATIO
A283 KORMOPHYTES, MACROPHYTES
BIOMASS
A284 KORMOPHYTES, MACROPHYTES
BIOMASS IN ROOTS
A285 KORMOPHYTES, MACROPHYTES
C/BIOMASS RATIO
A286 KORMOPHYTES, MACROPHYTES
CA/BIOMASS RATIO
A287 KORMOPHYTES, MACROPHYTES
CALORIC CONTENT
A288 KORMOPHYTES, MACROPHYTES
CARBOHYDRATE/BIOMASS RATIO
A289 KORMOPHYTES, MACROPHYTES
CHLOROPHYLL A/BIOMASS RATIO
A290 KORMOPHYTES, MACROPHYTES
CHLOROPHYLL B/BIOMASS RATIO
A291 KORMOPHYTES, MACROPHYTES
CRUDE FIBER/BIOMASS RATIO
A292 KORMOPHYTES, MACROPHYTES
CU/BIOMASS RATIO
A293 KORMOPHYTES, MACROPHYTES
DRY/WET WEIGHT RATIO
A294 KORMOPHYTES, MACROPHYTES
ETHER EXTRACT/BIOMASS RATIO
A295 KORMOPHYTES, MACROPHYTES
FE/BIOMASS RATIO
A296 KORMOPHYTES, MACROPHYTES
GROSS PHOTOSYNTHESIS
A297 KORMOPHYTES, MACROPHYTES
K/BIOMASS RATIO
A298 KORMOPHYTES, MACROPHYTES
MAX STANDING CROP
A299 KORMOPHYTES, MACROPHYTES
MG/BIOMASS RATIO
A300 KORMOPHYTES, MACROPHYTES
MN/BIOMASS RATIO
A301 KORMOPHYTES, MACROPHYTES
N/BIOMASS RATIO
A302 KORMOPHYTES, MACROPHYTES
NA/BIOMASS RATIO
A303 KORMOPHYTES, MACROPHYTES
P/BIOMASS RATIO
A304 KORMOPHYTES, MACROPHYTES
PRODUCTION
A305 KORMOPHYTES, MACROPHYTES
PROTEIN/BIOMASS RATIO
A306 KORMOPHYTES, MACROPHYTES
RATE OF NITROGEN UPTAKE
A307 KORMOPHYTES, MACROPHYTES
RATE OF PHOSPHORUS UPTAKE
A308 KORMOPHYTES, MACROPHYTES
S/BIOMASS RATIO
A309 KORMOPHYTES, MACROPHYTES
SI/BIOMASS RATIO
A310 KORMOPHYTES, MACROPHYTES
ZN/BIOMASS RATIO
A311 KORMOPHYTES, PLANTS
AL/BIOMASS RATIO
A312 KORMOPHYTES, PLANTS
ASH/BIOMASS RATIO
A313 KORMOPHYTES, PLANTS
AVERAGE WIDTH OF LEAVES
A314 KORMOPHYTES, PLANTS
BR/BIOMASS RATIO
A315 KORMOPHYTES, PLANTS
C/BIOMASS RATIO
A316 KORMOPHYTES, PLANTS
CA/BIOMASS RATIO
A317 KORMOPHYTES, PLANTS
CALORIC CONTENT
A318 KORMOPHYTES, PLANTS
CHANGE IN EQUIVAL.AIR TEMP/CHANGE IN HAG
A319 KORMOPHYTES, PLANTS
CL/BIOMASS RATIO
A320 KORMOPHYTES, PLANTS
COMPENSATION POINT
A321 KORMOPHYTES, PLANTS
CU/BIOMASS RATIO
A322 KORMOPHYTES, PLANTS
DEPTH CONTANING 63 % OF THE ROOTS
A323 KORMOPHYTES, PLANTS
DIFFERENCE OF EQUIVAL.AIR TEMP.BETWEEN TWO etcE
A324 KORMOPHYTES, PLANTS
DRAG COEFFICIENT OF THE LEAVES
A325 KORMOPHYTES, PLANTS
EXCHANGE COEFFICIENT FOR HEAT (MIN)
A326 KORMOPHYTES, PLANTS
EXTINCTION FACTOR FOR WIND
A327 KORMOPHYTES, PLANTS
FE/BIOMASS RATIO
A328 KORMOPHYTES, PLANTS
H/BIOMASS RATIO
A329 KORMOPHYTES, PLANTS
HEIGHT OF CANOPY
A330 KORMOPHYTES, PLANTS
I/BIOMASS RATIO
A331 KORMOPHYTES, PLANTS
K/BIOMASS RATIO
A332 KORMOPHYTES, PLANTS
LEAF AREA DENSITY
A333 KORMOPHYTES, PLANTS
LEAF AREA INDEX
A334 KORMOPHYTES, PLANTS
LEAF BIOMASS
A335 KORMOPHYTES, PLANTS
LEAF RES TO TRANSPIRATION
A336 KORMOPHYTES, PLANTS
MG/BIOMASS RATIO
A337 KORMOPHYTES, PLANTS
MIXING LENGTH
A338 KORMOPHYTES, PLANTS
MN/BIOMASS RATIO
A339 KORMOPHYTES, PLANTS
N/BIOMASS RATIO
A340 KORMOPHYTES, PLANTS
NA/BIOMASS RATIO
A341 KORMOPHYTES, PLANTS
NITROGEN FIXATION
A342 KORMOPHYTES, PLANTS
O/BIOMASS RATIO
A343 KORMOPHYTES, PLANTS
P/BIOMASS RATIO
A344 KORMOPHYTES, PLANTS
PB/BIOMASS RATIO
A345 KORMOPHYTES, PLANTS
PHOTOSYNTHESIS

L

V

Z

PART B

C

E

L

P

PART C

C1 ACCLIMATION TIME
CHANGE IN DEVELOPMENT TIME OF EGGS

C2 ACTINOMYCIN D
MORTALITY

C3 AG
MORTALITY

C4 AL
LEAF SYMPTONS

C5 AL
MORTALITY

C6 ALANAP
MORTALITY

C7 ALDRIN
HEART BEAT

C8 ALDRIN
MORTALITY

C9 ALKYL BENZEN SULFONATE
CHANGE IN GROWTH POTENTIAL

C10 ALKYL BENZEN SULFONATE
MORTALITY

C11 ALKYL POLYETHOXYLATE
CHANGE IN GROWTH POTENTIAL

C12 ALPHA OLEFIN SULFONATE
MORTALITY

C13 ALPHA-AMANITIN
MORTALITY

C14 AMMONIA
MORTALITY

C15 AMMONIUM
REPRESSION OF NITROGENASE ACTIVITY

C16 AS
MORTALITY

C17 ATRAZINE
MORTALITY

C438 AMMONIA
NITRIFICATION

C459 AMMONIA
MORTALITY

B

C18 B
DEFICIENCY SYMPTOMS

C19 B
MORTALITY

C20 B
SURVIVAL TIME

C21 BA
MORTALITY

C22 BCM
GROWTH INHIBITION

C23 BCM
MORTALITY

C24 BE
MORTALITY

C25 BENOMYL
GROWTH INHIBITION

C26 BENOMYL
MORTALITY

C27 BENSULIDE
MORTALITY

C28 BI
MORTALITY

C29 BO
DEFICIENCY SYMPTOMS

C30 BO
NORMAL GROWHT

C31 BO
TOXIC SYMPTOMS

C32 BORDEAUX MIXTURE
MORTALITY

C33 BOTULINAL
MORTALITY

C

C34 CA
MORTALITY

C35 CAFFEINE
MORTALITY

C36 CALCIUM ARSENATE
MORTALITY

C37 CAMPHECHLOR
MORTALITY

C38 CARBARYL
HEART BEAT

C39 CATION EXCHANGE CAPACITY
CHANGE IN UPTAKE OF CD

C40 CCL4
MORTALITY

C41 CD
BLOOD GLUCOSE

C42 CD
BLOOD LACTATE

C43 CD
DECREASE IN YIELD

C44 CD
HEMATOCRIT

C45 CD
HEMOGLOBIN

C46 CD
INCREASE OF YIELD

C47 CD
LIVER GLYCOGEN

C48 CD
MCHC

C49 CD
MEAN CORPUSCULAR HEMOGLOBIN

C50 CD
MEAN CORPUSCULAR VOLUME

C51 CD
MORTALITY

C52 CD
MUSCLE GLYCOGEN

C53 CD
PLASMA CA

C54 CD
PLASMA INORGANIC PHOSPHATE

C55 CD
PLASMA K

C56 CD
PLASMA MG

C57 CD
PLASMA OSMOLALITY

C58 CD
PLASMA PROTEIN

C59 CD
RBC

C60 CD
YIELD

C61 CE
MORTALITY

C62 CHLORAMBEN
MORTALITY

C63 CHLORAMINE
CHANGE IN HALF SATURATION CONSTANT FOR NIT etcN

C64 CHLORAMINE
CHANGE IN NITROGEN UPTAKE

C65 CHLORDANE
MORTALITY

C66 CHLORINE
CHANGE IN HALF SATURATION CONSTANT FOR NIT etcN

C67 CHLORINE
CHANGE IN NITROGEN UPTAKE

C68 CO
MORTALITY

C69 CO
OXYGEN CONSUMPTION

C70 CR
-

C71 CR
ACID PHOSPHATASE ACTIVITY

C72 CR
DESTRUCT. OF INTESTINAL EPITHELIUM

C73 CR
DIMINISHED GROWTH

C74 CR
G-6-PDH

C75 CR
GLUCOSE

D

E

F

G

H

P

Q

R

S

T

C386 TEMPERATURE
MORTALITY
C387 TEMPERATURE
OPTIMUM FOR GROWTH
C388 TEMPERATURE
UPPER LETHAL TEMPERATURE
C389 TETANUS
MORTALITY
C390 TETRACYCLINE HCL
MORTALITY
C391 TETRODOTOXIN
MORTALITY
C392 TH
MORTALITY
C393 THIOPHANATE-METHYL
GROWTH INHIBITION
C394 THIOPHANATE-METHYL
MORTALITY
C395 THIRAM
MORTALITY
C396 TL
MORTALITY
C397 TM
MORTALITY
C398 TRAFFIC INTENSITY
CHANGE IN PB CONTENT
C399 TRAFFIC INTENSITY
CHANGE IN PB CONTENT IN <<
C400 TRAFFIC INTENSITY
CHANGE IN PB CONTENT IN VEGETATION
C401 TRICHLOROMETHYLSULPHENYL BENZOATE
INHIBITION OF NITROGEN FIXATION
C402 TRIFLURALIN
MORTALITY
C403 TRINITROTULUENE
INHIBITION OF GROWTH
C404 TRINITROTULUENE
MORTALITY
C405 TWEEN 20 (?032)
DECREASE IN OXYGEN UPTAKE
C406 TWEEN 40 (?033)
DECREASE IN OXYGEN UPTAKE
C407 TWEEN 60 (?034)
DECREASE IN OXYGEN UPTAKE
C408 TWEEN 80 (?035)
DECREASE IN OXYGEN UPTAKE
C479 TEMPERATURE
GROWTH RATE

U

C409 U
MORTALITY

V

C410 V
MORTALITY

W

C411 W
MORTALITY
C439 WASTE WATER
REPRODUCTION
C472 WASTE WATER
GROWTH RATE

Y

C412 Y
MORTALITY
C413 YB
MORTALITY

Z

C414 ZN
CHANGE IN LARVAL DEVELOPMENT
C415 ZN
CHANGE IN UPTAKE OF CD
C416 ZN
DECREASE IN GROWTH RATE
C417 ZN
GROWTH
C418 ZN
HEMORRHAGING
C419 ZN
MORTALITY
C420 ZN
PHOTOSYNTHESIS
C421 ZN
RESPIRATION
C422 ZN
SURVIVAL TIME
C423 ZN
TOXIC SYMPTOMS
C424 ZN
VERTEBRAL DAMAGE
C425 ZR
MORTALITY
C467 ZN
OXYGEN IN BLOOD

C426 1-CHLOROPHTHALENE
MORTALITY
C427 1-NAPHTHOL
MORTALITY
C428 1,2-DIACETYLETHYLENE
INHIBITION OF NITROGEN FIXATION

C429 2,3,6-TBA
MORTALITY
C430 2,3,7,8-TCDD
MORTALITY
C431 2,4-D ACID
MORTALITY
C432 2,4-D DIMETHYL AMINE SALT
MORTALITY
C433 2,4,5-T
MORTALITY
C434 2,7-DCDD
MORTALITY

PART D

D1 BACKGROUND CONCENTRATION
ABUNDANCE IN AVERAGE CRUSTRAL ROCK
D2 BACKGROUND CONCENTRATION
AG
D3 BACKGROUND CONCENTRATION
AL
D4 BACKGROUND CONCENTRATION
AR
D5 BACKGROUND CONCENTRATION
AS
D6 BACKGROUND CONCENTRATION
ATP
D7 BACKGROUND CONCENTRATION
AU
D8 BACKGROUND CONCENTRATION
B
D9 BACKGROUND CONCENTRATION
BA
D10 BACKGROUND CONCENTRATION
BE
D11 BACKGROUND CONCENTRATION
BENZO(A)ANTHRACENE
D12 BACKGROUND CONCENTRATION
BENZO(A)PYRENE
D13 BACKGROUND CONCENTRATION
BI
D14 BACKGROUND CONCENTRATION
BIOSPHERE
D15 BACKGROUND CONCENTRATION
BIOTIN
D16 BACKGROUND CONCENTRATION
BR
D17 BACKGROUND CONCENTRATION
C
D18 BACKGROUND CONCENTRATION
CA
D19 BACKGROUND CONCENTRATION
CARBON DIOXIDE
D20 BACKGROUND CONCENTRATION
CARBON MONOXIDE
D21 BACKGROUND CONCENTRATION
CD
D22 BACKGROUND CONCENTRATION
CE
D23 BACKGROUND CONCENTRATION
CHLOROPHYLL A
D24 BACKGROUND CONCENTRATION
CH4
D25 BACKGROUND CONCENTRATION
CL
D26 BACKGROUND CONCENTRATION
CO
D27 BACKGROUND CONCENTRATION
CO3(2-) + HCO3(-)
D28 BACKGROUND CONCENTRATION
CR
D29 BACKGROUND CONCENTRATION
CS
D30 BACKGROUND CONCENTRATION
CU
D31 BACKGROUND CONCENTRATION
DDT
D32 BACKGROUND CONCENTRATION
DISSOLVED ORGANIC CARBON
D33 BACKGROUND CONCENTRATION
EU
D34 BACKGROUND CONCENTRATION
F
D35 BACKGROUND CONCENTRATION
FE
D36 BACKGROUND CONCENTRATION
FOLIC ACID
D37 BACKGROUND CONCENTRATION
GA
D38 BACKGROUND CONCENTRATION
GE
D39 BACKGROUND CONCENTRATION
H
D40 BACKGROUND CONCENTRATION
HCO3(-)
D41 BACKGROUND CONCENTRATION
HE
D42 BACKGROUND CONCENTRATION
HF
D43 BACKGROUND CONCENTRATION
HG
D44 BACKGROUND CONCENTRATION
HYDROGEN SULFIDE (H2S)
D45 BACKGROUND CONCENTRATION
H2O
D46 BACKGROUND CONCENTRATION
I
D47 BACKGROUND CONCENTRATION
IN
D48 BACKGROUND CONCENTRATION
K
D49 BACKGROUND CONCENTRATION
KR
D50 BACKGROUND CONCENTRATION
LA
D51 BACKGROUND CONCENTRATION
LI
D52 BACKGROUND CONCENTRATION
LU
D53 BACKGROUND CONCENTRATION
MG
D54 BACKGROUND CONCENTRATION
MN
D55 BACKGROUND CONCENTRATION
MO
D56 BACKGROUND CONCENTRATION
N
D57 BACKGROUND CONCENTRATION
NA
D58 BACKGROUND CONCENTRATION
NB
D59 BACKGROUND CONCENTRATION
NE
D60 BACKGROUND CONCENTRATION
NI
D61 BACKGROUND CONCENTRATION
NIACIN (NICOTINIC ACIDAMIDE)
D62 BACKGROUND CONCENTRATION
O2
D63 BACKGROUND CONCENTRATION
O3
D64 BACKGROUND CONCENTRATION
P
D65 BACKGROUND CONCENTRATION
PANTOTHENIC ACID
D66 BACKGROUND CONCENTRATION
PB
D67 BACKGROUND CONCENTRATION
POLYNUCLEAR AROMATIC HYDROCARBONS
D68 BACKGROUND CONCENTRATION
RA
D69 BACKGROUND CONCENTRATION
RB
D70 BACKGROUND CONCENTRATION
RN
D71 BACKGROUND CONCENTRATION
S
D72 BACKGROUND CONCENTRATION
SB
D73 BACKGROUND CONCENTRATION
SC
D74 BACKGROUND CONCENTRATION
SE
D75 BACKGROUND CONCENTRATION
SI
D76 BACKGROUND CONCENTRATION
SN
D77 BACKGROUND CONCENTRATION
SR
D78 BACKGROUND CONCENTRATION
TA
D79 BACKGROUND CONCENTRATION
TB
D80 BACKGROUND CONCENTRATION
TH
D81 BACKGROUND CONCENTRATION
THIAMINE
D82 BACKGROUND CONCENTRATION
TI
D83 BACKGROUND CONCENTRATION
TL
D84 BACKGROUND CONCENTRATION
U
D85 BACKGROUND CONCENTRATION
V

D86 BACKGROUND CONCENTRATION
VITAMIN B12
D87 BACKGROUND CONCENTRATION
W
D88 BACKGROUND CONCENTRATION
XE
D89 BACKGROUND CONCENTRATION
Y
D90 BACKGROUND CONCENTRATION
YB
D91 BACKGROUND CONCENTRATION
ZN
D92 BACKGROUND CONCENTRATION
ZR

C

D93 CONCENTRATION FACTOR
AG
D94 CONCENTRATION FACTOR
AL
D95 CONCENTRATION FACTOR
ALDRIN
D96 CONCENTRATION FACTOR
AM
D97 CONCENTRATION FACTOR
AS
D98 CONCENTRATION FACTOR
ATRAZINE
D99 CONCENTRATION FACTOR
AU
D100 CONCENTRATION FACTOR
AU-198
D101 CONCENTRATION FACTOR
B
D102 CONCENTRATION FACTOR
BA
D103 CONCENTRATION FACTOR
BA-140/LA-140
D104 CONCENTRATION FACTOR
BE
D105 CONCENTRATION FACTOR
BHC
D106 CONCENTRATION FACTOR
BR
D107 CONCENTRATION FACTOR
CA
D108 CONCENTRATION FACTOR
CACODYLIC ACID
D109 CONCENTRATION FACTOR
CAMPHECHLOR
D110 CONCENTRATION FACTOR
CARBARYL
D111 CONCENTRATION FACTOR
CD
D112 CONCENTRATION FACTOR
CD-109
D113 CONCENTRATION FACTOR
CD-115M
D114 CONCENTRATION FACTOR
CE
D115 CONCENTRATION FACTOR
CE-141
D116 CONCENTRATION FACTOR
CE-144
D117 CONCENTRATION FACTOR
CHLORDANE
D118 CONCENTRATION FACTOR
CHLORINATED NAPHTHALENE
D119 CONCENTRATION FACTOR
CL
D120 CONCENTRATION FACTOR
CO
D121 CONCENTRATION FACTOR
CO-58
D122 CONCENTRATION FACTOR
CO-60
D123 CONCENTRATION FACTOR
CR
D124 CONCENTRATION FACTOR
CR-51
D125 CONCENTRATION FACTOR
CS
D126 CONCENTRATION FACTOR
CS-133
D127 CONCENTRATION FACTOR
CS-134
D128 CONCENTRATION FACTOR
CS-137
D129 CONCENTRATION FACTOR
CU
D130 CONCENTRATION FACTOR
DDD
D131 CONCENTRATION FACTOR
DDE
D132 CONCENTRATION FACTOR
DDT
D133 CONCENTRATION FACTOR
DIELDRIN
D134 CONCENTRATION FACTOR
EDC-TAR
D135 CONCENTRATION FACTOR
ENDRIN
D136 CONCENTRATION FACTOR
EU
D137 CONCENTRATION FACTOR
F
D138 CONCENTRATION FACTOR
FE
D139 CONCENTRATION FACTOR
FE-59
D140 CONCENTRATION FACTOR
GA
D141 CONCENTRATION FACTOR
GE
D142 CONCENTRATION FACTOR
HCH
D143 CONCENTRATION FACTOR
HEPTACHLOR
D144 CONCENTRATION FACTOR
HEPTACHLOR AND -EPOXIDE
D145 CONCENTRATION FACTOR
HEXABROMOBENZENE
D146 CONCENTRATION FACTOR
HEXABROMOBIPHENYL
D147 CONCENTRATION FACTOR
HEXACHLOROBENZENE
D148 CONCENTRATION FACTOR
HG
D149 CONCENTRATION FACTOR
I
D150 CONCENTRATION FACTOR
I-131
D151 CONCENTRATION FACTOR
INSECTICIDES
D152 CONCENTRATION FACTOR
K
D153 CONCENTRATION FACTOR
K-40
D154 CONCENTRATION FACTOR
LA
D155 CONCENTRATION FACTOR
LI
D156 CONCENTRATION FACTOR
LU
D157 CONCENTRATION FACTOR
METHOXYCHLOR
D158 CONCENTRATION FACTOR
MG
D159 CONCENTRATION FACTOR
MIREX
D160 CONCENTRATION FACTOR
MN
D161 CONCENTRATION FACTOR
MN-54
D162 CONCENTRATION FACTOR
MO
D163 CONCENTRATION FACTOR
N
D164 CONCENTRATION FACTOR
N-NITROSOATRAZINE
D165 CONCENTRATION FACTOR
NA
D166 CONCENTRATION FACTOR
NAPHTHALENE
D167 CONCENTRATION FACTOR
NB
D168 CONCENTRATION FACTOR
NB-95
D169 CONCENTRATION FACTOR
NI
D170 CONCENTRATION FACTOR
ORGANOCHLORINES

D171 CONCENTRATION FACTOR
P
D172 CONCENTRATION FACTOR
PB
D173 CONCENTRATION FACTOR
PCB
D174 CONCENTRATION FACTOR
PHOTODIELDRIN
D175 CONCENTRATION FACTOR
PO
D176 CONCENTRATION FACTOR
PO-210
D177 CONCENTRATION FACTOR
PO-239
D178 CONCENTRATION FACTOR
POLYBROMINATED BIPHENYL
D179 CONCENTRATION FACTOR
PU
D180 CONCENTRATION FACTOR
PU-237
D181 CONCENTRATION FACTOR
PU-239
D182 CONCENTRATION FACTOR
RA
D183 CONCENTRATION FACTOR
RA-226
D184 CONCENTRATION FACTOR
RB
D185 CONCENTRATION FACTOR
RU
D186 CONCENTRATION FACTOR
RU-103
D187 CONCENTRATION FACTOR
RU-106
D188 CONCENTRATION FACTOR
S
D189 CONCENTRATION FACTOR
S-35
D190 CONCENTRATION FACTOR
SB
D191 CONCENTRATION FACTOR
SB-124
D192 CONCENTRATION FACTOR
SB-125
D193 CONCENTRATION FACTOR
SC
D194 CONCENTRATION FACTOR
SE
D195 CONCENTRATION FACTOR
SE-75
D196 CONCENTRATION FACTOR
SI
D197 CONCENTRATION FACTOR
SN
D198 CONCENTRATION FACTOR
SR
D199 CONCENTRATION FACTOR
SR-85
D200 CONCENTRATION FACTOR
SR-89
D201 CONCENTRATION FACTOR
SR-90
D202 CONCENTRATION FACTOR
TB
D203 CONCENTRATION FACTOR
TC-95
D204 CONCENTRATION FACTOR
TCDD
D205 CONCENTRATION FACTOR
TH
D206 CONCENTRATION FACTOR
TH-232
D207 CONCENTRATION FACTOR
TI
D208 CONCENTRATION FACTOR
TOXAPHENE
D209 CONCENTRATION FACTOR
U
D210 CONCENTRATION FACTOR
V
D211 CONCENTRATION FACTOR
W
D212 CONCENTRATION FACTOR
W-185
D213 CONCENTRATION FACTOR
XMC
D214 CONCENTRATION FACTOR
Y
D215 CONCENTRATION FACTOR
YB
D216 CONCENTRATION FACTOR
YT-90
D217 CONCENTRATION FACTOR
ZN
D218 CONCENTRATION FACTOR
ZN-65
D219 CONCENTRATION FACTOR
ZR
D220 CONCENTRATION FACTOR
ZR-95
D221 CONCENTRATION FACTOR
ZR-95/NB-95
D222 CONCENTRATION FACTOR
2,4-D
D223 CONCENTRATION FACTOR
2,4-DI-BROMOBIPHENYL
D224 CONCENTRATION FACTOR
2,4-DI-CHLOROBIPHENYL
D225 CONCENTRATION FACTOR
2,4,2`,5`-TETRA-CHLOROBIPHENYL
D226 CONCENTRATION FACTOR
2,4,5-T
D227 CONCENTRATION FACTOR
2,5,3`,4`-TETRA-CHLOROBIPHENYL
D228 CONCENTRATION FACTOR
2,5,4`-TRI-CHLOROBIPHENYL
D229 CONCENTRATION FACTOR
2,6-DI-BROMOBIPHENYL
D230 CONCENTRATION FACTOR
2,6-DI-CHLOROBIPHENYL
D231 CONCENTRATION FACTOR
2,6-DI-CHLOROBIPHENYL1
D232 CONCENTRATION FACTOR
2,6-DI-CHLOROBIPHENYL2
D233 CONCENTRATION FACTOR
2,6-DI-CHLOROBIPHENYL3
D234 CONCENTRATION FACTOR
3,4-DI-BROMOBIPHENYL
D235 CONCENTRATION FACTOR
3,4-DI-CHLOROBIPHENYL
D236 CONCENTRATION FACTOR
CS-137 CONCENTRATION FACTOR VERSUS TEMPERA etc
D237 CONCENTRATION FACTOR
CS-137 CONCENTRATION FACTOR VERSUS TEMPERA etc
D238 CONCENTRATION FACTOR
ZN CONCENTRATION FACTOR VERSUS TEMPERATURE etc
D239 CONCENTRATION FACTOR
ZN CONCENTRATION FACTOR VERSUS TEMPERATURE etc
D240 CONCENTRATION FACTOR
CS-137 CONCENTRATION FACTOR VERSUS TEMPERA etc
D241 CONCENTRATION FACTOR
CS-137 CONCENTRATION FACTOR VERSUS TEMPERA etc
D242 CONCENTRATION FACTOR
ZN CONCENTRATION FACTOR VERSUS TEMPERATURE etc
D243 CONCENTRATION FACTOR
ZN CONCENTRATION FACTOR VERSUS TEMPERATURE etc
D246 CONCENTRATION FACTOR
CS-137 CONCENTRATION FACTOR VERSUS TEMPERA etc
D247 CONCENTRATION FACTOR
CS-137 CONCENTRATION FACTOR VERSUS TEMPERA etc
D248 CONCENTRATION FACTOR
CO-60
D249 CONCENTRATION FACTOR
PARTITION COEFFICIENT VERSUS SOLUBILITY
D250 CONCENTRATION FACTOR
CD-109
D251 CONCENTRATION FACTOR
PU
D252 CONCENTRATION FACTOR
CS-137
D253 CONCENTRATION FACTOR
HG
D254 CONCENTRATIONS
AG
D255 CONCENTRATIONS
AL
D256 CONCENTRATIONS
ALDRIN
D257 CONCENTRATIONS
AS
D258 CONCENTRATIONS
ATRAZINE
D259 CONCENTRATIONS
B
D260 CONCENTRATIONS
BA

D

E

R

U

D415 UPTAKE
FE
D416 UPTAKE
HG
D417 UPTAKE
K
D418 UPTAKE
LI
D419 UPTAKE
MG
D420 UPTAKE
MN
D421 UPTAKE
MO
D422 UPTAKE
NA
D423 UPTAKE
NH3
D424 UPTAKE
NI
D425 UPTAKE
NO2
D426 UPTAKE
OCTACHLORODIBENZO-P-DIOXIN
D427 UPTAKE
P
D428 UPTAKE
PB
D429 UPTAKE
PCB
D430 UPTAKE
RB
D431 UPTAKE
SI
D432 UPTAKE
SN
D433 UPTAKE
SO2
D434 UPTAKE
SR
D435 UPTAKE
TI
D436 UPTAKE
V
D437 UPTAKE
ZN
D438 UPTAKE
ZR
D439 UPTAKE
2-MONOCHLOROBIPHENYL
D440 UPTAKE
2,2`-DICHLOROBIPHENYL
D441 UPTAKE
2,2`,4,4`,5,5`-HEXACHLOROBIPHENYL
D442 UPTAKE
2,2`,4,5,5`-PENTACHLOROBIPHENYL
D443 UPTAKE
2,2`,5-TRICHLOROBIPHENYL
D444 UPTAKE
2,2`,5,5`-TETRACHLOROBIPHENYL
D445 UPTAKE
2,3,4,5-TETRACHLOROBIPHENYL
D446 UPTAKE
2,3`-DICHLOROBIPHENYL
D447 UPTAKE
2,4-DICHLOROBIPHENYL
D448 UPTAKE
2,4,4`-TRICHLOROBIPHENYL
D449 UPTAKE
2,4`-DICHLOROBIPHENYL
D450 UPTAKE
2,6-DICHLOROBIPHENYL
D451 UPTAKE
2`,3,5-TRICHLOROBIPHENYL
D452 UPTAKE
3-MONOCHLOROBIPHENYL
D453 UPTAKE
3,3`,4-TRICHLOROBIPHENYL
D454 UPTAKE
3,3`,4,4`-TETRACHLOROBIPHENYL
D455 UPTAKE
3,3`4`-TRICHLOROBIPHENYL
D456 UPTAKE
3,4`-DICHLOROBIPHENYL
D457 UPTAKE
4-MONOCHLOROBIPHENYL
D458 UPTAKE
4,4`-DICHLOROBIPHENYL
D459 UPTAKE FROM SOIL WITH ADDED SLUDGE
AL
D460 UPTAKE FROM SOIL WITH ADDED SLUDGE
AS
D461 UPTAKE FROM SOIL WITH ADDED SLUDGE
B
D462 UPTAKE FROM SOIL WITH ADDED SLUDGE
BR
D463 UPTAKE FROM SOIL WITH ADDED SLUDGE
CA
D464 UPTAKE FROM SOIL WITH ADDED SLUDGE
CD
D465 UPTAKE FROM SOIL WITH ADDED SLUDGE
CL
D466 UPTAKE FROM SOIL WITH ADDED SLUDGE
CO
D467 UPTAKE FROM SOIL WITH ADDED SLUDGE
CR
D468 UPTAKE FROM SOIL WITH ADDED SLUDGE
CU
D469 UPTAKE FROM SOIL WITH ADDED SLUDGE
FE
D470 UPTAKE FROM SOIL WITH ADDED SLUDGE
HG
D471 UPTAKE FROM SOIL WITH ADDED SLUDGE
I
D472 UPTAKE FROM SOIL WITH ADDED SLUDGE
K
D473 UPTAKE FROM SOIL WITH ADDED SLUDGE
MG
D474 UPTAKE FROM SOIL WITH ADDED SLUDGE
MN
D475 UPTAKE FROM SOIL WITH ADDED SLUDGE
MO
D476 UPTAKE FROM SOIL WITH ADDED SLUDGE
NI
D477 UPTAKE FROM SOIL WITH ADDED SLUDGE
PB
D478 UPTAKE FROM SOIL WITH ADDED SLUDGE
SB
D479 UPTAKE FROM SOIL WITH ADDED SLUDGE
SC
D480 UPTAKE FROM SOIL WITH ADDED SLUDGE
SE
D481 UPTAKE FROM SOIL WITH ADDED SLUDGE
TA
D482 UPTAKE FROM SOIL WITH ADDED SLUDGE
TI
D483 UPTAKE FROM SOIL WITH ADDED SLUDGE
V
D484 UPTAKE FROM SOIL WITH ADDED SLUDGE
W
D485 UPTAKE FROM SOIL WITH ADDED SLUDGE
YB
D486 UPTAKE FROM SOIL WITH ADDED SLUDGE
ZN
D487 UPTAKE
CD FROM SOIL
D488 UPTAKE
CD FROM SOIL
D489 UPTAKE
CD FROM SOIL
D490 UPTAKE
MIREX
D491 UPTAKE
CO-60
D493 UPTAKE
HG
D494 UPTAKE
HG
D495 UPTAKE
HG
D496 UPTAKE
SE
D497 UPTAKE
HG
D498 UPTAKE
HG
D499 UPTAKE
HG
D503 UPTAKE
OZONE
D504 UPTAKE
HG
D505 UPTAKE
ADSORPTION OF HG
D506 UPTAKE
ADSORPTION OF CD

PART E

PART F

PART G

S

T

W

Abbreviations

+-	Indicating interval of determination. Not SD.
*	Multiplication
**	Iteration
---	Missing value, value not measured
AAS	Atom absorption spectophotometer
AB.GROUND	Above Ground
ABS	Absorption
ABS	Alkyl benzen sulfonate
ABS	Kelvin Degrees
ABSORP	Absorption
ACC	Acceleration
AE7.0	Alkyl polyethoxylate
AE7.4	Alkyl polyethoxylate
ALK	Alkalinity
ANIM	Animal
ANN	Annual
AOS	Alpha olefin sulfonate
APPL	Application
APPROX	Approximatly
ASSIM	Assimilation
ATT	Attenuation
AV	Average
AVIA	Aviable
B.C.	British Columbia
BACT	Bacteria
BEGIN	Beginning
BEL	Below
BETW	Between
BHC	Benzene hehachloride
BHC	Mixed isomers of 1,2,3,4,5,6-hexachloro-cyclohexane.
BI	Two generations per year
BIOL	Biological
BIOM	Biomass
BR.H.	Brest Height
CAL	Calories
CAL/CM2*MIN	Experimental condition of light intensity.
CALC	Calculated
CALIF	California
CAN	Canada
CAP	Phynylalanine hydroxylase
CAROT	Carotenoids
CC	Cm**3
CC	Millilitre
CDO	Control condition for cadmium experiment
CEC	Cation exchange capacity
CEL	Celcius
CELL	Cellular
CFL	Confidence limit
CH-P	Chlorella pyrenoidosa
CHEM	Chemically
CHEMO	Chemostat
CHL	Chlorophyll a
CHLA	Chlorophyll a
CHLOR	Chlorid
CI	Curie
CII	Copepodit stage 2
CL	Confidence limit
CLEAN	Cleaning
COEF	Coefficient
COEFF	Coefficient
COL	Colony
COMB	Compustion
COMM	Community
CON	Consumation
CONC	Concentration
COND	Conductivity
CONP	Consumption
CONST	Constant
CONSUMP	Consumption
CONT	Content
CONT	Continous
CONT	Counted
CONTAMIN	Contaminated
CR	Creek
CULT	Cultivable
CULTURE: xx K	XX Kelvin was used as temperature for cultivation culture
CUM	Cumulative
C1	Copepodit stage 1
D	(in condition) no. of days (measuring period
D	Days
DEF	Deficit
DEG	Degrees
DENM	Denmark
DERIV	Derivatives
DEVELOP	Development
DG	Degrees Days
DIA	Diameter
DIAM	Diameter
DIF	Different
DIFF	Different
DIFF	Diffusion
DIS	Dissolved
DISAPP	Disapperence
DISP	Dispersion
DISS	Dissolved
DISSOL	Dissolved
DK	Denmark
DM	Dry matter
DOM	Dissolved Organic Matter
DOM	Dominating
DUR	During
E	East
EFF	Effictive
EFFIC	Efficiency
ELIM	Elimination
ENGL	England
ENZ	Enzyme
EQUILBR	Equilibrium
ERODIBIL	Erodibilae
ERUPT	Eruptions (vulcanoes)
EST	Estimate
ESTU	Estuarie
EVAP	Evaporation
EXCH	Exchange
EXCL	Exclusive
EXCR	Excretion
EXCRET	Excretion
EXP	Expirimental
EXP. XX K	XX Kelvin was used as experimental condition
EXT	Extension
EXT	Extinction
F	Females
FACT	Factor
FERT	Fertilization
FIELD	Field data
FILT	Filtering
FILTRAT	Filtration
FL	Florida
FORM	Formation
FPS	Feet per second
FR	France
FR	From
FRACT	Fraction
FREUND	Freundlich isotherm, see footnote F057.
FTC	Foot light candles
FUNCT	Function
G.B.	Great Britain
G-6-P-D-H	Glucose-6-phosphate dehydrogenase
G-6-PDH	Glucose-6-phosphat dehydrogenase
G/G	Gram per gram (used about ratios)
GOT	Glutamate oxaloacetate transaminase
GPT	Glucose phosphate transferase
GPT	Glutamine pyruvate transaminase
GR	Growth rate
GR.RATE	Growth rate
H	Hour
HB	Hemaglobine
HRS	Hours
HUMID	Humidity
IK	Linear saturation point of photosynthesis curve.
IMP	Imperrious
IN	Inch
INCH	Inches

IND	Indiana
INGEST	Ingestion
INH	Inhibition
INIT	Initial concentration in experiment
INJECT	Injection
INORG	Inorganic
INSOL	Insolation
INTERV	Interval
INTRAV	Intravenous
JAP	Japaneese
JULIAN C.D.	Julian Calender Day
JUV	Juveniles
K	Kelvin degrees
K CAL	Kilo calories
K J	Kilo Joule
KLUX	Kilolux, unit of light
L	Lake
L	Litre
LA	Louisiana
LAB	Experiment done in laboratorium
LC50	CONC at witch 50 % of expiriment organisms dies.
LE	Less
LONG	Longitudinal
LY	Langleys
M	Males
M	Meter
M	Milli
M	Molar
M	Month
M/S	Meter/second
MAIN	Maintenance
MAT	Matter
MATC	Maximum acceptable toxicant concentration.
MAX	Maximum
MCH	Mean corpuscle hemoglobin (in blood)
MCV	Mean corpuscle volume
MCV	Mean corpuscle volume (in blood)
MEAN OF X	Mean of X determinations
MET	Metabolism
METAB	Metabolism
MIC. G	Microgram
MICH	Michigan
MIN	Minute
MINN	Minnesota
MISS	Missing
MOSM	Milli osmotic pressure
MULTI	More than 2 generations per year
MV	Milli Volt
MW	Milliwatt
MYH	Micro M/g of liver/hour
MYM	Micro M/g of liver/10 minutes
MYT	Micro M tyrosine/g of liver/hour
MYU	Micro units/ml
N	Nauplius
N	Nitrogen
N	North (about geographic names)
N=X	X numbers of samples
NC	North Carolina
NG	Nannogram
NO.	Number
NP	Net production
NV	North West
NWT	North Western Terretories in Canada
NZ	New Zealand
OBS	Observation period
OKL	Oklahoma
OPT	Optimum conditions
ORG	organic
ORG	Organism
OVA	Ovaries
O2	Oxygen
P	Pico
P	Production
P/R	Production/respiration = respiratory coefficient.
PART	Particle
PART	Particulate
PE-T	Pediastrum tetras
PHOTOSYNTH	Photosynthesis
PHYT	Phytoplankton
POLLUT	Pollutant
POT	Potiental
PPB	Parts per 10**(9
PPM	parts per 10**6
PROB	Probable
PROD	Production
PROD	Production
PROT	Protein
PT-50	-log(Poteintal toxicity)
PYR	Pyridine
R	Regression coefficient
R	River
RAD	Radiation
RBC	Red blood cells
REF	Reference
REL	Relation
REL	Relative
REL	Release
REMAIN	Remaining
REPROD	Reproduction
REQ	Required
REQ	Requirements
RESID	Residence
RESP	Respiration
REST	Resistance
REST	Resistance
S	Salinity
S	Second
S	South
SAP	Phenylalanine hydroxylase
SAT	Saturation
SC	South Carolina
SD	Standard deviation.
SE	Standard error of the mean
SE-P	Selenastrum capricornutum
SEC	Second
SED	Sediment
SEDIM	Sediment
SGPT	Serum glutamic pyruvic transaminase
SOUTH	Southern
SP	Number of spawns
SP	Species
SP	Specific
SPEC	Spectronic 20
SPH	Sphere
SSI	Spleen somatic ratio (g spleen/g body weight).
ST	Standing
STAGE C1	Copepodit stage 1
STRATO	Stratosphere
SW	South West
T	Tons
TECH	Technice
TEMP	Temperature
TENN	Tennesee
THR	Through
TOT	Total
TRANSP	Transport
TROP	Tropical
TWEEN	See footnotes F010, F018, F019 and F020.
U	Unit
U	10**(-6) m
UG	Microgram
UK	United Kingdom
UM	Micromol
UNI	One generation per year
UPT	Uptake
UTIL	Utilization
VAL	Value
VEL	Velocity
VELOC	Velocity
VERT	Vertical
VOL	Volume
VOLTINISM	Number of generations per per year, see UNI, BI and MULTI.
VS	Versus
W	Weeks
W	West
WAT	Water
WBC	White blood cells
WISC	Wisconsin
WT	Weight
WW	Wet Weight
Y	Years
YR	Year

PART A

Composition and Ecological Parameters of Living Organisms

ALGAE
BACTERIA
DETRITUS
FISH
FUNGI
INVERTEBRATES
KORMOPHYTES
VERTEBRATES
ZOO PLANKTON

A1

ALGAE

COMPOSITION
CELL SIZE
GROWTH RATE
HALF SATURATION CONSTANTS
PRODUCTION
RESPIRATION
SETTLING
UPTAKE
YIELD

TABLE A1

ALGAE : ASH/BIOMASS RATIO			
ITEM:	VALUE:	CONDITIONS:	REF.:
*ANABAENA SP.	5.19 % OF DRY WEIGHT	ALGAE FROM PONDS AND LAKES IN SOUTHEASTERN UNITED STATES	1660
*APHANIZOMENON SP.	7.21 % OF DRY WEIGHT	ALGAE FROM PONDS AND LAKES IN SOUTHEASTERN UNITED STATES	1660
*BACILLARIOPHYCEAE	58.5 % OF DRY WEIGHT	PLANKTON,8 SPECIES	1635
*CHARA SP.	43.4 % OF DRY WEIGHT	ALGAE FROM PONDS AND LAKES IN SOUTHEASTERN UNITED STATES	1660
*CLADOPHORA SP.	23.38 % OF DRY WEIGHT	ALGAE FROM PONDS AND LAKES IN SOUTHEASTERN UNITED STATES	1660
*EUGLENA SP.	4.12 % OF DRY WEIGHT	ALGAE FROM PONDS AND LAKES IN SOUTHEASTERN UNITED STATES	1660
*HYDRODICTYON SP.	17.94 % OF DRY WEIGHT	ALGAE FROM PONDS AND LAKES IN SOUTHEASTERN UNITED STATES	1660
*LYNGBYA SP.	17.20 % OF DRY WEIGHT	ALGAE FROM PONDS AND LAKES IN SOUTHEASTERN UNITED STATES	1660
*MICROCYSTIS SP.	6.2 % OF DRY WEIGHT	ALGAE FROM PONDS AND LAKES IN SOUTHEASTERN UNITED STATES	1660
*MOUGEOTIA SP.	14.54 % OF DRY WEIGHT	ALGAE FROM PONDS AND LAKES IN SOUTHEASTERN UNITED STATES	1660
*NITELLA SP.	19.11 % OF DRY WEIGHT	ALGAE FROM PONDS AND LAKES IN SOUTHEASTERN UNITED STATES	1660
*OEDOGONIUM SP.	12.69 % OF DRY WEIGHT	ALGAE FROM PONDS AND LAKES IN SOUTHEASTERN UNITED STATES	1660
*PHAEOPHYCEAE	28.8 % OF DRY WEIGHT	16 SPECIES	1635
*PITHOPHORA SP.	27.77 % OF DRY WEIGHT	ALGAE FROM PONDS AND LAKES IN SOUTHEASTERN UNITED STATES	1660
*RHIZOCLONIUM SP.	17.36 % OF DRY WEIGHT	ALGAE FROM PONDS AND LAKES IN SOUTHEASTERN UNITED STATES	1660
*SPIROGYRA SP.	13.06 % OF DRY WEIGHT	ALGAE FROM PONDS AND LAKES IN SOUTHEASTERN UNITED STATES	1660
*SPIROGYRA SP., GIANT	13.86 % OF DRY WEIGHT	ALGAE FROM PONDS AND LAKES IN SOUTHEASTERN UNITED STATES	1660

TABLE A2

ALGAE : ASSIMILATION			
ITEM:	VALUE:	CONDITIONS:	REF.:
*ALGAE FROM PACIFIC OCEAN	1.36 MICRO G C/(MICRO G CHLOROPHYLL * HOUR)	CHLOROPHYLL=0.155,INORG N=0.00, INORG P=0.26, SI=1.8 MG/M3	1584
*ALGAE FROM PACIFIC OCEAN	3.24 MICRO G C/(MICRO G CHLOROPHYLL * HOUR)	CHLOROPHYLL=0.099,INORG N=0.46, INORG P=0.24, SI=1.5 MG/M3	1584
*ALGAE FROM PACIFIC OCEAN	1.15 MICRO G C/(MICRO G CHLOROPHYLL * HOUR)	CHLOROPHYLL=0.092,INORG N=0.46, INORG P=0.33, SI=3.2 MG/M3	1584
*ALGAE FROM PACIFIC OCEAN	3.40 MICRO G C/(MICRO G CHLOROPHYLL * HOUR)	CHLOROPHYLL=0.135,INORG N=0.34, INORG P=0.23, SI=3.8 MG/M3	1584
*ALGAE FROM PACIFIC OCEAN	2.91 MICRO G C/(MICRO G CHLOROPHYLL * HOUR)	CHLOROPHYLL=0.174,INORG N=0.49, INORG P=0.22, SI=2.1 MG/M3	1584
*ALGAE FROM PACIFIC OCEAN	4.12 MICRO G C/(MICRO G CHLOROPHYLL * HOUR)	CHLOROPHYLL=0.224,INORG N=0.48, INORG P=0.27, SI=2.0 MG/M3	1584
*ALGAE FROM PACIFIC OCEAN	3.81 MICRO G C/(MICRO G CHLOROPHYLL * HOUR)	CHLOROPHYLL=0.142,INORG N=0.45, INORG P=0.30, SI=3.1 MG/M3	1584
*ALGAE FROM PACIFIC OCEAN	5.18 MICRO G C/(MICRO G CHLOROPHYLL * HOUR)	CHLOROPHYLL=0.158,INORG N=0.40, INORG P=0.24, SI=2.5 MG/M3	1584
*ALGAE FROM PACIFIC OCEAN	3.15 MICRO G C/(MICRO G CHLOROPHYLL * HOUR)	MEAN OF 8 NITROGEN POOR STATIONS	1584
*ALGAE FROM PACIFIC OCEAN	5.23 MICRO G C/(MICRO G CHLOROPHYLL * HOUR)	CHLOROPHYLL=0.201,INORG N=2.50, INORG P=0.35, SI=3.1 MG/M3	1584
*ALGAE FROM PACIFIC OCEAN	4.93 MICRO G C/(MICRO G CHLOROPHYLL * HOUR)	CHLOROPHYLL=0.211,INORG N=5.71, INORG P=0.27, SI=5.0 MG/M3	1584
*ALGAE FROM PACIFIC OCEAN	5.22 MICRO G C/(MICRO G CHLOROPHYLL * HOUR)	CHLOROPHYLL=0.151,INORG N=8.61, INORG P=0.75, SI=3.8 MG/M3	1584
*ALGAE FROM PACIFIC OCEAN	4.04 MICRO G C/(MICRO G CHLOROPHYLL * HOUR)	CHLOROPHYLL=0.102,INORG N=7.16, INORG P=0.62, SI=3.5 MG/M3	1584

TABLE A2

ALGAE : ASSIMILATION			
ITEM:	VALUE:	CONDITIONS:	REF.:
*ALGAE FROM PACIFIC OCEAN	3.53 MICRO G C/(MICRO G CHLOROPHYLL * HOUR)	CHLOROPHYLL=0.099,INORG N=7.81, INORG P=0.71, SI=--- MG/M3	1584
*ALGAE FROM PACIFIC OCEAN	5.84 MICRO G C/(MICRO G CHLOROPHYLL * HOUR)	CHLOROPHYLL=0.041,INORG N=8.11, INORG P=0.69, SI=1.6 MG/M3	1584
*ALGAE FROM PACIFIC OCEAN	4.98 MICRO G C/(MICRO G CHLOROPHYLL * HOUR)	CHLOROPHYLL=0.105,INORG N=7.42, INORG P=0.67, SI=3.6 MG/M3	1584
*ALGAE FROM PACIFIC OCEAN	6.19 MICRO G C/(MICRO G CHLOROPHYLL * HOUR)	CHLOROPHYLL=0.181,INORG N=6.09, INORG P=0.58, SI=3.2 MG/M3	1584
*ALGAE FROM PACIFIC OCEAN	4.55 MICRO G C/(MICRO G CHLOROPHYLL * HOUR)	CHLOROPHYLL=0.263,INORG N=8.35, INORG P=0.83, SI=5.2 MG/M3	1584
*ALGAE FROM PACIFIC OCEAN	4.95 MICRO G C/(MICRO G CHLOROPHYLL * HOUR)	MEAN OF 9 NITROGEN RICH STATIONS	1584
*ALGAE FROM PACIFIC OCEAN	10.6 MICROGRAN C/(MICROGRAM CHL*HOUR)	CHLOROPHYLL=0.155,INORG N=0.00, INORG P=0.26, SI=1.8 MG/M3	1584

TABLE A3

ALGAE : ATP-C/BIOMASS RATIO			
ITEM:	VALUE:	CONDITIONS:	REF.:
*TRICHODESMIUM SPP.	3.3 MICROGRAM ATP/COLONY	ATP-CARBON/COLONY,N.PACIFIC, N-FIXING ALGAE,N=6	967

TABLE A4

ALGAE : ATP/(CARBON+NITROGEN) RATIO			
ITEM:	VALUE:	CONDITIONS:	REF.:
*CHAETOCEROS LAUDERI	4.57*10**(-3) BY WEIGHT	CULTIVATED, MARINE	104
*CHLAMYDOMONAS MAGNUSII	1.80*10**(-3) BY WEIGHT	CULTIVATED, MARINE	104
*COSCINODISCUS SP.	3.19*10**(-3) BY WEIGHT	CULTIVATED, MARINE	104
*CRYPTOMONAS PSEUDOBALTICA	3.57*10**(-3) BY WEIGHT	CULTIVATED, MARINE	104
*HETEROTHRIX SP.	1.49*10**(-3) BY WEIGHT	CULTIVATED, MARINE	104
*PHAEODACTYLUM TRICORNUTUM	2.40*10**(-3) BY WEIGHT	CULTIVATED, MARINE	104
*TETRASELMIS STRICIA	1.53*10**(-3) BY WEIGHT	CULTIVATED, MARINE	104

TABLE A5

ALGAE : ATP/BIOMASS RATIO			
ITEM:	VALUE:	CONDITIONS:	REF.:
*ANABAENA FLOS-AQUAE	0.26 MG/G DRY WEIGHT	200 - 300 FTC,R=0.607	184
*CHAETOCEROS LAUDERI	0.269 MICROGRAM ATP/MM**3	CULTIVATED, MARINE	104
*CHLAMYDOMONAS MAGNUSII	0.402 MICROGRAM ATP/MM**3	CULTIVATED, MARINE	104
*CHLORELLA SP.	0.38 MG/G DRY WEIGHT	200 - 300 FTC,R=0.860	184
*COSCINODISCUS SP.	0.206 MICROGRAM ATP/MM**3	CULTIVATED, MARINE	104
*CRYPTOMONAS PSEUDOBALTICA	0.666 MICROGRAM ATP/MM**3	CULTIVATED, MARINE	104
*FRESHWATER ALGAE	0.96 MG/G DRY WEIGHT	ALGAE FROM ANDERSON-CUE LAKE, OLIGOTROPHIC,R=0.790	184
*FRESHWATER ALGAE	0.13 MG/G DRY WEIGHT	ALGAE FROM BIVINS ARM, HYPEREUTROPHIC,R=0.649	184
*FRESHWATER ALGAE	0.12 MG/G DRY WEIGHT	ALGAE FROM NEWNANS LAKE,EUTROPHIC, R=0.673	184

TABLE A5

ALGAE : ATP/BIOMASS RATIO

ITEM:	VALUE:	CONDITIONS:	REF.:
*HETEROTHRIX SP.	0.597 MICROGRAM ATP/MM**3	CULTIVATED, MARINE	104
*MARINE PHYTOPLANKTON	0.82 10**(-12) G/CELL	NO3 AS SOURCE OF N, STATIONARY PHASE,293 K, SOUTHERN CALIF	428
*MARINE PHYTOPLANKTON	1.0 10**(-12) G/CELL	NH4 AS SOURCE OF N, STATIONARY PHASE,293 K, SOUTHERN CALIF	428
*MARINE PHYTOPLANKTON	0.53 10**(-12) G/CELL	UREA AS SOURCE OF N, STATIONARY PHASE,293 K, SOUTHERN CALIF	428
*MICROCYSTIS AERUGINOSA	0.36 MG/G DRY WEIGHT	200 - 300 FTC,R=0.718	184
*PHAEODACTYLUM TRICORNUTUM	0.469 MICROGRAM ATP/MM**3	CULTIVATED, MARINE	104
*SELENASTRUM CAPRICORNUTUM	0.24 MG/G DRY WEIGHT	200 - 300 FTC,R=0.244	184
*TETRASELMIS STRICIA	0.159 MICROGRAM ATP/MM**3	CULTIVATED, MARINE	104
*THALASSIOSIRA SP.	3.9*10**(-15) GRAM/CELL	GROWTH RATE=0.041 1/H, CELL DENSITY=4.0*10**8 CELL/L, P-LIMITED CHEMO.	1243
*THALASSIOSIRA SP.	2.6*10**(-15) GRAM/CELL	GROWTH RATE=0.017 1/H, CELL DENSITY=6.7*10**8 CELL/L, P-LIMITED CHEMO.	1243
*TRICHODESMIUM SPP.	13 NG/COLONY	ATP/COLONY,N.PACIFIC,N-FIXING ALGAE,N=6	967

TABLE A6

ALGAE : ATP/C RATIO

ITEM:	VALUE:	CONDITIONS:	REF.:
*CHAETOCEROS LAUDERI	5.57*10**(-3) BY WEIGHT	CULTIVATED, MARINE	104
*CHLAMYDOMONAS MAGNUSII	2.13*10**(-3) BY WEIGHT	CULTIVATED, MARINE	104
*COSCINODISCUS SP.	3.89*10**(-3) BY WEIGHT	CULTIVATED, MARINE	104
*CRYPTOMONAS PSEUDOBALTICA	4.47*10**(-3) BY WEIGHT	CULTIVATED, MARINE	104
*HETEROTHRIX SP.	1.69*10**(-3) BY WEIGHT	CULTIVATED, MARINE	104
*MARINE PHYTOPLANKTON	3.57*10**(-3) BY WEIGHT	NO3 AS SOURCE OF N, LOG PHASE,293 K, SOUTHERN CALIF	428
*MARINE PHYTOPLANKTON	3.70*10**(-3) BY WEIGHT	NO3 AS SOURCE OF N, STATIONARY PHASE,293 K, SOUTHERN CALIF	428
*MARINE PHYTOPLANKTON	3.85*10**(-3) BY WEIGHT	NH4 AS SOURCE OF N, LOG PHASE,293 K, SOUTHERN CALIF	428
*MARINE PHYTOPLANKTON	4.35*10**(-3) BY WEIGHT	NH4 AS SOURCE OF N, STATIONARY PHASE,293 K, SOUTHERN CALIF	428
*MARINE PHYTOPLANKTON	3.85*10**(-3) BY WEIGHT	UREA AS SOURCE OF N, LOG PHASE,293 K, SOUTHERN CALIF	428
*MARINE PHYTOPLANKTON	3.45*10**(-3) BY WEIGHT	UREA AS SOURCE OF N, STATIONARY PHASE,293 K, SOUTHERN CALIF	428
*PERIDINIUM CINCTUM FA. WESTII	2374 BY WEIGHT	GROWN WITHOUT PHOSPHORUS	236
*PERIDINIUM CINCTUM FA. WESTII	297 BY WEIGHT	GROWN WITH PHOSPHORUS	236
*PERIDINIUM CINCTUM FA. WESTII	310 BY WEIGHT	NONE	236
*PHAEODACTYLUM TRICORNUTUM	3.00*10**(-3) BY WEIGHT	CULTIVATED, MARINE	104
*TETRASELMIS STRICIA	1.97*10**(-3) BY WEIGHT	CULTIVATED, MARINE	104
*THALASSIOSIRA PSEUDONANA	780 BY WEIGHT	P-LIMITED CHEMOSTAT,GROWTH=0.041 1/H, 295 K,0.07 CAL/CM2*MIN	1243
*THALASSIOSIRA PSEUDONANA	2088 BY WEIGHT	P-LIMITED CHEMOSTAT,GROWTH=0.017 1/H, 295 K,0.07 CAL/CM2*MIN	1243
*THALASSIOSIRA PSEUDONANA	7.19*10**(-3) BY WEIGHT	N-LIMITED CHEMOSTAT,GROWTH=0.041 1/H, 295 K,0.07 CAL/CM2*MIN	1243
*THALASSIOSIRA PSEUDONANA	5.55*10**(-3) BY WEIGHT	N-LIMITED CHEMOSTAT,GROWTH=0.017 1/H, 295 K,0.07 CAL/CM2*MIN	1243
*THALASSIOSIRA PSEUDONANA	3.24*10**(-3) BY WEIGHT	N-LIMITED CHEMOSTAT,GROWTH=0.0085 1/H,295 K,0.07 CAL/CM2*MIN	1243

TABLE A7

ALGAE : ATP/CHLOROPHYLL A RATIO

ITEM:	VALUE:	CONDITIONS:	REF.:
*CHLORELLA SP.	0.29 MG/MG	200 - 300 FTC,R=0.725	184

TABLE A7

ALGAE : ATP/CHLOROPHYLL A RATIO

ITEM:	VALUE:	CONDITIONS:	REF.:
*FRESHWATER ALGAE	0.69 MG/MG	ALGAE FROM ANDERSON-CUE LAKE, OLIGOTROPHIC,R=0.790	184
*FRESHWATER ALGAE	0.09 MG/MG	ALGAE FROM BIVINS ARM, HYPEREUTROPHIC,R=0.869	184
*FRESHWATER ALGAE	0.11 MG/MG	ALGAE FROM NEWNANS LAKE,EUTROPHIC, R=0.718	184
*MICROCYSTIS AERUGINOSA	0.09 MG/MG	200 - 300 FTC,R=0.683	184
*SELENASTRUM CAPRICORNATUM	0.35 MG/MG	200 - 300 FTC,R=0.539	184
*THALASSIOSIRA PSEUDONANA	0.037 BY WEIGHT	P-LIMITED CHEMOSTAT,GROWTH=0.041 1/H, 295 K,0.07 CAL/CM2*MIN	1243
*THALASSIOSIRA PSEUDONANA	0.038 BY WEIGHT	P-LIMITED CHEMOSTAT,GROWTH=0.017 1/H, 295 K,0.07 CAL/CM2*MIN	1243
*THALASSIOSIRA PSEUDONANA	0.2 BY WEIGHT	N-LIMITED CHEMOSTAT,GROWTH=0.041 1/H, 295 K,0.07 CAL/CM2*MIN	1243
*THALASSIOSIRA PSEUDONANA	0.25 BY WEIGHT	N-LIMITED CHEMOSTAT,GROWTH=0.017 1/H, 295 K,0.07 CAL/CM2*MIN	1243
*THALASSIOSIRA PSEUDONANA	0.33 BY WEIGHT	N-LIMITED CHEMOSTAT,GROWTH=0.0085 1/H,295 K,0.07 CAL/CM2*MIN	1243

TABLE A8

ALGAE : ATP/N RATIO

ITEM:	VALUE:	CONDITIONS:	REF.:
*CHAETOCEROS LAUDERI	26.2*10**(-3) BY WEIGHT	CULTIVATED, MARINE	104
*CHLAMYDOMONAS MAGNUSII	13.0*10**(-3) BY WEIGHT	CULTIVATED, MARINE	104
*COSCINODISCUS SP.	17.8*10**(-3) BY WEIGHT	CULTIVATED, MARINE	104
*CRYPTOMONAS PSEUDOBALTICA	15.3*10**(-3) BY WEIGHT	CULTIVATED, MARINE	104
*HETEROTHRIX SP.	12.7*10**(-3) BY WEIGHT	CULTIVATED, MARINE	104
*PHAEODACTYLUM TRICORNUTUM	12.5*10**(-3) BY WEIGHT	CULTIVATED, MARINE	104
*TETRASELMIS STRICIA	6.61*10**(-3) BY WEIGHT	CULTIVATED, MARINE	104
*THALASSIOSIRA PSEUDONANA	0.53*10**(-2) BY WEIGHT	P-LIMITED CHEMOSTAT,GROWTH=0.041 1/H, 295 K,0.07 CAL/CM2*MIN	1243
*THALASSIOSIRA PSEUDONANA	0.441*10**(-2) BY WEIGHT	P-LIMITED CHEMOSTAT,GROWTH=0.017 1/H, 295 K,0.07 CAL/CM2*MIN	1243
*THALASSIOSIRA PSEUDONANA	3.57*10**(-2) BY WEIGHT	N-LIMITED CHEMOSTAT,GROWTH=0.041 1/H, 295 K,0.07 CAL/CM2*MIN	1243
*THALASSIOSIRA PSEUDONANA	4.76*10**(-2) BY WEIGHT	N-LIMITED CHEMOSTAT,GROWTH=0.017 1/H, 295 K,0.07 CAL/CM2*MIN	1243
*THALASSIOSIRA PSEUDONANA	4*10**(-2) BY WEIGHT	N-LIMITED CHEMOSTAT,GROWTH=0.0085 1/H,295 K,0.07 CAL/CM2*MIN	1243

TABLE A9

ALGAE : ATP/P RATIO

ITEM:	VALUE:	CONDITIONS:	REF.:
*THALASSIOSIRA PSEUDONANA	0.115 BY WEIGHT	P-LIMITED CHEMOSTAT,GROWTH=0.041 1/H, 295 K,0.07 CAL/CM2*MIN	1243
*THALASSIOSIRA PSEUDONANA	0.123 BY WEIGHT	P-LIMITED CHEMOSTAT,GROWTH=0.017 1/H, 295 K,0.07 CAL/CM2*MIN	1243
*THALASSIOSIRA PSEUDONANA	0.175 BY WEIGHT	N-LIMITED CHEMOSTAT,GROWTH=0.041 1/H, 295 K,0.07 CAL/CM2*MIN	1243
*THALASSIOSIRA PSEUDONANA	0.182 BY WEIGHT	N-LIMITED CHEMOSTAT,GROWTH=0.017 1/H, 295 K,0.07 CAL/CM2*MIN	1243
*THALASSIOSIRA PSEUDONANA	0.114 BY WEIGHT	N-LIMITED CHEMOSTAT,GROWTH=0.0085 1/H,295 K,0.07 CAL/CM2*MIN	1243

TABLE A10

ALGAE : B/BIOMASS RATIO			
ITEM:	VALUE:	CONDITIONS:	REF.:
*CHARA SP.	6.7 MICROGRAM/L	ALGAE FROM PONDS AND LAKES IN SOUTHEASTERN UNITED STATES	1660
*CLADOPHORA SP.	84.6 MICROGRAM/L	ALGAE FROM PONDS AND LAKES IN SOUTHEASTERN UNITED STATES	1660
*EUGLENA SP.	3.8 MICROGRAM/L	ALGAE FROM PONDS AND LAKES IN SOUTHEASTERN UNITED STATES	1660
*LYNGBYA SP.	112 MICROGRAM/L	ALGAE FROM PONDS AND LAKES IN SOUTHEASTERN UNITED STATES	1660
*MICROCYSTIS SP.	3.6 MICROGRAM/L	ALGAE FROM PONDS AND LAKES IN SOUTHEASTERN UNITED STATES	1660
*MOUGEOTIA SP.	8 MICROGRAM/L	ALGAE FROM PONDS AND LAKES IN SOUTHEASTERN UNITED STATES	1660
*NITELLA SP.	9.8 MICROGRAM/L	ALGAE FROM PONDS AND LAKES IN SOUTHEASTERN UNITED STATES	1660
*OEDOGONIUM SP.	8.1 MICROGRAM/L	ALGAE FROM PONDS AND LAKES IN SOUTHEASTERN UNITED STATES	1660
*PITHOPHORA SP.	65 MICROGRAM/L	ALGAE FROM PONDS AND LAKES IN SOUTHEASTERN UNITED STATES	1660
*RHIZOCLONIUM SP.	1.8 MICROGRAM/L	ALGAE FROM PONDS AND LAKES IN SOUTHEASTERN UNITED STATES	1660
*SPIROGYRA SP.	4.2 MICROGRAM/L	ALGAE FROM PONDS AND LAKES IN SOUTHEASTERN UNITED STATES	1660
*SPIROGYRA SP., GIANT	4.3 MICROGRAM/L	ALGAE FROM PONDS AND LAKES IN SOUTHEASTERN UNITED STATES	1660

TABLE A11

ALGAE : BIOMASS			
ITEM:	VALUE:	CONDITIONS:	REF.:
*ALGAE	1600 KCAL/M**(-2)	MAX STANDING CROP, INITIAL VALUES, THERMAL ECOSYSTEM	1703
*CHLORELLA VULGARIS	CA. 3.4*10**(-11) G	GREEN ALGA	175
*CLADOPHORA SP.	34.6 DRY WEIGHT G/M2 (ANNUAL MEAN)	ANNUAL MEAN:T.INOR.P=1.1MG/L,NO3=2.7MG N/L,NH4=0.6MG N/L,RIVER ARROW	1262
*CLADOPHORA SP.	73.7 DRY WEIGHT G/M2 (ANNUAL MEAN)	ANNUAL MEAN:T.INOR.P=2.1MG/L,NO3=2.7MG N/L,NH4=0.4MG N/L,RIVER ARROW	1262
*CLADOPHORA SP.	5.0 DRY WEIGHT G/M2 (ANNUAL MEAN)	ANNUAL MEAN:T.INOR.P=0.7MG/L,NO3=4.9MG N/L,NH4=0.3MG N/L,RIVER COLE	1262
*CLADOPHORA SP.	50.1 DRY WEIGHT G/M2 (ANNUAL MEAN)	ANNUAL MEAN:T.INOR.P=1.8MG/L,NO3=4.3MG N/L,NH4=0.3MG N/L,RIVER COLE	1262
*CLADOPHORA SP.	60.3 DRY WEIGHT G/M2 (ANNUAL MEAN)	ANNUAL MEAN:T.INOR.P=1.5MG/L,NO3=4.3MG N/L,NH4=0.9MG N/L,RIVER BLYTHE	1262
*CLADOPHORA SP.	52.6 DRY WEIGHT G/M2 (ANNUAL MEAN)	ANNUAL MEAN:T.INOR.P=2.9MG/L,NO3=6.7MG N/L,NH4=0.9MG N/L,RIVER BLYTHE	1262
*CLADOPHORA SP.	54.8 DRY WEIGHT G/M2 (ANNUAL MEAN)	ANNUAL MEAN:T.INOR.P=2.8MG/L,NO3=6.6MG N/L,NH4=0.9MG N/L,RIVER BLYTHE	1262
*CLADOPHORA SP.	15.5 DRY WEIGHT G/M2 (ANNUAL MEAN)	ANNUAL MEAN:T.INOR.P=0.8MG/L,NO3=3.5MG N/L,NH4=0.4MG N/L,RIVER BLITHE	1262
*CLADOPHORA SP.	20.3 DRY WEIGHT G/M2 (ANNUAL MEAN)	ANNUAL MEAN:T.INOR.P=0.8MG/L,NO3=4.4MG N/L,NH4=0.4MG N/L,RIVER TEAN	1262
*CLADOPHORA SP.	23.0 DRY WEIGHT G/M2 (ANNUAL MEAN)	ANNUAL MEAN:T.INOR.P=1.1MG/L,NO3=4.7MG N/L,NH4=0.5MG N/L,RIVER TEAN	1262
*CLADOPHORA SP.	31.6 DRY WEIGHT G/M2 (ANNUAL MEAN)	ANNUAL MEAN:T.INOR.P=2.3MG/L,NO3=7.4MG N/L,NH4=0.7MG N/L,RIVER TEAN	1262
*CLADOPHORA SP.	14.5 DRY WEIGHT G/M2 (ANNUAL MEAN)	ANNUAL MEAN:T.INOR.P=2.3MG/L,NO3=7.4MG N/L,NH4=0.7MG N/L,RIVER RAY	1262
*CLADOPHORA SP.	66.4 DRY WEIGHT G/M2 (ANNUAL MEAN)	ANNUAL MEAN:T.INOR.P=4.6MG/L,NO3=12.5MG N/L,NH4=2.8MG N/L,RIVER RAY	1262
*CLADOPHORA SP.	5.0 DRY WEIGHT G/M2 (ANNUAL MEAN)	ANNUAL MEAN:T.INOR.P=1.3MG/L,NO3=7.0MG N/L,NH4=0.2MG N/L,R. GR. STOUR	1262

TABLE A11

ALGAE : BIOMASS

ITEM:	VALUE:	CONDITIONS:	REF.:
*CLADOPHORA SP.	37.4 DRY WEIGHT G/M2 (ANNUAL MEAN)	ANNUAL MEAN:T.INOR.P=1.7MG/L,NO3= 6.7MG N/L,NH4=0.05MG N/L,R. GR. STOU	1262
*CLADOPHORA SP.	74.7 DRY WEIGHT G/M2 (ANNUAL MEAN)	ANNUAL MEAN:T.INOR.P=1.6MG/L,NO3= 7.0MG N/L,NH4=0.2MG N/L,R. GR. STOUR	1262
*CLADOPHORA SP.	62.9 DRY WEIGHT G/M2 (ANNUAL MEAN)	ANNUAL MEAN:T.INOR.P=1.0MG/L,NO3= 4.8MG N/L,NH4=0.05MG N/L,RIVER DAREN	1262
*VALONIA VENTRICULOSA	68 G	THE LARGEST UNICELLULAR ALGA	175

TABLE A12

ALGAE : BIOTIN/BIOMASS RATIO

ITEM:	VALUE:	CONDITIONS:	REF.:
*MARINE PHYTOPLANKTON	3.2 10**(-16) G/CELL	NO3 AS SOURCE OF N, STATIONARY PHASE,293 K, SOUTHERN CALIF	428
*MARINE PHYTOPLANKTON	1.4 10**(-16) G/CELL	NH4 AS SOURCE OF N, STATIONARY PHASE,293 K, SOUTHERN CALIF	428
*MARINE PHYTOPLANKTON	1.9 10**(-16) G/CELL	UREA AS SOURCE OF N, STATIONARY PHASE,293 K, SOUTHERN CALIF	428

TABLE A13

ALGAE : C/BIOMASS RATIO

ITEM:	VALUE:	CONDITIONS:	REF.:
*AGMENELLUM QUADRUPLICATUM	42.0,44.9 % OF DRY WEIGHT	MARINE ALGAE, TWO DIFF. METHODS OF SAMPLING, SEE REF.	355
*ALGAE	.4-.5 FRACTION OF DRY WEIGHT	NONE	1571
*ALGAE	22.5 % OF DRY WEIGHT	MEAN ELEMENTARY COMPOSITION	157
*AMPHIDINIUM CARTERI	40.7,47.1 % OF DRY WEIGHT	MARINE ALGAE, TWO DIFF. METHODS OF SAMPLING, SEE REF.	355
*AMPHIPLEURA RUTILANS	46.42 % OF DRY WEIGHT	MARINE ALGAE, SEE REF.	355
*ANABAENA SP.	49.70 % OF DRY WEIGHT	ALGAE FROM PONDS AND LAKES IN SOUTHEASTERN UNITED STATES	1660
*ANAGAENOPSIS	49.57 % OF DRY WEIGHT	FRESHWATER ALGAE , SEE REF.	355
*APHANIZOMENON SP.	47.65 % OF DRY WEIGHT	ALGAE FROM PONDS AND LAKES IN SOUTHEASTERN UNITED STATES	1660
*CHAETOCEROS LAUDERI	49.3 MICROGRAM C/MM**3	CULTIVATED, MARINE	104
*CHAETOCEROS SP.	31.0 % OF DRY WEIGHT	MARINE ALGAE, SEE REF.	355
*CHARA SP.	29.3 % OF DRY WEIGHT	ALGAE FROM PONDS AND LAKES IN SOUTHEASTERN UNITED STATES	1660
*CHLAMYDOMONAS MAGNUSII	200 MICROGRAM C/MM**3	CULTIVATED, MARINE	104
*CHLAMYDOMONAS SP.	47.46 % OF DRY WEIGHT	FRESHWATER ALGAE , SEE REF.	355
*CHLORELLA PYRENOIDOSA	49.51 % OF DRY WEIGHT	FRESHWATER ALGAE , SEE REF.	355
*CHLORELLA PYRENOIDOSA	70.17 % OF DRY WEIGHT	FRESHWATER ALGAE , SEE REF.	355
*CHLORELLA PYRENOIDOSA	54.89 % OF DRY WEIGHT	FRESHWATER ALGAE , SEE REF.	355
*CHLORELLA VULGARIS	53.37 % OF DRY WEIGHT	FRESHWATER ALGAE , SEE REF.	355
*CLADOPHORA SP.	35.27 % OF DRY WEIGHT	ALGAE FROM PONDS AND LAKES IN SOUTHEASTERN UNITED STATES	1660
*COSCINODISCUS ANGSTII	840*10**(-6) MICROGRAM C/CELL	DIAMETER = 35 U	493
*COSCINODISCUS ANGSTII VAR. GRANULOMARGINATUS	3334*10**(-6) MICROGRAM C/CELL	DIAMETER = 87 U, CENTRIC	493
*COSCINODISCUS EXCENTRICUS	1644*10**(-6) MICROGRAM C/CELL	DIAMETER = 75 U	493
*COSCINODISCUS SP.	15.9,18.8 % OF DRY WEIGHT	MARINE ALGAE, TWO DIFF. METHODS OF SAMPLING, SEE REF.	355

TABLE A13

ALGAE : C/BIOMASS RATIO

ITEM:	VALUE:	CONDITIONS:	REF.:
*COSCINODISCUS SP.	53.5 MICROGRAM C/MM**3	CULTIVATED, MARINE	104
*CRYPTOMONAS PSEUDOBALTICA	167 MICROGRAM C/MM**3	CULTIVATED, MARINE	104
*DUNALIELLA SALINA	39.6,45.4 % OF DRY WEIGHT	MARINE ALGAE, TWO DIFF. METHODS OF SAMPLING, SEE REF.	355
*EGREGIA MENZIESII	49.97 % OF DRY WEIGHT	MARINE ALGAE, SEE REF.	355
*EUGLENA SP.	48.14 % OF DRY WEIGHT	ALGAE FROM PONDS AND LAKES IN SOUTHEASTERN UNITED STATES	1660
*EXUVIAELLA SP.	44.5,47.4 % OF DRY WEIGHT	MARINE ALGAE, TWO DIFF. METHODS OF SAMPLING, SEE REF.	355
*GIGARTINA AGARDHII	42.90 % OF DRY WEIGHT	MARINE ALGAE, SEE REF.	355
*GONYAULAX POLYEDRA	154.9 (SD=+-12.7) MICROGRAM C/ MM**3	N = 10	1204
*GYMNODINIUM SPLENDENS	129.2 (SD=+-5.3) MICROGRAM C/MM**3	N = 5	1204
*HETEROTHRIX SP.	351 MICROGRAM C/MM**3	CULTIVATED, MARINE	104
*HYDRODICTYON SP.	39.96 % OF DRY WEIGHT	ALGAE FROM PONDS AND LAKES IN SOUTHEASTERN UNITED STATES	1660
*LAUDERIA BOREALIS	85.0 (SD=+-13.8) MICROGRAM C/MM**3	CELL DIAMETER = 19 U, N=66	1204
*LAUDERIA BOREALIS	33.0 (SD=+-3.8) MICROGRAM C/MM**3	CELL DIAMETER = 36 U, N=6	1204
*LYNGBYA SP.	40.23 % OF DRY WEIGHT	ALGAE FROM PONDS AND LAKES IN SOUTHEASTERN UNITED STATES	1660
*MACROCYSTIS PYRIFERA	46.02 % OF DRY WEIGHT	MARINE ALGAE, SEE REF.	355
*MARINE PHYTOPLANKTON	220*10**(-12) G/CELL	NO3 AS SOURCE OF N, STATIONARY PHASE,293 K, SOUTHERN CALIF	428
*MARINE PHYTOPLANKTON	225*10**(-12) G/CELL	NH4 AS SOURCE OF N, STATIONARY PHASE,293 K, SOUTHERN CALIF	428
*MARINE PHYTOPLANKTON	154*10**(-12) G/CELL	UREA AS SOURCE OF N, STATIONARY PHASE,293 K, SOUTHERN CALIF	428
*MICROCYSTIS SP.	46.46 % OF DRY WEIGHT	ALGAE FROM PONDS AND LAKES IN SOUTHEASTERN UNITED STATES	1660
*MONOCHRYSIS LUTHERI	53.2,44.6 % OF DRY WEIGHT	MARINE ALGAE, TWO DIFF. METHODS OF SAMPLING, SEE REF.	355
*MOUGEOTIA SP.	40.74 % OF DRY WEIGHT	ALGAE FROM PONDS AND LAKES IN SOUTHEASTERN UNITED STATES	1660
*NAVICULA TORQUATUM	48.45 % OF DRY WEIGHT	MARINE ALGAE, SEE REF.	355
*NITELLA SP.	38.43 % OF DRY WEIGHT	ALGAE FROM PONDS AND LAKES IN SOUTHEASTERN UNITED STATES	1660
*NITZSCHIA CLOSTERIUM	54.98 % OF DRY WEIGHT	FRESHWATER ALGAE , SEE REF.	355
*OEDOGONIUM SP.	40.84 % OF DRY WEIGHT	ALGAE FROM PONDS AND LAKES IN SOUTHEASTERN UNITED STATES	1660
*OICOMONOS TERMO	51.76 % OF DRY WEIGHT	FRESHWATER ALGAE , SEE REF.	355
*PHAEODACTYLUM TRICORNUTUM	37.5,49.5 % OF DRY WEIGHT	MARINE ALGAE, TWO DIFF. METHODS OF SAMPLING, SEE REF.	355
*PHAEODACTYLUM TRICORNUTUM	160 MICROGRAM C/MM**3	CULTIVATED, MARINE	104
*PHAEOPHYCEAE	34.5 % OF DRY WEIGHT	MEAN ELEMENTARY COMPOSITION	157
*PITHOPHORA SP.	35.38 % OF DRY WEIGHT	ALGAE FROM PONDS AND LAKES IN SOUTHEASTERN UNITED STATES	1660
*PROROCENTRUM MICANS	213.8 (SD=+-17.7) MICROGRAM C/ MM**3	N = 11	1204
*RHIZOCLONIUM SP.	39.10 % OF DRY WEIGHT	ALGAE FROM PONDS AND LAKES IN SOUTHEASTERN UNITED STATES	1660
*SCENEDESMUS BASILENSIS	54.82 % OF DRY WEIGHT	FRESHWATER ALGAE , SEE REF.	355
*SCENEDESMUS OBLICUUS	54.87 % OF DRY WEIGHT	FRESHWATER ALGAE , SEE REF.	355
*SCENEDESMUS OBLIQUUS	54.55 % OF DRY WEIGHT	FRESHWATER ALGAE , SEE REF.	355
*SKELETONEMA COSTATUM	26.0,31.3 % OF DRY WEIGHT	MARINE ALGAE, TWO DIFF. METHODS OF SAMPLING, SEE REF.	355
*SPIROGYRA SP.	42.40 % OF DRY WEIGHT	ALGAE FROM PONDS AND LAKES IN SOUTHEASTERN UNITED STATES	1660
*STICHOCOCCUS BACILLARIS	52.65 % OF DRY WEIGHT	FRESHWATER ALGAE , SEE REF.	355
*STICHOCOCCUS BACILLARIS	57.14 % OF DRY WEIGHT	FRESHWATER ALGAE , SEE REF.	355
*STICHOCOCCUS BACILLARIS	53.82 % OF DRY WEIGHT	FRESHWATER ALGAE , SEE REF.	KETC
*SYRACOSPHAERA CARTERAE	34.6,35.2 % OF DRY WEIGHT	MARINE ALGAE, TWO DIFF. METHODS OF SAMPLING, SEE REF.	355

TABLE A13

ALGAE : C/BIOMASS RATIO			
ITEM:	VALUE:	CONDITIONS:	REF.:
*TETRASELMIS MACULATA	35.1,36.8 % OF DRY WEIGHT	MARINE ALGAE, TWO DIFF. METHODS OF SAMPLING, SEE REF.	355
*TETRASELMIS STRICIA	116 MICROGRAM C/MM**3	CULTIVATED, MARINE	104
*THALASSIOSIRA FLUVIATILIS	94*10**(-6) MICROGRAM C/CELL	DIAMETER = 11 U	493
*THALASSIOSIRA PSEUDONANA	8*10**(-6) MICROGRAM C/CELL	DIAMETER = 3.8 U	493
*THALASSIOSIRA SP.	2.5*10**(-13) MOLES/CELL	GROWTH RATE=0.041 1/H, CELL DENSITY=4.0*10**8 CELL/L, P-LIMITED CHEMO.	1243
*THALASSIOSIRA SP.	4.5*10**(-13) MOLES/CELL	GROWTH RATE=0.017 1/H, CELL DENSITY=6.7*10**8 CELL/L, P-LIMITED CHEMO.	1243
*TRICHODESMIUM SPP.	9.7 MICROGRAM	C/COLONY,N.PACIFIC,N-FIXING ALGAE, N=6	967
*ULVA SP.	44.36 % OF DRY WEIGHT	MARINE ALGAE, SEE REF.	355

TABLE A14

ALGAE : C/BIOTIN RATIO			
ITEM:	VALUE:	CONDITIONS:	REF.:
*MARINE PHYTOPLANKTON	5.9*10**5 BY WEIGHT	NO3 AS SOURCE OF N, LOG PHASE,293 K, SOUTHERN CALIF	428
*MARINE PHYTOPLANKTON	6.8*10**5 BY WEIGHT	NO3 AS SOURCE OF N, STATIONARY PHASE,293 K, SOUTHERN CALIF	428
*MARINE PHYTOPLANKTON	5.4*10**5 BY WEIGHT	NH4 AS SOURCE OF N, LOG PHASE,293 K, SOUTHERN CALIF	428
*MARINE PHYTOPLANKTON	1.6*10**6 BY WEIGHT	NH4 AS SOURCE OF N, STATIONARY PHASE,293 K, SOUTHERN CALIF	428
*MARINE PHYTOPLANKTON	5.9*10**5 BY WEIGHT	UREA AS SOURCE OF N, LOG PHASE,293 K, SOUTHERN CALIF	428
*MARINE PHYTOPLANKTON	8.2*10**5 BY WEIGHT	UREA AS SOURCE OF N, STATIONARY PHASE,293 K, SOUTHERN CALIF	428

TABLE A15

ALGAE : C/CHLOROPHYLL A RATIO			
ITEM:	VALUE:	CONDITIONS:	REF.:
*ALGAE	0.05 MG-C/MICROGRAM-CHL.	LAKE ONTARIO	1577
*ALGAE	0.05 MG-C/MICROGRAM-CHL.	POTOMAC RIVER	1153
*CHAETOCEROS DEBILIS	76 BY WEIGHT	SI-STARVED,72H,CONT.LIGHT:0.08LY/MIN,DILUTION=1/3-1/2 OF MAX.GR.RATE	631
*CHAETOCEROS DEBILIS	55 BY WEIGHT	SI-LIMITED,CONT.LIGHT:0.08LY/MIN, DILUTION=1/3-1/2 OF MAX.GR.RATE	631
*CHAETOCEROS DEBILIS	32 BY WEIGHT	NON-LIMITED,CONT.LIGHT:0.08LY/MIN, DILUTION=1/3-1/2 OF MAX.GR.RATE	631
*CHAETOCEROS DEBILIS	47 BY WEIGHT	NH4-LIMITED,CONT.LIGHT:0.08LY/MIN, DILUTION=1/3-1/2 OF MAX.GR.RATE	631
*CHAETOCEROS DEBILIS	43 BY WEIGHT	NH4-STARVED,72H,CONT.LIGHT:0.08LY/MIN,DILUTION=1/3-1/2 OF MAX.GR.RATE	631
*CHAETOCEROS DEBILIS	57.4-53.6 BY WEIGHT	NITROGEN-LIMITED,DILUTION RATE= 0.097-0.008/HR	1582
*CYCLOTELLA NANA	76-500 BY WEIGHT	NITRATE-LIMITED,DILUTION RATE= 0.077-0.009/HR	225
*MARINE PHYTOPLANKTON	90 BY WEIGHT	NO3 AS SOURCE OF N, LOG PHASE,293 K, SOUTHERN CALIF	428
*MARINE PHYTOPLANKTON	46 BY WEIGHT	NO3 AS SOURCE OF N, STATIONARY PHASE,293 K, SOUTHERN CALIF	428

TABLE A15

ALGAE : C/CHLOROPHYLL A RATIO			
ITEM:	VALUE:	CONDITIONS:	REF.:
*MARINE PHYTOPLANKTON	112 BY WEIGHT	NH4 AS SOURCE OF N, LOG PHASE,293 K, SOUTHERN CALIF	428
*MARINE PHYTOPLANKTON	58 BY WEIGHT	NH4 AS SOURCE OF N, STATIONARY PHASE,293 K, SOUTHERN CALIF	428
*MARINE PHYTOPLANKTON	96 BY WEIGHT	UREA AS SOURCE OF N, LOG PHASE,293 K, SOUTHERN CALIF	428
*MARINE PHYTOPLANKTON	77 BY WEIGHT	UREA AS SOURCE OF N, STATIONARY PHASE,293 K, SOUTHERN CALIF	428
*MARINE PHYTOPLANKTON	25 BY WEIGHT	VIGOROUSLY GROWING,EXCESS NITRATE IN WATER	39
*MARINE PHYTOPLANKTON	60 BY WEIGHT	UNHEALTHY PHYTOPLANKTON,NITRATE DEPLETED WATER	39
*MARINE PHYTOPLANKTON	30 BY WEIGHT	SEE REF.	1520
*SKELETONEMA COSTATUM	178 BY WEIGHT	SI-STARVED,72H,CONT.LIGHT:0.08LY/MIN,DILUTION=1/3-1/2 OF MAX.GR.RATE	631
*SKELETONEMA COSTATUM	65 BY WEIGHT	SI-LIMITED,CONT.LIGHT:0.08LY/MIN, DILUTION=1/3-1/2 OF MAX.GR.RATE	631
*SKELETONEMA COSTATUM	61 BY WEIGHT	NON-LIMITED,CONT.LIGHT:0.08LY/MIN, DILUTION=1/3-1/2 OF MAX.GR.RATE	631
*SKELETONEMA COSTATUM	72 BY WEIGHT	NH4-LIMITED,CONT.LIGHT:0.08LY/MIN, DILUTION=1/3-1/2 OF MAX.GR.RATE	631
*SKELETONEMA COSTATUM	98 BY WEIGHT	NH4-STARVED,72H,CONT.LIGHT:0.08LY/MIN,DILUTION=1/3-1/2 OF MAX.GR.RATE	631
*SKELETONEMA COSTATUM	70-220 BY WEIGHT	NITROGEN-LIMITED,BATCH CULTURE, 72-115HOURS STARVATION	686
*THALASSIOSIRA FLUVIATILIS	18-26 BY WEIGHT	NITRATE-LIMITED,BATCH CULTURE, 72-115HOURS STARVATION	673
*THALASSIOSIRA GRAVIDA	46 BY WEIGHT	SI-STARVED,72H,CONT.LIGHT:0.08LY/MIN,DILUTION=1/3-1/2 OF MAX.GR.RATE	631
*THALASSIOSIRA GRAVIDA	34 BY WEIGHT	SI-LIMITED,CONT.LIGHT:0.08LY/MIN, DILUTION=1/3-1/2 OF MAX.GR.RATE	631
*THALASSIOSIRA GRAVIDA	26 BY WEIGHT	NON-LIMITED,CONT.LIGHT:0.08LY/MIN, DILUTION=1/3-1/2 OF MAX.GR.RATE	631
*THALASSIOSIRA GRAVIDA	55 BY WEIGHT	NH4-LIMITED,CONT.LIGHT:0.08LY/MIN, DILUTION=1/3-1/2 OF MAX.GR.RATE	631
*THALASSIOSIRA GRAVIDA	106 BY WEIGHT	NH4-STARVED,72H,CONT.LIGHT:0.08LY/MIN,DILUTION=1/3-1/2 OF MAX.GR.RATE	631
*THALASSIOSIRA PSEUDONANA	28.6 BY WEIGHT	P-LIMITED CHEMOSTAT,GROWTH=0.041 1/H, 295 K,0.07 CAL/CM2*MIN	1243
*THALASSIOSIRA PSEUDONANA	66.7 BY WEIGHT	P-LIMITED CHEMOSTAT,GROWTH=0.017 1/H, 295 K,0.07 CAL/CM2*MIN	1243
*THALASSIOSIRA PSEUDONANA	28.7 BY WEIGHT	N-LIMITED CHEMOSTAT,GROWTH=0.041 1/H, 295 K,0.07 CAL/CM2*MIN	1243
*THALASSIOSIRA PSEUDONANA	44.3 BY WEIGHT	N-LIMITED CHEMOSTAT,GROWTH=0.017 1/H, 295 K,0.07 CAL/CM2*MIN	1243
*THALASSIOSIRA PSEUDONANA	91.4 BY WEIGHT	N-LIMITED CHEMOSTAT,GROWTH=0.0085 1/H,295 K,0.07 CAL/CM2*MIN	1243
*THALASSIOSIRA PSEUDONANA	28.7-91.4 BY WEIGHT	NITROGEN-LIMITED,DILUTION RATE= 0.041-0.008/HR	423
*THALASSIOSIRA PSEUDONANA	111-200 BY WEIGHT	NITRATE-LIMITED,BATCH CULTURE, 72-115HOURS STARVATION	673

TABLE A16

ALGAE : C/N RATIO			
ITEM:	VALUE:	CONDITIONS:	REF.:
*ALGAE	6.25 BY WEIGHT	MOSSDALE	1153
*ALGAE	20 - 5.88 BY WEIGHT	USED IN MODELS, 1976	55
*ALGAE	5 BY WEIGHT	USED IN MODELS, LAKE MICHIGAN	220

TABLE A16

ALGAE : C/N RATIO

ITEM:	VALUE:	CONDITIONS:	REF.:
*ALGAE	7.14 BY WEIGHT	USED IN MODELS, SAN JOAQUIN RIVER, CALIF.	1150
*ALGAE	7.14 BY WEIGHT	USED IN MODELS, AN JOAQUIN RIVER ESTUARY, CALIF.	1150
*ALGAE	5 BY WEIGHT	USED IN MODELS, POTOMAC ESTUARY	1150
*ALGAE	50/7 BY WEIGHT	USED IN MODELS, LAKE ERIE	1150
*ALGAE	5 BY WEIGHT	USED IN MODELS, LAKE ONTARIO	1150
*ALGAE	5.88 BY WEIGHT	USED IN MODELS, GRAYS HARBOR, WASHINGTON	78
*CHAETOCEROS DEBILIS	4.1 BY ATOMS	SI-STARVED,72H,CONT.LIGHT:0.08LY/MIN,DILUTION=1/3-1/2 OF MAX.GR.RATE	631
*CHAETOCEROS DEBILIS	5.4 BY ATOMS	SI-LIMITED,CONT.LIGHT:0.08LY/MIN, DILUTION=1/3-1/2 OF MAX.GR.RATE	631
*CHAETOCEROS DEBILIS	3.2 BY ATOMS	NON-LIMITED,CONT.LIGHT:0.08LY/MIN, DILUTION=1/3-1/2 OF MAX.GR.RATE	631
*CHAETOCEROS DEBILIS	6.4 BY ATOMS	NH4-LIMITED,CONT.LIGHT:0.08LY/MIN, DILUTION=1/3-1/2 OF MAX.GR.RATE	631
*CHAETOCEROS DEBILIS	4.0 BY ATOMS	NH4-STARVED,72H,CONT.LIGHT:0.08LY/MIN,DILUTION=1/3-1/2 OF MAX.GR.RATE	631
*CHAETOCEROS DEBILIS	9.8-12.9 BY ATOMS	NITROGEN-LIMITED,DILUTION RATE= 0.097-0.008/HR	1582
*CHAETOCEROS LAUDERI	5.6 BY ATOMS	GROWN UNDER OPTIMAL CONDITION	355
*CHAETOCEROS LAUDERI	4.79 BY WEIGHT	CULTIVATED, MARINE	104
*CHAETOCEROS SP.	6.6 BY ATOMS	GROWN UNDER OPTIMAL CONDITION	1220
*CHLAMYDOMONAS MAGNUSII	6.43 BY WEIGHT	CULTIVATED, MARINE	104
*COSCINODISCUS SP.	5.3 BY ATOMS	GROWN UNDER OPTIMAL CONDITION	355
*COSCINODISCUS SP.	4.56 BY WEIGHT	CULTIVATED, MARINE	104
*CRYPTOMONAS PSEUDOBALTICA	3.52 BY WEIGHT	CULTIVATED, MARINE	104
*CYCLOTELLA NANA	7.7-20.0 BY ATOMS	NITRATE-LIMITED,DILUTION RATE= 0.077-0.009/HR	225
*DITYLUM BRIGHTWELLII	5.2 BY ATOMS	GROWN UNDER OPTIMAL CONDITION	1522
*HETEROTHRIX SP.	7.51 BY WEIGHT	CULTIVATED, MARINE	104
*MARINE PHYTOPLANKTON	5.9BY WEIGHT	NO3 AS SOURCE OF N, LOG PHASE,293 K, SOUTHERN CALIF	428
*MARINE PHYTOPLANKTON	5.2BY WEIGHT	NO3 AS SOURCE OF N, STATIONARY PHASE,293 K, SOUTHERN CALIF	428
*MARINE PHYTOPLANKTON	3.9BY WEIGHT	NH4 AS SOURCE OF N, LOG PHASE,293 K, SOUTHERN CALIF	428
*MARINE PHYTOPLANKTON	4.3BY WEIGHT	NH4 AS SOURCE OF N, STATIONARY PHASE,293 K, SOUTHERN CALIF	428
*MARINE PHYTOPLANKTON	5.6BY WEIGHT	UREA AS SOURCE OF N, LOG PHASE,293 K, SOUTHERN CALIF	428
*MARINE PHYTOPLANKTON	7.0BY WEIGHT	UREA AS SOURCE OF N, STATIONARY PHASE,293 K, SOUTHERN CALIF	428
*MARINE PHYTOPLANKTON	3 BY WEIGHT	VIGOROUSLY GROWING,EXCESS NITRATE IN WATER	39
*MARINE PHYTOPLANKTON	4.5 BY WEIGHT	UNHEALTHY PHYTOPLANKTON,NITRATE DEPLETED WATER	39
*MARINE PHYTOPLANKTON	6+-2 BY WEIGHT	SEE REF.	1520
*PHAEODACTYLUM TRICORNUTUM	4.9 BY ATOMS	GROWN UNDER OPTIMAL CONDITION	355
*PHAEODACTYLUM TRICORNUTUM	4.23 BY WEIGHT	CULTIVATED, MARINE	104
*SKELETONEMA COSTATUM	13.2 BY ATOMS	SI-STARVED,72H,CONT.LIGHT:0.08LY/MIN,DILUTION=1/3-1/2 OF MAX.GR.RATE	631
*SKELETONEMA COSTATUM	6.8 BY ATOMS	SI-LIMITED,CONT.LIGHT:0.08LY/MIN, DILUTION=1/3-1/2 OF MAX.GR.RATE	631
*SKELETONEMA COSTATUM	5.1 BY ATOMS	NON-LIMITED,CONT.LIGHT:0.08LY/MIN, DILUTION=1/3-1/2 OF MAX.GR.RATE	631
*SKELETONEMA COSTATUM	9.5 BY ATOMS	NH4-LIMITED,CONT.LIGHT:0.08LY/MIN, DILUTION=1/3-1/2 OF MAX.GR.RATE	631
*SKELETONEMA COSTATUM	14.8 BY ATOMS	NH4-STARVED,72H,CONT.LIGHT:0.08LY/MIN,DILUTION=1/3-1/2 OF MAX.GR.RATE	631

TABLE A16

ALGAE : C/N RATIO			
ITEM:	VALUE:	CONDITIONS:	REF.:
*SKELETONEMA COSTATUM	4.8-9.4 BY ATOMS	NITROGEN-LIMITED,BATCH CULTURE, 72-115HOURS STARVATION	630
*SKELETONEMA COSTATUM	5.3 BY ATOMS	GROWN UNDER OPTIMAL CONDITION	1220
*TETRASELMIS STRICIA	3.78 BY WEIGHT	CULTIVATED, MARINE	104
*THALASSIOSIRA FLUVIATILIS	6.3-12.0 BY ATOMS	NITRATE-LIMITED,BATCH CULTURE, 72-115HOURS STARVATION	686
*THALASSIOSIRA GRAVIDA	4.6 BY ATOMS	SI-STARVED,72H,CONT.LIGHT:0.08LY/MIN,DILUTION=1/3-1/2 OF MAX.GR.RATE	631
*THALASSIOSIRA GRAVIDA	4.1 BY ATOMS	SI-LIMITED,CONT.LIGHT:0.08LY/MIN, DILUTION=1/3-1/2 OF MAX.GR.RATE	631
*THALASSIOSIRA GRAVIDA	2.7 BY ATOMS	NON-LIMITED,CONT.LIGHT:0.08LY/MIN, DILUTION=1/3-1/2 OF MAX.GR.RATE	631
*THALASSIOSIRA GRAVIDA	12.3 BY ATOMS	NH4-LIMITED,CONT.LIGHT:0.08LY/MIN, DILUTION=1/3-1/2 OF MAX.GR.RATE	631
*THALASSIOSIRA GRAVIDA	3.5 BY ATOMS	NH4-STARVED,72H,CONT.LIGHT:0.08LY/MIN,DILUTION=1/3-1/2 OF MAX.GR.RATE	631
*THALASSIOSIRA PSEUDONANA	5.2 BY ATOMS	GROWN UNDER OPTIMAL CONDITION	355
*THALASSIOSIRA PSEUDONANA	5.8 BY ATOMS	P-LIMITED CHEMOSTAT,GROWTH=0.041 1/H, 295 K,0.07 CAL/CM2*MIN	1243
*THALASSIOSIRA PSEUDONANA	10.7 BY ATOMS	P-LIMITED CHEMOSTAT,GROWTH=0.017 1/H, 295 K,0.07 CAL/CM2*MIN	1243
*THALASSIOSIRA PSEUDONANA	5.7 BY ATOMS	N-LIMITED CHEMOSTAT,GROWTH=0.041 1/H, 295 K,0.07 CAL/CM2*MIN	1243
*THALASSIOSIRA PSEUDONANA	10.0 BY ATOMS	N-LIMITED CHEMOSTAT,GROWTH=0.017 1/H, 295 K,0.07 CAL/CM2*MIN	1243
*THALASSIOSIRA PSEUDONANA	14.8 BY ATOMS	N-LIMITED CHEMOSTAT,GROWTH=0.0085 1/H,295 K,0.07 CAL/CM2*MIN	1243
*THALASSIOSIRA PSEUDONANA	5.7-14.8 BY ATOMS	NITROGEN-LIMITED,DILUTION RATE= 0.041-0.008/HR	423
*THALASSIOSIRA PSEUDONANA	4.8-10.0 BY ATOMS	NITRATE-LIMITED,BATCH CULTURE, 72-115HOURS STARVATION	673
*TRICHODESMIUM SPP.	4.1 BY WEIGHT	N.PACIFIC,N-FIXING ALGAE,N=6	967

TABLE A17

ALGAE : C/N/P RATIO			
ITEM:	VALUE:	CONDITIONS:	REF.:
*TRICHODESMIUM SPP.	81:20:1 BY WEIGHT	N.PACIFIC,N-FIXING ALGAE,N=6	967

TABLE A18

ALGAE : C/P RATIO			
ITEM:	VALUE:	CONDITIONS:	REF.:
*ASTERIONELLA JAPONICA	200 BY ATOMS	BATCH CULTURE, CELL VOL=12 % CARBON	1568
*CHAETOCEROS DEBILIS	33 BY ATOMS	SI-STARVED,72H,CONT.LIGHT:0.08LY/MIN,DILUTION=1/3-1/2 OF MAX.GR.RATE	631
*CHAETOCEROS DEBILIS	43 BY ATOMS	SI-LIMITED,CONT.LIGHT:0.08LY/MIN, DILUTION=1/3-1/2 OF MAX.GR.RATE	631
*CHAETOCEROS DEBILIS	39 BY ATOMS	NON-LIMITED,CONT.LIGHT:0.08LY/MIN, DILUTION=1/3-1/2 OF MAX.GR.RATE	631
*CHAETOCEROS DEBILIS	29 BY ATOMS	NH4-LIMITED,CONT.LIGHT:0.08LY/MIN, DILUTION=1/3-1/2 OF MAX.GR.RATE	631
*CHAETOCEROS DEBILIS	24 BY ATOMS	NH4-STARVED,72H,CONT.LIGHT:0.08LY/MIN,DILUTION=1/3-1/2 OF MAX.GR.RATE	631

TABLE A18

ALGAE : C/P RATIO

ITEM:	VALUE:	CONDITIONS:	REF.:
*CHAETOCEROS GRACILIS	1000 BY ATOMS	BATCH CULTURE, CELL VOL=12 % CARBON	1568
*MARINE PHYTOPLANKTON	20 BY WEIGHT	VIGOROUSLY GROWING,EXCESS NITRATE IN WATER	39
*MARINE PHYTOPLANKTON	33 BY WEIGHT	UNHEALTHY PHYTOPLANKTON,NITRATE DEPLETED WATER	39
*MARINE PHYTOPLANKTON	40+-15 BY WEIGHT	SEE REF.	1520
*MONOCHRYSIS LUTHERI	143 BY ATOMS	BATCH CULTURE, CELL VOL=12 % CARBON	1568
*PHAEODACTYLUM TRICORNUTUM	0.0024 BY ATOMS	BATCH CULTURE, DRY CELL=50 % CARBON	1568
*SKELETONEMA COSTATUM	104 BY ATOM	SI-STARVED,72H,CONT.LIGHT:0.08LY/MIN,DILUTION=1/3-1/2 OF MAX.GR.RATE	631
*SKELETONEMA COSTATUM	57 BY ATOM	SI-LIMITED,CONT.LIGHT:0.08LY/MIN, DILUTION=1/3-1/2 OF MAX.GR.RATE	631
*SKELETONEMA COSTATUM	51 BY ATOM	NON-LIMITED,CONT.LIGHT:0.08LY/MIN, DILUTION=1/3-1/2 OF MAX.GR.RATE	631
*SKELETONEMA COSTATUM	46 BY ATOM	NH4-LIMITED,CONT.LIGHT:0.08LY/MIN, DILUTION=1/3-1/2 OF MAX.GR.RATE	631
*SKELETONEMA COSTATUM	62 BY ATOMS	NH4-STARVED,72H,CONT.LIGHT:0.08LY/MIN,DILUTION=1/3-1/2 OF MAX.GR.RATE	631
*THALASSIOSIRA GRAVIDA	51 BY ATOMS	SI-STARVED,72H,CONT.LIGHT:0.08LY/MIN,DILUTION=1/3-1/2 OF MAX.GR.RATE	631
*THALASSIOSIRA GRAVIDA	40 BY ATOMS	SI-LIMITED,CONT.LIGHT:0.08LY/MIN, DILUTION=1/3-1/2 OF MAX.GR.RATE	631
*THALASSIOSIRA GRAVIDA	24 BY ATOMS	NON-LIMITED,CONT.LIGHT:0.08LY/MIN, DILUTION=1/3-1/2 OF MAX.GR.RATE	631
*THALASSIOSIRA GRAVIDA	24 BY ATOMS	NH4-LIMITED,CONT.LIGHT:0.08LY/MIN, DILUTION=1/3-1/2 OF MAX.GR.RATE	631
*THALASSIOSIRA GRAVIDA	18 BY ATOMS	NH4-STARVED,72H,CONT.LIGHT:0.08LY/MIN,DILUTION=1/3-1/2 OF MAX.GR.RATE	631
*THALASSIOSIRA PSEUDONANA	220 BY ATOMS	P-LIMITED CHEMOSTAT,GROWTH=0.041 1/H, 295 K,0.07 CAL/CM2*MIN	1243
*THALASSIOSIRA PSEUDONANA	665 BY ATOMS	P-LIMITED CHEMOSTAT,GROWTH=0.017 1/H, 295 K,0.07 CAL/CM2*MIN	1243
*THALASSIOSIRA PSEUDONANA	63.4 BY ATOMS	N-LIMITED CHEMOSTAT,GROWTH=0.041 1/H, 295 K,0.07 CAL/CM2*MIN	1243
*THALASSIOSIRA PSEUDONANA	84.3 BY ATOMS	N-LIMITED CHEMOSTAT,GROWTH=0.017 1/H, 295 K,0.07 CAL/CM2*MIN	1243
*THALASSIOSIRA PSEUDONANA	90.5 BY ATOMS	N-LIMITED CHEMOSTAT,GROWTH=0.0085 1/H,295 K,0.07 CAL/CM2*MIN	1243
*THALASSIOSIRA PSEUDONANA	63.4-90.5 BY ATOMS	NITROGEN-LIMITED,DILUTION RATE= 0.041-0.008/HR	423

TABLE A19

ALGAE : C/SI RATIO

ITEM:	VALUE:	CONDITIONS:	REF.:
*BACILLARIOPHYCEAE	5 BY WEIGHT	NONE	220
*CHAETOCEROS DEBILIS	19.6 BY ATOMS	SI-STARVED,72H,CONT.LIGHT:0.08LY/MIN,DILUTION=1/3-1/2 OF MAX.GR.RATE	631
*CHAETOCEROS DEBILIS	22.5 BY ATOMS	SI-LIMITED,CONT.LIGHT:0.08LY/MIN, DILUTION=1/3-1/2 OF MAX.GR.RATE	631
*CHAETOCEROS DEBILIS	5.8 BY ATOMS	NON-LIMITED,CONT.LIGHT:0.08LY/MIN, DILUTION=1/3-1/2 OF MAX.GR.RATE	631
*CHAETOCEROS DEBILIS	3.0 BY ATOMS	NH4-LIMITED,CONT.LIGHT:0.08LY/MIN, DILUTION=1/3-1/2 OF MAX.GR.RATE	631
*CHAETOCEROS DEBILIS	3.6 BY ATOMS	NH4-STARVED,72H,CONT.LIGHT:0.08LY/MIN,DILUTION=1/3-1/2 OF MAX.GR.RATE	631

TABLE A19

ALGAE : C/SI RATIO			
ITEM:	VALUE:	CONDITIONS:	REF.:
*MARINE PHYTOPLANKTON	1.67 BY WEIGHT	VIGOROUSLY GROWING,EXCESS NITRATE IN WATER	39
*MARINE PHYTOPLANKTON	1.0 BY WEIGHT	UNHEALTHY PHYTOPLANKTON,NITRATE DEPLETED WATER	39
*MARINE PHYTOPLANKTON	1.25 BY WEIGHT	SEE REF.	1520
*SKELETONEMA COSTATUM	36.5 BY ATOMS	SI-STARVED,72H,CONT.LIGHT:0.08LY/MIN,DILUTION=1/3-1/2 OF MAX.GR.RATE	631
*SKELETONEMA COSTATUM	27.7 BY ATOMS	SI-LIMITED,CONT.LIGHT:0.08LY/MIN, DILUTION=1/3-1/2 OF MAX.GR.RATE	631
*SKELETONEMA COSTATUM	9.4 BY ATOMS	NON-LIMITED,CONT.LIGHT:0.08LY/MIN, DILUTION=1/3-1/2 OF MAX.GR.RATE	631
*SKELETONEMA COSTATUM	7.8 BY ATOMS	NH4-LIMITED,CONT.LIGHT:0.08LY/MIN, DILUTION=1/3-1/2 OF MAX.GR.RATE	631
*SKELETONEMA COSTATUM	11.9 BY ATOMS	NH4-STARVED,72H,CONT.LIGHT:0.08LY/MIN,DILUTION=1/3-1/2 OF MAX.GR.RATE	631
*THALASSIOSIRA GRAVIDA	39.2 BY ATOMS	SI-STARVED,72H,CONT.LIGHT:0.08LY/MIN,DILUTION=1/3-1/2 OF MAX.GR.RATE	631
*THALASSIOSIRA GRAVIDA	24.4 BY ATOMS	SI-LIMITED,CONT.LIGHT:0.08LY/MIN, DILUTION=1/3-1/2 OF MAX.GR.RATE	631
*THALASSIOSIRA GRAVIDA	6.3 BY ATOMS	NON-LIMITED,CONT.LIGHT:0.08LY/MIN, DILUTION=1/3-1/2 OF MAX.GR.RATE	631
*THALASSIOSIRA GRAVIDA	7.6 BY ATOMS	NH4-LIMITED,CONT.LIGHT:0.08LY/MIN, DILUTION=1/3-1/2 OF MAX.GR.RATE	631
*THALASSIOSIRA GRAVIDA	3.5 BY ATOMS	NH4-STARVED,72H,CONT.LIGHT:0.08LY/MIN,DILUTION=1/3-1/2 OF MAX.GR.RATE	631

TABLE A20

ALGAE : C/VITAMIN B 1 RATIO			
ITEM:	VALUE:	CONDITIONS:	REF.:
*MARINE PHYTOPLANKTON	5.8*10**4 BY WEIGHT	NO3 AS SOURCE OF N, LOG PHASE,293 K, SOUTHERN CALIF	428
*MARINE PHYTOPLANKTON	2.7*10**4 BY WEIGHT	NO3 AS SOURCE OF N, STATIONARY PHASE,293 K, SOUTHERN CALIF	428
*MARINE PHYTOPLANKTON	6.0*10**4 BY WEIGHT	NH4 AS SOURCE OF N, LOG PHASE,293 K, SOUTHERN CALIF	428
*MARINE PHYTOPLANKTON	3.5*10**4 BY WEIGHT	NH4 AS SOURCE OF N, STATIONARY PHASE,293 K, SOUTHERN CALIF	428
*MARINE PHYTOPLANKTON	6.7*10**4 BY WEIGHT	UREA AS SOURCE OF N, LOG PHASE,293 K, SOUTHERN CALIF	428
*MARINE PHYTOPLANKTON	4.1*10**4 BY WEIGHT	UREA AS SOURCE OF N, STATIONARY PHASE,293 K, SOUTHERN CALIF	428

TABLE A21

ALGAE : C/VITAMIN B 12 RATIO			
ITEM:	VALUE:	CONDITIONS:	REF.:
*MARINE PHYTOPLANKTON	8.7*10**5 BY WEIGHT	NO3 AS SOURCE OF N, LOG PHASE,293 K, SOUTHERN CALIF	428
*MARINE PHYTOPLANKTON	6.8*10**5 BY WEIGHT	NO3 AS SOURCE OF N, STATIONARY PHASE,293 K, SOUTHERN CALIF	428
*MARINE PHYTOPLANKTON	7.5*10**5 BY WEIGHT	NH4 AS SOURCE OF N, LOG PHASE,293 K, SOUTHERN CALIF	428
*MARINE PHYTOPLANKTON	1.4*10**6 BY WEIGHT	NH4 AS SOURCE OF N, STATIONARY PHASE,293 K, SOUTHERN CALIF	428
*MARINE PHYTOPLANKTON	7.8*10**5 BY WEIGHT	UREA AS SOURCE OF N, LOG PHASE,293 K, SOUTHERN CALIF	428

TABLE A21

ALGAE : C/VITAMIN B 12 RATIO			
ITEM:	VALUE:	CONDITIONS:	REF.:
*MARINE PHYTOPLANKTON	7.5*10**5 BY WEIGHT	NH4 AS SOURCE OF N, LOG PHASE,293 K, SOUTHERN CALIF	428
*MARINE PHYTOPLANKTON	1.4*10**6 BY WEIGHT	NH4 AS SOURCE OF N, STATIONARY PHASE,293 K, SOUTHERN CALIF	428
*MARINE PHYTOPLANKTON	7.8*10**5 BY WEIGHT	UREA AS SOURCE OF N, LOG PHASE,293 K, SOUTHERN CALIF	428
*MARINE PHYTOPLANKTON	7.7*10**5 BY WEIGHT	UREA AS SOURCE OF N, STATIONARY PHASE,293 K, SOUTHERN CALIF	428

TABLE A22

ALGAE : CA/BIOMASS RATIO			
ITEM:	VALUE:	CONDITIONS:	REF.:
*ANABAENA SP.	0.36 % OF DRY WEIGHT	ALGAE FROM PONDS AND LAKES IN SOUTHEASTERN UNITED STATES	1660
*APHANIZOMENON SP.	0.73 % OF DRY WEIGHT	ALGAE FROM PONDS AND LAKES IN SOUTHEASTERN UNITED STATES	1660
*CHARA SP.	8.03 % OF DRY WEIGHT	ALGAE FROM PONDS AND LAKES IN SOUTHEASTERN UNITED STATES	1660
*CHLOROPHYTA	0.157 MG/(GRAM OF ACID-SOLUBLE FRACTION)	AVERAGE OF 7 MARINE SPECIES,PUERTO RICO	386
*CLADOPHORA SP.	1.69 % OF DRY WEIGHT	ALGAE FROM PONDS AND LAKES IN SOUTHEASTERN UNITED STATES	1660
*EUGLENA SP.	0.05 % OF DRY WEIGHT	ALGAE FROM PONDS AND LAKES IN SOUTHEASTERN UNITED STATES	1660
*HYDRODICTYON SP.	0.69 % OF DRY WEIGHT	ALGAE FROM PONDS AND LAKES IN SOUTHEASTERN UNITED STATES	1660
*LYNGBYA SP.	0.45 % OF DRY WEIGHT	ALGAE FROM PONDS AND LAKES IN SOUTHEASTERN UNITED STATES	1660
*MICROCYSTIS SP.	0.53 % OF DRY WEIGHT	ALGAE FROM PONDS AND LAKES IN SOUTHEASTERN UNITED STATES	1660
*MOUGEOTIA SP.	1.68 % OF DRY WEIGHT	ALGAE FROM PONDS AND LAKES IN SOUTHEASTERN UNITED STATES	1660
*NITELLA SP.	1.89 % OF DRY WEIGHT	ALGAE FROM PONDS AND LAKES IN SOUTHEASTERN UNITED STATES	1660
*OEDOGONIUM SP.	0.44 % OF DRY WEIGHT	ALGAE FROM PONDS AND LAKES IN SOUTHEASTERN UNITED STATES	1660
*PHAEOPHYCEAE	0.099 GRAM/(GRAM OF ACID-SOLUBLE FRACTION)	AVERAGE OF 4 MARINE SPECIES,PUERTO RICO	386
*PITHOPHORA SP.	3.82 % OF DRY WEIGHT	ALGAE FROM PONDS AND LAKES IN SOUTHEASTERN UNITED STATES	1660
*RHIZOCLONIUM SP.	0.52 % OF DRY WEIGHT	ALGAE FROM PONDS AND LAKES IN SOUTHEASTERN UNITED STATES	1660
*RHODOPHYTA	0.106 GRAM/(GRAM OF ACID-SOLUBLE FRACTION)	AVERAGE OF 9 MARINE SPECIES,PUERTO RICO	386
*SPIROGYRA SP.	0.57 % OF DRY WEIGHT	ALGAE FROM PONDS AND LAKES IN SOUTHEASTERN UNITED STATES	1660
*SPIROGYRA SP., GIANT	0.84 % OF DRY WEIGHT	ALGAE FROM PONDS AND LAKES IN SOUTHEASTERN UNITED STATES	1660

TABLE A23

ALGAE : CALORIC CONTENT			
ITEM:	VALUE:	CONDITIONS:	REF.:
*GYMNODINIUM SPLENDENS	5708 (SD=+-166.1) KCAL/KG ASH-FREE DRY WEIGHT	N=5	1204
*LAUDERIA BOREALIS	5418 (SD=+-57.3) KCAL/KG ASH-FREE DRY WEIGHT	N=7	1204

TABLE A23

ALGAE : CALORIC CONTENT			
ITEM:	VALUE:	CONDITIONS:	REF.:
*MARINE PHYTOPLANKTON	314 CAL/100 MG ASH-FREE DRY WEIGHT	ADRIATIC SEA	69
*PROROCENTRUM MICANS	5218 (SD=+-52.0) KCAL/KG ASH-FREE DRY WEIGHT	N=4	1204
*THALASSIOSIRA ROTULA	5479 (SD=+-57.3) KCAL/KG ASH-FREE DRY WEIGHT	N=2	1204

TABLE A24

ALGAE : CARBOHYDRATE/BIOMASS RATIO			
ITEM:	VALUE:	CONDITIONS:	REF.:
*AGMENELLUM QUADRUPLICATUM	38 % OF ASH FREE DRY WEIGHT	BLUE GREEN ALGAE,MARINE,ANTHRONE REACTION	1220
*AMPHIDINIUM CARTERI	39 % OF ASH FREE DRY WEIGHT	DINOPHYCEAE,MARINE,ANTHRONE REACTION	1220
*DUNALIELLA SALINA	32 % OF ASH FREE DRY WEIGHT	CHLOROPHYCEAE,MARINE,ANTHRONE REACTION	1220
*EXUVIAELLA SP.	42 % OF ASH FREE DRY WEIGHT	DINOPHYCEAE,MARINE,ANTHRONE REACTION	1220
*MARINE PHYTOPLANKTON	0.2 BY WEIGHT	VIGOROUSLY GROWING,EXCESS NITRATE IN WATER	39
*MARINE PHYTOPLANKTON	0.11 BY WEIGHT	UNHEALTHY PHYTOPLANKTON,NITRATE DEPLETED WATER	39
*MONOCHRYSIS LUTHERI	34 % OF ASH FREE DRY WEIGHT	CHRYSOPHYCEAE,MARINE,ANTHRONE REACTION	1220
*SKELETONEMA COSTATUM	33 % OF ASH FREE DRY WEIGHT	BACILLARIOPHYCEAE,MARINE,ANTHRONE REACTION	1220
*TETRASELMIS MACULATA	20 % OF ASH FREE DRY WEIGHT	PRASINOPHYCEAE,MARINE,ANTHRONE REACTION	1220

TABLE A25

ALGAE : CARBON (INTRACELLULAR)			
ITEM:	VALUE:	CONDITIONS:	REF.:
*BACILLARIOPHYCEAE	33 % OF DRY WEIGHT	RANGE=19-50%	1521
*BLUE GREEN ALGAE	36 % OF DRY WEIGHT	RANGE=28-45%	1521
*CHLOROPHYCEAE	43 % OF DRY WEIGHT	RANGE=35-48%	1521
*CHRYSOPHYCEAE	40 % OF DRY WEIGHT	RANGE=35-45%	1521
*DINOPHYCEAE	43 % OF DRY WEIGHT	RANGE=37-47%	1521

TABLE A26

ALGAE : CARBON FIXATION/NITROGEN FIXATION RATIO			
ITEM:	VALUE:	CONDITIONS:	REF.:
*TRICHODESMIUM SPP.	1.2-2.1:1 BY WEIGHT	N.PACIFIC,N-FIXING ALGAE	967

TABLE A27

ALGAE : CAROTENOIDS/CHLOROPHYLL A RATIO			
ITEM:	VALUE:	CONDITIONS:	REF.:
*MARINE PHYTOPLANKTON	0.95 BY WEIGHT	VIGOROUSLY GROWING,EXCESS NITRATE IN WATER	39

TABLE A27

ALGAE : CAROTENOIDS/CHLOROPHYLL A RATIO			
ITEM:	VALUE:	CONDITIONS:	REF.:
*MARINE PHYTOPLANKTON	1.15 BY WEIGHT	UNHEALTHY PHYTOPLANKTON,NITRATE DEPLETED WATER	39
*MARINE PHYTOPLANKTON	0.9 BY WEIGHT	SEE REF.	1520

TABLE A28

ALGAE : CELL DIAMETER			
ITEM:	VALUE:	CONDITIONS:	REF.:
*COCCOLITHUS HUXLEYI	5 MY	NONE	1219
*DITYLUM BRIGHTWELLII	30 MY	NONE	1219
*THALASSIOSIRA PSEUDONANA	2.5 - 10 MY	CLONE 3H	649
*THALASSIOSIRA PSEUDONANA	4.0 - 11 MY	CLONE 13-1	649

TABLE A29

ALGAE : CELL VOLUME			
ITEM:	VALUE:	CONDITIONS:	REF.:
*AGMENELLUM QUADRUPLICATUM	1.5 MICRO M**3	BLUE GREEN ALGAE,MARINE	1220
*AMPHIDINIUM CARTERI	740 MICRO M**3	DINOPHYCEAE,MARINE	1220
*AMPHIPHORA ORNATA	17650 MICRO M**3	NONE	1689
*ANABAENA FLOS-AQUA	80000 MICRO M**3	FRESHWATER PLANKTON,CALCULATED VOLUME	734
*ANABAENA FLOS-AQUAE	80000 MICRO M**3	COL.	1689
*ANABAENA SPP.	100 - 150 MICRO M**3	LAKE KINNERET	106
*ANKISTRODESMUS FALCATUS	250 MICRO M**3	NONE	1689
*ANKISTRODESMUS FALCATUS	250 MICRO M**3	FRESHWATER PLANKTON,CALCULATED VOLUME	734
*APHANOCAPSA DELICATISSIMA	4 MICRO M**3	NONE	1689
*APHANOTHECE CLATHRATA	10 MICRO M**3	NONE	1689
*APHANOTHECE NIDULANS	5 MICRO M**3	NONE	1689
*ASTERIONELLA FORMOSA	350 MICRO M**3	MICHIGAN	1689
*ASTERIONELLA FORMOSA	700 MICRO M**3	EUROPE	1689
*ASTERIONELLA FORMOSA	700 MICRO M**3	FRESHWATER PLANKTON,CALCULATED VOLUME	734
*ASTERIONELLA FORMOSA	480 MICRO M**3	LAKE TAHOE,SURFACE AREA/VOLUME=1.7	1596
*BOTRYOCOCCUS BRAUNII	10000 MICRO M**3	COL.	1689
*BOTRYOCOCCUSBRAUNII	10000 MICRO M**3	FRESHWATER PLANKTON,CALCULATED VOLUME, COL	734
*CERATIUM HIRUNDELLA	75000 MICRO M**3	LAKE KINNERET	106
*CERATIUM HIRUNDINELLA	4000 MICRO M**3	NONE	1689
*CERATIUM HIRUNDINELLA	70000 MICRO M**3	FRESHWATER PLANKTON,CALCULATED VOLUME	734
*CHAETOCEROS LAUDERI	2275 MICRO M**3	CULTIVATED, MARINE	104
*CHLAMYDOMONAS MAGNUSII	65.3 MICRO M**3	CULTIVATED, MARINE	104
*CHLAMYDOMONAS SUBCOMPLETA	250 MICRO M**3 MICRO M**3	NONE	1689
*CHLORELLA SP.	30 MICRO M**3	LAKE KINNERET	106
*CHLORELLA SP.	24 MICROM**3	COUNT BY MICROSCOPE	255
*CHLORELLA VULGARIS	200 MICRO M**3	NONE	1689
*CHLORELLA VULGARIS	200 MICRO M**3	FRESHWATER PLANKTON,CALCULATED VOLUME	734

TABLE A29

ALGAE : CELL VOLUME

ITEM:	VALUE:	CONDITIONS:	REF.:
*CHODATELLA SPP.	200 - 600 MICRO M**3	LAKE KINNERET	106
*CHROMULINA PYRIFORMIS	50 MICRO M**3	NONE	1689
*CHROOCOCCUS LIMNETICUS	400 MICRO M**3	COL.	1689
*CHROOCOCCUS SPP.	60 - 500 MICRO M**3	LAKE KINNERET	106
*CHROOCOCCUS TURGIDUS	1000 MICRO M**3	COL.	1689
*CHROOMONAS NORDSTEDTII	35 MICRO M**3	NONE	1689
*CLOSTERIUM ACICULARIS	4000 MICRO M**3	NONE	1689
*CLOSTERIUM ACICULARIS	4000 MICRO M**3	FRESHWATER PLANKTON,CALCULATED VOLUME	734
*CLOSTERIUM ACICULARIS VAR. SUBPRONUM	3000 MICRO M**3	LAKE KINNERET	106
*COELOSPHAERIUM NAEGELIANUM	15000 MICRO M**3	COL.	1689
*COMENSIS	400 MICRO M**3	NONE	1689
*COSCINODISCUS ANGSTII	26000 MICRO M**3	DIAMETER = 35 U	493
*COSCINODISCUS ANGSTII VAR. GRANULOMARGINATUS	160000 MICRO M**3	DIAMETER = 87 U, CENTRIC	493
*COSCINODISCUS EXCENTRICUS	63000 MICRO M**3	DIAMETER = 75 U	493
*COSCINODISCUS SP.	60400 MICRO M**3	CULTIVATED, MARINE	104
*COSMARIUM PHASEOLUSS	3000 MICRO M**3	NONE	1689
*COSMARIUM PHASEOLUSS	3000 MICRO M**3	FRESHWATER PLANKTON,CALCULATED VOLUME	734
*COSMARIUM RENIFORME	30000 MICRO M**3	NONE	1689
*COSMARIUM RENIFORME	30000 MICRO M**3	FRESHWATER PLANKTON,CALCULATED VOLUME	734
*COSMARIUM SP.	500 MICRO M**3	LAKE KINNERET	106
*CRYPTOMONAS EROSA	1000 MICRO M**3	MICHIGAN	1689
*CRYPTOMONAS EROSA	250 MICRO M**3	EUROPE	1689
*CRYPTOMONAS EROSA	250 MICRO M**3	FRESHWATER PLANKTON,CALCULATED VOLUME	734
*CRYPTOMONAS OVATA	250 MICRO M**3	NONE	1689
*CRYPTOMONAS OVATA	250 MICRO M**3	FRESHWATER PLANKTON,CALCULATED VOLUME	734
*CRYPTOMONAS PSEUDOBALTICA	410 MICRO M**3	CULTIVATED, MARINE	104
*CRYPTOMONAS SPP.	250 - 3600 - 6200 MICRO M**3	LAKE KINNERET	106
*CYCLOTELLA BODANICA	10000 MICRO M**3	NONE	1689
*CYCLOTELLA BODANICA	10000 MICRO M**3	FRESHWATER PLANKTON,CALCULATED VOLUME	734
*CYCLOTELLA COMENSIS	400 MICRO M**3	FRESHWATER PLANKTON,CALCULATED VOLUME	734
*CYCLOTELLA SP.	200 - 1000 MICRO M**3	LAKE KINNERET	106
*CYCLOTELLA STELLIGERA	77 MICRO M**3	LAKE TAHOE,SURFACE AREA/VOLUME=1.6	1596
*CYMATOPLEURA SOLEA	80000 MICRO M**3	FRESHWATER PLANKTON,CALCULATED VOLUME	734
*CYMATOPLEURA SOLEA	80000 MICRO M**3	NONE	1689
*DACTYLOCOCOCPSIS SMITHII	1500 MICRO M**3	COL.	1689
*DIATOMA VULGARE	4350 MICRO M**3	NONE	1689
*DINOBRYON BORGEI	1500 MICRO M**3	NONE	1689
*DINOBRYON DIVERGENS	800 MICRO M**3	NONE	1689
*DINOBRYON DIVERGENS PLUS D.SOCIALE	800 MICRO M**3	FRESHWATER PLANKTON,CALCULATED VOLUME	734
*DINOBRYON SOCIALE	800 MICRO M**3	NONE	1689
*DINOBRYON SP.	900 MICRO M**3	LAKE TAHOE,SURFACE AREA/VOLUME=0.5	1596
*DIPLOPSALIS ACUTA	11000 MICRO M**3	LAKE KINNERET	106
*DUNALIELLA SALINA	400 MICRO M**3	CHLOROPHYCEAE,MARINE	1220
*ELAKATOTHRIX GELATINOSA	90 - 300 MICRO M**3	LAKE KINNERET	106
*ERKENIA SUBAEQUICILIATA	30 MICRO M**3	LAKE KINNERET	106

TABLE A29

ALGAE : CELL VOLUME			
ITEM:	VALUE:	CONDITIONS:	REF.:
*EXUVIAELLA SP.	780 MICRO M**3	DINOPHYCEAE,MARINE	1220
*FRAGILARIA CROTONENSIS	200000 MICRO M**3	1 MM OF A THREAD	1689
*FRAGILARIA CROTONENSIS	200000 MICRO M**3	FRESHWATER PLANKTON,CALCULATED VOLUME, 1 MM OF A THREAD	734
*FRAGILLARIA CROTONENSIS	510 MICRO M**3	LAKE TAHOE,SURFACE AREA/VOLUME=1.5	1596
*GLOEOCAPSA RUPESTRIS	18 MICRO M**3	NONE	1689
*GLOEOCOCOCCUS SCHROETERI	5000 MICRO M**3	COL.	1689
*GLOEOCOCOCCUS SCHROETERI	5000 MICRO M**3	FRESHWATER PLANKTON,CALCULATED VOLUME, COL	734
*GOMPHOSPHAERIA LACUSTRIS	2000 MICRO M**3	COL.	1689
*GOMPHOSPHAERIA LACUSTRIS	2000 MICRO M**3	FRESHWATER PLANKTON,CALCULATED VOLUME, COL	734
*GYMNODINIUM FUSCUM	10000 MICRO M**3	NONE	1689
*GYMNODINIUM HELVETICUM	20000 MICRO M**3	NONE	1689
*GYMNODINIUM HELVETICUM	20000 MICRO M**3	FRESHWATER PLANKTON,CALCULATED VOLUME	734
*GYMNODINIUM ORDINATUM	400 MICRO M**3	NONE	1689
*HETEROTHRIX SP.	101 MICRO M**3	CULTIVATED, MARINE	104
*KEPHYRION SP.	50 MICRO M**3	FRESHWATER PLANKTON,CALCULATED VOLUME	734
*MALLOMONAS CAUDATA	12000 MICRO M**3	NONE	1689
*MALLOMONAS CAUDATA	10000 MICRO M**3	FRESHWATER PLANKTON,CALCULATED VOLUME	734
*MALLOMONAS URNAFORMIS	1200 MICRO M**3	NONE	1689
*MELOSIRA GRANULATA	60000 MICRO M**3	1 MM OF A THREAD	1689
*MELOSIRA GRANULATA	320 - 1100 - 1500 MICRO M**3	LAKE KINNERET	106
*MELOSIRA GRANULATA	60000 MICRO M**3	FRESHWATER PLANKTON,CALCULATED VOLUME, 1 MM OF A THREAD	734
*MELOSIRA ISLANDICA	80000 MICRO M**3	1 MM	1689
*MELOSIRA ISLANDICA	80000 MICRO M**3	FRESHWATER PLANKTON,CALCULATED VOLUME, 1 MM OF A THREAD	734
*MERISMOPEDIA TENUISSIMA	8 MICRO M**3	NONE	1689
*MICROCYSTIS AERUGINOSA	100000 MICRO M**3	COL.	1689
*MICROCYSTIS AERUGINOSA	100000 MICRO M**3	FRESHWATER PLANKTON,CALCULATED VOLUME, COL	734
*MICROCYSTIS FLOS-AQUAE	50 MICRO M**3	NONE	1689
*MICROCYSTIS SPP.	20 - 100 MICRO M**3	LAKE KINNERET	106
*MONOCHRYSIS LUTHERI	28 MICRO M**3	CHRYSOPHYCEAE,MARINE	1220
*NAVICULA SPP.	300 - 1400 MICRO M**3	LAKE KINNERET	106
*NITZSCHIA GRACILIS	240 MICRO M**3	NONE	1689
*OOCYSTIS SOLITARIA	400 MICRO M**3	NONE	1689
*OOCYSTIS SOLITARIA	400 MICRO M**3	FRESHWATER PLANKTON,CALCULATED VOLUME	734
*OOCYSTIS SP.	200 - 300 MICRO M**3	LAKE KINNERET	106
*OSCILLATORIA LIMNETICA	17500 MICRO M**3	1 MM OF A THREAD	1689
*OSCILLATORIA RUBESCENS	30000 MICRO M**3	1 MM OF A THREAD	1689
*OSCILLATORIA RUBESCENS	30000 MICRO M**3	FRESHWATER PLANKTON,CALCULATED VOLUME, 1 MM OF A THREAD	734
*PANDORINA MORUM	4000 MICRO M**3	COL.	1689
*PANDORINA MORUM	4000 MICRO M**3	FRESHWATER PLANKTON,CALCULATED VOLUME	734
*PEDIASTRUM SPP.	30 - 100 - 200 MICRO M**3	LAKE KINNERET	106
*PERIDINIUM CINCTUM	40000 MICRO M**3	NONE	1689
*PERIDINIUM CINCTUM FA. WESTII	75000 - 1250 MICRO M**300 MICRO M**3	LAKE KINNERET	106
*PERIDINIUM SPP.	5400 - 10600 MICRO M**3	LAKE KINNERET	106
*PERIDINIUM WILLEI	40000 MICRO M**3	NONE	1689

TABLE A29

ALGAE : CELL VOLUME

ITEM:	VALUE:	CONDITIONS:	REF.:
*PERIDINIUM WILLEI	40000 MICRO M**3	FRESHWATER PLANKTON,CALCULATED VOLUME	734
*PHAEODACTYLUM TRICORNUTUM	555 MICRO M**3	CULTIVATED, MARINE	104
*RADIOCYSTIS GEMINATA	7 MICRO M**3	LAKE KINNERET	106
*RHIZOCHRYSIS LIMNETICA	1200 MICRO M**3	NONE	1689
*RHODOMONAS LACUSTRIS	175 MICRO M**3	MICHIGAN	1689
*RHODOMONAS LACUSTRIS	200 MICRO M**3	EUROPE	1689
*RHODOMONAS LACUSTRIS	200 MICRO M**3	FRESHWATER PLANKTON,CALCULATED VOLUME	734
*RHODOMONAS MINUTA	200 MICRO M**3	NONE	1689
*RHODOMONAS MINUTA	100 - 250 MICRO M**3	LAKE KINNERET	106
*SCENEDESMUS QUADRICAUDA	1000 MICRO M**3	NONE	1689
*SCENEDESMUS QUADRICAUDA	1000 MICRO M**3	FRESHWATER PLANKTON,CALCULATED VOLUME	734
*SCENEDESMUS SPP.	30 - 100 - 200 MICRO M**3	LAKE KINNERET	106
*SELENASTRUM CAPRICORNUTUM	60 MICROM**3	PARTICLE COUNTER	255
*SELENASTRUM CAPRICORNUTUM	29 MICROM**3	PARTICLE COUNTER	255
*SELENASTRUM CAPRICORNUTUM	80 MICROM**3	PARTICLE COUNTER	255
*SELENASTRUM CAPRICORNUTUM	30 MICROM**3	PARTICLE COUNTER	255
*SKELETONEMA COSTATUM	1390 MICRO M**3	BACILLARIOPHYCEAE,MARINE	1220
*STAURASTRUM PARADOXUM	20000 MICRO M**3	NONE	1689
*STAURASTRUM PARADOXUM	20000 MICRO M**3	FRESHWATER PLANKTON,CALCULATED VOLUME	734
*STEPHANODISCUS ASTRAEA	2000 MICRO M**3	NONE	1689
*STEPHANODISCUS ASTRAEA	2000 MICRO M**3	FRESHWATER PLANKTON,CALCULATED VOLUME	734
*STEPHANODISCUS HANTZSCHII VAR. PUSILLUS	200 MICRO M**3	NONE	1689
*STEPHANODISCUS HANTZSCHII VAR. PUSILLUS	110 - 200 MICRO M**3	LAKE KINNERET	106
*STEPHANODISCUS HANTZSCHII VAR. PUSILLUS	200 MICRO M**3	FRESHWATER PLANKTON,CALCULATED VOLUME	734
*STEPHANODISCUS NIAGAREA	5000 MICRO M**3	NONE	1689
*SYNECHOCOCCUS AERUGINOSUS	350 MICRO M**3	NONE	1689
*SYNEDRA ACUS	250 MICRO M**3	NONE	1689
*SYNEDRA ACUS VAR. ANGUSTISSIMA	1000 MICRO M**3	NONE	1689
*SYNEDRA ACUS VAR. ANGUSTISSIMA	1000 MICRO M**3	FRESHWATER PLANKTON,CALCULATED VOLUME	734
*SYNEDRA CAPITATA	950 MICRO M**3	NONE	1689
*SYNEDRA DELICATISSIMA	300 MICRO M**3	NONE	1689
*SYNEDRA ULNA	50 MICRO M**3	NONE	1689
*TABELLARIA FENESTRATA	3000 MICRO M**3	MICHIGAN	1689
*TABELLARIA FENESTRATA	4000 MICRO M**3	EUROPE	1689
*TABELLARIA FENESTRATA	4000 MICRO M**3	FRESHWATER PLANKTON,CALCULATED VOLUME	734
*TETRAEDRON MINIMUM	40 MICRO M**3	NONE	1689
*TETRAEDRON MINIMUM	40 MICRO M**3	FRESHWATER PLANKTON,CALCULATED VOLUME	734
*TETRAEDRON SPP.	30 - 140 MICRO M**3	LAKE KINNERET	106
*TETRASELMIS MACULATA	310 MICRO M**3	PRASINOPHYCEAE,MARINE	1220
*TETRASELMIS STRICIA	376 MICRO M**3	CULTIVATED, MARINE	104
*THALASSIOSIRA FLUVIATILIS	1450 MICRO M**3	DIAMETER = 11 U	493
*THALASSIOSIRA PSEUDONANA	55 MICRO M**3	DIAMETER = 3.8 U	493
*TRACHELOMONAS HISPIDA	4200 MICRO M**3	NONE	1689
*TRACHELOMONAS VOLVOCINA	1800 MICRO M**3	NONE	1689
*ULOTHRIXX ZONATA	6000 MICRO M**3	1 MM	1689

TABLE A29

ALGAE : CELL VOLUME			
ITEM:	VALUE:	CONDITIONS:	REF.:
*UROGLENA AMERICANA	90000 MICRO M**3	COL.	1689
*UROGLENA AMERICANA	90000 MICROM**3	FRESHWATER PLANKTON, CALCULATED VOLUME, COL	734

TABLE A30

ALGAE : CHLOROPHYLL A/BIOMASS RATIO			
ITEM:	VALUE:	CONDITIONS:	REF.:
*ALGAE	100 CELL/ML=1.75 MICROGRAM-CHL/L	MOSSDALE	1153
*MARINE PHYTOPLANKTON	4.8 10**(-12) G/CELL	NO3 AS SOURCE OF N, STATIONARY PHASE,293 K, SOUTHERN CALIF	428
*MARINE PHYTOPLANKTON	3.9 10**(-12) G/CELL	NH4 AS SOURCE OF N, STATIONARY PHASE,293 K, SOUTHERN CALIF	428
*MARINE PHYTOPLANKTON	2.0 10**(-12) G/CELL	UREA AS SOURCE OF N, STATIONARY PHASE,293 K, SOUTHERN CALIF	428
*OSCILLATORIA REDEKEI	0.25 % OF DRY WEIGHT	GROWN AT 293K	523
*THALASSIOSIRA SP.	10.5*10**(-14) GRAM/CELL	GROWTH RATE=0.041 1/H, CELL DENSITY=4.0*10**8 CELL/L, P-LIMITED CHEMO.	1243
*THALASSIOSIRA SP.	8.1*10**(-14) GRAM/CELL	GROWTH RATE=0.017 1/H, CELL DENSITY=6.7*10**8 CELL/L, P-LIMITED CHEMO.	1243
*TRICHODESMIUM SPP.	34 NG/COLONY	CHL.A/COLONY,N.PACIFIC,N-FIXING ALGAE,N=6	967

TABLE A31

ALGAE : CU/BIOMASS RATIO			
ITEM:	VALUE:	CONDITIONS:	REF.:
*CHARA SP.	19 MICROGRAM/L	ALGAE FROM PONDS AND LAKES IN SOUTHEASTERN UNITED STATES	1660
*CLADOPHORA SP.	190 MICROGRAM/L	ALGAE FROM PONDS AND LAKES IN SOUTHEASTERN UNITED STATES	1660
*EUGLENA SP.	290 MICROGRAM/L	ALGAE FROM PONDS AND LAKES IN SOUTHEASTERN UNITED STATES	1660
*HYDRODICTYON SP.	114 MICROGRAM/L	ALGAE FROM PONDS AND LAKES IN SOUTHEASTERN UNITED STATES	1660
*LYNGBYA SP.	101 MICROGRAM/L	ALGAE FROM PONDS AND LAKES IN SOUTHEASTERN UNITED STATES	1660
*MICROCYSTIS SP.	37 MICROGRAM/L	ALGAE FROM PONDS AND LAKES IN SOUTHEASTERN UNITED STATES	1660
*MOUGEOTIA SP.	143 MICROGRAM/L	ALGAE FROM PONDS AND LAKES IN SOUTHEASTERN UNITED STATES	1660
*NITELLA SP.	39 MICROGRAM/L	ALGAE FROM PONDS AND LAKES IN SOUTHEASTERN UNITED STATES	1660
*OEDOGONIUM SP.	75 MICROGRAM/L	ALGAE FROM PONDS AND LAKES IN SOUTHEASTERN UNITED STATES	1660
*PITHOPHORA SP.	23 MICROGRAM/L	ALGAE FROM PONDS AND LAKES IN SOUTHEASTERN UNITED STATES	1660
*RHIZOCLONIUM SP.	75 MICROGRAM/L	ALGAE FROM PONDS AND LAKES IN SOUTHEASTERN UNITED STATES	1660
*SPIROGYRA SP.	47 MICROGRAM/L	ALGAE FROM PONDS AND LAKES IN SOUTHEASTERN UNITED STATES	1660
*SPIROGYRA SP., GIANT	34 MICROGRAM/L	ALGAE FROM PONDS AND LAKES IN SOUTHEASTERN UNITED STATES	1660

TABLE A32

ALGAE : C2H2 REDUCTION RATE			
ITEM:	VALUE:	CONDITIONS:	REF.:
*TRICHODESMIUM SPP.	251 PMOLES C2H4/(MICROGRAM ALGAL N)*HR	N.PACIFIC,N-FIXING ALGAE, INCUBATION IN LABORATORY	967

TABLE A32

ALGAE : C2H2 REDUCTION RATE			
ITEM:	VALUE:	CONDITIONS:	REF.:
*TRICHODESMIUM SPP.	620 PMOLES C2H4/HR	N.PACIFIC,N-FIXING ALGAE, MAX RATE	967

TABLE A33

ALGAE : C2H2 REDUCTION/N2 FIXATION RATIO			
ITEM:	VALUE:	CONDITIONS:	REF.:
*TRICHODESMIUM SPP.	1.9:1 (MOLE/MOLE)	N.PACIFIC,N-FIXING ALGAE, INCUBATION IN LABORATORY	967

TABLE A34

ALGAE : DARK UPTAKE			
ITEM:	VALUE:	CONDITIONS:	REF.:
*ALGAE FROM PACIFIC OCEAN	5.2 % OF PHOTOSYNTHESIS	CHLOROPHYLL=0.099,INORG N=0.46, INORG P=0.24, SI=1.5MG/M3	1584
*ALGAE FROM PACIFIC OCEAN	15.4 % OF PHOTOSYNTHESIS	CHLOROPHYLL=0.092,INORG N=0.46, INORG P=0.33, SI=3.2MG/M3	1584
*ALGAE FROM PACIFIC OCEAN	2.2 % OF PHOTOSYNTHESIS	CHLOROPHYLL=0.135,INORG N=0.34, INORG P=0.23, SI=3.8MG/M3	1584
*ALGAE FROM PACIFIC OCEAN	3.6 % OF PHOTOSYNTHESIS	CHLOROPHYLL=0.174,INORG N=0.49, INORG P=0.22, SI=2.1MG/M3	1584
*ALGAE FROM PACIFIC OCEAN	2.7 % OF PHOTOSYNTHESIS	CHLOROPHYLL=0.224,INORG N=0.48, INORG P=0.27, SI=2.0MG/M3	1584
*ALGAE FROM PACIFIC OCEAN	3.3 % OF PHOTOSYNTHESIS	CHLOROPHYLL=0.142,INORG N=0.45, INORG P=0.30, SI=3.1MG/M3	1584
*ALGAE FROM PACIFIC OCEAN	3.0 % OF PHOTOSYNTHESIS	CHLOROPHYLL=0.158,INORG N=0.40, INORG P=0.24, SI=2.5MG/M3	1584
*ALGAE FROM PACIFIC OCEAN	5.7 % OF PHOTOSYNTHESIS	MEAN OF 8 NITROGEN POOR STATIONS	1584
*ALGAE FROM PACIFIC OCEAN	1.5 % OF PHOTOSYNTHESIS	CHLOROPHYLL=0.201,INORG N=2.50, INORG P=0.35, SI=3.1	1584
*ALGAE FROM PACIFIC OCEAN	1.3 % OF PHOTOSYNTHESIS	CHLOROPHYLL=0.211,INORG N=5.71, INORG P=0.27, SI=5.0	1584
*ALGAE FROM PACIFIC OCEAN	2.8 % OF PHOTOSYNTHESIS	CHLOROPHYLL=0.151,INORG N=8.61, INORG P=0.75, SI=3.8	1584
*ALGAE FROM PACIFIC OCEAN	3.3 % OF PHOTOSYNTHESIS	CHLOROPHYLL=0.102,INORG N=7.16, INORG P=0.62, SI=3.5	1584
*ALGAE FROM PACIFIC OCEAN	8.0 % OF PHOTOSYNTHESIS	CHLOROPHYLL=0.099,INORG N=7.81, INORG P=0.71, SI=---	1584
*ALGAE FROM PACIFIC OCEAN	5.0 % OF PHOTOSYNTHESIS	CHLOROPHYLL=0.041,INORG N=8.11, INORG P=0.69, SI=1.6	1584
*ALGAE FROM PACIFIC OCEAN	13.6 % OF PHOTOSYNTHESIS	CHLOROPHYLL=0.105,INORG N=7.42, INORG P=0.67, SI=3.6	1584
*ALGAE FROM PACIFIC OCEAN	3.1 % OF PHOTOSYNTHESIS	CHLOROPHYLL=0.263,INORG N=8.35, INORG P=0.83, SI=5.2	1584
*ALGAE FROM PACIFIC OCEAN	4.8 % OF PHOTOSYNTHESIS	MEAN OF 8 NITROGEN RICH STATIONS	1584

TABLE A35

ALGAE : DRY/WET WEIGHT RATIO			
ITEM:	VALUE:	CONDITIONS:	REF.:
*AGMENELLUM QUADRUPLICATUM	0.563 MG DRY WEIGHT/MM**3	NONE	955
*AMPHIDINIUM CARTERI	0.346 MG DRY WEIGHT/MM**3	NONE	955

TABLE A35

ALGAE : DRY/WET WEIGHT RATIO			
ITEM:	VALUE:	CONDITIONS:	REF.:
*APHANIZOMENON FLOS-AQUAE F. GRACILE	0.35 MG DRY WEIGHT/MM**3	NONE	955
*CHAETOCEROS SP.	0.352 MG DRY WEIGHT/MM**3	NONE	955
*COENOCOCCUS PLANKTONICUS	0.508 MG DRY WEIGHT/MM**3	NONE	955
*COSCINODISCUS SP.	0.214 MG DRY WEIGHT/MM**3	NONE	955
*CYCLOTELLA MENEGHINIANA	0.310 MG DRY WEIGHT/MM**3	NONE	955
*DICTYOSPHAERIUM PULCHELLUM	0.611 MG DRY WEIGHT/MM**3	NONE	955
*DUNALIELLA SALINA	0.333 MG DRY WEIGHT/MM**3	NONE	955
*EXUVIAELLA SP.	0.341 MG DRY WEIGHT/MM**3	NONE	955
*MELOSIRA GRANULATA	1.02 MG DRY WEIGHT/MM**3	NONE	955
*MELOSIRA ITALICA SUBSP. SUBARTICA	0.857 MG DRY WEIGHT/MM**3	NONE	955
*MONOCHRYSIS LUTHERI	1.25 MG DRY WEIGHT/MM**3	NONE	955
*OSCILLATORIA AGARDHII VAR ISOTHRIX	0.425 MG DRY WEIGHT/MM**3	NONE	955
*OSCILLATORIA BOURRELLYI	0.620 MG DRY WEIGHT/MM**3	NONE	955
*PAULSCHULZIA TENERA	0.667 MG DRY WEIGHT/MM**3	NONE	955
*PHAEODACTYLUM TRICORNUTUM	0.181 MG DRY WEIGHT/MM**3	NONE	955
*SKELETONEMA COSTATUM	0.484 MG DRY WEIGHT/MM**3	NONE	955
*STEPHANODISCUS ASTRAEA	0.889 MG DRY WEIGHT/MM**3	NONE	955
*STEPHANODISCUS HANTZSCHII	0.230 MG DRY WEIGHT/MM**3	NONE	955
*SYRACOSPHAERA CARTERI	0.286 MG DRY WEIGHT/MM**3	NONE	955
*TETRASELMIS MACULATA	0.738 MG DRY WEIGHT/MM**3	NONE	955

TABLE A36

ALGAE : EXCRETION OF DISSOLVED ORGANIC CARBON			
ITEM:	VALUE:	CONDITIONS:	REF.:
*ANABAENA CYLINDRICA	2.0 - 4.0 % OF GROSS PHOTOSYNTHESIS	N=2, 12 H, STATIONARY PHASE	105
*ANABAENA CYLINDRICA	16.9 % OF GROSS PHOTOSYNTHESIS	12 H, OLD CULTURE, STATIONARY PHASE	105
*CHLORELLA SP.	1.5 - 1.8 % OF PHOTOSYNTHESIS	3 H, FULL SUNLIGHT, 305 K, 117500 CELL/ML, CHL A = 47 NG/ML	1378
*COCCOLITHUS HUXLEYI	2.1 - 5.2 % OF GROSS PHOTOSYNTHESIS	N=3, 12 H, STATIONARY PHASE	105
*COCCOLITHUS HUXLEYI	32 - 35 % OF GROSS PHOTOSYNTHESIS	12 H, OLD CULTURE, STATIONARY PHASE	105
*DICRATERIA SP.	4.1 - 4.6 % OF PHOTOSYNTHESIS	3 H, FULL SUNLIGHT, 305 K, 270000 CELL/ML, CHL A = 88 NG/ML	1378
*DUNALIELLA TERTIOLECTA	1.0 - 5.1 % OF GROSS PHOTOSYNTHESIS	N=7, 12 H, STATIONARY PHASE	105
*DUNALIELLA TERTIOLECTA	8 % OF GROSS PHOTOSYNTHESIS	12 H, OLD CULTURE, STATIONARY PHASE	105
*FRESHWATER PHYTOPLANKTON	1 - 70 % OF PHOTOSYNTHESIS	6 H, SWEDISH LAKES	1422
*FRESHWATER PHYTOPLANKTON	3 - 49 % OF PHOTOSYNTHESIS	4 H, LAKE WINDERMERE	1422
*FRESHWATER PHYTOPLANKTON	22 -67 % OF PHOTOSYNTHESIS	5 H, LAKE ONTARIO	1422
*FRESHWATER PHYTOPLANKTON	2 - 60 % OF PHOTOSYNTHESIS	4 H, BRITISH RESEVOIRS	1422
*GONYAULAX POLYEDRA	2.3 - 6.3 % OF GROSS PHOTOSYNTHESIS	N=8, 12 H, STATIONARY PHASE	105
*GONYAULAX POLYEDRA	32 % OF GROSS PHOTOSYNTHESIS	12 H, OLD CULTURE, STATIONARY PHASE	105
*GYMNODINIUM SPLENDENS	0.6 - 5.5 % OF GROSS PHOTOSYNTHESIS	N=10, 12 H, STATIONARY PHASE	105
*LAUDERIA SP.	13.0 - 19.0 % OF GROSS PHOTOSYNTHESIS	N=5, 12 H, STATIONARY PHASE	105
*MARINE PHYTOPLANKTON	10 - 75 % OF PHOTOSYNTHESIS	6 H, MENAI STRAIT	1422

TABLE A36

ALGAE : EXCRETION OF DISSOLVED ORGANIC CARBON			
ITEM:	VALUE:	CONDITIONS:	REF.:
*MARINE PHYTOPLANKTON	7 - 17 % OF PHOTOSYNTHESIS	6 H, COASTAL N PACIFIC	1422
*MARINE PHYTOPLANKTON	13 - 26 % OF PHOTOSYNTHESIS	6 H, OFFSHORE N PACIFIC	1422
*MARINE PHYTOPLANKTON	10 - 18 % OF PHOTOSYNTHESIS	6 H, GULF OF CALIFORNIA	1422
*MARINE PHYTOPLANKTON	20 - 27 % OF PHOTOSYNTHESIS	OFFSHORE N PACIFIC, 12 H	1422
*MARINE PHYTOPLANKTON	13 - 70 % OF PHOTOSYNTHESIS	WESTERN N ATLANTIC, 4 H	1422
*MARINE PHYTOPLANKTON	4 - 16 % OF PHOTOSYNTHESIS	NERITIC N ATLANTIC, 3 - 24 H	1422
*MARINE PHYTOPLANKTON	1 - 38 % OF PHOTOSYNTHESIS	INSHORE ANTARTIC, 4 H	1422
*MARINE PHYTOPLANKTON	0.- 46 % OF PHOTOSYNTHESIS	6 H, SCOTTISH SEA LOCHS	1422
*MARINE PHYTOPLANKTON	10 - 29 % OF PHOTOSYNTHESIS	2 H, INSHORE INDIAN OCEAN	1422
*MARINE PHYTOPLANKTON	0.2 % OF PHOTOSYNTHESIS	2 H, INSHORE N ATLANTIC	1422
*MARINE PHYTOPLANKTON	0.- 27 % OF PHOTOSYNTHESIS	2 H, COASTRAL N ATLANTIC	1422
*MARINE PHYTOPLANKTON	0 - 45 % OF PHOTOSYNTHESIS	6 H, OFFSHORE N ATLANTIC	1422
*MARINE PHYTOPLANKTON	9 - 26 % OF PHOTOSYNTHESIS	4 H, STRAIT OF GEORGIA	1422
*MARINE PHYTOPLANKTON	6.9 - 9.7 % OF GROSS PHOTOSYNTHESIS	N=3, 12 H, STATIONARY PHASE	105
*MONOCHRYSIS LUTHERI	29.8 % OF GROSS PHOTOSYNTHESIS	12 H, OLD CULTURE, STATIONARY PHASE	105
*NOSTOC MUSCORUM	2.0 - 5.0 % OF GROSS PHOTOSYNTHESIS	N=2, 12 H, STATIONARY PHASE	105
*NOSTOC MUSCORUM	39.2 % OF GROSS PHOTOSYNTHESIS	12 H, OLD CULTURE, STATIONARY PHASE	105
*SKELETONEMA COSTATUM	38 % OF PHOTOSYNTHESIS	5 H, LIGHT = 120000 LUX	1422
*SKELETONEMA COSTATUM	4.6 % OF PHOTOSYNTHESIS	48 H, LIGHT = 10000 LUX	1422
*SKELETONEMA COSTATUM	9.0 % OF PHOTOSYNTHESIS	48 H, LIGHT = 3000 LUX	1422
*SKELETONEMA COSTATUM	5.6 % OF PHOTOSYNTHESIS	96 H, LIGHT = 25000 LUX	1422
*SKELETONEMA COSTATUM	6.2 % OF PHOTOSYNTHESIS	96 H, LIGHT = 3000 LUX	1422
*TETRASELMIS CARTERIFORMIS	1.05 - 1.8 % OF PHOTOSYNTHESIS	3 H, FULL SUNLIGHT, 305 K, 21300 CELL/ML, CHL A = 93 NG/ML	1378
*TETRASELMIS GRACILIS	2.8 - 3.4 % OF PHOTOSYNTHESIS	3 H, FULL SUNLIGHT, 305 K, 43500 CELL/ML, CHL A = 107 NG/ML	1378
*THALASSIOSIRA FLUVIATILIS	13 % OF PHOTOSYNTHESIS	5 H, LIGHT = 120000 LUX	1422
*THALASSIOSIRA FLUVIATILIS	4.5 % OF PHOTOSYNTHESIS	48 H, LIGHT = 10000 LUX	1422
*THALASSIOSIRA FLUVIATILIS	6.2 % OF PHOTOSYNTHESIS	48 H, LIGHT = 3000 LUX	1422
*THALASSIOSIRA FLUVIATILIS	3.6 % OF PHOTOSYNTHESIS	96 H, LIGHT = 25000 LUX	1422
*THALASSIOSIRA FLUVIATILIS	4.9 % OF PHOTOSYNTHESIS	96 H, LIGHT = 3000 LUX	1422
*THALASSIOSIRA PSEUDONANA	27 % OF PHOTOSYNTHESIS	5 H, LIGHT = 120000 LUX	1422
*THALASSIOSIRA PSEUDONANA	4.0 % OF PHOTOSYNTHESIS	48 H,LIGHT = 10000 LUX	1422
*THALASSIOSIRA PSEUDONANA	4.6 % OF PHOTOSYNTHESIS	48 H, LIGHT = 3000 LUX	1422

TABLE A37

ALGAE : EXCRETION OF PHOSPHORUS			
ITEM:	VALUE:	CONDITIONS:	REF.:
*FRESHWATER ALGAE	0.840 1/HOUR	STEADY STATE SOLUTION TO CASTLE LAKE, AUTOLYSIS OF DISS ORGANIC P	1326
*SESTON	0.010 MICROGRAM P/HOUR	STONEHOUSE POND, FEB 22, 1973	454
*SESTON	0.4608 MICROGRAM P/HOUR	STONEHOUSE POND, APRIL 6, 1973	454
*SESTON	0.0002 MICROGRAM P/HOUR	STONEHOUSE POND, MAY 17, 1973	454
*SESTON	0.0512 MICROGRAM P/HOUR	STONEHOUSE POND, JUNE 17, 1973	454

TABLE A38

ALGAE : FAT/BIOMASS RATIO			
ITEM:	VALUE:	CONDITIONS:	REF.:
*AGMENELLUM QUADRUPLICATUM	16 % OF ASH FREE DRY WEIGHT	BLUE GREEN ALGAE, MARINE, SAPONIFIABLE FRACTION ONLY	1220

TABLE A38

ALGAE : FAT/BIOMASS RATIO			
ITEM:	VALUE:	CONDITIONS:	REF.:
*AMPHIDINIUM CARTERI	23 % OF ASH FREE DRY WEIGHT	DINOPHYCEAE,MARINE,SAPONIFIABLE FRACTION ONLY	1220
*DUNALIELLA SALINA	7 % OF ASH FREE DRY WEIGHT	CHLOROPHYCEAE,MARINE,SAPONIFIABLE FRACTION ONLY	1220
*EXUVIAELLA SP.	17 % OF ASH FREE DRY WEIGHT	DINOPHYCEAE,MARINE,SAPONIFIABLE FRACTION ONLY	1220
*MONOCHRYSIS LUTHERI	13 % OF ASH FREE DRY WEIGHT	CHRYSOPHYCEAE,MARINE,SAPONIFIABLE FRACTION ONLY	1220
*SKELETONEMA COSTATUM	7 % OF ASH FREE DRY WEIGHT	BACILLARIOPHYCEAE,MARINE, SAPONIFIABLE FRACTION ONLY	1220
*TETRASELMIS MACULATA	4 % OF ASH FREE DRY WEIGHT	PRASINOPHYCEAE,MARINE,SAPONIFIABLE FRACTION ONLY	1220

TABLE A39

ALGAE : FE/BIOMASS RATIO			
ITEM:	VALUE:	CONDITIONS:	REF.:
*ANABAENA SP.	80 MICROGRAM/L	ALGAE FROM PONDS AND LAKES IN SOUTHEASTERN UNITED STATES	1660
*APHANIZOMENON SP.	833 MICROGRAM/L	ALGAE FROM PONDS AND LAKES IN SOUTHEASTERN UNITED STATES	1660
*CHARA SP.	2520 MICROGRAM/L	ALGAE FROM PONDS AND LAKES IN SOUTHEASTERN UNITED STATES	1660
*CLADOPHORA SP.	2300 MICROGRAM/L	ALGAE FROM PONDS AND LAKES IN SOUTHEASTERN UNITED STATES	1660
*EUGLENA SP.	1545 MICROGRAM/L	ALGAE FROM PONDS AND LAKES IN SOUTHEASTERN UNITED STATES	1660
*HYDRODICTYON SP.	1963 MICROGRAM/L	ALGAE FROM PONDS AND LAKES IN SOUTHEASTERN UNITED STATES	1660
*LYNGBYA SP.	3866 MICROGRAM/L	ALGAE FROM PONDS AND LAKES IN SOUTHEASTERN UNITED STATES	1660
*MICROCYSTIS SP.	322 MICROGRAM/L	ALGAE FROM PONDS AND LAKES IN SOUTHEASTERN UNITED STATES	1660
*MOUGEOTIA SP.	60 MICROGRAM/L	ALGAE FROM PONDS AND LAKES IN SOUTHEASTERN UNITED STATES	1660
*NITELLA SP.	2180 MICROGRAM/L	ALGAE FROM PONDS AND LAKES IN SOUTHEASTERN UNITED STATES	1660
*OEDOGONIUM SP.	1645 MICROGRAM/L	ALGAE FROM PONDS AND LAKES IN SOUTHEASTERN UNITED STATES	1660
*PITHOPHORA SP.	2836 MICROGRAM/L	ALGAE FROM PONDS AND LAKES IN SOUTHEASTERN UNITED STATES	1660
*RHIZOCLONIUM SP.	1820 MICROGRAM/L	ALGAE FROM PONDS AND LAKES IN SOUTHEASTERN UNITED STATES	1660
*SPIROGYRA SP.	1368 MICROGRAM/L	ALGAE FROM PONDS AND LAKES IN SOUTHEASTERN UNITED STATES	1660
*SPIROGYRA SP., GIANT	1793 MICROGRAM/L	ALGAE FROM PONDS AND LAKES IN SOUTHEASTERN UNITED STATES	1660

TABLE A40

ALGAE : GROWTH RATE			
ITEM:	VALUE:	CONDITIONS:	REF.:
*AMPHIDINIUM CARTERI	8.8 HOURS	DOUBLING TIME, CONTINUOUS SATURATING LIGHT,291 K	466
*AMPHIDINIUM SP.	3.0 DAYS (GENERATION TIME) (F001)	2*10**(-3) G-ATOM N/L ADDED AS NO3,MARINE, BATCH,293 K	1692
*AMPHIDINIUM SP.	NO GROWTH (F001)	2*10**(-3) G-ATOM N/L ADDED AS ALA,MARINE, BATCH,293 K	1692

TABLE A40

ALGAE : GROWTH RATE

ITEM:	VALUE:	CONDITIONS:	REF.:
*AMPHIDINIUM SP.	NO GROWTH (F001)	2*10**(-3) G-ATOM N/L ADDED AS ASP,MARINE, BATCH,293 K	1692
*AMPHIDINIUM SP.	NO GROWTH (F001)	2*10**(-3) G-ATOM N/L ADDED AS GLU,MARINE, BATCH,293 K	1692
*AMPHIDINIUM SP.	NO GROWTH (F001)	2*10**(-3) G-ATOM N/L ADDED AS GLY,MARINE, BATCH,293 K	1692
*AMPHIDINIUM SP.	4.5 DAYS (GENERATION TIME) (F001)	2*10**(-3) G-ATOM N/L ADDED AS LYS,MARINE, BATCH,293 K	1692
*AMPHIDINIUM SP.	3.3 DAYS (GENERATION TIME)(F001)	2*10**(-3) G-ATOM N/L ADDED AS ORN,MARINE, BATCH,293 K	1692
*AMPHIDINIUM SP.	NO GROWTH (F001)	2*10**(-3) G-ATOM N/L ADDED AS SER,MARINE, BATCH,293 K	1692
*AMPHIDINIUM SP.	NO GROWTH (F001)	2*10**(-3) G-ATOM N/L ADDED AS THR,MARINE, BATCH,293 K	1692
*AMPHIDINIUM SP.	9.2 DAYS (GENERATION TIME) (F001)	2*10**(-3) G-ATOM N/L ADDED AS VAL,MARINE, BATCH,293 K	1692
*AMPHIDINIUM SP.	NO GROWTH (F001)	2*10**(-3) G-ATOM N/L ADDED AS MIX OF AMINO ACIDS,MARINE, BATCH,293 K	1692
*ANABAENA CYLINDRICA	0.96 DOUBLINGS/DAY	298K,SATURATING LIGHT,SYNTHETIC MEDIUM,BLUE GREEN ALGA	688
*ANABAENA CYLINDRICA	10.6 HOURS	DOUBLING TIME, CONTINUOUS SATURATING LIGHT,298 K	466
*ANACYSTIS NIDULANS	11.5 DOUBLINGS/DAY	314K,SATURATING LIGHT,SYNTHETIC MEDIUM,BLUE GREEN ALGA	688
*ANACYSTIS NIDULANS	2.0 HOURS	DOUBLING TIME, CONTINUOUS SATURATING LIGHT,314 K	466
*ASTERIONELLA FORMOSA	1.9 DOUBLINGS/DAY	291.5K,SATURATING LIGHT, NONSYNTHETIC MEDIUM,DIATOM	688
*ASTERIONELLA FORMOSA	9.6 HOURS	DOUBLING TIME, CONTINUOUS SATURATING LIGHT,293 K	466
*ASTERIONELLA JAPONICA	1.1 DOUBLINGS OF CELL NO/DAY	CULTURE IN LOG PHASE, 20 % SUNLIGHT, 75 UG-ATOM N AS NO3, 293-298 K	428
*ASTERIONELLA JAPONICA	0.9 DOUBLINGS OF CELL NO/DAY	CULTURE IN LOG PHASE, 20 % SUNLIGHT, 75 UG-ATOM N AS UREA, 293-298 K	428
*ASTERIONELLA JAPONICA	0.0 DOUBLINGS OF CELL NO/DAY	CULTURE IN LOG PHASE, 20 % SUNLIGHT, NO ADDED N, 293-298 K	428
*ASTERIONELLA JAPONICA	13.9 HOURS	DOUBLING TIME, CONTINUOUS SATURATING LIGHT,293-298 K	466
*BACILLARIOPHYCEAE	1.35 DOUBLINGS OF CELL NO/DAY	CULTURE IN LOG PHASE, 20 % SUNLIGHT, 75 UG-ATOM N/L AS NO3, 293-298 K	428
*BACILLARIOPHYCEAE	1.16 DOUBLINGS OF CELL NO/DAY	CULTURE IN LOG PHASE, 20 % SUNLIGHT, 75 UG-ATOM N/L AS NH4, 293-298 K	428
*BACILLARIOPHYCEAE	1.33 DOUBLINGS OF CELL NO/DAY	CULTURE IN LOG PHASE, 20 % SUNLIGHT, NO ADDED N, 293-298 K	428
*BACILLARIOPHYCEAE	0.29 DOUBLINGS OF CELL NO/DAY	CULTURE IN LOG PHASE, 20 % SUNLIGHT, NO ADDED N, 293-298 K	428
*CACHONINA SP.	2.6 DAYS (GENERATION TIME) (F001)	2*10**(-3) G-ATOM N/L ADDED AS NO3,MARINE, BATCH,293 K	1692
*CACHONINA SP.	NO GROWTH (F001)	2*10**(-3) G-ATOM N/L ADDED AS ASP,MARINE, BATCH,293 K	1692
*CACHONINA SP.	NO GROWTH (F001)	2*10**(-3) G-ATOM N/L ADDED AS GLU,MARINE, BATCH,293 K	1692
*CACHONINA SP.	NO GROWTH (F001)	2*10**(-3) G-ATOM N/L ADDED AS GLY,MARINE, BATCH,293 K	1692
*CACHONINA SP.	NO GROWTH (F001)	2*10**(-3) G-ATOM N/L ADDED AS LYS,MARINE, BATCH,293 K	1692
*CACHONINA SP.	NO GROWTH (F001)	2*10**(-3) G-ATOM N/L ADDED AS ORN,MARINE, BATCH,293 K	1692
*CACHONINA SP.	NO GROWTH (F001)	2*10**(-3) G-ATOM N/L ADDED AS SER,MARINE, BATCH,293 K	1692
*CACHONINA SP.	NO GROWTH (F001)	2*10**(-3) G-ATOM N/L ADDED AS THR,MARINE, BATCH,293 K	1692
*CACHONINA SP.	NO GROWTH (F001)	2*10**(-3) G-ATOM N/L ADDED AS VAL,MARINE, BATCH,293 K	1692

TABLE A40

ALGAE : GROWTH RATE			
ITEM:	VALUE:	CONDITIONS:	REF.:
*CACHONINA SP.	8.0 DAYS (GENERATION TIME) (F001)	2*10**(-3) G-ATOM N/L ADDED AS MIX OF AMINO ACIDS,MARINE, BATCH,293 K	1692
*CERATIUM TRIPOS	82.8 HOURS	DOUBLING TIME, CONTINUOUS SATURATING LIGHT,293 K	466
*CHAETOCEROS LAUDERI	33 HOURS	GENERATION TIME,GROWN UNDER OPTIMAL CONDITION	355
*CHAETOCEROS SP.	1.45 DOUBLINGS OF CELL NO/DAY	CULTURE IN LOG PHASE, 20 % SUNLIGHT, 75 UG-ATOM N AS NO3, 293-298 K	428
*CHAETOCEROS SP.	1.13 DOUBLINGS OF CELL NO/DAY	CULTURE IN LOG PHASE, 20 % SUNLIGHT, 75 UG-ATOM N AS NH4, 293-298 K	428
*CHAETOCEROS SP.	1.21 DOUBLINGS OF CELL NO/DAY	CULTURE IN LOG PHASE, 20 % SUNLIGHT, 75 UG-ATOM N AS UREA, 293-298 K	428
*CHAETOCEROS SP.	0.31 DOUBLINGS OF CELL NO/DAY	CULTURE IN LOG PHASE, 20 % SUNLIGHT, NO ADDED N, 293-298 K	428
*CHAETOCEROS SP.	2.5 DAYS (GENERATION TIME) (F001)	2*10**(-3) G-ATOM N/L ADDED AS NO3,MARINE, BATCH,293 K	1692
*CHAETOCEROS SP.	3.6 DAYS (GENERATION TIME) (F001)	2*10**(-3) G-ATOM N/L ADDED AS ALA,MARINE, BATCH,293 K	1692
*CHAETOCEROS SP.	NO GROWTH (F001)	2*10**(-3) G-ATOM N/L ADDED AS ASP,MARINE, BATCH,293 K	1692
*CHAETOCEROS SP.	NO GROWTH (F001)	2*10**(-3) G-ATOM N/L ADDED AS GLU,MARINE, BATCH,293 K	1692
*CHAETOCEROS SP.	8.2 DAYS (GENERATION TIME) (F001)	2*10**(-3) G-ATOM N/L ADDED AS GLY,MARINE, BATCH,293 K	1692
*CHAETOCEROS SP.	VERY SLOW ALGAL GROWTH (F001)	2*10**(-3) G-ATOM N/L ADDED AS LYS,MARINE, BATCH,293 K	1692
*CHAETOCEROS SP.	3.8 DAYS (GENERATION TIME) (F001)	2*10**(-3) G-ATOM N/L ADDED AS ORN,MARINE, BATCH,293 K	1692
*CHAETOCEROS SP.	NO GROWTH (F001)	2*10**(-3) G-ATOM N/L ADDED AS SER,MARINE, BATCH,293 K	1692
*CHAETOCEROS SP.	VERY SLOW ALGAL GROWTH (F001)	2*10**(-3) G-ATOM N/L ADDED AS THR,MARINE, BATCH,293 K	1692
*CHAETOCEROS SP.	4.2 DAYS (GENERATION TIME) (F001)	2*10**(-3) G-ATOM N/L ADDED AS MIX OF AMINO ACIDS,MARINE, BATCH,293 K	1692
*CHAETOCEROS SP.	17 HOURS	GENERATION TIME,GROWN UNDER OPTIMAL CONDITION	1220
*CHLAMYDOMONAS REINHARDTII	3.8 DOUBLINGS/DAY	298K,SATURATING LIGHT,SYNTHETIC MEDIUM,GREEN ALGA	688
*CHLAMYDOMONAS SP.	3.4 DAYS (GENERATION TIME) (F001)	2*10**(-3) G-ATOM N/L ADDED AS NO3,MARINE, BATCH,293 K	1692
*CHLAMYDOMONAS SP.	3.4 DAYS (GENERATION TIME) (F001)	2*10**(-3) G-ATOM N/L ADDED AS ALA,MARINE, BATCH,293 K	1692
*CHLAMYDOMONAS SP.	NO GROWTH (F001)	2*10**(-3) G-ATOM N/L ADDED AS GLU,MARINE, BATCH,293 K	1692
*CHLAMYDOMONAS SP.	NO GROWTH (F001)	2*10**(-3) G-ATOM N/L ADDED AS GLY,MARINE, BATCH,293 K	1692
*CHLAMYDOMONAS SP.	NO GROWTH (F001)	2*10**(-3) G-ATOM N/L ADDED AS LYS,MARINE, BATCH,293 K	1692
*CHLAMYDOMONAS SP.	NO GROWTH(F001)	2*10**(-3) G-ATOM N/L ADDED AS SER,MARINE, BATCH,293 K	1692
*CHLAMYDOMONAS SP.	2.1 DAYS (GENERATION TIME)(F001)	2*10**(-3) G-ATOM N/L ADDED AS MIX OF AMINO ACIDS,MARINE, BATCH,293 K	1692
*CHLORELLA ELLIPSOIDEA	3.6 DOUBLINGS/DAY	298K,SATURATING LIGHT,SYNTHETIC MEDIUM,GREEN ALGA	688
*CHLORELLA PYRENOIDOSA	60.2 HOURS	DOUBLING TIME, CONTINUOUS SATURATING LIGHT,283 K, PLANKTONIC STRAIN	466
*CHLORELLA PYRENOIDOSA	19.6 HOURS	DOUBLING TIME, CONTINUOUS SATURATING LIGHT,293 K, PLANKTONIC STRAIN	466
*CHLORELLA PYRENOIDOSA	7.75 HOURS	DOUBLING TIME, CONTINUOUS SATURATING LIGHT,298 K	466
*CHLORELLA PYRENOIDOSA	8.0 HOURS	DOUBLING TIME, CONTINUOUS SATURATING LIGHT,298 K, EMERSON STRAIN	466

TABLE A40

ALGAE : GROWTH RATE			
ITEM:	VALUE:	CONDITIONS:	REF.:
*CHLORELLA PYRENOIDOSA	2.6 HOURS	DOUBLING TIME, CONTINUOUS SATURATING LIGHT,312 K, 7-11-05 STRAIN	466
*CHLORELLA SP.	1.5 DAYS (GENERATION TIME) (F001)	2*10**(-3) G-ATOM N/L ADDED AS NO3,MARINE, BATCH,293 K	1692
*CHLORELLA SP.	NO GROWTH (F001)	2*10**(-3) G-ATOM N/L ADDED AS ALA,MARINE, BATCH,293 K	1692
*CHLORELLA SP.	NO GROWTH (F001)	2*10**(-3) G-ATOM N/L ADDED AS ASP,MARINE, BATCH,293 K	1692
*CHLORELLA SP.	NO GROWTH (F001)	2*10**(-3) G-ATOM N/L ADDED AS GLU,MARINE, BATCH,293 K	1692
*CHLORELLA SP.	NO GROWTH (F001)	2*10**(-3) G-ATOM N/L ADDED AS GLY,MARINE, BATCH,293 K	1692
*CHLORELLA SP.	NO GROWTH (F001)	2*10**(-3) G-ATOM N/L ADDED AS LYS,MARINE, BATCH,293 K	1692
*CHLORELLA SP.	5.5 DAYS (GENERATION TIME) (F001)	2*10**(-3) G-ATOM N/L ADDED AS ORN,MARINE, BATCH,293 K	1692
*CHLORELLA SP.	NO GROWTH (F001)	2*10**(-3) G-ATOM N/L ADDED AS SER,MARINE, BATCH,293 K	1692
*CHLORELLA SP.	11.0 DAYS (GENERATION TIME) (F001)	2*10**(-3) G-ATOM N/L ADDED AS THR,MARINE, BATCH,293 K	1692
*CHLORELLA SP.	NO GROWTH (F001)	2*10**(-3) G-ATOM N/L ADDED AS VAL,MARINE, BATCH,293 K	1692
*CHLORELLA SP.	23.5 DAYS (GENERATION TIME) (F001)	2*10**(-3) G-ATOM N/L ADDED AS MIX OF AMINO ACIDS,MARINE, BATCH,293 K	1692
*COCCOLITHUS HUXLEYI	1.46 DOUBLINGS/DAY	NO COCCOLITHS,SUBARCTIC PACIFIC, SPRING	1219
*COCCOLITHUS HUXLEYI	1.37 DOUBLINGS/DAY	WITH COCCOLITHS,SUBTROPICAL,SUMMER	1219
*COCCOLITHUS HUXLEYI	1.59 DOUBLINGS/DAY	WITH COCCOLITHS,TROPICAL UPWELLING	1219
*COCCOLITHUS HUXLEYI	1.61 DOUBLINGS/DAY	NO COCCOLITHS,ANTARCTIC UPWELLING, SUMMER	1219
*COCCOLITHUS HUXLEYI	0.80 DOUBLINGS/DAY	WITH COCCOLITHS,COASTAL ESTUARINE, SPRING	1219
*COCCOLITHUS SP.	2.0 DAYS (GENERATION TIME) (F001)	2*10**(-3) G-ATOM N/L ADDED AS NO3,MARINE, BATCH,293 K	1692
*COCCOLITHUS SP.	3.3 DAYS (GENERATION TIME) (F001)	2*10**(-3) G-ATOM N/L ADDED AS ALA,MARINE, BATCH,293 K	1692
*COCCOLITHUS SP.	NO GROWTH (F001)	2*10**(-3) G-ATOM N/L ADDED AS ASP,MARINE, BATCH,293 K	1692
*COCCOLITHUS SP.	NO GROWTH (F001)	2*10**(-3) G-ATOM N/L ADDED AS GLU,MARINE, BATCH,293 K	1692
*COCCOLITHUS SP.	3.5 DAYS (GENERATION TIME) (F001)	2*10**(-3) G-ATOM N/L ADDED AS GLY,MARINE, BATCH,293 K	1692
*COCCOLITHUS SP.	NO GROWTH (F001)	2*10**(-3) G-ATOM N/L ADDED AS LYS,MARINE, BATCH,293 K	1692
*COCCOLITHUS SP.	NO GROWTH (F001)	2*10**(-3) G-ATOM N/L ADDED AS ORN,MARINE, BATCH,293 K	1692
*COCCOLITHUS SP.	NO GROWTH (F001)	2*10**(-3) G-ATOM N/L ADDED AS SER,MARINE, BATCH,293 K	1692
*COCCOLITHUS SP.	NO GROWTH (F001)	2*10**(-3) G-ATOM N/L ADDED AS THR,MARINE, BATCH,293 K	1692
*COCCOLITHUS SP.	NO GROWTH (F001)	2*10**(-3) G-ATOM N/L ADDED AS VAL,MARINE, BATCH,293 K	1692
*COCCOLITHUS SP.	NO GROWTH (F001)	2*10**(-3) G-ATOM N/L ADDED AS MIX OF AMINO ACIDS,MARINE, BATCH,293 K	1692
*COSCINODISCUS SP.	29 HOURS	GENERATION TIME,GROWN UNDER OPTIMAL CONDITION	355-
*COSCINODISCUS SP.	30.0 HOURS	DOUBLING TIME, CONTINUOUS SATURATING LIGHT,291 K	466
*CRICOSPHAERA GALBANA	20.1 HOURS	DOUBLING TIME, CONTINUOUS SATURATING LIGHT,291 K	466
*CRICOSPHAERA SP.	2.2 DAYS (GENERATION TIME) (F001)	2*10**(-3) G-ATOM N/L ADDED AS NO3,MARINE, BATCH,293 K	1692
*CRICOSPHAERA SP.	3.7 DAYS (GENERATION TIME) (F001)	2*10**(-3) G-ATOM N/L ADDED AS ALA,MARINE, BATCH,293 K	1692

TABLE A40

ALGAE : GROWTH RATE

ITEM:	VALUE:	CONDITIONS:	REF.:
*CRICOSPHAERA SP.	NO GROWTH (F001)	2*10**(-3) G-ATOM N/L ADDED AS ASP,MARINE, BATCH,293 K	1692
*CRICOSPHAERA SP.	NO GROWTH (F001)	2*10**(-3) G-ATOM N/L ADDED AS GLU,MARINE, BATCH,293 K	1692
*CRICOSPHAERA SP.	NO GROWTH (F001)	2*10**(-3) G-ATOM N/L ADDED AS GLY,MARINE, BATCH,293 K	1692
*CRICOSPHAERA SP.	3.0 DAYS (GENERATION TIME) (F001)	2*10**(-3) G-ATOM N/L ADDED AS LYS,MARINE, BATCH,293 K	1692
*CRICOSPHAERA SP.	2.6 DAYS (GENERATION TIME) (F001)	2*10**(-3) G-ATOM N/L ADDED AS ORN,MARINE, BATCH,293 K	1692
*CRICOSPHAERA SP.	NO GROWTH (F001)	2*10**(-3) G-ATOM N/L ADDED AS SER,MARINE, BATCH,293 K	1692
*CRICOSPHAERA SP.	NO GROWTH (F001)	2*10**(-3) G-ATOM N/L ADDED AS THR,MARINE, BATCH,293 K	1692
*CRICOSPHAERA SP.	3.5 DAYS (GENERATION TIME) (F001)	2*10**(-3) G-ATOM N/L ADDED AS VAL,MARINE, BATCH,293 K	1692
*CRICOSPHAERA SP.	2.4 DAYS (GENERATION TIME) (F001)	2*10**(-3) G-ATOM N/L ADDED AS MIX OF AMINO ACIDS,MARINE, BATCH,293 K	1692
*CYCLOTELLA SP.	2.1 DAYS (GENERATION TIME) (F001)	2*10**(-3) G-ATOM N/L ADDED AS NO3,MARINE, BATCH,293 K	1692
*CYCLOTELLA SP.	NO GROWTH (F001)	2*10**(-3) G-ATOM N/L ADDED AS ALA,MARINE, BATCH,293 K	1692
*CYCLOTELLA SP.	NO GROWTH (F001)	2*10**(-3) G-ATOM N/L ADDED AS ASP,MARINE, BATCH,293 K	1692
*CYCLOTELLA SP.	NO GROWTH (F001)	2*10**(-3) G-ATOM N/L ADDED AS GLU,MARINE, BATCH,293 K	1692
*CYCLOTELLA SP.	NO GROWTH (F001)	2*10**(-3) G-ATOM N/L ADDED AS GLY,MARINE, BATCH,293 K	1692
*CYCLOTELLA SP.	NO GROWTH (F001)	2*10**(-3) G-ATOM N/L ADDED AS LYS,MARINE, BATCH,293 K	1692
*CYCLOTELLA SP.	NO GROWTH (F001)	2*10**(-3) G-ATOM N/L ADDED AS ORN,MARINE, BATCH,293 K	1692
*CYCLOTELLA SP.	NO GROWTH (F001)	2*10**(-3) G-ATOM N/L ADDED AS SER,MARINE, BATCH,293 K	1692
*CYCLOTELLA SP.	NO GROWTH (F001)	2*10**(-3) G-ATOM N/L ADDED AS THR,MARINE, BATCH,293 K	1692
*CYCLOTELLA SP.	NO GROWTH (F001)	2*10**(-3) G-ATOM N/L ADDED AS VAL,MARINE, BATCH,293 K	1692
*CYCLOTELLA SP.	NO GROWTH (F001)	2*10**(-3) G-ATOM N/L ADDED AS MIX OF AMINO ACIDS,MARINE, BATCH,293 K	1692
*CYLINDROTHECA CLOSTERIUM	1.13 DOUBLINGS OF CELL NO/DAY	CULTURE IN LOG PHASE, 20 % SUNLIGHT, 75 UG-ATOM N/L AS NO3, 293-298 K	428
*CYLINDROTHECA CLOSTERIUM	1.10 DOUBLINGS OF CELL NO/DAY	CULTURE IN LOG PHASE, 20 % SUNLIGHT, 75 UG-ATOM N/L AS NH4, 293-298 K	428
*CYLINDROTHECA CLOSTERIUM	1.10 DOUBLINGS OF CELL NO/DAY	CULTURE IN LOG PHASE, 20 % SUNLIGHT, 75 UG-ATOM N/L AS UREA, 293-298 K	428
*CYLINDROTHECA CLOSTERIUM	0.0 DOUBLINGS OF CELL NO/DAY	CULTURE IN LOG PHASE, 20 % SUNLIGHT, NO ADDED N, 293-298 K	428
*DITYLUM BRIGHTWELLII	1.22 DOUBLINGS/DAY	HEALTHY CELLS,SUBARCTIC PACIFIC, SPRING	1219
*DITYLUM BRIGHTWELLII	0.84 DOUBLINGS/DAY	SENESCENT CELLS,SUBTROPICAL,SUMMER	1219
*DITYLUM BRIGHTWELLII	1.90 DOUBLINGS/DAY	HEALTHY CELLS,TROPICAL UPWELLING	1219
*DITYLUM BRIGHTWELLII	1.66 DOUBLINGS/DAY	HEALTHY CELLS,ANTARCTIC UPWELLING, SUMMER	1219
*DITYLUM BRIGHTWELLII	1.48 DOUBLINGS/DAY	HEALTHY CELLS,COASTAL ESTUARINE, SPRING	1219
*DITYLUM BRIGHTWELLII	15 HOURS	GENERATION TIME,GROWN UNDER OPTIMAL CONDITION	1522
*DUNALIELLA SP.	2.2 DAYS (GENERATION TIME) (F001)	2*10**(-3) G-ATOM N/L ADDED AS NO3,MARINE, BATCH,293 K	1692
*DUNALIELLA SP.	NO GROWTH (F001)	2*10**(-3) G-ATOM N/L ADDED AS ALA,MARINE, BATCH,293 K	1692
*DUNALIELLA SP.	NO GROWTH (F001)	2*10**(-3) G-ATOM N/L ADDED AS ASP,MARINE, BATCH,293 K	1692

TABLE A40

ALGAE : GROWTH RATE			
ITEM:	VALUE:	CONDITIONS:	REF.:
*DUNALIELLA SP.	NO GROWTH (F001)	2*10**(-3) G-ATOM N/L ADDED AS GLU,MARINE, BATCH,293 K	1692
*DUNALIELLA SP.	NO GROWTH (F001)	2*10**(-3) G-ATOM N/L ADDED AS GLY,MARINE, BATCH,293 K	1692
*DUNALIELLA SP.	NO GROWTH (F001)	2*10**(-3) G-ATOM N/L ADDED AS LYS,MARINE, BATCH,293 K	1692
*DUNALIELLA SP.	NO GROWTH (F001)	2*10**(-3) G-ATOM N/L ADDED AS ORN,MARINE, BATCH,293 K	1692
*DUNALIELLA SP.	NO GROWTH (F001)	2*10**(-3) G-ATOM N/L ADDED AS SER,MARINE, BATCH,293 K	1692
*DUNALIELLA SP.	NO GROWTH (F001)	2*10**(-3) G-ATOM N/L ADDED AS THR,MARINE, BATCH,293 K	1692
*DUNALIELLA SP.	NO GROWTH (F001)	2*10**(-3) G-ATOM N/L ADDED AS VAL,MARINE, BATCH,293 K	1692
*DUNALIELLA SP.	NO GROWTH	2*10**(-3) G-ATOM N/L ADDED AS MIX OF AMINO ACIDS,MARINE, BATCH,293 K	1692
*DUNALIELLA TERTIOLECTA	0.018+-0.001 1/HOUR	NO3=4.46, NO2=0.20, NH4=0.19 UG-ATOM N/L, SPECIFIC GROWTH RATE,N=3	119
*DUNALIELLA TERTIOLECTA	0.045+-0.001 1/HOUR	NO3=4.03, NO2=0.29, NH4=0.31 UG-ATOM N/L, SPECIFIC GROWTH RATE,N=3	119
*DUNALIELLA TERTIOLECTA	0.057+-0.001 1/HOUR	NO3=6.60, NO2=0.16, NH4=0.14 UG-ATOM N/L, SPECIFIC GROWTH RATE,N=3	119
*DUNALIELLA TERTIOLECTA	0.065+-0.002 1/HOUR	NO3=19.53, NO2=0.05, NH4=0.45 UG-ATOM N/L, SPECIFIC GROWTH RATE,N=3	119
*DUNALIELLA TERTIOLECTA	0.076+-0.001 1/HOUR	NO3=16.79, NO2=0.02, NH4=0.00 UG-ATOM N/L, SPECIFIC GROWTH RATE,N=3	119
*DUNALIELLA TERTIOLECTA	0.074+-0.001 1/HOUR	NO3=19.00, NO2=0.00, NH4=0.10 UG-ATOM N/L, SPECIFIC GROWTH RATE,N=3	119
*DUNALIELLA TERTIOLECTA	0.073+-0.001 1/HOUR	NO3=17.16, NO2=0.24, NH4=0.47 UG-ATOM N/L, SPECIFIC GROWTH RATE,N=3	119
*DUNALIELLA TERTIOLECTA	0.074+-0.001 1/HOUR	NO3=16.20, NO2=0.22, NH4=5.13 UG-ATOM N/L, SPECIFIC GROWTH RATE,N=3	119
*DUNALIELLA TERTIOLECTA	0.074+-0.001 1/HOUR	NO3=12.75, NO2=0.04, NH4=13.10 UG-ATOM N/L, SPECIFIC GROWTH RATE,N=3	119
*DUNALIELLA TERTIOLECTA	0.073+-0.001 1/HOUR	NO3=6.42, NO2=0.28, NH4=23.80 UG-ATOM N/L, SPECIFIC GROWTH RATE,N=3	119
*DUNALIELLA TERTIOLECTA	0.073 1/HOUR	NO3=4.05, NO2=0.78, NH4=14.98 UG-ATOM N/L, SPECIFIC GROWTH RATE,N=3	119
*DUNALIELLA TERTIOLECTA	0.073+-0.002 1/HOUR	NO3=2.92, NO2=0.36, NH4=23.70 UG-ATOM N/L, SPECIFIC GROWTH RATE,N=3	119
*DUNALIELLA TERTIOLECTA	0.074 1/HOUR	NO3=2.88, NO2=0.06, NH4=27.50 UG-ATOM N/L, SPECIFIC GROWTH RATE,N=3	119
*DUNALIELLA TERTIOLECTA	0.073+-0.001 1/HOUR	NO3=7.91, NO2=0.14, NH4=13.50 UG-ATOM N/L, SPECIFIC GROWTH RATE,N=3	119
*EUGLENA GRACILIS	2.2 DOUBLINGS/DAY	298K,SATURATING LIGHT,SYNTHETIC MEDIUM,EUGLENOID ALGA	688
*HEMIAULUS SINENSIS	1.3 DOUBLINGS OF CELL NO/DAY	CULTURE IN LOG PHASE, 20 % SUNLIGHT, 75 UG-ATOM N/L AS NO3, 293-298 K	428
*HEMIAULUS SINENSIS	0.70 DOUBLINGS OF CELL NO/DAY	CULTURE IN LOG PHASE, 20 % SUNLIGHT, 75 UG-ATOM N/L AS NH4, 293-298 K	428
*HEMIAULUS SINENSIS	1.16 DOUBLINGS OF CELL NO/DAY	CULTURE IN LOG PHASE, 20 % SUNLIGHT, 75 UG-ATOM N/L AS UREA, 293-298 K	428
*HEMIAULUS SINENSIS	0.0 DOUBLINGS OF CELL NO/DAY	CULTURE IN LOG PHASE, 20 % SUNLIGHT, NO ADDED N, 293-298 K	428
*ISOCHRYSIS GALBANA	30.2 HOURS	DOUBLING TIME, CONTINUOUS SATURATING LIGHT,293 K	466

TABLE A40

ALGAE : GROWTH RATE

ITEM:	VALUE:	CONDITIONS:	REF.:
*ISOCHRYSIS SP.	1.5 DAYS (GENERATION TIME) (F001)	2*10**(-3) G-ATOM N/L ADDED AS NO3,MARINE, BATCH,293 K	1692
*ISOCHRYSIS SP.	8.2 DAYS (GENERATION TIME) (F001)	2*10**(-3) G-ATOM N/L ADDED AS ALA,MARINE, BATCH,293 K	1692
*ISOCHRYSIS SP.	NO GROWTH (F001)	2*10**(-3) G-ATOM N/L ADDED AS ASP,MARINE, BATCH,293 K	1692
*ISOCHRYSIS SP.	NO GROWTH (F001)	2*10**(-3) G-ATOM N/L ADDED AS GLU,MARINE, BATCH,293 K	1692
*ISOCHRYSIS SP.	3.6 DAYS (GENERATION TIME) (F001)	2*10**(-3) G-ATOM N/L ADDED AS GLY,MARINE, BATCH,293 K	1692
*ISOCHRYSIS SP.	5.1 DAYS (GENERATION TIME) (F001)	2*10**(-3) G-ATOM N/L ADDED AS LYS,MARINE, BATCH,293 K	1692
*ISOCHRYSIS SP.	5.0 DAYS (GENERATION TIME) (F001)	2*10**(-3) G-ATOM N/L ADDED AS ORN,MARINE, BATCH,293 K	1692
*ISOCHRYSIS SP.	6.6 DAYS (GENERATION TIME) (F001)	2*10**(-3) G-ATOM N/L ADDED AS SER,MARINE, BATCH,293 K	1692
*ISOCHRYSIS SP.	NO GROWTH (F001)	2*10**(-3) G-ATOM N/L ADDED AS THR,MARINE, BATCH,293 K	1692
*ISOCHRYSIS SP.	5.5 DAYS (GENERATION TIME) (F001)	2*10**(-3) G-ATOM N/L ADDED AS VAL,MARINE, BATCH,293 K	1692
*ISOCHRYSIS SP.	6.5 DAYS (GENERATION TIME)(F001)	2*10**(-3) G-ATOM N/L ADDED AS MIX OF AMINO ACIDS,MARINE, BATCH,293 K	1692
*LAMINARIA LONGICRURIS	0.21 - 0.29 1/MOUNTH	STRIPE LENGTH = 0-24 CM, BIOMASS = 122 G, NOVA SCOTIA	977
*LEPTOCYLINDRUS DANICUS	0.95 DOUBLINGS OF CELL NO/DAY	CULTURE IN LOG PHASE, 20 % SUNLIGHT, 75 UG-ATOM N AS NO3, 293-298 K	428
*LEPTOCYLINDRUS DANICUS	0.96 DOUBLINGS OF CELL NO/DAY	CULTURE IN LOG PHASE, 20 % SUNLIGHT, 75 UG-ATOM N AS NH4, 293-298 K	428
*LEPTOCYLINDRUS DANICUS	1.34 DOUBLINGS OF CELL NO/DAY	CULTURE IN LOG PHASE, 20 % SUNLIGHT, 75 UG-ATOM N AS UREA, 293-298 K	428
*LEPTOCYLINDRUS DANICUS	0.40 DOUBLINGS OF CELL NO/DAY	CULTURE IN LOG PHASE, 20 % SUNLIGHT, NO ADDED N, 293-298 K	428
*MARINE PHYTOPLANKTON	0.26 DOUBLINGS/DAY	AVERAGE VAL.,NUTRIENT-POOR WATER, SARGASSO SEA	427
*MARINE PHYTOPLANKTON	0.45 DOUBLINGS/DAY	AVERAGE VAL.,NUTRIENT-POOR WATER, FLORIDA STRAIT	427
*MARINE PHYTOPLANKTON	0.37 DOUBLINGS/DAY	AVERAGE VAL.,NUTRIENT-POOR WATER, OFF THE CAROLINAS	427
*MARINE PHYTOPLANKTON	0.35 DOUBLINGS/DAY	AVERAGE VAL.,NUTRIENT-POOR WATER, OFF MONTAUK PT.	427
*MARINE PHYTOPLANKTON	0.25-0.4 DOUBLINGS/DAY	AVERAGE VAL.,NUTRIENT-POOR WATER, OFF SOUTHERN CALIFORNIA,293K	427
*MARINE PHYTOPLANKTON	0.7 DOUBLINGS/DAY	AVERAGE VAL.,NUTRIENT-POOR WATER, OFF SOUTHERN CALIFORNIA,285-294K	427
*MARINE PHYTOPLANKTON	0.67 DOUBLINGS/DAY	AVERAGE VAL.,NUTRIENT-RICH WATER, PERU CURRENT,290-293K,APRIL 1966	427
*MARINE PHYTOPLANKTON	0.73 DOUBLINGS/DAY	AVERAGE VAL.,NUTRIENT-RICH WATER, PERU CURRENT,291-292K,JUNE 1969	427
*MARINE PHYTOPLANKTON	1.0 DOUBLINGS/DAY	AVERAGE VAL.,NUTRIENT-RICH WATER, OFF S.W.AFRICA	427
*MARINE PHYTOPLANKTON	> 1.0 DOUBLINGS/DAY	AVERAGE VAL.,NUTRIENT-RICH WATER, WESTERN ARABIAN SEA,300-301K	427
*MELOSIRA SP.	VERY SLOW ALGAL GROWTH (F001)	2*10**(-3) G-ATOM N/L ADDED AS NO3,MARINE, BATCH,293 K	1692
*MELOSIRA SP.	VERY SLOW ALGAL GROWTH (F001)	2*10**(-3) G-ATOM N/L ADDED AS ALA,MARINE, BATCH,293 K	1692
*MELOSIRA SP.	NO GROWTH (F001)	2*10**(-3) G-ATOM N/L ADDED AS ASP,MARINE, BATCH,293 K	1692
*MELOSIRA SP.	NO GROWTH (F001)	2*10**(-3) G-ATOM N/L ADDED AS GLU,MARINE, BATCH,293 K	1692
*MELOSIRA SP.	VERY SLOW ALGAL GROWTH (F001)	2*10**(-3) G-ATOM N/L ADDED AS GLY,MARINE, BATCH,293 K	1692
*MELOSIRA SP.	NO GROWTH (F001)	2*10**(-3) G-ATOM N/L ADDED AS LYS,MARINE, BATCH,293 K	1692

TABLE A40

ALGAE : GROWTH RATE

ITEM:	VALUE:	CONDITIONS:	REF.:
*MELOSIRA SP.	NO GROWTH (F001)	2*10**(-3) G-ATOM N/L ADDED AS ORN,MARINE, BATCH,293 K	1692
*MELOSIRA SP.	VERY SLOW ALGAL GROWTH (F001)	2*10**(-3) G-ATOM N/L ADDED AS SER,MARINE, BATCH,293 K	1692
*MELOSIRA SP.	VERY SLOW ALGAL GROWTH	2*10**(-3) G-ATOM N/L ADDED AS THR,MARINE, BATCH,293 K	1692
*MELOSIRA SP.	NO GROWTH (F001)	2*10**(-3) G-ATOM N/L ADDED AS VAL,MARINE, BATCH,293 K	1692
*MELOSIRA SP.	NO GROWTH (F001)	2*10**(-3) G-ATOM N/L ADDED AS MIX OF AMINO ACIDS,MARINE, BATCH,293 K	1692
*MONODUS SUBTERRANEUS	97.7 HOURS	DOUBLING TIME, CONTINUOUS SATURATING LIGHT,288 K	466
*MONODUS SUBTERRANEUS	37.8 HOURS	DOUBLING TIME, CONTINUOUS SATURATING LIGHT,293 K	466
*MONODUS SUBTERRANEUS	24.3 HOURS	DOUBLING TIME, CONTINUOUS SATURATING LIGHT,298 K	466
*MONODUS SUBTERRANEUS	42.7 HOURS	DOUBLING TIME, CONTINUOUS SATURATING LIGHT,303 K	466
*NAVICULA INCERTA	3.8 DAYS (GENERATION TIME) (F001)	2*10**(-3) G-ATOM N/L ADDED AS NO3,MARINE, BATCH,293 K	1692
*NAVICULA INCERTA	4.0 DAYS (GENERATION TIME) (F001)	2*10**(-3) G-ATOM N/L ADDED AS ALA,MARINE, BATCH,293 K	1692
*NAVICULA INCERTA	3.0 DAYS (GENERATION TIME) (F001)	2*10**(-3) G-ATOM N/L ADDED AS ASP,MARINE, BATCH,293 K	1692
*NAVICULA INCERTA	3.0 DAYS (GENERATION TIME)(F001)	2*10**(-3) G-ATOM N/L ADDED AS GLU,MARINE, BATCH,293 K	1692
*NAVICULA INCERTA	9.4 DAYS (GENERATION TIME)(F001)	2*10**(-3) G-ATOM N/L ADDED AS GLY,MARINE, BATCH,293 K	1692
*NAVICULA INCERTA	7.6 DAYS (GENERATION TIME) (F001)	2*10**(-3) G-ATOM N/L ADDED AS LYS,MARINE, BATCH,293 K	1692
*NAVICULA INCERTA	13.8 DAYS (GENERATION TIME)(F001)	2*10**(-3) G-ATOM N/L ADDED AS ORN,MARINE, BATCH,293 K	1692
*NAVICULA INCERTA	4.7 DAYS (GENERATION TIME) (F001)	2*10**(-3) G-ATOM N/L ADDED AS SER,MARINE, BATCH,293 K	1692
*NAVICULA INCERTA	3.8 DAYS (GENERATION TIME) (F001)	2*10**(-3) G-ATOM N/L ADDED AS THR,MARINE, BATCH,293 K	1692
*NAVICULA INCERTA	4.0 DAYS (GENERATION TIME) (F001)	2*10**(-3) G-ATOM N/L ADDED AS VAL,MARINE, BATCH,293 K	1692
*NAVICULA INCERTA	4.0 DAYS (GENERATION TIME)(F001)	2*10**(-3) G-ATOM N/L ADDED AS MIX OF AMINO ACIDS,MARINE, BATCH,293 K	1692
*NAVICULA MINIMA	1.4 DOUBLINGS/DAY	298K,SATURATING LIGHT,NONSYNTHETIC MEDIUM,DIATOM	688
*NAVICULA SALINARUM	11.3 DAYS (GENERATION TIME) (F001)	2*10**(-3) G-ATOM N/L ADDED AS NO3,MARINE, BATCH,293 K	1692
*NAVICULA SALINARUM	7.5 DAYS (GENERATION TIME) (F001)	2*10**(-3) G-ATOM N/L ADDED AS ALA,MARINE, BATCH,293 K	1692
*NAVICULA SALINARUM	NO GROWTH (F001)	2*10**(-3) G-ATOM N/L ADDED AS ASP,MARINE, BATCH,293 K	1692
*NAVICULA SALINARUM	NO GROWTH (F001)	2*10**(-3) G-ATOM N/L ADDED AS GLU,MARINE, BATCH,293 K	1692
*NAVICULA SALINARUM	NO GROWTH (F001)	2*10**(-3) G-ATOM N/L ADDED AS GLY,MARINE, BATCH,293 K	1692
*NAVICULA SALINARUM	NO GROWTH (F001)	2*10**(-3) G-ATOM N/L ADDED AS LYS,MARINE, BATCH,293 K	1692
*NAVICULA SALINARUM	NO GROWTH (F001)	2*10**(-3) G-ATOM N/L ADDED AS SER,MARINE, BATCH,293 K	1692
*NAVICULA SALINARUM	NO GROWTH (F001)	2*10**(-3) G-ATOM N/L ADDED AS THR,MARINE, BATCH,293 K	1692
*NAVICULA SALINARUM	NO GROWTH (F001)	2*10**(-3) G-ATOM N/L ADDED AS VAL,MARINE, BATCH,293 K	1692
*NAVICULA SALINARUM	8.0 DAYS (GENERATION TIME) (F001)	2*10**(-3) G-ATOM N/L ADDED AS MIX OF AMINO ACIDS,MARINE, BATCH,293 K	1692
*NITZSCHIA FRUSTULUM	3.4 DAYS (GENERATION TIME)(F001)	2*10**(-3) G-ATOM N/L ADDED AS NO3,MARINE, BATCH,293 K	1692
*NITZSCHIA FRUSTULUM	4.3 DAYS (GENERATION TIME)(F001)	2*10**(-3) G-ATOM N/L ADDED AS ALA,MARINE, BATCH,293 K	1692

TABLE A40

ALGAE : GROWTH RATE

ITEM:	VALUE:	CONDITIONS:	REF.:
*NITZSCHIA FRUSTULUM	5.8 DAYS (GENERATION TIME) (F001)	2*10**(-3) G-ATOM N/L ADDED AS ASP,MARINE, BATCH,293 K	1692
*NITZSCHIA FRUSTULUM	5.8 DAYS (GENERATION TIME)(F001)	2*10**(-3) G-ATOM N/L ADDED AS GLU,MARINE, BATCH,293 K	1692
*NITZSCHIA FRUSTULUM	7.3 DAYS (GENERATION TIME) (F001)	2*10**(-3) G-ATOM N/L ADDED AS GLY,MARINE, BATCH,293 K	1692
*NITZSCHIA FRUSTULUM	4.7 DAYS (GENERATION TIME) (F001)	2*10**(-3) G-ATOM N/L ADDED AS LYS,MARINE, BATCH,293 K	1692
*NITZSCHIA FRUSTULUM	5.2 DAYS (GENERATION TIME)(F001)	2*10**(-3) G-ATOM N/L ADDED AS ORN,MARINE, BATCH,293 K	1692
*NITZSCHIA FRUSTULUM	6.5 DAYS (GENERATION TIME)(F001)	2*10**(-3) G-ATOM N/L ADDED AS SER,MARINE, BATCH,293 K	1692
*NITZSCHIA FRUSTULUM	6.0 DAYS (GENERATION TIME)(F001)	2*10**(-3) G-ATOM N/L ADDED AS THR,MARINE, BATCH,293 K	1692
*NITZSCHIA FRUSTULUM	5.7 DAYS (GENERATION TIME) (F001)	2*10**(-3) G-ATOM N/L ADDED AS VAL,MARINE, BATCH,293 K	1692
*NITZSCHIA FRUSTULUM	6.5 DAYS (GENERATION TIME) (F001)	2*10**(-3) G-ATOM N/L ADDED AS MIX OF AMINO ACIDS,MARINE, BATCH,293 K	1692
*NITZSCHIA OVALIS	3.5 DAYS (GENERATION TIME) (F001)	2*10**(-3) G-ATOM N/L ADDED AS NO3,MARINE, BATCH,293 K	1692
*NITZSCHIA OVALIS	3.3 DAYS (GENERATION TIME)(F001)	2*10**(-3) G-ATOM N/L ADDED AS ALA,MARINE, BATCH,293 K	1692
*NITZSCHIA OVALIS	1.7 DAYS (GENERATION TIME) (F001)	2*10**(-3) G-ATOM N/L ADDED AS ASP,MARINE, BATCH,293 K	1692
*NITZSCHIA OVALIS	1.7 DAYS (GENERATION TIME) (F001)	2*10**(-3) G-ATOM N/L ADDED AS GLU,MARINE, BATCH,293 K	1692
*NITZSCHIA OVALIS	5.4 DAYS (GENERATION TIME) (F001)	2*10**(-3) G-ATOM N/L ADDED AS GLY,MARINE, BATCH,293 K	1692
*NITZSCHIA OVALIS	4.4 DAYS (GENERATION TIME) (F001)	2*10**(-3) G-ATOM N/L ADDED AS LYS,MARINE, BATCH,293 K	1692
*NITZSCHIA OVALIS	4.0 DAYS (GENERATION TIME) (F001)	2*10**(-3) G-ATOM N/L ADDED AS ORN,MARINE, BATCH,293 K	1692
*NITZSCHIA OVALIS	4.8 DAYS (GENERATION TIME) (F001)	2*10**(-3) G-ATOM N/L ADDED AS SER,MARINE, BATCH,293 K	1692
*NITZSCHIA OVALIS	9.5 DAYS (GENERATION TIME) (F001)	2*10**(-3) G-ATOM N/L ADDED AS THR,MARINE, BATCH,293 K	1692
*NITZSCHIA OVALIS	4.4 DAYS (GENERATION TIME) (F001)	2*10**(-3) G-ATOM N/L ADDED AS VAL,MARINE, BATCH,293 K	1692
*NITZSCHIA OVALIS	2.0 DAYS (GENERATION TIME) (F001)	2*10**(-3) G-ATOM N/L ADDED AS MIX OF AMINO ACIDS,MARINE, BATCH,293 K	1692
*NITZSCHIA SP.	1.52 DOUBLINGS OF CELL NO/DAY	CULTURE IN LOG PHASE, 20 % SUNLIGHT, 75 UG-ATOM N AS NO3, 293-298 K	428
*NITZSCHIA SP.	1.36 DOUBLINGS OF CELL NO/DAY	CULTURE IN LOG PHASE, 20 % SUNLIGHT, 75 UG-ATOM N AS NH4, 293-298 K	428
*NITZSCHIA SP.	1.48 DOUBLINGS OF CELL NO/DAY	CULTURE IN LOG PHASE, 20 % SUNLIGHT, 75 UG-ATOM N AS UREA, 293-298 K	428
*NITZSCHIA SP.	0.30 DOUBLINGS OF CELL NO/DAY	CULTURE IN LOG PHASE, 20 % SUNLIGHT, NO ADDED N, 293-298 K	428
*NITZSCHIA SP.	1.8 DAYS (GENERATION TIME) (F001)	2*10**(-3) G-ATOM N/L ADDED AS NO3,MARINE, BATCH,293 K	1692
*NITZSCHIA SP.	4.2 DAYS (GENERATION TIME) (F001)	2*10**(-3) G-ATOM N/L ADDED AS ALA,MARINE, BATCH,293 K	1692
*NITZSCHIA SP.	3.0 DAYS (GENERATION TIME) (F001)	2*10**(-3) G-ATOM N/L ADDED AS ASP,MARINE, BATCH,293 K	1692
*NITZSCHIA SP.	3.0 DAYS (GENERATION TIME) (F001)	2*10**(-3) G-ATOM N/L ADDED AS GLU,MARINE, BATCH,293 K	1692
*NITZSCHIA SP.	2.0 DAYS (GENERATION TIME) (F001)	2*10**(-3) G-ATOM N/L ADDED AS GLY,MARINE, BATCH,293 K	1692
*NITZSCHIA SP.	2.9 DAYS (GENERATION TIME) (F001)	2*10**(-3) G-ATOM N/L ADDED AS LYS,MARINE, BATCH,293 K	1692
*NITZSCHIA SP.	2.7 DAYS (GENERATION TIME) (F001)	2*10**(-3) G-ATOM N/L ADDED AS ORN,MARINE, BATCH,293 K	1692
*NITZSCHIA SP.	2.0 DAYS (GENERATION TIME)(F001)	2*10**(-3) G-ATOM N/L ADDED AS SER,MARINE, BATCH,293 K	1692

TABLE A40

ALGAE : GROWTH RATE

ITEM:	VALUE:	CONDITIONS:	REF.:
*NITZSCHIA SP.	3.0 DAYS (GENERATION TIME) (F001)	2*10**(-3) G-ATOM N/L ADDED AS THR,MARINE, BATCH,293 K	1692
*NITZSCHIA SP.	3.7 DAYS (GENERATION TIME) (F001)	2*10**(-3) G-ATOM N/L ADDED AS VAL,MARINE, BATCH,293 K	1692
*NITZSCHIA SP.	3.0 DAYS (GENERATION TIME) (F001)	2*10**(-3) G-ATOM N/L ADDED AS MIX OF AMINO ACIDS,MARINE, BATCH,293 K	1692
*PHAEODACTYLUM TRICORNUTUM	3.5 DAYS (GENERATION TIME) (F001)	2*10**(-3) G-ATOM N/L ADDED AS NO3,MARINE, BATCH,293 K	1692
*PHAEODACTYLUM TRICORNUTUM	1.5 DAYS (GENERATION TIME) (F001)	2*10**(-3) G-ATOM N/L ADDED AS NO3,MARINE, BATCH,293 K	1692
*PHAEODACTYLUM TRICORNUTUM	4.1 DAYS (GENERATION TIME) (F001)	2*10**(-3) G-ATOM N/L ADDED AS ALA,MARINE, BATCH,293 K	1692
*PHAEODACTYLUM TRICORNUTUM	1.0 DAYS (GENERATION TIME) (F001)	2*10**(-3) G-ATOM N/L ADDED AS ALA,MARINE, BATCH,293 K	1692
*PHAEODACTYLUM TRICORNUTUM	2.5 DAYS (GENERATION TIME) (F001)	2*10**(-3) G-ATOM N/L ADDED AS ASP,MARINE, BATCH,293 K	1692
*PHAEODACTYLUM TRICORNUTUM	2.2 DAYS (GENERATION TIME) (F001)	2*10**(-3) G-ATOM N/L ADDED AS ASP,MARINE, BATCH,293 K	1692
*PHAEODACTYLUM TRICORNUTUM	2.5 DAYS (GENERATION TIME) (F001)	2*10**(-3) G-ATOM N/L ADDED AS GLU,MARINE, BATCH,293 K	1692
*PHAEODACTYLUM TRICORNUTUM	2.2 DAYS (GENERATION TIME)(F001)	2*10**(-3) G-ATOM N/L ADDED AS GLU,MARINE, BATCH,293 K	1692
*PHAEODACTYLUM TRICORNUTUM	5.6 DAYS (GENERATION TIME)(F001)	2*10**(-3) G-ATOM N/L ADDED AS GLY,MARINE, BATCH,293 K	1692
*PHAEODACTYLUM TRICORNUTUM	2.1 DAYS (GENERATION TIME) (F001)	2*10**(-3) G-ATOM N/L ADDED AS GLY,MARINE, BATCH,293 K	1692
*PHAEODACTYLUM TRICORNUTUM	3.6 DAYS (GENERATION TIME) (F001)	2*10**(-3) G-ATOM N/L ADDED AS LYS,MARINE, BATCH,293 K	1692
*PHAEODACTYLUM TRICORNUTUM	2.3 DAYS (GENERATION TIME) (F001)	2*10**(-3) G-ATOM N/L ADDED AS LYS,MARINE, BATCH,293 K	1692
*PHAEODACTYLUM TRICORNUTUM	5.7 DAYS (GENERATION TIME) (F001)	2*10**(-3) G-ATOM N/L ADDED AS ORN,MARINE, BATCH,293 K	1692
*PHAEODACTYLUM TRICORNUTUM	1.7 DAYS (GENERATION TIME) (F001)	2*10**(-3) G-ATOM N/L ADDED AS ORN,MARINE, BATCH,293 K	1692
*PHAEODACTYLUM TRICORNUTUM	4.4 DAYS (GENERATION TIME) (F001)	2*10**(-3) G-ATOM N/L ADDED AS SER,MARINE, BATCH,293 K	1692
*PHAEODACTYLUM TRICORNUTUM	1.8 DAYS (GENERATION TIME) (F001)	2*10**(-3) G-ATOM N/L ADDED AS SER,MARINE, BATCH,293 K	1692
*PHAEODACTYLUM TRICORNUTUM	NO GROWTH (F001)	2*10**(-3) G-ATOM N/L ADDED AS THR,MARINE, BATCH,293 K	1692
*PHAEODACTYLUM TRICORNUTUM	NO GROWTH (F001)	2*10**(-3) G-ATOM N/L ADDED AS THR,MARINE, BATCH,293 K	1692
*PHAEODACTYLUM TRICORNUTUM	1.7 DAYS (GENERATION TIME) (F001)	2*10**(-3) G-ATOM N/L ADDED AS VAL,MARINE, BATCH,293 K	1692
*PHAEODACTYLUM TRICORNUTUM	3.0 DAYS (GENERATION TIME) (F001)	2*10**(-3) G-ATOM N/L ADDED AS VAL,MARINE, BATCH,293 K	1692
*PHAEODACTYLUM TRICORNUTUM	5.9 DAYS (GENERATION TIME) (F001)	2*10**(-3) G-ATOM N/L ADDED AS MIX OF AMINO ACIDS,MARINE, BATCH,293 K	1692
*PHAEODACTYLUM TRICORNUTUM	2.0 DAYS (GENERATION TIME) (F001)	2*10**(-3) G-ATOM N/L ADDED AS MIX OF AMINO ACIDS,MARINE, BATCH,293 K	1692
*PHAEODACTYLUM TRICORNUTUM	13 HOURS	GENERATION TIME,GROWN UNDER OPTIMAL CONDITION	355
*PHAEODACTYLUM TRICORNUTUM	2.4 DOUBLINGS/DAY	281K,NONSYNTHETIC MEDIUM,DIATOM	688
*PHAEODACTYLUM TRICORNUTUM	10.0 HOURS	DOUBLING TIME, CONTINUOUS SATURATING LIGHT,298 K	466
*PLATYMONAS SP.	1.6 DAYS (GENERATION TIME) (F001)	2*10**(-3) G-ATOM N/L ADDED AS NO3,MARINE, BATCH,293 K	1692
*PLATYMONAS SP.	1.2 DAYS (GENERATION TIME) (F001)	2*10**(-3) G-ATOM N/L ADDED AS ALA,MARINE, BATCH,293 K	1692
*PLATYMONAS SP.	3.6 DAYS (GENERATION TIME) (F001)	2*10**(-3) G-ATOM N/L ADDED AS ASP,MARINE, BATCH,293 K	1692
*PLATYMONAS SP.	3.6 DAYS (GENERATION TIME) (F001)	2*10**(-3) G-ATOM N/L ADDED AS GLU,MARINE, BATCH,293 K	1692
*PLATYMONAS SP.	1.7 DAYS (GENERATION TIME) (F001)	2*10**(-3) G-ATOM N/L ADDED AS GLY,MARINE, BATCH,293 K	1692

TABLE A40

ALGAE : GROWTH RATE

ITEM:	VALUE:	CONDITIONS:	REF.:
*PLATYMONAS SP.	NO GROWTH (F001)	2*10**(-3) G-ATOM N/L ADDED AS LYS,MARINE, BATCH,293 K	1692
*PLATYMONAS SP.	6.9 DAYS (GENERATION TIME) (F001)	2*10**(-3) G-ATOM N/L ADDED AS ORN,MARINE, BATCH,293 K	1692
*PLATYMONAS SP.	1.5 DAYS (GENERATION TIME) (F001)	2*10**(-3) G-ATOM N/L ADDED AS SER,MARINE, BATCH,293 K	1692
*PLATYMONAS SP.	27.5 DAYS (GENERATION TIME)(F001)	2*10**(-3) G-ATOM N/L ADDED AS THR,MARINE, BATCH,293 K	1692
*PLATYMONAS SP.	23.5 DAYS (GENERATION TIME)(F001)	2*10**(-3) G-ATOM N/L ADDED AS VAL,MARINE, BATCH,293 K	1692
*PLATYMONAS SP.	1.5 DAYS (GENERATION TIME) (F001)	2*10**(-3) G-ATOM N/L ADDED AS MIX OF AMINO ACIDS,MARINE, BATCH,293 K	1692
*PLATYMONAS SP.	30 HOUR	DOUBLING TIME, NEWPORT BAY	1137
*PLATYMONAS SUBCORDIFORMIS	2.2 DAYS (GENERATION TIME) (F001)	2*10**(-3) G-ATOM N/L ADDED AS NO3,MARINE, BATCH,293 K	1692
*PLATYMONAS SUBCORDIFORMIS	2.2 DAYS (GENERATION TIME)(F001)	2*10**(-3) G-ATOM N/L ADDED AS ALA,MARINE, BATCH,293 K	1692
*PLATYMONAS SUBCORDIFORMIS	2.5 DAYS (GENERATION TIME)(F001)	2*10**(-3) G-ATOM N/L ADDED AS ASP,MARINE, BATCH,293 K	1692
*PLATYMONAS SUBCORDIFORMIS	2.5 DAYS (GENERATION TIME) (F001)	2*10**(-3) G-ATOM N/L ADDED AS GLU,MARINE, BATCH,293 K	1692
*PLATYMONAS SUBCORDIFORMIS	1.4 DAYS (GENERATION TIME)(F001)	2*10**(-3) G-ATOM N/L ADDED AS GLY,MARINE, BATCH,293 K	1692
*PLATYMONAS SUBCORDIFORMIS	NO GROWTH (F001)	2*10**(-3) G-ATOM N/L ADDED AS LYS,MARINE, BATCH,293 K	1692
*PLATYMONAS SUBCORDIFORMIS	9.1 DAYS (GENERATION TIME)(F001)	2*10**(-3) G-ATOM N/L ADDED AS ORN,MARINE, BATCH,293 K	1692
*PLATYMONAS SUBCORDIFORMIS	1.7 DAYS (GENERATION TIME)(F001)	2*10**(-3) G-ATOM N/L ADDED AS SER,MARINE, BATCH,293 K	1692
*PLATYMONAS SUBCORDIFORMIS	3.2 DAYS (GENERATION TIME)(F001)	2*10**(-3) G-ATOM N/L ADDED AS THR,MARINE, BATCH,293 K	1692
*PLATYMONAS SUBCORDIFORMIS	6.5 DAYS (GENERATION TIME) (F001)	2*10**(-3) G-ATOM N/L ADDED AS VAL,MARINE, BATCH,293 K	1692
*PLATYMONAS SUBCORDIFORMIS	3.1 DAYS (GENERATION TIME)(F001)	2*10**(-3) G-ATOM N/L ADDED AS MIX OF AMINO ACIDS,MARINE, BATCH,293 K	1692
*PORPHYRIDIUM AERUGINEUM	1.4 DOUBLINGS/DAY	294K,SATURATING LIGHT,SYNTHETIC MEDIUM,RED ALGA	688
*PORPHYRIDIUM SP.	2.5 DAYS (GENERATION TIME)(F001)	2*10**(-3) G-ATOM N/L ADDED AS NO3,MARINE, BATCH,293 K	1692
*PROROCENTRUM MICANS	0.70 DOUBLINGS OF CELL NO/DAY	CULTURE IN LOG PHASE, 20 % SUNLIGHT, 75 UG-ATOM N/L AS NO3, 293-298 K	428
*PROROCENTRUM MICANS	0.63 DOUBLINGS OF CELL NO/DAY	CULTURE IN LOG PHASE, 20 % SUNLIGHT, 75 UG-ATOM N/L AS NH4, 293-298 K	428
*PROROCENTRUM MICANS	0.63 DOUBLINGS OF CELL NO/DAY	CULTURE IN LOG PHASE, 20 % SUNLIGHT, NO ADDED N, 293-298 K	428
*PROROCENTRUM MICANS	0.0 DOUBLINGS OF CELL NO/DAY	CULTURE IN LOG PHASE, 20 % SUNLIGHT, NO ADDED N, 293-298 K	428
*PROROCENTRUM MICANS	55.5 HOURS	DOUBLING TIME, CONTINUOUS SATURATING LIGHT,293 K	466
*SCENEDESMUS OBLIQUUS	2.2 DOUBLINGS/DAY	298K,SATURATING LIGHT,SYNTHETIC MEDIUM,GREEN ALGA	688
*SCHIZOTHRIX CALCICOLA	3.4 DOUBLINGS/DAY	303K,SATURATING LIGHT,SYNTHETIC MEDIUM,BLUE GREEN ALGA	688
*SKELETONEMA COSTATUM	13.1 HOURS	DOUBLING TIME, CONTINUOUS SATURATING LIGHT,291 K	466
*SKELETONEMA COSTATUM	1.3 DOUBLINGS OF CELL NO/DAY	CULTURE IN LOG PHASE, 20 % SUNLIGHT, 75 UG-ATOM N/L AS NO3, 293-298 K	428
*SKELETONEMA COSTATUM	0.88 DOUBLINGS OF CELL NO/DAY	CULTURE IN LOG PHASE, 20 % SUNLIGHT, 75 UG-ATOM N/L AS NH4, 293-298 K	428
*SKELETONEMA COSTATUM	1.44 DOUBLINGS OF CELL NO/DAY	CULTURE IN LOG PHASE, 20 % SUNLIGHT, 75 UG-ATOM N/L AS UREA, 293-298 K	428
*SKELETONEMA COSTATUM	0.25 DOUBLINGS OF CELL NO/DAY	CULTURE IN LOG PHASE, 20 % SUNLIGHT, NO ADDED N, 293-298 K	428

TABLE A40

ALGAE : GROWTH RATE

ITEM:	VALUE:	CONDITIONS:	REF.:
*SKELETONEMA COSTATUM	13 HOURS	GENERATION TIME,GROWN UNDER OPTIMAL CONDITION	1220
*SKELETONEMA SP.	5.5 DAYS (GENERATION TIME)(F001)	2*10**(-3) G-ATOM N/L ADDED AS NO3,MARINE, BATCH,293 K	1692
*SKELETONEMA SP.	NO GROWTH (F001)	2*10**(-3) G-ATOM N/L ADDED AS ALA,MARINE, BATCH,293 K	1692
*SKELETONEMA SP.	NO GROWTH (F001)	2*10**(-3) G-ATOM N/L ADDED AS ASP,MARINE, BATCH,293 K	1692
*SKELETONEMA SP.	NO GROWTH (F001)	2*10**(-3) G-ATOM N/L ADDED AS GLU,MARINE, BATCH,293 K	1692
*SKELETONEMA SP.	6.0 DAYS (GENERATION TIME)(F001)	2*10**(-3) G-ATOM N/L ADDED AS GLY,MARINE, BATCH,293 K	1692
*SKELETONEMA SP.	NO GROWTH (F001)	2*10**(-3) G-ATOM N/L ADDED AS LYS,MARINE, BATCH,293 K	1692
*SKELETONEMA SP.	NO GROWTH (F001)	2*10**(-3) G-ATOM N/L ADDED AS ORN,MARINE, BATCH,293 K	1692
*SKELETONEMA SP.	NO GROWTH (F001)	2*10**(-3) G-ATOM N/L ADDED AS SER,MARINE, BATCH,293 K	1692
*SKELETONEMA SP.	NO GROWTH (F001)	2*10**(-3) G-ATOM N/L ADDED AS THR,MARINE, BATCH,293 K	1692
*SKELETONEMA SP.	NO GROWTH (F001)	2*10**(-3) G-ATOM N/L ADDED AS VAL,MARINE, BATCH,293 K	1692
*SKELETONEMA SP.	NO GROWTH (F001)	2*10**(-3) G-ATOM N/L ADDED AS MIX OF AMINO ACIDS,MARINE, BATCH,293 K	1692
*STEPHANOPYXIS SP.	1.7 DAYS (GENERATION TIME) (F001)	2*10**(-3) G-ATOM N/L ADDED AS NO3,MARINE, BATCH,293 K	1692
*STEPHANOPYXIS SP.	NO GROWTH (F001)	2*10**(-3) G-ATOM N/L ADDED AS ALA,MARINE, BATCH,293 K	1692
*STEPHANOPYXIS SP.	NO GROWTH (F001)	2*10**(-3) G-ATOM N/L ADDED AS ASP,MARINE, BATCH,293 K	1692
*STEPHANOPYXIS SP.	NO GROWTH (F001)	2*10**(-3) G-ATOM N/L ADDED AS GLU,MARINE, BATCH,293 K	1692
*STEPHANOPYXIS SP.	NO GROWTH (F001)	2*10**(-3) G-ATOM N/L ADDED AS GLY,MARINE, BATCH,293 K	1692
*STEPHANOPYXIS SP.	NO GROWTH (F001)	2*10**(-3) G-ATOM N/L ADDED AS LYS,MARINE, BATCH,293 K	1692
*STEPHANOPYXIS SP.	NO GROWTH (F001)	2*10**(-3) G-ATOM N/L ADDED AS ORN,MARINE, BATCH,293 K	1692
*STEPHANOPYXIS SP.	NO GROWTH (F001)	2*10**(-3) G-ATOM N/L ADDED AS SER,MARINE, BATCH,293 K	1692
*STEPHANOPYXIS SP.	NO GROWTH (F001)	2*10**(-3) G-ATOM N/L ADDED AS THR,MARINE, BATCH,293 K	1692
*STEPHANOPYXIS SP.	NO GROWTH (F001)	2*10**(-3) G-ATOM N/L ADDED AS VAL,MARINE, BATCH,293 K	1692
*STEPHANOPYXIS SP.	NO GROWTH (F001)	2*10**(-3) G-ATOM N/L ADDED AS MIX OF AMINO ACIDS,MARINE, BATCH,293 K	1692
*TABELLARIA FLOCCULOSA VAR. FLOCCULOSA	1.4 DOUBLINGS/DAY	293K,SATURATING LIGHT,NONSYNTHETIC MEDIUM,DIATOM	688
*THALASSIOSIRA PSEUDONANA	15 HOURS	GENERATION TIME,GROWN UNDER OPTIMAL CONDITION	355
*THALASSIOSIRA ROTULA	38.9 (35.7 - 42.7) HOURS	GENERATION TIME, 600 LUX, 99 % CL., 285 K	1395
*THALASSIOSIRA ROTULA	37.7 (34.3 - 41.9) HOURS	GENERATION TIME, 1200 LUX, 99 % CL., 278 K	1395
*THALASSIOSIRA ROTULA	27.6 (26.1 - 29.2) HOURS	GENERATION TIME, 1200 LUX, 99 % CL., 285 K	1395
*THALASSIOSIRA ROTULA	20.6 (17.7 - 24.6) HOURS	GENERATION TIME, 1200 LUX, 99 % CL., 291 K	1395
*THALASSIOSIRA ROTULA	39.7 (34.6 - 46.7) HOURS	GENERATION TIME, 2400 LUX, 99 % CL., 277 K	1395
*THALASSIOSIRA ROTULA	28.5 (27.0 - 30.0) HOURS	GENERATION TIME, 2400 LUX, 99 % CL., 278 K	1395
*THALASSIOSIRA ROTULA	15.5 (14.9 - 16.3) HOURS	GENERATION TIME, 2400 LUX, 99 % CL., 285 K	1395
*THALASSIOSIRA ROTULA	10.8 (9.1 - 13.3) HOURS	GENERATION TIME, 2400 LUX, 99 % CL., 291 K	1395

TABLE A40

ALGAE : GROWTH RATE			
ITEM:	VALUE:	CONDITIONS:	REF.:
*THALASSIOSIRA ROTULA	9.2 (8.4 - 10.1) HOURS	GENERATION TIME, 2400 LUX, 99 % CL., 295 K	1395
*THALASSIOSIRA ROTULA	11.6 (9.6 - 14.4) HOURS	GENERATION TIME, 4000 LUX, 99 % CL., 285 K	1395
*THALASSIOSIRA ROTULA	9.7 (9.2 - 10.3) HOURS	GENERATION TIME, 5500 LUX, 99 % CL., 285 K	1395
*THALASSIOSIRA SP.	1.6 DAYS (GENERATION TIME)(F001)	2*10**(-3) G-ATOM N/L ADDED AS NO3,MARINE, BATCH,293 K	1692
*THALASSIOSIRA SP.	NO GROWTH (F001)	2*10**(-3) G-ATOM N/L ADDED AS ALA,MARINE, BATCH,293 K	1692
*THALASSIOSIRA SP.	NO GROWTH (F001)	2*10**(-3) G-ATOM N/L ADDED AS ALA,MARINE, BATCH,293 K	1692
*THALASSIOSIRA SP.	NO GROWTH (F001)	2*10**(-3) G-ATOM N/L ADDED AS ASP,MARINE, BATCH,293 K	1692
*THALASSIOSIRA SP.	NO GROWTH (F001)	2*10**(-3) G-ATOM N/L ADDED AS GLU,MARINE, BATCH,293 K	1692
*THALASSIOSIRA SP.	NO GROWTH (F001)	2*10**(-3) G-ATOM N/L ADDED AS GLY,MARINE, BATCH,293 K	1692
*THALASSIOSIRA SP.	NO GROWTH (F001)	2*10**(-3) G-ATOM N/L ADDED AS LYS,MARINE, BATCH,293 K	1692
*THALASSIOSIRA SP.	2.5 DAYS (GENERATION TIME)(F001)	2*10**(-3) G-ATOM N/L ADDED AS ORN,MARINE, BATCH,293 K	1692
*THALASSIOSIRA SP.	5.3 DAYS (GENERATION TIME)(F001)	2*10**(-3) G-ATOM N/L ADDED AS SER,MARINE, BATCH,293 K	1692
*THALASSIOSIRA SP.	NO GROWTH (F001)	2*10**(-3) G-ATOM N/L ADDED AS THR,MARINE, BATCH,293 K	1692
*THALASSIOSIRA SP.	NO GROWTH(F001)	2*10**(-3) G-ATOM N/L ADDED AS VAL,MARINE, BATCH,293 K	1692
*THALASSIOSIRA SP.	4.1 DAYS (GENERATION TIME)(F001)	2*10**(-3) G-ATOM N/L ADDED AS MIX OF AMINO ACIDS,MARINE, BATCH,293 K	1692
*TRICHODESMIUM SPP.	185 HOURS	DOUBLING TIME,FROM N ASSIMILATION, N.PACIFIC,N-FIXING ALGAE	967
*TRICHODESMIUM SPP.	520 HOURS	DOUBLING TIME,FROM C ASSIM. (TOT.PART.CARB.),N.PACIFIC, N-FIXING ALGAE	967
*TRICHODESMIUM SPP.	177 HOURS	DOUBLING TIME,BASED ON ATP CARBON, N.PACIFIC,N-FIXING ALGAE	967

TABLE A41

ALGAE : H/BIOMASS RATIO			
ITEM:	VALUE:	CONDITIONS:	REF.:
*ALGAE	4.6 % OF DRY WEIGHT	MEAN ELEMENTARY COMPOSITION	157
*PHAEOPHYCEAE	4.1 % OF DRY WEIGHT	MEAN ELEMENTARY COMPOSITION	157

TABLE A42

ALGAE : HALF SATURATION CONSTANT FOR AMMONIUM GROWTH			
ITEM:	VALUE:	CONDITIONS:	REF.:
*SELENASTRUM CAPRICORNUTUM	5.237 MICROGRAM N/L	293K,CHEMOST.,AMMONIUM GROWTH LIMIT.,R=0.975,MONOD EQ.	1319
*SELENASTRUM CAPRICORNUTUM	5.251 MICROGRAM N/L	297K,CHEMOST.,AMMONIUM GROWTH LIMIT.,R=0.996,MONOD EQ.	1319
*SELENASTRUM CAPRICORNUTUM	5.246 MICROGRAM N/L	300K,CHEMOST.,AMMONIUM GROWTH LIMIT.,R=0.989,MONOD EQ.	1319
*SELENASTRUM CAPRICORNUTUM	5.248 MICROGRAM N/L	301K,CHEMOST.,AMMONIUM GROWTH LIMIT.,R=0.799,MONOD EQ.	1319

TABLE A42

ALGAE : HALF SATURATION CONSTANT FOR AMMONIUM GROWTH

ITEM:	VALUE:	CONDITIONS:	REF.:
*SELENASTRUM CAPRICORNUTUM	5.505 MICROGRAM N/L	306K,CHEMOST.,AMMONIUM GROWTH LIMIT.,R=0.949,MONOD EQ.	1319
*SELENASTRUM CAPRICORNUTUM	5.237 MICROGRAM N/L	293 K, CHEMOSTAT	355

TABLE A43

ALGAE : HALF SATURATION CONSTANT FOR AMMONIUM UPTAKE

ITEM:	VALUE:	CONDITIONS:	REF.:
*ASTERIONELLA JAPONICA	1.0 MICROMOLES/L	NONE	425
*CHAETOCEROS DEBILIS	0.5 UG-ATOM/L	CHEMOSTAT, UPTAKE=0 FOR CONC = 0.2 UG-ATOM/L, NH4-LIMITATION	279
*CHAETOCEROS DEBILIS	2.2 UG-ATOM/L	CHEMOSTAT, UPTAKE=0 FOR CONC = 0.3 UG-ATOM/L, SI-LIMITATION	279
*CHAETOCEROS GRACILIS	0.4 MICROMOLES/L	NONE	425
*COCCOLITHUS HUXLEYI	0.1 MICROMOLES/L	NONE	425
*COSCINODISCUS LINEATUS	2.0 MICROMOLES/L	NONE	425
*COSCINODISCUS WAILESII	4.9 MICROMOLES/L	NONE	425
*CYCLOTELLA NANA	0.4 MICROMOLES/L	NONE	425
*DITYLUM BRIGHTWELLII	1.1 MICROMOLES/L	NONE	425
*DUNALIELLA TERTIOLECTA	0.6 MICROMOLES/L	NONE	425
*DUNALIELLA TERTIOLECTA	0.17 MICROMOLES/L	NONE	225
*GONYAULAX POLYEDRA	5.5 MICROMOLES/L	NONE	425
*GYMNODINIUM SPLENDENS	1.1 MICROMOLES/L	NONE	425
*LEPTOCYLINDRUS DANICUS	0.7 MICROMOLES/L	NONE	425
*MARINE PHYTOPLANKTON	0.15 MICROGRAM-ATOM/L	NORTH PACIFIC	429
*MONOCHRYSIS LUTHERI	0.29 MICROMOLES/L	NONE	225
*MONOCHRYSIS LUTHERI	0.4 MICROMOLES/L	NONE	425
*NERITIC BACILLARIOPHYCEAE	7.-120 MICROGRAM/L	NONE	431
*NERITIC OR LITTORAL FLAGELLATE	7.-77 MICROGRAM/L	NONE	431
*OCEANIC SPECIES	1.4-5.6 MICROGRAM/L	NONE	431
*OSCILLATORIA THIEBAUTII	6.74 MICROGRAM/L	SARGASSO SEA	231
*RHIZOSOLENIA ROBUSTA	7.5 MICROMOLES/L	NONE	425
*RHIZOSOLENIA STOLTERFOTHII	0.5 MICROMOLES/L	NONE	425
*SKELETONEMA COSTATUM	0.8 MICROMOLES/L	NONE	425
*SKELETONEMA COSTATUM	0.5 UG-ATOM/L	CHEMOSTAT, UPTAKE=0 FOR CONC = 0.2 UG-ATOM/L, NH4-LIMITATION	279
*SKELETONEMA COSTATUM	1.3 UG-ATOM/L	CHEMOSTAT, UPTAKE=0 FOR CONC = 0.2 UG-ATOM/L, SI-LIMITATION	279
*THALASSIOSIRA GRAVIDA	0.5 UG-ATOM/L	CHEMOSTAT, UPTAKE=0 FOR CONC = 0.2 UG-ATOM/L, NH4-LIMITATION	279
*THALASSIOSIRA GRAVIDA	0.3 UG-ATOM/L	CHEMOSTAT, UPTAKE=0 FOR CONC = 0.5 UG-ATOM/L, SI-LIMITATION	279

TABLE A44

ALGAE : HALF SATURATION CONSTANT FOR LIGHT

ITEM:	VALUE:	CONDITIONS:	REF.:
*ALGAE	0.004-0.006 KCAL/M**(2)*SEC	NONE	1571
*CHLOROPHYTE	0.0024 LY/MIN	N AS NITRATE, SUN ADAPTED, ESTAURINE PHYTOPLANKTON	76

TABLE A44

ALGAE : HALF SATURATION CONSTANT FOR LIGHT			
ITEM:	VALUE:	CONDITIONS:	REF.:
*CHLOROPHYTE	0.0017 LY/MIN	N AS NITRATE, SHADE ADAPTED, ESTAURINE PHYTOPLANKTON	76
*CHLOROPHYTE	0.0066 LY/MIN	N AS NITRATE AND NH4, SUN ADAPTED, ESTAURINE PHYTOPLANKTON	76
*CHLOROPHYTE	0.0026 LY/MIN	N AS NITRATE AND NH4, SHADE ADAPTED, ESTAURINE PHYTOPLANKTON	76
*COCCOLITHUS HUXLEYI	0.002 LY/MIN	NONE	1219
*DITYLUM BRIGHTWELLII	0.009 LY/MIN	NONE	1219
*SARGASSUM SP.	0.014 LY/MIN	INCUBATOR,COMM.RESP.=20-30% OF MAX PHOTOSYNT.RATE =0.45MG C/G	232
*SKELETONEMA COSTATUM	0.0003 LY/MIN	N AS NITRATE, SUN ADAPTED, ESTAURINE PHYTOPLANKTON	76
*SKELETONEMA COSTATUM	0.0004 LY/MIN	N AS NITRATE, SHADE ADAPTED, ESTAURINE PHYTOPLANKTON	76
*SKELETONEMA COSTATUM	0.003 LY/MIN	N AS NITRATE AND NH4, SUN ADAPTED, ESTAURINE PHYTOPLANKTON	76
*SKELETONEMA COSTATUM	0.007 LY/MIN	N AS NITRATE AND NH4, SHADE ADAPTED, ESTAURINE PHYTOPLANKTON	76

TABLE A45

ALGAE : HALF SATURATION CONSTANT FOR NITRATE GROWTH			
ITEM:	VALUE:	CONDITIONS:	REF.:
*NERITIC OR LITTORAL FLAGELLATE	8.4-130 MICROGRAM/L	FLAGELLATE	431

TABLE A46

ALGAE : HALF SATURATION CONSTANT FOR NITRATE UPTAKE			
ITEM:	VALUE:	CONDITIONS:	REF.:
*ANABAENA CYLINDRICA	70 MICROMOLES/L	NONE	643
*ASTERIONELLA JAPONICA	1.0 MICROMOLES/L	NONE	425
*BELLOCHIA SP.	0.1 - 0.9 MICROMOLES/L	NONE	355
*CHAETOCEROS GRACILIS	0.2 MICROMOLES/L	NONE	425
*COCCOCHLORIS STAGNINA	0.31 MICROMOLES/L	NONE	225
*COCCOLITHUS HUXLEYI	0.1 MICROMOLES/L	NONE	425
COCCOLITHUS HUXLEYI	0.1 MICROGRAM ATOM/L	NONE	1219
*COSCINODISCUS LINEATUS	2.6 MICROMOLES/L	NONE	425
*COSCINODISCUS WAILESII	3.6 MICROMOLES/L	NONE	425
*CYCLOTELLA NANA	0.4 - 1.9 MICROMOLES/L	NONE	230
*CYCLOTELLA NANA	1.8 MICROMOLES/L	NONE	964
*CYCLOTELLA NANA	0.35 MICROMOLES/L	NONE	225
*CYCLOTELLA NANA	0.5 MICROMOLES/L	NONE	425
*DITYLUM BRIGHTWELLII	0.6 MICROGRAM ATOM/L	NONE	1219
*DITYLUM BRIGHTWELLII	0.6 MICROMOLES/L	NONE	425
*DUNALIELLA SP.	0.28 (+-0.68) *10**(-6)M	288K,95% CONF.LIM.,LIGHT=9500LUX, BATCH CULTURE	1583
*DUNALIELLA SP.	0.95 (+-0.72) *10**(-6)M	298K,95% CONF.LIM.,LIGHT=9500LUX, BATCH CULTURE	1583
*DUNALIELLA TERTIOLECTA	1.4 MICROMOLES/L	NONE	425
*DUNALIELLA TERTIOLECTA	0.21 MICROMOLES/L	NONE	225
*FRAGILARIA PINNATA	0.6 - 1.6 MICROMOLES/L	NONE	230

TABLE A46

ALGAE : HALF SATURATION CONSTANT FOR NITRATE UPTAKE			
ITEM:	VALUE:	CONDITIONS:	REF.:
*GONYAULAX POLYEDRA	9.5 MICROMOLES/L	NONE	425
*GYMNODINIUM SPLENDENS	3.8 MICROMOLES/L	NONE	425
*GYMNODINIUM SPLENDENS	1.02 (+-0.99) *10**(-6)M	291K,95% CONF.LIM.,LIGHT=9500LUX, BATCH CULTURE	1583
*GYMNODINIUM SPLENDENS	6.55 (+-0.85) *10**(-6)M	298K,95% CONF.LIM.,LIGHT=9500LUX, BATCH CULTURE	1583
*ISOCHRYSIS GALBANA	0.1 MICROMOLES/L	NONE	425
*LEPTOCYLINDRUS DANICUS	1.25 MICROMOLES/L	NONE	425
*MONOCHRYSIS LUTHERI	0.42 MICROMOLES/L	NONE	225
*MONOCHRYSIS LUTHERI	0.6 MICROMOLES/L	NONE	425
*NERITIC BACILLARIOPHYCEAE	6.3-28 MICROGRAM/L	NONE	431
*OCEANIC SPECIES	1.4-7. MICROGRAM/L	NONE	431
*PHAEODACTYLUM TRICORNUTUM	2.6 MICROMOLES/L	NONE	813
*RHIZOSOLENIA ROBUSTA	3.0 MICROMOLES/L	NONE	425
*RHIZOSOLENIA STOLTERFOTHII	1.7 MICROMOLES/L	NONE	425
*SKELETONEMA COSTATUM	0.45 MICROMOLES/L	NONE	425
*THALASSIOSIRA PSEUDONANA	1.87 MICRO M	CLONE 3H	649
*THALASSIOSIRA PSEUDONANA	0.38 MICRO M	CLONE 13-1	649

TABLE A47

ALGAE : HALF SATURATION CONSTANT FOR NITRITE UPTAKE			
ITEM:	VALUE:	CONDITIONS:	REF.:
*ANABAENA CYLINDRICA	40 MICROMOLES/L	NONE	643
*CHLORELLA PYRENOIDOSA	25 MICROMOLES/L	NONE	425

TABLE A48

ALGAE : HALF SATURATION CONSTANT FOR NITROGEN GROWTH			
ITEM:	VALUE:	CONDITIONS:	REF.:
*ALGAE	0.025 MG/L	POTOMAC RIVER	1153
*ALGAE	0.025 MG/L	LAKE ONTARIO	1577
*ALGAE	0.1-0.2 MG/L	NONE	1571
*ALGAE	0.05-0.1 MG/L	NONE	1571
*PHYTOPLANKTON	25.0 MICROGRAM/L	SAN JOAQUIN(RIVER)	1149
*PHYTOPLANKTON	25.0 MICROGRAM/L	DELTA(ESTUARY)	1149
*PHYTOPLANKTON	25.0 MICROGRAM/L	POTOMAC(ESTUARY)	1149
*PHYTOPLANKTON	25.0 MICROGRAM/L	LAKE ERIE	1149
*PHYTOPLANKTON	25.0 MICROGRAM/L	LAKE ONTARIO	1149

TABLE A49

ALGAE : HALF SATURATION CONSTANT FOR NITROGEN UPTAKE			
ITEM:	VALUE:	CONDITIONS:	REF.:
*OSCILLATORIA AGARDHII GOMOMT	36 MICROMOLES/L	LIGHT INT.=0.1W/M**2,PH=7.1-7.6,N LIMIT.,NO3,BATCH CULT.	923

TABLE A49

ALGAE : HALF SATURATION CONSTANT FOR NITROGEN UPTAKE			
ITEM:	VALUE:	CONDITIONS:	REF.:
*OSCILLATORIA AGARDHII GOMOMT	35 MICROMOLES/L	LIGHT INT.=3.6W/M**2,PH=7.3-7.8,N LIMIT.,NO3,BATCH CULT.	923
*OSCILLATORIA AGARDHII GOMOMT	36 MICROMOLES/L	LIGHT INT.=5.5W/M**2,PH=7.1-7.6,N LIMIT.,NO3,BATCH CULT.	923
*THALASSIOSIRA PSEUDONANA	5.0*10**(-7) M	GROWTH RATE = 0.041 1/H, CHEMOSTAT, OLIGOTROF ISOLATE	1243
*THALASSIOSIRA PSEUDONANA	5.0*10**(-7) M	GROWTH RATE = 0.017 1/H, CHEMOSTAT, OLIGOTROF ISOLATE	1243
*THALASSIOSIRA PSEUDONANA	5.0*10**(-7) M	GROWTH RATE =0.0085 1/H, CHEMOSTAT, OLIGOTROF ISOLATE	1243

TABLE A50

ALGAE : HALF SATURATION CONSTANT FOR PHOSPHORUS GROWTH			
ITEM:	VALUE:	CONDITIONS:	REF.:
*ALGAE	0.005 MG/L	POTOMAC RIVER	1153
*ALGAE	0.002 MG/L	LAKE ONTARIO	1577
*ALGAE	0.02-0.05 MG/L	NONE	1571
*ALGAE	0.04-0.08 MG/L	NONE	1571
*CHAETOCEROS GRACILIS	25 MICROGRAM/L	NONE	1577
*PHYTOPLANKTON	5.0 MICROGRAM/L	POTOMAC(ESTUARY)	1149
*PHYTOPLANKTON	10 MICROGRAM/L	LAKE ERIE	1149
*PHYTOPLANKTON	2.0 MICROGRAM/L	LAKE ONTARIO	1149

TABLE A51

ALGAE : HALF SATURATION CONSTANT FOR PHOSPHORUS UPTAKE			
ITEM:	VALUE:	CONDITIONS:	REF.:
*ASTERIONELLA JAPONICA	0.12 MICROGRAM-AT.P/L	TROPICAL OCEAN,BATCH CULTURE	1581
*BIDDULPHIA SINENSIS	0.17 MICROGRAM-AT.P/L	TROPICAL ESTUARY,BATCH CULTURE	1288
*CERATAULINA BERGONII	3.33 MICROGRAM P/L	VOLUME=2712 U**3, P CONTENT= 1700*10**(-15) G-ATOM/CELL	456
*CERATIUM FURCA	0.15 MICROGRAM-AT.P/L	TROPICAL ESTUARY,BATCH CULTURE	355
*CHAETOCEROS CURVISETUS	0.778 - 1.11 MICROGRAM P/L	VOLUME=2170 U**3, P CONTENT= 100-310*10**(-15) G-ATOM/CELL	456
*CHAETOCEROS GRACILIS	3.7 MG/M3	NONE	468
*CHAETOCEROS SOCIALIS	0.389 - 1.11 MICROGRAM P/L	VOLUME=100 U**3, P CONTENT= 3.2-7.0*10**(-15) G-ATOM/CELL	456
*CHLORELLA PYRENOIDOSA	4. - 5. MICROMOLES/L	NONE	744
*COSCINODISCUS GIGAS	1.78 - 3.89 MICROGRAM P/L	VOLUME=5150805 U**3	456
*COSCINODISCUS GRANII	0.144 - 1.11 MICROGRAM P/L	VOLUME=172292 U**3, P CONTENT= 1000-4000*10**(-15) G-ATOM/CELL	456
*CYCLOTELLA NANA	0.58 MICROMOLES/L	NONE	496
*CYCLOTELLA NANA	1 MG/M3	NONE	468
*DINOBRYON CYLINDRICUM	0.8 MICROMOLES/L	NONE	355
*DINOBRYON CYLINDRICUM	0.72 MICROMOLES P	NEW JERSEY	904
*DINOBRYON SOCIALE VAR. AMERICANUM	0.42 MICROMOLES/L	NEW JERSEY, DIRECTLY FROM CULTURE	904
*DINOBRYON SOCIALE VAR. AMERICANUM	0.5 MICROMOLES/L	NONE	355
*DINOBRYON SOCIALE VAR. AMERICANUM	0.39 MICROMOLES P	NEW JERSEY, IN P-FREE MEDIUM FOR 2 D	904
*DITYLUM BRIGHTWELLII	3.33 MICROGRAM P/L	VOLUME=175472 U**3, P CONTENT= 10000*10**(-15) G-ATOM/CELL	456

TABLE A51

ALGAE : HALF SATURATION CONSTANT FOR PHOSPHORUS UPTAKE			
ITEM:	VALUE:	CONDITIONS:	REF.:
*EUGLENA GRACILIS	16 MICROMOLES/L	NONE	138
*EUGLENA GRACILIS	0.3 MICROGRAM-AT.P/L	CONTINUOUS CULTURE	249
*FRESHWATER ALGAE	0.010 - 0.012 MG ORTHO P/L	DETERMINED FROM WINTER PHOSPHATE CONC AND MAX PRIMARY PRODUCTION	355
*FRESHWATER PHYTOPLANKTON	0.24 MICRO M	CENTRAL LAKE KINNERET, 287.5 K, P= 0.14 MICRO M	600
*FRESHWATER PHYTOPLANKTON	0.79 MICRO M	SOUTHERN LAKE KINNERET, 290.0 K, P =0.10 MICRO M	600
*FRESHWATER PHYTOPLANKTON	0.21 MICRO M	SOUTHERN LAKE LINNERET, 301 K, P= 0.11 MICRO M	600
*FRESHWATER PHYTOPLANKTON	0.25 MICRO M	SOUTHERN LAKE KINNERET, 304 K, P= 0.037 MICRO M	600
*FRESHWATER PHYTOPLANKTON	0.18 MICRO M	JORDAN RIVER, UPSTREAM L. KINNERET, 289 K, P=0.44 MICRO M	600
*MARINE PHYTOPLANKTON	0.09 - 1.72 MICROGRAM-ATOM P/L	CHESAPEAKE BAY, NATURAL PHYTOPLANKTON, 5 STATIONS	355
*MICROCYSTIS AERUGINOSA	10 MICROGRAM/L(ESTIMATED)	NONE	1341
*NAVICULA ARENARIA	0.1 MICROGRAM-AT.P/L	ESTUARY,IMPROVED BATCH CULTURE	5
*NITZSCHIA ACTINASTROIDES	1.0 MICROMOLES/L	NONE	1077
*NITZSCHIA ACTINASTROIDES	0.014 MICROGRAM-AT.P/L	OLIGOTROPHIC LAKE,CONTINUOUS CULTURE	1077
*PEDIASTRUM DUPLEX	1.1 MICROMOLES/L	NONE	355
*PHAEODACTYLUM TRICORNUTUM	10 MICROGRAM/L	NONE	515
*SCENEDESMUS OBLIQUUS	48.4 (?) MICROGRAM-AT.P/L	FRESH WATER,CONTINUOUS CULTURE	396
*SCENEDESMUS OBLIQUUS	1.5 MG/M3	NONE	468
*SCENEDESMUS SP.	0.6 MICROMOLES/L	NONE	1321
*SCENEDESMUS SP.	0.03 MICROGRAM-AT.P/L	FRESH WATER,CONTINUOUS CULTURE	1767
*SELENASTRUM CAPRICORNUTUM	5 MG/M3	BATCH CULTURE	1602
*SKELETONEMA COSTATUM	0.244 - 1.2 MICROGRAM P/L	VOLUME=144 U**3, P CONTENT= 8.0-28.0*10**(-15) G-ATOM/CELL	456
*THALASSIOSIRA FLUVIATILIS	1.72 MICROMOLES/L	NONE	496
*THALASSIOSIRA FLUVIATILIS	1 MG/M3	NONE	468
*THALASSIOSIRA PSEUDONANA	7.0*10**(-7) M	GROWTH RATE = 0.035 1/H, CHEMOSTAT, OLIGOTROF ISOLATE	1243
*THALASSIOSIRA PSEUDONANA	6.0*10**(-7) M	GROWTH RATE = 0.041 1/H, CHEMOSTAT, OLIGOTROF ISOLATE	1243
*THALASSIOSIRA PSEUDONANA	7.0*10**(-7) M	GROWTH RATE = 0.017 1/H, CHEMOSTAT, OLIGOTROF ISOLATE	1243
*THALASSIOSIRA PSEUDONANA	7.0*10**(-7) M	GROWTH RATE = 0.041 1/H, CHEMOSTAT, ISOLATE FROM MORICHES BAY	1243

TABLE A52

ALGAE : HALF SATURATION CONSTANT FOR SILICIUM UPTAKE			
ITEM:	VALUE:	CONDITIONS:	REF.:
*DITYLUM BRIGHTWELLII	2.96 MICROMOLES/L	NONE	1203
*DITYLUM BRIGHTWELLII	2.96 MICROGRAM-AT SI/L	UPTAKE, NO UPTAKE FOR =2.96 UG-AT/ L, 25000 ERG/(CM2*SEC), 293 K ($47)	1203
*LICMOPHORA SP.	2.58 MICROMOLES/L	NONE	1203
*LICMOPHORA SP.	2.58 MICROGRAM-AT SI/L	UPTAKE, NO UPTAKE FOR =1.30 UG-AT/ L, 25000 ERG/(CM2*SEC), 293 K ($47)	1203
*NITZSCHIA ACTINASTROIDES	3.5 MICROMOLES/L	NONE	1077
*SKELETONEMA COSTATUM	0.80 MICROMOLES/L	NONE	1203
*SKELETONEMA COSTATUM	0.80 MICROGRAM-AT SI/L	UPTAKE, NO UPTAKE FOR =0.32 UG-AT/ L, 25000 ERG/(CM2*SEC), 293 K ($47)	1203

TABLE A52

ALGAE : HALF SATURATION CONSTANT FOR SILICIUM UPTAKE			
ITEM:	VALUE:	CONDITIONS:	REF.:
*THALASSIOSIRA DECIPIENS	3.37 MICROMOLES/L	NONE	1203
*THALASSIOSIRA DECIPIENS	3.37 MICROGRAM-AT SI/L	UPTAKE, NO UPTAKE FOR =1.33 UG-AT/L, 25000 ERG/(CM2*SEC), 293 K ($47)	1203
*THALASSIOSIRA PSEUDONANA	1.4 - 2.9 MICROMOLES/L	NONE	355
*THALASSIOSIRA PSEUDONANA	0.98 MICRO M	CLONE 3H	649
*THALASSIOSIRA PSEUDONANA	0.19 MICRO M	CLONE 13-1	649
*THALASSIOSIRA PSEUDONANA	1.39 MICROGRAM-AT SI/L	UPTAKE, NO UPTAKE FOR =0.67 UG-AT/L, 25000 ERG/(CM2*SEC), 293 K ($47)	1203

TABLE A53

ALGAE : HALF SATURATION CONSTANT FOR TOTAL NITROGEN GROWTH			
ITEM:	VALUE:	CONDITIONS:	REF.:
*SCENEDESMUS GRACILE	150 MICROGRAM/L	NONE	1153

TABLE A54

ALGAE : HALF SATURATION CONSTANT FOR TOTAL PHOSPHORUS GROWTH			
ITEM:	VALUE:	CONDITIONS:	REF.:
*SCENEDESMUS GRACILE	10 MICROGRAM/L	NONE	1153

TABLE A55

ALGAE : K/BIOMASS RATIO			
ITEM:	VALUE:	CONDITIONS:	REF.:
*ANABAENA SP.	1.20 % OF DRY WEIGHT	ALGAE FROM PONDS AND LAKES IN SOUTHEASTERN UNITED STATES	1660
*APHANIZOMENON SP.	0.68 % OF DRY WEIGHT	ALGAE FROM PONDS AND LAKES IN SOUTHEASTERN UNITED STATES	1660
*CHARA SP.	2.3 % OF DRY WEIGHT % OF DRY WEIGHT	ALGAE FROM PONDS AND LAKES IN SOUTHEASTERN UNITED STATES	1660
*CLADOPHORA SP.	6.08 % OF DRY WEIGHT	ALGAE FROM PONDS AND LAKES IN SOUTHEASTERN UNITED STATES	1660
*EUGLENA SP.	0.34 % OF DRY WEIGHT	ALGAE FROM PONDS AND LAKES IN SOUTHEASTERN UNITED STATES	1660
*HYDRODICTYON SP.	4.21 % OF DRY WEIGHT	ALGAE FROM PONDS AND LAKES IN SOUTHEASTERN UNITED STATES	1660
*LYNGBYA SP.	0.42 % OF DRY WEIGHT	ALGAE FROM PONDS AND LAKES IN SOUTHEASTERN UNITED STATES	1660
*MICROCYSTIS SP.	0.79 % OF DRY WEIGHT	ALGAE FROM PONDS AND LAKES IN SOUTHEASTERN UNITED STATES	1660
*MOUGEOTIA SP.	1.20 % OF DRY WEIGHT	ALGAE FROM PONDS AND LAKES IN SOUTHEASTERN UNITED STATES	1660
*NITELLA SP.	3.73 % OF DRY WEIGHT	ALGAE FROM PONDS AND LAKES IN SOUTHEASTERN UNITED STATES	1660
*OEDOGONIUM SP.	3.03 % OF DRY WEIGHT	ALGAE FROM PONDS AND LAKES IN SOUTHEASTERN UNITED STATES	1660
*PITHOPHORA SP.	3.06 % OF DRY WEIGHT	ALGAE FROM PONDS AND LAKES IN SOUTHEASTERN UNITED STATES	1660
*RHIZOCLONIUM SP.	1.90 % OF DRY WEIGHT	ALGAE FROM PONDS AND LAKES IN SOUTHEASTERN UNITED STATES	1660

TABLE A55

ALGAE : K/BIOMASS RATIO			
ITEM:	VALUE:	CONDITIONS:	REF.:
*SPIROGYRA SP.	0.92 % OF DRY WEIGHT	ALGAE FROM PONDS AND LAKES IN SOUTHEASTERN UNITED STATES	1660
*SPIROGYRA SP., GIANT	0.99 % OF DRY WEIGHT	ALGAE FROM PONDS AND LAKES IN SOUTHEASTERN UNITED STATES	1660

TABLE A56

ALGAE : LIPID/BIOMASS RATIO			
ITEM:	VALUE:	CONDITIONS:	REF.:
*MARINE PHYTOPLANKTON	0.15 BY WEIGHT	VIGOROUSLY GROWING,EXCESS NITRATE IN WATER	39
*MARINE PHYTOPLANKTON	0.35 BY WEIGHT	UNHEALTHY PHYTOPLANKTON,NITRATE DEPLETED WATER	39

TABLE A57

ALGAE : MAX GROWTH RATE			
ITEM:	VALUE:	CONDITIONS:	REF.:
*ACROSIPHONIA CENTRALIS	15 MG O2/(GRAM DRY WEIGHT*HOUR)	SPRING, 283 K, LIGHT=14.4 MW/CM2, BENTHIC, BALTIC SEA	829
*ALGAE	0.1 1/DAY	MOSSDALE	1153
*ALGAE	0.1 1/DAY	POTOMAC RIVER	1153
*ALGAE	0.58 1/DAY	TEMP. FACTOR IS 1 AT 273K,LAKE ONTARIO	1577
*ALGAE	1.-2. 1/DAY	TEMP. FACTOR IS 1 AT 293K	1571
*ALGAE	1.-3. 1/DAY	TEMP. FACTOR IS 1 AT 293K	1571
*ALGAE	0.63 1/DAY	277 K, NATURAL ASSOCIATION	354
*ALGAE	0.51 1/DAY	275.6 K, NATURAL ASSOCIATION	354
*ALGAE	0.01-1.2 1/DAY	NONE	958
*ALGAE	0.06-2.0 1/DAY	DETAILS SEE REF	720
*ALGAE	0.06-2.0 1/DAY	DETAILS SEE REF	1469
*CERAMIUM STRICTUM	17 MG O2/(GRAM DRY WEIGHT*HOUR)	AUTUMN, 288 K, LIGHT=16.3 MW/CM2, BENTHIC ALGAE, BALTIC SEA	829
*CERATAULINA BERGONII	1.47 DIVISIONS/DAY	VOLUME=2712 U**3, P CONTENT= 1700*10**(-15) G-ATOM/CELL	456
*CHAETOCEROS CURVISETUS	1.45 - 2.20 DIVISIONS/DAY	VOLUME=2170 U**3, P CONTENT= 100-310*10**(-15) G-ATOM/CELL	456
*CHAETOCEROS GRACILIS	0.107 1/HOUR	N LIMIT.,NH4,298K,CONT.LIGHT: 0.05LY/MIN,CIM(?048)=0.129B	1582
*CHAETOCEROS SOCIALIS	1.53 - 2.13 DIVISIONS/DAY	VOLUME=100 U**3, P CONTENT= 3.2-7.0*10**(-15) G-ATOM/CELL	456
*CHAETOMORPHA SP.	14.0 MICROMOLES C/(MG CHLOROPHYLL A*HOUR)	297 K, 1.3*10**17 QUANTA/(S*CM2), BUBBLED WITH AIR	205
*CHLAMYDOMONAS REINHARDTII	2.64 1/DAY	298 K	354
*CHLAMYDOMONAS REINHARDTII	2.64 1/DAY	298K	1481
*CHLORELLA ELLIPSOIDEA	3.14 1/DAY	298 K	354
*CHLORELLA ELLIPSOIDEA	1.2 1/DAY	288 K	354
*CHLORELLA ELLIPSOIDEA	3.14 1/DAY	298K	1554
*CHLORELLA PYRENOI	0.180 1/HOUR	N LIMIT.,NO3,308K,CONT.LIGHT: 0.05CAL/(CM**2)*MIN,CIM(?048)= 0.086A	1425
*CHLORELLA PYRENOIDOSA	1.36 1/DAY	292 K, NH4 LIMITING, CONTINUOUS CULT., FRESHWATER	1426
*CHLORELLA PYRENOIDOSA	1.45 1/DAY	292 K, NO3 LIMITING, CONTINUOUS CULT., FRESHWATER	1427

TABLE A57

ALGAE : MAX GROWTH RATE			
ITEM:	VALUE:	CONDITIONS:	REF.:
*CHLORELLA PYRENOIDOSA	1.95 1/DAY	298 K, P LIMITING, CONTINUOUS CULT., FRESHWATER	1768
*CHLORELLA PYRENOIDOSA	2.14 1/DAY	298 K, P LIMITING, CONTINUOUS CULT., FRESHWATER	1768
*CHLORELLA PYRENOIDOSA	1.84 1/DAY	301.5 K, NH4 LIMITING, CONTINUOUS CULT., FRESHWATER	1426
*CHLORELLA PYRENOIDOSA	2.22 1/DAY	301.5 K, NO3 LIMITING, CONTINUOUS CULT., FRESHWATER	1427
*CHLORELLA PYRENOIDOSA	3.94 1/DAY	308 K, NH4 LIMITING, CONTINUOUS CULT., FRESHWATER	1426
*CHLORELLA PYRENOIDOSA	4.32 1/DAY	308.5 K, NO3 LIMITING, CONTINUOUS CULT., FRESHWATER	1426
*CHLORELLA PYRENOIDOSA	4.26 1/DAY	312.2 K, NH4 LIMITING, CONTINUOUS CULT., FRESHWATER	1426
*CHLORELLA PYRENOIDOSA	5.65 1/DAY	312.2 K, NO3 LIMITING, CONTINUOUS CULT., FRESHWATER	1427
*CHLORELLA PYRENOIDOSA	1.96 1/DAY	298 K	354
*CHLORELLA PYRENOIDOSA	2.15 1/DAY	298 K	354
*CHLORELLA PYRENOIDOSA	0.2 1/DAY	283 K, HIGH-TEMPERATURE STRAIN	354
*CHLORELLA PYRENOIDOSA	1.1 1/DAY	288 K, HIGH-TEMPERATURE STRAIN	354
*CHLORELLA PYRENOIDOSA	2.4 1/DAY	293 K, HIGH-TEMPERATURE STRAIN	354
*CHLORELLA PYRENOIDOSA	9.9 MICROMOLES C/(MG CHLOROPHYLL A*HOUR)	297 K, 1.3*10**17 QUANTA/(S*CM2), BUBBLED WITH AIR	205
*CHLORELLA PYRENOIDOSA	0.180 1/HOUR	SPECIFIC GROWTH RATE, 308 K, SUBSTRATE NO3, LIGTH=0.05 CAL/(CM2*MIN)	1146
*CHLORELLA PYRENOIDOSA	1.96 1/DAY	298K	1089
*CHLORELLA PYRENOIDOSA	2.15 1/DAY	298K	1481
*CHLORELLA PYRENOIDOSA	0.2 1/DAY	283K	1482
*CHLORELLA PYRENOIDOSA	1.1 1/DAY	288K	1482
*CHLORELLA PYRENOIDOSA	2.4 1/DAY	293K	1482
*CHLORELLA PYRENOIDOSA	3.9 1/DAY	298K	1482
*CHLORELLA PYRENOIDOSA	0.9 MG DW/ML IN 65 HOURS	DETERMINATED AT END OF EXPONENTIAL FASE	1755
*CHLORELLA SP.	1.88 1/DAY	298 K, NO3 LIMITING, CONTINUOUS CULT., FRESHWATER	1716
*CHLORELLA VULGARIS	1.8 1/DAY	298 K	354
*CHLORELLA VULGARIS	1.8 1/DAY	298K	1481
*CLADOPHORA GLOMERATA	27 MG O2/(GRAM DRY WEIGHT*HOUR)	SUMMER, 293 K, LIGHT=18.4 MW/CM2, BENTHIC ALGAE, BALTIC SEA	829
*COCCOCHLORIS SP.	2.16 1/DAY	298 K, NO3 LIMITING, CONTINUOUS CULT., MARINE	225
*COCCOCHLORIS STAGNINA	0.086 1/HOUR	SPECIFIC GROWTH RATE, 298 K, SUBSTRATE NO3, LIGTH= 5.8 CAL/(CM2*H)	1146
*COCCOCHLORIS STAGNINA	0.086 1/HOUR	N LIMIT.,NO3,298K,CONT.LIGHT: 5.8CAL/(CM**2)*HR,CIM(?048)=0.184B	225
*COCCOLITHUS HUXLEYI	1.75 DOUBLINGS/DAY	OPTIMAL CONDITIONS	1219
*COSCINODISCUS GIGAS	0.51 - 0.73 DIVISIONS/DAY	VOLUME=5150805 U**3	456
*COSCINODISCUS GRANII	0.50 - 0.92 DIVISIONS/DAY	VOLUME=172292 U**3, P CONTENT= 1000-4000*10**(-15) G-ATOM/CELL	456
*CYCLOTELLA NANA	0.086 1/HOUR	N LIMIT.,NO3,298K,CONT.LIGHT: 5.8CAL/(CM**2)*HR,CIM(?048)=0.158B	225
*DELESSERIA SANGUINEA	8.8 MG O2/(GRAM DRY WEIGHT*HOUR)	SPRING, 283 K, LIGHT=10 MW/CM2, BENTHIC ALGAE, BALTIC SEA	829
*DELESSERIA SANGUINEA	4.0 MG O2/(GRAM DRY WEIGHT*HOUR)	SUMMER, 293 K, LIGHT=4.9 MW/CM2, BENTHIC ALGAE, BALTIC SEA	829
*DELESSERIA SANGUINEA	3.0 MG O2/(GRAM DRY WEIGHT*HOUR)	AUTUMN, 288 K, LIGHT=8.0 MW/CM2, BENTHIC ALGAE, BALTIC SEA	829
*DELESSERIA SANGUINEA	2.2 MG O2/(GRAM DRY WEIGHT*HOUR)	WINTER, 278 K, LIGHT=10 MW/CM2, BENTHIC ALGAE, BALTIC SEA	829
*DITYLUM BRIGHTWELLII	1.20 DIVISIONS/DAY	VOLUME=175472 U**3, P CONTENT= 10000*10**(-15) G-ATOM/CELL	456

TABLE A57

ALGAE : MAX GROWTH RATE

ITEM:	VALUE:	CONDITIONS:	REF.:
*DITYLUM BRIGHTWELLII	1.20 DIVISIONS/DAY	VOLUME=175472 U**3, P CONTENT= 10000*10**(-15) G-ATOM/CELL	456
*DITYLUM BRIGHTWELLII	2.32 DOUBLINGS/DAY	OPTIMAL CONDITIONS	1219
*DUMONTIA INCRASSATA	17 MG O2/(GRAM DRY WEIGHT*HOUR)	SPRING, 283 K, LIGHT=18.4 MW/CM2, BENTHIC ALGAE, BALTIC SEA	829
*DUMONTIA INCRASSATA	9 MG O2/(GRAM DRY WEIGHT*HOUR)	WINTER, 278 K, LIGHT=5.5 MW/CM2, BENTHIC ALGAE, BALTIC SEA	829
*DUNALIELLA SP.	1.93 (+-0.26) DOUBLINGS/DAY	288K,95% CONF.LIM.,LIGHT=9500LUX, BATCH CULTURE	1583
*DUNALIELLA SP.	1.85 (+-0.24) DOUBLINGS/DAY	298K,95% CONF.LIM.,LIGHT=9500LUX, BATCH CULTURE	1583
*DUNALIELLA TERTIOLECTA	0.80 1/DAY	288 K, FE LIMITING, CONTINUOUS CULT., MARINE	326
*DUNALIELLA TERTIOLECTA	1.83 1/DAY	298 K, NH4 LIMITING, CONTINUOUS CULT., MARINE	225
*DUNALIELL^ TERTIOLECTA	0.070 1/HOUR	N LIMIT.,NH4,298K,CONT.LIGHT: 5.8CAL/(CM**2)*HR,CIM(?048)=0.146B	225
*ECTOCARPUS CONFERVOIDES	14 MG O2/(GRAM DRY WEIGHT*HOUR)	SPRING, 283 K, LIGHT=26.7 MW/CM2, BENTHIC ALGAE, BALTIC SEA	829
*ENTEROMORPHA INTESTINALIS	30 MG O2/(GRAM DRY WEIGHT*HOUR)	SPRING, 283 K, LIGHT=17.5 MW/CM2, BENTHIC ALGAE, BALTIC SEA	829
*ENTEROMORPHA INTESTINALIS	28 MG O2/(GRAM DRY WEIGHT*HOUR)	WINTER, 278 K, LIGHT=13.5 MW/CM2, BENTHIC ALGAE, BALTIC SEA	829
*ENTEROMORPHA LINZA	22 MG O2/(GRAM DRY WEIGHT*HOUR)	AUTUMN, 288 K, LIGHT=18.4 MW/CM2, BENTHIC ALGAE, BALTIC SEA	829
*ENTEROMORPHA PROLIFERA	42 MG O2/(GRAM DRY WEIGHT*HOUR)	AUTUMN, 288 K, LIGHT=18.4 MW/CM2, BENTHIC ALGAE, BALTIC SEA	829
*ENTEROMORPHA PROLIFERA	13 MG O2/(GRAM DRY WEIGHT*HOUR)	SUMMER, 293 K, LIGHT=5.6 MW/CM2, BENTHIC ALGAE, BALTIC SEA	829
*ENTEROMORPHA SP.	79.6 MICROMOLES C/(MG CHLOROPHYLL A*HOUR)	297 K, 1.3*10**17 QUANTA/(S*CM2), BUBBLED WITH AIR	205
*FUCUS SERRATUS	5.4 MG O2/(GRAM DRY WEIGHT*HOUR)	WINTER, 278 K, LIGHT=43.6 MW/CM2, BENTHIC ALGAE, BALTIC SEA	829
*FUCUS SERRATUS	3.1 MG O2/(GRAM DRY WEIGHT*HOUR)	AUTUMN, 288 K, LIGHT=3.1 MW/CM2, BENTHIC ALGAE, BALTIC SEA	829
*FUCUS SERRATUS	2.4 MG O2/(GRAM DRY WEIGHT*HOUR)	SPRING, 283 K, LIGHT=23.6 MW/CM2, BENTHIC ALGAE, BALTIC SEA	829
*FUCUS VESICULOSUS	6.6 MG O2/(GRAM DRY WEIGHT*HOUR)	AUTUMN, 288 K, LIGHT=23.6 MW/CM2, BENTHIC ALGAE, BALTIC SEA	829
*FUCUS VESICULOSUS	6.0 MG O2/(GRAM DRY WEIGHT*HOUR)	SPRING, 283 K, LIGHT=28.4 MW/CM2, BENTHIC ALGAE, BALTIC SEA	829
*FUCUS VESICULOSUS	5.4 MG O2/(GRAM DRY WEIGHT*HOUR)	WINTER, 278 K, LIGHT=30.2 MW/CM2, BENTHIC ALGAE, BALTIC SEA	829
*FURCELLARIA FASTIGIATA	3.2 MG O2/(GRAM DRY WEIGHT*HOUR)	SPRING, 283 K, LIGHT=10 MW/CM2, BENTHIC ALGAE, BALTIC SEA	829
*FURCELLARIA FASTIGIATA	2.3 MG O2/(GRAM DRY WEIGHT*HOUR)	WINTER, 278 K, LIGHT=10 MW/CM2, BENTHIC ALGAE, BALTIC SEA	829
*FURCELLARIA FASTIGIATA	1.8 MG O2/(GRAM DRY WEIGHT*HOUR)	AUTUMN, 288 K, LIGHT=4.3 MW/CM2, BENTHIC ALGAE, BALTIC SEA	829
*GLENODINIUM SP.	61.7 MICROMOLES C/(MG CHLOROPHYLL A*HOUR)	297 K, 1.3*10**17 QUANTA/(S*CM2), BUBBLED WITH AIR	205
*GYMNODINIUM SPLENDENS	0.48 (+-0.06) DOUBLINGS/DAY	291K,95% CONF.LIM.,LIGHT=9500LUX, BATCH CULTURE	1583
*GYMNODINIUM SPLENDENS	0.83 (+-0.42) DOUBLINGS/DAY	298K,95% CONF.LIM.,LIGHT=9500LUX, BATCH CULTURE	1583
*LAMINARIA DIGITATA	3.9 MG O2/(GRAM DRY WEIGHT*HOUR)	AUTUMN, 288 K, LIGHT=6.3 MW/CM2, BENTHIC ALGAE, BALTIC SEA	829
*LAMINARIA DIGITATA	3.8 MG O2/(GRAM DRY WEIGHT*HOUR)	SPRING, 283 K, LIGHT=26.7 MW/CM2, BENTHIC ALGAE, BALTIC SEA	829
*LAMINARIA DIGITATA	1.3 MG O2/(GRAM DRY WEIGHT*HOUR)	WINTER, 278 K, LIGHT=1.3 MW/CM2, BENTHIC ALGAE, BALTIC SEA	829
*LAMINARIA SACCHARINA	2.0 MG O2/(GRAM DRY WEIGHT*HOUR)	WINTER, 278 K, LIGHT=2.8 MW/CM2, BENTHIC ALGAE, BALTIC SEA	829
*LAMINARIA SACCHARINA	2.0 MG O2/(GRAM DRY WEIGHT*HOUR)	AUTUMN, 288 K, LIGHT=23.6 MW/CM2, BENTHIC ALGAE, BALTIC SEA	829
*MARINE PHYTOPLANKTON	3.70 MG C/(M**3*H)	LOW NUTRIENT LEVEL, 15 MILES FROM COAST	118

TABLE A57

ALGAE : MAX GROWTH RATE

ITEM:	VALUE:	CONDITIONS:	REF.:
*MARINE PHYTOPLANKTON	1.20 MG C/(M**3*H)	LOW NUTRIENT LEVEL, CENTRAL NORTH PACIFIC	118
*MONOCHRYSIS LUTHERI	0.84 1/DAY	292 K, B12 LIMITING, CONTINUOUS CULT., MARINE	369
*NANNOCHLORIS ATOMUS	2.16 1/DAY	293 K	354
*NANNOCHLORIS ATOMUS	1.54 1/DAY	283 K	354
*NANNOCHLORIS ATOMUS	2.16 1/DAY	293K, FLAGELLATE	355
*NANNOCHLORIS ATOMUS	1.54 1/DAY	283K, FLAGELLATE	355
*NITZSCHIA ACTINASTROIDES	2.06 1/DAY	296 K, P LIMITING, CONTINUOUS CULT., MARINE	1474
*NITZSCHIA CLOSTERIUM	1.75 1/DAY	300 K	354
*NITZSCHIA CLOSTERIUM	1.55 1/DAY	292 K	354
*NITZSCHIA CLOSTERIUM	1.19 1/DAY	288.5 K	354
*NITZSCHIA CLOSTERIUM	0.67 1/DAY	283 K	354
*NITZSCHIA CLOSTERIUM	1.75 1/DAY	300K	1490
*NITZSCHIA CLOSTERIUM	1.55 1/DAY	292K	1490
*NITZSCHIA CLOSTERIUM	1.19 1/DAY	288.5K	1490
*NITZSCHIA CLOSTERIUM	0.67 1/DAY	283K	1490
*OSCILLATORIA AGARDHII	0.036 1/HOUR	SPECIFIC GROWTH RATE, 293 K, SUBSTRATE NO3, LIGTH SATURATED	1146
*OSCILLATORIA AGARDHII GOMOMT	0.017 1/HOUR	LIGHT =0.1W/M**2,PH=7.1-7.6,N LIMIT.,NO3,BATCH CULT.	923
*OSCILLATORIA AGARDHII GOMOMT	0.020 1/HOUR	LIGHT =3.6W/M**2,PH=7.3-7.8,N LIMIT.,NO3,BATCH CULT.	923
*OSCILLATORIA AGARDHII GOMOMT	0.020 1/HOUR	LIGHT =5.5W/M**2,PH=7.1-7.6,N LIMIT.,NO3,BATCH CULT.	923
*OSCILLATORIA AGARDHII GOMOMT	0.036 1/HOUR	N LIMIT.,NO3,293K,CONT.SAT.LIGHT, CIM(?048)=0.092A	933
*PETALONIA FASCIA	34 MG O2/(GRAM DRY WEIGHT*HOUR)	SPRING, 283 K, LIGHT=43.6 MW/CM2, BENTHIC ALGAE, BALTIC SEA	829
*PHYCODRYS RUBENS	4.8 MG O2/(GRAM DRY WEIGHT*HOUR)	SPRING, 283 K, LIGHT=10 MW/CM2, BENTHIC ALGAE, BALTIC SEA	829
*PHYCODRYS RUBENS	4.1 MG O2/(GRAM DRY WEIGHT*HOUR)	AUTUMN, 288 K, LIGHT=4.3 MW/CM2, BENTHIC ALGAE, BALTIC SEA	829
*PHYCODRYS RUBENS	3.7 MG O2/(GRAM DRY WEIGHT*HOUR)	SUMMER, 293 K, LIGHT=4.3 MW/CM2, BENTHIC ALGAE, BALTIC SEA	829
*PHYCODRYS RUBENS	2.0 MG O2/(GRAM DRY WEIGHT*HOUR)	WINTER, 278 K, LIGHT=1.4 MW/CM2, BENTHIC ALGAE, BALTIC SEA	829
*PHYLLOPHORA BRODIAEI	2.0 MG O2/(GRAM DRY WEIGHT*HOUR)	WINTER, 278 K, LIGHT=4.9 MW/CM2, BENTHIC ALGAE, BALTIC SEA	829
*PHYLLOPHORA BRODIAEI	1.9 MG O2/(GRAM DRY WEIGHT*HOUR)	SPRING, 283 K, LIGHT=9.12 MW/CM2, BENTHIC ALGAE, BALTIC SEA	829
*PHYLLOPHORA BRODIAEI	1.2 MG O2/(GRAM DRY WEIGHT*HOUR)	AUTUMN, 288 K, LIGHT=6.3 MW/CM2, BENTHIC ALGAE, BALTIC SEA	829
*PHYTOPLANKTON	2.0 1/DAY	SAN JOAQUIN(RIVER),293K	1149
*PHYTOPLANKTON	2.5 1/DAY	DELTA(ESTUARY),293K	1149
*PHYTOPLANKTON	2.0 1/DAY	POTOMAC(ESTUARY),293K	1149
*PHYTOPLANKTON	1.3 1/DAY	LAKE ERIE,293K	1149
*PHYTOPLANKTON	2.1 1/DAY	LAKE ONTARIO,293K	1149
*POLYSIPHONIA NIGRESCENS	5.3 MG O2/(GRAM DRY WEIGHT*HOUR)	AUTUMN, 288 K, LIGHT=14.5 MW/CM2, BENTHIC ALGAE, BALTIC SEA	829
*POLYSIPHONIA NIGRESCENS	4.3 MG O2/(GRAM DRY WEIGHT*HOUR)	SPRING, 283 K, LIGHT=18.5 MW/CM2, BENTHIC ALGAE, BALTIC SEA	829
*POLYSIPHONIA NIGRESCENS	2.3 MG O2/(GRAM DRY WEIGHT*HOUR)	WINTER, 278 K, LIGHT=26.7 MW/CM2, BENTHIC ALGAE, BALTIC SEA	829
*PORPHYRA LEUCOSTICTA	29 MG O2/(GRAM DRY WEIGHT*HOUR)	WINTER, 278 K, LIGHT=4.3 MW/CM2, BENTHIC ALGAE, BALTIC SEA	829
*SARGASSUM SP.	44.0 MICROMOLES C/(MG CHLOROPHYLL A*HOUR)	297 K, 1.3*10**17 QUANTA/(S*CM2), BUBBLED WITH AIR	205
*SCENEDESMUS OBLIQUUS	1.52 1/DAY	298 K	354
*SCENEDESMUS OBLIQUUS	1.52 1/DAY	298K	1481

TABLE A57

ALGAE : MAX GROWTH RATE			
ITEM:	VALUE:	CONDITIONS:	REF.:
*SCENEDESMUS QUADRICAUDA	2.02 1/DAY	298 K	354
*SCENEDESMUS QUADRICAUDA	2.02 1/DAY	298K	1089
*SCYTOSIPHON LOMENTARIA	12.5 MG O2/(GRAM DRY WEIGHT*HOUR)	SPRING, 283 K, LIGHT=26.7 MW/CM2, BENTHIC ALGAE, BALTIC SEA	829
*SELENASTRUM CAPRICORNUTUM	1.85 1/DAY	BATCH CULTURE,SPECIFIC GROWTH RATE	1602
*SELENASTRUM CAPRICORNUTUM	1.398 1/DAY	293K,CHEMOST.,AMMONIUM-N LIMIT.,R= 0.975,MONOD EQ.	1319
*SELENASTRUM CAPRICORNUTUM	1.544 1/DAY	297K,CHEMOST.,AMMONIUM-N LIMIT.,R= 0.996,MONOD EQ.	1319
*SELENASTRUM CAPRICORNUTUM	1.540 1/DAY	300K,CHEMOST.,AMMONIUM-N LIMIT.,R= 0.989,MONOD EQ.	1319
*SELENASTRUM CAPRICORNUTUM	1.596 1/DAY	301K,CHEMOST.,AMMONIUM-N LIMIT.,R= 0.799,MONOD EQ.	1319
*SELENASTRUM CAPRICORNUTUM	1.274 1/DAY	306K,CHEMOST.,AMMONIUM-N LIMIT.,R= 0.949,MONOD EQ.	1319
*SELENASTRUM CAPRICORNUTUM	1.365 1/DAY	293K,CHEMOST.,AMMONIUM-N LIMIT., MONOD EQ.,MOST PROB.SPEC.RATE	1319
*SELENASTRUM CAPRICORNUTUM	1.992 1/DAY	297K,CHEMOST.,AMMONIUM-N LIMIT., MONOD EQ.,MOST PROB.SPEC.RATE	1319
*SELENASTRUM CAPRICORNUTUM	1.412 1/DAY	300K,CHEMOST.,AMMONIUM-N LIMIT., MONOD EQ.,MOST PROB.SPEC.RATE	1319
*SELENASTRUM CAPRICORNUTUM	1.390 1/DAY	301K,CHEMOST.,AMMONIUM-N LIMIT., MONOD EQ.,MOST PROB.SPEC.RATE	1319
*SELENASTRUM CAPRICORNUTUM	1.274 1/DAY	306K,CHEMOST.,AMMONIUM-N LIMIT., MONOD EQ.,MOST PROB.SPEC.RATE	1319
*SELENASTRUM CAPRICORNUTUM	1.85 1/DAY	297 K, P LIMITING, CONTINUOUS CULT., FRESHWATER	1603
*SKELETONEMA COSTATUM	1.27 1/DAY	292 K, B12 LIMITING, CONTINUOUS CULT., MARINE	369
*SKELETONEMA COSTATUM	1.51 - 5.00 DIVISIONS/DAY	VOLUME=144 U**3, P CONTENT= 8.0-28.0*10**(-15) G-ATOM/CELL	456
*THALASSIOSIRA PSEUDONANA	3.6 DOUBLINGS/DAY	CLONE 3H	649
*THALASSIOSIRA PSEUDONANA	2.1 DOUBLINGS/DAY	CLONE 13-1	649
*THALASSIOSIRA PSEUDONANA	1.27 1/DAY	286.5 K, P LIMITING, CONTINUOUS CULT., MARINE	495
*THALASSIOSIRA PSEUDONANA	1.14 1/DAY	291 K, P LIMITING, CONTINUOUS CULT., MARINE	495
*THALASSIOSIRA PSEUDONANA	1.46 1/DAY	297 K, P LIMITING, CONTINUOUS CULT., MARINE	495
*THALASSIOSIRA PSEUDONANA	2.09 1/DAY	298 K, NO3 LIMITING, CONTINUOUS CULT., MARINE	225
*THALASSIOSIRA PSEUDONANA	2.77 1/DAY	293 K, SI LIMITING, CONTINUOUS CULT., MARINE	1202
*THALASSIOSIRA PSEUDONANA	1.34 1/DAY	289 K, B12 LIMITING, CONTINUOUS CULT., MARINE	1539
*THALASSIOSIRA PSEUDONANA	74.6 MICROMOLES C/(MG CHLOROPHYLL A*HOUR)	297 K, 1.3*10**17 QUANTA/(S*CM2), BUBBLED WITH AIR	205
*ULOTRIX SPECIOSA	35 MG O2/(GRAM DRY WEIGHT*HOUR)	SPRING, 283 K, LIGHT=23.6 MW/CM2, BENTHIC ALGAE, BALTIC SEA	829
*ULVOPSIS GREVILLEI	47 MG O2/(GRAM DRY WEIGHT*HOUR)	SPRING, 283 K, LIGHT=26.7 MW/CM2, BENTHIC ALGAE, BALTIC SEA	829
*ULVOPSIS GREVILLEI	13 MG O2/(GRAM DRY WEIGHT*HOUR)	WINTER, 278 K, LIGHT=14.5 MW/CM2, BENTHIC ALGAE, BALTIC SEA	829
*ZOOXANTHELLAE	16.5 MICROMOLES C/(MG CHLOROPHYLL A*HOUR)	297 K, 1.3*10**17 QUANTA/(S*CM2), BUBBLED WITH AIR	205

TABLE A58

ALGAE : MAX PHOTOSYNTHETIC EFFICIENCY			
ITEM:	VALUE:	CONDITIONS:	REF.:
*ANABAENA CYLINDRICA	1.7 MG C/(MG CHL)*HOUR*KLUX	CULTURE:293K;EXP.:303K, LIGHT = 10KLUX	44

TABLE A58

ALGAE : MAX PHOTOSYNTHETIC EFFICIENCY			
ITEM:	VALUE:	CONDITIONS:	REF.:
*CHAETOCEROS SP.	1.5 MG C/(MG CHL)*HOUR*KLUX	CULTURE:293K, LIGHT = 8KLUX;EXP.: 293K, LIGHT = 20KLUX	1096
*CHLORELLA ELLIPSOIDEA	5.5 MG C/(MG CHL)*HOUR*KLUX	CULTURE:293K;EXP.:303K, LIGHT = 20KLUX	44
*CHLORELLA VULGARIS	1.1 MG C/(MG CHL(A+B))*HOUR*KLUX	CULTURE:LIGHT INT.=3KLUX;EXP.: LIGHT INT.=6KLUX	1506
*CHLORELLA VULGARIS	3.8 MG C/((MG CHL A)*HOUR*KLUX)	CULTURE:LIGHT INT.=30KLUX;EXP.: LIGHT INT.=30KLUX	1506
*COCCOLITHUS HUXLEYI	> 2.2 MG C/(MG CHL(A+B))*HOUR*KLUX	CULTURE:287K, LIGHT = 38KLUX;EXP.: LIGHT INT.:>38KLUX	750
*CYCLOTELLA MENEGHINIANA	2.1-3.4 MG C/((MG CHL A) *HOUR*KLUX)	CULTURE:LIGHT INT.=3KLUX	760
*CYCLOTELLA MENEGHINIANA	3.6-4.4 MG C/((MG CHL A) *HOUR*KLUX)	CULTURE:LIGHT INT.=30KLUX	760
*HYMENOMONAS SP.	> 1.9 MG C/(MG CHL(A+B))*HOUR*KLUX	CULTURE:287K, LIGHT = 38KLUX;EXP.: LIGHT INT.:>38KLUX	750
*ICE FLORA	2.6 MG C/((MG CHL A)*HOUR*KLUX)	ANTARCTIC,271.4K	199
*ICE FLORA	0.4 MG C/((MG CHL A)*HOUR*KLUX)	ANTARCTIC,271.5K	197
*PHYTOPLANKTON	1.0-1.5 MG C/((MG CHL A) *HOUR*KLUX)	DEPTH=0M,ARCTIC,SUMMER	1504
*PHYTOPLANKTON	2.3 MG C/((MG CHL A)*HOUR*KLUX)	ANTARCTIC	199
*PHYTOPLANKTON	2.9-3.4 MG C/(MG CHL)*HOUR*KLUX	NORTHERN(GENERAL),NO VERTICAL STABILITY	1504
*PHYTOPLANKTON	1.4-2.0 MG C/((MG CHL A) *HOUR*KLUX)	DEPTH=10M,281-293K,WESTERN NORTH PACIFIC,SUMMER	1549
*PHYTOPLANKTON	4.0-4.2 MG C/(MG CHL)*HOUR*KLUX	DEPTH=0M,TEMPERATE,SUMMER	1504
*PHYTOPLANKTON	1.5 MG C/(MG CHL)*HOUR*KLUX	DEPTH=0M,TEMPERATE,WINTER	1504
*PHYTOPLANKTON	3-6 MG C/(MG CHL)*HOUR*KLUX	OYASHIO(GENERAL),SUMMER	713
*PHYTOPLANKTON	0.3-0.7 MG C/(MG CHL)*HOUR*KLUX	KUROSHIO(GENERAL),SUMMER	713
*PHYTOPLANKTON	2-6 MG C/(MG CHL)*HOUR*KLUX	BAYS AND COASTAL WATERS NEAR JAPAN (GENERAL)	713
*PHYTOPLANKTON	9.0-16.9 MG C/(MG CHL)*HOUR*KLUX	293-298K,TOKYO BAY,SKELETONEMA BLOOM,SUMMER	679
*PHYTOPLANKTON	8.0 MG C/(MG CHL)*HOUR*KLUX	TROPICAL(GENERAL)	1504
*PHYTOPLANKTON	1.1-5.2 MG C/((MG CHL A) *HOUR*KLUX)	296-300K,TROPICAL PACIFIC,AUTUMN	1549
*PHYTOPLANKTON	3.15 MG C/((MG CHL A)*HOUR*KLUX)	TROPICAL PACIFIC(GENERAL),SPRING, NITROGEN-POOR WATER	355
*PHYTOPLANKTON	2-6 MG C/(MG CHL)*HOUR*KLUX	EUTROPHIC LAKE(GENERAL)	713
*PHYTOPLANKTON	1-2 MG C/(MG CHL)*HOUR*KLUX	MESOTROPHIC LAKE(GENERAL)	713
*PHYTOPLANKTON	0.1-1 MG C/(MG CHL)*HOUR*KLUX	OLIGOTROPHIC LAKE(GENERAL)	713
*PHYTOPLANKTON	4.95 MG C/((MG CHL A)*HOUR*KLUX)	TROPICAL PACIFIC(GENERAL),SPRING, NITROGEN-RICH WATER	355
*PHYTOPLANKTON	3.0-4.0 MG C/(MG CHL)*HOUR*KLUX	DEPTH=10M,293K,TROPICAL ATLANTIC	1760
*SCENEDESMUS SP.	5.0 MG C/(MG CHL)*HOUR*KLUX	CULTURE:293K;EXP.:303K	44
*SKELETONEMA COSTATUM	6.2 MG C/(MG CHL)*HOUR*KLUX	CULTURE:293K, LIGHT = 8KLUX;EXP.: 293K, LIGHT = 20KLUX	1096
*SKELETONEMA COSTATUM p	3.6 MG C/(MG CHL)*HOUR*KLUX	CULTURE:293K;EXP.:293K, LIGHT = 13KLUX	43
*SYNEDRA SP.	2.0 MG C/(MG CHL)*HOUR*KLUX	CULTURE:293K;EXP.:293K, LIGHT = 15KLUX	44

TABLE A59

ALGAE : MAX RATE OF AMMONIUM UPTAKE			
ITEM:	VALUE:	CONDITIONS:	REF.:
*BIDDULPHIA AURITA	2.1*10**(-8) MICROMOLES / (CELL*HOUR)	NONE	954

TABLE A59

ALGAE : MAX RATE OF AMMONIUM UPTAKE			
ITEM:	VALUE:	CONDITIONS:	REF.:
*CHAETOCEROS DEBILIS	0.16 1/HOUR	CHEMOSTAT, UPTAKE=0 FOR CONC = 0.2 UG-ATOM/L, NH4-LIMITATION	279
*CHAETOCEROS DEBILIS	0.18 1/HOUR	CHEMOSTAT, UPTAKE=0 FOR CONC = 0.3 UG-ATOM/L, SI-LIMITATION	279
*DUNALIELLA TERTIOLECTA	2.6 - 10.6 *10**(-8) MICROMOLES / (CELL*HOUR)	NONE	225
*MARINE PHYTOPLANKTON	57 NG-ATOMS/(L*DAY)	NORTH PACIFIC	429
*MONOCHRYSIS LUTHERI	2.1 - 3.8 *10**(-8) MICROMOLES / (CELL*HOUR)	NONE	225
*OSCILLATORIA THIEBAUTII	0.59 PICOGRAM N/(CELL*HOUR)	SARGASSO SEA	231
*SKELETONEMA COSTATUM	0.16 1/HOUR	CHEMOSTAT, UPTAKE=0 FOR CONC = 0.2 UG-ATOM/L, NH4-LIMITATION	279
*SKELETONEMA COSTATUM	0.10 1/HOUR	CHEMOSTAT, UPTAKE=0 FOR CONC = 0.2 UG-ATOM/L, SI-LIMITATION	279
*THALASSIOSIRA GRAVIDA	0.16 1/HOUR	CHEMOSTAT, UPTAKE=0 FOR CONC = 0.2 UG-ATOM/L, NH4-LIMITATION	279
*THALASSIOSIRA GRAVIDA	0.03 1/HOUR	CHEMOSTAT, UPTAKE=0 FOR CONC = 0.5 UG-ATOM/L, SI-LIMITATION	279

TABLE A60

ALGAE : MAX RATE OF NITRATE UPTAKE			
ITEM:	VALUE:	CONDITIONS:	REF.:
*BIDDULPHIA AURITA	0.9*10**(-8) MICROMOLES / (CELL*HOUR)	NONE	954
*CHLOROPHYTE	4.78 MICROGRAM-ATOMS NO3-N/(10**6 CELLS*HOUR)	N AS NITRATE, SUN ADAPTED, ESTAURINE PHYTOPLANKTON	76
*CHLOROPHYTE	5.56 MICROGRAM-ATOMS NO3-N/(10**6 CELLS*HOUR)	N AS NITRATE, SHADE ADAPTED, ESTAURINE PHYTOPLANKTON	76
*CHLOROPHYTE	4.35 MICROGRAM-ATOMS NO3-N/(10**6 CELLS*HOUR)	N AS NITRATE AND NH4, SUN ADAPTED, ESTAURINE PHYTOPLANKTON	76
*CHLOROPHYTE	2.14 MICROGRAM-ATOMS NO3-N/(10**6 CELLS*HOUR)	N AS NITRATE AND NH4, SHADE ADAPTED, ESTAURINE PHYTOPLANKTON	76
*COCCOCHLORIS STAGNINA	4.0 - 9.0 *10**(-10) MICROMOLES / (CELL*HOUR)	NONE	225
*CYCLOTELLA NANA	4. - 9.*10**(-9) MICROMOLES / (CELL*HOUR)	NONE	230
*CYCLOTELLA NANA	0.3 - 1.6 *10**(-8) MICROMOLES / (CELL*HOUR)	NONE	225
*DITYLUM BRIGHTWELLII	1.2*10**(-6) MICROMOLES / (CELL*HOUR)	NONE	424
*DUNALIELLA TERTIOLECTA	2.2 - 3.7 *10**(-8) MICROMOLES / (CELL*HOUR)	NONE	225
*MONOCHRYSIS LUTHERI	1.4*10**(-8) MICROMOLES / (CELL*HOUR)	NONE	225
*SKELETONEMA COSTATUM	7.14 MICROGRAM-ATOMS NO3-N/(10**6 CELLS*HOUR)	N AS NITRATE, SUN ADAPTED, ESTAURINE PHYTOPLANKTON	76
*SKELETONEMA COSTATUM	8.33 MICROGRAM-ATOMS NO3-N/(10**6 CELLS*HOUR)	N AS NITRATE, SHADE ADAPTED, ESTAURINE PHYTOPLANKTON	76
*SKELETONEMA COSTATUM	2.78 MICROGRAM-ATOMS NO3-N/(10**6 CELLS*HOUR)	N AS NITRATE AND NH4, SUN ADAPTED, ESTAURINE PHYTOPLANKTON	76
*SKELETONEMA COSTATUM	1.52 MICROGRAM-ATOMS NO3-N/(10**6 CELLS*HOUR)	N AS NITRATE AND NH4, SHADE ADAPTED, ESTAURINE PHYTOPLANKTON	76

TABLE A61

ALGAE : MAX RATE OF PHOSPHORUS UPTAKE			
ITEM:	VALUE:	CONDITIONS:	REF.:
*ASTERIONELLA FORMOSA	8*10**(-9) MICROMOLES /(CELL*HOUR)	NONE	961

TABLE A61

ALGAE : MAX RATE OF PHOSPHORUS UPTAKE

ITEM:	VALUE:	CONDITIONS:	REF.:
*DINOBRYON CYLINDRICUM	2.01*10**(-9) MICRO MOLES/ (CELL*HOUR)	NEW JERSEY	904
*DINOBRYON SOCIALE VAR. AMERICANUM	13.9*10**(-9) MICRO MOLES/ (CELL*HOUR)	NEW JERSEY, IN P-FREE MEDIUM FOR 2 D	904
*DINOBRYON SOCIALE VAR. AMERICANUM	3.66*10**(-9) MICRO MOLES/ (CELL*HOUR)	NEW JERSEY, DIRECTLY FROM CULTURE	904
*EUGLENA GRACILIS	2.4*10**(-7) MICROMOLES / (CELL*HOUR)	NONE	138
*FRESHWATER PHYTOPLANKTON	0.040 MICROMOLES/HOUR	CENTRAL LAKE KINNERET, 287.5 K, P= 0.14 MICRO M	600
*FRESHWATER PHYTOPLANKTON	0.18 MICROMOLES/HOUR	SOUTHERN LAKE KINNERET, 290.0 K, P =0.10 MICRO M	600
*FRESHWATER PHYTOPLANKTON	0.066 MICROMOLES/HOUR	SOUTHERN LAKE LINNERET, 301 K, P= 0.11 MICRO M	600
*FRESHWATER PHYTOPLANKTON	1.11 MICROMOLES/HOUR	SOUTHERN LAKE KINNERET, 304 K, P= 0.037 MICRO M	600
*FRESHWATER PHYTOPLANKTON	0.060 MICROMOLES/HOUR	JORDAN RIVER, UPSTREAM L. KINNERET, 289 K, P=0.44 MICRO M	600
*MARINE PHYTOPLANKTON	0.02 - 2.95 MICROGRAM-ATOM P/(L*H)	CHESAPEAKE BAY, NATURAL PHYTOPLANKTON, 5 STATIONS	355
*PEDIASTRUM DUPLEX	4*10**(-8) MICROMOLES /(CELL*HOUR)	NONE	355
*PHAEODACTYLUM TRICORNUTUM	0.7 - 1.4 *10**(-9) MICROMOLES / (CELL*HOUR)	NONE	863
*SCENEDESMUS SP.	1.2*10**(-8) MICROMOLES / (CELL*HOUR)	NONE	1321

TABLE A62

ALGAE : MAX RATE OF PHOTOSYNTHESIS

ITEM:	VALUE:	CONDITIONS:	REF.:
*ANABAENA SP. AND OSCILLATORIA SP.	0.66 MG C/(MG CHLOROPHYLL A*HOUR)	283-304 K, PHOTOSYNTHETIC CAPACITY = 13.6 MG C/(M3*H)	355
*ASTERIONELLA SP.	1.54 MG C/(MG CHLOROPHYLL A*HOUR	291 K, PHOTOSYNTHETIC CAPACITY = 6.0 MG C/(M3*H)	355
*MELOSIRA SP.	2.28 MG C/(MG CHLOROPHYLL A*HOUR)	291 K, PHOTOSYNTHETIC CAPACITY = 19.5 MG C/(M3*H)	355
*MELOSIRA SP.	2.01 MG C/(MG CHLOROPHYLL A*HOUR)	301.5 K, PHOTOSYNTHETIC CAPACITY = 5.0 MG C/(M3*H)	355
*MICROCYSTIS SP.	6.40 MG C/(MG CHLOROPHYLL A*HOUR)	301.9 K, PHOTOSYNTHETIC CAPACITY = 10.5 MG C/(M3*H)	355
*NITZSCHIA SP.	3.96 MG C/(MG CHLOROPHYLL A*HOUR)	295-299 K, PHOTOSYNTHETIC CAPACITY = 8.9 MG C/(M3*H)	355
*TRICHODESMIUM SPP.	0.38 MICROGRAM C/(MICROGRAM CHL.A) *HR	N.PACIFIC,N-FIXING ALGAE	967

TABLE A63

ALGAE : MAX RATE OF SILICIUM UPTAKE

ITEM:	VALUE:	CONDITIONS:	REF.:
*DITYLUM BRIGHTWELLII	9.5*10**(-7) MICROMOLES / (CELL*HOUR)	NONE	1203
*DITYLUM BRIGHTWELLII	26.6 PICOGRAM SI/(CELL*HOUR)	UPTAKE, NO UPTAKE FOR =2.96 UG-AT/ L, 25000 ERG/(CM2*SEC), 293 K (F002)	1203
*LICMOPHORA SP.	7.7*10**(-8) MICROMOLES / (CELL*HOUR)	NONE	1203
*LICMOPHORA SP.	2.15 PICOGRAM SI/(CELL*HOUR)	UPTAKE, NO UPTAKE FOR =1.30 UG-AT/ L, 25000 ERG/(CM2*SEC), 293 K ($47)	1203

TABLE A63

ALGAE : MAX RATE OF SILICIUM UPTAKE			
ITEM:	VALUE:	CONDITIONS:	REF.:
*NAVICULA PELLICULOSA	1*10**(-8) MICROMOLES /(CELL*HOUR)	NONE	208
*NITZSCHIA ALBA	2.5 - 5.7 *10**(-9) MICROMOLES / (CELL*HOUR)	NONE	915
*SKELETONEMA COSTATUM	3.4*10**(-9) MICROMOLES / (CELL*HOUR)	NONE	1203
*SKELETONEMA COSTATUM	0.095 PICOGRAM SI/(CELL*HOUR)	UPTAKE, NO UPTAKE FOR =0.32 UG-AT/ L, 25000 ERG/(CM2*SEC), 293 K ($47)	1203
*THALASSIOSIRA DECIPIENS	1.46*10**(-7) MICROMOLES / (CELL*HOUR)	NONE	1203
*THALASSIOSIRA DECIPIENS	4.09 PICOGRAM SI/(CELL*HOUR)	UPTAKE, NO UPTAKE FOR =1.33 UG-AT/ L, 25000 ERG/(CM2*SEC), 293 K ($47)	1203
*THALASSIOSIRA PSEUDONANA	2.6*10**(-9) MICROMOLES / (CELL*HOUR)	NONE	1203
*THALASSIOSIRA PSEUDONANA	0.073 PICOGRAM SI/(CELL*HOUR)	UPTAKE, NO UPTAKE FOR =0.67 UG-AT/ L, 25000 ERG/(CM2*SEC), 293 K ($47)	1203

TABLE A64

ALGAE : MAX YIELD			
ITEM:	VALUE:	CONDITIONS:	REF.:
*SELENASTRUM CAPRICORNUTUM	805 MG BIOMASS/MG P UTILIZED	BATCH CULTURE	1602

TABLE A65

ALGAE : MAX YIELD COEFFICIENT			
ITEM:	VALUE:	CONDITIONS:	REF.:
*SELENASTRUM CAPRICORNUTUM	22.01 MG CELL/MG AMMONIUM-N	293K,CHEMOST.	1319
*SELENASTRUM CAPRICORNUTUM	61.20 MG CELL/MG AMMONIUM-N	297K,CHEMOST.	1319
*SELENASTRUM CAPRICORNUTUM	33.44 MG CELL/MG AMMONIUM-N	300K,CHEMOST.	1319
*SELENASTRUM CAPRICORNUTUM	29.94 MG CELL/MG AMMONIUM-N	301K,CHEMOST.	1319
*SELENASTRUM CAPRICORNUTUM	25.57 MG CELL/MG AMMONIUM-N	306K,CHEMOST.	1319

TABLE A66

ALGAE : MG/BIOMASS RATIO			
ITEM:	VALUE:	CONDITIONS:	REF.:
*ANABAENA SP.	0.42 % OF DRY WEIGHT	ALGAE FROM PONDS AND LAKES IN SOUTHEASTERN UNITED STATES	1660
*APHANIZOMENON SP.	0.21 % OF DRY WEIGHT	ALGAE FROM PONDS AND LAKES IN SOUTHEASTERN UNITED STATES	1660
*CHARA SP.	0.92 % OF DRY WEI¤HT	ALGAE FROM PONDS AND LAKES IN SOUTHEASTERN UNITED STATES	1660
*CLADOPHORA SP.	0.23 % OF DRY WEIGHT	ALGAE FROM PONDS AND LAKES IN SOUTHEASTERN UNITED STATES	1660
*EUGLENA SP.	0.07 % OF DRY WEIGHT	ALGAE FROM PONDS AND LAKES IN SOUTHEASTERN UNITED STATES	1660
*HYDRODICTYON SP.	0.17 % OF DRY WEIGHT	ALGAE FROM PONDS AND LAKES IN SOUTHEASTERN UNITED STATES	1660
*LYNGBYA SP.	0.14 % OF DRY WEIGHT	ALGAE FROM PONDS AND LAKES IN SOUTHEASTERN UNITED STATES	1660

TABLE A66

ALGAE : MG/BIOMASS RATIO			
ITEM:	VALUE:	CONDITIONS:	REF.:
*MICROCYSTIS SP.	0.17 % OF DRY WEIGHT	ALGAE FROM PONDS AND LAKES IN SOUTHEASTERN UNITED STATES	1660
*MOUGEOTIA SP.	0.57 % OF DRY WEIGHT	ALGAE FROM PONDS AND LAKES IN SOUTHEASTERN UNITED STATES	1660
*NITELLA SP.	0.95 % OF DRY WEIGHT	ALGAE FROM PONDS AND LAKES IN SOUTHEASTERN UNITED STATES	1660
*OEDOGONIUM SP.	0.16 % OF DRY WEIGHT	ALGAE FROM PONDS AND LAKES IN SOUTHEASTERN UNITED STATES	1660
*PITHOPHORA SP.	0.20 % OF DRY WEIGHT	ALGAE FROM PONDS AND LAKES IN SOUTHEASTERN UNITED STATES	1660
*RHIZOCLONIUM SP.	0.21 % OF DRY WEIGHT	ALGAE FROM PONDS AND LAKES IN SOUTHEASTERN UNITED STATES	1660
*SPIROGYRA SP.	0.45 % OF DRY WEIGHT	ALGAE FROM PONDS AND LAKES IN SOUTHEASTERN UNITED STATES	1660
*SPIROGYRA SP., GIANT	0.30 % OF DRY WEIGHT	ALGAE FRO¤ PONDS AND LAKES IN SOUTHEASTERN UNITED STATES	1660

TABLE A67

ALGAE : MIN CELL N			
ITEM:	VALUE:	CONDITIONS:	REF.:
*ANABAENA SP.	1*10**(-7) MICROMOLES/CELL	NONE	569
*ASTERIONELLA FORMOSA	6*10**(-7) MICROMOLES/CELL	NONE	569
*DINOBRYON SP.	1.8*10**(-8) MICROMOLES/CELL	NONE	569
*GYMNODINIUM SP.	3.9*10**(-7) MICROMOLES/CELL	NONE	569
*ISOCHRYSIS GALBANA	3*10**(-8) MICROMOLES/CELL	NONE	370

TABLE A68

ALGAE : MIN CELL P			
ITEM:	VALUE:	CONDITIONS:	REF.:
*ANABAENA SP.	2.5*10**(-9) MICROMOLES/CELL	NONE	569
*ASTERIONELLA FORMOSA	2*10**(-9) MICROMOLES/CELL	NONE	961
*ASTERIONELLA FORMOSA	1.5 - 3.0 *10**(-9) MICROMOLES/ CELL	NONE	1077
*ASTERIONELLA FORMOSA	3*10**(-8) MICROMOLES/CELL	NONE	569
*ASTERIONELLA JAPONICA	1.5*10**(-9) MICROMOLES/CELL	NONE	1077
*CHLORELLA PYRENOIDOSA	3*10**(-9) MICROMOLES/CELL	NONE	1077
*CYCLOTELLA NANA	0.9*10**(-9) MICROMOLES/CELL	NONE	496
*DINOBRYON SP.	0.5*10**(-9) MICROMOLES/CELL	NONE	569
*GYMNODINIUM SP.	1.1*10**(-8) MICROMOLES/CELL	NONE	569
*NITZSCHIA ACTINASTROIDES	3*10**(-9) MICROMOLES/CELL	DIAMETER = 35 MY	1077
*NITZSCHIA ACTINASTROIDES	4*10**(-9) MICROMOLES/CELL	DIAMETER = 50 - 55 MY	1077
*PHAEODACTYLUM TRICORNUTUM	2*10**(-9) MICROMOLES/CELL	NONE	1077
*SCENEDESMUS QUADRICAUDA	4.5*10**(-9) MICROMOLES/CELL	NONE	1077
*SCENEDESMUS SP.	1.7*10**(-9) MICROMOLES/CELL	NONE	1321
*THALASSIOSIRA FLUVIATILIS	12.2*10**(-9) MICROMOLES/CELL	NONE	496

TABLE A69

ALGAE : MIN CELL SI			
ITEM:	VALUE:	CONDITIONS:	REF.:
*ASTERIONELLA FORMOSA	2*10**(-6) MICROMOLES/CELL	NONE	703

TABLE A69

ALGAE : MIN CELL SI			
ITEM:	VALUE:	CONDITIONS:	REF.:
*ASTERIONELLA FORMOSA	1.8*10**(-6) MICROMOLES/CELL	NONE	569
*FRAGILARIA CROTONENSIS	4*10**(-6) MICROMOLES/CELL	NONE	569
*NAVICULA PELLICULOSA	0.5*10**(-7) MICROMOLES/CELL	NONE	208
*NITZSCHIA ALBA	3*10**(-7) MICROMOLES/CELL	NONE	915
*THALASSIOSIRA PSEUDONANA	2*10**(-8) MICROMOLES/CELL	NONE	PASS

TABLE A70

ALGAE : MINIMAL PHOSPHORUS REQUIREMENTS			
ITEM:	VALUE:	CONDITIONS:	REF.:
*ASTERIONELLA SP.	< 0.2 MICRO G P/(MM**3 CELL VOLUME)	NONE	1640
*FRAGILARIA SP.	0.2 - 0.35 MICRO G P/(MM**3 CELL VOLUME)	NONE	1640
*MICROCYSTIS SP.	> 0.5 MICRO G P/(MM**3 CELL VOLUME)	NONE	1640
*OSCILLATORIA SP.	> 0.5 MICRO G P/(MM**3 CELL VOLUME)	NONE	1640
*SCENEDESMUS SP.	> 0.5 MICRO G P/(MM**3 CELL VOLUME)	NONE	1640
*TABELLARIA SP.	0.45 - 0.6 MICRO G P/(MM**3 CELL VOLUME)	NONE	1640

TABLE A71

ALGAE : MN/BIOMASS RATIO			
ITEM:	VALUE:	CONDITIONS:	REF.:
*ANABAENA SP.	800 MICROGRAM/L	ALGAE FROM PONDS AND LAKES IN SOUTHEASTERN UNITED STATES	1660
*APHANIZOMENON SP.	167 MIGROGRAM/L	ALGAE FROM PONDS AND LAKES IN SOUTHEASTERN UNITED STATES	1660
*CHARA SP.	2926 MICROGRAM/L	ALGAE FROM PONDS AND LAKES IN SOUTHEASTERN UNITED STATES	1660
*CLADOPHORA SP.	1040 MICROGRAM/L	ALGAE FROM PONDS AND LAKES IN SOUTHEASTERN UNITED STATES	1660
*EUGLENA SP.	240 MICROGRAM/L	ALGAE FROM PONDS AND LAKES IN SOUTHEASTERN UNITED STATES	1660
*HYDRODICTYON SP.	1373 MICROGRAM/L	ALGAE FROM PONDS AND LAKES IN SOUTHEASTERN UNITED STATES	1660
*LYNGBYA SP.	5230 MICROGRAM/L	ALGAE FROM PONDS AND LAKES IN SOUTHEASTERN UNITED STATES	1660
*MICROCYSTIS SP.	2751 MICROGRAM/L	ALGAE FROM PONDS AND LAKES IN SOUTHEASTERN UNITED STATES	1660
*MOUGEOTIA SP.	1080 MICROGRAM/L	ALGAE FROM PONDS AND LAKES IN SOUTHEASTERN UNITED STATES	1660
*NITELLA SP.	2388 MICROGRAM/L	ALGAE FROM PONDS AND LAKES IN SOUTHEASTERN UNITED STATES	1660
*OEDOGONIUM SP.	1729 MICROGRAM/L	ALGAE FROM PONDS AND LAKES IN SOUTHEASTERN UNITED STATES	1660
*PITHOPHORA SP.	829 MICROGRAM/L	ALGAE FROM PONDS AND LAKES IN SOUTHEASTERN UNITED STATES	1660
*RHIZOCLONIUM SP.	1687 MICROGRAM/L	ALGAE FROM PONDS AND LAKES IN SOUTHEASTERN UNITED STATES	1660
*SPIROGYRA SP.	1641 MICROGRAM/L	ALGAE FROM PONDS AND LAKES IN SOUTHEASTERN UNITED STATES	1660
*SPIROGYRA SP., GIANT	1658 MICROGRAM/L	ALGAE FROM PONDS AND LAKES IN SOUTHEASTERN UNITED STATES	1660

TABLE A72

ALGAE : MORTALITY			
ITEM:	VALUE:	CONDITIONS:	REF.:
*FRESHWATER PHYTOPLANKTON	0.004 1/HOUR	STEADY STATE SOLUTION TO CASTLE LAKE.	1326

TABLE A73

ALGAE : N/BIOMASS			
ITEM:	VALUE:	CONDITIONS:	REF.:
*ALGAE	3.8 % OF DRY WEIGHT	MEAN ELEMENTARY COMPOSITION	157

TABLE A74

ALGAE : N/BIOMASS RATIO			
ITEM:	VALUE:	CONDITIONS:	REF.:
*ALGAE	.07-.09 FRACTION OF DRY WEIGHT	NONE	1571
*ANABAENA SP.	9.43 % OF DRY WEIGHT	ALGAE FROM PONDS AND LAKES IN SOUTHEASTERN UNITED STATES	1660
*APHANIZOMENON SP.	8.57 % OF DRY WEIGHT	ALGAE FROM PONDS AND LAKES IN SOUTHEASTERN UNITED STATES	1660
*CHAETOCEROS LAUDERI	10.3 MICROGRAM N/MM**3	CULTIVATED, MARINE	104
*CHARA SP.	2.46 % OF DRY WEIGHT	ALGAE FROM PONDS AND LAKES IN SOUTHEASTERN UNITED STATES	1660
*CHLAMYDOMONAS MAGNUSII	31.4 MICROGRAM N/MM**3	CULTIVATED, MARINE	104
*CHLOROPHYTA	9.54 MG/(GRAM OF ACID-SOLUBLE FRACTION)	AVERAGE OF 7 MARINE SPECIES,PUERTO RICO	386
*CLADOPHORA SP.	2.30 % OF DRY WEIGHT	ALGAE FROM PONDS AND LAKES IN SOUTHEASTERN UNITED STATES	1660
*COSCINODISCUS SP.	11.7 MICROGRAM N/MM**3	CULTIVATED, MARINE	104
*CRYPTOMONAS PSEUDOBALTICA	47.6 MICROGRAM N/MM**3	CULTIVATED, MARINE	104
*EUGLENA SP.	5.14 % OF DRY WEIGHT	ALGAE FROM PONDS AND LAKES IN SOUTHEASTERN UNITED STATES	1660
*HETEROTHRIX SP.	46.6 MICROGRAM N/MM**3	CULTIVATED, MARINE	104
*HYDRODICTYON SP.	3.87 % OF DRY WEIGHT	ALGAE FROM PONDS AND LAKES IN SOUTHEASTERN UNITED STATES	1660
*LYNGBYA SP.	5.01 % OF DRY WEIGHT	ALGAE FROM PONDS AND LAKES IN SOUTHEASTERN UNITED STATES	1660
*MARINE PHYTOPLANKTON	42 10**(-12) G/CELL	NO3 AS SOURCE OF N, STATIONARY PHASE,293 K, SOUTHERN CALIF	428
*MARINE PHYTOPLANKTON	53 10**(-12) G/CELL	NH4 AS SOURCE OF N, STATIONARY PHASE,293 K, SOUTHERN CALIF	428
*MARINE PHYTOPLANKTON	22 10**(-12) G/CELL	UREA AS SOURCE OF N, STATIONARY PHASE,293 K, SOUTHERN CALIF	428
*MICROCYSTIS SP.	8.08 % OF DRY WEIGHT	ALGAE FROM PONDS AND LAKES IN SOUTHEASTERN UNITED STATES	1660
*MOUGEOTIA SP.	1.77 % OF DRY WEIGHT	ALGAE FROM PONDS AND LAKES IN SOUTHEASTERN UNITED STATES	1660
*NITELLA SP.	2.70 % OF DRY WEIGHT	ALGAE FROM PONDS AND LAKES IN SOUTHEASTERN UNITED STATES	1660
*OEDOGONIUM SP.	2.64 % OF DRY WEIGHT	ALGAE FROM PONDS AND LAKES IN SOUTHEASTERN UNITED STATES	1660
*PHAEODACTYLUM TRICORNUTUM	37.6 MICROGRAM N/MM**3	CULTIVATED, MARINE	104
*PHAEOPHYCEAE	1.5 % OF DRY WEIGHT	MEAN ELEMENTARY COMPOSITION	157
*PHAEOPHYCEAE	8.87 MG/(GRAM OF ACID-SOLUBLE FRACTION)	AVERAGE OF 4 MARINE SPECIES,PUERTO RICO	386
*PITHOPHORA SP.	2.57 % OF DRY WEIGHT	ALGAE FROM PONDS AND LAKES IN SOUTHEASTERN UNITED STATES	1660

TABLE A74

ALGAE : N/BIOMASS RATIO

ITEM:	VALUE:	CONDITIONS:	REF.:
*RHIZOCLONIUM SP.	3.46 % OF DRY WEIGHT	ALGAE FROM PONDS AND LAKES IN SOUTHEASTERN UNITED STATES	1660
*RHODOPHYTA	10.50 MG/(GRAM OF ACID-SOLUBLE FRACTION)	AVERAGE OF 9 MARINE SPECIES,PUERTO RICO	386
*SPIROGYRA SP.	3.01 % OF DRY WEIGHT	ALGAE FROM PONDS AN¤ LAKES IN SOUTHEASTERN UNITED STATES	1660
*SPIROGYRA SP., GIANT	2.35 % OF DRY WEIGHT	ALGAE FROM PONDS AND LAKES IN SOUTHEASTERN UNITED STATES	1660
*TETRASELMIS STRICIA	34.6 MICROGRAM N/MM**3	CULTIVATED, MARINE	104
*THALASSIOSIRA SP.	4.3*10**(-14) MOLES/CELL	GROWTH RATE=0.041 1/H, CELL DENSITY=4.0*10**8 CELL/L, P-LIMITED CHEMO.	1243
*THALASSIOSIRA SP.	4.2*10**(-14) MOLES/CELL	GROWTH RATE=0.017 1/H, CELL DENSITY=6.7*10**8 CELL/L, P-LIMITED CHEMO.	1243
*TRICHODESMIUM SPP.	2.4 MICROGRAM N/COLONY	N/COLONY,N.PACIFIC,N-FIXING ALGAE, N=6	967

TABLE A75

ALGAE : N/CHLOROPHYLL A RATIO

ITEM:	VALUE:	CONDITIONS:	REF.:
*ALGAE	7 BY WEIGHT	USED IN MODELS, SAN JOAQUIN RIVER, CALIF.	1150
*ALGAE	7 BY WEIGHT	USED IN MODELS, SAN JOAQUIN RIVER ESTUARY, CALIF.	1150
*ALGAE	10 BY WEIGHT	USED IN MODELS, POTOMAC ESTUARY	1150
*ALGAE	7 BY WEIGHT	USED IN MODELS, LAKE ERIE	1150
*ALGAE	10 BY WEIGHT	USED IN MODELS, LAKE ONTARIO	1150
*ALGAE	0.01 MG-N/MICROGRAM-CHL.	POTOMAC RIVER	1153
*CHAETOCEROS DEBILIS	21.9 BY WEIGHT	SI-STARVED,72H,CONT.LIGHT:0.08LY/MIN,DILUTION=1/3-1/2 OF MAX.GR.RATE	631
*CHAETOCEROS DEBILIS	11.7 BY WEIGHT	SI-LIMITED,CONT.LIGHT:0.08LY/MIN, DILUTION=1/3-1/2 OF MAX.GR.RATE	631
*CHAETOCEROS DEBILIS	11.6 BY WEIGHT	NON-LIMITED,CONT.LIGHT:0.08LY/MIN, DILUTION=1/3-1/2 OF MAX.GR.RATE	631
*CHAETOCEROS DEBILIS	8.5 BY WEIGHT	NH4-LIMITED,CONT.LIGHT:0.08LY/MIN, DILUTION=1/3-1/2 OF MAX.GR.RATE	631
*CHAETOCEROS DEBILIS	12.6 BY WEIGHT	NH4-STARVED,72H,CONT.LIGHT:0.08LY/MIN,DILUTION=1/3-1/2 OF MAX.GR.RATE	631
*CHAETOCEROS DEBILIS	2.9-4.8 BY WEIGHT	NITROGEN-LIMITED,DILUTION RATE= 0.097-0.008/HR	1582
*CYCLOTELLA NANA	11.7-35.6 BY WEIGHT	NITRATE-LIMITED,DILUTION RATE= 0.077-0.009/HR	225
*MARINE PHYTOPLANKTON	15 BY WEIGHT	NO3 AS SOURCE OF N, LOG PHASE,293 K, SOUTHERN CALIF	428
*MARINE PHYTOPLANKTON	8.8 BY WEIGHT	NO3 AS SOURCE OF N, STATIONARY PHASE,293 K, SOUTHERN CALIF	428
*MARINE PHYTOPLANKTON	29 BY WEIGHT	NH4 AS SOURCE OF N, LOG PHASE,293 K, SOUTHERN CALIF	428
*MARINE PHYTOPLANKTON	14 BY WEIGHT	NH4 AS SOURCE OF N, STATIONARY PHASE,293 K, SOUTHERN CALIF	428
*MARINE PHYTOPLANKTON	17 BY WEIGHT	UREA AS SOURCE OF N, LOG PHASE,293 K, SOUTHERN CALIF	428
*MARINE PHYTOPLANKTON	11 BY WEIGHT	UREA AS SOURCE OF N, STATIONARY PHASE,293 K, SOUTHERN CALIF	428
*SKELETONEMA COSTATUM	15.7 BY WEIGHT	SI-STARVED,72H,CONT.LIGHT:0.08LY/MIN,DILUTION=1/3-1/2 OF MAX.GR.RATE	631
*SKELETONEMA COSTATUM	11.2 BY WEIGHT	SI-LIMITED,CONT.LIGHT:0.08LY/MIN, DILUTION=1/3-1/2 OF MAX.GR.RATE	631

TABLE A75

ALGAE : N/CHLOROPHYLL A RATIO			
ITEM:	VALUE:	CONDITIONS:	REF.:
*SKELETONEMA COSTATUM	14.0 BY WEIGHT	NON-LIMITED,CONT.LIGHT:0.08LY/MIN, DILUTION=1/3-1/2 OF MAX.GR.RATE	631
*SKELETONEMA COSTATUM	8.8 BY WEIGHT	NH4-LIMITED,CONT.LIGHT:0.08LY/MIN, DILUTION=1/3-1/2 OF MAX.GR.RATE	631
*SKELETONEMA COSTATUM	7.7 BY WEIGHT	NH4-STARVED,72H,CONT.LIGHT:0.08LY/ MIN,DILUTION=1/3-1/2 OF MAX.GR.RATE	631
*SKELETONEMA COSTATUM	28.9-5.6 BY WEIGHT	AMMONIUM-LIMITED,DILUTION RATE= 0.110-0.042/HR	630
*SKELETONEMA COSTATUM	11.5-20.0 BY WEIGHT	NITROGEN-LIMITED,BATCH CULTURE, 72-115HOURS STARVATION	686
*SKELETONEMA COSTATUM	74.6-10.7 BY WEIGHT	SILICATE-LIMITED,DILUTION RATE= 0.086-0.022/HR	630
*THALASSIOSIRA FLUVIATILIS	2.9-2.2 BY WEIGHT	NITRATE-LIMITED,BATCH CULTURE, 72-115HOURS STARVATION	673
*THALASSIOSIRA GRAVIDA	17.2 BY WEIGHT	SI-STARVED,72H,CONT.LIGHT:0.08LY/ MIN,DILUTION=1/3-1/2 OF MAX.GR.RATE	631
*THALASSIOSIRA GRAVIDA	9.8 BY WEIGHT	SI-LIMITED,CONT.LIGHT:0.08LY/MIN, DILUTION=1/3-1/2 OF MAX.GR.RATE	631
*THALASSIOSIRA GRAVIDA	11.4 BY WEIGHT	NON-LIMITED,CONT.LIGHT:0.08LY/MIN, DILUTION=1/3-1/2 OF MAX.GR.RATE	631
*THALASSIOSIRA GRAVIDA	5.3 BY WEIGHT	NH4-LIMITED,CONT.LIGHT:0.08LY/MIN, DILUTION=1/3-1/2 OF MAX.GR.RATE	631
*THALASSIOSIRA GRAVIDA	15.1 BY WEIGHT	NH4-STARVED,72H,CONT.LIGHT:0.08LY/ MIN,DILUTION=1/3-1/; OF MAX.GR.RATE	631
*THALASSIOSIRA PSEUDONANA	5.7 BY WEIGHT	P-LIMITED CHEMOSTAT,GROWTH=0.041 1/H, 295 K,0.07 CAL/CM2*MIN	1243
*THALASSIOSIRA PSEUDONANA	7.3 BY WEIGHT	P-LIMITED CHEMOSTAT,GROWTH=0.017 1/H, 295 K,0.07 CAL/CM2*MIN	1243
*THALASSIOSIRA PSEUDONANA	5.8 BY WEIGHT	N-LIMITED CHEMOSTAT,GROWTH=0.041 1/H, 295 K,0.07 CAL/CM2*MIN	1243
*THALASSIOSIRA PSEUDONANA	5.2 BY WEIGHT	N-LIMITED CHEMOSTAT,GROWTH=0.017 1/H, 295 K,0.07 CAL/CM2*MIN	1243
*THALASSIOSIRA PSEUDONANA	7.5 BY WEIGHT	N-LIMITED CHEMOSTAT,GROWTH=0.0085 1/H,295 K,0.07 CAL/CM2*MIN	1243
*THALASSIOSIRA PSEUDONANA	5.8-7.5 BY WEIGHT	NITROGEN-LIMITED,DILUTION RATE= 0.041-0.008/HR	423
*THALASSIOSIRA PSEUDONANA	23.3-20.0 BY WEIGHT	NITRATE-LIMITED,BATCH CULTURE, 72-115HOURS STARVATION	673

TABLE A76

ALGAE : N/P RATIO			
ITEM:	VALUE:	CONDITIONS:	REF.:
*CHAETOCEROS DEBILIS	8.0 BY ATOMS	SI-STARVED,72H,CONT.LIGHT:0.08LY/ MIN,DILUTION=1/3-1/2 OF MAX.GR.RATE	631
*CHAETOCEROS DEBILIS	7.9 BY ATOMS	SI-LIMITED,CONT.LIGHT:0.08LY/MIN, DILUTION=1/3-1/2 OF MAX.GR.RATE	631
*CHAETOCEROS DEBILIS	12.0 BY ATOMS	NON-LIMITED,CONT.LIGHT:0.08LY/MIN, DILUTION=1/3-1/2 OF MAX.GR.RATE	631
*CHAETOCEROS DEBILIS	4.6 BY ATOMS	NH4-LIMITED,CONT.LIGHT:0.08LY/MIN, DILUTION=1/3-1/2 OF MAX.GR.RATE	631
*CHAETOCEROS DEBILIS	6.0 BY ATOMS	NH4-STARVED,72H,CONT.LIGHT:0.08LY/ MIN,DILUTION=1/3-1/2 OF MAX.GR.RATE	631
*FRESHWATER ALGAE	10 - 20	MEAN ATOMIC RATIO	355
*FRESHWATER ALGAE	4.5 - 9	MEAN WEIGHT RATIO	355
*FRESHWATER ALGAE	7.9 (5.6 - 10.9)	OPTIMUM VALUE DETERMINED FROM BATCH C¤LTURES FROM 33 STATIONS IN LAKE	355
*FRESHWATER ALGAE	4.1 BY WEIGHT	L. VARESE,0.363 MG P/L,1.494 MG N/ L	355

TABLE A76

ALGAE : N/P RATIO

ITEM:	VALUE:	CONDITIONS:	REF.:
*FRESHWATER ALGAE	5.1 BY WEIGHT	L. LUGANO ,0.122 MG P/L,0.616 MG N/L	355
*FRESHWATER ALGAE	5.7 BY WEIGHT	L. OGGIONO ,0.161 MG P/L,0.91 MG N/L	355
*FRESHWATER ALGAE	5.6 BY WEIGHT	L. VIVERONE,0.117 MG P/L,0.652 MG N/L	355
*FRESHWATER ALGAE	16 BY WEIGHT	L. PUSIANO,0.068 MG P/L,1.116 MG N/L	355
*FRESHWATER ALGAE	6.7 BY WEIGHT	L.LUGANO,0.091 MG P/L,0.616 MG N/L	355
*FRESHWATER ALGAE	19 BY WEIGHT	L.ALSERIO,0.061 MG P/L,1.165 MG N/L	355
*FRESHWATER ALGAE	19 BY WEIGHT	L. COMO,0.035 MG P/L,0.67 MG N/L	355
*FRESHWATER ALGAE	15 BY WEIGHT	L. COMO (BRIENNO),0.053 MG P/L, 0.78 NG N/L	355
*FRESHWATER ALGAE	19 BY WEIGHT	L. COMO (LECCO),0.035 MG P/L,0.653 MG N/L	355
*FRESHWATER ALGAE	9.3 BY WEIGHT	L.AVIGLIANA,0.033 MG P/L,0.3 MG N/L	355
*FRESHWATER ALGAE	26 BY WEIGHT	L. GARLATE,0.029 MG P/L,0.746 MG N/L	355
*FRESHWATER ALGAE	24 BY WEIGHT	L. OLGINATE,0.029 MG P/L,0.707 MG N/L	355
*FRESHWATER ALGAE	36 BY WEIGHT	L. ISEO (NORTH),0.017 MG P/L,0.695 MG N/L	355
*FRESHWATER ALGAE	43 BY WEIGHT	L. MAGGIORE(PALLANZA),0.018 MG P/L,0.754 MG N/L	355
*FRESHWATER ALGAE	68 BY WEIGHT	L. COMABBIO,0.004 MG P/L,0.274 MG N/L	355
*FRESHWATER ALGAE	16 BY WEIGHT	L. CANDIA,0.005 MG P/L,0.083 MG N/L	355
*FRESHWATER ALGAE	90 BY WEIGHT	L.MAGGIORE(MEINA),0.019 MG P/L, 0.717 MG N/L	355
*FRESHWATER ALGAE	35 BY WEIGHT	L. ISEO(SOUTH),0.019 MG P/L,0.669 MG N/L	355
*FRESHWATER ALGAE	53 BY WEIGHT	L. GARDA (RIVE), 0.006 MG P/L. 0.322 MG N/L	355
*FRESHWATER ALGAE	104 BY WEIGHT	L. PIANO,0.01 MG P/L,1.045 MG N/L	355
*FRESHWATER ALGAE	20 BY WEIGHT	L.ENDINE,0.005 MG P/L,0.1 MG N/L	355
*FRESHWATER ALGAE	116 BY WEIGHT	L. GHIRLA,0.005 MG P/L,0.582 MG N/L	355
*FRESHWATER ALGAE	156 BY WEIGHT	L. MEZZOLA,0.003 MG P/L,0.468 MG N/L	355
*FRESHWATER ALGAE	63 BY WEIGHT	L. GARDA(SIRMIONE),0.005 MG P/L, 0.317 MG N/L	355
*FRESHWATER ALGAE	79 BY WEIGHT	L. GARDA(GARGNANO),0.004 MG P/L, 0.319 MG N/L	355
*FRESHWATER ALGAE	65 BY WEIGHT	L. IDRO,0.006 MG P/L,0.389 MG N/L	355
*FRESHWATER ALGAE	247 BY WEIGHT	L. GANNA,0.002 MG P/L,0.494 MG N/L	355
*FRESHWATER ALGAE	291 BY WEIGHT	L. SEGRINO,0.005 MG P/L,1.454 MG N/L	355
*FRESHWATER ALGAE	58 BY WEIGHT	L.MONTORFANO,0.004 MG P/L, 0.524 MG N/L	355
*FRESHWATER ALGAE	104 BY WEIGHT	L. MERGOZZO,0.004 MG P/L,0.524 MG N/L	355
*FRESHWATER ALGAE	189 BY WEIGHT	L.ANNONA,0.002 MG P/L,0.472 MG N/L	355
*FRESHWATER ALGAE	126 BY WEIGHT	L. MONATE,0.001 MG P/L,0.177 MG N/L	355
*MARINE PHYTOPLANKTON	12-15 BY ATOMS	VIGOROUSLY GROWING,EXCESS NITRATE IN WATER	39
*MARINE PHYTOPLANKTON	15-18 BY ATOMS	UNHEALTHY PHYTOPLANKTON,NITRATE DEPLETED WATER	39
*MARINE PHYTOPLANKTON	14.5 BY ATOMS	SEE REF.	1520
*SKELETONEMA COSTATUM	7.7 BY ATOMS	SI-STARVED,72H,CONT.LIGHT:0.08LY/MIN,DILUTION=1/3-1/2 OF MAX.GR.RATE	631

TABLE A76

ALGAE : N/P RATIO			
ITEM:	VALUE:	CONDITIONS:	REF.:
*SKELETONEMA COSTATUM	8.4 BY ATOMS	SI-LIMITED,CONT.LIGHT:0.08LY/MIN, DILUTION=1/3-1/2 OF MAX.GR.RATE	631
*SKELETONEMA COSTATUM	10.0 BY ATOMS	NON-LIMITED,CONT.LIGHT:0.08LY/MIN, DILUTION=1/3-1/2 OF MAX.GR.RATE	631
*SKELETONEMA COSTATUM	4.8 BY ATOMS	NH4-LIMITED,CONT.LIGHT:0.08LY/MIN, DILUTION=1/3-1/2 OF MAX.GR.RATE	631
*SKELETONEMA COSTATUM	4.2 BY ATOMS	NH4-STARVED,72H,CONT.LIGHT:0.08LY/ MIN,DILUTION=1/3-1/2 OF MAX.GR.RATE	631
*SKELETONEMA COSTATUM	15.2-5.1 BY ATOMS	AMMONIUM-LIMITED,DILUTION RATE= 0.110-0.042/HR	630
*SKELETONEMA COSTATUM	10.0-6.1 BY ATOMS	SILICATE-LIMITED,DILUTION RATE= 0.061-0.035/HR	330
*SKELETONEMA COSTATUM	10.0-9.8 BY ATOMS	SILICATE-LIMITED,DILUTION RATE= 0.086-0.022/HR	630
*THALASSIOSIRA GRAVIDA	11.1 BY ATOMS	SI-STARVED,72H,CONT.LIGHT:0.08LY/ MIN,DILUTION=1/3-1/2 OF MAX.GR.RATE	631
*THALASSIOSIRA GRAVIDA	9.7 BY ATOMS	SI-LIMITED,CONT.LIGHT:0.08LY/MIN, DILUTION=1/3-1/2 OF MAX.GR.RATE	631
*THALASSIOSIRA GRAVIDA	8.8 BY ATOMS	NON-LIMITED,CONT.LIGHT:0.08LY/MIN, DILUTION=1/3-1/2 OF MAX.GR.RATE	631
*THALASSIOSIRA GRAVIDA	2.0 BY ATOMS	NH4-LIMITED,CONT.LIGHT:0.08LY/MIN, DILUTION=1/3-1/2 OF MAX.GR.RATE	631
*THALASSIOSIRA GRAVIDA	5.1 BY ATOMS	NH4-STARVED,72H,CONT.LIGHT:0.08LY/ MIN,DILUTION=1/3-1/2 OF MAX.GR.RATE	631
*THALASSIOSIRA PSEUDONANA	37.2 BY WEIGHT	P-LIMITED CHEMOSTAT,GROWTH=0.041 1/H, 295 K,0.07 CAL/CM2*MIN	1243
*THALASSIOSIRA PSEUDONANA	61.1 BY WEIGHT	P-LIMITED CHEMOSTAT,GROWTH=0.017 1/H, 295 K,0.07 CAL/CM2*MIN	1243
*THALASSIOSIRA PSEUDONANA	9.7 BY WEIGHT	N-LIMITED CHEMOSTAT,GROWTH=0.041 1/H, 295 K,0.07 CAL/CM2*MIN	1243
*THALASSIOSIRA PSEUDONANA	8.3 BY WEIGHT	N-LIMITED CHEMOSTAT,GROWTH=0.017 1/H, 295 K,0.07 CAL/CM2*MIN	1243
*THALASSIOSIRA PSEUDONANA	6.4 BY MOLE	N-LIMITED CHEMOSTAT,GROWTH=0.0085 1/H,295 K,0.07 CAL/CM2*MIN	1243
*THALASSIOSIRA PSEUDONANA	9.7-6.4 BY ATOMS	NITROGEN-LIMITED,DILUTION RATE= 0.041-0.008/HR	423

TABLE A77

ALGAE : N/SI RATIO			
ITEM:	VALUE:	CONDITIONS:	REF.:
*CHAETOCEROS DEBILIS	4.8 BY ATOMS	SI-STARVED,72H,CONT.LIGHT:0.08LY/ MIN,DILUTION=1/3-1/2 OF MAX.GR.RATE	631
*CHAETOCEROS DEBILIS	4.1 BY ATOMS	SI-LIMITED,CONT.LIGHT:0.08LY/MIN, DILUTION=1/3-1/2 OF MAX.GR.RATE	631
*CHAETOCEROS DEBILIS	1.8 BY ATOMS	NON-LIMITED,CONT.LIGHT:0.08LY/MIN, DILUTION=1/3-1/2 OF MAX.GR.RATE	631
*CHAETOCEROS DEBILIS	0.5 BY ATOMS	NH4-LIMITED,CONT.LIGHT:0.08LY/MIN, DILUTION=1/3-1/2 OF MAX.GR.RATE	631
*CHAETOCEROS DEBILIS	0.9 BY ATOMS	NH4-STARVED,72H,CONT.LIGHT:0.08LY/ MIN,DILUTION=1/3-1/2 OF MAX.GR.RATE	631
*SKELETONEMA COSTATUM	2.8 BY ATOMS	SI-STARVED,72H,CONT.LIGHT:0.08LY/ MIN,DILUTION=1/3-1/2 OF MAX.GR.RATE	631
*SKELETONEMA COSTATUM	4.1 BY ATOMS	SI-LIMITED,CONT.LIGHT:0.08LY/MIN, DILUTION=1/3-1/2 OF MAX.GR.RATE	631
*SKELETONEMA COSTATUM	1.9 BY ATOMS	NON-LIMITED,CONT.LIGHT:0.08LY/MIN, DILUTION=1/3-1/2 OF MAX.GR.RATE	631
*SKELETONEMA COSTATUM	0.8 BY ATOMS	NH4-LIMITED,CONT.LIGHT:0.08LY/MIN, DILUTION=1/3-1/2 OF MAX.GR.RATE	631

TABLE A77

ALGAE : N/SI RATIO			
ITEM:	VALUE:	CONDITIONS:	REF.:
*SKELETONEMA COSTATUM	0.8 BY ATOMS	NH4-STARVED,72H,CONT.LIGHT:0.08LY/MIN,DILUTION=1/3-1/2 OF MAX.GR.RATE	631
*SKELETONEMA COSTATUM	3.8-0.58 BY ATOMS	AMMONIUM-LIMITED,DILUTION RATE=0.110-0.042/HR	630
*SKELETONEMA COSTATUM	3.3-5.7 BY ATOMS	SILICATE-LIMITED,DILUTION RATE=0.086-0.022/HR	630
*THALASSIOSIRA GRAVIDA	8.5 BY ATOMS	SI-STARVED,72H,CONT.LIGHT:0.08LY/MIN,DILUTION=1/3-1/2 OF MAX.GR.RATE	631
*THALASSIOSIRA GRAVIDA	6.0 BY ATOMS	SI-LIMITED,CONT.LIGHT:0.08LY/MIN,DILUTION=1/3-1/2 OF MAX.GR.RATE	631
*THALASSIOSIRA GRAVIDA	2.3 BY ATOMS	NON-LIMITED,CONT.LIGHT:0.08LY/MIN,DILUTION=1/3-1/2 OF MAX.GR.RATE	631
*THALASSIOSIRA GRAVIDA	0.6 BY ATOMS	NH4-LIMITED,CONT.LIGHT:0.08LY/MIN,DILUTION=1/3-1/2 OF MAX.GR.RATE	631
*THALASSIOSIRA GRAVIDA	1.4 BY ATOMS	NH4-STARVED,72H,CONT.LIGHT:0.08LY/MIN,DILUTION=1/3-1/2 OF MAX.GR.RATE	631

TABLE A78

ALGAE : NA/BIOMASS RATIO			
ITEM:	VALUE:	CONDITIONS:	REF.:
*ANABAENA SP.	0.18 % OF DRY WEIGHT	ALGAE FROM PONDS AND LAKES IN SOUTHEASTERN UNITED STATES	1660
*APHANIZOMENON SP.	0.19 % OF DRY WEIGHT	ALGAE FROM PONDS AND LAKES IN SOUTHEASTERN UNITED STATES	1660
*CHARA SP.	0.13 % OF DRY WEIGHT	ALGAE FROM PONDS AND LAKES IN SOUTHEASTERN UNITED STATES	1660
*CLADOPHORA SP.	0.18 % OF DRY WEIGHT	ALGAE FROM PONDS AND LAKES IN SOUTHEASTERN UNITED STATES	1660
*EUGLENA SP.	0.02 % OF DRY WEIGHT	ALGAE FROM PONDS AND LAKES IN SOUTHEASTERN UNITED STATES	1660
*HYDRODICTYON SP.	0.38 % OF DRY WEIGHT	ALGAE FROM PONDS AND LAKES IN SOUTHEASTERN UNITED STATES	1660
*LYNGBYA SP.	0.06 % OF DRY WEIGHT	ALGAE FROM PONDS AND LAKES IN SOUTHEASTERN UNITED STATES	1660
*MICROCYSTIS SP.	0.04 % OF DRY WEIGHT	ALGAE FROM PONDS AND LAKES IN SOUTHEASTERN UNITED STATES	1660
*MOUGEOTIA SP.	0.49 % OF DRY WEIGHT	ALGAE FROM PONDS AND LAKES IN SOUTHEASTERN UNITED STATES	1660
*NITELLA SP.	0.28 % OF DRY WEIGHT	ALGAE FROM PONDS AND LAKES IN SOUTHEASTERN UNITED STATES	1660
*OEDOGONIUM SP.	0.06 % OF DRY WEIGHT	ALGAE FROM PONDS AND LAKES IN SOUTHEASTERN UNITED STATES	1660
*PITHOPHORA SP.	0.07 % OF DRY WEIGHT	ALGAE FROM PONDS AND LAKES IN SOUTHEASTERN UNITED STATES	1660
*RHIZOCLONIUM SP.	0.09 % OF DRY WEIGHT	ALGAE FROM PONDS AND LAKES IN SOUTHEASTERN UNITED STATES	1660
*SPIROGYRA SP.	1.42 % OF DRY WEIGHT	ALGAE FROM PONDS AND LAKES IN SOUTHEASTERN UNITED STATES	1660
*SPIROGYRA SP., GIANT	1.43 % OF DRY WEIGHT	ALGAE FROM PONDS AND LAKES IN SOUTHEASTERN UNITED STATES	1660

TABLE A79

ALGAE : NITROGEN (INTRACELLULAR)			
ITEM:	VALUE:	CONDITIONS:	REF.:
*BACILLARIOPHYCEAE	4.9 % OF AVERAGE DRY WEIGHT	RANGE=2.7-5.9%	1521

TABLE A79

ALGAE : NITROGEN (INTRACELLULAR)			
ITEM:	VALUE:	CONDITIONS:	REF.:
*BLUE GREEN ALGAE	4.9 % OF AVERAGE DRY WEIGHT	RANGE=4.5-5.8%	1521
*CHLOROPHYCEAE	7.8 % OF AVERAGE DRY WEIGHT	RANGE=6.6-9.1%	1521
*CHRYSOPHYCEAE	4.4 % OF AVERAGE DRY WEIGHT	RANGE=3.3-5.0%	1521

TABLE A80

ALGAE : NITROGEN FIXATION			
ITEM:	VALUE:	CONDITIONS:	REF.:
*ANABAENA DOMINANCE	0 - 1 MICRO G N/(L*HOUR)	SMITH, ALASKA, 1963	1689
*ANABAENA SCHEREMETIEVI	1.12*10**(-9) MG N/(CELL*DAY)	CHERNOYE, USSR, 1937	1689
*BLUE GREEN ALGAE DOMINANCE	0.07 - 43 MICRO G N/(L*HOUR)	L. MENDOTA, WISC., 1960, 1967	1689
*BLUE GREEN ALGAE DOMINANCE	0 - 6 MICRO G N/(L*HOUR)	SANCTUARY, PENN., 1959	1689
*BLUE GREEN ALGAE DOMINANCE	0.005 - 1 MICRO G N/(L*HOUR)	WINGRA¤ WISC., 1961	1689
*FRESHWATER ALGAE	5 MICROGRAM N/(MG N*DAY)	LAKE TJEUKEMEER	693
*FRESHWATER ALGAE	33 % OF TOTAL INPUT	LAKE GEORGE	1633
*FRESHWATER ALGAE	2.8 - 70 MICRO G N/(MG CHLOROPHYLL*DAY)	ASSUMING 10 % N AND 1.5 % CHLOROPHYLL	550
*FRESHWATER ALGAE	3 MG N2/(G CARBOHYDRATE)	ASSUMING RESPIRATION RATE OF 24 MG O2/(MG CHLOROPHYLL*DAY)	550
*TRICHODESMIUM SPP.	0.27 MICROGRAM N/(MICROGRAM CHL.A) *HR	N.PACIFIC,N-FIXING ALGAE,USING C2H4:N2 CONVERSION RATIO OF 1.9: 1, MAX	967
*TRICHODESMIUM SPP.	134 PMOLES N2/(MICROGRAM ALGAL N) *HR	N.PACIFIC,N-FIXING ALGAE, INCUBATION IN LABORATORY	967

TABLE A81

ALGAE : NUTRIENT UTILIZATION CONSTANT			
ITEM:	VALUE:	CONDITIONS:	REF.:
*SELENASTRUM CAPRICORNUTUM	1.05 DAYS = KA	293K,CHEMOST.,R=0.934,EQ.(F003),KA =1/K1,K1=NUTRIENT UTIL.RATE	1319
*SELENASTRUM CAPRICORNUTUM	3.16 DAYS = KA	297K,CHEMOST.,R=0.908,EQ.(F003) ,KA =1/K1,K1=NUTRIENT UTIL.RATE	1319
*SELENASTRUM CAPRICORNUTUM	1.30 DAYS = KA	300K,CHEMOST.,R=0.470,EQ (F003) ,KA =1/K1,K1=NUTRIENT UTIL.RATE	1319
*SELENASTRUM CAPRICORNUTUM	1.36 DAYS = KA	301K,CHEMOST.,R=0.800,EQ.(F003),KA =1/K1,K1=NUTRIENT UTIL.RATE	1319
*SELENASTRUM CAPRICORNUTUM	0.558 DAYS = KB	293K,CHEMOST.,R=0.931,EQ.(F004), NET YIELD COEFF.=MAX.YIELD COEFF./ 2	1319
*SELENASTRUM CAPRICORNUTUM	2.640 DAYS = KB	297K,CHEMOST.,R=0.899,EQ.(F004), NET YIELD COEFF.=MAX.YIELD COEFF./ 2	1319
*SELENASTRUM CAPRICORNUTUM	0.825 DAYS = KB	300K,CHEMOST.,R=0.442,EQ.(F004), NET YIELD COEFF.=MAX.YIELD COEFF./ 2	1319
*SELENASTRUM CAPRICORNUTUM	0.750 DAYS = KB	301K,CHEMOST.,R=0.804,EQ.(F004), NET YIELD COEFF.=MAX.YIELD COEFF./ 2	1319
*SELENASTRUM CAPRICORNUTUM	0.84 DAYS = KA	306K,CHEMOST.,R=-0.459,EQ.(F003), KA=1/K1,K1=NUTRIENT UTIL.RATE	1319
*SELENASTRUM CAPRICORNUTUM	0.427 DAYS = KB	303K,CHEMOST.,R=-0.472,EQ.(F004), NET YIELD COEFF.=MAX.YIELD COEFF./ 2	1319

TABLE A82

ALGAE : O/BIOMASS RATIO			
ITEM:	VALUE:	CONDITIONS:	REF.:
*ALGAE	44.0 % OF DRY WEIGHT	MEAN ELEMENTARY COMPOSITION/ BIOMASS RATIO	157

TABLE A82

ALGAE : O/BIOMASS RATIO			
ITEM:	VALUE:	CONDITIONS:	REF.:
*PHAEOPHYCEAE	37.7 % OF DRY WEIGHT	MEAN ELEMENTARY COMPOSITION	157

TABLE A83

ALGAE : ORGANIC CONTENT			
ITEM:	VALUE:	CONDITIONS:	REF.:
*CHLOROPHYTA	0.304 GRAM/(GRAM OF ACID-SOLUBLE FRACTION)	AVERAGE OF 7 MARINE SPECIES,PUERTO RICO	386
*PHAEOPHYCEAE	0.313 GRAM/(GRAM OF ACID-SOLUBLE FRACTION)	AVERAGE OF 4 MARINE SPECIES,PUERTO RICO, TOTAL ORGANIC CONTENT	386
*RHODOPHYTA	0.406 GRAM/(GRAM OF ACID-SOLUBLE FRACTION)	AVERAGE OF 9 MARINE SPECIES,PUERTO RICO, TOTAL ORGANIC CONTENT	386

TABLE A84

ALGAE : OXYGEN DEMAND			
ITEM:	VALUE:	CONDITIONS:	REF.:
*BLUE GREEN ALGAE	2.8/1.23 MICROGRAM OXYGEN/ (MICROGRAM CHL.)*H	ILLUMINATED/UNILLUMINATED,288K,48H OLD,LOUGH NEAGH AT A DEPTH OF 0.M	523
*BLUE GREEN ALGAE	1.44/1. MICROGRAN OXYGEN/ (MICROGRAM CHL.)*H	ILLUMINATED/UNILLUMINATED,288K,48H OLD,LOUGH NEAGH AT A DEPTH OF 5.M	523
*BLUE GREEN ALGAE	1.39/0.87 MICROGRAM OXYGEN/ (MICROGRAM CHL.)*H	ILLUMINATED/UNILLUMINATED,288K,48H OLD,LOUNG NEAGH AT A DEPTH OF 10M	523
*BLUE GREEN ALGAE	2.2/1.7 MICROGRAM OXYGEN/ (MICROGRAM CHL.)*H	ILLUMINATED/UNILLUMINATED,288K,48H OLD,LAKE WASHIN. AT A DEPTH OF 4M	523

TABLE A85

ALGAE : P/BIOMASS RATIO			
ITEM:	VALUE:	CONDITIONS:	REF.:
*ALGAE	0.01 - 0.012 BY WEIGHT	NONE	1571
*ANABAENA SP.	0.77 % OF DRY WEIGHT	ALGAE FROM PONDS AND LAKES IN SOUTHEASTERN UNITED STATES	1660
*APHANIZOMENON SP.	1.17 % OF DRY WEIGHT	ALGAE FROM PONDS AND LAKES IN SOUTHEASTERN UNITED STATES	1660
*CHARA SP.	0.25 % OF DRY WEIGHT	ALGAE FROM PONDS AND LAKES IN SOUTHEASTERN UNITED STATES	1660
*CLADOPHORA SP.	0.56 % OF DRY WEIGHT	ALGAE FROM PONDS AND LAKES IN SOUTHEASTERN UNITED STATES	1660
*EUGLENA SP.	0.67 % OF DRY WEIGHT	ALGAE FROM PONDS AND LAKES IN SOUTHEASTERN UNITED STATES	1660
*HYDRODICTYON SP.	0.24 % OF DRY WEIGHT	ALGAE FROM PONDS AND LAKES IN SOUTHEASTERN UNITED STATES	1660
*LYNGBYA SP.	0.31 % OF DRY WEIGHT	ALGAE FROM PONDS AND LAKES IN SOUTHEASTERN UNITED STATES	1660
*MICROCYSTIS SP.	0.68 % OF DRY WEIGHT	ALGAE FROM PONDS AND LAKES IN SOUTHEASTERN UNITED STATES	1660
*MOUGEOTIA SP.	0.25 % OF DRY WEIGHT	ALGAE FROM PONDS AND LAKES IN SOUTHEASTERN UNITED STATES	1660
*NITELLA SP.	0.23 % OF DRY WEIGHT	ALGAE FROM PONDS AND LAKES IN SOUTHEASTERN UNITED STATES	1660
*OEDOGONIUM SP.	0.08 % OF DRY WEIGHT	ALGAE FROM PONDS AND LAKES IN SOUTHEASTERN UNITED STATES	1660

TABLE A85

ALGAE : P/BIOMASS RATIO			
ITEM:	VALUE:	CONDITIONS:	REF.:
*PITHOPHORA SP.	0.30 % OF DRY WEIGHT	ALGAE FROM PONDS AND LAKES IN SOUTHEASTERN UNITED STATES	1660
*RHIZOCLONIUM SP.	0.43 % OF DRY WEIGHT	ALGAE FROM PONDS AND LAKES IN SOUTHEASTERN UNITED STATES	1660
*SPIROGYRA SP.	0.20 % OF DRY WEIGHT	ALGAE FROM PONDS AND LAKES IN SOUTHEASTERN UNITED STATES	1660
*SPIROGYRA SP., GIANT	0.23 % OF DRY WEIGHT	ALGAE FROM PONDS AND LAKES IN SOUTHEASTERN UNITED STATES	1660
*THALASSIOSIRA SP.	1.1*10**(-15) MOLES/CELL	GROWTH RATE=0.041 1/H, CELL DENSITY=4.0*10**8 CELL/L, P-LIMITED CHEMO.	1243
*THALASSIOSIRA SP.	0.68*10**(-15) MOLES/CELL	GROWTH RATE=0.017 1/H, CELL DENSITY=6.7*10**8 CELL/L, P-LIMITED CHEMO.	1243
*TRICHODESMIUM SPP.	0.12 MICROGRAM	P/COLONY,N.PACIFIC,N-FIXING ALGAE, N=6	967

TABLE A86

ALGAE : P/CHLOROPHYLL A RATIO			
ITEM:	VALUE:	CONDITIONS:	REF.:
*ALGAE	0.001 MG-P/MICROGRAM-CHL.	POTOMAC RIVER	1153
*CHAETOCEROS DEBILIS	6.2 BY WEIGHT	SI-STARVED,72H,CONT.LIGHT:0.08LY/MIN,DILUTION=1/3-1/2 OF MAX.GR.RATE	631
*CHAETOCEROS DEBILIS	2.6 BY WEIGHT	SI-LIMITED,CONT.LIGHT:0.08LY/MIN, DILUTION=1/3-1/2 OF MAX.GR.RATE	631
*CHAETOCEROS DEBILIS	2.2 BY WEIGHT	NON-LIMITED,CONT.LIGHT:0.08LY/MIN, DILUTION=1/3-1/2 OF MAX.GR.RATE	631
*CHAETOCEROS DEBILIS	4.3 BY WEIGHT	NH4-LIMITED,CONT.LIGHT:0.08LY/MIN, DILUTION=1/3-1/2 OF MAX.GR.RATE	631
*CHAETOCEROS DEBILIS	4.8 BY WEIGHT	NH4-STARVED,72H,CONT.LIGHT:0.08LY/MIN,DILUTION=1/3-1/2 OF MAX.GR.RATE	631
*SKELETONEMA COSTATUM	4.6 BY WEIGHT	SI-STARVED,72H,CONT.LIGHT:0.08LY/MIN,DILUTION=1/3-1/2 OF MAX.GR.RATE	631
*SKELETONEMA COSTATUM	3.0 BY WEIGHT	SI-LIMITED,CONT.LIGHT:0.08LY/MIN, DILUTION=1/3-1/2 OF MAX.GR.RATE	631
*SKELETONEMA COSTATUM	2.8 BY WEIGHT	NON-LIMITED,CONT.LIGHT:0.08LY/MIN, DILUTION=1/3-1/2 OF MAX.GR.RATE	631
*SKELETONEMA COSTATUM	4.2 BY WEIGHT	NH4-LIMITED,CONT.LIGHT:0.08LY/MIN, DILUTION=1/3-1/2 OF MAX.GR.RATE	631
*SKELETONEMA COSTATUM	4.6 BY WEIGHT	NH4-STARVED,72H,CONT.LIGHT:0.08LY/MIN,DILUTION=1/3-1/2 OF MAX.GR.RATE	631
*SKELETONEMA COSTATUM	6.1-2.6 BY WEIGHT	AMMONIUM-LIMITED,DILUTION RATE= 0.110-0.042/HR	630
*SKELETONEMA COSTATUM	19.2-2.5 BY WEIGHT	SILICATE-LIMITED,DILUTION RATE= 0.086-0.022/HR	630
*THALASSIOSIRA GRAVIDA	2.1 BY WEIGHT	SI-STARVED,72H,CONT.LIGHT:0.08LY/MIN,DILUTION=1/3-1/2 OF MAX.GR.RATE	631
*THALASSIOSIRA GRAVIDA	2.3 BY WEIGHT	SI-LIMITED,CONT.LIGHT:0.08LY/MIN, DILUTION=1/3-1/2 OF MAX.GR.RATE	631
*THALASSIOSIRA GRAVIDA	3.0 BY WEIGHT	NON-LIMITED,CONT.LIGHT:0.08LY/MIN, DILUTION=1/3-1/2 OF MAX.GR.RATE	631
*THALASSIOSIRA GRAVIDA	6.1 BY WEIGHT	NH4-LIMITED,CONT.LIGHT:0.08LY/MIN, DILUTION=1/3-1/2 OF MAX.GR.RATE	631
*THALASSIOSIRA GRAVIDA	15.7 BY WEIGHT	NH4-STARVED,72H,CONT.LIGHT:0.08LY/MIN,DILUTION=1/3-1/2 OF MAX.GR.RATE	631
*THALASSIOSIRA PSEUDONANA	0.32 BY WEIGHT	P-LIMITED CHEMOSTAT,GROWTH=0.041 1/H, 295 K,0.07 CAL/CM2*MIN	1243
*THALASSIOSIRA PSEUDONANA	0.27 BY WEIGHT	P-LIMITED CHEMOSTAT,GROWTH=0.017 1/H, 295 K,0.07 CAL/CM2*MIN	1243

TABLE A86

ALGAE : P/CHLOROPHYLL A RATIO			
ITEM:	VALUE:	CONDITIONS:	REF.:
*THALASSIOSIRA PSEUDONANA	1.17 BY WEIGHT	N-LIMITED CHEMOSTAT,GROWTH=0.041 1/H, 295 K,0.07 CAL/CM2*MIN	1243
*THALASSIOSIRA PSEUDONANA	1.36 BY WEIGHT	N-LIMITED CHEMOSTAT,GROWTH=0.017 1/H, 295 K,0.07 CAL/CM2*MIN	1243
*THALASSIOSIRA PSEUDONANA	2.61 BY WEIGHT	N-LIMITED CHEMOSTAT,GROWTH=0.0085 1/H,295 K,0.07 CAL/CM2*MIN	1243
*THALASSIOSIRA PSEUDONANA	1.2-2.6 BY WEIGHT	NITROGEN-LIMITED,DILUTION RATE= 0.041-0.008/HR	423

TABLE A87

ALGAE : PHOSPHORUS (INTRACELLOLAR)			
ITEM:	VALUE:	CONDITIONS:	REF.:
*BACILLARIOPHYCEAE	1.1 % OF DRY WEIGHT	RANGE=0.4-2.0%	1521
*BLUE GREEN ALGAE	1.1 % OF DRY WEIGHT	RANGE=0.8-1.4%	1521
*CHLOROPHYCEAE	2.9 % OF DRY WEIGHT	RANGE=2.4-3.3%	1521
*CHRYSOPHYCEAE	2.1 % OF DRY WEIGHT	RANGE=1.2-3.0%	1521
*DINOPHYCEAE	4.4 % OF DRY WEIGHT	RANGE=3.3-5.0%	1521
*DINOPHYCEAE	1.0 % OF DRY WEIGHT	RANGE=0.6-1.1%	1521

TABLE A88

ALGAE : PHOTOSYNTHESIS			
ITEM:	VALUE:	CONDITIONS:	REF.:
*FUCUS VESICULOSUS	1.2 MG O2/(G DW*H)	288 K, 21.5 MW/CM2=SATURATION, THALLUS	829
*FUCUS VESICULOSUS	9.27 MG O2/(G DW*H)	288 K, 21.5 MW/CM2=SATURATION, APICES	829
*LAMINARIA DIGITATA	1.19 MG O2/(G DW*H)	288 K, 21.5 MW/CM2=SATURATION, THALLUS	829
*LAMINARIA DIGITATA	3.97 MG O2/(G DW*H)	288 K, 21.5 MW/CM2=SATURATION, SEVERAL CM FROM THALLUS MARGIN	829

TABLE A89

ALGAE : PHOTOSYNTHETIC EFFICIENCY			
ITEM:	VALUE:	CONDITIONS:	REF.:
*ANABAENA CYLINDRICA	0.25 MG C/(MG CHL)*HR*KLUX	INITIAL SLOPE,NATURAL POPULATION, DOM.SPECIE,283-303K,POND	44
*ASTERIONELLA JAPONICA	14.50 KILOLUX	SATURATION POINT, TROPICAL MARINE ALGAE, 2 H	1287
*ASTERIONELLA SP.	2.8 KLUX	IK, LINEAR SATURATION LEVEL OF PHOTOSYNTHESIS	1553
*BIDDULPHIA SINENSIS	17.75 KILOLUX	SATURATION POINT, TROPICAL MARINE ALGAE, 2 H	1287
*CERATIUM FURCA	13.00 KILOLUX	SATURATION POINT, TROPICAL MARINE ALGAE, 2 H	1287
*CHAETOCEROS LORENZIANUS	11.75 KILOLUX	SATURATION POINT, TROPICAL MARINE ALGAE, 2 H	1287
*CHAETOCEROS SP.	0.13 MG C/(MG CHL)*HR*KLUX	INITIAL SLOPE,CULTURE,293K	1096
*CHLORELLA ELLIPSOIDEA	0.50 MG C/(MG CHL)*HR*KLUX	INITIAL SLOPE,CULTURE,278-303K	44

TABLE A89

ALGAE : PHOTOSYNTHETIC EFFICIENCY

ITEM:	VALUE:	CONDITIONS:	REF.:
*CHLORELLA PYRENOIDOSA	0.48 MG C/((MG CHL A)*HR*KLUX)	INITIAL SLOPE,CULTURE	1505
*CHLORELLA VULGARIS	0.50 MG C/(MG CHL(A+B))*HR*KLUX	INITIAL SLOPE,CULTURE,293K	1506
*COCCOLITHUS HUXLEYI	0.13 MG C/(MG CHL(A+B))*HR*KLUX	INITIAL SLOPE,CULTURE	750
*COSCINODISCUS RADIATUS	11.50 KILOLUX	SATURATION POINT, TROPICAL MARINE ALGAE, 2 H	1287
*CRYPTOMONAS SP.	CA. 0.8 MG C/((MG CHL A)*HR*KLUX)	INITIAL SLOPE,NATURAL POPULATION, DOM.SPECIE,LAKE	714
*DICTYOCHA SP.	15.50 KILOLUX	SATURATION POINT, TROPICAL MARINE ALGAE, 2 H	1287
*DINOPHYSIS MILES	34.50 KILOLUX	SATURATION POINT, TROPICAL MARINE ALGAE, 2 H	1287
*FRESHWATER ALGAE	14 - 17 KERG/SEC	IK, LINEAR SATURATION LEVEL OF PHOTOSYNTHESIS	1639
*HYMENOMONAS SP.	0.10 MG C/(MG CHL(A+B))*HR*KLUX	INITIAL SLOPE,CULTURE	750
*MARINE NANNOPLANKTON	6.7+-0.6 MG C/(MG CHLA*HOUR)	0.1-1 UG-ATOM NO3/L, 288-291 K, CALIF CURRENT,MORNING	973
*MARINE NANNOPLANKTON	6.4+-0.4 MG C/(MG CHLA*HOUR)	0.1-1 UG-ATOM NO3/L, 288-291 K, CALIF CURRENT,EVENING	973
*MARINE NANNOPLANKTON	2.1+-0.5 MG C/(MG CHLA*HOUR)	0 NO3, 298-302 K, CALIF CURRENT, MORNING	973
*MARINE NANNOPLANKTON	1.1+-0.4 MG C/(MG CHLA*HOUR)	0 NO3, 298-302 K, CALIF CURRENT, EVENING	973
*MARINE NANNOPLANKTON	18.3+-2.6 MG C/(MG CHLA*HOUR)	5-12 UG-ATOM NO3/L, 298-301 K, CALIF CURRENT,MORNING	973
*MARINE NANNOPLANKTON	12.1+-2.8 MG C/(MG CHLA*HOUR)	5-12 UG-ATOM NO3/L, 298-301 K, CALIF CURRENT,EVENING	973
*MARINE NETPLANKTON	3.8+-0.9 MG C/(MG CHLA*HOUR)	0.1-1 UG-ATOM NO3/L, 288-291 K, CALIF CURRENT,MORNING	973
*MARINE NETPLANKTON	3.4+-0.6 MG C/(MG CHLA*HOUR)	0.1-1 UG-ATOM NO3/L, 288-291 K, CALIF CURRENT,EVENING	973
*MARINE NETPLANKTON	0.8+-0.4 MG C/(MG CHLA*HOUR)	0 NO3, 298-302 K, CALIF CURRENT, MORNING	973
*MARINE NETPLANKTON	2.6+-0.2 MG C/(MG CHLA*HOUR)	0 NO3, 298-302 K, CALIF CURRENT, EVENING	973
*MARINE NETPLANKTON	6.3+-1.1 MG C/(MG CHLA*HOUR)	5-12 UG-ATOM NO3/L, 298-301 K, CALIF CURRENT,MORNING	973
*MARINE NETPLANKTON	11.1+-1.6 MG C/(MG CHLA*HOUR)	5-12 UG-ATOM NO3/L, 298-301 K, CALIF CURRENT,EVENING	973
*PHYTOPLANKTON	0.36 MG C/((MG CHL A)*HR*KLUX)	INITIAL SLOPE,NATURAL POPULATION, 277.5-279K,ARCTIC	1504
*PHYTOPLANKTON	0.35 MG C/((MG CHL A)*HR*KLUX)	INITIAL SLOPE,NATURAL POPULATION, TROPICAL,TEMPERATE(SUMMER,WINTER)	1504
*PHYTOPLANKTON	0.35 MG C/((MG CHL A)*HR*KLUX)	INITIAL SLOPE,NATURAL POPULATION, NORTHERN AND ARCTIC(GENERAL)	1504
*PHYTOPLANKTON	0.05 MG C/((MG CHL A)*HR*KLUX)	INITIAL SLOPE,NATURAL POPULATION, KUROSHIO(GENERAL)	715
*PHYTOPLANKTON	0.35 MG C/((MG CHL A)*HR*KLUX)	INITIAL SLOPE,NATURAL POPULATION, OYASHIO(GENERAL)	715
*PHYTOPLANKTON	0.17 MG C/((MG CHL A)*HR*KLUX)	INITIAL SLOPE,NATURAL POPULATION, KUROSHIO & OYASHIO(GENERAL)	715
*PHYTOPLANKTON	0.42 MG C/((MG CHL A)*HR*KLUX)	INITIAL SLOPE,NATURAL POPULATION, ANTARCTIC	199
*PLANKTONIELLA SOL	14.50 KILOLUX	SATURATION POINT, TROPICAL MARINE ALGAE, 2 H	1287
*RHIZOSOLENIA STYLIFORMIS	23.50 KILOLUX	SATURATION POINT, TROPICAL MARINE ALGAE, 2 H	1287
*SCENEDESMUS SP.	0.65 MG C/(MG CHL)*HR*KLUX	INITIAL SLOPE,CULTURE,283-313K	44
*SKELETONEMA COSTATUM	0.52 MG C/((MG CHL A)*HR*KLUX)	INITIAL SLOPE,CULTURE	1505
*SKELETONEMA COSTATUM	0.44 MG C/(MG CHL)*HR*KLUX	INITIAL SLOPE,CULTURE,293K	1096
*SKELETONEMA COSTATUM	0.56 MG C/(MG CHL)*HR*KLUX	INITIAL SLOPE,NATURAL POPULATION, DOM.SPECIE,293K,TOKYO BAY	44
*SYNEDRA SP.	0.65 MG C/(MG CHL)*HR*KLUX	INITIAL SLOPE,NATURAL POPULATION, DOM.SPECIE,283-303K,POND	44
*TRICERATIUM FAVUS	17.25 KILOLUX	SATURATION POINT, TROPICAL MARINE ALGAE, 2 H	1287

TABLE A90

ALGAE : PRODUCTION/BIOMASS RATIO

ITEM:	VALUE:	CONDITIONS:	REF.:
*AGARUM CRIBROSUM	4.2	ANNUAL PRODUCTION, INITIAL BIOMASS, MEAN	977
*BOSSIELLA SP.	0.5 1/YEAR	YEARLY TURNOVER, SOUTHERN CALIFORNIA	1462
*CALLIARTHRON SP.	0.3 1/YEAR	YEARLY TURNOVER, SOUTHERN CALIFORNIA	1462
*CORALLINA SP.	0.5 1/YEAR	YEARLY TURNOVER, SOUTHERN CALIFORNIA	1462
*FRESHWATER LAKES	0.01 - 1.02 MG C ASSIMILATED/(MG BIOMASS C*DAY)	LAKE KINNERET	106
*FRESHWATER LAKES	0.28 - 4.11 MG C ASSIMILATED/(MG BIOMASS C*DAY)	LAKE MAGGIORE	106
*FRESHWATER LAKES	0.1 - 0.6 MG C ASSIMILATED/(MG BIOMASS C*DAY)	SWEDISH LAKES	106
*FRESHWATER LAKES	0.08 - 0.3 MG C ASSIMILATED/(MG BIOMASS C*DAY)	LAKE ERKEN	106
*FRESHWATER LAKES	0.67 - 1.5 MG C ASSIMILATED/(MG BIOMASS C*DAY)	POLISH LAKES	106
*FRESHWATER LAKES	0.2 - 0.44 MG C ASSIMILATED/(MG BIOMASS C*DAY)	CZECH RESEVOIRS	106
*FRESHWATER LAKES	0.1 - 0.8 MG C ASSIMILATED/(MG BIOMASS C*DAY)	DEVILS LAKE, N.D.	106
*FRESHWATER LAKES	0.2 - 1.2 MG C ASSIMILATED/(MG BIOMASS C*DAY)	UNTERSEE	106
*FRESHWATER LAKES	0.05 - 0.13 MG C ASSIMILATED/(MG BIOMASS C*DAY)	LAKE ZURICH	106
*FRESHWATER LAKES	0.31 - 0.46 MG C ASSIMILATED/(MG BIOMASS C*DAY)	OBERSEE	106
*FRESHWATER LAKES	0.39 MG C ASSIMILATED/(MG BIOMASS C*DAY)	ROTSEE	106
*FRESHWATER LAKES	0.26 - 11.3 MG C ASSIMILATED/(MG BIOMASS C*DAY)	LAKE NORRVIKEN	106
*FRESHWATER LAKES	0.2 - 3.4 MG C ASSIMILATED/(MG BIOMASS C*DAY)	TYSTRUP-BAVELSE LAKES	106
*LAMINARIA DIGITATA	9.8	ANNUAL PRODUCTION, INITIAL BIOMASS, MEAN	977
*LAMINARIA LONGICRURIS	7.2	ANNUAL PRODUCTION, INITIAL BIOMASS, MEAN	977
*LITHOTHAMNIOID ALGAE	1.5 1/YEAR	YEARLY TURNOVER, SOUTHERN CALIFORNIA	1462
*MELOBESIA SP.	2.0 1/YEAR	YEARLY TURNOVER, SOUTHERN CALIFORNIA	1462
*MISCELLANEOUS ARTICULATED ALGAE	0.5 1/YEAR	YEARLY TURNOVER, SOUTHERN CALIFORNIA	1462

TABLE A91

ALGAE : PROTEIN/BIOMASS RATIO

ITEM:	VALUE:	CONDITIONS:	REF.:
*AGMENELLUM QUADRUPLICATUM	44 % OF ASH FREE DRY WEIGHT	BLUE GREEN ALGAE,MARINE, NITROGEN*6.25	1220
*AMPHIDINIUM CARTERI	36 % OF ASH FREE DRY WEIGHT	DINOPHYCEAE,MARINE,NITROGEN*6.25	1220
*DUNALIELLA SALINA	58 % OF ASH FREE DRY WEIGHT	CHLOROPHYCEAE,MARINE,NITROGEN*6.25	1220
*EXUVIAELLA SP.	35 % OF ASH FREE DRY WEIGHT	DINOPHYCEAE,MARINE,NITROGEN*6.25	1220
*MARINE PHYTOPLANKTON	2.0 BY WEIGHT	VIGOROUSLY GROWING,EXCESS NITRATE IN WATER	39
*MARINE PHYTOPLANKTON	1.3 BY WEIGHT	UNHEALTHY PHYTOPLANKTON,NITRATE DEPLETED WATER	39
*MONOCHRYSIS LUTHERI	53 % OF ASH FREE DRY WEIGHT	CHRYSOPHYCEAE,MARINE,NITROGEN*6.25	1220
*SKELETONEMA COSTATUM	58 % OF ASH FREE DRY WEIGHT	BACILLARIOPHYCEAE,MARINE, NITROGEN*6.25	1220

TABLE A91

ALGAE : PROTEIN/BIOMASS RATIO			
ITEM:	VALUE:	CONDITIONS:	REF.:
*TETRASELMIS MACULATA	68 % OF ASH FREE DRY WEIGHT	PRASINOPHYCEAE,MARINE, NITROGEN*6.25	1220

TABLE A92

ALGAE : RA/BIOMASS RATIO			
ITEM:	VALUE:	CONDITIONS:	REF.:
*CHLOROPHYTA	0.0487 PICO CURIE/(GRAM OF ACID-SOLUBLE FRACTION)	AVERAGE OF 7 MARINE SPECIES,PUERTO RICO	386
*PHAEOPHYCEAE	0.095 PICO CURIE/(GRAM OF ACID-SOLUBLE FRACTION)	AVERAGE OF 4 MARINE SPECIES,PUERTO RICO	386
*RHODOPHYTA	0.047 PICO CURIE/(GRAM OF ACID-SOLUBLE FRACTION)	AVERAGE OF 9 MARINE SPECIES,PUERTO RICO	386

TABLE A93

ALGAE : RATE OF AMMONIUM UPTAKE			
ITEM:	VALUE:	CONDITIONS:	REF.:
*CHAETOCEROS DEBILIS	0.085 1/HOUR	SURGE UPTAKE, CHEMOSTAT, NH4-LIMITATION	279
*CHAETOCEROS DEBILIS	0.253 1/HOUR	SURGE UPTAKE, CHEMOSTAT, NH4-STARVATION	279
*CHAETOCEROS DEBILIS	0.167 1/HOUR	SURGE UPTAKE, CHEMOSTAT, SI-LIMITATION	279
*CHAETOCEROS DEBILIS	0.063 1/HOUR	SURGE UPTAKE, CHEMOSTAT, SI-STARVATION	279
*CHAETOCEROS DEBILIS	0.085 1/HOUR	INTERNALLY CONTROLLED UPTAKE, CHEMOSTAT,NH4-LIMITATION	279
*CHAETOCEROS DEBILIS	0.076 1/HOUR	INTERNALLY CONTROLLED UPTAKE, CHEMOSTAT,NH4-STARVATION	279
*CHAETOCEROS DEBILIS	0.125 1/HOUR	INTERNALLY CONTROLLED UPTAKE, CHEMOSTAT,SI-LIMITATION	279
*CHAETOCEROS DEBILIS	0.019 1/HOUR	INTERNALLY CONTROLLED UPTAKE, CHEMOSTAT,SI-STARVATION	279
*MARINE BACILLARIOPHYCEAE	0.0309 1/HOUR	STATIONARY PHASE,N ADDED AS 75 UG-ATOMS NH4-N, 293 K	1000
*MARINE BACILLARIOPHYCEAE	0.0318 1/HOUR	LOG PHASE,N ADDED AS 75 UG-ATOMS NH4-N, 293 K	1000
*PHYTOPLANKTON	0.1 1/DAY	STADY STATE SOLUTION TO MODEL OF CORPUS CHRISTI BAY	1242
*SKELETONEMA COSTATUM	0.268 1/HOUR	SURGE UPTAKE, CHEMOSTAT, NH4-LIMITATION	279
*SKELETONEMA COSTATUM	0.172 1/HOUR	SURGE UPTAKE, CHEMOSTAT, NH4-STARVATION	279
*SKELETONEMA COSTATUM	0.170 1/HOUR	SURGE UPTAKE, CHEMOSTAT, SI-LIMITATION	279
*SKELETONEMA COSTATUM	0.042 1/HOUR	SURGE UPTAKE, CHEMOSTAT, SI-STARVATION	279
*SKELETONEMA COSTATUM	0.124 1/HOUR	INTERNALLY CONTROLLED UPTAKE, CHEMOSTAT,NH4-LIMITATION	279
*SKELETONEMA COSTATUM	0.088 1/HOUR	INTERNALLY CONTROLLED UPTAKE, CHEMOSTAT,NH4-STARVATION	279
*SKELETONEMA COSTATUM	0.060 1/HOUR	INTERNALLY CONTROLLED UPTAKE, CHEMOSTAT,SI-LIMITATION	279
*SKELETONEMA COSTATUM	0.010 1/HOUR	INTERNALLY CONTROLLED UPTAKE, CHEMOSTAT,SI-STARVATION	279
*THALASSIOSIRA GRAVIDA	0.206 1/HOUR	SURGE UPTAKE, CHEMOSTAT, NH4-LIMITATION	279

TABLE A93

ALGAE : RATE OF AMMONIUM UPTAKE

ITEM:	VALUE:	CONDITIONS:	REF.:
*THALASSIOSIRA GRAVIDA	0.095 1/HOUR	SURGE UPTAKE, CHEMOSTAT, SI-LIMITATION	279
*THALASSIOSIRA GRAVIDA	0.110 1/HOUR	INTERNALLY CONTROLLED UPTAKE, CHEMOSTAT,NH4-LIMITATION	279
*THALASSIOSIRA GRAVIDA	0.040 1/HOUR	INTERNALLY CONTROLLED UPTAKE, CHEMOSTAT,SI-LIMITATION	279

TABLE A94

ALGAE : RATE OF NITRATE UPTAKE

ITEM:	VALUE:	CONDITIONS:	REF.:
*MARINE BACILLARIOPHYCEAE	0.0236 1/HOUR	STATIONARY PHASE,N ADDED AS 75 UG-ATOMS NO3-N, 293 K	1000
*MARINE BACILLARIOPHYCEAE	0.0278 1/HOUR	LOG PHASE,N ADDED AS 75 UG-ATOMS NO3-N, 293 K	1000
*PHYTOPLANKTON	0.4 1/DAY	STADY STATE SOLUTION TO MODEL OF CORPUS CHRISTI BAY	1242

TABLE A95

ALGAE : RATE OF NITROGEN UPTAKE

ITEM:	VALUE:	CONDITIONS:	REF.:
*DISSODINIUM SP.	0.005 PG-ATOMS N/(CELL*MIN)	INITAL 9UM N AS NO3,3087CELL/ML, 4 H,LIGHT, 296 K,LOG PHASE	117
*DISSODINIUM SP.	0.003 PG-ATOMS N/(CELL*MIN)	INITAL 6UM N AS NH4,3215CELL/ML, 4 H,LIGHT, 296 K,LOG PHASE	117
*DISSODINIUM SP.	0.003 PG-ATOMS N/(CELL*MIN)	INITAL 10UM N AS NO3,3147CELL/ML, 4 H,DARK, 296 K,LOG PHASE	117
*DISSODINIUM SP.	0.002 PG-ATOMS N/(CELL*MIN)	INITAL 6UM N AS NH4,2972CELL/ML, 4 H,DARK, 296 K,LOG PHASE	117
*PYROCYSTIS FUSIFORMIS	0.14 PG-ATOMS N/(CELL*MIN)	INITAL 9UM N AS NO3,195CELL/ML, 4 H,LIGHT, 296 K,LOG PHASE	117
*PYROCYSTIS FUSIFORMIS	0.12 PG-ATOMS N/(CELL*MIN)	INITAL 9UM N AS NH4,118CELL/ML, 4 H,LIGHT, 296 K,LOG PHASE	117
*PYROCYSTIS FUSIFORMIS	0.11 PG-ATOMS N/(CELL*MIN)	INITAL 10UM N AS NO3,110CELL/ML, 4 H,DARK, 296 K,LOG PHASE	117
*PYROCYSTIS FUSIFORMIS	0.11 PG-ATOMS N/(CELL*MIN)	INITAL 6UM N AS NH4,129CELL/ML, 4 H,DARK, 296 K,LOG PHASE	117
*PYROCYSTIS NOCTILUCA	0.07 PG-ATOMS N/(CELL*MIN)	INITAL 9UM N AS NO3,303CELL/ML, 4 H,LIGHT, 296 K,LOG PHASE	117
*PYROCYSTIS NOCTILUCA	0.06 PG-ATOMS N/(CELL*MIN)	INITAL 10UM N AS NH4,85CELL/ML, 4 H,LIGHT, 296 K,LOG PHASE	117
*PYROCYSTIS NOCTILUCA	0.05 PG-ATOMS N/(CELL*MIN)	INITAL 4UM N AS NO3,134CELL/ML, 4 H,DARK, 296 K,LOG PHASE	117
*PYROCYSTIS NOCTILUCA	0.04 PG-ATOMS N/(CELL*MIN)	INITAL 6UM N AS NH4,91CELL/ML, 4 H,DARK, 296 K,LOG PHASE	117
*PYROCYSTIS NOCTILUCA	0.08 PG-ATOMS N/(CELL*MIN)	INITAL 9UM N AS NO3,23CELL/ML, 4 H,LIGHT, 296 K,LOG PHASE	117
*PYROCYSTIS NOCTILUCA	0.05 PG-ATOMS N/(CELL*MIN)	INITAL 9UM N AS NH4,70CELL/ML, 4 H,LIGHT, 296 K,LOG PHASE	117
*PYROCYSTIS NOCTILUCA	0.06 PG-ATOMS N/(CELL*MIN)	INITAL 4UM N AS NO3,238CELL/ML, 4 H,DARK, 296 K,LOG PHASE	117
*PYROCYSTIS NOCTILUCA	0.03 PG-ATOMS N/(CELL*MIN)	INITAL 4UM N AS NH4,173CELL/ML, 4 H,DARK, 296 K,LOG PHASE	117

TABLE A96

ALGAE : RATE OF PHOSPHORUS UPTAKE

ITEM:	VALUE:	CONDITIONS:	REF.:
*FRESHWATER PHYTOPLANKTON	0.319 1/HOUR	STEADY STATE SOLUTION TO CASTLE LAKE.	1326

TABLE A96

ALGAE : RATE OF PHOSPHORUS UPTAKE			
ITEM:	VALUE:	CONDITIONS:	REF.:
*SESTON	0.0370 MICROGRAM P/HOUR	STONEHOUSE POND, FEB 22, 1973	454
*SESTON	0.0640 MICROGRAM P/HOUR	STONEHOUSE POND, APRIL 6, 1973	454
*SESTON	0.1438 MICROGRAM P/HOUR	STONEHOUSE POND, MAY 17, 1973	454
*SESTON	STONEHOUSE POND, JUNE 17, 1973	3.1820 MICROGRAM P/HOUR	454

TABLE A97

ALGAE : RATE OF UREA UPTAKE			
ITEM:	VALUE:	CONDITIONS:	REF.:
*MARINE BACILLARIOPHYCEAE	0.0327 1/HOUR	STATIONARY PHASE,N ADDED AS 75 UG-ATOMS UREA-N, 293 K	1000
*MARINE BACILLARIOPHYCEAE	0.0418 1/HOUR	LOG PHASE,N ADDED AS 75 UG-ATOMS UREA-N, 293 K	1000

TABLE A98

ALGAE : REFLECTANCE RATIO			
ITEM:	VALUE:	CONDITIONS:	REF.:
*ANABAENA SP.	0.620	625 TO 650 NM, CONC.OF ALGAE=4MG/L	558
*ANABAENA SP.	0.326	625 TO 650 NM, CONC.OF ALGAE=8MG/L	558
*CHLORELLA SP.	1.520	625 TO 650 NM, CONC.OF ALGAE=4MG/L	558
*CHLORELLA SP.	2.220	625 TO 650 NM, CONC.OF ALGAE=8MG/L	558
*MICROCYSTIS SP.	0.824	625 TO 650 NM, CONC.OF ALGAE=4MG/L	558
*MICROCYSTIS SP.	0.636	625 TO 650 NM, CONC.OF ALGAE=8MG/L	558
*SELENASTRUM SP.	1.153	625 TO 650 NM, CONC.OF ALGAE=4MG/L	558
*SELENASTRUM SP.	1.510	625 TO 650 NM, CONC.OF ALGAE=8MG/L	558

TABLE A99

ALGAE : RESPIRATION RATE			
ITEM:	VALUE:	CONDITIONS:	REF.:
*ALGAE	0.005 1/DAY-CELCIUS	MOSSDALE, ENDOGENOUS RESPIRATION	1153
*ALGAE	0.005 1/DAY-CELCIUS	POTOMAC RIVER, ENDOGENOUS RESPIRATION	1153
*ALGAE	0.1 1/DAY	TEMP FACTOR IS 1 AT 273K,LAKE ONTARIO, ENDOGENOUS RESPIRATION	1577
*ALGAE	0.05-0.2 1/DAY	TEMP.FACTOR IS 1 AT 293K, TEMP.>278K, ENDOGENOUS RESPIRATION	1571
*ALGAE	0.001-0.01 1/DAY	TEMP.FACTOR IS 1 AT 293K, TEMP.<278K, ENDOGENOUS RESPIRATION	1571
*ALGAE	0.031 KCAL/KCAL*DAY	INITIAL VALUES, THERMAL ECOSYSTEM, TRANSFER COEFF.	1703
*ANABAENA SP.	4.2-4.8 MICROLITER/(MG DRY WT.)*H	298K,NO OTHER INDICATION, ENDOGENOUS RESPIRATION	523
*ANABAENA VARIABILIS	8.4/1.7 MICROLITER/(MG DRY WT.)*H	GROWING CELL/STARVING CELL,298K, ENDOGENOUS RESPIRATION	523
*ANABAENA VARIABILIS	3.08 MICROLITER/(MG DRY WT.)*H	GROWING CELL,0.1M POTASSIUM PHOSPHATE BUFFER PH=7., ENDOGENOUS RESP	523
*ANABAENA VARIABILIS	12.3/3.08 MICROLITER/(MG DRY WT.) *H	GROWING CELL/STARVING CELL,307K, MIN. SALT MEDIUM, ENDOGENOUS RESPIRATION	523

TABLE A99

ALGAE : RESPIRATION RATE			
ITEM:	VALUE:	CONDITIONS:	REF.:
*ANACYSTIS NIDULANS	4-6 MICROLITER/(MG DRY WT.)*H	298K,NO OTHER INDICATION, ENDOGENOUS RESPIRATION	523
*ANACYSTIS NIDULANS	1.6/0.3 MICROLITER/(MG DRY WT.)*H	GROWING CELL/STARVING CELL,298K, ENDOGENOUS RESPIRATION	523
*ANACYSTIS NIDULANS	4.7/1.9 MICROLITER/(MG DRY WT.)*H	GROWING CELL/STARVING CELL,312K, ENDOGENOUS RESPIRATION	523
*ANACYSTIS NIDULANS	7.5/2.9 MICROLITER/(MG DRY WT.)*H	GROWING CELL/STARVING CELL,NONE TEMP INDICATION, ENDOGENOUS RESPIRATION	523
*ANACYSTIS NIDULANS	6.8 MICROLITER/(MG DRY WT.)*H	GROWING CELL,307K,0.1M POTASSIUM PHOSPHATE BUFFER PH=7., ENDOGENOUS RESP	523
*COSCINODISCUS EXCENTRICUS	0.075 1/DAY	289K, ENDOGENOUS RESPIRATION	1337
*COSCINODISCUS EXCENTRICUS	0.11 1/DAY	289K, ENDOGENOUS RESPIRATION	1337
*MIXED NATUREL	0.4-2.2 MICROGRAM/(MICROGRAM CHL.) *H	290-307K,LABORATORY EXPIRIMENT, ENDOGENOUS RESPIRATION	523
*MIXED NATUREL	0.2-4. MICROGRAM/(MICROGRAM CHL.) *H	296-308K,FIELD OBSERVATION, ENDOGENOUS RESPIRATION	523
*MIXED NATUREL	0.1-2.95 MICROGRAM/(MICROGRAM CHL.)*H	277-288K,FIELD OBSERVATION, ENDOGENOUS RESPIRATION	523
*NATURAL ASSOCIATION	0.03 1/DAY	275K, ENDOGENOUS RESPIRATION	1337
*NATURAL ASSOCIATION	0.12 1/DAY	291K, ENDOGENOUS RESPIRATION	1337
*NATURAL ASSOCIATION	0.024(+-0.012) 1/DAY	275K, ENDOGENOUS RESPIRATION	1337
*NATURAL ASSOCIATION	0.11(+-0.007) 1/DAY	291K, ENDOGENOUS RESPIRATION	1337
*NITZSCHIA CLOSTERIUM	0.035 1/DAY	279K, ENDOGENOUS RESPIRATION	1337
*NITZSCHIA CLOSTERIUM	0.17 1/DAY	308K, ENDOGENOUS RESPIRATION	1337
*NITZSCHIA CLOSTERIUM	0.08 1/DAY	293K, ENDOGENOUS RESPIRATION	1337
*NOSTOC MUSCORUM	4.4/1.1 MICROLITER/(MG DRY WT.)*H	GROWING CELL/STARVING CELL,298K, ENDOGENOUS RESPIRATION	523
*OSCILLATORIA REDEKEI	0.16-6.3 MICROGRAM/(MICROGRAM CHL.)*H	293K,LABORATORY EXPIRIMENT, ENDOGENOUS RESPIRATION	523
*PHORMIDIUM LURIDUM	4-6 MICROLITER/(MG DRY WT.)*H	298K,NO OTHER INDICATION, ENDOGENOUS RESPIRATION	523
*PHYTOPLANKTON	0.10 1/DAY	SAN JOAQUIN(RIVER),293 K, ENDOGENOUS RESPIRATION	1149
*PHYTOPLANKTON	0.10 1/DAY	DELTA(ESTUARY),293 K, ENDOGENOUS RESPIRATION	1149
*PHYTOPLANKTON	0.10 1/DAY	POTOMAC(ESTUARY),293 K, ENDOGENOUS RESPIRATION	1149
*PHYTOPLANKTON	0.08 1/DAY	LAKE ERIE,293 K, ENDOGENOUS RESPIRATION	1149
*PHYTOPLANKTON	0.10 1/DAY	LAKE ONTARIO,293 K, ENDOGENOUS RESPIRATION	1149

TABLE A100

ALGAE : S/BIOMASS RATIO			
ITEM:	VALUE:	CONDITIONS:	REF.:
*ALGAE	0.6 % OF DRY WEIGHT	MEAN ELEMENTARY COMPOSITION, INTERPOLATED	157
*ANABAENA SP.	0.53 % OF DRY WEIGHT	ALGAE FROM PONDS AND LAKES IN SOUTHEAS¤ERN UNITED STATES, END RESP	1660
*APHANIZOMENON SP.	1.18 % OF DRY WEIGHT	ALGAE FROM PONDS AND LAKES IN SOUTHEASTERN UNITED STATES	1660
*CHARA SP.	0.55 % OF DRY WEIGHT	ALGAE FROM PONDS AND LAKES IN SOUTHEASTERN UNITED STATES	1660
*CLADOPHORA SP.	1.58 % OF DRY WEIGHT	ALGAE FROM PONDS AND LAKES IN SOUTHEASTERN UNITED STATES	1660
*EUGLENA SP.	0.19 % OF DRY WEIGHT	ALGAE FROM PONDS AND LAKES IN SOUTHEASTERN UNITED STATES	1660

TABLE A100

ALGAE : S/BIOMASS RATIO

ITEM:	VALUE:	CONDITIONS:	REF.:
*HYDRODICTYON SP.	1.41 % OF DRY WEIGHT	ALGAE FROM PONDS AND LAKES IN SOUTHEASTERN UNITED STATES	1660
*LYNGBYA SP.	0.28 % OF DRY WEIGHT	ALGAE FROM PONDS AND LAKES IN SOUTHEASTERN UNITED STATES	1660
*MICROCYSTIS SP.	0.27 % OF DRY WEIGHT	ALGAE FROM PONDS AND LAKES IN SOUTHEASTERN UNITED STATES	1660
*MOUGEOTIA SP.	0.36 % OF DRY WEIGHT	ALGAE FROM PONDS AND LAKES IN SOUTHEASTERN UNITED STATES	1660
*NITELLA SP.	0.34 % OF DRY WEIGHT	ALGAE FROM PONDS AND LAKES IN SOUTHEASTERN UNITED STATES	1660
*OEDOGONIUM SP.	0.15 % OF DRY WEIGHT	ALGAE FROM PONDS AND LAKES IN SOUTHEASTERN UNITED STATES	1660
*PHAEOPHYCEAE	1.2 % OF DRY WEIGHT	MEAN ELEMENTARY COMPOSITION	157
*PITHOPHORA SP.	1.42 % OF DRY WEIGHT	ALGAE FROM PONDS AND LAKES IN SOUTHEASTERN UNITED STATES	1660
*RHIZOCLONIUM SP.	0.27	ALGAE FROM PONDS AND LAKES IN SOUTHEASTERN UNITED STATES	1660
*SPIROGYRA SP.	0.27 % OF DRY WEIGHT	ALGAE FROM PONDS AND LAKES IN SOUTHEASTERN UNITED STATES	1660
*SPIROGYRA SP., GIANT	0.24 % OF DRY WEIGHT	ALGAE FROM PONDS AND LAKES IN SOUTHEASTERN UNITED STATES	1660

TABLE A101

ALGAE : SATURATING LIGHT INTENSITY

ITEM:	VALUE:	CONDITIONS:	REF.:
*PHYTOPLANKTON	300 LANGLEYS/DAY	SAN JOAQUIN(RIVER)	1149
*PHYTOPLANKTON	300 LANGLEYS/DAY	DELTA(ESTUARY)	1149
*PHYTOPLANKTON	300 LANGLEYS/DAY	POTOMAC(ESTUARY)	1149
*PHYTOPLANKTON	350 LANGLEYS/DAY	LAKE ERIE	1149
*PHYTOPLANKTON	350 LANGLEYS/DAY	LAKE ONTARIO	1149

TABLE A102

ALGAE : SETTLING RATE

ITEM:	VALUE:	CONDITIONS:	REF.:
*ALGAE	0.1 M/DAY	LAKE ONTARIO	1577
*ALGAE	0.-2. M/DAY	NONE	1571
*ALGAE	0.3-0.5 M/DAY	NONE	958
*ALGAE	0.1-0.4 M/DAY	SEE DETAILS SEE REF	720
*ALGAE	0.1-0.27 M/DAY	SEE DETAILS SEE REF	1469
*ASTERIONELLA FORMOSA	0.26 M/DAY	1-4 CELLS/CHAIN, 24*10**3 CELL/ML, 9 D CULTURE, 293 K, LIVE	1457
*ASTERIONELLA FORMOSA	0.22 M/DAY	1-4 CELLS/CHAIN, 51*10**3 CELL/ML, 11 D CULTURE, 293 K, LIVE	1457
*ASTERIONELLA FORMOSA	0.34 M/DAY	1-4 CELLS/CHAIN, 116*10**3 CELL/ ML,15 D CULTURE, 293 K, LIVE	1457
*ASTERIONELLA FORMOSA	0.36 M/DAY	1-4 CELLS/CHAIN, 215*10**3 CELL/ ML,17 D CULTURE, 293 K, LIVE	1457
*ASTERIONELLA FORMOSA	0.40 M/DAY	1-4 CELLS/CHAIN, 240*10**3 CELL/ ML,19 D CULTURE, 293 K, LIVE	1457
*ASTERIONELLA FORMOSA	0.61 M/DAY	1-4 CELLS/CHAIN, 219*10**3 CELL/ ML,22 D CULTURE, 293 K, LIVE	1457
*ASTERIONELLA FORMOSA	0.26 M/DAY	5-8 CELLS/CHAIN, 24*10**3 CELL/ML, 9 D CULTURE, 293 K, LIVE	1457

TABLE A102

ALGAE : SETTLING RATE

ITEM:	VALUE:	CONDITIONS:	REF.:
*ASTERIONELLA FORMOSA	0.25 M/DAY	5-8 CELLS/CHAIN, 51*10**3 CELL/ML, 11 D CULTURE, 293 K, LIVE	1457
*ASTERIONELLA FORMOSA	0.40 M/DAY	5-8 CELLS/CHAIN, 116*10**3 CELL/ ML,15 D CULTURE, 293 K, LIVE	1457
*ASTERIONELLA FORMOSA	0.37 M/DAY	5-8 CELLS/CHAIN, 215*10**3 CELL/ ML,17 D CULTURE, 293 K, LIVE	1457
*ASTERIONELLA FORMOSA	0.43 M/DAY	5-8 CELLS/CHAIN, 240*10**3 CELL/ ML,19 D CULTURE, 293 K, LIVE	1457
*ASTERIONELLA FORMOSA	0.59 M/DAY	5-8 CELLS/CHAIN, 219*10**3 CELL/ ML,22 D CULTURE, 293 K, LIVE	1457
*ASTERIONELLA FORMOSA	0.33 M/DAY	9-12 CELLS/CHAIN, 24*10**3 CELL/ ML,9 D CULTURE, 293 K, LIVE	1457
*ASTERIONELLA FORMOSA	0.32 M/DAY	9-12 CELLS/CHAIN, 51*10**3 CELL/ ML,11 D CULTURE, 293 K, LIVE	1457
*ASTERIONELLA FORMOSA	0.27 M/DAY	9-12 CELLS/CHAIN, 116*10**3 CELL/ ML,15 D CULTURE, 293 K, LIVE	1457
*ASTERIONELLA FORMOSA	0.31 M/DAY	9-12 CELLS/CHAIN, 215*10**3 CELL/ ML,17 D CULTURE, 293 K, LIVE	1457
*ASTERIONELLA FORMOSA	0.50 M/DAY	9-12 CELLS/CHAIN, 240*10**3 CELL/ ML,19 D CULTURE, 293 K, LIVE	1457
*ASTERIONELLA FORMOSA	0.76 M/DAY	9-12 CELLS/CHAIN, 219*10**3 CELL/ ML,22 D CULTURE, 293 K, LIVE	1457
*ASTERIONELLA FORMOSA	0.32 M/DAY	ALIVE, 293 K	1457
*ASTERIONELLA FORMOSA	0.35 M/DAY	500 UG CUSO4/L, 293 K	1457
*ASTERIONELLA FORMOSA	0.44 M/DAY	IN FORMALIN, 293 K	1457
*ASTERIONELLA FORMOSA	0.52 M/DAY	HEAT KILLED, 293 K	1457
*ASTERIONELLA FORMOSA	0.42 M/DAY	IN LUGOLS SOLUTION, 293 K	1457
*ASTERIONELLA FORMOSA	0.23 M/DAY	LIVING CELLS,1-4 CELLS PER COLONY	1457
*ASTERIONELLA FORMOSA	0.29 M/DAY	LIVING CELLS,5-8 CELLS PER COLONY	1457
*ASTERIONELLA FORMOSA	0.34 M/DAY	LIVING CELLS,>8 CELLS PER COLONY	1457
*ASTERIONELLA FORMOSA	0.31 M/DAY	LIVING CELLS,TOTAL COLONY	1457
*ASTERIONELLA FORMOSA	0.38 M/DAY	500MICROGRAM CUSO4/L,1-4 CELLS	1457
*ASTERIONELLA FORMOSA	0.38 M/DAY	500MICROGRAM CUSO4/L,5-8 CELLS	1457
*ASTERIONELLA FORMOSA	0.36 M/DAY	500MICROGRAM CUSO4/L,>8 CELLS	1457
*ASTERIONELLA FORMOSA	0.38 M/DAY	500MICROGRAM CUSO4/L,TOTAL COLONY	1457
*ASTERIONELLA FORMOSA	0.45 M/DAY	1500MICROGRAM CUSO4/L,1-4 CELLS	1457
*ASTERIONELLA FORMOSA	0.44 M/DAY	1500MICROGRAM CUSO4/L,5-8 CELLS	1457
*ASTERIONELLA FORMOSA	0.31 M/DAY	1500MICROGRAM CUSO4/L,>8 CELLS	1457
*ASTERIONELLA FORMOSA	0.41 M/DAY	1500MICROGRAM CUSO4/L,TOTAL COLONY	1457
*ASTERIONELLA FORMOSA	0.20+-0.06 M/DAY	EXPONENTIAL PHASE,95% CONF.LIM., RADIUS=12.5MIC.M,293K,N=6	1600
*ASTERIONELLA FORMOSA	1.48+-1.05 M/DAY	STATIONARY PHASE,95% CONF.LIM., RADIUS=12.5MIC.M,293K,N=7	1600
*BACTERIASTRUM HYALINUM	0.39 - 1.27 M/DAY	AREA/VOL RATIO=0.29-0.33,CELL SIZE =10*30 UM, CENTRIC , CHAIN-FORMING	1453
*CHAETOCEROS LAUDERI	0.46 - 1.54 M/DAY	AREA/VOL RATIO=0.19-0.41,CELL SIZE =20*34 UM, CENTRIC , CHAIN-FORMING	1453
*COCCOLITHUS HUXLEYI	0.3 M/DAY	NO COCCOLITHS	1219
*COCCOLITHUS HUXLEYI	1.5 M/DAY	WITH COCCOLITHS	1219
*COSCINODISCUS LINEATUS	1.9 M/DAY	EXPONENTIAL PHASE,RADIUS=25MIC.M	431
*COSCINODISCUS LINEATUS	6.8 M/DAY	STATIONARY PHASE,RADIUS=25MIC.M	431
*CYCLOTELLA MENEGHINIANA	0.08+-0.10 M/DAY	EXPONENTIAL PHASE,95% CONF.LIM., RADIUS=1.0MIC.M,293K,N=4	1600
*CYCLOTELLA MENEGHINIANA	0.24+-0.31 M/DAY	STATIONARY PHASE,95% CONF.LIM., RADIUS=1.0MIC.M,293K,N=4	1600
*DITYLUM BRIGHTWELLII	1.3 M/DAY	MEAN OF 7 VALUES DUR.A 8HR LIGHT-16HR DARK CYCLUS	30
*DITYLUM BRIGHTWELLII	0.5 M/DAY	HEALTHY CELLS	1219
*DITYLUM BRIGHTWELLII	2.0 M/DAY	SENESCENT CELLS	1219

TABLE A102

ALGAE : SETTLING RATE			
ITEM:	VALUE:	CONDITIONS:	REF.:
*DITYLUM BRIGHTWELLII	0.6 M/DAY	EXPONENTIAL PHASE,RADIUS=75MIC.M	431
*DITYLUM BRIGHTWELLII	3.1 M/DAY	STATIONARY PHASE,RADIUS=75MIC.M	431
*MELOSIRA AGASSIZII	0.67+-0.48 M/DAY	EXPONENTIAL PHASE,95% CONF.LIM., RADIUS=27.4MIC.M,293K,N=3	1600
*MELOSIRA AGASSIZII	1.87+-0.38 M/DAY	STATIONARY PHASE,95% CONF.LIM., RADIUS=27.4MIC.M,293K,N=15	1600
*NITZSCHIA SERIATA	0.26 - 0.50 M/DAY	AREA/VOL RATIO=1.18-1.65,CELL SIZE =4*40 UM, PENNATE, CHAIN-FORMING	1451
*PHYTOPLANKTON	0-30 M/DAY	LIVING CELLS	1455
*PHYTOPLANKTON	0.1 M/DAY	LAKE ONTARIO	1149
*RHIZOSOLENIA ROBUSTA	1.1 M/DAY	EXPONENTIAL PHASE,RADIUS=42MIC.M	SMAY
*RHIZOSOLENIA ROBUSTA	4.7 M/DAY	STATIONARY PHASE,RADIUS=42MIC.M	SMAY
*RHIZOSOLENIA SETIGERA	0.19 - 0.44 M/DAY	CELL SIZE= 6*245 UM, CENTRIC, ELONGATE, NORMAL PRE-AUXOSPORE	1452
*RHIZOSOLENIA SETIGERA	0.22 - 0.63 M/DAY	AREA/VOLUME RATIO=0.62-0.75, SPINELESS PRE-AUXOSPORE, CENTRIC, ELONG	1452
*RHIZOSOLENIA SETIGERA	0.79 - 1.94 M/DAY	AREA/VOL RATIO=0.10-0.16,CELL SIZE =33*363 UM, POST-AUXOSPORE	1452
*RHIZOSOLENIA SETIGERA	0.63 - 0.95 M/DAY	TOTAL POPULATION	1452
*SCENEDESMUS QUADRICAUDA	0.27+-0.04 M/DAY	EXPONENTIAL PHASE,95% CONF.LIM., RADIUS=4.2MIC.M,293K,N=8	1600
*SCENEDESMUS QUADRICAUDA	0.89+-0.06 M/DAY	STATIONARY PHASE,95% CONF.LIM., RADIUS=4.2MIC.M,293K,N=14	1600
*SKELETONEMA COSTATUM	0.31 - 1.35 M/DAY	AREA/VOL RATIO=0.81-1.01,CELL SIZE =5*20 UM, CENTRIC , CHAIN-FORMING	1452
*TABELLARIA FLOCCULOSA	0.39 M/DAY	ALIVE, 293 K	1457
*TABELLARIA FLOCCULOSA	1.11 M/DAY	500 UG CUSO4/L, 293 K	1457
*TABELLARIA FLOCCULOSA	0.46 M/DAY	IN FORMALIN, 293 K	1457
*TABELLARIA FLOCCULOSA	1.50 M/DAY	HEAT KILLED, 293 K	1457
*TABELLARIA FLOCCULOSA	1.29 M/DAY	IN LUGOLS SOLUTION	1457
*TABELLARIA FLOCCULOSA	0.22 M/DAY	LIVING CELLS	1457
*THALASSIOSIRA NANA	0.10 - 0.28 M/DAY	AREA/VOL RATIO=0.88-1.20,CELL SIZE =4.3-5.2 UM, CENTRIC, UNICELLULAR	1451
*THALASSIOSIRA PSEUDONANA	0.15 M/DAY	CLONE 13-1, HEALTHY	649
*THALASSIOSIRA PSEUDONANA	0.85 M/DAY	CLONE 13-1, SENESCENT	649
*THALASSIOSIRA ROTULA	5.5 M/DAY	1200 LUX, 285 K	1395
*THALASSIOSIRA ROTULA	9.2 M/DAY	2400 LUX, 285 K	1395
*THALASSIOSIRA ROTULA	17.1 M/DAY	5500 LUX, 285 K	1395
*THALASSIOSIRA ROTULA	0.39 - 2.10 M/DAY	AREA/VOL RATIO=0.23-0.29,CELL SIZE =19-34 UM, CENTRIC , CHAIN-FORMING	1451

TABLE A103

ALGAE : SI/BIOMASS RATIO			
ITEM:	VALUE:	CONDITIONS:	REF.:
*ASTERIONELLA FORMOSA	139 (RANGE = 97-171) MICROGRAMSIO2/(10**6 CELLS)	GERMANY,N=5	956
*ASTERIONELLA FORMOSA	137 (RANGE = 100-175) MICROGRAMSIO2/(10**6 CELLS)	ENGLAND,N=28	956
*ASTERIONELLA FORMOSA	153 (RANGE = 140-165) MICROGRAMSIO2/(10**6 CELLS)	ENGLAND, N=2	956
*ASTERIONELLA FORMOSA	178 MICROGRAMSIO2/(10**6 CELLS)	USA	956
*ASTERIONELLA FORMOSA	50 (RANGE = 43-59) MICROGRAMSIO2/ (10**6 CELLS)	CULTURE	956
*CYCLOTELLA BODANICA	32000 MICROGRAMSIO2/(10**6 CELLS)	GERMANY	956

TABLE A103

ALGAE : SI/BIOMASS RATIO			
ITEM:	VALUE:	CONDITIONS:	REF.:
*CYCLOTELLA COMTA	750 MICROGRAMSIO2/(10**6 CELLS)	GERMANY, N=1	956
*CYCLOTELLA GLOMERATA	220 MICROGRAMSIO2/(10**6 CELLS)	GERMANY, N=1	956
*CYCLOTELLA SOCIALIS	1000 (RANGE = 850-1150) MICROGRAMSIO2/(10**6 CELLS)	GERMANY, N=2	956
*DIATOMA ELONGATUM	250 MICROGRAMSIO2/(10**6 CELLS)	GERMANY	956
*FRAGILARIA CROTONENSIS	189 (RANGE = 170-215) MICROGRAMSIO2/(10**6 CELLS)	GERMANY, N=4	956
*FRAGILARIA CROTONENSIS	188 (RANGE = 189-190) MICROGRAMSIO2/(10**6 CELLS)	ENGLAND, N=2	956
*MELOSIRA GRANULATA	130 MICROGRAMSIO2/(10**6 CELLS)	GERMANY, N=1	956
*MELOSIRA GRANULATA	620 MICROGRAMSIO2/(10**6 CELLS)	SUDAN, N=1	956
*MELOSIRA ISLANDICA SUBSP. HELVETICA	280 MICROGRAMSIO2/(10**6 CELLS)	GERMANY, N=1	956
*MELOSIRA ITALICA	320 MICROGRAMSIO2/(10**6 CELLS)	GERMANY, N=1	956
*MELOSIRA ITALICA SUBSP. SUBARTICA	237 (RANGE = 165-274) MICROGRAMSIO2/(10**6 CELLS)	ENGLAND, N=3	956
*STEPHANODISCUS ASTRAEA	8500 MICROGRAMSIO2/(10**6 CELLS)	GERMANY, N=1	956
*STEPHANODISCUS HANTZSCHII	41 (RANGE = 32-47) MICROGRAMSIO2/ (10**6 CELLS)	ENGLAND, N=3	956
*STEPHANODISCUS HANTZSCHII VAR. PUSILLUS	100 MICROGRAMSIO2/(10**6 CELLS)	GERMANY, N=1	956
*SYNEDRA ACUS VAR. ANGUSTISSIMA	1175 (RANGE = 1100-1250) MICROGRAMSIO2/(10**6 CELLS)	GERMANY, N=2	956
*SYNEDRA ACUS VAR. RADIANS	286 MICROGRAMSIO2/(10**6 CELLS)	ENGLAND, N=1	956
*TABELLARIA FLOCCULOSA VAR. ASTERIONELLOIDES	310 (RANGE = 250-420) MICROGRAMSIO2/(10**6 CELLS)	ENGLAND, N=5	956
*TABELLARIA FLOCCULOSA VAR. ASTERIONELLOIDES	395 (RANGE = 370-420) MICROGRAMSIO2/(10**6 CELLS)	GERMANY/SWITZERLAND, N=2	956

TABLE A104

ALGAE : SI/CHLOROPHYLL A RATIO			
ITEM:	VALUE:	CONDITIONS:	REF.:
*CHAETOCEROS DEBILIS	9.0 BY WEIGHT	SI-STARVED,72H,CONT.LIGHT:0.08LY/MIN,DILUTION=1/3-1/2 OF MAX.GR.RATE	631
*CHAETOCEROS DEBILIS	5.6 BY WEIGHT	SI-LIMITED,CONT.LIGHT:0.08LY/MIN, DILUTION=1/3-1/2 OF MAX.GR.RATE	631
*CHAETOCEROS DEBILIS	13.0 BY WEIGHT	NON-LIMITED,CONT.LIGHT:0.08LY/MIN, DILUTION=1/3-1/2 OF MAX.GR.RATE	631
*CHAETOCEROS DEBILIS	36.3 BY WEIGHT	NH4-LIMITED,CONT.LIGHT:0.08LY/MIN, DILUTION=1/3-1/2 OF MAX.GR.RATE	631
*CHAETOCEROS DEBILIS	28.1 BY WEIGHT	NH4-STARVED,72H,CONT.LIGHT:0.08LY/MIN,DILUTION=1/3-1/2 OF MAX.GR.RATE	631
*SKELETONEMA COSTATUM	11.4 BY WEIGHT	SI-STARVED,72H,CONT.LIGHT:0.08LY/MIN,DILUTION=1/3-1/2 OF MAX.GR.RATE	631
*SKELETONEMA COSTATUM	5.5 BY WEIGHT	SI-LIMITED,CONT.LIGHT:0.08LY/MIN, DILUTION=1/3-1/2 OF MAX.GR.RATE	631
*SKELETONEMA COSTATUM	15.0 BY WEIGHT	NON-LIMITED,CONT.LIGHT:0.08LY/MIN, DILUTION=1/3-1/2 OF MAX.GR.RATE	631
*SKELETONEMA COSTATUM	21.6 BY WEIGHT	NH4-LIMITED,CONT.LIGHT:0.08LY/MIN, DILUTION=1/3-1/2 OF MAX.GR.RATE	631
*SKELETONEMA COSTATUM	20.8 BY WEIGHT	NH4-STARVED,72H,CONT.LIGHT:0.08LY/MIN,DILUTION=1/3-1/2 OF MAX.GR.RATE	631
*SKELETONEMA COSTATUM	22.7-21.0 BY WEIGHT	AMMONIUM-LIMITED,DILUTION RATE= 0.110-0.042/HR	630
*SKELETONEMA COSTATUM	43.9-3.7 BY WEIGHT	SILICATE-LIMITED,DILUTION RATE= 0.086-0.022/HR	630

TABLE A104

ALGAE : SI/CHLOROPHYLL A RATIO			
ITEM:	VALUE:	CONDITIONS:	REF.:
*THALASSIOSIRA GRAVIDA	2.4 BY WEIGHT	SI-STARVED,72H,CONT.LIGHT:0.08LY/MIN,DILUTION=1/3-1/2 OF MAX.GR.RATE	631
*THALASSIOSIRA GRAVIDA	3.3 BY WEIGHT	SI-LIMITED,CONT.LIGHT:0.08LY/MIN, DILUTION=1/3-1/2 OF MAX.GR.RATE	631
*THALASSIOSIRA GRAVIDA	9.8 BY WEIGHT	NON-LIMITED,CONT.LIGHT:0.08LY/MIN, DILUTION=1/3-1/2 OF MAX.GR.RATE	631
*THALASSIOSIRA GRAVIDA	17.1 BY WEIGHT	NH4-LIMITED,CONT.LIGHT:0.08LY/MIN, DILUTION=1/3-1/2 OF MAX.GR.RATE	631
*THALASSIOSIRA GRAVIDA	50.7 BY WEIGHT	NH4-STARVED,72H,CONT.LIGHT:0.08LY/MIN,DILUTION=1/3-1/2 OF MAX.GR.RATE	631

TABLE A105

ALGAE : SI/P RATIO			
ITEM:	VALUE:	CONDITIONS:	REF.:
*CHAETOCEROS DEBILIS	1.7 BY ATOMS	SI-STARVED,72H,CONT.LIGHT:0.08LY/MIN,DILUTION=1/3-1/2 OF MAX.GR.RATE	631
*CHAETOCEROS DEBILIS	1.9 BY ATOMS	SI-LIMITED,CONT.LIGHT:0.08LY/MIN, DILUTION=1/3-1/2 OF MAX.GR.RATE	631
*CHAETOCEROS DEBILIS	6.7 BY ATOMS	NON-LIMITED,CONT.LIGHT:0.08LY/MIN, DILUTION=1/3-1/2 OF MAX.GR.RATE	631
*CHAETOCEROS DEBILIS	9.7 BY ATOMS	NH4-LIMITED,CONT.LIGHT:0.08LY/MIN, DILUTION=1/3-1/2 OF MAX.GR.RATE	631
*CHAETOCEROS DEBILIS	6.6 BY ATOMS	NH4-STARVED,72H,CONT.LIGHT:0.08LY/MIN,DILUTION=1/3-1/2 OF MAX.GR.RATE	631
*SKELETONEMA COSTATUM	2.8 BY ATOMS	SI-STARVED,72H,CONT.LIGHT:0.08LY/MIN,DILUTION=1/3-1/2 OF MAX.GR.RATE	631
*SKELETONEMA COSTATUM	2.0 BY ATOMS	SI-LIMITED,CONT.LIGHT:0.08LY/MIN, DILUTION=1/3-1/2 OF MAX.GR.RATE	631
*SKELETONEMA COSTATUM	5.4 BY ATOMS	NON-LIMITED,CONT.LIGHT:0.08LY/MIN, DILUTION=1/3-1/2 OF MAX.GR.RATE	631
*SKELETONEMA COSTATUM	5.9 BY ATOMS	NH4-LIMITED,CONT.LIGHT:0.08LY/MIN, DILUTION=1/3-1/2 OF MAX.GR.RATE	631
*SKELETONEMA COSTATUM	5.2 BY ATOMS	NH4-STARVED,72H,CONT.LIGHT:0.08LY/MIN,DILUTION=1/3-1/2 OF MAX.GR.RATE	631
*SKELETONEMA COSTATUM	4.0-8.8 BY ATOMS	AMMONIUM-LIMITED,DILUTION RATE= 0.110-0.042/HR	630
*SKELETONEMA COSTATUM	3.0-1.5 BY ATOMS	SILICATE-LIMITED,DILUTION RATE= 0.061-0.035/HR	330
*SKELETONEMA COSTATUM	3.0-1.7 BY ATOMS	SILICATE-LIMITED,DILUTION RATE= 0.086-0.022/HR	630
*THALASSIOSIRA GRAVIDA	1.3 BY ATOMS	SI-STARVED,72H,CONT.LIGHT:0.08LY/MIN,DILUTION=1/3-1/2 OF MAX.GR.RATE	631
*THALASSIOSIRA GRAVIDA	1.6 BY ATOMS	SI-LIMITED,CONT.LIGHT:0.08LY/MIN, DILUTION=1/3-1/2 OF MAX.GR.RATE	631
*THALASSIOSIRA GRAVIDA	3.8 BY ATOMS	NON-LIMITED,CONT.LIGHT:0.08LY/MIN, DILUTION=1/3-1/2 OF MAX.GR.RATE	631
*THALASSIOSIRA GRAVIDA	3.2 BY ATOMS	NH4-LIMITED,CONT.LIGHT:0.08LY/MIN, DILUTION=1/3-1/2 OF MAX.GR.RATE	631
*THALASSIOSIRA GRAVIDA	3.7 BY ATOMS	NH4-STARVED,72H,CONT.LIGHT:0.08LY/MIN,DILUTION=1/3-1/2 OF MAX.GR.RATE	631

TABLE A106

ALGAE : SPECIFIC GRAVITY			
ITEM:	VALUE:	CONDITIONS:	REF.:
*ANABAENA FLOS-AQUAE	1.0065	NORMAL CELLS	835

TABLE A106

ALGAE : SPECIFIC GRAVITY			
ITEM:	VALUE:	CONDITIONS:	REF.:
*ANABAENA FLOS-AQUAE	1.0085	CELLS EXPOSED TO A SUDDEN PRESSURE TO DESTROYE VACUOLES	835

TABLE A107

ALGAE : TEMPERATURE COEFFICIENT			
ITEM:	VALUE:	CONDITIONS:	REF.:
*ALGAE	1.066	GROWTH,LAKE ONTARIO	1577
*ALGAE	1.08	RESPIRATION,LAKE ONTARIO	1577

TABLE A108

ALGAE : TH/BIOMASS RATIO			
ITEM:	VALUE:	CONDITIONS:	REF.:
*CHLOROPHYTA	0.122 MICRO GRAM/(GRAM ACID SOLUBLE FRACTION)	AVERAGE OF 7 MARINE SPECIES,PUERTO RICO	386
*PHAEOPHYCEAE	0.17 MICRO GRAM/(GRAM ACID SOLUBLE FRACTION)	AVERAGE OF 4 MARINE SPECIES,PUERTO RICO	386
*RHODOPHYTA	0.205 MICRO GRAM/(GRAM ACID SOLUBLE FRACTION)	AVERAGE OF 9 MARINE SPECIES,PUERTO RICO	386

TABLE A109

ALGAE : THRESHOLD DENSITY			
ITEM:	VALUE:	CONDITIONS:	REF.:
*ALGAE	100 KCAL/M**(-2)	INITIAL VALUES, THERMAL ECOSYSTEM, LIMIT-EGESTION PROPORTION	1703

TABLE A110

ALGAE : U/BIOMASS RATIO			
ITEM:	VALUE:	CONDITIONS:	REF.:
*CHLOROPHYTA	0.88 MICRO GRAM/(GRAM ACID SOLUBLE FRACTION)	AVERAGE OF 7 MARINE SPECIES,PUERTO RICO	386
*PHAEOPHYCEAE	0.79 MICRO GRAM/(GRAM ACID SOLUBLE FRACTION)	AVERAGE OF 4 MARINE SPECIES,PUERTO RICO	386
*RHODOPHYTA	0.39 MICRO GRAM/(GRAM ACID SOLUBLE FRACTION)	AVERAGE OF 9 MARINE SPECIES,PUERTO RICO	386

TABLE A111

ALGAE : VITAMIN B 1/BIOMASS RATIO			
ITEM:	VALUE:	CONDITIONS:	REF.:
*MARINE PHYTOPLANKTON	82 10**(-16) G/CELL	NO3 AS SOURCE OF N, STATIONARY PHASE,293 K, SOUTHERN CALIF	428

TABLE A111

ALGAE : VITAMIN B 1/BIOMASS RATIO			
ITEM:	VALUE:	CONDITIONS:	REF.:
*MARINE PHYTOPLANKTON	64 10**(-16) G/CELL	NH4 AS SOURCE OF N, STATIONARY PHASE,293 K, SOUTHERN CALIF	428
*MARINE PHYTOPLANKTON	38 10**(-16) G/CELL	UREA AS SOURCE OF N, STATIONARY PHASE,293 K, SOUTHERN CALIF	428

TABLE A112

ALGAE : VITAMIN B 12/BIOMASS RATIO			
ITEM:	VALUE:	CONDITIONS:	REF.:
*MARINE PHYTOPLANKTON	3.2 10**(-16) G/CELL	NO3 AS SOURCE OF N, STATIONARY PHASE,293 K, SOUTHERN CALIF	428
*MARINE PHYTOPLANKTON	1.6 10**(-16) G/CELL	NH4 AS SOURCE OF N, STATIONARY PHASE,293 K, SOUTHERN CALIF	428
*MARINE PHYTOPLANKTON	2.0 10**(-16) G/CELL	UREA AS SOURCE OF N, STATIONARY PHASE,293 K, SOUTHERN CALIF	428

TABLE A113

ALGAE : ZN/BIOMASS RATIO			
ITEM:	VALUE:	CONDITIONS:	REF.:
*APHANIZOMENON SP.	120 MICROGRAM/L	ALGAE FROM PONDS AND LAKES IN SOUTHEASTERN UNITED STATES	1660
*CHARA SP.	89 MICROGRAM/L	ALGAE FROM PONDS AND LAKES IN SOUTHEASTERN UNITED STATES	1660
*CLADOPHORA SP.	10 MICROGRAM/L	ALGAE FROM PONDS AND LAKES IN SOUTHEASTERN UNITED STATES	1660
*EUGLENA SP.	73 MICROGRAM/L	ALGAE FROM PONDS AND LAKES IN SOUTHEASTERN UNITED STATES	1660
*HYDRODICTYON SP.	129 MICROGRAM/L	ALGAE FROM PONDS AND LAKES IN SOUTHEASTERN UNITED STATES	1660
*LYNGBYA SP.	171 MICROGRAM/L	ALGAE FROM PONDS AND LAKES IN SOUTHEASTERN UNITED STATES	1660
*MICROCYSTIS SP.	48 MICROGRAM/L	ALGAE FROM PONDS AND LAKES IN SOUTHEASTERN UNITED STATES	1660
*MOUGEOTIA SP.	520 MICROGRAM/L	ALGAE FROM PONDS AND LAKES IN SOUTHEASTERN UNITED STATES	1660
*NITELLA SP.	240 MICROGRAM/L	ALGAE FROM PONDS AND LAKES IN SOUTHEASTERN UNITED STATES	1660
*OEDOGONIUM SP.	119 MICROGRAM/L	ALGAE FROM PONDS AND LAKES IN SOUTHEASTERN UNITED STATES	1660
*PITHOPHORA SP.	29 MICROGRAM/L	ALGAE FROM PONDS AND LAKES IN SOUTHEASTERN UNITED STATES	1660
*RHIZOCLONIUM SP.	89 MICROGRAM/L	ALGAE FROM PONDS AND LAKES IN SOUTHEASTERN UNITED STATES	1660
*SPIROGYRA SP.	72 MICROGRAM/L	ALGAE FROM PONDS AND LAKES IN SOUTHEASTERN UNITED STATES	1660
*SPIROGYRA SP., GIANT	46 MICROGRAM/L	ALGAE FROM PONDS AND LAKES IN SOUTHEASTERN UNITED STATES	1660

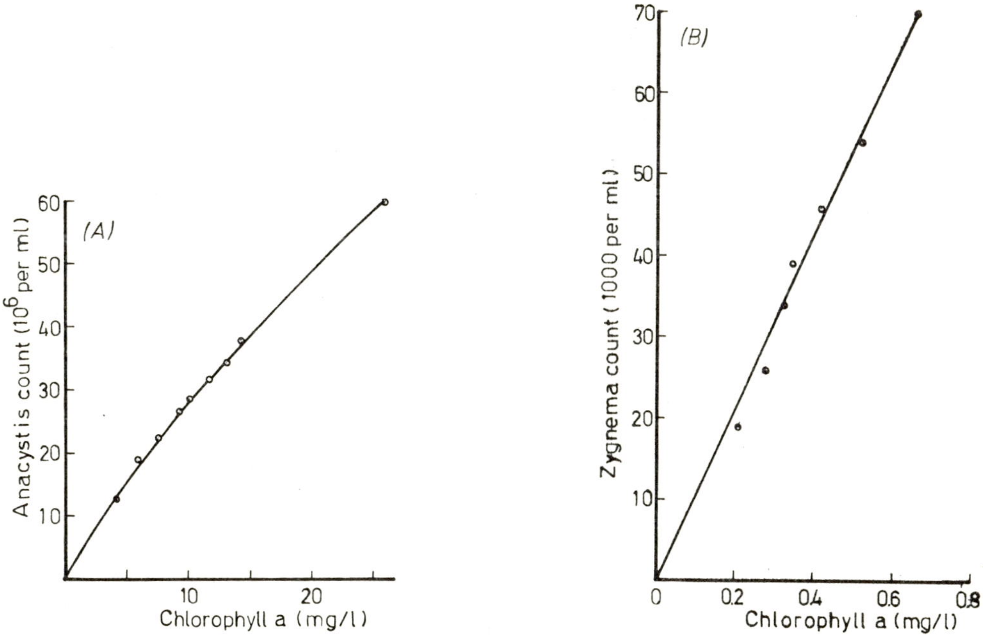

Chlorophyll A/Algae. A: Anacystis. B: Zygnema. C: Chlorella. D: Scenedesmus. A114

(ref: 826)

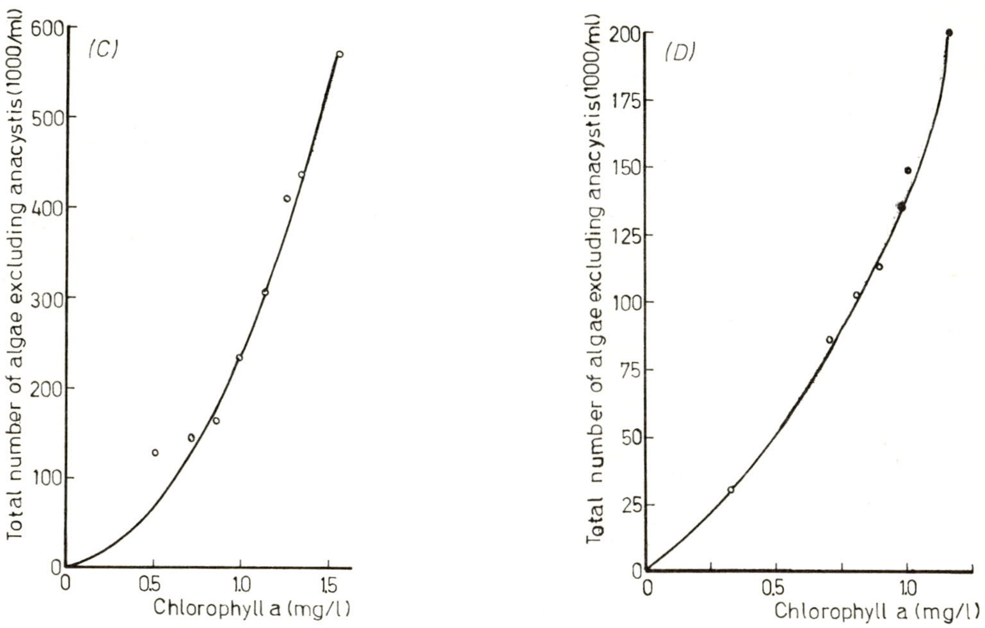

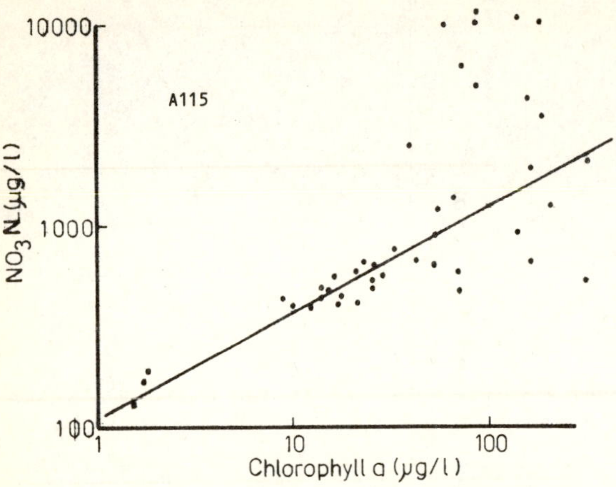

Max. algae biomass versus max. nitrate-N (winter).

(ref: 957)

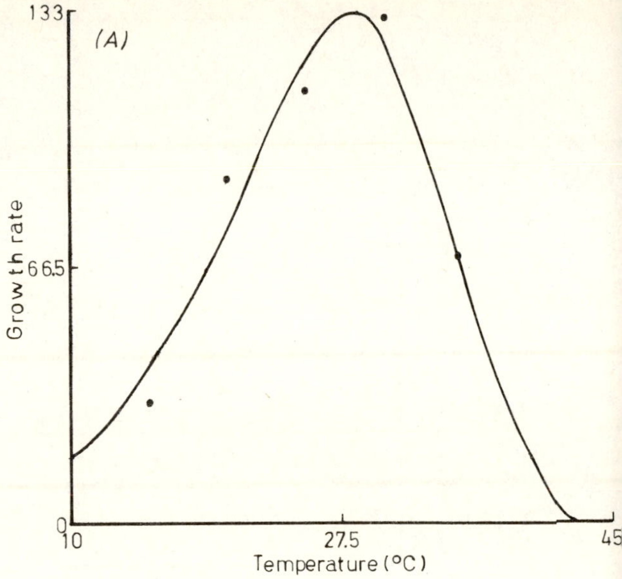

Growth rate versus temperature for A: Chlorella capricornutum. B: Ganoderma aplanatum. C: Lymnae stagnalis A116

(ref: 886)

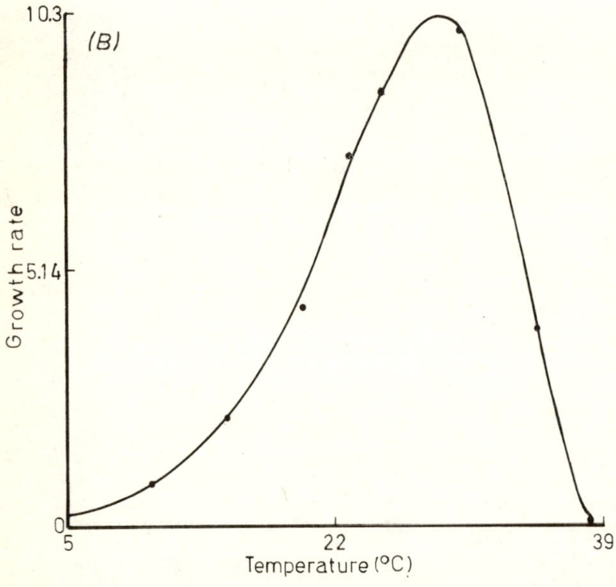

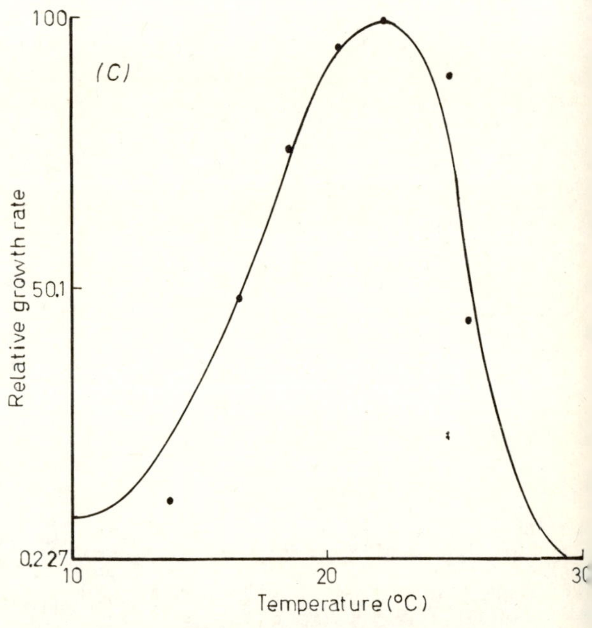

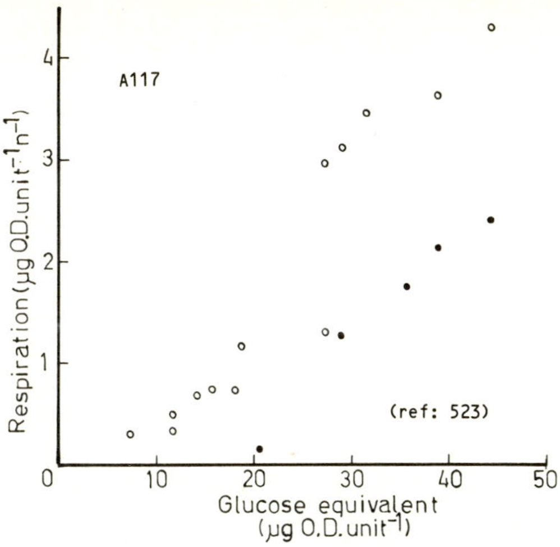

Respiration rate versus carbohydrate content for Oscillatoria redekei. ug OD/(units* hour) is plotted to glucose equivalent (μg OD/units) for cells put into dark and starved cells put into light Temperature 20 celcius.

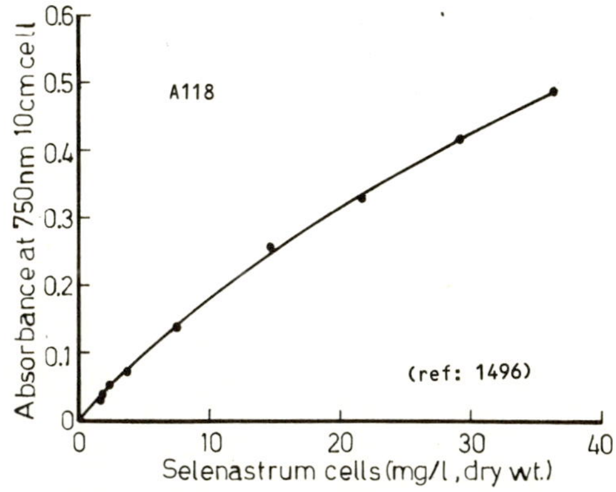

Absorbance vercus dry weight of Selenastrum capricornutum cells.

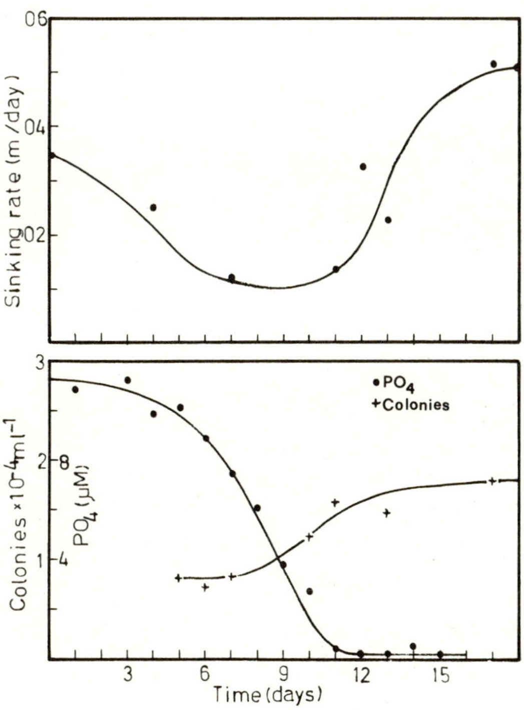

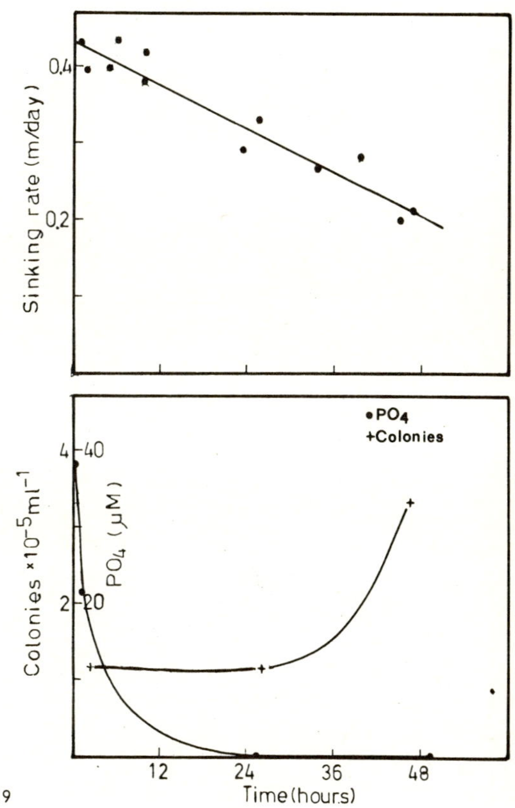

A119

Sinking rate (m/day) related to phosphate (extracellular, reactive) and population size for Scenedesmus quadricauda.
• PO4 μM, + 10**5 colonies/ml.

((ref: 1600)

Sinking rate (m/day) related to phosphate (extracellular, reactive) and population size for Asterionella formosa.
• PO4 μM, + 10**4 colonies/ml.

(ref: 1600

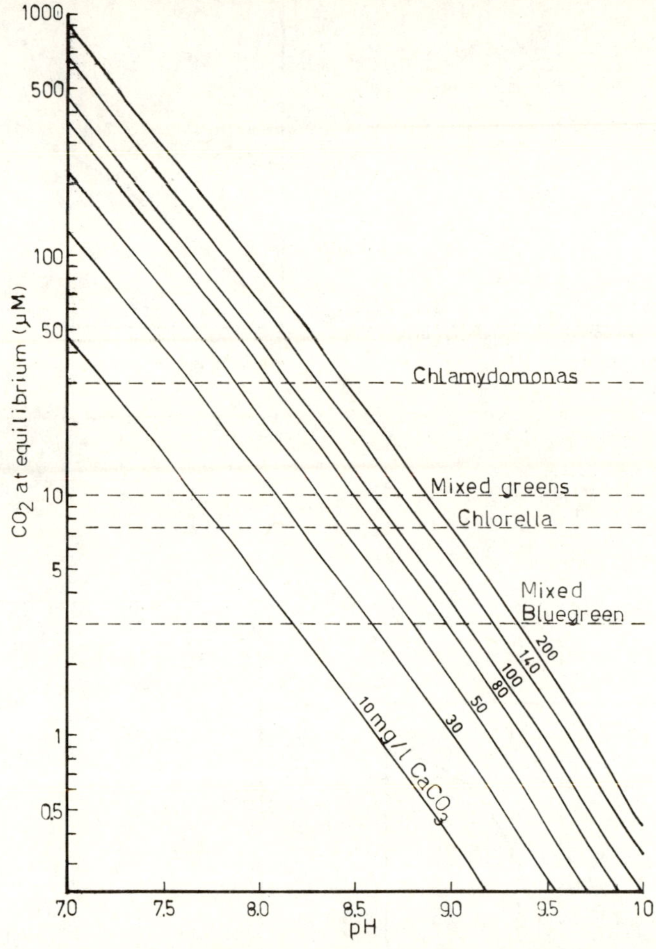

(ref: 826)

Alkalinity, pH, equilibrium carbon dioxide and the point at which carbon dioxide begins to limit different algal types.

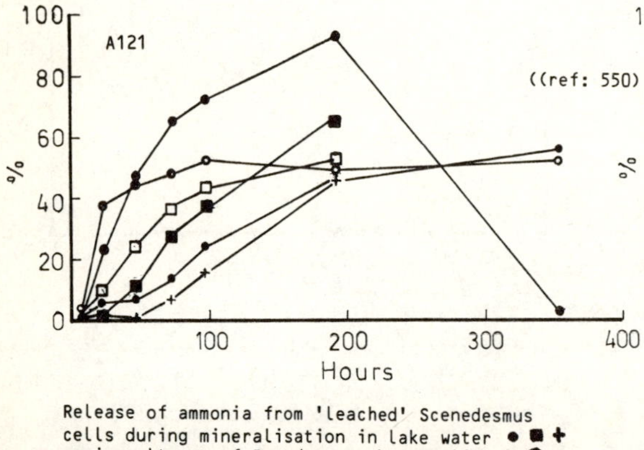

Release of ammonia from 'leached' Scenedesmus cells during mineralisation in lake water ● ■ + or in cultures of Pseudomonas boreopolis o □ • at three temperatures: 30 °C ● o 20 °C ■ □ 10 °C + •

((ref: 550)

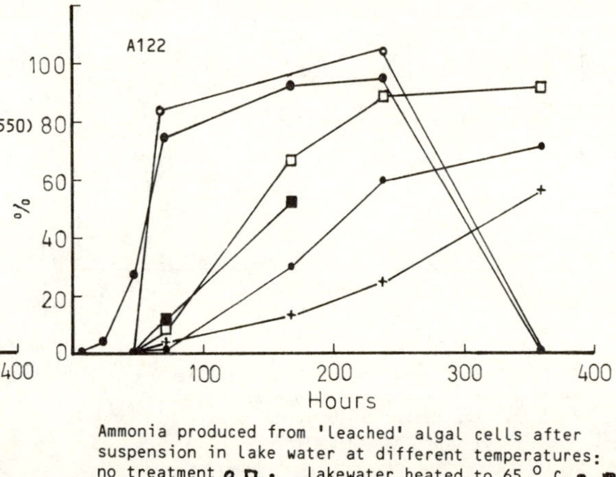

Ammonia produced from 'leached' algal cells after suspension in lake water at different temperatures: no treatment o □ • lakewater heated to 65 °C ● ■ + for 1 hour 30 °C ● o 20 °C ■ □ 10 °C + •

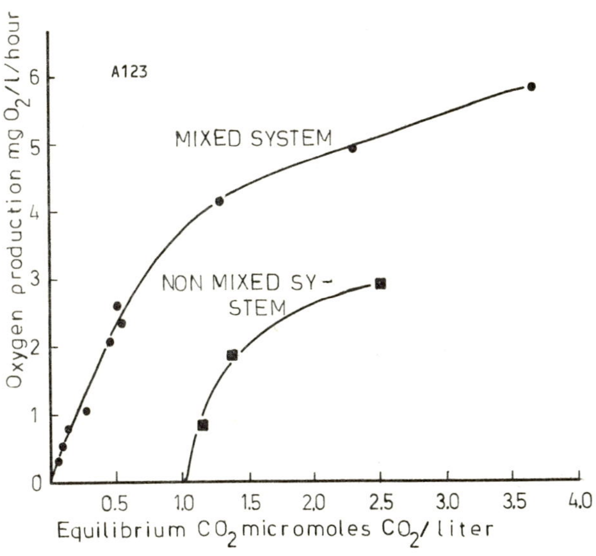

Photosynthesis versus carbon dioxide availability. Effect of mixing.

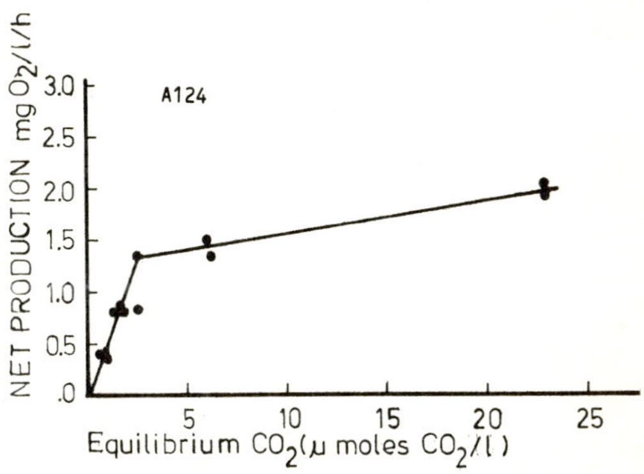

Photosynthesis versus carbon dioxide availability for mixed blue-green algae.

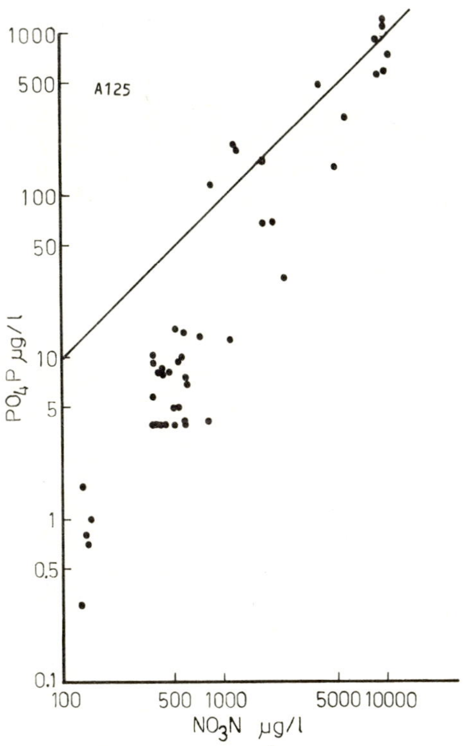

(ref: 957)

Ortho-P, max (winter) versus nitrate-N, max (winter) for several lakes in England.

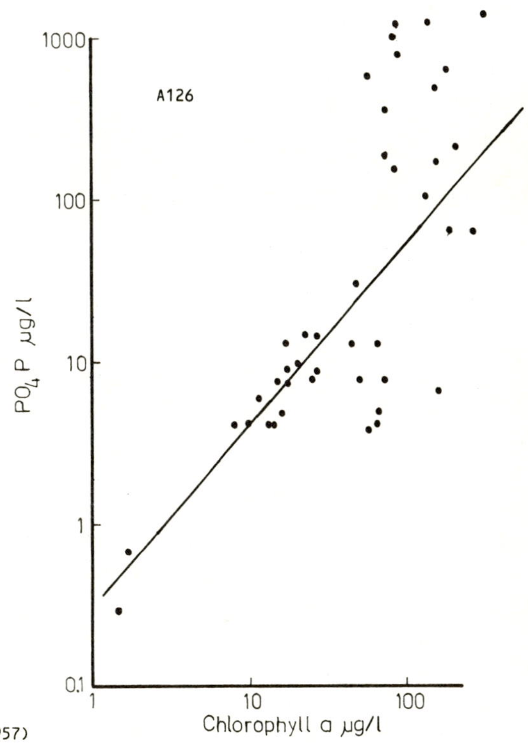

Algae biomass (max) versus ortho-P for several lakes in England.

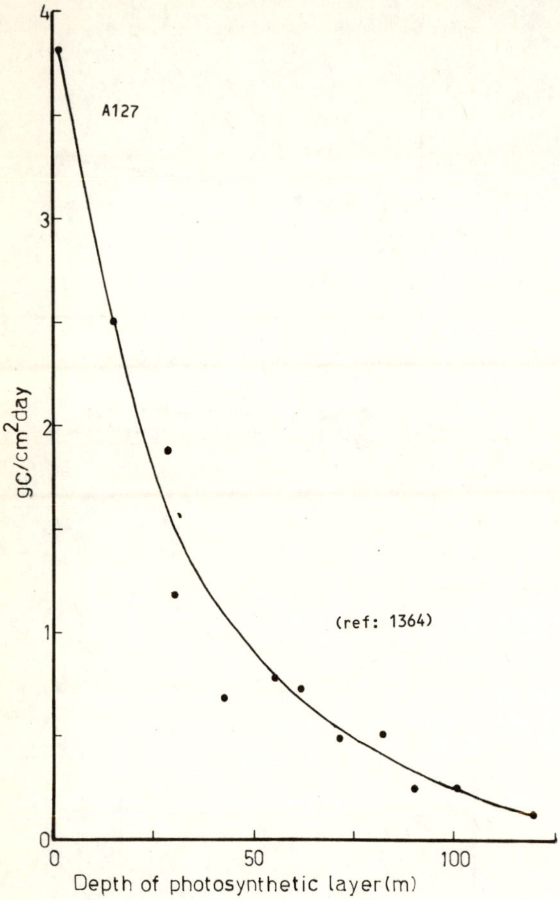

Productivity and depth of the euphotic zone relation.

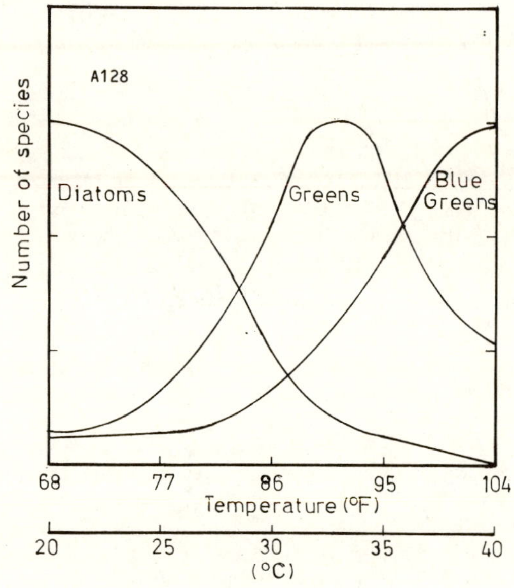

Algae shifts as the temperature rises.

(ref: 258)

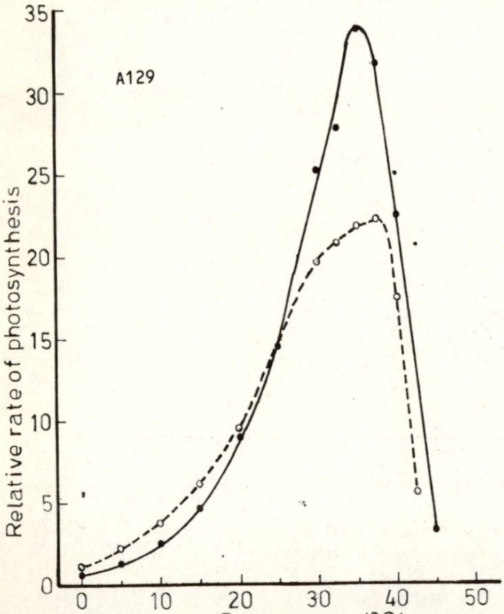

(ref: 258)

Photosynthesis of Nostoc muscorum • and Chlorella pyrenoidosa ○ versus temperature.

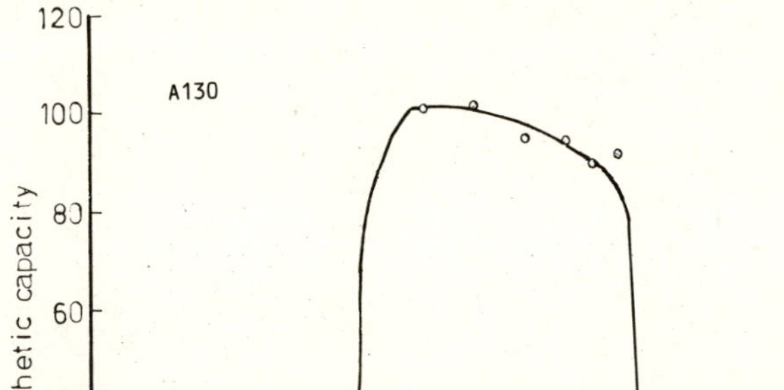

Photosynthesis of Nostoc muscorum versus pH.

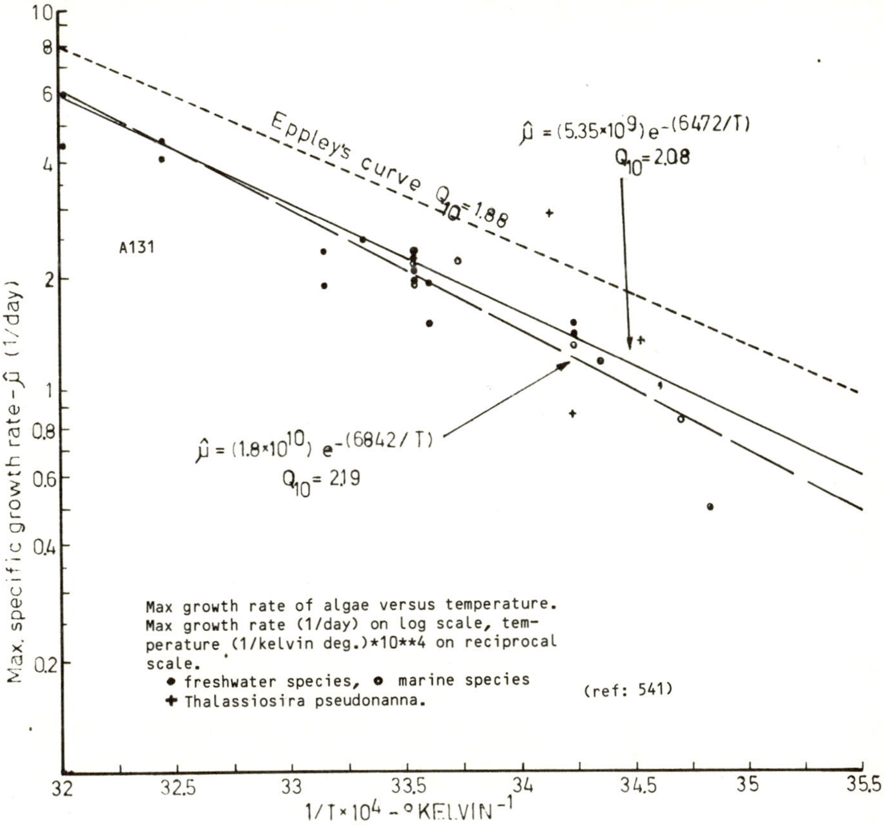

Max growth rate of algae versus temperature. Max growth rate (1/day) on log scale, temperature (1/kelvin deg.)*10**4 on reciprocal scale.
• freshwater species, ∘ marine species
✚ Thalassiosira pseudonanna. (ref: 541)

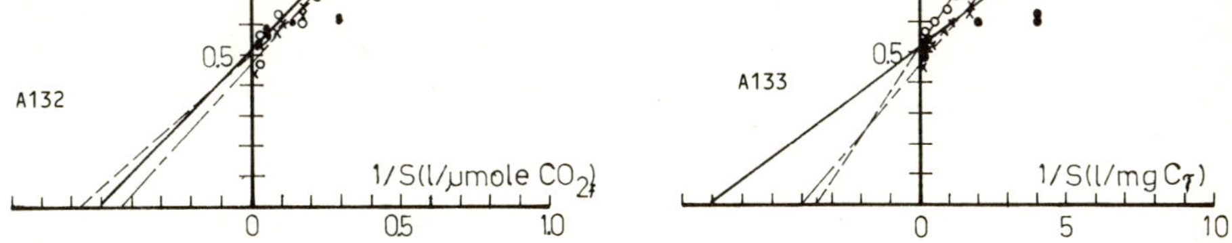

Growth rate of Scenedermus quadricauda versus carbon concentration (free Carbon dioxide resp. total carbon). Reciprocal plots.
pH range 7.10 - 7.21 •
pH range 7.44 - 7.61 ∘ (ref: 825)
pH range 7.25 - 7.39 ×

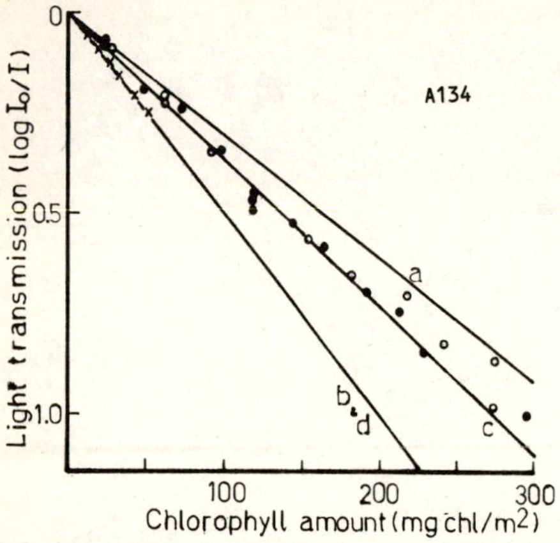

Chlorophyll versus light transmission for sulphur bacteria and algae. a: Chlorella. b: Skeletonema. c: Chromatium. ○ 20 klux. ● 500 klux.

(ref: 355)

TABLE A135

General Requirements for Vitamins among the Algae.
(++ = required in many species; + = few species; - = requirement rare; o = no known requirement)

REF.: 1281

Algae group	Biotin	Thiamine	Vitamin B12	Dominant Vitamin
Cyanophyceae	o	o	++	B12
Rhodophyceae	o	o	++	B12
Bacillariophyceae	o	+	++	B12
Xanthophyceae	o	o	o	none
Phaeophyceae	o	o	+	none
Chlorophyceae	o	+	++	B12
Chrysophyceae	-	++	+	Thiamine
Haptophyceae	-	++	+	Thiamine
Cryptophyceae	-	-	+	none
Dinophyceae	+	o	++	B12
Euglenophyceae	-	-	+	none

TABLE A136

Distribution of Photosynthetic Pigments among the Algae.
Algal groups are numbered from 1-10:
(1): Cyanophyceae (2): Chlorophyta (3): Xanthophyceae
(4): Chrysophyceae (5): Bacillariophyceae
(6): Cryptophyceae (7): Dinophyceae (8): Euglenophyceae
(9): Phaeophyceae (10): Rhodophyceae
REF.: 1689

Pigments	(1)	(2)	(3)	(4)	(5)	(6)	(7)	(8)	(9)	(10)
Chlorophyll(a)	+	+	+	+	+	+	+	+	+	+
Chlorophyll(b)	-	+	-	-	-	-	-	+	-	-
Chlorophyll(c)	-	-	-	-	+	+	+	-	+	-
Chlorophyll(d)	-	-	-	-	-	-	-	-	-	+
Alpha-carotene	-	+	-	-	-	+	-	-	-	+
Beta-carotene	+	+	+	+	+	-	+	+	+	+
Gamma-carotene	-	+	-	-	-	-	-	-	-	-
Epsilon-carotene	-	-	-	-	+	+	-	-	-	-
Lutein	+	+	-	+	-	-	-	-	+	+
Violaxanthin	-	+	+	-	-	-	-	-	+	+
Fucoxanthin	-	-	-	+	+	-	-	-	+	-
Neoxanthin	-	+	+	-	-	-	-	+	-	-
Astaxanthin	-	+	-	-	-	-	-	+	-	-
Diatoxanthin	-	-	-	+	+	-	-	-	+	-
Diadinoxanthin	-	-	-	+	+	-	+	-	-	-
Peridinin	-	-	-	-	-	-	+	-	-	-
Dinoxanthin	-	-	-	-	-	-	+	-	-	-
Teraxanthin	-	-	-	-	-	-	-	-	-	+
Antheraxanthin	-	-	-	-	-	-	-	+	-	-
Myxoxanthin	+	-	-	-	-	-	-	-	-	-
Myxoxanthophyll	+	-	-	-	-	-	-	-	-	-
Oscilloxanthin	+	-	-	-	-	-	-	-	-	-
Echinenone	+	-	-	-	-	-	-	+	-	-
Phycocyanin	+	-	-	-	-	+	-	-	-	+
Phycoerythrin	+	-	-	-	-	+	-	-	-	+

A2

BACTERIA and BACTERIOPHAGES

COMPOSITION
GROWTH RATE
HALF SATURATION CONSTANTS
MORTALITY
OXIDATION RATE
PRODUCTION
RESPIRATION

TABLE A137

BACTERIA : ASH/BIOMASS RATIO			
ITEM:	VALUE:	CONDITIONS:	REF.:
*BACTERIA	6.97 % OF DRY WEIGHT	21 SPECIES	1488

TABLE A138

BACTERIA : ATP/BIOMASS RATIO			
ITEM:	VALUE:	CONDITIONS:	REF.:
*AEROBACTER AEROGENES	0.28 *10**(-10) MICROGRAM ATP/CELL	GROWN AT OPTIMAL CONDITIONS	239
*BACILLUS CEREUS	1.1 *10**(-10) MICROGRAM ATP/CELL	GROWN AT OPTIMAL CONDITIONS	239
*BACILLUS COAGULANS	1.7 *10**(-10) MICROGRAM ATP/CELL	GROWN AT OPTIMAL CONDITIONS	239
*BACILLUS GLOBIGII	5.4 *10**(-10) MICROGRAM ATP/CELL	GROWN AT OPTIMAL CONDITIONS	239
*BACILLUS SUBTILIS	2.82 *10**(-10) MICROGRAM ATP/CELL	MEAN VALUE	348
*BREVIBACTERIUM HELVOLUM	0.37 *10**(-10) MICROGRAM ATP/CELL	GROWN AT OPTIMAL CONDITIONS	239
*ERWINA CAROTOVORA	0.44 *10**(-10) MICROGRAM ATP/CELL	GROWN AT OPTIMAL CONDITIONS	239
*ESCHERICHIA COLI	1.45 *18**(-10) MICROGRAM ATP/CELL	MEAN VALUE	348
*ESCHERICHIA COLI	1.0 *10**(-10) MICROGRAM ATP/CELL	GROWN AT OPTIMAL CONDITIONS	239
*FLAVOBACTERIUM ARBORESCENS	1.5 *10**(-10) MICROGRAM ATP/CELL	GROWN AT OPTIMAL CONDITIONS	239
*GAFFKYA TETRAGENA	0.61 *10**(-10) MICROGRAM ATP/CELL	GROWN AT OPTIMAL CONDITIONS	239
*KLEBSIELLA PNEUMONIAE	5.0 *10**(-10) MICROGRAM ATP/CELL	GROWN AT OPTIMAL CONDITIONS	239
*MICROCOCCUS LYSODEIKTICUS	1.3 *10**(-10) MICROGRAM ATP/CELL	GROWN AT OPTIMAL CONDITIONS	239
*MYCOBACTERIUM PHLEI	1.9 *10**(-10) MICROGRAM ATP/CELL	GROWN AT OPTIMAL CONDITIONS	239
*MYCOBACTERIUM SMEGMATIS	8.9 *10**(-10) MICROGRAM ATP/CELL	GROWN AT OPTIMAL CONDITIONS	239
*PROTEUS VULGARIS	1.8 *10**(-10) MICROGRAM ATP/CELL	GROWN AT OPTIMAL CONDITIONS	239
*PSEUDOMONAS AEROGINOSA	1.0 *10**(-10) MICROGRAM ATP/CELL	GROWN AT OPTIMAL CONDITIONS	239
*PSEUDOMONAS FLUORESCENS	1.6 *10**(-10) MICROGRAM ATP/CELL	MEAN VALUE	348
*PSEUDOMONAS FLURESCENS	3.1 *10**(-10) MICROGRAM ATP/CELL	GROWN AT OPTIMAL CONDITIONS	239
*SARCINA LUTEA	0.37 *10**(-10) MICROGRAM ATP/CELL	GROWN AT OPTIMAL CONDITIONS	239
*SERRATIA MARCESCENS	1.0 *10**(-10) MICROGRAM ATP/CELL	GROWN AT OPTIMAL CONDITIONS	239
*STAPHYLOCOCCUS AUERUS	0.64 *10**(-10) MICROGRAM ATP/CELL	GROWN AT OPTIMAL CONDITIONS	239

TABLE A139

BACTERIA : C/BIOMASS RATIO			
ITEM:	VALUE:	CONDITIONS:	REF.:
*BACTERIA	48.5 % OF DRY WEIGHT	MEAN ELEMENTARY COMPOSITION	157
*ESCHERICHIA COLI	12.14 %	ATOMIC COMPOSITION	1067
*MEAN DECOMPOSER	49 % OF DRY WEIGHT	MEAN ELEMENTARY COMPOSITION	157

TABLE A140

BACTERIA : CA/BIOMASS RATIO			
ITEM:	VALUE:	CONDITIONS:	REF.:
*ESCHERICHIA COLI	0.03 %	ATOMIC COMPOSITION	1067

TABLE A141

BACTERIA : CELL DRY WEIGHT			
ITEM:	VALUE:	CONDITIONS:	REF.:
*NITROSOMONAS SP.	10**(-12) G	NONE	310

TABLE A142

BACTERIA : CONCENTRATION FOR OPTIMAL GROWTH			
ITEM:	VALUE:	CONDITIONS:	REF.:
*NITRIFYING BACTERIA	1 MG CA/L	SEE REF.	830
*NITRIFYING BACTERIA	6 MG FE/L	SEE REF.	1024
*NITRIFYING BACTERIA	0.0002 MG MG/L	SEE REF.	1023
*NITRIFYING BACTERIA	0.003-0.02 MG P/L	SEE REF.	1022
*NITROBACTER SP.	310-3100 MG P/L	SEE REF	1626
*NITROBACTER SP.	0.3 MG FE/L	SEE REF.	1024
*NITROSOMONAS SP.	310-930 MG P/L	SEE REF.	1626
*NITROSOMONAS SP.	0.1 MG FE/L	SEE REF.	1024

TABLE A143

BACTERIA : COVALENT BOND DISTRIBUTION			
ITEM:	VALUE:	CONDITIONS:	REF.:
*ESCHERICHIA COLI	2862*10**7 EV	CONTRIBUTION TO TOTAL BOND ENERGY FROM H-C (SINGLE BOND)	1067
*ESCHERICHIA COLI	584*10**7 EV	CONTRIBUTION TO TOTAL BOND ENERGY FROM H-N (SINGLE BOND)	1067
*ESCHERICHIA COLI	230*10**7 EV	CONTRIBUTION TO TOTAL BOND ENERGY FROM H-O (SINGLE BOND)	1067
*ESCHERICHIA COLI	1128*10**7 EV	CONTRIBUTION TO TOTAL BOND ENERGY FROM C-C (SINGLE BOND)	1067
*ESCHERICHIA COLI	147*10**7 EV	CONTRIBUTION TO TOTAL BOND ENERGY FROM C-C (DOUBLE BOND)	1067
*ESCHERICHIA COLI	662*10**7 EV	CONTRIBUTION TO TOTAL BOND ENERGY FROM C-N (SINGLE BOND)	1067
*ESCHERICHIA COLI	116*10**7 EV	CONTRIBUTION TO TOTAL BOND ENERGY FROM C-N (DOUBLE BOND)	1067
*ESCHERICHIA COLI	377*10**7 EV	CONTRIBUTION TO TOTAL BOND ENERGY FROM C-O (SINGLE BOND)	1067
*ESCHERICHIA COLI	830*10**7 EV	CONTRIBUTION TO TOTAL BOND ENERGY FROM C-O (DOUBLE BOND)	1067

TABLE A144

BACTERIA : H/BIOMASS RATIO			
ITEM:	VALUE:	CONDITIONS:	REF.:
*BACTERIA	7.4 % OF DRY WEIGHT	MEAN ELEMENTARY COMPOSITION	157
*ESCHERICHIA COLI	9.94 %	ATOMIC COMPOSITION	1067
*MEAN DECOMPOSER	7 % OF DRY WEIGHT	MEAN ELEMENTARY COMPOSITION	157

TABLE A145

BACTERIA : HALF SATURATION CONSTANT FOR ACETATE UPTAKE			
ITEM:	VALUE:	CONDITIONS:	REF.:
*ANCALOMICROBIUM SP.	0.762 MICROGRAM/ML	OXIDATION LAGOON, 290 K	1498
*ANCALOMICROBIUM SP.	1.13 MICROGRAM/ML	OXIDATION LAGOON, 290 K	1498
*BACTERIA COMMUNITY	0.387 MICROGRAM/ML	OXIDATION LAGOON, 290 K	1498
*BACTERIA COMMUNITY	1.45 MICROGRAM/ML	OXIDATION LAGOON, 290 K	1498
*ELONGATED ROD	0.166 MICROGRAM/ML	OXIDATION LAGOON, 290 K	1498

TABLE A146

BACTERIA : HALF SATURATION CONSTANT FOR ACETIC ACID			
ITEM:	VALUE:	CONDITIONS:	REF.:
*SULPHATE REDUCING BACTERIA	5.7 MG/L	304 K, EXPERIMENTAL DATA ACCORDING TO EQ (~13)	1034

TABLE A147

BACTERIA : HALF SATURATION CONSTANT FOR AMMONIUM UPTAKE			
ITEM:	VALUE:	CONDITIONS:	REF.:
*MARINE PSEUDOMONAD	0.26 MILLIM	CHEMOSTAT, LATE LOG PHASE, 288 K	179

TABLE A148

BACTERIA : HALF SATURATION CONSTANT FOR GLUCOSE			
ITEM:	VALUE:	CONDITIONS:	REF.:
*ESCHERICHIA COLI	22 MICRO M	MINIMAL MEDIUM, 1942	1048
*ESCHERICHIA COLI	19.4 MICRO M	MINIMAL MEDIUM, 1942	1048
*ESCHERICHIA COLI	41.7 MICRO M	MINIMAL MEDIUM, 1958	1048
*ESCHERICHIA COLI	405 MICRO M	303 K, 1964, MINIMAL MEDIUM	1048
*ESCHERICHIA COLI	550 MICRO M	303 K, MINIMAL MEDIUM	1048
*ESCHERICHIA COLI	44 MICRO M	SEAWATER, 293 K	1048
*ESCHERICHIA COLI	0.378 MICRO M	303 K	1048

TABLE A149

BACTERIA : HALF SATURATION CONSTANT FOR LACTOSE			
ITEM:	VALUE:	CONDITIONS:	REF.:
*ESCHERICHIA COLI	50 MICRO M	SEAWATER, 293 K	1048
*ESCHERICHIA COLI	111 MICRO M	MINIMAL MEDIUM, 293 K	1048

TABLE A150

BACTERIA : HALF SATURATION CONSTANT FOR NITRATE UPTAKE			
ITEM:	VALUE:	CONDITIONS:	REF.:
*MARINE PSEUDOMONAD	0.26 MM	CHEMOSTAT, LATE LOG PHASE, 288 K	179

TABLE A151

BACTERIA : HALF SATURATION CONSTANT FOR NITROGEN OXIDATION			
ITEM:	VALUE:	CONDITIONS:	REF.:
*NITROBACTER SP.	8 MG N/L	305K	1207
*NITROBACTER SP.	5 MG N/L	298K	1207
*NITROSOMONAS SP.	10 MG N/L	303K	1207

TABLE A151

BACTERIA : HALF SATURATION CONSTANT FOR NITROGEN OXIDATION			
ITEM:	VALUE:	CONDITIONS:	REF.:
*NITROSOMONAS SP.	3.5 MG N/L	298K	1207
*NITROSOMONAS SP.	1.2 MG N/L	293K	1207

TABLE A152

BACTERIA : HALF SATURATION CONSTANT FOR NITROGEN UPTAKE			
ITEM:	VALUE:	CONDITIONS:	REF.:
*NITROSOMONAS EUROPAEA	56 MG (NH4(+) + NH3)-N/L	298K,PH=7	1536
*NITROSOMONAS EUROPAEA	5.6 MG (NH4(+) + NH3)-N/L	298K,PH=8	1536
*NITROSOMONAS EUROPAEA	2.0 MG (NH4(+) + NH3)-N/L	298K,PH=9	1536
*NITROSOMONAS EUROPAEA	0.32 MG NH3-N/L	298K,PH=7	1536
*NITROSOMONAS EUROPAEA	0.36 MG NH3-N/L	298K,PH=8	1536
*NITROSOMONAS EUROPAEA	0.81 MG NH3-N/L	298K,PH=9	1536

TABLE A153

BACTERIA : HALF SATURATION CONSTANT FOR OXYGEN UPTAKE			
ITEM:	VALUE:	CONDITIONS:	REF.:
*LARGER MARINE BACTERIA	6.2 MICROGRAM-ATOM/L	LENGTH=4.0 U, 293 K, O2 LIMITING	259
*NITROBACTER SP.	1.0 MG/L	303K	1207
*NITROBACTER SP.	0.5 MG/L	305K	1207
*NITROBACTER SP.	0.25 MG/L	291K	1207
*NITROSOMONAS SP.	0.5 MG/L	303K	1207
*NITROSOMONAS SP.	0.3 MG/L	293K	1207
*SMALL MARINE BACTERIA	0.022 MICROGRAM-ATOM/L	LENGTH=0.5 U, 293 K, O2 LIMITING	259

TABLE A154

BACTERIA : HALF SATURATION CONSTANT FOR PHOSPHORUS UPTAKE			
ITEM:	VALUE:	CONDITIONS:	REF.:
*ESCHERICHIA COLI	0.7 MICRO M	MINIMAL MEDIUM, 293 K, UPTAKE STUDY	1048
*ESCHERICHIA COLI	17.35 MICRO M	MINIMAL MEDIUM, 303 K	1048

TABLE A155

BACTERIA : K/BIOMASS RATIO			
ITEM:	VALUE:	CONDITIONS:	REF.:
*ESCHERICHIA COLI	0.05 %	ATOMIC COMPOSITION	1067

TABLE A156

BACTERIA : LIMIT FOR GROWTH			
ITEM:	VALUE:	p CONDITIONS:	REF.:
*ACONTIUM VELATUM	0	LOWER PH LIMIT	175

TABLE A156

BACTERIA : LIMIT FOR GROWTH

ITEM:	VALUE:	CONDITIONS:	REF.:
*BACTERIA	261 K	LOWER TEMPERATURE LIMIT	175
*BACTERIA	261 K	LOWER TEMPERATURE LIMIT	175
*DEEP SEA BACTERIA	1400 ATM.	UPPER HYDROSTATIC PRESSURE LIMIT	175
*DUNALIELLA SP.	SATURATED BRINES	UPPER SALINITY LIMIT	175
*HALOPHILIC BACTERIA	SATURATED BRINES	UPPER SALINITY LIMIT	175
*HETEROTROPHIC BACTERIA	DOUBLE DISTILLED WATER	LOWER SALINITY LIMIT	175
*IRON BACTERIA	+850 MV	UPPER EH LIMIT AT PH=3	175
*MICROORGANISMS	ESSENTIALLY 0 ATM.	LOWER HYDROSTATIC PRESSURE LIMIT	175
*PLECTONEMA NOSTOCORUM	13 ?	UPPER PH LIMIT	175
*SULPHATE REDUCING BACTERIA	377 K	UPPER TEMPERATURE LIMIT AT 1000 ATM.	175
*SULPHATE REDUCING BACTERIA	-350 TO -450 MV	LOWER EH LIMIT AT PH=8-9.5	175
*THIOBACILLUS THIOOXIDANS	0	LOWER PH LIMIT	175

TABLE A157

BACTERIA : MAX GROWTH RATE

ITEM:	VALUE:	CONDITIONS:	REF.:
*AMMONIA OXIDATION BACTERIA	0.33 1/DAY	MAX. SPECIFIC GROWTH RATE,AMMONIA OXID. IN ACTIVATED SLUDGE, 298 K	368
*AZOTOBACTER SP.	0.3-0.4 1/HR	SPEC.RATE	12
*BACTERIA	3.7 1/DAY	MAX SPECIFIC GROWTH RATE,AEROBIC TREATMENT OF WASTE WATER,298 K	92
*DENITRIFYING BACTERIA	1.8 1/DAY	MAX. SPECIFIC GROWTH RATE, DENITRIFICATION USING METHANOL,298 K	1508
*METHANE FERMENTING BACTERIA	0.13 1/DAY	MAX. SPECIFIC GROWTH RATE,METHANE FERM. OF ACETIC ACID, 298 K	896
*NITRITE OXIDATION BACTERIA	0.65 1/DAY	MAX. SPECIFIC GROWTH RATE,NITRITE OXIDATION IN ESTUARY WATER,298 K	842
*NITROBACTER SP.	1.39 1/DAY	305K	147
*NITROBACTER SP.	0.14 1/DAY	IN SLUDGE,293K	347
*NITROBACTER SP.	1.39 1/DAY	PURE CULTURE,293K	347
*NITROBACTER SP.	1.0 1/DAY	IN THAMES WATER,293K	347
*NITROSOMONAS SP.	1.0 1/DAY	ISOLATED FROM ACTIVATED SLUDGE, 300K,DOUBLING TIME=16HOURS	951
*NITROSOMONAS SP.	2.2 1/DAY	CLEAR MEDIUM,BATCH CULTURE,303K, DOUBLING TIME=8HOURS	1446
*NITROSOMONAS SP.	1.5 1/DAY	CLEAR MEDIUM,CONTINUOUS CULTURE, 303K,DOUBLING TIME=11HOURS	1446
*NITROSOMONAS SP.	0.7 1/DAY	293K,PH=7.5,PURE CULTURE AND IN THAMES RIVER WATER	368
*NITROSOMONAS SP.	0.33 1/DAY	293K,PH=7.5,ACTIVATED SLUDGE TREATED SEWAGE,VAL.USUAL.MUCH LESS	368
*NITROSOMONAS SP.	0.59 1/DAY	293K,ACTIVATED SLUDGE TREATED SEWAGE	596
*SULFOLOBUS SP.	4.9 HOURS	GENERATION TIME,IN NATURE,348K	1674
*SULFOLOBUS SP.	5.8 HOURS	GENERATION TIME,IN NATURE,348K	1674
*SULFOLOBUS SP.	6.4 HOURS	GENERATION TIME,IN NATURE,348K	1674
*SULFOLOBUS SP.	9.9 HOURS	GENERATION TIME,IN NATURE,353K	1674
*SULPHATE REDUCING BACTERIA	0.33 1/DAY	MAX SPECIFIC GROWTH RATE, USING ACETIC ACID, 298 K	1034

TABLE A158

BACTERIA : MAX RATE OF ACETIC ACID UPTAKE

ITEM:	VALUE:	CONDITIONS:	REF.:
*SULPHATE REDUCING BACTERIA	8.3 1/DAY	304 K, EXPERIMENTAL DATA ACCORDING TO EQ (!13)	1034

TABLE A159

BACTERIA : MAX RATE OF OF ACETATE UPTAKE			
ITEM:	VALUE:	CONDITIONS:	REF.:
*ANCALOMICROBIUM SP.	0.0257*10**3 MICROGRAM/(ML*MIN)	OXIDATION LAGOON, 290 K	1498
*ANCALOMICROBIUM SP.	0.374*10**3 MICROGRAM/(ML*MIN)	OXIDATION LAGOON, 290 K	1498
*BACTERIA COMMUNITY	2.21*10**3 MICROGRAM/(ML*MIN)	OXIDATION LAGOON, 290 K	1498
*BACTERIA COMMUNITY	10.9*10**3 MICROGRAM/(ML*MIN)	OXIDATION LAGOON, 290 K	1498
*ELONGATED ROD	0.388*10**3 MICROGRAM/(ML*MIN)	OXIDATION LAGOON, 290 K	1498
*FRESHWATER BACTERIA	17 (RANGE=2 -55) MICROGRAM/ (L*DAY)	UPPER KLAMATH LAKE I	1754

TABLE A160

BACTERIA : MAX RATE OF OF GLUCOSE UPTAKE			
ITEM:	VALUE:	CONDITIONS:	REF.:
*FRESHWATER BACTERIA	9 (RANGE= 2 -31) MICROGRAM/ (L*DAY)	UPPER KLAMATH LAKE I	1754

TABLE A161

BACTERIA : MAX RATE OF OF GLYCINE UPTAKE			
ITEM:	VALUE:	CONDITIONS:	REF.:
*FRESHWATER BACTERIA	7 (RANGE=0.04-23) MICROGRAM/ (L*DAY)	UPPER KLAMATH LAKE I	1754

TABLE A162

BACTERIA : MAX RATE OF OF GLYCOLATE UPTAKE			
ITEM:	VALUE:	CONDITIONS:	REF.:
*FRESHWATER BACTERIA	10 (RANGE=1 -37) MICROGRAM/ (L*DAY)	UPPER KLAMATH LAKE I	1754

TABLE A163

BACTERIA : MAX RATE OF PHOTOSYNTHESIS			
ITEM:	VALUE:	CONDITIONS:	REF.:
*CHLOROBIUM SP.	0.38 MG C/(MG CHLOROPHYLL A*HOUR)	283-304.5 K, PHOTOSYNTHETIC CAPACITY = 13.5 MG C/(M3*H)	355
*CHROMATIUM SP.	0.26 MG C/(MG CHLOROPHYLL A*HOUR)	291 K, PHOTOSYNTHETIC CAPACITY = 5.6 MG C/(M3*H)	355
*CHROMATIUM SP.	0.30 MG C/(MG CHLOROPHYLL A*HOUR)	291 K, PHOTOSYNTHETIC CAPACITY = 6.0 MG C/(M3*H)	355
*CHROMATIUM SP.	2.14 MG C/(MG CHLOROPHYLL A*HOUR)	295-299 K, PHOTOSYNTHETIC CAPACITY = 61.7 MG C/(M3*H)	355
*CHROMATIUM SP. AND GREEN SULFUR BACTERIA	0.17 MG C/(MG CHLOROPHYLL A*HOUR)	301.9 K, PHOTOSYNTHETIC CAPACITY = 172 MG C/(M3*H)	355
*PURPLE AND GREEN SULFUR BACTERIA	0.25 MG C/(MG CHLOROPHYLL A*HOUR)	301.5 K, PHOTOSYNTHETIC CAPACITY = 45.3 MG C/(M3*H)	355

TABLE A164

BACTERIA : MEAN GENERATION TIME			
ITEM:	VALUE:	CONDITIONS:	REF.:
*NITROSOMONAS SP.	11-58 HOURS	SEE REF.	310

TABLE A165

BACTERIA : MG/BIOMASS RATIO			
ITEM:	VALUE:	CONDITIONS:	REF.:
*ESCHERICHIA COLI	0.03 %	ATOMIC COMPOSITION	1067

TABLE A166

BACTERIA : MORTALITY			
ITEM:	VALUE:	CONDITIONS:	REF.:
*COLIFORM BACTERIA	0.049 1/HOUR	OHIO RIVER, SUMMER, 293 K	1048
*COLIFORM BACTERIA	0.045 1/HOUR	OHIO RIVER, WINTER, 278 K	1048
*COLIFORM BACTERIA	0.085 1/HOUR	UPPER ILLINOIS RIVER, JUNE-SEPT.	1048
*COLIFORM BACTERIA	0.105 1/HOUR	UPPER ILLINOIS RIVER, OCT AND MAY	1048
*COLIFORM BACTERIA	0.024 1/HOUR	UPPER ILLINOIS RIVER, DEC - MAR.	1048
*COLIFORM BACTERIA	0.043 1/HOUR	UPPER ILLINOIS RIVER, APR. AND NOV.	1048
*COLIFORM BACTERIA	0.085 1/HOUR	LOWER ILLINOIS RIVER, JUNE - SEPT.	1048
*COLIFORM BACTERIA	0.037 1/HOUR	LOWER ILLINOIS RIVER, OCT AND MAY	1048
*COLIFORM BACTERIA	0.026 1/HOUR	LOWER ILLINOIS RIVER, DEC - MAR.	1048
*COLIFORM BACTERIA	0.029 1/HOUR	LOWER ILLINOIS RIVER, APR. AND NOV.	1048
*COLIFORM BACTERIA	0.63 1/HOUR	SHALLOW TURBULENT STREAM	1048
*COLIFORM BACTERIA	0.020 1/HOUR	MISSOURI RIVER, WINTER	1048
*COLIFORM BACTERIA	0.043 1/HOUR	TENNESSEE RIVER, KNOXVILLE	1048
*COLIFORM BACTERIA	0.055 1/HOUR	TENNESSEE RIVER, CHATTANOOGA	1048
*COLIFORM BACTERIA	0.072 1/HOUR	SACRAMENTO RIVER, SUMMER	1048
*COLIFORM BACTERIA	0.049 1/HOUR	OHIO RIVER, SUMMER, 293 K	1048
*COLIFORM BACTERIA	0.045 1/HOUR	OHIO RIVER, WINTER, 278 K	1048
*COLIFORM BACTERIA	0.085 1/HOUR	UPPER ILLINOIS RIVER, JUNE-SEPT.	1048
*COLIFORM BACTERIA	0.105 1/HOUR	UPPER ILLINOIS RIVER, OCT AND MAY	1048
*COLIFORM BACTERIA	0.024 1/HOUR	UPPER ILLINOIS RIVER, DEC - MAR.	1048
*COLIFORM BACTERIA	0.043 1/HOUR	UPPER ILLINOIS RIVER, APR. AND NOV.	1048
*COLIFORM BACTERIA	0.085 1/HOUR	LOWER ILLINOIS RIVER, JUNE - SEPT.	1048
*COLIFORM BACTERIA	0.037 1/HOUR	LOWER ILLINOIS RIVER, OCT AND MAY	1048
*COLIFORM BACTERIA	0.026 1/HOUR	LOWER ILLINOIS RIVER, DEC - MAR.	1048
*COLIFORM BACTERIA	0.029 1/HOUR	LOWER ILLINOIS RIVER, APR. AND NOV.	1048
*COLIFORM BACTERIA	0.63 1/HOUR	SHALLOW TURBULENT STREAM	1048
*COLIFORM BACTERIA	0.020 1/HOUR	MISSOURI RIVER, WINTER	1048
*COLIFORM BACTERIA	0.043 1/HOUR	TENNESSEE RIVER, KNOXVILLE, SUMMER	1048
*COLIFORM BACTERIA	0.055 1/HOUR	TENNESSEE RIVER, CHATTANOOGA, SUMMER	1048
*COLIFORM BACTERIA	0.072 1/HOUR	SACRAMENTO RIVER, SUMMER	1048
*COLIFORM BACTERIA	0.23 1/HOUR	CUMBERLAND RIVER, SUMMER	1048
*COLIFORM BACTERIA	1.1 1/HOUR	GLATT RIVER	1048
*COLIFORM BACTERIA	0.021 1/HOUR	GROUNDWATER STREAM, 283 K	1048
*COLIFORM BACTERIA	0.017 1/HOUR	LEAF RIVER, MISS.	1048
*COLIFORM BACTERIA	0.00833 - 0.029 1/HOUR	WASTEWATER LAGOON, 280.9 - 298.5 K	1048
*COLIFORM BACTERIA	0.083 1/HOUR	MATURATION PONDS, 280.9 - 298.5 K	1048
*COLIFORM BACTERIA	0.07 1/HOUR	MATURATION PONDS	1048
*COLIFORM BACTERIA	0.042 1/HOUR	NORTH FORK KINGS RIVER, CALIF.	1567
*COLIFORM BACTERIA	0.0004 - 0.146 1/HOUR	ALL STREAMS	1567
*COLIFORM BACTERIA	0.02 - 0.083 1/HOUR	LAKE ONTARIO	1567

TABLE A166

BACTERIA : MORTALITY

ITEM:	VALUE:	CONDITIONS:	REF.:
*COLIFORM BACTERIA	0.02 1/HOUR	LAKE WASHINGTON	1567
*COLIFORM BACTERIA	0.042 - 0.125 1/HOUR	ALL STREAMS	1567
*COLIFORM BACTERIA	0.02 1/HOUR	BOISE RIVER, IDAHO	1567
*COLIFORM BACTERIA	0.02 1/HOUR	SAN FRANCISCO BAY ESTUARY	1567
*COLIFORM BACTERIA	0.02 - 0.333 1/HOUR	LONG ISLAND ESTUARY, NEW YORK	1567
*COLIFORM BACTERIA	0.481 CM**2/CAL	LIGTH INDUCEDE DEATH RATE, N=14, FIRST ORDER REACTION, FIELD DATA	1567
*COLIFORM BACTERIA	0.168 CM**2/CAL	LIGTH INDUCEDE DEATH RATE, N=24, FIRST ORDER REACTION, FIELD DATA	1567
*COLIFORM BACTERIA	0.136 CM**2/CAL	LIGTH INDUCEDE DEATH RATE, N=61, FIRST ORDER REACTION, LAB DATA	1567
*ESCHERICHIA COLI	0.362 CM**2/CAL	LIGTH INDUCEDE DEATH RATE, N= 4, FIRST ORDER REACTION, FIELD DATA	1567
*ESCHERICHIA COLI	0.354 CM**2/CAL	LIGTH INDUCEDE DEATH RATE, N= 4, FIRST ORDER REACTION, LAB DATA	1567
*FECAL COLIFORM	0.0105 1/HOUR	AVERAGE FOR STORMWATER RUNOFF, 283 K, 14 D	944
*FECAL COLIFORM	0.0654 1/HOUR	AVERAGE FOR STORMWATER RUNOFF, 293 K, FIRST DAY	944
*FECAL COLIFORM	0.0413 1/HOUR	AVERAGE FOR STORMWATER RUNOFF, 293 K, 2ND DAY	944
*FECAL COLIFORM	0.0259 1/HOUR	AVERAGE FOR STORMWATER RUNOFF, 293 K, 3RD DAY	944
*FECAL COLIFORM	0.0134 1/HOUR	AVERAGE FOR STORMWATER RUNOFF, 293 K, 4TH-7TH DAY	944
*FECAL COLIFORM	0.0076 1/HOUR	AVERAGE FOR STORMWATER RUNOFF, 293 K, 8TH-14TH DAY	944
*FECAL COLIFORM	0.0577 1/HOUR	STORMWATER RUNOFF, INITIAL DIE-AWAY	944
*FECAL COLIFORM	0.0481 1/HOUR	STORMWATER RUNOFF, 24 H AFTER INITIAL DIE-AWAY	944
*FECAL COLIFORM	0.0962 1/HOUR	STORMWATER RUNOFF, 48 H AFTER INITIAL DIE-AWAY	944
*FECAL COLIFORM	0.13 1/HOUR	UNCHLORINATED PRIMARY DOMESTIC EFFLUENT, JUNE	944
*FECAL COLIFORM	0.0668 1/HOUR	INDUSTRIAL AND PRIMARY TREATMENT DISCHARGE, JUNE	944
*FECAL COLIFORM	0.0419 1/HOUR	MANY SMALL SECONDARY TREATMENT EFFLUENTS, SUGAR BEET, JUNE	944
*FECAL COLIFORM	0.0542 1/HOUR	MANY SMALL SECONDARY TREATMENT EFFLUENTS, SUGAR BEET, AUG	944
*FECAL STREPTOCOCCI	0.048 - 0.123 CM**2/CAL	LIGTH INDUCEDE DEATH RATE, N=3 , FIRST ORDER REACTION, LAB DATA	1567
*FECAL STREPTOCOCCI	0.000 CM**2/CAL	LIGTH INDUCEDE DEATH RATE, N= 3, FIRST ORDER REACTION, FIELD DATA	1567
*FECAL STREPTOCOCCI	0.007 CM**2/CAL	LIGTH INDUCEDE DEATH RATE, N= 1, FIRST ORDER REACTION, FIELD DATA	1567
*FECAL STREPTOCOCCI	0.091 CM**2/CAL	LIGTH INDUCEDE DEATH RATE, N=12, FIRST ORDER REACTION, FIELD, INITIAL	1567
*FECAL STREPTOCOCCI	0.0625 1/HOUR	STORMWATER RUNOFF, INITIAL DIE-AWAY	944
*FECAL STREPTOCOCCI	0.0019 1/HOUR	AVERAGE STORMWATER RUNOFF, 283 K, 14 D	944
*FECAL STREPTOCOCCI	0.0125 1/HOUR	AVERAGE STORMWATER RUNOFF, 293 K, 2 D	944
*FECAL STREPTOCOCCI	0.0065 1/HOUR	AVERAGE STORMWATER RUNOFF, 293 K, 3RD DAY	944
*FECAL STREPTOCOCCI	0.0044 1/HOUR	AVERAGE STORMWATER RUNOFF, 293 K, 4TH-14TH D	944
*FRESHWATER BACTERIA	0.038 1/HOUR	STEADY STATE SOLUTION TO CASTLE LAKE.	1326
*SALMONELLA TYPHIMURIUM	1.480 CM**2/CAL	LIGTH INDUCEDE DEATH RATE, N= 2, FIRST ORDER REACTION, LAB DATA	1567
*SERRATIA MARCESCENS	0.002 CM**2/CAL	LIGTH INDUCEDE DEATH RATE, N=1 , FIRST ORDER REACTION, LAB DATA	1567

TABLE A166

BACTERIA : MORTALITY			
ITEM:	VALUE:	CONDITIONS:	REF.:
*TOTAL COLIFORM	0.0288 - 0.0096 1/HOUR	CLEAN RIVERS AND LAKES	944
*TOTAL COLIFORM	0.0384 - 0.0144 1/HOUR	POLLUTED RIVERS AND LAKES	944
*TOTAL COLIFORM	0.0384 1/HOUR	STORMWATER RUNOFF, INITIAL REGROWTH	944
*TOTAL COLIFORM	0.0211 1/HOUR	STORMWATER RUNOFF, INITIAL DIE-AWAY	944
*TOTAL COLIFORM	0.1053 1/HOUR	UNCHLORINATED PRIMARY DOMESTIC EFFLUENT, JUNE	944
*TOTAL COLIFORM	0.04 1/HOUR	UNCHLORINATED PRIMARY DOMESTIC EFFLUENT, OCT.	944
*TOTAL COLIFORM	0.0223 1/HOUR	INDUSTRIAL AND PRIMARY TREATMENT DISCHARGE, JUNE	944
*TOTAL COLIFORM	0.0182 1/HOUR	INDUSTRIAL AND PRIMARY TREATMENT DISCHARGE, OCT.	944
*TOTAL COLIFORM	0.0425 1/HOUR	MANY SMALL SECONDARY TREATMENT EFFLUENTS, SUGAR BEET, JUNE	944
*TOTAL COLIFORM	0.051 1/HOUR	MANY SMALL SECONDARY TREATMENT EFFLUENTS, SUGAR BEET, AUG	944
*TOTAL COLIFORM	0.0293 1/HOUR	MANY SMALL SECONDARY TREATMENT EFFLUENTS, SUGAR BEET, INITIAL OCT.	944
*TOTAL COLIFORM	0.0889 1/HOUR	ILLINOIS RIVER, FIRST 20 H AFTER PEAK, SUMMER	944
*TOTAL COLIFORM	0.0761 1/HOUR	ILLINOIS RIVER, 21-30 H AFTER PEAK, SUMMER	944
*TOTAL COLIFORM	0.0374 1/HOUR	ILLINOIS RIVER, 31-50 H AFTER PEAK, SUMMER	944

TABLE A167

BACTERIA : N/BIOMASS RATIO			
ITEM:	VALUE:	CONDITIONS:	REF.:
*BACTERIA	10.7 % OF DRY WEIGHT	MEAN ELEMENTARY COMPOSITION	157
*ESCHERICHIA COLI	3.04 %	ATOMIC COMPOSITION	1067
*MEAN DECOMPOSER	8 % OF DRY WEIGHT	MEAN ELEMENTARY COMPOSITION	157

TABLE A168

BACTERIA : NA/BIOMASS RATIO			
ITEM:	VALUE:	CONDITIONS:	REF.:
*ESCHERICHIA COLI	0.06 %	ATOMIC COMPOSITION	1067

TABLE A169

BACTERIA : NITROGEN FIXATION			
ITEM:	VALUE:	CONDITIONS:	REF.:
*AZOTOBACTER SP.	1.6*10**(-11) MG N/(CELL*DAY)	CHERNOYE, USSR, 1937,	1689
*HETEROTROPHIC BACTERIA	0.003 - 0.047 MICRO G N/(L*HOUR)	MARY, WISC.,11968	1689
*HETEROTROPHIC BACTERIA	0.083 - 0.308 MICRO G N/(L*HOUR)	MIZE, FLA., JULY 1968	1689
*HETEROTROPHIC BACTERIA	0.000 - 0.083 MICRO G N/(L*HOUR)	MIZE, FLA., AUGUST 1968	1689

TABLE A170

BACTERIA : NUMBERS			
ITEM:	VALUE:	CONDITIONS:	REF.:
*FRESHWATER BACTERIA	0.5*10**6/ML	TOTAL BAC.,SAPROPH.=300/ML,PRIM PROD=16 MG C/(M2*D),SCH@HSEE	1193
*FRESHWATER BACTERIA	0.5*10**6/ML	TOTAL BAC.,SAPROPH.=350/ML,PRIM PROD=17 MG C/(M2*D),SCHLUENSEE	1193
*FRESHWATER BACTERIA	1.0*10**6/ML	TOTAL BAC.,SAPROPH.=650/ML,PRIM PROD=35 MG C/(M2*D),GROSS. PL@NER SEE	1193
*FRESHWATER BACTERIA	1.1*10**6/ML	TOTAL BAC.,SAPROPH.=1000/ML,PRIM PROD=57 MG C/(M2*D), PLUSS-SEE	1193

TABLE A171

BACTERIA : O/BIOMASS RATIO			
ITEM:	VALUE:	CONDITIONS:	REF.:
*BACTERIA	27.0 % OF DRY WEIGHT	MEAN ELEMENTARY COMPOSITION	157
*ESCHERICHIA COLI	73.68 %	ATOMIC COMPOSITION	1067
*MEAN DECOMPOSER	30 % OF DRY WEIGHT	MEAN ELEMENTARY COMPOSITION/ BIOMASS RATIO	157

TABLE A172

BACTERIA : OXIDATION RATE			
ITEM:	VALUE:	CONDITIONS:	REF.:
*NITROSOMONAS SP.	0.67 MG N/(MG DRY CELLS)*HR	CELL DRY WT.=10**(-12)G,NH4-N CONC.=10MG/L,293K,RIVER TAME	310
*NITROSOMONAS SP.	0.89 MG N/(MG DRY CELLS)*HR	CELL DRY WT.=10**(-12)G,NH4-N CONC.=100MG/L,293K,RIVER TAME	310
*NITROSOMONAS SP.	1.20 MG N/(MG DRY CELLS)*HR	CELL DRY WT.=10**(-12)G,NH4-N CONC.=10MG/L,293K,RIVER TRENT	310
*NITROSOMONAS SP.	1.16 MG N/(MG DRY CELLS)*HR	CELL DRY WT.=10**(-12)G,NH4-N CONC.=10MG/L,293K,RIVER TRENT	310
*NITROSOMONAS SP.	1.08 MG N/(MG DRY CELLS)*HR	GROWTH RATE=1,YIELD=0.04,293K, CALC.FROM EQ.?046(A+B+C)	310
*NITROSOMONAS SP.	0.98 MG N/(MG DRY CELLS)*HR	293K	310
*NITROSOMONAS SP.	0.292 MG N/(MG DRY CELLS)*HR	CALC.FROM VAL.AT 293K,280K	310
*NITROSOMONAS SP.	0.325 MG N/(MG DRY CELLS)*HR	281K,FEBRUARY,RIVER TAME	310
*NITROSOMONAS SP.	0.286 MG N/(MG DRY CELLS)*HR	279.5K,MARCH,RIVER TAME	310
*NITROSOMONAS SP.	0.390 MG N/(MG DRY CELLS)*HR	283K,APRIL,RIVER TAME	310
*NITROSOMONAS SP.	0.610 MG N/(MG DRY CELLS)*HR	287.8K,JUNE,RIVER TAME	310
*NITROSOMONAS SP.	1.104 MG N/(MG DRY CELLS)*HR	294K,JULY,RIVER TAME	310
*NITROSOMONAS SP.	0.909 MG N/(MG DRY CELLS)*HR	292K,AUGUST,RIVER TAME	310
*NITROSOMONAS SP.	0.325 MG N/(MG DRY CELLS)*HR	281K,NOVEMBER,RIVER TAME	310
*NITROSOMONAS SP.	0.338 MG N/(MG DRY CELLS)*HR	281.5K,DECEMBER,RIVER TAME	310
*NITROSOMONAS SP.	0.292 MG N/(MG DRY CELLS)*HR	280K,FEBRUARY,RIVER TRENT	310
*NITROSOMONAS SP.	0.299 MG N/(MG DRY CELLS)*HR	280.3K,MARCH,RIVER TRENT	310
*NITROSOMONAS SP.	0.390 MG N/(MG DRY CELLS)*HR	283K,APRIL,RIVER TRENT	310
*NITROSOMONAS SP.	0.831 MG N/(MG DRY CELLS)*HR	291K,MAY,RIVER TRENT	310
*NITROSOMONAS SP.	0.662 MG N/(MG DRY CELLS)*HR	288.5K,JUNE,RIVER TRENT	310
*NITROSOMONAS SP.	1.273 MG N/(MG DRY CELLS)*HR	295.5K,JULY,RIVER TRENT	310
*NITROSOMONAS SP.	0.961 MG N/(MG DRY CELLS)*HR	292.5K,AUGUST,RIVER TRENT	310
*NITROSOMONAS SP.	0.753 MG N/(MG DRY CELLS)*HR	290K,SEPTEMBER,RIVER TRENT	310
*NITROSOMONAS SP.	0.364 MG N/(MG DRY CELLS)*HR	282.3K,NOVEMBER,RIVER TRENT	310

TABLE A172

BACTERIA : OXIDATION RATE			
ITEM:	VALUE:	CONDITIONS:	REF.:
*NITROSOMONAS SP.	0.370 MG N/(MG DRY CELLS)*HR	282.5K,DECEMBER,RIVER TRENT	310

TABLE A173

BACTERIA : P/BIOMASS RATIO			
ITEM:	VALUE:	CONDITIONS:	REF.:
*ESCHERICHIA COLI	0.60 %	ATOMIC COMPOSITION	1067

TABLE A174

BACTERIA : PRODUCTION/BIOMASS RATIO			
ITEM:	VALUE:	CONDITIONS:	REF.:
*BACTERIA	0.3 - 0.6	PER DAY, SOUTHERN SEAS OF USSR	567

TABLE A175

BACTERIA : RATE OF PHOSPHORUS UPTAKE			
ITEM:	VALUE:	CONDITIONS:	REF.:
*FRESHWATER BACTERIA	0.050 1/HOUR	STEADY STATE SOLUTION TO CASTLE LAKE.	1326

TABLE A176

BACTERIA : RATE OF PRODUCTION OF TETRAMETHYL-PB			
ITEM:	VALUE:	CONDITIONS:	REF.:
*BACTERIA FROM LAKE SEDIMENTS	125 UG MET-PB/W¤EK	FIRST WEEK, PB ADDED AS ME3PBOAC, 10 MG/50 G SEDIMENT	1744
*BACTERIA FROM LAKE SEDIMENTS	642 UG MET-PB/WEEK	SECOND WEEK, PB ADDED AS ME3PBOAC, 10 MG/50 G SEDIMENT	1744
*BACTERIA FROM LAKE SEDIMENTS	555 UG MET-PB/WEEK	THIRD WEEK, PB ADDED AS ME3PBOAC, 10 MG/50 G SEDIMENT	1744
*BACTERIA FROM LAKE SEDIMENTS	256 UG MET-PB/WEEK	FOURTH WEEK, PB ADDED AS ME3PBOAC, 10 MG/50 G SEDIMENT	1744

TABLE A177

BACTERIA : RESPIRATION RATE			
ITEM:	VALUE:	CONDITIONS:	REF.:
*AZOTOBACTER CHROOCOCCUM	500-1000 ML O2/(G WET WT.)*HR	301K	524
*BACILLUS FLUORESCENS NON LIQUEFACIENS	4100 ML O2/(G WET WT.)*HR	TEMP.NOT STATED	524
*BACILLUS MESENTERICUS VULGATUS	12.1 ML O2/(G WET WT.)*HR	289K	524
*BACTERIA	7*10**(-9) MM**3 O2/HR*INDIVIDUAL	GRASSLAND SOIL,293K,EQUIVALENT TO 575GCAL/G*DAY AT 289K	963

TABLE A178

BACTERIA : S/BIOMASS RATIO			
ITEM:	VALUE:	CONDITIONS:	REF.:
*BACTERIA	0.61 % OF DRY WEIGHT	MEAN ELEMENTARY COMPOSITION	157
*ESCHERICHIA COLI	0.32 %	ATOMIC COMPOSITION	1067
*MEAN DECOMPOSER	0.5 % OF DRY WEIGHT	MEAN ELEMENTARY COMPOSITION	157

TABLE A179

BACTERIA : TEMPERATURE COEFFICIENT			
ITEM:	VALUE:	CONDITIONS:	REF.:
*BACTERIA	1.047	BOD DECAY,TEMPARATURE FACTOR IS 1 AT 293K	1571
*BACTERIA	1.020	AMMONIA DECAY,TEMPARATURE FACTOR IS 1 AT 293K	1571
*BACTERIA	1.020	NITRITE DECAY,TEMPARATURE FACTOR IS 1 AT 293K	1571
*BACTERIA	1.040	DETRITUS DECAY,TEMPARATURE FACTOR IS 1 AT 293K	1571
*BACTERIA	1.040	COLIFORM DIEOFF,TEMPARATURE FACTOR IS 1 AT 293K	1571

TABLE A180

BACTERIA : WET WEIGHT			
ITEM:	VALUE:	CONDITIONS:	REF.:
*ESCHERICHIA COLI	9.6*10**(-13) G	NONE	175
*INFLUENZA BACILLUS	2.4*10**(-14) G	NONE	175
*PNEUMOCOCCUS BACTERIUM	1.9*10**(-15) G	NONE	175
*TYPHOID BACILLUS	1.7*10**(-12) G	NONE	175

TABLE A181

BACTERIA : YIELD			
ITEM:	VALUE:	CONDITIONS:	REF.:
*NITROBACTER SP.	0.02	YIELD=WT.OF CELLS/WT.OF NITROGEN OXIDIZED	1207
*NITROSOMONAS SP.	0.06	YIELD=WT.OF CELLS/WT.OF NITROGEN OXIDIZED	1207

TABLE A182

BACTERIOPHAGE : WET WEIGHT			
ITEM:	VALUE:	CONDITIONS:	REF.:
*T-EVEN BACTERIOPHAGE	3.3*10**(-16) G	NONE	175

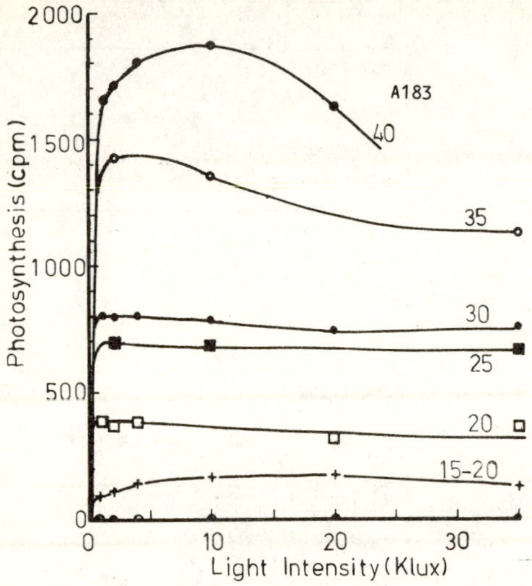

(ref: 355)

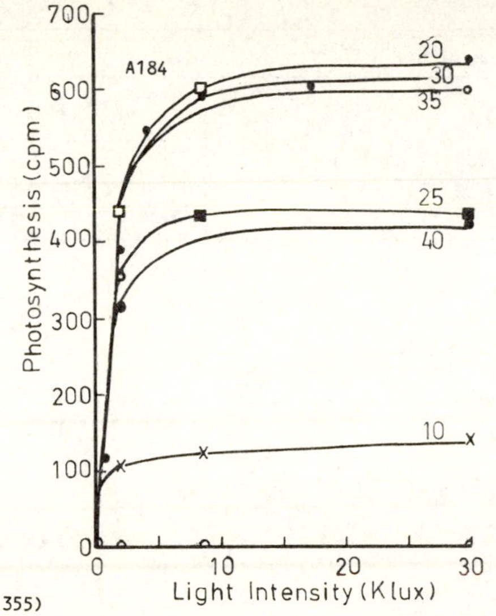

Photosythesis - Temperature curve of Chromatium and Chlorobium, left and rigth respectivily, at various in temperatures.

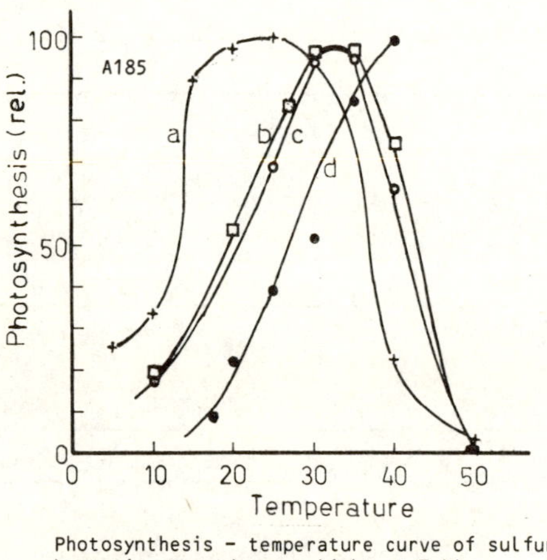

Photosynthesis - temperature curve of sulfur bacteria. Incandescent light at 7 klux.
a: Chromatium sp. from Lake Haruna.
b: Chromatium sp. from Yomuto hot S-spring.
c: Chlorobium sp. from Yomuto hot S-spring.
d: Chromatium-strain D cultivated.

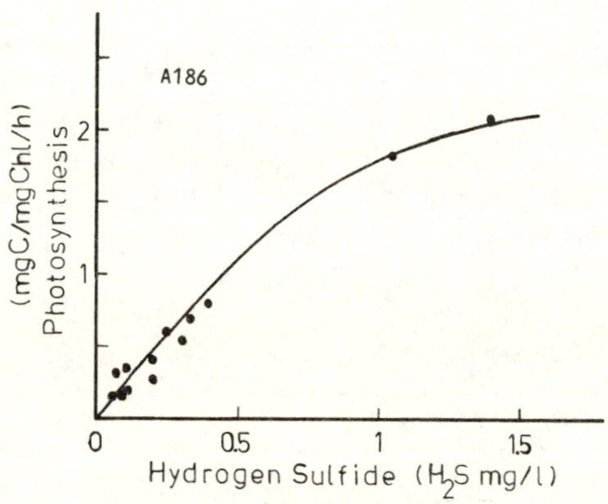

Photosynthesis versus hydrogensulphide concentrations for purple bacteria.

(ref: 355)

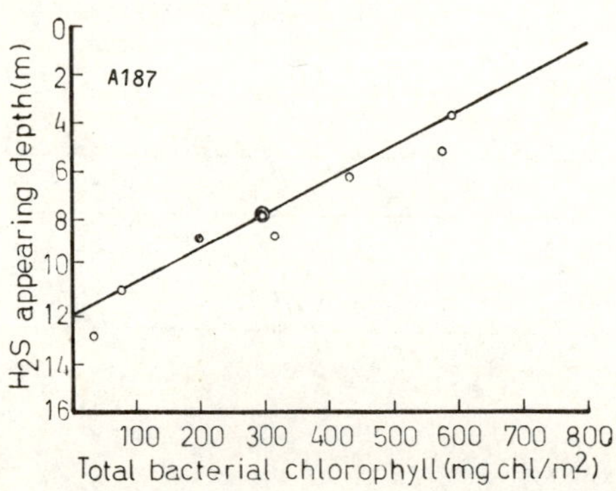

Bacterial chlorophyll versus the depth of the upper limit of the reducing environment during the growing period of photosynthetic bacteria in lakes.

A188

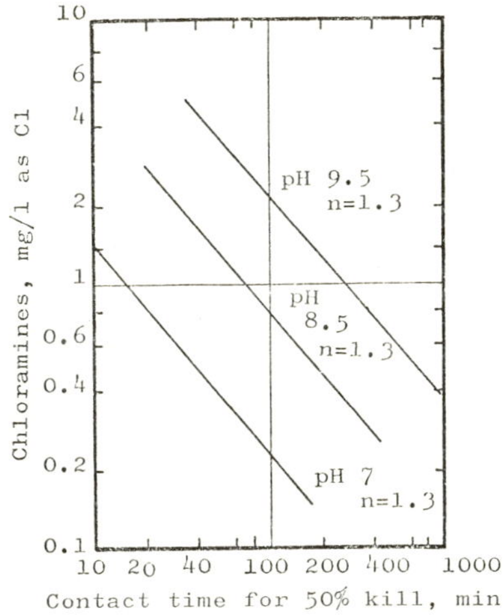

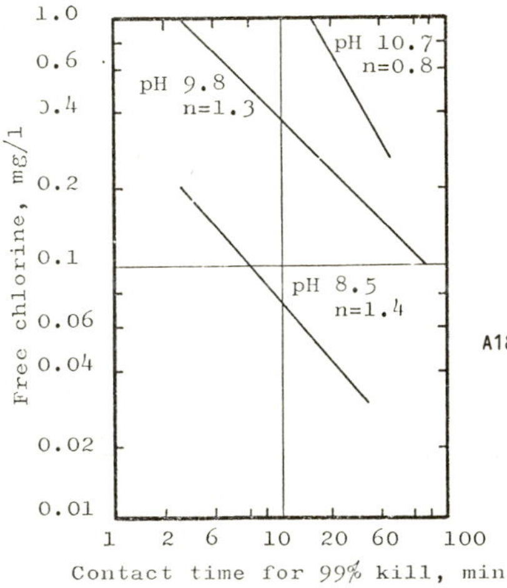

A189

(ref: 765)

A190

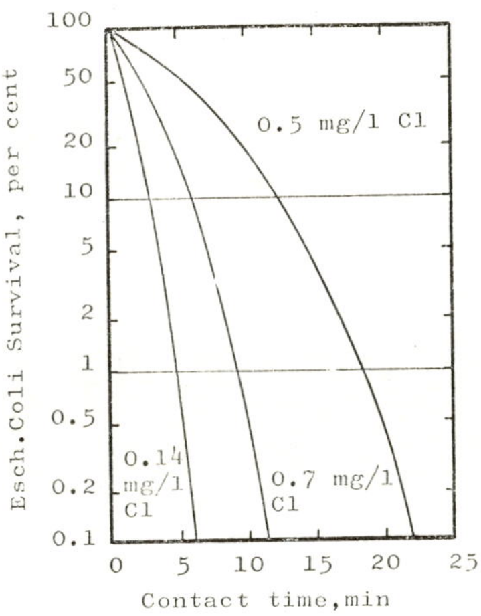

Effect of inhibitors A191

Non-competitive inhibitor	$K_{S,I}$ (mg/l)
Hg	2
Ag	5
CO	10
Cu	20
Ni	40
Cr^{6+}	200
CN^-	200-2000

Inhibition of biological oxidation.
MY = MYMAX*(K(s,i)/(K(s,i) + i).
i = inhibitor conventration.

(ref: 765)

Temperature effects on biological processes A192

Process	θ
Activated sludge (low loading)	1.00 - 1.01
Activated sludge (high loading)	1.02 - 1.03
Trickling filter	1.035
Aerated lagoons	1.06 - 1.08
Nitrification	1.143

Biological processes, temperature coefficient
K(T) = K(20)*θ**(T-20) ,Temperature in celcius.

(ref: 765)

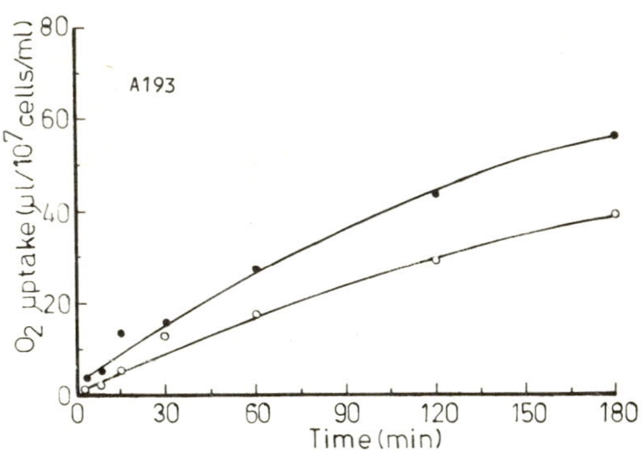

Oxygen utilization effected by temperature for Flavobacterium from Lake Ontario.
• 4 celcius. ○ 20 celcius. Oxygen uptake versus time.

(ref: 1297)

A194

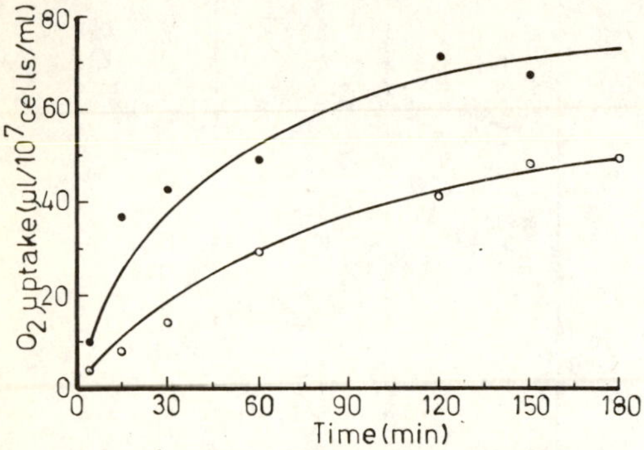

Oxygen utilization effected by temperature for Flavobacterium from Lake Superior.
• 4 celcius. o 20 celcius. Oxygen uptake versus time.

(ref: 1297)

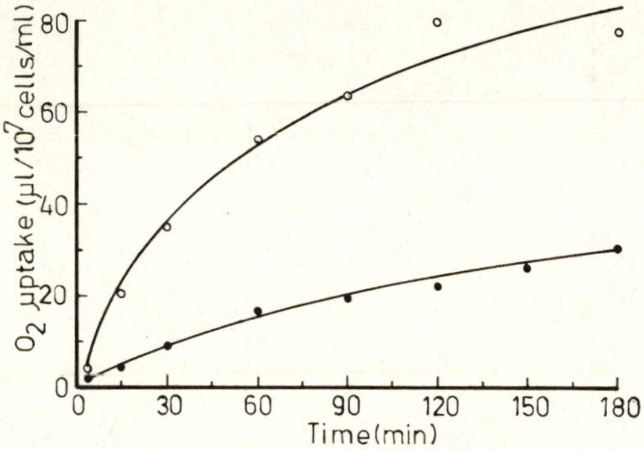

Oxygen utilization effected by temperature for E coli from St. Lawrence River.
• 4 celcius. o 20 celcius. Oxygen uptake versus time.

(ref: 1297)

TABLE A195

Predominating Bacteria that Form Hydrogen Sulphide from the Decomposition of Proteinaceous Organic Matter in Different Types of Lakes.

REF.: 866

Type of lake	Dominating Bacteria
Oligotrophic	Mycobacterium phlei, Mycobacterium filiforme
Mesotrophic	Bacterium nitrificans, Pseudomonas liquefaciens, Chromobacter aurantiacum
Eutrophic	Pseudomonas liquefaciens, bacterium delicatum
Bog lakes of high organic matter	Pseudomonas florescens, Bacillus pituitans
Saline Lakes and Estuaries	Mycobacterium luteum, Micrococcus nitrificans, Vibrio hydroslfureus
Saline Lakes and Estuaries	Achromobacter halophilum, Flavobacterium halophilum
Saline lakes and Estuaries	Bacterium albo-luteum

A3

DETRITUS

COMPOSITION
OXYGEN UPTAKE
SETTLING RATE

TABLE A196

DETRITUS : C/BIOMASS RATIO			
ITEM:	VALUE:	CONDITIONS:	REF.:
*DETRITUS	0.4-0.5 FRACTION OF DRY WEIGHT	NONE	1571

TABLE A197

DETRITUS : N/BIOMASS RATIO			
ITEM:	VALUE:	CONDITIONS:	REF.:
*DETRITUS	0.07-0.09 FRACTION OF DRY WEIGHT	NONE	1571

TABLE A198

DETRITUS : N/CHLOROPHYLL A RATIO			
ITEM:	VALUE:	CONDITIONS:	REF.:
*DETRITUS	0.01 MG-N/MG-CHL.	LAKE ONTARIO	1577

TABLE A199

DETRITUS : OXYGEN UPTAKE			
ITEM:	VALUE:	CONDITIONS:	REF.:
*DETRITUS	1.03 MG O(2)/(G*HOUR)	297 K, DET ASSOCIATED WITH THALASSIA, 5 D, PARTICLE DIAMETER = 0.1 MM	450
*DETRITUS	0.48 MG O(2)/(G*HOUR)	297 K,DET ASSOCIATED WITH THALASSIA, PART. DIAMETER=0.2 MM	450
*DETRITUS	0.3 MG O(2)/(G*HOUR)	297 K,DET ASSOCIATED WITH THALASSIA, PART. DIAMETER=0.28 MM	450
*DETRITUS	0.32 MG O(2)/(G*HOUR)	297 K,DET ASSOCIATED WITH THALASSIA, PART. DIAMETER=0.43 MM	450
*DETRITUS	0.28 MG O(2)/(G*HOUR)	297 K,DET ASSOCIATED WITH THALASSIA, PART. DIAMETER=1.4 MM	450

TABLE A200

DETRITUS : P/BIOMASS RATIO			
ITEM:	VALUE:	CONDITIONS:	REF.:
*DETRITUS	0.005-0.012	NONE	1571

TABLE A201

DETRITUS : P/CHLOROPHYLL A RATIO			
ITEM:	VALUE:	CONDITIONS:	REF.:
*DETRITUS	0.001 MG-P/MICROGRAM-CHL.	LAKE ONTARIO	1577

TABLE A202

DETRITUS : SETTLING RATE			
ITEM:	VALUE:	CONDITIONS:	REF.:
*DETRITUS	0.001 1/DAY	LAKE ONTARIO	1577

TABLE A202

DETRITUS : SETTLING RATE

ITEM:	VALUE:	CONDITIONS:	REF.:
*DETRITUS	0.-2. M/DAY	293K	1571
*FECAL PELLETS FROM ACARTIA CLAUSII	116.0 (72.0-210.0) M/DAY	FOOD=MICRO-FLAGELLATE,FECAL PEL.VOL.=0.21*10**6UM**3,N=7	1456
*FECAL PELLETS FROM EUPHAUSIA KROHNII	240.0 M/DAY	FOOD=NATURAL DIET,FECAL PEL.VOL.=(3-110)*10**6UM**3,N=113	479
*FECAL PELLETS FROM EUPHAUSIA PACIFICA	43.0 M/DAY	FOOD=SKELETONEMA COSTATUM	1188
*FECAL PELLETS FROM PONTELLA MEADII	54.0 (22.0-133.0) M/DAY	FOOD=SKELETONEMA COSTATUM,FECAL PEL.VOL.=8.2*10**6UM**3,N=22	1618
*FECAL PELLETS FROM PONTELLA MEADII	88.0 (18.0-153.0) M/DAY	FOOD=(?065),FECAL PEL.VOL.= 11.7*10**6UM**3,N=27	1618
*FECAL PELLETS FROM PONTELLA MEADII	51.0 (23.0-125.0) M/DAY	FOOD=DUNALIELLA SALINA,FECAL PEL.VOL.=3.7*10**6UM**3,N=28	1618
*FECAL PELLETS FROM PONTELLA MEADII	77.0 (15.0-117.0) M/DAY	FOOD=GONYAULAX MONILATA,FECAL PEL.VOL.=8.3*10**6UM**3,N=8	1618
*FECAL PELLETS FROM UNKNOWN SPECIES	666.0 (550.0-860.0) M/DAY	FECAL PEL.VOL.=38.6*10**6UM**3,N=6	1618
*FECAL PELLETS FROM UNKNOWN SPECIES	158.8 (36.0-376.0) M/DAY	FECAL PEL.VOL.=(0.2-2.5) *10**6UM**3,N=41	1454
*FECAL PELLETS FROM 3 SPECIES (F005)	53.0-411.0 M/DAY	FOOD=(F006),FECAL PEL.VOL.=(1.6-15.5)*10**6UM**3,N=75	479
*NITROGEN DETRITUS	0.05 1/DAY	USED IN MODELS, LAKE MICHIGAN	220
*NITROGEN DETRITUS	0.10 1/DAY	USED IN MODELS, POTOMAC ESTUARY	T(OC
*NITROGEN DETRITUS	0.028 1/DAY	USED IN MODELS, LAKE ERIE	T(OC
*NITROGEN DETRITUS	0.001 1/DAY	USED IN MODELS, LAKE ONTARIO	T(OC

A4

FISH

COMPOSITION
DEVELOPMENT OF LARVAE
EGESTION
FOOD
GROWTH RATE
HALF SATURATION CONSTANTS
MORTALITY
PRODUCTION
RESPIRATION

TABLE A203

FISH : ASSIMILATION			
ITEM:	VALUE:	CONDITIONS:	REF.:
*CTENOPHARYNGODON SP.	14.0 % OF INGESTION	NONE	1689
*FUNDULUS HETEROCLITUS	0.05 % OF INGESTED ZN	DIET = PLANT DETRITUS, MEASURED AFTER 24 H	1742
*FUNDULUS HETEROCLITUS	0.30 % OF INGESTED ZN	DIET = SHRIMP, MEASURED AFTER 24 H	1742
*PERCA FLUVIATILIS	35.8 % OF INGESTION	NONE	1689
*PLEURONECTES PLATESSA	0.10 % OF INGESTED ZN	DIET = MUSSEL OR NEREIS, MEASURED AFTER 4 D	1742
*PLEURONECTES PLATESSA	0.28 % OF INGESTED ZN	DIET = SHRIMP, MEASURED AFTER 4 D	1742
*PLEURONECTES PLATESSA	0.36 % OF INGESTED ZN	DIET = NERIES, MEASURED AFTER 4 D	1742

TABLE A204

FISH : C/BIOMASS RATIO			
ITEM:	VALUE:	CONDITIONS:	REF.:
*FISH	0.4-0.5	NONE	1571
*MOLA MOLA	48.2 % OF DRY WEIGHT	NONE	905
*PISCES	47.5 % OF DRY WEIGHT	MEAN ELEMENTARY COMPOSITION	157

TABLE A205

FISH : CALORIC CONTENT			
ITEM:	VALUE:	CONDITIONS:	REF.:
*FISH	1 KCAL/G WET WEIGHT	NONE	1661
*SOLEA SOLEA	0.626 - 1.073 K CALG WET WEIGHT	1.5 - 5.0 G	176

TABLE A206

FISH : COMPOSION OF SALMON EGGS			
ITEM:	VALUE:	CONDITIONS:	REF.:
*SALMON	W=66%,A=1.3%,F=3.6%,P=3.8%,PR=23.2%N=2.3%,C=0.2%	W=WATER,A=ASH,F=FATT,P=PHOSPHOLIPID,PR=PROTEIN,N=ORG.FREE N,C=C-HYD.	721

TABLE A207

FISH : CONVERSION EFFICIENCY			
ITEM:	VALUE:	CONDITIONS:	REF.:
*CLUPEA HARENGUS	6.99 %	FOOD AT SATURATION, 277.5 K	350
*CLUPEA HARENGUS	9.78 %	FOOD AT SATURATION, 279.5 K	350
*CLUPEA HARENGUS	12.25 %	FOOD = 1.3 % OF BODY WET WEIGHT, 279.2 K	350
*FISH	0.4-0.5 FRACTION OF INTAKE	NONE	1571

TABLE A208

FISH : DEVELOPMENT OF LARVAE			
ITEM:	VALUE:	CONDITIONS:	REF.:
*ACHIRUS LINEATUS	79.8 HOURS	HOURS AFTER HATCH WHEN YOLK IS ABSORBED, 297 K	695

TABLE A208

FISH : DEVELOPMENT OF LARVAE			
ITEM:	VALUE:	CONDITIONS:	REF.:
*ACHIRUS LINEATUS	59.7 HOURS	HOURS AFTER HATCH WHEN YOLK IS ABSORBED, 301 K	695
*ACHIRUS LINEATUS	58.2 HOURS	HOURS AFTER HATCH WHEN YOLK IS ABSORBED, 305 K	695
*ACHIRUS LINEATUS	20.5 HOURS	HOURS AFTER EYE PIGMENTATION WHEN YOLK IS ABSORBED, 297 K	695
*ACHIRUS LINEATUS	12.5 HOURS	HOURS AFTER EYE PIGMENTATION WHEN YOLK IS ABSORBED, 301 K	695
*ACHIRUS LINEATUS	12.0 HOURS	HOURS AFTER EYE PIGMENTATION WHEN YOLK IS ABSORBED, 305 K	695
*ACHIRUS LINEATUS	32 HOURS	GOOD SURVIVAL IF FED BEFORE THIS HOUR AFTER EYE PIGMENTATION, 297 K	695
*ACHIRUS LINEATUS	24 HOURS	GOOD SURVIVAL IF FED BEFORE THIS HOUR AFTER EYE PIGMENTATION, 301 K	695
*ACHIRUS LINEATUS	11.5 HOURS	HOURS AFTER YOLK ABSORPTION BEFORE WHICH LARVAE MUST BE FED, 297 K	695
*ACHIRUS LINEATUS	11.5 HOURS	HOURS AFTER YOLK ABSORPTION BEFORE WHICH LARVAE MUST BE FED, 301 K	695
*ANCHOA MITCHILLI	41.0 HOURS	HOURS AFTER HATCH WHEN YOLK IS ABSORBED, 297 K	695
*ANCHOA MITCHILLI	7.5 HOURS	HOURS AFTER EYE PIGMENTATION WHEN YOLK IS ABSORBED, 297 K	695
*ANCHOA MITCHILLI	40 HOURS	GOOD SURVIVAL IF FED BEFORE THIS HOUR AFTER EYE PIGMENTATION, 297 K	695
*ANCHOA MITCHILLI	32.5 HOURS	HOURS AFTER YOLK ABSORPTION BEFORE WHICH LARVAE MUST BE FED, 297 K	695
*ANCHOA MITCHILLI	36.6 HOURS	HOURS AFTER HATCH WHEN YOLK IS ABSORBED, 301 K	695
*ANCHOA MITCHILLI	9.5 HOURS	HOURS AFTER EYE PIGMENTATION WHEN YOLK IS ABSORBED, 301 K	695
*ANCHOA MITCHILLI	32 HOURS	GOOD SURVIVAL IF FED BEFORE THIS HOUR AFTER EYE PIGMENTATION, 301 K	695
*ANCHOA MITCHILLI	22.5 HOURS	HOURS AFTER YOLK ABSORPTION BEFORE WHICH LARVAE MUST BE FED, 301 K	695
*ANCHOA MITCHILLI	27.4 HOURS	HOURS AFTER HATCH WHEN YOLK IS ABSORBED, 305 K	695
*ANCHOA MITCHILLI	1.0 HOURS	HOURS AFTER EYE PIGMENTATION WHEN YOLK IS ABSORBED, 305 K	695
*ANCHOA MITCHILLI	16 HOURS	GOOD SURVIVAL IF FED BEFORE THIS HOUR AFTER EYE PIGMENTATION, 305 K	695
*ANCHOA MITCHILLI	15.0 HOURS	HOURS AFTER YOLK ABSORPTION BEFORE WHICH LARVAE MUST BE FED, 305 K	695
*ARCHOSARGUS RHOMBOIDALIS	52.9 HOURS	HOURS AFTER HATCH WHEN YOLK IS ABSORBED, 295 K	695
*ARCHOSARGUS RHOMBOIDALIS	50.0 HOURS	HOURS AFTER HATCH WHEN YOLK IS ABSORBED, 299 K	695
*ARCHOSARGUS RHOMBOIDALIS	40.9 HOURS	HOURS AFTER HATCH WHEN YOLK IS ABSORBED, 303 K	695
*ARCHOSARGUS RHOMBOIDALIS	10.5 HOURS	HOURS AFTER EYE PIGMENTATION WHEN YOLK IS ABSORBED, 295 K	695
*ARCHOSARGUS RHOMBOIDALIS	20.5 HOURS	HOURS AFTER EYE PIGMENTATION WHEN YOLK IS ABSORBED, 299 K	695
*ARCHOSARGUS RHOMBOIDALIS	16.0 HOURS	HOURS AFTER EYE PIGMENTATION WHEN YOLK IS ABSORBED, 303 K	695
*ARCHOSARGUS RHOMBOIDALIS	40 HOURS	GOOD SURVIVAL IF FED BEFORE THIS HOUR AFTER EYE PIGMENTATION, 295 K	695
*ARCHOSARGUS RHOMBOIDALIS	40 HOURS	GOOD SURVIVAL IF FED BEFORE THIS HOUR AFTER EYE PIGMENTATION, 299 K	695
*ARCHOSARGUS RHOMBOIDALIS	32 HOURS	GOOD SURVIVAL IF FED BEFORE THIS HOUR AFTER EYE PIGMANTATION, 303 K	695
*ARCHOSARGUS RHOMBOIDALIS	29.5 HOURS	HOURS AFTER YOLK ABSORPTION BEFORE WHICH LARVAE MUST BE FED, 295 K	695

TABLE A208

FISH : DEVELOPMENT OF LARVAE			
ITEM:	VALUE:	CONDITIONS:	REF.:
*ARCHOSARGUS RHOMBOIDALIS	19.5 HOURS	HOURS AFTER YOLK ABSORPTION BEFORE WHICH LARVAE MUST BE FED, 299 K	695
*ARCHOSARGUS RHOMBOIDALIS	16.0 HOURS	HOURS AFTER YOLK ABSORPTION BEFORE WHICH LARVAE MUST BE FED, 303 K	695

TABLE A209

FISH : EFFICIENCY			
ITEM:	VALUE:	CONDITIONS:	REF.:
*MARINE FISH	20 %	FISH PRODUCTION/PRIMARY PRODUCTION, UPWELLING AREA	1365
*MARINE FISH	15 %	FISH PRODUCTION/PRIMARY PRODUCTION, COASTAL AREA	1365
*MARINE FISH	10 %	FISH PRODUCTION/PRIMARY PRODUCTION, OCEANIC AREA	1365

TABLE A210

FISH : EGESTION			
ITEM:	VALUE:	CONDITIONS:	REF.:
*CTENOPHARYNGODON SP.	86.0 % OF INGESTION	NONE	1689
*PERCA FLUVIATILIS	64.2 % OF INGESTION	NONE	1689

TABLE A211

FISH : FOOD ITEMS			
ITEM:	VALUE:	CONDITIONS:	REF.:
*BAFFALOFISH	PLANT = 5 %	CALCULATED BY VOLUME	905
*BAFFALOFISH	DETRITUS = 40 %	CALCULATED BY VOLUME	905
*BAFFALOFISH	INVERTEBRATES = 5 %	CALCULATED BY VOLUME	905
*BAFFALOFISH	ZOOPLANKTON = 50 %	CALCULATED BY VOLUME	905
*BLACK BASS	INVERTEBRATES = 8 %	CALCULATED BY VOLUME	905
*BLACK BASS	FISH = 86 %	CALCULATED BY VOLUME	905
*BLACK BASS	TERRESTRIAL = 6 %	CALCULATED BY VOLUME	905
*BOWFIN	FISH = 100 %	CALCULATED BY VOLUME	905
*BULLHEADS	PLANT = 10 %	CALCULATED BY VOLUME	905
*BULLHEADS	DETRITUS = 25 %	CALCULATED BY VOLUME	905
*BULLHEADS	INVERTEBRATES = 50 %	CALCULATED BY VOLUME	905
*BULLHEADS	FISH = 15 %	CALCULATED BY VOLUME	905
*CARP	PLANT = 30 %	CALCULATED BY VOLUME	905
*CARP	DETRITUS = 40 %	CALCULATED BY VOLUME	905
*CARP	INVERTEBRATES = 20 %	CALCULATED BY VOLUME	905
*CARP	ZOOPLANKTON = 10 %	CALCULATED BY VOLUME	905
*CARPSUCKERS	PLANT = 15 %	CALCULATED BY VOLUME	905
*CARPSUCKERS	DETRITUS = 65 %	CALCULATED BY VOLUME	905
*CARPSUCKERS	INVERTEBRATES = 5 %	CALCULATED BY VOLUME	905
*CARPSUCKERS	ZOOPLANKTON = 15 %	CALCULATED BY VOLUME	905

TABLE A211

FISH : FOOD ITEMS			
ITEM:	VALUE:	CONDITIONS:	REF.:
*CATFISH	PLANT = 10 %	CALCULATED BY VOLUME	905
*CATFISH	INVERTEBRATES = 10 %	CALCULATED BY VOLUME	905
*CATFISH	FISH = 80 %	CALCULATED BY VOLUME	905
*CRAPPIES	PLANT = 5 %	CALCULATED BY VOLUME	905
*CRAPPIES	DETRITUS = 5 %	CALCULATED BY VOLUME	905
*CRAPPIES	INVERTEBRATES = 20 %	CALCULATED BY VOLUME	905
*CRAPPIES	ZOOPLANKTON = 15 %	CALCULATED BY VOLUME	905
*CRAPPIES	FISH = 55 %	CALCULATED BY VOLUME	905
*FRESHWATER DRUM	DETRITUS = 8 %	CALCULATED BY VOLUME	905
*FRESHWATER DRUM	INVERTEBRATES = 58 %	CALCULATED BY VOLUME	905
*FRESHWATER DRUM	FISH = 34 %	CALCULATED BY VOLUME	905
*GARS	FISH = 100 %	CALCULATED BY VOLUME	905
*GIZZARD SHAD	PLANT = 10 %	CALCULATED BY VOLUME	905
*GIZZARD SHAD	DETRITUS = 80 %	CALCULATED BY VOLUME	905
*GIZZARD SHAD	INVERTEBRATES = 5 %	CALCULATED BY VOLUME	905
*GIZZARD SHAD	ZOOPLANKTON = 5 %	CALCULATED BY VOLUME	905
*HADDOCK	FISH = 11 %	% WEIGHT OF STOMACH CONTENT, 31 - 35 CM	788
*HADDOCK	CRUSTACEA = 41 %	% WEIGHT OF STOMACH CONTENT, 31 - 35 CM	788
*HADDOCK	ANNELIDS = 11 %	% WEIGHT OF STOMACH CONTENT, 31 - 35 CM	788
*HADDOCK	MOLLUSKS = 10 %	% WEIGHT OF STOMACH CONTENT, 31 - 35 CM	788
*HADDOCK	ECHINODERMS = 27 %	% WEIGHT OF STOMACH CONTENT, 31 - 35 CM	788
*HOG SUCKERS	DETRITUS = 80 %	CALCULATED BY VOLUME	905
*HOG SUCKERS	INVERTEBRATES = 5 %	CALCULATED BY VOLUME	905
*HOG SUCKERS	ZOOPLANKTON = 15 %	CALCULATED BY VOLUME	905
*MADTOMS	DETRITUS = 27 %	CALCULATED BY VOLUME	905
*MADTOMS	INVERTEBRATES = 55 %	CALCULATED BY VOLUME	905
*MADTOMS	FISH = 18 %	CALCULATED BY VOLUME	905
*MINNOWS	PLANT = 20 %	CALCULATED BY VOLUME	905
*MINNOWS	INVERTEBRATES = 20 %	CALCULATED BY VOLUME	905
*MINNOWS	ZOOPLANKTON = 60 %	CALCULATED BY VOLUME	905
*PERCH	INVERTEBRATES = 20 %	CALCULATED BY VOLUME	905
*PERCH	ZOOPLANKTON = 20 %	CALCULATED BY VOLUME	905
*PERCH	FISH = 60 %	CALCULATED BY VOLUME	905
*PICKERELS	FISH = 100 %	CALCULATED BY VOLUME	905
*REDHORSES	INVERTEBRATES = 100 %	CALCULATED BY VOLUME	905
*SALMO GAIRDNERI	PLANT = 5 %	CALCULATED BY VOLUME	905
*SALMO GAIRDNERI	INVERTEBRATES = 60 %	CALCULATED BY VOLUME	905
*SALMO GAIRDNERI	ZOOPLANKTON = 15 %	CALCULATED BY VOLUME	905
*SALMO GAIRDNERI	FISH = 10 %	CALCULATED BY VOLUME	905
*SALMO GAIRDNERI	FISH = 10 %	CALCULATED BY VOLUME	905
*SALMO GAIRDNERI	TERRESTRIAL = 10 %	CALCULATED BY VOLUME	905
*SALVELINUS FONTINALIS	INVERTEBRATES = 90 %	CALCULATED BY VOLUME	905
*SALVELINUS FONTINALIS	ZOOPLANKTON = 5 %	CALCULATED BY VOLUME	905
*SALVELINUS FONTINALIS	TERRESTRIAL = 5 %	CALCULATED BY VOLUME	905
*SILVERSIDES	INVERTEBRATES = 20 %	CALCULATED BY VOLUME	905
*SILVERSIDES	ZOOPLANKTON = 80 %	CALCULATED BY VOLUME	905
*SUCKERS	PLANT = 15 %	CALCULATED BY VOLUME	905
*SUCKERS	DETRITUS = 65 %	CALCULATED BY VOLUME	905

TABLE A211

FISH : FOOD ITEMS			
ITEM:	VALUE:	CONDITIONS:	REF.:
*SUCKERS	INVERTEBRATES = 5 %	CALCULATED BY VOLUME	905
*SUCKERS	ZOOPLANKTON = 15 %	CALCULATED BY VOLUME	905
*SUNFISHES	PLANT = 10 %	CALCULATED BY VOLUME	905
*SUNFISHES	DETRITUS = 5 %	CALCULATED BY VOLUME	905
*SUNFISHES	INVERTEBRATES = 65 %	CALCULATED BY VOLUME	905
*SUNFISHES	FISH = 5 %	CALCULATED BY VOLUME	905
*SUNFISHES	TERRESTRIAL = 15 %	CALCULATED BY VOLUME	905
*TEMPERATE BASSES	INVERTEBRATES = 20 %	CALCULATED BY VOLUME	905
*TEMPERATE BASSES	ZOOPLANKTON = 10 %	CALCULATED BY VOLUME	905
*TEMPERATE BASSES	FISH = 70 %	CALCULATED BY VOLUME	905
*THREADFIN SHAD, ADULT	PLANT = 30 %	CALCULATED BY VOLUME	905
*THREADFIN SHAD, ADULT	DETRITUS = 5 %	CALCULATED BY VOLUME	905
*THREADFIN SHAD, ADULT	INVERTEBRATES = 15 %	CALCULATED BY VOLUME	905
*THREADFIN SHAD, ADULT	ZOOPLANKTON = 15 %	CALCULATED BY VOLUME	905
*THREADFIN SHAD, ADULT	ZOOPLANKTON = 55 %	CALCULATED BY VOLUME	905
*THREADFIN SHAD, YOUNG	PLANT = 30 %	CALCULATED BY VOLUME	905
*THREADFIN SHAD, YOUNG	DETRITUS = 50 %	CALCULATED BY VOLUME	905
*THREADFIN SHAD, YOUNG	INVERTEBRATES = 10 %	CALCULATED BY VOLUME	905
*THREADFIN SHAD, YOUNG	ZOOPLANKTON = 10 %	CALCULATED BY VOLUME	905
*WHITLING	FISH = 32 %	% WEIGHT OF STOMACH CONTENT, 21 - 25 CM	788
*WHITLING	CRUSTACEA = 43 %	% WEIGHT OF STOMACH CONTENT, 21 - 25 CM	788
*WHITLING	ANNELIDS = 16 %	% WEIGHT OF STOMACH CONTENT, 21 - 25 CM	788
*WHITLING	MOLLUSKS = 9 %	% WEIGHT OF STOMACH CONTENT, 21 - 25 CM	788

TABLE A212

FISH : GROWTH AND REPRODUCTION			
ITEM:	VALUE:	CONDITIONS:	REF.:
*CTENOPHARYNGODON SP.	1.9 % OF INGESTION	NONE	1689
*CTENOPHARYNGODON SP.	14.0 % OF INGESTION	NONE	1689
*PERCA FLUVIATILIS	19.5 % OF INGESTION	NONE	1689
*PERCA FLUVIATILIS	54.5 % OF ASSIMILATION	NONE	1689

TABLE A213

FISH : H/BIOMASS RATIO			
ITEM:	VALUE:	CONDITIONS:	REF.:
*PISCES	6.8 % OF DRY WEIGHT	MEAN ELEMENTARY COMPOSITION	157

TABLE A214

FISH : HALF SATURATION CONSTANT			
ITEM:	VALUE:	CONDITIONS:	REF.:
*CHANNEL CATFISH	3.1 % OF BODY WEIGHT CONSUMED/DAY	WEIGHT = 4 G, 303 K, MIXED DIET	905

TABLE A214

FISH : HALF SATURATION CONSTANT			
ITEM:	VALUE:	CONDITIONS:	REF.:
*FISH	50-2000 MG/M**2	GRAZING ON BENTHOS	1571
*FISH	0.05-0.1 MG/L	VALUE FOR GRAZING ON ZOOPLANKTON	1571
*MICROPTERUS SALMOIDES	4.6 % OF BODY WEIGHT CONSUMED/DAY	LENGHT = 24.5 CM, 294 K, FOOD = MINNOWS	905
*MUSKELLUNGE	5.6 % OF BODY WEIGHT CONSUMED/DAY	LENGTH = 17.0 CM, 17.0 G, 292.5 K, FOOD=MINNOWS	905
*RETICULATE SCULPIN	4.4 % OF BODY WEIGHT CONSUMED/DAY	WEIGHT = 1.2 G, 284.1 K, FOOD = MIDGE LARVAE	905
*SMALLMOUTH BASS	7.2 % OF BODY WEIGHT CONSUMED/DAY	LENGTH = 8.3 - 20.2 CM, 293.3 K, FOOD = MINNOWS	905
*SOCKEYE SALMON	3.9 % OF BODY WEIGHT CONSUMED/DAY	WEIGHT = 6.9 G, 283 K, MIXED DIET	905
*SOCKEYE SALMON	7.9 % OF BODY WEIGHT CONSUMED/DAY	WEIGHT = 7.1 G, 288 K, MIXED DIET	905

TABLE A215

FISH : MAX GROWTH RATE			
ITEM:	VALUE:	CONDITIONS:	REF.:
*CHANNEL CATFISH	3.4 % OF BODY WEIGHT/DAY	W¤IGHT = 4 G, 303 K, MIXED DIET	905
*FISH	0.02-0.03 1/DAY	TEMP. FACTOR IS 1 AT 293K	1571
*MICROPTERUS SALMOIDES	3.9 % OF BODY WEIGHT/DAY	LENGHT = 24.5 CM, 294 K, FOOD = MINNOWS	905
*MUSKELLUNGE	3.9 % OF BODY WEIGHT/DAY	LENGTH = 17.0 CM, 17.0 G, 292.5 K, FOOD=MINNOWS	905
*RETICULATE SCULPIN	1.7 % OF BODY WEIGHT/DAY	WEIGHT = 1.2 G, 284.1 K, FOOD = MIDGE LARVAE	905
*SMALLMOUTH BASS	4.7 % OF BODY WEIGHT/DAY	LENGTH = 8.3 - 20.2 CM, 293.3 K, FOOD = MINNOWS	905
*SOCKEYE SALMON	1.8 % OF BODY WEIGHT/DAY	WEIGHT = 6.9 G, 283 K, MIXED DIET	905
*SOCKEYE SALMON	4.2 % OF BODY WEIGHT/DAY	WEIGHT = 7.1 G, 288 K, MIXED DIET	905

TABLE A216

FISH : MORTALITY			
ITEM:	VALUE:	CONDITIONS:	REF.:
*ENGRAULIS JAPONICUS	1.6	AVERAGE MORTALITY VALUE, YEARLY	19
*ENGRAULIS RINGENS	1.15	AVERAGE MORTALITY VALUE, YEARLY	19
*EOPSETTA JORDANI	0.2	AVERAGE MORTALITY VALUE, YEARLY	19
*EOPSETTA JORDANI	0.24	AVERAGE MORTALITY VALUE, YEARLY	19
*FISH	0.001-0.005 1/DAY	293K	1571
*GADUS MACROCEPHALUS	0.6 - 0.9	AVERAGE MORTALITY VALUE, YEARLY	19
*LEPIDOPSETTA BILINEATA	0.25	AVERAGE MORTALITY VALUE, YEARLY	19
*LEPIDOPSETTA BILINEATA	0.5	AVERAGE MORTALITY VALUE, YEARLY	19
*MERLUCCIUS PRODUCTUS	0.3	AVERAGE MORTALITY VALUE, YEARLY	19
*MERLUCCIUS PRODUCTUS	0.4	AVERAGE MORTALITY VALUE, YEARLY	19
*SEBASTODES ALUTUS	0.1	AVERAGE MORTALITY VALUE, YEARLY.	CX(A
*THUNNUS ALBACORES	0.8	AVERAGE MORTALITY VALUE, YEARLY	19

TABLE A217

FISH : N/BIOMASS RATIO			
ITEM:	VALUE:	CONDITIONS:	REF.:
*BLUEGILL	16.7 % OF DRY WEIGHT	NONE	905

TABLE A217

FISH : N/BIOMASS RATIO			
ITEM:	VALUE:	CONDITIONS:	REF.:
*BLUEGILL	2.72 % OF WET WEIGHT	NONE	905
*CARP	2.6 % OF WET WEIGHT	NONE	905
*CHANNEL CATFISH	2.35 % OF DRY WEIGHT	NONE	905
*FISH	0.07-0.09 FRACTION OF DRY WEIGHT	NONE	1571
*FISH AND FISH LARVAE	8.3-10.7 % OF DRY WEIGHT	DRIED AT 333K,SARGASSO SEA	88
*GENERAL AVERAGE	16.3 % OF DRY WEIGHT	NONE	905
*LARGESCALE SUCKER	2.4 % OF WET WEIGHT	NONE	905
*MOLA MOLA	16.6 - 18.2 % OF DRY WEIGHT	NONE	905
*NORTHERN SQUAWFISH	2.5+-0.1 % OF WET WEIGHT	NONE	905
*SALMO GAIRDNERI	2.9 % OF WET WEIGHT	NONE	905
*SOLEA SOLEA	16.53 - 26.14 MG N/G	1.5 - 5.0 G	176

TABLE A218

FISH : N/BIOMASS RATION			
ITEM:	VALUE:	CONDITIONS:	REF.:
*PISCES	11.4 % OF DRY WEIGHT	MEAN ELEMENTARY COMPOSITION	157

TABLE A219

FISH : O/BIOMASS RATIO			
ITEM:	VALUE:	CONDITIONS:	REF.:
*PISCES	29.0 % OF DRY WEIGHT	MEAN ELEMENTARY COMPOSITION	157

TABLE A220

FISH : P/BIOMASS RATIO			
ITEM:	VALUE:	CONDITIONS:	REF.:
*BLUEGILL	4.75+-0.70 % OF DRY WEIGHT	NONE	905
*CARP	0.5+-0.01 % OF WET WEIGHT	NONE	905
*CHANNEL CATFISH	0.86 % OF WET WEIGHT	NONE	905
*COD	0.60 % OF WET WEIGHT	NONE	905
*EEL	0.68 % OF WET WEIGHT	NONE	905
*FISH	.01-.012	NONE	1571
*GENERAL AVERAGE FOR FISH FLESH	0.22 (RANGE=0.1 - 0.4) % OF WET WEIGHT	NONE	905
*HADDOCK	0.97 % OF WET WEIGHT	NONE	905
*HALIBUT	0.44 % OF WET WEIGHT	NONE	905
*HERRING	0.56 % OF WET WEIGHT	NONE	905
*LARGESCALE SUCKER	0.3 % OF WET WEIGHT	NONE	905
*MACKEREL	0.56 % OF WET WEIGHT	NONE	905
*NORTHERN SQUAWFISH	0.4 % OF WET WEIGHT	NONE	905
*SALMO GAIRDNERI	0.4 % OF WET WEIGHT	NONE	905
*SALMON	0.59 % OF WET WEIGHT	NONE	905
*TROUT	0.81 % OF WET WEIGHT	NONE	905

TABLE A220

FISH : P/BIOMASS RATIO			
ITEM:	VALUE:	CONDITIONS:	REF.:
*TURBOT	0.48 % OF WET WEIGHT	NONE	905

TABLE A221

FISH : PRODUCTION/BIOMASS RATIO			
ITEM:	VALUE:	CONDITIONS:	REF.:
*ALBURNUS ALBURNUS	1.92 ON YEARLY BASIS	PROD=915 KG DW/(HA*Y), R. THAMES, ENGL., MAX 2 Y	1661
*ASPIUS ASPIUS	0.55 ON YEARLY BASIS	PROD=1.54 KG DW/(HA*Y), VISTUAL R., POLAND	1661
*COTTUS CAROLINAE	5.5 ON YEARLY BASIS	PROD=77.0 KG DW/(HA*Y), STEELES RUN, 2ND ORDER STREAM, KY	1661
*COTTUS CAROLINAE	1.8 ON YEARLY BASIS	PROD=8.0 KG DW/(HA*Y), STEELES RUN, 3ND ORDER STREAM, KY	1661
*COTTUS COGNATUS	1.2 ON YEARLY BASIS	PROD=59.4 KG DW/(HA*Y), VALLY L. MINN.	1661
*ESOX LUCIUS	0.7 ON YEARLY BASIS	PROD=1.5 KG DW/(HA*Y), VISTULA R., POLAND	1661
*ESOX LUCIUS	0.32 ON YEARLY BASIS	PROD=0.75 KG DW/(HA*Y), RESERVOIR, CZECH.	1661
*ETHEOSTOMA SPECTABILE	2.4 ON YEARLY BASIS	PROD=52 KG DW/(HA*Y), STEELES RUN, 2ND ORDER STREAM, KY	1661
*ETHEOSTOMA SPECTABILE	1.7 ON YEARLY BASIS	PROD=23 KG DW/(HA*Y), STEELES RUN, 3ND ORDER STREAM, KY	1661
*FISH	LESS THAN 0.01	PER DAY, SOUTHERN SEAS OF USSR	567
*GOBIO GOBIO	1.94 ON YEARLY BASIS	PROD=289 KG DW/(HA*Y), R. THAMES, ENGL., MAX 2 Y	1661
*LEUCISCUS CEPHALUS	0.43 ON YEARLY BASIS	PROD=0.10 KG DW/(HA*Y), VISTULA R., POLAND	1661
*LEUCISCUS CEPHALUS	0.27 ON YEARLY BASIS	PROD=2.1 KG DW/(HA*Y), RESERVOIR, CZECH.	1661
*LEUCISCUS LEUCISCUS	1.75 ON YEARLY BASIS	PROD=52 KG DW/(HA*Y), R.THAMES, ENGL., 1 Y	1661
*ONCORHYNCHUS KISUTCH	2.4 ON YEARLY BASIS	PROD=86 KG DW/(HA*Y), N=3 STREAMS, JUVENILES, OREGON	1661
*PERCA FLUVIATILIS	0.41 ON YEARLY BASIS	PROD=0.04 KG DW/(HA*Y), VISTULA R., POLAND	1661
*PERCA FLUVIATILIS	0.35 ON YEARLY BASIS	PROD=7.3 KG DW/(HA*Y), RESERVOIR, CZECH., N=3 Y	1661
*RUTILUS RUTILUS	0.24 ON YEARLY BASIS	PROD=19.2 KG DW/(HA*Y), RESEVOIR, CZECH.	1661
*RUTILUS RUTILUS	1.12 ON YEARLY BASIS	PROD=281 KG DW/(HA*Y), R. THAMES, ENGL., MAX 2 Y	1661
*SALMO CLARKI	1.0 ON YEARLY BASIS	PROD=41 KG DW/(HA*Y), N=3 STREAMS, ONREGON	1661
*SALMO GAIRDNERI	2.0 ON YEARLY BASIS	PROD=47.6 KG DW/(HA*Y), WHISTLER BEND, OREG.	1661
*SALMO GAIRDNERI	2.4 ON YEARLY BASIS	PROD=132 KG DW/(HA*Y), JUVENILES, BOTHWELLS CR., ONTARIO	1661
*SALMO TRUTTA	0.5 ON YEARLY BASIS	PROD=38 KG DW/(HA*Y), R. YARTY, FRY PROD. IN TRIBS., ENGL	1661
*SALMO TRUTTA	2.0 ON YEARLY BASIS	PROD=544 KG DW/(HA*Y), HOROKIWI STREAM, N=6 PLACES, NZ	1661
*SALMO TRUTTA	1.0 ON YEARLY BASIS	PROD=106 KG DW/(HA*Y), BLACK EARTH CR., POLLUTED, WISC.	1661
*SALMO TRUTTA	1.2 ON YEARLY BASIS	PROD=328 KG DW/(HA*Y), BLACK EARTH R., POLL. ABATED, WISC.	1661
*SALMO TRUTTA	1.1 ON YEARLY BASIS	PROD=22 KG DW/(HA*Y), N=3 STREAMS, ENGL., NORMAL GROWTH	1661
*SALMO TRUTTA	0.5 ON YEARLY BASIS	PROD=23 KG DW/(HA*Y),GREAT DODGEN POT, SLOW GROWTH, ENGL.	1661
*SALMO TRUTTA	4.9 ON YEARLY BASIS	PROD=85 KG DW/(HA*Y), NURSERY STREAM, HINAU MAIN WATER, NZ	1661

TABLE A221

FISH : PRODUCTION/BIOMASS RATIO			
ITEM:	VALUE:	CONDITIONS:	REF.:
*SALMO TRUTTA	5.0 ON YEARLY BASIS	PROD=89 KG DW/(HA*Y), NURSERY STREAM, HINAKI MAIN WATER, NZ	1661
*SALMO TRUTTA	1.1 ON YEARLY BASIS	PROD=132 KG DW/(HA*Y), WALLA BR., N=2 Y, ENGL.	1661
*SALVELINUS FONTINALIS	0.71 ON YEARLY BASIS	PROD=2.2 KG DW/(HA*Y), MATAMEK LAKE, QUEBEE	1661
*SALVELINUS FONTINALIS	0.62 ON YEARLY BASIS	PROD=0.21 KG DW/(HA*Y), BILL L., QUEBEE	1661
*SALVELINUS FONTINALIS	1.3 ON YEARLY BASIS	PROD=145 KG DW/(HA*Y), VALLEY CR., POST-FLOOD, MINN	1661
*SALVELINUS FONTINALIS	1.6 ON YEARLY BASIS	PROD=116 KG DW/(HA*Y), LAWRENCE CR., N=6 Y, MINN.	1661
*SALVELINUS FONTINALIS	0.7 ON YEARLY BASIS	PROD=38 KG DW/(HA*Y), GALLIENNE CR., N=2 Y, QUEBEC	1661
*SALVELINUS FONTINALIS	1.6 ON YEARLY BASIS	PROD=63 KG DW/(HA*Y), KAILHOSRU CR., N=2 Y, QUEBEC	1661
*SALVELINUS FONTINALIS	1.05 ON YEARLY BASIS	PROD=25 KG DW/(HA*Y), SHERRY CR., N=3 Y, QUEBEC	1661
*SALVELINUS FONTINALIS	1.2 ON YEARLY BASIS	PROD=16 KG DW/(HA*Y), TCHINICAMAN R., N=2 Y, QUEBEC	1661
*STIZOSTEDION LUCIOPERCA	0.72 ON YEARLY BASIS	PROD=0.91 KG DW/(HA*Y), VISTULA R., POLAND	1661
*STIZOSTEDION LUCIOPERCA	0.57 ON YEARLY BASIS	PROD=5.5 KG DW/(HA*Y), L. BALATON, HUNGARY	1661
*STIZOSTEDION VITREUM VITREUM	0.43 ON YEARLY BASIS	PROD=2.1 KG DW/(HA*Y), WEST BLUE L., MANITOBA	1661
*TINCA TINCA	0.18 ON YEARLY BASIS	PROD=0.17 KG DW/(HA*Y), RESEVOIR, CZECH.	1661
*TOTAL FISH FAUNA, ALL TROPHIC LEVELS	0.85 ON YEARLY BASIS	PROD=244 KG DW/(HA*Y), RYBINSK R., USSR	1661
*TOTAL FISH FAUNA, BENTHIVORES	0.5 ON YEARLY BASIS	PROD=13 KG DW/(HA*Y), NAROCH L., USSR	1661
*TOTAL FISH FAUNA, BENTHIVORES	0.4 ON YEARLY BASIS	PROD=26 KG DW/(HA*Y), MYASTRO L., USSR	1661
*TOTAL FISH FAUNA, BENTHIVORES	0.5 ON YEARLY BASIS	PROD=60 KG DW/(HA*Y), BATORIN L., USSR	1661
*TOTAL FISH FAUNA, PISCIVORES	0.4 ON YEARLY BASIS	PROD=9 KG DW/(HA*Y), NAROCH L., USSR	1661
*TOTAL FISH FAUNA, PISCIVORES	0.3 ON YEARLY BASIS	PROD=14 KG DW/(HA*Y), MYASTRO L., USSR	1661
*TOTAL FISH FAUNA, PISCIVORES	0.4 ON YEARLY BASIS	PROD=14 KG DW/(HA*Y), BATORIN L., USSR	1661
*TOTAL FISH FAUNA, PLANKTIVORES	0.7 ON YEARLY BASIS	PROD=22 KG DW/(HA*Y), NAROCH L., USSR	1661
*TOTAL FISH FAUNA, PLANKTIVORES	0.7 ON YEARLY BASIS	PROD=24 KG DW/(HA*Y), MYASTRO L., USSR	1661
*TOTAL FISH FAUNA, PLANKTIVORES	0.8 ON YEARLY BASIS	PROD=9 KG DW/(HA*Y), BATORIN L., USSR	1661

TABLE A222

FISH : RESPIRATION RATE			
ITEM:	VALUE:	CONDITIONS:	REF.:
*CTENOPHARYNGODON SP.	12.2 % OF INGESTION	NONE	1689
*CTENOPHARYNGODON SP.	86.0 % OF ASSIMILATION	NONE	1689
*FISH	0.001-0.01 1/DAY	TEMP.FACTOR IS 1 AT 293K, TEMP>278K, ENDOGENOUS RESP	1571
*FISH	0.0001-0.001 1/DAY	ENDOGENOUS RESP., TEMP FACTOR IS 1 AT 293K,TEMP.<278K, ENDOGENOUS RESP	1571
*PERCA FLUVIATILIS	16.2 % OF INGESTION	NONE	1689
*PERCA FLUVIATILIS	45.5 % OF ASSIMILATION	NONE	1689

TABLE A222

FISH : RESPIRATION RATE			
ITEM:	VALUE:	CONDITIONS:	REF.:
*SALVELINUS FONTINALIS	0.52 ML O2/HOUR	OXYGEN CONC = 3.2 MG O2/L,WEIGHT = 5 G, 8 CM	965
*SALVELINUS FONTINALIS	0.85 ML O2/HOUR	OXYGEN CONC = 5.0 MG O2/L,WEIGHT = 5 G, 8 CM	965
*SALVELINUS FONTINALIS	1.00 ML O2/HOUR	OXYGEN CONC = 6.4 MG O2/L,WEIGHT = 5 G, 8 CM	965
*SALVELINUS FONTINALIS	2.9 ML O2/HOUR	OXYGEN CONC = 9.3 MG O2/L,WEIGHT = 5 G, 8 CM	965
*SALVELINUS FONTINALIS	10 ML O2/HOUR	OXYGEN CONC = 3.4 MG O2/L,WEIGHT = 135 G, 23 CM	965
*SALVELINUS FONTINALIS	12.6 ML O2/HOUR	OXYGEN CONC = 5.0 MG O2/L,WEIGHT = 135 G, 23 CM	965
*SALVELINUS FONTINALIS	17.2 ML O2/HOUR	OXYGEN CONC = 6.4 MG O2/L,WEIGHT = 135 G, 23 CM	965
*SALVELINUS FONTINALIS	23.5 ML O2/HOUR	OXYGEN CONC = 9.3 MG O2/L,WEIGHT = 135 G, 23 CM	965
*SALVELINUS FONTINALIS	47.7 ML O2/HOUR	OXYGEN CONC = 3.6 MG O2/L,WEIGHT = 1000 G, 45 GM	965
*SALVELINUS FONTINALIS	67.4 ML O2/HOUR	OXYGEN CONC = 5.0 MG O2/L,WEIGHT = 1000 G, 45 GM	965
*SALVELINUS FONTINALIS	92.1 ML O2/HOUR	OXYGEN CONC = 6.4 MG O2/L,WEIGHT = 1000 G, 45 GM	965
*SALVELINUS FONTINALIS	102 ML O2/HOUR	OXYGEN CONC = 9.3 MG O2/L,WEIGHT = 1000 G, 45 GM	965
*SALVELINUS FONTINALIS	0.17 ML O2/HOUR	WEIGHT = 5 G, 8 CM, BASAL METABOLISM, OXYGEN NOT LIMITING, 278 K	965
*SALVELINUS FONTINALIS	0.36 ML O2/HOUR	WEIGHT = 5 G, 8 CM, BASAL METABOLISM, OXYGEN NOT LIMITING, 283 K	965
*SALVELINUS FONTINALIS	0.60 ML O2/HOUR	WEIGHT = 5 G, 8 CM, BASAL METABOLISM, OXYGEN NOT LIMITING, 288 K	965
*SALVELINUS FONTINALIS	0.68 ML O2/HOUR	WEIGHT = 5 G, 8 CM, BASAL METABOLISM, OXYGEN NOT LIMITING, 293 K	965
*SALVELINUS FONTINALIS	0.97 ML O2/HOUR	WEIGHT = 5 G, 8 CM, ACTIVE METABOLISM, OXYGEN NOT LIMITING, 278 K	965
*SALVELINUS FONTINALIS	1.38 ML O2/HOUR	WEIGHT = 5 G, 8 CM, ACTIVE METABOLISM, OXYGEN NOT LIMITING, 283 K	965
*SALVELINUS FONTINALIS	1.9 ML O2/HOUR	WEIGHT = 5 G, 8 CM, ACTIVE METABOLISM, OXYGEN NOT LIMITING, 288 K	965
*SALVELINUS FONTINALIS	2.8 ML O2/HOUR	WEIGHT = 5 G, 8 CM, ACTIVE METABOLISM, OXYGEN NOT LIMITING, 293 K	965
*SALVELINUS FONTINALIS	3.4 ML O2/HOUR	WEIGHT = 135 G,23 CM, BASAL METABOLISM, OXYGEN NOT LIMITING, 278 K	965
*SALVELINUS FONTINALIS	7.0 ML O2/HOUR	WEIGHT = 135 G,23 CM, BASAL METABOLISM, OXYGEN NOT LIMITING, 283 K	965
*SALVELINUS FONTINALIS	9.7 ML O2/HOUR	WEIGHT = 135 G,23 CM, BASAL METABOLISM, OXYGEN NOT LIMITING, 288 K	965
*SALVELINUS FONTINALIS	11.3 ML O2/HOUR	WEIGHT = 135 G,23 CM, BASAL METABOLISM, OXYGEN NOT LIMITING, 293 K	965
*SALVELINUS FONTINALIS	20.7 ML O2/HOUR	WEIGHT = 135 G,23 CM, ACTIVE METABOLISM, OXYGEN NOT LIMITING, 278 K	965
*SALVELINUS FONTINALIS	23.9 ML O2/HOUR	WEIGHT = 135 G,23 CM, ACTIVE METABOLISM, OXYGEN NOT LIMITING, 283 K	965
*SALVELINUS FONTINALIS	29.1 ML O2/HOUR	WEIGHT = 135 G,23 CM, ACTIVE METABOLISM, OXYGEN NOT LIMITING, 288 K	965
*SALVELINUS FONTINALIS	31.6 ML O2/HOUR	WEIGHT = 135 G,23 CM, ACTIVE METABOLISM, OXYGEN NOT LIMITING, 293 K	965

TABLE A222

FISH : RESPIRATION RATE			
ITEM:	VALUE:	CONDITIONS:	REF.:
*SALVELINUS FONTINALIS	18.3 ML O2/HOUR	WEIGHT = 1 KG, 45 GM, BASAL METABOLISM, OXYGEN NOT LIMITING, 278 K	965
*SALVELINUS FONTINALIS	37.0 ML O2/HOUR	WEIGHT = 1 KG, 45 GM, BASAL METABOLISM, OXYGEN NOT LIMITING, 283 K	965
*SALVELINUS FONTINALIS	48.9 ML O2/HOUR	WEIGHT = 1 KG, 45 GM, BASAL METABOLISM, OXYGEN NOT LIMITING, 288 K	965
*SALVELINUS FONTINALIS	55.2 ML O2/HOUR	WEIGHT = 1 KG, 45 GM, BASAL METABOLISM, OXYGEN NOT LIMITING, 293 K	965
*SALVELINUS FONTINALIS	127 ML O2/HOUR	WEIGHT = 1 KG, 45 GM, ACTIVE METABOLISM, OXYGEN NOT LIMITING, 278 K	965
*SALVELINUS FONTINALIS	124 ML O2/HOUR	WEIGHT = 1 KG, 45 GM, ACTIVE METABOLISM, OXYGEN NOT LIMITING, 283 K	965
*SALVELINUS FONTINALIS	158 ML O2/HOUR	WEIGHT = 1 KG, 45 GM, ACTIVE METABOLISM, OXYGEN NOT LIMITING, 288 K	965
*SALVELINUS FONTINALIS	149 ML O2/HOUR	WEIGHT = 1 KG, 45 GM, ACTIVE METABOLISM, OXYGEN NOT LIMITING, 293 K	965

TABLE A223

FISH : S/BIOMASS RATIO			
ITEM:	VALUE:	CONDITIONS:	REF.:
*PISCES	1.0 % OF DRY WEIGHT	MEAN ELEMENTARY COMPOSITION	157

TABLE A224

FISH : WET/DRY WEIGHT RATIO			
ITEM:	VALUE:	CONDITIONS:	REF.:
*FISH	5	NONE	1661

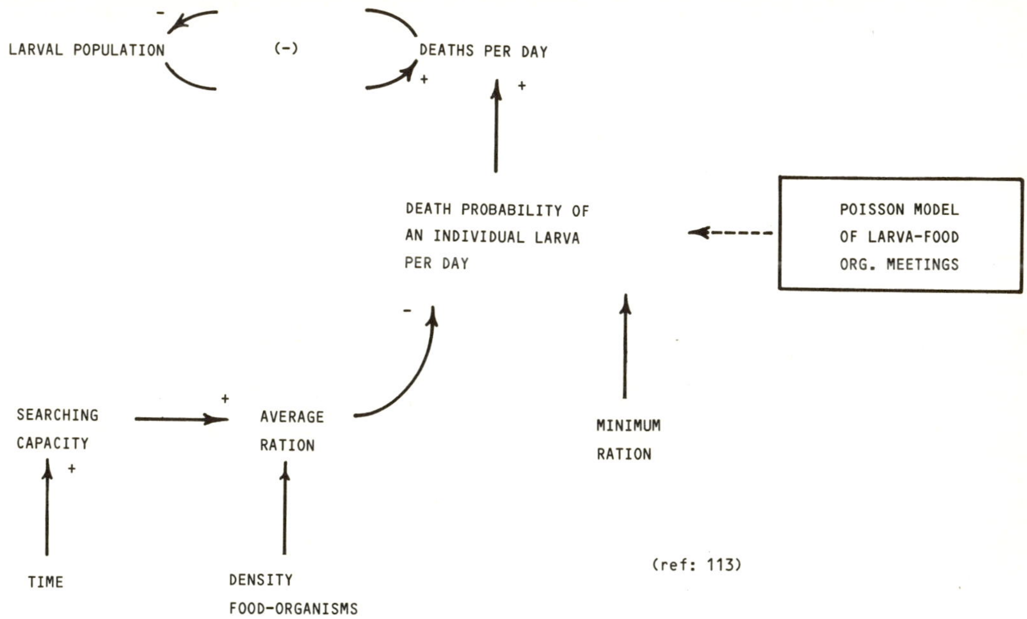

(ref: 113)

Fish model. The increasing ration model. A225

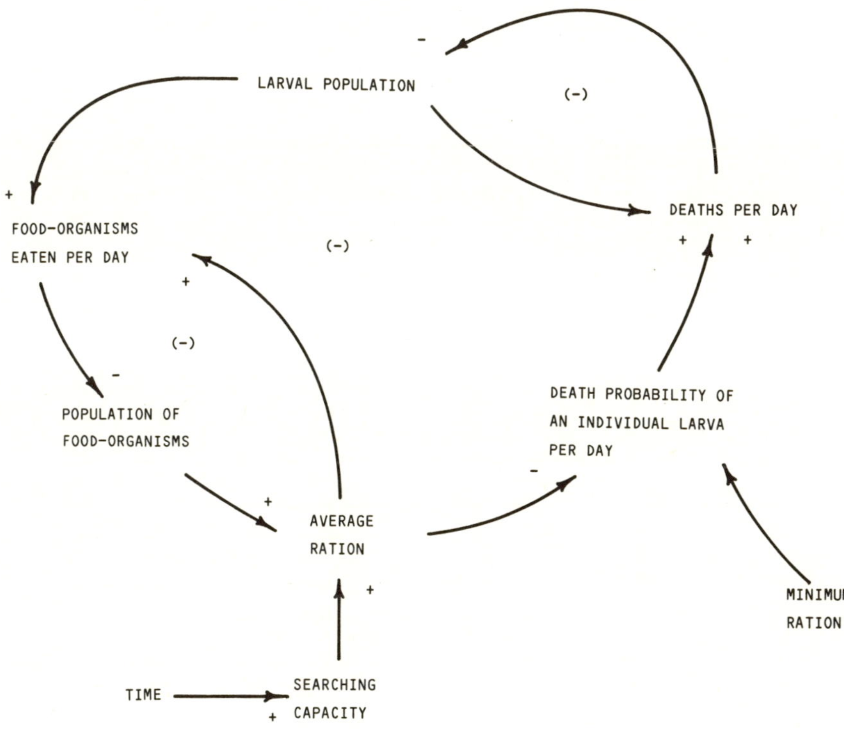

Fish model. Jones basic model.
The figure is composed of the increasing ration model and of the basis of the decreacing average ration model.

(ref: 113)

A226

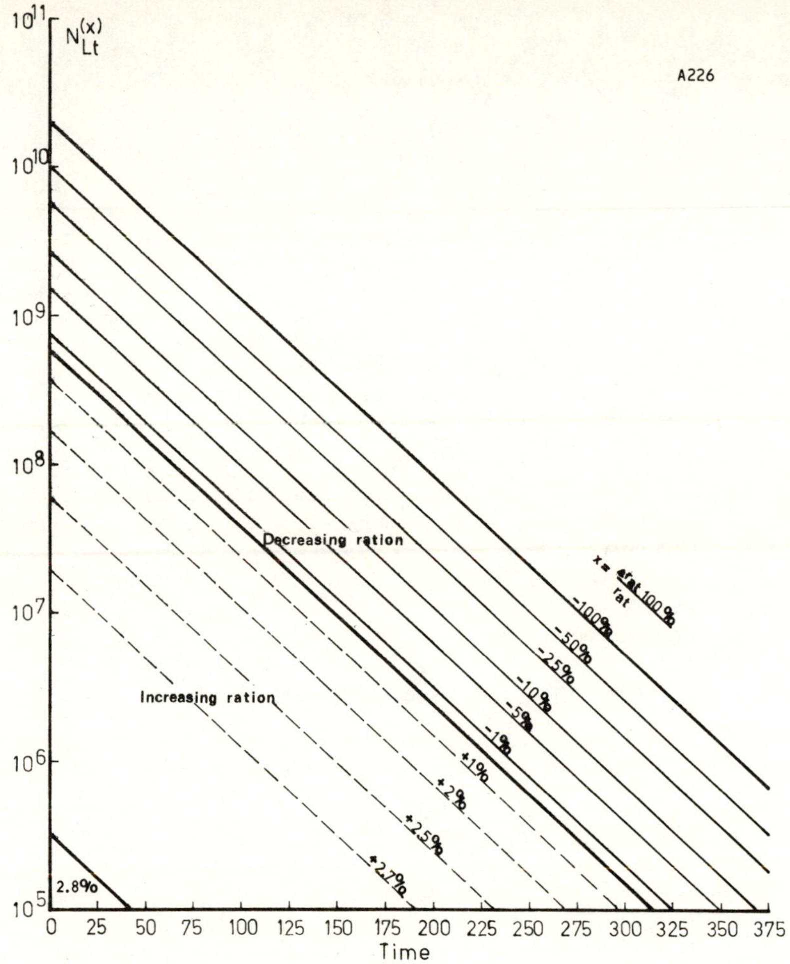

Growth of Fish population larvae, Nlt.
The lines decipts Nlt against time for various values of relative change in the average ration per day. The figures shows the ONE time step change in the ration as a function of Nlt.

(ref: 113)

A227

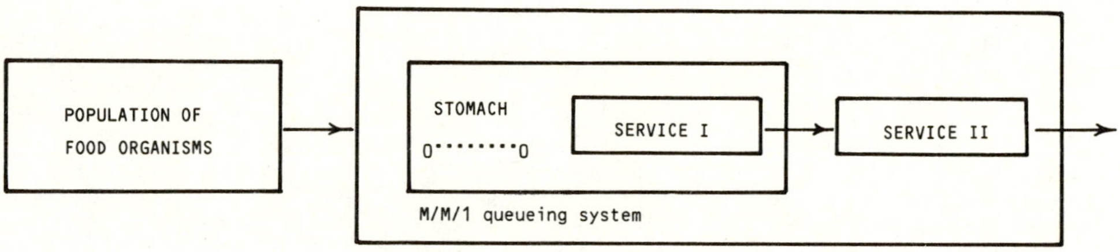

Fish model. Fish larva system.
The larva queueing system M/M/l in another biological interpretation. Service in the gut is considerd to take place in two series connected channals. Only service I is described in the queueing model. Only ONE food-organism can be served at a time in phase I, whereas several food-portions are served in phase II, which represent most of the gut.

A FISH LARVAE SYSTEM

(ref: 113)

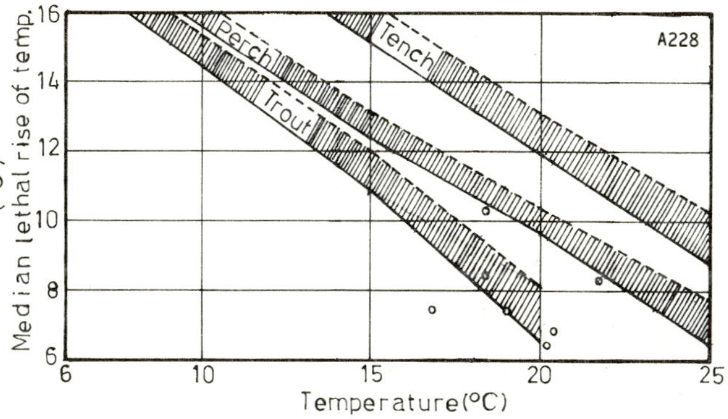

Median lethal rise of temperature for 3 species of fish versus acclimation temperature (= mean river temperature)
The mean lethal rise of temperature is understood as the mean difference of temperature between normal effluent (C degrees)
o mean temperature condition at generalizy stations ---- 100 min. medien
—— 1000 min. median

(ref: 215)

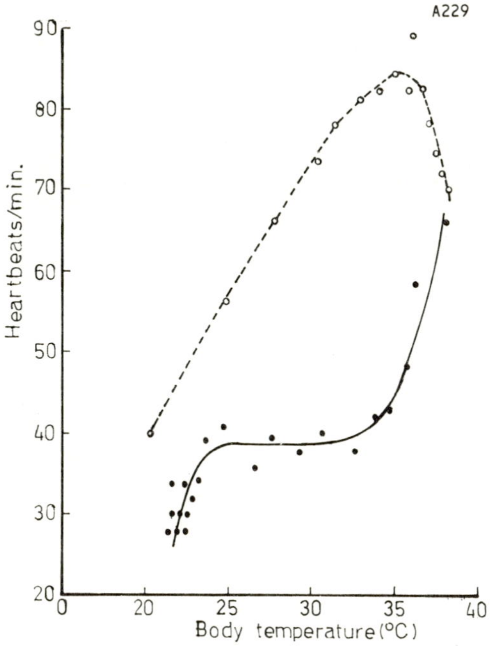

Relation of heart rate to body temperature in Galapapos marine ignana during heating o and cooling • . Weight 1.360 g.

(ref: 75)

Temperature adaption for Galapogos marine ignana.
T1 the difference between body and ambient temperature, is plotted to time (minutes), during heating (40 celcius) o and during cooling (20 celcius) •
water —— ,air ---

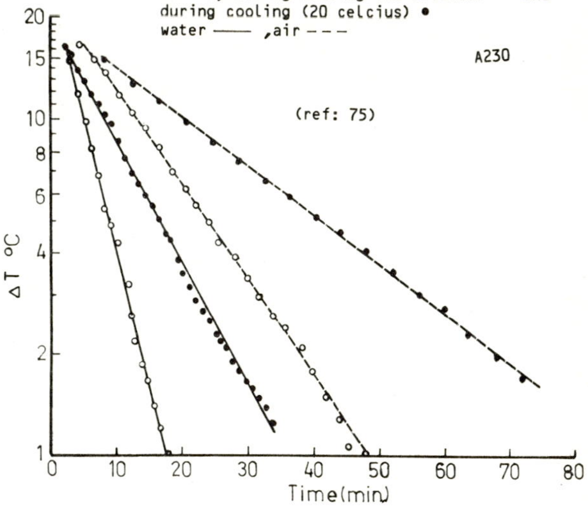

(ref: 75)

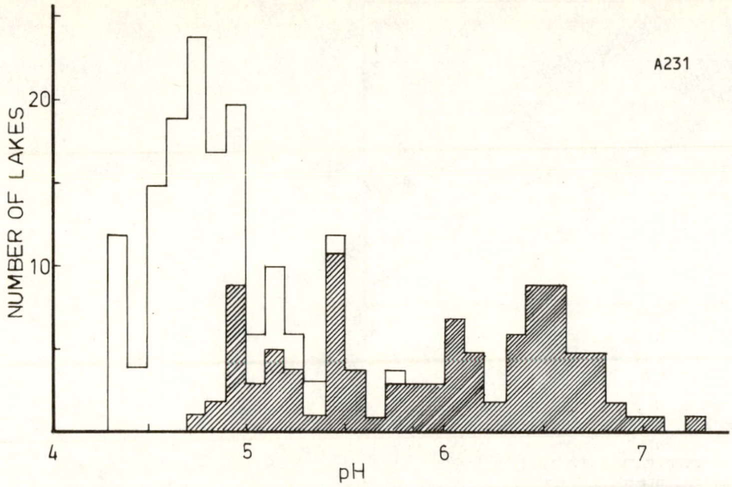

Effect of pH on fish populations. Determined by survey gill netting during the summer of 1975 in Adrirondack Mountain lakes. ☐ No fish present. ▨ Fish present. (ref: 1397)

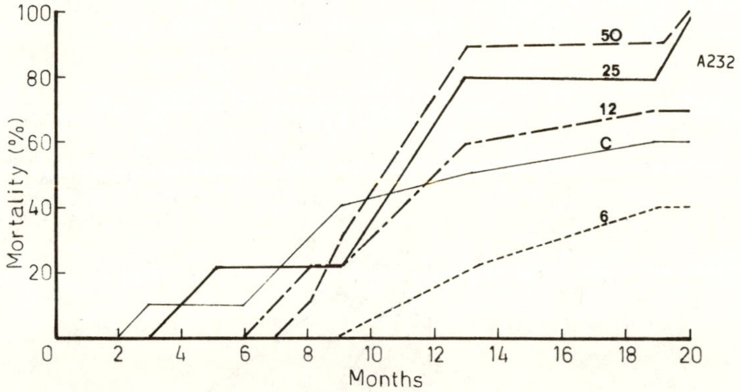

Mortality of brook trout by different concentrations of lime neutralized suspended iron. Juvenile brook trout during a 20 month exposure time. mg Fe/l : 0 - 50. (ref: 1389)

Noxious limits for fish (ref: 765) A233

Metal ion	Concentration (mg/l)
Hg^{2+}	0.1-0.9
Zn^{2+}	2
Cd^{2+}	3-20
Ni^{2+}	15-55
CN^-	0.03-0.25
CrO_4^{2-}	15-80
Cr^{3+}	15-80
Cu^{2+}	0.08-0.8

A5

FUNGI

COMPOSITION
RESPIRATION

TABLE A234

FUNGI : ASH/BIOMASS RATIO			
ITEM:	VALUE:	CONDITIONS:	REF.:
*FUNGI	6.32 % OF DRY WEIGHT	11 SPECIES	81

TABLE A235

FUNGI : C/BIOMASS RATIO			
ITEM:	VALUE:	CONDITIONS:	REF.:
*FUNGI	49.4 % OF DRY WEIGHT	MEAN ELEMENTARY COMPOSITION	157

TABLE A236

FUNGI : GROWTH RATE			
ITEM:	VALUE:	CONDITIONS:	REF.:
*BASISPORIUM GALLARUM	0.45 DOUBLINGS/DAY	298K,LIQUID SHAKEN CULTURE	175
*MONILINIA FRUCTICOLA	0.8 DOUBLINGS/DAY	298K,LIQUID SHAKEN CULTURE	175
*SACCHAROMYCES CEREVISIAE	12 DOUBLINGS/DAY	303K,LIQUID SHAKEN CULTURE	175

TABLE A237

FUNGI : H/BIOMASS RATIO			
ITEM:	VALUE:	CONDITIONS:	REF.:
*FUNGI	5.5 % OF DRY WEIGHT	MEAN ELEMENTARY COMPOSITION, INTERPOLATED	157

TABLE A238

FUNGI : N/BIOMASS RATIO			
ITEM:	VALUE:	CONDITIONS:	REF.:
*FUNGI	5.1 % OF DRY WEIGHT	MEAN ELEMENTARY COMPOSITION	157

TABLE A239

FUNGI : O/BIOMASS RATIO			
ITEM:	VALUE:	CONDITIONS:	REF.:
*FUNGI	34.0 % OF DRY WEIGHT	MEAN ELEMENTARY COMPOSITION	157

TABLE A240

FUNGI : RESPIRATION RATE			
ITEM:	VALUE:	CONDITIONS:	REF.:
*FUNGI	-	GRASSLAND SOIL,293K,EQUIVALENT TO 161GCAL/G*DAY AT 289K	963

TABLE A240

FUNGI : RESPIRATION RATE			
ITEM:	VALUE:	CONDITIONS:	REF.:
*NEUROSPORA CRASSA	6.4 ML O2/(G WET WT.)*HR	299K	524
*SACCHAROMYCES CEREVISIAE	8-14.5 ML O2/(G WET WT.)*HR	299K	524

TABLE A241

FUNGI : S/BIOMASS RATIO			
ITEM:	VALUE:	CONDITIONS:	REF.:
*FUNGI	0.4 % OF DRY WEIGHT	MEAN ELEMENTARY COMPOSITION, INTERPOLATED	157

TABLE A242

FUNGI : WET WEIGHT			
ITEM:	VALUE:	CONDITIONS:	REF.:
*SACCHAROMYCES CEREVISIAE	7.1*10**(-11) G	NONE	175

A6

INVERTEBRATES

COMPOSITION
DEVELOPMENT TIME
EGESTION
FILTRATION RATE
GROWTH RATE
HALF SATURATION CONSTANTS
MORTALITY
PRODUCTION
RESPIRATION

TABLE A243

INVERTEBRATES : ASH/BIOMASS RATIO			
ITEM:	VALUE:	CONDITIONS:	REF.:
*ACHETA DOMESTICUS	5.7 ASH % OF DRY WEIGHT	NONE	390
*AGRYPNIA OBSOLETA	4.72 (+-0.02) % OF DRY WEIGHT	TRICHOPTERA, 95 % CONF LIMIT, 12 H AT 873 K	1375
*ASELLUS AQUATICUS	28.62 (+-0.22) % OF DRY WEIGHT	ISOPODA, 95 % CONF LIMIT, 12 H AT 873 K	1375
*BRACHORIA INITIALIS	54.9 ASH % OF DRY WEIGHT	NONE	390
*CAMPONOTUS AMERICANUS	2.8 ASH % OF DRY WEIGHT	NONE	390
*CENTROPTILUM LUTEOLUM	8.14 (+-0.17) % OF DRY WEIGHT	EPHEMEROPTERA, 95 % CONF LIMIT, 12 H AT 873 K	1375
*CHAOBORUS FLAVICANS	6.93 (+-0.04) % OF DRY WEIGHT	DIPTERA, 95 % CONF LIMIT, 12 H AT 873 K	1375
*CHIRONOMUS PLUMOSUS	5.16 (+-0.11) % OF DRY WEIGHT	DIPTERA, 95 % CONF LIMIT, 12 H AT 873 K	1375
*CORDULIA AENEA	4.52 (+-0.07) % OF DRY WEIGHT	ODONATA, 95 % CONF LIMIT, 12 H AT 873 K	1375
*CYLISTICUS CONVEXUS	45.2 ASH % OF DRY WEIGHT	NONE	1309
*DRASSYLLUS VIRGINIANA	3.8 ASH % OF DRY WEIGHT	NONE	1072
*ERPOBDELLA OCTOCULATA	4.47 (+-0.08) % OF DRY WEIGHT	HIRUDINEA , 95 % CONF LIMIT, 12 H AT 873 K	1375
*EVATHRUS SODALIS	3.2 ASH % OF DRY WEIGHT	NONE	390
*GAMMARAQANTUS LACUSTRIS	20.69 (+-0.21) % OF DRY WEIGHT	AMPHIPODA, 95 % CONF LIMIT, 12 H AT 873 K	1375
*HEPTAGENIA FUSCOGRISEA	5.39 (+-0.13) % OF DRY WEIGHT	EPHEMEROPTERA, 95 % CONF LIMIT, 12 H AT 873 K	1375
*LEPTOPHLEBIA SP.	6.24 (+-0.07) % OF DRY WEIGHT	EPHEMEROPTERA, 95 % CONF LIMIT, 12 H AT 873 K	1375
*LIMNEPHILIDAE SP.	9.02 (+-0.12) % OF DRY WEIGHT	TRICHOPTERA, 95 % CONF LIMIT, 12 H AT 873 K	1375
*LUMBRICUS TERRESTRIS	23.0 ASH % OF DRY WEIGHT	NONE	485
*MESODON THYROIDES	22.8 ASH % OF DRY WEIGHT	NONE	390
*MYSIS RELICTA	8.61 (+-0.02) % OF DRY WEIGHT	MYSIDACEA, 95 % CONF LIMIT, 12 H AT 873 K	1375
*NOTONECTA GLAUCA	3.86 (+-0.08) % OF DRY WEIGHT	HETERPOTERA, 95 % CONF LIMIT, 12 H AT 873 K	1375
*OTOCRYPTOPS SEXSPINOSUS	7.2 ASH % OF DRY WEIGHT	NONE	390
*PALLACEA QUADRISPINOSA	30.37 (+-0.76) % OF DRY WEIGHT	AMPHIPODA, 95 % CONF LIMIT, 12 H AT 873 K	1375
*PARCOBLATTA SP.	4.0 ASH % OF DRY WEIGHT	NONE	1072
*SIALIS SP.	4.28 (+-0.04) % OF DRY WEIGHT	MEGALOPTERA, 95 % CONF LIMIT, 12 H AT 873 K	1375
*TOMOCERUS SP.	8.7 ASH % OF DRY WEIGHT	NONE	1072
*ZOOINVERTEBRATES AND ZOOPLANKTON	0.9	NONE	1661

TABLE A244

INVERTEBRATES : ASSIMILATION			
ITEM:	VALUE:	CONDITIONS:	REF.:
*ASELLUS AQUATICUS	30.3 % OF INGESTION	NONE	1689
*BLOOD WORM	60 % OF HG INTAKE	FOOD = ALGAE-DETRITUS	355
*BLOOD WORM	60 % OF HG INTAKE	FOOD = ALGAE-DETRITUS	355
*DIPLACODES TRIVIALIS	91 % OF INGESTION	46 MG ANIMAL, INGESTION = 61.5 CAL/(KCAL*ANIM.*D)	1209
*LANICE CONCHILEGA	70.6 - 77.2 %	BASED ON CALORIES	193
*LESTES SPONSA	36.6 % OF INGESTION	NONE	1689
*MEGANYCTIPHANES NORVEGICA	0.1 - 0.2 % OF INGESTED ZN	DIET = ARTEMIA	1742
*PHILAENUS SPUMARIUS	30 - 58 % OF INGESTION	3RD INSTAR, 0.39 MG, 296 K	1705

TABLE A244

INVERTEBRATES : ASSIMILATION			
ITEM:	VALUE:	CONDITIONS:	REF.:
*PHILAENUS SPUMARIUS	30 - 57 % OF INGESTION	4RD INSTAR, 1.16 MG, 298 K	1705

TABLE A245

INVERTEBRATES : C/BIOMASS RATIO			
ITEM:	VALUE:	CONDITIONS:	REF.:
*AGRYPNIA OBSOLETA	49.7 (+-1.3) % OF ASH FREE DRY WEIGHT	TRICHOPTERA, ORGANIC CARBON, 95 % CONF LIMIT	1375
*AGRYPNIA OBSOLETA	47.3 (+-1.4) % OF DRY WEIGHT	TRICHOPTERA, 95 % CONF LIMIT, TOTAL C	1375
*ANNELIDA	40.2 % OF DRY WEIGHT	MEAN ELEMENTARY COMPOSITION	157
*ARTHROPODA	44.6 % OF DRY WEIGHT	MEAN ELEMENTARY COMPOSITION,LAND	157
*ASELLUS AQUATICUS	44.7 (+-1.1) % OF ASH FREE DRY WEIGHT	ISOPODA, ORGANIC CARBON, 95 % CONF LIMIT	1375
*ASELLUS AQUATICUS	34.3 (+-0.9) % OF DRY WEIGHT	ISOPODA, 95 % CONF LIMIT, TOTAL C	1375
*BENTHOS	0.4 BY WEIGHT	USED IN MODELS, N FORK KINGS RIVER, CALIF.	1566
*BENTHOS	0.5 BY WEIGHT	USED IN MODELS, BOISE RIVER, IDAHO	241
*BENTHOS	0.4 BY WEIGHT	USED IN MODELS, N FORK KINGS RIVER, CALIF.	1566
*BENTHOS	0.5 BY WEIGHT	USED IN MODELS, BOISE RIVER, IDAHO	241
*BENTHOS	.4-.5	NONE	1571
*CENTROPTILUM LUTEOLUM	54.1 (+-1.9) % OF ASH FREE DRY WEIGHT	EPHEMEROPTERA, ORGANIC CARBON, 95 % CONF LIMIT	1375
*CENTROPTILUM LUTEOLUM	49.7 (+-1.8) % OF DRY WEIGHT	EPHEMEROPTERA, 95 % CONF LIMIT, TOTAL C	1375
*CHAOBORUS FLAVICANS	50.8 (+-(.7) % OF ASH FREE DRY WEIGHT	DIPTERA, ORGANIC CARBON, 95 % CONF LIMIT	1375
*CHAOBORUS FLAVICANS	47.3 (+-0.7) % OF DRY WEIGHT	DIPTERA, 95 % CONF LIMIT, TOTAL C	1375
*CHIRONOMUS PLUMOSUS	47.5 (+-0.9) % OF ASH FREE DRY WEIGHT	DIPTERA, ORGANIC CARBON, 95 % CONF LIMIT	1375
*CHIRONOMUS PLUMOSUS	45.1 (+-0.8) % OF DRY WEIGHT	DIPTERA, 95 % CONF LIMIT, TOTAL C	1375
*COELENTERATA	43.6 % OF DRY WEIGHT	MEAN ELEMENTARY COMPOSITION	157
*CORDULIA AENEA	49.6 (+-0.8) % OF ASH FREE DRY WEIGHT	ODONATA, ORGANIC CARBON, 95 % CONF LIMIT	1375
*CORDULIA AENEA	47.4 (+-0.8) % OF DRY WEIGHT	ODONATA, 95 % CONF LIMIT, TOTAL C	1375
*ECHINODERMATA	40.0 % OF DRY WEIGHT	MEAN ELEMENTARY COMPOSITION, INTERPOLATED	157
*ERPOBDELLA OCTOCULATA	50.6 (+-2.4) % OF ASH FREE DRY WEIGHT	HIRUDINEA , ORGANIC CARBON, 95 % CONF LIMIT	1375
*ERPOBDELLA OCTOCULATA	48.3 (+-2.3) % OF DRY WEIGHT	HIRUDINEA , 95 % CONF LIMIT, TOTAL C	1375
*GAMMARAQANTUS LACUSTRIS	51.1 (+-0.9) % OF ASH FREE DRY WEIGHT	AMPHIPODA, ORGANIC CARBON, 95 % CONF LIMIT	1375
*GAMMARAQANTUS LACUSTRIS	41.7 (+-0.8) % OF DRY WEIGHT	AMPHIPODA, 95 % CONF LIMIT, TOTAL C	1375
*HEPTAGENIA FUSCOGRISEA	55.1 (+-1.2) % OF ASH FREE DRY WEIGHT	EPHEMEROPTERA, ORGANIC CARBON, 95 % CONF LIMIT	1375
*HEPTAGENIA FUSCOGRISEA	52.2 (+-1.2) % OF DRY WEIGHT	EPHEMEROPTERA, 95 % CONF LIMIT, TOTAL C	1375
*INVERTEBRATE	51.8 (+-4.0) % OF ASH FREE DRY WEIGHT	STANDARD DERIVATION, MEAN OF 19 SPECIES	1375
*LEPTOPHLEBIA SP.	52.6 (+-1.3) % OF ASH FREE DRY WEIGHT	EPHEMEROPTERA, ORGANIC CARBON, 95 % CONF LIMIT	1375
*LEPTOPHLEBIA SP.	49.3 (+-1.2) % OF DRY WEIGHT	EPHEMEROPTERA, 95 % CONF LIMIT, TOTAL C	1375
*LIMNEPHILIDAE SP.	51.0 (+-0.8) % OF ASH FREE DRY WEIGHT	TRICHOPTERA, ORGANIC CARBON, 95 % CONF LIMIT	1375
*LIMNEPHILIDAE SP.	46.4 (+-0.7) % OF DRY WEIGHT	TRICHOPTERA, 95 % CONF LIMIT, TOTAL C	1375

TABLE A245

INVERTEBRATES : C/BIOMASS RATIO

ITEM:	VALUE:	CONDITIONS:	REF.:
*MOLLUSCA	39.9 % OF DRY WEIGHT	MEAN ELEMENTARY COMPOSITION,SOFT PARTS	157
*MYSIS RELICTA	54.6 (+-1.1) % OF ASH FREE DRY WEIGHT	MYSIDACEA, ORGANIC CARBON, 95 % CONF LIMIT	1375
*MYSIS RELICTA	50.0 (+-1.0) % OF DRY WEIGHT	MYSIDACEA, 95 % CONF LIMIT, TOTAL C	1375
*NEMATODA	40.0 % OF DRY WEIGHT	MEAN ELEMENTARY COMPOSITION, INTERPOLATED	157
*NOTONECTA GLAUCA	52.0 (+-1.0) % OF ASH FREE DRY WEIGHT	HETERPOTERA, ORGANIC CARBON, 95 % CONF LIMIT	1375
*NOTONECTA GLAUCA	50.0 (+-0.9) % OF DRY WEIGHT	HETERPOTERA, 95 % CONF LIMIT, TOTAL C	1375
*PALLACEA QUADRISPINOSA	47.0 (+-1.7) % OF ASH FREE DRY WEIGHT	AMPHIPODA, ORGANIC CARBON, 95 % CONF LIMIT	1375
*PALLACEA QUADRISPINOSA	35.4 (+-1.2) % OF DRY WEIGHT	AMPHIPODA, 95 % CONF LIMIT, TOTAL C	1375
*SIALIS SP.	51.1 (+-0.6) % OF ASH FREE DRY WEIGHT	MEGALOPTERA, ORGANIC CARBON, 95 % CONF LIMIT	1375
*SIALIS SP.	49.2 (+-0.6) % OF DRY WEIGHT	MEGALOPTERA, 95 % CONF LIMIT, TOTAL C	1375
*ZOOINVERTEBRATES AND ZOOPLANKTON	0.5 G C/GRAM DRY WEIGHT	NONE	1661

TABLE A246

INVERTEBRATES : CA/BIOMASS RATIO

ITEM:	VALUE:	CONDITIONS:	REF.:
*ARCTIIDAE	4.7 MG/(G ASH FREE DRY WT)	SOIL OR LITTER INVERT.	1311
*CAMPONOTUS PENNSYLVANICUS	2.9 MG/(G ASH FREE DRY WT)	SOIL OR LITTER INVERT.	1311
*CEUTHOPHILUS GRACILIPES	2.1 MG/(G ASH FREE DRY WT)	SOIL OR LITTER INVERT.	1311
*DIXIDESMUS ERASUS	294.9 MG/(G ASH FREE DRY WT)	SOIL OR LITTER INVERT.	1311
*GEOTRUPES SP.	0.8 MG/(G ASH FREE DRY WT)	SOIL OR LITTER INVERT.	1311
*LEIOBUNUM FLAVUM	7.2 MG/(G ASH FREE DRY WT)	SOIL OR LITTER INVERT.	1311
*LIGIDIUM BLUERIDGENSIS	108.9 MG/(G ASH FREE DRY WT)	SOIL OR LITTER INVERT.	1311
*MESOMPHIX VULGATUS	16.7 MG/(G ASH FREE DRY WT)	SOIL OR LITTER INVERT.	1311
*OTOCRYPTOPS SEXSPINOSUS	0.6 MG/(G ASH FREE DRY WT)	SOIL OR LITTER INVERT.	1311
*PACHYLOMERIDES AUDOUINI	2.1 MG/(G ASH FREE DRY WT)	SOIL OR LITTER INVERT.	1311
*SCIOMYZIDAE	2.0 MG/(G ASH FREE DRY WT)	SOIL OR LITTER INVERT.	1311
*SCYTONOTUS GRANULATUS	103.8 MG/(G ASH FREE DRY WT)	SOIL OR LITTER INVERT.	1311
*SPHAERODERUS STENOSTOMUS	0.5 MG/(G ASH FREE DRY WEIGHT)	SOIL OR LITTER INVERT.	1311

TABLE A247

INVERTEBRATES : CALORIC CONTENT

ITEM:	VALUE:	CONDITIONS:	REF.:
*ACHETA DOMESTICUS	5896 G CAL/(G ASH FREE DRY WT)	ASH = 5.7 % OF DRY WEIGHT	390
*AESCHNA CYANEA	5432 CAL/G DRY WEIGHT	GERMANY	235
*AESCHNA CYANEA	5837 CAL/G ASH FREE DRY WEIGHT	GERMANY	235
*AGRYPNIA OBSOLETA	21.5 (+-0.4) KJ/G ASH FREE DRY WEIGHT	TRICHOPTERA, 95 % CONF LIM, NO CORRECTION FOR NITROGEN	1375
*AGRYPNIA OBSOLETA	19.6 (+-1.1) KJ/G ASH FREE DRY WEIGHT	TRICHOPTERA, 95 % CONF LIM, CORRECTION FOR NITROGEN	1375
*AGRYPNIA OBSOLETA	43.3 (+-1.3) KJ/G ORGANIC CARBON	TRICHOPTERA, 95 % CONF LIM, NO CORECTION FOR NITROGEN	1375

TABLE A247

INVERTEBRATES : CALORIC CONTENT

ITEM:	VALUE:	CONDITIONS:	REF.:
*AGRYPNIA OBSOLETA	39.2 (+-2.9) KJ/G ORGANIC CARBON	TRICHOPTERA, 95 % CONF LIM, CORRECTION FOR NITROGEN	1375
*APHRODITA HASTATA	2638 (1914-3407) GCAL/G DRY WT.	INCL.ASH,WATER=80+-4%,20ANIMALS, 6SAMPLES,ALL SEASONS	1619
*ARACHNIDAE	5.820 KCAL/G ASH FREE DRY WEIGHT	NONE	521
*ARACHNIDAE	5.620 KCAL/G ASH FREE DRY WEIGHT	NONE	1447
*ARCTICA ISLANDICA	4031 (3684-4276) GCAL/G DRY WT.	SHELL-FREE,INCL.ASH,WATER=86+-5%, 72ANIMALS,11SAMPLES,ALL SEASONS	1619
*ASELLUS AQUATICUS	19.9 (+-0.6) KJ/G ASH FREE DRY WEIGHT	ISOPODA, 95 % CONF LIM, NO CORRECTION FOR NITROGEN	1375
*ASELLUS AQUATICUS	17.6 (+-0.6) KJ/G ASH FREE DRY WEIGHT	ISOPODA, 95 % CONF LIM, CORRECTION FOR NITROGEN	1375
*ASELLUS AQUATICUS	44.6 (+-2.5) KJ/G ORGANIC CARBON	ISOPODA, 95 % CONF LIM, NO CORECTION FOR NITROGEN	1375
*ASELLUS AQUATICUS	39.3 (+-1.9) KJ/G ORGANIC CARBON	ISOPODA, 95 % CONF LIM, CORRECTION FOR NITROGEN	1375
*ASELLUS AQUATICUS	3493 CAL/G DRY WEIGHT	GERMANY	235
*ASELLUS AQUATICUS	5521 CAL/G ASH FREE DRY WEIGHT	GERMANY	235
*ASTARTE UNDATA	4234 (3734-4521) GCAL/G DRY WT.	SHELL-FREE,INCL.ASH,WATER=87+-3%, 154ANIMALS,10SAMPLES,ALL SEASONS	1619
*BAETIS RHODANI	6044 CAL/G DRY WEIGHT	GERMANY	235
*BRACHORIA INITIALIS	5453 G CAL/(G ASH FREE DRY WT)	ASH = 7.2 % OF DRY WEIGHT	390
*CAMPONOTUS AMERICANUS	6247 G CAL/(G ASH FREE DRY WT)	ASH = 2.8 % OF DRY WEIGHT	390
*CENTROPTILUM LUTEOLUM	26.1 (+-0.4) KJ/G ASH FREE DRY WEIGHT	EPHEMEROPTERA, 95 % CONF LIM, NO CORRECTION FOR NITROGEN	1375
*CENTROPTILUM LUTEOLUM	23.7 (+-0.5) KJ/G ASH FREE DRY WEIGHT	EPHEMEROPTERA, 95 % CONF LIM, CORRECTION FOR NITROGEN	1375
*CENTROPTILUM LUTEOLUM	48.4 (+-1.7) KJ/G ORGANIC CARBON	EPHEMEROPTERA, 95 % CONF LIM, NO CORECTION FOR NITROGEN	1375
*CENTROPTILUM LUTEOLUM	43.8 (+-1.2) KJ/G ORGANIC CARBON	EPHEMEROPTERA, 95 % CONF LIM, CORRECTION FOR NITROGEN	1375
*CHAEPTOPSIS SP.	5.520 KCAL/G ASH FREE DRY WEIGHT	NONE	791
*CHAOBORUS FLAVICANS	22.7 (+-0.3) KJ/G ASH FREE DRY WEIGHT	DIPTERA, 95 % CONF LIM, NO CORRECTION FOR NITROGEN	1375
*CHAOBORUS FLAVICANS	20.4 (+-0.4) KJ/G ASH FREE DRY WEIGHT	DIPTERA, 95 % CONF LIM, CORRECTION FOR NITROGEN	1375
*CHAOBORUS FLAVICANS	44.7 (+-0.8) KJ/G ORGANIC CARBON	DIPTERA, 95 % CONF LIM, NO CORECTION FOR NITROGEN	1375
*CHAOBORUS FLAVICANS	40.1 (+-0.6) KJ/G ORGANIC CARBON	DIPTERA, 95 % CONF LIM, CORRECTION FOR NITROGEN	1375
*CHIRONOMUS PLUMOSUS	20.8 (+-2.2) KJ/G ASH FREE DRY WEIGHT	DIPTERA, 95 % CONF LIM, NO CORRECTION FOR NITROGEN	1375
*CHIRONOMUS PLUMOSUS	18.6 (+-2.0) KJ/G ASH FREE DRY WEIGHT	DIPTERA, 95 % CONF LIM, CORRECTION FOR NITROGEN	1375
*CHIRONOMUS PLUMOSUS	43.7 (+-4.6) KJ/G ORGANIC CARBON	DIPTERA, 95 % CONF LIM, NO CORECTION FOR NITROGEN	1375
*CHIRONOMUS PLUMOSUS	39.1 (+-4.1) KJ/G ORGANIC CARBON	DIPTERA, 95 % CONF LIM, CORRECTION FOR NITROGEN	1375
*COLUS STIMPSONI	4587 GCAL/G DRY WT.	SHELL-FREE,INCL.ASH,WATER=68%, 4ANIMALS,1SAMPLE,JULY TO SEPT.	1619
*CORDULIA AENEA	22.4 (+-0.8) KJ/G ASH FREE DRY WEIGHT	ODONATA, 95 % CONF LIM, NO CORRECTION FOR NITROGEN	1375
*CORDULIA AENEA	20.2 (+-0.9) KJ/G ASH FREE DRY WEIGHT	ODONATA, 95 % CONF LIM, CORRECTION FOR NITROGEN	1375
*CORDULIA AENEA	45.2 (+-1.2) KJ/G ORGANIC CARBON	ODONATA, 95 % CONF LIM, NO CORECTION FOR NITROGEN	1375
*CORDULIA AENEA	40.7 (+-2.4) KJ/G ORGANIC CARBON	ODONATA, 95 % CONF LIM, CORRECTION FOR NITROGEN	1375
*CRANGON SEPTEMSPINOSA	4272 (4087-4345) GCAL/G DRY WT.	INCL.ASH,WATER=74+-2%,43ANIMALS, 4SAMPLES,JAN.TO AUG.	1619
*CRANGON SEPTEMSPINOSA	4088 GCAL/G DRY WT.	INCL.ASH,WATER=76%,15ANIMALS, 1SAMPLE,NOVEMBER	1619
*CREMATOGASTER CLARA	6.130 KCAL/G ASH FREE DRY WEIGHT	NONE	791
*CRENOBIA ALPINA	6055 CAL/G DRY WEIGHT	GERMANY	235

TABLE A247

INVERTEBRATES : CALORIC CONTENT

ITEM:	VALUE:	CONDITIONS:	REF.:
*CRENOBIA ALPINA	6426 CAL/G ASH FREE DRY WEIGHT	GERMANY	235
*CYLISTICUS CONVEXUS	6123 (SE=+-9.7) G CAL/(G ASH FREE DRY WT)	ASH = 45.2 % OF DRY WEIGHT	1309
*DENTALIUM ENTALE	4560 GCAL/G DRY WT.	SHELL-FREE,INCL.ASH,WATER=73%, 100ANIMALS,1SAMPLE,APRIL	1619
*DRASSYLLUS VIRGINIANA	5794 (SE=+-89.4) G CAL/(G ASH FREE DRY WT)	ASH = 3.8 % OF DRY WEIGHT	1072
*DUGESIA GONOCEPHALA	6043 CAL/G DRY WEIGHT	GERMANY	235
*DUGESIA GONOCEPHALA	6453 CAL/G ASH FREE DRY WEIGHT	GERMANY	235
*ERPOBDELLA OCTOCULATA	22.1 (+-0.6) KJ/G ASH FREE DRY WEIGHT	HIRUDINEA , 95 % CONF LIM, NO CORRECTION FOR NITROGEN	1375
*ERPOBDELLA OCTOCULATA	19.8 (+-0.6) KJ/G ASH FREE DRY WEIGHT	HIRUDINEA , 95 % CONF LIM, CORRECTION FOR NITROGEN	1375
*ERPOBDELLA OCTOCULATA	43.9 (+-2.5) KJ/G ORGANIC CARBON	HIRUDINEA , 95 % CONF LIM, NO CORECTION FOR NITROGEN	1375
*ERPOBDELLA OCTOCULATA	40.8 (+-2.8) KJ/G ORGANIC CARBON	HIRUDINEA , 95 % CONF LIM, CORRECTION FOR NITROGEN	1375
*ERPOBDELLA OCTOCULATA	5607 CAL/G DRY WEIGHT	GERMANY	235
*ERPOBDELLA OCTOCULATA	5995 CAL/G ASH FREE DRY WEIGHT	GERMANY	235
*EUSIMULIUM SP.	4510 CAL/G DRY WEIGHT	GERMANY	235
*EUSIMULIUM SP.	5562 CAL/G ASH FREE DRY WEIGHT	GERMANY	235
*EVATHRUS SODALIS	5672 G CAL/(G ASH FREE DRY WT)	ASH = 3.2 % OF DRY WEIGHT	390
*GAMMARAQANTUS LACUSTRIS	24.1 (+-0.2) KJ/G ASH FREE DRY WEIGHT	AMPHIPODA, 95 % CONF LIM, NO CORRECTION FOR NITROGEN	1375
*GAMMARAQANTUS LACUSTRIS	21.7 (+-0.2) KJ/G ASH FREE DRY WEIGHT	AMPHIPODA, 95 % CONF LIM, CORRECTION FOR NITROGEN	1375
*GAMMARAQANTUS LACUSTRIS	47.1 (+-0.8) KJ/G ORGANIC CARBON	AMPHIPODA, 95 % CONF LIM, NO CORECTION FOR NITROGEN	1375
*GAMMARAQANTUS LACUSTRIS	45.5 (+-1.1) KJ/G ORGANIC CARBON	AMPHIPODA, 95 % CONF LIM, CORRECTION FOR NITROGEN	1375
*GAMMARUS FOSSARUM	3735 CAL/G DRY WEIGHT	GERMANY, 2 - 5 MM	235
*GAMMARUS FOSSARUM	3896 CAL/G DRY WEIGHT	GERMANY, 6 - 9 MM	235
*GAMMARUS FOSSARUM	3578 CAL/G DRY WEIGHT	GERMANY, 10 - 13 MM	235
*GAMMARUS FOSSARUM	5452 CAL/G ASH FREE DRY WEIGHT	GERMANY, 2 - 5 MM	235
*GAMMARUS FOSSARUM	5354 CAL/G ASH FREE DRY WEIGHT	GERMANY, 6 - 9 MM	235
*GAMMARUS FOSSARUM	5201 CAL/G ASH FREE DRY WEIGHT	GERMANY, 10 - 13 MM	235
*GLOSSIPHONIA COMPLANATA	5544 CAL/G DRY WEIGHT	GERMANY	235
*GLOSSIPHONIA COMPLANATA	5713 CAL/G ASH FREE DRY WEIGHT	GERMANY	235
*HEPTAGENIA FUSCOGRISEA	26.2 (+-0.6) KJ/G ASH FREE DRY WEIGHT	EPHEMEROPTERA, 95 % CONF LIM, NO CORRECTION FOR NITROGEN	1375
*HEPTAGENIA FUSCOGRISEA	24.1 (+-0.5) KJ/G ASH FREE DRY WEIGHT	EPHEMEROPTERA, 95 % CONF LIM, CORRECTION FOR NITROGEN	1375
*HEPTAGENIA FUSCOGRISEA	47.5 (+-1.7) KJ/G ORGANIC CARBON	EPHEMEROPTERA, 95 % CONF LIM, NO CORECTION FOR NITROGEN	1375
*HEPTAGENIA FUSCOGRISEA	43.6 (+-1.3) KJ/G ORGANIC CARBON	EPHEMEROPTERA, 95 % CONF LIM, CORRECTION FOR NITROGEN	1375
*HYAS ARANEUS	4958 GCAL/G DRY WT.	SHELL-FREE,INCL.ASH,WATER=81%, 1ANIMAL,1SAMPLE,MARCH	1619
*HYDROPSYCHE SP.	5348 CAL/G DRY WEIGHT	GERMANY	235
*HYDROPSYCHE SP.	5814 CAL/G ASH FREE DRY WEIGHT	GERMANY	235
*INVERTEBRATE	23.7 (+-2.6) KJ/G ASH FREE DRY WEIGHT	STANDARD DERIVATION, MEAN OF 19 SPECIES, NO CORRECTION FOR NITROGEN	1375
*INVERTEBRATE	21.5 (+-2.7) KJ/G ASH FREE DRY WEIGHT	STANDARD DERIVATION, MEAN OF 19 SPECIES, CORRECTION FOR NITROGEN	1375
*INVERTEBRATE	45.8 (+-1.7) KJ/G ORGANIC CARBON	STANDARD DERIVATION, MEAN OF 19 SPECIES, NO CORRECTION FOR NITROGEN	1375
*INVERTEBRATE	41.4 (+-1.9) KJ/G ORGANIC CARBON	STANDARD DERIVATION, MEAN OF 19 SPECIES, CORRECTION FOR NITROGEN	1375
*ISOPERLA GORTZI	5038 CAL/G DRY WEIGHT	GERMANY	235

TABLE A247

INVERTEBRATES : CALORIC CONTENT

ITEM:	VALUE:	CONDITIONS:	REF.:
*LEPTOCHEIRUS PINGUIS	2740 (2319-3348) GCAL/G DRY WT.	INCL.ASH,WATER=83+-3%,1500ANIMALS, 8SAMPLES,ALL SEASONS	1619
*LEPTOPHLEBIA MARGINATA	5647 CAL/G DRY WEIGHT	GERMANY	235
*LEPTOPHLEBIA SP.	24.4 (+-0.8) KJ/G ASH FREE DRY WEIGHT	EPHEMEROPTERA, 95 % CONF LIM, NO CORRECTION FOR NITROGEN	1375
*LEPTOPHLEBIA SP.	21.9 (+-0.8) KJ/G ASH FREE DRY WEIGHT	EPHEMEROPTERA, 95 % CONF LIM, CORRECTION FOR NITROGEN	1375
*LEPTOPHLEBIA SP.	46.4 (+-1.3) KJ/G ORGANIC CARBON	EPHEMEROPTERA, 95 % CONF LIM, NO CORECTION FOR NITROGEN	1375
*LEPTOPHLEBIA SP.	41.6 (+-1.4) KJ/G ORGANIC CARBON	EPHEMEROPTERA, 95 % CONF LIM, CORRECTION FOR NITROGEN	1375
*LIMNEPHILIDAE SP.	22.0 (+-0.3) KJ/G ASH FREE DRY WEIGHT	TRICHOPTERA, 95 % CONF LIM, NO CORRECTION FOR NITROGEN	1375
*LIMNEPHILIDAE SP.	20.5 (+-1.3) KJ/G ASH FREE DRY WEIGHT	TRICHOPTERA, 95 % CONF LIM, CORRECTION FOR NITROGEN	1375
*LIMNEPHILIDAE SP.	43.2 (+-0.4) KJ/G ORGANIC CARBON	TRICHOPTERA, 95 % CONF LIM, NO CORECTION FOR NITROGEN	1375
*LIMNEPHILIDAE SP.	40.2 (+-0.5) KJ/G ORGANIC CARBON	TRICHOPTERA, 95 % CONF LIM, CORRECTION FOR NITROGEN	1375
*LUMBRICUS TERRESTRIS	5745.7 G CAL/(G ASH FREE DRY WT)	ASH = 23.0 % OF DRY WEIGHT	485
*LUMBRINERIS FRAGILIS	4565 (4359-4810) GCAL/G DRY WT.	INCL.ASH,WATER=80+-3%,200ANIMALS, 12SAMPLES,ALL SEASONS	1619
*MEGANYCTIPHANES NORVEGICA	4945 (4559-5294) GCAL/G DRY WT.	INCL.ASH,WATER=81+-4%,855ANIMALS, 11SAMPLES,ALL SEASONS	1619
*MEGANYCTIPHANES NORVEGICA	5040 (4420-5460) GCAL/G DRY WT.	INCL.ASH,WATER=81+-5%,767ANIMALS, 12SAMPLES,ALL SEASONS	1619
*MESODON THYROIDES	5970.9 G CAL/(G ASH FREE DRY WT)	ASH = 22.8 % OF DRY WEIGHT	390
*MYRMICA SP.	6.309 KCAL/G DRY WEIGHT	EQUAL TO 1.92 KCAL/G FRESH WEIGHT	1244
*MYSIS RELICTA	25.7 (+-0.1) KJ/G ASH FREE DRY WEIGHT	MYSIDACEA, 95 % CONF LIM, NO CORRECTION FOR NITROGEN	1375
*MYSIS RELICTA	23.3 (+-0.2) KJ/G ASH FREE DRY WEIGHT	MYSIDACEA, 95 % CONF LIM, CORRECTION FOR NITROGEN	1375
*MYSIS RELICTA	47.3 (+-1.2) KJ/G ORGANIC CARBON	MYSIDACEA, 95 % CONF LIM, NO CORECTION FOR NITROGEN	1375
*MYSIS RELICTA	42.5 (+-1.0) KJ/G ORGANIC CARBON	MYSIDACEA, 95 % CONF LIM, CORRECTION FOR NITROGEN	1375
*MYSIS STENOLEPIS	4714 (4563-4824) GCAL/G DRY WT.	INCL.ASH,WATER=79+-3%,300ANIMALS, 3SAMPLES,FALL-WINTER	1619
*NEMOURA SP.	5488 CAL/G DRY WEIGHT	GERMANY	235
*NEPHTYS INCISA	3984 (3773-4156) GCAL/G DRY WT.	INCL.ASH,WATER=81+-3%,156ANIMALS, 12SAMPLES,ALL SEASONS	1619
*NOTONECTA GLAUCA	23.8 (+-0.4) KJ/G ASH FREE DRY WEIGHT	HETERPOTERA, 95 % CONF LIM, NO CORRECTION FOR NITROGEN	1375
*NOTONECTA GLAUCA	21.3 (+-0.5) KJ/G ASH FREE DRY WEIGHT	HETERPOTERA, 95 % CONF LIM, CORRECTION FOR NITROGEN	1375
*NOTONECTA GLAUCA	45.9 (+-1.7) KJ/G ORGANIC CARBON	HETERPOTERA, 95 % CONF LIM, NO CORECTION FOR NITROGEN	1375
*NOTONECTA GLAUCA	41.0 (+-1.6) KJ/G ORGANIC CARBON	HETERPOTERA, 95 % CONF LIM, CORRECTION FOR NITROGEN	1375
*NUCULANA SP.	4184 GCAL/G DRY WT.	SHELL-FREE,INCL.ASH,WATER=68%, 100ANIMALS,1SAMPLE,JUNE TO AUG.	1619
*ORCHELIMUM FIDICINIUM	5.432 KCAL/G DRY WEIGHT	NONE	547
*OTOCRYPTOPS SEXSPINOSUS	5344 G CAL/(G ASH FREE DRY WT)	ASH = 6.9 % OF DRY WEIGHT	390
*PALLACEA QUADRISPINOSA	21.3 (+-0.4) KJ/G ASH FREE DRY WEIGHT	AMPHIPODA, 95 % CONF LIM, NO CORRECTION FOR NITROGEN	1375
*PALLACEA QUADRISPINOSA	19.0 (+-0.4) KJ/G ASH FREE DRY WEIGHT	AMPHIPODA, 95 % CONF LIM, CORRECTION FOR NITROGEN	1375
*PALLACEA QUADRISPINOSA	45.3 (+-2.1) KJ/G ORGANIC CARBON	AMPHIPODA, 95 % CONF LIM, NO CORECTION FOR NITROGEN	1375
*PALLACEA QUADRISPINOSA	40.4 (+-1.8) KJ/G ORGANIC CARBON	AMPHIPODA, 95 % CONF LIM, CORRECTION FOR NITROGEN	1375
*PANDALUS MONTAGUI	4610 (4390-4815) GCAL/G DRY WT.	INCL.ASH,WATER=72+-3%,67ANIMALS, 9SAMPLES,ALL SEASONS	1619
*PARCOBLATTA SP.	6063 G CAL/(G ASH FREE DRY WT)	ASH = 8.7 % OF DRY WEIGHT	1072

TABLE A247

INVERTEBRATES : CALORIC CONTENT			
ITEM:	VALUE:	CONDITIONS:	REF.:
*PHILAENUS SPUMARIUS	5.807 KCAL/G DRY WEIGHT	NONE	1706
*PLECTROCNEMIA CONSPERSA	5703 CAL/G DRY WEIGHT	GERMANY	235
*PLECTROCNEMIA CONSPERSA	5958 CAL/G ASH FREE DRY WEIGHT	GERMANY	235
*PRAXILLELLA PRAETERMISSA	2365 GCAL/G DRY WT.	INCL.ASH,WATER=79%,300ANIMALS, 1SAMPLE,APRIL TO NOV.	1619
*PROKELISIA MARGINATA	5.380 KCAL/G ASH FREE DRY WEIGHT	NONE	791
*RADIX PEREGRA OVATA	4608 CAL/G DRY WEIGHT	GERMANY	235
*RADIX PEREGRA OVATA	5957 CAL/G ASH FREE DRY WEIGHT	GERMANY	235
*RHITHROGENA SEMICOLORATA	4955 CAL/G DRY WEIGHT	GERMANY	235
*RHITHROGENA SEMICOLORATA	5579 CAL/G ASH FREE DRY WEIGHT	GERMANY	235
*RHYACOPHILA FASCIATA	5869 CAL/G DRY WEIGHT	GERMANY	235
*RHYACOPHILA FASCIATA	6181 CAL/G ASH FREE DRY WEIGHT	GERMANY	235
*SARCOPHAGA BULLATA	5464 (SE=+-243) G CAL/(G ASH FREE DRY WT)	NONE	1448
*SCHISTOCERCA SP.	5.363 KCAL/G DRY WEIGHT	NONE	547
*SIALIS SP.	23.5 (+-0.3) KJ/G ASH FREE DRY WEIGHT	MEGALOPTERA, 95 % CONF LIM, NO CORRECTION FOR NITROGEN	1375
*SIALIS SP.	21.2 (+-0.4) KJ/G ASH FREE DRY WEIGHT	MEGALOPTERA, 95 % CONF LIM, CORRECTION FOR NITROGEN	1375
*SIALIS SP.	46.0 (+-0.8) KJ/G ORGANIC CARBON	MEGALOPTERA, 95 % CONF LIM, NO CORECTION FOR NITROGEN	1375
*SIALIS SP.	41.5 (+-1.0) KJ/G ORGANIC CARBON	MEGALOPTERA, 95 % CONF LIM, CORRECTION FOR NITROGEN	1375
*SPIRONTOCARIS SP.	4423 GCAL/G DRY WT.	INCL.ASH,WATER=73%,20ANIMALS, 1SAMPLE,OCTOBER	1619
*SPIRONTOCARIS SP.	4065 GCAL/G DRY WT.	INCL.ASH,WATER=78%,18ANIMALS, 1SAMPLE,APRIL	1619
*TENEBRIO MOLITOR	6.314 KCAL/G ASH FREE DRY WEIGHT	NONE	1447
*TOMOCERUS SP.	6063 G CAL/(G ASH FREE DRY WT)	ASH = 3.8 % OF DRY WEIGHT	1072
*TRIBOLIUM CASTANEUM	6.000 KCAL/G DRY WEIGHT	NONE	839
*TRIGONOTYLUS SP.	5.300 KCAL/G ASH FREE DRY WEIGHT	NONE	791
*TYROGLYPUS INTERNA	5808 +- 446 G CAL/(G ASH FREE DRY WT)	95 % CL.	1447
*UNCIOLA LEUCOPIS	2147 (1845-2512) GCAL/G DRY WT.	INCL.ASH,WATER=80+-7%,1800ANIMALS, 7SAMPLES,ALL SEASONS	1619
*VENERICARDIA BOREALIS	3678 (3400-4202) GCAL/G DRY WT.	SHELL-FREE,INCL.ASH,WATER=82+-2%, 188ANIMALS,10SAMPLES,ALL SEASONS	1619
*YOLDIA SAPOTILLA	4216 (3880-4555) GCAL/G DRY WT.	SHELL-FREE,INCL.ASH,WATER=78+-1%, 100ANIMALS,4SAMPLES,ALL SEASONS	1619
ZOOINVERTEBRATES AND ZOOPLANKTON	5 KCAL/G DRY WEIGHT	NONE	1661

TABLE A248

INVERTEBRATES : CONVERSION EFFICIENCY			
ITEM:	VALUE:	CONDITIONS:	REF.:
*BENTHOS	0.4-0.8	NONE	1571

TABLE A249

INVERTEBRATES : DEVELOPMENT TIME			
ITEM:	VALUE:	CONDITIONS:	REF.:
*FLY, EGGS	2 DAYS	INITIAL VALUES, THERMAL ECOSYSTEM, TO HATCHING	1703

TABLE A249

INVERTEBRATES : DEVELOPMENT TIME			
ITEM:	VALUE:	CONDITIONS:	REF.:
*FLY, PUPAE	6 DAYS	INITIAL VALUES, THERMAL ECOSYSTEM	1703
*FLY, 1,2 INSTAR LARVAE	2 DAYS	INITIAL VALUES, THERMAL ECOSYSTEM	1703
*FLY, 3 INSTAR LARVAE	4 DAYS	INITIAL VALUES, THERMAL ECOSYSTEM	1703

TABLE A250

INVERTEBRATES : DRY WEIGHT			
ITEM:	VALUE:	CONDITIONS:	REF.:
*ABLABESMYIA	0.154 MG	EMERGENT INSECTS, DRIED AT 343 K	595
*ABLABESMYIA SP.	0.005 MG	BENTHOS, DRIED AT 343 K, 0 - 1 MM	595
*ABLABESMYIA SP.	0.007 MG	BENTHOS, DRIED AT 343 K, 1 - 2 MM	595
*ABLABESMYIA SP.	0.010 MG	BENTHOS, DRIED AT 343 K, 2 - 3 MM	595
*ABLABESMYIA SP.	0.060 MG	BENTHOS, DRIED AT 343 K , MORE THAN 3 MM	595
*BHAOBORUS SP.	0.75	EMERGENT INSECTS, DRIED AT 343 K	595
*CAENIS SIMULANS	0.010 MG	BENTHOS, DRIED AT 343 K, 0 - 1 MM	595
*CAENIS SIMULANS	0.044 MG	BENTHOS, DRIED AT 343 K, 1 - 2 MM	595
*CAENIS SIMULANS	0.130 MG	BENTHOS, DRIED AT 343 K, 2 - 3 MM	595
*CAENIS SIMULANS	0.320 MM	BENTHOS, DRIED AT 343 K, 3 - 4 MM	595
*CAENIS SIMULANS	0.573 MG	BENTHOS, DRIED AT 343 K, 4 - 5 MM	595
*CAENIS SIMULANS	0.154 MG	EMERGENT INSECTS, DRIED AT 343 K	595
*CALLIBAETIS FERRUGINEUS	0.02 MM	BENTHOS, DRIED AT 343 K, 0 - 1 MM	595
*CALLIBAETIS FERRUGINEUS	0.05 MG	BENTHOS, DRIED AT 343 K, 1 - 2 MM	595
*CALLIBAETIS FERRUGINEUS	0.08 MG	BENTHOS, DRIED AT 343 K, 2 - 3 MM	595
*CALLIBAETIS FERRUGINEUS	0.14 MG	BENTHOS, DRIED AT 343 K, 3 - 4 MM	595
*CALLIBAETIS FERRUGINEUS	0.31 MG	BENTHOS, DRIED AT 343 K, 4 - 5 MM	595
*CALLIBAETIS FERRUGINEUS	0.63 MG	BENTHOS, DRIED AT 343 K, 5 - 6 MM	595
*CALLIBAETIS FERRUGINEUS	0.85 MG	BENTHOS, DRIED AT 343 K, 6 - 7 MM	595
*CALLIBAETIS FERRUGINEUS	1.68 MM	BENTHOS, DRIED AT 343 K, MORE THAN 7 MM	595
*CALLIBAETIS FERRUGINEUS	1.25 MG	EMERGENT INSECTS, DRIED AT 343 K	595
*CERATOPOGONIDAE	0.025 MG	BENTHOS, DRIED AT 343 K, 0 - 5 MM	595
*CERATOPOGONIDAE	0.14 MG	BENTHOS, DRIED AT 343 K, 5 - 10 MM	595
*CERATOPOGONIDAE	0.26 MG	BENTHOS, DRIED AT 343 K, 10 - 15 MM	595
*CHIRONOMUS SP.	0.007 MG	BENTHOS, DRIED AT 343 K, INSTAR 2	595
*CHIRONOMUS SP.	0.20 MG	BENTHOS, DRIED AT 343 K, INSTAR 3	595
*CHIRONOMUS SP.	1.50 MG	BENTHOS, DRIED AT 343 K, INSTAR 4	595
*CHIRONOMUS TENTANS	0.04 MG	BENTHOS, DRIED AT 343 K, INSTAR 2	595
*CHIRONOMUS TENTANS	0.45 MG	BENTHOS, DRIED AT 343 K, INSTAR 3	595
*CHIRONOMUS TENTANS	1.30 MG	BENTHOS, DRIED AT 343 K, INSTAR 4	595
*CHIRONOMUS TENTANS	2.10 MG	BENTHOS, DRIED AT 343 K, PUPAE	595
*CHIRONOMUS TENTANS	1.22 MG	EMERGENT INSECTS, BENTHOS, DRIED AT 343 K	595
*CORIXIDAE	0.60 MG	BENTHOS, DRIED AT 343 K, 0 - 10 MM	595
*CORYNONEURA SP.	0.007 MG	BENTHOS, DRIED AT 343 K, 1 - 2 MM	595
*DYTISCIDAE	0.04 MG	BENTHOS, DRIED AT 343 K, INSTAR 1	595
*DYTISCIDAE	0.17 MG	BENTHOS, DRIED AT 343 K, INSTAR 2	595
*DYTISCIDAE	0.97 MG	BENTHOS, DRIED AT 343 K, INSTAR 3	595
*GLYPTOTENDIPES BARBIPES	0.002 MG/ANIMAL	INSTAR 1, LAB, 297 K	822

TABLE A250

INVERTEBRATES : DRY WEIGHT			
ITEM:	VALUE:	CONDITIONS:	REF.:
*GLYPTOTENDIPES BARBIPES	0.025 MG	INSTAR 2, LAB, 297 K	822
*GLYPTOTENDIPES BARBIPES	0.160 MG	INSTAR 3, LAB, 297 K	822
*GLYPTOTENDIPES BARBIPES	0.650 MG	INSTAR 4, LAB, 297 K	822
*GLYPTOTENDIPES SP.	0.005 MG	BENTHOS, DRIED AT 343 K, INSTAR 1	595
*GLYPTOTENDIPES SP.	0.02 MG	BENTHOS, DRIED AT 343 K, INSTAR 2	595
*GLYPTOTENDIPES SP.	0.15 MG	BENTHOS, DRIED AT 343 K, INSTAR 3	595
*GLYPTOTENDIPES SP.	0.45 MG	BENTHOS, DRIED AT 343 K, INSTAR 4	595
*GLYPTOTENDIPES SP.	0.85 MG	EMERGENT INSECTS, DRIED AT 343 K	595
*HALIPLIDAE	0.47 MM	BENTHOS, DRIED AT 343 K, 0 - 5 MG	595
*HALIPLIDAE	0.94 MM	BENTHOS, DRIED AT 343 K, MORE THAN 5 MG	595
*HIRUDINEA	1.43 MG	BENTHOS, DRIED AT 343 K, 0 - 10 MM	595
*HIRUDINEA	5.70 MG	BENTHOS, DRIED AT 343 K, MORE THAN 10 MM	595
*HYALELLA AZTECA	0.015 MG	BENTHOS, DRIED AT 343 K, 0 - 1 MM	595
*HYALELLA AZTECA	0.036 MG	BENTHOS, DRIED AT 343 K, 1 - 2 MM	595
*HYALELLA AZTECA	0.080 MG	BENTHOS, DRIED AT 343 K, 2 - 3 MM	595
*HYALELLA AZTECA	0.133 MG	BENTHOS, DRIED AT 343 K, 3 - 4 MM	595
*HYALELLA AZTECA	0.20 MG	BENTHOS, DRIED AT 343 K, 4 - 5 MM	595
*HYALELLA AZTECA	0.41 MG	BENTHOS, DRIED AT 343 K, 5 - 6 MM	595
*HYALELLA AZTECA	0.60 MG	BENTHOS, DRIED AT 343 K, MORE THAN 6 MM	595
*HYDROPHILIDAE	1.04 MG	BENTHOS, DRIED AT 343 K, 0 - 5 MM	595
*HYDROPHILIDAE	2.14 MG	BENTHOS, DRIED AT 343 K, 5 - 10 MM	595
*HYDROPHILIDAE	4.42 MG	BENTHOS, DRIED AT 343 K, 10 - 15 MM	595
*MICROTENDIPES SP.	0.005 MG	BENTHOS, DRIED AT 343 K, INSTAR 1	595
*MICROTENDIPES SP.	0.01 MG	BENTHOS, DRIED AT 343 K, INSTAR 2	595
*MICROTENDIPES SP.	0.02 MG	BENTHOS, DRIED AT 343 K, INSTAR 3	595
*MICROTENDIPES SP.	0.10 MG	BENTHOS, DRIED AT 343 K, INSTAR 4	595
*MISCELLANEOUS DIPTERA	0.005 MG	BENTHOS, DRIED AT 343 K, 0 - 1 MM	595
*MISCELLANEOUS DIPTERA	0.007 MG	BENTHOS, DRIED AT 343 K, 1 - 2 MM	595
*MISCELLANEOUS DIPTERA	0.01 MG	BENTHOS, DRIED AT 343 K, 2 - 3 MM	595
*MISCELLANEOUS DIPTERA	0.02 MG	BENTHOS, DRIED AT 343 K, 3 - 4 MM	595
*MISCELLANEOUS DIPTERA	0.04 MG	BENTHOS, DRIED AT 343 K, 4 - 5 MM	595
*MISCELLANEOUS DIPTERA	0.065 MG	BENTHOS, DRIED AT 343 K, 5 - 6 MM	595
*MISCELLANEOUS DIPTERA	1.50 MG	BENTHOS, DRIED AT 343 K, MORE THAN 6 MM	595
*MISCELLANEOUS DIPTERA	0.100 MG	BENTHOS, DRIED AT 343 K, PUPAE	595
*MISCELLANEOUS DIPTERA	0.146 MG	EMERGENT INSECTS, DRIED AT 343 K	595
*MISCELLANEOUS TRICHOPTERA	0.06 MG	EMERGENT INSECTS, DRIED AT 343 K	595
*PHRYGANEA SP.	18.05 MG	EMERGENT INSECTS, DRIED AT 343 K	595
*POLYCENTROPUS SP.	0.91 MG	EMERGENT INSECTS, DRIED AT 343 K	595
*PROCLADIUS SP.	0.007 MG	BENTHOS, DRIED AT 343 K, 1 - 2 MM	595
*PROCLADIUS SP.	0.01 MG	BENTHOS, DRIED AT 343 K, 2 - 3 MM	595
*PROCLADIUS SP.	0.02 MG	BENTHOS, DRIED AT 343 K, 3 - 4 MM	595
*PROCLADIUS SP.	0.035 MG	BENTHOS, DRIED AT 343 K, MORE THAN 4 MM	595
*TANYTARSINI	0.007 MG	BENTHOS, DRIED AT 343 K, 1 - 2 MM	595
*TANYTARSINI	0.01 MG	BENTHOS, DRIED AT 343 K, 2 - 3 MM	595
*TANYTARSINI	0.02 MG	BENTHOS, DRIED AT 343 K, 3 - 4 MM	595
*TANYTARSINI	0.035 MG	BENTHOS, DRIED AT 343 K, MORE THAN 4 MM	595
*TRICHOPTERA	0.026 MG	BENTHOS, DRIED AT 343 K, 1 - 3 MM	595

TABLE A250

INVERTEBRATES : DRY WEIGHT			
ITEM:	VALUE:	CONDITIONS:	REF.:
*TRICHOPTERA	0.074 MG	BENTHOS, DRIED AT 343 K, 3 - 5 MM	595
*TRICHOPTERA	0.768 MG	BENTHOS, DRIED AT 343 K, MORE THAN 5 MM	595
*ZYGOPTERA	0.35 MG	BENTHOS, DRIED AT 343 K, 0 - 5 MM	595
*ZYGOPTERA	2.19 MG	BENTHOS, DRIED AT 343 K, 5 - 10 MM	595
*ZYGOPTERA	5.83 MG	BENTHOS, DRIED AT 343 K, 11 - 15 MM	595
*ZYGOPTERA	5.61 MG	EMERGENT INSECTS, DRIED AT 343 K	595

TABLE A251

INVERTEBRATES : EGESTION			
ITEM:	VALUE:	CONDITIONS:	REF.:
*ASELLUS AQUATICUS	69.7 % OF INGESTION	NONE	1689
*FLY, ADULT	0.300 FRACTION OF INGESTION	INITIAL VALUES, THERMAL ECOSYSTEM, LIMIT-EGESTION PROPORTION	1703
*FLY, 1,2 INSTAR	0.300 FRACTION OF INGESTION	INITIAL VALUES, THERMAL ECOSYSTEM, LIMIT-EGESTION PROPORTION	1703
*FLY, 3 INSTAR	0.300 FRACTION OF INGESTION	INITIAL VALUES, THERMAL ECOSYSTEM, LIMIT-EGESTION PROPORTION	1703
*LESTES SPONSA	63.4 % OF INGESTION	NONE	1689
*PHILAENUS SPUMARIUS	0.43 - 1.35 CAL/(ANIMAL*DAY)	3RD INSTAR, 0.39 MG, 296 K	1705
*PHILAENUS SPUMARIUS	1.12 - 3.48 CAL/(ANIMAL*DAY)	ADULT, 3.87 MG, 294 K, EXCRETING URIC ACID	1705
*PHILAENUS SPUMARIUS	0.41 CAL/(ANIMAL*DAY)	ADULT, 3.87 MG, 294 K, EXCRETING URIC ACID	1705
*PHILAENUS SPUMARIUS	0.25 CAL/(ANIMAL*DAY)	ADULT, 3.87 MG, 294 K, EXCRETING UREA	1705

TABLE A252

INVERTEBRATES : EMIGRATION RATE			
ITEM:	VALUE:	CONDITIONS:	REF.:
*FLY	0.007 KCAL/KCAL*DAY	INITIAL VALUES, THERMAL ECOSYSTEM, TRANSFER COEFF., INCL.DISSOL.ORG.MATTER	1703

TABLE A253

INVERTEBRATES : FILTRATION RATE			
ITEM:	VALUE:	CONDITIONS:	REF.:
*ARCTICA ISLANDICA	2.0 L/HR	277K	1737
*ARCTICA ISLANDICA	4.1 L/HR	287K	1737
*ARCTICA ISLANDICA	3.50 L/HR	283K	1737
*ARCTICA ISLANDICA	4.30 L/HR	293K	1737
*HIATELLA ARCTICA	0.0055 L/HR	278K	1403
*HIATELLA ARCTICA	0.0341 L/HR	288K	1403
*HIATELLA ARCTICA	0.0165 L/HR	283K	1403
*HIATELLA ARCTICA	0.0121 L/HR	293K	1403

TABLE A253

INVERTEBRATES : FILTRATION RATE			
ITEM:	VALUE:	CONDITIONS:	REF.:
*HYDROIDES NORVEGICA	900 ML/HR*G WET WT.	289-290K,ORG.WT.=12.3MG,2-3HOURS,N =44,GRAPHITE	321
*LANICE CONCHILEGA	105 ML/HR*G WET WT.	285K,ORG.WT.=248MG,5EXP.,40-DAY CONT.DET.,DUNALIELLA MARINA(?053)	193
*LANICE CONCHILEGA	175 ML/HR*G WET WT.	285K,ORG.WT.=77.1MG,4EXP.,46-DAY CONT.DET.,DUNALIELLA MARINA(?053)	193
*LANICE CONCHILEGA	180 ML/HR*G WET WT.	285K,ORG.WT.=72.4MG,3EXP.,20-DAY CONT.DET.,DUNALIELLA MARINA(?053)	193
*LANICE CONCHILEGA	610 ML/HR*G WET WT.	285K,ORG.WT.=18.5MG,2EXP.,24-DAY CONT.DET.,DUNALIELLA MARINA(?053)	193
*LANICE CONCHILEGA	1240 ML/HR*G WET WT.	285K,ORG.WT.=7.3MG,1EXP.,21-DAY CONT.DET.,DUNALIELLA MARINA(?053)	193
*MYTILUS EDULIS	0.620 L/HR	277K	1572
*MYTILUS EDULIS	1.330 L/HR	287K	1572
*MYTILUS EDULIS	0.353 L/HR	278K	1403
*MYTILUS EDULIS	1.751 L/HR	288K	1403
*MYTILUS EDULIS	1.100 L/HR	283K	1572
*MYTILUS EDULIS	1.240 L/HR	293K	1572
*MYTILUS EDULIS	1.468 L/HR	283K	1403
*MYTILUS EDULIS	1.796 L/HR	293K	1403
*MYTILUS EDULIS	1.751 L/HR	288K	1403
*MYTILUS EDULIS	1.798 L/HR	298K	1403
*MYTILUS EDULIS	1.796 L/HR	293K	1403
*MYTILUS EDULIS	0.106 L/HR	303K	1403
*MYXICOLA INFUNDIBULUM	100 ML/HR*G WET WT.	290-293K,ORG.WT.=2695MG,2-3HOURS,N =45,GRAPHITE	321
*OYSTERS	200 - 300 L/DAY	WET WEIGHT = 5 G	1061
*POMATOCEROS TRIQUETER	1400 ML/HR*G WET WT.	289-290K,ORG.WT.=18.5MG,2-3HOURS,N =45,GRAPHITE	321
*POMATOCEROS TRIQUETER	900 ML/HR*G WET WT.	285K,ORG.WT.=11.8MG,0.5-1HR,N=12, DUNALIELLA SP.(22*10**6CELLS/L)	840
*POMATOCEROS TRIQUETER	970 ML/HR*G WET WT.	291K,ORG.WT.=11.8MG,0.5-1HR,N=11, DUNALIELLA SP.(24*10**6CELLS/L)	840
*SABELLA PAVONINA	390 ML/HR*G WET WT.	291-293K,ORG.WT.=187MG,2-3HOURS,N= 50,GRAPHITE	321
*SALMACINA DYSTERI	2090 ML/HR*G WET WT.	290-291K,ORG.WT.=0.14MG,2-3HOURS,N =50,GRAPHITE	321
*SCHIZOBRANCHIA INSIGNIS	70 ML/HR*G WET WT. (MEAN)	285-286K,ORG.WT.=500-2000MG, 3HOURS,N=10,GRAPHITE	322
*SPIRORBIS BOREALIS	950 ML/HR*G WET WT.	290-292K,ORG.WT.=0.24MG,2-3HOURS,N =99,GRAPHITE	321

TABLE A254

INVERTEBRATES : GRAZING RATE			
ITEM:	VALUE:	CONDITIONS:	REF.:
*FLY, ADULT	0.751 KCAL/KCAL*DAY	INITIAL VALUES, THERMAL ECOSYSTEM, TRANSFER COEFF.	1703
*FLY, 1,2 INSTAR LARVAE	2.940 KCAL/KCAL*DAY	INITIAL VALUES, THERMAL ECOSYSTEM, TRANSFER COEFF.	1703
*FLY, 3 INSTAR LARVAE	1.456 KCAL/KCAL*DAY	INITIAL VALUES, THERMAL ECOSYSTEM, TRANSFER COEFF.	1703

TABLE A255

INVERTEBRATES : GROWTH AND REPRODUCTION			
ITEM:	VALUE:	CONDITIONS:	REF.:
*ASELLUS AQUATICUS	5.6 % OF INGESTION	NONE	1689

TABLE A255

INVERTEBRATES : GROWTH AND REPRODUCTION

ITEM:	VALUE:	CONDITIONS:	REF.:
*ASELLUS AQUATICUS	18.2 % OF ASSIMILATION	NONE	1689
*LESTES SPONSA	23.5 % OF INGESTION	NONE	1689
*LESTES SPONSA	64.0 % OF ASSIMILATION	NONE	1689

TABLE A256

INVERTEBRATES : GROWTH EFFICIENCY

ITEM:	VALUE:	CONDITIONS:	REF.:
*ALBURNUS ALBURNUS	8.8 % OF INGESTION	PROD = 30.5 KCAL/(M2*Y)	1012
*ALLOLOBOPHORA ROSEA	17.58 % OF INGESTION	PROD = 2.717 KCAL/(M2*Y)	1012
*ANATOPYNIA DYARI	15.2 % OF INGESTION	PROD = 37.4 KCAL/(M2*Y)	1012
*ASELLUS AQUATICUS	24.1 % OF INGESTION	PROD = 16.296 KCAL/(M2*Y)	1012
*ASELLUS MILIARIS	18.2 % OF INGESTION	PROD = 108.3 KCAL/(M2*Y)	1012
*CALOPSETRA DIVES	25.1 % OF INGESTION	PROD = 130.7 KCAL/(M2*Y)	1012
*CHORTHIPPUS PARALLELLUS	55.3 % OF INGESTION	PROD = 0.420 KCAL/(M2*Y)	1012
*CHORTHIPPUS PARALLELLUS	58.9 % OF INGESTION	PROD = 0.273 KCAL/(M2*Y)	1012
*DIPLACODES TRIVIALIS	48.1 % OF ASSIMILATION	46 MG ANIMAL, INGESTION = 61.5 CAL/(KCAL*ANIM.*D)	1209
*GAMMARUS PSEUDOLIMNEUS	15.0 % OF INGESTION	PROD = 183.0 KCAL/(M2*Y)	1012
*GOBIO GOBIO	6.6 % OF INGESTION	PROD = 2.6 KCAL/(M2*Y)	1012
*JAPONARIA LAMINATA ARMIGERA	46.5 % OF INGESTION	PROD = 26.9 KCAL/(M2*Y)	1012
*JAPONARIA LAMINATA ARMIGERA	50.0 % OF INGESTION	PROD = 1.5 KCAL/(M2*Y)	1012
*LEPTOTERNA DOLOBRATA	58.1 % OF INGESTION	PROD = 0.140 KCAL/(M2*Y)	1012
*LEPTOTERNA DOLOBRATA	52.3 % OF INGESTION	PROD = 0.096 KCAL/(M2*Y)	1012
*LEUCISCUS LEUCISCUS	5.96 % OF INGESTION	PROD = 1.5 KCAL/(M2*Y)	1012
*LIGIDIUM JAPONICA	18.0 % OF INGESTION	PROD = 3.5 KCAL/(M2*Y)	1012
*LIMNODRILUS HOFFMEISTERI CLARAPAREDE	24.9 % OF INGESTION	PROD = 160.5 KCAL/(M2*Y)	1012
*LITHOBIUS CRASSIPES	5.51 % OF INGESTION	PROD = 0.074 KCAL/(M2*Y)	1012
*LITHOBIUS FORTIFICATUS	4.08 % OF INGESTION	PROD = 0.085 KCAL/(M2*Y)	1012
*LITTORINA IRRORATA	14.0 % OF INGESTION	PROD = 40.6 KCAL/(M2*Y)	1012
*METAPENAEUS MONOCEROS	21.6 % OF INGESTION	GROSS EFF., FOOD = ESTUARINE DETRITUS	1286
*METAPENAEUS MONOCEROS	24.1 % OF ASSIMILATION	NET EFFIC., FOOD = ESTUARINE DETRITUS	1286
*MODIOLUS DEMISSUS	30.0 % OF INGESTION	PROD = 16.7 KCAL/(M2*Y)	1012
*ONYCHIURUS PROCAMPATUS	46.2 % OF INGESTION	PROD = 2.61 KCAL/(M2*Y)	1012
*ORCHELIMUM FIDICINIUM	36.7 % OF INGESTION	PROD = 10.8 KCAL/(M2*Y)	1012
*ORIBATID MITES	21.4 % OF INGESTION	PROD = 0.43 KCAL/(M2*Y)	1012
*ORTHOPTERA	15.6 % OF INGESTION	PROD = 4.0 KCAL/(M2*Y), 3 SPECIES	1012
*ORTHOPTERA	36.3 % OF INGESTION	PROD = 0.638 KCAL/(M2*Y), MOST MELANOPUS	1012
*ORTHOPTERA	38.5 % OF INGESTION	PROD = 0.373 KCAL/(M2*Y)	1012
*ORTHOPTERA	34.4 % OF INGESTION	PROD = 4.573 KCAL/(M2*Y)	1012
*PERCA FLUVIATILIS	7.2 % OF INGESTION	PROD = 1.8 KCAL/(M2*Y)	1012
*PHILAENUS SPUMARIUS	41.2 % OF INGESTION	PROD = 15963 KCAL/(M2*Y)	1012
*PHYSA PSEUDOLIMNEUS	27.9 % OF INGESTION	PROD = 170.6 KCAL/(M2*Y)	1012
*PIMPLA INSTIGATOR	65.4 % OF INGESTION	PROD = 0.017 KCAL/(M2*Y), LARVAE	1012
*PISIDIUM VIRGINICUM	47.3 % OF INGESTION	PROD = 81.8 KCAL/(M2*Y)	1012
*PTEROMALUS PUPARUM	56.9 % OF INGESTION	PROD = 0.074 KCAL/(M2*Y), LARVAE	1012
*PYRRHOSOMA NYMPHULA	58.2 % OF INGESTION	PROD = 4.436 KCAL/(M2*Y)	1012

TABLE A256

INVERTEBRATES : GROWTH EFFICIENCY			
ITEM:	VALUE:	CONDITIONS:	REF.:
*PYRRHOSOMA NYMPHULA	51.3 % OF INGESTION	PROD = 3.889 KCAL/(M2*Y)	1012
*RUTILUS RUTILUS	8.3 % OF INGESTION	PROD = 10.7 KCAL/(M2*Y)	1012
*SCROBICULARIA PLANA	24.3 % OF INGESTION	PROD = 17.5 KCAL/(M2*Y)	1012
*SCROBICULARIA PLANA	20.65 % OF INGESTION	PROD = 124.0 KCAL/(M2*Y)	1012
*TRICHOPTERA	23.7 % OF INGESTION	PROD = 20.9 KCAL/(M2*Y)	1012

TABLE A257

INVERTEBRATES : GROWTH RATE			
ITEM:	VALUE:	CONDITIONS:	REF.:
*CALANDRA ORYZAE	113.6 RATE	NET PRODUCTION RATE, LENGHT OF 1 GENERATION=58.1 D	891
*DAPHNIA PULEX	21.4 RATE	NET PRODUCTION RATE, LENGHT OF 1 GENERATION=10.6 D	891
*DAPHNIA PULEX	20.7 RATE	NET PRODUCTION RATE, LENGHT OF 1 GENERATION=15.4 D	891
*DUGESIA TIGRINA	2.0 RATE	NET PRODUCTION RATE, LENGHT OF 1 GENERATION=10.3 D	891
*EUROSTUS HILLERI	3.1 RATE	NET PRODUCTION RATE, LENGHT OF 1 GENERATION=115.6 D	891
*GIBBIUM PSYLLOIDES	75.2 RATE	NET PRODUCTION RATE, LENGHT OF 1 GENERATION=182.6 D	891
*GLYPTOTENDIPES BARBIPES	469 MG/(GRAM*DAY)	AGE = 1.6 D, 297 K, LAB, RELATIVE GROWTH	822
*GLYPTOTENDIPES BARBIPES	363 MG/(GRAM*DAY)	AGE = 5.5 D, 297 K, LAB, RELATIVE GROWTH	822
*GLYPTOTENDIPES BARBIPES	294 MG/(GRAM*DAY)	AGE = 8.7 D, 297 K, LAB, RELATIVE GROWTH	822
*GLYPTOTENDIPES BARBIPES	172 MG/(GRAM*DAY)	AGE = 14.7 D, 297 K, LAB, RELATIVE GROWTH	822
*GLYPTOTENDIPES BARBIPES	116 MG/(GRAM*DAY)	AGE = 20 D, 297 K, LAB, RELATIVE GROWTH	822
*GLYPTOTENDIPES BARBIPES	40 MG/(GRAM*DAY)	AGE = 24.5 D, 297 K, LAB, RELATIVE GROWTH	822
*GLYPTOTENDIPES BARBIPES	44 MG/(GRAM*DAY)	AGE = 32.1 D, 297 K, LAB, RELATIVE GROWTH	822
*GLYPTOTENDIPES BARBIPES	45 MG/(GRAM*DAY)	AGE = 42 MG/(GRAM*DAY)	822
*GLYPTOTENDIPES BARBIPES	25 MG/(GRAM*DAY)	AGE = 52 D, 297 K, LAB, RELATIVE GROWTH	822
*GLYPTOTENDIPES BARBIPES	400 MG/(GRAM*DAY)	INSTAR 1, LAB, 297 K	822
*GLYPTOTENDIPES BARBIPES	200 MG/(GRAM*DAY)	INSTAR 2, LAB, 297 K	822
*GLYPTOTENDIPES BARBIPES	100 MG/(GRAM*DAY)	INSTAR 3, LAB, 297 K	822
*GLYPTOTENDIPES BARBIPES	50 MG/(GRAM*DAY)	INSTAR 4, LAB, 297 K	822
*LASIODERMA SERRICORNE	13.2 RATE	NET PRODUCTION RATE, LENGHT OF 1 GENERATION=35.6 D	891
*MEZIUM AFFINE	64.2 RATE	NET PRODUCTION RATE, LENGHT OF 1 GENERATION=287.3 D	891
*MICROTUS AGRESTIS	5.9 RATE	NET PRODUCTION RATE, LENGHT OF 1 GENERATION=170.4 D	891
*NEZARA VIRIDULA	1.6 RATE	NET PRODUCTION RATE, LENGHT OF 1 GENERATION=10.3 D	891
*NIPTUS HOLOLEUCUS	2.3 RATE	NET PRODUCTION RATE, LENGHT OF 1 GENERATION=138.1 D	891
*OCTOPUS VULGARIS	1.11% PER DAY	288K,WEIGHT=530G,MATURE FEMALE, OPTIMAL CONDITION	975
*OCTOPUS VULGARIS	0.52% PER DAY	288K,WEIGHT=587G,MATURE FEMALE, OPTIMAL CONDITION	975
*OCTOPUS VULGARIS	0.19% PER DAY	288K,WEIGHT=664G,MATURE FEMALE, OPTIMAL CONDITION	975

TABLE A257

INVERTEBRATES : GROWTH RATE			
ITEM:	VALUE:	CONDITIONS:	REF.:
*OCTOPUS VULGARIS	1.35% PER DAY	288K, WEIGHT=500G, IMMATURE FEMALE, OPTIMAL CONDITION	975
*OCTOPUS VULGARIS	0.63% PER DAY	288K, WEIGHT=580G, IMMATURE FEMALE, OPTIMAL CONDITION	975
*OCTOPUS VULGARIS	0.79% PER DAY	288K, WEIGHT=675G, IMMATURE FEMALE, OPTIMAL CONDITION	975
*PAVONA CLIVOSA	2.5 CM/YEAR	RADIAL GROWTH RATE, CORAL	968
*PAVONA GIGANTEA	2 CM/YEAR	RADIAL GROWTH RATE, CORAL	968
*PEDICULUS HUMANUS	30.9 RATE	NET PRODUCTION RATE, LENGHT OF 1 GENERATION=30.9 D	891
*PHILAENUS SPUMARIUS	0.25 CAL/(ANIMAL*DAY)	3RD INSTAR, 0.39 MG, 296 K	1705
*PHILAENUS SPUMARIUS	0.74 CAL/(ANIMAL*DAY)	4RD INSTAR, 1.16 MG, 298 K	1705
*PHYLLOPERTHA HORTICOLA	2.9 RATE	NET PRODUCTION RATE, LENGHT OF 1 GENERATION=365.0 D	891
*PHYSA GYRINA	515.3 RATE	NET PRODUCTION RATE, LENGHT OF 1 GENERATION=142.0 D	891
*PHYSA GYRINA	642.7 RATE	NET PRODUCTION RATE, LENGHT OF 1 GENERATION=242.7 D	891
*POCILLOPORA DAMICORNIS	10 CM/YEAR	RADIAL GROWTH RATE, CORAL	968
*POCILLOPORA ROBUSTA	5 CM/YEAR	RADIAL GROWTH RATE, CORAL	968
*PORITES PANAMENSIS	1 CM/YEAR	RADIAL GROWTH RATE, CORAL	968
*PSAMMOCORA STELLATA	1 CM/YEAR	RADIAL GROWTH RATE, CORAL	968
*PTINUS FUR	10.8 RATE	NET PRODUCTION RATE, LENGHT OF 1 GENERATION=182.3 D	891
*PTINUS SEXPUNCTATUS	4.1 RATE	NET PRODUCTION RATE, LENGHT OF 1 GENERATION=227.5 D	891
*PTINUS TECTUS	116.1 RATE	NET PRODUCTION RATE, LENGHT OF 1 GENERATION=149.6 D	891
*STETHOMEZIUM SQUAMOSUM	41.0 RATE	NET PRODUCTION RATE, LENGHT OF 1 GENERATION=190.6 D	891
*TRIBOLIUM CASTANEUM	275.0 RATE	NET PRODUCTION RATE, LENGHT OF 1 GENERATION=80.0 D	891
*TRIGONOGENIUS GLOBULUS	49.8 RATE	NET PRODUCTION RATE, LENGHT OF 1 GENERATION=171.8 D	891

TABLE A258

INVERTEBRATES : H/BIOMASS RATIO			
ITEM:	VALUE:	CONDITIONS:	REF.:
*ANNELIDA	5.9 % OF DRY WEIGHT	MEAN ELEMENTARY COMPOSITION	157
*ARTHROPODA	7.3 % OF DRY WEIGHT	MEAN ELEMENTARY COMPOSITION,LAND	157
*COELENTERATA	4.5 % OF DRY WEIGHT	MEAN ELEMENTARY COMPOSITION, INTERPOLATED	157
*ECHINODERMATA	4.5 % OF DRY WEIGHT	MEAN ELEMENTARY COMPOSITION, INTERPOLATED	157
*MOLLUSCA	6.0 % OF DRY WEIGHT	MEAN ELEMENTARY COMPOSITION,SOFT PARTS	157
*NEMATODA	6.0 % OF DRY WEIGHT	MEAN ELEMENTARY COMPOSITION, INTERPOLATED	157

TABLE A259

INVERTEBRATES : HALF SATURATION CONSTANT FOR SEDIMENT			
ITEM:	VALUE:	CONDITIONS:	REF.:
*BENTHOS	50-2000 MG/M**2	GRAZING ON SEDIMENT	1571

TABLE A259

INVERTEBRATES : HALF SATURATION CONSTANT FOR SEDIMENT			
ITEM:	VALUE:	CONDITIONS:	REF.:
*BENTHOS	100 MG/M**2	USED IN MODELS, SAN FRANCISCO BAY ESTUARY	242
*BENTHOS	50 MG/M**2	USED IN MODELS, BOISE RIVER, IDAHO	241

TABLE A260

INVERTEBRATES : MAX GROWTH RATE			
ITEM:	VALUE:	CONDITIONS:	REF.:
*RHABDITIS MARINA	11.7 MICRO M/HOUR	298 K, 2.5 % SALINITET, FEMALE	1594
*RHABDITIS MARINA	11.3 MICRO M/HOUR	298 K, 2.5 % SALINITET, MALE	1594

TABLE A261

INVERTEBRATES : MAX STANDING CROP			
ITEM:	VALUE:	CONDITIONS:	REF.:
*FLY, ADULT	315 KCAL/M**(-2)	INITIAL VALUES, THERMAL ECOSYSTEM	1703
*FLY, 1,2 INSTAR LARVAE	150 KCAL/M**(-2)	INITIAL VALUES, THERMAL ECOSYSTEM	1703
*FLY, 3 INSTAR LARVAE	150 KCAL/M**(-2)	INITIAL VALUES, THERMAL ECOSYSTEM	1703

TABLE A262

INVERTEBRATES : MORTALITY			
ITEM:	VALUE:	CONDITIONS:	REF.:
*BENTHOS	0.003 1/DAY	USED IN MODELS, N FORK KINGS RIVER, CALIF.	1566
*BENTHOS	0.001 1/DAY	USED IN MODELS	1567
*BENTHOS	0.001 1/DAY	NONE	1571
*FLY, ADULT	0.148 KCAL/KCAL*DAY	INITIAL VALUES, THERMAL ECOSYSTEM, TRANSFER COEFF.	1703
*FLY, EGG	0.215 KCAL/KCAL*DAY	INITIAL VALUES, THERMAL ECOSYSTEM, TRANSFER COEFF.	1703
*FLY, PUPAE	0.056 KCAL/KCAL*DAY	INITIAL VALUES, THERMAL ECOSYSTEM, TRANSFER COEFF.	1703
*FLY, 1,2 INSTAR LARVAE	0.221 KCAL/KCAL*DAY	INITIAL VALUES, THERMAL ECOSYSTEM, TRANSFER COEFF.	1703
*FLY, 3 INSTAR LARVAE	0.122 KCAL/KCAL*DAY	INITIAL VALUES, THERMAL ECOSYSTEM, TRANSFER COEFF.,BASE RATE	1703

TABLE A263

INVERTEBRATES : N/BIOMASS RATIO			
ITEM:	VALUE:	CONDITIONS:	REF.:
*AGRYPNIA OBSOLETA	8.0 (+-1.9) % OF ASH FREE DRY WEIGHT	TRICHOPTERA, 95 % CONF LIMIT	1375
*ANNELIDA	9.9 % OF DRY WEIGHT	MEAN ELEMENTARY COMPOSITION	157
*ARTHROPODA	12.3 % OF DRY WEIGHT	MEAN ELEMENTARY COMPOSITION,LAND	157

TABLE A263

INVERTEBRATES : N/BIOMASS RATIO

ITEM:	VALUE:	CONDITIONS:	REF.:
*ASELLUS AQUATICUS	9.7 (+-1.4) % OF ASH FREE DRY WEIGHT	ISOPODA, 95 % CONF LIMIT	1375
*BENTHOS	0.08 BY WEIGHT	USED IN MODELS, N FORK KINGS RIVER, CALIF.	1566
*BENTHOS	0.09 BY WEIGHT	USED IN MODELS, BOISE RIVER, IDAHO	241
*BENTHOS	0.08 BY WEIGHT	USED IN MODELS, N FORK KINGS RIVER, CALIF.	1566
*BENTHOS	0.09 BY WEIGHT	USED IN MODELS, BOISE RIVER, IDAHO	241
*BENTHOS	0.07-0.09	NONE	1571
*CENTROPTILUM LUTEOLUM	9.9 (+-1.2) % OF ASH FREE DRY WEIGHT	EPHEMEROPTERA, 95 % CONF LIMIT	1375
*CHAOBORUS FLAVICANS	9.5 (+-0.4) % OF ASH FREE DRY WEIGHT	DIPTERA, 95 % CONF LIMIT	1375
*CHIRONOMUS PLUMOSUS	8.7 (+-1.4) % OF ASH FREE DRY WEIGHT	DIPTERA, 95 % CONF LIMIT	1375
*COELENTERATA	6.3 % OF DRY WEIGHT	MEAN ELEMENTARY COMPOSITION	157
*CORDULIA AENEA	9.0 (+-0.7) % OF ASH FREE DRY WEIGHT	ODONATA, 95 % CONF LIMIT	1375
*ECHINODERMATA	4.4 % OF DRY WEIGHT	MEAN ELEMENTARY COMPOSITION	157
*ERPOBDELLA OCTOCULATA	9.4 (+-0.6) % OF ASH FREE DRY WEIGHT	HIRUDINEA , 95 % CONF LIMIT	1375
*GAMMARAQANTUS LACUSTRIS	9.6 (+-0.7) % OF ASH FREE DRY WEIGHT	AMPHIPODA, 95 % CONF LIMIT	1375
*HEPTAGENIA FUSCOGRISEA	8.8 (+-0.8) % OF ASH FREE DRY WEIGHT	EPHEMEROPTERA, 95 % CONF LIMIT	1375
*INVERTEBRATE	9.0 (+-1.3) % OF ASH FREE DRY WEIGHT	STANDARD DERIVATION, MEAN OF 19 SPECIES	1375
*LEPTOPHLEBIA SP.	10.1 (+-0.8) % OF ASH FREE DRY WEIGHT	EPHEMEROPTERA, 95 % CONF LIMIT	1375
*LIMNEPHILIDAE SP.	6.2 (+-1.0) % OF ASH FREE DRY WEIGHT	TRICHOPTERA, 95 % CONF LIMIT	1375
*MOLLUSCA	8.5 % OF DRY WEIGHT	MEAN ELEMENTARY COMPOSITION,SOFT PARTS	157
*MYSIS RELICTA	10.0 (+-0.4) % OF ASH FREE DRY WEIGHT	MYSIDACEA, 95 % CONF LIMIT	1375
*NEMATODA	11.0 % OF DRY WEIGHT	MEAN ELEMENTARY COMPOSITION	157
*NOTONECTA GLAUCA	10.3 (+-0.6) % OF ASH FREE DRY WEIGHT	HETERPOTERA, 95 % CONF LIMIT	1375
*PALLACEA QUADRISPINOSA	9.4 (+-0.7) % OF ASH FREE DRY WEIGHT	AMPHIPODA, 95 % CONF LIMIT	1375
*SIALIS SP.	9.3 (+-1.1) % OF ASH FREE DRY WEIGHT	MEGALOPTERA, 95 % CONF LIMIT	1375

TABLE A264

INVERTEBRATES : NA/BIOMASS RATIO

ITEM:	VALUE:	CONDITIONS:	REF.:
*ARCTIIDAE	9.2 MG/(G ASH FREE DRY WT)	SOIL OR LITTER INVERT.	1311
*CAMPONOTUS PENNSYLVANICUS	9.4 MG/(G ASH FREE DRY WT)	SOIL OR LITTER INVERT.	1311
*CEUTHOPHILUS GRACILIPES	5.7 MG/(G ASH FREE DRY WT)	SOIL OR LITTER INVERT.	1311
*DIXIDESMUS ERASUS	3.7 MG/(G ASH FREE DRY WT)	SOIL OR LITTER INVERT.	1311
*GEOTRUPES SP.	3.4 MG/(G ASH FREE DRY WT)	SOIL OR LITTER INVERT.	1311
*LEIOBUNUM FLAVUM	5.7 MG/(G ASH FREE DRY WT)	SOIL OR LITTER INVERT.	1311
*LIGIDIUM BLUERIDGENSIS	5.1 MG/(G ASH FREE DRY WT)	SOIL OR LITTER INVERT.	1311
*MESOMPHIX VULGATUS	8.8 MG/(G ASH FREE DRY WT)	SOIL OR LITTER INVERT.	1311
*MESOMPHIX VULGATUS	2.2 MG/(G ASH FREE DRY WT)	SOIL OR LITTER INVERT.	1311
*OTOCRYPTOPS SEXSPINOSUS	9.2 MG/(G ASH FREE DRY WT)	SOIL OR LITTER INVERT.	1311

TABLE A264

INVERTEBRATES : NA/BIOMASS RATIO			
ITEM:	VALUE:	CONDITIONS:	REF.:
*PACHYLOMERIDES AUDOUINI	7.3 MG/(G ASH FREE DRY WT)	SOIL OR LITTER INVERT.	1311
*SCIOMYZIDAE	1.2 MG/(G ASH FREE DRY WT)	SOIL OR LITTER INVERT.	1311
*SCYTONOTUS GRANULATUS	1.4 MG/(G ASH FREE DRY WT)	SOIL OR LITTER INVERT.	1311
*SPHAERODERUS STENOSTOMUS	3.8 MG/(G ASH FREE DRY WT)	SOIL OR LITTER INVERT.	1311

TABLE A265

INVERTEBRATES : NUMBERS OF EGGS			
ITEM:	VALUE:	CONDITIONS:	REF.:
*AMPHIPHOLIS SQUAMATA	2-10 EGGS/INDIVIDUAL	OPHIUROID,DIRECT DEVELOPMENT	1396
*AMPHIURA ANGULARIS	5 EGGS/INDIVIDUAL	OPHIUROID,DIRECT DEVELOPMENT	1396
*AMPHIURA ANNULIFERA	2 EGGS/INDIVIDUAL	OPHIUROID,DIRECT DEVELOPMENT	1396
*AMPHIURA BELGICAE	10-200 EGGS/INDIVIDUAL	OPHIUROID,DIRECT DEVELOPMENT	1396
*AMPHIURA FILIFORMIS	CA 5000? EGGS/INDIVIDUAL	OPHIUROID,INDIRECT DEVELOPMENT	1396
*AMPHIURA MICROPLAX	100 EGGS/INDIVIDUAL	OPHIUROID,DIRECT DEVELOPMENT	1396
*AMPHIURA MONORIMA	5-10 EGGS/INDIVIDUAL	OPHIUROID,DIRECT DEVELOPMENT	1396
*ASTROCHLAMYS BRUNEUS	2000 EGGS/INDIVIDUAL	OPHIUROID,DIRECT DEVELOPMENT	1396
*KIRKS OPHIUROID	10-100 EGGS/INDIVIDUAL	OPHIUROID,DIRECT DEVELOPMENT	1396
*OPHIACANTHA DENSISPINA	20 EGGS/INDIVIDUAL	OPHIUROID,DIRECT DEVELOPMENT	1396
*OPHIACANTHA VIVIPARA	2-6 EGGS/INDIVIDUAL	OPHIUROID,DIRECT DEVELOPMENT	1396
*OPHIOCERES INCIPIENS	10-800 EGGS/INDIVIDUAL	OPHIUROID,DIRECT DEVELOPMENT	1396
*OPHIOCHONDRUS STELLIGER	5 EGGS/INDIVIDUAL	OPHIUROID,DIRECT DEVELOPMENT	1396
*OPHIOCOMA ECHINATA	717000 EGGS/INDIVIDUAL	OPHIUROID,INDIRECT DEVELOPMENT	1396
*OPHIOCOMA ERINACEUS	253000 EGGS/INDIVIDUAL	OPHIUROID,INDIRECT DEVELOPMENT	1396
*OPHIOLEBELLA BISCUTIFERA	3 EGGS/INDIVIDUAL	OPHIUROID,DIRECT DEVELOPMENT	1396
*OPHIOMITRELLA CLAVIGERA	10-20 EGGS/INDIVIDUAL	OPHIUROID,DIRECT DEVELOPMENT	1396
*OPHIOMITRELLA FALKLANDICA	20-50 EGGS/INDIVIDUAL	OPHIUROID,DIRECT DEVELOPMENT	1396
*OPHIOMITRELLA INGRATA	10 EGGS/INDIVIDUAL	OPHIUROID,DIRECT DEVELOPMENT	1396
*OPHIOMYXA BREVIRIMA	2-1200 EGGS/INDIVIDUAL	OPHIUROID,DIRECT DEVELOPMENT	1396
*OPHIONOTUS HEXACTIS	30-80 EGGS/INDIVIDUAL	OPHIUROID,DIRECT DEVELOPMENT	1396
*OPHIOPHOLIS ACULEATA	352000 EGGS/INDIVIDUAL	OPHIUROID,INDIRECT DEVELOPMENT	1396
*OPHIOPLOCUS ESMARKI	200 EGGS/INDIVIDUAL	OPHIUROID,DIRECT DEVELOPMENT	1396
*OPHIOSCOLEX NUTRIX	10-20 EGGS/INDIVIDUAL	OPHIUROID,DIRECT DEVELOPMENT	1396
*OPHIOSTEIRA ANTARCTICA	20-50 EGGS/INDIVIDUAL	OPHIUROID,DIRECT DEVELOPMENT	1396
*OPHIOZONELLA FALKLANDI	20-30 EGGS/INDIVIDUAL	OPHIUROID,DIRECT DEVELOPMENT	1396
*OPHIURA MERIDIONALIS	60-80 EGGS/INDIVIDUAL	OPHIUROID,DIRECT DEVELOPMENT	1396
*OPHIURA ROUCHI	100 EGGS/INDIVIDUAL	OPHIUROID,DIRECT DEVELOPMENT	1396
*OPHIUROLEPIS MARTENSI	50-60 EGGS/INDIVIDUAL	OPHIUROID,DIRECT DEVELOPMENT	1396
*PECTINURA CYLINDRICA	< 10 EGGS/INDIVIDUAL	OPHIUROID,DIRECT DEVELOPMENT	1396
*PECTINURA GRACILIS	< 10 EGGS/INDIVIDUAL	OPHIUROID,DIRECT DEVELOPMENT	1396
*STEGOPHIURA NODOSA	120 EGGS/INDIVIDUAL	OPHIUROID,DIRECT DEVELOPMENT	1396
*STEGOPHIURA SCULPTA	700 EGGS/INDIVIDUAL	OPHIUROID,DIRECT DEVELOPMENT	1396

TABLE A266

INVERTEBRATES : O/BIOMASS RATIO			
ITEM:	VALUE:	CONDITIONS:	REF.:
*ANNELIDA	34.0 % OF DRY WEIGHT	MEAN ELEMENTARY COMPOSITION	157

TABLE A266

INVERTEBRATES : O/BIOMASS RATIO			
ITEM:	VALUE:	CONDITIONS:	REF.:
*ARTHROPODA	32.3 % OF DRY WEIGHT	MEAN ELEMENTARY COMPOSITION,LAND	157
*COELENTERATA	27.1 % OF DRY WEIGHT	MEAN ELEMENTARY COMPOSITION	157
*ECHINODERMATA	45.6 % OF DRY WEIGHT	MEAN ELEMENTARY COMPOSITION	157
*MOLLUSCA	39.0 % OF DRY WEIGHT	MEAN ELEMENTARY COMPOSITION,SOFT PARTS	157
*NEMATODA	35.4 % OF DRY WEIGHT	MEAN ELEMENTARY COMPOSITION	157

TABLE A267

INVERTEBRATES : OVIPOSITION RATE			
ITEM:	VALUE:	CONDITIONS:	REF.:
*FLY	0.056 KCAL/KCAL*DAY	INITIAL VALUES, THERMAL ECOSYSTEM, TRANSFER COEFF.	1703

TABLE A268

INVERTEBRATES : P/BIOMASS RATIO			
ITEM:	VALUE:	CONDITIONS:	REF.:
*BENTHOS	0.015 BY WEIGHT	USED IN MODELS, N FORK KINGS RIVER, CALIF.	1566
*BENTHOS	0.015 BY WEIGHT	USED IN MODELS, BOISE RIVER, IDAHO	241
*BENTHOS	0.015 BY WEIGHT	USED IN MODELS, N FORK KINGS RIVER, CALIF.	1566
*BENTHOS	0.015 BY WEIGHT	USED IN MODELS, BOISE RIVER, IDAHO	241
*BENTHOS	0.01-0.012	NONE	1571

TABLE A269

INVERTEBRATES : PRODUCTION/BIOMASS RATIO			
ITEM:	VALUE:	CONDITIONS:	REF.:
*ACANTHINA SPINATA	0.3 1/YEAR	YEARLY TURNOVER, SOUTHERN CALIFORNIA MACROINVERTEBRATES	1462
*ACMAEA SP.	1.0 ON YEARLY BASIS	YEARLY TURNOVER, SOUTHERN CALIFORNIA MACROINVERTEBRATES	1462
*ALETES SQUAMIGERUS	0.3 ON YEARLY BASIS	YEARLY TURNOVER, SOUTHERN CALIFORNIA MACROINVERTEBRATES	1462
*ALLOPERLA MEDIANA	2.8 ON YEARLY BASIS	PROD=0.17 KG DW/(HA*Y), WALKER BRANCH, VOLTINISM=2 Y, TENN.	1661
*AMPHIPODA	0.008 - 0.04	PER DAY, SOUTHERN SEAS OF USSR	567
*ANCYLUS LACUSTRIS	3.3 ON YEARLY AND SHELL-FREE BASIS	PROD=0.09 KG DW/(HA*Y), R. THAMES, VOLTINISM=UNI, ENGL.	1661
*ANCYLUS LACUSTRIS	3.32 ON YEARLY BASIS	PRODUCTION=0.83G WET WT./YR,ALLAN CURVE METHOD	976
*ANODONTA ANATINA	0.20 ON YEARLY AND SHELL-FREE BASIS	PROD=22.0 KG DW/(HA*Y), R. THAMES, VOLTINISM=2+ Y, ENGL.	1661
*ANODONTA ANATINA	0.20 ON YEARLY BASIS	PRODUCTION=211.0G WET WT./YR,LESS SHELLS	976
*ANODONTA PISCINALIS	0.33 ON YEARLY AND SHELL-FREE BASIS	PROD=4.9 KG DW/(HA*Y), DANUBE R. POOLS, VOLTINISM=2+ Y, ROMANIA	1661
*ANTROPORA SP.	2.0 ON YEARLY BASIS	YEARLY TURNOVER, SOUTHERN CALIFORNIA MACROINVERTEBRATES	1462

TABLE A269

INVERTEBRATES : PRODUCTION/BIOMASS RATIO

ITEM:	VALUE:	CONDITIONS:	REF.:
*ARCHYNOPTERYX CURVATA	3.8 ON YEARLY BASIS	SAGEHEN CREEK, VOLTINISM=UNI, CALIF.	1661
*ASELLUS AQUATICUS	2.0 ON YEARLY BASIS	PROD=27.8 KG DW/(HA*Y), L. PAJEP MASKE., VOLTINISM=2 Y, SWEDEN	1661
*ASELLUS AQUATICUS	2.0 ON YEARLY BASIS	PROD=52.7 KG DW/(HA*Y), L. ERKEN, VOLTINISM=UNI, SWEDEN	1661
*ASELLUS AQUATICUS	3.2 ON YEARLY BASIS	PROD=1.38 KG DW/(HA*Y), R. THAMES, VOLTINISM=MULTI, ENGL.	1661
*ASELLUS AQUATICUS	4.7 ON YEARLY BASIS	PROD=9.4 KG DW/(HA*Y), EGLWYS NUNYDD R., VOLTINISM=BI, WALES, UK	1661
*ASELLUS AQUATICUS	3.17 ON YEARLY BASIS	PRODUCTION=13.30G WET WT./YR,ALLAN CURVE METHOD	976
*ASTRAEA UNDOSA	0.1 ON YEARLY BASIS	YEARLY TURNOVER, SOUTHERN CALIFORNIA MACROINVERTEBRATES	1462
*ASTROMETIS SERTULIFERA	0.3 ON YEARLY BASIS	YEARLY TURNOVER, SOUTHERN CALIFORNIA MACROINVERTEBRATES	1462
*ASTROPECTEN ARMATUS	0.3 ON YEARLY BASIS	YEARLY TURNOVER, SOUTHERN CALIFORNIA MACROINVERTEBRATES	1462
*ATHRIPSOIDES ANCYLUS	5.8 ON YEARLY BASIS	PROD=0.099 KG DW/(HA*Y), BRASHEEAS CR., VOLTINISM=UNI, KENTUCKY	1661
*BAETIS RHODANI	8.4 ON YEARLY BASIS	PROD=8.9 KG DW/(HA*Y), TROUT STREAMS, VOLTINISM=BI, CZECH.	1661
*BAETIS VAGANS	9.7 ON YEARLY BASIS	PROD=21 KG DW/(HA*Y), VALLEY CR., VOLTINISM=TRI, MINN.	1661
*BALANOPHYLLIA SP.	0.1 ON YEARLY BASIS	YEARLY TURNOVER, SOUTHERN CALIFORNIA MACROINVERTEBRATES	1462
*BALANUS SP.	1.0 ON YEARLY BASIS	YEARLY TURNOVER, SOUTHERN CALIFORNIA MACROINVERTEBRATES	1462
*BELLAMYA UNICOLOR	5.5 ON YEARLY AND SHELL-FREE BASIS	PROD=156.3 KG DW/(HA*Y), LAKE CHAD, VOLTINISM=UNI, MAX VALUE	1661
*BITHYNIA TENTACULATA	1.6 ON YEARLY AND SHELL-FREE BASIS	PROD=1.1 KG DW/(HA*Y), R. THAMES, VOLTINISM=2-3 Y, ENGL.	1661
*BITHYNIA TENTACULATA	1.63 ON YEARLY BASIS	PRODUCTION=10.60G WET WT./YR,ALLAN CURVE METHOD	976
*BURSA CALIFORNICUS	0.1 ON YEARLY BASIS	YEARLY TURNOVER, SOUTHERN CALIFORNIA MACROINVERTEBRATES	1462
*CHAMA PELLUCIDA	0.3 ON YEARLY BASIS	YEARLY TURNOVER, SOUTHERN CALIFORNIA MACROINVERTEBRATES	1462
*CHAOBORUS FLAVICANS	1.7 ON YEARLY BASIS	PROD=29.2 KG DW/(HA*Y), L. ESROM, VOLTINISM=UNI, DEMN	1661
*CHIONE UNDATELLA	0.2 ON YEARLY BASIS	YEARLY TURNOVER, SOUTHERN CALIFORNIA MACROINVERTEBRATES	1462
*CHIRONOMIDAE	4.97 ON YEARLY BASIS	PRODUCTION=8.65G WET WT./YR,7SPP., ALLAN CURVE METHOD	976
*CHIRONOMUS ANTHRACINUS	3.8 ON YEARLY BASIS	PROD=128.9 KG DW/(HA*Y), L. ESROM, VOLTINISM=2 Y, DENM.	1661
*CHIRONOMUS COMMUTATUS	1.49	MAX BIOMASS USED, YEARLY VALUE	894
*CHIRONOMUS PLUMOSUS	6.4 ON YEARLY BASIS	PROD=35.1 KG DW/(HA*Y), EGLWYS NUNYDD R., UK, VOLTINISM=BI	1661
*CHITONS	0.3 ON YEARLY BASIS	YEARLY TURNOVER, SOUTHERN CALIFORNIA MACROINVERTEBRATES	1462
*CHOROTERPES MEXICANUS	15.4 ON YEARLY BASIS	PROD=13.7 KG DW/(HA*Y), BRAZOS R. VOLTINISM=TRI, TEXAS	1661
*CHTHAMALUS SP.	2.0 ON YEARLY BASIS	YEARLY TURNOVER, SOUTHERN CALIFORNIA MACROINVERTEBRATES	1462
*CLADOTANYTARSUS MANCUS	5.2 ON YEARLY BASIS	PROD= 1.0 KG DW/(HA*Y), EGLWYS NUNYDD R., UK, VOLTINISM=BI	1661
*CLEOPATRA BULIMOIDES	3.5 ON YEARLY AND SHELL-FREE BASIS	PROD=119.0 KG DW/(HA*Y), LAKE CHAD, VOLTINISM=1+ Y, MAX VALUE	1661
*CONUS CALIFORNICUS	0.2 ON YEARLY BASIS	YEARLY TURNOVER, SOUTHERN CALIFORNIA MACROINVERTEBRATES	1462
*CORBICULA AFRICANA	2.9 ON YEARLY AND SHELL-FREE BASIS	PROD=106.3 KG DW/(HA*Y), LAKE CHAD, VOLTINISM=1-2 Y, MAX VALUE	1661
*CRANGONYX PSEUDOGRACILIS	5.8 ON YEARLY BASIS	PROD=0.22 KG DW/(HA*Y), R. THAMES, VOLTINISM=MULTI, ENGL.	1661
*CRANGONYX PSEUDOGRACILIS	5.83 ON YEARLY BASIS	PRODUCTION=2.10G WET WT./YR,ALLAN CURVE METHOD	976

TABLE A269

INVERTEBRATES : PRODUCTION/BIOMASS RATIO

ITEM:	VALUE:	CONDITIONS:	REF.:
*CRANGONYX RICHMONDENSIS	2.0 ON YEARLY BASIS	PROD=3.5 KG DW/(HA*Y), MARION L., VOLTINISM=1.5 Y, B.C.	1661
*CREPIDULA SP.	0.2 ON YEARLY BASIS	YEARLY TURNOVER, SOUTHERN CALIFORNIA MACROINVERTEBRATES	1462
*CYPRAEA SPADICEA	1.0 ON YEARLY BASIS	YEARLY TURNOVER, SOUTHERN CALIFORNIA MACROINVERTEBRATES	1462
*CYRNUS TRIMACULATUS	4.6 ON YEARLY BASIS	PROD=0.38 KG DW/(HA*Y), R. THAMES, VOLTINISM=UNI, ENGL.	1661
*CYRNUS TRIMACULATUS	4.63 ON YEARLY BASIS	PRODUCTION=3.70G WET WT./YR,ALLAN CURVE METHOD	976
*DELEATIDIUM SP.	3.5 ON YEARLY BASIS	Z42.3 KG DW/(HA*Y), SELWYN R., VOLTINISM=BI, NZ	1661
*DERMASTERIAS IMBRICATA	0.3 ON YEARLY BASIS	YEARLY TURNOVER, SOUTHERN CALIFORNIA MACROINVERTEBRATES	1462
*DIAPEROCCIA CALIFORNICA	0.3 ON YEARLY BASIS	YEARLY TURNOVER, SOUTHERN CALIFORNIA MACROINVERTEBRATES	1462
*ECDYONURUS SP.	8.4 ON YEARLY BASIS	PROD=14.1 KG DW/(HA*Y), TROUT STREAMS, VOLTINISM=UNI, CZECH.	1661
*EPHEMERA STRIGATA	2.3 ON YEARLY BASIS	PROD=7.1 KG DW/(HA*Y), YOSHINO R., VOLTINISM=UNI, JAPAN	1661
*EPHEMERELLA SUBVARIA	5.8 ON YEARLY BASIS	PROD=44.5 KG DW/(HA*Y), LUXEMBURG CR., VOLTINISM=UNI, MINN.	1661
*EPITONIUM SP.	0.2 ON YEARLY BASIS	YEARLY TURNOVER, SOUTHERN CALIFORNIA MACROINVERTEBRATES	1462
*ERPOBDELLA OCTOCULATA	1.5 ON YEARLY BASIS	PROD=36.0 KG DW/(HA*Y), WILFIN BECK, VOLTINISM=2 Y, N=3 Y, ENGL.	1661
*ERPOBDELLA OCTOCULATA	3.5 ON YEARLY BASIS	PROD=3.4 KG DW/(HA*Y), R. THAMES, VOLTINISM=2-3 Y, ENGL.	1661
*ERPOBDELLA OCTOCULATA	3.52 ON YEARLY BASIS	PRODUCTION=32.80G WET WT./YR,ALLAN CURVE METHOD	976
*EUPOMATUS SP.	1.0 ON YEARLY BASIS	YEARLY TURNOVER, SOUTHERN CALIFORNIA MACROINVERTEBRATES	1462
*FISSURELLA VOLCANO	0.3 ON YEARLY BASIS	YEARLY TURNOVER, SOUTHERN CALIFORNIA MACROINVERTEBRATES	1462
*GAMMARUS PULEX	3.35 ON YEARLY BASIS	YOUNG COHORT, PRODUCTION=2.6 G DW/(M2*Y)	729
*GAMMARUS PULEX	1.22 ON YEARLY BASIS	OLD COHORT, PRODUCTION=1.2 G DW/(M2*Y)	729
*GAMMARUS PULEX	6.7 ON YEARLY BASIS	PROD=0.14 KG DW/(HA*Y), R. THAMES, VOLTINISM=MULTI, ENGL.	1661
*GAMMARUS PULEX	6.66 ON YEARLY BASIS	PRODUCTION=1.40G WET WT./YR,ALLAN CURVE METHOD	976
*GAMMARUS TIGRINUS	6.2 ON YEARLY BASIS	PROD=130 KG DW/(HA*Y), TJEUKEMEER R., VOLTINISM=UNI, NETHERLANDS	1661
*GLOSSIPHONIA COMPLANATA	2.1 ON YEARLY BASIS	PROD=0.58 KG DW/(HA*Y), R. THAMES, VOLTINISM=2-3 Y, ENGL.	1661
*GLOSSIPHONIA COMPLANATA	2.07 ON YEARLY BASIS	PRODUCTION=5.57G WET WT./YR,ALLAN CURVE METHOD	976
*GLOSSIPHONIA HETEROCLITA	5.3 ON YEARLY BASIS	PROD=0.17 KG DW/(HA*Y), R. THAMES, VOLTINISM=UNI, ENGL.	1661
*GLOSSIPHONIA HETEROCLITA	5.25 ON YEARLY BASIS	PRODUCTION=1.68G WET WT./YR,ALLAN CURVE METHOD	976
*GLYPTOTENDIPES BARBIPES	8.5 ON YEARLY BASIS	PROD=74 KG DW/(HA*Y), WASTE LAGOON, ENTIRE LAG., VOLTINISM=MULTI, ORE.	1661
*GLYPTOTENDIPES BARBIPES	8.5 ON YEARLY BASIS	PROD=1616 KG DW/(HA*Y), WASTE LAGOON, NEAR SHORE, VOLTINISM=MULTI,ORE.	1661
*GLYPTOTENDIPES PARIPES	5.9 ON YEARLY BASIS	PROD=30.8 KG DW/(HA*Y), EGLWYS NUNYDD R., UK, VOLTINISM=BI	1661
*GLYPTOTENDIPES SP.	5.0 ON YEARLY BASIS	PROD=405 KG DW/(HA*Y), LOCH LEVEN, MAX 2 Y, VOLTINISM=MULTI, SCOTLAND	1661
*HALIOTIS RUFESCENS	0.05 ON YEARLY BASIS	YEARLY TURNOVER, SOUTHERN CALIFORNIA MACROINVERTEBRATES	1462
*HELOBDELLA STAGNALIS	3.0 ON YEARLY BASIS	PROD=11.9 KG DW/(HA*Y), EGLWYS NUNYDD R., VOLTINISM=BI, MAX 2 Y, UK	1661

TABLE A269

INVERTEBRATES : PRODUCTION/BIOMASS RATIO			
ITEM:	VALUE:	CONDITIONS:	REF.:
*HELOBDELLA STAGNALIS	3.5 ON YEARLY BASIS	PROD=8.7 KG DW/(HA*Y), EGLWYS NUNYDD R., VOLTINISM=BI, UK	1661
*HELOBDELLA STAGNALIS	4.8 ON YEARLY BASIS	PROD=3.5 KG DW/(HA*Y), R. THAMES, VOLTINISM=2-3 Y, ENGL.	1661
*HELOBDELLA STAGNALIS	4.84 ON YEARLY BASIS	PRODUCTION=21.7G WET WT./YR,ALLAN CURVE METHOD	976
*HENRICIA LEVIUSCULA	0.3 ON YEARLY BASIS	YEARLY TURNOVER, SOUTHERN CALIFORNIA MACROINVERTEBRATES	1462
*HEXAGENIA LIMBATA	4.4 ON YEARLY BASIS	PROD=1.2 KG DW/(HA*Y), TUTTLE CR. R., N=4 Y, VOLTINISM=UNI, KANSAS	1661
*HEXAGENIA LIMBATA AND H. BILINEATA	2.8 ON YEARLY BASIS	PROD=15.0 KG DW(HA*Y), CLARK L. VOLTINISM=1.7 Y, NEBRASKA	1661
*HYALELLA AZTECA	4.8 ON YEARLY BASIS	PROD=19.3 KG DW/(HA*Y), SUGARLOAF L., VOLTINISM=BI, MICH.	1661
*HYALELLA AZTECA	4.2 ON YEARLY BASIS	PROD=11.5 KG DW/(HA*Y), MARION L., VOLTINISM=UNI, B.C.	1661
*ILYODRILUS HAMMONIENSIS	0.7 ON YEARLY BASIS	PROD=11.7 KG DW/(HA*Y), LAKE ESROM, DENMARK	1661
*KELLETTIA KELLETTII	0.1 ON YEARLY BASIS	YEARLY TURNOVER, SOUTHERN CALIFORNIA MACROINVERTEBRATES	1462
*LIMNAEA PEREGER	11.52 ON YEARLY BASIS	PRODUCTION=48.40G WET WT./YR,ALLAN CURVE METHOD	976
*LIMNOCHIRONOMUS PULSUS	6.3 ON YEARLY BASIS	PROD=11.0 KG DW(HA*Y), EGLWYS NUNYDD R., UK, VOLTINISM=BI	1661
*LIMNOCHIRONOMUS SP.	3.6 ON YEARLY BASIS	PROD=5.9 KG DW/(H*Y), LOCH LEVEN, MAX 2 Y, VOLTINISM=MULTI, SCOTLAND	1661
*LIMNODRILUS SPP.	12.5 ON YEARLY BASIS	PROD=53.6 KG DW/(HA*Y), EGLWYS NUNYDD R., VOLTINISM=MULTI, WALES, UK	1661
*LINCKIA COLUMBIAE	0.3 ON YEARLY BASIS	YEARLY TURNOVER, SOUTHERN CALIFORNIA MACROINVERTEBRATES	1462
*LITHOPHAGA PLUMULA	0.2 ON YEARLY BASIS	YEARLY TURNOVER, SOUTHERN CALIFORNIA MACROINVERTEBRATES	1462
*LITTORINA SAXATILIS	4.1 ON YEARLY AND SHELL-FREE BASIS	PROD=32.5 KG DW/(HA*Y), NOVA SCOTIA, VOLTINISM=2+ Y, INTERTIDAL ESTU.	1661
*LOTTIA GIGANTEA	0.2 ON YEARLY BASIS	YEARLY TURNOVER, SOUTHERN CALIFORNIA MACROINVERTEBRATES	1462
*LOVENIA CORDIFORMIS	0.3 ON YEARLY BASIS	YEARLY TURNOVER, SOUTHERN CALIFORNIA MACROINVERTEBRATES	1462
*LYMNAEA PEREGRA	11.5 ON YEARLY AND SHELL-FREE BASIS	PROD=5.02 KG DW/(HA*Y), R. THAMES, VOLTINISM=MULTI, ENGL.	1661
*LYTECHINUS ANAMESUS	1.0 ON YEARLY BASIS	YEARLY TURNOVER, SOUTHERN CALIFORNIA MACROINVERTEBRATES	1462
*MACOMA BALTICA	1.5 ON YEARLY AND SHELL-FREE BASIS	PROD=19.3 KG DW/(H*Y), NVA SCOTIA, VOLTINISM=2+ Y, INTERTIDAL ESTU.	1661
*MEGATHURA CRENULATA	0.3 ON YEARLY BASIS	YEARLY TURNOVER, SOUTHERN CALIFORNIA MACROINVERTEBRATES	1462
*MELANIA TUBERCULATA	4.4 ON YEARLY AND SHELL-FREE BASIS	PROD=154.5 KG DW/(HA*Y), LAKE CHAD, VOLTINISM=1-2 Y, MAX VALUE	1661
*MEMBRANIPORA SP.	2.0 ON YEARLY BASIS	YEARLY TURNOVER, SOUTHERN CALIFORNIA MACROINVERTEBRATES	1462
*MESOPLANKTONIC GROUP	0.01 - 0.1	PER DAY, SOUTHERN SEAS OF USSR	567
*MICROTENDIPES CHLORIS	12.7 ON YEARLY BASIS	PROD=0.24 KG DW/(HA*Y), MIKOLAJSKIE L., VOLTINISM=MULTI, POLAND	1661
*MICROTENDIPES SP.	5.7 ON YEARLY BASIS	PROD=16.2 KG DW/(HA*Y), EGLWYS NUNYDD R., UK, VOLTINISM=BI	1661
*MISCELLANEOUS ANNELIDS	1.0 ON YEARLY BASIS	YEARLY TURNOVER, SOUTHERN CALIFORNIA MACROINVERTEBRATES	1462
*MISCELLANEOUS BARNACLES	1.0 ON YEARLY BASIS	YEARLY TURNOVER, SOUTHERN CALIFORNIA MACROINVERTEBRATES	1462
*MISCELLANEOUS BRACHIOPODS	1.0 ON YEARLY BASIS	YEARLY TURNOVER, SOUTHERN CALIFORNIA MACROINVERTEBRATES	1462
*MISCELLANEOUS BRYOZOANS	2.0 ON YEARLY BASIS	YEARLY TURNOVER, SOUTHERN CALIFORNIA MACROINVERTEBRATES	1462
*MISCELLANEOUS COELENTERATES	0.1 ON YEARLY BASIS	YEARLY TURNOVER, SOUTHERN CALIFORNIA MACROINVERTEBRATES	1462

TABLE A269

INVERTEBRATES : PRODUCTION/BIOMASS RATIO

ITEM:	VALUE:	CONDITIONS:	REF.:
*MISCELLANEOUS ECHINODERMS	0.3 ON YEARLY BASIS	YEARLY TURNOVER, SOUTHERN CALIFORNIA MACROINVERTEBRATES	1462
*MISCELLANEOUS MOLLUSEC	0.3 ON YEARLY BASIS	YEARLY TURNOVER, SOUTHERN CALIFORNIA MACROINVERTEBRATES	1462
*MITELLA POLYMERUS	1.0 ON YEARLY BASIS	YEARLY TURNOVER, SOUTHERN CALIFORNIA MACROINVERTEBRATES	1462
*MOLLUSCA	0.005 - 0.02	PER DAY, SOUTHERN SEAS OF USSR	567
*MURICEA SP.	0.1 ON YEARLY BASIS	YEARLY TURNOVER, SOUTHERN CALIFORNIA MACROINVERTEBRATES	1462
*MYA ARENARIA	2.5 ON YEARLY AND SHELL-FREE BASIS	PROD=116 KG DW/(HA*Y), NOVA SCOTIA, VOLTINISM=2+ Y, INTERTIDAL ESTU.	1661
*MYTILUS SP.	0.3 ON YEARLY BASIS	YEARLY TURNOVER, SOUTHERN CALIFORNIA MACROINVERTEBRATES	1462
*NEANTHES VIRENS	1.6 ON YEARLY BASIS	PROD=90.4 KG DW/(HA*Y), R. THAMES, VOLTINISM=2 Y, ESTUARY, ENGL.	1661
*NUTTALLINA CALIFORNICA	0.3 ON YEARLY BASIS	YEARLY TURNOVER, SOUTHERN CALIFORNIA MACROINVERTEBRATES	1462
*OLIGOCHAETA	5.5 - 9.3 ON YEARLY BASIS	PROD=12-97 KG DW/(HA*Y), PEAT LAND, CALIF.	1661
*OPHIUROIDEA	0.2 ON YEARLY BASIS	YEARLY TURNOVER, SOUTHERN CALIFORNIA MACROINVERTEBRATES	1462
*ORCONECTES VIRILIS	2.3 ON YEARLY BASIS	PROD=28.3 KG DW/(HA*Y), WEST LOST L., VOLTINISM=3.5 Y, MICH., 1962-3	1661
*ORCONECTES VIRILIS	1.1 ON YEARLY BASIS	PROD=8.2 KG DW/(HA*Y), NORTH TWIN L., VOLTINISM=3.5 Y, N=9 Y, MICH	1661
*ORCONECTES VIRILIS	0.94 ON YEARLY BASIS	PROD=22.3 KG DW/(HA*Y), WEST LOST L., VOLTINISM=3.5 Y, MICH, 1966-7	1661
*OSTREA LURIDA	0.1 ON YEARLY BASIS	YEARLY TURNOVER, SOUTHERN CALIFORNIA MACROINVERTEBRATES	1462
*PARACHIRONOMUS TENER	5.8 ON YEARLY BASIS	PROD= 1.1 KG DW/(HA*Y), EGLWYS NUNYDD R., UK, VOLTINISM=BI	1661
*PATIRIA MINIATA	0.3 ON YEARLY BASIS	YEARLY TURNOVER, SOUTHERN CALIFORNIA MACROINVERTEBRATES	1462
*PENITELLA CONRADI	0.3 ON YEARLY BASIS	YEARLY TURNOVER, SOUTHERN CALIFORNIA MACROINVERTEBRATES	1462
*PHIDOLOPORA SP.	0.5 ON YEARLY BASIS	YEARLY TURNOVER, SOUTHERN CALIFORNIA MACROINVERTEBRATES	1462
*PISATER SP.	0.3 ON YEARLY BASIS	YEARLY TURNOVER, SOUTHERN CALIFORNIA MACROINVERTEBRATES	1462
*PISIDIUM CASERTANUM	0.2 ON YEARLY AND SHELL-FREE BASIS	PROD=4.38 KG DW/(HA*Y), L. ESROM, VOLTINISM=2 Y, DENM.	1661
*PLANORBIS CONTORTUS	0.3 ON YEARLY AND SHELL-FREE BASIS	PROD=0.017 KG DW/(HA*Y), R. THAMES, VOLTINISM=UNI, ENGL	1661
*PLANORBIS CONTORTUS	0.32 ON YEARLY BASIS	PRODUCTION=0.16G WET WT./YR,ALLAN CURVE METHOD	976
*PLANORBIS VORTEX	6.5 ON YEARLY AND SHELL-FREE BASIS	PROD=0.11 KG DW/(HA*Y), R. THAMES, VOLTINISM=UNI, ENGL	1661
*PLANORBIS VORTEX	6.47 ON YEARLY BASIS	PRODUCTION=1.10G WET WT./YR,ALLAN CURVE METHOD	976
*PODODESMUS PERNOIDES	0.7 ON YEARLY BASIS	YEARLY TURNOVER, SOUTHERN CALIFORNIA MACROINVERTEBRATES	1462
*PONTOPOREIA AFFINIS	4.1 ON YEARLY BASIS	PROD=5.2 KG DW/(HA*Y), L. KRASNOYE, VOLTINISM=UNI, USSR	1661
*POTAMOPHYLAX CINGULATUS	4.4 ON YEARLY BASIS	PROD=7.7 KG DW/(HA*Y), STAMPES S., VOLTINISM=UNI, TISSUE ONLY, SWEDEN	1661
*POTAMOTHRIX HAMMONIENSIS	0.86 ON YEARLY BASIS	PROD=53.6 KG DW/(HA*Y), LAKE ESROM, VOLTINISM=4 Y, DK.	1661
*PROCLADIUS CHOREUS	6.5 ON YEARLY BASIS	PROD=34.1 KG DW/(HA*Y), EGLWYS NUNYDD R., UK, VOLTINISM=MULTI	1661
*PROCLADIUS PECTINATUS	1.9 ON YEARLY BASIS	PROD=5.2 KG DW/(HA*Y), L. ESROM, VOLTINISM=1 Y, DK.	1661
*PROCLADIUS RUFOVITTATUS	3.2 ON YEARLY BASIS	PROD=20.1 KG DW/(HA*Y), EGLWYS NUNYDD R., UK, VOLTINISM=UNI	1661
*PROTOTHARA STAMINEA	0.3 ON YEARLY BASIS	YEARLY TURNOVER, SOUTHERN CALIFORNIA MACROINVERTEBRATES	1462

TABLE A269

INVERTEBRATES : PRODUCTION/BIOMASS RATIO			
ITEM:	VALUE:	CONDITIONS:	REF.:
*PYRRHOSOMA NYMPHULA	3.2 ON YEARLY BASIS	PROD=7.1 KG DW/(HA*Y), PONDS IN ENGL., VOLTINISM=2 Y	1661
*RHITHROGENA SEMICOLORATA	8.4 ON YEARLY BASIS	PROD=20.8 KG DW/(HA*Y), TRUOT STREAMS, VOLTINISM=UNI, CZECH.	1661
*SCRUPOCELLARIA	2.0 ON YEARLY BASIS	YEARLY TURNOVER, SOUTHERN CALIFORNIA MACROINVERTEBRATES	1462
*SEMELE SP.	0.3 ON YEARLY BASIS	YEARLY TURNOVER, SOUTHERN CALIFORNIA MACROINVERTEBRATES	1462
*SERPULA SP.	1.0 ON YEARLY BASIS	YEARLY TURNOVER, SOUTHERN CALIFORNIA MACROINVERTEBRATES	1462
*SIALIS MITSUHASHII	3.9 ON YEARLY BASIS	PROD=2.5 KG DW/(HA*Y), L. TATSU-NAMA, VOLTINISM=3 Y, JAPAN	1661
*SIMULIUM BREVICAULE	6.2 ON YEARLY BASIS	PROD=0.60 KG DW/(H*Y), LISSURAGA R., VOLTINISM=MULTI, N=2 PLACES, FR.	1661
*SIMULIUM MONTICOLA	11.1 ON YEARLY BASIS	PROD=1.12 KG DW/(HA*Y), LISSURAGA R., VOLTINISM=MULTI, N=4 PLACES, FR.	1661
*SIMULIUM ORNATUM	12.9 ON YEARLY BASIS	PROD=1.29 KG DW/(H*Y), LISSURAGA R., N=2 PLACES, VOLTINISM=MULTI, FR.	1661
*SPHAERIUM CORNEUM	3.5 ON YEARLY AND SHELL-FREE BASIS	PROD=2.85 KG DW/(HA*Y), R. THAMES, VOLTINISM=2-3 Y, ENGL.	1661
*SPHAERIUM CORNEUM	3.53 ON YEARLY BASIS	PRODUCTION=27.50G WET WT./YR,ALLAN CURVE METHOD	976
*SPIRORBIS SP.	4.0 ON YEARLY BASIS	YEARLY TURNOVER, SOUTHERN CALIFORNIA MACROINVERTEBRATES	1462
*STENOPERLA PRASINA	2.5 ON YEARLY BASIS	PROD=20.8 KG DW/(HA*Y), SELWYN R., VOLTINISM=UNI, NZ	1661
*STICTOCHIRONOMUS SP.	9.8 ON YEARLY BASIS	PROD=102 KG DW/(HA*Y), LOCH LEVEN, MAX 2 Y, VOLTINISM=MULTI, SCOTLAND	1661
*STRONGYLOCENTROTUS FRANCISCANUS	1.0 ON YEARLY BASIS	YEARLY TURNOVER, SOUTHERN CALIFORNIA MACROINVERTEBRATES	1462
*STRONGYLOCENTROTUS PURPURATUS	1.0 ON YEARLY BASIS	YEARLY TURNOVER, SOUTHERN CALIFORNIA MACROINVERTEBRATES	1462
*SURCULITES SP.	0.2 ON YEARLY BASIS	YEARLY TURNOVER, SOUTHERN CALIFORNIA MACROINVERTEBRATES	1462
*TANYTARSUS HOLOCHLORUS	6.6 ON YEARLY BASIS	PROD=15.5 KG DW/(HA*Y), EGLWYS NUNYDD R., UK, VOLTINISM=BI	1661
*TANYTARSUS INOPERTUS	5.2 ON YEARLY BASIS	PROD= 9.1 KG DW/(HA*Y), EGLWYS NUNYDD R., UK, VOLTINISM=BI	1661
*TANYTARSUS LUGENS	7.6 ON YEARLY BASIS	PROD=24.0 KG DW/(HA*Y), EGLWYS NUNYDD R., UK, VOLTINISM=TRI	1661
*TEGULA FUNEBRALIS	0.1 ON YEARLY BASIS	YEARLY TURNOVER, SOUTHERN CALIFORNIA MACROINVERTEBRATES	1462
*TEREBRATALIA SP.	1.0 ON YEARLY BASIS	YEARLY TURNOVER, SOUTHERN CALIFORNIA MACROINVERTEBRATES	1462
*TEREBRATULINA SP.	1.0 ON YEARLY BASIS	YEARLY TURNOVER, SOUTHERN CALIFORNIA MACROINVERTEBRATES	1462
*TETRACLITA SP.	1.0 ON YEARLY BASIS	YEARLY TURNOVER, SOUTHERN CALIFORNIA MACROINVERTEBRATES	1462
*TUBIFICIDAE	1.00 ON YEARLY BASIS	PRODUCTION=71.0G WET WT./YR	976
*UNIO PICTORUM	0.14 ON YEARLY AND SHELL-FREE BASIS	PROD=8.8 KG DW/(HA*Y), R. THAMES, VOLTINISM=10-12 Y, ENGL.	1661
*UNIO PICTORUM	0.15 ON YEARLY AND SHELL-FREE BASIS	PROD=3.7 KG DW/(HA*Y), DANUBE R. POOLS, VOLTINISM=2+ Y, ROMANIA	1661
*UNIO PICTORUM	0.14 ON YEARLY BASIS	PRODUCTION=84.6G WET WT./YR	976
*UNIO TUMIDUS	0.13 ON YEARLY AND SHELL-FREE BASIS	PROD=3.3 KG DW/(HA*Y), R. THAMES, VOLTINISM=10-12 Y, ENGL.	1661
*UNIO TUMIDUS	0.25 ON YEARLY AND SHELL-FREE BASIS	PROD=3.4 KG DW/(HA*Y), DANUBE R. POOLS, VOLTINISM=2+ Y, ROMANIA	1661
*VIVIPARUS VIVIPARUS	1.6 ON YEARLY AND SHELL-FREE BASIS	PROD=4.46 KG DW/(HA*Y), R. THAMES, VOLTINISM=2-3 Y, ENGL.	1661
*VIVIPARUS VIVIPARUS	1.59 ON YEARLY BASIS	PRODUCTION=42.94G WET WT./YR,ALLAN CURVE METHOD	976

TABLE A270

INVERTEBRATES : RESPIRATION RATE			
ITEM:	VALUE:	CONDITIONS:	REF.:
*ARANEAE	3.0 MICROLITRE O2/(HOUR*ANIMAL)	GRASSLAND SOIL,292K,EQUIVALENT TO 27GCAL/(G*DAY) AT 289K	963

TABLE A270

INVERTEBRATES : RESPIRATION RATE

ITEM:	VALUE:	CONDITIONS:	REF.:
*ASELLUS AQUATICUS	24.7 % OF INGESTION	NONE	1689
*ASELLUS AQUATICUS	81.8 % OF ASSIMILATION	NONE	1689
*BENTHIC FORAMINIFERA	0.5 - 4.5 MICROLITRE/(MG WET WEIGHT*HOUR)	OPTIMUM LAB. CONDITIONS, 278 - 308 K	901
*BENTHOS	0.002 1/DAY	USED IN MODELS, N FORK KINGS RIVER, CALIF.	1566
*BENTHOS	0.001 1/DAY	USED IN MODELS, N FORK KINGS RIVER, CALIF.	1566
*BENTHOS	0.001 - 0.01 1/DAY	USED IN MODELS	1567
*BENTHOS	0.0001 - 0.001 1/DAY	USED IN MODELS	1567
*BENTHOS	0.001 1/DAY	USED IN MODELS, LAKE WASHINGTON	240
*BENTHOS	0.001 1/DAY	USED IN MODELS, SAN FRANCISCO BAY ESTUARY	242
*BENTHOS	0.01 1/DAY	USED IN MODELS, BOISE RIVER, IDAHO	241
*BENTHOS	0.002 1/DAY	USED IN MODELS, N FORK KINGS RIVER, CALIF., ACTIVE	1566
*BENTHOS	0.001 1/DAY	USED IN MODELS, N FORK KINGS RIVER, CALIF., INACTIVE	1566
*BENTHOS	0.001 - 0.01 1/DAY	USED IN MODELS, ACTIVE	1567
*BENTHOS	0.0001 - 0.001 1/DAY	USED IN MODELS, INACTIVE	1567
*BENTHOS	0.001 1/DAY	USED IN MODELS, LAKE WASHINGTON	240
*BENTHOS	0.001 1/DAY	USED IN MODELS, SAN FRANCISCO BAY ESTUARY	242
*BENTHOS	0.01 1/DAY	USED IN MODELS, BOISE RIVER, IDAHO	241
*BENTHOS	0.001-0.01 1/DAY	TEMP.FACTOR IS 1 AT 293K,TEMP>278K	1571
*BENTHOS	0.0001-0.001 1/DAY	TEMP.FACTOR IS 1 AT 293K, TEMP.<278K	1571
*CEUTHOPHILUS GRACILIPES	263 MICROLITREO2/(G WET WEIGHT*HOUR)	MEAN WET W. = 259 MG, 293 K, SOIL INVERT.	1310
*COGNETTIA COGNETTII	625 MICROLITREO2/(G WET WEIGHT*HOUR)	MEAN WET W. = 0.19 MG, 293 K, SOIL INVERT.	1150
*COLEOPTERA	60 MICROLITRE O2/(HOUR*ANIMAL)	GRASSLAND SOIL,286K,EQUIVALENT TO 39GCAL/(G*DAY) AT 289K	963
*COLLEMBOLA	0.55 MICROLITRE O2/(HOUR*ANIMAL)	GRASSLAND SOIL,289K,EQUIVALENT TO 144GCAL/(G*DAY) AT 289K	963
*CYLISTICUS CONVEXUS	205 MICROLITREO2/(G WET WEIGHT*HOUR)	MEAN WET W. = 35 MG,293 K, SOIL INVERT.	1309
*DIPLACODES TRIVIALIS	51.9 % OF ASSIMILATION	46 MG ANIMAL, INGESTION = 61.5 CAL/(KCAL*ANIM.*D)	1209
*DIPTERA	105 MICROLITRE O2/(HOUR*ANIMAL)	GRASSLAND SOIL,286K,EQUIVALENT TO 29GCAL/(G*DAY) AT 289K	963
*DIXIDESMUS ERASUS	140 MICROLITREO2/(G WET WEIGHT*HOUR)	MEAN WET W. = 78 MG, 293 K, SOIL INVERT.	1310
*ENCHYTRAEIDAE	0.117 MM**3 O2/HR*INDIVIDUAL	GRASSLAND SOIL,289K,EQUIVALENT TO 100GCAL/G*DAY AT 289K	963
*FLY, ADULT	0.470 KCAL/KCAL*DAY	INITIAL VALUES, THERMAL ECOSYSTEM, TRANSFER COEFF.	1703
*FLY, EGG	0.060 KCAL/KCAL*DAY	INITIAL VALUES, THERMAL ECOSYSTEM, TRANSFER COEFF.	1703
*FLY, PUPAE	0.030 KCAL/KCAL*DAY	INITIAL VALUES, THERMAL ECOSYSTEM, TRANSFER COEFF.	1703
*FLY, 1,2 INSTAR LARVAE	0.360 KCAL/KCAL*DAY	INITIAL VALUES, THERMAL ECOSYSTEM, TRANSFER COEFF.	1703
*FLY, 3 INSTAR LARVAE	0.330 KCAL/KCAL*DAY	INITIAL VALUES, THERMAL ECOSYSTEM, TRANSFER COEFF.	1703
*ISOPODA	9 MICROLITRE O2/(HOUR*ANIMAL)	GRASSLAND SOIL,292K,EQUIVALENT TO 36GCAL/(G*DAY) AT 289K	963
*ISOPTERA	199 MICROLITREO2/(G WET WEIGHT*HOUR)	MEAN WET W. = 2.1 MG, 293 K, SOIL INVERT.	1310
*LESTES SPONSA	13.2 % OF INGESTION	NONE	1689
*LESTES SPONSA	36.0 % OF ASSIMILATION	NONE	1689
*LIMNODRILUS HOFFMEISTERI	0.18 MICROLITRE /MG*HR	278K,RESPIRATION RATE,MEAN OF 3 EXP.	173

TABLE A270

INVERTEBRATES : RESPIRATION RATE			
ITEM:	VALUE:	CONDITIONS:	REF.:
*LIMNODRILUS HOFFMEISTERI	0.21 MICROLITRE /MG*HR	283K,RESPIRATION RATE,MEAN OF 3 EXP.	173
*LIMNODRILUS HOFFMEISTERI	0.31 MICROLITRE /MG*HR	288K,RESPIRATION RATE,MEAN OF 3 EXP.	173
*LIMNODRILUS HOFFMEISTERI	0.48 MICROLITRE /MG*HR	293K,RESPIRATION RATE,MEAN OF 3 EXP.	173
*LUMBRICIDAE	300 MICROLITRE O2/(HOUR*ANIMAL)	GRASSLAND SOIL,289K,EQUIVALENT TO 7GCAL/(G*DAY) AT 289K	963
*LUMBRICUS TERRESTRIS	36 MICROLITREO2/(G WET WEIGHT*HOUR)	MEAN WET W. = 2700 MG, 288 K, SOIL INVERT.	1383
*MESODON THYROIDES	91 MICROLITREO2/(G WET WEIGHT*HOUR)	MEAN WET W. = 2108 MG(INC SHELL), 293 K, SOIL INVERT.	390
*MITOPUS MORIO	610 - 960 MICROLITREO2/(G WET WEIGHT*HOUR)	MEAN WET W. = 3.7 - 56.1 MG, 289 K, SOIL INVERT.	1254
*MYRIAPODA	10 MICROLITRE O2/(HOUR*ANIMAL)	GRASSLAND SOIL,292K,EQUIVALENT TO 36GCAL/(G*DAY) AT 289K	963
*NEMATODA	582 - 354 MICROLITREO2/(G WET WEIGHT*HOUR)	MEAN WET W. = 0.000135-0.000196 MG,289 K, SOIL INVERT.	1125
*NEMATODA	1.2*10**(-3) MICROLITRE O2/ HR*INDIVIDUAL	GRASSLAND SOIL,289K,EQUIVALENT TO 144GCAL/G*DAY AT 289K	963
*ONYCHIURUS PROCAMPATUS	320 - 1000 MICROLITREO2/(G WET WEIGHT*HOUR)	MEAN WET W. = 0.01 - 0.10 MG, 288 K, SOIL INVERT.	648
*OPILIONES	10.5 MICROLITRE O2/(HOUR*ANIMAL)	GRASSLAND SOIL,292K,EQUIVALENT TO 53GCAL/(G*DAY) AT 289K	963
*ORBITED MITES	0.12 MICROLITRE O2/(HOUR*ANIMAL)	GRASSLAND SOIL,287K,EQUIVALENT TO 72GCAL/(G*DAY) AT 289K	963
*OTOCRYPTOPS SEXSPINOSUS	73 MICROLITREO2/(G WET WEIGHT*HOUR)	MEAN WET W. = 432 MG, 293 K, SOIL INVERT.	390
*PARAMECIUM SP.	0.5 ML O2/(G WET WT.)*HR	293K,CILIATE	524
*PARASITIC MITES	0.5 MICROLITRE O2/(HOUR*ANIMAL)	GRASSLAND SOIL,289K,EQUIVALENT TO 280GCAL/(G*DAY) AT 289K	963
*PARCOBLATTA SP.	158 MICROLITREO2/(G WET WEIGHT*HOUR)	MEAN WET W. = 173 MG, 293 K, SOIL INVERT.	1310
*PARHYALLA WHELPLEYI	2.95 MG O(2)/(G DRY WEIGHT*HOUR)	297 K, 92 H	450
*PELOSCOLEX MULTISETOSUS	0.39 MICROLITRE /MG*HR	278K,RESPIRATION RATE,MEAN OF 3 EXP.	173
*PELOSCOLEX MULTISETOSUS	0.35 MICROLITRE /MG*HR	283K,RESPIRATION RATE,MEAN OF 3 EXP.	173
*PELOSCOLEX MULTISETOSUS	0.42 MICROLITRE /MG*HR	288K,RESPIRATION RATE,MEAN OF 3 EXP.	173
*PELOSCOLEX MULTISETOSUS	0.56 MICROLITRE /MG*HR	293K,RESPIRATION RATE,MEAN OF 3 EXP.	173
*PHILAENUS SPUMARIUS	0.34 CAL/(ANIMAL*DAY)	3RD INSTAR, 0.39 MG, 296 K	1705
*PHILAENUS SPUMARIUS	0.76 CAL/(ANIMAL*DAY)	4RD INSTAR, 1.16 MG, 298 K	1705
*PHILAENUS SPUMARIUS	1.00 CAL/(ANIMAL*DAY)	ADULT, 3.87 MG, 294 K, EXCRETING URIC ACID	1705
*PHILAENUS SPUMARIUS	1.00 CAL/(ANIMAL*DAY)	ADULT, 3.87 MG, 294 K, EXCRETING UREA	1705
*PLATYNORTHRUS PELTIFER	925 MICROLITREO2/(G WET WEIGHT*HOUR)	MEAN WET W. = 0.063 MG, 288 K, SOIL INVERT.	109
*POPILIUS DISJUNCTUS	124 MICROLITREO2/(G WET WEIGHT*HOUR)	MEAN WET W. = 1630 MG, 293 K, SOIL INVERT.	1310
*PROTOZOA	10**(-2) MICROLITRE O2/ HR*INDIVIDUAL	GRASSLAND SOIL,295.5K,EQUIVALENT TO 14GCAL/G*DAY AT 289K	963
*PTYOIULUS IMPRESSUS	56 MICROLITREO2/(G WET WEIGHT*HOUR)	MEAN WET W. = 432 MG, 293 K, SOIL INVERT.	1310
*RHINCALANUS GIGAS	0.47 (SD=+-0.05) MICROL O2/(MG DRY WEIGHT*HOUR)	271.7 K, RESPIRATION ARTIC POND	1293
*RHINCALANUS GIGAS	3.07 (SD=+-0.40) MICROL O2/ (AMINAL*HOUR)	271.7 K, RESPIRATION ARTIC POND	1293
*SCHIZOCOSA SP.	251 MICROLITREO2/(G WET WEIGHT*HOUR)	MEAN WET W. = 23 MG, 293 K, SOIL INVERT.	1072
*SNAILS	1000 MM**3 O2/HR*INDIVIDUAL	GRASSLAND SOIL,301K,EQUIVALENT TO 29GCAL/G*DAY AT 289K	963
*SPHAERODERUS STENOSTOMUS	194 MICROLITREO2/(G WET WEIGHT*HOUR)	MEAN WET W. = 169 MG, 293 K, SOIL INVERT.	1310

TABLE A270

INVERTEBRATES : RESPIRATION RATE			
ITEM:	VALUE:	CONDITIONS:	REF.:
*TIPULA SP.	146 MICROLITREO2/(G WET WEIGHT*HOUR)	MEAN WET W. = 277 - 607 MG, 286 K, SOIL INVERT.	149
*TUBIFEX TUBIFEX	0.24 MICROLITRE /MG*HR	278K,RESPIRATION RATE,MEAN OF 3 EXP.	173
*TUBIFEX TUBIFEX	0.21 MICROLITRE /MG*HR	283K,RESPIRATION RATE,MEAN OF 3 EXP.	173
*TUBIFEX TUBIFEX	0.40 MICROLITRE /MG*HR	288K,RESPIRATION RATE,MEAN OF 3 EXP.	173
*TUBIFEX TUBIFEX	0.51 MICROLITRE /MG*HR	293K,RESPIRATION RATE,MEAN OF 3 EXP.	173

TABLE A271

INVERTEBRATES : RESPIRATORY COEFFICIENT			
ITEM:	VALUE:	CONDITIONS:	REF.:
*ASELLUS AQUATICUS	4.49	P/R, 0 - 116 D	1012
*CTENOPHARYNGODON IDELLA	5.90	P/R, ON PLANT FOOD, 6 M, 20-70 G	1012
*CTENOPHARYNGODON IDELLA	1.47	P/R, ON AMINAL FOOD, 6 M, 40-120 G	1012
*LESTES SPONSA	0.56	P/R, 0 - 37 D	1012
*RHIZOGLYPHUS ECHINOPUS	1.84	P/R, LARVAE ACTIVE, 0-11 D	1012
*RHIZOGLYPHUS ECHINOPUS	4.35	P/R, LARVAE RESTING, 0-13 D	1012
*RHIZOGLYPHUS ECHINOPUS	0.69	P/R, PROTONYMPHA ACTIVE, 0-16 D	1012
*RHIZOGLYPHUS ECHINOPUS	1.02	P/R, PROTONYMPHA RESTING, 0-17 D	1012
*RHIZOGLYPHUS ECHINOPUS	0.60	P/R, TRICHONYMPHA ACTIVE, 0-20 D	1012
*RHIZOGLYPHUS ECHINOPUS	0.80	P/R, TRICONYMPHA RESTING, 0-21 D	1012
*RHIZOGLYPHUS ECHINOPUS	0.67	P/R, AVERAGE NEW HATCHED ADULT, 0-22 D	1012
*RHIZOGLYPHUS ECHINOPUS	0.43	P/R, AVERAGE ADULT + EGGS, 0-72 D	1012
*SIMOCEPHALUS VECTULUS	0.37	P/R, FEMALE, 0-7 D, IMMATURE	1012
*SIMOCEPHALUS VECTULUS	0.54	P/R, FEMALE, 0-20 D, REPRODUCING	1012
*TRIBOLIUM CASTANEUM	2.40	P/R, PREPUPA, 0-24 D	1012
*TRIBOLIUM CASTANEUM	3.80	P/R, NEW HATCHED ADULT, AVERAGE, 0-36 D	1012
*TRIBOLIUM CASTANEUM	1.26	P/R, AVERAGE ADULT + EGGS, 0-116 D	1012

TABLE A272

INVERTEBRATES : S/BIOMASS RATIO			
ITEM:	VALUE:	CONDITIONS:	REF.:
*ANNELIDA	1.4 % OF DRY WEIGHT	MEAN ELEMENTARY COMPOSITION	157
*ARTHROPODA	0.44 % OF DRY WEIGHT	MEAN ELEMENTARY COMPOSITION,LAND	157
*COELENTERATA	1.9 % OF DRY WEIGHT	MEAN ELEMENTARY COMPOSITION	157
*ECHINODERMATA	0.5 % OF DRY WEIGHT	MEAN ELEMENTARY COMPOSITION	157
*MOLLUSCA	1.5 % OF DRY WEIGHT	MEAN ELEMENTARY COMPOSITION,SOFT PARTS	157
*NEMATODA	1.3 % OF DRY WEIGHT	MEAN ELEMENTARY COMPOSITION, INTERPOLATED	157

TABLE A273

INVERTEBRATES : SETTLING RATE			
ITEM:	VALUE:	CONDITIONS:	REF.:
*FORAMINIFERAL SHELLS	2.3 (2.1-2.4) CM/SEC	>250MY	102

TABLE A273

INVERTEBRATES : SETTLING RATE			
ITEM:	VALUE:	CONDITIONS:	REF.:
*FORAMINIFERAL SHELLS	1.0 (0.8-1.3) CM/SEC	177-250MY	102
*FORAMINIFERAL SHELLS	0.6 (0.56-0.73) CM/SEC	125-177MY	102
*FORAMINIFERAL SHELLS	0.3 (0.24-0.35) CM/SEC	62-125MY	102

TABLE A274

INVERTEBRATES : THRESHOLD DENSITY			
ITEM:	VALUE:	CONDITIONS:	REF.:
*FLY, ADULT	50 KCAL/M**(-2)	INITIAL VALUES, THERMAL ECOSYSTEM, LIMIT-EGESTION PROPORTION	1703
*FLY, ADULT	284 KCAL/M**(-2)	INITIAL VALUES, THERMAL ECOSYSTEM, LIMIT-EGESTION PROPORTION	1703
*FLY, 1,2 INSTAR LARVAE	50 KCAL/M**(-2)	INITIAL VALUES, THERMAL ECOSYSTEM, LIMIT-EGESTION PROPORTION	1703
*FLY, 1,2 INSTAR LARVAE	50 KCAL/M**(-2)	INITIAL VALUES, THERMAL ECOSYSTEM, LIMIT-EGESTION PROPORTION	1703
*FLY, 3 INSTAR LARVAE	50 KCAL/M**(-2)	INITIAL VALUES, THERMAL ECOSYSTEM, LIMIT-EGESTION PROPORTION	1703
*FLY, 3 INSTAR LARVAE	50 KCAL/M**(-2)	INITIAL VALUES, THERMAL ECOSYSTEM, LIMIT-EGESTION PROPORTION	1703

TABLE A275

INVERTEBRATES : TURNOVER TIME OF IODINE			
ITEM:	VALUE:	CONDITIONS:	REF.:
*AMPHIPODA	34.4 HOURS	LIVING IN FRESHWATER SEDIMENT	1011

TABLE A276

INVERTEBRATES : WET WEIGHT			
ITEM:	VALUE:	CONDITIONS:	REF.:
*TETRAHYMENA PYRIFORMIS	1.9*10**(-8) G	NONE	175
*TYPICAL PARAMECIUM	4.6*10**(-7) G	CILIATE	175

TABLE A277

INVERTEBRATES : WET/DRY WEIGHT RATIO			
ITEM:	VALUE:	CONDITIONS:	REF.:
*ZOOINVERTEBRATES AND ZOOPLANKTON	6 BY WEIGHT	NONE	1661

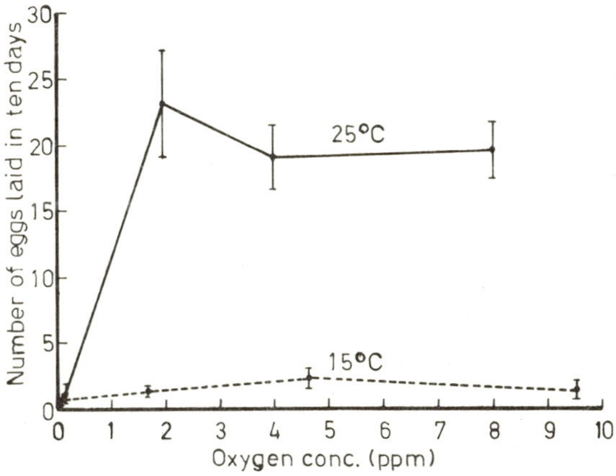

A278 Egg production versus oxygen concentration for Limnodrilus hoffmeisteri at 15 and 25 celcius. Means and standard errors are shown.

(ref: 47)

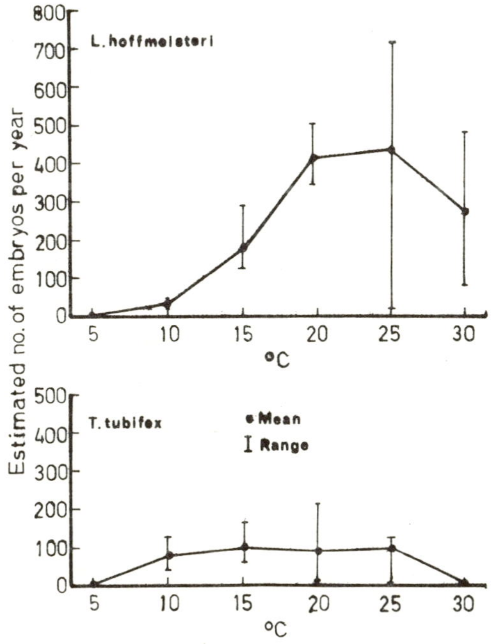

A279 Embryo production versus temperature for a: Limnodrilus hoffmeisteri, b: Tubifex tubifex. Mean and standard errors are shown.

(ref: 47)

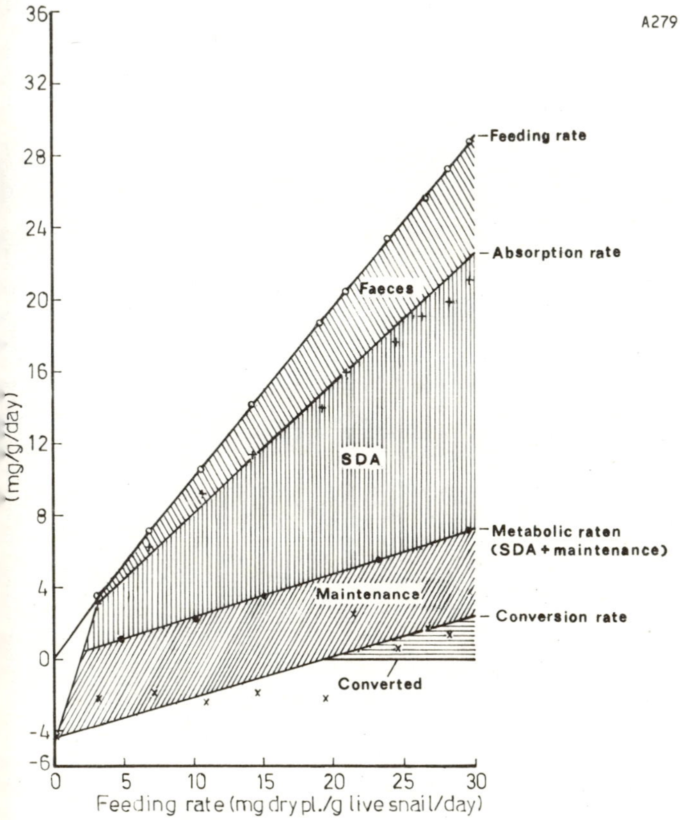

A280 Growth, metabolism and faeces versus feeding rate in young snails (Ceratophyllum) for 30 days at 28 celcius.
Feeding rate in mg dry plant/(g live snail* day) versus mg/(g*day).
Metabolic rate covers specific dynamic action (SDA) + maintenance.
o Feeding rate + adsorption rate
·etabolic rate x conversion rate

(ref: 1637)

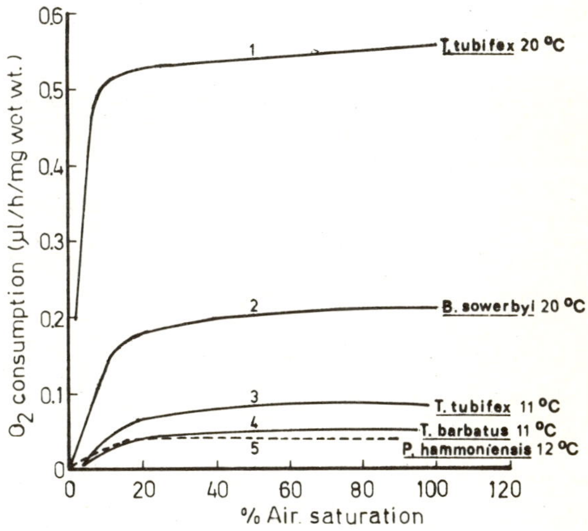

A281 Oxygen consumption versus oxygen concentration for tubicid worms.

(ref: 47)

A7

KORMOPHYTES, MACROPHYTES PLANTS TREES

COMPOSITION
GROWTH RATE
PRODUCTION
RATE OF NUTRIENT UPTAKE
RESPIRATION
ROOT AND LEAF
TRANSPIRATION
WIND PARAMETERS

TABLE A282

KORMOPHYTES, MACROPHYTES : ASH/BIOMASS RATIO			
ITEM:	VALUE:	CONDITIONS:	REF.:
*AMBLYSTEGIUM RIPARIUM	29.24 (SD=+-0.15) % OF DRY WEIGHT	MUSCI, N = 4, SAN MARCOS RIVER	608
*ANGIOSPERMS	17.2 % OF DRY WEIGHT	MEAN OF 15 SPECIES, SAN MARCOS RIVER	608
*CABOMBA CAROLINIANA	22.80 (SD=+-0.85) % OF DRY WEIGHT	N = 4, ANGIOSPERM, SAN MARCOS RIVER	608
*CERATOPHYLLUM DEMERSUM	8.78 (SD=+-0.20) % OF DRY WEIGHT	N = 4, ANGIOSPERM, SAN MARCOS RIVER	608
*ELODEA DENSA	16.77 (SD=+-0.97) % OF DRY WEIGHT	N = 4, ANGIOSPERM, SAN MARCOS RIVER	608
*EMERGENT	12 (RANGE =5 - 12) % OF DRY WEIGHT	NONE	1515
*FREE FLOATING	18.28 (SE=+-1.805) % OF DRY MATTER	AMAZON	696
*HETERANTHERA DUBIA	19.71 (SD=+-2.77) % OF DRY WEIGHT	N = 4, ANGIOSPERM, SAN MARCOS RIVER	608
*LUDWIGIA PALUSTRA	25.32 (SD=+-1.60) % OF DRY WEIGHT	N = 4, ANGIOSPERM, SAN MARCOS RIVER	608
*MYRIOPHYLLUM BRASILIENSE	10.08 (SD=+-1.17) % OF DRY WEIGHT	N = 4, ANGIOSPERM, SAN MARCOS RIVER	608
*MYRIOPHYLLUM HETEROPHYLLUM	8.89 (SD=+-0.40) % OF DRY WEIGHT	N = 4, ANGIOSPERM, SAN MARCOS RIVER	608
*NAJAS GUADALUPENSIS	18.28 (SD=+-1.37) % OF DRY WEIGHT	N = 4, ANGIOSPERM, SAN MARCOS RIVER	608
*PLANTS WITH FLOATING LEAVES	16 (RANGE =10 - 25) % OF DRY WEIGHT	NONE	1515
*POTAMOGETON CRISPUS	19.27 (SD=+-0.76) % OF DRY WEIGHT	N = 4, ANGIOSPERM, SAN MARCOS RIVER	608
*POTAMOGETON ILLINOEASIS	15.85 (SD=+-1.04) % OF DRY WEIGHT	N = 4, ANGIOSPERM, SAN MARCOS RIVER	608
*POTAMOGETON PECTINATUS	15.35 (SD=+-1.08) % OF DRY WEIGHT	N = 4, ANGIOSPERM, SAN MARCOS RIVER	608
*RHIZOCLONIUM HIEROGLYPHICUM	29.17 (SD=+-0.47) % OF DRY WEIGHT	N = 4, CHLOROPHYCEAE,SAN MARCOS RIVER	608
*RICCIA FLUVITANS	27.06 (SD=+-0.13) % OF DRY WEIGHT	N = 4, HEPATICAE,SAN MARCOS RIVER	608
*ROOTED	10.57 (SE=+-2.352) % OF DRY MATTER	AMAZON	696
*SAGITTARIA PLATYPHYLLA	22.60 (SD=+-3.07) % OF DRY WEIGHT	N = 4, ANGIOSPERM, SAN MARCOS RIVER	608
*SUBMERSED	21 (RANGE =9-25) % OF DRY WEIGHT	NONE	1515
*UTRICULARIA SUBULATA	19.92 (SD=+-0.17) % OF DRY WEIGHT	N = 4, ANGIOSPERM, SAN MARCOS RIVER	608
*VALLISNERIA AMERICANA	24.33 (SD=+-2.38) % OF DRY WEIGHT	N = 4, ANGIOSPERM, SAN MARCOS RIVER	608
*ZANNICHELLIA PALUSTRUS	10.60 (SD=+-0.37) % OF DRY WEIGHT	N = 4, ANGIOSPERM, SAN MARCOS RIVER	608

TABLE A283

KORMOPHYTES, MACROPHYTES : BIOMASS			
ITEM:	VALUE:	CONDITIONS:	REF.:
*EMERGENTS DOMINATING	0.4 - 10.7 G DRY WEIGHT/M**2	MAX BIOMASS,LADOGA L., USSR	1299
*EMERGENTS DOMINATING	0.1 - 33 G DRY WEIGHT/M**2	MAX BIOMASS,ONEGA L., USSR	1299
*EMERGENTS DOMINATING	440 - 830 G DRY WEIGHT/M**2	MAX BIOMASS,POLISH LAKES	1545
*EMERGENTS DOMINATING	850 G DRY WEIGHT/M**2	CAREX, MAX BIOMASS,MINNESOTA WETLANDS	107
*EMERGENTS DOMINATING	800 - 1100 G DRY WEIGHT/M**2	MAX BIOMASS,SURLINGHAM BROAD, ENG.	214
*EMERGENTS DOMINATING	1100 - 2200 G DRY WEIGHT/M**2	PHRAGMITES, AERIAL, MAX BIOMASS, OPATOVICKY P., CZEC.	384
*EMERGENTS DOMINATING	6000 - 8560 G DRY WEIGHT/M**2	PHRAGMITES, BELOW GROUND, MAX BIOMASS,OPATOVICKY P., CZEC.	384
*EMERGENTS DOMINATING	4640 G DRY WEIGHT/M**2	TYPHA, MAX BIOMASS,CEDAR CREEK, MINN.	165

TABLE A283

KORMOPHYTES, MACROPHYTES : BIOMASS			
ITEM:	VALUE:	CONDITIONS:	REF.:
*FLOATING	630 - 1472 G DRY WEIGHT/M**2	EICHHORINA, MAX BIOMASS,NEW ORLEANS, LA.	1235
*SUBMERGENTS DOMINATING	0.07 G DRY WEIGHT/M**2	MAX BIOMASS,TROUT L., SOFTWATER, WISC	1728
*SUBMERGENTS DOMINATING	1.73 G DRY WEIGHT/M**2	MAX BIOMASS,SWEENEY L., SOFTWATER, WISC	1727
*SUBMERGENTS DOMINATING	16.8 G DRY WEIGHT/M**2	MAX BIOMASS,WEBER L., SOFTWATER, WISC	1273
*SUBMERGENTS DOMINATING	32 G DRY WEIGHT/M**2	MAX BIOMASS,LOWES L., DYSTROPHIC, SCOTLAND	1489
*SUBMERGENTS DOMINATING	100 G DRY WEIGHT/M**2	MAX BIOMASS,SPIGGIE L., DYSTROPHIC, SCOTLAND	1489
*SUBMERGENTS DOMINATING	202 G DRY WEIGHT/M**2	MAX BIOMASS,L. MENDOTA, HARDWATER, WISC.	1331
*SUBMERGENTS DOMINATING	338 G DRY WEIGHT/M**2	SCIRPUS SUBTERMINALIS, MAX BIOMASS,LAWRENCE L., HARDWATER, MICH.	1325
*SUBMERGENTS DOMINATING	110 G DRY WEIGHT/M**2	CHARA, MAX BIOMASS,LAWRENCE L., HARDWATER, MICH	1325
*SUBMERGENTS DOMINATING	130 G DRY WEIGHT/M**2	ANNUALS, MAX BIOMASS,LAWRENCE L., HARDWATER, MICH	1325
*SUBMERGENTS DOMINATING	400 G DRY WEIGHT/M**2	MAX BIOMASS,CROISPOL, SCOTLAND	1489
*SUBMERGENTS DOMINATING	60 G DRY WEIGHT/M**2	RUPPIA, MAX BIOMASS,BORAX ., CALIF	1688
*SUBMERGENTS DOMINATING	500 G DRY WEIGHT/M**2	BERULA, MAX BIOMASS,R. IVEL, ENG.	1687
*SUBMERGENTS DOMINATING	100 - 400 G DRY WEIGHT/M**2	RANUNCULUS, MAX BIOMASS,R. TEST, ENG.	1199
*SUBMERGENTS DOMINATING	380 G DRY WEIGHT/M**2	POTAMOGETON, MAX BIOMASS,R. YARE, ENG.	1200
*SUBMERGENTS DOMINATING	700 - 7300 G DRY WEIGHT/M**2	THALASSIA, MAX BIOMASS,SALINE CHANNALS, PUERTO RICO	3553

TABLE A284

KORMOPHYTES, MACROPHYTES : BIOMASS IN ROOTS			
ITEM:	VALUE:	CONDITIONS:	REF.:
*ALISMA PLANTAGO AQUATICA	40 % OF TOTAL BIOMASS	EMERGENT	1680
*CERATOPHYLLUM SP.	< 5 % OF TOTAL BIOMASS	SUBMERSED	1680
*CHARA SP.	< 10 % OF TOTAL BIOMASS	SUBMERSED	1680
*CYPERUS FUSCUS	7 - 8 % OF TOTAL BIOMASS	EMERGENT	1680
*EICHHORNIA CRASSIPES	10 - 48 % OF TOTAL BIOMASS	SUBMERSED ROOTS, FLOATING	1680
*ELODEA CANADENSIS	2.6 % OF TOTAL BIOMASS	SUBMERSED	1680
*EQUISETUM FLUVIATILE	40 - 83 % OF TOTAL BIOMASS	EMERGENT	1680
*GLYCERIA MAXIMA	> 30 - 67 % OF TOTAL BIOMASS	EMERGENT	1680
*LITTORELLA UNIFLORA	46 - 55 % OF TOTAL BIOMASS	SUBMERSED	1680
*NUPHAR LUTEUM	50 - 80 % OF TOTAL BIOMASS	EMERGENT	1680
*NUPHAR PUMILUM	48 - 80 % OF TOTAL BIOMASS	EMERGENT	1680
*PHRAGMITES COMMUNIS	> 36 - 96 % OF TOTAL BIOMASS	EMERGENT	1680
*SCHOENOPLECTUS LACUSTRIS	> 46 - 90 % OF TOTAL BIOMASS	EMERGENT	1680
*SPARGANIUM SPP.	> 25 - 66 % OF TOTAL BIOMASS	EMERGENT	1680
*THALASSIA TESTUDIUNUM	75 - 85 % OF TOTAL BIOMASS	MARINE, SUBMERSED	1680
*TYPHA ANGUSTIFOLIA	> 32 - 59 % OF TOTAL BIOMASS	EMERGENT	1680
*TYPHA LATIFOLIA	43 - 50 % OF TOTAL BIOMASS	EMERGENT	1680
*TYPHA SP.	64 % OF TOTAL BIOMASS	HYBRID, EMERGENT	1680
*ZIZANIA AQUATICA	7 - 8 % OF TOTAL BIOMASS	FLOATING-LEAVED	1680

TABLE A285

KORMOPHYTES, MACROPHYTES : C/BIOMASS RATIO			
ITEM:	VALUE:	CONDITIONS:	REF.:
*AMBLYSTEGIUM RIPARIUM	27.12 (SD=+-0.24) % OF DRY WEIGHT	MUSCI, N = 5, SAN MARCOS RIVER	608

TABLE A285

KORMOPHYTES, MACROPHYTES : C/BIOMASS RATIO

ITEM:	VALUE:	CONDITIONS:	REF.:
*ANGIOSPERMS	31.7 % OF DRY WEIGHT	MEAN OF 15 SPECIES, SAN MARCOS RIVER	608
*CABOMBA CAROLINIANA	28.36 (SD=+-0.74) % OF DRY WEIGHT	N = 5, ANGIOSPERM, SAN MARCOS RIVER	608
*CERATOPHYLLUM DEMERSUM	34.34 (SD=+-0.81) % OF DRY WEIGHT	N = 5, ANGIOSPERMS, SAN MARCOS RIVER	608
*ELODEA DENSA	30.84 (SD=+-0.74) % OF DRY WEIGHT	N = 5, ANGIOSPERMS, SAN MARCOS RIVER	608
*HETERANTHERA DUBIA	27.18 (SD=+-0.59) % OF DRY WEIGHT	N = 5, ANGIOSPERMS, SAN MARCOS RIVER	608
*LUDWIGIA PALUSTRA	29.54 (SD=+-0.67) % OF DRY WEIGHT	N = 5, ANGIOSPERMS, SAN MARCOS RIVER	608
*MYRIOPHYLLUM BRASILIENSE	35.28 (SD=+-1.19) % OF DRY WEIGHT	N = 5, ANGIOSPERMS, SAN MARCOS RIVER	608
*MYRIOPHYLLUM HETEROPHYLLUM	37.61 (SD=+-0.43) % OF DRY WEIGHT	N = 5, ANGIOSPERMS, SAN MARCOS RIVER	608
*NAJAS GUADALUPENSIS	31.92 (SD=+-0.79) % OF DRY WEIGHT	N = 5, ANGIOSPERMS, SAN MARCOS RIVER	608
*POTAMOGETON CRISPUS	30.74 (SD=+-0.71) % OF DRY WEIGHT	N = 5, ANGIOSPERMS, SAN MARCOS RIVER	608
*POTAMOGETON ILLINOEASIS	32.73 (SD=+-0.76) % OF DRY WEIGHT	N = 5, ANGIOSPERMS, SAN MARCOS RIVER	608
*POTAMOGETON PECTINATUS	30.49 (SD=+-0.86) % OF DRY WEIGHT	N = 5, ANGIOSPERMS, SAN MARCOS RIVER	608
*RHIZOCLONIUM HIEROGLYPHICUM	25.84 (SD=+-0.41) % OF DRY WEIGHT	N = 5, CHLOROPHYCEAE, SAN MARCOS RIVER	608
*RICCIA FLUVITANS	28.66 (SD=+-0.24) % OF DRY WEIGHT	N = 5, HEPATICAE, SAN MARCOS RIVER	608
*SAGITTARIA PLATYPHYLLA	33.30 (SD=+-0.56) % OF DRY WEIGHT	N = 5, ANGIOSPERMS, SAN MARCOS RIVER	608
*UTRICULARIA SUBULATA	29.69 (SD=+-0.79) % OF DRY WEIGHT	N = 5, ANGIOSPERMS, SAN MARCOS RIVER	608
*VALLISNERIA AMERICANA	28.60 (SD=+-0.77) % OF DRY WEIGHT	N = 5, ANGIOSPERMS, SAN MARCOS RIVER	608
*ZANNICHELLIA PALUSTRUS	34.83 (SD=+-0.58) % OF DRY WEIGHT	N = 5, ANGIOSPERMS, SAN MARCOS RIVER	608

TABLE A286

KORMOPHYTES, MACROPHYTES : CA/BIOMASS RATIO

ITEM:	VALUE:	CONDITIONS:	REF.:
*BRASENIA SCHREBERI	1.79 (SD=+- 0.05) % OF DRY WEIGHT	PAR POND, SC, MAY 69, SOFT WAT, WATER=2.64 CA PPM, N=4	160
*CERATOPHYLLUM DEMERSUM	0.77 (SD=+- 0.06) % OF DRY WEIGHT	PAR POND, SC, MAY 69, SOFT WAT, WATER=2.64 CA PPM, N=4	160
*ELEOCHARIS ACICULARIS	0.53 (SD=+- 0.13) % OF DRY WEIGHT	PAR POND, SC, MAY 69, SOFT WAT, WATER=2.64 CA PPM, N=4	160
*ELEOCHARIS QUADRANGULATA	0.20 % OF DRY WEIGHT	PAR POND, SC, MAY 69, SOFT WAT, WATER=2.64 CA PPM, N=1	160
*FREE FLOATING	14.16 (SE=+-3.149) G/KG DRY WEIGHT	AMAZON	696
*HALOPHILA OVALIS	1.06 (SD=+- 0.01) % OF DRY WEIGHT	PAR POND, SC, MAY 69, SOFT WAT, WATER=2.64 CA PPM, N=5	160
*HYDROCOTYLE SP.	1.85 (SD=+- 0.26) % OF DRY WEIGHT	PAR POND, SC, MAY 69, SOFT WAT, WATER=2.64 CA PPM, N=4	160
*JUNCUS EFFUSUS	0.38 (SD=+- 0.11) % OF DRY WEIGHT	PAR POND, SC, MAY 69, SOFT WAT, WATER=2.64 CA PPM, N=5	160
*MYRIOPHYLLUM HETEROPHYLLUM	1.47 (SD=+- 0.11) % OF DRY WEIGHT	PAR POND, SC, MAY 69, SOFT WAT, WATER=2.64 CA PPM, N=4	160
*NAJAS GUADALUPENSIS	0.98 (SD=+- 0.14) % OF DRY WEIGHT	PAR POND, SC, MAY 69, SOFT WAT, WATER=2.64 CA PPM, N=4	160
*NELUMBO LUTEA	1.56 (SD=+- 0.18) % OF DRY WEIGHT	PAR POND, SC, MAY 69, SOFT WAT, WATER=2.64 CA PPM, N=2	160
*NUPHAR ADVENA	1.08 % OF DRY WEIGHT	PAR POND, SC, MAY 69, SOFT WAT, WATER=2.64 CA PPM, N=2	160

TABLE A286

KORMOPHYTES, MACROPHYTES : CA/BIOMASS RATIO

ITEM:	VALUE:	CONDITIONS:	REF.:
*PANICUM HEMITONIUM	0.38 (SD=+- 0.04) % OF DRY WEIGHT	PAR POND, SC, MAY 69, SOFT WAT, WATER=2.64 CA PPM, N=3	160
*PONTEDERIA CORDATA	0.96 % OF DRY WEIGHT	PAR POND, SC, MAY 69, SOFT WAT, WATER=2.64 CA PPM, N=1	160
*POTAMOGETON DIVERSIFOLIUS	1.14 % OF DRY WEIGHT	PAR POND, SC, MAY 69, SOFT WAT, WATER=2.64 CA PPM, N=1	160
*ROOTED	4.18 (SE=+-1.597) G/KG DRY WEIGHT	AMAZON	696
*SAGITTERIA LATIFOLIA	0.55 % OF DRY WEIGHT	PAR POND, SC, MAY 69, SOFT WAT, WATER=2.64 CA PPM, N=2	160
*SCIRPUS AMERICANUS	0.50 (SD=+- 0.01) % OF DRY WEIGHT	PAR POND, SC, MAY 69, SOFT WAT, WATER=2.64 CA PPM, N=5	160
*TYPHA LATIFOLIA	0.76 (SD=+- 0.18) % OF DRY WEIGHT	PAR POND, SC, MAY 69, SOFT WAT, WATER=2.64 CA PPM, N=4	160
*UTRICULARIA INFLATA	0.67 (SD=+- 0.07) % OF DRY WEIGHT	PAR POND, SC, MAY 69, SOFT WAT, WATER=2.64 CA PPM, N=4	160

TABLE A287

KORMOPHYTES, MACROPHYTES : CALORIC CONTENT

ITEM:	VALUE:	CONDITIONS:	REF.:
*EMERGENT	4480 (RANGE =4207 - 4987) CAL/G	NONE	1515
*FREE FLOATING	4.615 (SE=+-0.087) CAL/(G ASH FREE DRY WEIGHT)	AMAZON	696
*PLANTS WITH FLOATING LEAVES	4770 (RANGE =4560 - 5140) CAL/G	NONE	1515
*ROOTED	4.479 (SE=+-0.039) CAL/(G ASH FREE DRY WEIGHT)	AMAZON	696
*SUBMERSED	4580 (RANGE =4165 - 5201) CAL/G	NONE	1515

TABLE A288

KORMOPHYTES, MACROPHYTES : CARBOHYDRATE/BIOMASS RATIO

ITEM:	VALUE:	CONDITIONS:	REF.:
*EMERGENT	50 (RANGE =23 - 69) % OF ORGANIC MATTER	NONE	1515
*PLANTS WITH FLOATING LEAVES	42 (RANGE =31 - 64) % OF ORGANIC MATTER	NONE	1515
*SUBMERSED	51 (RANGE =-20 - 70) % OF ORGANIC MATTER	NONE	1515

TABLE A289

KORMOPHYTES, MACROPHYTES : CHLOROPHYLL A/BIOMASS RATIO

ITEM:	VALUE:	CONDITIONS:	REF.:
*BRASENIA SCHREBERI	3.26 (SD=+- 0.32) MG/G DRY WEIGHT	PAR POND, SOUTH CAROLINA, MAY 69, SOFT WATER	160
*HYDROCOTYLE SP.	5.13 (SD=+- 0.15) MG/G DRY WEIGHT	PAR POND, SOUTH CAROLINA, MAY 69, SOFT WATER	160
*JUNCUS EFFUSUS	1.26 (SD=+- 0.08) MG/G DRY WEIGHT	PAR POND, SOUTH CAROLINA, MAY 69, SOFT WATER	160
*MYRIOPHYLLUM HETEROPHYLLUM	7.74 (SD=+- 0.75) MG/G DRY WEIGHT	PAR POND, SOUTH CAROLINA, MAY 69, SOFT WATER	160
*NAJAS GUADALUPENSIS	8.52 (SD=+- 0.52) MG/G DRY WEIGHT	PAR POND, SOUTH CAROLINA, MAY 69, SOFT WATER	160

TABLE A289

KORMOPHYTES, MACROPHYTES : CHLOROPHYLL A/BIOMASS RATIO			
ITEM:	VALUE:	CONDITIONS:	REF.:
*NELUMBO LUTEA	3.62 (SD=+- 0.54) MG/G DRY WEIGHT	PAR POND, SOUTH CAROLINA, MAY 69, SOFT WATER	160
*NUPHAR ADVENA	3.86 (SD=+- 0.58) MG/G DRY WEIGHT	PAR POND, SOUTH CAROLINA, MAY 69, SOFT WATER	160
*NYMPHAEA ODORATA	3.38 (SD=+- 0.14) MG/G DRY WEIGHT	PAR POND, SOUTH CAROLINA, MAY 69, SOFT WATER	160
*PONTEDERIA CORDATA	3.26 (SD=+- 0.07) MG/G DRY WEIGHT	PAR POND, SOUTH CAROLINA, MAY 69, SOFT WATER	160
*POTAMOGETON DIVERSIFOLIUS	5.74 (SD=+- 0.13) MG/G DRY WEIGHT	PAR POND, SOUTH CAROLINA, MAY 69, SOFT WATER	160
*SCIRPUS AMERICANUS	2.32 (SD=+- 0.21) MG/G DRY WEIGHT	PAR POND, SOUTH CAROLINA, MAY 69, SOFT WATER	160
*TYPHA LATIFOLIA	2.61 (SD=+- 0.21) MG/G DRY WEIGHT	PAR POND, SOUTH CAROLINA, MAY 69, SOFT WATER	160
*UTRICULARIA INFLATA	6.19 (SD=+- 0.33) MG/G DRY WEIGHT	PAR POND, SOUTH CAROLINA, MAY 69, SOFT WATER	160

TABLE A290

KORMOPHYTES, MACROPHYTES : CHLOROPHYLL B/BIOMASS RATIO			
ITEM:	VALUE:	CONDITIONS:	REF.:
*BRASENIA SCHREBERI	0.10 (SD=+- 0.06) MG/G DRY WEIGHT	PAR POND, SOUTH CAROLINA, MAY 69, SOFT WATER	160
*HYDROCOTYLE SP.	1.23 (SD=+- 0.05) MG/G DRY WEIGHT	PAR POND, SOUTH CAROLINA, MAY 69, SOFT WATER	160
*JUNCUS EFFUSUS	0.23 (SD=+- 0.02) MG/G DRY WEIGHT	PAR POND, SOUTH CAROLINA, MAY 69, SOFT WATER	160
*MYRIOPHYLLUM HETEROPHYLLUM	2.21 (SD=+- 0.09) MG/G DRY WEIGHT	PAR POND, SOUTH CAROLINA, MAY 69, SOFT WATER	160
*NAJAS GUADALUPENSIS	2.51 (SD=+- 0.19) MG/G DRY WEIGHT	PAR POND, SOUTH CAROLINA, MAY 69, SOFT WATER	160
*NELUMBO LUTEA	0.82 (SD=+- 0.16) MG/G DRY WEIGHT	PAR POND, SOUTH CAROLINA, MAY 69, SOFT WATER	160
*NUPHAR ADVENA	0.79 (SD=+- 0.15) MG/G DRY WEIGHT	PAR POND, SOUTH CAROLINA, MAY 69, SOFT WATER	160
*NYMPHAEA ODORATA	0.43 (SD=+- 0.19) MG/G DRY WEIGHT	PAR POND, SOUTH CAROLINA, MAY 69, SOFT WATER	160
*PONTEDERIA CORDATA	0.77 (SD=+- 0.15) MG/G DRY WEIGHT	PAR POND, SOUTH CAROLINA, MAY 69, SOFT WATER	160
*POTAMOGETON DIVERSIFOLIUS	1.46 (SD=+- 0.01) MG/G DRY WEIGHT	PAR POND, SOUTH CAROLINA, MAY 69, SOFT WATER	160
*SCIRPUS AMERICANUS	0.41 (SD=+- 0.07) MG/G DRY WEIGHT	PAR POND, SOUTH CAROLINA, MAY 69, SOFT WATER	160
*TYPHA LATIFOLIA	0.65 (SD=+- 0.06) MG/G DRY WEIGHT	PAR POND, SOUTH CAROLINA, MAY 69, SOFT WATER	160
*UTRICULARIA INFLATA	0.65 (SD=+- 0.11) MG/G DRY WEIGHT	PAR POND, SOUTH CAROLINA, MAY 69, SOFT WATER	160

TABLE A291

KORMOPHYTES, MACROPHYTES : CRUDE FIBER/BIOMASS RATIO			
ITEM:	VALUE:	CONDITIONS:	REF.:
*EMERGENT	32 (RANGE =17.6 - 45.4) % OF ORGANIC MATTER	NONE	1515
*PLANTS WITH FLOATING LEAVES	27 (RANGE =14 - 38) % OF ORGANIC MATTER	NONE	1515
*SUBMERSED	27 (RANGE =14.0 - 61.6) % OF ORGANIC MATTER	NONE	1515

TABLE A292

KORMOPHYTES, MACROPHYTES : CU/BIOMASS RATIO

ITEM:	VALUE:	CONDITIONS:	REF.:
*BRASENIA SCHREBERI	32 (SD=+- 5) MG/KG DRY WEIGHT	PAR POND, SC, MAY 69, SOFT WAT, WATER=TRACE CU PPM, N=4	160
*CERATOPHYLLUM DEMERSUM	30 (SD=+- 7) MG/KG DRY WEIGHT	PAR POND, SC, MAY 69, SOFT WAT, WATER=TRACE CU PPM, N=4	160
*ELEOCHARIS ACICULARIS	42 (SD=+- 6) MG/KG DRY WEIGHT	PAR POND, SC, MAY 69, SOFT WAT, WATER=TRACE CU PPM, N=4	160
*ELEOCHARIS QUADRANGULATA	20 MG/KG DRY WEIGHT	PAR POND, SC, MAY 69, SOFT WAT, WATER=TRACE CU PPM, N=2	160
*HYDROCOTYLE SP.	53 (SD=+- 8) MG/KG DRY WEIGHT	PAR POND, SC, MAY 69, SOFT WAT, WATER=TRACE CU PPM, N=4	160
*MYRIOPHYLLUM HETEROPHYLLUM	44 (SD=+- 9) MG/KG DRY WEIGHT	PAR POND, SC, MAY 69, SOFT WAT, WATER=TRACE CU PPM, N=4	160
*NAJAS GUADALUPENSIS	48 (SD=+- 8) MG/KG DRY WEIGHT	PAR POND, SC, MAY 69, SOFT WAT, WATER=TRACE CU PPM, N=4	160
*NELUMBO LUTEA	40 (SD=+- 7) MG/KG DRY WEIGHT	PAR POND, SC, MAY 69, SOFT WAT, WATER=TRACE CU PPM, N=4	160
*NUPHAR ADVENA	35 MG/KG DRY WEIGHT	PAR POND, SC, MAY 69, SOFT WAT, WATER=TRACE CU PPM, N=2	160
*NYMPHAEA ODORATA	36 (SD=+- 7) MG/KG DRY WEIGHT	PAR POND, SC, MAY 69, SOFT WAT, WATER=TRACE CU PPM, N=5	160
*PANICUM HEMITONIUM	26 (SD=+- 2) MG/KG DRY WEIGHT	PAR POND, SC, MAY 69, SOFT WAT, WATER=TRACE CU PPM, N=3	160
*PONTEDERIA CORDATA	60 MG/KG DRY WEIGHT	PAR POND, SC, MAY 69, SOFT WAT, WATER=TRACE CU PPM, N=1	160
*POTAMOGETON DIVERSIFOLIUS	36 MG/KG DRY WEIGHT	PAR POND, SC, MAY 69, SOFT WAT, WATER=TRACE CU PPM, N=1	160
*SAGITTERIA LATIFOLIA	57 MG/KG DRY WEIGHT	PAR POND, SC, MAY 69, SOFT WAT, WATER=TRACE CU PPM, N=2	160
*TYPHA LATIFOLIA	37 (SD=+- 7) MG/KG DRY WEIGHT	PAR POND, SC, MAY 69, SOFT WAT, WATER=TRACE CU PPM, N=4	160
*UTRICULARIA INFLATA	47 (SD=+- 6) MG/KG DRY WEIGHT	PAR POND, SC, MAY 69, SOFT WAT, WATER=TRACE CU PPM, N=4	160

TABLE A293

KORMOPHYTES, MACROPHYTES : DRY/WET WEIGHT RATIO

ITEM:	VALUE:	CONDITIONS:	REF.:
*AMBLYSTEGIUM RIPARIUM	30.57 (SD=+-0.82) % OF FRESH WEIGHT	N=3, MUSCI, SAN MACRO RIVER	608
*ANGIOSPERMS	11.5	MEAN FOR 15 SPECIES, SAN MARCOS RIVER	608
*CABOMBA CAROLINIANA	7.47 (SD=+-0.56) % OF FRESH WEIGHT	N = 3,ANGIOSPERM, SAN MARCOS RIVER	608
*CERATOPHYLLUM DEMERSUM	8.90 (SD=+-0.36) % OF FRESH WEIGHT	N = 3,ANGIOSPERM, SAN MARCOS RIVER	608
*ELODEA DENSA	10.47 (SD=+-1.75) % OF FRESH WEIGHT	N = 3,ANGIOSPERM, SAN MARCOS RIVER	608
*EMERGENT	0.21 (RANGE = 0.15 - 0.30) FRACTION OF DRY WEIGHT	NONE	1515
*FREE FLOATING	7.15 (SE=+-0.520) % OF WET WEIGHT	AMAZON	696
*HETERANTHERA DUBIA	6.53 (SD=+-0.41) % OF FRESH WEIGHT	N = 3,ANGIOSPERM, SAN MARCOS RIVER	608
*LUDWIGIA PALUSTRA	13.50 (SD=+-0.50) % OF FRESH WEIGHT	N = 3,ANGIOSPERM, SAN MARCOS RIVER	608
*MYRIOPHYLLUM BRASILIENSE	20.00 (SD=+-0.20) % OF FRESH WEIGHT	N = 3,ANGIOSPERM, SAN MARCOS RIVER	608
*MYRIOPHYLLUM HETEROPHYLLUM	12.77 (SD=+-0.22) % OF FRESH WEIGHT	N = 3,ANGIOSPERM, SAN MARCOS RIVER	608
*NAJAS GUADALUPENSIS	10.83 (SD=+-1.22) % OF FRESH WEIGHT	N = 3,ANGIOSPERM, SAN MARCOS RIVER	608
*PLANTS WITH FLOATING LEAVES	0.09 (RANGE = 0.16 - 0.20) FRACTION OF DRY WEIGHT	NONE	1515
*POTAMOGETON CRISPUS	13.66 (SD=+-0.31) % OF FRESH WEIGHT	N = 3,ANGIOSPERM, SAN MARCOS RIVER	608

TABEL A293

KORMOPHYTES, MACROPHYTES : DRY/WET WEIGHT RATIO			
ITEM:	VALUE:	CONDITIONS:	REF.:
*POTAMOGETON ILLINOEASIS	14.20 (SD=+-0.26) % OF FRESH WEIGHT	N = 3,ANGIOSPERM, SAN MARCOS RIVER	608
*POTAMOGETON PECTINATUS	15.56 (SD=+-0.20) % OF FRESH WEIGHT	N = 3,ANGIOSPERM, SAN MARCOS RIVER	608
*RHIZOCLONIUM HIEROGLYPHICUM	27.43 (SD=+-1.80) % OF FRESH WEIGHT	N = 3, CHLOROPHYCEAE, SAN MARCOS RIVER	608
*RICCIA FLUVITANS	8.87 (SD=+-0.42) % OF FRESH WEIGHT	N = 3,ANGIOSPERM, SAN MARCOS RIVER	608
*ROOTED	19.81 (SE=+-2.937) % OF WET WEIGHT	AMAZON	696
*SAGITTARIA PLATYPHYLLA	6.40 (SD=+-0.26) % OF FRESH WEIGHT	N = 3,ANGIOSPERM, SAN MARCOS RIVER	608
*SUBMERSED	0.12 (RANGE 0.08 - 0.15) FRACTION OF DRY WEIGHT	NONE	1515
*UTRICULARIA SUBULATA	13.47 (SD=+-0.16) % OF FRESH WEIGHT	N = 3,ANGIOSPERM, SAN MARCOS RIVER	608
*VALLISNERIA AMERICANA	6.50 (SD=+-0.45) % OF FRESH WEIGHT	N = 3,ANGIOSPERM, SAN MARCOS RIVER	608
*ZANNICHELLIA PALUSTRUS	12.86 (SD=+-0.16) % OF FRESH WEIGHT	N = 3,ANGIOSPERM, SAN MARCOS RIVER	608

TABLE A294

KORMOPHYTES, MACROPHYTES : ETHER EXTRACT/BIOMASS RATIO			
ITEM:	VALUE:	CONDITIONS:	REF.:
*EMERGENT	2.1 (RANGE =0.4 - 6.1) % OF ORGANIC MATTER	NONE	1515
*PLANTS WITH FLOATING LEAVES	4.0 (RANGE =2.8 - 5.7) % OF ORGANIC MATTER	NONE	1515

TABLE A295

KORMOPHYTES, MACROPHYTES : FE/BIOMASS RATIO			
ITEM:	VALUE:	CONDITIONS:	REF.:
*BRASENIA SCHREBERI	500 (SD=+- 40) MG/KG DRY WEIGHT	PAR POND, SC, MAY 69, SOFT WAT, WATER=0.029 FE PPM, N=4	160
*CERATOPHYLLUM DEMERSUM	1053 (SD=+- 344) MG/KG DRY WEIGHT	PAR POND, SC, MAY 69, SOFT WAT, WATER=0.029 FE PPM, N=4	160
*ELEOCHARIS ACICULARIS	2920 (SD=+- 613) MG/KG DRY WEIGHT	PAR POND, SC, MAY 69, SOFT WAT, WATER=0.029 FE PPM, N=4	160
*ELEOCHARIS QUADRANGULATA	460 MG/KG DRY WEIGHT	PAR POND, SC, MAY 69, SOFT WAT, WATER=0.029 FE PPM, N=2	160
*HYDROCOTYLE SP.	1245 (SD=+- 326) MG/KG DRY WEIGHT	PAR POND, SC, MAY 69, SOFT WAT, WATER=0.029 FE PPM, N=4	160
*MYRIOPHYLLUM HETEROPHYLLUM	2000 (SD=+- 286) MG/KG DRY WEIGHT	PAR POND, SC, MAY 69, SOFT WAT, WATER=0.029 FE PPM, N=4	160
*NAJAS GUADALUPENSIS	712 (SD=+- 215) MG/KG DRY WEIGHT	PAR POND, SC, MAY 69, SOFT WAT, WATER=0.029 FE PPM, N=4	160
*NELUMBO LUTEA	126 (SD=+- 46) MG/KG DRY WEIGHT	PAR POND, SC, MAY 69, SOFT WAT, WATER=0.029 FE PPM, N=4	160
*NUPHAR ADVENA	740 MG/KG DRY WEIGHT	PAR POND, SC, MAY 69, SOFT WAT, WATER=0.029 FE PPM, N=2	160
*NYMPHAEA ODORATA	600 (SD=+- 128) MG/KG DRY WEIGHT	PAR POND, SC, MAY 69, SOFT WAT, WATER=0.029 FE PPM, N=5	160
*PANICUM HEMITONIUM	560 MG/KG DRY WEIGHT	PAR POND, SC, MAY 69, SOFT WAT, WATER=0.029 FE PPM, N=3	160
*PONTEDERIA CORDATA	200 MG/KG DRY WEIGHT	PAR POND, SC, MAY 69, SOFT WAT, WATER=0.029 FE PPM, N=1	160
*POTAMOGETON DIVERSIFOLIUS	2140 MG/KG DRY WEIGHT	PAR POND, SC, MAY 69, SOFT WAT, WATER=0.029 FE PPM, N=1	160
*SAGITTERIA LATIFOLIA	460 MG/KG DRY WEIGHT	PAR POND, SC, MAY 69, SOFT WAT, WATER=0.029 FE PPM, N=2	160

TABLE A295

KORMOPHYTES, MACROPHYTES : FE/BIOMASS RATIO			
ITEM:	VALUE:	CONDITIONS:	REF.:
*SCIRPUS AMERICANUS	133 (SD=+- 19) MG/KG DRY WEIGHT	PAR POND, SC, MAY 69, SOFT WAT, WATER=0.029 FE PPM, N=3	160
*TYPHA LATIFOLIA	120 (SD=+- 40) MG/KG DRY WEIGHT	PAR POND, SC, MAY 69, SOFT WAT, WATER=0.029 FE PPM, N=4	160
*UTRICULARIA INFLATA	2112 (SD=+- 222) MG/KG DRY WEIGHT	PAR POND, SC, MAY 69, SOFT WAT, WATER=0.029 FE PPM, N=4	160

TABLE A296

KORMOPHYTES, MACROPHYTES : GROSS PHOTOSYNTHESIS			
ITEM:	VALUE:	CONDITIONS:	REF.:
*LOTIC COMMUNITY	0.4 - 14 GRAM OXYGEN/(M2*DAY)	COMM. RESP = 4.2 - 20.2 G O(2)/(M2*D),P/R=0.1 - 1.1, ITCHEN RIVER	1165
*LOTIC COMMUNITY	3.2 - 17.6 GRAM OXYGEN/(M2*DAY)	COMM. RESP. = 6.7 - 15.4 G O(2)/(M2*D),P/R=1.13, IVEL RIVER	393
*LOTIC COMMUNITY	35.0 GRAM OXYGEN/(M2*DAY)	COMM. RESP = 5.0 G O(2)/(M2*D),P/R =7.0, SPRING, SILVER SPRINGS FL.	1165
*LOTIC COMMUNITY	8.0 GRAM OXYGEN/(M2*DAY)	COMM. RESP = 2.8 G O(2)/(M2*D),P/R =2.9, WINTER, SILVER SPRINGS FL.	1166
*LOTIC COMMUNITY	16.6 GRAM OXYGEN/(M2*DAY)	COMM. RESP = 14.6 G O(2)/(M2*D),P/R=1.1, HEADWATER AREA, SILVER SPR.	1165
*LOTIC COMMUNITY	0.3 - 9.8 GRAM OXYGEN/(M2*DAY)	COMM. RESP = 0.7 - 21.5 G O(2)/(M2*D),NEUSE RIVER, NC	694
*LOTIC COMMUNITY	6.8 GRAM OXYGEN/(M2*DAY)	COMM. RESP = 11.0 G O(2)/(M2*D),P/R=0.64, LIMESTONE BED, BLUE RIVER	374
*LOTIC COMMUNITY	21.3 GRAM OXYGEN/(M2*DAY)	COMM. RESP = 12.6 G O(2)/(M2*D),P/R=1.70, GRANITE BED, BLUE RIVER	374
*LOTIC COMMUNITY	3.0 GRAM OXYGEN/(M2*DAY)	COMM. RESP = 7.7 G O(2)/(M2*D),P/R =0.39, SAND BED,BLUE RIVER, OKL	374
*LOTIC COMMUNITY	2.5 - 27.4 GRAM OXYGEN/(M2*DAY)	COMM. RESP = 4.1 - 19.9 G O(2)/(M2*D),P/R=0.59 - 1.38, SAN MARCOS R.	608

TABLE A297

KORMOPHYTES, MACROPHYTES : K/BIOMASS RATIO			
ITEM:	VALUE:	CONDITIONS:	REF.:
*BRASENIA SCHREBERI	0.99 (SD=+- 0.01) % OF DRY WEIGHT	PAR POND, SC, MAY 69, SOFT WAT, WATER=1.61 K PPM, N=4	160
*CERATOPHYLLUM DEMERSUM	4.01 (SD=+- 0.59) % OF DRY WEIGHT	PAR POND, SC, MAY 69, SOFT WAT, WATER=1.61 K PPM, N=4	160
*ELEOCHARIS ACICULARIS	2.86 (SD=+- 0.71) % OF DRY WEIGHT	PAR POND, SC, MAY 69, SOFT WAT, WATER=1.61 K PPM, N=4	160
*ELEOCHARIS QUADRANGULATA	1.81 % OF DRY WEIGHT	PAR POND, SC, MAY 69, SOFT WAT, WATER=1.61 K PPM, N=1	160
*FREE FLOATING	36.75 (SE=+-4.239) G/KG DRY WEIGHT	AMAZON	696
*HYDROCOTYLE SP.	1.73 (SD=+- 0.42) % OF DRY WEIGHT	PAR POND, SC, MAY 69, SOFT WAT, WATER=1.61 K PPM, N=4	160
*JUNCUS EFFUSUS	0.89 (SD=+- 0.06) % OF DRY WEIGHT	PAR POND, SC, MAY 69, SOFT WAT, WATER=1.61 K PPM, N=5	160
*MYRIOPHYLLUM HETEROPHYLLUM	1.25 (SD=+- 0.08) % OF DRY WEIGHT	PAR POND, SC, MAY 69, SOFT WAT, WATER=1.61 K PPM, N=4	160
*NAJAS GUADALUPENSIS	3.49 (SD=+- 0.53) % OF DRY WEIGHT	PAR POND, SC, MAY 69, SOFT WAT, WATER=1.61 K PPM, N=4	160
*NELUMBO LUTEA	2.27 (SD=+- 0.25) % OF DRY WEIGHT	PAR POND, SC, MAY 69, SOFT WAT, WATER=1.61 K PPM, N=2	160
*NUPHAR ADVENA	1.88 % OF DRY WEIGHT	PAR POND, SC, MAY 69, SOFT WAT, WATER=1.61 K PPM, N=2	160

TABLE A297

KORMOPHYTES, MACROPHYTES : K/BIOMASS RATIO			
ITEM:	VALUE:	CONDITIONS:	REF.:
*NYMPHAEA ODORATA	1.28 (SD=+- 0.13) % OF DRY WEIGHT	PAR POND, SC, MAY 69, SOFT WAT, WATER=1.61 K PPM, N=5	160
*PANICUM HEMITONIUM	1.06 (SD=+- 0.13) % OF DRY WEIGHT	PAR POND, SC, MAY 69, SOFT WAT, WATER=1.61 K PPM, N=3	160
*PONTEDERIA CORDATA	2.58 % OF DRY WEIGHT	PAR POND, SC, MAY 69, SOFT WAT, WATER=1.61 K PPM, N=1	160
*POTAMOGETON DIVERSIFOLIUS	3.08 % OF DRY WEIGHT	PAR POND, SC, MAY 69, SOFT WAT, WATER=1.61 K PPM, N=1	160
*ROOTED	25.43 (SE=+-4.583) G/KG DRY WEIGHT	AMAZON	696
*SAGITTERIA LATIFOLIA	4.04 % OF DRY WEIGHT	PAR POND, SC, MAY 69, SOFT WAT, WATER=1.61 K PPM, N=2	160
*SCIRPUS AMERICANUS	2.83 (SD=+- 0.17) % OF DRY WEIGHT	PAR POND, SC, MAY 69, SOFT WAT, WATER=1.61 K PPM, N=5	160
*TYPHA LATIFOLIA	2.65 (SD=+- 0.50) % OF DRY WEIGHT	PAR POND, SC, MAY 69, SOFT WAT, WATER=1.61 K PPM, N=4	160
*UTRICULARIA INFLATA	1.98 (SD=+- 0.13) % OF DRY WEIGHT	PAR POND, SC, MAY 69, SOFT WAT, WATER=1.61 K PPM, N=4	160

TABLE A298

KORMOPHYTES, MACROPHYTES : MAX STANDING CROP			
ITEM:	VALUE:	CONDITIONS:	REF.:
*SUBMERSED MACROPHYTES IN FRESHWATER	520 G DRY WEIGHT/M2	RIVER IVEL, ENGLAND	393
*SUBMERSED MACROPHYTES IN FRESHWATER	320 G DRY WEIGHT/M2	RIVER IVEL, ENGLAND	1200
*SUBMERSED MACROPHYTES IN FRESHWATER	385 G DRY WEIGHT/M2	RIVER TEST, ENGLAND	1200
*SUBMERSED MACROPHYTES IN FRESHWATER	381 G DRY WEIGHT/M2	RIVER YARE, ENGLAND	1200
*SUBMERSED MACROPHYTES IN FRESHWATER	322 G DRY WEIGHT/M2	RIVER CHESS, ENGLAND	1200
*SUBMERSED MACROPHYTES IN FRESHWATER	123 G DRY WEIGHT/M2	MAPLE LODGE CHANNEL, ENGLAND, POLLUTED	1678
*SUBMERSED MACROPHYTES IN FRESHWATER	621 G DRY WEIGHT/M2	SILVER SPRINGS, FLORIDA	1166
*SUBMERSED MACROPHYTES IN FRESHWATER	809 G DRY WEIGHT/M2	SILVER SPRINGS, FLORIDA, MACROPHYTES AND ASSOCIATED AUFWUCHS	1166
*SUBMERSED MACROPHYTES IN FRESHWATER	526 G DRY WEIGHT/M2	THREE FLORIDA SPRINGS	1236
*SUBMERSED MACROPHYTES IN FRESHWATER	412 G DRY WEIGHT/M2	THREE FLORIDA RIVERS	1236
*SUBMERSED MACROPHYTES IN FRESHWATER	638 G DRY WEIGHT/M2	SAN MARCOS RIVER, TEXAS	608

TABLE A299

KORMOPHYTES, MACROPHYTES : MG/BIOMASS RATIO			
ITEM:	VALUE:	CONDITIONS:	REF.:
*BRASENIA SCHREBERI	0.26 (SD=+- 0.01) % OF DRY WEIGHT	PAR POND, SC, MAY 69, SOFT WAT, WATER=1.04 MG PPM, N=4	160
*CERATOPHYLLUM DEMERSUM	0.42 (SD=+- 0.10) % OF DRY WEIGHT	PAR POND, SC, MAY 69, SOFT WAT, WATER=1.04 MG PPM, N=4	160
*ELEOCHARIS ACICULARIS	0.33 (SD=+- 0.06) % OF DRY WEIGHT	PAR POND, SC, MAY 69, SOFT WAT, WATER=1.04 MG PPM, N=4	160
*ELEOCHARIS QUADRANGULATA	0.10 % OF DRY WEIGHT	PAR POND, SC, MAY 69, SOFT WAT, WATER=1.04 MG PPM, N=1	160

TABLE A299

KORMOPHYTES, MACROPHYTES : MG/BIOMASS RATIO			
ITEM:	VALUE:	CONDITIONS:	REF.:
*FREE FLOATING	3.90 (SE=+-0.602) G/KG DRY WEIGHT	AMAZON	696
*HYDROCOTYLE SP.	0.47 (SD=+- 0.09) % OF DRY WEIGHT	PAR POND, SC, MAY 69, SOFT WAT, WATER=1.04 MG PPM, N=4	160
*JUNCUS EFFUSUS	0.11 (SD=+- 0.01) % OF DRY WEIGHT	PAR POND, SC, MAY 69, SOFT WAT, WATER=1.04 MG PPM, N=5	160
*MYRIOPHYLLUM HETEROPHYLLUM	0.26 (SD=+- 0.05) % OF DRY WEIGHT	PAR POND, SC, MAY 69, SOFT WAT, WATER=1.04 MG PPM, N=4	160
*NAJAS GUADALUPENSIS	0.47 (SD=+- 0.06) % OF DRY WEIGHT	PAR POND, SC, MAY 69, SOFT WAT, WATER=1.04 MG PPM, N=4	160
*NELUMBO LUTEA	0.23 (SD=+- 0.02) % OF DRY WEIGHT	PAR POND, SC, MAY 69, SOFT WAT, WATER=1.04 MG PPM, N=2	160
*NUPHAR ADVENA	0.27 % OF DRY WEIGHT	PAR POND, SC, MAY 69, SOFT WAT, WATER=1.04 MG PPM, N=2	160
*NYMPHAEA ODORATA	0.14 (SD=+- 0.01) % OF DRY WEIGHT	PAR POND, SC, MAY 69, SOFT WAT, WATER=1.04 MG PPM, N=5	160
*PANICUM HEMITONIUM	0.25 (SD=+- 0.04) % OF DRY WEIGHT	PAR POND, SC, MAY 69, SOFT WAT, WATER=1.04 MG PPM, N=3	160
*PONTEDERIA CORDATA	0.15 % OF DRY WEIGHT	PAR POND, SC, MAY 69, SOFT WAT, WATER=1.04 MG PPM, N=1	160
*POTAMOGETON DIVERSIFOLIUS	0.19 % OF DRY WEIGHT	PAR POND, SC, MAY 69, SOFT WAT, WATER=1.04 MG PPM, N=1	160
*ROOTED	1.93 (SE=+-0.426) G/KG DRY WEIGHT	AMAZON	696
*SAGITTERIA LATIFOLIA	0.18 % OF DRY WEIGHT	PAR POND, SC, MAY 69, SOFT WAT, WATER=1.04 MG PPM, N=2	160
*SCIRPUS AMERICANUS	0.22 (SD=+- 0.02) % OF DRY WEIGHT	PAR POND, SC, MAY 69, SOFT WAT, WATER=1.04 MG PPM, N=5	160
*TYPHA LATIFOLIA	0.15 (SD=+- 0.03) % OF DRY WEIGHT	PAR POND, SC, MAY 69, SOFT WAT, WATER=1.04 MG PPM, N=4	160
*UTRICULARIA INFLATA	0.21 (SD=+- 0.02) % OF DRY WEIGHT	PAR POND, SC, MAY 69, SOFT WAT, WATER=1.04 MG PPM, N=4	160

TABLE A300

KORMOPHYTES, MACROPHYTES : MN/BIOMASS RATIO			
ITEM:	VALUE:	CONDITIONS:	REF.:
*BRASENIA SCHREBERI	265 (SD=+- 72) MG/KG DRY WEIGHT	PAR POND, SC, MAY 69, SOFT WAT, WATER=0.002 MN PPM, N=4	160
*CERATOPHYLLUM DEMERSUM	486 (SD=+- 62) MG/KG DRY WEIGHT	PAR POND, SC, MAY 69, SOFT WAT, WATER=0.002 MN PPM, N=4	160
*ELEOCHARIS ACICULARIS	192 (SD=+- 30) MG/KG DRY WEIGHT	PAR POND, SC, MAY 69, SOFT WAT, WATER=0.002 MN PPM, N=4	160
*ELEOCHARIS QUADRANGULATA	460 MG/KG DRY WEIGHT	PAR POND, SC, MAY 69, SOFT WAT, WATER=0.002 MN PPM, N=2	160
*HYDROCOTYLE SP.	196 (SD=+- 71) MG/KG DRY WEIGHT	PAR POND, SC, MAY 69, SOFT WAT, WATER=0.002 MN PPM, N=4	160
*MYRIOPHYLLUM HETEROPHYLLUM	473 (SD=+- 64) MG/KG DRY WEIGHT	PAR POND, SC, MAY 69, SOFT WAT, WATER=0.002 MN PPM, N=4	160
*NAJAS GUADALUPENSIS	201 (SD=+- 62) MG/KG DRY WEIGHT	PAR POND, SC, MAY 69, SOFT WAT, WATER=0.002 MN PPM, N=4	160
*NELUMBO LUTEA	607 (SD=+- 152) MG/KG DRY WEIGHT	PAR POND, SC, MAY 69, SOFT WAT, WATER=0.002 MN PPM, N=4	160
*NUPHAR ADVENA	300 MG/KG DRY WEIGHT	PAR POND, SC, MAY 69, SOFT WAT, WATER=0.002 MN PPM, N=2	160
*NYMPHAEA ODORATA	128 (SD=+- 13) MG/KG DRY WEIGHT	PAR POND, SC, MAY 69, SOFT WAT, WATER=0.002 MN PPM, N=5	160
*PANICUM HEMITONIUM	560 MG/KG DRY WEIGHT	PAR POND, SC, MAY 69, SOFT WAT, WATER=0.002 MN PPM, N=4	160
*PONTEDERIA CORDATA	200 MG/KG DRY WEIGHT	PAR POND, SC, MAY 69, SOFT WAT, WATER=0.002 MN PPM, N=1	160
*POTAMOGETON DIVERSIFOLIUS	160 MG/KG DRY WEIGHT	PAR POND, SC, MAY 69, SOFT WAT, WATER=0.002 MN PPM, N=1	160
*SAGITTERIA LATIFOLIA	460 MG/KG DRY WEIGHT	PAR POND, SC, MAY 69, SOFT WAT, WATER=0.002 MN PPM, N=2	160

TABLE A300

KORMOPHYTES, MACROPHYTES : MN/BIOMASS RATIO

ITEM:	VALUE:	CONDITIONS:	REF.:
*SCIRPUS AMERICANUS	292 (SD=+- 60) MG/KG DRY WEIGHT	PAR POND, SC, MAY 69, SOFT WAT, WATER=0.002 MN PPM, N=3	160
*TYPHA LATIFOLIA	412 (SD=+- 67) MG/KG DRY WEIGHT	PAR POND, SC, MAY 69, SOFT WAT, WATER=0.002 MN PPM, N=4	160
*UTRICULARIA INFLATA	480 (SD=+- 114) MG/KG DRY WEIGHT	PAR POND, SC, MAY 69, SOFT WAT, WATER=0.002 MN PPM, N=4	160

TABLE A301

KORMOPHYTES, MACROPHYTES : N/BIOMASS RATIO

ITEM:	VALUE:	CONDITIONS:	REF.:
*FREE FLOATING	22.4 (SE=+-1.950) G/KG DRY WEIGHT	AMAZON	696
*ROOTED	15.13 (SE=+-2.986) G/KG DRY WEIGHT	AMAZON	696

TABLE A302

KORMOPHYTES, MACROPHYTES : NA/BIOMASS RATIO

ITEM:	VALUE:	CONDITIONS:	REF.:
*BRASENIA SCHREBERI	0.66 (SD=+- 0.06) % OF DRY WEIGHT	PAR POND, SC, MAY 69, SOFT WAT, WATER=6.70 NA PPM, N=4	160
*CERATOPHYLLUM DEMERSUM	1.16 (SD=+- 0.16) % OF DRY WEIGHT	PAR POND, SC, MAY 69, SOFT WAT, WATER=6.70 NA PPM, N=4	160
*ELEOCHARIS ACICULARIS	0.54 (SD=+- 0.15) % OF DRY WEIGHT	PAR POND, SC, MAY 69, SOFT WAT, WATER=6.70 NA PPM, N=4	160
*ELEOCHARIS QUADRANGULATA	0.12 % OF DRY WIEGHT	PAR POND, SC, MAY 69, SOFT WAT, WATER=6.70 NA PPM, N=1	160
*FREE FLOATING	6.36 (SE=+-1.729) G/KG DRY WEIGHT	AMAZON	696
*HYDROCOTYLE SP.	0.98 (SD=+- 0.15) % OF DRY WEIGHT	PAR POND, SC, MAY 69, SOFT WAT, WATER=6.70 NA PPM, N=4	160
*JUNCUS EFFUSUS	0.40 (SD=+- 0.07) % OF DRY WEIGHT	PAR POND, SC, MAY 69, SOFT WAT, WATER=6.70 NA PPM, N=5	160
*MYRIOPHYLLUM HETEROPHYLLUM	1.87 (SD=+- 0.02) % OF DRY WEIGHT	PAR POND, SC, MAY 69, SOFT WAT, WATER=6.70 NA PPM, N=4	160
*NAJAS GUADALUPENSIS	0.61 (SD=+- 0.14) % OF DRY WEIGHT	PAR POND, SC, MAY 69, SOFT WAT, WATER=6.70 NA PPM, N=4	160
*NELUMBO LUTEA	0.28 (SD=+- 0.07) % OF DRY WEIGHT	PAR POND, SC, MAY 69, SOFT WAT, WATER=6.70 NA PPM, N=2	160
*NUPHAR ADVENA	1.47 % OF DRY WEIGHT	PAR POND, SC, MAY 69, SOFT WAT, WATER=6.70 NA PPM, N=2	160
*NYMPHAEA ODORATA	1.35 (SD=+- 0.06) % OF DRY WEIGHT	PAR POND, SC, MAY 69, SOFT WAT, WATER=6.70 NA PPM, N=5	160
*PANICUM HEMITONIUM	0.19 (SD=+- 0.02) % OF DRY WEIGHT	PAR POND, SC, MAY 69, SOFT WAT, WATER=6.70 NA PPM, N=3	160
*PONTEDERIA CORDATA	0.83 % OF DRY WEIGHT	PAR POND, SC, MAY 69, SOFT WAT, WATER=6.70 NA PPM, N=1	160
*POTAMOGETON DIVERSIFOLIUS	0.44 % OF DRY WEIGHT	PAR POND, SC, MAY 69, SOFT WAT, WATER=6.70 NA PPM, N=1	160
*ROOTED	2.32 (SE=+-2.032) G/KG DRY WEIGHT	AMAZON	696
*SAGITTERIA LATIFOLIA	0.14 % OF DRY WEIGHT	PAR POND, SC, MAY 69, SOFT WAT, WATER=6.70 NA PPM, N=2	160
*SCIRPUS AMERICANUS	0.09 (SD=+- 0.03) % OF DRY WEIGHT	PAR POND, SC, MAY 69, SOFT WAT, WATER=6.70 NA PPM, N=5	160
*TYPHA LATIFOLIA	0.28 (SD=+- 0.07) % OF DRY WEIGHT	PAR POND, SC, MAY 69, SOFT WAT, WATER=6.70 NA PPM, N=4	160
*UTRICULARIA INFLATA	1.52 (SD=+- 0.22) % OF DRY WEIGHT	PAR POND, SC, MAY 69, SOFT WAT, WATER=6.70 NA PPM, N=4	160

TABLE A303

KORMOPHYTES, MACROPHYTES : P/BIOMASS RATIO

ITEM:	VALUE:	CONDITIONS:	REF.:
*BRASENIA SCHREBERI	0.14 (SD=+- 0.01) % OF DRY WEIGHT	PAR POND, SC, MAY 69, SOFT WAT, WATER=0.008 P PPM, N=4	160
*CERATOPHYLLUM DEMERSUM	0.26 (SD=+- 0.02) % OF DRY WEIGHT	PAR POND, SC, MAY 69, SOFT WAT, WATER=0.008 P PPM, N=4	160
*ELEOCHARIS ACICULARIS	0.24 (SD=+- 0.07) % OF DRY WEIGHT	PAR POND, SC, MAY 69, SOFT WAT, WATER=0.008 P PPM, N=4	160
*ELEOCHARIS QUADRANGULATA	0.10 % OF DRY WEIGHT	PAR POND, SC, MAY 69, SOFT WAT, WATER=0.008 P PPM, N=1	160
*FREE FLOATING	2.11 (SE=+-0.295) G/KG DRY WEIGHT	AMAZON	696
*HYDROCOTYLE SP.	0.18 (SD=+- 0.02) % OF DRY WEIGHT	PAR POND, SC, MAY 69, SOFT WAT, WATER=0.008 P PPM, N=4	160
*JUNCUS EFFUSUS	0.27 (SD=+- 0.01) % OF DRY WEIGHT	PAR POND, SC, MAY 69, SOFT WAT, WATER=0.008 P PPM, N=5	160
*MYRIOPHYLLUM HETEROPHYLLUM	0.16 (SD=+- 0.02) % OF DRY WEIGHT	PAR POND, SC, MAY 69, SOFT WAT, WATER=0.008 P PPM, N=4	160
*NAJAS GUADALUPENSIS	0.15 (SD=+- 0.03) % OF DRY WEIGHT	PAR POND, SC, MAY 69, SOFT WAT, WATER=0.008 P PPM, N=4	160
*NELUMBO LUTEA	0.19 (SD=+- 0.05) % OF DRY WEIGHT	PAR POND, SC, MAY 69, SOFT WAT, WATER=0.008 P PPM, N=2	160
*NUPHAR ADVENA	0.40 % OF DRY WEIGHT	PAR POND, SC, MAY 69, SOFT WAT, WATER=0.008 P PPM, N=2	160
*NYMPHAEA ODORATA	0.18 (SD=+- 0.02) % OF DRY WEIGHT	PAR POND, SC, MAY 69, SOFT WAT, WATER=0.008 P PPM, N=5	160
*PANICUM HEMITONIUM	0.14 (SD=+- 0.02) % OF DRY WEIGHT	PAR POND, SC, MAY 69, SOFT WAT, WATER=0.008 P PPM, N=3	160
*PONTEDERIA CORDATA	0.24 % OF DRY WEIGHT	PAR POND, SC, MAY 69, SOFT WAT, WATER=0.008 P PPM, N=1	160
*POTAMOGETON DIVERSIFOLIUS	0.27 % OF DRY WEIGHT	PAR POND, SC, MAY 69, SOFT WAT, WATER=0.008 P PPM, N=1	160
*ROOTED	1.50 (SE=+-0.371) G/KG DRY WEIGHT	AMAZON	696
*SAGITTERIA LATIFOLIA	0.30 % OF DRY WEIGHT	PAR POND, SC, MAY 69, SOFT WAT, WATER=0.008 P PPM, N=2	160
*SCIRPUS AMERICANUS	0.18 (SD=+- 0.01) % OF DRY WEIGHT	PAR POND, SC, MAY 69, SOFT WAT, WATER=0.008 P PPM, N=5	160
*TYPHA LATIFOLIA	0.14 (SD=+- 0.02) % OF DRY WEIGHT	PAR POND, SC, MAY 69, SOFT WAT, WATER=0.008 P PPM, N=4	160
*UTRICULARIA INFLATA	0.12 (SD=+- 0.01) % OF DRY WEIGHT	PAR POND, SC, MAY 69, SOFT WAT, WATER=0.008 P PPM, N=4	160

TABLE A304

KORMOPHYTES, MACROPHYTES : PRODUCTION

ITEM:	VALUE:	CONDITIONS:	REF.:
*EMERGENTS DOMINATING	180 G DRY WEIGHT/(YEAR*M**2)	SPHAGNUM, ON HUMMOCKS, BLANCET BOG, ENGL.	263
*EMERGENTS DOMINATING	290 G DRY WEIGHT/(YEAR*M**2)	SPHAGNUM, IN POOLS, BLANKET BOG, ENGL.	263
*EMERGENTS DOMINATING	340 G DRY WEIGHT/(YEAR*M**2)	SPHAGNUM, ON 'LAWNS', BLANKET BOG, ENGL.	263
*EMERGENTS DOMINATING	738 G DRY WEIGHT/(YEAR*M**2)	CAREX, MINNESOTA WETLANDS	107
*EMERGENTS DOMINATING	2500 G DRY WEIGHT/(YEAR*M**2)	TYPHA, CEDAR CREEK, MINN.	165
*FLOATING	1500 - 4400 G DRY WEIGHT/ (YEAR*M**2)	EICHHORINA, NEW ORLEANS, LA.	1235
*SUBMERGENTS DOMINATING	595 G DRY WEIGHT/(YEAR*M**2)	SCIRPUS SUBTERMINALIS, L LAWRENCE, HARDWATER	1325
*SUBMERGENTS DOMINATING	155 G DRY WEIGHT/(YEAR*M**2)	CHARA, L LAWRENCE, HARDWATER	1325
*SUBMERGENTS DOMINATING	199 G DRY WEIGHT/(YEAR*M**2)	ANNUALS, L LAWRENCE, HARDWATER	1325
*SUBMERGENTS DOMINATING	64 G DRY WEIGHT/(YEAR*M**2)	RUPPIA, BORAX L, CALIF, SALINE	1688

TABLE A305

KORMOPHYTES, MACROPHYTES : PROTEIN/BIOMASS RATIO

ITEM:	VALUE:	CONDITIONS:	REF.:
*EMERGENT	13 (RANGE =3.5 - 26.7) % OF ORGANIC MATTER	NONE	1515

TABLE A305

KORMOPHYTES, MACROPHYTES : PROTEIN/BIOMASS RATIO			
ITEM:	VALUE:	CONDITIONS:	REF.:
*PLANTS WITH FLOATING LEAVES	26.5 (RANGE =18 - 44) % OF ORGANIC MATTER	NONE	1515
*SUBMERSED	22 (RANGE =7.5 - 34.7) % OF ORGANIC MATTER	NONE	1515

TABLE A306

KORMOPHYTES, MACROPHYTES : RATE OF NITROGEN UPTAKE			
ITEM:	VALUE:	CONDITIONS:	REF.:
*CERATOPHYLLUM DEMERSUM	0.591 MICROGRAM NO3-N/(MG PLANT N*HOUR)	20 CM PLANT, 4 H START AT 1040 CDT, LIGHT=192 LY/4 H, 297.4 K	1605
*CERATOPHYLLUM DEMERSUM	0.399 MICROGRAM NO3-N/(MG PLANT N*HOUR)	20 CM PLANT, 4 H START AT 1440 CDT, LIGHT=57 LY/4 H, 297.4 K	1605
*CERATOPHYLLUM DEMERSUM	0.320 MICROGRAM NO3-N/(MG PLANT N*HOUR)	20 CM PLANT, 4 H START AT 1840 CDT, LIGHT=1.6 LY/4 H, 296 K	1605
*CERATOPHYLLUM DEMERSUM	0.015 MICROGRAM NO3-N/(MG PLANT N*HOUR)	20 CM PLANT, 4 H START AT 2240 CDT, LIGHT=0 LY/4 H, 294.6 K	1605
*CERATOPHYLLUM DEMERSUM	0.010 MICROGRAM NO3-N/(MG PLANT N*HOUR)	20 CM PLANT, 4 H START AT 0240 CDT, LIGHT=0.1 LY/4 H, 294.9 K	1605
*CERATOPHYLLUM DEMERSUM	0.504 MICROGRAM NO3-N/(MG PLANT N*HOUR)	20 CM PLANT, 4 H START AT 0640 CDT, LIGHT=23 LY/4 H, 294.7 K	1605
*CERATOPHYLLUM DEMERSUM	0.257 MICROGRAM NH4-N/(MG PLANT N*HOUR)	20 CM PLANT, 4 H START AT 1040 CDT, LIGHT=192 LY/4 H, 297.4 K	1605
*CERATOPHYLLUM DEMERSUM	0.291 MICROGRAM NH4-N/(MG PLANT N*HOUR)	20 CM PLANT, 4 H START AT 1440 CDT, LIGHT=57 LY/4 H, 297.4 K	1605
*CERATOPHYLLUM DEMERSUM	0.298 MICROGRAM NH4-N/(MG PLANT N*HOUR)	20 CM PLANT, 4 H START AT 1840 CDT, LIGHT=1. LY/4 H, 296 K	1605
*CERATOPHYLLUM DEMERSUM	0.238 MICROGRAM NH4-N/(MG PLANT N*HOUR)	20 CM PLANT, 4 H START AT 2240 CDT, LIGHT=0 LY/4 H, 294.6 K	1605
*CERATOPHYLLUM DEMERSUM	0.237 MICROGRAM NH4-N/(MG PLANT N*HOUR)	20 CM PLANT, 4 H START AT 0240 CDT, LIGHT=0.1 LY/4 H, 294.9 K	1605
*CERATOPHYLLUM DEMERSUM	0.394 MICROGRAM NH4-N/(MG PLANT N*HOUR)	20 CM PLANT, 4 H START AT 0640 CDT, LIGHT=23 LY/4 H, 294.7 K	1605
*EICHHORNIA CRASSIPES	5.3 G N/(PLANT*DAY)	STATIC WATER, 10 % HOAGLANDS SOLUTION, NORTH CAROLINA	1333
*EICHHORNIA CRASSIPES	11.4 G N/(PLANT*DAY)	STATIC WATER, 25 % HOAGLANDS SOLUTION, NORTH CAROLINA	1333
*EICHHORNIA CRASSIPES	19.8 G N/(PLANT*DAY)	STATIC WATER, 50 % HOAGLANDS SOLUTION, NORTH CAROLINA	1333
*EICHHORNIA CRASSIPES	6.6 G N/(PLANT*DAY)	STATIC WATER, SEWAGE EFFLUENT, NORTH CAROLINA	1333
*EICHHORNIA CRASSIPES	9.9 G N/(PLANT*DAY)	FLOWING WAT., 10 % HOAGLANDS SOLUTION, NORTH CAROLINA	1333
*EICHHORNIA CRASSIPES	18.4 G N/(PLANT*DAY)	FLOWING WAT., 25 % HOAGLANDS SOLUTION, NORTH CAROLINA	1333
*EICHHORNIA CRASSIPES	20.8 G N/(PLANT*DAY)	FLOWING WAT., 50 % HOAGLANDS SOLUTION, NORTH CAROLINA	1333
*EICHHORNIA CRASSIPES	96 KG N/METRIC TON OF MACROPHYTE	MACROPHYTE IN AERATION POND	290

TABLE A307

KORMOPHYTES, MACROPHYTES : RATE OF PHOSPHORUS UPTAKE			
ITEM:	VALUE:	CONDITIONS:	REF.:
*EICHHORNIA CRASSIPES	1.1 G P/(PLANT*DAY)	STATIC WATER, 10 % HOAGLANDS SOLUTION, NORTH CAROLINA	1333
*EICHHORNIA CRASSIPES	2.1 G P/(PLANT*DAY)	STATIC WATER, 25 % HOAGLANDS SOLUTION, NORTH CAROLINA	1333
*EICHHORNIA CRASSIPES	3.1 G P/(PLANT*DAY)	STATIC WATER, 50 % HOAGLANDS SOLUTION, NORTH CAROLINA	1333

TABLE A307

KORMOPHYTES, MACROPHYTES : RATE OF PHOSPHORUS UPTAKE			
ITEM:	VALUE:	CONDITIONS:	REF.:
*EICHHORNIA CRASSIPES	1.6 G P/(PLANT*DAY)	STATIC WATER, SEWAGE EFFLUENT, NORTH CAROLINA	1333
*EICHHORNIA CRASSIPES	1.7 G P/(PLANT*DAY)	FLOWING WAT., 10 % HOAGLANDS SOLUTION, NORTH CAROLINA	1333
*EICHHORNIA CRASSIPES	2.5 G P/(PLANT*DAY)	FLOWING WAT., 25 % HOAGLANDS SOLUTION, NORTH CAROLINA	1333
*EICHHORNIA CRASSIPES	3.3 G P/(PLANT*DAY)	FLOWING WAT., 50 % HOAGLANDS SOLUTION, NORTH CAROLINA	1333
*EICHHORNIA CRASSIPES	18 KG PO4-P/METRIC TON OF MACROPHYTE	MACROPHYTE IN AERATION POND	290
*ZOSTERA MARINA	0.66 MICROGRAM PHOSPHATE-P/PLANT	LIGHT, 3000 MIN, FROM LOWER WATER TO UPPER WATER	1013
*ZOSTERA MARINA	0.16 MICROGRAM PHOSPHATE-P/PLANT	DARK, 3000 MIN, FROM LOWER WATER TO UPPER WATER	1013
*ZOSTERA MARINA	0.088 MICROGRAM PHOSPHATE-P/PLANT	LIGHT, 3000 MIN, FROM UPPER WATER TO LOWER WATER	1013
*ZOSTERA MARINA	0.0016 MICROGRAM PHOSPHATE-P/PLANT	DARK, 3000 MIN, FROM UPPER WATER TO LOWER WATER	1013
*ZOSTERA MARINA	0.88 MICROGRAM PHOSPHATE-P/PLANT	LIGHT, 3000 MIN, FROM UPPER WATER TO LEAVES AND STEMS	1013
*ZOSTERA MARINA	0.116 MICROGRAM PHOSPHATE-P/PLANT	DARK, 3000 MIN, FROM UPPER WATER TO LEAVES AND STEMS	1013
*ZOSTERA MARINA	0.16 MICROGRAM PHOSPHATE-P/PLANT	LIGHT, 3000 MIN, FROM LOWER WATER TO ROOTS AND RHIZOMES	1013
*ZOSTERA MARINA	0.088 MICROGRAM PHOSPHATE-P/PLANT	DARK, 3000 MIN, FROM LOWER WATER TO ROOTS AND RHIZOMES	1013
*ZOSTERA MARINA	0.964 MICROGRAM PHOSPHATE-P/PLANT	LIGHT, 3000 MIN, FROM UPPER WATER TO PLANT	1013
*ZOSTERA MARINA	0.216 MICROGRAM PHOSPHATE-P/PLANT	DARK, 3000 MIN, FROM UPPER WATER TO PLANT	1013
*ZOSTERA MARINA	0.34 MICROGRAM PHOSPHATE-P/PLANT	LIGHT, 3000 MIN, FROM LOWER WATER TO PLANT	1013
*ZOSTERA MARINA	0.098 MICROGRAM PHOSPHATE-P/PLANT	DARK, 3000 MIN, FROM LOWER WATER TO PLANT	1013
*ZOSTERA MARINA	0.392 MICROGRAM PHOSPHATE-P/PLANT	LIGHT, 3000 MIN, FROM UPPER WATER TO PLANT, NET TRANSPORT	1013
*ZOSTERA MARINA	0.057 MICROGRAM PHOSPHATE-P/PLANT	DARK, 3000 MIN, FROM UPPER WATER TO PLANT, NET TRANSPORT	1013
*ZOSTERA MARINA	0.912 MICROGRAM PHOSPHATE-P/PLANT	LIGHT, 3000 MIN, FROM LOWER WATER TO PLANT, NET TRANSPORT	1013
*ZOSTERA MARINA	0.256 MICROGRAM PHOSPHATE-P/PLANT	DARK, 3000 MIN, FROM LOWER WATER TO PLANT, NET TRANSPORT	1013

TABLE A308

KORMOPHYTES, MACROPHYTES : S/BIOMASS RATIO			
ITEM:	VALUE:	CONDITIONS:	REF.:
*BRASENIA SCHREBERI	0.11 (SD=+- 0.01) % OF DRY WEIGHT	PAR POND, SC, MAY 69, SOFT WAT, WATER=1.03 S PPM, N=4	160
*CERATOPHYLLUM DEMERSUM	0.30 (SD=+- 0.02) % OF DRY WEIGHT	PAR POND, SC, MAY 69, SOFT WAT, WATER=1.03 S PPM, N=4	160
*ELEOCHARIS ACICULARIS	0.28 (SD=+- 0.04) % OF DRY WEIGHT	PAR POND, SC, MAY 69, SOFT WAT, WATER=1.03 S PPM, N=4	160
*ELEOCHARIS QUADRANGULATA	0.15 % OF DRY WEIGHT	PAR POND, SC, MAY 69, SOFT WAT, WATER=1.03 S PPM, N=1	160
*HYDROCOTYLE SP.	0.16 (SD=+- 0.01) % OF DRY WEIGHT	PAR POND, SC, MAY 69, SOFT WAT, WATER=1.03 S PPM, N=4	160
*JUNCUS EFFUSUS	0.26 (SD=+- 0.01) % OF DRY WEIGHT	PAR POND, SC, MAY 69, SOFT WAT, WATER=1.03 S PPM, N=5	160
*MYRIOPHYLLUM HETEROPHYLLUM	0.24 (SD=+- 0.02) % OF DRY WEIGHT	PAR POND, SC, MAY 69, SOFT WAT, WATER=1.03 S PPM, N=4	160
*NAJAS GUADALUPENSIS	0.28 (SD=+- 0.01) % OF DRY WEIGHT	PAR POND, SC, MAY 69, SOFT WAT, WATER=1.03 S PPM, N=4	160

TABLE A308

KORMOPHYTES, MACROPHYTES : S/BIOMASS RATIO			
ITEM:	VALUE:	CONDITIONS:	REF.:
*NELUMBO LUTEA	0.16 (SD=+- 0.01) % OF DRY WEIGHT	PAR POND, SC, MAY 69, SOFT WAT, WATER=1.03 S PPM, N=2	160
*NUPHAR ADVENA	0.32 % OF DRY WEIGHT	PAR POND, SC, MAY 69, SOFT WAT, WATER=1.03 S PPM, N=2	160
*NYMPHAEA ODORATA	0.14 (SD=+- 0.02) % OF DRY WEIGHT	PAR POND, SC, MAY 69, SOFT WAT, WATER=1.03 S PPM, N=5	160
*PANICUM HEMITONIUM	0.23 (SD=+- 0.02) % OF DRY WEIGHT	PAR POND, SC, MAY 69, SOFT WAT, WATER=1.03 S PPM, N=3	160
*PONTEDERIA CORDATA	0.22 % OF DRY WEIGHT	PAR POND, SC, MAY 69, SOFT WAT, WATER=1.03 S PPM, N=1	160
*POTAMOGETON DIVERSIFOLIUS	0.50 % OF DRY WEIGHT	PAR POND, SC, MAY 69, SOFT WAT, WATER=1.03 S PPM, N=1	160
*SAGITTERIA LATIFOLIA	0.15 % OF DRY WEIGHT	PAR POND, SC, MAY 69, SOFT WAT, WATER=1.03 S PPM, N=2	160
*SCIRPUS AMERICANUS	0.59 (SD=+- 0.02) % OF DRY WEIGHT	PAR POND, SC, MAY 69, SOFT WAT, WATER=1.03 S PPM, N=5	160
*TYPHA LATIFOLIA	0.15 (SD=+- 0.02) % OF DRY WEIGHT	PAR POND, SC, MAY 69, SOFT WAT, WATER=1.03 S PPM, N=4	160
*UTRICULARIA INFLATA	0.26 (SD=+- 0.02) % OF DRY WEIGHT	PAR POND, SC, MAY 69, SOFT WAT, WATER=1.03 S PPM, N=4	160

TABLE A309

KORMOPHYTES, MACROPHYTES : SI/BIOMASS RATIO			
ITEM:	VALUE:	CONDITIONS:	REF.:
*FREE FLOATING	12.35 (SE=+-2.752) G/KG DRY WEIGHT	AMAZON	696
*ROOTED	17.2 (SE=+-4.785) G/KG DRY WEIGHT	AMAZON	696

TABLE A310

KORMOPHYTES, MACROPHYTES : ZN/BIOMASS RATIO			
ITEM:	VALUE:	CONDITIONS:	REF.:
*BRASENIA SCHREBERI	267 (SD=+- 94) MG/KG DRY WEIGHT	PAR POND, SC, MAY 69, SOFT WAT, WATER=0.008 ZN PPM, N=4	160
*CERATOPHYLLUM DEMERSUM	100 (SD=+- 24) MG/KG DRY WEIGHT	PAR POND, SC, MAY 69, SOFT WAT, WATER=0.008 ZN PPM, N=4	160
*ELEOCHARIS ACICULARIS	68 (SD=+- 7) MG/KG DRY WEIGHT	PAR POND, SC, MAY 69, SOFT WAT, WATER=0.008 ZN PPM, N=4	160
*ELEOCHARIS QUADRANGULATA	45 MG/KG DRY WEIGHT	PAR POND, SC, MAY 69, SOFT WAT, WATER=0.008 ZN PPM, N=2	160
*HYDROCOTYLE SP.	53 (SD=+- 8) MG/KG DRY WEIGHT	PAR POND, SC, MAY 69, SOFT WAT, WATER=0.008 ZN PPM, N=4	160
*HYDROCOTYLE SP.	31 (SD=+- 5) MG/KG DRY WEIGHT	PAR POND, SC, MAY 69, SOFT WAT, WATER=0.008 ZN PPM, N=3	160
*MYRIOPHYLLUM HETEROPHYLLUM	54 (SD=+- 14) MG/KG DRY WEIGHT	PAR POND, SC, MAY 69, SOFT WAT, WATER=0.008 ZN PPM, N=4	160
*NAJAS GUADALUPENSIS	48 (SD=+- 17) MG/KG DRY WEIGHT	PAR POND, SC, MAY 69, SOFT WAT, WATER=0.008 ZN PPM, N=4	160
*NELUMBO LUTEA	50 (SD=+- 8) MG/KG DRY WEIGHT	PAR POND, SC, MAY 69, SOFT WAT, WATER=0.008 ZN PPM, N=4	160
*NUPHAR ADVENA	50 MG/KG DRY WEIGHT	PAR POND, SC, MAY 69, SOFT WAT, WATER=0.008 ZN PPM, N=2	160
*NYMPHAEA ODORATA	32 (SD=+- 5) MG/KG DRY WEIGHT	PAR POND, SC, MAY 69, SOFT WAT, WATER=0.008 ZN PPM, N=5	160
*PONTEDERIA CORDATA	67 MG/KG DRY WEIGHT	PAR POND, SC, MAY 69, SOFT WAT, WATER=0.008 ZN PPM, N=1	160
*POTAMOGETON DIVERSIFOLIUS	60 MG/KG DRY WEIGHT	PAR POND, SC, MAY 69, SOFT WAT, WATER=0.008 ZN PPM, N=1	160

TABLE A310

KORMOPHYTES, MACROPHYTES : ZN/BIOMASS RATIO

ITEM:	VALUE:	CONDITIONS:	REF.:
*SAGITTERIA LATIFOLIA	46 MG/KG DRY WEIGHT	PAR POND, SC, MAY 69, SOFT WAT, WATER=0.008 ZN PPM, N=2	160
*TYPHA LATIFOLIA	30 (SD=+- 5) MG/KG DRY WEIGHT	PAR POND, SC, MAY 69, SOFT WAT, WATER=0.008 ZN PPM, N=4	160
*UTRICULARIA INFLATA	108 (SD=+- 27) MG/KG DRY WEIGHT	PAR POND, SC, MAY 69, SOFT WAT, WATER=0.008 ZN PPM, N=4	160

TABLE A311

KORMOPHYTES, PLANTS : AL/BIOMASS RATIO			
ITEM:	VALUE:	CONDITIONS:	REF.:
*ALFALFA	0.0025 %	ATOMIC COMPOSITION	109

TABLE A312

KORMOPHYTES, PLANTS : ASH/BIOMASS RATIO			
ITEM:	VALUE:	CONDITIONS:	REF.:
*ANGIOSPERMS	6.39 % OF DRY WEIGHT	LEAVES,7 SPECIES	81
*ANGIOSPERMS	0.57 % OF DRY WEIGHT	WOOD,9 SPECIES	81
*BRASENIA SCHREBERI	7.6% OF DRY WEIGHT	NONE	158
*BRYOPHYTES	3.28 % OF DRY WEIGHT	8 SPECIES	81
*CERATOPHYLLUM DEMERSUM	14.9 % OF DRY WEIGHT	NONE	158
*ELEOCHARIS ACICULARIS	11.2 % OF DRY WEIGHT	NONE	158
*FERNS	7.71 % OF DRY WEIGHT	6 SPECIES	81
*GYMNOSPERMS	2.41 % OF DRY WEIGHT	LEAVES AND TWIGS,3 SPECIES	81
*GYMNOSPERMS	0.166 % OF DRY WEIGHT	WOOD,5 SPECIES	81
*MYRIOPHYLLUM HETEROPHYLLUM	12.2 % OF DRY WEIGHT	NONE	158
*NAJAS GUADALUPENSIS	12.8 % OF DRY WEIGHT	NONE	158
*NELUMBO LUTEA	8.8 % OF DRY WEIGHT	NONE	158
*NUPHAR ADVENA	10.6 % OF DRY WEIGHT	NONE	158
*NYMPHAEA ODORATA	8.1 % OF DRY WEIGHT	NONE	158
*TYPHA LATIFOLIA	7.5 % OF DRY WEIGHT	NONE	158
*UTRICULARIA INFLATA	14.0 % OF DRY WEIGHT	NONE	158

TABLE A313

KORMOPHYTES, PLANTS : AVERAGE WIDTH OF LEAVES			
ITEM:	VALUE:	CONDITIONS:	REF.:
*CONIFEROUS FOREST	0.2 M	INPUT VALUE	556
*GRASS	0.005 M	INPUT VALUE	556
*MAIZE	0.05 M	INPUT VALUE	556

TABLE A314

KORMOPHYTES, PLANTS : BR/BIOMASS RATIO			
ITEM:	VALUE:	CONDITIONS:	REF.:
*ALFALFA	0.0001 %	ATOMIC COMPOSITION	109

TABLE A315

KORMOPHYTES, PLANTS : C/BIOMASS RATIO			
ITEM:	VALUE:	CONDITIONS:	REF.:
*ALFALFA	11.34 %	ATOMIC COMPOSITION	109

TABLE A315

KORMOPHYTES, PLANTS : C/BIOMASS RATIO			
ITEM:	VALUE:	CONDITIONS:	REF.:
*ANGIOSPERMS	45.4 % OF DRY WEIGHT	MEAN ELEMENTARY COMPOSITION	157
*MEAN PLANT	45 % OF DRY WEIGHT	MEAN ELEMENTARY COMPOSITION	157

TABLE A316

KORMOPHYTES, PLANTS : CA/BIOMASS RATIO			
ITEM:	VALUE:	CONDITIONS:	REF.:
*ALFALFA	0.58 %	ATOMIC COMPOSITION	109

TABLE A317

KORMOPHYTES, PLANTS : CALORIC CONTENT			
ITEM:	VALUE:	CONDITIONS:	REF.:
*ALPINE	5000 CAL/G DRY MATTER	SHOOTS OR LEAVES	584
*ALPINE	5102 CAL/G ASH FREE DRY WEIGHT	SHOOTS OR LEAVES	584
*ALPINE	4619 CAL/G DRY MATTER	ROOTS OR RHIZOMES	584
*ALPINE	4713 CAL/G ASH FREE DRY WEIGHT	ROOTS OR RHIZOMES	584
*BRASENIA SCHREBERI	4.03 KCAL/G DRY WEIGHT	NONE	158
*CERATOPHYLLUM DEMERSUM	3.91 KCAL/G DRY WEIGHT	NONE	158
*CONIFERS	4927 CAL/G DRY MATTER	SHOOTS OR LEAVES	1195
*CORNT	3.5 KCAL/G	PER G HARVEST WEIGHT	1623
*CRUDE SUGAR	3.7 KCAL/G	PER G HARVEST WEIGHT	1623
*CYMODOCEA SP.	3253 - 3228 CAL/G DRY MATTER	TROPICAL SEAGRAS, LEAF, GREEN	129
*CYMODOCEA SP.	3253 CAL/G DRY MATTER	TROPICAL SEAGRAS, LEAF, BROWN	129
*CYMODOCEA SP.	3266 - 3421 CAL/G DRY MATTER	TROPICAL SEAGRAS, RHIZOME	129
*CYMODOCEA SP.	2841 - 2841 CAL/G DRY MATTER	TROPICAL SEAGRAS, SHEATH	129
*CYMODOCEA SP.	3239 CAL/G DRY MATTER	SHOOTS OR LEAVES	129
*CYMODOCEA SP.	3999 CAL/G ASH FREE DRY WEIGHT	SHOOTS OR LEAVES	129
*DECIDUOUS FOREST	4759 CAL/G DRY MATTER	SHOOTS OR LEAVES	1195
*ELEOCHARIS ACICULARIS	4.26 KCAL/G DRY WEIGHT	NONE	158
*ENHALUS SP.	3124 - 3032 CAL/G DRY MATTER	TROPICAL SEAGRAS, LEAF, PROXIMAL HALF	129
*ENHALUS SP.	2515 CAL/G DRY MATTER	TROPICAL SEAGRAS, LEAF, DISTAL HALF	129
*ENHALUS SP.	3581 - 3522 CAL/G DRY MATTER	TROPICAL SEAGRAS, RHIZOME	129
*ENHALUS SP.	3109 CAL/G DRY MATTER	TROPICAL SEAGRAS, SHEATH	129
*ENHALUS SP.	2890 CAL/G DRY MATTER	SHOOTS OR LEAVES	129
*ENHALUS SP.	3802 CAL/G ASH FREE DRY WEIGHT	SHOOTS OR LEAVES	129
*HALODULE SP.	3246 - 3193 CAL/G DRY MATTER	TROPICAL SEAGRAS, LEAF	129
*HALODULE SP.	3018 - 3249 CAL/G DRY MATTER	TROPICAL SEAGRAS, RHIZOME	129
*HALODULE SP.	2507 - 2808 CAL/G DRY MATTER	TROPICAL SEAGRAS, SHEATH	129
*HALODULE SP.	3219 CAL/G DRY MATTER	SHOOTS OR LEAVES	129
*HALODULE SP.	4350 CAL/G ASH FREE DRY WEIGHT	SHOOTS OR LEAVES	129
*HALOPHILA MINOR	1407 - 1495 CAL/G DRY MATTER	TROPICAL SEAGRAS, WHOLE PLANT	129
*HALOPHILA OVALIS	2814 CAL/G DRY MATTER	SHOOTS OR LEAVES	129
*HALOPHILA OVALIS	6507 CAL/G ASH FREE DRY WEIGHT	SHOOTS OR LEAVES	129
*HALOPHILA OVALIS	2814 CAL/G DRY MATTER	TROPICAL SEAGRAS, LEAF	129

TABLE A317

KORMOPHYTES, PLANTS : CALORIC CONTENT

ITEM:	VALUE:	CONDITIONS:	REF.:
*HALOPHILA OVALIS	3124 CAL/G DRY MATTER	TROPICAL SEAGRAS, WHOLE PLANT	129
*HALOPHILA OVALIS	2603 CAL/G DRY MATTER	TROPICAL SEAGRAS, LEAF	129
*HALOPHILA OVALIS	2959 CAL/G DRY MATTER	TROPICAL SEAGRAS, RHIZOME	129
*HALOPHILA OVALIS	1954 CAL/G DRY MATTER	TROPICAL SEAGRAS, WHOLE PLANT	129
*HALOPHILA SPINULOSA	2814 CAL/G DRY MATTER	SHOOTS OR LEAVES	129
*HALOPHILA SPINULOSA	3629 CAL/G ASH FREE DRY WEIGHT	SHOOTS OR LEAVES	129
*HALOPHILA SPINULOSA	2814 CAL/G DRY MATTER	TROPICAL SEAGRAS, LEAF	129
*HYDROTRIDA CAROLINENSIS	4.06 KCAL/G DRY WEIGHT	NONE	158
*MYRIOPHYLLUM HETEROPHYLLUM	3.96 KCAL/G DRY WEIGHT	NONE	158
*NAJAS GUADALUPENSIS	3.92 KCAL/G DRY WEIGHT	NONE	158
*NELUMBO LUTEA	4.23 KCAL/G DRY WEIGHT	NONE	158
*NUPHAR ADVENA	4.32 KCAL/G DRY WEIGHT	NONE	158
*NYMPHAEA ODORATA	4.18 KCAL/G DRY WEIGHT	NONE	158
*OLD FIELD HERBS	4056 CAL/G DRY MATTER	SHOOTS OR LEAVES	544
*POTATOES	0.9 KCAL/G	PER G HARVEST WEIGHT	1623
*RED MANGROVE	4190 CAL/G DRY MATTER	SHOOTS OR LEAVES	544
*RED MANGROVE	4034 CAL/G DRY MATTER	ROOTS OR RHIZOMES	544
*RICE	3.6 KCAL/G	PER G HARVEST WEIGHT	1623
*SOYBEANS	4.0 KCAL/G	PER G HARVEST WEIGHT	1623
*SYRINGODIUM SP.	2178 CAL/G DRY MATTER	TROPICAL SEAGRAS, LEAF	129
*SYRINGODIUM SP.	2966 CAL/G DRY MATTER	TROPICAL SEAGRAS, RHIZOME	129
*SYRINGODIUM SP.	2792 CAL/G DRY MATTER	TROPICAL SEAGRAS, SHEATH	129
*SYRINGODIUM SP.	2178 CAL/G DRY MATTER	SHOOTS OR LEAVES	129
*SYRINGODIUM SP.	3630 CAL/G ASH FREE DRY WEIGHT	SHOOTS OR LEAVES	129
*THALASSIA SP.	1990 CAL/G DRY MATTER	SHOOTS OR LEAVES	201
*THALASSIA SP.	2618 CAL/G ASH FREE DRY WEIGHT	SHOOTS OR LEAVES	201
*TYPHA LATIFOLIA	4.26 KCAL/G DRY WEIGHT	NONE	158
*UTRICULARIA INFLATA	4.02 KCAL/G DRY WEIGHT	NONE	158
*WET TROPICAL FOREST	4065 CAL/G DRY MATTER	SHOOTS OR LEAVES	545
*WET TROPICAL FOREST	4041 CAL/G DRY MATTER	ROOTS OR RHIZOMES	545
*WHEAT	3.3 KCAL/G	PER G HARVEST WEIGHT	1623
*ZOSTERA SP.	3369 CAL/G DRY MATTER	SHOOTS OR LEAVES	1015
*ZOSTERA SP.	4211 CAL/G ASH FREE DRY WEIGHT	SHOOTS OR LEAVES	1015
*ZOSTERA SP.	2857 CAL/G DRY MATTER	ROOTS OR RHIZOMES	1015
*ZOSTERA SP.	3571 CAL/G ASH FREE DRY WEIGHT	ROOTS OR RHIZOMES	1015

TABLE A318

KORMOPHYTES, PLANTS : CHANGE IN EQUIVAL.AIR TEMP/CHANGE IN HAG

ITEM:	VALUE:	CONDITIONS:	REF.:
*CONIFEROUS FOREST	0.015 DEG.CELCIUS/M	EQUIVALENT HEAT FLUX=1 J/(M**2) *SEC, HAG=HEIGHT ABOVE GROUND	556
*CONIFEROUS FOREST	0.066 DEG.CELCIUS/M	EQUIVALENT HEAT FLUX=10 J/(M**2) *SEC, HAG=HEIGHT ABOVE GROUND	556
*CONIFEROUS FOREST	0.30 DEG.CELCIUS/M	EQUIVALENT HEAT FLUX=100 J/(M**2) *SEC, HAG=HEIGHT ABOVE GROUND	556
*GRASS	4.2 DEG.CELCIUS/M	EQUIVALENT HEAT FLUX=1 J/(M**2) *SEC, HAG=HEIGHT ABOVE GROUND	556
*GRASS	19.5 DEG.CELCIUS/M	EQUIVALENT HEAT FLUX=10 J/(M**2) *SEC, HAG=HEIGHT ABOVE GROUND	556
*GRASS	90. DEG.CELCIUS/M	EQUIVALENT HEAT FLUX=100 J/(M**2) *SEC, HAG=HEIGHT ABOVE GROUND	556

TABLE A318

KORMOPHYTES, PLANTS : CHANGE IN EQUIVAL.AIR TEMP/CHANGE IN HAG			
ITEM:	VALUE:	CONDITIONS:	REF.:
*MAIZE	0.12 DEG.CELCIUS/M	EQUIVALENT HEAT FLUX=1 J/(M**2) *SEC, HAG=HEIGHT ABOVE GROUND	556
*MAIZE	0.56 DEG.CELCIUS/M	EQUIVALENT HEAT FLUX¤10 J/(M**2) *SEC, HAG=HEIGHT ABOVE GROUND	556
*MAIZE	2.56 DEG.CELCIUS/M	EQUIVALENT HEAT FLUX=100 J/(M**2) *SEC, HAG=HEIGHT ABOVE GROUND	556

TABLE A319

KORMOPHYTES, PLANTS : CL/BIOMASS RATIO			
ITEM:	VALUE:	CONDITIONS:	REF.:
*ALFALFA	0.07 %	ATOMIC COMPOSITION	109
*CYMODOCEA SP. AND HALODULE SP.	17% OF ASH WEIGHT	TROPICAL SEAGRASSES,LEAVES	129
*CYMODOCEA SP. AND HALODULE SP.	25% OF ASH WEIGHT	TROPICAL SEAGRASSES,RHIZOMES	129
*ENHALUS SP. AND SYRINGODIUM SP.	26% OF ASH WEIGHT	TROPICAL SEAGRASSES,LEAVES	129
*ENHALUS SP. AND SYRINGODIUM SP.	27% OF ASH WEIGHT	TROPICAL SEAGRASSES,RHIZOMES	129
*HALOPHILA DECIPIENS	15% OF ASH WEIGHT	TROPICAL SEAGRASSES	129
*HALOPHILA SPINULOSA	32% OF ASH WEIGHT	TROPICAL SEAGRASSES	129

TABLE A320

KORMOPHYTES, PLANTS : COMPENSATION POINT			
ITEM:	VALUE:	CONDITIONS:	REF.:
*ARTEMISIA HERBA-ALBA	35.0-46.0 DEGREES CELCIUS	UPPER TEMP.AT COMP.POINT OF CO2 EXCHANGE,MAY 2,N=4,NON-IRRIGATED	874
*HAMMADA SCOPARIA	46.3-60.0 DEGREES CELCIUS	UPPER TEMP.AT COMP.POINT OF CO2 EXCHANGE,AUG.16,N=13, NON-IRRIGATED	874
*HAMMADA SCOPARIA	50.1-57.7 DEGREES CELCIUS	UPPER TEMP.AT COMP.POINT OF CO2 EXCHANGE,AUG.16,N=11,IRRIGATED	874
*PRUNUS ARMENIACA	42.7-54.5 DEGREES CELCIUS	UPPER TEMP.AT COMP.POINT OF CO2 EXCHANGE,JUNE 27,N=11, NON-IRRIGATED	874
*PRUNUS ARMENIACA	43.5-52.4 DEGREES CELCIUS	UPPER TEMP.AT COMP.POINT OF CO2 EXCHANGE,JULY 5,N=13,IRRIGATED	874
*REAUMURIA NEGEVENSIS	43.0-49.8 DEGREES CELCIUS	UPPER TEMP.AT COMP.POINT OF CO2 EXCHANGE,MAY 16,N=4,NON-IRRIGATED	874
*REAUMURIA NEGEVENSIS	44.5-50.6 DEGREES CELCIUS	UPPER TEMP.AT COMP.POINT OF CO2 EXCHANGE,AUG.16,N=4,IRRIGATED	874
*SALSOLA INERMIS	46.8-52.8 DEGREES CELCIUS	UPPER TEMP.AT COMP.POINT OF CO2 EXCHANGE,AUG.30,N=4,NON-IRRIGATED	874
*ZYGOPHYLLUM DUMOSUM	38.3-49.5 DEGREES CELCIUS	UPPER TEMP.AT COMP.POINT OF CO2 EXCHANGE,MAY 16,N=11, NON-IRRIGATED	874
*ZYGOPHYLLUM DUMOSUM	43.6-52.0 DEGREES CELCIUS	UPPER TEMP.AT COMP.POINT OF CO2 EXCHANGE,AUG.30,N=5,IRRIGATED	874

TABLE A321

KORMOPHYTES, PLANTS : CU/BIOMASS RATIO			
ITEM:	VALUE:	CONDITIONS:	REF.:
*ALFALFA	0.0003 %	ATOMIC COMPOSITION	109

TABLE A322

KORMOPHYTES, PLANTS : DEPTH CONTANING 63 % OF THE ROOTS			
ITEM:	VALUE:	CONDITIONS:	REF.:
*BARLEY	32.7 CM	DENSITY=50 PLANTS/M2, FRESH W.= 0.556 KG/M2	832
*BARLEY	70.4 CM	DENSITY=200 PLANTS/M2, FRESH W.= 0.742 KG/M2	832
*BARLEY	36.4 CM	DENSITY=800 PLANTS/M2, FRESH W.= 0.727 KG/M2	832
*BARLEY	57.1 CM	HIGH SOIL MOISTURE, FRESH W.=0.755 KG/M2	832
*BARLEY	38.4 CM	LOW SOIL MOISTURE, FRESH W.=0.963 KG/M2	832
*BRUSSEL SPROUTS	70.4 CM	BY ROOT NUMBER PROFILE, AGE=180 D, FRESH W.=8.64 KG/M2	1631
*CABBAGE	2.8 CM	SANDY LOAM, AGE=41 D, FRESH W.= 0.72 KG/M2	516
*CABBAGE	2.8 CM	SANDY LOAM, AGE=41 D, FRESH W.= 0.69 KG/M2	516
*CABBAGE	5.5 CM	SANDY LOAM, AGE=56 D, FRESH W.=2.6 KG/M2	516
*CABBAGE	7.1 CM	SANDY LOAM, AGE=57 D, FRESH W.=2.8 KG/M2	516
*CABBAGE	5.5 CM	SANDY LOAM, AGE=63 D, FRESH W.=2.6 KG/M2	516
*CABBAGE	6.2 CM	SANDY LOAM, AGE=64 D, FRESH W.=3.0 KG/M2	516
*CABBAGE	14.1 CM	SANDY LOAM, AGE=73 D, FRESH W.=5.6 KG/M2	516
*CABBAGE	15.1 CM	SANDY LOAM, AGE=73 D, FRESH W.=6.8 KG/M2	516
*CABBAGE	17.3 CM	SANDY LOAM, AGE=77 D, FRESH W.=7.6 KG/M2	516
*CABBAGE	41.5 CM	SANDY LOAM, AGE=77 D, FRESH W.=9.6 KG/M2	516
*CABBAGE	17.8 CM	SANDY LOAM, AGE=86 D, FRESH W.=8.9 KG/M2	516
*CABBAGE	6.4 CM	SANDY LOAM, AGE=86 D, FRESH W.=8.2 KG/M2	516
*CABBAGE	9.6 CM	SANDY LOAM, AGE=92 D, FRESH W.= 11.3 KG/M2	516
*CARROT	24.8 CM	HIGH FERTILITY, AGE=38 D, FRESH W. =0.9 KG/M2	1205
*CARROT	11.8 CM	HIGH FERTILITY, AGE=59 D, FRESH W. =5.5 KG/M2	1205
*CARROT	22.9 CM	HIGH FERTILITY, AGE = 79 D, FRESH W.=8.9 KG/M2	1205
*CARROT	24.8 CM	HIGH FERTILITY, AGE = 135 D, FRESH W.=12.1 KG/M2	1205
*CARROT	20.9 CM	LOW FERTILITY, AGE = 38 D, FRESH W.=1.5 KG/M2	1205
*CARROT	12.5 CM	LOW FERTILITY, AGE = 59 D, FRESH W.=3.6 KG/M2	1205
*CARROT	19.7 CM	LOW FERTILITY, AGE = 79 D, FRESH W.=6.6 KG/M2	1205
*CARROT	33.9 CM	LOW FERTILITY, AGE = 135 D, FRESH W.=9.9 KG/M2	1205
*CARROT	14.6 CM	PEAT,BY SOIL INJECTION OF P-32	602
*CARROT	23.8 CM	PEAT,BY SOIL INJECTION OF P-32	602
*CARROT	23.1 CM	PEAT,BY SOIL INJECTION OF P-32	602
*CARROT	38.8 CM	PEAT,BY SOIL INJECTION OF P-32, FRESH W.=8.25 KG/M2	602
*CELERIAC	19.6 CM	AGE = 108 D	678
*COTTON	93.5 CM	IRRIGATED, BY NUMBERS OF ROOTS	137
*COTTON	31.3 CM	IRRIGATED, BY WEIGHT OF ROOTS	137
*ENDIVE	24.1 CM	AGE = 65 D	678
*HYPARRHENIA RUFA	49.0 CM	AGE = 120 D, BY DRY WEIGHT	324

TABLE A322

KORMOPHYTES, PLANTS : DEPTH CONTANING 63 % OF THE ROOTS			
ITEM:	VALUE:	CONDITIONS:	REF.:
*KOHL-RABI	21.3 CM	AGE = 87 D	678
*LEEK	19.8 CM	AGE = 136 D	678
*LETTUCE	10.4 CM	HIGH FERTILITY, AGE=57 D, FRESH W. =4.3 KG/M2	1205
*LETTUCE	16.5 CM	LOW FERTILITY, AGE=78 D, FRESH W.= 1.6 KG/M2	1205
*LETTUCE	16.1 CM	LOW FERTILITY, AGE=75 D, FRESH W.= 3.1 KG/M2	1205
*LETTUCE	16.8 CM	IRRIGATED, AGE=42 D, FRESH W.=6.1 KG/M2	517
*LETTUCE	16.8 CM	IRRIGATED, AGE=42 D, FRESH W.=6.1 KG/M2	517
*LETTUCE	32.9 CM	NON-IRRIGATED, AGE=42 D, FRESH W.= 2.8 KG/M2	517
*LETTUCE	24.1 CM	NON-IRRIGATED, AGE=42 D, FRESH W.= 2.8 KG/M2	517
*MAIZE	44.0 CM	AGE =100 D, FRESH W.=1.26 KG/M2	475
*MAIZE	44.8 CM	CONTROL, AGE=100 D, FRESH W.=0.672 KG/M2	448
*MAIZE	63.1 CM	PHOSPHATE ADDED, AGE=100 D, FRESH W.=0.915 KG/M2	448
*MAIZE	16.2 CM	LOW N3 DEFOLIATIONS, 2 Y,BY INJECTION OF RB-86	1122
*MAIZE	21.6 CM	LOW N1 DEFOLIATIONS, 2 Y,BY INJECTION OF RB-86	1122
*MAIZE	14.0 CM	HIGH N3 DEFOLIATIONS, 2 Y,BY INJECTION OF RB-86	1122
*MAIZE	13.9 CM	HIGH N1 DEFOLIATIONS, 2 Y,BY INJECTION OF RB-86	1122
*MAIZE	14.1 CM	LOW N3 DEFOLIATIONS, 2 Y,BY ROOT DRY WEIGHT	1122
*OATS	40.6 CM	AGE=105 D, BY ROOT WEIGHT	137
*ONION	11.7 CM	SANDY LOAM, AGE=102 D, FRESH W.= 5.1 KG/M2	517
*ONION	11.4 CM	AGE = 106 D	678
*ONION	16.1 CM	SILT,BY SOIL INJECTION OF P-32	602
*ONION	10.9 CM	SILT,BY SOIL INJECTION OF P-32	602
*ONION	11.7 CM	SILT,BY SOIL INJECTION OF P-32	602
*ONION	5.9 CM	SILT,BY SOIL INJECTION OF P-32, FRESH W.=3.05 KG/M2	602
*ONION	5.9 CM	PEAT,BY SOIL INJECTION OF P-32	602
*ONION	5.8 CM	PEAT,BY SOIL INJECTION OF P-32	602
*ONION	9.5 CM	PEAT,BY SOIL INJECTION OF P-32	602
*ONION	19.1 CM	PEAT,BY SOIL INJECTION OF P-32, FRESH W.=6.31 KG/M2	602
*PARSLEY	27.1 CM	AGE = 188 D	678
*PASTURE	16.3 CM	AGE = 120 D, BY DRY WEIGHT	1407
*PEA	32.9 CM	AGE = 116 D	678
*PEA	64.5 CM	IRRIGATED, AGE=105 D, BY NUMBERS	137
*PEA	35.7 CM	IRRIGATED, AGE=105 D, BY WEIGHT	137
*PEPPER	10.8 CM	BY SOIL INJECTION OF P-32	602
*PEPPER	14.2 CM	BY SOIL INJECTION OF P-32	602
*PEPPER	14.3 CM	BY SOIL INJECTION OF P-32	602
*PEPPER	14.1 CM	BY SOIL INJECTION OF P-32, FRESH W.=2.70 KG/M2	602
*RYE-GRASS	13.5 CM	LOW N1 DEFOLIATIONS, 2 Y,BY ROOT DRY WEIGHT	1122
*RYE-GRASS	16.7 CM	HIGH N3 DEFOLIATIONS, 2 Y,BY ROOT DRY WEIGHT	1122
*RYE-GRASS	13.9 CM	HIGH N1 DEFOLIATIONS, 2 Y,BY ROOT DRY WEIGHT	1122

TABLE A322

KORMOPHYTES, PLANTS : DEPTH CONTANING 63 % OF THE ROOTS

ITEM:	VALUE:	CONDITIONS:	REF.:
*RYE-GRASS	20.5 CM	LOW N3 DEFOLIATIONS, 2 Y,BY ROOT NUMBER	1122
*RYE-GRASS	19.3 CM	LOW N1 DEFOLIATIONS, 2 Y,BY ROOT NUMBER	1122
*RYE-GRASS	20.1 CM	HIGH N3 DEFOLIATIONS, 2 Y,BY ROOT NUMBER	1122
*RYE-GRASS	21.6 CM	HIGH N1 DEFOLIATIONS, 2 Y,BY ROOT NUMBER	1122
*RYE-GRASS, SINGLE PLANTS	13.2 CM	AGE = 120 D, BY DRY WEIGHT	1298
*RYE-GRASS, SINGLE PLANTS	17.8 CM	AGE =120 D, BY TOTAL NUMBER	1298
*RYE-GRASS, SINGLE PLANTS	16.4 CM	AGE = 120 D, BY TOTAL LENGTH	1298
*SORGHUM SP.	144.6 CM	INCREASING MOISTURE (1), AGE=83 D	1097
*SORGHUM SP.	135.1 CM	INCREASING SOIL MOISTURE (2), AGE= 83 D	1097
*SORGHUM SP.	92.1 CM	INCREASING SOIL MOISTURE (3), AGE= 83 D	1097
*SORGHUM SP.	59.9 CM	INCREASING SOIL MOISTURE (4), AGE= 83 D	1097
*SPRING BARLEY	9.6 CM	CULTIVATION DEPTH = 0 CM,AGE=34 D, FRESH W.=0.114 KG/M2	1121
*SPRING BARLEY	11.1 CM	CULTIVATION DEPTH = 7.5 CM, AGE=34 D, FRESH W.=0.154 KG/M2	1121
*SPRING BARLEY	11.4 CM	CULTIVATION DEPTH = 15 CM, AGE=34 D, FRESH W.=0.148 KG/M2	1121
*SPRING BARLEY	11.4 CM	CULTIVATION DEPTH = 23 CM, AGE=34 D, FRESH W.=0.160 KG/M2	1121
*SPRING BARLEY	14.3 CM	CULTIVATION DEPTII = 0 CM, AGE-69 D, FRESH W.=0.890 KG/M2	1121
*SPRING BARLEY	20.2 CM	CULTIVATION DEPTH = 7.5 CM, AGE=69 D, FRESH W.=0.907 KG/M2	1121
*SPRING BARLEY	12.2 CM	CULTIVATION DEPTH = 15 CM, AGE=69 D, FRESH W.=0.929 KG/M2	1121
*SPRING BARLEY	14.6 CM	CULTIVATION DEPTH = 23 CM, AGE=69 D, FRESH W.=0.929 KG/M2	1121
*SPRING BARLEY	8.7 CM	CULTIVATION DEPTH = 0 CM, AGE=94 D	1121
*SPRING BARLEY	9.1 CM	CULTIVATION DEPTH = 7.5 CM, AGE=95 D	1121
*SPRING BARLEY	7.2 CM	CULTIVATION DEPTH = 15 CM, AGE=97 D	1121
*TOMATO	21.7 CM	AGE = 103 D	678
*TOMATO	46.1 CM	IRRIGATED, BY NUMBERS	137
*TOMATO	20.8 CM	IRRIGATED, BY WEIGHT	137
*VETCH	54.9 CM	IRRIGATED, AGE=105 D, BY NUMBERS OF ROOTS	137
*VETCH	45.7 CM	IRRIGATED, AGE=105 D, BY WEIGHT OF ROOTS	137
*WINTER WHEAT	38.0 CM	AGE=180 D, RANGER	1120
*WINTER WHEAT	43.0 CM	AGE=180 D, CULTURE 3	1120
*WINTER WHEAT	44.8 CM	AGE=180 D, CULTURE 25	1120

TABLE A323

KORMOPHYTES, PLANTS : DIFFERENCE OF EQUIVAL.AIR TEMP.BETWEEN TWO LEVELS

ITEM:	VALUE:	CONDITIONS:	REF.:
*CONIFEROUS FOREST	0.15 DEG.CELCIUS	EQUIVALENT HEAT FLUX=1 J/(M**2) *SEC, TWO LEVELS OF AIR TEMP	556
*CONIFEROUS FOREST	0.66 DEG.CELCIUS	EQUIVALENT HEAT FLUX=10 J/(M**2) *SEC, TWO LEVELS OF AIR TEMP	556
*CONIFEROUS FOREST	3.0 DEG.CELCIUS	EQUIVALENT HEAT FLUX=100 J/(M**2) *SEC, TWO LEVELS OF AIR TEMP	556

TABLE A323

KORMOPHYTES, PLANTS : DIFFERENCE OF EQUIVAL.AIR TEMP.BETWEEN TWO LEVELS			
ITEM:	VALUE:	CONDITIONS:	REF.:
*GRASS	0.84 DEG.CELCIUS	EQUIVALENT HEAT FLUX=1 J/(M**2) *SEC, TWO LEVELS OF AIR TEMP	556
*GRASS	3.9 DEG.CELCIUS	EQUIVALENT HEAT FLUX=10 J/(M**2) *SEC, TWO LEVELS OF AIR TEMP	556
*GRASS	18 DEG.CELCIUS	EQUIVALENT HEAT FLUX=100 J/(M**2) *SEC, TWO LEVELS OF AIR TEMP	556
*MAIZE	0.30 DEG.CELCIUS	EQUIVALENT HEAT FLUX=1 J/(M**2) *SEC, TWO LEVELS OF AIR TEMP	556
*MAIZE	1.4 DEG.CELCIUS	EQUIVALENT HEAT FLUX=10 J/(M**2) *SEC, TWO LEVELS OF AIR TEMP	556
*MAIZE	6.4 DEG.CELCIUS	EQUIVALENT HEAT FLUX=100 J/(M**2) *SEC	556

TABLE A324

KORMOPHYTES, PLANTS : DRAG COEFFICIENT OF THE LEAVES			
ITEM:	VALUE:	CONDITIONS:	REF.:
*CONIFEROUS FOREST	0.2	INPUT VALUE	556
*GRASS	0.2	INPUT VALUE	556
*MAIZE	0.2	INPUT VALUE	556

TABLE A325

KORMOPHYTES, PLANTS : EXCHANGE COEFFICIENT FOR HEAT (MIN)			
ITEM:	VALUE:	CONDITIONS:	REF.:
*CONIFEROUS FOREST	55*10**(-3) M**2/SEC	EQUIVALENT HEAT FLUX=1 J/(M**2) *SEC	556
*CONIFEROUS FOREST	122*10**(-3) M**2/SEC	EQUIVALENT HEAT FLUX=10 J/(M**2) *SEC	556
*CONIFEROUS FOREST	260*10**(-3) M**2/SEC	EQUIVALENT HEAT FLUX=100 J/(M**2) *SEC	556
*GRASS	0.19*10**(-3) M**2/SEC	EQUIVALENT HEAT FLUX=1 J/(M**2) *SEC	556
*GRASS	0.41*10**(-3) M**2/SEC	EQUIVALENT HEAT FLUX=10 J/(M**2) *SEC	556
*GRASS	0.89*10**(-3) M**2/SEC	EQUIVALENT HEAT FLUX=100 J/(M**2) *SEC	556
*MAIZE	6.7*10**(-3) M**2/SEC	EQUIVALENT HEAT FLUX=1 J/(M**2) *SEC	556
*MAIZE	14.5*10**(-3) M**2/SEC	EQUIVALENT HEAT FLUX=10 J/(M**2) *SEC	556
*MAIZE	31.2*10**(-3) M**2/SEC	EQUIVALENT HEAT FLUX=100 J/(M**2) *SEC	556

TABLE A326

KORMOPHYTES, PLANTS : EXTINCTION FACTOR FOR WIND			
ITEM:	VALUE:	CONDITIONS:	REF.:
*CONIFEROUS FOREST	1.88 M	CALCULATED VALUE	556
*GRASS	3.54	CALCULATED VALUE	556
*MAIZE	2.55 M	CALCULATED VALUE	556

TABLE A327

KORMOPHYTES, PLANTS : FE/BIOMASS RATIO			
ITEM:	VALUE:	CONDITIONS:	REF.:
*ALFALFA	0.0270 %	ATOMIC COMPOSITION	109

TABLE A328

KORMOPHYTES, PLANTS : H/BIOMASS RATIO			
ITEM:	VALUE:	CONDITIONS:	REF.:
*ALFALFA	8.72 % OF DRY WEIGHT	ATOMIC COMPOSITION	109
*ANGIOSPERMS	5.5 % OF DRY WEIGHT	MEAN ELEMENTARY COMPOSITION	157
*MEAN PLANT	5 % OF DRY WEIGHT	MEAN ELEMENTARY COMPOSITION	157

TABLE A329

KORMOPHYTES, PLANTS : HEIGHT OF CANOPY			
ITEM:	VALUE:	CONDITIONS:	REF.:
*CONIFEROUS FOREST	10. M	INPUT VALUE	556
*CONIFEROUS FOREST	4.33 M	CALC VALUE, (HEIGHT OF CAN)-(ZERO PLANE DISPLACEMENT	556
*GRASS	0.2 M	INPUT VALUE	556
*GRASS	0.053 M	CALC VALUE, (HEIGHT OF CAN)-(ZERO PLANE DISPLACEMENT	556
*MAIZE	2.5 M	INPUT VALUE	556
*MAIZE	0.84 M	CALC VALUE, (HEIGHT OF CAN)-(ZERO PLANE DISPLACEMENT	556

TABLE A330

KORMOPHYTES, PLANTS : I/BIOMASS RATIO			
ITEM:	VALUE:	CONDITIONS:	REF.:
*ALFALFA	0.0000 %	ATOMIC COMPOSITION	109

TABLE A331

KORMOPHYTES, PLANTS : K/BIOMASS RATIO			
ITEM:	VALUE:	CONDITIONS:	REF.:
*ALFALFA	0.17 %	ATOMIC COMPOSITION	109
*BRASENIA SCHREBERI	0.99 % OF DRY WEIGHT	NONE	158
*CERATOPHYLLUM DEMERSUM	4.01 % OF DRY WEIGHT	NONE	158
*CYMODOCEA SP. AND HALODULE SP.	5.0% OF ASH WEIGHT	TROPICAL SEAGRASSES,LEAVES	129
*CYMODOCEA SP. AND HALODULE SP.	10.0% OF ASH WEIGHT	TROPICAL SEAGRASSES,RHIZOMES	129
*ELEOCHARIS ACICULARIS	2.86 % OF DRY WEIGHT	NONE	158
*ENHALUS SP. AND SYRINGODIUM SP.	2.5% OF ASH WEIGHT	TROPICAL SEAGRASSES,LEAVES	129
*ENHALUS SP. AND SYRINGODIUM SP.	10.8% OF ASH WEIGHT	TROPICAL SEAGRASSES,RHIZOMES	129
*HALOPHILA DECIPIENS	1.7% OF ASH WEIGHT	TROPICAL SEAGRASSES	129
*HALOPHILA SPINULOSA	4.1% OF ASH WEIGHT	TROPICAL SEAGRASSES	129

TABLE A331

KORMOPHYTES, PLANTS : K/BIOMASS RATIO			
ITEM:	VALUE:	CONDITIONS:	REF.:
*MYRIOPHYLLUM HETEROPHYLLUM	1.25 % OF DRY WEIGHT	NONE	158
*NAJAS GUADALUPENSIS	3.49 % OF DRY WEIGHT	NONE	158
*NELUMBO LUTEA	2.27 % OF DRY WEIGHT	NONE	158
*NUPHAR ADVENA	1.88 % OF DRY WEIGHT	NONE	158
*NYMPHAEA ODORATA	1.28 % OF DRY WEIGHT	NONE	158
*TYPHA LATIFOLIA	2.65 % OF DRY WEIGHT	NONE	158
*UTRICULARIA INFLATA	1.98 % OF DRY WEIGHT	NONE	158

TABLE A332

KORMOPHYTES, PLANTS : LEAF AREA DENSITY			
ITEM:	VALUE:	CONDITIONS:	REF.:
*CONIFEROUS FOREST	0.2 M**2 LEAF/M**3 AIR	INPUT VALUE	556
*GRASS	25. M**2 LEAF/M**3 AIR	INPUT VALUE	556
*MAIZE	1.2 M**2 LEAF/M**3 AIR	INPUT VALUE	556

TABLE A333

KORMOPHYTES, PLANTS : LEAF AREA INDEX			
ITEM:	VALUE:	CONDITIONS:	REF.:
*CONIFEROUS FOREST	2. M**2 LEAF/M**2 GROUND	INPUT VALUE	556
*GRASS	5. M**2 LEAF/M**2 GROUND	INPUT VALUE	556
*MAIZE	3. M**2 LEAF/M**2 GROUND	INPUT VALUE	556

TABLE A334

KORMOPHYTES, PLANTS : LEAF BIOMASS			
ITEM:	VALUE:	CONDITIONS:	REF.:
*ABIES SPP.	14.9 METRIC T/HA	EVERGREEN,CONIFEROUS, 9 SAMPLES, JAPAN	1385
*BETULA ERMANII	2.7 METRIC T/HA	DECIDUOUS, BROADLEAVED, 10 SAMPLES, JAPAN	1385
*BETULA SPP.	2.7 METRIC T/HA	DECIDUOUS, BROADLEAVED, 21 SAMPLES, JAPAN	1385
*CAMELLIA JAPONICA	8.6 METRIC T/HA	EVERGREEN, BROADLEAVED, 4 SAMPLES, JAPAN	1385
*CASTANOPSIS CUSPIDATA	8.5 METRIC T/HA	EVERGREEN, BROADLEAVED, 11 SAMPLES, JAPAN	1385
*CHAMAECYPARIS OBTUSA	13.2 METRIC T/HA	EVERGREEN, CONIFEROUS, 7 SAMPLES, JAPAN	1385
*FAGUS CRENATA	3.1 METRIC T/HA	DECIDUOUS, BROADLEAVED, 86 SAMPLES, JAPAN	1385
*LARIX LEPTOLEPIS	3.9 METRIC T/HA	DECIDUOUS, CONIFEROUS, 15 SAMPLES, JAPAN	1385
*PINUS DENSIFLORA	6.6 METRIC T/HA	EVERGREEN, CONIFEROUS, 108 SAMPLES, JAPAN	1385
*PINUS THUNBERGII	9.9 METRIC T/HA	EVERGREEN, CONIFEROUS, 12 SAMPLES, JAPAN	1385

TABLE A335

KORMOPHYTES, PLANTS : LEAF RES TO TRANSPIRATION			
ITEM:	VALUE:	CONDITIONS:	REF.:
*ACER RUBRUM	0.20 MIN/CM	FREE WATER SURFACE=0.0367 MIN/CM, MIN RES OF OF THE BOUNDARY LAYER	268

TABLE A335

KORMOPHYTES, PLANTS : LEAF RES TO TRANSPIRATION			
ITEM:	VALUE:	CONDITIONS:	REF.:
*ADENOSTOMA FASCICULATUM	1.0 MIN/CM	FREE WATER SURFACE=0.0367 MIN/CM, MIN RES OF OF THE BOUNDARY LAYER	268
*ADENOSTOMA FASCICULATUM	0.71 MIN/CM	FREE WATER SURFACE=0.0367 MIN/CM, MIN RES OF OF THE BOUNDARY LAYER	268
*AMMOPHILA BREVILIGULATA	0.05 MIN/CM	FREE WATER SURFACE=0.0367 MIN/CM, MIN RES OF OF THE BOUNDARY LAYER	268
*ARCTOSTAPHYLOS UVA-URSI	0.0834 MIN/CM	FREE WATER SURFACE=0.0367 MIN/CM, MIN RES OF OF THE BOUNDARY LAYER	268
*AVICENNIA TOMENTOSA	0.04 MIN/CM	FREE WATER SURFACE=0.0367 MIN/CM, MIN RES OF OF THE BOUNDARY LAYER	268
*BETULA PAPYRIFERA	0.1 MIN/CM	FREE WATER SURFACE=0.0367 MIN/CM, MIN RES OF OF THE BOUNDARY LAYER	268
*CEANOTHUS CRASSIFOLIUS	0.35 MIN/CM	FREE WATER SURFACE=0.0367 MIN/CM, MIN RES OF OF THE BOUNDARY LAYER	268
*CHAMAEDAPHNE CALYCULATA	0.2 MIN/CM	FREE WATER SURFACE=0.0367 MIN/CM, MIN RES OF OF THE BOUNDARY LAYER	268
*COCCOLOBA UVIFERA	0.04 MIN/CM	FREE WATER SURFACE=0.0367 MIN/CM, MIN RES OF OF THE BOUNDARY LAYER	268
*CONOCARPUS ERECTUS	0.05 MIN/CM	FREE WATER SURFACE=0.0367 MIN/CM, MIN RES OF OF THE BOUNDARY LAYER	268
*GOSSYPIUM SP.	0.020 MIN/CM	FREE WATER SURFACE=0.0367 MIN/CM, MIN RES. OF BL., LOWER	268
*GOSSYPIUM SP.	0.148 MIN/CM	FREE WATER SURFACE=0.0367 MIN/CM, MIN RES. OF BL., UPPER	268
*GOSSYPIUM SP.	0.084 MIN/CM	FREE WATER SURFACE=0.0367 MIN/CM, MIN RES. OF BL., MEAN	268
*LAGUNCULARIA RACEMOSA	0.03 MIN/CM	FREE WATER SURFACE=0.0367 MIN/CM, MIN RES OF OF THE BOUNDARY LAYER	268
*PINUS RESINOSA	0.50 MIN/CM	FREE WATER SURFACE=0.0367 MIN/CM, MIN RES OF OF THE BOUNDARY LAYER	268
*PINUS STROBUS	0.67 MIN/CM	FREE WATER SURFACE=0.0367 MIN/CM, MIN RES OF OF THE BOUNDARY LAYER	268
*POPULUS GRANDIDENTATA	0.0667 MIN/CM	FREE WATER SURFACE=0.0367 MIN/CM, MIN RES OF OF THE BOUNDARY LAYER	268
*POPULUS TREMULOIDES	0.05 MIN/CM	FREE WATER SURFACE=0.0367 MIN/CM, MIN RES. OF BL., WET SITE	268
*POPULUS TREMULOIDES	0.0667 MIN/CM	FREE WATER SURFACE=0.0367 MIN/CM, MIN RES. OF BL., DRY SITE	268
*PTERIDIUM AQUILINUM	3.3 MIN/CM	FREE WATER SURFACE=0.0367 MIN/CM, MIN RES OF OF THE BOUNDARY LAYER	268
*QUERCUS DUMOSA	0.196 MIN/CM	FREE WATER SURFACE=0.0367 MIN/CM, MIN RES OF OF THE BOUNDARY LAYER	268
*RHIZOPHORA MANGLE	0.04 MIN/CM	FREE WATER SURFACE=0.0367 MIN/CM, MIN RES OF OF THE BOUNDARY LAYER	268
*THUJA OCCIDENTALIS	1.25 MIN/CM	FREE WATER SURFACE=0.0367 MIN/CM, MIN RES. OF BL., SHADE	268
*THUJA OCCIDENTALIS	1.25 MIN/CM	FREE WATER SURFACE=0.0367 MIN/CM, MIN RES. OF BL., SUNE BOUNDARY LAYER	268
*VITIS SP.	0.25 MIN/CM	FREE WATER SURFACE=0.0367 MIN/CM, MIN RES OF OF THE BOUNDARY LAYER	268

TABLE A336

KORMOPHYTES, PLANTS : MG/BIOMASS RATIO			
ITEM:	VALUE:	CONDITIONS:	REF.:
*ALFALFA	0.08 %	ATOMIC COMPOSITION	109

TABLE A337

KORMOPHYTES, PLANTS : MIXING LENGTH			
ITEM:	VALUE:	CONDITIONS:	REF.:
*CONIFEROUS FOREST	1.13 M	CALCULATED VALUE	556

TABLE A337

KORMOPHYTES, PLANTS : MIXING LENGTH			
ITEM:	VALUE:	CONDITIONS:	REF.:
*GRASS	0.016 M	CALCULATED VALUE	556
*MAIZE	0.23 M	CALCULATED VALUE	556

TABLE A338

KORMOPHYTES, PLANTS : MN/BIOMASS RATIO			
ITEM:	VALUE:	CONDITIONS:	REF.:
*ALFALFA	0.0004 %	ATOMIC COMPOSITION	109

TABLE A339

KORMOPHYTES, PLANTS : N/BIOMASS RATIO			
ITEM:	VALUE:	CONDITIONS:	REF.:
*ALFALFA	0.825 %	ATOMIC COMPOSITION	109
*ANGIOSPERMS	3.3 % OF DRY WEIGHT	MEAN ELEMENTARY COMPOSITION	157
*MEAN PLANT	3 % OF DRY WEIGHT	MEAN ELEMENTARY COMPOSITION	157

TABLE A340

KORMOPHYTES, PLANTS : NA/BIOMASS RATIO			
ITEM:	VALUE:	CONDITIONS:	REF.:
*ALFALFA	0.03 %	ATOMIC COMPOSITION	109
*CYMODOCEA SP. AND HALODULE SP.	11.8% OF ASH WEIGHT	TROPICAL SEAGRASSES,LEAVES	129
*CYMODOCEA SP. AND HALODULE SP.	14.1% OF ASH WEIGHT	TROPICAL SEAGRASSES,RHIZOMES	129
*ENHALUS SP. AND SYRINGODIUM SP.	15.5% OF ASH WEIGHT	TROPICAL SEAGRASSES,LEAVES	129
*ENHALUS SP. AND SYRINGODIUM SP.	20.0% OF ASH WEIGHT	TROPICAL SEAGRASSES,RHIZOMES	129
*HALOPHILA DECIPIENS	9.6% OF ASH WEIGHT	TROPICAL SEAGRASSES	129
*HALOPHILA SPINULOSA	18.6% OF ASH WEIGHT	TROPICAL SEAGRASSES	129

TABLE A341

KORMOPHYTES, PLANTS : NITROGEN FIXATION			
ITEM:	VALUE:	CONDITIONS:	REF.:
*ARACHIS HYPOGAEA	124 KG N/(HA*YEAR)	LEGUMINOUS PLANT	1141
*ARISTIDA ARGENTEA	0 - 3000 MICROGRAM N/(M**2*HOUR)	DURING DAYLIGHT, JAN 75, NON-LEGUME	670
*BURKEA AFRICANA	1200 - 2200 MICROGRAM N/(M**2*HOUR)	DURING DAYLIGHT, JAN 75, LEGUME	670
*CAJANUS CAJAN	224 KG N/(HA*YEAR)	LEGUMINOUS PLANT	1141
*CALOPOGONIUM MUCUNOIDES	202 KG N/(HA*YEAR)	LEGUMINOUS PLANT	1141
*CANAVALIA ENSIFORMIS	49 KG N/(HA*YEAR)	LEGUMINOUS PLANT	1141
*CASSIA MIMOSOIDES	240 - 9600 MICROGRAM N/(M**2*HOUR)	DURING DAYLIGHT, JAN 75, LEGUME	670
*CENTROSEMA PUBESCENS	259 KG N/(HA*YEAR)	LEGUMINOUS PLANT	1141

TABLE A341

KORMOPHYTES, PLANTS : NITROGEN FIXATION			
ITEM:	VALUE:	CONDITIONS:	REF.:
*CICER ARIETINUM	103 KG N/(HA*YEAR)	LEGUMINOUS PLANT	1141
*CYAMOPSIS TETRAGONOLOBUS	130 KG N/(HA*YEAR)	LEGUMINOUS PLANT	1141
*DESMODIUM INTORTUM + D. CANUM	897 KG N/(HA*YEAR)	LEGUMINOUS PLANT	1141
*DICHAPETALUM CYMOSUM	160 - 1480 MICROGRAM N/(M**2*HOUR)	DURING DAYLIGHT, JAN 75, NON-LEGUME	670
*DIGITARIA ERIANTHA	0 - 440 MICROGRAM N/(M**2*HOUR)	DURING DAYLIGHT, JAN 75, NON-LEGUME	670
*ELEPHANTORRHIZA SP.	360 - 2720 MICROGRAM N/(M**2*HOUR)	DURING DAYLIGHT, JAN 75, LEGUME	670
*ENTEROLOBIUM SAMAN	150 KG N/(HA*YEAR)	LEGUMINOUS PLANT	1141
*ERAGROSTIS PALLENS	160 - 3120 MICROGRAM N/(M**2*HOUR)	DURING DAYLIGHT, JAN 75, NON-LEGUME	670
*GLYCINE MAX	103 KG N/(HA*YEAR)	LEGUMINOUS PLANT	1141
*HYPERTHELIA DISSOLUTA	40 - 1880 MICROGRAM N/(M**2*HOUR)	DURING DAYLIGHT, JAN 75, NON-LEGUME	670
*INDEGOFERA SORDIDA	520 - 3760 MICROGRAM N/(M**2*HOUR)	DURING DAYLIGHT, JAN 75, LEGUME	670
*INDIGOFERIA OXYTROPIS	1920 - 77200 MICROGRAM N/(M**2*HOUR)	DURING DAYLIGHT, JAN 75, LEGUME	670
*LENS CULINARIS	101 KG N/(HA*YEAR)	LEGUMINOUS PLANT	1141
*LESPEDEZA SP.	95 KG N/(HA*YEAR)	LEGUMINOUS PLANT	1141
*LEUCAENA GLAUCA	277 KG N/(HA*YEAR)	LEGUMINOUS PLANT	1141
*LOTONOSIS BAINESII	62 KG N/(HA*YEAR)	LEGUMINOUS PLANT	1141
*LOTUS CORNICULATUS	116 KG N/(HA*YEAR)	LEGUMINOUS PLANT	1141
*LUPINUS SP.	176 KG N/(HA*YEAR)	LEGUMINOUS PLANT	1141
*MEDICAGO SP.	199 KG N/(HA*YEAR)	LEGUMINOUS PLANT	1141
*MELILOTUS SP.	199 KG N/(HA*YEAR)	LEGUMINOUS PLANT	1141
*MIKANEA CORDATA	120 KG N/(HA*YEAR)	LEGUMINOUS PLANT	1141
*OCHNA PULCHRA	800 - 3240 MICROGRAM N/(M**2*HOUR)	DURING DAYLIGHT, JAN 75, NON-LEGUME	670
*PARINARI CAPENSE	40 - 7240 MICROGRAM N/(M**2*HOUR)	DURING DAYLIGHT, JAN 75, NON-LEGUME	670
*PHASEOLUS AUREUS	61 KG N/(HA*YEAR)	LEGUMINOUS PLANT, MUNG	1141
*PHASEOLUSS ATRAPURPUREA	291 KG N/(HA*YEAR)	LEGUMINOUS PLANT	1141
*PHASEOLUSS AUREUS	202 KG N/(HA*YEAR)	LEGUMINOUS PLANT, GREEN GRAIN	1141
*PISUM SATIVUM	65 KG N/(HA*YEAR)	LEGUMINOUS PLANT	1141
*PUERARIA PHASEOLOIDES	99 KG N/(HA*YEAR)	LEGUMINOUS PLANT	1141
*PYGMAEOTHAMNUS ZEYHERI	80 - 1840 MICROGRAM N/(M**2*HOUR)	DURING DAYLIGHT, JAN 75, NON-LEGUME	670
*RHYNCHOSIA MONOPHYLLA	920 - 9080 MICROGRAM N/(M**2*HOUR)	DURING DAYLIGHT, JAN 75, LEGUME	670
*SESBANIA CANNABINA	542 KG N/(HA*YEAR)	LEGUMINOUS PLANT	1141
*STYLOSANTHES SP.	124 KG N/(HA*YEAR)	LEGUMINOUS PLANT	1141
*TEPHROSIA LUPINIIFOLIA	4240 - 23902 MICROGRAM N/(M**2*HOUR)	DURING DAYLIGHT, JAN 75, LEGUME	670
*TERMINALIA SERICEA	40 - 800 MICROGRAM N/(M**2*HOUR)	DURING DAYLIGHT, JAN 75, NON-LEGUME	670
*THALASSIA SP.	2.4-16.5 MICROGRAM N/MG PLANT-N*HR	IN LIGHT,WITH EPIPHYTES,REDFISH BAY	535
*TRIGONELLA FOENUMGRAECUM	342 KG N/(HA*YEAR)	LEGUMINOUS PLANT	1141
*VICIA FABA	210 KG N/(HA*YEAR)	LEGUMINOUS PLANT	1141
*VIGNA SINENSIS	198 KG N/(HA*YEAR)	LEGUMINOUS PLANT	1141
*XEROPHYTA RETINERVIS	0 - 0 MICROGRAM N/(M**2*HOUR)	DURING DAYLIGHT, JAN 75, NON-LEGUME	670

TABLE A342

KORMOPHYTES, PLANTS : O/BIOMASS RATIO			
ITEM:	VALUE:	CONDITIONS:	REF.:
*ALFALFA	77.90 %	ATOMIC COMPOSITION	109

TABLE A342

KORMOPHYTES, PLANTS : O/BIOMASS RATIO			
ITEM:	VALUE:	CONDITIONS:	REF.:
*ANGIOSPERMS	41.0 % OF DRY WEIGHT	MEAN ELEMENTARY COMPOSITION	157
*MEAN PLANT	40 % OF DRY WEIGHT	MEAN ELEMENTARY COMPOSITION	157

TABLE A343

KORMOPHYTES, PLANTS : P/BIOMASS RATIO			
ITEM:	VALUE:	CONDITIONS:	REF.:
*ALFALFA	0.71 %	ATOMIC COMPOSITION	109
*BRASENIA SCHREBERI	0.14% OF DRY WEIGHT	NONE	158
*CERATOPHYLLUM DEMERSUM	0.26% OF DRY WEIGHT	NONE	158
*CYMODOCEA SP. AND HALODULE SP.	0.6% OF ASH WEIGHT	TROPICAL SEAGRASSES,LEAVES	129
*CYMODOCEA SP. AND HALODULE SP.	0.6% OF ASH WEIGHT	TROPICAL SEAGRASSES,RHIZOMES	129
*ELEOCHARIS ACICULARIS	0.24% OF DRY WEIGHT	NONE	158
*ENHALUS SP. AND SYRINGODIUM SP.	0.7% OF ASH WEIGHT	TROPICAL SEAGRASSES,LEAVES	129
*ENHALUS SP. AND SYRINGODIUM SP.	0.8% OF ASH WEIGHT	TROPICAL SEAGRASSES,RHIZOMES	129
*HALOPHILA DECIPIENS	0.5% OF ASH WEIGHT	TROPICAL SEAGRASSES	129
*HALOPHILA SPINULOSA	0.4% OF ASH WEIGHT	TROPICAL SEAGRASSES	129
*MYRIOPHYLLUM HETEROPHYLLUM	0.16% OF DRY WEIGHT	NONE	158
*NAJAS GUADALUPENSIS	0.15% OF DRY WEIGHT	NONE	158
*NELUMBO LUTEA	0.19% OF DRY WEIGHT	NONE	158
*NUPHAR ADVENA	0.40% OF DRY WEIGHT	NONE	158
*NYMPHAEA ODORATA	0.18% OF DRY WEIGHT	NONE	158
*TYPHA LATIFOLIA	0.14% OF DRY WEIGHT	NONE	158
*UTRICULARIA INFLATA	0.12% OF DRY WEIGHT	NONE	158

TABLE A344

KORMOPHYTES, PLANTS : PB/BIOMASS RATIO			
ITEM:	VALUE:	CONDITIONS:	REF.:
*ALFALFA	0.0000 %	ATOMIC COMPOSITION	109

TABLE A345

KORMOPHYTES, PLANTS : PHOTOSYNTHESIS			
ITEM:	VALUE:	CONDITIONS:	REF.:
*AMARANTHUS EDULIS	58 (SD = +- 5) MG CO2/ (DM**2*HOUR)	NONE	399
*BETA VULGARIS	24 (SD = +- 2) MG CO2/ (DM**2*HOUR)	NONE	399
*GOSSYPIUM BARBADENSE	29 (SD = +- 2) MG CO2/ (DM**2*HOUR)	NONE	399
*GOSSYPIUM HIRSUTUM	30 (SD = +- 3) MG CO2/ (DM**2*HOUR)	NONE	399
*HELIANTHUS ANNUS	36 (SD = +- 4) MG CO2/ (DM**2*HOUR)	NONE	399
*SOLIDAGO VIRGAUREA	3.10 MG CO2/(DM**2*HOUR*10**4*ERG/ (CM**2*SEC))	MEAN OF 6, SHADED HABITAT, INITIAL SLOPE	134

TABLE A345

KORMOPHYTES, PLANTS : PHOTOSYNTHESIS			
IT¤M:	VALUE:	CONDITIONS:	REF.:
*SOLIDAGO VIRGAUREA	17.9 MG CO2/(DM**2*HOUR)	MEAN OF 6, SHADED HABITAT, SATURATION VALUE	134
*SOLIDAGO VIRGAUREA	2.31 MG CO2/(DM**2*HOUR*10**4*ERG/ (CM**2*SEC))	MEAN OF 6, EXPOSED HABITAT, INITIAL SLOPE	134
*SOLIDAGO VIRGAUREA	25.5 +-2.0 MG CO2/(DM**2*HOUR)	MEAN OF 6, EXPOSED HABITAT, SATURATION VALUE	134
*SORGHUM VULGARE	55 (SD = +- 6) MG CO2/ (DM**2*HOUR)	NONE	399
*ZEA MAYS	53 (SD = +- 6) MG CO2/ (DM**2*HOUR)	NONE	399

TABLE A346

KORMOPHYTES, PLANTS : PRODUCTION			
ITEM:	VALUE:	CONDITIONS:	REF.:
*BETULA MAXIMOWICZIANA, BRANCH	1.1 METRIC T/(HA*YEAR)	NET PRODUCTION, TEMP. DECIDUOUS BROADLEAVED FOREST, JAPAN, ESTIMAT	1385
*BETULA MAXIMOWICZIANA, BRANCH	1.0 METRIC T/(HA*YEAR)	NET PRODUCTION, TEMP. DECIDUOUS BROADLEAVED FOREST, JAPAN, ESTIMAT	1385
*BETULA MAXIMOWICZIANA, BRANCH	1.0 METRIC T/(HA*YEAR)	NET PRODUCTION, TEMP. DECIDUOUS BROADLEAVED FOREST, JAPAN, ESTIMAT	1385
*BETULA MAXIMOWICZIANA, LEAF	2.2 METRIC T/(HA*YEAR)	NET PRODUCTION, TEMP. DECIDUOUS BROADLEAVED FOREST, JAPAN, ESTIMAT	1385
*BETULA MAXIMOWICZIANA, LEAF	2.6 METRIC T/(HA*YEAR)	NET PRODUCTION, TEMP. DECIDUOUS BROADLEAVED FOREST, JAPAN, ESTIMAT	1385
*BETULA MAXIMOWICZIANA, LEAF	1.8 METRIC T/(HA*YEAR)	NET PRODUCTION, TEMP. DECIDUOUS BROADLEAVED FOREST, JAPAN, ESTIMAT	1385
*BETULA MAXIMOWICZIANA, STEM	2.9 METRIC T/(HA*YEAR)	NET PRODUCTION, TEMP. DECIDUOUS BROADLEAVED FOREST, JAPAN, ESTIMAT	1385
*BETULA MAXIMOWICZIANA, STEM	3.7 METRIC T/(HA*YEAR)	NET PRODUCTION, TEMP. DECIDUOUS BROADLEAVED FOREST, JAPAN, ESTIMAT	1385
*BETULA MAXIMOWICZIANA, STEM	2.8 METRIC T/(HA*YEAR)	NET PRODUCTION, TEMP. DECIDUOUS BROADLEAVED FOREST, JAPAN, ESTIMAT	1385
*BETULA MAXIMOWICZIANA, TOTAL STAND	6.3 METRIC T/(HA*YEAR)	NET PRODUCTION, TEMP. DECIDUOUS BROADLEAVED FOREST, JAPAN, ESTIMAT	1385
*BETULA MAXIMOWICZIANA, TOTAL STAND	7.5 METRIC T/(HA*YEAR)	NET PRODUCTION, TEMP. DECIDUOUS BROADLEAVED FOREST, JAPAN, ESTIMAT	1385
*BETULA MAXIMOWICZIANA, TOTAL STAND	6.1 METRIC T/(HA*YEAR)	NET PRODUCTION, TEMP. DECIDUOUS BROADLEAVED FOREST, JAPAN, ESTIMAT	1385
*BETULA MAXIMOWICZIANA, TOTAL TREE	6.2 METRIC T/(HA*YEAR)	NET PRODUCTION, TEMP. DECIDUOUS BROADLEAVED FOREST, JAPAN, ESTIMAT	1385
*BETULA MAXIMOWICZIANA, TOTAL TREE	7.2 METRIC T/(HA*YEAR)	NET PRODUCTION, TEMP. DECIDUOUS BROADLEAVED FOREST, JAPAN, ESTIMAT	1385
*BETULA MAXIMOWICZIANA, TOTAL TREE	5.5 METRIC T/(HA*YEAR)	NET PRODUCTION, TEMP. DECIDUOUS BROADLEAVED FOREST, JAPAN, ESTIMAT	1385
*BETULA MAXIMOWICZIANA, UNDERGROWTH	0.2 METRIC T/(HA*YEAR)	NET PRODUCTION, TEMP. DECIDUOUS BROADLEAVED FOREST, JAPAN, ESTIMAT	1385
*BETULA MAXIMOWICZIANA, UNDERGROWTH	0.3 METRIC T/(HA*YEAR)	NET PRODUCTION, TEMP. DECIDUOUS BROADLEAVED FOREST, JAPAN, ESTIMAT	1385
*BETULA MAXIMOWICZIANA, UNDERGROWTH	0.5 METRIC T/(HA*YEAR)	NET PRODUCTION, TEMP. DECIDUOUS BROADLEAVED FOREST, JAPAN, ESTIMAT	1385

TABLE A346

KORMOPHYTES, PLANTS : PRODUCTION

ITEM:	VALUE:	CONDITIONS:	REF.:
*FAGUS CRENATA, BRANCH	1.8 METRIC T/(HA*YEAR)	NET PRODUCTION, TEMP. DECIDUOUS BROADLEAVED FOREST, JAPAN, ESTIMAT	1385
*FAGUS CRENATA, LEAF	3.0 METRIC T/(HA*YEAR)	NET PRODUCTION, TEMP. DECIDUOUS BROADLEAVED FOREST, JAPAN, ESTIMAT	1385
*FAGUS CRENATA, STEM	2.5 METRIC T/(HA*YEAR)	NET PRODUCTION, TEMP. DECIDUOUS BROADLEAVED FOREST, JAPAN, ESTIMAT	1385
*FAGUS CRENATA, TOTAL TREE	7.3 METRIC T/(HA*YEAR)	NET PRODUCTION, TEMP. DECIDUOUS BROADLEAVED FOREST, JAPAN, ESTIMAT	1385
*POPULUS DAVIDIANA, BRANCH	0.9 METRIC T/(HA*YEAR)	NET PRODUCTION, TEMP. DECIDUOUS BROADLEAVED FOREST, JAPAN, ESTIMAT	1385
*POPULUS DAVIDIANA, LEAF	2.2 METRIC T/(HA*YEAR)	NET PRODUCTION, TEMP. DECIDUOUS BROADLEAVED FOREST, JAPAN, ESTIMAT	1385
*POPULUS DAVIDIANA, STEM	5.6 METRIC T/(HA*YEAR)	NET PRODUCTION, TEMP. DECIDUOUS BROADLEAVED FOREST, JAPAN, ESTIMAT	1385
*POPULUS DAVIDIANA, TOTAL STAND	12.4 METRIC T/(HA*YEAR)	NET PRODUCTION, TEMP. DECIDUOUS BROADLEAVED FOREST, JAPAN, ESTIMAT	1385
*POPULUS DAVIDIANA, TOTAL TREE	8.7 METRIC T/(HA*YEAR)	NET PRODUCTION, TEMP. DECIDUOUS BROADLEAVED FOREST, JAPAN, ESTIMAT	1385
*POPULUS DAVIDIANA, UNDERGROWTH	3.6 METRIC T/(HA*YEAR)	NET PRODUCTION, TEMP. DECIDUOUS BROADLEAVED FOREST, JAPAN, ESTIMAT	1385

TABLE A347

KORMOPHYTES, PLANTS : PRODUCTION, NET PRIMARY

ITEM:	VALUE:	CONDITIONS:	REF.:
*ABIES SACHALINENSIS	13.0 METRIC T/HA	SAME ALTITUDINAL ZONE , NET ABOVEGROUND PROD,JAPAN,AGE=26 Y	1385
*LARIX LEPTOLEPIS	16.5 METRIC T/HA	SAME ALTITUDINAL ZONE , NET ABOVEGROUND PROD,JAPAN,AGE=21 Y	1385
*PICEA ABIES	12.4 METRIC T/HA	SAME ALTITUDINAL ZONE , NET ABOVEGROUND PROD,JAPAN,AGE=47 Y	1385
*PICEA ABIES	11.7 METRIC T/HA	SAME ALTITUDINAL ZONE , NET ABOVEGROUND PROD,JAPAN,AGE=46 Y	1385
*PICEA ABIES	11.4 METRIC T/HA	SAME ALTITUDINAL ZONE , NET ABOVEGROUND PROD,JAPAN,AGE=46 Y	1385
*PICEA ABIES	7.3 METRIC T/HA	SAME ALTITUDINAL ZONE , NET ABOVEGROUND PROD,JAPAN,AGE=45 Y	1385
*POPULUS DAVIDIANA	8.7 METRIC T/HA	SAME ALTITUDINAL ZONE , NET ABOVEGROUND PROD.,JAPAN, AGE=40 Y	1385

TABLE A348

KORMOPHYTES, PLANTS : PROTEIN/BIOMASS RATIO

ITEM:	VALUE:	CONDITIONS:	REF.:
*BRASENIA SCHREBERI	10.9 % OF DRY WEIGHT	NONE	158
*CERATOPHYLLUM DEMERSUM	17.1 % OF DRY WEIGHT	NONE	158
*ELEOCHARIS ACICULARIS	14.1 % OF DRY WEIGHT	NONE	158
*ELEOCHARIS QUADRANGULATA	11.8 % OF DRY WEIGHT	CRUDE PROTEIN=NITROGEN*6.25	158
*HYDROTRIDA CAROLINENSIS	10.5 % OF DRY WEIGHT	NONE	158
*JUNCUS EFFUSUS	7.8 % OF DRY WEIGHT	CRUDE PROTEIN=NITROGEN*6.25	158

TABLE A348

KORMOPHYTES, PLANTS : PROTEIN/BIOMASS RATIO			
ITEM:	VALUE:	CONDITIONS:	REF.:
*MYRIOPHYLLUM HETEROPHYLLUM	13.5 % OF DRY WEIGHT	NONE	158
*NAJAS GUADALUPENSIS	14.4 % OF DRY WEIGHT	NONE	158
*NELUMBO LUTEA	12.1 % OF DRY WEIGHT	NONE	158
*NUPHAR ADVENA	21.6 % OF DRY WEIGHT	NONE	158
*NYMPHAEA ODORATA	14.6 % OF DRY WEIGHT	NONE	158
*ORONTIUM AQUATICUM	21.2 % OF DRY WEIGHT	CRUDE PROTEIN=NITROGEN*6.25	158
*PONTEDERIA CORDATA	8.8 % OF DRY WEIGHT	CRUDE PROTEIN=NITROGEN*6.25	158
*SCIRPUS AMERICANUS	7.6 % OF DRY WEIGHT	CRUDE PROTEIN=NITROGEN*6.25	158
*SCIRPUS VALIDUS	5.8 % OF DRY WEIGHT	CRUDE PROTEIN=NITROGEN*6.25	158
*TYPHA LATIFOLIA	4.0 % OF DRY WEIGHT	NONE	158
*UTRICULARIA INFLATA	11.4 % OF DRY WEIGHT	NONE	158

TABLE A349

KORMOPHYTES, PLANTS : RB/BIOMASS RATIO			
ITEM:	VALUE:	CONDITIONS:	REF.:
*ALFALFA	0.0005 %	ATOMIC COMPOSITION	109

TABLE A350

KORMOPHYTES, PLANTS : RESISTANCE TO HEAT			
ITEM:	VALUE:	CONDITIONS:	REF.:
*CONIFEROUS FOREST	182. SEC/M	EQUIVALENT HEAT FLUX=1 J/(M**2) *SEC, TRANS BETW CANOPY AND REF HEIGHT	556
*CONIFEROUS FOREST	82. SEC/M	EQUIVALENT HEAT FLUX=10 J/(M**2) *SEC, TRANS BETW CANOPY AND REF HEIGHT	556
*CONIFEROUS FOREST	38. SEC/M	EQUIVALENT HEAT FLUX=100 J/(M**2) *SEC, TRANS BETW CANOPY AND REF HEIGHT	556
*GRASS	1053 SEC/M	EQUIVALENT HEAT FLUX=1 J/(M**2) *SEC, TRANS BETW CANOPY AND REF HEIGHT	556
*GRASS	488 SEC/M	EQUIVALENT HEAT FLUX=10 J/(M**2) *SEC, TRANS BETW CANOPY AND REF HEIGHT	556
*GRASS	225 SEC/M	EQUIVALENT HEAT FLUX=100 J/(M**2) *SEC, TRANS BETW CANOPY AND REF HEIGHT	556
*MAIZE	373. SEC/M	EQUIVALENT HEAT FLUX=1 J/(M**2) *SEC, TRANS BETW CANOPY AND REF HEIGHT	556
*MAIZE	172. SEC/M	EQUIVALENT HEAT FLUX=10 J/(M**2) *SEC, TRANS BETW CANOPY AND REF HEIGHT	556
*MAIZE	80. SEC/M	EQUIVALENT HEAT FLUX=100 J/(M**2) *SEC, TRANS BETW CANOPY AND REF HEIGHT	556

TABLE A351

KORMOPHYTES, PLANTS : RESPIRATION RATE			
ITEM:	VALUE:	CONDITIONS:	REF.:
*AMARANTHUS EDULIS	5.2 (SD = +- 0.8) MG CO2/ (DM**2*HOUR)	IN DARK, NORMAL AIR	399

TABLE A351

KORMOPHYTES, PLANTS : RESPIRATION RATE			
ITEM:	VALUE:	CONDITIONS:	REF.:
*AMARANTHUS EDULIS	5.8 (SD = +- 1.0) MG CO2/ (DM**2*HOUR)	IN DARK, ZERO CO2	399
*AMARANTHUS EDULIS	0 MG CO2/(DM**2*HOUR)W	IN LIGHT, ZERO CO2	399
*BARLEY	LESS THAN 1 MG CO2/(G*HOUR)	SPECIFIC DARK RESPIRATION	AP(F
*BARLEY	5 - 20 % OF DAILY PHOTOSYNTHESIS	EXCEPT MAY WITH HIGHER VALUES, JAPAN, IN DARK	AP(F
*BETA VULGARIS	2.6 (SD = +- 0.4) MG CO2/ (DM**2*HOUR)	IN DARK, NORMAL AIR	399
*BETA VULGARIS	4.9 (SD = +- 0.8) MG CO2/ (DM**2*HOUR)	IN DARK, ZERO CO2	399
*BETA VULGARIS	2.4 (SD = +- 0.3) MG CO2/ (DM**2*HOUR)	IN LIGHT, ZERO CO2	399
*GOSSYPIUM BARBADENSE	3.5 (SD = +- 0.9) MG CO2/ (DM**2*HOUR)	IN DARK, NORMAL AIR	399
*GOSSYPIUM BARBADENSE	6.1 (SD = +- 0.9) MG CO2/ (DM**2*HOUR)	IN DARK, ZERO CO2	399
*GOSSYPIUM BARBADENSE	1.7 (SD = +- 1.1) MG CO2/ (DM**2*HOUR)	IN LIGHT, ZERO CO2	399
*GOSSYPIUM HIRSUTUM	3.2 (SD = +- 0.9) MG CO2/ (DM**2*HOUR)	IN DARK, NORMAL AIR	399
*GOSSYPIUM HIRSUTUM	5.6 (SD = +- 0.8) MG CO2/ (DM**2*HOUR)	IN DARK, ZERO CO2	399
*GOSSYPIUM HIRSUTUM	2.2 (SD = +- 0.3) MG CO2/ (DM**2*HOUR)	IN LIGHT, ZERO CO2	399
*HELIANTHUS ANNUS	5.0 (SD = +- 1.3) MG CO2/ (DM**2*HOUR)	IN DARK, NORMAL AIR	399
*HELIANTHUS ANNUS	6.1 (SD = +- 1.2) MG CO2/ (DM**2*HOUR)	IN DARK, ZERO CO2	399
*HELIANTHUS ANNUS	3.1 (SD = +- 0.9) MG CO2/ (DM**2*HOUR)	IN LIGHT, ZERO CO2	399
*SORGHUM VULGARE	1.6 (SD = +- 0.2) MG CO2/ (DM**2*HOUR)	IN DARK, NORMAL AIR	399
*SORGHUM VULGARE	2.8 (SD = +- 0.1) MG CO2/ (DM**2*HOUR)	IN DARK, ZERO CO2	399
*SORGHUM VULGARE	0 MG CO2/(DM**2*HOUR)	IN LIGHT, ZERO CO2	399
*ZEA MAYS	2.9 (SD = +- 0.3) MG CO2/ (DM**2*HOUR)	IN DARK, NORMAL AIR	399
*ZEA MAYS	4.1 (SD = +- 0.7) MG CO2/ (DM**2*HOUR)	IN DARK, ZERO CO2	399
*ZEA MAYS	0 MG CO2/(DM**2*HOUR)	IN LIGHT, ZERO CO2	399

TABLE A352

KORMOPHYTES, PLANTS : ROUGHNESS LENGTH			
ITEM:	VALUE:	CONDITIONS:	REF.:
*CONIFEROUS FOREST	1.27 M	CALCULATED VALUE	556
*GRASS	0.018 M	CALCULATED VALUE	556
*MAIZE	0.26 M	CALCULATED VALUE	556

TABLE A353

KORMOPHYTES, PLANTS : ROUGHNESS LENGTH/HEIGHT OF CANOPY			
ITEM:	VALUE:	CONDITIONS:	REF.:
*CONIFEROUS FOREST	0.127	CALCULATED VALUE	556
*GRASS	0.0916	CALCULATED VALUE	556

TABLE A353

KORMOPHYTES, PLANTS : ROUGHNESS LENGTH/HEIGHT OF CANOPY			
ITEM:	VALUE:	CONDITIONS:	REF.:
*MAIZE	0.105	CALCULATED VALUE	556

TABLE A354

KORMOPHYTES, PLANTS : S/BIOMASS RATIO			
ITEM:	VALUE:	CONDITIONS:	REF.:
*ALFALFA	0.10 %	ATOMIC COMPOSITION	109
*ANGIOSPERMS	0.44 % OF DRY WEIGHT	MEAN ELEMENTARY COMPOSITION	157
*MEAN PLANT	0.5 % OF DRY WEIGHT	MEAN ELEMENTARY COMPOSITION	157

TABLE A355

KORMOPHYTES, PLANTS : SI/BIOMASS RATIO			
ITEM:	VALUE:	CONDITIONS:	REF.:
*ALFALFA	0.0930 %	ATOMIC COMPOSITION	109

TABLE A356

KORMOPHYTES, PLANTS : SN/BIOMASS RATIO			
ITEM:	VALUE:	CONDITIONS:	REF.:
*ALFALFA	0.0000 %	ATOMIC COMPOSITION	109

TABLE A357

KORMOPHYTES, PLANTS : TRANSPIRATION			
ITEM:	VALUE:	CONDITIONS:	REF.:
*ALISMA PLANTAGO	1.7	RATIO MAX TRANSPIRATION/MIN TRANSPIRATION, AQUATIC	518
*AVENA SATIVA	170 G WATER/(G DRY MATTER)	TOTAL DW = 27.3,ROOT DW = 17.0, SHOOT DW = 10.3 G/POT,ISRAEL,N=4, OCT	815
*AVENA SATIVA	211 G WATER/(G DRY MATTER)	TOTAL DW = 105,ROOT DW = 20,SHOOT DW = 85 G/POT,ISRAEL,N=6,MAR	815
*AVENA SATIVA	112 G WATER/(G DRY MATTER)	TOTAL DW = 349,ROOT DW = 93,SHOOT DW = 256 G/POT,ISRAEL,N=6,AUG	815
*AVENA SATIVA	235 G WATER/(G DRY MATTER)	TOTAL DW = 175,ROOT DW = 52,SHOOT DW = 123 G/POT,ISRAEL,N=2,OCT	815
*AVENA SATIVA	233 G WATER/(G DRY MATTER)	TOTAL DW = 213,ROOT DW = 57,SHOOT DW = 156 G/POT,ISRAEL,N=2,OCT	815
*AVENA SATIVA	194 G WATER/(G DRY MATTER)	TOTAL DW = 367,ROOT DW = 87,SHOOT DW = 280 G/POT,ISRAEL,N=2,OCT-JAN	815
*AVENA SATIVA	221 G WATER/(G DRY MATTER)	TOTAL DW = 414,ROOT DW = 73,SHOOT DW = 341 G/POT,ISRAEL,N=2,OCT-FEB	815
*AVENA SATIVA	79 G WATER/(G DRY MATTER)	TOTAL DW = 254,ROOT DW = 79,SHOOT DW = 175 G/POT,ISRAEL,N=8,OCT-JAN	815
*AVENA STERILIS	124 G WATER/(G DRY MATTER)	TOTAL DW = 140,ROOT DW = 65,SHOOT DW = 75 G/POT,ISRAEL,N=10,MAR	815
*AVENA STERILIS	111 G WATER/(G DRY MATTER)	TOTAL DW = 72,ROOT DW = 29,SHOOT DW = 43 G/POT,ISRAEL,N=3,NOV-DEC	815

TABLE A357

KORMOPHYTES, PLANTS : TRANSPIRATION

ITEM:	VALUE:	CONDITIONS:	REF.:
*AVENA STERILIS	129 G WATER/(G DRY MATTER)	TOTAL DW = 141,ROOT DW = 45,SHOOT DW = 96 G/POT,ISRAEL,N=3,NOV-JAN	815
*AVENA STERILIS	154 G WATER/(G DRY MATTER)	TOTAL DW = 263,ROOT DW = 63,SHOOT DW = 201 G/POT,ISRAEL,N=3,NOV-MAR	815
*AVENA STERILIS	164 G WATER/(G DRY MATTER)	TOTAL DW = 309,ROOT DW = 93,SHOOT DW = 216 G/POT,ISRAEL,N=2,NOV-MAR	815
*BETA VULGARIS	3.34 (SD = +- 0.58) G WATER/ (DM**2*HOUR)	IN LIGHT	399
*BETA VULGARIS	0.636 (SD = +- 0.07) G WATER/ (DM**2*HOUR)	IN DARK	399
*EICHHORNIA CRASSIPES	1.5	RATIO MAX TRANSPIRATION/MIN TRANSPIRATION, AQUATIC	518
*ERUCARIA BOVEANA	207 G WATER/(G DRY MATTER)	TOTAL DW = 71,ROOT DW = 16,SHOOT DW = 55 G/POT,ISRAEL,N=2,DEC-MAR	815
*GOSSYPIUM BARBADENSE	2.58 (SD = +- 0.35) G WATER/ (DM**2*HOUR)	IN LIGHT	399
*GOSSYPIUM BARBADENSE	0.623 (SD = +- 0.12) G WATER/ (DM**2*HOUR)	IN DARK	399
*GOSSYPIUM HIRSUTUM	3.00 (SD = +- 0.47) G WATER/ (DM**2*HOUR)	IN LIGHT	399
*GOSSYPIUM HIRSUTUM	0.661 (SD = +- 0.08) G WATER/ (DM**2*HOUR)	IN DARK	399
*HELIANTHUS ANNUS	3.67 (SD = +- 0.36) G WATER/ (DM**2*HOUR)	IN LIGHT	399
*HELIANTHUS ANNUS	0.769 (SD = +- 0.17) G WATER/ (DM**2*HOUR)	IN DARK	399
*HORDEUM MURINUM	129 G WATER/(G DRY MATTER)	TOTAL DW = 78,ROOT DW = 31,SHOOT DW = 47 G/POT,ISRAEL,N=10,OCT-JAN	815
*HORDEUM MURINUM	144 G WATER/(G DRY MATTER)	TOTAL DW = 95,ROOT DW = 26,SHOOT DW = 69 G/POT,ISRAEL,N=2,OCT-DEC	815
*HORDEUM MURINUM	159 G WATER/(G DRY MATTER)	TOTAL DW = 166,ROOT DW = 38,SHOOT DW = 128 G/POT,ISRAEL,N=2,OCT-JAN	815
*HORDEUM MURINUM	178 G WATER/(G DRY MATTER)	TOTAL DW = 288,ROOT DW = 38,SHOOT DW = 128 G/POT,ISRAEL,N=2,OCT-FEB	815
*HORDEUM SATIVUM	173 G WATER/(G DRY MATTER)	TOTAL DW = 23.4,ROOT DW = 15.7, SHOOT DW = 7.7 G/POT,ISRAEL,N=4, OCT	815
*HORDEUM SATIVUM	104 G WATER/(G DRY MATTER)	TOTAL DW = 169,ROOT DW = 71,SHOOT DW = 98 G/POT,ISRAEL,N=8,OCT-JAN	815
*HORDEUM SATIVUM	187 G WATER/(G DRY MATTER)	TOTAL DW = 118,ROOT DW = 25,SHOOT DW = 93 G/POT,ISRAEL,N=6,MAR	815
*HORDEUM SATIVUM	118 G WATER/(G DRY MATTER)	TOTAL DW = 303,ROOT DW = 85,SHOOT DW = 218 G/POT,ISRAEL,N=7,AUG	815
*HORDEUM SATIVUM	217 G WATER/(G DRY MATTER)	TOTAL DW = 181,ROOT DW = 63,SHOOT DW = 118 G/POT,ISRAEL,N=3,OCT	815
*HORDEUM SATIVUM	191 G WATER/(G DRY MATTER)	TOTAL DW = 256,ROOT DW = 98,SHOOT DW = 167 G/POT,ISRAEL,N=3,OCT-NOV	815
*HORDEUM SATIVUM	187 G WATER/(G DRY MATTER)	TOTAL DW = 360,ROOT DW = 90,SHOOT DW = 270 G/POT,ISRAEL,N=3,OCT-DEC	815
*LAURUS NOBILIS	8.4	RATIO MAX TRANSPIRATION/MIN TRANSPIRATION, MESO- AND XEROPHYTES	518
*MALVA SYLVESTRIS	142 G WATER/(G DRY MATTER)	TOTAL DW = 139,ROOT DW = 34,SHOOT DW = 105 G/POT,ISRAEL,N=6,OCT-FEB	815
*MEDICAGO HISPIDA	157 G WATER/(G DRY MATTER)	TOTAL DW = 64,ROOT DW = 17,SHOOT DW = 47 G/POT,ISRAEL,N=10,OCT-JAN	815
*MEDICAGO HISPIDA	226 G WATER/(G DRY MATTER)	TOTAL DW = 34,ROOT DW = 7,SHOOT DW = 27 G/POT,ISRAEL,N=5,MAR	815
*MEDICAGO HISPIDA	192 G WATER/(G DRY MATTER)	TOTAL DW = 83,ROOT DW = 11,SHOOT DW = 72 G/POT,ISRAEL,N=3,OCT-MAR	815
*MEDICAGO HISPIDA	169 G WATER/(G DRY MATTER)	TOTAL DW = 60,ROOT DW = 12,SHOOT DW = 48 G/POT,ISRAEL,N=3,NOV-MAR	815
*NYMPHAEA MARLIACEA	3.2	RATIO MAX TRANSPIRATION/MIN TRANSPIRATION, AQUATIC	518
*PHALARIS MINOR	157 G WATER/(G DRY MATTER)	TOTAL DW = 83,ROOT DW = 30,SHOOT DW = 52 G/POT,ISRAEL,N=2,NOV-DEC	815
*PHALARIS MINOR	190 G WATER/(G DRY MATTER)	TOTAL DW = 116,ROOT DW = 27,SHOOT DW = 89 G/POT,ISRAEL,N=2,NOV-FEB	815

TABLE A357

KORMOPHYTES, PLANTS : TRANSPIRATION			
ITEM:	VALUE:	CONDITIONS:	REF.:
*PHALARIS MINOR	162 G WATER/(G DRY MATTER)	TOTAL DW = 283,ROOT DW = 105,SHOOT DW = 178 G/POT,ISRAEL,N=2,NOV-FEB	815
*POTAMOGETON NATANS	1.4	RATIO MAX TRANSPIRATION/MIN TRANSPIRATION, AQUATIC	518
*QUERCUS ILEX	4.6	RATIO MAX TRANSPIRATION/MIN TRANSPIRATION, MESO- AND XEROPHYTES	518
*REBOUDIA PINNATA	146 G WATER/(G DRY MATTER)	TOTAL DW = 42,ROOT DW = 7,SHOOT DW = 35 G/POT,ISRAEL,N=6,OCT-JAN	815
*REBOUDIA PINNATA	275 G WATER/(G DRY MATTER)	TOTAL DW = 60,ROOT DW = 6,SHOOT DW = 54 G/POT,ISRAEL,N=6,MAR	815
*REBOUDIA PINNATA	155 G WATER/(G DRY MATTER)	TOTAL DW = 65,ROOT DW = 15,SHOOT DW = 50 G/POT,ISRAEL,N=2,DEC-FEB	815
*REBOUDIA PINNATA	168 G WATER/(G DRY MATTER)	TOTAL DW = 41,ROOT DW = 4,SHOOT DW = 37 G/POT,ISRAEL,N=3,DEC-FEB	815
*SORGHUM VULGARE	3.26 (SD = +- 0.20) G WATER/(DM**2*HOUR)	IN LIGHT	399
*SORGHUM VULGARE	0.596 (SD = +- 0.10) G WATER/(DM**2*HOUR)	IN DARK	399
*STIPA CAPENSIS	181 G WATER/(G DRY MATTER)	TOTAL DW = 63,ROOT DW = 15,SHOOT DW = 48 G/POT,ISRAEL,N=2,OCT-JAN	815
*STIPA CAPENSIS	169 G WATER/(G DRY MATTER)	TOTAL DW = 134,ROOT DW = 24,SHOOT DW = 110 G/POT,ISRAEL,N=2,OCT-FEB	815
*STIPA CAPENSIS	165 G WATER/(G DRY MATTER)	TOTAL DW = 117,ROOT DW = 11,SHOOT DW = 106 G/POT,ISRAEL,N=2,OCT-MAR	815
*SYRINGA VULGARIS	4.9	RATIO MAX TRANSPIRATION/MIN TRANSPIRATION, MESO- AND XEROPHYTES	518
*TRIGONELLA ARABICA	222 G WATER/(G DRY MATTER)	TOTAL DW = 60,ROOT DW = 8,SHOOT DW = 92 G/POT,ISRAEL,N=6,MAR	815
*TRIGONELLA ARABICA	154 G WATER/(G DRY MATTER)	TOTAL DW = 26,ROOT DW = 4,SHOOT DW = 22 G/POT,ISRAEL,N=DEC-FEB	815
*TRIGONELLA ARABICA	131 G WATER/(G DRY MATTER)	TOTAL DW = 50,ROOT DW = 5.5,SHOOT DW = 44.5 G/POT,ISRAEL,N=3, DEC-MAR	815
*TRITICUM DURUM	194 G WATER/(G DRY MATTER)	TOTAL DW = 113,ROOT DW = 21,SHOOT DW = 92 G/POT,ISRAEL,N=6,MAR	815
*TRITICUM SATIVUM	129 G WATER/(G DRY MATTER)	TOTAL DW = 157,ROOT DW = 47,SHOOT DW = 110 G/POT,ISRAEL,N=6,OCT-JAN	815
*TRITICUM SATIVUM	220 G WATER/(G DRY MATTER)	TOTAL DW = 100,ROOT DW = 19,SHOOT DW = 81 G/POT,ISRAEL,N=6,MAR	815
*TRITICUM SATIVUM	217 G WATER/(G DRY MATTER)	TOTAL DW = 166,ROOT DW = 44,SHOOT DW = 122 G/POT,ISRAEL,N=3,OCT-NOV	815
*TRITICUM SATIVUM	190 G WATER/(G DRY MATTER)	TOTAL DW = 231,ROOT DW = 70,SHOOT DW = 161 G/POT,ISRAEL,N=3,OCT-DEC	815
*TRITICUM SATIVUM	131 G WATER/(G DRY MATTER)	TOTAL DW = 212,ROOT DW = 51,SHOOT DW = 161 G/POT,ISRAEL,N=3,NOV-FEB	815
*TRITICUM SATIVUM	100 G WATER/(G DRY MATTER)	TOTAL DW = 189,ROOT DW = 61,SHOOT DW = 128 G/POT,ISRAEL,N=3,DEC-MAR	815
*TRITICUM SATIVUM	166 G WATER/(G DRY MATTER)	TOTAL DW = 121,ROOT DW = 41,SHOOT DW = 80 G/POT,ISRAEL,N=3,JAN-MAR	815
*VIOLA TRICOLOR	7.6	RATIO MAX TRANSPIRATION/MIN TRANSPIRATION, MESO- AND XEROPHYTES	518
*ZEA MAYS	3.71 (SD = +- 0.51) G WATER/(DM**2*HOUR)	IN LIGHT	399
*ZEA MAYS	0.556 (SD = +- 0.10) G WATER/(DM**2*HOUR)	IN DARK	399

TABLE A358

KORMOPHYTES, PLANTS : TURBULENCE INTENSITY			
ITEM:	VALUE:	CONDITIONS:	REF.:
*CONIFEROUS FOREST	0.5	INPUT VALUE	556

TABLE A358

KORMOPHYTES, PLANTS : TURBULENCE INTENSITY			
ITEM:	VALUE:	CONDITIONS:	REF.:
*GRASS	0.5	INPUT VALUE	556
*MAIZE	0.5	INPUT VALUE	556

TABLE A359

KORMOPHYTES, PLANTS : VEIN LENGTH DENSITIES			
ITEM:	VALUE:	CONDITIONS:	REF.:
*ACER CAMPESTRIS	102 CM/CM**2	TERRESTRIAL PLANT	518
*FRAXINUS EXCELSIOR	88 CM/CM**2	TERRESTRIAL PLANT	518
*NELUMBO SP.	104 CM/CM**2	AERIAL LEAF, EDGE, AQUATIC PLANT	518
*NELUMBO SP.	91 CM/CM**2	AERIAL LEAF, MIDDLE, AQUATIC PLANT	518
*NELUMBO SP.	68 CM/CM**2	FLOATING LEAF, EDGE, AQUATIC PLANT	518
*NELUMBO SP.	71 CM/CM**2	FLOATING LEAF, MIDDLE, AQUATIC PLANT	518
*NUPHAR LUTEUM	7 CM/CM**2	SUBMERSED LEAF, AQUATIC PLANT	518
*NYMPHAEA MEXICANA	44 CM/CM**2	FLOATING LEAF, AQUATIC PLANT	518
*POTAMOGETON PRAELONGUS	14 CM/CM**2	SUBMERSED LEAF, AQUATIC PLANT	518
*QUERCUS CERRIS	86 CM/CM**2	TERRESTRIAL PLANT	518
*ROSA CANINA	108 CM/CM**2	TERRESTRIAL PLANT	518
*SALIX RUBRUM	52 CM/CM**2	TERRESTRIAL PLANT	518
*SALVINIA AURICULATA	14 CM/CM**2	FLOATING LEAF, AQUATIC PLANT	518

TABLE A360

KORMOPHYTES, PLANTS : VON KARMANS CONSTANT			
ITEM:	VALUE:	CONDITIONS:	REF.:
*CONIFEROUS FOREST	0.4	INPUT VALUE	556
*GRASS	0.4	INPUT VALUE	556
*MAIZE	0.4	INPUT VALUE	556

TABLE A361

KORMOPHYTES, PLANTS : WIND PARAMETER			
ITEM:	VALUE:	CONDITIONS:	REF.:
*CONIFEROUS FOREST	0.326	CALCULATED VALUE, FRICTION VEL/ WIND VEL AT TOP OF CANOPY	556
*CONIFEROUS FOREST	3.41	CALC VALUE, ((CANOPY HEIGHT)-(ZERO PLANE DISPLACM))/(ROUGHNESS LENGTH)	556
*GRASS	2.90	CALC VALUE, ((CANOPY HEIGHT)-(ZERO PLANE DISPLACM))/(ROUGHNESS LENGTH)	556
*GRASS	0.31	CALC VALUE,((LAI)*(AVERAGE WIDTH OF LEAVES)/(MIX LENGTH)	556
*GRASS	0.376	CALCULATED VALUE, FRICTION VEL/ WIND VEL AT TOP OF CANOPY	556
*MAIZE	3.21	CALC VALUE, ((CANOPY HEIGHT)-(ZERO PLANE DISPLACM))/(ROUGHNESS LENGTH)	556

TABLE A361

KORMOPHYTES, PLANTS : WIND PARAMETER			
ITEM:	VALUE:	CONDITIONS:	REF.:
*MAIZE	0.22	CALC VALUE,((LAI)*(AVERAGE WIDTH OF LEAVES)/(MIX LENGTH)	556
*MAIZE	0.343	CALCULATED VALUE, FRICTION VEL/ WIND VEL AT TOP OF CANOPY	556

TABLE A362

KORMOPHYTES, PLANTS : ZERO PLANE DISPLACEMENT			
ITEM:	VALUE:	CONDITIONS:	REF.:
*CONIFEROUS FOREST	5.67 M	CALCULATED VALUE	556
*GRASS	0.147 M	CALCULATED VALUE	556
*MAIZE	1.66 M	CALCULATED VALUE	556

TABLE A363

KORMOPHYTES, PLANTS : ZERO PLANE DISPLACEMENT/HEIGHT OF CANOPY			
ITEM:	VALUE:	CONDITIONS:	REF.:
*CONIFEROUS FOREST	0.567	CALCULATED VALUE	556
*GRASS	0.735	CALCULATED VALUE	556
*MAIZE	0.664	CALCULATED VALUE	556

TABLE A364

KORMOPHYTES, PLANTS : ZN/BIOMASS RATIO			
ITEM:	VALUE:	CONDITIONS:	REF.:
*ALFALFA	0.0004 %	ATOMIC COMPOSITION	109

TABLE A365

KORMOPHYTES, TREES : B/BIOMASS RATIO			
ITEM:	VALUE:	CONDITIONS:	REF.:
*ASPEN	0.91 G/COMPLETE TREE	TREE FRESH WET=232.1 KG,HEIGHT= 15.2 M,UNIV FOREST, STILLWATER, MAINE	1763
*BALSAM FIR	1.06 G/COMPLETE TREE	TREE FRESH WET=316.4 KG,HEIGHT= 13.7 M,UNIV FOREST, STILLWATER, MAINE	1763
*HEMLOCK	1.39 G/COMPLETE TREE	TREE FRESH WET=294.6 KG,HEIGHT= 13.4 M,UNIV FOREST, STILLWATER, MAINE	1763
*RED MAPLE	1.08 G/COMPLETE TREE	TREE FRESH WET=283.6 KG,HEIGHT= 14.7 M,UNIV FOREST, STILLWATER, MAINE	1763
*RED SPRUCE	0.98 G/COMPLETE TREE	TREE FRESH WET=275.9 KG,HEIGHT= 15.8 M,UNIV FOREST, STILLWATER, MAINE	1763
*WHITE BIRCH	1.33 G/COMPLETE TREE	TREE FRESH WET=371.9 KG,HEIGHT= 18.7 M,UNIV FOREST, STILLWATER, MAINE	1763
*WHITE PINE	0.89 G/COMPLETE TREE	TREE FRESH WET=376.9 KG,HEIGHT= 15.8 M,UNIV FOREST, STILLWATER, MAINE	1763

TABLE A366

KORMOPHYTES, TREES : COMPENSATION POINT			
ITEM:	VALUE:	CONDITIONS:	REF.:
*BEECH	7.5 %	PERCENT OF FULL WINTER SUNLIGHT IN MAINE	203
*EASTERN HEMLOCK	8.5 %	PERCENT OF FULL WINTER SUNLIGHT IN MAINE	203
*EASTERN WHITE PINE	10 %	PERCENT OF FULL WINTER SUNLIGHT IN MAINE	203
*ENGELMANN SPRUCE	11 %	PERCENT OF FULL WINTER SUNLIGHT IN MAINE	203
*LARCH	18 %	PERCENT OF FULL WINTER SUNLIGHT IN MAINE	203
*NORTHERN WHITE CEDAR	19 %	PERCENT OF FULL WINTER SUNLIGHT IN MAINE	203
*PONDEROSA PINE	31 %	PERCENT OF FULL WINTER SUNLIGHT IN MAINE	203
*RED OAK	14 %	PERCENT OF FULL WINTER SUNLIGHT IN MAINE	203
*SUGAR MAPLE	3.5 %	PERCENT OF FULL WINTER SUNLIGHT IN MAINE	203

TABLE A367

KORMOPHYTES, TREES : CU/BIOMASS RATIO			
ITEM:	VALUE:	CONDITIONS:	REF.:
*ASPEN	1.03 G/COMPLETE TREE	TREE FRESH WET=232.1 KG,HEIGHT= 15.2 M,UNIV FOREST, STILLWATER, MAINE	1763
*BALSAM FIR	1.71 G/COMPLETE TREE	TREE FRESH WET=316.4 KG,HEIGHT= 13.7 M,UNIV FOREST, STILLWATER, MAINE	1763
*HEMLOCK	0.84 G/COMPLETE TREE	TREE FRE¤H WET=294.6 KG,HEIGHT= 13.4 M,UNIV FOREST, STILLWATER, MAINE	1763
*RED MAPLE	1.32 G/COMPLETE TREE	TREE FRESH WET=283.6 KG,HEIGHT= 14.7 M,UNIV FOREST, STILLWATER, MAINE	1763

TABLE A367

KORMOPHYTES, TREES : CU/BIOMASS RATIO			
ITEM:	VALUE:	CONDITIONS:	REF.:
*RED SPRUCE	0.99 G/COMPLETE TREE	TREE FRESH WET=275.9 KG,HEIGHT= 15.8 M,UNIV FOREST, STILLWATER, MAINE	1763
*WHITE BIRCH	1.88 G/COMPLETE TREE	TREE FRESH WET=371.9 KG,HEIGHT= 18.7 M,UNIV FOREST, STILLWATER, MAINE	1763
*WHITE PINE	1.22 G/COMPLETE TREE	TREE FRESH WET=376.9 KG,HEIGHT= 15.8 M,UNIV FOREST, STILLWATER, MAINE	1763

TABLE A368

KORMOPHYTES, TREES : FE/BIOMASS RATIO			
ITEM:	VALUE:	CONDITIONS:	REF.:
*ASPEN	5.10 G/COMPLETE TREE	TREE FRESH WET=232.1 KG,HEIGHT= 15.2 M,UNIV FOREST, STILLWATER, MAINE	1763
*BALSAM FIR	9.64 G/COMPLETE TREE	TREE FRESH WET=316.4 KG,HEIGHT= 13.7 M,UNIV FOREST, STILLWATER, MAINE	1763
*HEMLOCK	7.84 G/COMPLETE TREE	TREE FRESH WET=294.6 KG,HEIGHT= 13.4 M,UNIV FOREST, STILLWATER, MAINE	1763
*PINUS RESINOSA, ALL AERIAL ORGANS	3.7466 KG/HA	TOTAL WEIGHT IN VEGETATION, HALLIDAY FIELD, ONT, CANADA	470
*PINUS RESINOSA, LIVING BRANCHES	0.2908 KG/HA	TOTAL WEIGHT IN VEGETATION, HALLIDAY FIELD, ONT, CANADA	470
*PINUS RESINOSA, NEEDLES	1.0717 KG/HA	TOTAL WEIGHT IN VEGETATION, HALLIDAY FIELD, ONT, CANADA	470
*PINUS RESINOSA, STEM BARK	0.0272 KG/HA	3-7 CM DIAM.,TOTAL WEIGHT IN VEGETATION,HALLIDAY FIELD, ONT, CANADA	470
*PINUS RESINOSA, STEM BARK	0.4646 KG/HA	7 CM-STUMP,TOTAL WEIGHT IN VEGETATION,HALLIDAY FIELD, ONT, CANADA	470
*PINUS RESINOSA, STEM WOOD	0.0096 KG/HA	3-7 CM DIAM.,TOTAL WEIGHT IN VEGETATION,HALLIDAY FIELD, ONT, CANADA	470
*PINUS RESINOSA, STEM WOOD	0.2604 KG/HA	7 CM-STUMP,TOTAL WEIGHT IN VEGETATION,HALLIDAY FIELD, ONT, CANADA	470
*PINUS RESINOSA, TWIGS	0.5602 KG/HA	TOTAL WEIGHT IN VEGETATION, HALLIDAY FIELD, ONT, CANADA	470
*PINUS RESINOSA,DEAD BRANCHES	1.0621 KG/HA	TOTAL WEIGHT IN VEGETATION, HALLIDAY FIELD, ONT, CANADA	470
*RED MAPLE	7.28 G/COMPLETE TREE	TREE FRESH WET=283.6 KG,HEIGHT= 14.7 M,UNIV FOREST, STILLWATER, MAINE	1763
*RED PINE, DEAD BRANCHES	0.5524 G/TREE	34 Y., N=10, HALLIDAY FIELD, ONT., CANADA	470
*RED PINE, NEEDLES	0.5574 G/TREE	34 Y., N=10, HALLIDAY FIELD, ONT., CANADA	470
*RED PINE, STEM BARK	0.0142 G/TREE	3-7 CM DIA.,34 Y., N=10, HALLIDAY FIELD, ONT., CANADA	470
*RED PINE, STEM WOOD	0.0050 G/TREE	3-7 CM DIA.,34 Y., N=10, HALLIDAY FIELD, ONT., CANADA	470
*RED PINE, TWIGS	0.2914 G/TREE	34 Y., N=10, HALLIDAY FIELD, ONT., CANADA	470
*RED SPRUCE	7.66 G/COMPLETE TREE	TREE FRESH WET=275.9 KG,HEIGHT= 15.8 M,UNIV FOREST, STILLWATER, MAINE	1763
*WHITE BIRCH	8.08 G/COMPLETE TREE	TREE FRESH WET=371.9 KG,HEIGHT= 18.7 M,UNIV FOREST, STILLWATER, MAINE	1763
*WHITE PINE	7.54 G/COMPLETE TREE	TREE FRESH WET=376.9 KG,HEIGHT= 15.8 M,UNIV FOREST, STILLWATER, MAINE	1763

TABLE A369

KORMOPHYTES, TREES : MN/BIOMASS RATIO

ITEM:	VALUE:	CONDITIONS:	REF.:
*ASPEN	9.60 G/COMPLETE TREE	TREE FRESH WET=232.1 KG,HEIGHT=15.2 M,UNIV FOREST, STILLWATER, MAINE	1763
*BALSAM FIR	69.30 G/COMPLETE TREE	TREE FRESH WET=316.4 KG,HEIGHT=13.7 M,UNIV FOREST, STILLWATER, MAINE	1763
*HEMLOCK	66.13 G/COMPLETE TREE	TREE FRESH WET=294.6 KG,HEIGHT=13.4 M,UNIV FOREST, STILLWATER, MAINE	1763
*PINUS RESINOSA, ALL AERIAL ORGANS	22.6457 KG/HA	TOTAL WEIGHT IN VEGETATION, HALLIDAY FIELD, ONT, CANADA	470
*PINUS RESINOSA, LIVING BRANCHES	1.3966 KG/HA	TOTAL WEIGHT IN VEGETATION, HALLIDAY FIELD, ONT, CANADA	470
*PINUS RESINOSA, NEEDLES	5.7170 KG/HA	TOTAL WEIGHT IN VEGETATION, HALLIDAY FIELD, ONT, CANADA	470
*PINUS RESINOSA, STEM BARK	0.2246 KG/HA	3-7 CM DIAM.,TOTAL WEIGHT IN VEGETATION,HALLIDAY FIELD, ONT, CANADA	470
*PINUS RESINOSA, STEM BARK	3.4499 KG/HA	7 CM-STUMP,TOTAL WEIGHT IN VEGETATION,HALLIDAY FIELD, ONT, CANADA	470
*PINUS RESINOSA, STEM WOOD	0.2560 KG/HA	3-7 CM DIAM.,TOTAL WEIGHT IN VEGETATION,HALLIDAY FIELD, ONT, CANADA	470
*PINUS RESINOSA, STEM WOOD	9.6777 KG/HA	7 CM-STUMP,TOTAL WEIGHT IN VEGETATION,HALLIDAY FIELD, ONT, CANADA	470
*PINUS RESINOSA, TWIGS	1.1018 KG/HA	TOTAL WEIGHT IN VEGETATION, HALLIDAY FIELD, ONT, CANADA	470
*PINUS RESINOSA,DEAD BRANCHES	0.8221 KG/HA	TOTAL WEIGHT IN VEGETATION, HALLIDAY FIELD, ONT, CANADA	470
*RED MAPLE	41.64 G/COMPLETE TREE	TREE FRESH WET=283.6 KG,HEIGHT=14.7 M,UNIV FOREST, STILLWATER, MAINE	1763
*RED PINE, DEAD BRANCHES	0.4276 G/TREE	34 Y., N=10, HALLIDAY FIELD, ONT., CANADA	470
*RED PINE, NEEDLES	2.9738 G/TREE	34 Y., N=10, HALLIDAY FIELD, ONT., CANADA	470
*RED PINE, STEM BARK	0.1169 G/TREE	3-7 CM DIA.,34 Y., N=10, HALLIDAY FIELD, ONT., CANADA	470
*RED PINE, STEM WOOD	0.1332 G/TREE	3-7 CM DIA.,34 Y., N=10, HALLIDAY FIELD, ONT., CANADA	470
*RED PINE, TWIGS	0.5731 G/TREE	34 Y., N=10, HALLIDAY FIELD, ONT., CANADA	470
*RED SPRUCE	61.03 G/COMPLETE TREE	TREE FRESH WET=275.9 KG,HEIGHT=15.8 M,UNIV FOREST, STILLWATER, MAINE	1763
*WHITE BIRCH	21.96 G/COMPLETE TREE	TREE FRESH WET=371.9 KG,HEIGHT=18.7 M,UNIV FOREST, STILLWATER, MAINE	1763
*WHITE PINE	16.42 G/COMPLETE TREE	TREE FRESH WET=376.9 KG,HEIGHT=15.8 M,UNIV FOREST, STILLWATER, MAINE	1763

TABLE A370

KORMOPHYTES, TREES : MO/BIOMASS RATIO

ITEM:	VALUE:	CONDITIONS:	REF.:
*ASPEN	0.30 G/COMPLETE TREE	TREE FRESH WET=232.1 KG,HEIGHT=15.2 M,UNIV FOREST, STILLWATER, MAINE	1763
*BALSAM FIR	0.34 G/COMPLETE TREE	TREE FRESH WET=316.4 KG,HEIGHT=13.7 M,UNIV FOREST, STILLWATER, MAINE	1763
*HEMLOCK	0.23 G/COMPLETE TREE	TREE FRESH WET=294.6 KG,HEIGHT=13.4 M,UNIV FOREST, STILLWATER, MAINE	1763

TABLE A370

KORMOPHYTES, TREES : MO/BIOMASS RATIO			
ITEM:	VALUE:	CONDITIONS:	REF.:
*RED MAPLE	0.35 G/COMPLETE TREE	TREE FRESH WET=283.6 KG,HEIGHT= 14.7 M,UNIV FOREST, STILLWATER, MAINE	1763
*RED SPRUCE	0.35 G/COMPLETE TREE	TREE FRESH WET=275.9 KG,HEIGHT= 15.8 M,UNIV FOREST, STILLWATER, MAINE	1763
*WHITE BIRCH	0.47 G/COMPLETE TREE	TREE FRESH WET=371.9 KG,HEIGHT= 18.7 M,UNIV FOREST, STILLWATER, MAINE p	1763
*WHITE PINE	0.18 G/COMPLETE TREE	TREE FRESH WET=376.9 KG,HEIGHT= 15.8 M,UNIV FOREST, STILLWATER, MAINE	1763

TABLE A371

KORMOPHYTES, TREES : NITROGEN DISTRIBUTION			
ITEM:	VALUE:	CONDITIONS:	REF.:
*PINE FOREST, IN BIOMASS	6.0 % OF TOTAL N	TEMPERATE BRITISH 55 Y FOREST	1198
*PINE FOREST, LEAF/WOOD RATIO	0.34	GALLERY FOREST	1198
*PINE FOREST, LEAVES	12.4 GRAM N/M**2	TEMPERATE BRITISH 55 Y FOREST	1198
*PINE FOREST, LITTER	40.9 GRAM N/M**2	TEMPERATE BRITISH 55 Y FOREST	1198
*PINE FOREST, ROOT/SHOOT RATIO	0.60	TEMPERATE BRITISH 55 Y FOREST	1198
*PINE FOREST, ROOTS	18.4 GRAM N/M**2	TEMPERATE BRITISH 55 Y FOREST	1198
*PINE FOREST, SOIL	730.8 GRAM N/M**2	TEMPERATE BRITISH 55 Y FOREST	1198
*PINE FOREST, TURNOVER	4.0	TEMPERATE BRITISH 55 Y FOREST	1198
*PINE FOREST, WOOD ABOVE GROUND	3.0 % OF TOTAL N	TEMPERATE BRITISH 55 Y FOREST	1198
*PINE FOREST, WOOD ABOVE GROUND	18.5 GRAM N/M**2	TEMPERATE BRITISH 55 Y FOREST	1198
*TROPICAL FOREST, IN BIOMASS	57.8 % OF TOTAL N	GALLERY FOREST	1198
*TROPICAL FOREST, LEAF/WOOD RATIO	0.76	GALLERY FOREST	1198
*TROPICAL FOREST, LEAVES	52.6 GRAM N/M**2	GALLERY FOREST	1198
*TROPICAL FOREST, LITTER	3.9 GRAM N/M**2	GALLERY FOREST	1198
*TROPICAL FOREST, ROOT/SHOOT RATIO	0.30	GALLERY FOREST	1198
*TROPICAL FOREST, ROOTS	28.2 GRAM N/M**2	GALLERY FOREST	1198
*TROPICAL FOREST, SOIL	85.3 GRAM N/M**2	GALLERY FOREST	1198
*TROPICAL FOREST, WOOD ABOVE GROUND	44.0 % OF TOTAL N	GALLERY FOREST	1198
*TROPICAL FOREST, WOOD ABOVE GROUND	41.2 GRAM N/M**2	GALLERY FOREST	1198

TABLE A372

KORMOPHYTES, TREES : PRODUCTION/BIOMASS RATIO			
ITEM:	VALUE:	CONDITIONS:	REF.:
*OAK-PINE FOREST	0.130 1/YEAR	ABOVE GROUND, NET PRODUCTION	1699
*OAK-PINE FOREST	0.1 1/YEAR	BELOW GROUND, NET PRODUCTION	1699
*OAK-PINE FOREST	0.833 1/YEAR	LEAVES, NET PRODUCTION	1699
*OAK-PINE FOREST	0.0806 1/YEAR	TOTAL WOOD TISSUE, NET PRODUCTION	1699

TABLE A373

KORMOPHYTES, TREES : ZN/BIOMASS			
ITEM:	VALUE:	CONDITIONS:	REF.:
*RED SPRUCE	3.69 G/COMPLETE TREE	TREE FRESH WET=275.9 KG,HEIGHT= 15.8 M,UNIV FOREST, STILLWATER, MAINE	1763

TABLE A374

KORMOPHYTES, TREES : ZN/BIOMASS RATIO

ITEM:	VALUE:	CONDITIONS:	REF.:
*ASPEN	5.62 G/COMPLETE TREE	TREE FRESH WET=232.1 KG,HEIGHT=15.2 M,UNIV FOREST, STILLWATER, MAINE	1763
*BALSAM FIR	4.65 G/COMPLETE TREE	TREE FRESH WET=316.4 KG,HEIGHT=13.7 M,UNIV FOREST, STILLWATER, MAINE	1763
*HEMLOCK	1.06 G/COMPLETE TREE	TREE FRESH WET=294.6 KG,HEIGHT=13.4 M,UNIV FOREST, STILLWATER, MAINE	1763
*RED MAPLE	7.66 G/COMPLETE TREE	TREE FRESH WET=283.6 KG,HEIGHT=14.7 M,UNIV FOREST, STILLWATER, MAINE	1763
*WHITE BIRCH	10.98 G/COMPLETE TREE	TREE FRESH WET=371.9 KG,HEIGHT=18.7 M,UNIV FOREST, STILLWATER, MAINE	1763
*WHITE PINE	4.91 G/COMPLETE TREE	TREE FRESH WET=376.9 KG,HEIGHT=15.8 M,UNIV FOREST, STILLWATER, MAINE	1763

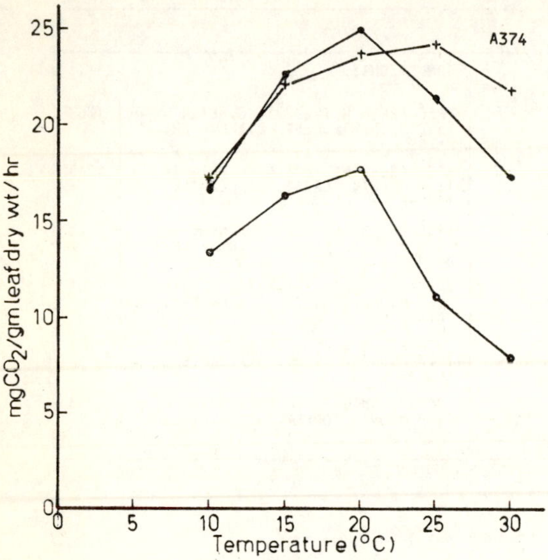

Photosynthesis (net) related to temperature for 3 populations of Thalichrum alpinum.
○ artic diploid ● artic triploid
+ alpine diploid.

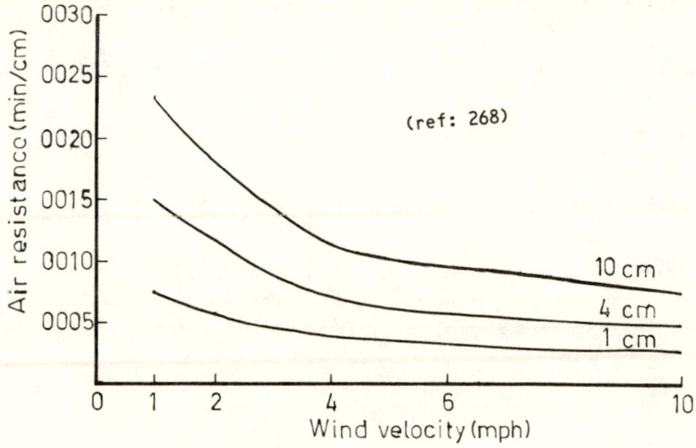

Air resistance to leaves relative to wind velocity and leaf size.

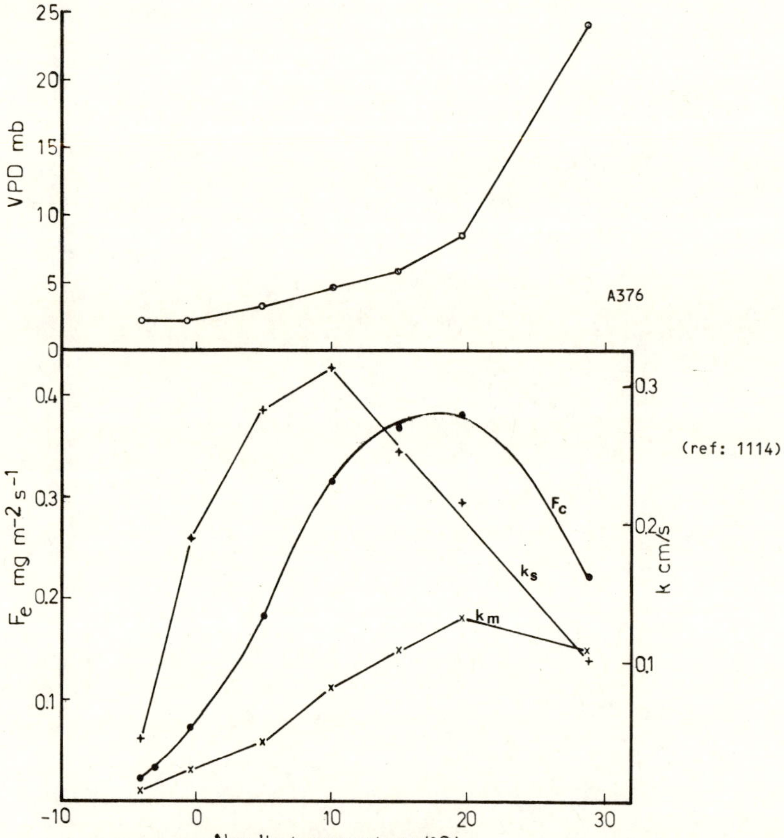

Needle temperature versus net photosynthesis (Fc), stomatal conductance (ks), mesophyll conductance (kM) and leaf-air vapour pressure difference (VPD) of a forest shoot of Picea sitchensis. Symbols used
Fc ● ,ks + ,kM × , VPD ○ .
Carbon dioxide concentration 320+-10 cm**3/m**3. Photonflux density 1000 μ E/(m**2*s)

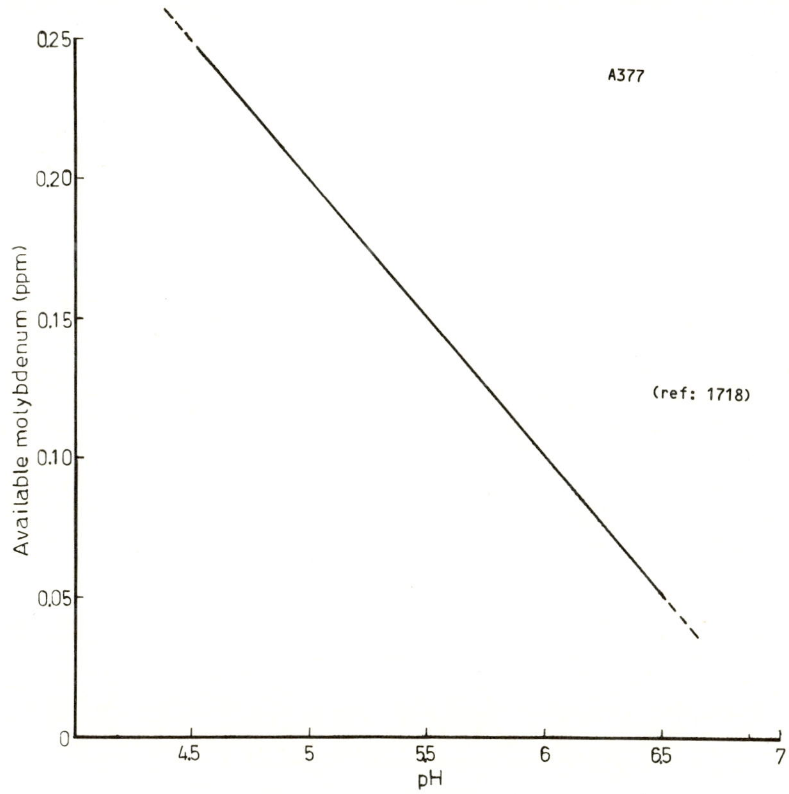

Molybdenum avaible (ppm) for clover grass versus pH.

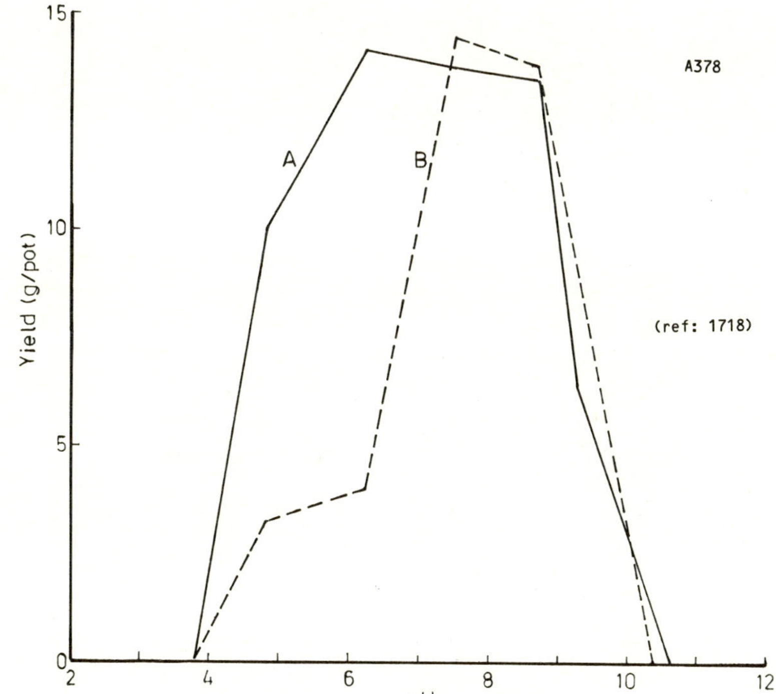

Yield versus pH for clover in a soil low in Mo. A: 125 g sodium molybdate/acre. b: without Mo.

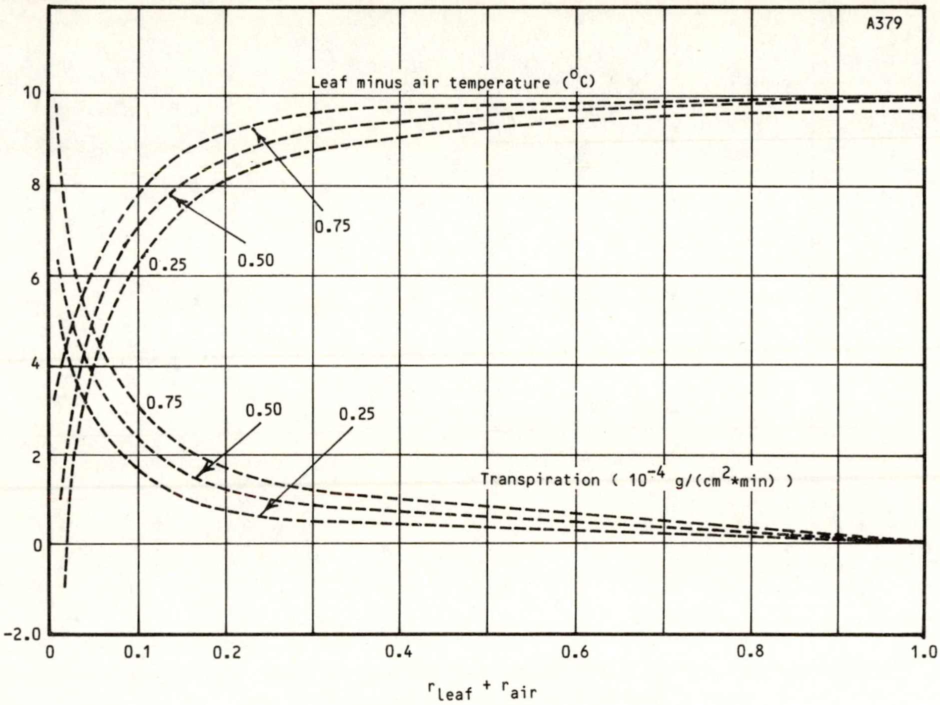

Leaf temperature and transpiration versus leaf and air reststance. Air temperature is 25 celcuis. Absorbed radiation 1.00 cal/(cm**2*min), wind speed 2 mph and leaf width is 4 cm. 3 different relative humidities is shown.

(ref: 268)

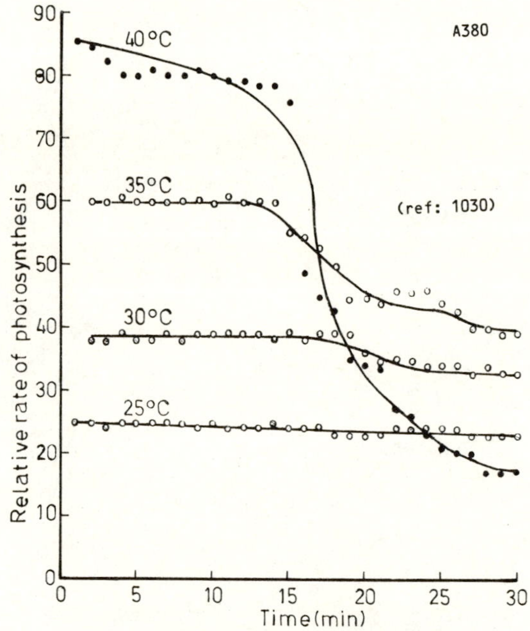

Photosynthesis (apperent, relative) for waterweed at different temperatures over a period of 30 minutes. 4 temperatures 40, 35, 30 and 25 celcius were used for the photosynthetic determinations of waterweed.

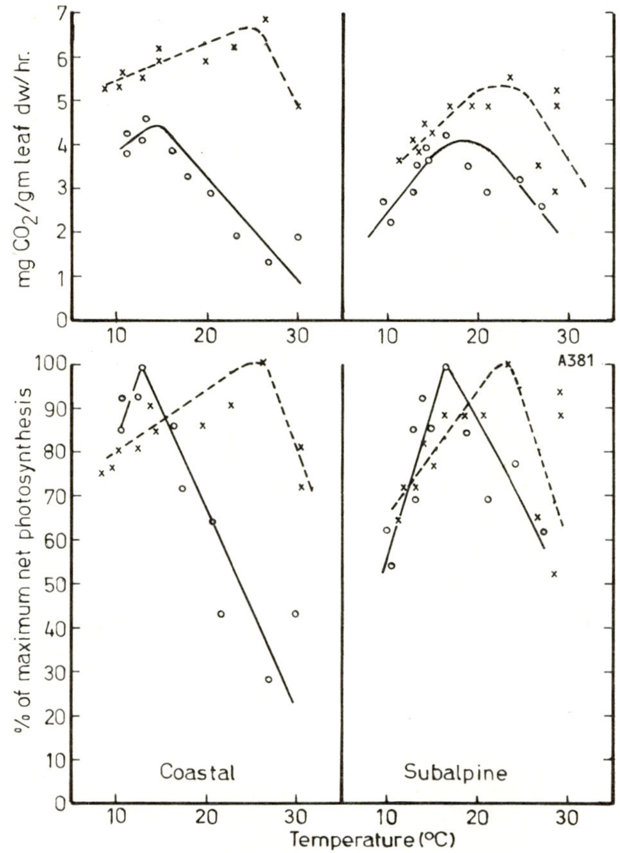

Photosynthetic acclimitation to warm temperatures of Polygonum bistortoides. Top curves are expressed on absolute basis and bottom curves are on a relative basis. First column coastral and second column subalpine populations. Circles indicate values measured immediatley upon removal from 10 days in the cold acclimation chamber. (o)indicate values measured after plants were kept at 30 celcius and full light for 25 hours.

(ref: 1059)

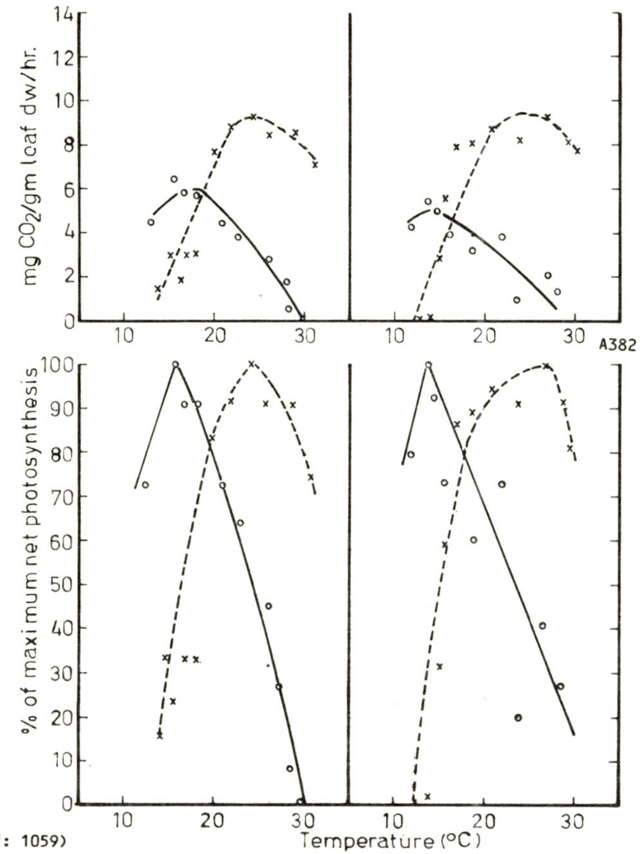

Photosynthetic acclimation of Encelia californica. Top curves are expresed on an absolute basis and bottom set on a relative basis. Circles indicate values measured immediately upon removal from 12 days in the cold chamber and (o)indicate values measured after the plants were kept at 30 celcius and full light in 23 hours.

A383

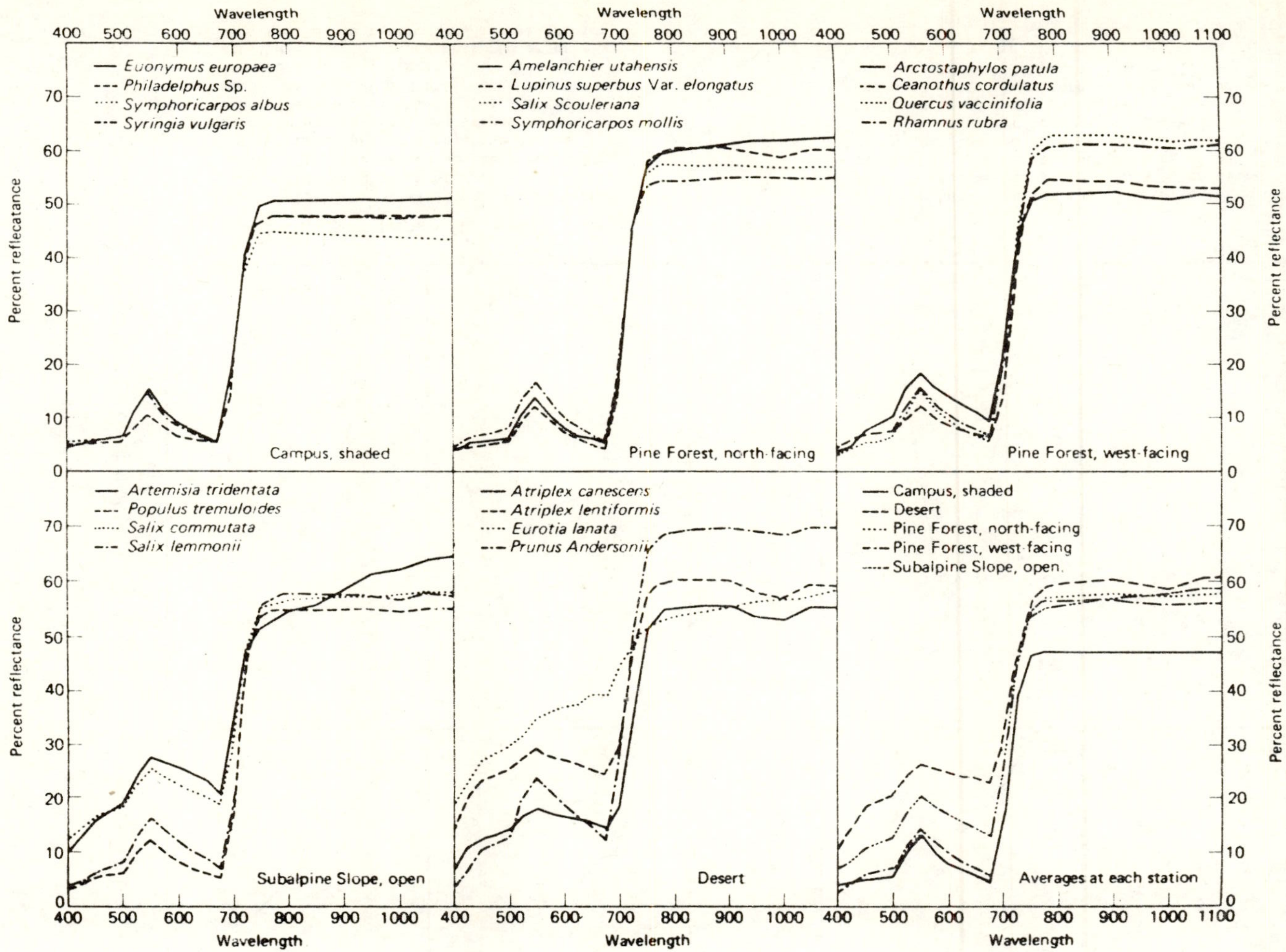

Reflectance from leaves versus wavelength for various species. (ref: 126)

A8

VERTEBRATES

COMPOSITION
GROWTH RATE
THERMAL CONDUCTIVITY

TABLE A385

VERTEBRATES : AL/BIOMASS RATIO			
ITEM:	VALUE:	CONDITIONS:	REF.:
*HOMO SAPIENS	0.0001 %	ATOMIC COMPOSITION	109

TABLE A386

VERTEBRATES : ASH/BIOMASS RATIO			
ITEM:	VALUE:	CONDITIONS:	REF.:
*ANNELIDA	22.4 % OF DRY WEIGHT	46 SPECIES	1635
*COELENTERATA	48.8 % OF DRY WEIGHT	16 SPECIES	1635
*CRUSTACEA	21.7 % OF DRY WEIGHT	25 SPECIES	1635
*ECHINODERMATA	60.8 % OF DRY WEIGHT	9 SPECIES	1635
*MOLLUSCA	11.0 % OF DRY WEIGHT	70 SPECIES	1635
*PISCES	5.2 % OF DRY WEIGHT	630 SPECIES	1635

TABLE A387

VERTEBRATES : ASSIMILATION			
ITEM:	VALUE:	CONDITIONS:	REF.:
*MICROPTERUS SALMOIDES	40 % OF HG INTAKE	FOOD = MOSQUITO FISH	701
*MICROPTERUS SALMOIDES	40 % OF HG INTAKE	FOOD = MOSQUITO FISH	701
*MOSQUITO FISH	12 % OF HG INTAKE	FOOD = BLOOD WORM	355
*MOSQUITO FISH	12 % OF HG INTAKE	FOOD = BLOOD WORM	355
*PREDATOR	85 % OF INGESTION	TYPICAL VALUE	355

TABLE A388

VERTEBRATES : BIOMASS AT DIFFERENT LIFE TIME			
ITEM:	VALUE:	CONDITIONS:	REF.:
*BONASA UMBELLUS	12 G WET WEIGHT	AGE CLASS (D) = 0, MALE	196
*BONASA UMBELLUS	21 G WET WEIGHT	AGE CLASS (D) = 7, MALE	196
*BONASA UMBELLUS	39 G WET WEIGHT	AGE CLASS (D) = 14, MALE	196
*BONASA UMBELLUS	95 G WET WEIGHT	AGE CLASS (D) = 28, MALE	196
*BONASA UMBELLUS	180 G WET WEIGHT	AGE CLASS (D) = 42, MALE	196
*BONASA UMBELLUS	295 G WET WEIGHT	AGE CLASS (D) = 56, MALE	196
*BONASA UMBELLUS	400 G WET WEIGHT	AGE CLASS (D) = 70, MALE	196
*BONASA UMBELLUS	479 G WET WEIGHT	AGE CLASS (D) = 84, MALE	196
*BONASA UMBELLUS	550 G WET WEIGHT	AGE CLASS (D) = 112, MALE	196
*BONASA UMBELLUS	579 G WET WEIGHT	AGE CLASS (D) = 140, MALE	196
*BONASA UMBELLUS	651 G WET WEIGHT	ADULT, MALE	196
*BONASA UMBELLUS	11 G WET WEIGHT	AGE CLASS (D) = 0, FEMALE	196
*BONASA UMBELLUS	20 G WET WEIGHT	AGE CLASS (D) = 7, FEMALE	196
*BONASA UMBELLUS	32 G WET WEIGHT	AGE CLASS (D) = 14, FEMALE	196
*BONASA UMBELLUS	81 G WET WEIGHT	AGE CLASS (D) = 28, FEMALE	196
*BONASA UMBELLUS	163 G WET WEIGHT	AGE CLASS (D) = 42, FEMALE	196
*BONASA UMBELLUS	260 G WET WEIGHT	AGE CLASS (D) = 56, FEMALE	196

TABLE A388

VERTEBRATES : BIOMASS AT DIFFERENT LIFE TIME

ITEM:	VALUE:	CONDITIONS	REF.:
*BONASA UMBELLUS	352 G WET WEIGHT	AGE CLASS (D) = 70, FEMALE	196
*BONASA UMBELLUS	428 G WET WEIGHT	AGE CLASS (D) = 84, FEMALE	196
*BONASA UMBELLUS	498 G WET WEIGHT	AGE CLASS (D) = 112, FEMALE	196
*BONASA UMBELLUS	511 G WET WEIGHT	AGE CLASS (D) = 140, FEMALE	196
*BONASA UMBELLUS	587 G WET WEIGHT	ADULT, FEMALE	196
*KOBUS LECHE	4.8 KG WET WEIGHT	AGE CLASS (YEARS) = 0, MALE	1346
*KOBUS LECHE	54.6 KG WET WEIGHT	AGE CLASS (YEARS) = 1, MALE	1346
*KOBUS LECHE	67.1 KG WET WEIGHT	AGE CLASS (YEARS) = 2, MALE	1346
*KOBUS LECHE	85.5 KG WET WEIGHT	AGE CLASS (YEARS) = 3, MALE	1346
*KOBUS LECHE	103.2 KG WET WEIGHT	AGE CLASS (YEARS) = 4, MALE	1346
*KOBUS LECHE	104.5 KG WET WEIGHT	AGE CLASS (YEARS) = 5, MALE	1346
*KOBUS LECHE	105.9 KG WET WEIGHT	AGE CLASS (YEARS) = 6, MALE	1346
*KOBUS LECHE	100.9 KG WET WEIGHT	AGE CLASS (YEARS) = 7, MALE	1346
*KOBUS LECHE	101.8 KG WET WEIGHT	AGE CLASS (YEARS) = 8, MALE	1346
*KOBUS LECHE	4.5 KG WET WEIGHT	AGE CLASS (YEARS) = 0, FEMALE	1346
*KOBUS LECHE	60.0 KG WET WEIGHT	AGE CLASS (YEARS) = 1, FEMALE	1346
*KOBUS LECHE	72.7 KG WET WEIGHT	AGE CLASS (YEARS) = 2, FEMALE	1346
*KOBUS LECHE	78.1 KG WET WEIGHT	AGE CLASS (YEARS) = 3, FEMALE	1346
*KOBUS LECHE	80.0 KG WET WEIGHT	AGE CLASS (YEARS) = 4, FEMALE	1346
*KOBUS LECHE	75.4 KG WET WEIGHT	AGE CLASS (YEARS) = 5, FEMALE	1346
*KOBUS LECHE	83.2 KG WET WEIGHT	AGE CLASS (YEARS) = 6, FEMALE	1346
*KOBUS LECHE	78.1 KG WET WEIGHT	AGE CLASS (YEARS) = 7, FEMALE	1346
*KOBUS LECHE	80.9 KG WET WEIGHT	AGE CLASS (YEARS) = 8, FEMALE	180
*LOXODONTA AFRICANA	90.0 KG WET WEIGHT	AGE CLASS (YEARS) = 1, BOTH SEXES	1250
*LOXODONTA AFRICANA	204.5 KG WET WEIGHT	AGE CLASS (YEARS) = 2, BOTH SEXES	1250
*LOXODONTA AFRICANA	318.1 KG WET WEIGHT	AGE CLASS (YEARS) = 3, BOTH SEXES	1250
*LOXODONTA AFRICANA	454.5 KG WET WEIGHT	AGE CLASS (YEARS) = 4, BOTH SEXES	1250
*LOXODONTA AFRICANA	613.6 KG WET WEIGHT	AGE CLASS (YEARS) = 5, BOTH SEXES	1250
*LOXODONTA AFRICANA	795.4 KG WET WEIGHT	AGE CLASS (YEARS) = 6, BOTH SEXES	1250
*LOXODONTA AFRICANA	1000.0 KG WET WEIGHT	AGE CLASS (YEARS) = 7, BOTH SEXES	1250
*LOXODONTA AFRICANA	1204.5 KG WET WEIGHT	AGE CLASS (YEARS) = 8, BOTH SEXES	1250
*LOXODONTA AFRICANA	1409.1 KG WET WEIGHT	AGE CLASS (YEARS) = 9, BOTH SEXES	1250
*LOXODONTA AFRICANA	1613.6 KG WET WEIGHT	AGE CLASS (YEARS) = 10, BOTH SEXES	1250
*LOXODONTA AFRICANA	4090.9 KG WET WEIGHT	AGE CLASS (YEARS) = 25, BOTH SEXES	1250
*MELEAGRIS GALLOPAVO	40 G WET WEIGHT	AGE CLASS (D) = 0, MALE	17
*MELEAGRIS GALLOPAVO	80 G WET WEIGHT	AGE CLASS (D) = 14, MALE	17
*MELEAGRIS GALLOPAVO	280 G WET WEIGHT	AGE CLASS (D) = 28, MALE	17
*MELEAG¤IS GALLOPAVO	560 G WET WEIGHT	AGE CLASS (D) = 42, MALE	17
*MELEAGRIS GALLOPAVO	850 G WET WEIGHT	AGE CLASS (D) = 56, MALE	17
*MELEAGRIS GALLOPAVO	1220 G WET WEIGHT	AGE CLASS (D) = 70, MALE	17
*MELEAGRIS GALLOPAVO	1640 G WET WEIGHT	AGE CLASS (D) = 84, MALE	17
*MELEAGRIS GALLOPAVO	2100 G WET WEIGHT	AGE CLASS (D) = 98, MALE	17
*MELEAGRIS GALLOPAVO	2600 G WET WEIGHT	AGE CLASS (D) = 112, MALE	17
*MELEAGRIS GALLOPAVO	3320 G WET WEIGHT	AGE CLASS (D) = 126, MALE	17
*MELEAGRIS GALLOPAVO	4050 G WET WEIGHT	AGE CLASS (D) = 140, MALE	17
*MELEAGRIS GALLOPAVO	4620 G WET WEIGHT	AGE CLASS (D) = 154, MALE	17
*MELEAGRIS GALLOPAVO	5100 G WET WEIGHT	AGE CLASS (D) = 168, MALE	17
*MELEAGRIS GALLOPAVO	5500 G WET WEIGHT	AGE CLASS (D) = 182, MALE	17
*MELEAGRIS GALLOPAVO	5780 G WET WEIGHT	AGE CLASS (D) = 196, MALE	17
*MELEAGRIS GALLOPAVO	5950 G WET WEIGHT	AGE CLASS (D) = 210, MALE	17

TABLE A388

VERTEBRATES : BIOMASS AT DIFFERENT LIFE TIME

ITEM:	VALUE:	CONDITIONS:	REF.:
*MELEAGRIS GALLOPAVO	6260 G WET WEIGHT	AGE CLASS (D) = 252, MALE	17
*MELEAGRIS GALLOPAVO	6350 G WET WEIGHT	AGE CLASS (D) = 280, MALE	17
*MELEAGRIS GALLOPAVO	40 G WET WEIGHT	AGE CLASS (D) = 0, FEMALE	17
*MELEAGRIS GALLOPAVO	80 G WET WEIGHT	AGE CLASS (D) = 14, FEMALE	17
*MELEAGRIS GALLOPAVO	250 G WET WEIGHT	AGE CLASS (D) = 28, FEMALE	17
*MELEAGRIS GALLOPAVO	480 G WET WEIGHT	AGE CLASS (D) = 42, FEMALE	17
*MELEAGRIS GALLOPAVO	690 G WET WEIGHT	AGE CLASS (D) = 56, FEMALE	17
*MELEAGRIS GALLOPAVO	960 G WET WEIGHT	AGE CLASS (D) = 70, FEMALE	17
*MELEAGRIS GALLOPAVO	1240 G WET WEIGHT	AGE CLASS (D) = 84, FEMALE	17
*MELEAGRIS GALLOPAVO	1580 G WET WEIGHT	AGE CLASS (D) = 98, FEMALE	17
*MELEAGRIS GALLOPAVO	1980 G WET WEIGHT	AGE CLASS (D) = 112, FEMALE	17
*MELEAGRIS GALLOPAVO	2520 G WET WEIGHT	AGE CLASS (D) = 126, FEMALE	17
*MELEAGRIS GALLOPAVO	3000 G WET WEIGHT	AGE CLASS (D) = 140, FEMALE	17
*MELEAGRIS GALLOPAVO	3320 G WET WEIGHT	AGE CLASS (D) = 154, FEMALE	17
*MELEAGRIS GALLOPAVO	3480 G WET WEIGHT	AGE CLASS (D) = 168, FEMALE	17
*MELEAGRIS GALLOPAVO	3620 G WET WEIGHT	AGE CLASS (D) = 182, FEMALE	17
*MELEAGRIS GALLOPAVO	3710 G WET WEIGHT	AGE CLASS (D) = 196, FEMALE	17
*MELEAGRIS GALLOPAVO	3770 G WET WEIGHT	AGE CLASS (D) = 210, FEMALE	17
*MELEAGRIS GALLOPAVO	3910 G WET WEIGHT	AGE CLASS (D) = 252, FEMALE	17
*MELEAGRIS GALLOPAVO	3960 G WET WEIGHT	AGE CLASS (D) = 280, FEMALE	17
*MUS MUSCULUS	1.25 G WET WEIGHT	AGE CLASS (W) = 0, MALE	17
*MUS MUSCULUS	4.03 G WET WEIGHT	AGE CLASS (W) = 1, MALE	17
*MUS MUSCULUS	6.54 G WET WEIGHT	AGE CLASS (W) = 2, MALE	17
*MUS MUSCULUS	8.76 G WET WEIGHT	AGE CLASS (W) = 3, MALE	17
*MUS MUSCULUS	12.13 G WET WEIGHT	AGE CLASS (W) = 4, MALE	17
*MUS MUSCULUS	14.94 G WET WEIGHT	AGE CLASS (W) = 5, MALE	17
*MUS MUSCULUS	17.60 G WET WEIGHT	AGE CLASS (W) = 6, MALE	17
*MUS MUSCULUS	20.06 G WET WEIGHT	AGE CLASS (W) = 7, MALE	17
*MUS MUSCULUS	21.59 G WET WEIGHT	AGE CLASS (W) = 8, MALE	17
*MUS MUSCULUS	24.78 G WET WEIGHT	AGE CLASS (W) = 12, MALE	17
*MUS MUSCULUS	27.76 G WET WEIGHT	AGE CLASS (W) = 16, MALE	17
*MUS MUSCULUS	29.91 G WET WEIGHT	AGE CLASS (W) = 20, MALE	17
*MUS MUSCULUS	31.07 G WET WEIGHT	AGE CLASS (W) = 24, MALE	17
*MUS MUSCULUS	33.07 G WET WEIGHT	AGE CLASS (W) = 28, MALE	17
*MUS MUSCULUS	34.07 G WET WEIGHT	AGE CLASS (W) = 32, MALE	17
*MUS MUSCULUS	35.49 G WET WEIGHT	AGE CLASS (W) = 36, MALE	17
*MUS MUSCULUS	1.26 G WET WEIGHT	AGE CLASS (W) = 0, FEMALE	17
*MUS MUSCULUS	4.04 G WET WEIGHT	AGE CLASS (W) = 1, FEMALE	17
*MUS MUSCULUS	6.52 G WET WEIGHT	AGE CLASS (W) = 2, FEMALE	17
*MUS MUSCULUS	8.72 G WET WEIGHT	AGE CLASS (W) = 3, FEMALE	17
*MUS MUSCULUS	11.94 G WET WEIGHT	AGE CLASS (W) = 4, FEMALE	17
*MUS MUSCULUS	14.73 G WET WEIGHT	AGE CLASS (W) = 5, FEMALE	17
*MUS MUSCULUS	16.77 G WET WEIGHT	AGE CLASS (W) = 6, FEMALE	17
*MUS MUSCULUS	18.42 G WET WEIGHT	AGE CLASS (W) = 7, FEMALE	17
*MUS MUSCULUS	19.81 G WET WEIGHT	AGE CLASS (W) = 8, FEMALE	17
*MUS MUSCULUS	22.99 G WET WEIGHT	AGE CLASS (W) = 12, FEMALE	17
*MUS MUSCULUS	26.11 G WET WEIGHT	AGE CLASS (W) = 16, FEMALE	17
*MUS MUSCULUS	29.03 G WET WEIGHT	AGE CLASS (W) = 20, FEMALE	17
*MUS MUSCULUS	31.00 G WET WEIGHT	AGE CLASS (W) = 24, FEMALE	17
*MUS MUSCULUS	33.00 G WET WEIGHT	AGE CLASS (W) = 28, FEMALE	17

TABLE A388

VERTEBRATES : BIOMASS AT DIFFERENT LIFE TIME

ITEM:	VALUE:	CONDITIONS:	REF.:
*MUS MUSCULUS	34.52 G WET WEIGHT	AGE CLASS (W) = 32, FEMALE	17
*MUS MUSCULUS	36.38 G WET WEIGHT	AGE CLASS (W) = 36, FEMALE	17
*ODOCOILEUS HEMIONUS COLUMBIANUS	3.1 KG WET WEIGHT	AGE CLASS (YEARS) = 0, MALE	180
*ODOCOILEUS HEMIONUS COLUMBIANUS	36.4 KG WET WEIGHT	AGE CLASS (YEARS) = 1, MALE	180
*ODOCOILEUS HEMIONUS COLUMBIANUS	54.9 KG WET WEIGHT	AGE CLASS (YEARS) = 2, MALE	180
*ODOCOILEUS HEMIONUS COLUMBIANUS	71.8 KG WET WEIGHT	AGE CLASS (YEARS) = 3, MALE	180
*ODOCOILEUS HEMIONUS COLUMBIANUS	74.5 KG WET WEIGHT	AGE CLASS (YEARS) = 4, MALE	180
*ODOCOILEUS HEMIONUS COLUMBIANUS	87.3 KG WET WEIGHT	AGE CLASS (YEARS) = 5, MALE	180
*ODOCOILEUS HEMIONUS COLUMBIANUS	96.4 KG WET WEIGHT	AGE CLASS (YEARS) = 6, MALE	180
*ODOCOILEUS HEMIONUS COLUMBIANUS	3.4 KG WET WEIGHT	AGE CLASS (YEARS) = 0, FEMALE	180
*ODOCOILEUS HEMIONUS COLUMBIANUS	35.4 KG WET WEIGHT	AGE CLASS (YEARS) = 1, FEMALE	180
*ODOCOILEUS HEMIONUS COLUMBIANUS	49.6 KG WET WEIGHT	AGE CLASS (YEARS) = 2, FEMALE	180
*ODOCOILEUS HEMIONUS COLUMBIANUS	56.7 KG WET WEIGHT	AGE CLASS (YEARS) = 3, FEMALE	180
*ODOCOILEUS HEMIONUS COLUMBIANUS	63.5 KG WET WEIGHT	AGE CLASS (YEARS) = 4, FEMALE	180
*ODOCOILEUS HEMIONUS COLUMBIANUS	41.4 KG WET WEIGHT	AGE CLASS (YEARS) = 5, FEMALE	180
*ODOCOILEUS HEMIONUS COLUMBIANUS	42.7 KG WET WEIGHT	AGE CLASS (YEARS) = 6, FEMALE	180
*PASSER DOMESTICUS	2 G WET WEIGHT	AGE CLASS (D) = 1, BOTH SEXES	1531
*PASSER DOMESTICUS	10 G WET WEIGHT	AGE CLASS (D) = 4, BOTH SEXES	1531
*PASSER DOMESTICUS	18 G WET WEIGHT	AGE CLASS (D) = 8, BOTH SEXES	1531
*PASSER DOMESTICUS	20 G WET WEIGHT	AGE CLASS (D) = 10, BOTH SEXES	1531
*PASSER DOMESTICUS	25 G WET WEIGHT	AGE CLASS (D) = 15, BOTH SEXES	1531
*PASSER DOMESTICUS	30.5 G WET WEIGHT	ADULT, MALE	1531
*PASSER DOMESTICUS	30.0 G WET WEIGHT	ADULT, FEMALE	1531
*RATTUS NORVEGINUS	6.08 G WET WEIGHT	AGE CLASS (W) = 0, MALE	17
*RATTUS NORVEGINUS	17.5 G WET WEIGHT	AGE CLASS (W) = 1, MALE	17
*RATTUS NORVEGINUS	36.9 G WET WEIGHT	AGE CLASS (W) = 2, MALE	17
*RATTUS NORVEGINUS	59.6 G WET WEIGHT	AGE CLASS (W) = 3, MALE	17
*RATTUS NORVEGINUS	92.9 G WET WEIGHT	AGE CLASS (W) = 4, MALE	17
*RATTUS NORVEGINUS	138 G WET WEIGHT	AGE CLASS (W) = 5, MALE	17
*RATTUS NORVEGINUS	188 G WET WEIGHT	AGE CLASS (W) = 6, MALE	17
*RATTUS NORVEGINUS	233 G WET WEIGHT	AGE CLASS (W) = 7, MALE	17
*RATTUS NORVEGINUS	274 G WET WEIGHT	AGE CLASS (W) = 8, MALE	17
*RATTUS NORVEGINUS	339 G WET WEIGHT	AGE CLASS (W) = 10, MALE	17
*RATTUS NORVEGINUS	393 G WET WEIGHT	AGE CLASS (W) = 12, MALE	17
*RATTUS NORVEGINUS	440 G WET WEIGHT	AGE CLASS (W) = 15, MALE	17
*RATTUS NORVEGINUS	490 G WET WEIGHT	AGE CLASS (W) = 20, MALE	17
*RATTUS NORVEGINUS	5.75 G WET WEIGHT	AGE CLASS (W) = 0, FEMALE	17
*RATTUS NORVEGINUS	16.2 G WET WEIGHT	AGE CLASS (W) = 1, FEMALE	17
*RATTUS NORVEGINUS	33.5 G WET WEIGHT	AGE CLASS (W) = 2, FEMALE	17
*RATTUS NORVEGINUS	53 G WET WEIGHT	AGE CLASS (W) = 3, FEMALE	17
*RATTUS NORVEGINUS	79.5 G WET WEIGHT	AGE CLASS (W) = 4, FEMALE	17
*RATTUS NORVEGINUS	113 G WET WEIGHT	AGE CLASS (W) = 5, FEMALE	17
*RATTUS NORVEGINUS	147 G WET WEIGHT	AGE CLASS (W) = 6, FEMALE	17
*RATTUS NORVEGINUS	172 G WET WEIGHT	AGE CLASS (W) = 7, FEMALE	17
*RATTUS NORVEGINUS	196 G WET WEIGHT	AGE CLASS (W) = 8, FEMALE	17
*RATTUS NORVEGINUS	227 G WET WEIGHT	AGE CLASS (W) = 10, FEMALE	17
*RATTUS NORVEGINUS	251 G WET WEIGHT	AGE CLASS (W) = 12, FEMALE	17
*RATTUS NORVEGINUS	274 G WET WEIGHT	AGE CLASS (W) = 15, FEMALE	17
*RATTUS NORVEGINUS	303 G WET WEIGHT	AGE CLASS (W) = 20, FEMALE	17
*RATTUS NORVEGINUS	335 G WET WEIGHT	AGE CLASS (W) = 30, FEMALE	17

TABLE A388

VERTEBRATES : BIOMASS AT DIFFERENT LIFE TIME			
ITEM:	VALUE:	CONDITIONS:	REF.:
*RATTUS NORVEGINUS	358 G WET WEIGHT	AGE CLASS (W) = 40, FEMALE	17
*SYLVILAGUS FLORIDANUS	25.6 G WET WEIGHT	AGE CLASS (W) = 0, BOTH SEXES	948
*SYLVILAGUS FLORIDANUS	58.3 G WET WEIGHT	AGE CLASS (W) = 1, BOTH SEXES	948
*SYLVILAGUS FLORIDANUS	68.5 G WET WEIGHT	AGE CLASS (W) = 2, BOTH SEXES	948
*SYLVILAGUS FLORIDANUS	92.4 G WET WEIGHT	AGE CLASS (W) = 3, BOTH SEXES	948
*SYLVILAGUS FLORIDANUS	134.5 G WET WEIGHT	AGE CLASS (W) = 4, BOTH SEXES	948
*SYLVILAGUS FLORIDANUS	234.5 G WET WEIGHT	AGE CLASS (W) = 5, BOTH SEXES	948
*SYLVILAGUS FLORIDANUS	274.5 G WET WEIGHT	AGE CLASS (W) = 6, BOTH SEXES	948
*SYLVILAGUS FLORIDANUS	374.5 G WET WEIGHT	AGE CLASS (W) = 7, BOTH SEXES	948
*SYLVILAGUS FLORIDANUS	435.6 G WET WEIGHT	AGE CLASS (W) = 8, BOTH SEXES	948
*SYLVILAGUS FLORIDANUS	414.0 G WET WEIGHT	AGE CLASS (W) = 9, BOTH SEXES	948
*SYLVILAGUS FLORIDANUS	497.4 G WET WEIGHT	AGE CLASS (W) = 10, BOTH SEXES	948
*SYLVILAGUS FLORIDANUS	517.5 G WET WEIGHT	AGE CLASS (W) = 11, BOTH SEXES	948
*SYLVILAGUS FLORIDANUS	659.7 G WET WEIGHT	AGE CLASS (W) = 12, BOTH SEXES	948
*SYLVILAGUS FLORIDANUS	765.3 G WET WEIGHT	AGE CLASS (W) = 13, BOTH SEXES	948
*SYLVILAGUS FLORIDANUS	785.4 G WET WEIGHT	AGE CLASS (W) = 14, BOTH SEXES	948
*SYLVILAGUS FLORIDANUS	945.0 G WET WEIGHT	AGE CLASS (W) = 15, BOTH SEXES	948
*SYLVILAGUS FLORIDANUS	975.2 G WET WEIGHT	AGE CLASS (W) = 16, BOTH SEXES	948
*SYLVILAGUS FLORIDANUS	981.0 G WET WEIGHT	AGE CLASS (W) = 19, BOTH SEXES	948
*SYLVILAGUS FLORIDANUS	1106.0 G WET WEIGHT	AGE CLASS (W) = 21, BOTH SEXES	948

TABLE A389

VERTEBRATES : BR/BIOMASS RATIO			
ITEM:	VALUE:	CONDITIONS:	REF.:
*HOMO SAPIENS	0.0002 %	ATOMIC COMPOSITION	109

TABLE A390

VERTEBRATES : C/BIOMASS RATIO			
ITEM:	VALUE:	CONDITIONS:	REF.:
*HOMO SAPIENS	19.37 %	ATOMIC COMPOSITION	109
*MAMMALIA	48.4 % OF DRY WEIGHT	MEAN ELEMENTARY COMPOSITION	157
*MEAN ANIMAL	45 % OF DRY WEIGHT	MEAN ELEMENTARY COMPOSITION	157

TABLE A391

VERTEBRATES : CA/BIOMASS RATIO			
ITEM:	VALUE:	CONDITIONS:	REF.:
*HOMO SAPIENS	1.38 %	ATOMIC COMPOSITION	109

TABLE A392

VERTEBRATES : CALORIC CONTENT			
ITEM:	VALUE:	CONDITIONS:	REF.:
*APODEMUS AGRARIUS	5.179 KCAL/G DRY WEIGHT	EQUAL TO 1.64 KCAL/G FRESH WEIGHT	552

TABLE A392

VERTEBRATES : CALORIC CONTENT

ITEM:	VALUE:	CONDITIONS:	REF.:
*APODEMUS FLAVICOLLIS	5.002 KCAL/G DRY WEIGHT	EQUAL TO 1.40 KCAL/G FRESH WEIGHT	552
*BUFO BUFO	5.003 KCAL/G DRY WEIGHT	EQUAL TO 1.25 KCAL/G FRESH WEIGHT	999
*CAPREOLUS CAPREOLUS	6.065 KCAL/G DRY WEIGHT	EQUAL TO 2.16 KCAL/G FRESH WEIGHT	547
*CLETHRIONOMYS GLAREOLUS	4.934 KCAL/G DRY WEIGHT	EQUAL TO 1.45 KCAL/G FRESH WEIGHT	552
*CYANOCITTA CRISTATA	5.090 KCAL/G DRY WEIGHT	NONE	1162
*HYLOCICHLA MUSTELLINA	7.725 KCAL/G DRY WEIGHT	BEFORE MIGRATION	1162
*HYLOCICHLA MUSTELLINA	5.990 KCAL/G DRY WEIGHT	AFTER MIGRATION	1162
*LEPUS EUROPAEUS	7.500 KCAL/G DRY WEIGHT	EQUAL TO 1.87 KCAL/G FRESH WEIGHT	1092
*MICROTUS ARVALIS	4.856 KCAL/G DRY WEIGHT	EQUAL TO 1.47 KCAL/G FRESH WEIGHT	552
*MICROTUS PENNSYLVANICUS	4.650 KCAL/G DRY WEIGHT	EQUAL TO 1.40 KCAL/G FRESH WEIGHT	547
*MUS MUSCULUS	5.675 KCAL/G DRY WEIGHT	EQUAL TO 1.70 KCAL/G FRESH WEIGHT	547
*MUS MUSCULUS	6.205 KCAL/G DRY WEIGHT	EQUAL TO 2.25 KCAL/G FRESH WEIGHT	1090
*ORYZOMYS PALUSTRIS	5.840 KCAL/G DRY WEIGHT	EQUAL TO 1.90 KCAL/G FRESH WEIGHT	331
*PARULA AMERICANA	7.680 KCAL/G DRY WEIGHT	BEFORE MIGRATION	1162
*PARULA AMERICANA	6.075 KCAL/G DRY WEIGHT	AFTER MIGRATION	1162
*PASSER DOMESTICUS	5.535 KCAL/G DRY WEIGHT	NONE	1162
*PASSER DOMESTICUS	5.512 KCAL/G DRY WEIGHT	EQUAL TO 1.91 KCAL/G FRESH WEIGHT	548
*PERDIX PERDIX	5.253 KCAL/G DRY WEIGHT	EQUAL TO 1.90 KCAL/G FRESH WEIGHT	1546
*PHOCA HISPIDA	2.3 - 5.3 KCAL/G WET WEIGHT	2 - 5.5 % ASH	1512
*PIRANGA OLIVACEA	7.575 KCAL/G DRY WEIGHT	BEFORE MIGRATION	1162
*PIRANGA OLIVACEA	6.645 KCAL/G DRY WEIGHT	AFTER MIGRATION	1162
*RANA ARVALIS	4.628 KCAL/G DRY WEIGHT	EQUAL TO 1.16 KCAL/G FRESH WEIGHT	999
*RICHMODENA CARDINALIS	5.530 KCAL/G DRY WEIGHT	NONE	1162
*SOREX ARANEUS	4.555 KCAL/G DRY WEIGHT	EQUAL TO 1.50 KCAL/G FRESH WEIGHT	552
*SOREX ARANEUS	4.991 KCAL/G DRY WEIGHT	EQUAL TO 1.53 KCAL/G FRESH WEIGHT	1091
*SOREX MINUTUS	5.025 KCAL/G DRY WEIGHT	EQUAL TO 1.63 KCAL/G FRESH WEIGHT	1091
*TELMATODYTES PALUSTRIS	6.170 KCAL/G ASH FREE DRY WEIGHT	NONE	791
*ZENOIDURA MACRAURA	2.61 KCAL/G FRESH WEIGHT	NONE	174

TABLE A393

VERTEBRATES : CL/BIOMASS RATIO

ITEM:	VALUE:	CONDITIONS:	REF.:
*HOMO SAPIENS	0.18 %	ATOMIC COMPOSITION	109

TABLE A394

VERTEBRATES : CONSUMPTION

ITEM:	VALUE:	CONDITIONS:	REF.:
*BALAENOPTERA BOREALIS	4.43 % OF BODY WET WEIGTH	EQUAL TO 900 KG FOOD PER DAY (WW)	803
*CLETHRIONOMYS GLAREOLUS AND APODEMUS FLAVICOLLIS	21 - 42*10**3 KCAL/(HA*YEAR)	FOOD AVAILABLE=2415-7080*10**3 KCAL/(HA*Y), PINEWOOD, POLAND	572
*CLETHRIONOMYS GLAREOLUS AND APODEMUS FLAVICOLLIS	20*10**3 KCAL/(HA*YEAR)	FOOD AVAILABLE=1024*10**3 KCAL/(HA*Y), CLADONIA, POLAND	572
*CLETHRIONOMYS GLAREOLUS AND APODEMUS FLAVICOLLIS	75 - 102*10**3 KCAL/(HA*YEAR)	FOOD AVAILABLE=13040*10**3 KCAL/(HA*Y), OAK-PPINE FOREST, POLAND	572
*CLETHRIONOMYS GLAREOLUS AND APODEMUS FLAVICOLLIS	75*10**3 KCAL/(HA*YEAR)	FOOD AVAILABLE=1950*10**3 KCAL/(HA*Y), BEECHWOOD, POLAND	572

TABLE A394

VERTEBRATES : CONSUMPTION			
ITEM:	VALUE:	CONDITIONS:	REF.:
*CLETHRIONOMYS GLAREOLUS, APODEMUS FLAVICOLLIS AND APODEMUS AGRARIUS	95*10**3 KCAL/(HA*YEAR)	FOOD AVAILABLE=2050*10**3 KCAL/(HA*Y), OAK-HORNBEAM FOREST, POLAND	572
*CLETHRIONOMYS RUTILUS AND TAMIASCIURUS SP.	179*10**3 KCAL/(HA*YEAR)	FOOD AVAILABLE=1320*10**3 KCAL/(HA*Y), TAIGA FOREST, ALASKA	572
*DIPODOMYS MERRIAMI AND LEPUS CALIFORNICUS	131*10**3 KCAL/(HA*YEAR)	FOOD AVAILABLE=2395*10**3 KCAL/(HA*Y), DESERT SHRUB, ARIZONA	572
*MICROTUS AGRESTIS	332 - 699*10**3 KCAL/(HA*YEAR)	FOOD AVAILABLE=1950*10**3 KCAL/(HA*Y), SPRUCE FOREST, SWEDEN	572
*MICROTUS OECONOMUS	375*10**3 KCAL/(HA*YEAR)	FOOD AVAILABLE=6650*10**3 KCAL/(HA*Y), FOREST ON PEAT-BOG, POLAND	572

TABLE A395

VERTEBRATES : CU/BIOMASS RATIO			
ITEM:	VALUE:	CONDITIONS:	REF.:
*HOMO SAPIENS	0.0004 %	ATOMIC COMPOSITION	109

TABLE A396

VERTEBRATES : EFFICIENCY			
ITEM:	VALUE:	CONDITIONS:	REF.:
*PHALLUSIA MAMMILLATA	90 (RANGE = 86 - 93) % OF FILTRATED AMOUNT	FOOD CONVERSION,PH=7.8,24 H,288 K, O2=5 MG/L,S=3.8%,FOOD=MONOCH. LUTH.	355

TABLE A397

VERTEBRATES : FE/BIOMASS RATIO			
ITEM:	VALUE:	CONDITIONS:	REF.:
*HOMO SAPIENS	0.0050 %	ATOMIC COMPOSITION	109

TABLE A398

VERTEBRATES : FILTRATION RATE			
ITEM:	VALUE:	CONDITIONS:	REF.:
*PHALLUSIA MAMMILLATA	3856 - 4730 ML/(HOUR*G ORG DRY WEIGHT)	MEAN=4380,FOOD=MONOCHRYSIS LUTHERI,24 H,288 K,PH=7.8,O2=5 MG/L,S=3.8%	355

TABLE A399

VERTEBRATES : GROWTH EFFICIENCY			
ITEM:	VALUE:	CONDITIONS:	REF.:
*ADENOTA KOB THOMASI	1.30 % OF INGESTION	PROD = 0.81 KCAL/(M2*Y)	1012

TABLE A399

VERTEBRATES : GROWTH EFFICIENCY			
ITEM:	VALUE:	CONDITIONS:	REF.:
*CLETHRIONOMYS RUTLUS	1.89 % OF INGESTION	PROD = 0.00024 KCAL/(M2*Y)	1012
*GLAUCOMYS SABRINUS YUKONENSIS	1.18 % OF INGESTION	PROD = 0.00002 KCAL/(M2*Y)	1012
*LEPTOTERNA DOLOBRATA	50.8 % OF INGESTION	PROD = 0.064 KCAL/(M2*Y)	1012
*LOXODONTA AFRICANA	1.46 % OF INGESTION	PROD = 0.34 KCAL/(M2*Y)	1012
*MICROTUS OECONOMUS	1.82 % OF INGESTION	PROD = 0.00024 KCAL/(M2*Y)	1012
*MICROTUS PENNSYLVANICUS	2.95 % OF INGESTION	PROD = 0.517 KCAL/(M2*Y)	1012
*MUSTELA RIXOSA ALLEGHNIENSIS	2.34 % OF INGESTION	PROD = 0.013 KCAL/(M2*Y)	1012
*PASSERCULUS SANDWICHENSIS	1.11 % OF INGESTION	PROD = 0.34 KCAL/(M2*Y)	1012
*PEROMYSCUS POLIONOTUS	1.79 % OF INGESTION	PROD = 0.12 KCAL/(M2*Y)	1012
*PITYMYS SUBTERRANEUS	1.57 % OF INGESTION	PROD = 0.0955 KCAL/(M2*Y)	1012
*TAMIASCIURUS HUDSONI PREBLEI	1.37 % OF INGESTION	PROD = 0.00020 KCAL/(M2*Y)	1012

TABLE A400

VERTEBRATES : GROWTH RATE			
ITEM:	VALUE:	CONDITIONS:	REF.:
*MICROTUS ORCADENSIS	12.2 RATE	NET PRODUCTATION RATE, LENGHT OF 1 GENERATION=332.4 D	891
*RATTUS NORVEGICUS	16.5 RATE	NET PRODUCTATION RATE, LENGHT OF 1 GENERATION=344.1 D	891
*RATTUS NORVEGICUS	16.8 RATE	NET PRODUCTATION RATE, LENGHT OF 1 GENERATION=236.5 D	891

TABLE A401

VERTEBRATES : H/BIOMASS RATIO			
ITEM:	VALUE:	CONDITIONS:	REF.:
*HOMO SAPIENS	9.31 %	ATOMIC COMPOSITION	109
*MAMMALIA	6.6 % OF DRY WEIGHT	MEAN ELEMENTARY COMPOSITION	157
*MEAN ANIMAL	6.5 % OF DRY WEIGHT	MEAN ELEMENTARY COMPOSITION	157

TABLE A402

VERTEBRATES : I/BIOMASS RATIO			
ITEM:	VALUE:	CONDITIONS:	REF.:
*HOMO SAPIENS	0.0001 %	ATOMIC COMPOSITION	109

TABLE A403

VERTEBRATES : K/BIOMASS RATIO			
ITEM:	VALUE:	CONDITIONS:	REF.:
*HOMO SAPIENS	0.22 %	ATOMIC COMPOSITION	109

TABLE A404

VERTEBRATES : MG/BIOMASS RATIO			
ITEM:	VALUE:	CONDITIONS:	REF.:
*HOMO SAPIENS	0.04 %	ATOMIC COMPOSITION	109

TABLE A405

VERTEBRATES : MN/BIOMASS RATIO			
ITEM:	VALUE:	CONDITIONS:	REF.:
*HOMO SAPIENS	0.0001 %	ATOMIC COMPOSITION	109

TABLE A406

VERTEBRATES : N/BIOMASS RATIO			
ITEM:	VALUE:	CONDITIONS:	REF.:
*HOMO SAPIENS	5.14 %	ATOMIC COMPOSITION	109
*MAMMALIA	12.8 % OF DRY WEIGHT	MEAN ELEMENTARY COMPOSITION	157
*MEAN ANIMAL	10 % OF DRY WEIGHT	MEAN ELEMENTARY COMPOSITION	157

TABLE A407

VERTEBRATES : NA/BIOMASS RATIO			
ITEM:	VALUE:	CONDITIONS:	REF.:
*HOMO SAPIENS	0.26 %	ATOMIC COMPOSITION	109

TABLE A408

VERTEBRATES : O/BIOMASS RATIO			
ITEM:	VALUE:	CONDITIONS:	REF.:
*HOMO SAPIENS	62.81 %	ATOMIC COMPOSITION	109
*MAMMALIA	18.6 % OF DRY WEIGHT	MEAN ELEMENTARY COMPOSITION	157
*MEAN ANIMAL	30 % OF DRY WEIGHT	MEAN ELEMENTARY COMPOSITION	157

TABLE A409

VERTEBRATES : P/BIOMASS RATIO			
ITEM:	VALUE:	CONDITIONS:	REF.:
*HOMO SAPIENS	0.63 %	ATOMIC COMPOSITION	109

TABLE A410

VERTEBRATES : PB/BIOMASS RATIO			
ITEM:	VALUE:	CONDITIONS:	REF.:
*HOMO SAPIENS	0.0001 %	ATOMIC COMPOSITION	109

TABLE A411

VERTEBRATES : RB/BIOMASS RATIO			
ITEM:	VALUE:	CONDITIONS:	REF.:
*HOMO SAPIENS	0.0009 %	ATOMIC COMPOSITION	109

TABLE A412

VERTEBRATES : S/BIOMASS RATIO			
ITEM:	VALUE:	CONDITIONS:	REF.:
*HOMO SAPIENS	0.64 %	ATOMIC COMPOSITION	109
*MAMMALIA	1.6 % OF DRY WEIGHT	MEAN ELEMENTARY COMPOSITION	157
*MEAN ANIMAL	1.3 % OF DRY WEIGHT	MEAN ELEMENTARY COMPOSITION	157

TABLE A413

VERTEBRATES : SI/BIOMASS RATIO			
ITEM:	VALUE:	CONDITIONS:	REF.:
*HOMO SAPIENS	0.0040 %	ATOMIC COMPOSITION	109

TABLE A414

VERTEBRATES : SN/BIOMASS RATIO			
ITEM:	VALUE:	CONDITIONS:	REF.:
*HOMO SAPIENS	0.0002 %	ATOMIC COMPOSITION	109

TABLE A415

VERTEBRATES : THERMAL CONDUCTIVITY			
ITEM:	VALUE:	CONDITIONS:	REF.:
*CARIBOU	4.9*10**(-3) CAL/(CM*MIN*DEG)	CONDUCTIVITY OF WINTER FUR, THICKNESS = 3.5 CM2	1399
*DALL SHEEP	8.1*10**(-3) CAL/(CM*MIN*DEG)	CONDUCTIVITY OF WINTER FUR, THICKNESS = 7 CM2	1399
*DOG	7.0*10**(-3) CAL/(CM*MIN*DEG)	CONDUCTIVITY OF WINTER FUR, THICKNESS = 4.0 CM2	1399
*POLAR BEAR	11.3*10**(-3) CAL/(CM*MIN*DEG)	CONDUCTIVITY OF WINTER FUR, THICKNESS = 6.5 CM2	1399
*RABBIT	5.2*10**(-3) CAL/(CM*MIN*DEG)	CONDUCTIVITY OF WINTER FUR, THICKNESS = 3.0 CM2	1399
*RED FOX	7.8*10**(-3) CAL/(CM*MIN*DEG)	CONDUCTIVITY OF WINTER FUR, THICKNESS = 4.5 CM2	1399
*SHREW	3.5*10**(-3) CAL/(CM*MIN*DEG)	CONDUCTIVITY OF WINTER FUR, THICKNESS = 0.5 CM2, SUMMER	1399
*SPERMOPHILUS UNDULATUS	5.2*10**(-3) CAL/(CM*MIN*DEG)	CONDUCTIVITY OF WINTER FUR, THICKNESS = 1.5 CM2	1399
*WEASEL	5.6*10**(-3) CAL/(CM*MIN*DEG)	CONDUCTIVITY OF WINTER FUR, THICKNESS = 0.8 CM2	1399
*WHITE FOX	5.8*10**(-3) CAL/(CM*MIN*DEG)	CONDUCTIVITY OF WINTER FUR, THICKNESS = 5 CM2	1399
*WOLF	9.0*10**(-3) CAL/(CM*MIN*DEG)	CONDUCTIVITY OF WINTER FUR, THICKNESS = 6.5 CM2	1399

TABLE A416

VERTEBRATES : ZN/BIOMASS RATIO			
ITEM:	VALUE:	CONDITIONS:	REF.:
*HOMO SAPIENS	0.0025 %	ATOMIC COMPOSITION	109

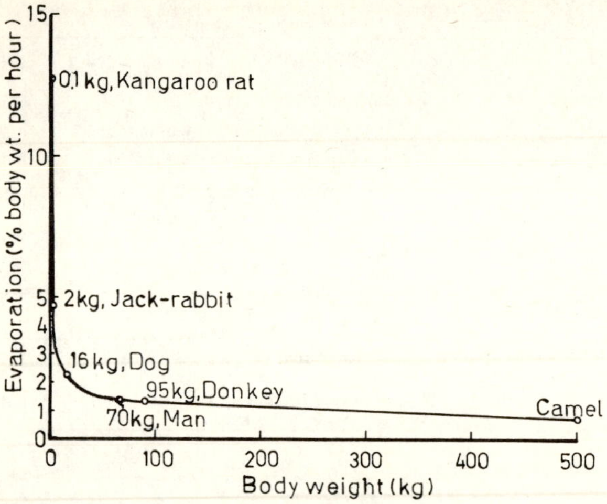

A417 Body size and evaporation relation necessary for the maintenance of a constant body temperature in a hot climate.

(ref: 1394)

A9

ZOOPLANKTON

COMPOSITION
DEVELOPMENT TIME
EXCRETION
FILTRATION RATE
GRAZING RATE
GROWTH
HALF SATURATION CONSTANTS
MORTALITY
PRODUCTION
RESPIRATION

TABLE A418

ZOOPLANKTON : AL/BIOMASS RATIO			
ITEM:	VALUE:	CONDITIONS:	REF.:
*CALANUS FINMARCHICUS	0.0000 %	ATOMIC COMPOSITION	1632

TABLE A419

ZOOPLANKTON : ASH/BIOMASS RATIO			
ITEM:	VALUE:	CONDITIONS:	REF.:
*AMPHIPODA	10-37 % OF DRY WEIGHT	DRIED AT 333K,NORTH PACIFIC	1184
*CHAETOGNATHA	4-5 % OF DRY WEIGHT	DRIED AT 333K,NORTH PACIFIC	1184
*COPEPODS	18-23 % OF DRY WEIGHT	DRIED AT 378K,CONTINENTAL SHELF OFF NEW YORK	308
*COPEPODS	2-6 % OF DRY WEIGHT	DRIED AT 333K,NORTH PACIFIC	1184
*CTENOPHORA	70-75 % OF DRY WEIGHT	DRIED AT 378K,CONTINENTAL SHELF OFF NEW YORK	308
*DAPHNIA CRISTATA	8.07 (+-0.03) % OF DRY WEIGHT	CLADOCERA , 95 % CONF LIMIT, 12 H AT 873 K	1375
*EUDIAPTOMUS GRACILIS	5.16 (+-0.05) % OF DRY WEIGHT	COPEPODA , 95 % CONF LIMIT, 12 H AT 873 K	1375
*EUPHAUSIACEA	19-22 % OF DRY WEIGHT	DRIED AT 378K,CONTINENTAL SHELF OFF NEW YORK	308
*EUPHAUSIACEA	8-9 % OF DRY WEIGHT	DRIED AT 333K,NORTH PACIFIC	1184
*LIMNOCALANUS MACRURUS	2.92 (+-0.05) % OF DRY WEIGHT	COPEPODA , 95 % CONF LIMIT, 12 H AT 873 K	1375
*MACROCYCLOPS ALBIDUS	4.66 (+-0.02) % OF DRY WEIGHT	COPEPODA , 95 % CONF LIMIT, 12 H AT 873 K	1375
*PTEROPODS	24-64 % OF DRY WEIGHT	DRIED AT 378K,CONTINENTAL SHELF OFF NEW YORK	308
*PTEROPODS	29-43 % OF DRY WEIGHT	DRIED AT 333K,NORTH PACIFIC	1184
*TUNICATA	71-77 % OF DRY WEIGHT	DRIED AT 378K,CONTINENTAL SHELF OFF NEW YORK	308

TABLE A420

ZOOPLANKTON : ASSIMILATION			
ITEM:	VALUE:	CONDITIONS:	REF.:
*CERIODAPHNIA RETICULATA	10.6 % OF INGESTION	NONE	1689
*DAPHNIA PULEX	23.9 % OF ENERGY CONSUMED	0-6 D OLD, FOOD = 25000 CELLS/ML, CHLAMYDOMONAS REINHARDTII	1328
*DAPHNIA PULEX	15.8 % OF ENERGY CONSUMED	0-6 D OLD, FOOD = 50000 CELLS/ML, CHLAMYDOMONAS REINHARDTII	1328
*DAPHNIA PULEX	8.4 % OF ENERGY CONSUMED	0-6 D OLD, FOOD = 75000 CELLS/ML, CHLAMYDOMONAS REINHARDTII	1328
*DAPHNIA PULEX	6.6 % OF ENERGY CONSUMED	0-6 D OLD, FOOD = 100000 CELLS/ML, CHLAMYDOMONAS REINHARDTII	1328
*DAPHNIA PULEX	31.7 % OF ENERGY CONSUMED	6-34 D OLD, FOOD = 25000 CELLS/ML, CHLAMYDOMONAS REINHARDTII	1328
*DAPHNIA PULEX	20.2 % OF ENERGY CONSUMED	6-34 D OLD, FOOD = 50000 CELLS/ML, CHLAMYDOMONAS REINHARDTII	1328
*DAPHNIA PULEX	16.8 % OF ENERGY CONSUMED	6-34 D OLD, FOOD = 75000 CELLS/ML, CHLAMYDOMONAS REINHARDTII	1328
*DAPHNIA PULEX	14.2 % OF ENERGY CONSUMED	6-34 D OLD, FOOD = 100000 CELLS/ ML, CHLAMYDOMONAS REINHARDTII	1328
*DAPHNIA PULEX	31 % OF ENERGY CONSUMED	0-40 D OLD, FOOD = 25000 CELLS/ML, CHLAMYDOMONAS REINHARDTII	1328
*DAPHNIA PULEX	20 % OF ENERGY CONSUMED	0-40 D OLD, FOOD = 50000 CELLS/ML, CHLAMYDOMONAS REINHARDTII	1328

TABLE A420

ZOOPLANKTON : ASSIMILATION

ITEM:	VALUE:	CONDITIONS:	REF.:
*DAPHNIA PULEX	16 % OF ENERGY CONSUMED	0-40 D OLD, FOOD = 75000 CELLS/ML, CHLAMYDOMONAS REINHARDTII	1328
*DAPHNIA PULEX	14 % OF ENERGY CONSUMED	0-40 D OLD, FOOD = 100000 CELLS/ ML, CHLAMYDOMONAS REINHARDTII	1328
*DAPHNIA PULEX	31.3 % OF CONSUMED	FOOD = 25*10**(-4) CELL/ML (CHLAMYDOMONAS), 40 D,AFTER 18 INSTARS	1327
*DAPHNIA PULEX	20.0 % OF CONSUMED	FOOD = 50*10**(-4) CELL/ML (CHLAMYDOMONAS), 40 D,AFTER 18 INSTARS	1327
*DAPHNIA PULEX	16.3 % OF CONSUMED	FOOD = 75*10**(-4) CELL/ML (CHLAMYDOMONAS), 40 D,AFTER 18 INSTARS	1327
*DAPHNIA PULEX	13.7 % OF CONSUMED	FOOD = 100*10**(-4) CELL/ML (CHLAMYDOMONAS), 40 D,AFTER 18 INSTARS	1327
*DAPHNIA PULEX	14 - 31 % OF INGESTION	NONE	1689
*DAPHNIA PULEX	71.0% OF INGESTED FOOD	FOOD=ANKISTRODESMUS FALCATUS AT K= 25	42
*DAPHNIA PULEX	100% OF INGESTED FOOD	FOOD=CHLORELLA VULGARIS AT K=25	42
*DAPHNIA PULEX	15.8% OF INGESTED FOOD	FOOD=ANACYSTIS NIDULANS AT K=25	42
*DAPHNIA PULEX	67.2% OF INGESTED FOOD	FOOD=SYNECHOCOCCUS ELONGATA AT K= 25	42
*DAPHNIA PULEX	85.3% OF INGESTED FOOD	FOOD=MERISMOPEDIA SP. AT K=25	42
*DAPHNIA PULEX	81.3% OF INGESTED FOOD	FOOD=SYNECHOCOCCUS CEDRORUM AT K= 25	42
*DAPHNIA PULEX	100% OF INGESTED FOOD	FOOD=ANABAENA FLOS-AQUA AT K=25	42
*DAPHNIA PULEX	77.8% OF INGESTED FOOD	FOOD=SYNECHOCYSTIS SP. AT K=25	42
*DAPHNIA PULEX	70.6% OF INGESTED FOOD	FOOD=GLOEOCAPSA ALPICOLA AT K=25	42
*DAPHNIA SCHODLERI	82+-2% OF INGESTION	283K,NO EGGS,FILTERING RATE= 0.55ML/ANIMAL*HR,N=5	647
*DAPHNIA SCHODLERI	94+-3% OF INGESTION	283K,WITH EGGS,FILTERING RATE= 0.50ML/ANIMAL*HR,N=5	647
*DAPHNIA SCHODLERI	78+-2% OF INGESTION	295K,NO EGGS,FILTERING RATE= 0.96ML/ANIMAL*HR,N=5	647
*DAPHNIA SCHODLERI	81+-2% OF INGESTION	295K,WITH EGGS,FILTERING RATE= 1.06ML/ANIMAL*HR,N=5	647
*LEPTODORA KINDTII	40.0 % OF INGESTION	NONE	1689
*MARINE ZOOPLANKTON	0.20 % OF INGESTED ZN	CALCULATED ON RELATIVE CONCENTRATION FACTOR	1742
*MESOCYCLOPS ALBIDUS	20 - 75 % OF INGESTION	NONE	1689
*SIMOCEPHALUS VECTULUS	72.4 % OF INGESTION	JUVENILE, FEMALE	1689
*SIMOCEPHALUS VECTULUS	31.7 % OF INGESTION	REPRODUCING FEMALE	1689
*ZOOPLANKTON	0.60 MG C/MG C	SAN JOAQUIN(RIVER)	1149
*ZOOPLANKTON	0.60 MG C/MG C	DELTA(ESTUARY)	1149
*ZOOPLANKTON	0.65 MG C/MG C	LAKE ERIE	1149
*ZOOPLANKTON	0.60 MG C/MG C	LAKE ONTARIO	1149

TABLE A421

ZOOPLANKTON : ATP/BIOMASS RATIO

ITEM:	VALUE:	CONDITIONS:	REF.:
*CALANUS FINMARCHICUS	1614.2 10**(-3)MICROGRAM/CALANUS (SD=507.6)	NO STARVATION,N=10,STAGE 4 AND 5	62
*CALANUS FINMARCHICUS	1271.1 10**(-3)MICROGRAM/CALANUS (SD=326.1)	STARVATION:8DAYS,N=10,STAGE 4 AND 5	62
*CALANUS FINMARCHICUS	975.6 10**(-3)MICROGRAM/CALANUS (SD=451.6)	STARVATION:14DAYS,N=10,STAGE 4 AND 5	62
*CALANUS FINMARCHICUS	1268.6 10**(-3)MICROGRAM/CALANUS (SD=544.0)	STARVATION:23DAYS,N=10,STAGE 4 AND 5	62

TABLE A421

ZOOPLANKTON : ATP/BIOMASS RATIO			
ITEM:	VALUE:	CONDITIONS:	REF.:
*CALANUS FINMARCHICUS	2.23% OF NONLIPID CARBON (SD=0.70)	NO STARVATION,STAGE 4 AND 5, NONLIPID C	62
*CALANUS FINMARCHICUS	2.07% OF NONLIPID CARBON (SD=0.70)	STARVATION:8DAYS,STAGE 4 AND 5, NONLIPID C	62
*CALANUS FINMARCHICUS	1.90% OF NONLIPID CARBON (SD=0.87)	STARVATION:14DAYS,STAGE 4 AND 5, NONLIPID C	62
*CALANUS FINMARCHICUS	2.04% OF NONLIPID CARBON (SD=0.78)	STARVATION:23DAYS,STAGE 4 AND 5, NONLIPID C	62

TABLE A422

ZOOPLANKTON : ATP/C RATIO			
ITEM:	VALUE:	CONDITIONS:	REF.:
*CALANUS FINMARCHICUS	0.87% OF TOTAL CARBON (SD=0.33)	NO STARVATION,STAGE 4 AND 5	62
*CALANUS FINMARCHICUS	0.92% OF TOTAL CARBON (SD=0.37)	STARVATION:8DAYS,STAGE 4 AND 5	62
*CALANUS FINMARCHICUS	0.86% OF TOTAL CARBON (SD=0.39)	STARVATION:14DAYS,STAGE 4 AND 5	62
*CALANUS FINMARCHICUS	0.91% OF TOTAL CARBON (SD=0.33)	STARVATION:23DAYS,STAGE 4 AND 5	62
*MESOCYCLOPS LEUCKARTI	0.0106 BY WEIGHT	ADULT FEMALE COPEPOD	236
*MESOCYCLOPS LEUCKARTI	0.00793 BY WEIGHT	COPEPODITE	236

TABLE A423

ZOOPLANKTON : ATP/CHLOROPHYLL A RATIO			
ITEM:	VALUE:	CONDITIONS:	REF.:
*ANABAENA FLOS-AQUAE	0.09 MG/MG	200 - 300 FTC,R=0.707	184

TABLE A424

ZOOPLANKTON : BR/BIOMASS RATIO			
ITEM:	VALUE:	CONDITIONS:	REF.:
*CALANUS FINMARCHICUS	0.0009 %	ATOMIC COMPOSITION	1632

TABLE A425

ZOOPLANKTON : C/BIOMASS RATIO			
ITEM:	VALUE:	CONDITIONS:	REF.:
*AMPHIPODA	26-48 % OF DRY WEIGHT	DRIED AT 333K,NORTH PACIFIC	1184
*CALANUS FINMARCHICUS	6.10 %	ATOMIC COMPOSITION	1632
*CALANUS FINMARCHICUS	187.4 MICROGRAM/CALANUS (SD=71.6)	NO STARVATION,N=9,STAGE 4 AND 5	62
*CALANUS FINMARCHICUS	153.0 MICROGRAM/CALANUS (SD=44.3)	STARVATION:8DAYS,N=10,STAGE 4 AND 5	62
*CALANUS FINMARCHICUS	118.1 MICROGRAM/CALANUS (SD=53.8)	STARVATION:14DAYS,N=10,STAGE 4 AND 5	62
*CALANUS FINMARCHICUS	153.3 MICROGRAM/CALANUS (SD=26.5)	STARVATION:23DAYS,N=12,STAGE 4 AND 5	62
*CHAETOGNATHA	22-34 % OF DRY WEIGHT	DRIED AT 333K,SARGASSO SEA	88

TABLE A425

ZOOPLANKTON : C/BIOMASS RATIO			
ITEM:	VALUE:	CONDITIONS:	REF.:
*CHAETOGNATHA	44-48 % OF DRY WEIGHT	DRIED AT 333K,NORTH PACIFIC	1184
*COPEPODS	35-48 % OF DRY WEIGHT	DRIED AT 333K,SARGASSO SEA	88
*COPEPODS	32-42 % OF DRY WEIGHT	DRIED AT 378K,CONTINENTAL SHELF OFF NEW YORK	308
*COPEPODS	39-66 % OF DRY WEIGHT	DRIED AT 333K,NORTH PACIFIC	1184
*CTENOPHORA	6.4 % OF DRY WEIGHT	DRIED AT 378K,CONTINENTAL SHELF OFF NEW YORK	308
*DAPHNIA CRISTATA	55.2 (+-0.9) % OF ASH FREE DRY WEIGHT	CLADOCERA , 95 % CONF LIMIT, ORGANIC C	1375
*DAPHNIA CRISTATA	50.7 (+-0.8) % OF DRY WEIGHT	CLADOCERA , 95 % CONF LIMIT, TOTAL C	1375
*EUDIAPTOMUS GRACILIS	52.6 (+-0.4) % OF ASH FREE DRY WEIGHT	COPEPODA , 95 % CONF LIMIT, ORGANIC C	1375
*EUDIAPTOMUS GRACILIS	49.8 (+-0.4) % OF DRY WEIGHT	COPEPODA , 95 % CONF LIMIT, TOTAL C	1375
*EUPHAUSIACEA	33-37 % OF DRY WEIGHT	DRIED AT 378K,CONTINENTAL SHELF OFF NEW YORK	308
*EUPHAUSIACEA	39-47 % OF DRY WEIGHT	DRIED AT 333K,NORTH PACIFIC	1184
*EUPHAUSIACEA AND MYSIDAE	35-43 % OF DRY WEIGHT	DRIED AT 333K,SARGASSO SEA	88
*FISH AND FISH LARVAE	33-42 % OF DRY WEIGHT	DRIED AT 333K,SARGASSO SEA	88
*HYDROMEDUSAE	5-10 % OF DRY WEIGHT	DRIED AT 333K,SARGASSO SEA	88
*LIMNOCALANUS MACRURUS	64.0 (+-0.6) % OF ASH FREE DRY WEIGHT	COPEPODA , 95 % CONF LIMIT, ORGANIC C	1375
*LIMNOCALANUS MACRURUS	62.1 (+-0.6) % OF DRY WEIGHT	COPEPODA , 95 % CONF LIMIT, TOTAL C	1375
*MACROCYCLOPS ALBIDUS	50.7 (+-0.6) % OF ASH FREE DRY WEIGHT	COPEPODA , 95 % CONF LIMIT, ORGANIC C	1375
*MACROCYCLOPS ALBIDUS	48.2 (+-0.5) % OF DRY WEIGHT	COPEPODA , 95 % CONF LIMIT, TOTAL C	1375
*POLYCHAETA	16-44 % OF DRY WEIGHT	DRIED AT 333K,SARGASSO SEA	88
*PTEROPODS	21-25 % OF DRY WEIGHT	DRIED AT 333K,SARGASSO SEA	88
*PTEROPODS	26-28 % OF DRY WEIGHT	DRIED AT 378K,CONTINENTAL SHELF OFF NEW YORK	308
*PTEROPODS	17-29 % OF DRY WEIGHT	DRIED AT 333K,NORTH PACIFIC	1184
*SIPHONOPHORA	3-16 % OF DRY WEIGHT	DRIED AT 333K,SARGASSO SEA	88
*TUNICATA	7-11 % OF DRY WEIGHT	DRIED AT 378K,CONTINENTAL SHELF OFF NEW YORK	308
*ZOOPLANKTON	10**4 NO./L=1.3 MG-C-L	MOSSDALE	1153
*ZOOPLANKTON	0.4-0.5	NONE	1571

TABLE A426

ZOOPLANKTON : CA/BIOMASS RATIO			
ITEM:	VALUE:	CONDITIONS:	REF.:
*CALANUS FINMARCHICUS	0.04 %	ATOMIC COMPOSITION	1632

TABLE A427

ZOOPLANKTON : CALORIC CONTENT			
ITEM:	VALUE:	CONDITIONS:	REF.:
*DAPHNIA CRISTATA	25.3 (+-0.6) KJ/G ASH FREE DRY WEIGHT	CLADOCERA , 95 % CONF LIM, NO CORRECTION FOR NITROGEN	1375
*DAPHNIA CRISTATA	23.4 (+-0.6) KJ/G ASH FREE DRY WEIGHT	CLADOCERA , 95 % CONF LIM, CORRECTION FOR NITROGEN	1375

TABLE A427

ZOOPLANKTON : CALORIC CONTENT

ITEM:	VALUE:	CONDITIONS:	REF.:
*DAPHNIA CRISTATA	45.8 (+-0.4) KJ/G ORGANIC CARBON	CLADOCERA , 95 % CONF LIM, NO CORECTION FOR NITROGEN	1375
*DAPHNIA CRISTATA	42.4 (+-0.7) KJ/G ORGANIC CARBON	CLADOCERA , 95 % CONF LIM, CORRECTION FOR NITROGEN	1375
*EUDIAPTOMUS GRACILIS	24.7 (+-0.3) KJ/G ASH FREE DRY WEIGHT	COPEPODA , 95 % CONF LIM, NO CORRECTION FOR NITROGEN	1375
*EUDIAPTOMUS GRACILIS	22.2 (+-0.2) KJ/G ASH FREE DRY WEIGHT	COPEPODA , 95 % CONF LIM, CORRECTION FOR NITROGEN	1375
*EUDIAPTOMUS GRACILIS	47.3 (+-2.1) KJ/G ORGANIC CARBON	COPEPODA , 95 % CONF LIM, NO CORECTION FOR NITROGEN	1375
*EUDIAPTOMUS GRACILIS	42.2 (+-0.7) KJ/G ORGANIC CARBON	COPEPODA , 95 % CONF LIM, CORRECTION FOR NITROGEN	1375
*LIMNOCALANUS MACRURUS	31.2 (+-0.4) KJ/G ASH FREE DRY WEIGHT	COPEPODA , 95 % CONF LIM, NO CORRECTION FOR NITROGEN	1375
*LIMNOCALANUS MACRURUS	29.7 (+-0.3) KJ/G ASH FREE DRY WEIGHT	COPEPODA , 95 % CONF LIM, CORRECTION FOR NITROGEN	1375
*LIMNOCALANUS MACRURUS	48.8 (+-0.4) KJ/G ORGANIC CARBON	COPEPODA , 95 % CONF LIM, NO CORECTION FOR NITROGEN	1375
*LIMNOCALANUS MACRURUS	46.4 (+-0.5) KJ/G ORGANIC CARBON	COPEPODA , 95 % CONF LIM, CORRECTION FOR NITROGEN	1375
*MACROCYCLOPS ALBIDUS	22.8 (+-0.3) KJ/G ASH FREE DRY WEIGHT	COPEPODA , 95 % CONF LIM, NO CORRECTION FOR NITROGEN	1375
*MACROCYCLOPS ALBIDUS	20.3 (+-0.4) KJ/G ASH FREE DRY WEIGHT	COPEPODA , 95 % CONF LIM, CORRECTION FOR NITROGEN	1375
*MACROCYCLOPS ALBIDUS	45.0 (+-0.4) KJ/G ORGANIC CARBON	COPEPODA , 95 % CONF LIM, NO CORECTION FOR NITROGEN	1375
*MACROCYCLOPS ALBIDUS	39.9 (+-0.4) KJ/G ORGANIC CARBON	COPEPODA , 95 % CONF LIM, CORRECTION FOR NITROGEN	1375

TABLE A428

ZOOPLANKTON : CL/BIOMASS RATIO

ITEM:	VALUE:	CONDITIONS:	REF.:
*CALANUS FINMARCHICUS	1.05 %	ATOMIC COMPOSITION	1632

TABLE A429

ZOOPLANKTON : CONVERSION EFFICIENCY

ITEM:	VALUE:	CONDITIONS:	REF.:
*ZOOPLANKTON	0.6	MOSSDALE	1153
*ZOOPLANKTON	0.5-0.8	NONE	1571
*ZOOPLANKTON,CARNIVOROUS	0.6	LAKE ONTARIO	1577
*ZOOPLANKTON,HERBIVOROUS	0.6	LAKE ONTARIO	1577
*ZOOPLANKTON,UPPER TROPHIC LEVEL 1	0.6	LAKE ONTARIO	1577
*ZOOPLANKTON,UPPER TROPHIC LEVEL 2	0.6	LAKE ONTARIO	1577

TABLE A430

ZOOPLANKTON : CU/BIOMASS RATIO

ITEM:	VALUE:	CONDITIONS:	REF.:
*CALANUS FINMARCHICUS	0.0004 %	ATOMIC COMPOSITION	1632

TABLE A431

ZOOPLANKTON : DEVELOPMENT TIME

ITEM:	VALUE:	CONDITIONS:	REF.:
*ACARTIA CLAUSII	31.2 (+-0.36) HOURS	MEDIAN DEVELOPMENT TIME TO NAUPLIUS STAGE 1, 293 K, AUTUMN, 95 % CL.	870
*ACARTIA CLAUSII	47.0 HOURS	MEDIAN DEVELOPMENT TIME TO NAUPLIUS STAGE 2, 293 K, AUTUMN	870
*ACARTIA CLAUSII	82.6 HOURS	MEDIAN DEVELOPMENT TIME TO NAUPLIUS STAGE 3, 293 K, AUTUMN	870
*ACARTIA CLAUSII	107.4 HOURS	MEDIAN DEVELOPMENT TIME TO NAUPLIUS STAGE 4, 293 K, AUTUMN	870
*ACARTIA CLAUSII	130.5 HOURS	MEDIAN DEVELOPMENT TIME TO NAUPLIUS STAGE 5, 293 K, AUTUMN	870
*ACARTIA CLAUSII	155.3 HOURS	MEDIAN DEVELOPMENT TIME TO NAUPLIUS STAGE 6, 293 K, AUTUMN	870
*ACARTIA CLAUSII	180.4 HOURS	MEDIAN DEVELOPMENT TIME TO COPEPODIT STAGE 1, 293 K, AUTUMN	870
*ACARTIA CLAUSII	204.7 HOURS	MEDIAN DEVELOPMENT TIME TO COPEPODIT STAGE 2, 293 K, AUTUMN	870
*ACARTIA CLAUSII	230.7 HOURS	MEDIAN DEVELOPMENT TIME TO COPEPODIT STAGE 3, 293 K, AUTUMN	870
*ACARTIA CLAUSII	254.0 HOURS	MEDIAN DEVELOPMENT TIME TO COPEPODIT STAGE 4, MALE, 293 K, AUTUMN	870
*ACARTIA CLAUSII	278.7 HOURS	MEDIAN DEVELOPMENT TIME TO COPEPODIT STAGE 5, MALE, 293 K, AUTUMN	870
*ACARTIA CLAUSII	312.0 HOURS	MEDIAN DEVELOPMENT TIME TO MALE, 293 K, AUTUMN	870
*ACARTIA CLAUSII	260.0 HOURS	MEDIAN DEVELOPMENT TIME TO COPEPODIT STAGE 4, FEMALE, 293 K, AUTUMN	870
*ACARTIA CLAUSII	287.0 HOURS	MEDIAN DEVELOPMENT TIME TO COPEPODIT STAGE 5, FEMALE, 293 K, AUTUMN	870
*ACARTIA CLAUSII	319.0 HOURS	MEDIAN DEVELOPMENT TIME TO FEMALE, 293 K, AUTUMN	870
*ACARTIA CLAUSII	145.3 HOURS	MEDIAN DEVELOPMENT TIME TO NAUPLIUS STAGE 6, 293 K, SPRING	870
*ACARTIA CLAUSII	171.4 HOURS	MEDIAN DEVELOPMENT TIME TO COPEPODIT STAGE 1, 293 K, SPRING	870
*ACARTIA CLAUSII	196.0 HOURS	MEDIAN DEVELOPMENT TIME TO COPEPODIT STAGE 2, 293 K, SPRING	870
*ACARTIA CLAUSII	218.9 HOURS	MEDIAN DEVELOPMENT TIME TO COPEPODIT STAGE 3, 293 K, SPRING	870
*ACARTIA CLAUSII	242.1 HOURS	MEDIAN DEVELOPMENT TIME TO COPEPODIT STAGE 4, MALE, 293 K, SPRING	870
*ACARTIA CLAUSII	263.0 HOURS	MEDIAN DEVELOPMENT TIME TO COPEPODIT STAGE 5, MALE, 293 K, SPRING	870
*ACARTIA CLAUSII	290.0 HOURS	MEDIAN DEVELOPMENT TIME TO MALE, 293 K, SPRING	870
*ACARTIA CLAUSII	239.6 HOURS	MEDIAN DEVELOPMENT TIME TO COPEPODIT STAGE 4, FEMALE, 293 K, SPRING	870
*ACARTIA CLAUSII	269.2 HOURS	MEDIAN DEVELOPMENT TIME TO COPEPODIT STAGE 5, FEMALE, 293 K, SPRING	870
*ACARTIA CLAUSII	300.9 HOURS	MEDIAN DEVELOPMENT TIME TO FEMALE, 293 K, SPRING	870
*ACARTIA CLAUSII	217.6 HOURS	MEDIAN DEVELOPMENT TIME TO NAUPLIUS STAGE 6, 288 K, SUMMER	870
*ACARTIA CLAUSII	260.9 HOURS	MEDIAN DEVELOPMENT TIME TO COPEPODIT STAGE 1, 288 K, SUMMER	870
*ACARTIA CLAUSII	299.0 HOURS	MEDIAN DEVELOPMENT TIME TO COPEPODIT STAGE 2, 288 K, SUMMER	870
*ACARTIA CLAUSII	339.0 HOURS	MEDIAN DEVELOPMENT TIME TO COPEPODIT STAGE 3, 288 K, SUMMER	870
*ACARTIA CLAUSII	363.6 HOURS	MEDIAN DEVELOPMENT TIME TO COPEPODIT STAGE 4, MALE, 288 K, SUMMER	870

TABLE A431

ZOOPLANKTON : DEVELOPMENT TIME

ITEM:	VALUE:	CONDITIONS:	REF.:
*ACARTIA CLAUSII	402.6 HOURS	MEDIAN DEVELOPMENT TIME TO COPEPODIT STAGE 5, MALE, 288 K, SUMMER	870
*ACARTIA CLAUSII	485.1 HOURS	MEDIAN DEVELOPMENT TIME TO MALE, 288 K, SUMMER	870
*ACARTIA CLAUSII	376.8 HOURS	MEDIAN DEVELOPMENT TIME TO COPEPODIT STAGE 4, FEMALE, 288 K, SUMMER	870
*ACARTIA CLAUSII	443.1 HOURS	MEDIAN DEVELOPMENT TIME TO COPEPODIT STAGE 5, FEMALE, 288 K, SUMMER	870
*ACARTIA CLAUSII	184.6 HOURS	MEDIAN DEVELOPMENT TIME TO NAUPLIUS STAGE 3, 283 K, SPRING	870
*ACARTIA CLAUSII	239.0 HOURS	MEDIAN DEVELOPMENT TIME TO NAUPLIUS STAGE 4, 283 K, SPRING	870
*ACARTIA CLAUSII	295.8 HOURS	MEDIAN DEVELOPMENT TIME TO NAUPLIUS STAGE 5, 283 K, SPRING	870
*ACARTIA CLAUSII	348.9 HOURS	MEDIAN DEVELOPMENT TIME TO NAUPLIUS STAGE 6, 283 K, SPRING	870
*ACARTIA CLAUSII	409.4 HOURS	MEDIAN DEVELOPMENT TIME TO COPEPODIT STAGE 1, 283 K, SPRING	870
*ACARTIA CLAUSII	475.4 HOURS	MEDIAN DEVELOPMENT TIME TO COPEPODIT STAGE 2, 283 K, SPRING	870
*ACARTIA CLAUSII	72.0 (+-0.52) HOURS	MEDIAN DEVELOPMENT TIME TO NAUPLIUS STAGE 1, 283 K, SPRING, 95 % CL.	870
*ACARTIA CLAUSII	45.8 (+-0.68) HOURS	MEDIAN DEVELOPMENT TIME TO NAUPLIUS STAGE 1, 288 K, SUMMER. 95 % CL.	870
*ACARTIA CLAUSII	29.3 (+-0.52) HOURS	MEDIAN DEVELOPMENT TIME TO NAUPLIUS STAGE 1, 293 K, SPRING, 95 % CL.	870
*CERIODAPHNIA SP.	1.8 (SE=+-0.09) DAYS	GRAZING ON 0.4*10**5 CELL ALGAE/L, 296 K, LAB, EGG DEVELOP	595
*CERIODAPHNIA SP.	1.9 (SE=+-0.08) DAYS	AFTER FIELD LIFE TABLES, 296 K, HIG NUTRIENT LEVEL, EGG DEVELOP	595
*CERIODAPHNIA SP.	1.8 (SE=+-0.11) DAYS	AFTER FIELD LIFE TABLES, 296 K, MEDIAN NUTRIENT LEVEL, EGG DEVELOP	595
*CERIODAPHNIA SP.	1.9 (SE=+-0.12) DAYS	AFTER FIELD LIFE TABLES, 296 K, LOW NUTRIENT LEVEL, EGG DEVELOP	595
*CERIODAPHNIA SP.	3.0 (SE=+-0.11) DAYS	GRAZING ON 4*10**5 CELL ALGAE/L, 296 K,LAB MATURATION TIME	595
*CERIODAPHNIA SP.	5.0 (SE=+-0.50) DAYS	GRAZING ON 0.4*10**5 CELL ALGAE/L, 296 K, LAB MATURATION TIME	595
*CERIODAPHNIA SP.	3.8 (SE=+-0.14) DAYS	AFTER FIELD LIFE TABLES, 296 K, HIG NUTRIENT LEVEL MATURATION TIME	595
*CERIODAPHNIA SP.	3.2 (SE=+-0.17) DAYS	AFTER FIELD LIFE TABLES, 296 K, MEDIAN NUTRIENT LEVEL MATURATION TIME	595
*CERIODAPHNIA SP.	3.1 (SE=+-0.24) DAYS	AFTER FIELD LIFE TABLES, 296 K, LOW NUTRIENT LEVEL MATURATION TIME	595
*KERATELLA VALGA	4 DAYS	290 K MATURATION TIME	848
*KERATELLA VALGA	1.5 DAYS	296 K MATURATION TIME	595
*LIMNOCALANUS MACRURUS	20 DAYS	NAUPLIUS STAGE 1, CHAR LAKE, FIELD DATA, ARTIC	1352
*LIMNOCALANUS MACRURUS	22 DAYS	NAUPLIUS STAGE 2, CHAR LAKE, FIELD DATA, ARTIC	1352
*LIMNOCALANUS MACRURUS	14 DAYS	NAUPLIUS STAGE 3, CHAR LAKE, FIELD DATA, ARTIC	1352
*LIMNOCALANUS MACRURUS	15 DAYS	NAUPLIUS STAGE 4, CHAR LAKE, FIELD DATA, ARTIC	1352
*LIMNOCALANUS MACRURUS	11 DAYS	NAUPLIUS STAGE 5, CHAR LAKE, FIELD DATA, ARTIC	1352
*LIMNOCALANUS MACRURUS	15 DAYS	NAUPLIUS STAGE 6, CHAR LAKE, FIELD DATA, ARTIC	1352
*LIMNOCALANUS MACRURUS	11 - 13 DAYS	COPEPODIT STAGE 1, CHAR LAKE,	1352

TABLE A431

ZOOPLANKTON : DEVELOPMENT TIME

ITEM:	VALUE:	CONDITIONS:	REF.:
*LIMNOCALANUS MACRURUS	19 - 20 DAYS	COPEPODIT STAGE 2, CHAR LAKE, FIELD DATA, ARTIC	1352
*LIMNOCALANUS MACRURUS	21 - 26 DAYS	COPEPODIT STAGE 3, CHAR LAKE, FIELD DATA, ARTIC	1352
*LIMNOCALANUS MACRURUS	20 - 25 DAYS	COPEPODIT STAGE 4, CHAR LAKE, FIELD DATA, ARTIC	1352
*LIMNOCALANUS MACRURUS	45 - 50 DAYS	COPEPODIT STAGE 5, CHAR LAKE, FIELD DATA, ARTIC	1352
*LIMNOCALANUS MACRURUS	65 -300 DAYS	COPEPODIT STAGE 6, CHAR LAKE, FIELD DATA, ARTIC	1352
*NEODIAPTOMUS SCHMACKERI	2 (RANGE 1 - 3) DAYS	EGG TO NAUPLIUS,301 - 303 K, LAB	380
*NEODIAPTOMUS SCHMACKERI	4 (RANGE 2 - 7) DAYS	NAUPLIUS TO COPEPODITE, 301 - 303 K, LAB	380
*NEODIAPTOMUS SCHMACKERI	26 (RANGE 22 - 29) DAYS	COPEPODITE TO ADULT, FEMALE, 301 - 303 K, LAB	380
*NEODIAPTOMUS SCHMACKERI	22 (RANGE 13 - 29) DAYS	COPEPODITE TO ADULT, MALE, 301 - 303 K, LAB	380
*NEODIAPTOMUS SCHMACKERI	32 (RANGE 25 - 37) DAYS	ONE COMPLETE GENERATION(EGG TO EGG), 301 - 303 K, LAB	380
*NEODIAPTOMUS SCHMACKERI	3 DAYS	NAUPLIUS TO COPEPODITE, 301 - 303 K, FIELD OBSERVATION	380
*NEODIAPTOMUS SCHMACKERI	25 DAYS	COPEPODITE TO ADULT, 301 - 303 K, FIELD OBSERVATION	380
*SIMOCEPHALUS SERRULATUS	2.0 DAYS	FIELD EXPERIMENT, 296 K, EGG DEVELOP	595
*SIMOCEPHALUS SERRULATUS	6.5 DAYS	FIELD EXPERIMENT, 296 K MATURATION TIME	595

TABLE A432

ZOOPLANKTON : DRY WEIGHT

ITEM:	VALUE:	CONDITIONS:	REF.:
*ACARTIA TONSA	12.9 MICROGRAM DRY WEIGHT/ANIMAL	FEMALE	318
*ACROPERUS HARPAE	6.0 MICROGRAM	SPECIFIC GRAVITY = 1.025, ADULT	595
*ACROPERUS HARPAE	3.0 MICROGRAM	SPECIFIC GRAVITY = 1.025,JUVENILE	595
*ALONA COSTATA	3.0 MICROGRAM	SPECIFIC GRAVITY = 1.025, ADULT	595
*ALONA COSTATA	1.0 MICROGRAM	SPECIFIC GRAVITY = 1.025, JUVENILE	595
*ALONA GUTTATA	1.8 MICROGRAM	SPECIFIC GRAVITY = 1.025 , ADULT	595
*ALONA GUTTATA	1.0 MICROGRAM	SPECIFIC GRAVITY = 1.025, JUVENILE	595
*ASCOMORPHA SP.	0.04 MICROGRAM	SPECIFIC GRAVITY = 1.025	595
*ASPLANCHNA GIRODI	1.0 MICROGRAM	SPECIFIC GRAVITY = 1.025	595
*ASPLANCHNA PRIODONTA	2.0 MICROGRAM	SPECIFIC GRAVITY = 1.025	595
*BOSMINA LONGIROSTRIS	1.8 MICROGRAM	SPECIFIC GRAVITY = 1.025, ADULT	595
*BOSMINA LONGIROSTRIS	1.0 MICROGRAM	SPECIFIC GRAVITY = 1.025, JUVENILE	595
*BRACHIONUS PLICATILIS	0.06 MICROGRAM	SPECIFIC GRAVITY = 1.025	595
*BRACHYURAN LARVAE	1.17 MG DRY WEIGHT/ANIMAL	MEGALOPS	827
*CALANOID COPEPODIDS	3.0 MICROGRAM	SPECIFIC GRAVITY = 1.025	595
*CALANOIDES CARINATUS	0.13 MG DRY WEIGHT/ANIMAL	NONE	827
*CALANUS DISCAUDATUS	26.15 MICROGRAM	LENGTH=2.04MM,COPEPODID STAGE 4	20
*CALANUS FINMARCHICUS	113.8 MICROGRAM DRY WEIGHT/ANIMAL	COPEPODIT 5	318
*CALANUS PACIFICUS	0.60 MICROGRAM	LENGTH=0.27MM,NAUPLII STAGE 3	20
*CALANUS PACIFICUS	0.179 MG DRY WEIGHT/ANIMAL	FEMALE	827
*CALANUS PACIFICUS	0.044 MG DRY WEIGHT/ANIMAL	COPEPODITE 5	827
*CALANUS PACIFICUS	0.026 MG DRY WEIGHT/ANIMAL	COPEPODITE 4	827
*CALANUS PACIFICUS	0.010 MG DRY WEIGHT/ANIMAL	COPEPODITE 2	827

TABLE A432

ZOOPLANKTON : DRY WEIGHT

ITEM:	VALUE:	CONDITIONS:	REF.:
*CALANUS PACIFICUS	0.001 MG DRY WEIGHT/ANIMAL	NAUPLII 1	827
*CENTROPAGES TYPICUS	37.3 MICROGRAM DRY WEIGHT/ANIMAL	FEMALE	318
*CERIODAPHNIA RETICULATA	4.2 MICROGRAM	SPECIFIC GRAVITY = 1.025, ADULT	595
*CERIODAPHNIA RETICULATA	2.2 MICROGRAM	SPECIFIC GRAVITY = 1.025, JUVENILE	595
*CHIRONOMUS TENTANS	0.005 MG	DRIED AT 343 K, INSTAR 1	595
*CHYDORUS SPHAERICUS	2.0 MICROGRAM	SPECIFIC GRAVITY = 1.025, ADULT	595
*CHYDORUS SPHAERICUS	1.0 MICROGRAM	SPECIFIC GRAVITY = 1.025, JUVENILE	595
*COMPTOCERUS RECTIROSTRIS	5.0 MICROGRAM	SPECIFIC GRAVITY = 1.025, ADULT	595
*COMPTOCERUS RECTIROSTRIS	2.0 MICROGRAM	SPECIFIC GRAVITY = 1.025, JUVENILE	595
*CYCLOPOID COPEPODIDS	3.0 MICROGRAM	SPECIFIC GRAVITY = 1.025	595
*CYCLOPOID NAUPLII	0.01 MICROGRAM	SPECIFIC GRAVITY = 1.025	595
*CYCLOPS VERICANS RUBELLUS	2.0 MICROGRAM	SPECIFIC GRAVITY = 1.025	595
*CYCLOPS VERNALIS	8.6 MICROGRAM	SPECIFIC GRAVITY = 1.025	595
*DAPHNIA PULEX	35.0 MICROGRAM	SPECIFIC GRAVITY = 1.025, ADULT	595
*DAPHNIA PULEX	8.0 MICROGRAM	SPECIFIC GRAVITY = 1.025, JUVENILE	595
*DIAPHANOSOMA BRACHYURUM	7.0 MICROGRAM	SPECIFIC GRAVITY = 1.025, ADULT	595
*DIAPHANOSOMA BRACHYURUM	3.0 MICROGRAM	SPECIFIC GRAVITY = 1.025, JUVENILE	595
*DIAPTOMUS LEPTOPUS	20 MICROGRAM	SPECIFIC GRAVITY = 1.025	595
*DIAPTOMUS PALLIDUS	8.1 MICROGRAM	SPECIFIC GRAVITY = 1.025	595
*EPILABIDOCERA AMPHITRITES	0.50 MG DRY WEIGHT/ANIMAL	NONE	827
*EUCYCLOPS AGILIS	8.0 MICROGRAM	SPECIFIC GRAVITY = 1.025	595
*EUPHAUSIA PACIFICA	6.36 MG DRY WEIGHT/ANIMAL	NONE	827
*EURYCERCUS LAMELLATUS	80 MICROGRAM	SPECIFIC GRAVITY = 1.025, ADULT	595
*EURYCERCUS LAMELLATUS	10 MICROGRAM	SPECIFIC GRAVITY = 1.025, JUVENILE	595
*GASTROPUS SP.	0.05 MICROGRAM	SPECIFIC GRAVITY = 1.025	595
*KERATELLA COCHLEARIS	0.05 MICROGRAM	SPECIFIC GRAVITY = 1.025	595
*KERATELLA HIEMALIS	0.05 MICROGRAM	SPECIFIC GRAVITY = 1.025	595
*LEYDIGIA ACANTHOCEROIDES	6.0 MICROGRAM	SPECIFIC GRAVITY = 1.025, ADULT	595
*LEYDIGIA ACANTHOCEROIDES	3.0 MICROGRAM	SPECIFIC GRAVITY = 1.025, JUVENILE	595
*LEYDIGIA QUADRANGULARIS	6.0 MICROGRAM	SPECIFIC GRAVITY = 1.025, ADULT	595
*LEYDIGIA QUADRANGULARIS	3.0 MICROGRAM	SPECIFIC GRAVITY = 1.025, JUVENILE	595
*MACROCYCLOPS ALBIDUS	40.0 MICROGRAM	SPECIFIC GRAVITY = 1.025	595
*MACROTHRIX LATICORNIS	2.0 MICROGRAM	SPECIFIC GRAVITY = 1.025, ADULT	595
*MACROTHRIX LATICORNIS	1.0 MICROGRAM	SPECIFIC GRAVITY = 1.025, JUVENILE	595
*MACROTHRIX ROSEA	2.0 MICROGRAM	SPECIFIC GRAVITY = 1.025, ADULT	595
*MACROTHRIX ROSEA	1.0 MICROGRAM	SPECIFIC GRAVITY = 1.025, JUVENILE	595
*MESOCYCLOPS EDAX	18 MICROGRAM	SPECIFIC GRAVITY = 1.025	595
*MESOCYCLOPS LEUCKARTI	15 MICROGRAM	SPECIFIC GRAVITY = 1.025	595
*NEMATOSCELIS ATLANTICA	0.91 MG DRY WEIGHT/ANIMAL	NONE	827
*PARACYCLOPS FIMBRIATUS POPPEI	3.0 MICROGRAM	SPECIFIC GRAVITY = 1.025	595
*PARATHEMISTO PACIFICA	1.65 MG DRY WEIGHT/ANIMAL	NONE	827
*PLEUROBRACHIA BACHEII	11.4 MG DRY WEIGHT/ANIMAL	NONE	827
*PLEURONCODES PLANIPES	432 MG DRY WEIGHT/ANIMAL	NONE	827
*PLEUROXUS DENTICULATUS	2.0 MICROGRAM	SPECIFIC GRAVITY = 1.025, ADULT	595
*PLEUROXUS DENTICULATUS	1.0 MICROGRAM	SPECIFIC GRAVITY = 1.025, JUVENILE	595
*PLEUROXUS PROCURVIS	4.0 MICROGRAM	SPECIFIC GRAVITY = 1.025, ADULT	595
*PLEUROXUS PROCURVIS	2.0 MICROGRAM	SPECIFIC GRAVITY = 1.025, JUVENILE	595
*POLYARTHRA SP.	0.05 MICROGRAM	SPECIFIC GRAVITY = 1.025	595
*PSEUDOCALANUS MINUTUS	21.5 MICROGRAM DRY WEIGHT/ANIMAL	FEMALE	318
*SAGITTA ELEGANS	1.43 MG DRY WEIGHT/ANIMAL	NONE	827

TABLE A432

ZOOPLANKTON : DRY WEIGHT

ITEM:	VALUE:	CONDITIONS:	REF.:
*SIMOCEPHALUS SERRULATUS	66.2 MICROGRAM	SPECIFIC GRAVITY = 1.025, LARGE ADULT	595
*SIMOCEPHALUS SERRULATUS	33.5 MICROGRAM	SPECIFIC GRAVITY = 1.025, SMALL ADULT	595
*SIMOCEPHALUS SERRULATUS	8.5 MICROGRAM	SPECIFIC GRAVITY = 1.025, JUVENILE	595
*SYNCHAETA PECTINATA	0.2 MICROGRAM	SPECIFIC GRAVITY = 1.025	595
*TOMOPTERIS SEPTENTRIONALIS	5.61 MG DRY WEIGHT/ANIMAL	NONE	827
*TORTANUS DISCAUDATUS	36.9 MICROGRAM	TOTAL LENGTH=1.97MM,FEMALE	20
*TROPOCYCLOPS PRASINUS	3.0 MICROGRAM	SPECIFIC GRAVITY OF VOLUMES = 1.025	595

TABLE A433

ZOOPLANKTON : DRY/WET WEIGHT RATIO

ITEM:	VALUE:	CONDITIONS:	REF.:
*AMPHIPODA	18.4-36.6 % OF WET WEIGHT	DRIED AT 333K,NORTH PACIFIC	1184
*CHAETOGNATHA	6.0-7.4 % OF WET WEIGHT	DRIED AT 333K,SARGASSO SEA	88
*CHAETOGNATHA	11.6-14.1 % OF WET WEIGHT	DRIED AT 333K,NORTH PACIFIC	1184
*COPEPODS	11.6-16.3 % OF WET WEIGHT	DRIED AT 333K,SARGASSO SEA	88
*COPEPODS	10.2-15.8 % OF WET WEIGHT	DRIED AT 378K,CONTINENTAL SHELF OFF NEW YORK	308
*COPEPODS	9.2-33.9 % OF WET WEIGHT	DRIED AT 333K,NORTH PACIFIC	1184
*CTENOPHORA	4.7-5.0 % OF WET WEIGHT	DRIED AT 378K,CONTINENTAL SHELF OFF NEW YORK	308
*EUPHAUSIACEA	19.0-20.0 % OF WET WEIGHT	DRIED AT 378K,CONTINENTAL SHELF OFF NEW YORK	308
*EUPHAUSIACEA	20.2-21.3 % OF WET WEIGHT	DRIED AT 333K,NORTH PACIFIC	1184
*EUPHAUSIACEA AND MYSIDAE	14.5-18.0 % OF WET WEIGHT	DRIED AT 333K,SARGASSO SEA	88
*FISH AND FISH LARVAE	11.9-16.0 % OF WET WEIGHT	DRIED AT 333K,SARGASSO SEA	88
*HYDROMEDUSAE	0.3-10.1 % OF WET WEIGHT	DRIED AT 333K,SARGASSO SEA	88
*POLYCHAETA	5.7-27.0 % OF WET WEIGHT	DRIED AT 333K,SARGASSO SEA	88
*PTEROPODS	22-32 % OF WET WEIGHT	DRIED AT 333K,SARGASSO SEA	88
*PTEROPODS	3.5-19 % OF WET WEIGHT	DRIED AT 378K,CONTINENTAL SHELF OFF NEW YORK	308
*PTEROPODS	25.0-36.4 % OF WET WEIGHT	DRIED AT 333K,NORTH PACIFIC	1184
*SIPHONOPHORA	0.3-6.1 % OF WET WEIGHT	DRIED AT 333K,SARGASSO SEA	88
*TUNICATA	4.0-4.1 % OF WET WEIGHT	DRIED AT 378K,CONTINENTAL SHELF OFF NEW YORK	308

TABLE A434

ZOOPLANKTON : EGESTION

ITEM:	VALUE:	CONDITIONS:	REF.:
*DAPHNIA PULEX	69 - 86 % OF INGESTION	NONE	1689
*LEPTODORA KINDTII	60.0 % OF INGESTION	NONE	1689
*SIMOCEPHALUS VECTULUS	27.6 % OF INGESTION	JUVENILE, FEMALE	1689
*SIMOCEPHALUS VECTULUS	68.3 % OF INGESTION	REPRODUCING FEMALE	1689

TABLE A435

ZOOPLANKTON : EGG DEVELOPMENT TIME

ITEM:	VALUE:	CONDITIONS:	REF.:
*CERIODAPHNIA SP.	1.8 (SE=+-0.04) DAYS	GRAZING ON 4*10**5 CELL ALGAE/L, 296 K,LAB, EGG DEVELOP	595

TABLE A436

ZOOPLANKTON : EXCRETION			
ITEM:	VALUE:	CONDITIONS:	REF.:
*PHRONIMA SEDENTARIA	5.5 +- 1.3 MICROGRAM NH4-N/(MG DRY WEIGHT*DAY)	12 H STARVATION BEFORE EXP., 286 K, RESP=17.3 UL O2/MG DW, 95 % CL	997
*PHRONIMA SEDENTARIA	5.3 MICROGRAN ORG-N/(MG DRY WEIGHT*DAY)	12 H STARVATION BEFORE EXP., 286 K, RESP=17.3 UL O2/MG DW	997
*PHRONIMA SEDENTARIA	10.8+- 0.3 MICROGRAM TOT-N/(MG DRY WEIGHT*DAY)	12 H STARVATION BEFORE EXP., 286 K, RESP=17.3 UL O2/MG DW	997
*SAGITTA SETOSA	30.7 +- 4.0 MICROGRAM NH4-N/(MG DRY WEIGHT*DAY)	16 H STAR. BEFORE EXP, 283-298 K, 6 H, 95 % CL	997
*SAGITTA SETOSA	20.1 MICROGRAM ORG-N/(MG DRY WEIGHT*DAY)	16 H STAR. BEFORE EXP, 283-298 K, 6 H	997
*SAGITTA SETOSA	70.8 +- 5.6 MICROGRAM NH4-N/(MG DRY WEIGHT*DAY)	16 H STAR. BEFORE EXP, 283-298 K, 6 H, 95 % CL	997
*SALPA FUSIFORMIS	2.0 +- 0.7 MICROGRAM NH4-N/(MG DRY WEIGHT*DAY)	3 H STARVATION BEFORE EXP., 288 K, RESP=7.7 UL O2/MG DW, 6 H, 95 % CL	997
*SALPA FUSIFORMIS	1.4 MICROGRAM ORG-N/(MG DRY WEIGHT*DAY)	3 H STARVATION BEFORE EXP., 288 K, RESP=7.7 UL O2/MG DW, 6 H	997
*SALPA FUSIFORMIS	3.4 +- 1.9 MICROGRAM TOT-N/(MG DRY WEIGHT*DAY)	3 H STARVATION BEFORE EXP., 288 K, RESP=7.7 UL O2/MG DW, 6 H, 95 % CL	997
*THALIA DEMOCRATICA	9.5 +- 0.8 MICROGRAM NH4-N/(MG DRY WEIGHT*DAY)	6 H STARVATION BEFORE EXP., 288 K, RESP=6.7 UL O2/MG DW, 6 H, 95 % CL	997
*THALIA DEMOCRATICA	4.1 MICROGRAM ORG-N/(MG DRY WEIGHT*DAY)	6 H STARVATION BEFORE EXP., 288 K, RESP=6.7 UL O2/MG DW, 6 H	997
*THALIA DEMOCRATICA	13.6+- 0.8 MICROGRAM TOT-N/(MG DRY WEIGHT*DAY)	6 H STARVATION BEFORE EXP., 288 K, RESP=6.7 UL O2/MG DW, 6 H, 95 % CL	997
*ZOOPLANKTON	2.6 % OF BODY N/DAY	WINTER	550
*ZOOPLANKTON	4.6 % OF BODY N/DAY	SUMMER	550
*ZOOPLANKTON	6.5 % OF BODY P/DAY	WINTER	550
*ZOOPLANKTON	10.3 % OF BODY P¤DAY	SUMMER	550

TABLE A437

ZOOPLANKTON : EXCRETION OF PHOSPHORUS			
ITEM:	VALUE:	CONDITIONS:	REF.:
*FRESHWATER ZOOPLANKTON	0.104 1/HOUR	STEADY STATE SOLUTION TO CASTLE LAKE.	1326
*FRESHWATER ZOOPLANKTON	0.0006 MICROGRAM P/(MG*HOUR)	STONEHOUSE POND, FEB 22, 1973	454
*FRESHWATER ZOOPLANKTON	0.0001 MICROGRAM P/(MG*HOUR)	STONEHOUSE POND, APRIL 6, 1973	454
*FRESHWATER ZOOPLANKTON	0.0172 MICROGRAM P/(MG*HOUR)	STONEHOUSE POND, MAY 17, 1973	454
*FRESHWATER ZOOPLANKTON	0.0203 MICROGRAM P/(MG*HOUR)	STONEHOUSE POND, JUNE 17, 1973	454

TABLE A438

ZOOPLANKTON : FE/BIOMASS RATIO			
ITEM:	VALUE:	CONDITIONS:	REF.:
*CALANUS FINMARCHICUS	0.0070 %	ATOMIC COMPOSITION	1632

TABLE A439

ZOOPLANKTON : FILTRATION RATE			
ITEM:	VALUE:	CONDITIONS:	REF.:
*ACARTIA CLAUSII	0.05 ML/(AMINAL*HOUR)	2 H, N=3, COULTER COUNTER TECHNIQUE	615

TABLE A439

ZOOPLANKTON : FILTRATION RATE

ITEM:	VALUE:	CONDITIONS:	REF.:
*ACARTIA CLAUSII	0.24 ML/(AMINAL*HOUR)	2 H, N=3, COULTER COUNTER TECHNIQUE	615
*ACARTIA CLAUSII	0.09 ML/(AMINAL*HOUR)	4 H, N=3, COULTER COUNTER TECHNIQUE	615
*ACARTIA CLAUSII	0.18 ML/(AMINAL*HOUR)	4 H, N=3, COULTER COUNTER TECHNIQUE	615
*ACARTIA CLAUSII	0.12 ML/(AMINAL*HOUR)	6 H, N=3, COULTER COUNTER TECHNIQUE	615
*ACARTIA CLAUSII	0.11 ML/(AMINAL*HOUR)	6 H, N=3, COULTER COUNTER TECHNIQUE	615
*ACARTIA CLAUSII	0.15 ML/(AMINAL*HOUR)	2 H, N=3, CHLOROPHYLL A TECHNIQUE	615
*ACARTIA CLAUSII	0.17 ML/(AMINAL*HOUR)	2 H, N=3, CHLOROPHYLL A TECHNIQUE	615
*ACARTIA CLAUSII	0.16 ML/(AMINAL*HOUR)	4 H, N=3, CHLOROPHYLL A TECHNIQUE	615
*ACARTIA CLAUSII	0.14 ML/(AMINAL*HOUR)	4 H, N=3, CHLOROPHYLL A TECHNIQUE	615
*ACARTIA CLAUSII	0.16 ML/(AMINAL*HOUR)	6 H, N=3, CHLOROPHYLL A TECHNIQUE	615
*ACARTIA CLAUSII	0.12 ML/(AMINAL*HOUR)	6 H, N=3, CHLOROPHYLL A TECHNIQUE	615
*ACARTIA CLAUSII	0.07 ML/(AMINAL*HOUR)	2 H, N=3, CARBON-14 TECHNIQUE	615
*ACARTIA CLAUSII	0.10 ML/(AMINAL*HOUR)	4 H, N=3, CARBON-14 TECHNIQUE	615
*ACARTIA CLAUSII	0.06 ML/(AMINAL*HOUR)	4 H, N=3, CARBON-14 TECHNIQUE	615
*ACARTIA CLAUSII	0.14 ML/(AMINAL*HOUR)	4 H, N=3, CARBON-14 TECHNIQUE	615
*ACARTIA CLAUSII	0.14 ML/(AMINAL*HOUR)	6 H, N=3, CARBON-14 TECHNIQUE	615
*ACARTIA CLAUSII	0.13 ML/(AMINAL*HOUR)	6 H, N=3, CARBON-14 TECHNIQUE	615
*ACARTIA TONSA	3.65 ML/COPEPOD*DAY (SD=4.23)	FOOD CONC.=474MICROGRAM C/L AS DETRITUS(?054),293K,SEAWATER,N=3	1354
*ACARTIA TONSA	5.28 ML/COPEPOD*DAY (SD=3.38)	FOOD CONC.=737MICROGRAM C/L AS DETRITUS(?054),293K,SEAWATER,N=3	1354
*ACARTIA TONSA	2.64 ML/COPEPOD*DAY (SD=0.89)	FOOD CONC.=1193MICROGRAM C/L AS DETRITUS(?054),293K,SEAWATER,N=3	1354
*ACARTIA TONSA	1.63 ML/COPEPOD*DAY (SD=1.75)	FOOD CONC.=1203MICROGRAM C/L AS DETRITUS(?054),293K,SEAWATER,N=3	1354
*BOSMINA LONGIROSTRIS	0.46 ML/(ANIM*DAY)	SEPT 1968, DROWNED BOG L., ONT., % OF TOTAL FILT.=85	607
*BOSMINA LONGIROSTRIS	0.45 ML/(ANIM*DAY)	SEPT 1969, DROWNED BOG L., ONT., % OF TOTAL FILT.=44.8	607
*BOSMINA LONGIROSTRIS	0.44 ML/(ANIMAL*DAY)	ANIMAL LENGTH = 0.4 - 0.6 MM	355
*BRACHIONUS CALYCIFLORUS	0.6-1.5 LITER/MG DRY WT.*DAY	NONE	707
*CALANUS FINMARCHICUS	0.05 LITER/MG DRY WT.*DAY	NONE	4
*CALANUS HELGOLANDICUS	200 - 500 ML/DAY	100 UG C ANIMAL, CULTURE	355
*CALANUS SP.	0.67-2.0 LITER/MG DRY WT.*DAY	NONE	1337
*CENTROPAGES HAMATUS	0.67-1.6 LITER/MG DRY WT.*DAY	NONE	36
*CERIODAPHNIA QUADRANGULA	4.6 ML/(ANIMAL*DAY)	ANIMAL LENGTH = 0.7 - 0.9 MM	355
*CHYDORUS SPHAERICUS	0.18 ML/(ANIMAL*DAY)	ANIMAL LENGTH = 0.1 - 0.2 MM	355
*DAPHNIA GALEATA	6.4 ML/(ANIMAL*DAY)	ANIMAL LENGTH = 1,5 - 1.7 MM	355
*DAPHNIA GALEATA	3.7 ML/(ANIMAL*DAY)	ANIMAL LENGTH = 1.5 - 1.7 MM	204
*DAPHNIA LONGISPINA	2.3 ML/(ANIMAL*DAY)	ANIMAL LENGTH = 0.7 - 1.2 MM	1104
*DAPHNIA MAGNA	0.74 LITER/MG DRY WT.*DAY	NONE	1361
*DAPHNIA MAGNA	0.2-0.3 LITER/MG DRY WT.*DAY	NONE	1007
*DAPHNIA PARVULA	1.6 ML/(ANIM*DAY)	SEPT 1969, DROWNED BOG L., ONT., % OF TOTAL FILT.=7.5	607
*DAPHNIA PARVULA	3.8 ML/(ANIMAL*DAY)	ANIMAL LENGTH = 0.7 - 1.2 MM	355
*DAPHNIA PULEX	2.20 10**6 CELLS/INDIV.*HR	FOOD=ANKISTRODESMUS FALCATUS AT K=25	42
*DAPHNIA PULEX	1.11 10**6 CELLS/INDIV.*HR	FOOD=CHLORELLA VULGARIS AT K=25	42
*DAPHNIA PULEX	1.65 10**6 CELLS/INDIV.*HR	FOOD=ANACYSTIS NIDULANS AT K=25	42
*DAPHNIA PULEX	0.33 10**6 CELLS/INDIV.*HR	FOOD=SYNECHOCOCCUS ELONGATA AT K=25	42
*DAPHNIA PULEX	0.51 10**6 CELLS/INDIV.*HR	FOOD=MERISMOPEDIA SP. AT K=25	42

TABLE A439

ZOOPLANKTON : FILTRATION RATE			
ITEM:	VALUE:	CONDITIONS:	REF.:
*DAPHNIA PULEX	1.12 10**6 CELLS/INDIV.*HR	FOOD=SYNECHOCOCCUS CEDRORUM AT K=25	42
*DAPHNIA PULEX	0.34 10**6 CELLS/INDIV.*HR	FOOD=ANABAENA FLOS-AQUA AT K=25	42
*DAPHNIA PULEX	1.01 10**6 CELLS/INDIV.*HR	FOOD=SYNECHOCYSTIS SP. AT K=25	42
*DAPHNIA PULEX	1.33 10**6 CELLS/INDIV.*HR	FOOD=GLOEOCAPSA ALPICOLA AT K=25	42
*DAPHNIA ROSEA	5.5 ML/(ANIMAL*DAY)	ANIMAL LENGTH = 1.3 - 1.6 MM	355
*DAPHNIA ROSEA	3.6 ML/(ANIMAL*DAY)	ANIMAL LENGTH = 1.3 - 1.6 MM	204
*DAPHNIA SCHODLERI	1.20 ML/ANIMAL*HR	FOOD=ANKISTRODESMUS SP.,MAX.VALUE, 293K,N=10	647
*DAPHNIA SCHODLERI	1.00 ML/ANIMAL*HR	FOOD=CHLAMYDOMONAS SP.,MAX.VALUE, 293K,N=4	647
*DAPHNIA SCHODLERI	1.10 ML/ANIMAL*HR	FOOD=FRUSTULIA SP.,MAX.VALUE,293K, N=2	647
*DAPHNIA SCHODLERI	1.20 ML/ANIMAL*HR	FOOD=FRUSTULIA AND ANKISTRODESMUS, MAX.VALUE,293K,N=4	647
*DAPHNIA SP.	1.06 LITER/MG DRY WT.*DAY	NONE	1752
*DAPHNIA SP.	0.2-1.6 LITER/MG DRY WT.*DAY	NONE	202
*DIAPHANOSOMA BRACHYURUM	1.2 ML/(ANIM*DAY)	SEPT 1969, DROWNED BOG L., ONT., % OF TOTAL FILT.=0.4	607
*DIAPHANOSOMA BRACHYURUM	1.6 ML/(ANIMAL*DAY)	ANI¤AL LENGTH = 0.9 - 1.4 MM	355
*DIAPTOMUS GRACILIS	1.92 - 1.96 ML/(ANIM*DAY)	FOOD=24.2-52.0CELLS*10**(-3)/ML, MELOSIRA AND ASTERIONELLA SP.	607
*DIAPTOMUS GRACILIS	0.68 ML/(ANIM*DAY)	FOOD=198.0CELLS*10**(-3)/ML, MELOSIRA AND ASTERIONELLA SP.	607
*DIAPTOMUS GRACILIS	0.61 ML/(ANIM*DAY)	FOOD=<30.0CELLS*10**(-3)/ML, CHLORELLA, 278 K	818
*DIAPTOMUS GRACILIS	1.51 ML/(ANIM*DAY)	FOOD=<30CELLS*10**(-3)/ML, CHLORELLA, 285 K	818
*DIAPTOMUS GRACILIS	2.40 ML/(ANIM*DAY)	FOOD=<30CELLS*10**(-3)/ML, CHLORELLA, 293 K	818
*DIAPTOMUS GRACILIS	0.94 ML/(ANIM*DAY)	FOOD=<30CELLS*10**(-3)/ML, SCENEDESMUS, 285 K	818
*DIAPTOMUS GRACILIS	1.32 ML/(ANIM*DAY)	FOOD=<30CELLS*10**(-3)/ML, SCENEDESMUS, 293 K	818
*DIAPTOMUS GRACILIS	1.76 ML/(ANIM*DAY)	FOOD=<30CELLS*10**(-3)/ML, DIPLOPHAERIA, 285 K	818
*DIAPTOMUS GRACILIS	2.54 ML/(ANIM*DAY)	FOOD=<30CELLS*10**(-3)/ML, DIPLOSPHAERIA, 293 K	818
*DIAPTOMUS GRACILIS	1.61 ML/(ANIM*DAY)	FOOD=<30CELLS*10*(-3)/ML, ANKISTRODESMUS, 285 K	818
*DIAPTOMUS GRACILIS	2.45 ML/(ANIM*DAY)	FOOD=<30CELLS*10**(-3)/ML, ANKISTRODESMUS, 293 K	818
*DIAPTOMUS GRACILIS	0.87 ML/(ANIM*DAY)	FOOD=<30CELLS*10**(-3)/ML, CARTERIA, 293 K	818
*DIAPTOMUS GRACILIS	1.96 ML/(ANIM*DAY)	FOOD=<30CELLS*10**(-3)/ML, NITZSCHIA, 293 K	818
*DIAPTOMUS GRACILIS	0.02 ML/(ANIM*DAY)	FOOD=<30CELLS*10**(-3)/ML, PEDIASTRUM, 293 K	818
*DIAPTOMUS GRACILIS	2.16 ML/(ANIM*DAY)	FOOD=<30CELLS*10**(-3)/ML, HAEMATOCOCCUS, 293 K	818
*DIAPTOMUS GRACILIS	0.19 ML/(ANIM*DAY)	FOOD=<30CELLS*10**(-3)/ML, BACTERIA, 293 K	818
*DIAPTOMUS GRACILOIDES	0.3 - 2.8 ML/(ANIM*DAY)	FOOD=NATURAL PHYTOPLANKTON	1104
*DIAPTOMUS GRACILOIDES	4.1 ML/(ANIM*DAY)	FOOD=13.6CELLS*10**(-3)/ML, SCENEDESMUS SP.	974
*DIAPTOMUS OREGONENSIS	2.5 - 1.4 ML/(ANIM*DAY)	FOOD=25.0-52.0CELLS*10**(-3)/ML, CHLAMYDOMONAS SP.	1329
*DIAPTOMUS OREGONENSIS	1.4 - 0.3 ML/(ANIM*DAY)	FOOD=52.0-198.0CELLS*10**(-3)/ML, CHLORELLA	1329
*DIAPTOMUS OREGONENSIS	1.4 - 0.00 ML/(ANIM*DAY)	FOOD=IN SITI PHYTOPLANKTON	607
*DIAPTOMUS OREGONENSIS	2.1 ML/(ANIM*DAY)	SEPT 1969, DROWNED BOG L., ONT., % OF TOTAL FILT.=0.9	607
*DIAPTOMUS SICILOIDES	2.0 ML/(ANIM*DAY)	FOOD=PANDORINA AND CHLAMYDOMONAS SP.	270

TABLE A439

ZOOPLANKTON : FILTRATION RATE

ITEM:	VALUE:	CONDITIONS:	REF.:
*DIAPTOMUS SICILOIDES	2.5 ML/(ANIM*DAY)	FOOD=PANDORINA AND CHLAMYDOMONAS SP.	1329
*HOLOPEDIUM GIBBERUM	9.4 ML/(ANIM*DAY)	SEPT 1969, DROWNED BOG L., ONT., % OF TOTAL FILT.=46.2	607
*LIMNOCALANUS MACRURUS	2.45 ML/(ANIM*DAY)	FOOD=SCENEDESMUS, > 278 K	819
*LIMNOCALANUS MACRURUS	1.24 ML/(ANIM*DAY)	FOOD=CHLAMYDOMONAS, > 278 K	819
*LIMNOCALANUS MACRURUS	0.1 ML/(ANIM*DAY)	FOOD=RHODOTURULA (YEAST), > 278 K	819
*NATURAL ASSOCIATION	0.8-1.1 LITER/MG DRY WT.*DAY	NONE	1337
*RHINCALANUS NASUTUS	0.3-2.2 LITER/MG DRY WT.*DAY	NONE	1079
*ZOOPLANKTON	0.13 L/MG*DAY*C	MOSSDALE	1153
*ZOOPLANKTON	0.13 L/(MG CARBON)*DAY	SAN JOAQUIN(RIVER),293K	1149
*ZOOPLANKTON	0.18 L/(MG CARBON)*DAY	DELTA(ESTUARY),293K	1149
*ZOOPLANKTON	0.25 L/(MG CARBON)*DAY	LAKE ERIE,293K	1149
*ZOOPLANKTON	1.2 L/(MG CARBON)*DAY	LAKE ONTARIO,293K	1149
*ZOOPLANKTON,CARNIVOROUS	0.06 1/MG*DAY-CELCIUS	LAKE ONTARIO	1577
*ZOOPLANKTON,HERBIVOROUS	0.06 1/MG*DAY	LAKE ONTARIO	1577
*ZOOPLANKTON,UPPER TROPHIC LEVEL 1	0.01 1/MG*DAY-CELCIUS	LAKE ONTARIO	1577
*ZOOPLANKTON,UPPER TROPHIC LEVEL 2	0.06 1/MG*DAY-CELCIUS	LAKE ONTARIO	1577

TABLE A440

ZOOPLANKTON : GRAZING RATE

ITEM:	VALUE:	CONDITIONS:	REF.:
*ACARTIA TONSA	0.06 MICROGRAM C/(MICROGRAM COP.C) *DAY (SD=0.07)	GRAZING RATE=3.65ML/COPEPOD*DAY(SD =4.23),293K,SEAWATER,N=3	1354
*ACARTIA TONSA	0.33 MICROGRAM C/(MICROGRAM COP.C) *DAY (SD=0.29)	GRAZING RATE=5.28ML/COPEPOD*DAY(SD =3.38),293K,SEAWATER,N=3	1354
*ACARTIA TONSA	0.66 MICROGRAM C/(MICROGRAM COP.C) *DAY (SD=0.22)	GRAZING RATE=2.64ML/COPEPOD*DAY(SD =0.89),293K,SEAWATER,N=3	1354
*ACARTIA TONSA	0.81 MICROGRAM C/(MICROGRAM COP.C) *DAY (SD=0.87)	GRAZING RATE=1.63ML/COPEPOD*DAY(SD =1.75),293K,SEAWATER,N=3	1354
*CALANUS HELGOLANDICUS	4.61 (RANGE=4.34-4.88) MICROGRAM C/ANIMAL	FOOD=LAUDERIA BOREALIS = 49 UG C/ L,N=2, STAGE NII TO CI	1204
*CALANUS HELGOLANDICUS	11.78 (RANGE=10.85-12.71) MICROGRAM C/ANIMAL	FOOD=LAUDERIA BOREALIS = 49 UG C/ L,N=2, STAGE CI TO CIII	1204
*CALANUS HELGOLANDICUS	47.28 (RANGE=47.35-49.21) MICROGRAM C/ANIMAL	FOOD=LAUDERIA BOREALIS = 49 UG C/ L,N=2, STAGE CIII TO CV	1204
*CALANUS HELGOLANDICUS	69.23 (RANGE=68.03-70.45) MICROGRAM C/ANIMAL	FOOD=LAUDERIA BOREALIS = 49 UG C/ L,N=2, STAGE CV TO ADULT	1204
*CALANUS HELGOLANDICUS	133.90 (RANGE=132.98-134.82) MICROGRAM C/ANIMAL	FOOD=LAUDERIA BOREALIS = 49 UG C/ L,N=2, STAGE TOTAL FROM NII TO ADULT	1204
*CALANUS HELGOLANDICUS	7.84 (RANGE=6.07-10.07) MICROGRAM C/ANIMAL	FOOD=LAUDERIA BOREALIS =100 UG C/ L, STAGE NII TO CI, N=3	1204
*CALANUS HELGOLANDICUS	21.55 (RANGE=18.28-23.34) MICROGRAM C/ANIMAL	FOOD=LAUDERIA BOREALIS =100 UG C/ L, STAGE CI TO CIII, N=3	1204
*CALANUS HELGOLANDICUS	98.00 (RANGE=94.10-100.38) MICROGRAM C/ANIMAL	FOOD=LAUDERIA BOREALIS =100 UG C/ L, STAGE CIII TO CV, N=3	1204
*CALANUS HELGOLANDICUS	40.84 MICROGRAM C/ANIMAL	FOOD=LAUDERIA BOREALIS =100 UG C/ L, STAGE CV TO ADULT, N=1	1204
*CALANUS PACIFICUS	0.34 MICROGRAM C/(ANIMAL*HOUR)	1010 CELL/ML=95 UG C/L, CELL DIA = 11 MY AS THALASSIOSIRA FLUVIATILIS	494
*CALANUS PACIFICUS	0.14 MICROGRAM C/(ANIMAL*HOUR)	200 CELL/ML=27 UG C/L, CELL DIA = 13 MY AS THALASSIOSIRA FLUVIATILIS	494
*CALANUS PACIFICUS	0.10 MICROGRAM C/(ANIMAL*HOUR)	23 CELL/ML= 19 UG C/L,CELLYDIA = 35 MY,COSCINODISCUS ANGSTII	494
*CALANUS PACIFICUS	0.11 MICROGRAM C/(ANIMAL*HOUR)	11 CELL/ML=18 UG C/L,CELL DIA = 75 MY AS COSCINODISCUS EXCENTRICUS	494

TABLE A440

ZOOPLANKTON : GRAZING RATE			
ITEM:	VALUE:	CONDITIONS:	REF.:
*CALANUS PACIFICUS	0.11 MICROGRAM C/(ANIMAL*HOUR)	4 CELL/ML=13 UG C/L, CELL DIA = 87 MY AS COSCINODISCUS ANGSTII	494
*DAPHNIA MAGNA	5.6 *10**6 CELLS/HR	E.COLI(0.9MY**3)= CONS=0.005MM**3/ HR*ANIM, MAX,	1007
*DAPHNIA MAGNA	0.50 *10**6 CELLS/HR	CHLORELLA VULG.(34MY**3)= CONS= 0.017MM**3/HR*ANIM, MAX,	1007
*DAPHNIA MAGNA	0.25 *10**6 CELLS/HR	SACCHAROMYCES CER.(66MY**3)= CONS= 0.016MM**3/HR*ANIM, MAX,	1007
*DAPHNIA MAGNA	0.0028 *10**6 CELLS/HR	TETRAHYMENA PYR.(18000MY**3)= CONS =0.051MM**3/HR*ANIM, MAX,	1007
*DAPHNIA PULEX	0.47 CAL	0-6 D OLD, FOOD = 25000 CELLS/ML, CHLAMYDOMONAS REINHARDTII	1328
*DAPHNIA PULEX	0.58 CAL	0-6 D OLD, FOOD = 50000 CELLS/ML, CHLAMYDOMONAS REINHARDTII	1328
*DAPHNIA PULEX	1.39 CAL	0-6 D OLD, FOOD = 75000 CELLS/ML, CHLAMYDOMONAS REINHARDTII	1328
*DAPHNIA PULEX	1.91 CAL	0-6 D OLD, FOOD = 100000 CELLS/ML, CHLAMYDOMONAS REINHARDTII	1328
*DAPHNIA PULEX	5.67 CAL	6-34 D OLD, FOOD = 25000 CELLS/ML, CHLAMYDOMONAS REINHARDTII	1328
*DAPHNIA PULEX	13.0 CAL	6-34 D OLD, FOOD = 50000 CELLS/ML, CHLAMYDOMONAS REINHARDTII	1328
*DAPHNIA PULEX	19.3 CAL	6-34 D OLD, FOOD = 75000 CELLS/ML, CHLAMYDOMONAS REINHARDTII	1328
*DAPHNIA PULEX	27.3 CAL	6-34 D OLD, FOOD = 100000 CELLS/ ML, CHLAMYDOMONAS REINHARDTII	1328
*DAPHNIA PULEX	6.1 CAL	0-40 D OLD, FOOD = 25000 CELLS/ML, CHLAMYDOMONAS REINHARDTII	1328
*DAPHNIA PULEX	13.6 CAL	0-40 D OLD, FOOD = 50000 CELLS/ML, CHLAMYDOMONAS REINHARDTII	1328
*DAPHNIA PULEX	20.7 CAL	0-40 D OLD, FOOD = 75000 CELLS/ML, CHLAMYDOMONAS REINHARDTII	1328
*DAPHNIA PULEX	29.2 CAL	0-40 D OLD, FOOD = 100000 CELLS/ ML, CHLAMYDOMONAS REINHARDTII	1328
*DAPHNIA SCHODLERI	10000 CELLS/ANIMAL*HR	FOOD=ANKISTRODESMUS SP.,MAX.VALUE, 293K,N=10	647
*DAPHNIA SCHODLERI	30000 CELLS/ANIMAL*HR	FOOD=CHLAMYDOMONAS SP.,MAX.VALUE, 293K,N=4	647
*DAPHNIA SCHODLERI	9800 CELLS/ANIMAL*HR	FOOD=FRUSTULIA SP.,MAX.VALUE,293K, N=2	647
*DAPHNIA SCHODLERI	10000 CELLS/ANIMAL*HR	FOOD=FRUSTULIA AND ANKISTRODESMUS, MAX.VALUE,293K,N=4	647
*FRESHWATER ZOOPLANKTON	0.015 1/HOUR	STEADY STATE SOLUTION TO CASTLE LAKE, ON PHYTOPLANKTON	1326
*FRESHWATER ZOOPLANKTON	2.8917 MICROGRAM P/(MG*HOUR)	FOOD = SESTON, STONEHOUSE POND, FEB 22, 1973	454
*FRESHWATER ZOOPLANKTON	88.960 MICROGRAM P/(MG*HOUR)	FOOD = SESTON, STONEHOUSE POND, APRIL 6, 1973	454
*FRESHWATER ZOOPLANKTON	0.2502 MICROGRAM P/(MG*HOUR)	FOOD = SESTON, STONEHOUSE POND, MAY 17, 1973	454
*FRESHWATER ZOOPLANKTON	88.960 MICROGRAM P/(MG*HOUR)	FOOD = SESTON, STONEHOUSE POND, JUNE 17, 1973	454

TABLE A441

ZOOPLANKTON : GROSS GROWTH EFFICIENCY			
ITEM:	VALUE:	CONDITIONS:	REF.:
*CALANUS FINMARCHICUS	35.7 %	FOOD = SKELETONEMA = 200 UG C/L, 283 K, BASED ON NITROGEN, CII TO CV	1204
*CALANUS FINMARCHICUS	34 %	FOOD=SEAWATER, HATCHING TO ADULTHOOD, 283 K, BASED ON NITROGEN	1204

TABLE A441

ZOOPLANKTON : GROSS GROWTH EFFICIENCY

ITEM:	VALUE:	CONDITIONS:	REF.:
*CALANUS FINMARCHICUS	18.9 - 34.6 %	FOOD=SEAWATER, BASED ON PHOSPHORUS, CV TO ADULT, 285 K	1204
*CALANUS FINMARCHICUS	17.2 %	FOOD=THALASSIOSIRA= 490 UG C/L, 279-291 K, CV AND ADULTS, BASED ON P	1204
*CALANUS HELGOLANDICUS	35.7 %	FOOD=SKELETONEMA= 200 UG C/L, 283 K, BASED ON NITROGEN, CII TO CV	1204
*CALANUS HELGOLANDICUS	21.4 - 37.7 %	FOOD=SEAWATEI, BASED ON NITROGEN, CV AND ADULT, 285 K	1204
*CALANUS HELGOLANDICUS	17.2 %	FOOD=THALASSIOSIRA= 490 UG C/L, 279-291 K, CV AND ADULT, BASED ON P	1204
*CALANUS HYPERBOREUS	13.0 - 17.3 %	FOOD=T. FLUVIATILIS=6.5 MG ASH-FREE DW/L,275-278 K, CV, ON ASH-FRE DW	1204
*CALANUS HYPERBOREUS	18.6 - 36.4 %	FOOD=T. FLUVIATILIS=1.7 MG ASG-FREE DW/L,275-278 K, CV, ON ASH-FRE DW	1204
*RHINCALANUS NASUTUS	30 %	FOOD=T. FLUVIATILIS=352 UG C/L, WHOLE LIFE, 283 K, BASED ON CARBON	1204
*RHINCALANUS NASUTUS	37 %	FOOD=T. FLUVIATILIS=196 UG C/L, WHOLE LIFE, 288 K, BASED ON CARBON	1204
*RHINCALANUS NASUTUS	34 %	FOOD=DITYLUM BRIGH.=200 UG C/L, WHOLE LIFE, 283 K, BASED ON CARBON	1204
*RHINCALANUS NASUTUS	45 %	FOOD=DITYLUM BRIGH.=148 UG C/L, WHOLE LIFE, 288 K, BASED ON CARBON	1204

TABLE A442

ZOOPLANKTON : GROWTH AND REPRODUCTION

ITEM:	VALUE:	CONDITIONS:	REF.:
*DAPHNIA PULEX	17.4 % OF CONSUMPTION	FOOD = 25*10**(-4) CELL/ML (CHLAMYDOMONAS), 40 D,AFTER 18 INSTARS	1327
*DAPHNIA PULEX	13.1 % OF CONSUMPTION	FOOD = 50*10**(-4) CELL/ML (CHLAMYDOMONAS), 40 D,AFTER 18 INSTARS	1327
*DAPHNIA PULEX	11.4 % OF CONSUMPTION	FOOD = 75*10**(-4) CELL/ML (CHLAMYDOMONAS), 40 D,AFTER 18 INSTARS	1327
*DAPHNIA PULEX	10.0 % OF CONSUMPTION	FOOD = 100*10**(-4) CELL/ML (CHLAMYDOMONAS), 40 D,AFTER 18 INSTARS	1327
*DAPHNIA PULEX	56.0 % OF ASSIMILATION	FOOD = 25*10**(-4) CELL/ML (CHLAMYDOMONAS), 40 D,AFTER 18 INSTARS	1327
*DAPHNIA PULEX	65.3 % OF ASSIMILATION	FOOD = 50*10**(-4) CELL/ML (CHLAMYDOMONAS), 40 D,AFTER 18 INSTARS	1327
*DAPHNIA PULEX	69.8 % OF ASSIMILATION	FOOD = 75*10**(-4) CELL/ML (CHLAMYDOMONAS), 40 D,AFTER 18 INSTARS	1327
*DAPHNIA PULEX	73.0 % OF ASSIMILATION	FOOD = 100*10**(-4) CELL/ML (CHLAMYDOMONAS), 40 D,AFTER 18 INSTARS	1327
*DAPHNIA PULEX	10 - 17 % OF INGESTION	NONE	1689
*DAPHNIA PULEX	56 - 73 % OF ASSIMILATION	NONE	1689
*LEPTODORA KINDTII	7.3 % OF ASSIMILATION	NONE	1689
*MESOCYCLOPS ALBIDUS	CA 25 % OF INGESTION	NONE	1689
*MESOCYCLOPS ALBIDUS	CA 50 % OF ASSIMILATION	NONE	1689
*SIMOCEPHALUS VECTULUS	52.9 % OF INGESTION	JUVENILE, FEMALE	1689
*SIMOCEPHALUS VECTULUS	73.1 % OF ASSIMILATION	REPRODUCING FEMALE	1689

TABLE A442

ZOOPLANKTON : GROWTH AND REPRODUCTION			
ITEM:	VALUE:	CONDITIONS:	REF.:
*SIMOCEPHALUS VECTULUS	20.5 % OF INGESTION	REPRODUCING FEMALE	1689
*SIMOCEPHALUS VECTULUS	64.7 % OF ASSIMILATION	NONE	1689

TABLE A443

ZOOPLANKTON : GROWTH EFFICIENCY			
ITEM:	VALUE:	CONDITIONS:	REF.:
*DAPHNIA PULEX	13.2 % OF ENERGY ASSIMILATED	0-6 D OLD, FOOD = 25000 CELLS/ML, CHLAMYDOMONAS REINHARDTII	1328
*DAPHNIA PULEX	9.1 % OF ENERGY ASSIMILATED	0-6 D OLD, FOOD = 50000 CELLS/ML, CHLAMYDOMONAS REINHARDTII	1328
*DAPHNIA PULEX	4.8 % OF ENERGY ASSIMILATED	0-6 D OLD, FOOD = 75000 CELLS/ML, CHLAMYDOMONAS REINHARDTII	1328
*DAPHNIA PULEX	3.9 % OF ENERGY ASSIMILATED	0-6 D OLD, FOOD = 100000 CELLS/ML, CHLAMYDOMONAS REINHARDTII	1328
*DAPHNIA PULEX	1.25 % OF ENERGY CONSUMED	6-34 D OLD, FOOD = 25000 CELLS/ML, CHLAMYDOMONAS REINHARDTII	1328
*DAPHNIA PULEX	0.78 % OF ENERGY CONSUMED	6-34 D OLD, FOOD = 50000 CELLS/ML, CHLAMYDOMONAS REINHARDTII	1328
*DAPHNIA PULEX	0.57 % OF ENERGY CONSUMED	6-34 D OLD, FOOD = 75000 CELLS/ML, CHLAMYDOMONAS REINHARDTII	1328
*DAPHNIA PULEX	0.43 % OF ENERGY CONSUMED	6-34 D OLD, FOOD = 100000 CELLS/ ML, CHLAMYDOMONAS REINHARDTII	1328
*DAPHNIA PULEX	3.95 % OF ENERGY ASSIMILATED	6-34 D OLD, FOOD = 25000 CELLS/ML, CHLAMYDOMONAS REINHARDTII	1328
*DAPHNIA PULEX	3.85 % OF ENERGY ASSIMILATED	6-34 D OLD, FOOD = 50000 CELLS/ML, CHLAMYDOMONAS REINHARDTII	1328
*DAPHNIA PULEX	3.41 % OF ENERGY ASSIMILATED	6-34 D OLD, FOOD = 75000 CELLS/ML, CHLAMYDOMONAS REINHARDTII	1328
*DAPHNIA PULEX	2.99 % OF ENERGY ASSIMILATED	6-34 D OLD, FOOD = 100000 CELLS/ ML, CHLAMYDOMONAS REINHARDTII	1328
*DAPHNIA PULEX	17.4 % OF ENERGY CONSUMED	0-40 D OLD, FOOD = 25000 CELLS/ML, CHLAMYDOMONAS REINHARDTII	1328
*DAPHNIA PULEX	13.1 % OF ENERGY CONSUMED	0-40 D OLD, FOOD = 50000 CELLS/ML, CHLAMYDOMONAS REINHARDTII	1328
*DAPHNIA PULEX	11.4 % OF ENERGY CONSUMED	0-40 D OLD, FOOD = 75000 CELLS/ML, CHLAMYDOMONAS REINHARDTII	1328
*DAPHNIA PULEX	10.0 % OF ENERGY CONSUMED	0-40 D OLD, FOOD = 100000 CELLS/ ML, CHLAMYDOMONAS REINHARDTII	1328
*DAPHNIA PULEX	56 % OF ENERGY ASSIMILATED	0-40 D OLD, FOOD = 25000 CELLS/ML, CHLAMYDOMONAS REINHARDTII	1328
*DAPHNIA PULEX	65 % OF ENERGY ASSIMILATED	0-40 D OLD, FOOD = 50000 CELLS/ML, CHLAMYDOMONAS REINHARDTII	1328
*DAPHNIA PULEX	70 % OF ENERGY ASSIMILATED	0-40 D OLD, FOOD = 75000 CELLS/ML, CHLAMYDOMONAS REINHARDTII	1328
*DAPHNIA PULEX	73 % OF ENERGY ASSIMILATED	0-40 D OLD, FOOD = 100000 CELLS/ ML, CHLAMYDOMONAS REINHARDTII	1328

TABLE A444

ZOOPLANKTON : GROWTH RATE			
ITEM:	VALUE:	CONDITIONS:	REF.:
*CANUELLA PERPLEXA	0.042 1/DAY	SPRING 1970, FIELD DATA, 280.6 K	650
*CERIODAPHNIA SP.	0.58 1/DAY	NEAR OPTIMAL LAB CONDITIONS, 296 K, GRAZING ON 4*10**5 ALGAE/L	595
*CERIODAPHNIA SP.	0.44 1/DAY	MEAN OF 18 DIFFERENT FIELD EXPERIMENT	595

TABLE A444

ZOOPLANKTON : GROWTH RATE

ITEM:	VALUE:	CONDITIONS:	REF.:
*CERIODAPHNIA SP.	7.2 DAYS	MEAN OF 18 DIFFERENT FIELD EXPERIMENT	355
*DAPHNIA PULEX	0.040 MICRO GRAM C/ (INDIVIDUAL*HOUR)	DETRITUS = 238.0 MG C/M3, DISSOLVED OXYGEN=2.8 MG/L	91
*DAPHNIA PULEX	0.032 MICRO GRAM C/ (INDIVIDUAL*HOUR)	DETRITUS = 194.7 MG C/M3, DISSOLVED OXYGEN=4.3 MG/L	91
*DAPHNIA PULEX	0.046 MICRO GRAM C/ (INDIVIDUAL*HOUR)	DETRITUS = 213.4 MG C/M3, DISSOLVED OXYGEN=4.5 MG/L	91
*DAPHNIA PULEX	0.043 MICRO GRAM C/ (INDIVIDUAL*HOUR)	DETRITUS = 194.7 MG C/M3, DISSOLVED OXYGEN=5.3 MG/L	91
*DAPHNIA PULEX	0.023 MICRO GRAM C/ (INDIVIDUAL*HOUR)	DETRITUS = 187.5 MG C/M3, DISSOLVED OXYGEN=5.6 MG/L	91
*DAPHNIA PULEX	0.74 1/DAY	FOOD=ANKISTRODESMUS FALCATUS AT K= 25,NET REPROD.RATE=2906.5	42
*DAPHNIA PULEX	0.09 1/DAY	FOOD=ANACYSTIS NIDULANS AT K=25, NET REPROD.RATE=2.1	42
*DAPHNIA PULEX	< 0.01 1/DAY	FOOD=SYNECHOCOCCUS ELONGATA AT K= 25,NET REPROD.RATE=0.9	42
*DAPHNIA PULEX	0 1/DAY	FOOD=MERISMOPEDIA SP. AT K=25,NET REPROD.RATE=0	42
*DAPHNIA PULEX	0.43 1/DAY	FOOD=SYNECHOCOCCUS CEDRORUM AT K= 25,NET REPROD.RATE=225.4	42
*DAPHNIA PULEX	0.21 1/DAY	FOOD=ANABAENA FLOS-AQUA AT K=25, NET REPROD.RATE=82.4	42
*DAPHNIA PULEX	0 1/DAY	FOOD=SYNECHOCYSTIS SP. AT K=25,NET REPROD.RATE=0	42
*DAPHNIA PULEX	0 1/DAY	FOOD=GLOEOCAPSA ALPICOLA AT K=25, NET REPROD.RATE=0	42
*DAPHNIA PULEX	0.57 1/DAY	FOOD=YEAST AT K=10,NET REPROD.RATE =1289.2	42
*DAPHNIA PULEX	0.04 1/DAY	NO FOOD,NET REPROD.RATE=1.6	42
*DAPHNIA PULEX	0.37 1/DAY	FOOD=ANKISTRODESMUS FALCATUS AT K= 1,NET REPROD.RATE=306.9	42
*DAPHNIA PULEX	0.26 1/DAY	FOOD=ANACYSTIS NIDULANS AT K=1,NET REPROD.RATE=39.9	42
*DAPHNIA PULEX	< 0.01 1/DAY	FOOD=SYNECHOCOCCUS ELONGATA AT K= 1,NET REPROD.RATE=0.1	42
*DAPHNIA PULEX	0 1/DAY	FOOD=MERISMOPEDIA SP. AT K=1,NET REPROD.RATE=0	42
*DAPHNIA PULEX	0.29 1/DAY	FOOD=SYNECHOCOCCUS CEDRORUM AT K= 1,NET REPROD.RATE=181.9	42
*DAPHNIA PULEX	0.16 1/DAY	FOOD=ANABAENA FLOS-AQUA AT K=1,NET REPROD.RATE=34.1	42
*DAPHNIA PULEX	0 1/DAY	FOOD=SYNECHOCYSTIS SP. AT K=1,NET REPROD.RATE=0	42
*DAPHNIA PULEX	0 1/DAY	FOOD=GLOEOCAPSA ALPICOLA AT K=1, NET REPROD.RATE=0	42
*DAPHNIA PULEX	0.050 MICRO GRAM C/ (INDIVIDUAL*HOUR)	DETRITUS = 228.5 MG C/M3, DISSOLVED OXYGEN=3.1 MG/L	91
*DAPHNIA ROSEA	76 NG P/MG*HR (SD=32)	FOOD=0.125*10**5CELLS RHODOTORULA/ ML,JUVENILE GROWTH CONST.,N=7	1013
*DAPHNIA ROSEA	118 NG P/MG*HR (SD=16)	FOOD=0.5*10**5CELLS RHODOTORULA/ ML,JUVENILE GROWTH CONST.,N=7	1013
*DAPHNIA ROSEA	119 NG P/MG*HR (SD=31)	FOOD=1*10**5CELLS RHODOTORULA/ML, JUVENILE GROWTH CONST.,N=14	1013
*DAPHNIA ROSEA	0.44 NG P/ANIMAL*HR (SD=0.075)	FOOD=0.125*10**5CELLS RHODOTORULA/ ML,ADULT GROWTH RATE,N=5	1013
*DAPHNIA ROSEA	0.184 NG P/ANIMAL*HR (SD=0.040)	FOOD=0.5*10**5CELLS RHODOTORULA/ ML,ADULT GROWTH RATE,N=6	1013
*DAPHNIA ROSEA	0.229 NG P/ANIMAL*HR (SD=0.050)	FOOD=1*10**5CELLS RHODOTORULA/ML, ADULT GROWTH RATE,N=10	1013
*DAPHNIA ROSEA	0.64 NG P/ANIMAL*HR (SD=0.16)	FOOD=0.125*10**5CELLS RHODOTORULA/ ML,REPROD.RATE,N=5	1013
*DAPHNIA ROSEA	1.74 NG P/ANIMAL*HR (SD=0.56)	FOOD=0.5*10**5CELLS RHODOTORULA/ ML,REPROD.RATE,N=6	1013
*DAPHNIA ROSEA	1.89 NG P/ANIMAL*HR (SD=0.47)	FOOD=1*10**5CELLS RHODOTORULA/ML, REPROD.RATE,N=10	1013

TABLE A444

ZOOPLANKTON : GROWTH RATE			
ITEM:	VALUE:	CONDITIONS:	REF.:
*EUCHLANIS DILATATA	0.76 1/DAY	MEAN VALUE, 295 K, LAB, LITTORAL ROTIFER	824
*HALICYCLOPS MAGNICEPS	0.056 1/DAY	SPRING 1969, FIELD DATA, 282.8 K	650
*KERATELLA VALGA	18 DAYS	290 K	355
*KERATELLA VALGA	6 DAYS	296 K	355
*PARONYCHOCAMPTUS NANUS	0.056 1/DAY	SPRING 1970, FIELD DATA, 280.6 K	650
*PARONYCHOCAMPTUS NANUS	0.042 1/DAY	SPRING 1969, FIELD DATA, 288 K	650
*PARONYCHOCAMPTUS NANUS	0.043 1/DAY	SPRING 1971, FIELD DATA, 288.4 K	650
*PARONYCHOCAMPTUS NANUS	0.048 1/DAY	AUTUM 1971, FIELD DATA	650
*PHILODINA ROSEOLA	0.196/DAY	288K,MEDIAN LIFE SPAN=28 DAYS	1389
*PHILODINA ROSEOLA	0.238/DAY	293K,MEDIAN LIFE SPAN=10.2 DAYS	1389
*PHILODINA ROSEOLA	0.322/DAY	298K,MEDIAN LIFE SPAN=5.9 DAYS	1389
*PHILODINA ROSEOLA	0.500/DAY	303K,MEDIAN LIFE SPAN=3.0 DAYS	1389
*SIMOCEPHALUS SERRULATUS	0.55 1/DAY	FEEDING ON 8*10**5 CELL ALGAE/L, LAB, 296 K	595
*SIMOCEPHALUS SERRULATUS	0.27 1/DAY	FIELD EXPERIMENT, 296 K	595
*SIMOCEPHALUS SERRULATUS	10 DAYS	FIELD EXPERIMENT, 296 K	355
*TACHIDIUS DISCIPEA	0.056 1/DAY	SPRING 1971, FIELD DATA, 280.8 K	650
*TACHIDIUS DISCIPEA	0.049 1/DAY	SPRING 1969, FIELD DATA, 282.8 K	650
*TACHIDIUS DISCIPEA	0.045 1/DAY	SPRING 1969, FIELD DATA, 283.5 K	650

TABLE A445

ZOOPLANKTON : H/BIOMASS RATIO			
ITEM:	VALUE:	CONDITIONS:	REF.:
*AMPHIPODA	4.4-7.6 % OF DRY WEIGHT	DRIED AT 333K,NORTH PACIFIC	1184
*CALANUS FINMARCHICUS	10.21 %	ATOMIC COMPOSITION	1632
*CHAETOGNATHA	7.2-7.6 % OF DRY WEIGHT	DRIED AT 333K,NORTH PACIFIC	1184
*COPEPODS	6.7-10.3 % OF DRY WEIGHT	DRIED AT 333K,NORTH PACIFIC	1184
*EUPHAUSIACEA	6.7-7.6 % OF DRY WEIGHT	DRIED AT 333K,NORTH PACIFIC	1184
*PTEROPODS	1.1-3.8 % OF DRY WEIGHT	DRIED AT 333K,NORTH PACIFIC	1184

TABLE A446

ZOOPLANKTON : HALF SATURATION CONSTANT FOR FEEDING ON ALGAE			
ITEM:	VALUE:	CONDITIONS:	REF.:
*BOSMINA COREGONI	0.¤ - 4.0 MG C/L	NONE	1387
*DAPHNIA MAGNA	9.6 MG C/L	FOOD = CHLORELLA VULGARIS = 0.124*10**(-6) MG C/CELL	1387
*DAPHNIA MAGNA	15.0 MG C/L	FOOD = CHLORELLA VULGARIS = 0.124*10**(-6) MG C/CELL	1387
*DAPHNIA ROSEA	0.16 MG C/L	FOOD = RHODOTORULA GLUTINIS = 0.776*10**(-8) MG C/CELL, 4.2 U LONG	1387
*DIAPTOMUS OREGONENSIS	1.6 MG C/L	FOOD= CHLAMYDOMONAS AND CHLORELLA = 0.124*10**(-6) MG C/CELL	1387

TABLE A447

ZOOPLANKTON : HALF SATURATION CONSTANT FOR GRAZING			
ITEM:	VALUE:	CONDITIONS:	REF.:
*ZOOPLANKTON	50 MICROGRAM CHLOR./L	SAN JOAQUIN(RIVER)	1149

TABLE A447

ZOOPLANKTON : HALF SATURATION CONSTANT FOR GRAZING			
ITEM:	VALUE:	CONDITIONS:	REF.:
*ZOOPLANKTON	50 MICROGRAM CHLOR./L	DELTA(ESTUARY)	1149
*ZOOPLANKTON	50 MICROGRAM CHLOR./L	LAKE ERIE	1149
*ZOOPLANKTON	10 MICROGRAM CHLOR./L	LAKE ONTARIO	1149
*ZOOPLANKTON	60 MICROGRAM CHL/L	MOSSDALE ON ALGAE	1153
*ZOOPLANKTON	60 MICROGRAM CHL/L	POTOMAC RIVER ON ALGAE	1153
*ZOOPLANKTON	0.4-0.6 MG/L	ON ALGAE	1571
*ZOOPLANKTON,HERBIVOROUS	10 MICROGRAM-CHL./L	LAKE ONTARIO ON ALGAE	1577

TABLE A448

ZOOPLANKTON : I/BIOMASS RATIO			
ITEM:	VALUE:	CONDITIONS:	REF.:
*CALANUS FINMARCHICUS	0.0002 %	ATOMIC COMPOSITION	1632

TABLE A449

ZOOPLANKTON : INGESTION TIME			
ITEM:	VALUE:	CONDITIONS:	REF.:
*CHAOBORUS TRIVITTATUS	24.3 (SE=+-5.6) SEC	TIME USED TO INGEST PREY = DAPHNIA = 1.0 MM, N=28, 4. INSTAR	1541
*CHAOBORUS TRIVITTATUS	74.2 (SE=+-10.8) SEC	TIME USED TO INGEST PREY = DAPHNIA = 1.4 MM, N=24, 4. INSTAR	1541
*CHAOBORUS TRIVITTATUS	210.5 (SE=+-44.1) SEC	TIME USED TO INGEST PREY = DAPHNIA = 1.8 MM, N=21, 4. INSTAR	1541
*CHAOBORUS TRIVITTATUS	5.0 (SE=+-0.2) SEC	TIME USED TO INGEST PREY = DIAPTOMUS = 1.0 MM, N=16, 4. INSTAR	1541
*CHAOBORUS TRIVITTATUS	11.4 (SE=+-1.1) SEC	TIME USED TO INGEST PREY = DIAPTOMUS = 1.4 MM, N=22, 4. INSTAR	1541
*CHAOBORUS TRIVITTATUS	103.6 (SE=+-28.0) SEC	TIME USED TO INGEST PREY = DIAPTOMUS = 1.8 MM, N=35, 4. INSTAR	1541

TABLE A450

ZOOPLANKTON : K/BIOMASS RATIO			
ITEM:	VALUE:	CONDITIONS:	REF.:
*CALANUS FINMARCHICUS	0.29 %	ATOMIC COMPOSITION	1632

TABLE A451

ZOOPLANKTON : LENGTH			
ITEM:	VALUE:	CONDITIONS:	REF.:
*ACANTHOCYCLOPS VERNALIS	1.08 - 1.17 MM	LATE COPEPODITS	361
*ACANTHOCYCLOPS VERNALIS	1.50 MM	ADULTS	361
*BOSMINA LONGIROSTRIS	0.40 MM	MAX.ADULT LENGTH,GATUN LAKE	1770

TABLE A451

ZOOPLANKTON : LENGTH

ITEM:	VALUE:	CONDITIONS:	REF.:
*BOSMINOPSIS DEITERSI	0.35 MM	MAX.ADULT LENGTH,GATUN LAKE	1770
*CERIODAPHNIA CORNUTA	0.40 MM	MAX.ADULT LENGTH,GATUN LAKE, UNHORNED OR HORNED MORPH	1770
*CHYDORUS EURYNOTUS	0.50 MM	MAX.ADULT LENGTH,GATUN LAKE	1770
*CYCLOPS VICINUS	1.08 - 1.17 MM	LATE COPEPODITS	361
*CYCLOPS VICINUS	1.50 MM	ADULTS	361
*DAPHNIA MIDDENDORFFIANA	2.32 - 2.92 MM	ADULTS	361
*DAPHNIA MIDDENDORFFIANA	0.32 - 1.08 MM	NEONATES	361
*DIAPHANOSOMA BRACHYURUM	0.90 MM	MAX.ADULT LENGTH,GATUN LAKE	1770
*DIAPTOMUS ALASKAENSIS	1.60 MM	ADULTS	361
*DIAPTOMUS GATUNENSIS	1.50 MM	MAX.ADULT LENGTH,GATUN LAKE,ADULTS	1770
*DIAPTOMUS GATUNENSIS	0.80 MM	MAX.ADULT LENGTH,GATUN LAKE, COPEPODIDS	1770
*EUBOSMINA TUBICENS	0.45 MM	MAX.ADULT LENGTH,GATUN LAKE	1770
*EUCYCLOPS AGILIS	1.50 MM	ADULTS	361
*EUCYCLOPS AGILIS	1.50 MM	MAX.ADULT LENGTH,GATUN LAKE,ADULTS	1770
*EUCYCLOPS AGILIS	0.80 MM	MAX.ADULT LENGTH,GATUN LAKE, COPEPODIDS	1770
*HETEROCOPE SEPTENTRIONALIS	0.66 - 0.80 MM	NAUPLII AND COPEPODIT 1	361
*HETEROCOPE SEPTENTRIONALIS	2.20 - 2.92 MM	COPEPODITS IV AND V	361
*MOINA MINUTUS	0.60 MM	MAX.ADULT LENGTH,GATUN LAKE	1770
*POLYARTEMIELLA HAZENI	1.23 MM	NAUPLII	361
*POLYARTEMIELLA HAZENI	1.67 - 3.42 MM	EARLY METANAUPLII	361
*POLYARTEMIELLA HAZENI	4.16 - 6.66 MM	LATE METANAUPLII	361
*TOTAL NAUPLII	0.25 MM	MAX.ADULT LENGTH,GATUN LAKE	1770

TABLE A452

ZOOPLANKTON : MAX FEEDING RATE

ITEM:	VALUE:	CONDITIONS:	REF.:
*CENTROPAGES TYPICUS	48% OF BODY WEIGHT/6HOURS	PREY=ARTEMIA NAUPLII,288K	37
*CENTROPAGES TYPICUS	22% OF BODY WEIGHT/6HOURS	PREY=ARTEMIA NAUPLII,281K	37
*LABIDOCERA JOLLAE	27% OF BODY WEIGHT/6HOURS	PREY=ANCHOVY FISH LARVAE	926
*METRIDIA LUCENS	24% OF BODY WEIGHT/6HOURS	PREY=ARTEMIA NAUPLII	613
*SAGITTA HISPIDA	16% OF BODY WEIGHT/6HOURS	PREY=ARTEMIA NAUPLII	1306
*TORTANUS DISCAUDATUS	17% OF BODY WEIGHT/6HOURS	PREY=CALANUS NAUPLII 3	20
*TORTANUS DISCAUDATUS	24% OF BODY WEIGHT/6HOURS	PREY=CALANUS NAUPLII 5	20

TABLE A453

ZOOPLANKTON : MAX GROWTH RATE

ITEM:	VALUE:	CONDITIONS:	REF.:
*EUCHLANIS DILATATA	1.4 1/DAYS	MAX VALUE, 295 K, LAB, LITTORAL ROTIFER	824
*ZOOPLANKTON	0.15-0.3 1/DAY	TEMP. FACTOR IS 1 AT 293K	1571

TABLE A454

ZOOPLANKTON : MEAN GENERATION TIME

ITEM:	VALUE:	CONDITIONS:	REF.:
*CERIODAPHNIA SP.	7.5 DAYS	4*10**5 CELL ALGAE/L, 296 K,NEAR OPTIMAL LAB, GENERATION TIME	595

TABLE A455

ZOOPLANKTON : MG/BIOMASS RATIO			
ITEM:	VALUE:	CONDITIONS:	REF.:
*CALANUS FINMARCHICUS	0.03 %	ATOMIC COMPOSITION	1632

TABLE A456

ZOOPLANKTON : MIN CONC FOR FEEDING			
ITEM:	VALUE:	CONDITIONS:	REF.:
*CALANUS PACIFICUS	0.016 MG C/L	NO FEEDING BELOW THIS VALUE	1387
*14 MARINE SPECIES	0.04 - 0.19 MG C/L	NO FEEDING BELOW THIS VALUE	1387

TABLE A457

ZOOPLANKTON : MN/BIOMASS RATIO			
ITEM:	VALUE:	CONDITIONS:	REF.:
*CALANUS FINMARCHICUS	0.0000 %	ATOMIC COMPOSITION	1632

TABLE A458

ZOOPLANKTON : MORTALITY			
ITEM:	VALUE:	CONDITIONS:	REF.:
*COPEPODS	0.003 - 0.005 1/DAY	USED IN MODEL, LAKE MICHIGAN	1567
*FRESHWATER ZOOPLANKTON	0.004 1/HOUR	STEADY STATE SOLUTION TO CASTLE LAKE.	1326
*SIMOCEPHALUS SERRULATUS	20 DAYS	FIELD EXPERIMENT, 296 K	595
*ZOOPLANKTON	0.075 1/DAY	MOSSDALE	1153
*ZOOPLANKTON	0.005 1/DAY	293K	1571
*ZOOPLANKTON	0.02 1/DAY	USED IN MODEL, NORTH FORK KINGS RIVER, CALIF	1567
*ZOOPLANKTON	0.005 1/DAY	USED IN MODEL	1567

TABLE A459

ZOOPLANKTON : N/BIOMASS RATIO			
ITEM:	VALUE:	CONDITIONS:	REF.:
*ALGAE	0.05 - 0.17 BY WEIGHT	USED IN MODELS, 1976	55
*ALGAE	0.2 BY WEIGHT	USED IN MODELS, LAKE MICHIGAN	220
*ALGAE	0.14 BY WEIGHT	USED IN MODELS, SAN JOAQUIN RIVER, CALIF.	1154
*ALGAE	0.14 BY WEIGHT	USED IN MODELS, SAN JOAQUIN RIVER ESTUARY, CALIF.	1154
*ALGAE	0.2 BY WEIGHT	USED IN MODELS, POTOMAC ESTUARY	1154
*ALGAE	7/50 BY WEIGHT	USED IN MODELS, LAKE ERIE	1154
*ALGAE	0.2 BY WEIGHT	USED IN MODELS, LAKE ONTARIO	1154
*AMPHIPODA	4.4-8.2 % OF DRY WEIGHT	DRIED AT 333K,NORTH PACIFIC	1184
*CALANUS FINMARCHICUS	1.52 %	ATOMIC COMPOSITION	1632

TABLE A459

ZOOPLANKTON : N/BIOMASS RATIO			
ITEM:	VALUE:	CONDITIONS:	REF.:
*CALANUS FINMARCHICUS	18.7 MICROGRAM/CALANUS (SD=4.2)	STARVATION:8DAYS,N=10,STAGE 4 AND 5	62
*CALANUS FINMARCHICUS	14.8 MICROGRAM/CALANUS (SD=4.4)	STARVATION:14DAYS,N=10,STAGE 4 AND 5	62
*CALANUS FINMARCHICUS	18.3 MICROGRAM/CALANUS (SD=2.7)	STARVATION:23DAYS,N=12,STAGE 4 AND 5	62
*CALANUS FINMARCHICUS	20.1 MICROGRAM/CALANUS (SD=5.6)	NO STARVATION,N=9,STAGE 4 AND 5	62
*CHAETOGNATHA	6.3-9.4 % OF DRY WEIGHT	DRIED AT 333K,SARGASSO SEA	88
*CHAETOGNATHA	10.7-11.1 % OF DRY WEIGHT	DRIED AT 333K,NORTH PACIFIC	1184
*COPEPODS	8.2-11.2 % OF DRY WEIGHT	DRIED AT 333K,SARGASSO SEA	88
*COPEPODS	4.7-7.1 % OF DRY WEIGHT	DRIED AT 378K,CONTINENTAL SHELF OFF NEW YORK	308
*COPEPODS	5.1-13.1 % OF DRY WEIGHT	DRIED AT 333K,NORTH PACIFIC	1184
*CTENOPHORA	0.2-1.1 % OF DRY WEIGHT	DRIED AT 378K,CONTINENTAL SHELF OFF NEW YORK	308
*DAPHNIA CRISTATA	7.4 (+-1.2) % OF ASH FREE DRY WEIGHT	CLADOCERA , 95 % CONF LIMIT	1375
*EUDIAPTOMUS GRACILIS	10.1 (+-0.4) % OF ASH FREE DRY WEIGHT	COPEPODA , 95 % CONF LIMIT	1375
*EUPHAUSIACEA	5.2-7.1 % OF DRY WEIGHT	DRIED AT 378K,CONTINENTAL SHELF OFF NEW YORK	308
*EUPHAUSIACEA	10.0-10.7 % OF DRY WEIGHT	DRIED AT 333K,NORTH PACIFIC	1184
*EUPHAUSIACEA AND MYSIDAE	9.4-10.5 % OF DRY WEIGHT	DRIED AT 333K,SARGASSO SEA	88
*HYDROMEDUSAE	1.4-6.9 % OF DRY WEIGHT	DRIED AT 333K,SARGASSO SEA	88
*LIMNOCALANUS MACRURUS	6.2 (+-0.3) % OF ASH FREE DRY WEIGHT	COPEPODA , 95 % CONF LIMIT	1375
*MACROCYCLOPS ALBIDUS	10.2 (+-1.0) % OF ASH FREE DRY WEIGHT	COPEPODA , 95 % CONF LIMIT	1375
*POLYCHAETA	4.4-11.2 % OF DRY WEIGHT	DRIED AT 333K,SARGASSO SEA	88
*PTEROPODS	2.7-4.2 % OF DRY WEIGHT	DRIED AT 333K,SARGASSO SEA	88
*PTEROPODS	2.2-5.0 % OF DRY WEIGHT	DRIED AT 378K,CONTINENTAL SHELF OFF NEW YORK	308
*PTEROPODS	1.5-6.0 % OF DRY WEIGHT	DRIED AT 333K,NORTH PACIFIC	1184
*SIPHONOPHORA	1.0-4.4 % OF DRY WEIGHT	DRIED AT 333K,SARGASSO SEA	88
*TUNICATA	0.3-1.5 % OF DRY WEIGHT	DRIED AT 378K,CONTINENTAL SHELF OFF NEW YORK	308
*ZOOPLANKTON	0.07-0.09	NONE % OF DRY WEIGHT	1571

TABLE A460

ZOOPLANKTON : N/P RATIO			
ITEM:	VALUE:	CONDITIONS:	REF.:
*CALANUS CRISTATUS	19 BY ATOMS	UNFILT.WATER,BASED ON RESPIRATION AND EXCRETION RATES	1547
*CALANUS FINMARCHICUS	10.8 BY ATOMS	FILT.WATER,BASED ON RESPIRATION AND EXCRETION RATES	211
*CALANUS FINMARCHICUS	11.0 BY ATOMS	FILT.WATER,BASED ON RESPIRATION AND EXCRETION RATES,TOTAL N/TOTAL P	210
*CALANUS HELGOLANDICUS	16.5 BY ATOMS	FILT.WATER,BASED ON RESPIRATION AND EXCRETION RATES,TOTAL N/TOTAL P	288
*CALANUS PLUMCHRUS	13 BY ATOMS	UNFILT.WATER,BASED ON RESPIRATION AND EXCRETION RATES	1547
*MIXED ZOOPLANKTON	7.0 BY ATOMS	UNFILT.WATER,BASED ON RESPIRATION AND EXCRETION RATES	625
*MIXED ZOOPLANKTON	9.98 BY ATOMS	UNFILT.WATER,BASED ON RESPIRATION AND EXCRETION RATES	987
*MIXED ZOOPLANKTON	10.33 BY ATOMS	UNFILT.WATER,BASED ON RESPIRATION AND EXCRETION RATES	897

TABLE A460

ZOOPLANKTON : N/P RATIO			
ITEM:	VALUE:	CONDITIONS:	REF.:
*MIXED ZOOPLANKTON	6.8 BY ATOMS	UNFILT.WATER,BASED ON RESPIRATION AND EXCRETION RATES	1080
*SAGITTA HISPIDA	11.3 BY ATOMS	FILT.WATER,BASED ON RESPIRATION AND EXCRETION RATES	87

TABLE A461

ZOOPLANKTON : NA/BIOMASS RATIO			
ITEM:	VALUE:	CONDITIONS:	REF.:
*CALANUS FINMARCHICUS	0.54 %	ATOMIC COMPOSITION	1632

TABLE A462

ZOOPLANKTON : O/N RATIO			
ITEM:	VALUE:	CONDITIONS:	REF.:
*ACARTIA CLAUSII	1.61 BY ATOMS	FILT.WATER,BASED ON RESPIRATION AND EXCRETION RATES	355
*BOREAL ZOOPLANKTON	6-200 BY ATOMS	UNFILT.WATER,BASED ON RESP.AND EXCRETION RATES,NINHYDRIN-N, 10SPECIES	276
*BOREAL,TEMPERATE,SUBTROPICAL AND TROPICAL ZOOPLANKTON	4-115 BY ATOMS	FILT.WATER,BASED ON RESPIRATION AND EXCRETION RATES,81 SPECIES	719
*CALANUS CRISTATUS	5.7 BY ATOMS	UNFILT.WATER,BASED ON RESPIRATION AND EXCRETION RATES	1547
*CALANUS HELGOLANDICUS	9.8-15.6 BY ATOMS	FILT.WATER,BASED ON RESPIRATION AND EXCRETION RATES,NINHYDRIN-N	287
*CALANUS PLUMCHRUS	6.8 BY ATOMS	UNFILT.WATER,BASED ON RESPIRATION AND EXCRETION RATES	1547
*MARINE ZOOPLANKTON	13.48 BY ATOMS	FOR RESPIRATION/EXCRETION OF NH4-N,22HOURS,288K,N=44	897
*MARINE ZOOPLANKTON	6.64 BY ATOMS	FOR RESPIRATION/EXCRETION OF TOTAL-N,22HOURS,288K,N=36	897
*MEGANYCTIPHANES NORVEGICA	4.77-12.13 BY ATOMS	FILT.WATER,BASED ON RESPIRATION AND EXCRETION RATES	355
*MIXED ZOOPLANKTON	7.7 BY ATOMS	UNFILT.WATER,BASED ON RESPIRATION AND EXCRETION RATES	625
*MIXED ZOOPLANKTON	41 BY ATOMS	UNFILT.WATER,BASED ON RESPIRATION AND EXCRETION RATES	987
*MIXED ZOOPLANKTON	13.48 BY ATOMS	UNFILT.WATER,BASED ON RESPIRATION AND EXCRETION RATES	897
*PHRONIMA SEDENTARIA	4.27 BY ATOMS	BASED ON RESPIRATION AND EXCRETION RATES	355
*SAGITTA HISPIDA	6.8 BY ATOMS	FILT.WATER,BASED ON RESPIRATION AND EXCRETION RATES	1307
*SAGITTA SETOSA	1.75 BY ATOMS	FILT.WATER,BASED ON RESPIRATION AND EXCRETION RATES	355
*TEMORA STYLIFERA	7-15 BY ATOMS	FILT.WATER,BASED ON RESPIRATION AND EXCRETION RATES	1131

TABLE A463

ZOOPLANKTON : O/P RATIO			
ITEM:	VALUE:	CONDITIONS:	REF.:
*CALANUS CRISTATUS	110 BY ATOMS	UNFILT.WATER,BASED ON RESPIRATION AND EXCRETION RATES	1547

TABLE A463

ZOOPLANKTON : O/P RATIO			
ITEM:	VALUE:	CONDITIONS:	REF.:
*CALANUS PLUMCHRUS	89 BY ATOMS	UNFILT.WATER,BASED ON RESPIRATION AND EXCRETION RATES	1547
*MARINE ZOOPLANKTON	142.4 BY ATOMS	FOR RESPIRATION/EXCRETION OF PO4-P,22HOURS,288K,N=41	897
*MARINE ZOOPLANKTON	66.8 BY ATOMS	FOR RESPIRATION/EXCRETION OF TOTAL-P,22HOURS,288K,N=40	897
*MIXED ZOOPLANKTON	54 BY ATOMS	UNFILT.WATER,BASED ON RESPIRATION AND EXCRETION RATES	625
*MIXED ZOOPLANKTON	222 BY ATOMS	UNFILT.WATER,BASED ON RESPIRATION AND EXCRETION RATES	987
*MIXED ZOOPLANKTON	142.4 BY ATOMS	UNFILT.WATER,BASED ON RESPIRATION AND EXCRETION RATES	897
*MIXED ZOOPLANKTON	72 BY ATOMS	UNFILT.WATER,BASED ON RESPIRATION AND EXCRETION RATES	1384

TABLE A464

ZOOPLANKTON : P/BIOMASS RATIO			
ITEM:	VALUE:	CONDITIONS:	REF.:
*CALANUS FINMARCHICUS	0.13 %	ATOMIC COMPOSITION	1632
*CHAETOGNATHA	0.5-0.7 % OF DRY WEIGHT	DRIED AT 333K,SARGASSO SEA	88
*COPEPODS	0.7-0.8 % OF DRY WEIGHT	DRIED AT 333K,SARGASSO SEA	88
*COPEPODS	0.4-0.8 % OF DRY WEIGHT	DRIED AT 378K,CONTINENTAL SHELF OFF NEW YORK	308
*CTENOPHORA	0.1-0.2 % OF DRY WEIGHT	DRIED AT 378K,CONTINENTAL SHELF OFF NEW YORK	308
*EUPHAUSIACEA	0.9-1.2 % OF DRY WEIGHT	DRIED AT 378K,CONTINENTAL SHELF OFF NEW YORK	308
*EUPHAUSIACEA AND MYSIDAE	1.4-1.6 % OF DRY WEIGHT	DRIED AT 333K,SARGASSO SEA	88
*FISH AND FISH LARVAE	0.9-1.8 % OF DRY WEIGHT	DRIED AT 333K,SARGASSO SEA	88
*HYDROMEDUSAE	0.1-0.4 % OF DRY WEIGHT	DRIED AT 333K,SARGASSO SEA	88
*POLYCHAETA	0.4-1.8 % OF DRY WEIGHT	DRIED AT 333K,SARGASSO SEA	88
*PTEROPODS	0.2-0.4 % OF DRY WEIGHT	DRIED AT 333K,SARGASSO SEA	88
*PTEROPODS	0.3-0.6 % OF DRY WEIGHT	DRIED AT 378K,CONTINENTAL SHELF OFF NEW YORK	308
*SIPHONOPHORA	< 0.1-0.2 % OF DRY WEIGHT	DRIED AT 333K,SARGASSO SEA	88
*TUNICATA	0.1-0.3 % OF DRY WEIGHT	DRIED AT 378K,CONTINENTAL SHELF OFF NEW YORK	308
*ZOOPLANKTON	.01-.012	NONE	1571

TABLE A465

ZOOPLANKTON : PB/BIOMASS RATIO			
ITEM:	VALUE:	CONDITIONS:	REF.:
*CALANUS FINMARCHICUS	0.0000 %	ATOMIC COMPOSITION	1632

TABLE A466

ZOOPLANKTON : PRODUCTION			
ITEM:	VALUE:	CONDITIONS:	REF.:
*ARCHTODIAPTOMUS SP.	0.0014 G DRY WEIGHT/(DAY*M**2)	FILTER ORG,ANNUAL, BIOMASS TURNOVER=162 D, SEVAN, S. USSR, 2 SPECIES	1735

TABLE A466

ZOOPLANKTON : PRODUCTION

ITEM:	VALUE:	CONDITIONS:	REF.:
*ASPLANCHNA PRIODONTA	0.0014 G DRY WEIGHT/(DAY*M**2)	PREDATOR MAY-OCT, BIOMASS TURNOVER =2.9 D, NAROCH L., USSR	1731
*ASPLANCHNA PRIODONTA	0.0061 G DRY WEIGHT/(DAY*M**2)	PREDATOR MAY-OCT, BIOMASS TURNOVER =2.5 D, MYASTRO L., USSR	1731
*ASPLANCHNA PRIODONTA	0.0105 G DRY WEIGHT/(DAY*M**2)	PREDATOR MAY-OCT, BIOMASS TURNOVER =3.2 D, BATORIN L., USSR	1731
*ASPLANCHNA SP.	0.0031 G DRY WEIGHT/(DAY*M**2)	PREDATOR ANNUAL, SEVERSON L., MINN., EUTROPHIC	272
*BOSMINA LONGIROSTRIS	0.007 G DRY WEIGHT/(DAY*M**2)	FILTER ORG,ANNUAL, SEVERSON L., MINN.	305
*BOSMINA LONGIROSTRIS	0.183 G DRY WEIGHT/(DAY*M**2)	FILTER ORG,MAY-NOV 1966, SANCTUARY L., PA. , PLUS B. COREGONI	305
*BOSMINA LONGIROSTRIS	0.067 G DRY WEIGHT/(DAY*M**2)	FILTER ORG,MAY-NOV 1967, SANCTURY L., PA., PLUS B. COREGONI	305
*BRACHIONUS SP.	0.0075 G DRY WEIGHT/(DAY*M**2)	FILTER ORG,ANNUAL, SEVERSON L., MINN., EUTROPHIC	272
*CERIODAPHNIA RETICULATA	0.031 G DRY WEIGHT/(DAY*M**2)	FILTER ORG,JULY-NOV, SANCTUARY L., PA.	305
*CHAOBORUS PUNCTIPENNIS	0.0001 G DRY WEIGHT/(DAY*M**2)	PREDATOR ANNUAL, SEVERSON L., MINN., EUTROPHIC	272
*CHYDORUS SPHAERICUS	0.004 G DRY WEIGHT/(DAY*M**2)	FILTER ORG,JULY-AUG 1966, SANCTURY L., PA. EUTROPHIC RES.	305
*CHYDORUS SPHAERICUS	0.047 G DRY WEIGHT/(DAY*M**2)	FILTER ORG,JULY-AUG 1967, SANCTURY L., PA.	305
*CLADOCERA	0.0026 G DRY WEIGHT/(DAY*M**2)	FILTER ORG,MAY-OCT, BIOMASS TURNOVER=13.7 D, NAROCH L. USSR	1730
*CLADOCERA	0.015 G DRY WEIGHT/(DAY*M**2)	FILTER ORG,MAY-OCT, BIOMASS TURNOVER=10.9 D, MYASTRO L. USSR	1730
*CLADOCERA	0.033 G DRY WEIGHT/(DAY*M**2)	FILTER ORG,MAY-OCT, BIOMASS TURNOVER=10.5 D, BATORIN L. USSR	1730
*CLADOCERA	0.0003 G DRY WEIGHT/(DAY*M**2)	PREDATOR MAY-OCT, BIOMASS TURNOVER =11.3 D, NAROCH L., USSR	1731
*CLADOCERA	0.0009 G DRY WEIGHT/(DAY*M**2)	PREDATOR MAY-OCT, BIOMASS TURNOVER =3.9 D, MYASTRO L., USSR	1731
*CLADOCERA	0.0002 G DRY WEIGHT/(DAY*M**2)	PREDATOR MAY-OCT, BIOMASS TURNOVER =10.8 D, BATORIN L., USSR	1731
*CYCLOPS SP.	0.0008 G DRY WEIGHT/(DAY*M**2)	PREDATOR MAY-OCT, BIOMASS TURNOVER =9.7 D, NAROCH L., USSR	1731
*CYCLOPS SP.	0.0023 G DRY WEIGHT/(DAY*M**2)	PREDATOR MAY-OCT, BIOMASS TURNOVER =19.4 D, MYASTRO L., USSR	1731
*CYCLOPS SP.	0.0094 G DRY WEIGHT/(DAY*M**2)	PREDATOR MAY-OCT, BIOMASS TURNOVER =14.2 D, BATORIN L., USSR	1731
*CYCLOPS STRENUUS	0.0004 G DRY WEIGHT/(DAY*M**2)	FILTER ORG,ANNUAL, BUTTERMERE, ENGL., OLIGOTROPHIC	1466
*CYCLOPS STRENUUS	0.0005 G DRY WEIGHT/(DAY*M**2)	FILTER ORG,ANNUAL, RYDAL WATER, ENGL., EUTROPHIC	1466
*CYCLOPS STRENUUS	0.0006 G DRY WEIGHT/(DAY*M**2)	FILTER ORG,ANNUAL, GRASMERE, ENGL., EUTROPHIC	1466
*CYCLOPS STRENUUS	0.0017 G DRY WEIGHT/(DAY*M**2)	FILTER ORG,ANNUAL, ESTHWAITE WATER, ENGL., EUTROPHIC	1466
*CYCLOPS STRENUUS	0.0007 G DRY WEIGHT/(DAY*M**2)	FILTER ORG,ANNUAL, BIOMASS TURNOVER=79.3 D, L. SEVAN, SOUTHERN USSR	1735
*DAPHNIA GALEATA MENDOTAE	0.407 G DRY WEIGHT/(DAY*M**2)	FILTER ORG,MAY-NOV 1966, SANCTURY L., PA., EUTROPHIC	305
*DAPHNIA GALEATA MENDOTAE	0.030 G DRY WEIGHT/(DAY*M**2)	FILTER ORG,MAY-NOV 1967, SANCTURY L., PA., EUTROPHIC	305
*DAPHNIA GALEATA MENDOTAE	0.114 G DRY WEIGHT/(DAY*M**2)	FILTER ORG,APRIL-SEP, CANYON FERRY ,MONT, EUTROPH, TURNOVER=10.0 D	1753
*DAPHNIA HYALINA	0.57 G DRY WEIGHT/(DAY*M**2)	FILTER ORG,ANNUAL,BIOMASS TURNOVER =21.3 D,EUTROPHIC, EGLWEYS RES.	514
*DAPHNIA HYALINA	0.32 G DRY WEIGHT/(DAY*M**2)	FILTER ORG,ANNUAL 1971,BIOMASS TURNOVER=15.9 D, EUTROPHIC, WALES	514
*DAPHNIA LONGISPINA	0.006 G DRY WEIGHT/(DAY*M**2)	FILTER ORG,ANNUAL, L. SEVAN, SOUTHERN USSR, BIOMASS TURNOVER= 58.9 D	1735
*DAPHNIA PARVULA	0.010 G DRY WEIGHT/(DAY*M**2)	FILTER ORG,ANNUAL, SEVERSON L. EUTROPHIC, MINN.	272

TABLE A466

ZOOPLANKTON : PRODUCTION

ITEM:	VALUE:	CONDITIONS:	REF.:
*DAPHNIA SCHODLERI	0.227 G DRY WEIGHT/(DAY*M**2)	FILTER ORG,APRIL-SEP, CANYON FERRY, MONT, EUTROPH, TURNOVER= 6.7 D	1753
*DIAPHANOSOMA LEUCHTENBERGIANUM	0.0067 G DRY WEIGHT/(DAY*M**2)	FILTER ORG,ANNUAL, SEVERSON L., MINN., EUTROPHIC	272
*DIAPTOMUS SICILOIDES	0.0342 G DRY WEIGHT/(DAY*M**2)	FILTER ORG,ANNUAL, SEVERSON L., MINN., EUTROPHIC	272
*EUDIAPTOMUS GRACILOIDES	0.0010 G DRY WEIGHT/(DAY*M**2)	FILTER ORG,MAY-OCT, BIOMASS TURNOVER=24.7 D, NAROCH L., USSR	1730
*EUDIAPTOMUS GRACILOIDES	0.0065 G DRY WEIGHT/(DAY*M**2)	FILTER ORG,MAY-OCT, BIOMASS TURNOVER=20.2 D, MYASTRO L. USSR	1730
*EUDIAPTOMUS GRACILOIDES	0.0070 G DRY WEIGHT/(DAY*M**2)	FILTER ORG,MAY-OCT, BIOMASS TURNOVER=15.4 D, BATORIN L. USSR	1730
*FILINIA LONGISETA	0.0011 G DRY WEIGHT/(DAY*M**2)	FILTER ORG,ANNUAL, SEVERSON L., MINN., EUTROPHIC	272
*KERATELLA COCHLEARIS	0.0007 G DRY WEIGHT/(DAY*M**2)	FILTER ORG,ANNUAL, SEVERSON L., MINN., EUTROPHIC	272
*KERATELLA QUADRATA	0.0021 G DRY WEIGHT/(DAY*M**2)	FILTER ORG,ANNUAL, SEVERSON L., MINN. EUTROPHIC	272
*LEPTODORA KINDTII	0.003 G DRY WEIGHT/(DAY*M**2)	PREDATOR MAY-NOV 1966, SANCTUARY L., PA.	305
*LEPTODORA KINDTII	0.013 G DRY WEIGHT/(DAY*M**2)	PREDATOR MAY-OCT 1967, SANCTUARY L., PA.	305
*MESOCYCLOPS EDAX	0.0046 G DRY WEIGHT/(DAY*M**2)	FILTER ORG,ANNUAL, SEVERSON L., MINN., EUTROPHIC	272
*POLYARTHRA SP.	0.0010 G DRY WEIGHT/(DAY*M**2)	FILTER ORG,ANNUAL, SEVERSON L., MINN., EUTROPHIC	272
*ROTIFERA	0.0027 G DRY WEIGHT/(DAY*M**2)	FILTER ORG,MAY-OCT., BIOMASS TURNOVER=1.7 D, NAROCH L., USSR	1730
*ROTIFERA	0.0024 G DRY WEIGHT/(DAY*M**2)	FILTER ORG,MAY-OCT, BIOMASS TURNOVER=3.9 D, MYASTRO L., USSR	1730
*ROTIFERA	0.0105 G DRY WEIGHT/(DAY*M**2)	FILTER ORG,MAY-OCT, BIOMASS TURNOVER=2.6 D, BATORIN L., USSR	1730
*SYNCHAETA SP.	0.00009 G DRY WEIGHT/(DAY*M**2)	PREDATOR ANNUAL, SEVERSON L., MINN., EUTROPHIC	272
*ZOOPLANKTON	963 MG DRY WEIGHT/M3	MEAN BIOMASS=67 MICRO G/L, LOW NUTRIENT, BASED ON 8 W, 1965	595
*ZOOPLANKTON	1316 MG DRY WEIGHT/M3	MEAN BIOMASS=104 MICRO G/L,MEDIAN NUTRIENT, BASED ON 8 W, 1965	595
*ZOOPLANKTON	5762 MG DRY WEIGHT/M3	MEAN BIOMASS=416 MICRO G/L,HIG NUTRIENT, BASED ON 8 W, 1965	595
*ZOOPLANKTON	3925 MG DRY WEIGHT/M3	MEAN BIOMASS=259 MICRO G/L,LOW NUTRIENT, BASED ON 12W, 1966	595
*ZOOPLANKTON	3514 MG DRY WEIGHT/M3	MEAN BIOMASS=141 MICRO G/L,MEDIAN NUTRIENT, BASED ON 12 W, 1966	595
*ZOOPLANKTON	4756 MG DRY WEIGHT/M3	MEAN BIOMASS=492 MICRO G/L,HIG NUTRIENT, BASED ON 12W, 1966	595
*ZOOPLANKTON	3412 MG DRY WEIGHT/M3	MEAN BIOMASS=124 MICRO G/L,LOW NUTRIENT, BASED ON 12W, 1967	595
*ZOOPLANKTON	3043 MG DRY WEIGHT/M3	MEAN BIOMASS=134 MICRO G/L,MEDIAN NUTRIENT, BASED ON 12 W, 1967	595
*ZOOPLANKTON	11823 MG DRY WEIGHT/M3	MEAN BIOMASS=1734 MICRO G/L,HIG NUTRIENT,BASED ON 12 W, 1967	595

TABLE A467

ZOOPLANKTON : PRODUCTION OF YOUNG

ITEM:	VALUE:	CONDITIONS:	REF.:
*DAPHNIA PULEX	16.5 % OF ENERGY CONSUMED	6-34 D OLD, FOOD = 25000 CELLS/ML, CHLAMYDOMONAS REINHARDTII	1328
*DAPHNIA PULEX	12.5 % OF ENERGY CONSUMED	6-34 D OLD, FOOD = 50000 CELLS/ML, CHLAMYDOMONAS REINHARDTII	1328
*DAPHNIA PULEX	10.9 % OF ENERGY CONSUMED	6-34 D OLD, FOOD = 75000 CELLS/ML, CHLAMYDOMONAS REINHARDTII	1328

TABLE A467

ZOOPLANKTON : PRODUCTION OF YOUNG

ITEM:	VALUE:	CONDITIONS:	REF.:
*DAPHNIA PULEX	10.0 % OF ENERGY CONSUMED	6-34 D OLD, FOOD = 100000 CELLS/ML, CHLAMYDOMONAS REINHARDTII	1328
*DAPHNIA PULEX	52.0 % OF ENERGY ASSIMILATED	6-34 D OLD, FOOD = 25000 CELLS/ML, CHLAMYDOMONAS REINHARDTII	1328
*DAPHNIA PULEX	61.8 % OF ENERGY ASSIMILATED	6-34 D OLD, FOOD = 50000 CELLS/ML, CHLAMYDOMONAS REINHARDTII	1328
*DAPHNIA PULEX	66.8 % OF ENERGY ASSIMILATED	6-34 D OLD, FOOD = 75000 CELLS/ML, CHLAMYDOMONAS REINHARDTII	1328
*DAPHNIA PULEX	70.5 % OF ENERGY ASSIMILATED	6-34 D OLD, FOOD = 100000 CELLS/ML, CHLAMYDOMONAS REINHARDTII	1328
*DAPHNIA PULEX	76.3 AVERAGE NUMBERS	6-34 D OLD, FOOD = 25000 CELLS/ML, CHLAMYDOMONAS REINHARDTII	1328
*DAPHNIA PULEX	132 AVERAGE NUMBERS	6-34 D OLD, FOOD = 50000 CELLS/ML, CHLAMYDOMONAS REINHARDTII	1328
*DAPHNIA PULEX	177 AVERAGE NUMBERS	6-34 D OLD, FOOD = 75000 CELLS/ML, CHLAMYDOMONAS REINHARDTII	1328
*DAPHNIA PULEX	223 AVERAGE NUMBERS	6-34 D OLD, FOOD = 100000 CELLS/ML, CHLAMYDOMONAS REINHARDTII	1328

TABLE A468

ZOOPLANKTON : PRODUCTION/ASSIMILATION

ITEM:	VALUE:	CONDITIONS:	REF.:
*ASELLOPSIS INTERMEDIA	56.7 %	CUM. C BUDGET TO 14 D	884
*ASELLOPSIS INTERMEDIA	35.9 %	CUM. C BUDGET TO 55 D	884
*ASELLOPSIS INTERMEDIA	17.9 %	CUM. C BUDGET TO 126 D	884
*ASELLOPSIS INTERMEDIA	11.8 %	CUM. C BUDGET TO 300 D	884
*CALANUS FINMARCHICUS	55.0 %	CUM N BUDGET , PRE-ADULT PHASE	286
*CALANUS FINMARCHICUS	22.7 %	CUM N BUDGET , ADULT PHASE, 250 EGGS/(FEMALE*35 D)	286
*CALANUS FINMARCHICUS	63.8 %	CUM N BUDGET , ADULT PHASE, 1500 EGGS/(FEMALE*35 D)	286
*CALANUS FINMARCHICUS	37.7 %	CUM N BUDGET , TOTAL LIFE, 250 EGGS/(FEMALE*35 D)	286
*CALANUS FINMARCHICUS	61.3 %	CUM N BUDGET , TOTAL LIFE, 1500 EGGS/(FEMALE*35 D)	286
*CALLIOPIUS LAEVIUSCULUS	43.4 %	CUM. C BUDGET, PRE-ADULT PHASE, 3 DIFF TEMP	319
*CALLIOPIUS LAEVIUSCULUS	24.1 %	CUM. C BUDGET, TOTAL LIFE, 3 DIFF TEMP	319
*EUPHAUSIA PACIFICA	25.8 %	CUM. C BUDGET OVER 70 D, LESS THAN 1 MG DRY WEIGHT	882
*EUPHAUSIA PACIFICA	11.8 %	CUM C BUDGET OVER 25 - 70 D, LESS THAN 1 MG DRY WEIGHT	882
*HYALELLA AZTECA	12.6 %	DAILY CALORIE BUDGET FOR LARGE ADULT	617
*MENIPPE MERCENARIA	28.4 %	DAILY CALORIE BUDGET, STAGE 1	1062
*MENIPPE MERCENARIA	38.4 %	DAILY CALORIE BUDGET, STAGE 2	1062
*MENIPPE MERCENARIA	46.2 %	DAILY CALORIE BUDGET, STAGE 3	1062
*MENIPPE MERCENARIA	46.8 %	DAILY CALORIE BUDGET, STAGE 4	1062
*MENIPPE MERCENARIA	59.3 %	DAILY CALORIE BUDGET, STAGE 5	1062
*MENIPPE MERCENARIA	43.9 %	DAILY CALORIE BUDGET, MEGALOPA	1062
*MENIPPE MERCENARIA	47.5 %	CUM. CALORIE BUDGET TO MEGALOPA	1062
*METAMYSIDOPSIS ELONGATA	38.0 %	CUM CALORIE BUDGET TO 45 D	262
*METAMYSIDOPSIS ELONGATA	44.0 %	CUM CALORIE BUDGET TO 53 D	262
*METAMYSIDOPSIS ELONGATA	43.0 %	CUM CALORIE BUDGET TO 68 D	262
*METAMYSIDOPSIS ELONGATA	30.0 %	CUM CALORIE BUDGET TO 103 D	262

TABLE A468

ZOOPLANKTON : PRODUCTION/ASSIMILATION

ITEM:	VALUE:	CONDITIONS:	REF.:
*MYSIS RELICTA	17.0 %	CUM CALORIE BUDGET TO , TOTAL LIFE	880
*TIGRIOPUS BREVICORNIS	17.2 %	CUM N BUDGET , PRE-ADULT PHASE	626
*TIGRIOPUS BREVICORNIS	29.3 %	CUM N BUDGET , ADULT PHASE	626
*TIGRIOPUS BREVICORNIS	26.5 %	CUM N BUDGET , TOTAL LIFE, FEMALE	626

TABLE A469

ZOOPLANKTON : PRODUCTION/BIOMASS RATIO

ITEM:	VALUE:	CONDITIONS:	REF.:
*CYCLOPS VICINUS	21 ON YEARLY BASIS	PROD=10.2 G DW/(M2*Y), EGLWYS NUNYDD R. WALES, UK	1661
*DAPHNIA HYALINA	23 ON YEARLY BASIS	PROD=35.3 G DW/(M2*Y), EGLWYS NUNYDD R., WALES, N=2 Y, UK	1661
*DAPHNIA LONGISPINA	2.5 ON YEARLY BASIS	PROD=0.56 G DW/(M2*Y), L. PORT-BIELH, FRANCE	1661
*EUDIAPTOMUS GRACILOIDES	6.8 ON YEARLY BASIS	PROD=7.7 G DW/(HA*Y), L. ESROM, DENMARK	1661
*EURYCERCUS LAMELLATUS	20 ON YEARLY BASIS	PROD=0.09 G DW/(HA*Y), L. NAROCH, USSR	1661
*MIXODIAPTOMUS LACINIATUS	2.5 ON YEARLY BASIS	PROD=1.3 G DW/(M2*Y), L. PORT-BIELH, FRANCE	1661
*THERMOCYCLOPS HYALINUS	28.5 ON YEARLY BASIS	PROD=16.1 G DW/(M2*Y), L. GEORGE, UGANDA	1661
*TOTAL CLADOCERA	22 ON YEARLY BASIS	PROD=19.2 G DW/(M2*Y), TALTOWISKO L., POLAND, MESOTROPHIC	1661
*TOTAL CLADOCERA	31 ON YEARLY BASIS	PROD=24.7 G DW/(M2*Y), MAKOLAJKIE L, POLAND, EUTROPHIC	1661
*TOTAL COPEPODA	1.2 ON YEARLY BASIS	PROD=0.27 G DW/(M2*Y), SNOWFLAKE L., ALBERTA	1661
*TOTAL COPEPODA	3.0 ON YEARLY BASIS	PROD=0.75 G DW/(M2*Y), TEARDROP L., ALBERTA	1661
*TOTAL COPEPODA	12 ON YEARLY BASIS	PROD=7.8 G DW/(M2*Y), TALTOWISKO L. POLAND, MESOTROPHIC	1661
*TOTAL COPEPODA	17.5 ON YEARLY BASIS	PROD=9.9 G DW/(M2*Y), MIKOLAJSKIE L., POLAND, EUTROPHIC	1661
*TOTAL ROTIFERA	42 ON YEARLY BASIS	PROD=6.2 G DW/(M2*Y), TALTOWISKO L. POLAND, MESOTROPHIC	1661
*TOTAL ROTIFERA	43 ON YEARLY BASIS	PROD=14.7 G DW/(M2*Y), MIKOLAJKIE L., POLAND, EUTROPHIC	1661
*TOTAL ROTIFERA	60 ON YEARLY BASIS	PROD=5.4 G DW/(M2*Y), RYBINSK R., USSR	1661
*TOTAL ROTIFERA	10.1 ON YEARLY BASIS	PROD=0.08 G DW/(M2*Y), L. ZELENETSKOYE, USSR, MAX 2 Y, COLD N. LAKE	1661
*TOTAL ROTIFERA	8.0 ON YEARLY BASIS	PROD=0.05 G DW/(M2*Y), L. AKULKINO, USSR, MAX 2 Y, COLD NORTH LAKE	1661
*TOTAL ROTIFERA, NON-PREDATOR	75 ON YEARLY BASIS	PROD=3.9 G DW/(M2*Y), THREE LAKES, USSR	1661
*TOTAL ROTIFERA, PREDATOR	64 ON YEARLY BASIS	PROD=4.6 G DW/(M2*Y), THREE LAKES, USSR	1661
*TOTAL ZOOPLANKTON, NON-PREDATOR	13 ON YEARLY BASIS	PROD=3.4 G DW/(M2*Y), LAKE KRIVOE, USSR	1661
*TOTAL ZOOPLANKTON, NON-PREDATOR	13 ON YEARLY BASIS	PROD=11 G DW/(M2*Y), LAKE KRUGLOE, USSR	1661
*TOTAL ZOOPLANKTON, NON-PREDATOR	18 ON YEARLY BASIS	PROD=17.7 G DW/(M2*Y), RED L., USSR	1661
*TOTAL ZOOPLANKTON, NON-PREDATOR	13.5 ON YEARLY BASIS	PROD=16.1 G DW/(M2*Y), L. BAIBAL, USSR	1661
*TOTAL ZOOPLANKTON, NON-PREDATOR	22 ON YEARLY BASIS	PROD=20.8 G DW/(M2*Y), N=9 LAKES IN USSR	1661
*TOTAL ZOOPLANKTON, PREDATOR	5 ON YEARLY BASIS	PROD=0.52 G DW/(M2*Y), L. KRIVOE, USSR	1661

TABLE A469

ZOOPLANKTON : PRODUCTION/BIOMASS RATIO			
ITEM:	VALUE:	CONDITIONS:	REF.:
*TOTAL ZOOPLANKTON, PREDATOR	11 ON YEARLY BASIS	PROD=2.0 G DW/(M2*Y), L. KRUGLOE, USSR	1661
*TOTAL ZOOPLANKTON, PREDATOR	15 ON YEARLY BASIS	PROD=2.0 G DW/(M2*Y), RED L., USSR	1661
*TOTAL ZOOPLANKTON, PREDATOR	2.4 ON YEARLY BASIS	PROD=1.6 G DW/(M2*Y), L. BAIKAL, USSR	1661
*TOTAL ZOOPLANKTON, ROTIFERA AND CRUSTACEA	7.6 ON YEARLY BASIS	PROD=0.70 G DW/(M2*Y), L. ZELENETZKOYE, USSR, MAX OF 2 Y	1661
*TOTAL ZOOPLANKTON, ROTIFERA AND CRUSTACEA	8.7 ON YEARLY BASIS	PROD=0.087 G DW/(M2*Y), L. AKULKINO, USSR, COLD N. LAKE	1661
*TOTAL ZOOPLANKTON, ROTIFERA, CLADOCERA AND COPEPODA	13 ON YEARLY BASIS	PROD=1.60 G DW/(M2*Y), L. KRUGLOE, USSR	1661
*TOTAL ZOOPLANKTON, ROTIFERA, COPEPODA AND CLADOCERA	14 ON YEARLY BASIS	PROD=1.98 G DW/(M2*Y), L. KRIVOE, USSR	1661
*TOTAL ZOOPLANKTON, ROTIFERA, COPEPODA AND CLADOCERA	22 ON YEARLY BASIS	PROD=33.2 G DW/(M2*Y), TALTOWISKO L., POLAND, MESOTROPHIC	1661
*TOTAL ZOOPLANKTON, ROTIFERA, COPEPODA AND CLADOCERA	27 ON YEARLY BASIS	PROD=49.3 G DW/(M2*Y), MIKOLAJSKIE L., POLAND, EUTROPHIC	1661

TABLE A470

ZOOPLANKTON : RB/BIOMASS RATIO			
ITEM:	VALUE:	CONDITIONS:	REF.:
*CALANUS FINMARCHICUS	0.0000 %	ATOMIC COMPOSITION	1632

TABLE A471

ZOOPLANKTON : RESPIRATION RATE			
ITEM:	VALUE:	CONDITIONS:	REF.:
*ACARTIA CLAUSII	13.0 K	ENDOGENOUS RESP., 283 MICROLITRE O2/(MG DW*HOUR)	944
*ACARTIA CLAUSII	12.4 MICROLITRE O2/(MG DW*HOUR)	ENDOGENOUS RESP., 286 K	944
*ACARTIA CLAUSII	15.8 MICROLITRE O2/(MG DW*HOUR)	ENDOGENOUS RESP., 290 K	944
*ACARTIA CLAUSII	18.9 MICROLITRE O2/(MG DW*HOUR)	ENDOGENOUS RESP., 293 K	944
*BRACHYURAN LARVAE	1.45+-0.19 MICROLITRE O2/(ANIMAL*HOUR)	DECAPODA,MEGALOPS,289K,MEAN OF 2 EXPERIMENTS	827
*CALANOIDES CARINATUS	0.472+-0.324 MICROLITRE O2/(ANIMAL*HOUR)	COPEPODA,289K,MEAN OF 13 EXPERIMENTS	827
*CALANUS FINMARCHICUS	4.2 ML O2/(MG DW*DAY)	ENDOGENOUS RESP., 293 K	354
*CALANUS FINMARCHICUS	2.3 ML O2/(MG DW*DAY)	ENDOGENOUS RESP., 288 K	354
*CALANUS FINMARCHICUS	1.4 ML O2/(MG DW*DAY)	ENDOGENOUS RESP., 283 K	354
*CALANUS FINMARCHICUS	1.3 ML O2/(MG DW*DAY)	ENDOGENOUS RESP., 277 K	354
*CALANUS FINMARCHICUS	4.2 ML O2/(MG DW*DAY)	293K, ENDOGENOUS RESP	1340
*CALANUS FINMARCHICUS	2.3 ML O2/(MG DW*DAY)	288K, ENDOGENOUS RESP	1340
*CALANUS FINMARCHICUS	1.4 ML O2/(MG DW*DAY)	283K, ENDOGENOUS RESP	1340
*CALANUS FINMARCHICUS	1.3 ML O2/(MG DW*DAY)	277K, ENDOGENOUS RESP	1340
*CALANUS PACIFICUS	0.590+-0.019 MICROLITRE O2/(ANIMAL*HOUR)	COPEPODA,FEMALE,288K,MEAN OF 3 EXPERIMENTS	827
*CALANUS PACIFICUS	0.259+-0.050 MICROLITRE O2/(ANIMAL*HOUR)	COPEPODA,COPEPODITE 5,288K,MEAN OF 5 EXPERIMENTS	827
*CALANUS PACIFICUS	0.086+-0.007 MICROLITRE O2/(ANIMAL*HOUR)	COPEPODA,COPEPODITE 4,288K,MEAN OF 3 EXPERIMENTS	827
*CALANUS PACIFICUS	0.044+-0.004 MICROLITRE O2/(ANIMAL*HOUR)	COPEPODA,COPEPODITE 2,288K,MEAN OF 2 EXPERIMENTS	827
*CALANUS PACIFICUS	0.004+-0.001 MICROLITRE O2/(ANIMAL*HOUR)	COPEPODA,NAUPLII 1,288K,MEAN OF 2 EXPERIMENTS	827

TABLE A471

ZOOPLANKTON : RESPIRATION RATE			
ITEM:	VALUE:	CONDITIONS:	REF.:
*CALANUS SP.	0.876+-0.105 MICROLITRE O2/ (ANIMAL*HOUR)	COPEPODA,289K,MEAN OF 3 EXPERIMENTS	827
*CENTROPAGES HAMATUS	2.27 MICROLITRE O2/(MG DW*HOUR)	ENDOGENOUS RESP., 279 K	944
*CENTROPAGES HAMATUS	4.54 MICROLITRE O2/(MG DW*HOUR)	ENDOGENOUS RESP., 283 K	944
*CENTROPAGES HAMATUS	6.34 MICROLITRE O2/(MG DW*HOUR)	ENDOGENOUS RESP., 286 K	944
*CENTROPAGES HAMATUS	7.56 MICROLITRE O2/(MG DW*HOUR)	ENDOGENOUS RESP., 290 K	944
*CERIODAPHNIA RETICULATA	1.8 % OF INGESTION	JUVENILE, FEMALE	1689
*CERIODAPHNIA SP. AND DAPHNIA SP.	2.7 MICROLITRE O2/(MG DW*HOUR)	ENDOGENOUS RESP., 277 K	944
*CERIODAPHNIA SP. AND DAPHNIA SP.	14.2 MICROLITRE O2/(MG DW*HOUR)	ENDOGENOUS RESP., 292 K	944
*CLADOCERA	14.2 ML O2/(MG DW*DAY)	ENDOGENOUS RESP., 291 K	354
*CLADOCERA	2.7 ML O2/(MG DW*DAY)	ENDOGENOUS RESP., 277 K	354
*CLADOCERA	14.2 ML O2/(MG DW*DAY)	291K, ENDOGENOUS RESP	131
*CLADOCERA	2.7 ML O2/(MG DW*DAY)	277K, ENDOGENOUS RESP	131
*COPEPODS	12.2 ML O2/(MG DW*DAY)	ENDOGENOUS RESP., 291 K	354
*COPEPODS	3.8 ML O2/(MG DW*DAY)	ENDOGENOUS RESP., 277 K	354
*COPEPODS	8.2 ML O2/(MG DW*DAY)	ENDOGENOUS RESP., 291 K	354
*COPEPODS	6.5 ML O2/(MG DW*DAY)	ENDOGENOUS RESP., 289 K	354
*COPEPODS	5.2 ML O2/(MG DW*DAY)	ENDOGENOUS RESP., 285 K	354
*COPEPODS	4.1 ML O2/(MG DW*DAY)	ENDOGENOUS RESP., 281 K	354
*COPEPODS	3.4 ML O2/(MG DW*DAY)	ENDOGENOUS RESP., 277 K	354
*COPEPODS	0.075 1/DAY	USED IN MODEL, SAN JOAQUIN RIVER, CALIF	1567
*COPEPODS	12.2 ML O2/(MG DW*DAY)	291K, ENDOGENOUS RESP	131
*COPEPODS	3.8 ML O2/(MG DW*DAY)	277K, ENDOGENOUS RESP	131
*COPEPODS	8.2 ML O2/(MG DW*DAY)	291K, ENDOGENOUS RESP	131
*COPEPODS	6.5 ML O2/(MG DW*DAY)	289K, ENDOGENOUS RESP	131
*COPEPODS	5.2 ML O2/(MG DW*DAY)	285K, ENDOGENOUS RESP	131
*COPEPODS	4.1 ML O2/(MG DW*DAY)	281K, ENDOGENOUS RESP	131
*COPEPODS	3.4 ML O2/(MG DW*DAY)	277K, ENDOGENOUS RESP	131
*CYCLOPS SP. AND DIAPTOMUS SP.	3.8 MICROLITRE O2/(MG DW*HOUR)	ENDOGENOUS RESP., 277 K	944
*CYCLOPS SP. AND DIAPTOMUS SP.	12.2 MICROLITRE O2/(MG DW*HOUR)	ENDOGENOUS RESP., 292 K	944
*DAPHNIA MAGNA	3.4 MICROLITRE O2/(MG DW*HOUR)	ENDOGENOUS RESP., 283 K	944
*DAPHNIA MAGNA	5.4 MICROLITRE O2/(MG DW*HOUR)	ENDOGENOUS RESP., 298 K	944
*DAPHNIA PULEX	78.6 % OF GROWTH + YOUNG PRODUCTION	0-40 D OLD, FOOD = 25000 CELLS/ML, CHLAMYDOMONAS REINHARDTII	1328
*DAPHNIA PULEX	53 % OF GROWTH + YOUNG PRODUCTION	0-40 D OLD, FOOD = 50000 CELLS/ML, CHLAMYDOMONAS REINHARDTII	1328
*DAPHNIA PULEX	43 % OF GROWTH + YOUNG PRODUCTION	0-40 D OLD, FOOD = 75000 CELLS/ML, CHLAMYDOMONAS REINHARDTII	1328
*DAPHNIA PULEX	37 % OF GROWTH + YOUNG PRODUCTION	0-40 D OLD, FOOD = 100000 CELLS/ ML, CHLAMYDOMONAS REINHARDTII	1328
*DAPHNIA PULEX	78.6 % OF GROWTH AND REPRODUCTION	FOOD = 25*10**(-4) CELL/ML (CHLAMYDOMONAS), 40 D,AFTER 18 INSTARS	1327
*DAPHNIA PULEX	53.0 % OF GROWTH AND REPRODUCTION	FOOD = 50*10**(-4) CELL/ML (CHLAMYDOMONAS), 40 D,AFTER 18 INSTARS	1327
*DAPHNIA PULEX	43.3 % OF GROWTH AND REPRODUCTION	FOOD = 75*10**(-4) CELL/ML (CHLAMYDOMONAS), 40 D,AFTER 18 INSTARS	1327
*DAPHNIA PULEX	37.0 % OF GROWTH AND REPRODUCTION	FOOD = 100*10**(-4) CELL/ML (CHLAMYDOMONAS), 40 D,AFTER 18 INSTARS	1327
*DAPHNIA PULEX	4 - 14 % OF INGESTION	NONE	1689
*DAPHNIA PULEX	27 - 44 % OF ASSIMILATION	NONE	1689
*DIAPTOMUS ARCTICUS	0.982 MICROLITRE O2/(MG DW*HOUR)	ENDOGENOUS RESP., 278 K	944
*DIAPTOMUS ARCTICUS	1.48 MICROLITRE O2/(MG DW*HOUR)	ENDOGENOUS RESP., 283 K	944

TABLE A471

ZOOPLANKTON : RESPIRATION RATE

ITEM:	VALUE:	CONDITIONS:	REF.:
*DIAPTOMUS ARCTICUS	2.06 MICROLITRE O2/(MG DW*HOUR)	ENDOGENOUS RESP., 288 K	944
*DIAPTOMUS ARCTICUS	2.93 MICROLITRE O2/(MG DW*HOUR)	ENDOGENOUS RESP., 293 K	944
*DIAPTOMUS ARCTICUS	4.38 MICROLITRE O2/(MG DW*HOUR)	ENDOGENOUS RESP., 298 K	944
*DIAPTOMUS CLACIPES	12.5 ML O2/(MG DW*DAY)	298K, ENDOGENOUS RESP	271
*DIAPTOMUS CLACIPES	8.5 ML O2/(MG DW*DAY)	293K, ENDOGENOUS RESP	271
*DIAPTOMUS CLACIPES	5.1 ML O2/(MG DW*DAY)	288K, ENDOGENOUS RESP	271
*DIAPTOMUS CLACIPES	2.4 ML O2/(MG DW*DAY)	283K, ENDOGENOUS RESP	271
*DIAPTOMUS CLACIPES	1.8 ML O2/(MG DW*DAY)	278K, ENDOGENOUS RESP	271
*DIAPTOMUS CLACIPES	2.14 MICROLITRE O2/(MG DW*HOUR)	ENDOGENOUS RESP., 278 K	944
*DIAPTOMUS CLACIPES	2.85 MICROLITRE O2/(MG DW*HOUR)	ENDOGENOUS RESP., 283 K	944
*DIAPTOMUS CLACIPES	4.64 MICROLITRE O2/(MG DW*HOUR)	ENDOGENOUS RESP., 288 K	944
*DIAPTOMUS CLACIPES	7.05 MICROLITRE O2/(MG DW*HOUR)	ENDOGENOUS RESP., 293 K	944
*DIAPTOMUS CLACIPES	10.6 MICROLITRE O2/(MG DW*HOUR)	ENDOGENOUS RESP., 298 K	944
*DIAPTOMUS CLACIPES	12.5 ML O2/(MG DW*DAY)	ENDOGENOUS RESP., 298 K	354
*DIAPTOMUS CLACIPES	8.5 ML O2/(MG DW*DAY)	ENDOGENOUS RESP., 293 K	354
*DIAPTOMUS CLACIPES	5.1 ML O2/(MG DW*DAY)	ENDOGENOUS RESP., 288 K	354
*DIAPTOMUS CLACIPES	2.4 ML O2/(MG DW*DAY)	ENDOGENOUS RESP., 283 K	354
*DIAPTOMUS CLACIPES	1.8 ML O2/(MG DW*DAY)	ENDOGENOUS RESP., 278 K	354
*DIAPTOMUS LEPTOPUS	2.33 MICROLITRE O2/(MG DW*HOUR)	ENDOGENOUS RESP., 278 K	944
*DIAPTOMUS LEPTOPUS	3.06 MICROLITRE O2/(MG DW*HOUR)	ENDOGENOUS RESP., 283 K	944
*DIAPTOMUS LEPTOPUS	5.07 MICROLITRE O2/(MG DW*HOUR)	ENDOGENOUS RESP., 288 K	944
*DIAPTOMUS LEPTOPUS	7.76 MICROLITRE O2/(MG DW*HOUR)	ENDOGENOUS RESP., 293 K	944
*DIAPTOMUS LEPTOPUS	11.7 MICROLITRE O2/(MG DW*HOUR)	ENDOGENOUS RESP., 298 K	944
*DIAPTOMUS LEPTOPUS	12.1 ML O2/(MG DW*DAY)	ENDOGENOUS RESP., 298 K	354
*DIAPTOMUS LEPTOPUS	7.4 ML O2/(MG DW*DAY)	ENDOGENOUS RESP., 293 K	354
*DIAPTOMUS LEPTOPUS	5.3 ML O2/(MG DW*DAY)	ENDOGENOUS RESP., 288 K	354
*DIAPTOMUS LEPTOPUS	2.8 ML O2/(MG DW*DAY)	ENDOGENOUS RESP., 283 K	354
*DIAPTOMUS LEPTOPUS	2.5 ML O2/(MG DW*DAY)	ENDOGENOUS RESP., 278 K	354
*DIAPTOMUS LEPTOPUS	12.5 ML O2/(MG DW*DAY)	298K, ENDOGENOUS RESP	271
*DIAPTOMUS LEPTOPUS	7.4 ML O2/(MG DW*DAY)	293K, ENDOGENOUS RESP	271
*DIAPTOMUS LEPTOPUS	5.3	288K, ENDOGENOUS RESP	271
*DIAPTOMUS LEPTOPUS	2.8 ML O2/(MG DW*DAY)	283K, ENDOGENOUS RESP	271
*DIAPTOMUS LEPTOPUS	2.5 ML O2/(MG DW*DAY)	278K, ENDOGENOUS RESP	271
*DIAPTOMUS OREGONENSIS	3.88 MICROLITRE O2/(MG DW*HOUR)	ENDOGENOUS RESP., 278 K	944
*DIAPTOMUS OREGONENSIS	4.71 MICROLITRE O2/(MG DW*HOUR)	ENDOGENOUS RESP., 283 K	944
*DIAPTOMUS OREGONENSIS	8.65 MICROLITRE O2/(MG DW*HOUR)	ENDOGENOUS RESP., 288 K	944
*DIAPTOMUS OREGONENSIS	13.8 MICROLITRE O2/(MG DW*HOUR)	ENDOGENOUS RESP., 293 K	944
*DIAPTOMUS OREGONENSIS	21.0 MICROLITRE O2/(MG DW*HOUR)	ENDOGENOUS RESP., 298 K	944
*DIAPTOMUS SICILOIDES	4.42 MICROLITRE O2/(MG DW*HOUR)	ENDOGENOUS RESP., 278 K	944
*DIAPTOMUS SICILOIDES	5.26 MICROLITRE O2/(MG DW*HOUR)	ENDOGENOUS RESP., 283 K	944
*DIAPTOMUS SICILOIDES	9.92 MICROLITRE O2/(MG DW*HOUR)	ENDOGENOUS RESP., 288 K	944
*DIAPTOMUS SICILOIDES	16.0 MICROLITRE O2/(MG DW*HOUR)	ENDOGENOUS RESP., 293 K	944
*DIAPTOMUS SICILOIDES	24.4 MICROLITRE O2/(MG DW*HOUR)	ENDOGENOUS RESP., 298 K	944
*DIAPTOMUS SICILOIDES	21 ML O2/(MG DW*DAY)	ENDOGENOUS RESP., 298 K	354
*DIAPTOMUS SICILOIDES	13.5 ML O2/(MG DW*DAY)	ENDOGENOUS RESP., 293 K	354
*DIAPTOMUS SICILOIDES	7.8 ML O2/(MG DW*DAY)	ENDOGENOUS RESP., 288 K	354
*DIAPTOMUS SICILOIDES	5.5 ML O2/(MG DW*DAY)	ENDOGENOUS RESP., 283 K	354
*DIAPTOMUS SICILOIDES	4.8 ML O2/(MG DW*DAY)	ENDOGENOUS RESP., 278 K	354
*DIAPTOMUS SICILOIDES	21 ML O2/(MG DW*DAY)	298K, ENDOGENOUS RESP	271
*DIAPTOMUS SICILOIDES	13.5 ML O2/(MG DW*DAY)	293K, ENDOGENOUS RESP	271

TABLE A471

ZOOPLANKTON : RESPIRATION RATE

ITEM:	VALUE:	CONDITIONS:	REF.:
*DIAPTOMUS SICILOIDES	7.8 ML O2/(MG DW*DAY)	288K, ENDOGENOUS RESP	271
*DIAPTOMUS SICILOIDES	5.5 ML O2/(MG DW*DAY)	283K, ENDOGENOUS RESP	271
*DIAPTOMUS SICILOIDES	4.8 ML O2/(MG DW*DAY)	278K, ENDOGENOUS RESP	271
*DIAPTOMUS SP.	4.3 ML O2/(MG DW*DAY)	ENDOGENOUS RESP., 298 K	354
*DIAPTOMUS SP.	3.0 ML O2/(MG DW*DAY)	ENDOGENOUS RESP., 293 K	354
*DIAPTOMUS SP.	2.1 ML O2/(MG DW*DAY)	ENDOGENOUS RESP., 288 K	354
*DIAPTOMUS SP.	1.7 ML O2/(MG DW*DAY)	ENDOGENOUS RESP., 283 K	354
*DIAPTOMUS SP.	1.1 ML O2/(MG DW*DAY)	ENDOGENOUS RESP., 278 K	354
*DIAPTOMUS SP.	4.3 ML O2/(MG DW*DAY)	298K, ENDOGENOUS RESP	271
*DIAPTOMUS SP.	3. ML O2/(MG DW*DAY)	293K, ENDOGENOUS RESP	271
*DIAPTOMUS SP.	2.1 ML O2/(MG DW*DAY)	288K, ENDOGENOUS RESP	271
*DIAPTOMUS SP.	1.7 ML O2/(MG DW*DAY)	283K, ENDOGENOUS RESP	271
*DIAPTOMUS SP.	1.1 ML O2/(MG DW*DAY)	278K, ENDOGENOUS RESP	271
*DIAPTOMUS SP. AND CYCLOPS SP.	3.3 MICROLITRE O2/(MG DW*HOUR)	ENDOGENOUS RESP., 277 K	944
*DIAPTOMUS SP. AND CYCLOPS SP.	4.3 MICROLITRE O2/(MG DW*HOUR)	ENDOGENOUS RESP., 281 K	944
*DIAPTOMUS SP. AND CYCLOPS SP.	5.7 MICROLITRE O2/(MG DW*HOUR)	ENDOGENOUS RESP., 285 K	944
*DIAPTOMUS SP. AND CYCLOPS SP.	6.6 MICROLITRE O2/(MG DW*HOUR)	ENDOGENOUS RESP., 289 K	944
*DIAPTOMUS SP. AND CYCLOPS SP.	8.07 MICROLITRE O2/(MG DW*HOUR)	ENDOGENOUS RESP., 292 K	944
*EPILABIDOCERA AMPHITRITES	52.6+-17.2 MICROLITRE O2/ (ANIMAL*HOUR)	COPEPODA,283K,MEAN OF 4 EXPERIMENTS	827
*EUPHAUSIA PACIFICA	7.57+-2.55 MICROLITRE O2/ (ANIMAL*HOUR)	EUPHAUSIACEA,281K,MEAN OF 5 EXPERIMENTS	827
*LEPTODORA KINDTII	92.7 % OF ASSIMIL¤TION	NONE	1689
*LEUCKARTIARA OCTONA	2.89+-1.34 MICROLITRE O2/ (ANIMAL*HOUR)	COELENTERATA,283K,MEAN OF 3 EXPERIMENTS	827
*MARINE ORGANISMS	12.6 (RANGE= 6.2 - 15.3) MG O2/(KG DRY WEIGHT*MIN)	FISH AND CRUSTACEANS,N=19, DW RANGE=0.0041-0.104 G, DEPTH=0-400 M	246
*MARINE ORGANISMS	4.50 (RANGE= 1.46 - 10) MG O2/(KG DRY WEIGHT*MIN)	FISH AND CRUSTACEANS,N=53, DW RANGE=0.0049-2.22 G,DEPTH=400-900 M	246
*MARINE ORGANISMS	1.20 (RANGE= 0.65-1.63) MG O2/(KG DRY WEIGHT*MIN)	FISH AND CRUSTACEANS,N= 8, DW RANGE=0.132-2.02 G,DEPTH=900-1300 M	246
*MESOCYCLOPS ALBIDUS	CA 20 % OF INGESTION	NONE	1689
*MESOCYCLOPS ALBIDUS	CA 50 % OF ASSIMILATION	NONE	1689
*MICROZOOPLANKTON	10**(-4) MICROLITRE O2/(HOUR*ANIM)	ANIMAL = 10**(-5) MG, PACIFIC OCEAN	837
*MICROZOOPLANKTON	5*10**(-1) MICROLITRE/(HOUR*ANIM)	ANIMAL = 0.5 MG, PACIFIC OCEAN	837
*MIXED COPEPODS	20.9+-21.2 MICROLITRE O2/ (ANIMAL*HOUR)	300K,MEAN OF 6 EXPERIMENTS	827
*MIXED ZOOPLANKTON	31.9+-22.2 MICROLITRE O2/ (ANIMAL*HOUR)	297K,MEAN OF 7 EXPERIMENTS	827
*NAUPLII ZOOPLANKTON	0.04 1/DAY	USED IN MODEL, LAKE MICHIGAN	1567
*NEMATOSCELIS ATLANTICA	4.78+-5.31 MICROLITRE O2/ (ANIMAL*HOUR)	EUPHAUSIACEA,289K,MEAN OF 8 EXPERIMENTS	827
*PARATHEMISTO PACIFICA	2.49+-1.08 MICROLITRE O2/ (ANIMAL*HOUR)	AMPHIPODA,283K,MEAN OF 7 EXPERIMENTS	827
*PHIALIDIUM GREGARIUM	1.94+-0.17 MICROLITRE O2/ (ANIMAL*HOUR)	COELENTERATA,286K,MEAN OF 9 EXPERIMENTS	827
*PLEUROBRACHIA BACHEII	2.19+-1.46 MICROLITRE O2/ (ANIMAL*HOUR)	CTENOPHORA,286K,MEAN OF 15 EXPERIMENTS	827
*PLEURONCODES PLANIPES	183.0+-12.4 MICROLITRE O2/ (ANIMAL*HOUR)	DECAPODA,296K,MEAN OF 5 EXPERIMENTS	827
*SAGITTA ELEGANS	0.896+-0.117 MICROLITRE O2/ (ANIMAL*HOUR)	CHAETOGNATHA,281K,MEAN OF 3 EXPERIMENTS	827
*SIMOCEPHALUS VECTULUS	19.5 % OF INGESTION	JUVENILE, FEMALE	1689
*SIMOCEPHALUS VECTULUS	26.9 % OF ASSIMILATION	JUVENILE, FEMALE	1689
*SIMOCEPHALUS VECTULUS	11.2 % OF INGESTION	REPRODUCING FEMALE	1689

TABLE A471

ZOOPLANKTON : RESPIRATION RATE			
ITEM:	VALUE:	CONDITIONS:	REF.:
*SIMOCEPHALUS VECTULUS	35.3 % OF ASSIMILATION	REPRODUCING FEMALE	1689
*STOMOTOCA ARTRA	10.0+-2.9 MICROLITRE O2/ (ANIMAL*HOUR)	COELENTERATA,286K,MEAN OF 36 EXPERIMENTS	827
*TEMORA LONGICORNIS	5.4 MICROLITRE O2/(MG DW*HOUR)	ENDOGENOUS RESP., 283 K	944
*TEMORA LONGICORNIS	5.4 MICROLITRE O2/(MG DW*HOUR)	ENDOGENOUS RESP., 286 K	944
*TEMORA LONGICORNIS	7.2 MICROLITRE O2/(MG DW*HOUR)	ENDOGENOUS RESP., 290 K	944
*TEMORA LONGICORNIS	10.3 MICROLITRE O2/(MG DW*HOUR)	ENDOGENOUS RESP., 293 K	944
*TOMOPTERIS SEPTENTRIONALIS	6.15+-1.82 MICROLITRE O2/ (ANIMAL*HOUR)	ANNELIDA,281K,MEAN OF 5 EXPERIMENTS	827
*ZOOPLANKTON	0.075 1/DAY	SAN JOAQUIN(RIVER),293 K, ENDOGENOUS RESPIRATION	1149
*ZOOPLANKTON	0.10 1/DAY	DELTA(ESTUARY),293 K, ENDOGENOUS RESPIRATION	1149
*ZOOPLANKTON	0.16 1/DAY	LAKE ERIE,293 K, ENDOGENOUS RESPIRATION	1149
*ZOOPLANKTON	0.02 1/DAY	LAKE ONTARIO,293 K, ENDOGENOUS RESPIRATION	1149
*ZOOPLANKTON	0.005 - 0.3 1/DAY	USED IN MODEL	1567
*ZOOPLANKTON	0.005 - 0.02 1/DAY	USED IN MODEL, NORTH FORK RIVER, CALIF	1567
*ZOOPLANKTON	0.001 - 0.10 1/DAY	USED IN MODEL	1567
*ZOOPLANKTON	0.01 1/DAY	USED IN MODEL, LAKE WASHINGTON	1567
*ZOOPLANKTON	0.10 1/DAY	USED IN MODEL, SAN JOAQUIN RIVER, DELTA ESTUARY, CALIF	1567
*ZOOPLANKTON	0.16 1/DAY	USED IN MODEL, LAKE ERIE	1567
*ZOOPLANKTON	0.02 1/DAY	USED IN MODEL, LAKE ONTARIO	1567
*ZOOPLANKTON	0.36 1/DAY	USED IN MODEL	1567
*ZOOPLANKTON	0.01 1/DAY	USED IN MODEL, BOISE RIVER, IDAHO	1567
*ZOOPLANKTON	0.01 1/DAY	USED IN MODEL, SAN FRANCISCO USED IN MODEL, BAY ESTUARY	1567
*ZOOPLANKTON	0.02-0.1 1/DAY	TEMP.FACTOR IS 1 AT 293K, TEMP>278K, ENDOGENOUS RESP	1571
*ZOOPLANKTON	0.001-0.01 1/DAY	TEMP.FACTOR IS 1 AT 293K, TEMP.<278K, ENDOGENOUS RESP	1571
*ZOOPLANKTON, ADULT	0.06 1/DAY	USED IN MODEL, LAKE MICHIGAN	1567
*ZOOPLANKTON,CARNIVOROUS	0.001 1/DAY-CELCIUS	LAKE ONTARIO, ENDOGENOUS RESP	1577
*ZOOPLANKTON,HERBIVOROUS	0.001 1/DAY-CELCIUS	LAKE ONTARIO, ENDOGENOUS RESP	1577
*ZOOPLANKTON,UPPER TROPHIC LEVEL 2	0.002 1/DAY-CELCIUS	LAKE ONTARIO, ENDOGENOUS RESP	1577

TABLE A472

ZOOPLANKTON : RESPIRATORY COEFFICIENT			
ITEM:	VALUE:	CONDITIONS:	REF.:
*COPEPOD	0.7	LIVING OFF FAT RESERVES	1218
*COPEPOD	1.0	FEEDING OFF PHYTOPLANKTON	1218

TABLE A473

ZOOPLANKTON : S/BIOMASS RATIO			
ITEM:	VALUE:	CONDITIONS:	REF.:
*CALANUS FINMARCHICUS	0.14 %	ATOMIC COMPOSITION	1632

TABLE A474

ZOOPLANKTON : SI/BIOMASS RATIO			
ITEM:	VALUE:	CONDITIONS:	REF.:
*CALANUS FINMARCHICUS	0.0070 %	ATOMIC COMPOSITION	1632

TABLE A475

ZOOPLANKTON : SN/BIOMASS RATIO			
ITEM:	VALUE:	CONDITIONS:	REF.:
*CALANUS FINMARCHICUS	0.0009 %	ATOMIC COMPOSITION	1632

TABLE A476

ZOOPLANKTON : VERTICAL MIGRATION			
ITEM:	VALUE:	CONDITIONS:	REF.:
*BOSMINA LONGIROSTRIS	20 M	MAXIMUM OBS AMPLITUDE, L. MICHIGAN	1010
*BOSMINA LONGIROSTRIS	12 M/HOUR	DESCENT, L. MICHIGAN	1010
*BOSMINA LONGIROSTRIS	19 M/HOUR	ASCENT, L. MICHIGAN	1010
*BOSMINA LONGIROSTRIS	1.8 M	MAXIMUM OBS AMPLITUDE, SILVER L. COLO.	1238
*CYCLOPS BICUSPIDATUS	3.2 M	MAXIMUM OBS AMPLITUDE, SILVER L., COLO.	1238
*DAPHNIA GALEATA MENDOTAE	10 M	MAXIMUM OBS AMPLITUDE, MICHIGAN L.	1676
*DAPHNIA GALEATA MENDOTAE	1.5 M	MAXIMUM OBS AMPLITUDE, L. MENDOTA, WISC.	1009
*DAPHNIA GALEATA MENDOTAE	1.4 M/HOUR	DESCENT AND ASCENT, MENDOTA, WISC.	1009
*DAPHNIA LONGISPINA	5.8 M	MAXIMUM OBS AMPLITUDE, GRAND L. COLO.	1238
*DAPHNIA LONGISPINA	31 M	MAXIMUM OBS AMPLITUDE, L. LUCERNE	1751
*DAPHNIA LONGISPINA	50 M	MAXIMUM OBS AMPLITUDE, L. VICTORIA	1751
*DAPHNIA PULEX	2.7 M	MAXIMUM OBS AMPLITUDE, SUMMIT, COLO.	1238
*DAPHNIA RETROCURVA	24.3 M	MAXIMUM OBS AMPLITUDE, L. MICHIGAN	1010
*DAPHNIA RETROCURVA	5.0 M/HOUR	DESCENT, L. MICHIGAN	1010
*DAPHNIA RETROCURVA	10.6 M/HOUR	ASCENT, L. MICHIGAN	1010
*DAPHNIA SCHODLERI	1.5 M	MAXIMUM OBS AMPLITUDE, L. MENDOTA	1009
*DAPHNIA SCHODLERI	1.4 M/HOUR	DESCENT AND ASCENT, L. MENDOTA	1009
*DIAPTOMUS SHOSHONE	8.8 M	MAXIMUM OBS AMPLITUDE, SUMMIT L.	1238
*HOLOPEDIUM GIBBERUM	13.7 M	MAXIMUM OBS AMPLITUDE, L. MICHIGAN	1010
*HOLOPEDIUM GIBBERUM	6.9 M/HOUR	ASCENT, L. MICHIGAN	1010
*LEPTODORA KINDTII	8.5 M	MAXIMUM OBS AMPLITUDE, L. MICHIGAN	1010
*LEPTODORA KINDTII	2.0 M/HOUR	DESCENT, L. MICHIGAN	1010
*LEPTODORA KINDTII	4.2 M/HOUR	ASCENT, L. MICHIGAN	1010
*LIMNOCALANUS MACRURUS	18.0 M/HOUR	ASCENT, L. MICHIGAN	1010
*LIMNOCALANUS MACRURUS	9.8 M/HOUR	DESCENT, L. MICHIGAN	1010
*LIMNOCALANUS MACRURUS	24 M	MAXIMUM OBS AMPLITUDE, L. MICHIGAN	1010
*MYSIS RELICTA	18 M/HOUR	ASCENT, L. MICHIGAN	1010
*MYSIS RELICTA	13.9 M/HOUR	DESCENT, L. MICHIGAN	1010
*MYSIS RELICTA	40 M	MAXIMUM OBS AMPLITUDE, L. MICHIGAN	1010
*MYSIS RELICTA	30 - 48 M/HOUR	DESCENT AND ASCENT, L. MICHIGAN	89
*MYSIS RELICTA	76 M	MAXIMUM OBS AMPLITUDE, L. MICHIGAN	89
*POLYPHEMUS PEDICULUS	13 M	MAXIMUM OBS AMPLITUDE, L. MICHIGAN	1010

TABLE A476

ZOOPLANKTON : VERTICAL MIGRATION			
ITEM:	VALUE:	CONDITIONS:	REF.:
*POLYPHEMUS PEDICULUS	2.0 M/HOUR	DESCENT, L. MICHIGAN	1010
*POLYPHEMUS PEDICULUS	3.6 M/HOUR	ASCENT, L. MICHIGAN	1010
*PONTOPOREIA AFFINIS	40 M	MAXIMUM OBS AMPLITUDE, L. MICHIGAN	1010
*PONTOPOREIA AFFINIS	13.9 M/HOUR	DESCENT, L. MICHIGAN	1010
*PONTOPOREIA AFFINIS	11.7 M/HOUR	ASCENT, L. MICHIGAN	1010

TABLE A477

ZOOPLANKTON : ZN/BIOMASS RATIO			
ITEM:	VALUE:	CONDITIONS:	REF.:
*CALANUS FINMARCHICUS	0.0000 %	ATOMIC COMPOSITION	1632

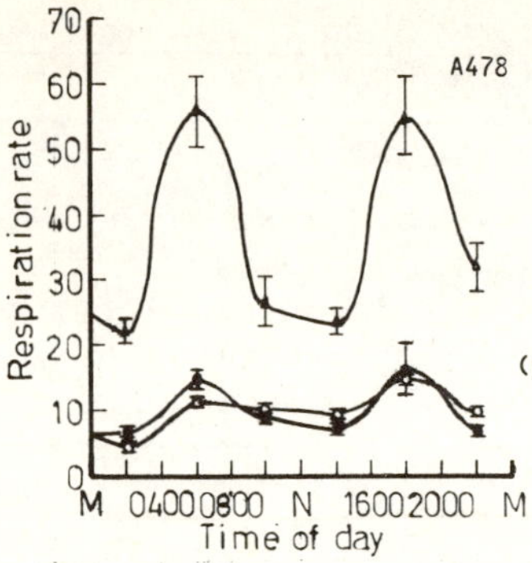

(ref: 383)

Zooplankton respiration versus time of day at 3 temperatures. All studies were conducted in continous darkness; rates corrected values, expected with a zooplankton biomass of 50 µg dry weight/ml. Principal species present (Lake Eunice) the copepods Diaptomus kenai (41.9 µg dry weight), Diaptomus typelli (13.5 ug dry weight) and the cladocerans Holopedium gibberumm and Daphnia rosea (6.9 µg dry weight)

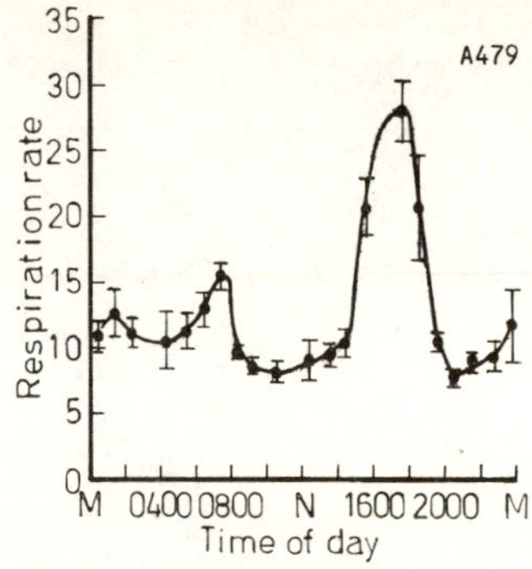

Zooplankton respiration versus time of day at 22 celcius in absence of light. Diaptomus kenai at an expected concentration of 50 µg dry weight/ml.

(ref: 383)

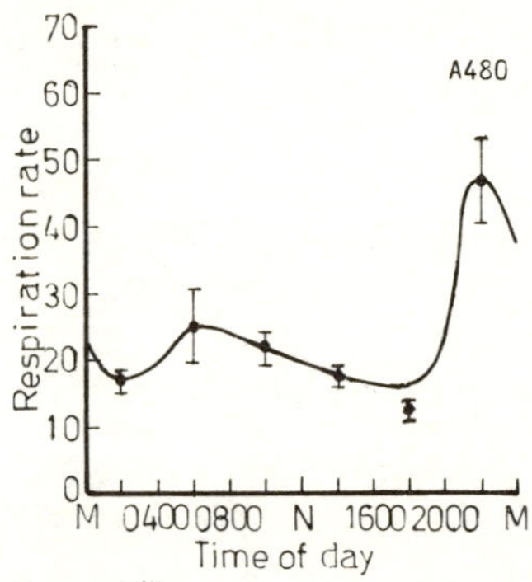

(ref: 383)

Zooplankton respiration versus time of day at 13 celcius in absence of light. Baline Lake zooplankton. (95 % Diapotomus ashlandii, mean dry weight 5.6 µg) at an expected concentration of 50 ug dry weight/ml.

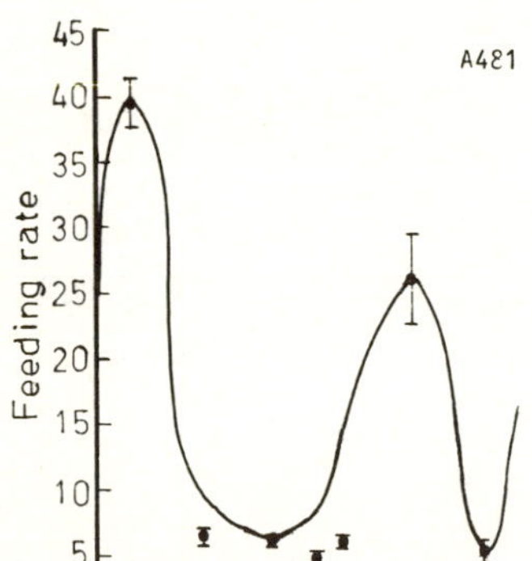

Zooplankton grazing rate versus time of at 21 celcius in absence of light. Composition see above (Lake Eunice). Converted to expected concentration of 50 µg dry weight/

(ref: 383)

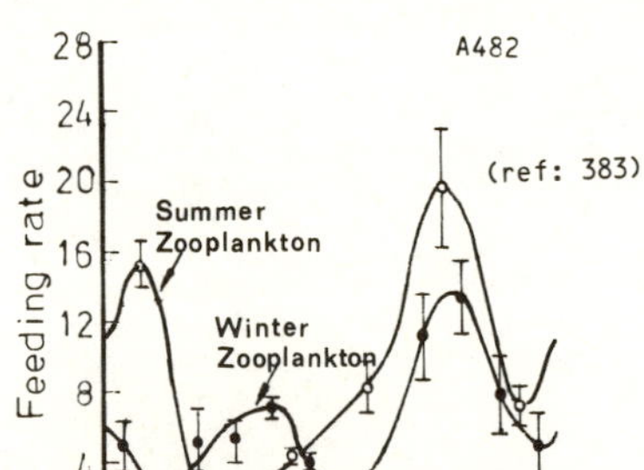

Zooplankton grazing rate versus time of day at winter and summer situations. Deer Lake zooplankton consisting primarily of Daphnia pulex (mean dry weight 8.5 µg) and Cyclops sculifer (mean dry weight 8.5 µg). Experiments were conducted at 10 celcius and 20 celcius respectively in abscence of light and all rates were corrected to values expected with a zooplankton concentration of 50 µg dry weight/ml.

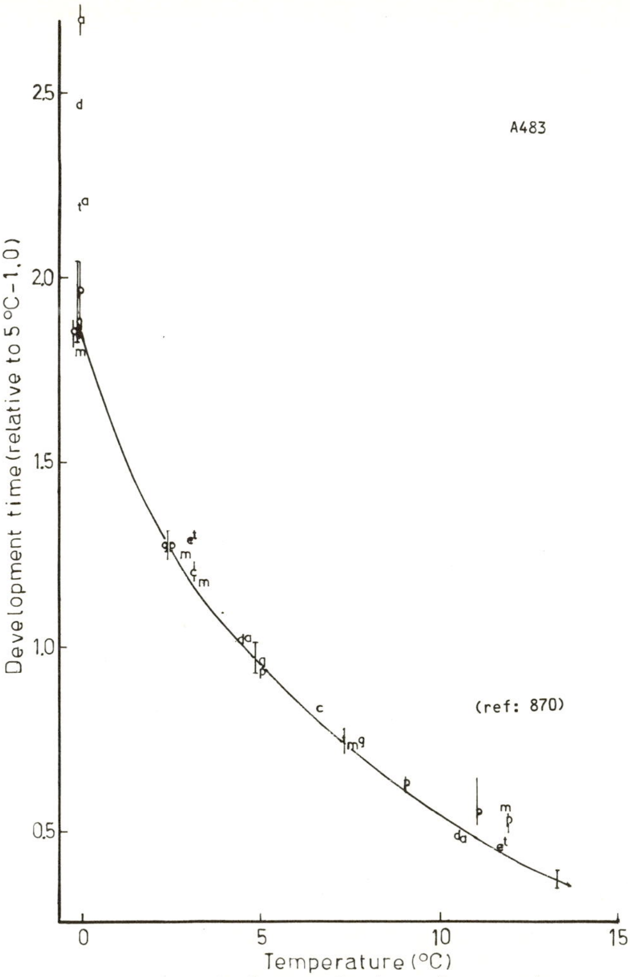

Egg development time for marine copepods versus temperature.a: Acartia clausi. b: Calanus hyperboreas. d: Turtanus chiscaudatus. e: Eurytemora hurundoides. g: Calanus glacialis. m: Metridia longa. p: Pseudocalanus minutus. A: Temora longicornis. I: Calanus finmarchicus.

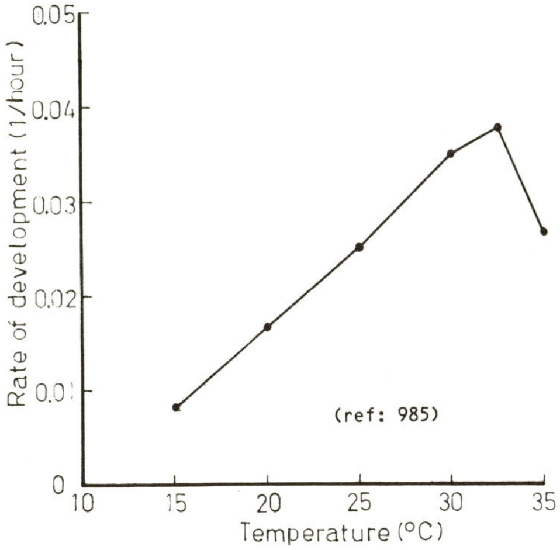

Development of eggs of Thermocyclops neglectus in relation to temperature. Development rate is expressed as the reciprocal development time in hours.

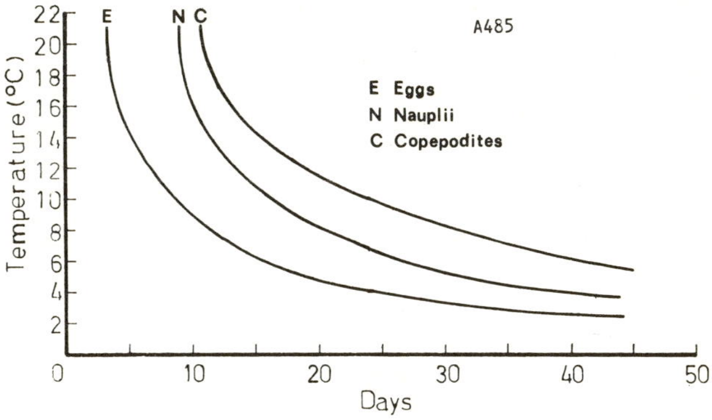

Development rate of different stages of Eudiaptomus graciloides in relation to temperature. E = Eggs. N = Nauplii. C = Copepodites.

(ref: 1105)

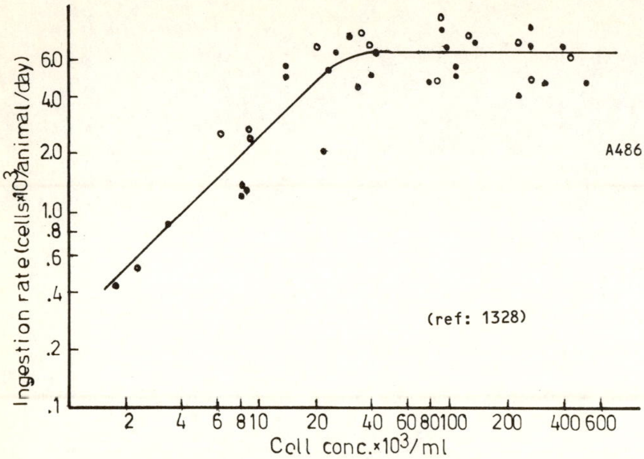

Ingestion rate of Diaphotomus oregonensis
in relation to food cell concentration
Cells*10**3/(animal*day) versus 10**3 cells/ml.
• Chlamydomonas ∘ Chlorella

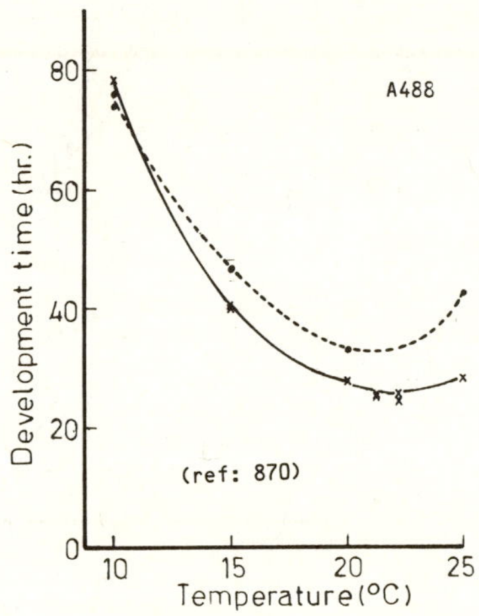

Egg development time for Acartia clausi
effected by temperature. 2 seasonal groups
are represented: solid line winter group.
brokenline: summer group. Bars indicates
95 % confidence interval.

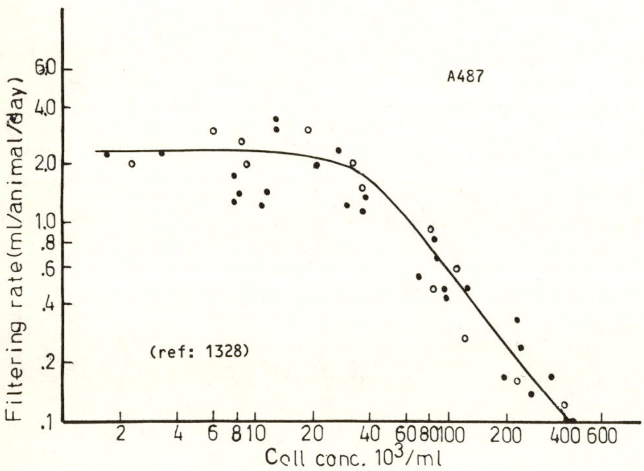

Filtering rate of Diaptomus oreogonensis
in relation to food cell concentration.
• Chlamydomonas ∘ Chlorella

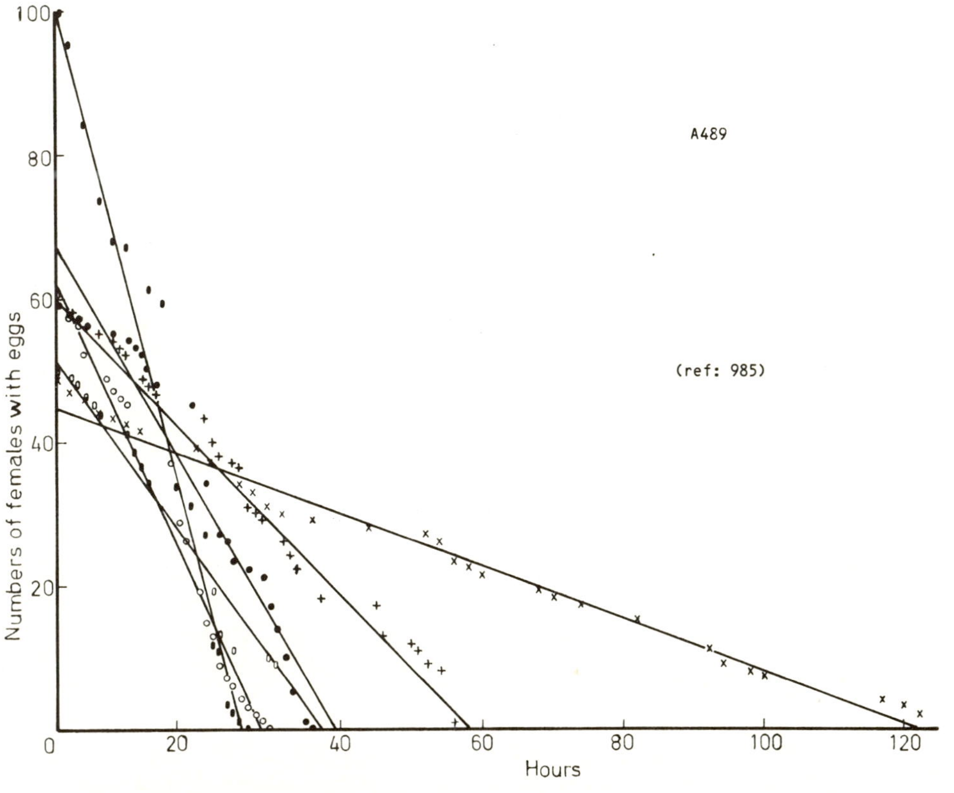

Development of eggs of Thermocyclops neglectus at various temperatures.
× 15 celcius. + 20 celcius. ● 25 celcius.
○ 30 celcius. • 32.5 celcius.
◦ 35 celcius

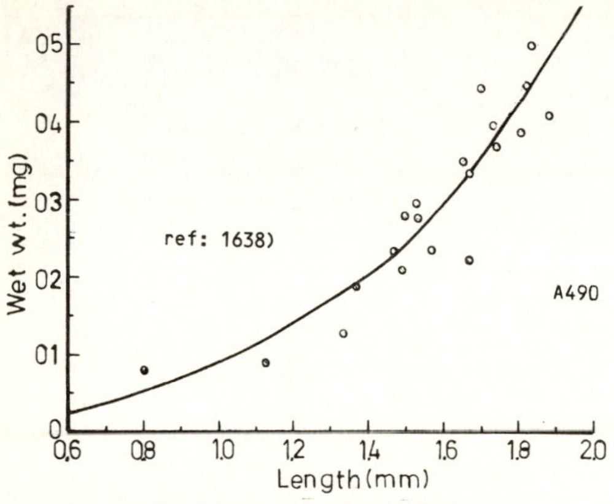

Wet weight (W) of Labidocera trispinosa versus metasome length (d). Procedure: washed in distilled water, weighed to +- 0.01 mg on Chan electrobalance.
W = 0.08255*d**2.7469

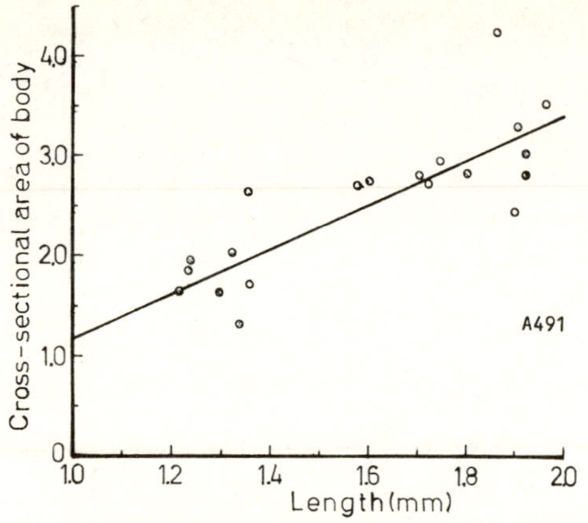

Length of metasome (d) for Labidocera trispinosa versus cross sectional area (A).
A = 0.034003*d**1.4206

(ref: 1638)

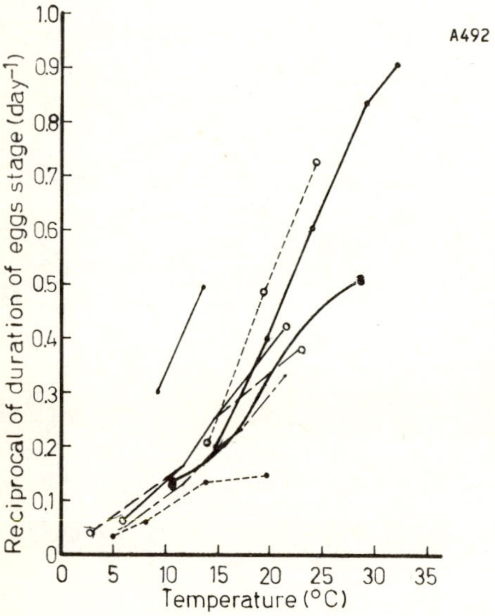

Development of eggs of Copepode in relation to temperature.
Keratella cochlearis
Mesocyclops leuckarti
Thermocyclops neglectus
Acantho salinus
Mesocyclops laciniatus
Acantho denticornis
Cyclops scutufer
Eudiaptomus gracilis

(ref: 988)

PART B

The Ecosphere and Chemical Compounds

BIOL. HALF LIFE TIME
CHEMICAL COMPOUNDS
ECOSPHERE
LIGHT
PESTICIDES
SOIL AND SEDIMENT
THERMODYNAMIC VALUES

B1

BIOLOGICAL HALF LIFE TIME

CHEMICAL COMPOUNDS
ELEMENTS

TABLE B1

BIOL. HALF-LIFE TIME : ALIPHATIC HYDROCARBONS			
ITEM:	VALUE:	CONDITIONS:	REF.:
*MUSSELS	3.5 DAYS	INITIAL =250 UG/KG DW, LAB. EXPERIMENT	471

TABLE B2

BIOL. HALF-LIFE TIME : AS			
ITEM:	VALUE:	CONDITIONS:	REF.:
*HOMO SAPIENS	30-60 HOURS	MAINLY URINARY EXCRETION	1702

TABLE B3

BIOL. HALF-LIFE TIME : BENZO(A)PYRENE			
ITEM:	VALUE:	CONDITIONS:	REF.:
*MYTILUS EDULIS	16 DAYS	INITIAL = 45 UG/KG DW, SEAWATER, 281 K, NATURAL CONTAMIN. ANIMALS	379
*RANGIA CUNEATA	2 - 5 DAYS	INITIAL =7.2 UG/KG WW, 24 H EXPOSURE	1110

TABLE B4

BIOL. HALF-LIFE TIME : CD			
ITEM:	VALUE:	CONDITIONS:	REF.:
*DOG	3-7 YEARS	SINGLE INTRAV.INJECT.OF 115-CD, ONLY URINARY EXCRETION	487
*DOG	419-659 DAYS	0.32-0.40MG 115-CD/KG,SINGLE INTRAV. INJECT.,FECAL EXCRETION	487
*DOG	260-500 DAYS	INTRAV.INJECT.OF 115-CD	487
*MOUSE	>100 DAYS	SINGLE INTRAPERITON.INJECT.OF 109-CD(17MIC.CI)	487
*MOUSE	40-300 DAYS	SINGLE INTRAPERITON.OR VEN.INJECT.,EXCL.1.DAY	487
*MOUSE	25-100 DAYS	SINGLE INTRAV.INJECT.,EXCL.FIRST RAPID CLEARENCE	487
*MOUSE	A COUPLE OF YEARS	0.25MG CD/KG DAY,5DAYS/W FOR 25W, SUBCUT.INJECT., MALE	487
*MOUSE	200 DAYS	SINGLE ORAL DOSE OF 109-CD, FEMALE	487
*RAT	300 DAYS	INTRAMUSC.INJECT.OF 109-CD (2MIC.CI/RAT), EXCL.INIT.EXCRET.PHASE	487
*SQUIRREL MONKEYS	>2 YEARS	4WEEKS AFTER ORAL EXPOSURE	487

TABLE B5

BIOL. HALF-LIFE TIME : CD-109			
ITEM:	VALUE:	CONDITIONS:	REF.:
*ACHETA DOMESTICUS	7 DAYS	BIOLOGICAL HALF-LIFE TIME, UPTAKE FROM FOOD	690
*GAMBUSIA AFFINIS	61 DAYS	BIOLOGICAL HALF-LIFE TIME, UPTAKE FROM FOOD	690

TABLE B5

BIOL. HALF-LIFE TIME : CD-109			
ITEM:	VALUE:	CONDITIONS:	REF.:
*LYCOSA SPP.	99 DAYS	BIOLOGICAL HALF-LIFE TIME, UPTAKE FROM FOOD	690
*LYSMATA SETICAUDATA	378 DAYS	EXPOSED TO CD IN 2 MONTHS	476
*LYSMATA SETICAUDATA	378 DAYS	EXPOSED TO CD IN 2 MONTHS	476
*MICROPTERUS SALMOIDES	21 DAYS	BIOLOGICAL HALF-LIFE TIME, UPTAKE FROM FOOD	690
*MYTILUS GALLOPROVINCIALIS	307 DAYS	FIELD ANIMALS	476
*MYTILUS GALLOPROVINCIALIS	1254 DAYS	LAB. ANIMALS	476
*MYTILUS GALLOPROVINCIALIS	307 DAYS	FIELD ANIMALS	476
*MYTILUS GALLOPROVINCIALIS	1254 DAYS	LAB. ANIMALS	476
*SIDMODON HISPIDUS	18 DAYS	BIOLOGICAL HALF-LIFE TIME, UPTAKE FROM FOOD	690
*SPIZELLA PASSERINA	99 DAYS	BIOLOGICAL HALF-LIFE TIME, UPTAKE FROM FOOD	690

TABLE B6

BIOL. HALF-LIFE TIME : CE-144			
ITEM:	VALUE:	CONDITIONS:	REF.:
*EUPHAUSIA PACIFICA	7.5 HOURS	BIOLOGICAL HALF-LIFE TIME, MEASURED OVER 2 MONTHS	480
*EUPHAUSIA PACIFICA	7.5 HOURS	BIOLOGICAL HALF-LIFE TIME, MEASURED OVER 2 MONTHS	480

TABLE B7

BIOL. HALF-LIFE TIME : CO			
ITEM:	VALUE:	CONDITIONS:	REF.:
*HOMO SAPIENS	800 DAYS	INTRAVEN.INJECT.OF COCL2,9-16% OF DOSE	1463

TABLE B8

BIOL. HALF-LIFE TIME : CO-57			
ITEM:	VALUE:	CONDITIONS:	REF.:
*MICROPOGON UNDULATUS	31 DAYS	BIOLOGICAL HALF-LIFE TIME	66

TABLE B9

BIOL. HALF-LIFE TIME : CO-58			
ITEM:	VALUE:	CONDITIONS:	REF.:
*MICROPOGON UNDULATUS	31 DAYS	BIOLOGICAL HALF-LIFE TIME	66

TABLE B10

BIOL. HALF-LIFE TIME : CO-60			
ITEM:	VALUE:	CONDITIONS:	REF.:
*AHNFELTIA PARADOXA, BLADE	20.7 DAYS	27 D, 288 K	1095

TABLE B10

BIOL. HALF-LIFE TIME : CO-60

ITEM:	VALUE:	CONDITIONS:	REF.:
*AHNFELTIA PARADOXA, STEM	57.6 DAYS	27 D, 288 K	1095
*CHONDRUS OCELLATUS, BLADE	18.3 DAYS	23 D, 288 K	1095
*CHONDRUS OCELLATUS, STEM	41.6 DAYS	23 D, 288 K	1095
*CLAM, SOFT TISSUE	30 DAYS	NONE	66
*CRANGON CRANGON	1.2 (SD=+-0.5) DAYS	UPTAKE FROM FOOD, FAST PHASE = 78 %	1671
*CRANGON CRANGON	9.8 (SD=+-2.7) DAYS	UPTAKE FROM FOOD, LONG LIVE COMPARTMENT = 22 %	1671
*LAMINARIA JAPONICA, BASE	12.8 DAYS	35 D, 288 K	1095
*LAMINARIA JAPONICA, STEM	36.7 DAYS	35 D, 288 K	1095
*LAMINARIA JAPONICA, TIP	24.1 DAYS	35 D, 288 K	1095
*MICROPOGON UNDULATUS	31 DAYS	NONE	66
*MICROPOGON UNDULATUS	31 DAYS	BIOLOGICAL HALF-LIFE TIME, AMBIENT TEMP AND SALINITY, INJECTION	66
*MICROPOGON UNDULATUS	31 DAYS	BIOLOGICAL HALF-LIFE TIME	66
*MYA ARENARIA, MOST SOFT TISSUE	3 - 15 DAYS	FAST PHASE	627
*MYA ARENARIA, MOST SOFT TISSUE	84 - 170 DAYS	SLOW PHASE	627
*MYA ARENARIA, MUSCLE	180 DAYS	NONE	627
*MYA ARENARIA, TOTAL SOFT TISSUE	240 DAYS	NONE	627
*MYTILUS EDULIS, INTACT	5,5 - 6 DAYS	FAST PHASE	1672
*MYTILUS EDULIS, INTACT	57 - 72 DAYS	SLOW PHASE	1672
*MYTILUS EDULIS, MOST SOFT TISSUE	4.7 - 17.7 DAYS	HALF-TIME OF ACCUMULATION	1241
*MYTILUS EDULIS, MUSCLE	17.7 DAYS	HALF-TIME OF ACCUMULATION	1241
*OYSTER, SOFT TISSUE	54 DAYS	NONE	66
*SARGASSUM THUNBERGII, BASE	79.6 DAYS	13 D, 288 K	1095
*SARGASSUM THUNBERGII, CENTRE	12.1 DAYS	13 D, 288 K	1095
*SARGASSUM THUNBERGII, TIP	7.5 DAYS	13 D, 288 K	1095

TABLE B11

BIOL. HALF-LIFE TIME : CR-51

ITEM:	VALUE:	CONDITIONS:	REF.:
*MICROPOGON UNDULATUS	70 DAYS	BIOLOGICAL HALF-LIFE TIME, AMBIENT TEMP AND SALINITY, INJECTION	66
*MICROPOGON UNDULATUS	0.5 DAYS	BIOLOGICAL HALF-LIFE TIME, FAST PHASE, 94 %	66
*MICROPOGON UNDULATUS	70 DAYS	BIOLOGICAL HALF-LIFE TIME, SLOW PHASE, 6 %	66

TABLE B12

BIOL. HALF-LIFE TIME : CS-134

ITEM:	VALUE:	CONDITIONS:	REF.:
*ARCTIC FOX	1.0 DAYS	FAST PHASE, 18 %, FEMALE, 4.9 KG	684
*ARCTIC FOX	17.5 DAYS	SLOW PHASE, 83 %, FEMALE, 4.9 KG	684
*COYOTE	1.5 DAYS	FAST PHASE, 20 %, FEMALE, 9.5 KG	684
*COYOTE	3.5 DAYS	FAST PHASE, 13 %, MALE, 11.5 KG	684
*COYOTE	26.0 DAYS	SLOW PHASE, 80 %, FEMALE, 9.5 KG	684
*COYOTE	22.0 DAYS	SLOW PHASE, 87 %, MALE, 11.5 KG	684

TABLE B12

BIOL. HALF-LIFE TIME : CS-134			
ITEM:	VALUE:	CONDITIONS:	REF.:
*PLEURONECTES PLATESSA	120 - 140 DAYS	30 G, BIOLOGICAL HALF-LIFE TIME, 283 K	745
*PLEURONECTES PLATESSA, MUSCLE	120 - 140 DAYS	30 G, BIOLOGICAL HALF-LIFE TIME, 283 K	745
*PLEURONECTES PLATESSA, MUSCLE	120 - 140 DAYS	30 G, BIOLOGICAL HALF-LIFE TIME, 283 K	745
*RAIA CLAVATA, MUSCLE	180 - 190 DAYS	80 G, BIOLOGICAL HALF-LIFE TIME, 283 K	745
*SILVER FOX	1.1 DAYS	FAST PHASE, 15 %, MALE, 5.3 KG	684
*SILVER FOX	25.3 DAYS	SLOW PHASE, 85 %, MALE, 5.3 KG	684
*WOLF	1.8 DAYS	FAST PHASE, 16 %, FEMALE, 31.3 KG	684
*WOLF	23.0 DAYS	SLOW PHASE, 84 %, FEMALE, 31.3 KG	684

TABLE B13

BIOL. HALF-LIFE TIME : CS-137			
ITEM:	VALUE:	CONDITIONS:	REF.:
*AIX SPONSA	17.3 (RANGE = 10.2 - 26.8) DAYS	FREE LIVING DUCKS, EFFECTIVE HALF-LIFE TIME	451
*BLENNIUS PHOLIS	20 - 30 DAYS	FAST PHASE	483
*BLENNIUS PHOLIS	203 +-18 DAYS	SLOW PHASE	483
*BLENNIUS PHOLIS	20 - 30 DAYS	FAST PHASE	482
*BLENNIUS PHOLIS	203 DAYS	SLOW PHASE	482
*BLENNIUS PHOLIS	20 - 30 DAYS	FAST PHASE	482
*BLENNIUS PHOLIS	203 DAYS	SLOW PHASE	482
*CANCER PAGURUS	4 DAYS	HALF-TIME OF ACCUMULATION	190
*CARCINUS MAENAS	10 DAYS	HALF-TIME OF ACCUMULATION	190
*CLAM, SOFT TISSUE	3 DAYS	FAST PHASE	66
*CLAM, SOFT TISSUE	12 DAYS	SLOW PHASE	66
*EUPHAUSIA PACIFICA	6 DAYS	BIOLOGICAL HALF-LIFE TIME, MEASURED OVER 2 M	480
*EUPHAUSIA PACIFICA	6 DAYS	BIOLOGICAL HALF-LIFE TIME, MEASURED OVER 2 M	480
*HOMO SAPIENS	0.3 YEAR	1956-1963, AUTHOR = RONDO, CITED IN REF.	916
*HOMO SAPIENS	0.1 YEAR	1961-1962, AUTHOR = ROSOFF, CITED IN REF.	916
*HOMO SAPIENS	0.2 YEAR	1961-1962, AUTHOR = COHN, CITED IN REF.	916
*HOMO SAPIENS	0.3 YEAR	1961-1962, AUTHOR = SUGURI, CITED IN REF.	916
*HOMO SAPIENS	1.9 YEAR	1964-1966, AUTHOR = RICHMOND, CITED IN REF.	916
*HOMO SAPIENS	1.6 YEAR	1964-1967, AUTHOR = GUSTAFSON, CITED IN REF.	916
*HOMO SAPIENS	1.4 YEAR	1964-1968, AUTHOR = SHUKLA, CITED IN REF.	916
*HOMO SAPIENS	1.5 YEAR	1965-1967, AUTHOR = MACDONALD, CITED IN REF.	916
*HOMO SAPIENS	1.3 YEAR	1965-1968, AUTHOR = HESP, CITED IN REF.	916
*HOMO SAPIENS	0.2 YEAR	1965-1968, AUTHOR = NAVERSTEN, CITED IN REF.	916
*HOMO SAPIENS	0.4 YEAR	1965-1968, AUTHOR = LLOYD, CITED IN REF.	916
*HOMO SAPIENS	1.5 YEAR	1965-1968, AUTHOR = SCOTT, CITED IN REF.	916
*HOMO SAPIENS	1.1 YEAR	1965-1968, AUTHOR = GODFREY, CITED IN REF.	916

TABLE B13

BIOL. HALF-LIFE TIME : CS-137

ITEM:	VALUE:	CONDITIONS:	REF.:
*HOMO SAPIENS	1.4 YEAR	1966-1968, AUTHOR = JOYET, CITED IN REF.	916
*HOMO SAPIENS	5.4 YEAR	1963-1970, AUTHOR = MACDONALD, CITED IN REF.	916
*HOMO SAPIENS	INFINITE	1963-1973, AUTHOR = RAHOLA, CITED IN REF.	916
*MICROPOGON UNDULATUS	6.2 DAYS	FAST PHASE, SKIN	65
*MICROPOGON UNDULATUS	26.2 DAYS	MEDIUM PHASE, SKIN	65
*MICROPOGON UNDULATUS	290 DAYS	SLOW PHASE, SKIN	65
*MICROPOGON UNDULATUS	34.8 DAYS	FAST PHASE, MUSCLE	65
*MICROPOGON UNDULATUS	94.7 DAYS	SLOW PHASE, MUSCLE	65
*MICROPOGON UNDULATUS	0.7 DAYS	FAST PHASE, LIVER	65
*MICROPOGON UNDULATUS	4.2 DAYS	MEDIUM I PHASE, LIVER	65
*MICROPOGON UNDULATUS	24.1 DAYS	MEDIUM II PHASE, LIVER	65
*MICROPOGON UNDULATUS	INFINITE	SLOW PHASE, LIVER	65
*MICROPOGON UNDULATUS	13.4 DAYS	FAST PHASE, GONAD	65
*MICROPOGON UNDULATUS	911 DAYS	SLOW PHASE, GONAD	65
*MICROPOGON UNDULATUS, MUSCLE	34.8 DAYS	FAST PHASE	65
*MICROPOGON UNDULATUS, MUSCLE	94.7 DAYS	SLOW PHASE	65
*MUGIL CHELO	20 - 30 DAYS	FAST PHASE	483
*MUGIL CHELO	154 +-24 DAYS	SLOW PHASE	483
*MUGIL CHELO	20 - 30 DAYS	FAST PHASE	482
*MUGIL CHELO	154 DAYS	SLOW PHASE	482
*MUGIL CHELO	20 - 30 DAYS	FAST PHASE	482
*MUGIL CHELO	154 DAYS	SLOW PHASE	482
*MYA ARENARIA, MOST SOFT TISSUE	1 - 4.6 DAYS	FAST PHASE	627
*MYA ARENARIA, MOST SOFT TISSUE	14 - 82 DAYS	SLOW PHASE	627
*MYA ARENARIA, MUSCLE	3.3 DAYS	FAST PHASE	627
*MYA ARENARIA, MUSCLE	33 DAYS	SLOW PHASE	627
*MYA ARENARIA, TOTAL SOFT TISSUE	40 DAYS	SLOW PHASE	627
*MYA ARENARIA, TOTAL SOFT TISSUE	3.6 DAYS	FAST PHASE	627
*MYTILUS GALLOPROVINCIALIS	2.5 - 2.9 DAYS	HALF LIFE TIME OF ACCUMULATION	189
*OYSTER, SOFT TISSUE	3 DAYS	FAST PHASE	65
*OYSTER, SOFT TISSUE	12 DAYS	SLOW PHASE	65
*PARALICHTHYS DENTATUS	5.3 DAYS	FAST PHASE, WHOLE BODY	65
*PARALICHTHYS DENTATUS	36.9 DAYS	SLOW PHASE, WHOLE BODY	65
*PARALICHTHYS DENTATUS	5.3 DAYS	FAST PHASE	65
*PARALICHTHYS DENTATUS	36.9 DAYS	LONG LIVED COMPARTMENT	65
*PERCH	200 DAYS	SLOW PHASE, ORALLY UPTAKE, 90 %	640
*PERCH	FEW DAYS	FAST PHASE, ORALLY UPTAKE, 5 - 10 %	640
*PLEURONECTES PLATESSA, MOST SOFT TISSUE	7 - 15 DAYS	FAST PHASE	748
*PLEURONECTES PLATESSA, MOST SOFT TISSUE	78 - 138 DAYS	SLOW PHASE	748
*PLEURONECTES PLATESSA, MUSCLE	136 DAYS	MARINE	748
*PORTUOAS PUBER	8 DAYS	HALF-TIME OF ACCUMULATION	190
*RAIA CLAVATA, MOST SOFT TISSUE	7 - 50 DAYS	FAST PHASE	748
*RAIA CLAVATA, MOST SOFT TISSUE	150 - 180 DAYS	SLOW PHASE	748
*RAIA CLAVATA, MUSCLE	180 DAYS	NONE	748
*ROACH	57 - 100 DAYS	SLOW PHASE, ORALLY UPTAKE, 90 %	640
*SALMO GAIRDNERI	20 - 80 DAYS	SLOW PHASE, ORALLY UPTAKE, 90 %	640
*SALMO TRUTTA	22.1 DAYS	UPTAKE FROM WATER, BIOLOGICAL HALF-LIFE, BLOOD CELLS	662

TABLE B13

BIOL. HALF-LIFE TIME : CS-137			
ITEM:	VALUE:	CONDITIONS:	REF.:
*SALMO TRUTTA	36.1 DAYS	UPTAKE FROM WATER, BIOLOGICAL HALF-LIFE, BLOOD PLASMA	662
*SALMO TRUTTA	22.6 DAYS	UPTAKE FROM WATER, BIOLOGICAL HALF-LIFE, WHOLE BLOOD	662
*SALMO TRUTTA	32.9 DAYS	UPTAKE FROM WATER, BIOLOGICAL HALF-LIFE, HEART	662
*SALMO TRUTTA	37.1 DAYS	UPTAKE FROM WATER, BIOLOGICAL HALF-LIFE, SPLEEN	662
*SALMO TRUTTA	36.7 DAYS	UPTAKE FROM WATER, BIOLOGICAL HALF-LIFE, LIVER	662
*SALMO TRUTTA	41.9 DAYS	UPTAKE FROM WATER, BIOLOGICAL HALF-LIFE, GUT	662
*SALMO TRUTTA	88.0 DAYS	UPTAKE FROM WATER, BIOLOGICAL HALF-LIFE, GONAD	662
*SALMO TRUTTA	38.2 DAYS	UPTAKE FROM WATER, BIOLOGICAL HALF-LIFE, KIDNEY	662
*SALMO TRUTTA	29.8 DAYS	UPTAKE FROM WATER, BIOLOGICAL HALF-LIFE, GILL FILAMENT	662
*SALMO TRUTTA	46.9 DAYS	UPTAKE FROM WATER, BIOLOGICAL HALF-LIFE, SKIN	662
*SALMO TRUTTA	99.0 DAYS	UPTAKE FROM WATER, BIOLOGICAL HALF-LIFE, MUSCLE	662
*SALMO TRUTTA	38.5 DAYS	UPTAKE FROM WATER, BIOLOGICAL HALF-LIFE, NERVE CORD	662
*SALMO TRUTTA	62.4 DAYS	UPTAKE FROM WATER, BIOLOGICAL HALF-LIFE, BONE	662
*SALMO TRUTTA	84.0 DAYS	UPTAKE FROM WATER, BIOLOGICAL HALF-LIFE, WHOLE FISH	662
*SALMO TRUTTA	41.4 DAYS	WHOLE BODY, UPTAKE FROM WATER, BIOLOGICAL HALF-LIFE, FISH= 1.5-16.5 G	662
*SALMO TRUTTA	37.6 DAYS	WHOLE BODY, UPTAKE FROM WATER, BIOLOGICAL HALF-LIFE, FISH= 8.0-40.0 G	662
*SALMO TRUTTA	79.9 DAYS	WHOLE BODY, UPTAKE FROM WATER, BIOLOGICAL HALF-LIFE, FISH=75-280 G	662
*SALMO TRUTTA	95 DAYS	WHOLE BODY, UPTAKE FROM WATER, BIOLOGICAL HALF-LIFE, FISH= 600-820 G	662

TABLE B14

BIOL. HALF-LIFE TIME : DDD			
ITEM:	VALUE:	CONDITIONS:	REF.:
*BEEF CATTLE	50 DAYS	DEPLETION FROM BODY FAT	1003
*LACTATING COWS	37 DAYS	DEPLETION FROM BODY FAT	489
*SHEEP	26 DAYS	DEPLETION FROM BODY FAT	1320
*SHEEP	4 WEEKS	DEPLETION FROM BODY FAT	704

TABLE B15

BIOL. HALF-LIFE TIME : DDE			
ITEM:	VALUE:	CONDITIONS:	REF.:
*BEEF CATTLE	213 DAYS	DEPLETION FROM BODY FAT	1003
*LACTATING COWS	70 DAYS	DEPLETION FROM BODY FAT	489
*SHEEP	223 DAYS	DEPLETION FROM BODY FAT	1320

TABLE B15

BIOL. HALF-LIFE TIME : DDE			
ITEM:	VALUE:	CONDITIONS:	REF.:
*SHEEP	14 WEEKS	DEPLETION FROM BODY FAT	704

TABLE B16

BIOL. HALF-LIFE TIME : DDT			
ITEM:	VALUE:	CONDITIONS:	REF.:
*BEEF CATTLE	103 DAYS	DEPLETION FROM BODY FAT	1003
*LACTATING COWS	140 DAYS	DEPLETION FROM BODY FAT	489
*SHEEP	90 DAYS	DEPLETION FROM BODY FAT	1320
*SHEEP	9 WEEKS	DEPLETION FROM BODY FAT	704

TABLE B17

BIOL. HALF-LIFE TIME : DI-2-ETHYLHEXYL PHTHALATE			
ITEM:	VALUE:	CONDITIONS:	REF.:
*PIMEPHALES PROMELAS	14.2 DAYS	1.24 G, UPTAKE FROM WATER IN 56 D, WATER = 1.9 UG/L	996
*PIMEPHALES PROMELAS	6.2 DAYS	1.24 G, UPTAKE FROM WATER IN 56 D, WATER = 2.5 UG/L	996
*PIMEPHALES PROMELAS	6.2 DAYS	1.24 G, UPTAKE FROM WATER IN 56 D, WATER = 4.6 UG/L	996
*PIMEPHALES PROMELAS	11.8 DAYS	1.24 G, UPTAKE FROM WATER IN 56 D, WATER = 8.1 UG/L	996
*PIMEPHALES PROMELAS	13.3 DAYS	1.24 G, UPTAKE FROM WATER IN 56 D, WATER = 14 UG/L	996
*PIMEPHALES PROMELAS	15.4 DAYS	1.24 G, UPTAKE FROM WATER IN 56 D, WATER = 30 UG/L	996
*PIMEPHALES PROMELAS	18.3 DAYS	1.24 G, UPTAKE FROM WATER IN 56 D, WATER = 62 UG/L	996

TABLE B18

BIOL. HALF-LIFE TIME : DIELDRIN			
ITEM:	VALUE:	CONDITIONS:	REF.:
*RAT	3.13 DAYS	INTAKE FROM FOOD, ADIPOSE TISSUE	389

TABLE B19

BIOL. HALF-LIFE TIME : ENDOSULFAN			
ITEM:	VALUE:	CONDITIONS:	REF.:
*MYTILUS EDULIS, DIGESTIVE GLAND	10.2 DAYS	UPTAKE FROM WATER = 0.1 MG/L	1345
*MYTILUS EDULIS, GILL	9.5 DAYS	UPTAKE FROM WATER = 0.1 MG/L	1345
*MYTILUS EDULIS, GONAD AND MANTLE	11.6 DAYS	UPTAKE FROM WATER = 0.1 MG/L	1345
*MYTILUS EDULIS, REMAINING TISSUE	6.0 DAYS	UPTAKE FROM WATER = 0.1 MG/L	1345
*MYTILUS EDULIS, WHOLE BODY	13.9 DAYS	UPTAKE FROM WATER = 0.1 MG/L	1345

TABLE B20

BIOL. HALF-LIFE TIME : ENDRIN			
ITEM:	VALUE:	CONDITIONS:	REF.:
*ICTALURUS PUNCTATUS, GASTROINTESTINAL TRACT	11 DAYS	AFTER 147 D UPTAKE FROM FOOD, 303 K, 17 G, FOOD=3 % OF WW/D	736
*ICTALURUS PUNCTATUS, KIDNEY	10 DAYS	AFTER 147 D UPTAKE FROM FOOD, 303 K, 17 G, FOOD=3 % OF WW/D	736
*ICTALURUS PUNCTATUS, LIVER	10 DAYS	AFTER 147 D UPTAKE FROM FOOD, 303 K, 17 G, FOOD=3 % OF WW/D	736
*ICTALURUS PUNCTATUS, REMAINING FISH	15 DAYS	AFTER 147 D UPTAKE FROM FOOD, 303 K, 17 G, FOOD=3 % OF WW/D	736
*ICTALURUS PUNCTATUS, WHOLE FISH	12 DAYS	AFTER 147 D UPTAKE FROM FOOD, 303 K, 17 G, FOOD=3 % OF WW/D	736

TABLE B21

BIOL. HALF-LIFE TIME : FE			
ITEM:	VALUE:	CONDITIONS:	REF.:
*HOMO SAPIENS	2040-2580 DAYS	PERSON ON 70KG,4-5G FE/70KG, EXCRETION=0.93MG/DAY	462

TABLE B22

BIOL. HALF-LIFE TIME : FE-55			
ITEM:	VALUE:	CONDITIONS:	REF.:
*MICROPOGON UNDULATUS	4.0 DAYS	BIOLOGICAL HALF-LIFE TIME, FAST PHASE, 4 %	66
*MICROPOGON UNDULATUS	215 DAYS	BIOLOGICAL HALF-LIFE TIME, SLOW PHASE, 96 %	66

TABLE B23

BIOL. HALF-LIFE TIME : FE-59			
ITEM:	VALUE:	CONDITIONS:	REF.:
*LITTORINA OBTUSATA	11.0 DAYS	INTAKE FROM FOOD (FUCUS S.), 283 K	1764
*MICROPOGON UNDULATUS	215 DAYS	BIOLOGICAL HALF-LIFE TIME, AMBIENT TEMP AND SALINITY, INJECTION	66
*MICROPOGON UNDULATUS	4.0 DAYS	BIOLOGICAL HALF-LIFE TIME, FAST PHASE, 4 %	66
*MICROPOGON UNDULATUS	215 DAYS	BIOLOGICAL HALF-LIFE TIME, SLOW PHASE, 96 %	66

TABLE B24

BIOL. HALF-LIFE TIME : HEXACHLOROBENZENE			
ITEM:	VALUE:	CONDITIONS:	REF.:
*LEPOMIS CYANELLUS	10.0 DAYS	CARCASS, UPTAKE FROM FOOD=100 PPM, ONE FEEDING, 296 K	1380
*LEPOMIS CYANELLUS	19.6 DAYS	LIVER, UPTAKE FROM FOOD=100 PPM, ONE FEEDING, 296 K	1380
*LEPOMIS CYANELLUS	8.0 DAYS	STOMACH AND PYLORIC CAECA, UPTAKE FROM FOOD=100 PPM, ONE FEEDING, 296 K	1380

TABLE B24

BIOL. HALF-LIFE TIME : HEXACHLOROBENZENE

ITEM:	VALUE:	CONDITIONS:	REF.:
*LEPOMIS CYANELLUS	8.1 DAYS	INTESTINE, UPTAKE FROM FOOD=100 PPM, ONE FEEDING, 296 K	1380
*LEPOMIS CYANELLUS	8.5 DAYS	MUSCLE, UPTAKE FROM FOOD=100 PPM, ONE FEEDING, 296 K	1380

TABLE B25

BIOL. HALF-LIFE TIME : HG

ITEM:	VALUE:	CONDITIONS:	REF.:
*CRASSOSTREA VIRGINICA	35.4 DAYS	SEAWAT DURING UPTAKE=10 PPB,12.1 PPM IN ANIM,80 D,298 K	306
*CRASSOSTREA VIRGINICA	19.9 DAYS	SEAWAT DURING UPTAKE=100 PPB,91.6 PPM IN ANIM,80 D,295 K	306
*CRASSOSTREA VIRGINICA	16.8 DAYS	SEAWAT DURING UPTAKE=10 PPB,5.4 PPM IN ANIM,25 D,295 K	306
*CRASSOSTREA VIRGINICA	9.3 DAYS	SEAWAT DURING UPTAKE=100 PPB,26.8 PPM IN ANIM,25 D,295 K	306
*GAMBUSIA AFFINIS	17.5 DAYS	UPTAKE FROM WATER AND FOOD	421
*HOMO SAPIENS	3-60 DAYS	ORAL INTAKE OF INORGANIC HG	487
*HOMO SAPIENS	70-90 DAYS	MONOMETHYL-HG	487
*HOMO SAPIENS	42+-3 DAYS	ORAL DOSE OF 0.1 UG HG/KG, ABSORP =15% OF DOSE,90D,URIN.EXCRETION	486
*LEBISTES RETICULATUS	38 - 75 DAYS	HIGH DESORPTION RATE FOR HIGH CONCENTRATION	861
*LEPOMIS MACROCHIRUS	38.5 DAYS	FAST PHASE = 40%, METHYL-HG, 297 K	206
*LEPOMIS MACROCHIRUS	130 DAYS	SLOW PHASE = 60%, METHYL-HG, 297 K	206
*LEPOMIS MACROCHIRUS	38.5 DAYS	FAST PHASE = 40%, METHYL-HG, 297 K	206
*LEPOMIS MACROCHIRUS	130 DAYS	SLOW PHASE = 60%, METHYL-HG, 297 K	206
*PIKE	5 - 10 %	FIRST FEW DAYS, FAST PHASE, 283 K, 300 G FISH	742
*PIKE	640 +- 120 DAYS	90-95 %, SLOW PHASE, 283 K	742
*PIKE	5 - 10 %	FIRST FEW DAYS, FAST PHASE, 283 K, 300 G FISH	742
*PIKE	640 +- 120 DAYS	90-95 %, SLOW PHASE, 283 K	742
*PUSA HISPIDA	500 DAYS	SLOW PHASE = 55 %, DAY 100-153	1595
*PUSA HISPIDA	20 DAYS	FAST PHASE = 45 %, DAY 30-100	1595
*SALMO GAIRDNERI	2-3 YEARS	283 K	206
*SALMO GAIRDNERI	2-3 YEARS	283 K	206
*SALMO GAIRDNERI	GREATER THAN 200 DAYS	INTRAGASTRIC DOSE OF METHYLHG, 0.5 MG HG/KG WW	522
*SALMO GAIRDNERI	GREATER THAN 200 DAYS	INTRAGASTRIC DOSE OF METHYLHG, 0.5 MG HG/KG WW	522

TABLE B26

BIOL. HALF-LIFE TIME : HG-203

ITEM:	VALUE:	CONDITIONS:	REF.:
*NEANTHES SUCCINEA	4.80 HOURS	POLYCHAETE, 7.9 % OF TOTAL BURDEN, FAST PHASE	959

TABLE B27

BIOL. HALF-LIFE TIME : HYDROCARBON

ITEM:	VALUE:	CONDITIONS:	REF.:
*MYTILUS EDULIS	11.4 DAYS	UPTAKE FROM WATER = 302 UG/L IN 80 DAYS, EXTRAC=N-HEXANE,JUNE-OCT	472

TABLE B27

BIOL. HALF-LIFE TIME : HYDROCARBON			
ITEM:	VALUE:	CONDITIONS:	REF.:
*MYTILUS EDULIS	16.1 DAYS	UPTAKE FROM WATER = 228 UG/L IN 40 DAYS, EXTRAC=N-HEXANE,NOV-JAN	472
*MYTILUS EDULIS	5.94 DAYS	UPTAKE FROM WATER = 381 UG/L IN 80 DAYS, EXTRAC=N-HEXANE,MARCH-JUNE	472

TABLE B28

BIOL. HALF-LIFE TIME : I-131			
ITEM:	VALUE:	CONDITIONS:	REF.:
*MICROPOGON UNDULATUS	24 DAYS	BIOLOGICAL HALF-LIFE TIME, AMBIENT TEMP AND SALINITY, INJECTION	66
*MICROPOGON UNDULATUS	0.25 DAYS	BIOLOGICAL HALF-LIFE TIME, FAST PHASE, 59 %	66
*MICROPOGON UNDULATUS	2.25 DAYS	BIOLOGICAL HALF-LIFE TIME, SLOW PHASE, 40 %	66
*SALMO GAIRDNERI	1 - 7 DAYS	287 K, BIOLOGICAL HALF-LIFE TIME	710

TABLE B29

BIOL. HALF-LIFE TIME : IN			
ITEM:	VALUE:	CONDITIONS:	REF.:
*RAT	44 DAYS	SUBCUTANOUSINJECTION	1461
*RAT	80 DAYS	INTRAMUSCULAR.INJECTION	1461

TABLE B30

BIOL. HALF-LIFE TIME : IN-114M			
ITEM:	VALUE:	CONDITIONS:	REF.:
*MICROPOGON UNDULATUS	224 DAYS	BIOLOGICAL HALF-LIFE TIME, AMBIENT TEMP AND SALINITY, INJECTION	66
*MICROPOGON UNDULATUS	3.5 DAYS	BIOLOGICAL HALF-LIFE TIME, FAST PHASE, 10 %	66
*MICROPOGON UNDULATUS	224 DAYS	BIOLOGICAL HALF-LIFE TIME, SLOW PHASE, 90 %	66

TABLE B31

BIOL. HALF-LIFE TIME : IN-131			
ITEM:	VALUE:	CONDITIONS:	REF.:
*MOUSE	21 DAYS	EFFECTIVE HALF-LIFE TIME	181

TABLE B32

BIOL. HALF-LIFE TIME : MN			
ITEM:	VALUE:	CONDITIONS:	REF.:
*HOMO SAPIENS	39 DAYS	FOR 70% OF DOSE,PARENTERAL(NO UPTAKE THR.INTESTINE)	969

TABLE B33

BIOL. HALF-LIFE TIME : MN-54			
ITEM:	VALUE:	CONDITIONS:	REF.:
*ANODONTA NUTTALLIANA	1300 DAYS	FRESHWATER	626

TABLE B34

BIOL. HALF-LIFE TIME : NA-24			
ITEM:	VALUE:	CONDITIONS:	REF.:
*CARP	10 - 25 DAYS	ORALLY UPTAKE	640
*PERCH	7 - 15 DAYS	ORALLY UPTAKE	640
*ROACH	7 - 11 DAYS	ORALLY UPTAKE	640

TABLE B35

BIOL. HALF-LIFE TIME : NAPHTALENE			
ITEM:	VALUE:	CONDITIONS:	REF.:
*MYTILUS EDULIS	1.3 DAYS	4 HOURS EXPOSURE, INITIAL = 7 MG/ KG DW	902

TABLE B36

BIOL. HALF-LIFE TIME : NB-95			
ITEM:	VALUE:	CONDITIONS:	REF.:
*MICROPOGON UNDULATUS	465 DAYS	BIOLOGICAL HALF-LIFE TIME, AMBIENT TEMP AND SALINITY, INJECTION	66
*MICROPOGON UNDULATUS	5.0 DAYS	BIOLOGICAL HALF-LIFE TIME, FAST PHASE, 54 %	66
*MICROPOGON UNDULATUS	465 DAYS	BIOLOGICAL HALF-LIFE TIME, SLOW PHASE, 46 %	66

TABLE B37

BIOL. HALF-LIFE TIME : PB			
ITEM:	VALUE:	CONDITIONS:	REF.:
*CRASSOSTREA VIRGINICA	13.2 DAYS	SEAWAT DURING UPTAKE=25 PPB,24 PPM IN ANIM,21 D,293 K	1280
*CRASSOSTREA VIRGINICA	33.3 DAYS	SEAWAT DURING UPTAKE=50 PPB,32 PPM IN ANIM,21 D,293 K	1280
*CRASSOSTREA VIRGINICA	52.0 DAYS	SEAWAT DURING UPTAKE=100 PPB,79 PPM IN ANIM,21 D,293 K	1280
*CRASSOSTREA VIRGINICA	143.0 DAYS	SEAWAT DURING UPTAKE=200 PPB,203 PPM IN ANIM,21 D,293 K	1280
*MYTILUS EDULIS	43.2 DAYS	SEAWAT DURING UPTAKE=5 PPB,33.7 PPM IN ANIM,40 D,285 K	1405
*MYTILUS EDULIS	68.5 DAYS	SEAWAT DURING UPTAKE=10 PPB,65.7 PPM IN ANIM,40 D,285 K	1405
*MYTILUS EDULIS	62.4 DAYS	SEAWAT DURING UPTAKE=50 PPB,395 PPM IN ANIM,40 D,285 K	1405
*MYTILUS EDULIS	48.1 DAYS	SEAWAT DURING UPTAKE=100 PPB,559 PPM IN ANIM,40 D,285 K	1405

TABLE B37

BIOL. HALF-LIFE TIME : PB			
ITEM:	VALUE:	CONDITIONS:	REF.:
*MYTILUS EDULIS	66.5 DAYS	SEAWAT DURING UPTAKE=200 PPB,1344 PPM IN ANIM,40 D,285 K	1405
*MYTILUS EDULIS	102.0 DAYS	SEAWAT DURING UPTAKE=500 PPB,2480 PPM IN ANIM,40 D,285 K	1405
*MYTILUS EDULIS	70.3 DAYS	SEAWAT DURING UPTAKE=1000 PPB,4972 PPM IN ANIM,40 D,285 K	1405
*MYTILUS EDULIS	66.8 DAYS	SEAWAT DURING UPTAKE=5000 PPB, 17625 PPM IN ANIM,40 D,285 K	1405

TABLE B38

BIOL. HALF-LIFE TIME : PB-210			
ITEM:	VALUE:	CONDITIONS:	REF.:
*RAT, BLOOD	44 (SD=+-4) DAYS	EFFECTIVE HALF-LIFE TIME, N=150, INJ. OF 25 U CI/KG, TERMINAL PHASE	1609
*RAT, FEMUR	109 (SD=+-6) DAYS	EFFECTIVE HALF-LIFE TIME, N=150, INJ. OF 25 U CI/KG, TERMINAL PHASE	1609
*RAT, GONADS	100 (SD=+-19) DAYS	EFFECTIVE HALF-LIFE TIME, N=150, INJ. OF 25 U CI/KG, TERMINAL PHASE	1609
*RAT, HEART	20 (SD=+-1) DAYS	EFFECTIVE HALF-LIFE TIME, N=150, INJ. OF 25 U CI/KG, TERMINAL PHASE	1609
*RAT, KIDNEY	27 (SD=+-1) DAYS	EFFECTIVE HALF-LIFE TIME, N=150, INJ. OF 25 U CI/KG, TERMINAL PHASE	1609
*RAT, LIVER	63 (SD=+-6) DAYS	EFFECTIVE HALF-LIFE TIME, N=150, INJ. OF 25 U CI/KG, TERMINAL PHASE	1609
*RAT, LUNGS	21 (SD=+-1) DAYS	EFFECTIVE HALF-LIFE TIME, N=150, INJ. OF 25 U CI/KG, TERMINAL PHASE	1609
*RAT, SKULL	86 (SD=+-6) DAYS	EFFECTIVE HALF-LIFE TIME, N=150, INJ. OF 25 U CI/KG, TERMINAL PHASE	1609
*RAT, SPINE	64 (SD=+-4) DAYS	EFFECTIVE HALF-LIFE TIME, N=150, INJ. OF 25 U CI/KG, TERMINAL PHASE	1609
*RAT, SPLEEN	18 (SD=+-1) DAYS	EFFECTIVE HALF-LIFE TIME, N=150, INJ. OF 25 U CI/KG, TERMINAL PHASE	1609

TABLE B39

BIOL. HALF-LIFE TIME : PENTACHLOROPHENOL			
ITEM:	VALUE:	CONDITIONS:	REF.:
*LEPOMIS CYANELLUS	12.5 DAYS	CARCASS, UPTAKE FROM FOOD=100 PPM, ONE FEEDING, 296 K	1380
*LEPOMIS CYANELLUS	12.4 DAYS	LIVER, UPTAKE FROM FOOD=100 PPM, ONE FEEDING, 296 K	1380
*LEPOMIS CYANELLUS	16.6 DAYS	STOMACH AND PYLORIC CAECA, UPTAKE FROM FOOD=100 PPM, ONE FEEDING, 296	1380
*LEPOMIS CYANELLUS	12.9 DAYS	INTESTINE, UPTAKE FROM FOOD=100 PPM, ONE FEEDING, 296 K	1380
*LEPOMIS CYANELLUS	8.9 DAYS	MUSCLE, UPTAKE FROM FOOD=100 PPM, ONE FEEDING, 296 K	1380

TABLE B40

BIOL. HALF-LIFE TIME : PHTHALIC ACID			
ITEM:	VALUE:	CONDITIONS:	REF.:
*RAT	10.1 HOURS	ORAL DOSE = 0.4 MG/KG, URINE=75 %, FECES=25 %, CALCULATED FROM REF	1715

TABLE B40

BIOL. HALF-LIFE TIME : PHTHALIC ACID			
ITEM:	VALUE:	CONDITIONS:	REF.:
*RAT	8.2 HOURS	ORAL DOSE = 4.0 MG/KG, URINE=75 %, FECES=25 %, CALCULATED FROM REF	1715

TABLE B41

BIOL. HALF-LIFE TIME : PO-210			
ITEM:	VALUE:	CONDITIONS:	REF.:
*HOMO SAPIENS	2.4 DAYS	EXCRETION WITH FECES, FAST PHASE	1409
*HOMO SAPIENS	52 DAYS	EXCRETION WITH FECES, SLOW PHASE	1409
*HOMO SAPIENS	33 DAYS	EXCRETION WITH URIN, 3 % OF TOTAL, VALUE FOR THE COMPARTMENT	1409
*MEGANYCTIPHANES NORVEGICA	9 DAYS	EXCRETION SOLUBLE PO-210, MARINE	355
*RAT, BLOOD	112 (SD=+-9) DAYS	EFFECTIVE HALF-LIFE TIME, N=150, INJ. OF 25 U CI/KG, TERMINAL PHASE	1609
*RAT, FEMUR	58 (SD=+-2) DAYS	EFFECTIVE HALF-LIFE TIME, N=150, INJ. OF 25 U CI/KG, TERMINAL PHASE	1609
*RAT, GONADS	72 (SD=+-10) DAYS	EFFECTIVE HALF-LIFE TIME, N=150, INJ. OF 25 U CI/KG, TERMINAL PHASE	1609
*RAT, HEART	60 (SD=+-4) DAYS	EFFECTIVE HALF-LIFE TIME, N=150, INJ. OF 25 U CI/KG, TERMINAL PHASE	1609
*RAT, KIDNEY	50 (SD=+-2) DAYS	EFFECTIVE HALF-LIFE TIME, N=150, INJ. OF 25 U CI/KG, TERMINAL PHASE	1609
*RAT, LIVER	67 (SD=+-7) DAYS	EFFECTIVE HALF-LIFE TIME, N=150, INJ. OF 25 U CI/KG, TERMINAL PHASE	1609
*RAT, LUNGS	56 (SD=+-7) DAYS	EFFECTIVE HALF-LIFE TIME, N=150, INJ. OF 25 U CI/KG, TERMINAL PHASE	1609
*RAT, SKULL	58 (SD=+-3) DAYS	EFFECTIVE HALF-LIFE TIME, N=150, INJ. OF 25 U CI/KG, TERMINAL PHASE	1609
*RAT, SPINE	41 (SD=+-2) DAYS	EFFECTIVE HALF-LIFE TIME, N=150, INJ. OF 25 U CI/KG, TERMINAL PHASE	1609
*RAT, SPLEEN	34 (SD=+-2) DAYS	EFFECTIVE HALF-LIFE TIME, N=150, INJ. OF 25 U CI/KG, TERMINAL PHASE	1609

TABLE B42

BIOL. HALF-LIFE TIME : PU-237			
ITEM:	VALUE:	CONDITIONS:	REF.:
*ICTALURUS PUNCTATUS	30.2 (SE=+-2.8) DAYS	PU-CITRATE, % UPTAKE = 10.5+-2.57, R=0.996	437
*ICTALURUS PUNCTATUS	24.3 (SE=+-4.5) DAYS	PU-FULVATE, % UPTAKE = 1.02+-0.18, R=0.972	437
*ICTALURUS PUNCTATUS	13.7 (SE=+-1.7) DAYS	PU-HYDROXIDE, % UPTAKE = 6.17+ -1.39, R=0.992	437
*ICTALURUS PUNCTATUS	30.2 (SE=+-2.8) DAYS	PU-CITRATE, % UPTAKE = 10.5+-2.57, R=0.996	437
*LYSMATA SETICAUDATA	42 DAYS	UPTAKE FROM SEAWATER, PU(+6)	476
*MYTILUS GALLOPROVINCIALIS	776 DAYS	UPTAKE FROM SEAWATER, PU(+6)	476
*MYTILUS GALLOPROVINCIALIS	39 DAYS	UPTAKE FROM PHYTOPLANKTON SUSPENSION, 7 DAYS OF UPTAKE, PU(+6), SEA.	476
*NERIES	79 DAYS	UPTAKE FROM SEAWATER, PU(+6), BETWEEN DAY 4 AND 35	476

TABLE B43

BIOL. HALF-LIFE TIME : SE			
ITEM:	VALUE:	CONDITIONS:	REF.:
*LUMBRICUS SP.	64 DAYS	ANNELIDA	529

TABLE B43

BIOL. HALF-LIFE TIME : SE			
ITEM:	VALUE:	CONDITIONS:	REF.:
*MOUSE	13 DAYS	NONE	529
*PHEASANT	10 DAYS	NONE	529

TABLE B44

BIOL. HALF-LIFE TIME : SE-75			
ITEM:	VALUE:	CONDITIONS:	REF.:
*LYSMATA SETICAUDATA	60 DAYS	UPTAKE OF SE(+4), 293 K, 0.8 G ANIMAL, LAB	478
*LYSMATA SETICAUDATA	58 DAYS	UPTAKE OF SE(+4), 293 K, 0.8 G ANIMAL, FIELD	478
*MYTILUS GALLOPROVINCIALIS	63 DAYS	UPTAKE OF SE(+4), 293 K, 8 G ANIMAL, LAB	478
*MYTILUS GALLOPROVINCIALIS	81 DAYS	UPTAKE OF SE(+4), 293 K, 8 G ANIMAL, FIELD	478

TABLE B45

BIOL. HALF-LIFE TIME : SN-113			
ITEM:	VALUE:	CONDITIONS:	REF.:
*MOUSE	25 DAYS	EFFECTIVE HALF-LIFE TIME	181

TABLE B46

BIOL. HALF-LIFE TIME : SR-85			
ITEM:	VALUE:	CONDITIONS:	REF.:
*MICROPOGON UNDULATUS	138 DAYS	BIOLOGICAL HALF-LIFE TIME, AMBIENT TEMP AND SALINITY, INJECTION	66
*MICROPOGON UNDULATUS	1.25 DAYS	BIOLOGICAL HALF-LIFE TIME, FAST PHASE, 21 %	66
*MICROPOGON UNDULATUS	138 DAYS	BIOLOGICAL HALF-LIFE TIME, SLOW PHASE, 79 %	66
*MICROPOGON UNDULATUS, BONE	223.0 DAYS	BIOLOGICAL HALF-LIFE TIME, 100 %	66
*MICROPOGON UNDULATUS, GILLS	6.0 DAYS	BIOLOGICAL HALF-LIFE TIME, FAST PHASE, 73 %	66
*MICROPOGON UNDULATUS, GILLS	91.0 DAYS	BIOLOGICAL HALF-LIFE TIME, SLOW PHASE, 27 %	66
*MICROPOGON UNDULATUS, GONAD	1.8 DAYS	BIOLOGICAL HALF-LIFE TIME, FAST PHASE, 84 %	66
*MICROPOGON UNDULATUS, GONAD	8.6 DAYS	BIOLOGICAL HALF-LIFE TIME, MEDIUM PHASE, 11 %	66
*MICROPOGON UNDULATUS, GONAD	103 DAYS	BIOLOGICAL HALF-LIFE TIME, SLOW PHASE, 4.3 %	66
*MICROPOGON UNDULATUS, HEART	8.0 DAYS	BIOLOGICAL HALF-LIFE TIME, 100 %	66
*MICROPOGON UNDULATUS, KIDNEY	4.8 DAYS	BIOLOGICAL HALF-LIFE TIME, FAST PHASE, 90 %	66
*MICROPOGON UNDULATUS, KIDNEY	94.0 DAYS	BIOLOGICAL HALF-LIFE TIME, SLOW PHASE, 10 %	66
*MICROPOGON UNDULATUS, LIVER	1.2 DAYS	BIOLOGICAL HALF-LIFE TIME, FAST PHASE, 84 %	66
*MICROPOGON UNDULATUS, LIVER	4.8 DAYS	BIOLOGICAL HALF-LIFE TIME, MEDIUM PHASE, 14 %	66

TABLE B46

BIOL. HALF-LIFE TIME : SR-85			
ITEM:	VALUE:	CONDITIONS:	REF.:
*MICROPOGON UNDULATUS, LIVER	136 DAYS	BIOLOGICAL HALF-LIFE TIME, SLOW PHASE, 2.6 %	66
*MICROPOGON UNDULATUS, MUSCLE	5.4 DAYS	BIOLOGICAL HALF-LIFE TIME, FAST PHASE, 89 %	66
*MICROPOGON UNDULATUS, MUSCLE	100 DAYS	BIOLOGICAL HALF-LIFE TIME, SLOW PHASE, 11 %	66
*MICROPOGON UNDULATUS, SCALES	120.0 DAYS	BIOLOGICAL HALF-LIFE TIME, 100 %	66
*MICROPOGON UNDULATUS, SKIN	3.6 DAYS	BIOLOGICAL HALF-LIFE TIME, FAST PHASE, 88 %	66
*MICROPOGON UNDULATUS, SKIN	51.0 DAYS	BIOLOGICAL HALF-LIFE TIME, SLOW PHASE, 12 %	66
*MICROPOGON UNDULATUS, SPLEEN	8.4 DAYS	BIOLOGICAL HALF-LIFE TIME, FAST PHASE, 71 %	66
*MICROPOGON UNDULATUS, SPLEEN	18.0 DAYS	BIOLOGICAL HALF-LIFE TIME, SLOW PHASE, 29 %	66
*MICROPOGON UNDULATUS, WHOLE BODY	1.25 DAYS	BIOLOGICAL HALF-LIFE TIME, FAST PHASE, 21 %	66
*MICROPOGON UNDULATUS, WHOLE BODY	138 DAYS	BIOLOGICAL HALF-LIFE TIME, SLOW PHASE, 79 %	66

TABLE B47

BIOL. HALF-LIFE TIME : SR-89			
ITEM:	VALUE:	CONDITIONS:	REF.:
*MICROPOGON UNDULATUS	1.25 DAYS	BIOLOGICAL HALF-LIFE TIME, FAST PHASE, 21 %	66
*MICROPOGON UNDULATUS	138 DAYS	BIOLOGICAL HALF-LIFE TIME, SLOW PHASE, 79 %	66

TABLE B48

BIOL. HALF-LIFE TIME : SR-90			
ITEM:	VALUE:	CONDITIONS:	REF.:
*LEPOMIS MACROCHIRUS	2 HOUR	IN BLOOD, 2 % OF TOTAL, FAST PHASE	1305
*LEPOMIS MACROCHIRUS	35 DAYS	IN BLOOD, 1 % OF TOTAL, MEDIUM PHASE	1305
*LEPOMIS MACROCHIRUS	INFINITE	IN BLOOD,97 % OF TOTAL, SLOW PHASE	1305
*MICROPOGON UNDULATUS	1.25 DAYS	BIOLOGICAL HALF-LIFE TIME, FAST PHASE, 21 %	66
*MICROPOGON UNDULATUS	138 DAYS	BIOLOGICAL HALF-LIFE TIME, SLOW PHASE, 79 %	66
*POMOXIS ANNULARIS	12 - 48 MINUTES	IN FISH FLESH, CLINCH RIVER, TENN.	1115

TABLE B49

BIOL. HALF-LIFE TIME : TC-95			
ITEM:	VALUE:	CONDITIONS:	REF.:
*HALIOTIS RUFESCENS	60 DAYS	UPTAKE ONLY FROM SEAWATER, CONC=50 NCI/L,MEAN OF 4, 284 K	1492

TABLE B50

BIOL. HALF-LIFE TIME : TCDD			
ITEM:	VALUE:	CONDITIONS:	REF.:
*MONKEY	20-30 DAYS	SEE REF.	1026

TABLE B50

BIOL. HALF-LIFE TIME : TCDD			
ITEM:	VALUE:	CONDITIONS:	REF.:
*RAT	20-30 DAYS	SEE REF.	1026
*RAT	11 DAYS	LIVER, MALE	488
*RAT	21 DAYS	BODY (EXCEPT LIVER), MALE	488
*RAT	12 DAYS	TOTAL BODY, MALE	488
*RAT	13 DAYS	LIVER, FEMALE	488
*RAT	17 DAYS	BODY (EXCEPT LIVER), FEMALE	488
*RAT	15 DAYS	TOTAL BODY, FEMALE	488

TABLE B51

BIOL. HALF-LIFE TIME : TL			
ITEM:	VALUE:	CONDITIONS:	REF.:
*HOMO SAPIENS	3-8 DAYS	NONE	1170

TABLE B52

BIOL. HALF-LIFE TIME : TOTAL AROMATIC CONTENT			
ITEM:	VALUE:	CONDITIONS:	REF.:
*MUSSELS	4 - 5 WEEKS	NATURAL CONTAMIN. ANIMALS	352

TABLE B53

BIOL. HALF-LIFE TIME : TOXAPHENE			
ITEM:	VALUE:	CONDITIONS:	REF.:
*RAT	8.3 DAYS	ORAL DOSE = 20 MG/KG, URINE = 70 % , FECES = 30 %	302

TABLE B54

BIOL. HALF-LIFE TIME : TRITIUM			
ITEM:	VALUE:	CONDITIONS:	REF.:
*HOMO SAPIENS	33 DAYS	FAST COMPARTMENT = 87 %	1055
*HOMO SAPIENS	2020 DAYS	SLOW COMPARTMENT = 13 %	1055
*HOMO SAPIENS	21 DAYS	FAST COMPARTMENT = 92 %	1055
*HOMO SAPIENS	280 DAYS	SLOW COMPARTMENT = 8 %	1055

TABLE B55

BIOL. HALF-LIFE TIME : ZN-65			
ITEM:	VALUE:	CONDITIONS:	REF.:
*ANODONTA NUTTALLIANA	650 DAYS	FRESHWATER	626

TABLE B55

BIOL. HALF-LIFE TIME : ZN-65			
ITEM:	VALUE:	CONDITIONS:	REF.:
*EUPHAUSIA PACIFICA	140 DAYS	BIOLOGICAL HALF-LIFE TIME, MEASURED OVER 2 M	480
*EUPHAUSIA PACIFICA	140 DAYS	BIOLOGICAL HALF-LIFE TIME, MEASURED OVER 2 M	480
*GAMBUSIA AFFINIS	2.47 DAYS	COMPARTMENT 1 = 8.72 %, 0.19 G, UPTAKE MAINLY FROM WATER	1725
*GAMBUSIA AFFINIS	13.6 DAYS	MEDIUM PHASE = 4.30 %, 0.19 G, UPTAKE MAINLY FROM WATER	1725
*GAMBUSIA AFFINIS	235 DAYS	SLOW PHASE 3 = 91.1 %, 0.19 G, UPTAKE MAINLY FROM WATER	1725
*HOMO SAPIENS	162 DAYS	UPTAKE FROM FOOD	474
*LITTORINA OBTUSATA	14.1 DAYS	INTAKE FROM WATER, 283 K	1764
*LITTORINA OBTUSATA	8.9 DAYS	INTAKE FROM FOOD (FUCUS S.), 283 K	1764
*MICROPOGON UNDULATUS	138 DAYS	BIOLOGICAL HALF-LIFE TIME, AMBIENT TEMP AND SALINITY, INJECTION	66
*MICROPOGON UNDULATUS	6.5 DAYS	BIOLOGICAL HALF-LIFE TIME, FAST PHASE, 10 %	66
*MICROPOGON UNDULATUS	138 DAYS	BIOLOGICAL HALF-LIFE TIME, SLOW PHASE, 90 %	66
*MYTILUS CALIFORNIANUS	76 +- 3.5 DAYS	BIOLOGICAL HALF-LIFE TIME	1761
*STARRY FLOUNDERS	139 DAYS	JUVENILE, DEPENDING ON INITIAL CONCENTRATION	1315

TABLE B56

BIOL. HALF-LIFE TIME : 2,3,2',3'-TCB			
ITEM:	VALUE:	CONDITIONS:	REF.:
*MOUSE	0.9 (0.7 - 12) DAYS	UPTAKE FROM FOOD = 10 PPM IN 20 D, CARCASS, INITIAL=0.13 UG/G WW	1052

TABLE B57

BIOL. HALF-LIFE TIME : 2,4,2',4'-TCB			
ITEM:	VALUE:	CONDITIONS:	REF.:
*MOUSE	9.2 (6.4 - 16) DAYS	UPTAKE FROM FOOD = 10 PPM IN 20 D, CARCASS, INITIAL=9.93 UG/G WW	1052
*MOUSE	7.8 (6.4 - 10) DAYS	UPTAKE FROM FOOD = 10 PPM IN 20 D, LIVER, INITIAL=4.94 UG/G WW	1052

TABLE B58

BIOL. HALF-LIFE TIME : 2,4,2',4'-TCB + PHENOBARBITAL			
ITEM:	VALUE:	CONDITIONS:	REF.:
*MOUSE	4.9 (3.7 - 7.0) DAYS	FOOD=10 TCB + 200 PPM PB IN 20 D, CARCASS, INITIAL=2.28 UG/G WW	1052

TABLE B59

BIOL. HALF-LIFE TIME : 2,5,2',5'-TCB			
ITEM:	VALUE:	CONDITIONS:	REF.:
*MOUSE	3.4 (2.2 - 7.3) DAYS	UPTAKE FROM FOOD = 10 PPM IN 20 D, CARCASS, INITIAL=4.36 UG/G WW	1052

TABLE B59

BIOL. HALF-LIFE TIME : 2,5,2',5'-TCB			
ITEM:	VALUE:	CONDITIONS:	REF.:
*MOUSE	2.6 (1.8 - 4.5) DAYS	UPTAKE FROM FOOD = 10 PPM IN 20 D, LIVER, INITIAL=0.84 UG/G WW	1052

TABLE B60

BIOL. HALF-LIFE TIME : 2,5,2',5'-TCB + PHENOBARBITAL			
ITEM:	VALUE:	CONDITIONS:	REF.:
*MOUSE	2.6 (1.6 - 6.2) DAYS	FOOD=10 TCB + 200 PPM PB IN 20 D, CARCASS, INITIAL=0.28 UG/G WW	1052

TABLE B61

BIOL. HALF-LIFE TIME : 3,4,3',4'-TCB			
ITEM:	VALUE:	CONDITIONS:	REF.:
*MOUSE	0.9 (0.7 - 1.5) DAYS	UPTAKE FROM FOOD = 10 PPM IN 20 D, CARCASS, INITIAL=0.25 UG/G WW	1052

TABLE B62

BIOL. HALF-LIFE TIME : 3,5,3',5'-TCB			
ITEM:	VALUE:	CONDITIONS:	REF.:
*MOUSE	2.1 (1.5 - 3.7) DAYS	UPTAKE FROM FOOD = 10 PPM IN 20 D, CARCASS, INITIAL=4.11 UG/G WW	1052
*MOUSE	2.2 (1.4 - 5.2) DAYS	UPTAKE FROM FOOD = 10 PPM IN 20 D, LIVER, INITIAL=0.53 UG/G WW	1052

TABLE B63

BIOL. HALF-LIFE TIME : 3,5,3',5'-TCB + PHENOBARBITAL			
ITEM:	VALUE:	CONDITIONS:	REF.:
*MOUSE	1.5 (1.2 - 2.1) DAYS	FOOD=10 TCB + 200 PPM PB IN 20 D, CARCASS, INITIAL=2.57 UG/G WW	1052

B2

CHEMICAL COMPOUNDS AND THE ECOSPHERE

ATMOSPHERE
CLIMATE
ELEMENT CYCLES
EVAPORATION RATE
HALF LIFE TIME
LOSS BY RUN OFF
PRODUCTIVITY
RESIDENCE TIME
SORPTION
WATER

TABLE B64

CHEMICAL COMPOUNDS : ENTHALPY, HYDROLYSIS OF PEPTIDE BONDS			
ITEM:	VALUE:	CONDITIONS:	REF.:
*GLYCINE-LEUCINE	-2110 +- 50 CAL/MOLE	BOND IN CARBOBENZOXYGLYCYL-L-LEUCINE	1529
*GLYCINE-PHENYLALANINE	-2550 +- 50 CAL/MOLE	BOND IN CARBOBENZOXYGLYCYL-L-PHENYLALANIN E	1529
*LYSINE-LYSINE	-1240 CAL/MOLE	BOND IN POLY-L-LYSINE	1529
*TYROSINE-GLYCINE	-1550 +- 150 CAL/MOLE	BOND IN BENZOYL-L-TYROSYLGLYCINAMIDE	1529
*TYROSINE-GLYCINE	-1335 +- 90 CAL/MOLE	BOND IN BENZOYL-L-TYROSYLGLYCINE	1529
*TYROSINE-GLYCINE	-1300 +- 150 CAL/MOLE	BOND IN L-TYROSYLGLYCINAMIDE	1529

TABLE B65

CHEMICAL COMPOUNDS : ENTROPY			
ITEM:	VALUE:	CONDITIONS:	REF.:
*ADENINE	36.1 CAL/(MOLE*DEG)	NONE.	1067
*ALANINE	30.883 CAL/(MOLE*DEG)	NONE.	1067
*ARGININE	59.9 CAL/(MOLE*DEG)	NONE.	1067
*ASPARAGINE	41.7 CAL/(MOLE*DEG)	NONE.	1067
*ASPARCTIC ACID	40.657 CAL/(MOLE*DEG)	NONE.	1067
*CO2	51.06 CAL/MOLE*DEG)	IN GAS PHASE	1067
*CYSTEINE	40.6 CAL/(MOLE*DEG)	NONE.	1067
*CYTOSINE	28.9 CAL/(MOLE*DEG)	NONE.	1067
*C6H12O6	-301.63 KCAL/MOLE	IN AQUEOUS SOLUTION	1067
*DEOXYRIBOSE	42.3 CAL/(MOLE*DEG)	NONE.	1067
*FATTY ACID	83.74 CAL/(MOLE*DEG)	NONE.	1067
*GLUCOSE	50.7 CAL/(MOLE*DEG)	NONE.	1067
*GLUTAMIC ACID	44.982 CAL/(MOLE*DEG)	NONE.	1067
*GLUTAMINE	46.631 CAL/(MOLE*DEG)	NONE.	1067
*GLYCERIN	48.87 CAL/(MOLE*DEG)	NONE.	1067
*GLYCINE	24.763 CAL/(MOLE*DEG)	NONE.	1067
*GUANINE	38.3 CAL/(MOLE*DEG)	NONE.	1067
*HISTIDINE	52.0 CAL/(MOLE*DEG)	NONE.	1067
*HNO3	35.0 CAL/MOLE*DEG)	DILUTE AQUEOUS SOLUTION	1067
*H2O	16.72 CAL/MOLE*DEG)	DILUTE AQUEOUS SOLUTION	1067
*H2SO4	4.1 CAL/MOLE*DEG)	DILUTE AQUEOUS SOLUTION	1067
*H3PO4	42.1 CAL/MOLE*DEG)	DILUTE AQUEOUS SOLUTION	1067
*ISOLEUCINE	49.632 CAL/(MOLE*DEG)	NONE.	1067
*LEUCINE	50.598 CAL/(MOLE*DEG)	NONE.	1067
*LYSINE	49.0 CAL/(MOLE*DEG)	NONE.	1067
*METHIONINE	55.317 CAL/(MOLE*DEG)	NONE.	1067
*NH3	26.3 CAL/MOLE*DEG)	DILUTE AQUEOUS SOLUTION	1067
*N2	45.77 CAL/MOLE*DEG)	IN GAS PHASE	1067
*O2	49.003 CAL/MOLE*DEG)	IN GAS PHASE	1067
*PHENYLALANINE	51.005 CAL/(MOLE*DEG)	NONE.	1067
*PHOSPHATE	26.41 CAL/(MOLE*DEG)	NONE., SATURATION	1067
*PROLINE	32.210 CAL/(MOLE*DEG)	NONE.	1067
*RIBOSE	42.3 CAL/(MOLE*DEG)	NONE.	1067
*SERINE	35.667 CAL/(MOLE*DEG)	NONE.	1067
*THREONINE	39.8 CAL/(MOLE*DEG)	NONE.	1067

TABLE B65

CHEMICAL COMPOUNDS : ENTROPY			
ITEM:	VALUE:	CONDITIONS:	REF.:
*THYMINE	32.8 CAL/(MOLE*DEG)	NONE.	1067
*TRYPTOPHAN	59.812 CAL/(MOLE*DEG)	NONE.	1067
*TYROSINE	51.096 CAL/(MOLE*DEG)	NONE.	1067
*URACIL	29.2 CAL/(MOLE*DEG)	NONE.	1067
*VALINE	42.719 CAL/(MOLE*DEG)	TABLE MADE BY REF.	1067

TABLE B66

CHEMICAL COMPOUNDS : EVAPORATION RATE			
ITEM:	VALUE:	CONDITIONS:	REF.:
*CCL2=CCL2	25.7 MIN	50 % EVAPORATION, DILUTE WATER SOL., INITIAL=1 PPM, 298 K	357
*CCL3CCL3	45 MIN	50 % EVAPORATION, DILUTE WATER SOL., INITIAL=1 PPM, 298 K	357
*CCL4	29 MIN	50 % EVAPORATION, DILUTE WATER SOL., INITIAL=1 PPM, 298 K	357
*CHCL=CCLCH2CL	49 MIN	50 % EVAPORATION, DILUTE WATER SOL., INITIAL=1 PPM, 298 K	357
*CHCL=CCL2	21.3 MIN	50 % EVAPORATION, DILUTE WATER SOL., INITIAL=1 PPM, 298 K	357
*CHCL=CHCH2CL	31 MIN	50 % EVAPORATION, DILUTE WATER SOL., INITIAL=1 PPM, 298 K	357
*CHCL=CHCH3	16 MIN	50 % EVAPORATION, DILUTE WATER SOL., INITIAL=1 PPM, 298 K	357
*CHCL=CHCL (CIS)	18 MIN	50 % EVAPORATION, DILUTE WATER SOL., INITIAL=1 PPM, 298 K	357
*CHCL=CHCL (TRANS)	24 MIN	50 % EVAPORATION, DILUTE WATER SOL., INITIAL=1 PPM, 298 K	357
*CHCL2CCL3	45 MIN	50 % EVAPORATION, DILUTE WATER SOL., INITIAL=1 PPM, 298 K	357
*CHCL2CHCL2	56 MIN	50 % EVAPORATION, DILUTE WATER SOL., INITIAL=1 PPM, 298 K	357
*CHCL3	21 MIN	50 % EVAPORATION, DILUTE WATER SOL., INITIAL=1 PPM, 298 K	357
*CH2=CCLCH2CL	20 MIN	50 % EVAPORATION, DILUTE WATER SOL., INITIAL=1 PPM, 298 K	357
*CH2=CCLCH3	29 MIN	50 % EVAPORATION, DILUTE WATER SOL., INITIAL=1 PPM, 298 K	357
*CH2=CCL2	22 MIN	50 % EVAPORATION, DILUTE WATER SOL., INITIAL=1 PPM, 298 K	357
*CH2=CHCH2CL	27 MIN	50 % EVAPORATION, DILUTE WATER SOL., INITIAL=1 PPM, 298 K	357
*CH2=CHCL	26 MIN	50 % EVAPORATION, DILUTE WATER SOL., INITIAL=1 PPM, 298 K	357
*CH2CLCCL2CH2CL	47 MIN	50 % EVAPORATION, DILUTE WATER SOL., INITIAL=1 PPM, 298 K	357
*CH2CLCCL3	43 MIN	50 % EVAPORATION, DILUTE WATER SOL., INITIAL=1 PPM, 298 K	357
*CH2CLCHCLCHCL2	51 MIN	50 % EVAPORATION, DILUTE WATER SOL., INITIAL=1 PPM, 298 K	357
*CH2CLCHCL2	21 MIN	50 % EVAPORATION, DILUTE WATER SOL., INITIAL=1 PPM, 298 K	357
*CH2CLCH2CL	29 MIN	50 % EVAPORATION, DILUTE WATER SOL., INITIAL=1 PPM, 298 K	357
*CH2CL2	20.7 MIN	50 % EVAPORATION, DILUTE WATER SOL., INITIAL=1 PPM, 298 K	357
*CH3CCL3	20 MIN	50 % EVAPORATION, DILUTE WATER SOL., INITIAL=1 PPM, 298 K	357
*CH3CHCL2	22 MIN	50 % EVAPORATION, DILUTE WATER SOL., INITIAL=1 PPM, 298 K	357
*CH3CH2CL	21 MIN	50 % EVAPORATION, DILUTE WATER SOL., INITIAL=1 PPM, 298 K	357

TABLE B66

CHEMICAL COMPOUNDS : EVAPORATION RATE			
ITEM:	VALUE:	CONDITIONS:	REF.:
*CH3CL	27 MIN	50 % EVAPORATION, DILUTE WATER SOL., INITIAL=1 PPM, 298 K	357
*PCB	1.74 MG/((CM**2)*HOUR)	SURFACE=12.28CM**2,373K,24 H, AROCLOR 1221	709
*PCB	0.874 MG/((CM**2)*HOUR)	SURFACE=12.28CM**2,373K,24 H, AROCLOR 1232	709
*PCB	0.338 MG/((CM**2)*HOUR)	SURFACE=12.28CM**2,373K,24 H, AROCLOR 1242	709
*PCB	0.053 MG/((CM**2)*HOUR)	SURFACE=12.28CM**2,373K,24 H, AROCLOR 1254	709
*PCB	0.009 MG/((CM**2)*HOUR)	SURFACE=12.28CM**2,373K,24 H, AROCLOR 1260	709
*PCB	0.013 MG/((CM**2)*HOUR)	SURFACE=12.28CM**2,373K,24 H, AROCLOR 1262	709
*PCB	0.152 MG/((CM**2)*HOUR)	SURFACE=12.28CM**2,373K,24 H, AROCLOR 1448	709
*PCB	0.007 MG/((CM**2)*HOUR)	SURFACE=12.28CM**2,373K,72 H, AROCLOR 4465	709
*PCB	0.004 MG/((CM**2)*HOUR)	SURFACE=12.28CM**2,373K,72 H, AROCLOR 5442	709
*PCB	0.004 MG/((CM**2)*HOUR)	SURFACE=12.28CM**2,373K,72 H, AROCLOR 5460	709
*PCB	1.74 MG/((CM**2)*HOUR)	SURFACE=12.28CM**2,373K,24 H, AROCLOR 1221	709
*PCB	0.874 MG/((CM**2)*HOUR)	SURFACE=12.28CM**2,373K,24 H, AROCLOR 1232	709
*PCB	0.338 MG/((CM**2)*HOUR)	SURFACE=12.28CM**2,373K,24 H, AROCLOR 1242	709
*PCB	0.053 MG/((CM**2)*HOUR)	SURFACE=12.28CM**2,373K,24 H, AROCLOR 1254	709
*PCB	0.009 MG/((CM**2)*HOUR)	SURFACE=12.28CM**2,373K,24 H, AROCLOR 1260	709
*PCB	0.013 MG/((CM**2)*HOUR)	SURFACE=12.28CM**2,373K,24 H, AROCLOR 1262	709
*PCB	0.152 MG/((CM**2)*HOUR)	SURFACE=12.28CM**2,373K,24 H, AROCLOR 1448	709
*PCB	0.007 MG/((CM**2)*HOUR)	SURFACE=12.28CM**2,373K,72 H, AROCLOR 4465	709
*PCB	0.004 MG/((CM**2)*HOUR)	SURFACE=12.28CM**2,373K,72 H, AROCLOR 5442	709
*PCB	0.004 MG/((CM**2)*HOUR)	SURFACE=12.28CM**2,373K,72 H, AROCLOR 5460	709

TABLE B67

CHEMICAL COMPOUNDS : HALF-LIFE TIME IN ATMOSPHERE			
ITEM:	VALUE:	CONDITIONS:	REF.:
*CCL2=CCL2	14.2 HOURS	INITIAL=10 PPM, HUMID.=35 %,UV= 2.6*NORMAL, 300 K, WITH 5 PPM NO	356
*CCL2=CCL2	8.3 HOURS	INITIAL=10 PPM, HUMID.=35 %,UV= 2.6*NORMAL, 300 K,WITH 16.8 PPM NO2	356
*CHCL=CCL2	3.5 HOURS	INITIAL=10 PPM, HUMID.=35 %,UV= 2.6*NORMAL, 300 K, WITH 5 PPM NO	356
*CHCL=CCL2	2.9 HOURS	INITIAL=10 PPM, HUMID.=35 %,UV= 2.6*NORMAL, 300 K,WITH 16.8 PPM NO2	356
*CHLOROBENZENE	8.7 HOURS	INITIAL=10 PPM, HUMID.=35 %,UV= 2.6*NORMAL, 300 K, WITH 5 PPM NO	356
*CH2CL2	LESS THAN 5 % IN 21 HOURS	INITIAL=10 PPM, HUMID.=35 %,UV= 2.6*NORMAL, 300 K, WITH 5 PPM NO	356
*CH2CL2	LESS THAN 5 % IN 7.5 HOURS	INITIAL=10 PPM, HUMID.=35 %,UV= 2.6*NORMAL, 300 K,WITH 16.8 PPM NO2	356

TABLE B67

CHEMICAL COMPOUNDS : HALF-LIFE TIME IN ATMOSPHERE

ITEM:	VALUE:	CONDITIONS:	REF.:
*CH3CCL3	LESS THAN 5 % IN 23.5 HOURS	INITIAL=10 PPM, HUMID.=35 %,UV= 2.6*NORMAL, 300 K, WITH 5 PPM NO	356
*CH3CCL3	LESS THEN 5 % IN 8 HOURS	INITIAL=10 PPM, HUMID.=35 %,UV= 2.6*NORMAL, 300 K,WITH 16.8 PPM NO2	356
*CIS-DICHLOROETHYLENE	3.0 HOURS	INITIAL=10 PPM, HUMID.=35 %,UV= 2.6*NORMAL, 300 K, WITH 5 PPM NO	356
*CIS-DICHLOROETHYLENE	2.8 HOURS	INITIAL=10 PPM, HUMID.=35 %,UV= 2.6*NORMAL, 300 K,WITH 16.8 PPM NO2	356
*CYCLOHEXANE	6.9 HOURS	INITIAL=10 PPM, HUMID.=35 %,UV= 2.6*NORMAL, 300 K, WITH 5 PPM NO	356
*CYCLOHEXANE	6.5 HOURS	INITIAL=10 PPM, HUMID.=35 %,UV= 2.6*NORMAL, 300 K,WITH 16.8 PPM NO2	356
*CYCLOHEXANE	7.5 HOURS	INITIAL=10 PPM, HUMID.=35 %,UV= 2.6*NORMAL, 300 K,WITH 5 PPM NO2	356
*CYCLOHEXENE	0.87 HOURS	INITIAL=10 PPM, HUMID.=35 %,UV= 2.6*NORMAL, 300 K, WITH 5 PPM NO	356
*CYCLOHEXENE	0.19 HOURS	INITIAL=10 PPM, HUMID.=35 %,UV= 2.6*NORMAL, 300 K,WITH 5 PPM NO2	356
*DIOXANE	3.4 HOURS	INITIAL=10 PPM, HUMID.=35 %,UV= 2.6*NORMAL, 300 K, WITH 5 PPM NO	356
*EPICHLOROHYDRIN	16.0 HOURS	INITIAL=10 PPM, HUMID.=35 %,UV= 2.6*NORMAL, 300 K, WITH 5 PPM NO	356
*ETHYL ACETATE	14.6 HOURS	INITIAL=10 PPM, HUMID.=35 %,UV= 2.6*NORMAL, 300 K, WITH 5 PPM NO	356
*ETHYL BENZENE	5.0 HOURS	INITIAL=10 PPM, HUMID.=35 %,UV= 2.6*NORMAL, 300 K, WITH 5 PPM NO	356
*ETHYLENE	2.9 HOURS	INITIAL=10 PPM, HUMID.=35 %,UV= 2.6*NORMAL, 300 K, WITH 5 PPM NO	356
*ETHYLENE	2.5 HOURS	INITIAL=10 PPM, HUMID.=35 %,UV= 2.6*NORMAL, 300 K,WITH 16.8 PPM NO2	356
*ISOBUTYL ALCOHOL	3.5 HOURS	INITIAL=10 PPM, HUMID.=35 %,UV= 2.6*NORMAL, 300 K, WITH 5 PPM NO	356
*M-XYLENE	2.9 HOURS	INITIAL= 8 PPM, HUMID.=35 %,UV= 2.6*NORMAL, 300 K, WITH 5 PPM NO	356
*METHYL ETHYL KETONE	9.8 HOURS	INITIAL=10 PPM, HUMID.=35 %,UV= 2.6*NORMAL, 300 K, WITH 5 PPM NO	356
*METHYL ISOBUTYL KETONE	3.5 HOURS	INITIAL=10 PPM, HUMID.=35 %,UV= 2.6*NORMAL, 300 K, WITH 5 PPM NO	356
*N-BUTYL ALCOHOL	6.5 HOURS	INITIAL=10 PPM, HUMID.=35 %,UV= 2.6*NORMAL, 300 K, WITH 5 PPM NO	356
*N-METHYLPYRROLE	0.16 HOURS	INITIAL=10 PPM, HUMID.=35 %,UV= 2.6*NORMAL, 300 K, WITH 5 PPM NO	356
*NITROMETHANE	9.2 HOURS	INITIAL=10 PPM, HUMID.=35 %,UV= 2.6*NORMAL, 300 K, WITH 5 PPM NO	356
*P-XYLENE	3.1 HOURS	INITIAL= 2 PPM, HUMID.=35 %,UV= 2.6*NORMAL, 300 K, WITH 5 PPM NO	356
*S-BUTYL ALCOHOL	4.0 HOURS	INITIAL=10 PPM, HUMID.=35 %,UV= 2.6*NORMAL, 300 K, WITH 5 PPM NO	356
*T-BUTYL ALCOHOL	34.5 HOURS	INITIAL=10 PPM, HUMID.=35 %,UV= 2.6*NORMAL, 300 K, WITH 5 PPM NO	356
*TOLUENE	6.8 HOURS	INITIAL=10 PPM, HUMID.=35 %,UV= 2.6*NORMAL, 300 K, WITH 5 PPM NO	356
*TRANS-DICHLOROETHYLENE	2.9 HOURS	INITIAL=10 PPM, HUMID.=35 %,UV= 2.6*NORMAL, 300 K, WITH 5 PPM NO	356
*TRANS-DICHLOROETHYLENE	2.8 HOURS	INITIAL=10 PPM, HUMID.=35 %,UV= 2.6*NORMAL, 300 K,WITH 16.8 PPM NO2	356
*TRANS-2-BUTENE	0.30 HOURS	INITIAL=10 PPM, HUMID.=35 %,UV= 2.6*NORMAL, 300 K, WITH 5 PPM NO	356
*TRANS-2-BUTENE	0.17 HOURS	INITIAL=10 PPM, HUMID.=35 %,UV= 2.6*NORMAL, 300 K, WITH 5 PPM NO2	356
*TRIOXANE	4.7 HOURS	INITIAL=10 PPM, HUMID.=35 %,UV= 2.6*NORMAL, 300 K, WITH 5 PPM NO	356
*VINYL CHLORIDE	4.3 HOURS	INITIAL=10 PPM, HUMID.=35 %,UV= 2.6*NORMAL, 300 K, WITH 5 PPM NO	356

TABLE B67

CHEMICAL COMPOUNDS : HALF-LIFE TIME IN ATMOSPHERE			
ITEM:	VALUE:	CONDITIONS:	REF.:
*VINYLIDENE CHLORIDE	2.1 HOURS	INITIAL=10 PPM, HUMID.=35 %,UV= 2.6*NORMAL, 300 K, WITH 5 PPM NO	356
*1-BUTENE OXIDE	15.9 HOURS	INITIAL=10 PPM, HUMID.=35 %,UV= 2.6*NORMAL, 300 K, WITH 5 PPM NO	356
*1-BUTENE OXIDE	15.2 HOURS	INITIAL=10 PPM, HUMID.=35 %,UV= 2.6*NORMAL, 300 K,WITH 5 PPM NO2	356
*1-METHOXY-2-PROPANOL	3.1 HOURS	INITIAL=10 PPM, HUMID.=35 %,UV= 2.6*NORMAL, 300 K, WITH 5 PPM NO	356
*1,1,2-TRICHLOROETHANE	15.9 HOURS	INITIAL=10 PPM, HUMID.=35 %,UV= 2.6*NORMAL, 300 K, WITH 5 PPM NO	356
*2,4,4-TRIMETHYL-1-PENTENE	0.60 HOURS	INITAL=10 PPM, HUMID.=35 %,UV= 2.6*NORMAL, 300 K, WITH 5 PPM NO	356

TABLE B68

CHEMICAL COMPOUNDS : HALF-LIFE TIME IN STRATOSPHERE			
ITEM:	VALUE:	CONDITIONS:	REF.:
*CHLORFLOURMETHAN, F-11	3*10**5 YEARS	ALTITUDE 10 KM	1359
*CHLORFLOURMETHAN, F-11	6.6 YEARS	ALTITUDE 20 KM	1359
*CHLORFLOURMETHAN, F-11	1.3 MONTHS	ALTITUDE 30 KM	1359
*CHLORFLOURMETHAN, F-11	4.7 DAYS	ALTITUDE 40 KM	1359
*CHLORFLOURMETHAN, F-11	2.1 DAYS	ALTITUDE 50 KM	1359
*CHLORFLOURMETHAN, F-11	15 HOURS	TOP OF ATMOSPHERE	1359
*CHLORFLOURMETHAN, F-12	66 YEARS	ALTITUDE 20 KM	1359
*CHLORFLOURMETHAN, F-12	4*10**6 YEARS	ALTITUDE 10 KM	1359
*CHLORFLOURMETHAN, F-12	11 MONTHS	ALTITUDE 30 KM	1359
*CHLORFLOURMETHAN, F-12	1.3 MONTHS	ALTITUDE 40 KM	1359
*CHLORFLOURMETHAN, F-12	16 DAYS	ALTITUDE 50 KM	1359
*CHLORFLOURMETHAN, F-12	3.5 DAYS	TOP OF ATMOSFERE	1359

TABLE B69

CHEMICAL COMPOUNDS : HALF-LIFE TIME IN WATER			
ITEM:	VALUE:	CONDITIONS:	REF.:
*SULPHUR	130 MIN	PH>11,ORDER OF REACTION WITH RESPECT TO SULPHIDE=1	15
*SULPHUR	24 MIN	PH=8.2,298K,ORDER OF REACTION WITH RESPECT TO SULPHIDE=I	15
*SULPHUR	175 MIN	PH=7.5-7.8,283K,ORDER OF REACTION WITH RESPECT TO SULPHIDE=1	15
*SULPHUR	336 MIN	PH=7.5-7.8,283K,ORDER OF REACTION WITH RESPECT TO SULPHIDE=1	15
*SULPHUR	861 MIN	PH=7.5-.78,283K,ORDER OF REACTION WITH RESPECT TO SULPHIDE=1	15
*SULPHUR	10,000 MIN	7.5-7.7,281-282K,NO ORDER OF REACTION	15
*SULPHUR	600 MIN	PH=7.6-7.7,281-282K,NO ODER OF REACTION	15
*SULPHUR	3000 MIN	PH=7.94,298K,ORDER OF REACTION WITH RESPECT TO SULPHIDE=1.34	15
*SULPHUR	180 MIN	PH=8,296-297K,ORDER OF REACTION WITH RESPECT TO SULPHIDE=1	15
*SULPHUR	280 MIN	PH=8,296-297K,ORDER OF REACTION WITH RESPECT TO SULPHIDE=1	15

TABLE B69

CHEMICAL COMPOUNDS : HALF-LIFE TIME IN WATER			
ITEM:	VALUE:	CONDITIONS:	REF.:
*SULPHUR	370 MIN	PH=8,283K,ORDER OF REACTION WITH RESPECT TO SULPHIDE=1	15
*SULPHUR	570 MIN	PH=8,283K,ORDER OF REACTION WITH RESPECT TO SULPHIDE=1	15

TABLE B70

CHEMICAL COMPOUNDS : ISOELECTRIC POINT			
ITEM	VALUE:	CONDITIONS:	REF.:
*ALBITE	2.0	PH VALUE OF ZERO POINT OF CHARGE	1526
*ALPHA-AL(OH)3	5.0	PH VALUE OF ZERO POINT OF CHARGE	1526
*ALPHA-AL2O3	9.1	PH VALUE OF ZERO POINT OF CHARGE	1526
*CUO	9.5	PH VALUE OF ZERO POINT OF CHARGE	1526
*FE(OH)3 (AMORPH)	8.5	PH VALUE OF ZERO POINT OF CHARGE	1526
*FE3O4	6.5	PH VALUE OF ZERO POINT OF CHARGE	1526
*GAMMA-FE2O3	6.7	PH VALUE OF ZERO POINT OF CHARGE	1526
*KAOLINITE	4.6	PH VALUE OF ZERO POINT OF CHARGE	1526
*MGO	12.4	PH VALUE OF ZERO POINT OF CHARGE	1526
*MNO2	2 - 4.5	PH VALUE OF ZERO POINT OF CHARGE	1526
*MONTMORILLONITE	2.5	PH VALUE OF ZERO POINT OF CHARGE	1526
*SIO2	2.0	PH VALUE OF ZERO POINT OF CHARGE	1526
*ZRSIO4	5	PH VALUE OF ZERO POINT OF CHARGE	1526

TABLE B71

CHEMICAL COMPOUNDS : LOSS BY RUNOFF			
ITEM:	VALUE:	CONDITIONS:	REF.:
*CARBON, FINE PARTICULATE ORGANIC	1.94 KG/HA	RUNOFF=92CM,OCT.67-MAY68, DEFORESTED	672
*CARBON, FINE PARTICULATE ORGANIC	1.13 KG/HA	RUNOFF=86CM,OCT.67-MAY68, UNDISTURBED	672
*CARBON, FINE PARTICULATE ORGANIC	2.58 KG/HA	RUNOFF=116CM,JUN.68-MAY69, DEFORESTED	672
*CARBON, FINE PARTICULATE ORGANIC	1.45 KG/HA	RUNOFF=90CM,JUN.68-MAY69, UNDISTURBED	672
*CARBON, FINE PARTICULATE ORGANIC	0.82 KG/HA	RUNOFF=36CM,JUN.69-OCT.69, DEFORESTED	672

TABLE B71

CHEMICAL COMPOUNDS : LOSS BY RUNOFF

ITEM:	VALUE:	CONDITIONS:	REF.:
*CARBON, FINE PARTICULATE ORGANIC	0.26 KG/HA	RUNOFF=17CM,JUN.69-OCT.69, UNDISTURBED	672
*DISSOLVED ORGANIC MATTER	9.84 KG/HA	RUNOFF=92CM,OCT.67-MAY68, DEFORESTED	672
*DISSOLVED ORGANIC MATTER	7.76 KG/HA	RUNOFF=86CM,OCT.67-MAY68, UNDISTURBED	672
*DISSOLVED ORGANIC MATTER	4.84 KG/HA	RUNOFF=116CM,JUN.68-MAY69, DEFORESTED	672
*DISSOLVED ORGANIC MATTER	8.51 KG/HA	RUNOFF=90CM,JUN.68-MAY69, UNDISTURBED	672
*DISSOLVED ORGANIC MATTER	5.89 KG/HA	RUNOFF=36CM,JUN.69-OCT.69, DEFORESTED	672
*DISSOLVED ORGANIC MATTER	4.18 KG/HA	RUNOFF=17CM,JUN.69-OCT.69, UNDISTURBED	672
*INORGANIC PARTICLES	47 G P/HA	PART.>1MM,9.OCT.67-22.MAY68, DEFORESTED	672
*INORGANIC PARTICLES	6 G P/HA	PART.>1MM,9.OCT.67-22.MAY68, UNDISTURBED	672
*INORGANIC PARTICLES	133 G P/HA	PART.>1MM,2.JUL.68-30.SEP.68, DEFORESTED	672
*INORGANIC PARTICLES	8.8 G P/HA	PART.>1MM,2.JUL.68-30.SEP.68, UNDISTURBED	672
*NITRATE	0.90 KG N/HA	CITRUS, 1973, 0.53 % OF SUPPLY, IN DRAINAGE, FLORIDA	979
*NITRATE	2.95 KG N/HA	CITRUS, 1974, 1.74 % OF SUPPLY, IN DRAINAGE, FLORIDA	979
*NITRATE	12.30 KG N/HA	CITRUS, 1975, 7.26 % OF SUPPLY, IN DRAINAGE, FLORIDA	979
*NITRATE	0.90 KG N/HA	CITRUS, 1973, 0.53 % OF SUPPLY, IN DRAINAGE, FLORIDA	979
*NITRATE	2.95 KG N/HA	CITRUS, 1974, 1.74 % OF SUPPLY, IN DRAINAGE, FLORIDA	979
*NITRATE	12.30 KG N/HA	CITRUS, 1975, 7.26 % OF SUPPLY, IN DRAINAGE, FLORIDA	979
*ORGANIC PARTICLES	37 G P/HA	PART.>1MM,9.OCT.67-22.MAY68, DEFORESTED	672
*ORGANIC PARTICLES	3.5 G P/HA	PART.>1MM,9.OCT.67-22.MAY68, UNDISTURBED	672
*ORGANIC PARTICLES	28 G P/HA	PART.>1MM,22.MAY68-2.JUL.68, DEFORESTED	672
*ORGANIC PARTICLES	1.7 G P/HA	PART.>1MM,22.MAY68-2.JUL.68, UNDISTURBED	672
*ORGANIC PARTICLES	7 G P/HA	PART.>1MM,2.JUL.68-30.SEP.68, DEFORESTED	672
*ORGANIC PARTICLES	0.6 G P/HA	PART.>1MM,2.JUL.68-30.SEP.68, UNDISTURBED	672
*ORGANIC PARTICLES	15 G P/HA	PART.>1MM,1.OCT.68-31.MAY69, DEFORESTED	672
*ORGANIC PARTICLES	1.2 G P/HA	PART.>1MM,1.OCT.68-31.MAY69, UNDISTURBED	672
*PHOSPHATE	0.82 KG P/HA	CITRUS, 1973, 4.44 % OF SUPPLY, IN DRAINAGE, FLORIDA	979
*PHOSPHATE	0.97 KG P/HA	CITRUS, 1974, 5.25 % OF SUPPLY, IN DRAINAGE, FLORIDA	979
*PHOSPHATE	0.51 KG P/HA	CITRUS, 1975, 0.81 % OF SUPPLY, IN DRAINAGE, FLORIDA	979
*PHOSPHATE	0.82 KG P/HA	CITRUS, 1973, 4.44 % OF SUPPLY, IN DRAINAGE, FLORIDA	979
*PHOSPHATE	0.97 KG P/HA	CITRUS, 1974, 5.25 % OF SUPPLY, IN DRAINAGE, FLORIDA	979
*PHOSPHATE	0.51 KG P/HA	CITRUS, 1975, 0.81 % OF SUPPLY, IN DRAINAGE, FLORIDA	979
*TOTAL DISSOLVED P PLUS FINE PARTICULATE P	15.62 G/HA	RUNOFF=92CM,OCT.67-MAY68, DEFORESTED	672
*TOTAL DISSOLVED P PLUS FINE PARTICULATE P	9.86 G/HA	RUNOFF=86CM,OCT.67-MAY68, UNDISTURBED	672
*TOTAL DISSOLVED P PLUS FINE PARTICULATE P	20.49 G/HA	RUNOFF=116CM,JUN.68-MAY69, DEFORESTED	672

TABLE B71

CHEMICAL COMPOUNDS : LOSS BY RUNOFF			
ITEM:	VALUE:	CONDITIONS:	REF.:
*TOTAL DISSOLVED P PLUS FINE PARTICULATE P	9.22 G/HA	RUNOFF=90CM,JUN.68-MAY69, UNDISTURBED	672

TABLE B72

CHEMICAL COMPOUNDS : OVERALL MASS TRANSFER COEFFICIENT, LIQUID			
ITEM:	VALUE:	CONDITIONS:	REF.:
*ALDRIN	3.72*10**(-3) METER/HOUR	IN WATER, 298 K	960
*AROCLOR 1242	0.057 METER/HOUR	IN WATER, 298 K	960
*AROCLOR 1248	0.072 METER/HOUR	IN WATER, 298 K	960
*AROCLOR 1254	0.067 METER/HOUR	IN WATER, 298 K	960
*AROCLOR 1260	0.067 METER/HOUR	IN WATER, 298 K	960
*BENZENE	0.144 METER/HOUR	IN WATER, 298 K	960
*BIPHENYL	0.092 METER/HOUR	IN WATER, 298 K	960
*CUMENE	0.119 METER/HOUR	IN WATER, 298 K	960
*DDT	9.34*10**(-3) METER/HOUR	IN WATER, 298 K	960
*DIELDRIN	5.33*10**(-5) METER/HOUR	IN WATER, 298 K	960
*HG	0.092 METER/HOUR	IN WATER, 298 K	960
*LINDANE	1.5*10**(-4) METER/HOUR	IN WATER, 298 K	960
*N-OCTANE	0.124 METER/HOUR	IN WATER, 298 K	960
*NAPHTHALENE	0.096 METER/HOUR	IN WATER, 298 K	960
*O-XYLENE	0.123 METER/HOUR	IN WATER, 298 K	960
*TOLUENE	0.133 METER/HOUR	IN WATER, 298 K	960
*2,2,4-TRIMETHYLPENTANE	0.124 METER/HOUR	IN WATER, 298 K	960

TABLE B73

CHEMICAL COMPOUNDS : RESIDENCE TIME IN STRATOSPHERE			
ITEM:	VALUE:	CONDITIONS:	REF.:
*AEROSOLS	1-2 YEARS	VARIES GREATLY,SINK:DIFFUSION INTO TROPOSPHERE	16
*CO2	2 YEARS	VARIES GREATLY,SINK:DIFFUSION INTO TROPOSPHERE	16
*OZONE	0.1-2 YEARS	VARIES GREATLY,SINK:REDUCTION TO OXYGEN	16
*WATER VAPOR	2 YEARS	VARIES GREATLY,SINK:DIFFUSION INTO TROPOPAUSE,FORM.OF NITRIC ACID	16

TABLE B74

CHEMICAL COMPOUNDS : RESIDENCE TIME IN TROPOSPHERE			
ITEM:	VALUE:	CONDITIONS:	REF.:
*AEROSOLS	LARGE AEROSOLS:10 DAYS, SMALL AEROSOLS:30 DAYS	VARIES GREATLY,SINK:RAIN,SNOW,HAIL	16
*AMMONIA	2-5 DAYS	VARIES GREATLY,SINK:AMMONIA+ SULF.DIOX. ->AMMONIUMSULF.,OX.TO NITRATE	16
*CARBON MONOXIDE	0.1-3 YEARS	VARIES GREATLY,SINK:PROB. OXIDATION TO CO2	16

TABLE B74

CHEMICAL COMPOUNDS : RESIDENCE TIME IN TROPOSPHERE			
ITEM:	VALUE:	CONDITIONS:	REF.:
*CO2	4 YEARS	VARIES GREATLY,SINK: BIOL.ABSORPTION,PHOTOSYNT., ABSORPTION IN OCEANS	16
*HYDROGEN SULPHIDE	2 DAYS	VARIES GREATLY,SINK:OXID.TO SULPHUR DIOXIDE,REMOVAL THROUGH RAINFALL	16
*METHANE	15-100 YEARS	VARIES GREATLY,SINK: PHOTOCHEM.REACT. (CARBON MONOXIDE MAJOR SINK?)	16
*NITRIC OXIDE	5 DAYS	VARIES GREATLY,SINK:OXID.TO NITRATE,SOURCE:COMBUST., BIOL.ACT.IN SOILS	16
*NITROGEN DIOXIDE	5 DAYS	VARIES GREATLY,SINK:OXIDATION TO NITRATE	16
*NITROUS OXIDE	1-4 YEARS	VARIES GREATLY,SINK: PHOTODISSIPAT.IN STRATO., BIOL.ACTION IN SOILS	16
*OZONE	1-3 MONTHS	VARIES GREATLY,SINK:REDUCTION TO OXYGEN	16
*SULPHUR DIOXIDE	4 DAYS	VARIES GREATLY,SINK:OXIDATION TO SULFATES	16
*WATER VAPOR	10 DAYS	VARIES GREATLY,SINK:TROPOPAUSE ACTS AS A COLD-TRAP	16

TABLE B75

CHEMICAL COMPOUNDS : SORPTION			
ITEM:	VALUE:	CONDITIONS:	REF.:
*ALKYL BENZENE SULFONATE ON ACTIVATED SLUDGE	0.0 MG/L	MISSING FROM FILTRATE,20MG SURFACT./L(INITIAL),21DAYS	332
*ALKYL BENZENE SULFONATE ON ACTIVATED SLUDGE	0.0 MG/L	MISSING FROM FILTRATE,20MG SURFACT./L(INITIAL),21DAYS	332
*ALKYL BENZENE SULFONATE ON ANABAENA CYLINDRICA	0.5 MG/L	MISS.FR.FILTRATE,20MG SURFACT./L (INITIAL),21 D,ALGAE-CONC.=58 MG/L	332
*ALKYL BENZENE SULFONATE ON ANABAENA CYLINDRICA	0.5 MG/L	MISS.FR.FILTRATE,20MG SURFACT./L (INITIAL),21 D,ALGAE-CONC.=58 MG/L	332
*ALKYL BENZENE SULFONATE ON ANABAENA VARIABILIS	0.9 MG/L	MISS.FR.FILTRATE,20MG SURFACT./L (INITIAL),21 D,ALGAE-CONC.=71 MG/L	332
*ALKYL BENZENE SULFONATE ON ANABAENA VARIABILIS	0.9 MG/L	MISS.FR.FILTRATE,20MG SURFACT./L (INITIAL),21 D,ALGAE-CONC.=71 MG/L	332
*ALKYL BENZENE SULFONATE ON ANACYSTIS NIDULANS	4.0 MG/L	MISS.FR.FILTRATE,20MG SURFACT./L (INITIAL),21 D,ALGAE-CONC.=121 MG/ L	332
*ALKYL BENZENE SULFONATE ON ANACYSTIS NIDULANS	4.0 MG/L	MISS.FR.FILTRATE,20MG SURFACT./L (INITIAL),21 D,ALGAE-CONC.=121 MG/ L	332
*ALKYL BENZENE S¤LFONATE ON ANKISTRODESMUS BRAUNII	0.5 MG/L	MISS.FR.FILTRATE,20MG SURFACT./L (INITIAL),21 D,ALGAE-CONC.=41 MG/L	332
*ALKYL BENZENE SULFONATE ON ANKISTRODESMUS BRAUNII	0.5 MG/L	MISS.FR.FILTRATE,20MG SURFACT./L (INITIAL),21 D,ALGAE-CONC.=41 MG/L	332
*ALKYL BENZENE SULFONATE ON CALOTHRIX PARIETINA	1.5 MG/L	MISS.FR.FILTRATE,20MG SURFACT./L (INITIAL),21 D,ALGAE-CONC.=47 MG/L	332
*ALKYL BENZENE SULFONATE ON CALOTHRIX PARIETINA	1.5 MG/L	MISS.FR.FILTRATE,20MG SURFACT./L (INITIAL),21 D,ALGAE-CONC.=47 MG/L	332
*ALKYL BENZENE SULFONATE ON CHLORELLA PYRENOIDOSA	0.0 MG/L	MISS.FR.FILTRATE,20MG SURFACT./L (INITIAL),21 D,ALGAE-CONC.=35 MG/L	332
*ALKYL BENZENE SULFONATE ON CHLORELLA PYRENOIDOSA	0.0 MG/L	MISS.FR.FILTRATE,20MG SURFACT./L (INITIAL),21 D,ALGAE-CONC.=35 MG/L	332
*ALKYL BENZENE SULFONATE ON CHLORELLA VULGARIS	0.2 MG/L	MISS.FR.FILT[ATE,20MG SURFACT./L (INITIAL),21 D,ALGAE-CONC.=40 MG/L	332
*ALKYL BENZENE SULFONATE ON CHLORELLA VULGARIS	0.2 MG/L	MISS.FR.FILTRATE,20MG SURFACT./L (INITIAL),21 D,ALGAE-CONC.=40 MG/L	332
*ALKYL BENZENE SULFONATE ON CYLINDROSPERMUM SP.	0.8 MG/L	MISS.FR.FILTRATE,20MG SURFACT./L (INITIAL),21 D,ALGAE-CONC.=55 MG/L	332
*ALKYL BENZENE SULFONATE ON CYLINDROSPERMUM SP.	0.8 MG/L	MISS.FR.FILTRATE,20MG SURFACT./L (INITIAL),21 D,ALGAE-CONC.=55 MG/L	332

TABLE B75

CHEMICAL COMPOUNDS : SORPTION

ITEM:	VALUE:	CONDITIONS:	REF.:
*ALKYL BENZENE SULFONATE ON GLOEOCAPSA ALPICOLA	0.5 MG/L	MISS.FR.FILTRATE,20MG SURFACT./L (INITIAL),21 D,ALGAE-CONC.=87 MG/L	332
*ALKYL BENZENE SULFONATE ON GLOEOCAPSA ALPICOLA	0.5 MG/L	MISS.FR.FILTRATE,20MG SURFACT./L (INITIAL),21 D,ALGAE-CONC.=87 MG/L	332
*ALKYL BENZENE SULFONATE ON NOSTOC MUSCORUM	0.9 MG/L	MISS.FR.FILTRATE,20MG SURFACT./L (INITIAL),21 D,ALGAE-CONC.=40 MG/L	332
*ALKYL BENZENE SULFONATE ON NOSTOC MUSCORUM	0.9 MG/L	MISS.FR.FILTRATE,20MG SURFACT./L (INITIAL),21 D,ALGAE-CONC.=40 MG/L	332
*ALKYL BENZENE SULFONATE ON OSCILLATORIA BORNETI	1.0 MG/L	MISS.FR.FILTRATE,20MG SURFACT./L (INITIAL),21 D,ALGAE-CONC.=50 MG/L	332
*ALKYL BENZENE SULFONATE ON OSCILLATORIA BORNETI	1.0 MG/L	MISS.FR.FILTRATE,20MG SURFACT./L (INITIAL),21 D,ALGAE-CONC.=50 MG/L	332
*ALKYL BENZENE SULFONATE ON OSCILLATORIA CHALYBIA	0.0 MG/L	MISS.FR.FILTRATE,20MG SURFACT./L (INITIAL),21 D,ALGAE-CONC.=30 MG/L	332
*ALKYL BENZENE SULFONATE ON OSCILLATORIA CHALYBIA	0.0 MG/L	MISS.FR.FILTRATE,20MG SURFACT./L (INITIAL),21 D,ALGAE-CONC.=30 MG/L	332
*ALKYL BENZENE SULFONATE ON OSCILLATORIA FORMOSA	0.0 MG/L	MISS.FR.FILTRATE,20MG SURFACT./L (INITIAL),21 D,ALGAE-CONC.=40 MG/L	332
*ALKYL BENZENE SULFONATE ON OSCILLATORIA FORMOSA	0.0 MG/L	MISS.FR.FILTRATE,20MG SURFACT./L (INITIAL),21 D,ALGAE-CONC.=40 MG/L	332
*ALKYL BENZENE SULFONATE ON OSCILLATORIA TENUIS	0.0 MG/L	MISS.FR.FILTRATE,20MG SURFACT./L (INITIAL),21 D,ALGAE-CONC.=72 MG/L	332
*ALKYL BENZENE SULFONATE ON OSCILLATORIA TENUIS	0.0 MG/L	MISS.FR.FILTRATE,20MG SURFACT./L (INITIAL),21 D,ALGAE-CONC.=72 MG/L	332
*ALKYL BENZENE SULFONATE ON PHORMIDIUM FAVEOLARUM	2.9 MG/L	MISS.FR.FILTRATE,20MG SURFACT./L (INITIAL),21 D,ALGAE-CONC.=90 MG/L	332
*ALKYL BENZENE SULFONATE ON PHORMIDIUM FAVEOLARUM	2.9 MG/L	MISS.FR.FILTRATE,20MG SURFACT./L (INITIAL),21 D,ALGAE-CONC.=90 MG/L	332
*ALKYL BENZENE SULFONATE ON POND WATER	1.5 MG/L	MISSING FROM FILTRATE,20MG SURFACT./L(INITIAL),21DAYS	332
*ALKYL BENZENE SULFONATE ON POND WATER	1.5 MG/L	MISSING FROM FILTRATE,20MG SURFACT./L(INITIAL),21DAYS	332
*ALKYL BENZENE SULFONATE ON RAW WASTEWATER	0.6 MG/L	MISSING FROM FILTRATE,20MG SURFACT./L(INITIAL),21DAYS	332
*ALKYL BENZENE SULFONATE ON RAW WASTEWATER	0.6 MG/L	MISSING FROM FILTRATE,20MG SURFACT./L(INITIAL),21DAYS	332
*ALKYL BENZENE SULFONATE ON SCENEDESMUS OBLIQUUS	0.4 MG/L	MISS.FR.FILTRATE,20MG SURFACT./L (INITIAL),21 D,ALGAE-CONC.=50 MG/L	332
*ALKYL BENZENE SULFONATE ON SCENEDESMUS OBLIQUUS	0.4 MG/L	MISS.FR.FILTRATE,20MG SURFACT./L (INITIAL),21 D,ALGAE-CONC.=50 MG/L	332
*ALKYL BENZENE SULFONATE ON SCENEDESMUS QUADRICAUDA	2.0 MG/L	MISS.FR.FILTRATE,20MG SURFACT./L (INITIAL),21 D,ALGAE-CONC.=20 MG/L	332
*ALKYL BENZENE SULFONATE ON SCENEDESMUS QUADRICAUDA	2.0 MG/L	MISS.FR.FILTRATE,20MG SURFACT./L (INITIAL),21 D,ALGAE-CONC.=20 MG/L	332
*ALKYL BENZENE SULFONATE ON SOIL	0.0 MG/L	MISSING FROM FILTRATE,20MG SURFACT./L(INITIAL),21DAYS	332
*ALKYL BENZENE SULFONATE ON SOIL	0.0 MG/L	MISSING FROM FILTRATE,20MG SURFACT./L(INITIAL),21DAYS	332
*ALKYL BENZENE SULFONATE ON ULOTHRIX FIMBRIATA	2.0 MG/L	MISS.FR.FILTRATE,20MG SURFACT./L (INITIAL),21 D,ALGAE-CONC.=90 MG/L	332
*ALKYL BENZENE SULFONATE ON ULOTHRIX FIMBRIATA	2.0 MG/L	MISS.FR.FILTRATE,20MG SURFACT./L (INITIAL),21 D,ALGAE-CONC.=90 MG/L	332
*ALKYL POLYETHOXYLATE ON ACTIVATED SLUDGE	8.0 MG/L	MISSING FROM FILTRATE,20MG SURFACT./L(INITIAL),21DAYS	332
*ALKYL POLYETHOXYLATE ON ACTIVATED SLUDGE	8.0 MG/L	MISSING FROM FILTRATE,20MG SURFACT./L(INITIAL),21DAYS	332
*ALKYL POLYETHOXYLATE ON ACTIVATED SLUDGE	8.5 MG/L	MISSING FROM FILTRATE,20MG SURFACT./L(INITIAL),21DAYS	332
*ALKYL POLYETHOXYLATE ON ACTIVATED SLUDGE	8.5 MG/L	MISSING FROM FILTRATE,20MG SURFACT./L(INITIAL),21DAYS	332
*ALKYL POLYETHOXYLATE ON ANABAENA CYLINDRICA	0.9 MG/L	MISS.FR.FILTRATE,20MG SURFACT./L (INITIAL),21 D,ALGAE-CONC.=58 MG/L	332
*ALKYL POLYETHOXYLATE ON ANABAENA CYLINDRICA	0.9 MG/L	MISS.FR.FILTRATE,20MG SURFACT./L (INITIAL),21 D,ALGAE-CONC.=58 MG/L	332
*ALKYL POLYETHOXYLATE ON ANABAENA CYLINDRICA	1.8 MG/L	MISS.FR.FILTRATE,20MG SURFACT./L (INITIAL),21 D,ALGAE-CONC.=58 MG/L	332
*ALKYL POLYETHOXYLATE ON ANABAENA CYLINDRICA	1.8 MG/L	MISS.FR.FILTRATE,20MG SURFACT./L (INITIAL),21 D,ALGAE-CONC.=58 MG/L	332

TABLE B75

CHEMICAL COMPOUNDS : SORPTION			
ITEM:	VALUE:	CONDITIONS:	REF.:
*ALKYL POLYETHOXYLATE ON ANABAENA VARIABILIS	2.7 MG/L	MISS.FR.FILTRATE,20MG SURFACT./L (INITIAL),21 D,ALGAE-CONC.=71 MG/L	332
*ALKYL POLYETHOXYLATE ON ANABAENA VARIABILIS	2.7 MG/L	MISS.FR.FILTRATE,20MG SURFACT./L (INITIAL),21 D,ALGAE-CONC.=71 MG/L	332
*ALKYL POLYETHOXYLATE ON ANABAENA VARIABILIS	4.2 MG/L	MISS.FR.FILTRATE,20MG SURFACT./L (INITIAL),21 D,ALGAE-CONC.=71 MG/L	332
*ALKYL POLYETHOXYLATE ON ANABAENA VARIABILIS	4.2 MG/L	MISS.FR.FILTRATE,20MG SURFACT./L (INITIAL),21 D,ALGAE-CONC.=71 MG/L	332
*ALKYL POLYETHOXYLATE ON ANACYSTIS NIDULANS	7.2 MG/L	MISS.FR.FILTRATE,20MG SURFACT./L (INITIAL),21 D,ALGAE-CONC.=121 MG/L	332
*ALKYL POLYETHOXYLATE ON ANACYSTIS NIDULANS	7.2 MG/L	MISS.FR.FILTRATE,20MG SURFACT./L (INITIAL),21 D,ALGAE-CONC.=121 MG/L	332
*ALKYL POLYETHOXYLATE ON ANACYSTIS NIDULANS	6.7 MG/L	MISS.FR.FILTRATE,20MG SURFACT./L (INITIAL),21 D,ALGAE-CONC.=121 MG/L	332
*ALKYL POLYETHOXYLATE ON ANACYSTIS NIDULANS	6.7 MG/L	MISS.FR.FILTRATE,20MG SURFACT./L (INITIAL),21 D,ALGAE-CONC.=121 MG/L	332
*ALKYL POLYETHOXYLATE ON ANKISTRODESMUS BRAUNII	8.9 MG/L	MISS.FR.FILTRATE,20MG SURFACT./L (INITIAL),21 D,ALGAE-CONC.=41 MG/L	332
*ALKYL POLYETHOXYLATE ON ANKISTRODESMUS BRAUNII	8.9 MG/L	MISS.FR.FILTRATE,20MG SURFACT./L (INITIAL),21 D,ALGAE-CONC.=41 MG/L	332
*ALKYL POLYETHOXYLATE ON ANKISTRODESMUS BRAUNII	11.0 MG/L	MISS.FR.FILTRATE,20MG SURFACT./L (INITIAL),21 D,ALGAE-CONC.=41 MG/L	332
*ALKYL POLYETHOXYLATE ON ANKISTRODESMUS BRAUNII	11.0 MG/L	MISS.FR.FILTRATE,20MG SURFACT./L (INITIAL),21 D,ALGAE-CONC.=41 MG/L	332
*ALKYL POLYETHOXYLATE ON CALOTHRIX PARIETINA	1.1 MG/L	MISS.FR.FILTRATE,20MG SURFACT./L (INITIAL),21 D,ALGAE-CONC.=47 MG/L	332
*ALKYL POLYETHOXYLATE ON CALOTHRIX PARIETINA	1.1 MG/L	MISS.FR.FILTRATE,20MG SURFACT./L (INITIAL),21 D,ALGAE-CONC.=47 MG/L	332
*ALKYL POLYETHOXYLATE ON CALOTHRIX PARIETINA	4.5 MG/L	MISS.FR.FILTRATE,20MG SURFACT./L (INITIAL),21 D,ALGAE-CONC.=47 MG/L	332
*ALKYL POLYETHOXYLATE ON CALOTHRIX PARIETINA	4.5 MG/L	MISS.FR.FILTRATE,20MG SURFACT./L (INITIAL),21 D,ALGAE-CONC.=47 MG/L	332
*ALKYL POLYETHOXYLATE ON CHLORELLA PYRENOIDOSA	0.7 MG/L	MISS.FR.FILTRATE,20MG SURFACT./L (INITIAL),21 D,ALGAE-CONC.=35 MG/L	332
*ALKYL POLYETHOXYLATE ON CHLORELLA PYRENOIDOSA	0.7 MG/L	MISS.FR.FILTRATE,20MG SURFACT./L (INITIAL),21 D,ALGAE-CONC.=35 MG/L	332
*ALKYL POLYETHOXYLATE ON CHLORELLA PYRENOIDOSA	1.3 MG/L	MISS.FR.FILTRATE,20MG SURFACT./L (INITIAL),21 D,ALGAE-CONC.=35 MG/L	332
*ALKYL POLYETHOXYLATE ON CHLORELLA PYRENOIDOSA	1.3 MG/L	MISS.FR.FILTRATE,20MG SURFACT./L (INITIAL),21 D,ALGAE-CONC.=35 MG/L	332
*ALKYL POLYETHOXYLATE ON CHLORELLA VULGARIS	5.5 MG/L	MISS.FR.FILTRATE,20MG SURFACT./L (INITIAL),21 D,ALGAE-CONC.=40 MG/L	332
*ALKYL POLYETHOXYLATE ON CHLORELLA VULGARIS	5.5 MG/L	MISS.FR.FILTRATE,20MG SURFACT./L (INITIAL),21 D,ALGAE-CONC.=40 MG/L	332
*ALKYL POLYETHOXYLATE ON CHLORELLA VULGARIS	6.1 MG/L	MISS.FR.FILTRATE,20MG SURFACT./L (INITIAL),21 D,ALGAE-CONC.=40 MG/L	332
*ALKYL POLYETHOXYLATE ON CHLORELLA VULGARIS	6.1 MG/L	MISS.FR.FILTRATE,20MG SURFACT./L (INITIAL),21 D,ALGAE-CONC.=40 MG/L	332
*ALKYL POLYETHOXYLATE ON CYLINDROSPERMUM SP.	5.6 MG/L	MISS.FR.FILTRATE,20MG SURFACT./L (INITIAL),21 D,ALGAE-CONC.=55 MG/L	332
*ALKYL POLYETHOXYLATE ON CYLINDROSPERMUM SP.	5.6 MG/L	MISS.FR.FILTRATE,20MG SURFACT./L (INITIAL),21 D,ALGAE-CONC.=55 MG/L	332
*ALKYL POLYETHOXYLATE ON CYLINDROSPERMUM SP.	6.9 MG/L	MISS.FR.FILTRATE,20MG SURFACT./L (INITIAL),21 D,ALGAE-CONC.=55 MG/L	332
*ALKYL POLYETHOXYLATE ON CYLINDROSPERMUM SP.	6.9 MG/L	MISS.FR.FILTRATE,20MG SURFACT./L (INITIAL),21 D,ALGAE-CONC.=55 MG/L	332
*ALKYL POLYETHOXYLATE ON GLOEOCAPSA ALPICOLA	4.9 MG/L	MISS.FR.FILTRATE,20MG SURFACT./L (INITIAL),21 D,ALGAE-CONC.=87 MG/L	332
*ALKYL POLYETHOXYLATE ON GLOEOCAPSA ALPICOLA	4.9 MG/L	MISS.FR.FILTRATE,20MG SURFACT./L (INITIAL),21 D,ALGAE-CONC.=87 MG/L	332
*ALKYL POLYETHOXYLATE ON GLOEOCAPSA ALPICOLA	5.6 MG/L	MISS.FR.FILTRATE,20MG SURFACT./L (INITIAL),21 D,ALGAE-CONC.=87 MG/L	332
*ALKYL POLYETHOXYLATE ON GLOEOCAPSA ALPICOLA	5.6 MG/L	MISS.FR.FILTRATE,20MG SURFACT./L (INITIAL),21 D,ALGAE-CONC.=87 MG/L	332

TABLE B75

CHEMICAL COMPOUNDS : SORPTION

ITEM:	VALUE:	CONDITIONS:	REF.:
*ALKYL POLYETHOXYLATE ON NOSTOC MUSCORUM	4.4 MG/L	MISS.FR.FILTRATE,20MG SURFACT./L (INITIAL),21 D,ALGAE-CONC.=40 MG/L	332
*ALKYL POLYETHOXYLATE ON NOSTOC MUSCORUM	4.4 MG/L	MISS.FR.FILTRATE,20MG SURFACT./L (INITIAL),21 D,ALGAE-CONC.=40 MG/L	332
*ALKYL POLYETHOXYLATE ON NOSTOC MUSCORUM	4.4 MG/L	MISS.FR.FILTRATE,20MG SURFACT./L (INITIAL),21 D,ALGAE-CONC.=40 MG/L	332
*ALKYL POLYETHOXYLATE ON NOSTOC MUSCORUM	4.4 MG/L	MISS.FR.FILTRATE,20MG SURFACT./L (INITIAL),21 D,ALGAE-CONC.=40 MG/L	332
*ALKYL POLYETHOXYLATE ON OSCILLATORIA BORNETI	5.5 MG/L	MISS.FR.FILTRATE,20MG SURFACT./L (INITIAL),21 D,ALGAE-CONC.=50 MG/L	332
*ALKYL POLYETHOXYLATE ON OSCILLATORIA BORNETI	5.5 MG/L	MISS.FR.FILTRATE,20MG SURFACT./L (INITIAL),21 D,ALGAE-CONC.=50 MG/L	332
*ALKYL POLYETHOXYLATE ON OSCILLATORIA BORNETI	4.0 MG/L	MISS.FR.FILTRATE,20MG SURFACT./L (INITIAL),21 D,ALGAE-CONC.=50 MG/L	332
*ALKYL POLYETHOXYLATE ON OSCILLATORIA BORNETI	4.0 MG/L	MISS.FR.FILTRATE,20MG SURFACT./L (INITIAL),21 D,ALGAE-CONC.=50 MG/L	332
*ALKYL POLYETHOXYLATE ON OSCILLATORIA CHALYBIA	5.3 MG/L	MISS.FR.FILTRATE,20MG SURFACT./L (INITIAL),21 D,ALGAE-CONC.=30 MG/L	332
*ALKYL POLYETHOXYLATE ON OSCILLATORIA CHALYBIA	5.3 MG/L	MISS.FR.FILTRATE,20MG SURFACT./L (INITIAL),21 D,ALGAE-CONC.=30 MG/L	332
*ALKYL POLYETHOXYLATE ON OSCILLATORIA CHALYBIA	2.8 MG/L	MISS.FR.FILTRATE,20MG SURFACT./L (INITIAL),21 D,ALGAE-CONC.=30 MG/L	332
*ALKYL POLYETHOXYLATE ON OSCILLATORIA CHALYBIA	2.8 MG/L	MISS.FR.FILTRATE,20MG SURFACT./L (INITIAL),21 D,ALGAE-CONC.=30 MG/L	332
*ALKYL POLYETHOXYLATE ON OSCILLATORIA FORMOSA	2.1 MG/L	MISS.FR.FILTRATE,20MG SURFACT./L (INITIAL),21 D,ALGAE-CONC.=40 MG/L	332
*ALKYL POLYETHOXYLATE ON OSCILLATORIA FORMOSA	2.1 MG/L	MISS.FR.FILTRATE,20MG SURFACT./L (INITIAL),21 D,ALGAE-CONC.=40 MG/L	332
*ALKYL POLYETHOXYLATE ON OSCILLATORIA FORMOSA	1.1 MG/L	MISS.FR.FILTRATE,20MG SURFACT./L (INITIAL),21 D,ALGAE-CONC.=40 MG/L	332
*ALKYL POLYETHOXYLATE ON OSCILLATORIA FORMOSA	1.1 MG/L	MISS.FR.FILTRATE,20MG SURFACT./L (INITIAL),21 D,ALGAE-CONC.=40 MG/L	332
*ALKYL POLYETHOXYLATE ON OSCILLATORIA TENUIS	3.5 MG/L	MISS.FR.FILTRATE,20MG SURFACT./L (INITIAL),21 D,ALGAE-CONC.=72 MG/L	332
*ALKYL POLYETHOXYLATE ON OSCILLATORIA TENUIS	3.5 MG/L	MISS.FR.FILTRATE,20MG SURFACT./L (INITIAL),21 D,ALGAE-CONC.=72 MG/L	332
*ALKYL POLYETHOXYLATE ON OSCILLATORIA TENUIS	1.6 MG/L	MISS.FR.FILTRATE,20MG SURFACT./L (INITIAL),21 D,ALGAE-CONC.=72 MG/L	332
*ALKYL POLYETHOXYLATE ON OSCILLATORIA TENUIS	1.6 MG/L	MISS.FR.FILTRATE,20MG SURFACT./L (INITIAL),21 D,ALGAE-CONC.=72 MG/L	332
*ALKYL POLYETHOXYLATE ON PHORMIDIUM FAVEOLARUM	0.4 MG/L	MISS.FR.FILTRATE,20MG SURFACT./L (INITIAL),21 D,ALGAE-CONC.=90 MG/L	332
*ALKYL POLYETHOXYLATE ON PHORMIDIUM FAVEOLARUM	0.4 MG/L	MISS.FR.FILTRATE,20MG SURFACT./L (INITIAL),21 D,ALGAE-CONC.=90 MG/L	332
*ALKYL POLYETHOXYLATE ON PHORMIDIUM FAVEOLARUM	5.0 MG/L	MISS.FR.FILTRATE,20MG SURFACT./L (INITIAL),21 D,ALGAE-CONC.=90 MG/L	332
*ALKYL POLYETHOXYLATE ON PHORMIDIUM FAVEOLARUM	5.0 MG/L	MISS.FR.FILTRATE,20MG SURFACT./L (INITIAL),21 D,ALGAE-CONC.=90 MG/L	332
*ALKYL POLYETHOXYLATE ON POND WATER	6.3 MG/L	MISSING FROM FILTRATE,20MG SURFACT./L(INITIAL),21DAYS	332
*ALKYL POLYETHOXYLATE ON POND WATER	6.3 MG/L	MISSING FROM FILTRATE,20MG SURFACT./L(INITIAL),21DAYS	332
*ALKYL POLYETHOXYLATE ON POND WATER	6.6 MG/L	MISSING FROM FILTRATE,20MG SURFACT./L(INITIAL),21DAYS	332
*ALKYL POLYETHOXYLATE ON POND WATER	6.6 MG/L	MISSING FROM FILTRATE,20MG SURFACT./L(INITIAL),21DAYS	332
*ALKYL POLYETHOXYLATE ON RAW WASTEWATER	1.0 MG/L	MISSING FROM FILTRATE,20MG SURFACT./L(INITIAL),21DAYS	332
*ALKYL POLYETHOXYLATE ON RAW WASTEWATER	1.0 MG/L	MISSING FROM FILTRATE,20MG SURFACT./L(INITIAL),21DAYS	332
*ALKYL POLYETHOXYLATE ON RAW WASTEWATER	6.5 MG/L	MISSING FROM FILTRATE,20MG SURFACT./L(INITIAL),21DAYS	332
*ALKYL POLYETHOXYLATE ON RAW WASTEWATER	6.5 MG/L	MISSING FROM FILTRATE,20MG SURFACT./L(INITIAL),21DAYS	332
*ALKYL POLYETHOXYLATE ON SCENEDESMUS OBLIQUUS	1.6 MG/L	MISS.FR.FILTRATE,20MG SURFACT./L (INITIAL),21 D,ALGAE-CONC.=50 MG/L	332
*ALKYL POLYETHOXYLATE ON SCENEDESMUS OBLIQUUS	1.6 MG/L	MISS.FR.FILTRATE,20MG SURFACT./L (INITIAL),21 D,ALGAE-CONC.=50 MG/L	332

TABLE B75

CHEMICAL COMPOUNDS : SORPTION

ITEM:	VALUE:	CONDITIONS:	REF.:
*ALKYL POLYETHOXYLATE ON SCENEDESMUS OBLIQUUS	5.7 MG/L	MISS.FR.FILTRATE,20MG SURFACT./L (INITIAL),21 D,ALGAE-CONC.=50 MG/L	332
*ALKYL POLYETHOXYLATE ON SCENEDESMUS OBLIQUUS	5.7 MG/L	MISS.FR.FILTRATE,20MG SURFACT./L (INITIAL),21 D,ALGAE-CONC.=50 MG/L	332
*ALKYL POLYETHOXYLATE ON SCENEDESMUS QUADRICAUDA	0.2 MG/L	MISS.FR.FILTRATE,20MG SURFACT./L (INITIAL),21 D,ALGAE-CONC.=20 MG/L	332
*ALKYL POLYETHOXYLATE ON SCENEDESMUS QUADRICAUDA	0.2 MG/L	MISS.FR.FILTRATE,20MG SURFACT./L (INITIAL),21 D,ALGAE-CONC.=20 MG/L	332
*ALKYL POLYETHOXYLATE ON SCENEDESMUS QUADRICAUDA	2.6 MG/L	MISS.FR.FILTRATE,20MG SURFACT./L (INITIAL),21 D,ALGAE-CONC.=20 MG/L	332
*ALKYL POLYETHOXYLATE ON SCENEDESMUS QUADRICAUDA	2.6 MG/L	MISS.FR.FILTRATE,20MG SURFACT./L (INITIAL),21 D,ALGAE-CONC.=20 MG/L	332
*ALKYL POLYETHOXYLATE ON SOIL	0.0 MG/L	MISSING FROM FILTRATE,20MG SURFACT./L(INITIAL),21DAYS	332
*ALKYL POLYETHOXYLATE ON SOIL	0.0 MG/L	MISSING FROM FILTRATE,20MG SURFACT./L(INITIAL),21DAYS	332
*ALKYL POLYETHOXYLATE ON SOIL	1.5 MG/L	MISSING FROM FILTRATE,20MG SURFACT./L(INITIAL),21DAYS	332
*ALKYL POLYETHOXYLATE ON SOIL	1.5 MG/L	MISSING FROM FILTRATE,20MG SURFACT./L(INITIAL),21DAYS	332
*ALKYL POLYETHOXYLATE ON ULOTHRIX FIMBRIATA	7.0 MG/L	MISS.FR.FILTRATE,20MG SURFACT./L (INITIAL),21 D,ALGAE-CONC.=90 MG/L	332
*ALKYL POLYETHOXYLATE ON ULOTHRIX FIMBRIATA	7.0 MG/L	MISS.FR.FILTRATE,20MG SURFACT./L (INITIAL),21 D,ALGAE-CONC.=90 MG/L	332
*ALKYL POLYETHOXYLATE ON ULOTHRIX FIMBRIATA	8.6 MG/L	MISS.FR.FILTRATE,20MG SURFACT./L (INITIAL),21 D,ALGAE-CONC.=90 MG/L	332
*ALKYL POLYETHOXYLATE ON ULOTHRIX FIMBRIATA	8.6 MG/L	MISS.FR.FILTRATE,20MG SURFACT./L (INITIAL),21 D,ALGAE-CONC.=90 MG/L	332
*LINEAR ALKYL SULFONATE ON ACTIVATED SLUDGE	4.9 MG/L	MISSING FROM FILTRATE,20MG SURFACT./L(INITIAL),21DAYS	332
*LINEAR ALKYL SULFONATE ON ACTIVATED SLUDGE	4.9 MG/L	MISSING FROM FILTRATE,20MG SURFACT./L(INITIAL),21DAYS	332
*LINEAR ALKYL SULFONATE ON ANABAENA CYLINDRICA	1.4 MG/L	MISS.FR.FILTRATE,20MG SURFACT./L (INITIAL),21 D,ALGAE-CONC.=58 MG/L	332
*LINEAR ALKYL SULFONATE ON ANABAENA CYLINDRICA	1.4 MG/L	MISS.FR.FILTRATE,20MG SURFACT./L (INITIAL),21 D,ALGAE-CONC.=58 MG/L	332
*LINEAR ALKYL SULFONATE ON ANABAENA VARIABILIS	1.3 MG/L	MISS.FR.FILTRATE,20MG SURFACT./L (INITIAL),21 D,ALGAE-CONC.=71 MG/L	332
*LINEAR ALKYL SULFONATE ON ANABAENA VARIABILIS	1.3 MG/L	MISS.FR.FILTRATE,20MG SURFACT./L (INITIAL),21 D,ALGAE-CONC.=71 MG/L	332
*LINEAR ALKYL SULFONATE ON ANACYSTIS NIDULANS	2.2 MG/L	MISS.FR.FILTRATE,20MG SURFACT./L (INITIAL),21 D,ALGAE-CONC.=121 MG/ L	332
*LINEAR ALKYL SULFONATE ON ANACYSTIS NIDULANS	2.2 MG/L	MISS.FR.FILTRATE,20MG SURFACT./L (INITIAL),21 D,ALGAE-CONC.=121 MG/ L	332
*LINEAR ALKYL SULFONATE ON ANKISTRODESMUS BRAUNII	3.5 MG/L	MISS.FR.FILTRATE,20MG SURFACT./L (INITIAL),21 D,ALGAE-CONC.=41 MG/L	332
*LINEAR ALKYL SULFONATE ON ANKISTRODESMUS BRAUNII	3.5 MG/L	MISS.FR.FILTRATE,20MG SURFACT./L (INITIAL),21 D,ALGAE-CONC.=41 MG/L	332
*LINEAR ALKYL SULFONATE ON CALOTHRIX PARIETINA	1.5 MG/L	MISS.FR.FILTRATE,20MG SURFACT./L (INITIAL),21 D,ALGAE-CONC.=47 MG/L	332
*LINEAR ALKYL SULFONATE ON CALOTHRIX PARIETINA	1.5 MG/L	MISS.FR.FILTRATE,20MG SURFACT./L (INITIAL),21 D,ALGAE-CONC.=47 MG/L	332
*LINEAR ALKYL SULFONATE ON CHLORELLA PYRENOIDOSA	0.0 MG/L	MISS.FR.FILTRATE,20MG SURFACT./L (INITIAL),21 D,ALGAE-CONC.=35 MG/L	332
*LINEAR ALKYL SULFONATE ON CHLORELLA PYRENOIDOSA	0.0 MG/L	MISS.FR.FILTRATE,20MG SURFACT./L (INITIAL),21 D,ALGAE-CONC.=35 MG/L	332
*LINEAR ALKYL SULFONATE ON CHLORELLA VULGARIS	0.6 MG/L	MISS.FR.FILTRATE,20MG SURFACT./L (INITIAL),21 D,ALGAE-CONC.=40 MG/L	332
*LINEAR ALKYL SULFONATE ON CHLORELLA VULGARIS	0.6 MG/L	MISS.FR.FILTRATE,20MG SURFACT./L (INITIAL),21 D,ALGAE-CONC.=40 MG/L	332
*LINEAR ALKYL SULFONATE ON CYLINDROSPERMUM SP.	0.0 MG/L	MISS.FR.FILTRATE,20MG SURFACT./L (INITIAL),21 D,ALGAE-CONC.=55 MG/L	332
*LINEAR ALKYL SULFONATE ON CYLINDROSPERMUM SP.	0.0 MG/L	MISS.FR.FILTRATE,20MG SURFACT./L (INITIAL),21 D,ALGAE-CONC.=55 MG/L	332
*LINEAR ALKYL SULFONATE ON GLOEOCAPSA ALPICOLA	0.3 MG/L	MISS.FR.FILTRATE,20MG SURFACT./L (INITIAL),21 D,ALGAE-CONC.=87 MG/L	332

TABLE B75

CHEMICAL COMPOUNDS : SORPTION

ITEM:	VALUE:	CONDITIONS:	REF.:
*LINEAR ALKYL SULFONATE ON GLOEOCAPSA ALPICOLA	0.3 MG/L	MISS.FR.FILTRATE,20MG SURFACT./L (INITIAL),21 D,ALGAE-CONC.=87 MG/L	332
*LINEAR ALKYL SULFONATE ON NOSTOC MUSCORUM	2.0 MG/L	MISS.FR.FILTRATE,20MG SURFACT./L (INITIAL),21 D,ALGAE-CONC.=40 MG/L	332
*LINEAR ALKYL SULFONATE ON NOSTOC MUSCORUM	2.0 MG/L	MISS.FR.FILTRATE,20MG SURFACT./L (INITIAL),21 D,ALGAE-CONC.=40 MG/L	332
*LINEAR ALKYL SULFONATE ON OSCILLATORIA BORNETI	0.1 MG/L	MISS.FR.FILTRATE,20MG SURFACT./L (INITIAL),21 D,ALGAE-CONC.=50 MG/L	332
*LINEAR ALKYL SULFONATE ON OSCILLATORIA BORNETI	0.1 MG/L	MISS.FR.FILTRATE,20MG SURFACT./L (INITIAL),21 D,ALGAE-CONC.=50 MG/L	332
*LINEAR ALKYL SULFONATE ON OSCILLATORIA CHALYBIA	0.0 MG/L	MISS.FR.FILTRATE,20MG SURFACT./L (INITIAL),21 D,ALGAE-CONC.=30 MG/L	332
*LINEAR ALKYL SULFONATE ON OSCILLATORIA CHALYBIA	0.0 MG/L	MISS.FR.FILTRATE,20MG SURFACT./L (INITIAL),21 D,ALGAE-CONC.=30 MG/L	332
*LINEAR ALKYL SULFONATE ON OSCILLATORIA FORMOSA	0.0 MG/L	MISS.FR.FILTRATE,20MG SURFACT./L (INITIAL),21 D,ALGAE-CONC.=40 MG/L	332
*LINEAR ALKYL SULFONATE ON OSCILLATORIA FORMOSA	0.0 MG/L	MISS.FR.FILTRATE,20MG SURFACT./L (INITIAL),21 D,ALGAE-CONC.=40 MG/L	332
*LINEAR ALKYL SULFONATE ON OSCILLATORIA TENUIS	1.1 MG/L	MISS.FR.FILTRATE,20MG SURFACT./L (INITIAL),21 D,ALGAE-CONC.=72 MG/L	332
*LINEAR ALKYL SULFONATE ON OSCILLATORIA TENUIS	1.1 MG/L	MISS.FR.FILTRATE,20MG SURFACT./L (INITIAL),21 D,ALGAE-CONC.=72 MG/L	332
*LINEAR ALKYL SULFONATE ON PHORMIDIUM FAVEOLARUM	1.9 MG/L	MISS.FR.FILTRATE,20MG SURFACT./L (INITIAL),21 D,ALGAE-CONC.=90 MG/L	332
*LINEAR ALKYL SULFONATE ON PHORMIDIUM FAVEOLARUM	1.9 MG/L	MISS.FR.FILTRATE,20MG SURFACT./L (INITIAL),21 D,ALGAE-CONC.=90 MG/L	332
*LINEAR ALKYL SULFONATE ON POND WATER	0.1 MG/L	MISSING FROM FILTRATE,20MG SURFACT./L(INITIAL),21DAYS	332
*LINEAR ALKYL SULFONATE ON POND WATER	0.1 MG/L	MISSING FROM FILTRATE,20MG SURFACT./L(INITIAL),21DAYS	332
*LINEAR ALKYL SULFONATE ON RAW WASTEWATER	6.0 MG/L	MISSING FROM FILTRATE,20MG SURFACT./L(INITIAL),21DAYS	332
*LINEAR ALKYL SULFONATE ON RAW WASTEWATER	6.0 MG/L	MISSING FROM FILTRATE,20MG SURFACT./L(INITIAL),21DAYS	332
*LINEAR ALKYL SULFONATE ON SCENEDESMUS OBLIQUUS	0.1 MG/L	MISS.FR.FILTRATE,20MG SURFACT./L (INITIAL),21 D,ALGAE-CONC.=50 MG/L	332
*LINEAR ALKYL SULFONATE ON SCENEDESMUS OBLIQUUS	0.1 MG/L	MISS.FR.FILTRATE,20MG SURFACT./L (INITIAL),21 D,ALGAE-CONC.=50 MG/L	332
*LINEAR ALKYL SULFONATE ON SCENEDESMUS QUADRICAUDA	1.7 MG/L	MISS.FR.FILTRATE,20MG SURFACT./L (INITIAL),21 D,ALGAE-CONC.=20 MG/L	332
*LINEAR ALKYL SULFONATE ON SCENEDESMUS QUADRICAUDA	1.7 MG/L	MISS.FR.FILTRATE,20MG SURFACT./L (INITIAL),21 D,ALGAE-CONC.=20 MG/L	332
*LINEAR ALKYL SULFONATE ON SOIL	1.4 MG/L	MISSING FROM FILTRATE,20MG SURFACT./L(INITIAL),21DAYS	332
*LINEAR ALKYL SULFONATE ON SOIL	1.4 MG/L	MISSING FROM FILTRATE,20MG SURFACT./L(INITIAL),21DAYS	332
*LINEAR ALKYL SULFONATE ON ULOTHRIX FIMBRIATA	2.6 MG/L	MISS.FR.FILTRATE,20MG SURFACT./L (INITIAL),21 D,ALGAE-CONC.=90 MG/L	332
*LINEAR ALKYL SULFONATE ON ULOTHRIX FIMBRIATA	2.6 MG/L	MISS.FR.FILTRATE,20MG SURFACT./L (INITIAL),21 D,ALGAE-CONC.=90 MG/L	332
*M-AMINOOPHENOL	12 % OF ADDED AMOUNT	0.5 MG/L PER 5 G BENTONITE (BENTONE 24), PH=7.7, 295 K	1037
*M-AMINOOPHENOL	0.8 % OF ADDED AMOUNT	0.5 MG/L PER 5 G BENTONITE (BENTONE 18C), PH=7.5, 295 K	1037
*M-CHLOROPHENOL	81.0 % OF ADDED AMOUNT	0.5 MG/L PER 5 G BENTONITE (BENTONE 24), PH=7.7, 295 K	1037
*M-CHLOROPHENOL	13 % OF ADDED AMOUNT	0.5 MG/L PER 5 G BENTONITE (BENTONE 18C), PH=7.6, 295 K	1037
*NONYL PHENOL POLYETHOXYLATE ON ACTIVATED SLUDGE	2.5 MG/L	MISSING FROM FILTRATE,20MG SURFACT./L(INITIAL),21DAYS	332
*NONYL PHENOL POLYETHOXYLATE ON ACTIVATED SLUDGE	2.5 MG/L	MISSING FROM FILTRATE,20MG SURFACT./L(INITIAL),21DAYS	332
*NONYL PHENOL POLYETHOXYLATE ON ANABAENA CYLINDRICA	3.6 MG/L	MISS.FR.FILTRATE,20MG SURFACT./L (INITIAL),21 D,ALGAE-CONC.=58 MG/L	332
*NONYL PHENOL POLYETHOXYLATE ON ANABAENA CYLINDRICA	3.6 MG/L	MISS.FR.FILTRATE,20MG SURFACT./L (INITIAL),21 D,ALGAE-CONC.=58 MG/L	332
*NONYL PHENOL POLYETHOXYLATE ON ANABAENA VARIABILIS	13.1 MG/L	MISS.FR.FILTRATE,20MG SURFACT./L (INITIAL),21 D,ALGAE-CONC.=71 MG/L	332

TABLE B75

CHEMICAL COMPOUNDS : SORPTION			
ITEM:	VALUE:	CONDITIONS:	REF.:
*NONYL PHENOL POLYETHOXYLATE ON ANABAENA VARIABILIS	13.1 MG/L	MISS.FR.FILTRATE,20MG SURFACT./L (INITIAL),21 D,ALGAE-CONC.=71 MG/L	332
*NONYL PHENOL POLYETHOXYLATE ON ANACYSTIS NIDULANS	5.1 MG/L	MISS.FR.FILTRATE,20MG SURFACT./L (INITIAL),21 D,ALGAE-CONC.=121 MG/L	332
*NONYL PHENOL POLYETHOXYLATE ON ANACYSTIS NIDULANS	5.1 MG/L	MISS.FR.FILTRATE,20MG SURFACT./L (INITIAL),21 D,ALGAE-CONC.=121 MG/L	332
*NONYL PHENOL POLYETHOXYLATE ON ANKISTRODESMUS BRAUNII	6.2 MG/L	MISS.FR.FILTRATE,20MG SURFACT./L (INITIAL),21 D,ALGAE-CONC.=41 MG/L	332
*NONYL PHENOL POLYETHOXYLATE ON ANKISTRODESMUS BRAUNII	6.2 MG/L	MISS.FR.FILTRATE,20MG SURFACT./L (INITIAL),21 D,ALGAE-CONC.=41 MG/L	332
*NONYL PHENOL POLYETHOXYLATE ON CALOTHRIX PARIETINA	2.1 MG/L	MISS.FR.FILTRATE,20MG SURFACT./L (INITIAL),21 D,ALGAE-CONC.=47 MG/L	332
*NONYL PHENOL POLYETHOXYLATE ON CALOTHRIX PARIETINA	2.1 MG/L	MISS.FR.FILTRATE,20MG SURFACT./L (INITIAL),21 D,ALGAE-CONC.=47 MG/L	332
*NONYL PHENOL POLYETHOXYLATE ON CHLORELLA PYRENOIDOSA	5.7 MG/L	MISS.FR.FILTRATE,20MG SURFACT./L (INITIAL),21 D,ALGAE-CONC.=35 MG/L	332
*NONYL PHENOL POLYETHOXYLATE ON CHLORELLA PYRENOIDOSA	5.7 MG/L	MISS.FR.FILTRATE,20MG SURFACT./L (INITIAL),21 D,ALGAE-CONC.=35 MG/L	332
*NONYL PHENOL POLYETHOXYLATE ON CHLORELLA VULGARIS	0.9 MG/L	MISS.FR.FILTRATE,20MG SURFACT./L (INITIAL),21 D,ALGAE-CONC.=40 MG/L	332
*NONYL PHENOL POLYETHOXYLATE ON CHLORELLA VULGARIS	0.9 MG/L	MISS.FR.FILTRATE,20MG SURFACT./L (INITIAL),21 D,ALGAE-CONC.=40 MG/L	332
*NONYL PHENOL POLYETHOXYLATE ON CYLINDROSPERMUM SP.	11.1 MG/L	MISS.FR.FILTRATE,20MG SURFACT./L (INITIAL),21 D,ALGAE-CONC.=55 MG/L	332
*NONYL PHENOL POLYETHOXYLATE ON CYLINDROSPERMUM SP.	11.1 MG/L	MISS.FR.FILTRATE,20MG SURFACT./L (INITIAL),21 D,ALGAE-CONC.=55 MG/L	332
*NONYL PHENOL POLYETHOXYLATE ON GLOEOCAPSA ALPICOLA	7.0 MG/L	MISS.FR.FILTRATE,20MG SURFACT./L (INITIAL),21 D,ALGAE-CONC.=87 MG/L	332
*NONYL PHENOL POLYETHOXYLATE ON GLOEOCAPSA ALPICOLA	7.0 MG/L	MISS.FR.FILTRATE,20MG SURFACT./L (INITIAL),21 D,ALGAE-CONC.=87 MG/L	332
*NONYL PHENOL POLYETHOXYLATE ON NOSTOC MUSCORUM	2.7 MG/L	MISS.FR.FILTRATE,20MG SURFACT./L (INITIAL),21 D,ALGAE-CONC.=40 MG/L	332
*NONYL PHENOL POLYETHOXYLATE ON NOSTOC MUSCORUM	2.7 MG/L	MISS.FR.FILTRATE,20MG SURFACT./L (INITIAL),21 D,ALGAE-CONC.=40 MG/L	332
*NONYL PHENOL POLYETHOXYLATE ON OSCILLATORIA BORNETI	1.9 MG/L	MISS.FR.FILTRATE,20MG SURFACT./L (INITIAL),21 D,ALGAE-CONC.=50 MG/L	332
*NONYL PHENOL POLYETHOXYLATE ON OSCILLATORIA BORNETI	1.9 MG/L	MISS.FR.FILTRATE,20MG SURFACT./L (INITIAL),21 D,ALGAE-CONC.=50 MG/L	332
*NONYL PHENOL POLYETHOXYLATE ON OSCILLATORIA CHALYBIA	6.5 MG/L	MISS.FR.FILTRATE,20MG SURFACT./L (INITIAL),21 D,ALGAE-CONC.=30 MG/L	332
*NONYL PHENOL POLYETHOXYLATE ON OSCILLATORIA CHALYBIA	6.5 MG/L	MISS.FR.FILTRATE,20MG SURFACT./L (INITIAL),21 D,ALGAE-CONC.=30 MG/L	332
*NONYL PHENOL POLYETHOXYLATE ON OSCILLATORIA FORMOSA	2.8 MG/L	MISS.FR.FILTRATE,20MG SURFACT./L (INITIAL),21 D,ALGAE-CONC.=40 MG/L	332
*NONYL PHENOL POLYETHOXYLATE ON OSCILLATORIA FORMOSA	2.8 MG/L	MISS.FR.FILTRATE,20MG SURFACT./L (INITIAL),21 D,ALGAE-CONC.=40 MG/L	332
*NONYL PHENOL POLYETHOXYLATE ON OSCILLATORIA TENUIS	7.0 MG/L	MISS.FR.FILTRATE,20MG SURFACT./L (INITIAL),21 D,ALGAE-CONC.=72 MG/L	332
*NONYL PHENOL POLYETHOXYLATE ON OSCILLATORIA TENUIS	7.0 MG/L	MISS.FR.FILTRATE,20MG SURFACT./L (INITIAL),21 D,ALGAE-CONC.=72 MG/L	332
*NONYL PHENOL POLYETHOXYLATE ON PHORMIDIUM FAVEOLARUM	0.8 MG/L	MISS.FR.FILTRATE,20MG SURFACT./L (INITIAL),21 D,ALGAE-CONC.=90 MG/L	332
*NONYL PHENOL POLYETHOXYLATE ON PHORMIDIUM FAVEOLARUM	0.8 MG/L	MISS.FR.FILTRATE,20MG SURFACT./L (INITIAL),21 D,ALGAE-CONC.=90 MG/L	332
*NONYL PHENOL POLYETHOXYLATE ON POND WATER	10.6 MG/L	MISSING FROM FILTRATE,20MG SURFACT./L(INITIAL),21DAYS	332
*NONYL PHENOL POLYETHOXYLATE ON POND WATER	10.6 MG/L	MISSING FROM FILTRATE,20MG SURFACT./L(INITIAL),21DAYS	332
*NONYL PHENOL POLYETHOXYLATE ON RAW WASTEWATER	7.5 MG/L	MISSING FROM FILTRATE,20MG SURFACT./L(INITIAL),21DAYS	332
*NONYL PHENOL POLYETHOXYLATE ON RAW WASTEWATER	7.5 MG/L	MISSING FROM FILTRATE,20MG SURFACT./L(INITIAL),21DAYS	332
*NONYL PHENOL POLYETHOXYLATE ON SCENEDESMUS OBLIQUUS	3.3 MG/L	MISS.FR.FILTRATE,20MG SURFACT./L (INITIAL),21 D,ALGAE-CONC.=50 MG/L	332
*NONYL PHENOL POLYETHOXYLATE ON SCENEDESMUS OBLIQUUS	3.3 MG/L	MISS.FR.FILTRATE,20MG SURFACT./L (INITIAL),21 D,ALGAE-CONC.=50 MG/L	332

TABLE B75

CHEMICAL COMPOUNDS : SORPTION			
ITEM:	VALUE:	CONDITIONS:	REF.:
*NONYL PHENOL POLYETHOXYLATE ON SCENEDESMUS QUADRICAUDA	2.9 MG/L	MISS.FR.FILTRATE,20MG SURFACT./L (INITIAL),21 D,ALGAE-CONC.=20 MG/L	332
*NONYL PHENOL POLYETHOXYLATE ON SCENEDESMUS QUADRICAUDA	2.9 MG/L	MISS.FR.FILTRATE,20MG SURFACT./L (INITIAL),21 D,ALGAE-CONC.=20 MG/L	332
*NONYL PHENOL POLYETHOXYLATE ON SOIL	0.0 MG/L	MISSING FROM FILTRATE,20MG SURFACT./L(INITIAL),21DAYS	332
*NONYL PHENOL POLYETHOXYLATE ON SOIL	0.0 MG/L	MISSING FROM FILTRATE,20MG SURFACT./L(INITIAL),21DAYS	332
*NONYL PHENOL POLYETHOXYLATE ON ULOTHRIX FIMBRIATA	14.8 MG/L	MISS.FR.FILTRATE,20MG SURFACT./L (INITIAL),21 D,ALGAE-CONC.=90 MG/L	332
*NONYL PHENOL POLYETHOXYLATE ON ULOTHRIX FIMBRIATA	14.8 MG/L	MISS.FR.FILTRATE,20MG SURFACT./L (INITIAL),21 D,ALGAE-CONC.=90 MG/L	332
*O-AMINOPHENOL	25.5 % OF ADDED AMOUNT	0.5 MG/L PER 5 G BENTONITE (BENTONE 24), PH=8.4, 295 K	1037
*O-AMINOPHENOL	29 % OF ADDED AMOUNT	0.5 MG/L PER 5 G BENTONITE (BENTONE 18C), PH=7.8, 295 K	1037
*O-BROMOPHENOL	85.6 % OF ADDED AMOUNT	0.5 MG/L PER 5 G BENTONITE (BENTONE 24), PH=7.6, 295 K	1037
*O-BROMOPHENOL	18 % OF ADDED AMOUNT	0.5 MG/L PER 5 G BENTONITE (BENTONE 18C), PH=7.4, 295 K	1037
*O-CHLOROPHENOL	76.6 % OF ADDED AMOUNT	0.5 MG/L PER 5 G BENTONITE (BENTONE 24), PH=7.8, 295 K	1037
*O-CHLOROPHENOL	15 % OF ADDED AMOUNT	0.5 MG/L PER 5 G BENTONITE (BENTONE 18C), PH=7.6, 295 K	1037
*O-CHLOROPHENOXYACETIC ACID	43.6 % OF ADDED AMOUNT	0.5 MG/L PER 5 G BENTONITE (BENTONE 24), PH=7.4, 295 K	1037
*O-CHLOROPHENOXYACETIC ACID	17.0 % OF ADDED AMOUNT	0.5 MG/L PER 5 G BENTONITE (BENTONE 18C), PH=7.3, 295 K	1037
*O-PHENYL PHENOL	96.4 % OF ADDED AMOUNT	0.5 MG/L PER 5 G BENTONITE (BENTONE 24), PH=7.3, 295 K	1037
*O-PHENYL PHENOL	47 % OF ADDED AMOUNT	0.5 MG/L PER 5 G BENTONITE (BENTONE 18C), PH=7.6, 295 K	1037
*P-CHLOROPHENOL	48.5 % OF ADDED AMOUNT	0.5 MG/L PER 5 G BENTONITE (BENTONE 24), PH=8.0, 295 K	1037
*P-CHLOROPHENOL	6 % OF ADDED AMOUNT	0.5 MG/L PER 5 G BENTONITE (BENTONE 18C), PH=7.7, 295 K	1037
*P-CHLOROPHENOXYACETIC ACID	45.6 % OF ADDED AMOUNT	0.5 MG/L PER 5 G BENTONITE (BENTONE 24), PH=7.4, 295 K	1037
*P-CHLOROPHENOXYACETIC ACID	11.0 % OF ADDED AMOUNT	0.5 MG/L PER 5 G BENTONITE (BENTONE 18C), PH=7.4, 295 K	1037
*2,4-DIBROMOPHENOL	97.9 % OF ADDED AMOUNT	0.5 MG/L PER 5 G BENTONITE (BENTONE 24), PH=7.6, 295 K	1037
*2,4-DIBROMOPHENOL	55.4 % OF ADDED AMOUNT	0.5 MG/L PER 5 G BENTONITE (BENTONE 18C), PH=7.4, 295 K	1037
*2,4-DICHLOROPHENOL	96.4 % OF ADDED AMOUNT	0.5 MG/L PER 5 G BENTONITE (BENTONE 24), PH=7.7, 295 K	1037
*2,4-DICHLOROPHENOL	38 % OF ADDED AMOUNT	0.5 MG/L PER 5 G BENTONITE (BENTONE 18C), PH=7.5, 295 K	1037
*2,4,5-TRICHLOROPHENOL	97.9 % OF ADDED AMOUNT	0.5 MG/L PER 5 G BENTONITE (BENTONE 24), PH=7.6, 295 K	1037
*2,4,5-TRICHLOROPHENOL	63.6 % OF ADDED AMOUNT	0.5 MG/L PER 5 G BENTONITE (BENTONE 18C), PH=7.4, 295 K	1037
*2,4,6-TRICHLOROPHENOL	91.8 % OF ADDED AMOUNT	0.5 MG/L PER 5 G BENTONITE (BENTONE 24), PH=7.5, 295 K	1037
*2,4,6-TRICHLOROPHENOL	55.0 % OF ADDED AMOUNT	0.5 MG/L PER 5 G BENTONITE (BENTONE 18C), PH=7.4, 295 K	1037
*2,5-DICHLOROPHENOL	93.9 % OF ADDED AMOUNT	0.5 MG/L PER 5 G BENTONITE (BENTONE 24), PH=7.9, 295 K	1037
*2,5-DICHLOROPHENOL	38 % OF ADDED AMOUNT	0.5 MG/L PER 5 G BENTONITE (BENTONE 18C), PH=7.5, 295 K	1037
*2,6-DICHLOROPHENOL	78.6 % OF ADDED AMOUNT	0.5 MG/L PER 5 G BENTONITE (BENTONE 24), PH=7.8, 295 K	1037
*2,6-DICHLOROPHENOL	32 % OF ADDED AMOUNT	0.5 MG/L PER 5 G BENTONITE (BENTONE 18C), PH=7.4, 295 K	1037
*2,6-DIMETHYLPHENOL	72.8 % OF ADDED AMOUNT	0.5 MG/L PER 5 G BENTONITE (BENTONE 24), PH=7.9, 295 K	1037
*2,6-DIMETHYLPHENOL	8 % OF ADDED AMOUNT	0.5 MG/L PER 5 G BENTONITE (BENTONE 18C), PH=7.6, 295 K	1037

TABLE B75

CHEMICAL COMPOUNDS : SORPTION			
ITEM:	VALUE:	CONDITIONS:	REF.:
*3,4-DICHLOROPHENOL	96.9 % OF ADDED AMOUNT	0.5 MG/L PER 5 G BENTONITE (BENTONE 24), PH=7.9, 295 K	1037
*3,4-DICHLOROPHENOL	34 % OF ADDED AMOUNT	0.5 MG/L PER 5 G BENTONITE (BENTONE 18C), PH=7.6, 295 K	1037
*3,4-DIMETHYLPHENOL	65.2 % OF ADDED AMOUNT	0.5 MG/L PER 5 G BENTONITE (BENTONE 24), PH=7.8, 295 K	1037
*3,4-DIMETHYLPHENOL	13 % OF ADDED AMOUNT	0.5 MG/L PER 5 G BENTONITE (BENTONE 18C), PH=7.7, 295 K	1037
*3,5-DICHLOROPHENOL	96.4 % OF ADDED AMOUNT	0.5 MG/L PER 5 G BENTONITE (BENTONE 24), PH=7.9, 295 K	1037
*3,5-DICHLOROPHENOL	42 % OF ADDED AMOUNT	0.5 MG/L PER 5 G BENTONITE (BENTONE 18C), PH=7.6, 295 K	1037

TABLE B076

Pools and Fluxes within the Lead Cycle.

REF.: 413

Flux	Value 10**12 g/year
World lead production (1966)	3.5
Northern hemisphere production	3.1
Lead burned as alkyles	0.31
River input of soluble lead to sea	0.24
River input of particulate lead to sea	0.5

TABLE B077

Trends to be Expected in the Development of Ecosystems.

REF.: 1160

System category	Ecosystem parameter	Developmental Stages	Mature Stages
Community energetics	Gross prod./community resp.(P/R ratio)	> or < 1	-> 1
Community Energetics	Gross prod./stand. crop biomass(P/B ratio)	high	low
Community energetics	Biomass supported/energy flow(B/E ratio)	low	high
Community energetics	Net community production(yield)	high	low
Community energetics	Food chains	linear, predom. grazing	Weblike, predom. detritus
Community chemicals	Total organic matter	small	large
Community chemicals	Inorganic nutrients	extrabiotic	intrabiotic
Diversity	Species diversity: variety component	low	high
Diversity	Species diversity: equitability component	low	high
Diversity	Biochemical diversity	low	high
Diversity	Pattern diversity	poorly organized	well-organized
Life history	Niche specialization	broad	narrow
Life history	Size of organism	small	large
Life history	Life cycles	short, simple	long, complex
Nutrient cycling	Mineral cycles	open	closed
Nutrient cycling	Exchange rate betw. organisms and environment	rapid	slow
Nutrient cycling	Role of detritus in nutrient regeneration	unimportant	important
Selection pressure	Growth form	rapid growth (r-selection)	feedback control (K-selection)
Selection pressure	production	quantity	quality
Overall homeostasis	Internal symbiosis	undeveloped	developed
Overall homeostasis	Nutrient conservation	poor	good
Overall homeostasis	Stability	poor	good
Overall homeostasis	Entropy	high	low
Overall homeostasis	Information	low	high

TABLE B078

The Proportions of Land Use for the World and for Major Regions.
Urban Land Use is not shown (1-2 %)
REF.: 355

Region	% cropland	% grazing land	% wasteland	% forest
World	11	19	40	30
Africa	9	21	45	25
USSR	10	17	34	39
North America	10	13	42	35
Latin America	5	20	27	48
South Asia	22	13	40	25
East Asia	10	15	63	12
Oceania	4	54	33	9
Europe	31	18	23	28

TABLE B079

Biosphere Exchange Rates

REF.: 355

Element	Turnover time in years
Water vapour	200E4
Oxygen	2000
Carbon dioxide	300

TABLE B080

General Ranges of Primary Productivity of Phytoplankton and Related Characteristics of Lakes of Different Trophic Categories.
REF.: 925

Trophic type	Mean Primary Prod. (mg C/m2/day)	Phytopl. density (cm3/m3)	Phytopl. Biomass (mg/m3)	Chloroph. a (mg/m3)	Dominant Phytoplankton	Light Extinct. Coeff. (1/m)	Tot. org. carbon (mg/l)	Tot. P (ug/l)	Tot. N (ug/l)	Tot. inorg. solids (mg/l)
Ultra-oligotrp	<50	<1	<50	0.01-0.5		0.03-0.8		<1-5	<1-250	2-15
Oligotrophic	50-300		20-100	0.3-3	Chryso+Cryptophyc	0.05-1.0	<1-3			
Oligo-mesotrp		1-3			Dino+Bacillariophyc			5-10	250-600	10-200
Mesotrophic	250-1000		100-300	2-15		0.1-2.0	<1-5			
Mesoeutrophic		3-5						10-30	500-1100	100-500
Eutrophic	>1000		>300	10-500	Bacillario+Cyanophyc	0.5-4.0	5-30			
Hypereutroph.		>10			Chloro+Euglenophyc			30->5000	500->15000	400-60000
Dystrophic	<50-500		<50-200	0.1-10		1.0-4.0	3-30	<1-10	<1-500	5-200

TABLE B081

Water Resources and Annual Water Balance of the Continents of the World.
REF.: 460

Component	Europe	Asia	Africa	North America	South America	Australia	Total
Area (1E6 km2)	9.8	45	30.3	20.7	17.8	8.7	132.3
Precipation (km3)	7165	32690	20780	13910	29355	6405	110E3
Total river runoff (km3)	3110	13190	4225	5960	10380	1965	38830
Underground runoff (km3)	1065	3410	1465	1740	3740	465	11885
Infiltration (km3)	5120	22910	18020	9690	22715	4905	83360
Evaporation (km3)	4055	19500	16555	7950	18975	4440	71475
% underground runoff of total	34	26	35	32	36	24	31

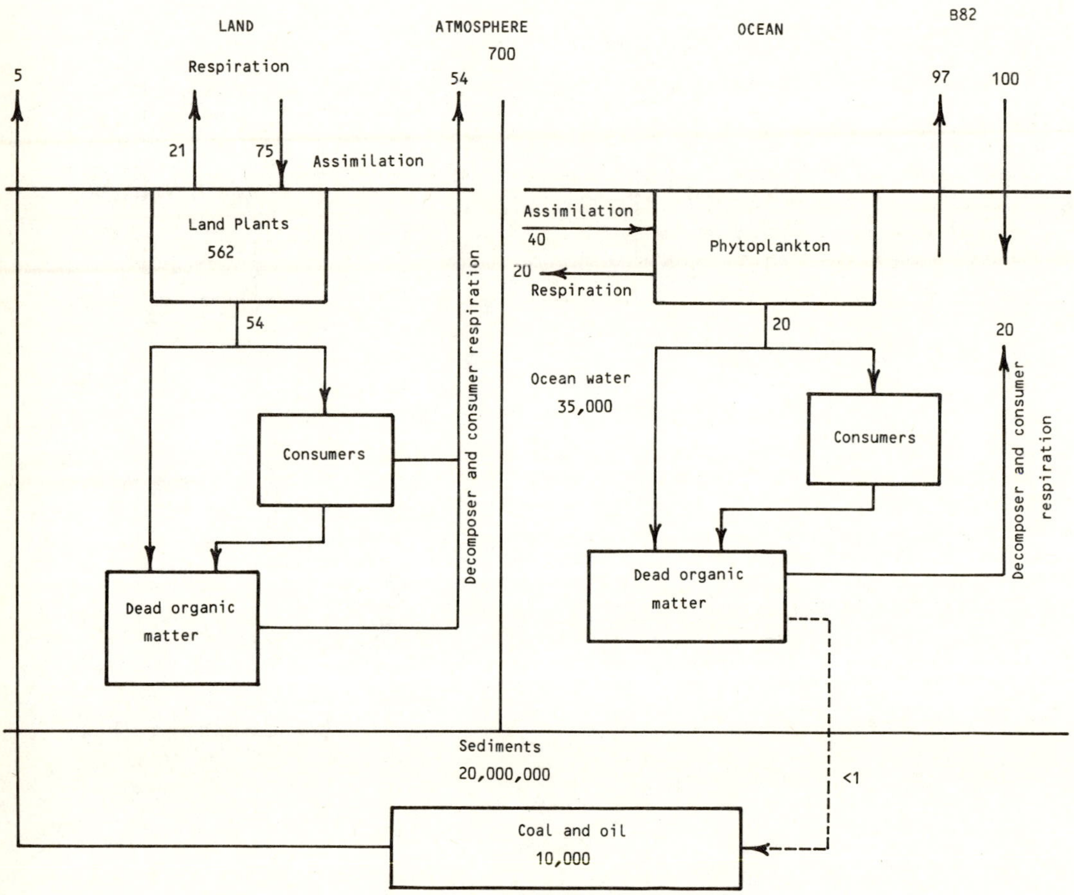

Carbon cycle, global. Values in compartments are in 10**9 tons and in fluxes 10**9 tons/ year.
b2

(ref: 268)

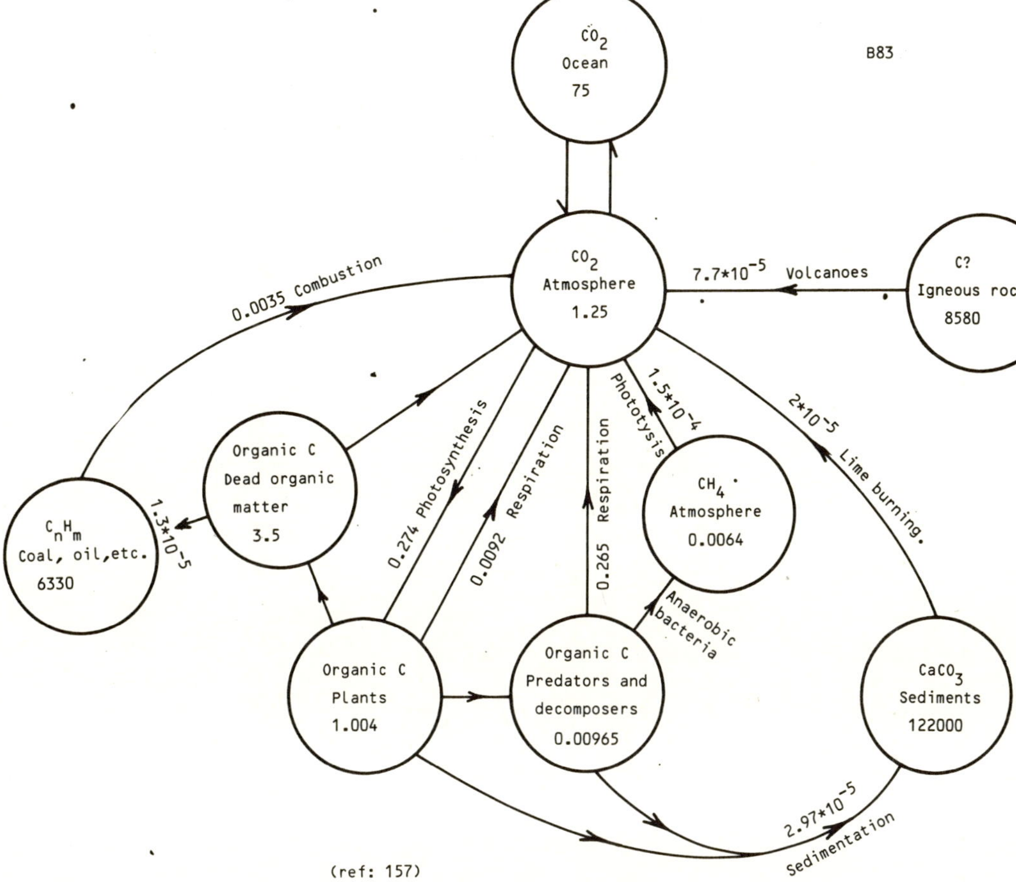

(ref: 157)

Carbon cycle, global. Values in compartments are in kg C/m**2, and in fluxes kg C/(m**2*year).

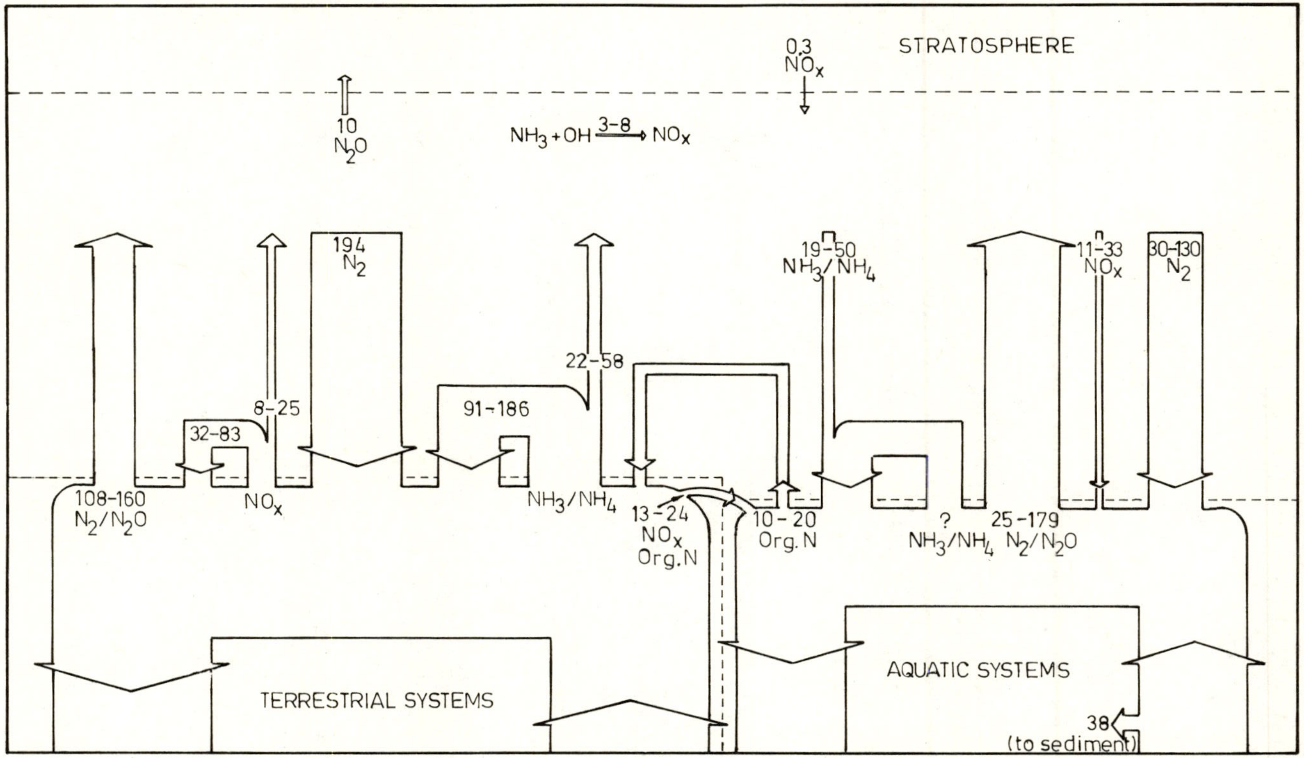

Cycle of nitrogen, global. Transfer in T g.year between the spheres. (T = 10**12).

(ref: 1347)

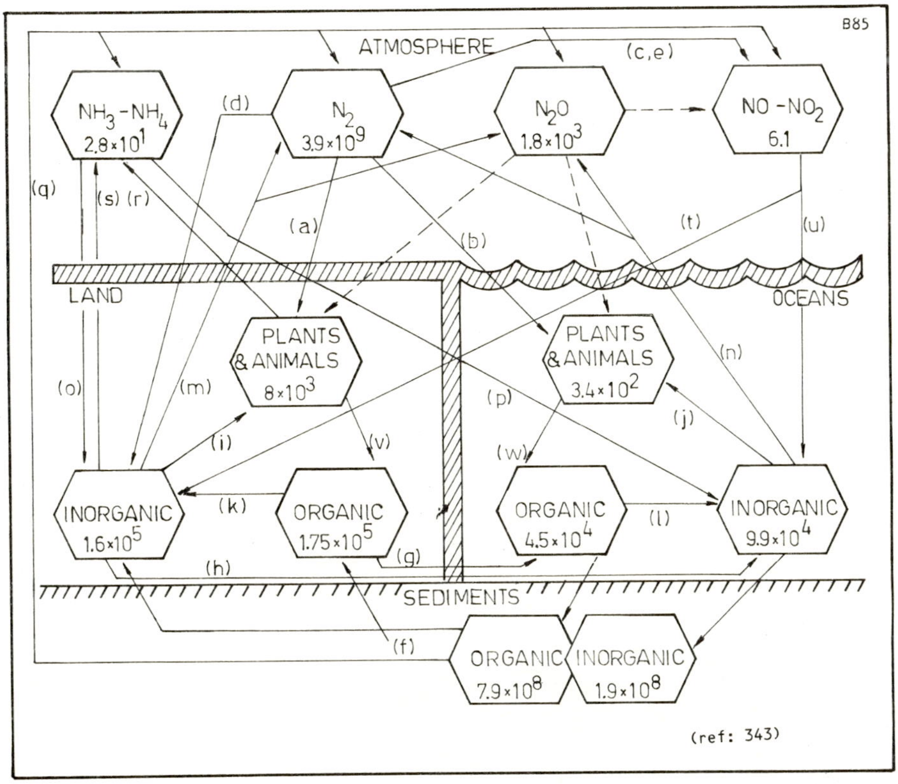

Cycle of nitrogen, global. Pools size in T g of N (T =10**12). transfer rates (T g N/year). a: N-fixation, land 99. b: N-fixation, ocean 30. c: N-fixation, atm. 7.4. d: N-fixation, industrial 40. e: N-fixation, combustion 18. f: weathering processes 4.4. g: Runoff, organic 22. h: Runoff, inorganic 13. i: Assimilation, land 1800. j: Assimilation, sea 1400. k: Minaralization, land1900. l: Mineralization, sea 1400. m: Denitrification, sea 40. o: NH4(+) rainout + washout land 65. p: NH4(+) rainout + washout sea 14. q: N from fossil fuels 3.5. r: NH4(+) volatilization, plants and animals 53. s* NH4(;) volitilization soil 22. t: NOx rainout + washout, land 25. u: NOx rainout + washout, sea 8.8. v: To organic pool, land 1900. w: To organic pool, sea 1400.

B86

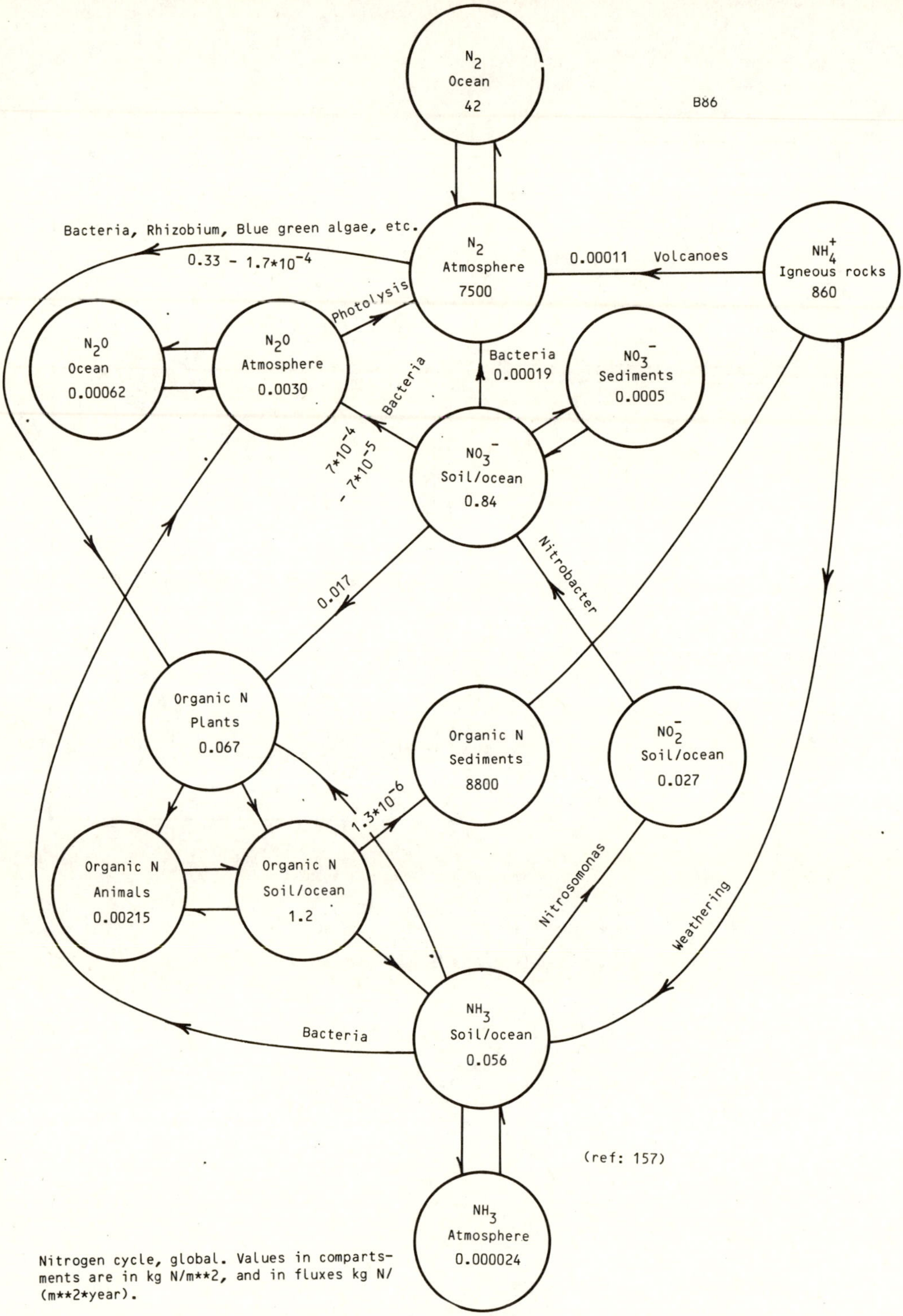

Nitrogen cycle, global. Values in comparts- ments are in kg N/m**2, and in fluxes kg N/ (m**2*year).

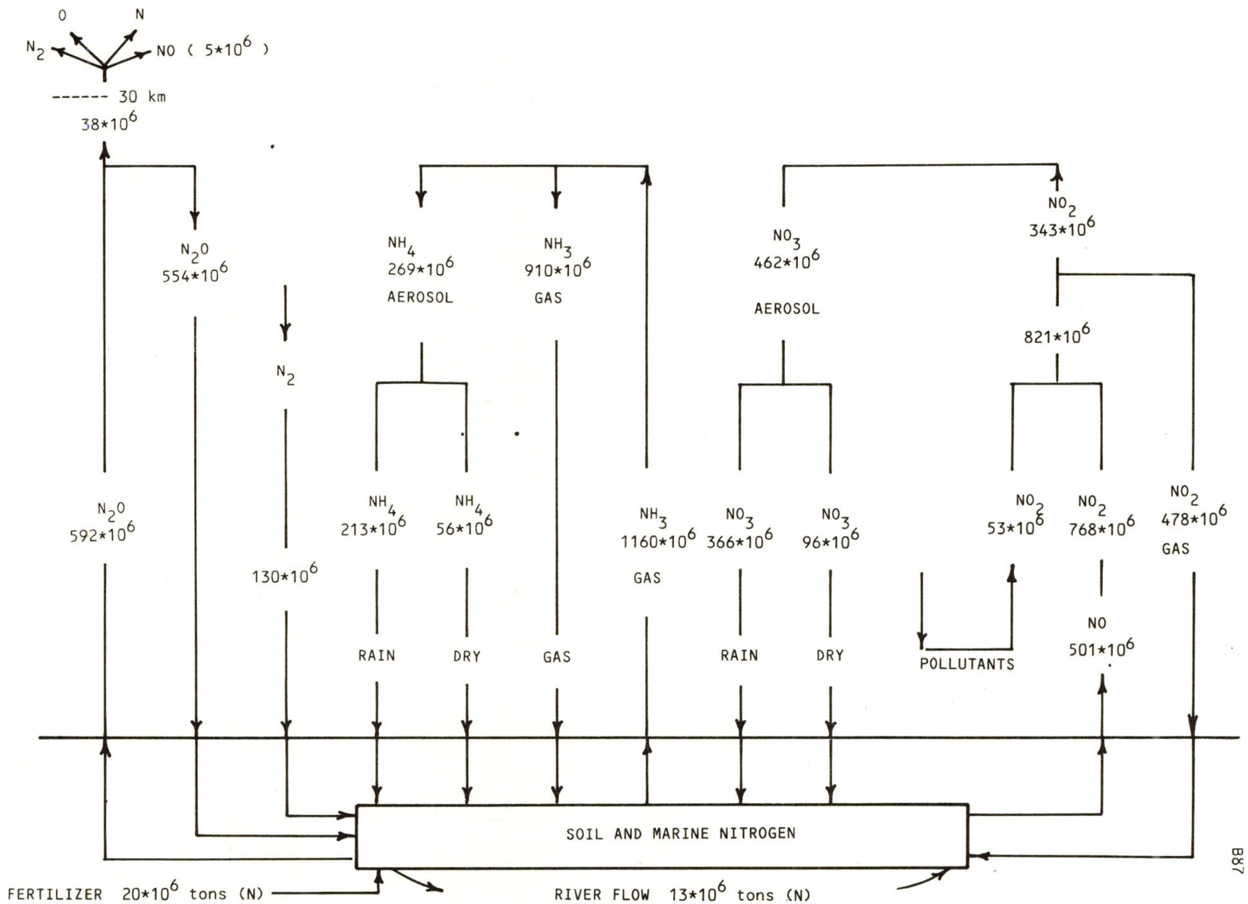

Nitrogen cycle, with special emphasis on atmosphere emissions and depositions. Units are in tons N/year.

(ref: 1347)

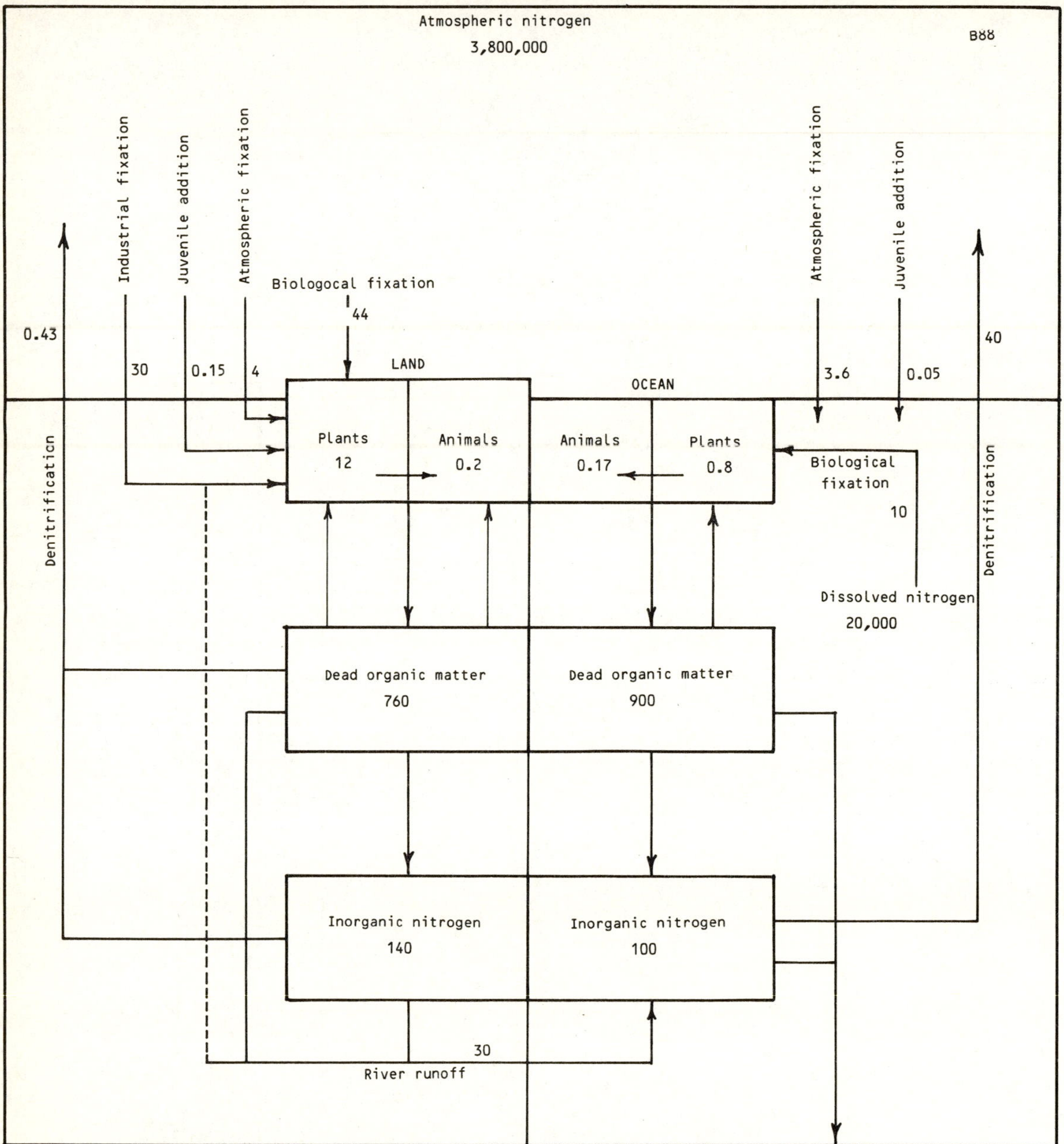

Nitrogen cycle, global. Values in compartsments are in 10**9 tons N, and fluxes in 10**6 tons N/year.Distribution of nitrogen in the biosphere and annual transfer can be estimated only within broad limits. The two quanties known with high confidence are the amount of nitrogen in atmosphere and the rate of industrial fixation. The apparent in the other figures shown here reflects chiefly an effort to preserve indicated or probable ratios among different inventories. Thus the figures for atmospheric fixation and biological fixation in the oceans could well by off by a factor 10. Because of the extensive use of industrially fixed nitrogen, the amounth of nitrogen avaiable to land plants may significantly exceed the nitrogen returned to the atmosphere by denitrifying bacteria in the soil. A portion of this excess fixed nitrogen is ultimately washed into the sea, but it is not included in the figure shown for river runoff. Similarly, the value for oceanic denitrification is no more than a rough estimate that is based on the assumption that the nitrogen cycle was in overall balance before mans interventation.Uptake of aminoacids by plants and uptake of detritus by animals are included.

(ref: 268)

B89

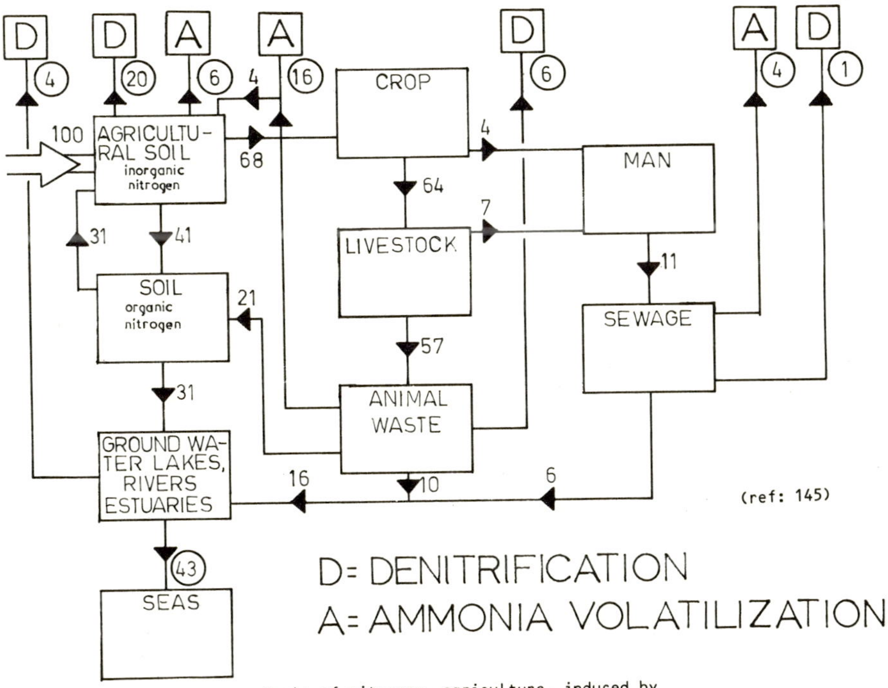

(ref: 145)

Cycle of nitrogen, agriculture, induced by input of 100 units of fertilizer. Encirled figures are final lossos. Taking into account the recycling of the soil N, application of 100 units makes 135 units avaiable due to use more than once. Loss to oceans is 43 %. D = denitrification. A = ammonia volitilization.

B90

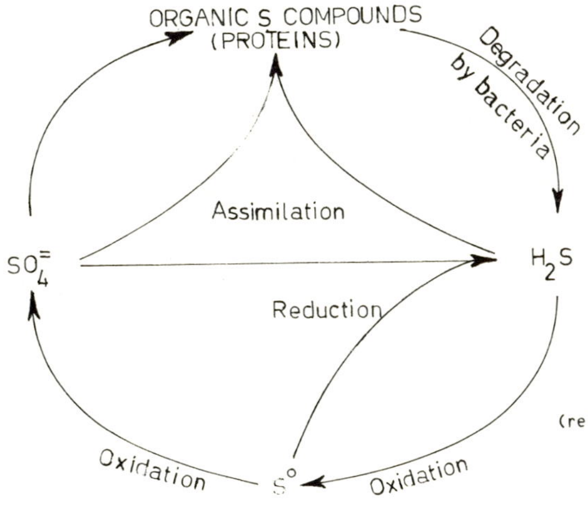

(ref: 1526)

Sulphur cycle.

B91

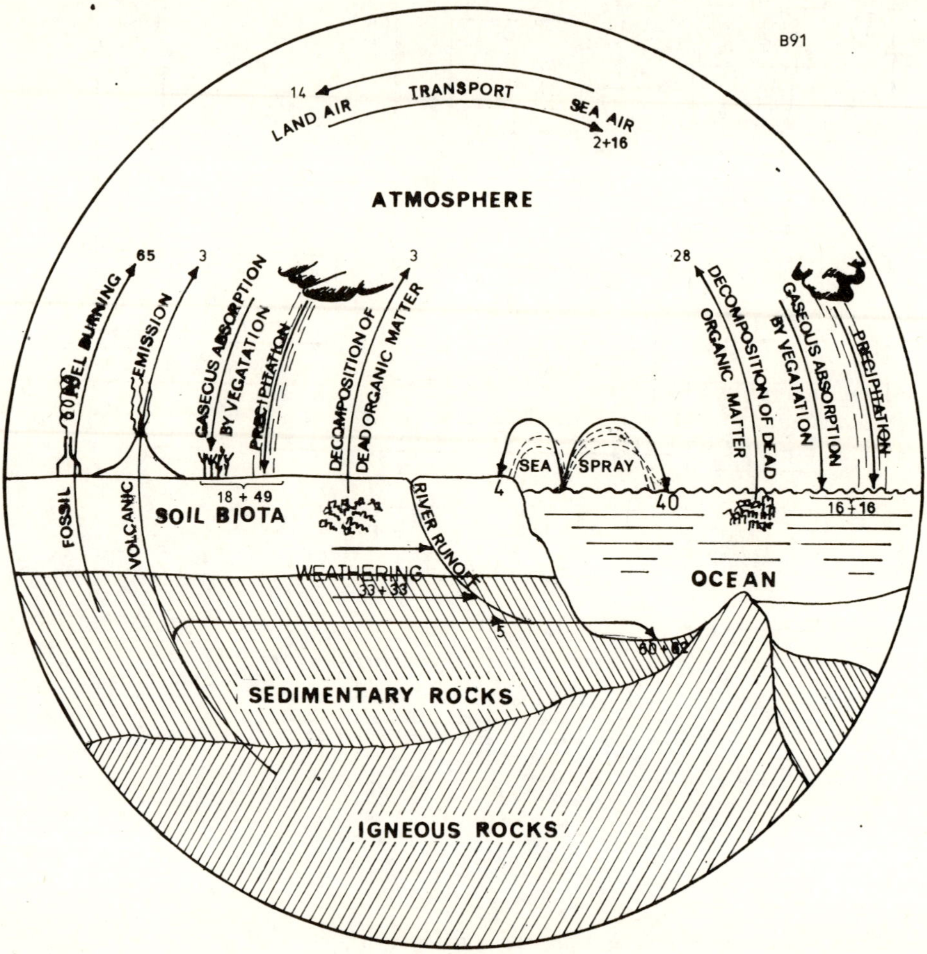

Cycle of sulphur, global. Transfers shown are in T g/year (T -10**12). Ligther typeface show transfer estimated to have prevailed before man influenced the cycle. Bold figures give estimates of what man since has added.

(ref: 144)

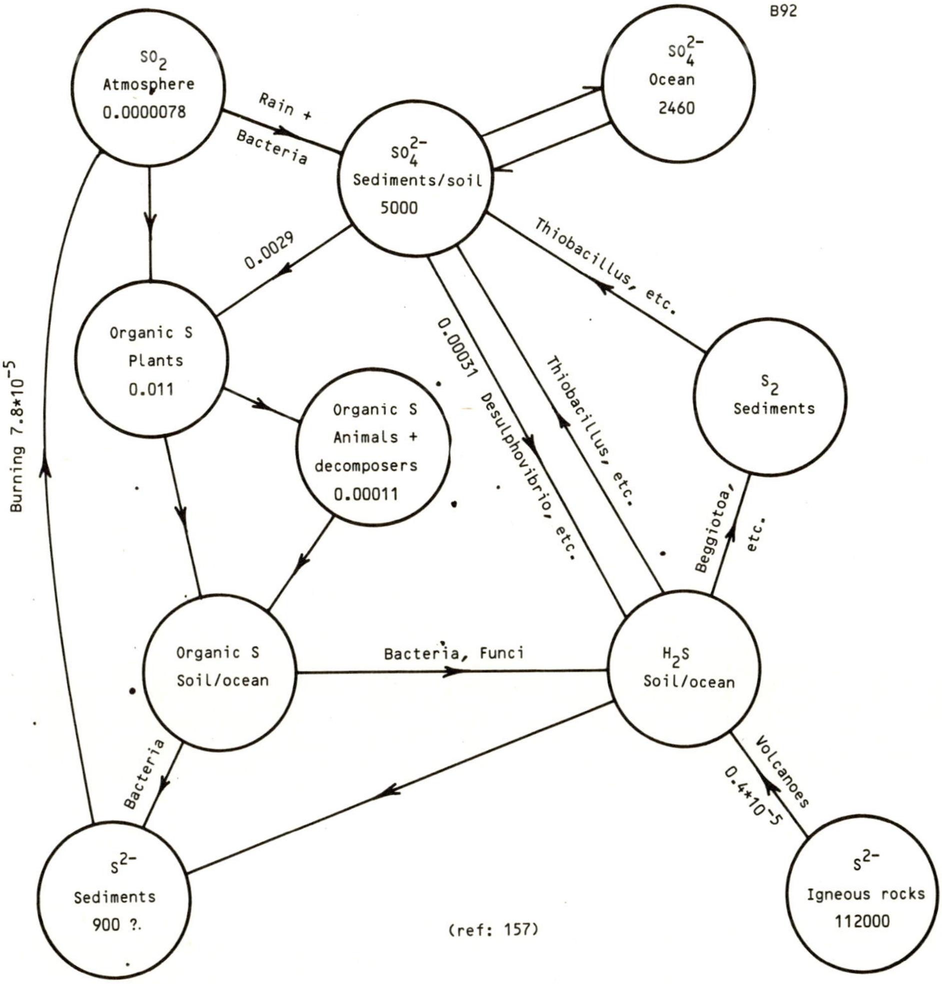

Sulphur cycle, global. Values in compartments are in kg S/m**2, and in fluxes in kg S/(m**2*year).

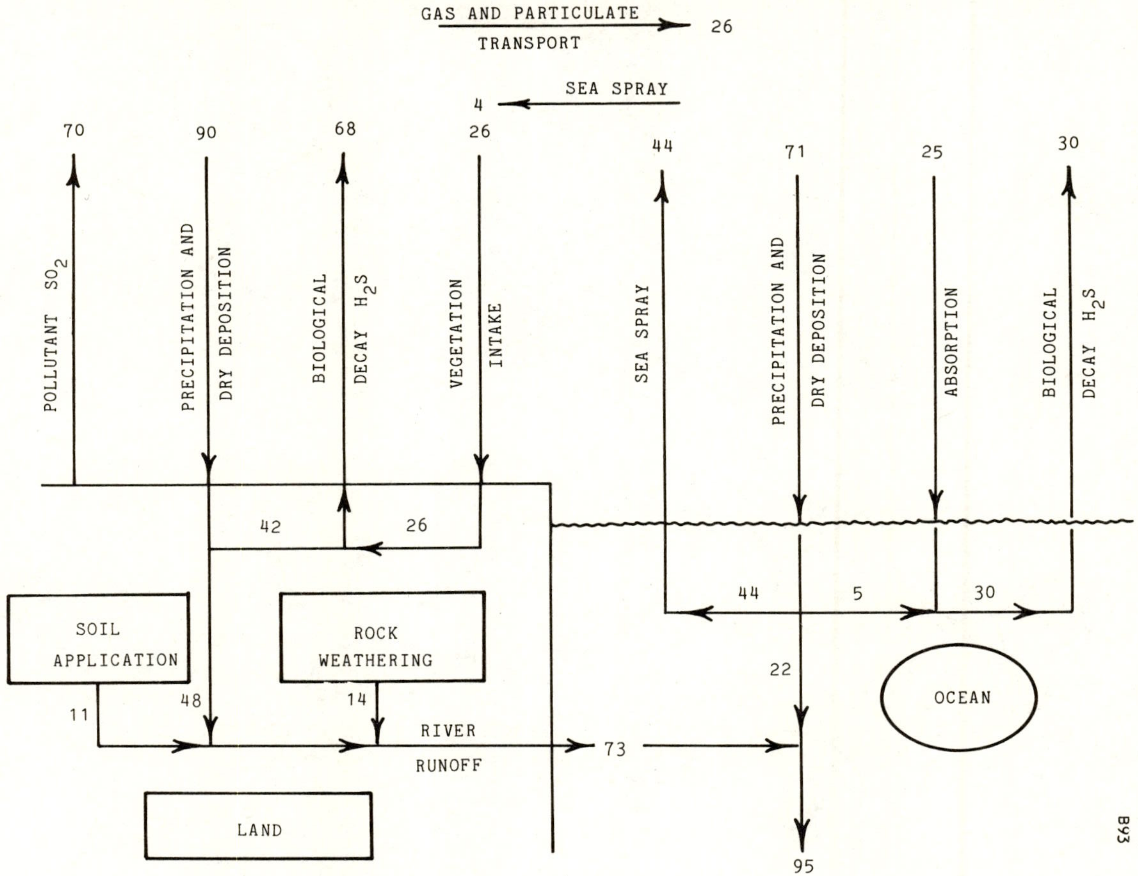

B93

Environmental sulphur circulation. Units 10**6 tons/year sulphur.

(ref: 1347)

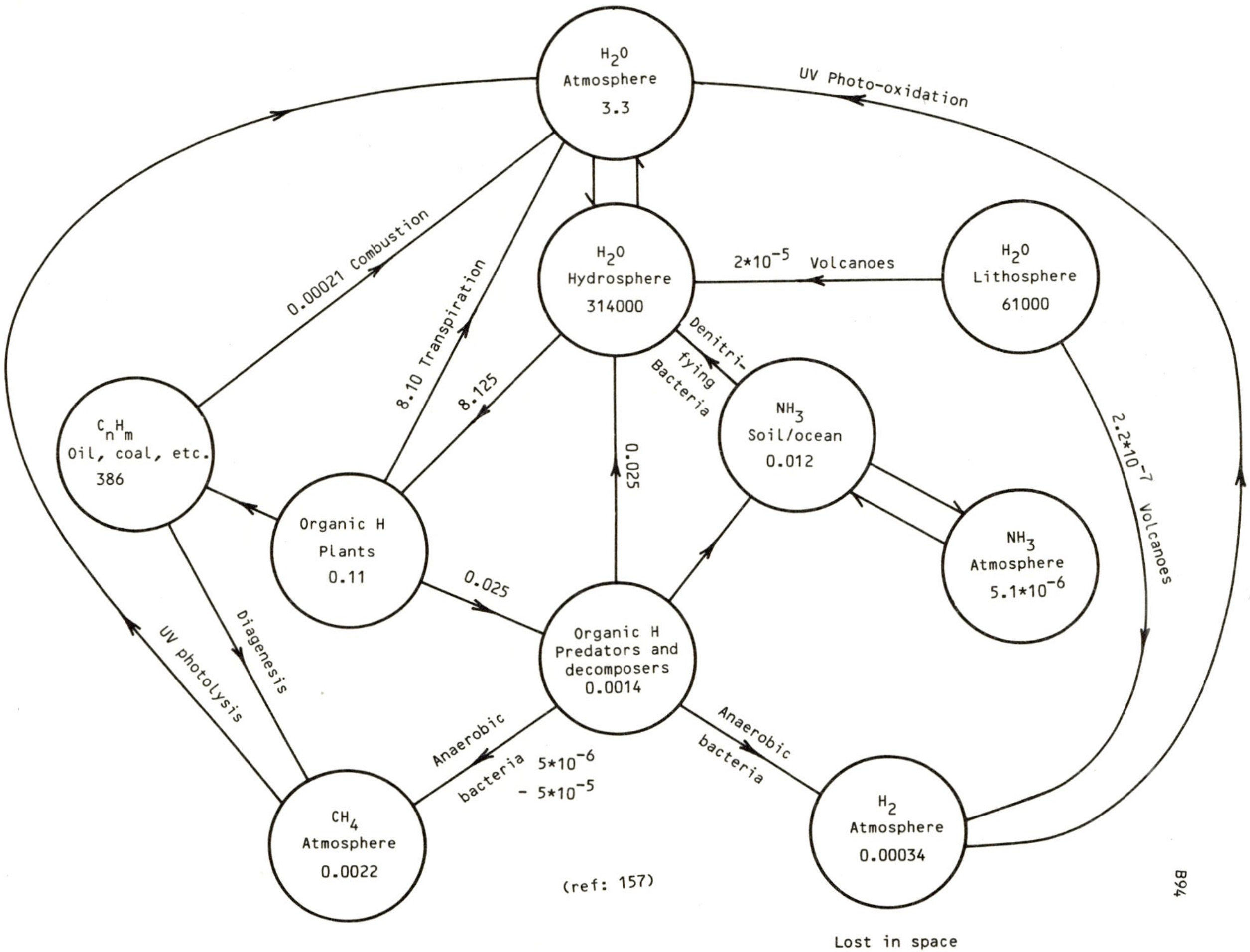

B94

Hydrogen cycle, global. Values in comparts-
ments are in kg H/m**2, and in fluxes kg H/
(m**2*year).

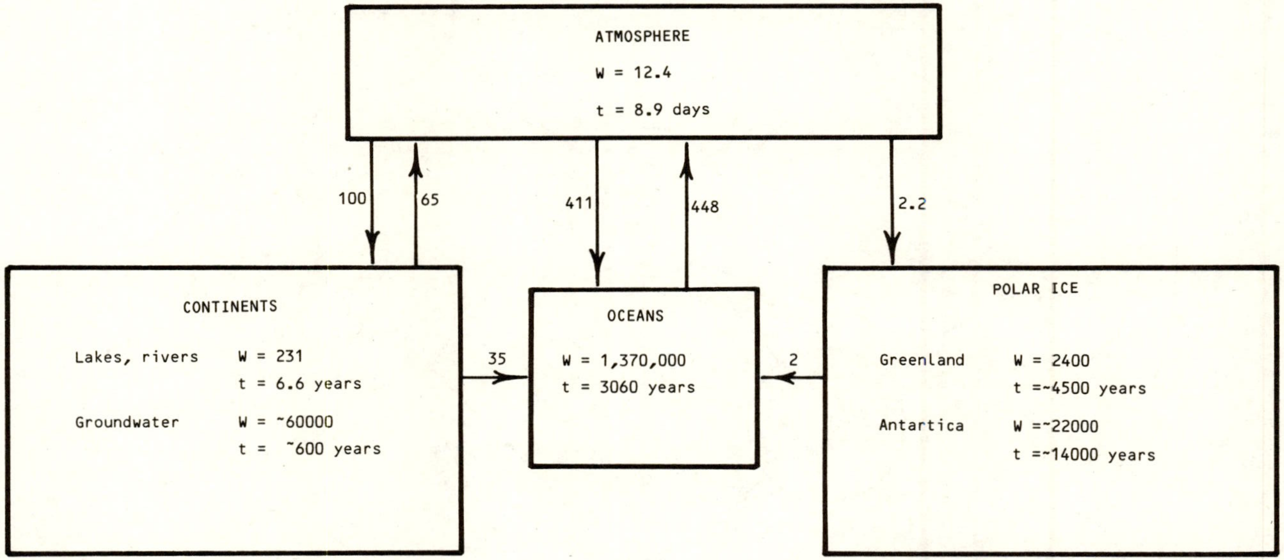

B95

Water cycle. W = water content in compartment in 10**3 km**3. T = retention time. Numbers on fluxes represents 10**3 km**3/year. Estimate of groundwater is to a depth of 5 km in the earths crust, much of this water is not activily exchanged.

(ref: 1347)

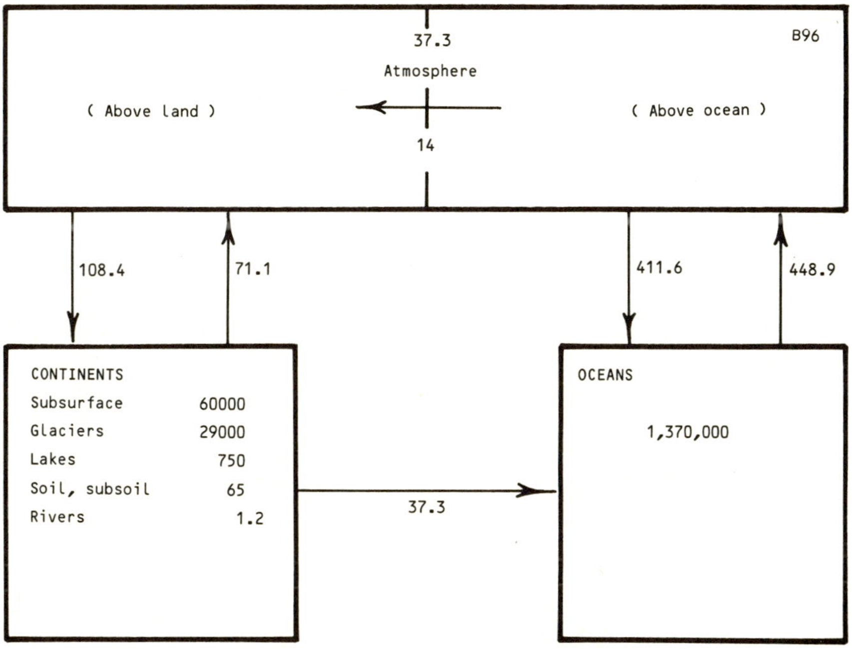

Water cycle. Numbers represent 10**3 km**3 in boxes and 10**3 km**3/year in flows. (ref: 1517)

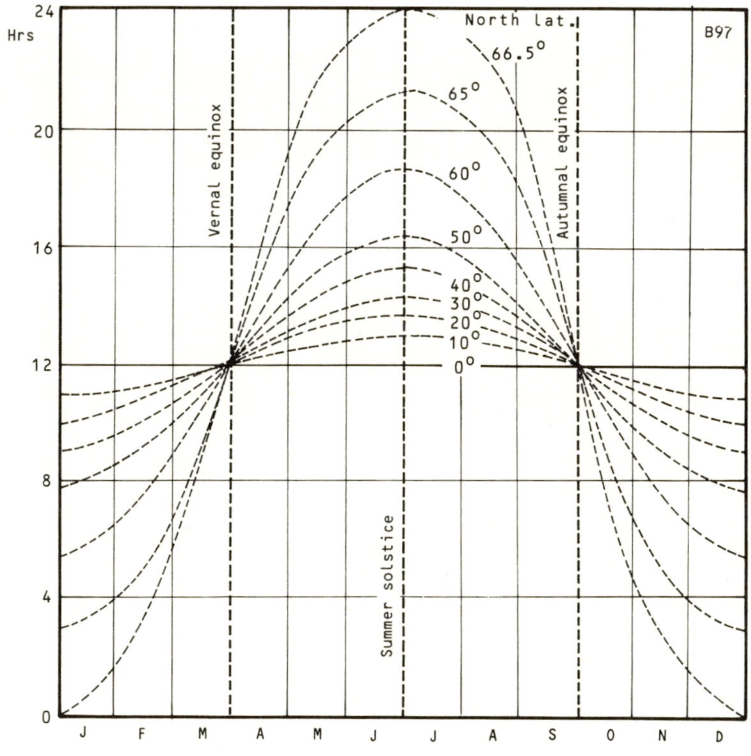

Duration of the day at various latitudes through tbe year. Vertical scale gives numbers of hours the sun is above the horizon. (Not corrected for the refraction of suns rays by the atmosphere). (ref: 793)

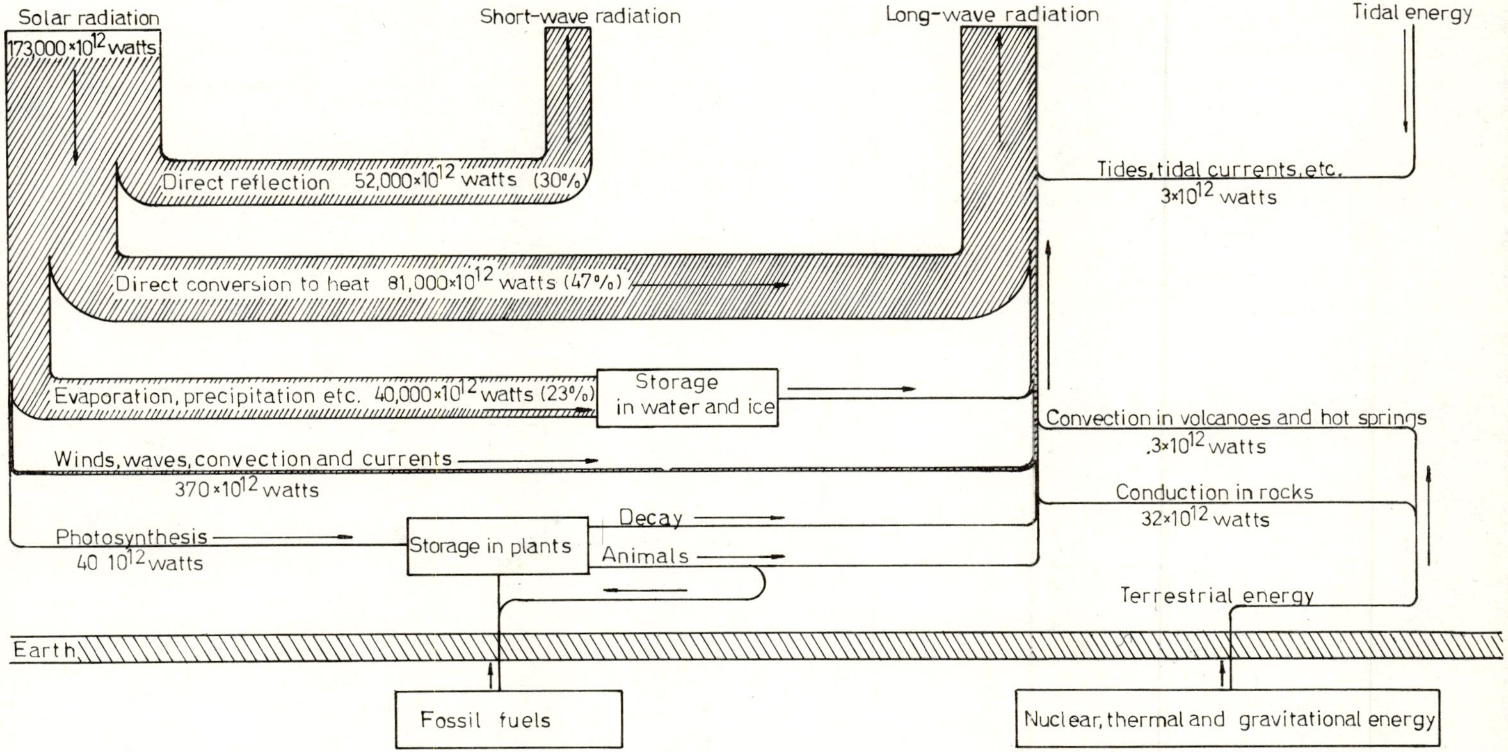

B98

Flow of energy, global. More than 99 % of input is solar radiation.

(ref: 700)

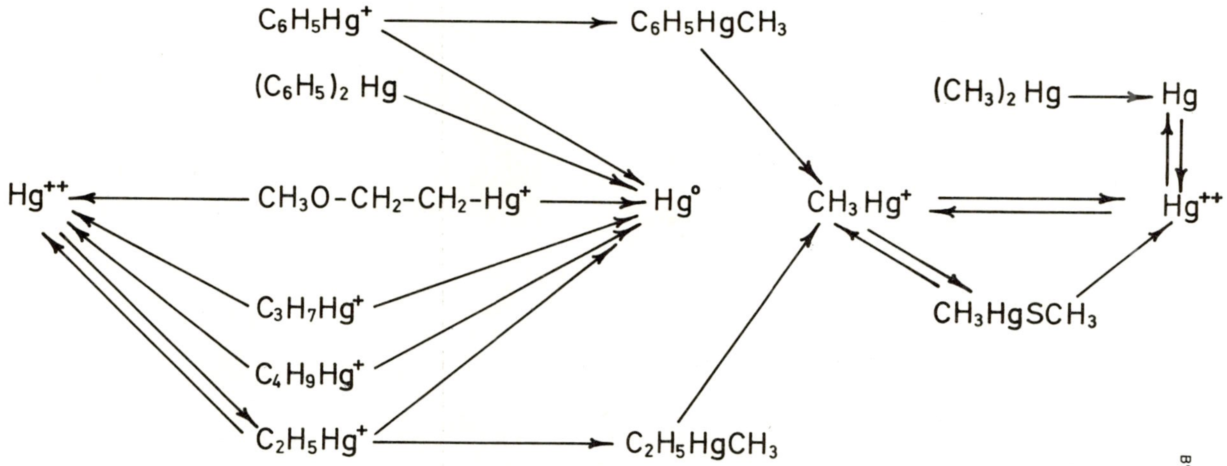

B99

(ref: 34)

Pathway of formation of mercury compounds. The enzymatic biological synthesis of monomethyl- and dimethylmercury. Methylcarbonium ion is chemically transferred to the mercuric ion to form dimethyl mercury
The reaction is produced in a neutral alkaline environment

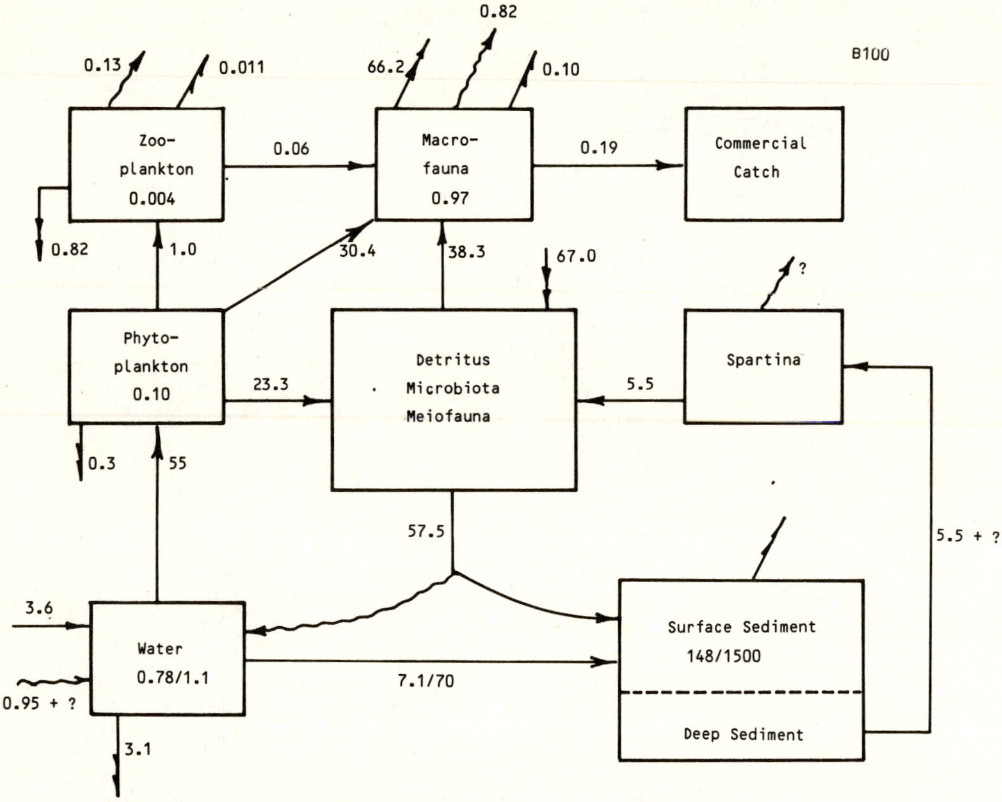

Zinc model. Newport River estuary : 13 km2 and a half-tide depth of 1.3 m. Values inside compartsments are in mg Zn/m**2 and arrows represent annual fluxes between compartments in mg Zn/(m**2*year).

Arrows:

Flushing (river volume only) and emigration

Excretion of soluble metals to water

Deposition of unassimilated materials to detritus

(ref; 1742)

B101

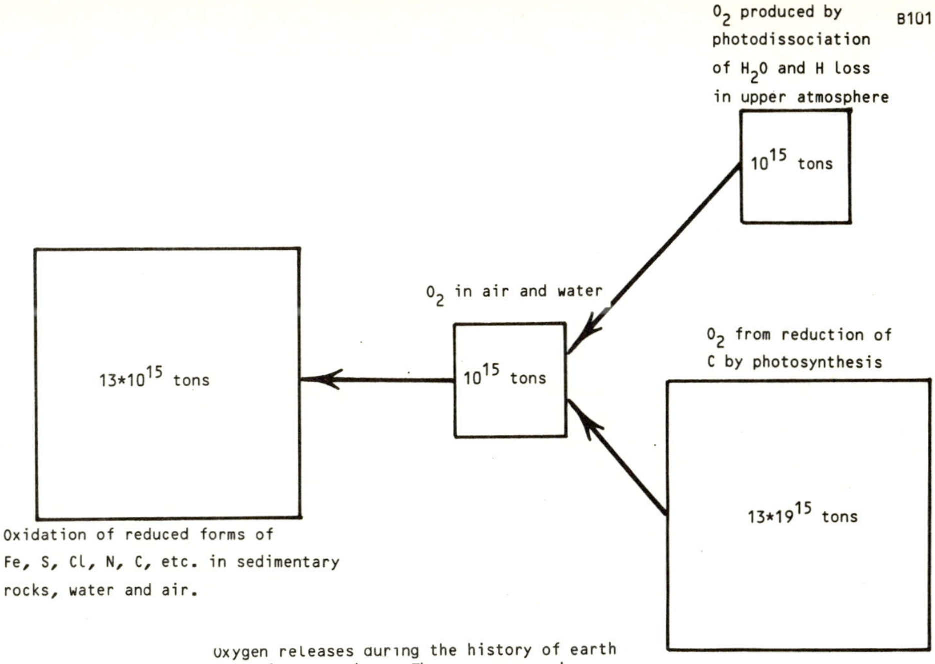

Oxygen releases during the history of earth into the atmosphere. The processes subsequently removing oxygen by oxidation are also shown.

(ref: 778)

B102

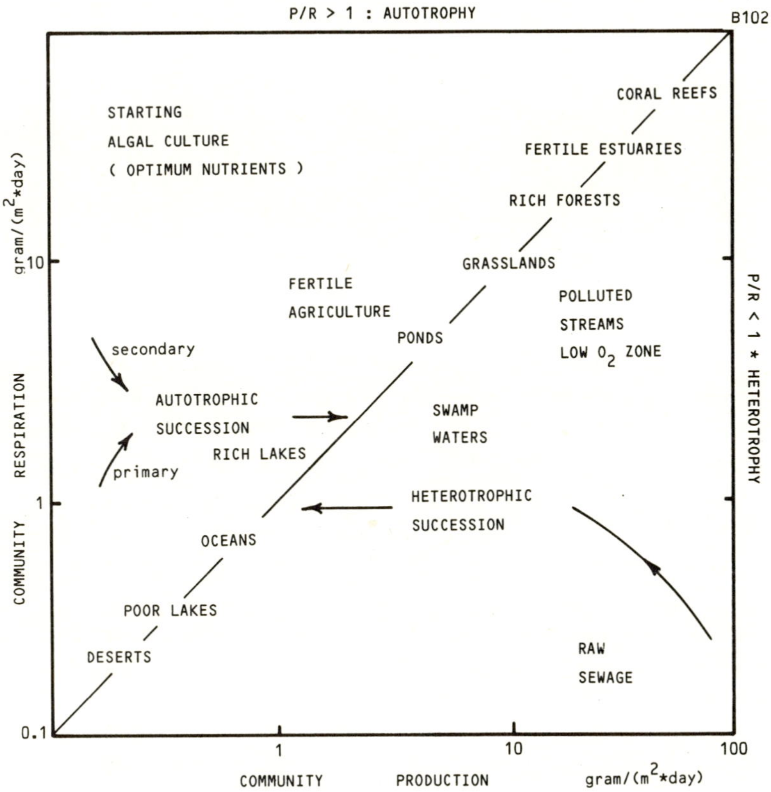

Production versus respiration. (ref: 1165)

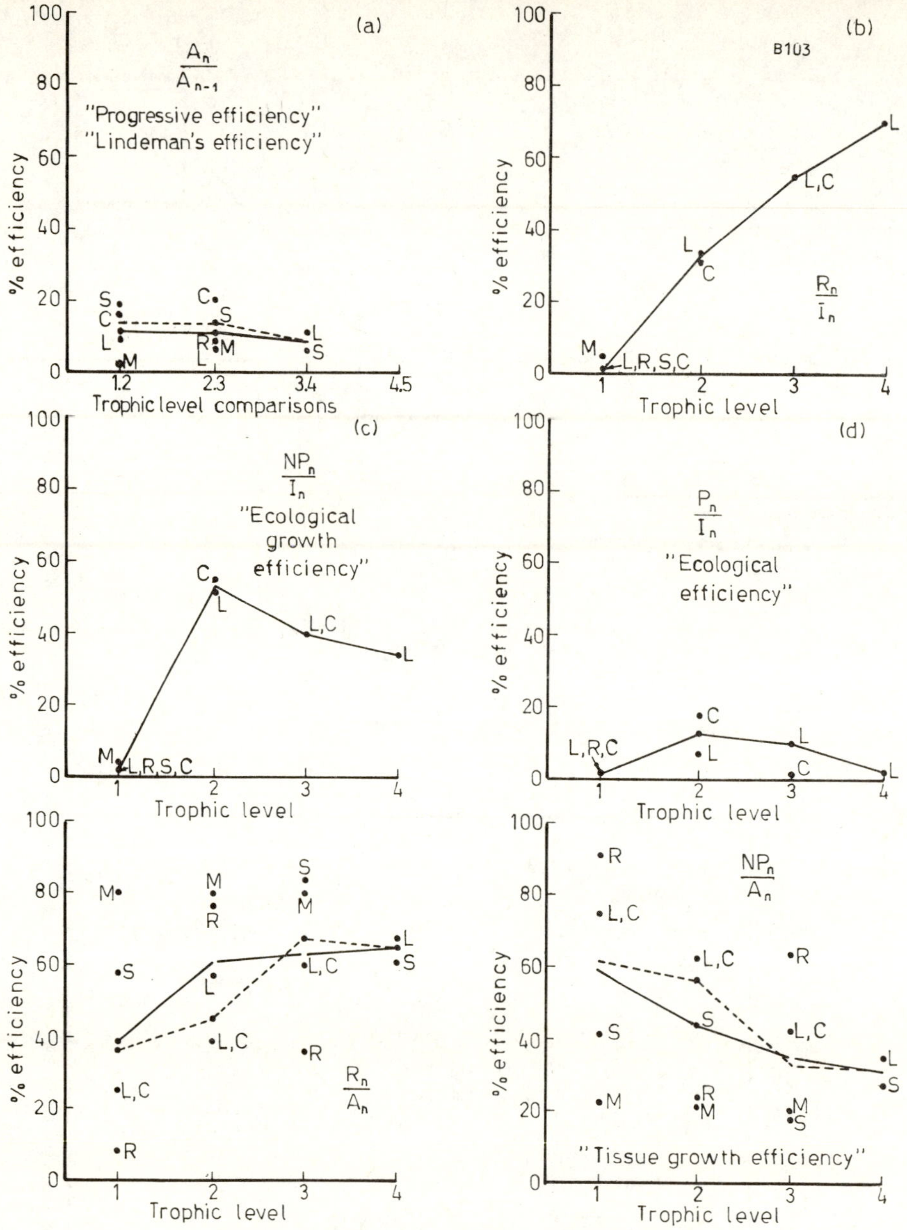

Efficiency ratios, the most commonly found in the litterature. the following symbols are used. L: Lake Mendota. C: Cedar bog lake. S: Silver Springs. R: Root spring. M: Salt Marsh. (ref: 268)

= average 5 communities.

= average 3 communities.

Fig(a) shows progressive efficiency (energy assimilated by trophic level n (A_n)/energy assimilated by trophic level n-1 (A_{n-1})) plotted to the trophic level.
Fig(b) shows the ratio of respiration rate (R_n) to ingestion rate plotted against trophic level (I_n).
Fig(c) shows a dovnward trend in the ratio of net production (NP_n) to ingestion (I_n) from trophic level 2.
Fig(d) gives the ratio of P_n (the rate of energy flow from level n to level n+1) to ingestion rate (I_n). When n=max value P_n -
Fig(e) gives the respiration rate (R_n) divided with assimilation rate (A_n) plotted to the trophic level.
Fig(f) gives net production relative to assimilation rate plotted to the trophic level.

B104

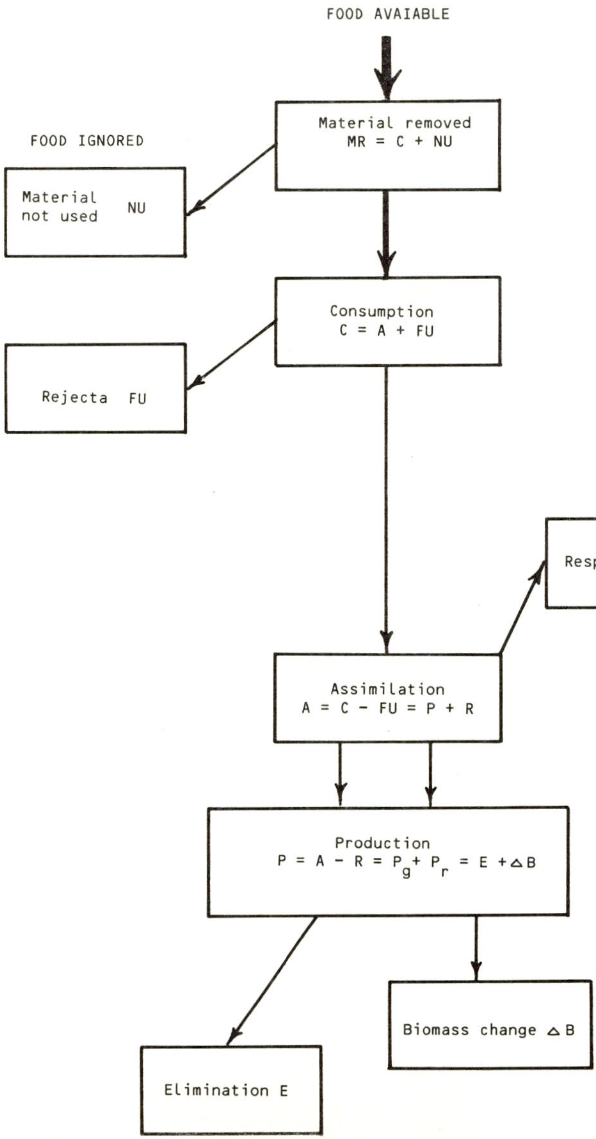

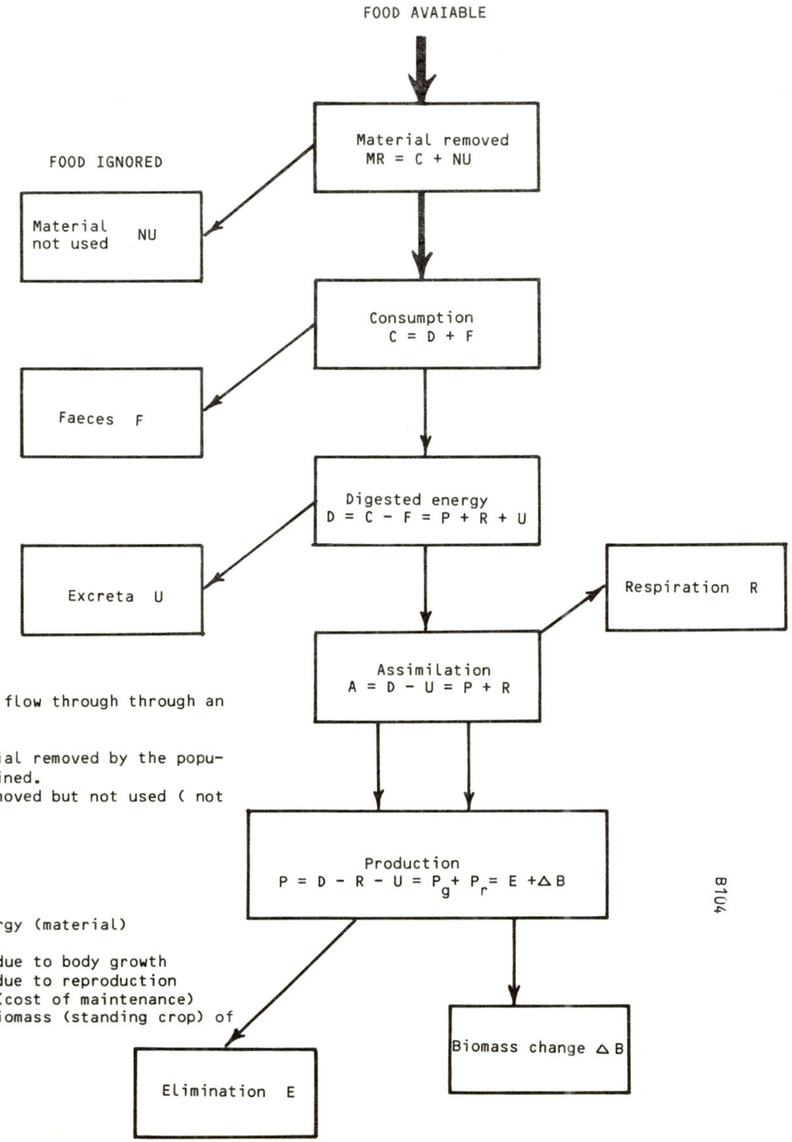

Matter and enrgy flow through through an ecological unit.

Symbols:

MR = Total material removed by the population examined.
NU = Material removed but not used (not consumed)
C = Consumption
FU = Rejecta
F = Egesta
U = Excreta
A = Assimilation
D = Digested energy (material)
P = Production
Pg = Production due to body growth
Pr = Production due to reproduction
R = Respiration (cost of maintenance)
B = Change in biomass (standing crop) of population
E = Elimination

(ref: 389)

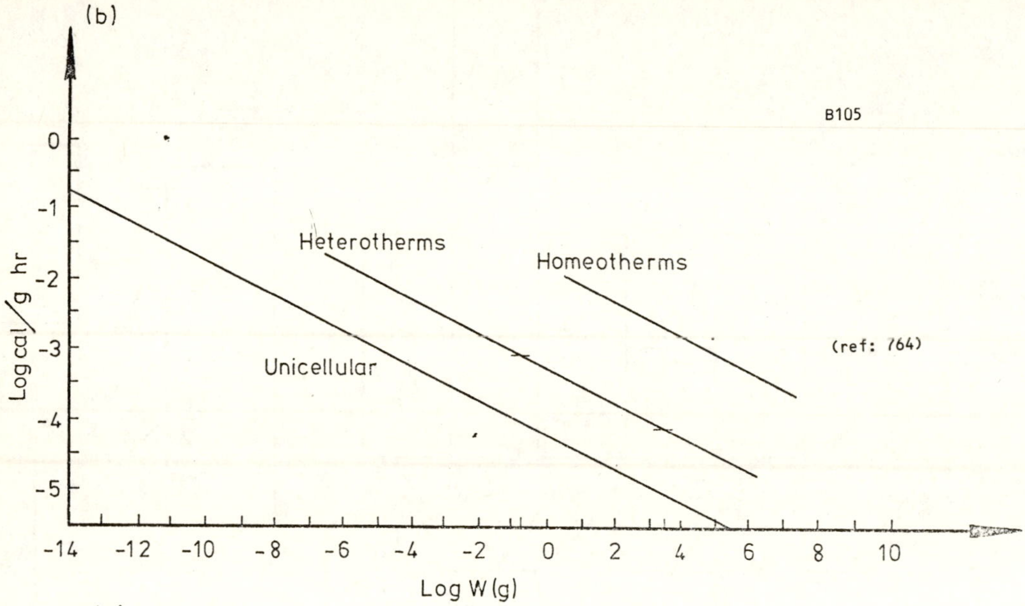

The relationship of metabolic rate to weight for various animals.

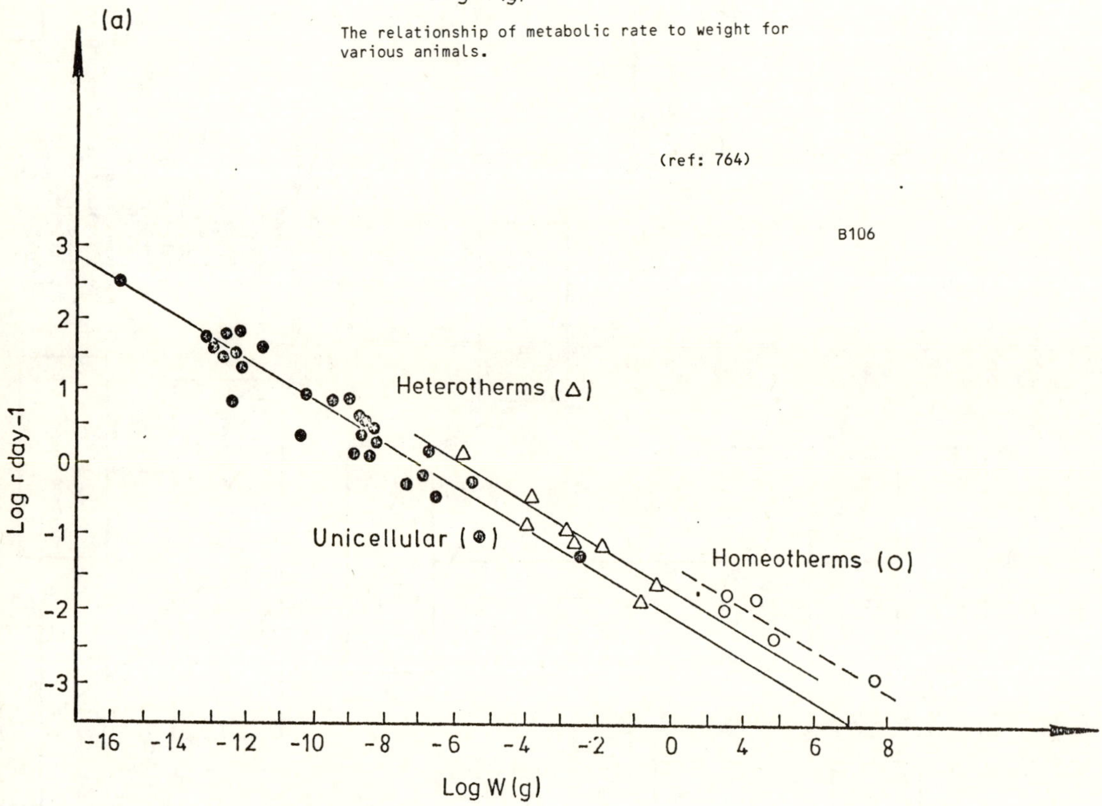

Intrinsic rate of natural increase to weight for various animals. Metabolic rate versus weight for varoius animals.

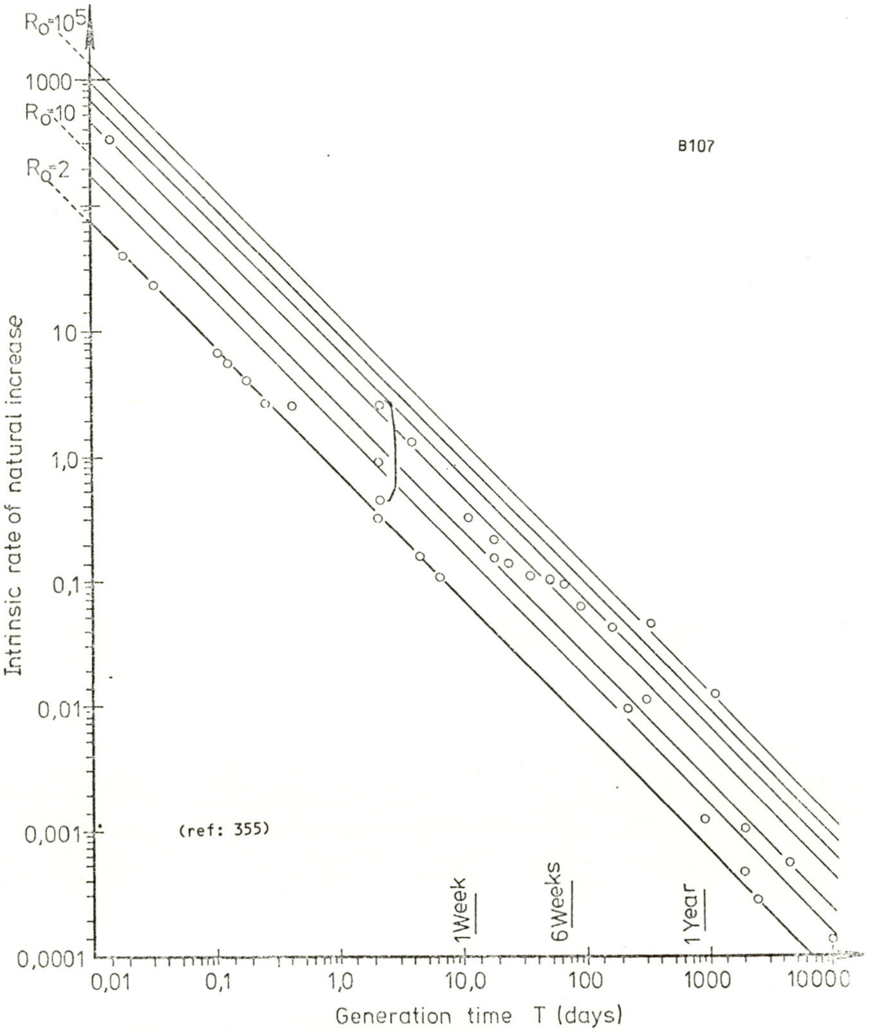

Intrinsic rate of of natural increase plotted to generation time with diagonal lines representing net reproduction rate from 2 to 10**5 for a variaty of organisms.

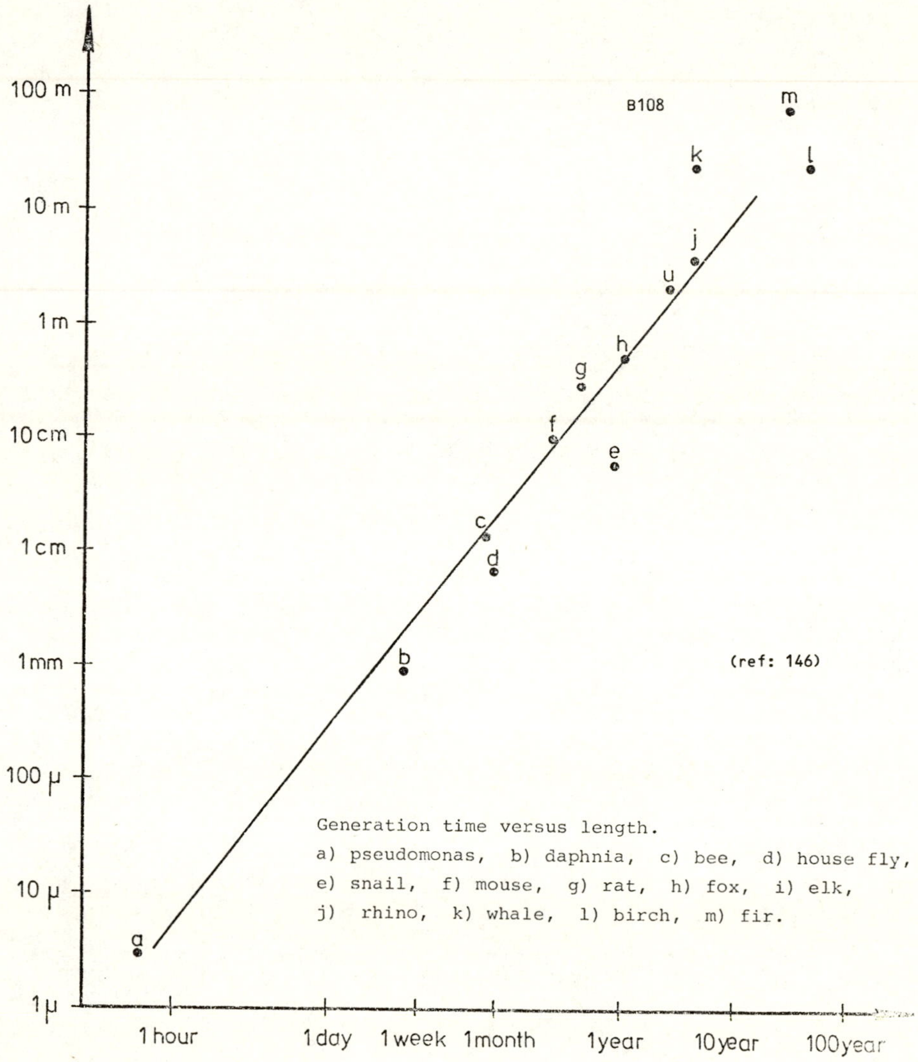

Length and generation time plotted on log-log scale.

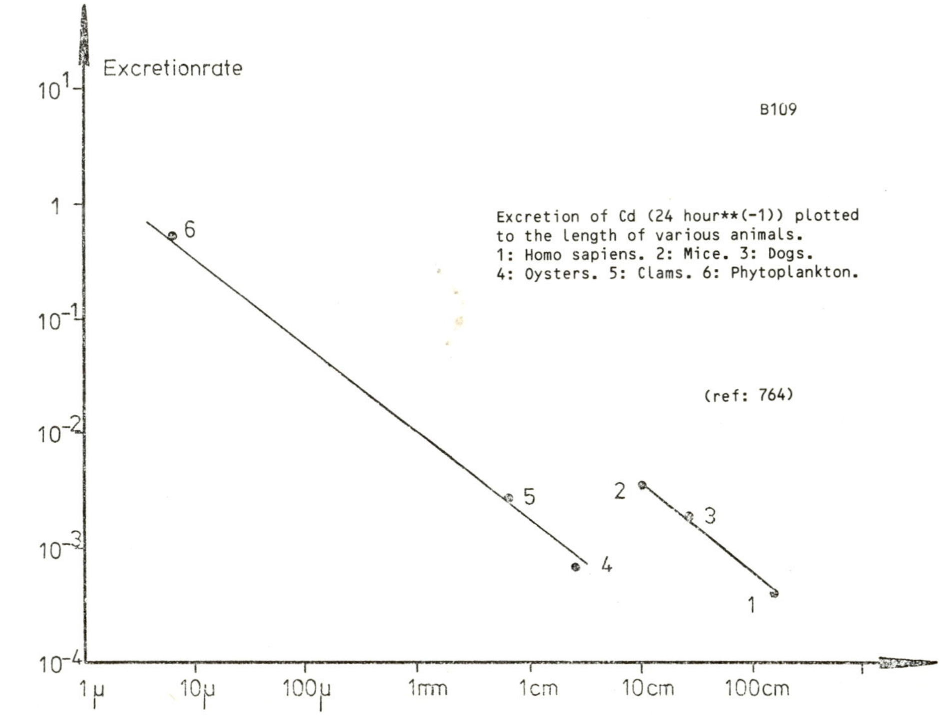
Excretionrate
B109
Excretion of Cd (24 hour**(-1)) plotted to the length of various animals.
1: Homo sapiens. 2: Mice. 3: Dogs.
4: Oysters. 5: Clams. 6: Phytoplankton.
(ref: 764)
10^1
1
10^-1
10^-2
10^-3
10^-4
1μ
10μ
100μ
1mm
1cm
10cm
100cm

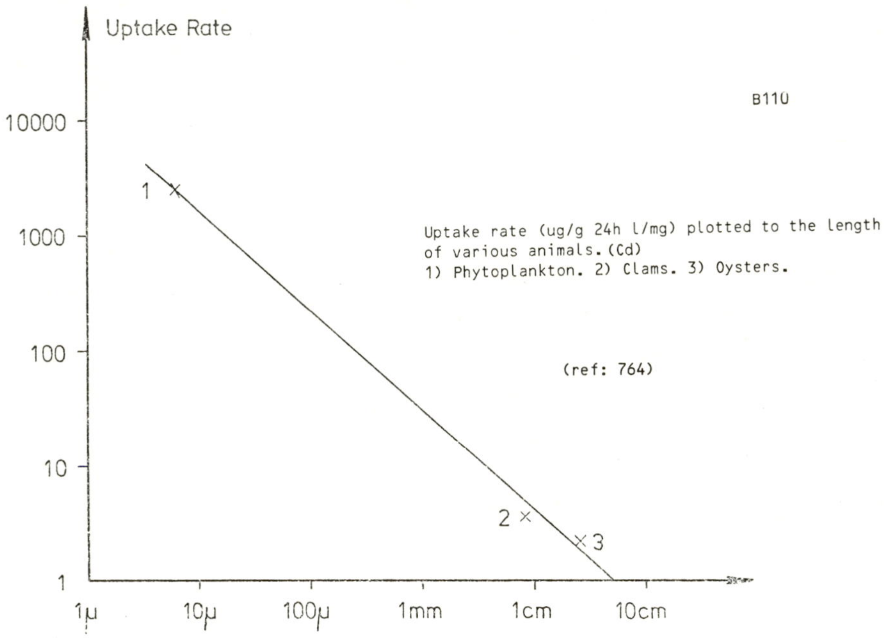
Uptake Rate
B110
Uptake rate (ug/g 24h l/mg) plotted to the length of various animals. (Cd)
1) Phytoplankton. 2) Clams. 3) Oysters.
(ref: 764)
10000
1000
100
10
1
1μ
10μ
100μ
1mm
1cm
10cm

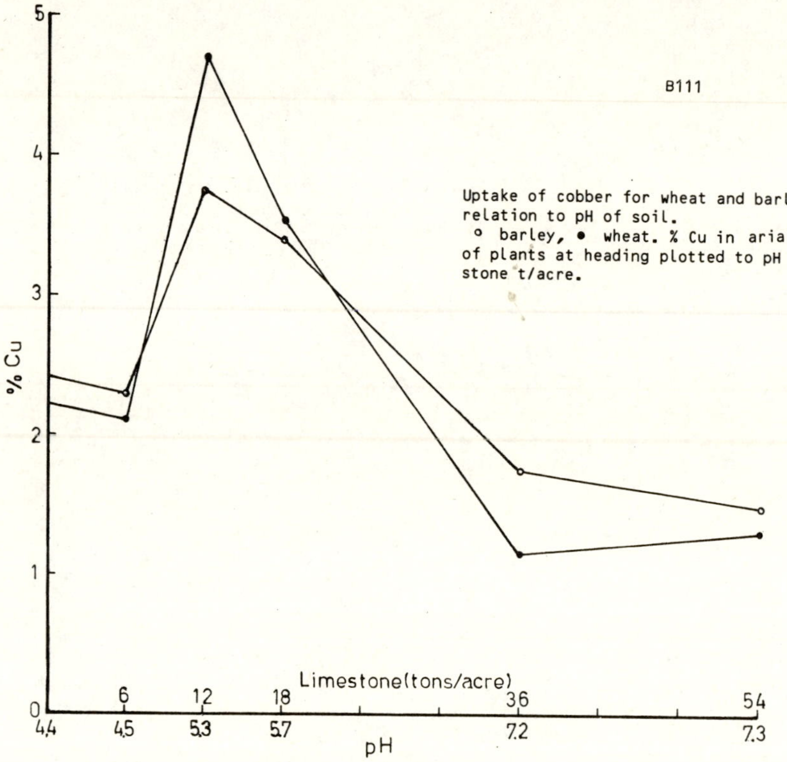

Uptake of cobber for wheat and barley in relation to pH of soil.
o barley, • wheat. % Cu in arial parts of plants at heading plotted to pH and limestone t/acre.

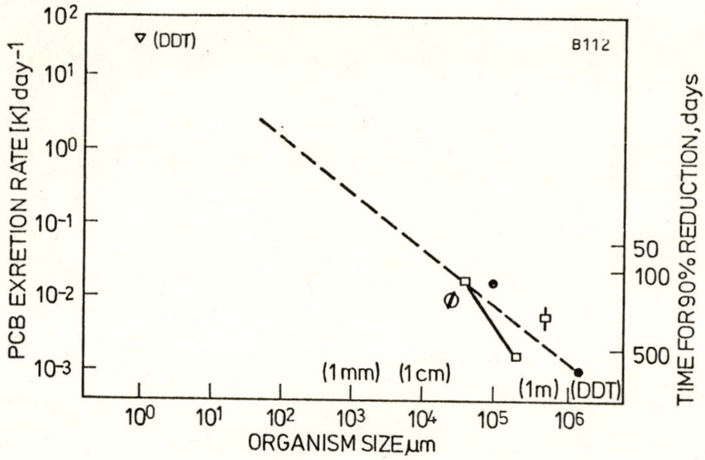

Ecretion of PCB and DDT versus organism size.

(ref: 1579)

B113

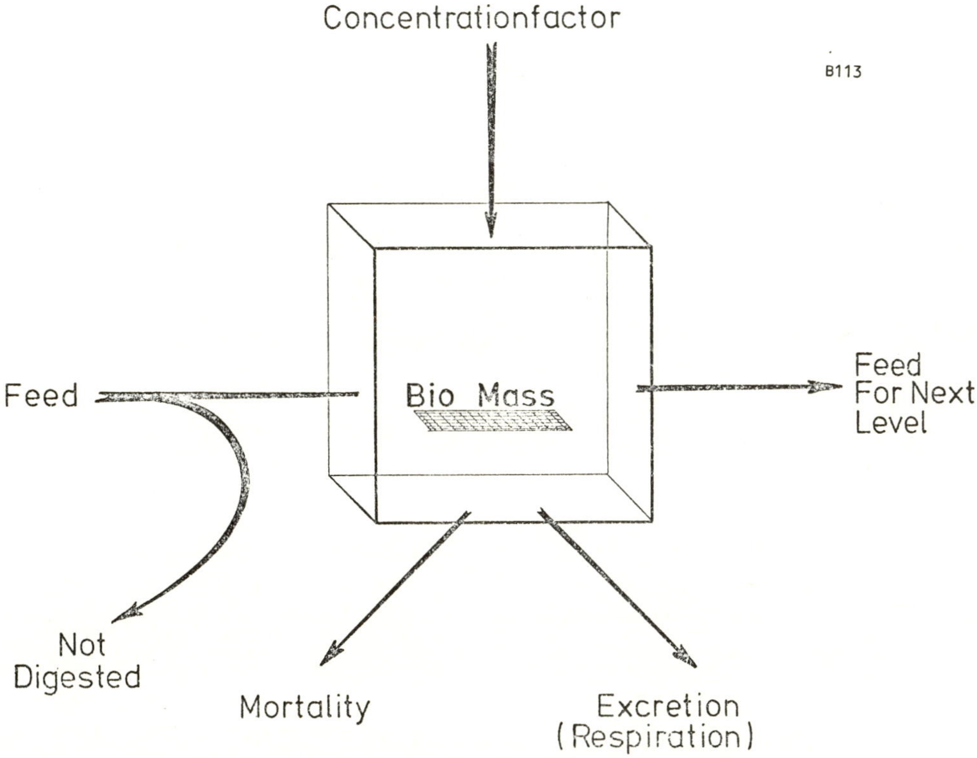

(ref: 763) Principles of modelling the concentration of a toxic substance in a trophic level.

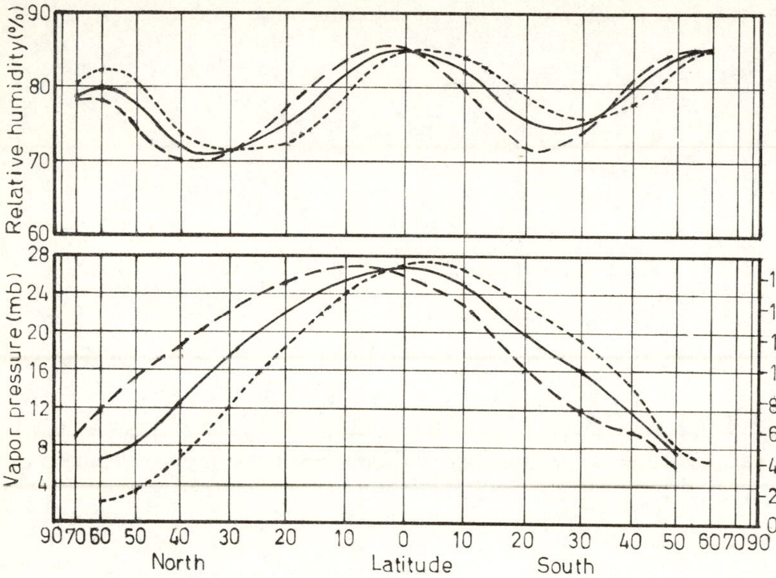

Relative humidity versus latitude.

Vapour pressure and mixing ratio versus latitude.

(ref: 644)

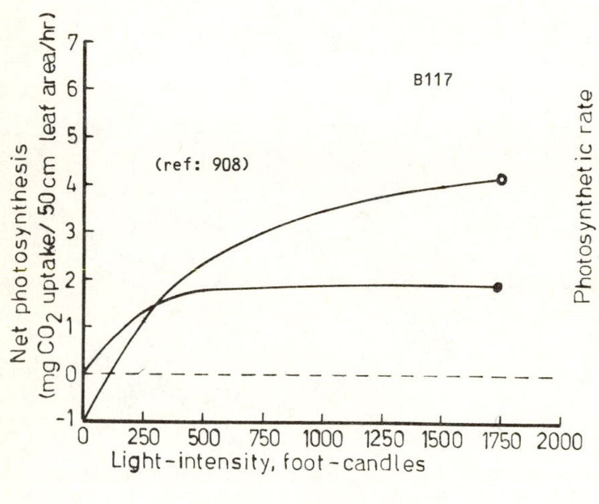

Photosynthesis versus ligth intensity for beech leaves, shade grown● and sun grown.○

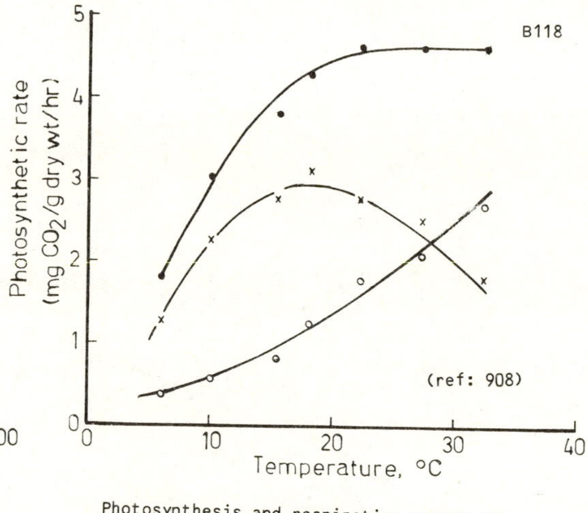

Photosynthesis and respiration versus temperature for spagnum moss. 750 foot-candle illumination ● Gross photosynthesis. ✖ Net photosynthesis ○ Respiration.

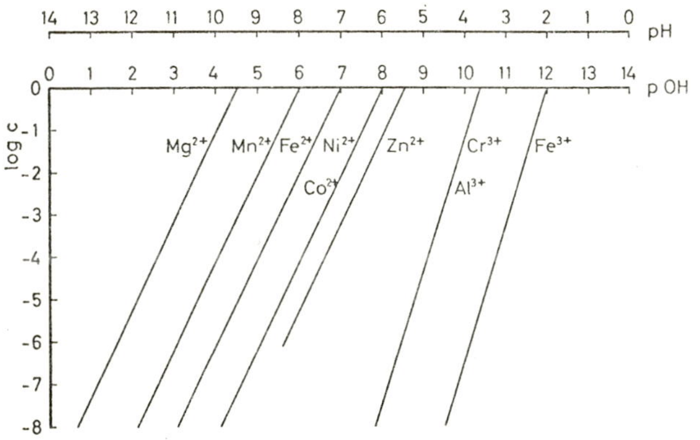

B114

Metal hydroxides, solubility versus pH and pOH.

(ref: 765)

B115

Complexes formation

Metal ion	log $K_{complex}$			
	Cyanide	Ammonia	Hexamine	EDTA
Fe^{2+}	36.0	-	-	14.3
Cu^{2+}	27.3	12.6	22.4	18.8
Zn^{2+}	16.9	9.1	16.2	16.5
Cl^{2+}	16.8	4.9	16.8	16.5
Ni^{2+}	22.0	8.5	19.3	18.6

Metal complexes with cyanide, ammonia, hexamine and EDTA.

TABLE B119

ECOSPHERE : AREA			
ITEM:	VALUE:	CONDITIONS:	REF.:
*ARABLE LAND	0.23*10**8 KM**2	NONE	157
*BROWN ALGAE ZONE	0.009*10**8 KM**2	DEPTH=0-20M	157
*CONIFEROUS FOREST	0.10*10**8 KM**2	NONE	157
*CONTINENTAL SHELF	0.27*10**8 KM**2	DEPTH=0-200M	157
*DECIDUOUS FOREST	0.049*10**8 KM**2	NONE	157
*DEEP SEAS	3.34*10**8 KM**2	DEPTH: >=200M	157
*DESERT	0.33*10**8 KM**2	NONE	157
*EARTH-SURFACE	5.11*10**14 M**2	NONE	1442
*FRESHWATER	0.0024*10**8 KM**2	NONE	157
*GRASSLAND	0.27*10**8 KM**2	NONE	157
*ICE	0.11*10**8 KM**2	NONE	157
*LAND	1.48*10**14 M**2	NONE	1442
*LITTORAL ZONE	0.000144*10**8 KM**2	NONE	157
*OTHER FOREST	0.144*10**8 KM**2	NONE	157
*TOTAL FOREST	0.44*10**8 KM**2	NONE	157
*TOTAL LAND	1.49*10**8 KM**2	NONE	157
*TOTAL OCEAN	3.61*10**8 KM**2	NONE	157
*TROPICAL FOREST	0.147*10**8 KM**2	NONE	157
*TUNDRA	0.11*10**8 KM**2	NONE	157
*WHOLE EARTH	5.10*10**8 KM**2	NONE	157

TABLE B120

ECOSPHERE : ATMOSPHERE			
ITEM:	VALUE:	CONDITIONS:	REF.:
*MASS OF ATMOSPHERE	5.3*10**21 G	VAL.PER UNIT AREA=1030G/CM**2	1442

TABLE B121

ECOSPHERE : ATMOSPHERIC TURBIDITY			
ITEM:	VALUE:	CONDITIONS:	REF.:
*DAVOS	0.026 UNIT OF SCHUEPP COEFFICIENT	1914-26	413
*DAVOS	0.046 UNIT OF SCHUEPP COEFFICIENT	1957-59	413
*JERUSALEM	0.043 UNIT OF SCHUEPP COEFFICIENT	1930-34	413
*JERUSALEM	0.050 UNIT OF SCHUEPP COEFFICIENT	1961-68	413
*MCMURDO	0.016 UNIT OF SCHUEPP COEFFICIENT	1949-52,NOVEMBER	413
*MCMURDO	0.026 UNIT OF SCHUEPP COEFFICIENT	1949-52,DECEMBER	413
*MCMURDO	0.022 UNIT OF SCHUEPP COEFFICIENT	1966,NOVEMBER	413
*MCMURDO	0.024 UNIT OF SCHUEPP COEFFICIENT	1966,DECEMBER	413
*MEXICO CITY	0.087 UNIT OF SCHUEPP COEFFICIENT	1911-28	413
*MEXICO CITY	0.181 UNIT OF SCHUEPP COEFFICIENT	1957-62	413
*MT.MONTEZUMA,CHILE	0.0171 UNIT OF SCHUEPP COEFFICIENT	1920-30	413
*MT.MONTEZUMA,CHILE	0.0076 UNIT OF SCHUEPP COEFFICIENT	1940-48	413
*TABLE MOUNTAIN,CALIFORNIA	0.0185 UNIT OF SCHUEPP COEFFICIENT	1925-30	413
*TABLE MOUNTAIN,CALIFORNIA	0.0081 UNIT OF SCHUEPP COEFFICIENT	1940-50	413
*WASHINGTON D.C.	0.105 UNIT OF SCHUEPP COEFFICIENT	1903-07	413

TABLE B121

ECOSPHERE : ATMOSPHERIC TURBIDITY			
ITEM:	VALUE:	CONDITIONS:	REF.:
*WASHINGTON D.C.	0.165 UNIT OF SCHUEPP COEFFICIENT	1962-66	413

TABLE B122

ECOSPHERE : AUTOTROPHIC RESPIRATION			
ITEM:	VALUE:	CONDITIONS:	REF.:
*ALFALFA FIELD	9200 KCAL/(YEAR**M**2)	ANNUAL VALUE, USA	1580
*FRESHWATER SPRING	12000 KCAL/(YEAR**M**2)	ANNUAL VALUE, SILVER SPRING, FLA.	1164
*MARINE SOUND	3200 KCAL/(YEAR**M**2)	ANNUAL VALUE, COASTAL, LONG ISLAND, N.Y.	1338
*OAK-PINE FOREST	6400 KCAL/(YEAR**M**2)	ANNUAL VALUE, MEDIUM AGED, NEW YORK, NO TIMBER HARVEST	1746
*PINE FOREST	4700 KCAL/(YEAR**M**2)	ANNUAL VALUE, YOUNG PLANTATION, MEAN FOR 0-50 Y, ENG.	1194
*RAIN FOREST	32000 KCAL/(YEAR**M**2)	ANNUAL VALUE, MATURE FOREST, PUERTO RICO	1163

TABLE B123

ECOSPHERE : BIOMASS			
ITEM:	VALUE:	CONDITIONS:	REF.:
*ARABLE LAND	0.67 KG DRY WEIGHT/M**2	TOTAL=0.15*10**14 KG,N=15	157
*CONIFEROUS FOREST	26.2 KG DRY WEIGHT/M**2	TOTAL=2.62*10**14 KG,N=18	157
*CORAL REEF, ENIWETOK	703 GRAM DRY WEIGHT/M**2	TYPICAL VALUE	355
*DECIDUOUS FOREST	12.2 KG DRY WEIGHT/M**2	TOTAL=0.60*10**14 KG,N=9	157
*DESERT	0.225 KG DRY WEIGHT/M**2	TOTAL=0.074*10**14 KG,ESTIMATED VALUES	157
*FRESHWATER, LAKE WISCONSIN	96 GRAM DRY WEIGHT/M**2	TYPICAL VALUE	355
*GRASSLAND	0.87 KG DRY WEIGHT/M**2	TOTAL=0.235*10**14 KG,N=6	157
*OCEAN WATER, ENGLISH CHANNAL	4 GRAM DRY WEIGHT/M**2	TYPICAL VALUE	355
*OLD FIELD, GEORGIA	500 GRAM DRY WEIGHT/M**2	TYPICAL VALUE	355
*OTHER FOREST	6.7 KG DRY WEIGHT/M**2	TOTAL=0.97*10**14 KG,ESTIMATED VALUES	157
*PART: BROWN ALGAE	3.4 KG DRY WEIGHT/M**2	TOTAL=0.0305*10**14 KG,N=8	157
*PART: LAND ANIMALS	0.014 KG DRY WEIGHT/M**2	TOTAL=0.021*10**14 KG	157
*PART: MARINE ANIMALS AND BACTERIA	0.0050 KG DRY WEIGHT/M**2	TOTAL=0.0183*10**14 KG	157
*PART: OCEAN PLANKTON	0.0067 KG DRY WEIGHT/M**2	TOTAL=0.024*10**14 KG,N=MANY	157
*PART: SOIL DECOMPOSERS	0.047 KG DRY WEIGHT/M**2	TOTAL=0.070*10**14 KG	157
*TOTAL DRY LAND	7.60 KG DRY WEIGHT/M**2	TOTAL=11.36*10**14 KG	157
*TOTAL OCEAN	0.0201 KG DRY WEIGHT/M**2	TOTAL=0.0728*10**14 KG	157
*TROPICAL FOREST	44.5 KG DRY WEIGHT/M**2	TOTAL=6.55*10**14 KG,ESTIMATED VALUES,N=1	157
*TROPICAL RAINFOREST, PANAMA	40000 GRAM DRY WEIGHT/M**2	TYPICAL VALUE	355
*TUNDRA	0.68 KG DRY WEIGHT/M**2	TOTAL=0.074*10**14 KG,ESTIMATED VALUES,N=1,SPHAGNUM	157
*WHOLE EARTH	2.25 KG DRY WEIGHT/M**2	TOTAL=11.43*10**14 KG	157

TABLE B124

ECOSPHERE : CARBON CONTENT			
ITEM:	VALUE:	CONDITIONS:	REF.:
*ARABLE LAND	0.30 KG/M**2	TOTAL=0.069*10**14 KG,N=15	157

TABLE B124

ECOSPHERE : CARBON CONTENT			
ITEM:	VALUE:	CONDITIONS:	REF.:
*CONIFEROUS FOREST	11.8 KG/M**2	TOTAL=1.18*10**14 KG,N=18	157
*DECIDUOUS FOREST	5.5 KG/M**2	TOTAL=0.27*10**14 KG,N=9	157
*DESERT	0.1 KG/M**2	TOTAL=0.033*10**14 KG,ESTIMATED VALUES	157
*GRASSLAND	0.39 KG/M**2	TOTAL=0.106*10**14 KG,N=6	157
*OTHER FOREST	3.0 KG/M**2	TOTAL=0.43*10**14 KG,ESTIMATED VALUES	157
*PART: BROWN ALGAE	1.0 KG/M**2	TOTAL=0.0090*10**14 KG,N=8	157
*PART: LAND ANIMALS	0.0064 KG/M**2	TOTAL=0.0095*10**14 KG	157
*PART: MARINE ANIMALS AND BACTERIA	0.0023 KG/M**2	TOTAL=0.0083*10**14 KG	157
*PART: OCEAN PLANKTON	0.0015 KG/M**2	TOTAL=0.0054*10**14 KG,N=MANY	157
*PART: SOIL DECOMPOSERS	0.021 KG/M**2	TOTAL=0.0315*10**14 KG	157
*TOTAL DRY LAND	3.42 KG/M**2	TOTAL=5.10*10**14 KG	157
*TOTAL OCEAN	0.0063 KG/M**2	TOTAL=0.0227*10**14 KG	157
*TROPICAL FOREST	20.0 KG/M**2	TOTAL=2.94*10**14 KG,ESTIMATED VALUE,N=1	157
*TUNDRA	0.3 KG/M**2	TOTAL=0.033*10**14 KG,ESTIMATED VALUES,N=1,SPHAGNUM	157
*WHOLE EARTH	1.004 KG/M**2	TOTAL=5.12*10**14 KG	157

TABLE B125

ECOSPHERE : CLASSIFICATION OF LAKES			
ITEM:	VALUE:	CONDITIONS:	REF.:
*CULTURAL EUTROPHIC	300 - 800 GRAM C/(M**2*YEAR)	NET PRIMARY PRODUCTIVITY	1351
*MESOTROPHIC	10 - 70 GRAM C/(M**2*YEAR)	NET PRIMARY PRODUCTIVITY	1351
*NATURAL EUTROPHIC	70 - 400 GRAM C/(M**2*YEAR)	NET PRIMARY PRODUCTIVITY	1351
*OLIGOTROPHIC	0.1 - 10 GRAM C/(M**2*YEAR)	NET PRIMARY PRODUCTIVITY	1351

TABLE B126

ECOSPHERE : CLIMATE			
ITEM:	VALUE:	CONDITIONS:	REF.:
*0.5 - 1.0 CEL. DEGREE	INCREASE IN ANNUAL MEAN TEMPERATURE	CLIMATE IN CITY COMPARED WITH SURROUNDING CONTRYSIDE	871
*1.0 - 2.0 CEL. DEGREE	INCREASE IN WINTER MINIMUM TEMPERATURE	CLIMATE IN CITY COMPARED WITH SURROUNDING CONTRYSIDE	871
*10 - 30 %	DECREASE IN EXTRAME GUSTS	CLIMATE IN CITY COMPARED WITH SURROUNDING CONTRYSIDE	871
*10 TIMES MORE	INCREASE IN PARTICULATE MATTER IN AIR	CLIMATE IN CITY COMPARED WITH SURROUNDING CONTRYSIDE	871
*100 %	INCREASE IN WINTER FOG	CLIMATE IN CITY COMPARED WITH SURROUNDING CONTRYSIDE	871
*15 - 20 %	DECREASE IN TOTAL INSOLATION	CLIMATE IN CITY COMPARED WITH SURROUNDING CONTRYSIDE	871
*2 - 3 %	DECREASE IN RELATIVE HUMIDITY	CLIMATE IN CITY COMPARED WITH SURROUNDING CONTRYSIDE	871
*20 - 30 %	DECREASE IN ANNUAL MEAN WIND SPEED	CLIMATE IN CITY COMPARED WITH SURROUNDING CONTRYSIDE	871
*30 %	DECREASE IN ULTRAVIOLET INSOLATION IN WINTER	CLIMATE IN CITY COMPARED WITH SURROUNDING CONTRYSIDE	871
*30 %	INCREASE IN SUMMER FOG	CLIMATE IN CITY COMPARED WITH SURROUNDING CONTRYSIDE	871

TABLE B126

ECOSPHERE : CLIMATE

ITEM:	VALUE:	CONDITIONS:	REF.:
*5 - 10 %	INCREASE IN CLOUD COVER	CLIMATE IN CITY COMPARED WITH SURROUNDING CONTRYSIDE	871
*5 - 10 %	INCREASE IN TOTAL PRECIPITATION	CLIMATE IN CITY COMPARED WITH SURROUNDING CONTRYSIDE	871
*5 - 15 %	DECREASE IN ULTRAVIOLET INSOLATION IN SUMMER	CLIMATE IN CITY COMPARED WITH SURROUNDING CONTRYSIDE	871
*5 - 20 %	INCREASE IN CALMS	CLIMATE IN CITY COMPARED WITH SURROUNDING CONTRYSIDE	871
*5 - 25 TIMES MORE	INCREASE IN GASEOUS POLLUTANTS	CLIMATE IN CITY COMPARED WITH SURROUNDING CONTRYSIDE	871
*5 %	DECREASE IN SNOWFALL	CLIMATE IN CITY COMPARED WITH SURROUNDING CONTRYSIDE	871

TABLE B127

ECOSPHERE : COMMUNITY STRUCTURE

ITEM:	VALUE:	CONDITIONS:	REF.:
*HABITAT DIVERSITY	DH = SQRT((SIGMA(F*P**2)-(SIGMA (F*P)**2/(SIGMA(F))/(SIGMA(F))	DH=HAB. DIV.,F=DENSITY OF FOLIAGE AT GIVEN DISTANCE P FROM FROM GROUND	1562

TABLE B128

ECOSPHERE : DENSITY

ITEM:	VALUE:	CONDITIONS:	REF.:
*DRY SAND	1.4 - 1.7 GRAM/CM**3	NONE	268
*ICE	1.7 - 2.3 GRAM/CM**3	NONE	268
*SOIL ORGANIC MATTER	1.3 GRAM/CM**3	NONE	268
*STILL AIR	1.25*10**(-3) GRAM/CM**3	NONE	268
*STILL WATER	1.0 GRAM/CM**3	NONE	268
*WET CLAY	1.7 - 2.2 GRAM/CM**3	NONE	268

TABLE B129

ECOSPHERE : DIURNAL DAMPING DEPTH

ITEM:	VALUE:	CONDITIONS:	REF.:
*CONCRETE	24.6 CM	NONE	268
*DRY CLAY	7.4 - 10.2 CM	NONE	268
*DRY SAND	5.8 - 10.5 CM	NONE	268
*DRY WOOD	7.4 - 8.3 CM	NONE	268
*ICE	14.1 - 20.5 CM	NONE	268
*IRON	265.5 CM	NONE	268
*ROCK	16.0 - 21.8 CM	NONE	268
*SILVER	215 CM	NONE	268
*SOIL ORGANIC MATTER	5.5 CM	NONE	268
*STILL AIR	69.7 - 75.7 CM	NONE	268
*STILL WATER	6.0 - 6.4 CM	NONE	268
*WET CLAY	13.5 - 18.5 CM	NONE	268

TABLE B129

ECOSPHERE : DIURNAL DAMPING DEPTH			
ITEM:	VALUE:	CONDITIONS:	REF.:
*WET SAND	16.6 CM	NONE	268

TABLE B130

ECOSPHERE : ENERGY FLOW, BIOMASS AND NUMBER			
ITEM:	VALUE:	CONDITIONS:	REF.:
*ACARTIA	2.5 KCAL/(M**2*DAY)	MARINE COPEPOD, BIOMASS=2.0G/M2, DENSITY=10**5NUMBERS/M2	1159
*LITTORINA SP.	1.0 KCAL/(M**2*DAY)	INTERTIDAL SNAIL, BIOMASS=10.0G/M2, DENSITY=200NUMBERS/M2	1159
*MICROTUS	0.7 KCAL/(M**2*DAY)	MEADOW MICE, BIOMASS=0.6G/M2, DENSITY=10**-2)NUMBERS/M2	1159
*ODOCOILEUS	0.5 KCAL/(M**2*DAY)	DEER, BIOMASS=1.1G/M2, DENSITY=10**(-5)NUMBERS/M2	1159
*ORCHELIMUM	0.4 KCAL/(M**2*DAY)	SALT MARSH GRASSHOPPERS, BIOMASS=1.0G/M2, DENSITY=10NUMBERS/M2	1159
*SOIL BACTERIA	1.0 KCAL/(M**2*DAY)	BIOMASS=0.001G/M2, DENSITY=10**12NUMBERS/M2	1159

TABLE B131

ECOSPHERE : GROSS PRIMARY PRODUCTION			
ITEM:	VALUE:	CONDITIONS:	REF.:
*ALFALFA FIELD	24000 KCAL/(YEAR**M**2)	ANNUAL VALUE, USA	1580
*BOG LAKE	0.3 G DRY ORG.MATTER/(M**2)*DAY	PHYTOPLANKTON ONLY,MINNESOTA, AVERAGE FOR 6MONTHS TO 1 Y	175
*CORAL REEF	18.2 G DRY ORG.MATTER/(M**2)*DAY	AVERAGE FOR 3 PACIFIC REEFS, AVERAGE FOR 6MONTHS TO 1 Y	175
*ESTUARIES	23 G DRY ORG.MATTER/(M**2)*DAY	TEXAS,SHORT FAVORABLE PERIODS	175
*EUTROPHIC LAKE	2.1 G DRY ORG.MATTER/(M**2)*DAY	SHALLOW LAKE,JAPAN,AVERAGE FOR 6MONTHS TO 1 Y	175
*FRESHWATER SPRING	20800 KCAL/(YEAR**M**2)	ANNUAL VALUE, SILVER SPRING, FLA.	1164
*LAGUNA MADRE	4.4 G DRY ORG.MATTER/(M**2)*DAY	TEXAS ESTUARIES,AVERAGE FOR 6MONTHS TO 1 Y	175
*LAKE ERIE	1.0 G DRY ORG.MATTER/(M**2)*DAY	WINTER,AVERAGE FOR 6MONTHS TO 1 Y	175
*LAKE ERIE	9.0 G DRY ORG.MATTER/(M**2)*DAY	SUMMER,AVERAGE FOR 6MONTHS TO 1 Y	175
*LAND	0.20 *10**14KG C/YR	NONE	1341
*LAND	0.24 *10**14 KG C/YR	NONE	465
*LAND	0.172 *10**14 KG C/YR	NONE	1076
*LONG ISLAND SOUND	3.2 G DRY ORG.MATTER/(M**2)*DAY	SHALLOW INSHORE WATERS,AVERAGE FOR 6MONTHS TO 1 Y	175
*MARINE SOUND	5700 KCAL/(YEAR**M**2)	ANNUAL VALUE, COASTAL, LONG ISLAND, N.Y.	1338
*MARINE TURTLE-GRASS FLATS	34 G DRY ORG.MATTER/(M**2)*DAY	FLORIDA,SHORT FAVORABLE PERIODS	175
*MARINE, COASTAL ZONES	2000 KCAL/(M**2*YEAR)	AREA=34.0*10**6 KM2,TOTAL G. PROD=6.8*10**16 KCAL/Y	1365
*MARINE, ESTUARIES AND REEFS	20000 KCAL/(M**2*YEAR)	AREA=2.0*10**6 KM2,TOTAL G. PROD=4.0*10**16 KCAL/Y	1161
*MARINE, OPEN OCEAN	1000 KCAL/(M**2*YEAR)	AREA=326.0*10**6 KM2,TOTAL G. PROD =32.6*10**16 KCAL/Y	1365
*MARINE, UPWELLING ZONES	6000 KCAL/(M**2*YEAR)	AREA=0.4*10**6 KM2,TOTAL G. PROD=0.2*10**16 KCAL/Y	1365
*MASS ALGAE CULTURE	43 G DRY ORG.MATTER/(M**2)*DAY	EXTRA CO2 ADDED,SHORT FAVORABLE PERIODS	175
*OAK-PINE FOREST	11500 KCAL/(YEAR**M**2)	ANNUAL VALUE, MEDIUM AGED, NEW YORK, NO TIMBER HARVEST	1746

TABLE B131

ECOSPHERE : GROSS PRIMARY PRODUCTION			
ITEM:	VALUE:	CONDITIONS:	REF.:
*OLIGOTROPHIC LAKE	0.7 G DRY ORG.MATTER/(M**2)*DAY	CLEAR DEEP LAKE,WISCONSIN,AVERAGE FOR 6MONTHS TO 1 Y	175
*PACIFIC OCEAN	0.2 G DRY ORG.MATTER/(M**2)*DAY	INFERTILE OPEN OCEAN,AVERAGE FOR 6MONTHS TO 1 Y	175
*PINE FOREST	12200 KCAL/(YEAR**M**2)	ANNUAL VALUE, YOUNG PLANTATION, MEAN FOR 0-50 Y, ENG.	1194
*POND	5.0 G DRY ORG.MATTER/(M**2)*DAY	FERTILIZED POND,N.C.,SHORT FAVORABLE PERIODS	175
*POND	9.0 G DRY ORG.MATTER/(M**2)*DAY	POND WITH TREATED SEWAGE WASTES, DENMARK,SHORT FAVORABLE PERIODS	175
*POND	27 G DRY ORG.MATTER/(M**2)*DAY	POND WITH UNTREATED WASTES,SOUTH DAKOTA,SHORT FAVORABLE PERIODS	175
*RAIN FOREST	45000 KCAL/(YEAR**M**2)	ANNUAL VALUE, MATURE FOREST, PUERTO RICO	1163
*RIVER	1.7 G DRY ORG.MATTER/(M**2)*DAY	TURBID RIVER,N.C.,SHORT FAVORABLE PERIODS	175
*SARGASSO SEA	0.5 G DRY ORG.MATTER/(M**2)*DAY	INFERTILE OPEN OCEAN,AVERAGE FOR 6MONTHS TO 1 Y	175
*SEA	0.44-2.03 *10**14 KG C/YR	NONE	1341
*SEA	0.235 *10**14 KG C/YR	NONE	1126
*SEA	=< 1.6 *10**14 KG C/YR	NONE	465
*SEA	0.25 *10**14 KG C/YR	NONE	1076
*SILVER SPRINGS	17.5 G DRY ORG.MATTER/(M**2)*DAY	FLORIDA,AVERAGE FOR 6MONTHS TO 1 Y	175
*SILVER SPRINGS	35 G DRY ORG.MATTER/(M**2)*DAY	FLORIDA,SHORT FAVORABLE PERIODS	175
*STREAM	57 G DRY ORG.MATTER/(M**2)*DAY	POLLUTED STREAM,INDIANA,SHORT FAVORABLE PERIODS	175
*TERRISTRIAL, BOREAL CONIFEROUS FOREST	3000 KCAL/(M**2*YEAR)	AREA=10.0*10**6 KM2,TOTAL G. PROD= 3.0*10**16 KCAL/Y	924
*TERRISTRIAL, BROADLEAVED EVERGREEN FOREST	20000 KCAL/(M**2*YEAR)	AREA=14.7*10**6 KM2,TOTAL G. PROD= 29.0*10**16 KCAL/Y, TROPIC + SUB-	924
*TERRISTRIAL, CULTIVATED LAND	8000 KCAL/(M**2*YEAR)	AREA=4.9*10**6 KM2,TOTAL G. PROD= 3.9*10**16 KCAL/Y, NO ENERGY SUBSIDY	924
*TERRISTRIAL, CULTIVATED LAND	12000 KCAL/(M**2*YEAR)	AREA=4.0*10**6 KM2,TOTAL G. PROD= 4.8*10**16 KCAL/Y, FUEL SUBSIDIZED	924
*TERRISTRIAL, DESERT AND TUNDRAS	200 KCAL/(M**2*YEAR)	AREA=40.0*10**6 KM2,TOTAL G. PROD= 0.8*10**16 KCAL/Y	924
*TERRISTRIAL, DRY FOREST	2500 KCAL/(M**2*YEAR)	AREA=9.4*10**6 KM2,TOTAL G. PROD= 2.4*10**16 KCAL/Y	924
*TERRISTRIAL, GRASSLANDS AND PASTURES	2500 KCAL/(M**2*YEAR)	AREA=42.0*10**6 KM2,TOTAL G. PROD= 10.5*10**16 KCAL/Y	924
*TERRISTRIAL, TEMPERATE FOREST	8000 KCAL/(M**2*YEAR)	AREA=4.9*10**6 KM2,TOTAL G. PROD= 3.9*10**16 KCAL/Y, MOIST	924
*TOTAL FOR BIOSPHERE	2000 KCAL/(M**2*YEAR)	AREA=500.0*10**6 KM2,TOTAL G. PROD =100.0*10**16 KCAL/Y, EXC. ICE CAPS	1161

TABLE B132

ECOSPHERE : HETEROTROPHIC RESPIRATION			
ITEM:	VALUE:	CONDITIONS:	REF.:
*ALFALFA FIELD	800 KCAL/(YEAR**M**2)	ANNUAL VALUE, USA	1580
*FRESHWATER SPRING	6800 KCAL/(YEAR**M**2)	ANNUAL VALUE, SILVER SPRING, FLA.	1164
*MARINE SOUND	2500 KCAL/(YEAR**M**2)	ANNUAL VALUE, COASTAL, LONG ISLAND, N.Y.	1338
*OAK-PINE FOREST	3000 KCAL/(YEAR**M**2)	ANNUAL VALUE, MEDIUM AGED, NEW YORK, NO TIMBER HARVEST	1746
*PINE FOREST	4600 KCAL/(YEAR**M**2)	ANNUAL VALUE, YOUNG PLANTATION, MEAN FOR 0-50 Y, ENG.	1194
*RAIN FOREST	13000 KCAL/(YEAR**M**2)	ANNUAL VALUE, MATURE FOREST, PUERTO RICO	1163

TABLE B133

ECOSPHERE : MEAN DAILY PRODUCTIVITY			
ITEM:	VALUE:	CONDITIONS:	REF.:
*PHYTOPLANKTON, LAKE BORAX	249 (RANGE = 10 - 524) MG C/ (DAY*M**2)	ANN PROD=91G C/(M2*Y), LARGE SHALLOW SALINE, 25 % OF TOTAL	1683
*PHYTOPLANKTON, LAKE CASTLE	98 (RANGE = 6 - 317) MG C/ (DAY*M**2)	ANN PROD=36G C/(M2*Y),DEEP, ALPINE, CALIF	1689
*PHYTOPLANKTON, LAKE CHAR	1.1 (RANGE = 0 - 35) MG C/ (DAY*M**2)	ANN PROD=4.1 G C/(M2*Y), 80 % BY BENTHIC FLORA, CAN, LAT 74, NWT	792
*PHYTOPLANKTON, LAKE CLEAR	438 (RANGE = 2 - 2440) MG C/ (DAY*M**2)	ANN PROD=260G C/(M2*Y), VERY LARGE, SHALLOW	540
*PHYTOPLANKTON, LAKE CROOKED	469 (RANGE = 142 - 1364) MG C/ (DAY*M**2)	ANN PROD=171G C/(M2*Y), LARGE, DEEP, HARDWATER, 1963	1685
*PHYTOPLANKTON, LAKE CROOKED	359 (RANGE = 23 - 870) MG C/ (DAY*M**2)	ANN PROD=131G C/(M2*Y), LARGE, DEEP, HARDWATER, 1964, IND.	1685
*PHYTOPLANKTON, LAKE ERIE	RANGE 30 - 4760 MG C/(DAY*M**2)	ANN PROD=310G C/(M2*Y), WESTERN STATIONS	1641
*PHYTOPLANKTON, LAKE ERIE	RANGE 120 - 1690 MG C/(DAY*M**2)	ANN PROD=210G C/(M2*Y), CENTRAL STATIONS	1641
*PHYTOPLANKTON, LAKE ERIE	RANGE 140 - 1440 MG C/(DAY*M**2)	ANN PROD=160G C/(M2*Y), EASTERN STATIONS	1641
*PHYTOPLANKTON, LAKE ERKEN	285 (RANGE = 40 - 2205) MG C/ (DAY*M**2)	ANN PROD=104G C/(M2*Y), LARGE DEEP, NAT. PRODUCTIVE	1350
*PHYTOPLANKTON, LAKE ESROM	370 (RANGE = 23 - 422) MG C/ (DAY*M**2)	ANN PROD=260G C/(M2*Y), LARGE, MODERATELY DEEP	782
*PHYTOPLANKTON, LAKE FREDERIKSBORG SLOTSS@	1030 (RANGE = 12 - 4160) MG C/ (DAY*M**2)	ANN PROD=376G C/(M2*Y), SHALLOW, ENRICHED, DEN.	1143
*PHYTOPLANKTON, LAKE FURES@	462 (RANGE = 0 - 1380) MG C/ (DAY*M**2)	ANN PROD=168G C/(M2*Y), LARGE, DEEP	782
*PHYTOPLANKTON, LAKE GOOSE	729 (RANGE = 166 - 1753) MG C/ (DAY*M**2)	ANN PROD=266G C/(M2*Y), SMALL KETTLE, IND.	1684
*PHYTOPLANKTON, LAKE GRANE LANGS@	248 (RANGE = 20 - 880) MG C/ (DAY*M**2)	ANN PROD=91G C/(M2*Y), DEEP, ACIDIC, HUMIC STAINED, DK	1143
*PHYTOPLANKTON, LAKE HURON	RANGE 150 - 700 MG C/(DAY*M**2)	ANN PROD=100 G C/(M2*Y), OFFSHORE STATIONS	1641
*PHYTOPLANKTON, LAKE LANAO	1700 (RANGE = 400 - 5000) MG C/ (DAY*M**2)	ANN PROD=620G C/(M2*Y), LARGE DEEP TROPICAL, PHILLIPPINES	917
*PHYTOPLANKTON, LAKE LAWRENCE	112.6 (RANGE = 5 - 497) MG C/ (DAY*M**2)	ANN PROD=41.1G C/(M2*Y), SMALL HARDWATER, 7 Y MEAN	1689
*PHYTOPLANKTON, LAKE LITTLE CROOKED	618 (RANGE = 263 - 1903) MG C/ (DAY*M**2)	ANN PROD=226G C/(M2*Y), SMALL, DEEP, KETTLE, 1963	(WET
*PHYTOPLANKTON, LAKE LITTLE CROOKED	598 (RANGE = 9 - 2451) MG C/ (DAY*M**2)	ANN PROD=218G C/(M2*Y), SMALL, DEEP, KETTLE, IND., 1964	1685
*PHYTOPLANKTON, LAKE LUNZER UNTRESEE	123 MG C/(M**2*DAY)	ANN PROD=45 G C/(M2*Y), SMALL ALPINE	783
*PHYTOPLANKTON, LAKE MARTIN	561 (RANGE = 27 - 1708) MG C/ (DAY*M**2)	ANN PROD=205G C/(M2*Y), DEEP, STAINED MARL, IND.	1686
*PHYTOPLANKTON, LAKE MERETTA	3.1 (RANGE = 0 - 170) MG C/ (DAY*M**2)	ANN PROD=11G C/(M2*Y), POLLUTED BY SEWAGE, NWT CANADA	792
*PHYTOPLANKTON, LAKE MICHIGAN	RANGE 70 - 1030 MG C/(DAY*M**2)	ANN PROD=130 G C/(M2*Y), OFFSHORE STATIONS	1641
*PHYTOPLANKTON, LAKE MINNETONKA	820 MG C/(DAY*M**2)	ANN PROD=300G C/(M2*Y), LARGE DEEP, COMPLEX BASIN, MINN.	1019
*PHYTOPLANKTON, LAKE OLIN	374 (RANGE = 89 - 996) MG C/ (DAY*M**2)	ANN PROD=137G C/(M2*Y), LARGE DEEP, MARL, IND.	1686
*PHYTOPLANKTON, LAKE OLIVER	336 (RANGE = 32 - 775) MG C/ (DAY*M**2)	ANN PROD=123G C/(M2*Y), LARGE DEEP, MARL, IND.	1686
*PHYTOPLANKTON, LAKE ONTARIO	RANGE 60 - 1400 MG C/(DAY*M**2)	ANN PROD=180 G C/(M2*Y), OFFSHORE STATIONS	1641
*PHYTOPLANKTON, LAKE PRETTY	440 (RANGE = 68 - 1850) MG C/ (DAY*M**2)	ANN PROD=161G C/(M2*Y), MODERATE-SIZE, DEEP MARL, IND.	1685
*PHYTOPLANKTON, LAKE PRETTY	305 (RANGE = 6 - 895) MG C/ (DAY*M**2)	ANN PROD=111G C/(M2*Y), MODERATE-SIZE, DEEP MARL, IND., 1964	1685
*PHYTOPLANKTON, LAKE RANSEREN	RANGE 23 - 66 MG C/(DAY*M**2)	SMALL LAPPLANDIC, SWEDEN	1350
*PHYTOPLANKTON, LAKE SMITH HOLE	194 (RANGE = 24 - 5960) MG C/ (DAY*M**2)	ANN PROD=71G C/(M2*Y), SHALLOW, HUMIC STAINED, IND.	1686
*PHYTOPLANKTON, LAKE STORE GRIBS@	230 (RANGE = 4 - 680) MG C/ (DAY*M**2)	ANN PROD=84G C/(M2*Y), DEEP, ACIDIC, HUMIC STAINED, DEN.	1143
*PHYTOPLANKTON, LAKE SUPERIOR	RANGE 50 - 260 MG C/(DAY*M**2)	MOST UNPRODUCTIVE OF LAURENTIAN GREAT LAKES	1285

TABLE B133

ECOSPHERE : MEAN DAILY PRODUCTIVITY			
ITEM:	VALUE:	CONDITIONS:	REF.:
*PHYTOPLANKTON, LAKE SYLVAN	1564 (RANGE = 9 - 4959) MG C/ (DAY*M**2)	ANN PROD=570G C/(M2*Y), COMPLEX BASIN, LARGE, SHALLOW, IND.	1684
*PHYTOPLANKTON, LAKE VICTORIA	1750 (RANGE = 1700 - 3800) MG C/ (DAY*M**2)	ANN PROD=640G C/(M2*Y), LARGE, TROPICAL	1552
*PHYTOPLANKTON, LAKE WALTER	210 (RANGE = 12 - 535) MG C/ (DAY*M**2)	ANN PROD=77G C/(M2*Y), BASIN B, MARL	1686
*PHYTOPLANKTON, LAKE WALTER	276 (RANGE = 30 - 1048) MG C/ (DAY*M**2)	ANN PROD=101G C/(M2*Y), BASIN C, MARL	1686
*PHYTOPLANKTON, LAKE WALTER	437 (RANGE = 98 - 1458) MG C/ (DAY*M**2)	ANN PROD=160G C/(M2*Y), BASIN D, MARL	1686
*PHYTOPLANKTON, LAKE WALTER	418 (RANGE = 102 - 1395) MG C/ (DAY*M**2)	ANN PROD=153G C/(M2*Y), BASIN A, MARL, IND.	1686
*PHYTOPLANKTON, LAKE WINTERGREEN	1012 (RANGE = 60 - 4160) MG C/ (DAY*M**2)	ANN PROD=369G C/(M2*Y), SHALLOW, HIG LOADING, MICH.	1689

TABLE B134

ECOSPHERE : MEAN TEMPERATURE AND HUMIDITY			
ITEM:	VALUE:	CONDITIONS:	REF.:
*AIR	300 K	BELEM, BRAZIL, ALTITUDE = 0 KM	1516
*AIR	294 K	TAKENGøN, INDONESIA, ALTITUDE = 1.2 KM	1516
*AIR	288 K	QUITO, ECUADOR, ALTITUDE = 2.9 KM	1516
*AIR	286 K	JAUJA, PERU, ALTITUDE = 3.4 KM	1516
*CONTINENTAL ARCTIC AIR MASS	227 K	VERY COLD, VERY DRY, SPECIFIC HUMIDITY = 0.1 G/KG	1516

TABLE B135

ECOSPHERE : N			
ITEM:	VALUE:	CONDITIONS:	REF.:
*ATMOSPHERE	5*10**21 G	TOTAL SUPPLY;PER UNIT AREA=10**3G/ CM**2	779

TABLE B136

ECOSPHERE : NET COMMUNITY PRODUCTION			
ITEM:	VALUE:	CONDITIONS:	REF.:
*ALFALFA FIELD	14400 KCAL/(YEAR**M**2)	ANNUAL VALUE, USA	1580
*FRESHWATER SPRING	2000 KCAL/(YEAR**M**2)	ANNUAL VALUE, SILVER SPRING, FLA.	1164
*MARINE SOUND	VERY LITTLE OR NONE	ANNUAL VALUE, COASTAL, LONG ISLAND, N.Y.	1338
*OAK-PINE FOREST	2000 KCAL/(YEAR**M**2)	ANNUAL VALUE, MEDIUM AGED, NEW YORK, NO TIMBER HARVEST	1746
*PINE FOREST	2900 KCAL/(YEAR**M**2)	ANNUAL VALUE, YOUNG PLANTATION, MEAN FOR 0-50 Y, ENG.	1194
*RAIN FOREST	VERY LITTLE OR NONE	ANNUAL VALUE, MATURE FOREST, PUERTO RICO	1163

TABLE B137

ECOSPHERE : NET PRIMARY PRODUCTION			
ITEM:	VALUE:	CONDITIONS:	REF.:
*AGRICULTURAL LAND	650 (NORMAL RANGE=100 - 4000) G DRY WEIGHT/(M**2*YEAR)	TOT NPP=9.1*10**9 D T/Y,BOIM.=1 DRY KG/M2,TOT BIOM.=14*10**9 D T	1700

TABLE B137

ECOSPHERE : NET PRIMARY PRODUCTION

ITEM:	VALUE:	CONDITIONS:	REF.:
*ALFALFA FIELD	15200 KCAL/(YEAR**M**2)	ANNUAL VALUE, USA	1580
*ATTACHED ALGAE AND ESTUARIES	2000 (NORMAL RANGE=500 - 4000) G DRY WEIGHT/(M**2*YEAR)	TOT NPP=4.0*10**9 D T/Y,BOIM.=1 DRY KG/M2,TOT BIOM.=2.0*10**9 D T	1700
*BOREAL FOREST	800 (NORMAL RANGE=400 - 2000) G DRY WEIGHT/(M**2*YEAR)	TOT NPP=9.6*10**9 D T/Y,BOIM.=20 DRY KG/M2,TOT BIOM.=240*10**9 D T	1700
*CONTINENTAL SHELF	350 (NORMAL RANGE=200 - 600) G DRY WEIGHT/(M**2*YEAR)	TOT NPP=9.5*10**9 D T/Y,BOIM.=0.01 DRY KG/M2,TOT BIOM.=0.3*10**9 D T	1700
*CORN	4500 KCAL/(M**2*YEAR)	USA, EDIBLE PORTION=1510 KCAL/ (M2*Y)	444
*CORN	1100 KCAL/(M**2*YEAR)	INDIA, EDIBLE PORTION=350 KCAL/ (M2*Y)	444
*CORN	2400 KCAL/(M**2*YEAR)	WORLD AVERAG, EDIBLE PORTION=810 KCAL/(M2*Y)	444
*DESERT SCRUB	70 (NORMAL RANGE=10 - 250) G DRY WEIGHT/(M**2*YEAR)	TOT NPP=1.3*10**9 D T/Y,BOIM.=0.7 DRY KG/M2,TOT BIOM.=13*10**9 D T	1700
*EXTREME DESERT, ROCK AND ICE	3 (NORMAL RANGE=0 - 10) G DRY WEIGHT/(M**2*YEAR)	TOT NPP=0.07*10**9 D T/Y,BOIM.= 0.02 DRY KG/M2,TOT BIOM.=0.5*10**9 D T	1700
*FRESHWATER SPRING	8800 KCAL/(YEAR**M**2)	ANNUAL VALUE, SILVER SPRING, FLA.	1164
*LAKE AND STREAM	500 (NORMAL RANGE=100 - 1500) G DRY WEIGHT/(M**2*YEAR)	TOT NPP=1.0*10**9 D T/Y,BOIM.=0.02 DRY KG/M2,TOT BIOM.=0.04*10**9 D T	1700
*MARINE SOUND	2500 KCAL/(YEAR**M**2)	ANNUAL VALUE, COASTAL, LONG ISLAND, N.Y.	1338
*OAK-PINE FOREST	5000 KCAL/(YEAR**M**2)	ANNUAL VALUE, MEDIUM AGED, NEW YORK, NO TIMBER HARVEST	1746
*OPEN OCEAN	125 (NORMAL RANGE=1 - 400) G DRY WEIGHT/(M**2*YEAR)	TOT NPP=41.5*10**9 D T/Y,BOIM.= 0.003 DRY KG/M2,TOT BIOM.= 1.0*10**9 D	1700
*PINE FOREST	7500 KCAL/(YEAR**M**2)	ANNUAL VALUE, YOUNG PLANTATION, MEAN FOR 0-50 Y, ENG.	1194
*RAIN FOREST	13000 KCAL/(YEAR**M**2)	ANNUAL VALUE, MATURE FOREST, PUERTO RICO	1163
*RICE	5500 KCAL/(M**2*YEAR)	JAPAN, EDIBLE PORTION=1840 KCAL/ (M2*Y)	444
*RICE	1700 KCAL/(M**2*YEAR)	BRAZIL, EDIBLE PORTION=580 KCAL/ (M2*Y)	444
*RICE	2300 KCAL/(M**2*YEAR)	WORLD AVERAGE, EDIBLE PORTION=760 KCAL/(M2*Y)	444
*SAVANNA	700 (NORMAL RANGE=200 - 2000) G DRY WEIGHT/(M**2*YEAR)	TOT NPP=10.5*10**9 D T/Y,BOIM.=4 DRY KG/M2,TOT BIOM.=60*10**9 D T	1700
*SOYBEANS	2400 KCAL/(M**2*YEAR)	CANADA, EDIBLE PORTION=800 KCAL/ (M2*Y)	444
*SOYBEANS	780 KCAL/(M**2*YEAR)	INDONESIA, EDIBLE PORTION=260 KCAL/(M2*Y)	444
*SOYBEANS	1400 KCAL/(M**2*YEAR)	WORLD AVERAGE, EDIBLE PORTION=480 KCAL/(M2*Y)	444
*SUGAR	12200 KCAL/(M**2*YEAR)	HAWAII, EDIBLE PORTION=4070 KCAL/ (M2*Y), FROM CANE	444
*SUGAR	7300 KCAL/(M**2*YEAR)	NETHERLANDS, EDIBLE PORTION=2440 KCAL/(M2*Y), FROM BEETS	444
*SUGAR	3700 KCAL/(M**2*YEAR)	CUBA, EDIBLE PORTION=1220 KCAL/ (M2*Y), FROM CANE	444
*SWAMP AND MARSH	2000 (NORMAL RANGE=800 - 4000) G DRY WEIGHT/(M**2*YEAR)	TOT NPP=4.0*10**9 D T/Y,BOIM.=12 DRY KG/M2,TOT BIOM.=24*10**9 D T	1700
*SWEET POTATOES AND YAMS	3600 KCAL/(M**2*YEAR)	JAPAN, EDIBLE PORTION=1800 KCAL/ (M2*Y)	444
*SWEET POTATOES AND YAMS	1100 KCAL/(M**2*YEAR)	INDONESIA, EDIBLE PORTION=570 KCAL/(M2*Y)	444
*SWEET POTATOES AND YAMS	1500 KCAL/(M**2*YEAR)	WORLD AVERAGE, EDIBLE PORTION=750 KCAL/(M2*Y)	444
*TEMPERATE FOREST	1300 (NORMAL RANGE=600 - 2500) G DRY WEIGHT/(M**2*YEAR)	TOT NPP=23.4*10**9 D T/Y,BOIM.=30 DRY KG/M2,TOT BIOM.=540*10**9 D T	1700
*TEMPERATE GRASSLAND	500 (NORMAL RANGE=150 - 1500) G DRY WEIGHT/(M**2*YEAR)	TOT NPP=4.5*10**9 D T/Y,BOIM.=4 DRY KG/M2,TOT BIOM.=60*10**9 D T	1700
*TOTAL FOR EARTH	320 G DRY WEIGHT/(M**2*YEAR)	TOT NPP=164*10**9 D T/Y,BOIM.=3.6 DRY KG/M2,TOT BIOM.=1855*10**9 D T	1700

TABLE B137

ECOSPHERE : NET PRIMARY PRODUCTION			
ITEM:	VALUE:	CONDITIONS:	REF.:
*TOTAL LAND	730 G DRY WEIGHT/(M**2*YEAR)	TOT NPP=109*10**9 D T/Y,BOIM.=12.5 DRY KG/M2,TOT BIOM.=1852*10**9 D T	1700
*TOTAL OCEAN	155 G DRY WEIGHT/(M**2*YEAR)	TOT NPP=55*10**9 D T/Y,BOIM.=0.009 DRY KG/M2,TOT BIOM.=3.3*10**9 D T	1700
*TROPICAL FOREST	2000 (NORMAL RANGE=1000 - 5000) G DRY WEIGHT/(M**2*YEAR)	TOT NPP=40.0*10**9 D T/Y,BOIM.=45 DRY KG/M2,TOT BIOM.=900*10**9 D T	1700
*TUNDRA AND ALPINE	140 (NORMAL RANGE=10 - 400) G DRY WEIGHT/(M**2*YEAR)	TOT NPP=1.1*10**9 D T/Y,BOIM.=0.6 DRY KG/M2,TOT BIOM.=5*10**9 D T	1700
*WHEAT	4400 KCAL/(M**2*YEAR)	NETHERLANDS, EDIBLE PORTION=1450 KCAL/(M2*Y)	444
*WHEAT	900 KCAL/(M**2*YEAR)	INDIA, EDIBLE PORTION=300 KCAL/(M2*Y)	444
*WHEAT	1300 KCAL/(M**2*YEAR)	WORLD AVERAGE, EDIBLE PORTION=430 KCAL/(M2*Y)	444
*WHITE POTATOES	4100 KCAL/(M**2*YEAR)	USA, EDIBLE PORTION=2040 KCAL/(M2*Y)	444
*WHITE POTATOES	1400 KCAL/(M**2*YEAR)	INDIA, EDIBLE PORTION=700 KCAL/(M2*Y)	444
*WHITE POTATOES	2200 KCAL/(M**2*YEAR)	WORLD AVERAGE, EDIBLE PORTION=1090 KCAL/(M2*Y)	444
*WOODLAND AND SHRUBLAND	600 (NORMAL RANGE=200 - 1200) G DRY WEIGHT/(M**2*YEAR)	TOT NPP=4.2*10**9 D T/Y,BOIM.=6 DRY KG/M2,TOT BIOM.=42*10**9 D T	1700

TABLE B138

ECOSPHERE : NET PRODUCTION			
ITEM:	VALUE:	CONDITIONS:	REF.:
*ARABLE LAND	0.21 *10**14 KG C/YR	IN DRY MATTER=0.46*10**14 KG/YR	157
*ARABLE LAND	0.91 KG C/M**2*YR	IN DRY MATTER=2.0 KG/M**2*YR	157
*BROWN ALGAE ZONE	0.009 *10**14 KG C/YR	IN DRY MATTER=0.026*10**14 KG/YR	157
*BROWN ALGAE ZONE	1.0 KG C/M**2*YR	IN DRY MATTER=2.9 KG/M**2*YR	157
*CONIFEROUS FOREST	0.14 *10**14 KG C/YR	IN DRY MATTER=0.28*10**14 KG/YR, VAL.+-50%	157
*CONIFEROUS FOREST	1.4 KG C/M**2*YR	IN DRY MATTER=2.8 KG/M**2*YR,VAL.+-50%	157
*CONTINENTAL SHELF	0.043 *10**14 KG C/YR	IN DRY MATTER=0.195*10**14 KG/YR	157
*CONTINENTAL SHELF	0.16 KG C/M**2*YR	IN DRY MATTER=0.73 KG/M**2*YR	157
*DECIDUOUS FOREST	0.025 *10**14 KG C/YR	IN DRY MATTER=0.06*10**14 KG/YR, VAL.+-50%	157
*DECIDUOUS FOREST	0.54 KG C/M**2*YR	IN DRY MATTER=1.2 KG/M**2*YR,VAL.+-50%	157
*DESERT	0.015 *10**14 KG C/YR	IN DRY MATTER=0.033*10**14 KG/YR	157
*DESERT	0.045 KG C/M**2*YR	IN DRY MATTER=0.1 KG/M**2*YR	157
*GRASSLAND	0.245 *10**14 KG C/YR	IN DRY MATTER=0.54*10**14 KG/YR, VAL.+-50%	157
*GRASSLAND	0.91 KG C/M**2*YR	IN DRY MATTER=2.0 KG/M**2*YR,VAL.+-50%	157
*LAND	0.163 *10**14 KG C/YR	NONE	1401
*LAND	0.151 *10**14 KG C/YR	NONE	1133
*LAND	0.21 *10**14 KG C/YR	NONE	465
*LAND	0.103 *10**14 KG C/YR	NONE	1076
*LAND	1.06 *10**14 KG C/YR	NONE	157
*OCEAN	0.24 *10**14 KG C/YR	IN DRY MATTER=1.070*10**14 KG/YR	157
*OCEAN	0.072 KG C/M**2*YR	IN DRY MATTER=0.32 KG/M**2*YR	157
*OTHER FOREST	0.078 *10**14 KG C/YR	IN DRY MATTER=0.17*10**14 KG/YR, VAL.+-50%	157
*OTHER FOREST	0.54 KG C/M**2*YR	IN DRY MATTER=1.2 KG/M**2*YR,VAL.+-50%	157

TABLE B138

ECOSPHERE : NET PRODUCTION			
ITEM:	VALUE:	CONDITIONS:	REF.:
*SEA	0 *10**14 KG C/YR	NONE	1401
*SEA	0.286 *10**14 KG C/YR	NONE	1133
*SEA	0.15 *10**14 KG C/YR	NONE	1126
*SEA	0.32 *10**14 KG C/YR	NONE	465
*SEA	0.53 *10**14 KG C/YR	NONE	1363
*SEA	0.25 *10**14 KG C/YR	NONE	1076
*SEA	0.29 *10**14 KG C/YR	NONE	157
*TOTAL DRY LAND	1.058 *10**14 KG C/YR	IN DRY MATTER=2.284*10**14 KG/YR	157
*TOTAL DRY LAND	0.071 KG C/M**2*YR	IN DRY MATTER=1.53 KG/M**2*YR	157
*TOTAL SEA	0.292 *10**14 KG C/YR	IN DRY MATTER=1.291*10**14 KG/YR	157
*TROPICAL FOREST	0.34 *10**14 KG C/YR	IN DRY MATTER=0.73*10**14 KG/YR, VAL.+-50%	157
*TROPICAL FOREST	2.3 KG C/M**2*YR	IN DRY MATTER=5.0 KG/M**2*YR,VAL.+-50%	157
*TUNDRA	0.005 *10**14 KG C/YR	IN DRY MATTER=0.011*10**14 KG/YR	157
*TUNDRA	0.045 KG C/M**2*YR	IN DRY MATTER=0.1 KG/M**2*YR	157
*WHOLE EARTH	1.35 *10**14 KG C/YR	IN DRY MATTER=3.575*10**14 KG/YR	157
*WHOLE EARTH	0.265 KG C/M**2*YR	IN DRY MATTER=0.70 KG/M**2*YR	157
*WHOLE SEA	0.081 KG C/M**2*YR	IN DRY MATTER=0.36 KG/M**2*YR	157

TABLE B139

ECOSPHERE : POOLS WITHIN THE CARBON CYCLE			
ITEM:	VALUE:	CONDITIONS:	REF.:
*ATMOSPHERE	58*10**15 MOLES	AS CO2	143
*BIOSPHERE	1.5*10**18 G	VAL.PER UNIT AREA=0.29G/CM**2	1442
*CARBONATE IN SEDIMENT	67000*10**18 GRAM C	TOTAL ON EARTH	1318
*CARBONATE SEDIMENTS	1700000*10**15 MOLES	AS CARBONATE	143
*CO2 AS CO3(-2) IN OCEAN	14.2*10**18 GRAM C	TOTAL ON EARTH	1318
*CO2 AS HCO3(-) IN OCEAN	114.7*10**18 GRAM C	TOTAL ON EARTH	1318
*CO2 AS H2CO3 IN OCEAN	0.8*10**18 GRAM C	TOTAL ON EARTH	1318
*CO2 CONSUMED IN PHOTOSYNTHESIS	1.1*10**17 G/YR	NONE	1442
*CO2 IN OCEAN WATER	1.4*10**19 G	VAL.PER UNIT AREA=27G/CM**2	1442
*CO2 IN THE ATMOSPHERE	2.35*10**18 GRAM C	TOTAL ON EARTH	1318
*CO2 INCREASE IN ATMOSPHERE	5*10**15 G/YR	0.3%/YR	1442
*CO2 PRODUCED GEOLOGICALLY	2.5*10**23 G	OVER 4.5 B.Y.,VAL.PER UNIT AREA= 5*10**4G/CM**2	1442
*CO2 RELEASED IN BURNING OF FOSSIL FUELS	1.5*10**16 G/YR	NONE	1442
*DEAD ORGANIC MATTER IN OCEAN	10.0*10**18 GRAM C	TOTAL ON EARTH	1318
*DEAD ORGANIC MATTER ON LAND	2.6*10**18 GRAM C	TOTAL ON EARTH	1318
*DECAYING ORGANIC MATTER	58*10**15 MOLES	ON LAND	143
*DECAYING ORGANIC MATTER	250*10**15 MOLES	IN OCEANS	143
*FOSSIL FUEL RESERVES	3*10**18 G	ECONOMICALLY EXPLOITABLE	1442
*FOSSIL FUELS	830*10**15 MOLES	AS ORGANIC CARBON	143
*LAND ORGANISMS	38*10**15 MOLES	AS ORGANIC CARBON	143
*LIVING MATTER ON LAND	0.3*10**18 GRAM C	TOTAL ON EARTH	1318
*LIVING ORGANIC MATTER IN OCEAN	0.03*10**18 GRAM C	TOTAL ON EARTH	1318
*MARINE ORGANISMS	0.830*10**15 MOLES	AS ORGANIC MATTER	143
*MASS OF ATMOSPHERIC CO2	2.38*10**18 G	VAL.PER UNIT AREA=0.45G/CM**2	1442

TABLE B139

ECOSPHERE : POOLS WITHIN THE CARBON CYCLE

ITEM:	VALUE:	CONDITIONS:	REF.:
*OCEAN DEEPS LAYERS	2900*10**15 MOLES	IN SOLUTION	143
*OCEAN SURFACE LAYERS	43*10**15 MOLES	IN SOLUTION	143
*ORGANIC CARBON IN SEDIMENT	25000*10**18 GRAM C	TOTAL ON EARTH	1318
*ORGANIC CARBON IN SEDIMENTARY ROCKS	5*10**21 G	VAL.PER UNIT AREA=10**3G/CM**2	1442
*TOTAL ACTIVE POOLS	3300*10**15 MOLES	ROUNDED TOTAL	143
*TOTAL COAL	15*10**18 G	NONE	1442
*TOTAL STORAGE POOLS	1700000*10**15 MOLES	ROUNDED TOTAL	143

TABLE B140

ECOSPHERE : POOLS WITHIN THE NITROGEN CYCLE

ITEM:	VALUE:	CONDITIONS:	REF.:
*ATMOSPHERE	270000000*10**12 MOLES	AS N2	342
*CRUSTAL ROCKS	1000000000*10**12 MOLES	AS ROCK-FORMING MINERALS	342
*DECAYING ORGANIC MATTER	120000*10**12 MOLES	AS ORGANIC N	342
*LAND ORGANISMS	870*10**12 MOLES	AS ORGANIC N	342
*MARINE ORGANISMS	69*10**12 MOLES	AS ORGANIC N	342
*OCEANS	1400000*10**12 MOLES	AS N2	342
*OCEANS SEDIMENTS	7100*10**12 MOLES	AS INORGANIC N	342
*SEDIMENTARY ROCKS	29000000*10**12 MOLES	AS ROCK-FORMING MINERALS	342
*SOILS	10000*10**12 MOLES	AS INORGANIC N	342
*TOTAL ACTIVE POOLS	140000*10**12 MOLES	ROUNDED TOTAL	342
*TOTAL STORAGE POOLS	1600000000*10**12 MOLES	ROUNDED TOTAL	342

TABLE B141

ECOSPHERE : POOLS WITHIN THE OXYGEN CYCLE

ITEM:	VALUE:	CONDITIONS:	REF.:
*ATMOSPHERE	150000*10**15 MOLES	AS O2	1360
*ATMOSPHERE	120*10**15 MOLES	AS CO2	1360
*ATMOSPHERE	700*10**15 MOLES	AS H2O	1360
*CRUSTAL ROCKS	7700000*10**15 MOLES	AS ROCK OXIDES	1360
*FRESH SURFACE WATER	6900*10**15 MOLES	AS H2O	1360
*GROUND WATER	460000*10**15 MOLES	AS H2O	1360
*LAND ORGANISMS	38*10**15 MOLES	AS ORGANIC O	1360
*MARINE ORGANISMS	1.83*10**15 MOLES	AS ORGANIC O	1360
*OCEANS	210*10**15 MOLES	AS O2	1360
*OCEANS	5900*10**15 MOLES	AS CO2	1360
*OCEANS	70000000*10**15 MOLES	AS H2O	1360
*ORGANIC MATTER	58*10**15 MOLES	IN SOILS, AS ORGANIC O	1360
*ORGANIC MATTER	250*10**15 MOLES	IN MARINA SEDIMENTS, AS ORGANIC O	1360
*SEDIMENTARY ROCKS	5100*10**15 MOLES	AS CARBONATES	1360
*SOIL MOISTURE	3700*10**15 MOLES	AS H2O	1360
*TOTAL ACTIVE POOLS	70000000*10**15 MOLES	ROUNDED TOTAL	1360
*TOTAL STORAGE POOLS	8200000*10**15 MOLES	ROUNDED TOTAL	1360

TABLE B142

ECOSPHERE : POOLS WITHIN THE WATER CYCLE			
ITEM:	VALUE:	CONDITIONS:	REF.:
*ATMOSPHERE	13000 KM**3	PERCENT OF TOTAL = 0.001	1302
*AVERAGE IN STREAM CHANNELS	1250 KM**3	PERCENT OF TOTAL = 0.0001	1302
*FRESHWATER LAKES	125000 KM**3	PERCENT OF TOTAL = 0.009	1302
*GROUND WATER	4170000 KM**3	PERCENT OF TOTAL = 0.31, WITHIN 0.8 KM DEPTH	1302
*GROUND WATER	4170000 KM**3	PERCENT OF TOTAL = 0.31, DEEP-LAYING	1302
*ICECAPS AND GLACIERS	29200000 KM**3	PERCENT OF TOTAL = 2.15	1302
*SALINE LAKES AND INLAND SEAS	104000 KM**3	PERCENT OF TOTAL = 0.008	1302
*SOIL MOISTURE AND INTERMEDIATE ZONE WATER	67000 KM**3	PERCENT OF TOTAL = 0.005	1302
*TOTAL LIQUID WATER IN LAND AREAS	8637000 KM**3	PERCENT OF TOTAL = 0.635	1302
*WORLD OCEANS	1322000000 KM**3	PERCENT OF TOTAL = 97.2	1302

TABLE B143

ECOSPHERE : PRIMARY PRODUCTION			
ITEM:	VALUE:	CONDITIONS:	REF.:
*ALGAL CULTURE	72 KCAL/(M**2*DAY)	EFFECIENCY OF PRIMARY PRODUCTION = 3.0 %	1167
*ARCTIC TUNDRA	1.8 KCAL/(M**2*DAY)	EFFECIENCY OF PRIMARY PRODUCTION = 0.08 %	1167
*ATLANTIC OCEAN	0-24 MG C/(M**2)*DAY	ICE ISLAND T3,MIDSUMMER	1445
*ATLANTIC OCEAN	0-6 MG C/(M**2)*DAY	STATION ALPHA,LATE SUMMER	1445
*ATLANTIC OCEAN	1300 MG C/(M**2)*DAY	W.BARENTS SEA,NEAR BEAR ISLAND, ARCTIC WATER,MAY	1445
*ATLANTIC OCEAN	275 MG C/(M**2)*DAY	W.BARENTS SEA,NEAR BEAR ISLAND, ATLANTIC WATER,MAY	1445
*ATLANTIC OCEAN	2400 MG C/(M**2)*DAY	NORWEGIAN SEA,NEAR SPITZBERGEN, N.ATLANTIC WATER,JUNE	1445
*ATLANTIC OCEAN	400-600 MG C/(M**2)*DAY	NORWEGIAN SEA,NEAR SPITZBERGEN, ARCTIC WATER,JUNE	1445
*ATLANTIC OCEAN	650-2700 MG C/(M**2)*DAY	N.N.ATLANTIC,FAROE-ICELAND RIDGE, SUMMER	1445
*ATLANTIC OCEAN	530-1300 MG C/(M**2)*DAY	N.N.ATLANTIC,NEAR ICELAND,SUMMER	1445
*ATLANTIC OCEAN	550 MG C/(M**2)*DAY	N.N.ATLANTIC,SOUTH OF GREENLAND, SUMMER	1445
*ATLANTIC OCEAN	150-250 MG C/(M**2)*DAY	N.N.ATLANTIC,IRMINGER SEA,SUMMER	1445
*ATLANTIC OCEAN	100-1500 MG C/(M**2)*DAY	NORTH SEA,ANNUAL RANGE	1445
*ATLANTIC OCEAN	220 MG C/(M**2)*DAY	NORTH SEA,N-E.COAST OF ENGLAND,MAY	1445
*ATLANTIC OCEAN	110 MG C/(M**2)*DAY	NORTH SEA,N-E.COAST OF ENGLAND, OCTOBER	1445
*ATLANTIC OCEAN	5 MG C/(M**2)*DAY	NORTH SEA,N-E.COAST OF ENGLAND, FEBRUARY	1445
*ATLANTIC OCEAN	300 MG C/(M**2)*DAY	DANISH COASTAL WATERS,MARCH BLOOM	1445
*ATLANTIC OCEAN	700 MG C/(M**2)*DAY	DANISH COASTAL WATERS,AUGUST(MAX)	1445
*ATLANTIC OCEAN	10 MG C/(M**2)*DAY	DANISH COASTAL WATERS,DECEMBER (MIN)	1445
*ATLANTIC OCEAN	100 MG C/(M**2)*DAY	EAST.N.ATLANTIC,15 MILES OFF OPORTO,SEPTEMBER	1445
*ATLANTIC OCEAN	150 MG C/(M**2)*DAY	EAST.N.ATLANTIC,200 MILES OFF OPORTO,SEPTEMBER	1445
*ATLANTIC OCEAN	30-40 MG C/(M**2)*DAY	MEDITERRANEAN SEA,2 MILES OFF FRENCH COAST,MIDSUMMER	1445
*ATLANTIC OCEAN	1930 MG C/(M**2)*DAY	WEST.N.ATLANTIC,CONTINENTAL SHELF AND INSHORE,SPRING FLOWERING	1445
*ATLANTIC OCEAN	560 MG C/(M**2)*DAY	WEST.N.ATLANTIC,CONTINENTAL SHELF AND INSHORE,MEAN RATE,SPRING	1445

TABLE B143

ECOSPHERE : PRIMARY PRODUCTION			
ITEM:	VALUE:	CONDITIONS:	REF.:
*ATLANTIC OCEAN	330 MG C/(M**2)*DAY	WEST.N.ATLANTIC,CONTINENTAL SHELF AND INSHORE,WEIGHTED ANNUAL MEAN	1445
*ATLANTIC OCEAN	890 MG C/(M**2)*DAY	SARGASSO SEA,APRIL BLOOM	1445
*ATLANTIC OCEAN	100-200 MG C/(M**2)*DAY	SARGASSO SEA,SUMMER	1445
*ATLANTIC OCEAN	100-200 MG C/(M**2)*DAY	CARIBBEAN SEA 10-20N	1445
*ATLANTIC OCEAN	60-160 MG C/(M**2)*DAY	EQUATORIAL ATLANTIC,W.BASIN,NEAR 20N,FEBRUARY	1445
*ATLANTIC OCEAN	230-300 MG C/(M**2)*DAY	EQUATORIAL ATLANTIC,W.BASIN,8 15'N,APRIL-MAY	1445
*ATLANTIC OCEAN	70-280 MG C/(M**2)*DAY	EQUATORIAL ATLANTIC,W.BASIN,8 15'S,MARCH	1445
*ATLANTIC OCEAN	20-130 MG C/(M**2)*DAY	EQUATORIAL ATLANTIC,W.BASIN,15 45'S,APRIL	1445
*ATLANTIC OCEAN	190-780 MG C/(M**2)*DAY	EQUATORIAL ATLANTIC,E.BASIN,NEAR 20N,FEBRUARY	1445
*ATLANTIC OCEAN	180-1480 MG C/(M**2)*DAY	EQUATORIAL ATLANTIC,E.BASIN,8 15'N,APRIL-MAY	1445
*ATLANTIC OCEAN	100-370 MG C/(M**2)*DAY	EQUATORIAL ATLANTIC,E.BASIN,8 15'S,MARCH	1445
*ATLANTIC OCEAN	90-420 MG C/(M**2)*DAY	EQUATORIAL ATLANTIC,E.BASIN,15 45'S,APRIL	1445
*ATLANTIC OCEAN	500-4000 MG C/(M**2)*DAY	S.W.ATLANTIC,WALVIS BAY AND BENGUELA CURRENT,DECEMBER	1445
*CORAL REEF	39 - 151 KCAL/(M**2*DAY)	EFFECIENCY OF PRIMARY PRODUCTION = 2.4 %	1167
*DESERT	1 (RANGE = 0 - 2) MT ORGANIC DRY/(HA*YEAR)	ARID	1679
*DESERT	0.4 KCAL/(M**2*DAY)	EFFECIENCY OF PRIMARY PRODUCTION = 0.05 %	1167
*FRESHWATER EMERGENT MACROPHYTE	38 (RANGE = 30 - 45) MT ORGANIC DRY/(HA*YEAR)	TEMPERATE	1679
*FRESHWATER EMERGENT MACROPHYTE	75 (RANGE = 65 - 85) MT ORGANIC DRY/(HA*YEAR)	TROPICAL	1679
*FRESHWATER PHYTOPLANKTON	2 (RANGE = 1 - 9) MT ORGANIC DRY/(HA*YEAR)	NONE	1679
*FRESHWATER SUBMERSED MACROPHYTE	17 (RANGE = 12 - 20) MT ORGANIC DRY/(HA*YEAR)	TROPICAL	1679
*FRESHWATER SUBMERSED MACROPHYTE	6 (RANGE = 1 - 7) MT ORGANIC DRY/(HA*YEAR)	TEMPERATE	1679
*INDIAN OCEAN	200-250 MG C/(M**2)*DAY	EQUATORIAL INDIAN OCEAN	1445
*MARINE EMERGENT MACROPHYTES	30 (RANGE = 25 - 85) MT ORGANIC DRY/(HA*YEAR)	SALT MARSH	1679
*MARINE PHYTOPLANKTON	2 (RANGE = 1 - 4.5) MT ORGANIC DRY/(HA*YEAR)	NONE	1679
*MARINE PHYTOPLANKTON	311 MG C/(M**2*DAY)	REGION PROVENCALE, MEDITER., CHLOROPHYLL A = 36 MG/M**2	738
*MARINE PHYTOPLANKTON	364 MG C/(M**2*DAY)	REGION COTE D'AZUR-CORSE, MEDITER., CHLOROPHYLL A = 49 MG/ M**2	738
*MARINE PHYTOPLANKTON	1.4 G (RANGE= 0.61 - 3.3) MG C/ (M**2*DAY)	SOUTHWEST AFRICA, N = 10	675
*MARINE PHYTOPLANKTON	4.1 G (RANGE= 1.10 - 9.2) MG C/ (M**2*DAY)	PERU, N = 8	675
*MARINE PHYTOPLANKTON	0.2 G (RANGE= 0.09 -0.35) MG C/ (M**2*DAY)	GULF OF MEXICO, 1968, N = 11	675
*MARINE SUBMERSED MACROPHYTE	29 (RANGE = 25 - 25) MT ORGANIC DRY/(HA*YEAR)	TEMPERATE	1679
*MARINE SUBMERSED MACROPHYTE	29 (RANGE = 30 - 40) MT ORGANIC DRY/(HA*YEAR)	TROPICAL	1679
*PACIFIC OCEAN	500-1000 MG C/(M**2)*DAY	EQUATORIAL PACIFIC,NEAR COAST OF ECUADOR,AUTUMN	1445
*PACIFIC OCEAN	410-800 MG C/(M**2)*DAY	EQUATORIAL PACIFIC,9N,90W,COSTA RICA DOME,NOVEMBER	1445
*PACIFIC OCEAN	10 MG C/(M**2)*DAY	EQUATORIAL PACIFIC,11N,115W,AUTUMN	1445
*PACIFIC OCEAN	100-250 MG C/(M**2)*DAY	EQUATORIAL PACIFIC,30N-30S,MARCH	1445

TABLE B143

ECOSPHERE : PRIMARY PRODUCTION			
ITEM:	VALUE:	CONDITIONS:	REF.:
*PACIFIC OCEAN	50-100 MG C/(M**2)*DAY	N.W.PACIFIC,SEA OF JAPAN,KUROSHIO CURRENT,SUMMER	1445
*PACIFIC OCEAN	250-500 MG C/(M**2)*DAY	N.W.PACIFIC,SEA OF JAPAN,OYASHIO SYSTEM,SUMMER	1445
*PACIFIC OCEAN	6-5100 MG C/(M**2)*DAY	N.W.PACIFIC,SEA OF OKHOTSK,NORTH OF JAPAN,MAY,RANGE	1445
*PACIFIC OCEAN	2000 MG C/(M**2)*DAY	N.W.PACIFIC,SEA OF OKHOTSK,NORTH OF JAPAN,MAY,MEAN	1445
*RAIN FOREST	50 (RANGE = 40 - 60) MT ORGANIC DRY/(HA*YEAR)	NONE	1679
*SUBTROPICAL DEEP OCEAN	2.9 KCAL/(M**2*DAY)	EFFECIENCY OF PRIMARY PRODUCTION = 0.09 %	1167
*SUGAR CANE FIELD	74 KCAL/(M**2*DAY)	EFFECIENCY OF PRIMARY PRODUCTION = 1.8 %	1167
*TEMPERATE ANNUALS	22 (RANGE = 19 - 25) MT ORGANIC DRY/(HA*YEAR)	NONE	1679
*TEMPERATE FOREST	12 (RANGE = 9 - 15) MT ORGANIC DRY/(HA*YEAR)	DECIDUOUS	1679
*TEMPERATE FOREST	28 (RANGE = 21 - 35) MT ORGANIC DRY/(HA*YEAR)	CONIFEROUS	1679
*TEMPERATE HERBS	20 (RANGE = 15 - 25) MT ORGANIC DRY/(HA*YEAR)	NONE	1679
*TROPICAL ANNUALS	30 (RANGE = 24 - 36) MT ORGANIC DRY/(HA*YEAR)	NONE	1679
*TROPICAL FOREST PLANTATION	28 KCAL/(M**2*DAY)	EFFECIENCY OF PRIMARY PRODUCTION = 0.7 %	1167
*TROPICAL MARINE MEADOW	20 - 141 KCAL/(M**2*DAY)	EFFECIENCY OF PRIMARY PRODUCTION = 2.0 %	1167
*TROPICAL RAINFOREST	131 KCAL/(M**2*DAY)	EFFECIENCY OF PRIMARY PRODUCTION = 3.5 %	1167

TABLE B144

ECOSPHERE : PRODUCTION			
ITEM:	VALUE:	CONDITIONS:	REF.:
*FONTINALIS SQUAMOSA	92 GRAM/M**2	PER YEAR ASSUMING ANNUAL BRANCHING PATTERN	337
*PERIPHYTON ON TYPHA	170 MG DRY WEIGHT/(D*M**2)	MAX BIOMASS = 1118 G	990
*POTAMOPHYLAX CINGULATUS	4500 CAL/(YEAR*M**2)	SOUTH SWEDISH STREAM	1192
*SALVELINUS FONTINALIS	50 KG WET WEIGHT/(HA*YEAR)	FLOOD YEAR,FLOOD-DAMAGED STREAM, MINN.	612
*SALVELINUS FONTINALIS	167 KG WET WEIGHT/(HA*YEAR)	RECOVERY YEAR, MINN., 1969, MAX VALUE	612
*SEAWEED	1750 G C/(M**2*YEAR)	IN THE SEAWEED ZONE, NOVA SCOTIA	977
*SEAWEED	648000 G C/(M SHORE LINE)	NOVA SCOTIA	977
*SIMULIUM SP.	1.05 KCAL/(M**2*YEAR)	NET PRODUCTION, LOTIC AREAS	1119

TABLE B145

ECOSPHERE : SECONDARY PRODUCTION			
ITEM:	VALUE:	CONDITIONS:	REF.:
*AFRICAN LAKES	0.4 - 50 KCAL/(M**2*YEAR)	MANS HARVEST, UNFERTILIZED WATER, MIXED CARNIVORES	1161
*GERMAN FISH PONDS	22 - 80 KCAL/(M**2*YEAR)	MANS HARVEST, UNFERTILIZED WATER, STOCED HERBIVORES	1161
*GERMAN FISH PONDS	202 - 336 KCAL/(M**2*YEAR)	MANS HARVEST, ARTIFICIALLY FERT. WATER, STOCED HERBIVORES	1161

TABLE B145

ECOSPHERE : SECONDARY PRODUCTION

ITEM:	VALUE:	CONDITIONS:	REF.:
*GREAT LAKES	0.2 - 1.6 KCAL/(M**2*YEAR)	MANS HARVEST, UNFERTILIZED WATER, MIXED CARNIVORES	1161
*HONG KONG	450 - 900 KCAL/(M**2*YEAR)	MANS HARVEST, OUTSIDE FOOD ADDED, HERBIVORES	1161
*MALAYA	785 KCAL/(M**2*YEAR)	MANS HARVEST, OUTSIDE FOOD ADDED, HERBIVORES	1161
*NORTH SEA	5.0 KCAL/(M**2*YEAR)	MANS HARVEST, UNFERTILIZED WATER, MIXED CARNIVORES	1161
*ONE ACRE POND, USA	450 KCAL/(M**2*YEAR)	MANS HARVEST, OUTSIDE FOOD ADDED, CARNIVORES	1161
*PERU UPWELLING AREA	335 KCAL/(M**2*YEAR)	MANS HARVEST, HEAVY NATURAL FERTILIZED WATER	1161
*PHILIPPINE MARINE PONDS	112 - 202 KCAL/(M**2*YEAR)	MANS HARVEST, ARTIFICIALLY FERT. WATER, STOCED HERBIVORES	1161
*SOUTH CHINA	202 - 3024 KCAL/(M**2*YEAR)	MANS HARVEST, OUTSIDE FOOD ADDED, HERBIVORES	1161
*US FISH PONDS	9 - 34 KCAL/(M**2*YEAR)	MANS HARVEST, UNFERTILIZED WATER, STOCED CARNIVORES	1161
*US FISH PONDS	45 - 112 KCAL/(M**2*YEAR)	MANS HARVEST, ARTIFICIALLY FERT. WATER, STOCED CARNIVORES	1161
*US SMALL LAKES	0.4 - 36 KCAL/(M**2*YEAR)	MANS HARVEST, UNFERTILIZED WATER, MIXED CARNIVORES	1161
*WORLD MARINE FISHERY	0.3 KCAL/(M**2*YEAR)	MANS HARVEST, UNFERTILIZED WATER, MIXED CARNIVORES	1161

TABLE B146

ECOSPHERE : SYNTHESIS OF 1 G OF BIOMASS

ITEM:	VALUE:	CONDITIONS:	REF.:
*CO2	410.35*10**4 MOLES	ACCORDING EQ: CO2+N2+H2O+H2SO4+H3PO4<->BIOMASS+O2	1067
*CO2	410.35*10**4 MOLES	ACCORDING EQ: CO2+HNO3+H2O+H2SO4+H3PO4<->BIOMASS+O2	1067
*CO2	410.35*10**4 MOLES	ACCORDING EQ: CO2+NH3+H2O+H2SO4+H3PO4<->BIOMASS+O2	1067
*CO2	25.53*10**4 MOLES	ACCORDING EQ: C6H12O6+NH3+H2SO4+H3PO4<->BIOMASS+CO2+H2O	1067
*C6H12O6	71.98*10**4 MOLES	ACCORDING EQ: C6H12O6+NH3+H2SO4+H3PO4<->BIOMASS+CO2+H2O	1067
*HNO3	93.28*10**4 MOLES	ACCORDING EQ: CO2+HNO3+H2O+H2SO4+H3PO4<->BIOMASS+O2	1067
*H2O	356.55*10**4 MOLES	ACCORDING EQ: CO2+HNO3+H2O+H2SO4+H3PO4<->BIOMASS+O2	1067
*H2O	398.20*10**4 MOLES	ACCORDING EQ: CO2+N2+H2O+H2SO4+H3PO4<->BIOMASS+O2	1067
*H2O	263.77*10**4 MOLES	ACCORDING EQ: CO2+NH3+H2O+H2SO4+H3PO4<->BIOMASS+O2	1067
*H2O	175.11*10**4 MOLES	ACCORDING EQ: C6H12O6+NH3+H2SO4+H3PO4<->BIOMASS+CO2+H2O	1067
*H2SO4	3.0*10**4 MOLES	ACCORDING EQ: CO2+N2+H2O+H2SO4+H3PO4<->BIOMASS+O2	1067
*H2SO4	3.0*10**4 MOLES	ACCORDING EQ: CO2+HNO3+H2O+H2SO4+H3PO4<->BIOMASS+O2	1067
*H2SO4	3.0*10**4 MOLES	ACCORDING EQ: CO2+NH3+H2O+H2SO4+H3PO4<->BIOMASS+O2	1067
*H2SO4	3.0*10**4 MOLES	ACCORDING EQ: C6H12O6+NH3+H2SO4+H3PO4<->BIOMASS+CO2+H2O	1067
*H3PO4	6.8*10**4 MOLES	ACCORDING EQ: CO2+N2+H2O+H2SO4+H3PO4<->BIOMASS+O2	1067
*H3PO4	6.8*10**4 MOLES	ACCORDING EQ: CO2+HNO3+H2O+H2SO4+H3PO4<->BIOMASS+O2	1067
*H3PO4	6.8*10**4 MOLES	ACCORDING EQ: CO2+NH3+H2O+H2SO4+H3PO4<->BIOMASS+O2	1067

TABLE B146

ECOSPHERE : SYNTHESIS OF 1 G OF BIOMASS			
ITEM:	VALUE:	CONDITIONS:	REF.:
*H3PO4	6.8*10**4 MOLES	ACCORDING EQ: C6H12O6+NH3+H2SO4+ H3PO4<->BIOMASS+CO2+H2O	1067
*NH3	93.28*10**4 MOLES	ACCORDING EQ: CO2+NH3+H2O+H2SO4+ H3PO4<->BIOMASS+O2	1067
*NH3	93.28*10**4 MOLES	ACCORDING EQ: C6H12O6+NH3+H2SO4+ H3PO4<->BIOMASS+CO2+H2O	1067
*N2	46.64*10**4 MOLES	ACCORDING EQ: CO2+N2+H2O+H2SO4+ H3PO4<->BIOMASS+O2	1067
*O2	506.60*10**4 MOLES	ACCORDING EQ: CO2+N2+H2O+H2SO4+ H3PO4<->BIOMASS+O2	1067
*O2	620.69*10**4 MOLES	ACCORDING EQ: CO2+HNO3+H2O+H2SO4+ H3PO4<->BIOMASS+O2	1067
*O2	434.38*10**4 MOLES	ACCORDING EQ: CO2+NH3+H2O+H2SO4+ H3PO4<->BIOMASS+O2	1067

TABLE B147

ECOSPHERE : WATER BALANCE FOR EARTH			
ITEM:	VALUE:	CONDITIONS:	REF.:
*EVAPORATION FROM AFRICA	43 CM/YEAR	NONE	192
*EVAPORATION FROM ASIA	31 CM/YEAR	NONE	192
*EVAPORATION FROM ATLANTIC OCEAN	124 CM/YEAR	NONE	192
*EVAPORATION FROM AUSTRALIA	42 CM/YEAR	NONE	192
*EVAPORATION FROM EUROPA	39 CM/YEAR	NONE	192
*EVAPORATION FROM INDIAN OCEAN	132 CM/YEAR	NONE	192
*EVAPORATION FROM NORTH AMERICA	32 CM/YEAR	NONE	192
*EVAPORATION FROM PACIFIC OCEAN	132 CM/YEAR	NONE	192
*EVAPORATION FROM SOUTH AMERICA	70 CM/YEAR	NONE	192
*EVAPORATION FROM TOTAL LAND AREAS	42 CM/YEAR	NONE	192
*EVAPORATION FROM WORLD OCEANS	126 CM/YEAR	NONE	192
*INFLOW TO ATLANTIC OCEAN	12 CM/YEAR	NONE	192
*INFLOW TO INDIAN OCEAN	7 CM/YEAR	NONE	192
*OUTFLOW FROM PACIFIC OCEAN	8 CM/YEAR	NONE	192
*OUTFLOW FROM WORLD OCEANS	0 CM/YEAR	NONE	192
*PRECIPITATION ON AFRICA	69 CM/YEAR	NONE	192
*PRECIPITATION ON ASIA	60 CM/YEAR	NONE	192
*PRECIPITATION ON ATLANTIC OCEAN	89 CM/YEAR	NONE	192
*PRECIPITATION ON AUSTRALIA	47 CM/YEAR	NONE	192
*PRECIPITATION ON EUROPA	64 CM/YEAR	NONE	192
*PRECIPITATION ON INDIAN OCEAN	117 CM/YEAR	NONE	192
*PRECIPITATION ON NORTH AMERICA	66 CM/YEAR	NONE	192
*PRECIPITATION ON PACIFIC OCEAN	133 CM/YEAR	NONE	192
*PRECIPITATION ON SOUTH AMERICA	163 CM/YEAR	NONE	192
*PRECIPITATION ON TOTAL LAND AREAS	73 CM/YEAR	NONE	192
*PRECIPITATION ON WORLD OCEANS	114 CM/YEAR	NONE	192
*RUNOFF FROM AFRICA	26 CM/YEAR	NONE	192
*RUNOFF FROM ASIA	29 CM/YEAR	NONE	192
*RUNOFF FROM ATLANTIC OCEAN	- 23 CM/YEAR	NONE	192
*RUNOFF FROM AUSTRALIA	5 CM/YEAR	NONE	192
*RUNOFF FROM EUROPA	25 CM/YEAR	NONE	192
*RUNOFF FROM INDIAN OCEAN	- 8 CM/YEAR	NONE	192
*RUNOFF FROM NORTH AMERICA	34 CM/YEAR	NONE	192

TABLE B147

ECOSPHERE : WATER BALANCE FOR EARTH			
ITEM:	VALUE:	CONDITIONS:	REF.:
*RUNOFF FROM PACIFIC OCEAN	- 7 CM/YEAR	NONE	192
*RUNOFF FROM SOUTH AMERICA	93 CM/YEAR	NONE	192
*RUNOFF FROM TOTAL LAND AREAS	31 CM/YEAR	NONE	192
*RUNOFF FROM WORLD OCEANS	- 12 CM/YEAR	NONE	192

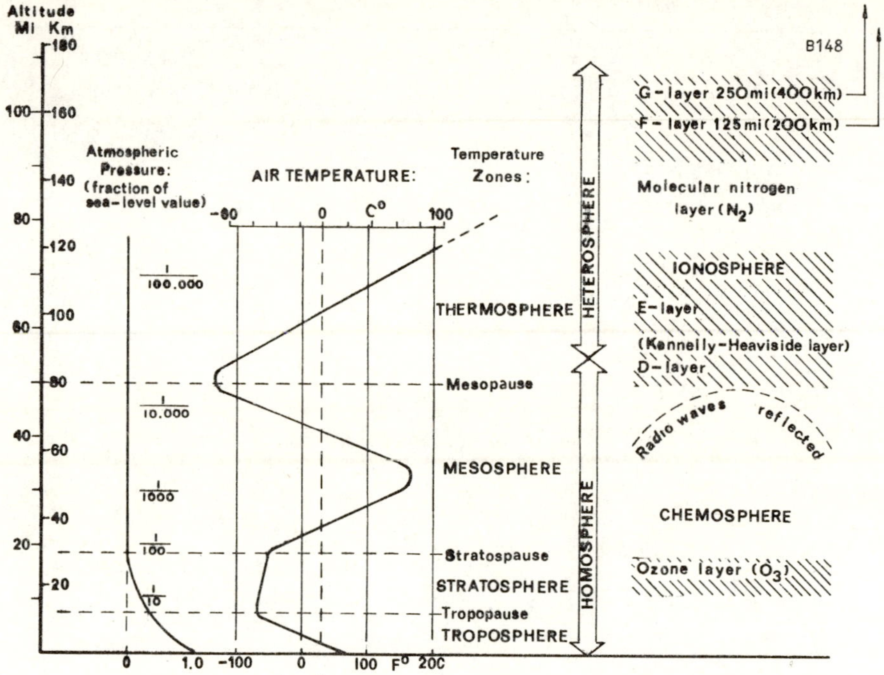

Atmosphere, structure. (ref: 1517)

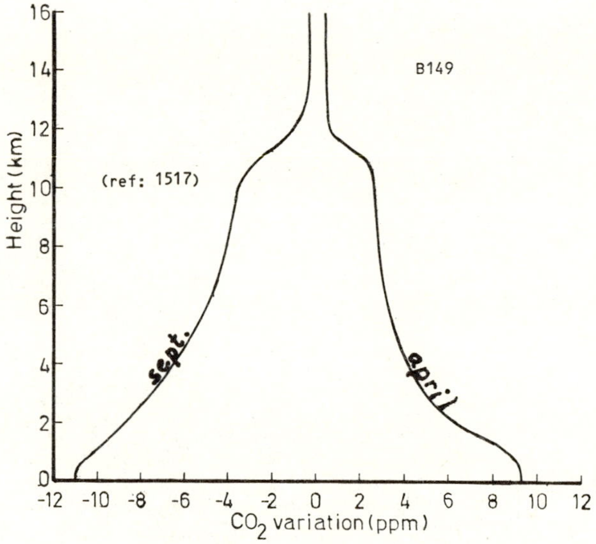

Carbon dioxide content of atmosphere, seasonal variation.

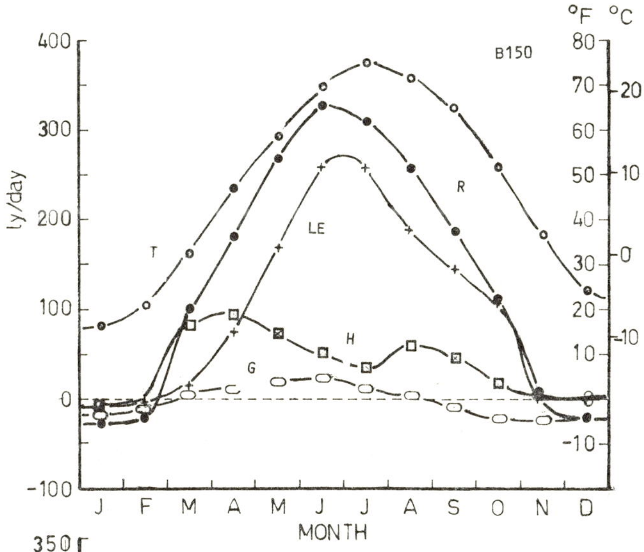

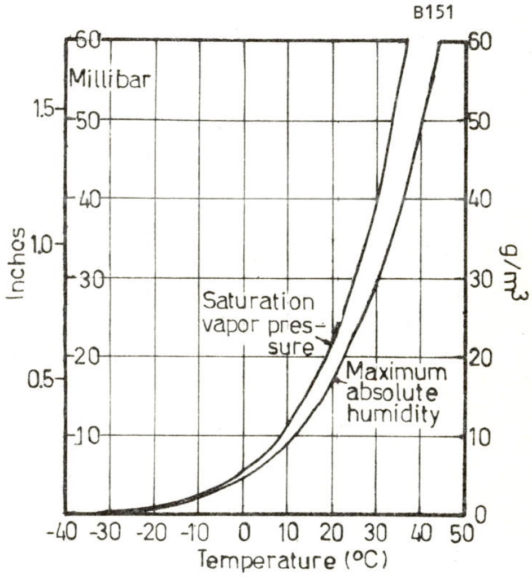

Maximum vapor pressure and absolute humidity.

(ref: 1517)

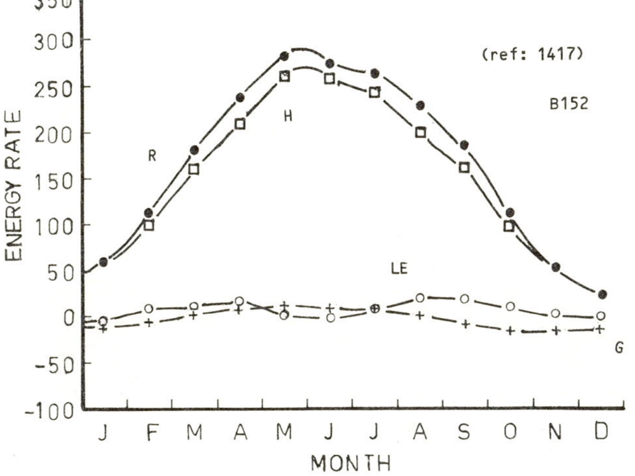

Heat balance, annual cycle in Madison (humid climate in middle lattitudes) and in Yuma (desert in subtropical zone). Symbols see B155

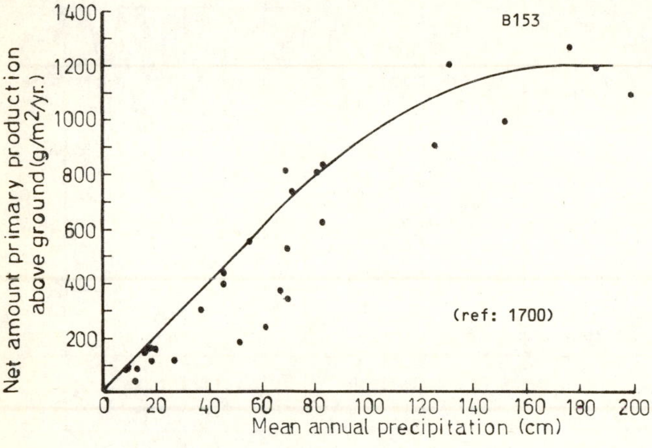

Net annual primary production to rainfall. Peak production of unstable communities are excluded.

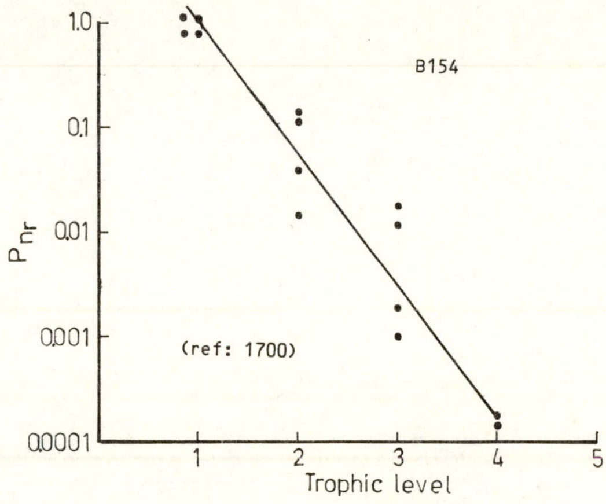

Net productivity of successive trophic levels in 4 aquatic communities relative to primary producers (= 1.0).

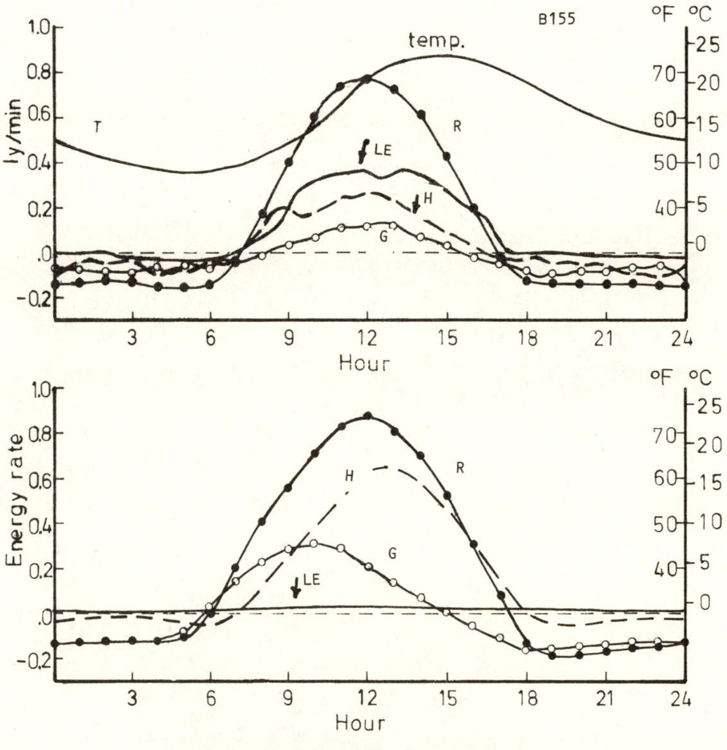

Heat balance, daily cycle in Hanrock Wisconsin (humid climate) and in El Mirage, California (desert climate). (ref: 1417)
T = Temperature. R = Radiation. LE = Latent heat flux (evaporation). H = Sensible heat flux. G = Heat of ground. R = H + LE + G.

B156

(ref: 691)

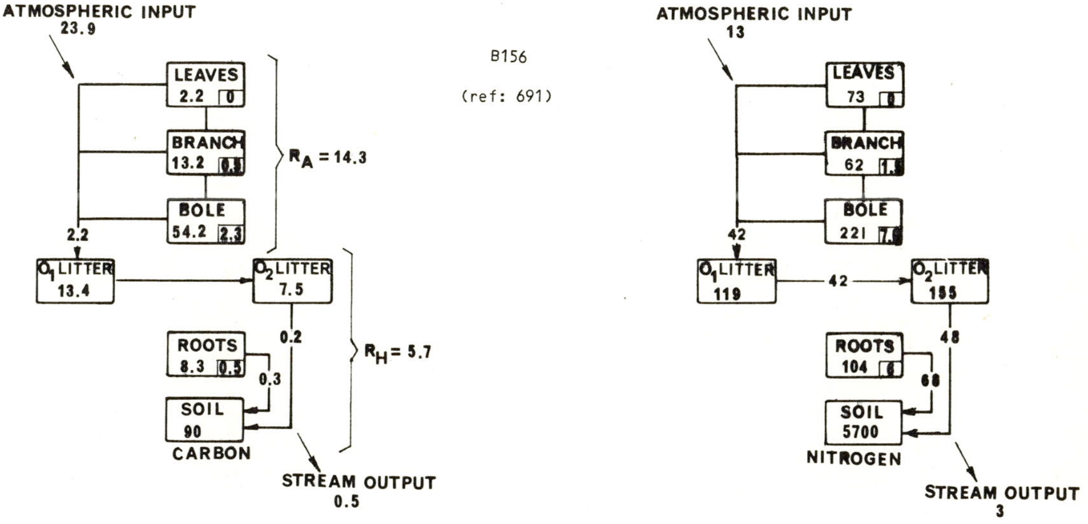

Carbon and Nitrogen cycles in a forest.
C (left) and N (right). Value in boxes
10**3 kg/ha, annual fluxes (on arrows) in
10**3 kg/(ha*year) and annual increments
(lower rigth corner of boxes) in
10**3 kg/ha. Rh = Heterotrophic respiration,
Ra = Autotrophic respiration. Eastern
Tennesee mixed deciduous forest.

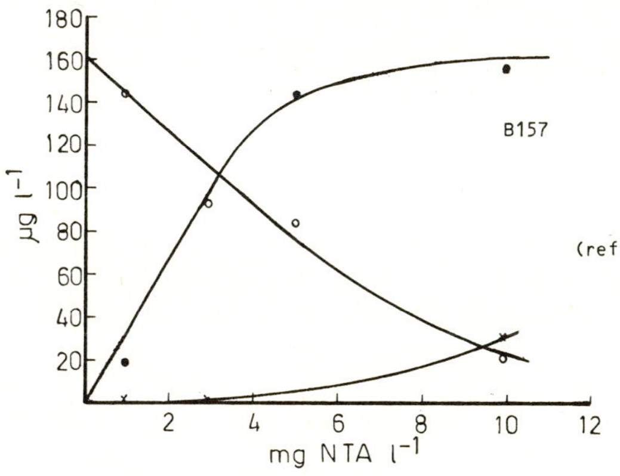

Leaching of Cu, Fe and Mn with NTA.
× Mn • Fe ○ Cu. 200 g wet sediment/l.

(ref: 1386)

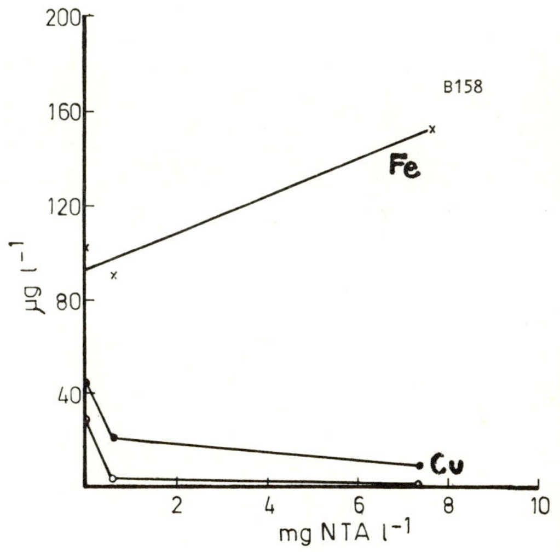

Leaching of Cu and Fe with NTA, 270 g sediment (wet)/l.

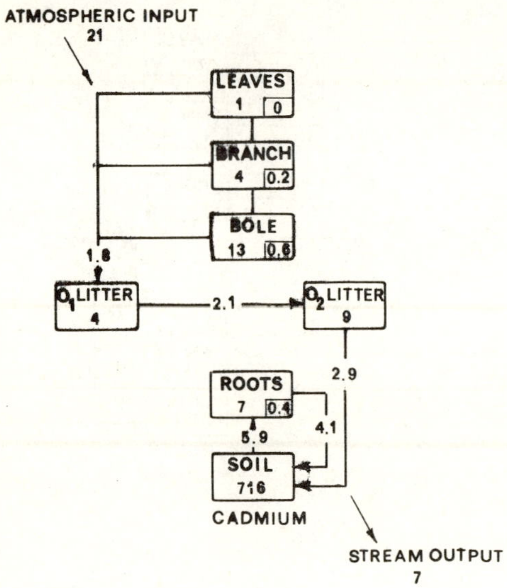

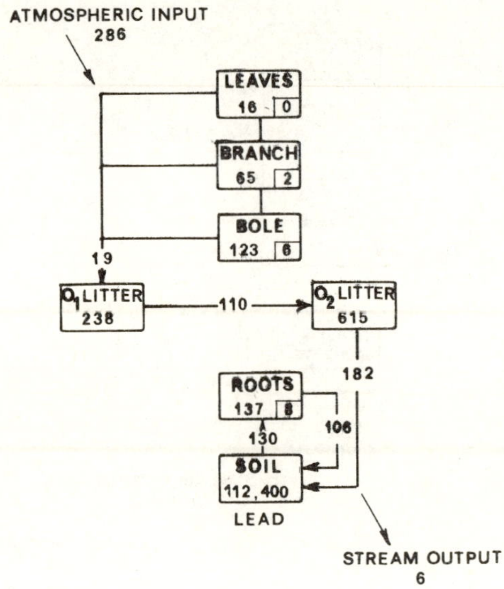

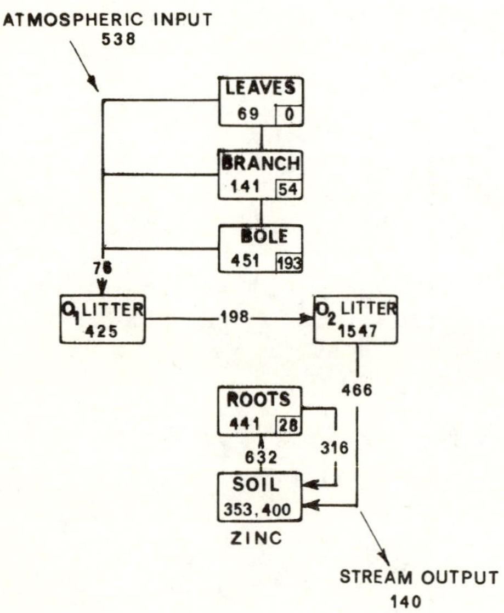

B159
(ref: 691)

Cadmium, Lead and Zinc cycles in a forest. Standing pool g/ha, annual fluxes (on arrows) g/(ha*year) and annual increments g/ha (lower right corner of boxes). Eastern Tennesee mixed deciduous forest.

SR-90 IN PERCH LAKE FOOD WEB

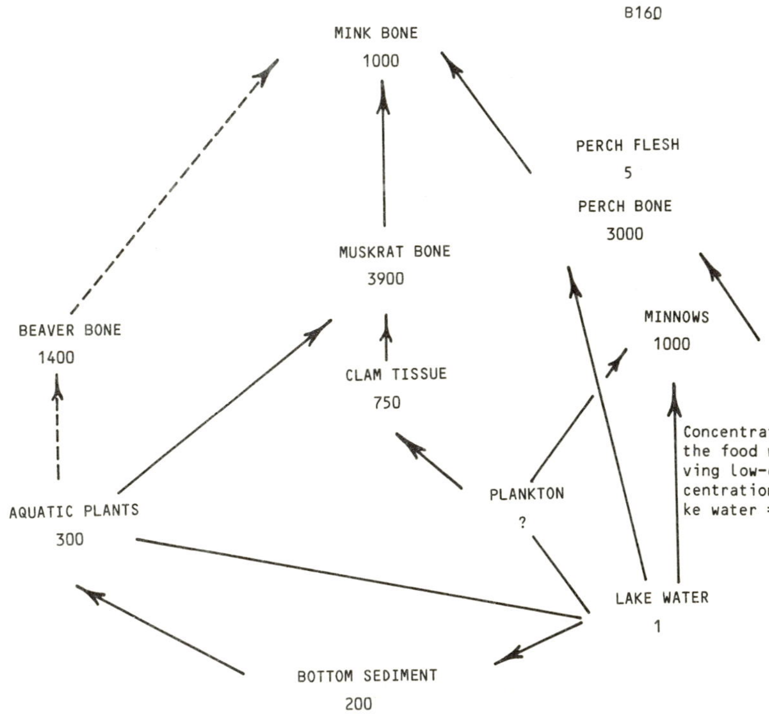

Concentrations of Sr-90 in various parts of the food web of a small Canadian lake reciving low-clevel atmoic wastes. Average concen centration factors are shown relative to lake water = 1.

(ref: 1185)

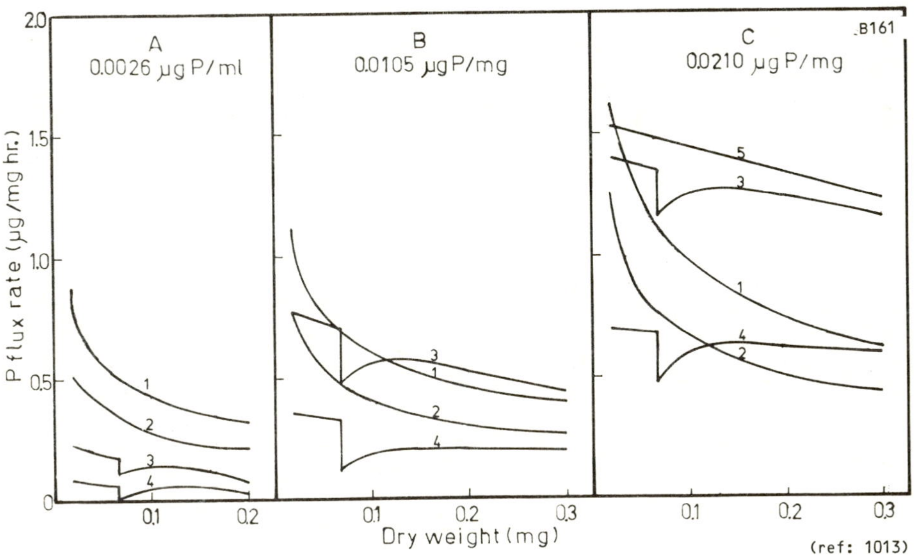

(ref: 1013)

Comparison of excretion rate of Daphnia rosea feeding on Rhodotorula, at 10 and 20 °C.

A: $0.125*10^5$ cells/ml.
1: excretion rate at 20 °C .
2: excretion rate at 10 °C.
3: ingestion - (growth + reproduction).
4: assimilation - (growth + reproduction).

B: $0.5*10^5$ cells/ml.
1. excretion at 20 °C.
2: excretion at 10 °C.
3: ingestion - (growth + reproduction).
4: assimilation - (growth + reproduction).

C: $1*10^5$ cells/ml.
1: excretion at 20 °C.
2. excretion at 10 °C.
3: ingestion - (growth + reproduction).
4: assimilation - (growth + reproduction).
5: ingestion.

B3

LIGHT

ABSORPTION
ALBEDO
PENETRATION
PRIMARY PRODUCTION
REFLECTION
SPECTRA

TABLE B162

LIGHT : ABSORPTION			
ITEM:	VALUE:	CONDITIONS:	REF.:
*MONOCHROMATIC LIGHT	91.1 %	EXP.ATTENUATION IN 1M WATER, WAVELENGTH=820NM,EXT.COEFF.=2.42	706
*MONOCHROMATIC LIGHT	89.4 %	EXP.ATTENUATION IN 1M WATER, WAVELENGTH=800NM,EXT.COEFF.=2.24	706
*MONOCHROMATIC LIGHT	90.1 %	EXP.ATTENUATION IN 1M WATER, WAVELENGTH=780NM,EXT.COEFF.=2.31	706
*MONOCHROMATIC LIGHT	91.4 %	EXP.ATTENUATION IN 1M WATER, WAVELENGTH=760NM,EXT.COEFF.=2.45	706
*MONOCHROMATIC LIGHT	88.5 %	EXP.ATTENUATION IN 1M WATER, WAVELENGTH=740NM,EXT.COEFF.=2.16	706
*MONOCHROMATIC LIGHT	64.5 %	EXP.ATTENUATION IN 1M WATER, WAVELENGTH=720NM,EXT.COEFF.=1.04	706
*MONOCHROMATIC LIGHT	45.0 %	EXP.ATTENUATION IN 1M WATER, WAVELENGTH=700NM,EXT.COEFF.=0.598	706
*MONOCHROMATIC LIGHT	36.6 %	EXP.ATTENUATION IN 1M WATER, WAVELENGTH=680NM,EXT.COEFF.=0.455	706
*MONOCHROMATIC LIGHT	31.0 %	EXP.ATTENUATION IN 1M WATER, WAVELENGTH=660NM,EXT.COEFF.=0.370	706
*MONOCHROMATIC LIGHT	26.6 %	EXP.ATTENUATION IN 1M WATER, WAVELENGTH=640NM,EXT.COEFF.=0.310	706
*MONOCHROMATIC LIGHT	23.5 %	EXP.ATTENUATION IN 1M WATER, WAVELENGTH=620NM,EXT.COEFF.=0.273	706
*MONOCHROMATIC LIGHT	19.0 %	EXP.ATTENUATION IN 1M WATER, WAVELENGTH=600NM,EXT.COEFF.=0.210	706
*MONOCHROMATIC LIGHT	7.0 %	EXP.ATTENUATION IN 1M WATER, WAVELENGTH=580NM,EXT.COEFF.=0.078	706
*MONOCHROMATIC LIGHT	3.9 %	EXP.ATTENUATION IN 1M WATER, WAVELENGTH=560NM,EXT.COEFF.=0.040	706
*MONOCHROMATIC LIGHT	3.0 %	EXP.ATTENUATION IN 1M WATER, WAVELENGTH=540NM,EXT.COEFF.=0.030	706
*MONOCHROMATIC LIGHT	1.6 %	EXP.ATTENUATION IN 1M WATER, WAVELENGTH=520NM,EXT.COEFF.=0.016	706
*MONOCHROMATIC LIGHT	0.77 %	EXP.ATTENUATION IN 1M WATER, WAVELENGTH=500NM,EXT.COEFF.=0.0075	706
*MONOCHROMATIC LIGHT	0.52 %	EXP.ATTENUATION IN 1M WATER, WAVELENGTH=480NM,EXT.COEFF.=0.0050	706
*MONOCHROMATIC LIGHT	0.52 %	EXP.ATTENUATION IN 1M WATER, WAVELENGTH=460NM,EXT.COEFF.=0.0054	706
*MONOCHROMATIC LIGHT	0.70 %	EXP.ATTENUATION IN 1M WATER, WAVELENGTH=440NM,EXT.COEFF.=0.0078	706
*MONOCHROMATIC LIGHT	0.92 %	EXP.ATTENUATION IN 1M WATER, WAVELENGTH=420NM,EXT.COEFF.=0.0088	706
*MONOCHROMATIC LIGHT	1.63 %	EXP.ATTENUATION IN 1M WATER, WAVELENGTH=400NM,EXT.COEFF.=0.0134	706
*MONOCHROMATIC LIGHT	2.10 %	EXP.ATTENUATION IN 1M WATER, WAVELENGTH=380NM,EXT.COEFF.=0.0255	706

TABLE B163

LIGHT : ALBEDO			
ITEM:	VALUE:	CONDITIONS:	REF.:
*BARE GROUND	10 - 20 %	% OF TOTAL INCOMING RADIATION, WITH DIFFUSE REFLECTION	934
*DARK CULTIVATED SOIL	7 - 10 %	% OF TOTAL INCOMING RADIATION, WITH DIFFUSE REFLECTION	934
*FOREST, GREEN	3 - 10 %	% OF TOTAL INCOMING RADIATION, WITH DIFFUSE REFLECTION	934
*FOREST, SNOW-COVERED GROUND	10 - 25 %	% OF TOTAL INCOMING RADIATION, WITH DIFFUSE REFLECTION	934
*FRESH SNOW	87 %	% OF TOTAL INCOMING RADIATION, WITH DIFFUSE REFLECTION	934
*FRESH SNOW COVER	75 - 95 %	% OF TOTAL INCOMING RADIATION, WITH DIFFUSE REFLECTION	934

TABLE B163

LIGHT : ALBEDO			
ITEM:	VALUE:	CONDITIONS:	REF.:
*GRASS, DRY	15 - 25 %	% OF TOTAL INCOMING RADIATION, WITH DIFFUSE REFLECTION	934
*GRASS, HIGH DRY	31 - 33 %	% OF TOTAL INCOMING RADIATION, WITH DIFFUSE REFLECTION	934
*GRASS, HIGH FRESH	26 %	% OF TOTAL INCOMING RADIATION, WITH DIFFUSE REFLECTION	934
*LIGHT SAND DUNES, SURF	30 - 60 %	% OF TOTAL INCOMING RADIATION, WITH DIFFUSE REFLECTION	934
*MEADOWS AND FIELDS	12 - 30 %	% OF TOTAL INCOMING RADIATION, WITH DIFFUSE REFLECTION	934
*OLD SNOW COVER	40 - 70 %	% OF TOTAL INCOMING RADIATION, WITH DIFFUSE REFLECTION	934
*SANDY SOIL	15 - 40 %	% OF TOTAL INCOMING RADIATION, WITH DIFFUSE REFLECTION	934
*SEA WATER SURFACE	3 - 10 %	% OF TOTAL INCOMING RADIATION, WITH DIFFUSE REFLECTION	934
*SNOW, OLD	46 %	% OF TOTAL INCOMING RADIATION, WITH DIFFUSE REFLECTION	934
*SNOW, SEVERAL DAYS OLD	70 %	% OF TOTAL INCOMING RADIATION, WITH DIFFUSE REFLECTION	934
*WOODS	5 - 20 %	% OF TOTAL INCOMING RADIATION, WITH DIFFUSE REFLECTION	934

TABLE B164

LIGHT : PENETRATION OF LIGHT			
ITEM:	VALUE:	CONDITIONS:	REF.:
*CLEAR ICE	72 %	THICKNESS = 43 CM, % TRANSMISSION OF SURFACE INSOLATION	10
*CLEAR ICE WITH VESTIGE SNOW	53 %	THICKNESS = 39 CM, % TRANSMISSION OF SURFACE INSOLATION	10
*COMPACTED OLD SNOW	1 - 5 %	THICKNESS = 17-20 CM, % TRANSMISSION OF SURFACE INSOLATION	10
*ICE WITH IRREGULAR SURFACE	58 %	THICKNESS = 29 CM, % TRANSMISSION OF SURFACE INSOLATION	10
*MILKY ICE WITH BUBBLES	54 %	THICKNESS = 29 CM, % TRANSMISSION OF SURFACE INSOLATION	10
*NEW SNOW	34 %	THICKNESS = 0.5 CM, % TRANSMISSION OF SURFACE INSOLATION	10
*NEW SNOW	20 %	THICKNESS = 5.0 CM, % TRANSMISSION OF SURFACE INSOLATION	10
*NEW SNOW	9 %	THICKNESS = 10 CM, % TRANSMISSION OF SURFACE INSOLATION	10
*NEW SNOW	8.8 - 6.7 %	THICKNESS = 17-20 CM, % TRANSMISSION OF SURFACE INSOLATION	10
*TRANSLUCENT ICE	11 - 18 %	THICKNESS = 25 CM, % TRANSMISSION OF SURFACE INSOLATION, SNOW ICE	10
*WET ICE WITH BUBBLES	41 %	THICKNESS = 39 CM, % TRANSMISSION OF SURFACE INSOLATION	10

TABLE B165

LIGHT : PLANETARY RADIATION			
ITEM:	VALUE:	CONDITIONS:	REF.:
*ABSORPTION OF ENERGY BY CLOUDS	3 % OF INCOMING RADIATION	INCOMING, TOTAL INPUT EQUAL TO 263 KILOLANGLEYS/Y	1417
*ABSORPTION OF ENERGY BY EARTH, LAND AND WATER SURFACE	50 % OF INCOMING RADIATION	INCOMING, TOTAL INPUT EQUAL TO 263 KILOLANGLEYS/Y	1417

TABLE B165

LIGHT : PLANETARY RADIATION			
ITEM:	VALUE:	CONDITIONS:	REF.:
*ABSORPTION OF ENERGY BY MOLECULES	15 % OF INCOMING RADIATION	INCOMING, TOTAL INPUT EQUAL TO 263 KILOLANGLEYS/Y	1417
*DIFFUSE REFLECTION DU TO RAYLEIGHT SCATTER AND DUST	5 % OF INCOMING RADIATION	INCOMING, TOTAL INPUT EQUAL TO 263 KILOLANGLEYS/Y	1417
*DIRECT REFLECTION FROM EARTHS SURFACE TO SPACE	6 % OF INCOMING RADIATION	INCOMING, TOTAL INPUT EQUAL TO 263 KILOLANGLEYS/Y	1417
*NET OUTGOING RADIATION FROM ATMOSPHERE	47 % OF INCOMING RADIATION	OUTGOING, TOTAL INPUT EQUAL TO 263 KILOLANGLEYS/Y	1417
*NET OUTGOING RADIATION FROM EARTH SURFACE	21 % OF INCOMING RADIATION	OUTGOING, TOTAL INPUT EQUAL TO 263 KILOLANGLEYS/Y	1417
*NET OUTGOING RADIATION FROM EARTH-ATMOSPHERE SYSTEM	68 % OF INCOMING RADIATION	OUTGOING, TOTAL INPUT EQUAL TO 263 KILOLANGLEYS/Y	1417
*RADIATION EMITTED BY ATMOSPHERE	137 % OF INCOMING RADIATION	OUTGOING, TOTAL INPUT EQUAL TO 263 KILOLANGLEYS/Y	1417
*RADIATION FROM ATMOSPHERE LOST TO SPACE	60 % OF INCOMING RADIATION	OUTGOING, TOTAL INPUT EQUAL TO 263 KILOLANGLEYS/Y	1417
*RADIATION FROM ATMOSPHERE RETURNED TO EARTH SURFACE	77 % OF INCOMING RADIATION	OUTGOING, TOTAL INPUT EQUAL TO 263 KILOLANGLEYS/Y	1417
*RADIATION FROM EARTH ABSORBED BY ATMOSPHERE	90 % OF INCOMING RADIATION	OUTGOING, TOTAL INPUT EQUAL TO 263 KILOLANGLEYS/Y	1417
*RADIATION FROM EARTH AURFACE LOST TO SPACE DIRECTLY	8 % OF INCOMING RADIATION	OUTGOING, TOTAL INPUT EQUAL TO 263 KILOLANGLEYS/Y	1417
*RADIATION FROM EARTHS LAND-AND-WATER SURFACE	98 % OF INCOMING RADIATION	OUTGOING, TOTAL INPUT EQUAL TO 263 KILOLANGLEYS/Y	1417
*REFLECTION FROM CLUDS TO SPACE	21 % OF INCOMING RADIATION	INCOMING, TOTAL INPUT EQUAL TO 263 KILOLANGLEYS/Y	1417
*TOTAL ABSORBED BY EARTH-ATMOSPHERE SYSTEM	68 % OF INCOMING RADIATION	INCOMING, TOTAL INPUT EQUAL TO 263 KILOLANGLEYS/Y	1417
*TOTAL AT TOP OF ATMOSPHERE	100 % OF INCOMING RADIATION	INCOMING, TOTAL INPUT EQUAL TO 263 KILOLANGLEYS/Y	1417
*TOTAL REFLECTION FROM EARTH TO SPACE	32 % OF INCOMING RADIATION	INCOMING, TOTAL INPUT EQUAL TO 263 KILOLANGLEYS/Y	1417

TABLE B166

LIGHT : REFLECTION			
ITEM:	VALUE:	CONDITIONS:	REF.:
*CAT FUR	40 (20 - 60) %	WAVE BAND=0.3-0.7 MICRONS	508
*CAT FUR	65 (60 - 70) %	WAVE BOND=0.7-3.0 MICRONS	508
*CAT FUR	5 %	WAVE BAND=8-12 MICRONS	508
*FUNGI	10 - 35 %	WAVE BAND=0.3-0.7 MICRONS	508
*FUNGI	60 (35 - 80) %	WAVE BOND=0.7-3.0 MICRONS	508
*FUNGI	10 %	WAVE BAND=8-12 MICRONS	508
*HOMO SAPIENS, SKIN	25 (5 - 45) %	WAVE BAND=0.3-0.7 MICRONS	508
*HOMO SAPIENS, SKIN	30 (3- 50) %	WAVE BOND=0.7-3.0 MICRONS	508
*LIVERWORT	7 %	WAVE BAND=0.3-0.7 MICRONS	508
*LIVERWORT	45 %	WAVE BOND=0.7-3.0 MICRONS	508
*LIVERWORT	4 %	WAVE BAND=8-12 MICRONS	508
*LIZARD	42 %	WAVE BAND=0.4-0.7 MICRONS, VENTRAL	508
*LIZARD	12 %	WAVE BAND=0.4-0.7 MICRONS, DORSAL	508
*LIZARD	30 %	WAVE BAND=0.7-1.1 MICRONS, VENTRAL	508
*LIZARD	9 %	WAVE BAND=0.7-1.1 MICRONS, DORSAL	508
*PIKA FUR, BACK	10 (5 - 15) %	WAVE BAND=0.3-0.7 MICRONS	508
*PIKA FUR, BACK	40 (20 - 60) %	WAVE BOND=0.7-3.0 MICRONS	508
*PIKA FUR, BACK	5 %	WAVE BAND=8-12 MICRONS	508
*POPULUS DELTOIDES, LEAF	7 %	WAVE BAND=0.3-0.7 MICRONS	508

TABLE B166

LIGHT : REFLECTION			
ITEM:	VALUE:	CONDITIONS:	REF.:
*POPULUS DELTOIDES, LEAF	35 %	WAVE BOND=0.7-3.0 MICRONS	508
*STELLAR JAY FEATHERS, BACK	3 %	WAVE BAND=0.3-0.7 MICRONS	508
*STELLAR JAY FEATHERS, BACK	40 (5 - 50) %	WAVE BOND=0.7-3.0 MICRONS	508

TABLE B167

LIGHT : REFLECTION FROM WATER SURFACE			
ITEM:	VALUE:	CONDITIONS:	REF.:
*DIFFUSE SKY LIGHT	17 %	SOLAR ELEVATION = 5 DEGREES	550
*DIFFUSE SKY LIGHT	15 %	SOLAR ELEVATION = 10 DEGREES	550
*DIFFUSE SKY LIGHT	14 %	SOLAR ELEVATION = 15 DEGREES	550
*DIFFUSE SKY LIGHT	13 %	SOLAR ELEVATION = 20 DEGREES	550
*DIFFUSE SKY LIGHT	11 %	SOLAR ELEVATION = 30 DEGREES	550
*DIFFUSE SKY LIGHT	9 %	SOLAR ELEVATION = 40 DEGREES	550
*DIFFUSE SKY LIGHT	7 %	SOLAR ELEVATION = 50 DEGREES	550
*DIRECT SUNLIGHT	58 %	SOLAR ELEVATION = 5 DEGREES	550
*DIRECT SUNLIGHT	35 %	SOLAR ELEVATION = 10 DEGREES	550
*DIRECT SUNLIGHT	21 %	SOLAR ELEVATION = 15 DEGREES	550
*DIRECT SUNLIGHT	13 %	SOLAR ELEVATION = 20 DEGREES	550
*DIRECT SUNLIGHT	6 %	SOLAR ELEVATION = 30 DEGREES	550
*DIRECT SUNLIGHT	3.5 %	SOLAR ELEVATION = 3.5 DEGREES	550
*DIRECT SUNLIGHT	2 %	SOLAR ELEVATION = 50 DEGREES	550

TABLE B168

Solar Spectral Irradiance Data. The wavelength l is in microns, the mean zero air mass spectral-irradiance H is in watts per square centimeter per micron, and P is the percentage of the solar constant associated with wavelengths shorter than wavelength l.
REF.: 777

l	H	P
0.22	0.003	0.02
0.225	.0042	0.03
0.23	.0052	0.05
0.24	.0058	0.09
0.245	.0064	0.11
0.25	.0064	0.13
0.255	0.01	0.16
0.26	0.013	0.2
0.265	0.02	0.27
0.27	0.025	0.34
0.275	0.022	0.43
0.28	0.024	0.51
0.285	0.034	0.62
0.29	0.052	0.77
0.295	0.063	0.98
0.3	0.061	1.23
0.305	0.067	1.43
0.31	0.087	1.69
0.315	0.082	1.97
0.32	0.085	2.26
0.325	0.102	2.6
0.33	0.115	3.02
0.335	0.111	3.4
0.34	0.111	3.8
0.345	0.117	4.21
0.35	0.118	4.63
0.355	0.116	5.04
0.36	0.116	5.47
0.365	0.129	5.89
0.37	0.133	6.36
0.38	0.123	7.29
0.385	0.115	7.72
0.39	0.112	8.13
0.395	0.12	8.54
0.4	0.154	9.03
0.405	0.188	9.65
0.41	0.194	10.3
0.415	0.192	11
0.42	0.192	11.7
0.425	0.189	12.4
0.43	0.178	13
0.435	0.182	13.7
0.44	0.203	14.4
0.445	0.215	15.1
0.46	0.216	17.5
0.465	0.215	18.2
0.47	0.217	19
0.475	0.22	19.8
0.48	0.216	20.6
0.485	0.203	21.3
0.49	0.199	22
0.495	0.204	22.8
0.5	0.198	23.5
0.505	0.197	24.2
0.51	0.196	24.9
0.515	0.189	25.6
0.52	0.187	26.3
0.525	0.192	26.9
0.53	0.195	27.6
0.535	0.197	28.3
0.54	0.198	29
0.545	0.198	29.8
0.55	0.195	30.5
0.555	0.192	31.2
0.56	0.19	31.8
0.565	0.189	32.5
0.57	0.187	33.2
0.575	0.187	33.9
0.58	0.187	34.5
0.585	0.185	35.2
0.59	0.184	35.9
0.595	0.183	36.5
0.6	0.181	37.2
0.61	0.177	38.4
0.62	0.174	39.7
0.63	0.17	40.9
0.64	0.166	42.1
0.65	0.162	43.3
0.66	0.159	44.5
0.67	0.155	45.6
0.68	0.151	46.7
0.69	0.148	47.8
0.7	0.144	48.8
0.71	0.141	49.8
0.72	0.137	50.8
0.73	0.134	51.8
0.74	0.13	52.7
0.75	0.127	53.7
0.8	.1127	57.9
0.85	.1003	61.7
0.9	.0895	65.1
0.95	.0803	68.1
1	.0725	70.9
1.1	.0606	75.7
1.2	.0501	79.6
1.3	.0406	82.9
1.4	.0328	85.5
1.5	.0267	87.6
1.6	0.011	89.4
1.7	.0182	90.83
1.8	.0152	92.03
1.9	.0127	93.02
2	.0118	93.87
2.1	.0092	94.58
2.2	.0078	95.2
2.3	.0068	95.71
2.4	.0058	96.18
2.5	.0051	96.57
2.6	.0044	96.9
2.7	.0039	97.21
2.8	.0034	97.47
2.9	0.003	97.72
3	.0027	97.9
3.1	.0023	98.08
3.2	.0021	98.24
3.3	.0019	98.39
3.4	.0017	98.52
3.5	.0015	98.63
3.6	.0014	98.74
3.7	.0012	98.85
3.8	.0011	98.91
3.9	0.001	98.99
4	.0009	99.05
4.1	.0009	99.13
4.2	.0008	99.18
4.3	.0007	99.33
4.4	.0007	99.29
4.5	.0006	99.23
4.6	.0006	99.38
4.7	.0005	99.41
4.8	.0005	99.45
4.9	.0004	99.48
5	.0004	99.51
6	.0021	99.74
7	.0001	99.86

TABLE B169

Units and Conversion Factors for Solar Radiation.

REF.: 1¤89

Unit	Unit
1 watt	1 joule/s
1 watt/cm2	1 joule/s/cm2
1 watt	14.3 cal/min
1 watt/cm2	14.3 langley/min
1 ly/min	0.0698 watt/cm2
1 ft-candle	10.764 lux
1 lux	0.0929 ft-candle
1 ly	2.11E15*A quanta/cm2 (A=wavelength in Angstroem)
1 ly/min	2.11E15*A quanta/cm2/min
1 ly/min	3.50E-9 einstein/cm2/min

TABLE B170

Units and Conversion Factors for Radient Flux Density and Irradiance Units. Wavelength L is in cm. Q denotes number of quanta.
REF.: 1689

Unit	Q/cm2/s	Einstein /cm2/s	erg/cm2/s	cal/cm2/s	cal/cm2/mn	cal/cm2/h	cal/dm2/h	kcal/m2/h	W/cm2	W/m2	BTU/ft2/mn
Q/cm2/s	1	1.66E-24	1.99E-16/L	4.75E-24/L	2.85E-22/L	1.71E-20/L	1.71E-18/L	1.71E-19/L	1.99E-23/L	1.99E-19/L	1.05E-21/L
E/cm2/s	6.02E23	1	1.20E8/L	2.86/L	1.72E2/L	1.03E4/L	1.03E6/L	1.03E5/L	12.0/L	1.20E5/L	6.34E2/L
rg/cm2/s	5.03E15*L	8.35E-9*L	1	2.39E-8	1.43E-6	8.6E-5	8.6E-3	8.6E-4	1E-7	1E-3	5.28E-6
cal/cm2/s	2.11E23*L	0.350*L	4.19E7	1	60	3.6E3	3.6E5	3.6E4	4.19	4.19E4	221
cal/cm2/mn	3.51E21*L	5.83E-3*L	6.98E5	1.67E-2	1	60	6000	600	6.98E-2	698	3.69
cal/cm2/h	5.85E19*L	9.71E-5*L	1.16E4	2.78E-4	1.67E-2	1	100	10	1.16E-3	11.63	6.17E-2
cal/dm2/h	5.85E17*L	9.71E-7*L	116	2.78E-6	1.67E-4	.01	1	.1	1.16E-5	.116	6.17E-4
kcal/m2/h	5.85E18*L	9.71E-6*L	1160	2.78E-5	1.67E-3	.1	10	1	1.16E-4	1.163	6.17E-3
W/cm2	5.03E22*L	8.35E-2*L	1E7	.239	14.33	860	8.6E-4	8.6E-3	1	1E4	52.8
W/m2	5.03E18*L	8.35E-6*L	1000	2.39E-5	1.43E-3	8.6E-2	8.6	.86	1E-4	1	5.28E-3
BTU/ft2/mn	0.95E21*L	1.58E-3*L	1.89E5	4.53E-3	0.27	16.2	1.62E3	162	1.89E-2	189	1

TABLE B171

The Electromagnetic Radiation Spectrum.

REF.: 1517

Radiation type	Wavelength (microns)	% of total solar energy
X rays and gamma rays	1/2000 - 1/100	9
Ultraviolet rays	0.2 - 0.4	small
Visible light rays	0.4 - 0.7	41
Infrared rays	0.7 - 3000	50

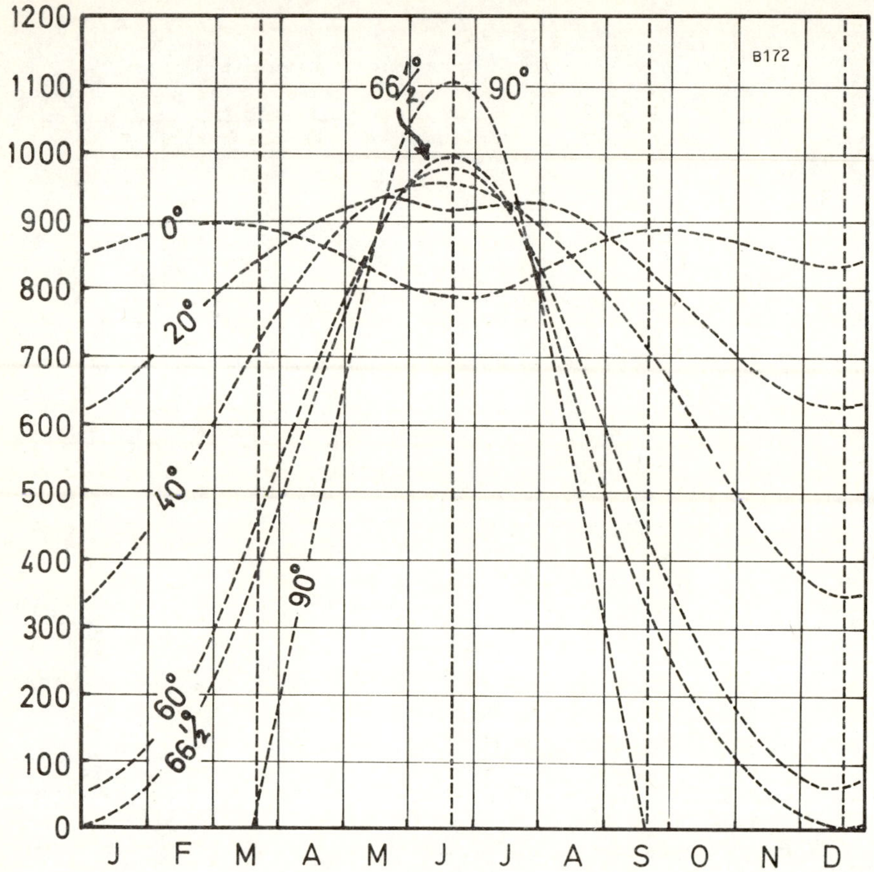

Sesonal variations of insolation versus latitude.

(ref: 1517)

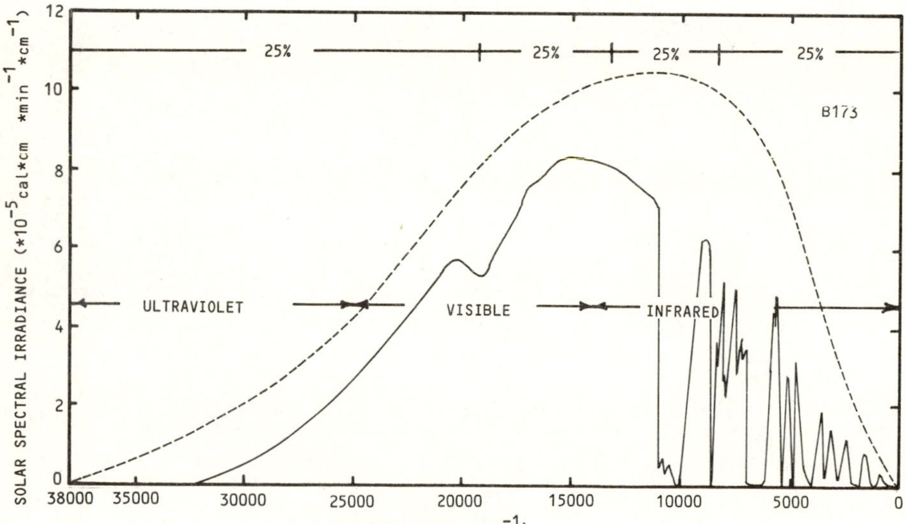

Solar spectrum at the mean solar distance from the earth as recived outside of earths atmosphere and at sea level on a surface perpendicular to the solar rays (solar constant of 2.00 cal/(cm**2*min**1*cm**1).
Extraterrestrial intensity
Intensity at sea level

(ref: 507)

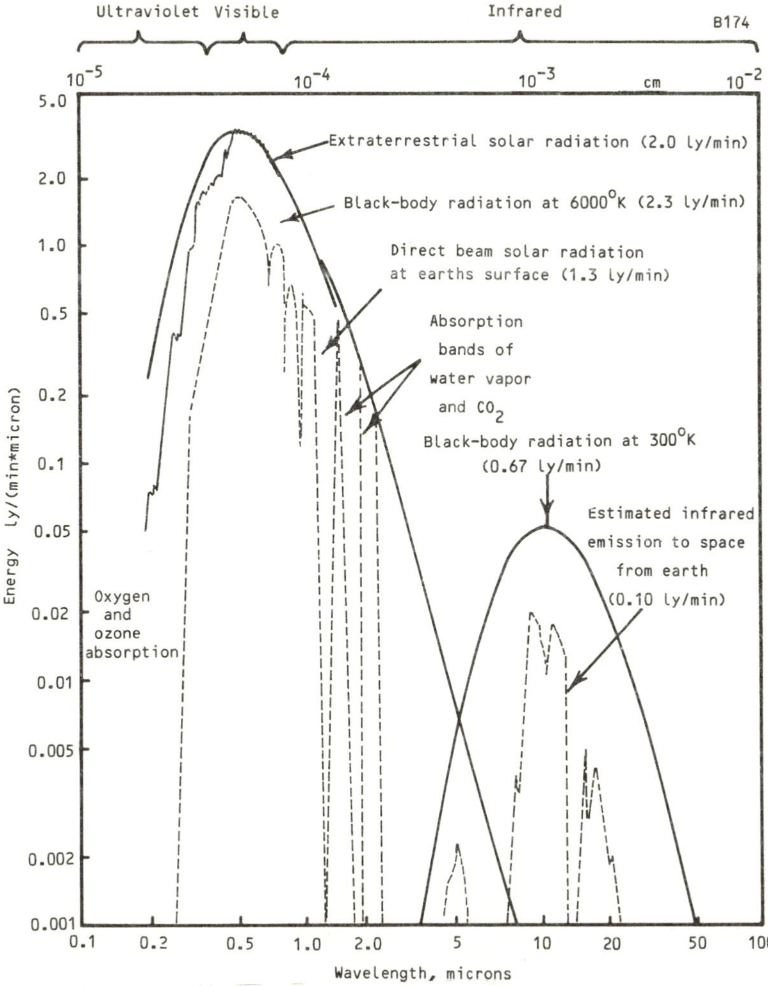

Solar spectra and earth radiation. (ref: 1517)

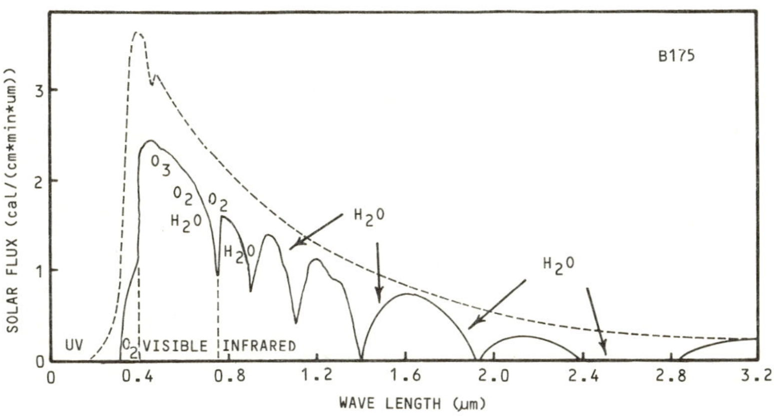

Solarspectrum at the surface of earth with major absorptions bands from oxygen, ozone and water vapour.
Extraterrestrial solar curve
Direct solar radiation at ground level
y-axis: solar flux (cal/(cm**2*min*um)
x-axis: wave length (um)

(ref: 507)

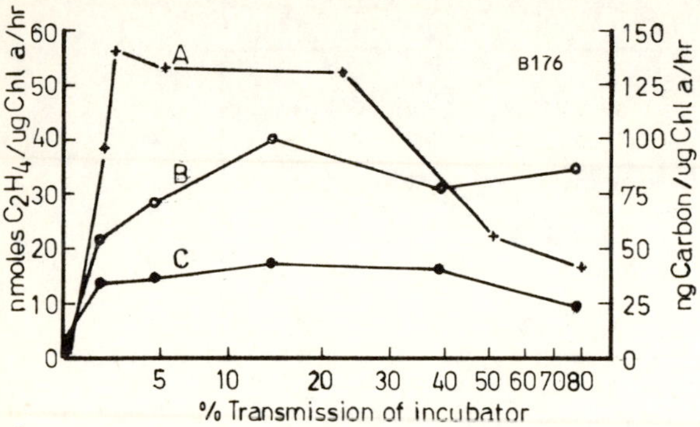

Light - acetylene reduction and carbon fixation by Richelia intracellularis/Rhizosolenia. A: Net incorporation of C-14 into particulate material. B: Ethylene production 1. experiment. C: Ethylene production 2. experiment. average sunlight on incubator was 1.3*10**17 quanta/(sec*cm**2) in 400-700 nm

(ref: 967)

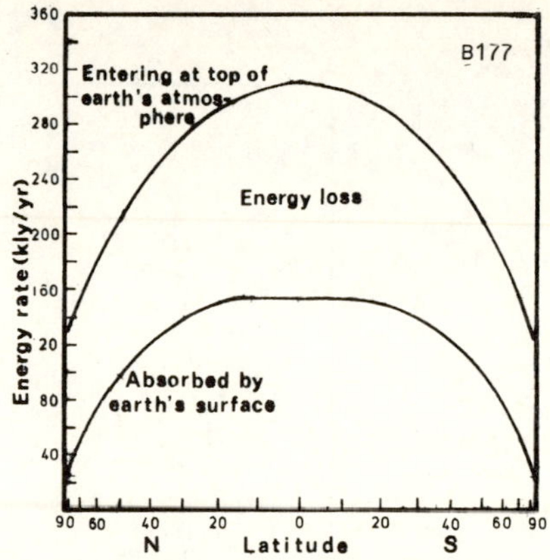

Solar radiation (mean annual) versus latitude. Both the energy entering the top of the atmosphere and the energy absorbed by the earth surface is shown. The difference is the energy loss.

(ref: 1417)

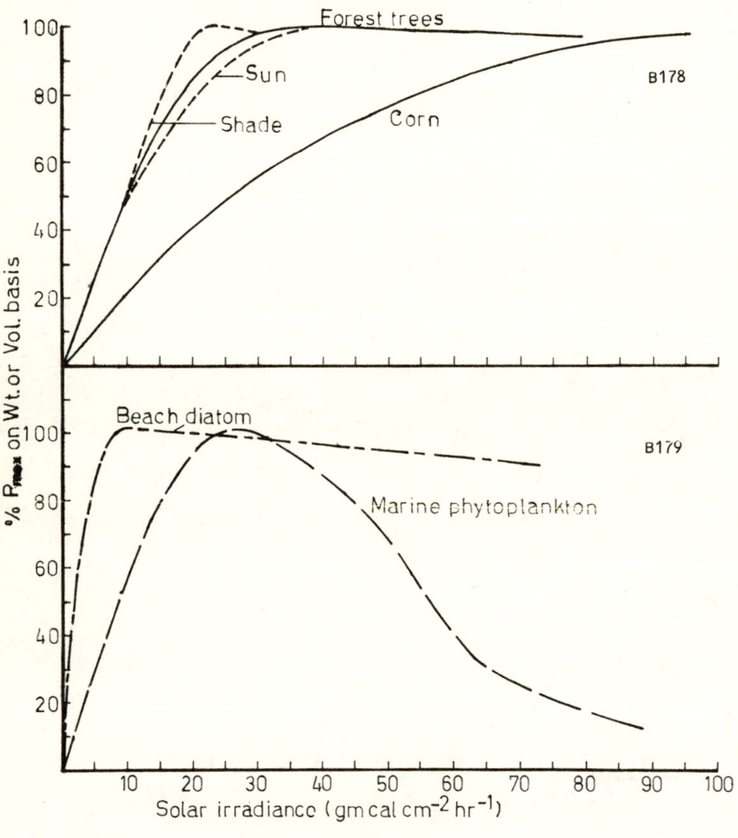

Light - photosynthesis curve for forest trees, corn, beach diatoms and marine phytoplankton. ——forest trees ——corn —·—beach diatoms — —marine phytoplankton.

(ref: 660)

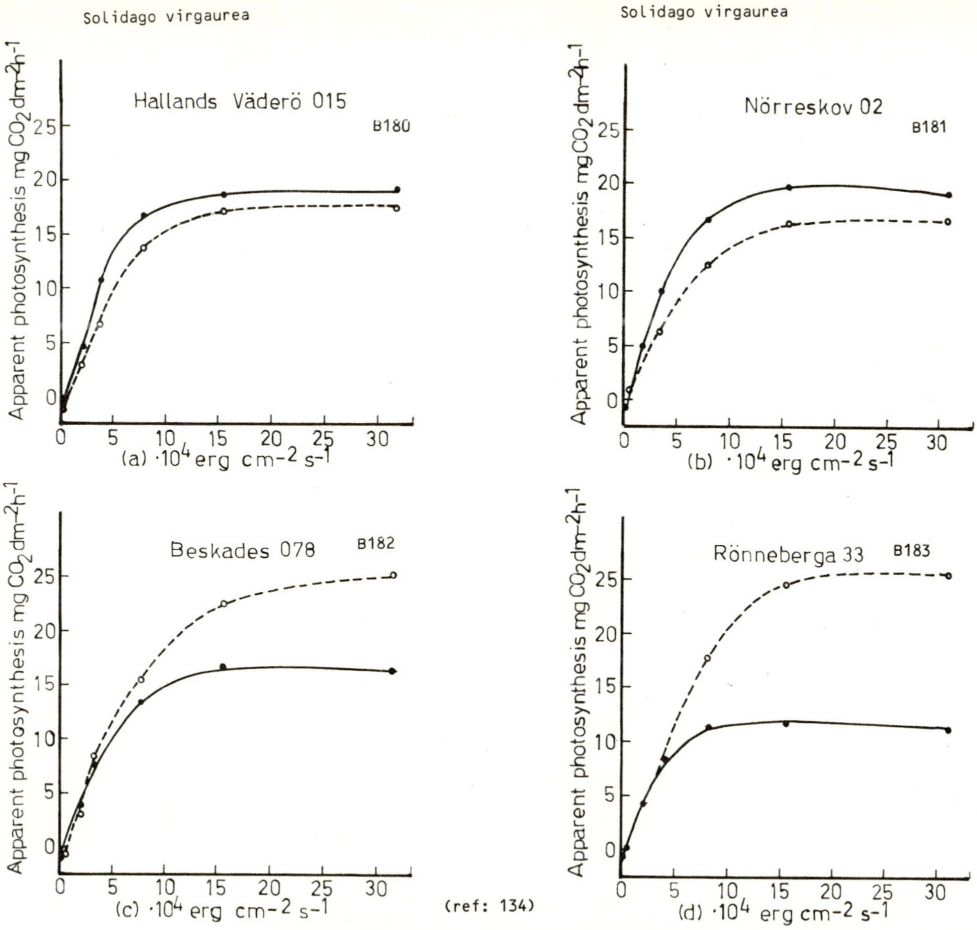

Photosynthesis versus light intensity in clone plants of four populations (a) (b) orginated from shahed habitats, (c) (d) from exposed habitats

cultivated at 3*10**4 ergs/(cm**2*s)

cultivated at 15*10**4 ergs/(cm**2*s)

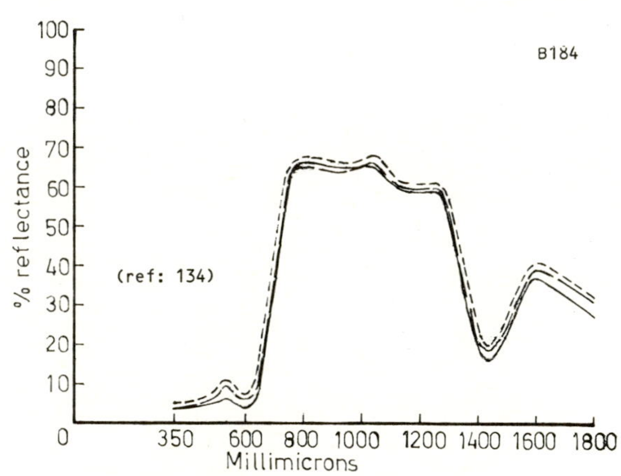

Reflectance of radient energy from upper leaflet surface in 3 populations of Thalictium alpinum. Significantly different (99 % level) at 600 nm and 1600 nm.

—— —— artic diploid ——— artic triploid

------ alpine diploid

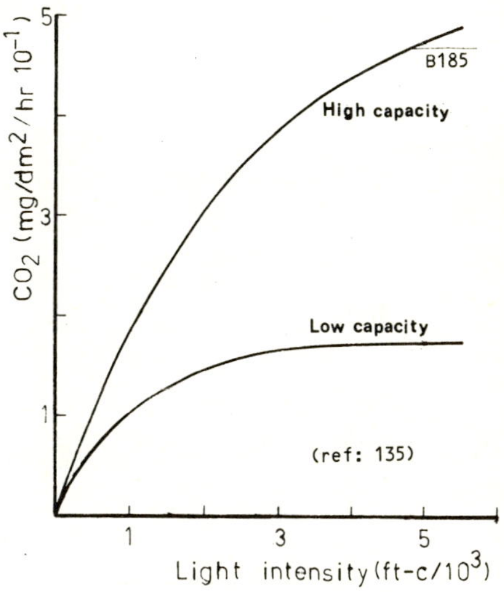

Photosynthetic capacity versus light for plants with hig and low capacity.

Population growth versus temperature and salinity. Growth is expressed as final concentration of cells/(initial concentration of cells). Salinity ppt (x-axis), Temperature in celcius (y-axis)

Photosynthesis, dependence on temperature and salinity.

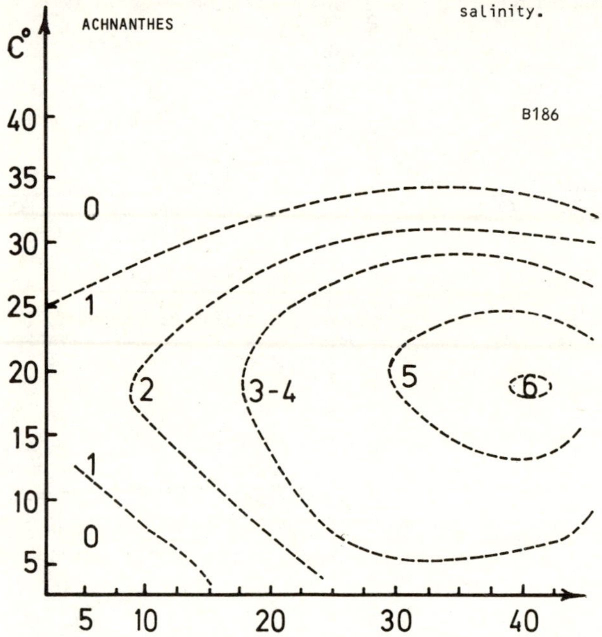

B186

(ref: 1528)

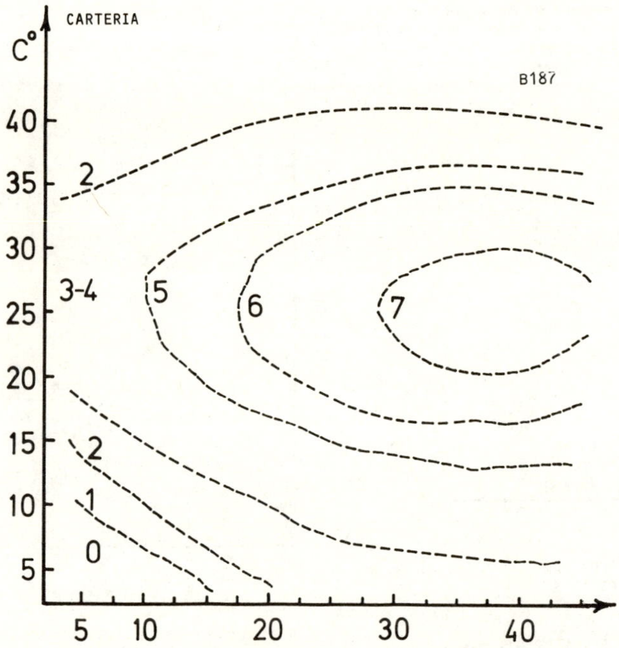

B187

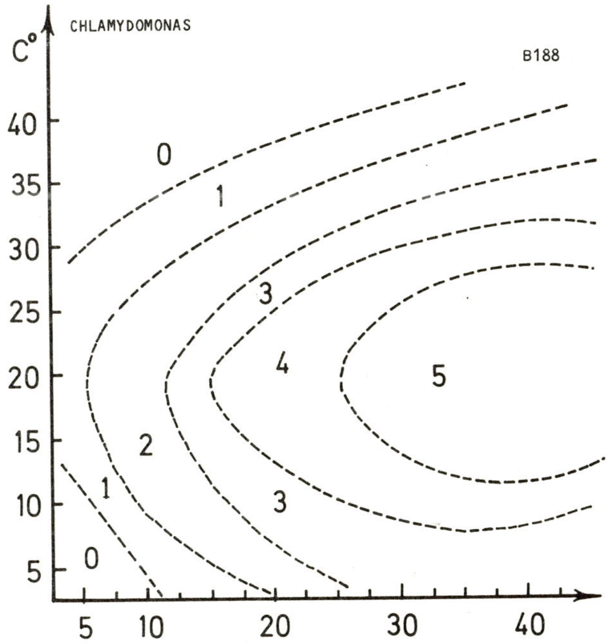
CHLAMYDOMONAS
C°
B188
40
35
30
25
20
15
10
5
0
1
2
3
4
5
3
1
0
5
10
20
30
40

(ref: 1528)

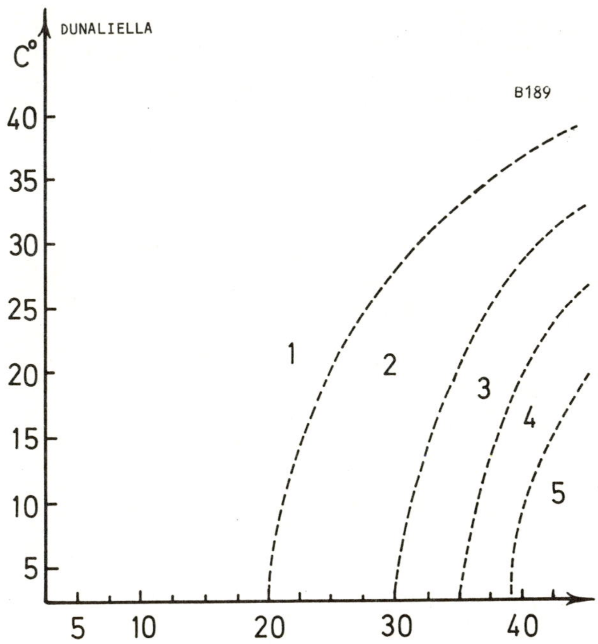
DUNALIELLA
C°
B189
40
35
30
25
20
15
10
5
1
2
3
4
5
5
10
20
30
40

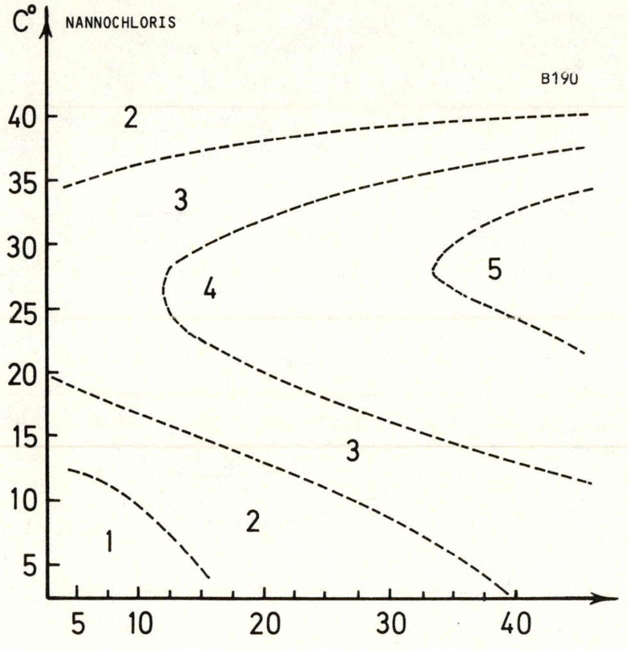
C°
NANNOCHLORIS
B190
2
3
4
5
3
2
1
40
35
30
25
20
15
10
5
5
10
20
30
40

(ref: 1528)

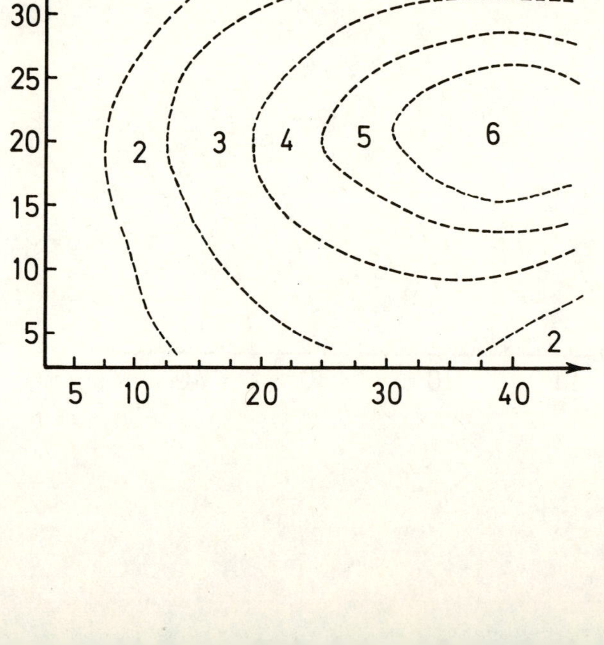
PHAEODACTYLUM
C°
B191
0
1
2
3
4
5
6
2
40
35
30
25
20
15
10
5
5
10
20
30
40

B4

PESTICIDES

DEGRADATION RATE
HALF LIFE TIME
LOSS BY RUN OFF
TRADE NAMES

TABLE B202

PESTICIDES : AERIAL FALLOUT			
ITEM:	VALUE:	CONDITIONS:	REF.:
*DDT	28 - 690*10**(-9) G DDT/(M**2*DAY)	SOUTHERN CALIFORNIA	1762

TABLE B203

PESTICIDES : DEGRADATION RATE			
ITEM:	VALUE:	CONDITIONS:	REF.:
*ALDRIN	40% OF INITIAL AMOUNT IN CULV SOIL	AFTER 20W,UNCULTIVATED:5% REMAINING AFTER 6W	1690
*ALDRIN	44-62% OF INITIAL AMOUNT IN CULV SOIL	AFTER 1YR,UNCULTIVATED:6.5-13% REMAINING AFTER 1YR	921
*ALDRIN	26-50% OF INITIAL AMOUNT IN CULV SOIL	AFTER 4MONTHS,UNCULTIVATED: 2.7-5.3% REMAINING AFTER 4MONTHS	920
*ALDRIN	90 WEEKS	IN SOIL, TIME USED TO REACH 70-100 % OF CONTROL	355
*AMIBEM	10 WEEKS	IN SOIL, TIME USED TO REACH 70-100 % OF CONTROL	355
*ATRAZINE	35 WEEKS	IN SOIL, TIME USED TO REACH 70-100 % OF CONTROL	355
*AZINPHOSMETHYL	95 % LOSS	LOAM SOIL,303K,6DAYS	733
*BARBAN	3 WEEKS	IN SOIL, TIME USED TO REACH 70-100 % OF CONTROL	355
*BENSULPHIDE	35 WEEKS	IN SOIL, TIME USED TO REACH 70-100 % OF CONTROL	355
*BHC	140 WEEKS	IN SOIL, TIME USED TO REACH 70-100 % OF CONTROL	355
*CDAA	8 WEEKS	IN SOIL, TIME USED TO REACH 70-100 % OF CONTROL	355
*CHLORDANE	300 WEEKS	IN SOIL, TIME USED TO REACH 70-100 % OF CONTROL	355
*CIPC	8 WEEKS	IN SOIL, TIME USED TO REACH 70-100 % OF CONTROL	355
*DALAPON	8 WEEKS	IN SOIL, TIME USED TO REACH 70-100 % OF CONTROL	355
*DDT	200 WEEKS	IN SOIL, TIME USED TO REACH 70-100 % OF CONTROL	355
*DICAMBA	8 WEEKS	IN SOIL, TIME USED TO REACH 70-100 % OF CONTROL	355
*DICHLOBENIL	18 WEEKS	IN SOIL, TIME USED TO REACH 70-100 % OF CONTROL	355
*DIELDRIN	140 WEEKS	IN SOIL, TIME USED TO REACH 70-100 % OF CONTROL	355
*DIPHENAMID	30 WEEKS	IN SOIL, TIME USED TO REACH 70-100 % OF CONTROL	355
*DIURON	30 WEEKS	IN SOIL, TIME USED TO REACH 70-100 % OF CONTROL	355
*EPTC	5 WEEKS	IN SOIL, TIME USED TO REACH 70-100 % OF CONTROL	355
*ETHION	12 % LOSS	SOIL DUST 1(?069),PH=6.9, ORG.MATTER=0.8%,200DAYS	733
*ETHION	36 % LOSS	SOIL DUST 2(?069),PH=7.3, ORG.MATTER=2.1%,200DAYS	733
*ETHION	23 % LOSS	SOIL DUST 3(?069),PH=7.3, ORG.MATTER=2.3%,200DAYS	733
*ETHION	36 % LOSS	SOIL DUST 4(?069),PH=7.6, ORG.MATTER=1.8%,200DAYS	733
*FENURON	30 WEEKS	IN SOIL, TIME USED TO REACH 70-100 % OF CONTROL	355
*HEPTACHLOR	15.0% OF INITIAL AMOUNT IN CULV SOIL	AFTER 3-4MONTHS,UNCULTIVATED:3.0% REMAINING AFTER 3-4MONTHS	1368
*HEPTACHLOR	26-50% OF INITIAL AMOUNT IN CULV SOIL	AFTER 4MONTHS,UNCULTIVATED: 2.7-5.3% REMAINING AFTER 4MONTHS	920

TABLE B203

PESTICIDES : DEGRADATION RATE			
ITEM:	VALUE:	CONDITIONS:	REF.:
*HEPTACHLOR	90 WEEKS	IN SOIL, TIME USED TO REACH 70-100 % OF CONTROL	LA(K
*IPA	5 WEEKS	IN SOIL, TIME USED TO REACH 70-100 % OF CONTROL	LA(K
*LINURON	18 WEEKS	IN SOIL, TIME USED TO REACH 70-100 % OF CONTROL	355
*MCPA	12 WEEKS	IN SOIL, TIME USED TO REACH 70-100 % OF CONTROL	355
*MONURON	35 WEEKS	IN SOIL, TIME USED TO REACH 70-100 % OF CONTROL	355
*PARATHION	85 % LOSS	SOIL DUST 2,3 AND 4(?069),PH= 7.3-7.6,ORG.MATTER=1.8-2.3%,10DAYS	733
*PARATHION	58 % LOSS	SOIL DUST 1(?069),PH=6.9, ORG.MATTER=0.8%,30DAYS	733
*PARATHION	99 % LOSS	SOIL DUST 1(?069),PH=6.9, ORG.MATTER=0.8%,50DAYS	733
*PHENTHOATE	50 % LOSS	SOIL DUST 2,3 AND 4(?069),PH= 7.3-7.6,ORG.MATTER=1.8-2.3%,10DAYS	733
*PHENTHOATE	95 % LOSS	SOIL DUST 2,3 AND 4(?069),PH= 7.3-7.6,ORG.MATTER=1.8-2.3%,30DAYS	733
*PHENTHOATE	38 % LOSS	SOIL DUST 1(?069),PH=6.9, ORG.MATTER=0.8%,75DAYS	733
*PHENTHOATE	> 99 % LOSS	SOIL DUST 1(?069),PH=6.9, ORG.MATTER=0.8%,150DAYS	733
*PICLORAM	80 WEEKS	IN SOIL, TIME USED TO REACH 70-100 % OF CONTROL	355
*PROMETRYN	10 WEEKS	IN SOIL, TIME USED TO REACH 70-100 % OF CONTROL	355
*PROPAZINE	80 WEEKS	IN SOIL, TIME USED TO REACH 70-100 % OF CONTROL	355
*SIMAZINE	40 WEEKS	IN SOIL, TIME USED TO REACH 70-100 % OF CONTROL	355
*TCA	12 WEEKS	IN SOIL, TIME USED TO REACH 70-100 % OF CONTROL	355
*TRIFLURALIN	23 WEEKS	IN SOIL, TIME USED TO REACH 70-100 % OF CONTROL	355
*2,3,5-T	21 WEEKS	IN SOIL, TIME USED TO REACH 70-100 % OF CONTROL	355
*2,3,6-TBA	47 WEEKS	IN SOIL, TIME USED TO REACH 70-100 % OF CONTROL	355
*2,4-D	5 WEEKS	IN SOIL, TIME USED TO REACH 70-100 % OF CONTROL	355

TABLE B204

PESTICIDES : HALF-LIFE TIME			
ITEM:	VALUE:	CONDITIONS:	REF.:
*CHLOROBROMURON	2.5 DAYS	IN ENSILAGE, PH=4.5, 300 K, INITIAL CONC = 0.12 PPM	802
*METHOXYCHLOR	7 DAYS	ELIMINATION FROM BEECH LEAVES IN WATER, PH=7.9	340
*METHOXYCHLOR	26 DAYS	ELIMINATION FROM MAPLE LEAVES IN WATER, PH=7.9	340
*METHOXYCHLOR	24 DAYS	ELIMINATION FROM OAK LEAVES IN WATER, PH=7.9	340
*PARATHION	5240+-260 HOURS	296 K,ESTUARINE WATER, SALINITY= 2.57%, HYDROLYSIS, PH=7.8	1668
*PARATHION	180+-15 HOURS	323 K,ESTUARINE WATER, SALINITY= 2.57%, HYDROLYSIS, PH=7.8	1668
*PARATHION	30.0+-0.9 HOURS	343 K,ESTUARINE WATER, SALINITY= 2.57%, HYDROLYSIS, PH=7.8	1668
*PARATHION	4.47+-0.3 HOURS	363 K,ESTUARINE WATER, SALINITY= 2.57%, HYDROLYSIS, PH=7.8	1668

TABLE B204

PESTICIDES : HALF-LIFE TIME			
ITEM:	VALUE:	CONDITIONS:	REF.:
*PARATHION	17.2+-0.3 HOURS	343 K,SEAWATER, SALINITY = 3.25 %, HYDROLYSIS, PH =8.0	1668
*PARATHION	19.3+-0.5 HOURS	343 K, SEAWATER, SALINITY =3.25 %, HYDROLYSIS, PH = 1.0	1668
*PARATHION	39.2+-1.3 HOURS	343 K, DISTILLED WATER, HYDROLYSIS, PH = 5.9	1668
*PARATHION	21.0+-0.7 HOURS	343 K, DISTILLED WATER,30 G NACL/ L, HYDROLYSIS, PH = 5.5	1668
*PARATHION	22.7+-0.6 HOURS	343 K, DISTILLED WATER + DIL NAOH, HYDROLYSIS, PH = 8.6	1668
*PARATHION	<0.5 HOURS	343 K, DISTILLED WATER + DIL NAOH, HYDROLYSIS, PH = 12	1668
*PARATHION	15.7+-0.2 HOURS	343 K, BUFFER, SALT 30 G/L, HYDROLYSIS, PH = 7.9	1668
*PARATHION	31.8+-0.7 HOURS	343 K, BUFFER, SALT 0.55 G/L, HYDROLYSIS, PH =7.5	1668
*PARATHION	5240+-260 HOURS	296 K,ESTUARINE WATER, SALINITY= 2.57%, HYDROLYSIS, PH=7.8	1668
*PARATHION	180+-15 HOURS	323 K,ESTUARINE WATER, SALINITY= 2.57%, HYDROLYSIS, PH=7.8	1668
*PARATHION	30.0+-0.9 HOURS	343 K,ESTUARINE WATER, SALINITY= 2.57%, HYDROLYSIS, PH=7.8	1668
*PARATHION	4.47+-0.3 HOURS	363 K,ESTUARINE WATER, SALINITY= 2.57%, HYDROLYSIS, PH=7.8	1668
*PARATHION	17.2+-0.3 HOURS	343 K,SEAWATER, SALINITY = 3.25 %, HYDROLYSIS, PH =8.0	1668
*PARATHION	19.3+-0.5 HOURS	343 K, SEAWATER, SALINITY =3.25 %, HYDROLYSIS, PH = 1.0	1668
*PARATHION	39.2+-1.3 HOURS	343 K, DISTILLED WATER, HYDROLYSIS, PH = 5.9	1668
*PARATHION	21.0+-0.7 HOURS	343 K, DISTILLED WATER,30 G NACL/ L, HYDROLYSIS, PH = 5.5	1668
*PARATHION	22.7+-0.6 HOURS	343 K, DISTILLED WATER + DIL NAOH, HYDROLYSIS, PH = 8.6	1668
*PARATHION	<0.5 HOURS	343 K, DISTILLED WATER + DIL NAOH, HYDROLYSIS, PH = 12	1668
*PARATHION	15.7+-0.2 HOURS	343 K, BUFFER, SALT 30 G/L, HYDROLYSIS, PH = 7.9	1668
*PARATHION	31.8+-0.7 HOURS	343 K, BUFFER, SALT 0.55 G/L, HYDROLYSIS, PH =7.5	1668
*PHOSPHAMIDON	FEW DAYS	SPRAYED BY AIRCRAFT, NEW BRUNSWICH, CAN., 3/6, 275-293 K, SUNNY	1629
*SEVIN	1.7 DAYS	DEGRADATION IN PLANT TISSUE	74

TABLE B205

PESTICIDES : HALF-LIFE TIME FOR PHOTOLYSIS			
ITEM:	VALUE:	CONDITIONS:	REF.:
*ATRAZINE	3*10**3 HOURS	LIGHT=3*10**13 PHOTONS/CM2,280-300 NM,LAT=30 N,LAB EXPERIMENT	1743
*ATRAZINE	3000 HOURS	LIGHT INT.=3*10**13PHOTONS/(CM**2) *SEC,280-300NM,HG-LAMP	1743
*CAPTAN	>2*10**3 HOURS	LIGHT=3*10**13 PHOTONS/CM2,280-300 NM,LAT=30 N,LAB EXPERIMENT	1743
*CAPTAN	> 2000 HOURS	LIGHT INT.=3*10**13PHOTONS/(CM**2) *SEC,280-300NM,HG-LAMP	1743
*CARBARYL	2*10**2 HOURS	LIGHT=3*10**13 PHOTONS/CM2,280-300 NM,LAT=30 N,LAB EXPERIMENT	1743
*CARBARYL	200 HOURS	LIGHT INT.=3*10**13PHOTONS/(CM**2) *SEC,280-300NM,HG-LAMP	1743
*CARBARYL	43 HOURS	LATITUDE=ON,SPRING,CALC.FOR MID-SEASON,DEPTH <10CM IN PURE WATER	1743

TABLE B205

PESTICIDES : HALF-LIFE TIME FOR PHOTOLYSIS

ITEM:	VALUE:	CONDITIONS:	REF.:
*CARBARYL	51 HOURS	LATITUDE=30N,SPRING,CALC.FOR MID-SEASON,DEPTH <10CM IN PURE WATER	1743
*CARBARYL	64 HOURS	LATITUDE=40N,SPRING,CALC.FOR MID-SEASON,DEPTH <10CM IN PURE WATER	1743
*CARBARYL	90 HOURS	LATITUDE=50N,SPRING,CALC.FOR MID-SEASON,DEPTH <10CM IN PURE WATER	1743
*CARBARYL	292 HOURS	LATITUDE=70N,SPRING,CALC.FOR MID-SEASON,DEPTH <10CM IN PURE WATER	1743
*CARBARYL	45 HOURS	LATITUDE=0N,SUMMER,CALC.FOR MID-SEASON,DEPTH <10CM IN PURE WATER	1743
*CARBARYL	46 HOURS	LATITUDE=30N,SUMMER,CALC.FOR MID-SEASON,DEPTH <10CM IN PURE WATER	1743
*CARBARYL	52 HOURS	LATITUDE=40N,SUMMER,CALC.FOR MID-SEASON,DEPTH <10CM IN PURE WATER	1743
*CARBARYL	61 HOURS	LATITUDE=50N,SUMMER,CALC.FOR MID-SEASON,DEPTH <10CM IN PURE WATER	1743
*CARBARYL	121 HOURS	LATITUDE=70N,SUMMER,CALC.FOR MID-SEASON,DEPTH <10CM IN PURE WATER	1743
*CARBARYL	41 HOURS	LATITUDE=0N,FALL,CALC.FOR MID-SEASON,DEPTH <10CM IN PURE WATER	1743
*CARBARYL	68 HOURS	LATITUDE=30N,FALL,CALC.FOR MID-SEASON,DEPTH <10CM IN PURE WATER	1743
*CARBARYL	102 HOURS	LATITUDE=40N,FALL,CALC.FOR MID-SEASON,DEPTH <10CM IN PURE WATER	1743
*CARBARYL	186 HOURS	LATITUDE=50N,FALL,CALC.FOR MID-SEASON,DEPTH <10CM IN PURE WATER	1743
*CARBARYL	1760 HOURS	LATITUDE=70N,FALL,CALC.FOR MID-SEASON,DEPTH <10CM IN PURE WATER	1743
*CARBARYL	43 HOURS	LATITUDE=0N,WINTER,CALC.FOR MID-SEASON,DEPTH <10CM IN PURE WATER	1743
*CARBARYL	103 HOURS	LATITUDE=30N,WINTER,CALC.FOR MID-SEASON,DEPTH <10CM IN PURE WATER	1743
*CARBARYL	200 HOURS	LATITUDE=40N,WINTER,CALC.FOR MID-SEASON,DEPTH <10CM IN PURE WATER	1743
*CARBARYL	571 HOURS	LATITUDE=50N,WINTER,CALC.FOR MID-SEASON,DEPTH <10CM IN PURE WATER	1743
*CARBARYL	43800 HOURS	LATITUDE=70N,WINTER,CALC.FOR MID-SEASON,DEPTH <10CM IN PURE WATER	1743
*DDT	280000 HOURS	MEDIUM=HYDROCARBON, LATITUDE 40, CALCULATED FROM EQUATION (~12)	1774
*DDT	>227000 HOURS	MEDIUM=WATER, LATITUDE 40, CALCULATED FROM EQUATION (~12)	1774
*DDT	280000 HOURS	IN HYDROCARBON,DIRECT PHOTOLYSIS, MIDDAY SUN(SUMMER),LAT.=40N	1743
*DDT	> 227000 HOURS	IN WATER,DIRECT PHOTOLYSIS,MIDDAY SUN(SUMMER),LAT.=40N	1743
*DIAZINON	1*10**3 HOURS	LIGHT=3*10**13 PHOTONS/CM2,280-300 NM,LAT=30 N,LAB EXPERIMENT	1743
*DIAZINON	1000 HOURS	LIGHT INT.=3*10**13PHOTONS/(CM**2) *SEC,280-300NM,HG-LAMP	1743
*MALATHION	2*10**4 HOURS	LIGHT=3*10**13 PHOTONS/CM2,280-300 NM,LAT=30 N,LAB EXPERIMENT	1743
*MALATHION	2*10**4 HOURS	LIGHT INT.=3*10**13PHOTONS/(CM**2) *SEC,280-300NM,HG-LAMP	1743

TABLE B205

PESTICIDES : HALF-LIFE TIME FOR PHOTOLYSIS			
ITEM:	VALUE:	CONDITIONS:	REF.:
*METHOPRENE	6 HOURS	ON GLASS, EXPOSED TO SUNLIGHT THROUGH GLASS, FEB, CALIF.	1289
*METHOXYCHLOR	4100 HOURS	MEDIUM=HYDROCARBON,LATITUDE 40 DEGRESS, CALCULATED FROM EQ (~12)	1774
*METHOXYCHLOR	690 HOURS	MEDIUM= WATER, LATITUDE 40, CALCULATED FROM EQUATION (~12)	1774
*METHOXYCHLOR	>300 HOURS	DISTILLED WATER, PH=6.3, CALCULATED FROM EQUATION (~12)	1774
*METHOXYCHLOR	2.2 HOURS	SUWANNEE RIVER, PH=4.7, CALCULATED FROM EQUATION (~12)	1774
*METHOXYCHLOR	5.4 HOURS	TOMBIGBEE RIVER, PH=7.6, CALCULATED FROM EQUATION (~12)	1774
*METHOXYCHLOR	2.9 HOURS	ALABAMA RIVER, PH=7.7, CALCULATED FROM EQUATION (~12)	1774
*METHOXYCHLOR	7.3 HOURS	DISTLLED WATER WITH 20 PPM HUMIC ACID, PH=5.2, CALCULATED FROM EQUATI	1774
*METHOXYCHLOR	3*10**1 HOURS	LIGHT=3*10**13 PHOTONS/CM2,280-300 NM,LAT=30 N,LAB EXPERIMENT	1743
*METHOXYCHLOR	30 HOURS	LIGHT INT.=3*10**13PHOTONS/(CM**2) *SEC,280-300NM,HG-LAMP	1743
*METHOXYCHLOR	4100 HOURS	IN HYDROCARBON,DIRECT PHOTOLYSIS, MIDDAY SUN(SUMMER),LAT.=40N	1743
*METHOXYCHLOR	690 HOURS	IN WATER,DIRECT PHOTOLYSIS,MIDDAY SUN(SUMMER),LAT.=40N	1743
*METHOXYCHLOR	> 300 HOURS	DISTILLED WATER,PH=6.3,MIDDAY SUN(MAY),LAT.=34N,40PPB METH	1743
*METHOXYCHLOR	2.2 HOURS	SUWANNEE RIVER,PH=4.7,MIDDAY SUN (MAY),LAT.=34N,40PPB METH	1743
*METHOXYCHLOR	5.4 HOURS	TOMBIGBEE RIVER,PH=7.6,MIDDAY SUN(MAY),LAT.=34N,40PPB METHOX.	1743
*METHOXYCHLOR	2.9 HOURS	ALABAMA RIVER,PH=7.7,MIDDAY SUN (MAY),LAT.=34N,40PPB METHOX.	1743
*METHOXYCHLOR	7.3 HOURS	20PPM"HUM.ACID"IN DEST.WAT.,PH= 5.2,MIDDAY SUN,MAY,LAT.=34N,40PPB METH	1743
*PARATHION	2*10**3 HOURS	LIGHT=3*10**13 PHOTONS/CM2,280-300 NM,LAT=30 N,LAB EXPERIMENT	1743
*PARATHION	2000 HOURS	LIGHT INT.=3*10**13PHOTONS/(CM**2) *SEC,280-300NM,HG-LAMP	1743
*TOXAPHENE	>2*10**4 HOURS	LIGHT=3*10**13 PHOTONS/CM2,280-300 NM,LAT=30 N,NO CHANGE UNDER EXP	1743
*TOXAPHENE	> 2*10**4 HOURS	LIGHT INT.=3*10**13PHOTONS/(CM**2) *SEC,280-300NM,HG-LAMP	1743
*2,4-D BEE	1*10**2	LIGHT=3*10**13 PHOTONS/CM2,BEE= BUTOXYETHYL ESTER OF 2,4-D,LAB EXP.	1743
*2,4-D BUTOXYETHYL	4.0 DAYS	IN HYDROCARBON,SEPTEMBER	1743
*2,4-D BUTOXYETHYL ESTER	100 HOURS	LIGHT INT.=3*10**13PHOTONS/(CM**2) *SEC,280-300NM,HG-LAMP	1743
*2,4-D BUTOXYETHYL ESTER	14 DAYS	IN WATER,SEPTEMBER	1743

TABLE B206

PESTICIDES : HALF-LIFE TIME FOR VOLATILIZATION			
ITEM:	VALUE:	CONDITIONS:	REF.:
*DDT	18.8 DAYS	P,P'-DDT, APPL = 1-1.3 KG/HA, COTTON	1100
*DDT	14.3 DAYS	O,P'-DDT, APPL = 1-1.3 KG/HA, COTTON	1100
*DDT	15.1 DAYS	P,P'-DDE, APPL = 1-1.3 KG/HA, COTTON	1100
*TOXAPHENE	15.1 DAYS	APPL = 2 - 2.7 KG/HA, COTTON	1100

TABLE B206

PESTICIDES : HALF-LIFE TIME FOR VOLATILIZATION			
ITEM:	VALUE:	CONDITIONS:	REF.:
*2,4-D METHYLESTER	22.0 DAYS	FROM A 1 M DEEP WATER LAYER, 298 K, HALF-LIFE TIME CALCULATED	1773
*2,4-D 1-BUTYLESTER	1.1 DAYS	FROM A 1 M DEEP WATER LAYER, 298 K, HALF-LIFE TIME CALCULATED	1773
*2,4-D 1-OCTYLESTER	11.5 DAYS	FROM A 1 M DEEP WATER LAYER, 298 K, HALF-LIFE TIME CALCULATED	1773
*2,4-D 2-BUYOXYETHYLESTER	895.0 DAYS	FROM A 1 M DEEP WATER LAYER, 298 K, HALF-LIFE TIME CALCULATED	1773

TABLE B207

PESTICIDES : HALF-LIFE TIME IN SOIL			
ITEM:	VALUE:	CONDITIONS:	REF.:
*AKTON	1.5 YEARS	10 LBS/ACRE, SPRAY, CORN, SULTAN SILT LOAM	519
*AKTON	1 YEAR	2 LBS/ACRE, GRANULES, CORN, SULTAN SILT LOAM	519
*AKTON	1.2 YEARS	2 LBS/ACRE, SPRAY, CORN, SULTAN SILT LOAM	519
*AKTON	32 WEEKS	LAB.	519
*ALDICARB	7 - 17 DAYS	OXIDATION, VALUE DEPENDING OF SOIL	1050
*ALDRIN	10 DAYS	IN IRRIGATED SOIL, COTTEN, GEZIRA, SUDAN	400
*ALDRIN	40 % REMAINING AFTER 14 YEARS	MAX VALUE	1099
*ALDRIN	28 % REMAINING AFTER 15 YEARS	MAX VALUE	1099
*AZINPHOSMETHYL	484 DAYS	STERILE SOIL, DRY, LAG PERIOD INCLUDED, 279 K	784
*AZINPHOSMETHYL	135 DAYS	STERILE SOIL, DRY, LAG PERIOD INCLUDED, 298 K	784
*AZINPHOSMETHYL	36 DAYS	STERILE SOIL, DRY, LAG PERIOD INCLUDED, 313 K	784
*AZINPHOSMETHYL	88 DAYS	STERILE SOIL, WET, LAG PERIOD INCLUDED, 279 K	784
*AZINPHOSMETHYL	29 DAYS	STERILE SOIL, WET, LAG PERIOD INCLUDED, 298 K	784
*AZINPHOSMETHYL	6 DAYS	STERILE SOIL, WET, LAG PERIOD INCLUDED, 313 K	784
*AZINPHOSMETHYL	484 DAYS	NATURAL SOIL, DRY, LAG PERIOD INCLUDED, 279 K	784
*AZINPHOSMETHYL	88 DAYS	NATURAL SOIL, DRY, LAG PERIOD INCLUDED, 298 K	784
*AZINPHOSMETHYL	32 DAYS	NATURAL SOIL, DRY, LAG PERIOD INCLUDED, 313 K	784
*AZINPHOSMETHYL	5 DAYS	313 K, POTATOES	1759
*AZINPHOSMETHYL	10 DAYS	CLAY AND SILT LOAM	733
*AZINPHOSMETHYL	15 DAYS	SILTY CLAY LOAM	733
*AZINPHOSMETHYL	25 DAYS	SILT LOAM	733
*AZINPHOSMETHYL	30 DAYS	SANDY LOAM	733
*AZINPHOSMETHYL	65 DAYS	SANDY LOAM	733
*AZINPHOSMETHYL	70 DAYS	FINE SANDY LOAM	733
*AZINPHOSMETHYL	80 DAYS	SANDY LOAM AND LOAM	733
*AZINPHOSMETHYL	29 DAYS	WET (50% OF SAT.) STERILIZED SOIL AT 298K	733
*AZINPHOSMETHYL	13 DAYS	NATURAL SOIL	733
*AZINPHOSMETHYL	12 DAYS	APPLIED AS AN EMULSION ON SOIL SURFACE	733
*AZINPHOSMETHYL	28 DAYS	GRANULAR FORM ROTOTILLING INTO SOIL	733

TABLE B207

PESTICIDES : HALF-LIFE TIME IN SOIL			
ITEM:	VALUE:	CONDITIONS:	REF.:
*BENOMYL	3 - 6 MONTHS	ON TURF, DELAWARE	79
*BENOMYL	6 - 12 MONTHS	BARE SOIL, DELAWARE	79
*BHC	10 % REMAINING AFTER 14 YEARS	MAX VALUE	1099
*BUX	1 - 2 WEEKS	HYDROLYSIS, LAB.	1050
*CARBARYL	8 DAYS	AGRICULTURAL SOIL	1050
*CARBARYL	64 DAYS	SANDY LOAM, APPL = 25.4 KG/HA	867
*CARBOFURAN	0.60 YEARS	IN SOIL, BETTER DRAINED, APPLICATION = 2.5 PPM, CLAY-MUCK, 1973	1717
*CARBOFURAN	0.92 YEARS	IN SOIL, POORLY DRAINED, APPLICATION = 2.5 PPM, CLAY-MUCK, 1973	1717
*CARBOFURAN	7.52 YEARS	IN SOIL, WELL DRAINED, APPLICATION = 2.5 PPM, CLAY-MUCK, 1974	1717
*CARBOFURAN	0.73 YEARS	IN SOIL, POORLY DRAINED, APPLICATION = 2.5 PPM, 1974, CLAY	1717
*CARBOFURAN	0.56 YEARS	IN SOIL, WELL DRAINED, APPLICATION = 2.5 PPM, CLAY-MUCK, 1973	1717
*CARBOFURAN	30 DAYS	HYDROLYSIS, FORMATION OF PHENOL	1050
*CHLORDANE	40 % REMAINING AFTER 14 YEARS	MAX VALUE	1099
*CHLOROBENZILATE	21 DAYS	IN LEON AND LAKELAND SANDS	979
*CHLOROBENZILATE	21 DAYS	IN LEON AND LAKELAND SANDS	979
*DDT	8 MONTHS	P,P'-DDT, SUBTROPICAL SOIL, FALL AND WINTER, APPL=5 KG/HA	1550
*DDT	14 DAYS	IN IRRIGATED SOIL, COTTEN, GEZIRA, SUDAN	400
*DDT	39 % REMAINING AFTER 17 YEARS	MAX VALUE	1099
*DICAMBA	3 WEEKS	ON LITTER, TEXAS	1069
*DICAMBA	2 WEEKS	ON NATIVE GRASSES, TEXAS	1069
*DIELDRIN	2.7 DAYS	LOSS BY VOLATILIZATION, GRASS, FIRST 5 D	1555
*DIELDRIN	7.5 MONTHS	P,P'-DDT, SUBTROPICAL SOIL, FALL AND WINTER, APPL=5 KG/HA	1550
*DIELDRIN	11 DAYS	IN IRRIGATED SOIL, COTTEN, GEZIRA, SUDAN	400
*DIELDRIN	31 % REMAINING AFTER 15 YEARS	MAX VALUE	1099
*DIOXATHION	55 DAYS	SOIL DUST 2(?069),PH=7.3, ORG.MATTER=2.1%	733
*DIOXATHION	45 DAYS	SOIL DUST 3(?069),PH=7.3, ORG.MATTER=2.3%	733
*DIOXATHION	30 DAYS	SOIL DUST 4(?069),PH=7.6, ORG.MATTER=1.8%	733
*ENDOSULFAN	7 DAYS	IN IRRIGATED SOIL, COTTEN, GEZIRA, SUDAN	400
*ENDRIN	41 % REMAINING AFTER 14 YEARS	MAX VALUE	1099
*ETHION	420 DAYS	NONE	733
*HEPTACHLOR	1.7 DAYS	LOSS BY VOLATILIZATION, GRASS, FIRST 5 D	1555
*HEPTACHLOR	16 % REMAINING AFTER 14 YEARS	MAX VALUE	1099
*MALATHION	3 DAYS	BASIC SILTY LOAM, ILLINOIS, APPL= 10 PPM, PH=6.2	1221
*MALATHION	7 DAYS	BASIC SILTY LOAM, ILLINOIS, APPL= 10 PPM, PH=8.2	1221
*MALATHION	4.3 DAYS	BASIC SILTY LOAM, ILLINOIS, APPL= 10 PPM, PH=7.2	1221
*MEOBAL	7 DAYS	FIELD CONDITION	1050
*METHOMYL	30 - 42 DAYS	NONE	1050
*NAPROAMIDE	54 DAYS	SOIL MOISTURE CONTENT = 10.0 %, 301 K, INITIAL = 4.5 KG/HA	1650
*NAPROAMIDE	63 DAYS	SOIL MOISTURE CONTENT = 7.5 %, 301 K, INITIAL = 4.5 KG/HA	1650

TABLE B207

PESTICIDES : HALF-LIFE TIME IN SOIL

ITEM:	VALUE:	CONDITIONS:	REF.:
*NAPROAMIDE	90 DAYS	SOIL MOISTURE CONTENT = 3.5 %, 301 K, INITIAL = 4.5 KG/HA	1650
*NAPROAMIDE	102 DAYS	SOIL MOISTURE CONTENT = 10.0 %, 287 K, INITIAL = 4.5 KG/HA	1650
*NAPROAMIDE	112 DAYS	SOIL MOISTURE CONTENT = 7.5 %, 287 K, INITIAL = 4.5 KG/HA	1650
*NAPROAMIDE	56 DAYS	SOIL MOISTURE CONTENT = 10.0 %, 301 K, INITIAL = 2.25 KG/HA	1650
*TERBACIL	225 DAYS	288 K	979
*TERBACIL	150 DAYS	303 K	979
*TERBACIL	225 DAYS	288 K	979
*TERBACIL	150 DAYS	303 K	979
*TOXAPHENE	45 % REMAINING AFTER 14 YEARS	MAX VALUE	1099
*TSUAMACIDE	7 DAYS	FIELD CONDITION	1050
*TSUAMACIDE	66 % IN 40 DAYS	HALF THE AMOUTH BOUND THIGTLY IN SOIL	1050
*2,4-D	3 DAYS	303 K	979
*2,4-D	10 - 14 DAYS	IN SOIL TREATED WITH 2,4-D IN SEVERAL YEARS	979
*2,4-D	3 WEEKS	ON LITTER, TEXAS	1069
*2,4-D	2 WEEKS	ON NAVITE GRASSES, TEXAS	1069
*2,4-D	3 DAYS	303 K	979
*2,4-D	10 - 14 DAYS	IN SOIL TREATED WITH 2,4-D IN SEVEREL YEARS	979
*2,4,5-T	3 WEEKS	ON LITTER, TEXAS	1069
*2,4,5-T	2 WEEKS	ON NATIVE GRASSES, TEXAS	1069

TABLE B208

PESTICIDES : HALF-LIFE TIME IN WATER

ITEM:	VALUE:	CONDITIONS:	REF.:
*ATRAZINE	5.1 DAYS	HYDROLYSIS, PH=1.3	41
*ATRAZINE	18.4 DAYS	HYDROLYSIS, PH=2.2	41
*ATRAZINE	66.4 DAYS	HYDROLYSIS, PH=3.1	41
*ATRAZINE	81.1 DAYS	HYDROLYSIS, PH=11.1	41
*ATRAZINE	15.2 DAYS	HYDROLYSIS, PH=11.9	41
*ATRAZINE	12.9 DAYS	HYDROLYSIS, PH=12.9	41
*CAPTAN	155 MIN	TOMBIGBEE RIVER,PH=7.0,301K	1743
*CARBARYL	99 MIN	PH = 10.0, 285 K,INITIAL CONC = 3*10**(-3) M, DARK	355
*CARBARYL	20 MIN	PH = 10.0, 298 K,INITIAL CONC = 3*10**(-3) M, DARK	355
*CARBARYL	8 MIN	PH = 10.0, 308 K,INITIAL CONC = 3*10**(-3) M, DARK	355
*CARBARYL	27 MIN	PH = 9.8, 298 K,INITIAL CONC = 3*10**(-3) M, DARK	355
*CARBARYL	58 MIN	PH = 9.5, 298 K,INITIAL CONC = 3*10**(-3) M, DARK	355
*CARBARYL	116 MIN	PH = 9.2, 298 K,INITIAL CONC = 3*10**(-3) M, DARK	355
*CARBARYL	173 MIN	PH = 9.0, 298 K,INITIAL CONC = 3*10**(-3) M, DARK	355
*CARBARYL	1 MONTH	PH = 8.0, 276.5 K, SEAWATER	800
*CARBARYL	4.8 DAYS	PH = 8.0, 290 K, SEAWATER	800
*CARBARYL	3.5 DAYS	PH = 8.0, 293 K, SEAWATER	800

TABLE B208

PESTICIDES : HALF-LIFE TIME IN WATER

ITEM:	VALUE:	CONDITIONS:	REF.:
*CARBARYL	1.0 DAYS	PH = 8.0, 301 K, SEAWATER	800
*CARBARYL	99 MIN	PH = 10.0, 285 K,INITIAL CONC = 3*10**(-3) M, DARK	355
*CARBARYL	20 MIN	PH = 10.0, 298 K,INITIAL CONC = 3*10**(-3) M, DARK	355
*CARBARYL	8 MIN	PH = 10.0, 308 K,INITIAL CONC = 3*10**(-3) M, DARK	355
*CARBARYL	27 MIN	PH = 9.8, 298 K,INITIAL CONC = 3*10**(-3) M, DARK	355
*CARBARYL	58 MIN	PH = 9.5, 298 K,INITIAL CONC = 3*10**(-3) M, DARK	355
*CARBARYL	116 MIN	PH = 9.2, 298 K,INITIAL CONC = 3*10**(-3) M, DARK	355
*CARBARYL	173 MIN	PH = 9.0, 298 K,INITIAL CONC = 3*10**(-3) M, DARK	355
*CARBARYL	1 MONTH	PH = 8.0, 276.5 K, SEAWATER	800
*CARBARYL	4.8 DAYS	PH = 8.0, 290 K, SEAWATER	800
*CARBARYL	3.5 DAYS	PH = 8.0, 293 K, SEAWATER	800
*CARBARYL	1.0 DAYS	PH = 8.0, 301 K, SEAWATER	800
*CARBARYL	3.2 HOURS	HYDROLYSIS, PH=9,300K	1743
*CARBARYL	1.3 DAYS	HYDROLYSIS, PH=8,300K	1743
*CARBARYL	13 DAYS	HYDROLYSIS, PH=7,300K	1743
*CARBARYL	4.4 MONTHS	HYDROLYSIS, PH=6,300K	1743
*CARBARYL	3.6 YEARS	HYDROLYSIS, PH=5,300K	1743
*DIAZINON	0.49 DAYS	HYDROLYSIS, PH=3.1	446
*DIAZINON	31 DAYS	HYDROLYSIS, PH=5.0	446
*DIAZINON	185 DAYS	HYDROLYSIS, PH=7.5	446
*DIAZINON	136 DAYS	HYDROLYSIS, PH=9.0	446
*DIAZINON	6 DAYS	HYDROLYSIS, PH=10.4	446
*DIAZOXON	0.017 DAY	HYDROLYSIS, PH=3.1	446
*DIAZOXON	1.27 DAYS	HYDROLYSIS, PH=5.0	446
*DIAZOXON	29 DAYS	HYDROLYSIS, PH=7.5	446
*DIAZOXON	18 DAYS	HYDROLYSIS, PH=9.0	446
*DIAZOXON	0.42 DAYS	HYDROLYSIS, PH=10.4	446
*DIMILIN	22.9 DAYS	PH = 7.7, 283 K, INITIAL = 0.1 PPM	1406
*DIMILIN	80.5 DAYS	PH =10.0, 283 K, INITIAL = 0.1 PPM	1406
*DIMILIN	28.7 DAYS	PH = 7.7, 297 K, INITIAL = 0.1 PPM	1406
*DIMILIN	8.31 DAYS	PH =10.0, 297 K, INITIAL = 0.1 PPM	1406
*DIMILIN	8.00 DAYS	PH = 7.7, 311 K, INITIAL = 0.09 PPM	1406
*DIMILIN	3.45 DAYS	PH =10.0, 311 K, INITIAL = 0.1 PPM	1406
*HEPTACHLOR	23.1 HOUR	299.8 K, FOR HYDROLYSIS, IN DISTILLED WATER	344
*METHOPRENE	30 HOURS	FRESHWATER POUND,INITIAL=0.001 PPM, FEBRUAR, IN SUNLIGHT, CALIF.	1400
*METHOPRENE	40 HOURS	FRESHWATER POUND, INITIAL=0.01 PPM,FEBRUAR, IN SUNLIGHT, CALIF	1400
*METHOXYCHLOR	58 HOURS	CONC.IN WATER=10**(-7)M,NO H2O2 ADDED,338K	1743
*METHOXYCHLOR	58 HOURS	HYDLS, WAT CONC=10**(-6)M,5% ACETONITRILE,NO H2O2 ADDED,338K	1743
*METHOXYCHLOR	< 1 HOUR	HYDLS, WAT CONC=10**(-6)M,5% ACETONITRILE,H2O2-CONC.=0.1M,338K	1743
*METHOXYCHLOR	< 1.7 HOUR	338K,HYDLS, WAT CONC=10**(-6)M,5% ACETONITRILE,H2O2-CONC.=8*10** (-2)M	1743
*METHOXYCHLOR	2.0 HOURS	338K,HYDLS, WAT CONC=10**(-6)M,5% ACETONITRILE,H2O2-CONC.=8*10** (-3)M	1743

TABLE B208

PESTICIDES : HALF-LIFE TIME IN WATER

ITEM:	VALUE:	CONDITIONS:	REF.:
*METHOXYCHLOR	17 HOURS	338K,HYDLS, WAT CONC=10**(-6)M,5% ACETONITRILE,H2O2-CONC.=8*10** (-4)M	1743
*METHOXYCHLOR	25 HOURS	338K,HYDLS, WAT CONC=10**(-6)M,5% ACETONITRILE,H2O2-CONC.=8*10** (-5)M	1743
*ZECTRAN	1 - 2 DAYS	PH = 9.5, 285 K	355
*ZECTRAN	27 DAYS	PH = 7.4, 285 K	355
*ZR515	100 DAYS	SALTWATER, 277.5 K, INITIAL = 830 PPM	1276
*ZR515	35 DAYS	FRESHWATER, 277.5 K, INITIAL = 763 PPM	1276
*2,4-D METHYL ESTER	1.1 HOUR	HYDROLYSIS IN WATER,PH=9,301K	1743
*2,4-D METHYL ESTER	44 HOURS	HYDROLYSIS IN WATER,PH=6,301K	1743
*2,4-D METHYLESTER	1.1 HOURS	HYDROLYSIS, 301 K, PH=9	1773
*2,4-D METHYLESTER	44 DAYS	HYDROLYSIS, 301 K, PH=6	1773
*2,4-D 1-BUTYL ESTER	5.2 HOURS	HYDROLYSIS IN WATER,PH=9,301K	1743
*2,4-D 1-BUTYL ESTER	220 HOURS	HYDROLYSIS IN WATER,PH=6,301K	1743
*2,4-D 1-BUTYLESTER	5.2 HOURS	HYDROLYSIS, 301 K, PH=9	1773
*2,4-D 1-BUTYLESTER	220 DAYS	HYDROLYSIS, 301 K, PH=6	1773
*2,4-D 1-OCTYL ESTER	5.2 HOURS	HYDROLYSIS IN WATER,PH=9,301K	1743
*2,4-D 1-OCTYL ESTER	220 HOURS	HYDROLYSIS IN WATER,PH=6,301K	1743
*2,4-D 1-OCTYLESTER	5.2 HOURS	HYDROLYSIS, 301 K, PH=9	1773
*2,4-D 1-OCTYLESTER	220 DAYS	HYDROLYSIS, 301 K, PH=6	1773
*2,4-D 2-BUTOXY-METHYLETHYL ESTER	4.4 HOURS	HYDROLYSIS IN WATER,PH=9,301K	1743
*2,4-D 2-BUTOXY-METHYLETHYL ESTER	180 HOURS	HYDROLYSIS IN WATER,PH=6,301K	1743
*2,4-D 2-BUTOXYETHYL ESTER	0.6 HOUR	HYDROLYSIS IN WATER,PH=9,301K	1743
*2,4-D 2-BUTOXYETHYL ESTER	26 HOURS	HYDROLYSIS IN WATER,PH=6,301K	1743
*2,4-D 2-BUTOXYETHYLESTER	0.6 HOURS	HYDROLYSIS, 301 K, PH=9	1773
*2,4-D 2-BUTOXYETHYLESTER	26 DAYS	HYDROLYSIS, 301 K, PH=6	1773
*2,4-D 2-BUTOXYMETHYLETHYLESTER	4.4 HOURS	HYDROLYSIS, 301 K, PH=9	1773
*2,4-D 2-BUTOXYMETHYLETHYLESTER	180 DAYS	HYDROLYSIS, 301 K, PH=6	1773
*2,4-D 2-OCTYL ESTER	37 HOURS	HYDROLYSIS IN WATER,PH=9,301K	1743
*2,4-D 2-OCTYL ESTER	1500 HOURS	HYDROLYSIS IN WATER,PH=6,301K	1743
*2,4-D 2-OCTYLESTER	37.0 HOURS	HYDROLYSIS, 301 K, PH=9	1773
*2,4-D 2-OCTYLESTER	1500 DAYS	HYDROLYSIS, 301 K, PH=6	1773
*2,4-D 2-PROPY#ESTER	17.0 HOURS	HYDROLYSIS, 301 K, PH=9	1773
*2,4-D 2-PROPYL ESTER	17 HOURS	HYDROLYSIS IN WATER,PH=9,301K	1743
*2,4-D 2-PROPYL ESTER	710 HOURS	HYDROLYSIS IN WATER,PH=6,301K	1743
*2,4-D 2-PROPYLESTER	710 DAYS	HYDROLYSIS, 301 K, PH=6	1773

TABLE B209

PESTICIDES : LOSS BY LEACHING

ITEM:	VALUE:	CONDITIONS:	REF.:
*LOSS FROM SOIL	2 % OF APPLICATION	APPLICATION = 3.36 KG/HA, POWDER, SOYABEANS	1514

TABLE B210

PESTICIDES : LOSS BY RUNOFF

ITEM:	VALUE:	CONDITIONS:	REF.:
*ALDRIN	5.2 % OF APPLICATION	SLOPE=1.5 %,CULTIVATED,APPLICATION =SIMULATED RAINFALL,SOIL=SIL,KY	583

TABLE B210

PESTICIDES : LOSS BY RUNOFF

ITEM:	VALUE:	CONDITIONS:	REF.:
*ATRAZINE	5.05 %	APPL = 2.2 KG/HA, ZEA MAYS	598
*ATRAZINE	4.83 %	APPL = 4.5 KG/HA, ZEA MAYS	598
*ATRAZINE	5.0 % OF APPLICATION	SLOPE=14 %,APP=SURFACE,CORN,SOIL= SICL, 5 MONTHS, PA	597
*ATRAZINE	4.8 % OF APPLICATION	SLOPE=14 %,CORN, APP=SURFACE, SOIL =SICL, 5 MONTHS, PA	597
*ATRAZINE	13.3 % OF APPLICATION	SLOPE=3 %,FALLOW,APP=INCORPORATED SIM. RAINFALL,SOIL=SL,1 H,ALA	57
*ATRAZINE	1.9 % OF APPLICATION (TO SEDIMENT)	SLOPE=3 %,FALLOW,APP=INCORPORATED SIM. RAINFALL,SOIL=SL,1 H, ALA	57
*ATRAZINE	11.4 % OF APPLICATION (TO WATER)	SLOPE=3 %,FALLOW,APP=INCORPORATED SIM. RAINFALL,SOIL=SL,1 H, ALA	57
*ATRAZINE	3.04 % OF APPLICATION (TO SEDIMENT)	SLPOPE=6 %,FALLOW,APP=INCORPORATED SIM. RAINFALL,SOIL=SL,1 H, ALA	57
*ATRAZINE	7.14 % OF APPLICATION (TO WATER)	SLOPE=6 %,FALLOW,APP=INCORPORATED SIM. RAINFALL, SOIL=SL,1 H, ALA	57
*ATRAZINE	1.95 % OF APPLICATION (TO SEDIMENT)	SLOPE=2 %,FALLOW,APP=INCORPORATED SIM. RAINFALL, SOIL=SL,1 H, ALA	57
*ATRAZINE	4.49 % OF APPLICATION (TO WATER)	SLOPE=2 %,FALLOW,APP=INCORPORATED SIM. RAINFALL, SOIL=SL,1 H, ALA	57
*ATRAZINE	2.17 % OF APPLICATION (TO SEDIMENT)	SLOPE=4 %,FALLOW,APP=INCORPORATED SIM. RAINFALL, SOIL=SCL, 1 H,ALA	57
*ATRAZINE	10.3 % OF APPLICATION (TO WATER)	SLOPE=4 %,FALLOW,APP=INCORPORATED SIM. RAINFALL, SOIL=SCL, 1 H, ALA	57
*ATRAZINE	2.5 % OF APPLICATION	SLOPE=14 %,CORN,APPLICATION= SURFACE, SOIL=SICL, 5 MONTHS, PA	597
*ATRAZINE	16 % OF APPLICATION	SLOPE=10-15 5,CORN,APP=SURFACE, SOIL=SIL,58 D,IA	1344
*ATRAZINE	5.7 - 7.3 % OF APPLICATION	SLOPE=6.5 %,FALLOW,APP=SIMULATED RAINFALL,SOIL=SL, 1 H, GA	1693
*BENOMYL	1.1 %	APPL = 20 LB/ACRE, RAIN=3.5 IN, 7 D, KEN.	1322
*CS13529	4.97 %	APPL = 2.2 KG/HA, ZEA MAYS	598
*CS13529	5.70 %	APPL = 2.2 KG/HA, ZEA MAYS	598
*DDT	6.3 % OF APPLICATION	SLOPE=1.5 %,CULTIVATED,APPLICATION =SIMULATED RAINFALL,SOIL=SIL, KY	583
*DDT	1.6 % OF APPLICATION	SLOPE=8 %,CONT. POTATOES,APP= SURFACE,SOIL=GL,24 MOUNTHS,ME	432
*DDT	1.0 % OF APPLICATION	SLOPE=8 %,ROT. POTATOES,APP= SURFACE,SOIL=GL,24 MONTHS, ME	432
*DDT	0.81 % OF APPLICATION	SLOPE=8 %,OATS,APPLICATION= SURFACE,SOIL=GL,12 MONTHS, ME	432
*DDT	2.83 % OF APPLICATION	SLOPE=3 %,COTTON,APP=FOLIAR, SOIL= LS & SL, 6 MONTHS, NC	162
*DICAMBA	APP. 2 % OF APPLICATION	SLOPE=3-8 %,FALLOW SOD, APP= SIMULATED RAINFALL,4 MONTHS,SOIL= CL,TX	1612
*DICHLOBENIL	2.33 % OF APPLICATION (TO SEDIMENT)	SLOPE=3 %,FALLOW,APP=INCORPORATED SIM. RAINFALL, SOIL=SL, 1 H, ALA	57
*DICHLOBENIL	2.55 % OF APPLICATION (TO SEDIMENT)	SLOPE=6 %,FALLOW,APP=INCORPARATED SIM. RAINFALL, SOIL=SL, 1 H, ALA	57
*DICHLOBENIL	2.07 % OF APPLICATION (TO WATER)	SLOPE=6 %,FALLOW,APP=INCORPORATED SIM. RAINFALL, SOIL=SL, 1 H, ALA	57
*DICHLOBENIL	0.36 % OF APPLICATION (TO SEDIMENT)	SLOPE=2 %,FALLOW,APP=INCORPORATED SIM. RAINFALL, SOIL=SL, 1 H, ALA	57
*DICHLOBENIL	2.36 % OF APPLICATION (TO WATER)	SLOPE=2 %,FALLOW,APP=INCORPORATED SIM. RAINFALL, SOIL=SL, 1 H, ALA	57
*DICHLOBENIL	2.55 % OF APPLICATION (TO SEDIMENT)	SLOPE=4 %,FALLOW,APP=INCORPORATED SIM. RAINFALL, SOIL=SL, 1 H, ALA	57
*DICHLOBENIL	5.99 % OF APPLICATION (TO WATER)	SLOPE=4 %,FALLOW,APP=INCORPORATED SIM. RAINFALL, SOIL=SL, 1 H, ALA	57
*DICHLOBENIL	7.61 % OF APPLICATION (TO WATER)	SLOPE=3 %,FALLOW,APP=INCORPORATED SIM. RAINFALL, SOIL=SL,1 H, ALA	57
*DIELDRIN	0.7 % OF APPLICATION	SLOPE=14 %,MOST CORN,APP= INCORPORATED,SOIL=SIL,26 MONTHS, OHIO	229

TABLE B210

PESTICIDES : LOSS BY RUNOFF

ITEM:	VALUE:	CONDITIONS:	REF.:
*DIELDRIN	0.007 % OF APPLICATION	SLOPE=14 %,MOST CORN,APP= INCORPORATED,SOIL=SIL,8 MONTHS, OHIO	229
*DIELDRIN	4.7 % OF APPLICATION	SLOPE=1.5 %,CULTIVATED,APP= SIMULATED RAINFALL,SOIL=SIL, KY	583
*DIURON	LESS THAN 0.042 % OF APPLICATION	APPL=0.84 KG/HA, COTTON, SURFACE RUNOFF, RAIN=9.7 CM, MISS., 1971	1723
*DIURON	0.025 % OF APPLICATION	APPL=0.84 KG/HA, COTTON, SURFACE RUNOFF, RAIN=9.2 CM, MISS., 1972	1723
*DIURON	0.199 % OF APPLICATION	APPL=0.84 KG/HA, COTTON, SURFACE RUNOFF, RAIN=15.5 CM, MISS, 1973	1723
*DIURON	MAX 0.05 % OF APPLICATION	SLOPE=0.2 %,COTTON,APP=SURFACE PONDED,SOIL=SICL,60 D, LA	1724
*ENDOSULFAN	0.35 % OF APPLICATION	SLOPE=8 %,CONT. POTATOES,APP= SURFACE,SOIL=GL,12 MONTHS, ME	432
*ENDOSULFAN	0.25 % OF APPLICATION	SLOPE=8 %,ROT. POTATOES,APP= SURFACE,SOIL=GL,12 MONTHS, ME	432
*ENDOSULFAN	0.01 % OF APPLICATION	SLOPE=8 %,OATS,APPLICATION=SURFACE ,SOIL=GL, 12 MONTHS, ME	432
*ENDRIN	0.9 % OF APPLICATION	SLOPE=8 %,CONT. POTATOES,APP= SURFACE, SOIL=GL, 24 MONTHS,ME	432
*ENDRIN	0.6 % OF APPLICATION	SLOPE=8 %,ROT. POTATOES,APP= SURFACE, SOIL=GL, 24 MONTHS,ME	432
*ENDRIN	0.1 % OF APPLICATION	SLOPE=-,SUGARCANE,APP=SURFACE,SOIL =SICL,9 MONTHS, LA	1721
*ENDRIN	0.4 % OF APPLICATION	SLOPE=-,SUGARCANE,APP=SURFACE,SOIL =SICL, 2 MONTHS, LA	1721
*FENAC	0.004 % OF APPLICATION	APPL=3.36 KG/HA, SUGARCANE, SURFACE RUNOFF, RAIN=9.2 CM, MISS, 1972	1723
*FENAC	2.8 % OF APPLICATION	APPL=3.36 KG/HA, SUGARCANE, SURFACE RUNOFF, RAIN=15.5 CM, MISS, 1973	1723
*FENAC	2.87 % OF APPLICATION	SLOPE=0.2 %,SUGAR CANE,APPLICATION =SURFACE,SOIL=SICL,60 D, LA	1724
*GS 14254	0.02 % OF APPLICATION	SLOPE=14 %,ALFAFA,APP=SURFACE, SOIL=SICL, 5 MONTHS, PA	599
*GS 14254	0.03 % OF APPLICATION	SLOPE=14 %,ALFAFA,APPLICATION= SURFACE, SOIL=SICL, 5 MONTHS, PA	599
*LINORUN	0.029 % OF APPLICATION	SLOPE=0.2 %,COTTON,APP=SURFACE PONDED, SOIL=SICL,60 D, LA	1724
*LINURON	0.04 % OF APPLICATION	APPL=2.24 KG/HA,SOYBEAN, SURFACE RUNOFF, RAIN=9.7 CM, MISS., 1971	1723
*LINURON	0.02 % OF APPLICATION	APPL=2.24 KG/HA,SOYBEAN, SURFACE RUNOFF, RAIN=9.2 CM, MISS., 1972	1723
*LINURON	0.29 % OF APPLICATION	APPL=2.24 KG/HA,SOYBEAN, SURFACE RUNOFF, RAIN=15.5 CM, MISS, 1973	1723
*MBC	0.63 %	APPL = 20 LB/ACRE, RAIN=3.5 IN, 7 D, KEN.	1322
*METHOXYCHLOR	0.4 % OF APPLICATION	SLOPE=-,GRASS, APPLICATION= SIMULATED RAINFALL,SOIL=SIL,14 MONTHS,OHIO	394
*PICLORAM	2.8 - 5.0 % OF APPLICATION	SLOPE=3-8 %,FALLOW SOD,APP= SIMULATED RAINFALL,SOIL=CL, 4 MONTHS, TX	1612
*PICLORAM	0.02 % OF APPLICATION	SLOPE=5 %,GRASS, APPLICATION= FOLIAR, SOIL=C, 24 MONTHS,TX	156
*PROMETRYN	0.5 % OF APPLICATION	SLOPE=1 %,COTTON, APP=SURFACE, SOIL=SL, 5 MONTHS, PA	599
*PROPACHLOR	2.3 % OF APPLICATION	SLOPE=10-15 %,APPLICATION=SURFACE, CORN, SOIL=SIL,8 D,IA	1344
*TERBACIL	82.72 G/HA	IN OPEN DRAIN, SHALLOW TILLAGE, APPL=4 LBS/HA, 14 D,74	979
*TERBACIL	80.94 G/HA	IN SUBMERGED DRAIN, SHALLOW TILLAGE, APPL=4 LBS/HA, 14 D,74	979
*TERBACIL	5.91 G/HA	IN SUBMERGED DRAIN, SHALLOW TILLAGE, APPL=4 LBS/HA, 14 D,75	979
*TERBACIL	26.88 G/HA	IN OPEN DRAIN, DEEP TILLAGE, APPL= 4 LBS/HA, 14 D,74	979

TABLE B210

PESTICIDES : LOSS BY RUNOFF			
ITEM:	VALUE:	CONDITIONS:	REF.:
*TERBACIL	24.19 G/HA	IN SUBMERGED DRAIN, DEEP TILLAGE, APPL=4 LBS/HA, 14 D,74	979
*TERBACIL	7.94 G/HA	IN SUBMERGED DRAIN, DEEP TIL. & LIME, APPL=4 LBS/HA, 14 D,75	979
*TERBACIL	37.08 G/HA	IN OPEN DRAIN, DEEP TIL. & LIME, APPL=4 LBS/HA, 14 D,74	979
*TERBACIL	20.81 G/HA	IN SUBMERGED DRAIN, DEEP TIL. & LIME, APPL=4 LBS/HA, 14 D,74	979
*TERBACIL	1.49 G/HA	IN SUBMERGED DRAIN, DEEP TIL. & LIME, APPL=4 LBS/HA, 14 D,75	979
*TERBACIL	35.3 G/HA	SHALLOW TILLAGE,IN SUBMERGED DRAIN, APPL=4 LBS/HA,14 D, CITRUS	979
*TERBACIL	47.3 G/HA	DEEP TILLAGE,IN SUBMERGED DRAIN, APPL=4 LBS/HA,14 D, CITRUS	979
*TERBACIL	8.9 G/HA	DEEP TIL. & LIME,IN SUBMERGED DRAIN, APPL=4 LBS/HA,14 D, CITRUS	979
*TERBACIL	483.9 G/HA	SHALLOW TILLAGE,IN SUBMERGED DRAIN, APPL=4 LBS/HA,14 D, CITRUS	979
*TERBACIL	494.5 G/HA	SHALLOW TILLAGE,IN OPEN DRAIN, APPL=4 LBS/HA,14 D, CITRUS	979
*TERBACIL	144.6 G/HA	DEEP TILLAGE,IN SUBMERGED DRAIN, APPL=4 LBS/HA,14 D, CITRUS	979
*TERBACIL	159.5 G/HA	DEEP TILLAGE,IN OPEN DRAIN, APPL=4 LBS/HA,14 D, CITRUS	979
*TERBACIL	124.4 G/HA	DEEP TILLAGE,IN SUBMERGED DRAIN, APPL=4 LBS/HA,14 D, CITRUS	979
*TERBACIL	221.7 G/HA	DEEP TIL. & LIME,IN OPEN DRAIN, APPL=4 LBS/HA,14 D, CITRUS	979
*TERBACIL	82.72 G/HA	IN OPEN DRAIN, SHALLOW TILLAGE, APPL=4 LBS/HA, 14 D,74	979
*TERBACIL	80.94 G/HA	IN SUBMERGED DRAIN, SHALLOW TILLAGE, APPL=4 LBS/HA, 14 D,74	979
*TERBACIL	5.91 G/HA	IN SUBMERGED DRAIN, SHALLOW TILLAGE, APPL=4 LBS/HA, 14 D,75	979
*TERBACIL	26.88 G/HA	IN OPEN DRAIN, DEEP TILLAGE, APPL=4 LBS/HA, 14 D,74	979
*TERBACIL	24.19 G/HA	IN SUBMERGED DRAIN, DEEP TILLAGE, APPL=4 LBS/HA, 14 D,74	979
*TERBACIL	7.94 G/HA	IN SUBMERGED DRAIN, DEEP TIL. & LIME, APPL=4 LBS/HA, 14 D,75	979
*TERBACIL	37.08 G/HA	IN OPEN DRAIN, DEEP TIL. & LIME, APPL=4 LBS/HA, 14 D,74	979
*TERBACIL	20.81 G/HA	IN SUBMERGED DRAIN, DEEP TIL. & LIME, APPL=4 LBS/HA, 14 D,74	979
*TERBACIL	1.49 G/HA	IN SUBMERGED DRAIN, DEEP TIL. & LIME, APPL=4 LBS/HA, 14 D,75	979
*TERBACIL	35.3 G/HA	SHALLOW TILLAGE,IN SUBMERGED DRAIN, APPL=4 LBS/HA,14 D, CITRUS	979
*TERBACIL	47.3 G/HA	DEEP TILLAGE,IN SUBMERGED DRAIN, APPL=4 LBS/HA,14 D, CITRUS	979
*TERBACIL	8.9 G/HA	DEEP TIL. & LIME,IN SUBMERGED DRAIN, APPL=4 LBS/HA,14 D, CITRUS	979
*TERBACIL	483.9 G/HA	SHALLOW TILLAGE,IN SUBMERGED DRAIN, APPL=4 LBS/HA,14 D, CITRUS	979
*TERBACIL	494.5 G/HA	SHALLOW TILLAGE,IN OPEN DRAIN, APPL=4 LBS/HA,14 D, CITRUS	979
*TERBACIL	144.6 G/HA	DEEP TILLAGE,IN SUBMERGED DRAIN, APPL=4 LBS/HA,14 D, CITRUS	979
*TERBACIL	159.5 G/HA	DEEP TILLAGE,IN OPEN DRAIN, APPL=4 LBS/HA,14 D, CITRUS	979
*TERBACIL	124.4 G/HA	DEEP TILLAGE,IN SUBMERGED DRAIN, APPL=4 LBS/HA,14 D, CITRUS	979
*TERBACIL	221.7 G/HA	DEEP TIL. & LIME,IN OPEN DRAIN, APPL=4 LBS/HA,14 D, CITRUS	979
*TOXAPHENE	0.36 % OF APPLICATION	SLOPE=3 %,COTTON,APP=FOLIAR, SOIL= LS & SL,6 MONTHS, NC	162
*TRIFLURALIN	0.004 % OF APPLICATION	APPL=1.40 KG/HA,SOYBEAN, SURFACE RUNOFF, RAIN=13.8 CM, MISS, 1971	1723

TABLE B210

PESTICIDES : LOSS BY RUNOFF

ITEM:	VALUE:	CONDITIONS:	REF.:
*TRIFLURALIN	NEGLIGIBLE	APPL=1.40 KG/HA,SOYBEAN, SURFACE RUNOFF, RAIN=26.5 CM, MISS, 1971	1723
*TRIFLURALIN	0.4 % OF APPLICATION	APPL=1.40 KG/HA,SOYBEAN, SURFACE RUNOFF, RAIN=27.5 CM, MISS, 1973	1723
*TRIFLURALIN	0.04 % OF APPLICATION	SLOPE=0.2 %,COTTON AND SOYBEANS, APP=INCORPORATED,SOIL=SICL,60 D,LA	1724
*2,4-D	15.26 G/HA	IN OPEN DRAIN, SHALLOW TILLAGE, APPL=3 LBS/HA, 14 D,74	979
*2,4-D	7.66 G/HA	IN SUBMERGED DRAIN, SHALLOW TILLAGE, APPL=3 LBS/HA, 14 D,74	979
*2,4-D	NONE	IN OPEN DRAIN, DEEP TILLAGE, APPL= 3 LBS/HA, 14 D,74	979
*2,4-D	3.19 G/HA	IN SUBMERGED DRAIN, DEEP TILLAGE, APPL=3 LBS/HA, 14 D,74	979
*2,4-D	20.55 G/HA	IN OPEN DRAIN, DEEP TIL. & LIME, APPL=3 LBS/HA, 14 D,74	979
*2,4-D	8.81 G/HA	IN SUBMERGED DRAIN, DEEP TIL. & LIME, APPL=3 LBS/HA, 14 D,74	979
*2,4-D	45.8 G/HA	SHALLOW TILLAGE,IN SUBMERGED DRAIN, APPL=3 LBS/HA,14 D, CITRUS	979
*2,4-D	91.2 G/HA	SHALLOW TILLAGE,IN OPEN DRAIN, APPL=3 LBS/HA,14 D, CITRUS	979
*2,4-D	19.1 G/HA	DEEP TILLAGE,IN SUBMERGED DRAIN, APPL=3 LBS/HA,14 D, CITRUS	979
*2,4-D	0 G/HA	DEEP TILLAGE,IN OPEN DRAIN, APPL=3 LBS/HA,14 D, CITRUS	979
*2,4-D	52.7 G/HA	DEEP TIL. & LIME,IN SUBMERGED DRAIN, APPL=3 LBS/HA,14 D, CITRUS	979
*2,4-D	122.9 G/HA	DEEP TIL. & LIME,IN OPEN DRAIN, APPL=3 LBS/HA,14 D, CITRUS	979
*2,4-D	0.33 %	1970, SANDY COASTAL PLAIN SOIL, APPL = 0.56 KG/HA	1694
*2,4-D	0.27 %	1971, SANDY COASTAL PLAIN SOIL, APPL = 0.56 KG/HA	1694
*2,4-D	0.04 %	1972, SANDY COASTAL PLAIN SOIL, APPL = 0.56 KG/HA	1694
*2,4-D	15.26 G/HA	IN OPEN DRAIN, SHALLOW TILLAGE, APPL=3 LBS/HA, 14 D,74	979
*2,4-D	7.66 G/HA	IN SUBMERGED DRAIN, SHALLOW TILLAGE, APPL=3 LBS/HA, 14 D,74	979
*2,4-D	NONE	IN OPEN DRAIN, DEEP TILLAGE, APPL= 3 LBS/HA, 14 D,74	979
*2,4-D	3.19 G/HA	IN SUBMERGED DRAIN, DEEP TILLAGE, APPL=3 LBS/HA, 14 D,74	979
*2,4-D	20.55 G/HA	IN OPEN DRAIN, DEEP TIL. & LIME, APPL=3 LBS/HA, 14 D,74	979
*2,4-D	8.81 G/HA	IN SUBMERGED DRAIN, DEEP TIL. & LIME, APPL=3 LBS/HA, 14 D,74	979
*2,4-D	45.8 G/HA	SHALLOW TILLAGE,IN SUBMERGED DRAIN, APPL=3 LBS/HA,14 D, CITRUS	979
*2,4-D	91.2 G/HA	SHALLOW TILLAGE,IN OPEN DRAIN, APPL=3 LBS/HA,14 D, CITRUS	979
*2,4-D	19.1 G/HA	DEEP TILLAGE,IN SUBMERGED DRAIN, APPL=3 LBS/HA,14 D, CITRUS	979
*2,4-D	0 G/HA	DEEP TILLAGE,IN OPEN DRAIN, APPL=3 LBS/HA,14 D, CITRUS	979
*2,4-D	52.7 G/HA	DEEP TIL. & LIME,IN SUBMERGED DRAIN, APPL=3 LBS/HA,14 D, CITRUS	979
*2,4-D	122.9 G/HA	DEEP TIL. & LIME,IN OPEN DRAIN, APPL=3 LBS/HA,14 D, CITRUS	979
*2,4-D-AMINE	APP. 5 % OF APPLICATION	SLOPE=7 %,CULTIVATED,APP=SIMULATED RAINFALL,1 H,SOIL=SL,GA	73
*2,4-D-AMINE	APP. 4 % OF APPLICATION	SLOPE=7 %,CULTIVATED,APP=SIMULATED RAINFALL,SOIL=SL,48 H, GA	73
*2,4-D-AMINE	APP. 5 % OF APPLICATION	SLOPE=7 %,CULTIVATED,APP=SIMULATED RAINFALL,SOIL=SL, 96 H, GA	73
*2,4-D-BUTYLETHER	APP. 35 % OF APPLICATION	SLOPE=7 %,CULTIVATED,APP=SIMULATED RAINFALL,1 H, SOIL=SL, GA	73

TABLE B210

PESTICIDES : LOSS BY RUNOFF			
ITEM:	VALUE:	CONDITIONS:	REF.:
*2,4-D-ISOCTYL	APP 40 % OF APPLICATION	SLOPE=7 %,CULTIVATED,APP=SIMULATED RAINFALL, 1 H,SOIL=SL, GA	73
*2,4,5-T	0.05 % OF APPLICATION	SLOPE=-,GRASS,APP=SIMULATED RAINFALL, SOIL=SIL, 14 MONTHS, OHIO	394
*2,4,5-T	APP. 2 % OF APPLICATION	SLOPE=3-8 %,FALLOW SOD,APP= SIMULATED RAINFALL,4 MONTHS,SOIL= CL, TX	1612
*2,4,5-T	0.02 % OF APPLICATION	SLOPE=5 %,GRASS, APPLICATION= FOLIAR,SOIL=C, 24 MONTHS,TX	156

TABLE B211

PESTICIDES : LOSS BY VOLATILIZATION			
ITEM:	VALUE:	CONDITIONS:	REF.:
*DDT	80 % IN 1 YEAR	TOTAL DDT, DESERT SOIL, APPL = 1-2 LB/ACRE	1658
*DDT	92 % IN 1 YEAR	P,P'-DDT, DESERT SOIL, APPL. = 1-2 LB/ACRE	1658
*DDT	88 % IN 5 MONTHS	APPL = 2.0 LB/ACRE IN JUNE, ARIZ., FALLOW SOIL	1659
*DDT	% IN 5 MONTHS	APPL = 2.0 LB/ACRE IN JUNE, ARIZ., 1RRIGATED SOIL	1659
*DDT	67 % IN 5 MONTHS	APPL =32.0 LB/ACRE IN JUNE, ARIZ., FALLOW SOIL	1659
*DDT	50 % IN 5 MONTHS	APPL = 3.0 LB/ACRE IN JUNE, ARIZ., IRRIGATED SOIL	1659
*DIELDRIN	4.0 G/(HA*DAY)	INCORPORATED, MAX. VALUE, APPLICATION=5.6 KG/HA IN A 7.5 CM DEEP LAYE	1556
*DIELDRIN	199 G/(HA/ GROWING SEASON)	MEAN OF GROWING SEASON, APPL=5.6 KG/HA,7.5 CM DEEP LAYER	1556
*DIELDRIN	654 G/(HA*DAY)	1 DAY AFTER APPL. WITH 5.6 KG/HA, FARM RIG, GRASS,OBSERVED VALUE	789
*DIELDRIN	325 G/(HA*DAY)	2 DAY AFTER APPL. WITH 5.6 KG/HA, FARM RIG, GRASS,OBSERVED VALUE	789
*DIELDRIN	282 G/(HA*DAY)	3 DAY AFTER APPL. WITH 5.6 KG/HA, FARM RIG, GRASS,OBSERVED VALUE	789
*DIELDRIN	53.7 G/(HA*DAY)	6 DAY AFTER APPL. WITH 5.6 KG/HA, FARM RIG, GRASS,OBSERVED VALUE	789
*DIELDRIN	40.0 G/(HA*DAY)	9 DAY AFTER APPL. WITH 5.6 KG/HA, FARM RIG, GRASS,OBSERVED VALUE	789
*DIELDRIN	9.2 G/(HA*DAY)	14 DAY AFTER APPL. WITH 5.6 KG/HA, FARM RIG, GRASS,OBSERVED VALUE	789
*DIELDRIN	6.2 G/(HA*DAY)	23 DAY AFTER APPL. WITH 5.6 KG/HA, FARM RIG, GRASS,OBSERVED VALUE	789
*HEPTACHLOR	5.0 G/(HA*DAY)	INCORPORATED, MAX. VALUE, APPLICATION=5.6 KG/HA IN A 7.5 CM DEEP LAYER	1556
*HEPTACHLOR	383 G/(HA/ GROWING SEASON)	INCORPORATED, MEAN OF GROWING SEASON, APPL=5.6 KG/HA, 7.5 CM DEEP	1556
*HEPTACHLOR	2554 G/(HA*DAY)	1 DAY AFTER APPL. WITH 5.6 KG/HA, FARM RIG, GRASS,OBSERVED VALUE	789
*HEPTACHLOR	335 G/(HA*DAY)	2 DAY AFTER APPL. WITH 5.6 KG/HA, FARM RIG, GRASS,OBSERVED VALUE	789
*HEPTACHLOR	132 G/(HA*DAY)	3 DAY AFTER APPL. WITH 5.6 KG/HA, FARM RIG, GRASS,OBSERVED VALUE	789
*HEPTACHLOR	33.5 G/(HA*DAY)	6 DAY AFTER APPL. WITH 5.6 KG/HA, FARM RIG, GRASS,OBSERVED VALUE	789
*HEPTACHLOR	24.1 G/(HA*DAY)	9 DAY AFTER APPL. WITH 5.6 KG/HA, FARM RIG, GRASS,OBSERVED VALUE	789
*HEPTACHLOR	6.9 G/(HA*DAY)	14 DAY AFTER APPL. WITH 5.6 KG/HA, FARM RIG, GRASS,OBSERVED VALUE	789

TABLE B211

PESTICIDES : LOSS BY VOLATILIZATION			
ITEM:	VALUE:	CONDITIONS:	REF.:
*HEPTACHLOR	7.4 G/(HA*DAY)	23 DAY AFTER APPL. WITH 5.6 KG/HA, FARM RIG, GRASS,OBSERVED VALUE	789
*TRIFLURALIN	0.0517 KG/(HA*DAY)	INCORPORATED, 14 KG/HA	1491
*TRIFLURALIN	4.0 KG/(HA*DAY)	SURFACE APPLICATION, 10 KG/HA	1491

TABLE B212

PESTICIDES : LOSS DURING APPLICATION			
ITEM:	VALUE:	CONDITIONS:	REF.:
*DYFONATE	17.6 % OF APPLICATION	APPL=3.54 LB/ACRE, SPRAY FROM TRACTOR,CALM,298 K,BY PAN ANALYS, 1969	821
*DYFONATE	24.1 % OF APPLICATION	APPL=3.54 LB/ACRE, SPRAY FROM TRACTOR,CALM,298 K,BY SOIL ANALYS,196 9	821
*DYFONATE	12.6 % OF APPLICATION	APPL=3.54 LB/ACRE, SPRAY FROM TRACTOR,CALM,298 K,BY PAN ANALYS, 1970	821
*DYFONATE	16.7 % OF APPLICATION	APPL=3.54 LB/ACRE, SPRAY FROM TRACTOR,CALM,298 K,BY SOIL ANALYS,197 0	821
*ZINOPHOS	39.5 % OF APPLICATION	APPL=3.54 LB/ACRE, SPRAY FROM TRACTOR,CALM,298 K,BY PAN ANALYS	821
*ZINOPHOS	37.3 % OF APPLICATION	APPL=3.54 LB/ACRE, SPRAY FROM TRACTOR,CALM,298 K,BY SOIL ANALYS	821

TABLE B213

PESTICIDES : LOSS FROM SOIL			
ITEM:	VALUE:	CONDITIONS:	REF.:
*DDT	18 % IN 5 YEARS	HOOD RIVER SOIL, SANDY LOAM, NO CULTIVATION, 284 K, PRECIP=30 INCH	821
*DDT	17 % IN 5 YEARS	MEDFORD SOIL, CLAY ADOBE, CULTIVATED, 284 K, PRECIP=30 INCHES	821

TABLE B214

PESTICIDES : QUANTUM YIELD FOR PHOTOLYSIS			
ITEM:	VALUE:	CONDITIONS:	REF.:
*CARBARYL	0.016	IN WATER,313NM,298K,DEGASSED,PH= 5.5	1743
*CARBARYL	0.0056	IN WATER,313NM,298K,AIR-SATURATED, PH=5.5	1743
*CARBARYL	0.0052	IN WATER,313NM,298K,AIR-SATURATED, PH=5.2	1743
*CARBARYL	0.0058	IN WATER,313NM,298K,AIR-SATURATED, PH=6.1	1743
*CARBARYL	0.0055	IN WATER,313NM,298K,AIR-SATURATED, PH=6.9	1743
*METHOXYCHLOR	0.12	IN N-HEXANE,METHOXYCHLOR CONC.= 0.003M	1743
*METHOXYCHLOR	0.32	IN 1.1M WATER IN ACETONITRILE, METHOXYCHLOR CONC.=0.003M	1743

TABLE B214

PESTICIDES : QUANTUM YIELD FOR PHOTOLYSIS			
ITEM:	VALUE:	CONDITIONS:	REF.:
*METHOXYCHLOR	0.3	IN WATER,METHOXYCHLOR CONC.=30PPB	1743
*2,4-D BUTOXYETHYL ESTER	0.056	PHOTOLYSIS IN WATER,313NM,301K,PH= 5.3	1743
*2,4-D BUTOXYETHYL ESTER	0.052	PHOTOLYSIS IN WATER,313NM,301K,PH= 6.6	1743
*2,4-D BUTOXYETHYL ESTER	0.056	PHOTOLYSIS IN WATER,313NM,301K,PH= 7.8	1743
*2,4-D BUTOXYETHYL ESTER	0.17	PHOTOLYSIS IN N-HEXANE,313NM,301K	1743
*2,4-D BUTOXYETHYL ESTER	0.17	PHOTOLYSIS IN N-HEXADECANE,313NM, 301K	1743
*2,4-D BUTOXYETHYL ESTER	0.0080	PHOTOLYSIS IN N-HEXANE,313NM,301K, ACETONE-SENSITIZED	1743
*2,4-D METHYL ESTER	0.031	PHOTOLYSIS IN WATER,313NM,301K,PH= 6.6	1743
*2,4-D METHYL ESTER	0.13	PHOTOLYSIS IN N-HEXADECANE,313NM, 301K	1743

TABLE B215

PESTICIDES : SOLUBILITY			
ITEM:	VALUE:	CONDITIONS:	REF.:
*ALDRIN	0.01-0.2 PPM	293-298K	389
*ASULAM	5000.0 PPM	PURITY = 40.3 %, IN WATER, SODIUM SALT	233
*BUTRALIN	LE IN WATER, SODIUM SALT THAN 1.0 PPM	PURITY = 61.9 %, IN WATER, EMULSIFIABLE CONCENTRATE	233
*CAMPHECHLOR	0.4-3.0 PPM	293-298K	389
*CHLORDANE	VERY LOW	293-298K	389
*DDT	0.001-0.04 PPM	293-298K	389
*DICAMBA	4500.0 PPM	PURITY = 49.5 %, IN WATER, DIMETHYLAMINE SALT	233
*DIELDRIN	0.1-0.25 PPM	293-298K	389
*ENDRIN	0.1-0.25 PPM	293-298K	389
*FLUOMETURON	85.0 PPM	PURITY = 83.5 %, IN WATER, WETTABLE POWDER	233
*HEPTACHLOR	VERY LOW	293-298K	389
*LEPTOPHOS	0.03 PPM	PURITY = 0.321 G/ML, IN WATER, WETTABLE POWDER	233
*LINDANE	7.0-10.0 PPM	293-298K	389
*METHAZOLE	1.5 PPM	PURITY = 88.5 %, IN WATER, WETTABLE POWDER	233
*NORFLURAZON	28.0 PPM	PURITY = 88.5 %, IN WATER, WETTABLE POWDER	233
*OXADIAZON	0.7 PPM	PURITY = 0.223 G/ML, IN WATER, EMULSIFIABLE CONCENTRATE	233
*PROFLURALIN	LE IN WATER, SODIUM SALT THAN 1.0 PPM	PURITY = 80.4 %, IN WATER, WETTABLE POWDER	233
*PROMETRYN	48.0 PPM	PURITY = 80.4 %, IN WATER, WETTABLE POWDER	233
*TH 6040	0.3 PPM	PURITY = 25.9 %, IN WATER, WETTABLE POWDER	233

TABLE B216

PESTICIDES : SPECIFIC SUNLIGHT ABSORPTION RATE			
ITEM:	VALUE:	CONDITIONS:	REF.:
*ATRAZINE	6.6*10**6 1/SEC	MIDDAY SUN(MIDSUMMER),LAT.=40N	1743

TABLE B216

PESTICIDES : SPECIFIC SUNLIGHT ABSORPTION RATE			
ITEM:	VALUE:	CONDITIONS:	REF.:
*CARBARYL	623*10**6 1/SEC	MIDDAY SUN(MIDSUMMER),LAT.=40N	1743
*DIAZINON	6.2*10**6 1/SEC	MIDDAY SUN(MIDSUMMER),LAT.=40N	1743
*MALATHION	1.2*10**6 1/SEC	MIDDAY SUN(MIDSUMMER),LAT.=40N	1743
*METHOXYCHLOR	0.87*10**6 1/SEC	MIDDAY SUN(MIDSUMMER),LAT.=40N	1743
*PARATHION	6930*10**6 1/SEC	MIDDAY SUN(MIDSUMMER),LAT.=40N	1743
*2,4-D ESTER	194*10**6 1/SEC	MIDDAY SUN(MIDSUMMER),LAT.=40N	1743

TABLE B217

PESTICIDES : TRADE AND CHEMICAL NAMES			
ITEM:	VALUE:	CONDITIONS:	REF.:
*AKTON	O-(2-CHLORO-1-(2,5-DICHLOROPHENYL) VINYL)-O,O-DIETHYL PHOSPHOROTHIOATE	NONE	519
*ALACHLOR	2-CHLORO-2',6'DIETHYL-N-(METHOXYMETHYL)-ACETANILDE	NONE	906
*ALDRIN	1,2,3,4,10,10-HEXACHLORO-1,4,4A,5,8,8A-HEXAHYDRO-1,4-ENDO-EXO-WWW	WWW = 5,8-DIMETHANONAPHTHALENE	857
*ALDRIN	1,2,3,4,10,10,HEXACHLORO-1,4,4A,5,8,8A-HEXAHYDRO-EXO-1,4-ENDO-5,8-ZZ	ZZ = DIMETHANONAPHATHALENE	906
*AROCLOR 1254	POLYCHLORINATED BIPHENYL (54 % CHLORINE)	NONE	857
*AROCLOR 1262	POLYCHLORINATED BIPHENYL (62 % CHLORINE)	NONE	857
*ASULAM	METHYL SULFANILYLCARBAMATE	NONE	233
*ATRAZINE	2-CHLORO-4-(ETHYLAMINO)-6-(ISOPROPYLAMINO)-1,3,5-TRIAZINE	NONE	906
*AZINPHOSMETHYL	O,O-DIMETHYL S-((4-OXO-1,2,3-BENZO-TRIAZINE-3(4H)-YL)METHYL) ZZ	ZZ = PHOSPHORODITHIOATE	906
*BARBAN	4-CHLORO-2-BUTYNYL M-CHLOROCARBANILATE	NONE	906
*BENOMYL	2-AMINO-BENZIMIDAZOLE	NONE	1322
*BENSULPHIDE	S-(O,O-DIISOPROPYL PHOSPHORODITHIOATE) ESTER OF ZZ	ZZ = N-(2-MERCAPTOETHYL) BENZENESULFONAMIDE	906
*BENTAZON	3-ISOPROPYL-1H-2,1,3-BENZOTHIADIAZIN-(4)3H-ONE 2,2-DIOXIDE	NONE	1514
*BHC	1,2,3,4,5,6-HEXACHLOROCYCLOHEXANE	SEVERAL ISOMERS AND A SPECIFIED % OF GAMMA ISOMER	906
*BHC	BHC	SEE BHC	906
*BROMACIL	5-BROMO-3-SEC-BUTYL-6-METHYLURACIL	NONE	906
*BUFENCARB	3-(1-METHYLBUTYL) PHENYL-6-METHYLCARBAMATE + ZZ	ZZ = 3-(1-ETHYLPROPYL)PHENYL METHYLCARBAMATE, (3:1)	906
*BUTRALIN	4-(1,1-DIMETHYLETHYL)-N-(1-METHYLPROPYL)-2,6-DINITROBENZENAMINE	NONE	233
*BUTYLATE	S-ETHYL DIISOBUTYLTHIOCARBAMATE	NONE	906
*CARBARYL	1-NAPHTHYL METHYLCARBAMATE	NONE	796
*CARBARYL	1-NAPHTHYL N-METHYLCARBAMATE	NONE	906
*CARBOFURAN	2,3-DIHYDRO-2,2-DIMETHYL-7-BEN-	NONE	906
*CDAA-T	N,N-DIALLYL-2-CHLOROACETEAMIDE	NONE	906
*CERESAN M	N-(ETHYLMERCURI)-P-TOLUENE SULFONANILIDE	NONE	857
*CHLORABEN	3-AMINO-2,5-DICHLOROBENZOIC ACID	NONE	906
*CHLORDANE	1,2,4,5,7,8,8-OCTACHLORO-2,3,3A,4,7,7A-HEXAHYDRO-4,7,METHANOINDENE	NONE	857
*CHLORDANE	1,2,4,5,6,7,8,8-OCTOCHLORO-2,3,3A,4,7,7A-HEXAHYDRO-4,7-METHANOINDANE	60% MINIMUM AND NOT OVER 40% OF RELATED COMPOUNDS	906

TABLE B217

PESTICIDES : TRADE AND CHEMICAL NAMES

ITEM:	VALUE:	CONDITIONS:	REF.:
*CHLOROBROMURON	3-(4-BROMO;3-CHLOROPHENYL) -1-METHOXY-1-METHYL-UREA	NONE	802
*CHLOROXURON	3-(P-(P-CHLOROPHENOXY)PHENYLL)-1, 1-DIMETHYLUREA	NONE	906
*CHLORPROPHAM	ISOPROPYL M-CHLOROCARBANILATE	NONE	906
*CHLORPYRIFOS	0,0-DIETHYL 0-(3,5,6-TRICHLORO-	NONE	906
*CO-RAL	0,0-DIETHYL-0, 3-CHLORO-4-METHYL-1-OXO-2H-1-BENZO PYRAN-7-YL WWW	WWW = PHOSPHOROTHIOATE	857
*DALAPON	2,2-DICHLOROPROPIONIC ACID	NONE	906
*DDE	1,1-DICHLORO-2,2-BIS- (P-CHLOROPHENYL) ETHYLENE	NONE	857
*DDT	1,1,1-TRICHLORO-2,2-BIS- (P-CHLOROPHENYL) ETHANE	NONE	857
*DDT	1,1,1-TRICHLORO-2,2-BIS (P-CHLOPHENYL)ETHANE	ISOMERIC MIX OF DICHLORODI-	906
*DIAZINON	0,0-DIETHYL 0-(2-ISOPROPYL 4-METHYL-6-PYRIMIDINYL) PHOSPHOROTHIOATE	NONE	857
*DIAZINON	0,0-DIETHYL 0-(2-ISOPROPYL-4-	ZZ = PHOSPHOROTHIOATE	906
*DIBROM	1,2-DIBROMO-2,2-DICHLOROETHYL DIMETHYL PHOSPHATE	NONE	1628
*DICAMBA	3,6-DICHLORO-O-ANISIC ACID	NONE	233
*DICAMBA	3,6-DICHLORO-O-ANISIC ACID	NONE	906
*DICHLOBENIL	2,6-DICHLOROBENZONITRILE ACID	NONE	906
*DICOFOL	1,1-BIS(CHLOROPHENYL)-2,2, 2-TRICHLOROETHANOL	NONE	906
*DIELDRIN	1,2,3,4,10,10,-HEXACHLORO-6, 7-EPOXY-1,4,4A,5,6,7,8, 8A-OCTAHYDRO-WWW	WWW = 1,4-ENDO-EXO-5, 8-DIMETHANONAPHTHALENE	857
*DIELDRIN	HEXACHLORO-EPOXY-OCTAHYDRO-ENDO-EX O-DIMETHANOL-NAPHTHALENE	85 % MINIMUM	906
*DIPHENAMID	N-DIMETHYL-2,2-DIPHENYLACETAMIDE	NONE	906
*DIQUAT	6,7-DIHYDRODIPYRIDO-(1,2-X:2', 1'-C)PYRAZIDINIUM DIBROMIDE	X = ALPHA	906
*DISULFOTON	0,0-DIETHYL S-(2-(ETHYLTHIO)ETHYL) PHOSPHORODITHIOATE	NONE	906
*DIURON	3-(3,4-DICHLOROPHENYL)-1, 1-DIMETHYLUREA	NONE	906
*DSMA	DISODIUM METHANEARSONATE	NONE	906
*DYFONATE	O-ETHYL-S-PHENYL-ETHYLPHOSPHONODIT HIOATE	NONE	821
*ENDOSULFAN	6,7,8,9,10,10-HEXACHLORO-1,5,5A,6, 9,9A-HEXAHYDRO-6,9-METHANO-ZZ	ZZ = 2,4,3-BENZO(E) -DIOXATHIEPIN-3-OXIDE	906
*ENDRIN	1,2,3,4,10,10-HEXACHLORO-6, 7-EPOXY-1,4,4A,5,6,7,8, 8A-OCTAHYDRO-WWW	WWW = 1,4-ENDO-ENDO-5, 8-DIMETHAONAPHTHALENE	857
*ENDRIN	HEXACHLOROEPOXYOCTAHYDRO-ENDO, ENDODIMETHANONAPHTHALENE	NONE	906
*EPN	O-ETHYL O-P-NITROPHENYL PHENYLPHOSPHONOTHIOATA	NONE	857
*EPTC	S-ETHYL N,N-DIPROPYLTHIOCARBAMATE	NONE	906
*FENAC	(2,3,6-TRICHLOROPHENYL)ACETIC ACID	NONE	906
*FENITROTHION	0,0-DIMETHYL 0-)4-NITRO-M-TOLYL) PHOSPHOROTHIOATA	NONE	857
*FENSULFOTHION	0,0-DITHYL 0-(4-(METHYLSULFINYL) PHENYL)PHOSPHOROTHIOATE	NONE	906
*FENURON	3-PHENYL-1,1-DIMETHYLUREA	NONE	906
*FLUOMETURON	1,1-DIMETHYL-3-(X,X, X-TRIFLURO-M-TOLYL)UREA	X=ALPHA	233
*FLUOMETURON	1,1-DIMETHYL-3-(X,X, X-TRIFLUORO-M-TOYL) UREA	X = ALPHA	906
*GS13529	2-CHLORO-4-ETHYLAMINO-6-TERT- BUTYLAMINO-S-TRIAZINE	NONE	598

TABLE B217

PESTICIDES : TRADE AND CHEMICAL NAMES			
ITEM:	VALUE:	CONDITIONS:	REF.:
*GUTHION	O,O-DIMETHYL S-(4-OXO-1,2,3-BENZOTRIAZIN-3(4H)-YLMETHYL) WWW	WWW = PHOSPHORODITHIOATE	857
*HCP	2,2'-METHYLENE-BIS-(3,4,6-TRI-CHLOROPHENOL)	NONE	1439
*HEPTACHLOR	1,4,5,6,7,8,8-HEPTACHLORO-3A,4,7,7A-TETRAHYDRO-4,7-METHANOINDENE	NONE	906
*HEXACHLOROPHENE	2,2'-METHYLENE-BIS-(3,4,6-TRI-CHLOROPHENOL)	NONE	1439
*LEPTOPHOS	O-(4-BROMO-2,5-DICHLOROPHENYL) O-METHYL PHENYLPHOSPHONOTHIOATE	NONE	233
*LINDANE	1,2,3,4,5,6-HEXACHLOROCYCLOHEXANE	GAMMA ISOMER OF NOT LESS THAN 99% PURITY	906
*LINURON	3-(3,4-DICHLOROPHENYL) -1-METHOXY-1-METHYLUREA	NONE	1723
*LINURON	3-(3,4-DICHLOROPHENYL) -1-METHOXY-1-METHYLUREA	NONE	906
*MALATHION	S-(1,2-BIS(ETHOXYCARBONYL)ETHYL) O,O-DIMETHYL PHOSPHORODITHIOATE	NONE	857
*MALATHION	O,O-DIMETHYL S-(1,2-DICARBETHOXYETHYL)PHOSPHORODITHIOATE	NONE	906
*MCPA	((4-CHLORO-O-TOXYL)OXY)ACETIC ACID	NONE	906
*METHAZOLE	2-(3,4-DICHLOROPHENYL)-4-METHYL-1,2,4OXADIAZOLIDINE-3,5-DIONE	NONE	233
*METHOPRENE	1, ISOPROPYL (2E,4E)-11-METHOXY-3,7,11-TRIMETHYL-2,4-DODECADIENOATE	NONE	1400
*METHOXYCHLOR	1,1,1-TRICHLORO-2,2-BIS (P-METHOXYPHENYL)ETHANOL	NONE	906
*METHYL PARATHION	O,O-DIMETHYL O-P-NITROPHENYL PHOSPHOROTHIOATE	NONE	906
*METHYLENE CHLORIDE	DICHLOROMETHANE	NONE	906
*MIREX	DODECACACHLOROOCTAHYDRO-1,2,4-METHENO-2H-CYCLOBUTA (CD) PENTALENE	NONE	19
*MONOCROTOPHOS	DIMETHYL PHOSPHATE OF -3-HYDROXY-N-METHYL-CIS-CROTONAMIDE	NONE	906
*MONURON	3-(P-CHLOROPHENYL)-1,1-DIMETHYLUREA	NONE	906
*MSMA	MONOSODIUM METHANEARSONATE	NONE	906
*NAPROAMIDE	2-(ALPHA-NAPHTHOXY)-N,N-DIETHYLPROPIONAMIDE	NONE	1650
*NITRALIN	4-METHYLSULFONYL-2,6-DINITRO-N,N-DIPROPYLANILINE	NONE	906
*NORFLURAZON	4-CHLORO-5-(METHYLAMONI)-2-(X,X,X-TRIFLURO-M-TOLYL)-3(2H)-ZZZ	X = ALPHA, ZZZ = PYRIDAZINONE	233
*OXADIAZON	2-TERT-BUTYL-4-(2,4 DICHLORO-5-ISOPROPOXYPHENYL)-X-1,3,4,-W	X = DELTA**2, W = OXADIAZLIN-5-ONE	233
*PARAQUAT	1,1'-DIMETHYL-4,4'-DIPYRIDINIUM ION	NONE	906
*PARATHION	O,O-DIETHYL O-P-NITROPHENYL PHOSPHOROTHIOATE	NONE	857
*PARATHION	O,O-DIETHYL O-P-NITROPHENYL PHOSPHOROTHIOATE	NONE	906
*PHORATE	O,O-DIETHYL S-ETHYLTHIOMETHYL PHOSPHORODITHIATE	NONE	906
*PHOSPHAMIDON	2-CHLORO-2-DIETHYLCARBAMOYL-1-METHYLVINYL DIMETHYL PHOSPHATE	NONE	857
*PHOSPHAMIDON	2-CHLORO-N,N-DIETHYL-3-HYDROXYCROTON AMIDE, DIMETHYLPHOSPHATE	NONE	1629
*PICLORAM	4-AMINO-3,5,6-TRICHLOROPICOLINIC ACID	NONE	906
*PROFLURALIN	N-(WWW)-X,X,X-TRIFLURO-2,6-DINITRO-N-PROPYL-P-TOLUIDINE	WWW = CYCLOPROPYLMETHYL, X = ALPHA	233
*PROMETRYN	2,4-BIS(ISOPROPYLAMINO)-6-(METHYLTHIO)-S-TRIAZINE	NONE	906

TABLE B217

PESTICIDES : TRADE AND CHEMICAL NAMES			
ITEM:	VALUE:	CONDITIONS:	REF.:
*PROMETRYN	2,4-BIS(ISOPROPYLAMINO)-6-(METHYLTHIO)-S-TRIAZINE	NONE	233
*PROPACHLOR	2-CHLORO-N-ISOPROPYLACETANILIDE	NONE	906
*PROPANIL	3',4'-DICHLOROPROPIONANILIDE	NONE	906
*PROPAZINE	2-CHLORO-4,6-BIS(ISOPROPYLAMINO) S-TRIAZINE	NONE	906
*PROPHAM	ISOPROPYL CARBANILATE	NONE	906
*PYRAZON	5-AMINO-4-CHLORO-2-PHENYL-3(2H) -PYRIDAZINONE	NONE	906
*SECBUMETON	N-ETHYL-6-METHOXY-N'-(1-METHYL-PROPYL)-1,3,5-TRIAZINE-2,4-DIAMINE	NONE	906
*SILVEX	2-(2,4,5-TRICHLOROPHENOXY) PROPIONIC ACID	NONE	906
*SIMAZINE	2-CHLORO-4,6-BIS(ETHYLAMINO) -S-TRIAZINE	NONE	906
*TCA	TRICHLOROACETIC ACID	NONE	906
*TDE	1,1-DICHLORO-2,2-BIS (P-CHLOROPHENYL)ETHANE	NONE	906
*TEMEPHOS	0,0'-(THIODI-4,1-PHENYLENE) BIS (0,0-DIMETHYL PHOSPHOROTHIOATE)	NONE	906
*TH 6040	1-(4-CHLOROPHENYL)-3-(2,6-DIFLUROBENZOYL)UREA	NONE	233
*TOXAPHENE	CHLORINATED CAMPHENE	CONTAINING 67 - 69 % CHLORINE	906
*TRICHLROFON	0,0-DIMETHYL (1-HYDROXY-2,2,2-TRICHLOROETHYL) PHOSPHONATE	NONE	857
*TRIFLURALIN	X,X,X-TRIFLUORO-2,6-DINITRO-N,N-DIPROPYL-P-TOLUIDINE	X = ALPHA	906
*XMC	3,5-XYLYL METHYLCARBAMATE	NONE	796
*ZECTRAN	4-DIMETHYLAMINO-3,5-XYLYL METHYL CARBAMATE	NONE	355
*ZINOPHOS	0,0-DIETHYL-0-2-PYRAZINYL PHOSPHOROTHIOATE	NONE	821
*ZR515	ISOPROPYL 11-METHOXY-3,7,11-TRIMETHYLDODECA-2,4-DIENOATE	ALTOSID	1276
*1,1,1-TRICHLOROETHANE	1,1,1-TRICHLOROETHANE	NONE	906
*2,3,5-T	(2,3,5-TRICHLOROPHENOXY)ACETIC ACID	NONE	906
*2,3,6-TBA	2,3,6-TRICHLOROBENZOIC ACID	NONE	906
*2,4-D	(2,4-DICHLOROPHENOXY)ACETIC ACID	NONE	906

TABLE B218

PESTICIDES : VOLATILITY			
ITEM:	VALUE:	CONDITIONS:	REF.:
*ALDRIN	2.3*10**(-5) MM HG	293-298K,VAPOR PRESSURE	389
*CAMPHECHLOR	1.0*10**(-6) MM HG	293-298K,VAPOR PRESSURE	389
*CHLORDANE	1.0*10**(-5) MM HG	293-298K,VAPOR PRESSURE	389
*DDT	1.9*10**(-7) MM HG	293-298K,VAPOR PRESSURE	389
*DIELDRIN	1.8*10**(-7) MM HG	293-298K,VAPOR PRESSURE	389
*ENDRIN	2.0*10**(-7) MM HG	293-298K,VAPOR PRESSURE	389
*HEPTACHLOR	3.0*10**(-4) MM HG	293-298K,VAPOR PRESSURE	389
*LINDANE	9.4*10**(-6) MM HG	293-298K,VAPOR PRESSURE	389

TABLE B219

Number of Species of Arthropods with Populations Resistant to Chlorinated Hydrocarbon Insecticides.

REF.: 389

Arthropod Group	DDT	Cyclodiene insecticides
Diptera	44	68
Lepidoptera	14	14
Hemiptera	10	15
Coleoptera	5	19
Acarina	15	12
Other orders	15	12
Total	91	135

TABLE B220

Properties of Soils used for Adsorption of Pesticides. The hamra and the hydromorphic grumusol soils belong to the same catena in Bet Dagan; the loess soil was sampled at the Gilat Experiment Farm near Beer Sheva.
REF.: 8

Soil type	% Coarse sand of total	% Sand of total	% silt of total	% clay of total	% CaCO3	Cation exchange capacity	Organic matter	pH
Hamra	7.5	75.2	3.6	13.1	2.2	13.8	0.8	7.8
Loess	2.2	53.3	34.7	9.7	15.4	15.7	0.9	7.9
Hydromorphic grumsol	8.9	38.3	21.3	31.4	4.3	33.4	1.2	7.9

B221

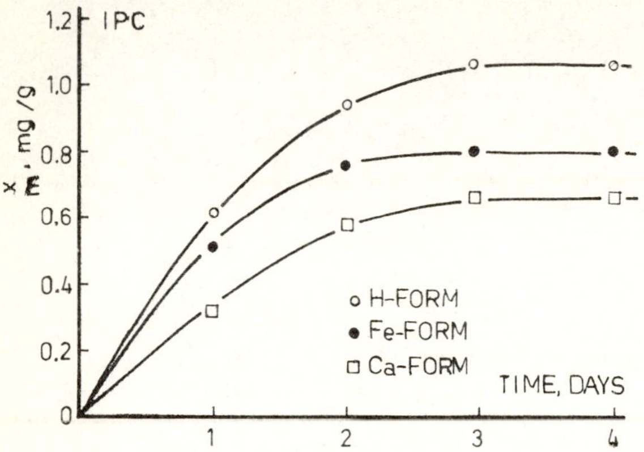

Adsorption of phenylamide pesticides on clay minerals.
Bentonite in H-form o , Fe-form ●
Ca-form ■ .

(ref: 401)

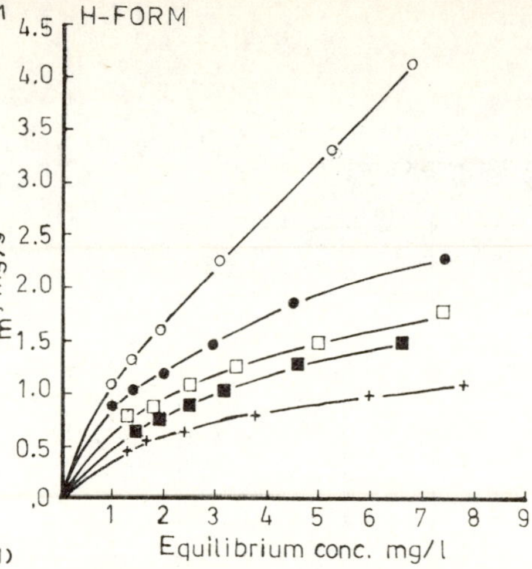

Adsorption of phenylamide pesticides on clay minerals.
Bentonite in H-form o Neburon ● Linuron
□ IPC ■ CIPC + Vinatex

B222

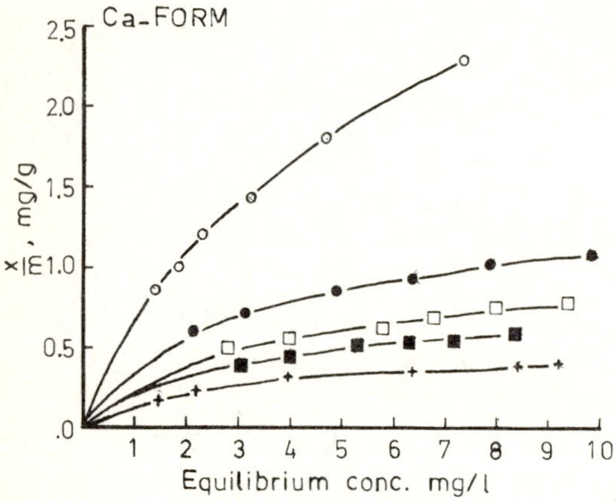

Bentonite in Ca-form. Symbol see above.

(ref: 401)

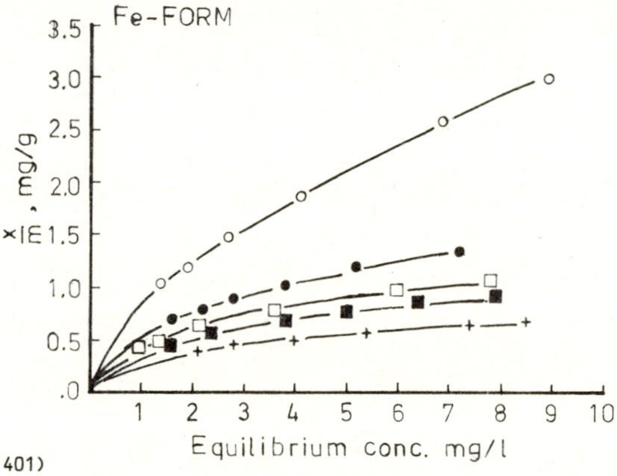

Bentonite in Fe-form. Symbols see above.

B223

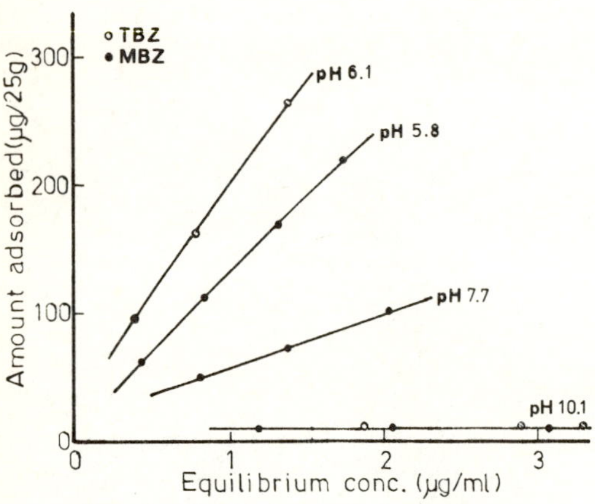

Adsorption isotherms for thiabendazole (TBZ) and 2-benzimidazole carbamic acid methyl ester (MBC) on loess soil at different pH and temperature 20 +- 1 celcius.
Properties of soils see table B220
● MBC o TBZ

(ref: 401)

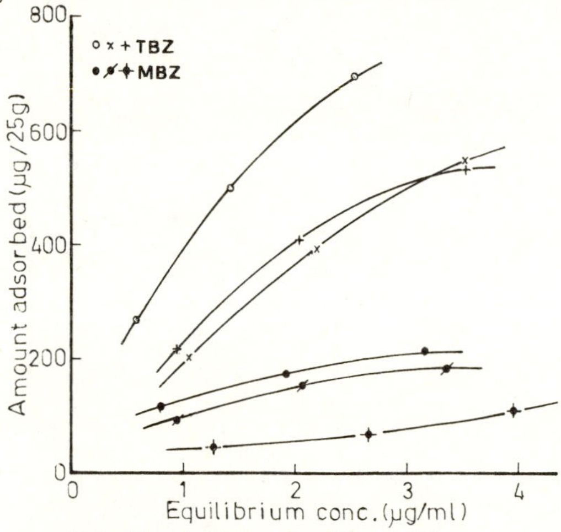

Adsorption isotherms for TBZ and MBC on 3 different soils.
x + o TBZ ✦ ✗ ● MBC
x,✦ hydromorphic grumusol +,✗ loess
hamra ●,o

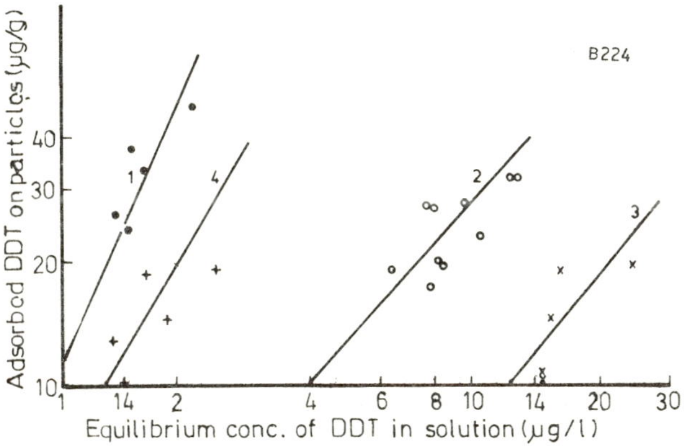

DDT adsorption isotherm on marine sediment, using 1: • = equilibrium concentration of DDT. 2: ○ = Equilibrium concentration of DDT in seawater corrected for blank results.
3: x = adsorbed DDT on suspended matter.
4: + = equilibrium concentration is seawater and adsorbed DDT on marine sediment.
As seen, there is a great difference in adsorption isotherms of DDT in cases, where the points of isotherms are obtained by different methods of calculation from experimental results.

(ref: 1257)

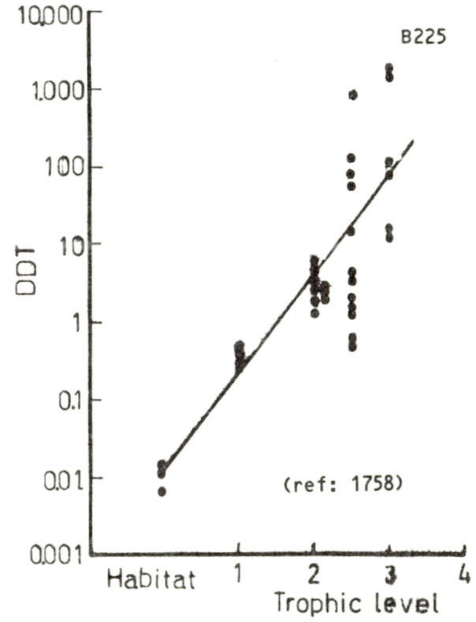

DDT concentration (ppm) versus trophic level in a detritus based food chain in Great Lakes.

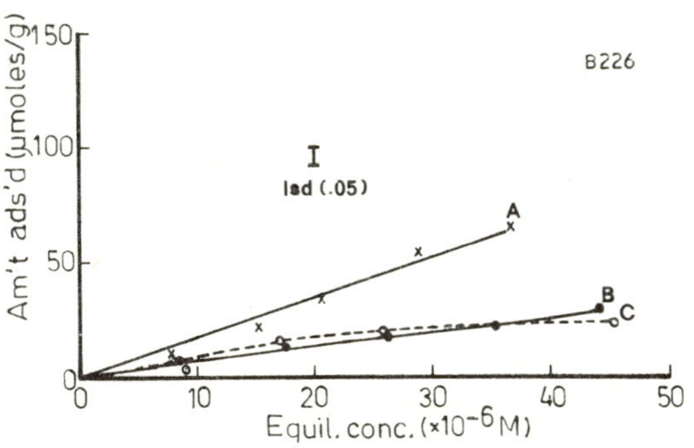

Adsorption isotherms for pesticides on montmorillonite. 24 +- 1 celcius, diionized water, Cu-form of montmorillonite.
A: Promethyn. B: Dicamba. C: Metuazole.

(ref: 233)

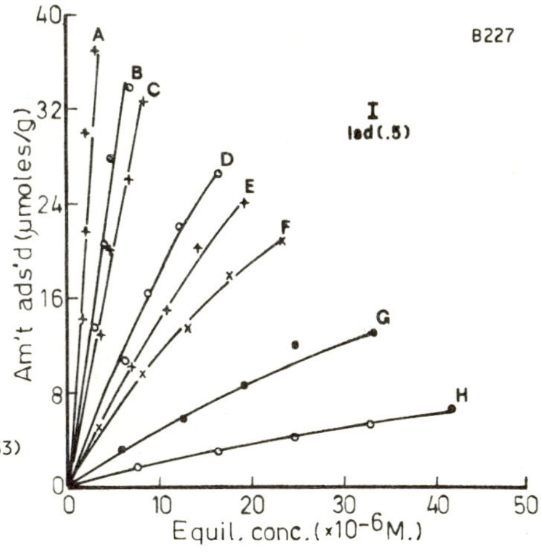

Adsorption isotherms for pesticides on organic matter.
24 +- 1 celcius, diionized water. Organic matter was prepared from featy muck soil (94 % combushible).
A: Promethyn. B: Leptophos. C: TH6040.
D: Oxadiazon. E: Metuazole. F: nor flurazon. G: Fluometuron. H: Asulam.

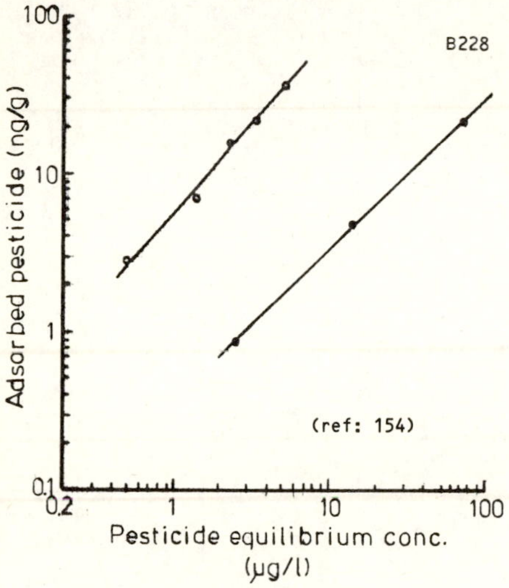

Adsorption isotherms for lindane and dieldrin on sand st 5°celcius. Equilibrium time 24 h, 50 cm**3 of pesticide water, 100 g of aquifer sand. o dieldrine • lindane.

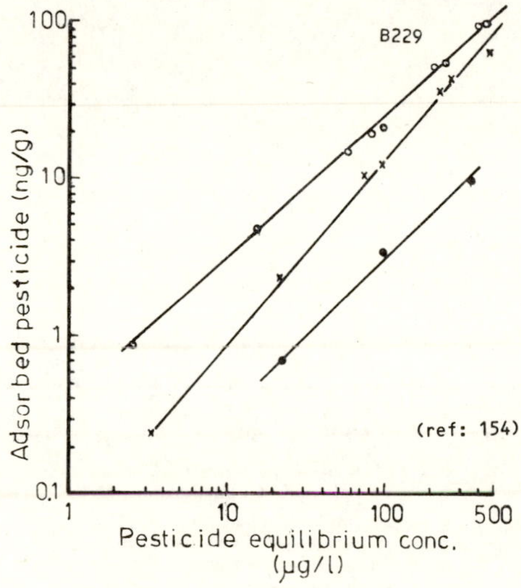

Adsorption isotherms for lindane on sand at 5°celcius. Equilibrium time: 24 h, 50 cm**3 lindane water with 100 g sand. o Aquifer sand. × Standard silica. 80 - 120 mesh. • Standard Ottawa sand 20 - 30 mesh.

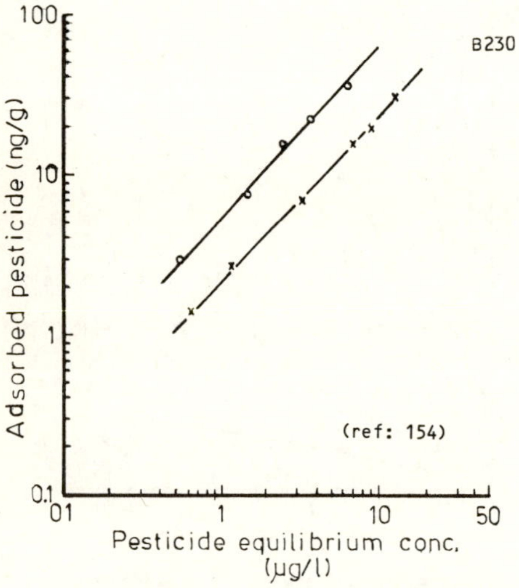

Adsorptions isotherms for dieldrin on sand at 5°celcius. Equilibrium time 24 h, 50 cm**3 of dieldrin water, 100 g of aquifer sand. o Aquifer sand. × Silica sand 80 - 120 mesh.

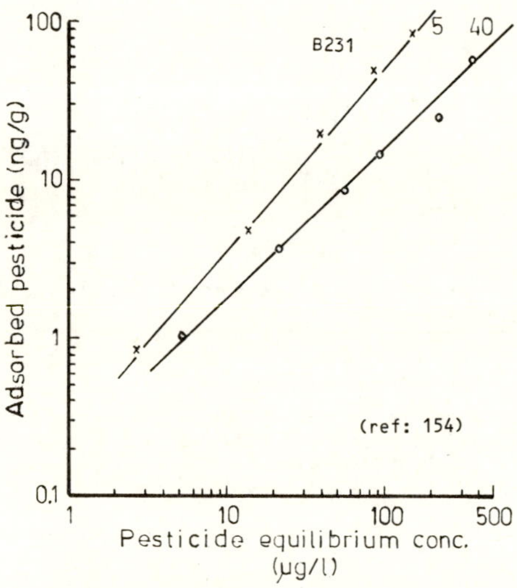

Adsorption isotherm of lindane on sand at 5°celcius and 40°celcius. Equilibrium time 24 hours, 50 cm**3 lindane water, 100 g aquifer sand.

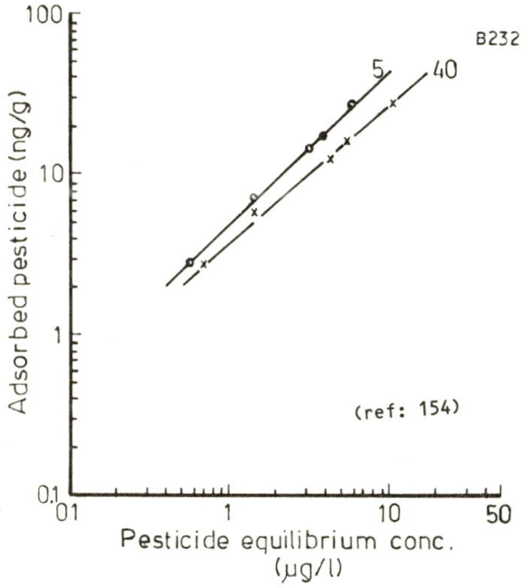

Adsorption isotherm of dieldrin on sand at 5°celcius and 40°celcius. Equilibrium time 24 hours, 50 cm**3 lindane water, 100 g aquifer sand.

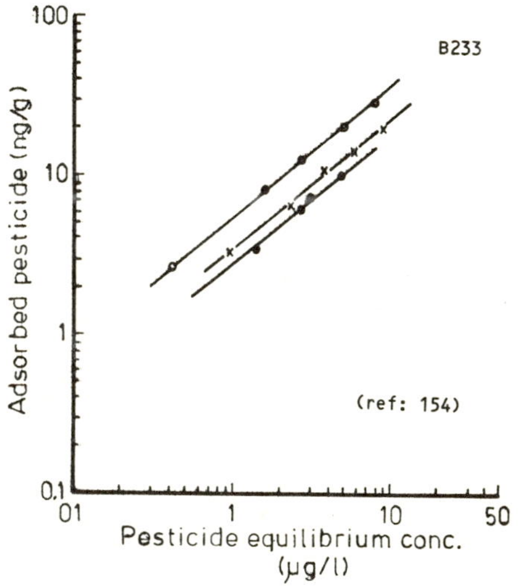

Adsorption isotherm of dieldrin on sand. Effect of lake water is shown. Equilibrium time 24 hours, 50 cm**3 dieldrin water, 100 g aquifer sand, temperature 5°celcius.
o distilled water, x Lake Mendota water, • Lake Mary water.

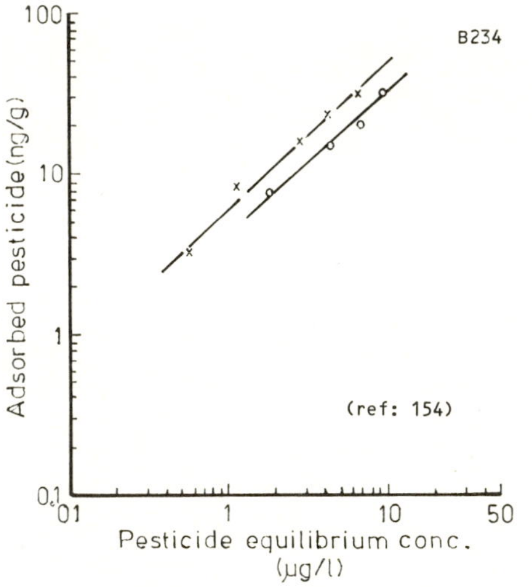

Adsorption of dieldrin on sand. Effect of particle size. 50 cm**3 dieldrin water, 100 g aquifer sand, equilibrium after 24 hours, temperature 5°celcius.
o 0.354 - 0.500 mm x 0.177 - 0.297 mm.

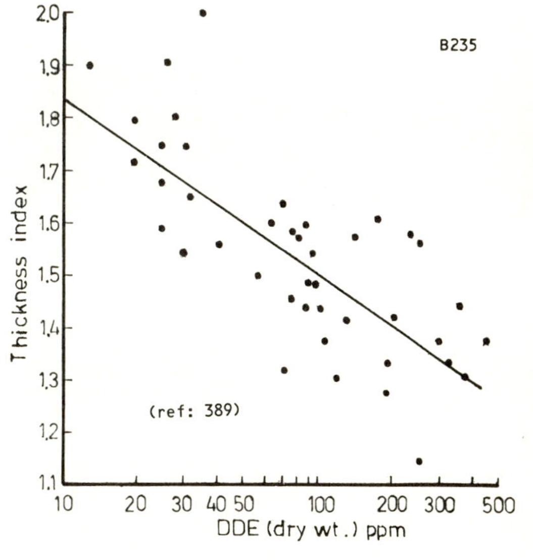

Eggshell thickness versus DDE residues in eggs for Alaskan peregrine falcons. Egg shell thickness expressed as weight in mg divided by length multiplied width in mm.

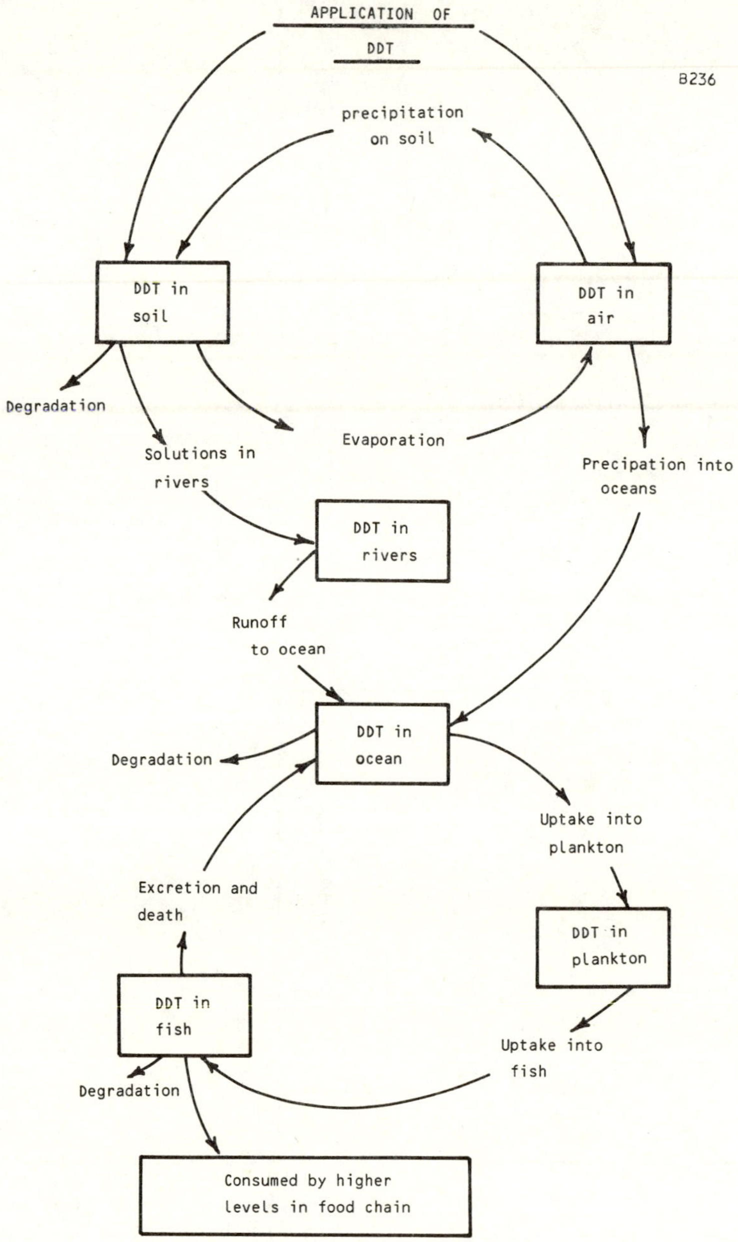

DDT model, flows in the environment.

(ref: 1295)

B237

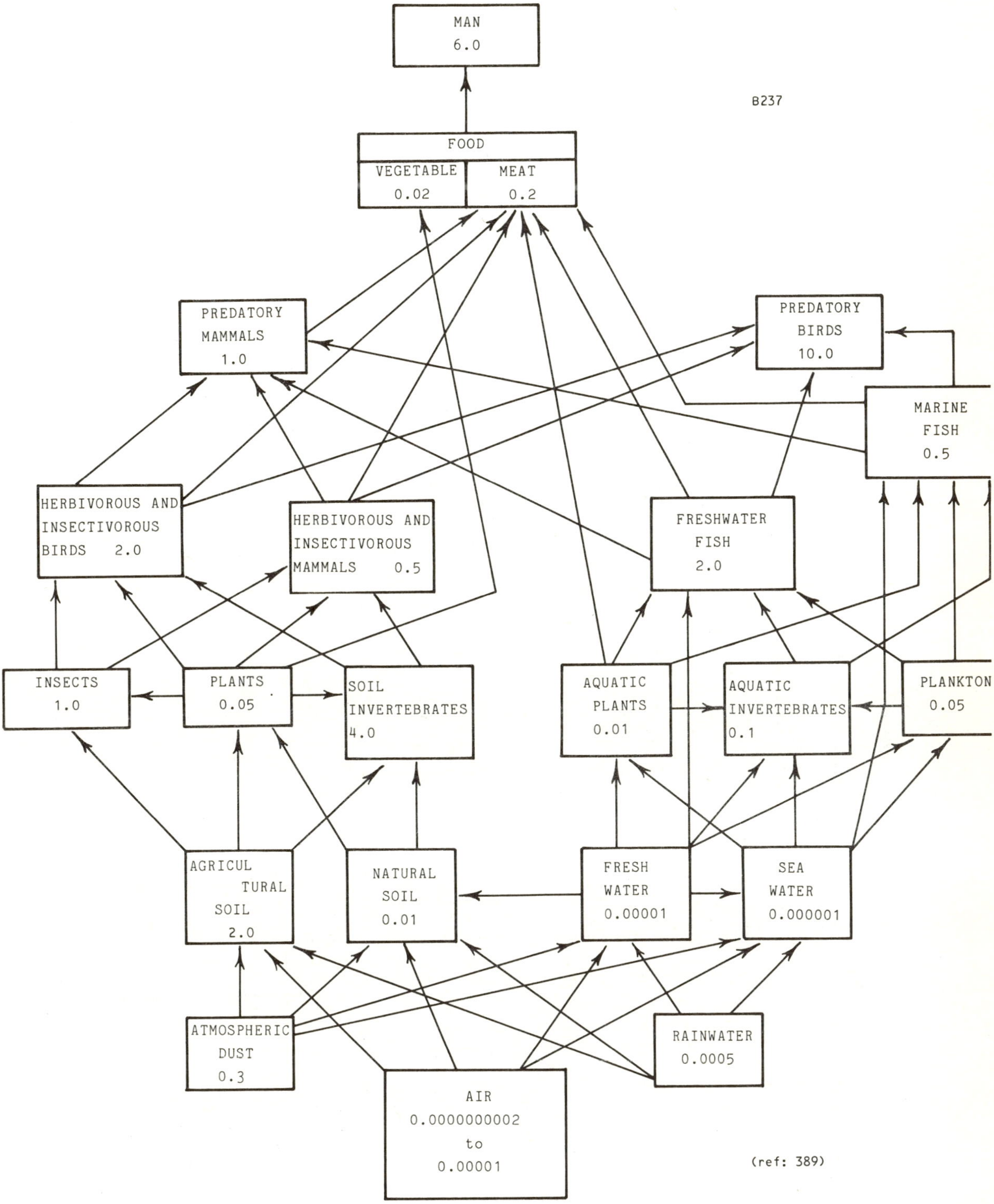

(ref: 389)

Typical concentrations of DDT in the environment. Concentration in mg/kg wet weight.

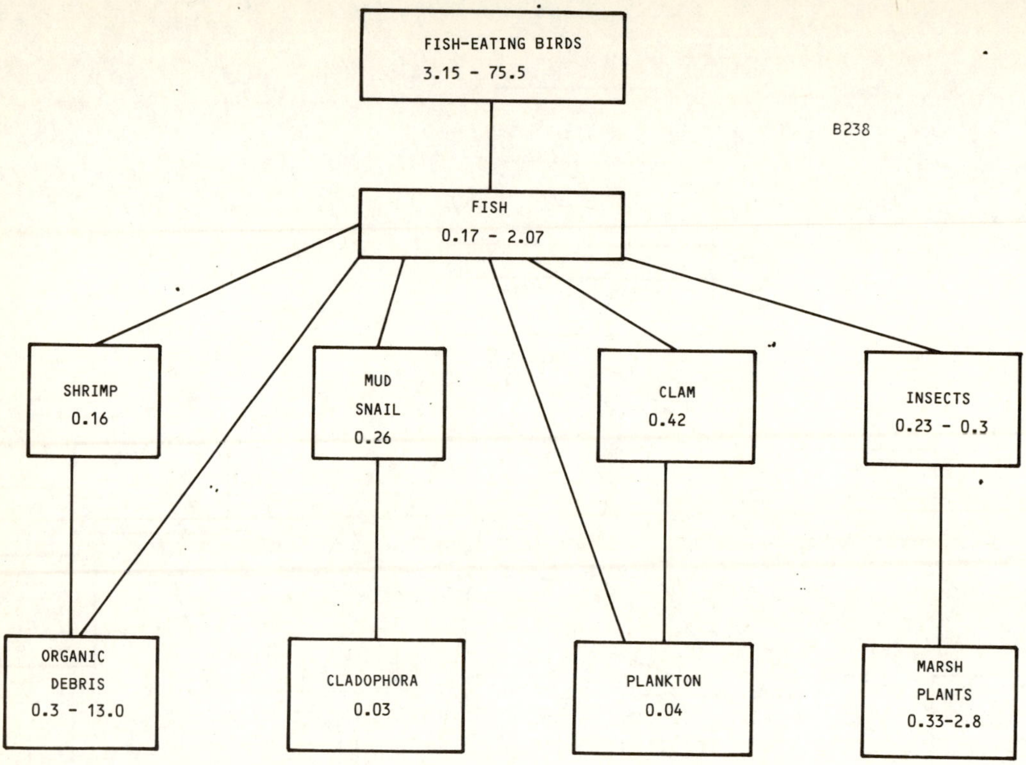

DDT accumulation through a food-chain. (ref: 389)

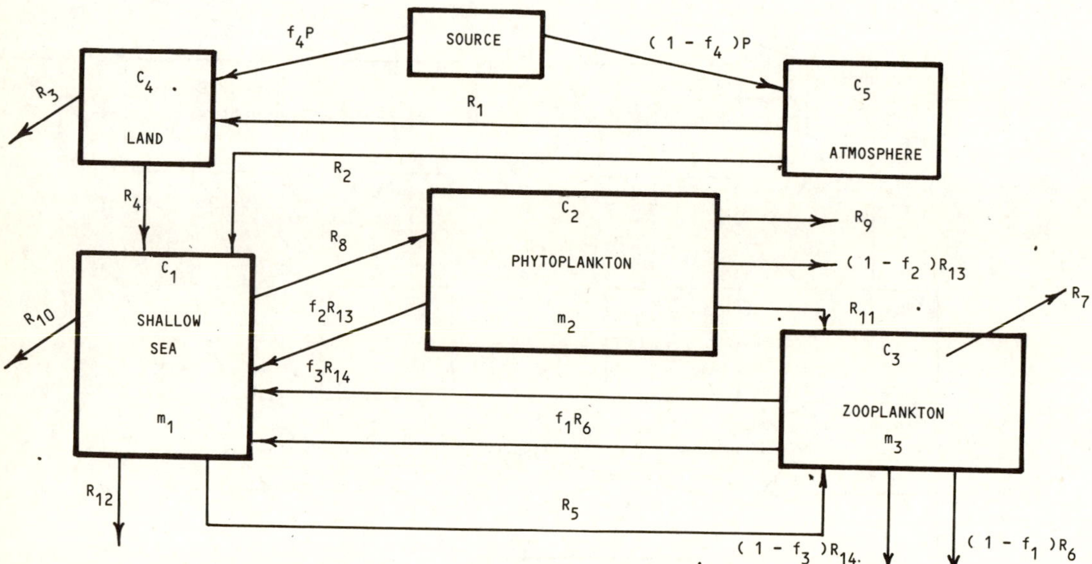

DDT accumulation through a food-chain.
b4
DDT - model. (Cramers circulation model).

R1 = k1*C5: Atmosphere to land transfer
R2 = k2*C5: Atmosphere to sea transfer
R3 = k3*C4: Degradation on land
R4 = k4*C4: Land to sea transfer
R5 = k5*m3*C1: Zooplankton absorption
R6 = k6*m3*C3: Zooplankton excretion
R7 = k7*m3*C3: Degradation on zooplankton
R8 = k8*m2*C1: Phytoplankton absorption
R9 = k9*m2*C2: Degradation in phytoplankton
R10 = k10*m1*C1: Degradation in shallow sea
R11 = mf*C2: Zooplankton feeding
R12 = mo*C1: Exchange between shallow and deep sea
R13 = md2*C1: Phytoplankton decay
R14 = md3*C3: Zooplankton decay
P: Rate of production of DDT.
f4: Fraction of DDT production going directly to land
f1,f2,f3: Fraction of decaying and excreted DDT going to shallow sea

(ref: 389)

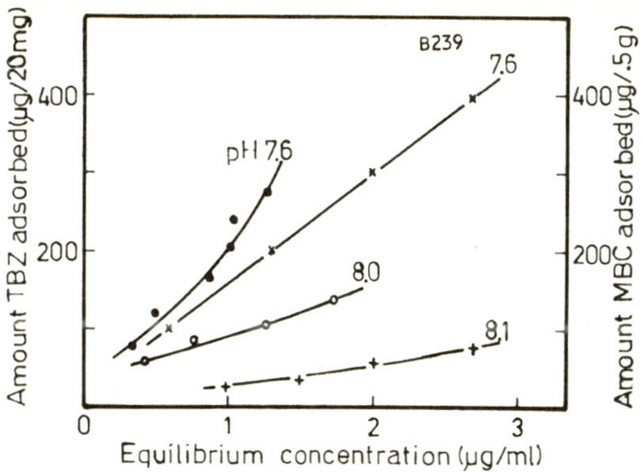

Adsorption isotherms for Thiabendazole (TBZ) and 2-benzimidazole carbonic acid methyl ester(MBC) on Ca x, • and Na-montmorilionite + , o Ionic strength 0.01 m of resp. ions, ug adsorbed per 20 mg adsorbant is plotted versus concentration (µg/ml).

(ref: 7)

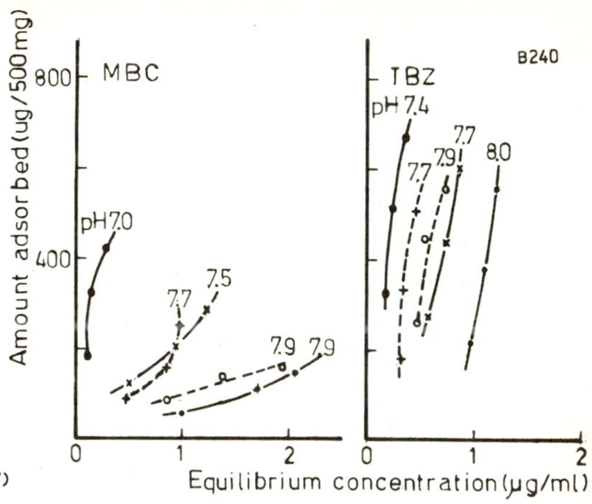

Adsorptions isotherms for thiabendazole (TBZ) and 2-benzimidazole carbonic acid ester methyl ester (MBC) on Ca-benthonite at different pH The equilibrium concentration (µg/ml) is plotted versus µg adsorbed per 0.5 g Ca-benthonite.

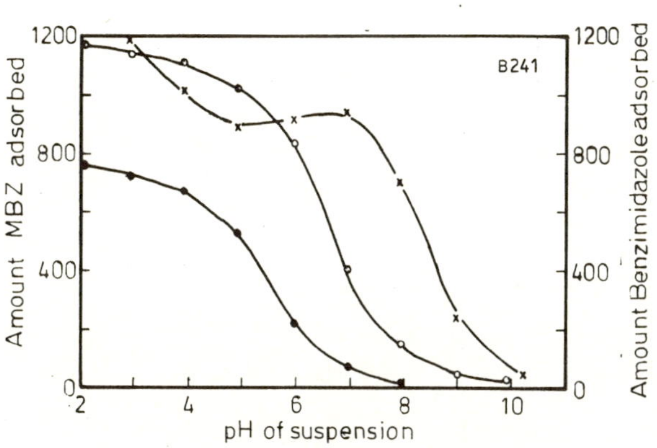

Adsorption of Thiabendazole x , Benzimidazole • and 2-benzimidazole carbonic acid methyl ester o on Ca-bentonite as function of pH (ionic strength 0.01 m CaCl2); µg adsorbed per 40 mg versus pH. (ref: 7)

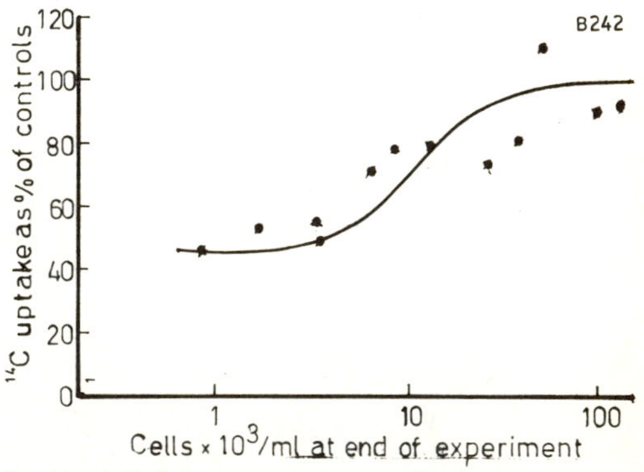

Growth of Skeletonema costatum at 10 ppb DDT. Net productivity of an algal culture in relation to cell density at the end of the experiment. DDT concentration was held constant so the data indicate that as DDT concentration relative to population size, increased, net productivity decresed.

(ref: 1758)

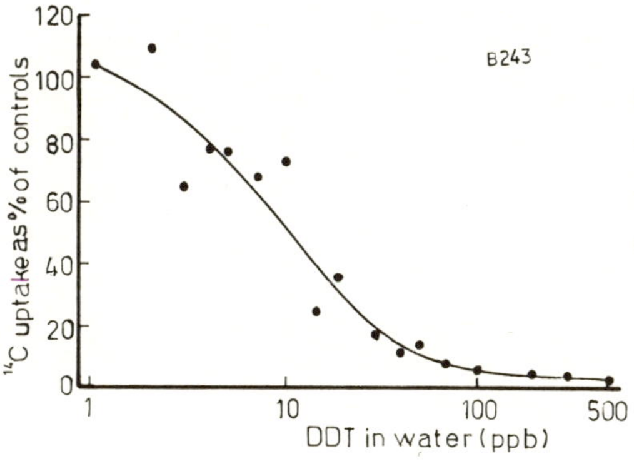

Productivity versus DDT concentration for marine phytoplankton.

(ref: 1758)

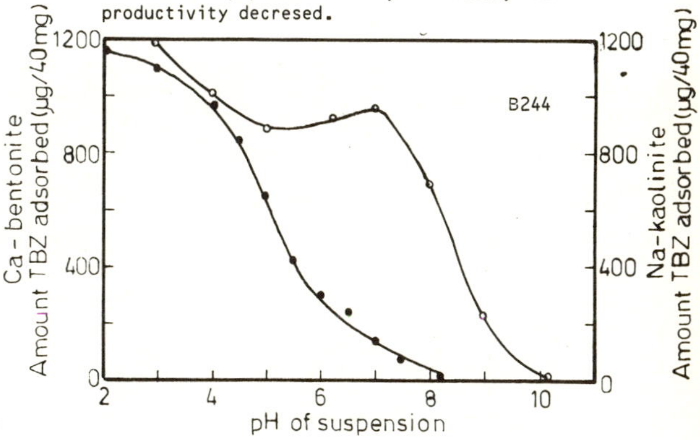

Adsorption isotherms for thiabendazole (TBZ) on Ca-benthonite o and Na-kaoline • as function of pH. µg/40 mg is plotted versus pH of suspension.

(ref: 7)

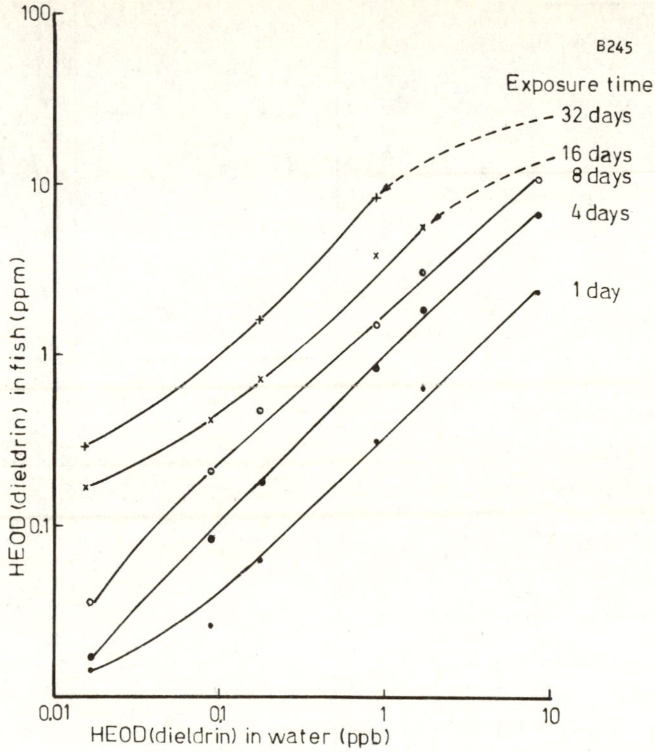

Dieldrin uptake by fish for different concentrations and exposure times. (ref: 389)

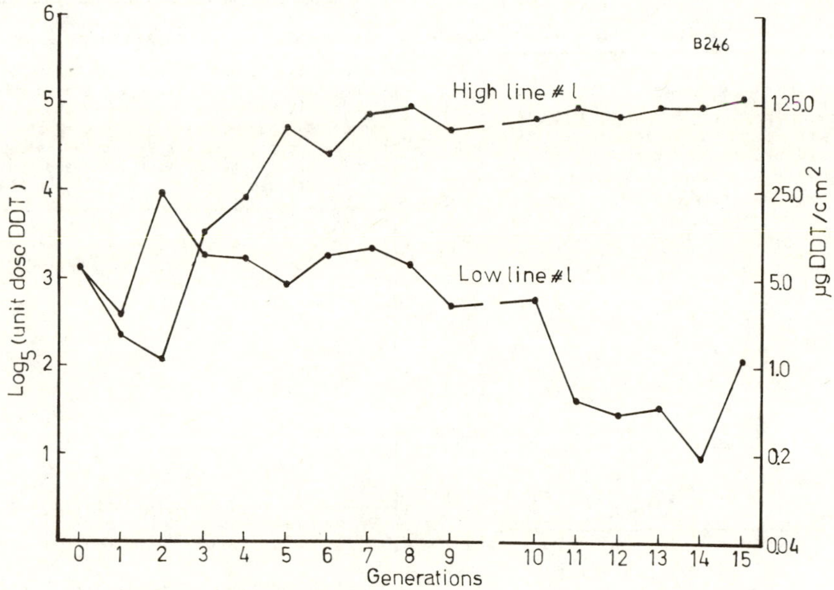

DDT resistance of fruit fly (Drossophilia melanogaster). Left hand scale is log of the dose tolerated and the rigth hand scale is the LC50 value. The two curves indicate populations exposed to directional selection for high respective low resistance. (ref: 97)

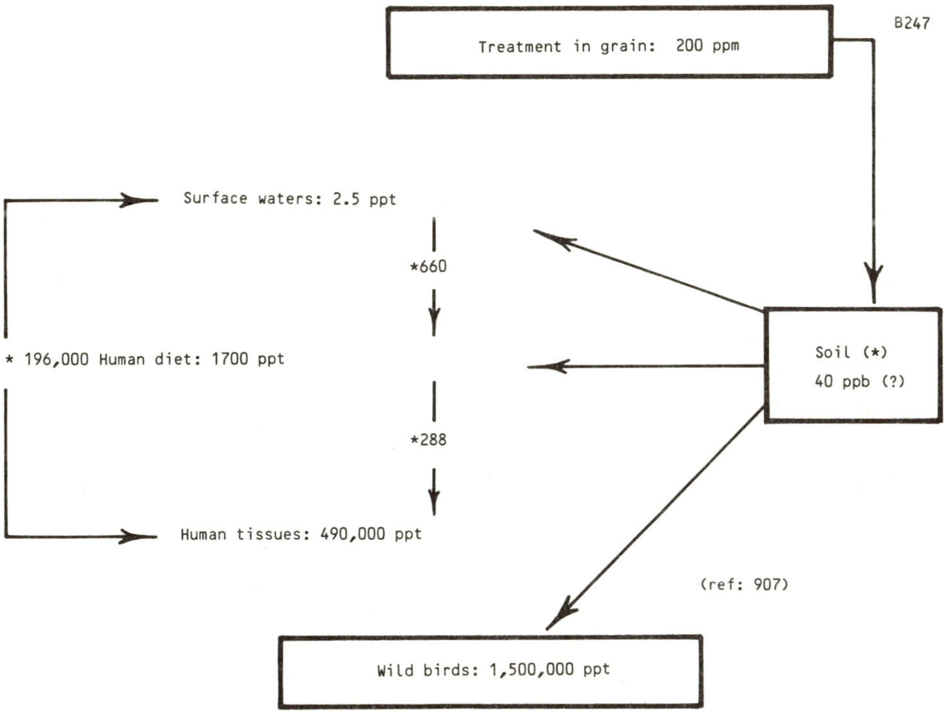

HCB - concentrations in the environment.
(*) According to data from the Wandeswil Research Institute (Switzerland) is the content of HCB in soil, 1-4 years after tre treatment from 18,000 to 170 ppb.

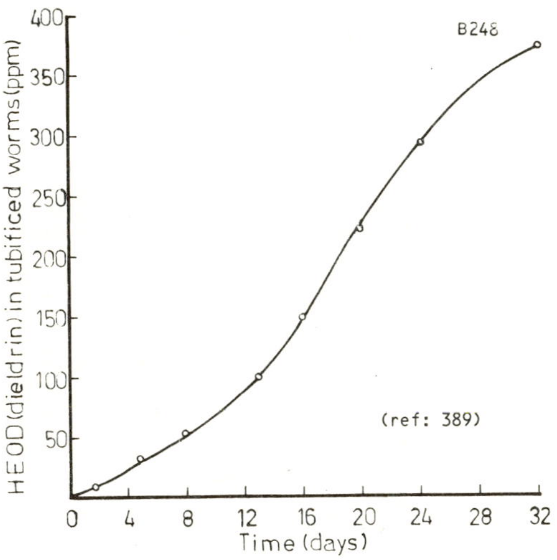

Dieldrin concentration (ppm) in tubificid worms, uptake from water plotted versus time exposure.

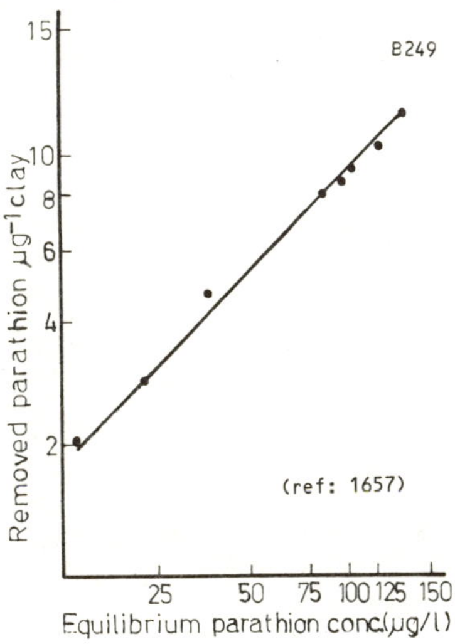

Adsorption isotherms of parathion on (hexane extracted) sediment. (Lake Mendota)

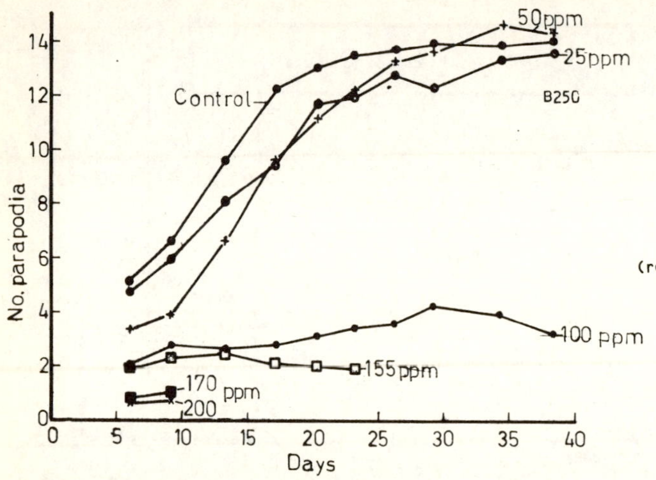

Growth rate of Ophryotrocha labronica, effect of phenol. The control series. Low concentration seems to have a growth promoting effect.

(ref: 1)

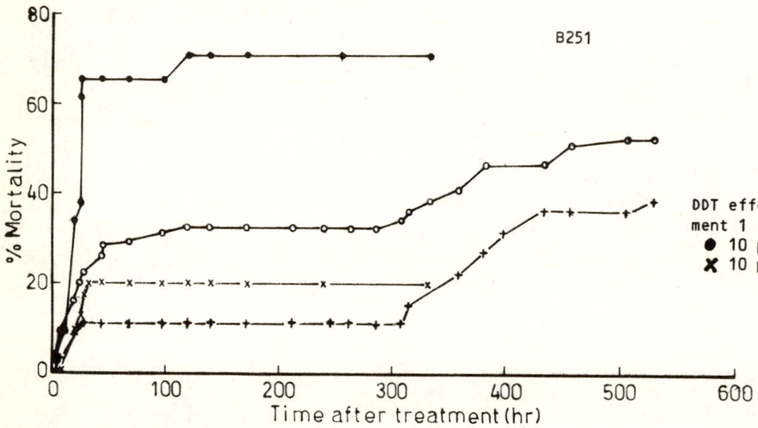

DDT effect on cumulative mortality. Experiment 1 (details see table): ✗ 1.0 DDT. ● 10 ppm DDT. Experiment 2: ✚ 1.0 DDT ✗ 10 ppm DDT. Contol deaths less than 3 %.

(ref: 281)

DDT effect on activity versus time for ● control, o 1 ppm DDT, ✗ 10 ppm DDT. Experiment 2. (details see table).

(ref: 281)

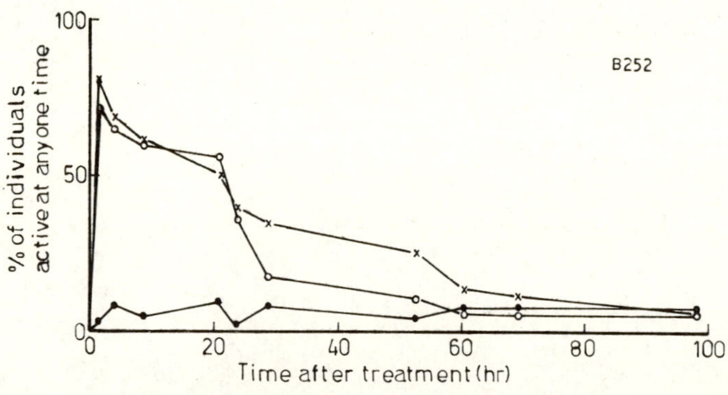

B5

SOIL AND SEDIMENTS

ACCUMULATIÓN RATE
OXYGEN RATE
RELEASE RATE
SULPHATE REDUCTION

TABLE B253

SEDIMENT : ACCUMULATION RATE			
ITEM:	VALUE:	CONDITIONS:	REF.:
*FRESHWATER SEDIMENT	0.3 CM/YEAR	LAKE ERIE, WESTERN BASIN, FROM SR-90 PROFILE	912
*FRESHWATER SEDIMENT	0.3 CM/YEAR	LAKE ERIE, WESTERN BASIN, FROM CS-137 PROFILE	912
*FRESHWATER SEDIMENT	0.6 CM/YEAR	LAKE ERIE, CENTRAL BASIN, FROM SR-90 PROFILE	912
*FRESHWATER SEDIMENT	0.6 CM/YEAR	LAKE ERIE, CENTRAL BASIN, FROM CS-137 PROFILE	912
*FRESHWATER SEDIMENT	0.05 CM/YEAR	LAKE ONTARIO, WESTERN PART, FROM SR-90 PROFILE	912
*FRESHWATER SEDIMENT	0.05 CM/YEAR	LAKE ONTARIO, WESTERN PART, FROM CS-137 PROFILE	912

TABLE B254

SEDIMENT : BINDING OF CU			
ITEM:	VALUE:	CONDITIONS:	REF.:
*LAKE SEDIMENT IN SUSPENSION	28 MG/G	100MG SEDIM./L,PH=6.7,RELEASE FROM SEDIM.:4.5MG CA/L AND 0.30MG MG/L	1386
*LAKE SEDIMENT IN SUSPENSION	28 MG/G	150MG SEDIM./L,PH=6.7,RELEASE FROM SEDIM.:5.7MG CA/L AND 0.43MG MG/L	1386
*LAKE SEDIMENT IN SUSPENSION	26 MG/G	200MG SEDIM./L,PH=6.8,RELEASE FROM SEDIM.:7.0MG CA/L AND 0.50MG MG/L	1386
*SEDIMENT IN SUSPENSION	44 MG/G	50MG SEDIM./L,PH=6.7,RELEASE FROM SEDIM.:5.5MG CA/L AND 0.28MG MG/L	1386

TABLE B255

SEDIMENT : OXYGEN UPTAKE			
ITEM:	VALUE:	CONDITIONS:	REF.:
*ESTUARINE MUD	1.5 (RANGE 1 - 2)GRAM 02/ (M**2*DAY)	298 K	1575
*FIBER SLUDGE	7 (RANGE 4 - 10) GRAM 02/ (M**2*DAY)	298 K,CELLULOSIC FIBER	1575
*FRESHWATER SEDIMENT	0.75 MG 02/(GRAM SEDIMENT*HOUR)	CHEMICAL UPTAKE = 59 %, ORG MATTER = 39.6 %, 281 K, ESROM L	618
*FRESHWATER SEDIMENT	1.48 MG 02/(GRAM SEDIMENT*HOUR)	CHEMICAL UPTAKE = 59 %, ORG MATTER = 40.0 %, 277 K, ESROM L	618
*FRESHWATER SEDIMENT	1.14 MG 02/(GRAM SEDIMENT*HOUR)	CHEMICAL UPTAKE = 31 %, ORG MATTER = 39.9 %, 275 K, ESROM L	618
*FRESHWATER SEDIMENT	0.74 MG 02/(GRAM SEDIMENT*HOUR)	CHEMICAL UPTAKE = 65 %, ORG MATTER = 72.0 %, 281 K, STORE GRIB L	618
*FRESHWATER SEDIMENT	1.34 MG 02/(GRAM SEDIMENT*HOUR)	CHEMICAL UPTAKE = 54 %, ORG MATTER = 45.0 %, 283 K, FRED. CASTLE L	618
*FRESHWATER SEDIMENT	0.10 MG 02/(GRAM SEDIMENT*HOUR)	CHEMICAL UPTAKE = 10 %, ORG MATTER = 21.3 %, 277 K, GRANE LANG L	618
*MARINE SEDIMENT	1.7 ML 02/(M**2*HOUR)	DEPHT=2750, CHEMICAL UPTAKE NOT DETERMINATED, 280 K, NE PACIFIC	1462
*MARINE SEDIMENT	4.2 ML 02/(M**2*HOUR)	DEPTH=2820 M, CHEMICAL UPTAKE = 3.1, 288 K, E TROP. PACIFIC	1462
*MARINE SEDIMENT	1.8 ML 02/(M**2*HOUR)	DEPTH=2900 M, CHEMICAL UPTAKE = 1.7, 288 K, E TROP. PACIFIC	1462
*MARINE SEDIMENT	0.7 ML 02/(M**2*HOUR)	DEPTH=1325 M, CHEMICAL UPTAKE = 0.4, 288 K, E TROP. PACIFIC	1462
*MARINE SEDIMENT	3.0 ML 02/(M**2*HOUR)	DEPTH=2755 M, CHEMICAL UPTAKE = 3.0, 281 K, NE PACIFIC	1462
*MARINE SEDIMENT	2.3 ML 02/(M**2*HOUR)	DEPTH=2790 M, CHEMICAL UPTAKE = 2.2, 281 K, NE PACIFIC	1462

TABLE B255

SEDIMENT : OXYGEN UPTAKE			
ITEM:	VALUE:	CONDITIONS:	REF.:
*MARINE SEDIMENT	2.3 ML O2/(M**2*HOUR)	DEPTH=2750 M, CHEMICAL UPTAKE = 1.8, 281 K, NE PACIFIC	1462
*MARINE SEDIMENT	2.4 ML O2/(M**2*HOUR)	DEPTH=2860 M, CHEMICAL UPTAKE = 2.4, 281 K, NE PACIFIC	1462
*MARINE SEDIMENT	3.0 ML O2/(M**2*HOUR)	DEPTH=5085 M, CHEMICAL UPTAKE = 3.0, 275 K, NE PACIFIC	1462
*MARINE SEDIMENT	1.7 ML O2/(M**2*HOUR)	DEPTH=4720 M, CHEMICAL UPTAKE = 1.1, 275 K, NE PACIFIC	1462
*MARINE SEDIMENT	1.2 ML O2/(M**2*HOUR)	DEPTH=5720 M, CHEMICAL UPTAKE = 1.1, 275 K, NE PACIFIC	1462
*MARINE SEDIMENT	1.5 ML O2/(M**2*HOUR)	DEPTH=5830 M, CHEMICAL UPTAKE = 1.5, 275 K, NE PACIFIC	1462
*MARINE SEDIMENT	2.8 ML O2/(M**2*HOUR)	DEPTH=5160 M, CHEMICAL UPTAKE = 2.3, 275 K, NE PACIFIC	1462
*MARINE SEDIMENT	1.9 ML O2/(M**2*HOUR)	DEPTH=4580 M, CHEMICAL UPTAKE = 1.9, 275 K, NE PACIFIC	1462
*MARINE SEDIMENT	0.8 ML O2/(M**2*HOUR)	DEPTH=7005 M, CHEMICAL UPTAKE = 0.2, 275 K, NE PACIFIC	1462
*MARINE SEDIMENT	3.3 ML O2/(M**2*HOUR)	DEPTH=6975 M, CHEMICAL UPTAKE = 2.3, 275 K, NE PACIFIC	1462
*MARINE SEDIMENT	0.9 ML O2/(M**2*HOUR)	DEPTH=4145 M, CHEMICAL UPTAKE = 0.9, 275 K, NE PACIFIC	1462
*MARINE SEDIMENT	1.7 ML O2/(M**2*HOUR)	DEPTH=3670 M, CHEMICAL UPTAKE = 1.7, 275 K, NE PACIFIC	1462
*MARINE SEDIMENT	1.2 ML O2/(M**2*HOUR)	DEPTH=3205 M, CHEMICAL UPTAKE = 0.8, 275 K, NE PACIFIC	1462
*MARINE SEDIMENT	2.0 ML O2/(M**2*HOUR)	DEPTH=3145 M, CHEMICAL UPTAKE = 2.0, 275 K, NE PACIFIC	1462
*MARINE SEDIMENT	1.5 ML O2/(M**2*HOUR)	DEPTH=2735 M, CHEMICAL UPTAKE = 1.4, 275 K, NE PACIFIC	1462
*MARINE SEDIMENT	5.2 ML O2/(M**2*HOUR)	DEPTH=2510 M, CHEMICAL UPTAKE = 3.7, 275 K, NE PACIFIC	1462
*MARINE SEDIMENT	0.5 ML O2/(M**2*HOUR)	DEPTH=1850 M, CHEMICAL UPTAKE = 0.0, 278 K, NEW ENGLAND SLOPE	1462
*MARINE SEDIMENT	2.4 ML O2/(M**2*HOUR)	DEPTH=1230 M, CHEMICAL UPTAKE = 1.04, 277 K, SAN DIEGO TROUGH	1462
*MINERAL SOILS	0.07 (RANGE 0.05 - 0.1) GRAM O2/(M**2*DAY)	298 K	1575
*MUNICIPAL SEWAGE SLUDGE	4 (RANGE 2 - 10) GRAM O2/(M**2*DAY)	298 K,OUTFALL VICINITY	1575
*MUNICIPAL SEWAGE SLUDGE	1.5 (RANGE 1 - 2) GRAM O2/(M**2*DAY)	298 K,AGED DOWNSTREAM OUTFALL	1575
*SANDY BOTTOM	0.5 (RANGE 0.2 - 1.0) GRAM O2/(M**2*DAY)	298 K	1575
*SEDIMENT FROM E.TROPICAL PACIFIC	3.8 ML/M**2*HR	DEPTH=275M,O2=0.8ML/L,CHEM.UPTAKE=71-77%,288K	1208
*SEDIMENT FROM E.TROPICAL PACIFIC	9.3 ML/M**2*HR	DEPTH=85M,O2=1.0ML/L,CHEM.UPTAKE=48-101%,288K	1208
*SEDIMENT FROM E.TROPICAL PACIFIC	4.2 ML/M**2*HR	DEPTH=2820M,O2=1.7ML/L,CHEM.UPTAKE=66-81%,288K	1208
*SEDIMENT FROM E.TROPICAL PACIFIC	1.8 ML/M**2*HR	DEPTH=2900M,O2=1.5-1.9ML/L, CHEM.UPTAKE=85-106%,288K	1208
*SEDIMENT FROM E.TROPICAL PACIFIC	0.7 ML/M**2*HR	DEPTH=1325M,O2=1.1ML/L,CHEM.UPTAKE=42-62%,288K	1208
*SEDIMENT FROM N.E.PACIFIC	1.7 ML/M**2*HR	DEPTH=2750M,O2=2.4ML/L,280K	1208
*SEDIMENT FROM N.E.PACIFIC	6.7 ML/M**2*HR	DEPTH=146M,O2=1.7-2.2ML/L, CHEM.UPTAKE=31-49%,280K	1208
*SEDIMENT FROM N.E.PACIFIC	3.0 ML/M**2*HR	DEPTH=105M,O2=2.2ML/L,CHEM.UPTAKE=87-112%,280K	1208
*SEDIMENT FROM N.E.PACIFIC	3.4 ML/M**2*HR	DEPTH=146M,O2=2.3-2.4ML/L, CHEM.UPTAKE=66-73%,280K	1208
*SEDIMENT FROM PUGET SOUND	6.7 ML/M**2*HR	DEPTH=180M,O2=3.5ML/L,CHEM.UPTAKE=50-59%,282K	1208
*SEDIMENT FROM PUGET SOUND	8.6 ML/M**2*HR	DEPTH=23M,O2=5.2-5.8ML/L,280K	1208
*SEDIMENT FROM PUGET SOUND	6.9 ML/M**2*HR	DEPTH=23M,O2=4.4-4.7ML/L,280K	1208
*SEDIMENT FROM PUGET SOUND	26.6 ML/M**2*HR	O2=2.53ML/L,DEPTH=74M,282K, TOT.UPTAKE	1208

TABLE B255

SEDIMENT : OXYGEN UPTAKE			
ITEM:	VALUE:	CONDITIONS:	REF.:
*SEDIMENT FROM PUGET SOUND	23.7 ML/M**2*HR	O2=2ML/L,DEPTH=74M,282K,TOT.UPTAKE	1208
*SEDIMENT FROM PUGET SOUND	18.9 ML/M**2*HR	O2=1.05ML/L,DEPTH=74M,282K, TOT.UPTAKE	1208
*SEDIMENT FROM PUGET SOUND	15.0 ML/M**2*HR	O2=0.56ML/L,DEPTH=74M,282K, TOT.UPTAKE	1208
*SEDIMENT FROM PUGET SOUND	4.7 ML/M**2*HR	O2=0.14ML/L,DEPTH=74M,282K, TOT.UPTAKE	1208
*SEDIMENT FROM PUGET SOUND	2.3 ML/M**2*HR	O2=0.07ML/L,DEPTH=74M,282K, TOT.UPTAKE	1208
*SEDIMENT FROM PUGET SOUND	0 ML/M**2*HR	O2=0ML/L,DEPTH=74M,282K,TOT.UPTAKE	1208
*SEDIMENT FROM PUGET SOUND	17.1 ML/M**2*HR	O2=2.53ML/L,DEPTH=74M,282K, RESPIRATION	1208
*SEDIMENT FROM PUGET SOUND	16.6 ML/M**2*HR	O2=2ML/L,DEPTH=74M,282K, RESPIRATION	1208
*SEDIMENT FROM PUGET SOUND	15.1 ML/M**2*HR	O2=1.05ML/L,DEPTH=74M,282K, RESPIRATION	1208
*SEDIMENT FROM PUGET SOUND	12.3 ML/M**2*HR	O2=0.56ML/L,DEPTH=74M,282K, RESPIRATION	1208
*SEDIMENT FROM PUGET SOUND	9.4 ML/M**2*HR	O2=0.42ML/L,DEPTH=74M,282K, RESPIRATION	1208
*SEDIMENT FROM PUGET SOUND	17.4 ML/M**2*HR	O2=3.44ML/L,DEPTH=74M,282K, CHEM.OXIDAT.	1208
*SEDIMENT FROM PUGET SOUND	11.1 ML/M**2*HR	O2=2.81ML/L,DEPTH=74M,282K, CHEM.OXIDAT.	1208
*SEDIMENT FROM PUGET SOUND	7.1 ML/M**2*HR	O2=2.04ML/L,DEPTH=74M,282K, CHEM.OXIDAT.	1208
*SEDIMENT FROM PUGET SOUND	5.6 ML/M**2*HR	O2=1.65ML/L,DEPTH=74M,282K, CHEM.OXIDAT.	1208
*SEDIMENT FROM PUGET SOUND	4.3 ML/M**2*HR	O2=1.23ML/L,DEPTH=74M,282K, CHEM.OXIDAT.	1208
*SEDIMENT FROM PUGET SOUND	2.3 ML/M**2*HR	O2=0.39ML/L,DEPTH=74M,282K, CHEM.OXIDAT.	1208
*SPHAEROTILUS	7 GRAM O2/(M**2*DAY)	298 K,10 G DRY WEIGHT/M2	1575

TABLE B256

SEDIMENT : RELEASE RATE			
ITEM:	VALUE:	CONDITIONS:	REF.:
*SEDIMENT	0.061 G BOD/(M**2*HOUR)	AEROBIC RATE	944
*SEDIMENT	0.154 G COD/(M**2*HOUR)	AEROBIC RATE	944
*SEDIMENT	0.125 G PO4-P/(M**2*HOUR)	AEROBIC RATE	944
*SEDIMENT	0.1083 G NH3-N/(M**2*HOUR)	AEROBIC RATE	944
*SEDIMENT	0.203 G BOD/(M**2*HOUR)	ANAEROBIC RATE	944
*SEDIMENT	0.495 G COD/(M**2*HOUR)	ANAEROBIC RATE	944
*SEDIMENT	0.250 G PO4-P/(M**2*HOUR)	ANAEROBIC RATE	944
*SEDIMENT	0.3125 G NH3-N/(M**2*HOUR)	ANAEROBIC RATE	944

TABLE B257

SEDIMENT : SULFATE REDUCTION			
ITEM:	VALUE:	CONDITIONS:	REF.:
*MARINE SEDIMENT	4.9 *10**(-9)M S/(CM**3)*DAY	ST.BARBARA BASIN,CALIFORNIA,UPPER 10CM	798
*MARINE SEDIMENT	1.4 *10**(-9)M S/(CM**3)*DAY	ST.BARBARA BASIN,CALIFORNIA,UPPER 3M	108

TABLE B257

SEDIMENT : SULFATE REDUCTION

ITEM:	VALUE:	CONDITIONS:	REF.:
*MARINE SEDIMENT	8.2 *10**(-9)M S/(CM**3)*DAY	BAY OF KIEL,GERMANY,SURFACE SEDIMENT,278-283K	638
*MARINE SEDIMENT	0.08-0.29 *10**(-9)M S/(CM**3)*DAY	BAY OF KIEL,GERMANY,10-30CM, 278-283K	638
*MARINE SEDIMENT	532-1444 *10**(-9)M S/(CM**3)*DAY	BARENTS SEA,LITTORAL SURFACE SEDIMENT	728
*MARINE SEDIMENT	29-68 (MEAN=49) *10**(-9)M S/(CM**3)*DAY	BLACK SEA,SURFACE SEDIMENT AT 200M,282K	1483
*MARINE SEDIMENT	0.8-38 (MEAN=14.3) *10**(-9)M S/(CM**3)*DAY	BLACK SEA,SURFACE SEDIMENT AT 2000M,282K	1483
*MARINE SEDIMENT	0 *10**(-9)M S/(CM**3)*DAY	BLACK SEA,BELOW 10CM AT 2000M,282K	1483
*MARINE SEDIMENT	55-135 *10**(-9)M S/(CM**3)*DAY	MODEL SYSTEM,LITTORAL SURFACE SEDIMENT,LIMFJORD,DENMARK,295K	767
*MARINE SEDIMENT	439-1013 *10**(-9)M S/(CM**3)*DAY	AARHUS BAY,DENMARK,SURFACE SEDIMENT	767

TABLE B258

SOIL : CATION EXCHANGE CAPACITY			
ITEM:	VALUE:	CONDITIONS:	REF.:
*ATTAPULGITE	0.2 EQ./KG	NONE	981
*CHLORITE	0.1-0.4 EQ./KG	NONE	981
*CLAY LOAMS	4.0 - 57.5 MEQ/(100 G DRY SOIL)	DETAILS SEE REF	752
*HALLOYSITE	0.05-0.5 EQ./KG	NONE	981
*HUMUS	1.7-2.9 EQ./KG	PH=4.5	157
*HUMUS	4.0-5.9 EQ./KG	PH=7.0	157
*ILLITE	10 - 40 MEQ/(100 G DRY SOIL)	DETAILS SEE REF	752
*KAOLINITE	0.03-0.15 EQ./KG	NONE	981
*KAOLINITE	3 - 15 MEQ/(100 G DRY SOIL)	DETAILS SEE REF	752
*LOAMS	7.5 - 15.9 MEQ/(100 G DRY SOIL)	DETAILS SEE REF	752
*MONTMORILLONITE	0.7-1.2 EQ./KG	NONE	981
*MONTMORILLONITE	100 - 150 MEQ/(100 G DRY SOIL)	DETAILS SEE REF	752
*SAND	2.0 - 3.5 MEQ/(100 G DRY SOIL)	DETAILS SEE REF	752
*SAND LOAMS	2.3 - 17.1 MEQ/(100 G DRY SOIL)	DETAILS SEE REF	752
*SILTY LOAMS	4.4 - 26.3 MEQ/(100 G DRY SOIL)	DETAILS SEE REF	752
*VERMICULITE	1.0-1.5 EQ./KG	NONE	981

B6

THERMO-DYNAMIC VALUES

BOND ENERGY
CONDUCTIVITY
ENERGY
ENTHALPY
ENTROPY
GIBBS FREE ENERGY
HEAT CAPACITY
HEAT TRANSFER
HENRY LAW CONSTANT
SOLUBILITY
VAPOR PRESSURE

TABLE B259

Units and Conversion Factors for Energy

REF.: 1689

Unit	erg	cal	kcal	Joule	Wh	KWh	BTU
Erg	1	2.39E-8	2.39E-11	1E-7	2.78E-11	2.78E-14	9.48E-11
cal	4.19E7	1	1E-3	4.19	1.16E-3	1.16E-6	3.97E-3
kcal	4.19E10	1000	1	4.19E3	1.16	1.16E-3	3.97
Joule	1E7	.239	2.39E-4	1	2.78E-4	2.78E-7	9.48E-4
Wh	3.6E10	8.60E2	8.60E-1	3.6E3	1	1E-3	3.41
KWh	3.6E13	8.60E5	8.60E2	3.6E6	1000	1	3.41E3
BTU	1.06E10	2.52E2	2.52E-1	1.06E3	2.93E-1	2.93E-4	1

TABLE B260

Units and Conversion Factors for Power.

REF.: 1689

Unit	erg/s	cal/s	cal/min	Watt	BTU/min
erg/s	1	2.39E-8	1.43E-6	1.00E-7	5.69E-9
cal/s	4.19E7	1	60	4.19	0.238
cal/min	6.97E5	1.67E-2	1	6.97E-2	3.97E-3
Watt	1.00E7	0.239	14.3	1	5.69E-2
BTU/min	1.76E8	4.20	252	17.6	1

TABLE B261

THERMODYNAMIC VALUES : BOND ENERGY			
ITEM:	VALUE:	CONDITIONS:	REF.:
*C-C (DOUBLE BOND)	145.8 KCAL/MOLE	AVERAGE, COVALENT BOND OF IMPORTANCE IN THE CHNOPS SYSTEM	291
*C-C (SINGLE BOND)	82.6 KCAL/MOLE	AVERAGE, COVALENT BOND OF IMPORTANCE IN THE CHNOPS SYSTEM	291
*C-C (TRIPLE BOND)	199.6 KCAL/MOLE	AVERAGE, COVALENT BOND OF IMPORTANCE IN THE CHNOPS SYSTEM	291
*C-H	98.7 KCAL/MOLE	AVERAGE, COVALENT BOND OF IMPORTANCE IN THE CHNOPS SYSTEM	291
*C-N (DOUBLE BOND)	147 KCAL/MOLE	AVERAGE, COVALENT BOND OF IMPORTANCE IN THE CHNOPS SYSTEM	291
*C-N (SINGLE BOND)	72.8 KCAL/MOLE	AVERAGE, COVALENT BOND OF IMPORTANCE IN THE CHNOPS SYSTEM	291
*C-N (TRIPLE BOND)	212.6 KCAL/MOLE	AVERAGE, COVALENT BOND OF IMPORTANCE IN THE CHNOPS SYSTEM	291
*C-O (DOUBLE BOND)	174 KCAL/MOLE	AVERAGE, COVALENT BOND OF IMPORTANCE IN THE CHNOPS SYSTEM	291
*C-O (SINGLE BOND)	85.5 KCAL/MOLE	AVERAGE, COVALENT BOND OF IMPORTANCE IN THE CHNOPS SYSTEM	291
*C-P (SINGLE BOND)	62 KCAL/MOLE	AVERAGE, COVALENT BOND OF IMPORTANCE IN THE CHNOPS SYSTEM	702
*C-S (SINGLE BOND)	65 KCAL/MOLE	AVERAGE, COVALENT BOND OF IMPORTANCE IN THE CHNOPS SYSTEM	291
*H-H	104.2 KCAL/MOLE	AVERAGE, COVALENT BOND OF IMPORTANCE IN THE CHNOPS SYSTEM	291
*H-N	93.4 KCAL/MOLE	AVERAGE, COVALENT BOND OF IMPORTANCE IN THE CHNOPS SYSTEM	291
*H-O	110.6 KCAL/MOLE	AVERAGE, COVALENT BOND OF IMPORTANCE IN THE CHNOPS SYSTEM	291
*H-P	77 KCAL/MOLE	AVERAGE, COVALENT BOND OF IMPORTANCE IN THE CHNOPS SYSTEM	291
*H-S	83.0 KCAL/MOLE	AVERAGE, COVALENT BOND OF IMPORTANCE IN THE CHNOPS SYSTEM	291
*N-N (DOUBLE BOND)	100 KCAL/MOLE	AVERAGE, COVALENT BOND OF IMPORTANCE IN THE CHNOPS SYSTEM	1229
*N-N (SINGLE BOND)	21 KCAL/MOLE	AVERAGE, COVALENT BOND OF IMPORTANCE IN THE CHNOPS SYSTEM	291
*N-N (TRIPLE BOND)	225.8 KCAL/MOLE	AVERAGE, COVALENT BOND OF IMPORTANCE IN THE CHNOPS SYSTEM	291
*N-O (DOUBLE BOND)	145 KCAL/MOLE	AVERAGE, COVALENT BOND OF IMPORTANCE IN THE CHNOPS SYSTEM	291
*N-O (SINGLE BOND)	53 KCAL/MOLE	AVERAGE, COVALENT BOND OF IMPORTANCE IN THE CHNOPS SYSTEM	291
*N-P (SINGLE BOND)	50 KCAL/MOLE	AVERAGE, COVALENT BOND OF IMPORTANCE IN THE CHNOPS SYSTEM	702
*N-P (TRIPLE BOND)	138 KCAL/MOLE	AVERAGE, COVALENT BOND OF IMPORTANCE IN THE CHNOPS SYSTEM	291
*O-O (DOUBLE BOND)	119.1 KCAL/MOLE	AVERAGE, COVALENT BOND OF IMPORTANCE IN THE CHNOPS SYSTEM	291
*O-O (SINGLE BOND)	41 KCAL/MOLE	AVERAGE, COVALENT BOND OF IMPORTANCE IN THE CHNOPS SYSTEM	291
*O-P (DOUBLE BOND)	120 KCAL/MOLE	AVERAGE, COVALENT BOND OF IMPORTANCE IN THE CHNOPS SYSTEM	291
*O-P (SINGLE BOND)	84 KCAL/MOLE	AVERAGE, COVALENT BOND OF IMPORTANCE IN THE CHNOPS SYSTEM	702
*O-S (DOUBLE BOND)	112 KCAL/MOLE	AVERAGE, COVALENT BOND OF IMPORTANCE IN THE CHNOPS SYSTEM	291
*P-P (SINGLE BOND)	48 KCAL/MOLE	AVERAGE, COVALENT BOND OF IMPORTANCE IN THE CHNOPS SYSTEM	291
*P-P (TRIPLE BOND)	116.7 KCAL/MOLE	AVERAGE, COVALENT BOND OF IMPORTANCE IN THE CHNOPS SYSTEM	291
*S-S (DOUBLE BOND)	84 KCAL/MOLE	AVERAGE, COVALENT BOND OF IMPORTANCE IN THE CHNOPS SYSTEM	291
*S-S (SINGLE BOND)	54 KCAL/MOLE	AVERAGE, COVALENT BOND OF IMPORTANCE IN THE CHNOPS SYSTEM	702

TABLE B262

THERMODYNAMIC VALUES : ENERGY			
ITEM:	VALUE:	CONDITIONS:	REF.:
*CITY	78 W/M**2	LOCAL ENERGY DISSIPATION RATE, 20000 PEOPLE/MILE**2,10KW/CAPITA	1442

TABLE B262

THERMODYNAMIC VALUES : ENERGY			
ITEM:	VALUE:	CONDITIONS:	REF.:
*EARTH	30 W/M**2	ENERGY STORAGE,TYPICAL VAL.,EITHER POS.OR NEG.	1442
*EARTH-SURFACE	220 W/M**2	SOLAR ENERGY ABSORBED,AVERAGE OVER ENTIRE EARTH	1442
*EARTH-SURFACE	220 W/M**2	INFRARED ENERGY EMITTED,AVERAGE	1442
*SUN	1400 W/M**2	SOLAR CONSTANT	1442
*U.S.A.	2.2*10**12 W	ENERGY CONSUMPTION RATE,1970	1442
*WHOLE WORLD	6.6*10**12 W	WORLD CONSUMPTION,1970	1442
*WHOLE WORLD	1*10**14 W	WORLD CONSUMPTION,POPULATION= 10*10**9,10KW(THERMAL)/CAPITA	1442

TABLE B263

THERMODYNAMIC VALUES : ENTHALPY OF INACTIVATION			
ITEM:	VALUE:	CONDITIONS:	REF.:
*AMYLASE MALT	41600 CAL/MOLE	PROTEIN, IN AQUEOUS SOLUTION	1501
*CATALASE	50500 CAL/MOLE	PROTEIN, IN AQUEOUS SOLUTION	1501
*EGG ALBUMIN	134000 CAL/MOLE	PROTEIN, IN AQUEOUS SOLUTION, PH = 7.7	1501
*EGG ALBUMIN	96800 CAL/MOLE	PROTEIN, IN AQUEOUS SOLUTION , PH = 3.4	1501
*EGG ALBUMIN	35200 CAL/MOLE	PROTEIN, IN AQUEOUS SOLUTION, PH = 1.35	1501
*EMULSIN	44900 CAL/MOLE	PROTEIN, IN AQUEOUS SOLUTION	1501
*ENTEROKINASE	42160 CAL/MOLE	PROTEIN, IN AQUEOUS SOLUTION	1501
*HEMOGLOBIN	75600 CAL/MOLE	PROTEIN, IN AQUEOUS SOLUTION	1501
*HEMOLYSIN	198000 CAL/MOLE	PROTEIN, IN AQUEOUS SOLUTION, FROM GOAT	1501
*INSULIN	35000 CAL/MOLE	PROTEIN, IN AQUEOUS SOLUTION	1501
*LEUCOSIN	84300 CAL/MOLE	PROTEIN, IN AQUEOUS SOLUTION	1501
*PEPSIN	55600 CAL/MOLE	PROTEIN, IN AQUEOUS SOLUTION	1501
*PEROXIDASE	185000 CAL/MOLE	PROTEIN, IN AQUEOUS SOLUTION, MILK	1501
*PROTEINASE	37860 CAL/MOLE	PROTEIN, IN AQUEOUS SOLUTION, PANCREATIC	1501
*RENNEN	89300 CAL/MOLE	PROTEIN, IN AQUEOUS SOLUTION	1501
*TETANOLYSIN	172000 CAL/MOLE	PROTEIN, IN AQUEOUS SOLUTION	1501
*TOBACCO MOSAIC VIRUS	40000 CAL/MOLE	PROTEIN, IN AQUEOUS SOLUTION	1501
*TRYPSIN	40200 CAL/MOLE	PROTEIN, IN AQUEOUS SOLUTION	1501
*TRYPSIN KINASE	44260 CAL/MOLE	PROTEIN, IN AQUEOUS SOLUTION	1501
*T1 BACTERIOPHAGE	95000 CAL/MOLE	PROTEIN, IN AQUEOUS SOLUTION	1501
*VIBRIOLYSIN	128000 CAL/MOLE	PROTEIN, IN AQUEOUS SOLUTION	1501
*YEAST INVERTASE	52400 CAL/MOLE	PROTEIN, IN AQUEOUS SOLUTION, PH = 5.7	1501
*YEAST INVERTASE	86400 CAL/MOLE	PROTEIN, IN AQUEOUS SOLUTION, PH = 5.2	1501
*YEAST INVERTASE	110400 CAL/MOLE	PROTEIN, IN AQUEOUS SOLUTION, PH = 4.0	1501
*YEAST INVERTASE	74400 CAL/MOLE	PROTEIN, IN AQUEOUS SOLUTION, PH = 3.0	1501

TABLE B264

THERMODYNAMIC VALUES : ENTROPY OF INACTIVATION			
ITEM:	VALUE:	CONDITIONS:	REF.:
*AMYLASE MALT	52.3 CAL/(MOLE*DEGREE)	PROTEIN, IN AQUEOUS SOLUTION	1501

TABLE B264

THERMODYNAMIC VALUES : ENTROPY OF INACTIVATION			
ITEM:	VALUE:	CONDITIONS:	REF.:
*CATALASE	81.0 CAL/(MOLE*DEGREE)	PROTEIN, IN AQUEOUS SOLUTION	1501
*EGG ALBUMIN	317.1 CAL/(MOLE*DEGREE)	PROTEIN, IN AQUEOUS SOLUTION, PH = 7.7	1501
*EGG ALBUMIN	223.7 CAL/(MOLE*DEGREE)	PROTEIN, IN AQUEOUS SOLUTION, PH = 3.4	1501
*EGG ALBUMIN	36.3 CAL/(MOLE*DEGREE)	PROTEIN, IN AQUEOUS SOLUTION, PH = 1.35	1501
*EMULSIN	65.3 CAL/(MOLE*DEGREE)	PROTEIN, IN AQUEOUS SOLUTION	1501
*ENTEROKINASE	52.8 CAL/(MOLE*DEGREE)	PROTEIN, IN AQUEOUS SOLUTION	1501
*HEMOGLOBIN	152.7 CAL/(MOLE*DEGREE)	PROTEIN, IN AQUEOUS SOLUTION	1501
*HEMOLYSIN	537.0 CAL/(MOLE*DEGREE)	PROTEIN, IN AQUEOUS SOLUTION, FROM GOAT	1501
*INSULIN	23.8 CAL/(MOLE*DEGREE)	PROTEIN, IN AQUEOUS SOLUTION	1501
*LEUCOSIN	185.0 CAL/(MOLE*DEGREE)	PROTEIN, IN AQUEOUS SOLUTION	1501
*PEPSIN	113.3 CAL/(MOLE*DEGREE)	PROTEIN, IN AQUEOUS SOLUTION	1501
*PEROXIDASE	466.0 CAL/(MOLE*DEGREE)	PROTEIN, IN AQUEOUS SOLUTION, MILK	1501
*PROTEINASE	40.6 CAL/(MOLE*DEGREE)	PROTEIN, IN AQUEOUS SOLUTION, PANCREATIC	1501
*RENNEN	208.0 CAL/(MOLE*DEGREE)	PROTEIN, IN AQUEOUS SOLUTION	1501
*TETANOLYSIN	459.0 CAL/(MOLE*DEGREE)	PROTEIN, IN AQUEOUS SOLUTION	1501
*TOBACCO MOSAIC VIRUS	18.0 CAL/(MOLE*DEGREE)	PROTEIN, IN AQUEOUS SOLUTION	1501
*TRYPSIN	44.7 CAL/(MOLE*DEGREE)	PROTEIN, IN AQUEOUS SOLUTION	1501
*TRYPSIN KINASE	57.6 CAL/(MOLE*DEGREE)	PROTEIN, IN AQUEOUS SOLUTION	1501
*T1 BACTERIOPHAGE	207.0 CAL/(MOLE*DEGREE)	PROTEIN, IN AQUEOUS SOLUTION	1501
*VIBRIOLYSIN	326.0 CAL/(MOLE*DEGREE)	PROTEIN, IN AQUEOUS SOLUTION	1501
*YEAST INVERTASE	84.7 CAL/(MOLE*DEGREE)	PROTEIN, IN AQUEOUS SOLUTION, PH = 5.7	1501
*YEAST INVERTASE	185.0 CAL/(MOLE*DEGREE)	PROTEIN, IN AQUEOUS SOLUTION, PH = 5.2	1501
*YEAST INVERTASE	262.5 CAL/(MOLE*DEGREE)	PROTEIN, IN AQUEOUS SOLUTION, PH = 4.0	1501
*YEAST INVERTASE	152.4 CAL/(MOLE*DEGREE)	PROTEIN, IN AQUEOUS SOLUTION, PH = 3.0	1501

TABLE B265

THERMODYNAMIC VALUES : GIBBS FREE ENERGY			
ITEM:	VALUE:	CONDITIONS:	REF.:
*FRUCTOSE-1-PHOSPHATE	-2.8 KCAL/MOLE	HYDROLYSIS OF SUGAR-PHOSPHATE BOND	1031
*FRUCTOSE-6-PHOSPHATE	-3.0 KCAL/MOLE	HYDROLYSIS OF SUGAR-PHOSPHATE BOND	1031
*GALACTOSE-6-PHOSPHATE	-3.0 KCAL/MOLE	HYDROLYSIS OF SUGAR-PHOSPHATE BOND	1031
*GLUCOSE-1-PHOSPHATE	-4.9 KCAL/MOLE	HYDROLYSIS OF SUGAR-PHOSPHATE BOND	1031
*GLUCOSE-6-PHOSPHATE	-3.0 KCAL/MOLE	HYDROLYSIS OF SUGAR-PHOSPHATE BOND	1031
*MANNOSE-6-PHOSPHATE	-2.7 KCAL/MOLE	HYDROLYSIS OF SUGAR-PHOSPHATE BOND	1031
*2-PHOSPHOGLYCERIC ACID	-4.1 KCAL/MOLE	HYDROLYSIS OF SUGAR-PHOSPHATE BOND	1031
*3-PHOSPHOGLYCERIC ACID	-3.0 KCAL/MOLE	HYDROLYSIS OF SUGAR-PHOSPHATE BOND	1031

TABLE B266

THERMODYNAMIC VALUES : HEAT CAPACITY			
ITEM:	VALUE:	CONDITIONS:	REF.:
*CONCRETE	500*10**(-3) CAL/(CM**3*DEG)	NONE	268

TABLE B266

THERMODYNAMIC VALUES : HEAT CAPACITY			
ITEM:	VALUE:	CONDITIONS:	REF.:
*DRY CLAY	100 - 400*10**(-3) CAL/(CM**3*DEG)	NONE	268
*DRY SAND	100 - 580*10**(-3) CAL/(CM**3*DEG)	NONE	268
*DRY WOOD	100 - 200*10**(-3) CAL/(CM**3*DEG)	NONE	268
*ICE	460*10**(-3) CAL/(CM**3*DEG)	NONE	268
*IRON	820*10**(-3) CAL/(CM**3*DEG)	NONE	268
*ROCK	430 - 580*10**(-3) CAL/(CM**3*DEG)	NONE	268
*SILVER	590*10**(-3) CAL/(CM**3*DEG)	NONE	268
*SOIL ORGANIC MATTER	600*10**(-3) CAL/(CM**3*DEG)	NONE	268
*STILL AIR	0.24 - 0.34*10**(-3) CAL/ (CM**3*DEG)	NONE	268
*STILL WATER	1000*10**(-3) CAL/(CM**3*DEG)	NONE	268
*WET CLAY	300 - 400*10**(-3) CAL/(CM**3*DEG)	NONE	268
*WET SAND	200 - 600*10**(-3) CAL/(CM**3*DEG)	NONE	268

TABLE B267

THERMODYNAMIC VALUES : HEAT OF COMBUSTION			
ITEM:	VALUE:	CONDITIONS:	REF.:
*ARTEMIS SP.	-6.737 KCAL/G ASH-FREE DW	BRINE SHRIMP, NAUPLII	1067
*BIPALIUM KEWENSE	-5.684 KCAL/G ASH-FREE DW	TERRISTRICAL FLATWORM	1067
*CALANUS HELGOLANDICUS	-5.400 KCAL/G ASH-FREE DW	COPEPOD	1067
*CHLOROHYDRA VIRIDISSIMA	-5.729 KCAL/G ASH-FREE DW	GREEN HYDRA	1067
*DUGESIA TIGRINA	-6.286 KCAL/G ASH-FREE DW	FLATWORM	1067
*EGGS	-5.75 KCAL/GRAM	ONLY THE PROTEIN FRACTION	1662
*EGGS	-9.50 KCAL/GRAM	ONLY THE FAT FRACTION	1662
*EGGS	-3.75 KCAL/GRAM	ONLY THE CARBOHYDRATE FRACTION	1662
*EXTRACTED BIRD FAT	-9.03 KCAL/G DRY WEIGHT	FAT RATIO = 100.0 % FAT OF DW.	1162
*FALL BIRDS	-8.08 KCAL/G DRY WEIGHT	FAT RATIO = 71.7 % FAT OF DW.	1162
*FALL BIRDS	-5.47 KCAL/G DRY WEIGHT	FAT RATIO = 0.0 % FAT OF DW., FAT EXTRACTED	1162
*FISH MEAT	-5.65 KCAL/GRAM	ONLY THE PROTEIN FRACTION	1662
*FISH MEAT	-9.50 KCAL/GRAM	ONLY THE FAT FRACTION	1662
*FRUITS	-5.20 KCAL/GRAM	ONLY THE PROTEIN FRACTION	1662
*FRUITS	-9.30 KCAL/GRAM	ONLY THE FAT FRACTION	1662
*FRUITS	-4.00 KCAL/GRAM	ONLY THE CARBOHYDRATE FRACTION	1662
*GELATIN	-5.27 KCAL/GRAM	ONLY THE PROTEIN FRACTION	1662
*GELATIN	-9.50 KCAL/GRAM	ONLY THE FAT FRACTION	1662
*GLUCOSE	-3.75 KCAL/GRAM	ONLY THE CARBOHYDRATE FRACTION	1662
*GLYCOGEN	-4.19 KCAL/GRAM	ONLY THE CARBOHYDRATE FRACTION	1662
*GOTTIDIA PYRAMIDATA	-4.397 KCAL/G ASH-FREE DW	BRANCHIOPOD	1067
*GRAIN	-5.80 KCAL/GRAM	ONLY THE PROTEIN FRACTION	1662
*GRAIN	-9.30 KCAL/GRAM	ONLY THE FAT FRACTION	1662
*GRAIN	-4.20 KCAL/GRAM	ONLY THE CARBOHYDRATE FRACTION	1662
*GREEN GRASS	-4.373 KCAL/ASH-FREE DRY WEIGHT	ANDROPOGON VIRINICUS OLD-FIELD COMMUNITY, GEORGIA	544
*GREEN HERBS	-4.288 KCAL/ASH-FREE DRY WEIGHT	ANDROPOGON VIRINICUS OLD-FIELD COMMUNITY, GEORGIA	544
*HYDRA LITTORALIS	-6.034 KCAL/G ASH-FREE DW	HYDRA	1067
*LEBISTES RETICULATUS	-5.823 KCAL/G ASH-FREE DW	GUPPIE	1067
*LEPTODORA KINDTII	-5.605 KCAL/G ASH-FREE DW	CLADOCERA	1067

TABLE B267

THERMODYNAMIC VALUES : HEAT OF COMBUSTION			
ITEM:	VALUE:	CONDITIONS:	REF.:
*LITTER	-4.139 KCAL/ASH-FREE DRY WEIGHT	ANDROPOGON VIRINICUS OLD-FIELD COMMUNITY, GEORGIA	544
*MILK	-5.65 KCAL/GRAM	ONLY THE PROTEIN FRACTION	1662
*MILK	-9.25 KCAL/GRAM	ONLY THE FAT FRACTION	1662
*MILK	-3.95 KCAL/GRAM	ONLY THE CARBOHYDRATE FRACTION	1662
*MUSHROOM	-5.00 KCAL/GRAM	ONLY THE PROTEIN FRACTION	1662
*MUSHROOM	-9.30 KCAL/GRAM	ONLY THE FAT FRACTION	1662
*MUSHROOM	-4.10 KCAL/GRAM	ONLY THE CARBOHYDRATE FRACTION	1662
*NONMIGRANTS BIRDS	-6.26 KCAL/G DRY WEIGHT	FAT RATIO = 21.2 % FAT OF DW.	1162
*NONMIGRANTS BIRDS	-5.44 KCAL/G DRY WEIGHT	FAT RATIO = 0.0 % FAT OF DW., FAT EXTRACTED	1162
*PHILENUS LEUCOPTHALMUS	-6.962 KCAL/G ASH-FREE DW	SPIT BUG	1067
*PYCNOPSYCHE GUTTIFER	-5.706 KCAL/G ASH-FREE DW	CADDIS FLY	1067
*PYCNOPSYCHE LEPIDO	-5.687 KCAL/G ASH-FREE DW	CADDIS FLY	1067
*ROOTS	-4.167 KCAL/ASH-FREE DRY WEIGHT	ANDROPOGON VIRINICUS OLD-FIELD COMMUNITY, GEORGIA	544
*SPRING BIRDS	-7.04 KCAL/G DRY WEIGHT	FAT RATIO = 44.1 % FAT OF DW.	1162
*SPRING BIRDS	-5.41 KCAL/G DRY WEIGHT	FAT RATIO = 0.0 % FAT OF DW., FAT EXTRACTED	1162
*STANDING DEAD VEGETATION	-4.290 KCAL/ASH-FREE DRY WEIGHT	ANDROPOGON VIRINICUS OLD-FIELD COMMUNITY, GEORGIA	544
*SUCCINA OVALIS	-5.415 KCAL/G ASH-FREE DW	AQUATIC SNAIL	1067
*SUCROSE	-3.95 KCAL/GRAM	ONLY THE CARBOHYDRATE FRACTION	1662
*TENEBRIO MOLITOR	-6.314 KCAL/G ASH-FREE DW	BEETLE	1067
*TETRAHYMENA PYRIFORMIS	-5.938 KCAL/G ASH-FREE DW	CILIATE	1067
*TRIGRIOPUS CALIFORNICUS	-5.515 KCAL/G ASH-FREE DW	COPEPOD	1067
*TYROGLYPHUS LINTNERI	-5.808 KCAL/G ASH-FREE DW	MITE	1067
*YEAST	-5.00 KCAL/GRAM	ONLY THE FAT FRACTION	1662
*YEAST	-9.30 KCAL/GRAM	ONLY THE CARBOHYDRATE FRACTION	1662
*YEAST	-4.20 KCAL/GRAM	ONLY THE CARBOHYDRATE FRACTION	1662

TABLE B268

THERMODYNAMIC VALUES : HEAT OF SOLUTION			
ITEM:	VALUE:	CONDITIONS:	REF.:
*ALANINE	2.25 KCAL/MOLE	SATURATION	1067
*ALANINE	2.04 KCAL/MOLE	INFINITE DILUTION	1067
*ARGININE	1.08 KCAL/MOLE	SATURATION	1067
*ARGININE	1.50 KCAL/MOLE	INFINITE DILUTION	1067
*ASPARAGINE	5.75 KCAL/MOLE	SATURATION	1067
*ASPARAGINE	5.75 KCAL/MOLE	INFINITE DILUTION	1067
*ASPARCTIC ACID	5.80 KCAL/MOLE	SATURATION	1067
*ASPARCTIC ACID	6.00 KCAL/MOLE	INFINITE DILUTION	1067
*GLUCOSE	2.96 KCAL/MOLE	SATURATION	1067
*GLUCOSE	2.63 KCAL/MOLE	INFINITE DILUTION	1067
*GLUTAMIC ACID	6.33 KCAL/MOLE	SATURATION	1067
*GLUTAMIC ACID	6.53 KCAL/MOLE	INFINITE DILUTION	1067
*GLYCERIN	5.46 KCAL/MOLE	SATURATION	1067
*GLYCERIN	-1.410 KCAL/MOLE	INFINITE DILUTION	1067
*GLUCINE	3.35 KCAL/MOLE	SATURATION	1067
*GLYCINE	3.75 KCAL/MOLE	INFINITE DILUTION	1067

TABLE B268

THERMODYNAMIC VALUES : HEAT OF SOLUTION			
ITEM:	VALUE:	CONDITIONS:	REF.:
*HISTIDINE	3.21 KCAL/MOLE	SATURATION	1067
*HISTIDINE	3.30 KCAL/MOLE	INFINITE DILUTION	1067
*ISOLEUCINE	0.84 KCAL/MOLE	SATURATION	1067
*LEUCINE	0.84 KCAL/MOLE	SATURATION	1067
*LEUCINE	0.78 KCAL/MOLE	INFINITE DILUTION	1067
*LYSINE	-3.50 KCAL/MOLE	SATURATION	1067
*LYSINE	-4.00 KCAL/MOLE	INFINITE DILUTION	1067
*METHIONINE	4.00 KCAL/MOLE	SATURATION	1067
*METHIONINE	4.00 KCAL/MOLE	INFINITE DILUTION	1067
*PHENYLALANINE	2.82 KCAL/MOLE	SATURATION	1067
*PHOSPHATE	15.63 KCAL/MOLE	SATURATION	1067
*PHOSPHATE	-2.0 KCAL/MOLE	INFINITE DILUTION	1067
*PROLINE	0.30 KCAL/MOLE	SATURATION	1067
*PROLINE	0.75 KCAL/MOLE	INFINITE DILUTION	1067
*SERINE	5.05 KCAL/MOLE	SATURATION	1067
*SERINE	5.18 KCAL/MOLE	INFINITE DILUTION	1067
*TRYPTOPHAN	1.36 KCAL/MOLE	SATURATION	1067
*TRYPTOPHAN	1.36 KCAL/MOLE	INFINITE DILUTION	1067
*VALINE	1.73 KCAL/MOLE	SATURATION	1067
*VALINE	1.43 KCAL/MOLE	INFINITE DILUTION	1067

TABLE B269

THERMODYNAMIC VALUES : HEAT TRANSFER, EARTH			
ITEM:	VALUE:	CONDITIONS:	REF.:
*CONVECTIVE HEAT LOSS	- 870 - 1100 KCAL/(M**2*DAY)	SYMBOL : QC	944
*CONVECTIVE HEAT LOSS OF GAIN	- 35 - 50 KCAL/(M**2*HOUR)	SYMBOL : QC	55
*EVAPORATION HEAT LOSS	25 - 900 KCAL/(M**2*HOUR)	SYMBOL : QE	55
*EVAPORATIVE HEAT LOSS	5500 - 20000 KCAL/(M**2*DAY)	SYMBOL : QE	944
*LONGWAVE ATMOSPHERIC RADIATION	225 - 360 KCAL/(M**2*HOUR)	SYMBOL : QA	55
*NET LONG WAVE RADIATION	220 - 345 KCAL/(M**2*HOUR)	SYMBOL : QAN = QA - QAR	55
*NET LONG WAVE RADIATION	6350 - 8150 KCAL/(M**2*DAY)	SYMBOL : QAN	944
*NET SHORTWAVE	25 - 375 KCAL/(M**2*HOUR)	SYMBOL : QSN = QS - QSR	55
*NET SURFACE RADIATION	225 - 400 KCAL/(M**2*HOUR)	SYMBOL : QBR	55
*NET SURFACE RADIATION	6500 - 9800 KCAL/(M**2*DAY)	SYMBOL : QBR	944
*REFLECTED LONG WAVE RADIATION	5 - 15 KCAL/(M**2*HOUR)	SYMBOL : QAR	55
*REFLECTED SHORTWAVE	5 - 25 KCAL/(M**2*HOUR)	SYMBOL : QSR	55
*REFLECTED SOLAR RADIATION	110 - 550 KCAL/(M**2*DAY)	SYMBOL : QR	944
*SHORTWAVE SOLAR RADIATION	30 - 300 KCAL/(M**2*HOUR)	SYMBOL : QS	55
*SOLAR RADIATION	1100 - 7600 KCAL/(M**2*DAY)	SYMBOL : QS	944

TABLE B270

THERMODYNAMIC VALUES : HENRY LAW CONSTANT			
ITEM:	VALUE:	CONDITIONS:	REF.:
*ALDRIN	1.44*10**(-5) ATM M**3/MOL	HENRY LAW CONSTANT=VAPOUR PRESSURE/SOLUBILITY IN WATER, 298 K	960

TABLE B270

THERMODYNAMIC VALUES : HENRY LAW CONSTANT			
ITEM:	VALUE:	CONDITIONS:	REF.:
*AROCLOR 1242	5.73*10**(-4) ATM M**3/MOL	HENRY LAW CONSTANT=VAPOUR PRESSURE/SOLUBILITY IN WATER, 298 K	960
*AROCLOR 1248	3.51*10**(-3) ATM M**3/MOL	HENRY LAW CONSTANT=VAPOUR PRESSURE/SOLUBILITY IN WATER, 298 K	960
*AROCLOR 1254	2.76*10**(-3) ATM M**3/MOL	HENRY LAW CONSTANT=VAPOUR PRESSURE/SOLUBILITY IN WATER, 298 K	960
*AROCLOR 1260	7.13*10**(-3) ATM M**3/MOL	HENRY LAW CONSTANT=VAPOUR PRESSURE/SOLUBILITY IN WATER, 298 K	960
*BENZENE	5.5*10**(-3) ATM M**3/MOL	HENRY LAW CONSTANT=VAPOUR PRESSURE/SOLUBILITY IN WATER, 298 K	960
*BIPHENYL	1.55*10**(-3) ATM M**3/MOL	HENRY LAW CONSTANT=VAPOUR PRESSURE/SOLUBILITY IN WATER, 298 K	960
*CUMENE	1.46*10**(-2) ATM M**3/MOL	HENRY LAW CONSTANT=VAPOUR PRESSURE/SOLUBILITY IN WATER, 298 K	960
*DDT	3.89*10**(-5) ATM M**3/MOL	HENRY LAW CONSTANT=VAPOUR PRESSURE/SOLUBILITY IN WATER, 298 K	960
*DIELDRIN	2.00*10**(-7) ATM M**3/MOL	HENRY LAW CONSTANT=VAPOUR PRESSURE/SOLUBILITY IN WATER, 298 K	960
*HG	1.14*10**(-2) ATM M**3/MOL	HENRY LAW CONSTANT=VAPOUR PRESSURE/SOLUBILITY IN WATER, 298 K	960
*LINDANE	4.93*10**(-7) ATM M**3/MOL	HENRY LAW CONSTANT=VAPOUR PRESSURE/SOLUBILITY IN WATER, 298 K	960
*N-OCTANE	3.21 ATM M**3/MOL	HENRY LAW CONSTANT=VAPOUR PRESSURE/SOLUBILITY IN WATER, 298 K	960
*NAPHTHALENE	1.18*10**(-3) ATM M**3/MOL	HENRY LAW CONSTANT=VAPOUR PRESSURE/SOLUBILITY IN WATER, 298 K	960
*O-XYLENE	5.27*10**(-3) ATM M**3/MOL	HENRY LAW CONSTANT=VAPOUR PRESSURE/SOLUBILITY IN WATER, 298 K	960
*TOLUENE	6.68*10**(-3) ATM M**3/MOL	HENRY LAW CONSTANT=VAPOUR PRESSURE/SOLUBILITY IN WATER, 298 K	960
*2,2,4-TRIMETHYLPENTANE	3.04 ATM M**3/MOL	HENRY LAW CONSTANT=VAPOUR PRESSURE/SOLUBILITY IN WATER, 298 K	960

TABLE B271

THERMODYNAMIC VALUES : MEAN TEMPERATURE AND HUMIDITY			
ITEM:	VALUE:	CONDITIONS:	REF.:
*CONTINENTAL POLAR AIR MASS	262 K	COLD, DRY, SPECIFIC HUMIDITY = 1.4 G/KG	1516
*CONTINENTAL TROPICAL AIR MASS	297 K	WARM, DRY, SPECIFIC HUMIDITY = 11 G/KG	1516
*MARITIME EQUATORIAL AIR MASS	300 K	WARM, VERY MOIST, SPECIFIC HUMIDITY = 19 G/KG	1516
*MARITIME POLAR AIR MASS	277 K	COOL, MOIST, SPECIFIC HUMIDITY = 4.4 G/KG	1516
*MARITIME TROPICAL AIR MASS	297 K	WARM, MOIST, SPECIFIC HUMIDITY = 17 G/KG	1516

TABLE B272

THERMODYNAMIC VALUES : SOLAR ENERGY DISTRIBUTION			
ITEM:	VALUE:	CONDITIONS:	REF.:
*FAR ULTRAVIOLET	0.02 % OF SOLAR SPECTRUM	WAVELENGTH = 0.1-0.2 U, ENERGY RANGE = 152.2-304.4 KCAL/EINSTEIN	777

TABLE B272

THERMODYNAMIC VALUES : SOLAR ENERGY DISTRIBUTION			
ITEM:	VALUE:	CONDITIONS:	REF.:
*MIDDLE INFRARED	2.10 % OF SOLAR SPECTRUM	WAVELENGTH = 3-30 U, ENERGY RANGE = 0.95-9.5 KCAL/EINSTEIN	777
*NEAR INTRARED	38.90 % OF SOLAR SPECTRUM	WAVELENGTH = 0.78-3 U, ENERGY RANGE = 9.5-36.7 KCAL/EINSTEIN	777
*ULTRAVIOLET	7.27 % OF SOLAR SPECTRUM	WAVELENGTH = 0.2-0.38 U, ENERGY RANGE = 75.3-152.2 KCAL/EINSTEIN	777
*VISIBLE	51.73 % OF SOLAR SPECTRUM	WAVELENGTH = 0.38-0.78 U, ENERGY RANGE = 36.7-75.3 KCAL/EINSTEIN	777

TABLE B273

THERMODYNAMIC VALUES : SOLUBILITY			
ITEM:	VALUE:	CONDITIONS:	REF.:
*ADENINE	0.0068 MOLES/LITER	IN WATER	1067
*ALANINE	1.665 MOLES/LITER	IN WATER	1067
*ALDRIN	0.2 MG/L	IN WATER, 298 K	960
*ANTHRACENE	75 MG/M3	IN WATER AT 298 K	355
*AROCLOR 1242	PCB	IN WATER, 298 K	960
*AROCLOR 1248	PCB	IN WATER, 298 K	960
*AROCLOR 1254	PCB	IN WATER, 298 K	960
*AROCLOR 1260	PCB	IN WATER, 298 K	960
*ASPARAGINE	0.186 MOLES/LITER	IN WATER	1067
*ASPARCTIC ACID	0.038 MOLES/LITER	IN WATER	1067
*BENZENE	1780 MG/L	IN WATER, 298 K	960
*BENZO(A)ANTHRACENE	10 MG/M3	IN WATER AT 298 K	355
*BIPHENYL	7.48 MG/L	IN WATER, 298 K	960
*CCL3F	110 MG/100 ML	293 K, IN WATER	1001
*CHRYSENE	6 MG/M3	IN WATER AT 298 K	355
*CUMENE	50 MG/L	IN WATER, 298 K	960
*CYTOSINE	0.064 MOLES/LITER	IN WATER	1067
*DDT	1.2*10**(-3) MG/L	IN WATER, 298 K	960
*DECACHLOROBIPHENYL	0.015 MG/L	IN WATER	709
*DECACHLOROBIPHENYL	0.015 MG/L	IN WATER	709
*DIELDRIN	0.25 MG/L	IN WATER, 298 K	960
*FATTY ACID	0.0000011 MOLES/LITER	IN WATER	1067
*FLUORANTHENE	265 MG/M3	IN WATER AT 298 K	355
*GLUCOSE	3.452 MOLES/LITER	IN WATER	1067
*GLUTAMIC ACID	0.058 MOLES/LITER	IN WATER	1067
*GLYCINE	2.900 MOLES/LITER	IN WATER	1067
*GUANINE	0.000033 MOLES/LITER	IN WATER	1067
*HG	3*10**(-2) MG/L	IN WATER, 298 K	960
*HISTIDINE	0.262 MOLES/LITER	IN WATER	1067
*ISOLEUCINE	0.303 MOLES/LITER	IN WATER	1067
*LEUCINE	0.181 MOLES/LITER	IN WATER	1067
*LINDANE	7.3 MG/L	IN WATER, 298 K	960
*N-OCTANE	0.66 MG/L	IN WATER, 298 K	960
*NAPHTALENE	1.5 MG/M3	IN WATER AT 298 K	355
*NAPHTHALENE	33 MG/L	IN WATER, 298 K	960
*NAPHTHALENE	12500 MG/M3	IN WATER AT 298 K	355
*O-XYLENE	175 MG/L	IN WATER, 298 K	960

TABLE B273

THERMODYNAMIC VALUES : SOLUBILITY			
ITEM:	VALUE:	CONDITIONS:	REF.:
*PHENANTHRENE	1600 MG/M3	IN WATER AT 298 K	355
*PHENYLALANINE	0.175 MOLES/LITER	IN WATER	1067
*PROLINE	6.572 MOLES/LITER	IN WATER	1067
*PYRENE	175 MG/M3	IN WATER AT 298 K	355
*SERINE	0.464 MOLES/LITER	IN WATER	1067
*THYMINE	0.032 MOLES/LITER	IN WATER	1067
*TOLUENE	515 MG/L	IN WATER, 298 K	960
*TRYPTOPHAN	0.055 MOLES/LITER	IN WATER	1067
*TYROSINE	0.0025 MOLES/LITER	IN WATER	1067
*URACIL	0.032 MOLES/LITER	IN WATER	1067
*VALINE	0.740 MOLES/LITER	IN WATER	1067
*2-MONOCHLOROBIPHENYL	5.9 MG/L	IN WATER	709
*2,2,4-TRIMETHYLPENTANE	2.44 MG/L	IN WATER, 298 K	960
*2,2'-DICHLOROBIPHENYL	1.50 MG/L	IN WATER	709
*2,2',3,3'-TETRACHLOROBIPHENYL	0.034 MG/L	IN WATER	709
*2,2',3,3',4,4',5,5'-OCTACHLOROBIPHENYL	0.0070 MG/L	IN WATER	709
*2,2',3,5'-TETRACHLOROBIPHENYL	0.170 MG/L	IN WATER	709
*2,2',4,4'-TETRACHLOROBIPHENYL	0.068 MG/L	IN WATER	709
*2,2',4,4',5,5'-HEXACHLOROBIPHENYL	0.0088 MG/L	IN WATER	709
*2,2',4,5,5'-PENTACHLOROBIPHENYL	0.031 MG/L	IN WATER	709
*2,2',5,5'-TETRACHLOROBIPHENYL	0.046 MG/L	IN WATER	709
*2,2'3,4,5'-PENTACHLOROBIPHENYL	0.022 MG/L	IN WATER	709
*2,3',4,4'-TETRACHLOROBIPHENYL	0.058 MG/L	IN WATER	709
*2,3',4',5-TETRACHLOROBIPHENYL	0.041 MG/L	IN WATER	709
*2,4-DICHLOROBIPHENYL	1.40 MG/L	IN WATER	709
*2,4,4'-TRICHLOROBIPHENYL	0.085 MG/L	IN WATER	709
*2,4'-DICHLOROBIPHENYL	1.88 MG/L	IN WATER	709
*2',3,4-TRICHLOROBIPHENYL	0.078 NG/L	IN WATER	709
*3-MONOCHLOROBIPHENYL	3.5 MG/L	IN WATER	709
*3,3',4,4'-TETRACHLOROBIPHENYL	0.175 MG/L	IN WATER	709
*4-MONOCHLOROBIPHENYL	1.19 MG/L	IN WATER	709

TABLE B273

THERMODYNAMIC VALUES : SOLUBILITY			
ITEM:	VALUE:	CONDITIONS:	REF.:
*4,4'-DICHLOROBIPHENYl	0.08 MG/L	IN WATER	709
*4,4'-DICHLOROBIPHENYL + HUMIC ACID EXTRACT	0.07 MG/L	IN WATER	709
*4,4'-DICHLOROBIPHENYL + TWEEN 80 (0.1%) (F010)	5.9 MG/L	IN WATER	709
*4,4'-DICHLOROBIPHENYL + TWEEN 80 (1%) (F010)	>= 10.0 MG/L	IN WATER	709

TABLE B274

THERMODYNAMIC VALUE : STANDARD HEAT OF FORMATION			
ITEM:	VALUE:	CONDITIONS:	REF.:
*ADENINE	+23.21 KCAL/MOLE	NONE.	1067
*ALANINE	-133.96 KCAL/MOLE	NONE.	1067
*ARGININE	-148.66 KCAL/MOLE	NONE.	1067
*ASPARAGINE	-188.50 KCAL/MOLE	NONE.	1067
*ASPARCTIC ACID	-232.47 KCAL/MOLE	NONE.	1067
*CYSTEINE	-127.2 KCAL/MOLE	NONE.	1067
*CYTOSINE	-111 KCAL/MOLE	NONE.	1067
*C6H12O6	63.1 CAL/MOLE*DEG)	DILUTE AQUEOUS SOLUTION	1067
*DEOXYRIBOSE	-214.4 KCAL/MOLE	NONE.	1067
*FATTY ACID	-220.5 KCAL/MOLE	NONE.	1067
*GLUCOSE	-304.26 KCAL/MOLE	NONE.	1067
*GLUTAMIC ACID	-240.05 KCAL/MOLE	NONE.	1067
*GLUTAMINE	-197.8 KCAL/MOLE	NONE.	1067
*GLYCERIN	-159.160 KCAL/MOLE	NONE.	1067
*GLYCINE	-126.22 KCAL/MOLE	NONE.	1067
*GUANINE	-43.72 KCAL/MOLE	NONE.	1067
*HISTIDINE	-54.2 KCAL/MOLE	NONE.	1067
*HNO3	-49.37 KCAL/MOLE	IN AQUEOUS SOLUTION	1067
*H2O	-68.32 KCAL/MOLE	IN AQUEOUS SOLUTION	1067
*H2SO4	-216.90 KCAL/MOLE	IN AQUEOUS SOLUTION	1067
*H3PO4	-308.20 KCAL/MOLE	IN AQUEOUS SOLUTION	1067
*ISOLEUCINE	-151.8 KCAL/MOLE	NONE.	1067
*LEUCINE	-151.9 KCAL/MOLE	NONE.	1067
*LYSINE	-162.2 KCAL/MOLE	NONE.	1067
*METHIONINE	-180.4 KCAL/MOLE	NONE.	1067
*NH3	-19.32 KCAL/MOLE	IN AQUEOUS SOLUTION	1067
*O2	-94.05 KCAL/MOLE	IN GAS PHASE	1067
*PHENYLALANINE	-111.9 KCAL/MOLE	NONE.	1067
*PHOSPHATE	-320.6 KCAL/MOLE	NONE.	1067
*PROLINE	-125.7 KCAL/MOLE	NONE.	1067
*RIBOSE	-253.9 KCAL/MOLE	NONE.	1067

TABLE B274

THERMODYNAMIC VALUES : STANDARD HEAT OF FORMATION			
ITEM:	VALUE:	CONDITIONS:	REF.:
*SERINE	-173.6 KCAL/MOLE	NONE.	1067
*THREONINE	-181.4 KCAL/MOLE	NONE.	1067
*THYMINE	-111.9 KCAL/MOLE	NONE.	1067
*TRYPTOPHAN	-99.8 KCAL/MOLE	NONE.	1067
*TYROSINE	-163.4 KCAL/MOLE	NONE.	1067
*URACIL	-110.2 KCAL/MOLE	NONE.	1067
*VALINE	-148.2 KCAL/MOLE	NONE.	1067

TABLE B275

THERMODYNAMIC VALUES : TERRESTRIAL ENERGY SOURCES			
ITEM:	VALUE:	CONDITIONS:	REF.:
*COSMIC RAYS	0.0015 CAL/(CM**2*YEAR)	NONE	1038
*ELECTRIC DISCHARGES	4 CAL/(CM**2*YEAR)	NONE	1038
*RADIOACTIVITY	0.8 CAL/(CM**2*YEAR)	TO 1.0 KM DEPHT	1038
*SUNLIGHT	260000 CAL/(CM**2*YEAR)	ALL WAVELENGTHS	1038
*SUNLIGHT	570 CAL/(CM**2*YEAR)	< 2500	1038
*SUNLIGHT	85 CAL/(CM**2*YEAR)	< 2000	1038
*SUNLIGHT	3.5 CAL/(CM**2*YEAR)	< 1500	1038
*VOLCANOES	0.13 CAL/(CM**2*YEAR)	NONE	1038

TABLE B276

THERMODYNAMIC VALUES : THERMAL CONDUCTIVITY			
ITEM:	VALUE:	CONDITIONS:	REF.:
*BALSA	81.7 - 144*10**(-3) CAL/ (CM*MIN*DEG)	NONE	268
*CONCRETE	1488 - 2232*10**(-3) CAL/ (CM*MIN*DEG)	NONE	268
*CYPRESS	166*10**(-3) CAL/(CM*MIN*DEG)	NONE	268
*DRY CLAY	12 - 90*10**(-3) CAL/(CM*MIN*DEG)	NONE	268
*DRY SAND	24 - 42*10**(-3) CAL/(CM*MIN*DEG)	NONE	268
*DRY WOOD	12 - 30*10**(-3) CAL/(CM*MIN*DEG)	NONE	268
*GLASS	1236 - 1488*10**(-3) CAL/ (CM*MIN*DEG)	NONE	268
*GRANITE	3222 - 6942*10**(-3) CAL/ (CM*MIN*DEG)	NONE	268
*ICE	300 - 420*10**(-3) CAL/ (CM*MIN*DEG)	NONE	268
*IRON	126000*10**(-3) CAL/(CM*MIN*DEG)	NONE	268
*MAHOGANY	224*10**(-3) CAL/(CM*MIN*DEG)	NONE	268
*MAPLE	273*10**(-3) CAL/(CM*MIN*DEG)	NONE	268
*MARBLE	3468 - 4956*10**(-3) CAL/ (CM*MIN*DEG)	NONE	268
*NEW SNOW	12 - 18*10**(-3) CAL/(CM*MIN*DEG)	NONE	268
*OAK	253*10**(-3) CAL/(CM*MIN*DEG)	NONE	268
*OLD SNOW	180 - 300*10**(-3) CAL/ (CM*MIN*DEG)	NONE	268
*ROCK	240 - 600*10**(-3) CAL/ (CM*MIN*DEG)	NONE	268

TABLE B276

THERMODYNAMIC VALUES : THERMAL CONDUCTIVITY			
ITEM:	VALUE:	CONDITIONS:	REF.:
*SANDSTONE	1980 - 3960*10**(-3) CAL/(CM*MIN*DEG)	NONE	268
*SILVER	60000*10**(-3) CAL/(CM*MIN*DEG)	NONE	268
*SOIL ORGANIC MATTER	40*10**(-3) CAL/(CM*MIN*DEG)	NONE	268
*STILL AIR	3.0 - 3.6*10**(-3) CAL/(CM*MIN*DEG)	NONE	268
*STILL WATER	78 - 90*10**(-3) CAL/(CM*MIN*DEG)	NONE	268
*VIRGINIA PINE	243*10**(-3) CAL/(CM*MIN*DEG)	NONE	268
*WET CLAY	120 - 360*10**(-3) CAL/(CM*MIN*DEG)	NONE	268
*WET SAND	120 - 360*10**(-3) CAL/(CM*MIN*DEG)	NONE	268
*WHITA PINE	193*10**(-3) CAL/(CM*MIN*DEG)	NONE	268

TABLE B277

THERMODYNAMIC VALUES : VAPOR PRESSURE			
ITEM:	VALUE:	CONDITIONS:	REF.:
*ALDRIN	6*10**(-6) MM HG	298 K	960
*AROCLOR 1242	4.06*10**(-4) MM HG	298 K	960
*AROCLOR 1248	4.94*10**(-4) MM HG	298 K	960
*AROCLOR 1254	7.71*10**(-5) MM HG	298 K	960
*AROCLOR 1260	4.05*10**(-5) MM HG	298 K	960
*BENZENE	95.2 MM HG	298 K	960
*BIPHENYL	0.057 MM HG	298 K	960
*CCL3F	667.4 MM HG	293K	1001
*CUMENE	4.6 MM HG	298 K	960
*DDT	1*10**(-7) MM HG	298 K	960
*DIELDRIN	1*10**(-7) MM HG	298 K	960
*HG	1.3*10**(-3) MM HG	298 K	960
*LINDANE	9.4*10**(-6) MM HG	298 K	960
*N-OCTANE	14.1 MM HG	298 K	960
*NAPHTHALENE	0.23 MM HG	298 K	960
*O-XYLENE	6.6 MM HG	298 K	960
*TOLUENE	28.4 MM HG	298 K	960
*2,2,4-TRIMETHYLPENTANE	49.3 MM HG	298 K	960

Henry constant (1 atm) for gasses as a function of temperature.

B428

Henry's Constant (x 10^{-3})

Gas	Temperature (°C) 0	5	10	15	20	25	30
Acetylene	0.72	0.84	0.96	1.08	1.21	1.33	1.46
Air (atm)	0.43	0.49	0.55	0.61	0.66	0.72	0.77
Carbon dioxide	0.73	0.88	1.04	1.22	1.42	1.64	1.86
Carbon monoxide	0.35	0.40	0.44	0.49	0.54	0.58	0.62
Hydrogen	0.58	0.61	0.64	0.66	0.68	0.70	0.73
Ethane	0.13	0.16	0.19	0.23	0.26	0.30	0.34
Hydrogen sulphide	26.80	31.50	36.70	42.30	48.30	54.50	60.90
Methane	0.22	0.26	0.30	0.34	0.38	0.41	0.45
Nitrous oxide	0.17	0.19	0.22	0.24	0.26	0.29	0.30
Nitrogen	0.53	0.60	0.67	0.74	0.80	0.87	0.92
Nitric oxide	-	1.17	1.41	1.66	1.98	2.25	2.59
Oxygen	0.25	0.29	0.33	0.36	0.40	0.44	0.48

Temperature (°C)	H · 10^{-4} Oxygen	Ozone
0	2.5	0.25
5	2.9	0.29
10	3.3	0.33
15	3.6	0.38
20	4.0	0.45
25	4.4	0.52
30	4.8	0.60
35	5.1	0.73
40	5.4	0.89
45	5.6	1.0
50	5.9	1.2

(ref: 765)

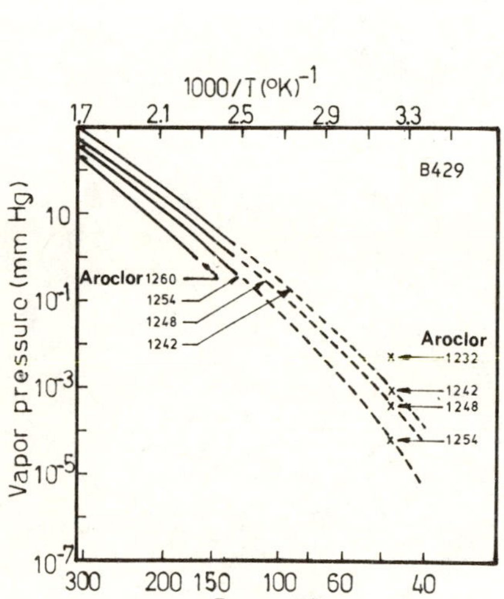

Vapour pressure of different types of PCB.

(ref: 709)

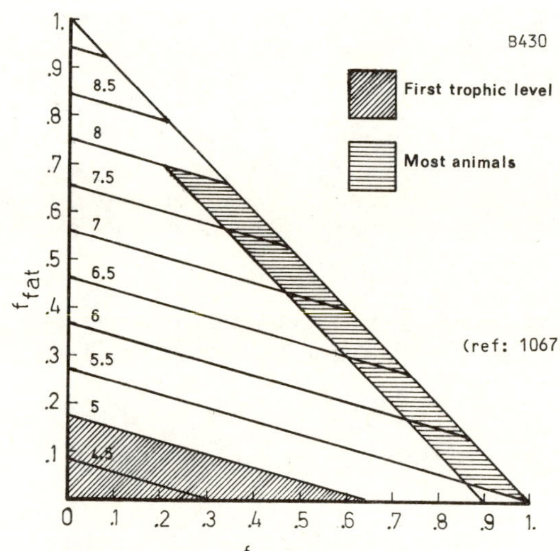

(ref: 1067)

Heat of combustion versus composition. First trophic level in the lower left corner. Most animal species in the lined portion of the graph.

C

BIOLOGICAL EFFECTS

OF
COMPOUNDS AND
ELEMENTS
ON
BIOLOGICAL
PROCESSES

TABLE C1

ACCLIMATION TIME : CHANGE IN DEVELOPMENT TIME OF EGGS			
ITEM:	VALUE:	CONDITIONS:	REF.:
*ACARTIA CLAUSI	27.22 HOURS AT 293 K	WINTER ACCLIMATED ANIMAL, 281 - 283 K	870
*ACARTIA CLAUSI	28.7 HOURS AT 293 K	WINTER ACCLIMATED ANIMAL, 9 D AT 293 K	870
*ACARTIA CLAUSI	30.3 HOURS AT 293 K	1. GENERATION FROM WINTER ACCLIMATED ANIMALS, WHOLE LIFE AT 293 K	870
*ACARTIA CLAUSI	32.67 HOURS AT 293 K	SUMMER ACCLIMATED ANIMALS, 291-293 K, FIELD TEMP.	870

TABLE C2

ACTINOMYCIN D : MORTALITY			
ITEM:	VALUE:	CONDITIONS:	REF.:
*MOUSE	50%	ORAL DOSE=0.7MG/KG(=5.58*10**(-7) MOL/KG),PT-50 =6.25 (F007)	953

TABLE C3

AG : MORTALITY			
ITEM:	VALUE:	CONDITIONS:	REF.:
*DAPHNIA MAGNA	50%	AG(NO3)2,EXPOS.TIME=4DAYS,CONC.IN WATER=30 UG/L	1004
*MOUSE	50%	COLLOID, ORAL DOSE=100MG/KG BODY	1611
*MOUSE	50%	NITRATE,INTRAPERITON.,DOSE=129MG/KG BODY	1611
*RAT	50%	OXIDE, ORAL DOSE=2820MG/KG BODY	1611

TABLE C4

AL : LEAF SYMPTONS			
ITEM:	VALUE:	CONDITIONS:	REF.:
*TOMATO LEAVES	NO SYMPTOMS BETWEEN 65 - 90 MG/KG DRY WEIGHT IN LEAVES	SYMPTOMS OUTSIDE THIS RANGE, STEAM STERILIZED SOIL	635

TABLE C5

AL : MORTALITY			
ITEM:	VALUE:	CONDITIONS:	REF.:
*DAPHNIA MAGNA	50%	ALCL3,EXPOS.TIME=2DAYS,CONC.IN WATER=3900 UG/L	122
*MOUSE	50%	SULFATE,ORAL,DOSE=770MG/KG BODY	1611
*RAT	50%	CHLORIDE,ORAL,DOSE=3700MG/KG BODY	1611
*RAT	50%	NITRATE,ORAL,DOSE=4300MG/KG BODY	1611

TABLE C6

ALANAP : MORTALITY			
ITEM:	VALUE:	CONDITIONS:	REF.:
*ICTALURUS PUNCTATUS	LESS THAN 10 %	10 PPM, 48 H, 14 G, 293 K, PH=8.2, ALK=80 PPM	1002

TABLE C7

ALDRIN : HEART BEAT			
ITEM:	VALUE:	CONDITIONS:	REF.:
*BIOMPHALARIA GLABRATA	150 BEATS/MIN	ALDRIN=1MICRO-MOL,AFTER 24H,299K,5 SNAILS	264
*BIOMPHALARIA GLABRATA	29 BEATS/MIN	ALDRIN=7MICRO-MOL,AFTER 24H,299K,5 SNAILS	264
*BIOMPHALARIA GLABRATA	24 BEATS/MIN	ALDRIN=281MICRO-MOL,AFTER 24H, 299K,5 SNAILS	264
*BIOMPHALARIA GLABRATA	0 BEATS/MIN	ALDRIN=300MICRO-MOL,AFTER 24H, 299K,5 SNAILS	264
*BIOMPHALARIA GLABRATA	136 BEATS/MIN	ALDRIN=1MICRO-MOL,24H,299K	264
*BIOMPHALARIA GLABRATA	18 BEATS/MIN	ALDRIN=14MICRO-MOL,24H,299K	264
*BIOMPHALARIA GLABRATA	25 BEATS/MIN	ALDRIN=28MICRO-MOL,24H,299K	264
*BIOMPHALARIA GLABRATA	24 BEATS/MIN	ALDRIN=56MICRO-MOL,24H,299K	264
*BIOMPHALARIA GLABRATA	24 BEATS/MIN	ALDRIN=112MICRO-MOL,24H,299K	264
*BIOMPHALARIA GLABRATA	20 BEATS/MIN	ALDRIN=168MICRO-MOL,24H,299K	264
*BIOMPHALARIA GLABRATA	24 BEATS/MIN	ALDRIN=225MICRO-MOL,24H,299K	264

TABLE C8

ALDRIN : MORTALITY			
ITEM:	VALUE:	CONDITIONS:	REF.:
*BOBWHITE	50%	CONC.IN FOOD=37MG/KG(33-41),95% CONF.LIMITS	389
*EARTHWORMS	95%	CONC.IN SOIL:>125 MG/KG	389
*JAPANESE QUAIL	50%	CONC.IN FOOD=34MG/KG(28-41),95% CONF.LIMITS	389
*MALLARD	50%	CONC.IN FOOD=155MG/KG(129-186),95% CONF.LIMITS	389
*PHEASANT	50%	CONC.IN FOOD=57MG/KG(50-64),95% CONF.LIMITS	389
*PHEASANT	50%	CONC.IN FOOD=14.0MG/KG	349
*QUAIL	50%	CONC.IN FOOD=4.0MG/KG	349
*RATS	50%	ACUTE ORAL TOXICITY=32-67MG/KG	389
*RATS	50%	ACUTE DERMAL TOXICITY=99MG/KG	389
*SLUGS	95%	CONC.IN SOIL:>250 MG/KG	389

TABLE C9

ALKYL BENZEN SULFONATE : CHANGE IN GROWTH POTENTIAL (F008)			
ITEM:	VALUE:	CONDITIONS:	REF.:
*ANABAENA CYLINDRICA	3	7DAYS,CONTROL=4 ,INITIAL CONC.=20MG SURFACTANT/L	332
*ANABAENA CYLINDRICA	1	14DAYS,CONTROL=4 ,INITIAL CONC.=20MG SURFACTANT/L	332
*ANABAENA CYLINDRICA	1	21DAYS,CONTROL=4 ,INITIAL CONC.=20MG SURFACTANT/L	332
*ANABAENA VARIABILIS	2	7DAYS,CONTROL=4 ,INITIAL CONC.=20MG SURFACTANT/L	332
*ANABAENA VARIABILIS	4	14DAYS,CONTROL=4(?071),INITIAL CONC.=20MG SURFACTANT/L	332
*ANABAENA VARIABILIS	4	21DAYS,CONTROL=4 ,INITIAL CONC.=20MG SURFACTANT/L	332
*ANACYSTIS NIDULANS	1	7DAYS,CONTROL=4 ,INITIAL CONC.=20MG SURFACTANT/L	332
*ANACYSTIS NIDULANS	3	14DAYS,CONTROL=4 ,INITIAL CONC.=20MG SURFACTANT/L	332

TABLE C9

ALKYL BENZEN SULFONATE : CHANGE IN GROWTH POTENTIAL (F008)			
ITEM:	VALUE:	CONDITIONS:	REF.:
*ANACYSTIS NIDULANS	3	21DAYS,CONTROL=4 ,INITIAL CONC.=20MG SURFACTANT/L	332
*ANKISTRODESMUS BRAUNII	4	7DAYS,CONTROL=4 ,INITIAL CONC.=20MG SURFACTANT/L	332
*ANKISTRODESMUS BRAUNII	4	14DAYS,CONTROL=4 ,INITIAL CONC.=20MG SURFACTANT/L	332
*ANKISTRODESMUS BRAUNII	4	21DAYS,CONTROL=4 ,INITIAL CONC.=20MG SURFACTANT/L	332
*CALOTHRIX PARIETINA	4	7,14 AND 21DAYS,CONTROL=4 INITIAL CONC.=20MG SURFACTANT/L	332
*CHLORELLA PYRENOIDOSA	2	7DAYS,CONTROL=4 ,INITIAL CONC.=20MG SURFACTANT/L	332
*CHLORELLA PYRENOIDOSA	4	14DAYS,CONTROL=4 ,INITIAL CONC.=20MG SURFACTANT/L	332
*CHLORELLA PYRENOIDOSA	4	21DAYS,CONTROL=4 ,INITIAL CONC.=20MG SURFACTANT/L	332
*CHLORELLA VULGARIS	2	7DAYS,CONTROL=4 ,INITIAL CONC.=20MG SURFACTANT/L	332
*CHLORELLA VULGARIS	4	14DAYS,CONTROL=4 ,INITIAL CONC.=20MG SURFACTANT/L	332
*CHLORELLA VULGARIS	4	21DAYS,CONTROL=4 ,INITIAL CONC.=20MG SURFACTANT/L	332
*CYLINDROSPERMUM SP.	3	7DAYS,CONTROL=4 INITIAL CONC.=20MG SURFACTANT/L	332
*CYLINDROSPERMUM SP.	1	14DAYS,CONTROL=4 ,INITIAL CONC.=20MG SURFACTANT/L	332
*CYLINDROSPERMUM SP.	2	21DAYS,CONTROL=4 ,INITIAL CONC.=20MG SURFACTANT/L	332
*GLOEOCAPSA ALPICOLA	3	7DAYS,CONTROL=4 ,INITIAL CONC.=20MG SURFACTANT/L	332
*GLOEOCAPSA ALPICOLA	4	14DAYS,CONTROL=4 ,INITIAL CONC.=20MG SURFACTANT/L	332
*GLOEOCAPSA ALPICOLA	4	21DAYS,CONTROL=4 ,INITIAL CONC.=20MG SURFACTANT/L	332
*NOSTOC MUSCORUM	4	7DAYS,CONTROL=4 ,INITIAL CONC.=20MG SURFACTANT/L	332
*NOSTOC MUSCORUM	3	14DAYS,CONTROL=4 ,INITIAL CONC.=20MG SURFACTANT/L	332
*NOSTOC MUSCORUM	4	21DAYS,CONTROL=4 ,INITIAL CONC.=20MG SURFACTANT/L	332
*OSCILLATORIA BORNETI	3	7DAYS,CONTROL=4 ,INITIAL CONC.=20MG SURFACTANT/L	332
*OSCILLATORIA BORNETI	3	14DAYS,CONTROL=4 ,INITIAL CONC.=20MG SURFACTANT/L	332
*OSCILLATORIA BORNETI	4	21DAYS,CONTROL=4 ,INITIAL CONC.=20MG SURFACTANT/L	332
*OSCILLATORIA CHALYBIA	2	7DAYS,CONTROL=4 ,INITIAL CONC.=20MG SURFACTANT/L	332
*OSCILLATORIA CHALYBIA	2	14DAYS,CONTROL=4 ,INITIAL CONC.=20MG SURFACTANT/L	332
*OSCILLATORIA CHALYBIA	3	21DAYS,CONTROL=4 ,INITIAL CONC.=20MG SURFACTANT/L	332
*OSCILLATORIA FORMOSA	2	7DAYS,CONTROL=4 ,INITIAL CONC.=20MG SURFACTANT/L	332
*OSCILLATORIA FORMOSA	2	14DAYS,CONTROL=4 ,INITIAL CONC.=20MG SURFACTANT/L	332
*OSCILLATORIA FORMOSA	3	21DAYS,CONTROL=4 ,INITIAL CONC.=20MG SURFACTANT/L	332
*OSCILLATORIA TENUIS	2	7DAYS,CONTROL=4 ,INITIAL CONC.=20MG SURFACTANT/L	332
*OSCILLATORIA TENUIS	1	14DAYS,CONTROL=4 ,INITIAL CONC.=20MG SURFACTANT/L	332
*OSCILLATORIA TENUIS	3	21DAYS,CONTROL=4 ,INITIAL CONC.=20MG SURFACTANT/L	332
*PHORMIDIUM FAVEOLARUM	2	7DAYS,CONTROL=4 ,INITIAL CONC.=20MG SURFACTANT/L	332
*PHORMIDIUM FAVEOLARUM	3	14DAYS,CONTROL=4 ,INITIAL CONC.=20MG SURFACTANT/L	332

TABLE C9

ALKYL BENZEN SULFONATE : CHANGE IN GROWTH POTENTIAL (F008)			
ITEM:	VALUE:	CONDITIONS:	REF.:
*PHORMIDIUM FAVEOLARUM	4	21DAYS,CONTROL=4 ,INITIAL CONC.=20MG SURFACTANT/L	332
*SCENEDESMUS OBLIQUUS	3	7DAYS,CONTROL=4 ,INITIAL CONC.=20MG SURFACTANT/L	332
*SCENEDESMUS OBLIQUUS	4	14DAYS,CONTROL=4 ,INITIAL CONC.=20MG SURFACTANT/L	332
*SCENEDESMUS OBLIQUUS	4	21DAYS,CONTROL=4 ,INITIAL CONC.=20MG SURFACTANT/L	332
*SCENEDESMUS QUADRICAUDA	3	7DAYS,CONTROL=4 ,INITIAL CONC.=20MG SURFACTANT/L	332
*SCENEDESMUS QUADRICAUDA	4	14DAYS,CONTROL=4 ,INITIAL CONC.=20MG SURFACTANT/L	332
*SCENEDESMUS QUADRICAUDA	4	21DAYS,CONTROL=4 ,INITIAL CONC.=20MG SURFACTANT/L	332
*ULOTHRIX FIMBRIATA	1	7DAYS,CONTROL=4 ,INITIAL CONC.=20MG SURFACTANT/L	332
*ULOTHRIX FIMBRIATA	4	14DAYS,CONTROL=4 ,INITIAL CONC.=20MG SURFACTANT/L	332
*ULOTHRIX FIMBRIATA	4	21DAYS,CONTROL=4 ,INITIAL CONC.=20MG SURFACTANT/L	332

TABLE C10

ALKYL BENZEN SULFONATE : MORTALITY			
ITEM:	VALUE:	CONDITIONS:	REF.:
*CYPRINUS CARPIO	50 %	ABS-COMM.+ENZ.=12.5MG/L,ABS-TECH.= 21MG/L,24H,294K,PH=7.5-7.7	947
*CYPRINUS CARPIO	50 %	ABS-COMM.+ENZ.=12.5MG/L,ABS-TECH.= 20MG/L,48H,294K	947
*CYPRINUS CARPIO	50 %	ABS-COMM.+ENZ.=12.5MG/L,ABS-TECH.= 18MG/L,96H,294K	947
*TILAPIA MELANOPLEURA	50 %	ABS-COMM.+ENZ.=32MG/L,ABS-TECH.= 37MG/L,ABS.##=25MG/L,24H,294K	947
*TILAPIA MELANOPLEURA	50 %	ABS-COMM.+ENZ.=30MG/L,ABS-TECH.= 37MG/L,ABS-COMM.=25MG/L,48H,294K	947
*TILAPIA MELANOPLEURA	50 %	ABS-COMM.+ENZ.=28MG/L,ABS-TECH.= 37MG/L,ABS-COMM.=23MG/L,96H,294K	947

TABLE C11

ALKYL POLYETHOXYLATE : CHANGE IN GROWTH POTENTIAL (F008)			
ITEM:	VALUE:	CONDITIONS:	REF.:
*ANABAENA CYLINDRICA	3	7DAYS,CONTROL=4 ,INITIAL CONC.=20 MG AE7.0/L	332
*ANABAENA CYLINDRICA	3	14DAYS,CONTROL=4 ,INITIAL CONC.=20 MG AE7.0/L	332
*ANABAENA CYLINDRICA	4	21DAYS,CONTROL=4 ,INITIAL CONC.=20 MG AE7.0/L	332
*ANABAENA CYLINDRICA	4	7DAYS,CONTROL=4 ,INITIAL CONC.=20 MG AE7.4/L	332
*ANABAENA CYLINDRICA	3	14DAYS,CONTROL=4 ,INITIAL CONC.=20 MG AE7.4/L	332
*ANABAENA CYLINDRICA	4	21DAYS,CONTROL=4 ,INITIAL CONC.=20 MG AE7.4/L	332
*ANABAENA VARIABILIS	1	7DAYS,CONTROL=4 ,INITIAL CONC.=20 MG AE7.0/L	332
*ANABAENA VARIABILIS	2	14DAYS,CONTROL=4 ,INITIAL CONC.=20 MG AE7.0/L	332

TABLE C11

ALKYL POLYETHOXYLATE : CHANGE IN GROWTH POTENTIAL (F008)			
ITEM:	VALUE:	CONDITIONS:	REF.:
*ANABAENA VARIABILIS	3	21DAYS,CONTROL=4 ,INITIAL CONC.=20 MG AE7.0/L	332
*ANABAENA VARIABILIS	1	7DAYS,CONTROL=4 ,INITIAL CONC.=20 MG AE7.4/L	332
*ANABAENA VARIABILIS	2	14DAYS,CONTROL=4 ,INITIAL CONC.=20 MG AE7.4/L	332
*ANABAENA VARIABILIS	4	21DAYS,CONTROL=4 ,INITIAL CONC.=20 MG AE7.4/L	332
*ANACYSTIS NIDULANS	1	7DAYS,CONTROL=4 ,INITIAL CONC.=20 MG AE7.0/'	332
*ANACYSTIS NIDULANS	1	14DAYS,CONTROL=4 ,INITIAL CONC.=20 MG AE7.0/L	332
*ANACYSTIS NIDULANS	3	21DAYS,CONTROL=4 ,INITIAL CONC.=20 MG AE7.0/L	332
*ANACYSTIS NIDULANS	2	7DAYS,CONTROL=4 ,INITIAL CONC.=20 MG AE7.4/L	332
*ANACYSTIS NIDULANS	2	14DAYS,CONTROL=4 ,INITIAL CONC.=20 MG AE7.4/L	332
*ANACYSTIS NIDULANS	3	21DAYS,CONTROL=4 ,INITIAL CONC.=20 MG AE7.4/L	332
*ANKISTRODESMUS BRAUNII	2	7DAYS,CONTROL=4 ,INITIAL CONC.=20 MG AE7.0/L	332
*ANKISTRODESMUS BRAUNII	3	14DAYS,CONTROL=4 ,INITIAL CONC.=20 MG AE7.0/L	332
*ANKISTRODESMUS BRAUNII	4	21DAYS,CONTROL=4 ,INITIAL CONC.=20 MG AE7.0/L	332
*ANKISTRODESMUS BRAUNII	2	7DAYS,CONTROL=4 ,INITIAL CONC.=20 MG AE7.4/L	332
*ANKISTRODESMUS BRAUNII	4	14DAYS,CONTROL=4 ,INITIAL CONC.=20 MG AE7.4/L	332
*ANKISTRODESMUS BRAUNII	4	21DAYS,CONTROL=4 ,INITIAL CONC.=20 MG AE7.4/L	332
*CALOTHRIX PARIETINA	4	7,14 AND 21DAYS,CONTROL=4 INITIAL CONC.=20 MG AE7.0/L	332
*CALOTHRIX PARIETINA	4	7,14 AND 21DAYS,CONTROL=4 INITIAL CONC.=20 MG AE7.4/L	332
*CHLORELLA PYRENOIDOSA	1	7DAYS,CONTROL=4 ,INITIAL CONC.=20 MG AE7.0/L	332
*CHLORELLA PYRENOIDOSA	3	14DAYS,CONTROL=4 ,INITIAL CONC.=20 MG AE7.0/L	332
*CHLORELLA PYRENOIDOSA	4	21DAYS,CONTROL=4 ,INITIAL CONC.=20 MG AE7.0/L	332
*CHLORELLA PYRENOIDOSA	1	7DAYS,CONTROL=4 ,INITIAL CONC.=20 MG AE7.4/L	332
*CHLORELLA PYRENOIDOSA	3	14DAYS,CONTROL=4 ,INITIAL CONC.=20 MG AE7.4/L	332
*CHLORELLA PYRENOIDOSA	4	21DAYS,CONTROL=4 ,INITIAL CONC.=20 MG AE7.4/L	332
*CHLORELLA VULGARIS	1	7DAYS,CONTROL=4 INITIAL CONC.=20 MG AE7.0/L	332
*CHLORELLA VULGARIS	1	14DAYS,CONTROL=4 ,INITIAL CONC.=20 MG AE7.0/L	332
*CHLORELLA VULGARIS	2	21DAYS,CONTROL=4 ,INITIAL CONC.=20 MG AE7.0/L	332
*CHLORELLA VULGARIS	1	7DAYS,CONTROL=4 ,INITIAL CONC.=20 MG AE7.4/L	332
*CHLORELLA VULGARIS	1	14DAYS,CONTROL=4 ,INITIAL CONC.=20 MG AE7.4/L	332
*CHLORELLA VULGARIS	2	21DAYS,CONTROL=4 ,INITIAL CONC.=20 MG AE7.4/L	332
*CYLINDROSPERMUM SP.	3	7DAYS,CONTROL=4 ,INITIAL CONC.=20 MG AE7.0/L	332
*CYLINDROSPERMUM SP.	2	14DAYS,CONTROL=4 ,INITIAL CONC.=20 MG AE7.0/L	332
*CYLINDROSPERMUM SP.	4	21DAYS,CONTROL=4 ,INITIAL CONC.=20 MG AE7.0/L	332
*CYLINDROSPERMUM SP.	3	7DAYS,CONTROL=4 ,INITIAL CONC.=20 MG AE7.4/L	332

TABLE C11

ALKYL POLYETHOXYLATE : CHANGE IN GROWTH POTENTIAL (F008)			
ITEM:	VALUE:	CONDITIONS:	REF.:
*CYLINDROSPERMUM SP.	2	14DAYS,CONTROL=4 ,INITIAL CONC.=20 MG AE7.4/L	332
*CYLINDROSPERMUM SP.	3	21DAYS,CONTROL=4 ,INITIAL CONC.=20 MG AE7.4/L	332
*GLOEOCAPSA ALPICOLA	3	7DAYS,CONTROL=4 ,INITIAL CONC.=20 MG AE7.0/L	332
*GLOEOCAPSA ALPICOLA	3	14DAYS,CONTROL=4 ,INITIAL CONC.=20 MG AE7.0/L	332
*GLOEOCAPSA ALPICOLA	4	21DAYS,CONTROL=4 ,INITIAL CONC.=20 MG AE7.0/L	332
*GLOEOCAPSA ALPICOLA	2	7DAYS,CONTROL=4 ,INITIAL CONC.=20 MG AE7.4/L	332
*GLOEOCAPSA ALPICOLA	2	14DAYS,CONTROL=4 ,INITIAL CONC.=20 MG AE7.4/L	332
*GLOEOCAPSA ALPICOLA	3	21DAYS,CONTROL=4 ,INITIAL CONC.=20 MG AE7.4/L	332
*NOSTOC MUSCORUM	4	7DAYS,CONTROL=4 ,INITIAL CONC.=20 MG AE7.0/L	332
*NOSTOC MUSCORUM	4	14DAYS,CONTROL=4 ,INITIAL CONC.=20 MG AE7.0/L	332
*NOSTOC MUSCORUM	4	21DAYS,CONTROL=4 ,INITIAL CONC.=20 MG AE7.0/L	332
*NOSTOC MUSCORUM	4	7DAYS,CONTROL=4 ,INITIAL CONC.=20 MG AE7.4/L	332
*NOSTOC MUSCORUM	4	14DAYS,CONTROL=4 ,INITIAL CONC.=20 MG AE7.4/L	332
*NOSTOC MUSCORUM	4	21DAYS,CONTROL=4 ,INITIAL CONC.=20 MG AE7.4/L	332
*OSCILLATORIA BORNETI	2	7DAYS,CONTROL=4 ,INITIAL CONC.=20 MG AE7.0/L	332
*OSCILLATORIA BORNETI	2	14DAYS,CONTROL=4 ,INITIAL CONC.=20 MG AE7.0/L	332
*OSCILLATORIA BORNETI	2	21DAYS,CONTROL=4 ,INITIAL CONC.=20 MG AE7.0/L	332
*OSCILLATORIA BORNETI	2	7DAYS,CONTROL=4 ,INITIAL CONC.=20 MG AE7.4/L	332
*OSCILLATORIA BORNETI	2	14DAYS,CONTROL=4 ,INITIAL CONC.=20 MG AE7.4/L	332
*OSCILLATORIA BORNETI	2	21DAYS,CONTROL=4 ,INITIAL CONC.=20 MG AE7.4/L	332
*OSCILLATORIA CHALYBIA	3	7DAYS,CONTROL=4 ,INITIAL CONC.=20 MG AE7.0/L	332
*OSCILLATORIA CHALYBIA	4	14DAYS,CONTROL=4 ,INITIAL CONC.=20 MG AE7.0/L	332
*OSCILLATORIA CHALYBIA	3	21DAYS,CONTROL=4 ,INITIAL CONC.=20 MG AE7.0/L	332
*OSCILLATORIA CHALYBIA	3	7DAYS,CONTROL=4 INITIAL CONC.=20 MG AE7.4/L	332
*OSCILLATORIA CHALYBIA	3	14DAYS,CONTROL=4 ,INITIAL CONC.=20 MG AE7.4/L	332
*OSCILLATORIA CHALYBIA	3	21DAYS,CONTROL=4 ,INITIAL CONC.=20 MG AE7.4/L	332
*OSCILLATORIA FORMOSA	3	7DAYS,CONTROL=4 ,INITIAL CONC.=20 MG AE7.0/L	332
*OSCILLATORIA FORMOSA	4	14DAYS,CONTROL=4 ,INITIAL CONC.=20 MG AE7.0/L	332
*OSCILLATORIA FORMOSA	4	21DAYS,CONTROL=4 ,INITIAL CONC.=20 MG AE7.0/L	332
*OSCILLATORIA FORMOSA	3	7DAYS,CONTROL=4 ,INITIAL CONC.=20 MG AE7.4/L	332
*OSCILLATORIA FORMOSA	4	14DAYS,CONTROL=4 ,INITIAL CONC.=20 MG AE7.4/L	332
*OSCILLATORIA FORMOSA	4	21DAYS,CONTROL=4 ,INITIAL CONC.=20 MG AE7.4/L	332
*OSCILLATORIA TENUIS	2	7DAYS,CONTROL=4 ,INITIAL CONC.=20 MG AE7.0/L	332
*OSCILLATORIA TENUIS	3	14DAYS,CONTROL=4 ,INITIAL CONC.=20 MG AE7.0/L	332

TABLE C11

ALKYL POLYETHOXYLATE : CHANGE IN GROWTH POTENTIAL (F008)

ITEM:	VALUE:	CONDITIONS:	REF.:
*OSCILLATORIA TENUIS	4	21DAYS,CONTROL=4 ,INITIAL CONC.=20 MG AE7.0/L	332
*OSCILLATORIA TENUIS	2	7DAYS,CONTROL=4 ,INITIAL CONC.=20 MG AE7.4/L	332
*OSCILLATORIA TENUIS	3	14DAYS,CONTROL=4 ,INITIAL CONC.=20 MG AE7.4/L	332
*OSCILLATORIA TENUIS	4	21DAYS,CONTROL=4 ,INITIAL CONC.=20 MG AE7.4/L	332
*PHORMIDIUM FAVEOLARUM	4	7DAYS,CONTROL=4 ,INITIAL CONC.=20 MG AE7.0/L	332
*PHORMIDIUM FAVEOLARUM	4	14DAYS,CONTROL=4 ,INITIAL CONC.=20 MG AE7.0/L	332
*PHORMIDIUM FAVEOLARUM	4	21DAYS,CONTROL=4 ,INITIAL CONC.=20 MG AE7.0/L	332
*PHORMIDIUM FAVEOLARUM	2	7DAYS,CONTROL=4 INITIAL CONC.=20 MG AE7.4/L	332
*PHORMIDIUM FAVEOLARUM	3	14DAYS,CONTROL=4 ,INITIAL CONC.=20 MG AE7.4/L	332
*PHORMIDIUM FAVEOLARUM	4	21DAYS,CONTROL=4 ,INITIAL CONC.=20 MG AE7.4/L	332
*SCENEDESMUS OBLIQUUS	3	7DAYS,CONTROL=4 ,INITIAL CONC.=20 MG AE7.0/L	332
*SCENEDESMUS OBLIQUUS	4	14DAYS,CONTROL=4 ,INITIAL CONC.=20 MG AE7.0/L	332
*SCENEDESMUS OBLIQUUS	4	21DAYS,CONTROL=4 ,INITIAL CONC.=20 MG AE7.0/L	332
*SCENEDESMUS OBLIQUUS	3	7DAYS,CONTROL=4 ,INITIAL CONC.=20 MG AE7.4/L	332
*SCENEDESMUS OBLIQUUS	4	14DAYS,CONTROL=4 ,INITIAL CONC.=20 MG AE7.4/L	332
*SCENEDESMUS OBLIQUUS	4	21DAYS,CONTROL=4 ,INITIAL CONC.=20 MG AE7.4/L	332
*SCENEDESMUS QUADRICAUDA	4	7DAYS,CONTROL=4 ,INITIAL CONC.=20 MG AE7.0/L	332
*SCENEDESMUS QUADRICAUDA	4	14DAYS,CONTROL=4 ,INITIAL CONC.=20 MG AE7.0/L	332
*SCENEDESMUS QUADRICAUDA	4	21DAYS,CONTROL=4 ,INITIAL CONC.=20 MG AE7.0/L	332
*SCENEDESMUS QUADRICAUDA	4	7DAYS,CONTROL=4 ,INITIAL CONC.=20 MG AE7.4/L	332
*SCENEDESMUS QUADRICAUDA	4	14DAYS,CONTROL=4 ,INITIAL CONC.=20 MG AE7.4/L	332
*SCENEDESMUS QUADRICAUDA	4	21DAYS,CONTROL=4 ,INITIAL CONC.=20 MG AE7.4/L	332
*ULOTHRIX FIMBRIATA	2	7DAYS,CONTROL=4 ,INITIAL CONC.=20 MG AE7.0/L	332
*ULOTHRIX FIMBRIATA	1	14DAYS,CONTROL=4 ,INITIAL CONC.=20 MG AE7.0/L	332
*ULOTHRIX FIMBRIATA	3	21DAYS,CONTROL=4 ,INITIAL CONC.=20 MG AE7.0/L	332
*ULOTHRIX FIMBRIATA	1	7DAYS,CONTROL=4 ,INITIAL CONC.=20 MG AE7.4/L	332
*ULOTHRIX FIMBRIATA	2	14DAYS,CONTROL=4 ,INITIAL CONC.=20 MG AE7.4/L	332
*ULOTHRIX FIMBRIATA	3	21DAYS,CONTROL=4 ,INITIAL CONC.=20 MG AE7.4/L	332

TABLE C12

ALPHA OLEFIN SULFONATE : MORTALITY

ITEM:	VALUE:	CONDITIONS:	REF.:
*CYPRINUS CARPIO	50 %	AOS-TECH.=3.2MG/L,24H,294K	947

TABLE C12

ALPHA OLEFIN SULFONATE : MORTALITY			
ITEM:	VALUE:	CONDITIONS:	REF.:
*CYPRINUS CARPIO	50 %	AOS-TECH.=3.MG/L,48H,294K	947
*CYPRINUS CARPIO	50 %	AOS-TECH.=3.MG/L,96H,294K	947
*TILAPIA MELANOPLEURA	50 %	AOS-TECH.=2MG/L,24H,294K	947
*TILAPIA MELANOPLEURA	50 %	AOS-TECH.=2MG/L,48H,294K	947
*TILAPIA MELANOPLEURA	50 %	AOS-TECH.=2MG/L,96H,294K	947

TABLE C13

ALPHA-AMANITIN : MORTALITY			
ITEM:	VALUE:	CONDITIONS:	REF.:
*MOUSE	50%	ORAL DOSE=0.3MG/KG(=3.28*10**(-7) MOL/KG),PT-50(F007)=6.48	953

TABLE C14

AMMONIA : MORTALITY			
ITEM:	VALUE:	CONDITIONS:	REF.:
*CHINOOK SALMON	0 %	AMMONIA (NH3)=0.018MG/L,6W	397
*CYPRINUS CARPIO	18 %	AMMONIA (NH3)=1.1MG/L,240H,289K	397
*SALMO GAIRDNERI	50 %	AMMONIA (NH3)=0.47MG/L,24H	397
*SALMO GAIRDNERI	15 %	AMMONIA (NH3)=0.22MG/L,24H	397
*SALMO GAIRDNERI	5 %	AMMONIA (NH3)=0.085MG/L,24H	397
*SALMO SALAR	50 %	AMMONIA (NH3)=0.28MG/L,24H	397
*SALMONOIDS	NO EFFECT	AMMONIA (NH3)(UNIONIZED)=0.025MG/L	397

TABLE C15

AMMONIUM : REPRESSION OF NITROGENASE ACTIVITY			
ITEM:	VALUE:	CONDITIONS:	REF.:
*AZOTOBACTER CHROOCOCCUM	50%	14MG NH4(+)-N/L	371
*AZOTOBACTER CHROOCOCCUM	100%	28MG NH4(+)-N/L	371

TABLE C16

AS : MORTALITY			
ITEM:	VALUE:	CONDITIONS:	REF.:
*DAPHNIA MAGNA	50%	NA2HASO4,EXPOS.TIME=2DAYS,CONC.IN WATER=7400 UG/L	122
*MOUSE	50%	ORAL DOSE=9MG/KG(=5.49*10**(-5) MOL/KG),PT-50(F007)=4.26 NAH2ASSO4	953
*RAT	50%	5-OXIDE,ORAL,DOSE=8MG/KG BODY	1611
*RAT	50%	3-OXIDE,ORAL,DOSE=45MG/KG BODY	1611
*RAT	50%	K-ARSENITE,ORAL,DOSE=1140MG/KG BODY	1611

TABLE C16

AS : MORTALITY			
ITEM:	VALUE:	CONDITIONS:	REF.:
*RAT	50%	NA-ARSENITE,ORAL,DOSE=41MG/KG BODY	1611
*RATS	50%	ACUTE ORAL TOXICITY=138 MG/KG , AS2o3,PESTICIDE	389
*RATS	50%	ACUTE ORAL TOXICITY=10-50 MG SODIUM ARSENITE/KG, PESTICIDE	389

TABLE C17

ATRAZINE : MORTALITY			
ITEM:	VALUE:	CONDITIONS:	REF.:
*RATS	50%	ACUTE ORAL TOXICITY=3080MG/KG, HERBICIDE	389

TABLE C18

B : DEFICIENCY SYMPTOMS			
ITEM:	VALUE:	CONDITIONS:	REF.:
*PLANTS	0.5 MG/KG AIR DRIED SOIL	NO SYMPT.OVER THIS VAL.,SANDY SOIL,SOIL-PH<6.5	77
*PLANTS	0.7 MG/KG AIR DRIED SOIL	NO SYMPT.OVER THIS VAL.,SANDY SOIL,SOIL-PH>6.5	77
*PLANTS	0.7 MG/KG AIR DRIED SOIL	NO SYMPT.OVER THIS VAL.,HEAVY MINERAL SOIL,SOIL-PH<6.5	77
*PLANTS	1.0 MG/KG AIR DRIED SOIL	NO SYMPT.OVER THIS VAL.,HEAVY MINERAL SOIL,SOIL-PH>6.5	77
*PLANTS	1.0 MG/KG AIR DRIED SOIL	NO SYMPT.OVER THIS VAL.,PEAT, SOIL-PH<6.5	77
*PLANTS	1.5 MG/KG AIR DRIED SOIL	NO SYMPT.OVER THIS VAL.,PEAT, SOIL-PH>6.5	77

TABLE C19

B : MORTALITY			
ITEM:	VALUE:	CONDITIONS:	REF.:
*GAMBUSIA AFFINIS	50%	NA2B4O7,EXPOS.TIME=2DAYS,CONC.IN WATER=8200000 UG/L	1004
*MOUSE	50%	BORAX,ORAL,DOSE=41MG/KG BODY	1611
*ONCORHYNCHUS KISUTCH	50 %	BORON IN FRESH WATER=113MICROGRAM/ ML,283 K, UNDERYEARLING	1588
*ONCORHYNCHUS KISUTC	50 %	BORON IN SEA WATER=12.2G/ML,283 K, UNDERYEARLING	1588
*RAT	50%	BORAX,ORAL,DOSE=4500MG/KG BODY	1611
*RAT	50%	BORIC ACID,ORAL,DOSE=2660MG/KG BODY	1611

TABLE C20

B : SURVIVAL TIME			
ITEM:	VALUE:	CONDITIONS:	REF.:
*ONCORHYNCHUS KISUTCH	33.4 HOUR	BORON=656MICROGRAM/ML,20 FISH, MEDIAN TIME	1588

TABLE C20

B : SURVIVAL TIME			
ITEM:	VALUE:	CONDITIONS:	REF.:
*ONCHRHYNCHUS KISUTCH	33.4 HOUR	BORON=656MICROGRAM/ML,20 FISH, MEDIAN TIME	1588
*ONCHRHYNCHUS KISUTCH	84. HOUR	BORON=328MICROGRAM/ML,20 FISH, MEDIAN TIME	1588

TABLE C21

BA : MORTALITY			
ITEM:	VALUE:	CONDITIONS:	REF.:
*DAPHNIA MAGNA	50%	BACL2,EXPOS.TIME=2DAYS,CONC.IN WATER=14500 UG/L	122
*MOUSE	50%	CHLORIDE,INTRAVEN.,DOSE=8MG/KG BODY	1611
*MOUSE	50%	NITRATE,INTRAVEN.,DOSE=8.5MG/KG BODY	1611
*MOUSE	50%	ACETATE,INTRAVEN.,DOSE=11MG/KG BODY	1611
*MOUSE	50%	ORAL DOSE=500MG/KG(=2.40*10**(-3) MOL/KG),PT-50(F007)=2.62, AS BACL2	953
*RAT	50%	CHLORIDE,ORAL,DOSE=150MG/KG BODY	1611
*RAT	50%	SULFONATE,ORAL,DOSE=3000MG/KG BODY	1611
*RAT	50%	CARBONATE,ORAL,DOSE=800MG/KG BODY	419

TABLE C22

BCM : GROWTH INHIBITION			
ITEM:	VALUE:	CONDITIONS:	REF.:
*CHLORELLA PYRENOIDOSA	50 %	48 H, FRESHWATER = 0.34 MG/L, 297 K, LOG PHASE, 10**7 CELLS/L	221

TABLE C23

BCM : MORTALITY			
ITEM:	VALUE:	CONDITIONS:	REF.:
*DAPHNIA MAGNA	50 %	48 H, FRESHWATER = 0.46 MG/L, 293 K, AGE=< 1 D, N=20	221
*SALMO GAIRDNERI	50 %	48 H, FRESHWATER = 1.8 MG/L, 288 K, AGE = 3 M, N=10	221

TABLE C24

BE : MORTALITY			
ITEM:	VALUE:	CONDITIONS:	REF.:
*LIBISTES RETICULATUS	50%	BERYLLIUM=22.3MG/L,24H,HARD WATER, 296K	355
*LIBISTES RETICULATUS	50%	BERYLLIUM=22.1MG/L,48H,HARD WATER, 296K	355
*LIBISTES RETICULATUS	50%	BERYLLIUM=20.3MG/L,96H,HARD WATER, 296K	355

TABLE C24

BE : MORTALITY			
ITEM:	VALUE:	CONDITIONS:	REF.:
*LIBISTES RETICULATUS	50%	BERYLLIUM=2.58MG/L,24H,SOFT WATER, 296K	355
*LIBISTES RETICULATUS	50%	BERYLLIUM=0.43MG/L,48H,SOFT WATER, 296K	355
*LIBISTES RETICULATUS	50%	BERYLLIUM=0.19MG/L,96H,SOFT WATER, 296K	355
*MOUSE	50%	ORAL DOSE=12MG/KG(=1.50*10**(-4) MOL/KG),PT-50(FOOT)=3.82, AS BECL2	953
*PIMEPHALES PROMELAS	50%	BECL2,EXPOS.TIME=4DAYS,CONC.IN WATER=150MICROGRAM/L,LIMNOL.	1004
*RAT	50%	SULFATE,ORAL,DOSE=82MG/KG BODY	1611
*RAT	50%	CHLORIDE,ORAL,DOSE=86MG/KG BODY	1611
*RAT	50%	FLUORIDE,ORAL,DOSE=98MG/KG BODY	1611
*RAT	50%	HYDROXIDE,INTRAVEN.,DOSE=2.3MG/KG BODY	1611

TABLE C25

BENOMYL : GROWTH INHIBITION			
ITEM:	VALUE:	CONDITIONS:	REF.:
*CHLORELLA PYRENOIDOSA	50 %	48 H, FRESHWATER = 1.4 MG/L, 297 K, LOG PHASE, 10**7 CELLS/L	221

TABLE C26

BENOMYL : MORTALITY			
ITEM:	VALUE:	CONDITIONS:	REF.:
*DAPHNIA MAGNA	50 %	48 H, FRESHWATER = 0.64 MG/L, 293 K, AGE=< 1 D, N=20	221
*LEBISTES RETICULATUS	50 %	48 H, FRESHWATER = 3.4 MG/L, 288 K, AGE=3 M, N=10	221
*RATS	50%	ACUTE ORAL TOXICITY=9590MG/KG	389
*SALMO GAIRDNERI	50 %	48 H, FRESHWATER = 0.48 MG/L, 288 K, AGE = 3 M, N=10	221

TABLE C27

BENSULIDE : MORTALITY			
ITEM:	VALUE:	CONDITIONS:	REF.:
*ICTALURUS PUNCTATUS	50 %	379 (327-402) PPB, 96 H, 14 G, 293 K,PH=8.2, ALK=80 PPM, 95 %CL	1002

TABLE C28

BI : MORTALITY			
ITEM:	VALUE:	CONDITIONS:	REF.:
*RAT	50%	DIHYDROXYPROPOXY,INTRAVEN.,DOSE= 13MG/KG BODY	1611

TABLE C29

BO : DEFICIENCY SYMPTOMS			
ITEM:	VALUE:	CONDITIONS:	REF.:
*AGATHIS AUSTRALIS	0.001 MG/L	TREE, 100 D IN GREENHOUSE, CONC IN CULTURE SOLUTION	1248

TABLE C30

BO : NORMAL GROWHT			
ITEM:	VALUE:	CONDITIONS:	REF.:
*AGATHIS AUSTRALIS	0.500 MG/L	TREE, 100 D IN GREENHOUSE, CONC IN CULTURE SOLUTION	1248

TABLE C31

BO : TOXIC SYMPTOMS			
ITEM:	VALUE:	CONDITIONS:	REF.:
*AGATHIS AUSTRALIS	5.000 MG/L	TREE, 100 D IN GREENHOUSE, CONC IN CULTURE SOLUTION	1248

TABLE C32

BORDEAUX MIXTURE : MORTALITY			
ITEM:	VALUE:	CONDITIONS:	REF.:
*RATS	50%	ACUTE ORAL TOXICITY=300MG/KG, FUNGICIDE	389

TABLE C33

BOTULINAL : MORTALITY			
ITEM:	VALUE:	CONDITIONS:	REF.:
*MOUSE	50%	ORAL DOSE=1.14*10**(-6)MG/KG, BOTULINAL A,PT-50(F007)=14.9	953
*MOUSE	50%	ORAL DOSE=3.2*10**(-7)MG/KG, BOTULINAL D,PT-50(F007)=15.49	953
*MOUSE	50%	ORAL DOSE=2.5*10**(-3)MG/KG, BOTULINAL E,PT-50(F007)=11.15	953

TABLE C34

CA : MORTALITY			
ITEM:	VALUE:	CONDITIONS:	REF.:
*MOUSE	50%	ORAL DOSE=1350MG/KG(=7.36*10**(-3) MOL/KG),PT-50(F007)=2.13, AS CACL2	953
*RAT	50%	ACETATE,ORAL,DOSE=4280MG/KG BODY	1611
*RAT	50%	CHLORIDE,ORAL,DOSE=4000MG/KG BODY	1611

TABLE C35

CAFFEINE : MORTALITY			
ITEM:	VALUE:	CONDITIONS:	REF.:
*MOUSE	50%	ORAL DOSE=250MG/KG(=1.29*10**(-3) MOL/KG),PT-50(?030)=2.89	953

TABLE C36

CALCIUM ARSENATE : MORTALITY			
ITEM:	VALUE:	CONDITIONS:	REF.:
*RATS	50%	ACUTE ORAL TOXICITY=10-40MG/KG, PESTICIDE	389

TABLE C37

CAMPHECHLOR : MORTALITY			
ITEM:	VALUE:	CONDITIONS:	REF.:
*BOBWHITE	50%	CONC.IN FOOD=828MG/KG(619-1102), 95% CONF.LIMITS	389
*EARTHWORMS	95%	CONC.IN SOIL:>15 MG/KG	389
*JAPANESE QUAIL	50%	CONC.IN FOOD=686MG/KG(523-1002), 95% CONF.LIMITS	389
*MALLARD	50%	CONC.IN FOOD=538MG/KG(474-614),95% CONF.LIMITS	389
*PHEASANT	50%	CONC.IN FOOD=542MG/KG(462-638),95% CONF.LIMITS	389
*PHEASANT	50%	CONC.IN FOOD=450.0MG/KG	349
*QUAIL	50%	CONC.IN FOOD=500.0MG/KG	349
*RATS	50%	ACUTE ORAL TOXICITY=40-150MG/KG, INSECTICIDE	389
*RATS	50%	ACUTE DERMAL TOXICITY=780-1075MG/KG,INSECTICIDE	389

TABLE C38

CARBARYL : HEART BEAT			
ITEM:	VALUE:	CONDITIONS:	REF.:
*BIOMPHALARIA GLABRATA	96 BEATS/MIN	CARBARYL=80MICRO-MOL,AFTER 24H, 299K,5 SNAILS	264
*BIOMPHALARIA GLABRATA	31 BEATS/MIN	CARBARYL=320MICRO-MOL,AFTER 24H, 299K,5 SNAILS	264
*BIOMPHALARIA GLABRATA	0 BEATS/MIN	CARBARYL=400MICRO-MOL,AFTER 24H, 299K,5 SNAILS	264

TABLE C39

CATION EXCHANGE CAPACITY : CHANGE IN UPTAKE OF CD			
ITEM:	VALUE:	CONDITIONS:	REF.:
*AVENA SATIVA	13.7 MG CD/KG DRY WEIGHT	CEC=17.1 MEQ/100 G SOIL,EXCH SOIL CD=2.94 PPM, DW=0.99 G	590
*AVENA SATIVA	12.9 MG CD/KG DRY WEIGHT	CEC=18.9 MEQ/100 G SOIL,EXCH SOIL CD=2.90 PPM, DW=1.34 G	590
*AVENA SATIVA	11.0 MG CD/KG DRY WEIGHT	CEC=23.1 MEQ/100 G SOIL,EXCH SOIL CD=2.55 PPM, DW=1.34 G	590
*AVENA SATIVA	9.5 MG CD/KG DRY WEIGHT	CEC=26.9 MEQ/100 G SOIL,EXCH SOIL CD=2.34 PPM, DW=1.66 G	590
*AVENA SATIVA	7.9 MG CD/KG DRY WEIGHT	CEC=30.5 MEQ/100 G SOIL,EXCH SOIL CD=1.95 PPM, DW=1.76 G	590

TABLE C40

CCL4 : MORTALITY			
ITEM:	VALUE:	CONDITIONS:	REF.:
*MOUSE	50%	ORAL DOSE=4620MG/KG(=3.0*10**(-2) MOL/KG),PT-50(F007)=1.52	953

TABLE C41

CD : BLOOD GLUCOSE			
ITEM:	VALUE:	CONDITIONS:	REF.:
*PLEURONECTES FLESUS	27.8.(+-1.2) MG/100 ML	CD=5PPB,4W,9-10 FISH ,283K	879
*PLEURONECTES FLESUS	29.2 (+-1.2) MG/100 ML	CD=5PPB,9W,9-10 FISH ,283K	879
*PLEURONECTES FLESUS	31.7 (+-1.2) MG/100 ML	CD=50PPB,4W,9-10 FISH ,283K	879
*PLEURONECTES FLESUS	41.0 (+-2.7) MG/100 ML	CD=50PPB,9W,9-10 FISH ,283K	879
*PLEURONECTES FLESUS	34.1 (+-1.4) MG/100 ML	CD=500PPB,4W,9-10 FISH ,283K	879
*PLEURONECTES FLESUS	28.4 (+-3.6) MG/100 ML	CD=500PPB,9W,9-10 FISH ,283K	879
*PLEURONECTES FLESUS	25.9 (+-0.8) MG/100 ML	CD=0.PPB,4W,9-10 FISH ,283K	879
*PLEURONECTES FLESUS	24.7 (+-0.8) MG/100 ML	CD=0.PPB,9W,9-10 FISH ,283K	879
*PLEURONECTES FLESUS	20. MG/100ML	CD=0.MG/L,15D,	879
*PLEURONECTES FLESUS	21. MG/100ML	CD=0.1MG/L,15D	879
*PLEURONECTES FLESUS	29. MG/100ML	CD=1.MG/L,15D	879
*PLEURONECTES FLESUS	35. MG/100ML	CD=10.MG/L,15D	879

TABLE C42

CD : BLOOD LACTATE			
ITEM:	VALUE:	CONDITIONS:	REF.:
*PLEURONECTES FLESUS	5. MG/100ML	CD=0.MG/100ML,15D	879
*PLEURONECTES FLESUS	3.9 MG/100ML	CD=0.1MG/L,15D	879
*PLEURONECTES FLESUS	4.8 MG/100ML	CD=1.MG/L,15D	879
*PLEURONECTES FLESUS	1.6 MG/100ML	CD=10.MG/L,15D	879

TABLE C43

CD : DECREASE IN YIELD			
ITEM:	VALUE:	CONDITIONS:	REF.:
*RICE	0 % OF CONTROL	CD ADDED TO SOIL=0.001% OF CDO	487
*RICE	8 % OF CONTROL	CD ADDED TO SOIL=0.003% OF CDO	487
*RICE	8 % OF CONTROL	CD ADDED TO SOIL=0.01% OF CDO	487
*RICE	7 % OF CONTROL	CD ADDED TO SOIL=0.03% OF CDO	487
*RICE	31 % OF CONTROL	CD ADDED TO SOIL=0.1% OF CDO	487
*RICE	68 % OF CONTROL	CD ADDED TO SOIL=0.3% OF CDO	487
*RICE	81 % OF CONTROL	CD ADDED TO SOIL=0.6% OF CDO	487
*RICE	99 % OF CONTROL	CD ADDED TO SOIL=1.0% OF CDO	487
*WHEAT	28 % OF CONTROL	CD ADDED TO SOIL=0.003% OF CDO	487
*WHEAT	84 % OF CONTROL	CD ADDED TO SOIL=0.01% OF CDO	487
*WHEAT	87 % OF CONTROL	CD ADDED TO SOIL=0.03% OF CDO	487
*WHEAT	97 % OF CONTROL	CD ADDED TO SOIL=0.1% OF CDO	487
*WHEAT	97 % OF CONTROL	CD ADDED TO SOIL=0.3% OF CDO	487
*WHEAT	98 % OF CONTROL	CD ADDED TO SOIL=0.6% OF CDO	487
*WHEAT	99 % OF CONTROL	CD ADDED TO SOIL=1.0% OF CDO	487

TABLE C44

CD : HEMATOCRIT			
ITEM:	VALUE:	CONDITIONS:	REF.:
*PLEURONECTES FLESUS	20.7 (+-0.9) %	CD=5PPB,4W,9-10 FISH ,283K	879

TABLE C44

CD : HEMATOCRIT			
ITEM:	VALUE:	CONDITIONS:	REF.:
*PLEURONECTES FLESUS	16.0 (+-0.8) %	CD=5PPB,9W,9-10 FISH ,283K	879
*PLEURONECTES FLESUS	18.8 (+-0.9) %	CD=50PPB,4W,9-10 FISH ,283K	879
*PLEURONECTES FLESUS	16,5 (+-0.8) %	CD=50PPB,9W,9-10 FISH ,283K	879
*PLEURONECTES FLESUS	15.9 (+-1.7) %	CD=500PPB,4W,9-10 FISH ,283K	879
*PLEURONECTES FLESUS	16.3 (+-1.6) %	CD=500PPB,9W,9-10 FISH ,283K	879
*PLEURONECTES FLESUS	21.8 (+-0.9) %	CD=0.PPB,4W,9-10 FISH ,283K	879
*PLEURONECTES FLESUS	22.1 (+-0.8) %	CD=0.PPB,9W,9-10 FISH ,283K	879
*PLEURONECTES FLESUS	21. %	CD=0.MG/L,15D,280K	879
*PLEURONECTES FLESUS	19. %	CD=0.1MG/L,15D,280K	879
*PLEURONECTES FLESUS	18 %	CD=1.MG/L,15D,280K	879
*PLEURONECTES FLESUS	17. %	CD=10.MG/L,15D,280K	879
*PSEUDOPLEURONECTES AMERICANUS	34. (+-1.) % PACKED RED CELLS	CD=5PPB.60D,32 FISH	216
*PSEUDOPLEURONECTES AMERICANUS	33. (+-1.) % PACKED RED CELLS	CD=10PPB,60D,32 FISH	216
*PSEUDOPLEURONECTES AMERICANUS	35. (+-1.) % PACKED RED CELLS	CD=0.PPB,60D,35 FISH	216

TABLE C45

CD : HEMOGLOBIN			
ITEM:	VALUE:	CONDITIONS:	REF.:
*PLEURONECTES FLESUS	6.1 (+-0.3) G/100 ML	CD=5PPB,4W,9-10 FISH ,283K	879
*PLEURONECTES FLESUS	4.4 (+-0.3) G/100 ML	CD=5PPB,9W,9-10 FISH ,283K	879
*PLEURONECTES FLESUS	5.5 (+-0.2) G/100 ML	CD=50PPB,4W,9-10 FISH ,283K	879
*PLEURONECTES FLESUS	4.5 (+-0.3) G/100 ML	CD=50PPB,9W,9-10 FISH ,283K	879
*PLEURONECTES FLESUS	4.9 (+-0.4) G/100 ML	CD=500PPB,4W,9-10 FISH ,283K	879
*PLEURONECTES FLESUS	4.6 (+-0.4) G/100 ML	CD=500PPB,9W,9-10 FISH ,283K	879
*PLEURONECTES FLESUS	6.3 (+-0.2) G/100 ML	CD=0.PPB,4W,9-10 FISH ,283K	879
*PLEURONECTES FLESUS	6.1 (+-0.2) G/100 ML	CD=0.PPB,9W,9-10 FISH ,283K	879
*PLEURONECTES FLESUS	5.8 %	CD=0.MG/L,15D,280K	879
*PLEURONECTES FLESUS	5.5 %	CD=0.1MG/L,15D,280K	879
*PLEURONECTES FLESUS	5.2 %	CD=1.MG/L,15D,280K	879
*PLEURONECTES FLESUS	4.8 %	CD=10.MG/L,15D,280K	879
*PSEUDOPLEURONECTES AMERICANUS	8.1 (+-0.3) G/100 ML WHOLE BLOOD	CD=5PPB,60D,31 FISH	216
*PSEUDOPLEURONECTES AMERICANUS	7.4 (+-0.3) G/100 ML WHOLE BLOOD	CD=10PPB,60D,32 FISH	216
*PSEUDOPLEURONECTES AMERICANUS	7.4 (+-0.3) G/100 ML WHOLE BLOOD	CD=0.PPB,60D,32 FISH	216

TABLE C46

CD : INCREASE OF YIELD			
ITEM:	VALUE:	CONDITIONS:	REF.:
*WHEAT	6 % OF CONTROL	CD ADDED TO SOIL=0.001% OF CDO	487

TABLE C47

CD : LIVER GLYCOGEN			
ITEM:	VALUE:	CONDITIONS:	REF.:
*PLEURONECTES FLESUS	2.2 MG/100MG WET. WT.	CD=0.MG/L,15D	879

TABLE C47

CD : LIVER GLYCOGEN			
ITEM:	VALUE:	CONDITIONS:	REF.:
*PLEURONECTES FLESUS	3. MG/100MG WET. WT.	CD=0.1MG/L,15D	879
*PLEURONECTES FLESUS	3.6 MG/100MG WET. WT.	CD=1.MG/L,15D	879
*PLEURONECTES FLESUS	3.8 MG/100MG WET. WT	CD=10.MG/L,15D	879

TABLE C48

CD : MCHC			
ITEM:	VALUE:	CONDITIONS:	REF.:
*PLEURONECTES FLESUS	28 %	CD=0.MG/L,15D,280K	879
*PLEURONECTES FLESUS	29 %	CD=0.1MG/L,15D,280K	879
*PLEURONECTES FLESUS	29.5 %	CD=1.MG/L,15D,280K	879
*PLEURONECTES FLESUS	27.5 %	CD=10.MG/L,15D,280K	879

TABLE C49

CD : MEAN CORPUSCULAR HEMOGLOBIN			
ITEM:	VALUE:	CONDITIONS:	REF.:
*PSEUDOPLEURONECTES AMERICANUS	27.9 (+-3.7) PICOGRAMS/CELL	CD=5PPB,60D,9 FISH	216
*PSEUDOPLEURONECTES AMERICANUS	26.2 (+-1.6) PICOGRAMS/CELL	CD=10PPB,60D,14 FISH	216
*PSEUDOPLEURONECTES AMERICANUS	26.2 (+-1.6) PICOGRAMS/CELL	CD=0.PPB,60D,14 FISH	216
*PSEUDOPLEURONECTES AMERICANUS	23.8 (+-1.) G/100 ML PACKED RED CELLS	CD=5PPB,60D,31 FISH	216
*PSEUDOPLEURONECTES AMERICANUS	22.7 (+-0.7) G/100 ML PACKED RED CELLS	CD=10PPB,60D,31 FISH	216
*PSEUDOPLEURONECTES AMERICANUS	22.7 (+-0.7) G/100 ML PACKED RED CELLS	CD=0.PPB,60D,31 FISH	216

TABLE C50

CD : MEAN CORPUSCULAR VOLUME			
ITEM:	VALUE:	CONDITIONS:	REF.:
*PSEUDOPLEURONECTES AMERICANUS	114. (+-6.1) MICRO**3/CELL	CD=5PPB,60D,9 FISH	216
*PSEUDOPLEURONECTES AMERICANUS	114. (+-5.) MICRO**3/CELL	CD=10PPB,60D,14 FISH	216
*PSEUDOPLEURONECTES AMERICANUS	121. (+-7.) MICRO**3/CELL	CD=0.PPB,60D,17 FISH	216

TABLE C51

CD : MORTALITY			
ITEM:	VALUE:	CONDITIONS:	REF.:
*CRANGON CRANGON	50%	CDCL2,EXPOS.TIME=2DAYS,CONC.IN WATER=3300-10000 UG/L,MARINE	1272
*CRANGON CRANGON	50%	CDCL2,EXPOS.TIME=2DAYS,CONC.IN WATER=1200MICROGRAM/L,BRACKISH	1103
*DAPHNIA MAGNA	50%	CDCL2,EXPOS.TIME=2DAYS,CONC.IN WATER=65 UG/L,LIMNOL.	122

TABLE C51

CD : MORTALITY			
ITEM:	VALUE:	CONDITIONS:	REF.:
*DOG	50%	4000MIN*MG/M**3(APPROX.),INHAL.OF CD OXIDE FUME FOR 10-30MIN	487
*GUINEA PIGS	50%	3500MIN*MG/M**3(APPROX.),INHAL.OF CD OXIDE FUME FOR 10-30MIN	487
*MONKEYS	50%	15000MIN*MG/M**3(APPROX.),INHAL.OF CD OXIDE FUME FOR 10-30MIN	487
*MOUSE	50%	<700MIN*MG/M**3 (500?),INHAL.OF CD OXIDE FUME FOR 10-30MIN	487
*NOEMACHEILUS BARBARTULUS	50 %	CADMIUM=2.MG/L,54D,HARD WATER	1477
*PHILODINA SP.	50 %	CDSO4= 6.4 +- 3.1 PPM, 24 H, 25 PPM CACO3, 293 K, PH=7.4-7.9	1373
*PHILODINA SP.	50 %	CDSO4= 6.4 +- 0.5 PPM, 48 H, 25 PPM CACO3, 293 K, PH=7.4-7.9	1373
*PHILODINA SP.	50 %	CDSO4= 2.0 +- 0.1 PPM, 96 H, 25 PPM CACO3, 293 K, PH=7.4-7.9	1373
*RABBITS	50%	2500MIN*MG/M**3(APPROX.),INHAL.OF CD OXIDE FUME FOR 10-30MIN	487
*RAT	50%	CHLORIDE,ORAL,DOSE=88MG/KG BODY	1611
*RAT	50%	OXIDE,ORAL,DOSE=72MG/KG BODY	1611
*RAT	50%	STEARATE,ORAL,DOSE=1225MG/KG BODY	1611
*RAT	50%	SUCCINATE,ORAL,DOSE=660MG/KG BODY	1611
*RATS	50%	500MIN*MG/M**3,N=160,INHAL.OF CD OXIDE FUME FOR 10-30MIN	487
*SALMO GAIRDNERI	50 %	CADMIUM=0.017MG/L,5.5D,HARD WATER	1477

TABLE C52

CD : MUSCLE GLYCOGEN			
ITEM:	VALUE:	CONDITIONS:	REF.:
*PLEURONECTES FLESUS	74. MICROGRAMS/100MG WET. WT.	CD=0.MG/L,15D	879
*PLEURONECTES FLESUS	44. MICROGRAMS/100MG WET. WT.	CD=0.1MG/L,15D	879
*PLEURONECTES FLESUS	54. MICROGRAMS/100MG WET. WT.	CD=1.MG/L,15D	879
*PLEURONECTES FLESUS	55. MICROGRAMS/100MG WET. WT.	CD=10.MG/L,15D	879

TABLE C53

CD : PLASMA CA			
ITEM:	VALUE:	CONDITIONS:	REF.:
*PLEURONECTES FLESUS	2.9 MILLI-MOL	CD=0.PPM,15D	879
*PLEURONECTES FLESUS	2.5 MILLI-MOL	CD=0.1PPM,15D	879
*PLEURONECTES FLESUS	2.4 MILLI-MOL	CD=1.PPM,15D	879
*PLEURONECTES FLESUS	2.2 MILLI-MOL	CD=10.PPM,15D	879

TABLE C54

CD : PLASMA INORGANIC PHOSPHATE			
ITEM:	VALUE:	CONDITIONS:	REF.:
*PLEURONECTES FLESUS	5.7 MILLI-MOL	CD=0.PPM,15D	879
*PLEURONECTES FLESUS	5.9 MILLI-MOL	CD=0.1PPM,15D	879

TABLE C54

CD : PLASMA INORGANIC PHOSPHATE			
ITEM:	VALUE:	CONDITIONS:	REF.:
*PLEURONECTES FLESUS	6.6 MILLI-MOL	CD=1.PPM,15D	879
*PLEURONECTES FLESUS	6.6 MILLI-MOL	CD=10.PPM,15D	879

TABLE C55

CD : PLASMA K			
ITEM:	VALUE:	CONDITIONS:	REF.:
*PLEURONECTES FLESUS	3.8 MILLI-MOL	CD=0.PPM,15D	879
*PLEURONECTES FLESUS	3.3 MILLI-MOL	CD=0.1PPM,15D	879
*PLEURONECTES FLESUS	3.1 MILLI-MOL	CD=1.PPM,15D	879
*PLEURONECTES FLESUS	3.1 MILLI-MOL	CD=10.PPM,15D	879

TABLE C56

CD : PLASMA MG			
ITEM:	VALUE:	CONDITIONS:	REF.:
*PLEURONECTES FLESUS	0.5 MILLI-MOL	CD=0.PPM,15D	879
*PLEURONECTES FLESUS	1.2 MILLI-MOL	CD=0.1PPM,15D	879
*PLEURONECTES FLESUS	1.1 MILLI-MOL	CD=10.PPM,15D	879
*PLEURONECTES FLESUS	1.4 MILLI-MOL	CD=1.PPM,15D	879

TABLE C57

CD : PLASMA OSMOLALITY			
ITEM:	VALUE:	CONDITIONS:	REF.:
*PSEUDOPLEURONECTES AMERICANUS	380. (+-7.) MOSM/KG H2O	CD=5PPB,60D,22 FISH	216
*PSEUDOPLEURONECTES AMERICANUS	401. (+-5.) MOSM/KG H2O	CD=10PPB,60D,19 FISH	216
*PSEUDOPLEURONECTES AMERICANUS	401. (+-0.7) MOSM/KG H2O	CD=0.PPM,60D,22 FISH	216

TABLE C58

CD : PLASMA PROTEIN			
ITEM:	VALUE:	CONDITIONS:	REF.:
*PSEUDOPLEURONECTES AMERICANUS	6.7 (+-0.2) G%	CD=5PPB,60D,30 FISH	216
*PSEUDOPLEURONECTES AMERICANUS	6.8 (+-0.2) G%	CD=10PPB,60D,31 FISH	216
*PSEUDOPLEURONECTES AMERICANUS	6.7 (+-0.2) G%	CD=0.PPB,60D,34 FISH	216

TABLE C59

CD : RBC			
ITEM:	VALUE:	CONDITIONS:	REF.:
*PSEUDOPLEURONECTES AMERICANUS	2.69 (+-0.12) MIL. CELLS/MM**3	CD=5PPB,60D,10 FISH	216

TABLE C59

CD : RBC			
ITEM:	VALUE:	CONDITIONS:	REF.:
*PSEUDOPLEURONECTES AMERICANUS	2.96 (+-0.18) MIL. CELLS/MM**3	CD=10PPB,60D,14 FISH	216
*PSEUDOPLEURONECTES AMERICANUS	2.9 (+-1.7) MIL. CELLS/MM**3	CD=0.PPB,60D,17 FISH	216

TABLE C60

CD : YIELD			
ITEM:	VALUE:	CONDITIONS:	REF.:
*AVENA SATIVA, GRAINS	27.28 DRY WT.(G/POT)	CONTROL =42.57DRY WT.(G/POT),CD IN SOIL=40PPM,GROWTH PERIOD=100D	775
*AVENA SATIVA, GRAINS	18.37 DRY WT.(G/POT)	CONTROL =42.57DRY WT.(G/POT),CD IN SOIL=200PPM,GROWTH PERIOD=100D	775
*AVENA SATIVA, HUSKS	8.31 DRY WT.(G/POT)	CONTROL =5.56DRY WT.(G/POT),CD IN SOIL=40PPM,GROWTH PERIOD=100D	775
*AVENA SATIVA, HUSKS	6.85 DRY WT.(G/POT)	CONTROL =5.56DRY WT.(G/POT),CD IN SOIL=200PPM,GROWTH PERIOD=100D	775
*AVENA SATIVA, LEAF	21.5 DRY WT.(G/POT)	CONTROL =20.84DRY WT.(G/POT),CD IN SOIL=40PPM,GROWTH PERIOD=100D	775
*AVENA SATIVA, LEAF	18.72 DRY WT.(G/POT)	CONTROL =20.84DRY WT.(G/POT),CD IN SOIL=200PPM,GROWTH PERIOD=100D	775
*AVENA SATIVA, ROOTS	3.18 DRY WT.(G/POT)	CONTROL =3.75DRY WT.(G/POT),CD IN SOIL=40PPM,GROWTH PERIOD=100D	775
*AVENA SATIVA, ROOTS	2.88 DRY WT.(G/POT)	CONTROL =3.75DRY WT.(G/POT),CD IN SOIL=200PPM,GROWTH PERIOD=100D	775
*AVENA SATIVA, STALK	39.17 DRY WT.(G/POT)	CONTROL =29.90DRY WT.(G/POT),CD IN SOIL=40PPM,GROWTH PERIOD=100D	775
*AVENA SATIVA, STALK	23.28 DRY WT.(G/POT)	CONTROL =29.90DRY WT.(G/POT),CD IN SOIL=200PPM,GROWTH PERIOD=100D	775
*BRASSICA OLERACEA, BOTRYTIS, LEAVES	3.17 DRY WT.(G/POT)	CONTROL =103.67DRY WT.(G/POT),CD IN SOIL=200PPM,GROWTH PERIOD=60D	775
*BRASSICA OLERACEA, BOTRYTIS, ROOT	1.47 DRY WT.(G/POT)	CONTROL =1.78DRY WT.(G/POT),CD IN SOIL=40PPM,GROWTH PERIOD=60D	775
*BRASSICA OLERACEA, BOTRYTIS, ROOT	0.17 DRY WT.(G/POT)	CONTROL =1.78DRY WT.(G/POT),CD IN SOIL=200PPM,GROWTH PERIOD=60D	775
*BRASSICA OLERACEA, ITALICA, LEAVES	122.5 DRY WT.(G/POT)	CONTROL =95.93DRY WT.(G/POT),CD IN SOIL=40PPM,GROWTH PERIOD=60D	775
*BRASSICA OLERACEA, ITALICA, LEAVES	35.20 DRY WT.(G/POT)	CONTROL =95.93DRY WT.(G/POT),CD IN SOIL=200PPM,GROWTH PERIOD=60D	775
*BRASSICA OLERACEA, ITALICA, ROOT	4.08 DRY WT.(G/POT)	CONTROL =2.46DRY WT.(G/POT),CD IN SOIL=40PPM,GROWTH PERIOD=60D	775
*BRASSICA OLERACEA, ITALICA, ROOT	2.09 DRY WT.(G/POT)	CONTROL =2.46DRY WT.(G/POT),CD IN SOIL=200PPM,GROWTH PERIOD=60D	775
*DAUCUS CAROTA, ROOT	0.81 DRY WT.(G/POT)	CONTROL =0.96DRY WT.(G/POT),GROWTH PERIOD=130D	775
*DAUCUS CAROTA, TOP	16.92 DRY WT.(G/POT)	CONTROL =18.95DRY WT.(G/POT),CD IN SOIL=40PPM,GROWTH PERIOD=130D	775
*DAUCUS CAROTA, TOP	1.5 DRY WT.(G/POT)	CONTROL =18.95DRY WT.(G/POT),CD IN SOIL=200PPM,GROWTH PERIOD=130D	775
*DAUCUS CAROTA, TUBERS	38.27 DRY WT.(G/POT)	CONTROL =41.72DRY WT.(G/POT),CD IN SOIL=40PPM,GROWTH PERIOD=130D	775
*LACTUCA SATIVA, LEAVES	18.82 DRY WT.(G/POT)	CONTROL =17.5DRY WT.(G/POT),CD IN SOIL=40PPM,GROWTH PERIOD=35D	775
*LACTUCA SATIVA, LEAVES	1.55 DRY WT.(G/POT)	CONTROL =17.5DRY WT.(G/POT),CD IN SOIL=200PPM,GROWTH PERIOD=35D	775
*LACTUCA SATIVA, ROOT	0.99 DRY WT.(G/POT)	CONTROL =0.91DRY WT.(G/POT),CD IN SOIL=40PPM,GROWTH PERIOD=35D	775
*LACTUCA SATIVA, ROOT	0.36 DRY WT.(G/POT)	CONTROL =0.91DRY WT.(G/POT),CD IN SOIL=200PPM,GROWTH PERIOD=35D	775
*PISUM SATIVUM, PODS	7.45 DRY WT.(G/POT)	CONTROL =10.59DRY WT.(G/POT),CD IN SOIL=40PPM,GROWTH PERIOD=95D	775
*PISUM SATIVUM, PODS	0.84 DRY WT.(G/POT)	CONTROL =10.59DRY WT.(G/POT),CD IN SOIL=200PPM,GROWTH PERIOD=95D	775
*PISUM SATIVUM, ROOT	1.8 DRY WT.(G/POT)	CONTROL =1.98DRY WT.(G/POT),CD IN SOIL=40PPM,GROWTH PERIOD=95D	775

TABLE C60

CD : YIELD

ITEM:	VALUE:	CONDITIONS:	REF.:
*PISUM SATIVUM, ROOT	0.56 DRY WT.(G/POT)	CONTROL =1.98DRY WT.(G/POT),CD IN SOIL=200PPM,GROWTH PERIOD=95D	775
*PISUM SATIVUM, SEEDS	16.98 DRY WT.(G/POT)	CONTROL =22.45DRY WT.(G/POT),CD IN SOIL=40PPM,GROWTH PERIOD=95D	775
*PISUM SATIVUM, SEEDS	0.21 DRY WT.(G/POT)	CONTROL =27.45DRY WT.(G/POT),CD IN SOIL=200PPM,GROWTH PERIOD=95D	775
*PISUM SATIVUM, VINES	14.43 DRY WT.(G/POT)	CONTROL =19.64DRY WT.(G/POT),CD IN SOIL=40PPM,GROWTH PERIOD=95D	775
*PISUM SATIVUM, VINES	2.47 DRY WT.(G/POT)	CONTROL =19.64DRY WT.(G/POT),CD IN SOIL=200PPM,GROWTH PERIOD=95D	775
*RAPHANUS SATIVUS, ROOT	0.13 DRY WT.(G/POT)	CONTROL =0.96DRY WT.(G/POT),CD IN SOIL=200PPM,GROWTH PERIOD=130D	775
*RAPHANUS SATIVUS, TOP	6.25 DRY WT.(G/POT)	CONTROL =8.24DRY WT.(G/POT),CD IN SOIL=40PPM,GROWTH PERIOD=45D	775
*RAPHANUS SATIVUS, TOP	1.47 DRY WT.(G/POT)	CONTROL =8.24DRY WT.(G/POT),CD IN SOIL=200PPM,GROWTH PERIOD=45D	775
*RAPHANUS SATIVUS, TUBERS	5.12 DRY WT.(G/POT)	CONTROL =7.1DRY WT.(G/POT),CD IN SOIL=40PPM,GROWTH PERIOD=45D	775
*RAPHANUS SATIVUS, TUBERS	0.48 DRY WT.(G/POT)	CONTROL =7.1DRY WT.(G/POT),CD IN SOIL=200PPM,GROWTH PERIOD=45D	775
*RAPHANUS SATIVUS, TUBERS	1.52 DRY WT.(G/POT)	CONTROL =41.72DRY WT.(G/POT),CD IN SOIL=200PPM,GROWTH PERIOD=130D	775
*SPINACIA OLERACEA, LEAVES	0.39 DRY WT.(G/POT)	CONTROL =10.96DRY WT.(G/POT),CD IN SOIL=40PPM,GROWTH PERIOD=55D	775
*SPINACIA OLERACEA, LEAVES	0.16 DRY WT.(G/POT)	CONTROL =10.96DRY WT.(G/POT),CD IN SOIL=200PPM,GROWTH PERIOD=55D	775
*SPINACIA OLERACEA, ROOT	0.02 DRY WT.(G/POT)	CONTROL =0.51DRY WT.(G/POT),CD IN SOIL=40PPM,GROWTH PERIOD=55D	775
*SPINACIA OLERACEA, ROOT	0.02 DRY WT.(G/POT)	CONTROL =0.51DRY WT.(G/POT),CD IN SOIL=200PPM,GROWTH PERIOD=55D	775

TABLE C61

CE : MORTALITY

ITEM:	VALUE:	CONDITIONS:	REF.:
*RAT	50%	NITRATE,ORAL,DOSE=4200MG/KG BODY	1611

TABLE C62

CHLORAMBEN : MORTALITY

ITEM:	VALUE:	CONDITIONS:	REF.:
*ICTALURUS PUNCTATUS	LESS THAN 10 %	10 PPM, 48 H, 14 G, 293 K, PH=8.2, ALK=80 PPM	1002

TABLE C63

CHLORAMINE : CHANGE IN HALF SATURATION CONSTANT FOR NITROGEN

ITEM:	VALUE:	CONDITIONS:	REF.:
*PHYTOPLANKTON	-0.11 MICRO M NITROGEN-N (1/L)	CONTROL = -1.46, NITRATE, NH2CL = 0.1 MG/L ,PH=8.2, 24 H ,R=0.50	1604

TABLE C64

CHLORAMINE : CHANGE IN NITROGEN UPTAKE

ITEM:	VALUE:	CONDITIONS:	REF.:
*PHYTOPLANKTON	0.004 MICRO M NITROGEN-N (L/H)	CONTROL = 0.123, NITRATE, NH2CL = 0.1 MG/L ,PH=8.2, 24 H ,R=0.50	1604

TABLE C65

CHLORDANE : MORTALITY			
ITEM:	VALUE:	CONDITIONS:	REF.:
*BOBWHITE	50%	CONC.IN FOOD=331MG/KG(197-479),95% CONF.LIMITS	389
*EARTHWORMS	95%	CONC.IN SOIL=5 MG/KG	389
*JAPANESE QUAIL	50%	CONC.IN FOOD=350MG/KG(305-403),95% CONF.LIMITS	389
*MALLARD	50%	CONC.IN FOOD=858MG/KG(629-1241), 95% CONF.LIMITS	389
*PHAESANT	50%	CONC.IN FOOD=430MG/KG(366-505),95% CONF.LIMITS	389
*PHEASANT	50%	CONC.IN FOOD=500.0MG/KG	349
*QUAIL	50%	CONC.IN FOOD=250.0MG/KG	349
*RATS	50%	ACUTE ORAL TOXICITY=457-590MG/KG, INSECTICIDE	389
*SLUGS	95%	CONC.IN SOIL:>10 MG/KG	389

TABLE C66

CHLORINE : CHANGE IN HALF SATURATION CONSTANT FOR NITROGEN			
ITEM:	VALUE:	CONDITIONS:	REF.:
*PHYTOPLANKTON	-22.62 MICRO M NITROGEN-N (1/L)	CONTROL = -1.46, NITRATE, CL2 = 0.1 MG/L ,PH=8.2, 24 H ,R=0.96	1604
*PHYTOPLANKTON	-5.01 MICRO M NITROGEN-N (1/L)	CONTROL = 2.78 , AMMONIA (NH3), CL2 = 0.1 MG/L ,PH=8.2, 24 H ,R= 0.77	1604
*PHYTOPLANKTON	2.96 MICRO M NITROGEN-N (1/L)	CONTROL = 0.99 , NITRATE, CL2 = 0.1 MG/L ,PH=8.2, 24 H ,R=0.99	1604
*PHYTOPLANKTON	11.55 MICRO M NITROGEN-N (1/L)	CONTROL = 4.40 , NITRATE, CL2 = 0.1 MG/L ,PH=8.2, 24 H ,R=0.21	1604
*PHYTOPLANKTON	20.48 MICRO M NITROGEN-N (1/L)	CONTROL = 9.38 , AMMONIA (NH3), CL2 = 0.1 MG/L ,PH=8.2, 24 H ,R= 0.52	1604

TABLE C67

CHLORINE : CHANGE IN NITROGEN UPTAKE			
ITEM:	VALUE:	CONDITIONS:	REF.:
*PHYTOPLANKTON	0.013 MICRO M NITROGEN-N (L/H)	CONTROL = 0.123, NITRATE, CL2 = 0.1 MG/L ,R=0.94	1604
*PHYTOPLANKTON	0.007 MICRO M NITROGEN-N (L/H)	CONTROL = 0.326, AMMONIA (NH3), CL2 = 0.1 MG/L ,PH=8.2, 24 H ,R= 0.77	1604
*PHYTOPLANKTON	0.010 MICRO M NITROGEN-N (L/H)	CONTROL = 0.038, NITRATE, CL2 = 0.1 MG/L ,PH=8.2, 24 H ,R=0.99	1604
*PHYTOPLANKTON	0.265 MICRO M NITROGEN-N (L/H)	CONTROL = 0.249, NITRATE, CL2 = 0.1 MG/L ,PH=8.2, 24 H ,R=0.21	1604
*PHYTOPLANKTON	0.822 MICRO M NITROGEN-N (L/H)	CONTROL = 0.581, AMMONIA (NH3), CL2 = 0.1 MG/LY.R=0.52	1604

TABLE C68

CO : MORTALITY			
ITEM:	VALUE:	CONDITIONS:	REF.:
*DAPHNIA MAGNA	50%	COCL2,EXPOS.TIME=2DAYS,CONC.IN WATER=1110MICROGRAM/L,LIMNOL.	122

TABLE C68

CO : MORTALITY			
ITEM:	VALUE:	CONDITIONS:	REF.:
*MOUSE	50%	CHLORIDE,ORAL,DOSE=80MG/KG BODY	1611
*RAT	50%	CHLORIDE,ORAL,DOSE=180MG/KG BODY	1611
*RAT	50%	OXIDE,ORAL,DOSE=1700MG/KG BODY	1611

TABLE C69

CO : OXYGEN CONSUMPTION			
ITEM:	VALUE:	CONDITIONS:	REF.:
*DIALYSIS CULTURE	0.226 ML/H	COBALT=0.MG/L,CULTIVATION HOAGLANDS SOLUTION,SEE PURVIS	1382
*DIALYSIS CULTURE	0.179 ML/H	COBALT=0.2MG/L,CULTIVATION HOAGLANDS SOLUTION,SEE PURVIS	1382
*DIALYSIS CULTURE	0.133 ML/H	COBALT=0.35MG/L,CULTIVATION HOAGLANDS SOLUTION,SEE PURVIS	1382

TABLE C70

CR : GLUCOSE			
ITEM:	VALUE:	CONDITIONS:	REF.:
*SALMO GAIRDNERI	3.4 G/100 ML	CR(VI)=10.PPM,32D,20 FISHES	1523

TABLE C71

CR : ACID PHOSPHATASE ACTIVITY			
ITEM:	VALUE:	CONDITIONS:	REF.:
*RUTILUS RUTILUS	28.46 MILLI-U/100 ML	CR(VI)=0MG/L,32D,283K,4*13 FISH	1523
*RUTILUS RUTILUS	41.18 MILLI-U/100 ML	CR(VI)=0.1MG/L,32D,283K,4*13 FISHES	1523
*RUTILUS RUTILUS	42.21 MILLI-U/100 ML	CR(VI)=1.0MG/L,32D,283K,4*13 FISHES	1523
*RUTILUS RUTILUS	26.94 MILLI-U/100 ML	CR(VI)=10MG/L,32D,283K,4*13 FISH	1523
*SALMO GAIRDNERI	28.46 MILLI-U/100 ML	CR(VI)=0.PPM,32D,20 FISH	1523
*SALMO GAIRDNERI	41.18 MILLI-U/100 ML	CR(VI)=0.1PPM,32D,20 FISH	1523
*SALMO GAIRDNERI	42.21 MILLI-U/100 ML	CR(VI)=1.PPM,32D,20 FISH	1523
*SALMO GAIRDNERI	26.94 MILLI-U/100 ML	CR(VI)=10.PPM,32D,20 FISH	1523

TABLE C72

CR : DESTRUCT. OF INTESTINAL EPITHELIUM			
ITEM:	VALUE:	CONDITIONS:	REF.:
*SALMO GAIRDNERI	EFFECT OBSERVED	CR(VI)=100.MG/L	491

TABLE C73

CR : DIMINISHED GROWTH			
ITEM:	VALUE:	CONDITIONS:	REF.:
*SALMO GAIRDNERI	EFFECT OBSERVED	CR(VI)=0.2 MG/L,YOUNG SALMON	1004

TABLE C74

CR : G-6-PDH			
ITEM:	VALUE:	CONDITIONS:	REF.:
*SALMO GAIRDNERI	16.46 MILLI-U/MG TISSUE	CR(VI)=0.PPM,32D,20 FISH	1523
*SALMO GAIRDNERI	16.68 MILLI-U/MG TISSUE	CR(VI)=0.1PPM,32D,20 FISH	1523
*SALMO GAIRDNERI	16.70 MILLI-U/MG TISSUE	CR(VI)=1.PPM,32D,20 FISH	1523
*SALMO GAIRDNERI	17.42 MILLI-U/MG TISSUE	CR(VI)=10.PPM,32D,20 FISH	1523

TABLE C75

CR : GLUCOSE (SEE ALSO TABLE C70)			
ITEM:	VALUE:	CONDITIONS:	REF.:
*RUTILUS RUTILUS	42.98 MG/100 ML	CR(VI)=0MG/L,32D,283K,4*13 FISH	1523
*RUTILUS RUTILUS	41.69 MG/100 ML	CR(VI)=0.1MG/L,32D,283K,4*13 FISH	1523
*RUTILUS RUTILUS	30.38 MG/100 ML	CR(VI)=1.0MG/L,32D,283K,4*13 FISH	1523
*RUTILUS RUTILUS	57.22 MG/100 ML	CR(VI)=10MG/L,32D,283K,4*13 FISH	1523
*SALMO GAIRDNERI	42.98 MG/100 ML	CR(VI)=0.PPM,32D,20 FISH	1523
*SALMO GAIRDNERI	41.69 MG/100 ML	CR(VI)=0.1PPM,32D,20 FISH	1523
*SALMO GAIRDNERI	30.38 MG/100 ML	CR(VI)=1.PPM,32D,20 FISH	1523
*SALMO GAIRDNERI	57.22 MG/100 ML	CR(VI)=10.PPM,32D,20 FISH	1523
*SALMO GAIRDNERI	3.6 MILLIMOL/L	CR(VI)=0MG/L,15-22D,281K,20 FISH	1523
*SALMO GAIRDNERI	8.4 MILLIMOL/L	CR(VI)=10MG/L,15-22D,281K,20 FISH	1523

TABLE C76

CR : GONAD WEIGHT			
ITEM:	VALUE:	CONDITIONS:	REF.:
*SALMO GAIRDNERI	0.84 %(G/100G FISH)	CR(VI)=0.PPM,32D,20 FISH ,MALE	1523
*SALMO GAIRDNERI	0.96 %(G/100G FISH)	CR(VI)=0.1PPM,32D,20 FISH ,MALE	1523
*SALMO GAIRDNERI	1.23 %(G/100G FISH)	CR(VI)=1.PPM,32D,20 FISH ,MALE	1523
*SALMO GAIRDNERI	1.51 %(G/100G FISH)	CR(VI)=10.PPM,32D,20 FISH ,MALE	1523
*SALMO GAIRDNERI	2.41 %(G/100G FISH)	CR(VI)=0.PPM,32D,20 FISH ,FEMALE	1523
*SALMO GAIRDNERI	1.94 %(G/100G FISH)	CR(VI)=0.1PPM,32D,20 FISH ,FEMALE	1523
*SALMO GAIRDNERI	2.79 %(G/100G FISH)	CR(VI)=1.PPM,32D,20 FISH ,FEMALE	1523
*SALMO GAIRDNERI	1.82 %(G/100G FISH)	CR(VI)=10.PPM,32D,20 FISH ,FEMALE	1523

TABLE C77

CR : GOT			
ITEM:	VALUE:	CONDITIONS:	REF.:
*SALMO GAIRDNERI	78. MILLI-U/MG TISSUE	CR(VI)=0.PPM,32D,20 FISH	1523
*SALMO GAIRDNERI	57. MILLI-U/MG TISSUE	CR(VI)=0.1PPM,32D,20 FISH	1523
*SALMO GAIRDNERI	76.6 MILLI-U/MG TISSUE	CR(VI)=1.PPM,32D,20 FISH	1523
*SALMO GAIRDNERI	30.9 MILLI-U/MG TISSUE	CR(VI)=10.PPM,32D,20 FISH	1523

TABLE C78

CR : GPT			
ITEM:	VALUE:	CONDITIONS:	REF.:
*SALMO GAIRDNERI	36.9 MILLI-U/MG TISSUE	CR(VI)=0.PPM,32D,20 FISH	1523

TABLE C78

CR : GPT			
ITEM:	VALUE:	CONDITIONS:	REF.:
*SALMO GAIRDNERI	26.3 MILLI-U/MG TISSUE	CR(VI)=0.1PPM,32D,20 FISH	1523
*SALMO GAIRDNERI	37.5 MILLI-U/MG TISSUE	CR(VI)=1.PPM,32D,20 FISH	1523
*SALMO GAIRDNERI	28.8 MILLI-U/MG TISSUE	CR(VI)=10.PPM,32D,20 FISH	1523

TABLE C79

CR : HEMATOCRIT			
ITEM:	VALUE:	CONDITIONS:	REF.:
*RUTILUS RUTILUS	14.7 %	CR(VI)=0MG/L,32D,283K,4*13 FISH	1523
*RUTILUS RUTILUS	17.1 %	CR(VI)=0.1MG/L,32D,283K,4*13 FISH	1523
*RUTILUS RUTILUS	15.1 %	CR(VI)=1.0MG/L,32D,283K,4*13 FISH	1523
*RUTILUS RUTILUS	13.5 %	CR(VI)=10MG/L,32D,283K,4*13 FISH	1523
*SALMO GAIRDNERI	INCREASED	CR(VI)=2.MG/L	492
*SALMO GAIRDNERI	14.7 %	CR(VI)=0.PPM,32D,20 FISH	1523
*SALMO GAIRDNERI	17.1 %	CR(VI)=0.1PPM,32D,20 FISH	1523
*SALMO GAIRDNERI	15.1 %	CR(VI)=1.PPM,32D,20 FISH	1523
*SALMO GAIRDNERI	13.5 %	CR(VI)=10.PPM,32D,20 FISH	1523
*SALMO GAIRDNERI	22.9 %	CR(VI)=0.PPM,15-22D,10 FISH	1523
*SALMO GAIRDNERI	29.6 %	CR(VI)=10.PPM,15-22D,10 FISH	1523
*SALMO GAIRDNERI	22.9 %	CR(VI)=0MG/L,15-22D,281K,20 FISH	1523
*SALMO GAIRDNERI	29.6 %	CR(VI)=10MG/L,15-22D,281K,20 FISH	1523

TABLE C80

CR : HEMOGLOBIN			
ITEM:	VALUE:	CONDITIONS:	REF.:
*RUTILUS RUTILUS	3.5 G/100 ML	CR(VI)=0MG/L,32D,283K,4*13 FISH	1523
*RUTILUS RUTILUS	3.2 G/100 ML	CR(VI)=0.1MG/L,32D,283K,4*13 FISH	1523
*RUTILUS RUTILUS	2.6 G/100 ML	CR(VI)=1.0MG/L,32D,283K,4*13 FISH	1523
*RUTILUS RUTILUS	3.4 G/100 ML	CR(VI)=10MG/L,32D,283K,4*13 FISH	1523
*SALMO GAIRDNERI	3.5 G/100 ML	CR(VI)=0.PPM,32D,20 FISH	1523
*SALMO GAIRDNERI	3.2 G/100 ML	CR(VI)=0.1PPM,32D,20 FISH	1523
*SALMO GAIRDNERI	2.8 G/100 ML	CR(VI)=1.PPM,32D,20 FISH	1523
*SALMO GAIRDNERI	12.9 MICROMOL/L	CR(VI)=0.PPM,15-22D,10 FISH	1523
*SALMO GAIRDNERI	17.6 MICROMOL/L	CR(VI)=10PPM,15-22D,10 FISH	1523
*SALMO GAIRDNERI	12.9 MICROMOL/L	CR(VI)=0MG/L,15-22D,281K,20 FISH	1523
*SALMO GAIRDNERI	17.6 MICROMOL/L	CR(VI)=10MG/L,15-22D,281K,20 FISH	1523

TABLE C81

CR : K IN BLOOD			
ITEM:	VALUE:	CONDITIONS:	REF.:
*SALMO GAIRDNERI	3.7 MILLIEQ/L	CR(VI)=0.PPM,15-22D,10 FISH	1523

TABLE C81

CR : K IN BLOOD			
ITEM:	VALUE:	CONDITIONS:	REF.:
*SALMO GAIRDNERI	4.2 MILLIEQ/L	CR(VI)=10.PPM,15-22D,10 FISH	1523
*SALMO GAIRDNERI	3.7 MILLIEQ/L	CR(VI)=0MG/L,15-22D,281K,20 FISH	1523
*SALMO GAIRDNERI	4.2 MILLIEQ/L	CR(VI)=10MG/L,15-22D,281K,20 FISH	1523

TABLE C82

CR : KIDNEY WEIGHT			
ITEM:	VALUE:	CONDITIONS:	REF.:
*SALMO GAIRDNERI	0.3 %(G/100G FISH)	CR(VI)=0.PPM,32D,20 FISH	1523
*SALMO GAIRDNERI	0.27 %(G/100G FISH)	CR(VI)=0.1PPM,32D,20 FISH	1523
*SALMO GAIRDNERI	0.27 %(G/100G FISH)	CR(VI)=1.PPM,32D,20 FISH	1523
*SALMO GAIRDNERI	0.33 %(G/100G FISH)	CR(VI)=10.PPM,32D,20 FISH	1523

TABLE C83

CR : LIVER WEIGHT			
ITEM:	VALUE:	CONDITIONS:	REF.:
*SALMO GAIRDNERI	0.92 %(G/100G FISH)	CR(VI)=0.PPM,281K,20 FISH ,2Y	1523
*SALMO GAIRDNERI	10.9 %(G/100G FISH)	CR(VI)=0.1PPM,281K,20 FISH ,2Y	1523
*SALMO GAIRDNERI	10.0 %(G/100G FISH)	CR(VI)=1.0PPM,281K,20 FISH ,2Y	1523
*SALMO GAIRDNERI	0.94 %(G/100G FISH)	CR(VI)=10.PPM,32D,20 FISH	1523

TABLE C84

CR : MORTALITY			
ITEM:	VALUE:	CONDITIONS:	REF.:
*ALGAE	100. %	CR(VI)=0.032-6.4MG/L	659
*CRANGON CRANGON	50%	CRO3,EXPOS.TIME=2DAYS,CONC.IN WATER= 100 MG/L,MARINE	1272
*DAPHNIA SP.	100. %	CR(VI)=0.05MG/L	273
*FISH	100. %	CR(VI)=17.-118.MG/L,DIFF. SPECIES	655
*LEPOMIS MACROCHIRUS	0 %	CR(VI)=45MG/L,20D,HARD WATER	366
*ONCORHYNCHUS KISUTCH	100 %	CR(VI)=31.8MG/L	683
*PHILODINA SP.	50 %	K2CR2O7=31.6 +- 2.5 PPM, 24 H, 25 PPM CACO3, 293 K, PH=7.4-7.9	1373
*PHILODINA SP.	50 %	K2CR2O7=31.2 +- 3.0 PPM, 48 H, 25 PPM CACO3, 293 K, PH=7.4-7.9	1373
*PHILODINA SP.	50 %	K2CR2O7= 3.1 +- 2.1 PPM, 96 H, 25 PPM CACO3, 293 K, PH=7.4-7.9	1373
*PIMEPHALES PROMELAS	50. %	CR(VI)=33MG/L,96H,HARD WATER	273
*PIMEPHALES PROMELAS	0. %	CR(VI)=1.MG/L,96H,HARD WATER	273
*RAT	50%	CHLORIDE,ORAL,DOSE=1870MG/KG BODY	1611
*RAT	50%	NITRATE,ORAL,DOSE=3250MG/KG BODY	1611
*SALMO GAIRDNERI	50. %	CR(VI)=69.MG/L,96H,SOFT WATER	273
*SALMO GAIRDNERI	0. %	CR(VI)=0.3MG/L,96H,SOFT WATER	273
*SALMO GAIRDNERI	50 %	CR(VI)=69MG/L,96H	98

TABLE C84

CR : MORTALITY			
ITEM:	VALUE:	CONDITIONS:	REF.:
*SALMO GAIRDNERI	0 %	CR(VI)=0.35MG/L,8M, CALCIUMCARBONATE=45MG/L,PH=7-8	98
*SALVELINUS FONTANALIS	50 %	CR(VI)=59MG/L,96H	98
*SALVELINUS FONTANALIS	0 %	CR(VI)=0.2MG/L,8M CALCIUMCARBONATE =45MG/L,PH=7-8	98
*SALVELINUS FONTINALIS	50. %	CR(VI)=50MG/L,96H,SOFT WATER	273
*SALVELINUS FONTINALIS	0. %	CR(VI)=0.3MG/L,96H,SOFT WATER	273

TABLE C85

CR : NA IN BLOOD			
ITEM:	VALUE:	CONDITIONS:	REF.:
*SALMO GAIRDNERI	3.6 MILLIEQ/L	CR(VI)=0.PPM,15-22D,10 FISH	1523
*SALMO GAIRDNERI	8.4 MILLIEQ/L	CR(VI)=10.PPM,15-22D,10 FISH	1523
*SALMO GAIRDNERI	151. MILLIEQ/L	CR(VI)=0MG/L,15-22D,281K,20 FISH	1523
*SALMO GAIRDNERI	115.1 MILLIEQ/L	CR(VI)=10MG/L,15-22D,281K,20 FISH	1523

TABLE C86

CR : NUMBER OF RED BLOOD CELLS			
ITEM:	VALUE:	CONDITIONS:	REF.:
*RUTILUS RUTILUS	1.19 10**6MICRO-L	CR(VI)=0MG/L,32D,283K,4*13 FISH	1523
*RUTILUS RUTILUS	1.24 10**6MICRO-L	CR(VI)=0.1MG/L,32D,283K,4*13 FISH	1523
*RUTILUS RUTILUS	1.13 10**6MICRO-L	CR(VI)=1.0MG/L,32D,283K,4*13 FISH	1523
*RUTILUS RUTILUS	1.11 10**6MICRO-L	CR(VI)=10MG/L,32D,283K,4*13 FISH	1523
*SALMO GAIRDNERI	1.19 10**6/MICRO-L	CR(VI)=0.PPM,32D,20 FISH	1523
*SALMO GAIRDNERI	1.24 10**6/MICRO-L	CR(VI)=0.1PPM,32D,20 FISH	1523
*SALMO GAIRDNERI	1.13 10**6/MICRO-L	CR(VI)=1.PPM,32D,20 FISH	1523
*SALMO GAIRDNERI	1.11 10**6/MICRO-L	CR(VI)=10.PPM,32D,20 FISH	1523

TABLE C87

CR : WHITE BLOOD CELLS			
ITEM:	VALU¤¤¤¤¤¤	CONDITIONS:	REF.:
*RUTILUS RUTILUS	98.1 10**3/MICRO-L	CR(VI)=0MG/L,32D,283K,4*13 FISH	1523
*RUTILUS RUTILUS	95.3 10**3/MICRO-L	CR(VI)=0.1MG/L,32D,283K,4*13 FISH	1523
*RUTILUS RUTILUS	82.8 10**3/MICRO-L	CR(VI)=1.0MG/L,32D,283K,4*13 FISH	1523
*RUTILUS RUTILUS	89.6 10**3/MICRO-L	CR(VI)=10MG/L,32D,283K,4*13 FISH	1523
*SALMO GAIRDNERI	98.1 MG/100 ML	CR(VI)=0.PPM,32D,20 FISH	1523
*SALMO GAIRDNERI	95.3 MG/100 ML	CR(VI)=0.1PPM,32D,20 FISH	1523
*SALMO GAIRDNERI	82.8 MG/100 ML	CR(VI)=1.PPM,32D,20 FISH	1523
*SALMO GAIRDNERI	89.6 MG/100 ML	CR(VI)=10.PPM,32D,20 FISH	1523

TABLE C88

CS : MORTALITY			
ITEM:	VALUE:	CONDITIONS:	REF.:
*MOUSE	50%	CARBONATE,INTRAVEN.,DOSE=940MG/KG BODY	1611
*RAT	50%	HYDROXIDE,ORAL,DOSE=1026MG/KG BODY	1611

TABLE C89

CU : CHANGE IN CHROMAGENS (F011)			
ITEM:	VALUE:	CONDITIONS:	REF.:
*BACTERIAL COLONIES	32.5% CHROMAGENS	CONTROL=19.4%,CU=2PPM,AFTER 14D, 299K,BRACKISH WATER(CLEAR L.)	1444
*BACTERIAL COLONIES	10.0 % CHROMAGENS	CONTROL=18.8%,CU=2PPM,AFTER 14D, 299K,FRESH WATER(L.HOUSTON)	1444

TABLE C90

CU : CHANGE IN GROWTH RATE			
ITEM:	VALUE:	CONDITIONS:	REF.:
*PARACENTROTUS LIVIDUS	INHIBITION OF GROWTH OF ARMS	30 UG/L	155

TABLE C91

CU : CHANGE IN NUTRIENT UPTAKE (F009)			
ITEM:	VALUE:	CONDITIONS:	REF.:
*BACTERIAL CELLS	0.2 PPM CO	CU=2.0PPM,CONT = 0.04PPM CO WITH WAT=0.04PPM CO,298K,BRACKISH	579
*BACTERIAL CELLS	0.7 PPM CR	CU=2.0PPM,CONT = 0.1PPM CR WITH WAT=0.4PPM CR,298K,BRACKISH	579
*BACTERIAL CELLS	4.0 PPM CU	CU=2.0PPM,CONT = 1.4PPM CU WITH WAT=1.2PPM CU,298K,BRACKISH	579
*BACTERIAL CELLS	162.0 PPM FE	CU=2.0PPM,CONT = 48.0PPM FE WITH WAT=16.0PPM FE,298K,BRACKISH	579
*BACTERIAL CELLS	0.02 PPM HG	CU=2.0PPM,CONT = 0.08PPM HG WITH WAT=0.008PPM HG,298K,BRACKISH	579
*BACTERIAL CELLS	2.2 PPM MN	CU=2.0PPM,CONT = 0.3PPM MN WITH WAT=0.09PPM MN,298K,BRACKISH	579
*BACTERIAL CELLS	9.5 PPM ZN	CU=2.0PPM,CONT = 0.8PPM ZN WITH WAT=0.5PPM ZN,298K,BRACKISH (?04)	579
*BACTERIAL CELLS	43.0 PPM AL-SI	CU=2.0PPM,CONT = 18.6PPM AL WITH WAT=13.0PPM AL,298K,BRACKISH	579
*BACTERIAL CELLS	13.3 PPM MG	CU=2.0PPM,CONT = 6.3PPM MG WITH WAT=6.2PPM MG,298K,BRACKISH	579
*BACTERIAL CELLS	29.0 PPM TI	CU=2.0PPM,CONT = 8.3PPM TI WITH WAT=1.1PPM TI,298K,BRACKISH	579
*BACTERIAL CELLS	18.8 PPM CA	CU=2.0PPM,CONT = 16.1PPM CA WITH WAT=21.8PPM CA,298K,BRACKISH	579
*BACTERIAL CELLS	21.0 PPM K	CU=2.0PPM,CONT = 13.8PPM K WITH WAT=9.7PPM K,298K,BRACKISH	579
*BACTERIAL CELLS	40.0 PPM NA	CU=2.0PPM,CONT = 13.6PPM NA WITH WAT=19.0PPM NA,298K,BRACKISH	579

TABLE C91

CU : CHANGE IN NUTRIENT UPTAKE (F009)

ITEM:	VALUE:	CONDITIONS:	REF.:
*BACTERIAL CELLS	0.4 PPM BR	CU=2.0PPM,CONT = 0.1PPM BR WITH WAT=0.2PPM BR,298K,BRACKISH	579
*BACTERIAL CELLS	13.4 PPM CL	CU=2.0PPM,CONT = 2.3PPM CL WITH WAT=14.1PPM CL,298K,BRACKISH	579
*BACTERIAL CELLS	0.6 PPM I	CU=2.0PPM,CONT = 0.1PPM I WITH WAT =0.1PPM I,298K,BRACKISH	579
*BACTERIAL CELLS	0.06 PPM CO	CU=2.0PPM,CONT = 0.08PPM CO WITH WAT=0.1PPM CO,298K,ASH R.	579
*BACTERIAL CELLS	0.3 PPM CR	CU=2.0PPM,CONT = 0.4PPM CR WITH WAT=0.2PPM CR,298K,ASH R.	579
*BACTERIAL CELLS	0.3 PPM CU	CU=2.0PPM,CONT = 0.7PPM CU WITH WAT=0.4PPM CU,298K,ASH R.	579
*BACTERIAL CELLS	23.3 PPM FE	CU=2.0PPM,CONT = 21.9PPM FE WITH WAT=16.9PPM FE,298K,ASH R.	579
*BACTERIAL CELLS	0.02 PPM HG	CU=2.0PPM,CONT = 0.02PPM HG WITH WAT=0.03PPM HG,298K,ASH R.	579
*BACTERIAL CELLS	0.05 PPM MN	CU=2.0PPM,CONT = 0.07PPM MN WITH WAT=0.07PPM MN,298K,ASH R.	579
*BACTERIAL CELLS	0.3 PPM ZN	CU=2.0PPM,CONT = 0.6PPM ZN WITH WAT=0.4PPM ZN,298K,ASH R.	579
*BACTERIAL CELLS	32.5 PPM AL-SI	CU=2.0PPM,CONT = 16.8PPM AL WITH WAT=13.0PPM AL,298K,ASH R.	579
*BACTERIAL CELLS	3.8 PPM MG	CU=2.0PPM,CONT = 4.9PPM MG WITH WAT=4.1PPM MG,298K,ASH R.	579
*BACTERIAL CELLS	1.9 PPM TI	CU=2.0PPM,CONT = 1.3PPM TI WITH WAT=0.9PPM TI,298K,ASH R.	579
*BACTERIAL CELLS	5.0 PPM CA	CU=2.0PPM,CONT = 4.2PPM CA WITH WAT=9.2PPM CA,298K,ASH R.	579
*BACTERIAL CELLS	6.9 PPM K	CU=2.0PPM,CONT = 4.6PPM K WITH WAT =6.1PPM K,298K,ASH R.	579
*BACTERIAL CELLS	2.4 PPM NA	CU=2.0PPM,CONT = 19.9PPM NA WITH WAT=7.7PPM NA,298K,ASH R.	579
*BACTERIAL CELLS	0.1 PPM BR	CU=2.0PPM,CONT = 0.1PPM BR WITH WAT=0.1PPM BR,298K,ASH R.	579
*BACTERIAL CELLS	2.3 PPM CL	CU=2.0PPM,CONT = 10.7PPM CL WITH WAT=3.8PPM CL,298K,ASH R.	579
*BACTERIAL CELLS	0.01 PPM I	CU=2.0PPM,CONT = 0.02PPM I WITH WAT=0.1PPM I,298K,ASH R.	579
*BACTERIAL CELLS	0.06 PPM CO	CU=2.0PPM,CONT = 0.03PPM CO WITH WAT=0.06PPM CO,298K,FRESH WAT	579
*BACTERIAL CELLS	0.75 PPM CR	CU=2.0PPM,CONT = 0.65PPM CR WITH WAT=0.25PPM CR,298K,FRESH WAT	579
*BACTERIAL CELLS	3.2 PPM CU	CU=2.0PPM,CONT = 1.2PPM CU WITH WAT=5.4PPM CU,298K,FRESH WAT	579
*BACTERIAL CELLS	33.0 PPM FE	CU=2.0PPM,CONT = 29.0PPM FE WITH WAT=23.0PPM FE,298K,FRESH WAT	579
*BACTERIAL CELLS	0.01 PPM HG	CU=2.0PPM,CONT = 0.02PPM HG WITH WAT=0.01PPM HG,298K,FRESH WAT	579
*BACTERIAL CELLS	0.23 PPM MN	CU=2.0PPM,CONT = 0.07PPM MN WITH WAT=0.03PPM MN,298K,FRESH WAT	579
*BACTERIAL CELLS	4.0 PPM ZN	CU=2.0PPM,CONT = 1.4PPM ZN WITH WAT=1.17PPM ZN,298K,FRESH WAT	579
*BACTERIAL CELLS	8.1 PPM AL-SI	CU=2.0PPM,CONT = 6.8PPM AL WITH WAT=3.0PPM AL,298K,FRESH WAT	579
*BACTERIAL CELLS	22.4 PPM MG	CU=2.0PPM,CONT = 26.9PPM MG WITH WAT=236.0PPM MG,298K,FRESH WAT	579
*BACTERIAL CELLS	2.7 PPM TI	CU=2.0PPM,CONT = 2.8PPM TI WITH WAT=13.8PPM TI,298K,FRESH WAT	579
*BACTERIAL CELLS	44.1 PPM CA	CU=2.0PPM,CONT = 43.2PPM CA WITH WAT=41.0PPM CA,298K,FRESH WAT	579
*BACTERIAL CELLS	12.3 PPM K	CU=2.0PPM,CONT = 41.0PPM K WITH WAT=80.0PPM K,298K,FRESH WAT	579
*BACTERIAL CELLS	84.8 PPM NA	CU=2.0PPM,CONT = 110.0PPM NA WITH WAT=1364.0PPM NA,298K,FRESH WAT	579
*BACTERIAL CELLS	0.5 PPM BR	CU=2.0PPM,CONT = 1.2PPM BR WITH WAT=9.0PPM BR,298K,FRESH WAT	579

TABLE C91

CU : CHANGE IN NUTRIENT UPTAKE			
ITEM:	VALUE:	CONDITIONS:	REF.:
*BACTERIAL CELLS	98.0 PPM CL	CU=2.0PPM,CONT = 123.0PPM CL WITH WAT=1434.0PPM CL,298K,FRESH WAT	579
*BACTERIAL CELLS	0.08 PPM I	CU=2.0PPM,CONT = 0.12PPM I WITH WAT=0.05PPM I,298K,FRESH WAT	579

TABLE C92

CU : CHANGE IN YIELD			
ITEM:	VALUE:	CONDITIONS:	REF.:
*MUSTARD	26.0 G WET WT./PLOT	SOIL-PH=5.7,CU IN SOIL=100PPM	1226
*MUSTARD	24.4 G WET WT./PLOT	SOIL-PH=6.4,CU IN SOIL=100PPM	1226
*MUSTARD	5.6 G WET WT./PLOT	SOIL-PH=5.7,CU IN SOIL=200PPM	1226
*MUSTARD	23.6 G WET WT./PLOT	SOIL-PH=6.4,CU IN SOIL=200PPM	1226
*MUSTARD	1.2 G WET WT./PLOT	SOIL-PH=5.7,CU IN SOIL=500PPM	1226
*MUSTARD	6.4 G WET WT./PLOT	SOIL-PH=6.4,CU IN SOIL=500PPM	1226
*OAT	19.4 G WET WT./PLOT	SOIL-PH=5.7,CU IN SOIL=100PPM	1226
*OAT	19.1 G WET WT./PLOT	SOIL-PH=6.4,CU IN SOIL=100PPM	1226
*OAT	17.1 G WET WT./PLOT	SOIL-PH=5.7,CU IN SOIL=200PPM	1226
*OAT	19.7 G WET WT./PLOT	SOIL-PH=6.4,CU IN SOIL=200PPM	1226
*OAT	7.3 G WET WT./PLOT	SOIL-PH=5.7,CU IN SOIL=500PPM	1226
*OAT	16.4 G WET WT./PLOT	SOIL-PH=6.4,CU IN SOIL=500PPM	1226

TABLE C93

CU : DECREASE IN DIVERSITY (F012)			
ITEM:	VALUE:	CONDITIONS:	REF.:
*WEED	1.48+-0.39 SHANNON DIVERSITY	SOIL-CU=140.6+-96PPM,N=8;CONTROL= 1.96+-0.27,SOIL-CU=17.8PPM,N=7, OHTA	1065
*WEED	1.70+-0.23 SHANNON DIVERSITY	SOIL-CU=92.6+-40PPM,N=8;CONTROL= 1.80+-0.09,SOIL-CU=10.8PPM,N=6, TSURU	1065
*WEED	0.78 SHANNON DIVERSITY	SOIL-CU=218.3PPM,N=4;CONTROL=1.51, SOIL-CU=15.0PPM,N=2,OHTA	1065
*WEED	1.38 SHANNON DIVERSITY	SOIL-CU=63.0PPM,N=4;CONTROL=1.39, SOIL-CU=14.0PPM,N=2,OHTA	1065

TABLE C94

CU : DECREASE IN GROWTH RATE			
ITEM:	VALUE:	CONDITIONS:	REF.:
*ASELLUS MERIDIANUS	0.017 MM/DAY	CONTROL=0.033MM/DAY, CU =0.1 MG/L 10 JUV. ,RIVER HAYLE,U.K.	178
*ASELLUS MERIDIANUS	0.022 MM/DAY	CONTROL=0.022MM/DAY, CU =0.1 MG/L 10 JUV. ,RIVER HAYLE,U.K.	178
*ASELLUS MERIDIANUS	0.006 MM/DAY	CONTROL=0.017MM/DAY, CU =0.1 MG/L 10 JUV. ,RIVER HAYLE,U.K.	178
*ASELLUS MERIDIANUS	0.011 MM/DAY	CONTROL=0.033MM/DAY, CU =0.1 MG/L 10 JUV. ,RIVER GANNAL,U.K.	178
*ASELLUS MERIDIANUS	0.006 MM/DAY	CONTROL=0.028MM/DAY, CU =0.1 MG/L 10 JUV. ,BREDWELL BROOK,U.K.	178

TABLE C94

CU : DECREASE IN GROWTH RATE

ITEM:	VALUE:	CONDITIONS:	REF.:
*ASELLUS MERIDIANUS	0.017 MM/DAY	CONTROL=0.044MM/DAY, CU =0.1 MG/L 10 JUV. ,RIVER GANNAL,U.K.	178
*ASELLUS MERIDIANUS	0.017 MM/DAY	CONTROL=0.033MM/DAY, CU =0.1 MG/L 10 JUV. ,RIVER HAYLE,U.K.	178
*ASELLUS MERIDIANUS	0.028 MM/DAY	CONTROL=0.033MM/DAY, CU =0.1 MG/L 10 JUV. ,RIVER HAYLE,U.K.	178
*ASELLUS MERIDIANUS	0.005 MM/DAY	CONTROL= 0.028MM/DAY, CU =0.1 MG/L 10 JUV. ,RIVER GANNAL,U.K.	178
*ASELLUS MERIDIANUS	0.002 MM/DAY	CONTROL=0.014MM/DAY, CU =0.1 MG/L 10 JUV. ,BREDWELL BROOK,U.K.	178

TABLE C95

CU : EGGS PER FEMALE

ITEM:	VALUE:	CONDITIONS:	REF.:
*PIMEPHALES PROMALES	1272	CU=0.033 MG/L, CONTROL= 1185	184
*PIMEPHALES PROMALES	735	CU=0.065 MG/L, CONTROL= 1185	184
*PIMEPHALES PROMALES	80	CU=0.12 MG/L, CONTROL= 1185	184
*PIMEPHALES PROMALES	0	CU=0.18 MG/L, CONTROL= 1185,NO EGGS AT CU=0.32 AND 0.57 MG/L	184

TABLE C96

CU : EGGS PER SPAWNING

ITEM:	VALUE:	CONDITIONS:	REF.:
*PIMEPHALES PROMALES	138	CU=0.033 MG/L, CONTROL= 174	184
*PIMEPHALES PROMALES	113	CU=0.065 MG/L, CONTROL= 174	184
*PIMEPHALES PROMALES	65	CU=0.12 MG/L, CONTROL= 174	184
*PIMEPHALES PROMALES	0	CU=0.18 MG/L, CONTROL= 174,NO EGGS AT CU=0.32 AND 0.57 MG/L	184

TABLE C97

CU : GROWTH

ITEM:	VALUE:	CONDITIONS:	REF.:
*AGATHIS AUSTRALIS	DEFICIENCY SYMPTOMS AT 0.001 MG/L	TREE, 100 D IN GREENHOUSE, CONC IN CULTURE SOLUTION	1248
*AGATHIS AUSTRALIS	NORMAL GROWTH AT 0.01 MG/L	TREE, 100 D IN GREENHOUSE, CONC IN CULTURE SOLUTION	1248
*AGATHIS AUSTRALIS	TOXIC SYMPTOMS AT 0.2 MG/L	TREE, 100 D IN GREENHOUSE, CONC IN CULTURE SOLUTION	1248

TABLE C98

CU : INCREASE IN GRAIN YIELD

ITEM:	VALUE:	CONDITIONS:	REF.:
*SPRING BARLEY	3.48+-0.16 T/HA AT 85% DRY MATTER	ADDED CUSO4=63-67KG/HA,LIGHT PEAT; CONTROL=2.55T/HA	218

TABLE C98

CU : INCREASE IN GRAIN YIELD

ITEM:	VALUE:	CONDITIONS:	REF.:
*SPRING BARLEY	1.18+-0.095 T/HA AT 85% DRY MATTER	ADDED CUSO4=11KG/HA,LOAMY SAND; CONTROL=0.70T/HA	218
*SPRING BARLEY	1.58+-0.095 T/HA AT 85% DRY MATTER	ADDED CUSO4=22KG/HA,LOAMY SAND; CONTROL=0.70T/HA	218
*SPRING BARLEY	1.55+-0.095 T/HA AT 85% DRY MATTER	ADDED CUSO4=31-34KG/HA,LOAMY SAND; CONTROL=0.70T/HA	218
*SPRING BARLEY	1.49+-0.095 T/HA AT 85% DRY MATTER	ADDED CUSO4=45KG/HA,LOAMY SAND; CONTROL=0.70T/HA	218
*SPRING BARLEY	1.85+-0.095 T/HA AT 85% DRY MATTER	ADDED CUSO4=63-67KG/HA,LOAMY SAND; CONTROL=0.70T/HA	218
*SPRING BARLEY	1.76+-0.095 T/HA AT 85% DRY MATTER	ADDED CUSO4=125-134KG/HA,LOAMY SAND;CONTROL=0.70T/HA	218
*SPRING WHEAT	2.68+-0.105 T/HA AT 85% DRY MATTER	ADDED CUSO4=31-34KG/HA,CALC.PEATY LOAM;CONTROL=1.96T/HA	218
*SPRING WHEAT	2.86+-0.105 T/HA AT 85% DRY MATTER	ADDED CUSO4=63-67KG/HA,CALC.PEATY LOAM;CONTROL=1.96T/HA	218
*SPRING WHEAT	2.75+-0.105 T/HA AT 85% DRY MATTER	ADDED CUSO4=125-134KG/HA, CALC.PEATY LOAM;CONTROL=1.96T/HA	218

TABLE C99

CU : INCREASE IN NUMBERS OF COLONY FORMING UNITS (F011)

ITEM:	VALUE:	CONDITIONS:	REF.:
*BACTERIAL COLONIES	5.25 (LOG-UNIT)	CONTROL=4.62,CU=2PPM,AFTER 14D, 299K,BRACKISH WATER(CLEAR L.)	1444
*BACTERIAL COLONIES	6.50 (LOG-UNIT)	CONTROL=5.25,CU=2PPM,AFTER 14D, 299K,FRESH WATER(L.HOUSTON)	1444

TABLE C100

CU : MORTALITY

ITEM:	VALUE:	CONDITIONS:	REF.:
*ASELLUS MERIDIANUS	50 %	COPPER=1.7MG/L,10 JUVENILES,293K, TEST ANIMALS FROM RIVER HAYLE,U.K	178
*ASELLUS MERIDIANUS	50 %	COPPER=1.9MG/L,10 JUVENILES,293K, TEST ANIMALS FROM RIVER GANNEL, U.K.	178
*ASELLUS MERIDIANUS	50 %	COPPER=1.2MG/L,10 JUVENILES,293K, ANIMALS FROM BREDWELL BROOK,U.K.	178
*CRANGON CRANGON	50%	CUSO4,EXPOS.TIME=2DAYS,CONC.IN WATER = 29.5 MG/L,MARINE	1272
*DAPHNIA MAGNA	50%	CUCL2,EXPOS.TIME=2DAYS,CONC.IN WATER=9.8MICROGRAM/L,LIMNOL.	122
*NOEMACHEILUS BARBARTULUS	50 %	COPPER=0.26MG/L63D,HARD WATER	1477
*PHILODINA SP.	50 %	CUSO4= 1.6 +- 4.2 PPM, 24 H, 25 PPM CACO3, 293 K, PH=7.4-7.9	1373
*PHILODINA SP.	50 %	CUSO4= 0.7 +- 3.3 PPM, 48 H, 25 PPM CACO3, 293 K, PH=7.4-7.9	1373
*PHILODINA SP.	50 %	CUSO4= 0.4 +- 2.2 PPM, 96 H, 25 PPM CACO3, 293 K, PH=7.4-7.9	1373
*RAT	50%	ACETATE,ORAL,DOSE=710MG/KG BODY	1611
*RAT	50%	CARBONATE,ORAL,DOSE=159MG/KG BODY	1611
*RAT	50%	CHLORIDE,ORAL,DOSE=140MG/KG BODY	1611
*RAT	50%	NITRATE,ORAL,DOSE=940MG/KG BODY	1611
*RAT	50%	1-OXIDE,ORAL,DOSE=470MG/KG BODY	1611
*RAT	50%	SULFATE,ORAL,DOSE=300MG/KG BODY	1611

TABLE C100

CU : MORTALITY			
ITEM:	VALUE:	CONDITIONS:	REF.:
*SALMO GAIRDNERI	50 %	COPPER=0.28MG/L,119D,HARD WATER	1477
*SALMO GAIRDNERI	50 %	COPPER(2+)=0.14MG/L,24H,PH=6.5	1423
*SALMO GAIRDNERI	50 %	COPPER(2+)=0.13MG/L,24H,PH=7.5	1423

TABLE C101

CU : PHOTOSYNTHESIS			
ITEM:	VALUE:	CONDITIONS:	REF.:
*AUFWUCHS COMMUNITIES	105 % OF CONTROL	CU= 0.001 MG CU/L, 3 H, GOLDEN HORN	1186
*AUFWUCHS COMMUNITIES	85 % OF CONTROL	CU= 0.01 MG CU/L, 3 H, GOLDEN HORN	1186
*AUFWUCHS COMMUNITIES	82 % OF CONTROL	CU= 0.1 MG CU/L, 3 H, GOLDEN HORN	1186
*AUFWUCHS COMMUNITIES	38 % OF CONTROL	CU= 1.0 MG CU/L, 3 H, GOLDEN HORN	1186
*AUFWUCHS COMMUNITIES	8 % OF CONTROL	CU= 10.0 MG CU/L, 3 H, GOLDEN HORN	1186

TABLE C102

CU : RESPIRATION			
ITEM:	VALUE:	CONDITIONS:	REF.:
*AUFWUCHS COMMUNITIES	100 % OF CONTROL	CU= 0.001 MG CU/L, 3 H, GOLDEN HORN	1186
*AUFWUCHS COMMUNITIES	98 % OF CONTROL	CU= 0.01 MG CU/L, 3 H, GOLDEN HORN	1186
*AUFWUCHS COMMUNITIES	90 % OF CONTROL	CU= 0.1 MG CU/L, 3 H, GOLDEN HORN	1186
*AUFWUCHS COMMUNITIES	80 % OF CONTROL	CU= 1.0 MG CU/L, 3 H, GOLDEN HORN	1186
*AUFWUCHS COMMUNITIES	50 % OF CONTROL	CU= 10.0 MG CU/L, 3 H, GOLDEN HORN	1186

TABLE C103

CU : SPAWNINGS PER FEMALE			
ITEM:	VALUE:	CONDITIONS:	REF.:
*PIMEPHALES PROMALES	8.9	CU=0.033 MG/L, CONTROL= 6.6	184
*PIMEPHALES PROMALES	6.5	CU=0.065 MG/L, CONTROL= 6.6	184
*PIMEPHALES PROMALES	0.8	CU=0.12 MG/L, CONTROL= 6.6	184
*PIMEPHALES PROMALES	0	CU=0.18 MG/L, CONTROL= 174,NO SPAWNINGS AT CU=0.32 AND 0.57 MG/L	184

TABLE C104

CU : SURVIVAL TIME			
ITEM:	VALUE:	CONDITIONS:	REF.:
*DAPHNIA MAGNA	>2880 MIN	CONTROL, DISSOLVED CU=0.02 MICRO M, PH=7.85-8.05,291 K, L. SUPEROIR	32
*DAPHNIA MAGNA	1740 MIN	TOT CU=0.2,DIS CU=0.2MICRO M,L. SUPEROIR,291 K	32

TABLE C104

CU : SURVIVAL TIME

ITEM:	VALUE:	CONDITIONS:	REF.:
*DAPHNIA MAGNA	480 MIN	TOT CU=0.32,DIS CU=0.33MICRO M,L. SUPEROIR,291 K	32
*DAPHNIA MAGNA	372 MIN	TOT CU=0.51,DIS CU=0.41MICRO M,L. SUPEROIR,291 K	32
*DAPHNIA MAGNA	168 MIN	TOT CU=0.8,DIS CU=0.61MICRO M,L. SUPEROIR,291 K	32
*DAPHNIA MAGNA	180 MIN	TOT CU=1.27,DIS CU=1.02MICRO M,L. SUPEROIR,291 K	32
*DAPHNIA MAGNA	84 MIN	TOT CU=2.01,DIS CU=1.54MICRO M,L. SUPEROIR,291 K	32
*DAPHNIA MAGNA	>2880 MIN	CONTROL, DISSOLVED CU =0.02MICRO M + 48 MICRO M NAHCO3	32
*DAPHNIA MAGNA	246 MIN	TOT CU=0.8,DIS CU=0.6 + 48 MICRO M NAHCO3, 291 K, L. SUPEROIR	32
*DAPHNIA MAGNA	120 MIN	TOT CU=1.27,DIS CU=0.98 + 48 MICRO M NAHCO3, 291 K, L. SUPEROIR	32
*DAPHNIA MAGNA	84 MIN	TOT CU=2.01,DIS CU=1.45 + 48 MICRO M NAHCO3, 291 K, L. SUPEROIR	32
*DAPHNIA MAGNA	>2880 MIN	CONTROL, DISSOLVED CU = 0.02 MICRO M + 210 MICRO M NA2HPO4, L. SUPEROIR	32
*DAPHNIA MAGNA	618 MIN	TOT CU=0.8,DIS CU=0.54 + 210 MICRO M NAHPO4, 291 K, L. SUPEROIR	32
*DAPHNIA MAGNA	486 MIN	TOT CU=1.27,DIS CU=0.76 + 210 MICRO M NAHPO4, 291 K, L. SUPEROIR	32
*DAPHNIA MAGNA	316 MIN	TOT CU=2.01,DIS CU=1.07 + 210 MICRO M NAHPO4, 291 K, L. SUPEROIR	32
*DAPHNIA MAGNA	>2880 MIN	CONTROL, DISSOLVED CU = 0.02 MICRO M + 110 MICRO M NA4P2O7, L. SUPEROIR	32
*DAPHNIA MAGNA	>2880 MIN	TOT CU=0.8,DIS CU=0.77 + 110 MICRO M NA4P2O7, 291 K, L. SUPEROIR	32
*DAPHNIA MAGNA	1.092 MIN	TOT CU=1.27,DIS CU=1.24 + 110 MICRO M NA4P2O7, 291 K, L. SUPEROIR	32
*DAPHNIA MAGNA	>2880 MIN	TOT CU=2.01,DIS CU=1.95 + 110 MICRO M NA4P2O7, 291 K, L. SUPEROIR	32
*NOEMACHEILUS BARBARTULUS	>64 DAYS	COPPER=0.002MG/L,285K,10 FISH	1477
*NOEMACHEILUS BARBARTULUS	>64 DAYS	0.12MG/L,285K,10 FISH	1477
*NOEMACHEILUS BARBARTULUS	35.1 DAYS	COPPER=0.29MG/L,285K,10 FISH	1477
*NOEMACHEILUS BARBARTULUS	8.3 DAYS	COPPER=0.39MG/L,285K,10 FISH	1477
*NOEMACHEILUS BARBARTULUS	18.4 DAYS	COPPER=0.49MG/L,285K,10 FISH	1477
*NOEMACHEILUS BARBARTULUS	4.5 DAYS	COPPER=0.71MG/L,285K,10 FISH	1477
*NOEMACHEILUS BARBARTULUS	3.7 DAYS	COPPER=0.76MG/L,285K,10 FISH	1477
*SALMO GAIRDNERI	260 MIN	COPPER(2+)=2.MG/L,PH=6.5	1423
*SALMO GAIRDNERI	340 MIN	COPPER(2+)=1.2MG/L,PH=6.5	1423
*SALMO GAIRDNERI	500 MIN	COPPER(2+)=0.8MG/L,PH=6.5	1423
*SALMO GAIRDNERI	620 MIN	COPPER(2+)=0.5MG/L,PH=6.5	1423
*SALMO GAIRDNERI	640 MIN	COPPER(2+)=0.35MG/L,PH=6.5	1423
*SALMO GAIRDNERI	1000 MIN	COPPER(2+)=0.2MG/L,PH=6.5	1423
*SALMO GAIRDNERI	330 MIN	COPPER(2+)=2.MG/L,PH=7.5	1423
*SALMO GAIRDNERI	400 MIN	COPPER(2+)=1.2MG/L,PH=7.5	1423
*SALMO GAIRDNERI	560 MIN	COPPER(2+)=0.8MG/L,PH=7.5	1423
*SALMO GAIRDNERI	680 MIN	COPPER(2+)=0.5MG/L,PH=7.5	1423
*SALMO GAIRDNERI	740 MIN	COPPER(2+)=0.35MG/L,PH=7.5	1423
*SALMO GAIRDNERI	900 MIN	COPPER(2+)=0.2MG/L,PH=7.5	1423

TABLE C105

CYANATE : INHIBITION OF NITROGEN FIXATION

ITEM:	VALUE:	CONDITIONS:	REF.:
*AZOTOBACTER SP.	INHIBITED	0.14MG N/L	54

TABLE C106

CYANIDE : DECREASE IN NUMBERS OF EGGS PER FEMALE			
ITEM:	VALUE:	CONDITIONS:	REF.:
*PIMEPHALES PROMELAS	2512 EGGS	5.9 MG HCN/M3,CONTROL=3464 EGGS,PH =8.07,256 D,298 K	927
*PIMEPHALES PROMELAS	1845 EGGS	11.14 MG HCN/M3,CONTROL=3464 EGGS, PH=8.07,256 D,298 K	927
*PIMEPHALES PROMELAS	1468 EGGS	17.9 MG HCN/M3,CONTROL=3464 EGGS, PH=8.07,256 D,298 K	927
*PIMEPHALES PROMELAS	1367 EGGS	24.7 MG HCN/M3,CONTROL=3464 EGGS, PH=8.07,256 D,298 K	927
*PIMEPHALES PROMELAS	1010 EGGS	32.8 MG HCN/M3,CONTROL=3464 EGGS, PH=8.07,256 D,298 K	927
*PIMEPHALES PROMELAS	1124 EGGS	40.5 MG HCN/M3,CONTROL=3464 EGGS, PH=8.07,256 D,298 K	927
*PIMEPHALES PROMELAS	72 EGGS	57.5 MG HCN/M3,CONTROL=3464 EGGS, PH=8.07,256 D,298 K	927
*PIMEPHALES PROMELAS	319 EGGS	66.8 MG HCN/M3,CONTROL=3464 EGGS, PH=8.07,256 D,298 K	927
*PIMEPHALES PROMELAS	243 EGGS	75.3 MG HCN/M3,CONTROL=3464 EGGS, PH=8.07,256 D,298 K	927
*PIMEPHALES PROMELAS	0 EGGS	88.9 MG HCN/M3,CONTROL=3464 EGGS, PH=8.07,256 D,298 K	927
*PIMEPHALES PROMELAS	0 EGGS	98.1 MG HCN/M3,CONTROL=3464 EGGS, PH=8.07,256 D,298 K	927

TABLE C107

CYANIDE : MORTALITY			
ITEM:	VALUE:	CONDITIONS:	REF.:
*MOUSE	50%	ORAL DOSE=3MG/KG(=1.11*10**(-4) MOL/KG),PT-50(FOOD)=3.95	953

TABLE C108

DALAPON : MORTALITY			
ITEM:	VALUE:	CONDITIONS:	REF.:
*ICTALURUS PUNCTATUS	LESS THAN 10 %	10 PPM, 48 H, 14 G, 293 K, PH=8.2, ALK=80 PPM	1002

TABLE C109

DAY LENGTH : CHANGE IN DIELDRIN UPTAKE			
ITEM:	VALUE:	CONDITIONS:	REF.:
*SECALE CEREALE	0.11 PPM DIELDRIN (WET WT.)	LONG DAY,GROWN IN SAND WITH 10PPM DIELDRIN	1691
*SECALE CEREALE	0.13 PPM DIELDRIN (WET WT.)	SHORT DAY,GROWN IN SAND WITH 10PPM DIELDRIN	1691

TABLE C110

DDD : MORTALITY			
ITEM:	VALUE:	CONDITIONS:	REF.:
*BOBWHITE	50%	CONC.IN FOOD=2178 MG/KG WW (1835-2584),95% CONF.LIMITS	389

TABLE C110

DDD : MORTALITY

ITEM:	VALUE:	CONDITIONS:	REF.:
*JAPANESE QUAIL	50%	CONC.IN FOOD=3165 MG/KG WW (2534-3978),95% CONF.LIMITS	389
*MALLARD	50%	CONC.IN FOOD=4814 MG/KG WW (3451-7054),95% CONF.LIMITS	389
*PHEASANT	50%	CONC.IN FOOD=579 MG/KG WW (499-668),95% CONF.LIMITS	389

TABLE C111

DDE : CHANGE IN THICKNESS OF EGG SHELL

ITEM:	VALUE:	CONDITIONS:	REF.:
*ANAS RUBRIPES	38 % DECREASE	PP-DDE=30PPM/MEAL(UP TO),6M,MEAN EGG CONTENT=135PPM(UP TO)	283
*FALCO SPARVERIUS	10 % DECREASE	PP-DDE=10PPM/MEAL,MORE THAN 1Y, MEAN EGG CONTENT=32 PPM	283
*MALLARD	13 % DECREASE	PP-DDE=40PPM/MEAL(UP TO),1-2Y,MEAN EGG CONTENT=NOT GIVEN	283
*STREPTOPELIA RISORIA	23 % DECREASE	PP-DDE=80*150MG/KG INJECTIONS,MEAN EGG CONTENT=80PPM/MEAL	283

TABLE C112

DDE : MORTALITY

ITEM:	VALUE:	CONDITIONS:	REF.:
*BOBWHITE	50%	CONC.IN FOOD=825 MG/KG WW (697-976),95% CONF.LIMITS	389
*JAPANESE QUAIL	50%	CONC.IN FOOD=1355 MG/KG WW (1111-1648),95% CONF.LIMITS	389
*MALLARD	50%	CONC.IN FOOD=3572 MG/KG WW (2811-4669),95% CONF.LIMITS	389
*PHEASANT	50%	CONC.IN FOOD=841 MG/KG WW (731-967),95% CONF.LIMITS	389

TABLE C113

DDT : BODY WEIGHT

ITEM:	VALUE:	CONDITIONS:	REF.:
*MOUSE	7.57+-0.86 G	INJECTION OF 50 MG DDT/ KG BODY WEIGHT,16 D,CONT=7.25+-0.76 G	820
*MOUSE	10.86+-1.43 G	INJECTION OF 50 MG DDT/ KG BODY WEIGHT,21 D,CONT=8.96+-0.85 G	820
*MOUSE	22.62+-3.55 G	INJECTION OF 50 MG DDT/ KG BODY WEIGHT,33 D,CONT=21.30+-2.34 G	820

TABLE C114

DDT : CHANGE IN DEVELOPMENT AND CONCENTRATION

ITEM:	VALUE:	CONDITIONS:	REF.:
*RANA TEMPORARIA	19.4 MICROGRAMS/SAMPLE	DDT=0.5PPM,FRESH SPAWN,5D BEFORE HATCHING,180 OVA	282

TABLE C114

DDT : CHANGE IN DEVELOPMENT AND CONCENTRATION			
ITEM:	VALUE:	CONDITIONS:	REF.:
*RANA TEMPORARIA	<0.01 MICROGRAMS/SAMPLE	DDT=0.5PPM,OVA BEGIN.TO ELONGATE, 3D BEFORE HATCHING,300 OVA	282
*RANA TEMPORARIA	<0.01 MICROGRAMS/SAMPLE	DDT=0.5PPM,HEAD-BODY DISTINGUISHABLE,2D BEFORE HATCHING,170 OVA	282

TABLE C115

DDT : CHANGE IN THICKNESS OF EGG SHELL			
ITEM:	VALUE:	CONDITIONS:	REF.:
*CHICKEN	9 % DECREASE	PP-DDT=10PPM/MEAL(UP TO),10W,MEAN EGG CONTENT=NOT GIVEN	283
*COLINUS VIRGINIANUS	4 % DECREASE	PP-DDT=30PPM/MEAL(UP TO),96D,MEAN EGG CONTENT=NOT GIVEN	283
*COTURNIX COTURNIX JAPONICA	6 % DECREASE	PP-DDT=100PP,45D,MEAN EGG CONTENT= 520PPM/MEAL	283
*COTURNIX COTURNIX JAPONICA	7 % DECREASE	PP-DDT=25PPM/MEAL,26W,MEAN EGG CONTENT=NOT GIVEN	283
*GALLUS DOMESTICUS	11 % DECREASE	PP-DDT=100PPM/MEAL,2M,PG=6PPM IN YOLK	283
*LONCHURA STRIATA	7 % DECREASE	PP-DDT=300MICROGRAMS/D MEAL,15W, MEAN EGG CONTENT=NOT GIVEN	283
*MALLARD	13 % DECREASE	PP-DDT=40PPM/MEAL(UP TO),1-2Y,MEAN EGG CONTENT=NOT GIVEN	283
*MALLARD	5 % DECREASE	PP-DDT=30PPM/MEAL(UP TO),96D,MEAN EGG CONTENT=NOT GIVEN	283
*MALLARD	9 % DECREASE	PP-DDT=1000MG/KG D,16D,MEAN EGG CONTENT=NOT GIVEN	283

TABLE C116

DDT : CONC. OF DDT IN HATCHING TADPOLES			
ITEM:	VALUE:	CONDITIONS:	REF.:
*RANA TEMPORARIA	<0.01 MICROGRAMS/SAMPLE	DDT=0.PPM,FRESH SPAWN,5D BEFORE HATCHING,180 OVA	282
*RANA TEMPORARIA	<0.01 MICROGRAMS/SAMPLE	DDT=0.PPM,OVA BEGIN. TO ELONGATE, 3D BEFORE HATCHING,300 OVA	282
*RANA TEMPORARIA	<0.01 MICROGRAMS/SAMPLE	DDT=0.PPM,HEAD-BODY DISTINGUISHABLE,2D BEFORE HATCHING,170 OVA	282

TABLE C117

DDT : DECREASE			
ITEM:	VALUE:	CONDITIONS:	REF.:
*COTURNIX COTURNIX JAPONICA	4 % DECREASE IN THICKNESS OF EGG SHELL	OP-DDT=27PPM, 45D,MEAN EGG CONTENT =27PPM/MEAL	283
*FALCO SPARVERIUS	17 % DECREASE IN THICHNESS OF EGG SHELL	PP-DDT+DIELDRIN=15+3PPM/MEAL(UP TO),1-2Y,MEAN EGG CONTENT=NOT GIVEN	283

TABLE C118

DDT : HEART BEAT			
ITEM:	VALUE:	CONDITIONS:	REF.:
*BIOMPHALARIA GLABRATA	92 BEATS/MIN	DDT=3MICRO-MOL,AFTER 24H,299K,5 SNAILS	264

TABLE C118

DDT : HEART BEAT			
ITEM:	VALUE:	CONDITIONS:	REF.:
*BIOMPHALARIA GLABRATA	30 BEATS/MIN	DDT=116MICRO-MOL,AFTER 24H,299K,5 SNAILS	264
*BIOMPHALARIA GLABRATA	0 BEATS/MIN	DDT=137MICRO-MOL,AFTER 24H,299K,5 SNAILS	264

TABLE C119

DDT : IMMOBILISATION			
ITEM:	VALUE:	CONDITIONS:	REF.:
*DAPHNIA SP.	100 %	DDT=0.002 MG/L, FOR 50 H	83

TABLE C120

DDT : INCREASE IN THYROID WEIGHT			
ITEM:	VALUE:	CONDITIONS:	REF.:
*COLOMBA LIVIA	3 MG/KG*DAY	DDT IN FEED=18-36MG/KG*D,DDT IN LIVER=3.1PPM, THINSHELLED EGGS	746

TABLE C121

DDT : MORTALITY			
ITEM:	VALUE:	CONDITIONS:	REF.:
*BOBWHITE	50%	CONC.IN FOOD=611 MG/KG WW (514-724),95% CONF.LIMITS	389
*BUFO BUFO	3. % (F013),DDT<0.01 MICROGRAMS/ SAMPLE	DDT=0.PPM,MEAN WT.=9.3MG,24H TREATMENT,2D OBS.	282
*BUFO BUFO	6. % (F013),DDT<0.01 MICROGRAMS/ SAMPLE	DDT=0.PPM,MEAN WT.=13.1MG,48H TREATMENT,2D OBS.	282
*BUFO BUFO	15. %	DDT=0.5PPM,MEAN WT.=9.8 OR 13.1MG, 2D AFTER TREATMENT	282
*BUFO BUFO	100. %	DDT=0.5PPM,64.6MG,48H TREATMENT	282
*BUFO BUFO	100. %	DDT=0.5PPM,MEAN WT.=79 OR 101MG,5D AFTER TREATMENT	282
*EARTHWORMS	95%	CONC.IN SOIL:>200 MG/KG	389
*JAPANESE QUAIL	50%	CONC.IN FOOD=568 MG/KG WW (470-687),95% CONF.LIMITS	389
*MALLARD	50%	CONC.IN FOOD=1869 MG/KG WW (1500-2372),95% CONF.LIMITS	389
*PHEASANT	50%	CONC.IN FOOD=311 MG/KG WW (256-374),95% CONF.LIMITS	389
*PHEASANT	50%	CONC.IN FOOD=300.0 MG/KG WW	349
*QUAIL	50%	CONC.IN FOOD=500.0 MG/KG WW	349
*RANA TEMPORARIA	100. % (F013),DDT=53.2 PPM	DDT=0.5PPM,MEAN WT.=37.6MG,48H TREATMENT	282
*RANA TEMPORARIA	100. % (F013),DDT=36.6 PPM	DDT=0.5PPM,MEAN WT.=150MG,48H TREATMENT	282
*RATS	50%	ACUTE ORAL TOXICITY=113-250MG/KG, INSECTICIDE	389
*RATS	50%	ACUTE DERMAL TOXICITY=2500-3000MG/ KG,INSECTICIDE	389
*SLUGS	95%	CONC.IN SOIL:>50 MG/KG	389

TABLE C122

DDT : MORTALITY AND CONCENTRATION (F013)

ITEM:	VALUE:		CONDITIONS:	REF.:
*BUFO BUFO	0. % SAMPLE	,DDT<0.01 MICROGRAMS/	DDT=0.05PPM,MEAN WT.=12.7MG,48H TREAT, FRANTIC BEHAVIOUR,2D OBS.	282
*BUFO BUFO	0. %	,DDT=6.9 PPM	DDT=0.005PPM,MEAN WT.=24.8MG,24H TREAT, FRANTIC BEHAVIOUR	282
*BUFO BUFO	0. %	,DDT=13.5 PPM	DDT=0.005PPM,MEAN WT.=25.2MG,48H TREAT, FRANTIC BEHAVIOUR	282
*BUFO BUFO	0. %	,DDT=65.2 PPM	DDT=0.05PPM,MEAN WT.=24.6MG,24H TREAT, FRANTIC BEHAVIOUR	282
*BUFO BUFO	0. %	,DDT=4.8 PPM	DDT=0.02PPM,MEAN WT.=71.2MG,24H TREAT, FRANTIC BEHAVIOUR	282
*BUFO BUFO	0. %	,DDT=3.5 PPM	DDT=0.02PPM,MEAN WT.=106MG,24H TREAT, FRANTIC BEHAVIOUR, ABNORMAL SNOUTS	282
*BUFO BUFO	0. %	,DDT=7.6 PPM	DDT=0.02PPM,MEAN WT.=92.MG,48H TREAT, FRANTIC-RESIGNED BEHV, ABNORMAL SN	282
*BUFO BUFO	5. %		DDT=0.005PPM,MEAN WT.=24.8 OR 25.2MG,5D AFTER TREAT, FRANTIC BEHAVIOUR	282
*BUFO BUFO	3. %	,DDT=4.3 PPM	DDT=0.02PPM,MEAN WT.=63.7MG,48H TREAT, FRANTIC BEHAVIOUR	282
*BUFO BUFO	5. %		DDT=0.02PPM,MEAN WT.=63.7 OR 71.6MG,5D AFTER TREAT, FRANTIC BEHAVIOUR	282
*BUFO BUFO	90. %		DDT=0.5PPM,MEAN WT.=20.1 OR 21.5MG,5D AFTER TREAT, SHUDDERING BEHAVIOUR	282
*BUFO BUFO	20. %	,DDT=134. PPM	DDT=0.5PPM,MEAN WT.=79MG,48H TREAT, SHUDDERING BEHAVIOUR	282
*BUFO BUFO	0. %	,DDT=139. PPM	DDT=0.05PPM,MEAN WT.=23.0MG,48H TREAT, RESIGNED BEHAVIOUR,7D OBS.	282
*BUFO BUFO	0. %	,DDT=326. PPM	DDT=0.5PPM,MEAN WT.=21.5MG,24H TREAT, RESIGNED-SHUDDERING BEHAVIOUR	282
*BUFO BUFO	0. %	,DDT=73.2 PPM	DDT=0.5PPM,MEAN WT.=101.MG,24H TREAT, RESIGNED-SHUDDERING BEHAVIOUR	282
*BUFO BUFO	0. %	,DDT=96. PPM	DDT=0.5PPM,MEAN WT.=64.6MG,24H TREAT, RESIGNED-SHUDDERING BEHAVIOUR,5D O	282
*BUFO BUFO	0. %	,DDT=478. PPM	DDT=0.5PPM,MEAN WT.=20.1MG,48H TREAT, SHUDDERING BEHAVIOUR	282
*RANA TEMPORARIA	0. %		DDT=0.02PPM,MEAN WT.=22.MG,24H TREAT, FRANTIC BEHAVIOUR,1D OBS.	282
*RANA TEMPORARIA	0. %	,DDT=2.4 PPM	DDT=0.005PPM,MEAN WT.=42.4MG,24H TREAT, FRANTIC BEHAVIOUR,1D OBS.	282
*RANA TEMPORARIA	0. %		DDT=0.02PPM,MEAN WT.=16.8MG,48H TREAT, FRANTIC BEHAVIOUR,1D OBS.	282
*RANA TEMPORARIA	0. %	DDT=2.4 PPM	DDT=0.02PPM,MEAN WT.=127MG,24H TREAT, FRANTIC-RESIGNED BEHV, ABNORMAL SN	282
*RANA TEMPORARIA	0. %	,DDT=75.5 PPM	DDT=).5PPM,MEAN WT.=14.6MG,24H TREAT, RESIGNED BEHV, DEFORMED,2D OBS.	282
*RANA TEMPORARIA	0. %		DDT=0.02PPM,MEAN WT.=15.9MG,24H TREAT, FRANTIC BEHV, DEFORMED,2D OBS.	282
*RANA TEMPORARIA	100. %		DDT=0.5PPM,MEAN WT.=347MG,5-8D AFTER TREAT, ABNORMAL SNOUTS	282
*RANA TEMPORARIA	94. %	,DDT=20.2 PPM	DDT=0.5PPM,MEAN WT.=347MG,48H TREAT, FRANTIC, ABNORMAL SNOUTS	282
*RANA TEMPORARIA	80. %		DDT=0.5PPM,MEAN WT.=17.8MG,48H TREAT, MORIBUND BEHAVIOUR	282
*RANA TEMPORARIA	80. %	,DDT=19.4 PPM	DDT=0.05PPM,MEAN WT.=29.4MG,48H TREAT, MORIBUND BEHAVIOUR,1D OBS.	282
*RANA TEMPORARIA	25. %	,DDT=53.2 PPM	DDT=0.5PPM,MEAN WT.=37.6MG,24H TREAT, MORIBUND BEHAVIOUR,1D OBS.	282
*RANA TEMPORARIA	90. %		DDT=0.5PPM,MEAN WT.=17.8 OR 19.5MG,1D AFTER TREAT, MORIBUND BEHAVIOUR	282
*RANA TEMPORARIA	6. %	,DDT=3.9 PPM	DDT=0.005PPM,MEAN WT.=31.2MG,48H TREATMENT,1D OBS.	282

TABLE C122

DDT : MORTALITY AND CONCENTRATION (F013)

ITEM:	VALUE:	CONDITIONS:	REF.:
*RANA TEMPORARIA	15. % ,DDT=13.1 PPM	DDT=0.05PPM,MEAN WT.=38.1MG,24H TREATMENT,1D OBS.	282
*RANA TEMPORARIA	29. %	DDT=0.02PPM,MEAN WT.=127 OR 134MG, 5D AFTER TREATMENT	282
*RANA TEMPORARIA	13. % DDT=36.6 PPM	DDT=0.5PPM,MEAN WT.=150MG,24H TREAT, RESIGNED-MORIBUND BEHAVIOUR	282
*RANA TEMPORARIA	0. % ,DDT=5.6 PPM	DDT=0.02PPM,MEAN WT.=134MG,48H TREAT, RESIGNED BEHV, ABNORMAL SNOUTS	282
*RANA TEMPORARIA	0. %	DDT=0.5PPM,MEAN WT.=19.5MG,24H TREAT, RESIGNED-MORIBUND BEHAVIOUR	282

TABLE C123

DDT : MORTALITY,FRANTIC-RESIGNED BEHAVIOUR,ABNORMAL SNOUTS

ITEM:	VALUE:	CONDITIONS:	REF.:
*RANA TEMPORARIA	5. %(F013),DDT=20.2 PPM	DDT=0.5PPM,MEAN WT.=347MG,24H TREAT, FRANTIC-RESIGNED BEHV, DEFORMED	282

TABLE C124

DDT : NO EFFECT AND CONCENTRATION (F013)

ITEM:	VALUE:	CONDITIONS:	REF.:
*BUFO BUFO(TADPOLES)	0. % ,DDT<0.01 MICROGRAMS/ SAMPLE	DDT=0.PPM,MEAN WT.=347MG,48H TREATMENT	282
*BUFO BUFO(TADPOLES)	0. % ,DDT<0.01 MICROGRAMS/ SAMPLE	DDT=0.005PPM,MEAN WT.=9.8MG,24H TREATMENT,2D OBS.	282
*BUFO BUFO(TADPOLES)	0. % ,DDT<0.01 MICROGRAMS/ SAMPLE	DDT=0.005PPM,MEAN WT.=12.5MG,48H TREATMENT,2D OBS.	282
*BUFO BUFO(TADPOLES)	0. % ,DDT=30. PPM	DDT=0.05PPM,MEAN WT.=10.MG,24H TREATMENT,2D OBS.	282
*BUFO BUFO(TADPOLES)	0. % ,DDT=82. PPM	DDT=0.5PPM,MEAN WT.=9.8MG,24H TREATMENT	282
*BUFO BUFO(TADPOLES)	0. % ,DDT=306. PPM	DDT=0.5PPM,MEAN WT.=13.1MG,48H TREATMENT	282
*BUFO BUFO(TADPOLES)	0. % ,DDT<0.01 MICROGRAMS/ SAMPLE	DDT=0.PPM,MEAN WT.=26.3MG,24H TREATMENT,7D OBS.	282
*BUFO BUFO(TADPOLES)	0. % ,DDT<0.01 MICROGRAMS/ SAMPLE	DDT=0.PPM,MEAN WT.=24.1MG,48D TREATMENT,7D OBS.	282
*BUFO BUFO(TADPOLES)	0. % ,DDT<0.01 MICROGRAMS/ SAMPLE	DDT=0.PPM,MEAN WT.=72.MG,24H TREATMENT,5D OBS.	282
*BUFO BUFO(TADPOLES)	0. % ,DDT<0.01 MICROGRAMS/ SAMPLE	DDT=0.PPM,MEAN WT.=72.2MG,48H TREATMENT,5D OBS.	282
*BUFO BUFO(TADPOLES)	0. % ,DDT=0.3 PPM	DDT=0.0008PPM,MEAN WT.=68.MG,24H TREATMENT,5D OBS.	282
*BUFO BUFO(TADPOLES)	0. % ,DDT=0.37 PPM	DDT=0.0008PPM,MEAN WT.=68.4MG,48H TREATMENT,5D OBS.	282
*BUFO BUFO(TADPOLES)	0. % ,DDT<0.01 MICROGRAMS/ SAMPLE	DDT=0.PPM,MEAN WT.=105MG,24H TREATMENT,5D OBS.	282
*BUFO BUFO(TADPOLES)	0. % ,DDT<0.01 MICROGRAMS/ SAMPLE	DDT=0.PPM,MEAN WT.=88.MG,48H TREATMENT,5D OBS.	282
*BUFO BUFO(TADPOLES)	0. % ,DDT=0.09 PPM	DDT=0.0008PPM,MEAN WT.=111.MG,24H TREATMENT,5D OBS.	282
*BUFO BUFO(TADPOLES)	0. % ,DDT=0.54 PPM	DDT=0.0008PPM,MEAN WT.=93.MG,48H TREATMENT,5D OBS.	282
*RANA TEMPORARIA	0. %	DDT=0.PPM,MEAN WT.=15.9MG,24H TREATMENT,2D OBS.	282

TABLE C124

DDT : NO EFFECT AND CONCENTRATION (F013)			
ITEM:	VALUE:	CONDITIONS:	REF.:
*RANA TEMPORARIA	0 %	DDT=0.0008PPM,MEAN WT.=14.8MG,24H TREATMENT,2D OBS.	282
*RANA TEMPORARIA	0 %	DDT=0.PPM,MEAN WT.=21.4MG,24H TREATMENT,2D OBS.	282
*RANA TEMPORARIA	0. %	DDT=0.PPM,MEAN WT.=27.9MG,48H TREATMENT,2D OBS.	282
*RANA TEMPORARIA	0. %	DDT=0.0008PPM,MEAN WT.=23.1MG,24H TREATMENT,2D OBS.	282
*RANA TEMPORARIA	0. %	DDT=0.0008PPM,MEAN WT.=26.9MG,48H TREATMENT,1D OBS.	282
*RANA TEMPORARIA	0. % ,DDT<0.01 MICROGRAMS/ SAMPLE	DDT=0.PPM,MEAN WT.=47.4MG,24H TREATMENT,1D OBS.	282
*RANA TEMPORARIA	0. % ,DDT<0.01 MICROGRAMS/ SAMPLE	DDT=0.PPM,MEAN WT.=44.7MG,48H TREATMENT,1D OBS.	282
*RANA TEMPORARIA	0. % ,DDT=0.23 PPM	DDT=0.PPM,MEAN WT.=160MG,24H TREATMENT,5D OBS.	282
*RANA TEMPORARIA	0. % ,DDT<0.01 MICROGRAMS/ SAMPLE	DDT=0.PPM,MEAN WT.=152MG,48H TREATMENT,5D OBS.	282
*RANA TEMPORARIA	0. % ,DDT=0.17 PPM	DDT=0.0008PPM,MEAN WT.=177MG,24H TREATMENT,5D OBS.	282
*RANA TEMPORARIA	0. % ,DDT=0.29 PPM	DDT=0.0008PPM,MEAN WT.=154MG,48H TREATMENT,5D OBS.	282
*RANA TEMPORARIA	0. % ,DDT<0.01 MICROGRAMS/ SAMPLE	DDT=0.PPM,MEAN WT.=347MG,24H TREATMENT,8D OBS.	282

TABLE C125

DEGRADATION : LITTER COMPOSITION			
ITEM:	VALUE:	CONDITIONS:	REF.:
*CA	0.43 % CA OF DRY WEIGHT	CHANGE IN COMPOSITION INITIAL = 0.35 % CA, 70 % DEGRADATED	1180
*K	0.25 % K OF DRY WEIGHT	CHANGE IN COMPOSITION INITIAL = 0.04 % K, 70 % DEGRADATED	1180
*MG	0.06 % MG OF DRY WEIGHT	CHANGE IN COMPOSITION INITIAL = 0.04 % MG, 70 % DEGRADATED	1180
*N	0.79 % N OF DRY WEIGHT	CHANGE IN COMPOSITION INITIAL = 0.96 % N, 70 % DEGRADATED	1180
*P	0.14 % P OF DRY WEIGHT	CHANGE IN COMPOSITION INITIAL = 0.04 % P, 70 % DEGRADATED	1180

TABLE C126

DEPH (DI-2-ETHYLHEXYL-PHTHALATE) : DECREASE IN HEART BEATS			
ITEM:	VALUE:	CONDITIONS:	REF.:
*GOLDFISH	87+-5 BEATS/MIN.	CONC.IN WATER=200PPM	755

TABLE C127

DEPH (DI-2-ETHYLHEXYL-PHTHALATE) : INHIBITION OF REPRODUCTION			
ITEM:	VALUE:	CONDITIONS:	REF.:
*DAPHNIA MAGNA	60%	CONC.IN WATER=3PPB	755
*DAPHNIA MAGNA	70%	CONC.IN WATER=10PPB	755
*DAPHNIA MAGNA	83%	CONC.IN WATER=30PPB	755

TABLE C128

DEPH (DI-2-ETHYLHEXYL-PHTHALATE) : NO EFFECT

ITEM:	VALUE:	CONDITIONS:	REF.:
*DOG	0	ORAL DOSE=0.06G/KG*DAY FOR 2YRS	52
*PIG	0	ORAL DOSE=0.06G/KG*DAY FOR 2YRS	52

TABLE C129

DETERGENTS : MORTALITY

ITEM:	VALUE:	CONDITIONS:	REF.:
*AEDES AEGYPTI	50. %	APH-11E0=500.PPM,24H,40 LARVAE	1627
*AEDES AEGYPTI	100. %	APH-11E0=1000.PPM,24H,40 LARVAE	1627
*AEDES AEGYPTI	100. %	PA-9E0=200.PPM,24H,40 LARVAE	1627
*AEDES AEGYPTI	50. %	PAES=11.PPM,24H,40 LARVAE	1627
*AEDES AEGYPTI	100. %	PAES=12.PPM,24H,40 LARVAE	1627
*AEDES AEGYPTI	50. %	RAS 311=4.PPM,24H,40 LARVAE	1627
*AEDES AEGYPTI	100. %	PAS 311=5.PPM,24H,40 LARVAE	1627
*AEDES AEGYPTI	100. %	TEEPOL CH53=170.PPM,24H,40 LARVAE	1627
*BIOMPHALARIA GLABRATA	100. %	APH-11E0=23.PPM,24H,2*3 SNAILS	1627
*BIOMPHALARIA GLABRATA	100. %	PA-9E0=11.PPM,24H,2*3 SNAILS	1627
*BIOMPHALARIA GLABRATA	100. %	PAES=12.PPM,24H,2*3 SNAILS	1627
*BIOMPHALARIA GLABRATA	100. %	PAS 311=4.PPM,24H,2*3 SNAILS	1627
*BIOMPHALARIA GLABRATA	100. %	TEEPOL CH53=7.PPM,24H,2*3 SNAILS	1627
*LEBISTES RETICULATUS	100. %	APH-11E0=57.PPM,24H,3 FISH , FEMALE	1627
*LEBISTES RETICULATUS	100. %	APH-11E0=64PPM,24H,3 FISH ,MALE	1627
*LEBISTES RETICULATUS	100. %	APH-11E0=52.PPM,24H,5 YOUNG FISH	1627
*LEBISTES RETICULATUS	100. %	PA-9E0=9.PPM,24H,3 FISH ,FEMALE	1627
*LEBISTES RETICULATUS	100. %	PA-9E0=11.PPM,24H,3 FISH ,MALE	1627
*LEBISTES RETICULATUS	100. %	PA-9E0=6.PPM,24H,5 YOUNG FISH	1627
*LEBISTES RETICULATUS	100. %	PAES=5.PPM,24H,3 FISH ,FEMALE	1627
*LEBISTES RETICULATUS	100. %	PAES=8.PPM,24H,3 FISH ,MALE	1627
*LEBISTES RETICULATUS	100. %	PAES=4.PPM,24H,5 YOUNG FISH	1627
*LEBISTES RETICULATUS	100. %	PAS 311=2.PPM,24H,3 FISH ,FEMALE	1627
*LEBISTES RETICULATUS	100. %	PAS 311=2.PPM,24H,3 FISH ,MALE	1627
*LEBISTES RETICULATUS	100. %	PAS 311=0.5PPM,24H,5 YOUNG FISH	1627
*LEBISTES RETICULATUS	100. %	TEEPOL CH53=4.PPM,24H,3 FISH , FEMALE	1627
*LEBISTES RETICULATUS	100. %	TEEPOL CH53=4.PPM,24H,3 FISH,MALE	1627

TABLE C130

DIBUTYL-PHTHALATE : DECREASE IN HEART BEATS

ITEM:	VALUE:	CONDITIONS:	REF.:
*GOLDFISH	130+-15 BEATS/MIN.	CONC.IN WATER=0.5PPM	755
*GOLDFISH	105+-10 BEATS/MIN.	CONC.IN WATER=1PPM	755
*GOLDFISH	60+-10 BEATS/MIN.	CONC.IN WATER=5PPM	755
*GOLDFISH	48+-10 BEATS/MIN.	CONC.IN WATER=12PPM	755
*GOLDFISH	120+-5 BEATS/MIN.	CONC.IN WATER=12PPM,200PPM ATROPIN ADDED	755

TABLE C131

DIBUTYL-PHTHALATE : MORTALITY			
ITEM:	VALUE:	CONDITIONS:	REF.:
*RAT	50%	ORAL DOSE=13.8ML/KG,24HRS	1303
*RAT	50%	PERITONEAL DOSE=8.3ML/KG,24HRS	1303
*RAT	50%	ORAL DOSE=11.9ML/KG,7DAYS	1303
*RAT	50%	PERITONEAL DOSE=6.8ML/KG,7DAYS	1303

TABLE C132

DICHLOBENIL : MORTALITY			
ITEM:	VALUE:	CONDITIONS:	REF.:
*RATS	50%	ACUTE ORAL TOXICITY=3160MG/KG, HERBICIDE	389

TABLE C133

DIELDRIN : DECREASE			
ITEM:	VALUE:	CONDITIONS:	REF.:
*ANAS PLATYRHYNCHOS	4 % DECREASE IN THICKNESS OF EGG SHELL	DIELDRIN=10PPM/MEAL(UP TO),1-2Y, MEAN EGG CONTENT=44PPM(UP TO)	283

TABLE C134

DIELDRIN : IMMOBILISATION			
ITEM:	VALUE:	CONDITIONS:	REF.:
*DAPHNIA SP.	100 %	DIELDRIN=0.25MG/L ,FOR 50 H	83

TABLE C135

DIELDRIN : MORTALITY			
ITEM:	VALUE:	CONDITIONS:	REF.:
*BOBWHITE	50%	CONC.IN FOOD=39MG/KG(37-41),95% CONF.LIMITS	389
*CRASSOSTREA VIRGINICA	50 %	FLOWING SEAWATER = 12.5 UG/L, 96 H	1041
*CYPRINODON VARIEGATUS	50 %	FLOWING SEAWATER = 23.6 UG/L, 96 H	1041
*EARTHWORMS	95%	CONC.IN SOIL:>100 MG/KG	389
*JAPANESE QUAIL	50%	CONC.IN FOOD=57MG/KG(53-61),95% CONF.LIMITS	389
*MALLARD	50%	CONC.IN FOOD=185MG/KG(152-217),95% CONF.LIMITS	389
*PALAEMONETES PUGIO	50 %	FLOWING SEAWATER = 11.4 UG/L, 96 H	1041
*PENAEUS DUORARUM	50 %	FLOWING SEAWATER = 0.9 UG/L, 96 H	1041
*PHEASANT	50%	CONC.IN FOOD=56MG/KG(51-60),95% CONF.LIMITS	389
*PHEASANT	50%	CONC.IN FOOD=50.0MG/KG	349
*QUAIL	50%	CONC.IN FOOD=35.0MG/KG	349
*RATS	50%	ACUTE ORAL TOXICITY=38-87MG/KG, INSECTICIDE	389

TABLE C135

DIELDRIN : MORTALITY			
ITEM:	VALUE:	CONDITIONS:	REF.:
*RATS	50%	ACUTE DERMAL TOXICITY=10-102MG/KG, INSECTICIDE	389

TABLE C136

DIMETHYLARSINIC ACID : MORTALITY			
ITEM:	VALUE:	CONDITIONS:	REF.:
*RATS	50%	ACUTE ORAL TOXICITY=1350MG/KG, PESTICIDE	389

TABLE C137

DIOCTYL-PHTHALATE : MORTALITY			
ITEM:	VALUE:	CONDITIONS:	REF.:
*RAT	50%	ORAL DOSE=45.1ML/KG,24HRS	1303
*RAT	50%	PERITONEAL DOSE=50.5ML/KG,24HRS	1303
*RAT	50%	ORAL DOSE=33.7ML/KG,7DAYS	1303
*RAT	50%	PERITONEAL DOSE=37.3ML/KG,7DAYS	1303

TABLE C138

DISODIUM METHYLARSONATE : MORTALITY			
ITEM:	VALUE:	CONDITIONS:	REF.:
*RATS	50%	ACUTE ORAL TOXICITY=750MG/KG, PESTICIDE	389

TABLE C139

DISSOLVED SUBSTANCES : CHANGE IN REAREATION FROM BUBBLES			
ITEM:	VALUE:	CONDITIONS:	REF.:
*REAREATION IN STREAMS	99 % OF NORMAL	SODIUM CHLORIDE = 1000 MG/L	1270
*REAREATION IN STREAMS	92 % OF NORMAL	GLYCINE = 1000 MG/L	1270
*REAREATION IN STREAMS	90 % OF NORMAL	GLUCOSE = 1000 MG/L	1270
*REAREATION IN STREAMS	87 % OF NORMAL	SOLUBLE STARCH = 1000 MG/L	1270
*REAREATION IN STREAMS	51 % OF NORMAL	PEPTONE = 1000 MG/L	1270
*REAREATION IN STREAMS	47 % OF NORMAL	NUTRIENT BROTH = 1000 MG/L	1270

TABLE C140

DIURON : MORTALITY			
ITEM:	VALUE:	CONDITIONS:	REF.:
*ICTALURUS PUNCTATUS	LESS THAN 10 %	10 PPM, 48 H, 14 G, 293 K, PH=8.2, ALK=80 PPM	1002

TABLE C140

DIURON : MORTALITY			
ITEM:	VALUE:	CONDITIONS:	REF.:
*RATS	50%	ACUTE ORAL TOXICITY=3400MG/KG, HERBICIDE	389

TABLE C141

DNBP : MORTALITY			
ITEM:	VALUE:	CONDITIONS:	REF.:
*ICTALURUS PUNCTATUS	50 %	118 (103-135) PPB, 96 H, 14 G, 293 K,PH=8.2, ALK=80 PPM, 95 %CL	1002

TABLE C142

DSMA : MORTALITY			
ITEM:	VALUE:	CONDITIONS:	REF.:
*ICTALURUS PUNCTATUS	LESS THAN 10 %	10 PPM, 48 H, 14 G, 293 K, PH=8.2, ALK=80 PPM	1002

TABLE C143

DY : MORTALITY			
ITEM:	VALUE:	CONDITIONS:	REF.:
*MOUSE	50%	CHLORIDE,INTRAPERITON.,DOSE=585MG/KG BODY	1611
*RAT	50%	NITRATE,ORAL,DOSE=3100MG/KG BODY	1611

TABLE C144

EDC-TAR : SURVIVAL			
ITEM:	VALUE:	CONDITIONS:	REF.:
*BALANUS BALANOIDES	100 %	EDC-TAR=0.PPM,20D,56 BANACLES	1356
*BALANUS BALANOIDES	95 %	EDC-TAR=1.PPM,20D,76 BANACLES	1356
*BALANUS BALANOIDES	100 %	EDC-TAR=2.PPM,20D,48 BANACLES	1356
*BALANUS BALANOIDES	100 %	EDC-TAR=5.PPM,20D,70 BANACLES	1356
*BALANUS BALANOIDES	45 %	EDC-TAR=10.PPM,20D,72 BANACLES	1356
*BALANUS BALANOIDES	6 %	EDC-TAR=25.PPM,20D,54 BANACLES	1356
*BALANUS BALANOIDES	0 %	EDC-TAR=50.PPM,20D,43 BANACLES	1356
*BALANUS BALANOIDES	0 %	EDC-TAR=100.PPM,20D,46 BANACLES	1356

TABLE C145

ENDOSULFAN : MORTALITY			
ITEM:	VALUE:	CONDITIONS:	REF.:
*RATS	50%	ACUTE ORAL TOXICITY=30-220MG/KG, INSECTICIDE	389

TABLE C146

ENDRIN : MORTALITY			
ITEM:	VALUE:	CONDITIONS:	REF.:
*EARTHWORMS	95%	CONC.IN SOIL=10 MG/KG	389
*ICTALURUS PUNCTATUS	50 %	0.8 (0.5-1.2) PPB, 96 H, 14 G, 293 K,PH=8.2, ALK=80 PPM, 95 %CL	1002
*PHEASANT	50%	CONC.IN FOOD=14.0MG/KG	349
*QUAIL	50%	CONC.IN FOOD=5.0MG/KG	349
*RATS	50%	ACUTE ORAL TOXICITY=7.5-7.8MG/KG, INSECTICIDE	389
*RATS	50%	ACUTE DERMAL TOXICITY=15MG/KG, INSECTICIDE	389
*SLUGS	95%	CONC.IN SOIL:>20 MG/KG	389

TABLE C147

EPINEPHRINE : MORTALITY			
ITEM:	VALUE:	CONDITIONS:	REF.:
*MOUSE	50%	ORAL DOSE=4MG/KG(=2.18*10**(-5) MOL/KG),PT-50(F007)=4.66	953

TABLE C148

EPTC : MORTALITY			
ITEM:	VALUE:	CONDITIONS:	REF.:
*ICTALURUS PUNCTATUS	LESS THAN 10 %	10 PPM, 48 H, 14 G, 293 K, PH=8.2, ALK=80 PPM	1002

TABLE C149

ER : MORTALITY			
ITEM:	VALUE:	CONDITIONS:	REF.:
*MOUSE	50%	CHLORIDE(CITRATECOMPLEX), INTRAPERITON.,DOSE=122MG/KG BODY	1611
*RAT	50%	NITRATE,INTRAPERITON.,DOSE=230MG/KG BODY	1611

TABLE C150

ETHYL MERCURIC CHLORIDE : MORTALITY			
ITEM:	VALUE:	CONDITIONS:	REF.:
*RATS	50%	ACUTE ORAL TOXICITY=30MG/KG, PESTICIDE	389

TABLE C151

EU : MORTALITY			
ITEM:	VALUE:	CONDITIONS:	REF.:
*MOUSE	50%	CHLORIDE,INTRAPERITON.,DOSE=500MG/KG BODY	1611

TABLE C151

EU : MORTALITY			
ITEM:	VALUE:	CONDITIONS:	REF.:
*RAT	50%	NITRATE,INTRAPERITON.,DOSE=210MG/ KG BODY	1611

TABLE C152

FE : DEFICIENCY SYMPTOmS			
ITEM:	VALUE:	CONDITIONS:	REF.:
*AGATHIS AUSTRALIS	0.001 MG/L	TREE, 100 D IN GREENHOUSE, CONC IN CULTURE SOLUTION	1248

TABLE C153

FE : GROWTH			
ITEM:	VALUE:	CONDITIONS:	REF.:
*AGATHIS AUSTRALIS	TOXIC SYMPTOMS AT 12.000 MG/L	TREE, 100 D IN GREENHOUSE, CONC IN CULTURE SOLUTION	1248

TABLE C154

FE : MORTALITY			
ITEM:	VALUE:	CONDITIONS:	REF.:
*MOUSE	50%	2-SULFATE,ORAL,DOSE=1170MG/KG BODY	1611
*RAT	50%	3-NITRATE,ORAL,DOSE=3250MG/KG BODY	1611
*RAT	50%	2-SULFATE,ORAL,DOSE=1480MG/KG BODY	1611
*RAT	50%	3-CHLORIDE,ORAL,DOSE=900MG/KG BODY	1611
*RAT	50%	2-CHLORIDE,ORAL,DOSE=984MG/KG BODY	1611
*RAT	50%	2-GLUCONATE,ORAL,DOSE=4600MG/KG BODY	1611

TABLE C155

FE : NORMAL GROWTH			
ITEM:	VALUE:	CONDITIONS:	REF.:
*AGATHIS AUSTRALIS	1.200 MG/L	TREE, 100 D IN GREENHOUSE, CONC IN CULTURE SOLUTION	1248

TABLE C156

FENAC : MORTALITY			
ITEM:	VALUE:	CONDITIONS:	REF.:
*RATS	50%	ACUTE ORAL TOXICITY=1780MG/KG, HERBICIDE	389

TABLE C157

FLUOMETURON : MORTALITY			
ITEM:	VALUE:	CONDITIONS:	REF.:
*ICTALURUS PUNCTATUS	LESS THAN 10 %	10 PPM, 48 H, 14 G, 293 K, PH=8.2, ALK=80 PPM	1002

TABLE C158

FLUORIDE : MORTALITY			
ITEM:	VALUE:	CONDITIONS:	REF.:
*MOUSE	50%	NA-FLOURIDE, ORAL DOSE=125MG/KG(= 2.98*10**(-3)MOL/KG),PT-50(F007)= 2.53	953
*RAINBOW TROUT	0.01 %	1 MG/L, FRESHWATER	657

TABLE C159

FLUOROACETATE : MORTALITY			
ITEM:	VALUE:	CONDITIONS:	REF.:
*CAT	50%	DOSE=0.2MG/KG,INTRAVEN.	1227
*CLAWED TOAD	50%	DOSE:>500MG/KG,INTRAPERITON.,SOUTH AFRICA	1227
*DOG	50%	DOSE=0.06MG/KG,INTRAVEN.	1227
*FROG	50%	DOSE=150MG/KG,SUBCUT.	1227
*GUINEA PIG	50%	DOSE=0.35MG/KG,INTRAPERITON.	1227
*HORSE	50%	DOSE=1.0MG/KG,ORAL	1227
*HUMAN BEING	50%	DOSE=2-10MG/KG,ORAL	1227
*MOUSE	50%	DOSE=7MG/KG,INTRAPERITON.	1227
*RABBIT	50%	DOSE=0.5MG/KG,INTRAVEN.	1227
*RAT	50%	DOSE=5MG/KG,INTRAPERITON.OR SUBCUT.	1227
*SHEEP	50%	DOSE=0.3MG/KG,ORAL	1227

TABLE C160

FOOD CONCENTRATION : CHANGE IN LIFE TIME			
ITEM:	VALUE:	CONDITIONS:	REF.:
*DAPHNIA HYALINA	8.1 (SD=+-2.88) WEEKS	FOOD=LAKE WATER FILTRED AT 0.6 U,N =17,MEAN LIFE DURATION TIME	194
*DAPHNIA HYALINA	9.6 (SD=+-3.88) WEEKS	FOOD=LAKE WATER THROUGH 3 U,N=14, MEAN LIFE DURATION TIME	194
*DAPHNIA HYALINA	9.8 (SD=+-3.62) WEEKS	FOOD=LAKE WATER FILTRED AT 3 U + 40000 CELLS/ML,N=20,MEAN LIFE TIME	194
*DAPHNIA HYALINA	7.8 (SD=+-2.03) WEEKS	FOOD=LAKE WATER THROUGH 3 U + 200000 CELLS/ML,N=18,MEAN LIFE TIME	194
*DAPHNIA HYALINA	6.1 (SD=+-1.28) WEEKS	FOOD=LAKE WATER,N=16,MEAN LIFE DURATION TIME	194
*DAPHNIA HYALINA	5.9 (SD=+-2.16) WEEKS	FOOD= 400000 CELLS /ML,N=17,MEAN LIFE DURATION TIME	194
*DAPHNIA HYALINA	6.6 (SD=+-1.88) WEEKS	FOOD= 600000 CELLS/ML,N=19,MEAN LIFE DURATION TIME	194

TABLE C161

FOOD CONCENTRATION : SURVIVAL OF LARVAE IN THE 16 FIRST DAYS AFTER HATCHING			
ITEM:	VALUE:	CONDITIONS:	REF.:
*ANCHOA MITCHILLI	63.6 %	FOOD = 5000 CHLORELLA/L, DW PER SURVIVOR = 849.9 UG	695

TABLE C162

FOOD CONCENTRATION : SURVIVAL OF LARVAE IN THE 16 FIRST DAYS AFTER HATCHING			
ITEM:	VALUE:	CONDITIONS:	REF.:
*ACHIRUS LINEATUS	0.0 %	LARVAE = 0.1 NO/L, FOOD = 50 CHLORELLA/L	695

TABLE C162

FOOD CONCENTRATION : SURVIVAL OF LARVAE IN THE 16 FIRST DAYS AFTER HATCHING

ITEM:	VALUE:	CONDITIONS:	REF.:
*ACHIRUS LINEATUS	5.6 %	LARVAE = 0.1 NO/L, FOOD = 53 CHLORELLA/L,DW PER SURVIVOR = 70 UG	695
*ACHIRUS LINEATUS	27.8 %	LARVAE = 0.1 NO/L, FOOD = 100 CHLORELLA/L,DW PER SURVIVOR = 142 UG	695
*ACHIRUS LINEATUS	11.1 %	LARVAE = 0.1 NO/L, FOOD = 106 CHLORELLA/L,DW PER SURVIVOR = 198 UG	695
*ACHIRUS LINEATUS	77.8 %	LARVAE = 0.1 NO/L, FOOD = 1240 CHLORELLA/L,DW PER SURVIVOR = 392 UG	695
*ACHIRUS LINEATUS	66.7 %	LARVAE = 0.1 NO/L, FOOD = 941 CHLORELLA/L,DW PER SURVIVOR = 377 UG	695
*ACHIRUS LINEATUS	0.0 %	LARVAE = 2.0 NO/L, FOOD = 74 CHLORELLA/L	695
*ACHIRUS LINEATUS	0.0 %	LARVAE = 2.0 NO/L, FOOD = 64 CHLORELLA/L	695
*ACHIRUS LINEATUS	10.0 %	LARVAE = 2.0 NO/L, FOOD = 125 CHLORELLA/L,DW PER SURVIVOR = 133 UG	695
*ACHIRUS LINEATUS	4.3 %	LARVAE = 2.0 NO/L, FOOD = 97 CHLORELLA/L,DW PER SURVIVOR = 30 UG	695
*ACHIRUS LINEATUS	58.6 %	LARVAE = 2.0 NO/L, FOOD = 1080 CHLORELLA/L,DW PER SURVIVOR = 253 UG	695
*ACHIRUS LINEATUS	14.3 %	LARVAE = 2.0 NO/L, FOOD = 1110 CHLORELLA/L,DW PER SURVIVOR = 152 UG	695
*ACHIRUS LINEATUS	0.0 %	LARVAE = 8.0 NO/L, FOOD = 60 CHLORELLA/L	695
*ACHIRUS LINEATUS	7.1 %	LARVAE = 8.0 NO/L, FOOD = 48 CHLORELLA/L,DW PER SURVIVOR = 84 UG	695
*ACHIRUS LINEATUS	4.6 %	LARVAE = 8.0 NO/L, FOOD = 94 CHLORELLA/L,DW PER SURVIVOR = 115 UG	695
*ACHIRUS LINEATUS	4.6 %	LARVAE = 8.0 NO/L, FOOD = 76 CHLORELLA/L,DW PER SURVIVOR = 138 UG	695
*ACHIRUS LINEATUS	56.1 %	LARVAE = 8.0 NO/L, FOOD = 982 CHLORELLA/L,DW PER SURVIVOR = 256 UG	695
*ACHIRUS LINEATUS	45.4 %	LARVAE = 8.0 NO/L, FOOD = 985 CHLORELLA/L,DW PER SURVIVOR = 214 UG	695
*ACHIRUS LINEATUS	0.0 %	LARVAE = 16.0 NO/L, FOOD = 72 CHLORELLA/L	695
*ACHIRUS LINEATUS	0.0 %	LARVAE = 16.0 NO/L, FOOD = 67 CHLORELLA/L	695
*ACHIRUS LINEATUS	3.2 %	LARVAE = 16.0 NO/L, FOOD = 119 CHLORELLA/L,DW PER SURVIVOR = 117 UG	695
*ACHIRUS LINEATUS	0.0 %	LARVAE = 16.0 NO/L, FOOD = 125 CHLORELLA/L	695
*ACHIRUS LINEATUS	15.4 %	LARVAE = 16.0 NO/L, FOOD = 984 CHLORELLA/L,DW PER SURVIVOR = 207 UG	695
*ACHIRUS LINEATUS	46.1 %	LARVAE = 16.0 NO/L, FOOD = 1228 CHLORELLA/L,DW PER SURVIVOR = 186 UG	695
*ANCHOA MITCHILLI	5.9 %	FOOD = 50 CHLORELLA/L, DW PER SURVIVOR = 181.7 UG	695
*ANCHOA MITCHILLI	4.7 %	FOOD = 100 CHLORELLA/L, DW PER SURVIVOR = 173.8 UG	695
*ANCHOA MITCHILLI	42.9 %	FOOD = 1000 CHLORELLA/L, DW PER SURVIVOR = 507.9 UG	695
*ARCHIRUS LINEATUS	1.6 %	FOOD = 50 CHLORELLA/L, DW PER SURVIVOR = 77.0 UG	695
*ARCHIRUS LINEATUS	8.2 %	FOOD = 100 CHLORELLA/L, DW PER SURVIVOR = 124.5 UG	695

TABLE C162

FOOD CONCENTRATION : SURVIVAL OF LARVAE IN THE 16 FIRST DAYS AFTER HATCHING			
ITEM:	VALUE:	CONDITIONS:	REF.:
*ARCHIRUS LINEATUS	47.5 %	FOOD = 1000 CHLORELLA/L, DW PER SURVIVOR = 254.7 UG	695

TABLE C163

GA : MORTALITY			
ITEM:	VALUE:	CONDITIONS:	REF.:
*GUINEA PIG	50%	NITRATE,ORAL,DOSE=3280MG/KG BODY	1611
*MOUSE	50%	CHLORIDE,INTRAPERITON.,DOSE=5MG/KG BODY	1611

TABLE C164

GD : MORTALITY			
ITEM:	VALUE:	CONDITIONS:	REF.:
*MOUSE	50%	CHLORIDE(CITRATECOMPLEX), INTRAPERITON.,DOSE=153MG/KG BODY	1611
*RAT	50%	NITRATE,INTRAPERITON.,DOSE=225MG/KG BODY	1611

TABLE C165

GE : MORTALITY			
ITEM:	VALUE:	CONDITIONS:	REF.:
*RAT	50%	OXIDE,INTRAPERITON.,DOSE=750MG/KG BODY	1611

TABLE C166

HCDD : MORTALITY			
ITEM:	VALUE:	CONDITIONS:	REF.:
*RAT	50%	ORAL DOSE=100MG/KG BODY WT.	1408

TABLE C167

HCDD : TERATOGENICITY			
ITEM:	VALUE:	CONDITIONS:	REF.:
*RAT	SOME EFFECT	DOSE=10-100MICROGRAM/KG*DAY,FROM DAY 6-15 IN PREGNANCY-PERIODE	1408
*RAT	NO EFFECT LEVEL	DOSE=0.1MICROGRAM/KG*DAY,FROM DAY 6-15 IN PREGNANCY-PERIODE	1408

TABLE C168

HEPTACHLOR : MORTALITY			
ITEM:	VALUE:	CONDITIONS:	REF.:
*BOBWHITE	50%	CONC.IN FOOD=92 MG/KG WW(76-113), 95% CONF.LIMITS	389

TABLE C168

HEPTACHLOR : MORTALITY			
ITEM:	VALUE:	CONDITIONS:	REF.:
*EARTHWORMS	95%	CONC.IN SOIL=20 MG/KG	389
*JAPANESE QUAIL	50%	CONC.IN FOOD=93 MG/KG WW(74-116), 95% CONF.LIMITS	389
*MALLARD	50%	CONC.IN FOOD=480 MG/KG WW(389-570) ,95% CONF.LIMITS	389
*PHEASANT	50%	CONC.IN FOOD=224 MG/KG WW(191-265) ,95% CONF.LIMITS	389
*PHEASANT	50%	CONC.IN FOOD=150.0-400.0 MG/KG WW	349
*QUAIL	50%	CONC.IN FOOD=125.0 MG/KG WW	349
*RATS	50%	ACUTE ORAL TOXICITY=60-100MG/KG	389
*RATS	50%	ACUTE DERMAL TOXICITY=195MG/KG	389

TABLE C169

HEXACHLORO-1,3-BUTADIEN : G-6-P			
ITEM:	VALUE:	CONDITIONS:	REF.:
*CARASSIUS AURATUS	130 MICRO-M/G OF LIVER/10MIN.	HCBD=0.3PPM(AV.CONC.),CONTR.= 137MYM,49D,6 FISH	903
*CARASSIUS AURATUS	115 MICRO-M/G OF LIVER/10MIN.	HCBD=3.PPM(AV.CONC.),CONTR.= 137MYM,49D,6 FISH	903
*CARASSIUS AURATUS	84 MICRO-M/G OF LIVER/10MIN.	HCBD=9.6PPM(AV.CONC.),CONTR.= 137MYM,49D,6 FISH	903

TABLE C170

HEXACHLORO-1,3-BUTADIEN : G-6-PDH			
ITEM:	VALUE:	CONDITIONS:	REF.:
*CARASSIUS AURATUS	1.7 MICRO-M/G OF LIVER/H	HCBD=0.3PPM(AV.CONC.),CONTR.= 1.6MYH,49D,6 FISH	903
*CARASSIUS AURATUS	2.1 MICRO-M/G OF LIVER/H	HCBD=3.PPM(AV.CONC.),CONTR.= 1.6MYH,49D,6 FISH	903
*CARASSIUS AURATUS	2.3 MICRO-M/G OF LIVER/H	HCBD=9.6PPM(AV.CONC.),CONTR.= 1.6MYH,49D,6 FISH	903

TABLE C171

HEXACHLORO-1,3-BUTADIEN : HEMATOCRIT			
ITEM:	VALUE:	CONDITIONS:	REF.:
*CARASSIUS AURATUS	40.8 %	HCBD=0.3PPM(AV.CONC.),CONTR.=46.3% ,49D,6 FISH	903
*CARASSIUS AURATUS	41.9 %	HCBD=3.PPM(AV.CONC.),CONTR.=46.3%, 49D,6 FISH	903
*CARASSIUS AURATUS	44.6 %	HCBD=9.6PPM(AV.CONC.),CONTR.=46.3% ,49D,6 FISH	903
*CARASSIUS AURATUS	42.8 %	HCBD=30.PPM(AV.CONC.),CONTR.=46.3% ,49D,6 FISH	903

TABLE C172

HEXACHLORO-1,3-BUTADIEN : HEMOGLOBIN			
ITEM:	VALUE:	CONDITIONS:	REF.:
*CARASSIUS AURATUS	8.9 G/100 ML	HCBD=0.3PPM(AV.CONC.),CONTR.= 10.2G/100ML,49D,6 FISH	903

TABLE C172

HEXACHLORO-1,3-BUTADIEN : HEMOGLOBIN			
ITEM:	VALUE:	CONDITIONS:	REF.:
*CARASSIUS AURATUS	9.1 G/100 ML	HCBD=3.PPM(AV.CONC.),CONTR.=10.2G/100ML,49D,6 FISH	903
*CARASSIUS AURATUS	9.4 G/100 ML	HCBD=9.6PPM(AV.CONC.),CONTR.=10.2G/100ML,49D,6 FISH	903
*CARASSIUS AURATUS	8.9 G/100 ML	HCBD=30.PPM(AV.CONC.),CONTR.=10.2G/100ML,49D,6 FISH	903

TABLE C173

HEXACHLORO-1,3-BUTADIEN : INCREASE IN BODY WEIGHT			
ITEM:	VALUE:	CONDITIONS:	REF.:
*CARASSIUS AURATUS	11 %	HCBD=0.3PPM(AV.CONC.),CONTR.=8%,49D,6 FISH	903
*CARASSIUS AURATUS	1 %	HCBD=3.PPM(AV.CONC.),CONTR.=8%,49D,6 FISH	903
*CARASSIUS AURATUS	4 %	HCBD=9.6PPM(AV.CONC.),CONTR.=8%,49D,6 FISH	903
*CARASSIUS AURATUS	-7 %	HCBD=30.PPM(AV.CONC.),CONTR.=8%($AU),49D,6 FISH	903
*CARASSIUS AURATUS	35 %	HCBD=0.3PPM(AV.CONC.),CONTR.=35%($AU),67D,6 FISH	903
*CARASSIUS AURATUS	27 %	HCBD=3.PPM(AV.CONC.),CONTR.=35%($AU),67D,6 FISH	903
*CARASSIUS AURATUS	28 %	HCBD=9.6PPM(AV.CONC.),CONTR.=35%($AU),67D,6 FISH	903
*CARASSIUS AURATUS	6 %	HCBD=30.PPM(AV.CONC.),CONTR.=35%($AU),67D,6 FISH	903

TABLE C174

HEXACHLORO-1,3-BUTADIEN : LIVER WEIGHT			
ITEM:	VALUE:	CONDITIONS:	REF.:
*CARASSIUS AURATUS	4.2 % OF BODY WEIGHT	HCBD=0.3PPM(AV.CONC.),CONTRL.=4.2%,49D,6 FISH	903
*CARASSIUS AURATUS	4.1 % OF BODY WEIGHT	HCBD=3.PPM(AV.CONC.),CONTRL.=4.2%,49D,6 FISH	903
*CARASSIUS AURATUS	5.4 % OF BODY WEIGHT	HCBD=9.6PPM(AV.CONC.),CONTRL.=4.2%,49D,6 FISH	903

TABLE C175

HEXACHLORO-1,3-BUTADIEN : MORTALITY			
ITEM:	VALUE:	CONDITIONS:	REF.:
*ASELLUS AQUATICUS	50 %	HCBD=0.16PPM,72H,290.5K,6*5 AMINALS	903
*ASELLUS AQUATICUS	50 %	HCBD=0.13PPM,96H,290.5K,6*5 AMINALS	903
*CARASSIUS AURATUS	50 %	HCBD=0.09PPM,96H,290.5K,5*5 FISH	903
*LYMNAEA STAGNALIS	50 %	HCBD=0.21PPM,24H,290.5K,5*10 SPECIMENS	903
*LYMNAEA STAGNALIS	50 %	HCBD=0.21PPM,48H,290.5K,5*10 SPECIMENS	903
*LYMNAEA STAGNALIS	50 %	HCBD=0.21PPM,72H,290.5K,5*10 SPECIMENS	903

TABLE C175

HEXACHLORO-1,3-BUTADIEN : MORTALITY			
ITEM:	VALUE:	CONDITIONS:	REF.:
*LYMNAEA STAGNALIS	50 %	HCBD=0.21PPM,96H,290.5K,5*10 SPECIMENS	903

TABLE C176

HEXACHLORO-1,3-BUTADIEN : PHENYLALANINE HYDROXYLASE			
ITEM:	VALUE:	CONDITIONS:	REF.:
*CARASSIUS AURATUS	5.5 MICRO-U/ML	HCBD=0.3PPM(AV.CONC.),CONTR.= 6.7MYU,67D,6 FISH	903
*CARASSIUS AURATUS	7.9 MICRO-U/ML	HCBD=3.PPM(AV.CONC.),CONTR.= 6.7MYU,67D,6 FISH	903
*CARASSIUS AURATUS	9.1 MICRO-U/ML	HCBD=9.6PPM(AV.CONC.),CONTR.= 6.7MYU,67D,6 FISH	903
*CARASSIUS AURATUS	5.8 MICRO-U/ML	HCBD=30.PPM(AV.CONC.),CONTR.= 6.7MYU,67D,6 FISH	903

TABLE C177

HEXACHLORO-1,3-BUTADIEN : PHENYLALANINE HYDROXYLASE ACTIVITY IN LIVER			
ITEM:	VALUE:	CONDITIONS:	REF.:
*CARASSIUS AURATUS	14.6 MICRO-M TYROSINE/G OF LIVER / H	HCBD=0.3PPM(AV.CONC.),CONTR.= 13.9MYT,67D,6 FISH	903
*CARASSIUS AURATUS	15.7 MICRO-M TYROSINE/G OF LIVER / H	HCBD=3.PPM(AV.CONC.),CONTR.= 13.9MYT,67D,6 FISH	903
*CARASSIUS AURATUS	18.1 MICRO-M TYROSINE/G OF LIVER / H	HCBD=9.6PPM(AV.CONC.),CONTR.= 13.9MYT,67D,6 FISH	903
*CARASSIUS AURATUS	21.1 MICRO-M TYROSINE/G OF LIVER / H	HCBD=30.PPM(AV.CONC.),CONTR.= 13.9MYT,67D,6 FISH	903

HEXACHLORO-1,3-BUTADIEN : SERUM GLUTAMIC PYRUVIC TRANSAMINASE			
ITEM:	VALUE:	CONDITIONS:	REF.:
*CARASSIUS AURATUS	4.2 MICRO-U/ML	HCBD=0.3PPM(AV.CONC.),CONTR.= 5.5MYU,67D,6 FISH	903
*CARASSIUS AURATUS	4.2 MICRO-U/ML	HCBD=3.PPM(AV.CONC.),CONTR.= 5.5MYU,67D,6 FISH	903
*CARASSIUS AURATUS	4.9 MICRO-U/ML	HCBD=9.6PPM(AV.CONC.),CONTR.= 5.5MYU,67D,6 FISH	903
*CARASSIUS AURATUS	5.0 MICRO-U/ML	HCBD=30.PPM(AV.CONC.),CONTR.= 5.5MYU,67D,6 FISH	903

TABLE C179

HEXACHLOROBENZENE : MORTALITY			
ITEM:	VALUE:	CONDITIONS:	REF.:
*RATS	50%	ACUTE ORAL TOXICITY=10000MG/KG, FUNGICIDE	389

TABLE C180

HF : MORTALITY			
ITEM:	VALUE:	CONDITIONS:	REF.:
*MOUSE	50%	CHLORIDE,INTRAPERITON.,DOSE=112MG/ KG BODY	1611

TABLE C180

HF : MORTALITY			
ITEM:	VALUE:	CONDITIONS:	REF.:
*MOUSE	50%	OXYCHLORIDE,INTRAPERITON.,DOSE= 113MG/KG BODY	1611
*RAT	50%	GLUCONATE,INTRAPERITON.,DOSE= 2000MG/KG BODY	1611

TABLE C181

HG : CHANGE IN CHROMAGENS (F003)			
ITEM:	VALUE:	CONDITIONS:	REF.:
*BACTERIAL COLONIES	3.8% CHROMAGENS	CONTROL=19.4%,CU=2PPM,AFTER 14D, 299K,BRACKISH WATER(CLEAR L.)	1444
*BACTERIAL COLONIES	12.5% CHROMAGENS	CONTROL=18.8% HG =2PPM,AFTER 14D, 299K,FRESH WATER(L.HOUSTON)	1444

TABLE C182

HG : CHANGE IN NUTRIENT UPTAKE (F004)			
ITEM:	VALUE:	CONDITIONS:	REF.:
*BACTERIAL CELLS	0.08 PPM CO	HG=0.04PPM, CONT =0.04PPM CO WITH WAT =0.04PPM CO,298K, BRACKISH	579
*BACTERIAL CELLS	0.3 PPM CR	HG=0.04PPM, CONT =0.1PPM CR WITH WAT =0.4PPM CR,298K, BRACKISH	579
*BACTERIAL CELLS	3.4 PPM CU	HG=0.04PPM, CONT =1.4PPM CU WITH WAT =1.2PPM CU,298K, BRACKISH	579
*BACTERIAL CELLS	100.0 PPM FE	HG=0.04PPM, CONT =48.0PPM FE WITH WAT =16.0PPM FE,298K, BRACKISH	579
*BACTERIAL CELLS	0.2 PPM HG	HG=0.04PPM, CONT =0.08PPM HG WITH WAT =0.008PPM HG,298K, BRACKISH	579
*BACTERIAL CELLS	1.0 PPM MN	HG=0.04PPM, CONT =0.3PPM MN WITH WAT =0.09PPM MN,298K, BRACKISH	579
*BACTERIAL CELLS	18.0 PPM ZN	HG=0.04PPM, CONT =0.8PPM ZN WITH WAT =0.5PPM ZN,298K, BRACKISH	579
*BACTERIAL CELLS	27.7 PPM AL-SI	HG=0.04PPM, CONT =18.6PPM AL WITH WAT =13.0PPM AL,298K, BRACKISH	579
*BACTERIAL CELLS	40.8 PPM MG	HG=0.04PPM, CONT =6.3PPM MG WITH WAT =6.2PPM MG,298K, BRACKISH	579
*BACTERIAL CELLS	11.0 PPM TI	HG=0.04PPM, CONT =8.3PPM TI WITH WAT =1.1PPM TI,298K, BRACKISH	579
*BACTERIAL CELLS	18.5 PPM CA	HG=0.04PPM, CONT =16.1PPM CA WITH WAT =21.8PPM CA,298K, BRACKISH	579
*BACTERIAL CELLS	23.0 PPM K	HG=0.04PPM, CONT =13.8PPM K WITH WAT =9.7PPM K,298K, BRACKISH	579
*BACTERIAL CELLS	22.0 PPM NA	HG=0.04PPM, CONT =13.6PPM NA WITH WAT =19.0PPM NA,298K, BRACKISH	579
*BACTERIAL CELLS	0.4 PPM BR	HG=0.04PPM, CONT =0.1PPM BR WITH WAT =0.2PPM BR,298K, BRACKISH	579
*BACTERIAL CELLS	6.6 PPM CL	HG=0.04PPM, CONT =2.3PPM CL WITH WAT =14.1PPM CL,298K, BRACKISH	579

TABLE C182

HG : CHANGE IN NUTRIENT UPTAKE (F004)

ITEM:	VALUE:	CONDITIONS:	REF.:
*BACTERIAL CELLS	0.2 PPM I	HG=0.04PPM, CONT =0.1PPM I WITH WAT =0.1PPM I,298K, BRACKISH	579
*BACTERIAL CELLS	0.05 PPM CO	HG=0.04PPM, CONT =0.08PPM CO WITH WAT =0.1PPM CO,298K, ASH R.	579
*BACTERIAL CELLS	0.5 PPM CR	HG=0.04PPM, CONT =0.4PPM CR WITH WAT =0.2PPM CR,298K, ASH R.	579
*BACTERIAL CELLS	0.9 PPM CU	HG=0.04PPM, CONT =0.7PPM CU WITH WAT =0.4PPM CU,298K, ASH R.	579
*BACTERIAL CELLS	26.9 PPM FE	HG=0.04PPM, CONT =21.9PPM FE WITH WAT =16.9PPM FE,298K, ASH R.	579
*BACTERIAL CELLS	0.04 PPM HG	HG=0.04PPM, CONT =0.02PPM HG WITH WAT =0.03PPM HG,298K, ASH R.	579
*BACTERIAL CELLS	0.08 PPM MN	HG=0.04PPM, CONT =0.07PPM MN WITH WAT =0.07PPM MN,298K, ASH R.	579
*BACTERIAL CELLS	0.5 PPM ZN	HG=0.04PPM, CONT =0.6PPM ZN WITH WAT =0.4PPM ZN,298K, ASH R.	579
*BACTERIAL CELLS	13.9 PPM AL-SI	HG=0.04PPM, CONT =16.8PPM AL.WATER =13.0PPM AL,298K, ASH R.	579
*BACTERIAL CELLS	2.9 PPM MG	HG=0.04PPM, CONT =4.9PPM MG WITH WAT =4.1PPM MG,298K, ASH R.	579
*BACTERIAL CELLS	1.3 PPM TI	HG=0.04PPM, CONT =1.3PPM TI WITH WAT =0.9PPM TI,298K, ASH R.	579
*BACTERIAL CELLS	10.4 PPM CA	HG=0.04PPM, CONT =4.2PPM CA WITH WAT =9.2PPM CA,298K, ASH R.	579
*BACTERIAL CELLS	5.1 PPM K	HG=0.04PPM, CONT =4.6PPM K WITH WAT =6.1PPM K,298K, ASH R.	579
*BACTERIAL CELLS	13.2 PPM NA	HG=0.04PPM, CONT =19.9PPM NA WITH WAT =7.7PPM NA,298K, ASH R.	579
*BACTERIAL CELLS	0.09 PPM BR	HG=0.04PPM, CONT =0. PPM BR WITH WAT =0.1PPM BR,298K, ASH R.	579
*BACTERIAL CELLS	11.7 PPM CL	HG=0.04PPM, CONT =10.7PPM CL WITH WAT =3.8PPM CL,298K, ASH R.	579
*BACTERIAL CELLS	0.03 PPM I	HG=0.04PPM, CONT =0.02PPM I WITH WAT =0.1PPM I,298K, ASH R.	579
*BACTERIAL CELLS	0.10 PPM CO	HG=0.04PPM, CONT =0.03PPM CO WITH WAT =0.06PPM CO,298K, FRESH WAT	579
*BACTERIAL CELLS	1.0 PPM CR	HG=0.04PPM, CONT =0.65PPM CR WITH WAT =0.25PPM CR,298K, FRESH WAT	579
*BACTERIAL CELLS	10.0 PPM CU	HG=0.04PPM, CONT =1.2PPM CU WITH WAT =5.4PPM CU,298K, FRESH WAT	579
*BACTERIAL CELLS	52.0 PPM FE	HG=0.04PPM, CONT =29.0PPM FE WITH WAT =23.0PPM FE,298K, FRESH WAT	579
*BACTERIAL CELLS	0.03 PPM HG	HG=0.04PPM, CONT =0.02PPM HG WITH WAT =0.01PPM HG,298K, FRESH WAT	579
*BACTERIAL CELLS	0.48 PPM MN	HG=0.04PPM, CONT =0.07PPM MN WITH WAT =0.03PPM MN,298K, FRESH WAT	579
*BACTERIAL CELLS	4.0 PPM ZN	HG=0.04PPM, CONT =1.4PPM ZN WITH WAT =1.17PPM ZN,298K, FRESH WAT	579
*BACTERIAL CELLS	28.0 PPM AL-SI	HG=0.04PPM, CONT =6.8PPM AL WITH WAT =3.0PPM AL,298K, FRESH WAT	579
*BACTERIAL CELLS	11.0 PPM MG	HG=0.04PPM, CONT =26.9PPM MG WITH WAT =236.0PPM MG,298K, FRESH WAT	579
*BACTERIAL CELLS	4.0 PPM TI	HG=0.04PPM, CONT =2.8PPM TI WITH WAT =13.8PPM TI,298K, FRESH WAT	579
*BACTERIAL CELLS	26.0 PPM CA	HG=0.04PPM, CONT =43.2PPM CA WITH WAT =41.0PPM CA,298K, FRESH WAT	579
*BACTERIAL CELLS	10.0 PPM K	HG=0.04PPM, CONT =41.0PPM K WITH WAT =80.0PPM K,298K, FRESH WAT	579

TABLE C182

HG : CHANGE IN NUTRIENT UPTAKE (F004)			
ITEM:	VALUE:	CONDITIONS:	REF.:
*BACTERIAL CELLS	36.0 PPM NA	HG=0.04PPM, CONT =110PPM NA WITH WAT =1364PPM NA,298K, FRESH WAT	579
*BACTERIAL CELLS	0.8 PPM BR	HG=0.04PPM, CONT =1.2PPM BR WITH WAT =9.0PPM BR,298K, FRESH WAT	579
*BACTERIAL CELLS	27.0 PPM CL	HG=0.04PPM, CONT =123PPM CL WITH WAT =1434PPM CL,298K, FRESH WAT	579
*BACTERIAL CELLS	0.08 PPM I	HG=0.04PPM, CONT =0.12PPM I WITH WAT =0.05PPM I,298K, FRESH WAT	579

TABLE C183

HG : CHANGE IN PHOTOSYNTHESIS			
ITEM:	VALUE:	CONDITIONS:	REF.:
*PHYTOPLANKTON	REDUCTION	0.1 UG/L AS ORGANOMERCURIAL FUNGICIDES	628

TABLE C184

HG : HEMATOCRIT			
IT¤M:	VALUE:	CONDITIONS:	REF.:
*PSEUDOPLEURONECTES AMERICANUS	29. (+-1.5) % PACKED RED CELLS	HG=5PPB,60D,25 FISH	216
*PSEUDOPLEURONECTES AMERICANUS	23. (+-1.4) % PACKED RED CELLS	HG=10PPB,60D,27 FISH	216
*PSEUDOPLEURONECTES AMERICANUS	28. (+-1.6) % PACKED RED CELLS	HG=0.PPB,60D,25 FISH	216

TABLE C185

HG : HEMOGLOBIN			
ITEM:	VALUE:	CONDITIONS:	REF.:
*PSEUDOPLEURONECTES AMERICANUS	5.4 (+-0.2) G/100 ML WHOLE BLOOD	HG=5PPB,60D,22 FISH	216
*PSEUDOPLEURONECTES AMERICANUS	4.2 (+-0.2) G/100 ML WHOLEEBLOOD	HG=10PPB,60D,27 FISH	216
*PSEUDOPLEURONECTES AMERICANUS	5.3 (+-0.3) G/100 ML WHOLE BLOOD	HG=0.PPB,60D,27 FISH	216

TABLE C186

HG : INCREASE IN NUMBERS OF COLONY FORMING UNITS (F003)			
ITEM:	VALUE:	CONDITIONS:	REF.:
*BACTERIAL COLONIES	6.94 (LOG-UNIT)	CONTROL=4.62 HG =2PPM,AFTER 14D, 299K,BRACKISH WATER(CLEAR L.)	1444
*BACTERIAL COLONIES	6.25 (LOG-UNIT)	CONTROL=5.25 HG =2PPM,AFTER 14D, 299K,FRESH WATER(L.HOUSTON)	1444

TABLE C187

HG : MEAN CORPUSCULAR HEMOGLOBIN			
ITEM:	VALUE:	CONDITIONS:	REF.:
*PSEUDOPLEURONECTES AMERICANUS	26.5 (+-1.5) PICOGRAMS/CELL	HG=5PPB,60D,11 FISH	216

TABLE C187

HG : MEAN CORPUSCULAR HEMOGLOBIN			
ITEM:	VALUE:	CONDITIONS:	REF.:
*PSEUDOPLEURONECTES AMERICANUS	23.5 (+-1.9) PICOGRAMS/CELL	HG=10PPB,60D,11 FISH	216
*PSEUDOPLEURONECTES AMERICANUS	30.7 (+-2.) PICOGRAMS/CELL	HG=0.PPB,60D,12 FISH	216
*PSEUDOPLEURONECTES AMERICANUS	18.9 (+-0.4) G/100 ML PACKED RED CELLS	HG=5PPB,60D,22 FISH	216
*PSEUDOPLEURONECTES AMERICANUS	17.8 (+-0.6) G/100 ML PACKED RED CELLS	HG=10PPB,60D,26 FISH	216
*PSEUDOPLEURONECTES AMERICANUS	19.4 (+-0.8) G/100 ML PACKED RED CELLS	HG=0.PPB,60D,29 FISH	216

TABLE C188

HG : MEAN CORPUSCULAR VOLUME			
ITEM:	VALUE:	CONDITIONS:	REF.:
*PSEUDOPLEURONECTES AMERICANUS	145. (+-12.) MICRO**3/CELL	HG=5PPB,60D,12 FISH	216
*PSEUDOPLEURONECTES AMERICANUS	127. (+-6.) MICRO**3/CELL	HG=10PPB,60D,10 FISH	216
*PSEUDOPLEURONECTES AMERICANUS	151. (+-10.5) MICRO**3/CELL	HG=0.PPB,60D,11 FISH	216

TABLE C189

HG : MORTALITY			
ITEM:	VALUE:	CONDITIONS:	REF.:
*CRANGON CRANGON	50%	HGCL2,EXPOS.TIME=2DAYS,CONC.IN WATER=3300-10000MICROGRAM/L, MARINE	1272
*DAPHNIA MAGNA	50%	HGCL2,EXPOS.TIME=2DAYS,CONC.IN WATER=5MICROGRAM/L,LIMNOL.	122
*MOUSE	50%	HYDROXYMETHYL,INTRAPERITON.,DOSE= 17MG/KG BODY	1135
*MOUSE	50%	HGCL2, ORAL DOSE=5MG/KG(=1.84*10** (-5)MOL/KG),PT-50(F007)=4.74	953
*PHILODINA SP.	50 %	HGCL2= 1.0 +- 2.2 PPM, 24 H, 25 PPM CACO3, 293 K, PH=7.4-7.9	1373
*PHILODINA SP.	50 %	HGCL2= 1.3 +- 2.1 PPM, 48 H, 25 PPM CACO3, 293 K, PH=7.4-7.9	1373
*PHILODINA SP.	50 %	HGCL2= 0.7 +- 1.9 PPM, 96 H, 25 PPM CACO3, 293 K, PH=7.4-7.9	1373
*RAT	50%	1-CHLORIDE,ORAL,DOSE=210MG/KG BODY	1611
*RAT	50%	2-CHLORIDE,ORAL,DOSE=37MG/KG BODY	1611
*RAT	50%	CHLOROETHYL,ORAL,DOSE=30MG/KG BODY	1611
*RAT	50%	PHENYLACETATE,ORAL,DOSE=22MG/KG BODY	1135
*RAT	50%	METHOXYETHYLACETATE,ORAL,DOSE= 10MG/KG BODY	1135
*RATS	50%	ACUTE ORAL TOXICITY=37 MG HGCL/KG, PESTICIDE	389
*SALMO GAIRDNERI	50 %	278 K, 0.40 MG HG/L, HGCL	962
*SALMO GAIRDNERI	50 %	283 K, 0.28 MG HG/L, HGCL	962
*SALMO GAIRDNERI	50 %	293 K, 0.22 MG HG/L, HGCL	962

TABLE C190

HG : PHOTOSYNTHESIS			
ITEM:	VALUE:	CONDITIONS:	REF.:
*AUFWUCHS COMMUNITIES	97 % OF CONTROL	HG= 0.001 MG HG/L, 3 H, GOLDEN HORN	1186

TABLE C190

HG : PHOTOSYNTHESIS			
ITEM:	VALUE:	CONDITIONS:	REF.:
*AUFWUCHS COMMUNITIES	93 % OF CONTROL	HG= 0.01 MG HG/L, 3 H, GOLDEN HORN	1186
*AUFWUCHS COMMUNITIES	72 % OF CONTROL	HG= 0.1 MG HG/L, 3 H, GOLDEN HORN	1186
*AUFWUCHS COMMUNITIES	16 % OF CONTROL	HG= 1.0 MG HG/L, 3 H, GOLDEN HORN	1186
*AUFWUCHS COMMUNITIES	0 % OF CONTROL	HG= 10.0 MG HG/L, 3 H, GOLDEN HORN	1186

TABLE C191

HG : PLASMA OSMOLALITY			
ITEM:	VALUE:	CONDITIONS:	REF.:
*PSEUDOPLEURONECTES AMERICANUS	348. (+-4.) MOSM/KG H2O	HG=5PPB,60D,11 FISH	216
*PSEUDOPLEURONECTES AMERICANUS	360. (+-4.) MOSM/KG H2O	HG=10PPB,60D,15 FISH	216
*PSEUDOPLEURONECTES AMERICANUS	362. (+-2.6) MOSM/KG H2O	HG=0.PPB,60D, 12 FISH	216

TABLE C192

HG : PLASMA PROTEIN			
ITEM:	VALUE:	CONDITIONS:	REF.:
*PSEUDOPLEURONECTES AMERICANUS	6.3 (+-0.2) G/100 ML	HG=5PPB,60D,25 FISH	216
*PSEUDOPLEURONECTES AMERICANUS	6. (+-0.2) G/100 ML	HG=10PPB,60D,27 FISH	216
*PSEUDOPLEURONECTES AMERICANUS	5.4 (+-0.2) G/100 ML	HG=0.PPB,60D, 26 FISH	216

TABLE C193

HG : RBC			
ITEM:	VALUE:	CONDITIONS:	REF.:
*PSEUDOPLEURONECTES AMERICANUS	2.1 (+-0.18) MIL. CELLS/MM**3	HG=5PPB,60D,12 FISH	216
*PSEUDOPLEURONECTES AMERICANUS	1.78 (+-0.16) MIL. CELLS/MM**3	HG=10PPB,60D,11 FISH	216
*PSEUDOPLEURONECTES AMERICANUS	1.87 (+-0.16) MIL. CELLS/MM**3	HG=0.PPB,60D,12 FISH	216

TABLE C194

HG : RESPIRATION			
ITEM:	VALUE:	CONDITIONS:	REF.:
*AUFWUCHS COMMUNITIES	94 % OF CONTROL	HG= 0.001 MG HG/L, 3 H, GOLDEN HORN	1186
*AUFWUCHS COMMUNITIES	87 % OF CONTROL	HG= 0.01 MG HG/L, 3 H, GOLDEN HORN	1186
*AUFWUCHS COMMUNITIES	70 % OF CONTROL	HG= 0.1 MG HG/L, 3 H, GOLDEN HORN	1186
*AUFWUCHS COMMUNITIES	60 % OF CONTROL	HG= 1.0 MG HG/L, 3 H, GOLDEN HORN	1186
*AUFWUCHS COMMUNITIES	10 % OF CONTROL	HG= 10.0 MG HG/L, 3 H, GOLDEN HORN	1186

TABLE C195

HG+CU : CHANGE IN CHROMAGENS (F003)			
ITEM:	VALUE:	CONDITIONS:	REF.:
*BACTERIAL COLONIES	12.5% CHROMAGENS	CONTROL=19.4%, CU+HG =2PPM,AFTER 14D,299K,BRACKISH WATER(CLEAR L.)	1444

TABLE C195

HG+CU : CHANGE IN CHROMAGENS (F003)			
ITEM:	VALUE:	CONDITIONS:	REF.:
*BACTERIAL COLONIES	15.6% CHROMAGENS	CONTROL=18.8%, CU+HG =2PPM,AFTER 14D,299K,FRESH WATER(L.HOUSTON)	1444

TABLE C196

HG+CU : INCREASE IN NUMBERS OF COLONY FORMING UNITS (F003)			
ITEM:	VALUE:	CONDITIONS:	REF.:
*BACTERIAL COLONIES	6.38 (LOG-UNIT)	CONTROL=4.62, CU+HG =2PPM,AFTER 14D,299K,BRACKISH WATER(CLEAR L.)	1444
*BACTERIAL COLONIES	6.70 (LOG-UNIT)	CONTROL=5.25, CU+HG =2PPM,AFTER 14D,299K,FRESH WATER(L.HOUSTON)	1444

TABLE C197

HO : MORTALITY			
ITEM:	VALUE:	CONDITIONS:	REF.:
*RAT	50%	NITRATE,ORAL,DOSE=3000MG/KG BODY	1611
*RAT	50%	CHLORIDE,INTRAPERITON.,DOSE=312MG/ KG BODY	1611
*RAT	50%	CHLORIDE(CITRATECOMPLEX), INTRAPERITON.,DOSE=117MG/KG BODY	1611

TABLE C198

HUMIDITY : CHANGE IN DIELDRIN UPTAKE			
ITEM:	VALUE:	CONDITIONS:	REF.:
*SECALE CEREALE	0.07 PPM DIELDRIN (WET WT.)	HIGH HUMIDITY,GROWN IN SAND WITH 10 PPM DIELDRIN	1691
*SECALE CEREALE	0.16 PPM DIELDRIN (WET WT.)	LOW HUMIDITY,GROWN IN SAND WITH 10 PPM DIELDRIN	1691

TABLE C199

IN : MORTALITY			
ITEM:	VALUE:	CONDITIONS:	REF.:
*MOUSE	50%	NITRATE,ORAL,DOSE=3350MG/KG BODY	1611
*RAT	50%	METALLIC,ORAL,DOSE=4200MG/KG BODY	1611
*RAT	50%	CHLORIDE,INTRAPERITON.,DOSE=1.8MG/ KG BODY	419

TABLE C200

IR : MORTALITY			
ITEM:	VALUE:	CONDITIONS:	REF.:
*MOUSE	50%	AS CHLORIDE, ,INTRAPERITON.,DOSE= 56MG/KG BODY	1611

TABLE C201

ISOBRANZAN : MORTALITY			
ITEM:	VALUE:	CONDITIONS:	REF.:
*RATS	50%	ACUTE ORAL TOXICITY=4.8-5.5MG-KG,	389

TABLE C202

K : MORTALITY			
ITEM:	VALUE:	CONDITIONS:	REF.:
*RAT	50%	ACETATE,ORAL,DOSE=3250MG/KG BODY	1611
*RAT	50%	CARBONATE,ORAL,DOSE=1870MG/KG BODY	1611
*RAT	50%	CHLORIDE,INTRAPERITON.,DOSE=825MG/ KG BODY	1611
*RAT	50%	FLOURIDE,ORAL,DOSE=245MG/KG BODY	1611
*RAT	50%	HYDROXIDE,ORAL,DOSE=365MG/KG BODY	1611

TABLE C203

K-ORTHOPHOSPHITE : INHIBITION OF NITROGEN FIXATION			
ITEM:	VALUE:	CONDITIONS:	REF.:
*AZOTOBACTER SP.	100% INHIBITED	0.01M,REVERSIBLE BY ADDITION OF NH4-ACETATE	195

TABLE C204

LA : MORTALITY			
ITEM:	VALUE:	CONDITIONS:	REF.:
*RAT	50%	METALLIC,INTRAVEN.,DOSE=3.5MG/KG BODY	1611
*RAT	50%	ACETATE,ORAL,DOSE=4400MG/KG BODY	1611
*RAT	50%	CHLORIDE,ORAL,DOSE=4200MG/KG BODY	1611
*RAT	50%	NITRATE,ORAL,DOSE=4500MG/KG BODY	1611
*RAT	50%	SULFATE,INTRAPERITON.,DOSE=275MG/ KG BODY	1611
*RAT	50%	OXIDE,ORAL,DOSE=10000MG/KG BODY	419

TABLE C205

LEAD ARSENATE : MORTALITY			
ITEM:	VALUE:	CONDITIONS:	REF.:
*RATS	50%	ACUTE ORAL TOXICITY=800MG/KG, PESTICIDE	389

TABLE C206

LI : MORTALITY			
ITEM:	VALUE:	CONDITIONS:	REF.:
*DAPHNIA MAGNA	50%	LICL,EXPOS.TIME=2DAYS,CONC.IN WATER=7200MICROGRAM/L,MARINE	1004

TABLE C206

LI : MORTALITY			
ITEM:	VALUE:	CONDITIONS:	REF.:
*GUINEA PIG	50%	FLOURIDE,ORAL,DOSE=200MG/KG BODY	1611
*RAT	50%	CARBONATE,ORAL,DOSE=710MG/KG BODY	1611

TABLE C207

LIGHT AND TEMPERATURE : CHANGE IN MN CONTENT			
ITEM:	VALUE:	CONDITIONS:	REF.:
*TOMATO, LEAF	1740 MG/KG DRY WT.	COOL DAY,UPPER LEAVES,SHADED	635
*TOMATO, LEAF	2400 MG/KG DRY WT.	COOL DAY,UPPER LEAVES,UNSHADED	635
*TOMATO, LEAF	2900 MG/KG DRY WT.	COOL DAY,LOWER LEAVES,SHADED	635
*TOMATO, LEAF	3400 MG/KG DRY WT.	COOL DAY,LOWER LEAVES,UNSHADED	635
*TOMATO, LEAF	2000 MG/KG DRY WT.	HOT DAY,UPPER LEAVES,SHADED	635
*TOMATO, LEAF	3090 MG/KG DRY WT.	HOT DAY,UPPER LEAVES,UNSHADED	635
*TOMATO, LEAF	3200 MG/KG DRY WT.	HOT DAY,LOWER LEAVES,SHADED	635
*TOMATO, LEAF	3940 MG/KG DRY WT.	HOT DAY,LOWER LEAVES,UNSHADED	635

TABLE C208

LIGHT INTENSITY : CHANGE IN MN CONTENT			
ITEM:	VALUE:	CONDITIONS:	REF.:
*TOMATO, LEAF	3350 MG/KG DRY WT.	SHADED,EXCHAN.MN(SOIL):0-15CM=78PPM;15-30CM=186PPM,TOXICITY SYMPTOMS	635
*TOMATO, LEAF	5650 MG/KG DRY WT.	UNSHADED,EXCHAN.MN(SOIL):0-15CM=96PPM;15-30CM=189PPM,SEVERE SYMPTOMS	635

TABLE C209

LIMESTONE : DECREASE IN B CONTENT			
ITEM:	VALUE:	CONDITIONS:	REF.:
*TOMATO LEAVES	25 MG/KG DRY WT.	ADDED LIMESTONE=5.6T/HA;CONTROL=55MG/KG DRY WT.	77
*TOMATO LEAVES	17 MG/KG DRY WT.	ADDED LIMESTONE=13.44T/HA;CONTROL=55MG/KG DRY WT.	77

TABLE C210

LIMESTONE DUST : DENSITY IN DUSTY AREA			
ITEM:	VALUE:	CONDITIONS:	REF.:
*ACER PENNSYLVANICUM	30 NO./HA	DENSITY IN CONTROL AREA=660NO./HA, BIG STONY CREEK,VA.,USA	749
*ACER RUBRUM	160 NO./HA	DENSITY IN CONTROL AREA=740NO./HA, BIG STONY CREEK,VA.,USA	749
*ACER SACCHARUM	845 NO./HA	DENSITY IN CONTROL AREA=10NO./HA, BIG STONY CREEK,VA.,USA	749
*AECULUS OCTANDRA	5 NO./HA	DENSITY IN CONTROL AREA=5NO./HA, BIG STONY CREEK,VA.,USA	749

TABLE C210

LIMESTONE DUST : DENSITY IN DUSTY AREA			
ITEM:	VALUE:	CONDITIONS:	REF.:
*CARYA SPP.	235 NO./HA	DENSITY IN CONTROL AREA=140NO./HA, BIG STONY CREEK,VA.,USA	749
*CASTANEA DENTATA	20 NO./HA	DENSITY IN CONTROL AREA=25NO./HA, BIG STONY CREEK,VA.,USA	749
*CERCIS CANADENSIS	915 NO./HA	DENSITY IN CONTROL AREA=75NO./HA, BIG STONY CREEK,VA.,USA	749
*CORNUS FLORIDA	1110 NO./HA	DENSITY IN CONTROL AREA=890NO./HA, BIG STONY CREEK,VA.,USA	749
*FRAXINUS AMERICANA	350 NO./HA	DENSITY IN CONTROL AREA=75NO./HA, BIG STONY CREEK,VA.,USA	749
*HAMAMELIS VIRGINIANA	195 NO./HA	DENSITY IN CONTROL AREA=1315NO./ HA,BIG STONY CREEK,VA.,USA	749
*JUNIPERUS VIRGINIANA	100 NO./HA	DENSITY IN CONTROL AREA=5NO./HA, BIG STONY CREEK,VA.,USA	749
*LIRIODENDRON TULIPIFERA	125 NO./HA	DENSITY IN CONTROL AREA=25NO./HA, BIG STONY CREEK,VA.,USA	749
*MAGNOLIA ACUMINATA	105 NO./HA	DENSITY IN CONTROL AREA=120NO./HA, BIG STONY CREEK,VA.,USA	749
*NYSSA SYLVATICA	85 NO./HA	DENSITY IN CONTROL AREA=325NO./HA, BIG STONY CREEK,VA.,USA	749
*OSTRYA VIRGINIANA	1075 NO./HA	DENSITY IN CONTROL AREA=320NO./HA, BIG STONY CREEK,VA.,USA	7¤9
*OXYDENDRON ARBOREUM	90 NO./HA	DENSITY IN CONTROL AREA=335NO./HA, BIG STONY CREEK,VA.,USA	749
*QUERCUS ALBA	15 NO./HA	DENSITY IN CONTROL AREA=-,BIG STONY CREEK,VA.,USA	749
*QUERCUS PRINUS	30 NO./HA	DENSITY IN CONTROL AREA=735NO./HA, BIG STONY CREEK,VA.,USA	749
*QUERCUS RUBRA	80 NO./HA	DENSITY IN CONTROL AREA=520NO./HA, BIG STONY CREEK,VA.,USA	749
*RHODODENDRON MAXIMUM	25 NO./HA	DENSITY IN CONTROL AREA=470NO./HA, BIG STONY CREEK,VA.,USA	749
*ROBINIA PSEUDOACACIA	85 NO./HA	DENSITY IN CONTROL AREA=110NO./HA, BIG STONY CREEK,VA.,USA	749
*SASSAFRAS ALBIDUM	175 NO./HA	DENSITY IN CONTROL AREA=855NO./HA, BIG STONY CREEK,VA.,USA	749
*TILIA AMERICANA	105 NO./HA	DENSITY IN CONTROL AREA=80NO./HA, BIG STONY CREEK,VA.,USA	749
*ULMUS RUBRA	55 NO./HA	DENSITY IN CONTROL AREA=5NO./HA, BIG STONY CREEK,VA.,USA	749
*VIBURNUM PRUNIFOLIUM	185 NO./HA	DENSITY IN CONTROL AREA=-,BIG STONY CREEK,VA.,USA	749

TABLE C211

LIMESTONE DUST : GROWTH RATE			
ITEM:	VALUE:	CONDITIONS:	REF.:
*TSUGA CANADENSIS	3.6 CM/YEAR	DUST=HEAVY,LOCATION 1,BIG STONY CREEK VALLEY,VIRGINIA,USA.	978
*TSUGA CANADENSIS	3.2 CM/YEAR	DUST=HEAVY,LOCATION 2,BIG STONY GREEK VALLAEY,VIRGINIA,USA.	978
*TSUGA CANADENSIS	5.4 CM/YEAR	DUST=HEAVY,LOCATION 3,BIG STONY CREEK VALLEY,VIRGINIA,USA.	978
*TSUGA CANADENSIS	4.5 CM/YEAR	DUST=HEAVY,LOCATION 4,BIG STONY CREEK VALLEY,VIRGINIA,USA.	978
*TSUGA CANADENSIS	5.6 CM/YEAR	DUST=HEAVY,LOCATION 5,BIG STONY CREEK VALLEY,VIRGINIA,USA.	978
*TSUGA CANADENSIS	8.1 CM/YEAR	DUST=NOT VISIBLE,LOCATION 1,BIG STONY CREEK VALLEY,VIRGINIA,USA.	978
*TSUGA CANADENSIS	7.6 CM/YEAR	DUST=NOT VISIBLE,LOCATION 2,BIG STONY CREEK VALLEY,VIRGINIA,USA.	978
*TSUGA CANADENSIS	7.2 CM/YEAR	DUST=NOT VISIBLE,LOCATION 3,BIG STONY CREEK VALLEY,VIRGINIA,USA.	978

TABLE C211

LIMESTONE DUST : GROWTH RATE			
ITEM:	VALUE:	CONDITIONS:	REF.:
*TSUGA CANADENSIS	8.6 CM/YEAR	DUST=NOT VISIBLE,LOCATION 4,BIG STONY CREEK VALLEY,VIRGINIA,USA.	978
*TSUGA CANADENSIS	12.1 CM/YEAR	DUST=NOT VISIBLE,LOCATION 5,BIG STONY CREEK VALLEY,VIRGINIA,USA.	978

TABLE C212

LIMESTONE DUST : NUMBER OF COLONIES OF BACTERIA (F014)			
ITEM:	VALUE:	CONDITIONS:	REF.:
*SASSAFRAS ALBIDUM	11.7 COLONIES/CM**2 LEAF	DUST=MODERATE,LEAF-PRINTS,BIG STONY CREEK VALLEY,VA.,USA.	978
*SASSAFRAS ALBIDUM	14. COLONIES/G LEAF	DUST=MODERATE,DILUTION-PLATES,BIG STONY CREEK VALLEY,VA.,USA.	978
*SASSAFRAS ALBIDUM	1.8 COLONIES/CM**2 LEAF	DUST=NOT VISIBLE,LEAF-PRINTS,BIG STONY CREEK VALLEY,VA.,USA.	978
*SASSAFRAS ALBIDUM	3. COLONIES/G LEAF	DUST=NOT VISIBLE,DILUTION-PLATES, BIG STONY CREEK VALLEY,VA.,USA.	978
*TSUGA CANADENSIS	8.1 COLONIES/CM**2 LEAF	DUST=HEAVY,LEAF-PRINTS,BIG STONY CREEK VALLEY,VA.,USA.	978
*TSUGA CANADENSIS	2.3 COLONIES/G LEAF	DUST=HEAVY,DILUTION-PLATES,BIG STONY CREEK VALLEY,VA.,USA.	978
*TSUGA CANADENSIS	14. COLONIES/CM**2 LEAF	DUST=NOT VISIBLE,LEAF-PRINTS,BIG STONY CREEK VALEY,VA.,USA.	978
*TSUGA CANADENSIS	7.5 COLONIES/G LEAF	DUST=NOT VISIBLE,DILUTION-PLATES, BIG STONY CREEK VALLEY,VA.,USA.	978
*VITIS VULPINA	0.2 COLONIES/CM**2 LEAF	DUST=HEAVY ENCRUSTED,LEAF-PRINTS, BIG CREEK STONY VALLEY,VA.USA.	978
*VITIS VULPINA	0. COLONIES/G LEAF	DUST=HEAVY ENCRUSTED,LEAF-PRINTS, BIG STONY CREEK VALLEY,VA.,USA.	978
*VITIS VULPINA	9.4 COLONIES/CM**2 LEAF	DUST=MODERATE,LEAF-PRINTS,BIG STONY CREEK VALLEY,VA.,USA.	978
*VITIS VULPINA	13.4 COLONIES/G LEAF	DUST=MODERATE,VITIS VULPINA,BIG STONY CREEK VALLEY,VA.,USA.	978
*VITIS VULPINA	4.3 COLONIES/CM**2 LEAF	DUST=NOT VISIBLE,LEAF-PRINTS,BIG STONY CREEK VALLEY,VA.,USA.	978
*VITIS VULPINA	6.2 COLONIES/G LEAF	DUST=NOT VISIBLE,DILUTION-PLATES, BIG STONY CREEK VALLEY,VA.,USA.	978

TABLE C213

LIMESTONE DUST : NUMBER OF COLONIES OF FUNGI (F015)			
ITEM:	VALUE:	CONDITIONS:	REF.:
*SASSAFRAS ALBIDUM	8.2 COLONIES/CM**2 LEAF	DUST=MODERATE,LEAF-PRINTS,BIG STONY CREEK VALLEY,VA.,USA.	978
*SASSAFRAS ALBIDUM	12.3 COLONIES/G LEAF	DUST=MODERATE,DILUTION-PLATES,BIG STONY CREEK VALLEY,VA.,USA.	978
*SASSAFRAS ALBIDUM	3.4 COLONIES/CM**2 LEAF	DUST=NOT VISIBLE,SASAFRAS ALBIDUM, BIG STONY CREEK VALLEY,VA.,USA.	978
*SASSAFRAS ALBIDUM	4.8 COLONIES/G LEAF	DUST=NOT VISIBLE,DILUTION-PLATES, BIG STONY CREEK VALLEY,VA.,USA.	978
*TSUGA CANADENSIS	19.5 COLONIES/CM**2 LEAF	DUST=HEAVY,LEAF-PRINTS,BIG STONY CREEK VALLEY,VA.,USA.	978
*TSUGA CANADENSIS	29.5 COLONIES/G LEAF	DUST=HEAVY,DILUTION-PLATES,BIG STONY CREEK VALLEY,VA.,USA.	978
*TSUGA CANADENSIS	15.6 COLONIES/CM**2 LEAF	DUST=NOT VISIBLE,LEAF-PRINTS,BIG STONY CREEK VALLEY,VA.,USA.	978
*TSUGA CANADENSIS	5.8 COLONIES/G LEAF	DUST=NOT VISIBLE,DILUTION-PLATES, BIG STONY CREEK VALLEY,VA.,USA.	978

TABLE C213

LIMESTONE DUST : NUMBER OF COLONIES OF FUNGI(F015)			
ITEM:	VALUE:	CONDITIONS:	REF.:
*VITIS VULPINA	0.7 COLONIES/CM**2 LEAF	DUST=HEAVY ENCRUSTED,LEAF-PRINTS, BIG STONY CREEK VALLEY,VA.,USA.	978
*VITIS VULPINA	2.2 COLONIES/G LEAF	DUST=HEAVY ENCRUSTED, DILUTION-PLATES,BIG STONY CREEK VALLEY,VA.,USA	978
*VITIS VULPINA	10.8 COLONIES/CM**2 LEAF	DUST=MODERATE,LEAF-PRINTS,BIG STONY CREEK VALLEY,VA.,USA.	978
*VITIS VULPINA	36.5 COLONIES/G LEAF	DUST=MODERATE,DILUTION-PLATES,BIG STONY CREEK VALLEY,VA.,USA.	978
*VITIS VULPINA	4.4 COLONIES/CM**2 LEAF	DUST=NOT VISIBLE,LEAF-PRINTS,BIG STONY CREEK VALLEY,VA.,USA.	978
*VITIS VULPINA	4.9 COLONIES/G LEAF	DUST=NOT VISIBLE,DILUTION-PLATES, BIG STONY CREEK VALLEY,VA.,USA.	978

TABLE C214

LIMESTONE DUST : OCCURRENCE OF CLADOSPORIUM (F015)			
ITEM:	VALUE:	CONDITIONS:	REF.:
*CLADOSPORIUM SP. ON VITIS VULPINA	26.6 %	DUST=HEAVY,LEAF-PRINTS,BIG STONY CREEK VALLEY,VA.,USA.	978

TABLE C215

LIMESTONE DUST : OCCURRENCE OF FUNGI (F016)			
ITEM:	VALUE:	CONDITIONS:	REF.:
*ALTERNARIA SP. ON SASSAFRAS ALBIDUM	4.2 %	NO DUST,DILUTION-PLATES,BIG STONY CREEK VALLEY,VA.,USA.	978
*ALTERNARIA SP. ON SASSAFRAS ALBIDUM	8.1 %	DUST=MODERATE,DILUTION-PLATES,BIG STONY CREEK VALLEY,VA.,USA.	978
*ALTERNARIA SP. ON SASSAFRAS ALBIDUM	4. %	NO DUST,DILUTION-PLATES,BIG STONY CREEK VALLEY,VA.,USA.	978
*ALTERNARIA SP. ON TSUGA CANADENSIS	28. %	NO DUST,LEAF-PRINTS,BIG STONY CREEK VALLEY,VA.,USA.	978
*ALTERNARIA SP. ON VITIS VULPINA	6.3 %	DUST=MODERATE,VITIS VULPINA,BIG STONY CREEK VALLEY,VA.,USA.	978
*ALTERNARIA SP. ON VITIS VULPINA	2.8 %	DUST=MODERATE,LEAF-PRINTS,BIG STONY CREEK VALLEY,VA.,USA.	978
*BLACK MYCELIUM ON SASSAFRAS ALBIDUM	4.2 %	DUST=MODERATE,DILUTION-PLATES,BIG STONY CREEK VALLEY,VA.,USA.	978
*BLACK MYCELIUM ON TSUGA CANADENSIS	6.8 %	DUST=HEAVY,LEAF-PRINTS,BIG STONY CREEK VALLEY,VA.,USA.	978
*BLACK MYCELIUM ON VITIS VULPINA	6.8 %	DUST=HEAVY,LEAF-PRINTS,BIG STONY CREEK VALLEY,VA.,USA.	978
*BLACK MYCELIUM ON VITIS VULPINA	4.2 %	NO DUST,LEAF-PRINTS,BIG STONY CREEK VALLEY,VA.,USA.	978
*BROWN MYCELIUM ON SASSAFRAS ALBIDUM	0.6 %	DUST=HEAVY,DILUTION-PLATES,BIG STONY CREEK VALLEY,VA.,USA.	978
*BROWN MYCELIUM ON VITIS VULPINA	20. %	DUST=HEAVY,LEAF-PRINTS,BIG STONY CREEK VALLEY,VA.,USA.	978
*BROWN MYCELIUM ON VITIS VULPINA	23.8 %	DUST=MODERATE,LEAF-PRINTS,BIG STONY CREEK VALLEY,VA.,USA.	978
*BROWN MYCELIUM ON VITIS VULPINA	24.9 %	LAEF-PRINTS,LEAF-PRINTS,BIG STONY CREEK VALLEY,VA.,USA.	978
*CANDIDA SP. ON CANDIDA SP. ON VITIS VULPINA	15.6 %	LEAD-PRINTS,LEAF-PRINTS,BIG STONY CREEK VALLEY,VA.,USA.	978
*CANDIDA SP. ON SASSAFRAS ALBIDUM	97.6 %	DUST=HEAVY,DILUTION-PLATES,BIG STONY CREEK VALLEY,VA.,USA.	978
*CANDIDA SP. ON SASSAFRAS ALBIDUM	12.2 %	NO DUST,DILUTION-PLATES,BIG STONY CREEK VALLEY,VA.,USA.	978

TABLE C215

LIMESTONE DUST : OCCURRENCE OF FUNGI (F016)

ITEM:	VALUE:	CONDITIONS:	REF.:
*CANDIDA SP. ON TSUGA CANADENSIS	2.3 %	DUST=HEAVY,DILUTION-PLATES,BIG STONY CREEK VALLEY,VA.,USA.	978
*CANDIDA SP. ON TSUGA CANADENSIS	6.8 %	NO DUST,DILUTION-PLATES,BIG STONY CREEK VALLEY,VA.,USA.	978
*CANDIDA SP. ON VITIS VULPINA	20. %	DUST=HEAVY,LEAF-PRINTS,BIG STONY CREEK VALLEY,VA.,USA.	978
*CLADOSPORIUM SP. ON TSUGA CANADENSIS	26.9 %	DUST=HEAVY,LEAF-PRINTS,BIG STONY CREEK VALLEY,VA.,USA.	978
*CLADOSPORIUMSP. ON SASSAFRAS ALBIDUM	1.2 %	DUST=HEAVY,DILUTION-PLATES, BIGTSTONY CREEK VALLEY,VA.,USA.	978
*CLADOSPORIUMSP. ON SASSAFRAS ALBIDUM	16.4 %	DUST=MODERATE,DILUTION-PLATES,BIG STONY CREEK VALLEY,VA.,USA.	978
*CLADOSPORIUMSP. ON SASSAFRAS ALBIDUM	18.5 %	NO DUST,DILUTION-PLATES,BIG STONY CREEK VALLEY,VA.,USA.	978
*CLADOSPORIUMSP. ON SASSAFRAS ALBIDUM	32.5 %	DUST=MODERATE,DILUTION-PLATES,BIG STONY CREEK VALLEY,VA.,USA.	978
*CLADOSPORIUMSP. ON SASSAFRAS ALBIDUM	8.7 %	NO DUST,DILUTION-PLATES,BIG STONY CREEK VALLEY,VA.,USA	978
*CLADOSPORIUMSP. ON TSUGA CANADENSIS	19.4 %	NO DUST,TSYGA CANADENSIS,BIG STONY CREEK VALLEY,VA.,USA.	978
*CLADOSPORIUMSP. ON TSUGA CANADENSIS	11.7 %	DUST=HEAVY,DILUTION-PLATES,BIG STONY CREEK VALLEY,VA.,USA.	978
*CLADOSPORIUMSP. ON TSUGA CANADENSIS	13.6 %	NO DUST,DILUTION-PLATES,BIG STONY CREEK VALLEY,VA.,USA.	978
*CLADOSPORIUMSP. ON VITIS VULPINA	2.3 %	DUST=MODERATE,LEAF-PRINTS,BIG STONY CREEK VALLEY,VA.,USA.	978
*COLLETOTRICHUM SP. ON SASSAFRAS ALBIDUM	8.4 %	NO DUST,DILUTION-PLATES,BIG STONY CREEK VALLEY,VA.,USA.	978
*COLLETOTRICHUM SP. ON SASSAFRAS ALBIDUM	6.2 %	NO DUST,DILUTION-PLATES,BIG STONY CREEK VALLEY,VA.,USA.	978
*COLLETOTRICHUM SP. ON TSUGA CANADENSIS	3.5 %	DUST=HEAVY,LEAF-PRINTS,BIG STONY CREEK VALLEY,VA.,USA.	978
*COLLETOTRICHUM SP. ON VITIS VULPINA	31.8 %	DUST=MODERATE,SASSAFRAS ALBIDUM, BIG STONY CREEK VALLEY,VA.,USA.	978
*COLLETOTRICHUM SP. ON VITIS VULPINA	21.6 %	NO DUST,LEAF-PRINTS,BIG STONY CREEK VALLEY,VA.,USA.	978
*CRYPTOCOCCUS SP. ON TSUGA CANADENSIS	28.2 %	DUST=HEAVY,LEAF-PRINTS,BIG STONY CREEK VALLEY,VA.,USA.	978
*CRYPTOCOCCUS SP. ON TSUGA CANADENSIS	22.8 %	NO DUST,LEAF-PRINTS,BIG STONY CREEK VALLEY,VA.,USA.	978
*CRYPTOCOCCUS SP. ON TSUGA CANADENSIS	10.1 %	DUST=HEAVY,DILUTION-PLATES,BIG STONY CREEK VALLEY,VA.,USA.	978
*CRYPTOCOCCUS SP. ON TSUGA CANADENSIS	13.7 %	NO DUST,DILUTION-PLATES,BIG STONY CREEK VALLEY,VA.,USA.	978
*CRYPTOCOCCUS SP. ON VITIS VULPINA	13.2 %	DUST=MODERATE,LEAF-PRINTS,BIG STONY CREEK VALLEY,VA.,USA.	978
*CRYPTOCOCCUS SP. ON VITIS VULPINA	25.2 %	NO DUST,LEAF-PRINTS,BIG STONY CREEK VALLEY,VA.,USA.	978
*CRYPTOCOCCUS SP. ON VITIS VULPINA	6. %	DUST=MODERATE,LEAF-PRINTS,BIG STONY CREEK VALLEY,VA.,USAA.	978
*CRYPTOCOCCUS SP. ON VITIS VULPINA	14.9 %	NO DUST,LEAF-PRINTS,BIG STONY CREEK VALLEY,VA.,USA.	978
*CURVULARIA SP. ON SASSAFRAS ALBIDUM	6.2 %	NO DUST,DILUTION-PLATES,BIG STONY CREEK VALLEY,VA.,USA.	978
*CURVULARIA SP. ON TSUGA CANADENSIS	3.3 %	DUST=HEAVY,DILUTION-PLATES,BIG STONY CREEK VALLEY,VA.,USA.	978
*CURVULARIA SP. ON TSUGA CANADENSIS	5.1 %	NO DUST,DILUTION-PLATES,BIG STONY CREEK VALLEY,VA.,USA.	978
*FUNGI ON VITIS VULPINA	5.3 %	DUST=MODERATE,LEAF-PRINTS,BIG STONY CREEK VALLEY,VA.,USA.	978
*FUNGI SP. ON TSUGA CANADENSIS	12.3 %	NO DUST,LEAF-PRINTS,BIG STONY CREEK VALLEY,VA.,USA.	978
*FUSARIUM SP. ON SASSAFRAS ALBIDUM	2. %	NO DUST,DILUTION-PLATES,BIG STONY CREEK VALLEY,VA.,USA.	978
*FUSARIUM SP. ON VITIS VULPINA	8.1 %	DUST=MODERATE,LEAF-PRINTS,BIG STONY CREEK VALLEY,VA.,USA.	978
*GREY MYCELIUM ON SASSAFRAS ALBIDUM	18. %	DUST=MODERATE,DILUTION-PLATES,BIG STINY CREEK VALLEY,VA.,USA.	978

TABLE C215

LIMESTONE DUST : OCCURRENCE OF FUNGI (F016)

ITEM:	VALUE:	CONDITIONS:	REF.:
*GREY MYCELIUM ON SASSAFRAS ALBIDUM	16.2 %	NO DUST,DILUTION-PLATES,BIG STONY CREEK VALLEY,VA.,USA.	978
*GREY MYCELIUM ON SASSAFRAS ALBIDUM	4.1 %	DUST=MODERATE,DILUTION-PLATES,BIG STONY CREEK VALLEY,VA.,USA.	978
*GREY MYCELIUM ON SASSAFRAS ALBIDUM	6.2 %	NO DUST,DILUTION-PLATES,BIG STONY CREEK VALLEY,VA.,USA.	978
*GREY MYCELIUM ON TSUGA CANADENSIS	6.8 %	DUST=HEAVY,LEAF-PRINTS,BIG STONY CREEK VALLEY,VA.,USA.	978
*GREY MYCELIUM ON TSUGA CANADENSIS	17.5 %	NO DUST,LEAF-PRINTS,BIG STONY CREKK VALLEY,VA.,USA.	978
*GREY MYCELIUM ON VITIS VULPINA	3.3 %	NO DUST,LEAF-PRINTS,BIG STONY CREEK VALLEY,VA.,USA.	978
*GREY MYCELIUM ON VITIS VULPINA	15.4 %	DUST=MODERATE,LEAF-PRINTS,BIG STONY CREEK VALLEY,VA.,USA.	978
*PERICONIELLA SP. ON SASSAFRAS ALBIDUM	6.4 %	DUST=MODERATE,DILUTION-PLATES,BIG STONY CREEK VALLEY,VA.,USA.	978
*PERICONIELLA SP. ON SASSAFRAS ALBIDUM	31.2 %	NO DUST,SASSAFRAS ALBIDUM,BIG STONY CREEK VALLEY,VA.,USA.	978
*PERICONIELLA SP. ON TSUGA CANADENSIS	12.8 %	DUST=HEAVY,LEAF-PRINTS,BIG STONY CREEK VALLEY,VA.,USA.	978
*PESTALOTIA SP. ON VITIS VULPINA	9. %	DUST=MODERATE,LEAF-PRINTS,BIG STONY CREEK VALLEY,VA.,USA.	978
*PESTALOTIA SP. ON VITIS VULPINA	5.5 %	NO DUST,LEAF-PRINTS,BIG STONY CREEK VALLEY,VA.,USA.	978
*PINICILLIUM SP. ON SASIS VULPINA	22. %	DUST=MODERATE,LEAF-PRINTS,BIG STONY CREEK VALLEY,VA.,USA.	978
*PINICILLIUM SP. ON SASIS VULPINA	42.5 %	NO DUST,LEAF-PRINTS,BIG STONY CREEK VALLEY,VA.,USA.	978
*PINICILLIUM SP. ON SASSAFRAS ALBIDUM	2.7 %	DUST=MODERATE,DILUTION-PLATES,BIG STONY CREEK VALLEY,VA.,USA.	978
*PINICILLIUM SP. ON SASSAFRAS ALBIDUM	4.2 %	NO DUST,DILUTION-PLATES,BIG STONY CREEK VALLEY,VA.,USA.	978
*PIRICAUDA SP. ON SASSAFRAS ALBIDUM	4.1 %	DUST=MODERATE,VITIS VULPINA,BIG STONY CREEK VALLEY,VA.,USA.	978
*PIRICAUDA SP. ON SASSAFRAS ALBIDUM	1.7 %	NO DUST,DILUTION-PLATES,BIG STONY CREEK VALLEY,VA.,USA.	978
*PIRICAUDA SP. ON SASSAFRAS ALBIDUM	4.2 %	DUST=MODERATE,DILUTION-PLATES,BIG STONY CREEK VALLEY,VA.,USA.	978
*PIRICAUDA SP. ON SASSAFRAS ALBIDUM	6.2 %	NO DUST,DILUTION-PLATES,BIG STONY CREEK VALLEY,VA.,USA.	978
*PIRICAUDA SP. ON TSUGA CANADENSIS	3.3 %	DUST=HEAVY,DILUTION-PLATES,BIG STONY CREEK VALLEY,VA.,USA.	978
*PIRICAUDA SP. ON TSUGA CANADENSIS	4. %	NO DUST,DILUTION-PLATES,BIG STONY CREEK VALLEY,VA.,USA.	978
*PIRICAUDA SP. ON VITIS VULPINA	2.5 %	LEAD-PRINTS,LEAF-PRINTS,BIG STONY CREEK VALLEY,VA.,USA.	978
*PIRICAUDA SP. ON VITIS VULPINA	9.2 %	NO DUST,LEAF-PRINTS,BIG STONY CREEK VALLEY,VA.,USA.	978
*PIRICAUDA SP. ON VITIS VULPINA	12.7 %	NO DUST,LEAF-PRINTS,BIG STONY CREEK VALLEY,VA.,USA.	978
*RHODOTORULA SP. ON SASSAFRAS ALBIDUM	0.6 %	DUST=HEAVY,DILUTION-PLATES,BIG STONY CREEK VALLEY,VA.,USA.	978
*RHODOTORULA SP. ON SASSAFRAS ALBIDUM	56.1 %	DUST=MODERATE,DILUTION-PLATES,BIG STONY CREEK VALLEY,VA.,USA.	978
*RHODOTORULA SP. ON SASSAFRAS ALBIDUM	12.1 %	DUST=MODERATE,DILUTION-PLATES,BIG STONY CREEK VALLEY,VA.,USA.	978
*RHODOTORULA SP. ON SASSAFRAS ALBIDUM	12.6 %	NO DUST,DILUTION-PLATES,BIG STONY CREEK VALLEY,VA.,USA.	978
*RHODOTORULA SP. ON TSUGA CANADENSIS	69.3 %	DUST=HEAVY,DILUTION-PLATES,BIG STONY CREEK VALLEY,VA.,USA.	978
*RHODOTORULA SP. ON TSUGA CANADENSIS	56.8 %	NO DUST,DILUTION-PLATES,BIG STONY CREEK VALLEY,VA.,USA.	978
*RHODOTORULA SP. ON VITIS VULPINA	26.6 %	DUST=HEAVY.VITIS VULPINALL,BIG STONY CREEK VALLEY,VA.,USA.	978
*RHODOTORULA SP. ON VITIS VULPINA	33.6 %	DUST=MODERATE,LEAF-PRINTS,BIG STONY CREEK VALLEY,VA.,USA.	978
*RHODOTORULA SP. ON VITIS VULPINA	4.1 %	NO DUST,LEAF-PRINTS,BIG STONY CREEK VALLEY,VA.,USA.	978

TABLE C215

LIMESTONE DUST : OCCURRENCE OF FUNGI (F016)			
ITEM:	VALUE:	CONDITIONS:	REF.:
*RHODOTORULA SP. ON VITIS VULPINA	17.7 %	NO DUST,LEAF-PRINTS,BIG STONY CREEK VALLEY,VA.,USA.	978
*SACCHAROMYCES SP. ON SASSAFRAS ALBIDUM	2.7 %	DUST=MODERATE,DILUTION-PLATES,BIG STONY CREEK VALLEY,VA.,USA.	978
*WHITE MYCELIUM ON VITIS VULPINA	8. %	DUST=MODERATE,LEAF-PRINTS,BIG STONY CREEK VALLEY,VA.,USA.	978
*WHITE MYCELIUM ON VITIS VULPINA	2.7 %	NO DUST,LEAF-PRINTS,BIG STONY CREEK VALLEY,VA.,USA.	978

TABLE C216

LIMESTONE DUST : OCCURRENCE OF SACCHAROYCES (F016)			
ITEM:	VALUE:	CONDITIONS:	REF.:
*SACCHAROMYCES SP. ON SASSAFRAS ALBIDUM	34.6 %	NO DUST,DILUTION-PLATES,BIG STONY CREEK VALLEY,VA.,USA.	978
*SACCHAROMYCES SP. ON SASSAFRAS ALBIDUM	28.4 %	DUST=MODERATE,DILUTION-PLATES,BIG STONY CREEK VALLEY,VA.,USA.	978
*SACCHAROMYCES SP. ON SASSAFRAS ALBIDUM	16.6 %	NO DUST,DILUTION-PLATES,BIG STONY CREEK VALLEY,VA.,USA.	978
*SACCHAROMYCES SP. ON TSUGA CANADENSIS	17.8 %	DUST=HEAVY,LEAF-PRINTS,BIG STONY CREEK VALLEY,VA.,USA.	978
*SACCHAROMYCES SP. ON VITIS VULPINA	5.8 %	DUST=MODERATE,LEAF-PRINTS,BIG STONY CREEK VALLEY,VA.,USA.	978

TABLE C217

LINDANE : IMMOBILISATION			
ITEM:	VALUE:	CONDITIONS:	REF.:
*DAPHNIA SP.	100 %	LINDANE=0.3 MG/L, FOR 50 H	83

TABLE C218

LINDANE : MORTALITY			
ITEM:	VALUE:	CONDITIONS:	REF.:
*BOBWHITE	50%	CONC.IN FOOD=882 MG/KG WW (755-1041),95% CONF.LIMITS	389
*EARTHWORMS	95%	CONC.IN SOIL:>65 MG/KG	389
*JAPANESE QUAIL	50%	CONC.IN FOOD=425 MG/KG WW (347-520),95% CONF.LIMITS	389
*MALLARD	50%	CONC.IN FOOD:>5000 MG/KG WW ,40% MORT.AT 5000 MG/KG WW	389
*PHEASANT	50%	CONC.IN FOOD=561 MG/KG WW (445-690),95% CONF.LIMITS	389
*RATS	50%	ACUTE ORAL TOXICITY=88-125MG/KG, INSECTICIDE	389
*RATS	50%	ACUTE DERMAL TOXICITY=900-1000MG/KG,INSECTICIDE	389

TABLE C219

LINEAR ALKYL BENZEN SULFONATE : CHANGE IN GROWTH POTENTIAL (F008)			
ITEM:	VALUE:	CONDITIONS:	REF.:
*ANABAENA CYLINDRICA	3	7DAYS,CONTROL=4 ,INITIAL CONC.=20MG SURFACTANT/L	332

TABLE C219

LINEAR ALKYL BENZEN SULFONATE : CHANGE IN GROWTH POTENTIAL (F008)			
ITEM:	VALUE:	CONDITIONS:	REF.:
*ANABAENA CYLINDRICA	1	14DAYS,CONTROL=4 ,INITIAL CONC.=20MG SURFACTANT/L	332
*ANABAENA CYLINDRICA	3	21DAYS,CONTROL=4 ,INITIAL CONC.=20MG SURFACTANT/L	332
*ANABAENA VARIABILIS	3	7DAYS,CONTROL=4 ,INITIAL CONC.=20MG SURFACTANT/L	332
*ANABAENA VARIABILIS	4	14DAYS,CONTROL=4 ,INITIAL CONC.=20MG SURFACTANT/L	332
*ANABAENA VARIABILIS	4	21DAYS,CONTROL=4 ,INITIAL CONC.=20MG SURFACTANT/L	332
*ANACYSTIS NIDULANS	1	7DAYS,CONTROL=4 ,INITIAL CONC.=20MG SURFACTANT/L	332
*ANACYSTIS NIDULANS	4	14DAYS,CONTROL=4 ,INITIAL CONC.=20MG SURFACTANT/L	332
*ANACYSTIS NIDULANS	4	21DAYS,CONTROL=4 ,INITIAL CONC.=20MG SURFACTANT/L	332
*ANKISTRODESMUS BRAUNII	4	7DAYS,CONTROL=4 ,INITIAL CONC.=20MG SURFACTANT/L	332
*ANKISTRODESMUS BRAUNII	4	14DAYS,CONTROL=4 ,INITIAL CONC.=20MG SURFACTANT/L	332
*ANKISTRODESMUS BRAUNII	4	21DAYS,CONTROL=4 ,INITIAL CONC.=20MG SURFACTANT/L	332
*CALOTHRIX PARIETINA	4	7,14 AND 21DAYS,CONTROL=4 , INITIAL CONC.=20MG SURFACTANT/L	332
*CHLORELLA PYRENOIDOSA	2	7DAYS,CONTROL=4 ,INITIAL CONC.=20MG SURFACTANT/L	332
*CHLORELLA PYRENOIDOSA	4	14DAYS,CONTROL=4 ,INITIAL CONC.=20MG SURFACTANT/L	332
*CHLORELLA PYRENOIDOSA	4	21DAYS,CONTROL=4 ,INITIAL CONC.=20MG SURFACTANT/L	332
*CHLORELLA VULGARIS	3	7DAYS,CONTROL=4 ,INITIAL CONC.=20MG SURFACTANT/L	332
*CHLORELLA VULGARIS	4	14DAYS,CONTROL=4 ,INITIAL CONC.=20MG SURFACTANT/L	332
*CHLORELLA VULGARIS	4	21DAYS,CONTROL=4 ,INITIAL CONC.=20MG SURFACTANT/L	332
*CYLINDROSPERMUM SP.	4	7DAYS,CONTROL=4 ,INITIAL CONC.=20MG SURFACTANT/L	332
*CYLINDROSPERMUM SP.	2	14DAYS,CONTROL=4 ,INITIAL CONC.=20MG SURFACTANT/L	332
*CYLINDROSPERMUM SP.	4	21DAYS,CONTROL=4 ,INITIAL CONC.=20MG SURFACTANT/L	332
*GLOEOCAPSA ALPICOLA	2	7DAYS,CONTROL=4 ,INITIAL CONC.=20MG SURFACTANT/L	332
*GLOEOCAPSA ALPICOLA	4	14DAYS,CONTROL=4 ,INITIAL CONC.=20MG SURFACTANT/L	332
*GLOEOCAPSA ALPICOLA	4	21DAYS,CONTROL=4 ,INITIAL CONC.=20MG SURFACTANT/L	332
*NOSTOC MUSCORUM	4	7DAYS,CONTROL=4 ,INITIAL CONC.=20MG SURFACTANT/L	332
*NOSTOC MUSCORUM	3	14DAYS,CONTROL=4 ,INITIAL CONC.=20MG SURFACTANT/L	332
*NOSTOC MUSCORUM	4	21DAYS,CONTROL=4 ,INITIAL CONC.=20MG SURFACTANT/L	332
*OSCILLATORI[illegible]RNETI	3	7DAYS,CONTROL=4 ,INITIAL CONC.=20MG SURFACTANT/L	332
*OSCILLATORIA BORNETI	4	14DAYS,CONTROL=4 ,INITIAL CONC.=20MG SURFACTANT/L	332
*OSCILLATORIA BORNETI	4	21DAYS,CONTROL=4 ,INITIAL CONC.=20MG SURFACTANT/L	332
*OSCILLATORIA CHALYBIA	3	7DAYS,CONTROL=4 ,INITIAL CONC.=20MG SURFACTANT/L	332
*OSCILLATORIA CHALYBIA	3	14DAYS,CONTROL=4 ,INITIAL CONC.=20MG SURFACTANT/L	332
*OSCILLATORIA CHALYBIA	4	21DAYS,CONTROL=4 ,INITIAL CONC.=20MG SURFACTANT/L	332
*OSCILLATORIA FORMOSA	4	7DAYS,CONTROL=4 ,INITIAL CONC.=20MG SURFACTANT/L	332

TABLE C219

LINEAR ALKYL BENZEN SULFONATE : CHANGE IN GROWTH POTENTIAL (F008)			
ITEM:	VALUE:	CONDITIONS:	REF.:
*OSCILLATORIA FORMOSA	3	14DAYS,CONTROL=4 ,INITIAL CONC.=20MG SURFACTANT/L	332
*OSCILLATORIA FORMOSA	4	21DAYS,CONTROL=4 ,INITIAL CONC.=20MG SURFACTANT/L	332
*OSCILLATORIA TENUIS	3	7DAYS,CONTROL=4 ,INITIAL CONC.=20MG SURFACTANT/L	332
*OSCILLATORIA TENUIS	2	14DAYS,CONTROL=4 ,INITIAL CONC.=20MG SURFACTANT/L	332
*OSCILLATORIA TENUIS	3	21DAYS,CONTROL=4 ,INITIAL CONC.=20MG SURFACTANT/L	332
*PHORMIDIUM FAVEOLARUM	4	7DAYS,CONTROL=4 ,INITIAL CONC.=20MG SURFACTANT/L	332
*PHORMIDIUM FAVEOLARUM	4	14DAYS,CONTROL=4 ,INITIAL CONC.=20MG SURFACTANT/L	332
*PHORMIDIUM FAVEOLARUM	4	21DAYS,CONTROL=4 ,INITIAL CONC.=20MG SURFACTANT/L	332
*SCENEDESMUS OBLIQUUS	4	7DAYS,CONTROL=4 ,INITIAL CONC.=20MG SURFACTANT/L	332
*SCENEDESMUS OBLIQUUS	3	14DAYS,CONTROL=4 ,INITIAL CONC.=20MG SURFACTANT/L	332
*SCENEDESMUS OBLIQUUS	4	21DAYS,CONTROL=4 ,INITIAL CONC.=20MG SURFACTANT/L	332
*SCENEDESMUS QUADRICAUDA	3	7DAYS,CONTROL=4 ,INITIAL CONC.=20MG SURFACTANT/L	332
*SCENEDESMUS QUADRICAUDA	4	14DAYS,CONTROL=4 ,INITIAL CONC.=20MG SURFACTANT/L	332
*SCENEDESMUS QUADRICAUDA	4	21DAYS,CONTROL=4 ,INITIAL CONC.=20MG SURFACTANT/L	332
*ULOTHRIX FIMBRIATA	1	7DAYS,CONTROL=4 ,INITIAL CONC.=20MG SURFACTANT/L	332
*ULOTHRIX FIMBRIATA	4	14DAYS,CONTROL=4 ,INITIAL CONC.=20MG SURFACTANT/L	332
*ULOTHRIX FIMBRIATA	4	21DAYS,CONTROL=4 ,INITIAL CONC.=20MG SURFACTANT/L	332

TABLE C220

LINEAR ALKYL BENZEN SULFONATE : MORTALITY			
ITEM:	VALUE:	CONDITIONS:	REF.:
*AEDES AEGYPTI	50. %	LAS 345 6.PPM,24H,40 LARVAE	1627
*AEDES AEGYPTI	100. %	LAS 345 8.PPM,24H,40 LARVAE	1627
*AEDES AEGYPTI	50. %	LAS 358 2.PPM,24H,40 LARVAE	1627
*AEDES AEGYPTI	100. %	LAS 358 3.PPM,24H,40 LARVAE	1627
*BIOMPHALARIA GLABRATA	100. %	LAS 345 3.PPM,24H,2*3 SNAILS	1627
*BIOMPHALARIA GLABRATA	100. %	LAS 358 1.PPM,24H,2*3 SNAILS	1627
*CYPRINUS CARPIO	50 %	LAS-COMM.+ENZ.=5.4MG/L,LAS-TECH.= 6.8,24H,,294K	947
*CYPRINUS CARPIO	50 %	LAS-COMM.+ENZ.=5.4MG/L,LAS-TECH.= 6.8MG/L,48H,294K	947
*CYPRINUS CARPIO	50 %	LAS-COMM.+ENZ.=5.4MG/L,LAS-TECH.= 5.MG/L,96H,294K	947
*LEBISTES RETICULATUS	100. %	LAS 345 2PPM,24H,3 FISHES, FEMALE	1627
*LEBISTES RETICULATUS	100. %	LAS 345 2.PPM,24H,3 FISHES, MALE	1627
*LEBISTES RETICULATUS	100. %	LAS 345 1.PPM,24H,5 YOUNG FISHES	1627
*LEBISTES RETICULATUS	100. %	LAS 358 1.PPM,24H,3 FISHES, FEMALE	1627
*LEBISTES RETICULATUS	100. %	LAS 358 1.PPM,24H,3 FISHES, MALE	1627

TABLE C220

LINEAR ALKYL BENZEN SULFONATE : MORTALITY			
ITEM:	VALUE:	CONDITIONS:	REF.:
*LEBISTES RETICULATUS	100. %	LAS 358 0.5PPM,24H,5 YOUNG FISHES	1627
*TILAPIA MELANOPLEURA	50 %	LAS-COMM.+ENZ.=26MG/L,LAS-TECH.= 12MG/L,LAS-COMM.=4.8MG/L,24H,294K	947
*TILAPIA MELANOPLEURA	50 %	LAS-COMM.+ENZ.=26MG/L,LAS-TECH.= 12MG/L,LAS-COMM.=4.8MG/L,48H,294K	947
*TILAPIA MELANOPLEURA	50 %	LAS-COMM.+ENZ.=23MG/L,LAS-TECH.= 12MG/L,LAS-COMM.=4.6MG/L,96H,294K	947

TABLE C221

LIPORIC ACID : INHIBITION OF NITROGEN FIXATION			
ITEM:	VALUE:	CONDITIONS:	REF.:
*CLOSTRIDIUM PASTEURIANUM	INHIBITED	0.1G/L,INHIB NULL BY ADDITION OF 10 UG BIOTIN/L	228

TABLE C222

LU : MORTALITY			
ITEM:	VALUE:	CONDITIONS:	REF.:
*MOUSE	50%	CHLORIDE(CITRATECOMPLEX), INTRAPERITON.,DOSE=135MG/KG BODY	1611
*RAT	50%	NITRATE,INTRAPERITON.,DOSE=335MG/ KG BODY	1611

TABLE C223

MALATHION : HEART BEAT			
ITEM:	VALUE:	CONDITIONS:	REF.:
*BIOMPHALARIA GLABRATA	136 BEATS/MIN	MALATHION=243MICRO-MOL,AFTER 24H, 299K,5 SNAILS	264
*BIOMPHALARIA GLABRATA	100 BEATS/MIN	MALATHION=375MICRO-MOL,AFTER 24H, 299K,5 SNAILS	264
*BIOMPHALARIA GLABRATA	32 BEATS/MIN	MALATHION=486MICRO-MOL,AFTER 24H, 299K,5 SNAILS	264
*BIOMPHALARIA GLABRATA	0 BEATS/MIN	MALATHION=520MICRO-MOL,AFTER 24H, 299K,5 SNAILS	264

TABLE C224

METHOXYCHLOR : MORTALITY			
ITEM:	VALUE:	CONDITIONS:	REF.:
*BOBWHITE	50%	CONC.IN FOOD:>5000 MG/KG WW	389
*CATOSTOMUS COMMERSONI	50 %	METHOXYCHLOR=0.0345PPM,288K,96H	1646
*JAPANESE QUAIL	50%	CONC.IN FOOD:>5000 MG/KG WW	389
*MALLARD	50%	CONC.IN FOOD:>5000 MG/KG WW	389
*PHEASANT	50%	CONC.IN FOOD:>5000 MG/KG WW	389
*PHEASANT	50%	CONC.IN FOOD=25000 MG/KG WW	349

TABLE C224

METHOXYCHLOR : MORTALITY			
ITEM:	VALUE:	CONDITIONS:	REF.:
*QUAIL	50%	CONC.IN FOOD=22000 MG/KG WW	349
*RATS	50%	ACUTE ORAL TOXICITY=6000MG/KG	389

TABLE C225

METHOXYETHYLMERCURY : MORTALITY			
ITEM:	VALUE:	CONDITIONS:	REF.:
*RATS	50%	ACUTE ORAL TOXICITY=570MG/KG	389

TABLE C226

METHYL THIOPHANATE : MORTALITY			
ITEM:	VALUE:	CONDITIONS:	REF.:
*RATS	50%	ACUTE ORAL TOXICITY=7500MG/KG	389

TABLE C227

METRIBUZIN : MORTALITY			
ITEM:	VALUE:	CONDITIONS:	REF.:
*ICTALURUS PUNCTATUS	LESS THAN 10 %	10 PPM, 48 H, 14 G, 293 K, PH=8.2, ALK=80 PPM	1002

TABLE C228

MG : MORTALITY			
ITEM:	VALUE:	CONDITIONS:	REF.:
*RAT	50%	ACETATE,ORAL,DOSE=18MG/KG BODY	1611
*RAT	50%	CHLORIDE,ORAL,DOSE=2800MG/KG BODY	1611
*RAT	50%	CHLORIDE,INTRAVEN.,DOSE=14MG/KG BODY	1611

TABLE C229

MN : CHANGE IN MN CONTENT			
ITEM:	VALUE:	CONDITIONS:	REF.:
*CUCUMBER, LEAF	77 MG/KG DRY WT.	NO SYMPTOMS,EXCHAN.MN(SOIL)=45PPM	635
*CUCUMBER, LEAF	600 MG/KG DRY WT.	NO SYMPTOMS,ADD.MN(SOIL)=200PPM	635
*CUCUMBER, LEAF	1350 MG/KG DRY WT.	BROWN SPOT.ON STEM AND PETIOLES OF LOW.LEAV.,ADD.MN(SOIL)=400PPM	635
*CUCUMBER, LEAF	1920 MG/KG DRY WT.	DARK BROWN SPOT.OF STEM AND PETIOLES OF MOST LEAVES,	635
*CUCUMBER, LEAF	1920 MG/KG DRY WT.	VEINS BROWN AND LAMINAE CHLOROTIC, ADD.MN(SOIL)=600PPM	635

TABLE C229

MN : CHANGE IN MN CONTENT			
ITEM:	VALUE:	CONDITIONS:	REF.:
*TOMATO, LEAF	110 MG/KG DRY WT.	NO SYMPTOMS,EXCHAN.MN(SOIL)=90PPM, PH=7.1	635
*TOMATO, LEAF	740 MG/KG DRY WT.	NO SYMPTOMS,ADD.MN(SOIL)=200PPM,PH =7.0	635
*TOMATO, LEAF	2240 MG/KG DRY WT.	CHLOROSIS OF LOW.LEAV.,VEINS PURPLE BROWN,ADD.MN(SOIL)=400PPM, PH=6.8	635
*TOMATO, LEAF	3040 MG/KG DRY WT.	DARK BROWN LESIONS ON STEM AND ON MAIN VEINS,LOW.LEAV.CHLOROTIC	635
*TOMATO, LEAF	3040 MG/KG DRY WT.	ADD.MN(SOIL)=600PPM,PH=6.9	635

TABLE C230

MN : GROWTH			
ITEM:	VALUE:	CONDITIONS:	REF.:
*AGATHIS AUSTRALIS	DEFICIENCY SYMOTOMS AT 0.001 MG/L	TREE, 100 D IN GREENHOUSE, CONC IN CULTURE SOLUTION	1248
*AGATHIS AUSTRALIS	NORMAL GROWTH AT 0.500 MG/L	TREE, 100 D IN GREENHOUSE, CONC IN CULTURE SOLUTION	1248
*AGATHIS AUSTRALIS	TOXIC SYMPTOMS AT 2.500 MG/L	TREE, 100 D IN GREENHOUSE, CONC IN CULTURE SOLUTION	1248

TABLE C231

MN : INCREASE IN GRAIN YIELD			
ITEM:	VALUE:	CONDITIONS:	REF.:
*BARLEY	4.71 T/HA AT 15% MOISTURE	ADDED MNSO4=5.6KG/HA,ORG.SOIL; CONTROL=3.25T/HA,1962,N.WALES	77
*OAT	2.59 T/HA AT 15% MOISTURE	ADDED MNSO4=5.6KG/HA,ORG.SOIL; CONTROL=1.13T/HA,1959,N.WALES	77
*OAT	2.88 T/HA AT 15% MOISTURE	ADDED MNSO4=5.6KG/HA,ORG.SOIL; CONTROL=0.98T/HA,1960,N.WALES	77
*OAT	4.85 T/HA AT 15% MOISTURE	ADDED MNSO4=5.6KG/HA,ZY CL(F017); CONTROL=3.48T/HA,1960,N.WALES	77
*OAT	1.71 T/HA AT 15% MOISTURE	ADDED MNSO4=5.6KG/HA,ORG.SOIL; CONTROL=1.01T/HA,1961,N.WALES	77

TABLE C232

MN : LEAF SYMPTONS			
ITEM:	VALUE:	CONDITIONS:	REF.:
*TOMATO LEAVES	3750 MG/KG DRY WT.	NO SYMPTONS AT 1600MG/KG DRY WT., STEAM STERILIZED SOIL	635

TABLE C233

MN : MORTALITY			
ITEM:	VALUE:	CONDITIONS:	REF.:
*DAPHNIA MAGNA	50%	MNCL2,EXPOS.TIME=2DAYS,CONC.IN WATER=9800MICROGRAM/L,LIMNOL.	122

TABLE C233

MN : MORTALITY			
ITEM:	VALUE:	CONDITIONS:	REF.:
*MOUSE	50%	SULFATE,INTRAPERITON.,DOSE=44MG/KG BODY	1611
*RAT	50%	ACETATE,ORAL,DOSE=3730MG/KG BODY	1611
*RAT	50%	K-PERMANGANATE,ORAL,DOSE=1090MG/KG BODY	1611

TABLE C234

MO : GROWTH			
ITEM:	VALUE:	CONDITIONS:	REF.:
*AGATHIS AUSTRALIS	DEFICIENCY SYMOTOMS AT 0.0001 MG/L	TREE, 100 D IN GREENHOUSE, CONC IN CULTURE SOLUTION	1248
*AGATHIS AUSTRALIS	NORMAL GROWTH AT 0.5000 MG/L	TREE, 100 D IN GREENHOUSE, CONC IN CULTURE SOLUTION	1248
*AGATHIS AUSTRALIS	NO TOXIC SYMPTOMS AT HIGH CONCENTRATIONS	TREE, 100 D IN GREENHOUSE, CONC IN CULTURE SOLUTION	1248

TABLE C235

MO : MORTALITY			
ITEM:	VALUE:	CONDITIONS:	REF.:
*PIMEPHALES PROMELAS	50%	MO03,EXPOS.TIME=4DAYS,CONC.IN WATER=70000MICROGRAM/L,LIMNOL.	1004
*RAT	50%	TRIOXIDE,ORAL,DOSE=125MG/KG BODY	1611
*RAT	50%	AMMONIUMMOLYBDATE,ORAL,DOSE=333MG/KG BODY	419

TABLE C236

MONURON : MORTALITY			
ITEM:	VALUE:	CONDITIONS:	REF.:
*ICTALURUS PUNCTATUS	LESS THAN 10 %	10 PPM, 48 H, 14 G, 293 K, PH=8.2, ALK=80 PPM	1002
*RATS	50%	ACUTE ORAL TOXICITY=3600MG/KG, HERBICIDE	389

TABLE C237

MORPHINE : MORTALITY			
ITEM:	VALUE:	CONDITIONS:	REF.:
*MOUSE	50%	ORAL DOSE=285MG/KG(=9.99*10**(-4) MOL/KG),PT-50(F007)=3.00	953

TABLE C238

MSMA : MORTALITY			
ITEM:	VALUE:	CONDITIONS:	REF.:
*ICTALURUS PUNCTATUS	LESS THAN 10 %	10 PPM, 48 H, 14 G, 293 K, PH=8.2, ALK=80 PPM	1002

TABLE C239

NA : MORTALITY			
ITEM:	VALUE:	CONDITIONS:	REF.:
*MOUSE	50%	SULFATE,INTRAPERITON.,DOSE=193MG/KG BODY	1611
*MOUSE	50%	ORAL DOSE=2600MG NACL/KG(=4.45*10**(-2)MOL/KG),PT-50(F007)=1.35	953
*RAT	50%	CARBONATE,ORAL,DOSE=4200MG/KG BODY	1611
*RAT	50%	CHLORIDE,ORAL,DOSE=3000MG/KG BODY	1611

TABLE C240

NA-FLOUROACETATE : MORTALITY			
ITEM:	VALUE:	CONDITIONS:	REF.:
*MOUSE	50%	ORAL DOSE=18MG/KG(=1.80*10**(-4)MOL/KG),PT-50(F007)=3.74	953

TABLE C241

NB : MORTALITY			
ITEM:	VALUE:	CONDITIONS:	REF.:
*MOUSE	50%	CHLORIDE,ORAL,DOSE=830MG/KG BODY	1611
*RAT	50%	CHLORID,ORAL,DOSE=41MG/KG BODY	419
*RAT	50%	K-NIOBATE,ORAL,DOSE=725MG/KG BODY	419

TABLE C242

ND : MORTALITY			
ITEM:	VALUE:	CONDITIONS:	REF.:
*MOUSE	50%	CHLORIDE,INTRAPERITON.,DOSE=347MG/KG BODY	1611
*MOUSE	50%	CHLORIDE(CITRATECOMPLEX),INTRAPERITON.,DOSE=140MG/KG BODY	1611
*RAT	50%	NITRATE,ORAL,DOSE=2750MG/KG BODY	1611

TABLE C243

NEBURON : MORTALITY			
ITEM:	VALUE:	CONDITIONS:	REF.:
*RATS	50%	ACUTE ORAL TOXICITY=11000MG/KG, HERBICIDE	389

TABLE C244

NI : CHANGE IN HATCHABILITY			
ITEM:	VALUE:	CONDITIONS:	REF.:
*PIMEPHALES PROMELAS EGGS	42.4% (0-97%)	NI=0.73MG/L,N=23;CONTROL=93.8%(86-97),N=10	1258

TABLE C244

NI : CHANGE IN HATCHABILITY			
ITEM:	VALUE:	CONDITIONS:	REF.:
*PIMEPHALES PROMELAS EGGS	94.2% (84-100%)	NI=0.38MG/L,N=13;CONTROL=93.8% (86-97),N=10	1258
*PIMEPHALES PROMELAS EGGS	95.8% (87-100%)	NI=0.18MG/L,N=12;CONTROL=93.8% (86-97),N=10	1258
*PIMEPHALES PROMELAS EGGS	95.9% (88-100%)	NI=0.082MG/L,N=11;CONTROL=93.8% (86-97),N=10	1258

TABLE C245

NI : CHANGE IN YIELD			
ITEM:	VALUE:	CONDITIONS:	REF.:
*MUSTARD	17.2 G WET WT.	SOIL-PH=5.7,NI ADDED TO SOIL= 50PPM,CONTROL=24.8G WET WT.	1226
*MUSTARD	0.8 G WET WT.	SOIL-PH=5.7,NI ADDED TO SOIL= 100PPM,CONTROL=24.8G WET WT.	1226
*MUSTARD	0.4 G WET WT.	SOIL-PH=5.7,NI ADDED TO SOIL= 250PPM,CONTROL=24.8G WET WT.	1226
*MUSTARD	23.2 G WET WT.	SOIL-PH=6.4,NI ADDED TO SOIL= 50PPM,CONTROL=25.6G WET WT.	1226
*MUSTARD	25.8 G WET WT.	SOIL-PH=6.4,NI ADDED TO SOIL= 100PPM,CONTROL=25.6G WET WT.	1226
*MUSTARD	8.0 G WET WT.	SOIL-PH=6.4,NI ADDED TO SOIL= 250PPM,CONTROL=25.6G WET WT.	1226
*OAT	17.1 G WET WT.	SOIL-PH=5.7,NI ADDED TO SOIL= 50PPM,CONTROL=20.3G WET WT.	1226
*OAT	5.9 G WET WT.	SOIL-PH=5.7,NI ADDED TO SOIL= 100PPM,CONTROL=20.3G WET WT.	1226
*OAT	2.5 G WET WT.	SOIL-PH=5.7,NI ADDED TO SOIL= 250PPM,CONTROL=20.3G WET WT.	1226
*OAT	16.8 G WET WT.	SOIL-PH=6.4,NI ADDED TO SOIL= 50PPM,CONTROL=19.8G WET WT.	1226
*OAT	14.6 G WET WT.	SOIL-PH=6.4,NI ADDED TO SOIL= 100PPM,CONTROL=19.8G WET WT.	1226
*OAT	13.8 G WET WT.	SOIL-PH=6.4,NI ADDED TO SOIL= 250PPM,CONTROL=19.8G WET WT.	1226
*OAT	1.4 G DRY WT.	SOIL-PH=5.3,ADD.SLUDGE=37.5T/HA, ADDED NI=97PPM,CONTROL=1.4G DRY WT.	1226
*OAT	1.3 G DRY WT.	SOIL-PH=5.3,ADD.SLUDGE=75T/HA, ADDED NI=194PPM,CONTROL=1.4G DRY WT.	1226
*OAT	1.0 G DRY WT.	SOIL-PH=5.3,ADDED SLUDGE=150T/HA, ADDED NI=389PPM,CONTROL=1.4G DRY WT.	1226
*OAT	1.6 G DRY WT.	SOIL-PH=6.8,ADD.SLUDGE=37.5T/HA, ADDED NI=97PPM,CONTROL=1.4G DRY WT.	1226
*OAT	1.6 G DRY WT.	SOIL-PH=6.8,ADD.SLUDGE=75T/HA, ADDED NI=194PPM,CONTROL=1.4G DRY WT.	1226
*OAT	1.6 G DRY WT.	SOIL-PH=6.8,ADDED SLUDGE=150T/HA, ADDED NI=389PPM,CONTROL=1.4G DRY WT.	1226

TABLE C246

NI : DECREASE IN EGGPRODUCTION			
ITEM:	VALUE:	CONDITIONS:	REF.:
*PIMEPHALES PROMELAS	218 EGGS	NI=1.6MG/L,MAL.=6,FEM.=6,SPAWN.=5, CONTROL=12824EGGS,M=5,F=8,SP=65	1258

TABLE C246

NI : DECREASE IN EGGPRODUCTION			
ITEM:	VALUE:	CONDITIONS:	REF.:
*PIMEPHALES PROMELAS	3489 EGGS	NI=0.73MG/L,MAL.=5,FEM.=7,SPAWN.=53,CONTROL=12824EGGS,M=5,F=8,SP=65	1258
*PIMEPHALES PROMELAS	9787 EGGS	NI=0.38MG/L,MAL.=5,FEM.=7,SPAWN.=60,CONTROL=12824EGGS,M=5,F=8,SP=65	1258
*PIMEPHALES PROMELAS	9240 EGGS	NI=0.18MG/L,MAL.=6,FEM.=7,SPAWN.=56,CONTROL=12824EGGS,M=5,F=8,SP=65	1258
*PIMEPHALES PROMELAS	5520 EGGS	NI=0.082MG/L,MAL.=6,FEM.=5,SPAWN.=39,CONTROL=12824EGGS,M=5,F=8,SP=65	1258

TABLE C247

NI : MORTALITY			
ITEM:	VALUE:	CONDITIONS:	REF.:
*CRANGON CRANGON	50%	NISO4,EXPOS.TIME=2DAYS,CONC.IN WATER=125000 UG/L,MARINE	1272
*DAPHNIA MAGNA	50%	NICL2,EXPOS.TIME=2DAYS,CONC.IN WATER=510 UG/L,LIMNOL.	122
*MOUSE	50%	FLOURIDE,INTRAVEN.,DOSE=130MG/KG BODY	1611
*RAT	50%	TETRACARBONYL,INTRAVEN.,DOSE=22MG/KG BODY	1611
*RAT	50%	NITRATE,ORAL,DOSE=1620MG/KG BODY	1611

TABLE C248

NICOTINIC ACID : MORTALITY			
ITEM:	VALUE:	CONDITIONS:	REF.:
*MOUSE	50%	ORAL DOSE=1860MG/KG(=1.51*10**(-2) MOL/KG),PT-50(F007)=1.82	953

TABLE C249

NITRALIN : MORTALITY			
ITEM:	VALUE:	CONDITIONS:	REF.:
*ICTALURUS PUNCTATUS	LESS THAN 10 %	10 PPM, 48 H, 14 G, 293 K, PH=8.2, ALK=80 PPM	1002

TABLE C250

NITROGEN : INCREASE IN BASEL AREA BY NITROGEN APPLICATION			
ITEM:	VALUE:	CONDITIONS:	REF.:
*PINUS NIGRA	4.48 M**2/HA	APPL= 84 KG N/HA, CONTROL=2.8 M2/HA, SCOTLAND, AGE=36 Y, NH4-SULPHATE	1036
*PINUS NIGRA	5.11 M**2/HA	APPL=168 KG N/HA, CONTROL=2.8 M2/HA, SCOTLAND, AGE=36 Y, NH4-SULPHATE	1036
*PINUS NIGRA	5.44 M**2/HA	APPL=336 KG N/HA, CONTROL=2.8 M2/HA, SCOTLAND, AGE=36 Y, NH4-SULPHATE	1036

TABLE C250

NITROGEN : INCREASE IN BASEL AREA BY NITROGEN APPLICATION

ITEM:	VALUE:	CONDITIONS:	REF.:
*PINUS NIGRA	5.79 M**2/HA	APPL=504 KG N/HA, CONTROL=2.8 M2/HA, SCOTLAND, AGE=36 Y, NH4-SULPHATE	1036

TABLE C251

NITROGEN : INHIBITION OF NITROGEN FIXATION

ITEM:	VALUE:	CONDITIONS:	REF.:
*AZOTOBACTER SP.	INHIBITED	AS NITRITE, MORE THAN 1.4MG N/L, STIMULATORY AT LOWER CONC.	54

TABLE C252

NITROUS ACIDE/OXIDE : INHIBITION OF NITROGEN FIXATION

ITEM:	VALUE:	CONDITIONS:	REF.:
*AZOTOBACTER SP.	INHIBITED	0.14MG N/L,REVERSIBLE,STIMULATORY AT LOWER CONC.	54

TABLE C253

NONYL PHENOL POLYETHOXYLATE : CHANGE IN GROWTH POTENTIAL (F008)

ITEM:	VALUE:	CONDITIONS:	REF.:
*ANABAENA CYLINDRICA	4	7DAYS,CONTROL=4 ,INITIAL CONC.=20 MG ABE9.5/L	332
*ANABAENA CYLINDRICA	4	14DAYS,CONTROL=4 ,INITIAL CONC.=20 MG ABE9.5/L	332
*ANABAENA CYLINDRICA	4	21DAYS,CONTROL=4 ,INITIAL CONC.=20 MG ABE9.5/L	332
*ANABAENA VARIABILIS	1	7DAYS,CONTROL=4 ,INITIAL CONC.=20 MG ABE9.5/L	332
*ANABAENA VARIABILIS	1	14DAYS,CONTROL=4 ,INITIAL CONC.=20 MG ABE9.5/L	332
*ANABAENA VARIABILIS	1	21DAYS,CONTROL=4 ,INITIAL CONC.=20 MG ABE9.5/L	332
*ANACYSTIS NIDULANS	3	7DAYS,CONTROL=4 ,INITIAL CONC.=20 MG ABE9.5/L	332
*ANACYSTIS NIDULANS	3	14DAYS,CONTROL=4 ,INITIAL CONC.=20 MG ABE9.5/L	332
*ANACYSTIS NIDULANS	4	21DAYS,CONTROL=4 ,INITIAL CONC.=20 MG ABE9.5/L	332
*ANKISTRODESMUS BRAUNII	1	7DAYS,CONTROL=4 ,INITIAL CONC.=20 MG ABE9.5/L	332
*ANKISTRODESMUS BRAUNII	4	14DAYS,CONTROL=4 ,INITIAL CONC.=20 MG ABE9.5/L	332
*ANKISTRODESMUS BRAUNII	4	21DAYS,CONTROL=4 ,INITIAL CONC.=20 MG ABE9.5/L	332
*CALOTHRIX PARIETINA	4	7,14 AND 21DAYS,CONTROL=4 INITIAL CONC.=20 MG ABE9.5/L	332
*CHLORELLA PYRENOIDOSA	1	7DAYS,CONTROL=4 ,INITIAL CONC.=20 MG ABE9.5/L	332
*CHLORELLA PYRENOIDOSA	3	14DAYS,CONTROL=4 ,INITIAL CONC.=20 MG ABE9.5/L	332
*CHLORELLA PYRENOIDOSA	4	21DAYS,CONTROL=4 ,INITIAL CONC.=20 MG ABE9.5/L	332

TABLE C253

NONYL PHENOL POLYETHOXYLATE : CHANGE IN GROWTH POTENTIAL (F008)

ITEM:	VALUE:	CONDITIONS:	REF.:
*CHLORELLA VULGARIS	1	7DAYS,CONTROL=4 ,INITIAL CONC.=20 MG ABE9.5/L	332
*CHLORELLA VULGARIS	2	14DAYS,CONTROL=4 ,INITIAL CONC.=20 MG ABE9.5/L	332
*CHLORELLA VULGARIS	2	21DAYS,CONTROL=4 ,INITIAL CONC.=20 MG ABE9.5/L	332
*CYLINDROSPERMUM SP.	3	7DAYS,CONTROL=4 ,INITIAL CONC.=20 MG ABE9.5/L	332
*CYLINDROSPERMUM SP.	3	14DAYS,CONTROL=4 ,INITIAL CONC.=20 MG ABE9.5/L	332
*CYLINDROSPERMUM SP.	3	21DAYS,CONTROL=4 ,INITIAL CONC.=20 MG ABE9.5/L	332
*GLOEOCAPSA ALPICOLA	2	7DAYS,CONTROL=4 ,INITIAL CONC.=20 MG ABE9.5/L	332
*GLOEOCAPSA ALPICOLA	2	14DAYS,CONTROL=4 ,INITIAL CONC.=20 MG ABE9.5/L	332
*GLOEOCAPSA ALPICOLA	3	21DAYS,CONTROL=4 ,INITIAL CONC.=20 MG ABE9.5/L	332
*NOSTOC MUSCORUM	4	7DAYS,CONTROL=4 ,INITIAL CONC.=20 MG ABE9.5/L	332
*NOSTOC MUSCORUM	3	14DAYS,CONTROL=4 ,INITIAL CONC.=20 MG ABE9.5/L	332
*NOSTOC MUSCORUM	4	21DAYS,CONTROL=4 ,INITIAL CONC.=20 MG ABE9.5/L	332
*OSCILLATORIA BORNETI	2	7DAYS,CONTROL=4 ,INITIAL CONC.=20 MG ABE9.5/L	332
*OSCILLATORIA BORNETI	2	14DAYS,CONTROL=4 ,INITIAL CONC.=20 MG ABE9.5/L	332
*OSCILLATORIA BORNETI	2	21DAYS,CONTROL=4 ,INITIAL CONC.=20 MG ABE9.5/L	332
*OSCILLATORIA CHALYBIA	2	7DAYS,CONTROL=4 ,INITIAL CONC.=20 MG ABE9.5/L	332
*OSCILLATORIA CHALYBIA	2	14DAYS,CONTROL=4 ,INITIAL CONC.=20 MG ABE9.5/L	332
*OSCILLATORIA CHALYBIA	2	21DAYS,CONTROL=4 ,INITIAL CONC.=20 MG ABE9.5/L	332
*OSCILLATORIA FORMOSA	2	7DAYS,CONTROL=4 ,INITIAL CONC.=20 MG ABE9.5/L	332
*OSCILLATORIA FORMOSA	3	14DAYS,CONTROL=4 ,INITIAL CONC.=20 MG ABE9.5/L	332
*OSCILLATORIA FORMOSA	4	21DAYS,CONTROL=4 ,INITIAL CONC.=20 MG ABE9.5/L	332
*OSCILLATORIA TENUIS	2	7DAYS,CONTROL=4 ,INITIAL CONC.=20 MG ABE9.5/L	332
*OSCILLATORIA TENUIS	2	14DAYS,CONTROL=4 ,INITIAL CONC.=20 MG ABE9.5/L	332
*OSCILLATORIA TENUIS	4	21DAYS,CONTROL=4 ,INITIAL CONC.=20 MG ABE9.5/L	332
*PHORMIDIUM FAVEOLARUM	4	7DAYS,CONTROL=4 ,INITIAL CONC.=20 MG ABE9.5/L	332
*PHORMIDIUM FAVEOLARUM	4	14DAYS,CONTROL=4 ,INITIAL CONC.=20 MG ABE9.5/L	332
*PHORMIDIUM FAVEOLARUM	4	21DAYS,CONTROL=4 ,INITIAL CONC.=20 MG ABE9.5/L	332
*SCENEDESMUS OBLIQUUS	2	7DAYS,CONTROL=4 ,INITIAL CONC.=20 MG ABE9.5/L	332
*SCENEDESMUS OBLIQUUS	4	14DAYS,CONTROL=4 ,INITIAL CONC.=20 MG ABE9.5/L	332
*SCENEDESMUS OBLIQUUS	4	21DAYS,CONTROL=4 ,INITIAL CONC.=20 MG ABE9.5/L	332
*SCENEDESMUS QUADRICAUDA	3	7DAYS,CONTROL=4 ,INITIAL CONC.=20 MG ABE9.5/L	332
*SCENEDESMUS QUADRICAUDA	4	14DAYS,CONTROL=4 ,INITIAL CONC.=20 MG ABE9.5/L	332
*SCENEDESMUS QUADRICAUDA	4	21DAYS,CONTROL=4 ,INITIAL CONC.=20 MG ABE9.5/L	332
*ULOTHRIX FIMBRIATA	2	7DAYS,CONTROL=4 ,INITIAL CONC.=20 MG ABE9.5/L	332

TABLE C253

NONYL PHENOL POLYETHOXYLATE : CHANGE IN GROWTH POTENTIAL (F008)			
ITEM:	VALUE:	CONDITIONS:	REF.:
*ULOTHRIX FIMBRIATA	2	14DAYS,CONTROL=4 ,INITIAL CONC.=20 MG ABE9.5/L	332
*ULOTHRIX FIMBRIATA	4	21DAYS,CONTROL=4 ,INITIAL CONC.=20 MG ABE9.5/L	332

TABLE C254

OCDD : MORTALITY			
ITEM:	VALUE:	CONDITIONS:	REF.:
*RAT	50%	ORAL DOSE:>1G/KG BODY WT.	1408

TABLE C255

OCDD : TERATOGENICITY			
ITEM:	VALUE:	CONDITIONS:	REF.:
*RAT	NO EFFECT	DOSE=100-500MG/KG*DAY,FROM DAY 6-15 IN PREGNANCY-PERIODE	1026

TABLE C256

OIL ANTIPOLLUTION PRODUCTS : MORTALITY			
ITEM:	VALUE:	CONDITIONS:	REF.:
*CARASSIUS AURATUS	50 %	BP MOD 9=1/160(V/V),72H,10 FISH	63
*CARASSIUS AURATUS	50 %	BP 1100=1/160(V/V),72H,10 FISH	63
*CARASSIUS AURATUS	50 %	DISPERSOC SD=1/150(V/V),72H,10 FISH	63
*CARASSIUS AURATUS	50 %	DREW. AMER. OIL SPILL EMULS. NO 1= 1/450(V/V),72H,10 FISH	63
*CARASSIUS AURATUS	50 %	ESSO COREXIT 7664=1/10(V/V),72H,10 FISH	63
*CARASSIUS AURATUS	50 %	ESSO COREXIT 8666=1/5-1/4(V/V), 72H,10 FISH	63
*CARASSIUS AURATUS	50 %	FINA SOL OSR/2=1/40(V/V),72H,10 FISH	63
*CARASSIUS AURATUS	50 %	FINA SOL SC=1/80(V/V),72H,10 FISH	63
*CARASSIUS AURATUS	50 %	GAMLEN OIL SPILL REMOVER=1/200(V/ V),72H,10 FISH	63
*CARASSIUS AURATUS	50 %	SHELL OIL DISPERSER=1/210(V/V), 72H,10 FISH	63
*CARASSIUS AURATUS	50 %	SLIP CLEAM M=1/200(V/V),72H,10 FISH	63
*CARASSIUS AURATUS	50 %	T C 66=1/160(V/V),72H,10 FISH	63
*CARASSIUS AURATUS	50 %	TAR SOLVENT ES=1/160(V/V),72H,10 FISH	63
*CARASSIUS AURATUS	50 %	ZEROL SEA EL=1/300(V/V),72H,10 FISH	63

TABLE C257

OIL POLLUTION : HEAVY OIL COVERRAGE			
ITEM:	VALUE:	CONDITIONS:	REF.:
*MYTILUS SPP.	DAMAGE ALONG TRANSECT 1 %	CAMPUS POINT,UCPB,CAL.,USA	473

TABLE C257

OIL POLLUTION : HEAVY OIL COVERRAGE			
ITEM:	VALUE:	CONDITIONS:	REF.:
*PHYLLOSPADIX TORREYI	DAMAGE ALONG TRANSECT 30-50 %	% EXPOSED BLADES DAMAGED, CARPINTERIA REEF,CAL.,USA	473
*PHYLLOSPADIX TORREYI	DAMAGE ALONG TRANSECT 50-60 %	% EXPOSED BLADES DAMAGED,SANTA BARBARA POINT,CAL.,USA	473
*PHYLLOSPADIX TORREYI	DAMAGE ALONG TRANSECT 50-60 %	% EXPOSED BLADES DAMAGED,SANTA BARBARA POINT,CAL.,USA	473

TABLE C258

OIL POLLUTION : LIGHT OIL COVERRAGE			
ITEM:	VALUE:	CONDITIONS:	REF.:
*CHTHAMALUS FISSUS	DAMAGE ALONG TRANSECT 1 %	COAL OIL POINT,CAL.,USA	473
*CHTHAMALUS FISSUS	DAMAGE ALONG TRANSECT 20 %	CAMPUS POINT,UCSB,CAL.,USA	473
*ENTERMORPHA INTESTINALIS	DAMAGE ALONG TRANSECT 1-5 %	THALLI WHITE,COAL OIL POINT,CAL., USA	473
*GIGARTINA LEPTORYNCHOS	DAMAGE ALONG TRANSECT 1-2 %	LOON POINT,MONTECITO,CAL.,USA	473
*PORPHYRA SP.	DAMAGE ALONG TRANSECT 1-5 %	THALLI WHITE,EL CAPITAN STATE PK, CAL.,USA	473
*ULVA ANGUSTA	DAMAGE ALONG TRANSECT 1-5 %	THALLI WHITE,HOBSON BAECH,CAL.,USA	473

TABLE C259

OIL POLLUTION : MODERATE OIL COVERRAGE			
ITEM:	VALUE:	CONDITIONS:	REF.:
*ENTERMORPHA INTESTINALIS	DAMAGE ALONG TRANSECT 10-20 %	THALLI WHITE,CAMPUS POINT,UCSB, CAL.,USA	473
*ENTERMORPHA INTESTINALIS	DAMAGE ALONG TRANSECT 20 %	THALLI WHITE,CARPINTERIA REEF, CAL.,USA	473
*POLLICIPES POLYMERUS	DAMAGE ALONG TRANSECT 1-5 %	CAMPUS POINT,UCBS,CAL.,USA	473
*POLLICIPES POLYMERUS	DAMAGE ALONG TRANSECT 1-5 %	CARPINTERIA REEF,CAL.,USA	473
*PORPHYRA SP.	DAMAGE ALONG TRANSECT 1-5 %	COIL OIL POINT,CAL.,USA	473

TABLE C260

OIL POLLUTION : TOTAL NUMBER AFTER POLLUTION,1969			
ITEM:	VALUE:	CONDITIONS:	REF.:
*PLANT SPECIES	20.	NUMBERS BEFORE 18(1956-7) 20 (1966-7),EL CAPITAN STATE PK., CAL.,USA	473
*PLANT SPECIES	31.	NUMBERS BEFORE 18(1956-7) 18 (1966-7),COIL OIL POINT,CAL.,USA	473
*PLANT SPECIES	12.	NUMBERS BEFORE 13(1956-7) 14 (1966-7),CAMPUS POINT,UCSB,CAL., USA	473
*PLANT SPECIES	21.	NUMBERS BEFORE 29(1956-7) 40 (1966-7),SANTA BARBARA POINT,CAL., USA	473
*PLANT SPECIES	14.	NUMBERS BEFORE 22(1956-7) 20 (1966-7),EUCALYPTUS LANE,CAL.,USA	473
*PLANT SPECIES	17.	NUMBERS BEFORE 27(1956-7) 26 (1966-7),LOON POINT,MONTECITO, CAL.,USA	473

TABLE C260

OIL POLLUTION : TOTAL NUMBER AFTER POLLUTION,1969			
ITEM:	VALUE:	CONDITIONS:	REF.:
*PLANT SPECIES	34.	NUMBERS BEFORE 14(1956-7) 29 (1966-7),CARPINTERIA REEF,CAL.,USA	473
*PLANT SPECIES	12.	NUMBERS 21(1956-7) 13(1966-7), HOBSON BEACH,CAL.,USA	473
*PLANT SPECIES	9.	NUMBERS BEFORE 19(1956-7) 17 (1966-7),PALM STREET,VENTURA, CAL.USA	473
*PLANT SPECIES	2.	NUMBERS BEFORE 18(1956-7) 14 (1966-7),LEO CARRILLO,ST BEACH, CAL.,USA	473

TABLE C261

OXYGEN : DECREASE IN GROWTH			
ITEM:	VALUE:	CONDITIONS:	REF.:
*SALVELINUS FONTINALIS	0 % OF MAX GROWTH	OXYGEN CONC =1.7 MG O2/L, 278 K, WEIGHT = 5 G, 8 CM	965
*SALVELINUS FONTINALIS	30 % OF MAX GROWTH	OXYGEN CONC =2.05 MG O2/L, 278 K, WEIGHT = 5 G, 8 CM	965
*SALVELINUS FONTINALIS	80 % OF MAX GROWTH	OXYGEN CONC =4.80 MG O2/L, 278 K, WEIGHT = 5 G, 8 CM	965
*SALVELINUS FONTINALIS	MAX GROWTH	OXYGEN CONC =6.9 MG O2/L, 278 K, WEIGHT = 5 G, 8 CM	965
*SALVELINUS FONTINALIS	0 % OF MAX GROWTH	OXYGEN CONC =1.8 MG O2/L, 283 K, WEIGHT = 5 G, 8 CM	965
*SALVELINUS FONTINALIS	30 % OF MAX GROWTH	OXYGEN CONC =2.23 MG O2/L, 283 K, WEIGHT = 5 G, 8 CM	965
*SALVELINUS FONTINALIS	80 % OF MAX GROWTH	OXYGEN CONC =5.09 MG O2/L, 283 K, WEIGHT = 5 G, 8 CM	965
*SALVELINUS FONTINALIS	MAX GROWTH	OXYGEN CONC =11.4 MG O2/L, 283 K, WEIGHT = 5 G, 8 CM	965
*SALVELINUS FONTINALIS	0 % OF MAX GROWTH	OXYGEN CONC =4.24 MG O2/L, 288 K, WEIGHT = 5 G, 8 CM	965
*SALVELINUS FONTINALIS	30 % OF MAX GROWTH	OXYGEN CONC =4.64 MG O2/L, 288 K, WEIGHT = 5 G, 8 CM	965
*SALVELINUS FONTINALIS	80 % OF MAX GROWTH	OXYGEN CONC =9.71 MG O2/L, 288 K, WEIGHT = 5 G, 8 CM	965
*SALVELINUS FONTINALIS	0 % OF MAX GROWTH	OXYGEN CONC =4.33 MG O2/L, 293 K, WEIGHT = 5 G, 8 CM	965
*SALVELINUS FONTINALIS	30 % OF MAX GROWTH	OXYGEN CONC =5.36 MG O2/L, 293 K, WEIGHT = 5 G, 8 CM	965
*SALVELINUS FONTINALIS	0 % OF MAX GROWTH	OXYGEN CONC =1.00 MG O2/L, 278 K, WEIGHT = 400 G	965
*SALVELINUS FONTINALIS	30 % OF MAX GROWTH	OXYGEN CONC =1.12 MG O2/L, 278 K, WEIGHT = 400 G	965
*SALVELINUS FONTINALIS	80 % OF MAX GROWTH	OXYGEN CONC =2.41 MG O2/L, 278 K, WEIGHT = 400 G	965
*SALVELINUS FONTINALIS	MAX GROWTH	OXYGEN CONC =3.13 MG O2/L, 278 K, WEIGHT = 400 G	965
*SALVELINUS FONTINALIS	0 % OF MAX GROWTH	OXYGEN CONC =1.74 MG O2/L, 283 K, WEIGHT = 400 G	965
*SALVELINUS FONTINALIS	30 % OF MAX GROWTH	OXYGEN CONC =1.83 MG O2/L, 283 K, WEIGHT = 400 G	965
*SALVELINUS FONTINALIS	80 % OF MAX GROWTH	OXYGEN CONC =3.64 MG O2/L, 283 K, WEIGHT = 400 G	965
*SALVELINUS FONTINALIS	MAX GROWTH	OXYGEN CONC =6.05 MG O2/L, 283 K, WEIGHT = 400 G	965
*SALVELINUS FONTINALIS	0 % OF MAX GROWTH	OXYGEN CONC =2.90 MG O2/L, 288 K, WEIGHT = 400 G	965
*SALVELINUS FONTINALIS	30 % OF MAX GROWTH	OXYGEN CONC =3.17 MG O2/L, 288 K, WEIGHT = 400 G	965
*SALVELINUS FONTINALIS	80 % OF MAX GROWTH	OXYGEN CONC =6.47 MG O2/L, 288 K, WEIGHT = 400 G	965

TABLE C261

OXYGEN : DECREASE IN GROWTH			
ITEM:	VALUE:	CONDITIONS:	REF.:
*SALVELINUS FONTINALIS	0 % OF MAX GROWTH	OXYGEN CONC =3.57 MG O2/L, 293 K, WEIGHT = 400 G	965
*SALVELINUS FONTINALIS	30 % OF MAX GROWTH	OXYGEN CONC =4.38 MG O2/L, 293 K, WEIGHT = 400 G	965

TABLE C262

OZONE : PHOTOSYNTHESIS			
ITEM:	VALUE:	CONDITIONS:	REF.:
*PINUS ELLIOTTII	2.6 MG/(G DRY WEIGHT*HOUR)	OZONE=0.PPHM,36D,302K,8 MONTH-OLD SEEDLINGS	72
*PINUS ELLIOTTII	3.1 MG/(G DRY WEIGHT*HOUR)	OZONE=5.PPHM,36D,302K,8 MONTH-OLD SEEDLINGS	72
*PINUS ELLIOTTII	2.0 MG/(G DRY WEIGHT*HOUR)	OZONE=15.PPHM,36D,302K,8 MONTH-OLD SEEDLINGS	72
*PINUS ELLIOTTII	0.9 MG/(G DRY WEIGHT*HOUR)	OZONE=0.PPHM,77D,302K,8 MONTH-OLD SEEDLINGS	72
*PINUS ELLIOTTII	1.6 MG/(G DRY WEIGHT*HOUR)	OZONE=5.PPHM,77D,302K,8 MONTH-OLD SEEDLINGS	72
*PINUS ELLIOTTII	0.9 MG/(G DRY WEIGHT*HOUR)	OZONE=15.PPHM,77D,302K,8 MONTH-OLD SEEDLINGS	72
*PINUS ELLIOTTII	1.9 MG/(G DRY WEIG¤T*HOUR)	OZONE=0.PPHM,88D,302K,8 MONTH-OLD SEEDLINGS	72
*PINUS ELLIOTTII	1.7 MG/(G DRY WEIGHT*HOUR)	OZONE=5.PPHM,88D,302K,8 MONTH-OLD SEEDLINGS	72
*PINUS ELLIOTTII	1.4 MG/(G DRY WEIGHT*HOUR)	OZONE=15.PPHM,88D,302K,8 MONTH-OLD SEEDLINGS	72
*PINUS SEROTINA	3.5 MG/(G DRY WEIGHT*HOUR)	OZONE=0.PPHM,36D,302K,8 MONTH-OLD SEEDLINGS	72
*PINUS SEROTINA	4.2 MG/(G DRY WEIGHT*HOUR)	OZONE=5.PPHM,36D,302K,8 MONTH-OLD SEEDLINGS	72
*PINUS SEROTINA	3.3 MG/(G DRY WEIGHT*HOUR)	OZONE=15.PPHM,36D,302K,8 MONTH-OLD SEEDLINGS	72
*PINUS SEROTINA	0.9 MG/(G DRY WEIGHT*HOUR)	OZONE=0.PPHM,77D,302K,8 MONTH-OLD SEEDLINGS	72
*PINUS SEROTINA	1.8 MG/(G DRY WEIGHT*HOUR)	OZONE=5.PPHM,77D,302K,8 MONTH-OLD SEEDLINGS	72
*PINUS SEROTINA	1.6 MG/(G DRY WEIGHT*HOUR)	OZONE=15.PPHM,77D,302K,8 MONTH-OLD SEEDLINGS	72
*PINUS SEROTINA	3.4 MG/(G DRY WEIGHT*HOUR)	OZONE=0.PPHM,88D,302K,8 MONTH-OLD SEEDLINGS	72
*PINUS SEROTINA	3.8 MG/(G DRY WEIGHT*HOUR)	OZONE=5.PPHM,88D,302K,8 MONTH-OLD SEEDLINGS	72
*PINUS SEROTINA	3.3 MG/(G DRY WEIGHT*HOUR)	OZONE=15.PPHM,88D,302K,8 MONTH-OLD SEEDLINGS	72
*PINUS STROBUS	4.9 MG/(G DRY WEIGHT*HOUR)	OZONE=0.PPHM,19D,302K,4 MONTH-OLD SEEDLINGS	72
*PINUS STROBUS	4.8 MG/(G DRY WEIGHT*HOUR)	OZONE=5.PPHM,19D,302K,4 MONTH-OLD SEEDLINGS	72
*PINUS STROBUS	4.4 MG/(G DRY WEIGHT*HOUR)	OZONE=15.PPHM,19D,302K,4 MONTH-OLD SEEDLINGS	72
*PINUS STROBUS	3.2 MG/(G DRY WEIGHT*HOUR)	OZONE=0.PPHM,36D,302K,4 MONTH-OLD SEEDLINGS	72
*PINUS STROBUS	3.0 MG/(G DRY WEIGHT*HOUR)	OZONE=5.PPHM,36D,302K,4 MONTH-OLD SEEDLINGS	72
*PINUS STROBUS	2.5 MG/(G DRY WEIGHT*HOUR)	OZONE=15.PPHM,36D,302K,4 MONTH-OLD SEEDLINGS	72
*PINUS STROBUS	2.1 MG/(G DRY WEIGHT*HOUR)	OZONE=0.PPHM,77D,302K,4 MONTH-OLD SEEDLINGS	72
*PINUS STROBUS	1.9 MG/(G DRY WEIGHT*HOUR)	OZONE=5.PPHM,77D,302K,4 MONTH-OLD SEEDLINGS	72
*PINUS STROBUS	2.0 MG/(G DRY WEIGHT*HOUR)	OZONE=15.PPHM,77D,302K,4 MONTH-OLD SEEDLINGS	72

TABLE C262

OZONE : PHOTOSYNTHESIS			
ITEM:	VALUE:	CONDITIONS:	REF.:
*PINUS TAEDA	3.7 MG/(G DRY WEIGHT*HOUR)	OZONE=0.PPHM,36D,302K,8 MONTH-OLD SEEDLINGS	72
*PINUS TAEDA	4. MG/(G DRY WEIGHT*HOUR)	OZONE=5.PPHM,36D,302K,8 MONTH-OLD SEEDLINGS	72
*PINUS TAEDA	3.2 MG/(G DRY WEIGHT*HOUR)	OZONE=15.PPHM,36D,302K,8 MONTH-OLD SEEDLINGS	72
*PINUS TAEDA	1.8 MG/(G DRY WEIGHT*HOUR)	OZONE=0.PPHM,77D,302K,8 MONTH-OLD SEEDLINGS	72
*PINUS TAEDA	1.8 MG/(G DRY WEIGHT*HOUR)	OZONE=5.PPHM,77D,302K,8 MONTH-OLD SEEDLINGS	72
*PINUS TAEDA	1.7 MG/(G DRY WEIGHT*HOUR)	OZONE=15.PPHM,77D,302K,8 MONTH-OLD SEEDLINGS	72
*PINUS TAEDA	4.3 MG/(G DRY WEIGHT*HOUR)	OZONE=0.PPHM,88D,302K,8 MONTH-OLD SEEDLINGS	72
*PINUS TAEDA	4.0 MG/(G DRY WEIGHT*HOUR)	OZONE=5.PPHM,88D,302K,8 MONTH-OLD SEEDLINGS	72
*PINUS TAEDA	4.1 MG/(G DRY WEIGHT*HOUR)	OZONE=15.PPHM,88D,302K,8 MONTH-OLD SEEDLINGS	72

TABLE C263

OZONE : RESPIRATION			
ITEM:	VALUE:	CONDITIONS:	REF.:
*PINUS ELLIOTTII	0.7 MG/(G DRY WEIGHT*HOUR)	OZONE=0.PPHM,36D,302K,8 MONTH-OLD SEEDLINGS	72
*PINUS ELLIOTTII	0.6 MG/(G DRY WEIGHT*HOUR)	OZONE=5.PPHM,36D,302K,8 MONTH-OLD SEEDLINGS	72
*PINUS ELLIOTTII	1.3 MG/(G DRY WEIGHT*HOUR)	OZONE=15.PPHM,36D,302K,8 MONTH-OLD SEEDLINGS	72
*PINUS ELLIOTTII	0.4 MG/(G DRY WEIGHT*HOUR)	OZONE=0.PPHM,77D,302K,8 MONTH-OLD SEEDLINGS	72
*PINUS ELLIOTTII	0.4 MG/(G DRY WEIGHT*HOUR)	OZONE=5.PPHM,77D,302K,8 MONTH-OLD SEEDLINGS	72
*PINUS ELLIOTTII	0.6 MG/(G DRY WEIGHT*HOUR)	OZONE=15.PPHM,77D,302K,8 MONTH-OLD SEEDLINGS	72
*PINUS ELLIOTTII	0.5 MG/(G DRY WEIGHT*HOUR)	OZONE=0.PPHM,88D,302K,8 MONTH-OLD SEEDLINGS	72
*PINUS ELLIOTTII	0.5 MG/(G DRY WEIGHT*HOUR)	OZONE=5.PPHM,88D,302K,8 MONTH-OLD SEEDLINGS	72
*PINUS ELLIOTTII	0.8 MG/(G DRY WEIGHT*HOUR)	OZONE=15.PPHM,88D,302K,8 MONTH-OLD SEEDLINGS	72
*PINUS SEROTINA	1.0 MG/(G DRY WEIGHT*HOUR)	OZONE=0.PPHM,36D,302K,8 MONTH-OLD SEEDLINGS	72
*PINUS SEROTINA	0.9 MG/(G DRY WEIGHT*HOUR)	OZONE=5.PPHM,36D,302K,8 MONTH-OLD SEEDLINGS	72
*PINUS SEROTINA	1.2 MG/(G DRY WEIGHT*HOUR)	OZONE=15.PPHM,36D,302K,8 MONTH-OLD SEEDLINGS	72
*PINUS SEROTINA	0.6 MG/(G DRY WEIGHT*HOUR)	OZONE=0.PPHM,77D,302K,8 MONTH-OLD SEEDLINGS	72
*PINUS SEROTINA	0.5 MG/(G DRY WEIGHT*HOUR)	OZONE=5.PPHM,77D,302K,8 MONTH-OLD SEEDLINGS	72
*PINUS SEROTINA	0.5 MG/(G DRY WEIGHT*HOUR)	OZONE=15.PPHM,77D,302K,8 MONTH-OLD SEEDLINGS	72
*PINUS SEROTINA	1.0 MG/(G DRY WEIGHT*HOUR)	OZONE=0.PPHM,88D,302K,8 MONTH-OLD SEEDLINGS	72
*PINUS SEROTINA	1.2 MG/(G DRY WEIGHT*HOUR)	OZONE=5.PPHM,88D,302K,8 MONTH-OLD SEEDLINGS	72
*PINUS SEROTINA	1.2 MG/(G DRY WEIGHT*HOUR)	OZONE=15.PPHM,88D,302K,8 MONTH-OLD SEEDLINGS	72
*PINUS STROBUS	0.8 MG/(G DRY WEIGHT*HOUR)	OZONE=0.PPHM,36D,302K,4 MONTH-OLD SEEDLINGS	72
*PINUS STROBUS	0.6 MG/(G DRY WEIGHT*HOUR)	OZONE=5.PPHM,36D,302K,4 MONTH-OLD SEEDLINGS	72

TABLE C263

OZONE : RESPIRATION

ITEM:	VALUE:	CONDITIONS:	REF.:
*PINUS STROBUS	1.4 MG/(G DRY WEIGHT*HOUR)	OZONE=15.PPM,36D,302K,4 MONTH-OLD SEEDLINGS	72
*PINUS STROBUS	0.6 MG/(G DRY WEIGHT*HOUR)	OZONE=0.PPHM,77D,302K,4 MONTH-OLD SEEDLINGS	72
*PINUS STROBUS	0.6 MG/(G DRY WEIGHT*HOUR)	OZONE=5.PPHM,77D,302K,4 MONTH-OLD SEEDLINGS	72
*PINUS STROBUS	0.9 MG/(G DRY WEIGHT*HOUR)	OZONE=15.PPHM,77D,302K,4 MONTH-OLD SEEDLINGS	72
*PINUS TAEDA	0.7 MG/(G DRY WEIGHT*HOUR)	OZONE=0.PPHM,36D,302K,8 MONTH-OLD SEEDLINGS	72
*PINUS TAEDA	0.7 MG/(G DRY WEIGHT*HOUR)	OZONE=5.PPHM,36D,302K,8 MONTH-OLD SEEDLINGS	72
*PINUS TAEDA	1.2 MG/(G DRY WEIGHT*HOUR)	OZONE=15.PPHM,36D,302K,8 MONTH-OLD SEEDLINGS	72
*PINUS TAEDA	0.6 MG/(G DRY WEIGHT*HOUR)	OZONE=0.PPHM,77D,302K,8 MONTH-OLD SEEDLINGS	72
*PINUS TAEDA	0.6 MG/(G DRY WEIGHT*HOUR)	OZONE=5.PPHM,77D,302K,8 MONTH-OLD SEEDLINGS	72
*PINUS TAEDA	0.5 MG/(G DRY WEIGHT*HOUR)	OZONE=15.PPHM,77D,302K,8 MONTH-OLD SEEDLINGS	72
*PINUS TAEDA	1.0 MG/(G DRY WEIGHT*HOUR)	OZONE=0.PPHM,88D,302K,8 MONTH-OLD SEEDLINGS	72
*PINUS TAEDA	1.0 MG/(G DRY WEIGHT*HOUR)	OZONE=5.PPHM,88D,302K,8 MONTH-OLD SEEDLINGS	72
*PINUS TAEDA	1.2 MG/(G DRY WEIGHT*HOUR)	OZONE=15.PPHM,88D,302K,8 MONTH-OLD SEEDLINGS	72

TABLE C264

PALYTOXIN : MORTALITY

ITEM:	VALUE:	CONDITIONS:	REF.:
*MOUSE	50%	ORAL DOSE=1.5*10**(-4)MG/KG(= 4.55*10**(-11)MOL/KG),PT-50(F007)= 10.34	953

TABLE C265

PARATHION : MORTALITY

ITEM:	VALUE:	CONDITIONS:	REF.:
*MOUSE	50%	ORAL DOSE=5.5MG/KG(=1.89*10**(-5) MOL/KG),PT-50(F007)=4.72	953
*RAT	50%	SINGLE ORAL DOSE=10000MICROGRAM/KG BODY WT.	355

TABLE C266

PB : GROWTH RATE

ITEM:	VALUE:	CONDITIONS:	REF.:
*ASELLUS MERIDIANUS	0.006 MM/DAY	CONTROL=0.033MM/DAY,LEAD=0.1MG/L, 10 JUVENILES,RIVER HAYLE,U.K.	178
*ASELLUS MERIDIANUS	0.022 MM/DAY	CONTROL=0.022MM/DAY,LEAD=0.1MG/L, 10 JUVENILES,RIVER HAYLE,U.K.	178
*ASELLUS MERIDIANUS	0.006 MM/DAY	CONTROL=0.017MM/DAY,LEAD=0.1MG/L, 10 JUVENILES,RIVER HAYLE,U.K.	178

TABLE C266

PB : GROWTH RATE

ITEM:	VALUE:	CONDITIONS:	REF.:
*ASELLUS MERIDIANUS	0.006 MM/DAY	CONTROL=0.028MM/DAY,LEAD=0.1MG/L, 10 JUVENILES,BREDWELL BROOK,U.K.	178
*ASELLUS MERIDIANUS	0.05 MM/DAY	CONTROL=0.044MM/DAY,LEAD=0.1MG/L, 10 JUVENILES,RIVER GANNAL,U.K.	178
*ASELLUS MERIDIANUS	0.006 MM/DAY	CONTROL=0.033MM/DAY,LEAD=0.1MG/L, 10 JUVENILES,RIVER HAYLE,U.K.	178
*ASELLUS MERIDIANUS	0.033 MM/DAY	CONTROL=0.033MM/DAY,LEAD=0.1MG/L, 10 JUVENILES,RIVER HAYLE,U.K.	178
*ASELLUS MERIDIANUS	0.033 MM/DAY	CONTROL=0.028MM/DAY,LEAD=0.1MG/L, 10 JUVENILES,RIVER GANNAL,U.K.	178
*ASELLUS MERIDIANUS	0.002 MM/DAY	CONTROL=0.014MM/DAY,LEAD=0.1MG/L, 10 JUVENILES,BREDWELL BROOK,U.K.	1835

TABLE C267

PB : LOSS IN REPRODUCTION

ITEM:	VALUE:	CONDITIONS:	REF.:
*DAPHNIA MAGNA	50 %	100 PPB IN WATER	122
*DAPHNIA MAGNA	16 %	30 PPB IN WATER	122

TABLE C268

PB : MORTALITY

ITEM:	VALUE:	CONDITIONS:	REF.:
*ASELLUS MERIDIANUS	50 %	LEAD=1.0MG/L,10 JUVENILES,293K, TEST ANIMALS FROM RIVER HAYLE,U.K.	178
*ASELLUS MERIDIANUS	50 %	LEAD=1.75MG/L,10 JUVENILES,293K, TEST ANIMALS FROM RIVER HAYLE,U.K.	178
*ASELLUS MERIDIANUS	50 %	LEAD=3.5MG/L,10 JUVENILES,293K, TEST ANIMALS FROM RIVER HAYLE,U.K.	178
*ASELLUS MERIDIANUS	50 %	LEAD=3.5MG/L,10 JUVENILES,293K, TEST ANIMALS FROM RIVER GANNEL, U.K.	178
*ASELLUS MERIDIANUS	50 %	LEAD=0.28MG/L,10 JUVENILES,293K, TEST ANIMALS FROM BREDWELL BROOK, U.K.	178
*DAPHNIA MAGNA	50%	PBCL2,EXPOS.TIME=2DAYS,CONC.IN WATER=450 UG/L,LIMNOL.	122
*PHILODINA SP.	50 %	PBCL2=47.5 +- 1.1 PPM, 24 H, 25 PPM CACO3, 293 K, PH=7.4-7.9	1373
*PHILODINA SP.	50 %	PBCL2=39.9 +- 1.5 PPM, 48 H, 25 PPM CACO3, 293 K, PH=7.4-7.9	1373
*PHILODINA SP.	50 %	PBCL2=40.8 +- 1.6 PPM, 96 H, 25 PPM CACO3, 293 K, PH=7.4-7.9	1373
*RAT	50%	ACETATE,INTRAVEN.,DOSE=120MG/KG BODY	1611
*RAT	50%	ARSENATE,ORAL,DOSE=100MG/KG BODY	1611
*RAT	50%	2-OXIDE,INTRAPERITON.,DOSE=400MG/ KG BODY	1611
*RAT	50%	SULFIDE,INTRAPERITON.,DOSE=1600MG/ KG BODY	1611
*RAT	50%	TRIETHYL,INTRAPERITON.,DOSE= 11.2MG/KG BODY	1611

TABLE C269

PB : PHOTOSYNTHESIS

ITEM:	VALUE:	CONDITIONS:	REF.:
*AUFWUCHS COMMUNITIES	95 % OF CONTROL	PB= 0.001 MG PB/L, 3 H	1186

TABLE C269

PB : PHOTOSYNTHESIS			
ITEM:	VALUE:	CONDITIONS:	REF.:
*AUFWUCHS COMMUNITIES	90 % OF CONTROL	PB= 0.02 MG PB/L, 3 H	1186
*AUFWUCHS COMMUNITIES	84 % OF CONTROL	PB= 0.1 MG PB/L, 3 H	1186
*AUFWUCHS COMMUNITIES	35 % OF CONTROL	1.0 MG PB/L, 3 H	1186
*AUFWUCHS COMMUNITIES	20 % OF CONTROL	PB= 10.0 MG PB/L	1186

TABLE C270

PB : RESPIRATION			
ITEM:	VALUE:	CONDITIONS:	REF.:
*AUFWUCHS COMMUNITIES	104 % OF CONTROL	PB= 0.001 MG PB/L, 3 H, GOLDEN HORN	1186
*AUFWUCHS COMMUNITIES	84 % OF CONTROL	PB= 0.01 MG PB/L, 3 H, GOLDEN HORN	1186
*AUFWUCHS COMMUNITIES	70 % OF CONTROL	PB= 0.1 MG PB/L, 3 H, GOLDEN HORN	1186
*AUFWUCHS COMMUNITIES	40 % OF CONTROL	PB= 1.0 MG PB/L, 3 H, GOLDEN HORN	1186
*AUFWUCHS COMMUNITIES	17 % OF CONTROL	PB= 10.0 MG PB/L, 3 H, GOLDEN HORN	1186

TABLE C271

PCB : BLOOD GLUCOSE			
ITEM:	VALUE:	CONDITIONS:	REF.:
*ANGUILLA ANGUILLA	82(+-8) MG/100 ML	CONTROL=82.7MG/100 ML,PCB=3*1000 MG/KG WW (EVERY SEC.DAY),1W	325
*ANGUILLA ANGUILLA	61.9(+-3.2) MG/100 ML	CONTROL=65.7MG/100 ML,PCB=6*1000 MG/KG WW (EVERY SEC.DAY),2W	325
*ANGUILLA ANGUILLA	62.5(+-2.2) MG/100 ML	CONTROL=62.6MG/100 ML,PBC=1*200 MG/KG WW (ONE TIME),1M	325
*ANGUILLA ANGUILLA	79.9(+-5.5) MG/100 ML	CONTROL=70.MG/100 ML,PCB=1*200 MG/KG WW (ONE TIME),4M	325

TABLE C272

PCB : BLOOD LACTATE			
ITEM:	VALUE:	CONDITIONS:	REF.:
*ANGUILLA ANGUILLA	19.7(+-1.9) MG/100 ML	CONTROL=14.3MG/100 ML,PCB=3*1000 MG/KG WW (EVERY SEC.DAY),1W	325
*ANGUILLA ANGUILLA	9.(+-3.6) MG/100 ML	CONTROL=4.9MG/100 ML,PCB=6*1000 MG/KG WW (EVERY SEC.DAY),2W	325
*ANGUILLA ANGUILLA	6.58(+-0.88) MG/100 ML	CONTROL=5.86MG/100 ML,PBC=1*200 MG/KG WW (ONE TIME),1M	325
*ANGUILLA ANGUILLA	7.78(+-1.35) MG/100 ML	CONTROL=9.51MG/100 ML,PCB=1*200 MG/KG WW (ONE TIME),4M	325

TABLE C273

PCB : BODY WEIGHT			
ITEM:	VALUE:	CONDITIONS:	REF.:
*MOUSE	11.54+-0.72 G	INJECTION OF 50 MG PCB/ KG BODY WEIGHT,21 D,CONT=8.96+-0.85 G	820

TABLE C273

PCB : BODY WEIGHT			
ITEM:	VALUE:	CONDITIONS:	REF.:
*MOUSE	22.76+-2.23 G	INJECTION OF 50 MG PCB/ KG BODY WEIGHT,33 D,CONT=21.30+-2.34 G	820
*MOUSE	8.47+-0.94 G	INJECTION OF 50 MG PCB/ KG WW AT BIRTH, CONT = 7.25+-0.76 G, 16 D	820

TABLE C274

PCB : ENZYME ACTIVITY IN BLOOD			
ITEM:	VALUE:	CONDITIONS:	REF.:
*SALMO GAIRDNERI	5.2(+-0.2) MICROGRAM PBG/(ML*HOUR)	CONTROL=5.6MICROGRAM PBG/(ML*HOUR) ,PCB=0.2G/FISH ONE TIME),3D	1681
*SALMO GAIRDNERI	5.3(+-0.2) MICROGRAM PBG/(ML*HOUR)	CONTROL=4.7MICROGRAM PBG/(ML*HOUR) ,PCB=0.2G/FISH ONE TIME),7D	1681
*SALMO GAIRDNERI	6.4(+-0.3) MICROGRAM PBG/(ML*HOUR)	CONTROL=5.9MICROGRAM PBG/(ML*HOUR) ,PCB=0.2G/FISH ONE TIME),14D	1681
*SALMO GAIRDNERI	5.4(+-0.2) MICROGRAM PBG/(ML*HOUR)	CONTROL=5.1MICROGRAM PBG/(ML*HOUR) ,PCB=0.2G/FISH ONE TIME),21D	1681
*SALMO GAIRDNERI	203.7(+-11.8) MICROGRAM PBG/ML*H	CONTROL=238.9MICROGRAM PBG/ML*H, PCB=0.2G/FISH ONE TIME),3D	1681
*SALMO GAIRDNERI	202.(+-11.1) MICROGRAM PBG/ML*H	CONTROL=180.MICROGRAM PBG/ML*H,PCB =0.2G/FISH ONE TIME),7D	1681
*SALMO GAIRDNERI	215.(+-16.) MICROGRAM PBG/ML*H	CONTROL=245.MICROGRAM PBG/ML*H,PCB =0.2G/FISH ONE TIME),14D	1681
*SALMO GAIRDNERI	182.4(+-13.2) MICROGRAM PBG/ML*H	CONTROL=188.9MICROGRAM PBG/ML*H, PCB=0.2G/FISH ONE TIME),21D	1681

TABLE C275

PCB : ENZYME ACTIVITY IN KIDNEY			
ITEM:	VALUE:	CONDITIONS:	REF.:
*SALMO GAIRDNERI	94.7(+-6.9) MICROGRAM PBG/G*TISSUE W.WT.*H	CONTROL=94.7MICROGRAM PBG/G*TISSUE W.WT.*H,PCB=0.2G/FISH ONE TIME), 3D	1681
*SALMO GAIRDNERI	88.5(+-2.) MICROGRAM PBG/G*TISSUE W.WT.*H	CONTROL=90.1MICROGRAM PBG/G*TISSUE W.WT.*H,PCB=0.2G/FISH ONE TIME), 7D	1681
*SALMO GAIRDNERI	92.7(+-4.1) MICROGRAM PBG/G*TISSUE W.WT.*H	CONTROL=102.9MICROGRAM PBG/ G*TISSUE W.WT.*H,PCB=0.2G/FISH ONE TIME),14D	1681
*SALMO GAIRDNERI	82.7(+-3.5) MICROGRAM PBG/G*TISSUE W.WT.*H	CONTROL=82.6MICROGRAM PBG/G*TISSUE W.WT.*H,PCB=0.2G/FISH ONE TIME), 21D	1681

TABLE C276

PCB : ENZYME ACTIVITY IN SPLEEN			
ITEM:	VALUE:	CONDITIONS:	REF.:
*SALMO GAIRDNERI	256.6(+-17.8) MICROGRAM PBG/ G*TISSUE W.WT.*H	CONTROL=197.4MICROGRAM PBG/ G*TISSUE W.WT.*H,PCB=0.2G/FISH ONE TIME),3D	1681
*SALMO GAIRDNERI	339.(+-39.5) MICROGRAM PBG/ G*TISSUE W.WT.*H	CONTROL=313.1MICROGRAM PBG/ G*TISSUE W.WT.*H,PCB=0.2G/FISH ONE TIME),7D	1681
*SALMO GAIRDNERI	399.6(+-28.6) MICROGRAM PBG/ G*TISSUE W.WT.*H	CONTROL=297.5MICROGRAM PBG/ G*TISSUE W.WT.*H,PCB=0.2G/FISH ONE TIME),14D	1681

TABLE C276

PCB : ENZYME ACTIVITY IN SPLEEN

ITEM:	VALUE:	CONDITIONS:	REF.:
*SALMO GAIRDNERI	341.(+25.4) MICROGRAM PBG/G*TISSUE W.WT.*H	CONTROL=244.8MICROGRAM PBG/ G*TISSUE W.WT.*H,PCB=0.2G/FISH ONE TIME),21D	1681

TABLE C277

PCB : GONAD SOMATIC INDEX

ITEM:	VALUE:	CONDITIONS:	REF.:
*ANGUILLA ANGUILLA	0.35(+-0.04)	CONTROL=0.38,PCB=3*1000MG/KG BODYWT.(EVERY SEC.DAY),1W	325
*ANGUILLA ANGUILLA	0.5(+-0.19)	CONTROL=0.32,PCB=6*1000MG/KG BODYWT.(EVERY SEC.DAY),2W	325
*ANGUILLA ANGUILLA	0.62(+-0.11)	CONTROL=0.64,PBC=1*200MG/KG BODYWT.(ONE TIME),1M	325
*ANGUILLA ANGUILLA	0.32(+-0.09)	CONTROL=0.43,PCB=1*200MG/KG BODYWT.(ONE TIME),4M	325

TABLE C278

PCB : GROWTH RATE

ITEM:	VALUE:	CONDITIONS:	REF.:
*ANGUILLA ANGUILLA	5.92(+-0.65) %	CONTROL=5.30%,PCB=3*1000MG/KG BODYWT.(EVERY SEC.DAY),1W	325
*ANGUILLA ANGUILLA	5.27(+-0.69) %	CONTROL=6.73%,PCB=6*1000MG/KG BODYWT.(EVERY SEC.DAY),2W	325
*ANGUILLA ANGUILLA	8.63(+-1.27) %	CONTROL=7.36%,PBC=1*200MG/KG BODYWT.(ONE TIME),1M	325
*ANGUILLA ANGUILLA	15.85(+-1.71) %	CONTROL=18.04%,PCB=1*200MG/KG BODYWT.(ONE TIME),4M	325
*DUNIALIELLA TERTIOLECTA	0.74 D**(-1)	PCB=0PPM,285K,120H,CONTROL=100%	459
*DUNIALIELLA TERTIOLECTA	0.75 D**(-1)	PCB=0.1PPM,285K,120H,CONTROL=101%	459
*DUNIALIELLA TERTIOLECTA	0.79 D**(-1)	PCB=10PMM,285K,120H,CONTROL=107%	459
*DUNIALIELLA TERTIOLECTA	1.22 D**(-1)	PCB=0PPM,290K,72H,CONTROL=100%	459
*DUNIALIELLA TERTIOLECTA	1.23 D**(-1)	PCB=0.1PPM,290K,72H,CONTROL=101%	459
*DUNIALIELLA TERTIOLECTA	1.18 D**(-1)	PCB=10PMM,290K,72H,CONTROL=97%	459
*DUNIALIELLA TERTIOLECTA	1.66 D**(-1)	PCB=0PPM,298K,72H,CONTROL=100%	459
*DUNIALIELLA TERTIOLECTA	1.66 D**(-1)	PCB=0.1PPM,298K,72H,CONTROL=100%	459
*DUNIALIELLA TERTIOLECTA	1.69 D**(-1)	PCB=10PMM,298K,72H,CONTROL=102%	459
*RHIZOSOLENIA SETIGERA	0.6 D**(-1)	PCB=0PPM,283K,192H,CONTROL=100%	459
*RHIZOSOLENIA SETIGERA	0.49 D**(-1)	PCB=0.1PPM,283K,192H,CONTROL=82%	459
*RHIZOSOLENIA SETIGERA	0.46 D**(-1)	PCB=1PPM,283K,192H,CONTROL=77%	459
*RHIZOSOLENIA SETIGERA	0.49 D**(-1)	PCB=10PPM,283K,192H,CONTROL=82%	459
*RHIZOSOLENIA SETIGERA	0.86 D**(-1)	PCB=0PPM,288H,144H,CONTROL=100%	459
*RHIZOSOLENIA SETIGERA	0.81 D**(-1)	PCB=0.1PPM,288K,144H,CONTROL=95%	459
*RHIZOSOLENIA SETIGERA	0.75 D**(-1)	PCB=1PPM,288K,144H,CONTROL=87%	459
*RHIZOSOLENIA SETIGERA	0.73 D**(-1)	PCB=10PPM,288K,144H,CONTROL=85%	459
*THALASSIOSIRA PSEUDONANA	1.43 D**(-1)	PCB=0PPM,285K,120H,CONTROL=100%	459
*THALASSIOSIRA PSEUDONANA	1.57 D**(-1)	PCB=0.1PPM,285K,120H,CONTROL=109%	459
*THALASSIOSIRA PSEUDONANA	0.6 D**(-1)	PCB=10PMM,285K,120H,CONTROL=42%	459
*THALASSIOSIRA PSEUDONANA	2.34 D**(-1)	PCB=0PPM,290K,72H,CONTROL=100%	459
*THALASSIOSIRA PSEUDONANA	2.15 D**(-1)	PCB=0.1PPM,290K,72H,CONTROL=84%	459

TABLE C278

PCB : GROWTH RATE			
ITEM:	VALUE:	CONDITIONS:	REF.:
*THALASSIOSIRA PSEUDONANA	1.31 D**(-1)	PCB=10PMM,290K,72H,CONTROL=56%	459
*THALASSIOSIRA PSEUDONANA	2.57 D**(-1)	PCB=0PPM,298K,72H,CONTROL=100%	459
*THALASSIOSIRA PSEUDONANA	2.59 D**(-1)	PCB=0.1PPM,298K,72H,CONTROL=101%	459
*THALASSIOSIRA PSEUDONANA	2.13 D**(-1)	PCB=10PMM,298K,72H,CONTROL=83%	459

TABLE C279

PCB : HEMATOCRIT			
ITEM:	VALUE:	CONDITIONS:	REF.:
*ANGUILLA ANGUILLA	33.2(+-2.4) %	CONTROL=34.%,PCB=3*1000MG/KG BODYWT.(EVERY SEC.DAY),1W	773
*ANGUILLA ANGUILLA	31.1(+-0.9) %	CONTROL=38.%,PBC=6*1000MG/KG BODYWT.(EVERY SEC.DAY),2W	773
*ANGUILLA ANGUILLA	34.1(+-1.4) %	CONTROL=36.1%,PBC=1*200MG/KG BODYWT.(ONE TIME),4W	773
*ANGUILLA ANGUILLA	31.7(+-1.2) %	CONTROL=30.4%,PCB=1*200MG/KG BODYWT.(ONE TIME),16W	773
*SALMO GAIRDNERI	43.4(+-4.8) %	CONTROL=49.7%,PCB=1*200MG/KG BODYWT.(ONE TIME),5D	773
*SALMO GAIRDNERI	48.7(+-3.2) %	CONTROL=46.3%,PCB=1*200MG/KG BODYWT.(ONE TIME),10D	773
*SALMO GAIRDNERI	48.5(+-2.9) %	CONTROL=55.2%,PCB=1*200MG/KG BODYWT.(ONE TIME),17D	773
*SALMO GAIRDNERI	39.4(+-1.9) %	CONTROL=43.9%,PCB=0.2G(ONE TIME), 3D	1681
*SALMO GAIRDNERI	38.4(+-2.2) %	CONTROL=37.9%,PCB=0.2G(ONE TIME), 7D	1681
*SALMO GAIRDNERI	34.4(+-4.2) %	CONTROL=41.9%,PCB=0.2G(ONE TIME), 14D	1681
*SALMO GAIRDNERI	32.9(+-37.5) %	CONTROL=37.5%,PCB=0.2G(ONE TIME), 21D	1681

TABLE C280

PCB : HEMOGLOBIN			
ITEM:	VALUE:	CONDITIONS:	REF.:
*ANGUILLA ANGUILLA	9.9(+-0.8) G/100 ML	CONTROL=9.3G/100 ML,PCB=3*1000MG/KG BODYWT.(EVERY SEC.DAY),1W	773
*ANGUILLA ANGUILLA	9.6(+-0.5) G/100 ML	CONTROL=10.3G/100 ML,PBC=6*1000MG/KG BODYWT.(EVERY SEC.DAY),2W	773
*ANGUILLA ANGUILLA	9.7(+-0.4) G/100 ML	CONTROL=10.5G/100 ML,PBC=1*200MG/KG BODYWT.(ONE TIME),4W	773
*ANGUILLA ANGUILLA	9.3(+-0.5) G/100 ML	CONTROL=8.6G/100 ML,PCB=1*200MG/KG BODYWT.(ONE TIME),16W	773
*SALMO GAIRDNERI	8.5(+-0.4) G/100 ML	CONTROL=10.G/100 ML,PCB=1*200MG/KG BODYWT.(ONE TIME),5D	773
*SALMO GAIRDNERI	9.4(+-0.6) G/100 ML	CONTROL=10.4G/100 ML,PCB=1*200MG/KG BODYWT.(ONE TIME),10D	773
*SALMO GAIRDNERI	10.(+-0.7) G/100 ML	CONTROL=11.4G/100 ML,PCB=1*200MG/KG BODYWT.(ONE TIME),17D	773
*SALMO GAIRDNERI	8.2(+-0.6) G/100 ML	CONTROL=8.3G/100 ML,PCB=0.2G(ONE TIME),3D	1681
*SALMO GAIRDNERI	9.(+-0.6) G/100 ML	CONTROL=7.8G/100 ML,PCB=0.2G(ONE TIME),7D	1681
*SALMO GAIRDNERI	7.6(+-0.9) G/100 ML	CONTROL=8.8G/100 ML,PCB=0.2G(ONE TIME),14D	1681
*SALMO GAIRDNERI	8.(+-0.6) G/100 ML	CONTROL=8.G/100 ML,PCB=0.2G(ONE TIME),21D	1681

TABLE C281

PCB : LIVER GLYCOGEN

ITEM:	VALUE:	CONDITIONS:	REF.:
*ANGUILLA ANGUILLA	8.(+-1.88) MG/100 MG	CONTROL=6.94MG/100 MG,PCB=3*1000MG/KG BODYWT.(EVERY SEC.DAY),1W	325
*ANGUILLA ANGUILLA	5.11(+-1.05) MG/100 MG	CONTROL=6.03MG/100 MG,PCB=6*1000MG/KG BODYWT.(EVERY SEC.DAY),2W	325
*ANGUILLA ANGUILLA	5.29(+-0.83) MG/100 MG	CONTROL=6.47MG/100 MG,PBC=1*200MG/KG BODYWT.(ONE TIME),1M	325
*ANGUILLA ANGUILLA	4.47(+-0.33) MG/100 MG	CONTROL=4.08MG/100 MG,PCB=1*200MG/KG BODYWT.(ONE TIME),4M	325

TABLE C282

PCB : LIVER PHOSPHOLIPIDS

ITEM:	VALUE:	CONDITIONS:	REF.:
*ANGUILLA ANGUILLA	2.27(0.21)	CONTROL=2.81,PCB=3*1000MG/KG BODYWT.(EVERY SEC.DAY),1W	325
*ANGUILLA ANGUILLA	2.49(+-0.18)	CONTROL=2.41,PCB=6*1000MG/KG BODYWT.(EVERY SEC.DAY),2W	325

TABLE C283

PCB : LIVER PROTEIN

ITEM:	VALUE:	CONDITIONS:	REF.:
*ANGUILLA ANGUILLA	13.9(+-0.9) MG/100 MG	CONTROL=14.1MG/100 MG,PCB=3*1000MG/KG BODYWT.(EVERY SEC.DAY),1W	325
*ANGUILLA ANGUILLA	15.2(+-1.1) MG/100 MG	CONTROL=14.1MG/100 MG,PCB=6*1000MG/KG BODYWT.(EVERY SEC.DAY),2W	325
*ANGUILLA ANGUILLA	19.(+-0.5) MG/100 MG	CONTROL=18.8MG/100 MG,PBC=1*200MG/KG BODYWT.(ONE TIME),1M	325
*ANGUILLA ANGUILLA	20.(+-1.) MG/100 MG	CONTROL=20.6MG/100 MG,PCB=1*200MG/KG BODYWT.(ONE TIME),4M	325

TABLE C284

PCB : LIVER SOMATIC INDEX

ITEM:	VALUE:	CONDITIONS:	REF.:
*ANGUILLA ANGUILLA	1.54(+-0.09)	CONTROL=1.43,PCB=3*1000MG/KG BODYWT.(EVERY SEC.DAY),1W	325
*ANGUILLA ANGUILLA	1.59(+-0.22)	CONTROL=1.42,PCB=6*1000MG/KG BODYWT.(EVERY SEC.DAY),2W	325
*ANGUILLA ANGUILLA	0.853(+-0.038)	CONTROL=0.814,PBC=1*200MG/KG BODYWT.(ONE TIME),1M	325
*ANGUILLA ANGUILLA	0.715(+-0.032)	CONTROL=0.758,PCB=1*200MG/KG BODYWT.(ONE TIME),4M	325

TABLE C285

PCB : LIVER TRIGLYCERIDES

ITEM:	VALUE:	CONDITIONS:	REF.:
*ANGUILLA ANGUILLA	6.06(+-2.7) MG/100 MG	CONTROL=5.73MG/100 MG,PCB=3*1000MG/KG BODYWT.(EVERY SEC.DAY),1W	325

TABLE C285

PCB : LIVER TRIGLYCERIDES			
ITEM:	VALUE:	CONDITIONS:	REF.:
*ANGUILLA ANGUILLA	8.26(+-1.09) MG/100 MG	CONTROL=5.12MG/100 MG,PCB= 6*1000MG/KG BODYWT.(EVERY SEC.DAY) ,2W	325
*ANGUILLA ANGUILLA	13.4(+-1.) MG/100 MG	CONTROL=12.6MG/100 MG,PBC=1*200MG/ KG BODYWT.(ONE TIME),1M	325
*ANGUILLA ANGUILLA	13.(+-2.7) MG/100 MG	CONTROL=12.6MG/100 MG,PCB=1*200MG/ KG BODYWT.(ONE TIME),4M	325

TABLE C286

PCB : LIVER WATER			
ITEM:	VALUE:	CONDITIONS:	REF.:
*ANGUILLA ANGUILLA	68.5(+-1.2) MG/100 MG	CONTROL=68.6MG/100 MG,PCB= 3*1000MG/KG BODYWT.(EVERY SEC.DAY) ,1W	325
*ANGUILLA ANGUILLA	66.5(+-0.6) MG/100 MG	CONTROL=68.3MG/100 MG,PCB= 6*1000MG/KG BODYWT.(EVERY SEC.DAY) ,2W	325
*ANGUILLA ANGUILLA	64.7(+-1.) MG/100 MG	CONTROL=63.2MG/100 MG,PBC=1*200MG/ KG BODYWT.(ONE TIME),1M	325
*ANGUILLA ANGUILLA	65.4(+-2.) MG/100 MG	CONTROL=65.6MG/100 MG,PCB=1*200MG/ KG BODYWT.(ONE TIME),4M	325

TABLE C287

PCB : MCH			
ITEM:	VALUE:	CONDITIONS:	REF.:
*ANGUILLA ANGUILLA	96.5(+-15.4) PICOGRAM/CELL	CONTROL=82PICOGRAM/CELL,PCB= 3*1000MG/KG BODYWT.(EVERY SEC.DAY) ,1W	773
*ANGUILLA ANGUILLA	97.7(+-12.6) PICOGRAM/CELL	CONTROL=85.4PICOGRAM/CELL,PBC= 6*1000MG/KG BODYWT.(EVERY SEC.DAY) ,2W	773
*ANGUILLA ANGUILLA	79.4(+-6.7) PICOGRAM/CELL	CONTROL=91.7PICOGRAM/CELL,PBC= 1*200MG/KG BODYWT.(ONE TIME),4W	773
*ANGUILLA ANGUILLA	67.2(+-4.1) PICOGRAM/CELL	CONTROL=70.6PICOGRAM/CELL,PCB= 1*200MG/KG BODYWT.(ONE TIME),16W	773
*SALMO GAIRDNERI	73.6(+-4.5) PICOGRAM/CELL	CONTROL=83.2PICOGRAM/CELL,PCB= 1*200MG/KG BODYWT.(ONE TIME),5D	773
*SALMO GAIRDNERI	81.7(+-3.8) PICOGRAM/CELL	CONTROL=84.8PICOGRAM/CELL,PCB= 1*200MG/KG BODYWT.(ONE TIME),10D	773
*SALMO GAIRDNERI	70.9(+-3.6) PICOGRAM/CELL	CONTROL=74.1PICOGRAM/CELL,PCB= 1*200MG/KG BODYWT.(ONE TIME),17D	773

TABLE C288

PCB : MCHC			
ITEM:	VALUE:	CONDITIONS:	REF.:
*ANGUILLA ANGUILLA	29.7(+-1.) %	CONTROL=27.4%,PCB=3*1000MG/KG BODYWT.(EVERY SEC.DAY),1W	773
*ANGUILLA ANGUILLA	29.8(+-1.3) %	CONTROL=27.1%,PBC=6*1000MG/KG BODYWT.(EVERY SEC.DAY),2W	773
*ANGUILLA ANGUILLA	28.5(+-1.3) %	CONTROL=29.3%,PBC=1*200MG/KG BODYWT.(ONE TIME),4W	773
*ANGUILLA ANGUILLA	29.2(+-0.8) %	CONTROL=28.5%,PCB=1*200MG/KG BODYWT.(ONE TIME),16W	773

TABLE C288

PCB : MCHC

ITEM:	VALUE:	CONDITIONS:	REF.:
*SALMO GAIRDNERI	21.7(+-1.) %	CONTROL=20.%,PCB=1*200MG/KG BODYWT.(ONE TIME),5D	773
*SALMO GAIRDNERI	19.3(+-0.4) %	CONTROL=22.5%,PCB=1*200MG/KG BODYWT.(ONE TIME),10D	773
*SALMO GAIRDNERI	20.5(+-0.5) %	CONTROL=20.7%,PCB=1*200MG/KG BODYWT.(ONE TIME),17D	773
*SALMO GAIRDNERI	20.6(+-0.8) %	CONTROL=18.5%,PCB=0.2G(ONE TIME), 3D	1681
*SALMO GAIRDNERI	23.5(+-0.8) %	CONTROL=20.4%,PCB=0.2G(ONE TIME), 7D	1681
*SALMO GAIRDNERI	22.2(+-0.6) %	CONTROL=21.1%,PCB=0.2G(ONE TIME), 14D	1681
*SALMO GAIRDNERI	24.7(+-0.8) %	CONTROL=21.6%,PCB=0.2G(ONE TIME), 21D	1681

TABLE C289

PCB : MCV

ITEM:	VALUE:	CONDITIONS:	REF.:
*ANGUILLA ANGUILLA	328.4(+-58.2) MICRO**3/CELL	CONTROL=296.9MICRO**3/CELL,PCB= 3*1000MG/KG BODYWT.(EVERY SEC.DAY) ,1W	773
*ANGUILLA ANGUILLA	327.(+-35.4) MICRO**3/CELL	CONTROL=317.9MICRO**3/CELL,PBC= 6*1000MG/KG BODYWT.(EVERY SEC.DAY) ,2W	773
*ANGUILLA ANGUILLA	285.2(+-30.6) MICRO**3/CELL	CONTROL=310.1MICRO**3/CELL,PBC= 1*200MG/KG BODYWT.(ONE TIME),4W	773
*ANGUILLA ANGUILLA	222.6(+-15.2) MICRO**3/CELL	CONTROL=249.8MICRO**3/CELL,PCB= 1*200MG/KG BODYWT.(ONE TIME),16W	773
*SALMO GAIRDNERI	356.1(+-21.) MICRO**3/CELL	CONTROL=417.3MICRO**3/CELL,PCB= 1*200MG/KG BODYWT.(ONE TIME),5D	773
*SALMO GAIRDNERI	424.3(+-21.4) MICRO**3/CELL	CONTROL=375.4MICRO**3/CELL.PCB= 1*200MG/KG BODYWT.(ONE TIME),10D	773
*SALMO GAIRDNERI	344.3(+-21.3) MICRO**3/CELL	CONTROL=359.2MICRO**3/CELL.PCB= 1*200MG/KG BODYWT.(ONE TIME),17D	773

TABLE C290

PCB : MEAN WEIGHTS (F022)

ITEM:	VALUE:	CONDITIONS:	REF.:
*ESCHERICHIA COLI	86.0 MG	PCB=0.1 MG/L, CONTROL= 44.5 MG, 24 H	808
*ESCHERICHIA COLI	90.4 MG	PCB=0.01 MG/L, CONTROL= 44.5 MG, 24 H	808
*ESCHERICHIA COLI	186.4 MG	PCB=0.1 MG/L, CONTROL= 160.8 MG, 24 H	808
*ESCHERICHIA COLI	153.2 MG	PCB=0.01 MG/L, CONTROL= 160.8 MG, 24 H	808

TABLE C291

PCB : MORTALITY

ITEM:	VALUE:	CONDITIONS:	REF.:
*CYPRINUS CARPIO	50 %	PCB AS TRICHLORATED FORM (KC300) = 1.45 PPM, 96 H, FRY	823

TABLE C291

PCB : MORTALITY			
ITEM:	VALUE:	CONDITIONS:	REF.:
*LEBISTES RETICULATUS	50 %	PCB AS TRICHLORATED FORM (KC300) = 0.90 PPM, 96 H, FRY	823
*LEBISTES RETICULATUS	50 %	PCB AS TRICHLORATED FORM (KC300) = 3.2 PPM, 96 H, ADULT	823

TABLE C292

PCB : MUSCLE GLYCOGEN			
ITEM:	VALUE:	CONDITIONS:	REF.:
*ANGUILLA ANGUILLA	0.22(+-0.01) MG/100 MG	CONTROL=0.2MG/100 MG,PCB=3*1000MG/KG BODYWT.(EVERY SEC.DAY),1W	325
*ANGUILLA ANGUILLA	0.2(+-0.05) MG/100 MG	CONTROL=0.26MG/100 MG,PCB=6*1000MG/KG BODYWT.(EVERY SEC.DAY),2W	325
*ANGUILLA ANGUILLA	0.125(+-0.018) MG/100 MG	CONTROL=0.147MG/100 MG,PBC=1*200MG/KG BODYWT.(ONE TIME),1M	325
*ANGUILLA ANGUILLA	0.192(+-0.026) MG/100 MG	CONTROL=0.189MG/100 MG,PCB=1*200MG/KG BODYWT.(ONE TIME),4M	325

TABLE C293

PCB : MUSCLE PHOSPHOLIPIDS			
ITEM:	VALUE:	CONDITIONS:	REF.:
*ANGUILLA ANGUILLA	0.66(+-0.06) MG/100 MG	CONTROL=0.63MG/100 MG,PCB=3*1000MG/KG BODYWT.(EVERY SEC.DAY),1W	325
*ANGUILLA ANGUILLA	0.51(+-0.01) MG/100 MG	CONTROL=0.54MG/100 MG,PCB=6*1000MG/KG BODYWT.(EVERY SEC.DAY),2W	325

TABLE C294

PCB : MUSCLE PROTEIN			
ITEM:	VALUE:	CONDITIONS:	REF.:
*ANGUILLA ANGUILLA	21.6(+-0.8) MG/100 MG	CONTROL=21.4MG/100 MG,PCB=3*1000MG/KG BODYWT.(EVERY SEC.DAY),1W	325
*ANGUILLA ANGUILLA	20.9(+-0.6) MG/100 MG	CONTROL=20.4MG/100 MG,PCB=6*1000MG/KG BODYWT.(EVERY SEC.DAY),2W	325
*ANGUILLA ANGUILLA	22.7(+-0.4) MG/100 MG	CONTROL=22.3MG/100 MG,PBC=1*200MG/KG BODYWT.(ONE TIME),1M	325
*ANGUILLA ANGUILLA	23.5(+-0.5) MG/100 MG	CONTROL=23.9MG/100 MG,PCB=1*200MG/KG BODYWT.(ONE TIME),4M	325

TABLE C295

PCB : MUSCLE TRIGLYCERIDES			
ITEM:	VALUE:	CONDITIONS:	REF.:
*ANGUILLA ANGUILLA	7.02(+-1.93) MG/100 MG	CONTROL=8.66MG/100 MG,PCB=3*1000MG/KG BODYWT.(EVERY SEC.DAY),1W	325

TABLE C295

PCB : MUSCLE TRIGLYCERIDES			
ITEM:	VALUE:	CONDITIONS:	REF.:
*ANGUILLA ANGUILLA	10.35(+-0.04) MG/100 MG	CONTROL=7.23MG/100 MG,PCB= 6*1000MG/KG BODYWT.(EVERY SEC.DAY) ,2W	325
*ANGUILLA ANGUILLA	18.9(+-3.3) MG/100 MG	CONTROL=21.6MG/100 MG,PBC=1*200MG/ KG BODYWT.(ONE TIME),1M	325
*ANGUILLA ANGUILLA	21.5(+-2.4) MG/100 MG	CONTROL=18.4MG/100 MG,PCB=1*200MG/ KG BODYWT.(ONE TIME),4M	325

TABLE C296

PCB : MUSCLE WATER			
ITEM:	VALUE:	CONDITIONS:	REF.:
*ANGUILLA ANGUILLA	70.9(+-2.) MG/100 MG	CONTROL=69.9MG/100 MG,PCB= 3*1000MG/KG BODYWT.(EVERY SEC.DAY) ,1W	325
*ANGUILLA ANGUILLA	69.7(+-0.4) MG/100 MG	CONTROL=73.1MG/100 MG,PCB= 6*1000MG/KG BODYWT.(EVERY SEC.DAY) ,2W	325
*ANGUILLA ANGUILLA	62.9(+-2.2) MG/100 MG	CONTROL=63.MG/100 MG,PBC=1*200MG/ KG BODYWT.(ONE TIME),1M	325
*ANGUILLA ANGUILLA	63.7(+-1.5) MG/100 MG	CONTROL=66.1MG/100 MG,PCB=1*200MG/ KG BODYWT.(ONE TIME),4M	325

TABLE C297

PCB : PLASMA CALCIUM			
ITEM:	VALUE:	CONDITIONS:	REF.:
*ANGUILLA ANGUILLA	2,67(+-0.05) MILLI-MOL	CONTROL=2.85MILLI-MOL,PBC=1*200MG/ KG BODYWT.(ONE TIME),1M	325
*ANGUILLA ANGUILLA	2.67(+-0.02) MILLI-MOL	CONTROL=2.75MILLI-MOL,PCB=1*200MG/ KG BODYWT.(ONE TIME),4M	325

TABLE C298

PCB : PLASMA CHLORIDE			
ITEM:	VALUE:	CONDITIONS:	REF.:
*ANGUILLA ANGUILLA	130.(+-2.) MILLI-MOL	CONTROL=132.MILLI-MOL,PCB= 3*1000MG/KG BODYWT.(EVERY SEC.DAY) ,1W	325
*ANGUILLA ANGUILLA	131.(+-1.) MILLI-MOL	CONTROL=135.MILLI-MOL,PCB= 6*1000MG/KG BODYWT.¤EVERY SEC.DAY) ,2W	325
*ANGUILLA ANGUILLA	123.(+-2.) MILLI-MOL	CONTROL=114.MILLI-MOL,PBC=1*200MG/ KG BODYWT.(ONE TIME),1M	325
*ANGUILLA ANGUILLA	106.(+-3.) MILLI-MOL	CONTROL=106.MILLI-MOL,PCB=1*200MG/ KG BODYWT.(ONE TIME),4M	325

TABLE C299

PCB : PLASMA CORTISOL			
ITEM:	VALUE:	CONDITIONS:	REF.:
*ANGUILLA ANGUILLA	8.(+-1.8) MICROGRAM/100 ML	CONTROL=8.8MICROGRAM/100 ML,PBC= 1*200MG/KG BODYWT.(ONE TIME),1M	325

TABLE C299

PCB : PLASMA CORTISOL			
ITEM:	VALUE:	CONDITIONS:	REF.:
*ANGUILLA ANGUILLA	3.8(+-1.5) MICROGRAM/100 ML	CONTROL=4.MICROGRAM/100 ML,PBC=1*200MG/KG BODYWT.(ONE TIME),1M	325

TABLE C300

PCB : PLASMA ESTERIFIED CHOLESTEROL			
ITEM:	VALUE:	CONDITIONS:	REF.:
*ANGUILLA ANGUILLA	3.13(+-0.07) MILLI-MOL	CONTROL=3.54MILLI-MOL,PCB=3*1000MG/KG BODYWT.(EVERY SEC.DAY),1W	325
*ANGUILLA ANGUILLA	3.87(+-03) MILLI-MOL	CONTROL=4.06MILLI-MOL,PCB=6*1000MG/KG BODYWT.(EVERY SEC.DAY),2W	325

TABLE C301

PCB : PLASMA FREE CHOLESTEROL			
ITEM:	VALUE:	CONDITIONS:	REF.:
*ANGUILLA ANGUILLA	0.79(+-0.41) MILLI-MOL	CONTROL=0.32MILLI-MOL,PCB=3*1000MG/KG BODYWT.(EVERY SEC.DAY),1W	325
*ANGUILLA ANGUILLA	0.47(+-0.29) MILLI-MOL	CONTROL=0.42MILLI-MOL,PCB=6*1000MG/KG BODYWT.(EVERY SEC.DAY),2W	325

TABLE C302

PCB : PLASMA INORGANIC PHOSPHATE			
ITEM:	VALUE:	CONDITIONS:	REF.:
*ANGUILLA ANGUILLA	1.29(+-0.08) MILLI-MOL	CONTROL=1.14MILLI-MOL,PCB=3*1000MG/KG BODYWT.(EVERY SEC.DAY),1W	325
*ANGUILLA ANGUILLA	1.55(+-0.19) MILLI-MOL	CONTROL=1.81MILLI-MOL,PCB=6*1000MG/KG BODYWT.(EVERY SEC.DAY),2W	325
*ANGUILLA ANGUILLA	1.43(+-0.06) MILLI-MOL	CONTROL=1.4MILLI-MOL,PBC=1*200MG/KG BODYWT.(ONE TIME),1M	325
*ANGUILLA ANGUILLA	1.43(+-0.04) MILLI-MOL	CONTROL=1.3MILLI-MOL,PCB=1*200MG/KG BODYWT.(ONE TIME),4M	325

TABLE C303

PCB : PLASMA MAGNESIUM			
ITEM:	VALUE:	CONDITIONS:	REF.:
*ANGUILLA ANGUILLA	0.94(+-0.03) MILLI-MOL	CONTROL=1.01MILLI-MOL,PBC=1*200MG/KG BODYWT.(ONE TIME),1M	325
*ANGUILLA ANGUILLA	1.07(+-0.02) MILLI-MOL	CONTROL=1.06MILLI-MOL,PCB=1*200MG/KG BODYWT.(ONE TIME),4M	325

TABLE C304

PCB : PLASMA PHOSPHOLIPIDS			
ITEM:	VALUE:	CONDITIONS:	REF.:
*ANGUILLA ANGUILLA	7.06(+-0.49) MILLI-MOL	CONTROL=8.1MILLI-MOL,PCB=3*1000MG/KG BODYWT.(EVERY SEC.DAY),1W	325

TABLE C304

PCB : PLASMA PHOSPHOLIPIDS

ITEM:	VALUE:	CONDITIONS:	REF.:
*ANGUILLA ANGUILLA	9.2(+-0.85) MILLI-MOL	CONTROL=6.78MILLI-MOL,PCB= 6*1000MG/KG BODYWT.(EVERY SEC.DAY) ,2W	325

TABLE C305

PCB : PLASMA POTASSIUM

ITEM:	VALUE:	CONDITIONS:	REF.:
*ANGUILLA ANGUILLA	6.2(+-1.) MILLI-MOL	CONTROL=5.3MILLI-MOL,PBC=1*200MG/ KG BODYWT.(ONE TIME),1M	325
*ANGUILLA ANGUILLA	2.9(+-0.1) MILLI-MOL	CONTROL=3.1MILLI-MOL,PCB=1*200MG/ KG BODYWT.(ONE TIME),4M	325

TABLE C306

PCB : PLASMA PROTEIN

ITEM:	VALUE:	CONDITIONS:	REF.:
*ANGUILLA ANGUILLA	5.75(+-0.17) G/100 ML	CONTROL=5.95G/100 ML,PCB=6*1000MG/ KG BODYWT.(EVERY SEC.DAY),2W	325
*ANGUILLA ANGUILLA	6.42(+-0.18) MG/100 ML	CONTROL=6.48MG/100 ML,PBC=1*200MG/ KG BODYWT.(ONE TIME),1M	325
*ANGUILLA ANGUILLA	6.34(+-0.95) MG/100 ML	CONTROL=5.98MG/100 ML,PCB=1*200MG/ KG BODYWT.(ONE TIME),4M	325

TABLE C307

PCB : PLASMA SODIUM

ITEM:	VALUE:	CONDITIONS:	REF.:
*ANGUILLA ANGUILLA	161.(+-1.) MILLI-MOL	CONTROL=159.MILLI-MOL,PBC=1*200MG/ KG BODYWT.(ONE TIME),1M	325
*ANGUILLA ANGUILLA	157.(+-2.) MILLI-MOL	CONTROL=155.MILLI-MOL,PCB=1*200MG/ KG BODYWT.(ONE TIME),4M	325

TABLE C308

PCB : PLASMA TRIGLIVER PROTEINCERIDES

ITEM:	VALUE:	CONDITIONS:	REF.:
*ANGUILLA ANGUILLA	4.66(+-0.72) MILLI-MOL	CONTROL=4.33MILLI-MOL,PCB= 3*1000MG/KG BODYWT.(EVERY SEC.DAY) ,1W	325
*ANGUILLA ANGUILLA	6.28(+-1.56) MILLI-MOL	CONTROL=5.1MILLI-MOL,PCB=6*1000MG/ KG BODYWT.(EVERY SEC.DAY),2W	325
*ANGUILLA ANGUILLA	5.14(+-0.54) MILLI-MOL	CONTROL=5.71MILLI-MOL,PBC=1*200MG/ KG BODYWT.(ONE TIME),1M	325
*ANGUILLA ANGUILLA	6.95(+-1.04) MILLI-MOL	CONTROL=7.4MILLI-MOL,PCB=1*200MG/ KG BODYWT.(ONE TIME),4M	325

TABLE C309

PCB : RBC

ITEM:	VALUE:	CONDITIONS:	REF.:
*ANGUILLA ANGUILLA	1.07(+-0.13) MIL./MICRO-L	CONTROL=1.21MIL./MICRO-L,PCB= 3*1000MG/KG BODYWT.(EVERY SEC.DAY) ,1W	773

TABLE C309

PCB : RBC			
ITEM:	VALUE:	CONDITIONS:	REF.:
*ANGUILLA ANGUILLA	1.01(+-0.11) MIL./MICRO-L	CONTROL=1.26MIL./MICRO-L,PBC=6*1000MG/KG BODYWT.(EVERY SEC.DAY),2W	773
*ANGUILLA ANGUILLA	1.29(+-0.11) MIL./MICRO-L	CONTROL=1.22MIL./MICRO-L,PBC=1*200MG/KG BODYWT.(ONE TIME),4W	773
*ANGUILLA ANGUILLA	1.42(+-0.1) MIL./MICRO-L	CONTROL=1.28MIL./MICRO-L,PCB=1*200MG/KG BODYWT.(ONE TIME),16W	773
*SALMO GAIRDNERI	1.22(+-0.09) MIL./MICRO-L	CONTROL=1.23MIL./MICRO-L,PCB=1*200MG/KG BODYWT.(ONE TIME),5D	773
*SALMO GAIRDNERI	1.15(+-0.08) MIL./MICRO-L	CONTROL=1.24MIL./MICRO-L,PCB=1*200MG/KG BODYWT.(ONE TIME),10D	773
*SALMO GAIRDNERI	1.46(+-0.11) MIL./MICRO-L	CONTROL=1.55MIL./MICRO-L,PCB=1*200MG/KG BODYWT.(ONE TIME),17D	773

TABLE C310

PCB : SPLEEN SOMATIC INDEX			
ITEM:	VALUE:	CONDITIONS:	REF.:
*ANGUILLA ANGUILLA	0.12(+-0.01) SPLEEN WT.*100/B.W.	CONTROL=0.11SSI,PCB=3*1000MG/KG BODYWT.(EVERY SEC.DAY),1W	773
*ANGUILLA ANGUILLA	0.09(+-0.03) SPLEEN WT.*100/B.W.	CONTROL=0.07SSI,PBC=6*1000MG/KG BODYWT.(EVERY SEC.DAY),2W	773
*ANGUILLA ANGUILLA	0.11(+-0.02) SPLEEN WT.*100/B.W.	CONTROL=0.07SPLEEN WT.*100/B.W., PBC=1*200MG/KG BODYWT.(ONE TIME), 4W	773
*ANGUILLA ANGUILLA	0.08(+-0.02) SPLEEN WT.*100/B.W.	CONTROL=0.09SPLEEN WT.*100/B.W., PCB=1*200MG/KG BODYWT.(ONE TIME), 16W	773
*SALMO GAIRDNERI	0.17(+-0.03) SP.WT.*100/BODYWT.	CONTROL=0.11SP.WT.*100/BODYWT.,PCB =0.2G(ONE TIME),3D	1681
*SALMO GAIRDNERI	0.2(+-0.05) SP.WT.*100/BODYWT.	CONTROL=0.21SP.WT.*100/BODYWT.,PCB =0.2G(ONE TIME),7D	1681
*SALMO GAIRDNERI	0.3(+-0.09) SP.WT.*100/BODYWT.	CONTROL=0.23SP.WT.*100/BODYWT.,PCB =0.2G(ONE TIME),14D	1681
*SALMO GAIRDNERI	0.24(+-0.05) SP.WT.*100/BODYWT.	CONTROL=0.13SP.WT.*100/BODYWT.,PCB =0.2G(ONE TIME),21D	1681
*SALMO GAIRDNERI	0.38(+-0.1) SPLEEN WT.*100/B.W.	CONTROL=0.23SPLEEN WT.*100/B.W., PCB=1*200MG/KG BODYWT.(ONE TIME), 5D	773
*SALMO GAIRDNERI	0.29(+-0.04) SPLEEN WT.*100/B.W.	CONTROL=0.25SPLEEN WT.*100/ B.W.,PCB=1*200MG/KG BODYWT.(ONE TIME),10D	773
*SALMO GAIRDNERI	0.26(+-0.04) SPLEEN WT.*100/B.W.	CONTROL=0.23SPLEEN WT.*100/B.W., PCB=1*200MG/KG BODYWT.(ONE TIME), 17D	773

TABLE C311

PH : CHANGE IN BIOMASS			
ITEM:	VALUE:	CONDITIONS:	REF.:
*BIOMASS IN STREAMS	186 MICROGRAM CHL A/100 CM**2	ALGAL BIOMASS, PH=4, SRING	656
*BIOMASS IN STREAMS	122 MICROGRAM CHL A/100 CM**2	ALGAL BIOMASS, PH=4.3 - 5.5, SPRING, NATURAL STREAM	656
*BIOMASS IN STREAMS	82 MICROGRAM CHL A/100 CM**2	ALGAL BIOMASS, PH=6, SRING	656
*BIOMASS IN STREAMS	140 MICROGRAM CHL A/100 CM**2	ALGAL BIOMASS, PH=4, SRING	656
*BIOMASS IN STREAMS	140 MICROGRAM CHL A/100 CM**2	ALGAL BIOMASS, PH=4.3 - 5.5, SUMMER, NATURAL STREAM	656
*BIOMASS IN STREAMS	63 MICROGRAM CHL A/100 CM**2	ALGAL BIOMASS, PH=6, SRING	656
*BIOMASS IN STREAMS	62 MICROGRAM CHL A/100 CM**2	ALGAL BIOMASS, PH=4, SRING	656

TABLE C311

PH : CHANGE IN BIOMASS

ITEM:	VALUE:	CONDITIONS:	REF.:
*BIOMASS IN STREAMS	43 MICROGRAM CHL A/100 CM**2	ALGAL BIOMASS, PH=4.3 - 5.5, FALL, NATURAL STREAM	656
*BIOMASS IN STREAMS	36 MICROGRAM CHL A/100 CM**2	ALGAL BIOMASS, PH=6, SRING	656

TABLE C312

PH : CHANGE IN DIVERSITY

ITEM:	VALUE:	CONDITIONS:	REF.:
*RIVER COMMUNITY	0.92 (SHANNON DIVERSITY)	PH=3.20, NUMBER OF INDIVIDUALS=21/10 FT2/NUMBER OF TAXA=2,STRIP MINE	1606
*RIVER COMMUNITY	3.10 (SHANNON DIVERSITY)	PH=6,NUMBER OF IND=467/10 FT2, NUMBER OF TAXA=30,CONTROL STATION	1606
*RIVER COMMUNITY	1.95 (SHANNON DIVERSITY)	PH=4.68,NO OF INDIVEDUALS=143,NO OF TAXA=17	1606
*RIVER COMMUNITY	1.43 (SHANNON DIVERSITY)	PH=4.90,NO OF IND=250/10 FT**2,NO OF TAXA=15,10 FT2=0.93 M2	1606
*RIVER COMMUNITY	1.94 (SHANNON DIVERSITY)	PH=5.00, NO OF IND=359/10 FT2,NO OF TAXA=13	1606

TABLE C313

PH : CHANGE IN HACTHING OF EGGS

ITEM:	VALUE:	CONDITIONS:	REF.:
*FISH EGGS	89 %	PH=10.0, 289 K,253-274 EGGS, BRACHYDANIO RERIO,72 H AFTER FERT	772
*FISH EGGS	95 %	PH=8.5, 289 K,253-274 EGGS, BRACHYDANIO RERIO,72 H AFTER FERT	772
*FISH EGGS	92 %	PH=7.0, 289 K,253-274 EGGS, BRACHYDANIO RERIO,72 H AFTER FERT	772
*FISH EGGS	83 %	PH=5.5, 289 K,253-274 EGGS, BRACHYDANIO RERIO, 72 H AFTER FERT	772
*FISH EGGS	20 %	PH=4.0, 289 K,253-274 EGGS, BRACHYDANIO RERIO, 72 H AFTER FERT	772
*FISH EGGS	39 %	PH=7.0 - 10.0, 289 K,253-274 EGGS, BRACHYDANIO RERIO,72 H AFTER F	772
*FISH EGGS	5 %	PH=5.5, 289 K,253-274 EGGS, BRACHYDANIO RERIO, 72 H AFTER FERT	772
*FISH EGGS	2 %	PH=4.0, 289 K,253-274 EGGS, BRACHYDANIO RERIO, 72 H AFTER FERT	772

TABLE C314

PH : CHANGE IN HATCHING OF EGGS

ITEM:	VALUE:	CONDITIONS:	REF.:
*JORDANELLA FLORIDAE	56.5 % OF HATCHED OF FERTILE EGGS	CONTROL=44.4%(PH=6.8),PH=6.0,AFTER 20D,299K	296
*JORDANELLA FLORIDAE	51.1 % OF HATCHED OF FERTILE EGGS	CONTROL=44.4%(PH=6.8),PH=5.5,AFTER 20D,299K	296
*JORDANELLA FLORIDAE	17.0 % OF HATCHED OF FERTILE EGGS	CONTROL=44.4%(PH=6.8),PH=5.0,AFTER 20D,299K	296

TABLE C315

PH : CHANGE IN UPTAKE OF PB

ITEM:	VALUE:	CONDITIONS:	REF.:
*PLANT ROOTS	3700 MG/G DRY WEIGHT	SOYBEAN, 12 H, PH = 4.5, LEAD NITRATE	45
*PLANT ROOTS	680 MG/G DRY WEIGHT	CORN, 12 H, PH = 4.5, LEAD NITRATE	45
*PLANT ROOTS	1010 MG/G DRY WEIGHT	BARLEY, 12 H, PH = 4.5, LEAD NITRATE	45
*PLANT ROOTS	910 MG/G DRY WEIGHT	RADISH, 12 H, PH = 4.5, LEAD NITRATE	45
*PLANT ROOTS	1450 MG/G DRY WEIGHT	BEAN, PH = 4.5, 4 H, LEAD NITRATE	45
*PLANT ROOTS	2950 MG/G DRY WEIGHT	SOYBEAN, 12 H, PH = 4.5, LEAD NITRATE	45
*PLANT ROOTS	4600 MG/G DRY WEIGHT	BEAN, PH = 4.5, 24 H, LEAD NITRATE	45
*PLANT ROOTS	5150 MG/G DRY WEIGHT	SOYBEAN, 12 H, PH = 5.5, LEAD NITRATE	45
*PLANT ROOTS	920 MG/G DRY WEIGHT	CORN, 12 H, PH = 5.5, LEAD NITRATE	45
*PLANT ROOTS	1240 MG/G DRY WEIGHT	BARLEY, 12 H, PH = 5.5, LEAD NITRATE	45
*PLANT ROOTS	1000 MG/G DRY WEIGHT	RADISH, 12 H, PH = 5.5, LEAD NITRATE	45
*PLANT ROOTS	1650 MG/G DRY WEIGHT	BEAN, PH = 5.5, 4 H, LEAD NITRATE	45
*PLANT ROOTS	2925 MG/G DRY WEIGHT	BEAN, 12 H, PH = 5.5, LEAD NITRATE	45
*PLANT ROOTS	5175 MG/G DRY WEIGHT	BEAN, PH = 5.5, 24 H, LEAD NITRATE	45
*PLANT ROOTS	4900 MG/G DRY WEIGHT	SOYBEAN, 12 H, PH = 7.0, LEAD NITRATE	45
*PLANT ROOTS	1200 MG/G DRY WEIGHT	CORN, 12 H, PH = 7.0, LEAD NITRATE	45
*PLANT ROOTS	1200 MG/G DRY WEIGHT	BARLEY, 12 H, PH = 7.0, LEAD NITRATE	45
*PLANT ROOTS	900 MG/G DRY WEIGHT	RADISH, 12 H, PH = 7.0, LEAD NITRATE	45
*PLANT ROOTS	837 MG/G DRY WEIGHT	BEAN, PH = 7.0, 4 H, LEAD NITRATE	45
*PLANT ROOTS	1875 MG/G DRY WEIGHT	BEAN, 12 H, PH = 7.0, LEAD NITRATE	45
*PLANT ROOTS	3350 MG/G DRY WEIGHT	BEAN, PH = 7.0, 24 H, LEAD NITRATE	45

TABLE C316

PH : DECREASE IN MEAN WEIGHT

ITEM:	VALUE:	CONDITIONS:	REF.:
*JORDANELLA FLORIDAE	0.42 GRAM	CONTROL=0.55GRAM (PH=6.8),PH=6.0, AFTER 45D, 299 K, FRY	296
*JORDANELLA FLORIDAE	0.26 GRAM	CONTROL=0.55GRAM (PH=6.8),PH=5.5, AFTER 45D, 299 K, FRY	296
*JORDANELLA FLORIDAE	0.075 GRAM	CONTROL=0.55GRAM (PH=6.8),PH=5.0, AFTER 45D, 299 K, FRY	296

TABLE C317

PH : DECREASE IN NUMBER OF SPAWNINGS

ITEM:	VALUE:	CONDITIONS:	REF.:
*JORDANELLA FLORIDAE	24	CONTROL=32(PH=6.8),PH=6.0,AFTER 20D,299K	296
*JORDANELLA FLORIDAE	15	CONTROL=32(PH=6.8),PH=5.5,AFTER 20D,299K	296
*JORDANELLA FLORIDAE	7	CONTROL=32(PH=6.8),PH=5.0,AFTER 20D,299K	296
*JORDANELLA FLORIDAE	1	CONTROL=32(PH=6.8),PH=4.5,AFTER 20D,299K	296

TABLE C318

PH : HATCHING OF EGGS

ITEM:	VALUE:	CONDITIONS:	REF.:
*FISH EGGS	14 % DYING LATER THAN 72 HOURS AFTER FERT	PH=7.0 -10.0,289 K,121-205 EGGS, BRACHYDANIO RERIO, LATER 72 H A F	772
*FISH EGGS	40 % DYING LATER THAN 72 HOURS AFTER FERT	PH=5.5, 289 K,121-205 EGGS, BRACHYDANIO RERIO,LATER 72 H A F	772
*FISH EGGS	52 % HATCHED OF FERTILE EGGS	PH=10.0, 289 K,253-274 EGGS, BRACHYDANIO RERIO	772
*FISH EGGS	49 % HATCHED OF FERTILE EGGS	PH=8.5, 289 K,253-274 EGGS, BRACHYDANIO RERIO	772
*FISH EGGS	48 % HATCHED OF FERTILE EGGS	PH=7.0, 289 K,253-274 EGGS, BRACHYDANIO RERIO	772
*FISH EGGS	20 % HATCHED OF FERTILE EGGS	PH=5.5, 289 K,253-274 EGGS, BRACHYDANIO RERIO	772
*FISH EGGS	4 % HATCHED OF FERTILE EGGS	PH=4.0, 289 K,253-274 EGGS, BRACHYDANIO RERIO	772
*FISH EGGS	98 % HATCHED OF FERTILE EGGS	PH=10.0, 289 K,132-133 EGGS, BRACHYDANIO RERIO,96 H AFTER FERT	772
*FISH EGGS	84 % HATCHED OF FERTILE EGGS	PH=8.5, 289 K,132-133 EGGS, BRACHYDANIO RERIO,96 H AFTER FERT	772
*FISH EGGS	75 % HATCHED OF FERTILE EGGS	PH=7.0, 289 K,132-133 EGGS, BRACHYDANIO RERIO,96 H AFTER FERT	772
*FISH EGGS	50 % HATCHED OF FERTILE EGGS	PH=5.5, 289 K,132-133 EGGS, BRACHYDANIO RERIO,96 H AFTER FERT	772
*FISH EGGS	13 % HATCHED OF FERTILE EGGS	PH=4.0, 289 K,132-133 EGGS, BRACHYDANIO RERIO,96 H AFTER FERT	772
*JORDANELLA FLORIDAE	78.6 % OF FERTILE EGGS	CONTROL=90.1%(PH=6.8),PH=6.0,AFTER 20D,299K	296
*JORDANELLA FLORIDAE	68.0 % OF FERTILE EGGS	CONTROL=90.1%(PH=6.8),PH=5.5,AFTER 20D,299K	296
*JORDANELLA FLORIDAE	55.2 % OF FERTILE EGGS	CONTROL=90.1%(PH=6.8),PH=5.0,AFTER 20D,299K	296

TABLE C319

PH : HATCHING OF EGGS PER SPAWING

ITEM:	VALUE:	CONDITIONS:	REF.:
*JORDANELLA FLORIDAE	1277 EGGS PER SPAWING	CONTROL=2467EGGS(PH=6.8),PH=6.0, AFTER 20D,299K	296
*JORDANELLA FLORIDAE	557 EGGS PER SPAWING	CONTROL=2467EGGS(PH=6.8),PH=5.5, AFTER 20D,299K	296
*JORDANELLA FLORIDAE	319 EGGS PER SPAWING	CONTROL=2467EGGS(PH=6.8),PH=5.0, AFTER 20D,299K	296
*JORDANELLA FLORIDAE	1 EGG	CONTROL=2467EGGS(PH=6.8),PH=4.5, AFTER 20D,299K	296

TABLE C320

PH : MORTALITY

ITEM:	VALUE:	CONDITIONS:	REF.:
*ACRONEURIA LYCORIAS	50%	PH=3.85, 30 DAYS EXPOSURE	90
*BOYERIA VINOSA	50%	PH=4.42, 30 DAYS EXPOSURE	90
*BRACHYCENTRUS AMERICANUS	50%	PH=2.45, 30 DAYS EXPOSURE	90
*EPHEMERELLA SUBVARIA	50%	PH=5.38, 30 DAYS EXPOSURE	90
*HYDROPSYCHE BETTENI	50%	PH=3.38, 30 DAYS EXPOSURE	90
*ISOGENUS FRONTALIS	50%	PH=4.50, 30 DAYS EXPOSURE	90
*OPHIOGOMPHUS RUPINSULENSIS	50%	PH=4.30, 30 DAYS EXPOSURE	90
*PTERONARCYS DORSATA	50%	PH=5.00, 30 DAYS EXPOSURE	90

TABLE C320

PH : MORTALITY			
ITEM:	VALUE:	CONDITIONS:	REF.:
*TAENIOPTERYX MAURA	50%	PH=3.71, 30 DAYS EXPOSURE	90

TABLE C321

PH : SUCCESFUL EMERGENCE			
ITEM:	VALUE:	CONDITIONS:	REF.:
*ACRONEURIA LYCORIAS	50 %	PH=5.0, 30 DAYS EXPOSURE	90
*BOYERIA VINOSA	50 %	PH=5.2, 30 DAYS EXPOSURE	90
*BRACHYCENTRUS AMERICANUS	50 %	PH=4.0, 30 DAYS EXPOSURE	90
*EPHEMERELLA SUBVARIA	50 %	PH=5.9, 30 DAYS EXPOSURE	90
*HYDROPSYCHE BETTENI	50 %	PH=4.7, 30 DAYS EXPOSURE	90
*ISOGENUS FRONTALIS	50 %	PH=6.6, 30 DAYS EXPOSURE	90
*OPHIOGOMPHUS RUPINSULENSIS	50 %	PH=5.2, 30 DAYS EXPOSURE	90
*PTERONARCYS DORSATA	50 %	PH=5.8, 30 DAYS EXPOSURE	90
*TAENIOPTERYX MAURA	50 %	PH=5.0, 30 DAYS EXPOSURE	90

TABLE C322

PH AND HCO3(-) : CHANGE IN CUPRIC ION/TOTAL SOLUBLE CU RATIO			
ITEM:	VALUE:	CONDITIONS:	REF.:
*CU	0.25	PH=7.0,1*10**(-3)M HCO3(-)	1606
*CU	0.11	PH=7.5,1*10**(-3)M HCO3(-)	1606
*CU	0.14	PH=7.0,2*10**(-3)M HCO3(-)	1606
*CU	0.06	PH=7.5,2*10**(-3)M HCO3(-)	1606
*CU	0.02	PH=8.0,2*10**(-3)M HCO3(-)	1606
*CU	0.09	PH=7.0,5*10**(-3)M HCO3(-)	1606
*CU	0.03	PH=7.5,5*10**(-3)M HCO3(-)	1606
*CU	0.009	PH=8.0,5*10**(-3)M HCO3(-)	1606

TABLE C323

PHENYL MERCURIC ACETATE : MORTALITY			
ITEM:	VALUE:	CONDITIONS:	REF.:
*RATS	50%	ACUTE ORAL TOXICITY=2080MG/KG	389

TABLE C324

PHOSPHATE : CHANGE IN NITROGEN FIXATION			
ITEM:	VALUE:	CONDITIONS:	REF.:
*PHYTOPLANKTON	21.8 *10**(-9) M C2H2/((UG CHL A) *HOUR)	MAINLY RHIZOSOLENIA SPP.,ADDED PO4 =0 MICRO MOL/L	967
*PHYTOPLANKTON	49.6 *10**(-9) M C2H2/((UG CHL A) *HOUR)	MAINLY RHIZOSOLENIA SPP.,ADDED PO4 =0.5 MICRO MOL/L	967

TABLE C324

PHOSPHATE : CHANGE IN NITROGEN FIXATION			
ITEM:	VALUE:	CONDITIONS:	REF.:
*PHYTOPLANKTON	41.5 *10**(-9) M C2H2/((UG CHL A) *HOUR)	MAINLY RHIZOSOLENIA SPP.,ADDED PO4 =1.0MICRO MOL/L	967
*PHYTOPLANKTON	39.4 *10**(-9) M C2H2/((UG CHL A) *HOUR)	MAINLY RHIZOSOLENIA SPP.,ADDED PO4 =2.0MICRO MOL/L	967
*PHYTOPLANKTON	52.8 *10**(-9) M C2H2/((UG CHL A) *HOUR)	MAINLY RHIZOSOLENIA SPP.,ADDED PO4 =5.0MICRO MOL/L	967
*PHYTOPLANKTON	37.0 *10**(-9) M C2H2/((UG CHL A) *HOUR)	MAINLY RHIZOSOLENIA SPP.,ADDED PO4 =10MICRO MOL/L	967
*PHYTOPLANKTON	29.7 *10**(-9) M C2H2/((UG CHL A) *HOUR)	MAINLY RHIZOSOLENIA SPP.,ADDED PO4 =20MICRO MOL/L	967
*PHYTOPLANKTON	19.1 *10**(-9) M C2H2/((UG CHL A) *HOUR)	MAINLY RHIZOSOLENIA SPP.,ADDED PO4 =50MICRO MOL/L	967

TABLE C325

PHOSPHORUS STARVATION : DECREASE IN CHLOROPHYLL A CONTENT			
ITEM:	VALUE:	CONDITIONS:	REF.:
*ANABAENA SP.	2.125 MG CHL.A/100MG ALGAE	NO STARVATION,PO4-P IN ALGAE= 0.070MG/100MG ALGAE	558
*ANABAENA SP.	0.350 MG CHL.A/100MG ALGAE	7DAYS IN P-FREE MEDIA,PO4-P IN ALGAE=0.0MG/100MG ALGAE	558
*ANABAENA SP.	0.049 MG CHL.A/100MG ALGAE	14DAYS IN P-FREE MEDIA,PO4-P IN ALGAE=0.0MG/100MG ALGAE	558
*CHLORELLA SP.	2.000 MG CHL.A/100MG ALGAE	NO STARVATION,PO4-P IN ALGAE= 0.139MG/100MG ALGAE	558
*CHLORELLA SP.	0.019 MG CHL.A/100MG ALGAE	7DAYS IN P-FREE MEDIA,PO4-P IN ALGAE=0.027MG/100MG ALGAE	558
*MICROCYSTIS SP.	1.215 MG CHL.A/100MG ALGAE	NO STARVATION,PO4-P IN ALGAE= 0.064MG/100MG ALGAE	558
*MICROCYSTIS SP.	0.625 MG CHL.A/100MG ALGAE	7DAYS IN P-FREE MEDIA,PO4-P IN ALGAE=0.0MG/100MG ALGAE	558
*MICROCYSTIS SP.	0.190 MG CHL.A/100MG ALGAE	14DAYS IN P-FREE MEDIA	558
*SELENASTRUM SP.	2.475 MG CHL.A/100MG ALGAE	NO STARVATION,PO4-P IN ALGAE= 0.192MG/100MG ALGAE	558
*SELENASTRUM SP.	0.762 MG CHL.A/100MG ALGAE	14DAYS IN P-FREE MEDIA,PO4-P IN ALGAE=0.121MG/100MG ALGAE	558

TABLE C326

PHOSPHORUS STARVATION : DECREASE IN EXTRACTABLE SOLUBLE PHOSPHORUS			
ITEM:	VALUE:	CONDITIONS:	REF.:
*ANABAENA SP.	0.0 MG PO4-P/100MG ALGAE	7DAYS IN P-FREE MEDIA,CONTROL= 0.070MG PO4-P/100MG ALGAE	558
*CHLORELLA SP.	0.027 MG PO4-P/100MG ALGAE	7DAYS IN P-FREE MEDIA,CONTROL= 0.139MG PO4-P/100MG ALGAE	558
*MICROCYSTIS SP.	0.0 MG PO4-P/100MG ALGAE	7DAYS IN P-FREE MEDIA,CONTROL= 0.064MG PO4-P/100MG ALGAE	558
*SELENASTRUM SP.	0.121 MG PO4-P/100MG ALGAE	14DAYS IN P-FREE MEDIA,CONTROL= 0.192MG PO4-P/100MG ALGAE	558

TABLE C327

PHTHALACID ESTER : MORTALITY			
ITEM:	VALUE:	CONDITIONS:	REF.:
*MOUSE	4.5%	DOSE=0.43-0.61MG/L,INHALATION,N=22	1303

TABLE C327

PHTHALACID ESTER : MORTALITY			
ITEM:	VALUE:	CONDITIONS:	REF.:
*MOUSE	50%	DOSE=0.7-0.93MG/L,INHALATION,N=22	1303
*MOUSE	100%	DOSE=1.9MG/L,INHALATION	1303

TABLE C328

PICLORAM : MORTALITY			
ITEM:	VALUE:	CONDITIONS:	REF.:
*RATS	50%	ACUTE ORAL TOXICITY=8200MG/KG, HERBICIDE	389

TABLE C329

POLYELECTROLYT : MORTALITY			
ITEM:	VALUE:	CONDITIONS:	REF.:
*DAPHNIA MAGNA	50 %	CALCON M-500=0.65MILLIGRAM/L,48H, 290-292K,5 DAPHNIA SP.	123
*DAPHNIA MAGNA	50 %	SUPERFLOC 330=0.34MILLIGRAM/L,96H, 290-292K,5 DAPHNIA SP.	123
*LIMNOCALANUS MACRURUS	50 %	CALCON M-500=2.MILLIGRAM/L,48H, 279-281K,5 COPEPODES	123
*LIMNOCALANUS MACRURUS	50 %	CALCON M-500=2.MILLIGRAM/L,96H, 279-281K,5 COPEPODES	123
*LIMNOCALANUS MACRURUS	50 %	SUPERFLOC 330=0.35MILLIGRAM/L,48H, 279-281K,5 COPEPODES	123
*LIMNOCALANUS MACRURUS	50 %	SUPERFLOC 330=0.29MILLIGRAM/L,96H, 290-292K,5 DAPHNIA SP.	123
*MYSIS RETICTA	50 %	CALCON M-500=4.MILLIGRAM/L,96H, 279-281K,5 MYSIDES	123
*MYSIS RETICTA	50 %	SUPERFLOC 330=0.5MILLIGRAM/L,96H, 279-281K,5 MYSIDES	123
*SALMO GAIRDNERI	50 %	CALCON M-500=6.5MILLIGRAM/L,48H, 279-281K,5 FISHES	123
*SALMO GAIRDNERI	50 %	CALCON M-500=6.15MILLIGRAM/L,96H, 279-281K,5 FISHES	123
*SALMO GAIRDNERI	50 %	SUPERFLOC 330=2.35MILLIGRAM/L,48H, 279-281K,5 FISHES	123
*SALMO GAIRDNERI	50 %	SUPERFLOC 330=2.12MILLIGRAM/L,96H, 279-281,5 FISHES	123
*SALVELINUS NAMAYCUSH	50 %	CALCON M-500>8.MILLIGRAM/L,48H, 279-281K,5 FISHES	123
*SALVELINUS NAMAYCUSH	50 %	CALCON M-500=5.7MILLIGRAM/L,96H, 279-281K,5 FISHES	123
*SALVELINUS NAMAYCUSH	50 %	SUPERFLOC 330=2.9MILLIGRAM/L,48H, 279-281K,5 FISHES	123
*SALVELINUS NAMAYCUSH	50 %	SUPERFLOC 330=2.85MILLIGRAM/L, 96H.279-281K,5 FISHES	123

TABLE C330

POLYELECTROLYTES : MORTALITY			
ITEM:	VALUE:	CONDITIONS:	REF.:
*DAPHNIA MAGNA	50 %	DOW AP-30=345MILLIGRAM/L,48H, 290-292K,5 DAPHNIA SP.	123

TABLE C330

POLYELECTROLYTES : MORTALITY

ITEM:	VALUE:	CONDITIONS:	REF.:
*DAPHNIA MAGNA	50 %	DOW AP-30=17MILLIGRAM/L,96H, 290-292K,5,DAPHNIA SP.	123
*DAPHNIA MAGNA	50 %	GENDRIV 162=42.MILLIGRAM/L,48H, 290-292K,5 DAPHNIA SP.	123
*DAPHNIA MAGNA	50 %	GENDRIV 162<6.2MILLIGRAM/L,96H, 290-292K,5 DAPHNIA SP.	123
*DAPHNIA MAGNA	50 %	MAGNIFLOC 521C= 3.7 mg/L, 48 H, 290-292 k. 5 DAPHNIA	123
*DAPHNIA MAGNA	50 %	MAGNIFLOC 521C= 2.1 MG/L, 96 H, 290-292 K. 5 DAPHNIA	123
*DAPHNIA MAGNA	50 %	MAGNIFLOC 905N> 50 MG/L, 48 H, 290- 292 K, 5 DAPHNIA	123
*DAPHNIA MAGNA	50 %	MAGNIFLOC 905N= 16.5 MG/L, 96 H, 290-292 K, 5 DAPHNIA	123
*SALMO GAIRDNERI	50 %	GENDRIV 162=218MILLIGRAM/L,96H, 287-289K,5 FISHES	123
*SALMO GAIRDNERI	50 %	MAGNIFLOC 521C= 8.7 MG/L, 48 H, 290-292 K, 5 FISH	123
*SALMO GAIRDNERI	50 %	MAGNIFLOC 521C= 8.7 MG/L, 96 H, 290-292 K, 5 FISH	123
*SALMO GAIRDNERI	50 %	MAGNIFLOC 905N > 8 MG/L, 96 H, 287-289 K, 5 FISH	123
*SALMO GAIRDNERI	50 %	MAGNIFLOC 905N > 8 MG/L, 48 H, 287-289 K, 5 FISH	123

TABLE C331

POLYOXYETHYLENE : MORTALITY

ITEM:	VALUE:	CONDITIONS:	REF.:
*ANGUILLA ANGUILLA	50 %	POLYOXYETHYLENE 14-MONOLAURATE = 240MG/L,24H,283-284K,10 FISH	1709
*ANGUILLA ANGUILLA	50 %	POLYOXYETHYLENE 14-MONOLAURATE = 170MG/L,48H,283-284K,10 FISH	1709
*ANGUILLA ANGUILLA	50 %	POLYOXYETHYLENE 14-MONOLAURATE = 120MG/L,96H,283-284K,10 FISH	1709
*SALMO SALAR	50 %	POLYOXYETHYLENE AMINES = 0.14-0.74MG/L,24H,288K,10 FISH	1709
*SALMO SALAR	50 %	POLYOXYETHYLENE AMINES = 0.11-0.66MG/L,48H,288K,10 FISH	1709
*SALMO SALAR	50 %	POLYOXYETHYLENE ESTERS = 27.5-300MG/L,24H,288K,10 FISH	1709
*SALMO SALAR	50 %	POLYOXYETHYLENE ESTERS = 13-25MG/ L,48H,288K,10 FISH	1709
*SALMO SALAR	50 %	POLYOXYETHYLENE ESTERS = 6.4MG/L, 96H,288K,10 FISH	1709
*SALMO SALAR	50 %	POLYOXYETHYLENE ESTERS = 21.5MG/L, 24H,283-284K,10 FISH	1709
*SALMO SALAR	50 %	POLYOXYETHYLENE ESTERS = 14.5MG/L, 48H,283-284K,10 FISH	1709
*SALMO SALAR	50 %	POLYOXYETHYLENE ESTERS = 3.5MG/L, 24H,283-284K,10 FISH	1709
*SALMO SALAR	50 %	POLYOXYETHYLENE ESTERS = 2.7MG/L, 48H,283-284K,10 FISH	1709
*SALMO SALAR	50 %	POLYOXYETHYLENE ESTERS = 1.5MG/L, 96H,283-284K,10 FISH	1709
*SALMO SALAR	50 %	POLYOXYETHYLENE GLYCOL > 1000MG/L, 24H,288K,10 FISH	1709
*SALMO SALAR	50 %	POLYOXYETHYLENE GLYCOL > 1000MG/L, 48H,288K,10 FISH	1709
*SALMO SALAR	50 %	POLYOXYETHYLENE GLYCOL > 1000MG/L, 48H,288K,10 FISH	1709
*SALMO SALAR	50 %	POLYOXYTHYLENE 14-MONOLAURATE = 32MG/L,24H	1708
*SALMO SALAR	50 %	POLYOXYETHYLENE 14-MONOLAURATE = 21.1MG/L,24H,283-284K,10 FISH	1709
*SALMO SALAR	50 %	POLYOXYETHYLENE 14-MONOLAURATE = 22MG/L,48H	1708

TABLE C331

POLYOXYETHYLENE : MORTALITY			
ITEM:	VALUE:	CONDITIONS:	REF.:
*SALMO SALAR	50 %	POLYOXYETHYLENE 14-MONOLAURATE = 14.5MG/L,48H,283-284K,10 FISH	1709
*SALMO SALAR	50 %	POLYOXYETHYLENE 14-MONOLAURATE = 15MG/L,96H	1708
*SALMO SALAR	50 %	POLYOXYETHYLENE 14-MONOLAURATE = 9.6MG/L,96H,283-284K,10 FISH	1709
*SALMO SALAR	50 %	POLYOXYETHYLENE 4-MONOLAURATE = 2.2MG/L,24H	1708
*SALMO SALAR	50 %	POLYOXYETHYLENE 4-MONOLAURATE = 3.2MG/L,24H,283-284K,10 FISH	1709
*SALMO SALAR	50 %	POLYOXYETHYLENE 4-MONOLAURATE = 2.7MG/L,48H,283-284K,10 FISH	1709
*SALMO SALAR	50 %	POLYOXYETHYLENE 4-MONOLAURATE = 1.5MG/L,96H,283-284K,10 FISH	1709
*SALMO SALAR	50 %	POLYOXYETHYLENE AMINES = 0.09-0.64MG/L,96H,288K,10 FISH	1709

TABLE C332

PR : MORTALITY			
ITEM:	VALUE:	CONDITIONS:	REF.:
*RAT	50%	NITRATE,ORAL,DOSE=3500MG/KG BODY	1611

TABLE C333

PROPANIL : MORTALITY			
ITEM:	VALUE:	CONDITIONS:	REF.:
*ICTALURUS PUNCTATUS	50 %	3796 (3475-4194) PPB, 14 G, 293 K, PH=8.2, ALK=80 PPM, 95 % CL., 96 H	1002

TABLE C334

PROPAZINE : MORTALITY			
ITEM:	VALUE:	CONDITIONS:	REF.:
*RATS	50%	ACUTE ORAL TOXICITY=5000MG/KG, HERBICIDE	389

TABLE C335

PROPHAM : MORTALITY			
ITEM:	VALUE:	CONDITIONS:	REF.:
*RATS	50%	ACUTE ORAL TOXICITY=5000MG/KG, HERBICIDE	389

TABLE C336

PU-CITRATE : MORTALITY			
ITEM:	VALUE:	CONDITIONS:	REF.:
*MOUSE	50%	ORAL DOSE=1750MG/KG(=4.02*10**(-3) MOL/KG),PT-50(F007)=2.40	953

TABLE C337

QUINTOZENE : MORTALITY			
ITEM:	VALUE:	CONDITIONS:	REF.:
*RATS	50%	ACUTE ORAL TOXICITY=12000MG/KG, FUNGICIDE	389

TABLE C338

RB : MORTALITY			
ITEM:	VALUE:	CONDITIONS:	REF.:
*RAT	50%	HYDROXIDE,ORAL,DOSE=586MG/KG BODY	1611
*RAT	50%	IODIDE,ORAL,DOSE=4708MG/KG BODY	1611

TABLE C339

RE : MORTALITY			
ITEM:	VALUE:	CONDITIONS:	REF.:
*RAT	50%	3-CHLORIDE,INTRAPERITON.,DOSE= 280MG/KG BODY	1611

TABLE C340

RH : MORTALITY			
ITEM:	VALUE:	CONDITIONS:	REF.:
*RAT	50%	3-CHLORIDE,INTRAVEN.,DOSE=198MG/KG BODY	1611

TABLE C341

RHINEWATER : GLUCOSE			
ITEM:	VALUE:	CONDITIONS:	REF.:
*SALMO GAIRDNERI	83.2(+-22.3) MG/100 ML	CONTROL=64.8MG/100 ML,6M,20 FISH	1266
*SALMO GAIRDNERI	46.5(+-11.7) MG/100 ML	CONTROL=37.4MG/100 ML,9M,20 FISH	1266

TABLE C342

RHINEWATER : GRANULOCYTES			
ITEM:	VALUE:	CONDITIONS:	REF.:
*SALMO GAIRDNERI	1.79(+-2.09) %	CONTROL=3.%,1M,30 FISH	1266
*SALMO GAIRDNERI	3.03(+-2.96) %	CONTROL=1.66%,3M,30 FISH	1266
*SALMO GAIRDNERI	2.(+-2.83) %	CONTROL=1.2%,6M,20 FISH	1266
*SALMO GAIRDNERI	13.6(+-9.4) %	CONTROL=5.01%,9M,20 FISH	1266

TABLE C343

RHINEWATER : HEMATOCRIT			
ITEM:	VALUE:	CONDITIONS:	REF.:
*SALMO GAIRDNERI	41.8(+-5.2) %	CONTROL=45.5%,1M,30 FISH	1266

TABLE C343

RHINEWATER : HEMATOCRIT			
ITEM:	VALUE:	CONDITIONS:	REF.:
*SALMO GAIRDNERI	42.3(+-5.1) %	CONTROL=45.5%,3M,30 FISH	1266
*SALMO GAIRDNERI	39.3(+-5.6) %	CONTROL=45.8%,6M,20 FISH	1266
*SALMO GAIRDNERI	38.6(+-4.) %	CONTROL=42.3%,9M,20 FISH	1266

TABLE C344

RHINEWATER : HEMOGLOBIN			
ITEM:	VALUE:	CONDITIONS:	REF.:
*SALMO GAIRDNERI	7.1(+-0.9) G/100 ML	CONTROL=7.9G/100 ML,6M,20 FISH	1266
*SALMO GAIRDNERI	8.2(+-1.1) G/100 ML	CONTROL=9.7G/100 ML,9M,20 FISH	1266

TABLE C345

ROTENONE : MORTALITY			
ITEM:	VALUE:	CONDITIONS:	REF.:
*MOUSE	50%	ORAL DOSE=2.8MG/KG(=7.10*10**(-6) MOL/KG),PT-50(F007)=5.15	953

TABLE C346

SAXITOXIN : MORTALITY			
ITEM:	VALUE:	CONDITIONS:	REF.:
*MOUSE	50%	ORAL DOSE=3.4*10**(-3)MG/KG(= 9.14*10**(-9)MOL/KG),PT-50(F007)= 8.04	953

TABLE C347

SB : MORTALITY			
ITEM:	VALUE:	CONDITIONS:	REF.:
*PIMEPHALES PROMELAS	50%	SBCL3,EXPOS.TIME=4DAYS,CONC.IN WATER=9000UG/L,LIMNOL.	1004
*RAT	50%	METALLIC,ORAL,DOSE=100MG/KG BODY	1611
*RAT	50%	5-CHLORIDE,ORAL,DOSE=675MG/KG BODY	1611
*RAT	50%	3-CHLORIDE,ORAL,DOSE=10000MG/KG BODY	419

TABLE C348

SC : MORTALITY			
ITEM:	VALUE:	CONDITIONS:	REF.:
*MOUSE	50%	CHLORIDE,INTRAPERITON.,DOSE=755MG/ KG BODY	1611

TABLE C349

SE : INCREASE IN SE-CONTENT			
ITEM:	VALUE:	CONDITIONS:	REF.:
*CORN	0.06 MICROGRAM/G (+-5.8%)	ADDED TO FIELD:120G SE/HA;CONTROL= 0.01UG/G	529
*CORN	0.1 MICROGRAM/G	ADDITION OF 75G SE/HA*YR FOR 5YRS; CONTROL=0.01UG/G	529

TABLE C350

SE : MORTALITY			
ITEM:	VALUE:	CONDITIONS:	REF.:
*DAPHNIA MAGNA	50%	NA2SE03,EXPOS.TIME=2DAYS,CONC.IN WATER=2500MICROGRAM/L,LIMNOL.	1004
*RAT	50%	H2SE03,ORAL,DOSE=7MG/KG BODY	1611
*RAT	50%	2-SULFIDE,ORAL,DOSE=38MG/KG BODY	1611

TABLE C351

SIMAZINE : MORTALITY			
ITEM:	VALUE:	CONDITIONS:	REF.:
*RATS	50%	ACUTE ORAL TOXICITY=5000MG/KG, HERBICIDE	389

TABLE C352

SIZE : CHANGE IN MINIMUM DIVISION TIME			
ITEM:	VALUE:	CONDITIONS:	REF.:
*LIVING ORGANISMS	30.9 M SEC	RADIUS = 10**(-3) CM	1067
*LIVING ORGANISMS	3.09 SEC	RADIUS = 10**(-2) CM	1067
*LIVING ORGANISMS	5.15 MIN	RADIUS = 10**(-1) CM	1067
*LIVING ORGANISMS	8.53 HOURS	RADIUS = 1 CM	1067
*LIVING ORGANISMS	35.3 DAYS	RADIUS = 10 CM	1067

TABLE C353

SM : MORTALITY			
ITEM:	VALUE:	CONDITIONS:	REF.:
*MOUSE	50%	3-CHLORIDE,INTRAPERITON.,DOSE= 365MG/KG BODY	1611
*RAT	50%	NITRATE,ORAL,DOSE=2900MG/KG BODY	1611

TABLE C354

SN : MORTALITY			
ITEM:	VALUE:	CONDITIONS:	REF.:
*DAPHNIA MAGNA	50%	SNCL2,EXPOS.TIME=2DAYS,CONC.IN WATER=55000MICROGRAM/L,LIMNOL.	122

TABLE C354

SN : MORTALITY			
ITEM:	VALUE:	CONDITIONS:	REF.:
*MOUSE	50%	2-CHLORIDE,INTRAPERITON.,DOSE= 41MG/KG BODY	1611
*MOUSE	50%	4-CHLORIDE,INTRAPERITON.,DOSE= 21MG/KG BODY	1611

TABLE C355

SODIUM CHLORATE : MORTALITY			
ITEM:	VALUE:	CONDITIONS:	REF.:
*RATS	50%	ACUTE ORAL TOXICITY=1200MG/KG, HERBICIDE	389

TABLE C356

SODIUM FLUORIDE : MORTALITY			
ITEM:	VALUE:	CONDITIONS:	REF.:
*RATS	50%	ACUTE ORAL TOXICITY=200MG/KG, PESTICIDE	389

TABLE C357

SODIUM FLUOSILICATE : MORTALITY			
ITEM:	VALUE:	CONDITIONS:	REF.:
*RATS	50%	ACUTE ORAL TOXICITY=125MG/KG, PESTICIDE	389

TABLE C358

SODIUM LAURYL SULPHATE : SURVIVAL TIME			
ITEM:	VALUE:	CONDITIONS:	REF.:
*CARASSIUS AURATUS	101 MIN	SLS=70PPM,ACCLIM. HARDNESS=300PPM, TREAT. SOL. HARDN.=300PPM,8 FISH	1610
*CARASSIUS AURATUS	112 MIN	SLS=70PPM,ACCLIM. HARDNESS=300PPM, TREAT. SOL. HARDN.=200PPM,8 FISH	1610
*CARASSIUS AURATUS	204 MIN	SLS=70PPM,ACCLIM. HARDNESS=300PPM, TREAT. SOL. HARDN.=60PPM,8 FISH	1610
*CARASSIUS AURATUS	>1440 MIN	SLS=70PPM,ACCLIM. HARDNESS=300PPM, TREAT. SOL. HARDN.=0PPM,8 FISH	1610
*CARASSIUS AURATUS	111 MIN	SLS=70PPM,ACCLIM. HARDNESS=200PPM, TREAT. SOL. HARDN.=300PPM,8 FISH	1610
*CARASSIUS AURATUS	121 MIN	SLS=70PPM,ACCLIM. HARDNESS=200PPM, TREAT. SOL. HARDN.=200PPM,8 FISH	1610
*CARASSIUS AURATUS	291 MIN	SLS=70PPM,ACCLIM. HARDNESS=200PPM, TREAT. SOL. HARDN.=60PPM,8 FISH	1610
*CARASSIUS AURATUS	>1440 MIN	SLS=70PPM,ACCLIM. HARDNESS=200PPM, TREAT. SOL. HARDN.=0PPM,8 FISH	1610
*CARASSIUS AURATUS	131 MIN	SLS=70PPM,ACCLIM. HARDNESS=60PPM, TREAT. SOL. HARDN.=300PPM,8 FISH	1610
*CARASSIUS AURATUS	156 MIN	SLS=70PPM,ACCLIM. HARDNESS=60PPM, TREAT. SOL. HARDN.=200PPM,8 FISH	1610

TABLE C358

SODIUM LAURYL SULPHATE : SURVIVAL TIME			
ITEM:	VALUE:	CONDITIONS:	REF.:
*CARASSIUS AURATUS	390 MIN	SLS=70PPM,ACCLIM. HARDNESS=60PPM, TREAT. SOL. HARDN.=60PPM,8 FISH	1610
*CARASSIUS AURATUS	>1440 MIN	SLS=70PPM,ACCLIM. HARDNESS=60PPM, TREAT. SOL. HARDN.=0PPM,8 FISH	1610
*CARASSIUS AURATUS	151 MIN	SLS=70PPM,ACCLIM. HARDNESS=0PPM, TREAT. SOL. HARDN.=300PPM,8 FISH	1610
*CARASSIUS AURATUS	173 MIN	SLS=70PPM,ACCLIM. HARDNESS=0PPM, TREAT. SOL. HARDN.=200PPM,8 FISH	1610
*CARASSIUS AURATUS	404 MIN	SLS=70PPM,ACCLIM. HARDNESS=0PPM, TREAT. SOL. HARDN.=60PPM,8 FISH	1610
*CARASSIUS AURATUS	>1440 MIN	SLS=70PPM,ACCLIM. HARDNESS=0PPM, TREAT. SOL. HARDN.=0PPM,8 FISH	1610
*CARASSIUS AURATUS	1440 MIN	SLS=70PPM,SODIUMCHLORIDE=1MILLI-M	1610
*CARASSIUS AURATUS	1440 MIN	SLS=70PPM,SODIUMCHLORIDE=3MILLI-M	1610
*CARASSIUS AURATUS	1440 MIN	SLS=70PPM,POTASSIUMCHLORIDE= 1MILLI-M	1610
*CARASSIUS AURATUS	1247(+-65) MIN	SLS=70PPM,POTASSIUMCHLORIDE= 3MILLI-M	1610
*CARASSIUS AURATUS	480(+-19) MIN	SLS=70PPM,MAGNIUMCHLORIDE=1MILLI-M	1610
*CARASSIUS AURATUS	106(+-5.5) MIN	SLS=70PPM,MAGNIUMCHLORIDE=3MILLI-M	1610
*CARASSIUS AURATUS	282(+-0.4) MIN	SLS=70PPM,CALCIUMCHLORIDE=1MILLI-M	1610
*CARASSIUS AURATUS	116(+-9.6) MIN	SLS=70PPM,CALCIUMCHLORIDE=3MILLI-M	1610
*SALMO GAIRDNERI	27 MIN	SLS=70PPM,ACCLIM. HARDNESS=300PPM, TREAT. SOL. HARDN.=300PPM,4 FISH	1610
*SALMO GAIRDNERI	43 MIN	SLS=70PPM,ACCLIM. HARDNESS=300PPM, TREAT. SOL. HARDN.=200PPM,4 FISH	1610
*SALMO GAIRDNERI	67 MIN	SLS=70PPM,ACCLIM. HARDNESS=300PPM, TREAT. SOL. HARDN.=60PPM,4 FISH	1610
*SALMO GAIRDNERI	216 MIN	SLS=70PPM,ACCLIM. HARDNESS=300PPM, TREAT. SOL. HARDN.=0PPM,4 FISH	1610
*SALMO GAIRDNERI	45 MIN	SLS=70PPM,ACCLIM. HARDNESS=200PPM, TREAT. SOL. HARDN.=300PPM,4 FISH	1610
*SALMO GAIRDNERI	52 MIN	SLS=70PPM,ACCLIM. HARDNESS=200PPM, TREAT. SOL. HARDN.=200PPM,4 FISH	1610
*SALMO GAIRDNERI	147 MIN	SLS=70PPM,ACCLIM. HARDNESS=200PPM, TREAT. SOL. HARDN.=60PPM,4 FISH	1610
*SALMO GAIRDNERI	1034 MIN	SLS=70PPM,ACCLIM. HARDNESS=200PPM, TREAT. SOL. HARDN.=0PPM,4 FISH	1610
*SALMO GAIRDNERI	57 MIN	SLS=70PPM,ACCLIM. HARDNESS60PPM, TREAT. SOL. HARDN.=300PPM,4 FISH	1610
*SALMO GAIRDNERI	77 MIN	SLS=70PPM,ACCLIM. HARDNESS=60PPM, TREAT. SOL. HARDN.=200PPM,4 FISH	1610
*SALMO GAIRDNERI	201 MIN	SLS=70PPM,ACCLIM. HARDNESS=60PPM, TREAT. SOL. HARDN.=60PPM,4 FISH	1610
*SALMO GAIRDNERI	1100 MIN	SLS=70PPM,ACCLIM. HARDNESS=60PPM, TREAT. SOL. HARDN.=0PPM,4 FISH	1610

TABLE C359

SR : MORTALITY			
ITEM:	VALUE:	CONDITIONS:	REF.:
*DAPHNIA MAGNA	50%	SRCL2,EXPOS.TIME=2DAYS,CONC.IN WATER=125000UG/L,LIMNOL.	122
*MOUSE	50%	CHLORIDE,INTRAVEN.,DOSE=148MG/KG BODY	1611
*RAT	50%	ACETATE,INTRAVEN.,DOSE=105MG/KG BODY	1611
*RAT	50%	BROMIDE,INTRAVEN.,DOSE=1000MG/KG BODY	1611
*RAT	50%	IODIDE,INTRAPERITON.,DOSE=800MG/KG BODY	1611

TABLE C359

SR : MORTALITY			
ITEM:	VALUE:	CONDITIONS:	REF.:
*RAT	50%	NITRATE,INTRAPERITON.,DOSE=540MG/KG BODY	1611

TABLE C360

STOCKING DENSITY : CHANGE IN SURVIVAL OF LARVAE IN THE 16 FIRST DAYS AFTER HATCHING			
ITEM:	VALUE:	CONDITIONS:	REF.:
*ANCHOA MITCHILLI	32.9 %	LARVAE = 0.5 NO/L, DW PER SURVIVOR = 578.0 UG	695
*ANCHOA MITCHILLI	31.3 %	LARVAE = 2.0 NO/L, DW PER SURVIVOR = 518.7 UG	695
*ANCHOA MITCHILLI	22.0 %	LARVAE = 8.0 NO/L, DW PER SURVIVOR = 402.7 UG	695
*ANCHOA MITCHILLI	31.0 %	LARVAE = 32.0 NO/L, DW PER SURVIVOR = 334.2 UG	695
*ARCHIRUS LINEATUS	31.5 %	LARVAE = 0.5 NO/L, DW PER SURVIVOR = 235.8 UG	695
*ARCHIRUS LINEATUS	14.5 %	LARVAE = 2.0 NO/L, DW PER SURVIVOR = 141.9 UG	695
*ARCHIRUS LINEATUS	19.6 %	LARVAE = 8.0 NO/L, DW PER SURVIVOR = 161.3 UG	695
*ARCHIRUS LINEATUS	10.8 %	LARVAE = 32.0 NO/L, DW PER SURVIVOR = 169.9 UG	695

TABLE C361

STREPTOMYCIN : MORTALITY			
ITEM:	VALUE:	CONDITIONS:	REF.:
*MOUSE	50%	ORAL DOSE=610MG/KG(=1.05*10**(-3) MOL/KG),PT-50(F007)=2.98	953

TABLE C362

STRYCHNINE : MORTALITY			
ITEM:	VALUE:	CONDITIONS:	REF.:
*MOUSE	50%	ORAL DOSE=0.98MG/KG(=2.93*10**(-6) MOL/KG),PT-50(F007)=5.53	953

TABLE C363

SULPHUR : CHLOROPHYLL			
ITEM:	VALUE:	CONDITIONS:	REF.:
*QUERCUS PETRAEA	15.9 MICROGRAM CHL A/CM**2 LEAF AREA	S=21.31MICROGRAMS/CM**2 LEAF AREA, 4 JUNE,ABERDARE VALLEY,S. WALES	1720
*QUERCUS PETRAEA	28.7 MICROGRAM CHL A/CM**2 LEAF AREA	S=27.63MICROGRAMS/CM**2 LEAF AREA, 1 AUGUST,ABERDARE VALLEY,S. WALES	1720
*QUERCUS PETRAEA	27.2 MICROGRAM CHL A/CM**2 LEAF AREA	S=42.45MICROGRAMS/CM**2 LEAF AREA, 16 AUGUST,ABERDARE VALLEY,S. WALES	1720
*QUERCUS PETRAEA	6.0 MICROGRAM CHL B/CM**2 LEAF AREA	S=21.31MICROGRAMS/CM**2 LEAF AREA, 4 JUNE,ABERDARE VALLEY,S. WALES	1720

TABLE C363

SULPHUR : CHLOROPHYLL			
ITEM:	VALUE:	CONDITIONS:	REF.:
*QUERCUS PETRAEA	9.4 MICROGRAM CHL B/CM**2 LEAF AREA	S=27.63MICROGRAMS/CM**2 LEAF AREA, 1 AUGUST,ABERDARE VALLEY,S. WALES	1720
*QUERCUS PETRAEA	11.3 MICROGRAM CHL B/CM**2 LEAF AREA	1S=42.45MICROGRAMS/CM**2 LEAF AREA,16 AUGUST,ABERDARE VALLEY,S. WALES	1720

TABLE C364

TA : MORTALITY			
ITEM:	VALUE:	CONDITIONS:	REF.:
*RAT	50%	K-TANTALFLOURIDE,ORAL,DOSE=2500MG/KG BODY	1611
*RAT	50%	CHLORIDE,ORAL,DOSE=958MG/KG BODY	1611
*RAT	50%	5-CHLORIDE,ORAL,DOSE=1900MG/KG BODY	1611

TABLE C365

TB : MORTALITY			
ITEM:	VALUE:	CONDITIONS:	REF.:
*MOUSE	50%	CHLORIDE(CITRATECOMPLEX), INTRAPERITON.,DOSE=121MG/KG BODY	1611
*RAT	50%	NITRATE,INTRAPERITON.,DOSE=260MG/KG BODY	1611

TABLE C366

TCA : MORTALITY			
ITEM:	VALUE:	CONDITIONS:	REF.:
*RATS	50%	ACUTE ORAL TOXICITY=3200-5000MG/KG,HERBICIDE	389

TABLE C367

TCDD : MORTALITY			
ITEM:	VALUE:	CONDITIONS:	REF.:
*POECILIA RETICULATUS	50 %	WATER = 0.1 PPB, 50 % AFTER 21.7 DAYS, 293 K	1136
*POECILIA RETICULATUS	50 %	WATER = 1.0 PPB, 50 % AFTER 11.6 DAYS, 293 K	1136
*POECILIA RETICULATUS	50 %	WATER =10.0 PPB, 50 % AFTER 18.2 DAYS, 293 K	1136

TABLE C368

TCDD : TERATOGENICITY			
ITEM:	VALUE:	CONDITIONS:	REF.:
*ALL INVESTIGATED ANIMALS	SOME EFFECT	DOSE=1-10UG/KG*DAY (?037),NO EFFECT ON MOTHER ANIMAL	1026

TABLE C368

TCDD : TERATOGENICITY			
ITEM:	VALUE:	CONDITIONS:	REF.:
*RAT	ZERO EFFECT LEVEL	DOSE=0.03UG/KG*DAY (?037),FROM DAY 6-15 IN PREGNANCY-PERIODE	1026

TABLE C369

TE : MORTALITY			
ITEM:	VALUE:	CONDITIONS:	REF.:
*MOUSE	50%	NA-TELLURITE,ORAL,DOSE=20MG/KG BODY	1611
*RAT	50%	NA-TELLURITE,ORAL,DOSE=83MG/KG BODY	419

TABLE C370

TEMPERATURE : CHANGE IN BOIL. HALF-LIFE TIME FOR CS-137			
ITEM:	VALUE:	CONDITIONS:	REF.:
*CARP	55 DAYS	293 K, 5 Y	640
*CARP	120 DAYS	283 K, 5 Y	640
*RAINBOW TROUT	20 DAYS	293 K, 1 Y	640
*RAINBOW TROUT	36 DAYS	280-281 K, 1 Y	640

TABLE C371

TEMPERATURE : CHANGE IN DEVELOPMENT TIME OF EGGS			
ITEM:	VALUE:	CONDITIONS:	REF.:
*THERMOCYCLOPS NEGLECTUS	58.7 (SE=+-1.71) HOURS	293 K, N = 11	198
*THERMOCYCLOPS NEGLECTUS	36.4 (SE=+-2.83) HOURS	298 K, N = 5	198
*THERMOCYCLOPS NEGLECTUS	27.0 (SE=+-1.99) HOURS	302 K, N = 15	198
*THERMOCYCLOPS NEGLECTUS	27.3 (SE=+-2.14) HOURS	304.5 K, N = 15	198
*THERMOCYCLOPS NEGLECTUS	37 HOURS	307 K, 22 % MORTALITY OF ADULTS	198
*THERMOCYCLOPS NEGLECTUS	NO HATCHNING	309.5 K, 100 % MORTALITY OF ADULTS	198

TABLE C372

TEMPERATURE : CHANGE IN DIELDRIN UPTAKE			
ITEM:	VALUE:	CONDITIONS:	REF.:
*SECALE CEREALE	0.11 PPM DIELDRIN (WET WT.)	75DEGREE F.,GROWN IN SAND WITH 10PPM DIELDRIN	1691
*SECALE CEREALE	0.14 PPM DIELDRIN (WET WT.)	65DEGREE F.,GROWN IN SAND WITH 10PPM DIELDRIN	1691

TABLE C373

TEMPERATURE : CHANGE IN EXCRETION OF NITROGEN FROM ZOOPLANKTON			
ITEM:	VALUE:	CONDITIONS:	REF.:
*PHRONIMA SEDENTARIA	5.1 +- 0.6 MICROGRAM TOTAL-N/(MG DRY WEIGHT*DAY)	278 K, 95 % CL.	997

TABLE C373

TEMPERATURE : CHANGE IN EXCRETION OF NITROGEN FROM ZOOPLANKTON			
ITEM:	VALUE:	CONDITIONS:	REF.:
*PHRONIMA SEDENTARIA	9.8 +- 2.3 MICROGRAM TOTAL-N/(MG DRY WEIGHT*DAY)	283.5 K, 95 % CL.	997
*PHRONIMA SEDENTARIA	10.4 +- 1.0 MICROGRAM TOTAL-N/(MG DRY WEIGHT*DAY)	288. K, 95 % CL.	997
*PHRONIMA SEDENTARIA	12.5 +- 1.2 MICROGRAM TOTAL-N/(MG DRY WEIGHT*DAY)	292.5 K, 95 % CL.	997
*PHRONIMA SEDENTARIA	2.4 +- 0.4 MICROGRAM NH4-N/(MG DRY WEIGHT*DAY)	278 K, 95 % CL.	997
*PHRONIMA SEDENTARIA	6.4 +- 2.4 MICROGRAM NH4-N/(MG DRY WEIGHT*DAY)	283.5 K, 95 % CL.	997
*PHRONIMA SEDENTARIA	7.5 +- 1.4 MICROGRAM NH4-N/(MG DRY WEIGHT*DAY)	288.5 K, 95 % CL.	997
*PHRONIMA SEDENTARIA	9.6 +- 0.4 MICROGRAM NH4-N/(MG DRY WEIGHT*DAY)	292.5 K, 95 % CL.	997

TABLE C374

TEMPERATURE : CHANGE IN FEEDING RATE			
ITEM:	VALUE:	CONDITIONS:	REF.:
*DAPHNIA SCHOEDLERI	4535 CELLS/ANIMAL*HR	278K,N=2,FOOD CONC.=10000CELLS/ML, ABOVE INCIPIENT LIMITING LEVEL	647
*DAPHNIA SCHOEDLERI	999 CELLS/ANIMAL*HR	278K,N=2,FOOD CONC.=5000CELLS/ML, BELOW INCIPIENT LIMITING LEVEL	647
*DAPHNIA SCHOEDLERI	9750 CELLS/ANIMAL*HR	283K,N=2,FOOD CONC.=10000CELLS/ML, ABOVE INCIPIENT LIMITING LEVEL	647
*DAPHNIA SCHOEDLERI	5126 CELLS/ANIMAL*HR	283K,N=2,FOOD CONC.=5000CELLS/ML, BELOW INCIPIENT LIMITING LEVEL	647
*DAPHNIA SCHOEDLERI	12267 CELLS/ANIMAL*HR	288K,N=2,FOOD CONC.=10000CELLS/ML, ABOVE INCIPIENT LIMITING LEVEL	647
*DAPHNIA SCHOEDLERI	9459 CELLS/ANIMAL*HR	288K,N=2,FOOD CONC.=5000CELLS/ML, BELOW INCIPIENT LIMITING LEVEL	647
*DAPHNIA SCHOEDLERI	1668 CELLS/ANIMAL*HR	303K,N=2,FOOD CONC.=10000CELLS/ML, ABOVE INCIPIENT LIMITING LEVEL	647
*DAPHNIA SCHOEDLERI	7490 CELLS/ANIMAL*HR	303K,N=2,FOOD CONC.=5000CELLS/ML, BELOW INCIPIENT LIMITING LEVEL	647

TABLE C375

TEMPERATURE : CHANGE IN FILTRATION RATE			
ITEM:	VALUE:	CONDITIONS:	REF.:
*MODIOLUS MODIOLUS	834 ML/HR	SIZE=7.6CM,FOOD=CHLAMYDOMONAS SP.(2*10**7CELLS/L,10*7.5MICRON), 277K	1737
*MODIOLUS MODIOLUS	1753 ML/HR	SIZE=7.6CM,FOOD=CHLAMYDOMONAS SP.(2*10**7CELLS/L,10*7.5MICRON), 285K	1737
*MODIOLUS MODIOLUS	2527 ML/HR	SIZE=7.6CM,FOOD=CHLAMYDOMONAS SP.(2*10**7CELLS/L,10*7.5MICRON), 293K	1737
*MYTILUS EDULIS	353 ML/HR	SIZE=4CM,FOOD=PLATYMONAS SUECICA (5*10**6CELLS/L,7.2*9.4MICRON), 278K	1403
*MYTILUS EDULIS	1468 ML/HR	SIZE=4CM,FOOD=PLATYMONAS SUECICA (5*10**6CELLS/L,7.2*9.4MICRON), 283K	1403
*MYTILUS EDULIS	1751 ML/HR	SIZE=4CM,FOOD=PLATYMONAS SUECICA (5*10**6CELLS/L,7.2*9.4MICRON), 288K	1403
*MYTILUS EDULIS	1796 ML/HR	SIZE=4CM,FOOD=PLATYMONAS SUECICA (5*10**6CELLS/L,7.2*9.4MICRON), 293K	1403

TABLE C375

TEMPERATURE : CHANGE IN FILTRATION RATE			
ITEM:	VALUE:	CONDITIONS:	REF.:
*MYTILUS EDULIS	1798 ML/HR	SIZE=4CM,FOOD=PLATYMONAS SUECICA (5*10**6CELLS/L,7.2*9.4MICRON), 298K	1403
*MYTILUS EDULIS	106 ML/HR	SIZE=4CM,FOOD=PLATYMONAS SUECICA (5*10**6CELLS/L,7.2*9.4MICRON), 303K	1403
*MYTILUS EDULIS	650 ML/HR	SIZE=4.5CM,FOOD=AQUADAG GRAPHITE (CA.2MICRON),278K	1572
*MYTILUS EDULIS	1100 ML/HR	SIZE=4.5CM,FOOD=AQUADAG GRAPHITE (CA.2MICRON),283K	1572
*MYTILUS EDULIS	1320 ML/HR	SIZE=4.5CM,FOOD=AQUADAG GRAPHITE (CA.2MICRON),288K	1572
*MYTILUS EDULIS	1370 ML/HR	SIZE=4.5CM,FOOD=AQUADAG GRAPHITE (CA.2MICRON),290K	1572
*MYTILUS EDULIS	1250 ML/HR	SIZE=4.5CM,FOOD=AQUADAG GRAPHITE (CA.2MICRON),293K	1572
*MYTILUS EDULIS	250 ML/HR	SIZE=4.5CM,FOOD=AQUADAG GRAPHITE (CA.2MICRON),298K	1572

TABLE C376

TEMPERATURE : CHANGE IN GROWTH RATE			
ITEM:	VALUE:	CONDITIONS:	REF.:
*PEPPER PLANTS	0.36 MM/DAY	96 DAYS OLD PLANT, 302.1 K	365
*PEPPER PLANTS	2.4 MM/DAY	96 DAYS OLD PLANT, 298.5 K	365
*PEPPER PLANTS	2.8 MM/DAY	96 DAYS OLD PLANT, 292.1 K	365
*PEPPER PLANTS	4.1 MM/DAY	96 DAYS OLD PLANT, 288.9 K	365
*PEPPER PLANTS	4.45 MM/DAY	96 DAYS OLD PLANT, 285.7 K	365
*PEPPER PLANTS	2.3 MM/DAY	96 DAYS OLD PLANT, 281.2 K	365
*PEPPER PLANTS	7.24 MM/DAY	24 DAYS OLD PLANT, 302.1 K	365
*PEPPER PLANTS	9.0 MM/DAY	24 DAYS OLD PLANT, 298.5 K	365
*PEPPER PLANTS	8 MM/DAY	24 DAYS OLD PLANT, 292.1 K	365
*PEPPER PLANTS	3 MM/DAY	24 DAYS OLD PLANT, 281.2 K	365
*PEPPER PLANTS	12.7 MM/DAY	24 DAYS OLD PLANT, 285.7 K	365

TABLE C377

TEMPERATURE : CHANGE IN RATE IN REPRODUCTION			
ITEM:	VALUE:	CONDITIONS:	REF.:
*CALANDRA ORYZAE	1 EGGS/(FEMALE*WEEK)	288.2 K,FEMALES/(FEMALE*W) = 0.005,TIME FROM EGG TO ADULT = 32.9	128
*CALANDRA ORYZAE	4 EGGS/(FEMALE*WEEK)	291.2 K,FEMALES/(FEMALE*W) = 0.14, TIME FROM EGG TO ADULT = 15.5 W	128
*CALANDRA ORYZAE	266 EGGS/(FEMALE*WEEK)	296 K,FEMALES/(FEMALE*W) = 0.43, TIME FROM EGG TO ADULT = 6.1 W	128
*CALANDRA ORYZAE	384 EGGS/(FEMALE*WEEK)	298.5 K,FEMALES/(FEMALE*W) = 0.61, TIME FROM EGG TO ADULT = 4.9 W	128
*CALANDRA ORYZAE	344 EGGS/(FEMALE*WEEK)	302.1 K,FEMALES/(FEMALE*W) = 0.77, TIME FROM EGG TO ADULT = 4.0 W	128
*CALANDRA ORYZAE	197 EGGS/(FEMALE*WEEK)	305.3 EGGS/(FEMALE*WEEK)	128
*CALANDRA ORYZAE	27 EGGS/(FEMALE*WEEK)	306.5 K,FEMALES/(FEMALE*W) = 0.12, TIME FROM EGG TO ADULT = 9.4 W	128
*CALANDRA ORYZAE	5 EGGS/(FEMALE*WEEK)	308.0 K,FEMALES/(FEMALE*W) = 0, TIME FROM EGG TO ADULT = INFINITE	128

TABLE C378

TEMPERATURE : CHANGE IN RATE OF DEVELOPMENT OF ZOOPLANKTON STAGES			
ITEM:	VALUE:	CONDITIONS:	REF.:
*CYCLOPS VICINUS	37.3 (SD=+-3.8) DAYS	FROM NAUPLIUS TO COPEPODIT STAGE 1, 278 K, NO. OF GROUPS=20	1082
*CYCLOPS VICINUS	4.6 (SD=+-1.0) DAYS	FROM COPEPODIT STAGE 1 TO 2, 278 K, NO. OF GROUPS=20	1082
*CYCLOPS VICINUS	4.0 (SD=+-1.2) DAYS	FROM COPEPODIT STAGE 2 TO 3, 278 K, NO. OF GROUPS=20	1082
*CYCLOPS VICINUS	5.1 (SD=+-0.8) DAYS	FROM COPEPODIT STAGE 3 TO 4, 278 K, NO. OF GROUPS=20	1082
*CYCLOPS VICINUS	3.6 (SD=+-2.2) DAYS	FROM COPEPODIT STAGE 4 TO 5, MALE, 278 K, NO. OF GROUPS=18	1082
*CYCLOPS VICINUS	10.4 (SD=+-4.9) DAYS	FROM COPEPODIT STAGE 4 TO 5, FEMALE, 278 K, NO. OF GROUPS=15	1082
*CYCLOPS VICINUS	8.3 (SD=+-1.8) DAYS	FROM COPEPODIT STAGE 5 TO ADULT, MALE, 278 K, NO. OF GROUPS=18	1082
*CYCLOPS VICINUS	13.1 (SD=+-2.9) DAYS	FROM COPEPODIT STAGE 5 TO ADULT, FEMALE, 278 K, NO. OF GROUPS=15	1082
*CYCLOPS VICINUS	25.3 (SD=+-2.3) DAYS	TOTAL COPEPODIT DEVELOPMENT, MALE, 278 K, NO. OF GROUPS=18	1082
*CYCLOPS VICINUS	37.8 (SD=+-6.4) DAYS	TOTAL COPEPODIT DEVELOPMENT, FEMALE, 278 K, NO. OF GROUPS=15	1082
*CYCLOPS VICINUS	14.5 (SD=+-2.0) DAYS	FROM NAUPLIUS TO COPEPODIT STAGE 1, 283 K, NO. OF GROUPS=20	1082
*CYCLOPS VICINUS	2.7 (SD=+-0.7) DAYS	FROM COPEPODIT STAGE 1 TO 2, 283 K, NO. OF GROUPS=20	1082
*CYCLOPS VICINUS	2.4 (SD=+-0.5) DAYS	FROM COPEPODIT STAGE 2 TO 3, 283 K, NO. OF GROUPS=20	1082
*CYCLOPS VICINUS	2.6 (SD=+-0.7) DAYS	FROM COPEPODIT STAGE 3 TO 4, 283 K, NO. OF GROUPS=20	1082
*CYCLOPS VICINUS	1.9 (SD=+-1.4) DAYS	FROM COPEPODIT STAGE 4 TO 5, MALE, 283 K, NO. OF GROUPS=19	1082
*CYCLOPS VICINUS	6.3 (SD=+-9.3) DAYS	FROM COPEPODIT STAGE 4 TO 5, FEMALE, 283 K, NO. OF GROUPS=19	1082
*CYCLOPS VICINUS	3.5 (SD=+-1.0) DAYS	FROM COPEPODIT STAGE 5 TO ADULT, MALE, 283 K, NO. OF GROUPS=19	1082
*CYCLOPS VICINUS	4.7 (SD=+-1.5) DAYS	FROM COPEPODIT STAGE 5 TO ADULT, FEMALE, 283 K, NO. OF GROUPS=19	1082
*CYCLOPS VICINUS	13.0 (SD=+-2.2) DAYS	TOTAL COPEPODIT DEVELOPMENT, MALE, 283 K, NO. OF GROUPS=19	1082
*CYCLOPS VICINUS	18.9 (SD=+-10.4) DAYS	TOTAL COPEPODIT DEVELOPMENT, FEMALE, 283 K, NO. OF GROUPS=19	1082
*CYCLOPS VICINUS	18.8 (SD=+-2.0) DAYS	FROM NAUPLIUS TO COPEPODIT STAGE 1, 288 K, NO. OF GROUPS=18	1082
*CYCLOPS VICINUS	1.7 (SD=+-1.1) DAYS	FROM COPEPODIT STAGE 1 TO 2, 288 K, NO. OF GROUPS=18	1082
*CYCLOPS VICINUS	1.5 (SD=+-0.9) DAYS	FROM COPEPODIT STAGE 2 TO 3, 288 K, NO. OF GROUPS=18	1082
*CYCLOPS VICINUS	1.3 (SD=+-0.5) DAYS	FROM COPEPODIT STAGE 3 TO 4, 288 K, NO. OF GROUPS=18	1082
*CYCLORS VICINUS	2.2 (SD=+-1.6) DAYS	FROM COPEPODIT STAGE 4 TO 5, MALE, 288 K, NO. OF GROUPS=12	1082
*CYCLOPS VICINUS	2.8 (SD=+-1.8) DAYS	FROM COPEPODIT STAGE 4 TO 5, FEMALE, 288 K, NO. OF GROUPS=14	1082
*CYCLOPS VICINUS	2.5 (SD=+-0.5) DAYS	FROM COPEPODIT STAGE 5 TO ADULT, MALE, 288 K, NO. OF GROUPS=12	1082
*CYCLOPS VICINUS	3.4 (SD=+-1.1) DAYS	FROM COPEPODIT STAGE 5 TO ADULT, FEMALE, 288 K, NO. OF GROUPS=14	1082
*CYCLOPS VICINUS	8.9 (SD=+-2.0) DAYS	TOTAL COPEPODIT DEVELOPMENT, MALE, 288 K, NO. OF GROUPS=12	1082
*CYCLOPS VICINUS	10.9 (SD=+-2.2) DAYS	TOTAL COPEPODIT DEVELOPMENT, FEMALE, 288 K, NO. OF GROUPS=14	1082
*CYCLOPS VICINUS	16.6 (SD=+-1.0) DAYS	FROM NAUPLIUS TO COPEPODIT STAGE 1, 293 K, NO. OF GROUPS=6	1082
*CYCLOPS VICINUS	1.6 (SD=+-0.6) DAYS	FROM COPEPODIT STAGE 1 TO 2, 293 K, NO. OF GROUPS=6	1082
*CYCLOPS VICINUS	1.0 (SD=+-0.5) DAYS	FROM COPEPODIT STAGE 2 TO 3, 293 K, NO. OF GROUPS=6	1082
*CYCLOPS VICINUS	1.7 (SD=+-1.3) DAYS	FROM COPEPODIT STAGE 3 TO 4, 293 K, NO. OF GROUPS=6	1082

TABLE C378

TEMPERATURE : CHANGE IN RATE OF DEVELOPMENT OF ZOOPLANKTON STAGES			
ITEM:	VALUE:	CONDITIONS:	REF.:
*CYCLOPS VICINUS	0.8 (SD=+-1.1) DAYS	FROM COPEPODIT STAGE 4 TO 5, MALE, 293 K, NO. OF GROUPS=6	1082
*CYCLOPS VICINUS	2.4 (SD=+-1.9) DAYS	FROM COPEPODIT STAGE 4 TO 5, FEMALE, 293 K, NO. OF GROUPS=6	1082
*CYCLOPS VICINUS	2.1 (SD=+-2.0) DAYS	FROM COPEPODIT STAGE 5 TO ADULT, MALE, 293 K, NO. OF GROUPS=6	1082
*CYCLOPS VICINUS	5.5 (SD=+-2.7) DAYS	FROM COPEPODIT STAGE 5 TO ADULT, FEMALE, 293 K, NO. OF GROUPS=6	1082
*CYCLOPS VICINUS	7.1 (SD=+-2.3) DAYS	TOTAL COPEPODIT DEVELOPMENT, MALE, 293 K, NO. OF GROUPS=6	1082
*CYCLOPS VICINUS	12.2 (SD=+-3.2) DAYS	TOTAL COPEPODIT DEVELOPMENT, FEMALE, 293 K, NO. OF GROUPS=6	1082
*EUDIAPTOMUS GRACILIS	36.2 (SD=+-4.4) DAYS	FROM NAUPLIUS TO COPEPODIT STAGE 1, 278 K, NO. OF GROUPS=16	1082
*EUDIAPTOMUS GRACILIS	9.8 (SD=+-1.4) DAYS	FROM COPEPODIT STAGE 1 TO 2, 278 K, NO. OF GROUPS=16	1082
*EUDIAPTOMUS GRACILIS	9.2 (SD=+-1.7) DAYS	FROM COPEPODIT STAGE 2 TO 3, 278 K, NO. OF GROUPS=14	1082
*EUDIAPTOMUS GRACILIS	15.1 (SD=+-5.4) DAYS	FROM COPEPODIT STAGE 3 TO 4, 278 K, NO. OF GROUPS=13	1082
*EUDIAPTOMUS GRACILIS	13.6 (SD=+-3.3) DAYS	FROM COPEPODIT STAGE 4 TO 5, MALE, 278 K, NO. OF GROUPS=8	1082
*EUDIAPTOMUS GRACILIS	18.6 (SD=+-6.6) DAYS	FROM COPEPODIT STAGE 4 TO 5, FEMALE, 278 K, NO. OF GROUPS=7	1082
*EUDIAPTOMUS GRACILIS	16.7 (SD=+-4.3) DAYS	FROM COPEPODIT STAGE 5 TO ADULT, MALE, 278 K, NO. OF GROUPS=8	1082
*EUDIAPTOMUS GRACILIS	17.3 (SD=+-4.6) DAYS	FROM COPEPODIT STAGE 5 TO ADULT, FEMALE, 278 K, NO. OF GROUPS=7	1082
*EUDIAPTOMUS GRACILIS	60.6 (SD=+-7.3) DAYS	TOTAL COPEPODIT DEVELOPMENT, MALE, 278 K, NO. OF GROUPS=8	1082
*EUDIAPTOMUS GRACILIS	67.6 (SD=+-8.8) DAYS	TOTAL COPEPODIT DEVELOPMENT, FEMALE, 278 K, NO. OF GROUPS=7	1082
*EUDIAPTOMUS GRACILIS	19.4 (SD=+-4.9) DAYS	FROM NAUPLIUS TO COPEPODIT STAGE 1, 283 K, NO. OF GROUPS=19	1082
*EUDIAPTOMUS GRACILIS	5.3 (SD=+-1.8) DAYS	FROM COPEPODIT STAGE 1 TO 2, 283 K, NO. OF GROUPS=18	1082
*EUDIAPTOMUS GRACILIS	6.1 (SD=+-4.4) DAYS	FROM COPEPODIT STAGE 2 TO 3, 283 K, NO. OF GROUPS=14	1082
*EUDIAPTOMUS GRACILIS	4.8 (SD=+-2.5) DAYS	FROM COPEPODIT STAGE 3 TO 4, 283 K, NO. OF GROUPS=11	1082
*EUDIAPTOMUS GRACILIS	7.8 (SD=+-2.2) DAYS	FROM COPEPODIT STAGE 4 TO 5, MALE, 283 K, NO. OF GROUPS=6	1082
*EUDIAPTOMUS GRACILIS	9.2 (SD=+-3.2) DAYS	FROM COPEPODIT STAGE 4 TO 5, FEMALE, 283 K, NO. OF GROUPS=8	1082
*EUDIAPTOMUS GRACILIS	10.5 (SD=+-4.0) DAYS	FROM COPEPODIT STAGE 5 TO ADULT, MALE, 283 K, NO. OF GROUPS=6	1082
*EUDIAPTOMUS GRACILIS	15.3 (SD=+-7.4) DAYS	FROM COPEPODIT STAGE 5 TO ADULT, FEMALE, 283 K, NO. OF GROUPS=8	1082
*EUDIAPTOMUS GRACILIS	31.0 (SD=+-6.7) DAYS	TOTAL COPEPODIT DEVELOPMENT, MALE, 283 K, NO. OF GROUPS=6	1082
*EUDIAPTOMUS GRACILIS	37.9 (SD=+-9.0) DAYS	TOTAL COPEPODIT DEVELOPMENT, FEMALE, 283 K, NO. OF GROUPS=8	1082
*EUDIAPTOMUS GRACILIS	15.2 (SD=+-0.9) DAYS	FROM NAUPLIUS TO COPEPODIT STAGE 1, 288 K, NO. OF GROUPS=9	1082
*EUDIAPTOMUS GRACILIS	3.0 (SD=+-1.8) DAYS	FROM COPEPODIT STAGE 1 TO 2, 288 K, NO. OF GROUPS=9	1082
*EUDIAPTOMUS GRACILIS	2.4 (SD=+-1.5) DAYS	FROM COPEPODIT STAGE 2 TO 3, 288 K, NO. OF GROUPS=9	1082
*EUDIAPTOMUS GRACILIS	3.3 (SD=+-2.1) DAYS	FROM COPEPODIT STAGE 3 TO 4, 288 K, NO. OF GROUPS=9	1082
*EUDIAPTOMUS GRACILIS	4.0 (SD=+-3.9) DAYS	FROM COPEPODIT STAGE 4 TO 5, MALE, 288 K, NO. OF GROUPS=9	1082
*EUDIAPTOMUS GRACILIS	4.5 (SD=+-6.8) DAYS	FROM COPEPODIT STAGE 4 TO 5, FEMALE, 288 K, NO. OF GROUPS=6	1082
*EUDIAPTOMUS GRACILIS	7.9 (SD=+-3.6) DAYS	FROM COPEPODIT STAGE 5 TO ADULT, MALE, 288 K, NO. OF GROUPS=9	1082
*EUDIAPTOMUS GRACILIS	5.2 (SD=+-4.5) DAYS	FROM COPEPODIT STAGE 5 TO ADULT, FEMALE, 288 K, NO. OF GROUPS=6	1082

TABLE C378

TEMPERATURE : CHANGE IN RATE OF DEVELOPMENT OF ZOOPLANKTON STAGES			
ITEM:	VALUE:	CONDITIONS:	REF.:
*EUDIAPTOMUS GRACILIS	20.7 (SD=+-5.3) DAYS	TOTAL COPEPODIT DEVELOPMENT, MALE, 288 K, NO. OF GROUPS=9	1082
*EUDIAPTOMUS GRACILIS	18.7 (SD=+-5.5) DAYS	TOTAL COPEPODIT DEVELOPMENT, FEMALE, 288 K, NO. OF GROUPS=6	1082
*EUDIAPTOMUS GRACILIS	8.9 (SD=+-1.2) DAYS	FROM NAUPLIUS TO COPEPODIT STAGE 1, 293 K, NO. OF GROUPS=5	1082
*EUDIAPTOMUS GRACILIS	3.1 (SD=+-1.0) DAYS	FROM COPEPODIT STAGE 1 TO 2, 293 K, NO. OF GROUPS=4	1082
*EUDIAPTOMUS GRACILIS	2.3 DAYS	FROM COPEPODIT STAGE 2 TO 3, 293 K, NO. OF GROUPS=3	1082
*EUDIAPTOMUS GRACILIS	2.5 (SD=+-1.3) DAYS	FROM COPEPODIT STAGE 3 TO 4, 293 K, NO. OF GROUPS=3	1082
*EUDIAPTOMUS GRACILIS	1.8 (SD=+-0.7) DAYS	FROM COPEPODIT STAGE 4 TO 5, MALE, 293 K, NO. OF GROUPS=3	1082
*EUDIAPTOMUS GRACILIS	0.8 (SD=+-0.7) DAYS	FROM COPEPODIT STAGE 4 TO 5, FEMALE, 293 K, NO. OF GROUPS=3	1082
*EUDIAPTOMUS GRACILIS	4.0 DAYS	FROM COPEPODIT STAGE 5 TO ADULT, MALE, 293 K, NO. OF GROUPS=1	1082
*EUDIAPTOMUS GRACILIS	2.1 (SD=+-0.9) DAYS	FROM COPEPODIT STAGE 5 TO ADULT, FEMALE, 293 K, NO. OF GROUPS=2	1082
*EUDIAPTOMUS GRACILIS	15.5 DAYS	TOTAL COPEPODIT DEVELOPMENT, MALE, 293 K, NO. OF GROUPS=1	1082
*EUDIAPTOMUS GRACILIS	12.0 (SD=+-1.4) DAYS	TOTAL COPEPODIT DEVELOPMENT, FEMALE, 293 K, NO. OF GROUPS=2	1082

TABLE C379

TEMPERATURE : CHANGE IN RATE OF THERMAL DENATURATION			
ITEM:	VALUE:	CONDITIONS:	REF.:
*PROTEIN	1.25*10**(-7) 1/SEC	310 K, 1.08 % REPLACED IN 24 H, 1 ORDER REAC., PROTEIN AS BIOMASS	1501
*PROTEIN	1.89*10**(-7) 1/SEC	311 K, 1.63 % REPLACED IN 24 H, 1 ORDER REAC., PROTEIN AS BIOMASS	1501
*PROTEIN	3.70*10**(-7) 1/SEC	312 K, 3.20 % REPLACED IN 24 H, 1 ORDER REAC., PROTEIN AS BIOMASS	1501
*PROTEIN	5.12*10**(-7) 1/SEC	313 K, 4.42 % REPLACED IN 24 H, 1 ORDER REAC., PROTEIN AS BIOMASS	1501
*PROTEIN	6.62*10**(-7) 1/SEC	314 K, 5.71 % REPLACED IN 24 H, 1 ORDER REAC., PROTEIN AS BIOMASS	1501
*PROTEIN	1.05*10**(-6) 1/SEC	315 K, 9.08 % REPLACED IN 24 H, 1 ORDER REAC., PROTEIN AS BIOMASS	1501
*PROTEIN	1.50*10**(-6) 1/SEC	316 K, 13.81 % REPLACED IN 24 H, 1 ORDER REAC., PROTEIN AS BIOMASS	1501
*PROTEIN	2.37*10**(-6) 1/SEC	317 K, 20.50 % REPLACED IN 24 H, 1 ORDER REAC., PROTEIN AS BIOMASS	1501
*PROTEIN	3.52*10**(-6) 1/SEC	318 K, 28.65 % REPLACED IN 24 H, 1 ORDER REAC., PROTEIN AS BIOMASS	1501
*PROTEIN	5.35*10**(-6) 1/SEC	319 K, 46.20 % REPLACED IN 24 H, 1 ORDER REAC., PROTEIN AS BIOMASS	1501
*PROTEIN	7.92*10**(-6) 1/SEC	320 K, 68.40 % REPLACED IN 24 H, 1 ORDER REAC., PROTEIN AS BIOMASS	1501
*PROTEIN	13.5*10**(-6) 1/SEC	321 K, 116.64 % REPLACED IN 24 H, 1 ORDER REAC., PROTEIN AS BIOMASS	1501
*PROTEIN	18.1*10**(-6) 1/SEC	322 K, 161.51 % REPLACED IN 24 H, 1 ORDER REAC., PROTEIN AS BIOMASS	1501
*PROTEIN	27.1*10**(-6) 1/SEC	323 K, 234.14 % REPLACED IN 24 H, 1 ORDER REAC., PROTEIN AS BIOMASS	1501

TABLE C380

TEMPERATURE : CHANGE IN RESPIRATION OF ZOOPLANKTON			
ITEM:	VALUE:	CONDITIONS:	REF.:
*PHRONIMA SEDENTARIA	9.1 +- 1.1 MICROLITRE O2/(MG DRY WEIGHT*DAY)	278 K, 95 % CL.	997

TABLE C380

TEMPERATURE : CHANGE IN RESPIRATION OF ZOOPLANKTON			
ITEM:	VALUE:	CONDITIONS:	REF.:
*PHRONIMA SEDENTARIA	14.9 +- 0.5 MICROLITRE O2/(MG DRY WEIGHT*DAY)	283.5 K, 95 % CL.	997
*PHRONIMA SEDENTARIA	16.5 +- 2.7 MICROLITRE O2/(MG DRY WEIGHT*DAY)	288.5 K, 95 % CL.	997
*PHRONIMA SEDENTARIA	19.4 +- 1.3 MICROLITRE O2/(MG DRY WEIGHT*DAY)	292.5 K, 95 % CL.	997

TABLE C381

TEMPERATURE : CHANGE IN UPTAKE OF CD			
ITEM:	VALUE:	CONDITIONS:	REF.:
*GLYCINE MAX	5.55 MG CD/KG DRY WEIGHT	288.5 K,SOIL = 10 PPM CD	590
*GLYCINE MAX	6.91 MG CD/KG DRY WEIGHT	294.1 K,SOIL = 10 PPM CD	590
*GLYCINE MAX	7.90 MG CD/KG DRY WEIGHT	299.6 K,SOIL = 10 PPM CD	590
*GLYCINE MAX	9.79 MG CD/KG DRY WEIGHT	305.2 K,SOIL = 10 PPM CD	590

TABLE C382

TEMPERATURE : CHANGE IN UPTAKE OF PB			
ITEM:	VALUE:	CONDITIONS:	REF.:
*PLANT ROOTS	59 MG/G DRY WEIGHT	CORN, 40 MIN, 277 K	45
*PLANT ROOTS	130 MG/G DRY WEIGHT	CORN, 120 MIN, 277 K	45
*PLANT ROOTS	205 MG/G DRY WEIGHT	SOYBEAN, 40 MIN, 277 K	45
*PLANT ROOTS	560 MG/G DRY WEIGHT	SOYBEAN, 120 MIN, 277 K	45
*PLANT ROOTS	197 MG/G DRY WEIGHT	BEAN, 40 MIN, 277 K	45
*PLANT ROOTS	450 MG/G DRY WEIGHT	BEAN, 120 MIN, 277 K	45
*PLANT ROOTS	94 MG/G DRY WEIGHT	CORN, 40 MIN, 297 K	45
*PLANT ROOTS	170 MG/G DRY WEIGHT	CORN, 120 MIN, 297 K	45
*PLANT ROOTS	38 MG/G DRY WEIGHT	SOYBEAN, 40 MIN, 297 K	45
*PLANT ROOTS	37 MG/G DRY WEIGHT	SOYBEAN, 120 MIN, 297 K	45
*PLANT ROOTS	38 MG/G DRY WEIGHT	BEAN, 40 MIN, 297 K	45
*PLANT ROOTS	32 MG/G DRY WEIGHT	BEAN, 120 MIN, 297 K	45

TABLE C383

TEMPERATURE : CHANGE IN WATER SOLUBLE AND EXCHANGEABLE MN			
ITEM:	VALUE:	CONDITIONS:	REF.:
*SOIL	10 PPM	HEATED TO 333K FOR 30MIN.,UNHEATED =8PPM	635
*SOIL	27 PPM	HEATED TO 355K FOR 30MIN.,UNHEATED =8PPM	635
*SOIL	60 PPM	HEATED TO 366K FOR 30MIN.,UNHEATED =8PPM	635
*SOIL	82 PPM	HEATED TO 373K FOR 30MIN.,UNHEATED =8PPM	635

TABLE C384

TEMPERATURE : FIRST MORTALITY			
ITEM:	VALUE:	CONDITIONS:	REF.:
*ONCORHYNCHUS KETA	2640 SEC	299.8 K, NO MORTALITY UNDER 297 K	355

TABLE C384

TEMPERATURE : FIRST MORTALITY			
ITEM:	VALUE:	CONDITIONS:	REF.:
*ONCORHYNCHUS KETA	15 SEC	302.5 K, NO MORTALITY UNDER 297 K	355
*ONCORHYNCHUS KETA	6 SEC	305.3 SEC, NO MORTALITY UNDER 297 K	355
*ONCORHYNCHUS KETA	4 SEC	310.9 K, NO MORTALITY UNDER 297 K	355
*ONCORHYNCHUS TSHAWYTSCHA	100 SEC	299.8 K, NO MORTALITY UNDER 294.2 K	355
*ONCORHYNCHUS TSHAWYTSCHA	4 SEC	305.3 K, NO MORTALITY UNDER 294.2 K	355
*THALEICHTHYS PACIFICUS	36 MIN	291 K, NO MORTALITY UNDER 288 K	355
*THALEICHTHYS PACIFICUS	15 MIN	294 K, NO MORTALITY UNDER 288 K	355
*THALEICHTHYS PACIFICUS	1.4 MIN	300 K, NO MORTALITY UNDER 288 K	355
*THALEICHTHYS PACIFICUS	0.3 MIN	305 K, NO MORTALITY UNDER 288 K	355
*THALEICHTHYS PACIFICUS	5.0 MIN	297 K, NO MORTALITY UNDER 288 K	355

TABLE C385

TEMPERATURE : LOWER LETHAL TEMPERATURE			
ITEM:	VALUE:	CONDITIONS:	REF.:
*BLACK BASS	274.6 K	NONE	905
*CATFISH	273 K	NONE	905
*FISH	273 K	AVERAGE VALUE	905
*MINNOW	273 K	NONE	905
*PICKRELS	273 K	NONE	905
*SUNFISH	275.5 K	NONE	905
*YELLOW PERCH	273 K	NONE	905

TABLE C386

TEMPERATURE : MORTALITY (F021)			
ITEM:	VALUE:	CONDITIONS:	REF.:
*AMEIURUS NEBULOSUS	TEMPERATURE FOR ZERO	MINIMUM TEMP.=276.7K,MAXIMUM TEMP. =309.5K,ACCLIMATION TEMP.=303K	168
*AMEIURUS NEBULOSUS	TEMPERATURE FOR ZERO	MINIMUM TEMP.=272K,MAXIMUM TEMP.= 306.4K,ACCLIMATION TEMP.=293K	168
*ICTALURUS LACUSTRIS	TEMPERATURE FOR ZERO	MINIMUM TEMP.=273K,MAXIMUM TEMP.= 303.3K,ACCLIMATION TEMP.=288K	637
*ICTALURUS LACUSTRIS	TEMPERATURE FOR ZERO	MINIMUM TEMP.=279K,MAXIMUM TEMP.= 306.5K,ACCLIMATION TEMP.=298K	637
*LEPOMIS MACROCHIRUS	TEMPERATURE FOR ZERO	MINIMUM TEMP.=275.5K,MAXIMUM TEMP. =303.7K,ACCLIMATION TEMP.=288K	637
*LEPOMIS MACROCHIRUS	TEMPERATURE FOR ZERO	MINIMUM TEMP.=278K,MAXIMUM TEMP.= 304.5K,ACCLIMATION TEMP.=293K	637
*MICROPTERUS SALMOIDES	TEMPERATURE FOR ZERO	MINIMUM TEMP.=5.5278.5K,MAXIMUM TEMP.=305.5K,ACCLIMATION TEMP.= 293K	637
*MICROPTERUS SALMOIDES	TEMPERATURE FOR ZERO	MINIMUM TEMP.=283.8K,MAXIMUM TEMP. =309.4K,ACCLIMATION TEMP.=303K	637
*ONCORHYNCHUS GORBUSCHA	TEMPERATURE FOR ZERO	MINIMUM TEMP.=NONE,MAXIMUM TEMP.= 294.3K,ACCLIMATION TEMP.=278K	168
*ONCORHYNCHUS GORBUSCHA	TEMPERATURE FOR ZERO	MINIMUM TEMP.=NONE,MAXIMUM TEMP.= 296.9K,ACCLIMATION TEMP.=293K	168
*ONCORHYNCHUS KETA	TEMPERATURE FOR ZERO	MINIMUM TEMP.=273.5K,MAXIMUM TEMP. =295.6K,ACCLIMATION TEMP.=283K	168
*ONCORHYNCHUS KETA	TEMPERATURE FOR ZERO	MINIMUM TEMP.=299.5K,MAXIMUM TEMP. =296.7K,ACCLIMATION TEMP.=293K	168

TABLE C386

TEMPERATURE : MORTALITY (F021)			
ITEM:	VALUE:	CONDITIONS:	REF.:
*ONCORHYNCHUS KISUTCH	TEMPERATURE FOR ZERO	MINIMUM TEMP.=273.2K,MAXIMUM TEMP. =295.9K,ACCLIMATION TEMP.=278K	168
*ONCORHYNCHUS KISUTCH	TEMPERATURE FOR ZERO	MINIMUM TEMP.=277.5K,MAXIMUM TEMP. =298K,ACCLIMATION TEMP.=293K	168
*ONCORHYNCHUS NERKA	TEMPERATURE FOR ZERO	MINIMUM TEMP.=273K,MAXIMUM TEMP.= 295.2K,ACCLIMATION TEMP.=278K	168
*ONCORHYNCHUS NERKA	TEMPERATURE FOR ZERO	MINIMUM TEMP.=277.7K,MAXIMUM TEMP. =297.8K,ACCLIMATION TEMP.=293K	168
*ONCORHYNCHUS TSCHAWYTSCHA	TEMPERATURE FOR ZERO	MINIMUM TEMP.=273,8K,MAXIMUM TEMP. =297.3K,ACCLIMATION TEMP.=283K	168
*ONCORHYNCHUS TSCHAWYTSCHA	TEMPERATURE FOR ZERO	MINIMUM TEMP.=277.5K,MAXIMUM TEMP. =298.1K,ACCLIMATION TEMP.=293K	168
*PERCA FLAVESCENS	TEMPERATURE FOR ZERO	MINIMUM TEMP.=274.1K,MAXIMUM TEMP. =298K,ACCLIMATION TEMP.=283K	636
*PERCA FLAVESCENS	TEMPERATURE FOR ZERO	MINIMUM TEMP.=276.7K,MAXIMUM TEMP. =302.7K,ACCLIMATION TEMP.=298K	636
*THERMOCYCLOPS NEGLECTUS	4%	308K,20HRS,50 ANIMALS	985
*THERMOCYCLOPS NEGLECTUS	8%	308K,24HRS,50 ANIMALS	985
*THERMOCYCLOPS NEGLECTUS	12%	308K,31HRS,50 ANIMALS	985
*THERMOCYCLOPS NEGLECTUS	20%	308K,33HRS,50 ANIMALS	985
*THERMOCYCLOPS NEGLECTUS	22%	308K,36HRS,50 ANIMALS	985
*THERMOCYCLOPS NEGLECTUS	22%	310.5K,2HRS,50 ANIMALS	985
*THERMOCYCLOPS NEGLECTUS	80%	310.5K,4HRS,50 ANIMALS	985
*THERMOCYCLOPS NEGLECTUS	98%	310.5K,6HRS,50 ANIMALS	985
*THERMOCYCLOPS NEGLECTUS	100%	310.5K,24HRS,50 ANIMALS	985
*THERMOCYCLOPS NEGLECTUS	100%	313K,2HRS,50 ANIMALS	985

TABLE C387

TEMPERATURE : OPTIMUM FOR GROWTH			
ITEM:	VALUE:	CONDITIONS:	REF.:
*BLACK BASS	300 K	NONE	905
*CATFISH	303 K	NONE	905
*CRAPPIES	296 K	NONE	905
*FISH	297.2 K	AVERAGE VALUE	905
*MINNOW	300 K	NONE	905
*PICKRELS	298.4 K	NONE	905
*SUNFISH	300.5 K	NONE	905
*YELLOW PERCH	287.2 K	NONE	905

TABLE C388

TEMPERATURE : UPPER LETHAL TEMPERATURE			
ITEM:	VALUE:	CONDITIONS:	REF.:
*BLACK BASS	309.5 K	NONE	905
*CATFISH	310.1 K	NONE	905
*CRAPPIES	305.5 K	NONE	905
*FISH	307.8 K	AVERAGE VALUE	905
*MINNOW	306.4 K	NONE	905
*PICKRELS	307.4 K	NONE	905

TABLE C388

TEMPERATURE : UPPER LETHAL TEMPERATURE			
ITEM:	VALUE:	CONDITIONS:	REF.:
*SUNFISH	308.7 K	NONE	905
*YELLOW PERCH	303.9 K	NONE	905

TABLE C389

TETANUS : MORTALITY			
ITEM:	VALUE:	CONDITIONS:	REF.:
*MOUSE	50%	ORAL DOSE=1.67*10**(-6)MG/KG(= 2.53*10**(-14)MOL/KG),PT-50(F007)= 13.6	953

TABLE C390

TETRACYCLINE HCL : MORTALITY			
ITEM:	VALUE:	CONDITIONS:	REF.:
*MOUSE	50%	ORAL DOSE=650MG/KG(=1.35*10**(-3) MOL/KG),PT-50(F007)=2.87	953

TABLE C391

TETRODOTOXIN : MORTALITY			
ITEM:	VALUE:	CONDITIONS:	REF.:
*MOUSE	50%	ORAL DOSE=1*10**(-2)MG/KG(= 3.13*10**(-8)MOL/KG),PT-50(F007)= 7.50	953

TABLE C392

TH : MORTALITY			
ITEM:	VALUE:	CONDITIONS:	REF.:
*MOUSE	50%	4-CHLORIDE,INTRAPERITON.,DOSE= 114MG/KG BODY	1611
*RAT	50%	NITRATE,INTRAPERITON.,DOSE=68MG/KG BODY	1611

TABLE C393

THIOPHANATE-METHYL : GROWTH INHIBITION			
ITEM:	VALUE:	CONDITIONS:	REF.:
*CHLORELLA PYRENOIDOSA	50 %	48 H, FRESHWATER = 8.5 MG/L, 297 K, LOG PHASE, 10**7 CELLS/L	221

TABLE C394

THIOPHANATE-METHYL : MORTALITY			
ITEM:	VALUE:	CONDITIONS:	REF.:
*DAPHNIA MAGNA	50 %	48 H, FRESHWATER = 16 MG/L, 293 K, AGE=< 1 D, N=20	221

TABLE C394

THIOPHANATE-METHYL : MORTALITY			
ITEM:	VALUE:	CONDITIONS:	REF.:
*LEBISTES RETICULATUS	50 %	48 H, FRESHWATER = 130 MG/L, 288 K, AGE=3 M, N=10	221
*SALMO GAIRDNERI	50 %	48 H, FRESHWATER = 7.8 MG/L, 288 K, AGE = 3 M, N=10	221

TABLE C395

THIRAM : MORTALITY			
ITEM:	VALUE:	CONDITIONS:	REF.:
*RATS	50%	ACUTE ORAL TOXICITY=275-865MG/KG, FUNGICIDE	389

TABLE C396

TL : MORTALITY			
ITEM:	VALUE:	CONDITIONS:	REF.:
*MOUSE	50%	CHLORIDE,ORAL,DOSE=24MG/KG BODY	1611
*MOUSE	50%	NITRATE,ORAL,DOSE=32.5MG/KG BODY	1611
*MOUSE	50%	3-ACETATE,ORAL,DOSE=32MG/KG BODY	1611
*MOUSE	50%	CHLORODE, ORAL DOSE=24MG/KG(= 1.0*10**(-4)MOL/KG),PT-50(F007)= 4.00	953
*RAT	50%	2-OXIDE,ORAL,DOSE=43.6MG/KG BODY	1611
*RAT	50%	SULFATE,ORAL,DOSE=15.8MG/KG BODY	1611

TABLE C397

TM : MORTALITY			
ITEM:	VALUE:	CONDITIONS:	REF.:
*MOUSE	50%	CHLORIDE,INTRAPERITON.,DOSE=485MG/ KG BODY	1611
*MOUSE	50%	NITRATE,INTRAPERITON.,DOSE=255MG/ KG BODY	1611

TABLE C398

TRAFFIC INTENSITY : CHANGE IN PB CONTENT			
ITEM:	VALUE:	CONDITIONS:	REF.:
*SOIL	87.30 MG PB/KG DRY WEIGHT	TRAFFIC= 21040 VEHICLES/D, DISTANCE FROM ROAD = 6 M	543
*SOIL	79.60 MG PB/KG DRY WEIGHT	TRAFFIC= 21040 VEHICLES/D, DISTANCE FROM ROAD = 12 M	543
*SOIL	47.44 MG PB/KG DRY WEIGHT	TRAFFIC= 21040 VEHICLES/D, DISTANCE FROM ROAD = 18 M	543
*SOIL	109.74 MG PB/KG DRY WEIGHT	TRAFFIC= 8120 VEHICLES/D, DISTANCE FROM ROAD = 3 M	543
*SOIL	47.25 MG PB/KG DRY WEIGHT	TRAFFIC= 8120 VEHICLES/D, DISTANCE FROM ROAD = 6 M	543

TABLE C398

TRAFFIC INTENSITY : CHANGE IN PB CONTENT			
ITEM:	VALUE:	CONDITIONS:	REF.:
*SOIL	27.37 MG PB/KG DRY WEIGHT	TRAFFIC= 8120 VEHICLES/D, DISTANCE FROM ROAD = 12 M	543
*SOIL	19.91 MG PB/KG DRY WEIGHT	TRAFFIC= 8120 VEHICLES/D, DISTANCE FROM ROAD = 18 M	543
*SOIL	22.43 MG PB/KG DRY WEIGHT	TRAFFIC= 1085 VEHICLES/D, DISTANCE FROM ROAD = 3 M	543
*SOIL	26.44 MG PB/KG DRY WEIGHT	TRAFFIC= 1085 VEHICLES/D, DISTANCE FROM ROAD = 6 M	543
*SOIL	19.92 MG PB/KG DRY WEIGHT	TRAFFIC= 1085 VEHICLES/D, DISTANCE FROM ROAD = 12 M	543
*SOIL	27.47 MG PB/KG DRY WEIGHT	TRAFFIC= 1085 VEHICLES/D, DISTANCE FROM ROAD = 18 M	543
*SOIL	126.74 MG PB/KG DRY WEIGHT	TRAFFIC= 23500 VEHICLES/D, DISTANCE FROM ROAD = 6 M	543
*SOIL	42.34 MG PB/KG DRY WEIGHT	TRAFFIC= 23500 VEHICLES/D, DISTANCE FROM ROAD = 12 M	543
*SOIL	32.31 MG PB/KG DRY WEIGHT	TRAFFIC= 23500 VEHICLES/D, DISTANCE FROM ROAD = 18 M	543
*SOIL	22.35 MG PB/KG DRY WEIGHT	TRAFFIC= 23500 VEHICLES/D, DISTANCE FROM ROAD = 50 M	543
*SOIL	22.35 MG PB/KG DRY WEIGHT	CONTROL AREA IN FOREST, 500 M FROM HIGHWAY	543

TABLE C399

TRAFFIC INTENSITY : CHANGE IN PB CONTENT IN VEGETATION			
ITEM:	VALUE:	CONDITIONS:	REF.:
*VIRGINIA PINE	49.85 MG PB/KG DRY WEIGHT	TRAFFIC = 21040 VEHICLES/D, VIR., DISTANCE TO ROAD = 6 M	543

TABLE C400

TRAFFIC INTENSITY : CHANGE IN PB CONTENT IN VEGETATION			
ITEM:	VALUE:	CONDITIONS:	REF.:
*APPLE	72.36 MG PB/KG DRY WEIGHT	TRAFFIC = 21040 VEHICLES/D, VIR., DISTANCE TO ROAD = 12 M	543
*APPLE	39.87 MG PB/KG DRY WEIGHT	TRAFFIC = 8120 VEHICLES/D, VIR., DISTANCE TO ROAD = 12 M	543
*BLACK LOCUST	74.96 MG PB/KG DRY WEIGHT	TRAFFIC = 21040 VEHICLES/D, VIR., DISTANCE TO ROAD = 6 M	543
*BLACK LOCUST	34.82 MG PB/KG DRY WEIGHT	TRAFFIC = 1085 VEHICLES/D, VIR., DISTANCE TO ROAD = 6 M	543
*BLACK LOCUST	154.61 MG PB/KG DRY WEIGHT	TRAFFIC = 23500 VEHICLES/D, VIR., DISTANCE TO ROAD = 6 M	543
*CHICORY	24.46 MG PB/KG DRY WEIGHT	TRAFFIC = 21040 VEHICLES/D, VIR., DISTANCE TO ROAD = 18 M	543
*CORALBERRY	71.98 MG PB/KG DRY WEIGHT	TRAFFIC = 21040 VEHICLES/D, VIR., DISTANCE TO ROAD = 12 M	543
*CORALBERRY	57.42 MG PB/KG DRY WEIGHT	TRAFFIC = 8120 VEHICLES/D, VIR., DISTANCE TO ROAD = 6 M	543
*CORALBERRY FRUIT	27.35 MG PB/KG DRY WEIGHT	TRAFFIC = 8120 VEHICLES/D, VIR., DISTANCE TO ROAD = 6 M	543
*CROWN VETCH	74.87 MG PB/KG DRY WEIGHT	TRAFFIC = 21040 VEHICLES/D, VIR., DISTANCE TO ROAD = 6 M	543
*CROWN VETCH	42.32 MG PB/KG DRY WEIGHT	TRAFFIC = 21040 VEHICLES/D, VIR., DISTANCE TO ROAD = 12 M	543
*CROWN VETCH	27.43 MG PB/KG DRY WEIGHT	TRAFFIC = 21040 VEHICLES/D, VIR., DISTANCE TO ROAD = 18 M	543

TABLE C400

TRAFFIC INTENSITY : CHANGE IN PB CONTENT IN VEGETATION

ITEM:	VALUE:	CONDITIONS:	REF.:
*DANDELION	39.93 MG PB/KG DRY WEIGHT	TRAFFIC = 21040 VEHICLES/D, VIR., DISTANCE TO ROAD = 18 M	543
*JAPANESE HONEYSUCKLE	29.86 MG PB/KG DRY WEIGHT	TRAFFIC = 21040 VEHICLES/D, VIR., DISTANCE TO ROAD = 6 M	543
*JAPANESE HONEYSUCKLE	32.39 MG PB/KG DRY WEIGHT	TRAFFIC = 8120 VEHICLES/D, VIR., DISTANCE TO ROAD = 3 M	543
*JAPANESE HONEYSUCKLE	17.48 MG PB/KG DRY WEIGHT	TRAFFIC = 8120 VEHICLES/D, VIR., DISTANCE TO ROAD = 6 M	543
*JAPANESE HONEYSUCKLE	25.00 MG PB/KG DRY WEIGHT	TRAFFIC = 8120 VEHICLES/D, VIR., DISTANCE TO ROAD = 12 M	543
*JAPANESE HONEYSUCKLE	32.31 MG PB/KG DRY WEIGHT	TRAFFIC = 1085 VEHICLES/D, VIR., DISTANCE TO ROAD = 3 M	543
*JAPANESE HONEYSUCKLE	19.97 MG PB/KG DRY WEIGHT	TRAFFIC = 1085 VEHICLES/D, VIR., DISTANCE TO ROAD = 6 M	543
*LESPEDEZA SERICEA	19.97 MG PB/KG DRY WEIGHT	TRAFFIC = 1085 VEHICLES/D, VIR., DISTANCE TO ROAD = 12 M	543
*MULTIFLORA ROSE	115.10 MG PB/KG DRY WEIGHT	TRAFFIC = 21040 VEHICLES/D, VIR., DISTANCE TO ROAD = 6 M	543
*MULTIFLORA ROSE	32.40 MG PB/KG DRY WEIGHT	TRAFFIC = 1085 VEHICLES/D, VIR., DISTANCE TO ROAD = 6 M	543
*MULTIFLORA ROSE FRUIT	17.49 MG PB/KG DRY WEIGHT	TRAFFIC = 1085 VEHICLES/D, VIR., DISTANCE TO ROAD = 6 M	543
*NARROW-LEAF PLANTAIN	27.50 MG PB/KG DRY WEIGHT	TRAFFIC = 21040 VEHICLES/D, VIR., DISTANCE TO ROAD = 18 M	543
*PANIC GRASS	33.50 MG PB/KG DRY WEIGHT	TRAFFIC = 8120 VEHICLES/D, VIR., DISTANCE TO ROAD = 12 M	543
*PB CONCENTRATION IN VEGETATION	19.99 MG PB/KG DRY WEIGHT	TRAFFIC = 8120 VEHICLES/D, VIR., DISTANCE TO ROAD = 6 M	543
*PB CONCENTRATION IN VEGETATION	32.44 MG PB/KG DRY WEIGHT	TRAFFIC = 21040 VEHICLES/D, VIR., DISTANCE TO ROAD = 6 M	543
*PB CONCENTRATION IN VEGETATION	29.98 MG PB/KG DRY WEIGHT	TRAFFIC = 21040 VEHICLES/D, VIR., DISTANCE TO ROAD = 18 M	543
*PB CONCENTRATION IN VEGETATION	20.97 MG PB/KG DRY WEIGHT	TRAFFIC = 8120 VEHICLES/D, VIR., DISTANCE TO ROAD = 3 M	543
*PB CONCENTRATION IN VEGETATION	17.50 MG PB/KG DRY WEIGHT	TRAFFIC = 8120 VEHICLES/D, VIR., DISTANCE TO ROAD = 18 M	543
*PB CONCENTRATION IN VEGETATION	12.49 MG PB/KG DRY WEIGHT	TRAFFIC = 1085 VEHICLES/D, VIR., DISTANCE TO ROAD = 3 M	543
*PB CONCENTRATION IN VEGETATION	14.97 MG PB/KG DRY WEIGHT	TRAFFIC = 1085 VEHICLES/D, VIR., DISTANCE TO ROAD = 12 M	543
*PB CONCENTRATION IN VEGETATION	69.76 MG PB/KG DRY WEIGHT	TRAFFIC = 23500 VEHICLES/D, VIR., DISTANCE TO ROAD = 6 M	543
*PB CONCENTRATION IN VEGETATION	29.88 MG PB/KG DRY WEIGHT	TRAFFIC = 23500 VEHICLES/D, VIR., DISTANCE TO ROAD = 12 M	543
*PB CONCENTRATION IN VEGETATION	19.95 MG PB/KG DRY WEIGHT	TRAFFIC = 23500 VEHICLES/D, VIR., DISTANCE TO ROAD = 18 M	543
*PB CONCENTRATION IN VEGETATION	24.85 MG PB/KG DRY WEIGHT	TRAFFIC = 23500 VEHICLES/D, VIR., DISTANCE TO ROAD = 50 M	543
*PERIWINKLE	35.73 MG PB/KG DRY WEIGHT	TRAFFIC = 8120 VEHICLES/D, VIR., DISTANCE TO ROAD = 3 M	543
*PERIWINKLE	24.95 MG PB/KG DRY WEIGHT	TRAFFIC = 8120 VEHICLES/D, VIR., DISTANCE TO ROAD = 6 M	543
*RUBUS	40.84 MG PB/KG DRY WEIGHT	TRAFFIC = 8120 VEHICLES/D, VIR., DISTANCE TO ROAD = 12 M	543
*SUMAC	24.88 MG PB/KG DRY WEIGHT	TRAFFIC = 8120 VEHICLES/D, VIR., DISTANCE TO ROAD = 6 M	543
*VIBURNUM	72.22 MG PB/KG DRY WEIGHT	TRAFFIC = 21040 VEHICLES/D, VIR., DISTANCE TO ROAD = 6 M	543
*VIBURNUM BERRIES	29.90 MG PB/KG DRY WEIGHT	TRAFFIC = 21040 VEHICLES/D, VIR., DISTANCE TO ROAD = 6 M	543

TABLE C401

TRICHLOROMETHYLSULPHENYL BENZOATE : INHIBITION OF NITROGEN FIXATION

ITEM:	VALUE:	CONDITIONS:	REF.:
*CLOSTRIDIUM PASTEURIANUM	INHIBITED	5MG/L, REVERSIBLE BY ADDITION OF NA-MOLYBDATE	228

TABLE C402

TRIFLURALIN : MORTALITY			
ITEM:	VALUE:	CONDITIONS:	REF.:
*ICTALURUS PUNCTATUS	50 %	417 (380-447) PPB, 96 H, 14 G, 293 K,PH=8.2, ALK=80 PPM, 95 %CL	1002

TABLE C403

TRINITROTULUENE : INHIBITION OF GROWTH			
ITEM:	VALUE:	CONDITIONS:	REF.:
*MICROCYSTIS AERUGINOSA	NONE	ALFA TNT=15MG/L, INHIBITION OVER THIS CONC	1465
*SELENASTRUM CAPRICORNUTUM	NONE	ALFA TNT=5MG/L, INHIBITION OVER THIS CONC	1465
*SELENASTRUM CAPRICORNUTUM	NONE	TNT COLORED COMPLEX=9MG/L	1465

TABLE C404

TRINITROTULUENE : MORTALITY			
ITEM:	VALUE:	CONDITIONS:	REF.:
*MICROCYSTIS AERUGINOSA	100 %	TNT COLORED COMPLEX=50MG/L	1465
*PIMEPHALES PROMELAS	50 %	ALFA TNT=2.58MG/L	1465
*PIMEPHALES PROMELAS	50 %	TNT COLORED COMPLEX=1.6 MG/L, PINK WATER, DEG PRODUCT OF TNT	1465

TABLE C405

TWEEN 20 (F018) : DECREASE IN OXYGEN UPTAKE			
ITEM:	VALUE:	CONDITIONS:	REF.:
*THIOBACILLUS FERROOXIDANS	35.6 MICROLITRE O2/HR*MG PROTEIN	AGENT CONC.=10PPM,PH=1.9,308K; CONTROL=71.7;STERILE CONTROL=5.6	1608
*THIOBACILLUS FERROOXIDANS	30.1 MICROLITRE O2/HR*MG PROTEIN	AGENT CONC.=100PPM,PH=1.9,308K; CONTROL=71.7;STERILE CONTROL=5.6	1608
*THIOBACILLUS FERROOXIDANS	18.2 MICROLITRE O2/HR*MG PROTEIN	AGENT CONC.=1000PPM,PH=1.9,308K; CONTROL=71.7;STERILE CONTROL=5.6	1608
*THIOBACILLUS FERROOXIDANS	18.1 MICROLITRE O2/HR*MG PROTEIN	AGENT CONC.=10000PPM,PH=1.9,308K; CONTROL=71.7;STERILE CONTROL=5.6	1608

TABLE C406

TWEEN 40 (F019) : DECREASE IN OXYGEN UPTAKE			
ITEM:	VALUE:	CONDITIONS:	REF.:
*THIOBACILLUS FERROOXIDANS	47.0 MICROLITRE O2/HR*MG PROTEIN	AGENT CONC.=10PPM,PH=1.9,308K; CONTROL=71.7;STERILE CONTROL=5.6	1608
*THIOBACILLUS FERROOXIDANS	35.0 MICROLITRE O2/HR*MG PROTEIN	AGENT CONC.=100PPM,PH=1.9,308K; CONTROL=71.7;STERILE CONTROL=5.6	1608
*THIOBACILLUS FERROOXIDANS	13.5 MICROLITRE O2/HR*MG PROTEIN	AGENT CONC.=1000PPM,PH=1.9,308K; CONTROL=71.7;STERILE CONTROL=5.6	1608
*THIOBACILLUS FERROOXIDANS	14.4 MICROLITRE O2/HR*MG PROTEIN	AGENT CONC.=10000PPM,PH=1.9,308K; CONTROL=71.7;STERILE CONTROL=5.6	1608

TABLE C407

TWEEN 60 (F020) : DECREASE IN OXYGEN UPTAKE			
ITEM:	VALUE:	CONDITIONS:	REF.:
*THIOBACILLUS FERROOXIDANS	39.5 MICROLITRE O2/HR*MG PROTEIN	AGENT CONC.=10PPM,PH=1.9,308K; CONTROL=71.7;STERILE CONTROL=5.6	1608

TABLE C407

TWEEN 60 (F020) : DECREASE IN OXYGEN UPTAKE			
ITEM:	VALUE:	CONDITIONS:	REF.:
*THIOBACILLUS FERROOXIDANS	36.9 MICROLITRE O2/HR*MG PROTEIN	AGENT CONC.=100PPM,PH=1.9,308K; CONTROL=71.7;STERILE CONTROL=5.6	1608
*THIOBACILLUS FERROOXIDANS	15.3 MICROLITRE O2/HR*MG PROTEIN	AGENT CONC.=1000PPM,PH=1.9,308K; CONTROL=71.7;STERILE CONTROL=5.6	1608
*THIOBACILLUS FERROOXIDANS	14.9 MICROLITRE O2/HR*MG PROTEIN	AGENT CONC.=10000PPM,PH=1.9,308K; CONTROL=71.7;STERILE CONTROL=5.6	1608

TABLE C408

TWEEN 80 (F010) : DECREASE IN OXYGEN UPTAKE			
ITEM:	VALUE:	CONDITIONS:	REF.:
*THIOBACILLUS FERROOXIDANS	45.1 MICROLITRE O2/HR*MG PROTEIN	AGENT CONC.=10PPM,PH=1.9,308K; CONTROL=71.7;STERILE CONTROL=5.6	1608
*THIOBACILLUS FERROOXIDANS	36.9 MICROLITRE O2/HR*MG PROTEIN	AGENT CONC.=100PPM,PH=1.9,308K; CONTROL=71.7;STERILE CONTROL=5.6	1608
*THIOBACILLUS FERROOXIDANS	18.4 MICROLITRE O2/HR*MG PROTEIN	AGENT CONC.=1000PPM,PH=1.9,308K; CONTROL=71.7;STERILE CONTROL=5.6	1608
*THIOBACILLUS FERROOXIDANS	17.4 MICROLITRE O2/HR*MG PROTEIN	AGENT CONC.=10000PPM,PH=1.9,308K; CONTROL=71.7;STERILE CONTROL=5.6	1608

TABLE C409

U : MORTALITY			
ITEM:	VALUE:	CONDITIONS:	REF.:
*MOUSE	50%	ACETATE,INTRAPERITON.,DOSE=400MG/ KG BODY	1611
*MOUSE	50%	4-OXIDE,INTRAPERITON.,DOSE=6MG/KG BODY	1611
*RAT	50%	NITRATE,INTRAPERITON.,DOSE=135MG/ KG BODY	1611

TABLE C410

V : MORTALITY			
ITEM:	VALUE:	CONDITIONS:	REF.:
*PIMEPHALES PROMELAS	50%	VOSO4,EXPOS.TIME=4DAYS,CONC.IN WATER=4800MICROGRAM/L,LIMNOL.	1004
*RAT	50%	2-CHLORIDE,ORAL,DOSE=540MG/KG BODY	1611
*RAT	50%	4-CHLORIDE,ORAL,DOSE=160MG/KG BODY	1611
*RAT	50%	3-CHLORIDE,ORAL,DOSE=350MG/KG BODY	1611

TABLE C411

W : MORTALITY			
ITEM:	VALUE:	CONDITIONS:	REF.:
*RAT	50%	METALLIC,INTRAPERITON.,DOSE= 5000MG/KG BODY	1611
*RAT	50%	NA-WOLFRAMATE,ORAL,DOSE=1190MG/KG BODY	1611

TABLE C412

Y : MORTALITY			
ITEM:	VALUE:	CONDITIONS:	REF.:
*RAT	50%	3-CHLORIDE,INTRAPERITON.,DOSE= 450MG/KG BODY	1611
*RAT	50%	3-NITRATE,INTRAPERITON.,DOSE= 350MG/KG BODY	1611
*RAT	50%	3-OXIDE,INTRAPERITON.,DOSE=500MG/ KG BODY	1611

TABLE C413

YB : MORTALITY			
ITEM:	VALUE:	CONDITIONS:	REF.:
*RAT	50%	2-NITRATE,ORAL,DOSE=3100MG/KG BODY	1611

TABLE C414

ZN : CHANGE IN LARVAL DEVELOPMENT			
ITEM:	VALUE:	CONDITIONS:	REF.:
*CRASSOSTREA GIGAS	NORMAL	ZN = 50 UG/L	167
*CRASSOSTREA GIGAS	DECREASE IN GROWTH	ZN = 100 - 150 UG/L	167
*CRASSOSTREA GIGAS	NO GROWTH	ZN =200 UG/L	167

TABLE C415

ZN : CHANGE IN UPTAKE OF CD			
ITEM:	VALUE:	CONDITIONS:	REF.:
*GLYCINE MAX	7.48 MG CD/KG DRY WEIGHT	SOIL = 0 PPM APPLIED, SOIL = 10 PPM CD, 288.5-305.2 K	590
*GLYCINE MAX	9.41 MG CD/KG DRY WEIGHT	SOIL = 5 PPM APPLIED, SOIL = 10 PPM CD, 288.5-305.2 K	590
*GLYCINE MAX	9.35 MG CD/KG DRY WEIGHT	SOIL = 10 PPM APPLIED, SOIL = 10 PPM CD, 288.5-305.2 K	590
*GLYCINE MAX	8.60 MG CD/KG DRY WEIGHT	SOIL = 25 PPM APPLIED, SOIL = 10 PPM CD, 288.5-305.2 K	590
*GLYCINE MAX	8.26 MG CD/KG DRY WEIGHT	SOIL = 50 PPM APPLIED, SOIL = 10 PPM CD, 288.5-305.2 K	590
*GLYCINE MAX	6.91 MG CD/KG DRY WEIGHT	SOIL = 100 PPM APPLIED, SOIL = 10 PPM CD, 288.5-305.2 K	590
*GLYCINE MAX	4.68 MG CD/KG DRY WEIGHT	SOIL = 225 PPM APPLIED, SOIL = 10 PPM CD, 288.5-305.2 K	590
*GLYCINE MAX	5.38 MG CD/KG DRY WEIGHT	SOIL = 400 PPM APPLIED, SOIL = 10 PPM CD, 288.5-305.2 K	590
*PHASEOLUS VULGARIS, LEAVES	43.7 MICRO CI/(MG DRY WEIGHT)	ZN=0 PPM, CD=3 PPB AS CD-109, 5 D, 283-293 K	690
*PHASEOLUS VULGARIS, LEAVES	64.4 MICRO CI/(MG DRY WEIGHT)	ZN=1 PPM, CD=3 PPB AS CD-109, 5 D, 283-293 K	690
*PHASEOLUS VULGARIS, LEAVES	156.3 MICRO CI/(MG DRY WEIGHT)	ZN=5 PPM, CD=3 PPB AS CD-109, 5 D, 283-293 K	690
*PHASEOLUS VULGARIS, LEAVES	180.6 MICRO CI/(MG DRY WEIGHT)	ZN=10 PPM, CD=3 PPB AS CD-109, 5 D, 283-293 K	690
*PHASEOLUS VULGARIS, LEAVES	50.0 MICRO CI/(MG DRY WEIGHT)	ZN=100 PPM, CD=3 PPB AS CD-109, 5 D, 283-293 K	690
*PHASEOLUS VULGARIS, ROOT	16.3 MICRO CI/(MG DRY WEIGHT)	ZN=0 PPM, CD=3 PPB AS CD-109, 5 D, 283-293 K	690

TABLE C415

ZN : CHANGE IN UPTAKE OF CD			
ITEM:	VALUE:	CONDITIONS:	REF.:
*PHASEOLUS VULGARIS, ROOT	13.6 MICRO CI/(MG DRY WEIGHT)	ZN=1 PPM, CD=3 PPB AS CD-109, 5 D, 283-293 K	690
*PHASEOLUS VULGARIS, ROOT	13.2 MICRO CI/(MG DRY WEIGHT)	ZN=5 PPM, CD=3 PPB AS CD-109, 5 D, 283-293 K	690
*PHASEOLUS VULGARIS, ROOT	8.3 MICRO CI/(MG DRY WEIGHT)	ZN=10 PPM, CD=3 PPB AS CD-109, 5 D, 283-293 K	690
*PHASEOLUS VULGARIS, ROOT	0.86 MICRO CI/(MG DRY WEIGHT)	ZN=100 PPM, CD=3 PPB AS CD-109, 5 D, 283-293 K	690
*PHASEOLUS VULGARIS, STEM	68.0 MICRO CI/(MG DRY WEIGHT)	ZN=0 PPM, CD=3 PPB AS CD-109, 5 D, 283-293 K	690
*PHASEOLUS VULGARIS, STEM	57.2 MICRO CI/(MG DRY WEIGHT)	ZN=1 PPM, CD=3 PPB AS CD-109, 5 D, 283-293 K	690
*PHASEOLUS VULGARIS, STEM	195.9 MICRO CI/(MG DRY WEIGHT)	ZN=5 PPM, CD=3 PPB AS CD-109, 5 D, 283-293 K	690
*PHASEOLUS VULGARIS, STEM	215.7 MICRO CI/(MG DRY WEIGHT)	ZN=10 PPM, CD=3 PPB AS CD-109, 5 D, 283-293 K	690
*PHASEOLUS VULGARIS, STEM	168.5 MICRO CI/(MG DRY WEIGHT)	ZN=100 PPM, CD=3 PPB AS CD-109, 5 D, 283-293 K	690

TABLE C416

ZN : DECREASE IN GROWTH RATE			
ITEM:	VALUE:	CONDITIONS:	REF.:
*CHLORELLA VULGARIS	100 % OF CONTROL	ZN=0PPM,PH=6.23,288.5K,AFTER 96 HR	1290
*CHLORELLA VULGARIS	39.86 % OF CONTROL	ZN=1.0PPM,PH=6.17,288.5K,AFTER 96 HR,CONTROL-PH=6.23	1290
*CHLORELLA VULGARIS	43.87 % OF CONTROL	ZN=1.8PPM,PH=6.14,288.5K,AFTER 96 HR,CONTROL-PH=6.23	1290
*CHLORELLA VULGARIS	51.28 % OF CONTROL	ZN=2.4PPM,PH=6.12,288.5K,AFTER 96 HR,CONTROL-PH=6.23	1290
*CHLORELLA VULGARIS	58.61 % OF CONTROL	ZN=3.2PPM,PH=6.07,288.5K,AFTER 96 HR,CONTROL-PH=6.23	1290
*CHLORELLA VULGARIS	67.96 % OF CONTROL	ZN=10.0PPM,PH=5.83,288.5K,AFTER 96 HR,CONTROL-PH=6.23	1290

TABLE C417

ZN : GROWTH			
ITEM:	VALUE:	CONDITIONS:	REF.:
*AGATHIS AUSTRALIS	DEFICIENCY SYMPTOMS AT 0.001 MG/L	TREE, 100 D IN GREENHOUSE, CONC IN CULTURE SOLUTION	1248
*AGATHIS AUSTRALIS	NORMAL GROWTH AT 0.050 MG/L	TREE, 100 D IN GREENHOUSE, CONC IN CULTURE SOLUTION	1248
*AGATHIS AUSTRALIS	TOXIC SYMPTOMS AT 0.500 MG/L	TREE, 100 D IN GREENHOUSE, CONC IN CULTURE SOLUTION	1248

TABLE C418

ZN : HEMORRHAGING			
ITEM:	VALUE:	CONDITIONS:	REF.:
*PHOXINUS PHOXINUS	EFFECT OBSERVED	ZN=0.2-6.8 MG/L	95

TABLE C419

ZN : MORTALITY			
ITEM:	VALUE:	CONDITIONS:	REF.:
*DAPHNIA MAGNA	50%	ZNCL2,EXPOS.TIME=2DAYS,CONC.IN WATER=100MICROGRAM/L,LIMNOL.	122

TABLE C419

ZN : MORTALITY

ITEM:	VALUE:	CONDITIONS:	REF.:
*NOEMACHEILUS BARBARTULUS	50 %	ZINC=2.5MG/L,5D,HARD WATER	1477
*PHILODINA SP.	50 %	ZNSO4= 1.1 +- 2.0 PPM, 24 H, 25 PPM CACO3, 293 K, PH=7.4-7.9	1373
*PHILODINA SP.	50 %	ZNSO4= 0.9 +- 2.3 PPM, 48 H, 25 PPM CACO3, 293 K, PH=7.4-7.9	1373
*PHILODINA SP.	50 %	ZNSO4= 1.3 +- 1.8 PPM, 96 H, 25 PPM CACO3, 293 K, PH=7.4-7.9	1373
*PHOXINUS PHOXINUS	50%	ZN(NO3)2,EXPOS.TIME=4DAYS,CONC.IN WATER=3200MICROGRAM/L,LIMNOL.	95
*RAT	50%	ACETATE,ORAL,DOSE=2460MG/KG BODY	1611
*RAT	50%	CHLORIDE,ORAL,DOSE=350MG/KG BODY	419
*SALMO GAIRDNERI	50 %	ZINC=9.MG/L,8.3H,288.5K,10 FISH	1479
*SALMO GAIRDNERI	50 %	ZINC=6.6MG/L,11.7H,288.5K,10 FISH	1479
*SALMO GAIRDNERI	50 %	ZINC=6.0MG/L,15H,288.5K,10 FISH	1479
*SALMO GAIRDNERI	50 %	ZINC=5.3MG/L,24H,288.5K,10 FISH	1479
*SALMO GAIRDNERI	50 %	ZINC=4.76MG/L,101H,288.5K,10 FISH	1479
*SALMO GAIRDNERI	50 %	ZINC=4.6MG/L,5D,HARD WATER	1477

TABLE C420

ZN : PHOTOSYNTHESIS

ITEM:	VALUE:	CONDITIONS:	REF.:
*AUFWUCHS COMMUNITIES	100 % OF CONTROL	ZN= 0.001 MG ZN/L, 3 H, GOLDEN HORN	1186
*AUFWUCHS COMMUNITIES	95 % OF CONTROL	ZN= 0.01 MG ZN/L, 3 H, GOLDEN HORN	1186
*AUFWUCHS COMMUNITIES	106 % OF CONTROL	ZN= 0.1 MG ZN/L, 3 H, GOLDEN HORN	1186
*AUFWUCHS COMMUNITIES	82 % OF CONTROL	ZN= 1.0 MG ZN/L, 3 H,GOLDEN HORN	1186
*AUFWUCHS COMMUNITIES	60 % OF CONTROL	ZN= 10.0 MG ZN/L, 3 H, GOLDEN HORN	1186

TABLE C421

ZN : RESPIRATION

ITEM:	VALUE:	CONDITIONS:	REF.:
*AUFWUCHS COMMUNITIES	104 % OF CONTROL	ZN= 0.001 MG ZN/L, 3 H, GOLDEN HORN	1186
*AUFWUCHS COMMUNITIES	98 % OF CONTROL	ZN= 0.01 MG ZN/L, 3 H, GOLDEN HORN	1186
*AUFWUCHS COMMUNITIES	95 % OF CONTROL	ZN= 0.1 MG ZN/L, 3 H, GOLDEN HORN	1186
*AUFWUCHS COMMUNITIES	88 % OF CONTROL	ZN= 1.0 MG ZN/L, 3 H, GOLDEN HORN	1186
*AUFWUCHS COMMUNITIES	80 % OF CONTROL	ZN= 10.0 MG ZN/L, 3 H, GOLDEN HORN	1186

TABLE C422

ZN : SURVIVAL TIME

ITEM:	VALUE:	CONDITIONS:	REF.:
*SALMO GAIRDNERI	220 MIN	ZINC=7MG/L,CALCIUMCARBONATE=12MG/L,PH=6.6-6.7	1479
*SALMO GAIRDNERI	300 MIN	ZINC=7MG/L,CALCIUMCARBONATE=50MG/L,PH=7.-7.2	1479

TABLE C422

ZN : SURVIVAL TIME			
ITEM:	VALUE:	CONDITIONS:	REF.:
*SALMO GAIRDNERI	470 MIN	ZINC=7MG/L,CALCIUMCARBONATE=320MG/L,PH=7.6-7.8	1479
*SALMO GAIRDNERI	660 MIN	ZINC=7MG/L,CALCIUMCARBONATE=504MG/L,PH=7.7-8.	1479

TABLE C423

ZN : TOXIC SYMPTOMS			
ITEM:	VALUE:	CONDITIONS:	REF.:
*AGATHIS AUSTRALIS	0.500 MG/L	TREE, 100 D IN GREENHOUSE, CONC IN CULTURE SOLUTION	1248

TABLE C424

ZN : VERTEBRAL DAMAGE			
ITEM:	VALUE:	CONDITIONS:	REF.:
*FISH	8 OF 13 SPECIMENS	ZN=0.2-0.3 MG/L,AFTER 270D	93

TABLE C425

ZR : MORTALITY			
ITEM:	VALUE:	CONDITIONS:	REF.:
*RAT	50%	4-CHLORIDE,ORAL,DOSE=1688MG/KG BODY	1611
*RAT	50%	NITRATE,ORAL,DOSE=853MG/KG BODY	1611
*RAT	50%	ZRO-ACETATE,ORAL,DOSE=4100MG/KG BODY	1611
*RAT	50%	ZROCL2,ORAL,DOSE=3500MG/KG BODY	1611
*RAT	50%	ZRO-NITRATE,ORAL,DOSE=2500MG/KG BODY	1611
*RAT	50%	ZRO-SULFATE,ORAL,DOSE=3500MG/KG BODY	1611

TABLE C426

1-CHLOROPHTHALENE : MORTALITY			
ITEM:	VALUE:	CONDITIONS:	REF.:
*RANA PIPIENS	50 %	1-CHLOROPHTHALENE=900MG/KG	1533

TABLE C427

1-NAPHTHOL : MORTALITY			
ITEM:	VALUE:	CONDITIONS:	REF.:
*RANA PIPIENS	50 %	1-NAPHTHOL=130MG/KG	1533

TABLE C428

1,2-DIACETYLETHYLENE : INHIBITION OF NITROGEN FIXATION			
ITEM:	VALUE:	CONDITIONS:	REF.:
*CLOSTRIDIUM PASTEURIANUM	INHIBITED	10MG/L,INHIB.NULLIFIED BY ADDITION OF 10MICROGRAM BIOTIN/L	228

TABLE C429

2,3,6-TBA : MORTALITY			
ITEM:	VALUE:	CONDITIONS:	REF.:
*RATS	50%	ACUTE ORAL TOXICITY=1500MG/KG, HERBICIDE	389

TABLE C430

2,3,7,8-TCDD : MORTALITY			
ITEM:	VALUE:	CONDITIONS:	REF.:
*CHICKEN	50%	SINGLE ORAL DOSE=25-50MICROGRAM/KG BODY WT.	568
*GUINEA PIG	50%	SINGLE ORAL DOSE=0.6-2.1MICROGRAM/ KG BODY WT.	568
*RABBIT	50%	SINGLE ORAL DOSE=10-115MICROGRAM/ KG BODY WT.	568
*RAT	50%	SINGLE ORAL DOSE=10-100MICROGRAM/ KG BODY WT.	568

TABLE C431

2,4-D ACID : MORTALITY			
ITEM:	VALUE:	CONDITIONS:	REF.:
*ICTALURUS PUNCTATUS	LESS THAN 10 %	10 PPM, 48 H, 14 G, 293 K, PH=8.2, ALK=80 PPM	1002

TABLE C432

2,4-D DIMETHYL AMINE SALT : MORTALITY			
ITEM:	VALUE:	CONDITIONS:	REF.:
*ICTALURUS PUNCTATUS	LESS THAN 10 %	10 PPM, 48 H, 14 G, 293 K, PH=8.2, ALK=80 PPM	1002

TABLE C433

2,4,5-T : MORTALITY			
ITEM:	VALUE:	CONDITIONS:	REF.:
*ICTALURUS PUNCTATUS	LESS THAN 10 %	10 PPM, 48 H, 14 G, 293 K, PH=8.2, ALK=80 PPM	1002

TABLE C434

2,7-DCDD : MORTALITY			
ITEM:	VALUE:	CONDITIONS:	REF.:
*RAT	50%	ORAL DOSE:>1G/KG BODY WT.	1408

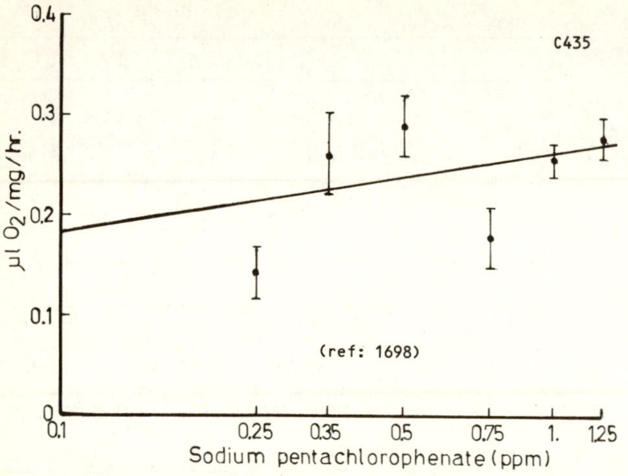

Oxygen consumption by Tubificidae versus pentachlorphenol concentration. pH = 7.5. Values equal to mean for 10 , bars represent S.D.

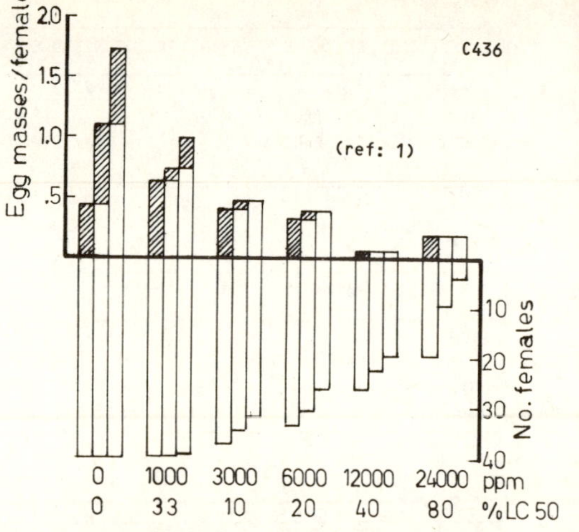

Reproduction and mortality of Ophryotrocha labronica, effect of the oil dispersant corexit 7664.
Each colum represents a 10-day period. Concentrations are expressed in ppm and in % of 48H LC-50.

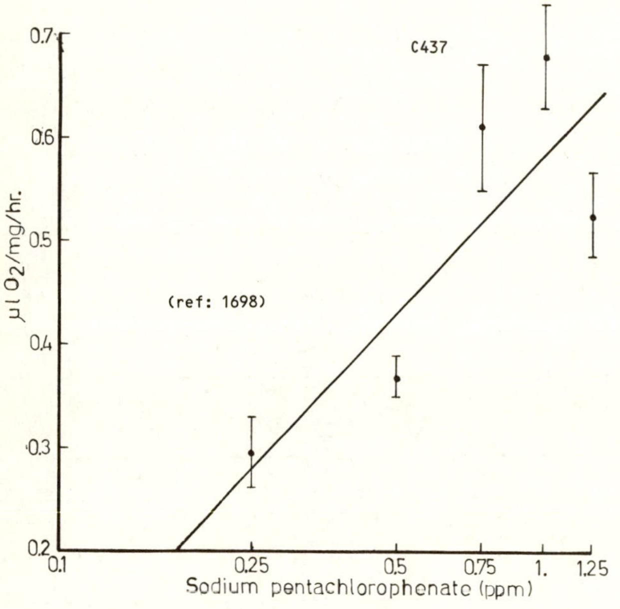

Oxygen consumption by Tubificidae versus pentachlorphenol concentration. pH = 8.5. Values equal to mean for 10 , bars represent S.D.

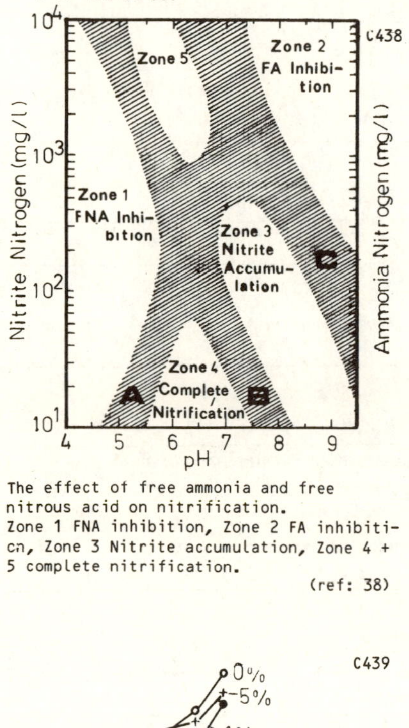

The effect of free ammonia and free nitrous acid on nitrification.
Zone 1 FNA inhibition, Zone 2 FA inhibition, Zone 3 Nitrite accumulation, Zone 4 + 5 complete nitrification.

(ref: 38)

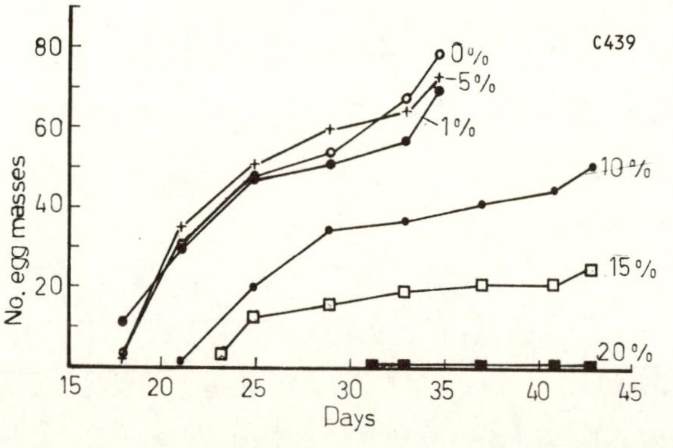

Reproduction of Ophryotrocha labronica, effect of sulphate pulp waste water. Salinity 33 ppt. Symbols see above.

(ref: 1)

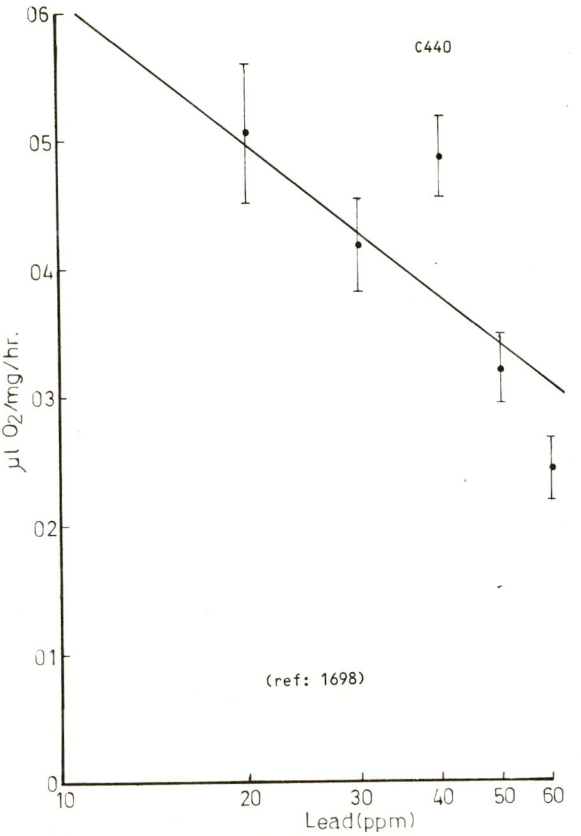

Oxygen consumption by Tubificidae versus lead concentration. pH = 8.5. Values equal to mean for 10 , bars represent S.D.

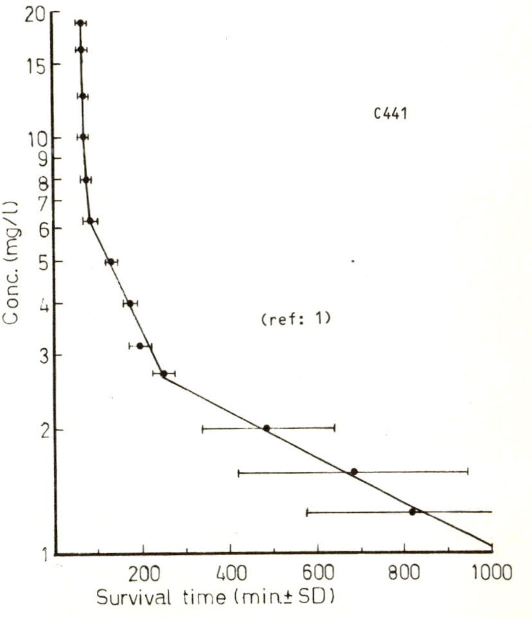

Survival time of rainbow trout in solutions of non-ionic surface active agent. Dobanol-45 was chosen as a typical non-ionic surfactant. The chain length is a mixture of linear C14 and C15 with an average of 7 ethylene oxide molecules attached to it.

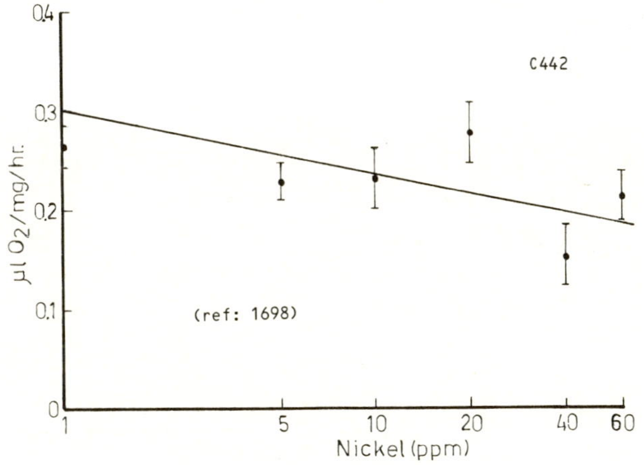

The microlitre of oxygen consumed by Tubificidae following exposure to varying concentrations of nickel chloride at pH 8.5. X is equal to the mean for 10 flasks; the vertical line represents the standard deviation of error.

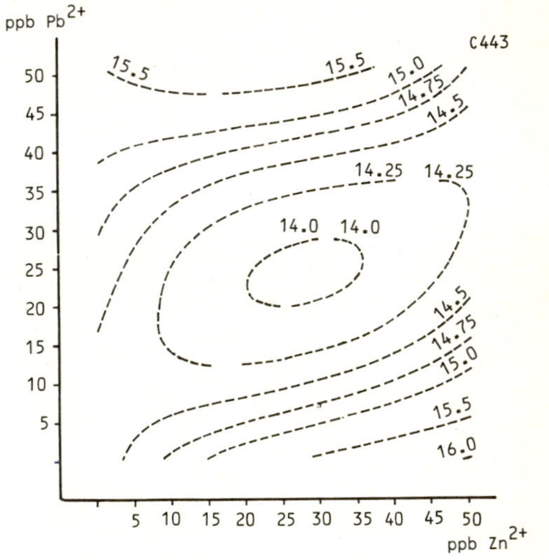

(ref: 96)

The influence of Pb and Zn on the larvae development until the megalopa stage (time in days) Experiment with R. harrisii. (crab larvae)

C444

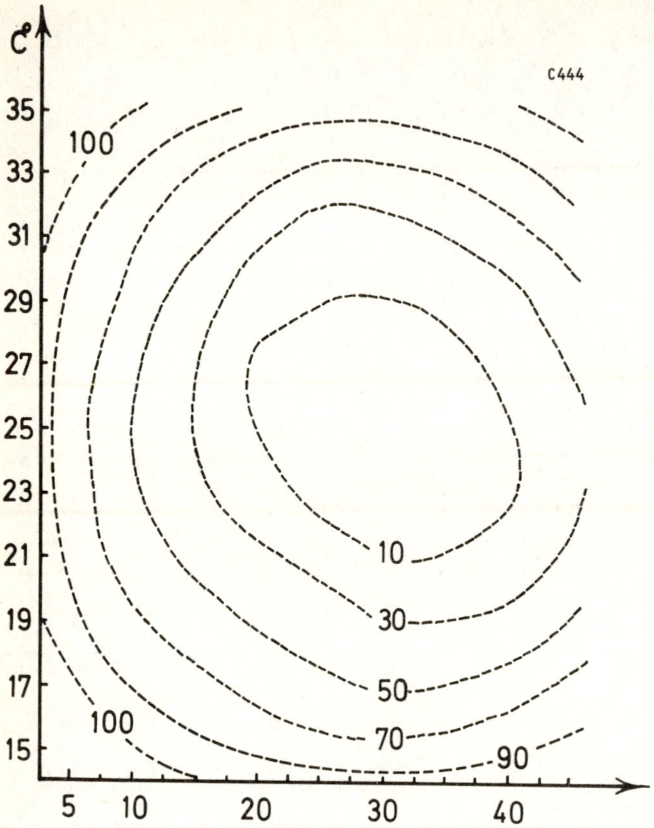

Mortality(%)of crab larvae at various combinations of temperature and salinity(‰) (ref: 1494)

C445

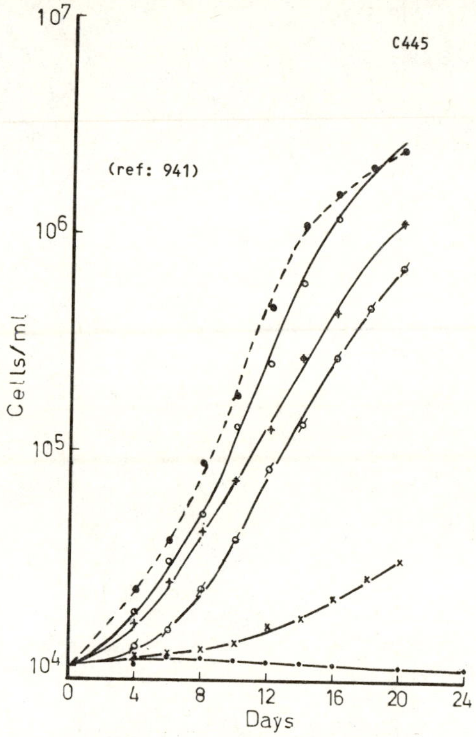

Growth of Chlamydomonas reinhardtii in media contaning methyl mercury. • control o 1 + 5 ø 10 × 20 • 40 ng Hg/ml. Temperature 16 +- 1 celcius under white light of 2.5*10**3 ergs/(cm**2*sec) on a 16:8 hours light-dark regime.

C446

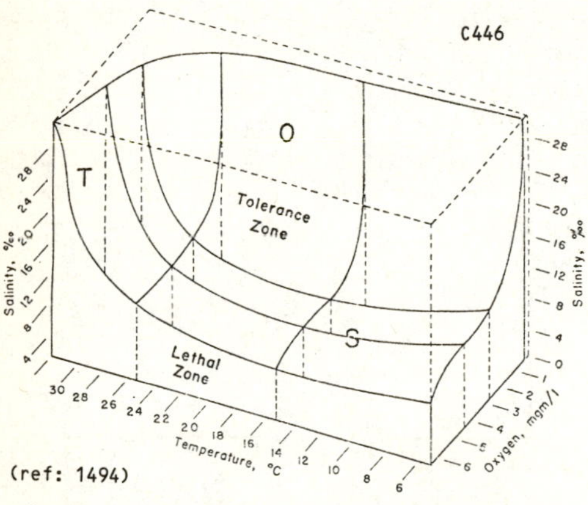

(ref: 1494)

LC-50 versus temperature, oxygen and salinity for lobster. 48H-LC50 test is used.

C447

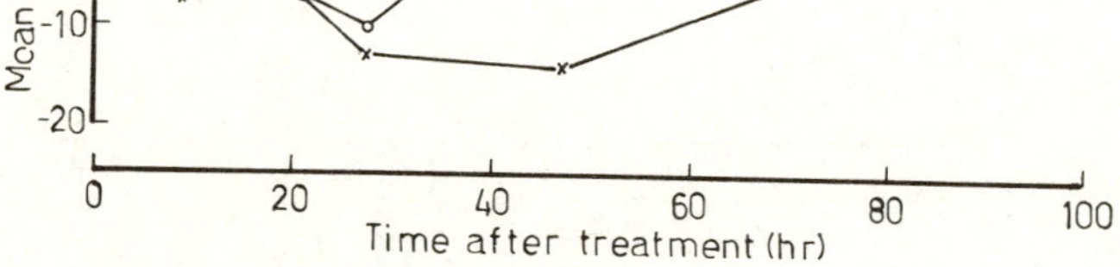

Change in body weight by different DDT concentrations for tadpoles. Means of four values are indicated. (ref: 355)
• control o 1.0 ppm DDT and × 10 ppm DDT.

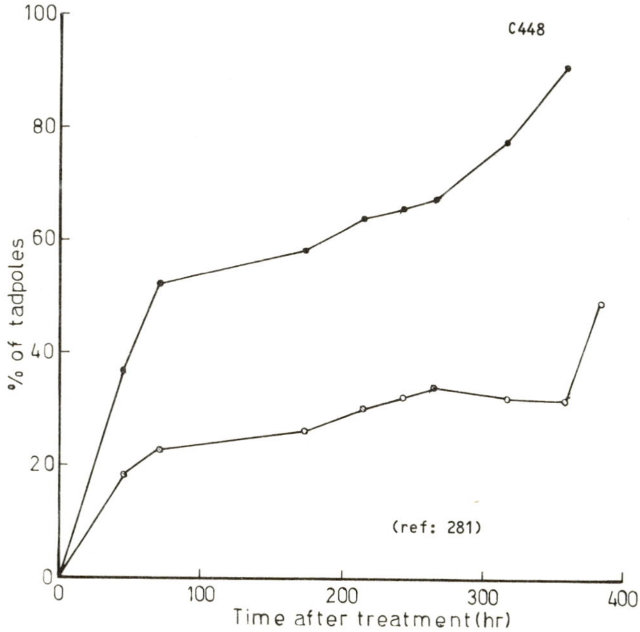

DDT effect on tadpoles (percentage). Experinent 2 (details see table)
○ 1.0 ppm DDT, ● 0.0 ppm DDT. % of tadpoles (up to and including XIX) with abnormal snouts.

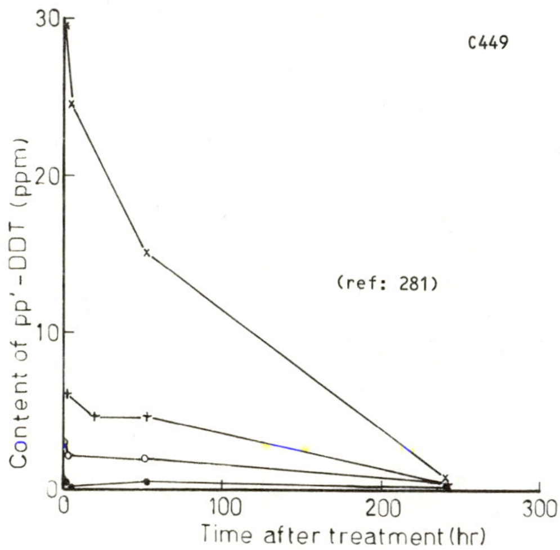

DDT-content in tadpoles of common frog. Experiment 1 (details see table).
● 0.01 ppm DDT, ○ 0.1 ppm DDT, + 1.0 ppm DDT, × 10 ppm DDT.
DDT concentration in ppm versus time in hours is shown.

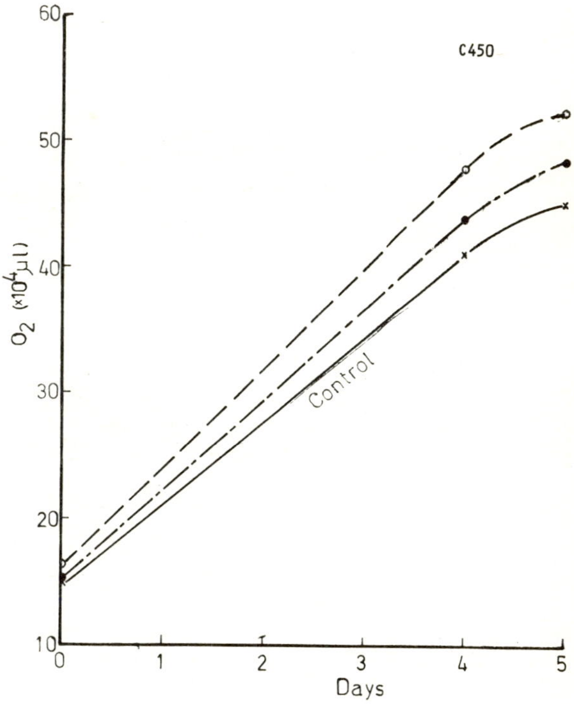

Photosynthesis of Enteromorpha influenced by wetting agents Tween 20● and Tween 60.○

(ref: 103)

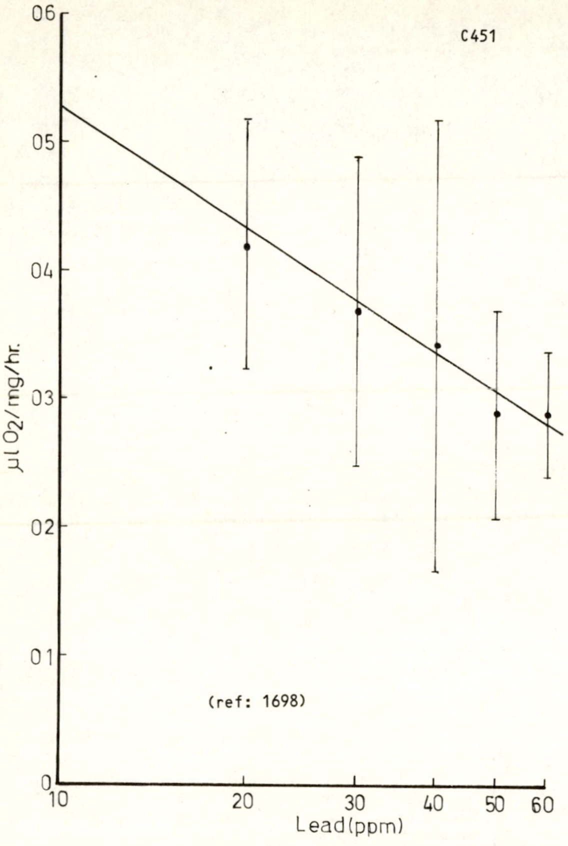

Oxygen consumption by Tubificidae versus lead concentration. pH = 6.5. Values equal to mean for 10 , bars represent S.D.

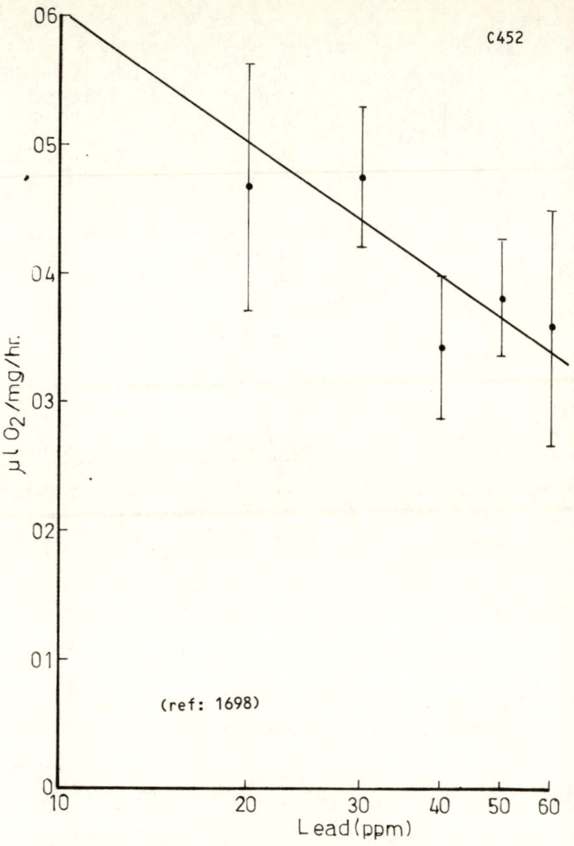

Oxygen consumption by Tubificidae versus lead concentration. pH = 7.5. Values equal to mean for 10 , bars represent S.D.

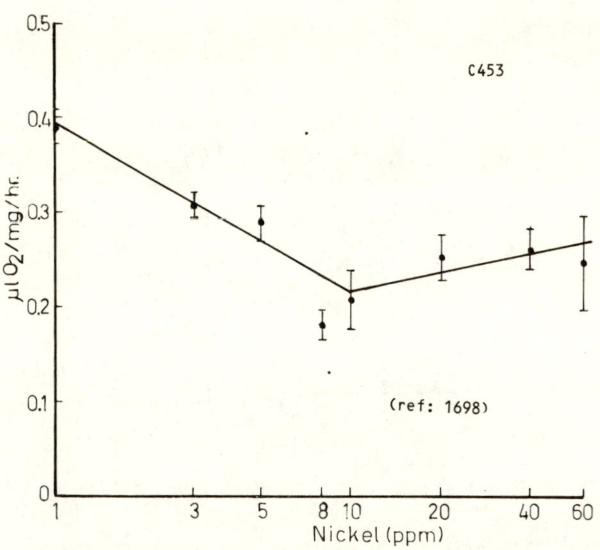

The microlitre of oxygen consumed by Tubificidae following exposure to varying concentrations of nickel chloride at pH 7.5. X is equal to the mean for 10 flasks; the vertical line represents the standard deviation of error.

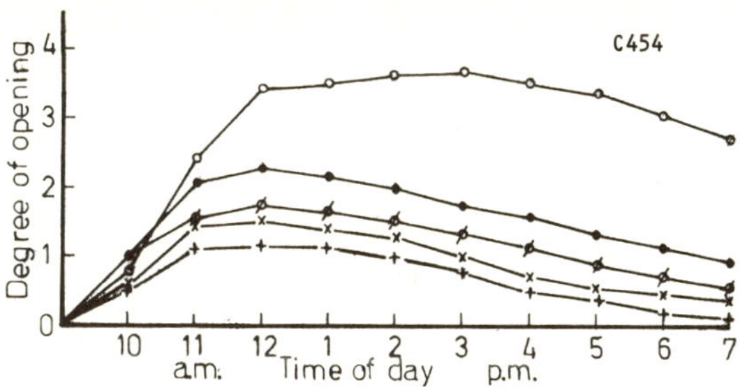

SO2- and CO2 concentration versus stomatal behaviour.
• 550 ppm CO2. & 0 ppm SO2. ⌀ 1000 ppm CO2. & 0 ppm SO2. x 550 ppm CO2 & 0.7 ppm SO2. + 1000 ppm CO2 & 0.7 ppm SO2.
○ control 320 - 330 ppm CO2 and 0 ppm So2

(ref: 970)

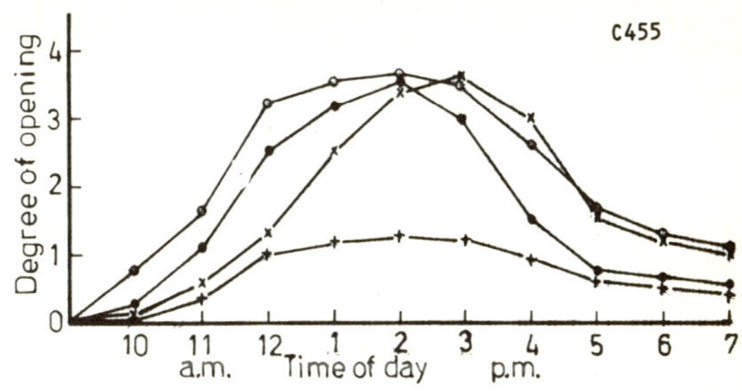

SO2 concentration versus stomatal behaviour in low light intensity.
○ 0.7 ppm SO2.,3500 lux. • 0 ppm SO2, 3500 lux. × 0.7 ppm SO2, 1800 lux. + 0 SO2, 180 lux. CO2 320 - 330 ppm.

(ref: 970)

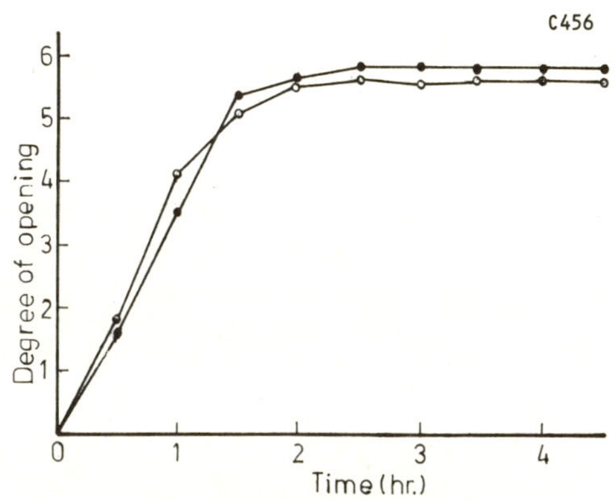

SO2 concentration versus stomatal behaviour.
• 0.7 ppm SO2. ○ Control with zero ppm SO2. Degree of opening versus time in hours is shown.

(ref: 970)

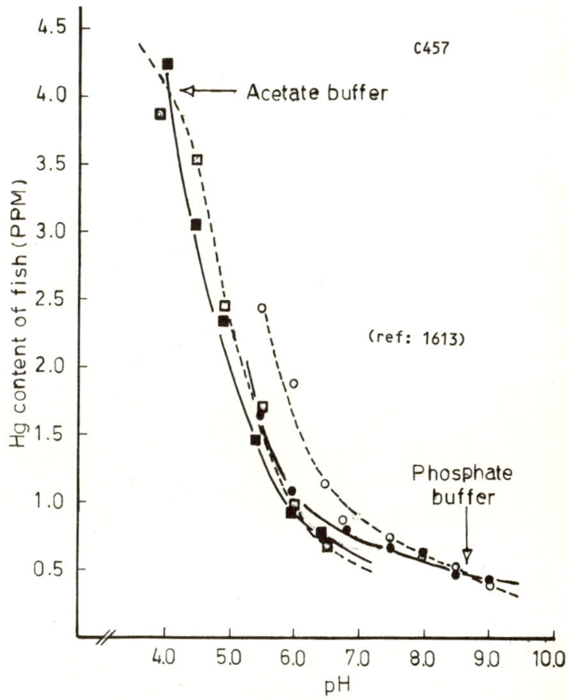

(ref: 1613)

Effect of pH on the mercury content of Fish.
Emerald shinner ■ Fathead minnows □
Emerald shinner ● Fathead minnows ○
Exposure to 1.5 ppm Hg from HgCl2 solutions.

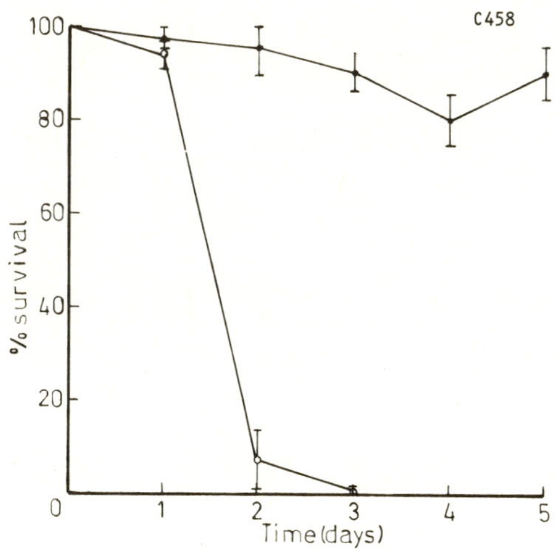

Survival of t. californicus with crude oil on water surface. Bars represent +- 1 S.D. Each point represent % of survival for 10 replicates. Time in days.
• Control. ○ Experimental.

(ref: 849)

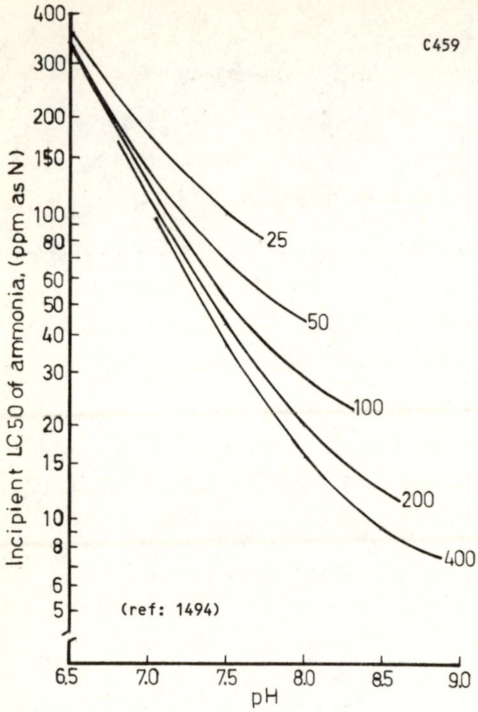

LC50 of ammonia for trouts versus pH at different HCO3(-) alkalinities.

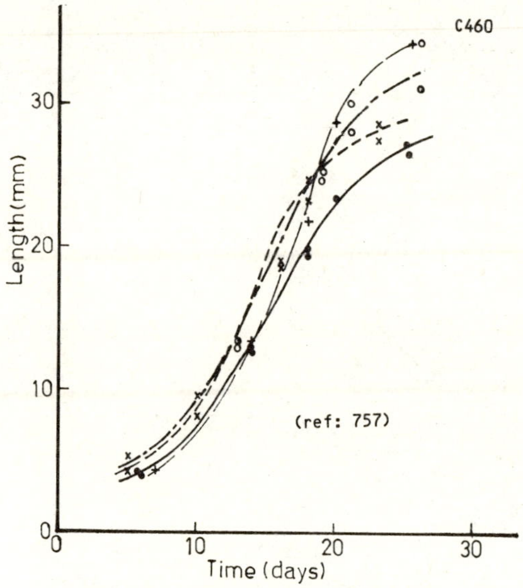

Growth rate at different concentrations of EDC-tar for Ophryotrocha labronica up to time for spawning. Temperature 21 +- 1 celcius. 0 ppm ● 0.5 ppm o 1 ppm x 2 ppm +

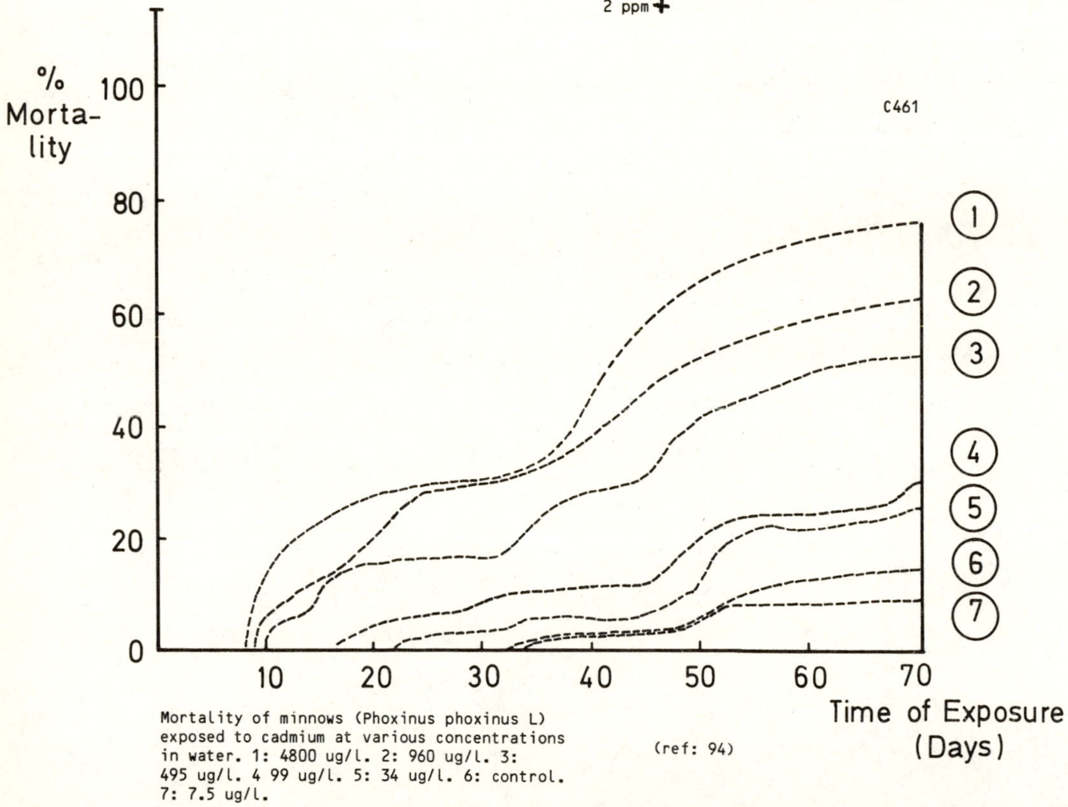

Mortality of minnows (Phoxinus phoxinus L) exposed to cadmium at various concentrations in water. 1: 4800 ug/l. 2: 960 ug/l. 3: 495 ug/l. 4 99 ug/l. 5: 34 ug/l. 6: control. 7: 7.5 ug/l.

(ref: 94)

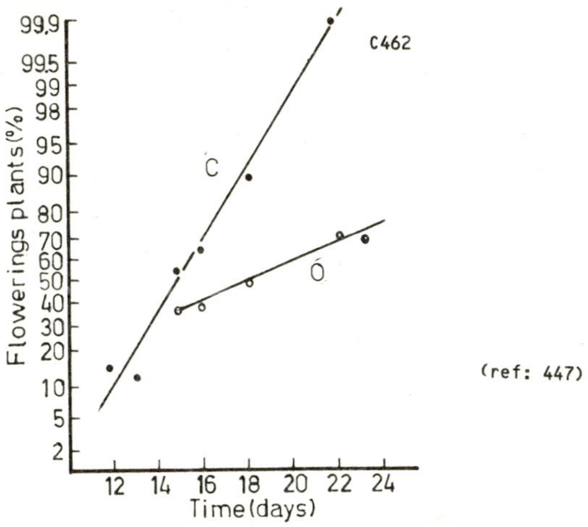

Ozone effect on floral initiation time and floral development in Lemna perpusilla exposed to ozone (0.1ppm) for 5 hour per day in 14 days. C = Control. O = Ozonated plants.

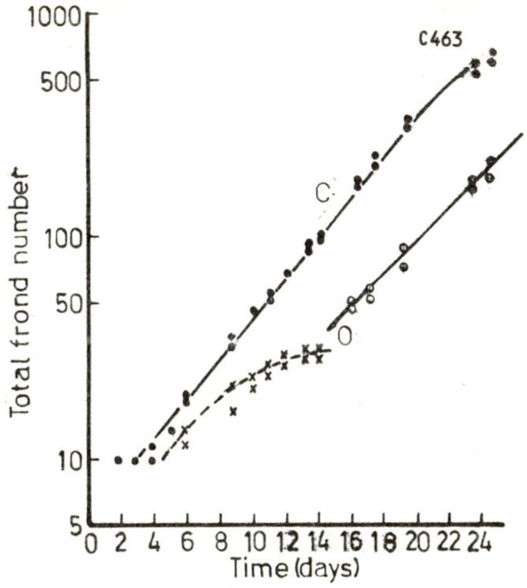

Ozone effect on frond number (total). Repression of frond multiplication of Lemna perpusilla exposed to ozone (0.1 ppm) for 5 hours per day for 14 days. C = Control. O = Ozonated plants.

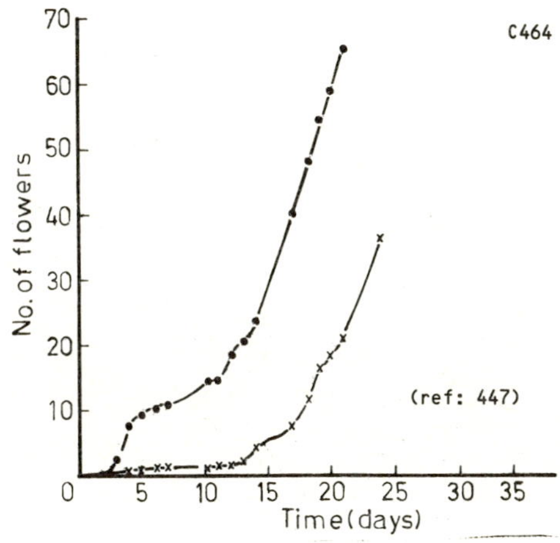

Ozone effect on number of flowers. Flower bud initiation before pinching in white sim carvation. ● growing in clean air and X air contaning 10 ppm of ozone. Note delay in floral initiation in plants growing in ozonated air. Rate of bud formation is quita similar once the lag phase is passed in ozonated plants.

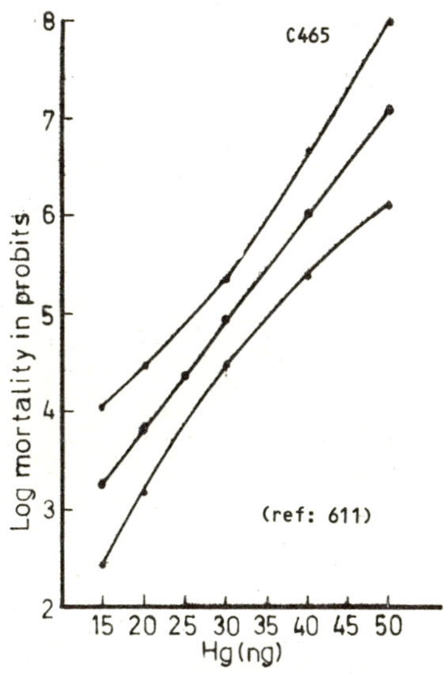

Mortality versus mercury concentration. Log killed celld of bacteria. Concentration in ug Hg/l. Errors of regression lines indicated by the curves.

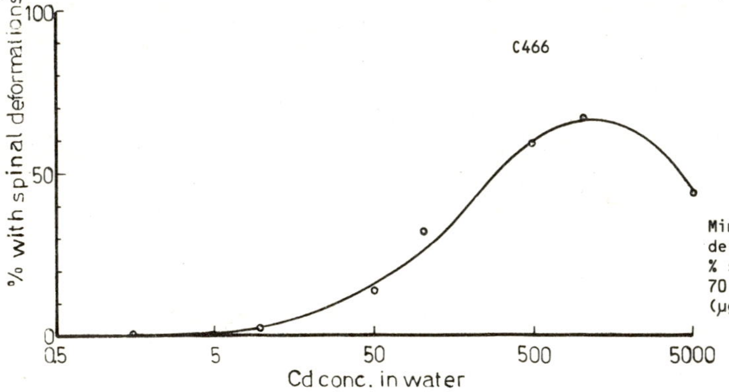

Minnows (phoxinus phoxinus L) with spinal deformations related to cadmium in the water. % surviving minnows with deformations after 70 days exposure to Cd-concentration (μg/l) is shown.

(ref: 94)

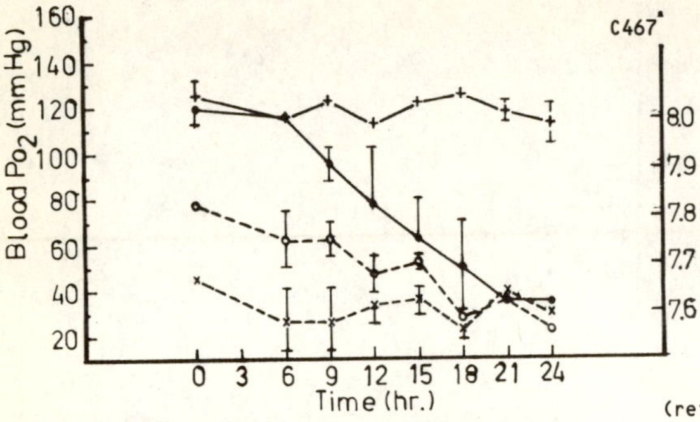

Change in partiel pressure of oxygen in blood and blood pH in rainbow trout exposed to 1.43 ppm zinc. Bars represent S.D.
+ control pressure ● test pressure
× control pH ○ test pH

(ref: 1416)

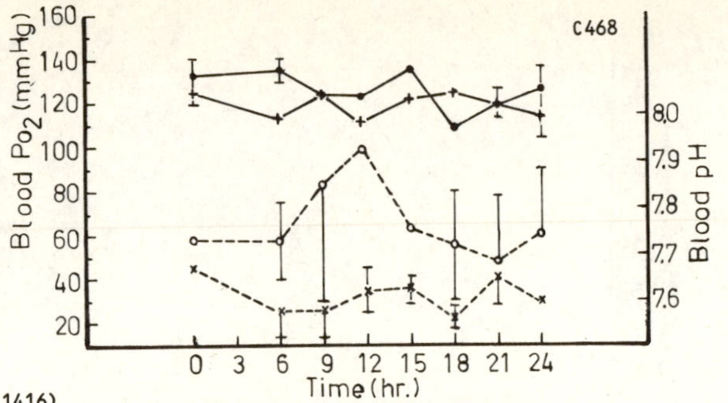

Change in partiel pressure of oxygen in blood and blood pH in rainbow trout exposed to 0.129 ppm cobber. Bars represent S.D.
+ control pressure ● test pressure
× control pH ○ test pH

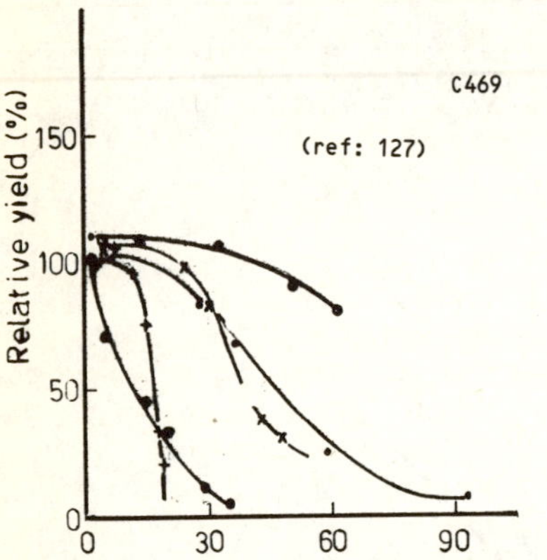

Yield versus cadmium content (µg/g) of diagnostic tissue. Relative yield in percentege is used.
● Soybean (left figure), + Field bean, × Wheat. • Corn. ○ Squash.

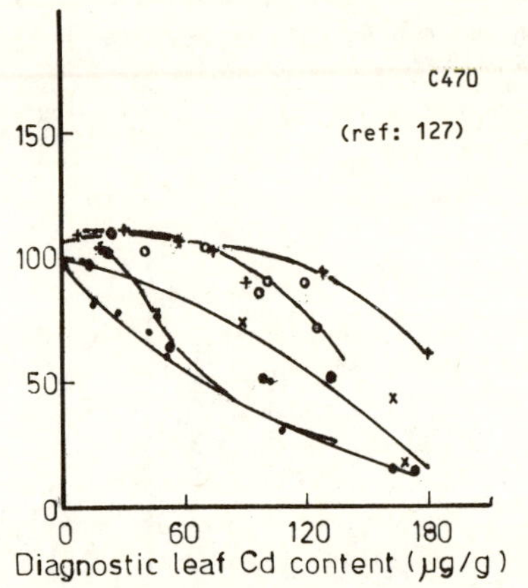

Yield versus cadmium content (µg/g) of diagnostic tissue. Relative yield in percentege is used.
● Lettuce + Cabbage. ○ Tomato.
× Spinach. • Carrot.

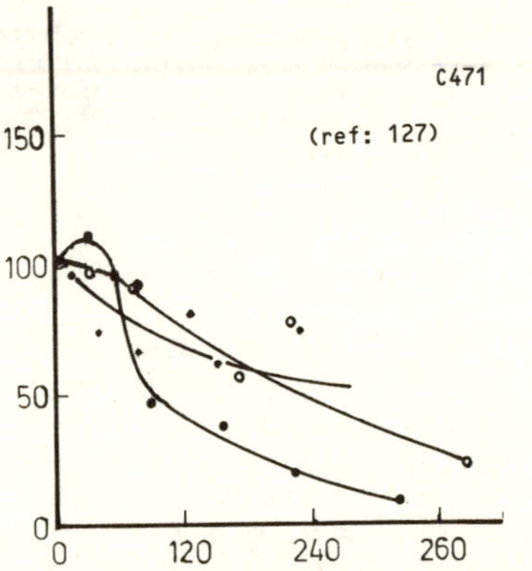

Yield versus cadmium content (µg/g) of diagnostic tissue. Relative yield in percentege is used.
● Curlycress. + Turnip. • Radish.

(ref: 1)

Growth rate and start of population of Ophyotrocha labronica, effect of sulphate pulp waste water. Salinity 33 ppt.
Solids dots indicate the presence of egg masses in the bowls. ○ 0 %. ● 1 %. + 5 % • 10 %. □ 15 %. ■ 20 % waste water.

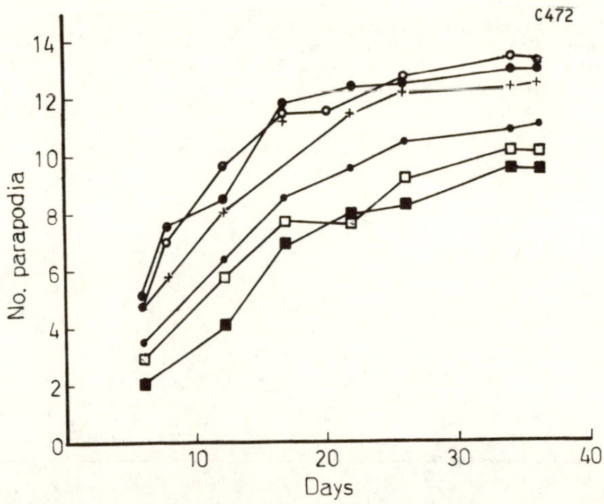

C473

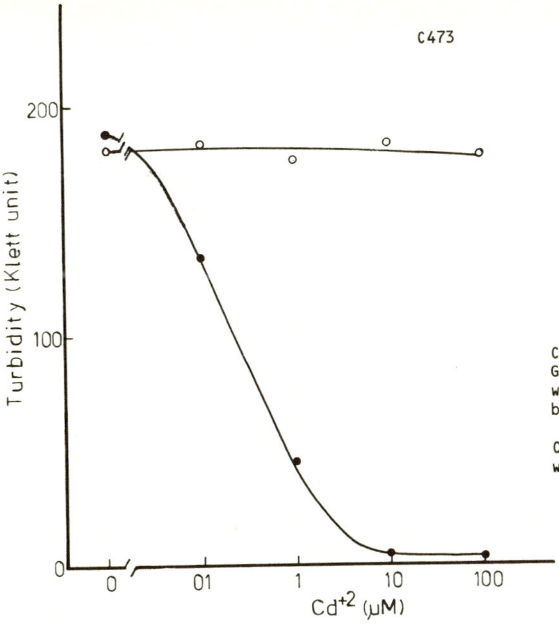

Cadmium resistance in Staphyllococcus aureus. Growth (turbidity) after 5 h incubation with aeration at 37 celcius in nutrient booth with and without added Cd.
o With a derivate of plasmid p1258 with Cd. ● with a derivate of plasmid p1258 with the cad 52 mutation in the cad A gene.

(ref: 1437)

C474

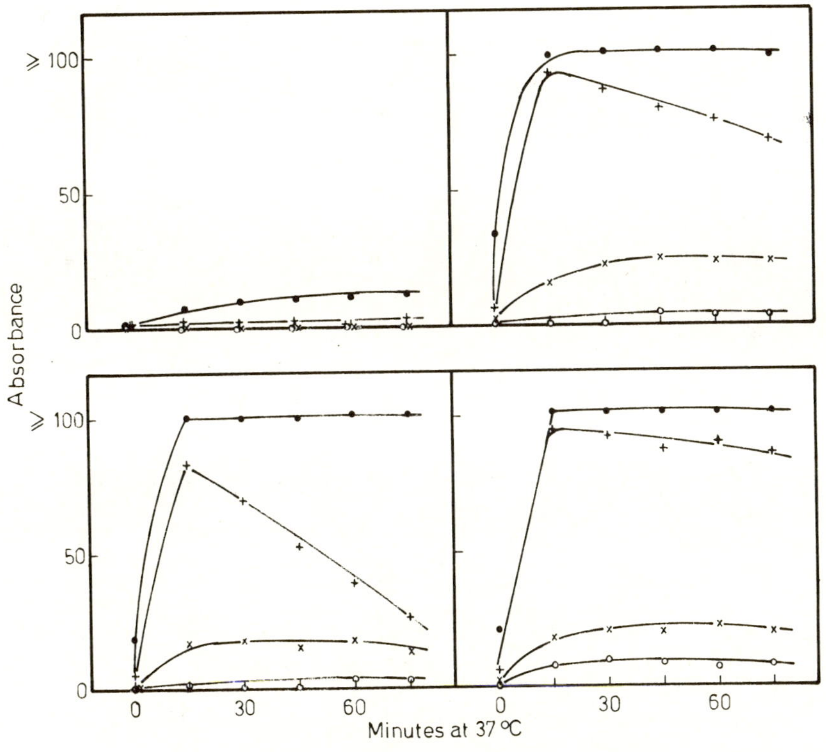

Plasmid determined decomposition of organo- mercurials. Absorption at 254 nm (Hg^{o}) was measured. Experiment was maded in standard BOD bottles in 37 ^{o} C water bath. Every 15 min the bottles were attached to flameless Hg AAS apparatus, so the Hg ^{o} generated were determined.

The bacterial strain of Pseudomonas aerugi nosa is written on the figure. PU21 is a plasmidless strain, the other strains has plasmids.

(ref: 1437)

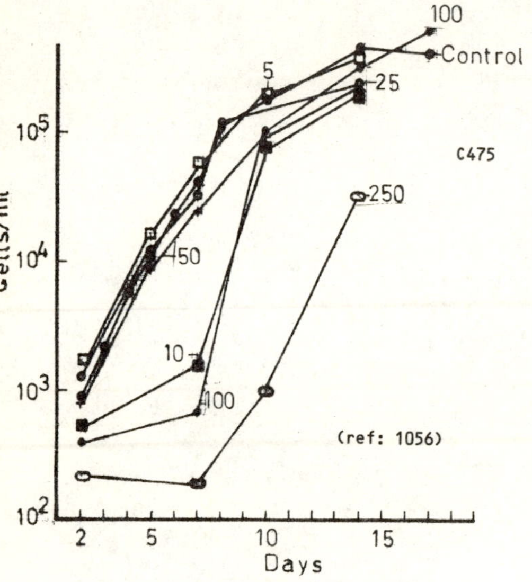

Effect of different concentrations (ppm) of mixture of weathered crude WC and emulsifier SC on the growth of Platymonas.

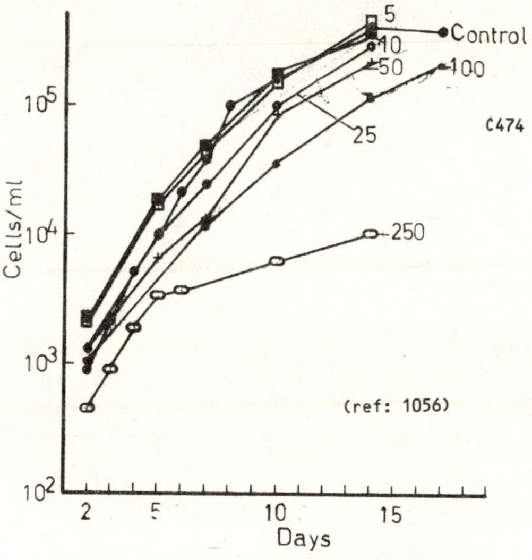

Growth curves of Platymonas in the presence of different concentrations (ppm) of mixture of crude C plus emulsifier OSR, compared with control.

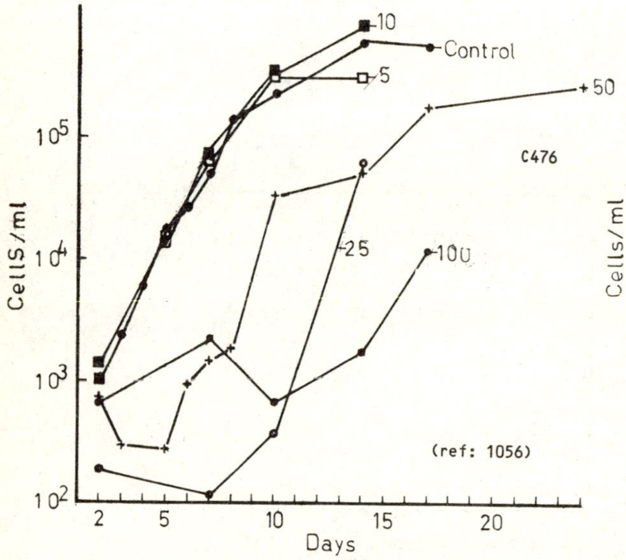

Effect of different concentrations (ppm) of emulsifier SC on the growth of Platymonas.

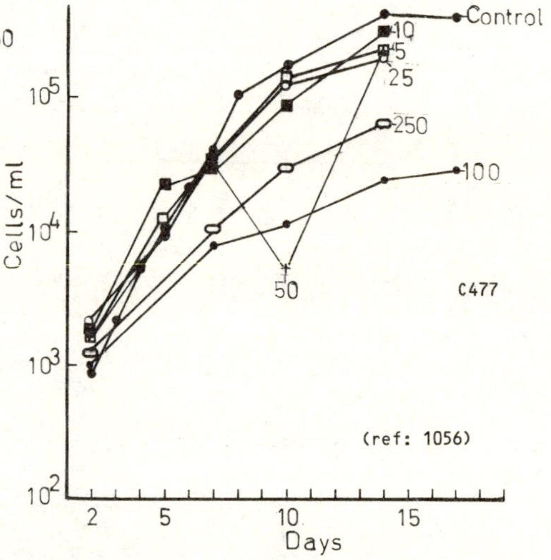

Effect of different concentrations (ppm) of weathered crude WC and Emulsifier OSR on the growth of Platymonas.

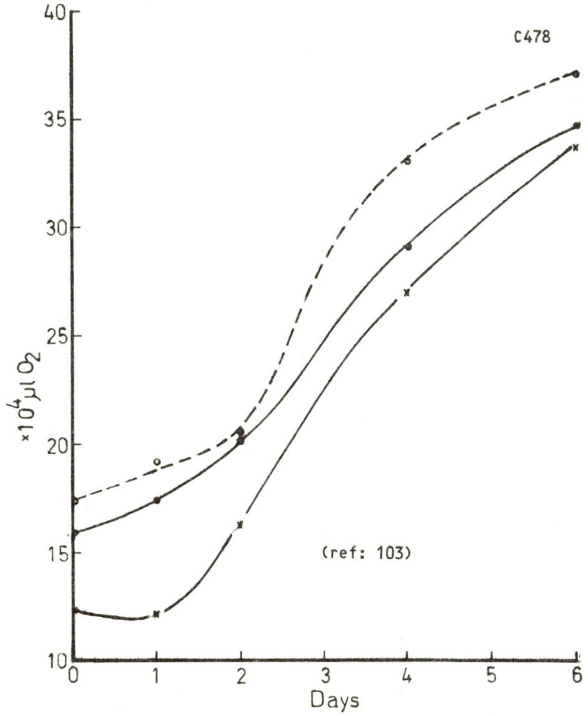

Photosynthesis influenced by the wetting agent Teepol. Chaetomorpha melagonium was used for the experiment, a multicellular marine green algae.
o dilution 1:12*10**6 × dilution 1*250000
• control.

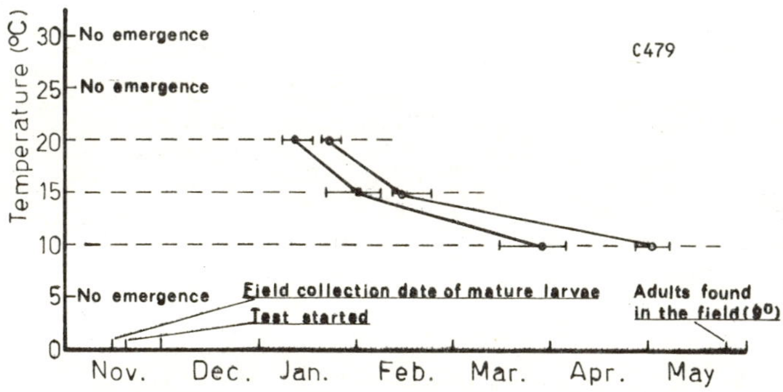

Temperature effect on the adult emergence of the stonefly, Pteronarcys dorsata.
• mean emergence date of males, o of females.

(ref: 1106)

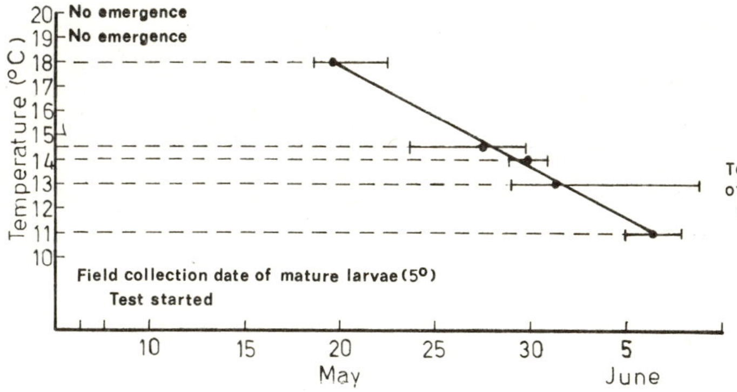

Temperature effect on the adult emergence of the mayfly, Ephemerella subvasia.
• mean emergence of subimago.

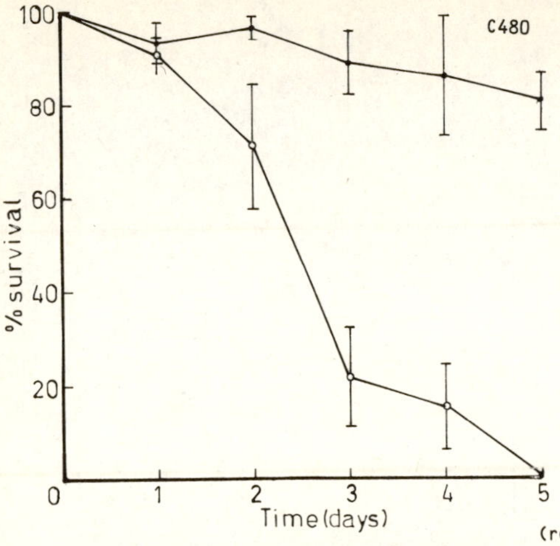

(ref: 849)

Survival of T. californicus with crude oil on water surface and bubbling air. Bars indicate 1 S.D. Each point represent % survival for 10 replicates.
• Control. ○ Expirimental.

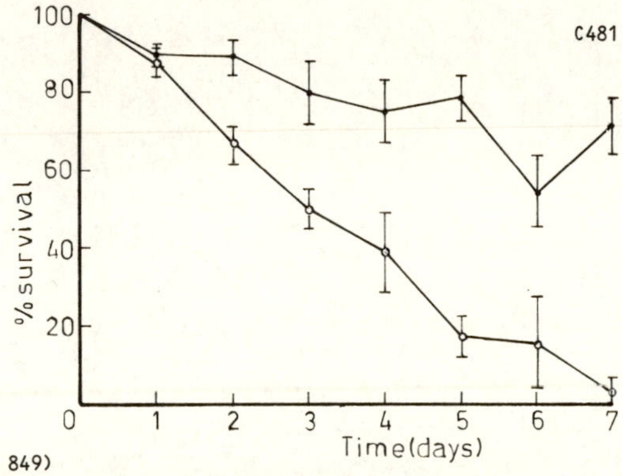

Survival of T. californicus with crude oil in dialysis membrane bag and bubbling air. Bars indicate +- 1 S.D. Each point represent % pf survival for 10 replicates.
• Control. ○ Expirimental.

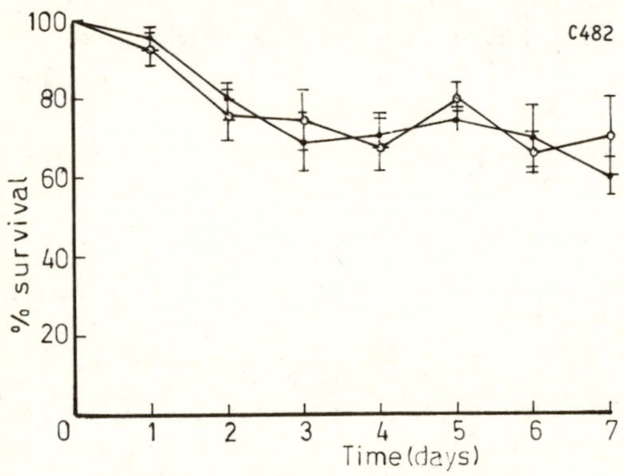

(ref: 849)

Survival of T. californicus with mineral oil on water surface and bubbling air. Bars indicate 1 S.D. Each point represent % survival for 10 replicates.
• Control. ○ Expirimental.

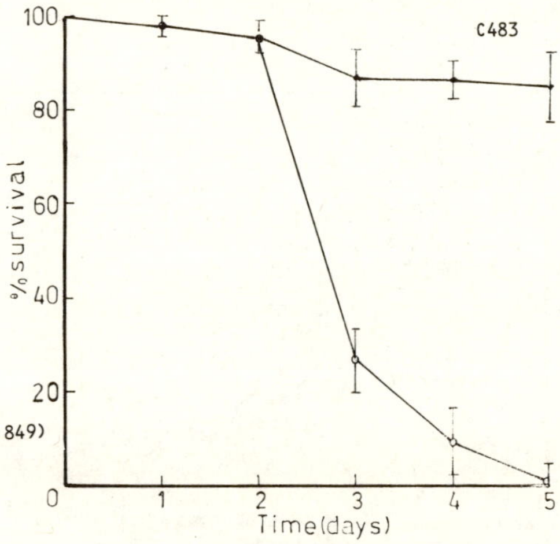

Survival of t. californicus with mineral oil on water surface. Bars represent +- 1 S.D. Each point represent % of survival for 10 replicates. Time in days.
• Control. ○ Experimental.

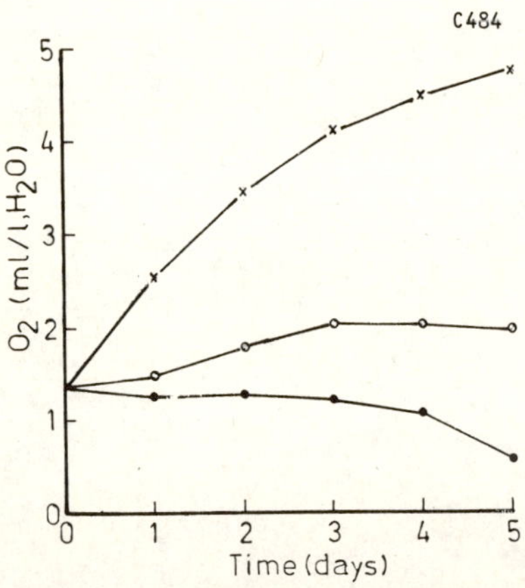

(ref: 849)

Oxygen diffusion rate influenced by crude oil and mineral oil. Diffusion rate across the air-water interface in ml O2/l H2O.
× Control. ○ Mineral oil. • Crude oil.

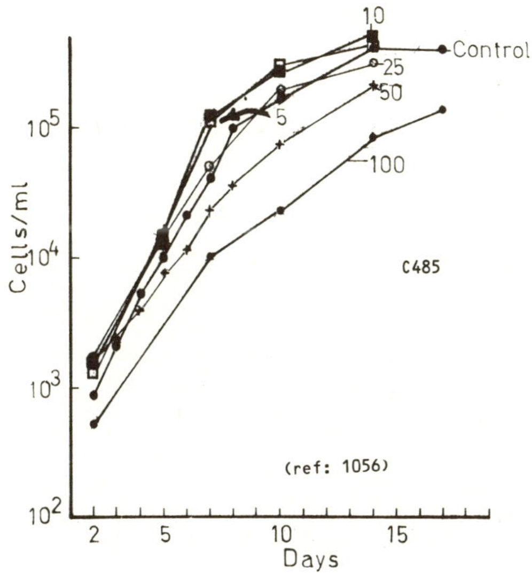

Effect of different concentration (ppm) of mixtures of weathered crude C and emulsifier SC on the growth of Platymonas.

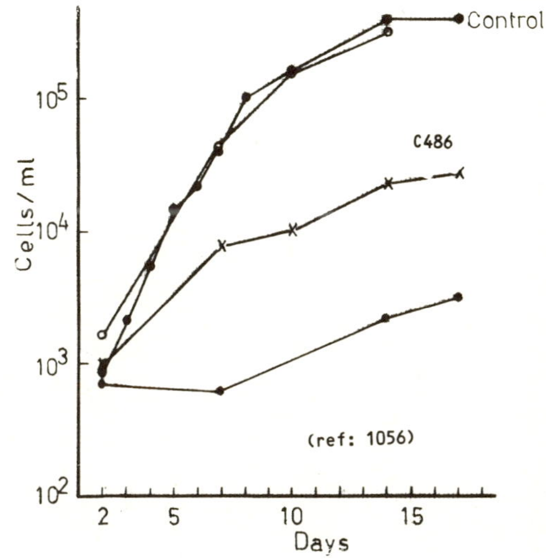

Effect of weathered crude WC (80 ppm) and emulsifoer OSR (20 ppm) X, weathered WC (100 ppm) • and of emulsifier OSR (25 ppm) alone o on the growth of Platymonas.

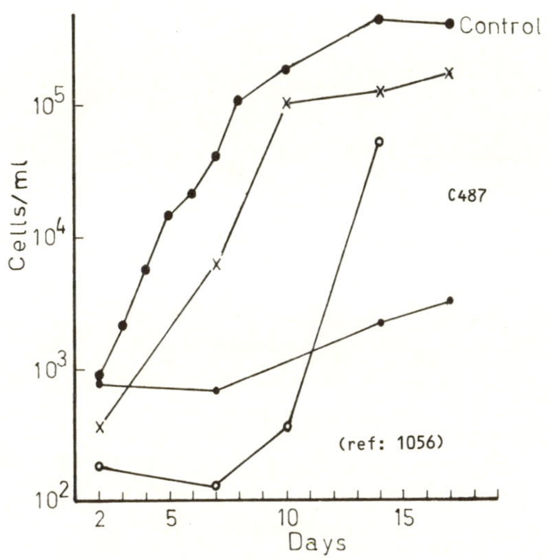

Effect of weathered crude WC (80 ppm) and emulsifier (20 ppm) X, of weather crude WC alone (100 ppm) • and of emulsifier SC alone (25 ppm) o on the growth of platymonas.

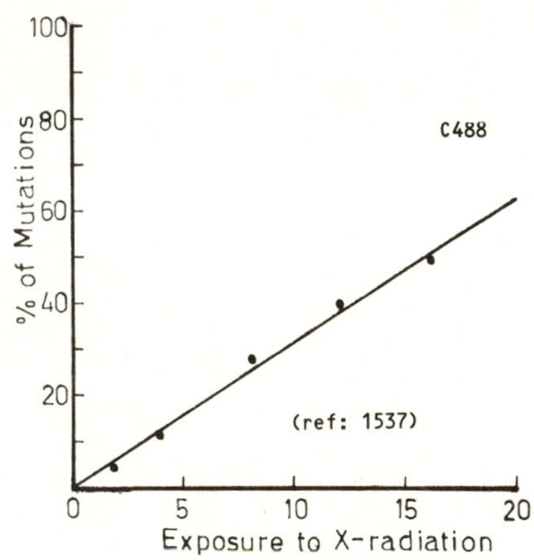

Mutagenic frequency to X-radiation for Aspergillus terrus. Abscissa in minutes exposure 8500 r/min.

Toxicity of cyanides		C489
	Toxic concentration (mg/l)	
Daphnia	1.8 : LD_{50} 48h	
Fish	0.1-1.0 : LD_{50} 24h	
Biological treatment	abt. 30 : 50% reduction, but strongly dependent on the culture	
Anaerobic sludge treatment	50 : 10% reduction	

(ref: 765)

Blood plasma parameters

C490

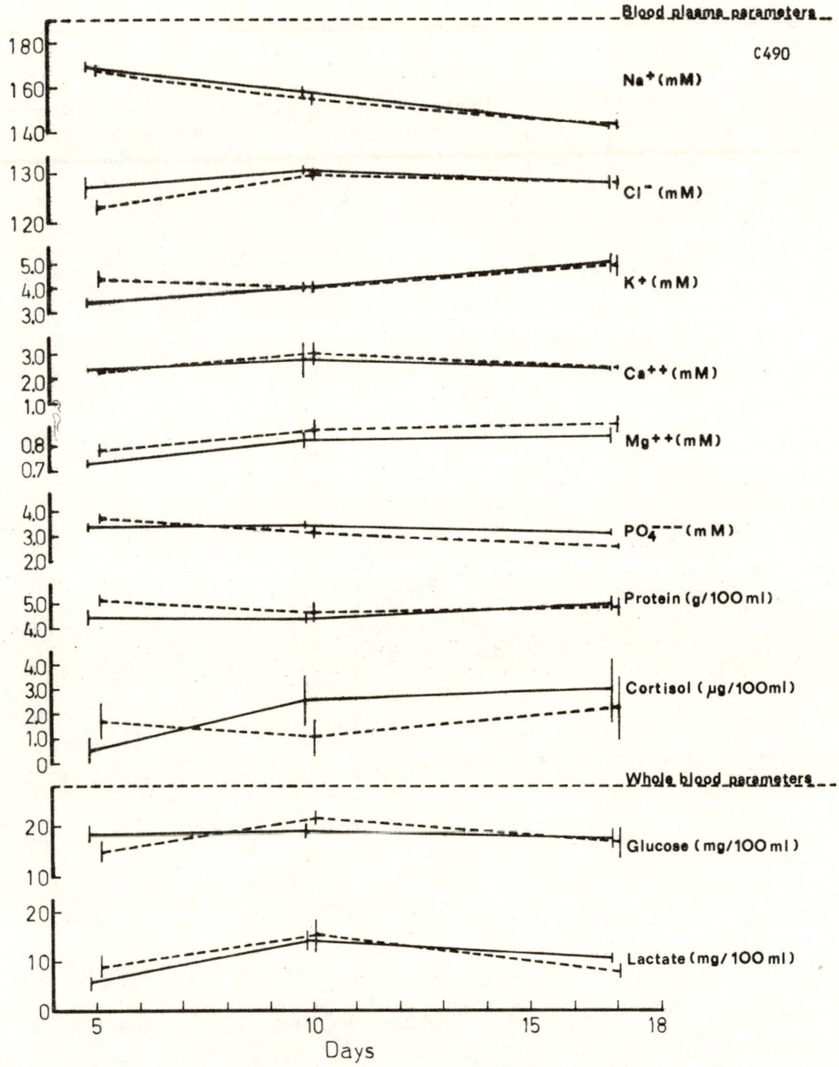

The effect of pcb on blood plasma and whole blood parameters in rainbow trout. Continous line = PCB treated. (200 mg PCB/kg wet weight). Broken line = control.

(ref: 922)

soild line control

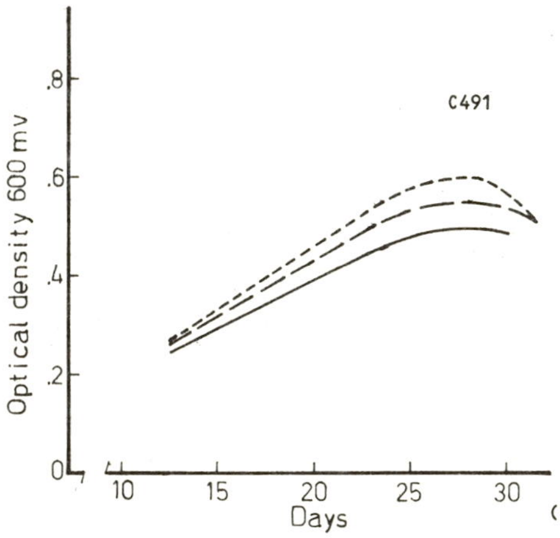

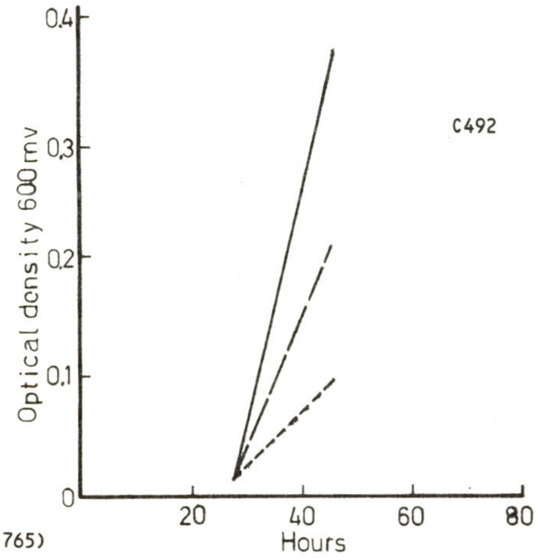

(ref: 1765)

Eudorina californica at 20 celcius
Small dashed line: 0.1 - 5.0 mg Cu(+2)/l
Large dashed line: 0.2 - 15 mg Cu(+2)/l

Anacystis nidulans at 28 celcius
Small dashed line: 0.1 mg Cu(+2)/l
Larger dashed line: 0.1 mg Cu(+2) and Ag(+)/l

solid line control

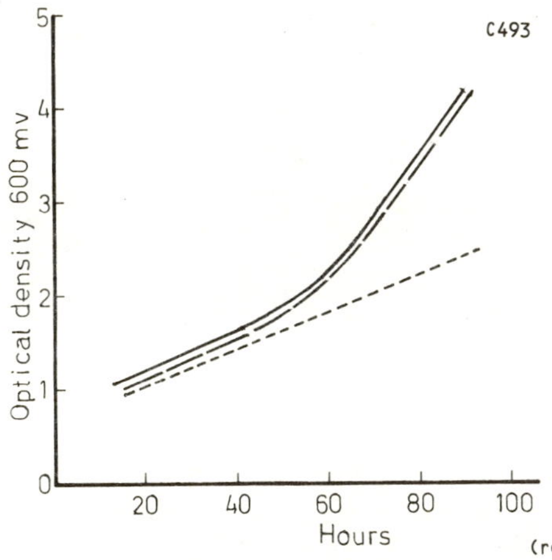

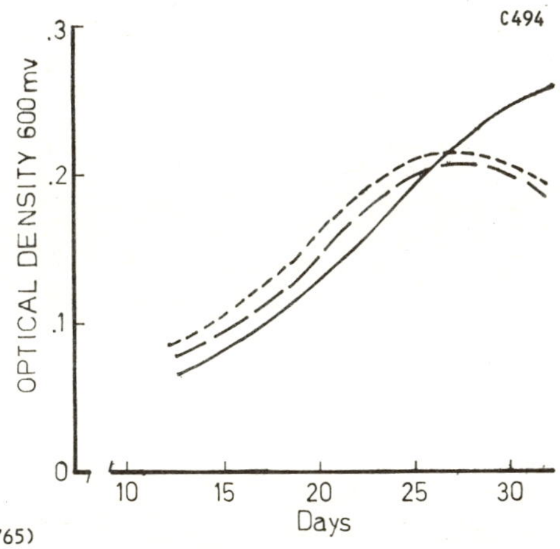

(ref: 1765)

Anabaena flosaqua
Small dashed line: 0.1 mg Cu(+2)/l
Large dashed line: 0.1 mg Cu(+2) and Ag(+)/l

Pediastrum biradiatum at 20 celcius.
Smaller dashed line 0.1 - 5.0 mg Cu(+2) and Ag(+)/l.
larger dashed line with 0.2 - 15 mg Cu(+2) and Ag(+)/l.

The effect of copper and silver on the growth of algae. Solid line is the control groups

C495

solid line control

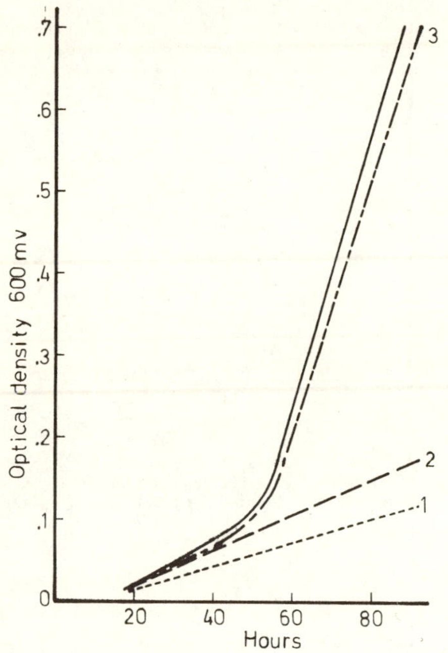

Chlorella vulgaris at 20 celcius,
3 0.1 mg Cu(+2)/l
2 0.2 mg Cu(+2)/l
1 0.3 mg Cu(+2)/l

(ref: 1765)

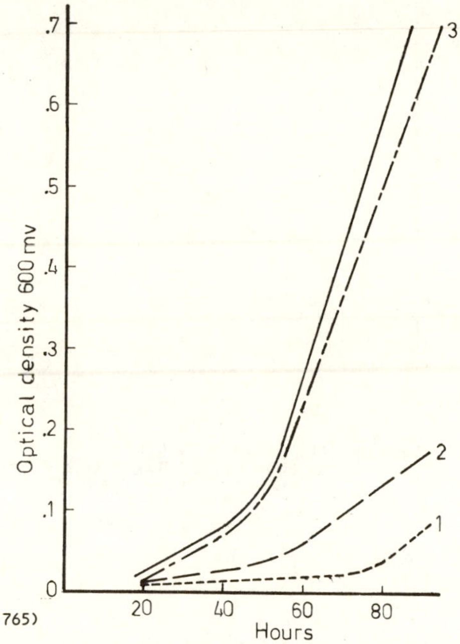

Chlorella vulgaris at 20 celcius,
3: 0.1 mg Cu(+2) and Ag(+)/l
2: 0.2 mg Cu(+2) and Ag(+)/l
1: 0.3 mg Cu(+2) and Ag(+)/l

solid line control

C496

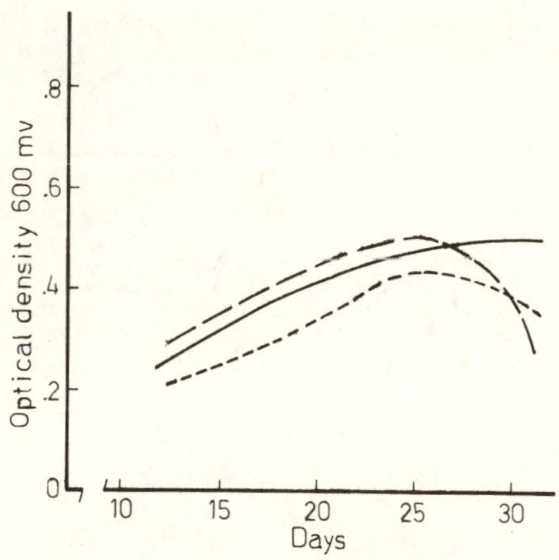

Eudorina californica at 20 celcius,
Smaller dashed line with 0.1 - 5.0 mg Cu(+2) and Ag(+)/l
Larger dashed line: 0.2-15 mg Cu(+2) and Ag(+)/l

(ref: 1765)

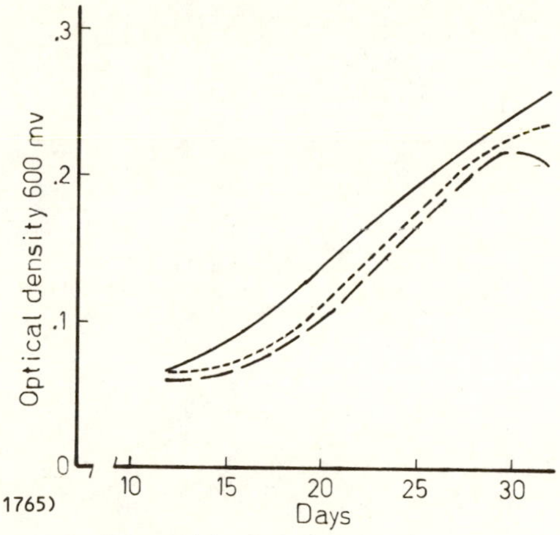

Pediastrum biradiatum at 20 celcius
(----) 0.1 - 5.0 mg Cu(+2)/l
(——) 0.2 - 15 mg Cu(+2)/l

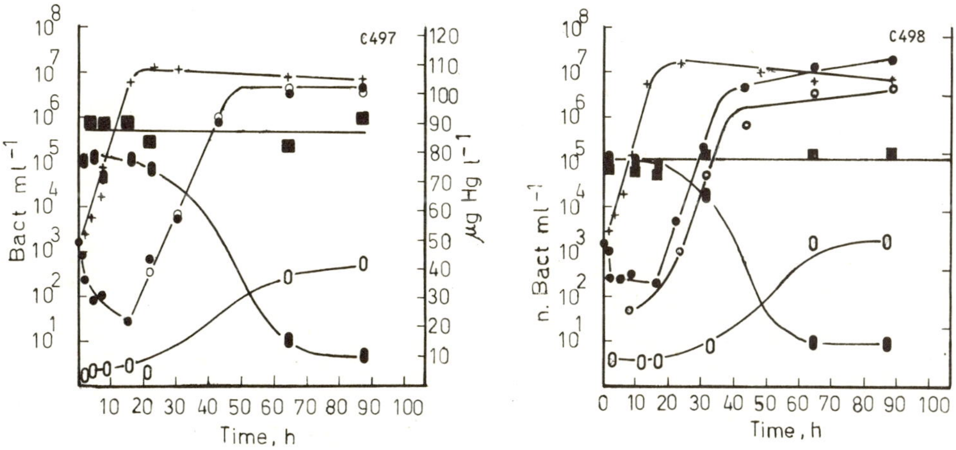

Effect of methyl mercury on the growth of bacteria. In sediment.
Total bacteria number in culture ●
Total bacteria number in control without Hg +
Methyl mercury resistant bacteria in culture ○
Organic mercury concentrations in culture ⬮
Mineral mercury in culture 0
Organic mercury in sterile control ■

(ref: 125)

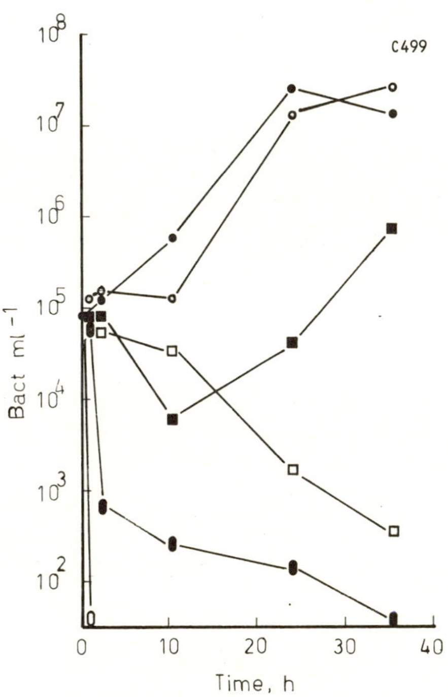

Effect of methyl mercury on the growth of bacteria. In sediment. ● without Hg ○ 10 μg Hg/l ■ 100 μg Hg/l □ 500 μg Hg/l ⬮ 1000 μg Hg/l 0 10000 μg Hg/l

(ref: 125)

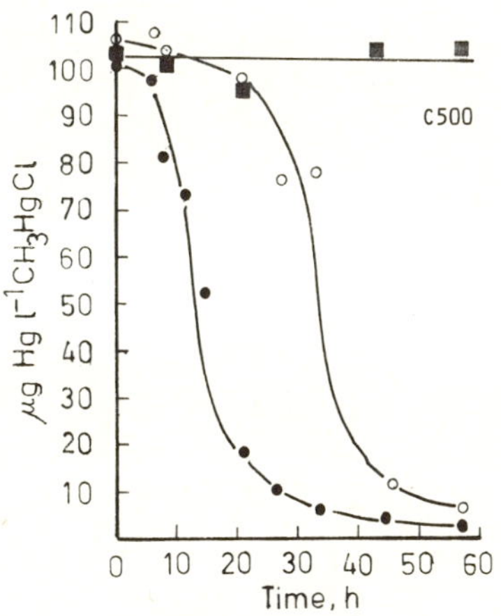

Effect of preincubation on the uptake of methyl mercury by bacteria. Disapperence of methyl mercury chloride initally present at 100 μg Hg/l in cultures, respectively, incubated with the same number of bacteria from 3 precultures containing ● 100 ○ 50 and ■ 0 ug Hg/l of methylmercury chloride.

(ref: 125)

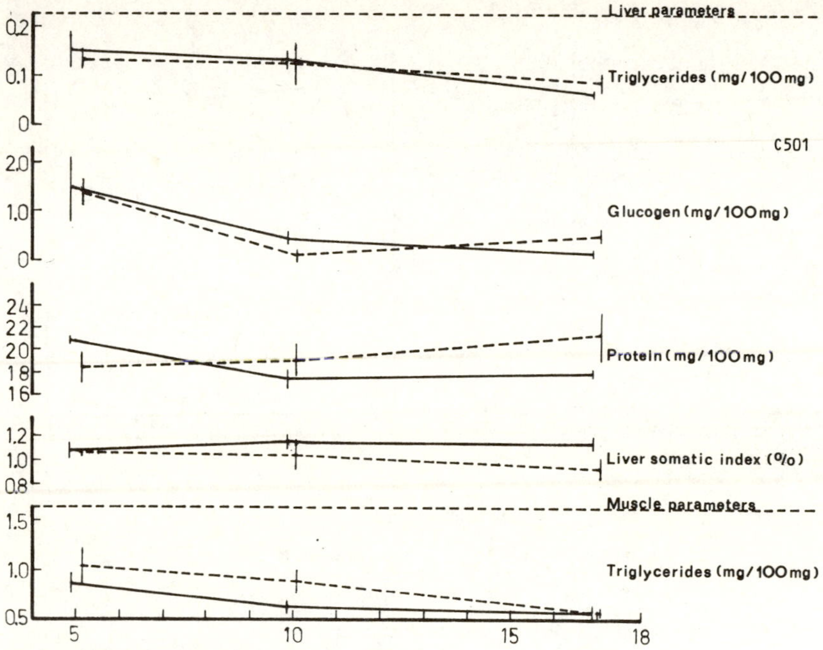

C501

The effect of PCB on liver and muscle parameters in rainbow trout.
Continous line = PCB treated (200 mg PCB/kg wet weight)
Broken line = control.

(ref: 922)

Time recorded for fish (tench) to die in relation to concentration of phenol (Nowacki et al., 1953)

C502

Phenol concentration in water (mg/l)	(Time in minutes) Fish are agitated	Fish swim on one side	Fish swim with belly upwards	Fish die
100	2	4	5	8
80	2	4	7	11
60	3	6	8	12
40	4	8	11	17
20	9	38	145	378
15	70	100	210	520
10	130	250	720	1030
8	170	310	950	1830
6	410	450	1730	3220
5		still alive after 4 days		

(ref: 765)

Toxic concentrations and LD_{50} values for fish

C503

Component	Formula	Toxicity limit (mg/l)	LD_{50} (mg/l)
Cyanides	HCN	0.05	0.2
Ammonia	NH_3	<0.5	0.5
Hydrogen sulphide	H_2S	0.5	3
Phenol	C_6H_5OH	3	10
Benzene	C_6H_6	2	18
Pyridine	C_5H_5N	100	>500
Chlorine	Cl_2	0.05	0.2

(ref: 765)

TABLE C504

Effects of Trace Amounts of Methyl Mercury Hydroxide (MMH) on the Growth of Tomato Seedlings after Thirteen Days. Six replications.
REF.: 606

Treatment (ppm) MMH	Mean shoot growth (cm)	% of Inhibition of Mean shoot growth	Mean seedlings wet weight (mg)	% Inhibition of mean wet weight	Mean seedlings dry weight (mg)	% Inhibition of mean dry weight
0.05	0.6	88.8	42	95.8	5	94.3
0.04	0.8	83.6	61	93.9	8	89.3
0.03	1.3	75.5	141	86.5	15	81
0.02	3.2	37.9	320	68.4	28	64
0.01	3.9	25.3	353	65.1	30	60.6
0	5.2	.	1013	.	77	.

TABLE C505

Growth response of Loblolly Pine (LP) and Red Maple (RM) versus Lead Concentration (Pb) in mole/l.
REF.: 333

Pb	heigth (cm), LP	heigth (cm), RM	stem dry wt (g), LP	stem dry wt (g), RM	root dry wt (g), LP	root dry wt (g), RM	Root/shoot ratio, LP	Root/shoot ratio, RM	Anthocyanin (rel.), LP	Anthocyanin (rel.), RM
0	7.26	10.02	0.3	0.76	0.8	1.89	2.8	2.86	0.99	3.18
2*10**(-4)	7.23	10.51	0.31	0.69	0.61	1.61	2.38	2.24	0.99	2.49
10**(-3)	5.03	5.71	0.2	0.48	0.46	0.84	2.58	2.15	0.73	7.18
2*10**(-3)	4.2	4.68	0.21	0.39	0.5	0.8	2.49	2.13	1.46	8.7
5*10**(-3)	3.18	3.33	0.17	0.29	0.33	0.43	1.76	1.57	1.42	11.75
Mean	5.38	6.87	0.24	0.53	0.54	1.11	2.4	2.18	1.12	6.66

TABLE C506

Growth response of Loblolly Pine (LP) and Red Maple (RM) versus Fluoride Concentration.
REF.: 333

Fluoride mole/l	heigth (cm) LP	heigth (cm) RM	stem dry wt (g) LP	stem dry wt (g) RM	root dry wt (g) LP	root dry wt (g) RM	root/shoot ratio LP	root/shoot ratio RM
0	6.74	9.31	0.29	0.76	0.76	1.97	3.02	2.79
2*10**(-4)	6.56	9.07	0.3	0.68	0.69	1.88	2.28	2.83
10**(-3)	6.68	7.99	0.3	0.68	0.7	2.45	2.89	3.71
2*10**(-3)	6.15	8.35	0.27	0.58	0.59	1.76	2.26	2.78
2*10**(-2)	3.74	3.91	0.19	0.31	0.32	0.55	1.77	1.84
Mean	5.97	7.73	0.27	0.6	0.61	1.72	2.44	2.79

TABLE C514

Difference in 96h LC50 between hard and soft water.

REF.: 1450

Toxicant	LC50 (mg/l) in hard water	LC50 (mg/l) in soft water	Hard:soft ratio
Hydrazine	3.85 (1.79 - 8.29)	0.61 (0.28 - 1.34)	6.31
Unsymmetrical dimethylhydrazine	10.1 (6.08-16.7)	26.5 (20.3-34.5)	0.38
Aerozine-50	2.25 (0.77-6.58)	1.17 (0.68-2.02)	1.92
Monomethyl hydrazine	3.26 (2.71-3.91)	2.58 (2.02-3.29)	1.26

TABLE C507

Effects of Methyl Mercury Hydroxide (MMH) on Corn and Bean Seedling Growth after Ten Days.
REF.: 606

Treatment (ppm)	Mean gain in shoot height (cm), corn	Mean gain in shoot heigth (cm), bean	% inhibition of mean shoot growth, corn	% inhibition of mean shoot growth, bean	Mean shoot wet wt (mg), corn	Mean shoot wet wt (mg), bean	% inhibition of mean wet wt, corn	% inhibition of mean wet wt, bean
25	0	0	100	100	96	80	95.3	96
5	2.6	0.4	79.2	96.6	256	147	87.5	92.5
1	11.2	0.8	56.6	91.8	571	191	72	90.3
0.5	123.3	2.1	27.7	79.4	1250	209	38.8	89.5
0.05	33.6	6.3	0.1	38.9	2269	816	0	58.8
0	34	10	.	.	2042	1978	.	.

TABLE C508

Comparison of Acute Toxicity of Organic Sulphur Compounds by Different Modes of Administration.
REF.: 953

Compound	Eye irritation, rabbits	IP LD50 (mg/kg), rats	Oral LD50 (mg/kg), rats	Inhalation LD50 (mg/m3), mice	Inhalation LD50 (mg/m3), rats	Skin LD50 (mg/kg), mice	Skin LD50 (mg/kg), rats	Skin LD50 (mg/kg), rabbits
Ethanethiol	slight	226	682	2770	4420	.	.	.
Propanethiol	moderate	515	1790	4010	7300	.	.	.
2-Methyl-1-propanethiol	very slight	917	7168	25000	25000	.	.	.
2-Methyl-2-propanethiol	slight	590	4729	16500	22200	.	.	.
Butanethiol	slight	399	1500	2400	4020	.	.	.
Hexanethiol	none	396	1254	528	1080	.	.	.
Methyl-heptanethiol	slight	12.9	83.5	47	51	19.1	1594	600
Benzenethiol	severe	9.8	46.2	28	33	6.5	300	134
Alpha-toluenethiol	slight	373	493	178	235	.	.	.

TABLE C509

Electronegativity of Carcinogenic Metals

REF.: 953

Element	Chemical carcinogenicity	Electronegativity
Cs	none	0.7
K	none	0.8
Ru	none	0.8
Na	none	0.9
Ba	none	0.9
Li	none	1
Ca	none	1
Sr	none	1
La	none	1.1
Y	suspected	1.2
Ha	none	1.2
Sc	suspected	1.3
Zr	positive	1.4
Be	positive	1.5
Cr	positive	1.5
Al	suspected	1.5
Ti	suspected	1.5
Zn	positive	1.6
Ga	suspected	1.6
V	none	1.6
Cd	positive	1.7
W	none	1.7
In	none	1.7
Co	positive	1.8

Element	Chemical carcinogenicity	Electronegativity
Ni	positive	1.8
Sn	positive	1.8
Pb	positive	1.8
Mn	suspected	1.8
Fe	suspected	1.8
Si	suspected	1.8
Mo	none	1.8
Tl	none	1.8
Ge	none	1.8
Cu	suspected	1.9
Te	none	1.9
Re	none	1.9
Hg	none	1.9
Sb	none	1.9
Bi	none	1.9
As	suspected	2
Te	none	2.1
Rh	suspected	2.2
Ru	none	2.2
Os	none	2.2
Ir	none	2.2
Pt	none	2.2
Se	positive	2.4
Ag	none	2.4
Au	none	2.4

TABLE C510

Fish Status for 1679 Lakes in Southern Norway Grouped according to pH

REF.: 656

pH	No. of lakes in pH range	% of lakes with no fish	% of lakes with sparce populations	% of lakes with good populations
<4.5	111	73	25	2
4.5-4.7	245	53	41	6
4.7-5.0	375	38	41	21
5.0-5.5	353	25	40	35
5.5-6.0	164	8	36	56
>6.0	431	1	13	86

OUT8
TABLE C511

Water Quality Standards for Domestic Water.

REF.: 355

Water use	Substances	Units	Int. WHO acceptable	Int. WHO allowable	European WHO	U.S.A.	Sweden	France	Bulgaria	Tanzania
Toxic effects	Lead	ug/l		50	100	50	20/50		100	100
Toxic effects	Arsenic	ug/l		50	50	50	10/50		50	50
Toxic effects	Selenium	ug/l		10	10	10	10/50		50	50
Toxic effects	Chromium	ug/l		50	50	50	20		50	50
Toxic effects	Cyanide	ug/l		200	50	10	10/20		10	200
Toxic effects	Cadmium	ug/l		10	10	10	10		50	50
Toxic effects	Barium	ug/l		1000	1000	1000			1000	1000
Toxic effects	Mercury	ug/l					1/5			
Toxic effects	Silver	ug/l				50				
Human Health	Fluoride	mg/l		1.5	0.7	0.8-1.7	1.5		0.7-1.0	8.0
Human Health	Nitrate	mg/l		30.0	50/100	45	30	44	30	(100)
General use	Color	mg pt/l	5	50		15	10		15	50
General use	Turbidity	mg SiO2/l	5	25		3	weak		30cm	30
General use	pH		7.0-8.5	6.5-9.2			6.0-8.0		6.5-8.5	6.5-9.2
General use	Tot. dis. matter	mg/l	500	1500			200	2000		2000
General use	Tot. hardness	mg CaCO3/l			500			300	450	600
General use	Calcium	mg/l	75	200					150	
General use	Magnesium	mg/l	50	150	125			125	50	
General use	Sulphate	mg/l	200	400	250	250	25/250	250	250	600
General use	Chloride	mg/l	200	600	600	250	25/250	250	250	800
General use	Iron	mg/l	0.3	1.0	1.0	0.3	0.2	0.2	0.2	1.0
General use	Manganese	mg/l	0.1	0.5	0.05	0.05	0.05	0.1	0.1	0.5
General use	Copper	mg/l	1.0	1.5	0.05	1.0	0.05/1.0	0.2	0.2	3.0
General use	Zinc	mg/l	5.0	15.0	5.0	5.0	0.3/5.0	3	3	15.0
General use	Phenol	ug/l	1	2	1	1			1	2

TABLE C512

Elements; Abundance and Biological Activity.
Symbols used:
a = elements formed by radioactive decay of uranium and thorium
Have short physical half-lives and their crustral abundance are too
low to be measured accurately.
b = very low, unmeasurable.
ra = radioactive
cs = carcinogenic, suspected only.
s = stimulatory
cp = carcinogenic, proven.
en = essential nutrient, established
ep = essential nutrient, probably or required under special conditions
t1 = toxic
t2 = very toxic
REF.: 1574

Element	Symbol	Atomic number	Crustal abundance, weight %	Abundance in hydrosphere, mg/l	Abundance in atmosphere, vol ppm	Biological activity	Threshold limit, mg/m3 in air in 8 h.
Actinium	Ac	89	manmade	manmade		ra	
Aluminium	Al	13	8	0.01		cs	
Americum	Am	95	manmade	manmade		ra	
Antimony	Sb	51	0.00002	0.0005		s t2	
Argon	Ar	18	0	0.6	9300		
Arsenic	As	33	0.00020	0.003		cs s t2	
Astatine	At	85	manmade	manmade		ra	
Barium	Ba	56	0.0380	0.03		s t1	0.5
Berkelium	Bk	97	manmade	manmade		ra	
Beryllium	Be	4	0.0002	b		cp (s) t2	0.002
Bismuth	Bi	83	4E-7	2E-5		t1	
Boron	B	5	0.0007	4.6			
Bromine	Br	35	0.00040	65		t1=Br2	
Cadmium	Cd	48	0.000018	0.001			0.2
Calcium	Ca	20	5.06	400		en	
Californium	Cf	98	manmade	manmade		ra	
Carbon	C	6	0.02	28	CO2=330	en	
Cerium	Ce	58	0.0083	0.0004		s	
Cesium	Cs	55	0.00016	0.0005			
Chlorine	Cl	17	0.019	18980		Cl(-)=en Cl2=t1	
Chromium	Cr	24	0.0096	5E-5		en cp s t1	0.1(CrO3)
Cobalt	Co	27	0.0028	0.0005		en cp	
Copper	Cu	29	0.0056	0.003		en s t1	
Curium	Cm	96	manmade	manmade		ra	
Dysprosium	Dy	66	0.00085	b		s	
Einsteinium	Es	99	manmade	manmade		ra	
Erbium	Er	68	0.00036	b		s	
Europium	Eu	63	0.00022	b			
Fermium	Fm	100	manmade	manmade		ra	
Fluorine	F	9	0.0460	1.3		ep s	
Francium	Fr	87	manmade	manmade		ra	
Gadolinium	Gd	64	0.00063	b			
Gallium	Ga	31	0.00063	b			
Germanium	Ge	32	0.00013	7E-5		s	
Gold	Au	79	2E-7	4E-6		s t1	
Hafnium	Hf	72	0.0004	b			
Helium	He	2	0	5E-6	5.2		
Holmium	Ho	67	0.00016	b		s	
Hydrogen	H	1	0.14	H2O	CH4=1.5 H2=0.5	en	
Indium	In	49	0.00002	0.02		s t2	
Iodine	I	53	0.00005	0.06		I(-)=en I2=t1	
Iridium	Ir	77	2E-8	b		t1	
Iron	Fe	26	5.80	0.01		en	
Krypton	Kr	36	0	0.0003	1		
Lanthanium	La	57	0.0050	0.0003			
Lead	Pb	82	0.0010	3E-5		t2 cp s	0.2
Lithium	Li	3	0.0020	0.17		s	
Lutetium	Lu	71	8E-5	b			
Magnesium	Mg	12	2.77	1350		en	
Manganese	Mn	25	0.100	0.002		en cs	5
Mendelevium	Md	101	manmade	manmade		ra	

TABLE C512 See page 585 for explanations

Mercury	Hg	50	2E-6	3E-5		s t2	
Molybdenium	Mo	42	0.00012	0.01		en	5-15
Neodymium	Nd	60	0.0044	b			
Neon	Ne	10	0	0.0001	18		
Neptumium	Np	93	manmade	manmade		ra	
Nickel	Ni	28	0.0072	0.002		ep cp s	
Niobium	Nb	28	0.0072	0.002		ep cp s	
Nitrogen	N	7	0.0020	0.5	780900	en	
Nobelium	No	102	manmade	manmade		ra	
Osmium	Os	76	2E-8	b			
Oxygen	O	8	45.2	H20	209500	en	
Palladium	Pd	46	3E-7	b		cs	
Phosphorus	P	15	0.1010	0.07		en	
Platinum	Pt	78	5E-7	b		t1	0.002
Plutonium	Pu	94	manmade	manmade		ra	
Polonium	Po	84	a	b		ra	
Potassium	K	19	1.68	380		en	
Praseodymium	Pr	59	0.0015				
Promethium	Pm	manmade	manmade				Protachtinium
Radium	Ra	88	a	1E-10		ra	
Radon	Rn	86	a	0.6E-15		ra	
Rhenium	Re	75	4E-8	b			
Rhodium	Rh	45	1E-8	b		cs	0.001
Rubidium	Rb	37	0.0070	0.12		s	
Rutherium	Ru	44	1E-8	b			
Samarium	Sm	62	0.00077	b			
Scandium	Sc	21	0.0022	4E-5		cs	
Selenium	Se	34	5E-6	0.004		cp s en t2	
Silicon	Si	14	27.2	3		ep cs	
Silver	Ag	47	8E-7	0.0003		cst1	0.01
Sodium	Na	11	2.32	10556		en	
Strontium	Sr	38	0.045	8			
Sulphur	S	16	0.030	885		en	
Tantalum	Ta	73	0.00024	b			5
Technetium	Tc	43	manmade	manmade		ra	

TABLE C513 TABLE C514 is on page 581

Maximum permissible metal equivalent (lb/acre) is based on 10 % of CEC, which in its turn is found as 0.5 + % clay + 2*% organic matter.
Maximum permissible metal equivalent (lb/acre) can be used to calculate max permissible sludge per acre by determinating metal equivalent in sludge from the following equation:
metal equiv. = Zn(+2) + 2*Cu(+2) + 4*Ni(+2). The unit of metals are Weight/(weight dry sludge matter). The table gives the the metal laodings as a function of the organic matter in the soil, calculated as tons organic matter/acre soil = % organic matter*10.

REF.: 506

Soil texture	Organic content = 5-10 T/acre	Organic content = 11-20 T/acre	Organic content = 21-30 T/acre	Organic content = 31-40 T/acre	Organic content = 41-50 T/acre	Organic content = 51-70 T/acre	Organic content > 70 T/acre
Sand	260	360	490	630	750	940	1140
Loamy sand	330	440	570	700	830	1020	1220
Sandy loam	420	520	650	780	910	1110	1300
Loam	590	680	810	940	1070	1200	1330
Silt loam	750	850	980	1110	1110	1370	1500
Silty clay loam	1240	1330	1460	1590	1720	1850	1980
Clay loam	1400	1500	1630	1760	1890	2020	2150
Clay	2050	2150	2280	2410	2510	2670	2800

D

CHEMICALS AND THE LIVING ORGANISMS

BACKGROUND VALUES
CONCENTRATION FACTOR
CONCENTRATIONS
DAILY INTAKE
EXCRETION AND UPTAKE

D1

BACKGROUND VALUES

IN
ATMOSPHERE
BIOSPHERE
HYDROSPHERE
LITOSPHERE

TABLE D1

BACKGROUND CONCENTRATION : ABUNDANCE IN AVERAGE CRUSTRAL ROCK			
ITEM:	VALUE:	CONDITIONS:	REF.:
*AG	0.000007 PERCENT BY WEIGHT	NONE	989
*AL	8.1 PERCENT BY WEIGHT	NONE	989
*AU	0.0000004 PERCENT BY WEIGHT	NONE	989
*CO	0.0025 PERCENT BY WEIGHT	NONE	989
*CR	0.010 PERCENT BY WEIGHT	NONE	989
*CU	0.0055 PERCENT BY WEIGHT	NONE	989
*FE	5.0 PERCENT BY WEIGHT	NONE	989
*HG	0.000008 PERCENT BY WEIGHT	NONE	989
*MG	2.1 PERCENT BY WEIGHT	NONE	989
*MN	0.10 PERCENT BY WEIGHT	NONE	989
*MO	0.00015 PERCENT BY WEIGHT	NONE	989
*NI	0.0075 PERCENT BY WEIGHT	NONE	989
*PB	0.0013 PERCENT BY WEIGHT	NONE	989
*PT	0.000001 PERCENT BY WEIGHT	NONE	989
*SB	0.00002 PERCENT BY WEIGHT	NONE	989
*SN	0.00020 PERCENT BY WEIGHT	NONE	989
*TI	0.44 PERCENT BY WEIGHT	NONE	989
*U	0.00018 PERCENT BY WEIGHT	NONE	989
*V	0.014 PERCENT BY WEIGHT	NONE	989
*W	0.00015 PERCENT BY WEIGHT	NONE	989
*ZN	0.0070 PERCENT BY WEIGHT	NONE	989

TABLE D2

BACKGROUND CONCENTRATION : AG			
ITEM:	VALUE:	CONDITIONS:	REF.:
*ATMOSPHERE	0.05-0.2 NG/M**3	SAN FRANCISCO BAY,9 STATIONS,1970	893
*RIVER WATER	0.13*10**(-3) PPM	RANGE=0.1*10**(-4)-0.35*10**(-2) PPM	157
*SEAWATER	0.0003 PPM	AGCL2(-),RESIDENCE TIME= 2100*10**3YRS	539
*SOIL	0.1 (0.01-5) PPM DRY SOIL	OVEN-DRIED	157

TABLE D3

BACKGROUND CONCENTRATION : AL			
ITEM:	VALUE:	CONDITIONS:	REF.:
*ATMOSPHERIC PARTICULATES	6.7+-7.2 NG/M**3	HAWAII,N>=56	550
*RIVER WATER	0.24 PPM	RANGE=0.01-2.5 PPM	157
*SEAWATER	0.01 PPM	RESIDENCE TIME=0.1*10**3YRS	539
*SOIL	71000 (10000-300000) PPM DRY SOIL	OVEN-DRIED	157

TABLE D4

BACKGROUND CONCENTRATION : AR			
ITEM:	VALUE:	CONDITIONS:	REF.:
*ATMOSPHERE	9300 UG/M**3	STP	790

TABLE D4

BACKGROUND CONCENTRATION : AR

ITEM:	VALUE:	CONDITIONS:	REF.:
*ATMOSPHERE	0.93% 10L.FRACTION	SEA-LEVEL DRY ATMOSPHERIC COMPOSITION	1445
*LOWER ATMOSPHERE	0.934% BY VOLUME	NONE	532
*RIVER WATER	0.6 PPM	NONE	157
*SEAWATER	0.6 PPM	NONE	539
*SEAWATER	0.3 ML/L	SAT.SEAWATER,SALINITY=3.43%,TEMP.= 283K	1445

TABLE D5

BACKGROUND CONCENTRATION : AS

ITEM:	VALUE:	CONDITIONS:	REF.:
*ATMOSPHERE	< 10 NG/M**3	NONE	893
*CLAMS, ALL SPECIES	3.13 (SD=+-3.85) (RANGE = 0.90 - 12.75) MG/KG WET WEIGHT	11 REFERENCES USED	1779
*CRAB, BLUE	12.16 (SD=+-1.79) (RANGE = 3.81 - 35.62) MG/KG WET WEIGHT	8 REFERENCES USED	1779
*GADUS MORHUA	2.14 (SD=+-1.26) (RANGE = 0.86 - 3.37) MG/KG WET WEIGHT	3 REFERENCES USED	1779
*ICTALURIDAE	1.15 (SD=+-1.46) (RANGE = 0.12 - 2.18) MG/KG WET WEIGHT	2 REFERENCES USED	1779
*LUTJANIDAE	0.06 MG/KG WET WEIGHT	1 REFERENCE USED	1779
*MELANOGRAMMUS AEGLEFINUS	2.56 (SD=+-2.77) (RANGE = 0.60 - 4.51) MG/KG WET WEIGHT	2 REFERENCES USED	1779
*OYSTER	9.12 (SD=+-15.59) (RANGE = 0.45 - 42.74) MG/KG WET WEIGHT	10 REFERENCES USED	1779
*PERCIDAE	0.31 (SD=+-0.26) (RANGE = 0.05 - 0.57) MG/KG WET WEIGHT	2 REFERENCES USED	1779
*PLEURONECTIDAE	2.45 (SD=+-1.92) (RANGE = 0.11 - 4.51) MG/KG WET WEIGHT	5 REFERENCES USED	1779
*POLLOCK	2.35 MG/KG WET WEIGHT	1 REFERENCE USED	1779
*RIVER WATER	0.4*10**(-3) PPM	RANGE:< 0.4*10**(-3)-0.23 PPM	157
*SCALLOPS, SEA	8.76 (SD=+-12.80) (RANGE = 0.39 - 33.56) MG/KG WET WEIGHT	7 REFERENCES USED	1779
*SEAWATER	0.003 PPM	ASO4H(2-)	539
*SHRIMP, BROWN	8.71 (SD=+-7.63) (RANGE = 1.77 - 23.75) MG/KG WET WEIGHT	8 REFERENCES USED	1779
*SILLAGINIDAE	0.73 MG/KG WET WEIGHT	1 REFERENCE USED	1779
*SOIL	6 (0.1-40) PPM DRY SOIL	OVEN-DRIED	157

TABLE D6

BACKGROUND CONCENTRATION : ATP

ITEM:	VALUE:	CONDITIONS:	REF.:
*ESTUARINE SEDIMENT	0.64 MICROGRAM/CM**3	DEPTH = 0 CM, WINTER, NEWPORT ESTUARY	453
*ESTUARINE SEDIMENT	0.16 MICROGRAM/CM**3	DEPTH = 5 CM, WINTER, NEWPORT ESTUARY	453
*ESTUARINE SEDIMENT	0.04 MICROGRAM/CM**3	DEPTH =15 CM, WINTER, NEWPORT ESTUARY	453
*ESTUARINE SEDIMENT	1.18 MICROGRAM/CM**3	DEPTH = 0 CM, SUMMER, NEWPORT ESTUARY	453
*ESTUARINE SEDIMENT	0.32 MICROGRAM/CM**3	DEPTH = 5 CM, SUMMER, NEWPORT ESTUARY	453

TABLE D7

BACKGROUND CONCENTRATION : AU			
ITEM:	VALUE:	CONDITIONS:	REF.:
*RIVER WATER	< 0.6*10**(-4) PPM	NONE	157
*SEAWATER	0.000011 PPM	AUCL4(-),RESIDENCE TIME= 560*10**3YRS	539

TABLE D8

BACKGROUND CONCENTRATION : B			
ITEM:	VALUE:	CONDITIONS:	REF.:
*RIVER WATER	0.013 PPM	RANGE=0.01-1 PPM	157
*SEAWATER	4.6 PPM	B(OH)3	539
*SOIL	10 (2-100) PPM DRY SOIL	OVEN-DRIED	157

TABLE D9

BACKGROUND CONCENTRATION : BA			
ITEM:	VALUE:	CONDITIONS:	REF.:
*RIVER WATER	0.054 PPM	RANGE=0.009-0.15 PPM	157
*SEAWATER	0.03 PPM	BA(++),RESIDENCE TIME=84*10**3YRS	539
*SOIL	500 (100-3000) PPM DRY SOIL	OVEN-DRIED	157

TABLE D10

BACKGROUND CONCENTRATION : BE			
ITEM:	VALUE:	CONDITIONS:	REF.:
*RIVER WATER	0.001 PPM	RANGE=0.1*10**(-3)-0.001 PPM	157
*SEAWATER	0.6*10**(-6) PPM	RESIDENCE TIME=0.15*10**3YRS	539
*SOIL	6 (0.1-40) PPM DRY SOIL	OVEN-DRIED	157

TABLE D11

BACKGROUND CONCENTRATION : BENZO(A)ANTHRACENE			
ITEM:	VALUE:	CONDITIONS:	REF.:
*ALGAE	60 MICROGRAM/KG WET WEIGHT	GREENLAND, SAMPLED AT 40 M	632
*ALGAE	60 MICROGRAM/KG WET WEIGHT	GREENLAND, SAMPLED AT SEABED	632
*ALGAE	2 MICROGRAM/KG WET WEIGHT	ITALY	632
*COD	15 MICROGRAM/KG WET WEIGHT	GREENLAND	632
*MOLLUSCS	60 MICROGRAM/KG WET WEIGHT	GREENLAND	632
*MOLLUSCS	2 MICROGRAM/KG WET WEIGHT	ITALY	632
*MUSSEL, BODY	55 MICROGRAM/KG WET WEIGHT	GREENLAND	632
*MUSSEL, BODY	130 AND 540 MICROGRAM/KG WET WEIGHT	ITALY	632
*MUSSEL, SHELL	18 MICROGRAM/KG WET WEIGHT	GREENLAND	632
*MUSSEL, SHELL	11 MICROGRAM/KG WET WEIGHT	ITALY	632

TABLE D11

BACKGROUND CONCENTRATION : BENZO(A)ANTHRACENE			
ITEM:	VALUE:	CONDITIONS:	REF.:
*PLANKTON	5 MICROGRAM/KG WET WEIGHT	GREENLAND	632
*PLANKTON	6 - 21 MICROGRAM/KG WET WEIGHT	ITALY	632
*PLANKTON	400 MICROGRAM/KG WET WEIGHT	FRENCH CHANNEL COAST	632
*SARDINE	65 MICROGRAM/KG WET WEIGHT	ITALY	632

TABLE D12

BACKGROUND CONCENTRATION : BENZO(A)PYRENE			
ITEM:	VALUE:	CONDITIONS:	REF.:
*MARINE SEDIMENT	5 MICROGRAM/KG DRY WEIGHT	GREENLAND, DEPTH = 0.2 M	632
*MARINE SEDIMENT	1000 - 3000 MICROGRAM/KG DRY WEIGHT	ITALY, DEPTH = 15 - 45 M, NAPLES, HIGLY INDUDTRIALIZED AREA	632
*MARINE SEDIMENT	7.5 MICROGRAM/KG DRY WEIGHT	ITALY, DEPTH = 13 M, NAPLES, HIGHLY INDUSTRIALIZED AREA	632
*MARINE SEDIMENT	10 - 530 MICROGRAM/KG DRY WEIGHT	ITALY, DEPTH = 2 - 65 M, NAPLES, HIGHLY INDUSTRILIZED AREA	632
*MARINE SEDIMENT	260 - 960 MICROGRAM/KG DRY WEIGHT	ITALY, DEPTH = 55 M, NAPLES, NEAR VOLCANIC POLLUTION	632
*MARINE SEDIMENT	1.4 MICROGRAM/KG DRY WEIGHT	ITALY, DEPTH = 120 M, NAPLES, NEAR VOLCANIC POLLUTION	632
*MARINE SEDIMENT	100 - 500 MICROGRAM/KG DRY WEIGHT	ITALY, NAPLES, ISLAND AFFECTED BY POLLUTION, DEPTH = 0 M	632
*MARINE SEDIMENT	400 MICROGRAM/KG DRY WEIGHT	FRENCH MEDITERRANEAN COAST, DEPTH = 14 M	632
*MARINE SEDIMENT	1500 MICROGRAM/KG DRY WEIGHT	FRENCH MEDITERRANEAN COAST, DEPTH = 1.6 M	632
*MARINE SEDIMENT	75 MICROGRAM/KG DRY WEIGHT	FRENCH MEDITERRANEAN COAST, DEPTH = 48 M	632
*MARINE SEDIMENT	TRACES	FRENCH MEDITERRANEAN COAST, DEPTH = 58 M	632
*MARINE SEDIMENT	400 MICROGRAM/KG DRY WEIGHT	FRENCH MEDITERRANEAN COAST, DEPTH = 82 M	632
*MARINE SEDIMENT	34 MICROGRAM/KG DRY WEIGHT	FRENCH MEDITERRANEAN COAST, ESTUARY, DEPTH = 0 M	632
*MARINE SEDIMENT	20 MICROGRAM/KG DRY WEIGHT	FRENCH MEDITERRANEAN COAST, ESTUARY, DEPTH = 1 M	632
*MARINE SEDIMENT	15 MICROGRAM/KG DRY WEIGHT	FRENCH MEDITERRANEAN COAST, ESTUARY, DEPTH = 4 M	632
*MARINE SEDIMENT	25 MICROGRAM/KG DRY WEIGHT	FRENCH MEDITERRANEAN COAST, ESTUARY, DEPTH = 5 M	632
*MARINE SEDIMENT	0 MICROGRAM/KG DRY WEIGHT	FRENCH MEDITERRANEAN COAST, ESTUARY, DEPTH = 102 M	632

TABLE D13

BACKGROUND CONCENTRATION : BI			
ITEM:	VALUE:	CONDITIONS:	REF.:
*SEAWATER	0.17*10**(-4) PPM	RESIDENCE TIME=45*10**3YRS	539

TABLE D14

BACKGROUND CONCENTRATION : BIOSPHERE			
ITEM:	VALUE:	CONDITIONS:	REF.:
*AL	4.12 MOLES/HECTARE	AVERAGE COMPOSITION OF LIVING MATTER	341

TABLE D14

BACKGROUND CONCENTRATION : BIOSPHERE			
ITEM:	VALUE:	CONDITIONS:	REF.:
*C	6560 MOLES/HECTARE	AVERAGE COMPOSITION OF LIVING MATTER	341
*CA	18.9 MOLES/HECTARE	AVERAGE COMPOSITION OF LIVING MATTER	341
*CL	2.80 MOLES/HECTARE	AVERAGE COMPOSITION OF LIVING MATTER	341
*FE	1.38 MOLES/HECTARE	AVERAGE COMPOSITION OF LIVING MATTER	341
*H	13100 MOLES/HECTARE	AVERAGE COMPOSITION OF LIVING MATTER	341
*K	11.7 MOLES/HECTARE	AVERAGE COMPOSITION OF LIVING MATTER	341
*MG	8.18 MOLES/HECTARE	AVERAGE COMPOSITION OF LIVING MATTER	341
*MN	0.765 MOLES/HECTARE	AVERAGE COMPOSITION OF LIVING MATTER	341
*N	71.6 MOLES/HECTARE	AVERAGE COMPOSITION OF LIVING MATTER	341
*NA	1.65 MOLES/HECTARE	AVERAGE COMPOSITION OF LIVING MATTER	341
*O	6540 MOLES/HECTARE	AVERAGE COMPOSITION OF LIVING MATTER	341
*P	3.35 MOLES/HECTARE	AVERAGE COMPOSITION OF LIVING MATTER	341
*S	4.44 MOLES/HECTARE	AVERAGE COMPOSITION OF LIVING MATTER	341
*SI	8.62 MOLES/HECTARE	AVERAGE COMPOSITION OF LIVING MATTER	341

TABLE D15

BACKGROUND CONCENTRATION : BIOTIN			
ITEM:	VALUE:	CONDITIONS:	REF.:
*FRESHWATER	0.0001 - 0.004 MG/M**3	LINDSEY POND, CONN.	99
*FRESHWATER	0.013 - 0.068 MG/M**3	SAGAMI, JAPAN	1178
*FRESHWATER	0.013 - 0.058 MG/M**3	DISSOLVED, TSUKUI, JAPAN, SURFACE	1179
*FRESHWATER	0.0005 - 0.0042 MG/M**3	PARTICULATE, TSUKUI, JAPAN, SURFACE	1179
*FRESHWATER	0.0021 - 0.050 MG/M**3	KASUMIGAURA, JAPAN	801
*FRESHWATER	<0.001 - 0.004 MG/M**3	SMALL SWISS PONDS	257
*FRESHWATER	0.004 MG/M**3	MEAN OF PRECIPITATION, MO., APRIL-NOV.	1213
*FRESHWATER	0.0008 MG/M**3	MEAN OF PRECIPATION, MO., NOV-MARCH	1213

TABLE D16

BACKGROUND CONCENTRATION : BR			
ITEM:	VALUE:	CONDITIONS:	REF.:
*RIVER WATER	0.021 PPM	RANGE=0.005-140 PPM	157
*SEAWATER	65 PPM	BR(-)	539
*SOIL	5 (1-10) PPM DRY SOIL	OVEN-DRIED	157

TABLE D17

BACKGROUND CONCENTRATION : C			
ITEM:	VALUE:	CONDITIONS:	REF.:
*ATMOSPHERE	164000 UG/M**3	STP	790

TABLE D17

BACKGROUND CONCENTRATION : C			
ITEM:	VALUE:	CONDITIONS:	REF.:
*RIVER WATER	11 PPM	HCO3(-),RANGE=6-19 PPM	157
*SEAWATER	28 PPM	CO3H(-),ORGANIC CARBON	539
*SOIL	20000 PPM DRY SOIL	OVEN-DRIED	157

TABLE D18

BACKGROUND CONCENTRATION : CA			
ITEM:	VALUE:	CONDITIONS:	REF.:
*ATMOSPHERIC PARTICULATES	160+-90 NG/M**3	HAWAII,N>=56	550
*RIVER WATER	15 PPM	RANGE=4-120 PPM	157
*RIVER, AFRICA	12.5 MG/L	NONE	1689
*RIVER, ASIA	18.4 MG/L	NONE	1689
*RIVER, AUSTRALIA	3.9 MG/L	NONE	1689
*RIVER, EUROPE	31.1 MG/L	NONE	1689
*RIVER, NORTH AMERICA	21.0 MG/L	NONE	1689
*RIVER, SOUTH AMERICA	7.2 MG/L	NONE	1689
*RIVER, WORLD	15. MG/L	NONE	1689
*SEAWATER	400 PPM	CA(2+),RESIDENCE TIME= 8000*10**3YRS	539
*SEAWATER	91 % OF TOTAL	AS FREE ION, CHLORINITY = 1.9 %, 298 K, PH = 8.15	504
*SEAWATER	8 % OF TOTAL	AS CASO4, CHLORINITY = 1.9 %, 298 K, PH = 8.15	504
*SEAWATER	1 % OF TOTAL	AS CA(HCO3)2, CHLORINITY = 1.9 %, 298 K, PH = 8.15	504
*SEAWATER	0.2 % OF TOTAL	AS CACO3(AQ), CHLORINITY = 1.9 %, 298 K, PH = 8.15	504
*SEAWATER	91 % OF TOTAL	AS FREE ION, CHLORINITY = 1.9 %, 298 K, PH = 8.15	504
*SEAWATER	8 % OF TOTAL	AS CASO4, CHLORINITY = 1.9 %, 298 K, PH = 8.15	504
*SEAWATER	1 % OF TOTAL	AS CA(HCO3)2, CHLORINITY = 1.9 %, 298 K, PH = 8.15	504
*SEAWATER	0.2 % OF TOTAL	AS CACO3(AQ), CHLORINITY = 1.9 %, 298 K, PH = 8.15	504
*SOIL	13700 (7000-500000) PPM DRY SOIL	OVEN-DRIED	157

TABLE D19

BACKGROUND CONCENTRATION : CARBON DIOXIDE			
ITEM:	VALUE:	CONDITIONS:	REF.:
*ATMOSPHERE	4:10**4 G/CM**2	TOTAL SUPPLY = 1.9*10**23 G**2	779
*ATMOSPHERE	0.45 G/CM**2	TOTAL IN ATMOS.(1975) = 2.25*10**18 G	779
*ATMOSPHERE	0.03% MOL.FRACTION	SEA-LEVEL DRY ATMOSPHERIC COMPOSITION	1445
*ATMOSPHERE	6*10**5 UG/M**3	STP ,RESIDENCE TIME=4YRS	790
*LOWER ATMOSPHERE	0.033% BY VOLUME	NONE	532
*SEAWATER	34-56 ML/L	SAT.SEAWATER,SALINITY=3.43%,TEMP.= 283K	1445

TABLE D20

BACKGROUND CONCENTRATION : CARBON MONOXIDE			
ITEM:	VALUE:	CONDITIONS:	REF.:
*ATMOSPHERE	1-20 UG/M**3	STP ,RESIDENCE TIME=0.3YRS	790

TABLE D21

BACKGROUND CONCENTRATION : CD

ITEM:	VALUE:	CONDITIONS:	REF.:
*ATMOSPHERE	1 NG/M**3	NOT CONTAMINATED AREA	893
*CLAMS, ALL SPECIES	0.22 (SD=+-0.17) (RANGE = 0.03 - 0.58) MG/KG WET WEIGHT	12 REFERENCES USED	1779
*CRAB, BLUE	3.62 (SD=+-6.31) (RANGE = 0.01 - 22.40) MG/KG WET WEIGHT	13 REFERENCES USED	1779
*FRESHWATER SEDIMENT	0.46 MG/KG DRY WEIGHT	RANGE OF 90 % = 0.10 - 1.20 PPM, 74 LAKES, PELITIC	464
*GADUS MORHUA	0.04 (SD=+-0.02) (RANGE = 0.03 - 0.06) MG/KG WET WEIGHT	3 REFERENCES USED	1779
*ICTALURIDAE	0.06 (SD=+-0.02) (RANGE = 0.05 - 0.08) MG/KG WET WEIGHT	3 REFERENCES USED	1779
*LUTJANIDAE	0.41 MG/KG WET WEIGHT	1 REFERENCE USED	1779
*MELANOGRAMMUS AEGLEFINUS	0.05 (SD=+-0.03) (RANGE = 0.03 - 0.08) MG/KG WET WEIGHT	3 REFERENCES USED	1779
*MERLUCCIUS SP.	0.04 (SD=+-0.03) (RANGE = 0.02 - 0.05) MG/KG WET WEIGHT	2 REFERENCES USED	1779
*OYSTER	1.88 (SD=+-1.29) (RANGE = 0.11 - 3.60) MG/KG WET WEIGHT	11 REFERENCES USED	1779
*PERCIDAE	0.55 (SD=+-0.62) (RANGE = 0.02 - 1.24) MG/KG WET WEIGHT	2 REFERENCES USED	1779
*PLEURONECTIDAE	0.03 (SD=+-0.01) (RANGE = 0.03 - 0.04) MG/KG WET WEIGHT	3 REFERENCES USED	1779
*POLLOCK	0.26X0.29) (RANGE = 0.05 - 0.46) MG/KG WET WEIGHT	2 REFERENCES USED	1779
*RIVER WATER	0.08 PPM	NONE	157
*SCALLOPS, SEA	2.77 (SD=+-4.45) (RANGE = 0.03 - 10.55) MG/KG WET WEIGHT	5 REFERENCES USED	1779
*SEAWATER	0.11*10**(-3) PPM	CD(2+),RESIDENCE TIME=500*10**3YRS	539
*SHALE	0.3 MG/KG DRY WEIGHT	AVERAGE VALUE	1616
*SHRIMP, BROWN	0.09 (SD=+-0.06) (RANGE = 0.04 - 0.24) MG/KG WET WEIGHT	8 REFERENCES USED	1779
*SILLAGINIDAE	0.01 MG/KG WET WEIGHT	1 REFERENCE USED	1779
*SOIL	0.06 (0.01-0.7) PPM DRY SOIL	OVEN-DRIED	157
*SYSTELLASPIS DEBLIS AND OPLOPHORUS SP.	13 MG/KG DRY WEIGHT	AZORES	898

TABLE D22

BACKGROUND CONCENTRATION : CE

ITEM:	VALUE:	CONDITIONS:	REF.:
*SEAWATER	0.4*10**(-3) PPM	RESIDENCE TIME=6.1*10**3YRS	539
*SEAWATER	0.058 MICROGRAM/L	TOTAL CONCENTRATION, JAPAN, CL = 1.873 %	1535
*SEAWATER	0.007 MICROGRAM/L	SOLUBLE CONCENTRATION, JAPAN, CL = 1.873 %	1535
*SOIL	50 PPM DRY SOIL	OVEN-DRIED	157

TABLE D23

BACKGROUND CONCENTRATION : CHLOROPHYLL A

ITEM:	VALUE:	CONDITIONS:	REF.:
*ESTUARINE SEDIMENT	7.45 MICROGRAM/CM**3	DEPTH = 0 CM, WINTER, NEWPORT ESTUARY	453
*ESTUARINE SEDIMENT	2.00 MICROGRAM/CM**3	DEPTH = 5 CM, WINTER, NEWPORT ESTUARY	453

TABLE D23

BACKGROUND CONCENTRATION : CHLOROPHYLL A			
ITEM:	VALUE:	CONDITIONS:	REF.:
*ESTUARINE SEDIMENT	0.60 MICROGRAM/CM**3	DEPTH =15 CM, WINTER, NEWPORT ESTUARY	453
*ESTUARINE SEDIMENT	3.77 MICROGRAM/CM**3	DEPTH = 0 CM, SUMMER, NEWPORT ESTUARY	453
*ESTUARINE SEDIMENT	2.01 MICROGRAM/CM**3	DEPTH = 5 CM, SUMMER, NEWPORT ESTUARY	453

TABLE D24

BACKGROUND CONCENTRATION : CH4			
ITEM:	VALUE:	CONDITIONS:	REF.:
*ATMOSPHERE	850-1100 UG/M**3	STP ,RESIDENCE TIME=100YRS	790
*LOWER ATMOSPHERE	0.00002% BY VOLUME	NONE	532
*SEAWATER	0-30 ML/L	SAT.SEAWATER,SALINITY=3.43%,283K, HIGH VAL.ONLY OVER ANOXIC BASIN	1445

TABLE D25

BACKGROUND CONCENTRATION : CL			
ITEM:	VALUE:	CONDITIONS:	REF.:
*RIVER WATER	7.8 PPM	RANGE=5-35 PPM	157
*RIVER, AFRICA	12.1 MG/L	NONE	1689
*RIVER, ASIA	8.7 MG/L	NONE	1689
*RIVER, AUSTRALIA	10. MG/L	NONE	1689
*RIVER, EUROPE	6.9 MG/L	NONE	1689
*RIVER, NORTH AMERICA	8. MG/L	NONE	1689
*RIVER, SOUTH AMERICA	4.9 MG/L	NONE	1689
*RIVER, WORLD	7.8 MG/L	NONE	1689
*SEAWATER	19000 PPM	CL(-)	539
*SEAWATER	100 % OF TOTAL	AS FREE ION, CHLORINITY = 1.9 %, 298 K, PH = 8.15	504
*SEAWATER	100 % OF TOTAL	AS FREE ION, CHLORINITY = 1.9 %, 298 K, PH = 8.15	504
*SOIL	100 PPM DRY SOIL	OVEN-DRIED	157

TABLE D26

BACKGROUND CONCENTRATION : CO			
ITEM:	VALUE:	CONDITIONS:	REF.:
*FRESHWATER SEDIMENT	17 MG/KG DRY WEIGHT	RANGE OF 90 % = 3 - 40 PPM, 74 LAKES, PELITIC	464
*RIVER WATER	0.9*10**(-3) PPM	RANGE:< 0.006 PPM	157
*SEAWATER	0.27*10**(-3) PPM	CO(2+),RESIDENCE TIME=18*10**3YRS	539
*SHALE	19 MG/KG DRY WEIGHT	AVERAGE VALUE	1616
*SOIL	8 (1-40) PPM DRY SOIL	OVEN-DRIED	157

TABLE D27

BACKGROUND CONCENTRATION : CO3(2-) + HCO3(-)			
ITEM:	VALUE:	CONDITIONS:	REF.:
*RIVER, AFRICA	43. MG/L	NONE	1689

TABLE D27

BACKGROUND CONCENTRATION : CO3(2-) + HCO3(-)			
ITEM:	VALUE:	CONDITIONS:	REF.:
*RIVER, ASIA	79. MG/L	NONE	1689
*RIVER, AUSTRALIA	31.6 MG/L	NONE	1689
*RIVER, EUROPE	95. MG/L	NONE	1689
*RIVER, NORTH AMERICA	68. MG/L	NONE	1689
*RIVER, SOUTH AMERICA	31. MG/L	NONE	1689
*RIVER, WORLD	58.4 MG/L	NONE	1689
*SEAWATER	9 % OF TOTAL	AS FREE ION, CHLORINITY = 1.9 %, 298 K, PH = 8.15	504
*SEAWATER	7 % OF TOTAL	AS CACO3, CHLORINITY = 1.9 %, 298 K, PH = 8.15	504
*SEAWATER	67 % OF TOTAL	AS MGCO3, CHLORINITY = 1.9 %, 298 K, PH = 8.15	504
*SEAWATER	17 % OF TOTAL	AS NA2CO3, CHLORINITY = 1.9 %, 298 K, PH = 8.15	504
*SEAWATER	9 % OF TOTAL	AS FREE ION, CHLORINITY = 1.9 %, 298 K, PH = 8.15	504
*SEAWATER	7 % OF TOTAL	AS CACO3, CHLORINITY = 1.9 %, 298 K, PH = 8.15	504
*SEAWATER	67 % OF TOTAL	AS MGCO3, CHLORINITY = 1.9 %, 298 K, PH = 8.15	504
*SEAWATER	17 % OF TOTAL	AS NA2CO3, CHLORINITY = 1.9 %, 298 K, PH = 8.15	504

TABLE D28

BACKGROUND CONCENTRATION : CR			
ITEM:	VALUE:	CONDITIONS:	REF.:
*CLAMS, ALL SPECIES	0.33 (SD=+-0.12) (RANGE = 0.17 - 0.44) MG/KG WET WEIGHT	6 REFERENCES USED	1779
*CRAB, BLUE	0.14 (SD=+-0.06) (RANGE = 0.10 - 0.17) MG/KG WET WEIGHT	2 REFERENCES USED	1779
*FRESHWATER SEDIMENT	82 MG/KG DRY WEIGHT	RANGE OF 90 % = 25 - 180PPM, 74 LAKES, PELITIC	464
*ICTALURIDAE	0.23 MG/KG WET WEIGHT	1 REFERENCE USED	1779
*LUTJANIDAE	0.12 MG/KG WET WEIGHT	1 REFERENCE USED	1779
*MELANOGRAMMUS AEGLEFINUS	0.11 (SD=+-0.12) (RANGE = 0.02 - 0.30) MG/KG WET WEIGHT	4 REFERENCES USED	1779
*MERLUCCIUS SP.	0.09 (SD=+-0.00) (RANGE = 0.09 - 0.09) MG/KG WET WEIGHT	2 REFERENCES USED	1779
*OYSTER	0.12 (SD=+-0.04) (RANGE = 0.09 - 0.14) MG/KG WET WEIGHT	2 REFERENCES USED	1779
*PERCIDAE	0.07 MG/KG WET WEIGHT	1 REFERENCE USED	1779
*PLEURONECTIDAE	0.09 (SD=+-0.08) (RANGE = 0.01 - 0.18) MG/KG WET WEIGHT	5 REFERENCES USED	1779
*POLLOCK	0.08 MG/KG WET WEIGHT	1 REFERENCE USED	1779
*RIVER WATER	0.18*10**(-3) PPM	RANGE=0.1*10**(-3)-0.06 PPM	157
*SCALLOPS, SEA	0.09 (SD=+-0.02) (RANGE = 0.05 - 0.11) MG/KG WET WEIGHT	5 REFERENCES USED	1779
*SEAWATER	0.5*10**(-4) PPM	RESIDENCE TIME=0.35*10**3YRS	539
*SHALE	90 MG/KG DRY WEIGHT	AVERAGE VALUE	1616
*SHRIMP, BROWN	0.07 (SD=+-0.05) (RANGE = 0.01 - 0.12) MG/KG WET WEIGHT	7 REFERENCES USED	1779
*SOIL	100 (5-3000) PPM DRY SOIL	OVEN-DRIED	157

TABLE D29

BACKGROUND CONCENTRATION : CS			
ITEM:	VALUE:	CONDITIONS:	REF.:
*RIVER WATER	0.2*10**(-3) PPM	RANGE=0.5*10**(-4)-0.2*10**(-3) PPM	157

TABLE D29

BACKGROUND CONCENTRATION : CS			
ITEM:	VALUE:	CONDITIONS:	REF.:
*SEAWATER	0.5*10**(-3) PPM	CS(+),RESIDENCE TIME=40*10**3YRS	539
*SOIL	6 (0.3-25) PPM DRY SOIL	OVEN-DRIED	157

TABLE D30

BACKGROUND CONCENTRATION : CU			
ITEM:	VALUE:	CONDITIONS:	REF.:
*ATMOSPHERIC PARTICULATES	1.9+-2.1 NG/M**3	HAWAII,N>=56	550
*FRESHWATER SEDIMENT	46 MG/KG DRY WEIGHT	RANGE OF 90 % = 20 - 80 PPM, 74 LAKES, PELITIC	464
*RIVER WATER	0.01 PPM	RANGE=0.6*10**(-3)-0.4 PPM	157
*SEAWATER	0.3*10**(-2) PPM	CU(2+),RESIDENCE TIME=50*10**3YRS	539
*SHALE	45 MG/KG DRY WEIGHT	AVERAGE VALUE	1616
*SOIL	20 (2-100) PPM DRY SOIL	OVEN-DRIED	157

TABLE D31

BACKGROUND CONCENTRATION : DDT			
ITEM:	VALUE:	CONDITIONS:	REF.:
*ALGAE IN OPEN OCEAN	0.1 MG/KG DRY WT.	TOTAL DDT=1.0*10**8T,DRY BIOMASS= 1.0*10**9 METRIC TONS	1747
*ALGAE ON CONTINENTAL SHELF	1.0 MG/KG DRY WT.	TOTAL DDT=3.0*10**8T,DRY BIOMASS= 0.3*10**9 METRIC TONS	1747
*ANIMALS IN OCEANS OTHER THAN FISH AND MAMMALIA	0.1 MG/KG DRY WT.	TOTAL DDT=3.02*10**8T,DRY BIOMASS= 3.02*10**9 METRIC TONS	1747
*ATTACHED ALGAE IN OCEAN	1.0 MG/KG DRY WT.	TOTAL DDT=20.0*10**8T,DRY BIOMASS= 2.0*10**9 METRIC TONS	1747
*BIRDS	1.0 MG/KG DRY WT.	TOTAL DDT=0.002*10**8T,DRY BIOMASS =0.00024*10**9 METRIC TONS	1747
*DOMESTIC MAMMALIA ON LAND	1.0 MG/KG DRY WT.	TOTAL DDT=1.7*10**8T,DRY BIOMASS= 0.17*10**9 METRIC TONS	1747
*FERAL MAMMALIA ON LAND	1.0 MG/KG DRY WT.	TOTAL DDT=0.09*10**8T,DRY BIOMASS= 0.009*10**9 METRIC TONS	1747
*FISH IN OCEANS	1.0 MG/KG DRY WT.	TOTAL DDT=6.5*10**8T,DRY BIOMASS= 0.65*10**9 METRIC TONS	1747
*HOMO SAPIENS	1.0 MG/KG DRY WT.	TOTAL DDT=3.0*10**8T,DRY BIOMASS= 0.30*10**9 METRIC TONS	1747
*MAMMALIA IN OCEANS	1.0 MG/KG DRY WT.	TOTAL DDT=0.55*10**8T,DRY BIOMASS= 0.055*10**9 METRIC TONS	1747
*PLANT IN LAKES AND STREAMS	0.010 MG/KG DRY WT.	TOTAL DDT=0.004*10**8T,DRY BIOMASS =0.04*10**9 METRIC TONS	1747
*PLANT IN SWAMPS AND MARSHES	0.001 MG/KG DRY WT.	TOTAL DDT=0.240*10**8T,DRY BIOMASS =24*10**9 METRIC TONS	1747
*PLANT IN TERRESTRIAL VEGETATION	0.0001 MG/KG DRY WT.	TOTAL DDT=1.814*10**8T,DRY BIOMASS =1814*10**9 METRIC TONS	1747
*PLANT ON AGRICULTURAL LAND	0.1 MG/KG DRY WT.	TOTAL DDT=14.000*10**8T,DRY BIOMASS=14*10**9 METRIC TONS	1747

TABLE D32

BACKGROUND CONCENTRATION : DISSOLVED ORGANIC CARBON			
ITEM:	VALUE:	CONDITIONS:	REF.:
*EAST CHINA SEA	1.25 (RANGE 0.96 - 1.42) MG C/L	MAY 1968, N=20	1173

TABLE D32

BACKGROUND CONCENTRATION : DISSOLVED ORGANIC CARBON

ITEM:	VALUE:	CONDITIONS:	REF.:
*NORTH PACIFIC	0.72 (RANGE 0.50 - 1.04) MG C/L	CALIF. - SAMOA, APRIL-MAY 1967, N= 6	1719
*NORTH PACIFIC	1.09 (RANGE 0.72 - 2.00) MG C/L	HAWAII - CODIAC, JAN 1966, N=10	1719
*NORTH PACIFIC	2.06 (RANGE 1.10 - 4.10) MG C/L	JUNE 1969 - JUNE 1970, N=8	1176
*NORTH-EQUATORIAL PACIFIC	1.14 (RANGE 0.81 - 1.46) MG C/L	155 DEG W, 50 DEG N - 15 DEG S, AUG-NOV 1969, N=15	1174
*NORTH-EQUATORIAL PACIFIC	1.12 (RANGE 0.88 - 1.49) MG C/L	APIA-TOKYO, AUG-NOV 1968, N=59	1174
*SAGAMI BAY	1.56 (RANGE 1.22 - 1.96) MG C/L	OCT 1970, N=30	1176
*SAGAMI BAY	1.31 (RANGE 1.04 - 1.69) MG C/L	OCT 1971, N=20	1176
*SAGAMI BAY	1.37 (RANGE 0.85 - 1.97) MG C/L	JAN - DEC 1971, N=30	1177
*SAGAMI BAY	1.24 (RANGE 0.73 - 2.31) MG C/L	JAN - DEC 1972, N=34	1177
*SAGAMI BAY	1.55 (RANGE 0.93 - 2.48) MG C/L	JAN - DEC 1973, N=12	1176
*SOUTH EAST PACIFIC	2.09 (RANGE 1.05 - 4.30) MG C/L	OFF PERU, MARCH-APRIL 1966, N=12	67
*SOUTH EAST PACIFIC	0.91 (RANGE 0.75 - 1.20) MG C/L	PERU CURRENT, APRIL 1966, N=12	67
*TOKYO BAY	3.61 (RANGE 2.26 - 5.32) MG C/L	JULY - AUG 1971, N=11	1176
*TOKYO BAY	3.07 (RANGE 2.40 - 3.70) MG C/L	AUG 1971, N=7	1176
*TOKYO BAY	1.94 (RANGE 1.29 - 3.01) MG C/L	SEPT 1971, N=10	1176
*TOKYO BAY	1.59 (RANGE 0.94 - 2.32) MG C/L	NOV - DEC 1971, N=13	1176
*TOKYO BAY	2.53 (RANGE 1.24 - 3.90) MG C/L	JUNE 1972, N=9	1176
*TOKYO BAY	3.63 (RANGE 1.63 - 4.36) MG C/L	JUNE 1973, N=9	1176
*TOKYO BAY	4.07 (RANGE 2.32 - 7.00) MG C/L	AUG 1973, N=27	1176
*TOKYO BAY	3.88 (RANGE 2.79 - 5.66) MG C/L	AUG 1973, N=30	1176
*TOKYO BAY	1.73 (RANGE 0.88 - 5.27) MG C/L	DEC 1973, N=32	1176
*WESTERN NORTH PACIFIC	1.10 (RANGE 0.83 - 1.68) MG C/L	JUNE 1966 - MAY 1968, N=22	1175
*WESTERN NORTH PACIFIC	0.88 (RANGE 0.73 - 1.13) MG C/L	JUME 1967, N=17	1172

TABLE D33

BACKGROUND CONCENTRATION : EU

ITEM:	VALUE:	CONDITIONS:	REF.:
*SEAWATER	0.0016 MICROGRAM/L	TOTAL CONCENTRATION, JAPAN, CL = 1.873 %	1535
*SEAWATER	0.0003 MICROGRAM/L	SOLUBLE CONCENTRATION, JAPAN, CL = 1.873 %	1535

TABLE D34

BACKGROUND CONCENTRATION : F

ITEM:	VALUE:	CONDITIONS:	REF.:
*RIVER WATER	0.09 PPM	NONE	157
*SEAWATER	1.3 PPM	NONE	539
*SOIL	200 (30-300) PPM DRY SOIL	OVEN-DRIED	157

TABLE D35

BACKGROUND CONCENTRATION : FE

ITEM:	VALUE:	CONDITIONS:	REF.:
*ATMOSPHERIC PARTICULATES	12+-9 NG/M**3	HAWAII,N>=56	550

TABLE D35

BACKGROUND CONCENTRATION : FE			
ITEM:	VALUE:	CONDITIONS:	REF.:
*FRESHWATER SEDIMENT	40100 MG/KG DRY WEIGHT	RANGE OF 90 % = 12000 - 69000 PPM, 74 LAKES, PELITIC	464
*RIVER WATER	0.67 PPM	RANGE=0.01-1.4 PPM	157
*RIVER, AFRICA	1.3 MG/L	NONE	1689
*RIVER, ASIA	0.01 MG/L	NONE	1689
*RIVER, AUSTRALIA	0.3 MG/L	NONE	1689
*RIVER, EUROPE	0.8 MG/L	NONE	1689
*RIVER, NORTH AMERICA	0.16 MG/L	NONE	1689
*RIVER, SOUTH AMERICA	1.4 MG/L	NONE	1689
*RIVER, WORLD	0.67 MG/L	NONE	1689
*SEAWATER	0.01 PPM	FE(OH)3,RESIDENCE TIME= 0.14*10**3YRS	539
*SHALE	46700 MG/KG DRY WEIGHT	AVERAGE VALUE	1616
*SOIL	38000 (7000-550000) PPM DRY SOIL	OVEN-DRIED	157

TABLE D36

BACKGROUND CONCENTRATION : FOLIC ACID			
ITEM:	VALUE:	CONDITIONS:	REF.:
*FRESHWATER	0.040 - 0.244 MG/M**3	KASUMIGAURA, JAPAN	801
*FRESHWATER	<0.01 - 0.48 MG/M**3	SMALL SWISS PONDS	257

TABLE D37

BACKGROUND CONCENTRATION : GA			
ITEM:	VALUE:	CONDITIONS:	REF.:
*RIVER WATER	< 0.001 PPM	NONE	157
*SEAWATER	0.3*10**(-4) PPM	RESIDENCE TIME=1.4*10**3YRS	539
*SOIL	30 (0.4-300) PPM DRY SOIL	OVEN-DRIED	157

TABLE D38

BACKGROUND CONCENTRATION : GE			
ITEM:	VALUE:	CONDITIONS:	REF.:
*SEAWATER	0.7*10**(-4) PPM	GE(OH)4,RESIDENCE TIME=7*10**3YRS	539
*SOIL	1 (1-50) PPM DRY SOIL	OVEN-DRIED	157

TABLE D39

BACKGROUND CONCENTRATION : H			
ITEM:	VALUE:	CONDITIONS:	REF.:
*ATMOSPHERE	250-365 UG/M**3	STP ,EXCLUSIVE H2O	790
*ATMOSPHERE	5*10**(-5)% MOL.FRACTION	SEA-LEVEL DRY ATMOSPHERIC COMPOSITION	1445

TABLE D39

BACKGROUND CONCENTRATION : H			
ITEM:	VALUE:	CONDITIONS:	REF.:
*ATMOSPHERE	36-90 UG/M**3	STP	790
*LOWER ATMOSPHERE	0.00005% BY VOLUME	NONE	532
*SEAWATER	108000 PPM	H20	539
*SEAWATER	3*10**(-3) ML/L	SAT.SEAWATER,SALINITY=3.43%,TEMP.= 283K	1445

TABLE D40

BACKGROUND CONCENTRATION : HCO3(-)			
ITEM:	VALUE:	CONDITIONS:	REF.:
*SEAWATER	69 % OF TOTAL	AS FREE ION, CHLORINITY = 1.9 %, 298 K, PH = 8.15	504
*SEAWATER	4 % OF TOTAL	AS CA(HCO3)2, CHLORINITY = 1.9 %, 298 K, PH = 8.15	504
*SEAWATER	19 % OF TOTAL	AS MG(HCO3)2, CHLORINITY = 1.9 %, 298 K, PH = 8.15	504
*SEAWATER	8 % OF TOTAL	AS NAHCO3, CHLORINITY = 1.9 %, 298 K, PH = 8.15	504
*SEAWATER	69 % OF TOTAL	AS FREE ION, CHLORINITY = 1.9 %, 298 K, PH = 8.15	504
*SEAWATER	4 % OF TOTAL	AS CA(HCO3)2, CHLORINITY = 1.9 %, 298 K, PH = 8.15	504
*SEAWATER	19 % OF TOTAL	AS MG(HCO3)2, CHLORINITY = 1.9 %, 298 K, PH = 8.15	504
*SEAWATER	8 % OF TOTAL	AS NAHCO3, CHLORINITY = 1.9 %, 298 K, PH = 8.15	504

TABLE D41

BACKGROUND CONCENTRATION : HE			
ITEM:	VALUE:	CONDITIONS:	REF.:
*ATMOSPHERE	5.2 MIC.G/M**3	STP ,RESIDENCE TIME=2*10**6YRS	790
*ATMOSPHERE	5*10**(-4)% MOL.FRACTION	SEA-LEVEL DRY ATMOSPHERIC COMPOSITION	1445
*LOWER ATMOSPHERE	0.00053% BY VOLUME	NONE	532
*SEAWATER	0.69*10**(-5) PPM	HE,RESIDENCE TIME=20000*10**3YRS	539

TABLE D42

BACKGROUND CONCENTRATION : HF			
ITEM:	VALUE:	CONDITIONS:	REF.:
*SEAWATER	< 0.8*10**(-5) PPM	NONE	539
*SOIL	6 PPM DRY SOIL	OVEN-DRIED	157

TABLE D43

BACKGROUND CONCENTRATION : HG			
ITEM:	VALUE:	CONDITIONS:	REF.:
*CLAMS, ALL SPECIES	0.08 (SD=+-0.05) (RANGE = 0.01 - 0.26) MG/KG WET WEIGHT	22 REFERENCES USED	1779

TABLE D43

BACKGROUND CONCENTRATION : HG			
ITEM:	VALUE:	CONDITIONS:	REF.:
*CRAB, BLUE	0.16 (SD=+-0.15) (RANGE = 0.07 - 0.48) MG/KG WET WEIGHT	6 REFERENCES USED	1779
*FISH	0.1 PPM	MARINE,EXCL.LARGE PREDATORS	1573
*FISH	0.5 PPM	FRESHWATER,PREDATORS,GREAT VARIATION	1573
*FRESHWATER SEDIMENT	0.49 MG/KG DRY WEIGHT	RANGE OF 90 % = 0.15 - 1.20 PPM, 74 LAKES, PELITIC	464
*GREY SEAL, LIVER	10-50 PPM	SEE REF.	681
*ICTALURIDAE	0.17 (SD=+-0.13) (RANGE = 0.04 - 0.50) MG/KG WET WEIGHT	11 REFERENCES USED	1779
*LUTJANIDAE	0.44 (SD=+-0.23) (RANGE = 0.28 - 0.60) MG/KG WET WEIGHT	2 REFERENCES USED	1779
*MELANOGRAMMUS AEGLEFINUS	0.07 (SD=+-0.06) (RANGE = 0.02 - 0.17) MG/KG WET WEIGHT	9 REFERENCES USED	1779
*MERLUCCIUS SP.	0.25 (SD=+-0.13) (RANGE = 0.09 - 0.45) MG/KG WET WEIGHT	10 REFERENCES USED	1779
*MOSS	0.4 MG/KG DRY WEIGHT	NO POLLUTION, GREAT SMOKY MOUNTAINS NATIONAL PARK	701
*MOSS	0.4 MG/KG DRY WEIGHT	NO POLLUTION, GREAT SMOKY MOUNTAINS NATIONAL PARK	701
*OYSTER	0.10 (SD=+-0.06) (RANGE = 0.03 - 0.23) MG/KG WET WEIGHT	9 REFERENCES USED	1779
*PELAGIC ORGANISMS	0.06 - 0.38 MG/KG DRY WEIGHT	AZORES	898
*PERCIDAE	1.18 (SD=+-2.32) (RANGE = 0.08 - 15.70) MG/KG WET WEIGHT	56 REFERENCES USED	1779
*PLANT	0.2 MG/KG AIR DRIED WEIGHT	NO POLLUTION, GREAT SMOKY MOUNTAINS NATIONAL PARK	701
*PLANT	0.2 MG/KG AIR DRIED WEIGHT	NO POLLUTION, GREAT SMOKY MOUNTAINS NATIONAL PARK	701
*PLEURONECTIDAE	0.22 (SD=+-0.20) (RANGE = 0.02 - 0.80) MG/KG WET WEIGHT	24 REFERENCES USED	1779
*POLLOCK	0.12 (SD=+-0.08) (RANGE = 0.05 - 0.23) MG/KG WET WEIGHT	4 REFERENCES USED	1779
*RIVER WATER	0.8*10**(-4) PPM	NONE	157
*SCALLOPS, SEA	0.04 (SD=+-0.008) (RANGE = 0.03 - 0.05) MG/KG WET WEIGHT	5 REFERENCES USED	1779
*SEAWATER	0.3*10**(-4) PPM	HGCL4(2-),RESIDENCE TIME= 42*10**3YRS	539
*SEAWATER	17 - 142 NG/L	AZORES, HIGHEST CONC IN DEEP WATER	898
*SHALE	0.4 MG/KG DRY WEIGHT	AVERAGE VALUE	1616
*SHRIMP, BROWN	0.70 (SD=+-2.02) (RANGE = 0.02 - 7.36) MG/KG WET WEIGHT	13 REFERENCES USED	1779
*THUNNUS SP.	0.51 (SD=+-0.30) (RANGE = 0.12 - 1.39) MG/KG WET WEIGHT	35 REFERENCES USED	1779

TABLE D44

BACKGROUND CONCENTRATION : HYDROGEN SULFIDE (H2S)			
ITEM:	VALUE:	CONDITIONS:	REF.:
*ATMOSPHERE	3-30 MIC.G/M**3	STP ,RESIDENCE TIME=0.11YRS	790
*SEAWATER	0-22 ML/L	SAT.SEAWATER,SALINITY=3.43%,283K, HIGH VAL.ONLY OVER ANOXIC BASIN	1445

TABLE D45

BACKGROUND CONCENTRATION : H2O			
ITEM:	VALUE:	CONDITIONS:	REF.:
*ATMOSPHERE	3*10**4 - 3*10**7 MIC.G/M**3	STP ,RESIDENCE TIME=0.027YRS	790

TABLE D45

BACKGROUND CONCENTRATION : H2O			
ITEM:	VALUE:	CONDITIONS:	REF.:
*ATMOSPHERE	3*10**5 G/CM**2	TOTAL SUPPLY = 1.5*10**24 G	779
*ATMOSPHERE	2 G/CM**2	TOTAL IN ATMOS.(1975) = 10**19 G	779

TABLE D46

BACKGROUND CONCENTRATION : I			
ITEM:	VALUE:	CONDITIONS:	REF.:
*RIVER WATER	0.002 PPM	NONE	157
*SEAWATER	0.06 PPM	I(-),IO3(-)?	539
*SOIL	5 PPM DRY SOIL	OVEN-DRIED	157

TABLE D47

BACKGROUND CONCENTRATION : IN			
ITEM:	VALUE:	CONDITIONS:	REF.:
*SEAWATER	<< 0.02 PPM	NONE	539

TABLE D48

BACKGROUND CONCENTRATION : K			
ITEM:	VALUE:	CONDITIONS:	REF.:
*ATMOSPHERIC PARTICULATES	120+-60 NG/M**3	HAWAII,N>=56	550
*RIVER WATER	2.3 PPM	RANGE=1.4-10 PPM	157
*RIVER, ASIA	3.8 MG/L	NONE	1689
*RIVER, AUSTRALIA	1.4 MG/L	NONE	1689
*RIVER, EUROPE	1.7 MG/L	NONE	1689
*RIVER, NORTH AMERICA	1.4 MG/L	NONE	1689
*RIVER, SOUTH AMERICA	2. MG/L	NONE	1689
*RIVER, WORLD	2.3 MG/L	NONE	1689
*SEAWATER	380 PPM	K(+),RESIDENCE TIME=11000*10**3YRS	539
*SEAWATER	1 % OF TOTAL	AS K2SO4, CHLORINITY = 1.9 %, 298 K, PH = 8.15	504
*SEAWATER	99 % OF TOTAL	AS FREE ION, CHLORINITY = 1.9 %, 298 K, PH = 8.15	504
*SOIL	14000 (400-30000) PPM DRY SOIL	OVEN-DRIED	157

TABLE D49

BACKGROUND CONCENTRATION : KR			
ITEM:	VALUE:	CONDITIONS:	REF.:
*ATMOSPHERE	1.1 MIC.G/M**3	STP	790

TABLE D49

BACKGROUND CONCENTRATION : KR			
ITEM:	VALUE:	CONDITIONS:	REF.:
*ATMOSPHERE	10**(-4)% MOL.FRACTION	SEA-LEVEL DRY ATMOSPHERIC COMPOSITION	1445
*LOWER ATMOSPHERE	0.00012% BY VOLUME	NONE	532
*SEAWATER	0.25*10**(-2) PPM	KR	539

TABLE D50

BACKGROUND CONCENTRATION : LA			
ITEM:	VALUE:	CONDITIONS:	REF.:
*SEAWATER	0.12*10**(-4) PPM	RESIDENCE TIME=0.44*10**3YRS	539
*SOIL	30 (1-5000) PPM DRY SOIL	OVEN-DRIED	157

TABLE D51

BACKGROUND CONCENTRATION : LI			
ITEM:	VALUE:	CONDITIONS:	REF.:
*FRESHWATER SEDIMENT	57 MG/KG DRY WEIGHT	RANGE OF 90 % = 15 - 200 PPM, 74 LAKES, PELITIC	464
*RIVER WATER	0.11*10**(-2) PPM	RANGE=0.7*10**(-4)-0.04 PPM	157
*SEAWATER	0.18 PPM	LI(+),RESIDENCE TIME= 20000*10**3YRS	539
*SHALE	66 MG/KG DRY WEIGHT	AVERAGE VALUE	1616
*SOIL	30 (7-200) PPM DRY SOIL	OVEN-DRIED	157

TABLE D52

BACKGROUND CONCENTRATION : LU			
ITEM:	VALUE:	CONDITIONS:	REF.:
*SEAWATER	0.0007 MICROGRAM/L	TOTAL CONCENTRATION, JAPAN, CL = 1.873 %	1535
*SEAWATER	0.0003 MICROGRAM/L	SOLUBLE CONCENTRATION, JAPAN, CL = 1.873 %	1535

TABLE D53

BACKGROUND CONCENTRATION : MG			
ITEM:	VALUE:	CONDITIONS:	REF.:
*ATMOSPHERIC PARTICULATES	410+-240 NG/M**3	HAWAII,N>=56	550
*RIVER WATER	4.1 PPM	RANGE=1.5-6 PPM	157
*RIVER, AFRICA	3.8 MG/L	NONE	1689
*RIVER, ASIA	5.6 MG/L	NONE	1689
*RIVER, AUSTRALIA	2.7 MG/L	NONE	1689
*RIVER, EUROPE	5.6 MG/L	NONE	1689
*RIVER, NORTH AMERICA	5. MG/L	NONE	1689
*RIVER, SOUTH AMERICA	1.5 MG/L	NONE	1689

TABLE D53

BACKGROUND CONCENTRATION : MG			
ITEM:	VALUE:	CONDITIONS:	REF.:
*RIVER, WORLD	4.1 MG/L	NONE	1689
*SEAWATER	1350 PPM	MG(2+),RESIDENCE TIME= 45000*10**3YRS	539
*SEAWATER	87 % OF TOTAL	AS FREE ION, CHLORINITY = 1.9 %, 298 K, PH = 8.15	504
*SEAWATER	1 % OF TOTAL	AS MG(HCO3)2, CHLORINITY = 1.9 %, 298 K, PH = 8.15	504
*SEAWATER	11 % OF TOTAL	AS MGSO4, CHLORINITY = 1.9 %, 298 K, PH = 8.15	504
*SEAWATER	0.3 % OF TOTAL	AS MGCO3(AQ), CHLORINITY = 1.9 %, 298 K, PH = 8.15	504
*SOIL	5000 (600-6000) PPM DRY SOIL	OVEN-DRIED	157

TABLE D54

BACKGROUND CONCENTRATION : MN			
ITEM:	VALUE:	CONDITIONS:	REF.:
*ATMOSPHERIC PARTICULATES	0.19+-0.14 NG/M**3	HAWAII,N>=56	550
*FRESHWATER SEDIMENT	882 MG/KG DRY WEIGHT	RANGE OF 90 % = 100 - 1800 PPM, 74 LAKES, PELITIC	464
*RIVER WATER	0.012 PPM	RANGE=0.2*10**(-4)-0.13 PPM	157
*SEAWATER	0.002 PPM	MN(2+),RESIDENCE TIME=1.4*10**3YRS	539
*SHALE	850 MG/KG DRY WEIGHT	AVERAGE VALUE	1616
*SOIL	850 (100-4000) PPM DRY SOIL	OVEN-DRIED	157

TABLE D55

BACKGROUND CONCENTRATION : MO			
ITEM:	VALUE:	CONDITIONS:	REF.:
*BEAN SPROUTS	0.61 MG/KG DRY WEIGHT	NORMAL LEAVES	1718
*CABBAGE, LEAF	0.42 MG/KG DRY WEIGHT	NORMAL LEAVES	1718
*CABBAGE, LEAF	0.09 MG/KG DRY WEIGHT	DEFICIENT LEAVES	1718
*CAULIFLOWER, LEAVES	0.10 - 3.0 MG/KG DRY WEIGHT	NORMAL LEAVES, HIGH MN IN LEAVES	1718
*CAULIFLOWER, LEAVES	0.02 - 0.08 MG/KG DRY WEIGHT	DEFICIENT LEAVES, HIGH MN IN LEAVES	1718
*CAULIFLOWER, LEAVES	1.0 MG/KG DRY WEIGHT	NORMAL LEAVES, MN TOXICITY COMPLICATING FACTOR	1718
*CAULIFLOWER, LEAVES	0.4 - 0.5 MG/KG DRY WEIGHT	DEFICIENT LEAVES, MN TOXICITY COMPLICATING FACTOR	1718
*CITRUS, LEAVES	0.03 - 0.08 MG/KG DRY WEIGHT	DEFICIENT LEAVES, SYMPTOMS PRONOUNCED BELOW SOIL PH = 5.6	1718
*CITRUS, LEAVES	0.02 - 0.03 MG/KG DRY WEIGHT	DEFICIENT LEAVES, SEVERLY AFFECTED LEAVES, IN WATER CULTURE	1718
*HOPS, LEAVES	0.04 MG/KG DRY WEIGHT	NORMAL LEAVES	1718
*HOPS, LEAVES	0.02 MG/KG DRY WEIGHT	DEFICIENT LEAVES, WHITE INTERVEINAL BREAKDOWN OF LEAVES	1718

TABLE D55

BACKGROUND CONCENTRATION : MO			
ITEM:	VALUE:	CONDITIONS:	REF.:
*LETTUCE, LEAVES	0.10 MG/KG DRY WEIGHT	NORMAL LEAVES, VERY LARGE YIELD INCREASE	1718
*LETTUCE, LEAVES	0.06 MG/KG DRY WEIGHT	DEFICIENT LEAVES, VERY LARGE YIELD INCREASE	1718
*RIVER WATER	0.35*10**(-4) PPM	RANGE:< 0.007 PPM	157
*SEAWATER	0.01 PPM	MOO4(2-),RESIDENCE TIME= 500*10**3YRS	539
*SOIL	2 (0.2-5) PPM DRY SOIL	OVEN-DRIED	157

TABLE D56

BACKGROUND CONCENTRATION : N			
ITEM:	VALUE:	CONDITIONS:	REF.:
*AMBIENT CONCENTRATION	0.25 PPM	COMPOUND=N2O,ATMOSPHERIC MASS AS N =1500*10**6TONS	1348
*AMBIENT CONCENTRATION	4 PPB AS NO2	COMPOUND=NO/NO2,LAND,ATMOSPHERIC MASS AS N=12*10**6TONS	1348
*AMBIENT CONCENTRATION	1 PPB	COMPOUND=NO/NO2,OCEAN,ATMOSPHERIC MASS AS N=3*10**6TONS	1348
*AMBIENT CONCENTRATION	6 PPB	COMPOUND=NH3,ATMOSPHERIC MASS AS N =30*10**6TONS	1348
*AMBIENT CONCENTRATION	0.2 UG/M**3	COMPOUND=NO3,ATMOSPHERIC MASS AS N =0.2*10**6TONS	1348
*AMBIENT CONCENTRATION	1.0 UG/M**3	COMPOUND=NH4,ATMOSPHERIC MASS AS N =4.1*10**6TONS	1348
*ATMOSPHERE	9.73*10**10 UG/M**3	STP N2	790
*ATMOSPHERE	78.09% MOL.FRACTION	SEA-LEVEL DRY ATMOSPHERIC COMPOSITION N2	1445
*ATMOSPHERE	8*10**2 G/CM**2	TOTAL IN ATMOS.(1975) = 4*10**21 G, N2	779
*ATMOSPHERE	0-15 UG/M**3	STP ,RESIDENCE TIME=SHORT, NH3	790
*ATMOSPHERE	0-6 UG/M**3	STP ,RESIDENCE TIME=SHORT, NO2	790
*ATMOSPHERE	9.73*10**10 UG/M**3	STP , N2	790
*ATMOSPHERE	500-1200 UG/M**3	STP ,RESIDENCE TIME=4YRS, N2O	790
*LOWER ATMOSPHERE	78.084% BY VOLUME	N2	532
*LOWER ATMOSPHERE	0.00005% BY VOLUME	N2O	532
*RIVER WATER	0.23 PPM	NO3(-),RANGE=0.01-0.8 PPM	157
*RIVER, AFRICA	0.8 MG/L	NO3(-)	1689
*RIVER, ASIA	0.7 MG/L	NO3(-)	1689
*RIVER, AUSTRALIA	0.05 MG/L	NO3(-)	1689
*RIVER, EUROPE	3.7 MG/L	NO3(-)	1689
*RIVER, NORTH AMERICA	1. MG/L	NO3(-)	1689
*RIVER, SOUTH AMERICA	0.7 MG/L	NO3(-)	1689
*RIVER, WORLD	1. MG/L	NO3(-)	1689
*SEAWATER	0.5 PPM	ORGANIC N,NO3(-),NH4(+),RESIDENCE TIME=2.5*10**3YRS	539
*SEAWATER	11.56 ML/L	SAT.SEAWATER,SALINITY=3.43%,TEMP.= 283K N2	1445
*SOIL	1000 (200-2500) PPM DRY SOIL	OVEN-DRIED	157

TABLE D57

BACKGROUND CONCENTRATION : NA			
ITEM:	VALUE:	CONDITIONS:	REF.:
*ATMOSPHERIC PARTICULATES	3500+-2000 NG/M**3	HAWAII,N>=56	550

TABLE D57

BACKGROUND CONCENTRATION : NA			
ITEM:	VALUE:	CONDITIONS:	REF.:
*RIVER WATER	6.3 PPM	RANGE=3-25 PPM	157
*RIVER, AFRICA	11. MG/L	NONE	1689
*RIVER, ASIA	5.5 MG/L	NONE	1689
*RIVER, AUSTRALIA	2.9 MG/L	NONE	1689
*RIVER, EUROPE	5.4 MG/L	NONE	1689
*RIVER, NORTH AMERICA	9. MG/L	NONE	1689
*RIVER, SOUTH AMERICA	4. MG/L	NONE	1689
*RIVER, WORLD	6.3 MG/L	NONE	1689
*SEAWATER	10500 PPM	NA(+),RESIDENCE TIME= 260000*10**3YRS	539
*SEAWATER	99 % OF TOTAL	AS FREE ION, CHLORINITY = 1.9 %, 298 K, PH = 8.15	504
*SEAWATER	1 % OF TOTAL	AS NA2SO4, CHLORINITY = 1.9 %, 298 K, PH = 8.15	504
*SOIL	6300 (750-7500) PPM DRY SOIL	OVEN-DRIED	157

TABLE D58

BACKGROUND CONCENTRATION : NB			
ITEM:	VALUE:	CONDITIONS:	REF.:
*SEAWATER	0.1*10**(-4) PPM	RESIDENCE TIME=0.3*10**3YRS	539

TABLE D59

BACKGROUND CONCENTRATION : NE			
ITEM:	VALUE:	CONDITIONS:	REF.:
*ATMOSPHERE	18 MIC.G/M**3	STP	790
*ATMOSPHERE	2*10**(-3)% MOL.FRACTION	SEA-LEVEL DRY ATMOSPHERIC COMPOSITION	1445
*LOWER ATMOSPHERE	0.00182% BY VOLUME	NONE	532
*SEAWATER	0.14*10**(-3) PPM	NE	539
*SEAWATER	1.7*10**(-4) ML/L	SAT.SEAWATER,SALINITY=3.43%,TEMP.= 283K	1445

TABLE D60

BACKGROUND CONCENTRATION : NI			
ITEM:	VALUE:	CONDITIONS:	REF.:
*FRESHWATER SEDIMENT	103 MG/KG DRY WEIGHT	RANGE OF 90 % = 30 - 250 PPM, 74 LAKES, PELITIC	464
*RIVER WATER	0.01 PPM	RANGE=0.2*10**(-3)-0.02 PPM	157
*SEAWATER	0.54*10**(-2) PPM	NI(2+),RESIDENCE TIME=18*10**3YRS	539
*SHALE	68 MG/KG DRY WEIGHT	AVERAGE VALUE	1616
*SOIL	40 (10-1000) PPM DRY SOIL	OVEN-DRIED	157

TABLE D61

BACKGROUND CONCENTRATION : NIACIN (NICOTINIC ACIDAMIDE)			
ITEM:	VALUE:	CONDITIONS:	REF.:
*FRESHWATER	0.15 - 0.89 MG/M**3	LINDSEY POND, CONN.	99
*FRESHWATER	0.30 - 3.3 MG/M**3	KASUMIGAURA, JAPAN	801
*FRESHWATER	<0.001 - 3.0 MG/M**3	SMALL SWISS PONDS	257
*FRESHWATER	2.0 MG/M**3	MEAN OF PRECIPITATION, MO., APRIL-NOV.	1213
*FRESHWATER	0.42 MG/M**3	MEAN OF PRECIPATION, MO., NOV-MARCH	1213

TABLE D62

BACKGROUND CONCENTRATION : O2			
ITEM:	VALUE:	CONDITIONS:	REF.:
*ATMOSPHERE	2.99*10**10 MIC.G/M**3	STP	790
*ATMOSPHERE	20.95% MOL.FRACTION	SEA-LEVEL DRY ATMOSPHERIC COMPOSITION	1445
*ATMOSPHERE	5*10**3 G/CM**2	TOTAL SUPPLY = 2.7*10**22 G	779
*ATMOSPHERE	2*10**2 G/CM**2	TOTAL IN ATMOS.(1975) = 10**21 G	1421
*LOWER ATMOSPHERE	20.946% BY VOLUME	NONE	532
*SEAWATER	857000 PPM	H20,02,SO4(2-)	539
*SEAWATER	6.44 ML/L	SAT.SEAWATER,SALINITY=3.43%,TEMP.= 283K	1445
*SOIL	490000 PPM DRY SOIL	OVEN-DRIED	157

TABLE D63

BACKGROUND CONCENTRATION : O3			
ITEM:	VALUE:	CONDITIONS:	REF.:
*ATMOSPHERE	10**(-6)% MOL.FRACTION	SEA-LEVEL DRY ATMOSPHERIC COMPOSITION	1445
*ATMOSPHERE	0-100 MIC.G/M**3	STP ,RESIDENCE TIME=2YRS	790

TABLE D64

BACKGROUND CONCENTRATION : P			
ITEM:	VALUE:	CONDITIONS:	REF.:
*RIVER WATER	0.005 PPM	RANGE=0.001-0.3(-12) PPM	157
*SEAWATER	0.07 PPM	HPO4(2-)	539
*SEAWATER	2*10**(-9) PPM	NONE	539
*SOIL	650 PPM DRY SOIL	OVEN-DRIED	157

TABLE D65

BACKGROUND CONCENTRATION : PANTOTHENIC ACID			
ITEM:	VALUE:	CONDITIONS:	REF.:
*FRESHWATER	0.01 - 0.26 MG/M**3	KASUMIGAURA, JAPAN	801

TABLE D65

BACKGROUND CONCENTRATION : PANTOTHENIC ACID			
ITEM:	VALUE:	CONDITIONS:	REF.:
*FRESHWATER	<0.01 - 0.034 MG/M**3	SMALL SWISS PONDS	257

TABLE D66

BACKGROUND CONCENTRATION : PB			
ITEM:	VALUE:	CONDITIONS:	REF.:
*ATMOSPHERE	0.5 NG/M**3	NONE	893
*ATMOSPHERIC PARTICULATES	3.0+-2.8 NG/M**3	HAWAII,N>=56	550
*CLAMS, ALL SPECIES	1.60 (SD=+-2.28) (RANGE = 0.15 - 7.37) MG/KG WET WEIGHT	9 REFERENCES USED	1779
*COASTRAL, NEAR FLORIDA	19.9 - 41.8 MG PB/M**3	DISSOLVED PB	633
*CRAB, BLUE	0.54 (SD=+-0.18) (RANGE = 0.30 - 0.81) MG/KG WET WEIGHT	5 REFERENCES USED	1779
*CREECK IN FLORIDA	200 MG PB/M**3	DISSOLVED PB	633
*FRESHWATER SEDIMENT	31 MG/KG DRY WEIGHT	RANGE OF 90 % = 8 - 75 PPM, 74 LAKES, PELITIC	464
*GADUS MORHUA	0.28 (SD=+-0.07) (RANGE = 0.20 - 0.34) MG/KG WET WEIGHT	3 REFERENCES USED	1779
*ICTALURIDAE	0.64 (SD=+-0.06) (RANGE = 0.59 - 0.68) MG/KG WET WEIGHT	2 REFERENCES USED	1779
*LUTJANIDAE	0.48 MG/KG WET WEIGHT	1 REFERENCE USED	1779
*MARINE, ENGLISH CHANNAL	0.2 MG PB/M**3	DISSOLVED PB	1278
*MARINE, IRISH SEA	0.1 MG PB/M**3	DISSOLVED PB	1278
*MARINE, NEAR ENGLAND	0.3 - 3.6 MG PB/M**3	DISSOLVED PB	213
*MARINE, NORTH SEA	0.2 MG PB/M**3	DISSOLVED PB	1278
*MARINE, 300 KM FROM CALIFORNIA	36 MG PB/M**3	DISSOLVED PB	633
*MELANOGRAMMUS AEGLEFINUS	0.33 (SD=+-0.17) (RANGE = 0.20 - 0.53) MG/KG WET WEIGHT	3 REFERENCES USED	1779
*MERLUCCIUS SP.	0.44 (SD=+-0.10) (RANGE = 0.38 - 0.51) MG/KG WET WEIGHT	2 REFERENCES USED	1779
*OYSTER	24.92 (SD=+-38.97) (RANGE = 0.20 - 100) MG/KG WET WEIGHT	12 REFERENCES USED	1779
*PERCIDAE	0.55 (SD=+-0.07) (RANGE = 0.50 - 0.60) MG/KG WET WEIGHT	2 REFERENCES USED	1779
*PLEURONECTIDAE	0.36 (SD=+-0.17) (RANGE = 0.20 - 0.53) MG/KG WET WEIGHT	3 REFERENCES USED	1779
*POLLOCK	1.64 MG/KG WET WEIGHT	1 REFERENCE USED	1779
*RIVER WATER	0.005 PPM	RANGE=0.6*10**(-3)-0.12 PPM	157
*RIVER, ENGLAND	4.5 - 8.7 MG PB/M**3	DISSOLVED PB	405
*RIVER, USA	4 MG PB/M**3	DISSOLVED PB	633
*SCALLOPS, SEA	0.32 (SD=+-0.12) (RANGE = 0.15 - 0.47) MG/KG WET WEIGHT	6 REFERENCES USED	1779
*SEAWATER	0.3*10**(-4) PPM	PB(2+),RESIDENCE TIME=2*10**3YRS	539
*SHALE	20 MG/KG DRY WEIGHT	AVERAGE VALUE	1616
*SHRIMP, BROWN	0.46 (SD=+-0.06) (RANGE = 0.39 - 0.52) MG/KG WET WEIGHT	4 REFERENCES USED	1779
*SOIL	10 (2-200) PPM DRY SOIL	OVEN-DRIED	157

TABLE D67

BACKGROUND CONCENTRATION : POLYNUCLEAR AROMATIC HYDROCARBONS			
ITEM:	VALUE:	CONDITIONS:	REF.:
*LAKE WATER	10 NG/L	LAKE CONSTANCE, CENTRIFUGED WATER	355

TABLE D67

BACKGROUND CONCENTRATION : POLYNUCLEAR AROMATIC HYDROCARBONS			
ITEM:	VALUE:	CONDITIONS:	REF.:
*RIVER WATER	800 NG/L	RIVER MAIN, CENTRIFUGED WATER	355
*TAP WATER	7 NG/L	MAINZ	355
*WELL WATER	25 NG/L	SOUTH OF FRANKFURT	355

TABLE D68

BACKGROUND CONCENTRATION : RA			
ITEM:	VALUE:	CONDITIONS:	REF.:
*RIVER WATER	3.9*10**(-10) PPM	NONE	157
*SEAWATER	6*10**(-11) PPM	NONE	539
*SOIL	8*10**(-7) (3-20*10**(-7)) PPM DRY SOIL	OVEN-DRIED	157

TABLE D69

BACKGROUND CONCENTRATION : RB			
ITEM:	VALUE:	CONDITIONS:	REF.:
*RIVER WATER	0.15*10**(-2) PPM	RANGE:< 0.008	157
*SEAWATER	0.12 PPM	RB(+),RESIDENCE TIME=270*10**3YRS	539
*SOIL	100 (20-600) PPM DRY SOIL	OVEN-DRIED	157

TABLE D70

BACKGROUND CONCENTRATION : RN			
ITEM:	VALUE:	CONDITIONS:	REF.:
*ATMOSPHERE	6*10**(-8)% MOL.FRACTION	SEA-LEVEL DRY ATMOSPHERIC COMPOSITION	1445
*RIVER WATER	1.7*10**(-15) PPM	NONE	157
*SEAWATER	6*10**(-16) PPM	RN	539

TABLE D71

BACKGROUND CONCENTRATION : S			
ITEM:	VALUE:	CONDITIONS:	REF.:
*ATMOSPHERE	3-50 UG/M**3	STP	790
*ATMOSPHERE	0-50 UG/M**3	STP ,RESIDENCE TIME=0.014YRS, SO2	790
*RIVER WATER	3.7 PPM	RANGE=0.9-30 PPM	157
*RIVER, AFRICA	13.5 MG/L	SO4(2-)	1689
*RIVER, ASIA	8.4 MG/L	SO4(2-)	1689
*RIVER, AUSTRALIA	2.6 MG/L	SO4(2-)	1689
*RIVER, EUROPE	24. MG/L	SO4(2-)	1689
*RIVER, NORTH AMERICA	20. MG/L	SO4(2-)	1689
*RIVER, SOUTH AMERICA	4.8 MG/L	SO4(2-)	1689

TABLE D71

BACKGROUND CONCENTRATION : S			
ITEM:	VALUE:	CONDITIONS:	REF.:
*RIVER, WORLD	11.2 MG/L	SO4(2-)	1689
*SEAWATER	885 PPM	SO4(2-)	539
*SEAWATER	54 % OF TOTAL	AS FREE ION, CHLORINITY = 1.9 %, 298 K, PH = 8.15, OF SO4 POOL	504
*SEAWATER	22 % OF TOTAL	AS MGS, CHLORINITY = 1.9 %, 298 K, PH = 8.15, OF SO4 POOL	504
*SEAWATER	0.5 % OF TOTAL	AS K2S, CHLORINITY = 1.9 %, 298 K, PH = 8.15, OF SO4 POOL	504
*SEAWATER	3 % OF TOTAL	AS CAS, CHLORINITY = 1.9 %, 298 K, PH = 8.15, OF SO4 POOL	504
*SEAWATER	21 % OF TOTAL	AS NA2S, CHLORINITY = 1.9 %, 298 K, PH = 8.15, OF SO4 POOL	504
*SOIL	700 (30-900) PPM DRY SOIL	OVEN-DRIED	157
*TROPOSPHERE	0.28 UG SO2-S/M**3	AS SO2: 0.2PPB	1348
*TROPOSPHERE	0.28 UG H2S-S/M**3	AS H2S: 0.2PPB	1348
*TROPOSPHERE	0.7 UG SO4-S/M**3	AS SO4: 2 UG/M**3	1348

TABLE D72

BACKGROUND CONCENTRATION : SB			
ITEM:	VALUE:	CONDITIONS:	REF.:
*PELAGIC ORGANISMS	0.1 MG/KG DRY WEIGHT	AZORES	898
*SEAWATER	0.33*10**(-3) PPM	RESIDENCE TIME=350*10**3YRS	539
*SOIL	2 - 10 PPM OF DRY SOIL	OVEN-DRIED	157

TABLE D73

BACKGROUND CONCENTRATION : SC			
ITEM:	VALUE:	CONDITIONS:	REF.:
*SEAWATER	< 0.4*10**(-5) PPM	RESIDENCE TIME=5.6*10**3YRS	539
*SOIL	7 (10-25) PPM DRY SOIL	OVEN-DRIED	157

TABLE D74

BACKGROUND CONCENTRATION : SE			
ITEM:	VALUE:	CONDITIONS:	REF.:
*RIVER WATER	< 0.02 PPM	NONE	157
*SEAWATER	0.9*10**(-4) PPM	NONE	539
*SOIL	0.2 (0.01-2) PPM DRY SOIL	OVEN-DRIED	157

TABLE D75

BACKGROUND CONCENTRATION : SI			
ITEM:	VALUE:	CONDITIONS:	REF.:
*RIVER WATER	6.5 PPM	RANGE=2-12 PPM , AS SI	157
*RIVER, AFRICA	23.2 MG/L	AS SIO2	1689
*RIVER, ASIA	11.7 MG/L	AS SIO2	1689
*RIVER, AUSTRALIA	3.9 MG/L	AS SIO2	1689
*RIVER, EUROPE	7.5 MG/L	AS SIO2	1689
*RIVER, NORTH AMERICA	9. MG/L	AS SIO2	1689
*RIVER, SOUTH AMERICA	11.9 MG/L	AS SIO2	1689
*RIVER, WORLD	13.1 MG/L	AS SIO2	1689
*SEAWATER	3 PPM	SI(OH)4,RESIDENCE TIME=8*10**3YRS , AS SI	539
*SOIL	330000 (250000-350000) PPM DRY SOIL	OVEN-DRIED , AS SI	157

TABLE D76

BACKGROUND CONCENTRATION : SN			
ITEM:	VALUE:	CONDITIONS:	REF.:
*RIVER WATER	0.4*10**(-4) PPM	NONE	157
*SEAWATER	0.003 PPM	RESIDENCE TIME=100*10**3YRS	539
*SOIL	10 (2-200) PPM DRY SOIL	OVEN-DRIED	157

TABLE D77

BACKGROUND CONCENTRATION : SR			
ITEM:	VALUE:	CONDITIONS:	REF.:
*ATMOSPHERIC PARTICULATES	2.4+-1.5 NG/M**3	HAWAII,N>=56	550
*FRESHWATER SEDIMENT	216 MG/KG DRY WEIGHT	RANGE OF 90 % = 55 - 600 PPM, 74 LAKES, PELITIC	464
*RIVER WATER	0.08 PPM	RANGE=0.003-0.8 PPM	157
*SEAWATER	8.1 PPM	SR(9+),RESIDENCE TIME= 19000*10**3YRS	539
*SHALE	300 MG/KG DRY WEIGHT	AVERAGE VALUE	1616
*SOIL	300 (50-1000) PPM DRY SOIL	OVEN-DRIED	157

TABLE D78

BACKGROUND CONCENTRATION : TA			
ITEM:	VALUE:	CONDITIONS:	REF.:
*SEAWATER	< 0.25*10**(-5) PPM	NONE	539

TABLE D79

BACKGROUND CONCENTRATION : TB			
ITEM:	VALUE:	CONDITIONS:	REF.:
*SEAWATER	0.0042 MICROGRAM/L	TOTAL CONCENTRATION, JAPAN, CL = 1.873 %	1535

TABLE D79

BACKGROUND CONCENTRATION : TB			
ITEM:	VALUE:	CONDITIONS:	REF.:
*SEAWATER	0.0008 MICROGRAM/L	SOLUBLE CONCENTRATION, JAPAN, CL = 1.873 %	1535

TABLE D80

BACKGROUND CONCENTRATION : TH			
ITEM:	VALUE:	CONDITIONS:	REF.:
*MARINE ALGAE	10**(-6) - 10**(-8) G TH/G ASH WEIGHT	ADRIATIC SEA, FILTRED WATER	1524
*RIVER WATER	0.0*10**(-4) PPM	NONE	157
*SEAWATER	0.5*10**(-4) PPM	RESIDENCE TIME=0.35*10**3YRS	539
*SEAWATER	8*10**(-13) G TH/L	ADRIATIC SEA, FILTRED WATER	1524
*SEAWATER	8*10**(-10) G TH/L	ADRIATIC SEA, UNFILTRED WATER	1524
*SOIL	5 (0.1-12) PPM DRY SOIL	OVEN-DRIED	157

TABLE D81

BACKGROUND CONCENTRATION : THIAMINE			
ITEM:	VALUE:	CONDITIONS:	REF.:
*FRESHWATER	0.008 - 0.077 MG/M**3	LINDSEY POND, CONN.	99
*FRESHWATER	0.05 - 12 MG/M**3	NORTHERN GERMAN LAKES	588
*FRESHWATER	0.001 - 0.38 MG/M**3	SAGAMI, JAPAN	1178
*FRESHWATER	0.075 - 0.436 MG/M**3	DISSOLVED, TSUKUI, JAPAN, SURFACE	1179
*FRESHWATER	0.031 - 0.159 MG/M**3	PARTICULATE, TSUKUI, JAPAN, SURFACE	1179
*FRESHWATER	<0.001 - 1.0 MG/M**3	SMALL SWISS PONDS	257

TABLE D82

BACKGROUND CONCENTRATION : TI			
ITEM:	VALUE:	CONDITIONS:	REF.:
*RIVER WATER	0.86*10**(-2) PPM	RANGE:< 0.11 PPM	157
*SEAWATER	0.001 PPM	RESIDENCE TIME=0.16*10**3YRS	539
*SOIL	5000 (1000-10000) PPM DRY SOIL	OVEN-DRIED	157

TABLE D83

BACKGROUND CONCENTRATION : TL			
ITEM:	VALUE:	CONDITIONS:	REF.:
*SEAWATER	< 0.1*10**(-4) PPM	TL(+)	539
*SOIL	0.1 PPM DRY SOIL	OVEN-DRIED	157

TABLE D84

BACKGROUND CONCENTRATION : U			
ITEM:	VALUE:	CONDITIONS:	REF.:
*RIVER WATER	0.001 PPM	RANGE=0.2*10**(-4)-0.05 PPM	157

TABLE D84

BACKGROUND CONCENTRATION : U			
ITEM:	VALUE:	CONDITIONS:	REF.:
*SEAWATER	0.003 PPM	UO2(CO3)3(4-),RESIDENCE TIME= 500*10**3YRS	539
*SOIL	1 (0.9-9) PPM DRY SOIL	OVEN-DRIED	157

TABLE D85

BACKGROUND CONCENTRATION : V			
ITEM:	VALUE:	CONDITIONS:	REF.:
*ATMOSPHERIC PARTICULATES	0.16+-0.11 NG/M**3	HAWAII,N>=56	550
*RIVER WATER	0.001 PPM	RANGE:< 0.007 PPM	157
*SEAWATER	0.002 PPM	VO5H3(2-),RESIDENCE TIME= 10*10**3YRS	539
*SOIL	100 (20-500) PPM DRY SOIL	OVEN-DRIED	157

TABLE D86

BACKGROUND CONCENTRATION : VITAMIN B12			
ITEM:	VALUE:	CONDITIONS:	REF.:
*FRESHWATER	0.06 - 0.075 MG/M**3	LINDSEY POND, CONN.	99
*FRESHWATER	0.005 - 0.85 MG/M**3	SAGAMI, JAPAN	1178
*FRESHWATER	0.0005 - 0.0042 MG/M**3	DISSOLVED, TSUKUI, JAPAN, SURFACE	1179
*FRESHWATER	0.0019 - 0.0203 MG/M**3	PARTICULATE, TSUKUI, JAPAN, SURFACE	1179
*FRESHWATER	0.005 - 0.028 MG/M**3	KASUMIGAURA, JAPAN	801
*FRESHWATER	0.001 MG/M**3	MEAN OF PRECIPITATION, MO., APRIL-NOV.	1213
*FRESHWATER	0.0004 MG/M**3	MEAN OF PRECIPATION, MO., NOV-MARCH	1213

TABLE D87

BACKGROUND CONCENTRATION : W			
ITEM:	VALUE:	CONDITIONS:	REF.:
*SEAWATER	0.1*10**(-3) PPM	WO4(2-),RESIDENCE TIME=1*10**3YRS	539

TABLE D88

BACKGROUND CONCENTRATION : XE			
ITEM:	VALUE:	CONDITIONS:	REF.:
*ATMOSPHERE	0.086 UG/M**3	STP	790
*ATMOSPHERE	8*10**(-6)% MOL.FRACTION	SEA-LEVEL DRY ATMOSPHERIC COMPOSITION	1445
*LOWER ATMOSPHERE	0.00009% BY VOLUME	NONE	532
*SEAWATER	0.52*10**(-4) PPM	XE	539

TABLE D89

BACKGROUND CONCENTRATION : Y			
ITEM:	VALUE:	CONDITIONS:	REF.:
*SEAWATER	0.3*10**(-3) PPM	RESIDENCE TIME=7.5*10**3YRS	539

TABLE D89

BACKGROUND CONCENTRATION : Y			
ITEM:	VALUE:	CONDITIONS:	REF.:
*SOIL	50 (25-250) PPM DRY SOIL	OVEN-DRIED	157

TABLE D90

BACKGROUND CONCENTRATION : YB			
ITEM:	VALUE:	CONDITIONS:	REF.:
*SEAWATER	0.0056 MICROGRAM/L	TOTAL CONCENTRATION, JAPAN, CL = 1.873 %	1535
*SEAWATER	0.0021 MICROGRAM/L	SOLUBLE CONCENTRATION, JAPAN, CL = 1.873 %	1535

TABLE D91

BACKGROUND CONCENTRATION : ZN			
ITEM:	VALUE:	CONDITIONS:	REF.:
*FRESHWATER SEDIMENT	123 MG/KG DRY WEIGHT	RANGE OF 90 % = 45 - 220 PPM, 74 LAKES, PELITIC	464
*RIVER WATER	0.01 PPM	RANGE=0.2*10**(-3)-1 PPM	157
*SEAWATER	0.01 PPM	ZN(2+),RESIDENCE TIME=180*10**3YRS	539
*SHALE	95 MG/KG DRY WEIGHT	AVERAGE VALUE	1616
*SOIL	50 (10-300) PPM DRY SOIL	OVEN-DRIED	157

TABLE D92

BACKGROUND CONCENTRATION : ZR			
ITEM:	VALUE:	CONDITIONS:	REF.:
*RIVER WATER	0.26*10**(-2) PPM	RANGE=0.5*10**(-4)-0.022 PPM	157
*SEAWATER	0.22*10**(-4) PPM	NONE	539
*SOIL	300 (60-2000) PPM DRY SOIL	OVEN-DRIED	157

D2

CONCENTRATION FACTOR

ELEMENTS
AND
CHEMICAL
COMPOUNDS

TABLE D93

CONCENTRATION FACTOR : AG			
ITEM:	VALUE:	CONDITIONS:	REF.:
*BROWN ALGAE	240 BASED ON WET WEIGHT	NONE	157
*DAPHNIA MAGNA	9.8 BASED ON WET WEIGHT	AG=0.5 MG/L,1 W	1563
*DAPHNIA MAGNA	32 BASED ON WET WEIGHT	AG=0.5 MG/L,3 W	1563
*DAPHNIA MAGNA	36 BASED ON WET WEIGHT	AG=0.5 MG/L,5 W	1563
*DAPHNIA MAGNA	24 BASED ON WET WEIGHT	AG=0.5 MG/L,7 W	1563
*DAPHNIA MAGNA	26 BASED ON WET WEIGHT	AG=0.5 MG/L,10 W	1563
*DAPHNIA MAGNA	5 BASED ON WET WEIGHT	AG=5.0 MG/L,1 W	1563
*DAPHNIA MAGNA	20 BASED ON WET WEIGHT	AG=5.0 MG/L,3 W	1563
*DAPHNIA MAGNA	28 BASED ON WET WEIGHT	AG=5.0 MG/L,5 W	1563
*DAPHNIA MAGNA	11 BASED ON WET WEIGHT	AG=5.0 MG/L,7 W	1563
*DAPHNIA MAGNA	9 BASED ON WET WEIGHT	AG=5.0 MG/L,10 W	1563
*FUCUS VESICULOSUS	5*10**3 BASED ON DRY WEIGHT	18 OBS.,ALL AREAS AROUND THE BRITISH ISLES	1278
*LIGUMIA SP. AND MARGARITIFERA SP.	3 BASED ON WET WEIGHT	AG=0.5 MG/L,1 W,BODY WEIGHT =5.67 G	1563
*LIGUMIA SP. AND MARGARITIFERA SP.	3.4 BASED ON WET WEIGHT	AG=0.5 MG/L,3 W,BODY WEIGHT =3.05 G	1563
*LIGUMIA SP. AND MARGARITIFERA SP.	3.6 BASED ON WET WEIGHT	AG=0.5 MG/L,5 W,BODY WEIGHT =4.11 G	1563
*LIGUMIA SP. AND MARGARITIFERA SP.	8.0 BASED ON WET WEIGHT	AG=0.5 MG/L,7 W,BODY WEIGHT =4.60 G	1563
*LIGUMIA SP. AND MARGARITIFERA SP.	6.6 BASED ON WET WEIGHT	AG=0.5 MG/L,10 W,BODY WEIGHT =3.12 G	1563
*LIGUMIA SP. AND MARGARITIFERA SP.	0.5 BASED ON WET WEIGHT	AG=5.0 MG/L,1 W,BODY WEIGHT =6.70 G	1563
*LIGUMIA SP. AND MARGARITIFERA SP.	1.9 BASED ON WET WEIGHT	AG=5.0 MG/L,3 W,BODY WEIGHT =2.67 G	1563
*LIGUMIA SP. AND MARGARITIFERA SP.	13.2 BASED ON WET WEIGHT	AG=5.0 MG/L,5 W,BODY WEIGHT =2.06 G	1563
*LIGUMIA SP. AND MARGARITIFERA SP.	22.0 BASED ON WET WEIGHT	AG=5.0 MG/L,7 W,BODY WEIGHT =5.16 G	1563
*LIGUMIA SP. AND MARGARITIFERA SP.	9.8 BASED ON WET WEIGHT	AG=5.0 MG/L,10 W,BODY WEIGHT =4.24 G	1563
*PATELLA VULGATA	3.1*10**4 BASED ON DRY WEIGHT	17 OBS.,ALL AREAS AROUND THE BRITISH ISLES	1278
*PHYTOPLANKTON	620-15000 BASED ON WET WEIGHT	NONE	311
*PIMEPHALES PROMELAS	3.2 BASED ON WET WEIGHT	AG=0.5 MG/L,1 W,BODY WEIGHT =1.14 G	1563
*PIMEPHALES PROMELAS	3.2 BASED ON WET WEIGHT	AG=0.5 MG/L,3 W,BODY WEIGHT =1.03 G	1563
*PIMEPHALES PROMELAS	2.4 BASED ON WET WEIGHT	AG=0.5 MG/L,5 W,BODY WEIGHT =1.16 G	1563
*PIMEPHALES PROMELAS	2.4 BASED ON WET WEIGHT	AG=0.5 MG/L,7 W,BODY WEIGHT =0.85 G	1563
*PIMEPHALES PROMELAS	4.0 BASED ON WET WEIGHT	AG=0.5 MG/L,10 W,BODY WEIGHT =1.20 G	1563
*PIMEPHALES PROMELAS	0.7 BASED ON WET WEIGHT	AG=5.0 MG/L,1 W,BODY WEIGHT =0.89 G	1563
*PIMEPHALES PROMELAS	2.7 BASED ON WET WEIGHT	AG=5.0 MG/L,3 W,BODY WEIGHT =0.94 G	1563
*PIMEPHALES PROMELAS	4.2 BASED ON WET WEIGHT	AG=5.0 MG/L,HG=5.0 MG HG/M3,BODY WEIGHT =0.80 G	1563
*PIMEPHALES PROMELAS	4.2 BASED ON WET WEIGHT	AG=5.0 MG/L,7 W,BODY WEIGHT =0.87 G	1563
*PIMEPHALES PROMELAS	6.2 BASED ON WET WEIGHT	AG=5.0 MG/L,10 W,BODY WEIGHT =1.19 G	1563
*PLANKTON	210 BASED ON WET WEIGHT	NONE	157
*PORPHYRA UMBILICALIS	1.9*10**3 BASED ON DRY WEIGHT	20 OBS.,ALL AREAS AROUND THE BRITISH ISLES	1278
*SCENEDESMUS SP.	52 BASED ON DRY WEIGHT	AG=0.5 MG/L,1 W	1563
*SCENEDESMUS SP.	182 BASED ON DRY WEIGHT	AG=0.5 MG/L,3 W	1563

TABLE D93

CONCENTRATION FACTOR : AG

ITEM:	VALUE:	CONDITIONS:	REF.:
*SCENEDESMUS SP.	240 BASED ON WET WEIGHT	AG=0.5 MG/L,5 W	1563
*SCENEDESMUS SP.	200 BASED ON WET WEIGHT	AG=0.5 MG/L,7 W	1563
*SCENEDESMUS SP.	96 BASED ON WET WEIGHT	AG=0.5 MG/L,10 W	1563
*SCENEDESMUS SP.	10 BASED ON WET WEIGHT	AG=5.0 MG/L,1 W	1563
*SCENEDESMUS SP.	XX BASED ON WET WEIGHT	AG=5.0 MG/L,3 W	1563
*SCENEDESMUS SP.	50 BASED ON WET WEIGHT	AG=5.0 MG/L,5 W	1563
*SCENEDESMUS SP.	72 BASED ON WET WEIGHT	AG=5.0 MG/L,7 W	1563
*SCENEDESMUS SP.	150 BASED ON WET WEIGHT	AG=5.0 MG/L,10 W	1563

TABLE D94

CONCENTRATION FACTOR : AL

ITEM:	VALUE:	CONDITIONS:	REF.:
*BROWN ALGAE	1550 BASED ON WET WEIGHT	NONE	157
*PLANKTON	25000 BASED ON WET WEIGHT	NONE	157

TABLE D95

CONCENTRATION FACTOR : ALDRIN

ITEM:	VALUE:	CONDITIONS:	REF.:
*ALFALFA	0.011 BASED ON WET WEIGHT	CONC.IN SOIL=0.84MG/KG DRY WT.	389
*BEETLE	3.83 BASED ON WET WEIGHT	CONC.IN SOIL=0.06MG/KG DRY WT., U.S.,SITES=2	389
*BUFFALOFISH	30000 BASED ON WET WEIGHT	CONC.IN WATER=0.007UG/L	389
*BUFFALOFISH	3913 BASED ON WET WEIGHT	CONC.IN WATER=0.023UG/L	389
*CARROT, ROOT	0.021 BASED ON WET WEIGHT	CONC.IN CLAY SOIL=0.48MG/KG DRY WT.	389
*CARROT, ROOT	0.001 BASED ON WET WEIGHT	CONC.IN MUCK SOIL=8.36MG/KG DRY WT.	389
*CARROT, ROOT	0.396 BASED ON WET WEIGHT	CONC.IN SILT LOAM SOIL=1.34MG/KG DRY WT.	389
*CARROT, ROOT	0.340 BASED ON WET WEIGHT	CONC.IN SOIL=0.94MG/KG DRY WT.	389
*CATFISH	1428 BASED ON WET WEIGHT	CONC.IN WATER=0.007UG/L	389
*CATFISH	4444 BASED ON WET WEIGHT	CONC.IN WATER=0.009UG/L	389
*CATFISH	952 BASED ON WET WEIGHT	CONC.IN WATER=0.021UG/L	389
*CATFISH	1590 BASED ON WET WEIGHT	CONC.IN WATER=0.044UG/L	389
*CRAYFISH	84286 BASED ON WET WEIGHT	CONC.IN WATER=0.014UG/L,U.S.	389
*CUCUMBER FRUIT	0.031 BASED ON WET WEIGHT	CONC.IN SOIL=3.7MG/KG DRY WT.	389
*EARTHWORMS	0.6 BASED ON WET WEIGHT	CONC.IN SOIL(ARABLE)=0.10MG/KG DRY WT.,G.B.,SITES=7	389
*EARTHWORMS	0.57 BASED ON WET WEIGHT	CONC.IN SOIL(ARABLE)=0.37MG/KG DRY WT.,G.B.,SITES=10	389
*EARTHWORMS	0.38 BASED ON WET WEIGHT	CONC.IN SOIL=0.24MG/KG DRY WT., G.B.,SITES=2	389
*EARTHWORMS	3.33 BASED ON DRY WEIGHT	CONC.IN SOIL=0.03MG/KG DRY WT., U.S.,SITES=67	389
*EARTHWORMS	1.17 BASED ON WET WEIGHT	CONC.IN SOIL=0.06MG/KG DRY WT., U.S.,SITES=1	389
*EARTHWORMS	4.0 BASED ON DRY WEIGHT	CONC.IN SOIL=0.008MG/KG DRY WT., U.S.,SITES=1	389
*EARTHWORMS EXCL. LUMBRICUS TERRESTRIS	0.91 BASED ON WET WEIGHT	CONC.IN SOIL=0.72MG/KG DRY WT., G.B.,SITES=1	389

TABLE D95

CONCENTRATION FACTOR : ALDRIN			
ITEM:	VALUE:	CONDITIONS:	REF.:
*GRYLLUS ASSIMILIS	0.17 BASED ON WET WEIGHT	CONC.IN SOIL=0.06MG/KG DRY WT., U.S.,SITES=2	389
*LETTUCE, FOLIAGE	0.013 BASED ON WET WEIGHT	CONC.IN SILT LOAM SOIL=2.3MG/KG DRY WT.	389
*LUMBRICUS TERRESTRIS	0.07 BASED ON WET WEIGHT	CONC.IN SOIL=0.72MG/KG DRY WT., G.B.,SITES=1	389
*OYSTER	10 BASED ON WET WEIGHT	CONC.IN WATER:<1.0UG/L,CONC.IN SEDIMENT:<0.01MG/KG,U.S.	389
*OYSTER	10 BASED ON WET WEIGHT	CONC.IN WATER:<1.0UG/L,CONC.IN SEDIMENT:<0.001MG/KG,U.S.	389
*PEAS, FOLIAGE	0.083 BASED ON WET WEIGHT	CONC.IN SAND=2.3MG/KG DRY WT.	389
*PEAS, ROOT	22.39 BASED ON WET WEIGHT	CONC.IN SAND=2.3MG/KG DRY WT.	389
*POTATO TUBERS	0.074 BASED ON WET WEIGHT	CONC.IN SOIL=0.94MG/KG DRY WT.	389
*SLUGS	16.67 BASED ON WET WEIGHT	CONC.IN SOIL=0.006MG/KG DRY WT., U.S.,SITES=4	389
*SOYBEAN	>6.0 BASED ON WET WEIGHT	CONC.IN SOIL:<0.0001MG/KG DRY WT.	389
*SUGAR BEET ROOT	0.057 BASED ON WET WEIGHT	CONC.IN SILT LOAM SOIL=2.3MG/KG DRY WT.	389
*SUGAR BEET ROOT	0.284 BASED ON WET WEIGHT	CONC.IN LOAM SOIL=0.35MG/KG DRY WT.	389
*TURNIPS	0.022 BASED ON WET WEIGHT	CONC.IN SILT LOAM SOIL=2.3MG/KG DRY WT.	389
*WHEAT, FOLIAGE	0.008 BASED ON WET WEIGHT	CONC.IN SILT LOAM SOIL=1.8MG/KG DRY WT.	389

TABLE D96

CONCENTRATION FACTOR : AM			
ITEM:	VALUE:	CONDITIONS:	REF.:
*BARLEY, GRAIN	1.4*10**(-6) BASED ON DRY PLANT AND DRY SOIL	NEVADA, LAB. SITE 11241 AM, 293-301 K	1402
*BARLEY, GRAIN	3*10**(-6) BASED ON DRY PLANT AND DRY SOIL	NEVADA, LAB. SITE 13241 AM, 293-301 K	1402
*BARLEY, LOWER VEGETATION	17*10**(-6) BASED ON DRY PLANT AND DRY SOIL	NEVADA, LAB. SITE 11,239,249 PU, 293-301 K	1402
*BARLEY, LOWER VEGETATION	58*10**(-6) BASED ON DRY PLANT AND DRY SOIL	NEVADA, LAB. SITE 13241 AM, 293-301 K	1402
*BARLEY, UPPER VEGETATION	156*10**(-6) BASED ON DRY PLANT AND DRY SOIL	NEVADA, LAB. SITE 11241 AM, 293-301 K	1402
*BARLEY, UPPER VEGETATION	93*10**(-6) BASED ON DRY PLANT AND DRY SOIL	NEVADA, LAB. SITE 13241 AM, 293-301 K	1402
*NEREIS DIVERSICOLOR	0.0003 (SD=+-0.0004) CONC IN ANIM/(CONC IN SEDIMENT)	WET WEIGHT BOTH SED. AND ANIM., BRAVO CRATER	84
*NEREIS DIVERSICOLOR	0.0006 (SD=+-0.0006) CONC IN ANIM/(CONC IN SEDIMENT)	WET WEIGHT BOTH SED. AND ANIM., WINDSCALE	84
*NEREIS DIVERSICOLOR	0.0012 (SD=+-0.0001) CONC IN ANIM/(CONC IN SEDIMENT)	DRY WEIGHT BOTH SED. AND ANIM., WINDSCALE,N=22, 5 D	84
*NEREIS DIVERSICOLOR	0.0020 (SD=+-0.0001) CONC IN ANIM/(CONC IN SEDIMENT)	DRY WEIGHT BOTH SED. AND ANIM., WINDSCALE,N=25, 15 D	84
*NEREIS DIVERSICOLOR	0.0024 (SD=+-0.0002) CONC IN ANIM/(CONC IN SEDIMENT)	DRY WEIGHT BOTH SED. AND ANIM., WINDSCALE,N=23, 40 D	84
*NEREIS DIVERSICOLOR	0.0043 (SD=+-0.0008) CONC IN ANIM/(CONC IN SEDIMENT)	DRY WEIGHT BOTH SED. AND ANIM., WINDSCALE,N=1, 225 D	84
*NEREIS DIVERSICOLOR	0.0004 (SD=+-0.0002) CONC IN ANIM/(CONC IN SEDIMENT)	DRY WEIGHT BOTH SED. AND ANIM., BIKINI ATOLL,N=9, 5 D	84
*NEREIS DIVERSICOLOR	0.0007 (SD=+-0.0001) CONC IN ANIM/(CONC IN SEDIMENT)	DRY WEIGHT BOTH SED. AND ANIM., BIKINI ATOLL,N=10, 15 D	84
*NEREIS DIVERSICOLOR	0.0013 (SD=+-0.0002) CONC IN ANIM/(CONC IN SEDIMENT)	DRY WEIGHT BOTH SED. AND ANIM., BIKINI ATOLL,N=6, 40 D	84
*NEREIS DIVERSICOLOR	0.0010 (SD=+-0.0001) CONC IN ANIM/(CONC IN SEDIMENT)	DRY WEIGHT BOTH SED. AND ANIM., BIKINI ATOLL,N=31, 40 D	84

TABLE D96

CONCENTRATION FACTOR : AS

ITEM:	VALUE:	CONDITIONS:	REF.:
*BROWN ALGAE	2500 BASED ON WET WEIGHT	NONE	157
*BROWN ALGAE	200-6000 BASED ON WET WEIGHT	NONE	994
*CERATOPHYLLUM DEMERSUM	13000 - 20000 BASED ON DRY WEIGHT	WAIKATO RIVER, JAPAN, 524 - 971MG AS/KG DW	1304
*COMPSOPOGON HOOKERI	7000 BASED ON DRY WEIGHT	WAIKATO RIVER, JAPAN, 550MG AS/KG DW	1304
*ENTEROMORPHA COMPRESSA	4000 BASED ON DRY WEIGHT	ENGLAND, 12 MG AS/KG DW	1304
*ENTEROMORPHA NANA	800 - 2800 BASED ON DRY WEIGHT	WAIKATO RIVER, JAPAN, 20 - 40MG AS/KG DW	1304
*FUCUS SERRATUS	600 BASED ON DRY WEIGHT	ENGLAND, 1.7MG AS/KG DW	1304
*ICTALURUS PUNCTATUS	0.4 BASED ON DRY WEIGHT	296 K, LAB. SYSTEM, 8 D, WILLACY SANDY LOAM,	1406
*LAGAROSIPHON MAJOR	4600 - 5300 BASED ON DRY WEIGHT	WAIKATO RIVER, JAPAN, 32-405MG AS/KG DW	1304
*LEMNA MINOR	3 BASED ON DRY WEIGHT	296 K, LAB. SYSTEM, 20 D, WILLACY SANDY LOAM	1406
*MACROCYSTIS PYRIFERA	19000 BASED ON DRY WEIGHT	NEW ZEALAND, 58MG AS/KG DW	1304
*OEDOGONIUM CARDIACUM	17 BASED ON DRY WEIGHT	296 K, LAB. SYSTEM, 20 D, HIDALGO CLAY LOAM	1406
*OEDOGONIUM CARDIACUM	5 BASED ON DRY WEIGHT	296 K, LAB. SYSTEM, 20 D, LAREDO SILT LOAM	1406
*OEDOGONIUM CARDIACUM	3 BASED ON DRY WEIGHT	296 K, LAB. SYSTEM, 20 D, WILLACY SANDY LOAM	1406
*PHYSA SP.	23.0 BASED ON DRY WEIGHT	296 K, LAB. SYSTEM, 8 D, HIDALGO CLAY LOAM,	1406
*PHYSA SP.	7 BASED ON DRY WEIGHT	296 K, LAB. SYSTEM, 8 D, WILLACY SANDY LOAM,	1406
*PHYSA SP.	21 BASED ON DRY WEIGHT	296 K, LAB. SYSTEM, 20 D, LAREDO SILT LOAM	1406
*PHYSA SP.	2 BASED ON DRY WEIGHT	296 K, LAB. SYSTEM, 20 D, WILLACY SANDY LOAM	1406
*PROCAMBARUS CLARKI, CHITIN	0.7 BASED ON DRY WEIGHT	296 K, LAB. SYSTEM, 8 D, LAREDO SILT LOAM,	1406
*PROCAMBARUS CLARKI, CHITIN	5.3 BASED ON DRY WEIGHT	296 K, LAB. SYSTEM, 8 D, WILLACY SANDY LOAM,	1406
*PROCAMBARUS CLARKI, CHITIN	0.1 BASED ON DRY WEIGHT	296 K, LAB. SYSTEM, 20 D, HIDALGO CLAY LOAM	1406
*PROCAMBARUS CLARKI, CHITIN	2.5 BASED ON DRY WEIGHT	296 K, LAB. SYSTEM, 20 D, LAREDO SILT LOAM	1406
*PROCAMBARUS CLARKI, CHITIN	0.9 BASED ON DRY WEIGHT	296 K, LAB. SYSTEM, 20 D, WILLACY SANDY LOAM	1406
*PROCAMBARUS CLARKI, SOFT TISSUE	4.2 BASED ON DRY WEIGHT	296 K, LAB. SYSTEM, 8 D, HIDALGO CLAY LOAM,	1406
*PROCAMBARUS CLARKI, SOFT TISSUE	5.2 BASED ON DRY WEIGHT	296 K, LAB. SYSTEM, 8 D, LAREDO SILT LOAM,	1406
*PROCAMBARUS CLARKI, SOFT TISSUE	4.7 BASED ON DRY WEIGHT	296 K, LAB. SYSTEM, 8 D, WILLACY SANDY LOAM,	1406
*PROCAMBARUS CLARKI, SOFT TISSUE	16.0 BASED ON DRY WEIGHT	296 K, LAB. SYSTEM, 20 D, LAREDO SILT LOAM	1406
*SALMO GAIRDNERI, EGG	2.7 BASED ON WET WEIGHT	0.05 MG AS(3+)/L, 160-200 EGGS, 279.5 K, 33 D	317
*SALMO GAIRDNERI, EGG	5.5 BASED ON WET WEIGHT	0.1 MG AS(3+)/L, 160-200 EGGS, 279.5 K, 33 D	317
*SALMO GAIRDNERI, EGG	0.9 BASED ON WET WEIGHT	0.5 MG AS(3+)/L, 160-200 EGGS, 279.5 K, 33 D	317
*SALMO GAIRDNERI, EGG	0.2 BASED ON WET WEIGHT	1.0 MG AS(3+)/L, 160-200 EGGS, 279.5 K, 33 D	317
*SALMO GAIRDNERI, EGG	0.1 BASED ON WET WEIGHT	5.0 MG AS(3+)/L, 160-200 EGGS, 279.5 K, 15 D	317
*SALMO GAIRDNERI, EGG	0.05 BASED ON WET WEIGHT	10.0 MG AS(3+)/L, 160-200 EGGS, 279.5 K, 15 D	317
*SALMO GAIRDNERI, EGG	0.03 BASED ON WET WEIGHT	50.0 MG AS(3+)/L, 160-200 EGGS. 279.5 K, 15 D	317
*SALMO GAIRDNERI, EGG	18.5 BASED ON WET WEIGHT	0.05 MG AS(5+)/L, 160-200 EGGS, 279.5 K, 33 D	317

TABLE D97

CONCENTRATION FACTOR : AS			
ITEM:	VALUE:	CONDITIONS:	REF.:
*SALMO GAIRDNERI, EGG	4.0 BASED ON WET WEIGHT	0.1 MG AS(5+)/L, 160-200 EGGS, 279.5 K, 33 D	317
*SALMO GAIRDNERI, EGG	0.8 BASED ON WET WEIGHT	0.5 MG AS(5+)/L, 160-200 EGGS, 279.5 K, 33 D	317
*SALMO GAIRDNERI, EGG	0.08 BASED ON WET WEIGHT	1.0 MG AS(5+)/L, 160-200 EGGS, 279.5 K, 33 D	317
*SALMO GAIRDNERI, EGG	0.01 BASED ON WET WEIGHT	5.0 MG AS(5+)/L, 160-200 EGGS, 279.5 K, 15 D	317
*SALMO GAIRDNERI, EGG	0.007 BASED ON WET WEIGHT	10.0 MG AS(5+)/L, 160-200 EGGS, 279.5 K, 15 D	317
*SALMO GAIRDNERI, EGG	0.02 BASED ON WET WEIGHT	50.0 MG AS(5+)/L, 160-200 EGGS, 279.5 K, 15 D	317
*TYPHA ORIENTALIS	100 BASED ON DRY WEIGHT	WAIKATO RIVER, JAPAN, 8MG AS/KG DW	1304

TABLE D98

CONCENTRATION FACTOR : ATRAZINE			
ITEM:	VALUE:	CONDITIONS:	REF.:
*GAMBUSIA AFFINIS	17 BASED ON WET WEIGHT	MODEL ECOSYSTEM, ORGANISM=197 PPB, WATER=5.0-3.8 PPB, 18 D	804
*HELISOMA SP.	8 BASED ON WET WEIGHT	MODEL ECOSYSTEM, ORGANISM=98 PPB, WATER=5.0-3.8 PPB, 18 D	804
*OEDOGONIUM CARDIACUM	9 BASED ON WET WEIGHT	MODEL ECOSYSTEM, ORGANISM=104 PPB, WATER=5.0-3.8 PPB, 18 D	804
*SCENEDESMUS ACUTUS	31.8 BASED ON DRY WEIGHT	WATER = 0.5*10**(-7) MOL, 303 K, PH=7-8, 8 H	141
*SCENEDESMUS ACUTUS	90 BASED ON DRY WEIGHT	WATER = 4.65*10**(-9) MOL/L, 303 K, PH=7-8, 8 H, LIVING CELLS	141
*SCENEDESMUS ACUTUS	60 BASED ON DRY WEIGHT	WATER = 4.65*10**(-8) MOL/L, 303 K, PH=7-8, 8 H, LIVING CELLS	141
*SCENEDESMUS ACUTUS	35 BASED ON DRY WEIGHT	WATER = 4.65*10**(-7) MOL/L, 303 K, PH=7-8, 8 H, LIVING CELLS	141
*SCENEDESMUS ACUTUS	10 BASED ON DRY WEIGHT	WATER = 4.65*10**(-6) MOL/L, 303 K, PH=7-8, 8 H, LIVING CELLS	141
*SCENEDESMUS ACUTUS	12 BASED ON DRY WEIGHT	WATER = 4.65*10**(-9) MOL/L, 303 K, PH=7-8, 8 H, DEAD CELLS	141
*SCENEDESMUS ACUTUS	9 BASED ON DRY WEIGHT	WATER = 4.65*10**(-8) MOL/L, 303 K, PH=7-8, 8 H, DEAD CELLS	141
*SCENEDESMUS ACUTUS	9 BASED ON DRY WEIGHT	WATER = 4.65*10**(-7) MOL/L, 303 K, PH=7-8, 8 H, DEAD CELLS	141
*SCENEDESMUS ACUTUS	9 BASED ON DRY WEIGHT	WATER = 4.65*10**(-6) MOL/L, 303 K, PH=7-8, 8 H, DEAD CELLS	141

TABLE D99

CONCENTRATION FACTOR : AU			
ITEM:	VALUE:	CONDITIONS:	REF.:
*BROWN ALGAE	270 BASED ON WET WEIGHT	NONE	157

TABLE D100

CONCENTRATION FACTOR : AU-198			
ITEM:	VALUE:	CONDITIONS:	REF.:
*ANODONTA NUTTALLIANA, SOFT PARTS	7 BASED ON WET WEIGHT	NO FOOD, 292 K, PH=7.5, FRESHWAT=2.3 PART. AND 16 SOLUB P CI/ML, 7 D	629

TABLE D100

CONCENTRATION FACTOR : AU-198			
ITEM:	VALUE:	CONDITIONS:	REF.:
*ANODONTA NUTTALLIANA, SOFT PARTS	7 BASED ON WET WEIGHT	NO FOOD,292 K,PH=7.5, FRESHWAT=8.9 PART. AND 25 SOLUB P CI/ML, 21 D	629
*ASTACUS SP., MUSCLE	LESS THAN 1 BASED ON WET WEIGHT	NO FOOD,292 K,PH=7.5, FRESHWAT=2.3 PART. AND 16 SOLUB P CI/ML, 7 D	629
*ASTACUS SP., MUSCLE	LESS THAN 1 BASED ON WET WEIGHT	NO FOOD,292 K,PH=7.5, FRESHWAT=8.9 PART. AND 25 SOLUB P CI/ML, 21 D	629
*ASTACUS SP., SOFT PARTS	11 BASED ON WET WEIGHT	NO FOOD,292 K,PH=7.5, FRESHWAT=2.3 PART. AND 16 SOLUB P CI/ML, 7 D	629
*ASTACUS SP., SOFT PARTS	12 BASED ON WET WEIGHT	NO FOOD,292 K,PH=7.5, FRESHWAT=WW	629
*ASTACUS SP., VISCERA	14 BASED ON WET WEIGHT	NO FOOD,292 K,PH=7.5, FRESHWAT=2.3 PART. AND 16 SOLUB P CI/ML, 7 D	629
*ASTACUS SP., VISCERA	14 BASED ON WET WEIGHT	NO FOOD,292 K,PH=7.5, FRESHWAT=8.9 PART. AND 25 SOLUB P CI/ML, 21 D	629
*CANCER PRODUCTUS, MUSCLE	4 BASED ON WET WEIGHT	NO FOOD,285 K,PH=8.2, SEAWATER=3.3 PART. AND 12 SOLUB P CI/ML, 20 D	629
*CANCER PRODUCTUS, SORT PARTS	12 BASED ON WET WEIGHT	NO FOOD,285 K,PH=8.2, SEAWATER=3.3 PART. AND 12 SOLUB P CI/ML, 20 D	629
*CANCER PRODUCTUS, VISCERA	20 BASED ON WET WEIGHT	NO FOOD,285 K,PH=8.2, SEAWATER=3.5 PART AND 6.9 SOLUB P CI/ML, 6 D	629
*CANCER PRODUCTUS, VISCERA	16 BASED ON WET WEIGHT	NO FOOD,285 K,PH=8.2, SEAWATER=3.3 PART. AND 12 SOLUB P CI/ML, 20 D	629
*CARASSIUS AURATUS, MUSCLE	LESS THAN 1 BASED ON WET WEIGHT	NO FOOD,292 K,PH=7.5, FRESHWAT=2.3 PART. AND 16 SOLUB P CI/ML, 7 D	629
*CARASSIUS AURATUS, NONMUSCLE	9 BASED ON WET WEIGHT	NO FOOD,292 K,PH=7.5, FRESHWAT=2.3 PART. AND 16 SOLUB P CI/ML, 7 D	629
*CARASSIUS AURATUS, WHOLE BODY	9 BASED ON WET WEIGHT	NO FOOD,292 K,PH=7.5, FRESHWAT=8.9 PART. AND 25 SOLUB P CI/ML, 21 D	629
*GILLICHTHYS MIRABILIS, MUSCLE	LESS THAN 1 BASED ON WET WEIGHT	NO FOOD,285 K,PH=8.2, SEAWATER=3.3 PART. AND 12 SOLUB P CI/ML, 20 D	629
*GILLICHTHYS MIRABILIS, NONMUSCLE	2 BASED ON WET WEIGHT	NO FOOD,285 K,PH=8.2, SEAWATER=3.3 PART. AND 12 SOLUB P CI/ML, 20 D	629
*GILLICHTHYS MIRABILIS, NONMUSCLE	1 BASED ON WET WEIGHT	NO FOOD,285 K,PH=8.2, SEAWATER=3.3 PART. AND 12 SOLUB P CI/ML, 20 D	629
*GILLICHTHYS MIRABILIS, WHOLE BODY	1 BASED ON WET WEIGHT	NO FOOD,285 K,PH=8.2, SEAWATER= 3.5 PART . AND 6.9 SOLUB P CI/ML, 6 D	629
*MYTILUS EDULIS, SOFT PARTS	11 BASED ON WET WEIGHT	NO FOOD,285 K,PH=8.2, SEAWATER=3.5 PART AND 6.9 SOLUB P CI/ML, 6 D	629
*SAXIDONUS GIGANTEUS, SOFT PARTS	3 BASED ON WET WEIGHT	NO FOOD,285 K,PH=8.2, SEAWATER=3.3 PART. AND 12 SOLUB P CI/ML, 20 D	629

TABLE D101

CONCENTRATION FACTOR : B			
ITEM:	VALUE:	CONDITIONS:	REF.:
*BROWN ALGAE	6.6 BASED ON WET WEIGHT	NONE	157
*CORAL	1.72*10**(-2) BASED ON CA CONTENT (F023)	DEEP OCEAN CORAL, MEAN OF 13 DETERMINATIONS	939
*CORAL	2.03*10**(-2) BASED ON CA CONTENT (F023)	SHALLOW OPEN OCEAN CORAL, MEAN OF 8 DETERMINATIONS	939
*CORAL	1.50*10**(-2) BASED ON CA CONTENT (F023)	SHALLOW COASTAL OCEAN, MEAN OF 13 DETERMINATIONS	939

TABLE D102

CONCENTRATION FACTOR : BA			
ITEM:	VALUE:	CONDITIONS:	REF.:
*BROWN ALGAE	260 BASED ON WET WEIGHT	NONE	157

TABLE D102

CONCENTRATION FACTOR : BA			
ITEM:	VALUE:	CONDITIONS:	REF.:
*CORAL	1.34 BASED ON CA CONTENT (F023)	DEEP OCEAN CORAL, MEAN OF 13 DETERMINATIONS	939
*CORAL	1.11 BASED ON CA CONTENT (F023)	SHALLOW OPEN OCEAN CORAL, MEAN OF 8 DETERMINATIONS	939
*CORAL	2.38 BASED ON CA CONTENT (F023)	SHALLOW COASTAL OCEAN, MEAN OF 13 DETERMINATIONS	939
*PLANKTON	120 BASED ON WET WEIGHT	NONE	157

TABLE D103

CONCENTRATION FACTOR : BA-140/LA-140			
ITEM:	VALUE:	CONDITIONS:	REF.:
*ANODONTA NUTTALLIANA, SOFT PARTS	1.6 BASED ON WET WEIGHT	NO FOOD, 292 K, PH=7.5, FRESHWAT=0.69 PART. AND 110 SOLUB P CI/ML, 7 D	629
*ANODONTA NUTTALLIANA, SOFT PARTS	3.2 BASED ON WET WEIGHT	NO FOOD, 292 K, PH=7.5, FRESHWAT=3.5 PART. AND 280 SOLUB P CI/ML, 21 D	629
*ASTACUS SP., MUSCLE	0.4 BASED ON WET WEIGHT	NO FOOD, 292 K, PH=7.5, FRESHWAT=0.69 PART. AND 110 SOLUB P CI/ML, 7 D	629
*ASTACUS SP., MUSCLE	1.7 BASED ON WET WEIGHT	NO FOOD, 292 K, PH=7.5, FRESHWAT=3.5 PART. AND 280 SOLUB P CI/ML, 21 D	629
*ASTACUS SP., SOFT PARTS	9 BASED ON WET WEIGHT	NO FOOD, 292 K, PH=7.5, FRESHWAT=0.69 PART. AND 110 SOLUB P CI/ML, 7 D	629
*ASTACUS SP., SOFT PARTS	25 BASED ON WET WEIGHT	NO FOOD, 292 K, PH=7.5, FRESHWAT=90 PART. AND 0 SOLUB P CI/L, 7 D	629
*ASTACUS SP., VISCERA	10 BASED ON WET WEIGHT	NO FOOD, 292 K, PH=7.5, FRESHWAT=0.69 PART. AND 110 SOLUB P CI/ML, 7 D	629
*ASTACUS SP., VISCERA	30 BASED ON WET WEIGHT	NO FOOD, 292 K, PH=7.5, FRESHWAT=3.5 PART. AND 280 SOLUB P CI/ML, 21 D	629
*CANCER PRODUCTUS, MUSCLE	1.3 BASED ON WET WEIGHT	NO FOOD, 285 K, PH=8.2, SEAWATER=0.67 PART AND 230 SOLUB P CI/ML, 20 D	629
*CANCER PRODUCTUS, SORT PARTS	7 BASED ON WET WEIGHT	NO FOOD, 285 K, PH=8.2, SEAWATER=0.67 PART AND 230 SOLUB P CI/ML, 20 D	629
*CANCER PRODUCTUS, VISCERA	5 BASED ON WET WEIGHT	NO FOOD, 285 K, PH=8.2, SEAWATER=1.10 PART. AND 220 SOLUB P CI/ML, 6 D	629
*CANCER PRODUCTUS, VISCERA	9 BASED ON WET WEIGHT	NO FOOD, 285 K, PH=8.2, SEAWATER=0.67 PART AND 230 SOLUB P CI/ML, 20 D	629
*CARASSIUS AURATUS, MUSCLE	LESS THAN 1 BASED ON WET WEIGHT	NO FOOD, 292 K, PH=7.5, FRESHWAT=0.69 PART. AND 110 SOLUB P CI/ML, 7 D	629
*CARASSIUS AURATUS, NONMUSCLE	3.3 BASED ON WET WEIGHT	NO FOOD, 292 K, PH=7.5, FRESHWAT=0.69 PART. AND 110 SOLUB P CI/ML, 7 D	629
*CARASSIUS AURATUS, WHOLE BODY	3 BASED ON WET WEIGHT	NO FOOD, 292 K, PH=7.5, FRESHWAT=3.5 PART. AND 280 SOLUB P CI/ML, 21 D	629
*GILLICHTHYS MIRABILIS, MUSCLE	LESS THAN 1 BASED ON WET WEIGHT	NO FOOD, 285 K, PH=8.2, SEAWATER=1.10 PART. AND 220 SOLUB P CI/ML, 6 D	629
*GILLICHTHYS MIRABILIS, MUSCLE	0.9 BASED ON WET WEIGHT	NO FOOD, 285 K, PH=8.2, SEAWATER=0.67 PART AND 230 SOLUB P CI/ML, 20 D	629
*GILLICHTHYS MIRABILIS, NONMUSCLE	1.3 BASED ON WET WEIGHT	NO FOOD, 285 K, PH=8.2, SEAWATER=1.10 PART. AND 220 SOLUB P CI/ML, 6 D	629
*GILLICHTHYS MIRABILIS, NONMUSCLE	0.8 BASED ON WET WEIGHT	NO FOOD, 285 K, PH=8.2, SEAWATER=0.67 PART AND 230 SOLUB P CI/ML, 20 D	629
*GILLICHTHYS MIRABILIS, WHOLE BODY	1.2 BASED ON WET WEIGHT	NO FOOD, 285 K, PH=8.2, SEAWATER=0.67 PART AND 230 SOLUB P CI/ML, 20 D	629

TABLE D103

CONCENTRATION FACTOR : BA-140/LA-140

ITEM:	VALUE:	CONDITIONS:	REF.:
*GILLICHTHYS MIRABILIS, WHOLE BODY	0.8 BASED ON WET WEIGHT	NO FOOD, 285 K, PH=8.2, SEAWATER=120 PART AND 0 SOLUB P CI/L, 6 D	629
*MYTILUS EDULIS, SOFT PARTS	0.9 BASED ON WET WEIGHT	NO FOOD, 285 K, PH=8.2, SEAWATER= 1.10 PART. AND 220 SOLUB P CI/ML, 6 D	629
*SAXIDONUS GIGANTEUS, SOFT PARTS	0.7 BASED ON WET WEIGHT	NO FOOD, 285 K, PH=8.2, SEAWATER= 0.67 PART AND 230 SOLUB P CI/ML, 20 D	629
*SUSPENDED SOLIDS	530 (RANGE = 320 - 780) BASED ON WET WEIGHT	DANUBE, 1973, NUCLEAR POWER PLANT SINCE 1967	658

TABLE D104

CONCENTRATION FACTOR : BE

ITEM:	VALUE:	CONDITIONS:	REF.:
*BROWN ALGAE	1500 BASED ON WET WEIGHT	NONE	994

TABLE D105

CONCENTRATION FACTOR : BHC

ITEM:	VALUE:	CONDITIONS:	REF.:
*CABBAGE, LEAF	0.218 BASED ON WET WEIGHT	CONC.IN SANDY LOAM SOIL=1.70MG/KG DRY WT.	389
*CARROT	0.021 BASED ON WET WEIGHT	CONC.IN MUCK SOIL=0.34MG/KG DRY WT.	389
*CARROT	0.263 BASED ON WET WEIGHT	CONC.IN SANDY SOIL=0.09MG/KG DRY WT.	389
*CARROT, ROOT	3.53 BASED ON WET WEIGHT	CONC.IN SANDY LOAM SOIL=1.70MG/KG DRY WT.	389
*CRASSOSTREA VIRGINICA	18 BASED ON WET WEIGHT	CONC.IN WATER=0.33UG/L,U.S.	389
*CRAYFISH	33 BASED ON WET WEIGHT	CONC.IN WATER=0.03UG/L,U.S.	389
*EARTHWORMS	5.0 BASED ON WET WEIGHT	CONC.IN SOIL(ARABLE)=0.01MG/KG DRY WT.,G.B.,SITES=10	389
*EARTHWORMS	3.75 BASED ON WET WEIGHT	CONC.IN SOIL(ORCHARD)=0.08MG/KG DRY WT.,G.B.,SITES=2	389
*EARTHWORMS EXCL. LUMBRICUS TERRESTRIS	4.25 BASED ON WET WEIGHT	CONC.IN SOIL=0.004MG/KG DRY WT., G.B.,SITES=1	389
*FERAL PIGEON, TISSUE	2.0 BASED ON WET WEIGHT	CONC.IN FOOD=72MG/KG	389
*GYRINIDAE	33 BASED ON WET WEIGHT	CONC.IN WATER=0.03UG/L,U.S.	389
*LUMBRICUS TERRESTRIS	1.60 BASED ON WET WEIGHT	CONC.IN SOIL=0.004MG/KG DRY WT., G.B.,SITES=1	389
*OYSTER	10 BASED ON WET WEIGHT	CONC.IN WATER:<1.0UG/L,CONC.IN SEDIMENT:<0.01,U.S.	389
*OYSTER	10 BASED ON WET WEIGHT	CONC.IN WATER:<1.0UG/L,CONC.IN SEDIMENT:<0.001MG/KG,U.S.	389
*PEA VINES	0.894 BASED ON WET WEIGHT	CONC.IN SANDY LOAM SOIL=1.70MG/KG DRY WT.	389
*PEAS, FOLIAGE	36.94 BASED ON WET WEIGHT	CONC.IN SAND=0.62MG/KG DRY WT.	389
*PEAS, ROOT	133.23 BASED ON WET WEIGHT	CONC.IN SAND=0.62MG/KG DRY WT.	389
*PLANKTON	7 BASED ON WET WEIGHT	CONC.IN WATER=0.03UG/L,U.S.	389
*PLANKTON	2500 BASED ON WET WEIGHT	CONC.IN WATER=0.040UG/L,U.S.	389
*POTATO TUBERS	0.207 BASED ON WET WEIGHT	CONC.IN SANDY LOAM SOIL=3.0MG/KG DRY WT.	389
*SOYBEAN	1.0 BASED ON WET WEIGHT	CONC.IN SOIL=0.006MG/KG DRY WT.	389
*TENDIPENDIDAE,LARVAE	33 BASED ON WET WEIGHT	CONC.IN WATER=0.03UG/L,U.S.	389

TABLE D106

CONCENTRATION FACTOR : BR			
ITEM:	VALUE:	CONDITIONS:	REF.:
*BROWN ALGAE	2.8 BASED ON WET WEIGHT	NONE	157

TABLE D107

CONCENTRATION FACTOR : CA			
ITEM:	VALUE:	CONDITIONS:	REF.:
*BRASSICA OLERACEA	13 BASED ON WET WEIGHT	CONC.IN SOIL SOLUTION=750PPM	157
*BROWN ALGAE	7.2 BASED ON WET WEIGHT	NONE	157
*BROWN ALGAE	0.5-10 BASED ON WET WEIGHT	NONE	994
*CANCER IRRORATUS	128.0 (SD=+-4.1) BASED ON WET WEIGHT	SEAWATER CONC=400 PPM, NOVA SCOTIA, AAS	984
*DAPHNIA MAGNA	6.9 BASED ON DRY WEIGHT	ON DRY WT.BASIS RELATIVE TO THE ALGAE CONSUMED	1630
*PHYTOPLANKTON	12-85 BASED ON WET WEIGHT	NONE	311
*PLANKTON	5 BASED ON WET WEIGHT	NONE	157
*PLANKTON	10 BASED ON WET WEIGHT	NONE	994
*ZOOPLANKTON	10.5 BASED ON WET WEIGHT	SEAWATER CONCENTRATION 400.0 MG/M3	985
*32 FRESHWATER PLANT	265 BASED ON DRY WEIGHT	ALGAE,BRYOPHYTES AND VASCULAR HYDROPHYTES	1597

TABLE D108

CONCENTRATION FACTOR : CACODYLIC ACID			
ITEM:	VALUE:	CONDITIONS:	REF.:
*DAPHNIA MAGNA	51 BASED ON DRY WEIGHT	296 K, LAB. SYSTEM, 8 D, HIDALGO CLAY LOAM,	1406
*DAPHNIA MAGNA	147 BASED ON DRY WEIGHT	296 K, LAB. SYSTEM, 8 D, LAREDO SILT LOAM,	1406
*ICTALURUS PUNCTATUS	0.8 BASED ON DRY WEIGHT	296 K, LAB. SYSTEM, 8 D, HIDALGO CLAY LOAM,	1406
*ICTALURUS PUNCTATUS	2.8 BASED ON DRY WEIGHT	296 K, LAB. SYSTEM, 8 D, LAREDO SILT LOAM,	1406
*ICTALURUS PUNCTATUS	0.7 BASED ON DRY WEIGHT	296 K, LAB. SYSTEM, 8 D, WILLACY SANDY LOAM,	1406
*ICTALURUS PUNCTATUS	8.5 BASED ON DRY WEIGHT	296 K, LAB. SYSTEM, 20 D, HIDALGO CLAY LOAM	1406
*ICTALURUS PUNCTATUS	275 BASED ON DRY WEIGHT	296 K, LAB. SYSTEM, 20 D, LAREDO SILT LOAM	1406
*ICTALURUS PUNCTATUS	2.1 BASED ON DRY WEIGHT	296 K, LAB. SYSTEM, 20 D, WILLACY SANDY LOAM	1406
*LEMNA MINOR	52 BASED ON DRY WEIGHT	296 K, LAB. SYSTEM, 8 D, HIDALGO CLAY LOAM,	1406
*LEMNA MINOR	88 BASED ON DRY WEIGHT	296 K, LAB. SYSTEM, 8 D, LAREDO SILT LOAM,	1406
*LEMNA MINOR	58 BASED ON DRY WEIGHT	296 K, LAB. SYSTEM, 8 D, WILLACY SANDY LOAM,	1406
*LEMNA MINOR	1244 BASED ON DRY WEIGHT	296 K, LAB. SYSTEM, 20 D, HIDALGO CLAY LOAM	1406
*LEMNA MINOR	3466 BASED ON DRY WEIGHT	296 K, LAB. SYSTEM, 20 D, LAREDO SILT LOAM	1406
*LEMNA MINOR	89 BASED ON DRY WEIGHT	296 K, LAB. SYSTEM, 20 D, WILLACY SANDY LOAM	1406
*OEDOGONIUM CARDIACUM	80 BASED ON DRY WEIGHT	296 K, LAB. SYSTEM, 8 D, HIDALGO CLAY LOAM,	1406
*OEDOGONIUM CARDIACUM	298 BASED ON DRY WEIGHT	296 K, LAB. SYSTEM, 8 D, LAREDO SILT LOAM,	1406

TABLE D108

CONCENTRATION FACTOR : CACODYLIC ACID			
ITEM:	VALUE:	CONDITIONS:	REF.:
*OEDOGONIUM CARDIACUM	101 BASED ON DRY WEIGHT	296 K, LAB. SYSTEM, 8 D, WILLACY SANDY LOAM,	1406
*OEDOGONIUM CARDIACUM	2483 BASED ON DRY WEIGHT	296 K, LAB. SYSTEM, 20 D, HIDALGO CLAY LOAM	1406
*OEDOGONIUM CARDIACUM	26959 BASED ON DRY WEIGHT	296 K, LAB. SYSTEM, 20 D, LAREDO SILT LOAM	1406
*OEDOGONIUM CARDIACUM	163 BASED ON DRY WEIGHT	296 K, LAB. SYSTEM, 20 D, WILLACY SANDY LOAM	1406
*PHYSA SP.	26 BASED ON DRY WEIGHT	296 K, LAB. SYSTEM, 8 D, HIDALGO CLAY LOAM	1406
*PHYSA SP.	25 BASED ON DRY WEIGHT	296 K, LAB. SYSTEM, 8 D, LAREDO SILT LOAM,	1406
*PHYSA SP.	22 BASED ON DRY WEIGHT	296 K, LAB. SYSTEM, 8 D, WILLACY SANDY LOAM,	1406
*PHYSA SP.	1050 BASED ON DRY WEIGHT	296 K, LAB. SYSTEM, 20 D, LAREDO SILT LOAM	1406
*PHYSA SP.	4.3 BASED ON DRY WEIGHT	296 K, LAB. SYSTEM, 20 D, WILLACY SANDY LOAM	1406
*PROCAMBARUS CLARKI, SOFT TISSUE	0.5 BASED ON DRY WEIGHT	296 K, LAB. SYSTEM, 8 D, HIDALGO CLAY LOAM,	1406
*PROCAMBARUS CLARKI, SOFT TISSUE	2.0 BASED ON DRY WEIGHT	296 K, LAB. SYSTEM, 8 D, LAREDO SILT LOAM,	1406
*PROCAMBARUS CLARKI, SOFT TISSUE	2.0 BASED ON DRY WEIGHT	296 K, LAB. SYSTEM, 8 D, WILLACY SANDY LOAM,	1406
*PROCAMBARUS CLARKI, SOFT TISSUE	5.9 BASED ON DRY WEIGHT	296 K, LAB. SYSTEM, 20 D, HIDALGO CLAY LOAM	1406
*PROCAMBARUS CLARKI, SOFT TISSUE	13.6 BASED ON DRY WEIGHT	296 K, LAB. SYSTEM, 20 D, LAREDO SILT LOAM	1406
*PROCAMBARUS CLARKI, SOFT TISSUE	3.0 BASED ON DRY WEIGHT	296 K, LAB. SYSTEM, 20 D, WILLACY SANDY LOAM	1406

TABLE D109

CONCENTRATION FACTOR : CAMPHECHLOR			
ITEM:	VALUE:	CONDITIONS:	REF.:
*AQUATIC PLANT	18 BASED ON WET WEIGHT	CONC.IN WATER=4.8NG/L	389
*AQUATIC PLANT	512 BASED ON WET WEIGHT	CONC.IN WATER=0.41NG/L	389
*CRAYFISH	20 BASED ON WET WEIGHT	CONC.IN WATER:<1.0UG/L,U.S.	389
*FISH	2000-2500 BASED ON WET WEIGHT	CONC.IN WATER=1.0-4.0UG/L	389
*FRESHWATER INVERTEBRATES	2270 BASED ON WET WEIGHT	CONC.IN WATER=0.63UG/L,U.S.	389
*GYRINIDAE	20 BASED ON WET WEIGHT	CONC.IN WATER:<1.0UG/L,U.S.	389
*PLANKTON	5 BASED ON WET WEIGHT	CONC.IN WATER:<1.0UG/L,U.S.	389
*SALMO GAIRDNERI	18829 BASED ON WET WEIGHT	CONC.IN WATER=0.41UG/L	389
*TENDIPENDIDAE,LARVAE	20 BASED ON WET WEIGHT	CONC.IN WATER:<1.0UG/L,U.S.	389

TABLE D110

CONCENTRATION FACTOR : CARBARYL			
ITEM:	VALUE:	CONDITIONS:	REF.:
*ICTALURUS PUNCTATUS	140 BASED ON DRY WEIGHT	MODEL ECOSYSTEM, 22 D, 296 K,WATER =1.66-9.45 PPB,SOIL=2.43 PPM	796
*LEMNA MINOR	3600 BASED ON DRY WEIGHT	MODEL ECOSYSTEM, 22 D, 296 K,WATER =1.66-9.45 PPB,SOIL=2.43 PPM	796
*OEDOGONIUM CARDIACUM	4000 BASED ON DRY WEIGHT	MODEL ECOSYSTEM, 22 D, 296 K,WATER =1.66-9.45 PPB,SOIL=2.43 PPM	796

TABLE D110

CONCENTRATION FACTOR : CARBARYL			
ITEM:	VALUE:	CONDITIONS:	REF.:
*PHYSA SP.	300 BASED ON DRY WEIGHT	MODEL ECOSYSTEM, 22 D, 296 K,WATER =1.66-9.45 PPB,SOIL=2.43 PPM	796
*PROCAMBARUSSP.	260 BASED ON DRY WEIGHT	MODEL ECOSYSTEM, 22 D, 296 K,WATER =1.66-9.45 PPB,SOIL=2.43 PPM	796

TABLE D111

CONCENTRATION FACTOR : CD			
ITEM:	VALUE:	CONDITIONS:	REF.:
*ADSORPTION ON FISH FAECALS	100 - 300 BASED ON DRY WEIGHT	ADDED CD =2.1 MG/M3, 2 H, RELATIVE TO DRY WEIGHT	502
*ADSORPTION ON FISH FEACALS	200 - 1000 BASED ON DRY WEIGHT	ADDED CD =2.1 MG/M3, 24 H, RELATIVE TO DRY WEIGHT	502
*ADSORPTION ON HUMIC ACID	14000 (+-2000) BASED ON DRY WEIGHT	ADDED CD =2.1 - 4.3 MG/M3, 2 H, RELATIVE TO DRY WEIGHT	502
*ADSORPTION ON HUMIC ACID	18000 (+-3000) BASED ON DRY WEIGHT	ADDED CD = 2.1 - 4.3 MG/M3, 24 H, RELATIVE TO DRY WEIGHT	502
*ADSORPTION ON KAOLIN	250 (+=30) BASED ON DRY WEIGHT	ADDED CD =7.3 MG/M3, 2 H,RELATIVE TO DRY WEIGHT	502
*ADSORPTION ON KAOLIN	380 (+-50) BASED ON DRY WEIGHT	ADDED CD =7.3 MG/M3, 24 H, RELATIVE TO DRY WEIGHT	502
*ADSORPTION ON PLANT MATERIAL	800 BASED ON DRY WEIGHT	ADDED CD =2.1 MG/M3, 2 H, RELATIVE TO DRY WEIGHT	502
*ADSORPTION ON PLANT MATERIAL	1000 BASED ON DRY WEIGHT	ADDED CD =2.1 MG/M3, 24 H, RELATIVE TO DRY WEIGHT	502
*ADSORPTION ON SILICA	430 (+-40) BASED ON DRY WEIGHT	ADDED CD =7.3 MG/M3, 2 H,RELATIVE TO DRY WEIGHT	502
*ADSORPTION ON SILICA	1000 (+-100) BASED ON DRY WEIGHT	ADDED CD =7.3 MG/M3, 24 H,RELATIVE TO DRY WEIGHT	502
*ADSORPTION ON SUSPENDED SOLIDS	25000 BASED ON DRY WEIGHT	ADDED CD =2.1 MG/M3, SUSPENDED SOLIDS = 34 MG/L, 1 H	502
*ADSORPTION ON SUSPENDED SOLIDS	13000 BASED ON DRY WEIGHT	ADDED CD =5.5 MG/M3, SUSPENDED SOLIDS = 34 MG/L, 1 H	502
*ADSORPTION ON SUSPENDED SOLIDS	8000 BASED ON DRY WEIGHT	ADDED CD =13.0 MG/M3, SUSPENDED SOLIDS = 34 MG/L, 1 H	502
*ADSORPTION ON SUSPENDED SOLIDS	2600 BASED ON DRY WEIGHT	ADDED CD =56 MG/M3, SUSPENDED SOLIDS =34 MG/L, 1 H	502
*ADSORPTION ON SUSPENDED SOLIDS	17000 BASED ON DRY WEIGHT	ADDED CD =2.1 MG/M3, 2 H, RIVER IVEL	502
*ADSORPTION ON SUSPENDED SOLIDS	40000 BASED ON DRY WEIGHT	ADDED CD =2.1 MG/M3, 24 H, RIVER IVEL	502
*ADSORPTION ON SUSPENDED SOLIDS	3000 BASED ON DRY WEIGHT	ADDED CD =2.1 MG/M3, 2 H, RIVER HIZ	502
*ADSORPTION ON SUSPENDED SOLIDS	6000 BASED ON DRY WEIGHT	ADDED CD =2.1 MG/M3, 24 H, RIVER HIZ	502
*ADSORPTION ON SUSPENDED SOLIDS	14000 BASED ON DRY WEIGHT	ADDED CD =2.1 MG/M3, 2 H, PIX BROOK	502
*ADSORPTION ON SUSPENDED SOLIDS	26000 BASED ON DRY WEIGHT	ADDED CD =2.1 MG/M3, 24 H, PIX BROOK	502
*ADSORPTION ON SUSPENDED SOLIDS	6000 - 25000 BASED ON DRY WEIGHT	ADDED CD =2.1 MG/M3, 2 H, STEVENAGE BROOK	502
*ADSORPTION ON SUSPENDED SOLIDS	42000 BASED ON DRY WEIGHT	EDTA = 0 MG/M3,ADDED CD =2.1 MG/ M3,SUSPENDED SOLIDS=88 MG/L	502
*ADSORPTION ON SUSPENDED SOLIDS	20000 BASED ON DRY WEIGHT	EDTA = 2.8 MG/M3,ADDED CD =2.1 MG/ M3,SUSPENDED SOLIDS=88 MG/L	502
*ADSORPTION ON SUSPENDED SOLIDS	24000 BASED ON DRY WEIGHT	EDTA = 27 MG/M3,ADDED CD =2.1 MG/ M3,SUSPENDED SOLIDS=88 MG/L	502
*ADSORPTION ON SUSPENDED SOLIDS	6900 BASED ON DRY WEIGHT	EDTA = 300 MG/M3,ADDED CD =2.1 MG/ M3,SUSPENDED SOLIDS=88 MG/L	502
*BROWN ALGAE	890 BASED ON WET WEIGHT	NONE	157
*CARASSIUS AURATUS	136 BASED ON WET WEIGHT	WATER = 3 UG/L	1093
*CARASSIUS AURATUS	400 BASED ON WET WEIGHT	WATER = 1 UG/L	1093

TABLE D111

CONCENTRATION FACTOR : CD			
ITEM:	VALUE:	CONDITIONS:	REF.:
*CLADOPHORA GLOMERATA	18*10**3 BASED ON DRY WEIGHT	WATER FROM MAIN DUCK ISL. (L.ONTARIO)=0.1+-0.1, (CONC. IN NG/ML+-S.D.)	806
*FUCUS VESICULOSUS	2.7*10**3 BASED ON DRY WEIGHT	18 OBS.,ALL AREAS AROUND THE BRITISH ISLES	1278
*MYTILUS EDULIS	220 BASED ON WET WEIGHT	STEADY STATE CF., WATER=0.050 MG CD/L	217
*MYTILUS GALLOPROVINCIALIS	400 BASED ON WET WEIGHT	WATER = 4.34 PPM	971
*PLANKTON	910 BASED ON WET WEIGHT	NONE	157
*ZOOPLANKTON	6000 BASED ON WET WEIGHT	SEAWATER CONCENTRATION 0.0001 MG/ M3	985
*1200 BASED ON DRY WEIGHT	ADSORPTION ON SUSPENDED SOLIDS	ADDED CD =177 MG/M3, SUSPENDED SOLIDS =34 MG/L, 1 H	502
*32 FRESHWATER PLANT	1620 BASED ON DRY WEIGHT	ALGAE,BRYOPHYTES AND VASCULAR HYDROPHYTES	1597

TABLE D112

CONCENTRATION FACTOR : CD-109			
ITEM:	VALUE:	CONDITIONS:	REF.:
*LYSMATA SETICAUDATA	600 BASED ON DRY WEIGHT	UPTAKE FROM WATER, 2 M	476
*MYTILUS GALLOPROVINCIALIS	130 BASED ON DRY WEIGHT	UPTAKE FROM WATER, 2 M	476

TABLE D113

CONCENTRATION FACTOR : CD-115M			
ITEM:	VALUE:	CONDITIONS:	REF.:
*CRUSTACEAN	GREATER THAN 250 BASED ON WET WEIGHT	MARINE	671
*ECHINODERM	110 BASED ON WET WEIGHT	MARINE	671
*FISH, VISCERA	6 - 10 BASED ON WET WEIGHT	MARINE	671

TABLE D114

CONCENTRATION FACTOR : CE			
ITEM:	VALUE:	CONDITIONS:	REF.:
*HIZIKIA FUSIFORME, WHOLE BODY	2740 BASED ON WET WEIGHT	REL. TO SOLUBLE CONC.,TOT CONC=58 NG/L, SOLUB=7 NG/L,SEAWAT,JAP.	1535
*MERETRIX LUSORIA, SOFT PART	520 BASED ON WET WEIGHT	REL. TO TOTAL CONC.,TOT CONC=58 NG/L, SOLUB=7 NG/L,SEAWAT,JAP.	1535
*PARALICHTHYS OLIVACEUS, MUSCLE	200 BASED ON WET WEIGHT	REL. TO SOLUBLE CONC.,TOT CONC=58 NG/L, SOLUB=7 NG/L,SEAWAT,JAP.	1535
*SARGASSUM FULVELLUM, WHOLE BODY	290 BASED ON WET WEIGHT	REL. TO TOTAL CONC.,TOT CONC=58 NG/L, SOLUB=7 NG/L,SEAWAT,JAP.	1535
*SERIOLA QUINQUERADIATA, MUSCLE	340 BASED ON WET WEIGHT	REL. TO SOLUBLE CONC.,TOT CONC=58 NG/L, SOLUB=7 NG/L,SEAWAT,JAP.	1535
*ULVA PERTUSA, WHOLE BODY	520 BASED ON WET WEIGHT	REL. TO TOTAL CONC.,TOT CONC=58 NG/L, SOLUB=7 NG/L,SEAWAT,JAP.	1535
*UNDARIA PINNATIFIDA, WHOLE BODY	530 BASED ON WET WEIGHT	REL. TO SOLUBLE CONC.,TOT CONC=58 NG/L, SOLUB=7 NG/L,SEAWAT,JAP.	1535
*32 FRESHWATER PLANT	7100 BASED ON DRY WEIGHT	ALGAE,BRYOPHYTES AND VASCULAR HYDROPHYTES	1597

TABLE D115

CONCENTRATION FACTOR : CE-141			
ITEM:	VALUE:	CONDITIONS:	REF.:
*ANODONTA NUTTALLIANA, SOFT PARTS	6 BASED ON WET WEIGHT	NO FOOD,292 K,PH=7.5, FRESHWAT=90 PART. AND 0 SOLUB P CI/L, 7 D	629
*ANODONTA NUTTALLIANA, SOFT PARTS	3 BASED ON WET WEIGHT	NO FOOD,292 K,PH=7.5, FRESHWAT=370 PART. AND 0 SOLUB P CI/L, 21 D	629
*ASTACUS SP., MUSCLE	LESS THAN 1 BASED ON WET WEIGHT	NO FOOD,292 K,PH=7.5, FRESHWAT=90 PART. AND 0 SOLUB P CI/L, 7 D	629
*ASTACUS SP., MUSCLE	LESS THAN 1 BASED ON WET WEIGHT	NO FOOD,292 K,PH=7.5, FRESHWAT=370 PART. AND 0 SOLUB P CI/L, 21 D	629
*ASTACUS SP., SOFT PARTS	110 BASED ON WET WEIGHT	NO FOOD,292 K,PH=7.5, FRESHWAT=90 PART. AND 0 SOLUB P CI/L, 7 D	629
*ASTACUS SP., SOFT PARTS	38 BASED ON WET WEIGHT	NO FOOD,292 K,PH=7.5, FRESHWAT=2.3 PART AND 1200 SOLUB P CI/ML, 7 D	629
*ASTACUS SP., VISCERA	130 BASED ON WET WEIGHT	NO FOOD,292 K,PH=7.5, FRESHWAT=90 PART. AND 0 SOLUB P CI/L, 7 D	629
*ASTACUS SP., VISCERA	46 BASED ON WET WEIGHT	NO FOOD,292 K,PH=7.5, FRESHWAT=370 PART. AND 0 SOLUB P CI/L, 21 D	629
*CANCER PRODUCTUS, MUSCLE	LESS THAN 1 BASED ON WET WEIGHT	NO FOOD,285 K,PH=8.2, SEAWATER=30 PART. AND 0 SOLUB P CI/L, 20 D	629
*CANCER PRODUCTUS, SORT PARTS	140 BASED ON WET WEIGHT	NO FOOD,285 K,PH=8.2, SEAWATER=30 PART. AND 0 SOLUB P CI/L, 20 D	629
*CANCER PRODUCTUS, VISCERA	61 BASED ON WET WEIGHT	NO FOOD,285 K,PH=8.2, SEAWATER=120 PART AND 0 SOLUB P CI/L, 6 D	629
*CANCER PRODUCTUS, VISCERA	200 BASED ON WET WEIGHT	NO FOOD,285 K,PH=8.2, SEAWATER=30 PART. AND 0 SOLUB P CI/L, 20 D	629
*CARASSIUS AURATUS, MUSCLE	LESS THAN 1 BASED ON WET WEIGHT	NO FOOD,292 K,PH=7.5, FRESHWAT=90 PART. AND 0 SOLUB P CI/L, 7 D	629
*CARASSIUS AURATUS, NONMUSCLE	7 BASED ON WET WEIGHT	NO FOOD,292 K,PH=7.5, FRESHWAT=90 PART. AND 0 SOLUB P CI/L, 7 D	629
*CARASSIUS AURATUS, WHOLE BODY	7 BASED ON WET WEIGHT	NO FOOD,292 K,PH=7.5, FRESHWAT=370 PART. AND 0 SOLUB P CI/L, 21 D	629
*FISH, MUSCLE	150 (RANGE = 30 - 560) BASED ON WET WEIGHT	DANUBE, 1973, NUCLEAR POWER PLANT SINCE 1967	658
*FRESHWATER SEDIMENT	750 (RANGE = 290 - 1080) BASED ON WET WEIGHT	DANUBE, 1973, NUCLEAR POWER PLANT SINCE 1967	658
*GILLICHTHYS MIRABILIS, MUSCLE	LESS THAN 1 BASED ON WET WEIGHT	NO FOOD,285 K,PH=8.2, SEAWATER=120 PART AND 0 SOLUB P CI/L, 6 D	629
*GILLICHTHYS MIRABILIS, MUSCLE	LESS THAN 1 BASED ON WET WEIGHT	NO FOOD,285 K,PH=8.2, SEAWATER=30 PART. AND 0 SOLUB P CI/L, 20 D	629
*GILLICHTHYS MIRABILIS, NONMUSCLE	0.6 BASED ON WET WEIGHT	NO FOOD,285 K,PH=8.2, SEAWATER=120 PART AND 0 SOLUB P CI/L, 6 D	629
*GILLICHTHYS MIRABILIS, NONMUSCLE	57 BASED ON WET WEIGHT	NO FOOD,285 K,PH=8.2, SEAWATER=30 PART. AND 0 SOLUB P CI/L, 20 D	629
*GILLICHTHYS MIRABILIS, WHOLE BODY	5 BASED ON WET WEIGHT	NO FOOD,285 K,PH=8.2, SEAWATER=30 PART. AND 0 SOLUB P CI/L, 20 D	629
*GILLICHTHYS MIRABILIS, WHOLE BODY	52 BASED ON WET WEIGHT	NO FOOD,285 K,PH=8.2, SEAWATER=34 PARTT AND 1300 SOLUB P CI/ML, 6 D	629
*MYTILUS EDULIS, SOFT PARTS	5 BASED ON WET WEIGHT	NO FOOD,285 K,PH=8.2, SEAWATER=120 PART AND 0 SOLUB P CI/L, 6 D	629
*SAXIDONUS GIGANTEUS, SOFT PARTS	60 BASED ON WET WEIGHT	NO FOOD,285 K,PH=8.2, SEAWATER=30 PART. AND 0 SOLUB P CI/L, 20 D	629
*SUSPENDED SOLIDS	1020 (RANGE = 340 - 1860) BASED ON WET WEIGHT	DANUBE, 1973, NUCLEAR POWER PLANT SINCE 1967	658
*WATER PLANT	270 (RANGE = 50 - 810) BASED ON WET WEIGHT	DANUBE, 1973, NUCLEAR POWER PLANT SINCE 1967	658

TABLE D116

CONCENTRATION FACTOR : CE-144			
ITEM:	VALUE:	CONDITIONS:	REF.:
*ALGAE	GREATER THAN 100 BASED ON WET WEIGHT	MARINE	671
*ALGAE	2000 - 10000 BASED ON WET WEIGHT	LAB VALUE	25

TABLE D116

CONCENTRATION FACTOR : CE-144

ITEM:	VALUE:	CONDITIONS:	REF.:
*ALGAE	GREATER THAN 100 BASED ON WET WEIGHT	MARINE	671
*CORALLINA OFFICINALIS	10000 - 20000 BASED ON DRY WEIGHT	SEAWATER = 1 - 2 P CI/L	24
*CRUSTACEAN, EXOSKELETON	2 BASED ON WET WEIGHT	MARINE	671
*CRUSTACEAN, EXOSKELETON	2 BASED ON WET WEIGHT	MARINE	671
*CRUSTACEAN, MUSCLE	25 BASED ON WET WEIGHT	MARINE	671
*CRUSTACEAN, MUSCLE	25 BASED ON WET WEIGHT	MARINE	671
*CRUSTACEAN, VISCERA	200 BASED ON WET WEIGHT	MARINE	671
*CRUSTACEAN, VISCERA	200 BASED ON WET WEIGHT	MARINE	671
*ECHINODERM, DIGESTIVE TRACT	GREATER THAN 250 BASED ON WET WEIGHT	MARINE	671
*ECHINODERM, DIGESTIVE TRACT	GREATER THAN 250 BASED ON WET WEIGHT	MARINE	671
*ECHINODERM, GONAD	2.5 BASED ON WET WEIGHT	MARINE	671
*ECHINODERM, GONAD	2.5 BASED ON WET WEIGHT	MARINE	671
*FISH	LESS THAN 200 BASED ON WET WEIGHT	LAB VALUE	25
*FISH, BONE	390 (RANGE = 140 - 760) BASED ON WET WEIGHT	DANUBE, 1973, NUCLEAR POWER PLANT SINCE 1967	658
*FISH, GILL	GREATER THAN 70 BASED ON WET WEIGHT	MARINE	671
*FISH, GILL	GREATER THAN 70 BASED ON WET WEIGHT	MARINE	671
*FISH, MUSCLE	60 (RANGE = 20 - 200) BASED ON WET WEIGHT	DANUBE, 1973, NUCLEAR POWER PLANT SINCE 1967	658
*FISH, MUSCLE	0.3 BASED ON WET WEIGHT	MARINE	671
*FISH, MUSCLE	0.3 BASED ON WET WEIGHT	MARINE	671
*FRESHWATER SEDIMENT	1010 (RANGE = 260 - 2180) BASED ON WET WEIGHT	DANUBE, 1973, NUCLEAR POWER PLANT SINCE 1967	658
*INVERTEBRATE	500 - 2000 BASED ON WET WEIGHT	LAB VALUE	25
*LAMPSILIS RADIATA	900 BASED ON WET WEIGHT	91 DAYS, FRESHWATER	639
*MARINE MOLLUSCS AND CRUSTACEANS	100 - 300 BASED ON WET WEIGHT	EDIBLE PART	918
*MARINE PHYTOPLANKTON	300 - 3000 BASED ON WET WEIGHT	30 W, 6 SPECIES	1324
*MOLLUSCS, ADDUCTOR MUSCLE	5 BASED ON WET WEIGHT	MARINE	671
*MOLLUSCS, ADDUCTOR MUSCLE	5 BASED ON WET WEIGHT	MARINE	671
*MOLLUSCS, VISCERA	GREATER THAN 40 BASED ON WET WEIGHT	MARINE	671
*MOLLUSCS, VISCERA	GREATER THAN 40 BASED ON WET WEIGHT	MARINE	671
*NITELLA OPACA	26000 BASED ON WET WEIGHT	EQUILIBRIUM AFTER 100 DAYS	603
*SUSPENDED SILT	86700 (RANGE = 16900 - 132000) BASED ON DRY WEIGHT	SEAWATER, BOMBAY	1261
*SUSPENDED SOLIDS	1330 (RANGE = 60 - 8160) BASED ON WET WEIGHT	DANUBE, 1973, NUCLEAR POWER PLANT SINCE 1967	658
*WATER PLANT	310 (RANGE = 50 - 630) BASED ON WET WEIGHT	DANUBE, 1973, NUCLEAR POWER PLANT SINCE 1967	658

TABLE D117

CONCENTRATION FACTOR : CHLORDANE

ITEM:	VALUE:	CONDITIONS:	REF.:
*ALFALFA FOLIAGE	302.4 BASED ON WET WEIGHT	CONC.IN SOIL=0.25MG/KG DRY WT.	389
*ALGAE	2 BASED ON WET WEIGHT	CONC.IN WATER=6.6NG/L	389
*CHUB	1.2 BASED ON WET WEIGHT	CONC.IN WATER=6.6UG/L	389
*EARTHWORMS	5.0 BASED ON WET WEIGHT	CONC.IN SOIL=0.02MG/KG DRY WT., U.S.,SITES=67	389

TABLE D117

CONCENTRATION FACTOR : CHLORDANE			
ITEM:	VALUE:	CONDITIONS:	REF.:
*GONIDEA SP.	1000 BASED ON WET WEIGHT	CONC.IN WATER=0.007UG/L,U.S.	389
*OYSTER	10 BASED ON WET WEIGHT	CONC.IN WATER:<1.0UG/L,FC: <0.001MG/KG,U.S.	389
*OYSTER	10 BASED ON WET WEIGHT	CONC.IN WATER:<1.0UG/L,CONC.IN SEDIMENT:<0.01MG/KG,U.S.	389
*SOYBEAN	0.005 BASED ON WET WEIGHT	CONC.IN SOIL=0.020MG/KG DRY WT.	389
*SUGAR BEET ROOT	0.182 BASED ON WET WEIGHT	CONC.IN LOAM SOIL=1.233MG/KG DRY WT.	389
*SWEET POTATO TUBERS	0.004 BASED ON WET WEIGHT	CONC.IN SOIL=0.28MG/KG DRY WT.	389
*VASCULAR PLANT	0.46 BASED ON WET WEIGHT	CONC.IN WATER=6.6NG/L	389

TABLE D118

CONCENTRATION FACTOR : CHLORINATED NAPHTHALENE			
ITEM:	VALUE:	CONDITIONS:	REF.:
*CHLOROCOCCUM SP.	25 BASED ON WET WT.	CONC.IN WATER=50PPB,24HRS,HALOWAX 1000(F024)	355
*CHLOROCOCCUM SP.	32 BASED ON WET WT.	CONC.IN WATER=100PPB,24HRS,HALOWAX 1000(F024)	355
*CHLOROCOCCUM SP.	60 BASED ON WET WT.	CONC.IN WATER=5PPB,24HRS,HALOWAX 1013(F025)	355
*CHLOROCOCCUM SP.	60 BASED ON WET WT.	CONC.IN WATER=10PPB,24HRS,HALOWAX 1013(F025)	355
*CHLOROCOCCUM SP.	84 BASED ON WET WT.	CONC.IN WATER=25PPB,24HRS,HALOWAX 1013(F025)	355
*CHLOROCOCCUM SP.	86 BASED ON WET WT.	CONC.IN WATER=50PPB,24HRS,HALOWAX 1013(F025)	355
*CHLOROCOCCUM SP.	120 BASED ON WET WT.	CONC.IN WATER=100PPB,24HRS,HALOWAX 1013(F025)	355
*CHLOROCOCCUM SP.	140 BASED ON WET WT.	CONC.IN WATER=5PPB,24HRS,HALOWAX 1014(F026)	355
*CHLOROCOCCUM SP.	110 BASED ON WET WT.	CONC.IN WATER=10PPB,24HRS,HALOWAX 1014(F026)	355
*CHLOROCOCCUM SP.	124 BASED ON WET WT.	CONC.IN WATER=25PPB,24HRS,HALOWAX 1014(F026)	355
*CHLOROCOCCUM SP.	110 BASED ON WET WT.	CONC.IN WATER=50PPB,24HRS,HALOWAX 1014(F026)	355
*CHLOROCOCCUM SP.	120 BASED ON WET WT.	CONC.IN WATER=100PPB,24HRS,HALOWAX 1014(F026)	355

TABLE D119

CONCENTRATION FACTOR : CL			
ITEM:	VALUE:	CONDITIONS:	REF.:
*BRASSICA OLERACEA	38 BASED ON WET WEIGHT	CONC.IN SOIL SOLUTION=29PPM	157
*BROWN ALGAE	0.062 BASED ON WET WEIGHT	NONE	157
*DAPHNIA MAGNA	17.5 BASED ON DRY WEIGHT	ON DRY WT.BASIS RELATIVE TO THE ALGAE CONSUMED	1630
*PLANKTON	1 BASED ON WET WEIGHT	NONE	157

TABLE D120

CONCENTRATION FACTOR : CO			
ITEM:	VALUE:	CONDITIONS:	REF.:
*AGARUM CRIBROSUM	200 BASED ON DRY WEIGHT	SEAWATER=5*10**(-7) G/L, VOSTOK BAY, SEA OF JAPAN	1367

TABLE D120

CONCENTRATION FACTOR : CO

ITEM:	VALUE:	CONDITIONS:	REF.:
*AHNFELTIA PARADOXA	910 BASED ON WET WEIGHT	SEAWATER = 0.14 UG/L, JAPAN	1095
*BROWN ALGAE	650 BASED ON WET WEIGHT	NONE	157
*BROWN ALGAE	450 BASED ON WET WEIGHT	NONE	994
*CHONDRUS OCELLATUS	860 BASED ON WET WEIGHT	SEAWATER = 0.14 UG/L, JAPAN	1095
*CHONDRUS YENDOI	150 BASED ON DRY WEIGHT	SEAWATER=5*10**(-7) G/L, VOSTOK BAY, SEA OF JAPAN	1367
*CHORDA FILUM	540 BASED ON DRY WEIGHT	SEAWATER=5*10**(-7) G/L, VOSTOK BAY, SEA OF JAPAN	1367
*CODIUM YESSOENSIS	60 BASED ON DRY WEIGHT	SEAWATER=5*10**(-7) G/L, VOSTOK BAY, SEA OF JAPAN	1367
*COSTARIA COSTATA	100 BASED ON DRY WEIGHT	SEAWATER=5*10**(-7) G/L, VOSTOK BAY, SEA OF JAPAN	1367
*EISENIA BICYCLIS	200 BASED ON WET WEIGHT	SEAWATER = 0.14 UG/L, JAPAN	1095
*ENTEROMORPHA PROLIFERA	1400 BASED ON DRY WEIGHT	SEAWATER=5*10**(-7) G/L, VOSTOK BAY, SEA OF JAPAN	1367
*FISH	50 BASED ON WET WEIGHT	MARINE ORGANISMS, TYPICAL VALUE	538
*FISH	90 BASED ON WET WEIGHT	FRESHWATER ORGANISMS, TYPICAL VALUE	1028
*FRESHWATER PLANKTON	37000	CONC.IN WATER=0.03UG/L,SCHOEHSEE	573
*HIZIKIA FUSIFORME	120 BASED ON WET WEIGHT	SEAWATER = 0.14 UG/L, JAPAN	1095
*LAMINARIA CHICHORIOIDES	500 BASED ON DRY WEIGHT	SEAWATER=5*10**(-7) G/L, VOSTOK BAY, SEA OF JAPAN	1367
*LAMINARIA JAPONICA	70 BASED ON WET WEIGHT	SEAWATER = 0.14 UG/L, JAPAN	1095
*MACROPHYTES	1367 BASED ON WET WEIGHT	FRESHWATER ORGANISMS, TYPICAL VALUE	1028
*MOLLUSCS	300 BASED ON WET WEIGHT	FRESHWATER ORGANISMS, TYPICAL VALUE	993
*MOLLUSCS, BIVALVE	30 - 1730 BASED ON WET WEIGHT	BOMBAY	116
*PELVETIA WRIGHTII	300 BASED ON DRY WEIGHT	SEAWATER=5*10**(-7) G/L, VOSTOK BAY, SEA OF JAPAN	1367
*PHYLLOSPADIX IWATENSIS	460 BASED ON DRY WEIGHT	SEAWATER=5*10**(-7) G/L, VOSTOK BAY, SEA OF JAPAN	1367
*PHYTOPLANKTON	15900-152000 BASED ON DRY WEIGHT	NONE	311
*PHYTOPLANKTON	100000 BASED ON DRY WEIGHT	MARINE	952
*PLANKTON	4600 BASED ON WET WEIGHT	NONE	157
*POLYSIPHONIA JAPONICA	740 BASED ON DRY WEIGHT	SEAWATER=5*10**(-7) G/L, VOSTOK BAY, SEA OF JAPAN	1367
*PTILOTA FILICINA	900 BASED ON DRY WEIGHT	SEAWATER=5*10**(-7) G/L, VOSTOK BAY, SEA OF JAPAN	1367
*RHODOMELA LARIX	750 BASED ON DRY WEIGHT	SEAWATER=5*10**(-7) G/L, VOSTOK BAY, SEA OF JAPAN	1367
*SARGASSUM PALLIDUM	300 BASED ON DRY WEIGHT	SEAWATER=5*10**(-7) G/L, VOSTOK BAY, SEA OF JAPAN	1367
*SARGASSUM RINGGOLDIANUM	260 BASED ON WET WEIGHT	SEAWATER = 0.14 UG/L, JAPAN	1095
*SARGASSUM THUNBERGII	1890 BASED ON WET WEIGHT	SEAWATER = 0.14 UG/L, JAPAN	1095
*SESTON	4.2 BASED ON WET WEIGHT	DEPTH = 8 M, SCH@HSEE	573
*SESTON	3.8 BASED ON WET WEIGHT	DEPTH = 17 M, SCH@HSEE	573
*ULVA FENESTRATA	200 BASED ON DRY WEIGHT	SEAWATER=5*10**(-7) G/L, VOSTOK BAY, SEA OF JAPAN	1367
*ULVA PERTUSA	340 BASED ON WET WEIGHT	SEAWATER = 0.14 UG/L, JAPAN	1095
*UNDARIA PINNATIFIDA	190 BASED ON WET WEIGHT	SEAWATER = 0.14 UG/L, JAPAN	1095
*ZOOPLANKTON	3200 BASED ON WET WEIGHT	SEAWATER CONCENTRATION 0.0005 MG/ M3	985
*ZOSTERA ASIATICA	770 BASED ON DRY WEIGHT	SEAWATER=5*10**(-7) G/L, VOSTOK BAY, SEA OF JAPAN	1367
*32 FRESHWATER PLANT	4425 BASED ON DRY WEIGHT	ALGAE,BRYOPHYTES AND VASCULAR HYDROPHYTES	1597

TABLE D121

CONCENTRATION FACTOR : CO-58

ITEM:	VALUE:	CONDITIONS:	REF.:
*FISH, BONE	220 (RANGE = 100 - 310) BASED ON WET WEIGHT	DANUBE, 1973, NUCLEAR POWER PLANT SINCE 1967	658

TABLE D121

CONCENTRATION FACTOR : CO-58			
ITEM:	VALUE:	CONDITIONS:	REF.:
*FISH, MUSCLE	130 (RANGE = 30 - 310) BASED ON WET WEIGHT	DANUBE, 1973, NUCLEAR POWER PLANT SINCE 1967	658
*FRESHWATER SEDIMENT	890 (RANGE = 310 - 2420) BASED ON WET WEIGHT	DANUBE, 1973, NUCLEAR POWER PLANT SINCE 1967	658
*MYTILUS EDULIS, BYSSUS	1500 BASED ON DRY WEIGHT	LAB., MARINE	1429
*MYTILUS EDULIS, SHELL	190 BASED ON DRY WEIGHT	LAB., MARINE	1429
*SUSPENDED SOLIDS	2860 (RANGE = 290 - 15440) BASED ON WET WEIGHT	DANUBE, 1973, NUCLEAR POWER PLANT SINCE 1967	658
*WATER PLANT	530 (RANGE = 15 - 1800) BASED ON WET WEIGHT	DANUBE, 1973, NUCLEAR POWER PLANT SINCE 1967	658

TABLE D122

CONCENTRATION FACTOR : CO-60			
ITEM:	VALUE:	CONDITIONS:	REF.:
*AHNFELTIA PARADOXA	290 BASED ON WET WEIGHT	SEAWATER = 0.14 UG/L, JAPAN	1095
*CHONDRUS OCELLATUS	833 BASED ON WET WEIGHT	SEAWATER = 0.14 UG/L, JAPAN	1095
*CLIBANARIUS VIRESCENS, WHOLE BODY	520 BASED ON DRY WEIGHT	15 - 30 D, MARINE	1428
*CRASSOSTREA GIGAS, BODY FLUID	20 BASED ON WET WEIGHT	STEADY STATE CF., INITIAL CF = 20	298
*CRASSOSTREA GIGAS, DIGESTIVE GLAND + STOMACH	140 BASED ON WET WEIGHT	STEADY STATE CF., INITIAL CF = 60	298
*CRASSOSTREA GIGAS, GILLS + LABIAL PALPS	96 BASED ON WET WEIGHT	STEADY STATE CF., INITIAL CF = 42	298
*CRASSOSTREA GIGAS, INTACT ANIMAL	300 BASED ON WET WEIGHT	STEADY STATE CF., INITIAL CF = 90	298
*CRASSOSTREA GIGAS, MANTLE	82 BASED ON WET WEIGHT	STEADY STATE CF., INITIAL CF = 38	298
*CRASSOSTREA GIGAS, MUSCLE	41 BASED ON WET WEIGHT	STEADY STATE CF., INITIAL CF = 33	298
*CRASSOSTREA GIGAS, SHELL	610 BASED ON WET WEIGHT	STEADY STATE CF., INITIAL CF = 200	298
*CRASSOSTREA GIGAS, SOFT TISSUES	51 BASED ON WET WEIGHT	STEADY STATE CF., INITIAL CF = 30	298
*CRASSOSTREA GIGAS, VISCERAL REMAINS	38 BASED ON WET WEIGHT	STEADY STATE CF., INITIAL CF = 19	298
*EISENIA BICYCLIS	34 BASED ON WET WEIGHT	SEAWATER = 0.14 UG/L, JAPAN	1095
*FISH, BONE	310 (RANGE = 140 - 420) BASED ON WET WEIGHT	DANUBE, 1973, NUCLEAR POWER PLANT SINCE 1967	658
*FISH, MUSCLE	130 (RANGE = 10 - 380) BASED ON WET WEIGHT	DANUBE, 1973, NUCLEAR POWER PLANT SINCE 1967	658
*FRESHWATER SEDIMENT	2830 (RANGE = 540 - 6070) BASED ON WET WEIGHT	DANUBE, 1973, NUCLEAR POWER PLANT SINCE 1967	658
*HIZIKIA FUSIFORME	127 BASED ON WET WEIGHT	SEAWATER = 0.14 UG/L, JAPAN	1095
*LAMINARIA JAPONICA	138 BASED ON WET WEIGHT	SEAWATER = 0.14 UG/L, JAPAN	1095
*LAMPSILIS RADIATA	790 BASED ON WET WEIGHT	91 DAYS, FRESHWATER	639
*LEANDER PACIFICUS, CARAPACE	18 BASED ON DRY WEIGHT	15 - 30 D, MARINE	1428
*LEANDER PACIFICUS, FLESH	7 BASED ON DRY WEIGHT	15 - 30 D, MARINE	1428
*MYTILUS EDULIS, SHELL	190 BASED ON DRY WEIGHT	15 - 30 D, MARINE	1428
*MYTILUS EDULIS, SOFT PARTS	140 BASED ON DRY WEIGHT	15 - 30 D, MARINE	1428
*PARTICLE FRACTIONS	1.6 % OF TOTAL	SIZE=2.0-0.5 MM,WEIGHT=3.2 %,AREA= 0.2 %,MARINE, JAPAN	1095
*PARTICLE FRACTIONS	3.2 % OF TOTAL	SIZE=0.5-0.25 MM,WEIGHT=8.5 %,AREA =1.4 %,MARINE, JAPAN	1095
*PARTICLE FRACTIONS	10.8 % OF TOTAL	SIZE=0.25-0.125 MM,WEIGHT=23 %, AREA=10.1 %,MARINE, JAPAN	1095
*PARTICLE FRACTIONS	10.5 % OF TOTAL	SIZE=0.125-0.075 MM,WEIGHT=16.9 %, AREA=10.4 %,MARINE, JAPAN	1095
*PARTICLE FRACTIONS	73.9 % OF TOTAL	SIZE BELOW 0.075 MM,WEIGHT=48.4 %, AREA=77.9 %,MARINE, JAPAN	1095
*SARGASSUM THUNBERGII	2574 BASED ON WET WEIGHT	SEAWATER = 0.14 UG/L, JAPAN	1095
*SEDIMENT	169 BASED ON DRY WEIGHT	SILTY SAND SEDIMENT, MARINE, URAZOKO BAY, JAPAN	1095

TABLE D122

CONCENTRATION FACTOR : CO-60

ITEM:	VALUE:	CONDITIONS:	REF.:
*SEDIMENT	231 BASED ON DRY WEIGHT	SANDY SILT SEDIMENT, MARINE, URAZOKO BAY, JAPAN	1095
*SEDIMENT	26 BASED ON DRY WEIGHT	SAND SEDIMENT, MARINE, URAZOKO BAY, JAPAN	1095
*SEDIMENT	36 BASED ON DRY WEIGHT	SAND SEDIMENT, MARINE, URAZOKO BAY, JAPAN	1095
*SEDIMENT	20 BASED ON DRY WEIGHT	SAND SEDIMENT, MARINE, URAZOKO BAY, JAPAN	1095
*SEDIMENT	16 BASED ON DRY WEIGHT	SAND SEDIMENT, MARINE, URAZOKO BAY, JAPAN	1095
*STYELA PLICATA, TEST	400 BASED ON DRY WEIGHT	15 - 30 D, MARINE	1428
*STYELA PLICATA, VISCERA	850 BASED ON DRY WEIGHT	15 - 30 D, MARINE	1428
*SUSPENDED SOLIDS	5380 (RANGE = 470 - 12040) BASED ON WET WEIGHT	DANUBE, 1973, NUCLEAR POWER PLANT SINCE 1967	658
*ULVA PERTUSA	439 BASED ON WET WEIGHT	SEAWATER = 0.14 UG/L, JAPAN	1095
*UNDARIA PINNATIFIDA	210 BA¤ED ON WET WEIGHT	SEAWATER = 0.14 UG/L, JAPAN	1095
*WATER PLANT	220 (RANGE = 25 - 520) BASED ON WET WEIGHT	DANUBE, 1973, NUCLEAR POWER PLANT SINC¤ 1967	658

TABLE D123

CONCENTRATION FACTOR : CR

ITEM:	VALUE:	CONDITIONS:	REF.:
*AGARUM CRIBROSUM	590000 BASED ON DRY WEIGHT	SEAWATER=2.0*10**(-9) G/L, VOSTOK BAY, SEA OF JAPAN	1367
*BROWN ALGAE	6500 BASED ON WET WEIGHT	NONE	157
*BROWN ALGAE	300-10000 BASED ON WET WEIGHT	NONE	994
*CHONDRUS YENDOI	450000 BASED ON DRY WEIGHT	SEAWATER=2.0*10**(-9) G/L, VOSTOK BAY, SEA OF JA¤AN	1367
*CHORDA FILUM	310000 BASED ON DRY WEIGHT	SEAWATER=2.0*10**(-9) G/L, VOSTOK BAY, SEA OF JAPAN	1367
*CODIUM YESSOENSIS	1095000 BASED ON DRY WEIGHT	SEAWATER=2.0*10**(-9) G/L, VOSTOK BAY, SEA OF JAPAN	1367
*COSTARIA COSTATA	175000 BASED ON DRY WEIGHT	SEAWATER=2.0*10**(-9) G/L, VOSTOK BAY, SEA OF JAPAN	1367
*ENTEROMORPHA PROLIFERA	1170000 BASED ON DRY WEIGHT	SEAWATER=2.0*10**(-9) G/L, VOSTOK BAY, SEA OF JAPAN	1367
*FISH	10 BASED ON WET WEIGHT	FRESHWATER ORGANISMS, TYPICAL VALUE	1028
*HETEROCHORDARIA ABIETINA	435000 BASED ON DRY WEIGHT	SEAWATER=2.0*10**(-9) G/L, VOSTOK BAY, SEA OF JAPAN	1367
*LAMINARIA CHICHORIOIDES	335000 BASED ON DRY WEIGHT	SEAWATER=2.0*10**(-9) G/L, VOSTOK BAY, SEA OF JAPAN	1367
*LAMINARIA JAPONICA	145000 BASED ON DRY WEIGHT	SEAWATER=2.0*10**(-9) G/L, VOSTOK BAY, SEA OF JAPAN	1367
*MACROPHYTES	2880 BASED ON WET WEIGHT	MARINE ORGANISMS, TYPICAL VALUE	589
*MOLLUSCS	21800 BASED ON WET WEIGHT	MARINE ORGANISMS, TYPICAL VALUE	1280
*MOLLUSCS	267 BASED ON WET WEIGHT	FRESHWATER ORGANISMS, TYPICAL VALUE	993
*PELVETIA WRIGHTII	325000 BASED ON DRY WEIGHT	SEAWATER=2.0*10**(-9) G/L, VOSTOK BAY, SEA OF JAPAN	1367
*PHYLLOSPADIX IWATENSIS	440000 BASED ON DRY WEIGHT	SEAWATER=2.0*10**(-9) G/L, VOSTOK BAY, SEA OF JAPAN	1367
*PHYTOPLANKTON	17900-81000 BASED ON WET WEIGHT	NONE	311
*PHYTOPLANKTON	7800 BASED ON WET WEIGHT	MARINE ORGANISMS, TYPICAL VALUE	986
*PLANKTON	17000 BASED ON WET WEIGHT	NONE	157
*POLYSIPHONIA JAPONICA	1120000 BASED ON DRY WEIGHT	SEAWATER=2.0*10**(-9) G/L, VOSTOK BAY, SEA OF JAPAN	1367
*PTILOTA FILICINA	300000 BASED ON DRY WEIGHT	SEAWATER=2.0*10**(-9) G/L, VOSTOK BAY, SEA OF JAPAN	1367

TABLE D123

CONCENTRATION FACTOR : CR			
ITEM:	VALUE:	CONDITIONS:	REF.:
*RHODOMELA LARIX	785000 BASED ON DRY WEIGHT	SEAWATER=2.0*10**(-9) G/L, VOSTOK BAY, SEA OF JAPAN	1367
*SARGASSUM PALLIDUM	195000 BASED ON DRY WEIGHT	SEAWATER=2.0*10**(-9) G/L, VOSTOK BAY, SEA OF JAPAN	1367
*ULVA FENESTRATA	172000 BASED ON DRY WEIGHT	SEAWATER=2.0*10**(-9) G/L, VOSTOK BAY, SEA OF JAPAN	1367
*ULVARIA SPENDENS	55000 BASED ON DRY WEIGHT	SEAWATER=2.0*10**(-9) G/L, VOSTOK BAY, SEA OF JAPAN	1367
*ZOSTERA ASIATICA	215000 BASED ON DRY WEIGHT	SEAWATER=2.0*10**(-9) G/L, VOSTOK BAY, SEA OF JAPAN	1367
*32 FRESHWATER PLANT	695 BASED ON DRY WEIGHT	ALGAE,BRYOPHYTES AND VASCULAR HYDROPHYTES	1597

TABLE D124

CONCENTRATION FACTOR : CR-51			
ITEM:	VALUE:	CONDITIONS:	REF.:
*HERMIONE HYSTRIX	15 BASED ON WET WEIGHT	19 D	248
*LAMPSILIS RADIATA	440 BASED ON WET WEIGHT	91 DAYS, FRESHWATER	639

TABLE D125

CONCENTRATION FACTOR : CS			
ITEM:	VALUE:	CONDITIONS:	REF.:
*AMERIURUS NEBULOSUS	212 BASED ON WET WEIGHT	WINTERGREEN LAKE, STABLE CS, MAY 1971	437
*BOWFIN	268 BASED ON WET WEIGHT	WINTERGREEN LAKE, STABLE CS, MAY 1971	437
*BROWN ALGAE	33 BASED ON WET WEIGHT	NONE	157
*BROWN ALGAE	1-100 BASED ON WET WEIGHT	NONE	994
*CERATOPHYLLUM SP.	420 BASED ON WET WEIGHT	WINTERGREEN LAKE, STABLE CS, MAY 1971	437
*HYBRID SUNFISH	333 BASED ON WET WEIGHT	WINTERGREEN LAKE, STABLE CS, SEP 1970	437
*HYBRID SUNFISH	152 BASED ON WET WEIGHT	WINTERGREEN LAKE, STABLE CS, MAY 1971	437
*LAKE CHUBSUCKER	262 BASED ON WET WEIGHT	WINTERGREEN LAKE, STABLE CS, SEP 1970	437
*LAKE CHUBSUCKER	292 BASED ON WET WEIGHT	WINTERGREEN LAKE, STABLE CS, MAY 1971	437
*LARGEMOUTH BASS	723 BASED ON WET WEIGHT	WINTERGREEN LAKE, STABLE CS, SEP 1970	437
*LARGEMOUTH BASS	348 BASED ON WET WEIGHT	WINTERGREEN LAKE, STABLE CS, MAY 1971	437
*LARGEMOUTH BASS	3188 BASED ON WET WEIGHT	WINTERGREEN LAKE, STABLE CS, MAY 1971	437
*NUPHAR SP.	416 BASED ON WET WEIGHT	WINTERGREEN LAKE, STABLE CS, MAY 1971	437
*PHYTOPLANKTON	2000-42000 BASED ON WET WEIGHT	NONE	311
*PHYTOPLANKTON	1595 BASED ON WET WEIGHT	WINTERGREEN LAKE, STABLE CS, SEP 1970	437
*PHYTOPLANKTON	112 BASED ON WET WEIGHT	WINTERGREEN LAKE, STABLE CS, MAY 1971	437
*PLANKTON	1-5 BASED ON WET WEIGHT	NONE	994
*SALMO TRUTTA	1.02*10**3 BASED ON WET WEIGHT	WHOLE FISH, STABLE CS, FISH= 6.6 G, SOFT WATER AREA	662

TABLE D125

CONCENTRATION FACTOR : CS

ITEM:	VALUE:	CONDITIONS:	REF.:
*SALMO TRUTTA	BASED ON WET WEIGHT	WHOLE FISH, STABLE CS, FISH=23.0 G, SOFT WATER AREA	662
*SALMO TRUTTA	6.80*10**3 BASED ON WET WEIGHT	WHOLE FISH, STABLE CS, FISH= 354 G, SOFT WATER AREA	662
*SALMO TRUTTA	7.80*10**3 BASED ON WET WEIGHT	WHOLE FISH, STABLE CS, FISH= 850 G, SOFT WATER AREA	662
*SALMO TRUTTA	7.8*10**3 BASED ON WET WEIGHT	WHOLE FISH, STABLE CS, FISH= 850 G, SOFT WATER AREA	662
*YELLOW PERCH	2009 BASED ON WET WEIGHT	WINTERGREEN LAKE, STABLE CS, SEP 1970	437
*YELLOW PERCH	1016 BASED ON WET WEIGHT	WINTERGREEN LAKE, STABLE CS, MAY 1971	437
*ZOOPLANKTON	248 BASED ON WET WEIGHT	WINTERGREEN LAKE, STABLE CS, SEP 1970	437
*32 FRESHWATER PLANT	480 BASED ON DRY WEIGHT	ALGAE,BRYOPHYTES AND VASCULAR HYDROPHYTES	1597

TABLE D126

CONCENTRATION FACTOR : CS-133

ITEM:	VALUE:	CONDITIONS:	REF.:
*GAMMARUS LACUSTRIS	24 BASED ON WET WEIGHT	0.06 D AFTER START, EAST TWIN LAKE, REL. TO FILTERED WATER	591
*GAMMARUS LACUSTRIS	243 BASED ON WET WEIGHT	3 D AFTER START, EAST TWIN LAKE, REL. TO FILTERED WATER	591
*GAMMARUS LACUSTRIS	994 BASED ON WET WEIGHT	39 D AFTER START, EAST TWIN LAKE, REL. TO FILTERED WATER	591
*GAMMARUS LACUSTRIS	1030 BASED ON WET WEIGHT	393 D AFTER START, EAST TWIN LAKE, REL. TO FILTERED WATER	591
*GAMMARUS LACUSTRIS	184 BASED ON WET WEIGHT	393 D AFTER START, EAST TWIN LAKE, REL. TO TOTAL CONC.	591
*GAMMARUS LACUSTRIS	24 BASED ON WET WEIGHT	0.06 D AFTER START, EAST TWIN LAKE, REL. TO FILTERED WATER	591
*GAMMARUS LACUSTRIS	243 BASED ON WET WEIGHT	3 D AFTER START, EAST TWIN LAKE, REL. TO FILTERED WATER	591
*GAMMARUS LACUSTRIS	994 BASED ON WET WEIGHT	39 D AFTER START, EAST TWIN LAKE, REL. TO FILTERED WATER	591
*GAMMARUS LACUSTRIS	1030 BASED ON WET WEIGHT	393 D AFTER START, EAST TWIN LAKE, REL. TO FILTERED WATER	591
*GAMMARUS LACUSTRIS	184 BASED ON WET WEIGHT	393 D AFTER START, EAST TWIN LAKE, REL. TO TOTAL CONC.	591
*LEMNA TRISULCA	140 BASED ON WET WEIGHT	3 D AFTER START, EAST TWIN LAKE, REL. TO FILTERED WATER	591
*LEMNA TRISULCA	324 BASED ON WET WEIGHT	39 D AFTER START, EAST TWIN LAKE, REL. TO FILTERED WATER	591
*LEMNA TRISULCA	1264 BASED ON WET WEIGHT	393 D AFTER START, EAST TWIN LAKE, REL. TO FILTERED WATER	591
*LEMNA TRISULCA	228 BASED ON WET WEIGHT	393 D AFTER START, EAST TWIN LAKE, REL. TO TOTAL CONC.	591
*LEMNA TRISULCA	140 BASED ON WET WEIGHT	3 D AFTER START, EAST TWIN LAKE, REL. TO FILTERED WATER	591
*LEMNA TRISULCA	324 BASED ON WET WEIGHT	39 D AFTER START, EAST TWIN LAKE, REL. TO FILTERED WATER	591
*LEMNA TRISULCA	1264 BASED ON WET WEIGHT	393 D AFTER START, EAST TWIN LAKE, REL. TO FILTERED WATER	591
*LEMNA TRISULCA	228 BASED ON WET WEIGHT	393 D AFTER START, EAST TWIN LAKE, REL. TO TOTAL CONC.	591
*NUPHAR POLYSEPALUM	18 BASED ON WET WEIGHT	3 D AFTER START, EAST TWIN LAKE, REL. TO FILTERED WATER	591
*NUPHAR POLYSEPALUM	101 BASED ON WET WEIGHT	39 D AFTER START, EAST TWIN LAKE, REL. TO FILTERED WATER	591
*NUPHAR POLYSEPALUM	280 BASED ON WET WEIGHT	393 D AFTER START, EAST TWIN LAKE, REL. TO FILTERED WATER	591

TABLE D126

CONCENTRATION FACTOR : CS-133

ITEM:	VALUE:	CONDITIONS:	REF.:
*NUPHAR POLYSEPALUM	50 BASED ON WET WEIGHT	393 D AFTER START, EAST TWIN LAKE, REL. TO TOTAL CONC.	591
*NUPHAR POLYSEPALUM	18 BASED ON WET WEIGHT	3 D AFTER START, EAST TWIN LAKE, REL. TO FILTERED WATER	591
*NUPHAR POLYSEPALUM	101 BASED ON WET WEIGHT	39 D AFTER START, EAST TWIN LAKE, REL. TO FILTERED WATER	591
*NUPHAR POLYSEPALUM	280 BASED ON WET WEIGHT	393 D AFTER START, EAST TWIN LAKE, REL. TO FILTERED WATER	591
*NUPHAR POLYSEPALUM	50 BASED ON WET WEIGHT	393 D AFTER START, EAST TWIN LAKE, REL. TO TOTAL CONC.	591
*SALMO GAIRDNERI	85 BASED ON WET WEIGHT	3 D AFTER START, EAST TWIN LAKE, REL. TO FILTERED WATER	591
*SALMO GAIRDNERI	873 BASED ON WET WEIGHT	39 D AFTER START, EAST TWIN LAKE, REL. TO FILTERED WATER	591
*SALMO GAIRDNERI	2447 BASED ON WET WEIGHT	393 D AFTER START, EAST TWIN LAKE, REL. TO FILTERED WATER	591
*SALMO GAIRDNERI	430 BASED ON WET WEIGHT	393 D AFTER START, EAST TWIN LAKE, REL. TO TOTAL CONC.	591
*SALMO GAIRDNERI	85 BASED ON WET WEIGHT	3 D AFTER START, EAST TWIN LAKE, REL. TO FILTERED WATER	591
*SALMO GAIRDNERI	873 BASED ON WET WEIGHT	39 D AFTER START, EAST TWIN LAKE, REL. TO FILTERED WATER	591
*SALMO GAIRDNERI	2447 BASED ON WET WEIGHT	393 D AFTER START, EAST TWIN LAKE, REL. TO FILTERED WATER	591
*SALMO GAIRDNERI	430 BASED ON WET WEIGHT	393 D AFTER START, EAST TWIN LAKE, REL. TO TOTAL CONC.	591
*SEDIMENT	86 ML/CM**2 (BASED ON SURFACE OF SEDIMENT)	0.06 D AFTER START, EAST TWIN LAKE, REL. TO FILTERED WATER	591
*SEDIMENT	639 ML/CM**2 (BASED ON SURFACE OF SEDIMENT)	3 D AFTER START, EAST TWIN LAKE, REL. TO FILTERED WATER	591
*SEDIMENT	1945 ML/CM**2 (BASED ON SURFACE OF SEDIMENT)	39 D AFTER START, EAST TWIN LAKE, REL. TO FILTERED WATER	591
*SEDIMENT	7570 ML/CM**2 (BASED ON SURFACE OF SEDIMENT)	393 D AFTER START, EAST TWIN LAKE, REL. TO FILTERED WATER	591
*SEDIMENT	1357 ML/CM**2 (BASED ON SURFACE OF SEDIMENT)	393 D AFTER START, EAST TWIN LAKE, REL. TO TOTAL CONC.	591
*SEDIMENT	86 ML/CM**2 (BASED ON SURFACE OF SEDIMENT)	0.06 D AFTER START, EAST TWIN LAKE, REL. TO FILTERED WATER	591
*SEDIMENT	639 ML/CM**2 (BASED ON SURFACE OF SEDIMENT)	3 D AFTER START, EAST TWIN LAKE, REL. TO FILTERED WATER	591
*SEDIMENT	1945 ML/CM**2 (BASED ON SURFACE OF SEDIMENT)	39 D AFTER START, EAST TWIN LAKE, REL. TO FILTERED WATER	591
*SEDIMENT	7570 ML/CM**2 (BASED ON SURFACE OF SEDIMENT)	393 D AFTER START, EAST TWIN LAKE, REL. TO FILTERED WATER	591
*SEDIMENT	1357 ML/CM**2 (BASED ON SURFACE OF SEDIMENT)	393 D AFTER START, EAST TWIN LAKE, REL. TO TOTAL CONC.	591
*ZOOPLANKTON	9.3 BASED ON WET WEIGHT	0.06 D AFTER START, EAST TWIN LAKE, REL. TO FILTERED WATER	591
*ZOOPLANKTON	63 BASED ON WET WEIGHT	3 D AFTER START, EAST TWIN LAKE, REL. TO FILTERED WATER	591
*ZOOPLANKTON	228 BASED ON WET WEIGHT	39 D AFTER START, EAST TWIN LAKE, REL. TO FILTERED WATER	591
*ZOOPLANKTON	112 BASED ON WET WEIGHT	393 D AFTER START, EAST TWIN LAKE, REL. TO FILTERED WATER	591
*ZOOPLANKTON	20 BASED ON WET WEIGHT	393 D AFTER START, EAST TWIN LAKE, REL. TO TOTAL CONC.	591
*ZOOPLANKTON	9.3 BASED ON WET WEIGHT	0.06 D AFTER START, EAST TWIN LAKE, REL. TO FILTERED WATER	591
*ZOOPLANKTON	63 BASED ON WET WEIGHT	3 D AFTER START, EAST TWIN LAKE, REL. TO FILTERED WATER	591
*ZOOPLANKTON	228 BASED ON WET WEIGHT	39 D AFTER START, EAST TWIN LAKE, REL. TO FILTERED WATER	591
*ZOOPLANKTON	112 BASED ON WET WEIGHT	393 D AFTER START, EAST TWIN LAKE, REL. TO FILTERED WATER	591
*ZOOPLANKTON	20 BASED ON WET WEIGHT	393 D AFTER START, EAST TWIN LAKE, REL. TO TOTAL CONC.	591

TABLE D127

CONCENTRATION FACTOR : CS-134

ITEM:	VALUE:	CONDITIONS:	REF.:
*FISH, BONE	240 (RANGE = 60 - 500) BASED ON WET WEIGHT	DANUBE, 1973, NUCLEAR POWER PLANT SINCE 1967	658
*FISH, MUSCLE	200 (RANGE = 30 - 380) BASED ON WET WEIGHT	DANUBE, 1973, NUCLEAR POWER PLANT SINCE 1967	658
*FRESHWATER PHYTOPLANKTON	4000 - 13000 BASED ON DRY WEIGHT	EQUILIBRIUM AFTER 36 H	534
*FRESHWATER PHYTOPLANKTON	4000 - 13000 BASED ON DRY WEIGHT	EQUILIBRIUM AFTER 36 H	534
*FRESHWATER SEDIMENT	3040 (RANGE = 840 - 5360) BASED ON WET WEIGHT	DANUBE, 1973, NUCLEAR POWER PLANT SINCE 1967	658
*MARINE ALGAE	1.2 - 2.0 BASED ON WET WEIGHT	1 D	671
*MARINE ALGAE	1.2 - 2.0 BASED ON WET WEIGHT	1 D	671
*MARINE CRUSTACEAN, SOFT PARTS	15 BASED ON WET WEIGHT	4 D	671
*MARINE CRUSTACEAN, SOFT PARTS	15 BASED ON WET WEIGHT	4 D	671
*ROOTED AQUATIC PLANT	400 BASED ON DRY WEIGHT	EQUILIBRIUM AFTER 36 H	534
*ROOTED AQUATIC PLANT	400 BASED ON DRY WEIGHT	EQUILIBRIUM AFTER 36 H	534
*SUSPENDED SOLIDS	13950 (RANGE = 1000 - 58210) BASED ON WET WEIGHT	DANUBE, 1973, NUCLEAR POWER PLANT SINCE 1967	658
*WATER PLANT	320 (RANGE = 10 - 1350) BASED ON WET WEIGHT	DANUBE, 1973, NUCLEAR POWER PLANT SINCE 1967	658

TABLE D128

CONCENTRATION FACTOR : CS-137

ITEM:	VALUE:	CONDITIONS:	REF.:
*ALGAE	LESS THAN 40 BASED ON WET WEIGHT	LAB VALUE	25
*AMERIURUS NEBULOSUS	842 BASED ON WET WEIGHT	WINTERGREEN LAKE, MAY 1971	437
*ASTROPECTEN ARANCIACUS	8 BASED ON DRY WEIGHT	EASTERN MEDITERRANEAN, DRY WEIGHT = 43.4 G	525
*BOOPS BOOPS	208 BASED ON DRY WEIGHT	EASTERN MEDITERRANEAN, DRY WEIGHT = 91.7 G	525
*BOWFIN	440 BASED ON WET WEIGHT	WINTERGREEN LAKE, MAY 1971	437
*BRISSOPSIS LYRIFERA	261 BASED ON DRY WEIGHT	EASTERN MEDITERRANEAN, DRY WEIGHT = 333.0 G	525
*BRISSOPSIS LYRIFERA	339 BASED ON DRY WEIGHT	EASTERN MEDITERRANEAN, DRY WEIGHT = 153.0 G	525
*CERATOPHYLLUM SP.	302 BASED ON WET WEIGHT	WINTERGREEN LAKE, MAY 1971	437
*CHARYDIS LONGICOLLIS	57 BASED ON DRY WEIGHT	EASTERN MEDITERRANEAN, DRY WEIGHT = 90.5 G	525
*CRASSOSTREA GIGAS, BODY FLUID	3.1 BASED ON WET WEIGHT	STEADY STATE CF., INITIAL CF = 3.1	298
*CRASSOSTREA GIGAS, DIGESTIVE GLAND + STOMACH	11.3 BASED ON WET WEIGHT	STEADY STATE CF., INITIAL CF = 10.7	298
*CRASSOSTREA GIGAS, GILLS + LABIAL PALPS	10.1 BASED ON WET WEIGHT	STEADY STATE CF., INITIAL CF = 9.5	298
*CRASSOSTREA GIGAS, INTACT ANIMAL	7.7 BASED ON WET WEIGHT	STEADY STATE CF., INITIAL CF = 5.3	298
*CRASSOSTREA GIGAS, MANTLE	9.4 BASED ON WET WEIGHT	STEADY STATE CF., INITIAL CF = 8.6	298
*CRASSOSTREA GIGAS, MUSCLE	6.6 BASED ON WET WEIGHT	STEADY STATE CF., INITIAL CF = 5.4	298
*CRASSOSTREA GIGAS, SHELL	9.2 BASED ON WET WEIGHT	STEADY STATE CF., INITIAL CF = 7.2	298
*CRASSOSTREA GIGAS, SOFT TISSUES	9.4 BASED ON WET WEIGHT	STEADY STATE CF., INITIAL CF = 8.8	298
*CRASSOSTREA GIGAS, VISCERAL REMAINS	6.9 BASED ON WET WEIGHT	STEADY STATE CF., INITIAL CF = 6.4	298
*ENTEROMORPHA SP.	19 BASED ON WET WEIGHT	CONC. IN BODY = 4900 MG/KG WW	M(VI
*ESKMEALS SILT	449.0 BASED ON WET WEIGHT	YEARLY MEAN, CUMBERLAND COASTRAL AREA, 1969	1240
*FISH	LESS THAN 40 BASED ON WET WEIGHT	LAB VALUE	25
*FISH, BONE	210 (RANGE = 20 - 640) BASED ON WET WEIGHT	DANUBE, 1973, NUCLEAR POWER PLANT SINCE 1967	658
*FISH, MUSCLE	170 (RANGE = 35 - 350) BASED ON WET WEIGHT	DANUBE, 1973, NUCLEAR POWER PLANT SINCE 1967	658

TABLE D128

CONCENTRATION FACTOR : CS-137

ITEM:	VALUE:	CONDITIONS:	REF.:
*FRESHWATER SEDIMENT	3160 (RANGE = 1340 - 5950) BASED ON WET WEIGHT	DANUBE, 1973, NUCLEAR POWER PLANT SINCE 1967	658
*FRESHWATER SEDIMENT	3500 BASED ON DRY WEIGHT	LAKE ERIE, WESTERN BASIN	912
*FRESHWATER SEDIMENT	10000 BASED ON DRY WEIGHT	LAKE ERIE, CENTRAL BASIN	912
*FRESHWATER SEDIMENT	10000 BASED ON DRY WEIGHT	LAKE ONTARIO, WESTERN PART	912
*FUCUS SERRATUS	46 BASED ON WET WEIGHT	CONC. IN BODY = 10550 MG/KG WW	M(VI
*FUCUS SERRATUS	32 BASED ON WET WEIGHT	NONE	186
*FUCUS SERRATUS	53 BASED ON WET WEIGHT	NONE	1561
*FUCUS VESICULOSUS	46 BASED ON WET WEIGHT	CONC. IN BODY = 10140 MG/KG WW	M(VI
*FUCUS VESICULOSUS	28 - 78 BASED ON WET WEIGHT	NONE	1044
*FUCUS VESICULOSUS	111 BASED ON WET WEIGHT	NONE	1045
*FUCUS VESICULOSUS	51 - 74 BASED ON WET WEIGHT	NONE	1047
*FUCUS VESICULOSUS	27 BASED ON WET WEIGHT	NONE	1046
*FUCUS VESICULOSUS	74 BASED ON WET WEIGHT	NONE	186
*HYBRID SUNFISH	525 BASED ON WET WEIGHT	WINTERGREEN LAKE, SEP 1970	437
*HYBRID SUNFISH	429 BASED ON WET WEIGHT	WINTERGREEN LAKE, MAY 1971	437
*INSHORE SAND	36.3 BASED ON WET WEIGHT	YEARLY MEAN, CUMBERLAND COASTRAL AREA, 1969	1240
*INVERTEBRATE	LESS THAN 40 BASED ON WET WEIGHT	LAB VALUE	25
*LAKE CHUBSUCKER	875 BASED ON WET WEIGHT	WINTERGREEN LAKE, SEP 1970	437
*LAKE CHUBSUCKER	642 BASED ON WET WEIGHT	WINTERGREEN LAKE, MAY 1971	437
*LAMPSILIS RADIATA	220 BASED ON WET WEIGHT	91 DAYS, FRESHWATER	639
*LARGEMOUTH BASS	1671 BASED ON WET WEIGHT	WINTERGREEN LAKE, SEP 1970	437
*LARGEMOUTH BASS	1426 BASED ON WET WEIGHT	WINTERGREEN LAKE, MAY 1971	437
*LARGEMOUTH BASS	701 BASED ON WET WEIGHT	WINTERGREEN LAKE, MAY 1971	437
*LITTORINA LITTOREA, SHELL	13 BASED ON WET WEIGHT	NONE	M(VI
*LITTORINA LITTOREA, SOFT PARTS	43 BASED ON WET WEIGHT	CONC. IN BODY = 3320 MG/KG WW	M(VI
*LITTORINA LITTOREA, SOFT PARTS	18 - 72 BASED ON WET WEIGHT	NONE	186
*MARINE ALGAE	53 BASED ON DRY WEIGHT	EASTERN MEDITERRANEAN, DRY WEIGHT = 200.0 G	525
*MARINE ALGAE	138 BASED ON DRY WEIGHT	EASTERN MEDITERRANEAN, DRY WEIGHT = 185.4 G	525
*MARINE ALGAE	30 BASED ON WET WEIGHT	BROWN ALGAE	1267
*MARINE ALGAE	10 BASED ON WET WEIGHT	GREEN ALGAE	1267
*MARINE MOLLUSCS AND CRUSTACEANS	4 - 10 BASED ON WET WEIGHT	EDIBLE PART	918
*METAPENAEUS MONOCEROS	57 BASED ON DRY WEIGHT	EASTERN MEDITERRANEAN, DRY WEIGHT = 127 G	525
*MUGIL CAPITO	106 BASED ON DRY WEIGHT	EASTERN MEDITERRANEAN, DRY WEIGHT = 238.0 G	525
*MUGIL CEPHALUS	106 BASED ON DRY WEIGHT	EASTERN MEDITERRANEAN, DRY WEIGHT = 238.0 G	525
*MYTILUS EDULIS, SHELL	9 BASED ON WET WEIGHT	NONE	190
*MYTILUS EDULIS, SOFT PARTS	47 BASED ON WET WEIGHT	CONC. IN BODY = 3040 MG/KG WW	190
*MYTILUS EDULIS, SOFT PARTS	12 - 44 BASED ON WET WEIGHT	NONE	995
*NEPHTYS SP.	3.5 BASED ON WET WEIGHT	YEARLY MEAN, CUMBERLAND COASTRAL AREA, 1969	1240
*NUCELLA LAPILLUS	57 BASED ON WET WEIGHT	NONE	995
*NUCELLA LAPILLUS, SHELL	31 BASED ON WET WEIGHT	NONE	190
*NUCELLA LAPILLUS, SOFT PARTS	99 BASED ON WET WEIGHT	CONC. IN BODY = 2890 MG/KG WW	190
*NUCULA SP.	14.3 BASED ON WET WEIGHT	YEARLY MEAN, CUMBERLAND COASTRAL AREA, 1969	1240
*NUPHAR SP.	333 BASED ON WET WEIGHT	WINTERGREEN LAKE, MAY 1971	437
*PARAPENAEUS LONGIROSTRIS	102 BASED ON DRY WEIGHT	EASTERN MEDITERRANEAN, DRY WEIGHT = 50.0 G	525
*PARTICLE FRACTIONS	0.3 % OF TOTAL	SIZE=2.0-0.5 MM,WEIGHT=3.2 %,AREA= 0.2 %,MARINE, JAPAN	1095

TABLE D128

CONCENTRATION FACTOR : CS-137

ITEM:	VALUE:	CONDITIONS:	REF.:
*PARTICLE FRACTIONS	3.9 % OF TOTAL	SIZE=0.5-0.25 MM,WEIGHT=8.5 %,AREA =1.4 %,MARINE, JAPAN	1095
*PARTICLE FRACTIONS	16.2 % OF TOTAL	SIZE=0.25-0.125 MM,WEIGHT=23 %, AREA=10.1 %,MARINE, JAPAN	1095
*PARTICLE FRACTIONS	15.5 % OF TOTAL	SIZE=0.125-0.075 MM,WEIGHT=16.9 %, AREA=10.4 %,MARINE, JAPAN	1095
*PARTICLE FRACTIONS	64.1 % OF TOTAL	SIZE BELOW 0.075 MM,WEIGHT=48.4 %, AREA=77.9 %,MARINE, JAPAN	1095
*PENAEUS JAPONICUS	35 BASED ON DRY WEIGHT	EASTERN MEDITERRANEAN, DRY WEIGHT = 133.0 G	525
*PENAEUS JAPONICUS	57 BASED ON DRY WEIGHT	EASTERN MEDITERRANEAN, DRY WEIGHT = 208.0 G	525
*PENNATULA RUBRA	71 BASED ON DRY WEIGHT	EASTERN MEDITERRANEAN, DRY WEIGHT = 97.2 G	525
*PHYTOPLANKTON	833 BASED ON WET WEIGHT	WINTERGREEN LAKE, SEP 1970	437
*PHYTOPLANKTON	151 BASED ON WET WEIGHT	WINTERGREEN LAKE, MAY 1971	437
*PLEURONECTES PLATESSA, FLESH	26.7 BASED ON WET WEIGHT	YEARLY MEAN, CUMBERLAND COASTRAL AREA, 1969	1240
*PLEURONECTES PLATESSA, GUT CONTENT	39.0 BASED ON WET WEIGHT	YEARLY MEAN, CUMBERLAND COASTRAL AREA, 1969	1240
*SALMO TRUTTA	5.5 BASED ON WET WEIGHT	WHOLE FISH, FISH = 1.5 - 16.5 G, UPTAKE FROM WATER = 1 U CI/L	662
*SALMO TRUTTA	4.4 BASED ON WET WEIGHT	WHOLE FISH, FISH = 8.0 - 40.0 G, UPTAKE FROM WATER = 1 U CI/L	662
*SALMO TRUTTA	3.6 BASED ON WET WEIGHT	WHOLE FISH, FISH = 75.0-280.0 G, UPTAKE FROM WATER = 1 U CI/L	662
*SALMO TRUTTA	1.8 BASED ON WET WEIGHT	WHOLE FISH, FISH = 600 -820.0 G, UPTAKE FROM WATER = 1 U CI/L	662
*SALMO TRUTTA	3.90*10**3 BASED ON WET WEIGHT	WHOLE FISH, SOFT WATER AREA	662
*SALMO TRUTTA	3.90*10**3 BASED ON WET WEIGHT	WHOLE FISH, SOFT WATER AREA	662
*SALMO TRUTTA, BLOOD CELLS	1.99 BASED ON WET WEIGHT	STEADY STATE CF., UPTAKE FROM WATER= 1 U CI/L, 420 D, ALK=224 PPM	662
*SALMO TRUTTA, BLOOD PLASMA	0.37 BASED ON WET WEIGHT	STEADY STATE CF., UPTAKE FROM WATER= 1 U CI/L, 420 D, ALK=224 PPM	662
*SALMO TRUTTA, BONE	3.4 BASED ON WET WEIGHT	STEADY STATE CF., UPTAKE FROM WATER= 1 U CI/L, 420 D, ALK=224 PPM	662
*SALMO TRUTTA, GILL FILAMENT	4.9 BASED ON WET WEIGHT	STEADY STATE CF., UPTAKE FROM WATER= 1 U CI/L, 420 D, ALK=224 PPM	662
*SALMO TRUTTA, GONAD	8.0 BASED ON WET WEIGHT	STEADY STATE CF., UPTAKE FROM WATER= 1 U CI/L, 420 D, ALK=224 PPM	662
*SALMO TRUTTA, GUT	8.2 BASED ON WET WEIGHT	STEADY STATE CF., UPTAKE FROM WATER= 1 U CI/L, 420 D, ALK=224 PPM	662
*SALMO TRUTTA, HEART	10.9 BASED ON WET WEIGHT	STEADY STATE CF., UPTAKE FROM WATER= 1 U CI/L, 420 D, ALK=224 PPM	662
*SALMO TRUTTA, KIDNEY	5.5 BASED ON WET WEIGHT	STEADY STATE CF., UPTAKE FROM WATER= 1 U CI/L, 420 D, ALK=224 PPM	662
*SALMO TRUTTA, LIVER	4.2 BASED ON WET WEIGHT	STEADY STATE CF., UPTAKE FROM WATER= 1 U CI/L, 420 D, ALK=224 PPM	662
*SALMO TRUTTA, MUSCLE	8.2 BASED ON WET WEIGHT	STEADY STATE CF., UPTAKE FROM WATER= 1 U CI/L, 420 D, ALK=224 PPM	662
*SALMO TRUTTA, NERVE CORD	7.0 BASED ON WET WEIGHT	STEADY STATE CF., UPTAKE FROM WATER= 1 U CI/L, 420 D, ALK=224 PPM	662
*SALMO TRUTTA, SKIN	2.9 BASED ON WET WEIGHT	STEADY STATE CF., UPTAKE FROM WATER= 1 U CI/L, 420 D, ALK=224 PPM	662
*SALMO TRUTTA, SPLEEN	5.8 BASED ON WET WEIGHT	STEADY STATE CF., UPTAKE FROM WATER= 1 U CI/L, 420 D, ALK=224 PPM	662
*SALMO TRUTTA, WHOLE BLOOD	1.08 BASED ON WET WEIGHT	STEADY STATE CF., UPTAKE FROM WATER= 1 U CI/L, 420 D, ALK=224 PPM	662

TABLE D128

CONCENTRATION FACTOR : CS-137

ITEM:	VALUE:	CONDITIONS:	REF.:
*SALMO TRUTTA, WHOLE FISH	5.1 BASED ON WET WEIGHT	STEADY STATE CF., UPTAKE FROM WATER= 1 U CI/L, 420 D, ALK=224 PPM	662
*SARDINA PILCHARDUS	110 BASED ON DRY WEIGHT	EASTERN MEDITERRANEAN, DRY WEIGHT = 127.0 G	525
*SARDINELLA AURITA	133 BASED ON DRY WEIGHT	EASTERN MEDITERRANEAN, DRY WEIGHT = 113.9 G	525
*SARDINELLA AURITA	133 BASED ON DRY WEIGHT	EASTERN MEDITERRANEAN, DRY WEIGHT = 204.6 G	525
*SARDINELLA AURITA	61 BASED ON DRY WEIGHT	EASTERN MEDITERRANEAN, DRY WEIGHT = 198.3 G	525
*SARDINELLA AURITA	51 BASED ON DRY WEIGHT	EASTERN MEDITERRANEAN, DRY WEIGHT = 198.2 G	525
*SCOMBER JAPONICUS	102 BASED ON DRY WEIGHT	EASTERN MEDITERRANEAN, DRY WEIGHT = 238.6 G	525
*SEDIMENT	48 BASED ON DRY WEIGHT	SILTY SAND SEDIMENT, MARINE, URAZOKO BAY, JAPAN	1095
*SEDIMENT	66 BASED ON DRY WEIGHT	SANDY SILT SEDIMENT, MARINE, URAZOKO BAY, JAPAN	1095
*SEDIMENT	32 BASED ON DRY WEIGHT	SAND SEDIMENT, MARINE, URAZOKO BAY, JAPAN	1095
*SEDIMENT	16 BASED ON DRY WEIGHT	SAND SEDIMENT, MARINE, URAZOKO BAY, JAPAN	1095
*SEDIMENT	5 BASED ON DRY WEIGHT	SAND SEDIMENT, MARINE, URAZOKO BAY, JAPAN	1095
*SEDIMENT	15 BASED ON DRY WEIGHT	SAND SEDIMENT, MARINE, URAZOKO BAY, JAPAN	1095
*SEPIA OFFCINALIS	65 BASED ON DRY WEIGHT	EASTERN MEDITERRANEAN, DRY WEIGHT = 245.5 G, SHELL	525
*SEPIA OFFCINALIS	47 BASED ON DRY WEIGHT	EASTERN MEDITERRANEAN, DRY WEIGHT = 245.5 G, EXCLUDING SHELL	525
*SIGANUS LURIDUS	41 BASED ON DRY WEIGHT	EASTERN MEDITERRANEAN, DRY WEIGHT = 88.4 G	525
*SPHYRAENA SP.	184 BASED ON DRY WEIGHT	EASTERN MEDITERRANEAN, DRY WEIGHT = 100.2 G	525
*SUSPENDED SILT	870 (RANGE = 422 - 1570) BASED ON DRY WEIGHT	SEAWATER, BOMBAY	1261
*SUSPENDED SOLIDS	18780 (RANGE = 1210 - 93140) BASED ON WET WEIGHT	DANUBE, 1973, NUCLEAR POWER PLANT SINCE 1967	658
*TRIFOLIUM PERENNE, AERIAL PART	RANGE 0.54 - 166.22 BASED ON DRY WEIGHT	CONT. WHEN SOWN, CULTURE SOL.= 4*10**(-4) U CI/ML	101
*TRIFOLIUM PERENNE, AERIAL PART	RANGE 0.51 - 26.36 BASED ON DRY WEIGHT	CONT. WHEN SHOETING, CULTURE SOL.= 4*10**(-4) U CI/ML, SOIL ONLY	101
*TRIFOLIUM PERENNE, AERIAL PART	RANGE 0.15 - 7.50 BASED ON DRY WEIGHT	CONT. WHEN SHOETING, CULTURE SOL.= 4*10**(-4) U CI/ML, IRRIGATED	101
*TRIFOLIUM PERENNE, AERIAL PART	RANGE 5.36 - 13.88 BASED ON DRY WEIGHT	CONT. WHEN FLOWERING, CULTURE SOL. =4*10**(-4) U CI/ML, SOIL ONLY	101
*TRIFOLIUM PERENNE, AERIAL PART	RANGE 1.45 - 27.95 BASED ON DRY WEIGHT	CONT. WHEN FLOWERING, CULTURE SOL. =4*10**(-4) U CI/ML, OVERHEAD IRRIG	101
*TRIFOLIUM PERENNE, AERIAL PART	RANGE 0.07 - 5.10 BASED ON DRY WEIGHT	CONT. WHEN SOWN, CULTURE SOL.= 4*10**(-2) U CI/ML	101
*TRIFOLIUM PERENNE, AERIAL PART	RANGE 0.06 - 3.73 BASED ON DRY WEIGHT	CONT. WHEN SHOETING, CULTURE SOL.= 4*10**(-2) U CI/ML, SOIL ONLY	101
*TRIFOLIUM PERENNE, AERIAL PART	RANGE 0.02 - 16.82 BASED ON DRY WEIGHT	CONT. WHEN SHOETING, CULTURE SOL.= 4*10**(-2) U CI/ML, IRRIGATED	101
*TRIFOLIUM PERENNE, AERIAL PART	RANGE 0.07 - 2.09 BASED ON DRY WEIGHT	CONT. WHEN FLOWERING, CULTURE SOL. =4*10**(-2) U CI/ML, SOIL ONLY	101
*TRIFOLIUM PERENNE, AERIAL PART	RANGE 0.12 - 1.39 BASED ON DRY WEIGHT	CONT. WHEN FLOWERING, CULTURE SOL. =4*10**(-2) U CI/ML, OVERHEAD IRRIG	101
*TRIFOLIUM PERENNE, AERIAL PART	RANGE 0.02 - 1.04 BASED ON DRY WEIGHT	CONT. WHEN SOWN, CULTURE SOL.= 4*10**(-1) U CI/ML	101
*TRIFOLIUM PERENNE, AERIAL PART	RANGE 0.04 - 3.74 BASED ON DRY WEIGHT	CONT. WHEN SHOETING, CULTURE SOL.= 4*10**(-1) U CI/ML, SOIL ONLY	101
*TRIFOLIUM PERENNE, AERIAL PART	RANGE 0.01 - 4.27 BASED ON DRY WEIGHT	CONT. WHEN SHOETING, CULTURE SOL.= 4*10**(-1) U CI/ML, IRRIGATED	101
*TRIFOLIUM PERENNE, AERIAL PART	RANGE 0.12 - 0.76 BASED ON DRY WEIGHT	CONT. WHEN FLOWERING, CULTURE SOL. =4*10**(-1) U CI/ML, SOIL ONLY	101

TABLE D128

CONCENTRATION FACTOR : CS-137

ITEM:	VALUE:	CONDITIONS:	REF.:
*TRIFOLIUM PERENNE, AERIAL PART	RANGE 0.10 - 1.69 BASED ON DRY WEIGHT	CONT. WHEN FLOWERING, CULTURE SOL. =4*10**(-1) U CI/ML, OVERHEAD IRRIG	101
*TRIFOLIUM PERENNE, ROOT	RANGE 1.47 - 58.34 BASED ON DRY WEIGHT	CONT. WHEN SOWN, CULTURE SOL.= 4*10**(-4) U CI/ML	101
*TRIFOLIUM PERENNE, ROOT	RANGE 1.83 - 20.51 BASED ON DRY WEIGHT	CONT. WHEN SHOETING, CULTURE SOL.= 4*10**(-4) U CI/ML, SOIL ONLY	101
*TRIFOLIUM PERENNE, ROOT	RANGE 0.56 - 3.67 BASED ON DRY WEIGHT	CONT. WHEN SHOETING, CULTURE SOL.= 4*10**(-4) U CI/ML, IRRIGATED	101
*TRIFOLIUM PERENNE, ROOT	RANGE 2.11 - 139.55 BASED ON DRY WEIGHT	CONT. WHEN FLOWERING, CULTURE SOL. =4*10**(-4) U CI/ML, SOIL ONLY	101
*TRIFOLIUM PERENNE, ROOT	RANGE 1.18 - 8.23 BASED ON DRY WEIGHT	CONT. WHEN FLOWERING, CULTURE SOL. =4*10**(-4) U CI/ML, IRRIGATED	101
*TRIFOLIUM PERENNE, ROOT	RANGE 0.14 - 1.98 BASED ON DRY WEIGHT	CONT. WHEN SOWN, CULTURE SOL.= 4*10**(-2) U CI/ML	101
*TRIFOLIUM PERENNE, ROOT	RANGE 0.15 - 5.10 BASED ON DRY WEIGHT	CONT. WHEN SHOETING, CULTURE SOL.= 4*10**(-2) U CI/ML, SOIL ONLY	101
*TRIFOLIUM PERENNE, ROOT	RANGE 0.04 - 0.63 BASED ON DRY WEIGHT	CONT. WHEN SHOETING, CULTURE SOL.= 4*10**(-2) U CI/ML, IRRIGATED	101
*TRIFOLIUM PERENNE, ROOT	RANGE 0.08 - 3.28 BASED ON DRY WEIGHT	CONT. WHEN FLOWERING, CULTURE SOL. =4*10**(-2) U CI/ML, SOIL ONLY	101
*TRIFOLIUM PERENNE, ROOT	RANGE 0.27 - 1.06 BASED ON DRY WEIGHT	CONT. WHEN FLOWERING, CULTURE SOL. =4*10**(-2) U CI/ML, IRRIGATED	101
*TRIFOLIUM PERENNE, ROOT	RANGE 0.02 - 5.10 BASED ON DRY WEIGHT	CONT. WHEN SOWN, CULTURE SOL.= 4*10**(-1) U CI/ML	101
*TRIFOLIUM PERENNE, ROOT	RANGE 0.28 - 4.11 BASED ON DRY WEIGHT	CONT. WHEN SHOETING, CULTURE SOL.= 4*10**(-1) U CI/ML, SOIL ONLY	101
*TRIFOLIUM PERENNE, ROOT	RANGE 0.07 - 1.14 BASED ON DRY WEIGHT	CONT. WHEN SHOETING, CULTURE SOL.= 4*10**(-1) U CI/ML, IRRIGATED	101
*TRIFOLIUM PERENNE, ROOT	RANGE 0.36 - 3.92 BASED ON DRY WEIGHT	CONT. WHEN FLOWERING, CULTURE SOL. =4*10**(-1) U CI/ML, SOIL ONLY	101
*TRIFOLIUM PERENNE, ROOT	RANGE 0.04 - 1.74 BASED ON DRY WEIGHT	CONT. WHEN FLOWERING, CULTURE SOL. =4*10**(-1) U CI/ML, IRRIGATED	101
*UPENEUS MOLUCCENSIS	37 BASED ON DRY WEIGHT	RED SEA, DRY WEIGHT = 121.0 G	525
*WATER PLANT	180 (RANGE = 10 - 1000) BASED ON WET WEIGHT	DANUBE, 1973, NUCLEAR POWER PLANT SINCE 1967	658
*YELLOW PERCH	1043 BASED ON WET WEIGHT	WINTERGREEN LAKE, SEP 1970	437
*YELLOW PERCH	1182 BASED ON WET WEIGHT	WINTERGREEN LAKE, MAY 1971	437
*ZOOPLANKTON	845 BASED ON WET WEIGHT	WINTERGREEN LAKE, SEP 1970	437

TABLE D129

CONCENTRATION FACTOR : CU

ITEM:	VALUE:	CONDITIONS:	REF.:
*AGARUM CRIBROSUM	900 BASED ON DRY WEIGHT	SEAWATER=2.5*10**(-7) G/L, VOSTOK BAY, SEA OF JAPAN	1367
*BLIDINGIA MINIMA	0.71 - 1.83*10**4 BASED ON DRY WEIGHT	CALCULATED ON BASIS OF DISSOLVED METAL CONCENTRATION IN SEAWATER	1412
*BRASSICA OLERACEA	275 BASED ON WET WEIGHT	CONC.IN SOIL SOLUTION=0.04PPM	157
*BROWN ALGAE	920 BASED ON WET WEIGHT BASED ON WET WEIGHT	NONE	157
*BROWN ALGAE	100	NONE	994
*CANCER IRRORATUS	42275 (SD=+-7232) BASED ON WET WEIGHT	SEAWATER CONC=0.47*10**(-3) PPM, NOVA SCOTIA, AAS	984
*CHONDRUS YENDOI	2300 BASED ON DRY WEIGHT	SEAWATER=2.5*10**(-7) G/L, VOSTOK BAY, SEA OF JAPAN	1367
*CHORDA FILUM	560 BASED ON DRY WEIGHT	SEAWATER=2.5*10**(-7) G/L, VOSTOK BAY, SEA OF JAPAN	1367
*CLADOPHORA GLOMERATA	2.2*10**3 BASED ON DRY WEIGHT	WATER FROM DEADMAN BAY (L.ONTARIO) =5.7+-1.1, (CONC. IN NG/ML+-S.D.)	806
*CLADOPHORA GLOMERATA	1.9*10**3 BASED ON DRY WEIGHT	WATER FROM MAIN DUCK ISL. (L.ONTARIO)=2.9+-0.6, (CONC. IN NG/ML+-S.D.)	806

TABLE D129

CONCENTRATION FACTOR : CU

ITEM:	VALUE:	CONDITIONS:	REF.:
*CODIUM YESSOENSIS	500 BASED ON DRY WEIGHT	SEAWATER=2.5*10**(-7) G/L, VOSTOK BAY, SEA OF JAPAN	1367
*CORAL	2.76 BASED ON CA CONTENT ($34)	DEEP OCEAN CORAL, MEAN OF 13 DETERMINATIONS	939
*CORAL	1.79 BASED ON CA CONTENT ($34)	SHALLOW OPEN OCEAN CORAL, MEAN OF 8 DETERMINATIONS	939
*CORAL	2.22 BASED ON CA CONTENT ($34)	SHALLOW COASTAL OCEAN, MEAN OF 13 DETERMINATIONS	939
*ENTEROMORPHA LINZA	0.56 - 0.77*10**4 BASED ON DRY WEIGHT	CALCULATED ON BASIS OF DISSOLVED METAL CONCENTRATION IN SEAWATER	1412
*ENTEROMORPHA PROLIFERA	2800 BASED ON DRY WEIGHT	SEAWATER=2.5*10**(-7) G/L, VOSTOK BAY, SEA OF JAPAN	1367
*FISH	127 BASED ON WET WEIGHT	MARINE ORGANISMS, TYPICAL VALUE	538
*FISH	60 BASED ON WET WEIGHT	FRESHWATER ORGANISMS, TYPICAL VALUE	1028
*FRESHWATER PLANKTON	60000 BASED ON DRY WEIGHT	CONC.IN WATER=1.0UG/L,SCHOEHSEE	573
*FUCUS VESICULOSUS	0.36 - 0.74*10**4 BASED ON DRY WEIGHT	CALCULATED ON BASIS OF DISSOLVED METAL CONCENTRATION IN SEAWATER	1412
*FUCUS VESICULOSUS	0.25*10**4 BASED ON DRY WEIGHT	CALCULATED ON BASIS OF DISSOLVED METAL CONCENTRATION IN SEAWATER	1412
*FUCUS VESICULOSUS	0.45*10**4 BASED ON DRY WEIGHT	CALCULATED ON BASIS OF DISSOLVED METAL CONCENTRATION IN SEAWATER	1412
*FUCUS VESICULOSUS	0.45*10**4 BASED ON DRY WEIGHT	CALCULATED ON BASIS OF DISSOLVED METAL CONCENTRATION IN SEAWATER	1412
*FUCUS VESICULOSUS	0.45 - 2.7*10**4 BASED ON DRY WEIGHT	CALCULATED ON BASIS OF DISSOLVED METAL CONCENTRATION IN SEAWATER	1412
*FUCUS VESICULOSUS	4.5*10**3 BASED ON DRY WEIGHT	18 OBS.,ALL AREAS AROUND THE BRITISH ISLES	1278
*HETEROCHORDARIA ABIETINA	800 BASED ON DRY WEIGHT	SEAWATER=2.5*10**(-7) G/L, VOSTOK BAY, SEA OF JAPAN	1367
*LAMINARIA CHICHORIOIDES	1600 BASED ON DRY WEIGHT	SEAWATER=2.5*10**(-7) G/L, VOSTOK BAY, SEA OF JAPAN	1367
*LAMINARIA JAPONICA	360 BASED ON DRY WEIGHT	SEAWATER=2.5*10**(-7) G/L, VOSTOK BAY, SEA OF JAPAN	1367
*MACROPHYTES	2890 BASED ON WET WEIGHT	MARINE ORGANISMS, TYPICAL VALUE	589
*MACROPHYTES	158 BASED ON WET WEIGHT	FRESHWATER ORGANISMS, TYPICAL VALUE	1028
*MOLLUSCS	3800 BASED ON WET WEIGHT	MARINE ORGANISMS, TYPICAL VALUE	1280
*MOLLUSCS	1500 BASED ON WET WEIGHT	FRESHWATER ORGANISMS, TYPICAL VALUE	993
*MYTILUS GALLOPROVINCIALIS	800 BASED ON WET WEIGHT	WATER = 9.25 PPM	971
*PATELLA VULGATA	7.6*10**3 BASED ON DRY WEIGHT	17 OBS.,ALL AREAS AROUND THE BRITISH ISLES	1278
*PELVETIA WRIGHTII	440 BASED ON DRY WEIGHT	SEAWATER=2.5*10**(-7) G/L, VOSTOK BAY, SEA OF JAPAN	1367
*PHYLLOSPADIX IWATENSIS	1000 BASED ON DRY WEIGHT	SEAWATER=2.5*10**(-7) G/L, VOSTOK BAY, SEA OF JAPAN	1367
*PHYTOPLANKTON	360-9300 BASED ON WET WEIGHT	NONE	311
*PHYTOPLANKTON	2800 BASED ON WET WEIGHT	MARINE ORGANISMS, TYPICAL VALUE	986
*PLANKTON	17000 BASED ON WET WEIGHT	NONE	157
*POLYSIPHONIA JAPONICA	2800 BASED ON DRY WEIGHT	SEAWATER=2.5*10**(-7) G/L, VOSTOK BAY, SEA OF JAPAN	1367
*PORPHYRA UMBILICALIS	6.3*10**3 BASED ON DRY WEIGHT	20 OBS.,ALL AREAS AROUND THE BRITISH ISLES	1278
*PTILOTA FILICINA	5200 BASED ON DRY WEIGHT	SEAWATER=2.5*10**(-7) G/L, VOSTOK BAY, SEA OF JAPAN	1367
*RHODOMELA LARIX	1800 BASED ON DRY WEIGHT	SEAWATER=2.5*10**(-7) G/L, VOSTOK BAY, SEA OF JAPAN	1367
*SARGASSUM PALLIDUM	1700 BASED ON DRY WEIGHT	SEAWATER=2.5*10**(-7) G/L, VOSTOK BAY, SEA OF JAPAN	1367
*SESTON	1.0 BASED ON WET WEIGHT	DEPTH = 8 M, SCH@HSEE	573
*SESTON	0.8 BASED ON WET WEIGHT	DEPTH = 17 M, SCH@HSEE	573
*ULVA FENESTRATA	8400 BASED ON DRY WEIGHT	SEAWATER=2.5*10**(-7) G/L, VOSTOK BAY, SEA OF JAPAN	1367

TABLE D129

CONCENTRATION FACTOR : CU			
ITEM:	VALUE:	CONDITIONS:	REF.:
*ULVA SP.	0.47 - 0.86*10**4 BASED ON DRY WEIGHT	CALCULATED ON BASIS OF DISSOLVED METAL CONCENTRATION IN SEAWATER	1412
*ULVA SP.	0.12*10**4 BASED ON DRY WEIGHT	CALCULATED ON BASIS OF DISSOLVED METAL CONCENTRATION IN SEAWATER	1412
*ULVARIA SPENDENS	5700 BASED ON DRY WEIGHT	SEAWATER=2.5*10**(-7) G/L, VOSTOK BAY, SEA OF JAPAN	1367
*ZOOPLANKTON	1630 BASED ON ZOOPLANKTON WET WEIGHT CONCENTRATION	SEAWATER CONCENTRATION 1630 MG/M3	985
*ZOOPLANKTON	1800 BASED ON WET WEIGHT	MARINE ORGANISMS, TYPICAL VALUE	986
*ZOSTERA ASIATICA	600 BASED ON DRY WEIGHT	SEAWATER=2.5*10**(-7) G/L, VOSTOK BAY, SEA OF JAPAN	1367

TABLE D130

CONCENTRATION FACTOR : DDD			
ITEM:	VALUE:	CONDITIONS:	REF.:
*AMERIURUS NEBULOSUS	1107-1771 BASED ON WET WEIGHT	CONC.IN WATER=14UG/L	389
*BLACK CRAPPIE	386-8214 BASED ON WET WEIGHT	CONC.IN WATER=14UG/L	389
*BLUEGILL	471-714 BASED ON WET WEIGHT	CONC.IN WATER=14UG/L	389
*LARGEMOUTH BASS	1407-1785 BASED ON WET WEIGHT	CONC.IN WATER=14UG/L	389
*SACRAMENTO BLACKFISH	778-1257 BASED ON WET WEIGHT	CONC.IN WATER=14UG/L	389
*WHITE CATFISH	2172-9214 BASED ON WET WEIGHT	CONC.IN WATER=14UG/L	389

TABLE D131

CONCENTRATION FACTOR : DDE			
ITEM:	VALUE:	CONDITIONS:	REF.:
*MYTILUS SPP.	310000 BASED ON WET WEIGHT	P,P-DDE, SEAWATER=5.1 NG/L, CALIF., ROYAL PALMS, 19/9 72	1343
*MYTILUS SPP.	45000 BASED ON WET WEIGHT	P,P-DDE, SEAWATER=0.11 NG/M3, CALIF., FRISCO BAY, 1975	1343

TABLE D132

CONCENTRATION FACTOR : DDT			
ITEM:	VALUE:	CONDITIONS:	REF.:
*AGRIOLIMAX RETICULATUS	8.26 BASED ON WET WEIGHT	SOIL = 2.30 PPM, TERRESTRIAL GASTROPOD	328
*AGRIOLIMAX RETICULATUS	2.33 BASED ON WET WEIGHT	SOIL = 17.20 PPM, TERRESTRIAL GASTROPOD	328
*ALFALFA FOLIAGE	0.081 BASED ON WET WEIGHT	CONC.IN SOIL=1.39MG/KG DRY WT.	389
*ALGAE	21250 BASED ON WET WEIGHT	CONC.IN WATER=0.004NG/L	389
*ALGAE	0.35 BASED ON WET WEIGHT	CONC.IN WATER=5.80NG/L	389
*ALGAE	0.06 BASED ON WET WEIGHT	CONC.IN WATER=1000.0NG/L	389
*ALGAE	126400 BASED ON WET WEIGHT	CONC.IN WATER=0.20NG/L	389
*ALGAE	500 BASED ON WET WEIGHT	CONC.IN WATER=0.016NG/L	389
*ALGAE	30 BASED ON WET WEIGHT	CONC.IN WATER=0.33NG/L	389
*ALGAE	MAX VALUE = 3.3*10**1, WET WEIGHT	BIOMAGNIFICATION FACTOR	906
*AMPHIDINIUM CARTERI	2.2*10**5 BASED ON WET WEIGHT	MARINE ALGAE, CELL C = 1226 UG, EQUILIBRIUM = 0.5 PPT	293

TABLE D132

CONCENTRATION FACTOR : DDT

ITEM:	VALUE:	CONDITIONS:	REF.:
*AMPHIDINIUM CARTERI	1.4*10**5 BASED ON WET WEIGHT	MARINE ALGAE, CELL C = 961 UG, EQUILIBRIUM = 0.5 PPT	293
*AMPHIDINIUM CARTERI	1.2*10**5 BASED ON WET WEIGHT	MARINE ALGAE, CELL C = 961 UG, EQUILIBRIUM = 0.8 PPT	293
*AMPHIPODA	65000 BASED ON WET WEIGHT	CONC.IN WATER=0.20UG/L,U.S.	389
*ANACYSTIS NIDULANS	0.07 BASED ON WET WEIGHT	CONC.IN WATER=1000.0NG/L	389
*ANKISTRODESMUS AMALLOIDES	21300 BASED ON WET WEIGHT	30 MIN, 100 CELLS/ML, WATER CONC = 0.72 PPB	1117
*ANKISTRODESMUS AMALLOIDES	43600 BASED ON WET WEIGHT	60 MIN, 100 CELLS/ML, WATER CONC = 0.72 PPB	1117
*ANKISTRODESMUS AMALLOIDES	53100 BASED ON WET WEIGHT	120 MIN, 100 CELLS/ML, WATER CONC = 0.72 PPB	1117
*ANKISTRODESMUS AMALLOIDES	61600 BASED ON WET WEIGHT	180 MIN, 100 CELLS/ML, WATER CONC = 0.72 PPB	1117
*APLYSIA SP.	178000 BASED ON WET WEIGHT	CONC.IN WATER=0.01UG/L,U.S.	389
*AQUATIC PLANT	MAX VALUE = 1*10**5, WET WEIGHT	BIOMAGNIFICATION FACTOR	906
*AQUATIC VEGETATION	326 BASED ON WET WEIGHT	CONC.IN WATER=20.0NG/L	389
*AQUATIC VEGETATION	375 BASED ON WET WEIGHT	CONC.IN WATER=200.0NG/L	389
*AQUIPECTEN IRRADIANS	27000 BASED ON WET WEIGHT	CONC.IN WATER=0.001(ESTIM.),U.S.	389
*ARION SP.	3.17 BASED ON WET WEIGHT	SOIL = 2.30 PPM, TERRESTRIAL GASTROPOD	329
*BALD EAGLE, MUSCLE	0.36 BASED ON WET WEIGHT	CONC.IN FOOD=800MG/KG	389
*BEETLE	1.0 BASED ON WET WEIGHT	CONC.IN SOIL(ARABLE)=0.31MG/KG DRY WT.,G.B.,SITES=10	389
*BEETLE	0.43 BASED ON WET WEIGHT	CONC.IN SOIL(ORCHARD)=9.75MG/KG DRY WT.,G.B.,SITES=2	389
*BEETLE	2.14 BASED ON WET WEIGHT	CONC.IN SOIL(ARABLE)=0.58MG/KG DRY WT.,G.B.,SITES=10	389
*BEETLE	0.52 BASED ON WET WEIGHT	CONC.IN SOIL(ORCHARD)=9.75MG/KG DRY WT.,G.B.,SITES=2	389
*BEETLE	4.33 BASED ON WET WEIGHT	CONC.IN SOIL=0.06MG/KG DRY WT., U.S.,SITES=2	389
*BEETLE LARVAE	35.33 BASED ON WET WEIGHT	CONC.IN SOIL=0.015MG/KG DRY WT., U.S.,SITES=2	389
*BLACK BULLHEAD	155 BASED ON WET WEIGHT	CONC.IN WATER=20UG/L	389
*BOBWHITE QUAIL, TISSUE	0.03 BASED ON WET WEIGHT	CONC.IN FOOD=50MG/KG	389
*BOBWHITE QUAIL, TISSUE	0.04 BASED ON WET WEIGHT	CONC.IN FOOD=100MG/KG	389
*BOBWHITE QUAIL, TISSUE	0.08 BASED ON WET WEIGHT	CONC.IN FOOD=250MG/KG	389
*BRACHIDONTES DEMISSUS PLICATULUS	157000 BASED ON WET WEIGHT	CONC.IN WATER=0.001UG/L(ESTIM.), U.S.	389
*BRACHIDONTES RECURVUS	24000 BASED ON WET WEIGHT	CONC.IN WATER=1.0UG/L,U.S.	389
*BULLHEAD TROUT	100-200 BASED ON WET WEIGHT	CONC.IN WATER=20UG/L	389
*CALLINECTES SAPIDUS	79000 BASED ON WET WEIGHT	CONC.IN WATER=0.001UG/L(ESTIM.), U.S.	389
*CAMPETOMA SP.	43 BASED ON WET WEIGHT	CONC.IN WATER=3.28UG/L,CANADA	389
*CARROT	0.007 BASED ON WET WEIGHT	CONC.IN SANDY SOIL=4.99MG/KG DRY WT.	389
*CARROT	0.0003 BASED ON WET WEIGHT	CONC.IN MUCK SOIL=12.61MG/KG DRY WT.	389
*CARROT, ROOT	0.128 BASED ON WET WEIGHT	CONC.IN SANDY LOAM SOIL=24.8MG/KG DRY WT.	389
*CERATOPHYLLUM DEMERSUM	2013 BASED ON WET WEIGHT	CONC.IN WATER=0.77NG/L	389
*CERATOPHYLLUM DEMERSUM	13550 BASED ON WET WEIGHT	CONC.IN WATER=0.20NG/L	389
*CHARA SP.	13050 BASED ON WET WEIGHT	CONC.IN WATER=0.20NG/L	389
*CHARA SP.	1636 BASED ON WET WEIGHT	CONC.IN WATER=0.77NG/L	389
*CHICKEN, MUSCLE	0.21 BASED ON WET WEIGHT	CONC.IN FOOD=10.00MG/KG	389
*CHICKEN, MUSCLE	0.21 BASED ON WET WEIGHT	CONC.IN FOOD=50.00MG/KG	389
*CHICKEN, MUSCLE	0.16 BASED ON WET WEIGHT	CONC.IN FOOD=250.00MG/KG	389
*CHIRONOMIDAE	26800 BASED ON WET WEIGHT	CONC.IN WATER=0.20UG/L,U.S.	389

TABLE D132

CONCENTRATION FACTOR : DDT

ITEM:	VALUE:	CONDITIONS:	REF.:
*CHUB	5 BASED ON WET WEIGHT	CONC.IN WATER=5.8UG/L	389
*CLADIUM JAMAICENSIS	471 BASED ON WET WEIGHT	CONC.IN WATER=0.017NG/L	389
*CLADOPHORA GRACILIS	1660 BASED ON WET WEIGHT	CONC.IN WATER=0.05NG/L	389
*CLADOPHORA SP.	12701 BASED ON WET WEIGHT	CONC.IN WATER=0.77NG/L	389
*COWBIRD, TISSUE	0.25 BASED ON WET WEIGHT	CONC.IN FOOD=500MG/KG	389
*CRAB	MAX VALUE = 1.44*10**2, WET WEIGHT	BIOMAGNIFICATION FACTOR	906
*CRAB	144 BASED ON WET WEIGHT	CONC.IN WATER=50.0UG/L,U.S.	389
*CRASSOSTREA GIGAS	20000 BASED ON WET WEIGHT	CONC.IN WATER=1.0UG/L,U.S.	389
*CRASSOSTREA GIGAS	70000 BASED ON WET WEIGHT	CONC.IN WATER=0.1UG/L,U.S.	389
*CRASSOSTREA VIRGINICA	26000 BASED ON WET WEIGHT	CONC.IN WATER=1.0UG/L,U.S.	389
*CRASSOSTREA VIRGINICA	70000 BASED ON WET WEIGHT	CONC.IN WATER=0.001UG/L(ESTIM.), U.S.	389
*CRASSOSTREA VIRGINICA	70000 BASED ON WET WEIGHT	CONC.IN WATER=0.10UG/L,U.S.	389
*CRASSOSTREA VIRGINICA	30000 BASED ON WET WEIGHT	CONC.IN WATER=1.0UG/L,U.S.	389
*CRASSOSTREA VIRGINICA	15100 BASED ON WET WEIGHT	CONC.IN WATER=10.0UG/L,U.S.	389
*CRAYFISH	25 BASED ON WET WEIGHT	CONC.IN WATER=0.08UG/L,U.S.	389
*CRAYFISH	MAX VALUE = 9.7*10**1, WET WEIGHT	BIOMAGNIFICATION FACTOR	906
*CROAKERS	20000 BASED ON WET WEIGHT	CONC.IN WATER=0.1UG/L	389
*DEROCERAS OR LIMAX	2.25 BASED ON DRY WEIGHT	SOIL = 15.30 PPM, TERRESTRIAL GASTROPOD	528
*DEROCERAS OR LIMAX	2.95 BASED ON DRY WEIGHT	SOIL = 15.30 PPM, TERRESTRIAL GASTROPOD	528
*DEROCERAS OR LIMAX	5.53 BASED ON DRY WEIGHT	SOIL = 3.39 PPM, TERRESTRIAL GASTROPOD	528
*DEROCERAS OR LIMAX	17.93 BASED ON DRY WEIGHT	SOIL = 2.94 PPM, TERRESTRIAL GASTROPOD	528
*EARTHWORMS	0.64 BASED ON WET WEIGHT	CONC.IN SOIL=17.50MG/KG DRY WT., N.ZEALAND,SITES=1	389
*EARTHWORMS	1.58 BASED ON WET WEIGHT	CONC.IN SOIL=7.77MG/KG DRY WT., G.B.,SITES=2	389
*EARTHWORMS	2.73 BASED ON WET WEIGHT	CONC.IN SOIL=1.96MG/KG DRY WT., G.B.,SITES=1	389
*EARTHWORMS	2.73 BASED ON WET WEIGHT	CONC.IN SOIL(ARABLE)=0.32MG/KG DRY WT.,G.B.,SITES=7	389
*EARTHWORMS	2.01 BASED ON WET WEIGHT	CONC.IN SOIL(ORCHARD)=9.75MG/KG DRY WT.,G.B.,SITEA=2	389
*EARTHWORMS	5.86 BASED ON WET WEIGHT	CONC.IN SOIL(ARABLE)=0.58MG/KG DRY WT.,G.B.,SITES=10	389
*EARTHWORMS	2.73 BASED ON WET WEIGHT	CONC.IN SOIL(ORCHARD)=9.75MG/KG DRY WT.,G.B.,SITES=2	389
*EARTHWORMS	0.15 BASED ON WET WEIGHT	CONC.IN SOIL(FOREST)=1.23MG/KG DRY WT.,CANADA,	389
*EARTHWORMS	0.02 BASED ON WET WEIGHT	CONC.IN SOIL=12.50MG/KG DRY WT., U.S.,SITES=1	389
*EARTHWORMS	9.44 BASED ON WET WEIGHT	CONC.IN SOIL=1.80MG/KG DRY WT., U.S.,SITES=3	389
*EARTHWORMS	9.04 BASED ON DRY WEIGHT	CONC.IN SOIL=1.36MG/KG DRY WT., U.S.,SITES=67	389
*EARTHWORMS	2.16 BASED ON DRY WEIGHT	CONC.IN SOIL=0.06MG/KG DRY WT., U.S.,SITES=1	389
*EARTHWORMS	6.0 BASED ON DRY WEIGHT	CONC.IN SOIL=0¤01MG/KG DRY WT., U.S.,SITES=20	389
*EARTHWORMS	0.71 BASED ON WET WEIGHT	CONC.IN SOIL=11.43MG/KG DRY WT., G.B.,SITES=1	389
*EARTHWORMS	3.16 BASED ON WET WEIGHT	CONC.IN SOIL=3.74MG/KG DRY WT., G.B.,SITES=1	389
*EARTHWORMS	18.90 BASED ON DRY WEIGHT	U.S.,SITES=3	389
*EARTHWORMS	9.8 BASED ON DRY WEIGHT	CONC.IN SOIL=0.98MG/KG DRY WT., U.S.,SITES=67	389
*EARTHWORMS EXCL. LUMBRICUS TERRESTRIS	0.78 BASED ON WET WEIGHT	CONC.IN SOIL=83.2MG/KG DRY WT., U.S.,SITES=1	389

TABLE D132

CONCENTRATION FACTOR : DDT

ITEM:	VALUE:	CONDITIONS:	REF.:
*EARTHWORMS EXCL. LUMBRICUS TERRESTRIS	0.60 BASED ON WET WEIGHT	CONC.IN SOIL=117.33MG/KG DRY WT., U.S.,SITES=1	389
*EARTHWORMS EXCL. LUMBRICUS TERRESTRIS	2.40 BASED ON WET WEIGHT	CONC.IN SOIL=10.65MG/KG DRY WT., U.S.,SITES=2	389
*EARTHWORMS EXCL. LUMBRICUS TERRESTRIS	0.79 BASED ON WET WEIGHT	CONC.IN SOIL=0.94MG/KG DRY WT., G.B.,SITES=1	389
*EASTERN OYSTER/CLAM	MAX VALUE = 7*10**4, WET WEIGHT	BIOMAGNIFICATION FACTOR	906
*ELEOCHARIS CELLULOSA	533 BASED ON WET WEIGHT	CONC.IN WATER=0.015NG/L	389
*ENSIS DIRECTUS	29000 BASED ON WET WEIGHT	CONC.IN WATER=0.001UG/L(ESTIM.), U.S.	389
*EUGLENA GRACILIS	0.14 BASED ON WET WEIGHT	CONC.IN WATER=1000.0NG/L	389
*FERAL PIGEON, MUSCLE	0.45 BASED ON WET WEIGHT	CONC.IN FOOD=72MG/KG	389
*FISH	10-100 BASED ON WET WEIGHT	CONC.IN WATER=10UG/L	389
*FISH	829300 BASED ON WET WEIGHT	CONC.IN WATER=0.015UG/L	389
*FISH	35000 BASED ON WET WEIGHT	CONC.IN WATER=0.11UG/L	389
*FISH	3333-21333 BASED ON WET WEIGHT	CONC.IN WATER=0.30UG/L	389
*FISH	130-1450 BASED ON WET WEIGHT	5 SPP.,CONC.IN WATER=30-40UG/L	389
*FISH	MAX VALUE = 8.29*10**5,WET WEIGHT	BIOMAGNIFICATION FACTOR	906
*GONIDEA SP.	1640000 BASED ON WET WEIGHT	CONC.IN WATER=0.005UG/L,U.S.	389
*GRYLLUS ASSIMILIS	1.0 BASED ON WET WEIGHT	CONC.IN SOIL=0.06MG/KG DRY WT., U.S.,SITES=2	389
*GYRINIDAE	11488 BASED ON WET WEIGHT	CONC.IN WATER=0.08UG/L,U.S.	389
*HIBISCUS PALUSTRIS	1403 BASED ON WET WEIGHT	CONC.IN WATER=0.77NG/L	389
*HIRUDINEA SP.	31400 BASED ON WET WEIGHT	CONC.IN WATER=0.20UG/L,U.S.	389
*HOMARUS AMERICANUS	216000 BASED ON WET WEIGHT	CONC.IN WATER=0.001UG/L(ESTIM.), U.S.	389
*KESTREL, TISSUE	18.38 BASED ON WET WEIGHT	CONC.IN FOOD=2.8MG/KG,DIED	389
*KESTREL, TISSUE	25.0 BASED ON WET WEIGHT	CONC.IN FOOD=2.8MG/KG,SACRIFICED	389
*LEMNA MINOR	13550 BASED ON WET WEIGHT	CONC.IN WATER=0.20NG/L	389
*LEMNA SP.	2974 BASED ON WET WEIGHT	CONC.IN WATER=0.77NG/L	389
*LUMBRICUS TERRESTRIS	0.69 BASED ON WET WEIGHT	CONC.IN SOIL=83.2MG/KG DRY WT., U.S.,SITES=1	389
*LUMBRICUS TERRESTRIS	2.03 BASED ON WET WEIGHT	CONC.IN SOIL=31.06MG/KG DRY WT., U.S.,SITES=1	389
*LUMBRICUS TERRESTRIS	1.24 BASED ON WET WEIGHT	CONC.IN SOIL=10.65MG/KG DRY WT., U.S.,SITES=2	389
*LUMBRICUS TERRESTRIS	2.53 BASED ON WET WEIGHT	CONC.IN SOIL=6.88MG/KG DRY WT., G.B.,SITES=1	389
*LUMBRICUS TERRESTRIS	0.67 BASED ON WET WEIGHT	CONC.IN SOIL=11.43MG/KG DRY WT., G.B.,SITES=1	389
*LUMBRICUS TERRESTRIS	1.17 BASED ON WET WEIGHT	CONC.IN SOIL=0.94MG/KG DRY WT., G.B.,SITES=1	389
*MARSH SNAIL	19000 BASED ON WET WEIGHT	CONC.IN WATER=0.001UG/L(ESTIM.), U.S.	¤389
*MERCENARIA MERCENARIA	6000 BASED ON WET WEIGHT	CONC.IN WATER=1.0UG/L,U.S.	389
*MERCENARIA MERCENARIA	26000 BASED ON WET WEIGHT	CONC.IN WATER=0.001UG/L(ESTIM.), U.S.	389
*MERCENARIA MERCENARIA	70000 BASED ON WET WEIGHT	CONC.IN WATER=0.10UG/L,U.S.	389
*MERCENARIA MERCENARIA	3000 BASED ON WET WEIGHT	CONC.IN WATER=1.0UG/L,U.S.	389
*MERCENARIA MERCENARIA	8400 BASED ON WET WEIGHT	CONC.IN WATER=0.05UG/L,CONC.IN SEDIMENT=0.14MG/KG(ESTIM.),U.S.	389
*MOSS	30 BASED ON WET WEIGHT	CONC.IN WATER=0.33NG/L	389
*MYA ARENARIA	62000 BASED ON WET WEIGHT	CONC.IN WATER=0.001UG/L(ESTIM.), U.S.	389
*MYRIOPHYLLUM SP.	948 BASED ON WET WEIGHT	CONC.IN WATER=0.77NG/L	389
*MYRIOPHYLLUM SP.	8350 BASED ON WET WEIGHT	CONC.IN WATER=0.20NG/L	389
*MYTILUS EDULIS	98000 BASED ON WET WEIGHT	CONC.IN WATER=0.001UG/L(ESTIM.), U.S.	389
*MYTILUS SPP.	90000 BASED ON WET WEIGHT	P,P-DDT, SEAWATER=0.54 NG/L, FRANCE, VALRAS-PLAGE, 9/7 74	1343

TABLE D132

CONCENTRATION FACTOR : DDT			
ITEM:	VALUE:	CONDITIONS:	REF.:
*MYTILUS SPP.	40000 BASED ON WET WEIGHT	P,P-DDT, SEAWATER=0.35 NG/M3, FRANCE, STES-MARIES, 10/7 74	1343
*MYTILUS SPP.	210000 BASED ON WET WEIGHT	P,P-DDT, SEAWATER=0.25 NG/M3, FRANCE, CARRO, 11/7 74	1343
*MYTILUS SPP.	690000 BASED ON WET WEIGHT	P,P-DDT, SEAWATER=1.3 NG/M3, FRANCE, MARSEILLE, 12/7 74	1343
*NASSARIUS OBSOLETUS	76000 BASED ON WET WEIGHT	CONC.IN WATER=0.001UG/L(ESTIM.), U.S.	389
*NASSARIUS OBSOLETUS	5200 BASED ON WET WEIGHT	CONC.IN WATER=0.05UG/L,U.S.	389
*NOTONECTIDAE	16000 BASED ON WET WEIGHT	CONC.IN WATER=0.20UG/L,U.S.	389
*OSTREA EDULIS	15000 BASED ON WET WEIGHT	CONC.IN WATER=1.0UG/L,U.S.	389
*OSTREA EQUESTRIS	23000 BASED ON WET WEIGHT	CONC.IN WATER=1.0UG/L,U.S.	389
*OYSTER	14670 BASED ON WET WEIGHT	CONC.IN WATER=1.0UG/L,U.S.	389
*OYSTER	70000 BASED ON WET WEIGHT	CONC.IN WATER=0.1UG/L,U.S.	389
*OYSTER	75 BASED ON WET WEIGHT	CONC.IN WATER=0.28UG/L,U.S.	389
*OYSTER	450 BASED ON WET WEIGHT	CONC.IN WATER:<1.0UG/L,CONC.IN SEDIMENT=0.082MG/KG,U.S.	389
*OYSTER	40 BASED ON WET WEIGHT	CONC.IN WATER:<1.0UG/L,CONC.IN SEDIMENT=0.01MG/KG,U.S.	389
*PALAEMONETES VULGARIS	61000 BASED ON WET WEIGHT	CONC.IN WATER=0.001UG/L(ESTIM.), U.S.	389
*PHEASANT, MUSCLE	0.48 BASED ON WET WEIGHT	CONC.IN FOOD=50MG/KG,DIED	389
*PHEASANT, MUSCLE	0.07 BASED ON WET WEIGHT	CONC.IN FOOD=50MG/KG,SACRIFICED	389
*PHEASANT, TISSUE	2.91 BASED ON WET WEIGHT	CONC.IN FOOD=10MG/KG	389
*PINFISH	40000 BASED ON WET WEIGHT	CONC.IN WATER=0.1UG/L	389
*PINFISH	12000 BASED ON WET WEIGHT	CONC.IN WATER=1.0UG/L	389
*PLANKTON	63 BASED ON WET WEIGHT	CONC.IN WATER=0.08UG/L,U.S.	389
*PLANKTON	250 BASED ON WET WEIGHT	CONC.IN WATER=20.0UG/L,U.S.	389
*PLANKTON	16666 BASED ON WET WEIGHT	CONC.IN WATER=0.30UG/L,U.S.	389
*PLANKTON	MAX VALUE = 1.67*10**4, WET WEIGHT	BIOMAGNIFICATION FACTOR	906
*PLANKTON	130 BASED ON WET WEIGHT	CONC.IN WATER=0.115UG/L,U.S.	389
*PLANKTON	70 BASED ON WET WEIGHT	CONC.IN WATER=1.0UG/L(ESTIM.),U.S.	389
*PLANKTON	800 BASED ON WET WEIGHT	CONC.IN WATER=0.05UG/L,U.S.	389
*PLANORBIDAE	5150 BASED ON WET WEIGHT	CONC.IN WATER=0.20UG/L,U.S.	389
*PONTEDERIA CORDATA	636 BASED ON WET WEIGHT	CONC.IN WATER=0.77NG/L	389
*PONTEDERIA LANCEOLATA	240 BASED ON WET WEIGHT	CONC.IN WATER=0.025NG/L	389
*POTAMOGETON CRISPUS	961 BASED ON WET WEIGHT	CONC.IN WATER=0.77NG/L	389
*POTAMOGETON FOLIOSUS	1727 BASED ON WET WEIGHT	CONC.IN WATER=0.77NG/L	389
*POTAMOGETON PECTINATUS	326 BASED ON WET WEIGHT	CONC.IN WATER=20.0NG/L	389
*POTAMOGETON PECTINATUS	1338 BASED ON WET WEIGHT	CONC.IN WATER=0.77NG/L	389
*POTAMOGETON PECTINATUS	9300 BASED ON WET WEIGHT	CONC.IN WATER=0.20NG/L	389
*POTAMOGETON SP.	160 BASED ON WET WEIGHT	CONC.IN WATER=10.0NG/L	389
*POTATO TUBERS	0.066 BASED ON WET WEIGHT	CONC.IN SANDY LOAM SOIL=24.8MG/KG DRY WT.	389
*RUTABAGA	0.036 BASED ON WET WEIGHT	CONC.IN SANDY LOAM SOIL=24.8MG/KG DRY WT.	389
*SAGITTARIA LATIFOLIA	6273 BASED ON WET WEIGHT	CONC.IN WATER=0.77NG/L	389
*SALMO GAIRDNERI	207 BASED ON WET WEIGHT	CONC.IN WATER=20UG/L	389
*SALVELINUS FONTINALIS	710 BASED ON WET WEIGHT	CONC.IN WATER=24UG/L	389
*SCENEDESMUS OBLIQUUS	0.05 BASED ON WET WEIGHT	CONC.IN WATER=1000.0NG/L	389
*SCIRPUS SP.	3650 BASED ON WET WEIGHT	CONC.IN WATER=0.20NG/L	389
*SCIRPUS VALIDUS	675 BASED ON WET WEIGHT	CONC.IN WATER=0.77NG/L	389
*SEA HARE	MAX VALUE = 1.78*10**5, WET WEIGHT	BIOMAGNIFICATION FACTOR	906
*SEA SQUID	MAX VALUE = 1*10**6, WET WEIGHT	BIOMAGNIFICATION FACTOR	906
*SEA SQUIRT	1600000 BASED ON WET WEIGHT	CONC.IN WATER=0.01UG/L,U.S.	389

TABLE D132

CONCENTRATION FACTOR : DDT			
ITEM:	VALUE:	CONDITIONS:	REF.:
*SEA SQUIRT	200000 BASED ON WET WEIGHT	CONC.IN WATER=0.10UG/L,U.S.	389
*SEA SQUIRT	200 BASED ON WET WEIGHT	CONC.IN WATER=100.0UG/L,U.S.	389
*SHRIMP	MAX VALUE = 2.8*10**3, WET WEIGHT	BIOMAGNIFICATION FACTOR	906
*SHRIMP	478 BASED ON WET WEIGHT	CONC.IN WATER=0.115UG/L,U.S.	389
*SHRIMP	280 BASED ON WET WEIGHT	CONC.IN WATER=0.5UG/L,U.S.	389
*SHRIMP	3200 BASED ON WET WEIGHT	CONC.IN WATER=0.05UG/L,U.S.	389
*SLUGS	3.31 BASED ON WET WEIGHT	CONC.IN SOIL=1.96MG/KG DRY WT., G.B.,SITES=1	389
*SLUGS	2.25 BASED ON WET WEIGHT	CONC.IN SOIL(ORCHARD)=9.75MG/KG DRY WT.,G.B.,SITES=2	389
*SLUGS	11.69 BASED ON WET WEIGHT	CONC.IN SOIL=7.10MG/KG DRY WT., G.B.,SITES=1	389
*SLUGS	2.41 BASED ON WET WEIGHT	CONC.IN SOIL(ORCHARD)=9.75MG/KG DRY WT.,G.B.,SITES=2	389
*SLUGS	6.33 BASED ON DRY WEIGHT	CONC.IN SOIL=5.53MG/KG DRY WT., U.S.,SITES=4	389
*SLUGS	6.54 BASED ON WET WEIGHT	CONC.IN SOIL=6.88MG/KG DRY WT., G.B.,SITES=1	389
*SLUGS	0.30 BASED ON WET WEIGHT	CONC.IN GRASS ROOT=20.90MG/KG DRY WT.,G.B.,SITES=1	389
*SLUGS	3.86 BASED ON WET WEIGHT	CONC.IN SOIL=3.74MG/KG DRY WT., G.B.,SITES=1	389
*SNAIL	MAX VALUE = 1.48*10**3, WET WEIGHT	BIOMAGNIFICATION FACTOR	906
*SNAIL	0.14 BASED ON DRY WEIGHT	SOIL = 15.30 PPM, TERRESTRIAL GASTROPOD	528
*SNAIL	1.05 BASED ON DRY WEIGHT	SOIL = 2.94 PPM, TERRESTRIAL GASTROPOD	528
*SNAIL	0.83 BASED ON DRY WEIGHT	SOIL = 2.94 PPM, TERRESTRIAL GASTROPOD	528
*SNAIL	1480 BASED ON WET WEIGHT	CONC.IN WATER=50.0UG/L,U.S.	389
*SOYBEAN	0.074 BASED ON WET WEIGHT	CONC.IN SOIL=0.57MG/KG DRY WT.	389
*SPARGANIUM EURYCARPUM	740 BASED ON WET WEIGHT	CONC.IN WATER=0.77NG/L	389
*SPARGANIUM EURYCARPUM	6450 BASED ON WET WEIGHT	CONC.IN WATER=0.20NG/L	389
*SPARTINA SP., MAT	4549 BASED ON WET WEIGHT	CONC.IN WATER=1.44NG/L(APPROX.)	389
*SPIROGYRA SP.	325 BASED ON WET WEIGHT	CONC.IN WATER=0.77NG/L	389
*SUGAR BEET ROOT	0.067 BASED ON WET WEIGHT	CONC.IN LOAM SOIL=1.537MG/KG DRY WT.	389
*SWEET POTATO TUBERS	0.008 BASED ON WET WEIGHT	CONC.IN SOIL=1.19MG/KG DRY WT.	389
*SYRACOSHHAERA CARTERAE	2.3*10**5 BASED ON WET WEIGHT	MARINE ALGAE, CELL C = 301 UG, EQUILIBRIUM = 2.0 PPT	293
*SYRACOSHHAERA CARTERAE	2.9*10**5 BASED ON WET WEIGHT	MARINE ALGAE, CELL C = 151 UG, EQUILIBRIUM = 2.6 PPT	293
*TENDIPENDIDAE,LARVAE	11488 BASED ON WET WEIGHT	CONC.IN WATER=0.08UG/L,U.S.	389
*THALASSIOSIRA FLUVIATILIS	1.4*10**5 BASED ON WET WEIGHT	MARINE ALGAE, CELL C = 730 UG, EQUILIBRIUM = 2.4 PPT	293
*THALASSIOSIRA FLUVIATILIS	16*10**5 BASED ON WET WEIGHT	MARINE ALGAE, CELL C = 365 UG, EQUILIBRIUM = 3.3 PPT	293
*TROUT	200 BASED ON WET WEIGHT	CONC.IN WATER=20UG/L	389
*TYPHA LATIFOLIA	1300 BASED ON WET WEIGHT	CONC.IN WATER=0.20NG/L	389
*UCA SP.	68000 BASED ON WET WEIGHT	CONC.IN WATER=0.001UG/L(ESTIM.), U.S.	389
*UTRICULARIA VULGARIS	3299 BASED ON WET WEIGHT	CONC.IN WATER=0.77NG/L	389
*UTRICULARIA VULGARIS	29200 BASED ON WET WEIGHT	CONC.IN WATER=0.20NG/L	389
*VASCULAR PLANT	0.52 BASED ON WET WEIGHT	CONC.IN WATER=5.8NG/L	389

TABLE D133

CONCENTRATION FACTOR : DIELDRIN			
ITEM:	VALUE:	CONDITIONS:	REF.:
*ALFALFA	0.481 BASED ON WET WEIGHT	CONC.IN LOAM SOIL=0.054MG/KG DRY WT.	389

TABLE D133

CONCENTRATION FACTOR : DIELDRIN

ITEM:	VALUE:	CONDITIONS:	REF.:
*ALGAE	4091	CONC.IN WATER=0.011NG/L	389
*ALGAE	7480 BASED ON WET WEIGHT	300 K, FRESHWATER, MERCALF MODEL ECOSYSTEM	1379
*AMPHIDINIUM CARTERI	982 BASED ON DRY WEIGHT	1.7 PPB DIELDRIN, 2 H, 02.3 % REMOVED FROM SOL., 1.67*10**6 CELL/ML	1323
*AMPHIDINIUM CARTERI	982 BASED ON DRY WEIGHT	1.7 PPB DIELDRIN, 2 H, 02.3 % REMOVED FROM SOL., 1.67*10**6 CELL/ML	1323
*ANKISTRODESMUS AMALLOIDES	11700 BASED ON WET WEIGHT	30 MIN, 100 CELLS/ML, WATER CONC = 0.72 PPB	1117
*ANKISTRODESMUS AMALLOIDES	19100 BASED ON WET WEIGHT	60 MIN, 100 CELLS/ML, WATER CONC = 0.72 PPB	1117
*ANKISTRODESMUS AMALLOIDES	27000 BASED ON WET WEIGHT	120 MIN, 100 CELLS/ML, WATER CONC = 0.72 PPB	1117
*ANKISTRODESMUS AMALLOIDES	32000 BASED ON WET WEIGHT	180 MIN, 100 CELLS/ML, WATER CONC = 0.72 PPB	1117
*BEETLE	0.50 BASED ON WET WEIGHT	CONC.IN SOIL(ARABLE)=0.15MG/KG DRY WT.,G.B.,SITES=10	389
*BEETLE	0.30 BASED ON WET WEIGHT	CONC.IN SOIL(ARABLE)=0.37MG/KG DRY WT.,G.B.,SITES=10	389
*BEETLE	20.64 BASED ON WET WEIGHT	CONC.IN SOIL=0.25MG/KG DRY WT., U.S.,SITES=2	389
*BUFFALOFISH	30000 BASED ON WET WEIGHT	CONC.IN WATER=0.007UG/L	389
*BUFFALOFISH	3913 BASED ON WET WEIGHT	CONC.IN WATER=0.023UG/L	389
*CALLINECTES SAPIDUS	13000 BASED ON WET WEIGHT	CONC.IN WATER=0.001UG/L(ESTIM.), U.S.	389
*CARROT	0.003 BASED ON WET WEIGHT	CONC.IN MUCK SOIL=1.53MG/KG DRY WT.	389
*CARROT	0.039 BASED ON WET WEIGHT	CONC.IN SANDY SOIL=1.165MG/KG DRY WT.	389
*CARROT, ROOT	0.229 BASED ON WET WEIGHT	CONC.IN CLAY SOIL=0.48MG/KG DRY WT.	389
*CARROT, ROOT	0.005 BASED ON WET WEIGHT	CONC.IN MUCK SOIL=3.90MG/KG DRY WT.	389
*CATFISH	1428 BASED ON WET WEIGHT	CONC.IN WATER=0.007UG/L	389
*CATFISH	4444 BASED ON WET WEIGHT	CONC.IN WATER=0.009UG/L	389
*CATFISH	952 BASED ON WET WEIGHT	CONC.IN WATER=0.021UG/L	389
*CATFISH	1590 BASED ON WET WEIGHT	CONC.IN WATER=0.044UG/L	389
*CHEUMATOPSYCHE SP.,LARVAE	5133 BASED ON WET WEIGHT	CONC.IN WATER=11.5UG/L,U.S.	389
*CHICKEN, TISSUE	40.80 BASED ON WET WEIGHT	CONC.IN FOOD=0.25MG/KG	389
*CHICKEN, TISSUE	47.60 BASED ON WET WEIGHT	CONC.IN FOOD=0.75MG/KG	389
*CHICKEN, TISSUE	0.06 BASED ON WET WEIGHT	CONC.IN FOOD=10.00MG/KG	389
*CHICKEN, TISSUE	0.05 BASED ON WET WEIGHT	CONC.IN FOOD=20.00MG/KG	389
*CORBICULA MANILENSIS	1015 BASED ON WET WEIGHT	300 K, FRESHWATER, METCALF MODEL ECOSYSTEM	1379
*CORN, ROOT	11.13 BASED ON WET WEIGHT	CONC.IN SOIL=0.84MG/KG DRY WT.	389
*CORN, ROOT	13.45 BASED ON WET WEIGHT	CONC.IN SOIL=4.12MG/KG DRY WT.	389
*CORN, SHOOT	0.286 BASED ON WET WEIGHT	CONC.IN SOIL=0.84MG/KG DRY WT.	389
*CORN, SHOOT	0.201 BASED ON WET WEIGHT	CONC.IN SOIL=4.12MG/KG DRY WT.	389
*CORYDALIS CORNUTUS, LARVAE	68 BASED ON WET WEIGHT	CONC.IN WATER=17.0UG/L,U.S.	389
*CRASSOSTREA VIRGINICA	3500 BASED ON WET WEIGHT	CONC.IN WATER=1.0UG/L,U.S.	389
*CRAYFISH	18 BASED ON WET WEIGHT	CONC.IN WATER=29.0UG/L,U.S.	389
*CUCUMBER FRUIT	0.031 BASED ON WET WEIGHT	CONC.IN SOIL=3.7MG/KG DRY WT.	389
*CUCUMBER FRUIT	0.031 BASED ON WET WEIGHT	CONC.IN SOIL=1.4MG/KG DRY WT.	389
*CUTWORMS	2.83 BASED ON WET WEIGHT	G.B.,SITES=1	389
*CYCLOTELLA NANA	4810 BASED ON DRY WEIGHT	1.7 PPB DIELDRIN, 2 H, 13.0 % REMOVED FROM SOL., 0.83*10**6 CELL/ML	1323
*CYCLOTELLA NANA	4810 BASED ON DRY WEIGHT	1.7 PPB DIELDRIN, 2 H, 13.0 % REMOVED FROM SOL., 0.83*10**6 CELL/ML	1323

TABLE D133

CONCENTRATION FACTOR : DIELDRIN

ITEM:	VALUE:	CONDITIONS:	REF.:
*DAPHNIA SP.	2145 BASED ON WET WEIGHT	300 K, FRESHWATER, MERCALF MODEL ECOSYSTEM	1379
*EARTHWORMS	4.86 BASED ON WET WEIGHT	CONC.IN SOIL(ARABLE)=0.07MG/KG DRY WT.,G.B.,SITES=7	389
*EARTHWORMS	1.16 BASED ON WET WEIGHT	CONC.IN SOIL(ARABLE)=0.37MG/KG DRY WT.,G.B.,SITES=10	389
*EARTHWORMS	4.64 BASED ON WET WEIGHT	CONC.IN SOIL=0.28MG/KG DRY WT., G.B.,SITES=2	389
*EARTHWORMS	9.9 BASED ON DRY WEIGHT	CONC.IN SOIL=0.10MG/KG DRY WT., U.S.,SITES=67	389
*EARTHWORMS	5.68 BASED ON DRY WEIGHT	CONC.IN SOIL=0.25MG/KG DRY WT., U.S.,SITES=1	389
*EARTHWORMS	5.85 BASED ON DRY WEIGHT	CONC.IN SOIL=0.103MG/KG DRY WT., U.S.,SITES=1	389
*EARTHWORMS	1.97 BASED ON WET WEIGHT	CONC.IN SOIL=0.63MG/KG DRY WT., U.S.,SITES=4	389
*EARTHWORMS	126.60 BASED ON DRY WEIGHT	CONC.IN SOIL=0.006MG/KG DRY WT., U.S.,SITES=67	389
*EARTHWORMS	7.50 BASED ON DRY WEIGHT	CONC.IN SOIL=0.40MG/KG DRY WT., U.S.,SITES=2	389
*EARTHWORMS EXCL. LUMBRICUS TERRESTRIS	4.82 BASED ON WET WEIGHT	CONC.IN SOIL=0.64MG/KG DRY WT., G.B.,SITES=1	389
*ELODEA SP.	1280 BASED ON WET WEIGHT	300 K, FRESHWATER, MERCALF MODEL ECOSYSTEM	1379
*ENSIS DIRECTUS	4000 BASED ON WET WEIGHT	CONC.IN WATER=0.001UG/L(ESTIM.), U.S.	389
*FISH	10-100 BASED ON WET WEIGHT	CONC.IN WATER=10UG/L	389
*FISH	6145 BASED ON WET WEIGHT	300 K, FRESHWATER, MERCALF MODEL ECOSYSTEM	1379
*GRYLLUS ASSIMILIS	0.88 BASED ON WET WEIGHT	CONC.IN SOIL=0.25MG/KG DRY WT., U.S.,SITES=2	389
*HOMARUS AMERICANUS	12000 BASED ON WET WEIGHT	CONC.IN WATER=0.001UG/L(ESTIM.), U.S.	389
*HYDROPSYCHE,LARVAE	5133 BASED ON WET WEIGHT	CONC.IN WATER=11.5U¤/L,U.S.	389
*ISOCHRYSIS GALBANA	8238 BASED ON DRY WEIGHT	1.7 PPB DIELDRIN, 2 H, 15.5 % REMOVED FROM SOL., 1.40*10**6 CELL/ML	1323
*ISOCHRYS¤S GALBANA	8238 BASED ON DRY WEIGHT	1.7 PPB DIELDRIN, 2 H, 15.5 % REMOVED FROM SOL., 1.40*10**6 CELL/ML	1323
*LAMB, TISSUE	1.60 BASED ON WET WEIGHT	CONC.IN FOOD=0.25MG/KG	389
*LAMB, TISSUE	0.80 BASED ON WET WEIGHT	CONC.IN FOOD=0.75MG/KG	389
*LAMB, TISSUE	0.70 BASED ON WET WEIGHT	CONC.IN FOOD=2.25MG/KG	389
*LEGUME FOLIAGE	0.05 BASED ON WET WEIGHT	CONC.IN SOIL=0.05MG/KG DRY WT.	389
*LEIOSTOMUS XANTHURUS	6000 BASED ON WET WEIGHT	UPTAKE FROM WATER	1041
*LETTUCE, FOLIAGE	0.013 BASED ON WET WEIGHT	CONC.IN SILT LOAM SOIL=2.3MG/KG DRY WT.	389
*LUMBRICUS TERRESTRIS	2.50 BASED ON WET WEIGHT	CONC.IN SOIL=0.64MG/KG DRY WT., G.B.,SITES=1	389
*NASSARIUS OBSOLETUS	11000 BASED ON WET WEIGHT	CONC.IN WATER=0.001UG/L(ESTIM.), U.S.	389
*OLISTHODISCUS LUTEUS	4900 BASED ON DRY WEIGHT	1.7 PPB DIELDRIN, 2 H, 13.0 % REMOVED FROM SOL., 0.12*10**6 CELL/ML	1323
*OLISTHODISCUS LUTEUS	4900 BASED ON DRY WEIGHT	1.7 PPB DIELDRIN, 2 H, 13.0 % REMOVED FROM SOL., 0.12*10**6 CELL/ML	1323
*PALAEMONETES VULGARIS	236000 BASED ON WET WEIGHT	CONC.IN WATER=0.001UG/L(ESTIM.), U.S.	389
*PHEASANT, MUSCLE	0.05 BASED ON WET WEIGHT	CONC.IN FOOD=50MG/KG	389
*PHYSA SP.	7978 BASED ON WET WEIGHT	CONC.IN WATER=6.0UG/L,U.S.	389
*PHYSA SP.	114935 BASED ON WET WEIGHT	300 K, FRESHWATER, MERCALF MODEL ECOSYSTEM	1379
*PIG, TISSUE	1.60 BASED ON WET WEIGHT	CONC.IN FOOD=0.25MG/KG	389
*PIG, TISSUE	3.70 BASED ON WET WEIGHT	CONC.IN FOOD=0.75MG/KG	389

TABLE D133

CONCENTRATION FACTOR : DIELDRIN

ITEM:	VALUE:	CONDITIONS:	REF.:
*PIG, TISSUE	1.90 BASED ON WET WEIGHT	CONC.IN FOOD=2.25MG/KG	389
*PLANKTON	278 BASED ON WET WEIGHT	CONC.IN WATER=0.018UG/L,U.S.	389
*RANGIA CUNEATA	54 BASED ON DRY WEIGHT	UPTAKE FROM FOOD = 11.16 UG/G WW IN ALGAE, 48 H, N=6	1251
*RANGIA CUNEATA	54 BASED ON DRY WEIGHT	UPTAKE FROM FOOD = 11.16 UG/G WW IN ALGAE, 48 H, N=6	1251
*RAT, TISSUE	15.10 BASED ON WET WEIGHT	CONC.IN FOOD=1.0MG/KG	389
*RAT, TISSUE	6.70 BASED ON WET WEIGHT	CONC.IN FOOD=10MG/KG	389
*SHRIMP	278 BASED ON WET WEIGHT	CONC.IN WATER=0.018UG/L,U.S.	389
*SIMULIUM VITTATUM,LARVAE	1786 BASED ON WET WEIGHT	CONC.IN WATER=11.5UG/L,U.S.	389
*SKELETONEMA COSTATUM	15882 BASED ON DRY WEIGHT	1.7 PPB DIELDRIN, 2 H, 42.0 % REMOVED FROM SOL., 1.19*10**6 CELL/ML	1323
*SKELETONEMA COSTATUM	15882 BASED ON DRY WEIGHT	1.7 PPB DIELDRIN, 2 H, 42.0 % REMOVED FROM SOL., 1.19*10**6 CELL/ML	1323
*SLUGS	7.50 BASED ON WET WEIGHT	CONC.IN SOIL(ARABLE)=0.04MG/KG DRY WT.,G.B.,SITES=1	389
*SLUGS	6.75 BASED ON WET WEIGHT	CONC.IN SOIL=0.20MG/KG DRY WT., G.B.,SITES=1	389
*SLUGS	0.81 BASED ON WET WEIGHT	CONC.IN SOIL(ARABLE)=0.37MG/KG DRY WT.,G.B.,SITES=10	389
*SLUGS	571.11 BASED ON DRY WEIGHT	CONC.IN SOIL=0.009MG/KG DRY WT., U.S.,SITES=4	389
*STEER, TISSUE	3.20 BASED ON WET WEIGHT	CONC.IN FOOD=0.25MG/KG	389
*STEER, TISSUE	4.70 BASED ON WET WEIGHT	CONC.IN FOOD=0.75MG/KG	389
*STEER, TISSUE	3.90 BASED ON WET WEIGHT	CONC.IN FOOD=2.25MG/KG	389
*SUGAR BEET ROOT	0.057 BASED ON WET WEIGHT	CONC.IN SILT LOAM SOIL=2.3MG/KG DRY WT.	389
*SUGAR BEET ROOT	0.284 BASED ON WET WEIGHT	CONC.IN LOAM SOIL=0.35MG/KG DRY WT.	389
*SWEET CLOVER	0.481 BASED ON WET WEIGHT	CONC.IN LOAM SOIL=0.054MG/KG DRY WT.	389
*SWEET POTATO TUBERS	0.024 BASED ON WET WEIGHT	CONC.IN SOIL=0.17MG/KG DRY WT.	389
*TETRASELMIS CHUII	8588 BASED ON DRY WEIGHT	1.7 PPB DIELDRIN, 2 H, 16.0 % REMOVED FROM SOL., 0.13*10**6 CELL/ML	1323
*TETRASELMIS CHUII	8588 BASED ON DRY WEIGHT	1.7 PPB DIELDRIN, 2 H, 16.0 % REMOVED FROM SOL., 0.13*10**6 CELL/ML	1323
*TROUT	3300 BASED ON WET WEIGHT	CONC.IN WATER=2.3UG/L	389
*TUBIFICID WORMS	4563 BASED ON WET WEIGHT	CONC.IN WATER=80.0UG/L,U.S.	389
*TURNIPS	0.022 BASED ON WET WEIGHT	CONC.IN SILT LOAM SOIL=2.3MG/KG DRY WT.	389
*UCA ININAX	247 BASED ON WET WEIGHT	300 K, FRESHWATER, MERCALF MODEL ECOSYSTEM	1379
*UCA SP.	15000 BASED ON WET WEIGHT	CONC.IN WATER=0.001UG/L(ESTIM.), U.S.	389
*WHEAT	0.15 BASED ON WET WEIGHT	CONC.IN SOIL=1.13MG/KG DRY WT.	389
*WHEAT	0.038 BASED ON WET WEIGHT	CONC.IN SOIL=28.39MG/KG DRY WT.	389
*WHEAT, FOLIAGE	0.008 BASED ON WET WEIGHT	CONC.IN SILT LOAM SOIL=1.8MG/KG DRY WT.	389
*WHITE-TAILED DEER, MUSCLE	0.02 BASED ON WET WEIGHT	CONC.IN FOOD=5MG/KG	389
*WHITE-TAILED DEER, MUSCLE	0.05 BASED ON WET WEIGHT	CONC.IN FOOD=25MG/KG	389

TABLE D134

CONCENTRATION FACTOR : EDC-TAR

ITEM:	VALUE:	CONDITIONS:	REF.:
*LEANDER ADSPERSUS	159 BASED ON WET WEIGHT	SEAWATER,S=3.3-3.4%,283-287K, 16DAYS,N=14,CONC.IN WATER=1PPM	757

TABLE D134

CONCENTRATION FACTOR : EDC-TAR			
ITEM:	VALUE:	CONDITIONS:	REF.:
*LEANDER ADSPERSUS	220 BASED ON WET WEIGHT	SEAWATER,S=3.3-3.4%,283-287K, 16DAYS.N=14,CONC.IN WATER=0.1PPM	757
*LEANDER ADSPERSUS	2900 BASED ON WET WEIGHT	SEAWATER,S=3.3-3.4%,283-287K, 16DAYS,N=14,CONC.IN WATER=0.01PPM	757
*MYTILUS EDULIS	37 BASED ON WET WEIGHT	SEAWATER,S=3.3-3.4%,283-287K, 16DAYS,N=14,CONC.IN WATER=10PPM	757
*MYTILUS EDULIS	59 BASED ON WET WEIGHT	SEAWATER,S=3.3-3.4%,283-287K, 16DAYS,N=14,CONC.IN WATER=1PPM	757
*MYTILUS EDULIS	1030 BASED ON WET WEIGHT	SEAWATER,S=3.3-3.4%,283-287K, 16DAYS,N=14,CONC.IN WATER=0.1PPM	757
*PLEURONECTES PLATESSA	15 BASED ON WET WEIGHT	SEAWATER,S=3.3-3.4%,283-287K, 16DAYS,N=14,CONC.IN WATER=1PPM	757

TABLE D135

CONCENTRATION FACTOR : ENDRIN			
ITEM:	VALUE:	CONDITIONS:	REF.:
*ALGAE	0.67 BASED ON WET WEIGHT	CONC.IN WATER=10.5NG/L	389
*AQUATIC VEGETATION	27 BASED ON WET WEIGHT	CONC.IN WATER=7.4NG/L	389
*CHUB	4.7 BASED ON WET WEIGHT	CONC.IN WATER=10.5UG/L	389
*CONIDEA SP.	1182 BASED ON WET WEIGHT	CONC.IN WATER=0.011UG/L,U.S.	389
*EARTHWORMS	3.57 BASED ON WET WEIGHT	CONC.IN SOIL=0.49MG/KG DRY WT., U.S.,SITES=67	389
*EARTHWORMS	5.8 BASED ON WET WEIGHT	CONC.IN SOIL=0.12MG/KG DRY WT., U.S.,SITES=2	389
*FISH	70000 BASED ON WET WEIGHT	CONC.IN WATER=0.10UG/L	389
*OYSTER	10 BASED ON WET WEIGHT	CONC.IN WATER:<1.0UG/L,CONC.IN SEDIMENT:<0.001MG/KG,U.S.	389
*OYSTER	10 BASED ON WET WEIGHT	CONC.IN WATER:<1.0UG/L,CONC.IN SEDIMENT:<0.01MG/KG,U.S.	389
*PIMEPHALES PROMELAS	10000 BASED ON WET WEIGHT	CONC.IN WATER=0.015UG/L	389
*SLUGS	40.62 BASED ON DRY WEIGHT	CONC.IN SOIL=1.20MG/KG DRY WT., U.S.,SITES=4	389
*VASCULAR PLANT	0.57 BASED ON WET WEIGHT	CONC.IN WATER=10.5NG/L	389
*WHEAT, FOLIAGE	0.020 BASED ON WET WEIGHT	CONC.IN SILT LOAM SOIL=1.918MG/KG DRY WT.	389

TABLE D136

CONCENTRATION FACTOR : EU			
ITEM:	VALUE:	CONDITIONS:	REF.:
*ENGRAULIS JAPONICUS, WHOLE BODY	92 BASED ON WET WEIGHT	REL. TO TOTAL CONC.,TOT CONC=1.6 NG/L, SOLUB=0.3 NG/L,SEAWAT,JAP.	1535
*ENGRAULIS JAPONICUS, WHOLE BODY	730 BASED ON WET WEIGHT	REL. TO SOLUBLE CONC.,TOT CONC=1.6 NG/L, SOLUB=0.3 NG/L,SEAWAT,JAP.	1535
*HIZIKIA FUSIFORME, WHOLE BODY	100 BASED ON WET WEIGHT	REL. TO TOTAL CONC.,TOT CONC=1.6 NG/L, SOLUB=0.3 NG/L,SEAWAT,JAP.	1535
*HIZIKIA FUSIFORME, WHOLE BODY	800 BASED ON WET WEIGHT	REL. TO SOLUBLE CONC.,TOT CONC=1.6 NG/L, SOLUB=0.3 NG/L,SEAWAT,JAP.	1535
*MERETRIX LUSORIA, SOFT PART	860 BASED ON WET WEIGHT	REL. TO TOTAL CONC.,TOT CONC=1.6 NG/L, SOLUB=0.3 NG/L,SEAWAT,JAP.	1535
*MERETRIX LUSORIA, SOFT PART	6870 BASED ON WET WEIGHT	REL. TO SOLUBLE CONC.,TOT CONC=1.6 NG/L, SOLUB=0.3 NG/L,SEAWAT,JAP.	1535

TABLE D136

CONCENTRATION FACTOR : EU			
ITEM:	VALUE:	CONDITIONS:	REF.:
*PARALICHTHYS OLIVACEUS, MUSCLE	17 BASED ON WET WEIGHT	REL. TO TOTAL CONC.,TOT CONC=1.6 NG/L, SOLUB=0.3 NG/L,SEAWAT,JAP.	1535
*PARALICHTHYS OLIVACEUS, MUSCLE	130 BASED ON WET WEIGHT	REL. TO SOLUBLE CONC.,TOT CONC=1.6 NG/L, SOLUB=0.3 NG/L,SEAWAT,JAP.	1535
*SARGASSUM FULVELLUM, WHOLE BODY	230 BASED ON WET WEIGHT	REL. TO TOTAL CONC.,TOT CONC=1.6 NG/L, SOLUB=0.3 NG/L,SEAWAT,JAP.	1535
*SARGASSUM FULVELLUM, WHOLE BODY	1800 BASED ON WET WEIGHT	REL. TO SOLUBLE CONC.,TOT CONC=1.6 NG/L, SOLUB=0.3 NG/L,SEAWAT,JAP.	1535
*SERIOLA QUINQUERADIATA, MUSCLE	58 BASED ON WET WEIGHT	REL. TO TOTAL CONC.,TOT CONC=1.6 NG/L, SOLUB=0.3 NG/L,SEAWAT,JAP.	1535
*SERIOLA QUINQUERADIATA, MUSCLE	470 BASED ON WET WEIGHT	REL. TO SOLUBLE CONC.,TOT CONC=1.6 NG/L, SOLUB=0.3 NG/L,SEAWAT,JAP.	1535
*ULVA PERTUSA, WHOLE BODY	320 BASED ON WET WEIGHT	REL. TO TOTAL CONC.,TOT CONC=1.6 NG/L, SOLUB=0.3 NG/L,SEAWAT,JAP.	1535
*ULVA PERTUSA, WHOLE BODY	2570 BASED ON WET WEIGHT	REL. TO SOLUBLE CONC.,TOT CONC=1.6 NG/L, SOLUB=0.3 NG/L,SEAWAT,JAP.	1535
*UNDARIA PINNATIFIDA, WHOLE BODY	38 BASED ON WET WEIGHT	REL. TO TOTAL CONC.,TOT CONC=1.6 NG/L, SOLUB=0.3 NG/L,SEAWAT,JAP.	1535
*UNDARIA PINNATIFIDA, WHOLE BODY	300 BASED ON WET WEIGHT	REL. TO SOLUBLE CONC.,TOT CONC=1.6 NG/L, SOLUB=0.3 NG/L,SEAWAT,JAP.	1535

TABLE D137

CONCENTRATION FACTOR : F			
ITEM:	VALUE:	CONDITIONS:	REF.:
*BROWN ALGAE	0.86 BASED ON WET WEIGHT	NONE	157

TABLE D138

CONCENTRATION FACTOR : FE			
ITEM:	VALUE:	CONDITIONS:	REF.:
*AGARUM CRIBROSUM	2900 BASED ON DRY WEIGHT	SEAWATER=37.5*10**(-6) G/L, VOSTOK BAY, SEA OF JAPAN	1367
*BRASSICA OLERACEA	120 BASED ON WET WEIGHT	CONC.IN SOIL SOLUTION=0.25PPM	157
*BROWN ALGAE	17000 BASED ON WET WEIGHT	NONE	157
*BROWN ALGAE	20000-35000 BASED ON WET WEIGHT	NONE	994
*CANCER IRRORATUS	5165 (SD=+-726) BASED ON WET WEIGHT	SEAWATER CONC=11.62*10**(-3) PPM, NOVA SCOTIA, AAS	984
*CHONDRUS YENDOI	1600 BASED ON DRY WEIGHT	SEAWATER=37.5*10**(-6) G/L, VOSTOK BAY, SEA OF JAPAN	1367
*CHORDA FILUM	100 BASED ON DRY WEIGHT	SEAWATER=37.5*10**(-6) G/L, VOSTOK BAY, SEA OF JAPAN	1367
*CODIUM YESSOENSIS	1000 BASED ON DRY WEIGHT	SEAWATER=37.5*10**(-6) G/L, VOSTOK BAY, SEA OF JAPAN	1367
*COSTARIA COSTATA	250 BASED ON DRY WEIGHT	SEAWATER=37.5*10**(-6) G/L, VOSTOK BAY, SEA OF JAPAN	1367
*ENTEROMORPHA PROLIFERA	1300 BASED ON DRY WEIGHT	SEAWATER=37.5*10**(-6) G/L, VOSTOK BAY, SEA OF JAPAN	1367
*FISH	190 BASED ON WET WEIGHT	FRESHWATER ORGANISMS, TYPICAL VALUE	1028
*FRESHWATER PLANKTON	63000 BASED ON WET WEIGHT	CONC.IN WATER=15UG/L,SCHOEHSEE	573
*HETEROCHORDARIA ABIETINA	2000 BASED ON DRY WEIGHT	SEAWATER=37.5*10**(-6) G/L, VOSTOK BAY, SEA OF JAPAN	1367
*LAMINARIA CHICHORIOIDES	460 BASED ON DRY WEIGHT	SEAWATER=37.5*10**(-6) G/L, VOSTOK BAY, SEA OF JAPAN	1367
*LAMINARIA JAPONICA	70 BASED ON DRY WEIGHT	SEAWATER=37.5*10**(-6) G/L, VOSTOK BAY, SEA OF JAPAN	1367

TABLE D138

CONCENTRATION FACTOR : FE			
ITEM:	VALUE:	CONDITIONS:	REF.:
*MACROPHYTES	3642 BASED ON WET WEIGHT	FRESHWATER ORGANISMS, TYPICAL VALUE	1028
*MOLLUSCS	14400 BASED ON WET WEIGHT	MARINE ORGANISMS, TYPICAL VALUE	1280
*PATELLA VULGATA	2.8*10**5 BASED ON DRY WEIGHT	17 OBS.,ALL AREAS AROUND THE BRITISH ISLES	1278
*PELVETIA WRIGHTII	940 BASED ON DRY WEIGHT	SEAWATER=37.5*10**(-6) G/L, VOSTOK BAY, SEA OF JAPAN	1367
*PHYLLOSPADIX IWATENSIS	3500 BASED ON DRY WEIGHT	SEAWATER=37.5*10**(-6) G/L, VOSTOK BAY, SEA OF JAPAN	1367
*PHYTOPLANKTON	76700-492000 BASED ON WET WEIGHT	NONE	311
*PHYTOPLANKTON	28300 BASED ON WET WEIGHT	MARINE ORGANISMS, TYPICAL VALUE	986
*PHYTOPLANKTON	90000 BASED ON DRY WEIGHT	MARINE	952
*PLANKTON	87000 BASED ON WET WEIGHT	NONE	157
*PLANKTON	2000-140000 BASED ON WET WEIGHT	NONE	994
*POLYSIPHONIA JAPONICA	24000 BASED ON DRY WEIGHT	SEAWATER=37.5*10**(-6) G/L, VOSTOK BAY, SEA OF JAPAN	1367
*PORPHYRA UMBILICALIS	3.6*10**4 BASED ON DRY WEIGHT	20 OBS.,ALL AREAS AROUND THE BRITISH ISLES	1278
*PTILOTA FILICINA	6000 BASED ON DRY WEIGHT	SEAWATER=37.5*10**(-6) G/L, VOSTOK BAY, SEA OF JAPAN	1367
*RHODOMELA LARIX	7300 BASED ON DRY WEIGHT	SEAWATER=37.5*10**(-6) G/L, VOSTOK BAY, SEA OF JAPAN	1367
*SARGASSUM PALLIDUM	300 BASED ON DRY WEIGHT	SEAWATER=37.5*10**(-6) G/L, VOSTOK BAY, SEA OF JAPAN	1367
*SESTON	6.1 BASED ON WET WEIGHT	DEPTH = 8 M, SCH@HSEE	573
*SESTON	6.5 BASED ON WET WEIGHT	DEPTH = 17 M, SCH@HSEE	573
*ULVA FENESTRATA	2200 BASED ON DRY WEIGHT	SEAWATER=37.5*10**(-6) G/L, VOSTOK BAY, SEA OF JAPAN	1367
*ULVARIA SPENDENS	4500 BASED ON DRY WEIGHT	SEAWATER=37.5*10**(-6) G/L, VOSTOK BAY, SEA OF JAPAN	1367
*ZOOPLANKTON	144000 BASED ON WET WEIGHT	SEAWATER CONCENTRATION 0.01 MG/M3	985
*ZOOPLANKTON	114600 BASED ON WET WEIGHT	MARINE ORGANISMS, TYPICAL VALUE	986
*ZOSTERA ASIATICA	230 BASED ON DRY WEIGHT	SEAWATER=37.5*10**(-6) G/L, VOSTOK BAY, SEA OF JAPAN	1367
*32 FRESHWATER PLANT	4935 BASED ON DRY WEIGHT	ALGAE,BRYOPHYTES AND VASCULAR HYDROPHYTES	1597

TABLE D139

CONCENTRATION FACTOR : FE-59			
ITEM:	VALUE:	CONDITIONS:	REF.:
*FISH, MUSCLE	540 BASED ON WET WEIGHT	DANUBE, 1973, NUCLEAR POWER PLANT SINCE 1967	658
*FRESHWATER SEDIMENT	660 BASED ON WET WEIGHT	DANUBE, 1973, NUCLEAR POWER PLANT SINCE 1967	658
*MARINE BENTHIC ALGAE	100 - 300 BASED ON DRY WEIGHT	EDTA ADDED	1559

TABLE D140

CONCENTRATION FACTOR : GA			
ITEM:	VALUE:	CONDITIONS:	REF.:
*BROWN ALGAE	4200 BASED ON WET WEIGHT	NONE	157
*PLANKTON	12000 BASED ON WET WEIGHT	NONE	157

TABLE D141

CONCENTRATION FACTOR : GE			
ITEM:	VALUE:	CONDITIONS:	REF.:
*BROWN ALGAE	15-200 BASED ON WET WEIGHT	NONE	994
*32 FRESHWATER PLANT	305 BASED ON DRY WEIGHT	ALGAE,BRYOPHYTES AND VASCULAR HYDROPHYTES	1597

TABLE D142

CONCENTRATION FACTOR : HCH			
ITEM:	VALUE:	CONDITIONS:	REF.:
*SALMO GAIRDNERI	0.1 BASED ON WET WEIGHT	UPTAKE FROM FOOD, CONC IN FISH FAT, FOOD: HCH =10 MG/KG, 2 WEEKS	224
*SALMO GAIRDNERI	0.26 BASED ON WET WEIGHT	UPTAKE FROM FOOD, CONC IN FISH FAT, FOOD: HCH =50 MG/KG, 2 WEEKS	224
*SALMO GAIRDNERI	0.38 BASED ON WET WEIGHT	UPTAKE FROM FOOD, CONC IN FISH FAT, FOOD: HCH = 250 MG/KG, 2 WEEKS	224
*SALMO GAIRDNERI	0.47 BASED ON WET WEIGHT	UPTAKE FROM FOOD, CONC IN FISH FAT, FOOD: HCH = 10 MG/KG, 4 WEEKS	224
*SALMO GAIRDNERI	0.42 BASED ON WET WEIGHT	UPTAKE FROM FOOD, CONC IN FISH FAT, FOOD: HCH =50MG/KG, 4 W	224
*SALMO GAIRDNERI	0.35 BASED ON WET WEIGHT	UPTAKE FROM FOOD, CONC IN FISH FAT, FOOD: HCH =250MG/KG, 4 W	224
*SALMO GAIRDNERI	0.34 BASED ON WET WEIGHT	UPTAKE FROM FOOD, CONC IN FISH FAT, FOOD: HCH =1250MG/KG, 4 W	224
*SALMO GAIRDNERI	0.44 BASED ON WET WEIGHT	UPTAKE FROM FOOD, CONC IN FISH FAT, FOOD: HCH =10MG/KG, 8 W	224
*SALMO GAIRDNERI	0.39 BASED ON WET WEIGHT	UPTAKE FROM FOOD, CONC IN FISH FAT, FOOD: HCH =50MG/KG, 8 W	224
*SALMO GAIRDNERI	0.40 BASED ON WET WEIGHT	UPTAKE FROM FOOD, CONC IN FISH FAT, FOOD: HCH =250MG/KG, 8 W	224
*SALMO GAIRDNERI	0.43 BASED ON WET WEIGHT	UPTAKE FROM FOOD, CONC IN FISH FAT, FOOD: HCH =1250MG/KG, 8 W	224
*SALMO GAIRDNERI	0.54 BASED ON WET WEIGHT	UPTAKE FROM FOOD, CONC IN FISH FAT, FOOD: HCH =10MG/KG, 12 W	224
*SALMO GAIRDNERI	0.37 BASED ON WET WEIGHT	UPTAKE FROM FOOD, CONC IN FISH FAT, FOOD: HCH =50MG/KG, 12 W	224
*SALMO GAIRDNERI	0.70 BASED ON WET WEIGHT	UPTAKE FROM FOOD, CONC IN FISH FAT, FOOD: HCH =250MG/KG, 12 W	224
*SALMO GAIRDNERI	0.41 BASED ON WET WEIGHT	UPTAKE FROM FOOD, CONC IN FISH FAT, FOOD: HCH =1250MG/KG, 12 W	224

TABLE D143

CONCENTRATION FACTOR : HEPTACHLOR			
ITEM:	VALUE:	CONDITIONS:	REF.:
*ALFALFA	0.036 BASED ON WET WEIGHT	CONC.IN SOIL=0.78MG/KG DRY WT.	389
*ALFALFA FOLIAGE	2.863 BASED ON WET WEIGHT	CONC.IN SOIL=0.234MG/KG DRY WT.	389
*ALFALFA ROOT	1.752 BASED ON WET WEIGHT	CONC.IN SOIL=0.234MG/KG DRY WT.	389
*BLUEGILL	314 BASED ON WET WEIGHT	CONC.IN WATER=50UG/L	389
*BLUEGILL	1130 BASED ON WET WEIGHT	CONC.IN WATER=50UG/L	389
*CARROT	0.089 BASED ON WET WEIGHT	CONC.IN SANDY SOIL=0.44MG/KG DRY WT.	389
*CARROT	0.003 BASED ON WET WEIGHT	CONC.IN MUCK SOIL=1.40MG/KG DRY WT.	389
*CARROT, ROOT	0.737 BASED ON WET WEIGHT	CONC.IN SILT LOAM SOIL=1.33MG/KG DRY WT.	389
*CARROT, ROOT	0.735 BASED ON WET WEIGHT	CONC.IN SOIL=0.49MG/KG DRY WT.	389

TABLE D143

CONCENTRATION FACTOR : HEPTACHLOR			
ITEM:	VALUE:	CONDITIONS:	REF.:
*CHICKEN, LIVER	0.13 BASED ON WET WEIGHT	CONC.IN FOOD=0.05MG/KG	389
*CUCUMBER FRUIT	0.721 BASED ON WET WEIGHT	CONC.IN SOIL=3.8MG/KG DRY WT.	389
*EARTHWORMS	3.50 BASED ON DRY WEIGHT	CONC.IN SOIL=0.02MG/KG DRY WT., U.S.,SITES=67	389
*LETTUCE	0.016 BASED ON WET WEIGHT	CONC.IN SILT LOAM SOIL=2.5MG/KG DRY WT.	389
*OAT, SEED	0.02 BASED ON WET WEIGHT	CONC.IN SOIL=1.0MG/KG DRY WT.	389
*POTATO TUBERS	0.102 BASED ON WET WEIGHT	CONC.IN SOIL=0.49MG/KG DRY WT.	389
*RUTABAGA, ROOT	0.075 BASED ON WET WEIGHT	CONC.IN SOIL=0.320MG/KG DRY WT.	389
*SOYBEAN	0.11 BASED ON WET WEIGHT	CONC.IN SOIL=1.0MG/KG DRY WT.	389
*SUGAR BEET ROOT	0.056 BASED ON WET WEIGHT	CONC.IN SILT LOAM SOIL=2.5MG/KG DRY WT.	389
*SWEET POTATO TUBERS	0.02 BASED ON WET WEIGHT	CONC.IN SOIL=0.02MG/KG DRY WT.	389
*TURNIPS	0.036 BASED ON WET WEIGHT	CONC.IN SILT LOAM SOIL=2.5MG/KG DRY WT.	389
*WHEAT	0.057 BASED ON WET WEIGHT	CONC.IN SOIL=1.94MG/KG DRY WT.	389
*WHEAT	0.045 BASED ON WET WEIGHT	CONC.IN SOIL=9.74MG/KG DRY WT.	389
*WOODCOCK, TISSUE	4.55 BASED ON WET WEIGHT	CONC.IN FOOD=2.86MG/KG	389
*WOODCOCK, TISSUE	2.62 BASED ON WET WEIGHT	CONC.IN FOOD=0.65MG/KG	389

TABLE D144

CONCENTRATION FACTOR : HEPTACHLOR AND -EPOXIDE			
ITEM:	VALUE:	CONDITIONS:	REF.:
*CRASSOSTREA VIRGINICA	7 BASED ON WET WEIGHT	CONC.IN WATER=0.28UG/L,U.S.	389
*CRAYFISH	17 BASED ON WET WEIGHT	CONC.IN WATER=0.06UG/L,U.S.	389
*GYRINIDAE	5200 BASED ON WET WEIGHT	CONC.IN WATER=0.06UG/L,U.S.	389
*OYSTER	10 BASED ON WET WEIGHT	CONC.IN WATER:<1.0UG/L,CONC.IN SEDIMENT:<0.002MG/KG,U.S.	389
*OYSTER	20 BASED ON WET WEIGHT	CONC.IN WATER:<1.0UG/L,CONC.IN SEDIMENT:<0.02MG/KG,U.S.	389
*PLANKTON	18 BASED ON WET WEIGHT	CONC.IN WATER=0.06UG/L,U.S.	389
*TENDIPENDIDAE,LARVAE	5200 BASED ON WET WEIGHT	CONC.IN WATER=0.06UG/L,U.S.	389

TABLE D145

CONCENTRATION FACTOR : HEXABROMOBENZENE			
ITEM:	VALUE:	CONDITIONS:	REF.:
*SALMO SALAR	1 BASED ON WET WEIGHT	UPTAKE FROM WATER, STEADY STATE CF., 288 K, 48 H, 6.24 G FISH	1776
*SALMO SALAR	1 BASED ON WET WEIGHT	UPTAKE FROM FOOD, STEADY STATE CF., 288 K, 48 H, 6.24 G FISH	1776

TABLE D146

CONCENTRATION FACTOR : HEXABROMOBIPHENYL			
ITEM:	VALUE:	CONDITIONS:	REF.:
*SALMO SALAR	1.73 BASED ON WET WEIGHT	JUVENILE, UPTAKE FROM WATER, 48 H, 5.29 G FISH	1778

TABLE D146

CONCENTRATION FACTOR : HEXABROMOBIPHENYL			
ITEM:	VALUE:	CONDITIONS:	REF.:
*SALMO SALAR	0.114 BASED ON WET WEIGHT	JUVENILE, UPTAKE FROM FOOD, 48 H, 5.29 G FISH	1778

TABLE D147

CONCENTRATION FACTOR : HEXACHLOROBENZENE			
ITEM:	VALUE:	CONDITIONS:	REF.:
*SALMO SALAR	690 BASED ON WET WEIGHT	UPTAKE FROM WATER, STEADY STATE CF., 288 K, 48 H, 6.24 G FISH	1776
*SALMO SALAR	0.174 BASED ON WET WEIGHT	UPTAKE FROM FOOD, STEADY STATE CF., 288 K, 48 H, 6.24 G FISH	1776

TABLE D148

CONCENTRATION FACTOR : HG			
ITEM:	VALUE:	CONDITIONS:	REF.:
*AEDES AEGYPTI	167.3 (SD=20.1)	300K,PHENYL-203-HG-ACETATE,24HRS INCUBATION,LARVAL AGE=3 DAYS (F027)	355
*AEDES AEGYPTI	238.1 (SD=18.2)	300K,PHENYL-203-HG-ACETATE,48HRS INCUBATION,LARVAL AGE=3 DAYS (F027)	355
*AEDES AEGYPTI	381.2 (SD=17.1)	300K,PHENYL-203-HG-ACETATE,72HRS INCUBATION,LARVAL AGE=3 DAYS (F027)	355
*AEDES AEGYPTI	165.0 (SD=4.0)	300K,PHENYL-203-HG-ACETATE,96HRS INCUBATION,LARVAL AGE=3 DAYS (F027)	355
*AEDES AEGYPTI	291.9 (SD=6.7)	300K,PHENYL-203-HG-ACETATE,24HRS INCUBATION,LARVAL AGE=5 DAYS (F027)	355
*AEDES AEGYPTI	334.5 (SD=25.3)	300K,PHENYL-203-HG-ACETATE,48HRS INCUBATION,LARVAL AGE=5 DAYS (F027)	355
*AEDES AEGYPTI	491.8 (SD=31.3)	300K,PHENYL-203-HG-ACETATE,72HRS INCUBATION,LARVAL AGE=5 DAYS (F027)	355
*AEDES AEGYPTI	325.0 (SD=13.0)	300K,PHENYL-203-HG-ACETATE,96HRS INCUBATION,LARVAL AGE=5 DAYS (F027)	355
*AEDES AEGYPTI	385.3 (SD=23.5)	300K,PHENYL-203-HG-ACETATE,24HRS INCUBATION,LARVAL AGE=7 DAYS (F027)	355
*AEDES AEGYPTI	447.2 (SD=53.7)	300K,PHENYL-203-HG-ACETATE,48HRS INCUBATION,LARVAL AGE=7 DAYS (F027)	355
*AEDES AEGYPTI	579.1 (SD=26.4)	300K,PHENYL-203-HG-ACETATE,48HRS INCUBATION,LARVAL AGE=7 DAYS (?057)	355
*AEDES AEGYPTI	372.1 (SD=7.0)	300K,PHENYL-203-HG-ACETATE,96HRS INCUBATION,LARVAL AGE=7 DAYS (F027)	355
*AEDES AEGYPTI	20.5 (SD=1.2)	300K,203-HG-NITRATE,24HRS INCUBATION,LARVAL AGE=3 DAYS (F027)	355
*AEDES AEGYPTI	24.1 (SD=2.1)	300K,203-HG-NITRATE,48HRS INCUBATION,LARVAL AGE=3 DAYS (F027)	355
*AEDES AEGYPTI	42.7 (SD=1.7)	300K,203-HG-NITRATE,72HRS INCUBATION,LARVAL AGE=3 DAYS (F027)	355
*AEDES AEGYPTI	95.1 (SD=4.7)	300K,203-HG-NITRATE,96HRS INCUBATION,LARVAL AGE=3 DAYS (F027)	355

TABLE D148

CONCENTRATION FACTOR : HG

ITEM:	VALUE:	CONDITIONS:	REF.:
*AEDES AEGYPTI	35.1 (SD=1.8)	300K,203-HG-NITRATE,24HRS INCUBATION,LARVAL AGE=5 DAYS (F027)	355
*AEDES AEGYPTI	46.7 (SD=2.8)	300K,203-HG-NITRATE,48HRS INCUBATION,LARVAL AGE=5 DAYS (F027)	355
*AEDES AEGYPTI	63.1 (SD=4.3)	300K,203-HG-NITRATE,72HRS INCUBATION,LARVAL AGE=5 DAYS (F027)	355
*AEDES AEGYPTI	114.3 (SD=4.8)	300K,203-HG-NITRATE,96HRS INCUBATION,LARVAL AGE=5 DAYS (F027)	355
*AEDES AEGYPTI	50.1 (SD=3.2)	300K,203-HG-NITRATE,24HRS INCUBATION,LARVAL AGE=7 DAYS (F027)	355
*AEDES AEGYPTI	79.8 (SD=6.1)	300K,203-HG-NITRATE,48HRS INCUBATION,LARVAL AGE=7 DAYS (F027)	355
*AEDES AEGYPTI	102.1 (SD=3.1)	300K,203-HG-NITRATE,72HRS INCUBATION,LARVAL AGE=7 DAYS (F027)	355
*AEDES AEGYPTI	130.2 (SD=3.3)	300K,203-HG-NITRATE,96HRS INCUBATION,LARVAL AGE=7 DAYS (F027)	355
*ALGAE	200-1200 BASED ON DRY WEIGHT	SEE REF.	610
*BACILLUS LICHENIFORMIS	6.7 (SD=0.5)	296K,PHENYL-203-HG-ACETATE,5HRS INCUBATION (F028)	355
*BACILLUS LICHENIFORMIS	8.1 (SD=1.3)	296K,PHENYL-203-HG-ACETATE,24HRS INCUBATION (F028)	355
*BACILLUS LICHENIFORMIS	8.2 (SD=1.8)	296K,PHENYL-203-HG-ACETATE,48HRS INCUBATION (F028)	355
*BACILLUS LICHENIFORMIS	9.4 (SD=0.4)	296K,PHENYL-203-HG-ACETATE,72HRS INCUBATION (F028)	355
*BACILLUS LICHENIFORMIS	8.8 (SD=1.9)	296K,PHENYL-203-HG-ACETATE,96HRS INCUBATION (F028)	355
*BACILLUS LICHENIFORMIS	4.3 (SD=0.7)	310K,PHENYL-203-HG-ACETATE,5HRS INCUBATION (F028)	355
*BACILLUS LICHENIFORMIS	31.3 (SD=3.5)	310K,PHENYL-203-HG-ACETATE,24HRS INCUBATION (F028)	355
*BACILLUS LICHENIFORMIS	53.9 (SD=10.7)	310K,PHENYL-203-HG-ACETATE,48HRS INCUBATION (F028)	355
*BACILLUS LICHENIFORMIS	39.4 (SD=10.6)	310K,PHENYL-203-HG-ACETATE,72HRS INCUBATION (F028)	355
*BACILLUS LICHENIFORMIS	19.0 (SD=3.2)	310K,PHENYL-203-HG-ACETATE,96HRS INCUBATION (F028)	355
*BACILLUS LICHENIFORMIS	8.8 (SD=0.5)	318K,PHENYL-203-HG-ACETATE,5HRS INCUBATION (F028)	355
*BACILLUS LICHENIFORMIS	42.8 (SD=6.3)	318K,PHENYL-203-HG-ACETATE,24HRS INCUBATION (F028)	355
*BACILLUS LICHENIFORMIS	51.4 (SD=9.0)	318K,PHENYL-203-HG-ACETATE,48HRS INCUBATION (F028)	355
*BACILLUS LICHENIFORMIS	55.5 (SD=5.6)	318K,PHENYL-203-HG-ACETATE,72HRS INCUBATION (F028)	355
*BACILLUS LICHENIFORMIS	22.4 (SD=3.9)	318K,PHENYL-203-HG-ACETATE,96HRS INCUBATION (F028)	355
*BACILLUS LICHENIFORMIS	9.1 (SD=2.0)	296K,203-HG-NITRATE,5HRS INCUBATION (F028)	355
*BACILLUS LICHENIFORMIS	17.7 (SD=3.1)	296K,203-HG-NITRATE,24HRS INCUBATION (F028)	355
*BACILLUS LICHENIFORMIS	25.7 (SD=5.2)	296K,203-HG-NITRATE,48HRS INCUBATION (F028)	355
*BACILLUS LICHENIFORMIS	33.6 (SD=5.3)	296K,203-HG-NITRATE,72HRS INCUBATION (F028)	355
*BACILLUS LICHENIFORMIS	28.6 (SD=5.7)	296K,203-HG-NITRATE,96HRS INCUBATION (F028)	355
*BACILLUS LICHENIFORMIS	13.6 (SD=2.8)	310K,203-HG-NITRATE,5HRS INCUBATION (F028)	355
*BACILLUS LICHENIFORMIS	61.6 (SD=3.7)	310K,203-HG-NITRATE,24HRS INCUBATION (F028)	355

TABLE D148

CONCENTRATION FACTOR : HG

ITEM:	VALUE:	CONDITIONS:	REF.:
*BACILLUS LICHENIFORMIS	70.9 (SD=6.7)	310K,203-HG-NITRATE,48HRS INCUBATION (F028)	355
*BACILLUS LICHENIFORMIS	78.0 (SD=2.0)	310K,203-HG-NITRATE,72HRS INCUBATION (F028)	355
*BACILLUS LICHENIFORMIS	16.8 (SD=7.2)	310K,203-HG-NITRATE,96HRS INCUBATION (F028)	355
*BACILLUS LICHENIFORMIS	34.7 (SD=6.5)	318K,203-HG-NITRATE,5HRS INCUBATION (F028)	355
*BACILLUS LICHENIFORMIS	69.9 (SD=10.8)	318K,203-HG-NITRATE,24HRS INCUBATION (F028)	355
*BACILLUS LICHENIFORMIS	51.9 (SD=3.0)	318K,203-HG-NITRATE,48HRS INCUBATION (F028)	355
*BACILLUS LICHENIFORMIS	42.2 (SD=5.8)	318K,203-HG-NITRATE,72HRS INCUBATION (F028)	355
*BACILLUS LICHENIFORMIS	24.0 (SD=3.5)	318K,203-HG-NITRATE,96HRS INCUBATION (F028)	355
*BROWN ALGAE	250 BASED ON WET WEIGHT	NONE	157
*CRASSOSTREA VIRGINICA	1400 - 2800 BASED ON WET WEIGHT	WATER = 10 - 100 UG.L, 45 D	307
*DAPHNIA MAGNA	65 BASED ON WET WEIGHT	HG=2.0 MG HG/M3,1 W	1563
*DAPHNIA MAGNA	33 BASED ON WET WEIGHT	HG=2.0 MG HG/M3,3 W	1563
*DAPHNIA MAGNA	55 BASED ON WET WEIGHT	HG=2.0 MG HG/M3,5 W	1563
*DAPHNIA MAGNA	55 BASED ON WET WEIGHT	HG=2.0 MG HG/M3,7 W	1563
*DAPHNIA MAGNA	50 BASED ON WET WEIGHT	HG=2.0 MG HG/M3,10 W	1563
*DAPHNIA MAGNA	45 BASED ON WET WEIGHT	HG=5.0 MG HG/M3,1 W	1563
*DAPHNIA MAGNA	46 BASED ON WET WEIGHT	HG=5.0 MG HG/M3,3 W	1563
*DAPHNIA MAGNA	38 BASED ON WET WEIGHT	HG=5.0 MG HG/M3,5 W	1563
*DAPHNIA MAGNA	25 BASED ON WET WEIGHT	HG=5.0 MG HG/M3,7 W	1563
*DAPHNIA MAGNA	20 BASED ON WET WEIGHT	HG=5.0 MG HG/M3,10 W	1563
*ESOX LUCIUS, AXIAL MUSCLE	3000 BASED ON WET WEIGHT	MIN VALUE, SWEDEN	776
*FISH	3000 BASED ON WET WEIGHT	FRESHWATER,METHYL-HG(+)	1135
*LEBISTES RETICULATUS	108.4	203-HG-NITRATE,24HRS INCUBATION, BASED ON WET WEIGHT	355
*LEBISTES RETICULATUS	282.2	203-HG-NITRATE,48HRS INCUBATION, BASED ON WET WEIGHT	355
*LEBISTES RETICULATUS	184.1	203-HG-NITRATE,72HRS INCUBATION, BASED ON WET WEIGHT	355
*LEBISTES RETICULATUS	169.1	203-HG-NITRATE,96HRS INCUBATION, BASED ON WET WEIGHT	355
*LEBISTES RETICULATUS	33.3	PHENYL-203-HG-ACETATE,24HRS INCUBATION,BASED ON WET WEIGHT	355
*LEBISTES RETICULATUS	91.6	PHENYL-203-HG-ACETATE,48HRS INCUBATION,BASED ON WET WEIGHT	355
*LEBISTES RETICULATUS	169.1	PHENYL-203-HG-ACETATE,72HRS INCUBATION,BASED ON WET WEIGHT	355
*LEBISTES RETICULATUS	18.3	PHENYL-203-HG-ACETATE,96HRS INCUBATION,BASED ON WET WEIGHT	355
*LIGUMIA SP. AND MARGARITIFERA SP.	12.5 BASED ON WET WEIGHT	HG=2.0 MG HG/M3,1 W,BODY WEIGHT = 6.17 G	1563
*LIGUMIA SP. AND MARGARITIFERA SP.	37.5 BASED ON WET WEIGHT	HG=2.0 MG HG/M3,3 W,BODY WEIGHT = 4.13 G	1563
*LIGUMIA SP. AND MARGARITIFERA SP.	30.0 BASED ON WET WEIGHT	HG=2.0 MG HG/M3,5 W,BODY WEIGHT = 4.31 G	1563
*LIGUMIA SP. AND MARGARITIFERA SP.	67.5 BASED ON WET WEIGHT	HG=2.0 MG HG/M3,7 W,BODY WEIGHT = 4.43 G	1563
*LIGUMIA SP. AND MARGARITIFERA SP.	55.0 BASED ON WET WEIGHT	HG=2.0 MG HG/M3|10 W,BODY WEIGHT = 4.61 G	1563
*LIGUMIA SP. AND MARGARITIFERA SP.	20.0 BASED ON WET WEIGHT	HG=5.0 MG HG/M3,1 W,BODY WEIGHT = 9.61 G	1563
*LIGUMIA SP. AND MARGARITIFERA SP.	67.0 BASED ON WET WEIGHT	HG=5.0 MG HG/M3,3 W,BODY WEIGHT = 2.86 G	1563
*LIGUMIA SP. AND MARGARITIFERA SP.	10.0 BASED ON WET WEIGHT	HG=5.0 MG HG/M3,5 W,BODY WEIGHT = 4.27 G	1563

TABLE D148

CONCENTRATION FACTOR : HG

ITEM:	VALUE:	CONDITIONS:	REF.:
*LIGUMIA SP. AND MARGARITIFERA SP.	30.0 BASED ON WET WEIGHT	HG=5.0 MG HG/M3,7 W,BODY WEIGHT = 3.42 G	1563
*LIGUMIA SP. AND MARGARITIFERA SP.	35 BASED ON WET WEIGHT	HG=5.0 MG HG/M3,10 W,BODY WEIGHT = 5.09 G	1563
*LIGUMIA SP. AND MARGARITIFERA SP.	160 BASED ON WET WEIGHT	HG=5.0 MG HG/M3,1 W,BODY WEIGHT = 0.75 G	1563
*LONGFIN SMELT	3400 BASED ON WET WEIGHT	SIZE = 9.3 CM, LAKE WASHINGTON, WATER CONC = 0.020 MG/M3, MUSCLE	1392
*MYTILUS GALLOPROVINCIALIS	15625 BASED ON WET WEIGHT	WATER = 41.2 PPM	971
*ORYZIAS LATIPES	1600 BASED ON WET WEIGHT	WATER=10 PPB, MEASURED ON EGGS, PH =6.9, 299 K, HGCL	652
*ORYZIAS LATIPES	1900 BASED ON WET WEIGHT	WATER=15 PPB, MEASURED ON EGGS, PH =6.9, 299 K, HGCL	652
*ORYZIAS LATIPES	2700 BASED ON WET WEIGHT	WATER=20 PPB, MEASURED ON EGGS, PH =6.9, 299 K, HGCL	652
*ORYZIAS LATIPES	1900 BASED ON WET WEIGHT	WATER=30 PPB, MEASURED ON EGGS, PH =6.9, 299 K, HGCL	652
*PALAEMONETES VULGARIS, EXOSKELETON	7.54 BASED ON WET WEIGHT	METHYLMERCURY CLORIDE, WATER = 1.5 PPB, 24 H, 293 K	1301
*PALAEMONETES VULGARIS, EXOSKELETON	8.51 BASED ON WET WEIGHT	MERCURIC CLORIDE, WATER = 1.5 PPB, 24 H, 293 K	1301
*PALAEMONETES VULGARIS, EXOSKELETON	11.3 BASED ON WET WEIGHT	METHYLMERCURY CLORIDE, WATER = 1.5 PPB, 72 H, 293 K	1301
*PALAEMONETES VULGARIS, FOREGUT	5.58 BASED ON WET WEIGHT	METHYLMERCURY CLORIDE, WATER = 1.5 PPB, 24 H, 293 K	1301
*PALAEMONETES VULGARIS, FOREGUT	4.30 BASED ON WET WEIGHT	MERCURIC CLORIDE, WATER = 1.5 PPB, 24 H, 293 K	1301
*PALAEMONETES VULGARIS, FOREGUT	8.7 BASED ON WET WEIGHT	METHYLMERCURY CLORIDE, WATER = 1.5 PPB, 72 H, 293 K	1301
*PALAEMONETES VULGARIS, GILLS	24.44 BASED ON WET WEIGHT	METHYLMERCURY CLORIDE, WATER = 1.5 PPB, 24 H, 293 K	1301
*PALAEMONETES VULGARIS, GILLS	147.54 BASED ON WET WEIGHT	MERCURIC CLORIDE, WATER = 1.5 PPB, 24 H, 293 K	1301
*PALAEMONETES VULGARIS, GILLS	16.3 BASED ON WET WEIGHT	METHYLMERCURY CLORIDE, WATER = 1.5 PPB, 72 H, 293 K	1301
*PALAEMONETES VULGARIS, REMSINDER	2.19 BASED ON WET WEIGHT	METHYLMERCURY CLORIDE, WATER = 1.5 PPB, 24 H, 293 K	1301
*PALAEMONETES VULGARIS, REMSINDER	2.41 BASED ON WET WEIGHT	MERCURIC CLORIDE, WATER = 1.5 PPB, 24 H, 293 K	1301
*PALAEMONETES VULGARIS, REMSINDER	5.3 BASED ON WET WEIGHT	METHYLMERCURY CLORIDE, WATER = 1.5 PPB, 72 H, 293 K	1301
*PALAEMONETES VULGARIS, VENTRAL NERVE CORD	113.22 BASED ON WET WEIGHT	METHYLMERCURY CLORIDE, WATER = 1.5 PPB, 24 H, 293 K	1301
*PALAEMONETES VULGARIS, VENTRAL NERVE CORD	29.23 BASED ON WET WEIGHT	MERCURIC CLORIDE, WATER = 1.5 PPB, 24 H, 293 K	1301
*PALAEMONETES VULGARIS, VENTRAL NERVE CORD	7.6 BASED ON WET WEIGHT	METHYLMERCURY CLORIDE, WATER = 1.5 PPB, 72 H, 293 K	1301
*PALAEMONETES VULGARIS, WHOLE BODIES	5.60 BASED ON WET WEIGHT	METHYLMERCURY CLORIDE, WATER = 1.5 PPB, 24 H, 293 K	1301
*PALAEMONETES VULGARIS, WHOLE BODIES	5.70 BASED ON WET WEIGHT	MERCURIC CLORIDE, WATER = 1.5 PPB, 24 H, 293 K	1301
*PALAEMONETES VULGARIS, WHOLE BODIES	7.90 BASED ON WET WEIGHT	METHYLMERCURY CLORIDE, WATER = 1.5 PPB, 72 H, 293 K	1301
*PHYTOPLANKTON	70-4800 BASED ON WET WEIGHT	NONE	311
*PTYCHOCHEILUS OREGONENSIS	8500 ON WET WEIGHT BASIS	SIZE = 30 CM, LAKE WASHINGTON, WATER CONC = 0.020 MG/M3, MUSCLE	1392
*PTYCHOCHEILUS OREGONENSIS	40000 ON WET WEIGHT BASIS	SIZE = 50 CM, LAKE WASHINGTON, WATER CONC = 0.020 MG/M3, MUSCLE	1392
*SALMO GAIRDNERI	4 BASED ON BASED ON WET WEIGHT	WATER = 0.1 MG HG/L AS HGCL, 278 K	962
*SALMO GAIRDNERI	10 BASED ON BASED ON WET WEIGHT	WATER = 0.1 MG HG/L AS HGCL, 283 K	962
*SALMO GAIRDNERI	22 BASED ON BASED ON WET WEIGHT	WATER = 0.1 MG HG/L AS HGCL, 293 K	962
*SALMO GAIRDNERI	80 - 100 BASED ON WET WEIGHT	293 K, 0.1 MG HG/L AS PHENYLMERCURIC ACETATE	962
*SCENEDESMUS SP.	43 BASED ON WET WEIGHT	HG= 2.0MG/M3, 1 WEEK	1563
*SCENEDESMUS SP.	73 BASED ON WET WEIGHT	HG=2.0 MG/M3, 3 W	1563

TABLE D148

CONCENTRATION FACTOR : HG			
ITEM:	VALUE:	CONDITIONS:	REF.:
*SCENEDESMUS SP.	113 BASED ON WET WEIGHT	HG= 2.0 MG/M3, 5 W	1563
*SCENEDESMUS SP.	180 BASED ON WET WEIGHT	HG= 2.0 MG/M3, 7 W	1563
*SCENEDESMUS SP.	180 BASED ON WET WEIGHT	HG=2.0 MG/M3, 10 W	1563
*SCENEDESMUS SP.	66 BASED ON WET WEIGHT	HG=5.0 MG/M3, 1 W	1563
*SCENEDESMUS SP.	100 BASED ON WET WEIGHT	HG= 5.0 MG/M3, 3 W	1563
*SCENEDESMUS SP.	120 BASED ON WET WEIGHT	HG= 5.0 MG/M3, 5 W	1563
*SCENEDESMUS SP.	144 BASED ON WET WEIGHT	HG= 5.0 MG/M3, 7 W	1563
*SCENEDESMUS SP.	146 BASED ON WET WEIGHT	HG= 5.0 MG/M3, 10 W	1563
*SOCKEYE	1750 ON WET WEIGHT BASIS	SIZE = 9.0 CM, LAKE WASHINGTON, WATER CONC = 0.020 MG/M3, MUSCLE	1392
*ZOOPLANKTON	650 ON WET WEIGHT BASIS	LAKE WASHINGTON, WATER CONC = 0.020 MG/M3	1392
*32 FRESHWATER PLANT	5915 BASED ON DRY WEIGHT	ALGAE,BRYOPHYTES AND VASCULAR HYDROPHYTES	1597

TABLE D149

CONCENTRATION FACTOR : I			
ITEM:	VALUE:	CONDITIONS:	REF.:
*BROWN ALGAE	6200 BASED ON WET WEIGHT	NONE	157
*BROWN ALGAE	3000-10000 BASED ON WET WEIGHT	NONE	994
*PLANKTON	1200 BASED ON WET WEIGHT	NONE	157
*32 FRESHWATER PLANT	370 BASED ON DRY WEIGHT	ALGAE,BRYOPHYTES AND VASCULAR HYDROPHYTES	1597

TABLE D150

CONCENTRATION FACTOR : I-131			
ITEM:	VALUE:	CONDITIONS:	REF.:
*ANODONTA NUTTALLIANA, SOFT PARTS	0.5 BASED ON WET WEIGHT	NO FOOD,292 K,PH=7.5, FRESHWAT=1.7 PART. AND 250 SOLUB P CI/ML, 7 D	629
*ANODONTA NUTTALLIANA, SOFT PARTS	0.7 BASED ON WET WEIGHT	NO FOOD,292 K,PH=7.5, FRESHWAT=5.7 PART. AND 280 P CI/ML, 21 D	629
*ASTACUS SP., MUSCLE	0.1 BASED ON WET WEIGHT	NO FOOD,292 K,PH=7.5, FRESHWAT=1.7 PART. AND 250 SOLUB P CI/ML, 7 D	629
*ASTACUS SP., MUSCLE	LESS THAN 1 BASED ON WET WEIGHT	NO FOOD,292 K,PH=7.5, FRESHWAT=5.7 PART. AND 280 P CI/ML, 21 D	629
*ASTACUS SP., SOFT PARTS	15 BASED ON WET WEIGHT	NO FOOD,292 K,PH=7.5, FRESHWAT= 0.69 PART. AND 110 SOLUB P CI/ML, 7 D	629
*ASTACUS SP., VISCERA	17 BASED ON WET WEIGHT	NO FOOD,292 K,PH=7.5, FRESHWAT=5.7 PART. AND 280 P CI/ML, 21 D	629
*CANCER PRODUCTUS, MUSCLE	0.5 BASED ON WET WEIGHT	NO FOOD,285 K,PH=8.2, SEAWATER=1.3 PART. AND 450 P CI/ML, 20 D	629
*CANCER PRODUCTUS, SORT PARTS	19 BASED ON WET WEIGHT	NO FOOD,285 K,PH=8.2, SEAWATER=1.3 PART. AND 450 P CI/ML, 20 D	629
*CANCER PRODUCTUS, VISCERA	2 BASED ON WET WEIGHT	NO FOOD,285 K,PH=8.2, SEAWATER=1.1 PART. AND 440 SOLUB P CI/ML, 6 D	629
*CANCER PRODUCTUS, VISCERA	21 BASED ON WET WEIGHT	NO FOOD,285 K,PH=8.2, SEAWATER=1.3 PART. AND 450 P CI/ML, 20 D	629
*CARASSIUS AURATUS, MUSCLE	0.2 BASED ON WET WEIGHT	NO FOOD,292 K,PH=7.5, FRESHWAT=1.7 PART. AND 250 SOLUB P CI/ML, 7 D	629
*CARASSIUS AURATUS, NONMUSCLE	1 BASED ON WET WEIGHT	NO FOOD,292 K,PH=7.5, FRESHWAT=1.7 PART. AND 250 SOLUB P CI/ML, 7 D	629
*CARASSIUS AURATUS, WHOLE BODY	0.9 BASED ON WET WEIGHT	NO FOOD,292 K,PH=7.5, FRESHWAT=5.7 PART. AND 280 P CI/ML, 21 D	629

TABLE D150

CONCENTRATION FACTOR : I-131			
ITEM:	VALUE:	CONDITIONS:	REF.:
*FISH, MUSCLE	520 (RANGE = 250 - 780) BASED ON WET WEIGHT	DANUBE, 1973, NUCLEAR POWER PLANT SINCE 1967	658
*FRESHWATER SEDIMENT	6800 (RANGE = 510 - 13080) BASED ON WET WEIGHT	DANUBE, 1973, NUCLEAR POWER PLANT SINCE 1967	658
*GILLICHTHYS MIRABILIS, MUSCLE	LESS THAN 1 BASED ON WET WEIGHT	NO FOOD,285 K,PH=8.2, SEAWATER=1.3 PART. AND 450 P CI/ML, 20 D	629
*GILLICHTHYS MIRABILIS, NONMUSCLE	0.7 BASED ON WET WEIGHT	NO FOOD,285 K,PH=8.2, SEAWATER=1.1 PART. AND 440 SOLUB P CI/ML, 6 D	629
*GILLICHTHYS MIRABILIS, NONMUSCLE	0.9 BASED ON WET WEIGHT	NO FOOD,285 K,PH=8.2, SEAWATER=1.3 PART. AND 450 P CI/ML, 20 D	629
*GILLICHTHYS MIRABILIS, WHOLE BODY	0.7 BASED ON WET WEIGHT	NO FOOD,285 K,PH=8.2, SEAWATER=1.3 PART. AND 450 P CI/ML, 20 D	629
*GILLICHTHYS MIRABILIS, WHOLE BODY	0.8 BASED ON WET WEIGHT	NO FOOD,285 K,PH=8.2, SEAWATER= 1.10 PART. AND 220 SOLUB P CI/ML, 6 D	629
*MYTILUS EDULIS, SOFT PARTS	0.8 BASED ON WET WEIGHT	NO FOOD,285 K,PH=8.2, SEAWATER=1.1 PART. AND 440 SOLUB P CI/ML, 6 D	629

TABLE D151

CONCENTRATION FACTOR : INSECTICIDES			
ITEM:	VALUE:	CONDITIONS:	REF.:
*SLUGS	1-100 BASED ON WET WEIGHT	CONC.FACT.=CONC.IN SLUGS(WET WT.)/ CONC.IN SOIL(DRY WT.)	389

TABLE D152

CONCENTRATION FACTOR : K			
ITEM:	VALUE:	CONDITIONS:	REF.:
*BRASSICA OLERACEA	115 BASED ON WET WEIGHT	CONC.IN SOIL SOLUTION=5.4PPM	157
*BROWN ALGAE	34 BASED ON WET WEIGHT	NONE	157
*BROWN ALGAE	3-50 BASED ON WET WEIGHT	NONE	994
*CANCER IRRORATUS	7.76 (SD=+-0.81) BASED ON WET WEIGHT	SEAWATER CONC=8.0 PPM, NOVA SCOTIA, AAS	984
*PHYTOPLANKTON	1400-7200 BASED ON WET WEIGHT	NONE	311

TABLE D153

CONCENTRATION FACTOR : K-40			
ITEM:	VALUE:	CONDITIONS:	REF.:
*FISH, BONE	1560 (RANGE = 970 - 2450) BASED ON WET WEIGHT	DANUBE, 1973, NUCLEAR POWER PLANT SINCE 1967	658
*FISH, MUSCLE	700 (RANGE = 570 - 1580) BASED ON WET WEIGHT	DANUBE, 1973, NUCLEAR POWER PLANT SINCE 1967	658
*FRESHWATER SEDIMENT	1300 (RANGE = 650 - 2080) BASED ON WET WEIGHT	DANUBE, 1973, NUCLEAR POWER PLANT SINCE 1967	658
*SUSPENDED SOLIDS	1070 (RANGE = 520 - 2010) BASED ON WET WEIGHT	DANUBE, 1973, NUCLEAR POWER PLANT SINCE 1967	658
*WATER PLANT	320 (RANGE = 220 - 490) BASED ON WET WEIGHT	DANUBE, 1973, NUCLEAR POWER PLANT SINCE 1967	658

TABLE D154

CONCENTRATION FACTOR : LA			
ITEM:	VALUE:	CONDITIONS:	REF.:
*BROWN ALGAE	8300 BASED ON WET WEIGHT	NONE	157

TABLE D155

CONCENTRATION FACTOR : LI			
ITEM:	VALUE:	CONDITIONS:	REF.:
*BROWN ALGAE	8? BASED ON WET WEIGHT	NONE	157
*CORAL	0.83 BASED ON CA CONTENT (F023)	DEEP OCEAN CORAL, MEAN OF 13 DETERMINATIONS	939
*CORAL	0.63 BASED ON CA CONTENT (F023)	SHALLOW OPEN OCEAN CORAL, MEAN OF 8 DETERMINATIONS	939
*CORAL	0.64 BASED ON CA CONTENT (F023)	SHALLOW COASTAL OCEAN, MEAN OF 13 DETERMINATIONS	939

TABLE D156

CONCENTRATION FACTOR : LU			
ITEM:	VALUE:	CONDITIONS:	REF.:
*ENGRAULIS JAPONICUS, WHOLE BODY	75 BASED ON WET WEIGHT	REL. TO TOTAL CONC.,TOT CONC=0.7 NG/L, SOLUB=0.3 NG/L,SEAWAT,JAP.	1535
*ENGRAULIS JAPONICUS, WHOLE BODY	200 BASED ON WET WEIGHT	REL. TO SOLUBLE CONC.,TOT CONC=0.7 NG/L, SOLUB=0.3 NG/L,SEAWAT,JAP.	1535
*HIZIKIA FUSIFORME, WHOLE BODY	280 BASED ON WET WEIGHT	REL. TO TOTAL CONC.,TOT CONC=0.7 NG/L, SOLUB=0.3 NG/L,SEAWAT,JAP.	1535
*HIZIKIA FUSIFORME, WHOLE BODY	730 BASED ON WET WEIGHT	REL. TO SOLUBLE CONC.,TOT CONC=0.7 NG/L, SOLUB=0.3 NG/L,SEAWAT,JAP.	1535
*MERETRIX LUSORIA, SOFT PART	300 BASED ON WET WEIGHT	REL. TO TOTAL CONC.,TOT CONC=0.7 NG/L, SOLUB=0.3 NG/L,SEAWAT,JAP.	1535
*MERETRIX LUSORIA, SOFT PART	770 BASED ON WET WEIGHT	REL. TO SOLUBLE CONC.,TOT CONC=0.7 NG/L, SOLUB=0.3 NG/L,SEAWAT,JAP.	1535
*PARALICHTHYS OLIVACEUS, MUSCLE	13 BASED ON WET WEIGHT	REL. TO TOTAL CONC.,TOT CONC=0.7 NG/L, SOLUB=0.3 NG/L,SEAWAT,JAP.	1535
*PARALICHTHYS OLIVACEUS, MUSCLE	33 BASED ON WET WEIGHT	REL. TO SOLUBLE CONC.,TOT CONC=5.6 NG/L, SOLUB=2.1 NG/L,SEAWAT,JAP.	1535
*SARGASSUM FULVELLUM, WHOLE BODY	350 BASED ON WET WEIGHT	REL. TO TOTAL CONC.,TOT CONC=0.7 NG/L, SOLUB=0.3 NG/L,SEAWAT,JAP.	1535
*SARGASSUM FULVELLUM, WHOLE BODY	930 BASED ON WET WEIGHT	REL. TO SOLUBLE CONC.,TOT CONC=0.7 NG/L, SOLUB=0.3 NG/L,SEAWAT,JAP.	1535
*SERIOLA QUINQUERADIATA, MUSCLE	13 BASED ON WET WEIGHT	REL. TO TOTAL CONC.,TOT CONC=0.7 NG/L, SOLUB=0.3 NG/L,SEAWAT,JAP.	1535
*SERIOLA QUINQUERADIATA, MUSCLE	33 BASED ON WET WEIGHT	REL. TO SOLUBLE CONC.,TOT CONC=5.6 NG/L, SOLUB=2.1 NG/L,SEAWAT,JAP.	1535
*ULVA PERTUSA, WHOLE BODY	300 BASED ON WET WEIGHT	REL. TO TOTAL CONC.,TOT CONC=0.7 NG/L, SOLUB=0.3 NG/L,SEAWAT,JAP.	1535
*ULVA PERTUSA, WHOLE BODY	800 BASED ON WET WEIGHT	REL. TO SOLUBLE CONC.,TOT CONC=0.7 NG/L, SOLUB=0.3 NG/L,SEAWAT,JAP.	1535
*UNDARIA PINNATIFIDA, WHOLE BODY	110 BASED ON WET WEIGHT	REL. TO TOTAL CONC.,TOT CONC=0.7 NG/L, SOLUB=0.3 NG/L,SEAWAT,JAP.	1535
*UNDARIA PINNATIFIDA, WHOLE BODY	300 BASED ON WET WEIGHT	REL. TO SOLUBLE CONC.,TOT CONC=0.7 NG/L, SOLUB=0.3 NG/L,SEAWAT,JAP.	1535

TABLE D157

CONCENTRATION FACTOR : METHOXYCHLOR			
ITEM:	VALUE:	CONDITIONS:	REF.:
*CHICKEN, MUSCLE AND TISSUE	0.01(MEAN) BASED ON WET WEIGHT	CONC.IN FOOD=1000MG/KG	389
*CHICKEN, TISSUE	0.006 BASED ON WET WEIGHT	CONC.IN FOOD=100.00MG/KG	389
*EARTHWORMS EXCL. LUMBRICUS TERRESTRIS	4.17 BASED ON WET WEIGHT	CONC.IN SOIL=2.66MG/KG DRY WT., U.S.,SITES=2	389
*LUMBRICUS TERRESTRIS	2.11 BASED ON WET WEIGHT	CONC.IN SOIL=2.66MG/KG DRY WT., U.S.,SITES=2	389
*OYSTER	10 BASED ON WET WEIGHT	CONC.IN WATER:<1.0UG/L,CONC.IN SEDIMENT:<0.01MG/KG,U.S.	389

TABLE D158

CONCENTRATION FACTOR : MG

ITEM:	VALUE:	CONDITIONS:	REF.:
*BRASSICA OLERACEA	17 BASED ON WET WEIGHT	CONC.IN SOIL SOLUTION=22PPM	157
*BROWN ALGAE	0.96 BASED ON WET WEIGHT	NONE	157
*CANCER IRRORATUS	2.24 (SD=+-0.103) BASED ON WET WEIGHT	SEAWATER CONC=1350 PPM, NOVA SCOTIA, AAS	984
*PHYTOPLANKTON	230-660 BASED ON WET WEIGHT	NONE	311
*PLANKTON	0.59 BASED ON WET WEIGHT	NONE	157
*ZOOPLANKTON	1.3 BASED ON WET WEIGHT	SEAWATER CONCENTRATION 1350.0 MG/M3	985

TABLE D159

CONCENTRATION FACTOR : MIREX

ITEM:	VALUE:	CONDITIONS:	REF.:
*CHLOROCOCCUM SP.	10000 BASED ON WET WEIGHT	LAB., WATER = 10 PPTR	685

TABLE D160

CONCENTRATION FACTOR : MN

ITEM:	VALUE:	CONDITIONS:	REF.:
*AGARUM CRIBROSUM	86600 BASED ON DRY WEIGHT	SEAWATER=6.4*10**(-7) G/L, VOSTOK BAY, SEA OF JAPAN	1367
*BRASSICA OLERACEA	460 BASED ON WET WEIGHT	CONC.IN SOIL SOLUTION=0.072PPM	157
*BROWN ALGAE	6500 BASED ON WET WEIGHT	NONE	157
*BROWN ALGAE	6500 BASED ON WET WEIGHT	NONE	994
*CANCER IRRORATUS	12699 (SD=+-1502) BASED ON WET WEIGHT	SEAWATER CONC=0.84*10**(-3) PPM, NOVA SCOTIA, AAS	984
*CHONDRUS YENDOI	10000 BASED ON DRY WEIGHT	SEAWATER=6.4*10**(-7) G/L, VOSTOK BAY, SEA OF JAPAN	1367
*CHORDA FILUM	2200 BASED ON DRY WEIGHT	SEAWATER=6.4*10**(-7) G/L, VOSTOK BAY, SEA OF JAPAN	1367
*CODIUM YESSOENSIS	24000 BASED ON DRY WEIGHT	SEAWATER=6.4*10**(-7) G/L, VOSTOK BAY, SEA OF JAPAN	1367
*COSTARIA COSTATA	7000 BASED ON DRY WEIGHT	SEAWATER=6.4*10**(-7) G/L, VOSTOK BAY, SEA OF JAPAN	1367
*ENTEROMORPHA PROLIFERA	71000 BASED ON DRY WEIGHT	SEAWATER=6.4*10**(-7) G/L, VOSTOK BAY, SEA OF JAPAN	1367
*FISH	373 BASED ON WET WEIGHT	MARINE ORGANISMS, TYPICAL VALUE	538
*FISH	23 BASED ON WET WEIGHT	FRESHWATER ORGANISMS, TYPICAL VALUE	1028
*FRESHWATER PLANKTON	29000 BASED ON WET WEIGHT	CONC.IN WATER=4.5UG/L,SCHOEHSEE	573
*HETEROCHORDARIA ABIETINA	4600 BASED ON DRY WEIGHT	SEAWATER=6.4*10**(-7) G/L, VOSTOK BAY, SEA OF JAPAN	1367
*LAMINARIA CHICHORIOIDES	8300 BASED ON DRY WEIGHT	SEAWATER=6.4*10**(-7) G/L, VOSTOK BAY, SEA OF JAPAN	1367
*LAMINARIA JAPONICA	1700 BASED ON DRY WEIGHT	SEAWATER=6.4*10**(-7) G/L, VOSTOK BAY, SEA OF JAPAN	1367
*MACROPHYTES	1450 BASED ON WET WEIGHT	FRESHWATER ORGANISMS, TYPICAL VALUE	1028
*MOLLUSCS	2300 BASED ON WET WEIGHT	MARINE ORGANISMS, TYPICAL VALUE	1280
*PELVETIA WRIGHTII	4600 BASED ON DRY WEIGHT	SEAWATER=6.4*10**(-7) G/L, VOSTOK BAY, SEA OF JAPAN	1367
*PHYLLOSPADIX IWATENSIS	8000 BASED ON DRY WEIGHT	SEAWATER=6.4*10**(-7) G/L, VOSTOK BAY, SEA OF JAPAN	1367
*PHYTOPLANKTON	3800 BASED ON WET WEIGHT	MARINE ORGANISMS, TYPICAL VALUE	986

TABLE D160

CONCENTRATION FACTOR : MN

ITEM:	VALUE:	CONDITIONS:	REF.:
*PLANKTON	9400 BASED ON WET WEIGHT	NONE	157
*PLANKTON	750 BASED ON WET WEIGHT	NONE	994
*POLYSIPHONIA JAPONICA	9000 BASED ON DRY WEIGHT	SEAWATER=6.4*10**(-7) G/L, VOSTOK BAY, SEA OF JAPAN	1367
*PTILOTA FILICINA	178000 BASED ON DRY WEIGHT	SEAWATER=6.4*10**(-7) G/L, VOSTOK BAY, SEA OF JAPAN	1367
*RHODOMELA LARIX	37000 BASED ON DRY WEIGHT	SEAWATER=6.4*10**(-7) G/L, VOSTOK BAY, SEA OF JAPAN	1367
*SARGASSUM PALLIDUM	7000 BASED ON DRY WEIGHT	SEAWATER=6.4*10**(-7) G/L, VOSTOK BAY, SEA OF JAPAN	1367
*SESTON	5.4 BASED ON WET WEIGHT	DEPTH = 8 M, SCH@HSEE	573
*SESTON	1.6 BASED ON WET WEIGHT	DEPTH = 17 M, SCH@HSEE	573
*ULVA FENESTRATA	44000 BASED ON DRY WEIGHT	SEAWATER=6.4*10**(-7) G/L, VOSTOK BAY, SEA OF JAPAN	1367
*ULVARIA SPENDENS	47000 BASED ON DRY WEIGHT	SEAWATER=6.4*10**(-7) G/L, VOSTOK BAY, SEA OF JAPAN	1367
*UNIO MANCUS	80000 BASED ON DRY WEIGHT	LAKE VARSEE	1027
*UNIO MANCUS	100000 BASED ON DRY WEIGHT	LAKE MAGGIORE	1027
*ZOOPLANKTON	1650 BASED ON WET WEIGHT	SEAWATER CONCENTRATION 0.002 MG/M3	985
*ZOOPLANKTON	3900 BASED ON WET WEIGHT	MARINE ORGANISMS, TYPICAL VALUE	986
*ZOSTERA ASIATICA	18600 BASED ON DRY WEIGHT	SEAWATER=6.4*10**(-7) G/L, VOSTOK BAY, SEA OF JAPAN	1367

TABLE D161

CONCENTRATION FACTOR : MN-54

ITEM:	VALUE:	CONDITIONS:	REF.:
*FISH, MUSCLE	230 (RANGE = 60 - 440) BASED ON WET WEIGHT	DANUBE, 1973, NUCLEAR POWER PLANT SINCE 1967	658
*FRESHWATER SEDIMENT	1520 (RANGE = 570 - 2450) BASED ON WET WEIGHT	DANUBE, 1973, NUCLEAR POWER PLANT SINCE 1967	658
*LAMPSILIS RADIATA	2400 BASED ON WET WEIGHT	91 DAYS, FRESHWATER	639
*SUSPENDED SOLIDS	1030 (RANGE = 500 - 1890) BASED ON WET WEIGHT	DANUBE, 1973, NUCLEAR POWER PLANT SINCE 1967	658
*UNIO MANCUS	40000 BASED ON DRY WEIGHT	LAKE VARSEE	1027
*UNIO MANCUS	300000 BASED ON DRY WEIGHT	LAKE MAGGIORE	1027
*WATER PLANT	240 (RANGE = 65 - 430) BASED ON WET WEIGHT	DANUBE, 1973, NUCLEAR POWER PLANT SINCE 1967	658

TABLE D162

CONCENTRATION FACTOR : MO

ITEM:	VALUE:	CONDITIONS:	REF.:
*AGARUM CRIBROSUM	60 BASED ON DRY WEIGHT	SEAWATER=11.0*10**(-7) G/L, VOSTOK BAY, SEA OF JAPAN	1367
*BROWN ALGAE	11 BASED ON WET WEIGHT	NONE	157
*CHONDRUS YENDOI	100 BASED ON DRY WEIGHT	SEAWATER=11.0*10**(-7) G/L, VOSTOK BAY, SEA OF JAPAN	1367
*CHORDA FILUM	200 BASED ON DRY WEIGHT	SEAWATER=11.0*10**(-7) G/L, VOSTOK BAY, SEA OF JAPAN	1367
*COSTARIA COSTATA	20 BASED ON DRY WEIGHT	SEAWATER=11.0*10**(-7) G/L, VOSTOK BAY, SEA OF JAPAN	1367
*ENTEROMORPHA PROLIFERA	250 BASED ON DRY WEIGHT	SEAWATER=11.0*10**(-7) G/L, VOSTOK BAY, SEA OF JAPAN	1367

TABLE D162

CONCENTRATION FACTOR : MO			
ITEM:	VALUE:	CONDITIONS:	REF.:
*FRESHWATER PLANKTON	20000 BASED ON WET WEIGHT	CONC.IN WATER=0.21UG/L,SCHOEHSEE	573
*HETEROCHORDARIA ABIETINA	10 BASED ON DRY WEIGHT	SEAWATER=11.0*10**(-7) G/L, VOSTOK BAY, SEA OF JAPAN	1367
*LAMINARIA CHICHORIOIDES	10 BASED ON DRY WEIGHT	SEAWATER=11.0*10**(-7) G/L, VOSTOK BAY, SEA OF JAPAN	1367
*LAMINARIA JAPONICA	70 BASED ON DRY WEIGHT	SEAWATER=11.0*10**(-7) G/L, VOSTOK BAY, SEA OF JAPAN	1367
*PELVETIA WRIGHTII	100 BASED ON DRY WEIGHT	SEAWATER=11.0*10**(-7) G/L, VOSTOK BAY, SEA OF JAPAN	1367
*PHYLLOSPADIX IWATENSIS	20 BASED ON DRY WEIGHT	SEAWATER=11.0*10**(-7) G/L, VOSTOK BAY, SEA OF JAPAN	1367
*PLANKTON	25 BASED ON WET WEIGHT	NONE	157
*POLYSIPHONIA JAPONICA	270 BASED ON DRY WEIGHT	SEAWATER=11.0*10**(-7) G/L, VOSTOK BAY, SEA OF JAPAN	1367
*PTILOTA FILICINA	350 BASED ON DRY WEIGHT	SEAWATER=11.0*10**(-7) G/L, VOSTOK BAY, SEA OF JAPAN	1367
*RHODOMELA LARIX	80 BASED ON DRY WEIGHT	SEAWATER=11.0*10**(-7) G/L, VOSTOK BAY, SEA OF JAPAN	1367
*SARGASSUM PALLIDUM	100 BASED ON DRY WEIGHT	SEAWATER=11.0*10**(-7) G/L, VOSTOK BAY, SEA OF JAPAN	1367
*SESTON	0.8 BASED ON WET WEIGHT	DEPTH = 8 M, SCH@HSEE	573
*SESTON	0.7 BASED ON WET WEIGHT	DEPTH = 17 M, SCH@HSEE	573
*ULVA FENESTRATA	50 BASED ON DRY WEIGHT	SEAWATER=11.0*10**(-7) G/L, VOSTOK BAY, SEA OF JAPAN	1367
*ULVARIA SPENDENS	50 BASED ON DRY WEIGHT	SEAWATER=11.0*10**(-7) G/L, VOSTOK BAY, SEA OF JAPAN	1367
*ZOSTERA ASIATICA	180 BASED ON DRY WEIGHT	SEAWATER=11.0*10**(-7) G/L, VOSTOK BAY, SEA OF JAPAN	1367

TABLE D163

CONCENTRATION FACTOR : N			
ITEM:	VALUE:	CONDITIONS:	REF.:
*BRASSICA OLERACEA	49 BASED ON WET WEIGHT	CONC.IN SOIL SOLUTION=215PPM	157
*BROWN ALGAE	7500 BASED ON WET WEIGHT	NONE	157
*PLANKTON	19000 BASED ON WET WEIGHT	NONE	157

TABLE D164

CONCENTRATION FACTOR : N-NITROSOATRAZINE			
ITEM:	VALUE:	CONDITIONS:	REF.:
*GAMBUSIA AFFINIS	16 BASED ON WET WEIGHT	MODEL ECOSYSTEM, ORGANISM=186 PPB, WATER=5.0-3.8 PPB, 18 D	804
*GAMBUSIA AFFINIS	22 BASED ON WET WEIGHT	MODEL ECOSYSTEM, ORGANISM=188 PPB, WATER=5.0-3.8 PPB, 18 D	804
*GAMBUSIA AFFINIS	31 BASED ON WET WEIGHT	MODEL ECOSYSTEM, ORGANISM=301 PPB, WATER=5.0-3.8 PPB, 18 D	804

TABLE D165

CONCENTRATION FACTOR : NA			
ITEM:	VALUE:	CONDITIONS:	REF.:
*BRASSICA OLERACEA	19 BASED ON WET WEIGHT	CONC.IN SOIL SOLUTION=39PPM	157

TABLE D165

CONCENTRATION FACTOR : NA			
ITEM:	VALUE:	CONDITIONS:	REF.:
*BROWN ALGAE	0.78 BASED ON WET WEIGHT	NONE	157
*BROWN ALGAE	1 BASED ON WET WEIGHT	NONE	994
*CANCER IRRORATUS	0.84 (SD=+-0.052) BASED ON WET WEIGHT	SEAWATER CONC=10500 PPM, NOVA SCOTIA, AAS	984
*DAPHNIA MAGNA	4.5	ON DRY WT. BASIS RELATIVE TO THE ALGAE CONSUMED	1630
*PHYTOPLANKTON	105-2600 BASED ON WET WEIGHT	NONE	311
*PLANKTON	0.14 BASED ON WET WEIGHT	NONE	157
*PSEUDEMYS SP.	300 BASED ON DRY WEIGHT	IN-VITRO, FRESHWATER = 575 MG/L	381

TABLE D166

CONCENTRATION FACTOR : NAPHTHALENE			
ITEM:	VALUE:	CONDITIONS:	REF.:
*RANGIA CUNEATA	2.3 BASED ON WET WT.	CONC.IN WATER=0.84PPM,24 HRS	1111
*RANGIA CUNEATA	8.5 BASED ON WET WT.	CONC.IN WATER=0.34PPM,24 HRS , 1-METHYLNAPHTHALENE	1111
*RANGIA CUNEATA	8.1 BASED ON WET WT.	CONC.IN WATER=0.48PPM,24 HRS , 2-METHYLNAPHTHALENE	1111
*RANGIA CUNEATA	17.1 BASED ON WET WT.	CONC.IN WATER=0.24PPM,24 HRS , DIMETHYLNAPHTHALENE	1111
*RANGIA CUNEATA	26.7 BASED ON WET WT.	CONC.IN WATER=0.03PPM,24 HRS , TRIMETHYLNAPHTHALENE	1111

TABLE D167

CONCENTRATION FACTOR : NB			
ITEM:	VALUE:	CONDITIONS:	REF.:
*BROWN ALGAE	450-1000 BASED ON WET WEIGHT	NONE	994
*32 FRESHWATER PLANT	7640 BASED ON DRY WEIGHT	ALGAE,BRYOPHYTES AND VASCULAR HYDROPHYTES	1597

TABLE D168

CONCENTRATION FACTOR : NB-95			
ITEM:	VALUE:	CONDITIONS:	REF.:
*FISH, MUSCLE	180 (RANGE = 40 - 460) BASED ON WET WEIGHT	DANUBE, 1973, NUCLEAR POWER PLANT SINCE 1967	658
*FRESHWATER SEDIMENT	1850 (RANGE = 710 - 3970) BASED ON WET WEIGHT	DANUBE, 1973, NUCLEAR POWER PLANT SINCE 1967	658
*SUSPENDED SOLIDS	870 (RANGE = 80 - 2700) BASED ON WET WEIGHT	DANUBE, 1973, NUCLEAR POWER PLANT SINCE 1967	658
*WATER PLANT	100 (RANGE = 20 - 250) BASED ON WET WEIGHT	DANUBE, 1973, NUCLEAR POWER PLANT SINCE 1967	658

TABLE D169

CONCENTRATION FACTOR : NI			
ITEM:	VALUE:	CONDITIONS:	REF.:
*AGARUM CRIBROSUM	320 BASED ON DRY WEIGHT	SEAWATER=7.6*10**(-7) G/L, VOSTOK BAY, SEA OF JAPAN	1367

TABLE D169

CONCENTRATION FACTOR : NI			
ITEM:	VALUE:	CONDITIONS:	REF.:
*BROWN ALGAE	140 BASED ON WET WEIGHT	NONE	157
*BROWN ALGAE	500 BASED ON WET WEIGHT	NONE	994
*CHONDRUS YENDOI	20 BASED ON DRY WEIGHT	SEAWATER=7.6*10**(-7) G/L, VOSTOK BAY, SEA OF JAPAN	1367
*CHORDA FILUM	50 BASED ON DRY WEIGHT	SEAWATER=7.6*10**(-7) G/L, VOSTOK BAY, SEA OF JAPAN	1367
*CODIUM YESSOENSIS	30 BASED ON DRY WEIGHT	SEAWATER=7.6*10**(-7) G/L, VOSTOK BAY, SEA OF JAPAN	1367
*COSTARIA COSTATA	110 BASED ON DRY WEIGHT	SEAWATER=7.6*10**(-7) G/L, VOSTOK BAY, SEA OF JAPAN	1367
*ENTEROMORPHA PROLIFERA	30 BASED ON DRY WEIGHT	SEAWATER=7.6*10**(-7) G/L, VOSTOK BAY, SEA OF JAPAN	1367
*FISH	235 BASED ON WET WEIGHT	MARINE ORGANISMS, TYPICAL VALUE	538
*FISH	85 BASED ON WET WEIGHT	FRESHWATER ORGANISMS, TYPICAL VALUE	1028
*FUCUS VESICULOSUS	2.8*10**3 BASED ON DRY WEIGHT	18 OBS.,ALL AREAS AROUND THE BRITISH ISLES	1278
*HETEROCHORDARIA ABIETINA	40 BASED ON DRY WEIGHT	SEAWATER=7.6*10**(-7) G/L, VOSTOK BAY, SEA OF JAPAN	1367
*LAMINARIA CHICHORIOIDES	60 BASED ON DRY WEIGHT	SEAWATER=7.6*10**(-7) G/L, VOSTOK BAY, SEA OF JAPAN	1367
*LAMINARIA JAPONICA	100 BASED ON DRY WEIGHT	SEAWATER=7.6*10**(-7) G/L, VOSTOK BAY, SEA OF JAPAN	1367
*MACROPHYTES	1050 BASED ON WET WEIGHT	MARINE ORGANISMS, TYPICAL VALUE	589
*MOLLUSCS	4000 BASED ON WET WEIGHT	MARINE ORGANISMS, TYPICAL VALUE	1280
*MOLLUSCS	650 BASED ON WET WEIGHT	FRESHWATER ORGANISMS, TYPICAL VALUE	993
*PELVETIA WRIGHTII	30 BASED ON DRY WEIGHT	SEAWATER=7.6*10**(-7) G/L, VOSTOK BAY, SEA OF JAPAN	1367
*PHYLLOSPADIX IWATENSIS	40 BASED ON DRY WEIGHT	SEAWATER=7.6*10**(-7) G/L, VOSTOK BAY, SEA OF JAPAN	1367
*PHYTOPLANKTON	570 BASED ON WET WEIGHT	MARINE ORGANISMS, TYPICAL VALUE	986
*PLANKTON	1700 BASED ON WET WEIGHT	NONE	157
*PTILOTA FILICINA	200 BASED ON DRY WEIGHT	SEAWATER=7.6*10**(-7) G/L, VOSTOK BAY, SEA OF JAPAN	1367
*RHODOMELA LARIX	350 BASED ON DRY WEIGHT	SEAWATER=7.6*10**(-7) G/L, VOSTOK BAY, SEA OF JAPAN	1367
*ULVA FENESTRATA	200 BASED ON DRY WEIGHT	SEAWATER=7.6*10**(-7) G/L, VOSTOK BAY, SEA OF JAPAN	1367
*ULVARIA SPENDENS	780 BASED ON DRY WEIGHT	SEAWATER=7.6*10**(-7) G/L, VOSTOK BAY, SEA OF JAPAN	1367
*ZOOPLANKTON	2500 BASED ON ZOOPLANKTON WET WEIGHT CONCENTRATION	SEAWATER CONCENTRATION 0.002 MG/M3	985
*ZOOPLANKTON	560 BASED ON WET WEIGHT	MARINE ORGANISMS, TYPICAL VALUE	986
*ZOSTERA ASIATICA	230 BASED ON DRY WEIGHT	SEAWATER=7.6*10**(-7) G/L, VOSTOK BAY, SEA OF JAPAN	1367

TABLE D170

CONCENTRATION FACTOR : ORGANOCHLORINES			
ITEM:	VALUE:	CONDITIONS:	REF.:
*AQUATIC PLANT	2222 BASED ON WET WEIGHT	CONC.IN WATER=0.45NG/L	389
*AQUATIC PLANT	3143 BASED ON WET WEIGHT	CONC.IN WATER=0.35NG/L	389
*AQUATIC PLANT	3478 BASED ON WET WEIGHT	CONC.IN WATER=0.23NG/L	389
*AQUATIC PLANT	101000 BASED ON WET WEIGHT	CONC.IN WATER=0.30NG/L	389

TABLE D171

CONCENTRATION FACTOR : P			
ITEM:	VALUE:	CONDITIONS:	REF.:
*BRASSICA OLERACEA	1600 BASED ON WET WEIGHT	CONC.IN SOIL SOLUTION=0.70PPM	157

TABLE D171

CONCENTRATION FACTOR : P

ITEM:	VALUE:	CONDITIONS:	REF.:
*BROWN ALGAE	10000 BASED ON WET WEIGHT	NONE	157
*BROWN ALGAE	10000 BASED ON WET WEIGHT	NONE	994
*PLANKTON	15000 BASED ON WET WEIGHT	NONE	157
*ZOSTERA MARINA	87 BASED ON DRY WEIGHT	3000 MIN, RHIZOME, LIGHT, ADDED TO LOWER COMPARTMENT	1013
*ZOSTERA MARINA	1011	LEAF BASE, DRY WEIGHT BASIS, 3000 MIN, ADDED TO LOWER COMPARTMENT	1013
*ZOSTERA MARINA	743	ROOT, DRY WEIGHT BASIS, 3000 MIN, LIGHT, ADDED TO LOWER COMPARTMENT	1013
*ZOSTERA MARINA	96	STEM, DRY WEIGHT BASIS, 3000 MIN, DARK, ADDED TO LOWER COMPARTMENT	1013
*ZOSTERA MARINA	215	LEAF BASE, DARK, DRY WEIGHT BASIS, 3000 MIN,ADDED TO LOWER COMPARTMEN	1013
*ZOSTERA MARINA	213	LEAF TIP, DRY WEIGHT BASIS, DARK, 3000 MIN, ADDED TO LOWER COMPARTMENT	1013
*ZOSTERA MARINA	382	LEAF MIDDLE,LIGHT,DRY W. BASIS, 3000 MIN, ADDED TO UPPER COMPARTMENT	1013
*ZOSTERA MARINA	4823	ROOT, LIGHT, DRY W. BASIS, 3000 MIN, ADDED TO UPPER COMPARTMENT	1013
*ZOSTERA MARINA	1806	LEAF TIP,LIGHT, DRY W. BASIS, 3000 MIN, ADDED TO UPPER COMPARTMENT	1013
*ZOSTERA MARINA	47	STEM,DARK, DRY W. BASIS, 3000 MIN, ADDED TO UPPER COMPARTMENT	1013
*ZOSTERA MARINA	455	LEAF BASE, DARK, DRY W. BASIS, 3000 MIN, ADDED TO UPPER COMPARTMENT	1013
*32 FRESHWATER PLANT	5480 BASED ON DRY WEIGHT	ALGAE,BRYOPHYTES AND VASCULAR HYDROPHYTES	1597

TABLE D172

CONCENTRATION FACTOR : PB

ITEM:	VALUE:	CONDITIONS:	REF.:
*BLIDINGIA MINIMA	2.7 - 8.2*10**4 BASED ON DRY WEIGHT	CALCULATED ON BASIS OF DISSOLVED METAL CONCENTRATION IN SEAWATER	1412
*BROWN ALGAE	70000 BASED ON WET WEIGHT	NONE	157
*CRASSOSTREA GIGAS	75 ON WET WEIGHT BASIS	BIG BEEF, WATER CONC = 1.3 MG/M3	1392
*ENTEROMORPHA LINZA	2.0 - 4.5*10**4 BASED ON DRY WEIGHT	CALCULATED ON BASIS OF DISSOLVED METAL CONCENTRATION IN SEAWATER	1412
*FUCUS VESICULOSUS	1.3 - 2.4*10**4 BASED ON DRY WEIGHT	CALCULATED ON BASIS OF DISSOLVED METAL CONCENTRATION IN SEAWATER	1412
*FUCUS VESICULOSUS	0.087*10**4 BASED ON DRY WEIGHT	CALCULATED ON BASIS OF DISSOLVED METAL CONCENTRATION IN SEAWATER	1412
*FUCUS VESICULOSUS	0.23*10**4 BASED ON DRY WEIGHT	CALCULATED ON BASIS OF DISSOLVED METAL CONCENTRATION IN SEAWATER	1412
*FUCUS VESICULOSUS	0.24*10**4 BASED ON DRY WEIGHT	CALCULATED ON BASIS OF DISSOLVED METAL CONCENTRATION IN SEAWATER	1412
*FUCUS VESICULOSUS	2.4*10**3 (MICROGRAMS/G DRY MAT.)/ (MICROGRANMS/G SEAWATER)	13 OBS.,ALL AREAS AROUND THE BRITISH ISLES	1278
*LONGFIN SMELT	75 ON WET WEIGHT BASIS	SIZE = 4 - 12 CM, LAKE WASHINGTON, WATER CONC = 2 MG/M3, MUSCLE	1392
*MYTILUS EDULIS	150 ON WET WEIGHT BASIS	BIG BEEF, WATER CONC = 1.3 MG/M3	1392
*MYTILUS GALLOPROVINCIALIS	323 BASED ON WET WEIGHT	WATER = 6.40 PPM	971
*PLANKTON	41000 BASED ON WET WEIGHT	NONE	157
*PTYCHOCHEILUS OREGONENSIS	1700 ON WET WEIGHT BASIS	SIZE = 30 CC, LAKE WASHINGTON, WATER CONC = 2 MG/M3, MUSCLE	1392
*PTYCHOCHEILUS OREGONENSIS	650 ON WET WEIGHT BASIS	SIZE = 50 CM, LAKE WASHINGTON, WATER CONC = 2 MG/M3, MUSCLE	1392

TABLE D172

CONCENTRATION FACTOR : PB			
ITEM:	VALUE:	CONDITIONS:	REF.:
*SOCKEYE	65 ON WET WEIGHT BASIS	SIZE = 5 - 17 CM, LAKE WASHINGTON, WATER CONC = 2 MG/M3, MUSCLE	1392
*ULVA SP.	1.7 - 4.9*10**4 BASED ON DRY WEIGHT	CALCULATED ON BASIS OF DISSOLVED METAL CONCENTRATION IN SEAWATER	1412
*ZOOPLANKTON	197000 BASED ON WET WEIGHT	SEAWATER CONCENTRATION 0.000003 MG/M3	985
*ZOOPLANKTON	1500 ON WET WEIGHT BASIS	LAKE WASHINGTON, WATER CONC = 2 MG/M3	1392

TABLE D173

CONCENTRATION FACTOR : PCB			
ITEM:	VALUE:	CONDITIONS:	REF.:
*AMBLOPLITES RUPESTRIS	117*10**3 BASED ON WET WEIGHT	WATER=3.0 PPB, RATIO TISSUE/SED = 0.12 ON WW, HUDSON RIVER	1094
*CHAOBORUS PUNCTIPENNIS	22000 BASED ON DRY WEIGHT	WATER=1.3 PPB, 1 D, 294 K, PH=7.4, UPTAKE FROM WATER, NO FOOD	1381
*CHAOBORUS PUNCTIPENNIS	23000 BASED ON DRY WEIGHT	WATER=1.3 PPB, 4 D, 294 K, PH=7.4, UPTAKE FROM WATER, NO FOOD	1381
*CHAOBORUS PUNCTIPENNIS	23800 BASED ON DRY WEIGHT	WATER=1.3 PPB. 7 D, 294 K, PH=7.4, UPTAKE FROM WATER, NO FOOD	1381
*CHAOBORUS PUNCTIPENNIS	24800 BASED ON DRY WEIGHT	WATER=1.3 PPB, 14 D, 294 K, PH= 7.4,UPTAKE FROM WATER, NO FOOD	1381
*CORYDALUS CORNUTUS	1400 BASED ON DRY WEIGHT	WATER=1.1 PPB, 1 D, 294 K, PH=7.4, UPTAKE FROM WATER, NO FOOD	1381
*CORYDALUS CORNUTUS	4600 BASED ON DRY WEIGHT	WATER=1.1 PPB, 4 D, 294 K, PH=7.4, UPTAKE FROM WATER, NO FOOD	1381
*CORYDALUS CORNUTUS	5700 BASED ON DRY WEIGHT	WATER=1.1 PPB, 7 D, 294 K, PH=7.4, UPTAKE FROM WATER, NO FOOD	1381
*CORYDALUS CORNUTUS	6600 BASED ON DRY WEIGHT	WATER=1.1 PPB, 14 D, 294 K, PH= 7.4,UPTAKE FROM WATER, NO FOOD	1381
*CORYDALUS CORNUTUS	6800 BASED ON DRY WEIGHT	WATER=1.1 PPB, 21 D, 294 K, PH= 7.4,UPTAKE FROM WATER, NO FOOD	1381
*CULEX TARSALIS	12600 BASED ON DRY WEIGHT	WATER=1.5 PPB, 1 D, 294 K, PH=7.4, UPTAKE FROM WATER, NO FOOD	1381
*CULEX TARSALIS	18000 BASED ON DRY WEIGHT	WATER=1.5 PPB, 4 D, 294 K, PH=7.4, UPTAKE FROM WATER, NO FOOD	1381
*CULEX TARSALIS	20000 BASED ON DRY WEIGHT	WATER=1.5 PPB, 7 D, 294 K, PH=7.4, UPTAKE FROM WATER, NO FOOD	1381
*DAPHNIA MAGNA	24700 BASED ON DRY WEIGHT	WATER=1.1 PPB, 1 D, 294 K, PH=7.4, UPTAKE FROM WATER, NO FOOD	1381
*DAPHNIA MAGNA	47000 BASED ON DRY WEIGHT	WATER=1.1 PPB, 4 D, 294 K, PH=7.4, UPTAKE FROM WATER, NO FOOD	1381
*GAMMARUS PSEUDOLIMNAEUS	17000 BASED ON DRY WEIGHT	WATER=1.6 PPB, 1 D, 294 K, PH=7.4, UPTAKE FROM WATER, NO FOOD	1381
*GAMMARUS PSEUDOLIMNAEUS	24000 BASED ON DRY WEIGHT	WATER=1.6 PPB, 4 D, 294 K, PH=7.4, UPTAKE FROM WATER, NO FOOD	1381
*GAMMARUS PSEUDOLIMNAEUS	26000 BASED ON DRY WEIGHT	WATER=1.6 PPB, 7 D, 294 K, PH=7.4, UPTAKE FROM WATER, NO FOOD	1381
*GAMMARUS PSEUDOLIMNAEUS	27500 BASED ON DRY WEIGHT	WATER=1.6 PPB, 14 D, 294 K, PH= 7.4,UPTAKE FROM WATER, NO FOOD	1381
*GAMMARUS PSEUDOLIMNAEUS	27000 BASED ON DRY WEIGHT	WATER=1.6 PPB, 21 D, 294 K, PH= 7.4,UPTAKE FROM WATER, NO FOOD	1381
*MYTILUS SPP.	270000 BASED ON WET WEIGHT	SEAWATER=0.94 NG/M3, FRANCE, VALRAS-PLAGE, 9/7 74	1343
*MYTILUS SPP.	470000 BASED ON WET WEIGHT	SEAWATER=0.95 NG/M3, FRANCE, STES-MARIES, 10/7 74	1343
*MYTILUS SPP.	300000 BASED ON WET WEIGHT	SEAWATER=0.68 NG/M3, FRANCE, CARRO, 11/7 74	1343
*MYTILUS SPP.	690000 BASED ON WET WEIGHT	SEAWATER=1.6 NG/M3, FRANCE, MARSEILLE, 12/7 74	1343
*MYTILUS SPP.	190000 BASED ON WET WEIGHT	SEAWATER=1.2 NG/L, CALIF., ROYAL PALMS, 19/9 72	1343

TABLE D173

CONCENTRATION FACTOR : PCB

ITEM:	VALUE:	CONDITIONS:	REF.:
*MYTILUS SPP.	69000 BASED ON WET WEIGHT	SEAWATER=0.98 NG/M3, CALIF., FRISCO BAY, 1975	1343
*NOTROPIS CORNUTUS	7.0*10**3 BASED ON WET WEIGHT	WATER=1.0 PPB, RATIO TISSUE/SED = 0.98 ON WW, HUDSON RIVER	1094
*NOTROPIS CORNUTUS	78*10**3 BASED ON WET WEIGHT	WATER=1.0 PPB, RATIO TISSUE/SED = 4.09 ON WW, HUDSON RIVER	1094
*ORCONECTES NAIS	570 BASED ON DRY WEIGHT	WATER=1.2 PPB, 1 D, 294 K, PH=7.4, UPTAKE FROM WATER, NO FOOD	1381
*ORCONECTES NAIS	1700 BASED ON DRY WEIGHT	WATER=1.2 PPB, 4 D, 294 K, PH=7.4, UPTAKE FROM WATER, NO FOOD	1381
*ORCONECTES NAIS	3400 BASED ON DRY WEIGHT	WATER=1.2 PPB, 7 D, 294 K, PH=7.4, UPTAKE FROM WATER, NO FOOD	1381
*ORCONECTES NAIS	4500 BASED ON DRY WEIGHT	WATER=1.2 PPB, 14 D, 294 K, PH=7.4,UPTAKE FROM WATER, NO FOOD	1381
*ORCONECTES NAIS	5100 BASED ON DRY WEIGHT	WATER=1.2 PPB, 21 D, 294 K, PH=7.4,UPTAKE FROM WATER, NO FOOD	1381
*PALAEMONETES KADIAKENSIS	10300 BASED ON DRY WEIGHT	WATER=1.3 PPB, 1 D, 294 K, PH=7.4, UPTAKE FROM WATER, NO FOOD	1381
*PALAEMONETES KADIAKENSIS	12300 BASED ON DRY WEIGHT	WATER=1.3 PPB, 4 D, 294 K, PH=7.4, UPTAKE FROM WATER, NO FOOD	1381
*PALAEMONETES KADIAKENSIS	13700 BASED ON DRY WEIGHT	WATER=1.3 PPB, 7 D, 294 K, PH=7.4, UPTAKE FROM WATER, NO FOOD	1381
*PALAEMONETES KADIAKENSIS	14200 BASED ON DRY WEIGHT	WATER=1.3 PPB, 14 D, 294 K, PH=7.4,UPTAKE FROM WATER, NO FOOD	1381
*PALAEMONETES KADIAKENSIS	16600 BASED ON DRY WEIGHT	WATER=1.3 PPB, 21 D, 294 K, PH=7.4,UPTAKE FROM WATER, NO FOOD	1381
*PALAEMONETES PUGIO	14000 BASED ON WET WEIGHT	SEAWATER = 1.3 UG/L, 290-301 K,16 D, MORTALITY = 40 %	1130
*PALAEMONETES PUGIO	6700 BASED ON WET WEIGHT	SEAWATER = 4.0 UG/L, 290-301 K,16 D, MORTALITY = 45 %	1130
*PALAEMONETES PUGIO	3700 BASED ON WET WEIGHT	SEAWATER = 12.5 UG/L, 290-301 K,16 D, MORTALITY = 55 %	1130
*PALAEMONETES PUGIO	7600 BASED ON WET WEIGHT	AROCLOR 1254, FLOWING SEAWAT=0.17 UG/L,S=2.6 %, 7 D, 290-301 K	1130
*PALAEMONETES PUGIO	8700 BASED ON WET WEIGHT	AROCLOR 1254, FLOWING SEAWAT=0.62 UG/L,S=2.6 %, 7 D, 290-301 K	1130
*PALAEMONETES PUGIO	3200 BASED ON WET WEIGHT	AROCLOR 1254, FLOWING SEAWAT=1.0 UG/L,S=2.6 %, 7 D, 290-301 K	1130
*PALAEMONETES PUGIO	11000 BASED ON WET WEIGHT	AROCLOR 1254, FLOWING SEAWAT=2.3 UG/L,S=2.6 %, 7 D, 290-301 K	1130
*PALAEMONETES PUGIO	7000 BASED ON WET WEIGHT	AROCLOR 1254, FLOWING SEAWAT=2.7 UG/L,S=2.6 %, 7 D, 290-301 K	1130
*PALAEMONETES PUGIO	4800 BASED ON WET WEIGHT	AROCLOR 1254, FLOWING SEAWAT=3.2 UG/L,S=2.6 %, 7 D, 290-301 K	1130
*PALAEMONETES PUGIO	8100 BASED ON WET WEIGHT	AROCLOR 1254, FLOWING SEAWAT=3.2 UG/L,S=2.6 %, 7 D, 290-301 K	1130
*PALAEMONETES PUGIO	5600 BASED ON WET WEIGHT	AROCLOR 1254, FLOWING SEAWAT=5.2 UG/L,S=2.6 %, 7 D, 290-301 K	1130
*PALAEMONETES PUGIO	3000 BASED ON WET WEIGHT	AROCLOR 1254, FLOWING SEAWAT=5.3 UG/L,S=2.6 %, 7 D, 290-301 K	1130
*PALAEMONETES PUGIO	5700 BASED ON WET WEIGHT	AROCLOR 1254, FLOWING SEAWAT=5.3 UG/L,S=2.6 %, 7 D, 290-301 K	1130
*PALAEMONETES PUGIO	7100 BASED ON WET WEIGHT	AROCLOR 1254, FLOWING SEAWAT=9.1 UG/L,S=2.6 %, 7 D, 290-301 K	1130
*PALAEMONETES PUGIO	7600 BASED ON WET WEIGHT	SEAWATER = 0.17 UG/L, 290-301 K,7 D, MORTALITY = 8 %	1130
*PALAEMONETES PUGIO	8700 BASED ON WET WEIGHT	SEAWATER = 0.62 UG/L, 290-301 K,7 D, MORTALITY = 4 %	1130
*PALAEMONETES PUGIO	7100 BASED ON WET WEIGHT	SEAWATER = 9.1 UG/L, 290-301 K,7 D, MORTALITY = 60 %	1130
*PHYSA SP.	1.9*10**3 BASED ON WET WEIGHT	WATER=1.0 PPB, RATIO TISSUE/SED = 3.63 ON WW, HUDSON RIVER	1094
*PHYSA SP.	45*10**3 BASED ON WET WEIGHT	WATER=2.2 PPB, RATIO TISSUE/SED = 6.52 ON WW, HUDSON RIVER	1094
*PHYSA SP.	27*10**3 BASED ON WET WEIGHT	WATER=1.0 PPB, RATIO TISSUE/SED = 11.82 ON WW, HUDSON RIVER	1094
*PTERONARCYS DORSATA	2100 BASED ON DRY WEIGHT	WATER=2.8 PPB, 1 D, 294 K, PH=7.4, UPTAKE FROM WATER, NO FOOD	1381

TABLE D173

CONCENTRATION FACTOR : PCB			
ITEM:	VALUE:	CONDITIONS:	REF.:
*PTERONARCYS DORSATA	2500 BASED ON DRY WEIGHT	WATER=2.8 PPB, 4 D, 294 K, PH=7.4, UPTAKE FROM WATER, NO FOOD	1381
*PTERONARCYS DORSATA	2800 BASED ON DRY WEIGHT	WATER=2.8 PPB, 7 D, 294 K, PH=7.4, UPTAKE FROM WATER, NO FOOD	1381
*PTERONARCYS DORSATA	2900 BASED ON DRY WEIGHT	WATER=2.8 PPB, 14 D, 294 K, PH= 7.4,UPTAKE FROM WATER, NO FOOD	1381
*PTERONARCYS DORSATA	2800 BASED ON DRY WEIGHT	WATER=2.8 PPB, 21 D, 294 K, PH= 7.4,UPTAKE FROM WATER, NO FOOD	1381
*SALMO SALAR	282 BASED ON WET WEIGHT	JUVENILE, UPTAKE FROM WATER, 48 H, 5.29 G FISH, AROCLOR 1254	1778
*SALMO SALAR	0.358 BASED ON WET WEIGHT	JUVENILE, UPTAKE FROM FOOD, 48 H, 5.29 G FISH, AROCLOR 1254	1778
*YELLOW PERCH	17.0*10**3 BASED ON WET WEIGHT	WATER=1.0 PPB, RATIO TISSUE/SED = 0.41 ON WW, HUDSON RIVER	1094

TABLE D174

CONCENTRATION FACTOR : PHOTODIELDRIN			
ITEM:	VALUE:	CONDITIONS:	REF.:
*ANKISTRODESMUS AMALLOIDES	4700 BASED ON WET WEIGHT	30 MIN, 100 CELLS/ML, WATER CONC = 0.72 PPB	1117
*ANKISTRODESMUS AMALLOIDES	3800 BASED ON WET WEIGHT	60 MIN, 100 CELLS/ML, WATER CONC = 0.72 PPB	1117
*ANKISTRODESMUS AMALLOIDES	4900 BASED ON WET WEIGHT	120 MIN, 100 CELLS/ML, WATER CONC = 0.72 PPB	1117
*ANKISTRODESMUS AMALLOIDES	4700 BASED ON WET WEIGHT	180 MIN, 100 CELLS/ML, WATER CONC = 0.72 PPB	1117

TABLE D175

CONCENTRATION FACTOR : PO			
ITEM:	VALUE:	CONDITIONS:	REF.:
*PRIMARY CONSUMERS	48 - 1495 BASED ON WET WEIGHT	ECALGRAIN BAY, MARINE	575
*PRIMARY CONSUMERS, ASCIDIANS	785 BASED ON WET WEIGHT	ECALGRAIN BAY, MARINE	575
*PRIMARY CONSUMERS, BALANUS BALANOIDES	503 BASED ON WET WEIGHT	ECALGRAIN BAY, MARINE	575
*PRIMARY CONSUMERS, LITTORINA LITTORALIS	205 BASED ON WET WEIGHT	ECALGRAIN BAY, MARINE	575
*PRIMARY CONSUMERS, SEDENTARY HERBIVOROUS SPECIES	119 BASED ON WET WEIGHT	ECALGRAIN BAY, MARINE	575
*PRIMARY CONSUMERS, SEDENTARY LIMNIVOROUS SPECIES	119 BASED ON WET WEIGHT	ECALGRAIN BAY, MARINE	575
*PRIMARY CONSUMERS, SESILE MACROPHAGE SPECIES	165 BASED ON WET WEIGHT	ECALGRAIN BAY, MARINE	575
*PRIMARY CONSUMERS, SESILE MICROPHAGE SPECIES	1082 BASED ON WET WEIGHT	ECALGRAIN BAY, MARINE	575
*PRIMARY CONSUMERS, SPONGES	1370 BASED ON WET WEIGHT	ECALGRAIN BAY, MARINE	575
*PRIMARY PRODUCERS, ALGAE	322 BASED ON WET WEIGHT	ECALGRAIN BAY, MARINE	575
*PRIMARY PRODUCERS, FUCUS SERRATUS	523 BASED ON WET WEIGHT	ECALGRAIN BAY, MARINE	575
*PRIMARY PRODUCERS, LICHENS	3756 BASED ON WET WEIGHT	ECALGRAIN BAY, MARINE	575
*SECONDARY CONSUMERS	38 - 452 BASED ON WET WEIGHT	ECALGRAIN BAY, MARINE	575
*SECONDARY CONSUMERS, ASTERINA GIBBOSA	452 BASED ON WET WEIGHT	ECALGRAIN BAY, MARINE	575
*SECONDARY CONSUMERS, BURROWING PREDACEOUS SPECIES	315 BASED ON WET WEIGHT	ECALGRAIN BAY, MARINE	575
*SECONDARY CONSUMERS, MOTILE PREDACEOUS SPECIES	63 BASED ON WET WEIGHT	ECALGRAIN BAY, MARINE	575

TABLE D175

CONCENTRATION FACTOR : PO

ITEM:	VALUE:	CONDITIONS:	REF.:
*SECONDARY CONSUMERS, NUCELLA LAPILLUS	70 BASED ON WET WEIGHT	ECALGRAIN BAY, MARINE	575
*SECONDARY CONSUMERS, SEDENTARY PREDACEOUS SPECIES	261 BASED ON WET WEIGHT	ECALGRAIN BAY, MARINE	575
*TERTIARY CONSUMERS	1 - 73 BASED ON WET WEIGHT	ECALGRAIN BAY, MARINE	575
*TERTIARY CONSUMERS, BENTHIC MOTILE PREDATORS	47 BASED ON WET WEIGHT	ECALGRAIN BAY, MARINE	575
*TERTIARY CONSUMERS, PELAGIC PREDATORS	8 BASED ON WET WEIGHT	ECALGRAIN BAY, MARINE	575

TABLE D176

CONCENTRATION FACTOR : PO-210

ITEM:	VALUE:	CONDITIONS:	REF.:
*CELERY	2.9*10**(-5)	BASED ON DRY PLANT AND DRY SOIL CONC AT END OF EXP., MATURE CROP	1664
*CORN	0.35*10**(-5)	BASED ON DRY PLANT AND DRY SOIL CONC AT END OF EXP., MATURE CROP	1664
*ONIONS	5.1*10**(-5)	BASED ON DRY PLANT AND DRY SOIL CONC AT END OF EXP., MATURE CROP	1664
*PEAS	3.5*10**(-5)	BASED ON DRY PLANT AND DRY SOIL CONC AT END OF EXP., MATURE CROP	1664
*POTATOES	62*10**(-5)	BASED ON DRY PLANT AND DRY SOIL CONC AT END OF EXP., MATURE CROP	1664
*RADISH	58*10**(-5)	BASED ON DRY PLANT AND DRY SOIL CONC AT END OF EXP., MATURE CROP	1664
*SPINACH	80*10**(-5)	BASED ON DRY PLANT AND DRY SOIL CONC AT END OF EXP., MATURE CROP	1664
*TOMATO	0.12*10**(-5)	BASED ON DRY PLANT AND DRY SOIL CONC AT END OF EXP., MATURE CROP	1664
*WHEAT	0.9*10**(-5)	BASED ON DRY PLANT AND DRY SOIL CONC AT END OF EXP., MATURE CROP	1664

TABLE D177

CONCENTRATION FACTOR : PO-239

ITEM:	VALUE:	CONDITIONS:	REF.:
*BASS, BACKBONE	160 BASED ON WET WEIGHT	MARINE VERTEBRATE	1138
*BASS, GUT	36 BASED ON WET WEIGHT	MARINE VERTEBRATE	1138
*BASS, LIVER	14 BASED ON WET WEIGHT	MARINE VERTEBRATE	1138
*BASS, MUSCLE	4 BASED ON WET WEIGHT	MARINE VERTEBRATE	1138
*BLUE FIN TUNA, BACKBONE	90 BASED ON WET WEIGHT	MARINE VERTEBRATE	1138
*BLUE FIN TUNA, MUSCLE	2 BASED ON WET WEIGHT	MARINE VERTEBRATE	1138
*BLUE FISH, BACKBONE	570 BASED ON WET WEIGHT	MARINE VERTEBRATE	1138
*BLUE FISH, LIVER	26 BASED ON WET WEIGHT	MARINE VERTEBRATE	1138
*BLUE FISH, MUSCLE	5 BASED ON WET WEIGHT	MARINE VERTEBRATE	1138
*BLUE SHARK, BACKBONE	21 BASED ON WET WEIGHT	MARINE VERTEBRATE	1138
*BLUE SHARK, LIVER	175 BASED ON WET WEIGHT	MARINE VERTEBRATE	1138
*BLUE SHARK, MUSCLE	4 BASED ON WET WEIGHT	MARINE VERTEBRATE	1138
*BONITO, WHOLE ANIMAL	3 BASED ON WET WEIGHT	MARINE VERTEBRATE	1138
*DUSKY SHARK, BACKBONE	49 BASED ON WET WEIGHT	MARINE VERTEBRATE	1138
*FLOUNDER, BACKBONE	50 BASED ON WET WEIGHT	MARINE VERTEBRATE	1138
*FLOUNDER, GUT	1060 BASED ON WET WEIGHT	MARINE VERTEBRATE	1138

TABLE D177

CONCENTRATION FACTOR : PO-239			
ITEM:	VALUE:	CONDITIONS:	REF.:
*FLOUNDER, LIVER	57 BASED ON WET WEIGHT	MARINE VERTEBRATE	1138
*FLOUNDER, MUSCLE	1 BASED ON WET WEIGHT	MARINE VERTEBRATE	1138
*MAKO SHARK, BACKBONE	29 BASED ON WET WEIGHT	MARINE VERTEBRATE	1138
*MAKO SHARK, MUSCLE	13 BASED ON WET WEIGHT	MARINE VERTEBRATE	1138
*TAUTOG, BACKBONE	510 BASED ON WET WEIGHT	MARINE VERTEBRATE	1138
*TAUTOG, GUT	730 BASED ON WET WEIGHT	MARINE VERTEBRATE	1138
*TAUTOG, LIVER	64 BASED ON WET WEIGHT	MARINE VERTEBRATE	1138

TABLE D178

CONCENTRATION FACTOR : POLYBROMINATED BIPHENYL			
ITEM:	VALUE:	CONDITIONS:	REF.:
*SALMO SALAR	48 BASED ON WET WEIGHT	JUVENILE, UPTAKE FROM WATER, 48 H, 5.29 G FISH, BP-6	1778
*SALMO SALAR	1.0 BASED ON WET WEIGHT	JUVENILE, UPTAKE FROM FOOD, 48 H, 5.29 G FISH, BP-6	1778

TABLE D179

CONCENTRATION FACTOR : PU			
ITEM:	VALUE:	CONDITIONS:	REF.:
*BARLEY, GRAIN	0.2*10**(-6) BASED ON DRY PLANT AND DRY SOIL	NEVADA, LAB. SITE 11,239,249 PU, 293-301 K	1402
*BARLEY, GRAIN	0.4*10**(-6) BASED ON DRY PLANT AND DRY SOIL	NEVADA, LAB. SITE 13,239,249 PU, 293-301 K	1402
*BARLEY, LOWER VEGETATION	9.2*10**(-6) BASED ON DRY PLANT AND DRY SOIL	NEVADA, LAB. SITE 13,239,249 PU, 293-301 K	1402
*BARLEY, LOWER VEGETATION	63*10**(-6) BASED ON DRY PLANT AND DRY SOIL	NEVADA, LAB. SITE 11241 AM, 293-301 K	1402
*BARLEY, ROOT	36000 BASED ON PLANT DRY WEIGHT AND DRY SOIL	SOIL = 10 U CI/G SOIL, RITZVILLE SILT LOAM	1710
*BARLEY, ROOT	60000 BASED ON PLANT DRY WEIGHT AND DRY SOIL	SOIL = 0.5 U CI/G SOIL, RITZVILLE SILT LOAM	1710
*BARLEY, ROOT	110000 BASED ON PLANT DRY WEIGHT AND DRY SOIL	SOIL = 0.05 U CI/G SOIL, RITZVILLE SILT LOAM	1710
*BARLEY, SHOOTS	5500 BASED ON PLANT DRY WEIGHT AND DRY SOIL	SOIL = 10 U CI/G SOIL, RITZVILLE SILT LOAM	1710
*BARLEY, SHOOTS	19000 - 21000 BASED ON PLANT DRY WEIGHT AND DRY SOIL	SOIL = 0.5 U CI/G SOIL, RITZVILLE SILT LOAM	1710
*BARLEY, SHOOTS	14000 - 56000 BASED ON PLANT DRY WEIGHT AND DRY SOIL	SOIL = 0.05 BASED ON PLANT DRY WEIGHT AND DRY SOIL	1710
*BARLEY, UPPER VEGETATION	23*10**(-6) BASED ON DRY PLANT AND DRY SOIL	NEVADA, LAB. SITE 11,239,249 PU, 293-301 K	1402
*BARLEY, UPPER VEGETATION	8.8*10**(-6) BASED ON DRY PLANT AND DRY SOIL	NEVADA, LAB. SITE 13,239,249 PU, 293-301 K	1402
*CRASSOSTREA GIGAS	210	CONC.IN WATER=0.0119PCI/L,SHELL, ST.VAAST-LA-HOUGUE,FRANCE (F029)	576
*CRASSOSTREA GIGAS	30	CONC.IN WATER=0.0119PCI/L,SOFT PARTS,ST.VAAST-LA-HOUGUE,FRANCE (F029)	576
*CRASSOSTREA GIGAS	170	CONC.IN WATER=0.0119PCI/L,ENTIRE ANIMAL,ST.VAAST-LA-HOUGUE,FRANCE (F029)	576
*CREPIDULA FORNICATA	2225	CONC.IN WATER=0.0119PCI/L,SHELL, ST.VAAST-LA-HOUGUE,FRANCE (F029)	576
*CREPIDULA FORNICATA	30	CONC.IN WATER=0.0119PCI/L,SOFT PARTS,ST.VAAST-LA-HOUGUE,FRANCE (F029)	576

TABLE D179

CONCENTRATION FACTOR : PU

ITEM:	VALUE:	CONDITIONS:	REF.:
*CREPIDULA FORNICATA	1580	CONC.IN WATER=0.0119PCI/L,ENTIRE ANIMAL,ST.VAAST-LA-HOUGUE,FRANCE (F029)	576
*ENSIS ENSIS	110	CONC.IN WATER=0.0119PCI/L,SHELL, ST.VAAST-LA-HOUGUE,FRANCE (F029)	576
*ENSIS ENSIS	20	CONC.IN WATER=0.0119PCI/L,SOFT PARTS,ST.VAAST-LA-HOUGUE,FRANCE (F029)	576
*ENSIS ENSIS	60	CONC.IN WATER=0.0119PCI/L,ENTIRE ANIMAL,ST.VAAST-LA-HOUGUE,FRANCE (F029)	576
*LITTORINA LITTORALIS	80	CONC.IN WATER=0.0119PCI/L,SHELL, ST.VAAST-LA-HOUGUE,FRANCE (F029)	576
*LITTORINA LITTORALIS	70	CONC.IN WATER=0.0119PCI/L,SOFT PARTS.ST.VAAST-LA-HOUGUE,FRANCE (F029)	576
*LITTORINA LITTORALIS	80	CONC.IN WATER=0.0119PCI/L,ENTIRE ANIMAL,ST.VAAST-LA-HOUGUE,FRANCE (F029)	576
*MYTILUS EDULIS	330	CONC.IN WATER=0.0119PCI/L,SHELL, ST.VAAST-LA-HOUGUE,FRANCE (F029)	576
*MYTILUS EDULIS	40	CONC.IN WATER=0.0119PCI/L,SOFT PARTS,ST.VAAST-LA-HOUGUE,FRANCE (F029)	576
*MYTILUS EDULIS	190	CONC.IN WATER=0.0119PCI/L,ENTIRE ANIMAL,ST.VAAST-LA-HOUGUE,FRANCE (F029)	576
*NEREIS DIVERSICOLOR	0.0015 (SD=+-0.0001) CONC IN ANIM/(CONC IN SEDIMENT)	WET WEIGHT BOTH SED. AND ANIM., BRAVO CRATER	84
*NEREIS DIVERSICOLOR	0.0012 (SD=+-0.0001) CONC IN ANIM/(CONC IN SEDIMENT)	WET WEIGHT BOTH SED. AND ANIM., WINDSCALE	84
*NEREIS DIVERSICOLOR	0.0018 (SD=+-0.0002) CONC IN ANIM/(CONC IN SEDIMENT)	DRY WEIGHT BOTH SED. AND ANIM., WINDSCALE,N=22, 5 D	84
*NEREIS DIVERSICOLOR	0.0035 (SD=+-0.0003) CONC IN ANIM/(CONC IN SEDIMENT)	DRY WEIGHT BOTH SED. AND ANIM., WINDSCALE,N=25, 15 D	84
*NEREIS DIVERSICOLOR	0.0035 (SD=+-0.0005) CONC IN ANIM/(CONC IN SEDIMENT)	DRY WEIGHT BOTH SED. AND ANIM., WINDSCALE,N=23, 40 D	84
*NEREIS DIVERSICOLOR	0.0060 (SD=+-0.0010) CONC IN ANIM/(CONC IN SEDIMENT)	DRY WEIGHT BOTH SED. AND ANIM., WINDSCALE,N=1, 225 D	84
*NEREIS DIVERSICOLOR	0.0012 (SD=+-0.0001) CONC IN ANIM/(CONC IN SEDIMENT)	DRY WEIGHT BOTH SED. AND ANIM., BIKINI ATOLL,N=9, 5 D	84
*NEREIS DIVERSICOLOR	0.0047 (SD=+-0.0005) CONC IN ANIM/(CONC IN SEDIMENT)	DRY WEIGHT BOTH SED. AND ANIM., BIKINI ATOLL,N=10, 15 D	84
*NEREIS DIVERSICOLOR	0.0028 (SD=+-0.0002) CONC IN ANIM/(CONC IN SEDIMENT)	DRY WEIGHT BOTH SED. AND ANIM., BIKINI ATOLL,N=6, 40 D	84
*NEREIS DIVERSICOLOR	0.0049 (SD=+-0.0005) CONC IN ANIM/(CONC IN SEDIMENT)	DRY WEIGHT BOTH SED. AND ANIM., BIKINI ATOLL,N=31, 40 D	84
*NEREIS DIVERSICOLOR	0.0008 (SD=+-0.00006) CONC IN ANIM/(CONC IN SEDIMENT)	DRY WEIGHT BOTH SED. AND ANIM., BIKINI ATOLL,N=25, 15 D	84
*NEREIS DIVERSICOLOR	0.0021 (SD=+-0.0001) CONC IN ANIM/(CONC IN SEDIMENT)	DRY WEIGHT BOTH SED. AND ANIM., BIKINI ATOLL,N=13, 40 D	84
*NUCELLA LAPILLUS	440	CONC.IN WATER=0.0119PCI/L,SHELL, ST.VAAST-LA-HOUGUE,FRANCE (F029)	576
*NUCELLA LAPILLUS	50	CONC.IN WATER=0.0119PCI/L,SOFT PARTS.ST.VAAST-LA-HOUGUE,FRANCE (F029)	576
*NUCELLA LAPILLUS	400	CONC.IN WATER=0.0119PCI/L,ENTIRE ANIMAL,ST.VAAST-LA-HOUGUE,FRANCE (F029)	576
*OSTREA EDULIS	290	CONC.IN WATER=0.0119PCI/L,SHELL, ST.VAAST-LA-HOUGUE,FRANCE (F029)	576
*OSTREA EDULIS	40	CONC.IN WATER=0.0119PCI/L,SOFT PARTS,ST.VAAST-LA-HOUGUE,FRANCE (F029)	576
*OSTREA EDULIS	250	CONC.IN WATER=0.0119PCI/L,ENTIRE ANIMAL,ST.VAAST-LA-HOUGUE,FRANCE (F029)	576
*PATELLA VULGATA	180	CONC.IN WATER=0.0119PCI/L,SHELL, ST.VAAST-LA-HOUGUE,FRANCE (F029)	576
*PATELLA VULGATA	50	CONC.IN WATER=0.0119PCI/L,SOFT PARTS.ST.VAAST-LA-HOUGUE,FRANCE (F029)	576

TABLE D179

CONCENTRATION FACTOR : PU			
ITEM:	VALUE:	CONDITIONS:	REF.:
*PATELLA VULGATA	120	CONC.IN WATER=0.0119PCI/L,ENTIRE ANIMAL,ST.VAAST-LA-HOUGUE,FRANCE (F029)	576
*VENERUPIS DECUSSATA	160	CONC.IN WATER=0.0119PCI/L,SHELL, ST.VAAST-LA-HOUGUE,FRANCE (F029)	576
*VENERUPIS DECUSSATA	50	CONC.IN WATER=0.0119PCI/L,SOFT PARTS.ST.VAAST-LA-HOUGUE,FRANCE (F029)	576
*VENERUPIS DECUSSATA	120	CONC.IN WATER=0.0119PCI/L,ENTIRE ANIMAL,ST.VAAST-LA-HOUGUE,FRANCE (F029)	576
*WORM	0.17 CONC IN ANIM/(CONC IN SEDIMENT)	WET WEIGHT BOTH SED. AND ANIM., MASS. COAST	84

TABLE D180

CONCENTRATION FACTOR : PU-237			
ITEM:	VALUE:	CONDITIONS:	REF.:
*LYSMATA SETICAUDATA	94 - 160 BASED ON WET WEIGHT	7 - 17 DAYS	476

TABLE D181

CONCENTRATION FACTOR : PU-239			
ITEM:	VALUE:	CONDITIONS:	REF.:
*ALGAE	10000 BASED ON WET WEIGHT	AT ROCKY FLATS PLANT, USA	1589
*BULLHEAD MINNOW	0.89 - 7.5 BASED ON WET WEIGHT	AT ROCKY FLATS PLANT, USA	1589
*CRAYFISH	1000 BASED ON WET WEIGHT	AT ROCKY FLATS PLANT, USA	1589
*LIMANDA LIMANDA	30 BASED ON WET WEIGHT	ORGANISM FROM IRISH SEA	661
*LITTORINA LITTOREA	2*10**3 BASED ON WET WEIGHT	ORGANISM FROM IRISH SEA	661
*MYTILUS EDULIS	2*10**3 BASED ON WET WEIGHT	ORGANISM FROM IRISH SEA	661
*PLEURONECTES PLATESSA	30 BASED ON WET WEIGHT	ORGANISM FROM IRISH SEA	661
*PORPHYRA SP.	3*10**3 BASED ON WET WEIGHT	ORGANISM FROM IRISH SEA	661
*SALMO SALAR	1 BASED ON WET WEIGHT	ORGANISM FROM IRISH SEA	661
*SATIUM LEGUMINOSAE	0.546 BASED ON DRY WEIGHT	N=4, COLLOID PU SIZE =0.2-0.6 U, PU IN 2 CN DEPTH, CHEALATED BY DTPA	932
*SATIUM LEGUMINOSAE	0.797 BASED ON DRY WEIGHT	N=4, COLLOID PU SIZE =0.6-1.0 U, PU IN 2 CN DEPTH, CHEALATED BY DTPA	932
*SATIUM LEGUMINOSAE	2.480 BASED ON DRY WEIGHT	N=4, COLLOID PU SIZE = > 1.00 U, PU IN 2 CN DEPTH, CHEALATED BY DTPA	932
*SATIUM LEGUMINOSAE	0.517 BASED ON DRY WEIGHT	N=4, COLLOID PU SIZE =0.2-0.6 U, PU IN 8 CN DEPTH, CHEALATED BY DTPA	932
*SATIUM LEGUMINOSAE	0.458 BASED ON DRY WEIGHT	N=4, COLLOID PU SIZE =0.6-1.0 U, PU IN 8 CN DEPTH, CHEALATED BY DTPA	932
*SATIUM LEGUMINOSAE	0.532 BASED ON DRY WEIGHT	N=4, COLLOID PU SIZE = > 1.00 U, PU IN 8 CN DEPTH, CHEALATED BY DTPA	932
*SATIUM LEGUMINOSAE	7.43*10**(-4) BASED ON DRY WEIGHT	N=4, COLLOID PU SIZE =0.2-0.6 U, PU IN 8 CN DEPTH, UNCHEALATED	932
*SATIUM LEGUMINOSAE	6.98*10**(-4) BASED ON DRY WEIGHT	N=4, COLLOID PU SIZE =0.6-1.0 U, PU IN 8 CN DEPTH, UNCHEALATED	932
*SATIUM LEGUMINOSAE	6.20*10**(-4) BASED ON DRY WEIGHT	N=4, COLLOID PU SIZE = > 1.00 U, PU IN 2 CN DEPTH, UNCHEALATED	932
*SEDIMENT	7*10**4 BASED ON DRY SEDIMENT	WATER = 0.48 P CI/L, 1 KM FROM OUTFALL, IRISH SEA	661

TABLE D181

CONCENTRATION FACTOR : PU-239

ITEM:	VALUE:	CONDITIONS:	REF.:
*SEDIMENT	22*10**4 BASED ON DRY SEDIMENT	WATER = 0.37 P CI/L, 4.5 KM FROM OUTFALL, IRISH SEA	661
*SEDIMENT	3.5*10**4 BASED ON DRY SEDIMENT	WATER = 0.65 P CI/L, 4.5 KM FROM OUTFALL, IRISH SEA	661
*SEDIMENT	3*10**4 BASED ON DRY SEDIMENT	WATER = 0.31 P CI/L, 9 KM FROM OUTFALL, IRISH SEA	661
*SEDIMENT	3*10**4 BASED ON DRY SEDIMENT	WATER = 0.46 P CI/L, 9 KM FROM OUTFALL, IRISH SEA	661
*SEDIMENT	9*10**4 BASED ON DRY SEDIMENT	WATER = 0.31 P CI/L, 9 KM FROM OUTFALL, IRISH SEA	661
*SEDIMENT	2.0*10**4 BASED ON DRY SEDIMENT	WATER = 1.02 P CI/L, 9 KM FROM OUTFALL, IRISH SEA	661
*SEDIMENT	0.6*10**4 BASED ON DRY SEDIMENT	WATER = 0.035 P CI/L, 75 KM FROM OUTFALL, IRISH SEA	661
*SEDIMENT	0.7*10**4 BASED ON DRY SEDIMENT	WATER = 0.05 P CI/L, 90 KM FROM OUTFALL, IRISH SEA	661
*SEDIMENT	2*10**4 BASED ON DRY SEDIMENT	WATER = 0.03 P CI/L, 110 KM FROM OUTFALL, IRISH SEA	661
*SEDIMENT	0.1*10**4 BASED ON DRY SEDIMENT	WATER = 0.040 P CI/L, 110 KM FROM OUTFALL, IRISH SEA	661
*SESTON	10000 - 100000 BASED ON WET WEIGHT	AT ROCKY FLATS PLANT, USA	1589
*SUSPENDED SILT	89600 (RANGE = 48400 - 131000) BASED ON DRY WEIGHT	SEAWATER, BOMBAY	1261
*VASCULAR PLANT	0.1 - 0.01 BASED ON WET WEIGHT	AT ROCKY FLATS PLANT, USA	1589
*ZOOPLANKTON	1000 BASED ON WET WEIGHT	AT ROCKY FLATS PLANT, USA	1589

TABLE D182

CONCENTRATION FACTOR : RA

ITEM:	VALUE:	CONDITIONS:	REF.:
*BROWN ALGAE	370 BASED ON WET WEIGHT	NONE	157
*BROWN ALGAE	100 BASED ON WET WEIGHT	NONE	994
*PLANKTON	4500 BASED ON WET WEIGHT	NONE	157
*PLANKTON	2750 BASED ON WET WEIGHT	NONE	994

TABLE D183

CONCENTRATION FACTOR : RA-226

ITEM:	VALUE:	CONDITIONS:	REF.:
*FISH, BONE	510 (RANGE = 170 - 810) BASED ON WET WEIGHT	DANUBE, 1973, NUCLEAR POWER PLANT SINCE 1967	658
*FISH, MUSCLE	70 (RANGE = 25 - 110) BASED ON WET WEIGHT	DANUBE, 1973, NUCLEAR POWER PLANT SINCE 1967	658
*FRESHWATER SEDIMENT	2070 (RANGE = 1390 - 2570) BASED ON WET WEIGHT	DANUBE, 1973, NUCLEAR POWER PLANT SINCE 1967	658
*SUSPENDED SOLIDS	2410 (RANGE = 480 - 10160) BASED ON WET WEIGHT	DANUBE, 1973, NUCLEAR POWER PLANT SINCE 1967	658
*WATER PLANT	190 (RANGE = 25 - 690) BASED ON WET WEIGHT	DANUBE, 1973, NUCLEAR POWER PLANT SINCE 1967	658

TABLE D184

CONCENTRATION FACTOR : RB

ITEM:	VALUE:	CONDITIONS:	REF.:
*BROWN ALGAE	15 BASED ON WET WEIGHT	NONE	157

TABLE D184

CONCENTRATION FACTOR : RB			
ITEM:	VALUE:	CONDITIONS:	REF.:
*BROWN ALGAE	5-50 BASED ON WET WEIGHT	NONE	994
*PHYTOPLANKTON	6600-35400 BASED ON WET WEIGHT	NONE	311
*32 FRESHWATER PLANT	1230 BASED ON DRY WEIGHT	ALGAE,BRYOPHYTES AND VASCULAR HYDROPHYTES	1597

TABLE D185

CONCENTRATION FACTOR : RU			
ITEM:	VALUE:	CONDITIONS:	REF.:
*BROWN ALGAE	15-2000 BASED ON WET WEIGHT	NONE	994
*PLANKTON	600-3000 BASED ON WET WEIGHT	NONE	994
*32 FRESHWATER PLANT	1700 BASED ON DRY WEIGHT	ALGAE,BRYOPHYTES AND VASCULAR HYDROPHYTES	1597

TABLE D186

CONCENTRATION FACTOR : RU-103			
ITEM:	VALUE:	CONDITIONS:	REF.:
*ANODONTA NUTTALLIANA, SOFT PARTS	0.4 BASED ON WET WEIGHT	NO FOOD,292 K,PH=7.5, FRESHWAT=40 PART. AND 3100 SOLUB P CI/L, 7 D	629
*ANODONTA NUTTALLIANA, SOFT PARTS	1.1 BASED ON WET WEIGHT	NO FOOD,292 K,PH=7.5, FRESHWAT=330 PART. AND 5500 SOLUB P CI/L, 21 D	629
*ASTACUS SP., MUSCLE	LESS THAN 1 BASED ON WET WEIGHT	NO FOOD,292 K,PH=7.5, FRESHWAT=40 PART. AND 3100 SOLUB P CI/L, 7 D	629
*ASTACUS SP., MUSCLE	LESS THAN 1 BASED ON WET WEIGHT	NO FOOD,292 K,PH=7.5, FRESHWAT=330 PART. AND 5500 SOLUB P CI/L, 21 D	629
*ASTACUS SP., SOFT PARTS	6 BASED ON WET WEIGHT	NO FOOD,292 K,PH=7.5, FRESHWAT=0 PART. AND 3600 SOLUB P CI/L, 7 D	629
*ASTACUS SP., VISCERA	7 BASED ON WET WEIGHT	NO FOOD,292 K,PH=7.5, FRESHWAT=330 PART. AND 5500 SOLUB P CI/L, 21 D	629
*CANCER PRODUCTUS, MUSCLE	0.3 BASED ON WET WEIGHT	NO FOOD,285 K,PH=8.2, SEAWATER=40 PART. AND 4000 SOLUB P CI/L, 20 D	629
*CANCER PRODUCTUS, SORT PARTS	3 BASED ON WET WEIGHT	NO FOOD,285 K,PH=8.2, SEAWATER=40 PART. AND 4000 SOLUB P CI/L, 20 D	629
*CANCER PRODUCTUS, VISCERA	3 BASED ON WET WEIGHT	NO FOOD,285 K,PH=8.2, SEAWATER=90 PART. AND 2800 SOLUB P CI/L, 6 D	629
*CANCER PRODUCTUS, VISCERA	3 BASED ON WET WEIGHT	NO FOOD,285 K,PH=8.2, SEAWATER=40 PART. AND 4000 SOLUB P CI/L, 20 D	629
*CARASSIUS AURATUS, MUSCLE	LESS THAN 1 BASED ON WET WEIGHT	NO FOOD,292 K,PH=7.5, FRESHWAT=40 PART. AND 3100 SOLUB P CI/L, 7 D	629
*CARASSIUS AURATUS, NONMUSCLE	0.8 BASED ON WET WEIGHT	NO FOOD,292 K,PH=7.5, FRESHWAT=40 PART. AND 3100 SOLUB P CI/L, 7 D	629
*CARASSIUS AURATUS, WHOLE BODY	0.7 BASED ON WET WEIGHT	NO FOOD,292 K,PH=7.5, FRESHWAT=330 PART. AND 5500 SOLUB P CI/L, 21 D	629
*FISH, MUSCLE	170 (RANGE = 40 - 270) BASED ON WET WEIGHT	DANUBE, 1973, NUCLEAR POWER PLANT SINCE 1967	658
*FRESHWATER SEDIMENT	450 (RANGE = 40 - 1190) BASED ON WET WEIGHT	DANUBE, 1973, NUCLEAR POWER PLANT SINCE 1967	658
*GILLICHTHYS MIRABILIS, MUSCLE	LESS THAN 1 BASED ON WET WEIGHT	NO FOOD,285 K,PH=8.2, SEAWATER=90 PART. AND 2800 SOLUB P CI/L, 6 D	629
*GILLICHTHYS MIRABILIS, MUSCLE	LESS THAN 1 BASED ON WET WEIGHT	NO FOOD,285 K,PH=8.2, SEAWATER=40 PART. AND 4000 SOLUB P CI/L, 20 D	629
*GILLICHTHYS MIRABILIS, NONMUSCLE	0.8 BASED ON WET WEIGHT	NO FOOD,285 K,PH=8.2, SEAWATER=90 PART. AND 2800 SOLUB P CI/L, 6 D	629
*GILLICHTHYS MIRABILIS, NONMUSCLE	0.5 BASED ON WET WEIGHT	NO FOOD,285 K,PH=8.2, SEAWATER=40 PART. AND 4000 SOLUB P CI/L, 20 D	629
*GILLICHTHYS MIRABILIS, WHOLE BODY	0.7 BASED ON WET WEIGHT	NO FOOD,285 K,PH=8.2, SEAWATER=40 PART. AND 4000 SOLUB P CI/L, 20 D	629

TABLE D186

CONCENTRATION FACTOR : RU-103			
ITEM:	VALUE:	CONDITIONS:	REF.:
*GILLICHTHYS MIRABILIS, WHOLE BODY	0.4 BASED ON WET WEIGHT	NO FOOD,285 K,PH=8.2, SEAWATER=0 PART. AND 7500 SOLUB P CI/L, 6 D	629
*MYTILUS EDULIS, SOFT PARTS	1.8 BASED ON WET WEIGHT	NO FOOD,285 K,PH=8.2, SEAWATER=90 PART. AND 2800 SOLUB P CI/L, 6 D	629
*SAXIDONUS GIGANTEUS, SOFT PARTS	0.4 BASED ON WET WEIGHT	NO FOOD,285 K,PH=8.2, SEAWATER=40 PART. AND 4000 SOLUB P CI/L, 20 D	629
*SUSPENDED SOLIDS	870 (RANGE = 360 - 2420) BASED ON WET WEIGHT	DANUBE, 1973, NUCLEAR POWER PLANT SINCE 1967	658
*WATER PLANT	100 (RANGE = 25 - 250) BASED ON WET WEIGHT	DANUBE, 1973, NUCLEAR POWER PLANT SINCE 1967	658

TABLE D187

CONCENTRATION FACTOR : RU-106			
ITEM:	VALUE:	CONDITIONS:	REF.:
*ALGAE	100 - 200 BASED ON WET WEIGHT	LAB VALUE	25
*CORALLINA OFFICINALIS	100 - 200 BASED ON DRY WEIGHT	SEAWATER = 1 - 2 P CI/L	24
*ESKMEALS SILT	7529.9 BASED ON WET WEIGHT	YEARLY MEAN, CUMBERLAND COASTRAL AREA, 1969	1240
*FISH	1 BASED ON WET WEIGHT	LAB VALUE	25
*FISH, BONE	580 (RANGE = 300 - 1430) BASED ON WET WEIGHT	DANUBE, 1973, NUCLEAR POWER PLANT SINCE 1967	658
*FISH, MUSCLE	70 (RANGE = 15 - 170) BASED ON WET WEIGHT	DANUBE, 1973, NUCLEAR POWER PLANT SINCE 1967	658
*FRESHWATER SEDIMENT	1190 (RANGE = 260 - 2790) BASED ON WET WEIGHT	DANUBE, 1973, NUCLEAR POWER PLANT SINCE 1967	658
*INSHORE SAND	1217.0 BASED ON WET WEIGHT	YEARLY MEAN, CUMBERLAND COASTAL AREA, 1969	1240
*INVERTEBRATE	20 - 50 BASED ON WET WEIGHT	LAB VALUE	25
*MARINE MOLLUSCS AND CRUSTACEANS	4 - 11 BASED ON WET WEIGHT	GILLS	918
*MARINE MOLLUSCS AND CRUSTACEANS	2 - 10 BASED ON WET WEIGHT	HEPATOPANCREAS	918
*MARINE MOLLUSCS AND CRUSTACEANS	3 - 28 BASED ON WET WEIGHT	SHELL	918
*NEPHTYS SP.	507.2 BASED ON WET WEIGHT	YEARLY MEAN, CUMBERLAND COASTRAL AREA, 1969	1240
*NUCULA SP.	3247.8 BASED ON WET WEIGHT	YEARLY MEAN, CUMBERLAND COASTRAL AREA, 1969	1240
*PLEURONECTES PLATESSA, GUT CONTENT	663.7 BASED ON WET WEIGHT	YEARLY MEAN, CUMBERLAND COASTRAL AREA, 1969	1240
*SUSPENDED SILT	4860 (RANGE = 1750 - 8720) BASED ON DRY WEIGHT	SEAWATER, BOMBAY	1261
*SUSPENDED SOLIDS	4730 (RANGE = 140 - 10760) BASED ON WET WEIGHT	DANUBE, 1973, NUCLEAR POWER PLANT SINCE 1967	658
*WATER PLANT	30 (RANGE = 10 - 90) BASED ON WET WEIGHT	DANUBE, 1973, NUCLEAR POWER PLANT SINCE 1967	658

TABLE D188

CONCENTRATION FACTOR : S			
ITEM:	VALUE:	CONDITIONS:	REF.:
*BRASSICA OLERACEA	21 BASED ON WET WEIGHT	CONC.IN SOIL SOLUTION=205PPM	157
*BROWN ALGAE	3.4 BASED ON WET WEIGHT	NONE	157
*BROWN ALGAE	10 BASED ON WET WEIGHT	NONE	994
*PLANKTON	1.7 BASED ON WET WEIGHT	NONE	157
*32 FRESHWATER PLANT	165 BASED ON DRY WEIGHT	ALGAE,BRYOPHYTES AND VASCULAR HYDROPHYTES	1597

TABLE D189

CONCENTRATION FACTOR : S-35			
ITEM:	VALUE:	CONDITIONS:	REF.:
*LAMPSILIS RADIATA	240 BASED ON WET WEIGHT	91 DAYS, FRESHWATER	639

TABLE D190

CONCENTRATION FACTOR : SB			
ITEM:	VALUE:	CONDITIONS:	REF.:
*PHYTOPLANKTON	280-3200 BASED ON WET WEIGHT	NONE	311
*PLANKTON	50 BASED ON WET WEIGHT	NONE	994

TABLE D191

CONCENTRATION FACTOR : SB-124			
ITEM:	VALUE:	CONDITIONS:	REF.:
*ANODONTA NUTTALLIANA, SOFT PARTS	1.7 BASED ON WET WEIGHT	NO FOOD,292 K,PH=7.5, FRESHWAT=0 PART. AND 3600 SOLUB P CI/L, 7 D	629
*ANODONTA NUTTALLIANA, SOFT PARTS	1.7 BASED ON WET WEIGHT	NO FOOD,292 K,PH=7.5, FRESHWAT=0 PART. AND 8800 SOLUB P CI/L, 21 D	629
*ASTACUS SP., MUSCLE	LESS THAN 1 BASED ON WET WEIGHT	NO FOOD,292 K,PH=7.5, FRESHWAT=0 PART. AND 3600 SOLUB P CI/L, 7 D	629
*ASTACUS SP., MUSCLE	LESS THAN 1 BASED ON WET WEIGHT	NO FOOD,292 K,PH=7.5, FRESHWAT=0 PART. AND 8800 SOLUB P CI/L, 21 D	629
*ASTACUS SP., SOFT PARTS	LESS THAN 1 BASED ON WET WEIGHT	NO FOOD,292 K,PH=7.5, FRESHWAT=0 PART. AND 3600 SOLUB P CI/L, 7 D	629
*ASTACUS SP., SOFT PARTS	1.2 BASED ON WET WEIGHT	NO FOOD,292 K,PH=7.5, FRESHWAT=1.7 PART. AND 250 SOLUB P CI/ML, 7 D	629
*ASTACUS SP., VISCERA	LESS THAN 1 BASED ON WET WEIGHT	NO FOOD,292 K,PH=7.5, FRESHWAT=0 PART. AND 3600 SOLUB P CI/L, 7 D	629
*ASTACUS SP., VISCERA	1.4 BASED ON WET WEIGHT	NO FOOD,292 K,PH=7.5, FRESHWAT=0 PART. AND 8800 SOLUB P CI/L, 21 D	629
*CANCER PRODUCTUS, MUSCLE	LESS THAN 1 BASED ON WET WEIGHT	NO FOOD,285 K,PH=8.2, SEAWATER=70 PART AND 13000 SOLUB P CI/L, 20 D	629
*CANCER PRODUCTUS, SORT PARTS	0.7 BASED ON WET WEIGHT	NO FOOD,285 K,PH=8.2, SEAWATER=70 PART AND 13000 SOLUB P CI/L, 20 D	629
*CANCER PRODUCTUS, VISCERA	0.7 BASED ON WET WEIGHT	NO FOOD,285 K,PH=8.2, SEAWATER=0 PART. AND 7500 SOLUB P CI/L, 6 D	629
*CANCER PRODUCTUS, VISCERA	0.8 BASED ON WET WEIGHT	NO FOOD,285 K,PH=8.2, SEAWATER=70 PART AND 13000 SOLUB P CI/L, 20 D	629
*CARASSIUS AURATUS, MUSCLE	LESS THAN 1 BASED ON WET WEIGHT	NO FOOD,292 K,PH=7.5, FRESHWAT=0 PART. AND 3600 SOLUB P CI/L, 7 D	629
*CARASSIUS AURATUS, NONMUSCLE	1.7 BASED ON WET WEIGHT	NO FOOD,292 K,PH=7.5, FRESHWAT=0 PART. AND 3600 SOLUB P CI/L, 7 D	629
*CARASSIUS AURATUS, WHOLE BODY	1.5 BASED ON WET WEIGHT	NO FOOD,292 K,PH=7.5, FRESHWAT=0 PART. AND 8800 SOLUB P CI/L, 21 D	629
*FISH, BONE	810 (RANGE = 350 - 1100) BASED ON WET WEIGHT	DANUBE, 1973, NUCLEAR POWER PLANT SINCE 1967	658
*FISH, MUSCLE	90 (RANGE = 70 - 120) BASED ON WET WEIGHT	DANUBE, 1973, NUCLEAR POWER PLANT SINCE 1967	658
*FRESHWATER SEDIMENT	220 (RANGE = 180 - 270) BASED ON WET WEIGHT	DANUBE, 1973, NUCLEAR POWER PLANT SINCE 1967	658
*GILLICHTHYS MIRABILIS, MUSCLE	LESS THAN 1 BASED ON WET WEIGHT	NO FOOD,285 K,PH=8.2, SEAWATER=0 PART. AND 7500 SOLUB P CI/L, 6 D	629
*GILLICHTHYS MIRABILIS, MUSCLE	LESS THAN 1 BASED ON WET WEIGHT	NO FOOD,285 K,PH=8.2, SEAWATER=70 PART AND 13000 SOLUB P CI/L, 20 D	629
*GILLICHTHYS MIRABILIS, NONMUSCLE	0.9 BASED ON WET WEIGHT	NO FOOD,285 K,PH=8.2, SEAWATER=0 PART. AND 7500 SOLUB P CI/L, 6 D	629
*GILLICHTHYS MIRABILIS, NONMUSCLE	0.4 BASED ON WET WEIGHT	NO FOOD,285 K,PH=8.2, SEAWATER=70 PART AND 13000 SOLUB P CI/L, 20 D	629

TABLE D191

CONCENTRATION FACTOR : SB-124			
ITEM:	VALUE:	CONDITIONS:	REF.:
*GILLICHTHYS MIRABILIS, WHOLE BODY	0.9 BASED ON WET WEIGHT	NO FOOD,285 K,PH=8.2, SEAWATER=70 PART AND 13000 SOLUB P CI/L, 20 D	629
*GILLICHTHYS MIRABILIS, WHOLE BODY	0.4 BASED ON WET WEIGHT	NO FOOD,285 K,PH=8.2, SEAWATER=1.1 PART. AND 440 SOLUB P CI/ML, 6 D	629
*MYTILUS EDULIS, SOFT PARTS	0.7 BASED ON WET WEIGHT	NO FOOD,285 K,PH=8.2, SEAWATER=0 PART. AND 7500 SOLUB P CI/L, 6 D	629
*SAXIDONUS GIGANTEUS, SOFT PARTS	LESS THAN 1 BASED ON WET WEIGHT	NO FOOD,285 K,PH=8.2, SEAWATER=70 PART AND 13000 SOLUB P CI/L, 20 D	629
*SUSPENDED SOLIDS	780 (RANGE = 760 - 800) BASED ON WET WEIGHT	DANUBE, 1973, NUCLEAR POWER PLANT SINCE 1967	658
*WATER PLANT	30 BASED ON WET WEIGHT	DANUBE, 1973, NUCLEAR POWER PLANT SINCE 1967	658

TABLE D192

CONCENTRATION FACTOR : SB-125			
ITEM:	VALUE:	CONDITIONS:	REF.:
*FISH, BONE	470 BASED ON WET WEIGHT	DANUBE, 1973, NUCLEAR POWER PLANT SINCE 1967	658
*FISH, MUSCLE	100 (RANGE = 30 - 170) BASED ON WET WEIGHT	DANUBE, 1973, NUCLEAR POWER PLANT SINCE 1967	658
*FRESHWATER SEDIMENT	560 (RANGE = 230 - 1000) BASED ON WET WEIGHT	DANUBE, 1973, NUCLEAR POWER PLANT SINCE 1967	658
*SUSPENDED SOLIDS	4300 (RANGE = 480 - 28000) BASED ON WET WEIGHT	DANUBE, 1973, NUCLEAR POWER PLANT SINCE 1967	658
*WATER PLANT	140 (RANGE = 60 - 330) BASED ON WET WEIGHT	DANUBE, 1973, NUCLEAR POWER PLANT SINCE 1967	658

TABLE D193

CONCENTRATION FACTOR : SC			
ITEM:	VALUE:	CONDITIONS:	REF.:
*BROWN ALGAE	1500-2600 BASED ON WET WEIGHT	NONE	994
*PHYTOPLANKTON	160000-995000 BASED ON WET WEIGHT	NONE	311

TABLE D194

CONCENTRATION FACTOR : SE			
ITEM:	VALUE:	CONDITIONS:	REF.:
*PHYTOPLANKTON	900-5500 BASED ON WET WEIGHT	NONE	311

TABLE D195

CONCENTRATION FACTOR : SE-75			
ITEM:	VALUE:	CONDITIONS:	REF.:
*LYSMATA SETICAUDATA, EXOSKELETON	30.0 BASED ON WET WEIGHT	51 D, SEAWATER, UPTAKE FROM WATER	477
*LYSMATA SETICAUDATA, EYES	15.2 BASED ON WET WEIGHT	51 D, SEAWATER, UPTAKE FROM WATER	477
*LYSMATA SETICAUDATA, MUSCLE	17.8 BASED ON WET WEIGHT	51 D, SEAWATER, UPTAKE FROM WATER	477

TABLE D195

ITEM:	VALUE:	CONDITIONS:	REF.:
CONCENTRATION FACTOR : SE-75			
*LYSMATA SETICAUDATA, VISCERA	25.7 BASED ON WET WEIGHT	51 D, SEAWATER, UPTAKE FROM WATER	477
*LYSMATA SETICAUDATA, WHOLE BODY	21.9 BASED ON WET WEIGHT	51 D, SEAWATER, UPTAKE FROM WATER	477
*MYTILUS GALLOPROVINCIALIS	46 (SD=+-10) BASED ON WET WEIGHT	WET WEIGHT = 2.1 G, UPTAKE OF SE(+ 4), 13 D, N=3	478
*MYTILUS GALLOPROVINCIALIS	29 (SD=+-8) BASED ON WET WEIGHT	WET WEIGHT =13.2 G, UPTAKE OF SE(+ 4), 13 D, N=5	478
*MYTILUS GALLOPROVINCIALIS	13 (SD=+-2.5) BASED ON WET WEIGHT	WET WEIGHT =21.8 G, UPTAKE OF SE(+ 4), 13 D, N=4	478
*MYTILUS GALLOPROVINCIALIS, GILLS	284 BASED ON WET WEIGHT	65 D, SEAWATER, UPTAKE FROM WATER	477
*MYTILUS GALLOPROVINCIALIS, MANTLE	80.7 BASED ON WET WEIGHT	65 D, SEAWATER, UPTAKE FROM WATER	477
*MYTILUS GALLOPROVINCIALIS, MUSCLE	186 BASED ON WET WEIGHTT	65 D, SEAWATER, UPTAKE FROM WATER	477
*MYTILUS GALLOPROVINCIALIS, SHELL	127 BASED ON WET WEIGHT	65 D, SEAWATER, UPTAKE FROM WATER	477
*MYTILUS GALLOPROVINCIALIS, VISCERA	494 BASED ON WET WEIGHTT	65 D, SEAWATER, UPTAKE FROM WATER	477
*MYTILUS GALLOPROVINCIALIS, WHOLE BODY	91.2 BASED ON WET WEIGHT	65 D, SEAWATER, UPTAKE FROM WATER	477

TABLE D196

ITEM:	VALUE:	CONDITIONS:	REF.:
CONCENTRATION FACTOR : SI			
*BROWN ALGAE	120 BASED ON WET WEIGHT	NONE	157
*PLANKTON	17000 BASED ON WET WEIGHT	NONE	157

TABLE D197

ITEM:	VALUE:	CONDITIONS:	REF.:
CONCENTRATION FACTOR : SN			
*BROWN ALGAE	92 BASED ON WET WEIGHT	NONE	157
*PLANKTON	2900 BASED ON WET WEIGHT	NONE	157

TABLE D198

ITEM:	VALUE:	CONDITIONS:	REF.:
CONCENTRATION FACTOR : SR			
*BROWN ALGAE	44 BASED ON WET WEIGHT	NONE	157
*BROWN ALGAE	1-40 BASED ON WET WEIGHT	NONE	994
*CANCER IRRORATUS	84.2 (SD=+-2.65) BASED ON WET WEIGHT	SEAWATER CONC=380 PPM, NOVA SCOTIA, AAS	984
*CORAL	1.25 BASED ON CA CONTENT (F023)	DEEP OCEAN CORAL, MEAN OF 13 DETERMINATIONS	939
*CORAL	1.22 BASED ON CA CONTENT (F023)	SHALLOW OPEN OCEAN CORAL, MEAN OF 8 DETERMINATIONS	939
*CORAL	1.10 BASED ON CA CONTENT (F023)	SHALLOW COASTAL OCEAN, MEAN OF 13 DETERMINATIONS	939
*PLANKTON	8 BASED ON WET WEIGHT	NONE	157
*PLANKTON	9 BASED ON WET WEIGHT	NONE	994
*ZOOPLANKTON	5.1 BASED ON ZOOPLANKTON WET WEIGHT CONCENTRATION	SEAWATER CONCENTRATION 8.0 MG/M3	985
*32 FRESHWATER PLANT	475 BASED ON DRY WEIGHT	ALGAE,BRYOPHYTES AND VASCULAR HYDROPHYTES	1597

TABLE D199

CONCENTRATION FACTOR : SR-85			
ITEM:	VALUE:	CONDITIONS:	REF.:
*FRESHWATER PHYTOPLANKTON	200 - 700 BASED ON DRY WEIGHT	EQUILIBRIUM AFTER 36 H	534
*FRESHWATER PHYTOPLANKTON	200 - 700 BASED ON DRY WEIGHT	EQUILIBRIUM AFTER 36 H	534
*ROOTED AQUATIC PLANT	600 BASED ON DRY WEIGHT	EQUILIBRIUM AFTER 36 H	534
*ROOTED AQUATIC PLANT	600 BASED ON DRY WEIGHT	EQUILIBRIUM AFTER 36 H	534

TABLE D200

CONCENTRATION FACTOR : SR-89			
ITEM:	VALUE:	CONDITIONS:	REF.:
*FISH, BONE	40 (RANGE = 10 - 60) BASED ON WET WEIGHT	DANUBE, 1973, NUCLEAR POWER PLANT SINCE 1967	658
*FISH, MUSCLE	20 (RANGE = 10 - 60) BASED ON WET WEIGHT	DANUBE, 1973, NUCLEAR POWER PLANT SINCE 1967	658
*FRESHWATER SEDIMENT	140 (RANGE = 20 - 350) BASED ON WET WEIGHT	DANUBE, 1973, NUCLEAR POWER PLANT SINCE 1967	658
*MARINE ALGAE	5.3 - 6.7 BASED ON WET WEIGHT	1 D	671
*MARINE ALGAE	5.3 - 6.7 BASED ON WET WEIGHT	1 D	671
*MARINE CRUSTACEAN, EXOSKELETON	15 BASED ON WET WEIGHT	4 D	671
*MARINE CRUSTACEAN, EXOSKELETON	15 BASED ON WET WEIGHT	4 D	671
*MARINE ECHINODERM	0.9 BASED ON WET WEIGHT	1 D	671
*MARINE ECHINODERM	0.9 BASED ON WET WEIGHT	1 D	671
*SUSPENDED SOLIDS	1250 (RANGE = 1 - 8340) BASED ON WET WEIGHT	DANUBE, 1973, NUCLEAR POWER PLANT SINCE 1967	658
*WATER PLANT	130 (RANGE = 2 - 510) BASED ON WET WEIGHT	DANUBE, 1973, NUCLEAR POWER PLANT SINCE 1967	658

TABLE D201

CONCENTRATION FACTOR : SR-90			
ITEM:	VALUE:	CONDITIONS:	REF.:
*CARP	33 BASED ON WET WEIGHT	SEMIMIGRATORY FISH, CASPIAN AND AZOV SEAS	1598
*DETRITUS	LESS THAN 1000 BASED ON DRY WEIGHT	MARINE DETRITUS, INDIAN OCEAN	1222
*FISH, BONE	360 (RANGE = 19 - 1090) BASED ON WET WEIGHT	DANUBE, 1973, NUCLEAR POWER PLANT SINCE 1967	658
*FISH, MUSCLE	20 (RANGE = 5 - 40) BASED ON WET WEIGHT	DANUBE, 1973, NUCLEAR POWER PLANT SINCE 1967	658
*FRESHWATER SEDIMENT	810 (RANGE = 60 - 1660) BASED ON WET WEIGHT	DANUBE, 1973, NUCLEAR POWER PLANT SINCE 1967	658
*FRESHWATER SEDIMENT	45 BASED ON DRY WEIGHT	LAKE ERIE, WESTERN BASIN	912
*FRESHWATER SEDIMENT	80 BASED ON DRY WEIGHT	LAKE ERIE, CENTRAL BASIN	912
*FRESHWATER SEDIMENT	120 BASED ON DRY WEIGHT	LAKE ONTARIO, WESTERN PART	912
*KUTUM	2.5 BASED ON WET WEIGHT	MIGRATORY FISH, CASPIAN AND AZOV SEAS	1598
*ROACH	40 BASED ON WET WEIGHT	SEMIMIGRATORY FISH, CASPIAN AND AZOV SEAS	1598
*SUSPENDED SILT	15.8 (RANGE = 4.3 - 31.0) BASED ON DRY WEIGHT	SEAWATER, BOMBAY	1261
*SUSPENDED SOLIDS	2460 (RANGE = 10 - 10560) BASED ON WET WEIGHT	DANUBE, 1973, NUCLEAR POWER PLANT SINCE 1967	658
*ULVA LACTUCA	0.32 BASED ON WET WEIGHT	EQUILIBRIUM AFTER 3 H, PH=8, CARRIER FREE	603
*WATER PLANT	20 (RANGE = 3 - 50) BASED ON WET WEIGHT	DANUBE, 1973, NUCLEAR POWER PLANT SINCE 1967	658

TABLE D202

CONCENTRATION FACTOR : TB			
ITEM:	VALUE:	CONDITIONS:	REF.:
*ENGRAULIS JAPONICUS, WHOLE BODY	150 BASED ON WET WEIGHT	REL. TO TOTAL CONC.,TOT CONC=4.2 NG/L, SOLUB=0.8 NG/L,SEAWAT,JAP.	1535
*ENGRAULIS JAPONICUS, WHOLE BODY	1040 BASED ON WET WEIGHT	REL. TO SOLUBLE CONC.,TOT CONC=4.2 NG/L, SOLUB=0.8 NG/L,SEAWAT,JAP.	1535
*HIZIKIA FUSIFORME, WHOLE BODY	110 BASED ON WET WEIGHT	REL. TO TOTAL CONC.,TOT CONC=4.2 NG/L, SOLUB=0.8 NG/L,SEAWAT,JAP.	1535
*HIZIKIA FUSIFORME, WHOLE BODY	780 BASED ON WET WEIGHT	REL. TO SOLUBLE CONC.,TOT CONC=4.2 NG/L, SOLUB=0.8 NG/L,SEAWAT,JAP.	1535
*MERETRIX LUSORIA, SOFT PART	460 BASED ON WET WEIGHT	REL. TO TOTAL CONC.,TOT CONC=4.2 NG/L, SOLUB=0.8 NG/L,SEAWAT,JAP.	1535
*MERETRIX LUSORIA, SOFT PART	3180 BASED ON WET WEIGHT	REL. TO SOLUBLE CONC.,TOT CONC=4.2 NG/L, SOLUB=0.8 NG/L,SEAWAT,JAP.	1535
*PARALICHTHYS OLIVACEUS, MUSCLE	5 BASED ON WET WEIGHT	REL. TO TOTAL CONC.,TOT CONC=4.2 NG/L, SOLUB=0.8 NG/L,SEAWAT,JAP.	1535
*PARALICHTHYS OLIVACEUS, MUSCLE	38 BASED ON WET WEIGHT	REL. TO SOLUBLE CONC.,TOT CONC=4.2 NG/L, SOLUB=0.8 NG/L,SEAWAT,JAP.	1535
*SARGASSUM FULVELLUM, WHOLE BODY	320 BASED ON WET WEIGHT	REL. TO TOTAL CONC.,TOT CONC=4.2 NG/L, SOLUB=0.8 NG/L,SEAWAT,JAP.	1535
*SARGASSUM FULVELLUM, WHOLE BODY	2230 BASED ON WET WEIGHT	REL. TO SOLUBLE CONC.,TOT CONC=4.2 NG/L, SOLUB=0.8 NG/L,SEAWAT,JAP.	1535
*SERIOLA QUINQUERADIATA, MUSCLE	13 BASED ON WET WEIGHT	REL. TO TOTAL CONC.,TOT CONC=4.2 NG/L, SOLUB=0.8 NG/L,SEAWAT,JAP.	1535
*SERIOLA QUINQUERADIATA, MUSCLE	883 BASED ON WET WEIGHT	REL. TO SOLUBLE CONC.,TOT CONC=4.2 NG/L, SOLUB=0.8 NG/L,SEAWAT,JAP.	1535
*ULVA PERTUSA, WHOLE BODY	180 BASED ON WET WEIGHT	REL. TO TOTAL CONC.,TOT CONC=4.2 NG/L, SOLUB=0.8 NG/L,SEAWAT,JAP.	1535
*ULVA PERTUSA, WHOLE BODY	1210 BASED ON WET WEIGHT	REL. TO SOLUBLE CONC.,TOT CONC=4.2 NG/L, SOLUB=0.8 NG/L,SEAWAT,JAP.	1535
*UNDARIA PINNATIFIDA, WHOLE BODY	54 BASED ON WET WEIGHT	REL. TO TOTAL CONC.,TOT CONC=4.2 NG/L, SOLUB=0.8 NG/L,SEAWAT,JAP.	1535
*UNDARIA PINNATIFIDA, WHOLE BODY	380 BASED ON WET WEIGHT	REL. TO SOLUBLE CONC.,TOT CONC=4.2 NG/L, SOLUB=0.8 NG/L,SEAWAT,JAP.	1535

TABLE D203

CONCENTRATION FACTOR : TC-95			
ITEM:	VALUE:	CONDITIONS:	REF.:
*HALIOTIS RUFESCENS, COLUMNAR MUSCLE	0.5 BASED ON WET WEIGHT	SEAWATER WITH 50 N CI/L, N=2-4, 284 K, 3 D	1492
*HALIOTIS RUFESCENS, COLUMNAR MUSCLE	0.8 BASED ON WET WEIGHT	SEAWATER WITH 50 N CI/L, N=2-4, 284 K, 10 D	1492
*HALIOTIS RUFESCENS, COLUMNAR MUSCLE	1.3 BASED ON WET WEIGHT	SEAWATER WITH 50 N CI/L, N=2-4, 284 K, 25 D	1492
*HALIOTIS RUFESCENS, DIGESTIVE GLAND	28 BASED ON WET WEIGHT	SEAWATER WITH 50 N CI/L, N=2-4, 284 K, 3 D	1492
*HALIOTIS RUFESCENS, DIGESTIVE GLAND	80 BASED ON WET WEIGHT	SEAWATER WITH 50 N CI/L, N=2-4, 284 K, 10 D	1492
*HALIOTIS RUFESCENS, DIGESTIVE GLAND	271 BASED ON WET WEIGHT	SEAWATER WITH 50 N CI/L, N=2-4, 284 K, 25 D	1492
*HALIOTIS RUFESCENS, GILL	117 BASED ON WET WEIGHT	SEAWATER WITH 50 N CI/L, N=2-4, 284 K, 3 D	1492
*HALIOTIS RUFESCENS, GILL	309 BASED ON WET WEIGHT	SEAWATER WITH 50 N CI/L, N=2-4, 284 K, 10 D	1492
*HALIOTIS RUFESCENS, GILL	253 BASED ON WET WEIGHT	SEAWATER WITH 50 N CI/L, N=2-4, 284 K, 25 D	1492
*HALIOTIS RUFESCENS, GONAD	1.5 BASED ON WET WEIGHT	SEAWATER WITH 50 N CI/L, N=2-4, 284 K, 3 D	1492
*HALIOTIS RUFESCENS, GONAD	3 BASED ON WET WEIGHT	SEAWATER WITH 50 N CI/L, N=2-4, 284 K, 10 D	1492
*HALIOTIS RUFESCENS, GONAD	5 BASED ON WET WEIGHT	SEAWATER WITH 50 N CI/L, N=2-4, 284 K, 25 D	1492
*HALIOTIS RUFESCENS, HEART	2.2 BASED ON WET WEIGHT	SEAWATER WITH 50 N CI/L, N=2-4, 284 K, 10 D	1492

TABLE D203

CONCENTRATION FACTOR : TC-95			
ITEM:	VALUE:	CONDITIONS:	REF.:
*HALIOTIS RUFESCENS, HEART	4 BASED ON WET WEIGHT	SEAWATER WITH 50 N CI/L, N=2-4, 284 K, 25 D	1492
*HALIOTIS RUFESCENS, HEART	6 BASED ON WET WEIGHT	SEAWATER WITH 50 N CI/L, N=2-4, 284 K, WW D	1492
*HALIOTIS RUFESCENS, KIDNEY	28 BASED ON WET WEIGHT	SEAWATER WITH 50 N CI/L, N=2-4, 284 K, 3 D	1492
*HALIOTIS RUFESCENS, KIDNEY	67 BASED ON WET WEIGHT	SEAWATER WITH 50 N CI/L, N=2-4, 284 K, 10 D	1492
*HALIOTIS RUFESCENS, KIDNEY	147 BASED ON WET WEIGHT	SEAWATER WITH 50 N CI/L, N=2-4, 284 K, 25 D	1492

TABLE D204

CONCENTRATION FACTOR : TCDD			
ITEM:	VALUE:	CONDITIONS:	REF.:
*ALGAE	=< 15000 BASED ON WET WEIGHT	SEE REF.	727
*DAPHNIA MAGNA	7800 BASED ON DRY WEIGHT	WATER=1330 PPT, SOIL=7.45 PPM, 4 D, METCALF MODEL, CF TO WAT. CONC	726
*DAPHNIA MAGNA	31200 BASED ON DRY WEIGHT	WATER=239 PPT, SOIL=3.17 PPM, 4 D, METCALF MODEL, CF TO WAT. CONC	726
*DAPHNIA MAGNA	22900 BASED ON DRY WEIGHT	WATER=7.13 PPT, SOIL=0.10 PPM, 4 D, METCALF MODEL, CF TO WAT. CONC	726
*DAPHNIA MAGNA	26800 BASED ON DRY WEIGHT	WATER=0.66 PPT, SOIL=0.01 PPM, 4 D, METCALF MODEL, CF TO WAT. CONC	726
*DAPHNIA MAGNA	18100 BASED ON DRY WEIGHT	WATER=0.26 PPT, SOIL=0.001 PPM, 4 D, METCALF MODEL, CF TO WAT. CONC	726
*DAPHNIA MAGNA	48000 BASED ON DRY WEIGHT	WATER=0.05 PPT, SOIL=0.0001 PPM, 4 D, METCALF MODEL, CF TO WAT. CONC	726
*FISH	=< 15000 BASED ON WET WEIGHT	SEE REF.	727
*GAMBUSIA AFFINIS	1000 BASED ON DRY WEIGHT	WATER=1330 PPT, SOIL=7.45 PPM, 4 D, METCALF MODEL, CF TO WAT. CONC	726
*GAMBUSIA AFFINIS	9200 BASED ON DRY WEIGHT	WATER=239 PPT, SOIL=3.17 PPM, 4 D, METCALF MODEL, CF TO WAT. CONC	726
*GAMBUSIA AFFINIS	11300 BASED ON DRY WEIGHT	WATER=48 PPT, SOIL=0.53 PPM, 4 D, METCALF MODEL, CF TO WAT. CONC	726
*GAMBUSIA AFFINIS	23300 BASED ON DRY WEIGHT	WATER=18 PPT, SOIL=0.29 PPM, 4 D, METCALF MODEL, CF TO WAT. CONC	726
*GAMBUSIA AFFINIS	12900 BASED ON DRY WEIGHT	WATER=7 PPT, SOIL=0.15 PPM, 4 D, METCALF MODEL, CF TO WAT. CONC	726
*GAMBUSIA AFFINIS	61600 BASED ON DRY WEIGHT	WATER=7.13 PPT, SOIL=0.10 PPM, 4 D, METCALF MODEL, CF TO WAT. CONC	726
*GAMBUSIA AFFINIS	63300 BASED ON DRY WEIGHT	WATER=0.66 PPT, SOIL=0.01 PPM, 4 D, METCALF MODEL, CF TO WAT. CONC	726
*GAMBUSIA AFFINIS	22700 BASED ON DRY WEIGHT	WATER=0.26 PPT, SOIL=0.001 PPM, 4 D, METCALF MODEL, CF TO WAT. CONC	726
*GAMBUSIA AFFINIS	24000 BASED ON DRY WEIGHT	WATER=0.05 PPT, SOIL=0.0001 PPM, 4 D, METCALF MODEL, CF TO WAT. CONC	726
*ICTALURUS PUNCTATUS	3000 BASED ON DRY WEIGHT	WATER=239 PPT, SOIL=3.17 PPM, 4 D, METCALF MODEL, CF TO WAT. CONC	726
*ICTALURUS PUNCTATUS	2300 BASED ON DRY WEIGHT	WATER=48 PPT, SOIL=0.53 PPM, 4 D, METCALF MODEL, CF TO WAT. CONC	726
*ICTALURUS PUNCTATUS	6700 BASED ON DRY WEIGHT	WATER=18 PPT, SOIL=0.29 PPM, 4 D, METCALF MODEL, CF TO WAT. CONC	726
*ICTALURUS PUNCTATUS	11400 BASED ON DRY WEIGHT	WATER=7 PPT, SOIL=0.15 PPM, 4 D, METCALF MODEL, CF TO WAT. CONC	726
*ICTALURUS PUNCTATUS	14400 BASED ON DRY WEIGHT	WATER=7.13 PPT, SOIL=0.10 PPM, 4 D, METCALF MODEL, CF TO WAT. CONC	726
*ICTALURUS PUNCTATUS	27900 BASED ON DRY WEIGHT	WATER=0.66 PPT, SOIL=0.01 PPM, 4 D, METCALF MODEL, CF TO WAT. CONC	726
*ICTALURUS PUNCTATUS	4600 BASED ON DRY WEIGHT	WATER=0.26 PPT, SOIL=0.001 PPM, 4 D, METCALF MODEL, CF TO WAT. CONC	726
*ICTALURUS PUNCTATUS	2000 BASED ON DRY WEIGHT	WATER=0.05 PPT, SOIL=0.0001 PPM, 4 D, METCALF MODEL, CF TO WAT. CONC	726

TABLE D204

CONCENTRATION FACTOR : TCDD			
ITEM:	VALUE:	CONDITIONS:	REF.:
*INSECTA	=< 15000 BASED ON WET WEIGHT	SEE REF.	727
*LEMNA MINOR	4300 BASED ON DRY WEIGHT	WATER=7.13 PPT, SOIL=0.10 PPM, 4 D, METCALF MODEL, CF TO WAT. CONC	726
*LEMNA MINOR	5000 BASED ON DRY WEIGHT	WATER=0.66 PPT, SOIL=0.01 PPM, 4 D, METCALF MODEL, CF TO WAT. CONC	726
*LEMNA MINOR	1200 BASED ON DRY WEIGHT	WATER=0.26 PPT, SOIL=0.001 PPM, 4 D, METCALF MODEL, CF TO WAT. CONC	726
*LEMNA MINOR	4000 BASED ON DRY WEIGHT	WATER=0.05 PPT, SOIL=0.0001 PPM, 4 D, METCALF MODEL, CF TO WAT. CONC	726
*OEDOGONIUM CARDIACUM	5000 BASED ON DRY WEIGHT	WATER=1330 PPT, SOIL=7.45 PPM, 4 D, METCALF MODEL, CF TO WAT. CONC	726
*OEDOGONIUM CARDIACUM	10500 BASED ON DRY WEIGHT	WATER=239 PPT, SOIL=3.17 PPM, 4 D, METCALF MODEL, CF TO WAT. CONC	726
*OEDOGONIUM CARDIACUM	8100 BASED ON DRY WEIGHT	WATER=48 PPT, SOIL=0.53 PPM, 4 D, METCALF MODEL, CF TO WAT. CONC	726
*OEDOGONIUM CARDIACUM	12800 BASED ON DRY WEIGHT	WATER=48 PPT, SOIL=0.53 PPM, 4 D, METCALF MODEL, CF TO WAT. CONC	726
*OEDOGONIUM CARDIACUM	18600 BASED ON DRY WEIGHT	WATER=7 PPT, SOIL=0.15 PPM, 4 D, METCALF MODEL, CF TO WAT. CONC	726
*OEDOGONIUM CARDIACUM	11100 BASED ON DRY WEIGHT	WATER=7.13 PPT, SOIL=0.10 PPM, 4 D, METCALF MODEL, CF TO WAT. CONC	726
*OEDOGONIUM CARDIACUM	7600 BASED ON DRY WEIGHT	WATER=0.66 PPT, SOIL=0.01 PPM, 4 D, METCALF MODEL, CF TO WAT. CONC	726
*OEDOGONIUM CARDIACUM	5400 BASED ON DRY WEIGHT	WATER=0.26 PPT, SOIL=0.001 PPM, 4 D, METCALF MODEL, CF TO WAT. CONC	726
*OEDOGONIUM CARDIACUM	2000 BASED ON DRY WEIGHT	WATER=0.05 PPT, SOIL=0.0001 PPM, 4 D, METCALF MODEL, CF TO WAT. CONC	726
*PHYSA SP.	1400 BASED ON DRY WEIGHT	WATER=1330 PPT, SOIL=7.45 PPM, 4 D, METCALF MODEL, CF TO WAT. CONC	726
*PHYSA SP.	11600 BASED ON DRY WEIGHT	WATER=239 PPT, SOIL=3.17 PPM, 4 D, METCALF MODEL, CF TO WAT. CONC	726
*PHYSA SP.	41000 BASED ON DRY WEIGHT	WATER=48 PPT, SOIL=0.53 PPM, 4 D, METCALF MODEL, CF TO WAT. CONC	726
*PHYSA SP.	16100 BASED ON DRY WEIGHT	WATER=18 PPT, SOIL=0.29 PPM, 4 D, METCALF MODEL, CF TO WAT. CONC	726
*PHYSA SP.	47100 BASED ON DRY WEIGHT	WATER=7 PPT, SOIL=0.15 PPM, 4 D, METCALF MODEL, CF TO WAT. CONC	726
*PHYSA SP.	17500 BASED ON DRY WEIGHT	WATER=7.13 PPT, SOIL=0.10 PPM, 4 D, METCALF MODEL, CF TO WAT. CONC	726
*PHYSA SP.	14700 BASED ON DRY WEIGHT	WATER=0.66 PPT, SOIL=0.01 PPM, 4 D, METCALF MODEL, CF TO WAT. CONC	726
*PHYSA SP.	5400 BASED ON DRY WEIGHT	WATER=0.26 PPT, SOIL=0.001 PPM, 4 D, METCALF MODEL, CF TO WAT. CONC	726
*PHYSA SP.	24000 BASED ON DRY WEIGHT	WATER=0.05 PPT, SOIL=0.0001 PPM, 4 D, METCALF MODEL, CF TO WAT. CONC	726
*SNAIL	=< 15000 BASED ON WET WEIGHT	SEE REF.	727

TABLE D205

CONCENTRATION FACTOR : TH			
ITEM:	VALUE:	CONDITIONS:	REF.:
*BROWN ALGAE	10 BASED ON WET WEIGHT	NONE	994
*MYA ARENARIA	18.6 BASED ON WET WEIGHT	WATER = 50 MG/M3, 80 D, THALLIUM (I) SULFATE	1777
*MYA ARENARIA	17.6 BASED ON WET WEIGHT	WATER =100 MG/M3, 80 D, THALLIUM (I) SULFATE	1777
*MYTILUS EDULIS	12.4 BASED ON WET WEIGHT	WATER = 50 MG/M3, 40 D, THALLIUM (I) SULFATE	1777
*MYTILUS EDULIS	10.9 BASED ON WET WEIGHT	WATER =100 MG/M3, 40 D, THALLIUM (I) SULFATE	1777

TABLE D206

CONCENTRATION FACTOR : TH-232			
ITEM:	VALUE:	CONDITIONS:	REF.:
*FISH, BONE	190 BASED ON WET WEIGHT	DANUBE, 1973, NUCLEAR POWER PLANT SINCE 1967	658

TABLE D206

CONCENTRATION FACTOR : TH-232			
ITEM:	VALUE:	CONDITIONS:	REF.:
*FISH, MUSCLE	80 (RANGE = 40 - 110) BASED ON WET WEIGHT	DANUBE, 1973, NUCLEAR POWER PLANT SINCE 1967	658
*FRESHWATER SEDIMENT	3160 (RANGE = 810 - 5950) BASED ON WET WEIGHT	DANUBE, 1973, NUCLEAR POWER PLANT SINCE 1967	658
*SUSPENDED SOLIDS	1900 (RANGE = 280 - 5000) BASED ON WET WEIGHT	DANUBE, 1973, NUCLEAR POWER PLANT SINCE 1967	658
*WATER PLANT	125 (RANGE = 90 - 180) BASED ON WET WEIGHT	DANUBE, 1973, NUCLEAR POWER PLANT SINCE 1967	658

TABLE D207

CONCENTRATION FACTOR : TI			
ITEM:	VALUE:	CONDITIONS:	REF.:
*AGARUM CRIBROSUM	23000 BASED ON DRY WEIGHT	SEAWATER=1.0*10**(-7) G/L, VOSTOK BAY, SEA OF JAPAN	1367
*BROWN ALGAE	3000 BASED ON WET WEIGHT	NONE	157
*CHONDRUS YENDOI	500 BASED ON DRY WEIGHT	SEAWATER=1.0*10**(-7) G/L, VOSTOK BAY, SEA OF JAPAN	1367
*CHORDA FILUM	1200 BASED ON DRY WEIGHT	SEAWATER=1.0*10**(-7) G/L, VOSTOK BAY, SEA OF JAPAN	1367
*CODIUM YESSOENSIS	3200 BASED ON DRY WEIGHT	SEAWATER=1.0*10**(-7) G/L, VOSTOK BAY, SEA OF JAPAN	1367
*COSTARIA COSTATA	900 BASED ON DRY WEIGHT	SEAWATER=1.0*10**(-7) G/L, VOSTOK BAY, SEA OF JAPAN	1367
*ENTEROMORPHA PROLIFERA	1900 BASED ON DRY WEIGHT	SEAWATER=1.0*10**(-7) G/L, VOSTOK BAY, SEA OF JAPAN	1367
*HETEROCHORDARIA ABIETINA	6000 BASED ON DRY WEIGHT	SEAWATER=1.0*10**(-7) G/L, VOSTOK BAY, SEA OF JAPAN	1367
*LAMINARIA CHICHORIOIDES	900 BASED ON DRY WEIGHT	SEAWATER=1.0*10**(-7) G/L, VOSTOK BAY, SEA OF JAPAN	1367
*LAMINARIA JAPONICA	200 BASED ON DRY WEIGHT	SEAWATER=1.0*10**(-7) G/L, VOSTOK BAY, SEA OF JAPAN	1367
*PELVETIA WRIGHTII	800 BASED ON DRY WEIGHT	SEAWATER=1.0*10**(-7) G/L, VOSTOK BAY, SEA OF JAPAN	1367
*PHYLLOSPADIX IWATENSIS	15700 BASED ON DRY WEIGHT	SEAWATER=1.0*10**(-7) G/L, VOSTOK BAY, SEA OF JAPAN	1367
*PHYTOPLANKTON	2700 BASED ON WET WEIGHT	MARINE ORGANISMS, TYPICAL VALUE	986
*PLANKTON	20000 BASED ON WET WEIGHT	NONE	157
*POLYSIPHONIA JAPONICA	37000 BASED ON DRY WEIGHT	SEAWATER=1.0*10**(-7) G/L, VOSTOK BAY, SEA OF JAPAN	1367
*PTILOTA FILICINA	44000 BASED ON DRY WEIGHT	SEAWATER=1.0*10**(-7) G/L, VOSTOK BAY, SEA OF JAPAN	1367
*RHODOMELA LARIX	35000 BASED ON DRY WEIGHT	SEAWATER=1.0*10**(-7) G/L, VOSTOK BAY, SEA OF JAPAN	1367
*SARGASSUM PALLIDUM	5300 BASED ON DRY WEIGHT	SEAWATER=1.0*10**(-7) G/L, VOSTOK BAY, SEA OF JAPAN	1367
*ULVA FENESTRATA	9400 BASED ON DRY WEIGHT	SEAWATER=1.0*10**(-7) G/L, VOSTOK BAY, SEA OF JAPAN	1367
*ULVARIA SPENDENS	18000 BASED ON DRY WEIGHT	SEAWATER=1.0*10**(-7) G/L, VOSTOK BAY, SEA OF JAPAN	1367
*ZOSTERA ASIATICA	11000 BASED ON DRY WEIGHT	SEAWATER=1.0*10**(-7) G/L, VOSTOK BAY, SEA OF JAPAN	1367

TABLE D208

CONCENTRATION FACTOR : TOXAPHENE			
ITEM:	VALUE:	CONDITIONS:	REF.:
*ASPERGILLUS SP.	1.7*10**4 BASED ON WET WEIGHT	EQUILIBRIUM CF., 301 K, 24 H	1212

TABLE D208

CONCENTRATION FACTOR : TOXAPHENE			
ITEM:	VALUE:	CONDITIONS:	REF.:
*BACILLUS SUBTILIS	3.4*10**3 BASED ON WET WEIGHT	EQUILIBRIUM CF., 301 K, 24 H	1212
*CHLORELLA PYRENOIDOSA	1.7*10**4 BASED ON WET WEIGHT	EQUILIBRIUM CF., 288 K, 200 H, 170 FTC	1212
*FLAVOBACTERIUM HARRISONII	5.2*10**3 BASED ON WET WEIGHT	EQUILIBRIUM CF., 301 K, 24 H	1212
*SESTON	6.6*10**3 BASED ON WET WEIGHT	EQUILIBRIUM CF., 301 K, 24 H	1212

TABLE D209

CONCENTRATION FACTOR : U			
ITEM:	VALUE:	CONDITIONS:	REF.:
*BROWN ALGAE	10 BASED ON WET WEIGHT	NONE	994
*CORAL	1.35 BASED ON CA CONTENT (F023)	DEEP OCEAN CORAL, MEAN OF 8 DETERMINATIONS	939
*CORAL	1.36 BASED ON CA CONTENT (F023)	SHALLOW OPEN OCEAN CORAL, MEAN OF 8 DETERMINATIONS	939
*CORAL	1.08 BASED ON CA CONTENT (F023)	SHALLOW COASTAL OCEAN, MEAN OF 10 DETERMINATIONS	939

TABLE D210

CONCENTRATION FACTOR : V			
ITEM:	VALUE:	CONDITIONS:	REF.:
*AGARUM CRIBROSUM	1100 BASED ON DRY WEIGHT	SEAWATER=4.4*10**(-7) G/L, VOSTOK BAY, SEA OF JAPAN	1367
*BROWN ALGAE	250 BASED ON WET WEIGHT	NONE	157
*CHONDRUS YENDOI	800 BASED ON DRY WEIGHT	SEAWATER=4.4*10**(-7) G/L, VOSTOK BAY, SEA OF JAPAN	1367
*CHORDA FILUM	370 BASED ON DRY WEIGHT	SEAWATER=4.4*10**(-7) G/L, VOSTOK BAY, SEA OF JAPAN	1367
*CODIUM YESSOENSIS	330 BASED ON DRY WEIGHT	SEAWATER=4.4*10**(-7) G/L, VOSTOK BAY, SEA OF JAPAN	1367
*HETEROCHORDARIA ABIETINA	300 BASED ON DRY WEIGHT	SEAWATER=4.4*10**(-7) G/L, VOSTOK BAY, SEA OF JAPAN	1367
*LAMINARIA CHICHORIOIDES	80 BASED ON DRY WEIGHT	SEAWATER=4.4*10**(-7) G/L, VOSTOK BAY, SEA OF JAPAN	1367
*LAMINARIA JAPONICA	120 BASED ON DRY WEIGHT	SEAWATER=4.4*10**(-7) G/L, VOSTOK BAY, SEA OF JAPAN	1367
*PELVETIA WRIGHTII	300 BASED ON DRY WEIGHT	SEAWATER=4.4*10**(-7) G/L, VOSTOK BAY, SEA OF JAPAN	1367
*PHYLLOSPADIX IWATENSIS	1100 BASED ON DRY WEIGHT	SEAWATER=4.4*10**(-7) G/L, VOSTOK BAY, SEA OF JAPAN	1367
*PLANKTON	620 BASED ON WET WEIGHT	NONE	157
*POLYSIPHONIA JAPONICA	3500 BASED ON DRY WEIGHT	SEAWATER=4.4*10**(-7) G/L, VOSTOK BAY, SEA OF JAPAN	1367
*PTILOTA FILICINA	1300 BASED ON DRY WEIGHT	SEAWATER=4.4*10**(-7) G/L, VOSTOK BAY, SEA OF JAPAN	1367
*RHODOMELA LARIX	600 BASED ON DRY WEIGHT	SEAWATER=4.4*10**(-7) G/L, VOSTOK BAY, SEA OF JAPAN	1367
*SARGASSUM PALLIDUM	130 BASED ON DRY WEIGHT	SEAWATER=4.4*10**(-7) G/L, VOSTOK BAY, SEA OF JAPAN	1367
*ULVA FENESTRATA	120 BASED ON DRY WEIGHT	SEAWATER=4.4*10**(-7) G/L, VOSTOK BAY, SEA OF JAPAN	1367
*ULVARIA SPENDENS	400 BASED ON DRY WEIGHT	SEAWATER=4.4*10**(-7) G/L, VOSTOK BAY, SEA OF JAPAN	1367

TABLE D211

CONCENTRATION FACTOR : W			
ITEM:	VALUE:	CONDITIONS:	REF.:
*BROWN ALGAE	87 BASED ON WET WEIGHT	NONE	157

TABLE D211

CONCENTRATION FACTOR : W			
ITEM:	VALUE:	CONDITIONS:	REF.:
*PLANT	2-18 BASED ON WET WEIGHT	CONC.IN SOIL=2.7UG/G	1670

TABLE D212

CONCENTRATION FACTOR : W-185			
ITEM:	VALUE:	CONDITIONS:	REF.:
*ANODONTA NUTTALLIANA, SOFT PARTS	0.5 BASED ON WET WEIGHT	NO FOOD,292 K,PH=7.5, FRESHWAT=2.3 PART AND 1200 SOLUB P CI/ML, 7 D	629
*ANODONTA NUTTALLIANA, SOFT PARTS	0.8 BASED ON WET WEIGHT	NO FOOD,292 K,PH=7.5, FRESHWAT=34 PART AND 1900 SOLUB P CI/ML, 21 D	629
*ASTACUS SP., MUSCLE	LESS THAN 1 BASED ON WET WEIGHT	NO FOOD,292 K,PH=7.5, FRESHWAT=2.3 PART AND 1200 SOLUB P CI/ML, 7 D	629
*ASTACUS SP., MUSCLE	LESS THAN 1 BASED ON WET WEIGHT	NO FOOD,292 K,PH=7.5, FRESHWAT=34 PART AND 1900 SOLUB P CI/ML, 21 D	629
*ASTACUS SP., SOFT PARTS	0.7 BASED ON WET WEIGHT	NO FOOD,292 K,PH=7.5, FRESHWAT=2.3 PART AND 1200 SOLUB P CI/ML, 7 D	629
*ASTACUS SP., SOFT PARTS	4 BASED ON WET WEIGHT	NO FOOD,292 K,PH=7.5, FRESHWAT=2.3 PART. AND 16 SOLUB P CI/ML, 7 D	629
*ASTACUS SP., VISCERA	0.9 BASED ON WET WEIGHT	NO FOOD,292 K,PH=7.5, FRESHWAT=2.3 PART AND 1200 SOLUB P CI/ML, 7 D	629
*ASTACUS SP., VISCERA	5 BASED ON WET WEIGHT	NO FOOD,292 K,PH=7.5, FRESHWAT=34 PART AND 1900 SOLUB P CI/ML, 21 D	629
*CANCER PRODUCTUS, MUSCLE	0.8 BASED ON WET WEIGHT	NO FOOD,285 K,PH=8.2, SEAWATER=20 PART AND 1500 SOLUB P CI/ML, 20 D	629
*CANCER PRODUCTUS, SORT PARTS	10 BASED ON WET WEIGHT	NO FOOD,285 K,PH=8.2, SEAWATER=20 PART AND 1500 SOLUB P CI/ML, 20 D	629
*CANCER PRODUCTUS, VISCERA	0.5 BASED ON WET WEIGHT	NO FOOD,285 K,PH=8.2, SEAWATER=34 PARTT AND 1300 SOLUB P CI/ML, 6 D	629
*CANCER PRODUCTUS, VISCERA	15 BASED ON WET WEIGHT	NO FOOD,285 K,PH=8.2, SEAWATER=20 PART AND 1500 SOLUB P CI/ML, 20 D	629
*CARASSIUS AURATUS, MUSCLE	LESS THAN 1 BASED ON WET WEIGHT	NO FOOD,292 K,PH=7.5, FRESHWAT=2.3 PART AND 1200 SOLUB P CI/ML, 7 D	629
*CARASSIUS AURATUS, NONMUSCLE	0.7 BASED ON WET WEIGHT	NO FOOD,292 K,PH=7.5, FRESHWAT=2.3 PART AND 1200 SOLUB P CI/ML, 7 D	629
*CARASSIUS AURATUS, WHOLE BODY	0.6 BASED ON WET WEIGHT	NO FOOD,292 K,PH=7.5, FRESHWAT=34 PART AND 1900 SOLUB P CI/ML, 21 D	629
*GILLICHTHYS MIRABILIS, MUSCLE	LESS THAN 1 BASED ON WET WEIGHT	NO FOOD,285 K,PH=8.2, SEAWATER=34 PARTT AND 1300 SOLUB P CI/ML, 6 D	629
*GILLICHTHYS MIRABILIS, MUSCLE	0.2 BASED ON WET WEIGHT	NO FOOD,285 K,PH=8.2, SEAWATER=20 PART AND 1500 SOLUB P CI/ML, 20 D	629
*GILLICHTHYS MIRABILIS, NONMUSCLE	0.8 BASED ON WET WEIGHT	NO FOOD,285 K,PH=8.2, SEAWATER=34 PARTT AND 1300 SOLUB P CI/ML, 6 D	629
*GILLICHTHYS MIRABILIS, NONMUSCLE	0.5 BASED ON WET WEIGHT	NO FOOD,285 K,PH=8.2, SEAWATER=20 PART AND 1500 SOLUB P CI/ML, 20 D	629
*GILLICHTHYS MIRABILIS, WHOLE BODY	0.6 BASED ON WET WEIGHT	NO FOOD,285 K,PH=8.2, SEAWATER=20 PART AND 1500 SOLUB P CI/ML, 20 D	629
*GILLICHTHYS MIRABILIS, WHOLE BODY	0.5 BASED ON WET WEIGHT	NO FOOD,285 K,PH=8.2, SEAWATER=3.5 PART AND 6.9 SOLUB P CI/ML, 6 D	629
*MYTILUS EDULIS, SOFT PARTS	1.3 BASED ON WET WEIGHT	NO FOOD,285 K,PH=8.2, SEAWATER=34 PARTT AND 1300 SOLUB P CI/ML, 6 D	629
*SAXIDONUS GIGANTEUS, SOFT PARTS	0.4 BASED ON WET WEIGHT	NO FOOD,285 K,PH=8.2, SEAWATER=20 PART AND 1500 SOLUB P CI/ML, 20 D	804

TABLE D213

CONCENTRATION FACTOR : XMC			
ITEM:	VALUE:	CONDITIONS:	REF.:
*ICTALURUS PUNCTATUS	550 BASED ON DRY WEIGHT	MODEL ECOSYSTEM, 22 D, 296 K,WATER =2.0-35.8 PPB,SOIL=2.78 PPM	796

TABLE D213

CONCENTRATION FACTOR : XMC			
ITEM:	VALUE:	CONDITIONS:	REF.:
*LEMNA MINOR	2300 BASED ON DRY WEIGHT	MODEL ECOSYSTEM, 22 D, 296 K,WATER =2.0-35.8 PPB,SOIL=2.78 PPM	796
*OEDOGONIUM CARDIACUM	3050 BASED ON DRY WEIGHT	MODEL ECOSYSTEM, 22 D, 296 K,WATER =2.0-35.8 PPB,SOIL=2.78 PPM	796
*PHYSA SP.	260 BASED ON DRY WEIGHT	MODEL ECOSYSTEM, 22 D, 296 K,WATER =2.0-35.8 PPB,SOIL=2.78 PPM	796
*PROCAMBARUSSP.	104 BASED ON DRY WEIGHT	MODEL ECOSYSTEM, 22 D, 296 K,WATER =2.0-35.8 PPB,SOIL=2.78 PPM	796

TABLE D214

CONCENTRATION FACTOR : Y			
ITEM:	VALUE:	CONDITIONS:	REF.:
*BROWN ALGAE	100-1000 BASED ON WET WEIGHT	NONE	994
*32 FRESHWATER PLANT	6880	ALGAE,BRYOPHYTES AND VASCULAR HYDROPHYTES	1597

TABLE D215

CONCENTRATION FACTOR : YB			
ITEM:	VALUE:	CONDITIONS:	REF.:
*ENGRAULIS JAPONICUS, WHOLE BODY	210 BASED ON WET WEIGHT	REL. TO TOTAL CONC.,TOT CONC=5.6 NG/L, SOLUB=2.1 NG/L,SEAWAT,JAP.	1535
*ENGRAULIS JAPONICUS, WHOLE BODY	710 BASED ON WET WEIGHT	REL. TO SOLUBLE CONC.,TOT CONC=5.6 NG/L, SOLUB=2.1 NG/L,SEAWAT,JAP.	1535
*HIZIKIA FUSIFORME, WHOLE BODY	94 BASED ON WET WEIGHT	REL. TO TOTAL CONC.,TOT CONC=5.6 NG/L, SOLUB=2.1 NG/L,SEAWAT,JAP.	1535
*HIZIKIA FUSIFORME, WHOLE BODY	320 BASED ON WET WEIGHT	REL. TO SOLUBLE CONC.,TOT CONC=5.6 NG/L, SOLUB=2.1 NG/L,SEAWAT,JAP.	1535
*MERETRIX LUSORIA, SOFT PART	750 BASED ON WET WEIGHT	REL. TO TOTAL CONC.,TOT CONC=5.6 NG/L, SOLUB=2.1 NG/L,SEAWAT,JAP.	1535
*MERETRIX LUSORIA, SOFT PART	2590 BASED ON WET WEIGHT	REL. TO SOLUBLE CONC.,TOT CONC=5.6 NG/L, SOLUB=2.1 NG/L,SEAWAT,JAP.	1535
*PARALICHTHYS OLIVACEUS, MUSCLE	39 BASED ON WET WEIGHT	REL. TO TOTAL CONC.,TOT CONC=5.6 NG/L, SOLUB=2.1 NG/L,SEAWAT,JAP.	1535
*PARALICHTHYS OLIVACEUS, MUSCLE	130 BASED ON WET WEIGHT	REL. TO SOLUBLE CONC.,TOT CONC=5.6 NG/L, SOLUB=2.1 NG/L,SEAWAT,JAP.	1535
*SARGASSUM FULVELLUM, WHOLE BODY	140 BASED ON WET WEIGHT	REL. TO TOTAL CONC.,TOT CONC=5.6 NG/L, SOLUB=2.1 NG/L,SEAWAT,JAP.	1535
*SARGASSUM FULVELLUM, WHOLE BODY	460 BASED ON WET WEIGHT	REL. TO SOLUBLE CONC.,TOT CONC=5.6 NG/L, SOLUB=2.1 NG/L,SEAWAT,JAP.	1535
*SERIOLA QUINQUERADIATA, MUSCLE	54 BASED ON WET WEIGHT	REL. TO TOTAL CONC.,TOT CONC=5.6 NG/L, SOLUB=2.1 NG/L,SEAWAT,JAP.	1535
*SERIOLA QUINQUERADIATA, MUSCLE	190 BASED ON WET WEIGHT	REL. TO SOLUBLE CONC.,TOT CONC=5.6 NG/L, SOLUB=2.1 NG/L,SEAWAT,JAP.	1535
*ULVA PERTUSA, WHOLE BODY	230 BASED ON WET WEIGHT	REL. TO TOTAL CONC.,TOT CONC=5.6 NG/L, SOLUB=2.1 NG/L,SEAWAT,JAP.	1535
*ULVA PERTUSA, WHOLE BODY	800 BASED ON WET WEIGHT	REL. TO SOLUBLE CONC.,TOT CONC=5.6 NG/L, SOLUB=2.1 NG/L,SEAWAT,JAP.	1535
*UNDARIA PINNATIFIDA, WHOLE BODY	85 BASED ON WET WEIGHT	REL. TO TOTAL CONC.,TOT CONC=5.6 NG/L, SOLUB=2.1 NG/L,SEAWAT,JAP.	1535
*UNDARIA PINNATIFIDA, WHOLE BODY	290 BASED ON WET WEIGHT	REL. TO SOLUBLE CONC.,TOT CONC=5.6 NG/L, SOLUB=2.1 NG/L,SEAWAT,JAP.	1535

TABLE D216

CONCENTRATION FACTOR : YT-90			
ITEM:	VALUE:	CONDITIONS:	REF.:
*ULVA LACTUCA	550 BASED ON WET WEIGHT	EQUILIBRIUM AFTER 3 H, PH=8, CARRIER FREE	603

TABLE D217

CONCENTRATION FACTOR : ZN

ITEM:	VALUE:	CONDITIONS:	REF.:
*AGARUM CRIBROSUM	200 BASED ON DRY WEIGHT	SEAWATER=66.6*10**(-7) G/L, VOSTOK BAY, SEA OF JAPAN	1367
*ARTHROBACTER MARINUS	200 BASED ON DRY WEIGHT	TOT.CONC.=350PPB,ADDED ZN=0PPB,293 K,	786
*ARTHROBACTER MARINUS	97 BASED ON DRY WEIGHT	TOT.CONC.=1050PPB,ADDED ZN=654PPB, 293 K,	786
*ARTHROBACTER MARINUS	55 BASED ON DRY WEIGHT	TOT.CONC.=3650PPB,ADDED ZN= 3269PPB,293 K,	786
*ARTHROBACTER MARINUS	48 BASED ON DRY WEIGHT	TOT.CONC.=6960PPB,ADDED ZN= 6538PPB,293 K,	786
*ARTHROBACTER MARINUS	72 BASED ON DRY WEIGHT	TOT.CONC.=13390PPB,ADDED ZN= 13076PPB,293 K,	786
*ARTHROBACTER MARINUS	202 BASED ON DRY WEIGHT	TOT.CONC.=18350PPB,ADDED ZN= 19614PPB,293 K,	786
*BRASSICA OLERACEA	41 BASED ON WET WEIGHT	CONC.IN SOIL SOLUTION=0.2PPM	157
*BROWN ALGAE	3400 BASED ON WET WEIGHT	NONE	157
*BROWN ALGAE	100-13000 BASED ON WET WEIGHT	NONE	994
*CANCER IRRORATUS	16599 (SD=+-1368) BASED ON WET WEIGHT	SEAWATER CONC=1.77*10**(-3) PPM, NOVA SCOTIA, AAS	984
*CARASSIUS AURATUS	75 BASED ON WET WEIGHT	WATER = 0.67 - 0.22 PPM	1093
*CARCINUS SP.	520 BASED ON WET WEIGHT	6=N, UPTAKE FROM WATER=23.3 - 117.1 UG/L. 96 D	1316
*CARCINUS SP.	450 BASED ON WET WEIGHT	6=N, UPTAKE FROM WATER AND FOOD, WATER=23.3-117.1 UG/L, 96 D	1316
*CHONDRUS YENDOI	260 BASED ON DRY WEIGHT	SEAWATER=66.6*10**(-7) G/L, VOSTOK BAY, SEA OF JAPAN	1367
*CHORDA FILUM	60 BASED ON DRY WEIGHT	SEAWATER=66.6*10**(-7) G/L, VOSTOK BAY, SEA OF JAPAN	1367
*CLADOPHORA GLOMERATA	2.9*10**3 BASED ON DRY WEIGHT	WATER FROM DEADMAN BAY (L.ONTARIO) =7.7+-3.1, (CONC. IN NG/ML+-S.D.)	806
*CLADOPHORA GLOMERATA	1*10**3 BASED ON DRY WEIGHT	WATER FROM MAIN DUCK ISL. (L.ONTARIO)=1.7+-1.2, (CONC. IN NG/ML+-S.D.)	806
*CLADOPHORA GLOMERATA	49*10**3 BASED ON DRY WEIGHT	WATER FROM DEADMAN BAY (L.ONTARIO) =0.6+-0.2, (CONC. IN NG/ML+-S.D.)	806
*CODIUM YESSOENSIS	540 BASED ON DRY WEIGHT	SEAWATER=66.6*10**(-7) G/L, VOSTOK BAY, SEA OF JAPAN	1367
*CORAL	0.71 BASED ON CA CONTENT (F023)	SHALLOW OPEN OCEAN CORAL, MEAN OF 13 DETERMINATIONS	939
*CORAL	5.04 BASED ON CA CONTENT (F023)	SHALLOW OPEN OCEAN CORAL, MEAN OF 8 DETERMINATIONS	939
*CORAL	0.40 BASED ON CA CONTENT (F023)	SHALLOW OPEN OCEAN CORAL, MEAN OF 13 DETERMINATIONS	939
*COSTARIA COSTATA	60 BASED ON DRY WEIGHT	SEAWATER=66.6*10**(-7) G/L, VOSTOK BAY, SEA OF JAPAN	1367
*CRASSOSTREA GIGAS	14000+-9700 BASED ON WET WEIGHT	WILLAPA BAY	1419
*CRASSOSTREA GRYPHOIDES	5500 - 13000 BASED ON WET WEIGHT	BOMBAY	116
*ENTEROMORPHA PROLIFERA	500 BASED ON DRY WEIGHT	SEAWATER=66.6*10**(-7) G/L, VOSTOK BAY, SEA OF JAPAN	1367
*FISH	533 BASED ON WET WEIGHT	MARINE ORGANISMS, TYPICAL VALUE	538
*FISH	228 BASED ON WET WEIGHT	FRESHWATER ORGANISMS, TYPICAL VALUE	1028
*FRESHWATER PLANKTON	61000 BASED ON WET WEIGHT	CONC.IN WATER=1.8UG/L,SCHOEHSEE	573
*FUCUS SERRATUS	1.1*10**4 BASED ON WET WEIGHT	STEADY STATE CF., 283 K, ZN=5-25 MG/M3, MARCH	1764
*FUCUS SERRATUS	7.4*10**3 BASED ON WET WEIGHT	STEADY STATE CF., 283 K, ZN=5-25 MG/M3, JULY	1764
*FUCUS SERRATUS	7.2*10**3 BASED ON WET WEIGHT	STEADY STATE CF., 283 K, ZN=5-25 MG/M3, NOVEMBER	1764
*FUCUS VESICULOSUS	2*10**4 BASED ON DRY WEIGHT	18 OBS.,ALL AREAS AROUND THE BRITISH ISLES	1278
*GOBIUS SP.	1400 BASED ON WET WEIGHT	5=N, UPTAKE FROM WATER=23.3 - 117.1 UG/L. 96 D	1316
*GOBIUS SP.	900 BASED ON WET WEIGHT	4=N, UPTAKE FROM WATER AND FOOD, WATER=23.3-117.1 UG/L, 96 D	1316

TABLE D217

CONCENTRATION FACTOR : ZN			
ITEM:	VALUE:	CONDITIONS:	REF.:
*HETEROCHORDARIA ABIETINA	150 BASED ON DRY WEIGHT	SEAWATER=66.6*10**(-7) G/L, VOSTOK BAY, SEA OF JAPAN	1367
*LAMINARIA CHICHORIOIDES	150 BASED ON DRY WEIGHT	SEAWATER=66.6*10**(-7) G/L, VOSTOK BAY, SEA OF JAPAN	1367
*LAMINARIA JAPONICA	70 BASED ON DRY WEIGHT	SEAWATER=66.6*10**(-7) G/L, VOSTOK BAY, SEA OF JAPAN	1367
*LYSMATA SETICAUDATA	2020 BASED ON WET WEIGHT	6=N, UPTAKE FROM WATER=23.3 - 117.1 UG/L. 96 D	1316
*LYSMATA SETICAUDATA	3200 BASED ON WET WEIGHT	6=N, UPTAKE FROM WATER AND FOOD, WATER=23.3-117.1 UG/L, 96 D	1316
*MACROPHYTES	7000 BASED ON WET WEIGHT	MARINE ORGANISMS, TYPICAL VALUE	589
*MACROPHYTES	318 BASED ON WET WEIGHT	FRESHWATER ORGANISMS, TYPICAL VALUE	1028
*MOLLUSCS	27300 BASED ON WET WEIGHT	MARINE ORGANISMS, TYPICAL VALUE	1280
*MOLLUSCS	2258 BASED ON WET WEIGHT	FRESHWATER ORGANISMS, TYPICAL VALUE	993
*MYTILUS EDULIS	75 BASED ON WET WEIGHT	STEADY STATE CF., WATER=0.750 MG ZN/L	217
*MYTILUS GALLOPROVINCIALIS	580 BASED ON WET WEIGHT	6=N, UPTAKE FROM WATER=23.3 - 117.1 UG/L. 96 D	1316
*PATELLA VULGATA	1.9*10**4 BASED ON DRY WEIGHT	17 OBS.,ALL AREAS AROUND THE BRITISH ISLES	1278
*PELVETIA WRIGHTII	20 BASED ON DRY WEIGHT	SEAWATER=66.6*10**(-7) G/L, VOSTOK BAY, SEA OF JAPAN	1367
*PHYLLOSPADIX IWATENSIS	200 BASED ON DRY WEIGHT	SEAWATER=66.6*10**(-7) G/L, VOSTOK BAY, SEA OF JAPAN	1367
*PHYTOPLANKTON	8900-75600 BASED ON WET WEIGHT	NONE	311
*PHYTOPLANKTON	5500 BASED ON WET WEIGHT	MARINE ORGANISMS, TYPICAL VALUE	986
*PLANKTON	65000 BASED ON WET WEIGHT	NONE	157
*PLANKTON	1000 BASED ON WET WEIGHT	NONE	994
*POLYSIPHONIA JAPONICA	300 BASED ON DRY WEIGHT	SEAWATER=66.6*10**(-7) G/L, VOSTOK BAY, SEA OF JAPAN	1367
*PORPHYRA UMBILICALIS	1*10**4 BASED ON DRY WEIGHT	20 OBS.,ALL AREAS AROUND THE BRITISH ISLES	1278
*PTILOTA FILICINA	480 BASED ON DRY WEIGHT	SEAWATER=66.6*10**(-7) G/L, VOSTOK BAY, SEA OF JAPAN	1367
*RHODOMELA LARIX	230 BASED ON DRY WEIGHT	SEAWATER=66.6*10**(-7) G/L, VOSTOK BAY, SEA OF JAPAN	1367
*SARGASSUM PALLIDUM	40 BASED ON DRY WEIGHT	SEAWATER=66.6*10**(-7) G/L, VOSTOK BAY, SEA OF JAPAN	1367
*SESTON	1.5 BASED ON WET WEIGHT	DEPTH = 8 M, SCH@HSEE	573
*SESTON	1.3 BASED ON WET WEIGHT	DEPTH = 17 M, SCH@HSEE	573
*ULVA FENESTRATA	30 BASED ON DRY WEIGHT	SEAWATER=66.6*10**(-7) G/L, VOSTOK BAY, SEA OF JAPAN	1367
*ULVARIA SPENDENS	100 BASED ON DRY WEIGHT	SEAWATER=66.6*10**(-7) G/L, VOSTOK BAY, SEA OF JAPAN	1367
*ZOOPLANKTON	5100 BASED ON WET WEIGHT	SEAWATER CONCENTRATION 0.01 MG/M3	985
*ZOOPLANKTON	8800 BASED ON WET WEIGHT	MARINE ORGANISMS, TYPICAL VALUE	986
*32 FRESHWATER PLANT	4600 BASED ON DRY WEIGHT	ALGAE,BRYOPHYTES AND VASCULAR HYDROPHYTES	1597

TABLE D218

CONCENTRATION FACTOR : ZN-65			
ITEM:	VALUE:	CONDITIONS:	REF.:
*ALGAE	290 BASED ON WET WEIGHT	MARINE	671
*ALGAE	290 BASED ON WET WEIGHT	MARINE	671
*CARCINUS SP.	130 BASED ON WET WEIGHT	4=N, UPTAKE FROM WATER=23.3 - 117.1 UG/L. 96 D	1316

TABLE D218

CONCENTRATION FACTOR : ZN-65			
ITEM:	VALUE:	CONDITIONS:	REF.:
*CARCINUS SP.	210 BASED ON WET WEIGHT	5=N, UPTAKE FROM WATER AND FOOD, WATER=23.3-117.1 UG/L, 96 D	1316
*CHIRONOMID LARVAE	1700 BASED ON WET WEIGHT	LAKE MALAR, SWEDEN, WATER = 3*10** (-6) UCI/ML	609
*CRASSOSTREA GIGAS	9900+-2700 BASED ON WET WEIGHT	WILLAPA BAY	1419
*CRASSOSTREA GIGAS	0.56 - 1.2*10**4 BASED ON DRY WEIGHT	13 MOUNTHS, WATER = 1.04-19.63*10**(-2) P CI/ML, 1962	1374
*CRASSOSTREA GIGAS	0.86 - 1.1*10**4 BASED ON DRY WEIGHT	13 MOUNTHS, WATER = 1.04-19.63*10**(-2) P CI/ML, 1964	1374
*CRUSTACEAN, VISCERA	GREATER THAN 500 BASED ON WET WEIGHT	MARINE	671
*CRUSTACEAN, VISCERA	GREATER THAN 500 BASED ON WET WEIGHT	MARINE	671
*ECHINODERM, DIGESTIVE TRACT	GREATER THAN 200 BASED ON WET WEIGHT	MARINE	671
*ECHINODERM, DIGESTIVE TRACT	GREATER THAN 200 BASED ON WET WEIGHT	MARINE	671
*FISH	2 - 20 BASED ON WET WEIGHT	42 D	794
*FISH, MUSCLE	510 (RANGE = 190 - 780) BASED ON WET WEIGHT	DANUBE, 1973, NUCLEAR POWER PLANT SINCE 1967	658
*FISH, VISCERA	22 BASED ON WET WEIGHT	MARINE	671
*FISH, VISCERA	22 BASED ON WET WEIGHT	MARINE	671
*FRESHWATER SEDIMENT	480 (RANGE = 230 - 740) BASED ON WET WEIGHT	DANUBE, 1973, NUCLEAR POWER PLANT SINCE 1967	658
*FRESHWATER SEDIMENT	20000 BASED ON WET WEIGHT	LAKE MALAR, SWEDEN, WATER = 3*10** (-6) UCI/ML	609
*GOBIUS SP.	10 BASED ON WET WEIGHT	4=N, UPTAKE FROM WATER=23.3 - 117.1 UG/L. 96 D	1316
*GOBIUS SP.	25 BASED ON WET WEIGHT	4=N, UPTAKE FROM WATER AND FOOD, WATER=23.3-117.1 UG/L, 96 D	1316
*LAMPSILIS RADIATA	4100 BASED ON WET WEIGHT	91 DAYS, FRESHWATER	639
*LEECHES	400 BASED ON WET WEIGHT	LAKE MALAR, SWEDEN, WATER = 3*10** (-6) UCI/ML	609
*LYSMATA SETICAUDATA	380 BASED ON WET WEIGHT	2=N, UPTAKE FROM WATER=23.3 - 117.1 UG/L. 96 D	1316
*LYSMATA SETICAUDATA	380 BASED ON WET WEIGHT	4=N, UPTAKE FROM WATER AND FOOD, WATER=23.3-117.1 UG/L, 96 D	1316
*MARINE ALGAE	1200 - 13000 BASED ON DRY WEIGHT	MAX CF. AT 500 FLC AND 291-298 K, NOT RELATED TO WATER CONC.	86
*MOLLUSCS, VICERA	68 BASED ON WET WEIGHT	MARINE	671
*MOLLUSCS, VICERA	68 BASED ON WET WEIGHT	MARINE	671
*MYTILUS GALLOPROVINCIALIS	170 BASED ON WET WEIGHT	6=N, UPTAKE FROM WATER=23.3 - 117.1 UG/L. 96 D	1316
*PIKE	1250 BASED ON WET WEIGHT	LAKE MALAR, SWEDEN, WATER = 3*10** (-6) UCI/ML	609
*SNAIL	590 BASED ON WET WEIGHT	LAKE MALAR, SWEDEN, WATER = 3*10** (-6) UCI/ML	609
*SUSPENDED SOLIDS	3030 (RANGE = 90 - 8820) BASED ON WET WEIGHT	DANUBE, 1973, NUCLEAR POWER PLANT SINCE 1967	658
*WATER PLANT	150 BASED ON WET WEIGHT	DANUBE, 1973, NUCLEAR POWER PLANT SINCE 1967	658

TABLE D219

CONCENTRATION FACTOR : ZR			
ITEM:	VALUE:	CONDITIONS:	REF.:
*BROWN ALGAE	350-1000 BASED ON WET WEIGHT	NONE	994
*PHYTOPLANKTON	1500-3000 BASED ON WET WEIGHT	NONE	311
*PLANKTON	1500-3000 BASED ON WET WEIGHT	NONE	994

TABLE D219

CONCENTRATION FACTOR : ZR			
ITEM:	VALUE:	CONDITIONS:	REF.:
*32 FRESHWATER PLANT	6230 BASED ON DRY WEIGHT	ALGAE,BRYOPHYTES AND VASCULAR HYDROPHYTES	1597

TABLE D220

CONCENTRATION FACTOR : ZR-95			
ITEM:	VALUE:	CONDITIONS:	REF.:
*FISH, BONE	500 (RANGE = 180 - 1220) BASED ON WET WEIGHT	DANUBE, 1973, NUCLEAR POWER PLANT SINCE 1967	658
*FISH, MUSCLE	160 (RANGE = 40 - 440) BASED ON WET WEIGHT	DANUBE, 1973, NUCLEAR POWER PLANT SINCE 1967	658
*FRESHWATER SEDIMENT	1430 (RANGE = 340 - 3570) BASED ON WET WEIGHT	DANUBE, 1973, NUCLEAR POWER PLANT SINCE 1967	658
*SUSPENDED SOLIDS	1250 (RANGE = 180 - 3600) BASED ON WET WEIGHT	DANUBE, 1973, NUCLEAR POWER PLANT SINCE 1967	658
*WATER PLANT	150 (RANGE = 65 - 460) BASED ON WET WEIGHT	DANUBE, 1973, NUCLEAR POWER PLANT SINCE 1967	658

TABLE D221

CONCENTRATION FACTOR : ZR-95/NB-95			
ITEM:	VALUE:	CONDITIONS:	REF.:
*ESKMEALS SILT	34637.7 BASED ON WET WEIGHT	YEARLY MEAN, CUMBERLAND COASTRAL AREA, 1969	1240
*INSHORE SAND	5268.7 BASED ON WET WEIGHT	YEARLY MEAN, CUMBERLAND COASTRAL AREA, 1969	1240
*NEPHTYS SP.	733.8 BASED ON WET WEIGHT	YEARLY MEAN, CUMBERLAND COASTRAL AREA, 1969	1240
*NUCULA SP.	15886.0 BASED ON WET WEIGHT	YEARLY MEAN, CUMBERLAND COASTRAL AREA, 1969	1240
*PLEURONECTES PLATESSA, GUT CONTENT	2670.6 BASED ON WET WEIGHT	YEARLY MEAN, CUMBERLAND COASTRAL AREA, 1969	1240

TABLE D222

CONCENTRATION FACTOR : 2,4-D			
ITEM:	VALUE:	CONDITIONS:	REF.:
*SCENEDESMUS ACUTUS	2.2 BASED ON DRY WEIGHT	WATER = 0.5*10**(-7) MOL, 303 K, PH=7-8, 8 H	141

TABLE D223

CONCENTRATION FACTOR : 2,4-DI-BROMOBIPHENYL			
ITEM:	VALUE:	CONDITIONS:	REF.:
*SALMO SALAR	1343 BASED ON WET WEIGHT	UPTAKE FROM WATER, STEADY STATE CF., 288 K, 48 H, 6.24 G FISH	1776
*SALMO SALAR	0.318 BASED ON WET WEIGHT	UPTAKE FROM FOOD, STEADY STATE CF., 288 K, 48 H, 6.24 G FISH	1776

TABLE D224

CONCENTRATION FACTOR : 2,4-DI-CHLOROBIPHENYL			
ITEM:	VALUE:	CONDITIONS:	REF.:
*SALMO SALAR	894 BASED ON WET WEIGHT	UPTAKE FROM WATER, STEADY STATE CF., 288 K, 48 H, 6.24 G FISH	1776

TABLE D224

CONCENTRATION FACTOR : 2,4-DI-CHLOROBIPHENYL			
ITEM:	VALUE:	CONDITIONS:	REF.:
*SALMO SALAR	0.054 BASED ON WET WEIGHT	UPTAKE FROM FOOD, STEADY STATE CF., 288 K, 48 H, 6.24 G FISH	1776

TABLE D225

CONCENTRATION FACTOR : 2,4,2`,5`-TETRA-CHLOROBIPHENYL			
ITEM:	VALUE:	CONDITIONS:	REF.:
*SALMO SALAR	389 BASED ON WET WEIGHT	UPTAKE FROM WATER, STEADY STATE CF., 288 K, 48 H, 6.24 G FISH	1776
*SALMO SALAR	0.414 BASED ON WET WEIGHT	UPTAKE FROM FOOD, STEADY STATE CF., 288 K, 48 H, 6.24 G FISH	1776

TABLE D226

CONCENTRATION FACTOR : 2,4,5-T			
ITEM:	VALUE:	CONDITIONS:	REF.:
*SCENEDESMUS ACUTUS	3.3 BASED ON DRY WEIGHT	WATER = 0.5*10**(-7) MOL, 303 K, PH=7-8, 8 H	141

TABLE D227

CONCENTRATION FACTOR : 2,5,3`,4`-TETRA-CHLOROBIPHENYL			
ITEM:	VALUE:	CONDITIONS:	REF.:
*SALMO SALAR	389 BASED ON WET WEIGHT	UPTAKE FROM WATER, STEADY STATE CF., 288 K, 48 H, 6.24 G FISH	1776
*SALMO SALAR	0.586 BASED ON WET WEIGHT	UPTAKE FROM FOOD, STEADY STATE CF., 288 K, 48 H, 6.24 G FISH	1776

TABLE D228

CONCENTRATION FACTOR : 2,5,4`-TRI-CHLOROBIPHENYL			
ITEM:	VALUE:	CONDITIONS:	REF.:
*SALMO SALAR	385 BASED ON WET WEIGHT	UPTAKE FROM WATER, STEADY STATE CF., 288 K, 48 H, 6.24 G FISH	1776
*SALMO SALAR	0.494 BASED ON WET WEIGHT	UPTAKE FROM FOOD, STEADY STATE CF., 288 K, 48 H, 6.24 G FISH	1776

TABLE D229

CONCENTRATION FACTOR : 2,6-DI-BROMOBIPHENYL			
ITEM:	VALUE:	CONDITIONS:	REF.:
*SALMO SALAR	1267 BASED ON WET WEIGHT	UPTAKE FROM WATER, STEADY STATE CF., 288 K, 48 H, 6.24 G FISH	1776
*SALMO SALAR	0.179 BASED ON WET WEIGHT	UPTAKE FROM FOOD, STEADY STATE CF., 288 K, 48 H, 6.24 G FISH	1776

TABLE D230

CONCENTRATION FACTOR : 2,6-DI-CHLOROBIPHENYL			
ITEM:	VALUE:	CONDITIONS:	REF.:
*SALMO SALAR	919 BASED ON WET WEIGHT	UPTAKE FROM WATER, STEADY STATE CF., 288 K, 48 H, 6.24 G FISH	1776
*SALMO SALAR	0.137 BASED ON WET WEIGHT	UPTAKE FROM FOOD, STEADY STATE CF., 288 K, 48 H, 6.24 G FISH	1776

TABLE D231

CONCENTRATION FACTOR : 2,6-DI-CHLOROBIPHENYL1			
ITEM:	VALUE:	CONDITIONS:	REF.:
*SALMO SALAR	425 BASED ON WET WEIGHT	UPTAKE FROM WATER, STEADY STATE CF., 288 K, 48 H, 6.24 G FISH	1776
*SALMO SALAR	0.449 BASED ON WET WEIGHT	UPTAKE FROM FOOD, STEADY STATE CF., 288 K, 48 H, 6.24 G FISH	1776

TABLE D232

CONCENTRATION FACTOR : 2,6-DI-CHLOROBIPHENYL2			
ITEM:	VALUE:	CONDITIONS:	REF.:
*SALMO SALAR	314 BASED ON WET WEIGHT	UPTAKE FROM WATER, STEADY STATE CF., 288 K, 48 H, 6.24 G FISH	1776
*SALMO SALAR	0.587 BASED ON WET WEIGHT	UPTAKE FROM FOOD, STEADY STATE CF., 288 K, 48 H, 6.24 G FISH	1776

TABLE D233

CONCENTRATION FACTOR : 2,6-DI-CHLOROBIPHENYL3			
ITEM:	VALUE:	CONDITIONS:	REF.:
*SALMO SALAR	111 BASED ON WET WEIGHT	UPTAKE FROM WATER, STEADY STATE CF., 288 K, 48 H, 6.24 G FISH	1776
*SALMO SALAR	0.571 BASED ON WET WEIGHT	UPTAKE FROM FOOD, STEADY STATE CF., 288 K, 48 H, 6.24 G FISH	1776

TABLE D234

CONCENTRATION FACTOR : 3,4-DI-BROMOBIPHENYL			
ITEM:	VALUE:	CONDITIONS:	REF.:
*SALMO SALAR	1343 BASED ON WET WEIGHT	UPTAKE FROM WATER, STEADY STATE CF., 288 K, 48 H, 6.24 G FISH	1776
*SALMO SALAR	0 BASED ON WET WEIGHT	UPTAKE FROM FOOD, STEADY STATE CF., 288 K, 48 H, 6.24 G FISH	1776

TABLE D235

CONCENTRATION FACTOR : 3,4-DI-CHLOROBIPHENYL			
ITEM:	VALUE:	CONDITIONS:	REF.:
*SALMO SALAR	358 BASED ON WET WEIGHT	UPTAKE FROM WATER, STEADY STATE CF., 288 K, 48 H, 6.24 G FISH	1776

TABLE D235

CONCENTRATION FACTOR : 3,4-DI-CHLOROBIPHENYL			
ITEM:	VALUE:	CONDITIONS:	REF.:
*SALMO SALAR	0 BASED ON WET WEIGHT	UPTAKE FROM FOOD, STEADY STATE CF., 288 K, 48 H, 6.24 G FISH	1776

Concentration factor of Cs-137 versus temperature and salinity. Concentration factor dpm cs-137 per gram dry weight algae/ dpm Cs-137 per ml sea water.
Temperature (y-axis) in celcius.
Salinity (x-axis) in ppt.

(ref: 1528)

The values on the following figures is given as the ln (natural logaritm) to the normal concentrations factors. This means that the value 0 on the figures is equal to the normal concentration factor-1; the value 1 on the figures is equal to the normal concentration factor 2.71 etc.

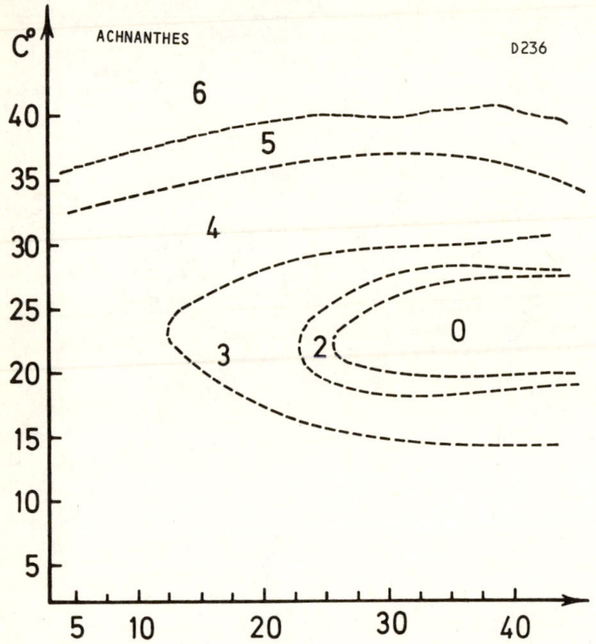

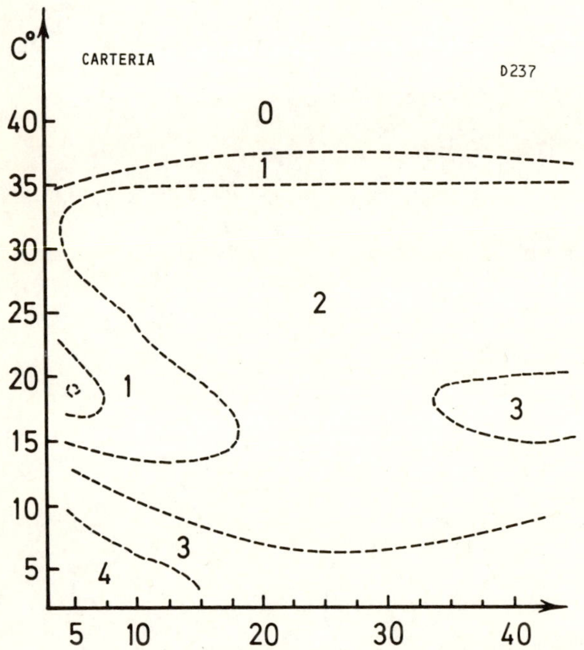

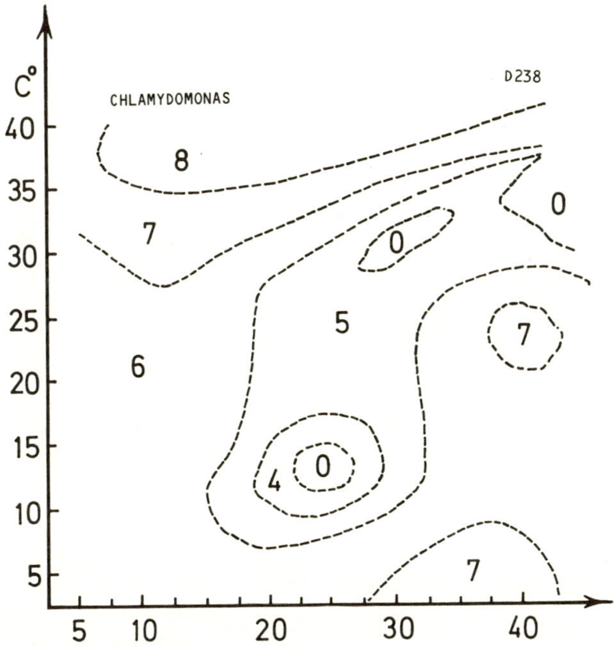
C°
CHLAMYDOMONAS
D238
40
35
30
25
20
15
10
5
5
10
20
30
40
8
7
0
0
5
7
6
4
0
7

see p. 700.

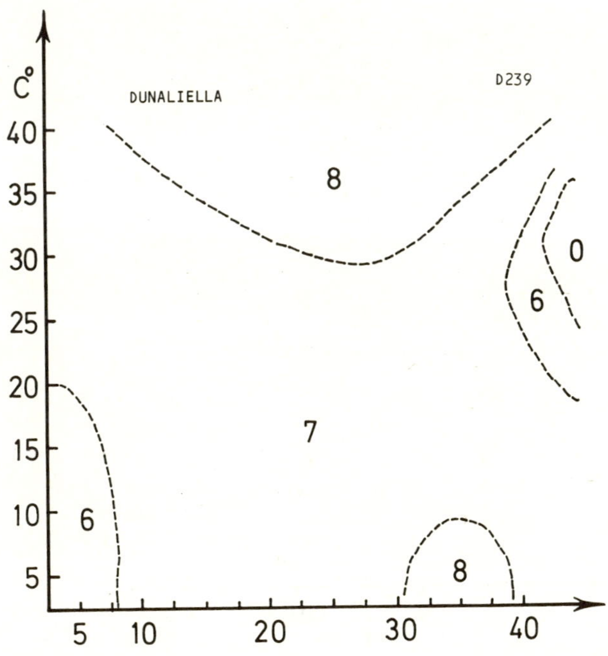
C°
DUNALIELLA
D239
40
35
30
25
20
15
10
5
5
10
20
30
40
8
0
6
7
6
8

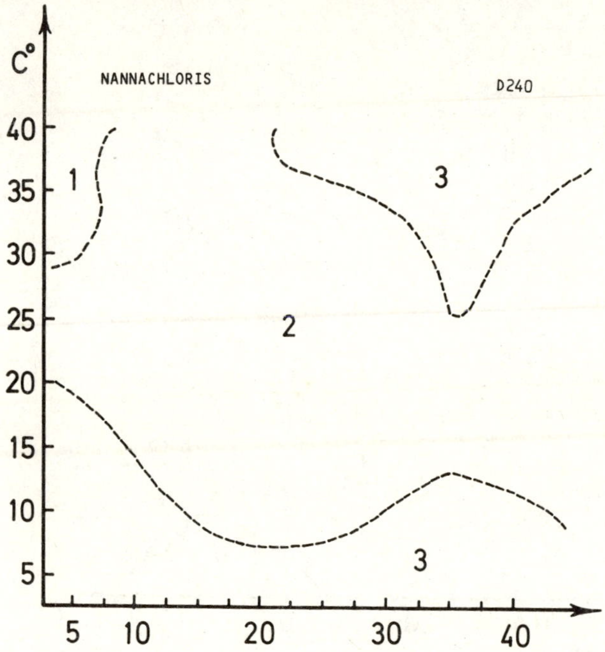
C°
NANNACHLORIS
D240
40
35
30
25
20
15
10
5
1
2
3
3
5
10
20
30
40

see p. 700.

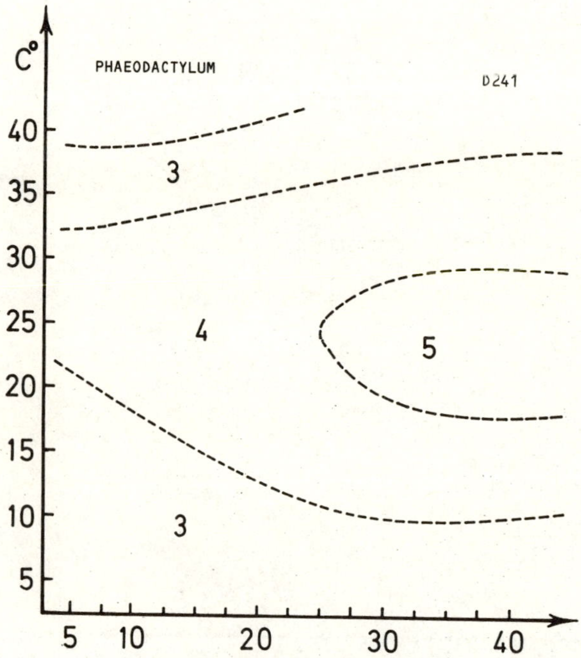
C°
PHAEODACTYLUM
D241
40
35
30
25
20
15
10
5
3
4
5
3
5
10
20
30
40

Zinc concentration factors (dpm ^{65}Zn per g dry wt algae/ dpm ^{65}Zn per ml sea water) for each algae on temperature (^{o}C) and salinity (ppt).

(ref: 1528)

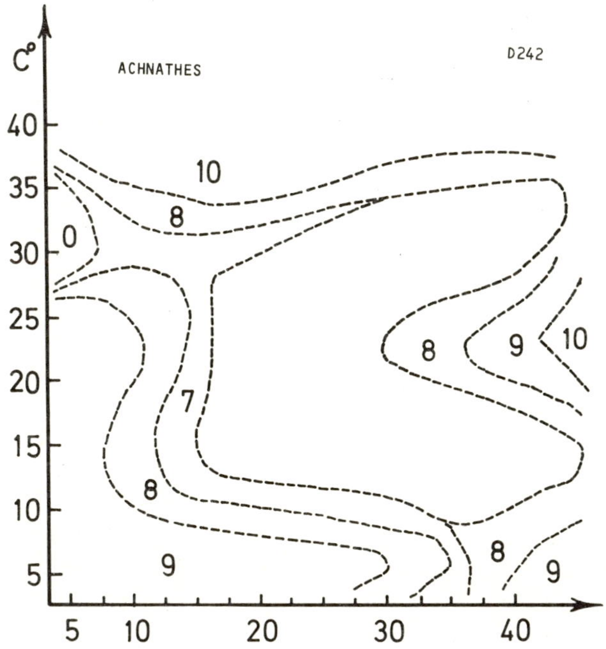

see p. 700.

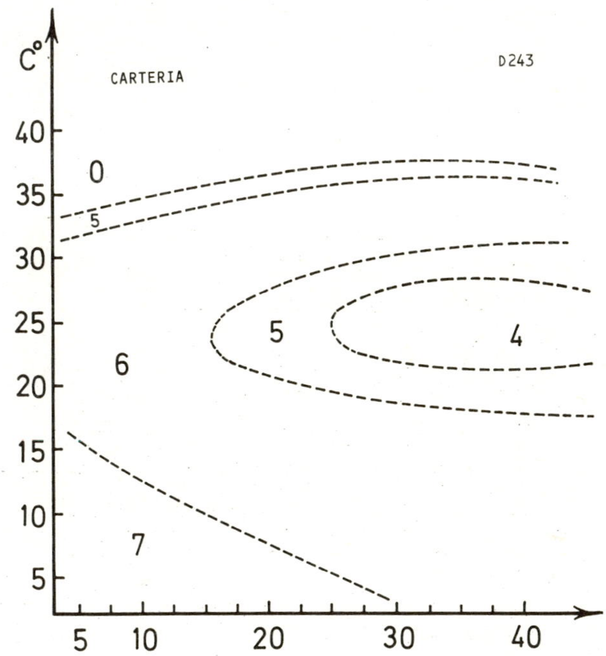

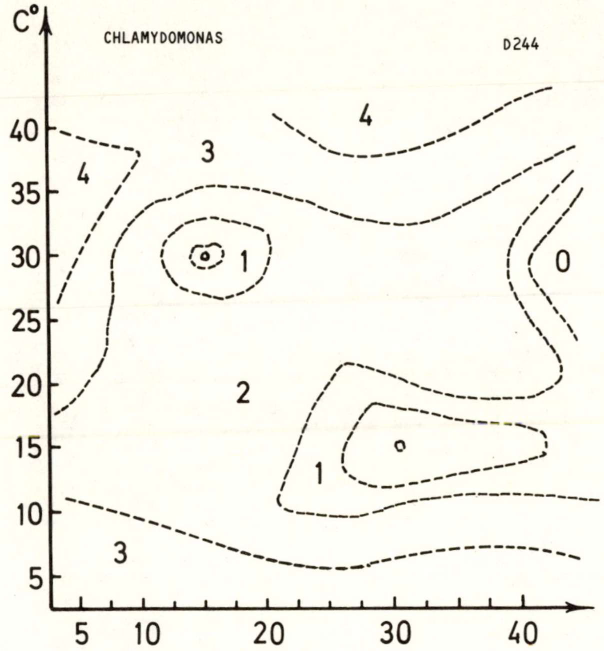
C°
CHLAMYDOMONAS
D244
4
3
4
1
0
2
1
3
40
35
30
25
20
15
10
5
5
10
20
30
40

see p. 700.

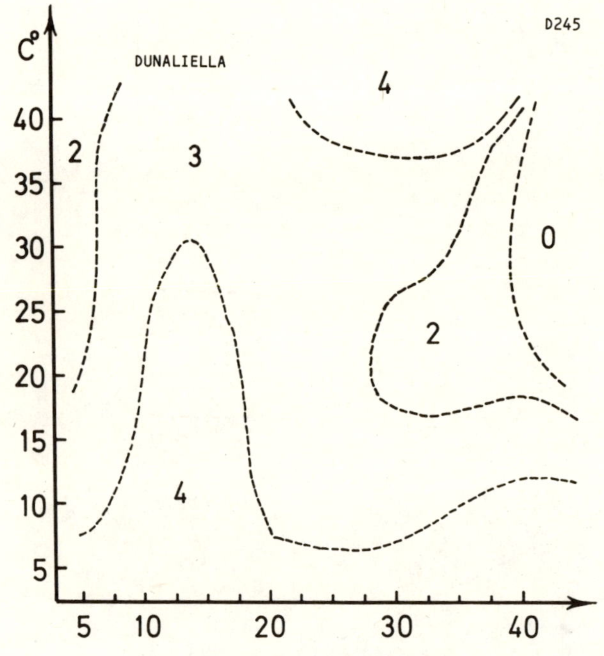
D245
C°
DUNALIELLA
4
2
3
0
2
4
40
35
30
25
20
15
10
5
5
10
20
30
40

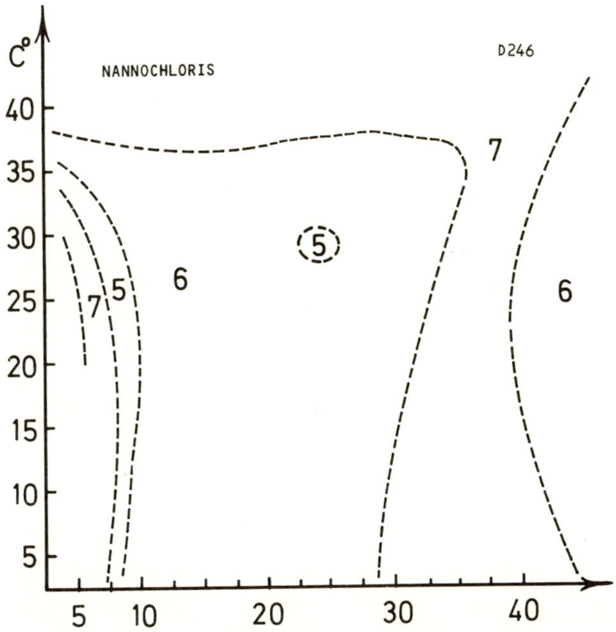
C°
NANNOCHLORIS
D246
40
35
30
25
20
15
10
5
7
5
6
(5)
7
6
5
10
20
30
40

see p. 700.

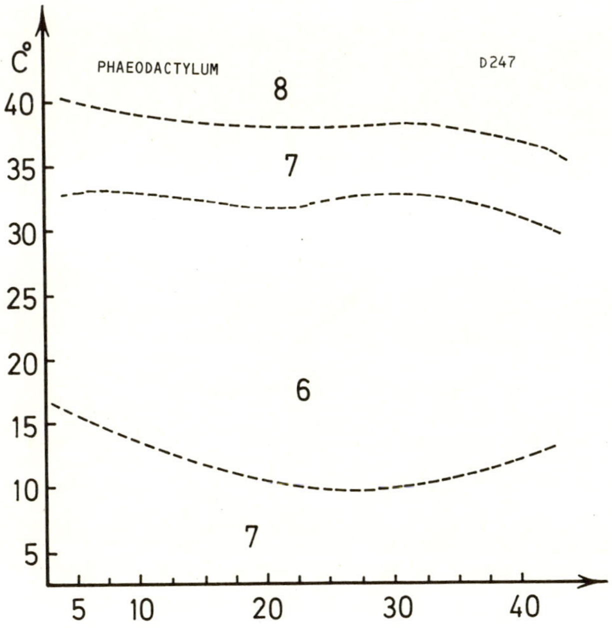
C°
PHAEODACTYLUM
D247
8
7
6
7
40
35
30
25
20
15
10
5
5
10
20
30
40

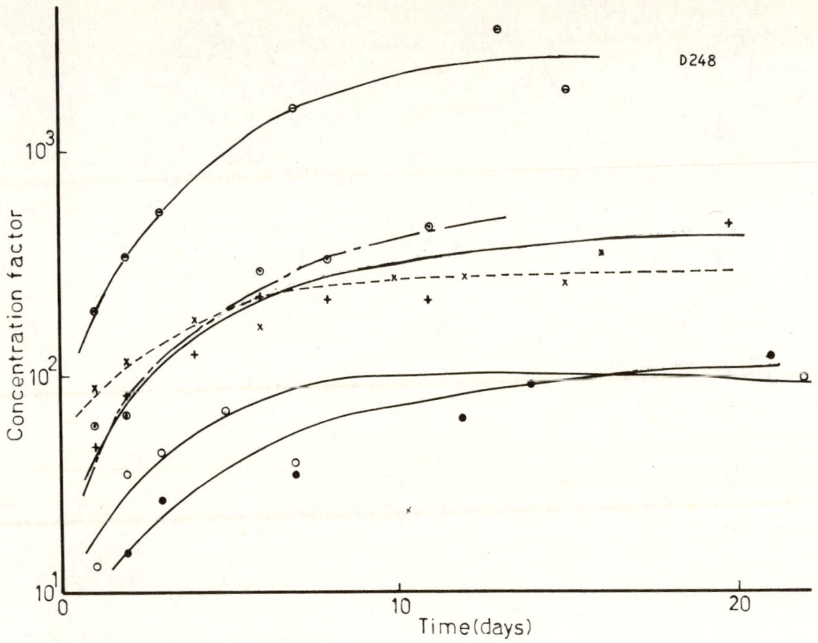

Concentration factor for Co-60 versus time for several kinds of seaweed. (ref: 1671)
S. thunberfii Ulva Annfelha
Chondrus Eisenia Laminaria.

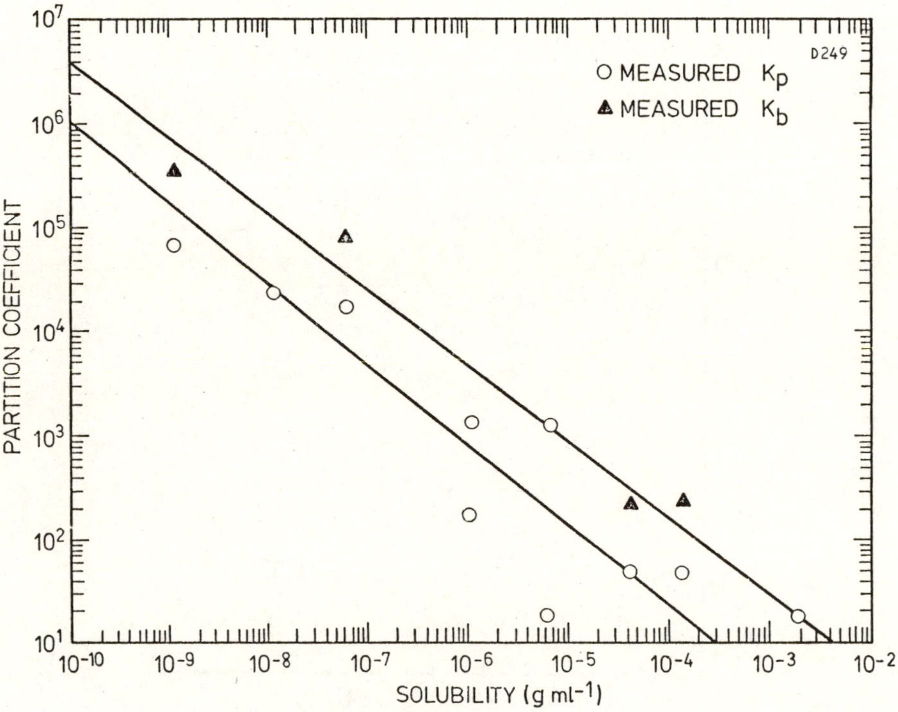

Solubility versus partition coefficient on sediment from Coyote Creek (Kp) and bacteria , mixed populations (Kb). This plot is valid for a number of organic compounds. (ref: 888)

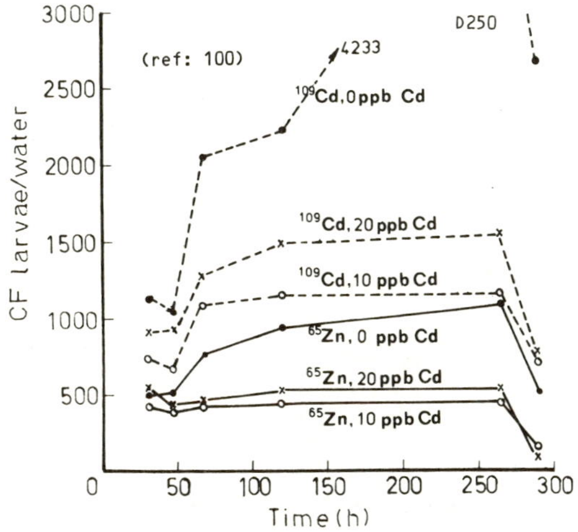

Uptake of Cd-109 by larvae of chironomides as a function of time. The concentration factor is plotted to the time in hours. • 0 ppb Cd × 20 ppb Cd o 10 ppb Cd in the water with respectively without Zn.

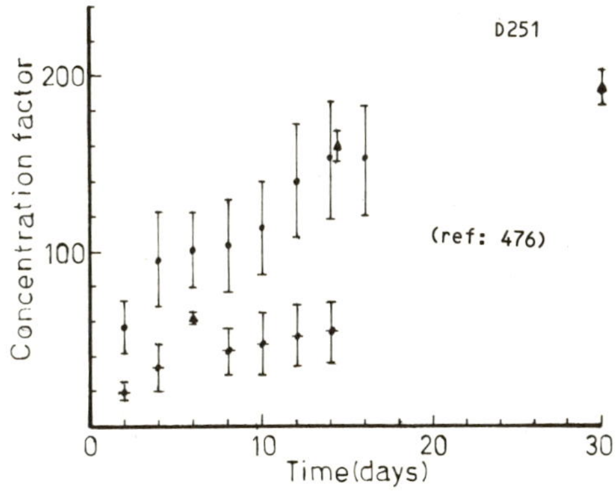

Concentration factor for plutonium from water by Neries diversicolor (worms) versus time. The water was previously conditionated by worms. Points indicate mean values, bars S.D. • plutonium-237 (+6) ▲ plutonium-239 (+4) • plutonium-237 (+6)

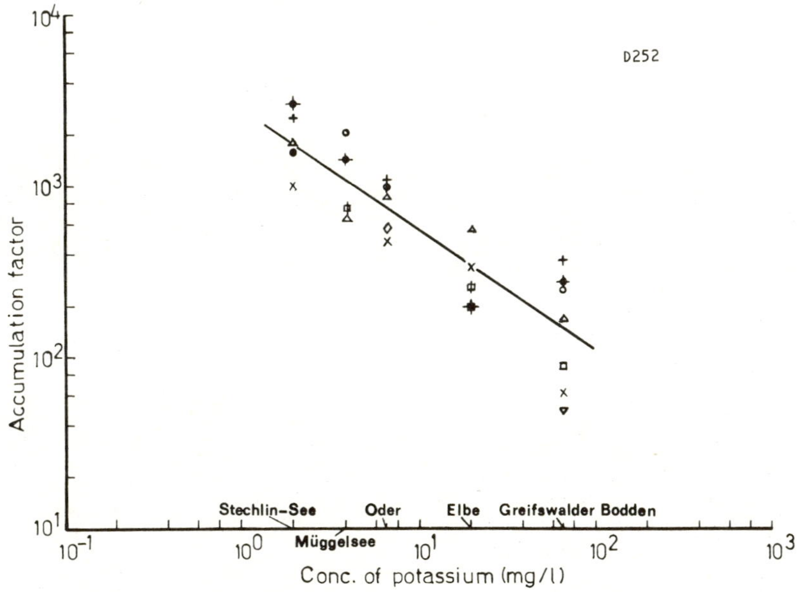

Concentration factor of Cs-137 for fish. Dependence on potassium concentration of water. + Esox lucius, o Lucioperca lucioperca, × Leuciscus rutilus, ✦ Perca fluviatilis, ▽ Clupea harengus, □ Pleuronectes flesus, △ Anguilla anguilla, • Coregonus albula typica, ♦ Aspius ropax, ◇ Tinca vulgaris, ⌑ Abranis brama, ⊕ Leucissceus idus.

(ref: 434)

TABLE D253

Mercury Content of Tomato Seedlings Grown in Methyl Mercury Hydroxide (MMH), and its Effect on Growth.
REF.: 606

Treatment (ppm) MMH	Hg in roots (ppm wet wt)	Concentration factor	Hg in Terminals (ppm wet wt)	Concentration factor	Mean gain in shoot length (cm)	% Inhibition of shoot length	Mean terminal wet wt (mg)	% Inhibition of terminal wet wt	Days of experiment
.06	19.8	497	3.4	57.1	90.9	73.3	69.5		2
.03	27.1	903	2.3	77.3	72.8		124.7	48	2
.006	8.2	1367	1.5	250	.8	27.2	150.8	37.1	2
.003	1.	333	.6	200	1	9.9	228	4.9	2
<.001	.5		.1		1.1		239.8		2
.06	39.1	651	1	16	.5	88.9	173.9	91.8	10
.03	10.2	341	.7	22	1.4	68.8	699.3	66.9	10
.006	1.6	267	.8	132	4	11	625.8	71.3	10
.003	.2	57	.2	63	5.1	0	2895.6	0	10
0	0	0	0	0	4.5		2109		10

D3

CONCENTRATIONS

IN
ANIMALS
PLANTS
SOIL
WATER

TABLE D254

CONCENTRATIONS : AG			
ITEM:	VALUE:	CONDITIONS:	REF.:
*WATER	0.085-5.3 UG /L	RIVERS,U.S.A.	893
*ZOOPLANKTON	< 0.96 MG/KG DRY WT.(MEAN)	19ST.,N.Y.BIGHT,CONC.IN PHYTOPL. (LONG ISLAND):<0.61MG/KG DRY WT.MEAN	565

TABLE D255

CONCENTRATIONS : AL			
ITEM:	VALUE:	CONDITIONS:	REF.:
*ATMOSPHERE	74-483 NG/M**3	ENGLAND,7 STATIONS	893
*ATMOSPHERE	250-2000 NG/M**3	SAN FRANCISCO BAY,U.S.	893
*ATMOSPHERE	0.57+-0.17 NG/M**3	SOUTHPOLE,OCT.1970	893
*RAIN	160 UG /L	ENGLAND,YEARLY MEAN,1 STATION	893
*WATER	11-213 UG/L	RIVERS IN U.S.A.	893

TABLE D256

CONCENTRATIONS : ALDRIN			
ITEM:	VALUE:	CONDITIONS:	REF.:
*SEDIMENT	0.17 MG/KG	CONC.IN WATER=0.014UG/L,U.S.	389
*SEDIMENT	<0.01 MG/KG	CONC.IN WATER:<1.0UG/L,U.S.	389
*SEDIMENT	<0.001 MG/KG	CONC.IN WATER:<1.0UG/L,U.S.	389

TABLE D257

CONCENTRATIONS : AS			
ITEM:	VALUE:	CONDITIONS:	REF.:
*COD, LIVER	3.4 MI UG/G	HUMAN FOOD	893
*COD, MUSCLE	0.5 MI UG /G	HUMAN FOOD	893
*RAIN	0.3-1 MI UG/L	NONE	893
*SEDIMENT	<5 - 30 PPM	WATER CONC.= < 10 MG/M3,GEOLOGIGAL AGE= PERMIAN	48
*SEDIMENT	<5 - 30 PPM	WATER CONC.=< 10 MG/M3,GEOLOGIGAL AGE= CARBONIFEROUS	48
*SHELLFISH	40-170 UG /G	HUMAN FOOD	893
*VEGETABLES	0.1 UG/G	HUMAN FOOD	893
*WATER	0.2-0.4 UG /L	RIVERS,SWEDEN	893

TABLE D258

CONCENTRATIONS : ATRAZINE			
ITEM:	VALUE:	CONDITIONS:	REF.:
*OAT, ROOT	0.660 MG DEISOPROPYLHYDROXY-ATRAZINE/KG DRY WEIGHT	SOIL TREATED WITH ATRAZINE = 4.5 KG/HA, 293-303 K, 2 W	0004

TABLE D258

CONCENTRATIONS : ATRAZINE

ITEM:	VALUE:	CONDITIONS:	REF.:
*OAT, ROOT	0.564 MG HYDROXY-ATRAZINE/KG FRESH WEIGHT	SOIL TREATED WITH ATRAZINE = 4.5 KG/HA, 293-303 K, 2 W	817
*OAT, ROOT	1.030 MG HYDROXY-ATRAZINE/KG DRY WEIGHT	SOIL TREATED WITH ATRAZINE = 4.5 PPM, LAB, 293-303 K, 2 W	817
*OAT, ROOT	0.464 MG HYDROXY-ATRAZINE/KG DRY WEIGHT	SOIL TREATED WITH ATRAZINE = 4.5 KG/HA, 293-303 K, 2 W	817
*OAT, ROOT	0.195 MG HYDROXY-ATRAZINE/KG DRY WEIGHT	SOIL TREATED WITH ATRAZINE = 0.5 PPM, LAB, 293-303 K, 2 W	817
*OAT, ROOT	0.670 MG HYDROXY-ATRAZINE/KG DRY WEIGHT	SOIL TREATED WITH ATRAZINE = 1.0 PPM, LAB, 293-303 K, 2 W	817
*OAT, ROOT	0.950 MG HYDROXY-ATRAZINE/KG DRY WEIGHT	SOIL TREATED WITH ATRAZINE = 2.0 PPM, LAB, 293-303 K, 2 W	817
*OAT, ROOT	0.634 MG DEETYLHYDROXY-ATRAZINE/KG DRY WEIGHT	SOIL TREATED WITH ATRAZINE = 4.5 KG/HA, 293-303 K, 2 W	817
*OAT, ROOT	0.312 MG DEETYLHYDROXY-ATRAZINE/KG DRY WEIGHT	SOIL TREATED WITH ATRAZINE = 4.5 KG/HA, 293-303 K, 2 W	817
*OAT, ROOT	0.653 MG DEISOPROPYLHYDROXY-ATRAZINE/KG DRY WEIGHT	SOIL TREATED WITH ATRAZINE = 4.5 KG/HA, 293-303 K, 2 W	817
*OAT, SHOOTS	0.193 MG HYDROXY-ATRAZINE/KG FRESH WEIGHT	SOIL TREATED WITH ATRAZINE = 4.5 KG/HA, 293-303 K, 2 W	817
*OAT, SHOOTS	0.236 MG HYDROXY-ATRAZINE/KG FRESH WEIGHT	SOIL TREATED WITH ATRAZINE = 0.5 PPM, LAB, 293-303 K, 2 W	817
*OAT, SHOOTS	0.330 MG HYDROXY-ATRAZINE/KG FRESH WEIGHT	SOIL TREATED WITH ATRAZINE = 1.0 PPM, LAB, 293-303 K, 2 W	817
*OAT, SHOOTS	0.514 MG HYDROXY-ATRAZINE/KG FRESH WEIGHT	SOIL TREATED WITH ATRAZINE = 2.0 PPM, LAB, 293-303 K, 2 W	817
*OAT, SHOOTS	0.409 MG DESISOPROPYLHYDROXY-ATRAZINE/KG WET WEIGHT	SOIL TREATED WITH ATRAZINE = 4.5 KG/HA, 293-303 K, 2 W	817
*OAT, SHOOTS	0.392 MG DESISOPROPYLHYDROXY-ATRAZINE/KG WET WEIGHT	SOIL TREATED WITH ATRAZINE = 4.5 PPM, LAB, 293-303 K, 2 W	817

TABLE D259

CONCENTRATIONS : B

ITEM:	VALUE:	CONDITIONS:	REF.:
*CABBAGE, PLANT	21. PPM DRY MATTER	15 PLANTS,RURAL AREA	1283
*CABBAGE, PLANT	28.8 PPM DRY MATTER	15 PLANTS,URBAN AREA	1283
*DRINKING WATER	40-60 UG/L	STOCKHOLM,1969	893
*ONCHORHYNCHUS KISUTCH	354 MICROGRAM/G	BORON IN WATER=656MICROGRAM/ML, 33.4H,20 FISH	1588
*ONCHORHYNCHUS KISUTCH	391 MICROGRAM/G	BORON IN WATER=656MICROGRAM/ML, 33.4H,20 FISH	1588
*ONCHORHYNCHUS KISUTCH	224 MICROGRAM/G	BORON IN WATER=328MICROGRAM/ML, 84H,20 FISH	1588
*RAIN	2 UG/L	SWEDEN,1953	893
*RAIN	80 UG/L	ENGLAND,1968	893
*RAIN	2-4 UG/L	FRANCE,AREA WITH NO INDUSTRY	893
*RAIN	10 UG/L	FLORIDA,1954	893
*WATER	5-67 UG/L	RIVERS,SWEDEN	893
*WATER	510-760 UG/L	RIVER:TRENT,U.K.	893

TABLE D260

CONCENTRATIONS : BA

ITEM:	VALUE:	CONDITIONS:	REF.:
*ATMOSPHERE	2-1500 NG/M**3	URBAN,U.S.A.	893

TABLE D260

CONCENTRATIONS : BA			
ITEM:	VALUE:	CONDITIONS:	REF.:
*WATER	54 UG/L	RIVERS,WORLD MEAN	893

TABLE D261

CONCENTRATIONS : BE			
ITEM:	VALUE:	CONDITIONS:	REF.:
*ATMOSPHERE	1-2 NG/M**3	URBAN,U.S.A.	893
*ATMOSPHERE	0.13 NG/M**3	NOT URBAN,U.S.A.	893
*DRINKING WATER	0.01-0.7 UG/L	U.S.A.	893
*WATER	1 UG/L	SURFACE WATER	893

TABLE D262

CONCENTRATIONS : BHC			
ITEM:	VALUE:	CONDITIONS:	REF.:
*CHICKEN, BRAIN	0.50 MG/KG	CONC.IN FOOD=10.00MG/KG	389
*CHICKEN, FAT	1.20 MG/KG	CONC.IN FOOD=1.00MG/KG	389
*CHICKEN, FAT	1.18 MG/KG	CONC.IN FOOD=1.00MG/KG	389
*CHICKEN, FAT	21.0 MG/KG	CONC.IN FOOD=10.00MG/KG	389
*CHICKEN, FAT	20.38 MG/KG	CONC.IN FOOD=10.00MG/KG	389
*CHICKEN, LIVER	0.026 MG/KG	CONC.IN FOOD=0.05MG/KG	389
*CHICKEN, LIVER	0.039 MG/KG	CONC.IN FOOD=0.15MG/KG	389
*CHICKEN, LIVER	0.082 MG/KG	CONC.IN FOOD=0.45MG/KG	389
*FERAL PIGEON, TISSUE	144.0 MG/KG	CONC.IN FOOD=72MG/KG	389
*SEDIMENT	<0.0002 MG/KG	CONC.IN WATER=0.03UG/L,U.S.	389
*SEDIMENT	0.027 MG/KG	CONC.IN WATER=0.040UG/L,U.S.	389
*SEDIMENT	<0.01 MG/KG	CONC.IN WATER:<1.0UG/L,U.S.	389
*SEDIMENT	<0.001 MG/KG	CONC.IN WATER:<1.0UG/L,U.S.	389

TABLE D263

CONCENTRATIONS : BI			
ITEM:	VALUE:	CONDITIONS:	REF.:
*ATMOSPHERE	1-38 NG/M**3	URBAN,U.S.A.	893
*ATMOSPHERE	0.1-0.4 NG/M**3	NOT URBAN,U.S.A.	893

TABLE D264

CONCENTRATIONS : BOD			
ITEM:	VALUE:	CONDITIONS:	REF.:
*AGRICULTURAL CROPLAND	7 MG/L	GENERAL CHARACTERISTICS	364
*ANIMAL FEEDLOT, RUNOFF	1000-11000 MG/L	GENERAL CHARACTERISTICS	364

TABLE D264

CONCENTRATIONS : BOD			
ITEM:	VALUE:	CONDITIONS:	REF.:
*MUNICIPAL SEWAGE	200 (100-300) MG/L	TYPICAL UNTREATED	364
*MUNICIPAL SEWAGE	135 (70-200) MG/L	TYPICAL TREATED,PRIMARY	364
*MUNICIPAL SEWAGE	25 (15-45) MG/L	TYPICAL TREATED,SECONDARY	364
*PRECIPITATION	12-13 MG/L	GENERAL CHARACTERISTICS	364
*URBAN LAND,DRAINAGE	12-160 MG/L	GENERAL CHARACTERISTICS	364

TABLE D265

CONCENTRATIONS : BR			
ITEM:	VALUE:	CONDITIONS:	REF.:
*STIGEOCLONIUM PACHYDERUM	72.8 % OF THE AMOUTH ADDED TO MEDIUM	MEDIUM = 1 MG/L, 45 MIN, 298 K, 40 W	1628

TABLE D266

CONCENTRATIONS : CAMPHECHLOR			
ITEM:	VALUE:	CONDITIONS:	REF.:
*SEDIMENT	<0.01 MG/KG	CONC.IN WATER:<1.0MIC.G/L,U.S.	389
*SEDIMENT	0.65 MG/KG	CONC.IN WATER=0.63MIC.G/L,U.S.	389

TABLE D267

CONCENTRATIONS : CARBOHYDRATE			
ITEM:	VALUE:	CONDITIONS:	REF.:
*CHLORELLA SP.	23.5 % OF DRY WEIGHT	UNFAVORABLE ENVIRONMENT	1039
*CHLORELLA SP.	33.4 % OF DRY WEIGHT	FAVORABLE ENVIRONMENT	1039
*OYSTER	28.2 % OF DRY WEIGHT	NONE	461
*SACCHAROMYCES CEREVISIAE	37 % OF DRY WEIGHT	NONE	574
*SEA URCHIN	3.4 % OF DRY WEIGHT	40 H LARVA	461
*SEA URCHIN	5.4 % OF DRY WEIGHT	UNFERTILIZED EGG	461
*SILKWORM	1.8 % OF DRY WEIGHT	LARVA	461
*SILKWORM	6.5 % OF DRY WEIGHT	IMAGO	461
*TORULOPSIS UTILIS	30 % OF DRY WEIGHT	NONE	574

TABLE D268

CONCENTRATIONS : CCL3F			
ITEM:	VALUE:	CONDITIONS:	REF.:
*TROPOSPHERE	50 PPT	1950-71,GLOBAL PRODUCT.=1.48*10**6 METRIC T,LOAD=1.01*10**6 METRIC T	1712
*TROPOSPHERE	60 PPT	1972,GLOBAL PRODUCT.=0.27*10**6 METRIC T,LOAD=1.21*10**6 METRIC T	1712
*TROPOSPHERE	80 PPT	1974,GLOBAL PRODUCT.=0.27*10**6 METRIC T,LOAD=1.62*10**6 METRIC T	1712

TABLE D269

CONCENTRATIONS : CD

ITEM:	VALUE:	CONDITIONS:	REF.:
*AQUIPECTEN IRRADIANS	19 % MORE THAN CONTROL	SEAWATER, 10 UG CD/L ADDED AS CDCL2, 21 D	398
*ATMOSPHERE	5 NG/M**3	STOCKHOLM,WEEKLY MEAN	893
*AVENA SATIVA, GRAINS	20.8 PPM DRY WT.	CD IN CONTROL PLANTS=3.9PPM,CD IN SOIL=40PPM	775
*AVENA SATIVA, GRAINS	33.6 PPM DRY WT.	CD IN CONTROL PLANTS=3.9PPM,CD IN SOIL=200PPM	775
*AVENA SATIVA, HUSKS	37.8 PPM DRY WT.	CD IN CONTROL PLANTS=4.5PPM,CD IN SOIL=40PPM	775
*AVENA SATIVA, HUSKS	96 PPM DRY WT.	CD IN CONTROL PLANTS=4.5PPM,CD IN SOIL=200PPM	775
*AVENA SATIVA, LEAF	45.4 PPM DRY WT.	CD IN CONTROL PLANTS=3.9PPM,CD IN SOIL=40PPM	775
*AVENA SATIVA, LEAF	177 PPM DRY WT.	CD IN CONTROL PLANTS=3.9PPM,CD IN SOIL=200PPM	775
*AVENA SATIVA, ROOT	144.8 PPM DRY WT.	CD IN CONTROL PLANTS=3.6PPM,CD IN SOIL=40PPM	775
*AVENA SATIVA, ROOT	663.2 PPM DRY WT.	CD IN CONTROL PLANTS=3.6PPM,CD IN SOIL=200PPM	775
*AVENA SATIVA, STALK	47.4 PPM DRY WT.	CD IN CONTROL PLANTS=3.9PPM,CD IN SOIL=40PPM	775
*AVENA SATIVA, STALK	116.5 PPM DRY WT.	CD IN CONTROL PLANTS=3.9PPM,CD IN SOIL=200PPM	775
*BEVERAGES	0.040 UG/G WET WT.(MAX)	U.S.A.,1968-1970,N=30, ATOM.ABS.AND/OR POLAROGRAPHY	487
*BOUND IN STREAM	95 % OF ADDED	CD RELEASED IN A STREAM AS CDCL2, BOUND IN 100 FIRST M	701
*BOUND IN STREAM	95 % OF ADDED	CD RELEASED IN A STREAM AS CDCL2, BOUND IN 100 FIRST M	701
*BRACHYDANIO RERIO	12.7 MG/KG WET WEIGHT	FOOD= 10 PPM CD AS ACETATE, FEMALE, 300 K, ALK=25 MG CA/L	1308
*BRACHYDANIO RERIO	5.10 MG/KG WET WEIGHT	FOOD= 10 PPM CD AS ACETATE, MALE, 300 K, ALK=25 MG CA/L	1308
*BRASSICA OLERACEA VAR. BOTRYTIS, ROOT	202.5 PPM DRY WT.	CD IN CONTROL PLANTS=1.8PPM,CD IN SOIL=40PPM	775
*BRASSICA OLERACEA VAR. BOTRYTIS, ROOT	1356.7 PPM DRY WT.	CD IN CONTROL PLANTS=1.8PPM,CD IN SOIL=200PPM	775
*BRASSICA OLERACEA VAR. ITALICA, LEAVES	36 PPM DRY WT.	CD IN CONTROL PLANTS=2.7PPM,CD IN SOIL=40PPM	775
*BRASSICA OLERACEA VAR. ITALICA, LEAVES	268.5 PPM DRY WT.	CD IN CONTROL PLANTS=2.7PPM,CD IN SOIL=200PPM	775
*BRASSICA OLERACEA VAR. ITALICA, ROOT	153 PPM DRY WT.	CD IN CONTROL PLANTS=6.5PPM,CD IN SOIL=40PPM	775
*BRASSICA OLERACEA VAR. ITALICA, ROOT	1647.4 PPM DRY WT.	CD IN CONTROL PLANTS=6.5PPM,CD IN SOIL=200PPM	775
*BRASSICA OLERACEA VAR. OLERACEA, LEAF	18.5 PPM DRY WT.	CD IN CONTROL PLANTS=4.8PPM,CD IN SOIL=40PPM	775
*BRASSICA OLERACEA VAR. OLERACEA, LEAF	198.6 PPM DRY WT.	CD IN CONTROL PLANTS=4.8PPM,CD IN SOIL=200PPM	775
*BULK PRECIPITATION	0.6 MICROGRAMS/L MEAN WT.	14JULY-7NOVEMBER-1971,NEW HAMPSHIRE,USA	1393
*CRASSOSTREA VIRGINICA	11.83 MG CD/KG DRY WEIGHT	0 W, 5 PPB IN UNFILT. SEAWATER, CONTROL = 11.83 PPM, 283 K	1771
*CRASSOSTREA VIRGINICA	21.75 MG CD/KG DRY WEIGHT	4 W, 5 PPB IN UNFILT. SEAWATER, CONTROL = 13.10 PPM, 283 K	1771
*CRASSOSTREA VIRGINICA	22.00 MG CD/KG DRY WEIGHT	8 W, 5 PPB IN UNFILT. SEAWATER, CONTROL = 15.57 PPM, 283 K	1771
*CRASSOSTREA VIRGINICA	24.90 MG CD/KG DRY WEIGHT	12 W, 5 PPB IN UNFILT. SEAWATER, CONTROL = 19.35 PPM, 283 K	1771
*CRASSOSTREA VIRGINICA	28.55 MG CD/KG DRY WEIGHT	16 W, 5 PPB IN UNFILT. SEAWATER, CONTROL = 13.42 PPM, 283 K	1771
*CRASSOSTREA VIRGINICA	27.59 MG CD/KG DRY WEIGHT	20 W, 5 PPB IN UNFILT. SEAWATER, CONTROL = 13.60 PPM, 283 K	1771
*CRASSOSTREA VIRGINICA	33.77 MG CD/KG DRY WEIGHT	24 W, 5 PPB IN UNFILT. SEAWATER, CONTROL = 16.75 PPM, 283 K	1771
*CRASSOSTREA VIRGINICA	39.22 MG CD/KG DRY WEIGHT	28 W, 5 PPB IN UNFILT. SEAWATER, CONTROL = 10.85 PPM, 283 K	1771

TABLE D269

CONCENTRATIONS : CD

ITEM:	VALUE:	CONDITIONS:	REF.:
*CRASSOSTREA VIRGINICA	63.27 MG CD/KG DRY WEIGHT	32 W, 5 PPB IN UNFILT. SEAWATER, CONTROL = 12.19 PPM, 283 K	1771
*CRASSOSTREA VIRGINICA	85.83 MG CD/KG DRY WEIGHT	36 W, 5 PPB IN UNFILT. SEAWATER, CONTROL = 17.47 PPM, 283 K	1771
*CRASSOSTREA VIRGINICA	104.91 MG CD/KG DRY WEIGHT	40 W, 5 PPB IN UNFILT. SEAWATER, CONTROL = 12.55 PPM, 283 K	1771
*CRASSOSTREA VIRGINICA	352 % MORE THAN CONTROL	SEAWATER, 10 UG CD/L ADDED AS CDCL2, 21 D	398
*CRATAEGUS MONOGYNA, UNWASHED LEAVES	6.04 PPM DRY WT.	AVONMOUTH,BRISTOL,GB	936
*DAIRY PRODUCTS	0.050 UG/G WET WT.(MAX)	U.S.A.,1968-1970,N=30,, ATOM.ABS.AND/OR POLAROGRAPHY	487
*DAUCUS CAROTA, ROOT	1737.5 PPM DRY WT.	CD IN CONTROL PLANTS=8.5PPM,CD IN SOIL=40PPM	775
*DAUCUS CAROTA, TOPS	79.3 PPM DRY WT.	CD IN CONTROL PLANTS=6.6PPM,CD IN SOIL=40PPM	775
*DAUCUS CAROTA, TUBERS	26.8 PPM DRY WT.	CD IN CONTROL PLANTS=2.4PPM,CD IN SOIL=40PPM	775
*DRINKING WATER	0.1-1 UG/L	SWEDEN	893
*FISH AND POULTRY, MEAT	0.045 UG/G WET WT.(MAX)	U.S.A.,1968-1970,N=30,, ATOM.ABS.AND/OR POLAROGRAPHY	487
*FODDER RAPE	637 NG/G D.M.	ADDED SEWAGE SLUDGE(D.M.)=0T/HA, SOIL-PH=4.8	31
*FODDER RAPE	845 NG/G D.M.	ADDED SEWAGE SLUDGE(D.M.)=6.5T/HA, SOIL-PH=4.8	31
*FODDER RAPE	906 NG/G D.M.	ADDED SEWAGE SLUDGE(D.M.)=19T/HA, SOIL-PH=4.8	31
*FODDER RAPE	1021 NG/G D.M.	ADDED SEWAGE SLUDGE(D.M.)=58T/HA, SOIL-PH=4.8	31
*FODDER RAPE	719 NG/G D.M.	ADDED SEWAGE SLUDGE(D.M.)=175T/HA, SOIL-PH=4.8	31
*FODDER RAPE	121 NG/G D.M.	ADDED SEWAGE SLUDGE(D.M.)=0T/HA, SOIL-PH=6.0	31
*FODDER RAPE	171 NG/G D.M.	ADDED SEWAGE SLUDGE(D.M.)=6.5T/HA, SOIL-PH=6.0	31
*FODDER RAPE	220 NG/G D.M.	ADDED SEWAGE SLUDGE(D.M.)=19T/HA, SOIL-PH=6.0	31
*FODDER RAPE	412 NG/G D.M.	ADDED SEWAGE SLUDGE(D.M.)=58T/HA, SOIL-PH=6.0	31
*FODDER RAPE	604 NG/G D.M.	ADDED SEWAGE SLUDGE(D.M.)=175T/HA, SOIL-PH=6.0	31
*FODDER RAPE	105 NG/G D.M.	ADDED SEWAGE SLUDGE(D.M.)=0T/HA, SOIL-PH=7.2	31
*FODDER RAPE	110 NG/G D.M.	ADDED SEWAGE SLUDGE(D.M.)=6.5T/HA, SOIL-PH=7.2	31
*FODDER RAPE	160 NG/G D.M.	ADDED SEWAGE SLUDGE(D.M.)=19T/HA, SOIL-PH=7.2	31
*FODDER RAPE	236 NG/G D.M.	ADDED SEWAGE SLUDGE(D.M.)=58T/HA, SOIL-PH=7.2	31
*FODDER RAPE	384 NG/G D.M.	ADDED SEWAGE SLUDGE(D.M.)=175T/HA, SOIL-PH=7.2	31
*FRUITS	0.225 UG/G WET WT.(MAX)	U.S.A.,1968-1970,N=30,, ATOM.ABS.AND/OR POLAROGRAPHY	487
*GARDEN FRUITS	0.070 UG/G WET WT.(MAX)	U.S.A.,1968-1970,N=30,, ATOM.ABS.AND/OR POLAROGRAPHY	487
*GLYCINE MAX, LEAF	0.4 MG/KG DRY WEIGHT	DIAGNOSTIC PART, NEW LEAF, SOIL= 0.0 UG CD/G, PH=7.7, ZN=47 PPM	127
*GLYCINE MAX, LEAF	3.7 MG/KG DRY WEIGHT	DIAGNOSTIC PART, NEW LEAF, SOIL= 2.5 UG CD/G, PH=7.7, ZN=47 PPM	127
*GLYCINE MAX, LEAF	5.5 MG/KG DRY WEIGHT	DIAGNOSTIC PART, NEW LEAF, SOIL= 5.0 UG CD/G, PH=7.7, ZN=47 PPM	127
*GLYCINE MAX, LEAF	16.0 MG/KG DRY WEIGHT	DIAGNOSTIC PART, NEW LEAF, SOIL= 10.0 UG CD/G, PH=7.7, ZN=47 PPM	127
*GLYCINE MAX, LEAF	22.0 MG/KG DRY WEIGHT	DIAGNOSTIC PART, NEW LEAF, SOIL= 20.0 UG CD/G, PH=7.7, ZN=47 PPM	127
*GLYCINE MAX, LEAF	29.0 MG/KG DRY WEIGHT	DIAGNOSTIC PART, NEW LEAF, SOIL= 40.0 UG CD/G, PH=7.7, ZN=47 PPM	127

TABLE D269

CONCENTRATIONS : CD			
ITEM:	VALUE:	CONDITIONS:	REF.:
*GLYCINE MAX, LEAF	36.0 MG/KG DRY WEIGHT	DIAGNOSTIC PART, NEW LEAF, SOIL= 80.0 UG CD/G, PH=7.7, ZN=47 PPM	127
*GLYCINE MAX, LEAF	35.0 MG/KG DRY WEIGHT	DIAGNOSTIC PART, NEW LEAF, SOIL= 160.0 UG CD/G, PH=7.7, ZN=47 PPM	127
*GLYCINE MAX, LEAF AND PETIOLE	<0.06 MG/KG	CD (SOIL TOT OR EXTRAC. IN .1N HCL)=0.66-0.1MG/KG,SLUGDE= 0.	669
*GLYCINE MAX, LEAF AND PETIOLE	0.63 MG/KG	CD (SOIL TOT OR EXTRAC. IN .1N HCL)=2.67-2.94MG/KG,SLUGDE= 6.4	669
*GLYCINE MAX, LEAF AND PETIOLE	2.89 MG/KG	CD (SOIL TOT OR EXTRAC. IN .1N HCL)=6.94-8.11MG/KG,SLUGDE= 12.7	669
*GLYCINE MAX, LEAF AND PETIOLE	6.28 MG/KG	CD (SOIL TOT OR EXTRAC. IN .1N HCL)=11.35-13.33MG/KG,SLUGDE= 25.4	669
*GLYCINE MAX, LEAF AND PETIOLE	0.27 MG/KG	CD (SOIL TOT OR EXTRAC. IN .1N HCL)=0.57-0.19MG/KG,SLUGDE= 0.	669
*GLYCINE MAX, LEAF AND PETIOLE	1.87 MG/KG	CD (SOIL TOT OR EXTRAC. IN .1N HCL)=4.37-3.63MG/KG,SLUGDE= 6.4	669
*GLYCINE MAX, LEAF AND PETIOLE	4.19 MG/KG	CD (SOIL TOT OR EXTRAC. IN .1N HCL)=11.11-9.63MG/KG,SLUGDE= 12.7	669
*GLYCINE MAX, LEAF AND PETIOLE	13.6 MG/KG	CD (SOIL TOT OR EXTRAC. IN .1N HCL)=18.13-15.27MG/KG,SLUGDE= 25.4	669
*GLYCINE MAX, LEAF AND PETIOLE	0.19 MG/KG	CD (SOIL TOT OR EXTRAC. IN .1N HCL)=0.51-0.16MG/KG,SLUGDE= 0.	669
*GLYCINE MAX, LEAF AND PETIOLE	0.85 MG/KG	CD (SOIL TOT OR EXTRAC. IN .1N HCL)=4.91-3.86MG/KG,SLUGDE= 6.4	669
*GLYCINE MAX, LEAF AND PETIOLE	1.76 MG/KG	CD (SOIL TOT OR EXTRAC. IN .1N HCL)=10.10-8.51MG/KG,SLUGDE= 12.7	669
*GLYCINE MAX, LEAF AND PETIOLE	3.85 MG/KG	CD (SOIL TOT OR EXTRAC. IN .1N HCL)=16.7-13.66MG/KG,SLUGDE= 25.4	669
*GLYCINE MAX, LEAF AND PETIOLE	0.22 MG/KG	CD (SOIL TOT OR EXTRAC. IN .1N HCL)<0.25-0.2MG/KG,SLUGDE= 0.	669
*GLYCINE MAX, LEAF AND PETIOLE	0.89 MG/KG	CD (SOIL TOT OR EXTRAC. IN .1N HCL)=3.96-4.03MG/KG,SLUGDE= 6.4	669
*GLYCINE MAX, LEAF AND PETIOLE	2.73 MG/KG	CD (SOIL TOT OR EXTRAC. IN .1N HCL)=7.57-8.14MG/KG,SLUGDE= 12.7	669
*GLYCINE MAX, LEAF AND PETIOLE	5.78 MG/KG	CD (SOIL TOT OR EXTRAC. IN .1N HCL)=16.77-14.69MG/KG,SLUGDE= 25.4	669
*GLYCINE MAX, LEAF AND PETIOLE	0.25 MG/KG	CD (SOIL TOT OR EXTRAC. IN .1N HCL)<0.25-0.15MG/KG,SLUGDE= 0.	669
*GLYCINE MAX, LEAF AND PETIOLE	0.64 MG/KG	CD (SOIL TOT OR EXTRAC. IN .1N HCL)=4.56-4.43MG/KG,SLUGDE= 6.4	669
*GLYCINE MAX, LEAF AND PETIOLE	1.22 MG/KG	CD (SOIL TOT OR EXTRAC. IN .1N HCL)=9.3-8.74MG/KG,SLUGDE= 12.7	669
*GLYCINE MAX, LEAF AND PETIOLE	2.46 MG/KG	CD (SOIL TOT OR EXTRAC. IN .1N HCL)=14.56-14.13MG/KG.SLUGDE= 25.4	669
*GLYCINE MAX, SEED	0.6 MG/KG DRY WEIGHT	EDIABLE PART, SOIL=0.0 UG CD/G, PH =7.7, ZN=47 PPM	127
*GLYCINE MAX, SEED	3.9 MG/KG DRY WEIGHT	EDIABLE PART, SOIL=2.5 UG CD/G, PH =7.7, ZN=47 PPM	127
*GLYCINE MAX, SEED	6.8 MG/KG DRY WEIGHT	EDIABLE PART, SOIL=5.0 UG CD/G, PH =7.7, ZN=47 PPM	127
*GLYCINE MAX, SEED	11.0 MG/KG DRY WEIGHT	EDIABLE PART, SOIL=10.0 UG CD/G, PH=7.7, ZN=47 PPM	127
*GLYCINE MAX, SEED	17.0 MG/KG DRY WEIGHT	EDIABLE PART, SOIL=20.0 UG CD/G, PH=7.7, ZN=47 PPM	127

TABLE D269

CONCENTRATIONS : CD

ITEM:	VALUE:	CONDITIONS:	REF.:
*GLYCINE MAX, SEED	29.0 MG/KG DRY WEIGHT	EDIABLE PART, SOIL=40.0 UG CD/G, PH=7.7, ZN=47 PPM	127
*GLYCINE MAX, SEED	29.0 MG/KG DRY WEIGHT	EDIABLE PART, SOIL=80.0 UG CD/G, PH=7.7, ZN=47 PPM	127
*GLYCINE MAX, SEED	21.0 MG/KG DRY WEIGHT	EDIABLE PART, SOIL=160.0 UG CD/G, PH=7.7, ZN=47 PPM	127
*GLYCINE MAX, SEED TOTAL L.	0.027 MG/KG	CD (SOIL TOT OR EXTRAC. IN 0.1N HCL)=0.66-0.1MG/KG,SLUGDE= 0.	669
*GLYCINE MAX, SEED TOTAL L.	0.41 MG/KG	CD (SOIL TOT OR EXTRAC. IN 0.1N HCL)=2.67-2.94MG/KG,SLUGDE= 6.4	669
*GLYCINE MAX, SEED TOTAL L.	0.90 MG/KG	CD (SOIL TOT OR EXTRAC. IN 0.1N HCL)=6.94-8.11MG/KG,SLUGDE= 12.7	669
*GLYCINE MAX, SEED TOTAL L.	1.55 MG/KG	CD (SOIL TOT OR EXTRAC. IN 0.1N HCL)=11.35-13.33MG/KG,SLUGDE =25.4	669
*GLYCINE MAX, SEED TOTAL L.	0.29 MG/KG	CD (SOIL TOT OR EXTRAC. IN 0.1N HCL)=0.57-0.19MG/KG,SLUGDE= 0.	669
*GLYCINE MAX, SEED TOTAL L.	0.49 MG/KG	CD (SOIL TOT OR EXTRAC. IN 0.1N HCL)=4.37-3.63MG/KG,SLUGDE= 6.4	669
*GLYCINE MAX, SEED TOTAL L.	0.94 MG/KG	CD (SOIL TOT OR EXTRAC. IN 0.1N HCL)=11.11-9.63MG/KG,SLUGDE= 12.7	669
*GLYCINE MAX, SEED TOTAL L.	3.0 MG/KG	CD (SOIL TOT OR EXTRAC. IN 0.1N HCL)=18.13-15.27MG/KG,SLUGDE =25.4	669
*GLYCINE MAX, SEED TOTAL L.	0.18 MG/KG	CD (SOIL TOT OR EXTRAC. IN 0.1N HCL)=0.51-0.16MG/KG,SLUGDE= 0.	669
*GLYCINE MAX, SEED TOTAL L.	0.31 MG/KG	CD (SOIL TOT OR EXTRAC. IN 0.1N HCL)=4.91-3.86MG/KG,SLUGDE= 6.4	669
*GLYCINE MAX, SEED TOTAL L.	0.55 MG/KG	CD (SOIL TOT OR EXTRAC. IN 0.1N HCL)=10.10-8.51MG/KG,SLUGDE= 12.7	669
*GLYCINE MAX, SEED TOTAL L.	0.75 MG/KG	CD (SOIL TOT OR EXTRAC. IN 0.1N HCL)=16.7-13.66MG/KG,SLUGDE= 25.4	669
*GLYCINE MAX, SEED TOTAL L.	0.31 MG/KG	CD (SOIL TOT OR EXTRAC. IN 0.1N HCL)<0.25-0.2MG/KG,SLUGDE= 0.	669
*GLYCINE MAX, SEED TOTAL L.	0.31 MG/KG	CD (SOIL TOT OR EXTRAC. IN 0.1N HCL)=3.96-4.03MG/KG,SLUGDE= 6.4	669
*GLYCINE MAX, SEED TOTAL L.	0.57 MG/KG	CD (SOIL TOT OR EXTRAC. IN 0.1N HCL)=7.57-8.14MG/KG,SLUGDE= 12.7	669
*GLYCINE MAX, SEED TOTAL L.	0.92 MG/KG	CD (SOIL TOT OR EXTRAC. IN 0.1N HCL)=16.77-14.69MG/KG,SLUGDE =25.4	669
*GLYCINE MAX, SEED TOTAL L.	0.13 MG/KG	CD (SOIL TOT OR EXTRAC. IN 0.1N HCL)<0.25-0.15MG/KG,SLUGDE= 0.	669
*GLYCINE MAX, SEED TOTAL L.	0.23 MG/KG	CD (SOIL TOT OR EXTRAC. IN 0.1N HCL)=4.56-4.43MG/KG,SLUGDE= 6.4	669
*GLYCINE MAX, SEED TOTAL L.	0.35 MG/KG	CD (SOIL TOT OR EXTRAC. IN 0.1N HCL)=9.3-8.74MG/KG,SLUGDE= 12.7	669
*GLYCINE MAX, SEED TOTAL L.	0.35 MG/KG	CD (SOIL TOT OR EXTRAC. IN 0.1N HCL)=14.56-14.13MG/KG.SLUGDE =25.4	669
*GRAIN AND CEREAL PRODUCTS	0.070 UG /G WET WT.(MAX)	U.S.A.,1968-1970,N=30,, ATOM.ABS.AND/OR POLAROGRAPHY	487
*GUINEA PIG, RENAL CORTEX	10.8 UG/G DRY WT.(SD=2.8)	1 MONTH OF CIGARETTE SMOKE,N=22; CONTROL=8.4MIC.G/G D.W.(SD=1.2),N= 15	487
*HOMARUS AMERICANUS	25 % MORE THAN CONTROL	SEAWATER, 10 UG CD/L ADDED AS CDCL2, 21 D	398
*HOMO SAPIENS, BLOOD OF EXPOSED WORKERS	27 NG/G	SWEDEN,PRESENTLY EXPOSED,N=22, SPECTROGRAPHIC,1971	487

TABLE D269

CONCENTRATIONS : CD

ITEM:	VALUE:	CONDITIONS:	REF.:
*HOMO SAPIENS, BLOOD OF EXPOSED WORKERS	8.6 NG/G	SWEDEN,NOT EXPOSED FOR 10-20YEARS, N=35,SPECTROGRAPHIC,1971	487
*HOMO SAPIENS, BLOOD OF EXPOSED WORKERS	41 NG/G	W.GERMANY,PRESENTLY EXPOSED,N=22, ATOM.ABS.AFTER EXTRACTION	487
*HOMO SAPIENS, BLOOD OF NOT EXPOSED PERSONS	5 (2-8) NG/G	SWEDEN,N=6,NEUTRON ACTIVATION,1961	487
*HOMO SAPIENS, BLOOD OF NOT EXPOSED PERSONS	<2 NG/G	SWEDEN,N=150,SPECTROGRAPHIC,1971	487
*HOMO SAPIENS, BLOOD OF NOT EXPOSED PERSONS	3.5 NG/G	W.GERMANY,N=18,ATOM.ABS.AFTER EXTRACTION,1972	487
*HOMO SAPIENS, BLOOD OF NOT EXPOSED PERSONS	9.5 NG/G (SE=2.1)	BELGIUM,N=24,ATOM.ABS.AFTER EXTRACTION,1972	487
*HOMO SAPIENS, BLOOD OF NOT EXPOSED PERSONS	<5 NG/G	U.S.A.,N=243,ATOM.ABS.AFTER EXTRACTION,1968	487
*HOMO SAPIENS, BLOOD OF NOT EXPOSED PERSONS	4.9 NG/G	U.K.,CHILDREN,N=89,ATOM.ABS.AFTER EXTRACTION,1972	487
*HOMO SAPIENS, SERUM OF EXPOSED WORKERS	1.9 NG/G	W.GERMANY,PRESENTLY EXPOSED,N=17, ATOM.ABS.AFTER EXTRACTION,1969	487
*HOMO SAPIENS, SERUM OF EXPOSED WORKERS	13.7 NG/G	W.GERMANY,PRESENTLY EXPOSED,N=22, ATOM.ABS.AFTER EXTRACTION	487
*HOMO SAPIENS, SERUM OF NOT EXPOSED PERSONS	3.3 NG/G	W.GERMANY,N=15,ATOM.ABS.AF¤ER EXTRACTION,1968	487
*HOMO SAPIENS, SERUM OF NOT EXPOSED PERSONS	12 NG/G	U.S.A.,N=27,ATOM.ABS.,1970	487
*LACTUCA SATIVA, LEAF	51.1 PPM DRY WT.	CD IN CONTROL PLANTS=4.8PPM,CD IN SOIL=40PPM	775
*LACTUCA SATIVA, LEAF	667.7 PPM DRY WT.	CD IN CONTROL PLANTS=4.8,CD IN SOIL=200PPM	775
*LACTUCA SATIVA, ROOT	294.7 PPM DRY WT.	CD IN CONTROL PLANTS=2.3PPM,CD IN SOIL=40PPM	775
*LACTUCA SATIVA, ROOT	1628.2 PPM DRY WT.	CD IN CONTROL PLANTS=2.3PPM,CD IN SOIL=200PPM	775
*LEAFY VEGETABLES	0.110 UG/G WET WT.(MAX)	U.S.A.,1968-1970,N=30, ATOM.ABS.AND/OR POLAROGRAPHY	487
*LEGUME VEGETABLES	0.035 UG/G WET WT.(MAX)	U.S.A.,1968-1970,N=30, ATOM.ABS.AND/OR POLAROGRAPHY	487
*MILK	0.0015-0.004 UG/G WET WT.	U.S.A.	487
*MILK	0.009 UG/G WET WT.	WESTERN GERMANY	487
*MILK	0.01 UG/G WET WT.	CZECHOSLOVAKIA	487
*MILK	0.003 UG/G WET WT.	JAPAN (NONPOLLUTED AREA)	487
*MOUSE, BLOOD	23.8 NG/G (SD=8.6)	3.WEEK,INJECT.OF 0.025MG CD/KG BODY/D (5D/WEEK)DUR.WEEK 1-3,N=4	487
*MOUSE, BLOOD	14.9 NG/G (SD=1.4)	6.WEEK,INJECT.OF 0.025MG CD/KG BODY/D (5D/WEEK)DUR.WEEK 1-3,N=4	487
*MOUSE, BLOOD	48.4 NG/G (SD=19.5)	21.WEEK,INJECT.OF 0.025MG CD/KG BODY/D (5D/WEEK)DUR.WEEK 7-21,N=4	487
*MOUSE, BLOOD	25.9 NG/G (SD=6.4)	24.WEEK,INJECT.OF 0.025MG CD/KG BODY/D (5D/WEEK)DUR.WEEK 7-21,N=8	487
*MOUSE, KIDNEY	162(141-172) UG/G WET WT.	INJECT.OF 0.25MG CD/KG BODY/D(5D/WEEK) FOR 6 MONTHS,N=4	487
*MOUSE, KIDNEY	172(158-194) UG/G WET WT.	INJECT.OF 0.5MG CD/KG BODY/D(5D/WEEK) FOR 6 MONTHS,N=4	487
*MOUSE, LIVER	281(271-288) UG/G WET WT.	INJECT.OF 0.25MG CD/KG BODY/D(5D/WEEK) FOR 6 MONTHS,N=4	487
*MOUSE, LIVER	258(190-288) UG/G WET WT.	INJECT.OF 0.5MG CD/KG BODY/D(5D/WEEK) FOR 6 MONTHS,N=4	487
*MOUSE, PANCEAS	71(67-74) UG/G WET WT.	INJECT.OF 0.25MG CD/KG BODY/D(5D/WEEK) FOR 6 MONTHS,N=4	487
*MOUSE, PANCEAS	114(111-121) UG/G WET WT.	INJECT.OF 0.5MG CD/KG BODY/D(5D/WEEK) FOR 6 MONTHS,N=4	487
*MOUSE, SPLEEN	10.1(9.5-10.6) UG/G WET WT.	INJECT.OF 0.25MG CD/KG BODY/D(5D/WEEK) FOR 6 MONTHS,N=4	487
*MOUSE, SPLEEN	19.5(18.4-21.4) UG/G WET WT.	INJECT.OF 0.5MG CD/KG BODY/D(5D/WEEK) FOR 6 MONTHS,N=4	487
*MOUSE, URINE	0.80 NG/24HR	3.WEEK,INJECT.OF 0.025MG CD/KG BODY/D (5D/WEEK)DUR.WEEK 1-3,N=4	487

TABLE D269

CONCENTRATIONS : CD			
ITEM:	VALUE:	CONDITIONS:	REF.:
*MOUSE, URINE	0.80 NG/24HR	6.WEEK,INJECT.OF 0.025MG CD/KG BODY/D (5D/WEEK)DUR.WEEK 1-3,N=4	487
*MOUSE, URINE	4.65 NG/24HR	21.WEEK,INJECT.OF 0.025MG CD/KG BODY/D (5D/WEEK)DUR.WEEK 7-21,N=4	487
*MOUSE, URINE	6.34 NG/24HR	24.WEEK,INJECT.OF 0.025MG CD/KG BODY/D (5D/WEEK)DUR.WEEK 7-21,N=8	487
*MOUSE, WHOLE BODY	11.0 UG	3.WEEK,INJECT.OF 0.025MG CD/KG BODY/D (5D/WEEK)DUR.WEEK 1-3,N=4	487
*MOUSE, WHOLE BODY	11.9 UG	6.WEEK,INJECT.OF 0.025MG CD/KG BODY/D (5D/WEEK)DUR.WEEK 1-3,N=4	487
*MOUSE, WHOLE BODY	61.1 UG	21.WEEK,INJECT.OF 0.025MG CD/KG BODY/D (5D/WEEK)DUR.WEEK 7-21,N=4	487
*MOUSE, WHOLE BODY	65.2 UG	24.WEEK,INJECT.OF 0.025MG CD/KG BODY/D (5D/WEEK)DUR.WEEK 7-21,N=8	487
*MULINIA LATERALIS	3.6 (SD=+-0.21) UG CD/G TISSUE DRY WEIGHT	UPTAKE FROM WATER = 5 UG/L, 283 K, 14 D, S=3 %, N=10	355
*MULINIA LATERALIS	8.9 (SD=+-1.03) UG CD/G TISSUE DRY WEIGHT	UPTAKE FROM WATER = 5 UG/L, 293 K, 14 D, S=3 %, N=10	355
*MULINIA LATERALIS	10.33 (SD=+-0.27) UG CD/G TISSUE DRY WEIGHT	UPTAKE FROM WATER = 20 UG/L, 283 K, 14 D, S=3 %, N=10	355
*MULINIA LATERALIS	20.46 UG CD/G TISSUE DRY WEIGHT	UPTAKE FROM WATER = 20 UG/L, 293 K, 14 D, S=3 %, N=10	355
*MYA ARENARIA	2.2 (SD=+-0.31) UG CD/G TISSUE DRY WEIGHT	UPTAKE FROM WATER = 5 UG/L, 283 K, 14 D, S=3 %, N=10	355
*MYA ARENARIA	4.2 (SD=+-0.25) UG CD/G TISSUE DRY WEIGHT	UPTAKE FROM WATER = 5 UG/L, 293 K, 14 D, S=3 %, N=10	355
*MYA ARENARIA	16.76 (SD=+-1.23) UG CD/G TISSUE DRY WEIGHT	UPTAKE FROM WATER = 20 UG/L, 283 K, 14 D, S=3 %, N=10	355
*MYA ARENARIA	29.04 (SD=+-0.34) UG CD/G TISSUE DRY WEIGHT	UPTAKE FROM WATER = 20 UG/L, 293 K, 14 D, S=3 %, N=10	355
*MYTILUS EDULIS	9.3 (SD=+-2.48) UG CD/G TISSUE DRY WEIGHT	UPTAKE FROM WATER = 5 UG/L, 283 K, 14 D, S=3 %, N=10	355
*MYTILUS EDULIS	9.4 (SD=+-0.91) UG CD/G TISSUE DRY WEIGHT	UPTAKE FROM WATER = 5 UG/L, 293 K, 14 D, S=3 %, N=10	355
*MYTILUS EDULIS	50.55 (SD=+-8.57) UG CD/G TISSUE DRY WEIGHT	UPTAKE FROM WATER = 20 UG/L, 283 K, 14 D, S=3 %, N=10	355
*MYTILUS EDULIS	60.35 (SD=+-11.8) UG CD/G TISSUE DRY WEIGHT	UPTAKE FROM WATER = 20 UG/L, 293 K, 14 D, S=3 %, N=10	3¤5
*NUCULA PROXIMA	2.0 (SD=+-1.13) UG CD/G TISSUE DRY WEIGHT	UPTAKE FROM WATER = 20 UG/L, 283 K, 14 D, S=3 %, N=10	355
*NUCULA PROXIMA	4.56 (SD=+-1.19) UG CD/G TISSUE DRY WEIGHT	UPTAKE FROM WATER = 20 UG/L, 293 K, 14 D, S=3 %, N=10	355
*OIL,FATS AND SHORTENING	0.085 UG/G WET WT.(MAX)	U.S.A.,1968-1970,N=30,, ATOM.ABS.AND/OR POLAROGRAPHY	487
*ORYZA SATIVA, GRAIN	0.1 MG/KG DRY WEIGHT	EDIBLE PART, SOIL=0.0 UG CD/G, PH=7.7, ZN=47 PPM	127
*ORYZA SATIVA, GRAIN	0.1 MG/KG DRY WEIGHT	EDIBLE PART, SOIL=2.5 UG CD/G, PH=7.7, ZN=47 PPM	127
*ORYZA SATIVA, GRAIN	0.2 MG/KG DRY WEIGHT	EDIBLE PART, SOIL=5.0 UG CD/G, PH=7.7, ZN=47 PPM	127
*ORYZA SATIVA, GRAIN	0.2 MG/KG DRY WEIGHT	EDIBLE PART, SOIL=10.0 UG CD/G, PH =7.7, ZN=47 PPM	127
*ORYZA SATIVA, GRAIN	0.3 MG/KG DRY WEIGHT	EDIBLE PART, SOIL=20.0 UG CD/G, PH =7.7, ZN=47 PPM	127
*ORYZA SATIVA, GRAIN	0.6 MG/KG DRY WEIGHT	EDIBLE PART, SOIL=40.0 UG CD/G, PH =7.7, ZN=47 PPM	127
*ORYZA SATIVA, GRAIN	0.7 MG/KG DRY WEIGHT	EDIBLE PART, SOIL=80.0 UG CD/G, PH =7.7, ZN=47 PPM	127
*ORYZA SATIVA, GRAIN	1.1 MG/KG DRY WEIGHT	EDIBLE PART, SOIL=160.0 UG CD/G, PH=7.7, ZN=47 PPM	127
*ORYZA SATIVA, LEAF	0.1 MG/KG DRY WEIGHT	DIAGNOSTIC PART, 2 TOP LEAVES, SOIL=0.0 UG CD/G, PH=7.7, ZN=47 PPM	127
*ORYZA SATIVA, LEAF	0.1 MG/KG DRY WEIGHT	DIAGNOSTIC PART, 2 TOP LEAVES, SOIL=2.5 UG CD/G, PH=7.7, ZN=47 PPM	127
*ORYZA SATIVA, LEAF	0.1 MG/KG DRY WEIGHT	DIAGNOSTIC PART, 2 TOP LEAVES, SOIL=5.0 UG CD/G, PH=7.7, ZN=47 PPM	127

TABLE D269

CONCENTRATIONS : CD

ITEM:	VALUE:	CONDITIONS:	REF.:
*ORYZA SATIVA, LEAF	0.1 MG/KG DRY WEIGHT	DIAGNOSTIC PART, 2 TOP LEAVES, SOIL=10.0 UG CD/G, PH=7.7, ZN=47 PPM	127
*ORYZA SATIVA, LEAF	0.1 MG/KG DRY WEIGHT	DIAGNOSTIC PART, 2 TOP LEAVES, SOIL=20.0 UG CD/G, PH=7.7, ZN=47 PPM	127
*ORYZA SATIVA, LEAF	0.1 MG/KG DRY WEIGHT	DIAGNOSTIC PART, 2 TOP LEAVES, SOIL=40.0 UG CD/G, PH=7.7, ZN=47 PPM	127
*ORYZA SATIVA, LEAF	0.3 MG/KG DRY WEIGHT	DIAGNOSTIC PART, 2 TOP LEAVES, SOIL=80.0 UG CD/G, PH=7.7, ZN=47 PPM	127
*ORYZA SATIVA, LEAF	0.7 MG/KG DRY WEIGHT	DIAGNOSTIC PART, 2 TOP LEAVES, SOIL=160.0 UG CD/G, PH=7.7, ZN=47 PPM	127
*PATELLA VULGATA	2.5*10**4 (M¤CROGRAMS/G DRY MAT.)/ (MICROGRAMS/G SEAWATER)	17 OBS.,ALL AREAS AROUND THE BRITISH ISLES	1278
*PHASEOLUS VULGARIS, DRY BEAN	0.05 MG/KG DRY WEIGHT	EDIBLE PART, SOIL=0.0 UG CD/G, PH= 7.7, ZN=47 PPM	127
*PHASEOLUS VULGARIS, DRY BEAN	0.2 MG/KG DRY WEIGHT	EDIBLE PART, SOIL=2.5 UG CD/G, PH= 7.7, ZN=47 PPM	127
*PHASEOLUS VULGARIS, DRY BEAN	0.4 MG/KG DRY WEIGHT	EDIBLE PART, SOIL=5.0 UG CD/G, PH= 7.7, ZN=47 PPM	127
*PHASEOLUS VULGARIS, DRY BEAN	0.7 MG/KG DRY WEIGHT	EDIBLE PART, SOIL=10.0 UG CD/G, PH =7.7, ZN=47 PPM	127
*PHASEOLUS VULGARIS, DRY BEAN	1.0 MG/KG DRY WEIGHT	EDIBLE PART, SOIL=20.0 UG CD/G, PH =7.7, ZN=47 PPM	127
*PHASEOLUS VULGARIS, DRY BEAN	1.7 MG/KG DRY WEIGHT	EDIBLE PART, SOIL=40.0 UG CD/G, PH =7.7, ZN=47 PPM	127
*PHASEOLUS VULGARIS, DRY BEAN	2.3 MG/KG DRY WEIGHT	EDIBLE PART, SOIL=80.0 UG CD/G, PH =7.7, ZN=47 PPM	127
*PHASEOLUS VULGARIS, DRY BEAN	3.0 MG/KG DRY WEIGHT	EDIBLE PART, SOIL=160.0 UG CD/G, PH=7.7, ZN=47 PPM	127
*PHASEOLUS VULGARIS, LEAF	0.6 MG/KG DRY WEIGHT	DIAGNOSTIC PART, NEW LEAF, SOIL= 0.0 UG CD/G, PH=7.7, ZN=47 PPM	127
*PHASEOLUS VULGARIS, LEAF	1.8 MG/KG DRY WEIGHT	DIAGNOSTIC PART, NEW LEAF, SOIL= 2.5 UG CD/G, PH=7.7, ZN=47 PPM	127
*PHASEOLUS VULGARIS, LEAF	4.3 MG/KG DRY WEIGHT	DIAGNOSTIC PART, NEW LEAF, SOIL= 5.0 UG CD/G, PH=7.7, ZN=47 PPM	127
*PHASEOLUS VULGARIS, LEAF	10.0 MG/KG DRY WEIGHT	DIAGNOSTIC PART, NEW LEAF, SOIL= 10.0 UG CD/G, PH=7.7, ZN=47 PPM	127
*PHASEOLUS VULGARIS, LEAF	12.0 MG/KG DRY WEIGHT	DIAGNOSTIC PART, NEW LEAF, SOIL= 20.0 UG CD/G, PH=7.7, ZN=47 PPM	127
*PHASEOLUS VULGARIS, LEAF	14.0 MG/KG DRY WEIGHT	DIAGNOSTIC PART, NEW LEAF, SOIL= 40.0 UG CD/G, PH=7.7, ZN=47 PPM	127
*PHASEOLUS VULGARIS, LEAF	19.0 MG/KG DRY WEIGHT	DIAGNOSTIC PART, NEW LEAF, SOIL= 80.0 UG CD/G, PH=7.7, ZN=47 PPM	127
*PHASEOLUS VULGARIS, LEAF	19.0 MG/KG DRY WEIGHT	DIAGNOSTIC PART, NEW LEAF, SOIL= 160.0 UG CD/G, PH=7.7, ZN=47 PPM	127
*PHASEOLUS VULGARIS, LEAVES AND STEMS	72.9 P CI/MG DRY WEIGHT	WATER=225 PPM N,CD-109=3 PPB,7 D, PH=6.5, 293 K,TOT PROD=577 MG DW	690
*PHASEOLUS VULGARIS, LEAVES AND STEMS	49.0 P CI/MG DRY WEIGHT	WATER=225 PPM N,CD-109=3 PPB,7 D, PH=6.5, 293 K,TOT PROD=444 MG DW	690
*PHASEOLUS VULGARIS, LEAVES AND STEMS	29.8 P CI/MG DRY WEIGHT	WATER=225 PPM N,CD-109=3 PPB,7 D, PH=6.5, 293 K,TOT PROD=468 MG DW	690
*PHASEOLUS VULGARIS, LOWER LEAVES	48.4 P CI/MG DRY WEIGHT	WAT=225 PPM N,CD-109=3 PPB, 7 D,PH =6.5,293 K,PROD=106 MG DW	690
*PHASEOLUS VULGARIS, LOWER LEAVES	29.1 P CI/MG DRY WEIGHT	WATER=338 PPM N,CD-109=3 PPB, 7 D, PH=6.5,293 K,PROD=98 MG DW	690
*PHASEOLUS VULGARIS, LOWER LEAVES	10.0 P CI/MG DRY WEIGHT	WATER=1463 PPM N,CD-109=3 PPB,7 D, PH=6.5,293 K,PROD=110 MG DW	690
*PHASEOLUS VULGARIS, ROOT	3707 P CI/MG DRY WEIGHT	WAT=225 PPM N,CD-109=3 PPB, 7 D,PH =6.5,293 K,PROD=5.1 MG DW	690
*PHASEOLUS VULGARIS, ROOT	4553 P CI/MG DRY WEIGHT	WATER=338 PPM N,CD-109=3 PPB, 7 D, PH=6.5,293 K,PROD=65.1 MG DW	690
*PHASEOLUS VULGARIS, ROOT	4325 P CI/MG DRY WEIGHT	WATER=1463 PPM N,CD-109=3 PPB,7 D, PH=6.5,293 K,PROD=62.0 MG DW	690
*PHASEOLUS VULGARIS, STEMS	116.2 P CI/MG DRY WEIGHT	WAT=225 PPM N,CD-109=3 PPB, 7 D,PH =6.5,293 K,PROD=135 MG DW	690

TABLE D269

CONCENTRATIONS : CD

ITEM:	VALUE:	CONDITIONS:	REF.:
*PHASEOLUS VULGARIS, STEMS	93.1 P CI/MG DRY WEIGHT	WATER=338 PPM N,CD-109=3 PPB, 7 D, PH=6.5,293 K,PROD=104 MG DW	690
*PHASEOLUS VULGARIS, STEMS	72.6 P CI/MG DRY WEIGHT	WATER=1463 PPM N,CD-109=3 PPB,7 D, PH=6.5,293 K,PROD=113 MG DW	690
*PHASEOLUS VULGARIS, TOP LEAVES	68.3 P CI/MG DRY WEIGHT	WAT=225 PPM N,CD-109=3 PPB, 7 D,PH =6.5,293 K,PROD=145 MG DW	690
*PHASEOLUS VULGARIS, TOP LEAVES	40.5 P CI/MG DRY WEIGHT	WATER=338 PPM N,CD-109=3 PPB, 7 D, PH=6.5,293 K,PROD=80 MG DW	690
*PHASEOLUS VULGARIS, TOP LEAVES	25.5 P CI/MG DRY WEIGHT	WATER=1463 PPM N,CD-109=3 PPB,7 D, PH=6.5,293 K,PROD=82 MG DW	690
*PISUM SATIVUM, PODS	9.5 PPM DRY WT.	CD IN CONTROL PLANTS=5.7PPM,CD IN SOIL=40PPM	775
*PISUM SATIVUM, PODS	28.2 PPM DRY WT.	CD IN CONTROL PLANTS=5.7PPM,CD IN SOIL=200PPM	775
*PISUM SATIVUM, ROOT	409.7 PPM DRY WT.	CD IN CONTROL PLANTS=8.6PPM,CD IN SOIL=40PPM	775
*PISUM SATIVUM, ROOT	1571.2 PPM DRY WT.	CD IN CONTROL PLANTS=8.6PPM,CD IN SOIL=200PPM	775
*PISUM SATIVUM, SEED	10.1 PPM DRY WT.	CD IN CONTROL PLANTS=5.4PPM,CD IN SOIL=40PPM	775
*PISUM SATIVUM, SEED	19.7 PPM DRY WT.	CD IN CONTROL PLANTS=5.4PPM,CD IN SOIL=200PPM	775
*PISUM SATIVUM, VINES	37.2 PPM DRY WT.	CD IN CONTROL PLANTS=7.4PPM,CD IN SOIL=40PPM	775
*PISUM SATIVUM, VINES	116.9 PPM DRY WT.	CD IN CONTROL PLANTS=7.4PPM,CD IN SOIL=200PPM	775
*PORPHYRA UMBILICALIS	6.6*10**2 (MICROGRAMS/G DRY MAT.)/ (MICROGRAMS/G SEAWATER)	20 OBS.,ALL AREAS AROUND THE BRITISH ISLES	1278
*POTATO	0.001 UG/G WET WT.	U.S.A.	487
*POTATO	0.039 UG/G WET WT.	WESTERN GERMANY	487
*POTATO	0.09 UG/G WET WT.	CZECHOSLOVAKIA	487
*POTATO	0.017 UG/G WET WT.	ROUMANIA	487
*POTATO	0.038 UG/G WET WT.	JAPAN (NONPOLLUTED AREA)	487
*POTATO	0.08 UG/G WET WT.	U.K.	487
*POTATOES	0.080 UG/G WET WT.(MAX)	U.S.A.,1969-70,N=30,,ATOM.ABS.AND/ OR POLAROGRAPHY	487
*PSEUDOPLEURONECTES AMERICANUS, BLOOD	<0.3 PPM WET. WT.	CD=0PPB,60D,3-6 FISH	216
*PSEUDOPLEURONECTES AMERICANUS, BLOOD	<0.2 PPM WET. WT.	CD=5PPB,60D,3-6 FISH	216
*PSEUDOPLEURONECTES AMERICANUS, BLOOD	<0.2 PPM WET. WT.	CD=10PPB,60D,3.6 FISH	216
*PSEUDOPLEURONECTES AMERICANUS, GILL	<0.3 PPM WET. WT.	CD=0PPB,60D,3-6 FISH	216
*PSEUDOPLEURONECTES AMERICANUS, GILL	<0.3 PPM WET. WT.	CD=5PPB,60D,3-6 FISH	216
*PSEUDOPLEURONECTES AMERICANUS, GILL	<0.3 PPM WET. WT.	CD=10PPB,60D,3-6 FISH	216
*QUERCUS ROBUR, UNWASHED LEAVES	6.82 PPM DRY WT.	AVONMOUTH,BRISTOL,GB	936
*RAPHANUS SATIVUS, TOPS	264.7 PPM DRY WT.	CD IN CONTROL PLANTS=9.8PPM,CD IN SOIL=40PPM	775
*RAPHANUS SATIVUS, TOPS	398 PPM DRY WT.	CD IN CONTROL PLANTS=9.8PPM,CD IN SOIL=40PPM	775
*RAPHANUS SATIVUS, TUBERS	54.6 PPM DRY WT.	CD IN CONTROL PLANTS=3.6PPM,CD IN SOIL=40PPM	775
*RAT, KIDNEY	4.4 % OF TOTAL INTRACELLULAR AMOUNT	NUCLEI,10 RATS,24 H AFTER INJEC OF 277 UG CD(+2)/275 G	1366
*RAT, KIDNEY	2.2 % OF TOTAL INTRACELLULAR AMOUNT	MITOCHONDRIA,10 RATS,24 H AFTER INJEC OF 277 UG CD(+2)/275 G	1366
*RAT, KIDNEY	2.1 % OF TOTAL INTRACELLULAR AMOUNT	LYSOSOMES,10 RATS,24 H AFTER INJEC OF 277 UG CD(+2)/275 G	1366
*RAT, KIDNEY	1.6 % OF TOTAL INTRACELLULAR AMOUNT	MICROSOMES,10 RATS,24 H AFTER INJEC. OF 277 UG CD(+2)/275 G	1366
*RAT, KIDNEY	89.7 % OF TOTAL INTRACELLULAR AMOUNT	SOLUBLE FRAC.,10 RATS,24 H AFTER INJEC. OF 277 UG CD(+2)/275 G	1366

TABLE D269

CONCENTRATIONS : CD			
ITEM:	VALUE:	CONDITIONS:	REF.:
*RAT, LIVER	3.5 % OF TOTAL INTRACELLULAR AMOUNT	NUCLEI,10 RATS ,24 H AFTER INJECTION OF 277 MICRO G CD(+2)	1366
*RAT, LIVER	2.7 % OF TOTAL INTRACELLULAR AMOUNT	MITOCHONDRIA,10 RATS,24 H AFTER INJEC OF 277 UG CD(+2)/275 G	1366
*RAT, LIVER	1 % OF TOTAL INTRACELLULAR AMOUNT	LYSOSOMES,10 RATS,24 H AFTER INJEC OF 277 UG CD(+2)/275 G	1366
*RAT, LIVER	2.3 % OF TOTAL INTRACELLULAR AMOUNT	MICROSOMES,10 RATS,24 H AFTER INJEC OF 277 UG CD(+2)/275 G	1366
*RAT, LIVER	90.5 % OF TOTAL INTRACELLULAR AMOUNT	SOLUBLE FRAC.,10 RATS,24 H AFTER INJEC OF 277 UG CD(+2)/275 G	1366
*ROOT VEGETABLES	0.080 UG/G WET WT.(MAX)	U.S.A.,1968-1970,N=30,, ATOM.ABS.AND/OR POLAROGRAPHY	487
*SALIX ALBA, UNWASHED LEAVES	7.85 PPM DRY WT.	AVONMOUTH,BRISTOL,GB	936
*SEDIMENT	2 PPM	WATER CONC = <1 MICROGRAM/L, YEARLY MEAN, CARNON RIVER	SCE(
*SEDIMENT	5 PPM	WATER CONC = 2 MICROGRAM/L, YEARLY MEAN, RED RIVER	SCE(
*SEDIMENT	2 PPM	WATER CONC = 1 MICROGRAM/L, YEARLY MEAN, RED RIVER	SCE(
*SEDIMENT	12 PPM	WATER CONC = 3 MICROGRAM/L, YEARLY MEAN, GANNEL RIVER	SCE(
*SEDIMENT	2 PPM	WATER CONC = 1 MICROGRAM/L, YEARLY MEAN, GANNEL RIVER	SCE(
*SEDIMENT	2 PPM	WATER CONC = <1 MICROGRAM/L, YEARLY MEAN, FOWEY RIVER	SCE(
*SEDIMENT	2 PPM	WATER CONC = <1 MICROGRAM/L, YEARLY MEAN, FOWEY RIVER	SCE(
*SEDIMENT	6.7 PPM	MAX WATER CONC = 195 MICROGRAM/L, 1 YEARS MEASUREMENT, CARNON R	SCE(
*SEDIMENT	2 PPM	MAX WATER CONC = 1 MICROGRAM/L, 1 YEARS MEASUREMENT, CARNON R	SCE(
*SEDIMENT	5 PPM	MAX WATER CONC = 4 MICROGRAM/L, 1 YEARS MEASUREMENT, RED R	SCE(
*SEDIMENT	2 PPM	MAX WATER CONC = 1 MICROGRAM/L, 1 YEARS MEASUREMENT, RED R	SCE(
*SEDIMENT	12 PPM	MAX WATER CONC = 6 MICROGRAM/L, MAX CONC, GANNEL RIVER	SCE(
*SEDIMENT	2 PPM	MAX WATER CONC = 2 MICROGRAM/L, MAX CONC, GANNEL RIVER	SCE(
*SEDIMENT	2 PPM	MAX WATER CONC = 2 MICROGRAM/L, 1 YEARS MEASUREMENT, FOWEY R	SCE(
*SEDIMENT	2 PPM	MAX WATER CONC = 1 MICROGRAM/L, 1 YEARS MEASUREMENT, FOWEY R	SCE(
*SEDIMENT	0.37 (RANGE 0.2 - 0.7) MG/KG DRY WEIGHT	UNPOLLUTED AREA, ESTUARY, URR WATER	1557
*SEDIMENT	0.9 (RANGE 0.8 - 1.3) MG/KG DRY WEIGHT	UNPOLLUTED AREA, ESTUARY, URR WATER	1558
*SEDIMENT	1.0 (RANGE 0.7 - 1.2) MG/KG DRY WEIGHT	UNPOLLUTED AREA, ESTUARY, URR WATER	1558
*SEDIMENT	0.4 (RANGE 0.3 - 0.5) MG/KG DRY WEIGHT	UNPOLLUTED AREA, RIVER,ILLINOIS RIVER	993
*SOIL (SEDIMENT)	1.4 PPM DRY SOIL	RANGE 0.4-2.4PPM,RIVER RHEIDOL, CARDIGANSHIRE,GB	327
*SPINACIA OLERACEA, LEAVES	207.5 PPM DRY WT.	CD IN CONTROL PLANTS=12.2PPM,CD IN SOIL=40PPM	775
*SPINACIA OLERACEA, LEAVES	239.3 PPM DRY WT.	CD IN CONTROL PLANTS=12.2PPM,CD IN SOIL=200PPM	775
*SPINACIA OLERACEA, ROOT	213.9 PPM DRY WT.	CD IN CONTROL PLANTS=8.5PPM,CD IN SOIL=40PPM	775
*SPINACIA OLERACEA, ROOT	490.3 PPM DRY WT.	CD IN CONTROL PLANTS=8.5PPM,CD IN SOIL=200PPM	775
*SUGAR AND ADJUNCTS	0.055 UG/G WET WT.(MAX)	U.S.A.,1968-1970,N=30,, ATOM.ABS.AND/OR POLAROGRAPHY	487
*TOMATO	0 UG/G WET WT.	U.S.A.	487
*TOMATO	0.015 UG/G WET WT.	WESTERN GERMANY	487
*TOMATO	0.013 UG/G WET WT.	RUMANIA	487

TABLE D269

CONCENTRATIONS : CD

ITEM:	VALUE:	CONDITIONS:	REF.:
*TOMATO	0.032 UG/G WET WT.	JAPAN (NONPOLLUTED AREA)	487
*TOMATO	0.06 UG/G WET WT.	U.K.	487
*TRITICUM AESTIVUM, GRAIN	0.1 MG/KG DRY WEIGHT	EDIBLE PART, SOIL=0.0 UG CD/G, PH=7.7, ZN=47 PPM	127
*TRITICUM AESTIVUM, GRAIN	1.5 MG/KG DRY WEIGHT	EDIBLE PART, SOIL=2.5 UG CD/G, PH=7.7, ZN=47 PPM	127
*TRITICUM AESTIVUM, GRAIN	2.8 MG/KG DRY WEIGHT	EDIBLE PART, SOIL=5.0 UG CD/G, PH=7.7, ZN=47 PPM	127
*TRITICUM AESTIVUM, GRAIN	5.8 MG/KG DRY WEIGHT	EDIBLE PART, SOIL=10.0 UG CD/G, PH=7.7, ZN=47 PPM	127
*TRITICUM AESTIVUM, GRAIN	8.7 MG/KG DRY WEIGHT	EDIBLE PART, SOIL=20.0 UG CD/G, PH=7.7, ZN=47 PPM	127
*TRITICUM AESTIVUM, GRAIN	10.0 MG/KG DRY WEIGHT	EDIBLE PART, SOIL=40.0 UG CD/G, PH=7.7, ZN=47 PPM	127
*TRITICUM AESTIVUM, GRAIN	13.0 MG/KG DRY WEIGHT	EDIBLE PART, SOIL=80.0 UG CD/G, PH=7.7, ZN=47 PPM	127
*TRITICUM AESTIVUM, GRAIN	14.0 MG/KG DRY WEIGHT	EDIBLE PART, SOIL=160.0 UG CD/G, PH=7.7, ZN=47 PPM	127
*TRITICUM AESTIVUM, LEAF	0.1 MG/KG DRY WEIGHT	DIAGNOSTIC PART, 2 TOP LEAVES, SOIL=0.0 UG CD/G, PH=7.7, ZN=47 PPM	127
*TRITICUM AESTIVUM, LEAF	2.7 MG/KG DRY WEIGHT	DIAGNOSTIC PART, 2 TOP LEAVES, SOIL=2.5 UG CD/G, PH=7.7, ZN=47 PPM	127
*TRITICUM AESTIVUM, LEAF	6.4 MG/KG DRY WEIGHT	DIAGNOSTIC PART, 2 TOP LEAVES, SOIL=5.0 UG CD/G, PH=7.7, ZN=47 PPM	127
*TRITICUM AESTIVUM, LEAF	12.0 MG/KG DRY WEIGHT	DIAGNOSTIC PART, 2 TOP LEAVES, SOIL=10.0 UG CD/G, PH=7.7, ZN=47 PPM	127
*TRITICUM AESTIVUM, LEAF	24.0 MG/KG DRY WEIGHT	DIAGNOSTIC PART, 2 TOP LEAVES, SOIL=20.0 UG CD/G, PH=7.7, ZN=47 PPM	127
*TRITICUM AESTIVUM, LEAF	31.0 MG/KG DRY WEIGHT	DIAGNOSTIC PART, 2 TOP LEAVES, SOIL=40.0 UG CD/G, PH=7.7, ZN=47 PPM	127
*TRITICUM AESTIVUM, LEAF	41.0 MG/KG DRY WEIGHT	DIAGNOSTIC PART, 2 TOP LEAVES, SOIL=80.0 UG CD/G, PH=7.7, ZN=47 PPM	127
*TRITICUM AESTIVUM, LEAF	48.0 MG/KG DRY WEIGHT	DIAGNOSTIC PART, 2 TOP LEAVES, SOIL=160.0 UG CD/G, PH=7.7, ZN=47 PPM	127
*WATER	< 1 UG/L	729 RIVERS AND LAKES,U.S.A.	893
*WHEAT FLOUR	0.07 UG/G WET WT.	U.S.A.	487
*WHEAT FLOUR	0.047 UG/G WET WT.	WESTERN GERMANY	487
*WHEAT FLOUR	0.02 UG/G WET WT.	CZECHOSLOVAKIA	487
*WHEAT FLOUR	0.025 UG/G WET WT.	JAPAN (NONPOLLUTED AREA)	487
*WHEAT FLOUR	0.05-0.10 UG/G WET WT.	CANADA	487
*ZEA MAYS, GRAIN	0.05 MG/KG DRY WEIGHT	EDIBLE PART, SOIL=0.0 UG CD/G, PH=7.7, ZN=47 PPM	127
*ZEA MAYS, GRAIN	0.05 MG/KG DRY WEIGHT	EDIBLE PART, SOIL=2.5 UG CD/G, PH=7.7, ZN=47 PPM	127
*ZEA MAYS, GRAIN	0.5 MG/KG DRY WEIGHT	EDIBLE PART, SOIL=5.0 UG CD/G, PH=7.7, ZN=47 PPM	127
*ZEA MAYS, GRAIN	1.4 MG/KG DRY WEIGHT	EDIBLE PART, SOIL=10.0 UG CD/G, PH=7.7, ZN=47 PPM	127
*ZEA MAYS, GRAIN	2.4 MG/KG DRY WEIGHT	EDIBLE PART, SOIL=20.0 UG CD/G, PH=7.7, ZN=47 PPM	127
*ZEA MAYS, GRAIN	3.4 MG/KG DRY WEIGHT	EDIBLE PART; SOIL=40.0 UG CD/G, PH=7.7, ZN=47 PPM	127
*ZEA MAYS, GRAIN	3.2 MG/KG DRY WEIGHT	EDIBLE PART, SOIL=80.0 UG CD/G, PH=7.7, ZN=47 PPM	127
*ZEA MAYS, GRAIN	6.6 MG/KG DRY WEIGHT	EDIBLE PART, SOIL=160.0 UG CD/G, PH=7.7, ZN=47 PPM	127
*ZEA MAYS, LEAF	3.9 MG/KG DRY WEIGHT	DIAGNOSTIC PART, 3+4 LEAF FROM TOP, SOIL=0.0 UG CD/G, PH=7.7, ZN=47 P	127

TABLE D269

CONCENTRATIONS : CD

ITEM:	VALUE:	CONDITIONS:	REF.:
*ZEA MAYS, LEAF	11.0 MG/KG DRY WEIGHT	DIAGNOSTIC PART, 3+4 LEAF FROM TOP, SOIL=2.5 UG CD/G, PH=7.7, ZN =47 P	127
*ZEA MAYS, LEAF	23.0 MG/KG DRY WEIGHT	DIAGNOSTIC PART, 3+4 LEAF FROM TOP, SOIL=5.0 UG CD/G, PH=7.7, ZN =47 P	127
*ZEA MAYS, LEAF	27.0 MG/KG DRY WEIGHT	DIAGNOSTIC PART, 3+4 LEAF FROM TOP, SOIL=10.0 UG CD/G, PH=7.7, ZN=47	127
*ZEA MAYS, LEAF	40.0 MG/KG DRY WEIGHT	DIAGNOSTIC PART, 3+4 LEAF FROM TOP, SOIL=20.0 UG CD/G, PH=7.7, ZN=47	127
*ZEA MAYS, LEAF	58.0 MG/KG DRY WEIGHT	DIAGNOSTIC PART, 3+4 LEAF FROM TOP, SOIL=40.0 UG CD/G, PH=7.7, ZN=47	127
*ZEA MAYS, LEAF	71.0 MG/KG DRY WEIGHT	DIAGNOSTIC PART, 3+4 LEAF FROM TOP, SOIL=80.0 UG CD/G, PH=7.7, ZN=47	127
*ZEA MAYS, LEAF	94.0 MG/KG DRY WEIGHT	DIAGNOSTIC PART, 3+4 LEAF FROM TOP, SOIL=160.0 UG CD/G, PH=7.7, ZN=47	127
*ZOOPLANKTON	2.08 MG/KG DRY WT.(MEAN)	19ST.,N.Y.BIGHT,CONC.IN PHYTOPL. (LONG ISLAND)=1.6MG/KG DRY WT. (MEAN)	565

TABLE D270

CONCENTRATIONS : CD IN LIVER/CD IN KIDNEY

ITEM:	VALUE:	CONDITIONS:	REF.:
*DOG	0.17	0.5UG CD/G WATER,AFTER 4 YEARS	487
*DOG	0.19	2.5UG CD/G WATER,AFTER 4 YEARS	487
*DOG	0.24	5.0UG CD/G WATER,AFTER 4 YEARS	487
*DOG	0.30	10.0UG CD/G WATER,AFTER 4 YEARS	487
*RAT	0.20	0.5UG CD/G WATER,AFTER 6 MONTHS	487
*RAT	0.38	2.5UG CD/G WATER,AFTER 6 MONTHS	487
*RAT	0.43	5.0UG CD/G WATER,AFTER 6 MONTHS	487
*RAT	0.67	10.0UG CD/G WATER,AFTER 6 MONTHS	487
*RAT	0.14	0.1UG CD/G WATER,AFTER 12 MONTHS	487
*RAT	0.19	0.5UG CD/G WATER,AFTER 12 MONTHS	487
*RAT	0.24	2.5UG CD/G WATER,AFTER 12 MONTHS	487
*RAT	0.31	5.0UG CD/G WATER,AFTER 12 MONTHS	487
*RAT	0.47	10.0UG CD/G WATER,AFTER 12 MONTHS	487

TABLE D271

CONCENTRATIONS : CHLORDANE

ITEM:	VALUE:	CONDITIONS:	REF.:
*SEDIMENT	<0.001 MG/KG	CONC.IN WATER:<1.0UG/L,U.S.	389
*SEDIMENT	<0.01 MG/KG	CONC.IN WATER:<1.0UG/L,U.S.	389

TABLE D272

CONCENTRATIONS : CO

ITEM:	VALUE:	CONDITIONS:	REF.:
*ATMOSPHERE	0.08-4 NG/M**3	ENGLAND,7 STATIONS	893

TABLE D272

CONCENTRATIONS : CO			
ITEM:	VALUE:	CONDITIONS:	REF.:
*ATMOSPHERE	6 NG/M**3	URBAN,U.S.A.	893
*ATMOSPHERE	0.00084+-0.00027 NG/M**3	SOUTHPOLE,1970	893
*COCKSFOOT	0.16 MG/KG DRY WT.	ADDED COSO4=2.2KG/HA,PH=5.6; CONTROL=0.03MG/KG DRY WT.	1048
*COCKSFOOT	0.25 MG/KG DRY WT.	ADDED COSO4=4.4KG/HA,PH=5.6; CONTROL=0.03MG/KG DRY WT.	1048
*COCKSFOOT	0.10 MG/KG DRY WT.	ADDED COSO4=2.2KG/HA,PH=7.0; CONTROL=0.02MG/KG DRY WT.	1048
*COCKSFOOT	0.15 MG/KG DRY WT.	ADDED COSO4=4.4KG/HA,PH=7.0; CONTROL=0.02MG/KG DRY WT.	1048
*FOOD	0.01-0.35 MG/KG DRY WT.	VEGETABLES	893
*FOOD	0.1-0.3 MG/KG DRY WT.	ANIMAL	893
*HYPOLIMNION	0.07 UG/L	CONC.IN EPILIMNION=0.03MIC.G/L, CONC.FACTOR=2.3,SCHOEHSEE	573
*RAIN	0.01 UG/L	SWEDEN,3 STATIONS	893
*RAIN	0.25 UG/L	ENGLAND,YEARLY MEAN	893
*RED CLOVER	0.31 MG/KG DRY WT.	ADDED COSO4=2.2KG/HA,PH=5.6; CONTROL=0.06MG/KG DRY WT.	1048
*RED CLOVER	0.53 MG/KG DRY WT.	ADDED COSO4=4.4KG/HA,PH=5.6; CONTROL=0.06MG/KG DRY WT.	1048
*RED CLOVER	0.17 MG/KG DRY WT.	ADDED COSO4=2.2KG/HA,PH=7.0; CONTROL=0.04MG/KG DRY WT.	1048
*RED CLOVER	0.38 MG/KG DRY WT.	ADDED COSO4=4.4KG/HA,PH=7.0; CONTROL=0.04MG/KG DRY WT.	1048
*SEDIMENT	6.6 MG/KG DRY WEIGHT	UNPOLLUTED AREA, ESTUARY, URR WATER	1557
*SEDIMENT	5.7 (RANGE 3.8 - 7.5) MG/KG DRY WEIGHT	UNPOLLUTED AREA, ESTUARY, URR WATER	1558
*SEDIMENT	7.9 (RANGE 5.3 - 11.9) MG/KG DRY WEIGHT	UNPOLLUTED AREA, ESTUARY, URR WATER	1558
*SEDIMENT	6 (RANGE 4 - 8) MG/KG DRY WEIGHT	UNPOLLUTED AREA, RIVER,ILLINOIS RIVER	993
*SEDIMENT	8.3 MIC.G/G DRY WT.	CONC.IN EPILIMNION=0.03MIC.G/L, CONC.FACTOR=280000,SCHOEHSEE	573
*TIMOTHY	0.19 MG/KG DRY WT.	ADDED LIMESTONE=0.63T/HA,SOIL-PH= 4.7,N.WALES	40
*TIMOTHY	0.11 MG/KG DRY WT.	ADDED LIMESTONE=2.5T/HA,SOIL-PH= 5.0,N.WALES	40
*TIMOTHY	0.10 MG/KG DRY WT.	ADDED LIMESTONE=5.0T/HA,SOIL-PH= 5.2,N.WALES	40
*TIMOTHY	0.08 MG/KG DRY WT.	ADDED LIMESTONE=15.0T/HA,SOIL-PH= 6.0,N.WALES	40
*VEGETATION	0.16 MG/KG DRY WT.	CONC.IN PODZOL:F=1,H=1,A1A2=2,B1= 2,B2=8,B3=5 (F030)	1048
*WATER	1-36 UG/L	RIVERS,U.S.A.	893
*WATER	0.50 UG/L	BALTIC SEA,GOTLANDSDYBET:250M, 1970-71	893
*WATER	3.75 UG/L	LAKE,SWEDEN	893
*WHITE CLOVER	0.40 MG/KG DRY WT.	ADDED LIMESTONE=0.63T/HA,SOIL-PH= 4.7,N.WALES	40
*WHITE CLOVER	0.36 MG/KG DRY WT.	ADDED LIMESTONE=2.5T/HA,SOIL-PH= 5.0,N.WALES	40
*WHITE CLOVER	0.30 MG/KG DRY WT.	ADDED LIMESTONE=5.0T/HA,SOIL-PH= 5.2,N.WALES	40
*WHITE CLOVER	0.29 MG/KG DRY WT.	ADDED LIMESTONE=15.0T/HA,SOIL-PH= 6.0,N.WALES	40

TABLE D273

CONCENTRATIONS : COD			
ITEM:	VALUE:	CONDITIONS:	REF.:
*AGRICULTURAL CROPLAND	80 MG/L	GENERAL CHARACTERISTICS	364

TABLE D273

CONCENTRATIONS : COD			
ITEM:	VALUE:	CONDITIONS:	REF.:
*ANIMAL FEEDLOT, RUNOFF	3100-41000 MG/L	GENERAL CHARACTERISTICS	364
*MUNICIPAL SEWAGE	500 (250-750) MG/L	TYPICAL UNTREATED	364
*MUNICIPAL SEWAGE	330 (165-500) MG/L	TYPICAL TREATED,PRIMARY	364
*MUNICIPAL SEWAGE	55 (25-80) MG/L	TYPICAL TREATED,SECONDARY	364
*PRECIPITATION	9-16 MG/L	GENERAL CHARACTERISTICS	364
*URBAN LAND,DRAINAGE	85-110 MG/L	GENERAL CHARACTERISTICS	364

TABLE D274

CONCENTRATIONS : COMPOUNDS IN DRY BIOMASS			
ITEM:	VALUE:	CONDITIONS:	REF.:
*ADENINE	1.40*10**(-4) MOLES/G DRY BIOMASS	NONE	1067
*ALANINE	4.54*10**(-4) MOLES/G DRY BIOMASS	NONE	1067
*ARGININE	2.52*10**(-4) MOLES/G DRY BIOMASS	NONE	1067
*ASPARAGINE	1.01*10**(-4) MOLES/G DRY BIOMASS	NONE	1067
*ASPARTIC ACID	2.01*10**(-4) MOLES/G DRY BIOMASS	NONE	1067
*CYSTEINE	1.01*10**(-4) MOLES/G DRY BIOMASS	NONE	1067
*CYSTOSINE	1.40*10**(-4) MOLES/G DRY BIOMASS	NONE	1067
*DEOXYRIBOSE	0.96*10**(-4) MOLES/G DRY BIOMASS	NONE	1067
*FATTY ACID	2.80*10**(-4) MOLES/G DRY BIOMASS	NONE	1067
*GLUTAMIC ACID	3.53*10**(-4) MOLES/G DRY BIOMASS	NONE	1067
*GLUTAMINE	2.01*10**(-4) MOLES/G DRY BIOMASS	NONE	1067
*GLYCERIN	1.40*10**(-4) MOLES/G DRY BIOMASS	NONE	1067
*GLYCINE	¤.03*10**(-4) MOLES/G DRY BIOMASS	NONE	1067
*GUANINE	1.40*10**(-4) MOLES/G DRY BIOMASS	NONE	1067
*HEXOSE	10.26*10**(-4) MOLES/G DRY BIOMASS	NONE	1067
*HISTIDINE	0.50*10**(-4) MOLES/G DRY BIOMASS	NONE	1067
*ISOLEUCINE	2.52*10**(-4) MOLES/G DRY BIOMASS	NONE	1067
*LEUCINE	4.03*10**(-4) MOLES/G DRY BIOMASS	NONE	1067
*LYSINE	4.03*10**(-4) MOLES/G DRY BIOMASS	NONE	1067
*METHIONINE	2.01*10**(-4) MOLES/G DRY BIOMASS	NONE	1067
*OTHER	2.08*10**(-4) MOLES/G DRY BIOMASS	NONE	1067
*PHENYLALANINE	1.51*10**(-4) MOLES/G DRY BIOMASS	NONE	1067
*PHOSPHATE	6.80*10**(-4) MOLES/G DRY BIOMASS	NONE	1067
*PROLINE	2.52*10**(-4) MOLES/G DRY BIOMASS	NONE	1067
*RIBOSE	4.47*10**(-4) MOLES/G DRY BIOMASS	NONE	1067
*SERINE	3.02*10**(-4) MOLES/G DRY BIOMASS	NONE	1067
*THREONINE	2.52*10**(-4) MOLES/G DRY BIOMASS	NONE	1067
*THYMINE	0.24*10**(-4) MOLES/G DRY BIOMASS	NONE	1067
*TRYPTOPHAN	0.50*10**(-4) MOLES/G DRY BIOMASS	NONE	1067
*TYROSINE	1.01*10**(-4) MOLES/G DRY BIOMASS	NONE	1067
*URACIL	1.15*10**(-4) MOLES/G DRY BIOMASS	NONE	1067
*VALINE	3.02*10**(-4) MOLES/G DRY BIOMASS	NONE	1067

TABLE D275

CONCENTRATIONS : CR			
ITEM:	VALUE:	CONDITIONS:	REF.:
*ALISMA PLANTAGO AQUATICA, OLD LEAVES	4.92 MG/KG DRY WT.	800M ABOVE POLL.ENTRY(1),CR IN WATER=NOT DETECT.,WHOLE LEAVES	1224

TABLE D275

CONCENTRATIONS : CR			
ITEM:	VALUE:	CONDITIONS:	REF.:
*ALISMA PLANTAGO AQUATICA, OLD LEAVES	2.71 MG/KG DRY WT.	800M ABOVE POLL.ENTRY(1),CR IN WATER=NOT DETECT.,PROCESSED LEAVES	1224
*ALISMA PLANTAGO AQUATICA, OLD LEAVES	3.36 MG/KG DRY WT.	150M BELOW POLL.ENTRY(1),CR IN WATER=0.030MG/L,WHOLE LEAVES	1224
*ALISMA PLANTAGO AQUATICA, OLD LEAVES	1.48 MG/KG DRY WT.	150M BELOW POLL.ENTRY(1),CR IN WATER=0.030MG/L,PROCESSED LEAVES	1224
*ALISMA PLANTAGO AQUATICA, OLD LEAVES	2.45 MG/KG DRY WT.	800 AND 50M BELOW POLL.(1)AND(2), CR IN WATER=0.050MG/L,WHOLE LEAVES	1224
*ALISMA PLANTAGO AQUATICA, OLD LEAVES	2.07 MG/KG DRY WT.	800 AND 50M BELOW POLL.(1)AND(2), CR IN WATER=0.050MG/L,PROCESSED LEAVES	1224
*ALISMA PLANTAGO AQUATICA, YOUNG LEAVES	2.65 MG/KG DRY WT.	800M ABOVE POLL.ENTRY(1),CR IN WATER=NOT DETECT.,WHOLE LEAVES	1224
*ALISMA PLANTAGO AQUATICA, YOUNG LEAVES	1.90 MG/KG DRY WT.	800M ABOVE POLL.ENTRY(1),CR IN WATER=NOT DETECT.,PROCESSED LEAVES	1224
*ALISMA PLANTAGO AQUAT¤CA, YOUNG LEAVES	14.10 MG/KG DRY WT.	150M BELOW POLL.ENTRY(1),CR IN WATER=0.030MG/L,WHOLE LEAVES	1224
*ALISMA PLANTAGO AQUATICA, YOUNG LEAVES	7.93 MG/KG DRY WT.	150M BELOW POLL.ENTRY(1),CR IN WATER=0.030MG/L,PROCESSED LEAVES	1224
*ALISMA PLANTAGO AQUATICA, YOUNG LEAVES	5.90 MG/KG DRY WT.	800 AND 50M BELOW POLL.(1)AND(2), CR IN WATER=0.050MG/L,WHOLE LEAVES	1224
*ALISMA PLANTAGO AQUATICA, YOUNG LEAVES	5.03 MG/KG DRY WT.	800 AND 50M BELOW POLL.(1)AND(2), CR IN WATER=0.050MG/L,PROCESSED LEAVES	1224
*ATMOSPHERE	0.0053+-0.003 NG/M**3	SOUTHPOLE,1970	893
*ATMOSPHERE	1-55 NG/M**3	URBAN,U.S.A.	893
*ATMOSPHERE	1-12 NG/M**3	NOT URBAN,U.S.A.	893
*ATMOSPHERE	0.8-11.5 NG/M**3	ENGLAND,7 STATIONS	893
*ATMOSPHERE	130-480 NG/M**3	SWEDEN,MEAN FROM 8 STATIONS	893
*BARLEY, LEAVES	2.1 MG/KG DRY WT.	SOIL-PH=5.6,ADDED CR-ALUM.=220PPM	1226
*BARLEY, LEAVES	2.2 MG/KG DRY WT.	SOIL-PH=5.6,ADDED K2CR2O7=200PPM	1226
*BARLEY, LEAVES	2.7 MG/KG DRY WT.	SOIL-PH=6.4,ADDED CR-ALUM.=200PPM, CONTROL=0.8MG/KG DRY WT.	1226
*BARLEY, LEAVES	2.3 MG/KG DRY WT.	SOIL-PH=6.4,ADDED K2CR2O7=200PPM, CONTROL=0.8MG/KG DRY WT.	1226
*BARLEY, LEAVES	2.6 MG/KG DRY WT.	SOIL-PH=7.8,ADDED CR-ALUM.=200PPM, CONTROL=0.5MG/KG DRY WT.	1226
*BARLEY, LEAVES	3.2 MG/KG DRY WT.	SOIL-PH=7.8,ADDED K2CR2O7=200PPM, CONTROL=0.5MG/KG DRY WT.	1226
*BARLEY, ROOT	46 MG/KG DRY WT.	SOIL-PH=5.6,ADDED CR-ALUM.=200PPM	1226
*BARLEY, ROOT	137 MG/KG DRY WT.	SOIL-PH=5.6,ADDED K2CR2O7=200PPM	1226
*BARLEY, ROOT	78 MG/KG DRY WT.	SOIL-PH=6.4,ADDED CR-ALUM.=200PPM, CONTROL=2.6MG/KG DRY WT.	1226
*BARLEY, ROOT	115 MG/KG DRY WT.	SOIL-PH=6.4,ADDED K2CR2O7=200PPM, CONTROL=2.6MG/KG DRY WT.	1226
*BARLEY, ROOT	146 MG/KG DRY WT.	SOIL-PH=7.8,ADDED CR-ALUM.=200PPM, CONTROL=1.9MG/KG DRY WT.	1226
*BARLEY, ROOT	168 MG/KG DRY WT.	SOIL-PH=7.8,ADDED K2CR2O7=200PPM, CONTROL=1.9MG/KG DRY WT.	1226
*FOOD	0.2-1 MG/KG DRY WT.	VEGETABLES	893
*LIVING SPRUCE NEEDLES	0.2 - 1.5 MG/KG DRY WEUGHT	LITTER=2.3-6.7, UNDERLAYING SOIL= 23-81MG/KG DW,SOUTH SWEDEN	1129
*RAIN	2.9 UG/L	ENGLAND,YEARLY MEAN,1 STATION	893
*RAIN	0.1-0.4 UG/L	SWEDEN,3 STATIONS	893
*RIVERS	0.7-84. PPB	15 RIVERS,NORTH AMERICA, CR(VI)	587
*SEDIMENT	9.9 MG/KG DRY WEIGHT	UNPOLLUTED AREA, ESTUARY, URR WATER	1557
*SEDIMENT	3.7 (RANGE 1.6 - 5.7) MG/KG DRY WEIGHT	UNPOLLUTED AREA, ESTUARY, URR WATER	1558
*SEDIMENT	5.7 (RANGE 3.9 - 8.4) MG/KG DRY WEIGHT	UNPOLLUTED AREA, ESTUARY, URR WATER	1558
*SEDIMENT	6 (RANGE 3 - 7) MG/KG DRY WEIGHT	UNPOLLUTED AREA, RIVER,ILLINOIS RIVER	993

TABLE D275

CONCENTRATIONS : CR			
ITEM:	VALUE:	CONDITIONS:	REF.:
*SEDIMENT	10.43 (5.2-14.0) MG/KG DRY WT.	800M ABOVE POLL.ENTRY(1),CR IN WATER=NOT DETECT.,POLLUTED RIVER	1224
*SEDIMENT	53.00 (43.58-59.00) MG/KG DRY WT.	150M BELOW POLL.ENTRY(1),CR IN WATER=0.030MG/L,POLLUTED RIVER	1224
*SEDIMENT	64.74 (62.67-66.60) MG/KG DRY WT.	800 AND 50M BELOW POLL.(1)AND(2), CR IN WATER=0.050MG/L,POLLUTED RIVER	1224
*SURFACE WATER	9.-88. PPB	HOLLAND, CR(VI)	1523
*VEGETATION	1.3 MG/KG DRY WT.	CONC.IN PODZOL:F=5,H=8,A1A2=80,B1=80,B2=100,B3=100,(?043)	1048
*ZOOPLANKTON	< 7.92 MG/KG DRY WT.(MEAN)	19ST.,N.Y.BIGHT,CONC.IN PHYTOPL. (LONG ISLAND):<2.88MG/KG DRY WT.MEAN	565

TABLE D276

CONCENTRATIONS : CR-51			
ITEM:	VALUE:	CONDITIONS:	REF.:
*DUNALIELLA SP.	94.2 % OF TOTAL	LAB, 1 H, SOLUBLE FRACTION, WITHOUT ALGAE = 94.2 %	482
*DUNALIELLA SP.	3.4 % OF TOTAL	LAB, 1 H, INSOLUBLE FRACTION, WITHOUT ALGAE = 5.8 %	482
*DUNALIELLA SP.	97.4 % OF TOTAL	LAB,24 H, SOLUBLE FRACTION, WITHOUT ALGAE = 93.6 %	482
*DUNALIELLA SP.	2.6 % OF TOTAL	LAB, 24 H, INSOLUBLE FRACTION, WITHOUT ALGAE = 6.4 %	482

TABLE D277

CONCENTRATIONS : CU			
ITEM:	VALUE:	CONDITIONS:	REF.:
*ALISMA PLANTAGO AQUATICA, OLD LEAVES	129.58 MG/KG DRY WT.	800M ABOVE POLL.ENTRY(1),CU IN WATER=0.032MG/L,WHOLE LEAVES	1224
*ALISMA PLANTAGO AQUATICA, OLD LEAVES	68.66 MG/KG DRY WT.	800M ABOVE POLL.ENTRY(1),CU IN WATER=0.032MG/L,PROCESSED LEAVES	1224
*ALISMA PLANTAGO AQUATICA, OLD LEAVES	77.10 MG/KG DRY WT.	150M BELOW POLL.ENTRY(1),CU IN WATER=0.113MG/L,WHOLE LEAVES	1224
*ALISMA PLANTAGO AQUATICA, OLD LEAVES	51.54 MG/KG DRY WT.	150M BELOW POLL.ENTRY(1),CU IN WATER=0.113MG/L,PROCESSED LEAVES	1224
*ALISMA PLANTAGO AQUATICA, OLD LEAVES	113.90 MG/KG DRY WT.	800 AND 50M BELOW POLL.(1)AND(2), CU IN WATER=0.085MG/L,WHOLE LEAVES	1224
*ALISMA PLANTAGO AQUATICA, OLD LEAVES	81.58 MG/KG DRY WT.	800 AND 50M BELOW POLL.(1)AND(2), CU IN WATER=0.085MG/L,PROCESSED LEAVES	1224
*ALISMA PLANTAGO AQUATICA, YOUNG LEAVES	155.67 MG/KG DRY WT.	800M ABOVE POLL.ENTRY(1),CU IN WATER=0.032MG/L,WHOLE LEAVES	1224
*ALISMA PLANTAGO AQUATICA, YOUNG LEAVES	70.05 MG/KG DRY WT.	800M ABOVE POLL.ENTRY(1),CU IN WATER=0.032MG/L,PROCESSED LEAVES	1224
*ALISMA PLANTAGO AQUATICA, YOUNG LEAVES	128.51 MG/KG DRY WT.	150M BELOW POLL.ENTRY(1),CU IN WATER=0.113MG/L,WHOLE LEAVES	1224
*ALISMA PLANTAGO AQUATICA, YOUNG LEAVES	75.67 MG/KG DRY WT.	150M BELOW POLL.ENTRY(1),CU IN WATER=0.113MG/L,PROCESSED LEAVES	1224
*ALISMA PLANTAGO AQUATICA, YOUNG LEAVES	136.31 MG/KG DRY WT.	800 AND 50M BELOW POLL.(1)AND(2), CU IN WATER=0.085MG/L,WHOLE LEAVES	1224
*ALISMA PLANTAGO AQUATICA, YOUNG LEAVES	77.60 MG/KG DRY WT.	800 AND 50M BELOW POLL.(1)AND(2), CU IN WATER=0.085MG/L,PROCESSED LEAVES	1224
*ANABAENA VARIABILIS	4.7 UG/CELL	ADD.=0.0MG CU++,AG+/L, INOCULUM=1X,CU IN MEDIA=8.4 MIC.G/40 ML	1765

TABLE D277

CONCENTRATIONS : CU			
ITEM:	VALUE:	CONDITIONS:	REF.:
*ANABAENA VARIABILIS	9.5 UG/CELL	ADD.=0.1MG CU++,AG+/L,INOCULUM=1X, CU IN MEDIA=29.0 MIC.G/40 ML	1765
*ANABAENA VARIABILIS	3.4 UG/CELL	ADD.=0.0MG CU++,AG+/L,INOCULUM=5X, CU IN MEDIA=11.0 MIC.G/40 ML	1765
*ANABAENA VARIABILIS	7.2 UG/CELL	ADD.=0.1MG CU++,AG+/L,INOCULUM=5X, CU IN MEDIA=29.0 MIC.G/40 ML	1765
*ATMOSPHERE	0.036+-0.019 NG/M**3	SOUTHPOLE	893
*ATMOSPHERE	90 NG/M**3	U.S.A.,1960-65,MEAN FOR 103 STATES,MAX=10000NG/M**3	893
*ATMOSPHERE	1-46 NG/M**3	URBAN,ENGLAND	893
*BUSYCON CANALICULATUM, DIGESTIVE GLAND	400 - 900 MICROGRAM CU/G WET WEIGHT	SUMMER, NARRAGANSETT BAY	110
*BUSYCON CANALICULATUM, DIGESTIVE GLAND	65 - 120 MICROGRAM CU/G WET WEIGHT	WINTER, NARRAGANSETT BAY	110
*CABBAGE, PLANT	4.1 PPM DRY MATTER	15 PLANTS,RURAL AREA	1283
*CABBAGE, PLANT	3.6 PPM DRY MATTER	15 PLANTS,URBAN AREA	1283
*CONIFEROUS TREES, NEEDLES	240 (SD=+-26) MG/KG ASH	SILICIC ROCK=50+-4,SOIL=110+-10 PPM, FINLAND,N=58	950
*CONIFEROUS TREES, NEEDLES	59 (SD=+-12) MG/KG ASH	ULTRABASIC ROCK = 27+-4,SOIL=22+-5 PPM, FINLAND,N=10	950
*CONIFEROUS TREES, NEEDLES	130 (SD=+-34) MG/KG ASH	CALCAREOUS ROCK=34+-7,SOIL=42+-10 PPM, FINLAND,N=9	950
*CONIFEROUS TREES, TWIGS	510 (SD=+-71) MG/KG ASH	SILICIC ROCK=50+-4,SOIL=110+-10 PPM, FINLAND,N=58	950
*CONIFEROUS TREES, TWIGS	82 (SD=+-10) MG/KG ASH	ULTRABASIC ROCK = 27+-4,SOIL=22+-5 PPM, FINLAND,N=10	950
*CONIFEROUS TREES, TWIGS	290 (SD=+-48) MG/KG ASH	CALCAREOUS ROCK=34+-7,SOIL=42+-10 PPM, FINLAND,N=9	950
*DECIDUOUS TREES AND SHRUBS, LEAVES	150 (SD=+-21) MG/KG ASH	SILICIC ROCK=50+-4,SOIL=110+-10 PPM, FINLAND,N=27	950
*DECIDUOUS TREES AND SHRUBS, LEAVES	94 (SD=+-10) MG/KG ASH	ULTRABASIC ROCK = 27+-4,SOIL=22+-5 PPM, FINLAND,N=14	950
*DECIDUOUS TREES AND SHRUBS, LEAVES	99 (SD=+-26) MG/KG ASH	CALCAREOUS ROCK=34+-7,SOIL=42+-10 PPM, FINLAND,N=16	950
*DECIDUOUS TREES AND SHRUBS, TWIGS	270 (SD=+-47) MG/KG ASH	SILICIC ROCK=50+-4,SOIL=110+-10 PPM, FINLAND,N=27	950
*DECIDUOUS TREES AND SHRUBS, TWIGS	130 (SD=+-20) MG/KG ASH	ULTRABASIC ROCK = 27+-4,SOIL=22+-5 PPM, FINLAND,N=14	950
*DECIDUOUS TREES AND SHRUBS, TWIGS	96 (SD=+-16) MG/KG ASH	CALCAREOUS ROCK=34+-7,SOIL=42+-10 PPM, FINLAND,N=16	950
*FOOD	1.49 UG/G	FISH AND SHELLFISH(EXCL.OYSTER), U.S.A.	893
*FOOD	1.17 UG/G	VEGETABLES,U.S.A.	893
*FOOD	4.63 UG/G	FAT(EXCL.LECITHIN),U.S.A.	893
*FOOD	14.82 UG/G	NUTS,U.S.A.	893
*FOOD	3.92 UG/G	MEAT (INCL.LIVER=11MIC.G/G),U.S.A.	893
*FOOD	1.76 UG/G	MILK PRODUCTS,U.S.A.	893
*HERBAGE	10.1 MG/KG DRY WT.	NO CU ADDED,SOUTH WALES	40
*HERBAGE	11.0 MG/KG DRY WT.	ADDED CUSO4=11.2KG/HA(EQ.CU),SOUTH WALES	40
*HERBAGE	12.0 MG/KG DRY WT.	ADDED CUSO4=22.4KG/HA(EQ.CU),SOUTH WALES	40
*HERBAGE	12.4 MG/KG DRY WT.	ADDED CU-OXIDE=11.2KG/HA(EQ.CU), SOUTH WALES	40
*HERBAGE	12.0 MG/KG DRY WT.	ADDED CU-OXIDE=22.4KG/HA(EQ.CU), SOUTH WALES	40
*HOMO SAPIENS, BLOOD	0.65 (SD=0.27) MICRO M	CU IN TAP WATER,1ST FLUSH=0.10, RUNNING WAT=0.04 MICRO M CU,N=30, GB	1060
*HUMAN ALA DEHYDRATASE ACTIVITY	40.3 (SD=5.5) MICRO MOL ALA/MIN/ML RBC	CU IN TAP WATER,1ST FLUSH=0.10, RUNNING WAT=0.04 MICRO M CU,N=30, GB	1060
*HYPOLIMNION	0.9 UG/L	CONC.IN EPILIMNION=1.0MIC.G/L, CONC.FACTOR=0.9,SCHOEHSEE	573

TABLE D277

CONCENTRATIONS : CU			
ITEM:	VALUE:	CONDITIONS:	REF.:
*LIVING SPRUCE NEEDLES	1.4 - 3.5 MG/KG DRY WEUGHT	LITTER=5.1-13.5, UNDERLAYING SOIL= 37-48MG/KG DW,SOUTH SWEDEN	1129
*NOEMACHEILUS BARBATULUS	7.9 MICROGRAM/G DRY MATTER	COPPER=0.76MG/L,CONTROL=3.MG/L, 285K,10 FISH	1477
*NOEMACHEILUS BARBATULUS	5.2 MICROGRAM/G DRY MATTER	COPPER=0.71MG/L,CONTROL=3.MG/L, 285K,10 FISH	1477
*NOEMACHEILUS BARBATULUS	6.2 MICROGRAM/G DRY MATTER	COPPER=0.49MG/L,CONTROL=3.MG/L, 285K,10 FISH	1477
*NOEMACHEILUS BARBATULUS	10.3 MICROGRAM/G DRY MATTER	COPPER=0.38MG/L,CONTROL=3.MG/L, 285K,10 FISH	1477
*NOEMACHEILUS BARBATULUS	7.8 MICROGRAM/G DRY MATTER	COPPER=0.29MG/L,CONTROL=3.MG/L, 285K,10 FISH	1477
*NOEMACHEILUS BARBATULUS	3.7 MICROGRAM/G DRY MATTER	COPPER=0.12MG/L,CONTROL=3.MG/L, 285K,10 FISH	1477
*NOEMACHEILUS BARBATULUS, BRAIN	53 MICROGRAM/G DRY MATTER	COPPER=0.76MG/L,NO CONTROL,285K,10 FISH	1477
*NOEMACHEILUS BARBATULUS, BRAIN	12.1 MICROGRAM/G DRY MATTER	COPPER=0.71MG/L,NO CONTROL,285K,10 FISH	1477
*NOEMACHEILUS BARBATULUS, BRAIN	4.2 MICROGRAM/G DRY MATTER	COPPER=0.49MG/L,NO CONTROL,285K,10 FISH	1477
*NOEMACHEILUS BARBATULUS, BRAIN	43.8 MICROGRAM/G DRY MATTER	COPPER=0.38MG/L,NO CONTROL,285K,10 FISH	1477
*NOEMACHEILUS BARBATULUS, BRAIN	25.9 MICROGRAM/G DRY MATTER	COPPER=0.29MG/L,NO CONTROL,285K,10 FISH	1477
*NOEMACHEILUS BARBATULUS, GILL	164 MICROGRAM/G DRY MATTER	COPPER=0.76MG/L,CONTROL=26.4MG/L, 285K,10 FISH	1477
*NOEMACHEILUS BARBATULUS, GILL	92 MICROGRAM/G DRY MATTER	COPPER=0.71MG/L,CONTROL=26.4MG/L, 285K,10 FISH	1477
*NOEMACHEILUS BARBATULUS, GILL	208 MICROGRAM/G DRY MATTER	COPPER=0.49MG/L,CONTROL=26.4MG/L, 285K,10 FISH	1477
*NOEMACHEILUS BARBATULUS, GILL	157 MICROGRAM/G DRY MATTER	COPPER=0.38MG/L,CONTROL=26.4MG/L, 285K,10 FISH	1477
*NOEMACHEILUS BARBATULUS, GILL	489 MICROGRAM/G DRY MATTER	COPPER=0.29MG/L,CONTROL=26.4MG/L, 285K,10 FISH	1477
*NOEMACHEILUS BARBATULUS, GILL	21.4 MICROGRAM/G DRY MATTER	COPPER=0.12MG/L,CONTROL=26.4MG/L, 285K,10 FISH	1477
*NOEMACHEILUS BARBATULUS, LIVER	115 MICROGRAM/G DRY MATTER	COPPER=0.76MG/L,CONTROL=53MG/L, 285K,10 FISH	1477
*NOEMACHEILUS BARBATULUS, LIVER	56 MICROGRAM/G DRY MATTER	COPPER=0.71MG/L,CONTROL=53MG/L, 285K,10 FISH	1477
*NOEMACHEILUS BARBATULUS, LIVER	92 MICROGRAM/G DRY MATTER	COPPER=0.49MG/L,CONTROL=53MG/L, 285K,10 FISH	1477
*NOEMACHEILUS BARBATULUS, LIVER	153 MICROGRAM/G DRY MATTER	COPPER=0.38MG/L,CONTROL=53MG/L, 285K,10 FISH	1477
*NOEMACHEILUS BARBATULUS, LIVER	163 MICROGRAM/G DRY MATTER	COPPER=0.29MG/L,CONTROL=53MG/L, 285K,10 FISH	1477
*NOEMACHEILUS BARBATULUS, LIVER	77 MICROGRAM/G DRY MATTER	COPPER=0.12MG/L,CONTROL=53MG/L, 285K,10 FISH	1477
*NOEMACHEILUS BARBATULUS, MUSCLE	8.5 MICROGRAM/G DRY MATTER	COPPER=0.76MG/L,CONTROL=5.1MG/L, 285K,10 FISH	1477
*NOEMACHEILUS BARBATULUS, MUSCLE	45.7 MICROGRAM/G DRY MATTER	COPPER=0.71MG/L,CONTROL=5.1MG/L, 285K,10 FISH	1477
*NOEMACHEILUS BARBATULUS, MUSCLE	33.5 MICROGRAM/G DRY MATTER	COPPER=0.49MG/L,CONTROL=5.1MG/L, 285K,10 FISH	1477
*NOEMACHEILUS BARBATULUS, MUSCLE	53 MICROGRAM/G DRY MATTER	COPPER=0.38MG/L,CONTROL=5.1MG/L, 285K,10 FISH	1477
*NOEMACHEILUS BARBATULUS, MUSCLE	8.2 MICROGRAM/G DRY MATTER	COPPER=0.29MG/L,CONTROL=5.1MG/L, 285K,10 FISH	1477
*NOEMACHEILUS BARBATULUS, MUSCLE	10.1 MICROGRAM/G DRY MATTER	COPPER=0.12MG/L,CONTROL=5.1MG/L, 285K,10 FISH	1477
*NOEMACHEILUS BARBATULUS, OPERCULUM	21.6 MICROGRAM/G DRY MATTER	COPPER=0.76MG/L,NO CONTROL,285K,10 FISH	1477
*NOEMACHEILUS BARBATULUS, OPERCULUM	69 MICROGRAM/G DRY MATTER	COPPER=0.71MG/L,NO CONTROL,285K,10 FISH	1477
*NOEMACHEILUS BARBATULUS, OPERCULUM	85 MICROGRAM/G DRY MATTER	COPPER=0.49MG/L,NO CONTROL,285K,10 FISH	1477
*NOEMACHEILUS BARBATULUS, OPERCULUM	44 MICROGRAM/G DRY MATTER	COPPER=0.38MG/L,NO CONTROL,285K,10 FISH	1477

TABLE D277

CONCENTRATIONS : CU

ITEM:	VALUE:	CONDITIONS:	REF.:
*NOEMACHEILUS BARBATULUS, OPERCULUM	31.5 MICROGRAM/G DRY MATTER	COPPER=0.29MG/L,NO CONTROL,285K,10 FISH	1477
*NOEMARCEILUS BARBATULUS, EYE	18.6 MICROGRAM/G DRY MATTER	COPPER=0.76MG/L,CONTROL=19.6MG/L, 285K,10FISH	1477
*NOEMARCEILUS BARBATULUS, EYE	21.8 MICROGRAM/G DRY MATTER	COPPER=0.71MG/L,CONTROL=19.6MG/L, 285K,10FISH	1477
*NOEMARCEILUS BARBATULUS, EYE	44.8 MICROGRAM/G DRY MATTER	COPPER=0.49MG/L,CONTROL=19.6MG/L, 285K,10FISH	1477
*NOEMARCEILUS BARBATULUS, EYE	36.7 MICROGRAM/G DRY MATTER	COPPER=0.38MG/L,CONTROL=19.6MG/L, 285K,10FISH	1477
*NOEMARCEILUS BARBATULUS, EYE	20.8 MICROGRAM/G DRY MATTER	COPPER=0.29MG/L,CONTROL=19.6MG/L, 285K,10FISH	1477
*NOEMARCEILUS BARBATULUS, EYE	3.4 MICROGRAM/G DRY MATTER	COPPER=0.12MG/L,CONTROL=19.6MG/L, 285K,10FISH	1477
*PIG FAECES	1421.8 PPM	FEED CONC.=328.1PPM,WT. OF PIG= 45KG,CU SUPPL. IN FEED	1349
*PIG FAECES	625. PPM	FEED CONC.=41.3PPM,WT. OF PIG=61KG	1349
*PIG FAECES	82.5 PPM	FEED CONC.=13.6PPM,WT. OF PIG=76KG	1349
*PIG FAECES	93.8 PPM	FEED CONC.=11.4PPM,WT. OF PIG=53KG	1349
*PIG FAECES	135. PPM	FEED CONC.=11.4,WT. OF PIG=64KG	1349
*PIG FAECES	150.5 PPM	FEED CONC.=25.9PPM,WT. OF PIG=73KG	1349
*PIG URINE	0.25 PPM	FEED CONC.=328.1PPM,WT. OF PIG= 45KG,CU SUPPL. IN FEED	1349
*PIG URINE	0.2 PPM	FEED CONC.=41.3PPM,WT. OF PIG=61KG	1349
*PIG URINE	0.78 PPM	FEED CONC.=13.6PPM,WT. OF PIG=76KG	1349
*PIG URINE	0.2 PPM	FEED CONC.=11.4PPM,WT. OF PIG=53KG	1349
*PIG URINE	0.125 PPM	FEED CONC.=11.4PPM,WT. OF PIG=64KG	1349
*PIG URINE	0.13 PPM	FEED CONC.=25.9PPM,WT. OF PIG=73KG	1349
*RAIN	23 UG/L	ENGLAND,1971,REL.UNCONTAMINATED,CU IN ATM.=22NG/M**3	893
*RAIN	3-8.5 UG/L	SWEDEN	893
*RED CLOVER	10.6 MG/KG DRY WT.	ADDED CUSO4=22KG/HA,PH=5.4;CONTROL =7.7MG/KG DRY WT.	1048
*RED CLOVER	12.1 MG/KG DRY WT.	ADDED CUSO4=22KG/HA,PH=6.2;CONTROL =3.8MG/KG DRY WT.	1048
*SEDIMENT	89 PPM	WATER CONC = 4 MICROGRAM/L, YEARLY MEAN, CARNON RIVER	48
*SEDIMENT	1100 PPM	WATER CONC = 30 MICROGRAM/L, YEARLY MEAN, RED RIVER	48
*SEDIMENT	374 PPM	WATER CONC = 5 MICROGRAM/L, YEARLY MEAN, RED RIVER	48
*SEDIMENT	333 PPM	WATER CONC = 11 MICROGRAM/L, YEARLY MEAN, GANNEL RIVER	48
*SEDIMENT	32 PPM	WATER CONC = 2 MICROGRAM/L, YEARLY MEAN, GANNEL RIVER	48
*SEDIMENT	26 PPM	WATER CONC = 2 MICROGRAM/L, YEARLY MEAN, FOWEY RIVER	48
*SEDIMENT	66 PPM	WATER CONC = 7 MICROGRAM/L, YEARLY MEAN, FOWEY RIVER	48
*SEDIMENT	291 PPM	WATER CONC = 3 MICROGRAM/L, YEARLY MEAN, CARNON RIVER	48
*SEDIMENT	89 PPM	MAX WATER CONC = 8 MICROGRAM/L, 1 YEARS MEASUREMENT, CARNON R	48
*SEDIMENT	1100 PPM	MAX WATER CONC = 39 MICROGRAM/L, 1 YEARS MEASUREMENT, RED R	48
*SEDIMENT	374 PPM	MAX WATER CONC = 8 MICROGRAM/L, 1 YEARS MEASUREMENT, RED R	48
*SEDIMENT	333 PPM	MAX WATER CONC = 15 MICROGRAM/L, MAX CONC, GANNEL RIVER	48
*SEDIMENT	32 PPM	MAX WATER CONC = 3 MICROGRAM/L, MAX CONC, GANNEL RIVER	48
*SEDIMENT	26 PPM	MAX WATER CONC = 2 MICROGRAM/L, 1 YEARS MEASUREMENT, FOWEY R	48

TABLE D277

CONCENTRATIONS : CU

ITEM:	VALUE:	CONDITIONS:	REF.:
*SEDIMENT	66 PPM	MAX WATER CONC = 12 MICROGRAM/L, 1 YEARS MEASUREMENT, FOWEY R	48
*SEDIMENT	291 PPM	MAX WATER CONC = 4 MICROGRAM/L, 1 YEARS MEASUREMENT, CARNON R	48
*SEDIMENT	4.2 PPM	UNPOLLUTED AREA, ESTUARY, URR WATER	1557
*SEDIMENT	6.9 (RANGE 1.9 - 12.2) MG/KG DRY WEIGHT	UNPOLLUTED AREA, ESTUARY, URR WATER	1558
*SEDIMENT	11.2 (RANGE 9.2 - 16.8) MG/KG DRY WEIGHT	UNPOLLUTED AREA, ESTUARY, URR WATER	1558
*SEDIMENT	7.7 (RANGE 3.5 - 11.2) MG/KG DRY WEIGHT	UNPOLLUTED AREA, RIVER, ILLINOIS RIVER	993
*SEDIMENT	140.26 (126.20-185.67) MG/KG DRY WT.	800M ABOVE POLL.ENTRY(1),CU IN WATER=0.032MG/L,POLLUTED RIVER	1224
*SEDIMENT	244.31 (221.50-290.18) MG/KG DRY WT.	150M BELOW POLL.ENTRY(1),CU IN WATER=0.113MG/L,POLLUTED RIVER	1224
*SEDIMENT	286.34 (274.19-295.28) MG/KG DRY WT.	800 AND 50M BELOW POLL.(1)AND(2), CU IN WATER=0.085MG/L,POLLUTED RIVER	1224
*SEDIMENT	95 UG/G DRY WT.	CONC.IN EPILIMNION=1.0UG/L, CONC.FACTOR=95000,SCHOEHSEE	573
*SOIL (SEDIMENT)	19. PPM DRY SOIL	RANGE 5-13PPM,RIVER RHEIDOL, CARDIGANSHIRE,GB	327
*SUSPENDED SOLIDS IN WATER	0.77 FRACTION OF TOTAL CU	SOLUBLE CU = 56 MG/M3,RIVER TAME	1511
*SUSPENDED SOLIDS IN WATER	0.76 FRACTION OF TOTAL CU	SOLUBLE CU = 66 MG/M3,RIVER TAME	1511
*SUSPENDED SOLIDS IN WATER	0.77 FRACTION OF TOTAL CU	SOLUBLE CU = 48 MG/M3,RIVER TAME	1511
*SUSPENDED SOLIDS IN WATER	0.74 FRACTION OF TOTAL CU	SOLUBLE CU = 54 MG/M3,RIVER TAME	1511
*SUSPENDED SOLIDS IN WATER	0.83 FRACTION OF TOTAL CU	SOLUBLE CU =200 MG/M3,RIVER TAME	1511
*SUSPENDED SOLIDS IN WATER	0.47 FRACTION OF TOTAL CU	SOLUBLE CU = 85 MG/M3,RIVER TAME	1511
*SUSPENDED SOLIDS IN WATER	0.88 FRACTION OF TOTAL CU	SOLUBLE CU = 20 MG/M3,RIVER LEE	1511
*SUSPENDED SOLIDS IN WATER	0.57 FRACTION OF TOTAL CU	SOLUBLE CU = 13 MG/M3,STEVENAGE BROOK	1511
*SUSPENDED SOLIDS IN WATER	0.43 FRACTION OF TOTAL CU	SOLUBLE CU = 31 MG/M3,PIX BROOK	1511
*TIMOTHY	1.9 MG/KG DRY WT.	ADDED LIMESTONE=0.63T/HA,SOIL-PH= 4.7,N.WALES	40
*TIMOTHY	1.8 MG/KG DRY WT.	ADDED LIMESTONE=2.5T/HA,SOIL-PH= 5.0,N.WALES	40
*TIMOTHY	2.4 MG/KG DRY WT.	ADDED LIMESTONE=5.0T/HA,SOIL-PH= 5.2,N.WALES	40
*TIMOTHY	2.0 MG/KG DRY WT.	ADDED LIMESTONE=15.0T/HA,SOIL-PH= 6.0,N.WALES	40
*VEGETATION	5.0 MG/KG DRY WT.	CONC.IN PODZOL:F=10,H=6,A1A2:<10, B1=20,B2=10,B3=10, (F030)	1048
*WATER	0.83-105 UG/L	RIVERS,U.S.A.,AVERAGE=5.3UG/L	893
*WHITE CLOVER	2.5 MG/KG DRY WT.	ADDED LIMESTONE=0.63T/HA,SOIL-PH= 4.7,N.WALES	40
*WHITE CLOVER	2.6 MG/KG DRY WT.	ADDED LIMESTONE=2.5T/HA,SOIL-PH= 5.0,N.WALES	40
*WHITE CLOVER	2.5 MG/KG DRY WT.	ADDED LIMESTONE=5.0T/HA,SOIL-PH= 5.2,N.WALES	40
*WHITE CLOVER	2.6 MG/KG DRY WT.	ADDED LIMESTONE=15.0T/HA,SOIL-PH= 6.0,N.WALES	40
*ZOOPLANKTON	< 16.22 MG/KG DRY WT.(MEAN)	19ST.,N.Y.BIGHT,CONC.IN PHYTOPL. (LONG ISLAND):<9.07MG/KG DRY WT.MEAN	565

TABLE D278

CONCENTRATIONS : DBT

ITEM:	VALUE:	CONDITIONS:	REF.:
*WILD FISH (!W)	0.37 PPM	AFTER 1D IN AQUARIUM,DBT IN WATER= 0.01PPM	754

TABLE D278

CONCENTRATIONS : DBT			
ITEM:	VALUE:	CONDITIONS:	REF.:
*WILD FISH (!W)	0.35 PPM	AFTER 8D IN AQUARIUM,DBT IN WATER= 0.01PPM	754
*WILD FISH (!W)	0.17 PPM	AFTER 32D IN AQUARIUM,DBT IN WATER =0.01PPM	754
*WILD FISH (!W)	NO MEASURABLE AMOUNT OF DBT	AFTER 35D IN AQUARIUM,DBT IN WATER =0.01PPM-32D,0.PPM-3D	754

TABLE D279

CONCENTRATIONS : DDD+DDT			
ITEM:	VALUE:	CONDITIONS:	REF.:
*CASMERODIUS ALBUS, BRAIN	2.1 PPM MEAN WET WT.	5 OBS.,%LIPID=6.8,DIELDRIN=4.36, ENDRIN<0.18,AUDUBON CANYON RANCH, USA	439
*CASMERODIUS ALBUS, BREAST MUSCLE	3.7 PPM MEAN WET WT.	5 OBS.,%LIPID=2.4,AUDUBON CANYON RANCH,CAL.,USA	439
*CASMERODIUS ALBUS, LIVER	6.1 PPM MEAN WET WT.	5 OBS.,%LIPID=3.5,HG=6.08,AUDUBON CANYON RANCH,CAL.,USA	439

TABLE D280

CONCENTRATIONS : DDE			
ITEM:	VALUE:	CONDITIONS:	REF.:
*BIRD, EGG YOLK LIPID	94. PPM WET WEIGHT	24 APRIL,%LIPID=100,AUDUBON CANYON RANCH,CAL.,USA (!!!)	439
*CASMERODIUS ALBUS, BRAIN	59.7 PPM MEAN WET WT.	5 OBS.,%LIPID=6.8,DIELDRIN=4.36, ENDRIN<0.18,AUDUBON CAYON RANCH, USA	439
*CASMERODIUS ALBUS, BREAST MUSCLE	93.9 PPM MEAN WET WT.	5 OBS.,%LIPID=2.4,AUDUBON CANYON RANCH,CAL.,USA	439
*CASMERODIUS ALBUS, LIVER	124.3 PPM MEAN WET WT.	5 OBS.,%LIPID=3.5,HG=6.08,AUDUBON CANYON RANCH,CAL.,USA	439
*EGRET, EGG YOLK LIPID	248. PPM WET WEIGHT	NO DATE.AUDUBON CANYON RANCH,CAL., USA	439
*HERON (PROBABLE), EGG YOLK LIPID	287. PPM WET WEIGHT	30MARCH,%LIPID=100.AUDUBON CANYON RANCH,CAL.,USA	439
*HERON (PROBABLE), EMBRYO	1.2 PPM WET WT.	NO DATE,%LIPID=3.6.AUDUBON CANYON RANCH,CAL.,USA	439
*HERON, EGG YOLK LIPID	108. PPM WET WEIGHT	20 APRIL,%LIPID=100.AUDUBON CANYON RANCH,CAL.,USA	439
*NESTLING,EGRET(AGE UNKNOWN)	1.1 PPM WET WT.	13 MAY,%LIPID=1.2,AUDUBON CANYON RANCH,CAL.,USA	439
*NESTLING,EGRET(1 WEEK)	0.55 PPM WET WT.	7 MAY,%LIPID=1.2,AUDUBON CANYON RANCH,CAL.,USA	439
*NESTLING,EGRET(1.5-2 WEEKS)	1.1 PPM WET WT.	11 MAY,%LIPID=1.2,AUDUBON CANYON RANCH,CAL.,USA	439
*NESTLING,EGRET(1-2 WEEKS)	3.2 PPM WET WT.	15-1969-70,%LIPID=1.7,AUDUBON CANYON RANCH,CAL.,USA	439
*NESTLING,EGRET(2-2.5 WEEKS)	2.1 PPM WET WT.	11 MAY,%LIPID=2.1,AUDUBON CANYON RANCH,CAL.,USA	439
*NESTLING,EGRET(5-6 DAYS)	4. PPM,WET WT.	11 MAY,%LIPID=2.2,AUDUBON CANYON RANCH,CAL.,USA	439
*NESTLING,HERON(2 WEEKS)	0.49 PPM WET WT.	24 APRIL,%LIPID=1.6.AUDUBON CANYON RANCH,CAL.,USA	439
*NESTLING,HERON(3 WEEKS)	1.18 PPM WET WT.	24 APRIL,%LIPID=1..AUDUBON CANYON RANCH,CAL.,USA	439
*NESTLING,HERON(4.5 WEEKS)	0.16 PPM WET WT.	24 APRIL,%LIPID=1.2,AUDUBON CANYON RANCH,CAL.,USA	439
*PAGOPHILUS GROENLANDICUS, BLUBBER	3.18 (SD=+-1.12) PPM (ON WET WEIGHT BASIS)	95.5 % FAT, 4 ANIMALS, ST LAWRENCE, CANADA	785

TABLE D280

CONCENTRATIONS : DDE			
ITEM:	VALUE:	CONDITIONS:	REF.:
*PAGOPHILUS GROENLANDICUS, BLUBBER	1.10 (SD=+-0.43) PPM (ON WET WEIGHT BASIS)	83,0 % FAT, 15 SAMPLES,1973,GULF OF ST LAWRENCE	484
*PAGOPHILUS GROENLANDICUS, BLUBBER	2.44 (SD=+-01.45) PPM (ON WET WEIGHT BASIS)	97.8 % F¤T, 10 SAMPLES,1971,GULF OF ST LAWRENCE	484
*PAGOPHILUS GROENLANDICUS, BRAIN	0.03 (SD=+-0.03) PPM (ON WET WEIGHT BASIS)	11.5 % FAT, 3 ANIMALS, ST LAWRENCE, CANADA	785
*PAGOPHILUS GROENLANDICUS, BRAIN	0.01 (SD=+-0.01) PPM (ON WET WEIGHT BASIS)	6.6 % FAT, 14 SAMPLES, 1973,GULF OF ST LAWRENCE	484
*PAGOPHILUS GROENLANDICUS, BRAIN	7.5 (SD=+-1.3) PPM (ON WET WEIGHT BASIS)	7.5 % FAT, 10 SAMPLES, 1971,GULF OF ST LAWRENCE	484
*PAGOPHILUS GROENLANDICUS, KIDNEY	14.4 (SD=+-9.2) PPM (ON WET WEIGHT BASIS)	14.4 % FAT, 13 SAMPLES,1973,GULF OF ST LAWRENCE	484
*PAGOPHILUS GROENLANDICUS, LIVER	0.02 (SD=+-0.01) PPM (ON WET WEIGHT BASIS)	1.8 % FAT, 4 AINMALS, GULF OF ST LAWRENCE, CANADA	785
*PAGOPHILUS GROENLANDICUS, LIVER	0.11 (SD=+-0.14) PPM (ON WET WEIGHT BASIS)	4.4 % FAT, 16 SAMPLES, 1973,GULF OF ST LAWRENCE	484
*PAGOPHILUS GROENLANDICUS, LIVER	0.15 (SD=+-0.19) PPM (ON WET WEIGHT BASIS)	7.0 % FAT, 11 SAMPLES, 1971,GULF OF ST LAWRENCE	484
*PAGOPHILUS GROENLANDICUS, MUSCLE	0.01 (SD=+-0.00) PPM (ON WET WEIGHT BASIS)	1.1 % FAT, 4 ANIMALS, ST LAWRENCE, CANADA	785

TABLE D281

CONCENTRATIONS : DDT			
ITEM:	VALUE:	CONDITIONS:	REF.:
*BALD EAGLE, BRAIN	0.70 MG/KG	CONC.IN FOOD=5MG/KG	389
*BALD EAGLE, BRAIN	80.0 MG/KG	CONC.IN FOOD=800MG/KG	389
*BALD EAGLE, FAT	35.70 MG/KG	CONC.IN FOOD=5MG/KG	389
*BALD EAGLE, LIVER	1.90 MG/KG	CONC.IN FOOD=5MG/KG	389
*BALD EAGLE, MUSCLE	291.0 MG/KG	CONC.IN FOOD=800MG/KG	389
*BIVALVES	0.23 +- 0.05 MG/KG WET WEIGHT	INITIAL WATER = 1 UG/L, 4 H, SMALL POND, FINLAND	1376
*BIVALVES	1.02 +- 0.09 MG/KG WET WEIGHT	INITIAL WATER = 1 UG/L, 8 D, SMALL POND, FINLAND	1376
*BLADDERWORT	0.40 +- 0.02 MG/KG WET WEIGHT	INITIAL WATER = 1 UG/L, 4 H, SMALL POND, FINLAND	1376
*BLADDERWORT	1.35 +- 0.04 MG/KG WET WEIGHT	INITIAL WATER = 1 UG/L, 8 D, SMALL POND, FINLAND	1376
*BOBWHITE QUAIL, BRAIN	0.40 MG/KG	CONC.IN FOOD=25MG/KG	389
*BOBWHITE QUAIL, BRAIN	0.57 MG/KG	CONC.IN FOOD=50MG/KG	389
*BOBWHITE QUAIL, BRAIN	1.02 MG/KG	CONC.IN FOOD=100MG/KG	389
*BOBWHITE QUAIL, BRAIN	1.94 MG/KG	CONC.IN FOOD=200MG/KG	389
*BOBWHITE QUAIL, BRAIN	8.75 MG/KG	CONC.IN FOOD=400MG/KG	389
*BOBWHITE QUAIL, BRAIN	14.20 MG/KG	CONC.IN FOOD=800MG/KG	389
*BOBWHITE QUAIL, TISSUE	1.60 MG/KG	CONC.IN FOOD=50MG/KG	389
*BOBWHITE QUAIL, TISSUE	4.10 MG/KG	CONC.IN FOOD=100MG/KG	389
*BOBWHITE QUAIL, TISSUE	19.80 MG/KG	CONC.IN FOOD=250MG/KG	389
*CADDISFLY LARVAE	0.27 +- 0.05 MG/KG WET WEIGHT	INITIAL WATER = 1 UG/L, 4 H, SMALL POND, FINLAND	1376
*CADDISFLY LARVAE	1.33 +- 0.08 MG/KG WET WEIGHT	INITIAL WATER = 1 UG/L, 8 D, SMALL POND, FINLAND	1376
*CHICKEN, BRAIN	0.60 MG/KG	CONC.IN FOOD=10.00MG/KG	389
*CHICKEN, BRAIN	3.70 MG/KG	CONC.IN FOOD=50.00MG/KG	389
*CHICKEN, BRAIN	18.30 MG/KG	CONC.IN FOOD=250.00MG/KG	389
*CHICKEN, FAT	73.0 MG/KG	CONC.IN FOOD=10.00MG/KG	389
*CHICKEN, FAT	518.0 MG/KG	CONC.IN FOOD=50.COMG/KG	389

TABLE D281

CONCENTRATIONS : DDT

ITEM:	VALUE:	CONDITIONS:	REF.:
*CHICKEN, FAT	2670.0 MG/KG	CONC.IN FOOD=250.00MG/KG	389
*CHICKEN, FAT	80.73 MG/KG	CONC.IN FOOD=10.00MG/KG	389
*CHICKEN, LIVER	4.80 MG/KG	CONC.IN FOOD=10.00MG/KG	389
*CHICKEN, LIVER	46.0 MG/KG	CONC.IN FOOD=50.00MG/KG	389
*CHICKEN, LIVER	198.0 MG/KG	CONC.IN FOOD=150.00MG/KG	389
*CHICKEN, LIVER	0.16 MG/KG	CONC.IN FOOD=0.05MG/KG	389
*CHICKEN, LIVER	0.41 MG/KG	CONC.IN FOOD=0.15MG/KG	389
*CHICKEN, LIVER	0.53 MG/KG	CONC.IN FOOD=0.45MG/KG	389
*CHICKEN, LIVER	2.12 MG/KG	CONC.IN FOOD=10.00MG/KG	389
*CHICKEN, MUSCLE	2.10 MG/KG	CONC.IN FOOD=10.00MG/KG	389
*CHICKEN, MUSCLE	10.50 MG/KG	CONC.IN FOOD=50.00MG/KG	389
*CHICKEN, MUSCLE	39.80 MG/KG	CONC.IN FOOD=250.00MG/KG	389
*CORMORANT	26.4 MG/KG WET WEIGHT	WATER=0.00005 PPM, TOTAL DDT, WHOLE ORGANISNS BASIS, ESTAURY	1749
*COWBIRD, BRAIN	66.0 MG/KG	CONC.IN FOOD=500MG/KG	389
*COWBIRD, TISSUE	123.0 MG/KG	CONC.IN FOOD=500MG/KG	389
*CRUCIAN CARP, GILLS	0.16 +- 0.03 MG/KG WET WEIGHT	INITIAL WATER = 1 UG/L, 4 H, SMALL POND, FINLAND	1376
*CRUCIAN CARP, GILLS	0.39 +- 0.04 MG/KG WET WEIGHT	INITIAL WATER = 1 UG/L, 8 D, SMALL POND, FINLAND	1376
*CRUCIAN CARP, LIVER	0.06 +- 0.02 MG/KG WET WEIGHT	INITIAL WATER = 1 UG/L, 4 H, SMALL POND, FINLAND	1376
*CRUCIAN CARP, LIVER	0.06 +- 0.01 MG/KG WET WEIGHT	INITIAL WATER = 1 UG/L, 8 D, SMALL POND, FINLAND	1376
*CRUCIAN CARP, MUSCLE	0.15 +- 0.03 MG/KG WET WEIGHT	INITIAL WATER = 1 UG/L, 4 H, SMALL POND, FINLAND	1376
*CRUCIAN CARP, MUSCLE	0.08 +- 0.01 MG/KG WET WEIGHT	INITIAL WATER = 1 UG/L, 8 D, SMALL POND, FINLAND	1376
*DAPHNIA MAGNA	128 MG/KG WET WEIGHT	WATER= 8.0 PPB, 26 H, 294.1 K, TOTAL DDT,FIRST INSTAR, 9 UG ANIMAL	300
*DAPHNIA MAGNA	266 MG/KG WET WEIGHT	WATER= 16.0 PPB, 26 H, 294.1 K, TOTAL DDT,FIRST INSTAR, 9 UG ANIMAL	300
*DAPHNIA MAGNA	1150 MG/KG WET WEIGHT	WATER= 50.0 PPB, 26 H, 294.1 K, TOTAL DDT,FIRST INSTAR, 9 UG ANIMAL	300
*DAPHNIA MAGNA	1260 MG/KG WET WEIGHT	WATER= 100 PPB, 26 H, 294.1 K, TOTAL DDT,FIRST INSTAR, 9 UG ANIMAL	300
*DAPHNIA MAGNA	2364 MG/KG WET WEIGHT	WATER= 200 PPB, 26 H, 294.1 K, TOTAL DDT,FIRST INSTAR, 9 UG ANIMAL	300
*DAPHNIA MAGNA	4280 MG/KG WET WEIGHT	WATER= 1100 PPB, 26 H, 294.1 K, TOTAL DDT,FIRST INSTAR, 9 UG ANIMAL	300
*ESCHERICHIA COLI	11.2 MG/KG	DDT IN MEDIUM=0.1 MG/L, 24 H,LAB	808
*ESCHERICHIA COLI	2.2 MG/KG	DDT IN MEDIUM=0.01 MG/L, 24 H,LAB	808
*ESCHERICHIA COLI	DDT=2.1;DDD=6.0;DDE=2.3;TOTAL DDT= 11.2 MG/KG	DDT IN MEDIUM=0.1 MG/L, 24 H, LAB	808
*ESCHERICHIA COLI	DDT=0;DDD=1.5;DDE=0.5;TOTAL DDT= 2.2 MG/KG	DDT IN MEDIUM=0.01 MG/L, 24 H, LAB	808
*FERAL PIGEON, LIVER	32.80 MG/KG	CONC.IN FOOD=72MG/KG	389
*FERAL PIGEON, MUSCLE	32.33 MG/KG	CONC.IN FOOD=72MG/KG	389
*FISH HAWK, EGG	13.8 MG/KG WET WEIGHT	WATER=0.00005 PPM, TOTAL DDT, WHOLE ORGANISNS BASIS, ESTAURY	1749
*GAMBUSIA AFFINIS	36 MICROGRAM/KG WET WEIGHT	FISH = 70.0 MG, INITIAL WATER = 41 PPT, 48 H, 293 K	1085
*GAMBUSIA AFFINIS	34 MICROGRAM/KG WET WEIGHT	FISH = 150 MG , INITIAL WATER = 41 PPT, 48 H, 293 K	1085
*GAMBUSIA AFFINIS	28 MICROGRAM/KG WET WEIGHT	FISH = 220 MG , INITIAL WATER = 41 PPT, 48 H, 293 K	1085
*GAMBUSIA AFFINIS	20 MICROGRAM/KG WET WEIGHT	FISH = 380 MG , INITIAL WATER = 41 PPT, 48 H, 293 K	1085

TABLE D281

CONCENTRATIONS : DDT

ITEM:	VALUE:	CONDITIONS:	REF.:
*GAMBUSIA AFFINIS	9 MICROGRAM/KG WET WEIGHT	FISH = 900 MG , INITIAL WATER = 41 PPT, 48 H, 293 K	1085
*GUINEA PIG, FAT	1.59 MG/KG	CONC.IN FOOD=10MG/KG	389
*HERON	3.57 MG/KG WET WEIGHT	WATER=0.00005 PPM, TOTAL DDT, WHOLE ORGANISNS BASIS, ESTAURY	1749
*HERRING GULL	6.00 MG/KG WET WEIGHT	WATER=0.00005 PPM, TOTAL DDT, WHOLE ORGANISNS BASIS, ESTAURY	1749
*HOMO SAPIENS	100 MG/KG WET WEIGHT	INTAKE FROM FOOD = 3.5 MG P, P`-DDT/(MAN*D), STEADY STATE	389
*HOMO SAPIENS	353 MG/KG WET WEIGHT	INTAKE FROM FOOD = 35 MG P,P`-DDT/ (MAN*D), STEADY STATE	389
*KESTREL, BRAIN	256.80 MG/KG	CONC.IN FOOD=2.8MG/KG,DIED	389
*KESTREL, BRAIN	14.90 MG/KG	CONC.IN FOOD=2.8MG/KG,SACRIFICED	389
*KESTREL, LIVER	190.80 MG/KG	CONC.IN FOOD=2.8MG/KG,DIED	389
*KESTREL, LIVER	24.0 MG/KG	CONC.IN FOOD=2.8MG/KG,SACRIFICED	389
*KESTREL, TISSUE	51.45 MG/KG	CONC.IN FOOD=2.8MG/KG,DIED	389
*KESTREL, TISSUE	70.0 MG/KG	CONC.IN FOOD=2.8MG/KG,SACRIFICED	389
*MERGANSER	22.8 MG/KG WET WEIGHT	WATER=0.00005 PPM, TOTAL DDT, WHOLE ORGANISNS BASIS, ESTAURY	1749
*MOSS	0.36 +- 0.01 MG/KG WET WEIGHT	INITIAL WATER = 1 UG/L, 4 H, SMALL POND, FINLAND	1376
*MOSS	0.73 +- 0.02 MG/KG WET WEIGHT	INITIAL WATER = 1 UG/L, 8 D, SMALL POND, FINLAND	1376
*NEEDLEFISH	2.07 MG/KG WET WEIGHT	WATER=0.00005 PPM, TOTAL DDT, WHOLE ORGANISNS BASIS, ESTAURY	1749
*NEWT	0.07 +- 0.02 MG/KG WET WEIGHT	INITIAL WATER = 1 UG/L, 4 H, SMALL POND, FINLAND	1376
*NEWT	1.17 +- 0.05 MG/KG WET WEIGHT	INITIAL WATER = 1 UG/L, 8 D, SMALL POND, FINLAND	1376
*NOTONECTA SP.	0.08 +- 0.01 MG/KG WET WEIGHT	INITIAL WATER = 1 UG/L, 4 H, SMALL POND, FINLAND	1376
*NOTONECTA SP.	0.21 +- 0.02 MG/KG WET WEIGHT	INITIAL WATER = 1 UG/L, 8 D, SMALL POND, FINLAND	1376
*PAGOPHILUS GROENLANDICUS, BLUBBER	0.85 (SD=+-0.12) PPM (ON WET WEIGHT BASIS)	95.5 % FAT, 4 ANIMALS, ST LAWRENCE, CANADA	785
*PAGOPHILUS GROENLANDICUS, BLUBBER	0.19 (SD=+-0.14) PPM (ON WET WEIGHT BASIS)	83.0 % FAT, 15 SAMPLES,1973,GULF OF ST LAWRENCE	484
*PAGOPHILUS GROENLANDICUS, BLUBBER	0.50 (SD=+-0.32) PPM (ON WET WEIGHT BASIS)	97.8 % FAT, 10 SAMPLES,1971,GULF OF ST LAWRENCE	484
*PAGOPHILUS GROENLANDICUS, BRAIN	NON DETECTABLE	11.5 % FAT, 3 ANIMALS, ST LAWRENCE, CANADA	785
*PAGOPHILUS GROENLANDICUS, BRAIN	NON DETECTABLE	6.6 % FAT, 14 SAMPLES, 1973,GULF OF ST LAWRENCE	484
*PAGOPHILUS GROENLANDICUS, BRAIN	0.06 (SD=+-0.19) PPM (ON WET WEIGHT BASIS)	7.5 % FAT, 10 SAMPLES, 1971,GULF OF ST LAWRENCE	484
*PAGOPHILUS GROENLANDICUS, KIDNEY	0.03 (SD=+-0.03) PPM (ON WET WEIGHT BASIS)	14.4 % FAT, 13 SAMPLES,1973,GULF OF ST LAWRENCE	484
*PAGOPHILUS GROENLANDICUS, LIVER	NON DETECTABLE	1.8 % FAT, 4 AINMALS, GULF OF ST LAWRENCE, CANADA	785
*PAGOPHILUS GROENLANDICUS, LIVER	0.01 (SD=+-0.01) PPM (ON WET WEIGHT BASIS)	4.4 % FAT, 16 SAMPLES, 1973,GULF OF ST LAWRENCE	484
*PAGOPHILUS GROENLANDICUS, LIVER	0.07 (SD=+-0.05	7.0 % FAT, 11 SAMPLES, 1971,GULF OF ST LAWRENCE	484
*PERCH, GILLS	1.67 +- 0.06 MG/KG WET WEIGHT	INITIAL WATER = 1 UG/L, 4 H, SMALL POND, FINLAND	1376
*PERCH, GILLS	1.72 +- 0.06 MG/KG WET WEIGHT	INITIAL WATER = 1 UG/L, 8 D, SMALL POND, FINLAND	1376
*PERCH, LIVER	5.63 +- 0.16 MG/KG WET WEIGHT	INITIAL WATER = 1 UG/L, 4 H, SMALL POND, FINLAND	1376
*PERCH, LIVER	2.00 +- 0.06 MG/KG WET WEIGHT	INITIAL WATER = 1 UG/L, 8 D, SMALL POND, FINLAND	1376
*PERCH, MES. ADIP.	0.39 +- 0.09 MG/KG WET WEIGHT	INITIAL WATER = 1 UG/L, 4 H, SMALL POND, FINLAND	1376
*PERCH, MES. ADIP.	16.80 +- 0.40 MG/KG WET WEIGHT	INITIAL WATER = 1 UG/L, 8 D, SMALL POND, FINLAND	1376

TABLE D281

CONCENTRATIONS : DDT

ITEM:	VALUE:	CONDITIONS:	REF.:
*PERCH, MUSCLE	0.19 +- 0.03 MG/KG WET WEIGHT	INITIAL WATER = 1 UG/L, 4 H, SMALL POND, FINLAND	1376
*PERCH, MUSCLE	0.22 +- 0.02 MG/KG WET WEIGHT	INITIAL WATER = 1 UG/L, 8 D, SMALL POND, FINLAND	1376
*PHEASANT, MUSCLE	23.90 MG/KG	CONC.IN FOOD=50MG/KG,DIED	389
*PHEASANT, MUSCLE	3.60 MG/KG	CONC.IN FOOD=50MG/KG,SACRIFICED	389
*PHEASANT, TISSUE	29.10 MG/KG	CONC.IN FOOD=10MG/KG	389
*PICKEREL	1.33 MG/KG WET WEIGHT	WATER=0.00005 PPM, TOTAL DDT, WHOLE ORGANISNS BASIS, ESTAURY	1749
*PLANKTON	0.04 MG/KG WET WEIGHT	WATER=0.00005 PPM, TOTAL DDT, WHOLE ORGANISNS BASIS, ESTAURY	1749
*RABBIT, FAT	18.78 MG/KG	CONC.IN FOOD=10MG/KG	389
*RABBIT, LIVER	0.77 MG/KG	CONC.IN FOOD=10MG/KG	389
*RAT, FAT	56.43 MG/KG	CONC.IN FOOD=10MG/KG	389
*RAT, LIVER	1.96 MG/KG	CONC.IN FOOD=10MG/KG	389
*SEDIMENT	0.0022 MG/KG	CONC.IN WATER=0.08UG/L,U.S.	389
*SEDIMENT	1.80 MG/KG	CONC.IN WATER=0.30UG/L,U.S.	389
*SEDIMENT	1.10 MG/KG	CONC.IN WATER=TRACE,U.S.	389
*SEDIMENT	0.0022 MG/KG	CONC.IN WATER=0.08UG/L,U.S.	389
*SEDIMENT	0.047 MG/KG	CONC.IN WATER=3.28UG/L,CANADA	389
*SEDIMENT	0.74 +- 0.03 MG/KG WET WEIGHT	INITIAL WATER = 1 UG/L, 4 H, SMALL POND, FINLAND	1376
*SEDIMENT	0.94 +- 0.03 MG/KG WET WEIGHT	INITIAL WATER = 1 UG/L, 8 D, SMALL POND, FINLAND	1376
*SEDIMENT	0.20 MG/KG	CONC.IN WATER=0.115UG/L,U.S.	389
*SEDIMENT	0.14 MG/KG	CONC.IN WATER=0.05UG/L,U.S.	389
*SEDIMENT	0.082 MG/KG	CONC.IN WATER:<1.0UG/L,U.S.	389
*SEDIMENT	0.01 MG/KG	CONC.IN WATER:<1.0UG/L,U.S.	389
*SESTON, 3 MY	36.1 +- 62.7 MG/KG WET WEIGHT	INITIAL WATER = 1 UG/L, 4 H, SMALL POND, FINLAND	1376
*SESTON, 3 MY	3.7 +- 17.7 MG/KG WET WEIGHT	INITIAL WATER = 1 UG/L, 8 D, SMALL POND, FINLAND	1376
*SESTON, 8 MY	9.0 +- 4.0 MG/KG WET WEIGHT	INITIAL WATER = 1 UG/L, 4 H, SMALL POND, FINLAND	1376
*SESTON, 8 MY	2.9 +- 1.3 MG/KG WET WEIGHT	INITIAL WATER = 1 UG/L, 8 D, SMALL POND, FINLAND	1376
*SHEEP, FAT	8.13 MG/KG	CONC.IN FOOD=10MG/KG	389
*SHEEP, LIVER	0.44 MG/KG	CONC.IN FOOD=10MG/KG	389
*SHEEPHEAD MINNOW	0.94 MG/KG WET WEIGHT	WATER=0.00005 PPM, TOTAL DDT, WHOLE ORGANISNS BASIS, ESTAURY	1749
*SILVERSIDE MINNOW	0.23 MG/KG WET WEIGHT	WATER=0.00005 PPM, TOTAL DDT, WHOLE ORGANISNS BASIS, ESTAURY	1749
*TERN	3.91 MG/KG WET WEIGHT	WATER=0.00005 PPM, TOTAL DDT, WHOLE ORGANISNS BASIS, ESTAURY	1749

TABLE D282

CONCENTRATIONS : DDT (TOTAL)

ITEM:	VALUE:	CONDITIONS:	REF.:
*BIRD, EGG YOLK LIPID	108. PPM WET WEIGHT	24 APRIL,%LIPID=100,AUDUBON CANYON RANCH,CAL.,USA	439
*BREVOORTIA TYRANNUS	0.343 MG/KG WET WEIGHT	LONG ISLAND(MARINE),ESTIM.PESTICID CONC. IN WATER=10**(-6)NG/L	463
*BREVOORTIA TYRANNUS, BODY	1.530 MG/KG WET WEIGHT	LONG ISLAND(MARINE).ESTIM.PESTICID CONC. IN WATER=10**(-6)NG/L	463
*BREVOORTIA TYRANNUS, HEAD AND BODYCAVITY	2.330 MG/KG WET WEIGHT	LONG ISLAND(MARINE),ESTIM.PESTICID CONC. IN WATER=10**(-6)NG/L	463

TABLE D282

CONCENTRATIONS : DDT (TOTAL)			
ITEM:	VALUE:	CONDITIONS:	REF.:
*CALLINECTES SAPIDUS	0.010-0.136 MG/KG WET WEIGHT	LONG ISLAND(MARINE),ESTIM.PESTICID CONC. IN WATER=10**(-6)NG/L	463
*CIRCUS HUDSONIUS EGGS	38.200 MG/KG WET WEIGHT	LONG ISLAND,FOOD=RALLUS,STERNA AND RYNCHOPS	463
*EGRET, EGG YOLK LIPID	285. PPM WET WEIGHT	NO DATE,%LIPID=100,AUDUBON CANYON RANCH,CAL.,USA	439
*ENSIS DIRECTUS	0.029 MG/KG WET WEIGHT	LONG ISLAND(MARINE),ESTIM.PESTICID CONC. IN WATER=10**(-6)NG/L	463
*FUNDULUS MAJALIS	0.412 MG/KG WET WEIGHT	LONG ISLAND(MARINE),ESTIM.PESTICID CONC. IN WATER=10**(-6)NG/L	463
*HERON (PROBABLE), EGG YOLK LIPID	320 PPM WET WEIGHT	30MARCH,%LIPID=100,AUDUBON CANYON RANCH,CAL.,USA	439
*HERON (PROBABLE), EMBRYO	1.4 PPM WET WT.	NO DATE,%LIPID=3.6,AUDUBON CANYON RANCH,CAL.,USA	439
*HERON, EGG YOLK LIPID	122. PPM WET WEIGHT	20 APRIL,%LIPID=100,AUDUBON CANYON RANCH,CAL.,USA	439
*HOMARUS AMERICANUS	0.216 MG/KG WET WEIGHT	LONG ISLAND(MARINE),ESTIM.PESTICID CONC. IN WATER=10**(-6)NG/L	463
*MELAMPUS BIDENATUS	0.019 MG/KG WET WEIGHT	LONG ISLAND(MARINE),ESTIM.PESTICID CONC. IN WATER=10**(-6)NG/L	463
*MENIDIA MENIDIA	0.149 MG/KG WET WEIGHT	LONG ISLAND(MARINE),ESTIM.PESTICID CONC. IN WATER=10**(-6)NG/L	463
*NASSARIUS OBSOLETUS	0.076 MG/KG WET WEIGHT	LONG ISLAND(MARINE),ESTIM.PESTICID CONC. IN WATER=10**(-6)NG/L	463
*NESTLING,EGRET(AGE UNKNOWN)	1.2 PPM WET WT.	13 MAY,%LIPID=1.2,AUDUBON CANYON RANCH,CAL.,USA	439
*NESTLING,EGRET(1 WEEK)	0.59 PPM WET WT.	7 MAY,%LIPID=1.2,AUDUBON CANYON RANCH,CAL.,USA	439
*NESTLING,EGRET(1.5-2 WEEKS)	1.2 PPM WET WT.	11 MAY,%LIPID=1.2,AUDUBON CANYON RANCH,CAL.,USA	439
*NESTLING,EGRET(1-2 WEEKS)	3.3 PPM WET WT.	15 MAY,%LIPID=1.7,AUDUBON CANYON RANCH,CAL.,USA	439
*NESTLING,EGRET(2-2.5 WEEKS)	2.2 PPM WET WT.	11 MAY,%LIPID=2.1,AUDUBON CANYON RANCH,CAL.,USA	439
*NESTLING,EGRET(5-6 DAYS)	4.1 PPM WET WT.	11 MAY,%LIPID=2.2,AUDUBON CANYON RANCH,CAL.,USA	439
*NESTLING,HERON(2 WEEKS)	0.56 PPM WET WT.	24 APRIL,%LIPID=1.6,AUDUBON CANYON RANCH,CAL.,USA	439
*NESTLING,HERON(3 WEEKS)	1.33 PPM WET WT.	24 APRIL,%LIPID=1.,AUDUBON CANYON RANCH,CAL.,USA	439
*NESTLING,HERON(4-5 WEEKS)	0.18 PPM WET WT.	24 APRIL,%LIPID=1.2,AUDUBON CANYON RANCH,CAL.,USA	439
*OPSANUS TAU	0.123 MG/KG WET WEIGHT	LONG ISLAND(MARINE),ESTIM.PESTICID CONC. IN WATER=10**(-6)NG/L	463
*PAGOPHILUS GROENLANDICUS, BLUBBER	4.40 (SD=+-1.27) PPM (ON WET WEIGHT BASIS)	95.5 % FAT, 4 ANIMALS, ST LAWRENCE, CANADA	785
*PAGOPHILUS GROENLANDICUS, BLUBBER	1.34 (SD=+-0.59) PPM (ON WET WEIGHT BASIS)	83,0 % FAT, 15 SAMPLES,1973,GULF OF ST LAWRENCE	484
*PAGOPHILUS GROENLANDICUS, BLUBBER	1.41 (SD=+-0.66) PPM (ON WET WEIGHT BASIS)	97.8 % FAT, 10 SAMPLES,1971,GULF OF ST LAWRENCE	484
*PAGOPHILUS GROENLANDICUS, BRAIN	0.02 (SD=+-0.02) PPM (ON WET WEIGHT BASIS)	6.6 % FAT, 14 SAMPLES, 1973,GULF OF ST LAWRENCE	484
*PAGOPHILUS GROENLANDICUS, BRAIN	0.09 (SD=+-0.17) PPM (ON WET WEIGHT BASIS)	7.5 % FAT, 10 SAMPLES, 1971,GULF OF ST LAWRENCE	484
*PAGOPHILUS GROENLANDICUS, KIDNEY	0.49 (SD=+-0.82) PPM (ON WET WEIGHT BASIS)	14.4 % FAT, 13 SAMPLES,1973,GULF OF ST LAWRENCE	484
*PAGOPHILUS GROENLANDICUS, LIVER	0.16 (SD=+-0.21) PPM (ON WET WEIGHT BASIS)	4.4 % FAT, 16 SAMPLES, 1973,GULF OF ST LAWRENCE	484
*PAGOPHILUS GROENLANDICUS, LIVER	0.28 (SD=+-0.17) PPM (ON WET WEIGHT BASIS)	7.0 % FAT, 11 SAMPLES, 1971,GULF OF ST LAWRENCE	484
*PAGOPHILUS GROENLANDICUS, MUSCLE	0.01 (SD=+-0.00) PPM (ON WET WEIGHT BASIS)	1.1 % FAT, 4 ANIMALS, ST LAWRENCE, CANADA	785
*PALAEMONETES VULGARIS	0.061 MG/KG WET WEIGHT	LONG ISLAND(MARINE),ESTIM.PESTICID CONC. IN WATER=10**(-6)NG/L	463
*POMATOMUS SALTATRIX	0.199 MG/KG WET WEIGHT	LONG ISLAND(MARINE),ESTIM.PESTICID CONC. IN WATER=10**(-6)NG/L	463
*PSEUDOPLEURONECTES AMERICANUS	0.059 MG/KG WET WEIGHT	LONG ISLAND(MARINE),ESTIM.PESTICID CONC. IN WATER=10**(-6)NG/L	463

TABLE D282

CONCENTRATIONS : DDT (TOTAL)

ITEM:	VALUE:	CONDITIONS:	REF.:
*RALLUS LONGIROSTRIS	1.430 MG/KG WET WEIGHT	LONG ISLAND,FOOD=MELAMPUS AND UCA	463
*RALLUS LONGIROSTRIS, EGG	1.400 MG/KG WET WEIGHT	LONG ISLAND,FOOD=MELAMPUS AND UCA	463
*ROCCUS AMERICANUS	1.990 MG/KG WET WEIGHT	LONG ISLAND(MARINE),ESTIM.PESTICID CONC. IN WATER=10**(-6)NG/L	463
*RYNCHOPS NIGRA NIGRA	13.900 MG/KG WET WEIGHT	LONG ISLAND,FOOD=MENIDIA AND FUNDULUS	463
*RYNCHOPS NIGRA NIGRA EGGS	5.400 MG/KG WET WEIGHT	LONG ISLAND,FOOD=MENIDIA AND FUNDULUS	463
*SCOMBER SCOMBRUS	0.177 MG/KG WET WEIGHT	LONG ISLAND(MARINE),ESTIM.PESTICID CONC. IN WATER=10**(-6)NG/L	463
*STERNA HIRUNDO	3.760 MG/KG WET WEIGHT	LONG ISLAND,FOOD=MENIDIA AND FUNDULUS	463
*STERNA HIRUNDO, EGG	5.700 MG/KG WET WEIGHT	LONG ISLAND,FOOD=MENIDIA AND FUNDULUS	463
*UCA SP.	0.081-0.083 MG/KG WET WEIGHT	LONG ISLAND(MARINE),ESTIM.PESTICID CONC. IN WATER=10**(-6)NG/L	463
*UCA SP.	0.054 MG/KG WET WEIGHT	LONG ISLAND(MARINE),ESTIM.PESTICID CONC. IN WATER=10**(-6)NG/L	463

TABLE D283

CONCENTRATIONS : DI-2-ETHYLHEXYL PHTHALATE

ITEM:	VALUE:	CONDITIONS:	REF.:
*PIMEPHALES PROMELAS	1.2 MICROGRAM/GRAM WET WEIGHT	1.24 G, 56 D, UPTAKE FROM WATER, CONC = 1.9 UG/L	996
*PIMEPHALES PROMELAS	2.2 MICROGRAM/GRAM WET WEIGHT	1.24 G, 56 D, UPTAKE FROM WATER, CONC = 2.5 UG/L	996
*PIMEPHALES PROMELAS	4.1 MICROGRAM/GRAM WET WEIGHT	1.24 G, 56 D, UPTAKE FROM WATER, CONC = 4.6 UG/L	996
*PIMEPHALES PROMELAS	3.6 MICROGRAM/GRAM WET WEIGHT	1.24 G, 56 D, UPTAKE FROM WATER, CONC = 8.1 UG/L	996
*PIMEPHALES PROMELAS	5.0 MICROGRAM/GRAM WET WEIGHT	1.24 G, 56 D, UPTAKE FROM WATER, CONC = 14 UG/L	996
*PIMEPHALES PROMELAS	8.6 MICROGRAM/GRAM WET WEIGHT	1.24 G, 56 D, UPTAKE FROM WATER, CONC = 30 UG/L	996
*PIMEPHALES PROMELAS	9.6 MICROGRAM/GRAM WET WEIGHT	1.24 G, 56 D, UPTAKE FROM WATER, CONC = 62 UG/L	996
*SEDIMENT	200 NG/G	CONC.IN WATER=300NG/G,LAKE SUPERIOR,ONTARIO	1303
*WALLEYE	800 NG/G	CONC.IN WATER=300NG/G,LAKE SUPERIOR,ONTARIO	1303

TABLE D284

CONCENTRATIONS : DIELDRIN

ITEM:	VALUE:	CONDITIONS:	REF.:
*CALLINECTES SAPIDUS	25.3 (SD=+-4.6) MICROGRAM/KG WET WEIGHT	FOOD=3 G WITH 103 UG/KG/D, 10 D, WATER=0, PH=8.2, 294 K, CONT=7.5 PPM	1252
*CALLINECTES SAPIDUS	96.6 (SD=+-7.8) MICROGRAM/KG WET WEIGHT	FOOD=3 G WITH 181 UG/KG/D, 10 D, WATER=0, PH=8.2, 294 K, CONT=4.7 PPM	1252
*CALLINECTES SAPIDUS	32.8 (SD=+-5.5) MICROGRAM/KG WET WEIGHT	FOOD=3 G WITH 193 UG/KG/D, 10 D, WATER=0, PH=8.2, 294 K, CONT=3.4 PPM	1252
*CALLINECTES SAPIDUS	47 % OF INTAKE	INCORPORATION FROM FOOD = 103 UG/KG, WATER = 0 PPB	1252
*CALLINECTES SAPIDUS	68 % OF INTAKE	INCORPORATION FROM FOOD = 181 UG/KG, WATER = 0 PPB	1252

TABLE D284

CONCENTRATIONS : DIELDRIN

ITEM:	VALUE:	CONDITIONS:	REF.:
*CALLINECTES SAPIDUS	78 % OF INTAKE	INCORPORATION FROM FOOD = 193 UG/KG, WATER = 0 PPB	1252
*CHICKEN, LIVER	0.12 MG/KG WET WEIGHT	CONC.IN FOOD=0.05MG/KG	389
*CHICKEN, LIVER	0.27 MG/KG WET WEIGHT	CONC.IN FOOD=0.15MG/KG	389
*CHICKEN, LIVER	0.50 MG/KG WET WEIGHT	CONC.IN FOOD=0.45MG/KG	389
*CHICKEN, LIVER	0.10 MG/KG WET WEIGHT	CONC.IN FOOD=0.05MG/KG	389
*CHICKEN, LIVER	0.20 MG/KG WET WEIGHT	CONC.IN FOOD=0.15MG/KG	389
*CHICKEN, LIVER	0.35 MG/KG WET WEIGHT	CONC.IN FOOD=0.45MG/KG	389
*CHICKEN, TISSUE	10.20 MG/KG WET WEIGHT	CONC.IN FOOD=0.25MG/KG	389
*CHICKEN, TISSUE	35.70 MG/KG WET WEIGHT	CONC.IN FOOD=0.75MG/KG	389
*CHICKEN, TISSUE	0.60 MG/KG WET WEIGHT	CONC.IN FOOD=10.00MG/KG	389
*CHICKEN, TISSUE	1.0 MG/KG WET WEIGHT	CONC.IN FOOD=20.0MG/KG	389
*COTURNIX QUAIL, BRAIN	0.72 MG/KG WET WEIGHT	CONC.IN FOOD=2MG/KG	389
*COTURNIX QUAIL, BRAIN	10.13 MG/KG WET WEIGHT	CONC.IN FOOD=10MG/KG	389
*COTURNIX QUAIL, BRAIN	13.19 MG/KG WET WEIGHT	CONC.IN FOOD=50MG/KG	389
*COTURNIX QUAIL, BRAIN	12.49 MG/KG WET WEIGHT	CONC.IN FOOD=250MG/KG	389
*COTURNIX QUAIL, LIVER	22.02 MG/KG WET WEIGHT	CONC.IN FOOD=10MG/KG	389
*COTURNIX QUAIL, LIVER	59.83 MG/KG WET WEIGHT	CONC.IN FOOD=50MG/KG	389
*COTURNIX QUAIL, LIVER	23.08 MG/KG WET WEIGHT	CONC.IN FOOD=250MG/KG	389
*DUNALIELLA PEIRCEI	1.27 MICROGRAM/G DRY WEIGHT	24 H, WATER = 10 UG/L, N=10	1251
*DUNALIELLA PEIRCEI	0.91 MICROGRAM/G DRY WEIGHT	48 H, WATER = 10 UG/L, N=10	1251
*DUNALIELLA PEIRCEI	0.84 MICROGRAM/G DRY WEIGHT	72 H, WATER = 10 UG/L, N=10	1251
*DUNALIELLA PEIRCEI	0.77 MICROGRAM/G DRY WEIGHT	96 H, WATER = 10 UG/L, N=10	1251
*DUNALIELLA PEIRCEI	1.27 MICROGRAM/G DRY WEIGHT	24 H, WATER = 10 UG/L, N=10	1251
*DUNALIELLA PEIRCEI	0.91 MICROGRAM/G DRY WEIGHT	48 H, WATER = 10 UG/L, N=10	1251
*DUNALIELLA PEIRCEI	0.84 MICROGRAM/G DRY WEIGHT	72 H, WATER = 10 UG/L, N=10	1251
*DUNALIELLA PEIRCEI	0.77 MICROGRAM/G DRY WEIGHT	96 H, WATER = 10 UG/L, N=10	1251
*ICTALURUS PUNCTATUS, DORSAL MUSCLE	45 MICROGRAM/KG WET WEIGHT	UPTAKE FROM WATER=75 NG/L,7 D, 150-225 MM	1421
*ICTALURUS PUNCTATUS, DORSAL MUSCLE	81 MICROGRAM/KG WET WEIGHT	UPTAKE FROM WATER=75 NG/L,21 D, 150-225 MM	1421
*ICTALURUS PUNCTATUS, DORSAL MUSCLE	175 MICROGRAM/KG WET WEIGHT	UPTAKE FROM WATER=75 NG/L,28 D, 150-225 MM	1421
*ICTALURUS PUNCTATUS, DORSAL MUSCLE	74 MICROGRAM/KG WET WEIGHT	UPTAKE FROM WATER=75 NG/L,7 D, 350-400 MM	1421
*ICTALURUS PUNCTATUS, DORSAL MUSCLE	179 MICROGRAM/KG WET WEIGHT	UPTAKE FROM WATER=75 NG/L,21 D, 350-400 MM	1421
*ICTALURUS PUNCTATUS, DORSAL MUSCLE	274 MICROGRAM/KG WET WEIGHT	UPTAKE FROM WATER=75 NG/L,28 D, 350-400 MM	1421
*ICTALURUS PUNCTATUS, DORSAL MUSCLE	179 MICROGRAM/KG WET WEIGHT	UPTAKE FROM FOOD = 2 PPM, 7 D, 150-225 MM	1421
*ICTALURUS PUNCTATUS, DORSAL MUSCLE	506 MICROGRAM/KG WET WEIGHT	UPTAKE FROM FOOD = 2 PPM, 21 D, 150-225 MM	1421
*ICTALURUS PUNCTATUS, DORSAL MUSCLE	544 MICROGRAM/KG WET WEIGHT	UPTAKE FROM FOOD = 2 PPM, 28 D, 150-225 MM	1421
*ICTALURUS PUNCTATUS, DORSAL MUSCLE	246 MICROGRAM/KG WET WEIGHT	UPTAKE FROM FOOD = 2 PPM, 7 D, 350-400 MM	1421
*ICTALURUS PUNCTATUS, DORSAL MUSCLE	796 MICROGRAM/KG WET WEIGHT	UPTAKE FROM FOOD = 2 PPM, 21 D, 350-400 MM	1421
*ICTALURUS PUNCTATUS, DORSAL MUSCLE	1243 MICROGRAM/KG WET WEIGHT	UPTAKE FROM FOOD = 2 PPM, 28 D, 350-400 MM	1421
*ICTALURUS PUNCTATUS, DORSAL MUSCLE	303 MICROGRAM/KG WET WEIGHT	UPTAKE FROM FOOD(2 PPM) AND WATER(75 NG/L), 7 D, 150-225 MM	1421
*ICTALURUS PUNCTATUS, DORSAL MUSCLE	606 MICROGRAM/KG WET WEIGHT	UPTAKE FROM FOOD(2 PPM) AND WATER(75 NG/L), 21 D, 150-225 MM	1421
*ICTALURUS PUNCTATUS, DORSAL MUSCLE	898 MICROGRAM/KG WET WEIGHT	UPTAKE FROM FOOD(2 PPM) AND WATER(75 NG/L), 28 D, 150-225 MM	1421
*ICTALURUS PUNCTATUS, DORSAL MUSCLE	280 MICROGRAM/KG WET WEIGHT	UPTAKE FROM FOOD(2 PPM) AND WATER(75 NG/L), 7 D, 350-400 MM	1421

TABLE D284

CONCENTRATIONS : DIELDRIN

ITEM:	VALUE:	CONDITIONS:	REF.:
*ICTALURUS PUNCTATUS, DORSAL MUSCLE	1199 MICROGRAM/KG WET WEIGHT	UPTAKE FROM FOOD(2 PPM) AND WATER(75 NG/L), 21 D, 350-400 MM	1421
*ICTALURUS PUNCTATUS, DORSAL MUSCLE	2418 MICROGRAM/KG WET WEIGHT	UPTAKE FROM FOOD(2 PPM) AND WATER(75 NG/L), 28 D, 350-400 MM	1421
*LAMB, TISSUE	0.40 MG/KG WET WEIGHT	CONC.IN FOOD=0.25MG/KG	389
*LAMB, TISSUE	0.60 MG/KG WET WEIGHT	CONC.IN FOOD=0.75MG/KG	389
*LAMB, TISSUE	1.70 MG/KG WET WEIGHT	CONC.IN FOOD=2.25MG/KG	389
*PAGOPHILUS GROENLANDICUS, BLUBBER	0.150 (SD=+-0.025) PPM (ON WET WEIGHT BASIS)	95.5 % FAT, 4 ANIMALS, ST LAWRENCE, CANADA	785
*PAGOPHILUS GROENLANDICUS, BLUBBER	0.050 (SD=+-0.046) PPM (ON WET WEIGHT BASIS)	83.0 % FAT, 15 SAMPLES,1973,GULF OF ST LAWRENCE	484
*PAGOPHILUS GROENLANDICUS, BLUBBER	TRACE	97.8 % FAT, 10 SAMPLES,1971,GULF OF ST LAWRENCE	484
*PAGOPHILUS GROENLANDICUS, BRAIN	0.001 (SD=+-0.001) PPM (ON WET WEIGHT BASIS)	11.5 % FAT, 3 ANIMALS, ST LAWRENCE, CANADA	785
*PAGOPHILUS GROENLANDICUS, BRAIN	TRACE	6.6 % FAT, 14 SAMPLES, 1973,GULF OF ST LAWRENCE	484
*PAGOPHILUS GROENLANDICUS, BRAIN	TRACE	7.5 % FAT, 10 SAMPLES, 1971,GULF OF ST LAWRENCE	484
*PAGOPHILUS GROENLANDICUS, KIDNEY	0.012 (SD=+-0.019) PPM (ON WET WEIGHT BASIS)	14.4 % FAT, 13 SAMPLES,1973,GULF OF ST LAWRENCE	484
*PAGOPHILUS GROENLANDICUS, LIVER	0.001 (SD=+-0.001) PPM (ON WET WEIGHT BASIS)	1.8 % FAT, 4 AINMALS	785
*PAGOPHILUS GROENLANDICUS, LIVER	0.002 (SD=+-0.002) PPM (ON WET WEIGHT BASIS)	4.4 % FAT, 16 SAMPLES, 1973,GULF OF ST LAWRENCE	484
*PAGOPHILUS GROENLANDICUS, LIVER	TRACE	7.0 % FAT, 11 SAMPLES, 1971,GULF OF ST LAWRENCE	484
*PAGOPHILUS GROENLANDICUS, MUSCLE	0.001 (SD=+-0.001) PPM (ON WET WEIGHT BASIS)	1.1 % FAT, 4 ANIMALS, ST LAWRENCE, CANADA	785
*PHEASANT, MUSCLE	2.70 MG/KG WET WEIGHT	CONC.IN FOOD=50MG/KG	389
*PIG, TISSUE	0.40 MG/KG WET WEIGHT	CONC.IN FOOD=0.25MG/KG	389
*PIG, TISSUE	2.80 MG/KG WET WEIGHT	CONC.IN FOOD=0.75MG/KG	389
*PIG, TISSUE	4.30 MG/KG WET WEIGHT	CONC.IN FOOD=2.25MG/KG	389
*RAT, BLOOD	0.254 MG/KG WET WEIGHT	UPTAKE FROM FOOD WITH 50 PPM DIELDRIN, STEADY STATE, 180 D	389
*RAT, BRAIN	0.24 MG/KG WET WEIGHT	CONC.IN FOOD=10MG/KG	389
*RAT, FAT	183 MG/KG WET WEIGHT	UPTAKE FROM FOOD WITH 50 PPM DIELDRIN, STEADY STATE, 180 D	389
*RAT, FAT	15.85 MG/KG WET WEIGHT	CONC.IN FOOD=10MG/KG	389
*RAT, LIVER	7.85 MG/KG WET WEIGHT	UPTAKE FROM FOOD WITH 50 PPM DIELDRIN, STEADY STATE, 180 D	389
*RAT, LIVER	0.85 MG/KG WET WEIGHT	CONC.IN FOOD=10MG/KG	389
*RAT, TISSUE	15.10 MG/KG WET WEIGHT	CONC.IN FOOD=1.0MG/KG	389
*RAT, TISSUE	67.50 MG/KG WET WEIGHT	CONC.IN FOOD=10MG/KG	389
*RHIZOPHORA MANGLE, HYPOCOTYL	0.01 MG/KG WET WT.	APPLICATION=0.06KG/HA,PH=6.3-6.7,S =2.5%,6WEEKS	1647
*RHIZOPHORA MANGLE, HYPOCOTYL	0.02 MG/KG WET WT.	APPLICATION=0.11KG/HA,PH=6.3-6.7,S =2.5%,6WEEKS	1647
*RHIZOPHORA MANGLE, HYPOCOTYL	0.02 MG/KG WET WT.	APPLICATION=0.28KG/HA,PH=6.3-6.7,S =2.5%,6WEEKS	1647
*RHIZOPHORA MANGLE, HYPOCOTYL	0.04 MG/KG WET WT.	APPLICATION=0.56KG/HA,PH=6.3-6.7,S =2.5%,6WEEKS	1647
*RHIZOPHORA MANGLE, HYPOCOTYL	0.08 MG/KG WET WT.	APPLICATION=1.12KG/HA,PH=6.3-6.7,S =2.5%,6WEEKS	1647
*RHIZOPHORA MANGLE, HYPOCOTYL	0.08 MG/KG WET WT.	APPLICATION=2.80KG/HA,PH=6.3-6.7,S =2.5%,6WEEKS	1647
*RHIZOPHORA MANGLE, HYPOCOTYL	0.08 MG/KG WET WT.	APPLICATION=5.60KG/HA,PH=6.3-6.7,S =2.5%,6WEEKS	1647
*RHIZOPHORA MANGLE, HYPOCOTYL	0.16 MG/KG WET WT.	APPLICATION=11.20KG/HA,PH=6.3-6.7, S=2.5%,6WEEKS	1647
*RHIZOPHORA MANGLE, LEAVES	0.07 MG/KG WET WT.	APPLICATION=0.06KG/HA,PH=6.3-6.7,S =2.5%,6WEEKS	1647
*RHIZOPHORA MANGLE, LEAVES	0.11 MG/KG WET WT.	APPLICATION=0.11KG/HA,PH=6.3-6.7,S =2.5%,6WEEKS	1647

TABLE D284

CONCENTRATIONS : DIELDRIN

ITEM:	VALUE:	CONDITIONS:	REF.:
*RHIZOPHORA MANGLE, LEAVES	0.11 MG/KG WET WT.	APPLICATION=0.28KG/HA,PH=6.3-6.7,S =2.5%,6WEEKS	1647
*RHIZOPHORA MANGLE, LEAVES	0.07 MG/KG WET WT.	APPLICATION=0.56KG/HA,PH=6.3-6.7,S =2.5%,6WEEKS	1647
*RHIZOPHORA MANGLE, LEAVES	0.11 MG/KG WET WT.	APPLICATION=1.12KG/HA,PH=6.3-6.7,S =2.5%,6WEEKS	1647
*RHIZOPHORA MANGLE, LEAVES	0.07 MG/KG WET WT.	APPLICATION=2.80KG/HA,PH=6.3-6.7,S =2.5%,6WEEKS	1647
*RHIZOPHORA MANGLE, LEAVES	0.09 MG/KG WET WT.	APPLICATION=5.60KG/HA,PH=6.3-6.7,S =2.5%,6WEEKS	1647
*RHIZOPHORA MANGLE, LEAVES	0.10 MG/KG WET WT.	APPLICATION=11.20KG/HA,PH=6.3-6.7, S=2.5%,6WEEKS	1647
*RHIZOPHORA MANGLE, ROOT	0.03 MG/KG WET WT.	APPLICATION=1.12KG/HA,PH=6.3-6.7,S =2.5%,6WEEKS	1647
*RHIZOPHORA MANGLE, ROOT	0.04 MG/KG WET WT.	APPLICATION=2.80KG/HA,PH=6.3-6.7,S =2.5%,6WEEKS	1647
*RHIZOPHORA MANGLE, ROOT	0.04 MG/KG WET WT.	APPLICATION=5.60KG/HA,PH=6.3-6.7,S =2.5%,6WEEKS	1647
*RHIZOPHORA MANGLE, ROOT	0.06 MG/KG WET WT.	APPLICATION=11.20KG/HA,PH=6.3-6.7, S=2.5%,6WEEKS	1647
*SEDIMENT	0.38 MG/KG WET WEIGHT	CONC.IN WATER=29.0 MICROGRAM/L, U.S.	389
*SEDIMENT	0.89 MG/KG WET WEIGHT	CONC.IN WATER=0.018 MICROGRAM/L, U.S.	389
*SEDIMENT	0.089 MG/KG WET WEIGHT	CONC.IN WATER=0.018 MICROGRAM/L, U.S.	389
*SEDIMENT	0.0002 MG/KG WET WEIGHT	CONC.IN WATER=TRACE,U.S.	389
*STEER, TISSUE	0.80 MG/KG WET WEIGHT	CONC.IN FOOD=0.25MG/KG	389
*STEER, TISSUE	3.50 MG/KG WET WEIGHT	CONC.IN FOOD=0.75MG/KG	389
*STEER, TISSUE	8.70 MG/KG WET WEIGHT	CONC.IN FOOD=2.25MG/KG	389
*WHITE-TAILED DEER, BRAIN	0.35 MG/KG WET WEIGHT	CONC.IN FOOD=5MG/KG	389
*WHITE-TAILED DEER, BRAIN	0.20 MG/KG WET WEIGHT	CONC.IN FOOD=5MG/KG	389
*WHITE-TAILED DEER, BRAIN	0.40 MG/KG WET WEIGHT	CONC.IN FOOD=15MG/KG	389
*WHITE-TAILED DEER, BRAIN	4.41 MG/KG WET WEIGHT	CONC.IN FOOD=25MG/KG	389
*WHITE-TAILED DEER, BRAIN	2.50 MG/KG WET WEIGHT	CONC.IN FOOD=45MG/KG	389
*WHITE-TAILED DEER, LIVER	0.98 MG/KG WET WEIGHT	CONC.IN FOOD=5MG/KG	389
*WHITE-TAILED DEER, LIVER	9.13 MG/KG WET WEIGHT	CONC.IN FOOD=25MG/KG	389
*WHITE-TAILED DEER, MUSCLE	0.10 MG/KG WET WEIGHT	CONC.IN FOOD=5MG/KG	389
*WHITE-TAILED DEER, MUSCLE	1.19 MG/KG WET WEIGHT	CONC.IN FOOD=25MG/KG	389

TABLE D285

CONCENTRATIONS : DINITRAMINE

ITEM:	VALUE:	CONDITIONS:	REF.:
*CARP, BILE	22.7 (RANGE = 20.1 - 27.0) MG/KG WET WEIGHT	12 H, UPTAKE FROM WATER=1 MG/L, 285 K, PH=7.8, HARDNESS=220 MG/L	1182
*CARP, MUSCLE	3.96 (RANGE = 1.50 - 6.73) MG/KG WET WEIGHT	12 H, UPTAKE FROM WATER=1 MG/L, 285 K, PH=7.8, HARDNESS=220 MG/L	1182
*CARP, PLASMA	24.4 (RANGE = 22.4 - 25.1) MG/KG WET WEIGHT	12 H, UPTAKE FROM WATER=1 MG/L, 285 K, PH=7.8, HARDNESS=220 MG/L	1182
*CHANNAL CATFISH, BILE	13.4 MG/KG WET WEIGHT	12 H, UPTAKE FROM WATER=1 MG/L, 285 K, PH=7.8, HARDNESS=220 MG/L	1182
*CHANNAL CATFISH, MUSCLE	46.1 (RANGE = 29.9 - 62.3) MG/KG WET WEIGHT	12 H, UPTAKE FROM WATER=1 MG/L, 285 K, PH=7.8, HARDNESS=220 MG/L	1182
*CHANNAL CATFISH, PLASMA	58.5 (RANGE = 54.0 - 63.0) MG/KG WET WEIGHT	12 H, UPTAKE FROM WATER=1 MG/L, 285 K, PH=7.8, HARDNESS=220 MG/L	1182

TABLE D286

CONCENTRATIONS : EDC-TAR

ITEM:	VALUE:	CONDITIONS:	REF.:
*LEANDER ADSPERSUS	160 PPM	EDC-TAR IN WATER= 1 PPM,283-287 K, 16 DAYS OF EXPOSURE	758

TABLE D286

CONCENTRATIONS : EDC-TAR

ITEM:	VALUE:	CONDITIONS:	REF.:
*LEANDER ADSPERSUS	8 PPM	EDC-TAR IN WATER=0.1 PPM,283-287 K,16 DAYS OF EXPOSURE	758
*LEANDER ADSPERSUS	3 PPM	EDC-TAR IN WATER=0.01 PPM,283-287 K,16 DAYS OF EXPOSURE	758
*MYTILUS EDULIS	13 PPM	EDC-TAR IN WATER=1 PPM,283-287 K, 16 DAYS OF EXPOSURE	758
*MYTILUS EDULIS	270 PPM	EDC-TAR IN WATER=10 PPM,283-287 K, 16 DAYS OF EXPOSURE	758
*MYTILUS EDULIS	26 PPM	EDC-TAR IN WATER=10 PPM,283-287 K, 16 DAYS OF EXPOSURE	758
*MYTILUS EDULIS	6 PPM	EDC-TAR IN WATER=0.1 PPM,283-287 K,16 DAYS OF EXPOSURE	758

TABLE D287

CONCENTRATIONS : ENDOSULFAN

ITEM:	VALUE:	CONDITIONS:	REF.:
*CHALAMYS OPERCULARIS, ADDUCTOR MUSCLE	0.88 MG/KG WET WEIGHT	WATER = 0.1 MG/L, 2 W	1345
*CHALAMYS OPERCULARIS, DIGESTIVE GLAND	6.1 MG/KG WET WEIGHT	WATER = 0.1 MG/L, 2 W	1345
*CHALAMYS OPERCULARIS, GILL	0.46 MG/KG WET WEIGHT	WATER = 0.1 MG/L, 2 W	1345
*CHALAMYS OPERCULARIS, GONAD	3.8 MG/KG WET WEIGHT	WATER = 0.1 MG/L, 2 W	1345
*CHALAMYS OPERCULARIS, MANTLE	1.34 MG/KG WET WEIGHT	WATER = 0.1 MG/L, 2 W	1345
*MYTILUS EDULIS, DIGESTIVE GLAND	6.2 MG/KG WET WEIGHT	WATER = 0.1 MG/L, 2 W	1345
*MYTILUS EDULIS, FOOT	0.75 MG/KG WET WEIGHT	WATER = 0.1 MG/L, 2 W	1345
*MYTILUS EDULIS, GILL	2.1 MG/KG WET WEIGHT	WATER = 0.1 MG/L, 2 W	1345
*MYTILUS EDULIS, GONAD	0.66 MG/KG WET WEIGHT	WATER = 0.1 MG/L, 2 W	1345
*MYTILUS EDULIS, MANTLE	0.66 MG/KG WET WEIGHT	WATER = 0.1 MG/L, 2 W	1345
*MYTILUS EDULIS, POSTERIOR ADDUCTOR MUSCLE	0.75 MG/KG WET WEIGHT	WATER = 0.1 MG/L, 2 W	1345

TABLE D288

CONCENTRATIONS : ENDRIN

ITEM:	VALUE:	CONDITIONS:	REF.:
*SEDIMENT	0.0006 MG/KG	CONC.IN WATER=TRACE,U.S.	389
*SEDIMENT	<0.001 MG/KG WET WEIGHT	CONC.IN WATER:<1.0UG/L,U.S.	389
*SEDIMENT	<0.01 MG/KG WET WEIGHT	CONC.IN WATER:<1.0UG/L,U.S.	389

TABLE D289

CONCENTRATIONS : FE

ITEM:	VALUE:	CONDITIONS:	REF.:
*ALISMA PLANTAGO AQUATICA, OLD LEAVES	869.07 MG/KG DRY WT.	800M ABOVE POLL.ENTRY(1),FE IN WATER=0.708MG/L,WHOLE LEAVES	1224
*ALISMA PLANTAGO AQUATICA, OLD LEAVES	596.23 MG/KG DRY WT.	800M ABOVE POLL.ENTRY(1),FE IN WATER=0.708MG/L,PROCESSED LEAVES	1224
*ALISMA PLANTAGO AQUATICA, OLD LEAVES	837.54 MG/KG DRY WT.	150M BELOW POLL.ENTRY(1),FE IN WATER=0.893MG/L,WHOLE LEAVES	1224

TABLE D289

CONCENTRATIONS : FE			
ITEM:	VALUE:	CONDITIONS:	REF.:
*ALISMA PLANTAGO AQUATICA, OLD LEAVES	599.09 MG/KG DRY WT.	150M BELOW POLL.ENTRY(1),FE IN WATER=0.893MG/L,PROCESSED LEAVES	1224
*ALISMA PLANTAGO AQUATICA, OLD LEAVES	795.83 MG/KG DRY WT.	800 AND 50M BELOW POLL.(1)AND(2), FE IN WATER=0.888MG/L,WHOLE LEAVES	1224
*ALISMA PLANTAGO AQUATICA, OLD LEAVES	549.81 MG/KG DRY WT.	800 AND 50M BELOW POLL.(1)AND(2), FE IN WATER=0.888MG/L,PROCESSED LEAVES	1224
*ALISMA PLANTAGO AQUATICA, YOUNG LEAVES	1145.85 MG/KG DRY WT.	800M ABOVE POLL.ENTRY(1),FE IN WATER=0.708MG/L,WHOLE LEAVES	1224
*ALISMA PLANTAGO AQUATICA, YOUNG LEAVES	863.99 MG/KG DRY WT.	800M ABOVE POLL.ENTRY(1),FE IN WATER=0.708MG/L,PROCESSED LEAVES	1224
*ALISMA PLANTAGO AQUATICA, YOUNG LEAVES	1115.78 MG/KG DRY WT.	150M BELOW POLL.ENTRY(1),FE IN WATER=0.893MG/L,WHOLE LEAVES	1224
*ALISMA PLANTAGO AQUATICA, YOUNG LEAVES	728.12 MG/KG DRY WT.	150M BELOW POLL.ENTRY(1),FE IN WATER=0.893MG/L,PROCESSED LEAVES	1224
*ALISMA PLANTAGO AQUATICA, YOUNG LEAVES	966.42 MG/KG DRY WT.	800 AND 50M BELOW POLL.(1)AND(2), FE IN WATER=0.888MG/L,WHOLE LEAVES	1224
*ALISMA PLANTAGO AQUATICA, YOUNG LEAVES	878.56 MG/KG DRY WT.	800 AND 50M BELOW POLL.(1)AND(2), FE IN WATER=0.888MG/L,PROCESSED LEAVES	1224
*ATMOSPHERE	0.84+-0.21 NG/M**3	SOUTHPOLE,1970	893
*ATMOSPHERE	90-650 NG/M**3	ENGLAND,7 STATIONS	893
*ATMOSPHERE	200-11000 NG/M**3	SWEDEN,MEAN FROM 12 STATIONS	893
*ATMOSPHERE	80-700 NG/M**3	NOT URBAN,U.S.A.	893
*ATMOSPHERE	300-4200 NG/M**3	URBAN,U.S.A.	893
*FUCUS VESICULOSUS	2.4*10**4 BASED ON DRY WEIGHT	18 OBS.,ALL AREAS AROUND THE BRITISH ISLES	1278
*HYPOLIMNION	425 MICROGRAM/L	CONC.IN EPILIMNION=15 MICROGRAM/L, CONC.FACTOR=28,SCHOEHSEE	573
*LIVING SPRUCE NEEDLES	29 - 111 MG/KG DRY WEUGHT	LITTER=574-2850, UNDERLAYING SOIL= 19000-24000MG/KG DW,SOUTH SWEDEN	1129
*RAIN	200 MICROGRAM/L	ENGLAND,YEARLY MEAN	893
*SEDIMENT	6.7 PPM OF DRY WEIGHT	WATER CONC = 91 MICROGRAM/L, YEARLY MEAN, CARNON RIVER	48
*SEDIMENT	3.0 PPM OF DRY WEIGHT	WATER CONC = 14 MICROGRAM/L, YEARLY MEAN, CARNON RIVER	48
*SEDIMENT	7.7 PPM OF DRY WEIGHT	WATER CONC = 83 MICROGRAM/L, YEARLY MEAN, RED RIVER	48
*SEDIMENT	5.1 PPM OF DRY WEIGHT	WATER CONC = 22 MICROGRAM/L, YEARLY MEAN, RED RIVER	48
*SEDIMENT	7.1 PPM OF DRY WEIGHT	WATER CONC = 380 MICROGRAM/L, YEARLY MEAN, GANNEL RIVER	48
*SEDIMENT	4.2 PPM OF DRY WEIGHT	WATER CONC = 79 MICROGRAM/L, YEARLY MEAN, GANNEL RIVER	48
*SEDIMENT	3.6 PPM OF DRY WEIGHT	WATER CONC = 20 MICROGRAM/L, YEARLY MEAN, FOWEY RIVER	48
*SEDIMENT	2.6 PPM OF DRY WEIGHT	WATER CONC = 33 MICROGRAM/L, YEARLY MEAN, FOWEY RIVER	48
*SEDIMENT	727 PPM OF DRY WEIGHT	WATER CONC = 167 MICROGRAM/L, YEARLY MEAN, CARNON RIVER	48
*SEDIMENT	3.0 PPM OF DRY WEIGHT	MAX WATER CONC = 26 MICROGRAM/L, 1 YEARS MEASUREMENT, CARNON R	48
*SEDIMENT	7.7 PPM OF DRY WEIGHT	MAX WATER CONC = 140 MICROGRAM/L, 1 YEARS MEASUREMENT, RED R	48
*SEDIMENT	5.1 PPM OF DRY WEIGHT	MAX WATER CONC = 50 MICROGRAM/L, 1 YEARS MEASUREMENT, RED R	48
*SEDIMENT	7.1 PPM OF DRY WEIGHT	MAX WATER CONC = 485 MICROGRAM/L, MAX CONC, GANNEL RIVER	48
*SEDIMENT	4.2 PPM OF DRY WEIGHT	MAX WATER CONC = 120 MICROGRAM/L, MAX CONC, GANNEL RIVER	48
*SEDIMENT	3.6 PPM OF DRY WEIGHT	MAX WATER CONC = 28 MICROGRAM/L, 1 YEARS MEASUREMENT, FOWEY R	48
*SEDIMENT	2.6 PPM OF DRY WEIGHT	MAX WATER CONC = 45 MICROGRAM/L, 1 YEARS MEASUREMENT, FOWEY R	48
*SEDIMENT	727 PPM OF DRY WEIGHT	MAX WATER CONC = 280 MICROGRAM/L, 1 YEARS MEASUREMENT, CARNON R	48

TABLE D289

CONCENTRATIONS : FE

ITEM:	VALUE:	CONDITIONS:	REF.:
*SEDIMENT	649.31 (596.66-683.36) MG/KG DRY WT.	800M ABOVE POLL.ENTRY(1),FE IN WATER=0.708MG/L,POLLUTED RIVER	1224
*SEDIMENT	683.96 (628.77-714.18) MG/KG DRY .WT.	150M BELOW POLL.ENTRY(1),FE IN WATER=0.893MG/L,POLLUTED RIVER	1224
*SEDIMENT	1593.91 (1548.25-1655.80) MG/KG DRY WT.	800 AND 50M BELOW POLL.(1)AND(2), FE IN WATER=0.888MG/L,POLLUTED RIVER	1224
*SEDIMENT	58000 MICROGRAM/G DRY WT.	CONC.IN EPILIMNION=15 MICROGRAM/L, CONC.FACTOR=3.9*10**6,SCHOEHSEE	573
*WATER	0.8-63 MICROGRAM/L	BALTIC SEA,8 STATIONS,1970-71	893
*WATER	31-1670 MICROGRAM/L	RIVERS,U.S.A.	893

TABLE D290

CONCENTRATIONS : GAMMA BHC

ITEM:	VALUE:	CONDITIONS:	REF.:
*SOIL	0.1 PPM	MIRPUR,DACCA,10G	1101
*SOIL	0.5 PPM	JOYDABPUR(A),DACCA,10G	1101
*SOIL	0.1 PPM	JOYDABPUR(B),DACCA,10G	1101
*SOIL	0.45 PPM	DEMRA(A),DACCA,10G	1101
*SOIL	0.5 PPM	DEMRA(B),DACCA,10G	1101
*SOIL	0.43 PPM	DEMRA(C),DACCA 10G	1101

TABLE D291

CONCENTRATIONS : HEPTACHLOR

ITEM:	VALUE:	CONDITIONS:	REF.:
*CHICKEN, LIVER	0.15 MG/KG WET WEIGHT	CONC.IN FOOD=0.15MG/KG	389
*CHICKEN, LIVER	0.31 MG/KG WET WEIGHT	CONC.IN FOOD=0.45MG/KG	389
*WOODCOCK, TISSUE	1.70 MG/KG WET WEIGHT	CONC.IN FOOD=0.65MG/KG	389
*WOODCOCK, TISSUE	13.0 MG/KG WET WEIGHT	CONC.IN FOOD=2.86MG/KG	389

TABLE D292

CONCENTRATIONS : HEPTACHLOR AND -EPOXIDE

ITEM:	VALUE:	CONDITIONS:	REF.:
*SEDIMENT	0.0008 MG/KG WET WEIGHT	CONC.IN WATER=0.06UG/L,U.S.	389
*SEDIMENT	<0.002 MG/KG WET WEIGHT	CONC.IN WATER:<1.0UG/L,U.S.	389
*SEDIMENT	<0.02 MG/KG WET WEIGHT	CONC.IN WATER:<1.0UG/L,U.S.	389

TABLE D293

CONCENTRATIONS : HEXACHLOROPHENE

ITEM:	VALUE:	CONDITIONS:	REF.:
*CAMBARINAE	1338 MICROGRAM/KG WET WEIGHT	SEDIMENT= 9.3-377.0 UG/KG DW, IMMATURE, BUFFALO CREEK	1439

TABLE D293

CONCENTRATIONS : HEXACHLOROPHENE

ITEM:	VALUE:	CONDITIONS:	REF.:
*HYDROPHILUS TRIANGULARIS	1500 MICROGRAM/KG WET WEIGHT	SEDIMENT= 9.3-377.0 UG/KG DW, BUFFALO CREEK	1439
*SEDIMENT	377.0 MICROGRAM/KG DRY WEIGHT	SURFACE WATER=44.3 PPM, BOTTOM WATER=30.0 PPM, BUFFALO CREEK	1439
*SEDIMENT	106.8 MICROGRAM/KG DRY WEIGHT	SURFACE WATER=18.0 PPM, BOTTOM WATER=22.0 PPM, BUFFALO CREEK	1439
*SEDIMENT	81.4 MICROGRAM/KG DRY WEIGHT	SURFACE WATER=21.2 PPM, BOTTOM WATER=22.0 PPM, BUFFALO CREEK	1439
*SEDIMENT	110.9 MICROGRAM/KG DRY WEIGHT	SURFACE WATER=31.2 PPM, BOTTOM WATER=21.6 PPM, BUFFALO CREEK	1439
*SEDIMENT	30.5 MICROGRAM/KG DRY WEIGHT	SURFACE WATER=37.8 PPM, BOTTOM WATER=41.4 PPM, BUFFALO CREEK	1439
*SEDIMENT	19.5 MICROGRAM/KG DRY WEIGHT	SURFACE WATER=16.4 PPM, BOTTOM WATER=21.2 PPM, BUFFALO CREEK	1439
*SEDIMENT	25.4 MICROGRAM/KG DRY WEIGHT	SURFACE WATER=41.5 PPM, BOTTOM WATER=19.2 PPM, BUFFALO CREEK	1439
*SEDIMENT	44.9 MICROGRAM/KG DRY WEIGHT	SURFACE WATER=37.6 PPM, BOTTOM WATER=15.2 PPM, BUFFALO CREEK	1439
*SEDIMENT	159.2 MICROGRAM/KG DRY WEIGHT	SURFACE WATER=28.9 PPM, BOTTOM WATER=48.3 PPM, BUFFALO CREEK	1439
*SEDIMENT	9.3 MICROGRAM/KG DRY WEIGHT	SURFACE WATER=20.0 PPM, BOTTOM WATER=20.8 PPM, BUFFALO CREEK	1439
*SEDIMENT	27.1 MICROGRAM/KG DRY WEIGHT	SURFACE WATER=21.6 PPM, BOTTOM WATER=26.8 PPM, BUFFALO CREEK	1439
*SEDIMENT	84.8 MICROGRAM/KG DRY WEIGHT	SURFACE WATER=34.8 PPM, BOTTOM WATER=16.4 PPM, BUFFALO CREEK	1439
*SIGARA SP.	27800 MICROGRAM/KG WET WEIGHT	SEDIMENT= 9.3-377.0 UG/KG DW, BUFFALO CREEK	1439
*TUBIFEX TUBIFEX	335 MICROGRAM/KG WET WEIGHT	SEDIMENT= 9.3-377.0 UG/KG DW, BUFFALO CREEK	1439

TABLE D294

CONCENTRATIONS : HG

ITEM:	VALUE:	CONDITIONS:	REF.:
*ATMOSPHERE	0.001-50 NG/M**3	NONE	893
*AVENA SATIVA, GRAINS	0.013 MG/KG DRY WEIGHT	CONTROL=0.009 PPM,SOIL=4 PPM,PH= 5.1, SILT LOAM,CEC=.383 MEQ/G, 293 K	774
*AVENA SATIVA, GRAINS	0.020 MG/KG DRY WEIGHT	CONTROL=0.009 PPM,SOIL=20 PPM,PH= 5.1,SILT LOAM,CEC=.383 MEQ/G, 293 K	774
*AVENA SATIVA, HUSKS	0.229 MG/KG DRY WEIGHT	CONTROL=0.107 PPM,SOIL=4 PPM,PH= 5.1, SILT LOAM,CEC=.383 MEQ/G, 293 K	774
*AVENA SATIVA, HUSKS	0.266 MG/KG DRY WEIGHT	CONTROL=0.107 PPM,SOIL=20 PPM,PH= 5.1,SILT LOAM,CEC=.383 MEQ/G, 293 K	774
*AVENA SATIVA, LEAVES	0.193 MG/KG DRY WEIGHT	CONTROL=0.176 PPM,SOIL=4 PPM,PH= 5.1, SILT LOAM,CEC=.383 MEQ/G, 293 K	774
*AVENA SATIVA, LEAVES	0.199 MG/KG DRY WEIGHT	CONTROL=0.176 PPM,SOIL=20 PPM,PH= 5.1,SILT LOAM,CEC=.383 MEQ/G, 293 K	774
*AVENA SATIVA, ROOT	0.157 MG/KG DRY WEIGHT	CONTROL=0.151 PPM,SOIL=4 PPM,PH= 5.1, SILT LOAM,CEC=.383 MEQ/G, 293 K	774
*AVENA SATIVA, ROOT	0.426 MG/KG DRY WEIGHT	CONTROL=0.151 PPM,SOIL=20 PPM,PH= 5.1,SILT LOAM,CEC=.383 MEQ/G, 293 K	774
*AVENA SATIVA, STALKS	0.008 MG/KG DRY WEIGHT	CONTROL=0.011 PPM,SOIL=4 PPM,PH= 5.1, SILT LOAM,CEC=.383 MEQ/G, 293 K	774
*AVENA SATIVA, STALKS	0.026 MG/KG DRY WEIGHT	CONTROL=0.011 PPM,SOIL=20 PPM,PH= 5.1,SILT LOAM,CEC=.383 MEQ/G, 293 K	774

TABLE D294

CONCENTRATIONS : HG

ITEM:	VALUE:	CONDITIONS:	REF.:
*BRASSICA OLERACEA VAR. BOTRYTIS, LEAVES	0.068 MG/KG DRY WEIGHT	CONTROL=0.079 PPM,SOIL=4 PPM,PH= 5.1, SILT LOAM,CEC=.383 MEQ/G, 293 K	774
*BRASSICA OLERACEA VAR. BOTRYTIS, LEAVES	0.061 MG/KG DRY WEIGHT	CONTROL=0.079 PPM,SOIL=20 PPM,PH= 5.1,SILT LOAM,CEC=.383 MEQ/G, 293 K	774
*BRASSICA OLERACEA VAR. BOTRYTIS, ROOT	0.197 MG/KG DRY WEIGHT	CONTROL=0.019 PPM,SOIL=4 PPM,PH= 5.1, SILT LOAM,CEC=.383 MEQ/G, 293 K	774
*BRASSICA OLERACEA VAR. BOTRYTIS, ROOT	2.447 MG/KG DRY WEIGHT	CONTROL=0.019 PPM,SOIL=20 PPM,PH= 5.1,SILT LOAM,CEC=.383 MEQ/G, 293 K	774
*BRASSICA OLERACEA VAR. ITALICA, LEAVES	0.078 MG/KG DRY WEIGHT	CONTROL=0.063 PPM,SOIL=4 PPM,PH= 5.1, SILT LOAM,CEC=.383 MEQ/G, 293 K	774
*BRASSICA OLERACEA VAR. ITALICA, LEAVES	0.029 MG/KG DRY WEIGHT	CONTROL=0.063 PPM,SOIL=20 PPM,PH= 5.1,SILT LOAM,CEC=.383 MEQ/G, 293 K	774
*BRASSICA OLERACEA VAR. ITALICA, ROOT	0.505 MG/KG DRY WEIGHT	CONTROL=0.171 PPM,SOIL=4 PPM,PH= 5.1, SILT LOAM,CEC=.383 MEQ/G, 293 K	774
*BRASSICA OLERACEA VAR. ITALICA, ROOT	1.870 MG/KG DRY WEIGHT	CONTROL=0.171 PPM,SOIL=20 PPM,PH= 5.1,SILT LOAM,CEC=.383 MEQ/G, 293 K	774
*BULK PRECIPITATION	0.06 MICROGRAMS/L MEAN WT.	30JULY-7NOVEMBER-1971,NEW HAMPSHIRE,USA	1393
*CALF, BRAIN	29 MICROGRAM/G WET WEIGHT	4.7MG HG/KG DAY,ORAL, ETHYL-HG-P-TOLUENE,38D,N=1	486
*CALF, BRAIN	12 MICROGRAM/G WET WEIGHT	47MG HG/KG DAY,ORAL, ETHYL-HG-P-TOLUENE,9D,N=1	486
*CALF, BRAIN	3.2 MICROGRAM/G WET WEIGHT	120MG HG/KG,ORAL, ETHYL-HG-P-TOLUENE,3D,N=1	486
*CALF, KIDNEY	60 MICROGRAM/G WET WEIGHT	4.7MG HG/KG DAY,ORAL, ETHYL-HG-P-TOLUENE,38D,N=1	486
*CALF, KIDNEY	62 MICROGRAM/G WET WEIGHT	47MG HG/KG DAY,ORAL, ETHYL-HG-P-TOLUENE,9D,N=1	486
*CALF, KIDNEY	29 MICROGRAM/G WET WEIGHT	120MG HG/KG,ORAL, ETHYL-HG-P-TOLUENE,3D,N=1	486
*CALF, LIVER	50 MICROGRAM/G WET WEIGHT	4.7MG HG/KG DAY,ORAL, ETHYL-HG-P-TOLUENE,38D,N=1	486
*CALF, LIVER	58 MICROGRAM/G WET WEIGHT	47MG HG/KG DAY,ORAL, ETHYL-HG-P-TOLUENE,9D,N=1	486
*CALF, LIVER	46 MICROGRAM/G WET WEIGHT	120MG HG/KG,ORAL, ETHYL-HG-P-TOLUENE,3D,N=1	486
*CALF, WHOLE BLOOD	7 MICROGRAM/G WET WEIGHT	4.7MG HG/KG DAY,ORAL, ETHYL-HG-P-TOLUENE,38D,N=1	486
*CALF, WHOLE BLOOD	23 MICROGRAM/G WET WEIGHT	47MG HG/KG DAY,ORAL, ETHYL-HG-P-TOLUENE,9D,N=1	486
*CALF, WHOLE BLOOD	5 MICROGRAM/G WET WEIGHT	120MG HG/KG,ORAL, ETHYL-HG-P-TOLUENE,3D,N=1	486
*CALIDRIS ALPINA, LIVER	1.9 PPM DRY WT.	12 BIRDS(3 FIRST WINTER,9 ADULT), 20.9-10.10-71-72,BRITISH ISLES	1216
*CALIDRIS ALPINA, LIVER	1.7 PPM DRY WT.	7 BIRDS(1 FIRST WINTER,6 ADULT), 25-27.10.72,BRITISH ISLES	1216
*CALIDRIS ALPINA, LIVER	3.3 PPM DRY WT.	3 BIRDS(2 FIRST WINTER,1 ADULT), 9.11.72,BRITISH ISLES	1216
*CALIDRIS ALPINA, LIVER	6.9 PPM DRY WT.	9 BIRDS,10.1.70,BRITISH ISLES	1216
*CALIDRIS ALPINA, LIVER	10.5 PPM DRY WT.	1 BIRD(1 ADULT),19.3.72,BRITISH ISLES	1216
*CALIDRIS CANUTUS, LIVER	1 PPM DRY WT.	2 BIRDS(FIRST WINTER),29.8.72, BRITISH ISLES	1216
*CALIDRIS CANUTUS, LIVER	1.3 PPM DRY WT.	4 BIRDS(3 FIRST WINTER,1 ADULT), 26.9-10.10.72,BRITISH ISLES	1216
*CALIDRIS CANUTUS, LIVER	7.3 PPM DRY WT.	2 BIRDS(2 ADULT),11.1.70,BRITISH ISLES	1216
*CALIDRIS CANUTUS, LIVER	9.8 PPM DRY WT.	7 BIRDS(7 ADULT),19.2.72,BRITISH ISLES	1216
*CALIDRIS CANUTUS, LIVER	14.4 PPM DRY WT.	6 BIRDS(6 ADULT),19.3.72,BRITISH ISLES	1216

TABLE D294

CONCENTRATIONS : HG			
ITEM:	VALUE:	CONDITIONS:	REF.:
*CALIDRIS CANUTUS, LIVER	18.5 PPM DRY WT.	2 BIRDS(2 FIRST WINTER),26.3.72, BRITISH ISLES	1216
*CAT, BRAIN	9.2 MICROGRAM/G WET WEIGHT	1.5MG HG/KG DAY,ORAL, METHYL-HG-SH-METHYL,22D,N=4	486
*CAT, BRAIN	28 MICROGRAM/G WET WEIGHT	1MG HG/KG DAY,ORAL,METHYL-HG-OH, 37-45D,N=2	486
*CAT, BRAIN	9 MICROGRAM/G WET WEIGHT	0.2-0.4MG HG/KG DAY,ORAL, METHYL-HG-OH,76-126D,N=2	486
*CAT, LIVER	62 MICROGRAM/G WET WEIGHT	1.5MG HG/KG DAY,ORAL, METHYL-HG-SH-METHYL,22D,N=4	486
*CAT, LIVER	100 MICROGRAM/G WET WEIGHT	1MG HG/KG DAY,ORAL,METHYL-HG-OH, 37-45D,N=2	486
*CAT, LIVER	31 MICROGR¤M/G WET WEIGHT	0.2-0.4MG HG/KG DAY,ORAL, METHYL-HG-OH,76-126D,N=2	486
*CAT, WHOLE BLOOD	13 MICROGRAM/G WET WEIGHT	1.5MG HG/KG DAY,ORAL, METHYL-HG-SH-METHYL,22D,N=4	486
*CAT, WHOLE BLOOD	61 MICROGRAM/G WET WEIGHT	1MG HG/KG DAY,ORAL,METHYL-HG-OH, 37-45D,N=2	486
*CAT, WHOLE BLOOD	19 MICROGRAM/G WET WEIGHT	0.2-0.4MG HG/KG DAY,ORAL, METHYL-HG-OH,76-126D,N=2	486
*DAUCUS CAROTA VAR. SATIVA, ROOT	0.428 MG/KG DRY WEIGHT	CONTROL=0.163 PPM,SOIL=4 PPM,PH= 5.1, SILT LOAM,CEC=.383 MEQ/G, 293 K	774
*DAUCUS CAROTA VAR. SATIVA, ROOT	1.058 MG/KG DRY WEIGHT	CONTROL=0.163 PPM,SOIL=20 PPM,PH= 5.1,SILT LOAM,CEC=.383 MEQ/G, 293 K	774
*DAUCUS CAROTA VAR. SATIVA, TOPS	0.061 MG/KG DRY WEIGHT	CONTROL=0.044 PPM,SOIL=4 PPM,PH= 5.1, SILT LOAM,CEC=.383 MEQ/G, 293 K	774
*DAUCUS CAROTA VAR. SATIVA, TOPS	0.072 MG/KG DRY WEIGHT	CONTROL=0.044 PPM,SOIL=20 PPM,PH= 5.1,SILT LOAM,CEC=.383 MEQ/G, 293 K	774
*DAUCUS CAROTA VAR. SATIVA, TUBERS	0.053 MG/KG DRY WEIGHT	CONTROL=0.044 PPM,SOIL=4 PPM,PH= 5.1, SILT LOAM,CEC=.383 MEQ/G, 293 K	774
*DAUCUS CAROTA VAR. SATIVA, TUBERS	0.039 MG/KG DRY WEIGHT	CONTROL=0.044 PPM,SOIL=20 PPM,PH= 5.1,SILT LOAM,CEC=.383 MEQ/G, 293 K	774
*FISH	0.27 MG/KG	HG SEDIMENT<0.04-0.62DRY WT.,HG WATER<0.2 MICROG/L,225 FISH	1655
*FOOD	0.01-0.1 MG/KG	SWEDEN,FISH FROM SEA, UNCONTAMINATED	893
*FOOD	0.05-0.2 MG/KG	SWEDEN,FISH FROM LAKES, UNCONTAMINATED	893
*KIDNEY OF CAT	20 MICROGRAM/G WET WEIGHT	1.5MG HG/KG DAY,ORAL, METHYL-HG-SH-METHYL,22D,N=4	486
*KIDNEY OF CAT	62 MICROGRAM/G WET WEIGHT	1MG HG/KG DAY,ORAL,METHYL-HG-OH, 37-45D,N=2	486
*KIDNEY OF CAT	19 MICROGRAM/G WET WEIGHT	0.2-0.4MG HG/KG DAY,ORAL, METHYL-HG-OH,76-126D,N=2	486
*LACTUCA SATIVA VAR. CRISPA, LEAVES	0.017 MG/KG DRY WEIGHT	CONTROL=0.031 PPM,SOIL=4 PPM,PH= 5.1, SILT LOAM,CEC=.383 MEQ/G, 293 K	774
*LACTUCA SATIVA VAR. CRISPA, LEAVES	0.045 MG/KG DRY WEIGHT	CONTROL=0.031 PPM,SOIL=20 PPM,PH= 5.1,SILT LOAM,CEC=.383 MEQ/G, 293 K	774
*LACTUCA SATIVA VAR. CRISPA, ROOT	0.175 MG/KG DRY WEIGHT	CONTROL=0.112 PPM,SOIL=4 PPM,PH= 5.1, SILT LOAM,CEC=.383 MEQ/G, 293 K	774
*LACTUCA SATIVA VAR. CRISPA, ROOT	0.387 MG/KG DRY WEIGHT	CONTROL=0.112 PPM,SOIL=20 PPM,PH= 5.1,SILT LOAM,CEC=.383 MEQ/G, 293 K	774
*LEBISTES RETICULATUS, BODY	5.2 % OF TOTAL	WATER = 1*10**(-7) M HG-ETHYLCL, 3-5 D, 298 K,REST IN SOL=77.2 %	443
*LEBISTES RETICULATUS, FIN	0.8 % OF TOTAL	WATER = 1*10**(-7) M HG-ETHYLCL, 3-5 D, 298 K,REST IN SOL=77.2 %	443
*LEBISTES RETICULATUS, HEAD	5.6 % OF TOTAL	WATER = 1*10**(-7) M HG-ETHYLCL, 3-5 D, 298 K,REST IN SOL=77.2 %	443
*LEBISTES RETICULATUS, INTERNAL ORGANS	9.8 % OF TOTAL	WATER = 1*10**(-7) M HG-ETHYLCL, 3-5 D, 298 K,REST IN SOL=77.2 %	443

TABLE D294

CONCENTRATIONS : HG

ITEM:	VALUE:	CONDITIONS:	REF.:
*LEBISTES RETICULATUS, TAIL	1.4 % OF TOTAL	WATER = 1*10**(-7) M HG-ETHYLCL, 3-5 D, 298 K,REST IN SOL=77.2 %	443
*MOUSE, BRAIN	4 MICROGRAM/G WET WEIGHT	2.5MG HG/KG DAY,SUBCUT., METHYL-HG-ACETATE,10D,N=3	486
*MOUSE, BRAIN	6 MICROGRAM/G WET WEIGHT	5MG HG/KG DAY,SUBCUT., METHYL-HG-ACETATE,10D,N=3	486
*MOUSE, BRAIN	0.02 MICROGRAM/G WET WEIGHT	0.03MG HG/KG,INTRAVEN., METHYL-HG-OH,6D,N=10	486
*MOUSE, BRAIN	0.2 MICROGRAM/G WET WEIGHT	0.3MG HG/KG,INTRAVEN., METHYL-HG-OH,6D,N=10	486
*MOUSE, BRAIN	0.6 MICROGRAM/G WET WEIGHT	1.0MG HG/KG,INTRAVEN., METHYL-HG-OH,6D,N=10	486
*MOUSE, BRAIN	3.3 MICROGRAM/G WET WEIGHT	5.0MG HG/KG,INTRAVEN., METHYL-HG-OH,6D,N=10	486
*MOUSE, BRAIN	0.37 MICROGRAM/G WET WEIGHT	1.0MG HG/KG,INTRAVEN., METHYL-HG-CL,22D,N=2	486
*MOUSE, BRAIN	0.2 MICROGRAM/G WET WEIGHT	3MG HG/KG,SUBCUT., N-PROPYL-HG-ACETATE,7D,N=3	486
*MOUSE, BRAIN	0.14 MICROGRAM/G WET WEIGHT	3MG HG/KG,SUBCUT., ISO-PROPYL-HG-ACETATE,7D,N=3	486
*MOUSE, BRAIN	0.23 MICROGRAM/G WET WEIGHT	3MG HG/KG,SUBCUT., N-BUTYL-HG-ACETATE,7D,N=3	486
*MOUSE, BRAIN	0.24 MICROGRAM/G WET WEIGHT	3MG HG/KG,SUBCUT., ISO-BUTYL-HG-ACETATE,7D,N=3	486
*MOUSE, BRAIN	0.12 MICROGRAM/G WET WEIGHT	3MG HG/KG,SUBCUT., N-AMYL-HG-ACETATE,7D,N=3	486
*MOUSE, KIDNEY	40 MICROGRAM/G WET WEIGHT	2.5MG HG/KG DAY,SUBCUT., METHYL-HG-ACETATE,10D,N=3	486
*MOUSE, KIDNEY	60 MICROGRAM/G WET WEIGHT	5MG HG/KG DAY,SUBCUT., METHYL-HG-ACETATE,10D,N=3	486
*MOUSE, KIDNEY	0.02 MICROGRAM/G WET WEIGHT	0.03MG HG/KG,INTRAVEN., METHYL-HG-OH,6D,N=10	486
*MOUSE, KIDNEY	0.2 MICROGRAM/G WET WEIGHT	0.3MG HG/KG,INTRAVEN., METHYL-HG-OH,6D,N=10	486
*MOUSE, KIDNEY	0.6 MICROGRAM/G WET WEIGHT	1.0MG HG/KG,INTRAVEN., METHYL-HG-OH,6D,N=10	486
*MOUSE, KIDNEY	4.0 MICROGRAM/G WET WEIGHT	5.0MG HG/KG,INTRAVEN., METHYL-HG-OH,6D,N=10	486
*MOUSE, KIDNEY	2.7 MICROGRAM/G WET WEIGHT	1.0MG HG/KG,INTRAVEN., METHYL-HG-CL,22D,N=2	486
*MOUSE, KIDNEY	32 MICROGRAM/G WET WEIGHT	3MG HG/KG,SUBCUT., N-PROPYL-HG-ACETATE,7D,N=3	486
*MOUSE, KIDNEY	22 MICROGRAM/G WET WEIGHT	3MG HG/KG,SUBCUT., ISO-PROPYL-HG-ACETATE,7D,N=3	486
*MOUSE, KIDNEY	19 MICROGRAM/G WET WEIGHT	3MG HG/KG,SUBCUT., N-BUTYL-HG-ACETATE,7D,N=3	486
*MOUSE, KIDNEY	21 MICROGRAM/G WET WEIGHT	3MG HG/KG,SUBCUT., ISO-BUTYL-HG-ACETATE,7D,N=3	486
*MOUSE, KIDNEY	17 MICROGRAM/G WET WEIGHT	3MG HG/KG,SUBCUT., N-AMYL-HG-ACETATE,7D,N=3	486
*MOUSE, LIVER	20 MICROGRAM/G WET WEIGHT	2.5MG HG/KG DAY,SUBCUT., METHYL-HG-ACETATE,10D,N=3	486
*MOUSE, LIVER	30 MICROGRAM/G WET WEIGHT	5MG HG/KG DAY,SUBCUT., METHYL-HG-ACETATE,10D,N=3	486
*MOUSE, LIVER	0.08 MICROGRAM/G WET WEIGHT	0.03MG HG/KG,INTRAVEN., METHYL-HG-OH,6D,N=10	486
*MOUSE, LIVER	0.6 MICROGRAM/G WET WEIGHT	0.3MG HG/KG,INTRAVEN., METHYL-HG-OH,6D,N=10	486
*MOUSE, LIVER	2.1 MICROGRAM/G WET WEIGHT	1.0MG HG/KG,INTRAVEN., METHYL-HG-OH,6D,N=10	486
*MOUSE, LIVER	9.8 MICROGRAM/G WET WEIGHT	5.0MG HG/KG,INTRAVEN., METHYL-HG-OH,6D,N=10	486
*MOUSE, LIVER	0.7 MICROGRAM/G WET WEIGHT	1.0MG HG/KG,INTRAVEN., METHYL-HG-CL,22D,N=2	486
*MOUSE, LIVER	15 MICROGRAM/G WET WEIGHT	3MG HG/KG,SUBCUT., N-PROPYL-HG-ACETATE,7D,N=3	486
*MOUSE, LIVER	16 MICROGRAM/G WET WEIGHT	3MG HG/KG,SUBCUT., ISO-PROPYL-HG-ACETATE,7D,N=3	486

TABLE D294

CONCENTRATIONS : HG (F031)

ITEM:	VALUE:	CONDITIONS:	REF.:
*MOUSE, LIVER	11 MICROGRAM/G WET WEIGHT	3MG HG/KG,SUBCUT., N-BUTYL-HG-ACETATE,7D,N=3	486
*MOUSE, LIVER	16 MICROGRAM/G WET WEIGHT	3MG HG/KG,SUBCUT., ISO-BUTYL-HG-ACETATE,7D,N=3	486
*MOUSE, LIVER	6.3 MICROGRAM/G WET WEIGHT	3MG HG/KG,SUBCUT., N-AMYL-HG-ACETATE,7D,N=3	486
*MOUSE, WHOLE BLOOD	5 MICROGRAM/G WET WEIGHT	2.5MG HG/KG DAY,SUBCUT., METHYL-HG-ACETATE,10D,N=3	486
*MOUSE, WHOLE BLOOD	9 MICROGRAM/G WET WEIGHT	5MG HG/KG DAY,SUBCUT., METHYL-HG-ACETATE,10D,N=3	486
*MOUSE, WHOLE BLOOD	0.02 MICROGRAM/G WET WEIGHT	0.03MG HG/KG,INTRAVEN., METHYL-HG-OH,6D,N=10	486
*MOUSE, WHOLE BLOOD	0.2 MICROGRAM/G WET WEIGHT	0.3MG HG/KG,INTRAVEN., METHYL-HG-OH,6D,N=10	486
*MOUSE, WHOLE BLOOD	0.5 MICROGRAM/G WET WEIGHT	1.0MG HG/KG,INTRAVEN., METHYL-HG-OH,6D,N=10	486
*MOUSE, WHOLE BLOOD	3.7 MICROGRAM/G WET WEIGHT	5.0MG HG/KG,INTRAVEN., METHYL-HG-OH,6D,N=10	486
*MOUSE, WHOLE BLOOD	0.5 MICROGRAM/G WET WEIGHT	1.0MG HG/KG,INTRAVEN., METHYL-HG-CL,22D,N=2	486
*MOUSE, WHOLE BLOOD	1.1 MICROGRAM/G WET WEIGHT	3MG HG/KG,SUBCUT., N-PROPYL-HG-ACETATE,7D,N=3	486
*MOUSE, WHOLE BLOOD	0.33 MICROGRAM/G WET WEIGHT	3MG HG/KG,SUBCUT., ISO-PROPYL-HG-ACETATE,7D,N=3	486
*MOUSE, WHOLE BLOOD	0.86 MICROGRAM/G WET WEIGHT	3MG HG/KG,SUBCUT., N-BUTYL-HG-ACETATE,7D,N=3	486
*MOUSE, WHOLE BLOOD	0.80 MICROGRAM/G WET WEIGHT	3MG HG/KG,SUBCUT., ISO-BUTYL-HG-ACETATE,7D,N=3	486
*MOUSE, WHOLE BLOOD	0.46 MICROGRAM/G WET WEIGHT	3MG HG/KG,SUBCUT., N-AMYL-HG-ACETATE,7D,N=3	486
*PAGOPHILUS GROENLANDICUS (ADULT), ADRENAL	14.20 PPM	TOTAL HG,OR.DOSE=0.25MG/KG BODY*DAY FOR 60D	1355
*PAGOPHILUS GROENLANDICUS (ADULT), ADRENAL	10.75 PPM	METHYL-HG,OR.DOSE=0.25MG/KG BODY*DAY FOR 60D	1355
*PAGOPHILUS GROENLANDICUS (ADULT), BLOOD	0.16 MG/L (SD=0.06)	METHYL-HG,OR.DOSE=0.25MG/KG BODY*DAY FOR 60D ;CONTROL= 0.15+-0.06	1355
*PAGOPHILUS GROENLANDICUS (ADULT), BLOOD	0.14 MG/L (SD=0.003)	TOTAL HG,OR.DOSE=0.25MG/KG BODY*DAY FOR 60D ;CONTROL= 0.11+-0.004	1355
*PAGOPHILUS GROENLANDICUS (ADULT), BLOOD	9.93 PPM	TOTAL HG,OR.DOSE=0.25MG/KG BODY*DAY FOR 60D ;CONTROL= 0.12PPM	1355
*PAGOPHILUS GROENLANDICUS (ADULT), BLOOD	8.85 PPM	METHYL-HG,OR.DOSE=0.25MG/KG BODY*DAY FOR 60D ;CONTROL= 0.10PPM	1355
*PAGOPHILUS GROENLANDICUS (ADULT), BLUBBER	0.28 PPM	TOTAL HG,OR.DOSE=0.25MG/KG BODY*DAY FOR 60D ;CONTROL= N.D.	1355
*PAGOPHILUS GROENLANDICUS (ADULT), BLUBBER	0.58 PPM	METHYL-HG,OR.DOSE=0.25MG/KG BODY*DAY FOR 60D ;CONTROL= 0.05PPM	1355
*PAGOPHILUS GROENLANDICUS (ADULT), BRAIN	14.80 PPM	TOTAL HG,OR.DOSE=0.25MG/KG BODY*DAY FOR 60D ;CONTROL= 0.45PPM	1355
*PAGOPHILUS GROENLANDICUS (ADULT), CLAWS	21.90 PPM	TOTAL HG,OR.DOSE=0.25MG/KG BODY*DAY FOR 60D ;CONTROL= 5.10PPM	1355
*PAGOPHILUS GROENLANDICUS (ADULT), GONAD	13.00 PPM	TOTAL HG,OR.DOSE=0.25MG/KG BODY*DAY FOR 60D ;CONTROL= N.D.	1355
*PAGOPHILUS GROENLANDICUS (ADULT), GONAD	10.08 PPM	METHYL-HG,OR.DOSE=0.25MG/KG BODY*DAY FOR 60D ;CONTROL= N.D.	1355
*PAGOPHILUS GROENLANDICUS (ADULT), HAIR	8.21 PPM	TOTAL HG,OR.DOSE=0.25MG/KG BODY*DAY FOR 60D ;CONTROL= 9.17PPM	1355
*PAGOPHILUS GROENLANDICUS (ADULT), HAIR	7.30 PPM	METHYL-HG,OR.DOSE=0.25MG/KG BODY*DAY FOR 60D ;CONTROL= 1.27PPM	1355
*PAGOPHILUS GROENLANDICUS (ADULT), HEART	12.90 PPM	TOTAL HG,OR.DOSE=0.25MG/KG BODY*DAY FOR 60D ;CONTROL= 0.25PPM	1355

TABLE D294

CONCENTRATIONS : HG (F031)

ITEM:	VALUE:	CONDITIONS:	REF.:
*PAGOPHILUS GROENLANDICUS (ADULT), HEART	18.04 PPM	METHYL-HG,OR.DOSE=0.25MG/KG BODY*DAY FOR 60D	1355
*PAGOPHILUS GROENLANDICUS (ADULT), KIDNEY	69.50 PPM	TOTAL HG,OR.DOSE=0.25MG/KG BODY*DAY FOR 60D ;CONTROL= 5.98PPM	1355
*PAGOPHILUS GROENLANDICUS (ADULT), KIDNEY	51.60 PPM	METHYL-HG,OR.DOSE=0.25MG/KG BODY*DAY FOR 60D ;CONTROL= 0.27PPM	1355
*PAGOPHILUS GROENLANDICUS (ADULT), LIVER	64.00 PPM	TOTAL HG,OR.DOSE=0.25MG/KG BODY*DAY FOR 60D ;CONTROL= 25.8PPM	1355
*PAGOPHILUS GROENLANDICUS (ADULT), LIVER	18.48 PPM	METHYL-HG,OR.DOSE=0.25MG/KG BODY*DAY FOR 60D ;CONTROL= 0.16PPM	1355
*PAGOPHILUS GROENLANDICUS (ADULT), LUNG	10.60 PPM	TOTAL HG,OR.DOSE=0.25MG/KG BODY*DAY FOR 60D ;CONTROL= 0.12PPM	1355
*PAGOPHILUS GROENLANDICUS (ADULT), MUSCLE	27.60 PPM	TOTAL HG,OR.DOSE=0.25MG/KG BODY*DAY FOR 60D ;CONTROL= 0.78PPM	1355
*PAGOPHILUS GROENLANDICUS (ADULT), MUSCLE	35.20 PPM	METHYL-HG,OR.DOSE=0.25MG/KG BODY*DAY FOR 60D ;CONTROL= 0.56PPM	1355
*PAGOPHILUS GROENLANDICUS (ADULT), SMALL INTESTINE	17.20 PPM	TOTAL HG,OR.DOSE=0.25MG/KG BODY*DAY FOR 60D ;CONTROL= 0.44PPM	1355
*PAGOPHILUS GROENLANDICUS (ADULT), SMALL INTESTINE	0.50 PPM	METHYL-HG,OR.DOSE=0.25MG/KG BODY*DAY FOR 60D ;CONTROL= 0.09PPM	1355
*PAGOPHILUS GROENLANDICUS (ADULT), SPLEEN	20.40 PPM	TOTAL HG,OR.DOSE=0.25MG/KG BODY*DAY FOR 60D ;CONTROL= 9.92PPM	1355
*PAGOPHILUS GROENLANDICUS (IMMATURE), ADRENAL	25.00 PPM	TOTAL HG,OR.DOSE=0.25MG/KG BODY*DAY FOR 90D ;CONTROL= N.D.	1355
*PAGOPHILUS GROENLANDICUS (IMMATURE), ADRENAL	26.00 PPM	TOTAL HG,OR.DOSE=25.0MG/KG BODY*DAY FOR 20DAYS	1355
*PAGOPHILUS GROENLANDICUS (IMMATURE), ADRENAL	34.90 PPM	TOTAL HG,OR.DOSE=25.0MG/KG BODY*DAY FOR 26DAYS	1355
*PAGOPHILUS GROENLANDICUS (IMMATURE), ADRENAL	21.70 PPM	METHYL-HG,OR.DOSE=0.25MG/KG BODY*DAY FOR 90D ;CONTROL= N.D.	1355
*PAGOPHILUS GROENLANDICUS (IMMATURE), ADRENAL	7.50 PPM	METHYL-HG,OR.DOSE=25.0MG/KG BODY*DAY FOR 20DAYS	1355
*PAGOPHILUS GROENLANDICUS (IMMATURE), ADRENAL	10.40 PPM	METHYL-HG,OR.DOSE=25.0MG/KG BODY*DAY FOR 26DAYS	1355
*PAGOPHILUS GROENLANDICUS (IMMATURE), BLOOD	0.09 MG/L (SD=0.07)	METHYL-HG,OR.DOSE=0.25MG/KG BODY*DAY FOR 90D ;CONTROL= 0.10+-0.08	1355
*PAGOPHILUS GROENLANDICUS (IMMATURE), BLOOD	0.10 MG/L (SD=0.07)	METHYL-HG,OR.DOSE=25.0MG/KG BODY*DAY FOR 90D ;CONTROL= 0.10+-0.08	1355
*PAGOPHILUS GROENLANDICUS (IMMATURE), BLOOD	0.10 MG/L (SD=0.06)	METHYL-HG,OR.DOSE=25.0MG/KG BODY*DAY FOR 60D ;CONTROL= 0.15+-0.06	1355
*PAGOPHILUS GROENLANDICUS (IMMATURE), BLOOD	0.07 MG/L (SD=0.04)	TOTAL HG,OR.DOSE=0.25MG/KG BODY*DAY FOR 90D ;CONTROL= 0.09+-0.04	1355
*PAGOPHILUS GROENLANDICUS (IMMATURE), BLOOD	0.08 MG/L (SD=0.05)	TOTAL HG,OR.DOSE=25.0MG/KG BODY*DAY FOR 90D ;CONTROL= 0.09+-0.04	1355
*PAGOPHILUS GROENLANDICUS (IMMATURE), BLOOD	0.07 MG/L (SD=0.04)	TOTAL HG,OR.DOSE=25.0MG/KG BODY*DAY FOR 60D ;CONTROL= 0.11+-0.004	1355
*PAGOPHILUS GROENLANDICUS (IMMATURE), BLOOD	13.10 PPM	TOTAL HG,OR.DOSE=0.25MG/KG BODY*DAY FOR 90D ;CONTROL= 0.10PPM	1355
*PAGOPHILUS GROENLANDICUS (IMMATURE), BLOOD	28.80 PPM	TOTAL HG,OR.DOSE=25.0MG/KG BODY*DAY FOR 20DAYS	1355
*PAGOPHILUS GROENLANDICUS (IMMATURE), BLOOD	30.30 PPM	TOTAL HG,OR.DOSE=25.0MG/KG BODY*DAY FOR 26DAYS	1355
*PAGOPHILUS GROENLANDICUS (IMMATURE), BLOOD	12.50 PPM	METHYL-HG,OR.DOSE=0.25MG/KG BODY*DAY FOR 90D ;CONTROL= 0.10PPM	1355

TABLE D294

CONCENTRATIONS : HG (F031)

ITEM:	VALUE:	CONDITIONS:	REF.:
*PAGOPHILUS GROENLANDICUS (IMMATURE), BLOOD	19.00 PPM	METHYL-HG,OR.DOSE=25.0MG/KG BODY*DAY FOR 20DAYS	1355
*PAGOPHILUS GROENLANDICUS (IMMATURE), BLOOD	23.70 PPM	METHYL-HG,OR.DOSE=25.0MG/KG BODY*DAY FOR 26DAYS	1355
*PAGOPHILUS GROENLANDICUS (IMMATURE), BLUBBER	0.20 PPM	TOTAL HG,OR.DOSE=0.25MG/KG BODY*DAY FOR 90D ;CONTROL= N.D.	1355
*PAGOPHILUS GROENLANDICUS (IMMATURE), BLUBBER	1.71 PPM	TOTAL HG,OR.DOSE=25.0MG/KG BODY*DAY FOR 20DAYS	1355
*PAGOPHILUS GROENLANDICUS (IMMATURE), BLUBBER	0.94 PPM	TOTAL HG,OR.DOSE=25.0MG/KG BODY*DAY FOR 26DAYS	1355
*PAGOPHILUS GROENLANDICUS (IMMATURE), BLUBBER	0.55 PPM	METHYL-HG,OR.DOSE=0.25MG/KG BODY*DAY FOR 90D ;CONTROL= N.D.	1355
*PAGOPHILUS GROENLANDICUS (IMMATURE), BLUBBER	0.59 PPM	METHYL-HG,OR.DOSE=25.0MG/KG BODY*DAY FOR 20DAYS	1355
*PAGOPHILUS GROENLANDICUS (IMMATURE), BLUBBER	1.00 PPM	METHYL-HG,OR.DOSE=25.0MG/KG BODY*DAY FOR 26DAYS	1355
*PAGOPHILUS GROENLANDICUS (IMMATURE), BRAIN	21.80 PPM	TOTAL HG,OR.DOSE=0.25MG/KG BODY*DAY FOR 90D ;CONTROL= 0.36PPM	1355
*PAGOPHILUS GROENLANDICUS (IMMATURE), BRAIN	42.90 PPM	TOTAL HG,OR.DOSE=25.0MG/KG BODY*DAY FOR 20DAYS	1355
*PAGOPHILUS GROENLANDICUS (IMMATURE), BRAIN	23.80 PPM	TOTAL HG,OR.DOSE=25.0MG/KG BODY*DAY FOR 26DAYS	1355
*PAGOPHILUS GROENLANDICUS (IMMATURE), CLAWS	22.30 PPM	TOTAL HG,OR.DOSE=0.25MG/KG BODY*DAY FOR 90D ;CONTROL= 2.36PPM	1355
*PAGOPHILUS GROENLANDICUS (IMMATURE), CLAWS	81.80 PPM	TOTAL HG,OR.DOSE=25.0MG/KG BODY*DAY FOR 20DAYS	1355
*PAGOPHILUS GROENLANDICUS (IMMATURE), CLAWS	38.80 PPM	TOTAL HG,OR.DOSE=25.0MG/KG BODY*DAY FOR 26DAYS	1355
*PAGOPHILUS GROENLANDICUS (IMMATURE), GONAD	27.00 PPM	TOTAL HG,OR.DOSE=25.0MG/KG BODY*DAY FOR 20DAYS	1355
*PAGOPHILUS GROENLANDICUS (IMMATURE), GONAD	33.00 PPM	TOTAL HG,OR.DOSE=25.0MG/KG BODY*DAY FOR 26DAYS	1355
*PAGOPHILUS GROENLANDICUS (IMMATURE), GONAD	11.00 PPM	METHYL-HG,OR.DOSE=0.25MG/KG BODY*DAY FOR 90D	1355
*PAGOPHILUS GROENLANDICUS (IMMATURE), GONAD	21.50 PPM	METHYL-HG,OR.DOSE=25.0MG/KG BODY*DAY FOR 20DAYS	1355
*PAGOPHILUS GROENLANDICUS (IMMATURE), GONAD	29.10 PPM	METHYL-HG,OR.DOSE=25.0MG/KG BODY*DAY FOR 26DAYS	1355
*PAGOPHILUS GROENLANDICUS (IMMATURE), HAIR	1.66 PPM	TOTAL HG,OR.DOSE=0.25MG/KG BODY*DAY FOR 90D ;CONTROL= 1.26PPM	1355
*PAGOPHILUS GROENLANDICUS (IMMATURE), HAIR	322.00 PPM	TOTAL HG,OR.DOSE=25.0MG/KG BODY*DAY FOR 20DAYS	1355
*PAGOPHILUS GROENLANDICUS (IMMATURE), HAIR	434.00 PPM	TOTAL HG,OR.DOSE=25.0MG/KG BODY*DAY FOR 26DAYS	1355
*PAGOPHILUS GROENLANDICUS (IMMATURE), HAIR	4.10 PPM	METHYL-HG,OR.DOSE=0.25MG/KG BODY*DAY FOR 90D ;CONTROL= 1.50PPM	1355
*PAGOPHILUS GROENLANDICUS (IMMATURE), HAIR	428.00 PPM	METHYL-HG,OR.DOSE=25.0MG/KG BODY*DAY FOR 20DAYS	1355
*PAGOPHILUS GROENLANDICUS (IMMATURE), HAIR	350.00 PPM	METHYL-HG,OR.DOSE=25.0MG/KG BODY*DAY FOR 26DAYS	1355
*PAGOPHILUS GROENLANDICUS (IMMATURE), HEART	15.90 PPM	TOTAL HG,OR.DOSE=0.25MG/KG BODY*DAY FOR 90D ;CONTROL= 0.38PPM	1355
*PAGOPHILUS GROENLANDICUS (IMMATURE), HEART	46.30 PPM	TOTAL HG,OR.DOSE=25.0MG/KG BODY*DAY FOR 20DAYS	1355
*PAGOPHILUS GROENLANDICUS (IMMATURE), HEART	50.90 PPM	TOTAL HG,OR.DOSE=25.0MG/KG BODY*DAY FOR 26DAYS	1355
*PAGOPHILUS GROENLANDICUS (IMMATURE), HEART	40.20 PPM	METHYL-HG,OR.DOSE=25.0MG/KG BODY*DAY FOR 20DAYS	1355
*PAGOPHILUS GROENLANDICUS (IMMATURE), HEART	41.50 PPM	METHYL-HG,OR.DOSE=25.0MG/KG BODY*DAY FOR 26DAYS	1355
*PAGOPHILUS GROENLANDICUS (IMMATURE), KIDNEY	50.60 PPM	TOTAL HG,OR.DOSE=0.25MG/KG BODY*DAY FOR 90D ;CONTROL= 0.92PPM	1355

TABLE D294

CONCENTRATIONS : HG (F031)

ITEM:	VALUE:	CONDITIONS:	REF.:
*PAGOPHILUS GROENLANDICUS (IMMATURE), KIDNEY	48.30 PPM	TOTAL HG,OR.DOSE=25.0MG/KG BODY*DAY FOR 20DAYS	1355
*PAGOPHILUS GROENLANDICUS (IMMATURE), KIDNEY	62.50 PPM	TOTAL HG,OR.DOSE=25.0MG/KG BODY*DAY FOR 26DAYS	1355
*PAGOPHILUS GROENLANDICUS (IMMATURE), KIDNEY	10.50 PPM	METHYL-HG,OR.DOSE=0.25MG/KG BODY*DAY FOR 90D ;CONTROL= 0.20PPM	1355
*PAGOPHILUS GROENLANDICUS (IMMATURE), KIDNEY	11.70 PPM	METHYL-HG,OR.DOSE=25.0MG/KG BODY*DAY FOR 20DAYS	1355
*PAGOPHILUS GROENLANDICUS (IMMATURE), KIDNEY	20.50 PPM	METHYL-HG,OR.DOSE=25.0MG/KG BODY*DAY FOR 26DAYS	1355
*PAGOPHILUS GROENLANDICUS (IMMATURE), LIVER	82.50 PPM	TOTAL HG,OR.DOSE=0.25MG/KG BODY*DAY FOR 90D ;CONTROL= 1.72PPM	1355
*PAGOPHILUS GROENLANDICUS (IMMATURE), LIVER	134.00 PPM	TOTAL HG,OR.DOSE=25.0MG/KG BODY*DAY FOR 20DAYS	1355
*PAGOPHILUS GROENLANDICUS (IMMATURE), LIVER	142.00 PPM	TOTAL HG,OR.DOSE=25.0MG/KG BODY*DAY FOR 26DAYS	1355
*PAGOPHILUS GROENLANDICUS (IMMATURE), LIVER	75.70 PPM	METHYL-HG,OR.DOSE=0.25MG/KG BODY*DAY FOR 90D ;CONTROL= 0.20PPM	1355
*PAGOPHILUS GROENLANDICUS (IMMATURE), LIVER	127.00 PPM	METHYL-HG,OR.DOSE=25.0MG/KG BODY*DAY FOR 20DAYS	1355
*PAGOPHILUS GROENLANDICUS (IMMATURE), LIVER	125.00 PPM	METHYL-HG,OR.DOSE=25.0MG/KG BODY*DAY FOR 26DAYS	1355
*PAGOPHILUS GROENLANDICUS (IMMATURE), LUNG	18.20 PPM	TOTAL HG,OR.DOSE=0.25MG/KG BODY*DAY FOR 90D ;CONTROL= 0.26PPM	1355
*PAGOPHILUS GROENLANDICUS (IMMATURE), LUNG	24.80 PPM	TOTAL HG,OR.DOSE=25.0MG/KG BODY*DAY FOR 20DAYS	1355
*PAGOPHILUS GROENLANDICUS (IMMATURE), LUNG	30.80 PPM	TOTAL HG,OR.DOSE=25.0MG/KG BODY*DAY FOR 26DAYS	1355
*PAGOPHILUS GROENLANDICUS (IMMATURE), LUNG	10.60 PPM	METHYL-HG,OR.DOSE=0.25MG/KG BODY*DAY FOR 90D ;CONTROL= N.D.	1355
*PAGOPHILUS GROENLANDICUS (IMMATURE), LUNG	7.30 PPM	METHYL-HG,OR.DOSE=25.0MG/KG BODY*DAY FOR 20DAYS	1355
*PAGOPHILUS GROENLANDICUS (IMMATURE), LUNG	5.70 PPM	METHYL-HG,OR.DOSE=25.0MG/KG BODY*DAY FOR 26DAYS	1355
*PAGOPHILUS GROENLANDICUS (IMMATURE), MUSCLE	42.70 PPM	TOTAL HG,OR.DOSE=0.25MG/KG BODY*DAY FOR 90D ;CONTROL= 0.54PPM	1355
*PAGOPHILUS GROENLANDICUS (IMMATURE), MUSCLE	96.70 PPM	TOTAL HG,OR.DOSE=25.0MG/KG BODY*DAY FOR 20DAYS	1355
*PAGOPHILUS GROENLANDICUS (IMMATURE), MUSCLE	125.00 PPM	TOTAL HG,OR.DOSE=25.0MG/KG BODY*DAY FOR 26DAYS	1355
*PAGOPHILUS GROENLANDICUS (IMMATURE), MUSCLE	40.20 PPM	METHYL-HG,OR.DOSE=0.25MG/KG BODY*DAY FOR 90D ;CONTROL= 0.50PPM	1355
*PAGOPHILUS GROENLANDICUS (IMMATURE), MUSCLE	91.50 PPM	METHYL-HG,OR.DOSE=25.0MG/KG BODY*DAY FOR 20DAYS	1355
*PAGOPHILUS GROENLANDICUS (IMMATURE), MUSCLE	115.00 PPM	METHYL-HG,OR.DOSE=25.0MG/KG BODY*DAY FOR 26DAYS	1355
*PAGOPHILUS GROENLANDICUS (IMMATURE), SMALL INTESTINE	17.10 PPM	TOTAL HG,OR.DOSE=0.25MG/KG BODY*DAY FOR 90D ;CONTROL= 0.37PPM	1355
*PAGOPHILUS GROENLANDICUS (IMMATURE), SMALL INTESTINE	1660.00 PPM	TOTAL HG,OR.DOSE=25.0MG/KG BODY*DAY FOR 20DAYS	1355
*PAGOPHILUS GROENLANDICUS (IMMATURE), SMALL INTESTINE	1730.00 PPM	TOTAL HG,OR.DOSE=25.0MG/KG BODY*DAY FOR 26DAYS	1355
*PAGOPHILUS GROENLANDICUS (IMMATURE), SMALL INTESTINE	0.50 PPM	METHYL-HG,OR.DOSE=0.25MG/KG BODY*DAY FOR 90D ;CONTROL= N.D.	1355
*PAGOPHILUS GROENLANDICUS (IMMATURE), SMALL INTESTINE	1367.00 PPM	METHYL-HG,OR.DOSE=25.0MG/KG BODY*DAY FOR 26DAYS	1355
*PAGOPHILUS GROENLANDICUS (IMMATURE), SPLEEN	20.00 PPM	TOTAL HG,OR.DOSE=0.25MG/KG BODY*DAY FOR 90D ;CONTROL= 0.33PPM	1355
*PAGOPHILUS GROENLANDICUS (IMMATURE), SPLEEN	81.80 PPM	TOTAL HG,OR.DOSE=25.0MG/KG BODY*DAY FOR 20DAYS	1355

TABLE D294

CONCENTRATIONS : HG (F031)

ITEM:	VALUE:	CONDITIONS:	REF.:
*PAGOPHILUS GROENLANDICUS (IMMATURE), SPLEEN	68.60 PPM	TOTAL HG,OR.DOSE=25.0MG/KG BODY*DAY FOR 26DAYS	1355
*PAGOPHILUS GROENLANDICUS (IMMATURE), SPLEEN	14.50 PPM	METHYL-HG,OR.DOSE=0.25MG/KG BODY*DAY FOR 90D ;CONTROL= N.D.	1355
*PAGOPHILUS GROENLANDICUS (IMMATURE), SPLEEN	60.30 PPM	METHYL-HG,OR.DOSE=25.0MG/KG BODY*DAY FOR 20DAYS	1355
*PAGOPHILUS GROENLANDICUS (IMMATURE), SPLEEN	51.50 PPM	METHYL-HG,OR.DOSE=25.0MG/KG BODY*DAY FOR 26DAYS	1355
*PIG, BRAIN	3.7 MICROGRAM/G WET WEIGHT	1.0MG HG/KG DAY,INTRAMUSC., METHYL-HG-ACETATE,7D,N=1	486
*PIG, BRAIN	13 MICROGRAM/G WET WEIGHT	5.0MG HG/KG DAY,INTRAMUSC., METHYL-HG-ACETATE,7D,N=1	486
*PIG, KIDNEY	12 MICROGRAM/G WET WEIGHT	1.0MG HG/KG DAY,INTRAMUSC., METHYL-HG-ACETATE,7D,N=1	486
*PIG, KIDNEY	57 MICROGRAM/G WET WEIGHT	5.0MG HG/KG DAY,INTRAMUSC., METHYL-HG-ACETATE,7D,N=1	486
*PIG, LIVER	22 MICROGRAM/G WET WEIGHT	1.0MG HG/KG DAY,INTRAMUSC., METHYL-HG-ACETATE,7D,N=1	486
*PIG, LIVER	50 MICROGRAM/G WET WEIGHT	5.0MG HG/KG DAY,INTRAMUSC., METHYL-HG-ACETATE,7D,N=1	486
*PIG, WHOLE BLOOD	2.0 MICROGRAM/G WET WEIGHT	1.0MG HG/KG DAY,INTRAMUSC., METHYL-HG-ACETATE,7D,N=1	486
*PIG, WHOLE BLOOD	7.0 MICROGRAM/G WET WEIGHT	5.0MG HG/KG DAY,INTRAMUSC., METHYL-HG-ACETATE,7D,N=1	486
*PIMEPHALES PROMELAS	65 BASED ON WET WEIGHT	HG=2.0 MG HG/M3,1 W,BODY WEIGHT = 0.84 G	1563
*PIMEPHALES PROMELAS	100 BASED ON WET WEIGHT	HG=2.0 MG HG/M3,3 W,BODY WEIGHT = 0.72 G	1563
*PIMEPHALES PROMELAS	XX BASED ON WET WEIGHT	HG=2.0 MG HG/M3,5 W,BODY WEIGHT = 0.95 G	1563
*PIMEPHALES PROMELAS	138 BASED ON WET WEIGHT	HG=2.0 MG HG/M3,7 W,BODY WEIGHT = 0.85 G	1563
*PIMEPHALES PROMELAS	158 BASED ON WET WEIGHT	HG=2.0 MG HG/M3,10 W,BODY WEIGHT = 1.03 G	1563
*PIMEPHALES PROMELAS	146 BASED ON WET WEIGHT	HG=5.0 MG HG/M3,3 W,BODY WEIGHT = 0.86 G	1563
*PIMEPHALES PROMELAS	126 BASED ON WET WEIGHT	HG=5.0 MG HG/M3,5 W,BODY WEIGHT = 1.01 G	1563
*PIMEPHALES PROMELAS	101 BASED ON WET WEIGHT	AG=0.5 MG AG/L,7 W,BODY WEIGHT = 0.84 G	1563
*PIMEPHALES PROMELAS	145 BASED ON WET WEIGHT	HG=5.0 MG HG/M3,10 W,BODY WEIGHT = 0.88 G	1563
*PISUM SATIVUM, PODS	0.011 MG/KG DRY WEIGHT	CONTROL=0.005 PPM,SOIL=4 PPM,PH= 5.1, SILT LOAM,CEC=.383 MEQ/G, 293 K	774
*PISUM SATIVUM, PODS	0.042 MG/KG DRY WEIGHT	CONTROL=0.005 PPM,SOIL=20 PPM,PH= 5.1,SILT LOAM,CEC=.383 MEQ/G, 293 K	774
*PISUM SATIVUM, ROOT	0.160 MG/KG DRY WEIGHT	CONTROL=0.011 PPM,SOIL=4 PPM,PH= 5.1, SILT LOAM,CEC=.383 MEQ/G, 293 K	774
*PISUM SATIVUM, ROOT	1.415 MG/KG DRY WEIGHT	CONTROL=0.011 PPM,SOIL=20 PPM,PH= 5.1,SILT LOAM,CEC=.383 MEQ/G, 293 K	774
*PISUM SATIVUM, SEEDS	0.002 MG/KG DRY WEIGHT	CONTROL=0.001 PPM,SOIL=4 PPM,PH= 5.1, SILT LOAM,CEC=.383 MEQ/G, 293 K	774
*PISUM SATIVUM, SEEDS	0.003 MG/KG DRY WEIGHT	CONTROL=0.001 PPM,SOIL=20 PPM,PH= 5.1,SILT LOAM,CEC=.383 MEQ/G, 293 K	774
*PISUM SATIVUM, VINES	0.187 MG/KG DRY WEIGHT	CONTROL=0.110 PPM,SOIL=4 PPM,PH= 5.1, SILT LOAM,CEC=.383 MEQ/G, 293 K	774
*PISUM SATIVUM, VINES	0.085 MG/KG DRY WEIGHT	CONTROL=0.110 PPM,SOIL=20 PPM,PH= 5.1,SILT LOAM,CEC=.383 MEQ/G, 293 K	774
*PSEUDOPLEURONECTES AMERICANUS, BLOOD	<0.4 PPM WET. WT.	HG=0PPB,60D,3-6 FISH	216
*PSEUDOPLEURONECTES AMERICANUS, BLOOD	2.9 PPM WET. WT.	HG=5PPB,60D,3-6 FISH	216

TABLE D294

CONCENTRATIONS : HG

ITEM:	VALUE:	CONDITIONS:	REF.:
*PSEUDOPLEURONECTES AMERICANUS, BLOOD	3.8 PPM WET. WT.	HG=10PPB,60D,3-6 FISH	216
*PSEUDOPLEURONECTES AMERICANUS, GILL	<0.14 PPM WET. WT.	HG=0PPB,60D,3-6 FISH	216
*PSEUDOPLEURONECTES AMERICANUS, GILL	20.6 PPM WET. WT.	HG=5PPB,60D,3-6 FISH	216
*PSEUDOPLEURONECTES AMERICANUS, GILL	42.8 PPM WET. WT.	HG=10PPB,60D,3-6 FISH	216
*RABBIT, BRAIN	0.01 MICROGRAM/G WET WEIGHT	2MG HG/KG,SUBCUT., PHENYL-HG-ACETATE,40D,N=2	486
*RABBIT, KIDNEY	6 MICROGRAM/G WET WEIGHT	2MG HG/KG,SUBCUT., PHENYL-HG-ACETATE,40D,N=2	486
*RABBIT, LIVER	0.6 MICROGRAM/G WET WEIGHT	2MG HG/KG,SUBCUT., PHENYL-HG-ACETATE,40D,N=2	486
*RABBIT, WHOLE BLOOD	0.01 MICROGRAM/G WET WEIGHT	2MG HG/KG,SUBCUT., PHENYL-HG-ACETATE,40D,N=2	486
*RAIN	0.1 UG/L	SWEDEN,60MIC.G HG/M**2*YR	893
*RAPHANUS SATIVUS, TOPS	0.218 MG/KG DRY WEIGHT	CONTROL=0.237 PPM,SOIL=4 PPM,PH= 5.1, SILT LOAM,CEC=.383 MEQ/G, 293 K	774
*RAPHANUS SATIVUS, TOPS	0.585 MG/KG DRY WEIGHT	CONTROL=0.237 PPM,SOIL=20 PPM,PH= 5.1,SILT LOAM,CEC=.383 MEQ/G, 293 K	774
*RAPHANUS SATIVUS, TUBERS	0.026 MG/KG DRY WEIGHT	CONTROL=0.013 PPM,SOIL=4 PPM,PH= 5.1, SILT LOAM,CEC=.383 MEQ/G, 293 K	774
*RAPHANUS SATIVUS, TUBERS	0.663 MG/KG DRY WEIGHT	CONTROL=0.013 PPM,SOIL=20 PPM,PH= 5.1,SILT LOAM,CEC=.383 MEQ/G, 293 K	774
*RAT, BRAIN	0.02 MICROGRAM/G WET WEIGHT	0.1MG HG/KG,INTRAVEN., METHYL-HG-CL,32D,N=10	486
*RAT, BRAIN	0.19 MICROGRAM/G WET WEIGHT	0.05MG HG/KG DAY,SUBCUT., METHYL-HG-OH,13D,N=4	486
*RAT, BRAIN	4 MICROGRAM/G WET WEIGHT	0.65MG HG/KG DAY,SUBCUT., METHYL-HG-DICYANDIAMIDE,42D,N=6	486
*RAT, BRAIN	0.01 MICROGRAM/G WET WEIGHT	0.04MG HG/KG,INTRAVEN., METHYL-HG-OH,6D,N=3	486
*RAT, BRAIN	0.14 MICROGRAM/G WET WEIGHT	0.4MG HG/KG,INTRAVEN., METHYL-HG-OH,6D,N=3	486
*RAT, BRAIN	1.5 MICROGRAM/G WET WEIGHT	4.0MG HG/KG,INTRAVEN., METHYL-HG-OH,6D,N=3	486
*RAT, BRAIN	21 MICROGRAM/G WET WEIGHT	40MG HG/KG,SUBCUT.,METHYL-HG-OH, 3D,N=20	486
*RAT, BRAIN	0.04 MICROGRAM/G WET WEIGHT	0.1MG HG/KG,INTRAVEN., METHYL-HG-OH,4D,N=5	486
*RAT, BRAIN	0.13 MICROGRAM/G WET WEIGHT	0.5MG HG/KG,INTRAVEN., METHYL-HG-OH,4D,N=5	486
*RAT, BRAIN	0.17 MICROGRAM/G WET WEIGHT	0.5MG HG/KG,INTRAVEN., METHYL-HG-OH,16D,N=3	486
*RAT, BRAIN	0.7 MICROGRAM/G WET WEIGHT	3MG HG/KG,INTRAMUSC.,ETHYL-HGCL, 7D,N=6	486
*RAT, BRAIN	1.4 MICROGRAM/G WET WEIGHT	10MG HG/KG,SUBCUT.,ETHYL-HGCL,8D,N =3	486
*RAT, BRAIN	1.4 MICROGRAM/G WET WEIGHT	10MG HG/KG,SUBCUT., ETHYL-HG-CYSTEINE,8D,N=3	486
*RAT, BRAIN	0.51 MICROGRAM/G WET WEIGHT	0.2MG HG/KG,INTRAVEN., PHENYL-HG-ACETATE,1D,N=4	486
*RAT, BRAIN	0.03 MICROGRAM/G WET WEIGHT	0.05MG HG/KG DAY,SUBCUT., PHENYL-HG-OH,18D,N=4	486
*RAT, BRAIN	<1 MICROGRAM/G WET WEIGHT	0.65MG HG/KG DAY,SUBCUT., PHENYL-HG-ACETATE,42D,N=6	486
*RAT, BRAIN	0.06 MICROGRAM/G WET WEIGHT	3MG HG/KG,ORAL,PHENYL-HG-ACETATE, 2D,N=2	486
*RAT, BRAIN	0.004 MICROGRAM/G WET WEIGHT	0.05MG HG/KG,INTRAVEN., PHENYL-HG-OH,4D,N=3	486
*RAT, BRAIN	0.017 MICROGRAM/G WET WEIGHT	0.5MG HG/KG,INTRAVEN., PHENYL-HG-OH,4D,N=3	486
*RAT, BRAIN	0.018 MICROGRAM/G WET WEIGHT	0.5MG HG/KG,PHENYL-HG-OH,16D,N=3	486

TABLE D294

CONCENTRATIONS : HG

ITEM:	VALUE:	CONDITIONS:	REF.:
*RAT, BRAIN	0.16 MICROGRAM/G WET WEIGHT	10MG HG/KG,SUBCUT.,PHENYL-HGCL,8D, N=3	486
*RAT, BRAIN	0.47 MICROGRAM/G WET WEIGHT	5MG HG/KG,INTRAVEN.,PHENYL-HG-OH, 3D,N=3	486
*RAT, BRAIN	0.5 MICROGRAM/G WET WEIGHT	25MG HG/KG,SUBCUT.,PHENYL-HG-OH, 3D,N=20	486
*RAT, BRAIN	0.009 MICROGRAM/G WET WEIGHT	0.5MG HG/KG DAY,SUBCUT., METHOXYETHYL-HG-OH,12D,N=4	486
*RAT, BRAIN	0.018 MICROGRAM/G WET WEIGHT	0.5MG HG/KG,INTRAVEN., METHOXYETHYL-HG-OH,4D,N=4	486
*RAT, BRAIN	0.022 MICROGRAM/G WET WEIGHT	0.5MG HG/KG,INTRAVEN., METHOXYETHYL-HG-OH,4D,N=4	486
*RAT, BRAIN	0.009 MICROGRAM/G WET WEIGHT	0.05MG HG/KG,INTRAVEN., METHOXYETHYL-HG-OH,4D,N=4	486
*RAT, BRAIN	0.15 MICROGRAM/G WET WEIGHT	3MG HG/KG,INTRAVEN., METHOXYETHYL-HG-OH,3D,N=3	486
*RAT, BRAIN	0.036 MICROGRAM/G WET WEIGHT	0.3MG HG/KG,INTRAVEN., METHOXYETHYL-HG-OH,3D,N=3	486
*RAT, BRAIN	0.027 MICROGRAM/G WET WEIGHT	0.03MG HG/KG,INTRAVEN., METHOXYETHYL-HG-OH,3D,N=3	486
*RAT, BRAIN	0.076 MICROGRAM/G WET WEIGHT	3MG HG/KG,INTRAVEN., METHOXYETHYL-HG-OH,12D,N=3	486
*RAT, BRAIN	0.007 MICROGRAM/G WET WEIGHT	0.3MG HG/KG,INTRAVEN., METHOXYETHYL-HG-OH,12D,N=3	486
*RAT, BRAIN	0.002 MICROGRAM/G WET WEIGHT	0.03MG HG/KG,INTRAVEN., METHOXYETHYL-HG-OH,12D,N=3	486
*RAT, BRAIN	0.2 MICROGRAM/G WET WEIGHT	20MG HG/KG,SUBCUT., METHOXYETHYL-HG-OH,3D,N=20	486
*RAT, KIDNEY	0.26 MICROGRAM/G WET WEIGHT	0.1MG HG/KG,INTRAVEN., METHYL-HG-CL,32D,N=10	486
*RAT, KIDNEY	4.6 MICROGRAM/G WET WEIGHT	0.05MG HG/KG DAY,SUBCUT., METHYL-HG-OH,13D,N=4	486
*RAT, KIDNEY	51 MICROGRAM/G WET WEIGHT	0.65MG HG/KG DAY,SUBCUT., METHYL-HG-DICYANDIAMIDE,42D,N=6	486
*RAT, KIDNEY	0.49 MICROGRAM/G WET WEIGHT	0.04MG HG/KG,INTRAVEN., METHYL-HG-OH,6D,N=3	486
*RAT, KIDNEY	2.2 MICROGRAM/G WET WEIGHT	0.4MG HG/KG,INTRAVEN., METHYL-HG-OH,6D,N=3	486
*RAT, KIDNEY	22 MICROGRAM/G WET WEIGHT	4.0MG HG/KG,INTRAVEN., METHYL-HG-OH,6D,N=3	486
*RAT, KIDNEY	98 MICROGRAM/G WET WEIGHT	40MG HG/KG,SUBCUT.,METHYL-HG-OH, 3D,N=20	486
*RAT, KIDNEY	0.59 MICROGRAM/G WET WEIGHT	0.1MG HG/KG,INTRAVEN., METHYL-HG-OH,4D,N=5	486
*RAT, KIDNEY	1.7 MICROGRAM/G WET WEIGHT	0.5MG HG/KG,INTRAVEN., METHYL-HG-OH,4D,N=5	486
*RAT, KIDNEY	2.6 MICROGRAM/G WET WEIGHT	0.5MG HG/KG,INTRAVEN., METHYL-HG-OH,16D,N=3	486
*RAT, KIDNEY	69 MICROGRAM/G WET WEIGHT	3MG HG/KG,INTRAMUSC.,ETHYL-HGCL, 7D,N=6	486
*RAT, KIDNEY	110 MICROGRAM/G WET WEIGHT	10MG HG/KG,SUBCUT.,ETHYL-HGCL,8D,N =3	486
*RAT, KIDNEY	95 MICROGRAM/G WET WEIGHT	10MG HG/KG,SUBCUT., ETHYL-HG-CYSTEINE,8D,N=3	486
*RAT, KIDNEY	9.1 MICROGRAM/G WET WEIGHT	0.2MG HG/KG,INTRAVEN., PHENYL-HG-ACETATE,1D,N=4	486
*RAT, KIDNEY	33 MICROGRAM/G WET WEIGHT	0.05MG HG/KG DAY,SUBCUT., PHENYL-HG-OH,18D,N=4	486
*RAT, KIDNEY	90 MICROGRAM/G WET WEIGHT	0.65MG HG/KG DAY,SUBCUT., PHENYL-HG-ACETATE,42D,N=6	486
*RAT, KIDNEY	27 MICROGRAM/G WET WEIGHT	3MG HG/KG,ORAL,PHENYL-HG-ACETATE, 2D,N=2	486
*RAT, KIDNEY	1.7 MICROGRAM/G WET WEIGHT	0.05MG HG/KG,INTRAVEN., PHENYL-HG-OH,4D,N=3	486
*RAT, KIDNEY	15 MICROGRAM/G WET WEIGHT	0.5MG HG/KG,INTRAVEN., PHENYL-HG-OH,4D,N=3	486
*RAT, KIDNEY	16 MICROGRAM/G WET WEIGHT	0.5MG HG/KG,PHENYL-HG-OH,16D,N=3	486

TABLE D294

CONCENTRATIONS : HG

ITEM:	VALUE:	CONDITIONS:	REF.:
*RAT, KIDNEY	18 MICROGRAM/G WET WEIGHT	10MG HG/KG,SUBCUT.,PHENYL-HGCL,8D, N=3	486
*RAT, KIDNEY	42 MICROGRAM/G WET WEIGHT	5MG HG/KG,INTRAVEN.,PHENYL-HG-OH, 3D,N=3	486
*RAT, KIDNEY	52 MICROGRAM/G WET WEIGHT	25MG HG/KG,SUBCUT.,PHENYL-HG-OH, 3D,N=20	486
*RAT, KIDNEY	27 MICROGRAM/G WET WEIGHT	0.5MG HG/KG DAY,SUBCUT., METHOXYETHYL-HG-OH,12D,N=4	486
*RAT, KIDNEY	17 MICROGRAM/G WET WEIGHT	0.5MG HG/KG,INTRAVEN., METHOXYETHYL-HG-OH,4D,N=4	486
*RAT, KIDNEY	12 MICROGRAM/G WET WEIGHT	0.5MG HG/KG,INTRAVEN., METHOXYETHYL-HG-OH,4D,N=4	486
*RAT, KIDNEY	2.4 MICROGRAM/G WET WEIGHT	0.05MG HG/KG,INTRAVEN., METHOXYETHYL-HG-OH,4D,N=4	486
*RAT, KIDNEY	19 MICROGRAM/G WET WEIGHT	3MG HG/KG,INTRAVEN., METHOXYETHYL-HG-OH,3D,N=3	486
*RAT, KIDNEY	7.6 MICROGRAM/G WET WEIGHT	0.3MG HG/KG,INTRAVEN., METHOXYETHYL-HG-OH,3D,N=3	486
*RAT, KIDNEY	1.2 MICROGRAM/G WET WEIGHT	0.03MG HG/KG,INTRAVEN., METHOXYETHYL-HG-OH,3D,N=3	486
*RAT, KIDNEY	0.73 MICROGRAM/G WET WEIGHT	3MG HG/KG,INTRAVEN., METHOXYETHYL-HG-OH,12D,N=3	486
*RAT, KIDNEY	3.4 MICROGRAM/G WET WEIGHT	0.3MG HG/KG,INTRAVEN., METHOXYETHYL-HG-OH,12D,N=3	486
*RAT, KIDNEY	0.42 MICROGRAM/G WET WEIGHT	0.03MG HG/KG,INTRAVEN., METHOXYETHYL-HG-OH,12D,N=3	486
*RAT, KIDNEY	29 MICROGRAM/G WET WEIGHT	20MG HG/KG,SUBCUT., METHOXYETHYL-HG-OH,3D,N=20	486
*RAT, LIVER	0.08 MICROGRAM/G WET WEIGHT	0.1MG HG/KG,INTRAVEN., METHYL-HG-CL,32D,N=10	486
*RAT, LIVER	0.92 MICROGRAM/G WET WEIGHT	0.05MG HG/KG DAY,SUBCUT., METHYL-HG-OH,13D,N=4	486
*RAT, LIVER	16 MICROGRAM/G WET WEIGHT	0.65MG HG/KG DAY,SUBCUT., METHYL-HG-DICYANDIAMIDE,42D,N=6	486
*RAT, LIVER	0.04 MICROGRAM/G WET WEIGHT	0.04MG HG/KG,INTRAVEN., METHYL-HG-OH,6D,N=3	486
*RAT, LIVER	0.41 MICROGRAM/G WET WEIGHT	0.4MG HG/KG,INTRAVEN., METHYL-HG-OH,6D,N=3	486
*RAT, LIVER	4.2 MICROGRAM/G WET WEIGHT	4.0MG HG/KG,INTRAVEN., METHYL-HG-OH,6D,N=3	486
*RAT, LIVER	77 MICROGRAM/G WET WEIGHT	40MG HG/KG,SUBCUT.,METHYL-HG-OH, 3D,N=20	486
*RAT, LIVER	0.17 MICROGRAM/G WET WEIGHT	0.1MG HG/KG,INTRAVEN., METHYL-HG-OH,4D,N=5	486
*RAT, LIVER	0.52 MICROGRAM/G WET WEIGHT	0.5MG HG/KG,INTRAVEN., METHYL-HG-OH,4D,N=5	486
*RAT, LIVER	0.48 MICROGRAM/G WET WEIGHT	0.5MG HG/KG,INTRAVEN., METHYL-HG-OH,16D,N=3	486
*RAT, LIVER	5.3 MICROGRAM/G WET WEIGHT	3MG HG/KG,INTRAMUSC.,ETHYL-HGCL, 7D,N=6	486
*RAT, LIVER	11 MICROGRAM/G WET WEIGHT	10MG HG/KG,SUBCUT.,ETHYL-HGCL,8D,N =3	486
*RAT, LIVER	12 MICROGRAM/G WET WEIGHT	10MG HG/KG,SUBCUT., ETHYL-HG-CYSTEINE,8D,N=3	486
*RAT, LIVER	0.52 MICROGRAM/G WET WEIGHT	0.2MG HG/KG,INTRAVEN., PHENYL-HG-ACETATE,1D,N=4	486
*RAT, LIVER	0.36 MICROGRAM/G WET WEIGHT	0.05MG HG/KG DAY,SUBCUT., PHENYL-HG-OH,18D,N=4	486
*RAT, LIVER	1.3 MICROGRAM/G WET WEIGHT	0.65MG HG/KG DAY,SUBCUT., PHENYL-HG-ACETATE,42D,N=6	486
*RAT, LIVER	0.58 MICROGRAM/G WET WEIGHT	3MG HG/KG,ORAL,PHENYL-HG-ACETATE, 2D,N=2	486
*RAT, LIVER	0.081 MICROGRAM/G WET WEIGHT	0.05MG HG/KG,INTRAVEN., PHENYL-HG-OH,4D,N=3	486
*RAT, LIVER	0.33 MICROGRAM/G WET WEIGHT	0.5MG HG/KG,INTRAVEN., PHENYL-HG-OH,4D,N=3	486
*RAT, LIVER	0.083 MICROGRAM/G WET WEIGHT	0.5MG HG/KG,PHENYL-HG-OH,16D,N=3	486

TABLE D294

CONCENTRATIONS : HG			
ITEM:	VALUE:	CONDITIONS:	REF.:
*RAT, LIVER	1.3 MICROGRAM/G WET WEIGHT	10MG HG/KG,SUBCUT.,PHENYL-HGCL,8D, N=3	486
*RAT, LIVER	11 MICROGRAM/G WET WEIGHT	5MG HG/KG,INTRAVEN.,PHENYL-HG-OH, 3D,N=3	486
*RAT, LIVER	19 MICROGRAM/G WET WEIGHT	25MG HG/KG,SUBCUT.,PHENYL-HG-OH, 3D,N=20	486
*RAT, LIVER	0.25 MICROGRAM/G WET WEIGHT	0.5MG HG/KG DAY,SUBCUT., METHOXYETHYL-HG-OH,12D,N=4	486
*RAT, LIVER	0.30 MICROGRAM/G WET WEIGHT	0.5MG HG/KG,INTRAVEN., METHOXYETHYL-HG-OH,4D,N=4	486
*RAT, LIVER	0.24 MICROGRAM/G WET WEIGHT	0.5MG HG/KG,INTRAVEN., METHOXYETHYL-HG-OH,4D,N=4	486
*RAT, LIVER	0.061 MICROGRAM/G WET WEIGHT	0.05MG HG/KG,INTRAVEN., METHOXYETHYL-HG-OH,4D,N=4	486
*RAT, LIVER	4.0 MICROGRAM/G WET WEIGHT	3MG HG/KG,INTRAVEN., METHOXYETHYL-HG-OH,3D,N=3	486
*RAT, LIVER	0.17 MICROGRAM/G WET WEIGHT	0.3MG HG/KG,INTRAVEN., METHOXYETHYL-HG-OH,3D,N=3	486
*RAT, LIVER	0.047 MICROGRAM/G WET WEIGHT	0.03MG HG/KG,INTRAVEN., METHOXYETHYL-HG-OH,3D,N=3	486
*RAT, LIVER	0.51 MICROGRAM/G WET WEIGHT	3MG HG/KG,INTRAVEN., METHOXYETHYL-HG-OH,12D,N=3	486
*RAT, LIVER	0.021 MICROGRAM/G WET WEIGHT	0.3MG HG/KG,INTRAVEN., METHOXYETHYL-HG-OH,12D,N=3	486
*RAT, LIVER	0.005 MICROGRAM/G WET WEIGHT	0.03MG HG/KG,INTRAVEN., METHOXYETHYL-HG-OH,12D,N=3	486
*¤AT, LIVER	44 MICROGRAM/G WET WEIGHT	20MG HG/KG,SUBCUT., METHOXYETHYL-HG-OH,3D,N=20	486
*RAT, WHOLE BLOOD	0.05 MICROGRAM/G WET WEIGHT	0.1MG HG/KG,INTRAVEN., METHYL-HG-CL,32D,N=10	486
*RAT, WHOLE BLOOD	3 MICROGRAM/G WET WEIGHT	0.05MG HG/KG DAY,SUBCUT., METHYL-HG-OH,13D,N=4	486
*RAT, WHOLE BLOOD	CA. 40 MICROGRAM/G WET WEIGHT	0.65MG HG/KG DAY,SUBCUT., METHYL-HG-DICYANDIAMIDE,42D,N=6	486
*RAT, WHOLE BLOOD	0.14 MICROGRAM/G WET WEIGHT	0.04MG HG/KG,INTRAVEN., METHYL-HG-OH,6D,N=3	486
*RAT, WHOLE BLOOD	1.7 MICROGRAM/G WET WEIGHT	0.4MG HG/KG,INTRAVEN., METHYL-HG-OH,6D,N=3	486
*RAT, WHOLE BLOOD	19 MICROGRAM/G WET WEIGHT	4.0MG HG/KG,INTRAVEN., METHYL-HG-OH,6D,N=3	486
*RAT, WHOLE BLOOD	290 MICROGRAM/G WET WEIGHT	40MG HG/KG,SUBCUT.,METHYL-HG-OH, 3D,N=20	486
*RAT, WHOLE BLOOD	0.43 MICROGRAM/G WET WEIGHT	0.1MG HG/KG,INTRAVEN., METHYL-HG-OH,4D,N=5	486
*RAT, WHOLE BLOOD	2.2 MICROGRAM/G WE¤ WEIGHT	0.5MG HG/KG,INTRAVEN., METHYL-HG-OH,4D,N=5	486
*RAT, WHOLE BLOOD	1.8 MICROGRAM/G WET WEIGHT	0.5MG HG/KG,INTRAVEN., METHYL-HG-OH,16D,N=3	486
*RAT, WHOLE BLOOD	14 MICROGRAM/G WET WEIGHT	3MG HG/KG,INTRAMUSC.,ETHYL-HGCL, 7D,N=6	486
*RAT, WHOLE BLOOD	19 MICROGRAM/G WET WEIGHT	10MG HG/KG,SUBCUT.,ETHYL-HGCL,8D,N =3	486
*RAT, WHOLE BLOOD	23 MICROGRAM/G WET WEIGHT	10MG HG/KG,SUBCUT., ETHYL-HG-CYSTEINE,8D,N=3	486
*RAT, WHOLE BLOOD	0.22 MICROGRAM/G WET WEIGHT	0.2MG HG/KG,INTRAVEN., PHENYL-HG-ACETATE,1D,N=4	486
*RAT, WHOLE BLOOD	0.039 MICROGRAM/G WET WEIGHT	0.05MG HG/KG DAY,SUBCUT., PHENYL-HG-OH,18D,N=4	486
*RAT, WHOLE BLOOD	CA. 0.5 MICROGRAM/G WET WEIGHT	0.65MG HG/KG DAY,SUBCUT., PHENYL-HG-ACETATE,42D,N=6	486
*RAT, WHOLE BLOOD	0.74 MICROGRAM/G WET WEIGHT	3MG HG/KG,ORAL,PHENYL-HG-ACETATE, 2D,N=2	486
*RAT, WHOLE BLOOD	0.006 MICROGRAM/G WET WEIGHT	0.05MG HG/KG,INTRAVEN., PHENYL-HG-OH,4D,N=3	486
*RAT, WHOLE BLOOD	0.04 MICROGRAM/G WET WEIGHT	0.5MG HG/KG,INTRAVEN., PHENYL-HG-OH,4D,N=3	486
*RAT, WHOLE BLOOD	0.047 MICROGRAM/G WET WEIGHT	0.5MG HG/KG,PHENYL-HG-OH,16D,N=3	486

TABLE D294

CONCENTRATIONS : HG

ITEM:	VALUE:	CONDITIONS:	REF.:
*RAT, WHOLE BLOOD	0.29 MICROGRAM/G WET WEIGHT	10MG HG/KG,SUBCUT.,PHENYL-HGCL,8D, N=3	486
*RAT, WHOLE BLOOD	4.4 MICROGRAM/G WET WEIGHT	5MG HG/KG,INTRAVEN.,PHENYL-HG-OH, 3D,N=3	486
*RAT, WHOLE BLOOD	6 MICROGRAM/G WET WEIGHT	25MG HG/KG,SUBCUT.,PHENYL-HG-OH, 3D,N=20	486
*RAT, WHOLE BLOOD	0.033 MICROGRAM/G WET WEIGHT	0.5MG HG/KG DAY,SUBCUT., METHOXYETHYL-HG-OH,12D,N=4	486
*RAT, WHOLE BLOOD	0.068 MICROGRAM/G WET WEIGHT	0.5MG HG/KG,INTRAVEN., METHOXYETHYL-HG-OH,4D,N=4	486
*RAT, WHOLE BLOOD	0.054 MICROGRAM/G WET WEIGHT	0.5MG HG/KG,INTRAVEN., METHOXYETHYL-HG-OH,4D,N=4	486
*RAT, WHOLE BLOOD	0.009 MICROGRAM/G WET WEIGHT	0.05MG HG/KG,INTRAVEN., METHOXYETHYL-HG-OH,4D,N=4	486
*RAT, WHOLE BLOOD	1.1 MICROGRAM/G WET WEIGHT	3MG HG/KG,INTRAVEN., METHOXYETHYL-HG-OH,3D,N=3	486
*RAT, WHOLE BLOOD	0.082 MICROGRAM/G WET WEIGHT	0.3MG HG/KG,INTRAVEN., METHOXYETHYL-HG-OH,3D,N=3	486
*RAT, WHOLE BLOOD	0.044 MICROGRAM/G WET WEIGHT	0.03MG HG/KG,INTRAVEN., METHOXYETHYL-HG-OH,3D,N=3	486
*RAT, WHOLE BLOOD	0.066 MICROGRAM/G WET WEIGHT	3MG HG/KG,INTRAVEN., METHOXYETHYL-HG-OH,12D,N=3	486
*RAT, WHOLE BLOOD	0.004 MICROGRAM/G WET WEIGHT	0.3MG HG/KG,INTRAVEN., METHOXYETHYL-HG-OH,12D,N=3	486
*RAT, WHOLE BLOOD	0.001 MICROGRAM/G WET WEIGHT	0.03MG HG/KG,INTRAVEN., METHOXYETHYL-HG-OH,12D,N=3	486
*RAT, WHOLE BLOOD	11 MICROGRAM/G WET WEIGHT	20MG HG/KG,SUBCUT., METHOXYETHYL-HG-OH,3D,N=20	486
*SAIMIRI SCIUREUS, BRAIN	18 MICROGRAM/G WET WEIGHT	0.3-0.7MG HG/KG DAY,ORAL, METHYL-HG-OH,35-36D,N=2, KILL.1AND9D AFT.EXPOS.	486
*SAIMIRI SCIUREUS, BRAIN	14 MICROGRAM/G WET WEIGHT	0.3-0.7MG HG/KG DAY,ORAL, METHYL-HG-OH,28D,N=1,KILL.4D AFT.EXPOS.STOP	486
*SAIMIRI SCIUREUS, BRAIN	2.5 MICROGRAM/G WET WEIGHT	0.3-0.7MG HG/KG DAY,ORAL, METHYL-HG-OH,21D,N=2,KILL.85D AFT.EXPOS.STOP	486
*SAIMIRI SCIUREUS, KIDNEY	6.0 MICROGRAM/G WET WEIGHT	0.3-0.7MG HG/KG DAY,ORAL, METHYL-HG-OH,35-36D,N=2, KILL.1AND9D AFT.EXPOS.	486
*SAIMIRI SCIUREUS, KIDNEY	12 MICROGRAM/G WET WEIGHT	0.3-0.7MG HG/KG DAY,ORAL, METHYL-HG-OH,28D,N=1,KILL.4D AFT.EXPOS.STOP	486
*SAIMIRI SCIUREUS, KIDNEY	4.9 MICROGRAM/G WET WEIGHT	0.3-0.7MG HG/KG DAY,ORAL, METHYL-HG-OH,21D,N=2,KILL.85D AFT.EXPOS.STOP	486
*SAIMIRI SCIUREUS, LIVER	9.8 MICROGRAM/G WET WEIGHT	0.3-0.7MG HG/KG DAY,ORAL, METHYL-HG-OH,35-36D,N=2, KILL.1AND9D AFT.EXPOS.	486
*SAIMIRI SCIUREUS, LIVER	24 MICROGRAM/G WET WEIGHT	0.3-0.7MG HG/KG DAY,ORAL, METHYL-HG-OH,28D,N=1,KILL.4D AFT.EXPOS.STOP	486
*SAIMIRI SCIUREUS, LIVER	2.7 MICROGRAM/G WET WEIGHT	0.3-0.7MG HG/KG DAY,ORAL, METHYL-HG-OH,21D,N=2,KILL.85D AFT.EXPOS.STOP	486
*SAIMIRI SCIUREUS, WHOLE BLOOD	1.5 MICROGRAM/G WET WEIGHT	0.3-0.7MG HG/KG DAY,ORAL, METHYL-HG-OH,35-36D,N=2, KILL.1AND9D AFT.EXPOS.	486
*SAIMIRI SCIUREUS, WHOLE BLOOD	1.9 MICROGRAM/G WET WEIGHT	0.3-0.7MG HG/KG DAY,ORAL, METHYL-HG-OH,28D,N=1,KILL.4D AFT.EXPOS.STOP	486
*SAIMIRI SCIUREUS, WHOLE BLOOD	0.3 MICROGRAM/G WET WEIGHT	0.3-0.7MG HG/KG DAY,ORAL, METHYL-HG-OH,21D,N=2,KILL.85D AFT.EXPOS.STOP	486
*SARCOSAPROPHAGOUS FLIES	0.62 MG/KG	FOOD = FISH = 0.47 MG/KG, FULL-GROWN FLY LARVAE	1140
*SARCOSAPROPHAGOUS FLIES	6.90 MG/KG	FOOD = FISH = 1.90 MG/KG, FULL-GROWN FLY LARVAE	1140
*SEDIMENT	0.07 (RANGE 0.02 - 0.33) MG/KG DRY WEIGHT	UNPOLLUTED AREA, ESTUARY, URR WATER	1557
*SEDIMENT	0.07 (RANGE 0.03 - 0.17) MG/KG DRY WEIGHT	UNPOLLUTED AREA, ESTUARY, URR WATER	1558

TABLE D294

CONCENTRATIONS : HG			
ITEM:	VALUE:	CONDITIONS:	REF.:
*SEDIMENT	0.06 (RANGE 0.02 - 0.12) MG/KG DRY WEIGHT	UNPOLLUTED AREA, ESTUARY, URR WATER	1558
*SPINACIA OLERACEA, LEAVES	0.339 MG/KG DRY WEIGHT	CONTROL=0.094 PPM,SOIL=4 PPM,PH= 5.1, SILT LOAM,CEC=.383 MEQ/G, 293 K	774
*SPINACIA OLERACEA, LEAVES	0.695 MG/KG DRY WEIGHT	CONTROL=0.094 PPM,SOIL=20 PPM,PH= 5.1,SILT LOAM,CEC=.383 MEQ/G, 293 K	774
*SPINACIA OLERACEA, ROOT	1.022 MG/KG DRY WEIGHT	CONTROL=0.095 PPM,SOIL=4 PPM,PH= 5.1, SILT LOAM,CEC=.383 MEQ/G, 293 K	774
*SPINACIA OLERACEA, ROOT	1.067 MG/KG DRY WEIGHT	CONTROL=0.095 PPM,SOIL=20 PPM,PH= 5.1,SILT LOAM,CEC=.383 MEQ/G, 293 K	774
*TRINGA TOTANUS, LIVER	1.6 PPM DRY WT.	3 BIRDS(KILLED),29.8.72,BRITISH ISLES	1216
*TRINGA TOTANUS, LIVER	2.7 PPM DRY WT.	3 BIRDS(FOUND DEAD),4-29.11.72, BRITISH ISLES	1216
*TRINGA TOTANUS, LIVER	12.3 PPM DRY WT.	5 BIRDS(2 KILLED,3 FOUND DEAD), 5.2-3.3-72,BRITISH ISLES	1216
*WATER	0.01-0.1 MIC.G/L	NORMAL VAL.FOR RIVERS AND LAKES	893
*WATER	0.02-0.07 MIC.G/L	GROUNDWATER	893
*WATER	0.12 MIC.G/L	BALTIC SEA,SURFACE WATER	893

TABLE D295

CONCENTRATIONS : HUMUS			
ITEM:	VALUE:	CONDITIONS:	REF.:
*CARBON	50 - 60 % OF DRY WEIGHT	ROUGHLY ESTIMATE	550
*H	5 % OF DRY WEIGHT	ROUGHLY ESTIMATE	550
*NITROGEN	3 - 5 % OF DRY WEIGHT	ROUGHLY ESTIMATE	550
*OXYGEN	30 - 40 % OF DRY WEIGHT	ROUGHLY ESTIMATE	550

TABLE D296

CONCENTRATIONS : HYDROCARBONS			
ITEM:	VALUE:	CONDITIONS:	REF.:
*MYTILUS EDULIS	353 MG/KG WET WEIGHT	UPTAKE FROM WATER = 302 UG/L, 80 DAYS, EXTRAC=N-HEXANE,JUNE-AUG	472
*MYTILUS EDULIS	225 MG/KG WET WEIGHT	UPTAKE FROM WATER = 302 UG/L, 30 DAYS, EXTRAC=N-HEXANE,NOV-DEC	472
*MYTILUS EDULIS	407 MG/KG WET WEIGHT	UPTAKE FROM WATER = 302 UG/L, 55 DAYS, EXTRAC=N-HEXANE,MARCH-MAY	472

TABLE D297

CONCENTRATIONS : IN			
ITEM:	VALUE:	CONDITIONS:	REF.:
*ATMOSPHERE	< 0.025-0.08 NG/M**3	ENGLAND	893
*RAIN	< 0.59 MIC.G/L	ENGLAND,YEARLY MEAN	893

TABLE D298

CONCENTRATIONS : INSECTICIDES			
ITEM:	VALUE:	CONDITIONS:	REF.:
*BUFFALO	0.032 MG/KG DRY WEIGHT	SOUTH DAK., FRESHWATER, WATER = 0.230 MG/M3, TOTAL INSECT., N=10	564

TABLE D298

CONCENTRATIONS : INSECTICIDES			
ITEM:	VALUE:	CONDITIONS:	REF.:
*BUFFALO	0.032 MG/KG DRY WEIGHT	SOUTH DAK., FRESHWATER, WATER = 0.230 MG/M3, TOTAL INSECT., N=10	564
*BULLHEAD	0.031 MG/KG DRY WEIGHT	SOUTH DAK., FRESHWATER, WATER = 0.230 MG/M3, TOTAL INSECT., N=10	564
*BULLHEAD	0.031 MG/KG DRY WEIGHT	SOUTH DAK., FRESHWATER, WATER = 0.230 MG/M3, TOTAL INSECT., N=10	564
*CARP	0.092 MG/KG DRY WEIGHT	SOUTH DAK., FRESHWATER, WATER = 0.230 MG/M3, TOTAL INSECT., N=10	564
*CARP	0.092 MG/KG DRY WEIGHT	SOUTH DAK., FRESHWATER, WATER = 0.230 MG/M3, TOTAL INSECT., N=10	564
*CRAPPIE	0.016 MG/KG DRY WEIGHT	SOUTH DAK., FRESHWATER, WATER = 0.230 MG/M3, TOTAL INSECT., N=10	564
*CRAPPIE	0.016 MG/KG DRY WEIGHT	SOUTH DAK., FRESHWATER, WATER = 0.230 MG/M3, TOTAL INSECT., N=10	564
*SEDIMENT	0.0016 MG/KG WET WEIGHT	SOUTH DAK., FRESHWATER, WATER = 0.230 MG/M3, TOTAL INSECT., N=10	564
*SEDIMENT	0.0016 MG/KG WET WEIGHT	SOUTH DAK., FRESHWATER, WATER = 0.230 MG/M3, TOTAL INSECT., N=10	564

TABLE D299

CONCENTRATIONS : K-40			
ITEM:	VALUE:	CONDITIONS:	REF.:
*PARTICLE FRACTIONS	2.0 % OF TOTAL	SIZE=2.0-0.5 MM,WEIGHT=3.2 %,AREA= 0.2 %,MARINE, JAPAN	1095
*PARTICLE FRACTIONS	8.5 % OF TOTAL	SIZE=0.5-0.25 MM,WEIGHT=8.5 %,AREA =1.4 %,MARINE, JAPAN	1095
*PARTICLE FRACTIONS	23.1 % OF TOTAL	SIZE=0.25-0.125 MM,WEIGHT=23 %, AREA=10.1 %,MARINE, JAPAN	1095
*PARTICLE FRACTIONS	20.1 % OF TOTAL	SIZE=0.125-0.075 MM,WEIGHT=16.9 %, AREA=10.4 %,MARINE, JAPAN	1095
*PARTICLE FRACTIONS	46.3 % OF TOTAL	SIZE BELOW 0.075 MM,WEIGHT=48.4 %, AREA=77.9 %,MARINE, JAPAN	1095

TABLE D300

CONCENTRATIONS : LI			
ITEM:	VALUE:	CONDITIONS:	REF.:
*WATER	0.1-400 MIC.G/L	RIVERS,MEAN=1UG/L	893

TABLE D301

CONCENTRATIONS : LIPID			
ITEM:	VALUE:	CONDITIONS:	REF.:
*CHLORELLA SP.	63.4 % OF DRY WEIGHT	UNFAVORABLE ENVIRONMENT	1039
*CHLORELLA SP.	20.2 % OF DRY WEIGHT	FAVORABLE ENVIRONMENT	1039
*EEL	63.8 % OF DRY WEIGHT	SILVER	1487
*EEL	39.4 % OF DRY WEIGHT	YELLOW	1487
*HERRING	49.5 % OF DRY WEIGHT	NONE	1487
*MUSSEL	11.9 % OF DRY WEIGHT	NONE	1487
*MYCOPLASMA GALLISEPTICUM	11 % OF DRY WEIGHT	NONE	1067

TABLE D301

CONCENTRATIONS : LIPID			
ITEM:	VALUE:	CONDITIONS:	REF.:
*OYSTER	11.1 % OF DRY WEIGHT	NONE	461
*PLAICE	9.4 % OF DRY WEIGHT	NONE	1487
*SACCHAROMYCES CEREVISIAE	2 % OF DRY WEIGHT	NONE	574
*SEA URCHIN	17.4 % OF DRY WEIGHT	40 H LARVA	461
*SEA URCHIN	31.2 % OF DRY WEIGHT	UNFERTILIZED EGG	461
*SILKWORM	13.3 % OF DRY WEIGHT	LARVA	461
*SILKWORM	24.3 % OF DRY WEIGHT	IMAGO	461
*TORULOPSIS UTILIS	3 % OF DRY WEIGHT	NONE	574

TABLE D302

CONCENTRATIONS : METHOXYCHLOR			
ITEM:	VALUE:	CONDITIONS:	REF.:
*CHICKEN, FAT	54.10 MG/KG	CONC.IN FOOD=1000MG/KG	389
*CHICKEN, FAT	0.10 MG/KG	CONC.IN FOOD=8.00MG/KG	389
*CHICKEN, FAT	1.26 MG/KG	CONC.IN FOOD=100.00MG/KG	389
*CHICKEN, FAT	1.40 MG/KG	CONC.IN FOOD=100MG/KG	389
*CHICKEN, LIVER	0.68 MG/KG	CONC.IN FOOD=1000MG/KG	389
*CHICKEN, MUSCLE	0.18 MG/KG	CONC.IN FOOD=1000MG/KG	389
*CHICKEN, TISSUE	18.90 MG/KG	CONC.IN FOOD=1000MG/KG	389
*CHICKEN, TISSUE	0.56 MG/KG	CONC.IN FOOD=100.00MG/KG	389
*RHIZOPHORA MANGLE, HYPOCOTYL	0.03 MG/KG WET WT.	APPLICATION=0.28KG/HA,PH=6.3-6.7,S =2.5%,6WEEKS	1647
*RHIZOPHORA MANGLE, HYPOCOTYL	0.26 MG/KG WET WT.	APPLICATION=0.56KG/HA,PH=6.3-6.7,S =2.5%,6WEEKS	1647
*RHIZOPHORA MANGLE, HYPOCOTYL	0.26 MG/KG WET WT.	APPLICATION=1.12KG/HA,PH=6.3-6.7,S =2.5%,6WEEKS	1647
*RHIZOPHORA MANGLE, HYPOCOTYL	0.29 MG/KG WET WT.	APPLICATION=2.80KG/HA,PH=6.3-6.7,S =2.5%,6WEEKS	1647
*RHIZOPHORA MANGLE, HYPOCOTYL	0.31 MG/KG WET WT.	APPLICATION=5.60KG/HA,PH=6.3-6.7,S =2.5%,6WEEKS	1647
*RHIZOPHORA MANGLE, HYPOCOTYL	0.49 MG/KG WET WT.	APPLICATION=11.20KG/HA,PH=6.3-6.7, S=2.5%,6WEEKS	1647
*RHIZOPHORA MANGLE, ROOT	0.02 MG/KG WET WT.	APPLICATION=2.80KG/HA,PH=6.3-6.7,S =2.5%,6WEEKS	1647
*RHIZOPHORA MANGLE, ROOT	0.06 MG/KG WET WT.	APPLICATION=5.60KG/HA,PH=6.3-6.7,S =2.5%,6WEEKS	1647
*RHIZOPHORA MANGLE, ROOT	0.07 MG/KG WET WT.	APPLICATION=11.20KG/HA,PH=6.3-6.7, S=2.5%,6WEEKS	1647
*SEDIMENT	<0.01 MG/KG	CONC.IN WATER:<1.0 MICROGRAM/L, U.S.	389

TABLE D303

CONCENTRATIONS : MN			
ITEM:	VALUE:	CONDITIONS:	REF.:
*ALISMA PLANTAGO AQUATICA, OLD LEAVES	2591.60 MG/KG DRY WT.	800M ABOVE POLL.ENTRY(1),MN IN WATER=0.038MG/L,WHOLE LEAVES	1224
*ALISMA PLANTAGO AQUATICA, OLD LEAVES	2553.65 MG/KG DRY WT.	800M ABOVE POLL.ENTRY(1),MN IN WATER=0.038MG/L,PROCESSED LEAVES	1224
*ALISMA PLANTAGO AQUATICA, OLD LEAVES	127.00 MG/KG DRY WT.	150M BELOW POLL.ENTRY(1),MN IN WATER=0.075MG/L,WHOLE LEAVES	1224

TABLE D303

CONCENTRATIONS : MN

ITEM:	VALUE:	CONDITIONS:	REF.:
*ALISMA PLANTAGO AQUATICA, OLD LEAVES	124.31 MG/KG DRY WT.	150M BELOW POLL.ENTRY(1),MN IN WATER=0.075MG/L,PROCESSED LEAVES	1224
*ALISMA PLANTAGO AQUATICA, OLD LEAVES	99.42 MG/KG DRY WT.	800 AND 50M BELOW POLL.(1)AND(2), MN IN WATER=0.099MG/L,WHOLE LEAVES	1224
*ALISMA PLANTAGO AQUATICA, OLD LEAVES	95.92 MG/KG DRY WT.	800 AND 50M BELOW POLL.(1)AND(2), MN IN WATER=0.099MG/L,PROCESSED LEAVES	1224
*ALISMA PLANTAGO AQUATICA, YOUNG LEAVES	1306.32 MG/KG DRY WT.	800M ABOVE POLL.ENTRY(1),MN IN WATER=0.038MG/L,WHOLE LEAVES	1224
*ALISMA PLANTAGO AQUATICA, YOUNG LEAVES	1513.98 MG/KG DRY WT.	800M ABOVE POLL.ENTRY(1),MN IN WATER=0.038MG/L,PROCESSED LEAVES	1224
*ALISMA PLANTAGO AQUATICA, YOUNG LEAVES	165.00 MG/KG DRY WT.	150M BELOW POLL.ENTRY(1),MN IN WATER=0.075MG/L,WHOLE LEAVES	1224
*ALISMA PLANTAGO AQUATICA, YOUNG LEAVES	171.99 MG/KG DRY WT.	150M BELOW POLL.ENTRY(1),MN IN WATER=0.075MG/L,PROCESSED LEAVES	1224
*ALISMA PLANTAGO AQUATICA, YOUNG LEAVES	188.12 MG/KG DRY WT.	800 AND 50M BELOW POLL.(1)AND(2), MN IN WATER=0.099MG/L,WHOLE LEAVES	1224
*ALISMA PLANTAGO AQUATICA, YOUNG LEAVES	160.20 MG/KG DRY WT.	800 AND 50M BELOW POLL.(1)AND(2), MN IN WATER=0.099MG/L,PROCESSED LEAVES	1224
*ATMOSPHERE	0.01+-0.005 NG/M**3	SOUTHPOLE,1970	893
*ATMOSPHERE	100 NG/M**3	U.S.A.,MEAN FROM 103 STATIONS, 1960-65,MAX=9980NG/M**3	893
*ATMOSPHERE	10-1440 NG/M**3	URBAN,U.S.A.	893
*ATMOSPHERE	5-47 NG/M**3	NOT URBAN,U.S.A.	893
*ATMOSPHERE	3-41 NG/M**3	ENGLAND,7 STATIONS	893
*ATMOSPHERE	440-3000 NG/M**3	SWEDEN,MEAN FROM 6 STATIONS	893
*CONIFEROUS TREES, NEEDLES	14000 (SD=+-1300) MG/KG ASH	SILICIC ROCK=610+-59, SOIL=1500+ -150 PPM, FINLAND,N=64	950
*CONIFEROUS TREES, NEEDLES	16000 (SD=+-3800) MG/KG ASH	ULTRABASIC ROCK=1400+-110,SOIL= 1200+-150 PPM,FINLAND,N=10	950
*CONIFEROUS TREES, NEEDLES	7800 (SD=+-1200) MG/KG ASH	CALCAREOUS ROCK=2300+-350,SOIL= 1800+-370 PPM, FINLAND,N=9	950
*CONIFEROUS TREES, TWIGS	10000 (SD=+-1100) MG/KG ASH	SILICIC ROCK=610+-59, SOIL=1500+ -150 PPM, FINLAND,N=64	950
*CONIFEROUS TREES, TWIGS	5200 (SD=AND-1100) MG/KG ASH	ULTRABASIC ROCK=1400+-110,SOIL= 1200+-150 PPM,FINLANDN=10	950
*CONIFEROUS TREES, TWIGS	6100 (SD=+-1400) MG/KG ASH	CALCAREOUS ROCK=2300+-350,SOIL= 1800+-370 PPM, FINLAND,N=9	950
*DECIDUOUS TREES AND SHRUBS, LEAVES	12000 (SD=AND-1200) MG/KG ASH	SILICIC ROCK=610+-59, SOIL=1500+ -150 PPM, FINLAND,N=80	950
*DECIDUOUS TREES AND SHRUBS, LEAVES	15000 (SD=+-3200) MG/KG ASH	ULTRABASIC ROCK=1400+-110,SOIL= 1200+-150 PPM,FINLAND,N=14	950
*DECIDUOUS TREES AND SHRUBS, LEAVES	5000 (SD=+-130) MG/KG ASH	CALCAREOUS ROCK=2300+-350,SOIL= 1800+-370 PPM, FINLAND,N=23	950
*DECIDUOUS TREES AND SHRUBS, TWIGS	9000 (SD1+-950) MG/KG ASH	SILICIC ROCK=610+-59, SOIL=1500+ -150 PPM, FINLAND,N=80	950
*DECIDUOUS TREES AND SHRUBS, TWIGS	11000 (SD=+-2800) MG/KG ASH	ULTRABASIC ROCK=1400+-110,SOIL= 1200+-150 PPM,FINLAND,N=14	950
*DECIDUOUS TREES AND SHRUBS, TWIGS	7400 (SD=+-160) MG/KG ASH	CALCAREOUS ROCK=2300+-350,SOIL= 1800+-370 PPM, FINLAND,N=23	950
*FOOD	0.5-1.5 MG/KG DRY WT.	VEGETABLES	893
*FOOD	63-129 MICROGRAM/L	MILK,U.S.A.	893
*FOOD	150-190 MG/KG DRY WT.	TEA	893
*FUCUS VESICULOSUS	2.3*10**4 (MICROGRAM/G DRY MAT)/ (MICROGRAMS/G SEAWATER)	18 OBS.,ALL AREAS AROUNDRTHESBRITISH ISLES	1278
*HYPOLIMNION	590 MICROG¤AM/L	CONC.IN EPILIMNION=4.5 MICROGRAM/ L,CONC.FACTOR=130,SCHOEHSEE	573
*LIVING SPRUCE NEEDLES	71 - 1090 MG/KG DRY WEUGHT	LITTER=131-318, UNDERLAYING SOIL= 290-323MG/KG DW,SOUTH SWEDEN	1129
*PATELLA VULGATA	1.1*10**4 (MICROGRAMS/G DRY MAT.)/ (MICROGRAMS/G SEAWATER)	17 OBS.,ALL AREAS AROUND THE BRITISH ISLES	1278
*PEDIASTRUM TETRAS	7.61 MG/KG DRY WT.	MN IN MEDIUM=0.0PPM,294K,LIGHT= 2500LUX	1682
*PEDIASTRUM TETRAS	27.40 MG/KG DRY WT.	MN IN MEDIUM=0.2PPM,294K,LIGHT= 2500LUX	1682

TABLE D303

CONCENTRATIONS : MN

ITEM:	VALUE:	CONDITIONS:	REF.:
*PEDIASTRUM TETRAS	39.99 MG/KG DRY WT.	MN IN MEDIUM=0.4PPM,294K,LIGHT= 2500LUX	1682
*PEDIASTRUM TETRAS	45.05 MG/KG DRY WT.	MN IN MEDIUM=0.5PPM,294K,LIGHT= 2500LUX	1682
*PORPHYRA UMBILICALIS	7.*10**3 (MICROGRAMS/G DRY MAT.)/ (MICROGRAMS/G SEAWATER)	20 OBS.,ALL AREAS AROUND THE BRITSH ISLES	1278
*RAIN	8.1 MICROGRAM/L	ENGLAND,YEARLY MEAN,1 STATION	893
*RAIN	6-17 MICROGRAM/L	SWEDEN,3 STATIONS	893
*SEDIMENT	490 PPM OF DRY WEIGHT	WATER CONC = 6 MICROGRAM/L, YEARLY MEAN, CARNON RIVER	48
*SEDIMENT	1688 PPM OF DRY WEIGHT	WATER CONC = 781 MICROGRAM/L, YEARLY MEAN, RED RIVER	48
*SEDIMENT	635 PPM OF DRY WEIGHT	WATER CONC = 69 MICROGRAM/L, YEARLY MEAN, RED RIVER	48
*SEDIMENT	3970 PPM OF DRY WEIGHT	WATER CONC = 367 MICROGRAM/L, YEARLY MEAN, GANNEL RIVER	48
*SEDIMENT	527 PPM OF DRY WEIGHT	WATER CONC = 16 MICROGRAM/L, YEARLY MEAN, GANNEL RIVER	48
*SEDIMENT	520 PPM OF DRY WEIGHT	WATER CONC = 8 MICROGRAM/L, YEARLY MEAN, FOWEY RIVER	48
*SEDIMENT	245 PPM OF DRY WEIGHT	WATER CONC = 9 MICROGRAM/L, YEARLY MEAN, FOWEY RIVER	48
*SEDIMENT	3245 PPM OF DRY WEIGHT	WATER CONC = 244 MICROGRAM/L, YEARLY MEAN, CARNON RIVER	48
*SEDIMENT	490 PPM OF DRY WEIGHT	MAX WATER CONC = 9 MICROGRAM/L, 1 YEARS MEASUREMENT, CARNON R	48
*SEDIMENT	1688 PPM OF DRY WEIGHT	MAX WATER CONC = 3000 MICROGRAM/L, 1 YEARS MEASUREMENT, RED R	48
*SEDIMENT	635 PPM OF DRY WEIGHT	MAX WATER CONC = 200 MICROGRAM/L, 1 YEARS MEASUREMENT, RED R	48
*SEDIMENT	3970 PPM OF DRY WEIGHT	MAX WATER CONC = 614 MICROGRAM/L, MAX CONC, GANNEL RIVER	48
*SEDIMENT	527 PPM OF DRY WEIGHT	MAX WATER CONC = 22 MICROGRAM/L, MAX CONC, GANNEL RIVER	48
*SEDIMENT	520 PPM OF DRY WEIGHT	MAX WATER CONC = 11 MICROGRAM/L, 1 YEARS MEASUREMENT, FOWEY R	48
*SEDIMENT	245 PPM OF DRY WEIGHT	MAX WATER CONC = 15 MICROGRAM/L, 1 YEARS MEASUREMENT, FOWEY R	48
*SEDIMENT	3245 PPM OF DRY WEIGHT	MAX WATER CONC = 515 MICROGRAM/L, 1 YEARS MEASUREMENT, CARNON R	48
*SEDIMENT	128.3 (RANGE 81.2 - 265.9) MG/KG DRY WEIGHT	UNPOLLUTED AREA, ESTUARY, URR WATER	1557
*SEDIMENT	333 (RANGE 221 - 587) MG/KG DRY WEIGHT	UNPOLLUTED AREA, ESTUARY, URR WATER	1558
*SEDIMENT	601 (RANGE 3.8 - 1156) MG/KG DRY WEIGHT	UNPOLLUTED AREA, ESTUARY, URR WATER	1558
*SEDIMENT	45.29 (36.92-51.26) MG/KG DRY WT.	800M ABOVE POLL.ENTRY(1),MN IN WATER=0.038MG/L,POLLUTED RIVER	1224
*SEDIMENT	50.39 (42.20-59.53) MG/KG DRY WT.	150M BELOW POLL.ENTRY(1),MN IN WATER=0.075MG/L,POLLUTED RIVER	1224
*SEDIMENT	130.11 (113.72-144.88) MG/KG DRY WT.	800 AND 50M BELOW POLL.(1)AND(2), MN IN WATER=0.099MG/L,POLLUTED RIVER	1224
*SEDIMENT	1600 MICROGRAM/G DRY WT.	CONC.IN EPILIMNION=4.5 MICROGRAM/ L,CONC.FACTOR=355000,SCHOEHSEE	573
*TOMATO, LEAF	5300 MG/KG DRY WT.	ADDED SUPERPHOSPHATE=2.4KG/M**3 SOIL,350PPM EXCHANGEABLE MN	635
*TOMATO, LEAF	5000 MG/KG DRY WT.	ADDED SUPERPHOSPHATE=9.6KG/M**3 SOIL,350PPM EXCHANGEABLE MN	635
*TOMATO, LEAF	19.0 MG/KG DRY WT.	0.50PPM MN,3.02M.EQ.KH2PO4/L CULTURE SOLUTION	635
*TOMATO, LEAF	21.5 MG/KG DRY WT.	0.50PPM MN,9.07M.EQ.KH2PO4/L CULTURE SOLUTION	635
*TOMATO, LEAF	185 MG/KG DRY WT.	5PPM MN,3.02M.EQ.KH2PO4/L CULTURE SOLUTION	635
*TOMATO, LEAF	876 MG/KG DRY WT.	25PPM MN,3.02M.EQ.KH2PO4/L CULTURE SOLUTION	635

TABLE D303

CONCENTRATIONS : MN			
ITEM:	VALUE:	CONDITIONS:	REF.:
*TOMATO, LEAF	762 MG/KG DRY WT.	25PPM MN,9.07M.EQ.KH2PO4/L CULTURE SOLUTION	635
*TOMATO, ROOT	99.7 MG/KG DRY WT.	0.50PPM MN,3.02M.EQ.KH2PO4/L CULTURE SOLUTION	635
*TOMATO, ROOT	95.9 MG/KG DRY WT.	0.50PPM MN,9.07M.EQ.KH2PO4/L CULTURE SOLUTION	635
*TOM¤TO, ROOT	990 MG/KG DRY WT.	5PPM MN,3.02M.EQ.KH2PO4/L CULTURE SOLUTION	635
*TOMATO, ROOT	3066 MG/KG DRY WT.	25PPM MN,3.02M.EQ.KH2PO4/L CULTURE SOLUTION	635
*TOMATO, ROOT	2371 MG/KG DRY WT.	25PPM MN,9.07M.EQ.KH2PO4/L CULTURE SOLUTION	635
*TOMATO, SHOOT	19.9 MG/KG DRY WT.	0.50PPM MN,3.02M.EQ.KH2PO4/L CULTURE SOLUTION	635
*TOMATO, SHOOT	23.5 MG/KG DRY WT.	0.50PPM MN,9.07M.EQ.KH2PO4/L CULTURE SOLUTION	635
*TOMATO, SHOOT	177 MG/KG DRY WT.	5PPM MN,3.02M.EQ.KH2PO4/L CULTURE SOLUTION	635
*TOMATO, SHOOT	832 MG/KG DRY WT.	25PPM MN,3.02M.EQ.KH2PO4/L CULTURE SOLUTION	635
*TOMATO, SHOOT	738 MG/KG DRY WT.	25PPM MN,9.07M.EQ.KH2PO4/L CULTURE SOLUTION	635
*TOMATO, WHOLE PLANT	37.1 MG/KG DRY WT.	0.50PPM MN,3.02M.EQ.KH2PO4/L CULTURE SOLUTION	635
*TOMATO, WHOLE PLANT	38.3 MG/KG DRY WT.	0.50PPM MN,9.07M.EQ.KH2PO4/L CULTURE SOLUTION	635
*TOMATO, WHOLE PLANT	339 MG/KG DRY WT.	5PPM MN,3.02M.EQ.KH2PO4/L CULTURE SOLUTION	635
*TOMATO, WHOLE PLANT	1320 MG/KG DRY WT.	25PPM MN,3.02M.EQ.KH2PO4/L CULTURE SOLUTION	635
*TOMATO, WHOLE PLANT	1153 MG/KG DRY WT.	25PPM MN,9.07M.EQ.KH2PO4/L CULTURE SOLUTION	635
*VEGETATION	380 MG/KG DRY WT.	CONC.IN PODZOL:F=150,H=100,A1A2=100,B1=500,B2=2000,B3=300,(F030)	1048
*WATER	0.0-160 MICROGRAM/L	GROUND- AND SURFACE WATER,U.S.A., 1961	893

TABLE D304

CONCENTRATIONS : MN-54			
ITEM:	VALUE:	CONDITIONS:	REF.:
*PARTICLE FRACTIONS	2.2 % OF TOTAL	SIZE=2.0-0.5 MM,WEIGHT=3.2 %,AREA=0.2 %,MARINE, JAPAN	1095
*PARTICLE FRACTIONS	3.2 % OF TOTAL	SIZE=0.5-0.25 MM,WEIGHT=8.5 %,AREA =1.4 %,MARINE, JAPAN	1095
*PARTICLE FRACTIONS	19.6 % OF TOTAL	SIZE=0.25-0.125 MM,WEIGHT=23 %,AREA=10.1 %,MARINE, JAPAN	1095
*PARTICLE FRACTIONS	12.5 % OF TOTAL	SIZE=0.125-0.075 MM,WEIGHT=16.9 %,AREA=10.4 %,MARINE, JAPAN	1095
*PARTICLE FRACTIONS	62.9 % OF TOTAL	SIZE BELOW 0.075 MM,WEIGHT=48.4 %,AREA=77.9 %,MARINE, JAPAN	1095

TABLE D305

CONCENTRATIONS : MO			
ITEM:	VALUE:	CONDITIONS:	REF.:
*ATMOSPHERE	10-30 NG/M**3	URBAN,U.S.A.	893
*ATMOSPHERE	0.1-3.2 NG/M**3	NOT URBAN,U.S.A.	893

TABLE D305

CONCENTRATIONS : MO			
ITEM:	VALUE:	CONDITIONS:	REF.:
*CONIFEROUS TREES, NEEDLES	2.3 (SD=+-0.6) MG/KG ASH	SILICIC ROCK=3.3, SOIL=3.4+-0.2 PPM, FINLAND,N=64	950
*CONIFEROUS TREES, NEEDLES	15 (SD=+-8) MG/KG ASH	ULTRABASIC ROCK=3.0 PPM, SOIL=3.0 PPM, FINLAND,N=10	950
*CONIFEROUS TREES, NEEDLES	2.2 (SD=+-1) MG/KG ASH	CALCAREOUS ROCK=3.2 PPM, SOIL=3.0 PPM, FINLAND,N=9	950
*CONIFEROUS TREES, NEEDLES	7.4 (SD=+-4) MG/KG ASH	CALCAREOUS ROCK=3.2 PPM, SOIL=3.0 PPM, FINLAND,N=9	950
*CONIFEROUS TREES, TWIGS	1.6 (SD=+-0.3) MG/KG ASH	SILICIC ROCK=3.3, SOIL=3.4+-0.2 PPM, FINLAND,N=64	950
*CONIFEROUS TREES, TWIGS	12 (SD=+-6) MG/KG ASH	ULTRABASIC ROCK=3.0 PPM, SOIL=3.0 PPM, FINLAND,N=10	950
*DECIDUOUS TREES AND SHRUBS, LEAVES	5.4 (SD=+-1) MG/KG ASH	SILICIC ROCK=3.3, SOIL=3.4+-0.2 PPM, FINLAND,N=75	950
*DECIDUOUS TREES AND SHRUBS, LEAVES	28 (SD=+-9) MG/KG ASH	ULTRABASIC ROCK=3.0 PPM, SOIL=3.0 PPM, FINLAND,N=14	950
*DECIDUOUS TREES AND SHRUBS, LEAVES	7.7 (SD=+-2.0) MG/KG ASH	CALCAREOUS ROCK=3.2 PPM, SOIL=3.0 PPM, FINLAND,N=23	950
*DECIDUOUS TREES AND SHRUBS, TWIGS	6.8 (SD=+-1.5) MG/KG ASH	SILICIC ROCK=3.3, SOIL=3.4+-0.2 PPM, FINLAND,N=75	950
*DECIDUOUS TREES AND SHRUBS, TWIGS	23 (SD=+-6) MG/KG ASH	ULTRABASIC ROCK=3.0 PPM, SOIL=3.0 PPM, FINLAND,N=14	950
*DECIDUOUS TREES AND SHRUBS, TWIGS	14.0 (SD=+-4) MG/KG ASH	CALCAREOUS ROCK=3.2 PPM, SOIL=3.0 PPM, FINLAND,N=23	950
*DRINKING WATER	1-29 MICROGRAM/L	SCHWEIZ	893
*FOOD	0.44 MG/KG DRY WT.	RICE	893
*FOOD	0.21 MG/KG DRY WT.	POTATO	893
*FOOD	2-4 MG/KG DRY WT.	SHEEP	893
*FOOD	0.14 MG/KG DRY WT.	CHICKEN	893
*HYPOLIMNION	0.30 MICROGRAM/L	CONC.IN EPILIMNION=0.21UG/L, CONC.FACTOR=1.4,SCHOEHSEE	573
*RED CLOVER	18.6 MG/KG DRY WT.	ADDED NAMO04=1.1KG/HA,PH=6.1; CONTROL=0.7MG/KG DRY WT.	1048
*RED CLOVER	35.6 MG/KG DRY WT.	ADDED NAMO04=1.1KG/HA,PH=6.5; CONTROL=3.4MG/KG DRY WT.	1048
*RYE-GRASS	2.0 MG/KG DRY WT.	ADDED NAMO04=1.1KG/HA,PH=6.1; CONTROL=0.8MG/KG DRY WT.	1048
*RYE-GRASS	3.4 MG/KG DRY WT.	ADDED NAMO04=1.1KG/HA,PH=6.5; CONTROL=1.1MG/KG DRY WT.	1048
*SEDIMENT	1.4 MICROGRAM/G DRY WT.	CONC.IN EPILIMNION=0.21 MICROGRAM/ L,CONC.FACTOR=7000,SCHOEHSEE	573
*WATER	0.68 MICROGRAM/L	SURFACE WATER,U.S.A.,MEAN	893
*WHITE CLOVER	0.4 MG/KG DRY WT.	ADDED LIMESTONE=0.63T/HA,SOIL-PH= 4.7,N.WALES	40
*WHITE CLOVER	0.4 MG/KG DRY WT.	ADDED LIMESTONE=2.5T/HA,SOIL-PH= 5.0,N.WALES	40
*WHITE CLOVER	0.6 MG/KG DRY WT.	ADDED LIMESTONE=5.0T/HA,SOIL-PH= 5.2,N.WALES	40
*WHITE CLOVER	1.0 MG/KG DRY WT.	ADDED LIMESTONE=15.0T/HA,SOIL-PH= 6.0,N.WALES	40
*WHITE CLOVER	2.3 MG/KG DRY WT.	ADDED LIMESTONE=0.63T/HA,ADDED NA2MO04=280G/HA,SOIL-PH=4.7, N.WALES	40
*WHITE CLOVER	2.3 MG/KG DRY WT.	ADDED LIMESTONE=2.5T/HA,ADDED NA2MO04=280G/HA,SOIL-PH=5.0, N.WALES	40
*WHITE CLOVER	4.0 MG/KG DRY WT.	ADDED LIMESTONE=5.0T/HA,ADDED NA2MO04=280G/HA,SOIL-PH=5.2, N.WALES	40
*WHITE CLOVER	6.3 MG/KG DRY WT.	ADDED LIMESTONE=15.0T/HA,ADDED NA2MO04=280G/HA,SOIL-PH=6.0, N.WALES	40

TABLE D306

CONCENTRATIONS : MO-99			
ITEM:	VALUE:	CONDITIONS:	REF.:
*NITELLA FLEXILIS	8.8 MICROGRAM/KG	WATER = 0.014 UG/L, UPTAKE FROM WATER, FERN LAKE	1433

TABLE D306

CONCENTRATIONS : MO-99			
ITEM:	VALUE:	CONDITIONS:	REF.:
*NITELLA FLEXILIS	130 MICROGRAM/KG	WATER = 3.3 MG/L, LAB.	1433

TABLE D307

CONCENTRATIONS : NI			
ITEM:	VALUE:	CONDITIONS:	REF.:
*ATMOSPHERE	1-118 NG/M**3	URBAN,U.S.A.	893
*ATMOSPHERE	0.6-12 NG/M**3	NOT URBAN,U.S.A.	893
*ATMOSPHERE	80-300 NG/M**3	SWEDEN,MEAN FROM 2 STATIONS,MAX= 300-800NG/M**3	893
*ATMOSPHERE	1-45 NG/M**3	ENGLAND,7 STATIONS	893
*DRINKING WATER	5.3 MICROGRAM/L (0.8-34)	GERMANY	893
*DRINKING WATER	1.1 MICROGRAM/L (0.8-1.5)	U.S.A.	893
*FOOD	0.3-3.3 MG/KG DRY WT.	VEGETABLES	893
*LIVING SPRUCE NEEDLES	0.7 - 3.1 MG/KG DRY WETGHT	LITTER=2.7-5.6, UNDERLAYING SOIL= 18-28MG/KG DW,SOUTH SWEDEN	1129
*OAT	8 MG/KG DRY WT.	SOIL-PH=5.3,NO NI ADDED	1226
*OAT	90 MG/KG DRY WT.	SOIL-PH=5.3,ADDED SLUDGE=37.5T/HA, ADDED NI=97PPM	1226
*OAT	120 MG/KG DRY WT.	SOIL-PH=5.3,ADDED SLUDGE=75T/HA, ADDED NI=194PPM	1226
*OAT	210 MG/KG DRY WT.	SOIL-PH=5.3,ADDED SLUDGE=150T/HA, ADDED NI=389PPM	1226
*OAT	4 MG/KG DRY WT.	SOIL-PH=6.8,NO NI ADDED	1226
*OAT	28 MG/KG DRY WT.	SOIL-PH=6.8,ADDED SLUDGE=37.5T/HA, ADDED NI=97PPM	1226
*OAT	50 MG/KG DRY WT.	SOIL-PH=6.8,ADDED SLUDGE=75T/HA, ADDED NI=194PPM	1226
*OAT	70 MG/KG DRY WT.	SOIL-PH=6.8,ADDED SLUDGE=150T/HA, ADDED NI=389PPM	1226
*PATELLA VULGATA	3.6*10**3 (MICROGRAMS/G DRY MAT.)/ (MICROGRAMS/G SEAWATER)	17 OBS.,ALL AREAS AROUND THE BRITISH ISLES	1278
*PORPHYRA UMBILICALIS	1.1*10**3 (MICROGRAMS/G DRY MAT.)/ (MICROGRAMS/G SEAWATER)	20 OBS.,ALL AREAS AROUND THE BRITISH ISLES	1278
*RAIN	6 MICROGRAM/L	ENGLAND,1 STATION,YEARLY MEAN	893
*RAIN	3 MICROGRAM/L	SWEDEN,3 STATIONS	893
*SEDIMENT	7.2 (RANGE 4.2 15.0) MG/KG DRY WEIGHT	UNPOLLUTED AREA, ESTUARY, URR WATER	1557
*SEDIMENT	9.2 (RANGE 6.1 - 14.3) MG/KG DRY WEIGHT	UNPOLLUTED AREA, ESTUARY, URR WATER	1558
*SEDIMENT	18.7 (RANGE 9.0 - 34.5) MG/KG DRY WEIGHT	UNPOLLUTED AREA, ESTUARY, URR WATER	1558
*SEDIMENT	16.0 (RANGE 10 - 22) MG/KG DRY WEIGHT	UNPOLLUTED AREA, RIVER,ILLINOIS RIVER	993
*SPRING WHEAT	2.5 MG/KG DRY WT.	SOIL-PH=5.1,NI ADDED TO SOIL=0PPM	1226
*SPRING WHEAT	2.2 MG/KG DRY WT.	SOIL-PH=5.5,NI ADDED TO SOIL=0PPM	1226
*SPRING WHEAT	1.0 MG/KG DRY WT.	SOIL-PH=6.5,NI ADDED TO SOIL=0PPM	1226
*SPRING WHEAT	4.5 MG/KG DRY WT.	SOIL-PH=5.1,NI ADDED TO SOIL=5PPM	1226
*SPRING WHEAT	2.5 MG/KG DRY WT.	SOIL-PH=5.5,NI ADDED TO SOIL=5PPM	1226
*SPRING WHEAT	0.75 MG/KG DRY WT.	SOIL-PH=6.5,NI ADDED TO SOIL=5PPM	1226
*SPRING WHEAT	0.5 MG/KG DRY WT.	SOIL-PH=7.5,NI ADDED TO SOIL=5PPM	1226
*SPRING WHEAT	3.0 MG/KG DRY WT.	SOIL-PH=5.1,NI ADDED TO SOIL=10PPM	1226
*SPRING WHEAT	3.7 MG/KG DRY WT.	SOIL-PH=5.5,NI ADDED TO SOIL=10PPM	1226
*SPRING WHEAT	2.2 MG/KG DRY WT.	SOIL-PH=6.5,NI ADDED TO SOIL=10PPM	1226
*SPRING WHEAT	0.5 MG/KG DRY WT.	SOIL-PH=7.5,NI ADDED TO SOIL=10PPM	1226

TABLE D307

CONCENTRATIONS : NI			
ITEM:	VALUE:	CONDITIONS:	REF.:
*SPRING WHEAT	8.0 MG/KG DRY WT.	SOIL-PH=5.1,NI ADDED TO SOIL=20PPM	1226
*SPRING WHEAT	4.7 MG/KG DRY WT.	SOIL-PH=5.5,NI ADDED TO SOIL=20PPM	1226
*SPRING WHEAT	2.0 MG/KG DRY WT.	SOIL-PH=6.5,NI ADDED TO SOIL=20PPM	1226
*SPRING WHEAT	10.0 MG/KG DRY WT.	SOIL-PH=5.1,NI ADDED TO SOIL=40PPM	1226
*SPRING WHEAT	6.5 MG/KG DRY WT.	SOIL-PH=5.5,NI ADDED TO SOIL=40PPM	1226
*SPRING WHEAT	2.75 MG/KG DRY WT.	SOIL-PH=6.5,NI ADDED TO SOIL=40PPM	1226
*SPRING WHEAT	0.75 MG/KG DRY WT.	SOIL-PH=7.5,NI ADDED TO SOIL=40PPM	1226
*SPRING WHEAT	74.0 MG/KG DRY WT.	SOIL-PH=5.1,NI ADDED TO SOIL=80PPM	1226
*SPRING WHEAT	17.2 MG/KG DRY WT.	SOIL-PH=5.5,NI ADDED TO SOIL=80PPM	1226
*SPRING WHEAT	3.0 MG/KG DRY WT.	SOIL-PH=6.5,NI ADDED TO SOIL=80PPM	1226
*SPRING WHEAT	1.25 MG/KG DRY WT.	SOIL-PH=7.5,NI ADDED TO SOIL=80PPM	1226
*SPRING WHEAT	105 MG/KG DRY WT.	SOIL-PH=5.5,NI ADDED TO SOIL= 160PPM	1226
*SPRING WHEAT	8.25 MG/KG DRY WT.	SOIL-PH=6.5,NI ADDED TO SOIL= 160PPM	1226
*SPRING WHEAT	3.0 MG/KG DRY WT.	SOIL-PH=7.5,NI ADDED TO SOIL= 160PPM	1226
*VEGETATION	1.1 MG/KG DRY WT.	CONC.IN PODZOL:F=8,H=5,A1A2=20,B1= 15,B2=20,B3=30,(?043)	1048
*WATER	0.2 MICROGRAM/L	BALTIC SEA,GOTLANDSDYBET:230M, 1970-71	893
*WATER	7.75 MICROGRAM/L	BALTIC SEA,FLADEN:50M,1970-71	893
*WATER	5-56 MICROGRAM/L	RIVERS,U.S.A.	893
*ZOOPLANKTON	< 2.75 MG/KG DRY WT.(MEAN)	19ST.,N.Y.BIGHT,CONC.IN PHYTOPL. (LONG ISLAND)=2.62MG/KG DRY WT. (MEAN)	565

TABLE D308

CONCENTRATIONS : NITRATE-N			
ITEM:	VALUE:	CONDITIONS:	REF.:
*AGRICULTURAL CROPLAND	0.4 MG/L WATER	GENERAL CHARACTERISTICS	364
*ANIMAL FEEDLOT, RUNOFF	10-23 MG/L WATER	GENERAL CHARACTERISTICS	364
*FORESTED LAND	0.1-1.3 MG/L WATER	GENERAL CHARACTERISTICS	364
*PRECIPITATION	0.14-1.1 MG/L WATER	GENERAL CHARACTERISTICS	364

TABLE D309

CONCENTRATIONS : NUCLEIC ACID			
ITEM:	VALUE:	CONDITIONS:	REF.:
*MYCOPLASMA GALLISEPTICUM	12 % OF DRY WEIGHT	NONE	1067
*SACCHAROMYCES CEREVISIAE	12 % OF DRY WEIGHT	NONE	574
*TORULOPSIS UTILIS	4.6 % OF DRY WEIGHT	NONE	574

TABLE D310

CONCENTRATIONS : O,P`DDT			
ITEM:	VALUE:	CONDITIONS:	REF.:
*SOIL	1.0 PPM	MIRPUR,DACCA,10G	1101

TABLE D310

CONCENTRATIONS : O,P'DDT			
ITEM:	VALUE:	CONDITIONS:	REF.:
*SOIL	0.1 PPM	JOUDABPUR(A),DACCA,10G	1101
*SOIL	0.5 PPM	JOYDABPUR(B),DACCA,10G	1101
*SOIL	0.3 PPM	DENRA(A),DACCA,10G	1101
*SOIL	0.5 PPM	DEMRA(B),DACCA,10G	1101
*SOIL	0.4 PPM	DEMRA(C),DACCA,10G	1101

TABLE D311

CONCENTRATIONS : OCTACHLORODIBENZO-P-DIOXIN			
ITEM:	VALUE:	CONDITIONS:	REF.:
*RAT, ADIPOSE TISSUE	0.04 (SE=+-0.04) MICROGRAM	INGESTED =120.7 UG, 2 W, 106 G ANIM.	1715
*RAT, FECES	13.9 (SE=+-0.7) MICROGRAM	INGESTED = 22.7 UG, 2 W, 106 G ANIM.	1715
*RAT, FECES	45.2 (SE=+-1.7) MICROGRAM	INGESTED =120.7 UG, 2 W, 106 G ANIM.	1715
*RAT, LIVER	0.46 (SE=+-0.09) MICROGRAM	INGESTED = 22.7 UG, 2 W, 106 G ANIM.	1715
*RAT, LIVER	1.35 (SE=+-0.19) MICROGRAM	INGESTED =120.7 UG, 2 W, 106 G ANIM.	1715

TABLE D312

CONCENTRATIONS : ORGANIC HG			
ITEM:	VALUE:	CONDITIONS:	REF.:
*ANCHOVY	76 % OF TOTAL HG	MONTEREY BAY, CALIF, TOTAL=38 NG/G WW, FIRST YEAR CLASS	841
*ANCHOVY	88 % OF TOTAL HG	MONTEREY BAY, CALIF, TOTAL HG = 75 NG/G WW, 3. YEAR CLASS	841
*ANCHOVY	100 % OF TOTAL HG	MONTEREY BAY, CALIF, TOTAL HG = 47 NG/G WW, 4. YEAR CLASS	841
*PHYTOPLANKTON	67 % OF TOTAL HG	MONTEREY BAY, CALIF, 24 MAY, TOTAL =5.7 NG/G WW	841
*PHYTOPLANKTON	34 % OF TOTAL HG	MONTEREY BAY, CALIF, 8 JUNE, TOTAL =5.9 NG/G WW	841
*PHYTOPLANKTON	12 % OF TOTAL HG	MONTEREY BAY, CALIF, 28 JUN, TOTAL =15.2 NG/G WW	841
*PHYTOPLANKTON	42 % OF TOTAL HG	MONTEREY BAY, CALIF, 15 AUG, TOTAL =10.6 NG/G WW	841

TABLE D313

CONCENTRATIONS : P			
ITEM:	VALUE:	CONDITIONS:	REF.:
*TOMATO, LEAF	0.44% DRY WT.	ADDED SUPERPHOSPHATE=2.4KG/M**3 SOIL	635
*TOMATO, LEAF	0.96% DRY WT.	ADDED SUPERPHOSPHATE=9.6KG/M**3 SOIL	635

TABLE D314

CONCENTRATIONS : P,P DDT			
ITEM:	VALUE:	CONDITIONS:	REF.:
*SOIL	3.0 PPM	MIRPUR,DACCA,10G	1101

TABLE D314

CONCENTRATIONS : P,P'DDT			
ITEM:	VALUE:	CONDITIONS:	REF.:
*SOIL	1.8 PPM	JOYDABPUR(A),DACCA,10G	1101
*SOIL	2.0 PPM	JOYDABPUR(B),DACCA,10G	1101
*SOIL	1.3 PPM	DEMRA(A),DACCA,10G	1101
*SOIL	1.6 PPM	DEMRA(B),DACCA,10G	1101
*SOIL	2.0 PPM	DEMRA(C),DACCA,10G	1101

TABLE D315

CONCENTRATIONS : PARTICULATE PU-237			
ITEM:	VALUE:	CONDITIONS:	REF.:
*RIVER WATER	34 %	RETAINED BY A 1.2 UM FILTER, PU(+ 6)	1087
*RIVER WATER	48 %	RETAINED BY A 0.45 UM FILTER, PU(+ 6)	1087
*RIVER WATER	57 %	RETAINED BY A 0.22 UM FILTER, PU(+ 6)	1087
*RIVER WATER	56 %	RETAINED BY A 0.10 UM FILTER, PU(+ 6)	1087
*SEAWATER	34 %	RETAINED BY A 1.2 UM FILTER, PU(+ 6)	1087
*SEAWATER	40 %	RETAINED BY A 0.45 UM FILTER, PU(+ 6)	1087
*SEAWATER	41 %	RETAINED BY A 0.22 UM FILTER, PU(+ 6)	1087
*SEAWATER	40 %	RETAINED BY A 0.10 UM FILTER, PU(+ 6)	1087

TABLE D316

CONCENTRATIONS : PB			
ITEM:	VALUE:	CONDITIONS:	REF.:
*ALISMA PLANTAGO AQUATICA, OLD LEAVES	4.92 MG/KG DRY WT.	800M ABOVE POLL.ENTRY(1),PB IN WATER=NOT DETECT.,WHOLE LEAVES	1224
*ALISMA PLANTAGO AQUATICA, OLD LEAVES	3.03 MG/KG DRY WT.	800M ABOVE POLL.ENTRY(1),PB IN WATER=NOT DETECT.,PROCESSED LEAVES	1224
*ALISMA PLANTAGO AQUATICA, OLD LEAVES	38.10 MG/KG DRY WT.	150M BELOW POLL.ENTRY(1),PB IN WATER=0.068MG/L WATER,WHOLE LEAVES	1224
*ALISMA PLANTAGO AQUATICA, OLD LEAVES	21.72 MG/KG DRY WT.	150M BELOW POLL.ENTRY(1),PB IN WATER=0.068MG/L WATER,PROCESSED LEAVES	1224
*ALISMA PLANTAGO AQUATICA, OLD LEAVES	7.36 MG/KG DRY WT.	800 AND 50M BELOW POLL.(1)AND(2), PB IN WATER=0.060MG/L WATER,WHOLE LEAVE	1224
*ALISMA PLANTAGO AQUATICA, OLD LEAVES	4.72 MG/KG DRY WT.	800 AND 50M BELOW POLL.(1)AND(2), PB IN WATER=0.060MG/L WATER, PROCESSED L	1224
*ALISMA PLANTAGO AQUATICA, YOUNG LEAVES	14.37 MG/KG DRY WT.	800M ABOVE POLL.ENTRY(1),PB IN WATER=NOT DETECT.,WHOLE LEAVES	1224
*ALISMA PLANTAGO AQUATICA, YOUNG LEAVES	14.25 MG/KG DRY WT.	800M ABOVE POLL.ENTRY(1),PB IN WATER=NOT DETECT.,PROCESSED LEAVES	1224
*ALISMA PLANTAGO AQUATICA, YOUNG LEAVES	23.61 MG/KG DRY WT.	150M BELOW POLL.ENTRY(1),PB IN WATER=0.068MG/L WATER,WHOLE LEAVES	1224
*ALISMA PLANTAGO AQUATICA, YOUNG LEAVES	18.77 MG/KG DRY WT.	150M BELOW POLL.ENTRY(1),PB IN WATER=0.068MG/L WATER,PROCESSED LEAVES	1224

TABLE D316

CONCENTRATIONS : PB

ITEM:	VALUE:	CONDITIONS:	REF.:
*ALISMA PLANTAGO AQUATICA, YOUNG LEAVES	18.25 MG/KG DRY WT.	800 AND 50M BELOW POLL.(1)AND(2), PB IN WATER=0.060MG/L WATER,WHOLE LEAVE	1224
*ALISMA PLANTAGO AQUATICA, YOUNG LEAVES	9.81 MG/KG DRY WT.	800 AND 50M BELOW POLL.(1)AND(2), PB IN WATER=0.060MG/L WATER, PROCESSED L	1224
*ATMOSPHERE	3.4-21.2 MICROGRAM/M**3	STOCKHOLM	893
*ATMOSPHERE	400-7600 NG/M**3	BIG CITIES,1960-70	893
*BRYUM PSEUDOTRIQUETRUM	3806 PPM	PB IN SOIL=1640PPM,SLATER MINE, FOOLOW,DERBYSHIRE,GB	1430
*BRYUM PSEUDOTRIQUETRUM	16670 PPM	PB IN SOIL=11104PPM,BLEA BECK JUNCTION,W.YORKSHIRE,GB	1430
*BRYUM PSEUDOTRIQUETRUM	1079 PPM	PB IN SOIL=1150PPM,SAND HAW MINE, W.YORKSHIRE,GB	1430
*BUCCINUM UNDATUM	41 MG PB/KG DRY WEIGHT	SEDIMENT = 182 PPM, FIRTH OF CLYDE	593
*BUCCINUM UNDATUM	5.4 MG PB/KG DRY WEIGHT	SEDIMENT =50 PPM, IRISH SEA	593
*BULK PRECIPITATION	13.4 MICROGRAMS/L MEAN WT.	14JULY-7NOVEMBER-1971,NEW HAMPSHIRE,USA	1393
*CABBAGE, PLANT	0.81 PPM DRY MATTER	15 PLANTS,RURAL AREA	1283
*CABBAGE, PLANT	0.73 PPM DRY MATTER	15 PLANTS,URBAN AREA	1283
*CAMPTOTHECIUM LUTESCENS	164 PPM	PB IN SOIL=80PPM,BLEA BECK HEAD, W.YORKSHIRE,GB	1430
*CLADOPHORA GLOMERATA	CONC FAC (=CONC IN ORG/CONC IN WATER)=16*10**3 (MICROGRAM/G DW) /(NG/ML)	WATER FROM DEADMAN BAY (L.ONTARIO) =4.2+-1.9, (CONC. IN NG/ML+-S.D.)	806
*CLADOPHORA GLOMERATA	CONC FAC (=CONC IN ORG/CONC IN WATER)=20*10**3 (MICROGRAM/G DW) /(NG/ML)	WATER FROM MAIN DUCK ISL. (L.ONTARIO)=0.5+-0.3, (CONC. IN NG/ML+-S.D.)	806
*CORNICULARIA MURICATA	12045 PPM	PB IN SOIL=2570PPM,BLEA BECK HEAD, W.YORKSHIRE,GB	1430
*CRANGON ALLMANI	32 MG PB/KG DRY WEIGHT	SEDIMENT = 182 PPM, FIRTH OF CLYDE	593
*CRANGON ALLMANI	17 MG PB/KG DRY WEIGHT	SEDIMENT = 50 PPM, IRISH SEA	593
*CRATAEGUS MONOGYNA, UNWASHED LEAVES	509 PPM DRY WT.	AVONMOUTH,BRISTOL,GB	936
*DICRANELLA VARIA	1325 PPM	PB IN SOIL=975PPM,SPEED MINE, FOOLOW,DERBYSHIRE,GB	1430
*DICRANELLA VARIA	10460 PPM	PB IN SOIL=7625PPM,SLATER MINE, FOOLOW,DERBYSHIRE,GB	1430
*DICRANELLA VARIA	11950 PPM	PB IN SOIL=8490PPM,PASTYREMINE, FOOLOW,DERBYSHIRE,GB	1430
*DICRANELLA VARIA	540 PPM	PB IN SOIL=1108PPM,BLYTHE WOOD, ALPORT,DERBYSHIRE,GB	1430
*DICRANELLA VARIA	7 PPM	PB IN SOIL=4PPM,BOTANIC GARDEN, COTTINGHAM,W.YORKSHIRE,GB	1430
*DICRANELLA VARIA	42300 PPM	PB IN SOIL=1870PPM,SLATER MINE, FOOLOW,DERBYSHIRE,GB	1430
*DICRANELLA VARIA	60450 PPM	PB IN SOIL=2400PPM,PASTYREMINE, FOOLOW,DERBYSHIRE,GB	1430
*ENCALYPTA VULGARIS	2423 PPM	PI IN SOIL =2423PPM,ALPORT CUPOLA SLAG HEAP,DERBYSHIRE,GB	1430
*FOOD	0.2-5.7 MICROGRAM/G	KIDNEY	893
*FOOD	0.2-2.8 MICROGRAM/G	LIVER	893
*HOMO SAPIENS, BLOOD	1.46 (SD=0.48) MICRO M	PB IN TAP WATER,1ST FLUSH=3.16, RUNNING WAT=1.86 MICRO M PB,N=27, GB	1060
*HUMAN ALA DEHYDRATASE ACTIVITY	28.1 (SD=7.5) MICRO MOL ALA/MIN/ML RBC	PB IN TAP WATER,1ST FLUSH=3.16, RUNNING WAT=1.86 MICRO M PB,N=27, GB	1060
*NUCULA SULCATA	50 MG PB/KG DRY WEIGHT	SEDIMENT = 182 PPM, FIRTH OF CLYDE	593
*PANDALUS MONTAGUI	31 MG PB/KG DRY WEIGHT	SEDIMENT = 182 PPM, FIRTH OF CLYDE	593
*PANDALUS MONTAGUI	24 MG PB/KG DRY WEIGHT	SEDIMENT = 50 PPM, IRISH SEA	593
*PATELLA VULGATA	5.5*10**3 (MICROGRAMS/G DRY MAT.)/ (MICROGRAMS/G SEAWATER)	12 OBS.,ALL AREAS AROUND THE BRITISH ISLES	1278
*PEDIASTRUM TETRAS	2.749 MG/KG DRY WT.	PB IN MEDIUM=0.0PPM,294K,LIGHT= 2500LUX	1682

TABLE D316

CONCENTRATIONS : PB

ITEM:	VALUE:	CONDITIONS:	REF.:
*PEDIASTRUM TETRAS	2.650 MG/KG DRY WT.	PB IN MEDIUM=0.2PPM,294K,LIGHT= 2500LUX	1682
*PEDIASTRUM TETRAS	4.753 MG/KG DRY WT.	PB IN MEDIUM=0.5PPM,294K,LIGHT= 2500LUX	1682
*PEDIASTRUM TETRAS	3.336 MG/KG DRY WT.	PB IN MEDIUM=1.0PPM,294K,LIGHT= 2500LUX	1682
*PEDIASTRUM TETRAS	3.747 MG/KG DRY WT.	PB IN MEDIUM=1.5PPM,294K,LIGHT= 2500LUX	1682
*PELTIGERA,CANINA	9943 PPM	PB IN SOIL=1654PPM,BLEA BECK HEAD, W.YORKSHIRE,GB	1430
*PELTIGERA,CANINA	5385 PPM	PB IN SOIL=2950PPM,ALPORT CUPOLA SLAG HEAP,DERBYSHIRE,GB	1430
*PHILONOTIS FONTANA(NEW)	432 PPM	PB IN SOIL=940PPM,BLEA BECK JUNCTION,W.YORKSHIRE,GB	1430
*PHILONOTIS FONTANA(NEW)	237 PPM	PB IN SOIL=1640PPM,SLATER MINE, FOOLOW,DERBYSHIRE,GB	1430
*PHILONOTIS FONTANA(OLD)	10015 PPM	PB IN SOIL=940PPM,BLEA BECK JUNCTION,W.YORKSHIRE,GB	1430
*PHILONOTIS FONTANA(OLD)	8750 PPM	PB IN SOIL=1640PPM,SLATER MINE, FOOLOW,DERBYSHIRE,GB	1430
*PORPHYRA UMBILICALIS	2.*10**3 (MICROGRAMS/G DRY MAT.)/ (MICROGRAMS/G SEAWATER)	15 OBS.,ALL AREAS AROUND THE BRITISH ISLES	1278
*QUERCUS ROBUR, UNWASHED LEAVES	753 PPM DRY WT.	AVONMOUTH,BRISTOL,GB	936
*RADISH, ROOT	2.0 MG/KG DRY WEIGHT	AIR = 0.03 UG/M3, SOIL = 27 PPM (HNO3-EXT.)	339
*RADISH, ROOT	1.8 MG/KG DRY WEIGHT	AIR = 0.96 UG/M3, SOIL = 27 PPM (HNO3-EXT.)	339
*RADISH, ROOT	1.9 MG/KG DRY WEIGHT	AIR = 0.09 UG/M3, SOIL = 17 PPM (HNO3-EXT.)	339
*RADISH, ROOT	2.0 MG/KG DRY WEIGHT	AIR = 1.45 UG/M3, SOIL = 17 PPM (HNO3-EXT.)	339
*RADISH, ROOT	0.8 MG/KG DRY WEIGHT	AIR = 1.07 UG/M3, SOIL = 27 PPM (HNO3-EXT.)	339
*RADISH, ROOT	0.8 MG/KG DRY WEIGHT	AIR = 1.71 UG/M3, SOIL = 45 PPM (HNO3-EXT.)	339
*RADISH, ROOT	0.7 MG/KG DRY WEIGHT	AIR = 2.32 UG/M3, SOIL = 82 PPM (HNO3-EXT.)	339
*RAIN	90 UG/L	SWEDEN,1972	893
*RAT, KIDNEY	34.2 % OF TOTAL INTRACELLULAR AMOUNT	NUCLEI,36 RATS,24 H AFTER INJEC OF 18 UG PB(+2)/275 G	1366
*RAT, KIDNEY	26.1 % OF TOTAL INTRACELLULAR AMOUNT	MITICHONDRIA,36 RATS,24 H AFTER INJEC OF 18 UG PB(+2)/275 G	1366
*RAT, KIDNEY	5.3 % OF TOTAL INTRACELLULAR AMOUNT	LYSOSOMES,36 RATS,24 H AFTER INJEC OF 18 UG PB(+2)/275 G	1366
*RAT, KIDNEY	5.0 % OF TOTAL INTRACELLULAR AMOUNT	MICROSOMES,36 RATS,24 H AFTER INJEC OF 18 UG PB(+2)/275 G	1366
*RAT, KIDNEY	32.1 % OF TOTAL INTRACELLULAR AMOUNT	SOLUBLE FRAC.,36 RATS,24 H AFTER INJEC OF 18 UG PB(+2)/275 G	1366
*RAT, LIVER	28.3 % OF TOTAL INTRACELLULAR AMOUNT	NUCLEI,36 RATS,24 H AFTER INJECTION OF 18 MICRO G PB(+2)	1366
*RAT, LIVER	9.5 % OF TOTAL INTRACELLULAR AMOUNT	MITOCHONDRIA,36 RATS,24 H AFTER INJEC OF 18 UG PB(+2)/275 G	1366
*RAT, LIVER	10.2 % OF TOTAL INTRACELLULAR AMOUNT	LYSOSOMES,36 RATS,24 H AFTER INJEC OF 18 UG PB(+2)/275 G	1366
*RAT, LIVER	10.9 % OF TOTAL INTRACELLULAR AMOUNT	MICROSOMES,36 RATS,24 H AFTER INJEC OF 18 UG PB(+2)/275 G	1366
*RAT, LIVER	39.2 % OF TOTAL INTRACELLULAR AMOUNT	SOLUBLE FRAC.,36 RATS,24 H AFTER INJEC OF 18 UG PB(+2)/275 G	1366
*SALIX ALBA, UNWASHED LEAVES	260 PPM DRY WT.	AVONMOUTH,BRISTOL,GB	936
*SEDIMENT	89 PPM OF DRY WEIGHT	WATER CONC = <1 MICROGRAM/L, YEARLY MEAN, CARNON RIVER	48
*SEDIMENT	367 PPM OF DRY WEIGHT	WATER CONC = 7 MICROGRAM/L, YEARLY MEAN, RED RIVER	48
*SEDIMENT	65 PPM OF DRY WEIGHT	WATER CONC = 2 MICROGRAM/L, YEARLY MEAN, RED RIVER	48
*SEDIMENT	4360 PPM OF DRY WEIGHT	WATER CONC = 73 MICROGRAM/L, YEARLY MEAN, GANNEL RIVER	48

TABLE D316

CONCENTRATIONS : PB

ITEM:	VALUE:	CONDITIONS:	REF.:
*SEDIMENT	82 PPM OF DRY WEIGHT	WATER CONC = 3 MICROGRAM/L, YEARLY MEAN, GANNEL RIVER	48
*SEDIMENT	98 PPM OF DRY WEIGHT	WATER CONC = 2 MICROGRAM/L, YEARLY MEAN, FOWEY RIVER	48
*SEDIMENT	46 PPM OF DRY WEIGHT	WATER CONC = 2 MICROGRAM/L, YEARLY MEAN, FOWEY RIVER	48
*SEDIMENT	2750 PPM OF DRY WEIGHT	WATER CONC = 1107 MICROGRAM/L, YEARLY MEAN, CARNON RIVER	48
*SEDIMENT	89 PPM OF DRY WEIGHT	MAX WATER CONC = 1.5 MICROGRAM/L, 1 YEARS MEASUREMENT, CARNON R	48
*SEDIMENT	367 PPM OF DRY WEIGHT	MAX WATER CONC = 15 MICROGRAM/L, 1 YEARS MEASUREMENT, RED R	48
*SEDIMENT	65 PPM OF DRY WEIGHT	MAX WATER CONC = 3 MICROGRAM/L, 1 YEARS MEASUREMENT, RED R	48
*SEDIMENT	4360 PPM OF DRY WEIGHT	MAX WATER CONC MICROGRAM/L, MAX CONC, GANNEL RIVER	48
*SEDIMENT	82 PPM OF DRY WEIGHT	MAX WATER CONC = 8 MICROGRAM/L, MAX CONC, GANNEL RIVER	48
*SEDIMENT	98 PPM OF DRY WEIGHT	MAX WATER CONC = 3 MICROGRAM/L, 1 YEARS MEASUREMENT, FOWEY R	48
*SEDIMENT	46 PPM OF DRY WEIGHT	MAX WATER CONC = 4 MICROGRAM/L, 1 YEARS MEASUREMENT, FOWEY R	48
*SEDIMENT	2750 PPM OF DRY WEIGHT	MAX WATER CONC = 2000 MICROGRAM/L, 1 YEARS MEASUREMENT, CARNON R	48
*SEDIMENT	31.2 (RANGE 21.3-65.7) MG/KG DRY WEIGHT	UNPOLLUTED AREA, ESTUARY, URR WATER	1557
*SEDIMENT	21.5 (RANGE 11.5 - 34.1) MG/KG DRY WEIGHT	UNPOLLUTED AREA, ESTUARY, URR WATER	1558
*SEDIMENT	23.2 (RANGE 12.9 - 32.9) MG/KG DRY WEIGHT	UNPOLLUTED AREA, ESTUARY, URR WATER	1558
*SEDIMENT	17 (RANGE 13 - 27) MG/KG DRY WEIGHT	UNPOLLUTED AREA, RIVER, ILLINOIS RIVER	993
*SEDIMENT	72.11 (64.36-78.65) MG/KG DRY WT.	800M ABOVE POLL.ENTRY(1),PB IN WATER=NOT DETECT.,POLLUTED RIVER	1224
*SEDIMENT	79.93 (75.50-84.57) MG/KG DRY WT.	150M BELOW POLL.ENTRY(1),PB IN WATER=0.068MG/L WATER,POLLUTED RIVER	1224
*SEDIMENT	53.08 (49.22-59.50) MG/KG DRY WT.	800 AND 50M BELOW POLL.(1)AND(2), PB IN WATER=0.060MG/L WATER, POLLUTED RI	1224
*SOIL (SEDIMENT)	42. PPM DRY SOIL	RANGE 12-72PPM,RIVER RHEIDOL, CARDIGANSHIRE,GB	327
*STREET DUST	1840+-510PPM DRY WEIGHT	SD, AREA=LAMBETH(=CENTRAL), N=15	375
*STREET DUST	1330+-660PPM DRY WEIGHT	SD, AREA=ISLINGTON(=CENTRAL), N=15	375
*STREET DUST	1040+-300PPM DRY WEIGHT	SD, AREA=WIMBLEDON(=MIDURBAN), N=16	375
*STREET DUST	1040+-450PPM DRY WEIGHT	SD, AREA=EDGWARE	375
*STREET DUST	920+-500PPM DRY WEIGHT	SD, AREA=HAROLD WOOD(=OUTER URBAN), N=18	375
*STREET DUST	35+-15PPM DRY WEIGHT	SD, AREA=WALTON ON THE HILL(RURAL), N=7	375
*STREET DUST	1530+-600PPM DRY WEIGHT	SD, ROAD TYPE=MAIN SD, ROADS, GUTTER	375
*STREET DUST	1460+-520PPM DRY WEIGHT	SD, ROAD TYPE=MAIN SD, ROADS, PAVEMENT	375
*STREET DUST	1030+-450PPM DRY WEIGHT	SD, ROAD TYPE=SIDE SD, ROADS, GUTTER	375
*STREET DUST	900+-420PPM DRY WEIGHT	SD, ROAD TYPE=SIDE SD, ROADS, PAVEMENT	375
*TOMATO, ROOT	51 MG/KG DRY WEIGHT	SOIL = 164 PPM	1071
*TOMATO, ROOT	418 MG/KG DRY WEIGHT	SUBSTRATE = SAND = 1 PPM PB	1071
*VEGETATION	7.6 MG/KG DRY WT.	CONC.IN PODZOL:F=200,H=100,A1A2=20,B1=15,B2=20,B3=10,(F030)	1048
*WATER	< 0.5 UG/L	SWEDEN	893
*WATER	19-21 UG/L	WASTE WATER FROM STOCKHOLM	893

TABLE D316

CONCENTRATIONS : PB

ITEM:	VALUE:	CONDITIONS:	REF.:
*ZALOPHUS CALIFORNIANUS, AORTA	2.2 PPM DRY WT.	6 SEA LIONS FROM SAN LUIS OBISPO COUNTY,CAL.,USA	163
*ZALOPHUS CALIFORNIANUS, CERABRAL HEMISPHERE	3.8 PPM DRY WT.	6 SEA LIONS FROM SAN LUIS OBISPO COUNTY,CAL.,USA	163
*ZALOPHUS CALIFORNIANUS, CEREBELLUM	2.7 PPM DRY WT.	6 SEA LIONS FROM SAN LUIS OBISPO COUNTY,CAL.,USA	163
*ZALOPHUS CALIFORNIANUS, FAT	0.3 PPM DRY WT.	6 SEA LIONS FROM SAN LUIS OBISPO COUNTY,CAL.,USA	163
*ZALOPHUS CALIFORNIANUS, FEMUR	20.6 PPM DRY WT.	6 SEA LIONS FROM SAN LUIS OBISPO COUNTY,CAL.,USA	163
*ZALOPHUS CALIFORNIANUS, GONAD	3 PPM DRY WT.	6 SEA LIONS FROM SAN LUIS OBISPO COUNTY,CAL.,USA	163
*ZALOPHUS CALIFORNIANUS, HAIR-SKIN	1.6 PPM DRY WT.	6 SEA LIONS FROM SAN LUIS OBISPO COUNTY,CAL.,USA	163
*ZALOPHU¤ CALIFORNIANUS, HUMERRUS	34.2 PPM DRY WT.	6 SEA LIONS FROM SAN LUIS OBISPO COUNTY,CAL.,USA	163
*ZALOPHUS CALIFORNIANUS, KIDNEY CORTEX	2.4 PPM DRY WT.	6 SEA LIONS FROM SAN LUIS OBISPO COUNTY,CAL.,USA	163
*ZALOPHUS CALIFORNIANUS, KIDNEY MEDULLA	2.0 PPM DRY WT.	6 SEA LIONS FROM SAN LUIS OBISPO COUNTY,CAL.,USA	163
*ZALOPHUS CALIFORNIANUS, LARGE INTESTINE	1.5 PPM DRY WT.	6 SEA LIONS FROM SAN LUIS OBISPO COUNTY,CAL.,USA	163
*ZALOPHUS CALIFORNIANUS, LIVER	1.3 PPM DRY WT.	6 SEA LIONS FROM SAN LUIS OBISPO COUNTY,CAL.,USA	163
*ZALOPHUS CALIFORNIANUS, LUNG	2.0 PPM DRY WT.	6 SEA LIONS FROM SAN LUIS OBISPO COUNTY,CAL.,USA	163
*ZALOPHUS CALIFORNIANUS, MUSCLE	1.1 PPM DRY WT.	6 SEA LIONS FROM SAN LUIS OBISPO COUNTY,CAL.,USA	163
*ZALOPHUS CALIFORNIANUS, RIB	8.7 PPM DRY WT.	6 SEA LIONS FROM SAN LUIS OBISPO COUNTY,CAL.,USA	163
*ZALOPHUS CALIFORNIANUS, SMALL INTESTINE	3.4 PPM DRY WT.	6 SEA LIONS FROM SAN LUIS OBISPO COUNTY,CAL.,USA	163
*ZALOPHUS CALIFORNIANUS, SPLEEN	1.3 PPM DRY WT.	6 SEA LIONS FROM SAN LUIS OBISPO COUNTY,CAL.,USA	163
*ZALOPHUS CALIFORNIANUS, STOMAC	0.5 PPM DRY WT.	6 SEA LIONS FROM SAN LUIS OBISPO COUNTY,CAL.,USA	163
*ZALOPHUS CALIFORNIANUS, TOOTH	12.9 PPM DRY WT.	6 SEA LIONS FROM SAN LUIS OBISPO COUNTY,CAL.,USA	163
*ZOOPLANKTON	25.03 MG/KG DRY WT.(MEAN)	19ST.,N.Y.BIGHT,CONC.IN PHYTOPL. (LONG ISLAND)=16.8MG/KG DRY WT. (MEAN)	565

TABLE D317

CONCENTRATIONS : PCB

ITEM:	VALUE:	CONDITIONS:	REF.:
*BIRD, EGG YOLK LIPID	241. PPM WET WEIGHT	24 APRIL,%LIPID=100,AUDUBON CANYON RANCH,CAL.,USA	439
*BUFFALO	0.06 MG/KG DRY WEIGHT	SOUTH DAK., FRESHWATER, WATER = 0.5 MG/M3, N=10	564
*BUFFALO	0.06 MG/KG DRY WEIGHT	SOUTH DAK., FRESHWATER, WATER = 0.5 MG/M3, N=10	564
*BULLHEAD	0.11 MG/KG DRY WEIGHT	SOUTH DAK., FRESHWATER, WATER = 0.5 MG/M3, N=10	564
*BULLHEAD	0.11 MG/KG DRY WEIGHT	SOUTH DAK., FRESHWATER, WATER = 0.5 MG/M3, N=10	564
*CARP	0.11 MG/KG DRY WEIGHT	SOUTH DAK., FRESHWATER, WATER = 0.5 MG/M3, N=10	564
*CARP	0.11 MG/KG DRY WEIGHT	SOUTH DAK., FRESHWATER, WATER = 0.5 MG/M3, N=10	564
*CASMERODIUS ALBUS, BRAIN	9.7 PPM MEAN WET WT.	5 OBS.,%LIPID=6.8,DIELDRIN=4.36, ENDRIN<0.18,AUDUBON CANYON RANCH, USA	439

TABLE D317

CONCENTRATIONS : PCB

ITEM:	VALUE:	CONDITIONS:	REF.:
*CASMERODIUS ALBUS, BREAST MUSCLE	16. PPM MEAN WET WT.	5 OBS.,%LIPID=2.4,AUDUBON CANYON RANCH,CAL.,USA	439
*CASMERODIUS ALBUS, LIVER	37.3 PPM MEAN WET WT.	5 OBS.,%LIPID=3.5,HG=6.08,AUDUBON CANYON RANCH,CAL.,USA	439
*CHLORELLA PYRENOIDOSA	32 MG/KG WET WEIGHT	WATER INITIAL = 10 PPB, 1 H, 295 K, 165 FTC, TETRACHLORO-PCB	1624
*CHLORELLA PYRENOIDOSA	70 MG/KG WET WEIGHT	WATER INITIAL = 10 PPB, 1 H, 295 K, 165 FTC, HEXACHLORO-PCB	1624
*CHLORELLA PYRENOIDOSA	16 MG/KG WET WEIGHT	WATER INITIAL = 10 PPB, 1 H, 295 K, 165 FTC, OCTACHLORO-PCB	1624
*CHLORELLA PYRENOIDOSA	52 MG/KG WET WEIGHT	WATER INITIAL = 10 PPB, 1 H, 295 K, 165 FTC, DECACHLORO-PCB	1624
*CHLORELLA PYRENOIDOSA, DEAD CELLS	62 MG/KG WET WEIGHT	WATER INITIAL = 10 PPB, 1 H, 295 K, 165 FTC, TETRACHLORO-PCB	1624
*CHLORELLA PYRENOIDOSA, DEAD CELLS	156 MG/KG WET WEIGHT	WATER INITIAL = 10 PPB, 1 H, 295 K, 165 FTC, HEXACHLORO-PCB	1624
*CRAPPIE	LESS THAN 0.05 MG/KG DRY WEIGHT	SOUTH DAK., FRESHWATER, WATER = 0.5 MG/M3, N=10	564
*CRAPPIE	LESS THAN 0.05 MG/KG DRY WEIGHT	SOUTH DAK., FRESHWATER, WATER = 0.5 MG/M3, N=10	564
*EGRET, EGG YOLK LIPID	40.3 PPM WET WEIGHT	NO DATE,%LIPID=100,AUDUBON CANYON RANCH,CAL.,USA	439
*ESCHERICHIA COLI	13.6 MG/KG	PCB IN MEDIUM=0.1 MG/L WATER, 24 H, LAB	808
*ESCHERICHIA COLI	4.4 MG/KG	PCB IN MEDIUM=0.01 MG/KG, 24 H, LAB	808
*HERON (PROBABLE), EMBRYO	1.2 PPM WET WT.	NO DATE,%LIPID=3.6,AUDUBON CANYON RANCH,CAL.,USA	439
*HERON, EGG YOLK LIPID	50.6 PPM WET WEIGHT	20 APRIL,%LIPID=100,AUDUBON CANYON RANCH,CAL.,USA	439
*MYTILUS EDULIS	85 NG/G WET WT.	CONC.IN SEDIMENT=1.8NG/G DRY WT., SOUTHERN PUGET SOUND	1073
*MYTILUS EDULIS	95 NG/G WET WT.	CONC.IN SEDIMENT=70NG/G DRY WT., SOUTHERN PUGET SOUND	1073
*MYTILUS EDULIS	210 NG/G WET WT.	CONC.IN SEDIMENT=330NG/G DRY WT., SOUTHERN PUGET SOUND	1073
*MYTILUS EDULIS	31 NG/G WET WT.	CONC.IN SEDIMENT=1.0NG/G DRY WT., SOUTHERN PUGET SOUND	1073
*MYTILUS EDULIS	72 NG/G WET WT.	CONC.IN SEDIMENT=6.3NG/G DRY WT., SOUTHERN PUGET SOUND	1073
*MYTILUS EDULIS	38 NG/G WET W¤.	CONC.IN SEDIMENT=17NG/G DRY WT., SOUTHERN PUGET SOUND	1073
*MYTILUS EDULIS	16 NG/G WET WT.	CONC.IN SEDIMENT=1.5NG/G DRY WT., SOUTHERN PUGET SOUND	1073
*MYTILUS EDULIS	50 NG/G WET WT.	CONC.IN SEDIMENT=2.6NG/G DRY WT., SOUTHERN PUGET SOUoD	1073
*MYTILUS EDULIS	27 NG/G WET WT.	CONC.IN SEDIMENT=1.1NG/G DRY WT., SOUTHERN PUGET SOUND	1073
*MYTILUS EDULIS	30 NG/G WET WT.	CONC.IN SEDIMENT=2.1NG/G DRY WT., SOUTHERN PUGET SOUND	1073
*MYTILUS EDULIS	14 NG/G WET WT.	CONC.IN SEDIMENT=0.7NG/G DRY WT., SOUTHERN PUGET SOUND	1073
*MYTILUS EDULIS	24 NG/G WET WT.	CONC.IN SEDIMENT=1.5NG/G DRY WT., SOUTHERN PUGET SOUND	1073
*MYTILUS EDULIS	11 NG/G WET WT.	CONC.IN SEDIMENT=3.1NG/G DRY WT., SOUTHERN PUGET SOUND	1073
*MYTILUS EDULIS	16 NG/G WET WT.	CONC.IN SEDIMENT=0.7NG/G DRY WT., SOUTHERN PUGET SOUND	1073
*MYTILUS EDULIS	10 NG/G WET WT.	CONC.IN SEDIMENT=5.7NG/G DRY WT., SOUTHERN PUGET SOUND	1073
*MYTILUS EDULIS	11 NG/G WET WT.	CONC.IN SEDIMENT=1.8NG/G DRY WT., SOUTHERN PUGET SOUND	1073
*MYTILUS EDULIS	14 NG/G WET WT.	CONC.IN SEDIMENT=10NG/G DRY WT., SOUTHERN PUGET SOUND	1073
*MYTILUS EDULIS	40 NG/G WET WT.	CONC.IN SEDIMENT=7.3NG/G DRY WT., SOUTHERN PUGET SOUND	1073
*NESTLING,EGRET(AGE UNKNOWN)	4.6 PPM WET WT.	13 MAY,%LIPID=1.2,AUDUBON CANYON RANCH,CAL.,USA	439

TABLE D317

CONCENTRATIONS : PCB			
ITEM:	VALUE:	CONDITIONS:	REF.:
*NESTLING,EGRET(1 WEEK)	1.75 PPM WET WT.	7 MAY,%LIPID=1.2,AUDUBON CANYON RANCH,CAL.,USA	439
*NESTLING,EGRET(1.5-2 WEEKS)	1.7 PPM WET WT.	11 MAY,%LIPID=1.2,AUDUBON CANYON RANCH,CAL.,USA	439
*NESTLING,EGRET(1-2 WEEKS)	1.8 PPM WET WT.	15 MAY,%LIPID=1.7,AUDUBON CANYON RANCH,CAL.,USA	439
*NESTLING,EGRET(2-2.5 WEEKS)	0.7 PPM WET WT.	11 MAY,%LIPID=2.1,AUDUBON CANYON RANCH,CAL.,USA	439
*NESTLING,EGRET(5-6 DAYS)	1.3 PPM WET WT.	11 MAY,%LIPID=2.2,AUDUBON CANYON RANCH,CAL.,USA	439
*NESTLING,HERON(2 WEEKS)	1.35 PPM WET WT.	24 APRIL,%LIPID=1.6,AUDUBON CANYON RANCH,CAL.,USA	439
*NESTLING,HERON(3 WEEKS)	1.82 PPM WET WT.	24 APRIL,%LIPID=1.,AUDUBON CANYON RANCH,CAL.,USA	439
*NESTLING,HERON(4-5 WEEKS)	0.49 PPM WET WT.	24 APRIL,%LIPID=1.2,AUDUBON CANYON RANCH,CAL.,USA	439
*PAGOPHILUS GROENLANDICUS, BLUBBER	6.05 (SD=+-2.05) PPM (ON WET WEIGHT BASIS)	95.5 % FAT, 4 ANIMALS, ST LAWRENCE, CANADA	785
*PAGOPHILUS GROENLANDICUS, BLUBBER	1.08 (SD=+-0.53) PPM (ON WET WEIGHT BASIS)	83,0 % FAT, 15 SAMPLES,1973,GULF OF ST LAWRENCE	484
*PAGOPHILUS GROENLANDICUS, BLUBBER	2.46 (SD=+-1.38) PPM (ON WET WEIGHT BASIS)	97.8 % FAT, 10 SAMPLES,1971,GULF OF ST LAWRENCE	484
*PAGOPHILUS GROENLANDICUS, BRAIN	0.09 (SD=+-0.0) PPM (ON WET WEIGHT BASIS)	11.5 % FAT, 3 ANIMALS, ST LAWRENCE, CANADA	785
*PAGOPHILUS GROENLANDICUS, BRAIN	0.03 (SD=+-0.04) PPM (ON WET WEIGHT BASIS)	6.6 % FAT, 14 SAMPLES, 1973,GULF OF ST LAWRENCE	484
*PAGOPHILUS GROENLANDICUS, BRAIN	0.05 (SD=+-0.13) PPM (ON WET WEIGHT BASIS)	7.5 % FAT, 10 SAMPLES, 1971,GULF OF ST LAWRENCE	484
*PAGOPHILUS GROENLANDICUS, KIDNEY	0.43 (SD=+-0.44) PPM (ON WET WEIGHT BASIS)	14.4 % FAT, 13 SAMPLES,1973,GULF OF ST LAWRENCE	484
*PAGOPHILUS GROENLANDICUS, LIVER	0.07 (SD=+-0.01) PPM (ON WET WEIGHT BASIS)	1.8 % FAT, 4 AINMALS	785
*PAGOPHILUS GROENLANDICUS, LIVER	0.16 (SD=+-0.18) PPM (ON WET WEIGHT BASIS)	4.4 % FAT, 16 SAMPLES, 1973,GULF OF ST LAWRENCE	484
*PAGOPHILUS GROENLANDICUS, LIVER	0.14 (SD=+-0.10) PPM (ON WET WEIGHT BASIS)	7.0 % FAT, 11 SAMPLES, 1971,GULF OF ST LAWRENCE	484
*PAGOPHILUS GROENLANDICUS, MUSCLE	0.45 (SD=+-0.01) PPM (ON WET WEIGHT BASIS)	1.1 % FAT, 4 ANIMALS, ST LAWRENCE, CANADA	785
*SEDIMENT	0.0064 MG/KG WET WEIGHT	SOUTH DAK., FRESHWATER, WATER = 0.5 MG/M3, N=10	564
*SEDIMENT	0.0064 MG/KG WET WEIGHT	SOUTH DAK., FRESHWATER, WATER = 0.5 MG/M3, N=10	564

TABLE D318

CONCENTRATIONS : PO-210			
ITEM:	VALUE:	CONDITIONS:	REF.:
*CYSTOSERIA OSMUNDACEA	70 +- 4 P CI/KG	1.2 F CI/CM2, SEAWATER= 0.023 P CI/KG, POINT LOMA,CALIF,73	676
*CYSTOSERIA OSMUNDACEA	50 +- 3 P CI/KG	1.6 F CI/CM2, SEAWATER= 0.023 P CI/KG, POINT LOMA,CALIF,73	676
*CYSTOSERIA OSMUNDACEA	70 +- 4 P CI/KG	1.2 F CI/CM2, SEAWATER= 0.023 P CI/KG, POINT LOMA,CALIF,73	676
*CYSTOSERIA OSMUNDACEA	50 +- 3 P CI/KG	1.6 F CI/CM2, SEAWATER= 0.023 P CI/KG, POINT LOMA,CALIF,73	676
*EISENIA ARBOREA	66 +- 4 P CI/KG	2.5 F CI/CM2, SEAWATER= 0.023 P CI/KG, POINT LOMA,CALIF,73	676
*EISENIA ARBOREA	58 +- 3 P CI/KG	2.0 F CI/CM2, SEAWATER= 0.023 P CI/KG, POINT LOMA,CALIF,73	676
*EISENIA ARBOREA	66 +- 4 P CI/KG	2.5 F CI/CM2, SEAWATER= 0.023 P CI/KG, POINT LOMA,CALIF,73	676
*EISENIA ARBOREA	58 +- 3 P CI/KG	2.0 F CI/CM2, SEAWATER= 0.023 P CI/KG, POINT LOMA,CALIF,73	676

TABLE D318

CONCENTRATIONS : PO-210

ITEM:	VALUE:	CONDITIONS:	REF.:
*LAMINARIA SACCHARINA	52 +- 3 P CI/KG	2.4 F CI/CM2, SEAWATER= 0.023 P CI/KG, POINT LOMA,CALIF,73	676
*LAMINARIA SACCHARINA	122 +- 7 P CI/KG	5.5 F CI/CM2, SEAWATER= 0.023 P CI/KG, POINT LOMA,CALIF,73	676
*LAMINARIA SACCHARINA	52 +- 3 P CI/KG	2.4 F CI/CM2, SEAWATER= 0.023 P CI/KG, POINT LOMA,CALIF,73	676
*LAMINARIA SACCHARINA	122 +- 7 P CI/KG	5.5 F CI/CM2, SEAWATER= 0.023 P CI/KG, POINT LOMA,CALIF,73	676
*MACROCYSTIS PYRIFERA	60 +- 2 P CI/KG	1.7 F CI/CM2, SEAWATER= 0.023 P CI/KG, POINT LOMA,CALIF,73, YOUNG BLA	676
*MACROCYSTIS PYRIFERA	64 +- 2 P CI/KG	1.8 F CI/CM2, SEAWATER= 0.023 P CI/KG, POINT LOMA,CALIF,73, YOUNG BLA	676
*MACROCYSTIS PYRIFERA	105 +- 5 P CI/KG WET WEIGHT	TIP, 1.9 F CI/CM2, SEAWATER=0.023 P CI/KG, POINT LOMA,CAL.	676
*MACROCYSTIS PYRIFERA	126 +- 10 P CI/KG WET WEIGHT	1 CM FROM TIP, 2.4 F CI/CM2, SEAWATER=0.023 P CI/KG, POINT LOMA,CAL.	676
*MACROCYSTIS PYRIFERA	75 +- 4 P CI/KG WET WEIGHT	2 CM FROM TIP, 1.8 F CI/CM2, SEAWATER=0.023 P CI/KG, POINT LOMA,CAL.	676
*MACROCYSTIS PYRIFERA	105 +- 5 P CI/KG WET WEIGHT	3 CM FROM TIP, 2.5 F CI/CM2, SEAWATER=0.023 P CI/KG, POINT LOMA,CAL.	676
*MACROCYSTIS PYRIFERA	142 +- 11 P CI/KG WET WEIGHT	4 CM FROM TIP, 3.2 F CI/CM2, SEAWATER=0.023 P CI/KG, POINT LOMA,CAL.	676
*MACROCYSTIS PYRIFERA	173 +- 17 P CI/KG WET WEIGHT	5 CM FROM TIP, 5.1 F CI/CM2, SEAWATER=0.023 P CI/KG, POINT LOMA,CAL.	676
*MACROCYSTIS PYRIFERA	183 +- 15 P CI/KG WET WEIGHT	6 CM FROM TIP, 4.8 F CI/CM2, SEAWATER=0.023 P CI/KG, POINT LOMA,CAL.	676
*MACROCYSTIS PYRIFERA	260 +- 80 P CI/KG WET WEIGHT	7 CM FROM TIP, 7.9 F CI/CM2, SEAWATER=0.023 P CI/KG, POINT LOMA,CAL.	676
*MACROCYSTIS PYRIFERA	320 +- 22 P CI/KG WET WEIGHT	8 CM FROM TIP, 7.6 F CI/CM2, SEAWATER=0.023 P CI/KG, POINT LOMA,CAL.	676
*MACROCYSTIS PYRIFERA	410 +- 30 P CI/KG WET WEIGHT	9 CM FROM TIP, 8.9 F CI/CM2, SEAWATER=0.023 P CI/KG, POINT LOMA,CAL.	676
*MACROCYSTIS PYRIFERA	283 +- 20 P CI/KG WET WEIGHT	10 CM FROM TIP, 7.0 F CI/CM2, SEAWATER=0.023 P CI/KG, POINT LOMA,CAL.	676
*MACROCYSTIS PYRIFERA	60 +- 2 P CI/KG	1.7 F CI/CM2, SEAWATER= 0.023 P CI/KG, POINT LOMA,CALIF,73, YOUNG BLA	676
*MACROCYSTIS PYRIFERA	64 +- 2 P CI/KG	1.8 F CI/CM2, SEAWATER= 0.023 P CI/KG, POINT LOMA,CALIF,73, YOUNG BLA	676
*MACROCYSTIS PYRIFERA	105 +- 5 P CI/KG WET WEIGHT	TIP, 1.9 F CI/CM2, SEAWATER=0.023 P CI/KG, POINT LOMA,CAL.	676
*MACROCYSTIS PYRIFERA	126 +- 10 P CI/KG WET WEIGHT	1 CM FROM TIP, 2.4 F CI/CM2, SEAWATER=0.023 P CI/KG, POINT LOMA,CAL.	676
*MACROCYSTIS PYRIFERA	75 +- 4 P CI/KG WET WEIGHT	2 CM FROM TIP, 1.8 F CI/CM2, SEAWATER=0.023 P CI/KG, POINT LOMA,CAL.	676
*MACROCYSTIS PYRIFERA	105 +- 5 P CI/KG WET WEIGHT	3 CM FROM TIP, 2.5 F CI/CM2, SEAWATER=0.023 P CI/KG, POINT LOMA,CAL.	676
*MACROCYSTIS PYRIFERA	142 +- 11 P CI/KG WET WEIGHT	4 CM FROM TIP, 3.2 F CI/CM2, SEAWATER=0.023 P CI/KG, POINT LOMA,CAL.	676
*MACROCYSTIS PYRIFERA	173 +- 17 P CI/KG WET WEIGHT	5 CM FROM TIP, 5.1 F CI/CM2, SEAWATER=0.023 P CI/KG, POINT LOMA,CAL.	676
*MACROCYSTIS PYRIFERA	183 +- 15 P CI/KG WET WEIGHT	6 CM FROM TIP, 4.8 F CI/CM2, SEAWATER=0.023 P CI/KG, POINT LOMA,CAL.	676
*MACROCYSTIS PYRIFERA	260 +- 80 P CI/KG WET WEIGHT	7 CM FROM TIP, 7.9 F CI/CM2, SEAWATER=0.023 P CI/KG, POINT LOMA,CAL.	676

TABLE D318

CONCENTRATIONS : PO-210

ITEM:	VALUE:	CONDITIONS:	REF.:
*MACROCYSTIS PYRIFERA	320 +- 22 P CI/KG WET WEIGHT	8 CM FROM TIP, 7.6 F CI/CM2, SEAWATER=0.023 P CI/KG, POINT LOMA,CAL.	676
*MACROCYSTIS PYRIFERA	410 +- 30 P CI/KG WET WEIGHT	9 CM FROM TIP, 8.9 F CI/CM2, SEAWATER=0.023 P CI/KG, POINT LOMA,CAL.	676
*MACROCYSTIS PYRIFERA	283 +- 20 P CI/KG WET WEIGHT	10 CM FROM TIP, 7.0 F CI/CM2, SEAWATER=0.023 P CI/KG, POINT LOMA,CAL.	676
*PTERYGOPHORA CALIFORNICA	96 +- 7 P CI/KG	2.6 F CI/CM2, SEAWATER= 0.023 P CI/KG, POINT LOMA,CALIF,73	676
*PTERYGOPHORA CALIFORNICA	138 +- 10 P CI/KG	2.3 F CI/CM2, SEAWATER= 0.023 P CI/KG, POINT LOMA,CALIF,73	676
*PTERYGOPHORA CALIFORNICA	96 +- 7 P CI/KG	2.6 F CI/CM2, SEAWATER= 0.023 P CI/KG, POINT LOMA,CALIF,73	676
*PTERYGOPHORA CALIFORNICA	138 +- 10 P CI/KG	2.3 F CI/CM2, SEAWATER= 0.023 P CI/KG, POINT LOMA,CALIF,73	676
*SCHIZYMENIA EPIPHYTICA	290 +- 26 P CI/KG	7.1 F CI/CM2, SEAWATER= 0.023 P CI/KG, POINT LOMA,CALIF,73	676
*SCHIZYMENIA EPIPHYTICA	144 +- 9 P CI/KG	2.2 F CI/CM2, SEAWATER= 0.023 P CI/KG, POINT LOMA,CALIF,73	676
*SCHIZYMENIA EPIPHYTICA	290 +- 26 P CI/KG	7.1 F CI/CM2, SEAWATER= 0.023 P CI/KG, POINT LOMA,CALIF,73	676
*SCHIZYMENIA EPIPHYTICA	144 +- 9 P CI/KG	2.2 F CI/CM2, SEAWATER= 0.023 P CI/KG, POINT LOMA,CALIF,73	676

TABLE D319

CONCENTRATIONS : PROTEIN

ITEM:	VALUE:	CONDITIONS:	REF.:
*CHLORELLA SP.	13.1 % OF DRY WEIGHT	UNFAVORABLE ENVIRONMENT	1039
*CHLORELLA SP.	46.4 % OF DRY WEIGHT	FAVORABLE ENVIRONMENT	1039
*EEL	33.6 % OF DRY WEIGHT	SILVER EEL	1487
*EEL	58.0 % OF DRY WEIGHT	YELLOW EEL	1487
*HERRING	45.7 % OF DRY WEIGHT	NONE	1487
*MUSSEL	73.7 % OF DRY WEIGHT	NONE	1487
*MYCOPLASMA GALLISEPTICUM	80 % OF DRY WEIGHT	NONE	1067
*OYSTER	51.2 % OF DRY WEIGHT	NONE	461
*PLAICE	79.5 % OF DRY WEIGHT	NONE	1487
*SACCHAROMYCES CEREVISIAE	40 % OF DRY WEIGHT	NONE	574
*SEA URCHIN	60.6 % OF DRY WEIGHT	40 H LARVA	461
*SEA URCHIN	66.9 % OF DRY WEIGHT	UNFERTILIZED EGG	461
*SILKWORM	55.5 % OF DRY WEIGHT	05870000	461
*SILKWORM	63.4 % OF DRY WEIGHT	IMAGO	461
*TORULOPSIS UTILIS	54.4 % OF DRY WEIGHT	NONE	574

TABLE D320

CONCENTRATIONS : PU-239

ITEM:	VALUE:	CONDITIONS:	REF.:
*MACROCYSTIS PYRIFERA	0.32 +- 0.05 P CI/KG WET WEIGHT	TIP, 7 A CI/CM2, SEAWATER=0.00025 P CI/KG, POINT LOMA,CAL.	676
*MACROCYSTIS PYRIFERA	0.48 +- 0.06 P CI/KG WET WEIGHT	1 CM FROM TIP, 11 A CI/CM2, SEAWATER=0.00025 P CI/KG, POINT LOMA,CAL.	676

TABLE D320

CONCENTRATIONS : PU-239			
ITEM:	VALUE:	CONDITIONS:	REF.:
*MACROCYSTIS PYRIFERA	0.54 +- 0.05 P CI/KG WET WEIGHT	2 CM FROM TIP, 15 A CI/CM2, SEAWATER=0.00025 P CI/KG, POINT LOMA,CAL.	676
*MACROCYSTIS PYRIFERA	0.50 +- 0.07 P CI/KG WET WEIGHT	3 CM FROM TIP, 13 A CI/CM2, SEAWATER=0.00025 P CI/KG, POINT LOMA,CAL.	676
*MACROCYSTIS PYRIFERA	0.60 +- 0.06 P CI/KG WET WEIGHT	4 CM FROM TIP, 15 A CI/CM2, SEAWATER=0.00025 P CI/KG, POINT LOMA,CAL.	676
*MACROCYSTIS PYRIFERA	0.68 +- 0.07 P CI/KG WET WEIGHT	5 CM FROM TIP, 19 A CI/CM2, SEAWATER=0.00025 P CI/KG, POINT LOMA,CAL.	676
*MACROCYSTIS PYRIFERA	0.61 +- 0.06 P CI/KG WET WEIGHT	6 CM FROM TIP, 19 A CI/CM2, SEAWATER=0.00025 P CI/KG, POINT LOMA,CAL.	676
*MACROCYSTIS PYRIFERA	1.00 +- 0.08 P CI/KG WET WEIGHT	7 CM FROM TIP, 31 A CI/CM2, SEAWATER=0.00025 P CI/KG, POINT LOMA,CAL.	676
*MACROCYSTIS PYRIFERA	0.44 +-0.06 P CI/KG WET WEIGHT	8 CM FROM TIP, 26 A CI/CM2, SEAWATER=0.00025 P CI/KG, POINT LOMA,CAL.	676
*MACROCYSTIS PYRIFERA	1.44 +- 0.13 P CI/KG WET WEIGHT	9 CM FROM TIP, 36 A CI/CM2, SEAWATER=0.00025 P CI/KG, POINT LOMA,CAL.	676
*MACROCYSTIS PYRIFERA	1.90 +- 0.16 P CI/KG WET WEIGHT	10 CM FROM TIP, 40 A CI/CM2, SEAWATER=0.00025 P CI/KG, POINT LOMA,CAL	676
*MACROCYSTIS PYRIFERA	0.32 +- 0.05 P CI/KG WET WEIGHT	TIP, 7 A CI/CM2, SEAWATER=0.00025 P CI/KG, POINT LOMA,CAL.	676
*MACROCYSTIS PYRIFERA	0.48 +- 0.06 P CI/KG WET WEIGHT	1 CM FROM TIP, 11 A CI/CM2, SEAWATER=0.00025 P CI/KG, POINT LOMA,CAL.	676
*MACROCYSTIS PYRIFERA	0.54 +- 0.05 P CI/KG WET WEIGHT	2 CM FROM TIP, 15 A CI/CM2, SEAWATER=0.00025 P CI/KG, POINT LOMA,CAL.	676
*MACROCYSTIS PYRIFERA	0.50 +- 0.07 P CI/KG WET WEIGHT	3 CM FROM TIP, 13 A CI/CM2, SEAWATER=0.00025 P CI/KG, POINT LOMA,CAL.	676
*MACROCYSTIS PYRIFERA	0.60 +- 0.06 P CI/KG WET WEIGHT	4 CM FROM TIP, 15 A CI/CM2, SEAWATER=0.00025 P CI/KG, POINT LOMA,CAL.	676
*MACROCYSTIS PYRIFERA	0.68 +- 0.07 P CI/KG WET WEIGHT	5 CM FROM TIP, 19 A CI/CM2, SEAWATER=0.00025 P CI/KG, POINT LOMA,CAL.	676
*MACROCYSTIS PYRIFERA	0.61 +- 0.06 P CI/KG WET WEIGHT	6 CM FROM TIP, 19 A CI/CM2, SEAWATER=0.00025 P CI/KG, POINT LOMA,CAL.	676
*MACROCYSTIS PYRIFERA	1.00 +- 0.08 P CI/KG WET WEIGHT	7 CM FROM TIP, 31 A CI/CM2, SEAWATER=0.00025 P CI/KG, POINT LOMA,CAL.	676
*MACROCYSTIS PYRIFERA	0.44 +-0.06 P CI/KG WET WEIGHT	8 CM FROM TIP, 26 A CI/CM2, SEAWATER=0.00025 P CI/KG, POINT LOMA,CAL.	676
*MACROCYSTIS PYRIFERA	1.44 +- 0.13 P CI/KG WET WEIGHT	9 CM FROM TIP, 36 A CI/CM2, SEAWATER=0.00025 P CI/KG, POINT LOMA,CAL.	676
*MACROCYSTIS PYRIFERA	1.90 +- 0.16 P CI/KG WET WEIGHT	10 CM FROM TIP, 40 A CI/CM2, SEAWATER=0.00025 P CI/KG, POINT LOMA,CAL	676

TABLE D321

CONCENTRATIONS : SB			
ITEM:	VALUE:	CONDITIONS:	REF.:
*ATMOSPHERE	42-85 NG/M**3	URBAN,U.S.A.	893
*ATMOSPHERE	1-2 NG/M**3	NOT URBAN,U.S.A.	893
*ATMOSPHERE	1 NG/M**3	NORTH ENGLAND,1971	893
*ATMOSPHERE	1.7+-0.6 PG/M**3	SOUTHPOLE,1970	893

TABLE D321

CONCENTRATIONS : SB			
ITEM:	VALUE:	CONDITIONS:	REF.:
*RAIN	1.8 UG/L	NORTH ENGLAND	893
*RAIN	0.1-0.2 UG/L	SWEDEN	893
*WATER	0.05-0.08 UG/L	RIVERS,SWEDEN	893

TABLE D322

CONCENTRATIONS : SE			
ITEM:	VALUE:	CONDITIONS:	REF.:
*ATMOSPHERE	1 NG/M**3	U.S.A.,1964-65	893
*ATMOSPHERE	0.48-3.0 NG/M**3	ENGLAND,7 CITIES,1972	893
*ATMOSPHERE	0.0056+-0.0012 NG/M**3	SOUTHPOLE,1972	893
*DRINKING WATER	0.061 UG/L	STOCKHOLM,SWEDEN	893
*RAIN	0.34 UG/L	ENGLAND,1971,REL.UNCONTAMINATED,SE IN ATM.=1NG/M**3	893
*RAT, KIDNEY	44.1 % OF TOTAL INTRACELLULAR AMOUNT	NUCLEI,9 RATS,24 H AFTER INJEC. OF 6 UG SE/275 G	1366
*RAT, KIDNEY	10.2 % OF TOTAL INTRACELLULAR AMOUNT	MITOCHONDRIA,9 RATS,24 H AFTER INJEC. OF 6 UG SE/275 G	1366
*RAT, KIDNEY	10.6 % OF TOTAL INTRACELLULAR AMOUNT	LYSOSOMES,9 RATS,24 H AFTER INJEC. OF 6 UG SE/275 G	1366
*RAT, KIDNEY	18.3 % OF TOTAL INTRACELLULAR AMOUNT	MICROSOMES,9 RATS,24 H AFTER INJEC. OF 6 UG SE/275 G	1366
*RAT, KIDNEY	16.8 % OF TOTAL INTRACELLULAR AMOUNT	SOLUBLE FRAC.,9 RATS,24 H AFTER INJEC. OF 6 UG SE/275 G	1366
*RAT, LIVER	41.8 % OF TOTAL INTRACELLULAR AMOUNT	NUCLEI,9 RATS ON 250-300 G,12 H AFTER INJECTION OF 6 MICRO G SEO3 (-2)	1366
*RAT, LIVER	8.2 % OF TOTAL INTRACELLULAR AMOUNT	MITOCHONDRIA,9 RATS,24 H AFTER INJEC. OF 6 UG SE/275 G	1366
*RAT, LIVER	13.4 % OF TOTAL INTRACELLULAR AMOUNT	LYSOSOMES,9 RATS,24 H AFTER INJEC. OF 6 UG SE/275 G	1366
*RAT, LIVER	21 % OF TOTAL INTRACELLULAR AMOUNT	MICROSOMES,9 RATS,24 H AFTER INJEC. OF 6 UG SE/275 G	1366
*RAT, LIVER	15.6 % OF TOTAL INTRACELLULAR AMOUNT	SOLUBLE FRAC.,9 RATS,24 H AFTER INJEC. OF 6 UG SE/275 G	1366
*WATER	0.10-0.15 UG/L	SWEDEN,WATER-SPRING	893

TABLE D323

CONCENTRATIONS : SN			
ITEM:	VALUE:	CONDITIONS:	REF.:
*ATMOSPHERE	10-30 NG/M**3	URBAN,U.S.A.	893
*ATMOSPHERE	0.2-1.8 NG/M**3	NOT URBAN,U.S.A.	893
*RAIN	10 UG/L	SWEDEN,3 STATIONS	893
*WATER	< 0.3-11 UG/L	RIVERS,U.S.A	893

TABLE D324

CONCENTRATIONS : SR			
ITEM:	VALUE:	CONDITIONS:	REF.:
*FOOD	9.56 UG/G	U.S.A.,FISH AND SHELLFISH	893

TABLE D324

CONCENTRATIONS : SR

ITEM:	VALUE:	CONDITIONS:	REF.:
*FOOD	1.20 UG/G	U.S.A.,MEAT	893
*FOOD	0.50 UG/G	U.S.A.,MILK	893
*FOOD	2.46 UG/G	U.S.A.,VEGETABLES	893
*WATER	11-802 UG/L	RIVERS AND LAKES,U.S.A.,1958-67	893

TABLE D325

CONCENTRATIONS : SUSPENDED SOLIDS

ITEM:	VALUE:	CONDITIONS:	REF.:
*MUNICIPAL SEWAGE	200 (100-350) MG/L WATER	TYPICAL UNTREATED	364
*MUNICIPAL SEWAGE	80 (40-120) MG/L WATER	TYPICAL TREATED,PRIMARY	364
*MUNICIPAL SEWAGE	15 (10-30) MG/L WATER	TYPICAL TREATED,SECONDARY	364
*PRECIPITATION	11-13 MG/L WATER	GENERAL CHARACTERISTICS	364
*URBAN LAND,DRAINAGE	5-7340 MG/L WATER	GENERAL CHARACTERISTICS	364

TABLE D326

CONCENTRATIONS : TA

ITEM:	VALUE:	CONDITIONS:	REF.:
*ASCIDIELLA ASPERSA	0.19 MG/KG DRY WEIGHT	ASSUMED SEAWATER CONC=0.02 MG/M3, SOUTHAMPTON WATER	207
*BUCCINUM UNDATUM	0.019 MG/KG DRY WEIGHT	ASSUMED SEAWATER CONC=0.02 MG/M3, WHITSTABLE, THAMES ESTUARY	207
*CARCINUS MAENAS, MANTLE	0.030 MG/KG DRY WEIGHT	ASSUMED SEAWATER CONC=0.02 MG/M3, SOUTHAMPTON WATER	207
*CARCINUS MAENAS, MUSCLE	0.024 MG/KG DRY WEIGHT	ASSUMED SEAWATER CONC=0.02 MG/M3, SOUTHAMPTON WATER	207
*CARDIUM EDULE	0.043 MG/KG DRY WEIGHT	ASSUMED SEAWATER CONC=0.02 MG/M3, WHITSTABLE, THAMES ESTUARY	207
*CIONA INTESTINALIS, WITH TEST	0.24 MG/KG DRY WEIGHT	ASSUMED SEAWATER CONC=0.02 MG/M3, SOUTHAMPTON WATER	207
*CIONA INTESTINALIS, WITHOUT TEST	0.096 MG/KG DRY WEIGHT	ASSUMED SEAWATER CONC=0.02 MG/M3, SOUTHAMPTON WATER	207
*CRASSOSTREA ANGULATA	0.009 MG/KG DRY WEIGHT	ASSUMED SEAWATER CONC=0.02 MG/M3, WHITSTABLE, THAMES ESTUARY	207
*HOLOTHURIA FORSKALI	0.032 MG/KG DRY WEIGHT	ASSUMED SEAWATER CONC=0.02 MG/M3, ENG. CHANNAL, OFF PLYMOUTH	207
*MERCENARIA MERCENARIA, DIGESTIVE GLAND	0.02 MG/KG DRY WEIGHT	ASSUMED SEAWATER CONC=0.02 MG/M3, SOUTHAMPTON WATER	207
*MERCENARIA MERCENARIA, FOOT	0.19 MG/KG DRY WEIGHT	ASSUMED SEAWATER CONC=0.02 MG/M3, SOUTHAMPTON WATER	207
*MYTILUS EDULIS, MANTLE	0.01 MG/KG DRY WEIGHT	ASSUMED SEAWATER CONC=0.02 MG/M3, SOUTHAMPTON WATER	207
*MYTILUS EDULIS, MUSCLE	<0.01 MG/KG DRY WEIGHT	ASSUMED SEAWATER CONC=0.02 MG/M3, SOUTHAMPTON WATER	207
*NEOMYSIS INTEGER	2.3 MG/KG DRY WEIGHT	ASSUMED SEAWATER CONC=0.02 MG/M3, TOTTON MARSHES, TEST R, ENG.	207
*PORANIA PULVILLUS	0.010 MG/KG DRY WEIGHT	ASSUMED SEAWATER CONC=0.02 MG/M3, ENG. CHANNAL, OFF PLYMOUTH	207

TABLE D327

CONCENTRATIONS : TC-99

ITEM:	VALUE:	CONDITIONS:	REF.:
*SOYBEAN	1.4 MICROGRAM/G DRY WEIGHT	SOIL = 0.01 UG/G DW	1711

TABLE D327

CONCENTRATIONS : TC-99

ITEM:	VALUE:	CONDITIONS:	REF.:
*SOYBEAN	6.7 MICROGRAM/G DRY WEIGHT	SOIL = 0.1 UG/G DW	1711
*SOYBEAN	380 MICROGRAM/G DRY WEIGHT	SOIL = 1.0 UG/G DW	1711
*SOYBEAN	380 MICROGRAM/G DRY WEIGHT	SOIL = 5.0 UG/G DW	1711
*SOYBEAN, BUD	0.00534 MICROGRAM/G DRY WEIGHT	SOIL = 0.001 UG/G DW	1711
*SOYBEAN, BUD	0.0134 MICROGRAM/G DRY WEIGHT	SOIL = 0.01 UG/G DW	1711
*SOYBEAN, BUD	0.255 MICROGRAM/G DRY WEIGHT	SOIL = 0.1 UG/G DW	1711
*SOYBEAN, COTYLEDON	0.38 MICROGRAM/G DRY WEIGHT	SOIL = 0.001 UG/G DW	1711
*SOYBEAN, COTYLEDON	20.7 MICROGRAM/G DRY WEIGHT	SOIL = 0.01 UG/G DW	1711
*SOYBEAN, COTYLEDON	50.4 MICROGRAM/G DRY WEIGHT	SOIL = 0.1 UG/G DW	1711
*SOYBEAN, COTYLEDON	484 MICROGRAM/G DRY WEIGHT	SOIL = 1.0 UG/G DW	1711
*SOYBEAN, COTYLEDON	342 MICROGRAM/G DRY WEIGHT	SOIL = 5.0 UG/G DW	1711
*SOYBEAN, LEAVES	0.186 MICROGRAM/G DRY WEIGHT	SOIL = 0.001 UG/G DW	1711
*SOYBEAN, LEAVES	0.964 MICROGRAM/G DRY WEIGHT	SOIL = 0.01 UG/G DW	1711
*SOYBEAN, LEAVES	8.46 MICROGRAM/G DRY WEIGHT	SOIL = 0.1 UG/G DW	1711
*SOYBEAN, STEM	0.0306 MICROGRAM/G DRY WEIGHT	SOIL = 0.001 UG/G DW	1711
*SOYBEAN, STEM	0.230 MICROGRAM/G DRY WEIGHT	SOIL = 0.01 UG/G DW	1711
*SOYBEAN, STEM	1.25 MICROGRAM/G DRY WEIGHT	SOIL = 0.1 UG/G DW	1711
*SOYBEAN, STEM	202 MICROGRAM/G DRY WEIGHT	SOIL = 1.0 UG/G DW	1711
*SOYBEAN, STEM	640 MICROGRAM/G DRY WEIGHT	SOIL = 5.0 UG/G DW	1711
*WHEAT	0.34 MICROGRAM/G DRY WEIGHT	SOIL = 0.001 UG.G DW	1711
*WHEAT	1.5 MICROGRAM/G DRY WEIGHT	SOIL = 0.01 UG/G DW	1711
*WHEAT	17 MICROGRAM/G DRY WEIGHT	SOIL = 0.1 UG/G DW	1711
*WHEAT, BLADE	0.476 MICROGRAM/G DRY WEIGHT	SOIL = 0.001 UG/G DW	1711
*WHEAT, BLADE	3.12 MICROGRAM/G DRY WEIGHT	SOIL = 0.01 UG/G DW	1711
*WHEAT, BLADE	39.7 MICROGRAM/G DRY WEIGHT	SOIL = 0.1 UG/G DW	1711
*WHEAT, CULM	0.00279 MICROGRAM/G DRY WEIGHT	SOIL = 0.001 UG/G DW	1711
*WHEAT, CULM	0.0235 MICROGRAM/G DRY WEIGHT	SOIL = 0.01 UG/G DW	1711
*WHEAT, CULM	0.130 MICROGRAM/G DRY WEIGHT	SOIL = 0.1 UG/G DW	1711
*WHEAT, HEAD	0.000437 MICROGRAM/G DRY WEIGHT	SOIL = 0.001 UG/G DW	1711
*WHEAT, HEAD	0.0201 MICROGRAM/G DRY WEIGHT	SOIL = 0.01 UG/G DW	1711
*WHEAT, HEAD	0.0230 MICROGRAM/G DRY WEIGHT	SOIL = 0.1 UG/G DW	1711
*WHEAT, HEAD	1.19 MICROGRAM/G DRY WEIGHT	SOIL = 0.001 UG/G DW	1711

TABLE D328

CONCENTRATIONS : TDE

ITEM:	VALUE:	CONDITIONS:	REF.:
*PAGOPHILUS GROENLANDICUS, BLUBBER	0.18 (SD=+-0.12) PPM (ON WET WEIGHT BASIS)	95.5 % FAT, 4 ANIMALS, ST LAWRENCE, CANADA	785
*PAGOPHILUS GROENLANDICUS, BLUBBER	0.13 (SD=+-0.12) PPM (ON WET WEIGHT BASIS)	83,0 % FAT, 15 SAMPLES,1973,GULF OF ST LAWRENCE	484
*PAGOPHILUS GROENLANDICUS, BLUBBER	0.50 (SD=+-0.32) PPM (ON WET WEIGHT BASIS)	97.8 % FAT, 10 SAMPLES,1971,GULF OF ST LAWRENCE	484
*PAGOPHILUS GROENLANDICUS, BRAIN	NON DETECTABLE	11.5 % FAT, 3 ANIMALS, ST LAWRENCE, CANADA	785
*PAGOPHILUS GROENLANDICUS, BRAIN	0.01 (SD=+-0.01) PPM (ON WET WEIGHT BASIS)	6.6 % FAT, 14 SAMPLES, 1973,GULF OF ST LAWRENCE	484
*PAGOPHILUS GROENLANDICUS, BRAIN	TRACE	7.5 % FAT, 10 SAMPLES, 1971,GULF OF ST LAWRENCE	484
*PAGOPHILUS GROENLANDICUS, KIDNEY	0.04 (SD=+-0.06) PPM (ON WET WEIGHT BASIS)	14.4 % FAT, 13 SAMPLES,1973,GULF OF ST LAWRENCE	484
*PAGOPHILUS GROENLANDICUS, LIVER	0.01 (SD=+-0.01) PPM (ON WET WEIGHT BASIS)	1.8 % FAT, 4 AINMALS, GULF OF ST LAWRENCE, CANADA	785

TABLE D328

CONCENTRATIONS : TDE			
ITEM:	VALUE:	CONDITIONS:	REF.:
*PAGOPHILUS GROENLANDICUS, LIVER	0.01 (SD=+-0.01) PPM (ON WET WEIGHT BASIS)	4.4 % FAT, 16 SAMPLES, 1973, GULF OF ST LAWRENCE	484
*PAGOPHILUS GROENLANDICUS, LIVER	0.03 (SD=+-0.13) PPM (ON WET WEIGHT BASIS)	7.0 % FAT, 11 SAMPLES, 1971, GULF OF ST LAWRENCE	484
*PAGOPHILUS GROENLANDICUS, MUSCLE	NON DETECTABLE	1.1 % FAT, 4 ANIMALS, ST LAWRENCE, CANADA	785

TABLE D329

CONCENTRATIONS : TELODRIN			
ITEM:	VALUE:	CONDITIONS:	REF.:
*COTURNIX QUAIL, BRAIN	1.60 MG/KG	CONC.IN FOOD=2MG/KG	389
*COTURNIX QUAIL, BRAIN	1.40 MG/KG	CONC.IN FOOD=10MG/KG	389

TABLE D330

CONCENTRATIONS : TI			
ITEM:	VALUE:	CONDITIONS:	REF.:
*ATMOSPHERE	10-70 NG/M**3	URBAN, U.S.A.	893
*ATMOSPHERE	3-18 NG/M**3	NOT URBAN, U.S.A.	893
*ATMOSPHERE	30 NG/M**3	SWEDEN, TIO2	893
*FOOD	1.8 MG/KG DRY WT.	POTATO	893
*FOOD	1.2 MG/KG DRY WT.	CABBAGE	893
*FOOD	12 MG/KG DRY WT.	SPINACH	893
*FOOD	2-4 MG/KG DRY WT.	SHEEP, HORSE AND HARE	893
*VEGETATION	21 MG/KG DRY WT.	CONC.IN PODZOL:F=250,H=500,A1A2=8000,B1=5000,B2=8000,B3=6000, (?043)	1048

TABLE D331

CONCENTRATIONS : TOTAL SOLIDS			
ITEM:	VALUE:	CONDITIONS:	REF.:
*URBAN LAND, DRAINAGE	194-8620 MG/L WATER	GENERAL CHARACTERISTICS	364

TABLE D332

CONCENTRATIONS : TOTAL-N			
ITEM:	VALUE:	CONDITIONS:	REF.:
*AGRICULTURAL CROPLAND	9 MG/L WATER	GENERAL CHARACTERISTICS	364
*ANIMAL FEEDLOT, RUNOFF	920-2100 MG/L WATER	GENERAL CHARACTERISTICS	364
*FORESTED LAND	0.3-1.8 MG/L WATER	GENERAL CHARACTERISTICS	364
*MUNICIPAL SEWAGE	40 MG/L WATER	TYPICAL UNTREATED	364
*MUNICIPAL SEWAGE	35 MG/L WATER	TYPICAL TREATED, PRIMARY	364
*MUNICIPAL SEWAGE	30 MG/L WATER	TYPICAL TREATED, SECONDARY	364

TABLE D332

CONCENTRATIONS : TOTAL-N

ITEM:	VALUE:	CONDITIONS:	REF.:
*PRECIPITATION	1.2-1.3 MG/L WATER	GENERAL CHARACTERISTICS	364
*URBAN LAND, DRAINAGE	3 MG/L WATER	GENERAL CHARACTERISTICS	364

TABLE D333

CONCENTRATIONS : TOTAL-P

ITEM:	VALUE:	CONDITIONS:	REF.:
*AGRICULTURAL CROPLAND	0.02-1.7 MG/L WATER	GENERAL CHARACTERISTICS	364
*ANIMAL FEEDLOT, RUNOFF	290-360	GENERAL CHARACTERISTICS	364
*FORESTED LAND	0.01-0.11 MG/L WATER	GENERAL CHARACTERISTICS	364
*MUNICIPAL SEWAGE	10 MG/L WATER	TYPICAL UNTREATED	364
*MUNICIPAL SEWAGE	7.5 MG/L WATER	TYPICAL TREATED, PRIMARY	364
*MUNICIPAL SEWAGE	5.0 MG/L WATER	TYPICAL TREATED, SECONDARY	364
*PRECIPITATION	0.02-0.04 MG/L WATER	GENERAL CHARACTERISTICS	364
*URBAN LAND, DRAINAGE	0.2-1.1 MG/L WATER	GENERAL CHARACTERISTICS	364

TABLE D334

CONCENTRATIONS : U-238

ITEM:	VALUE:	CONDITIONS:	REF.:
*PARTICLE FRACTIONS	5.1 % OF TOTAL	SIZE=2.0-0.5 MM, WEIGHT=3.2 %, AREA= 0.2 %, MARINE, JAPAN	1095
*PARTICLE FRACTIONS	10.6 % OF TOTAL	SIZE=0.5-0.25 MM, WEIGHT=8.5 %, AREA =1.4 %, MARINE, JAPAN	1095
*PARTICLE FRACTIONS	19.4 % OF TOTAL	SIZE=0.25-0.125 MM, WEIGHT=23 %, AREA=10.1 %, MARINE, JAPAN	1095
*PARTICLE FRACTIONS	17.5 % OF TOTAL	SIZE=0.125-0.075 MM, WEIGHT=16.9 %, AREA=10.4 %, MARINE, JAPAN	1095
*PARTICLE FRACTIONS	47.4 % OF TOTAL	SIZE BELOW 0.075 MM, WEIGHT=48.4 %, AREA=77.9 %, MARINE, JAPAN	1095

TABLE D335

CONCENTRATIONS : V

ITEM:	VALUE:	CONDITIONS:	REF.:
*ATMOSPHERE	0.0015+-0.0006 NG/M**3	SOUTHPOLE, 1970	893
*ATMOSPHERE	1-46 NG/M**3	URBAN, U.S.A.	893
*ATMOSPHERE	0.5-24 NG/M**3	NOT URBAN, U.S.A.	893
*ATMOSPHERE	4.2-32.2 NG/M**3	ENGLAND, 7 CITIES, 1972	893
*DRINKING WATER	0.5 UG/L	SWEDEN	893
*FOOD	0.01 UG/G WET WT.	CALFLIVER, SWEDEN	893
*FOOD	0.0001 UG/G WET WT.	MILK, SWEDEN	893
*FOOD	0.00082 UG/G WET WT.	POTATO, SWEDEN	893
*FOOD	0.23 UG/G	AM. POLISHED RICE	893
*FOOD	0.82 UG/G	JAP. POLISHED RICE	893
*RAIN	4.1 UG/L	ENGLAND, 1971, REL. UNCONTAMINATED, V IN ATM.=9.6NG/M**3	893

TABLE D335

CONCENTRATIONS : V

ITEM:	VALUE:	CONDITIONS:	REF.:
*RAT, KIDNEY	31.5 % OF TOTAL INTRACELLULAR AMOUNT	NUCLEI,12 RATS,24 H AFTER INJEC OF 10 UG V(+4)/275 G	1366
*RAT, KIDNEY	29.6 % OF TOTAL INTRACELLULAR AMOUNT	MITOCHONDRIA,12 RATS,24 H AFTER INJEC OF 10 UG V(+4)/275 G	1366
*RAT, KIDNEY	6.6 % OF TOTAL INTRACELLULAR AMOUNT	LYSOSOMES,12 RATS,24 H AFTER INJEC OF 10 UG V(+4)/275 G	1366
*RAT, KIDNEY	4.3 % OF TOTAL INTRACELLULAR AMOUNT	MICROSOMES,12 RATS,24 H AFTER INJEC OF 10 UG V(+4)/275 G	1366
*RAT, KIDNEY	28.1 % OF TOTAL INTRACELLULAR AMOUNT	SOLUBLE FRAC.,12 RATS,24 H AFTER INJEC OF 10 UG V(+4)/275 G	1366
*RAT, LIVER	44.3 % OF TOTAL INTRACELLULAR AMOUNT	NUCLEI,24 H AFTER ADMINISTRATIONM, 12 RATS ON 250-300 G,70 UCI V(+4)	1366
*RAT, LIVER	21.4 % OF TOTAL INTRACELLULAR AMOUNT	MITOCHONDRIA,24 H AFTER ADM.,12 RATS ON 250-300 G,70 UCI V(+4)	1366
*RAT, LIVER	7.2 % OF TOTAL INTRACELLULAR AMOUNT	LYSOSOMES,24 H AFTER ADMINISTRATION,12 RATS ON 250-300 G,70 UCI V(+4)	1366
*RAT, LIVER	4 % OF TOTAL INTRACELLULAR AMOUNT	MICROSOMES,12 RATS ON 250-300G, 24 H AFTER ADMIN. OFA700U CI V(+4)	1366
*RAT, LIVER	23.1 % OF TOTAL INTRACELLULAR AMOUNT	SOLUBLE FRACTION,12 RATS,24 H AFTER ADMIN.70 MICRO CI V(+4)	1366
*VEGETATION	1.0 MG/KG DRY WT.	CONC.IN PODZOL:F=20,H=10,A1A2=150, B1=200,B2=100,B3=100, (F030)	1048
*WATER	1 UG/L	SURFACE WATER,MEAN	893

TABLE D336

CONCENTRATIONS : W

ITEM:	VALUE:	CONDITIONS:	REF.:
*ATMOSPHERE	1 NG/M**3	NOT URBAN,ENGLAND,1971	893
*ATMOSPHERE	5.8 NG/M**3	CHICAGO,1969	893
*ATMOSPHERE	0.4 NG/M**3	NOT URBAN,U.S.A.,1969	893
*ATMOSPHERE	0.0015+-0.0006 NG/M**3	SOUTHPOLE,1970	893
*RAIN	0.35 UG/L	NOT URBAN,ENGLAND,1971,W IN ATM.= 1NG/M**3	893
*WATER	0.03 UG/L	RIVER,SWEDEN,1965	893

TABLE D337

CONCENTRATIONS : ZN

ITEM:	VALUE:	CONDITIONS:	REF.:
*ALISMA PLANTAGO AQUATICA, OLD LEAVES	97.46 MG/KG DRY WT.	800M ABOVE POLL.ENTRY(1),ZN IN WATER=0.056MG/L WATER,PROCESSED LEAVES	1224
*ALISMA PLANTAGO AQUATICA, OLD LEAVES	87.99 MG/KG DRY WT.	150M BELOW POLL.ENTRY(1),ZN IN WATER=0.190MG/L WATER,WHOLE LEAVES	1224
*ALISMA PLANTAGO AQUATICA, OLD LEAVES	52.64 MG/KG DRY WT.	150M BELOW POLL.ENTRY(1),ZN IN WATER=0.190MG/L WATER,PROCESSED LEAVES	1224
*ALISMA PLANTAGO AQUATICA, OLD LEAVES	117.63 MG/KG DRY WT.	800 AND 50M BELOW POLL.(1)AND(2), ZN IN WATER=0.107MG/L WATER,WHOLE LEAVE	1224
*ALISMA PLANTAGO AQUATICA, OLD LEAVES	98.22 MG/KG DRY WT.	800 AND 50M BELOW POLL.(1)AND(2), ZN IN WATER=0.107MG/L WATER, PROCESSED L	1224
*ALISMA PLANTAGO AQUATICA, YOUNG LEAVES	119.36 MG/KG DRY WT.	800M ABOVE POLL.ENTRY(1),ZN IN WATER=0.056MG/L WATER,WHOLE LEAVES	1224

TABLE D337

CONCENTRATIONS : ZN

ITEM:	VALUE:	CONDITIONS:	REF.:
*ALISMA PLANTAGO AQUATICA, YOUNG LEAVES	98.25 MG/KG DRY WT.	800M ABOVE POLL.ENTRY(1),ZN IN WATER=0.056MG/L WATER,PROCESSED LEAVES	1224
*ALISMA PLANTAGO AQUATICA, YOUNG LEAVES	199.36 MG/KG DRY WT.	150M BELOW POLL.ENTRY(1¤¤¤o IN WATER=0.190MG/L WATER,WHOLE LEAVES	1224
*ALISMA PLANTAGO AQUATICA, YOUNG LEAVES	119.82 MG/KG DRY WT.	150M BELOW POLL.ENTRY(1),ZN IN WATER=0.190MG/L WATER,PROCESSED LEAVES	1224
*ALISMA PLANTAGO AQUATICA, YOUNG LEAVES	192.45 MG/KG DRY WT.	800 AND 50M BELOW POLL.(1)AND(2), ZN IN WATER=0.107MG/L WATER,WHOLE LEAVE	1224
*ALISMA PLANTAGO AQUATICA, YOUNG LEAVES	145.87 MG/KG DRY WT.	800 AND 50M BELOW POLL.(1)AND(2), ZN IN WATER=0.107MG/L WATER, PROCESSED L	1224
*ATMOSPHERE	0.03+-0.01 NG/M**3	SOUTHPOLE,1970	893
*ATMOSPHERE	10-840 NG/M**3	URBAN,U.S.A.	893
*ATMOSPHERE	13-200 NG/M**3	NOT URBAN,U.S.A.	893
*ATMOSPHERE	57-345 NG/M**3	ENGLAND,7 STATIONS	893
*ATMOSPHERE	390 NG/M**3 (MEAN)	SWEDEN,MAX=1100NG/M**3	893
*BRYUM PSEUDOTRIQUETRUM	14045 PPM	ZN IN SOIL=1840PPM,SLATER MINE, FOOLOW,DERBYSHIRE,GB	1430
*BRYUM PSEUDOTRIQUETRUM	1544 PPM	ZN IN SOIL=1910PPM,BLEA BECK JUNCTION,W.YORKSHIRE,GB	1430
*BRYUM PSEUDOTRIQUETRUM	435 PPM	ZN IN SOIL=815PPM,SAND HAW MINE, W.YORKSHIRE,GB	1430
*CABBAGE, PLANT	20.9 PPM DRY MATTER	15 PLANTS,RURAL AREA	1283
*CABBAGE, PLANT	30.2 PPM DRY MATTER	15 PLANTS,URBAN AREA	1283
*CAMPTOTHECIUM LUTESCENS	4220 PPM	ZN IN SOIL=13220PPM,BLEA BECK HEAD,W.YORKSHIRE,GB	1430
*CONIFEROUS TREES, NEEDLES	1500 (SD=+-250) MG/KG ASH	SILICIC ROCK=180+-17,SOIL=320+-50 PPM , FINLAND,N=34	950
*CONIFEROUS TREES, NEEDLES	950 (SD=+-250) MG/KG ASH	ULTRABASIC ROCK=190+-29,SOIL=270+ -49 PPM, FINLAND,N=10	950
*CONIFEROUS TREES, NEEDLES	860 (SD=+-330) MG/KG ASH	CALCAREOUS ROCK=230+-44,SOIL=220+ -64 PPM, FINLAND,N=8	950
*CONIFEROUS TREES, TWIGS	1600 (SD=+-220) MG/KG ASH	SILICIC ROCK=180+-17,SOIL=320+-50 PPM , FINLAND,N=34	950
*CONIFEROUS TREES, TWIGS	1200 (SD=+-310) MG/KG ASH	ULTRABASIC ROCK=190+-29,SOIL=270+ -49 PPM, FINLAND,N=10	950
*CONIFEROUS TREES, TWIGS	1200 (SD=+-400) MG/KG ASH	CALCAREOUS ROCK=230+-44,SOIL=220+ -64 PPM, FINLAND,N=8	950
*CORNICULARIA MURICATA	5640 PPM	ZN IN SOIL=13220PPM,BLEA BECK HEAD,W.YORKSHIRE,GB	1430
*CRATAEGUS MONOGYNA, UNWASHED LEAVES	1300 PPM DRY WT.	AVONMOUTH,BRISTOL,GB	936
*DAPHNIA MAGNA	1 % OF TOTAL ZN AVAIBLE	UPTAKE FROM WATER	56
*DECIDUOUS TREES AND SHRUBS, LEAVES	2100 (SD=+-330) MG/KG ASH	SILICIC ROCK=180+-17,SOIL=320+-50 PPM , FINLAND,N=36	950
*DECIDUOUS TREES AND SHRUBS, LEAVES	1200 (SD=+-330) MG/KG ASH	ULTRABASIC ROCK=190+-29,SOIL=270+ -49 PPM, FINLAND,N=12	950
*DECIDUOUS TREES AND SHRUBS, LEAVES	1000 (SD=+-470) MG/KG ASH	CALCAREOUS ROCK=230+-44,SOIL=220+ -64 PPM, FINLAND,N=14	950
*DECIDUOUS TREES AND SHRUBS, TWIGS	4100 (SD=+-340) MG/KG ASH	SILICIC ROCK=180+-17,SOIL=320+-50 PPM , FINLAND,N=36	950
*DECIDUOUS TREES AND SHRUBS, TWIGS	2700 (SD=+-540) MG/KG ASH	ULTRABASIC ROCK=190+-29,SOIL=270+ -49 PPM, FINLAND,N=12	950
*DECIDUOUS TREES AND SHRUBS, TWIGS	1700 (SD=+-540) MG/KG ASH	CALCAREOUS ROCK=230+-44,SOIL=220+ -64 PPM, FINLAND,N=14	950
*DICRANELLA VARIA	5985 PPM	ZN IN SOIL=2816PPM,SPEED MINE, FOOLOW,DERBYSHIRE,GB	1430
*DICRANELLA VARIA	5250 PPM	ZN IN SOIL=2810PPM,SLATER MINE, FOOLOW,DERBYSHIRE,GB	1430
*DICRANELLA VARIA	2470 PPM	ZN IN SOIL=2700PPM,PASTYREMINE, FOOLOW,DERBYSHIRE,GB	1430
*DICRANELLA VARIA	4540 PPM	ZN IN SOIL=2645PPM,BLYTHE WOOD, ALPORT,DERBYSHIRE,GB	1430

TABLE D337

CONCENTRATIONS : ZN			
ITEM:	VALUE:	CONDITIONS:	REF.:
*DICRANELLA VARIA	18 PPM	ZN IN SOIL=20PPM,BOTANIC GARDEN, COTTINGHAM,W.YORKSHIRE,GB	1430
*DICRANELLA VARIA	6150 PPM	ZN IN SOIL=2810PPM,SLATER MINE, FOOLOW,DERBYSHIRE,GB	1430
*DICRANELLA VARIA	3600 PPM	ZN IN SOIL=2700PPM,PASTYREMINE, FOOLOW,DERBYSHIRE,GB	1430
*ENCALYPTA VULGARIS	1667 PPM	ZN IN SOIL=3960PPM,ALPORT CUPOLA SLAG HEAP,DERBYSHIRE,GB	1430
*FOOD	10-100 MG/KG DRY WT.	VEGETABLES	893
*FOOD	50-200 MG/KG DRY WT.	ANIMALS	893
*GLYCINE MAX, LEAF	57 MG/KG DRY WEIGHT	DIAGNOSTIC PART, NEW LEAF, SOIL= 0.0 UG CD/G, PH=7.7, ZN=47 PPM	127
*GLYCINE MAX, LEAF	58 MG/KG DRY WEIGHT	DIAGNOSTIC PART, NEW LEAF, SOIL= 2.5 UG CD/G, PH=7.7, ZN=47 PPM	127
*GLYCINE MAX, LEAF	49 MG/KG DRY WEIGHT	DIAGNOSTIC PART, NEW LEAF, SOIL= 5.0 UG CD/G, PH=7.7, ZN=47 PPM	127
*GLYCINE MAX, LEAF	53 MG/KG DRY WEIGHT	DIAGNOSTIC PART, NEW LEAF, SOIL= 10.0 UG CD/G, PH=7.7, ZN=47 PPM	127
*GLYCINE MAX, LEAF	48 MG/KG DRY WEIGHT	DIAGNOSTIC PART, NEW LEAF, SOIL= 20.0 UG CD/G, PH=7.7, ZN=47 PPM	127
*GLYCINE MAX, LEAF	38 MG/KG DRY WEIGHT	DIAGNOSTIC PART, NEW LEAF, SOIL= 40.0 UG CD/G, PH=7.7, ZN=47 PPM	127
*GLYCINE MAX, LEAF	36 MG/KG DRY WEIGHT	DIAGNOSTIC PART, NEW LEAF, SOIL= 80.0 UG CD/G, PH=7.7, ZN=47 PPM	127
*GLYCINE MAX, LEAF	28 MG/KG DRY WEIGHT	DIAGNOSTIC PART, NEW LEAF, SOIL= 160.0 UG CD/G, PH=7.7, ZN=47 PPM	127
*GLYCINE MAX, LEAF AND PETIOLE	49 MG/KG	ZN(SOIL TOTAL OR EXTRAC. IN 0.1N HCL)=53-13MG/KG,SLUGDE=0.MM	669
*GLYCINE MAX, LEAF AND PETIOLE	107 MG/KG	ZN(SOIL TOTAL OR EXTRAC. IN 0.1N HCL)=132-41MG/KG,SLUGDE=6.4MM	669
*GLYCINE MAX, LEAF AND PETIOLE	162 MG/KG	ZN(SOIL TOTAL OR EXTRAC. IN 0.1N HCL)=239-148MG/KG,SLUGDE=12.7MM	669
*GLYCINE MAX, LEAF AND PETIOLE	203 MG/KG	ZN(SOIL TOTAL OR EXTRAC. IN 0.1N HCL)=336-298MG/KG,SLUGDE=25.4MM	669
*GLYCINE MAX, LEAF AND PETIOLE	50 MG/KG	ZN(SOIL TOTAL OR EXTRAC. IN 0.1N HCL)=74-12MG/KG,SLUGDE=0.MM	669
*GLYCINE MAX, LEAF AND PETIOLE	146 MG/KG	ZN(SOIL TOTAL OR EXTRAC. IN 0.1N HCL)=143-70MG/KG,SLUGDE=6.4MM	669
*GLYCINE MAX, LEAF AND PETIOLE	222 MG/KG	ZN(SOIL TOTAL OR EXTRAC. IN 0.1N HCL)=265-193MG/KG,SLUGDE=12.7MM	669
*GLYCINE MAX, LEAF AND PETIOLE	444 MG/KG	ZN(SOIL TOTAL OR EXTRAC. IN 0.1N HCL)=390-316MG/KG,SLUGDE=25.4MM	669
*GLYCINE MAX, LEAF AND PETIOLE	42 MG/KG	ZN(SOIL TOTAL OR EXTRAC. IN 0.1N HCL)=80-13MG/KG,SLUGDE=0.MM	669
*GLYCINE MAX, LEAF AND PETIOLE	81 MG/KG	ZN(SOIL TOTAL OR EXTRAC. IN 0.1N HCL)=175-74MG/KG,SLUGDE=6.4MM	669
*GLYCINE MAX, LEAF AND PETIOLE	120 MG/KG	ZN(SOIL TOTAL OR EXTRAC. IN 0.1N HCL)=248-144MG/KG,SLUGDE=12.7MM	669
*GLYCINE MAX, LEAF AND PETIOLE	188 MG/KG	ZN(SOIL TOTAL OR EXTRAC. IN 0.1N HCL)=356-243MG/KG,SLUGDE=25.4MM	669
*GLYCINE MAX, LEAF AND PETIOLE	49 MG/KG	ZN(SOIL TOTAL OR EXTRAC. IN 0.1N HCL)=69-13MG/KG,SLUGDE=0.MM	669
*GLYCINE MAX, LEAF AND PETIOLE	103 MG/KG	ZN(SOIL TOTAL OR EXTRAC. IN 0.1N HCL)=136-63MG/KG,SLUGDE=6.4MM	669
*GLYCINE MAX, LEAF AND PETIOLE	132 MG/KG	ZN(SOIL TOTAL OR EXTRAC. IN 0.1N HCL)=203-126MG/KG,SLUGDE=12.7MM	669
*GLYCINE MAX, LEAF AND PETIOLE	143 MG/KG	ZN(SOIL TOTAL OR EXTRAC. IN 0.1N HCL)=370-269MG/KG,SLUGDE=25.4MM	669
*GLYCINE MAX, LEAF AND PETIO¤E	56 MG/KG	ZN(SOIL TOTAL OR EXTRAC. IN 0.1N HCL)=71-12MG/KG,SLUGDE=0.MM	669
*GLYCINE MAX, LEAF AND PETIOLE	85 MG/KG	ZN(SOIL TOTAL OR EXTRAC. IN 0.1N HCL)=178-69MG/KG,SLUGDE=6.4MM	669
*GLYCINE MAX, LEAF AND PETIOLE	106 MG/KG	ZN(SOIL TOTAL OR EXTRAC. IN 0.1N HCL)=232-145MG/KG,SLUGDE=12.7MM	669
*GLYCINE MAX, LEAF AND PETIOLE	130 MG/KG	ZN(SOIL TOTAL OR EXTRAC. IN 0.1N HCL)=339-136MG/KG,SLUGDE=25.4MM	669

TABLE D337

CONCENTRATIONS : ZN

ITEM:	VALUE:	CONDITIONS:	REF.:
*GLYCINE MAX, SEED	61 MG/KG DRY WEIGHT	EDIABLE PART, SOIL=0.0 UG CD/G, PH =7.7, ZN=47 PPM	127
*GLYCINE MAX, SEED	72 MG/KG DRY WEIGHT	EDIABLE PART, SOIL=2.5 UG CD/G, PH =7.7, ZN=47 PPM	127
*GLYCINE MAX, SEED	63 MG/KG DRY WEIGHT	EDIABLE PART, SOIL=5.0 UG CD/G, PH =7.7, ZN=47 PPM	127
*GLYCINE MAX, SEED	68 MG/KG DRY WEIGHT	EDIABLE PART, SOIL=10.0 UG CD/G, PH=7.7, ZN=47 PPM	127
*GLYCINE MAX, SEED	55 MG/KG DRY WEIGHT	EDIABLE PART, SOIL=20.0 UG CD/G, PH=7.7, ZN=47 PPM	127
*GLYCINE MAX, SEED	54 MG/KG DRY WEIGHT	EDIABLE PART, SOIL=40.0 UG CD/G, PH=7.7, ZN=47 PPM	127
*GLYCINE MAX, SEED	54 MG/KG DRY WEIGHT	EDIABLE PART, SOIL=80.0 UG CD/G, PH=7.7, ZN=47 PPM	127
*GLYCINE MAX, SEED	37 MG/KG DRY WEIGHT	EDIABLE PART, SOIL=160.0 UG CD/G, PH=7.7, ZN=47 PPM	127
*GLYCINE MAX, SEED	38 MG/KG DRY WEIGHT	DIAGNOSTIC PART, 3+4 LEAF FROM TOP, SOIL=0.0 UG CD/G, PH=7.7, ZN =47 P	127
*GLYCINE MAX, SEED TOTAL	58 MG/KG	ZN(SOIL TOTAL OR EXTRAC. IN 0.1N HCL)=53-13MG/KG,SLUGDE=0.MM	669
*GLYCINE MAX, SEED TOTAL	72 MG/KG	ZN(SOIL TOTAL OR EXTRAC. IN 0.1N HCL)=132-41MG/KG,SLUGDE=6.4MM	669
*GLYCINE MAX, SEED TOTAL	81 MG/KG	ZN(SOIL TOTAL OR EXTRAC. IN 0.1N HCL)=239-148MG/KG,SLUGDE=12.7MM	669
*GLYCINE MAX, SEED TOTAL	85 MG/KG	ZN(SOIL TOTAL OR EXTRAC. IN 0.1N HCL)=336-298MG/KG,SLUGDE=25.4MM	669
*GLYCINE MAX, SEED TOTAL	51 MG/KG	ZN(SOIL TOTAL OR EXTRAC. IN 0.1N HCL)=74-12MG/KG,SLUGDE=0.MM	669
*GLYCINE MAX, SEED TOTAL	72 MG/KG	ZN(SOIL TOTAL OR EXTRAC. IN 0.1N HCL)=143-70MG/KG,SLUGDE=6.4MM	669
*GLYCINE MAX, SEED TOTAL	82 MG/KG	ZN(SOIL TOTAL OR EXTRAC. IN 0.1N HCL)=265-193MG/KG,SLUGDE=12.7MM	669
*GLYCINE MAX, SEED TOTAL	94 MG/KG	ZN(SOIL TOTAL OR EXTRAC. IN 0.1N HCL)=390-316MG/KG,SLUGDE=25.4MM	669
*GLYCINE MAX, SEED TOTAL	55 MG/KG	ZN(SOIL TOTAL OR EXTRAC. IN 0.1N HCL)=80-13MG/KG,SLUGDE=0.MM	669
*GLYCINE MAX, SEED TOTAL	71 MG/KG	ZN(SOIL TOTAL OR EXTRAC. IN 0.1N HCL)=175-74MG/KG,SLUGDE=6.4MM	669
*GLYCINE MAX, SEED TOTAL	78 MG/KG	ZN(SOIL TOTAL OR EXTRAC. IN 0.1N HCL)=248-144MG/KG,SLUGDE=12.7MM	669
*GLYCINE MAX, SEED TOTAL	86 MG/KG	ZN(SOIL TOTAL OR EXTRAC. IN 0.1N HCL)=356-243MG/KG,SLUGDE=25.4MM	669
*GLYCINE MAX, SEED TOTAL	49 MG/KG	ZN(SOIL TOTAL OR EXTRAC. IN 0.1N HCL)=69-13MG/KG,SLUGDE=0.MM	669
*GLYCINE MAX, SEED TOTAL	63 MG/KG	ZN(SOIL TOTAL OR EXTRAC. IN 0.1N HCL)=136-63MG/KG,SLUGDE=6.4MM	669
*GLYCINE MAX, SEED TOTAL	65 MG/KG	ZN(SOIL TOTAL OR EXTRAC. IN 0.1N HCL)=203-126MG/KG,SLUGDE=12.4MM	669
*GLYCINE MAX, SEED TOTAL	78 MG/KG	ZN(SOIL TOTAL OR EXTRAC. IN 0.1N HCL)=370-269MG/KG,SLUGDE=25.4MM	669
*GLYCINE MAX, SEED TOTAL	54 MG/KG	ZN(SOIL TOTAL OR EXTRAC. IN 0.1N HCL)=71-12MG/KG,SLUGDE=0.MM	669
*GLYCINE MAX, SEED TOTAL	59 MG/KG	ZN(SOIL TOTAL OR EXTRAC. IN 0.1N HCL)=178-69MG/KG,SLUGDE=6.4MM	669
*GLYCINE MAX, SEED TOTAL	62 MG/KG	ZN(SOIL TOTAL OR EXTRAC. IN 0.1N HCL)=232-145MG/KG,SLUGDE=12.7MM	669
*GLYCINE MAX, SEED TOTAL	66 MG/KG	ZN(SOIL TOTAL OR EXTRAC. IN 0.1N HCL)=339-236MG/KG,SLUGDE=25.4MM	669
*GUINEA PIG, RENAL CORTEX	151 UG/G DRY WT.(SD=19)	1 MONTH OF CIGARETTE SMOKE,N=22, CONTROL=133UG/G D.W.(SD=9),N=15	487
*HYPOLIMNION	1.9 UG/L	CONC.IN EPILIMNION=1.9UG/L, CONC.FACTOR=1.1,SCHOEHSEE	573
*LIVING SPRUCE NEEDLES	27 - 188 MG/KG DRY WEUGHT	LITTER=60-126, UNDERLAYING SOIL= 28-41MG/KG DW,SOUTH SWEDEN	1129
*NEREIS DIVERSICOLOR	130 - 350 MICROGRAM/G DRY WEIGHT	RANGE IN SEDIMENNT = 100 - 300 UG/ G DW, ENGLAND	188

TABLE D337

CONCENTRATIONS : ZN

ITEM:	VALUE:	CONDITIONS:	REF.:
*ORYZA SATIVA, GRAIN	39 MG/KG DRY WEIGHT	EDIABLE PART, SOIL=0.0 UG CD/G, PH =7.7, ZN=47 PPM	127
*ORYZA SATIVA, GRAIN	43 MG/KG DRY WEIGHT	EDIABLE PART, SOIL=2.5 UG CD/G, PH =7.7, ZN=47 PPM	127
*ORYZA SATIVA, GRAIN	37 MG/KG DRY WEIGHT	EDIABLE PART, SOIL=5.0 UG CD/G, PH =7.7, ZN=47 PPM	127
*ORYZA SATIVA, GRAIN	29 MG/KG DRY WEIGHT	EDIABLE PART, SOIL=10.0 UG CD/G, PH=7.7, ZN=47 PPM	127
*ORYZA SATIVA, GRAIN	27 MG/KG DRY WEIGHT	EDIABLE PART, SOIL=20.0 UG CD/G, PH=7.7, ZN=47 PPM	127
*ORYZA SATIVA, GRAIN	28 MG/KG DRY WEIGHT	EDIABLE PART, SOIL=40.0 UG CD/G, PH=7.7, ZN=47 PPM	127
*ORYZA SATIVA, GRAIN	24 MG/KG DRY WEIGHT	EDIABLE PART, SOIL=80.0 UG CD/G, PH=7.7, ZN=47 PPM	127
*ORYZA SATIVA, GRAIN	21 MG/KG DRY WEIGHT	EDIABLE PART, SOIL=160.0 UG CD/G, PH=7.7, ZN=47 PPM	127
*ORYZA SATIVA, LEAF	15 MG/KG DRY WEIGHT	DIAGNOSTIC PART, 2 TOP LEAVES, SOIL=0.0 UG CD/G, PH=7.7, ZN=47 PPM	127
*ORYZA SATIVA, LEAF	16 MG/KG DRY WEIGHT	DIAGNOSTIC PART, 2 TOP LEAVES, SOIL=2.5 UG CD/G, PH=7.7, ZN=47 PPM	127
*ORYZA SATIVA, LEAF	17 MG/KG DRY WEIGHT	DIAGNOSTIC PART, 2 TOP LEAVES, SOIL=5.0 UG CD/G, PH=7.7, ZN=47 PPM	127
*ORYZA SATIVA, LEAF	16 MG/KG DRY WEIGHT	DIAGNOSTIC PART, 2 TOP LEAVES, SOIL=10.0 UG CD/G, PH=7.7, ZN=47 PPM	127
*ORYZA SATIVA, LEAF	14 MG/KG DRY WEIGHT	DIAGNOSTIC PART, 2 TOP LEAVES, SOIL=20.0 UG CD/G, PH=7.7, ZN=47 PPM	127
*ORYZA SATIVA, LEAF	16 MG/KG DRY WEIGHT	DIAGNOSTIC PART, 2 TOP LEAVES, SOIL=40.0 UG CD/G, PH=7.7, ZN=47 PPM	127
*ORYZA SATIVA, LEAF	20 MG/KG DRY WEIGHT	DIAGNOSTIC PART, 2 TOP LEAVES, SOIL=80.0 UG CD/G, PH=7.7, ZN=47 PPM	127
*ORYZA SATIVA, LEAF	13 MG/KG DRY WEIGHT	DIAGNOSTIC PART, 2 TOP LEAVES, SOIL=160.0 UG CD/G, PH=7.7, ZN=47 PPM	127
*PELTIGERA,CANINA	452 PPM	ZN IN SOIL=1800PPM,BLEA BECK HEAD, W.YORKSHIRE,GB	1430
*PELTIGERA,CANINA	5650 PPM	ZN IN SOIL=3960PPM,ALPORT CUPOLA SLAG HEAP,DERBYSHIRE,GB	1430
*PHASEOLUS VULGARIS, DRY BEAN	23 MG/KG DRY WEIGHT	EDIBLE PART, SOIL=0.0 UG CD/G, PH= 7.7, ZN=47 PPM	127
*PHASEOLUS VULGARIS, DRY BEAN	25 MG/KG DRY WEIGHT	EDIBLE PART, SOIL=2.5 UG CD/G, PH= 7.7, ZN=47 PPM	127
*PHASEOLUS VULGARIS, DRY BEAN	21 MG/KG DRY WEIGHT	EDIBLE PART, SOIL=5.0 UG CD/G, PH= 7.7, ZN=47 PPM	127
*PHASEOLUS VULGARIS, DRY BEAN	21 MG/KG DRY WEIGHT	EDIBLE PART, SOIL=10.0 UG CD/G, PH =7.7, ZN=47 PPM	127
*PHASEOLUS VULGARIS, DRY BEAN	20 MG/KG DRY WEIGHT	EDIBLE PART, SOIL=20.0 UG CD/G, PH =7.7, ZN=47 PPM	127
*P¤ASEOLUS VULGARIS, DRY BEAN	15 MG/KG DRY WEIGHT	EDIBLE PART, SOIL=40.0 UG CD/G, PH =7.7, ZN=47 PPM	127
*PHASEOLUS VULGARIS, DRY BEAN	13 MG/KG DRY WEIGHT	EDIBLE PART, SOIL=80.0 UG CD/G, PH =7.7, ZN=47 PPM	127
*PHASEOLUS VULGARIS, DRY BEAN	12 MG/KG DRY WEIGHT	EDIBLE PART, SOIL=160.0 UG CD/G, PH=7.7, ZN=47 PPM	127
*PHASEOLUS VULGARIS, LEAF	47 MG/KG DRY WEIGHT	DIAGNOSTIC PART, NEW LEAF, SOIL= 0.0 UG CD/G, PH=7.7, ZN=47 PPM	127
*PHASEOLUS VULGARIS, LEAF	44 MG/KG DRY WEIGHT	DIAGNOSTIC PART, NEW LEAF, SOIL= 2.5 UG CD/G, PH=7.7, ZN=47 PPM	127
*PHASEOLUS VULGARIS, LEAF	39 MG/KG DRY WEIGHT	DIAGNOSTIC PART, NEW LEAF, SOIL= 5.0 UG CD/G, PH=7.7, ZN=47 PPM	127
*PHASEOLUS VULGARIS, LEAF	31 MG/KG DRY WEIGHT	DIAGNOSTIC PART, NEW LEAF, SOIL= 10.0 UG CD/G, PH=7.7, ZN=47 PPM	127
*PHASEOLUS VULGARIS, LEAF	29 MG/KG DRY WEIGHT	DIAGNOSTIC PART, NEW LEAF, SOIL= 20.0 UG CD/G, PH=7.7, ZN=47 PPM	127

TABLE D337

CONCENTRATIONS : ZN

ITEM:	VALUE:	CONDITIONS:	REF.:
*PHASEOLUS VULGARIS, LEAF	29 MG/KG DRY WEIGHT	DIAGNOSTIC PART, NEW LEAF, SOIL= 40.0 UG CD/G, PH=7.7, ZN=47 PPM	127
*PHASEOLUS VULGARIS, LEAF	21 MG/KG DRY WEIGHT	DIAGNOSTIC PART, NEW LEAF, SOIL= 80.0 UG CD/G, PH=7.7, ZN=47 PPM	127
*PHASEOLUS VULGARIS, LEAF	21 MG/KG DRY WEIGHT	DIAGNOSTIC PART, NEW LEAF, SOIL= 160.0 UG CD/G, PH=7.7, ZN=47 PPM	127
*PHILONOTIS FONTANA(NEW)	297 PPM	ZN IN SOIL=1350PPM,SLATER MINE, FOOLOW,DERBYSHIRE,GB	1430
*PHILONOTIS FONTANA(NEW)	204 PPM	ZN IN SOIL=1840PPM,SLATER MINE, FOOLOW,DERBYSHIRE,GB	1430
*PHILONOTIS FONTANA(OLD)	2420 PPM	ZN IN SOIL=1350PPM,BLEA BECK JUNCTION,W.YORKSHIRE,GB	1430
*PHILONOTIS FONTANA(OLD)	8750 PPM	ZN IN SOIL=1840PPM,SLATER MINE, FOOLOW,DERBYSHIRE,GB	1430
*PLANTAGO AQUATICA, OLD LEAVES	125.11 MG/KG DRY WT.	800M ABOVE POLL.ENTRY(1),ZN IN WATER=0.056MG/L WATER,WHOLE LEAVES	1224
*QUERCUS ROBUR, UNWASHED LEAVES	1617 PPM DRY WT.	AVONMOUTH,BRISTOL,GB	936
*RAIN	85 UG/L	ENGLAND,1 STATION,YEARLY MEAN	893
*RAIN	15.5-84 UG/L	SWEDEN,3 STATIONS	893
*RED CLOVER	75 MG/KG DRY WT.	ADDED ZNSO4=22KG/HA,PH=6.1;CONTROL =33MG/KG DRY WT.	1048
*RED CLOVER	44 MG/KG DRY WT.	ADDED ZNSO4=22KG/HA,PH=6.5;CONTROL =29MG/KG DRY WT.	1048
*SALIX ALBA, UNWASHED LEAVES	1699 PPM DRY WT.	AVONMOUTH,BRISTOL,GB	936
*SEDIMENT	320 PPM OF DRY WEIGHT	WATER CONC = 10 MICROGRAM/L, YEARLY MEAN, CARNON RIVER	48
*SEDIMENT	1235 PPM OF DRY WEIGHT	WATER CONC = 110 MICROGRAM/L, YEARLY MEAN, RED RIVER	48
*SEDIMENT	445 PPM OF DRY WEIGHT	WATER CONC = 33 MICROGRAM/L, YEARLY MEAN, RED RIVER	48
*SEDIMENT	2610 PPM OF DRY WEIGHT	WATER CONC = 338 MICROGRAM/L, YEARLY MEAN, GANNEL RIVER	48
*SEDIMENT	247 PPM OF DRY WEIGHT	WATER CONC = 10 MICROGRAM/L, YEARLY MEAN, GANNEL RIVER	48
*SEDIMENT	161 PPM OF DRY WEIGHT	WATER CONC = 3 MICROGRAM/L, YEARLY MEAN, FOWEY RIVER	48
*SEDIMENT	65 PPM OF DRY WEIGHT	WATER CONC = 6 MICROGRAM/L, YEARLY MEAN, FOWEY RIVER	48
*SEDIMENT	3 PPM OF DRY WEIGHT	WATER CONC = 3 MICROGRAM/L, YEARLY MEAN, CARNON RIVER	48
*SEDIMENT	320 PPM OF DRY WEIGHT	MAX WATER CONC = 12 MICROGRAM/L, 1 YEARS MEASUREMENT, CARNON R	48
*SEDIMENT	1235 PPM OF DRY WEIGHT	MAX WATER CONC = 198 MICROGRAM/L, 1 YEARS MEASUREMENT, RED R	48
*SEDIMENT	445 PPM OF DRY WEIGHT	MAX WATER CONC = 105 MICROGRAM/L, 1 YEARS MEASUREMENT, RED R	48
*SEDIMENT	2610 PPM OF DRY WEIGHT	MAX WATER CONC = 1000 MICROGRAM/L, MAX CONC, GANNEL RIVER	48
*SEDIMENT	247 PPM OF DRY WEIGHT	MAX WATER CONC = 35 MICROGRAM/L, MAX CONC, GANNEL RIVER	48
*SEDIMENT	161 PPM OF DRY WEIGHT	MAX WATER CONC = 8 MICROGRAM/L, 1 YEARS MEASUREMENT, FOWEY R	48
*SEDIMENT	65 PPM OF DRY WEIGHT	MAX WATER CONC = 12 MICROGRAM/L, 1 YEARS MEASUREMENT, FOWEY R	48
*SEDIMENT	3 PPM OF DRY WEIGHT	MAX WATER CONC = 5 MICROGRAM/L, 1 YEARS MEASUREMENT, CARNON R	48
*SEDIMENT	24.8 (RANGE 17.2 - 42.0) MG/KG DRY WEIGHT	UNPOLLUTED AREA, ESTUARY, URR WATER	1557
*SEDIMENT	41.2 (RANGE 24.8 - 65.0) MG/KG DRY WEIGHT	UNPOLLUTED AREA, ESTUARY, URR WATER	1558
*SEDIMENT	67.0 (RANGE 43.0 - 97.5) MG/KG DRY WEIGHT	UNPOLLUTED AREA, ESTUARY, URR WATER	1558
*SEDIMENT	30 (RANGE 18 - 41) MG/KG DRY WEIGHT	UNPOLLUTED AREA, RIVER,ILLINOIS RIVER	993
*SEDIMENT	105.84 (96.33-112.61) MG/KG DRY WT.	800M ABOVE POLL.ENTRY(1),ZN IN WATER=0.056MG/L WATER,POLLUTED RIVER	1224

TABLE D337

CONCENTRATIONS : ZN			
ITEM:	VALUE:	CONDITIONS:	REF.:
*SEDIMENT	345.77 (328.51-365.64) MG/KG DRY WT.	150M BELOW POLL.ENTRY(1),ZN IN WATER=0.190MG/L WATER,POLLUTED RIVER	1224
*SEDIMENT	100.29 (96.12-105.72) MG/KG DRY WT.	800 AND 50M BELOW POLL.(1)AND(2), ZN IN WATER=0.107MG/L WATER, POLLUTED RI	1224
*SEDIMENT	350 UG/G DRY WT.	CONC.IN EPILIMNION=1.8UG/L, CONC.FACTOR=195000,SCHOEHSEE	573
*SOIL (SEDIMENT)	129. PPM DRY SOIL	RANGE 66-192PPM,RIVER RHEIDOL, CARDIGANSHIRE,GB	327
*TRITICUM AESTIVUM, GRAIN	45 MG/KG DRY WEIGHT	EDIABLE PART, SOIL=0.0 UG CD/G, PH =7.7, ZN=47 PPM	127
*TRITICUM AESTIVUM, GRAIN	47 MG/KG DRY WEIGHT	EDIABLE PART, SOIL=2.5 UG CD/G, PH =7.7, ZN=47 PPM	127
*TRITICUM AESTIVUM, GRAIN	42 MG/KG DRY WEIGHT	EDIABLE PART, SOIL=5.0 UG CD/G, PH =7.7, ZN=47 PPM	127
*TRITICUM AESTIVUM, GRAIN	48 MG/KG DRY WEIGHT	EDIABLE PART, SOIL=10.0 UG CD/G, PH=7.7, ZN=47 PPM	127
*TRITICUM AESTIVUM, GRAIN	41 MG/KG DRY WEIGHT	EDIABLE PART, SOIL=20.0 UG CD/G, PH=7.7, ZN=47 PPM	127
*TRITICUM AESTIVUM, GRAIN	44 MG/KG DRY WEIGHT	EDIABLE PART, SOIL=40.0 UG CD/G, PH=7.7, ZN=47 PPM	127
*TRITICUM AESTIVUM, GRAIN	32 MG/KG DRY WEIGHT	EDIABLE PART, SOIL=80.0 UG CD/G, PH=7.7, ZN=47 PPM	127
*TRITICUM AESTIVUM, GRAIN	29 MG/KG DRY WEIGHT	EDIABLE PART, SOIL=160.0 UG CD/G, PH=7.7, ZN=47 PPM	127
*TRITICUM AESTIVUM, LEAF	44 MG/KG DRY WEIGHT	DIAGNOSTIC PART, 3 TOP LEAF, SOIL= 0.0 UG CD/G, PH=7.7, ZN=47 PPM	127
*TRITICUM AESTIVUM, LEAF	38 MG/KG DRY WEIGHT	DIAGNOSTIC PART, 3 TOP LEAF, SOIL= 2.5 UG CD/G, PH=7.7, ZN=47 PPM	127
*TRITICUM AESTIVUM, LEAF	35 MG/KG DRY WEIGHT	DIAGNOSTIC PART, 3 TOP LEAF, SOIL= 5.0 UG CD/G, PH=7.7, ZN=47 PPM	127
*TRITICUM AESTIVUM, LEAF	41 MG/KG DRY WEIGHT	DIAGNOSTIC PART, 3 TOP LEAF, SOIL= 10.0 UG CD/G, PH=7.7, ZN=47 PPM	127
*TRITICUM AESTIVUM, LEAF	34 MG/KG DRY WEIGHT	DIAGNOSTIC PART, 3 TOP LEAF, SOIL= 20.0 UG CD/G, PH=7.7, ZN=47 PPM	127
*TRITICUM AESTIVUM, LEAF	37 MG/KG DRY WEIGHT	DIAGNOSTIC PART, 3 TOP LEAF, SOIL= 40.0 UG CD/G, PH=7.7, ZN=47 PPM	127
*TRITICUM AESTIVUM, LEAF	31 MG/KG DRY WEIGHT	DIAGNOSTIC PART, 3 TOP LEAF, SOIL= 80.0 UG CD/G, PH=7.7, ZN=47 PPM	127
*TRITICUM AESTIVUM, LEAF	24 MG/KG DRY WEIGHT	DIAGNOSTIC PART, 3 TOP LEAF, SOIL= 160.0 UG CD/G, PH=7.7, ZN=47 PPM	127
*WATER	1.67 UG/L	BALTIC SEA,LANDSORTDYBET:400M, 1970-71	893
*WATER	39.19 UG/L	BALTIC SEA,FLADEN:10M,1970-71	893
*WATER	16-96 UG/L	RIVERS,U.S.A.,MEAN	893
*ZEA MAYS, GRAIN	32 MG/KG DRY WEIGHT	EDIABLE PART, SOIL=0.0 UG CD/G, PH =7.7, ZN=47 PPM	127
*ZEA MAYS, GRAIN	38 MG/KG DRY WEIGHT	EDIABLE PART, SOIL=2.5 UG CD/G, PH =7.7, ZN=47 PPM	127
*ZEA MAYS, GRAIN	47 MG/KG DRY WEIGHT	EDIABLE PART, SOIL=5.0 UG CD/G, PH =7.7, ZN=47 PPM	127
*ZEA MAYS, GRAIN	47 MG/KG DRY WEIGHT	EDIABLE PART, SOIL=10.0 UG CD/G, PH=7.7, ZN=47 PPM	127
*ZEA MAYS, GRAIN	45 MG/KG DRY WEIGHT	EDIABLE PART, SOIL=20.0 UG CD/G, PH=7.7, ZN=47 PPM	127
*ZEA MAYS, GRAIN	54 MG/KG DRY WEIGHT	EDIABLE PART, SOIL=40.0 UG CD/G, PH=7.7, ZN=47 PPM	127
*ZEA MAYS, GRAIN	41 MG/KG DRY WEIGHT	EDIABLE PART, SOIL=80.0 UG CD/G, PH=7.7, ZN=47 PPM	127
*ZEA MAYS, GRAIN	53 MG/KG DRY WEIGHT	EDIABLE PART, SOIL=160.0 UG CD/G, PH=7.7, ZN=47 PPM	127
*ZEA MAYS, LEAF	42 MG/KG DRY WEIGHT	DIAGNOSTIC PART, 3+4 LEAF FROM TOP, SOIL=2.5 UG CD/G, PH=7.7, ZN =47 P	127
*ZEA MAYS, LEAF	39 MG/KG DRY WEIGHT	DIAGNOSTIC PART, 3+4 LEAF FROM TOP, SOIL=5.0 UG CD/G, PH=7.7, ZN =47 P	127

TABLE D337

CONCENTRATIONS : ZN

ITEM:	VALUE:	CONDITIONS:	REF.:
*ZEA MAYS, LEAF	33 MG/KG DRY WEIGHT	DIAGNOSTIC PART, 3+4 LEAF FROM TOP, SOIL=10.0 UG CD/G, PH=7.7, ZN=47	127
*ZEA MAYS, LEAF	26 MG/KG DRY WEIGHT	DIAGNOSTIC PART, 3+4 LEAF FROM TOP, SOIL=20.0 UG CD/G, PH=7.7, ZN=47	127
*ZEA MAYS, LEAF	27 MG/KG DRY WEIGHT	DIAGNOSTIC PART, 3+4 LEAF FROM TOP, SOIL=40.0 UG CD/G, PH=7.7, ZN=47	127
*ZEA MAYS, LEAF	24 MG/KG DRY WEIGHT	DIAGNOSTIC PART, 3+4 LEAF FROM TOP, SOIL=80.0 UG CD/G, PH=7.7, ZN=47	127
*ZEA MAYS, LEAF	22 MG/KG DRY WEIGHT	DIAGNOSTIC PART, 3+4 LEAF FROM TOP, SOIL=160.0 UG CD/G, PH=7.7, ZN=47	127
*ZOOPLANKTON	162.53 MG/KG DRY WT.(MEAN)	19ST.,N.Y.BIGHT,CONC.IN PHYTOPL. (LONG ISLAND)=63.93MG/KG DRY WT.MEAN	565

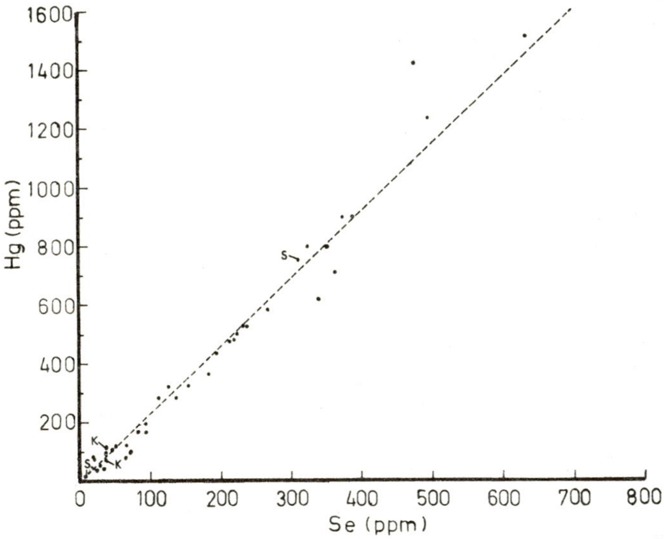

D338 Selenium and mercury correlation in seals:
in liver s spleen k kidney.

(ref: 355)

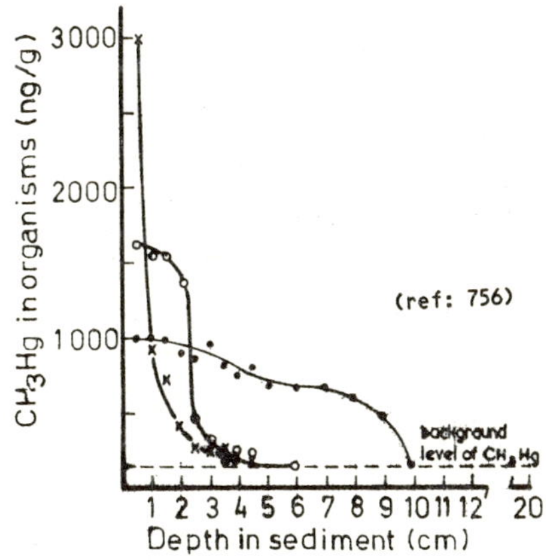

D339 Methyl mercury in fish, when the layer of inorganic mercury is situated in different depths.
Series with only microorganisms ×
Series with Tubificidae o
Series with Anodonta •

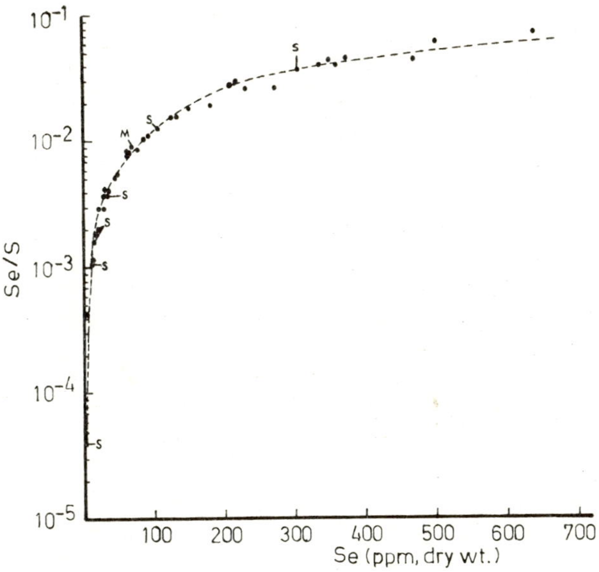

D340 Selenium/sulphur ratios versus total Se content for various organs of the seal:
• liver s spleen m muscle.

(ref: 355)

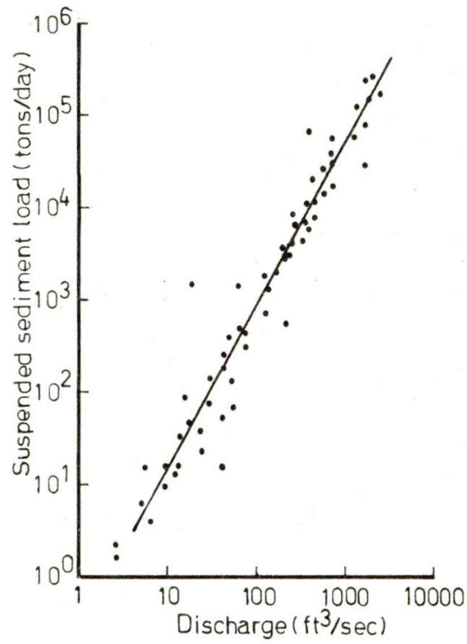

D341 Suspended sediment load versus discharge as measured at a single gauging station.

(ref: 910)

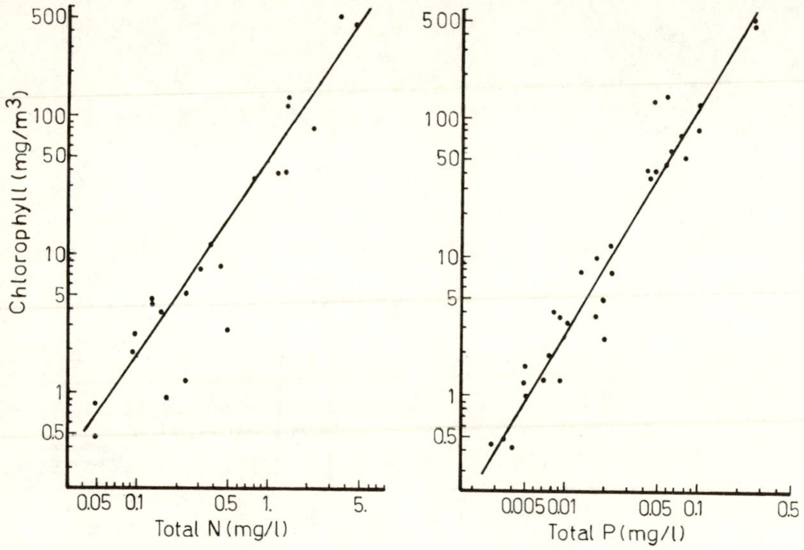

D342 Chlorophyll A and nitrogen resp. phosphorus relationship in japanese waters. (ref: 1369)

D343

Odour threshold of phenols before and after chlorination

Compounds	Odour threshold concentration (mg/l)	Odour threshold concentration after chlorination (mg/l)
Phenol	18	0.001
Cresols	0.0025	0.0001-0.001
Thymol	0.23	0.05-0.1
Naphthol	70	0.5-1.0
	40	-
Creosote (technical)	0.12	0.01-0.05

(ref: 765)

TABLE D344

Mercury Content of Aphids after Sixteen days feeding on Tomato Seedlings Grown in Methyl Mercury Hydroxide.
REF.: 606

Treatment (ppm)	Hg (ppm wet wt)	Mean wet wt / aphid (mg)	% inhibition of wet wt	Mean no. of aphids / seedling	% inhibition of aphids no.	Mean reproduction / adult female	% inhibition of fecundity	Days to become reproductive	% inhibition of development rate
0.04	3.62	0.2	13.8	2.67	99	0	100	.	100
0.02	1.61	0.175	24.5	57.63	68	4	41	14	29
0.004	0.38	0.124	46.5	118	35	4.7	31	12	17
0	0.56	0.232	.	180.9	.	6.8	.	10	.

TABLE D345

Copper Uptake from Soil by Barley in Suffolk. Concentration in Mean Grain Content (mg/kg dry weight).
REF.: 218

Copper sulphate applied in 1952 (kg/ha)	Grain conc. in 1952	Grain conc. in 1953	Grain conc. in 1954	Grain conc. in 1955	Grain conc. in 1956	Mean
0	1.6	1.6	1.1	1.2	1.8	1.4
11.2	1.9	1.2	1	1.2	2.5	1.6
22.4	1.5	1.4	2	2.6	2.4	2
33.6	2	1.8	2	3.2	3.4	2.5
44.8	2.7	2	1.3	2	3	2.2
67.2	1.8	2.6	2.8	3.8	3.7	2.9
134.4	2.5	3.3	2.4	3.6	4.4	3.2

TABLE D346

Distributions of Cd, Pb and Zn in the Major Vegetative Components (g/ha) and Soils (kg/ha) in an eastern Tennessee mixed deciduous forest.

REF.: 691

Component	Species	Cd	Pb	Zn
Leaves	Quercus prinus	1	15	71
Leaves	Quercus spp.-Carya spp.	1	18	81
Leaves	Pinus echinata	2	17	55
Leaves	Liriodendron tulipifera	1	14	65
Branches	Quercus prinus	4	58	120
Branches	Quercus spp.-Carya spp.	5	85	120
Branches	Pinus echinata	5	47	150
Branches	Liriodendron tulipifera	3	70	140
Bole	Quercus prinus	8	87	240
Bole	Quercus spp.-Carya spp.	14	133	540
Bole	Pinus echinata	13	100	380
Bole	Liriodendron tulipifera	9	88	360
01 litter	Quercus prinus	3	190	290
01 litter	Quercus spp.-Carya spp.	5	220	420
01 litter	Pinus echinata	3	340	610
01 litter	Liriodendron tulipifera	5	200	380
02 litter	Quercus prinus	15	940	2000
02 litter	Quercus spp.-Carya spp.	4	630	2200
02 litter	Pinus echinata	9	580	930
02 litter	Liriodendron tulipifera	8	320	980
soil	Quercus prinus	0.7	90	353
soil	Quercus spp.-Carya spp.	0.7	90	353
soil	Pinus echinata	0.7	90	353
soil	Liriodendron tulipifera	0.9	370	358
<0.5 cm dia. Roots	Quercus prinus	5	73	190
<0.5 cm dia. Roots	Quercus spp.-Carya spp.	3	200	280
<0.5 cm dia. Roots	Pinus echinata	3	58	320
<0.5 cm dia. Roots	Liriodendron tulipifera	6	95	470
0.5 - 2.0 cm dia. Roots	Quercus prinus	2	20	88
0.5 - 2.0 cm dia. Roots	Quercus spp.-Carya spp.	2	46	110
0.5 - 2.0 cm dia. Roots	Pinus echinata	2	11	88
0.5 - 2.0 cm dia. Roots	Liriodendron tulipifera	1	17	77
> 2.0 cm dia. Roots	Quercus prinus	1	6	37
> 2.0 cm dia. Roots	Quercus spp.-Carya spp.	1	11	43
> 2.0 cm dia. Roots	Pinus echinata	1	5	31
> 2.0 cm dia. Roots	Liriodendron tulipifera	1	7	43

TABLE D347

The Passage of Six Heavy Metals from Medium to Bacteries, and from Bacteries to Tubificid Worms. The Bacteries being used as the sole source of food for the worms.

REF.: 1223

Sample	Cr	Cu	Mn	Fe	Pb	Zn
Worms as collected, ug/g dry weight	19.1	80.71	13.04	1286	151.4	529.8
Worms fed cells grown in medium, no metals added, ug/g dry weight	3.92	236.2	13.19	737.7	178.9	261.9
Worms fed cells grown in medium + 1 ug/ml metal, ug/g dry weight	14.08	397	17.92	1242	559.2	685.1
Worms fed cells grown in medium + 2 ug/ml metal, ug/g dry weight	29.86	621.2	24.97	1944	568.3	868.3
Bacterial cells - no metals added, ug/g dry weight	109.1	213.2	16.53	289.3	119	206.6
Bacterial cells + 1 ug/ml metals, ug/g dry weight	983	765.9	292.8	2253	409.6	1648
Bacterial cells + 2 ug/ml metals, ug/g dry weight	2850	1068	558	3450	720.8	2010
Bacterial medium - no metals added, ug/ml	.	0.24	0.03	0.06	0.08	0.14
Bacterial medium + 1 ug/ml metal, ug/ml	1	1.2	0.78	0.85	1.05	1.28
Bacterial medium + 2 ug/ml metal, ug/ml	2.05	2.4	1.66	1.82	1.9	2

TABLE D348

Ratios and Enrichment factors based on National Air Sampling Network. Data for 1965. Ratio = concentration of metal (ug/m3)/(concentration of metal in suspended particles (ug/m3)). Enrichment factor = ratio in town/(ratio in background). Background = Matagorda.
REF.: 355

Element	Ratio for Huston	Ratio for Matagorda	Enrichment factor for Huston
Cd	78E-6	1E-5	7.8
Cr	.0015	.0004	3.66
Fe	.0078	.0049	1.59
Pb	.0062	.0005	12.76
Ni	70E-6	46E-6	1.52

D4

CONCENTRATION, PERMISSIBLE AND DAYLY INTAKE

OF
CHEMICAL COMPOUNDS
ELEMENTS

TABLE D349

CONCENTRATIONS PERMISSIBLE : ALDRIN			
ITEM:	VALUE:	CONDITIONS:	REF.:
*CITRUS FRUIT	0.05 MG/KG	MAX.CONC.PERMITTED IN HUMAN FOOD	500
*FOOD	0.1 MG/KG	MAX.CONC.PERMITTED IN HUMAN FOOD (F032)	500
*FRUITS (OTHER THAN CITRUS)	0.1 MG/KG	MAX.CONC.PERMITTED IN HUMAN FOOD	500
*POTATOES	0.2 MG/KG	MAX.CONC.PERMITTED IN HUMAN FOOD	500
*PUBLIC WATER SUPPLIES	17 NG/L	SURFACE WATER	1124
*RICE, ROUGH	0.02 MG/KG	MAX.CONC.PERMITTED IN HUMAN FOOD	500

TABLE D350

CONCENTRATIONS PERMISSIBLE : CARBAMATES			
ITEM:	VALUE:	CONDITIONS:	REF.:
*PUBLIC WATER SUPPLIES	100 NG/L	SURFACE WATER	1124

TABLE D351

CONCENTRATIONS PERMISSIBLE : CHLORDANE			
ITEM:	VALUE:	CONDITIONS:	REF.:
*COTTONSEED OIL, CRUDE	0.1 MG/KG	MAX.CONC.PERMITTED IN HUMAN FOOD	500
*COTTONSEED OIL, EDIBLE	0.02 MG/KG	MAX.CONC.PERMITTED IN HUMAN FOOD	500
*FOOD	0.3 MG/KG	MAX.CONC.PERMITTED IN HUMAN FOOD (F033)	500
*FOOD	0.2 MG/KG	MAX.CONC.PERMITTED IN HUMAN FOOD (F034)	500
*FOOD	0.02 MG/KG	MAX.CONC.PERMITTED IN HUMAN FOOD (F035)	500
*FOOD	0.05 MG/KG	MAX.CONC.PERMITTED IN HUMAN FOOD (F036)	500
*FOOD	0.1 MG/KG	MAX.CONC.PERMITTED IN HUMAN FOOD (F037)	500
*FOOD	0.1 MG/KG	MAX.CONC.PERMITTED IN HUMAN FOOD (F038)	500
*FOOD	0.02 MG/KG	MAX.CONC.PERMITTED IN HUMAN FOOD (F039)	500
*FOOD	0.5 MG/KG	MAX.CONC.PERMITTED IN HUMAN FOOD (F040)	500
*PUBLIC WATER SUPPLIES	3 NG/L	SURFACE WATER	1124
*SOYBEAN OIL,EDIBLE	0.02 MG/KG	MAX.CONC.PERMITTED IN HUMAN FOOD	500

TABLE D352

CONCENTRATIONS PERMISSIBLE : DDT			
ITEM:	VALUE:	CONDITIONS:	REF.:
*FOOD	7 MG/KG	MAX.CONC.PERMITTED IN HUMAN FOOD (F041)	500
*FOOD	1 MG/KG	MAX.CONC.PERMITTED IN HUMAN FOOD (F042)	500
*FOOD	3.5 MG/KG	MAX.CONC.PERMITTED IN HUMAN FOOD (F043)	500
*PUBLIC WATER SUPPLIES	42 NG/L	SURFACE WATER	1124

TABLE D353

CONCENTRATIONS PERMISSIBLE : DIELDRIN			
ITEM:	VALUE:	CONDITIONS:	REF.:
*CITRUS FRUIT	0.05 MG/KG	MAX.CONC.PERMITTED IN HUMAN FOOD	500
*FOOD	0.1 MG/KG	MAX.CONC.PERMITTED IN HUMAN FOOD (F032)	500
*FRUITS (OTHER THAN CITRUS)	0.1 MG/KG	MAX.CONC.PERMITTED IN HUMAN FOOD	500
*POTATOES	0.2 MG/KG	MAX.CONC.PERMITTED IN HUMAN FOOD	500
*PUBLIC WATER SUPPLIES	17 NG/L	SURFACE WATER	1124
*RICE, ROUGH	0.02 MG/KG	MAX.CONC.PERMITTED IN HUMAN FOOD	500

TABLE D354

CONCENTRATIONS PERMISSIBLE : ENDRIN			
ITEM:	VALUE:	CONDITIONS:	REF.:
*COTTONSEED AND - OIL,CRUDE	0.1 MG/KG	MAX.CONC.PERMITTED IN HUMAN FOOD	500
*FOOD	0.02 MG/KG	MAX.CONC.PERMITTED IN HUMAN FOOD (F044)	500
*FOOD	0.02 MG/KG	MAX.CONC.PERMITTED IN HUMAN FOOD (F045)	500
*PUBLIC WATER SUPPLIES	1 NG/L	SURFACE WATER	1124

TABLE D355

CONCENTRATIONS PERMISSIBLE : HEPTACHLOR			
ITEM:	VALUE:	CONDITIONS:	REF.:
*PINEAPPLE, EDIBLE PART	0.01 MG/KG	MAX.CONC.PERMITTED IN HUMAN FOOD	500
*PUBLIC WATER SUPPLIES	18 NG/L	SURFACE WATER	1124

TABLE D356

CONCENTRATIONS PERMISSIBLE : HEPTACHLOR EPOXIDE			
ITEM:	VALUE:	CONDITIONS:	REF.:
*PUBLIC WATER SUPPLIES	18 NG/L	SURFACE WATER	1124

TABLE D357

CONCENTRATIONS PERMISSIBLE : LINDANE			
ITEM:	VALUE:	CONDITIONS:	REF.:
*PUBLIC WATER SUPPLIES	56 NG/L	SURFACE WATER	1124

TABLE D358

CONCENTRATIONS PERMISSIBLE : METHOXYCHLOR			
ITEM:	VALUE:	CONDITIONS:	REF.:
*PUBLIC WATER SUPPLIES	35 NG/L	SURFACE WATER	1124

TABLE D359

CONCENTRATIONS PERMISSIBLE : ORG.PHOSPHATES			
ITEM:	VALUE:	CONDITIONS:	REF.:
*PUBLIC WATER SUPPLIES	100 NG/L	SURFACE WATER	1124

TABLE D360

CONCENTRATIONS PERMISSIBLE : TOXAPHENE:2,4-D PLUS 2,4,5-T AND 2,4,5-TP			
ITEM:	VALUE:	CONDITIONS:	REF.:
*PUBLIC WATER SUPPLIES	100 NG/L	SURFACE WATER	1124

ABLE D361

DAILY INTAKE FROM : FOOD

ITEM:	VALUE:	CONDITIONS:	REF.:
*AL	18-36 MG/24HRS	NORMAL INTAKE	893
*B	0.3-41 MG/24HRS	MOSTLY VEGETABLES	893
*BA	0.735 MG/24HRS	U.S.A.	893
*BE	0.012 MG/24HRS	U.S.A.	893
*BI	0.02 MG/24HRS	U.S.A.	893
*CD	4-60 UG/DAY	HOMO SAPIENS U.S.A.,DITHIZONE,1961	487
*CD	48 UG/DAY	HOMO SAPIENS WESTERN GERMANY, ATOM.ABS.AFTER EXTRACTION,1969	487
*CD	38-64 UG/DAY	HOMO SAPIENS ROUMANIA,DITHIZONE, 1970	487
*CD	60 UG/DAY	HOMO SAPIENS CZECHOSLOVAKIA, DITHIZONE,ISOTOPE DILUT.OR.ATOM.ABS.,1970	487
*CD	59 UG/DAY	HOMO SAPIENS JAPAN(NONPOLL.AR.), DITHIZONE OR ATOM.ABS.AFT.EXTRACT.,1970	487
*CD	92 UG/DAY	HOMO SAPIENS U.S.A.,ATOM.ABS.,1971	487
*CD	50 UG/DAY	HOMO SAPIENS U.S.A.(1969), DITHIZONE OR ATOM.ABS.	487
*CD	38 UG/DAY	HOMO SAPIENS U.S.A.(1970), DITHIZONE OR ATOM.ABS.	487
*CD	47 UG/DAY	HOMO SAPIENS JAPAN(NONPOLL.AR.), 1972	487
*CD	0.05 MG/24HRS	NONE	893
*CO	0.39 MG/24HRS	U.S.A.,INCL.DRINKING WATER	893
*CR	0.245 MG/24HRS	U.S.A.	893
*FE	15 MG/24HRS	SWEDEN	893
*HG	0.01-0.02 MG/24HRS	U.S.A.	893
*LI	2 MG/24HRS	NONE	893
*MN	4.40 MG/24HRS	INCL.WATER,U.S.A.	893
*MO	0.335 MG/24HRS	INCL.WATER,U.S.A.	893
*NI	0.6 MG/24HRS	U.S.A.,INCL.WATER	893
*SB	0.1 MG/24HRS	U.S.A.	893
*SE	0.06-0.15 MG/24HRS	U.S.A.	893
*TI	1.4 MG/24HRS	U.S.A.	893
*V	0.116 MG/24HRS	U.S.A.	893
*ZN	14.5 MG/24HRS	INCL.WATER,U.S.A.	893

TABLE D362

DAILY INTAKE, ACCEPTABLE : HOMO SAPIENS			
ITEM:	VALUE:	CONDITIONS:	REF.:
*ALDRIN	0.0001 MG/KG BODY WT.	NONE	500
*CHLORDANE	0.001 MG/KG BODY WT.	NONE	500
*DDT	0.005 MG/KG BODY WT.	NONE	500
*DIELDRIN	0.0001 MG/KG BODY WT.	NONE	500
*ENDRIN	0.0002 MG/KG BODY WT.	NONE	500
*HEPTACHLOR	0.0005 MG/KG BODY WT.	NONE	500

D5

EXCRETION AND UPTAKE

OF
CHEMICAL COMPOUNDS
ELEMENTS
FROM/TO
THE
ENVIRONMENT

TABLE D363

EXCRETION : AL			
ITEM:	VALUE:	CONDITIONS:	REF.:
*HOMO SAPIENS	17/18 OF DOSE	WITH FECES,ORAL INTAKE=18MG/DAY, 30DAYS	1599
*HOMO SAPIENS	1/17 OF DOSE	WITH URINE,ORAL INTAKE=18MG/DAY, 30DAYS	1599

TABLE D364

EXCRETION : BA			
ITEM:	VALUE:	CONDITIONS:	REF.:
*RAT	7% OF DOSE	WITH URINE,INJECT.OF BASO4,24HRS	1228
*RAT	20% OF DOSE	WITH FECES,INJECT.OF BASO4,24HRS	1228

TABLE D365

EXCRETION : BE			
ITEM:	VALUE:	CONDITIONS:	REF.:
*ANIMAL	94% OF DOSE	WITH FECES,ORAL INTAKE	183
*ANIMAL	1.6% OF DOSE	WITH URINE,ORAL INTAKE	183

TABLE D366

EXCRETION : CA-45			
ITEM:	VALUE:	CONDITIONS:	REF.:
*FUCUS VESICULOSUS	90 %	IN FIRST FEW HOURS, 295 K	1540

TABLE D367

EXCRETION : CD			
ITEM:	VALUE:	CONDITIONS:	REF.:
*ACHETA DOMESTICUS	0.9 +- 0.3 1/DAY	GRASSLAND FOOD CHAIN, PREDATOR, CD OXIDE, FAST PHASE = 83 %	689
*ACHETA DOMESTICUS	0.09 +- 0.04 1/DAY	GRASSLAND FOOD CHAIN, PREDATOR, CD OXIDE, SLOW PHASE = 17 %	689
*ACHETA DOMESTICUS	0.8 +- 0.3 1/DAY	GRASSLAND FOOD CHAIN, PREDATOR, CD (NO3)2, FAST PHASE = 83 %	689
*ACHETA DOMESTICUS	0.11 +- 0.05 1/DAY	GRASSLAND FOOD CHAIN, PREDATOR, CD (NO3)2, SLOW PHASE = 17 %	689
*ANIMAL	90-98% OF DOSE	WITH FECES,ORAL INTAKE	487
*GOAT, FECES	5.6% OF DOSE	SINGLE INTRAV.INJECT.OF 109-CD, AFTER 5DAYS	487
*GOAT, URINE	0.5% OF ABS.AMOUNT	SINGLE ORAL DOSE OF 109-CD,1DAY AFTER EXPOSURE	487
*GOAT, URINE	1.0% OF ABS.AMOUNT	SINGLE ORAL DOSE OF 109-CD,2DAYS AFTER EXPOSURE	487
*GOAT, URINE	0.1% OF ABS.AMOUNT	SINGLE ORAL DOSE OF 109-CD,3-7DAYS AFTER EXPOSURE	487
*GOAT, URINE	0.025% OF DOSE	SINGLE INTRAV.INJECT.OF 109-CD, 1DAY AFTER INJECT.	487

TABLE D367

EXCRETION : CD

ITEM:	VALUE:	CONDITIONS:	REF.:
*GOAT, URINE	0.004% OF DOSE	SINGLE INTRAV.INJECT.OF 109-CD, 2DAYS AFTER INJECT.	487
*GOAT, URINE	0.002% OF DOSE	SINGLE INTRAV.INJECT.OF 109-CD, 3-7DAYS AFTER INJECT.	487
*HOMO SAPIENS, KIDNEY	0.005% BODY BURDEN PER DAY	BIOL.HALF-TIME=38YEARS	487
*HOMO SAPIENS, KIDNEY	0.01% BODY BURDEN PER DAY	BIOL.HALF-TIME=19YEARS	487
*HOMO SAPIENS, KIDNEY	0.02% BODY BURDEN PER DAY	BIOL.HALF-TIME=9.5YEARS	487
*HOMO SAPIENS, KIDNEY	0.05% BODY BURDEN PER DAY	BIOL.HALF-TIME=3.8YEARS	487
*HOMO SAPIENS, URINE OF EXPOSED WORKERS	530-1120 UG/24HR	FRANCE,N=2,CDO DUST,NEPHELOMETRIC, 1954	487
*HOMO SAPIENS, URINE OF EXPOSED WORKERS	12-487 UG/L	ENGLAND,N=12,CDO DUST,DITHIZONE, 1955	487
*HOMO SAPIENS, URINE OF EXPOSED WORKERS	14-428 UG/L	ENGLAND,N=26,CDO DUST,DITHIZONE, 1957	487
*HOMO SAPIENS, URINE OF EXPOSED WORKERS	9-36 UG/L	ENGLAND,N=10,CDO DUST,NOT DIRECTLY EXPOSED,DITHIZONE,1957	487
*HOMO SAPIENS, URINE OF EXPOSED WORKERS	16-425 UG/L	ENGLAND,N=27,CDO FUME,DITHIZONE, 1957	487
*HOMO SAPIENS, URINE OF EXPOSED WORKERS	0-15.9 UG/L	JAPAN,N=34,CD STEARATE,DITHIZONE, 1965	487
*HOMO SAPIENS, URINE OF EXPOSED WORKERS	3-140 UG/L	JAPAN,N=13,CDO FUME,DITHIZONE,1967	487
*HOMO SAPIENS, URINE OF EXPOSED WORKERS	0-168 UG/L	ENGLAND,N=56,CDO DUST,DITHIZONE, 1969	487
*HOMO SAPIENS, URINE OF EXPOSED WORKERS	0.8-6.2 UG/G CREATININE	W.GERMANY,N=18,CDO FUME CDS,AAS, AFTER MIBK-APDC EXTRACTION,1969	487
*HOMO SAPIENS, URINE OF NOT EXPOSED PERSONS	1.59 (<0.5-10.8) UG/L	U.S.A.,ADULT,N=154,SPECTROGR.AFTER DITHIZONE EXTRACTION,1963	487
*HOMO SAPIENS, URINE OF NOT EXPOSED PERSONS	3.1 (0-15.9) UG/L	JAPAN,ADULT,N=30,DITHIZONE,1965	487
*HOMO SAPIENS, URINE OF NOT EXPOSED PERSONS	0.47 UG/L (SD=0.25)	JAPAN(GIFU),4-6 YRS,N=46,AAS, AFTER EXTRACTION,1971	487
*HOMO SAPIENS, URINE OF NOT EXPOSED PERSONS	0.65 UG/L (SD=0.45)	JAPAN(GIFU),9-10 YRS,N=40,AAS, AFTER EX¤RACTION,1971	487
*HOMO SAPIENS, URINE OF NOT EXPOSED PERSONS	0.72 UG/L (SD=0.5)	JAPAN(GIFU),14-15 YRS,N=41,AAS, AFTER EXTRACTION,1971	487
*HOMO SAPIENS, URINE OF NOT EXPOSED PERSONS	0.99 UG/L (SD=0.63)	JAPAN(GIFU),20-29 YRS,N=56,AAS, AFTER EXTRACTION,1971	487
*HOMO SAPIENS, URINE OF NOT EXPOSED PERSONS	1.13 UG/L (SD=1.06)	JAPAN(GIFU),30-39 YRS,N=37,AAS, AFTER EXTRACTION,1971	487
*HOMO SAPIENS, URINE OF NOT EXPOSED PERSONS	1.76 UG/L (SD=1.33)	JAPAN(GIFU),40-49 YRS,N=40,AAS, AFTER EXTRACTION,1971	487
*HOMO SAPIENS, URINE OF NOT EXPOSED PERSONS	1.75 UG/L (SD=1.38)	JAPAN(GIFU),50-59 YRS,N=43,AAS, AFTER EXTRACTION,1971	487
*HOMO SAPIENS, URINE OF NOT EXPOSED PERSONS	1.0 UG/L	W.GERMANY,ADULTS,N=14,DITHIZONE, 1969	487
*HOMO SAPIENS, URINE OF NOT EXPOSED PERSONS	0.98 (0.34-1.57) UG/24HR	W.GERMANY,ADULTS,N=15,AAS, AFTER MIBK-APDC EXTRACTION,1968	487
*HOMO SAPIENS, URINE OF NOT EXPOSED PERSONS	1.25 UG/G CREATININE	W.GERMANY,INCL.CHILDREN,N=169,AAS, AFTER MIBK-APDC EXTRACT.,1969	487
*HOMO SAPIENS, URINE OF NOT EXPOSED PERSONS	0.95 UG/G CREATININE (SD=0.8)	BELGIUM,ADULTS,N=44,AAS, AFTER EXTRACTION,1972	487
*HOMO SAPIENS, URINE OF NOT EXPOSED PERSONS	0.39 (0.05-0.77) UG/24HR	SWEDEN,20-47 = AGE,N=10,AAS, AFTER MIBK-APDC EXTRACTION	487
*HOMO SAPIENS, URINE OF NOT EXPOSED PERSONS	0.62 (0.1-2.0) UG/G CREATININE	SWEDEN,50-59 = AGE,N=88,AAS, AFTER EXTRACTION,1973	487
*HOMO SAPIENS, URINE OF NOT EXPOSED PERSONS	0.21 (0.1-0.3) UG/G CREATININE	SWEDEN,15-16 = AGE,N=10,AAS, AFTER EXTRACTION	487
*HOMO SAPIENS, URINE OF NOT EXPOSED PERSONS	1.7 (0.4-3.7) UG/G CREATININE	SWEDEN,34-63 YRS,CONTAM.AREA,N=10, ATOM.ABS.AFTER EXTRACTION,1972	487
*LYCOSA SP.	0.007 1/DAY	GRASSLAND FOOD CHAIN, PREDATOR	689
*RABBIT, FECES	1.6% OF DOSE/DAY	0.25MG CD/KG DAY,5DAYS/W FOR 29W, SUBCUT.INJ.,AFTER 11WEEKS	487
*RABBIT, FECES	2.8% OF DOSE/DAY	0.25MG CD/KG DAY,5DAYS/W FOR 29W, SUBCUT.INJ.,AFTER 17WEEKS	487

TABLE D367

EXCRETION : CD			
ITEM:	VALUE:	CONDITIONS:	REF.:
*RABBIT, FECES	11.2% OF DOSE/DAY	0.25MG CD/KG DAY,5DAYS/W FOR 29W, SUBCUT.INJ.,AFTER 23WEEKS	487
*RABBIT, FECES	6.6% OF DOSE/DAY	0.25MG CD/KG DAY,5DAYS/W FOR 29W, SUBCUT.INJ.,AFTER 29WEEKS	487
*RABBIT, URINE	<1% OF DOSE	0.65MG 115-CD/KG DAY, SUBCUT.INJECT.,0-7WEEKS	487
*RABBIT, URINE	>100*VALUE FOR 0-7 WEEKS	0.65MG 115-CD/KG DAY, SUBCUT.INJECT.,8-10WEEKS	487
*RAT, BILES	0.005% OF DOSE/HR	0.5MG 109-CD/KG,SUBCUT.INJECT., DURING 3DAYS	487
*RAT, BILES	5% OF FECAL EXCRETION/DAY	0.5MG 109-CD/KG,SUBCUT.INJECT., DURING 3DAYS	487
*RAT, FECES	7.3% OF DOSE	0.63MG 115-CD/KG,SINGLE INTRAV. INJEC.,AFTER 24HR	487
*RAT, FECES	18.5% OF DOSE	0.63MG 115-CD/KG,SINGLE INTRAV.INJECT.,AFTER 72HR	487
*RAT, FECES	5% OF DOSE	2.2UG 109-CD,SINGLE SUBCUT.INJECT.,AFTER 4DAYS	487
*RAT, URINE	1.25% OF ABS.AMOUNT	SINGLE ORAL DOSE,1DAY AFTER EXPOSURE	487
*RAT, URINE	0.65% OF ABS.AMOUNT	SINGLE ORAL DOSE,6DAYS AFTER EXPOSURE	487
*RAT, URINE	0.003-0.007% OF DOSE /DAY	SINGLE SUBCUT.INJECT.,AFTER 4DAYS	487

TABLE D368

EXCRETION : CE-141			
ITEM:	VALUE:	CONDITIONS:	REF.:
*MEGANYCTIPHANES NORVEGICA	99 %	LOSS OF TOTAL BODY BURDEN BY 1 MOLT, UPTAKE FROM WATER	481
*MEGANYCTIPHANES NORVEGICA	99 %	LOSS OF TOTAL BODY BURDEN BY 1 MOLT, UPTAKE FROM WATER	481

TABLE D369

EXCRETION : CO			
ITEM:	VALUE:	CONDITIONS:	REF.:
*HOMO SAPIENS	3-27% OF DOSE	WITH FECES,ORAL INTAKE,23 WOMEN, 3WEEKS	622
*HOMO SAPIENS	67% OF DOSE	WITH URINE,ORAL INTAKE,23 WOMEN, 3WEEKS	622

TABLE D370

EXCRETION : CU			
ITEM:	VALUE:	CONDITIONS:	REF.:
*ANIMAL	70% OF DOSE	WITH FECES,ORAL INTAKE	462
*ANIMAL	< 1% OF DOSE	WITH URINE,ORAL INTAKE	462

TABLE D371

EXCRETION : DIELDRIN			
ITEM:	VALUE:	CONDITIONS:	REF.:
*ICTALURUS PUNCTATUS, DORSAL MUSCLE	152 MICROGRAM/KG WET WEIGHT	UPTAKE FROM WATER=75 NG/L,INITIAL= 176 PPM,7 D,150-225 MM	1421

TABLE D371

EXCRETION : DIELDRIN			
ITEM:	VALUE:	CONDITIONS:	REF.:
*ICTALURUS PUNCTATUS, DORSAL MUSCLE	78 MICROGRAM/KG WET WEIGHT	UPTAKE FROM WATER=75 NG/L,INITIAL= 176 PPM,14 D,150-225 MM	1421
*ICTALURUS PUNCTATUS, DORSAL MUSCLE	47 MICROGRAM/KG WET WEIGHT	UPTAKE FROM WATER=75 NG/L,INITIAL= 176 PPM,21 D,150-225 MM	1421
*ICTALURUS PUNCTATUS, DORSAL MUSCLE	24 MICROGRAM/KG WET WEIGHT	UPTAKE FROM WATER=75 NG/L,INITIAL= 176 PPM,28 D,150-225 MM	1421
*ICTALURUS PUNCTATUS, DORSAL MUSCLE	162 MICROGRAM/KG WET WEIGHT	UPTAKE FROM WATER=75 NG/L,INITIAL= 186 PPM,7 D,350-400 MM	1421
*ICTALURUS PUNCTATUS, DORSAL MUSCLE	84 MICROGRAM/KG WET WEIGHT	UPTAKE FROM WATER=75 NG/L,INITIAL= 186 PPM,14 D,350-400 MM	1421
*ICTALURUS PUNCTATUS, DORSAL MUSCLE	59 MICROGRAM/KG WET WEIGHT	UPTAKE FROM WATER=75 NG/L,INITIAL= 186 PPM,21 D,350-400 MM	1421
*ICTALURUS PUNCTATUS, DORSAL MUSCLE	28 MICROGRAM/KG WET WEIGHT	UPTAKE FROM WATER=75 NG/L,INITIAL= 186 PPM,28 D,350-400 MM	1421
*ICTALURUS PUNCTATUS, DORSAL MUSCLE	217 MICROGRAM/KG WET WEIGHT	UPTAKE FROM FOOD = 2 PPM, INITIAL= 287 PPM,7 D,150-225 MM	1421
*ICTALURUS PUNCTATUS, DORSAL MUSCLE	168 MICROGRAM/KG WET WEIGHT	UPTAKE FROM FOOD = 2 PPM, INITIAL= 287 PPM,14 D,150-225 MM	1421
*ICTALURUS PUNCTATUS, DORSAL MUSCLE	152 MICROGRAM/KG WET WEIGHT	UPTAKE FROM FOOD = 2 PPM, INITIAL= 287 PPM,21 D,150-225 MM	1421
*ICTALURUS PUNCTATUS, DORSAL MUSCLE	108 MICROGRAM/KG WET WEIGHT	UPTAKE FROM FOOD = 2 PPM, INITIAL= 287 PPM,28 D,150-225 MM	1421
*ICTALURUS PUNCTATUS, DORSAL MUSCLE	229 MICROGRAM/KG WET WEIGHT	UPTAKE FROM FOOD = 2 PPM, INITIAL= 302 PPM,7 D,350-400 MM	1421
*ICTALURUS PUNCTATUS, DORSAL MUSCLE	184 MICROGRAM/KG WET WEIGHT	UPTAKE FROM FOOD = 2 PPM, INITIAL= 302 PPM,14 D,350-400 MM	1421
*ICTALURUS PUNCTATUS, DORSAL MUSCLE	142 MICROGRAM/KG WET WEIGHT	UPTAKE FROM FOOD = 2 PPM, INITIAL= 302 PPM,21 D,350-400 MM	1421
*ICTALURUS PUNCTATUS, DORSAL MUSCLE	109 MICROGRAM/KG WET WEIGHT	UPTAKE FROM FOOD = 2 PPM, INITIAL= 302 PPM,28 D,350-400 MM	1421
*ICTALURUS PUNCTATUS, DORSAL MUSCLE	241 MICROGRAM/KG WET WEIGHT	UP. FROM FOOD(2 PPM) AND WATE(75 NG/L),INIT=365 PPM7 D,150-225 MM	1421
*ICTALURUS PUNCTATUS, DORSAL MUSCLE	217 MICROGRAM/KG WET WEIGHT	UP. FROM FOOD(2 PPM) AND WATE(75 NG/L),INIT=365 PPM14 D,150-225 MM	1421
*ICTALURUS PUNCTATUS, DORSAL MUSCLE	195 MICROGRAM/KG WET WEIGHT	UP. FROM FOOD(2 PPM) AND WATE(75 NG/L),INIT=365 PPM21 D,150-225 MM	1421
*ICTALURUS PUNCTATUS, DORSAL MUSCLE	172 MICROGRAM/KG WET WEIGHT	UP. FROM FOOD(2 PPM) AND WATE(75 NG/L),INIT=365 PPM28 D,150-225 MM	1421
*ICTALURUS PUNCTATUS, DORSAL MUSCLE	358 MICROGRAM/KG WET WEIGHT	UP. FROM FOOD(2 PPM) AND WAT (75 NG/L),INIT=634 PPM7 D,350-400 MM	1421
*ICTALURUS PUNCTATUS, DORSAL MUSCLE	221 MICROGRAM/KG WET WEIGHT	UP. FROM FOOD(2 PPM) AND WAT (75 NG/L),INIT=634 PPM14 D,350-400 MM	1421
*ICTALURUS PUNCTATUS, DORSAL MUSCLE	194 MICROGRAM/KG WET WEIGHT	UP. FROM FOOD(2 PPM) AND WAT (75 NG/L),INIT=634 PPM21 D,350-400 MM	1421
*ICTALURUS PUNCTATUS, DORSAL MUSCLE	170 MICROGRAM/KG WET WEIGHT	UP. FROM FOOD(2 PPM) AND WAT (75 NG/L),INIT=634 PPM28 D,350-400 MM	1421

TABLE D372

EXCRETION : DINITRAMINE			
ITEM:	VALUE:	CONDITIONS:	REF.:
*CARP, BILE	13.8 MG/KG WET WEIGHT	24 H AFTER TRANSFERENCE TO UNCONT. WAT., INIT=22.7 MG/KG WET WEIGHT	1182
*CARP, MUSCLE	4.98 MG/KG WET WEIGHT	24 H AFTER TRANSFERENCE TO UNCONT. WAT., INIT=39.6 MG/KG WET WEIGHT	1182
*CARP, PLASMA	5.1 MG/KG WET WEIGHT	24 H AFTER TRANSFERENCE TO UNCONT. WAT., INIT=24.4 MG/KG WET WEIGHT	1182
*CHANNAL CATFISH, BILE	14.5 MG/KG WET WEIGHT	24 H AFTER TRANSFERENCE TO UNCONT. WAT., INIT=13.4 MG/KG WET WEIGHT	1182
*CHANNAL CATFISH, MUSCLE	20.0 MG/KG WET WEIGHT	24 H AFTER TRANSFERENCE TO UNCONT. WAT., INIT=46.1 MG/KG WET WEIGHT	1182
*CHANNAL CATFISH, PLASMA	19.6 MG/KG WET WEIGHT	24 H AFTER TRANSFERENCE TO UNCONT. WAT., INIT=58.5 MG/KG WET WEIGHT	1182

TABLE D373

EXCRETION : ETHYLENTHIOUREA			
ITEM:	VALUE:	CONDITIONS:	REF.:
*GUINEA PIG	17.75 (SE=+-1.72) % OF ADMINATRATED	0 - 6 H AFTER DOSE, EXCRETION WITH URINE, N=6	1123
*GUINEA PIG	27.18 (SE=+-0.66) % OF ADMINATRATED	6 - 24 H AFTER DOSE, EXCRETION WITH URINE, N=6	1123
*GUINEA PIG	2.48 (SE=+-0.43) % OF ADMINATRATED	48 - 72 H AFTER DOSE, EXCRETION WITH URINE, N=6	1123
*GUINEA PIG	TRACE	48 - 72 H AFTER DOSE, EXCRETION WITH URINE, N=6	1123
*GUINEA PIG	17.75 (SE=+-1.72) % OF ADMINATRATED	0 - 6 H AFTER DOSE, EXCRETION WITH URINE, N=6	1123
*GUINEA PIG	27.18 (SE=+-0.66) % OF ADMINATRATED	6 - 24 H AFTER DOSE, EXCRETION WITH URINE, N=6	1123
*GUINEA PIG	2.48 (SE=+-0.43) % OF ADMINATRATED	48 - 72 H AFTER DOSE, EXCRETION WITH URINE, N=6	1123
*GUINEA PIG	TRACE	48 - 72 H AFTER DOSE, EXCRETION WITH URINE, N=6	1123
*RAT	18.60 (SE=+-2.38) % OF ADMINATRATED	0 - 6 H AFTER DOSE, EXCRETION WITH URINE, N=6	1123
*RAT	42.68 (SE=+-2.36) % OF ADMINATRATED	6 - 24 H AFTER DOSE, EXCRETION WITH URINE, N=6	1123
*RAT	3.65 (SE=+-0.50) % OF ADMINATRATED	24 - 48 H AFTER DOSE, EXCRETION WITH URINE, N=6	1123
*RAT	0.42 (SE=+-0.12) % OF ADMINATRATED	48 -72 H AFTER DOSE, EXCRETION WITH URINE, N=6	1123
*RAT	TRACE	72 - 96 H AFTER DOSE, EXCRETION WITH URINE, N=6	1123
*RAT	18.60 (SE=+-2.38) % OF ADMINATRATED	0 - 6 H AFTER DOSE, EXCRETION WITH URINE, N=6	1123
*RAT	42.68 (SE=+-2.36) % OF ADMINATRATED	6 - 24 H AFTER DOSE, EXCRETION WITH URINE, N=6	1123
*RAT	3.65 (SE=+-0.50) % OF ADMINATRATED	24 - 48 H AFTER DOSE, EXCRETION WITH URINE, N=6	1123
*RAT	0.42 (SE=+-0.12) % OF ADMINATRATED	48 -72 H AFTER DOSE, EXCRETION WITH URINE, N=6	1123
*RAT	TRACE	72 - 96 H AFTER DOSE, EXCRETION WITH URINE, N=6	1123

TABLE D374

EXCRETION : FE			
ITEM:	VALUE:	CONDITIONS:	REF.:
*HOMO SAPIENS	90-95% OF DOSE	WITH FECES,ORAL INTAKE,NORMAL PERSON	462
*HOMO SAPIENS	80-90% OF DOSE	WITH FECES,ORAL INTAKE,PERSON WITH FE-DEFICIT	462

TABLE D375

EXCRETION : GE			
ITEM:	VALUE:	CONDITIONS:	REF.:
*DOG	90% OF DOSE	WITH URINE,INTRAVEN.INJECT.,72HRS	183
*RABBIT	75% OF DOSE	WITH URINE,INTRAVEN.INJECT.	183
*RABBIT	9% OF DOSE	WITH FECES,INTRAVEN.INJECT.	183

TABLE D376

EXCRETION : HG			
ITEM:	VALUE:	CONDITIONS:	REF.:
*HOMO SAPIENS, FECES	90% OF TOTAL ELIMINATION	ORAL DOSE OF METHYL-HG	486

TABLE D376

EXCRETION : HG			
ITEM:	VALUE:	CONDITIONS:	REF.:
*HOMO SAPIENS, FECES	34% OF DOSE	ORAL DOSE OF METHYL-HG,AFTER 49D	486
*HOMO SAPIENS, FECES	1.9% OF DOSE/DAY	ORAL DOSE OF METHYL-HG,DUR.WEEK 1	486
*HOMO SAPIENS, FECES	0.09MG/DAY	0.05-0.1MG HG-VAPOR/M**3,N= 15WORKERS,RESP.UPTAKE	486
*HOMO SAPIENS, FECES	0.14MG/DAY	0.11-0.2MG HG-VAPOR/M**3,N= 10WORKERS,RESP.UPTAKE	486
*HOMO SAPIENS, URINE	3% OF DOSE	ORAL DOSE OF METHYL-HG,AFTER 49D	486
*HOMO SAPIENS, URINE	0.01% OF DOSE/DAY	ORAL DOSE OF METHYL-HG,DUR.WEEK 1	486
*HOMO SAPIENS, URINE	0.12MG/DAY	0.05-0.1MG HG-VAPOR/M**3,N= 15WORKERS,RESP.UPTAKE	486
*HOMO SAPIENS, URINE	0.19MG/DAY	0.11-0.2MG HG-VAPOR/M**3,N= 10WORKERS,RESP.UPTAKE	486
*MOUSE	FECES/URINE=4	INTRAV.INJECT.OF METHYL-HG-SALT, DUR.21D	486
*PIG, FECES	10% OF DOSE	INJECT.OF METHYL-HG	486
*PIG, URINE	2% OF DOSE	INJECT.OF METHYL-HG,AFTER 14D	486
*RABBIT, FECES	27% OF DOSE	INTRAV.INJECT.OF 1MG HGCL2/KG, AFTER 14D	486
*RABBIT, URINE	39% OF DOSE	INTRAV.INJECT.OF 1MG HGCL2/KG, AFTER 14D	486
*RAT	20% OF INJECT.AMOUNT	INTRAV.INJECT.,HG-VAPOR,AFTER 30SEC.	486
*RAT	FECES/URINE=4	RESP.INTAKE OF ELEMENT.HG(1.4MG/ M**3)FOR 5HRS,DUR.THE FIRST 2D	486
*RAT	FECES/URINE: 4->2 (>2)	RESP.INTAKE OF ELEMENT.HG(1.4MG/ M**3)FOR 5HRS,AFTER 2D	486
*RAT, FECES	10% OF DOSE	INTRAV.INJECT.OF HG-ACETATE	486
*RAT, FECES	30% OF DOSE	INTRAV.INJECT.OF PHENYL-HG-ACETATE	486
*RAT, FECES	80% OF DOSE	ORAL DOSE OF 0.5MG HG-ACETATE/KG, AFTER 48HRS	486
*RAT, FECES	18% OF DOSE	INTRAV.INJECT.OF 0.5MG HG-ACETATE/ KG,AFTER 48HRS	486
*RAT, FECES	30% OF DOSE	SUBCUT.INJECT.OF 0.5MG HGCL2/KG DAY,7DAYS/WEEK,AFTER 14D	486
*RAT, URINE	1% OF DOSE	ORAL DOSE OF 0.5MG HG-ACETATE/KG, AFTER 48HRS	486
*RAT, URINE	10% OF DOSE	INTRAV.INJECT.OF 0.5MG HG-ACETATE/ KG,AFTER 48HRS	486
*RAT, URINE	70% OF DOSE	SUBCUT.INJECT.OF 0.5MG HGCL2/KG DAY,7DAYS/WEEK,AFTER 14D	486

TABLE D377

EXCRETION : IN			
ITEM:	VALUE:	CONDITIONS:	REF.:
*ANIMAL	40% OF DOSE	INJECT.,URINARY AND FECAL EXCRETION,30DAYS	1461
*ANIMAL	100% OF DOSE	ORAL INTAKE,4DAYS	1461

TABLE D378

EXCRETION : LI			
ITEM:	VALUE:	CONDITIONS:	REF.:
*ANIMAL	45-75% OF AMOUNT IN BODY	WITH URINE,24HRS	1139

TABLE D379

EXCRETION : MIREX			
ITEM:	VALUE:	CONDITIONS:	REF.:
*JAPANESE QUAIL	25 % OF ORAL DOSIS	CUMULATIV REL. WITH FECES, MALE, 1 D AFTER DOSIS, IN BODY = 73 %	730
*JAPANESE QUAIL	10 % OF ORAL DOSIS	CUMULATIV REL. WITH FECES, FEMALE, 1 D AFTER DOSIS, IN BODY = 86 %	730
*JAPANESE QUAIL	26 % OF ORAL DOSIS	CUMULATIV REL. WITH FECES, MALE, 3 D AFTER DOSIS, IN BODY = 71 %	730
*JAPANESE QUAIL	11 % OF ORAL DOSIS	CUMULATIV REL. WITH FECES, FEMALE, 3 D AFTER DOSIS, IN BODY = 76 %	730
*JAPANESE QUAIL	0.09 % OF ORAL DOSIS	CUMULATIV REL. WITH WHITE, FEMALE, 3 D AFTER DOSIS, IN BODY = 76 %	730
*JAPANESE QUAIL	13 % OF ORAL DOSIS	CUMULATIV REL. WITH YOLK, FEMALE, 3 D AFTER DOSIS, IN BODY = 76 %	730
*JAPANESE QUAIL	26 % OF ORAL DOSIS	CUMULATIV REL. WITH FECES, MALE, 7 D AFTER DOSIS, IN BODY = 68 %	730
*JAPANESE QUAIL	0.14 % OF ORAL DOSIS	CUMULATIV REL. WITH WHITE, FEMALE, 7 D AFTER DOSIS, IN BODY = 38 %	730
*JAPANESE QUAIL	12 % OF ORAL DOSIS	CUMULATIV REL. WITH FECES, FEMALE, 7 D AFTER DOSIS, IN BODY = 38 %	730
*JAPANESE QUAIL	48 % OF ORAL DOSIS	CUMULATIV REL. WITH YOLK, FEMALE, 7 D AFTER DOSIS, IN BODY = 38 %	730
*JAPANESE QUAIL	26 % OF ORAL DOSIS	CUMULATIV REL. WITH FECES, MALE, 28 D AFTER DOSIS, IN BODY = 52 %	730
*JAPANESE QUAIL	12 % OF ORAL DOSIS	CUMULATIV REL. WITH FECES, FEMALE, 28 D AFTER DOSIS, IN BODY = 20 %	730
*JAPANESE QUAIL	12 % OF ORAL DOSIS	CUMULATIV REL. WITH YOLK, FEMALE, 28 D AFTER DOSIS, IN BODY = 20 %	730
*JAPANESE QUAIL	26 % OF ORAL DOSIS	CUMULATIV REL. WITH FECES, MALE, 84 D AFTER DOSIS, IN BODY = 53 %	730
*JAPANESE QUAIL	12 % OF ORAL DOSIS	CUMULATIV REL. WITH FECES, FEMALE, 84 D AFTER DOSIS, IN BODY = 2 %	730
*JAPANESE QUAIL	85 % OF ORAL DOSIS	CUMULATIV REL. WITH YOLK, FEMALE, 84 D AFTER DOSIS, IN BODY = 2 %	730

TABLE D380

EXCRETION : MN			
ITEM:	VALUE:	CONDITIONS:	REF.:
*RAT	90% OF DOSE	WITH FECES,PARENTERAL(NO UPTAKE THR.INTESTINE)	561

TABLE D381

EXCRETION : MO			
ITEM:	VALUE:	CONDITIONS:	REF.:
*HOMO SAPIENS	24-29% OF DOSE	WITH URINE,INJECTION,10DAYS	1158
*HOMO SAPIENS	1-6.8% OF DOSE	WITH FECES,INJECTION,10DAYS	1158

TABLE D382

EXCRETION : N			
ITEM:	VALUE:	CONDITIONS:	REF.:
*CALANUS SP.	37.5 % OF DAILY RATION	FEEDING ON SPRING DIATOMS,LOST AS FAECAL PELLETS	355

TABLE D382

EXCRETION : N

ITEM:	VALUE:	CONDITIONS:	REF.:
*CALANUS SP.	35.7 % OF DAILY RATION	FEEDING ON SPRING DIATOMS,LOST AS SOLUBLE EXCRETION	355
*CALANUS SP.	26.8 % OF DAILY RATION	FEEDING ON SPRING DIATOMS,AVIABLE FOR GROWTH	355

TABLE D383

EXCRETION : NAPHTHALENE

ITEM:	VALUE:	CONDITIONS:	REF.:
*RANGIA CUNEATA	79% IN 24HRS	TISSUE CONC.=1.9PPM AFTER 24HRS EXPOS.	1111
*RANGIA CUNEATA	34 % IN 24HRS	TISSUE CONC.=2.9PPM AFTER 24HRS EXPOS.,1-METHYLNAPHTHALENE	1111
*RANGIA CUNEATA	51% IN 24HRS	TISSUE CONC.=3.9PPM AFTER 24HRS EXPOS.,2-METHYLNAPHTHALENE	1111
*RANGIA CUNEATA	32% IN 24HRS	TISSUE CONC.=4.1PPM AFTER 24HRS EXPOS.,DIMETHYLNAPHTHALENE	1111
*RANGIA CUNEATA	50% IN 24HRS	TISSUE CONC.=0.8PPM AFTER 24HRS EXPOS.,TRIMETHYLNAPHTHALENE	1111

TABLE D384

EXCRETION : P

ITEM:	VALUE:	CONDITIONS:	REF.:
*ACARTIA	1.3MG P/(G BODY WT)*DAY	SUMMER, C II-IV, SOLUBLE P	355
*ACARTIA	1.0MG P/(G BODY WT)*DAY	SUMMER, C V-V1, SOLUBLE P	355
*ACARTIA	1.4MG P/(G BODY WT)*DAY	SUMMER, NAUPLII, SOLUBLE P	355
*ACARTIA SPP.	11MG P/(G BODY WT)*DAY	SPRING, SOLUBLE P	355
*ACARTIA SPP.	2.4MG P/(G BODY WT)*DAY	SPRING, SOLUBLE P	355
*ACARTIA SPP.	6.6MG P/(G BODY WT)*DAY	AUTUMN, SOLUBLE P	355
*CALANUS FINMARCHICUS	2.2MG P/(G BODY WT)*DAY	SPRING, SOLUBLE P	355
*CALANUS FINMARCHICUS	0.6MG P/(G BODY WT)*DAY	AUTUMN, SOLUBLE P	355
*CALANUS SP.	23.0 % OF DAILY RATION	FEEDING ON SPRING DIATOMS,LOST AS FAECAL PELLETS	355
*CALANUS SP.	59.8 % OF DAILY RATION	FEEDING ON SPRING DIATOMS,LOST AS SOLUBLE EXCRETION	355
*CALANUS SP.	17.2 % OF DAILY RATION	FEEDING ON SPRING DIATOMS, AVAILABLE FOR GROWTH	355
*OITHONA SIMILIS	0.9MG P/(G BODY WT)*DAY	SUMMER, SOLUBLE P	355
*PSEUDOCALANUS MILUTUS	1.5MG P/(G BODY WT)*DAY	SUMMER, SOLUBLE P	355
*SAGITTA HISPIDA	2.4MG P/(G BODY WT)*DAY	SOLUBLE P	355
*TEMORA LONGICORNIS	1.3MG P/(G BODY WT)*DAY	SUMMER, SOLUBLE P	355
*ZOOPLANKTON, MIXED	3.7MG P/(G BODY WT)*DAY	SUMMER, SOLUBLE P	355

TABLE D385

EXCRETION : PB

ITEM:	VALUE:	CONDITIONS:	REF.:
*HOMO SAPIENS	50 - 80 %	ORAL INTAKE, FOR A LONG PERIOD	1587

TABLE D385

EXCRETION : PB			
ITEM:	VALUE:	CONDITIONS:	REF.:
*HOMO SAPIENS	99 % OF TOTAL EXCRETED	INORGANIC PB	1649
*RABBIT	90 - 95 % IN 60 DAYS	ORAL INTAKE	1587
*SHEEP	5 - 10 % IN 10 DAYS	ORAL INTAKE	1587

TABLE D386

EXCRETION : SB			
ITEM:	VALUE:	CONDITIONS:	REF.:
*RAT	93-98% OF DOSE	WITH FECES,ORAL INTAKE OF SBCL3 (609UG/L)	909

TABLE D387

EXCRETION : SE			
ITEM:	VALUE:	CONDITIONS:	REF.:
*PIG	15-65% OF DOSE	WITH FECES,ORAL INTAKE	462

TABLE D388

EXCRETION : SN			
ITEM:	VALUE:	CONDITIONS:	REF.:
*RAT	99 % OF DOSE	WITH FECES,ORAL INTAKE,48HRS	382

TABLE D389

EXCRETION : SR			
ITEM:	VALUE:	CONDITIONS:	REF.:
*HOMO SAPIENS	0.81 MG/DAY	WITH FECES,ORAL INTAKE=1.37MG/DAY FOR 30DAYS,WOMAN	1599
*HOMO SAPIENS	0.24 MG/DAY	WITH URINE,WOMAN,ORAL INTAKE= 1.37MG/DAY FOR 30DAYS	1599

TABLE D390

EXCRETION : SR-89			
ITEM:	VALUE:	CONDITIONS:	REF.:
*TILAPIA MOSSAMBICA	50 %	REMAINING AFTER 1 DAY, UPTAKE BY EATING CAPSULES	150

TABLE D391

EXCRETION : TE			
ITEM:	VALUE:	CONDITIONS:	REF.:
*ANIMAL	75% OF DOSE	WITH FECES,ORAL INTAKE	183

TABLE D391

EXCRETION : TE			
ITEM:	VALUE:	CONDITIONS:	REF.:
*ANIMAL	10.7-16.4% OF DOSE	INTRAVEN.INJECT.,1HR	183

TABLE D392

EXCRETION : TI			
ITEM:	VALUE:	CONDITIONS:	REF.:
*ANIMAL	97% OF DOSE	WITH FECES,ORAL INTAKE	183

TABLE D393

EXCRETION : V			
ITEM:	VALUE:	CONDITIONS:	REF.:
*HOMO SAPIENS	99.9% OF DOSE	WITH FECES,ORAL INTAKE	1228

TABLE D394

EXCRETION : W			
ITEM:	VALUE:	CONDITIONS:	REF.:
*RAT	86-90% OF DOSE	WITH URINE,INTRAMUSC.INJECT.,24HRS	382
*RAT	5-9% OF DOSE	WITH FECES,INTRAMUSC.INJECT.,24HRS	382

TABLE D395

EXCRETION : Y-91			
ITEM:	VALUE:	CONDITIONS:	REF.:
*TILAPIA MOSSAMBICA	2 %	REMAINING AFTER 2 D, UPTAKE BY EATING CAPSULES	151

TABLE D396

EXCRETION : ZN			
ITEM:	VALUE:	CONDITIONS:	REF.:
*HOMO SAPIENS	42% OF DOSE	WITH FECES,ORAL INTAKE=13.2MG/D	462
*HOMO SAPIENS	8% OF DOSE	WITH URINE,ORAL INTAKE=13.2MG/D	462

TABLE D397

EXCRETION : ZR-95			
ITEM:	VALUE:	CONDITIONS:	REF.:
*PORPHYRA UMBILICALIS	50 % OF INITIAL	REMAINING IN ITEM AFTER 6 DAYS	469

TABLE D397

EXCRETION : ZR-95			
ITEM:	VALUE:	CONDITIONS:	REF.:
*PORPHYRA UMBILICALIS	96 % OF INITIAL	REMAINING IN ITEM AFTER 65 DAYS	469

TABLE D398

REMOVAL : PB

ITEM:	VALUE:	CONDITIONS:	REF.:
*CRATAEGUS MONOGYNA, LEAF	1 % OF SOLUBLE METAL	WASHED WITH DEIONISED WATER, BRISTOL,GB	936
*CRATAEGUS MONOGYNA, LEAF	1 % OF TOTAL METAL	WASHED WITH DEIONISED WATER, BRISTOL,GB	936
*CRATAEGUS MONOGYNA, LEAF	3.08 % OF TOTAL METAL	WASHED WITH A DETERGENT,BRISTOL,GB	936
*CRATAEGUS MONOGYNA, LEAF	1.07 % OF SOLUBLE METAL	BOILED,BRISTOL,GB	936
*CRATAEGUS MONOGYNA, LEAF	1.15 % OF TOTAL METAL	BOILED,BRISTOL,GB	936
*CRATAEGUS MONOGYNA, LEAF	12.17 TOTAL/SOLUBLE	WASHED WITH DEIONISED WATER, BRISTOL,GB	936
*CRATAEGUS MONOGYNA, LEAF	4.53 TOTAL/SOLUBLE	WASHED WITH A DETERGENT,BRISTOL,GB	936
*CRATAEGUS MONOGYNA, LEAF	1.51 TOTAL/SOLUBLE	BOILED,BRISTOL,GB	936
*QUERCUS ROBUR, LEAF	2.2 % OF SOLUBLE METAL	WASHED WITH DEIONISED WATER, BRISTOL,GB	936
*QUERCUS ROBUR, LEAF	1.25 % OF TOTAL METAL	WASHED WITH DEIONISED WATER, BRISTOL,GB	936
*QUERCUS ROBUR, LEAF	2.91 % OF TOTAL METAL	WASHED WITH A DETERGENT,BRISTOL,GB	936
*QUERCUS ROBUR, LEAF	1 % OF SOLUBLE METAL	BOILED,BRISTOL,GB	936
*QUERCUS ROBUR, LEAF	1 % OF TOTAL METAL	BOILED,BRISTOL,GB	936
*QUERCUS ROBUR, LEAF	6.847TOTAL/SOLUBLE	WASHED WITH DEIONISED WATER, BRISTOL,GB	936
*QUERCUS ROBUR, LEAF	1.43 TOTAL/SOLUBLE	BOILED,BRISTOL,GB	936
*SALIX ALBA, LEAF	1.72 % OF SOLUBLE METAL	WASHED WITH DEIONISED WATER, BRISTOL,GB	936
*SALIX ALBA, LEAF	1.15 % OF TOTAL METAL	WASHED WITH DEIONISED WATER, BRISTOL,GB	936
*SALIX ALBA, LEAF	1 % OF TOTAL METAL	WASHED WITH A DETERGENT,BRISTOL,GB	936
*SALIX ALBA, LEAF	1.66 % OF SOLUBLE METAL	BOILED,BRISTOL,GB	936
*SALIX ALBA, LEAF	1.59 % OF TOTAL METAL	BOILED,BRISTOL,GB	936
*SALIX ALBA, LEAF	8.0 TOTAL/SOLUBLE	WASHED WITH DEIONISED WATER, BRISTOL,GB	936
*SALIX ALBA, LEAF	1.37 TOTAL/SOLUBLE	BOILED,BRISTOL,GB	936

TABLE D399

REMOVAL : ZN

ITEM:	VALUE:	CONDITIONS:	REF.:
*CRATAEGUS MONOGYNA, LEAF	1.34 % OF SOLUBLE METAL	WASHED WITH DEIONISED WATER, BRISTOL,GB	936
*CRATAEGUS MONOGYNA, LEAF	1.96 % OF TOTAL METAL	WASHED WITH DEIONISED WATER, BRISTOL,GB	936
*CRATAEGUS MONOGYNA, LEAF	1 % OF SOLUBLE METAL	WASHED WITH A DETERGENT,BRISTOL,GB	936
*CRATAEGUS MONOGYNA, LEAF	1 % OF TOTAL METAL	WASHED WITH A DETERGENT,BRISTOL,GB	936
*CRATAEGUS MONOGYNA, LEAF	1.75 % OF SOLUBLE METAL	BOILED,BRISTOL,GB	936
*CRATAEGUS MONOGYNA, LEAF	1.34 % OF TOTAL METAL	BOILED,BRISTOL,GB	936
*CRATAEGUS MONOGYNA, LEAF	2.57 TOTAL/SOLUBLE	WASHED WITH DEIONISED WATER, BRISTOL,GB	936
*CRATAEGUS MONOGYNA, LEAF	3.42 TOTAL/SOLUBLE	WASHED WITH A DETERGENT,BRISTOL,GB	936
*CRATAEGUS MONOGYNA, LEAF	0.99 TOTAL/SOLUBLE	BOILED,BRISTOL,GB	936
*QUERCUS ROBUR, LEAF	6.49 % OF SOLUBLE METAL	WASHED WITH DEIONISED WATER, BRISTOL,GB	936
*QUERCUS ROBUR, LEAF	5.38 % OF TOTAL METAL	WASHED WITH DEIONISED WATER, BRISTOL,GB	936
*QUERCUS ROBUR, LEAF	4.12 % OF SOLUBLE METAL	WASHED WITH A DETERGENT,BRISTOL,GB	936
*QUERCUS ROBUR, LEAF	3.3 % OF TOTAL METAL	WASHED WITH A DETERGENT,BRISTOL,GB	936
*QUERCUS ROBUR, LEAF	1 % OF SOLUBLE METAL	BOILED,BRISTOL,GB	936

TABLE D399

REMOVAL : ZN

ITEM:	VALUE:	CONDITIONS:	REF.:
*QUERCUS ROBUR, LEAF	1 % OF TOTAL METAL	BOILED,BRISTOL,GB	936
*QUERCUS ROBUR, LEAF	1.44 TOTAL/SOLUBLE (!MM)	WASHED WITH DEIONISED WATER, BRISTOL,GB	936
*QUERCUS ROBUR, LEAF	1.95 TOTAL/SOLUBLE (!MM)	WASHED WITH A DETERGENT,BRISTOL,GB	936
*QUERCUS ROBUR, LEAF	1.34 TOTAL/SOLUBLE (!MM)	BOILED,BRISTOL,GB	936
*SALIX ALBA, LEAF	1 % OF SOLUBLE METAL	WASHED WITH DEIONISED WATER, BRISTOL,GB	936
*SALIX ALBA, LEAF	1 % OF TOTAL METAL	WASHED WITH DEIONISED WATER, BRISTOL,GB	936
*SALIX ALBA, LEAF	1.7 % OF SOLUBLE METAL	WASHED WITH A DETERGENT,BRISTOL,GB	936
*SALIX ALBA, LEAF	1.61 % OF TOTAL METAL	WASHED WITH A DETERGENT,BRISTOL,GB	936
*SALIX ALBA, LEAF	1.64 % OF SOLUBLE METAL	BOILED,BRISTOL,GB	936
*SALIX ALBA, LEAF	1.54 % OF TOTAL METAL	BOILED,BRISTOL,GB	936
*SALIX ALBA, LEAF	1.87 TOTAL/SOLUBLE (!MM)	WASHED WITH DEIONISED WATER, BRISTOL,GB	936
*SALIX ALBA, LEAF	1.25 TOTAL/SOLUBLE (!MM)	WASHED WITH A DETERGENT,BRISTOL,GD	936
*SALIX ALBA, LEAF	1.02 TOTAL/SOLUBLE (!MM)	BOILED,BRISTOL,GB	936

TABLE D400

UPTAKE : AG			
ITEM:	VALUE:	CONDITIONS:	REF.:
*ABIES GRANDIS	0.12 KG/(ACRE*100 YEAR)	USED FOR GROWTH IN 100 Y	470
*ANGUILLA JAPONICA	1.2 - 4.2*10**(-7) MG/HOUR	THROUGH GILLS	663

TABLE D401

UPTAKE : AL			
ITEM:	VALUE:	CONDITIONS:	REF.:
*ABIES GRANDIS	9.0 KG/(ACRE*100 YEARS)	USED FOR GROWTH IN 100 Y	470

TABLE D402

UPTAKE : B			
ITEM:	VALUE:	CONDITIONS:	REF.:
*ABIES GRANDIS	0.180 KG/(ACRE*100 YEARS)	USED FOR GROWTH IN 100 Y	470

TABLE D403

UPTAKE : BA			
ITEM:	VALUE:	CONDITIONS:	REF.:
*ABIES GRANDIS	7.4 KG/(ACRE*100 YEARS)	USED FOR GROWTH IN 100 Y	470

TABLE D404

UPTAKE : BIPHENYL			
ITEM:	VALUE:	CONDITIONS:	REF.:
*RAT	99.5% RETAINED	ORAL DOSE=5MG/KG	709
*RAT	98.8% RETAINED	ORAL DOSE=50MG/KG	709
*RAT	98.2% RETAINED	ORAL DOSE=100MG/KG	709

TABLE D405

UPTAKE : CA			
ITEM:	VALUE:	CONDITIONS:	REF.:
*ABIES GRANDIS	333.0 KG/(ACRE*100 YEARS)	USED FOR GROWTH IN 100 Y	470

TABLE D406

UPTAKE : CD			
ITEM:	VALUE:	CONDITIONS:	REF.:
*ACER SACCHARUM, LEAF AND TWIG	60. MG/TREE	SINGLE GROWING SEASON, CD FROM ATM.	1464

TABLE D406

UPTAKE : CD

ITEM:	VALUE:	CONDITIONS:	REF.:
*CORN, ROOT	1200 MG/KG DW	SOLUTION CONC = 100 MG/L WATER, 12 D	845
*CORN, ROOT	CONC 2 - 4 TIMES CONCENTRATION IN THE SHOOT	GROWN IN SOLUTION, 12 D	845
*CORN, SHOOT	500 MG/KG DW	SOLUTION CONC = 100 MG/L WATER, 12 D	845
*CORN, SHOOT	48 MG/KG DW	LOAMY SAND SOIL, 10 MG/KG DW, 4 W	845
*CORN, SHOOT	8.4 MG/KG DW	SILTY CLAY LOAM SOIL, 10 MG/KG DW, 4 W	845
*FODDER RAPE	11.2 UG/POT	ADDED CD IN SEWAGE SLUDGE=0UG/POT, SOIL-PH=4.8	31
*FODDER RAPE	17.7 UG/POT	ADDED CD IN SEWAGE SLUDGE=255UG/POT, SOIL-PH=4.8	31
*FODDER RAPE	20.0 UG/POT	ADDED CD IN SEWAGE SLUDGE=764UG/POT, SOIL-PH=4.8	31
*FODDER RAPE	23.1 UG/POT	ADDED CD IN SEWAGE SLUDGE=2291UG/POT, SOIL-PH=4.8	31
*FODDER RAPE	17.3 UG/POT	ADDED CD IN SEWAGE SLUDGE=6874UG/POT, SOIL-PH=4.8	31
*FODDER RAPE	3.4 UG/POT	ADDED CD IN SEWAGE SLUDGE=0UG/POT, SOIL-PH=6.0	31
*FODDER RAPE	5.5 UG/POT	ADDED CD IN SEWAGE SLUDGE=255UG/POT, SOIL-PH=6.0	31
*FODDER RAPE	7.9 UG/POT	ADDED CD IN SEWAGE SLUDGE=764UG/POT, SOIL-PH=6.0	31
*FODDER RAPE	17.6 UG/POT	ADDED CD IN SEWAGE SLUDGE=2291UG/POT, SOIL-PH=6.0	31
*FODDER RAPE	30.4 UG/POT	ADDED CD IN SEWAGE SLUDGE=6874UG/POT, SOIL-PH=6.0	31
*FODDER RAPE	3.0 UG/POT	ADDED CD IN SEWAGE SLUDGE=0UG/POT, SOIL-PH=7.2	31
*FODDER RAPE	3.7 UG/POT	ADDED CD IN SEWAGE SLUDGE=255UG/POT, SOIL-PH=7.2	31
*FODDER RAPE	5.8 UG/POT	ADDED CD IN SEWAGE SLUDGE=764UG/POT, SOIL-PH=7.2	31
*FODDER RAPE	10.9 UG/POT	ADDED CD IN SEWAGE SLUDGE=2291UG/POT, SOIL-PH=7.2	31
*FODDER RAPE	19.2 UG/POT	ADDED CD IN SEWAGE SLUDGE=6874UG/POT, SOIL-PH=7.2	31
*GOAT, INTESTINE	2% OF DOSE	ORAL DOSE OF 109-CD	487
*LYCOSA SP.	0.08 1/DAY	GRASSLAND FOOD CHAIN, PREDATOR	689
*PTERONARCELLA BADIA	19.0 MICROGRAM/(G*DAY)	WATER CONC = 9.0 MG/L WATER, FIRST 24 H	261
*PTERONARCELLA BADIA	0.005 MICROGRAM/(G*DAY)	WATER CONC = 9.0 MG/L WATER, DAY 1 - 4	261
*PTERONARCYS CALIFORNICA	1.90 MICROGRAM/(G*DAY)	WATER CONC = 5.0 MG/L WATER, FIRST 24 H	261
*PTERONARCYS CALIFORNICA	0.001 MICROGRAM/(G*DAY)	WATER CONC = 5.0 MG/L WATER, DAY 1 - 4	261
*PTERONARCYS CALIFORNICA	1.45 MICROGRAM/(G*DAY)	WATER CONC = 3.0 MG/L WATER, FIRST 24 H	261
*PTERONARCYS CALIFORNICA	0.06 MICROGRAM/(G*DAY)	WATER CONC = 3.0 MG/L WATER, DAY 1 - 4	261
*PTERONEMOBIUS FASCIATUS	0.55 1/DAY	GRASSLAND FOOD CHAIN, PRIMARY CONSUMER	689
*RICE, BRAN	0.79 UG/G	CD ADDED TO SOIL=0.001% OF CDO, PH-DEPENDENT	487
*RICE, BRAN	0.84 UG/G	CD ADDED TO SOIL=0.003% OF CDO, PH-DEPENDENT	487
*RICE, BRAN	1.60 UG/G	CD ADDED TO SOIL=0.01% OF CDO, PH-DEPENDENT	487
*RICE, BRAN	2.68 UG/G	CD ADDED TO SOIL=0.03% OF CDO, PH-DEPENDENT	487
*RICE, BRAN	2.94 UG/G	CD ADDED TO SOIL=0.1% OF CDO, PH-DEPENDENT	487
*RICE, BRAN	3.19 UG/G	CD ADDED TO SOIL=0.3% OF CDO, PH-DEPENDENT	487

TABLE D406

UPTAKE : CD

ITEM:	VALUE:	CONDITIONS:	REF.:
*RICE, BRAN	3.94 UG/G	CD ADDED TO SOIL=0.6% OF CDO, PH-DEPENDENT	487
*RICE, BRAN	0.44 UG/G	CD ADDED TO SOIL=0% OF CDO (CONTROL),PH-DEPENDENT	487
*RICE, POLISHED	0.16 UG/G	CD ADDED TO SOIL=0% OF CDO (CONTROL),PH-DEPENDENT	487
*RICE, POLISHED	0.28 UG/G	CD ADDED TO SOIL=0.001% OF CDO, PH-DEPENDENT	487
*RICE, POLISHED	0.40 UG/G	CD ADDED TO SOIL=0.003% OF CDO, PH-DEPENDENT	487
*RICE, POLISHED	0.78 UG/G	CD ADDED TO SOIL=0.01% OF CDO, PH-DEPENDENT	487
*RICE, POLISHED	1.37 UG/G	CD ADDED TO SOIL=0.03% OF CDO, PH-DEPENDENT	487
*RICE, POLISHED	1.62 UG/G	CD ADDED TO SOIL=0.1% OF CDO, PH-DEPENDENT	487
*RICE, POLISHED	1.94 UG/G	CD ADDED TO SOIL=0.3% OF CDO, PH-DEPENDENT	487
*RICE, POLISHED	1.73 UG/G	CD ADDED TO SOIL=0.6% OF CDO, PH-DEPENDENT	487
*RICE, POLISHED	0.59 UG/G	CD ADDED TO SOIL=0% OF CDO, PH-DEPENDENT	487
*RICE, UNPOLISHED	4.98 UG/G	CD ADDED TO SOIL=1.0% OF CDO, PH-DEPENDENT	487
*SEDIMENT	320.0 MICROGRAM/(G*DAY)	WATER CONC = 5.0 MG/L WATER, FIRST 24 H	261
*SEDIMENT	84.0 MICROGRAM/(G*DAY)	WATER CONC = 5.0 MG/L WATER, DAY 1 - 4	261
*SEDIMENT	220.0 MICROGRAM/(G*DAY)	WATER CONC = 3.0 MG/L WATER, FIRST 24 H	261
*SEDIMENT	64.0 MICROGRAM/(G*DAY)	WATER CONC = 3.0 MG/L WATER, DAY 1 - 4	261
*SOIL	INCREASE FROM 0.7 TO 9.4 MG CD/KG SOIL	POTS PLACED 500 METERS FROM A CD EMISSION SOURCE, AEROSOL FALLOUT	845
*WHEAT	0.067 UG/G DRY WT.	ADD.SLUDGE=0T/HA,ADD.CD IN SOIL= 0UG/G, SOIL-PH=4.8,CAO=0%	487
*WHEAT	0.119 UG/G DRY WT.	ADD.SLUDGE=6.5T/HA,ADD.CD IN SOIL= 0.031UG/G, SOIL-PH=4.8,CAO=0%	487
*WHEAT	0.170 UG/G DRY WT.	ADD.SLUDGE=19T/HA,ADD.CD IN SOIL= 0.094UG/G, SOIL-PH=5.1,CAO=0%	487
*WHEAT	0.257 UG/G DRY WT.	ADD.SLUDGE=58T/HA,ADD.CD IN SOIL= 0.28UG/G, SOIL-PH=5.3,CAO=0%	487
*WHEAT	0.124 UG/G DRY WT.	ADD.SLUDGE=175T/HA,ADD.CD IN SOIL= 0.84UG/G, SOIL-PH=5.8,CAO=0%	487
*WHEAT	0.045 UG/G DRY WT.	ADD.SLUDGE=0T/HA,ADD.CD IN SOIL= 0UG/G, SOIL-PH=6.1,CAO=0.1%	487
*WHEAT	0.086 UG/G DRY WT.	ADD.SLUDGE=6.5T/HA,ADD.CD IN SOIL= 0.031UG/G, SOIL-PH=6.1,CAO=0.1%	487
*WHEAT	0.123 UG/G DRY WT.	ADD.SLUDGE=19T/HA,ADD.CD IN SOIL= 0.094UG/G, SOIL-PH=6.2,CAO=0.1%	487
*WHEAT	0.134 UG/G DRY WT.	ADD.SLUDGE=58T/HA,ADD.CD IN SOIL= 0.28UG/G, SOIL-PH=6.2,CAO=0.1%	487
*WHEAT	0.147 UG/G DRY WT.	ADD.SLUDGE=175T/HA,ADD.CD IN SOIL= 0.84UG/G, SOIL-PH=5.9,CAO=0.1%	487
*WHEAT	0.029 UG/G DRY WT.	ADD.SLUDGE=0T/HA,ADD.CD IN SOIL= 0UG/G, SOIL-PH=7.4,CAO=0.2%	487
*WHEAT	0.033 UG/G DRT WT	ADD.SLUDGE=6.5T/HA,ADD.CD IN SOIL= 0.031UG/G, SOIL-PH=7.2,CAO=0.2%	487
*WHEAT	0.050 UG/G DRY WT.	ADD.SLUDGE=19T/HA,ADD.CD IN SOIL= 0.094UG/G, SOIL-PH=7.2,CAO=0.2%	487
*WHEAT	0.086 UG/G DRY WT	ADD.SLUDGE=58T/HA,ADD.CD IN SOIL= 0.28UG/G, SOIL-PH=6.8,CAO=0.2%	487
*WHEAT	0.122 UG/G DRY WT.	ADD.SLUDGE=175T/HA,ADD.CD IN SOIL= 0.84UG/G, SOIL-PH=6.5,CAO=0.2%	487
*WHEAT, WHOLE GRAIN	8.27 UG/G	CD ADDED TO SOIL=0.001% OF CDO, PH-DEPENDENT	487
*WHEAT, WHOLE GRAIN	15.5 UG/G	CD ADDED TO SOIL=0.003% OF CDO, PH-DEPENDENT	487

TABLE D406

UPTAKE : CD			
ITEM:	VALUE:	CONDITIONS:	REF.:
*WHEAT, WHOLE GRAIN	29.9 UG/G	CD ADDED TO SOIL=0.01% OF CDO, PH-DEPENDENT	487
*WHEAT, WHOLE GRAIN	41.4 UG/G	CD ADDED TO SOIL=0.03% OF CDO, PH-DEPENDENT	487
*WHEAT, WHOLE GRAIN	60.7 UG/G	CD ADDED TO SOIL=0.1% OF CDO, PH-DEPENDENT	487
*WHEAT, WHOLE GRAIN	48.6 UG/G	CD ADDED TO SOIL=0.3% OF CDO, PH-DEPENDENT	487
*WHEAT, WHOLE GRAIN	90.8 UG/G	CD ADDED TO SOIL=0.6% OF CDO, PH-DEPENDENT	487
*WHEAT, WHOLE GRAIN	139.0 UG/G	CD ADDED TO SOIL=1.0% OF CDO, PH-DEPENDENT	487

TABLE D407

UPTAKE : CO			
ITEM:	VALUE:	CONDITIONS:	REF.:
*ABIES GRANDIS	0.007 KG/(ACRE*100 YEARS)	USED FOR GROWTH IN 100 Y	470
*SOIL	7.38*10**4 KG/KM**2 YEAR	CO FROM ATM.	723

TABLE D408

UPTAKE : CO-60			
ITEM:	VALUE:	CONDITIONS:	REF.:
*AZOLLA SP.	0.15 - 0.05 MICRO CURIE/ (GRAM*HOUR)	AT STEADY STATE, N=2	68
*CABOMBA SP.	0.03 - 0.67 MICRO CURIE/ (GRAM*HOUR)	AT STEADY STATE, N=2	68
*CERATOPHYLLUM SP.	0.98 - 0.69 MICRO CURIE/ (GRAM*HOUR)	AT STEADY STATE, N=2	68
*DIEMICTILUS SP.	0.01 - 0.02 MICRO CURIE/ (GRAM*HOUR)	AT STEADY STATE, N=2	68
*ELODEA SP.	1.58 - 0.30 MICRO CURIE/ (GRAM*HOUR)	AT STEADY STATE, N=2	68
*HALIOTIS DISCUS	47.1 (SD=+-3.3) %	ABS. RATE FROM FOOD,ZFOOD= LAMINARIA, 1.75 G FOOD, 2 D, N=6	1095
*HALIOTIS DISCUS	30.5 (SD=+-7.6) %	ABS. RATE FROM FOOD, FOOD= UNDULARIA, 2.7 G FOOD, 2 D, N=6	1095
*HALIOTIS DISCUS	25.5 (SD=+-6.3) %	ABS. RATE FROM FOOD, FOOD=EISENIA, 2.1 G FOOD, 2 D, N=4	1095
*HALIOTIS DISCUS	46.7 (SD=+-6.0) %	ABS. RATE FROM FOOD, FOOD=ULVA, 1.2 G FOOD, 2 D, N=4	1095
*HELISOMA SP.	0.21 - 0.03 MICRO CURIE/ (GRAM*HOUR)	AT STEADY STATE, N=2	68
*LEMNA SP.	0.69 - 0.38 MICRO CURIE/ (GRAM*HOUR)	AT STEADY STATE, N=2	68
*NAJAS SP.	1.28 - 1.08 MICRO CURIE/ (GRAM*HOUR)	AT STEADY STATE, N=2	68
*PLANKTON	0.007 - 0.12 MICRO CURIE/ (GRAM*HOUR)	AT STEADY STATE, N=2	68
*RAT	0.60 %	UPTAKE FROM FOOD	23
*VALLISNERIA SP.	0.32 - 0.48 MICRO CURIE/ (GRAM*HOUR)	AT STEADY STATE, N=2	68

TABLE D409

UPTAKE : CR			
ITEM:	VALUE:	CONDITIONS:	REF.:
*ABIES GRANDIS	0.015 KG/(ACRE*100 YEARS)	USED FOR GROWTH IN 100 Y	470

TABLE D409

UPTAKE : CR

ITEM:	VALUE:	CONDITIONS:	REF.:
*ACER SACCHARUM, LEAF AND TWIG	140. MG/TREE	SINGLE GROWING SEASON,CR FROM ATM.	1464
*ALISMA PLANTAGO AQUATICA, OLD PLANT	44.92% (F046)	800M ABOVE POLL.ENTRY(1),REMOVED FROM SURFACE	1224
*ALISMA PLANTAGO AQUATICA, OLD PLANT	55.95% (F046)	150M BELOW POLL.ENTRY(1),REMOVED FROM SURFACE	1224
*ALISMA PLANTAGO AQUATICA, OLD PLANT	15.51% (F046)	800 AND 50M BELOW POLL.(1)AND(2) RESP.,REMOVED FROM SURFACE	1224
*ALISMA PLANTAGO AQUATICA, YOUNG PLANT	28.30% (F046)	800M ABOVE POLL.ENTRY(1),REMOVED FROM SURFACE	1224
*ALISMA PLANTAGO AQUATICA, YOUNG PLANT	43.76% (F046)	150M BELOW POLL.ENTRY(1),REMOVED FROM SURFACE	1224
*ALISMA PLANTAGO AQUATICA, YOUNG PLANT	14.75% (F046)	800 AND 50M BELOW POLL.(1)AND(2) RESP.,REMOVED FROM SURFACE	1224
*ANGUILLA JAPONICA	0.8 - 1.9*10**(-8) MG/HOUR	THROUGH GILLS	663

TABLE D410

UPTAKE : CS

ITEM:	VALUE:	CONDITIONS:	REF.:
*PLEURONECTES PLATESSA	50 %	% OF BODY BURDEN OBTAINED FROM FOOD, MIN VALUE	745
*RAJA CLAVATA	80 %	% OF BODY BURDEN OBTAINED FROM FOOD, MIN VALUE	745

TABLE D411

UPTAKE : CU

ITEM:	VALUE:	CONDITIONS:	REF.:
*ABIES GRANDIS	0.760 KG/(ACRE*100 YEARS)	USED FOR GROWTH IN 100 Y	470
*ALISMA PLANTAGO AQUATICA, OLD PLANT	47.01% (F046)	800M ABOVE POLL.ENTRY(1),REMOVED FROM SURFACE	1224
*ALISMA PLANTAGO AQUATICA, OLD PLANT	33.15% (F046)	150M BELOW POLL.ENTRY(1),REMOVED FROM SURFACE	1224
*ALISMA PLANTAGO AQUATICA, OLD PLANT	28.42% (F046)	800 AND 50M BELOW POLL.(1)AND(2) RESP.,REMOVED FROM SURFACE	1224
*ALISMA PLANTAGO AQUATICA, YOUNG PLANT	55.00% (F046)	800M ABOVE POLL.ENTRY(1),REMOVED FROM SURFACE	1224
*ALISMA PLANTAGO AQUATICA, YOUNG PLANT	41.12% (F046)	150M BELOW POLL.ENTRY(1),REMOVED FROM SURFACE	1224
*ALISMA PLANTAGO AQUATICA, YOUNG PLANT	43.07% (F046)	800 AND 50M BELOW POLL.(1)AND(2) RESP.,REMOVED FROM SURFACE	1224
*MENHADEN	9 %	ASSIMILATION EFFICIENCY, INGESTED= 1.0 UG/D, JUVENILE	301
*PINFISH	2.9 %	ASSIMILATION EFFICIENCY, INGESTED= 0.14 UG/D, JUVENILE	301
*SPOT	1 %	ASSIMILATION EFFICIENCY, INGESTED= 5.14 UG/D, JUVENILE	301

TABLE D412

UPTAKE : DDT

ITEM:	VALUE:	CONDITIONS:	REF.:
*HOMO SAPIENS	1.0 % OF TOTAL INTAKE	FROM SUGARS AND ADJUNCTS	376

TABLE D412

UPTAKE : DDT			
ITEM:	VALUE:	CONDITIONS:	REF.:
*HOMO SAPIENS	1.5 % OF TOTAL INTAKE	FROM ROOT VEGETABLES	376
*HOMO SAPIENS	2.0 % OF TOTAL INTAKE	FROM LEGUME VEGETABLES	376
*HOMO SAPIENS	2.0 % OF TOTAL INTAKE	FROM POTATOES	376
*HOMO SAPIENS	3.1 % OF TOTAL INTAKE	FROM OILS AND FATS	376
*HOMO SAPIENS	3.3 % OF TOTAL INTAKE	FROM LEAFY VEGETABLES	376
*HOMO SAPIENS	7.8 % OF TOTAL INTAKE	FROM GRAINS-CERALS	376
*HOMO SAPIENS	9.9 % OF TOTAL INTAKE	FROM GARDEN FRUITS	376
*HOMO SAPIENS	13.4 % OF TOTAL INTAKE	FROM FRUITS	376
*HOMO SAPIENS	14.4 % OF TOTAL INTAKE	FROM DAIRY PRODUCTS	376
*HOMO SAPIENS	40.8 % OF TOTAL INTAKE	FROM MEATS, FISH AND POULTRY	376

TABLE D413

UPTAKE : DIELDRIN			
ITEM:	VALUE:	CONDITIONS:	REF.:
*LEPTIDIUS FLORIDANUS	0.010 MG/KG DRY WEIGHT/DAY	FROM FOOD = 213 PPM, SEAWATER	422
*LEPTIDIUS FLORIDANUS	0.191 MG/KG DRY WEIGHT/DAY	ONLY FROM WATER = 0.5 PPB, SEAWATER	422

TABLE D414

UPTAKE : ENDRIN			
ITEM:	VALUE:	CONDITIONS:	REF.:
*GAMBUSIA AFFINIS, BRAIN	198.6 PPM/H	WATER = 1000 PPB, UPTAKE FROM WATER, NOT RESISTANT FISH, FISH DIED	438
*GAMBUSIA AFFINIS, BRAIN	28.76 PPM/H	WATER = 1000 PPB, UPTAKE FROM WATER, RESISTANT FISH	438
*GAMBUSIA AFFINIS, LIVER	213.6 PPM/H	WATER = 1000 PPB, UPTAKE FROM WATER, NOT RESISTANT FISH, FISH DIED	438
*GAMBUSIA AFFINIS, LIVER	176.7 PPM/H	WATER = 1000 PPB, UPTAKE FROM WATER, RESISTANT FISH	438

TABLE D415

UPTAKE : FE			
ITEM:	VALUE:	CONDITIONS:	REF.:
*ABIES GRANDIS	1.8 KG/(ACRE*100 YEARS)	USED FOR GROWTH IN 100 Y	470
*ALISMA PLANTAGO AQUATICA, OLD PLANT	31.39% (F046)	800M ABOVE POLL.ENTRY(1),REMOVED FROM SURFACE	1224
*ALISMA PLANTAGO AQUATICA, OLD PLANT	28.46% (F046)	150M BELOW POLL.ENTRY(1),REMOVED FROM SURFACE	1224
*ALISMA PLANTAGO AQUATICA, OLD PLANT	30.91% (F046)	800 AND 50M BELOW POLL.(1)AND(2) RESP.,REMOVED FROM SURFACE	1224
*ALISMA PLANTAGO AQUATICA, YOUNG PLANT	24.60% (F046)	800M ABOVE POLL.ENTRY(1),REMOVED FROM SURFACE	1224
*ALISMA PLANTAGO AQUATICA, YOUNG PLANT	34.74% (F046)	150M BELOW POLL.ENTRY(1),REMOVED FROM SURFACE	1224
*ALISMA PLANTAGO AQUATICA, YOUNG PLANT	9.09% (F046)	800 AND 50M BELOW POLL.(1)AND(2) RESP.,REMOVED FROM SURFACE	1224

TABLE D415

UPTAKE : FE			
ITEM:	VALUE:	CONDITIONS:	REF.:
*MENHADEN	1 %	ASSIMILATION EFFICIENCY, INGESTED= 920 UG/D, JUVENILE	301
*PINFISH	0.6 %	ASSIMILATION EFFICIENCY, INGESTED= 240 UG/D, JUVENILE	301
*SPOT	0.3 %	ASSIMILATION EFFICIENCY, INGESTED= 1400 UG/D, JUVENILE	301

TABLE D416

UPTAKE : HG			
ITEM:	VALUE:	CONDITIONS:	REF.:
*ANGUILLA JAPONICA	0.3 - 3.2*10**(-6) MG/HOUR	THROUGH GILLS	663
*CAT	90% OF DOSE	ORAL INTAKE OF METHYL-HG, UPT.THR.INTESTINE	486
*GUINEA PIG	6% OF DOSE (MAX.)	16MG HGCL2/ML,AFTER 5HRS, UPT.THR.SKIN	486
*HOMO SAPIENS	78-85% OF INHALED AMOUNT	RESPIRATORY INTAKE OF ELEMENTAL HG (50-350UG/M**3)	486
*HOMO SAPIENS	8% OF DOSE (MINIMUM)	ORAL INTAKE OF HGCL2,AFTER 20MIN.-1HR,UPT.THR.INTESTINE	486
*HOMO SAPIENS	15% OF DOSE	PERORAL INTAKE OF PROTEIN-HG-NITRATE,UPT.THR.INTEST INE,N=7	486
*HOMO SAPIENS	CA. 100%	ORAL INTAKE OF METHYL-HG-SALT AND -PROTEINATE,UPT.THR.INTESTINE	486
*MONKEY	90% OF DOSE	ORAL INTAKE OF METHYL-HG, UPT.THR.INTESTINE	486
*MOUSE	75-100% OF INHALED AMOUNT	RESPIRATORY INTAKE OF ELEMENTAL HG	486
*MOUSE	2% OF DAILY INTAKE	ORAL INTAKE OF HGCL2, UPT.THR.INTESTINE	486
*RAT	50-100% OF INHALED AMOUNT	RESPIRATORY INTAKE OF ELEMENTAL HG	486
*RAT	<0.01% OF DOSE	ORAL INTAKE OF ELEMENTAL HG, UPT.THR.INTESTINE	486
*RAT	90% OF DOSE	ORAL INTAKE OF METHYL-HG, UPT.THR.INTESTINE	486
*RAT	20% OF DOSE	ORAL INTAKE OF 0.5MG HG-ACETATE/ KG,AFTER 48HRS,UPT.THR.INTESTINE	486
*RAT	40% OF DOSE	ORAL,0.5MG PHENYL-HG-ACETATE/KG, 48HRS,UPT.THR.INTESTINE	486

TABLE D417

UPTAKE : K			
ITEM:	VALUE:	CONDITIONS:	REF.:
*ABIES GRANDIS	148.0 KG/(ACRE*100 YEARS)	USED FOR GROWTH IN 100 Y	470

TABLE D418

UPTAKE : LI			
ITEM:	VALUE:	CONDITIONS:	REF.:
*ABIES GRANDIS	0.082 KG/(ACRE*100 YEARS)	USED FOR GROWTH IN 100 Y	470

TABLE D419

UPTAKE : MG			
ITEM:	VALUE:	CONDITIONS:	REF.:
*ABIES GRANDIS	64.0 KG/(ACRE*100 YEARS)	USED FOR GROWTH IN 100 Y	470

TABLE D420

UPTAKE : MN			
ITEM:	VALUE:	CONDITIONS:	REF.:
*ABIES GRANDIS	8.0 KG/(ACRE*100 YEARS)	USED FOR GROWTH IN 100 Y	470
*ALISMA PLANTAGO AQUATICA, OLD PLANT	1.46% (F046)	800M ABOVE POLL.ENTRY(1),REMOVED FROM SURFACE	1224
*ALISMA PLANTAGO AQUATICA, OLD PLANT	2.12% (F046)	150M BELOW POLL.ENTRY(1),REMOVED FROM SURFACE	1224
*ALISMA PLANTAGO AQUATICA, OLD PLANT	3.52% (F046)	800 AND 50M BELOW POLL.(1)AND(2) RESP.,REMOVED FROM SURFACE	1224
*ALISMA PLANTAGO AQUATICA, YOUNG PLANT	(F046)	800M ABOVE POLL.ENTRY(1),REMOVED FROM SURFACE	1224
*ALISMA PLANTAGO AQUATICA, YOUNG PLANT	(F046)	150M BELOW POLL.ENTRY(1),REMOVED FROM SURFACE	1224
*ALISMA PLANTAGO AQUATICA, YOUNG PLANT	14.84% (F046)	800 AND 50M BELOW POLL.(1)AND(2) RESP.,REMOVED FROM SURFACE	1224
*HOMARUS VULGARIS	0.4 MICROGRAM/DAY	500 G ANIMAL, UPTAKE FROM SEAWATER	187
*MENHADEN	3 %	ASSIMILATION EFFICIENCY, INGESTED= 11 UG/D, JUVENILE	301
*PINFISH	7 %	ASSIMILATION EFFICIENCY, INGESTED= 2.3 UG/D, JUVENILE	301
*SPOT	0.7 %	ASSIMILATION EFFICIENCY, INGESTED= 29 UG/D, JUVENILE	301

TABLE D421

UPTAKE : MO			
ITEM:	VALUE:	CONDITIONS:	REF.:
*ABIES GRANDIS	0.003 KG/(ACRE*100 YEARS)	USED FOR GROWTH IN 100 Y	470

TABLE D422

UPTAKE : NA			
ITEM:	VALUE:	CONDITIONS:	REF.:
*ABIES GRANDIS	3.0 KG/(ACRE*100 YEARS)	USED FOR GROWTH IN 100 Y	470

TABLE D423

UPTAKE : NH3			
ITEM:	VALUE:	CONDITIONS:	REF.:
*SOYBEAN	2.*10**3 KG/KM**2 YEAR	NH3 FROM ATM.	708

TABLE D424

UPTAKE : NI			
ITEM:	VALUE:	CONDITIONS:	REF.:
*ABIES GRANDIS	0.028 KG/(ACRE*100 YEARS)	USED FOR GROWTH IN 100 Y	470
*ACER SACCHARUM, LEAF AND TWIG	820. MG/TREE	SINGLE GROWING SEASON,NI FROM ATM.	1464

TABLE D425

UPTAKE : NO2			
ITEM:	VALUE:	CONDITIONS:	REF.:
*ALFALFA	3.58*10**4 KG/KM**2 YEAR	NO2 FROM ATM.	668

TABLE D426

UPTAKE : OCTACHLORODIBENZO-P-DIOXIN

ITEM:	VALUE:	CONDITIONS:	REF.:
*RAT	7% OF DOSE	UPT.THR.STOMACH-INTESTINE,ORAL DOSE=100UG/DAY FOR 21DAYS	1134

TABLE D427

UPTAKE : P

ITEM:	VALUE:	CONDITIONS:	REF.:
*ZOSTERA MARINA	11 MICROGRAM P/(GRAM PLANT)	WHOLE PLANT, LEAVES,	355
*ZOSTERA MARINA	22 MICROGRAM P/(GRAM PLANT)	WHOLE PLANT, ROOT AND RHIZOMES,	355
*ZOSTERA MARINA	9 MICROGRAM P/(GRAM PLANT)	DETACHED LEAVES,	355
*ZOSTERA MARINA	12 MICROGRAM P/(GRAM PLANT)	DETACHED ROOT AND RHIZOMES,	355
*ZOSTERA MARINA	320 MICROGRAM P/(GRAM PLANT)	WHOLE PLANT, LEAVES,	355
*ZOSTERA MARINA	280 MICROGRAM P/(GRAM PLANT)	WHOLE PLANT, ROOT AND RHIZOMES,	355
*ZOSTERA MARINA	520 MICROGRAM P/(GRAM PLANT)	DETACHED LEAVES,	355
*ZOSTERA MARINA	500 MICROGRAM P/(GRAM PLANT)	DETACHED ROOT AND RHIZOMES,	355

TABLE D428

UPTAKE : PB

ITEM:	VALUE:	CONDITIONS:	REF.:
*ABIES GRANDIS	1.3 KG/(ACRE*100 YEARS)	USED FOR GROWTH IN 100 Y	470
*ACER SACCHARUM, LEAF AND TWIG	5800 MG/TREE	SINGLE GROWING SEASON,PB FROM ATM.	1464
*ALISMA PLANTAGO AQUATICA, OLD PLANT	38.42% (F046)	800M ABOVE POLL.ENTRY(1),REMOVED FROM SURFACE	1224
*ALISMA PLANTAGO AQUATICA, OLD PLANT	42.99% (F046)	150M BELOW POLL.ENTRY(1),REMOVED FROM SURFACE	1224
*ALISMA PLANTAGO AQUATICA, OLD PLANT	35.87% (F046)	800 AND 50M BELOW POLL.(1)AND(2) RESP.,REMOVED FROM SURFACE	1224
*ALISMA PLANTAGO AQUATICA, YOUNG PLANT	0.86% (F046)	800M ABOVE POLL.ENTRY(1),REMOVED FROM SURFACE	1224
*ALISMA PLANTAGO AQUATICA, YOUNG PLANT	20.50% (F046)	150M BELOW POLL.ENTRY(1),REMOVED FROM SURFACE	1224
*ALISMA PLANTAGO AQUATICA, YOUNG PLANT	46.25% (F046)	800 AND 50M BELOW POLL.(1)AND(2) RESP.,REMOVED FROM SURFACE	1224
*CORN, OLD LEAVES	10000 MG/KG DW	HYDROPONIC SOLUTION CONC = 100 MG/ L WATER, 6 D,PHOSPHATE DEFICIENT PLANT	845
*CORN, OLD LEAVES	2000 MG/KG DW	HYDROPONIC SOLUTION CONC = 100 MG/ L WATER, 6 D, ADEQUATE PHOSPHORUS	845
*CORN, ROOT	10600 MG/KG DW	HYDROPONIC SOLUTION CONC = 100 MG/ L WATER, 3 D	845
*CORN, TOP	3900 MG/KG DW	HYDROPONIC SOLUTION CONC = 100 MG/ L WATER, 3 D	845
*CORN, TOP	200 - 400 MG/KG DW	SOIL CONC = 200 MG/KG DW, 4 W	845
*CORN, YOUNG LEAVES	900 MG/KG DW	HYDROPONIC SOLUTION CONC = 100 MG/ L WATER, 6 D, ADEQUATE PHOSPHORUS	845
*CORN, YOUNG LEAVES	6700 MG/KG DW	HYDROPONIC SOLUTION CONC = 100 MG/ L WATER, 6 D, PHOSPHATE DEFICIT PLANTS	845
*HOMO SAPIENS	5 - 17 % OF INTAKE	ORAL INTAKE	664
*PLANT, LEAF	EXTREMELY SMALL	UPTAKE FROM AIR FALLOUT ON LEAVES	845
*RABBIT	0.8 - 1.0 % OF INTAKE	ORAL INTAKE	664
*RADISH, ROOT	19.8 MG/KG DRY WEIGHT	FROM AIR, CONTROL IN GREENHOUSE = 19.3 PPM, PB IN SOIL = 29.9 PPM	868

TABLE D428

UPTAKE : PB

ITEM:	VALUE:	CONDITIONS:	REF.:
*RADISH, ROOT	28.8 MG/KG DRY WEIGHT	FROM AIR, CONTROL IN GREENHOUSE = 25.5 PPM, PB IN SOIL = 165 PPM	868
*RADISH, ROOT	35.5 MG/KG DRY WEIGHT	FROM AIR, CONTROL IN GREENHOUSE = 3¤.8 PPM, PB IN SOIL = 299 PPM	868
*RADISH, TOP	25.5 MG/KG DRY WEIGHT	FROM AIR, CONTROL IN GREENHOUSE = 12.9 PPM, PB IN SOIL = 29.9 PPM	868
*RADISH, TOP	30.4 MG/KG DRY WEIGHT	FROM AIR, CONTROL IN GREENHOUSE = 17.8 PPM, PB IN SOIL = 165 PPM	868
*RADISH, TOP	38.8 MG/KG DRY WEIGHT	FROM AIR, CONTROL IN GREENHOUSE = 24.2 PPM, PB IN SOIL = 299 PPM	868
*SHEEP	1.3 % OF INTAKE	ORAL INTAKE	664

TABLE D429

UPTAKE : PCB

ITEM:	VALUE:	CONDITIONS:	REF.:
*RHIZOPHORA MANGLE, HYPOCOTYL	0.35 MG/KG WET WT.	APPLICATION=0.56KG/HA,PH=6.3-6.7,S =2.5%,6W, AROCLOR 1242 (?073)	1647
*RHIZOPHORA MANGLE, HYPOCOTYL	0.79 MG/KG WET WT.	APPLICATION=1.12KG/HA,PH=6.3-6.7,S =2.5%,6W, AROCLOR 1242 (?073)	1647
*RHIZOPHORA MANGLE, HYPOCOTYL	0.92 MG/KG WET WT.	APPLICATION=2.80KG/HA,PH=6.3-6.7,S =2.5%,6W, AROCLOR 1242 (?073)	1647
*RHIZOPHORA MANGLE, HYPOCOTYL	0.53 MG/KG WET WT.	APPLICATION=5.60KG/HA,PH=6.3-6.7,S =2.5%,6W, AROCLOR 1242 (?073)	1647
*RHIZOPHORA MANGLE, HYPOCOTYL	1.50 MG/KG WET WT.	APPLICATION=11.20KG/HA,PH=6.3-6.7, S=2.5%,6W, AROCLOR 1242 (?073)	1647
*RHIZOPHORA MANGLE, LEAVES	0.46 MG/KG WET WT.	APPLICATION=0.56KG/HA,PH=6.3-6.7,S =2.5%,6W, AROCLOR 1242 (?073)	1647
*RHIZOPHORA MANGLE, LEAVES	1.11 MG/KG WET WT.	APPLICATION=1.12KG/HA,PH=6.3-6.7,S =2.5%,6W, AROCLOR 1242 (?073)	1647
*RHIZOPHORA MANGLE, LEAVES	0.92 MG/KG WET WT.	APPLICATION=2.80KG/HA,PH=6.3-6.7,S =2.5%,6W, AROCLOR 1242 (?073)	1647
*RHIZOPHORA MANGLE, LEAVES	0.80 MG/KG WET WT.	APPLICATION=5.60KG/HA,PH=6.3-6.7,S =2.5%,6W, AROCLOR 1242 (?073)	1647
*RHIZOPHORA MANGLE, LEAVES	0.90 MG/KG WET WT.	APPLICATION=11.20KG/HA,PH=6.3-6.7, S=2.5%,6W, AROCLOR 1242 (?073)	1647
*RHIZOPHORA MANGLE, ROOT	0.10 MG/KG WET WT.	APPLICATION=5.60KG/HA,PH=6.3-6.7,S =2.5%,6W, AROCLOR 1242 (?073)	1647
*RHIZOPHORA MANGLE, ROOT	0.10 MG/KG WET WT.	APPLICATION=11.20KG/HA,PH=6.3-6.7, S=2.5%,6W, AROCLOR 1242 (?073)	1647

TABLE D430

UPTAKE : RB

ITEM:	VALUE:	CONDITIONS:	REF.:
*ABIES GRANDIS	0.89 KG/(ACRE*100 YEARS)	USED FOR GROWTH IN 100 Y	470

TABLE D431

UPTAKE : SI

ITEM:	VALUE:	CONDITIONS:	REF.:
*ABIES GRANDIS	24.0 KG/(ACRE*100 YEARS)	USED FOR GROWTH IN 100 Y	470

TABLE D432

UPTAKE : SN			
ITEM:	VALUE:	CONDITIONS:	REF.:
*ABIES GRANDIS	0.003 KG/(ACRE*100 YEARS)	USED FOR GROWTH IN 100 Y	470

TABLE D433

UPTAKE : SO2			
ITEM:	VALUE:	CONDITIONS:	REF.:
*ALFALFA	1.27*10**6 KG/KM**2 YEAR	SO2 FROM ATM.	668

TABLE D434

UPTAKE : SR			
ITEM:	VALUE:	CONDITIONS:	REF.:
*ABIES GRANDIS	6.7 KG/(ACRE*100 YEARS)	USED FOR GROWTH IN 100 Y	470

TABLE D435

UPTAKE : TI			
ITEM:	VALUE:	CONDITIONS:	REF.:
*ABIES GRANDIS	1.1 KG/(ACRE*100 YEARS)	USED FOR GROWTH IN 100 Y	470

TABLE D436

UPTAKE : V			
ITEM:	VALUE:	CONDITIONS:	REF.:
*ABIES GRANDIS	0.0002 KG/(ACRE*100 YEARS)	USED FOR GROWTH IN 100 Y	470

TABLE D437

UPTAKE : ZN			
ITEM:	VALUE:	CONDITIONS:	REF.:
*ABIES GRANDIS	0.549 KG/(ACRE*100 YEARS)	USED FOR GROWTH IN 100 Y	470
*ALISMA PLANTAGO AQUATICA, OLD PLANT	22.10% (F046)	800M ABOVE POLL.ENTRY(1),REMOVED FROM SURFACE	1224
*ALISMA PLANTAGO AQUATICA, OLD PLANT	40.18% (F046)	150M BELOW POLL.ENTRY(1),REMOVED FROM SURFACE	1224
*ALISMA PLANTAGO AQUATICA, OLD PLANT	16.50% (F046)	800 AND 50M BELOW POLL.(1)AND(2) RESP.,REMOVED FROM SURFACE	1224
*ALISMA PLANTAGO AQUATICA, YOUNG PLANT	17.69% (F046)	800M ABOVE POLL.ENTRY(1),REMOVED FROM SURFACE	1224
*ALISMA PLANTAGO AQUATICA, YOUNG PLANT	39.90% (F046)	150M BELOW POLL.ENTRY(1),REMOVED FROM SURFACE	1224
*ALISMA PLANTAGO AQUATICA, YOUNG PLANT	24.20% (F046)	800 AND 50M BELOW POLL.(1)AND(2) RESP.,REMOVED FROM SURFACE	1224

TABLE D437

UPTAKE : ZN			
ITEM:	VALUE:	CONDITIONS:	REF.:
*ANGUILLA JAPONICA	0.7 - 4.1*10**(-6) MG/HOUR	THROUGH GILLS	663
*ARTHROBACTER MARINUS	31.7 PPB (EQ.)	TOT.CONC.=350PPB,ADDED ZN=0PPB, 293K	786
*ARTHROBACTER MARINUS	40.9 PPB (EQ.)	TOT.CONC.=1050PPB,ADDED ZN=654PPB, 293K	786
*ARTHROBACTER MARINUS	62.2 PPB (EQ.)	TOT.CONC.=3650PPB,ADDED ZN= 3269PPB,293K	786
*ARTHROBACTER MARINUS	90.9 PPB (EQ.)	TOT.CONC.=6960PPB,ADDED ZN= 6538PPB,293K	786
*ARTHROBACTER MARINUS	286.4 PPB (EQ.)	TOT.CONC.=13390PPB,ADDED ZN= 13076PPB,293K	786
*ARTHROBACTER MARINUS	480.8 PPB (EQ.)	TOT.CONC.=18350PPB,ADDED ZN= 19614PPB,293K	786
*MENHADEN	36 %	ASSIMILATION EFFICIENCY, INGESTED= 4.6 UG/D, JUVENILE	301
*PINFISH	19 %	ASSIMILATION EFFICIENCY, INGESTED= 3.5 UG/D, JUVENILE	301
*SPOT	2 %	ASSIMILATION EFFICIENCY, INGESTED= 34 UG/D, JUVENILE	301

TABLE D438

UPTAKE : ZR			
ITEM:	VALUE:	CONDITIONS:	REF.:
*ABIES GRANDIS	0.12 KG/(ACRE*100 YEARS)	USED FOR GROWTH IN 100 Y	470

TABLE D439

UPTAKE : 2-MONOCHLOROBIPHENYL			
ITEM:	VALUE:	CONDITIONS:	REF.:
*RAT	98.9% RETAINED	ORAL DOSE=5MG/KG	709
*RAT	98.0% RETAINED	ORAL DOSE=50MG/KG	709
*RAT	98.4% RETAINED	ORAL DOSE=100MG/KG	709

TABLE D440

UPTAKE : 2,2`-DICHLOROBIPHENYL			
ITEM:	VALUE:	CONDITIONS:	REF.:
*RAT	95.0% RETAINED	ORAL DOSE=5MG/KG	709
*RAT	97.4% RETAINED	ORAL DOSE=50MG/KG	709
*RAT	98.2% RETAINED	ORAL DOSE=100MG/KG	709

TABLE D441

UPTAKE : 2,2`,4,4`,5,5`-HEXACHLOROBIPHENYL			
ITEM:	VALUE:	CONDITIONS:	REF.:
*RAT	95.3% RETAINED	ORAL DOSE=5MG/KG	709

TABLE D442

UPTAKE : 2,2`,4,5,5`-PENTACHLOROBIPHENYL			
ITEM:	VALUE:	CONDITIONS:	REF.:
*RAT	95.1% RETAINED	ORAL DOSE=5MG/KG	709

TABLE D443

UPTAKE : 2,2`,5-TRICHLOROBIPHENYL			
ITEM:	VALUE:	CONDITIONS:	REF.:
*RAT	95.1% RETAINED	ORAL DOSE=50MG/KG	709
*RAT	94.3% RETAINED	ORAL DOSE=100MG/KG	709

TABLE D444

UPTAKE : 2,2`,5,5`-TETRACHLOROBIPHENYL			
ITEM:	VALUE:	CONDITIONS:	REF.:
*RAT	96.0% RETAINED	ORAL DOSE=5MG/KG	709

TABLE D445

UPTAKE : 2,3,4,5-TETRACHLOROBIPHENYL			
ITEM:	VALUE:	CONDITIONS:	REF.:
*RAT	97.3% RETAINED	ORAL DOSE=5MG/KG	709

TABLE D446

UPTAKE : 2,3`-DICHLOROBIPHENYL			
ITEM:	VALUE:	CONDITIONS:	REF.:
*RAT	96.8% RETAINED	ORAL DOSE=50MG/KG	709
*RAT	97.9% RETAINED	ORAL DOSE=100MG/KG	709

TABLE D447

UPTAKE : 2,4-DICHLOROBIPHENYL			
ITEM:	VALUE:	CONDITIONS:	REF.:
*RAT	95.7% RETAINED	ORAL DOSE=50MG/KG	709
*RAT	97.1% RETAINED	ORAL DOSE=100MG/KG	709

TABLE D448

UPTAKE : 2,4,4`-TRICHLOROBIPHENYL			
ITEM:	VALUE:	CONDITIONS:	REF.:
*RAT	92.6% RETAINED	ORAL DOSE=50MG/KG	709

TABLE D448

UPTAKE : 2,4,4`-TRICHLOROBIPHENYL			
ITEM:	VALUE:	CONDITIONS:	REF.:
*RAT	94.0% RETAINED	ORAL DOSE=100MG/KG	709

TABLE D449

UPTAKE : 2,4`-DICHLOROBIPHENYL			
ITEM:	VALUE:	CONDITIONS:	REF.:
*RAT	97.2% RETAINED	ORAL DOSE=5MG/KG	709
*RAT	95.5% RETAINED	ORAL DOSE=50MG/KG	709
*RAT	97.6% RETAINED	ORAL DOSE=100MG/KG	709

TABLE D450

UPTAKE : 2,6-DICHLOROBIPHENYL			
ITEM:	VALUE:	CONDITIONS:	REF.:
*RAT	95.0% RETAINED	ORAL DOSE=50MG/KG	709
*RAT	94.9% RETAINED	ORAL DOSE=100MG/KG	709

TABLE D451

UPTAKE : 2`,3,5-TRICHLOROBIPHENYL			
ITEM:	VALUE:	CONDITIONS:	REF.:
*RAT	95.6% RETAINED	ORAL DOSE=50MG/KG	709
*RAT	95.0% RETAINED	ORAL DOSE=100MG/KG	709

TABLE D452

UPTAKE : 3-MONOCHLOROBIPHENYL			
ITEM:	VALUE:	CONDITIONS:	REF.:
*RAT	98.0% RETAINED	ORAL DOSE=5MG/KG	709
*RAT	96.6% RETAINED	ORAL DOSE=50MG/KG	709
*RAT	97.3% RETAINED	ORAL DOSE=100MG/KG	709

TABLE D453

UPTAKE : 3,3`,4-TRICHLOROBIPHENYL			
ITEM:	VALUE:	CONDITIONS:	REF.:
*RAT	90.3% RETAINED	ORAL DOSE=100MG/KG	709

TABLE D454

UPTAKE : 3,3`,4,4`-TETRACHLOROBIPHENYL			
ITEM:	VALUE:	CONDITIONS:	REF.:
*RAT	91.8% RETAINED	ORAL DOSE=5MG/KG	709

TABLE D455

UPTAKE : 3,3'4'-TRICHLOROBIPHENYL			
ITEM:	VALUE:	CONDITIONS:	REF.:
*RAT	93.5% RETAINED	ORAL DOSE=50MG/KG	709

TABLE D456

UPTAKE : 3,4'-DICHLOROBIPHENYL			
ITEM:	VALUE:	CONDITIONS:	REF.:
*RAT	95.7% RETAINED	ORAL DOSE=5MG/KG	709
*RAT	94.3% RETAINED	ORAL DOSE=50MG/KG	709
*RAT	97.1% RETAINED	ORAL DOSE=100MG/KG	709

TABLE D457

UPTAKE : 4-MONOCHLOROBIPHENYL			
ITEM:	VALUE:	CONDITIONS:	REF.:
*RAT	97.3% RETAINED	ORAL DOSE=5MG/KG	709
*RAT	96.3% RETAINED	ORAL DOSE=50MG/KG	709
*RAT	96.4% RETAINED	ORAL DOSE=100MG/KG	709

TABLE D458

UPTAKE : 4,4'-DICHLOROBIPHENYL			
ITEM:	VALUE:	CONDITIONS:	REF.:
*RAT	95.4% RETAINED	ORAL DOSE=5MG/KG	709
*RAT	95.6% RETAINED	ORAL DOSE=50MG/KG	709
*RAT	97.4% RETAINED	ORAL DOSE=100MG/KG	709

TABLE D459

UPTAKE FROM SOIL WITH ADDED SLUDGE : AL			
ITEM:	VALUE:	CONDITIONS:	REF.:
*ALLIUM CEPA	18 MG/KG DRY WEIGHT	PH=7.1, 2 Y, 100 T DW SL AC, CONT= 7 PPM, SOIL=39 G/KG, SLUDGE=42 G/KG	499
*BRASSICA OLERACEA VAR. CAPITATA	24 MG/KG DRY WEIGHT	PH=7.1, 2 Y, 100 T DW SL AC, CONT= 24 PPM, SOIL=39 G/KG, SLUDGE=42 G/KG	499
*DAUCUS CAROTA VAR. SATIVA	35 MG/KG DRY WEIGHT	PH=7.1, 2 Y, 100 T DW SL AC, CONT= 14 PPM, SOIL=39 G/KG, SLUDGE=42 G/KG	499
*ECHINOCHLOA CRUSGALLI VAR. FRUMENTACEA	28 MG/KG DRY WEIGHT	PH=7.1, 2 Y, 100 T DW SL AC, CONT= . PPM, SOIL=39 G/KG, SLUDGE=42 G/KG	499
*LYCOPERSICON ESCULENTUM	15 MG/KG DRY WEIGHT	PH=7.1, 2 Y, 100 T DW SL AC, CONT= 13 PPM, SOIL=39 G/KG, SLUDGE=42 G/KG	499
*PHASEOLUS VULGARIS	19 MG/KG DRY WEIGHT	PH=7.1, 2 Y, 100 T DW SL AC, CONT= 17 PPM, SOIL=39 G/KG, SLUDGE=42 G/KG	499
*SOLANUM TUBEROSUM	11 MG/KG DRY WEIGHT	PH=7.1, 2 Y, 100 T DW SL AC, CONT= 10 PPM, SOIL=39 G/KG, SLUDGE=42 G/KG	499

TABLE D460

UPTAKE FROM SOIL WITH ADDED SLUDGE : AS			
ITEM:	VALUE:	CONDITIONS:	REF.:
*FODDER RAPE	61 (% INCREASE IN PLANT)/(% INCREASE IN SOIL)	BACKGROUND LEVEL = 12.3 PPM,SLUDGE LEVEL= 6.6 PPM	27

TABLE D461

UPTAKE FROM SOIL WITH ADDED SLUDGE : B			
ITEM:	VALUE:	CONDITIONS:	REF.:
*ALLIUM CEPA	9 MG/KG DRY WEIGHT	PH=7.1, 1 Y, 100 T DW SL AC, CONT= 6 PPM, SOIL=14 PPM, SLUDGE=63 PPM	499
*ALLIUM CEPA	24 MG/KG DRY WEIGHT	PH=7.1, 2 Y, 100 T DW SL AC, CONT= 15 PPM, SOIL=14 , SLUDGE=63 PPM	499
*BRASSICA OLERACEA VAR. CAPITATA	25 MG/KG DRY WEIGHT	PH=7.1, 1 Y, 100 T DW SL AC, CONT= 23 PPM, SOIL=14 PPM, SLUDGE=63 PPM	499
*BRASSICA OLERACEA VAR. CAPITATA	35 MG/KG DRY WEIGHT	PH=7.1, 2 Y, 100 T DW SL AC, CONT= 35 PPM, SOIL=14 , SLUDGE=63 PPM	499
*DAUCUS CAROTA VAR. SATIVA	22 MG/KG DRY WEIGHT	PH=7.1, 2 Y, 100 T DW SL AC, CONT= 1.1 PPM, SOIL=14 , SLUDGE=63 PPM	499
*ECHINOCHLOA CRUSGALLI VAR. FRUMENTACEA	9 MG/KG DRY WEIGHT	PH=7.1, 1 Y, 100 T DW SL AC, CONT= 17 PPM, SOIL=14 PPM, SLUDGE=63 PPM	499
*ECHINOCHLOA CRUSGALLI VAR. FRUMENTACEA	12 MG/KG DRY WEIGHT	PH=7.1, 2 Y, 100 T DW SL AC, CONT= 10 PPM, SOIL=14 , SLUDGE=63 PPM	499
*FODDER RAPE	0.8 (% INCREASE IN PLANT)/(% INCREASE IN SOIL)	BACKGROUND LEVEL = 0.59 PPM,SLUDGE LEVEL= 7.3 PPM	27
*LYCOPERSICON ESCULENTUM	12 MG/KG DRY WEIGHT	PH=7.1, 2 Y, 100 T DW SL AC, CONT= 9 PPM, SOIL=14 , SLUDGE=63 PPM	499
*PHASEOLUS VULGARIS	28 MG/KG DRY WEIGHT	PH=7.1, 1 Y, 100 T DW SL AC, CONT= 16 PPM, SOIL=14 PPM, SLUDGE=63 PPM	499
*PHASEOLUS VULGARIS	41 MG/KG DRY WEIGHT	PH=7.1, 2 Y, 100 T DW SL AC, CONT= 18 PPM, SOIL=14 , SLUDGE=63 PPM	499
*SOLANUM TUBEROSUM	9 MG/KG DRY WEIGHT	PH=7.1, 1 Y, 100 T DW SL AC, CONT= 4 PPM, SOIL=14 PPM, SLUDGE=63 PPM	499
*SOLANUM TUBEROSUM	10 MG/KG DRY WEIGHT	PH=7.1, 2 Y, 100 T DW SL AC, CONT= 6 PPM, SOIL=14 , SLUDGE=63 PPM	499

TABLE D462

UPTAKE FROM SOIL WITH ADDED SLUDGE : BR			
ITEM:	VALUE:	CONDITIONS:	REF.:
*ALLIUM CEPA	11.7 MG/KG DRY WEIGHT	PH=7.1, 1 Y, 100 T DW SL AC, CONT= 5.0 PPM, SOIL=3.5, SLUDGE=10 PPM	499
*ALLIUM CEPA	5.4 MG/KG DRY WEIGHT	PH=7.1, 1 Y, 100 T DW SL AC, CONT= 1.8 PPM, SOIL=3.5, SLUDGE=10 PPM	499
*ALLIUM CEPA	3.0 MG/KG DRY WEIGHT	PH=7.1, 2 Y, 100 T DW SL AC, CONT= 2.9 PPM, SOIL=3.5 PPM, SLUDGE=10 PPM	499
*BRASSICA OLERACEA VAR. CAPITATA	11 MG/KG DRY WEIGHT	PH=7.1, 1 Y, 100 T DW SL AC, CONT= 6.5 PPM, SOIL=3.5, SLUDGE=10 PPM	499
*BRASSICA OLERACEA VAR. CAPITATA	11.1 MG/KG DRY WEIGHT	PH=7.1, 2 Y, 100 T DW SL AC, CONT= 8.4 PPM, SOIL=3.5 PPM, SLUDGE=10 PPM	499
*DAUCUS CAROTA VAR. SATIVA	4.5 MG/KG DRY WEIGHT	PH=7.1, 1 Y, 100 T DW SL AC, CONT= 4.9 PPM, SOIL=3.5, SLUDGE=10 PPM	499
*DAUCUS CAROTA VAR. SATIVA	6.0 MG/KG DRY WEIGHT	PH=7.1, 2 Y, 100 T DW SL AC, CONT= 4.0 PPM, SOIL=3.5 PPM, SLUDGE=10 PPM	499
*ECHINOCHLOA CRUSGALLI VAR. FRUMENTACEA	10 MG/KG DRY WEIGHT	PH=7.1, 1 Y, 100 T DW SL AC, CONT= 13 PPM, SOIL=3.5, SLUDGE=10 PPM	499
*ECHINOCHLOA CRUSGALLI VAR. FRUMENTACEA	8.2 MG/KG DRY WEIGHT	PH=7.1, 2 Y, 100 T DW SL AC, CONT= 8.8 PPM, SOIL=3.5 PPM, SLUDGE=10 PPM	499
*LYCOPERSICON ESCULENTUM	5.6 MG/KG DRY WEIGHT	PH=7.1, 2 Y, 100 T DW SL AC, CONT= 3.7 PPM, SOIL=3.5 PPM, SLUDGE=10 PPM	499
*PHASEOLUS VULGARIS	6.2 MG/KG DRY WEIGHT	PH=7.1, 1 Y, 100 T DW SL AC, CONT= 5.5 PPM, SOIL=3.5, SLUDGE=10 PPM	499
*PHASEOLUS VULGARIS	3.5 MG/KG DRY WEIGHT	PH=7.1, 2 Y, 100 T DW SL AC, CONT= 5.1 PPM, SOIL=3.5 PPM, SLUDGE=10 PPM	499
*SOLANUM TUBEROSUM	6.1 MG/KG DRY WEIGHT	PH=7.1, 1 Y, 100 T DW SL AC, CONT= 6.2 PPM, SOIL=3.5, SLUDGE=10 PPM	499
*SOLANUM TUBEROSUM	5.2 MG/KG DRY WEIGHT	PH=7.1, 2 Y, 100 T DW SL AC, CONT= 3.3 PPM, SOIL=3.5 PPM, SLUDGE=10 PPM	499

TABLE D463

UPTAKE FROM SOIL WITH ADDED SLUDGE : CA			
ITEM:	VALUE:	CONDITIONS:	REF.:
*ALLIUM CEPA	13200 MG/KG DRY WEIGHT	PH=7.1, 1 Y, 100 T DW SL AC, CONT= 4210 PPM, SOIL=3.4 G/KG, SL=71 G/KG	499
*ALLIUM CEPA	6360 MG/KG DRY WEIGHT	PH=7.1, 2 Y, 100 T DW SL AC, CONT= 1920 PPM, SOIL=3.4 G/KG, SL=71 G/KG	499
*BRASSICA OLERACEA VAR. CAPITATA	7630 MG/KG DRY WEIGHT	PH=7.1, 1 Y, 100 T DW SL AC, CONT= 4500 PPM, SOIL=3.4 G/KG, SL=71 G/KG	499
*BRASSICA OLERACEA VAR. CAPITATA	19500 MG/KG DRY WEIGHT	PH=7.1, 2 Y, 100 T DW SL AC, CONT= 7700 PPM, SOIL=3.4 G/KG, SL=71 G/KG	499
*DAUCUS CAROTA VAR. SATIVA	2430 MG/KG DRY WEIGHT	PH=7.1, 1 Y, 100 T DW SL AC, CONT= 1650 PPM, SOIL=3.4 G/KG, SL=71 G/KG	499
*DAUCUS CAROTA VAR. SATIVA	2710 MG/KG DRY WEIGHT	PH=7.1, 2 Y, 100 T DW SL AC, CONT= 2060 PPM, SOIL=3.4 G/KG, SL=71 G/KG	499
*ECHINOCHLOA CRUSGALLI VAR. FRUMENTACEA	2650 MG/KG DRY WEIGHT	PH=7.1, 1 Y, 100 T DW SL AC, CONT= 3700 PPM, SOIL=3.4 G/KG, SL=71 G/KG	499
*ECHINOCHLOA CRUSGALLI VAR. FRUMENTACEA	3950 MG/KG DRY WEIGHT	PH=7.1, 2 Y, 100 T DW SL AC, CONT= 3780 PPM, SOIL=3.4 G/KG, SL=71 G/KG	499
*LYCOPERSICON ESCULENTUM	600 MG/KG DRY WEIGHT	PH=7.1, 1 Y, 100 T DW SL AC, CONT= 780 PPM, SOIL=3.4 G/KG, SL=71 G/KG	499
*LYCOPERSICON ESCULENTUM	1780 MG/KG DRY WEIGHT	PH=7.1, 2 Y, 100 T DW SL AC, CONT= 786 PPM, SOIL=3.4 G/KG, SL=71 G/KG	499

TABLE D463

UPTAKE FROM SOIL WITH ADDED SLUDGE : CA			
ITEM:	VALUE:	CONDITIONS:	REF.:
*PHASEOLUS VULGARIS	7050 MG/KG DRY WEIGHT	PH=7.1, 1 Y, 100 T DW SL AC, CONT= 2670 PPM, SOIL=3.4 G/KG, SL=71 G/ KG	499
*PHASEOLUS VULGARIS	6000 MG/KG DRY WEIGHT	PH=7.1, 2 Y, 100 T DW SL AC, CONT= 4610 PPM, SOIL=3.4 G/KG, SL=71 G/ KG	499
*SOLANUM TUBEROSUM	426 MG/KG DRY WEIGHT	PH=7.1, 1 Y, 100 T DW SL AC, CONT= 237 PPM, SOIL=3.4 G/KG, SL=71 G/KG	499
*SOLANUM TUBEROSUM	342 MG/KG DRY WEIGHT	PH=7.1, 2 Y, 100 T DW SL AC, CONT= 83 PPM, SOIL=3.4 G/KG, SLUDGE=71 G/KG	499

TABLE D464

UPTAKE FROM SOIL WITH ADDED SLUDGE : CD			
ITEM:	VALUE:	CONDITIONS:	REF.:
*ALLIUM CEPA	2.2 MG/KG DRY WEIGHT	PH=7.1, 1 Y, 100 T DW SL AC, CONT= 0.6 PPM, SOIL=0.1 PPM, SLUDGE=44 PPM	499
*ALLIUM CEPA	0.9 MG/KG DRY WEIGHT	PH=7.1, 2 Y, 100 T DW SL AC, CONT= 0.3 PPM, SOIL=0.1 PPM, SLUDGE=44 PPM	499
*BRASSICA OLERACEA VAR. CAPITATA	0.7 MG/KG DRY WEIGHT	PH=7.1, 1 Y, 100 T DW SL AC, CONT= 0.2 PPM, SOIL=0.1 PPM, SLUDGE=44 PPM	499
*BRASSICA OLERACEA VAR. CAPITATA	0.9 MG/KG DRY WEIGHT	PH=7.1, 2 Y, 100 T DW SL AC, CONT= 0.3 PPM, SOIL=0.1 PPM, SLUDGE=44 PPM	499
*DAUCUS CAROTA VAR. SATIVA	0.5 MG/KG DRY WEIGHT	PH=7.1, 1 Y, 100 T DW SL AC, CONT= 1.1 PPM, SOIL=0.1 PPM, SLUDGE=44 PPM	499
*DAUCUS CAROTA VAR. SATIVA	0.8 MG/KG DRY WEIGHT	PH=7.1, 2 Y, 100 T DW SL AC, CONT= 1.9 PPM, SOIL=0.1 PPM, SLUDGE=44 PPM	499
*ECHINOCHLOA CRUSGALLI VAR. FRUMENTACEA	0.4 MG/KG DRY WEIGHT	PH=7.1, 1 Y, 100 T DW SL AC, CONT= 0.2 PPM, SOIL=0.1 PPM, SLUDGE=44 PPM	499
*ECHINOCHLOA CRUSGALLI VAR. FRUMENTACEA	0.4 MG/KG DRY WEIGHT	PH=7.1, 2 Y, 100 T DW SL AC, CONT= 0.2 PPM, SOIL=0.1 PPM, SLUDGE=44 PPM	499
*FODDER RAPE	0.0 (% INCREASE IN PLANT)/(% INCREASE IN SOIL)	BACKGROUND LEVEL = 1.2 PPM,SLUDGE LEVEL= 11.0 PPM	27
*LYCOPERSICON ESCULENTUM	0.3 MG/KG DRY WEIGHT	PH=7.1, 1 Y, 100 T DW SL AC, CONT= 0.1 PPM, SOIL=0.1 PPM, SLUDGE=44 PPM	499
*LYCOPERSICON ESCULENTUM	0.3 MG/KG DRY WEIGHT	PH=7.1, 2 Y, 100 T DW SL AC, CONT= 0.1 PPM, SOIL=0.1 PPM, SLUDGE=44 PPM	499
*PHASEOLUS VULGARIS	0.1 MG/KG DRY WEIGHT	PH=7.1, 1 Y, 100 T DW SL AC, CONT= 0.1 PPM, SOIL=0.1 PPM, SLUDGE=44 PPM	499
*PHASEOLUS VULGARIS	0.1 MG/KG DRY WEIGHT	PH=7.1, 2 Y, 100 T DW SL AC, CONT= 0.1 PPM, SOIL=0.1 PPM, SLUDGE=44 PPM	499
*SOLANUM TUBEROSUM	0.5 MG/KG DRY WEIGHT	PH=7.1, 1 Y, 100 T DW SL AC, CONT= 0.3 PPM, SOIL=0.1 PPM, SLUDGE=44 PPM	499
*SOLANUM TUBEROSUM	0.3 MG/KG DRY WEIGHT	PH=7.1, 2 Y, 100 T DW SL AC, CONT= 0.3 PPM, SOIL=0.1 PPM, SLUDGE=44 PPM	499

TABLE D465

UPTAKE FROM SOIL WITH ADDED SLUDGE : CL			
ITEM:	VALUE:	CONDITIONS:	REF.:
*ALLIUM CEPA	4280 MG/KG DRY WEIGHT	PH=7.1, 1 Y, 100 T DW SL AC, CONT= 1650 PPM, SOIL=82 PPM, SLUDGE=5670 PPM	499

TABLE D465

UPTAKE FROM SOIL WITH ADDED SLUDGE : CL

ITEM:	VALUE:	CONDITIONS:	REF.:
*ALLIUM CEPA	1120 MG/KG DRY WEIGHT	PH=7.1, 2 Y, 100 T DW SL AC, CONT= 1260 PPM, SOIL=82 PPM, SL = 5670 PPM	499
*BRASSICA OLERACEA VAR. CAPITATA	3220 MG/KG DRY WEIGHT	PH=7.1, 1 Y, 100 T DW SL AC, CONT= 3210 PPM, SOIL=82 PPM, SL=5670 PPM	499
*BRASSICA OLERACEA VAR. CAPITATA	4400 MG/KG DRY WEIGHT	PH=7.1, 2 Y, 100 T DW SL AC, CONT= 4000 PPM, SOIL=82 PPM, SL = 5670 PPM	499
*DAUCUS CAROTA VAR. SATIVA	2040 MG/KG DRY WEIGHT	PH=7.1, 1 Y, 100 T DW SL AC, CONT= 3220 PPM, SOIL=82 PPM, SL=5670 PPM	499
*DAUCUS CAROTA VAR. SATIVA	2500 MG/KG DRY WEIGHT	PH=7.1, 2 Y, 100 T DW SL AC, CONT= 1880 PPM, SOIL=82 PPM, SL = 5670 PPM	499
*ECHINOCHLOA CRUSGALLI VAR. FRUMENTACEA	5780 MG/KG DRY WEIGHT	PH=7.1, 1 Y, 100 T DW SL AC, CONT= 5260 PPM, SOIL=82 PPM, SL=5670 PPM	499
*ECHINOCHLOA CRUSGALLI VAR. FRUMENTACEA	6230 MG/KG DRY WEIGHT	PH=7.1, 2 Y, 100 T DW SL AC, CONT= 4100 PPM, SOIL=82 PPM, SL = 5670 PPM	499
*LYCOPERSICON ESCULENTUM	3160 MG/KG DRY WEIGHT	PH=7.1, 1 Y, 100 T DW SL AC, CONT= 3200 PPM, SOIL=82 PPM, SL=5670 PPM	499
*LYCOPERSICON ESCULENTUM	2470 MG/KG DRY WEIGHT	PH=7.1, 2 Y, 100 T DW SL AC, CONT= 2390 PPM, SOIL=82 PPM, SL = 5670 PPM	499
*PHASEOLUS VULGARIS	2460 MG/KG DRY WEIGHT	PH=7.1, 1 Y, 100 T DW SL AC, CONT= 2270 PPM, SOIL=82 PPM, SL=5670 PPM	499
*PHASEOLUS VULGARIS	1970 MG/KG DRY WEIGHT	PH=7.1, 2 Y, 100 T DW SL AC, CONT= 1980 PPM, SOIL=82 PPM, SL = 5670 PPM	499
*SOLANUM TUBEROSUM	2750 MG/KG DRY WEIGHT	PH=7.1, 1 Y, 100 T DW SL AC, CONT= 870 PPM, SOIL=82 PPM, SL=5670 PPM	499
*SOLANUM TUBEROSUM	2620 MG/KG DRY WEIGHT	PH=7.1, 2 Y, 100 T DW SL AC, CONT= 2800 PPM, SOIL=82 PPM, SL = 5670 PPM	499

TABLE D466

UPTAKE FROM SOIL WITH ADDED SLUDGE : CO

ITEM:	VALUE:	CONDITIONS:	REF.:
*FODDER RAPE	5.7 (% INCREASE IN PLANT)/(% INCREASE IN SOIL)	BACKGROUND LEVEL = 14.2 PPM,SLUDGE LEVEL= 14.6 PPM	27

TABLE D467

UPTAKE FROM SOIL WITH ADDED SLUDGE : CR

ITEM:	VALUE:	CONDITIONS:	REF.:
*ALLIUM CEPA	5.9 MG/KG DRY WEIGHT	PH=7.1, 1 Y, 100 T DW SL AC, CONT= 1.1 PPM, SOIL=17.3 PPM, SLUDGE=302 PPM	499
*ALLIUM CEPA	0.2 MG/KG DRY WEIGHT	PH=7.1, 2 Y, 100 T DW SL AC, CONT= 0.1 PPM, SOIL=17.3 PPM, SLUDGE=302 PPM	499
*BRASSICA OLERACEA VAR. CAPITATA	3.5 MG/KG DRY WEIGHT	PH=7.1, 1 Y, 100 T DW SL AC, CONT= 0.2 PPM, SOIL=17.3 PPM, SLUDGE=302 PPM	499
*BRASSICA OLERACEA VAR. CAPITATA	0.2 MG/KG DRY WEIGHT	PH=7.1, 2 Y, 100 T DW SL AC, CONT= 0.2 PPM, SOIL=17.3 PPM, SLUDGE=302 PPM	499
*DAUCUS CAROTA VAR. SATIVA	0.9 MG/KG DRY WEIGHT	PH=7.1, 1 Y, 100 T DW SL AC, CONT= 2.8 PPM, SOIL=17.3 PPM, SLUDGE=302 PPM	499
*DAUCUS CAROTA VAR. SATIVA	0.4 MG/KG DRY WEIGHT	PH=7.1, 2 Y, 100 T DW SL AC, CONT= 0.1 PPM, SOIL=17.3 PPM, SLUDGE=302 PPM	499

TABLE D467

UPTAKE FROM SOIL WITH ADDED SLUDGE : CR			
ITEM:	VALUE:	CONDITIONS:	REF.:
*ECHINOCHLOA CRUSGALLI VAR. FRUMENTACEA	0.2 MG/KG DRY WEIGHT	PH=7.1, 1 Y, 100 T DW SL AC, CONT= 0.5 PPM, SOIL=17.3 PPM, SLUDGE=302 PPM	499
*ECHINOCHLOA CRUSGALLI VAR. FRUMENTACEA	0.2 MG/KG DRY WEIGHT	PH=7.1, 2 Y, 100 T DW SL AC, CONT= 0.3 PPM, SOIL=17.3 PPM, SLUDGE=302 PPM	499
*FODDER RAPE	0.8 (% INCREASE IN PLANT)/(% INCREASE IN SOIL)	BACKGROUND LEVEL = 36.1 PPM,SLUDGE LEVEL= 61.0 PPM	27
*LYCOPERSICON ESCULENTUM	0.2 MG/KG DRY WEIGHT	PH=7.1, 1 Y, 100 T DW SL AC, CONT= 0.2 PPM, SOIL=17.3 PPM, SLUDGE=302 PPM	499
*LYCOPERSICON ESCULENTUM	0.3 MG/KG DRY WEIGHT	PH=7.1, 2 Y, 100 T DW SL AC, CONT= 0.1 PPM, SOIL=17.3 PPM, SLUDGE=302 PPM	499
*PHASEOLUS VULGARIS	0.6 MG/KG DRY WEIGHT	PH=7.1, 1 Y, 100 T DW SL AC, CONT= 3.5 PPM, SOIL=17.3 PPM, SLUDGE=302 PPM	499
*PHASEOLUS VULGARIS	0.2 MG/KG DRY WEIGHT	PH=7.1, 2 Y, 100 T DW SL AC, CONT= 0.2 PPM, SOIL=17.3 PPM, SLUDGE=302 PPM	499
*SOLANUM TUBEROSUM	0.11 MG/KG DRY WEIGHT	PH=7.1, 1 Y, 100 T DW SL AC, CONT= 0.07 PPM, SOIL=17.3 PPMSL = 302 PPM	499
*SOLANUM TUBEROSUM	0.3 MG/KG DRY WEIGHT	PH=7.1, 2 Y, 100 T DW SL AC, CONT= 0.2 PPM, SOIL=17.3 PPM, SLUDGE=302 PPM	499

TABLE D468

UPTAKE FROM SOIL WITH ADDED SLUDGE : CU			
ITEM:	VALUE:	CONDITIONS:	REF.:
*ALLIUM CEPA	1.7 MG/KG DRY WEIGHT	PH=7.1, 1 Y, 100 T DW SL AC, CONT= 3.4 PPM, SOIL=47 PPM, SLUDGE=734 PPM p	499
*ALLIUM CEPA	1.5 MG/KG DRY WEIGHT	PH=7.1, 2 Y, 100 T DW SL AC, CONT= 1.0 PPM, SOIL=47 PPM, SLUDGE=734 PPM	499
*BRASSICA OLERACEA VAR. CAPITATA	4.2 MG/KG DRY WEIGHT	PH=7.1, 1 Y, 100 T DW SL AC, CONT= 3.0 PPM, SOIL=47 PPM, SLUDGE=734 PPM	499
*BRASSICA OLERACEA VAR. CAPITATA	2.2 MG/KG DRY WEIGHT	PH=7.1, 2 Y, 100 T DW SL AC, CONT= 0.6 PPM, SOIL=47 PPM, SLUDGE=734 PPM	499
*DAUCUS CAROTA VAR. SATIVA	2.6 MG/KG DRY WEIGHT	PH=7.1, 2 Y, 100 T DW SL AC, CONT= 1.3 PPM, SOIL=47 PPM, SLUDGE=734 PPM	499
*ECHINOCHLOA CRUSGALLI VAR. FRUMENTACEA	3.9 MG/KG DRY WEIGHT	PH=7.1, 1 Y, 100 T DW SL AC, CONT= 2.4 PPM, SOIL=47 PPM, SLUDGE=734 PPM	499
*ECHINOCHLOA CRUSGALLI VAR. FRUMENTACEA	2.0 MG/KG DRY WEIGHT	PH=7.1, 2 Y, 100 T DW SL AC, CONT= 1.2 PPM, SOIL=47 PPM, SLUDGE=734 PPM	499
*FODDER RAPE	0.4 (% INCREASE IN PLANT)/(% INCREASE IN SOIL)	BACKGROUND LEVEL =25.5 PPM,SLUDGE LEVEL= 90.5 PPM	27
*LYCOPERSICON ESCULENTUM	2.5 MG/KG DRY WEIGHT	PH=7.1, 1 Y, 100 T DW SL AC, CONT= 2.2 PPM, SOIL=47 PPM, SLUDGE=734 PPM	499
*LYCOPERSICON ESCULENTUM	1.7 MG/KG DRY WEIGHT	PH=7.1, 2 Y, 100 T DW SL AC, CONT= 1.6 PPM, SOIL=47 PPM, SLUDGE=734 PPM	499
*PHASEOLUS VULGARIS	3.4 MG/KG DRY WEIGHT	PH=7.1, 1 Y, 100 T DW SL AC, CONT= 3.2 PPM, SOIL=47 PPM, SLUDGE=734 PPM	499
*PHASEOLUS VULGARIS	2.2 MG/KG DRY WEIGHT	PH=7.1, 2 Y, 100 T DW SL AC, CONT= 1.8 PPM, SOIL=47 PPM, SLUDGE=734 PPM	499
*SOLANUM TUBEROSUM	5.7 MG/KG DRY WEIGHT	PH=7.1, 1 Y, 100 T DW SL AC, CONT= 3.1 PPM, SOIL=47 PPM, SLUDGE=734 PPM	499

TABLE D468

UPTAKE FROM SOIL WITH ADDED SLUDGE : CU			
ITEM:	VALUE:	CONDITIONS:	REF.:
*SOLANUM TUBEROSUM	4.6 MG/KG DRY WEIGHT	PH=7.1, 2 Y, 100 T DW SL AC, CONT= 2.5 PPM, SOIL=47 PPM, SLUDGE=734 PPM	499

TABLE D469

UPTAKE FROM SOIL WITH ADDED SLUDGE : FE			
ITEM:	VALUE:	CONDITIONS:	REF.:
*ALLIUM CEPA	264 MG/KG DRY WEIGHT	PH=7.1, 1 Y, 100 T DW SL AC, CONT= 100 PPM, SOIL=25 G/KG, SLUDGE=61 G/KG	499
*ALLIUM CEPA	96 MG/KG DRY WEIGHT	PH=7.1, 2 Y, 100 T DW SL AC, CONT= 40 PPM, SOIL=25 G/KG, SLUDGE=61 G/KG	499
*BRASSICA OLERACEA VAR. CAPITATA	177 MG/KG DRY WEIGHT	PH=7.1, 1 Y, 100 T DW SL AC, CONT= 78 PPM, SOIL=25 G/KG, SLUDGE=61 G/KG	499
*BRASSICA OLERACEA VAR. CAPITATA	90 MG/KG DRY WEIGHT	PH=7.1, 2 Y, 100 T DW SL AC, CONT= 136 PPM, SOIL=25 G/KG, SLUDGE=61 G/KG	499
*DAUCUS CAROTA VAR. SATIVA	146 MG/KG DRY WEIGHT	PH=7.1, 1 Y, 100 T DW SL AC, CONT= 143 PPM, SOIL=25 G/KG, SLUDGE=61 G/KG	499
*DAUCUS CAROTA VAR. SATIVA	34 MG/KG DRY WEIGHT	PH=7.1, 2 Y, 100 T DW SL AC, CONT= 71 PPM, SOIL=25 G/KG, SLUDGE=61 G/KG	499
*ECHINOCHLOA CRUSGALLI VAR. FRUMENTACEA	100 MG/KG DRY WEIGHT	PH=7.1, 1 Y, 100 T DW SL AC, CONT= 49 PPM, SOIL=25 G/KG, SLUDGE=61 G/KG	499
*ECHINOCHLOA CRUSGALLI VAR. FRUMENTACEA	164 MG/KG DRY WEIGHT	PH=7.1, 2 Y, 100 T DW SL AC, CONT= 74 PPM, SOIL=25 G/KG, SLUDGE=61 G/KG	499
*LYCOPERSICON ESCULENTUM	170 MG/KG DRY WEIGHT	PH=7.1, 1 Y, 100 T DW SL AC, CONT= 173 PPM, SOIL=25 G/KG, SLUDGE=61 G/KG	499
*LYCOPERSICON ESCULENTUM	85 MG/KG DRY WEIGHT	PH=7.1, 2 Y, 100 T DW SL AC, CONT= 20 PPM, SOIL=25 G/KG, SLUDGE=61 G/KG	499
*PHASEOLUS VULGARIS	97 MG/KG DRY WEIGHT	PH=7.1, 1 Y, 100 T DW SL AC, CONT= 108 PPM, SOIL=25 G/KG, SLUDGE=61 G/KG	499
*PHASEOLUS VULGARIS	111 MG/KG DRY WEIGHT	PH=7.1, 2 Y, 100 T DW SL AC, CONT= 84 PPM, SOIL=25 G/KG, SLUDGE=61 G/KG	499
*SOLANUM TUBEROSUM	70 MG/KG DRY WEIGHT	PH=7.1, 1 Y, 100 T DW SL AC, CONT= 49 PPM, SOIL=25 G/KG, SLUDGE=61 G/KG	499
*SOLANUM TUBEROSUM	56 MG/KG DRY WEIGHT	PH=7.1, 2 Y, 100 T DW SL AC, CONT= 35 PPM, SOIL=25 G/KG, SLUDGE=61 G/KG	499

TABLE D470

UPTAKE FROM SOIL WITH ADDED SLUDGE : HG			
ITEM:	VALUE:	CONDITIONS:	REF.:
*ALLIUM CEPA	0.5 MG/KG DRY WEIGHT	PH=7.1, 1 Y, 100 T DW SL AC, CONT= 0.3 PPM, SOIL=0.1 PPM, SLUDGE=11.7 PPM	499
*ALLIUM CEPA	0.2 MG/KG DRY WEIGHT	PH=7.1, 2 Y, 100 T DW SL AC, CONT= 0.1 PPM, SOIL=0.1 PPM, SLUDGE=11.7 PPM	499
*BRASSICA OLERACEA VAR. CAPITATA	0.4 MG/KG DRY WEIGHT	PH=7.1, 1 Y, 100 T DW SL AC, CONT= 0.1 PPM, SOIL=0.1 PPM, SLUDGE=11.7 PPM	499

TABLE D470

UPTAKE FROM SOIL WITH ADDED SLUDGE : HG

ITEM:	VALUE:	CONDITIONS:	REF.:
*BRASSICA OLERACEA VAR. CAPITATA	0.2 MG/KG DRY WEIGHT	PH=7.1, 2 Y, 100 T DW SL AC, CONT= 0.3 PPM, SOIL=0.1 PPM, SLUDGE=11.7 PPM	499
*DAUCUS CAROTA VAR. SATIVA	0.1 MG/KG DRY WEIGHT	PH=7.1, 1 Y, 100 T DW SL AC, CONT= 0.1 PPM, SOIL=0.1 PPM, SLUDGE=11.7 PPM	499
*DAUCUS CAROTA VAR. SATIVA	0.04 MG/KG DRY WEIGHT	PH=7.1, 2 Y, 100 T DW SL AC, CONT= 0.1 PPM, SOIL=0.1 PPM, SLUDGE=11.7 PPM	499
*ECHINOCHLOA CRUSGALLI VAR. FRUMENTACEA	1.0 MG/KG DRY WEIGHT	PH=7.1, 1 Y, 100 T DW SL AC, CONT= 0.7 PPM, SOIL=0.1 PPM, SLUDGE=11.7 PPM	499
*ECHINOCHLOA CRUSGALLI VAR. FRUMENTACEA	0.5 MG/KG DRY WEIGHT	PH=7.1, 2 Y, 100 T DW SL AC, CONT= 0.4 PPM, SOIL=0.1 PPM, SLUDGE=11.7 PPM	499
*FODDER RAPE	0.01 (% INCREASE IN PLANT)/(% INCREASE IN SOIL)	BACKGROUND LEVEL = 0.018 PPM, SLUDGE LEVEL= 12.0 PPM	27
*LYCOPERSICON ESCULENTUM	0.4 MG/KG DRY WEIGHT	PH=7.1, 1 Y, 100 T DW SL AC, CONT= 0.1 PPM, SOIL=0.1 PPM, SLUDGE=11.7 PPM	499
*LYCOPERSICON ESCULENTUM	0.2 MG/KG DRY WEIGHT	PH=7.1, 2 Y, 100 T DW SL AC, CONT= 0.1 PPM, SOIL=0.1 PPM, SLUDGE=11.7 PPM	499
*PHASEOLUS VULGARIS	0.1 MG/KG DRY WEIGHT	PH=7.1, 1 Y, 100 T DW SL AC, CONT= 0.2 PPM, SOIL=0.1 PPM, SLUDGE=11.7 PPM	499
*PHASEOLUS VULGARIS	0.1 MG/KG DRY WEIGHT	PH=7.1, 2 Y, 100 T DW SL AC, CONT= 0.3 PPM, SOIL=0.1 PPM, SLUDGE=11.7 PPM	499
*SOLANUM TUBEROSUM	0.1 MG/KG DRY WEIGHT	PH=7.1, 1 Y, 100 T DW SL AC, CONT= 0.1 PPM, SOIL=0.1 PPM, SLUDGE=11.7 PPM	499
*SOLANUM TUBEROSUM	0.1 MG/KG DRY WEIGHT	PH=7.1, 2 Y, 100 T DW SL AC, CONT= 0.1 PPM, SOIL=0.1 PPM, SLUDGE=11.7 PPM	499

TABLE D471

UPTAKE FROM SOIL WITH ADDED SLUDGE : I

ITEM:	VALUE:	CONDITIONS:	REF.:
*ALLIUM CEPA	1.3 MG/KG DRY WEIGHT	PH=7.1, 1 Y, 100 T DW SL AC, CONT= 0.7 PPM, SOIL=1.2 PPM	499
*BRASSICA OLERACEA VAR. CAPITATA	2.9 MG/KG DRY WEIGHT	PH=7.1, 1 Y, 100 T DW SL AC, CONT= 0.3 PPM, SOIL=1.2 PPM	499
*DAUCUS CAROTA VAR. SATIVA	1.0 MG/KG DRY WEIGHT	PH=7.1, 1 Y, 100 T DW SL AC, CONT= 0.9 PPM, SOIL=1.2 PPM	499
*ECHINOCHLOA CRUSGALLI VAR. FRUMENTACEA	11.7 MG/KG DRY WEIGHT	PH=7.1, 1 Y, 100 T DW SL AC, CONT= . PPM, SOIL=1.2 PPM	499
*LYCOPERSICON ESCULENTUM	1.7 MG/KG DRY WEIGHT	PH=7.1, 1 Y, 100 T DW SL AC, CONT= 0.4 PPM, SOIL=1.2 PPM	499
*PHASEOLUS VULGARIS	0.7 MG/KG DRY WEIGHT	PH=7.1, 1 Y, 100 T DW SL AC, CONT= 0.7 PPM, SOIL=1.2 PPM	499
*SOLANUM TUBEROSUM	0.5 MG/KG DRY WEIGHT	PH=7.1, 1 Y, 100 T DW SL AC, CONT= 0.04 PPM, SOIL=1.2 PPM	499

TABLE D472

UPTAKE FROM SOIL WITH ADDED SLUDGE : K

ITEM:	VALUE:	CONDITIONS:	REF.:
*ALLIUM CEPA	11000 MG/KG DRY WEIGHT	PH=7.1, 2 Y, 100 T DW SL AC, CONT= 14000 PPM, SOIL=14 G/KG, SL=2.18 G/KG	499

TABLE D472

UPTAKE FROM SOIL WITH ADDED SLUDGE : K

ITEM:	VALUE:	CONDITIONS:	REF.:
*BRASSICA OLERACEA VAR. CAPITATA	28400 MG/KG DRY WEIGHT	PH=7.1, 2 Y, 100 T DW SL AC, CONT= 17400 PPM, SOIL=14 G/KG, SL=2.18 G/KG	499
*DAUCUS CAROTA VAR. SATIVA	19300 MG/KG DRY WEIGHT	PH=7.1, 2 Y, 100 T DW SL AC, CONT= 18400 PPM, SOIL=14 G/KG, SL=2.18 G/KG	499
*ECHINOCHLOA CRUSGALLI VAR. FRUMENTACEA	15500 MG/KG DRY WEIGHT	PH=7.1, 2 Y, 100 T DW SL AC, CONT= 13100 PPM, SOIL=14 G/KG, SL=2.18 G/KG	499
*LYCOPERSICON ESCULENTUM	21900 MG/KG DRY WEIGHT	PH=7.1, 2 Y, 100 T DW SL AC, CONT= 17000 PPM, SOIL=14 G/KG, SL=2.18 G/KG	499
*PHASEOLUS VULGARIS	20600 MG/KG DRY WEIGHT	PH=7.1, 2 Y, 100 T DW SL AC, CONT= 24600 PPM, SOIL=14 G/KG, SL=2.18 G/KG	499
*SOLANUM TUBEROSUM	31400 MG/KG DRY WEIGHT	PH=7.1, 2 Y, 100 T DW SL AC, CONT= 20000 PPM, SOIL=14 G/KG, SL=2.18 G/KG	499

TABLE D473

UPTAKE FROM SOIL WITH ADDED SLUDGE : MG

ITEM:	VALUE:	CONDITIONS:	REF.:
*ALLIUM CEPA	630 MG/KG DRY WEIGHT	PH=7.1, 2 Y, 100 T DW SL AC, CONT= 720 PPM, SOIL=7.47 G/KG, SL=8.57 G/KG	499
*BRASSICA OLERACEA VAR. CAPITATA	1050 MG/KG DRY WEIGHT	PH=7.1, 2 Y, 100 T DW SL AC, CONT= 2850 PPM, SOIL=7.47 G/KG, SL=8.57 G/KG	499
*DAUCUS CAROTA VAR. SATIVA	880 MG/KG ¤RY WEIGHT	PH=7.1, 2 Y, 100 T DW SL AC, CONT= 1260 PPM, SOIL=7.47 G/KG, SL=8.57 G/KG	499
*ECHINOCHLOA CRUSGALLI VAR. FRUMENTACEA	4650 MG/KG DRY WEIGHT	PH=7.1, 2 Y, 100 T DW SL AC, CONT= 2100 PPM, SOIL=7.47 G/KG, SL=8.57 G/KG	499
*LYCOPERSICON ESCULENTUM	960 MG/KG DRY WEIGHT	PH=7.1, 2 Y, 100 T DW SL AC, CONT= 900 PPM, SOIL=7.47 G/KG, SL=8.57 G/KG	499
*PHASEOLUS VULGARIS	1670 MG/KG DRY WEIGHT	PH=7.1, 2 Y, 100 T DW SL AC, CONT= 2360 PPM, SOIL=7.47 G/KG, SL=8.57 G/KG	499
*SOLANUM TUBEROSUM	870 MG/KG DRY WEIGHT	PH=7.1, 2 Y, 100 T DW SL AC, CONT= 640 PPM, SOIL=7.47 G/KG, SL=8.57 G/KG	499

TABLE D474

UPTAKE FROM SOIL WITH ADDED SLUDGE : MN

ITEM:	VALUE:	CONDITIONS:	REF.:
*FODDER RAPE	18 (% INCREASE IN PLANT)/(% INCREASE IN SOIL)	BACKGROUND LEVEL = 476 PPM,SLUDGE LEVEL= 480 PPM	27

TABLE D475

UPTAKE FROM SOIL WITH ADDED SLUDGE : MO

ITEM:	VALUE:	CONDITIONS:	REF.:
*ALLIUM CEPA	1.8 MG/KG DRY WEIGHT	PH=7.1, 1 Y, 100 T DW SL AC, CONT= 0.7 PPM, SLUDGE = 21 PPM	499

TABLE D475

UPTAKE FROM SOIL WITH ADDED SLUDGE : MO			
ITEM:	VALUE:	CONDITIONS:	REF.:
*BRASSICA OLERACEA VAR. CAPITATA	0.4 MG/KG DRY WEIGHT	PH=7.1, 1 Y, 100 T DW SL AC, CONT= 1.0 PPM, SLUDGE = 21 PPM	499
*BRASSICA OLERACEA VAR. CAPITATA	4.9 MG/KG DRY WEIGHT	PH=7.1, 2 Y, 100 T DW SL AC, CONT= 0.8 PPM, SLUDGE = 21 PPM	499
*DAUCUS CAROTA VAR. SATIVA	0.2 MG/KG DRY WEIGHT	PH=7.1, 1 Y, 100 T DW SL AC, CONT= 0.2 PPM, SLUDGE = 21 PPM	499
*DAUCUS CAROTA VAR. SATIVA	0.6 MG/KG DRY WEIGHT	PH=7.1, 2 Y, 100 T DW SL AC, CONT= 0.2 PPM, SLUDGE = 21 PPM	499
*ECHINOCHLOA CRUSGALLI VAR. FRUMENTACEA	2.3 MG/KG DRY WEIGHT	PH=7.1, 1 Y, 100 T DW SL AC, CONT= 0.3 PPM, SLUDGE = 21 PPM	499
*FODDER RAPE	1.6 (% INCREASE IN PLANT)/(% INCREASE IN SOIL)	BACKGROUND LEVEL = 0.53 PPM,SLUDGE LEVEL= 7.4 PPM	27
*LYCOPERSICON ESCULENTUM	0.9 MG/KG DRY WEIGHT	PH=7.1, 1 Y, 100 T DW SL AC, CONT= 0.5 PPM, SLUDGE = 21 PPM	499
*PHASEOLUS VULGARIS	13 MG/KG DRY WEIGHT	PH=7.1, 1 Y, 100 T DW SL AC, CONT= 1.0 PPM, SLUDGE = 21 PPM	499
*PHASEOLUS VULGARIS	30 MG/KG DRY WEIGHT	PH=7.1, 2 Y, 100 T DW SL AC, CONT= 2.0 PPM, SLUDGE = 21 PPM	499
*SOLANUM TUBEROSUM	0.8 MG/KG DRY WEIGHT	PH=7.1, 1 Y, 100 T DW SL AC, CONT= 0.2 PPM, SLUDGE = 21 PPM	499

TABLE D476

UPTAKE FROM SOIL WITH ADDED SLUDGE : NI			
ITEM:	VALUE:	CONDITIONS:	REF.:
*ALLIUM CEPA	1.4 MG/KG DRY WEIGHT	PH=7.1, 2 Y, 100 T DW SL AC, CONT= 0.6 PPM, SOIL=17 PPM, SLUDGE =191 PPM	499
*BRASSICA OLERACEA VAR. CAPITATA	4.8 MG/KG DRY WEIGHT	PH=7.1, 1 Y, 100 T DW SL AC, CONT= 1.9 PPM, SOIL=17 PPM, SLUDGE=191 PPM	499
*BRASSICA OLERACEA VAR. CAPITATA	1.9 MG/KG DRY WEIGHT	PH=7.1, 2 Y, 100 T DW SL AC, CONT= 2.1 PPM, SOIL=17 PPM, SLUDGE =191 PPM	499
*DAUCUS CAROTA VAR. SATIVA	1.1 MG/KG DRY WEIGHT	PH=7.1, 2 Y, 100 T DW SL AC, CONT= 1.2 PPM, SOIL=17 PPM, SLUDGE =191 PPM	499
*ECHINOCHLOA CRUSGALLI VAR. FRUMENTACEA	3.1 MG/KG DRY WEIGHT	PH=7.1, 1 Y, 100 T DW SL AC, CONT= 1.4 PPM, SOIL=17 PPM, SLUDGE=191 PPM	499
*ECHINOCHLOA CRUSGALLI VAR. FRUMENTACEA	2.3 MG/KG DRY WEIGHT	PH=7.1, 2 Y, 100 T DW SL AC, CONT= 1.8 PPM, SOIL=17 PPM, SLUDGE =191 PPM	499
*FODDER RAPE	1.6 (% INCREASE IN PLANT)/(% INCREASE IN SOIL)	BACKGROUND LEVEL = 28.2 PPM,SLUDGE LEVEL=343.3 PPM	27
*LYCOPERSICON ESCULENTUM	1.4 MG/KG DRY WEIGHT	PH=7.1, 1 Y, 100 T DW SL AC, CONT= 0.5 PPM, SOIL=17 PPM, SLUDGE=191 PPM	499
*LYCOPERSICON ESCULENTUM	0.9 MG/KG DRY WEIGHT	PH=7.1, 2 Y, 100 T DW SL AC, CONT= 0.4 PPM, SOIL=17 PPM, SLUDGE =191 PPM	499
*PHASEOLUS VULGARIS	6.7 MG/KG DRY WEIGHT	PH=7.1, 1 Y, 100 T DW SL AC, CONT= 3.9 PPM, SOIL=17 PPM, SLUDGE=191 PPM	499
*PHASEOLUS VULGARIS	5.5 MG/KG DRY WEIGHT	PH=7.1, 2 Y, 100 T DW SL AC, CONT= 4.4 PPM, SOIL=17 PPM, SLUDGE =191 PPM	499
*SOLANUM TUBEROSUM	1.1 MG/KG DRY WEIGHT	PH=7.1, 1 Y, 100 T DW SL AC, CONT= 0.6 PPM, SOIL=17 PPM, SLUDGE=191 PPM	499
*SOLANUM TUBEROSUM	0.9 MG/KG DRY WEIGHT	PH=7.1, 2 Y, 100 T DW SL AC, CONT= 1.0 PPM, SOIL=17 PPM, SLUDGE =191 PPM	499

TABLE D477

UPTAKE FROM SOIL WITH ADDED SLUDGE : PB			
ITEM:	VALUE:	CONDITIONS:	REF.:
*FODDER RAPE	0.7 (% INCREASE IN PLANT)/(% INCREASE IN SOIL)	BACKGROUND LEVEL = 25.7 PPM,SLUDGE LEVEL= 293 PPM	27

TABLE D478

UPTAKE FROM SOIL WITH ADDED SLUDGE : SB

ITEM:	VALUE:	CONDITIONS:	REF.:
*ALLIUM CEPA	11.5 MG/KG DRY WEIGHT	PH=7.1, 1 Y, 100 T DW SL AC, CONT= 0.8 PPM, SOIL=0.8 PPM, SLUDGE=2.9 PPM	499
*BRASSICA OLERACEA VAR. CAPITATA	0.9 MG/KG DRY WEIGHT	PH=7.1, 1 Y, 100 T DW SL AC, CONT= 1.4 PPM, SOIL=0.8 PPM, SLUDGE=2.9 PPM	499
*DAUCUS CAROTA VAR. SATIVA	0.9 MG/KG DRY WEIGHT	PH=7.1, 1 Y, 100 T DW SL AC, CONT= 0.9 PPM, SOIL=0.8 PPM, SLUDGE=2.9 PPM	499
*ECHINOCHLOA CRUSGALLI VAR. FRUMENTACEA	0.3 MG/KG DRY WEIGHT	PH=7.1, 1 Y, 100 T DW SL AC, CONT= 1.1 PPM, SOIL=0.8 PPM, SLUDGE=2.9 PPM	499
*LYCOPERSICON ESCULENTUM	0.5 MG/KG DRY WEIGHT	PH=7.1, 1 Y, 100 T DW SL AC, CONT= 2.5 PPM, SOIL=0.8 PPM, SLUDGE=2.9 PPM	499
*PHASEOLUS VULGARIS	1.0 MG/KG DRY WEIGHT	PH=7.1, 1 Y, 100 T DW SL AC, CONT= 0.4 PPM, SOIL=0.8 PPM, SLUDGE=2.9 PPM	499
*SOLANUM TUBEROSUM	0.7 MG/KG DRY WEIGHT	PH=7.1, 1 Y, 100 T DW SL AC, CONT= 0.4 PPM, SOIL=0.8 PPM, SLUDGE=2.9 PPM	499

TABLE D479

UPTAKE FROM SOIL WITH ADDED SLUDGE : SC

ITEM:	VALUE:	CONDITIONS:	REF.:
*ALLIUM CEPA	0.02 MG/KG DRY WEIGHT	PH=7.1, 1 Y, 100 T DW SL AC, CONT= 0.01 PPM, SOIL=8.7 PPM, SLUDGE=3.2 PPM	499
*BRASSICA OLERACEA VAR. CAPITATA	0.01 MG/KG DRY WEIGHT	PH=7.1, 1 Y, 100 T DW SL AC, CONT= 0.03 PPM, SOIL=8.7 PPM, SLUDGE=3.2 PPM	499
*DAUCUS CAROTA VAR. SATIVA	0.00 MG/KG DRY WEIGHT	PH=7.1, 1 Y, 100 T DW SL AC, CONT= 0.01 PPM, SOIL=8.7 PPM, SLUDGE=3.2 PPM	499
*ECHINOCHLOA CRUSGALLI VAR. FRUMENTACEA	0.02 MG/KG DRY WEIGHT	PH=7.1, 1 Y, 100 T DW SL AC, CONT= 0.00 PPM, SOIL=8.7 PPM, SLUDGE=3.2 PPM	499
*LYCOPERSICON ESCULENTUM	0.01 MG/KG DRY WEIGHT	PH=7.1, 1 Y, 100 T DW SL AC, CONT= 0.03 PPM, SOIL=8.7 PPM, SLUDGE=3.2 PPM	499
*PHASEOLUS VULGARIS	0.02 MG/KG DRY WEIGHT	PH=7.1, 1 Y, 100 T DW SL AC, CONT= 0.01 PPM, SOIL=8.7 PPM, SLUDGE=3.2 PPM	499
*SOLANUM TUBEROSUM	0.01 MG/KG DRY WEIGHT	PH=7.1, 1 Y, 100 T DW SL AC, CONT= 0.004 PPM, SOIL=8.7 PPM, SLUDGE= 3.2 PP	499

TABLE D480

UPTAKE FROM SOIL WITH ADDED SLUDGE : SE

ITEM:	VALUE:	CONDITIONS:	REF.:
*ALLIUM CEPA	0.00 MG/KG DRY WEIGHT	PH=7.1, 1 Y, 100 T DW SL AC, CONT= 0.00 PPM, SOIL=0.3 PPM, SLUDGE=1.4 PPM	499
*ALLIUM CEPA	0.02 MG/KG DRY WEIGHT	PH=7.1, 2 Y, 100 T DW SL AC, CONT= 0.02 PPM, SOIL=0.3 PPM, SLUDGE=1.4 PPM	499
*BRASSICA OLERACEA VAR. CAPITATA	0.03 MG/KG DRY WEIGHT	PH=7.1, 1 Y, 100 T DW SL AC, CONT= 0.01 PPM, SOIL=0.3 PPM, SLUDGE=1.4 PPM	499
*BRASSICA OLERACEA VAR. CAPITATA	0.04 MG/KG DRY WEIGHT	PH=7.1, 2 Y, 100 T DW SL AC, CONT= 0.02 PPM, SOIL=0.3 PPM, SLUDGE=1.4 PPM	499

TABLE D480

UPTAKE FROM SOIL WITH ADDED SLUDGE : SE			
ITEM:	VALUE:	CONDITIONS:	REF.:
*DAUCUS CAROTA VAR. SATIVA	0.01 MG/KG DRY WEIGHT	PH=7.1, 1 Y, 100 T DW SL AC, CONT= 0.00 PPM, SOIL=0.3 PPM, SLUDGE=1.4 PPM	499
*DAUCUS CAROTA VAR. SATIVA	0.04 MG/KG DRY WEIGHT	PH=7.1, 2 Y, 100 T DW SL AC, CONT= 0.01 PPM, SOIL=0.3 PPM, SLUDGE=1.4 PPM	499
*ECHINOCHLOA CRUSGALLI VAR. FRUMENTACEA	0.02 MG/KG DRY WEIGHT	PH=7.1, 1 Y, 100 T DW SL AC, CONT= 0.02 PPM, SOIL=0.3 PPM, SLUDGE=1.4 PPM	499
*ECHINOCHLOA CRUSGALLI VAR. FRUMENTACEA	0.03 MG/KG DRY WEIGHT	PH=7.1, 2 Y, 100 T DW SL AC, CONT= 0.03 PPM, SOIL=0.3 PPM, SLUDGE=1.4 PPM	499
*FODDER RAPE	NO UPTAKE (% INCREASE IN PLANT)/(% INCREASE IN SOIL)	BACKGROUND LEVEL = 0.238,SLUDGE LEVEL7.3 PPM	27
*LYCOPERSICON ESCULENTUM	0.01 MG/KG DRY WEIGHT	PH=7.1, 1 Y, 100 T DW SL AC, CONT= 0.01 PPM, SOIL=0.3 PPM, SLUDGE=1.4 PPM	499
*LYCOPERSICON ESCULENTUM	0.02 MG/KG DRY WEIGHT	PH=7.1, 2 Y, 100 T DW SL AC, CONT= 0.02 PPM, SOIL=0.3 PPM, SLUDGE=1.4 PPM	499
*PHASEOLUS VULGARIS	0.04 MG/KG DRY WEIGHT	PH=7.1, 1 Y, 100 T DW SL AC, CONT= 0.02 PPM, SOIL=0.3 PPM, SLUDGE=1.4 PPM	499
*PHASEOLUS VULGARIS	0.02 MG/KG DRY WEIGHT	PH=7.1, 2 Y, 100 T DW SL AC, CONT= 0.01 PPM, SOIL=0.3 PPM, SLUDGE=1.4 PPM	499
*SOLANUM TUBEROSUM	0.02 MG/KG DRY WEIGHT	PH=7.1, 1 Y, 100 T DW SL AC, CONT= 0.01 PPM, SOIL=0.3 PPM, SLUDGE=1.4 PPM	499
*SOLANUM TUBEROSUM	0.03 MG/KG DRY WEIGHT	PH=7.1, 2 Y, 100 T DW SL AC, CONT= 0.02 PPM, SOIL=0.3 PPM, SLUDGE=1.4 PPM	499

TABLE D481

UPTAKE FROM SOIL WITH ADDED SLUDGE : TA			
ITEM:	VALUE:	CONDITIONS:	REF.:
*ALLIUM CEPA	0.6 MG/KG DRY WEIGHT	PH=7.1, 1 Y, 100 T DW SL AC, CONT= 0.3 PPM, SOIL=0.6 PPM, SLUDGE=0.6 PPM	499
*BRASSICA OLERACEA VAR. CAPITATA	0.06 MG/KG DRY WEIGHT	PH=7.1, 1 Y, 100 T DW SL AC, CONT= 0.03 PPM, SOIL=0.6 PPM, SLUDGE=0.6 PPM	499
*DAUCUS CAROTA VAR. SATIVA	0.06 MG/KG DRY WEIGHT	PH=7.1, 1 Y, 100 T DW SL AC, CONT= 0.07 PPM, SOIL=0.6 PPM, SLUDGE=0.6 PPM	499
*ECHINOCHLOA CRUSGALLI VAR. FRUMENTACEA	0.03 MG/KG DRY WEIGHT	PH=7.1, 1 Y, 100 T DW SL AC, CONT= 0.07 PPM, SOIL=0.6 PPM, SLUDGE=0.6 PPM	499
*LYCOPERSICON ESCULENTUM	0.01 MG/KG DRY WEIGHT	PH=7.1, 1 Y, 100 T DW SL AC, CONT= 0.01 PPM, SOIL=0.6 PPM, SLUDGE=0.6 PPM	499
*SOLANUM TUBEROSUM	0.06 MG/KG DRY WEIGHT	PH=7.1, 1 Y, 100 T DW SL AC, CONT= 0.04 PPM, SOIL=0.6 PPM, SLUDGE=0.6 PPM	499

TABLE D482

UPTAKE FROM SOIL WITH ADDED SLUDGE : TI			
ITEM:	VALUE:	CONDITIONS:	REF.:
*ALLIUM CEPA	3.9 MG/KG DRY WEIGHT	PH=7.1, 2 Y, 100 T DW SL AC, CONT= 0.5 PPM, SOIL=3292 PPM, SL=2270 PPM	499
*BRASSICA OLERACEA VAR. CAPITATA	6.7 MG/KG DRY WEIGHT	PH=7.1, 2 Y, 100 T DW SL AC, CONT= 3.2 PPM, SOIL=3292 PPM, SL=2270 PPM	499

TABLE D482

UPTAKE FROM SOIL WITH ADDED SLUDGE : TI

ITEM:	VALUE:	CONDITIONS:	REF.:
*DAUCUS CAROTA VAR. SATIVA	6.1 MG/KG DRY WEIGHT	PH=7.1, 2 Y, 100 T DW SL AC, CONT= . PPM, SOIL=3292 PPM, SL=2270 PPM	499
*ECHINOCHLOA CRUSGALLI VAR. FRUMENTACEA	2.0 MG/KG DRY WEIGHT	PH=7.1, 2 Y, 100 T DW SL AC, CONT= . PPM, SOIL=3292 PPM, SL=2270 PPM	499
*LYCOPERSICON ESCULENTUM	3.1 MG/KG DRY WEIGHT	PH=7.1, 2 Y, 100 T DW SL AC, CONT= 2.0 PPM, SOIL=3292 PPM, SL=2270 PPM	0000
*PHASEOLUS VULGARIS	1.6 MG/KG DRY WEIGHT	PH=7.1, 2 Y, 100 T DW SL AC, CONT= 12 PPM, SOIL=3292 PPM, SL=2270 PPM	499
*SOLANUM TUBEROSUM	2.5 MG/KG DRY WEIGHT	PH=7.1, 2 Y, 100 T DW SL AC, CONT= 3.0 PPM, SOIL=3292 PPM, SL=2270 PPM	499

TABLE D483

UPTAKE FROM SOIL WITH ADDED SLUDGE : V

ITEM:	VALUE:	CONDITIONS:	REF.:
*ALLIUM CEPA	0.4 MG/KG DRY WEIGHT	PH=7.1, 1 Y, 100 T DW SL AC, CONT= 0.2 PPM, SOIL=67 PPM, SLUDGE=29 PPM	499
*BRASSICA OLERACEA VAR. CAPITATA	0.3 MG/KG DRY WEIGHT	PH=7.1, 1 Y, 100 T DW SL AC, CONT= 0.4 PPM, SOIL=67 PPM, SLUDGE=29 PPM	499
*DAUCUS CAROTA VAR. SATIVA	0.1 MG/KG DRY WEIGHT	PH=7.1, 1 Y, 100 T DW SL AC, CONT= 0.4 PPM, SOIL=67 PPM, SLUDGE=29 PPM	499
*ECHINOCHLOA CRUSGALLI VAR. FRUMENTACEA	0.2 MG/KG DRY WEIGHT	PH=7.1, 1 Y, 100 T DW SL AC, CONT= 0.04 PPM, SOIL=67 PPM, SLUDGE=29 PPM	499
*LYCOPERSICON ESCULENTUM	0.02 MG/KG DRY WEIGHT	PH=7.1, 1 Y, 100 T DW SL AC, CONT= 0.12 PPM, SOIL=67 PPM, SLUDGE=29 PPM	499
*PHASEOLUS VULGARIS	0.1 MG/KG DRY WEIGHT	PH=7.1, 1 Y, 100 T DW SL AC, CONT= 0.06 PPM, SOIL=67 PPM, SLUDGE=29 PPM	499
*SOLANUM TUBEROSUM	0.04 MG/KG DRY WEIGHT	PH=7.1, 1 Y, 100 T DW SL AC, CONT= 4.8 PPM, SOIL=67 PPM, SLUDGE=29 PPM	499

TABLE D484

UPTAKE FROM SOIL WITH ADDED SLUDGE : W

ITEM:	VALUE:	CONDITIONS:	REF.:
*ALLIUM CEPA	0.4 MG/KG DRY WEIGHT	PH=7.1, 1 Y, 100 T DW SL AC, CONT= 0.2 PPM, SOIL=2.1 PPM, SLUDGE=4 PPM	499
*BRASSICA OLERACEA VAR. CAPITATA	1.0 MG/KG DRY WEIGHT	PH=7.1, 1 Y, 100 T DW SL AC, CONT= 0.4 PPM, SOIL=2.1 PPM, SLUDGE=4 PPM	499
*DAUCUS CAROTA VAR. SATIVA	0.4 MG/KG DRY WEIGHT	PH=7.1, 1 Y, 100 T DW SL AC, CONT= 0.4 PPM, SOIL=2.1 PPM, SLUDGE=4 PPM	499
*ECHINOCHLOA CRUSGALLI VAR. FRUMENTACEA	0.3 MG/KG DRY WEIGHT	PH=7.1, 1 Y, 100 T DW SL AC, CONT= 0.3 PPM, SOIL=2.1 PPM, SLUDGE=4 PPM	499
*LYCOPERSICON ESCULENTUM	0.5 MG/KG DRY WEIGHT	PH=7.1, 1 Y, 100 T DW SL AC, CONT= 0.4 PPM, SOIL=2.1 PPM, SLUDGE=4 PPM	499
*PHASEOLUS VULGARIS	0.5 MG/KG DRY WEIGHT	PH=7.1, 1 Y, 100 T DW SL AC, CONT= 0.2 PPM, SOIL=2.1 PPM, SLUDGE=4 PPM	499
*SOLANUM TUBEROSUM	0.4 MG/KG DRY WEIGHT	PH=7.1, 1 Y, 100 T DW SL AC, CONT= 0.04 PPM, SOIL=2.1 PPM, SLUDGE=4 PPM	499

TABLE D485

UPTAKE FROM SOIL WITH ADDED SLUDGE : YB			
ITEM:	VALUE:	CONDITIONS:	REF.:
*ALLIUM CEPA	0.2 MG/KG DRY WEIGHT	PH=7.1, 1 Y, 100 T DW SL AC, CONT= 0.2 PPM, SOIL=2.0 PPM, SLUDGE=0.2 PPM	499
*BRASSICA OLERACEA VAR. CAPITATA	0.1 MG/KG DRY WEIGHT	PH=7.1, 1 Y, 100 T DW SL AC, CONT= 0.01 PPM, SOIL=2.0 PPM, SLUDGE=0.2 PPM	499
*DAUCUS CAROTA VAR. SATIVA	0.1 MG/KG DRY WEIGHT	PH=7.1, 1 Y, 100 T DW SL AC, CONT= 0.01 PPM, SOIL=2.0 PPM, SLUDGE=0.2 PPM	499
*ECHINOCHLOA CRUSGALLI VAR. FRUMENTACEA	0.1 MG/KG DRY WEIGHT	PH=7.1, 1 Y, 100 T DW SL AC, CONT= 0.1 PPM, SOIL=2.0 PPM, SLUDGE=0.2 PPM	499
*LYCOPERSICON ESCULENTUM	0.2 MG/KG DRY WEIGHT	PH=7.1, 1 Y, 100 T DW SL AC, CONT= 0.1 PPM, SOIL=2.0 PPM, SLUDGE=0.2 PPM	499
*PHASEOLUS VULGARIS	0.1 MG/KG DRY WEIGHT	PH=7.1, 1 Y, 100 T DW SL AC, CONT= 0.04 PPM, SOIL=2.0 PPM, SLUDGE=0.2 PPM	499
*SOLANUM TUBEROSUM	0.1 MG/KG DRY WEIGHT	PH=7.1, 1 Y, 100 T DW SL AC, CONT= . PPM, SOIL=2.0 PPM, SLUDGE=0.2 PPM	499

TABLE D486

UPTAKE FROM SOIL WITH ADDED SLUDGE : ZN			
ITEM:	VALUE:	CONDITIONS:	REF.:
*FODDER RAPE	0.8 (% INCREASE IN PLANT)/(% INCREASE IN SOIL)	BACKGROUND LEVEL = 97.9 PPM,SLUDGE LEVEL= 368.8 PPM	499

D487

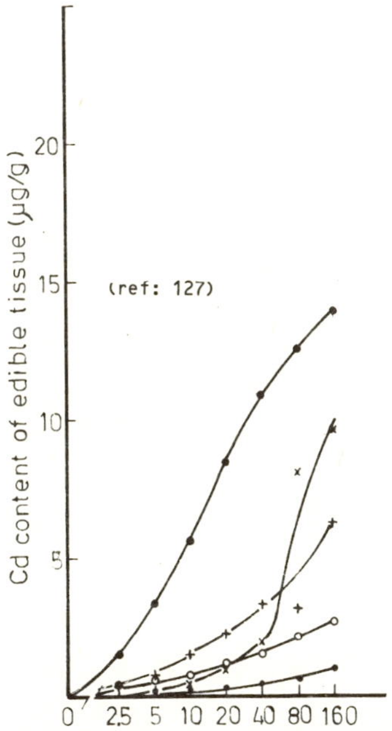

Cadmium content (µg/g) of edible plant tissue in relation to cadmium in soil (µg/g).
o field bean + corn × squash ● wheat
• rice

D488

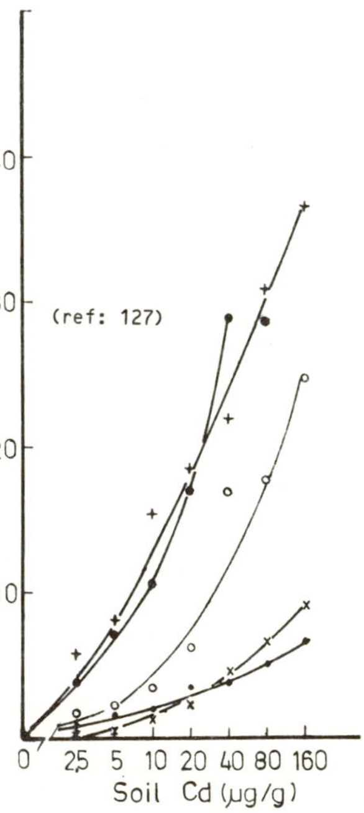

Cadmium content (µg/g) of edible plant tissue in relation to cadmium in soil (µg/g).
● soybean o radish + carrot • tomato
× cabbage

D489

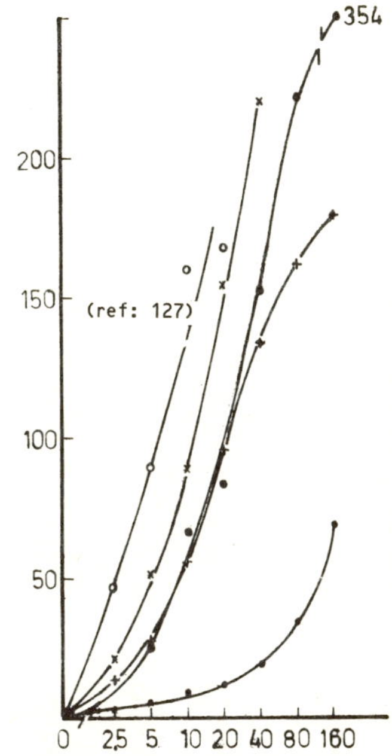

Cadmium content (µg/g) of edible plant tissue in relation to cadmium in soil (µg/g).
• turnip tuber ● turnip top × curlycress
o spinach + lettuce

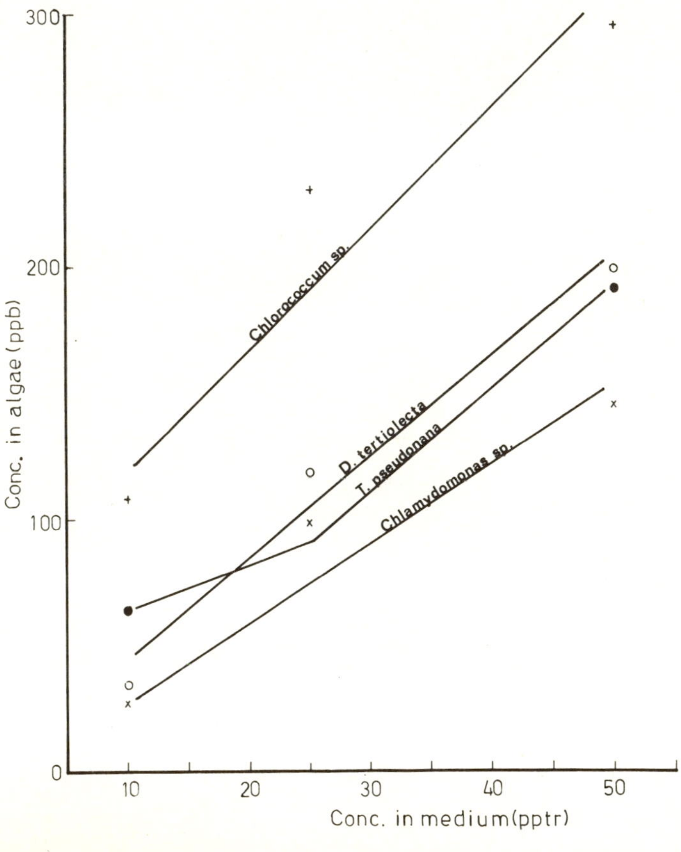

(ref: 222)

D490

Uptake of mirex by algae. Salinity 15 ppt.
+ Chlorococcum sp. o Dunaliella tertiolecta
● Thalassiosira pseudonana
× Chlamydomonas sp.

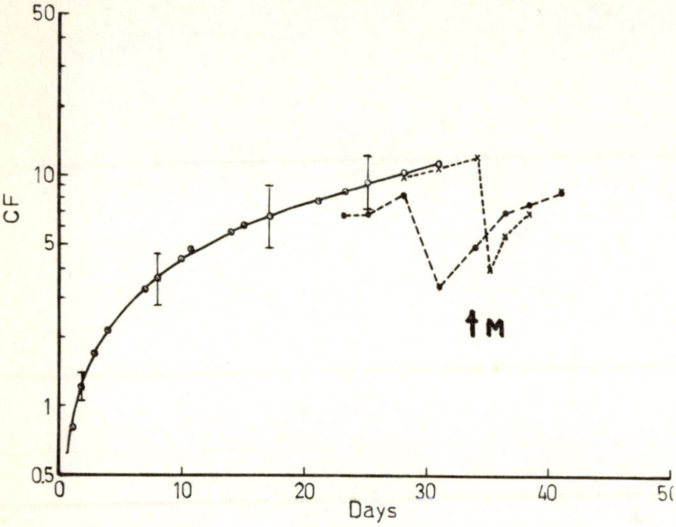

D491 Uptake of Co-60 by shrimps. o represent the mean and the vertical bars the S.D. of 8 or 6 individuals. x and • data for two individual shrimps. CF indicates the concentration factor. ↑ M means moulted.

(ref: 1671)

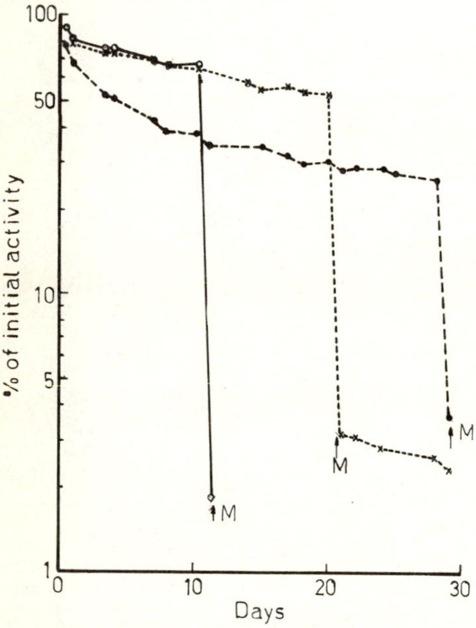

D492 Excretion of Co-60 by shrimps after 21 days uptake of Co-60 from seawater. The result of three shrimps are indicated: x and • . Notice that the shrimp with the highest initial rate of loss o had moulted two days before the end of the uptake perido. ↑ M means moulted.

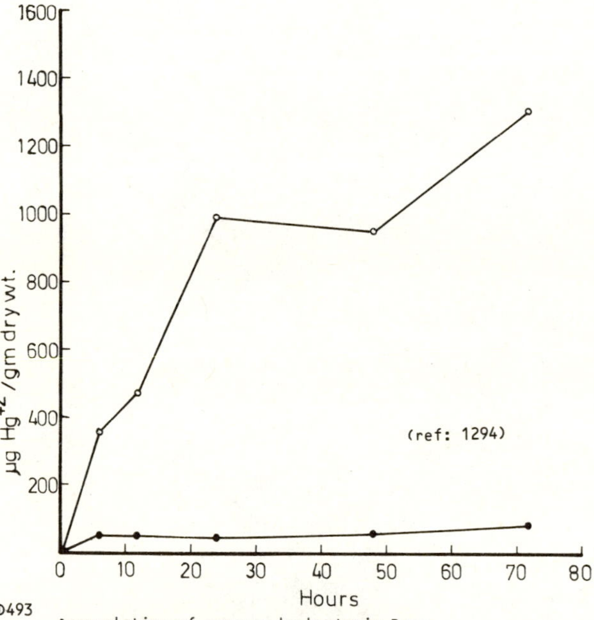

(ref: 1294)

D493 Accumulation of mercury by bacteria Pseudomonas flourescus and sediment from Ottawa River water.
Ottawa river water: Eh: + 345 - 430 mV, pH: 7.70 - 7.85. pCl: 3.68 - 3.80, Conductivity: 21 - 22 umhos, Turbidity 8 - 10 FTU, Complexing capacity 21 μM Hg(II), Sediment claytype (kaolinite and illite) 25 m**2/(gram dry weight), Organic content: 10.1 %, Grain size: 0.35 mm, Cation capacity: 18.58 meqv per 100 g dry weight. Pseudomonas flourescens: stationary phase at 30 celcius, 40+- 20 m**2/(gram dry weight).
o P. flourescens, • Sediment.

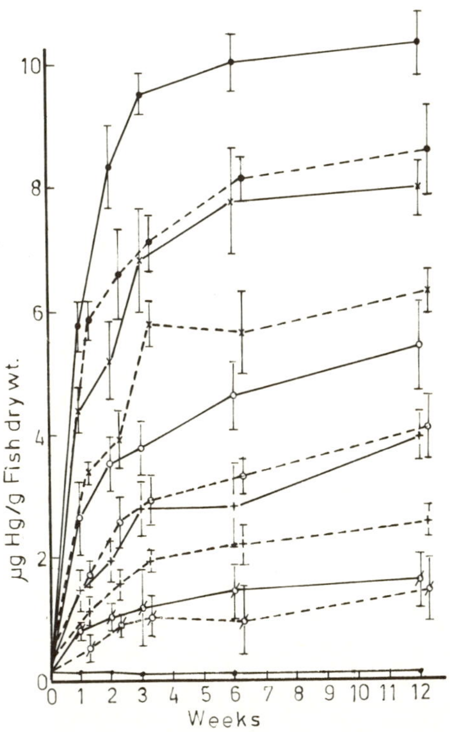

(ref: 941)

D494 Uptake of methyl mercury by Salmo gairdneri from different mercury concentrations in water. Continous flow at 14 celcius. control • 0.25 ø 0.5 + 1.0 o 2.5 x 5.0 ⊙ . Hg concentration in fish with muscus ——— . Means of 6 samples +- S.D.

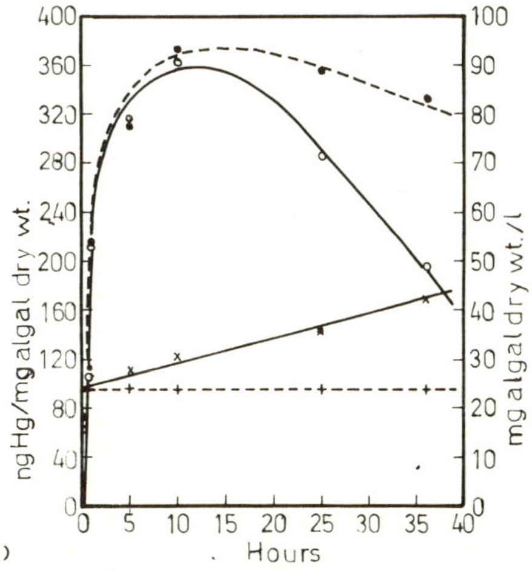

D495 Uptake of methyl mercury by Chlamydomonas reinhardtii. Living o and killed • cultures. Cultures were exposed to an initial concentration of 20 ng Hg/ml. Temperature 16 +- 1 celcius, white light 2.5*10**3 ergs/(cm**2*sec) on a 16:8 hours light:dark regime. Algal biomass (mg algal dry weight/l) in living x and killed + cultures.

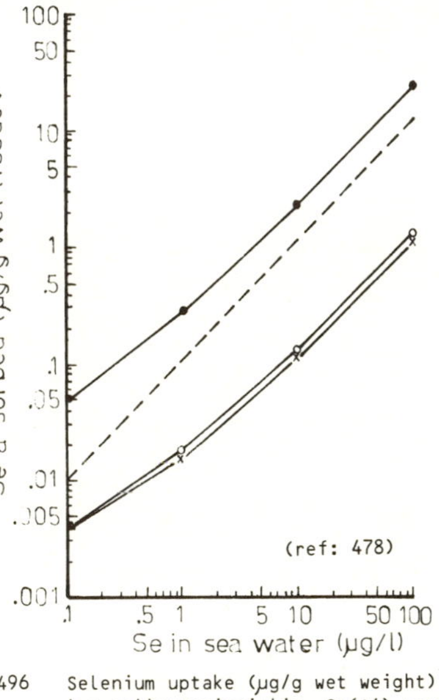

D496 Selenium uptake (µg/g wet weight) by Mytilus galloprovincialis. Se(+4) concentration in seawater is indicated. Points represent mean of 5 individiuals with an average wet weight of 3.3 g.
o whole animals. • soft parts x shell. Dashed line indicates slope of line if Se uptake is directly proportional to Se concentration in water.

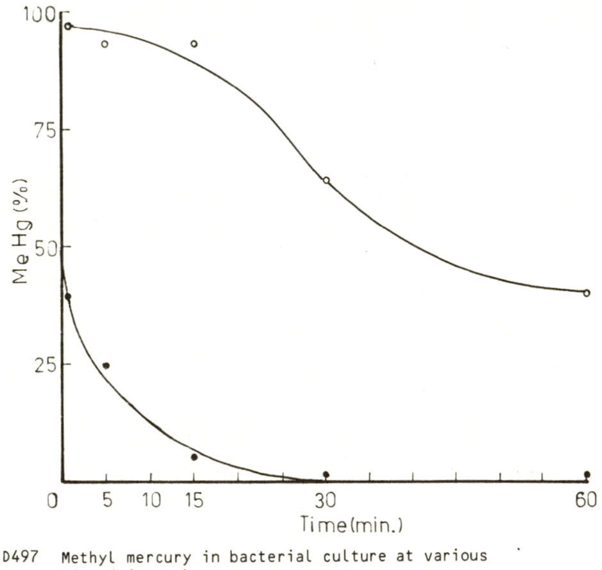

D497 Methyl mercury in bacterial culture at various time (minutes).
y-axis, % of methyl mercury found in the culture. o KI not added. • KI added.

(ref: 611,

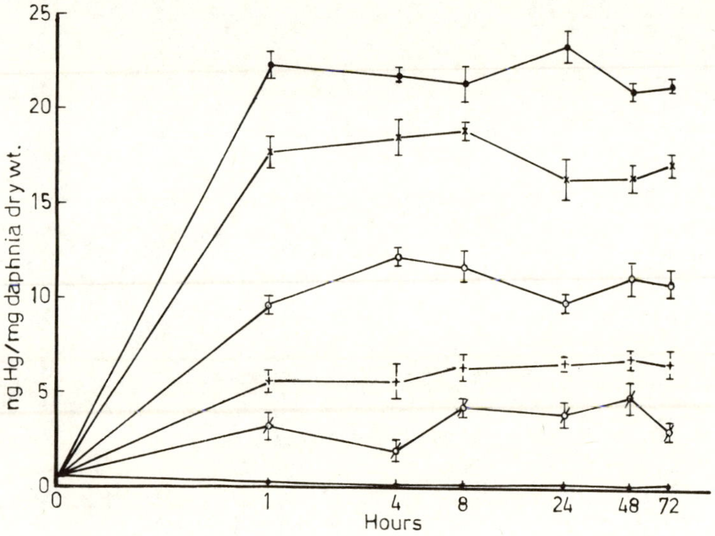

D498 Uptake of methyl mercury by Daphnia pulex from water contaning 0 • , 0.25 ø , 0.5 + , 1.0 o , 2.5 x , 5.0 ● ng Hg/l. Temperature 14 celcuis. Values indicated are average of 3 samples , bars S.D.

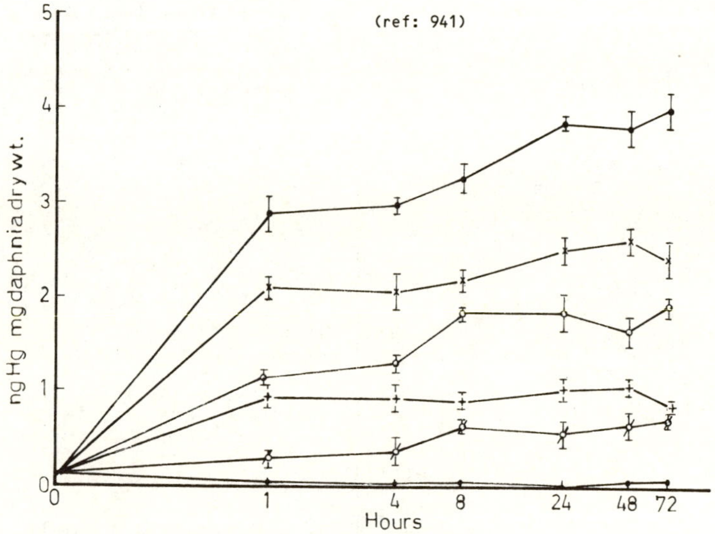

D499 Uptake of methyl mercury by Daphnia pulex from Chlamydomonas reinhardtii, which had previously been exposed to different initial concentrations of methyl mercury.
• control ø 0.25 + 0.5 o 1.0 x 2.5
● 5.0 ng Hg/ml. Values are average of 3 samples +- S.D.

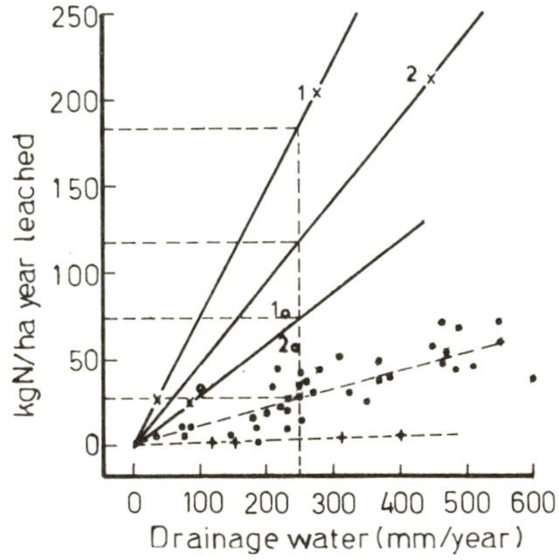

(ref: 847)

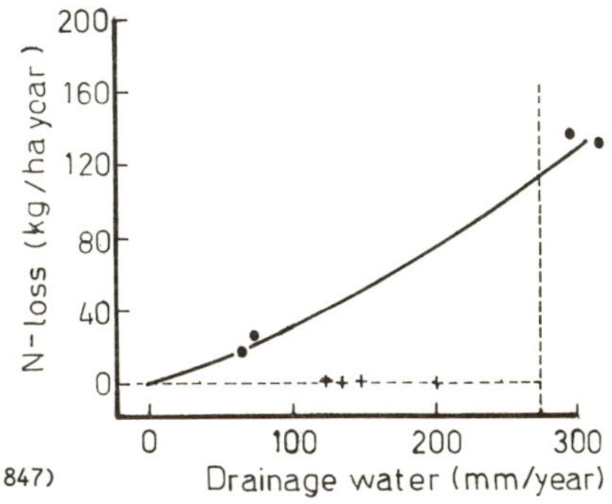

D500 Leaching of nitrogen from soil.
x peas, 1 fodder beats after peas,
2 winter wheat after peas, o old clover,
1,2 red clover or potatoes 2. year,
• summer crops without legumes, + grass land.

D501 Leaching of nitrogen from soil.
• clover, + grass.

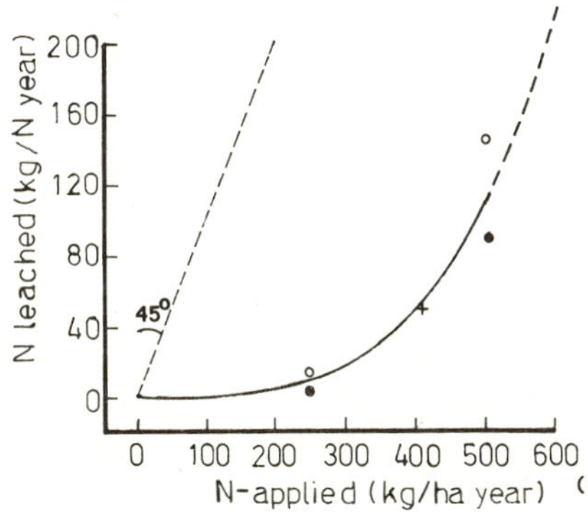

(ref: 1590)

D502 N-losses on grassland plotted to the annual use of fertilizer nitrogen.
o 30 March 1970 to 29 March 1971 (280 mm drainage).
● 20 March 1971 to 20 March 1972 (270 mm drainage).
✗ Hapselse Bay catchment area.

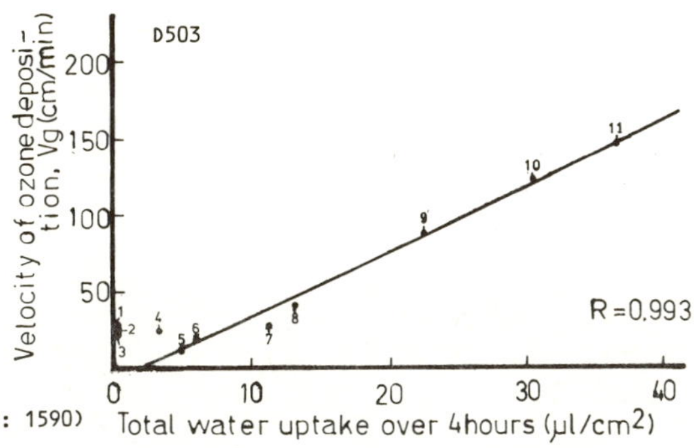

Ozone absorption and water uptake correlation by plant shoots. The plants are from left to right.
1: Acangh paper. 2: Osteasuermum. 3: Petunio. 4: Quercus. 5: Cameilla (old growth). 6: Ginkgo. 7: Bougainuillea. 8: Cameilla (new growth). 9: Chrysanthernum. 10: Osleospernum. 11: Petunia.

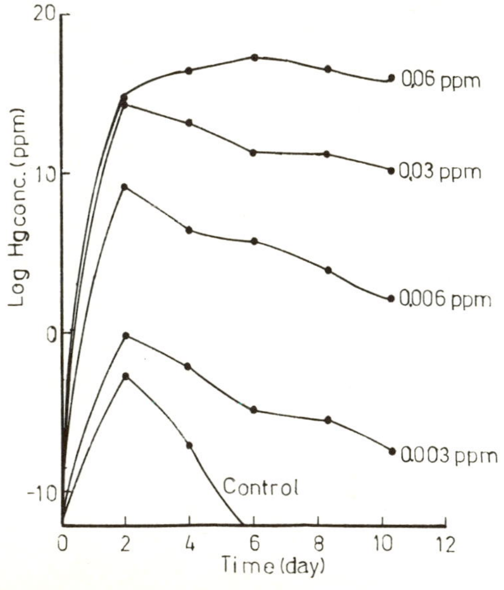

(ref: 606)

D504
Mercury concentration in tomato seedling root, exposed to 5 levels of CH_2HgOH concentrations in nutrient solutions over a 10 days period.

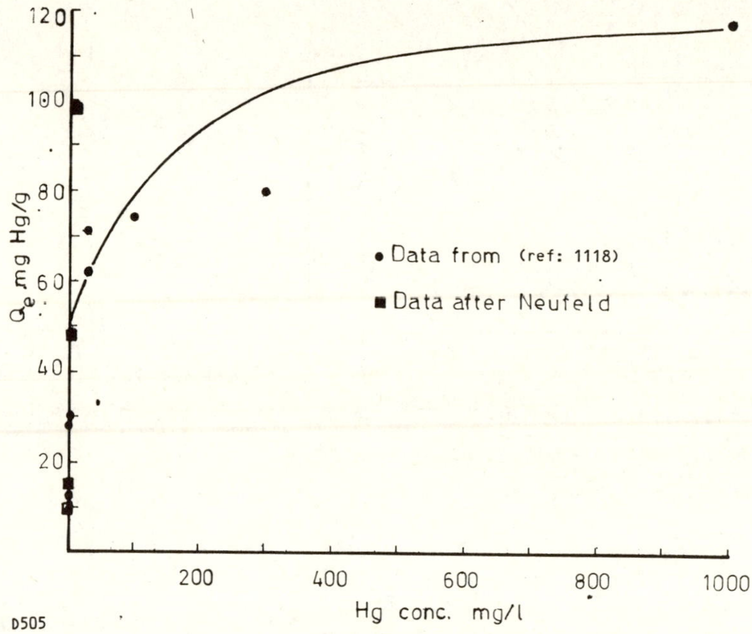

D505

Adsorption of mercury on sludge (activated)

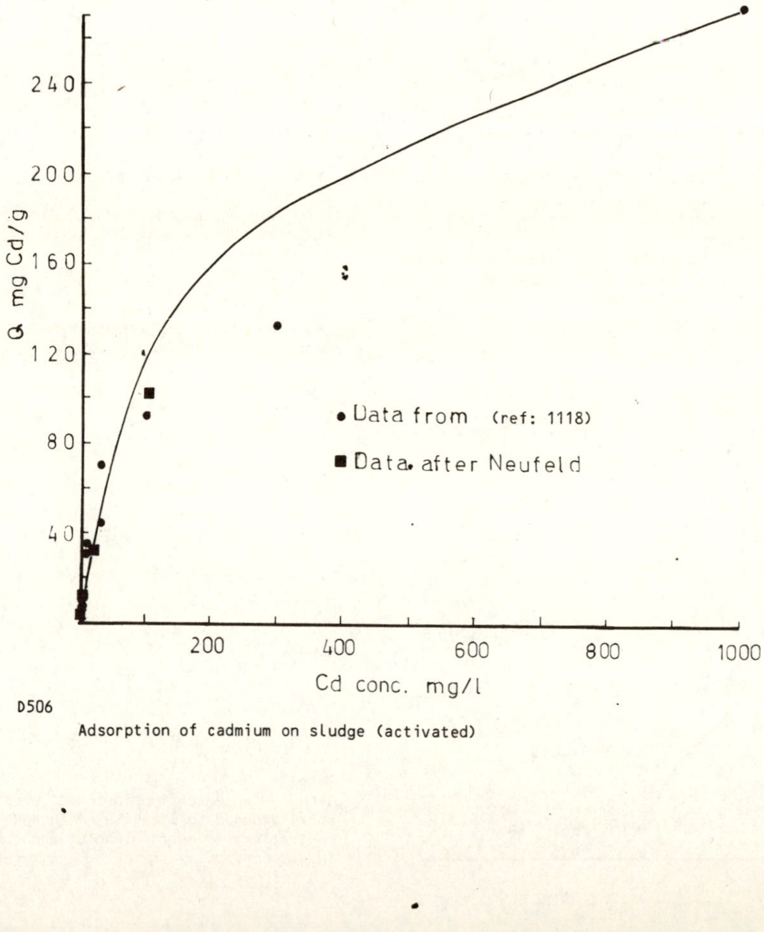

D506

Adsorption of cadmium on sludge (activated)

TABLE D507

Absorption of Elements Across the Human Gut Wall.

REF.: 157

Element	% Absorbed	Excretion largely with
B	>70	urine
Br	>70	urine
Cl	>70	urine
Cr(VI)	>70	urine
F	>70	urine
Ge	>70	urine
I	>70	urine
Mo(VI)	>70	urine
N	>70	urine
Se	>70	urine
Te	>70	urine
V(V)	>70	urine
W(VI)	>70	urine
Cs	>70	urine
Fr	>70	urine
K	>70	urine
Li	>70	urine
Na	>70	urine
Rb	>70	urine
Tl(I)	>70	urine
Ag(I)	>70	faeces
As	5-70	urine
P	5-70	urine
S	5-70	urine
Sb	5-70	urine
Si	5-70	urine
Co	5-70	urine
Cu	5-70	urine
Hg	5-70	urine
Mg	5-70	urine
Ni	5-70	urine
U(VI)	5-70	urine
Ba	5-70	faeces

Element	% Absorbed	Excretion largely with
Ca	5-70	faeces
Pb	5-70	faeces
Ra	5-70	faeces
Sn	5-70	faeces
Sr	5-70	faeces
Au	<5	faeces
Nb	<5	faeces
Pa	<5	faeces
Pd	<5	faeces
Pt	<5	feaces
Ta	<5	faeces
Al	<5	faeces
Am	<5	faeces
Be	<5	feaces
Bi	<5	faeces
Cd	<5	faeces
Ce	<5	faeces
Fe	<5	feaces
Ga	<5	faeces
In	<5	faeces
La	<5	faeces
Lanthanides	<5	faeces
Mn	<5	faeces
Np	<5	faeces
Pu	<5	faeces
Ru	<5	faeces
Sc	<5	faeces
Ti	<5	faeces
Y	<5	faeces
Zn	<5	faeces
Zr	<5	faeces

TABLE D508

Percentage Retention of Inhaled Particles in different Regions of the Respiratory Tract. Retention in %.
REF.: 953

Region	Size = 20 um	Size = 6 um	Size = 2 um	Size = 0.6 um	Size = 0.2 um
Mouth	15	0	0	0	0
Pharynx	8	0	0	0	0
Trachea	10	1	0	0	0
Pulmonary bronchi	12	2	0	0	8
Secondary bronchi	19	4	1	0	0
Tertiary bronchi	17	9	2	0	0
Quaternary bronchi	6	7	2	1	1
Terminal bronchioles	6	19	6	4	6
Respiratory bronchioles	0	11	5	3	4
Alveolar ducts	0	25	25	8	11
Alveolar sacs	0	5	0	0	0
Total	93	83	41	16	22

TABLE D509

Approximate Deposition of Inhaled Insoluble Particulates. Deposition in % in region.
REF.: 953

Material	Particle size (um)	Upper respiratory	Deep pulmonary	Approximate total
Gas	<0.1	0	>60	60
Fumes	.1	0	60	60
Fumes	.2	0	53	53
Fumes	.4	0	76	76
Fumes	.6	0	80	80
Smoke	1	3	53	56
Smoke	2	60	26	86
Dust	3	70	20	90
Dust	5	85	15	100
Dust	10	90	<10	100
Dust	20	100	0	100

TABLE D510

Release of Trace Metals From Digested Sludge Particulates after Dilution with Seawater. Temperature 288 K.
REF.: 1353

Element	Total Concentration of Trace Metals before Dilution (mg/l)	Initial Concentration of Trace Metals in Residue (mg/kg)	% of Trace Metals Released in 1 Hour (1:50)	% of Trace Metals Released in 1 Hour (1:100)	% of Trace Metals Released in 1 Hour (1:200)	% of Trace Metals Released in 5 Weeks (1:50)	% of Trace Metals Released in 5 Weeks (1:100)	% of Trace Metals Released in 5 Weeks (1:200)
Cd	5.62	234	24.8	37.2	20.5	93	95	96
Cu	61.9	2579	1	3.3	7.1	5	5.6	9
Cr	59.7	2441	0	6	14.4	2	2	3.8
Fe	326	13298	0	0	0	0	0	0
Mn	2.75	111	0.9	4.5	7.7	31.8	34.8	35.8
Ni	14.86	603	3.4	2.5	10.9	49	58	64
Pb	21.5	875	48.6	45.1	53.1	37.8	35.4	35.4
Zn	115.2	4256	8.6	9.8	14.5	18	24.4	38.7

TABLE D511

Release of Trace Metals from Suspended Solids in Primary Effluent after Dilution with Seawater at 5:1. Temperature 288 K.
REF.: 1353

Element	Total Concentration of Trace Metals after Dilution (ug/l)	Initial Concentration (mg/kg)	Concentration after 90 min (mg/kg)	Concentration after 28 days of Equilibrium (mg/kg)	% of Trace Metals Released after 90 min	% of Trace Metals Released after Reaching Equilibrium	% of Trace Metals Released after Reaching Equilibrium (T=298K,1:10)
Cd	1.5	75	60.5	4	1923	94.7	60.1
Cu	48.4	2719	2702	933	6	67	43.9
Cr	45	2695	2695	2695	0	0	.
Fe	110	5999	6688	6666	0	0	.
Mn	11.3	51.6	51	50	0	0	6
Ni	31	960	924	413	3.75	57	69.8
Pb	3.15	155.7	133.7	73.3	14	52.9	19
Zn	19.5	1140	1178	640	0	43.86	25.7

TABLE D512

Release of Trace Metals from Suspended Solids in Mixture of Primary and Secondary Effluent after Dilution with Seawater 5:1. Temp. 288 K
REF.: 1353

Element	Total Concentration of Trace Metals after Dilution (ug/l)	Initial Concentration (mg/kg)	Concentration after 90 min (mg/kg)	Concentration after 28 days at Equilibrium (mg/kg)	% Trace Metals Released after 90 min	% Trace Metals Released after Reaching Equilibrium	% Trace Metals Released after Reaching Equilibrium (T=298K,1:10)
Cd	1.8	87.3	72	6.15	17.5	92.5	57.5
Cu	30	2182	2000	667	8.34	69.4	53
Cr	42.3	3296	3296	3296	0	0	.
Fe	69.3	5177	5344	5283	0	0	.
Mn	7.6	51.4	52	50	0	0	8.42
Ni	37	896	848	333	5.4	62.8	44.6
Pb	2.1	140.3	120	58.3	14.5	58.4	42.8
Zn	19.8	1481	1456	917	1.7	38.1	17.6

TABLE D513

Release of Trace Metals from Suspended Solids in Dry Weather Flow from Los Angeles River after Dilution with Seawater at 2:1. Temp 288 K.
REF.: 1353

Element	Total Concentration of Trace Metals after Dilution (ug/l)	Initial Concentration (mg/kg)	Concentration after 90 min (mg/kg)	Concentration after 28 days Equilibrium (mg/kg)	% of Trace Metals Released After 90 min	% of Trace Metals Released after Reaching Equilibrium	% of Trace Metals Released after Reaching Equilibrium (T=298K,1:2)
Cu	7	1063	837	362	21.3	65.9	60
Cr	2	453	453	453	0	0	.
Fe	53.3	12167	12167	12167	0	0	.
Mn	3.4	498	498	475	0	0	0
Ni	6	407	181	113	55.5	72.2	21.5
Pb	3.5	516	475	430	729	16.7	31.9
Zn	4.8	905	814.5	362	10	60	21.9

E

EQUATIONS FOR BIOLOGICAL PROCESSES

RELATIONS BODY WEIGTH/SIZE
BIOLOGICAL PROCESSES
CORRELATIONS/PARAMETERS

TABLE E1

EQUATION : ABUNDANCE AS A FUNCTION OF SALINITY			
ITEM:	VALUE:	CONDITIONS:	REF.:
*EUTERPINA ACUTIFRONS	LN(Y)=1.98+26.3*S-36.4*S**2	Y=ABUNDANCE OF COPEPOD,S=SALINITY=3.4%,PELORUS SOUND,NEW ZEALAND	1066

TABLE E2

EQUATION : ACCUMULATION OF CD (F054)			
ITEM:	VALUE:	CONDITIONS:	REF.:
*KIDNEY CORTEX	DX=-A*X*DT+B*DT	X=ORG.BURDEN OF CD,A=EXCRET.RATE,B=CD ABS./TIME UNIT,T=TIME	487
*KIDNEY CORTEX	X=K1*(1-E**(-K2*T))	X=ORG.BURDEN OF CD,K1=FRACT.OF ORG.BURDEN REMAIN.AT TIME 1	487
*KIDNEY CORTEX	X=K1*(1-E**(-K2*T)	K2=FRACT.OF BODY BURDEN REMAIN.AT TIME 2,T=TIME	487

TABLE E3

EQUATION : ACTIVE UPTAKE OF IONS			
ITEM:	VALUE:	CONDITIONS:	REF.:
*LIVING CELLS	E=R*T*LN(CI/CO)	E=MIN.ENERGY NEEDED,R=GAS CONST.,T=ABSOLUTE TEMP.,	157
*LIVING CELLS	E=R*T*LN(CI/CO)	CI=CONC.INSIDE THE CELL,CO=CONC.OUTSIDE THE CELL	157

TABLE E4

EQUATION : ACUTE LETHAL EXPOSURE TO RADIATION VS INTERPHASE CHROMOSOME VOLUME			
ITEM:	VALUE:	CONDITIONS:	REF.:
*SEEDS PLANTS	LOG(Y) = 1.69422 - 0.93025*LOG(X)	Y=RAD (KILOROENTGENS), X=CHROMOSOME VOL (UM**3)	1486

TABLE E5

EQUATION : ALGAE			
ITEM:	VALUE:	CONDITIONS:	REF.:
*ALGAL STANDING CROP VS DISSOLVED CALCIUM	LOG(Y) = 2.31 + 3.99*LOG(X)	Y=ALGAL ST CROP (UG/L), X=DIS CA (U EQ/L), R=0.78, UK, FRESHWATER LAKES	554
*BIOMASS VS VOLUME	LOG(C) = 0.610 + 0.892*LOG(V + 0.1*B)	C=CARBON (PICO G), V=PLASMA VOL (UM**3), B=VACUOLE VOL (UM**3)	148
*CELL CARBON AS A FUNCTION OF CHLOROPHYLL A	PC=37+29*X	PC=CELL CARBON,X=CHL.A,R=0.73,95% CONF.LIM.,ST.1, MARINE PHYT,N=13	430
*CELL CARBON AS A FUNCTION OF CHLOROPHYLL A	PC=2+55*X	PC=CELL CARBON,X=CHL.A,R=0.88,95% CONF.LIM.,ST.2, MARINE PHYT,N=15	430
*CELL CARBON AS A FUNCTION OF CHLOROPHYLL A	PC=-1+80*X	PC=CELL CARBON,X=CHL.A,R=0.89,95% CONF.LIM.,ST.3, MARINE PHYT,N=15	430
*CELL CARBON AS A FUNCTION OF CHLOROPHYLL A	PC=4+40*X	PC=CELL CARBON,X=CHL.A,R=0.79,95% CONF.LIM., MARINE PHYT,N=16	1772
*CELL CARBON AS A FUNCTION OF CHLOROPHYLL A	PC=11+19*X	PC=CELL CARBON,X=CHL.A,R=0.54,95% CONF.LIM., MARINE PHYT,N=16	674

TABLE E5

EQUATION : ALGAE			
ITEM:	VALUE:	CONDITIONS:	REF.:
*CHLOROPHYLL VERSUS TOTAL PHOSPHORUS	LOG CHL = - 1.09 + 1.46*LOG (P-TOTAL)	143 LAKES, SUMMER LEVEL, PHYTOPLANKTON, R=0.95	787
*CORRELATION BETWEEN CHLOROPHYLL A AND TOTAL IRON	Y=1.782*X - 20.3	Y=LOG(CHLOROPHYLL A),X=LOG(FE),R= 0.842,AT END OF BATCH, AS-F	256
*CORRELATION BETWEEN CHLOROPHYLL A AND TOTAL IRON	Y=1.171*X - 6.5	Y=LOG(CHLOROPHYLL A),X=LOG(FE), HAEMATOCOCCUS PLUVIALIS, R=0.824	256
*CORRELATION BETWEEN CHLORO¤HYLL A AND TOTAL IRON	Y=1.163*X - 4.8	Y=LOG(CHLOROPHYLL A),X=LOG(FE),R= 0.875,AT END OF BATCH, SC-Q	256
*CORRELATION BETWEEN CHLOROPHYLL A AND TOTAL IRON	Y=1.303*X - 8.5	Y=LOG(CHLOROPHYLL A),X=LOG(FE),R= 0.870,AT END OF BATCH, SYNEDRA ACUS	256
*CORRELATION BETWEEN CHLOROPHYLL A AND TOTAL PHOSPHATE	Y=1.916*X - 1.93	Y=LOG(CHLOROPHYLL A),X=LOG(P),R= 0.826,AT END OF BATCH, AS-F	256
*CORRELATION BETWEEN CHLOROPHYLL A AND TOTAL PHOSPHATE	Y=1.317*X - 6.90	Y=LOG(CHLOROPHYLL A),X=LOG(P), HA-P, R=0.932,AT END OF BATCH	256
*CORRELATION BETWEEN CHLOROPHYLL A AND TOTAL PHOSPHATE	Y=1.51*X - 2.90	Y=LOG(CHLOROPHYLL A),X=LOG(P),R= 0.872,AT END OF BATCH, SC-Q	256
*CORRELATION BETWEEN CHLOROPHYLL A AND TOTAL PHOSPHATE	Y=1.407*X - 9.23	Y=LOG(CHLOROPHYLL A),X=LOG(P),R= 0.788,AT END OF BATCH, SYNEDRA ACUS	256
*DRY WEIGHT AS A FUNCTION OF OPTICAL DENSITY	Y = 971 - 7206*D + 57200*D**2	Y=DRY WEIGHT (MG/L), D=OPTICAL DENSITY, 600 NM, SPECTRONIC 20, CH-P	1144
*INCREASE OF PHOTOSYNTHETIC EFFICIENCY WITH DECREASED LIGHT	GA = EXP(0.685*IZ + 0.4)	GA= PHOTOSYNTHETIC EFFICIENCY,IZ = LIGHT AT DEPTH Z, PHYT	1326
*LIGHT INHIBITION AS A FUNCTION OF UV AND IR AT SURFACE(O) AND DEPTH	P = LN(UV+IR)(Z) + 5)/((LN(UV+IR) (O) + 5)*(1.54))	P=LIGTH INH, UV=ULTRA VIOLET,IR= INFRARED, DEPTH AS INDEX, PHYT	1326
*MAXIMUM DAILY PRODUCTION VS LIGHT INTENSITY AND ALGAE CONCENTRATION	G24 = CA*LZ/(LZ + 500)*0.048	G24=MAX PROD (MG C/(M3*D)),CA= ALGAE (G(N+P)/M3),LZ=AVIA LIGTH, IN WATER	114
*MAXIMUM GROWTH RATE AS A FUNCTION OF TEMPERATURE	LOG(V) = 0.0275*T - 0.070	V=GROWTH RATE(DOUBLINGS/DAY),T= TEMP (DEGRESS CELCIUS),273-313 K, PHYT	427
*MN CONTENT IN ALGA VS MEDIUM	Y=73.34*X+8.68	Y=MN CONTENT IN ALGA,X=MN CONC.IN MEDIUM,294K,LIGHT=2500LUX, PE-T	1682
*MORTALITY OF PHYTOPLANKTON	J = 0.005*Q/T	Q=BIOMASS,T = TIME (H)| J = FLUX (UG/(L*H))	1326
*NUTRIENT CONTENT PER CELL	Q=UPTAKE RATE/CELL DIVISION RATE	EXP.GROWTH PHASE,NUTRIENT CONTENT PER CELL(=Q) REMAINS CONST., PHYT	1218
*ORGANIC CARBON AS A FUNCTION OF CHLOROPHYLL A	POC=151+81*X	POC=ORG C,X=CHL.A,R=0.65,ST.1, MARINE PHYT,N=13	430
*ORGANIC CARBON AS A FUNCTION OF CHLOROPHYLL A	POC=55+140*X	POC=ORG C,X=CHL.A,R=0.70,ST.2, MARINE PHYT,N=15	430
*ORGANIC CARBON AS A FUNCTION OF CHLOROPHYLL A	POC=23+285*X	POC=ORG C,X=CHL.A,R=0.92,ST.3, MARINE PHYT,N=15	430
*ORGANIC CARBON AS A FUNCTION OF CHLOROPHYLL A	POC=91+206*X	POC=ORG C,X=CHL.A,R=0.59, MARINE PHYT,N=16	1772
*ORGANIC CARBON AS A FUNCTION OF CHLOROPHYLL A	POC=73+32*X	POC=ORG C,X=CHL.A,R=0.63, MARINE PHYT,N=16	674
*PARTICULATE NITROGEN AS A FUNCTION OF ASH-FREE DRY WEIGHT	Y = 0.142 + 27*X	Y= PART N (MG/L), X=DW (MG/L), R= 0.94	169
*PHOTOSYNTHESIS AS A FUNCTION OF SECCHI DEPTH	A/A(MAX) = 1.75*S**0.7	A=PHOTOSYN.,AM=MAX OF A, S=SECCHI D.(M), ESTIMATE, IN LAKES	1639
*RELATIVE GROWTH AS A FUNCTION OF LIGHT INTENSITY	RG = I/IOPT*EXP(1-I/IOPT)	I=LIGHT INTENSITY,IOPT=OPTIMAL I, RG=RELATIVE GROWTH	1132
*RESPIRATION	RESP=RESPK*F(T)*PHYT	(F047)	764
*RESPIRATION	RESP=RESPK(CA/CMAX)*F(T)*PHYT	AS C (F047)	764
*SPECIFIC GROWTH RATE AS A FUNCTION OF TEMPERATURE	GR = 0.006*T - 0.035	GR=GROWTH RATE (1/H),T=WATER TEMP (CEL), 279-301 K	944
*TOTAL ORGANIC CONTENT VS CELL SIZE	LOG(C) = 0.758*LOG(V) - 0.422	C=CARBON PER CELL (PG C), V=VOLUME (UM**3), DIATOMS	1518
*TOTAL ORGANIC CONTENT VS CELL SIZE	LOG(C) = 0.866*LOG(V) - 0.460	C=CARBON PER CELL (PG C), V=VOLUME (UM**3), PHYT (NOT DIATOMS)	1518

TABLE E6

EQUATION : ALGAL STANDING CROP VS CULTIVABLE PART OF DRAINAGE AREA			
ITEM:	VALUE:	CONDITIONS:	REF.:
*FRESHWATER VS WATERSHEDS	Y = 0.172*X - 0.98	Y=ALGAL ST CROP (UG DW/L),X= CULT DRAINAGE AREA (%),R=0.90, UK	554

TABLE E7

EQUATION : BIOMASS AS A FUNCTION OF DIAMETER AT BREAST HEIGHT

ITEM:	VALUE:	CONDITIONS:	REF.:
*PSEUDOTSUGA MENZIESSII	LOG(Y)=0.64294+2.39643*LOG(DBH)	Y=BIOMASS(KG),DBH=DIAMETER AT BREAST HEIGHT(CM),R**2=0.873, FOLIAGE	946
*PSEUDOTSUGA MENZIESSII	LOG(Y)=1.23727+2.24162*LOG(DBH)	Y=BIOMASS(KG),DBH=DIAM.AT BREAST HEIGHT(CM),R**2=0.903,TOTAL CROWN	946
*PSEUDOTSUGA MENZIESSII	LOG(Y)=-0.85223+2.30609*LOG(DBH)	Y=BIOMASS(KG),DBH=BR.H.DIAM.(CM), R**2=0.986,TOT.TREE BIOM.AB.GROUND	946

TABLE E8

EQUATION : BIOMASS AS A FUNCTION OF TIME

ITEM:	VALUE:	CONDITIONS:	REF.:
*BIRDS	B = 0.055*LOG(T)	B=BIOMASS (KG/HA), T=AGE OF ABANDONED AGRICULTURAL FIELD	1011
*BIRDS	B = 21.15 - 6.18*LOG(T)	B= INDIVIDUAL AVER BIOMASS (G/ BIRD), T=AGE OF ABANDONED AGRIC. FIELD (Y)	1011

TABLE E9

EQUATION : BODY LENGTH

ITEM:	VALUE:	CONDITIONS:	REF.:
*APLODINOTUS GRUNNIENS	L = 57.639 + 56.348*S	L=TOTAL LENGTH (MM), S=SPINE RADIUS (MM), TUTTLE CREEK, KANSAS	834
*ELEGINOPS MACLOVINUS	L = 46.57 + 65.66*R	L=TOT LENGTH (MM), R=SCALE TOT LENGTH (MM), R=0.9559, DESEADO R ESTUARY	555
*ICTALURUS PUNCTATUS	L = 7.50 + 87.96*S + 8.02*S**2	L=TOTAL LENGTH (MM), S=SPINE RADIUS (MM), TUTTLE CREEK, KANSAS	833
*LOXECHINUS ALBUS	Y = 3.75*X**0.79	Y=DIAMETER (MM), X=HEIGHT (MM), CHILEAN COAST, BOTH SEXES	580
*POLLACHIUS VIRENS	LN(R) = -1.6987 + 0.6781*LN(L)	R=OTOLITH RADIUS (MM), L=LENGTH (CM), NORTH AND NORWEGIAN SEA	254

TABLE E10

EQUATION : BODY LENGTH AS A FUNCTION OF MEAN FOOD CONCENTRATION

ITEM:	VALUE:	CONDITIONS:	REF.:
*ANCHOA MITCHILLI	L = 15.05*(1 - EXP(-0.18*LN(P))	L=LENGTH (MM), P=FOOD (CHLORELLA/ L),N=29, R=0.82, 16 D AFTER HATCH	695

TABLE E11

EQUATION : BODY LENGTH VS AGE GROUP

ITEM:	VALUE:	CONDITIONS:	REF.:
*CHRYSOPHRYS MAJOR	LT = 484.50*(1 - EXP(-0.18*(T + 0.47)))	LT=FORK L (MM) AT AGE T (Y), CENTRAL WATERS OF HIROSHIMA PREFECTURE	865
*COD, NORTH SEA	(LT - 36.5) = +-19.25*(|T-2.04|) **0.7225	LT = LENGTH (CM) AT AGE T (YEAR-CLASS)	1292
*ELEGINOPS MACLOVINUS	L = 108.29*T + 13.8	L=LENGTH (MM), T=TIME (Y), DESEADO RIVER ESTUARY	555

TABLE E11

EQUATION : BODY LENGTH VS AGE GROUP			
ITEM:	VALUE:	CONDITIONS:	REF.:
*GADUS MORHUA	L =110.8*(1 - EXP(-0.333*(T-0.87)))	L=LENGTH (CM), T=AGE GROUP (Y), SOUTHERN NORTH SEA	315
*GADUS MORHUA	L = 118.7*(1- EXP(-0.269*(T-0.87)))	L=LENGTH (CM), T=AGE GROUP (Y), NOTHERN NORTH SEA	315
*GENYPTERUS CAPENSIS	LT = 120.13*(1 - EXP(-0.1314*(T+ 0.1858)))	LT=LENGTH (MM) AT AGE T (Y), WEST COAST STOCKS, WALVIS BAY	1230
*GENYPTERUS CAPENSIS	LT = 148.51*(1 - EXP(-0.1120*(T+ 0.7404)))	LT=LENGTH (MM) AT AGE T (Y), SOUTHERN STOCKS, WALVIS BAY	1230
*HADDOCK	(LT - 23.25) = +-5.447*(|T-1.8|) **0.79	LT = LENGTH (CM) AT AGE T (YEAR-CLASS)	1292
*KATSUWONUS PELAMIS	L = 103.6*(1 - EXP(-0.302* (T-0.016)))	L=LENGTH (CM), T=AGE GROUP (Y),BY VERTEBRAL RING READING, TAIWAN	245
*KATSUWONUS PELAMIS	L = 103.8*(1 - EXP(-0.432* (T-0.016)))	L=LENGTH (CM), T=AGE GROUP (Y),BY INCREMENT METHOD, TAIWAN	245
*LABEO ROHITA	L = 1015*(1 - EXP(-0.276*(T+0.333)))	L=LENGTH (MM), T=AGE GROUP (Y)	816
*MACRURONUS MAGELLANICUS	LT = 66.8*(1 - EXP(-0.35984* (T-0.070851)))	LT = LENGTH (CM) AT AGE T (YEAR-CLLASS)	6
*MARLUCCIUS ANGUSTIMANUS	LT = 32.7*(1 - EXP(-0.35*(T - 0.16)))	LT = LENGTH (CM) AT AGE T (YEAR-CLLASS)	991
*MERLUCCIUS ANGUSTIMANUS	LT = 32.7*(1-EXP(-0.35(T-0.16))	LT=LENGTH (CM) AT TIME T (YEAR), SOUTH GOLF OF CALIF.	991
*PANULIRUS LAEVICAUDA	L = 21.3*(1 - EXP(-0.708*T))	L=TOTAL LENGTH (CM), T=AGE GROUP (Y), CEARA STATE, BRAZIL	732
*PANULIRUS LONGIPES CYGNUS	LT = 113.47*(1 - EXP(-0.459* (T-1.045)))	LT=LENGTH (MM) AT AGE T (Y), COASTAL REEFS	250
*PIMELODUS MACULATUS	L = 56.5*(1 - EXP(-0.1938*T))	L=LENGTH (CM), T=AGE GROUP (Y), MALE, BRAZIL	452
*PIMELODUS MACULATUS	L = 45.5*(1 - EXP(-0.2104*T))	L=LENGTH (CM), T=AGE GROUP (Y), FEMALE, BRAZIL	452
*PLACIE	(LT - 39) = +-3.802*(|T-8.2|) **0.6776	LT = LENGTH (CM) AT AGE T (YEAR-CLASS)	1292
*POLLACHIUS VIRENS	L = 109.4*(1 - EXP(-0.160*T))	L = LENGTH (CM), T=AGE GROUP (Y), NORTH VS NORWEGIAN SEA	254
*PSENOPSIS ANOMALA	L = 243.I*(1 - EXP(-0.34*(T+1.674)))	L = LENGTH (MM), T=AGE GROUP (Y), KII CHANNEL, JAPAN	1371
*PSEUDOTOLITHUS SENEGALENSIS	LT = 81*(1 - EXP(-0.33*(T - 0)))	LT = LENGTH (CM) AT AGE T (YEAR-CLLASS)	82
*PSEUDOTOLITHUS TYPUS	LT = 103*(1 - EXP(-0.29*(T - 0.15)))	LT = LENGTH (CM) AT AGE T (YEAR-CLLASS)	82
*SALMO SALAR	LT = 120.25*(1 - EXP(-0.439*(T + 0.335)))	LT=LENGTH (CM) AT AGE T=TIME (YEAR-CLASS), BALTIC	1593
*SALMO SALAR	LT = 156.03*(1 - EXP(-0.2876*(T + 0.456)))	LT=LENGTH (CM) AT AGE T=TIME (YEAR-CLASS), BALTIC	680
*SCOMBEROMORUS CAVALLA	L = 840M(1 - EXP(-0.35M(T + 2.5)))	L=LENGTH (MM), T=AGE GROUP (Y), FLORIDA, MALE	85
*SCOMBEROMORUS MACULATUS	L = 515*(1 - EXP(-0.48*(T+1.12)))	L=LENGTH (MM),W=AGE GROUP (Y), FLORIDA	1274
*SERRANUS ALEXANDRINUS	(LT - 24.33) = +-7.143*(|T-1.93|) **0.82897	LT = LENGTH (CM) AT AGE T (YEAR-CLASS)	1292
*SOLE, NORTH SEA	(LT - 9.5) = +-12.1*(|T-1|) **0.4298	LT = LENGTH (CM) AT AGE T (YEAR-CLASS)	1292
*TAUTOGOLABRUS ADSPERSUS	LT = 284.77*(1 - EXP(-0.1979* (T-0.1044)))	LT=LENGTH (MM) AT AGE T (Y), WEWEANTIC R. ESTUARY, MASS	1418
*THYMALLUS THYMALLUS	LT = 57.5*(1 - EXP(-0.223*(T - 0.115)))	LT=LENGTH (CM) AT AGE T=TIME (YEAR-CLASS), LLYN TEGID, 1946-1952	1750
*THYMALLUS THYMALLUS	LT = 45.6*(1 - EXP(-0.386*(T - 0.222)))	LT=LENGTH (CM) AT AGE T=TIME (YEAR-CLASS), LLYN TEGID, 1962-1968	1750
*THYMALLUS THYMALLUS	LT = 38.9*(1 - EXP(-0.424*(T - 0.149)))	LT=LENGTH (CM) AT AGE T=TIME (YEAR-CLASS), UPPPER DEE	1750
*THYMALLUS THYMALLUS	LT = 41.7*(1 - EXP(-0.484*(T - 0.281)))	LT=LENGTH (CM) AT AGE T=TIME (YEAR-CLASS), CORWEN	1750
*THYMALLUS THYMALLUS	LT = 57.5*(1-EXP(-0.233*T+ 0.233*0.115))	LT=LENGTH (CM) AT TIME T (YEAR), LLYN TEGID VS RIVER DEE	1750
*TRACHURUS JAPONICUS	LT = 424.9*(1 - EXP(-0.248*(T + 809)))	LT=FORK L (CM) AT AGE T (Y), WATERS OFF CHEJU ISLAND	33

TABLE E11

EQUATION : BODY LENGTH VS AGE GROUP			
ITEM:	VALUE:	CONDITIONS:	REF.:
*TRACHURUS TRACHURUS	LT = 56.41*(1 - EXP(-0.11*(T + 0.5461)))	LT=LENGTH (CM) AT AGE T (Y), CUBA	1564
*TRACHURUS TRACHURUS	LT = 55.5*(1 - EXP(-0.11*(T-1.96)))	LT=LENGTH (CM) AT AGE T (Y), SOUTH-WEST AFRICAN WATERS	1677

TABLE E12

EQUATION : BODY WEIGHT			
ITEM:	VALUE:	CONDITIONS:	REF.:
*ARMADILLIDIUM VULGARE	P = P0/EXP((K/0.4329)*T)	P=WEIGHT ,P0=P AT BIRTH, T=TIME (D), K=COEFFICIENT OF GROWTH	698
*LOXECHINUS ALBUS	Y = 0.0004*X**2.966	Y=DIAMETER (MM), X=WEIGHT (G), CHILEAN COAST, FEMALE	580
*SALMO TRUTTA	Y = 0.1522*X + 0.78	Y= WW OF GONADS (G), X= WW OF FISH (G), FEMALE, 1967-1969	1591
*SALMO TRUTTA	Y = 0.2378*X - 42.54	Y= WW OF GONADS (G), X= WW OF FISH (G), FEMALE, 1970-1971	1591
*SALMO TRUTTA	Y = 0.03316*X - 0.65	Y= WW OF GONADS (G), X= WW OF FISH (G), MALE	1591
*SOLEA SOLEA	C = 1.49*W - 0.00393*L**3	C=TOTAL CALORIC CONT (K CAL), W= WET W (G), L=LENGTH (CM)	176

TABLE E13

EQUATION : BODY WEIGHT VS AGE GROUP			
ITEM:	VALUE:	CONDITIONS:	REF.:
*CHRYSOPHRYS MAJOR	W = 2368.22*(1 - EXP(-0.18*(T + 0.47)))**2.84	W=WEIGHT (G) AT AGE T (Y), CENTRAL WATERS OF HIROSHIMA PREFECTURE	865
*COD, NORTH SEA	(WT - 51) = +-377*(|T-1|)**1.7482	WT = WEIGHT (G) AT AGE T (YEAR-CLASS	1292
*GADUS MORHUA	W =14150*(1 - EXP(-0.333*(T-0.87)))	W= WEIGHT (G), T=AGE GROUP (Y), SOUTHERN NORTH SEA	315
*GADUS MORHUA	W = 17390*(1- EXP(-0.269*(T-0.87)))	W= WEIGHT (G), T=AGE GROUP (Y), NOTHERN NORTH SEA	315
*HADDOCK	(WT - 124.2) = +-85.07*(|T-2.03|) **1.1152	WT = WEIGHT (G) AT AGE T (YEAR-CLASS	1292
*MACRURONUS MAGELLANICUS	WT = 804.1*(1 - EXP(-0.35984* (T-0.070851)))**3	WT = WEIGHT (G) AT AGE T (YEAR-CLASS)	6
*PLACIE	(WT - 920) = +-146.3*(|T-11.2|) **0.82617	WT = WEIGHT (G) AT AGE T (YEAR-CLASS)	1292
*PSENOPSIS ANOMALA	W = 486.7*(1 - EXP((-0.34*(T+ 1.674)))*2.8912	W = WEIGHT (G), T=AGE GROUP (Y), KII CHANNAL, JAPAN	1371
*SALMO SALAR	WT = 16504*(1 - EXP(-0.437*(T + 0.334)))	WT= WEIGHT (G) AT AGE T=TIME (YEAR-CLASS), BALTIC	1593
*SALMO SALAR	WT = 25230*(1 - EXP(-0.5397*(T - 0.082)))	WT= WEIGHT (G) AT AGE T=TIME (YEAR-CLASS), BALTIC	680
*SERRANUS ALEXANDRINUS	(WT - 71.9) = +-129.8*(|T-1|) **1.4084	WT = WEIGHT (G) AT AGE T (YEAR-CLASS)	1292
*SOLE, NORTH SEA	(WT - 144.9) = +-788*(|T-3|) **0.8483	WT = WEIGHT (G) AT AGE T (YEAR-CLASS)	1292
*TRACHURUS JAPONICUS	WT = 1.000*(1 - EXP(-0.248*(T + 809)))**3	WT=WET WEIGHT (G) AT AGE T (Y), WATERS OFF CHEJU ISLAND	33
*TRACHURUS TRACHURUS	W = 1307.2*(1 - EXP(-0.11*(T - 1.96)))**3	W=WEIGHT (G) AT AGE T (Y), SOUTH-WEST AFRICAN WATERS	1677

TABLE E14

EQUATION : BODY WEIGHT VS BODY LENGTH			
ITEM:	VALUE:	CONDITIONS:	REF.:
*ACRONEURIA SP.	W=0.027*L**3	PLECOPTERA,W=WEIGHT,L=LENGTH	388

TABLE E14

EQUATION : BODY WEIGHT VS BODY LENGTH

ITEM:	VALUE:	CONDITIONS:	REF.:
*AGARUM CRIBROSUM	Y = 0.0005*X**2.7740	Y=WW (G), X=LENGTH (CM), R=0.88, DEPTH=12 M, FOX POINT, 1968	977
*AGARUM CRIBROSUM	Y = 0.0099*X**2.2273	Y=WW (G), X=LENGTH (CM), R=0.856, DEPTH=12 M, FOX POINT, 1970	977
*AGENEIOSUS VALENCIENNESI	LOG(W) = 2.211 + 3.02*LOG(L)	W=WET W. (G), L=LENGTH (MM), BRAZIL, MALE, R=0.99	1436
*AGENEIOSUS VALENCIENNESI	LOG(W) = 3.042 + 3.042*LOG(L)	W=WET W. (G), L=LENGTH (MM), BRAZIL, FEMALE, R=0.99	1436
*AGENEIOSUS VALENCIENNESI	LOG(W) = 2.248 + 3.04*LOG(L)	W=WET W. (G), L=LENGTH (MM), BRAZIL, BOTH SEXES, R=0.99	1436
*ALOSA PSEUDOHARENGUS	LOG(W) = -5.28911 + 3.06370*LOG(L)	W=WET W (G), L=LENGTH (MM), CLAYTOR L., VIRGINIA	139
*ANTOCHA SP.	W=0.002*L**3	DIPTERA,W=WEIGHT,L=LENGTH	388
*APHELOCHIRUS VITTATUS	W=0.071*L**3	HEMIPTERA,W=WEIGHT,L=LENGTH	388
*APLODINOTUS GRUNNIENS	LOG(W) = -5.248 + 3.123*LOG(L)	L=TOTAL LENGTH (MM), W=WET W. (G), TUTTLE CREEK, KANSAS	834
*ATHERIX SP.	W=0.025*L**3	DIPTERA,W=WEIGHT,L=LENGTH	388
*BAETIS SP.	W=0.012*L**3	EPHEMEROPTERA,W=WEIGHT,L=LENGTH	388
*BALAENOPTERA ACUTOROSTRATA	W = 0.049574*W**2.31	W=WET WEIGHT (TONNES), L-LENGTH (M), ADJUSTED FOR FLUID LOSS	942
*BALAENOPTERA ACUTOROSTRATA	W = 0.0076*L**3.23	W=WET WEIGHT (TONNES),L=LENGTH (M) , ANTARCTIC, N=22	942
*BALAENOPTERA BOREALIS	W = 0.025763*L**2.43	W=WET WEIGHT (TONNES), L-LENGTH (M), ADJUSTED FOR FLUID LOSS	942
*BALAENOPTERA BOREALIS	W = 0.0242*L**2.43	W=WET WEIGHT (TONNES),L=LENGTH (M) , JAPAN, N=16	942
*BALAENOPTERA BRYDEI	W = 0.012965*W**2.74	W=WET WEIGHT (TONNES), L-LENGTH (M), ADJUSTED FOR FLUID LOSS	942
*BALAENOPTERA BRYDEI	W = 0.0122*L**2.74	W=WET WEIGHT (TONNES),L=LENGTH (M) , JAPAN, N=27	942
*BALAENOPTERA MUSCULUS	W = 0.002899*L**3.25	W=WET WEIGHT (TONNES), L-LENGTH (M), ADJUSTED FOR FLUID LOSS	942
*BALAENOPTERA MUSCULUS	W = 0.0046*L**3.09	W=WET WEIGHT (TONNES),L=LENGTH (M) , ANTARCTIC, N=43	942
*BALAENOPTERA MUSCULUS BREVICAUDA	W = 0.0004*L**3.97	W=WET WEIGHT (TONNES),L=LENGTH (M) , ANTARCTIC, N=5	942
*BALAENOPTERA PHYSALUS	W = 0.007996*L**2.90	W=WET WEIGHT (TONNES), L-LENGTH (M), ADJUSTED FOR FLUID LOSS	942
*BALAENOPTERA PHYSALUS	W = 0.0238*L**2.53	W=WET WEIGHT (TONNES),L=LENGTH (M) , ANTARCTIC, N=34	942
*BALAENOPTERA PHYSALUS	W = 0.0015*L**3.46	W=WET WEIGHT (TONNES),L=LENGTH (M) , N. HEMISPHERE, N=8	942
*BOSMINA LONGIROSTRIS	W**(-3)=0.56*L-0.01	W=WET WT.(MG),L=LENGTH(MM)	1424
*BOSMINA LONGIROSTRIS	W**(-3)=0.06*L+0.035	W=WET WT.(MG),L=LENGTH(MM)	1187
*BOSMINA SP.	W=0.124*L**2.181	W=WET WT.(MG),L=LENGTH(MM)	1234
*BRACHYCENTRUS SP.	W=0.051*L**3	TRICHOPTERA,W=WEIGHT,L=LENGTH	388
*BRANCHINECTA GIGAS	W = 0.108*L**3.27	W=DRY WEIGHT (UG),L=LENGTH (MM), MALE, N=70, 14-60 MM	316
*BRANCHINECTA GIGAS	W = 0.118*L**3.20	W=DRY WEIGHT (UG),L=LENGTH (MM), FEMALE, N=65, 11.4-72 MM	316
*BRANCHINECTA GIGAS	W = 7.56*L**1.45	W=DRY WEIGHT (UG),L=LENGTH (MM), LARVAE, N=14, 1.1-4.9 MM	316
*BRANCHINECTA MACKINI	W = 0.011*L**2.36	W=DRY WEIGHT (UG),L=LENGTH (MM), MALE, N=35, 6.2-24.7 MM	316
*BRANCHIN¤CTA MACKINI	W = 0.003*L**2.57	W=DRY WEIGHT (UG),L=LENGTH (MM), FEMALE, N=32, 6.1-25 MM	316
*BRANCHINECTA MACKINI	W = 0.192*L**1.51	W=DRY WEIGHT (UG),L=LENGTH (MM), LARVAE, N=26, 0.5-10.7 MM	316
*CARCINUS MEDITERRANEUS	W = 0.1819*L**3.16	W=WET W. (G), L=LENGTH (CM), MAY-JUNE, FRANCE, MALE	359
*CARCINUS MEDITERRANEUS	W = 0.1814*L**3.08	W=WET W. (G), L=LENGTH (CM), MAY-JUNE, FRANCE, FEMALE	359
*CARPIODES CYPRINUS	LOG(W) = -4.06042 + 2.78559*LOG(L)	W=WET W (G), L=LENGTH (MM), FOUR MILE CREEK, OHIO	1745

TABLE E14

EQUATION : BODY WEIGHT VS BODY LENGTH

ITEM:	VALUE:	CONDITIONS:	REF.:
*CARPIODES VELIFER	LOG(W) = -4.05206 + 2.80837*LOG(L)	W=WET W (G), L=LENGTH (MM), FOUR MILE CREEK, OHIO	1745
*CATACLYSTA MIDAS	W=0.04*L**3	LEPIDOPTERA,W=WEIGHT,L=LENGTH	388
*CERIODAPHNIA PULCHELLA	W**(-3)=0.57*L-0.02	W=WET WT.(MG),L=LENGTH(MM)	1187
*CHIRONOMUS SP.	W=0.0026*L**3	DIPTERA,W=WEIGHT,L=LENGTH	388
*CHONDROSTOMA NASUS PRESPENSE	LOG(W) = 2.98472*LOG(L) - 1.75947	W=WET W. (G), L=LENGTH (MM), PRESPA L.	1434
*COLOLABIS SAIRA	W = 0.00008309*L**3.19519	W=WET W. (G), L=LENGTH (MM), USA	1532
*COPEPODS	W**(-3)=0.34*L-0.03	W=WET WT.(MG),L=LENGTH(MM)	1424
*COPEPODS	W=0.055*L**2.73	W=WET WT.(MG),L=LENGTH(MM)	1424
*CYCLOPS SP.	Y = 0.01610*X - 0.00080	Y = DRY W. (UG), X=LENGTH (MM), STONEHOUSE POND	454
*DAPHNIA HYALINA	W**(-3)=0.32*L-0.08	W=WET WT.(MG),L=LENGTH(MM)	1424
*DAPHNIA HYALINA	W**(-3)=0.46*L-0.12	W=WET WT.(MG),L=LENGTH(MM)	1187
*DAPHNIA LONGISPINA	W**(-3)=0.46*L-0.12	W=WET WT.(MG),L=LENGTH(MM)	1187
*DAPHNIA PULICARIA	LOG(W) = 3.14*LOG(L) + 0.59	W=DRY WEIGHT (UG), L=LENGTH (MM)	1470
*DAPHNIA ROSEA	W=0.0053*L**2.17	W=DRY WT.(MG),L=LENGTH FR.TAIL-BASE TO ANTERIOR EDGE OF EYE,NONGRAVID	1013
*DAPHNIA ROSEA	W=0.0053*L**2.17	CORRECTION FACT.:ADD 2 UG PER EGG OR EMBRYO IN THE BROOD POUCH	1013
*DAPHNIA SP.	Y = 0.02794*X - 0.01365	Y = DRY W. (UG), X=LENGTH (MM), STONEHOUSE POND	454
*DAPHNIA SP.	W=0.052*L**3.012	W=WET WT.(MG),L=LENGTH(MM)	1234
*DIAPHANOSOMA SP.	W=0.092*L**2.449	W=WET WT.(MG),L=LENGTH(MM)	1234
*DIAPTOMUS NEVADENSIS	W = 0.001*L**3.61	W=DRY WEIGHT (UG),L=LENGTH (MM), FEMALE, N=51, 1.9-3.8 MM	316
*DIAPTOMUS NEVADENSIS	W = 0.107*L**1.79	W=DRY WEIGHT (UG),L=LENGTH (MM), LARVAE, N=20, 0.4-1.9 MM	316
*DIAPTOMUS SPP.	Y = 0.01629*X + 0.00061	Y = DRY W. (UG), X=LENGTH (MM), STONEHOUSE POND	454
*DINARTHRODES SP.	W=0.027*L**3	TRICHOPTERA,W=WEIGHT,L=LENGTH	388
*ECDYONURUS YOSHIDAE	W=0.042*L**3	EPHEMEROPTERA,W=WEIGHT,L=LENGTH	388
*ELEGINOPS MACLOVINUS	W = 1.25*10**(-5)*L**3.061	W=WET W. (G), L=LENGTH (MM), R= 0.9996, DESEADO RIVER ESTUARY	555
*ELMIS SP.	W=0.0062*L**3	COLEOPTERA,LARVA,W=WEIGHT,L=LENGTH	388
*ELMIS SP.	W=0.07*L**3	COLEOPTERA,IMAGO,W=WEIGHT,L=LENGTH	388
*ELOPS SAURUS	LOG(W) = 5.3295 + 3.1123*LOG(L)	W=WET W. (G), L= FORK LENGTH (MM), JUVENILE, LOUISIANA COASTAL STREAM	1414
*EPEORUS HIEMALIS	W=0.027*L**3	EPHEMEROPTERA,W=WEIGHT,L=LENGTH	388
*EPHEMERA LINEATA	W=0.01*L**3	EPHEMEROPTERA,W=WEIGHT,L=LENGTH	388
*EPHEMERELLA TRISPINA	W=0.03*L**3	EPHEMEROPTERA,W=WEIGHT,L=LENGTH	388
*EPIOPHLEBIA SP.	W=0.029*L**3	ODONATA,W=WEIGHT,L=LENGTH	388
*ERIOCERA SP.	W=0.026*L**3	DIPTERA,W=WEIGHT,L=LENGTH	388
*ESCHRICHTIUS ROBUSTUS	W = 0.0051*L**3.28	W=WET WEIGHT (TONNES),L=LENGTH (M) , N. PACIFIC, N=8	942
*EUBALAENA GLACIALIS SIEBOLDII	W = 0.013200*L**3.06	W=WET WEIGHT (TONNES), L-LENGTH (M), ADJUSTED FOR FLUID LOSS	942
*EUBALAENA GLACIALIS SIEBOLDII	W = 0.0124*L**3.06	W=WET WEIGHT (TONNES),L=LENGTH (M) , N. PACIFIC, N=21	942
*EUBRANCHIPUS BUNDYI	W = 0.550*L**2.50	W=DRY WEIGHT (UG),L=LENGTH (MM), MALE, N=44, 3.1-10.7 MM	316
*EUBRANCHIPUS BUNDYI	W = 0.513*L**2.42	W=DRY WEIGHT (UG),L=LENGTH (MM), FEMALE, N=37, 3.2-11.2 MM	316
*EUBRANCHIPUS BUNDYI	W = 0.001*L**3.39	W=DRY WEIGHT (UG),L=LENGTH (MM), MALE, N=42, 1.9-3.3 MM	316
*EUBRIANAX SP.	W=0.052*L**3	COLEOPTERA,W=WEIGHT,L=LENGTH	388
*FUNDULUS HETEROCLITUS	LOG(W) = 3.066*LOG(L) - 4.943	R=0.968, W=WET W. (G), L=LENGTH (CM), PATUXENT RIVER	1156

TABLE E14

EQUATION : BODY WEIGHT VS BODY LENGTH

ITEM:	VALUE:	CONDITIONS:	REF.:
*FUNDULUS MAJALIS	LOG(W) = 3.076*LOG(L) - 5.022	R=0.938, W=WET W. (G), L=LENGTH (CM), PATUXENT RIVER	1156
*GOERA JAPONICA	W=0.028*L**3	TRICHOPTERA,W=WEIGHT,L=LENGTH	388
*GOMPHUS SP.	W=0.043*L**3	ODONATA,W=WEIGHT,L=LENGTH	388
*HETEROPNEUSTES FOSSILIS	LOG(W) = -5.6440 + 3.1646*LOG(L)	W=WET W. (G), L=LENGTH (MM), DARBHANGA, INDIA, JUVENILE, R= 0.97	1570
*HETEROPNEUSTES FOSSILIS	LOG(W) = -5.0817 + 2.9418*LOG(L)	W=WET W. (G), L=LENGTH (MM), DARBHANGA, INDIA, FEMALE, R=0.975	1570
*HETEROPNEUSTES FOSSILIS	LOG(W) = -4.2147 + 2.5271*LOG(L)	W=WET W. (G), L=LENGTH (MM), DARBHANGA, INDIA, MALE, R=0.91	1570
*HOLOPEDIUM SP.	Y = 0.00686*X + 0.01206	Y = DRY W. (UG), X=LENGTH (MM), STONEHOUSE POND	454
*HYDROPSYCHE SP.	W=0.019*L**3	TRICHOPTERA,W=WEIGHT,L=LENGTH	388
*ICTALURUS PUNCTATUS	LOG(W) = -5.419 + 3.115*LOG(L)	L=TOTAL LENGTH (MM), W=WET W. (G), TUTTLE CREEK, KANSAS	833
*ISONYCHIA JAPONICA	W=0.023*L**3	EPHEMEROPTERA,W=WEIGHT,L=LENGTH	388
*LAMINARIA DIGITATA	Y = 0.0041*X**2.5927	Y=WW (G), X=LENGTH (CM), R=0.863, DEPTH=5 M, FOX POINT, 1968	977
*LAMINARIA DIGITATA	Y = 0.0128*X**2.2651	Y=WW (G), X=LENGTH (CM), R=0.892, DEPTH=12 M, FOX POINT, 1968	977
*LAMINARIA DIGITATA	Y = 0.0187*X**2.2587	Y=WW (G), X=LENGTH (CM), R=0.869, DEPTH=5 M, FOX POINT, 1970	977
*LAMINARIA DIGITATA	Y = 0.0013*X**2.8148	Y=WW (G), X=LENGTH (CM), R=0.969, DEPTH=12 M, FOX POINT, 1970	977
*LAMINARIA LONGICRURIS	Y = 0.0980*X**1.7234	Y=WW (G), X=LENGTH (CM), R=0.964, DEPTH=5 M,SRRAWBERRY ISL, 1968	977
*LAMINARIA LONGICRURIS	Y = 0.0121*X**1.9925	Y=WW (G), X=LENGTH (CM), R=0.948, DEPTH=5 M, FOX POINT, 1968	977
*LAMINARIA LONGICRURIS	Y = 0.0255*X**1.9180	Y=WW (G), X=LENGTH (CM), R=0.952, DEPTH=12 M, FOX POINT, 1968	977
*LAMINARIA LONGICRURIS	Y = 0.0038*X**2.4256	Y=WW (G), X=LENGTH (CM), R=0.985, DEPTH=5 M, LUKE ISL., 1970	977
*LAMINARIA LONGICRURIS	Y = 0.0010*X**2.7358	Y=WW (G), X=LENGTH (CM), R=0.961, DEPTH=10 M, LUKE ISL., 1970	977
*LAMINARIA LONGICRURIS	Y = 0.0018*X**2.5338	Y=WW (G), X=LENGTH (CM), R=0.96, DEPTH=5 M, FOX POINT, 1970	977
*LAMINARIA LONGICRURIS	Y = 0.0025*X**2.4150	Y=WW (G), X=LENGTH (CM), R=0.923, DEPTH=12 M, FOX POINT, 1970	977
*LOXECHINUS ALBUS	Y = 0.0006*X**2.835	Y=DIAMETER (MM), X=WEIGHT (G), CHILEAN COAST, MALE	580
*LUCIOLA CRUCIATA	W=0.03*L**3	COLEOPTERA,W=WEIGHT,L=LENGTH	388
*MACRONEMA SP.	W=0.02*L**3	TRICHOPTERA,W=WEIGHT,L=LENGTH	388
*MEGAPTERA NOVAEAGLIAE	W = 0.016473*L**2.95	W=WET WEIGHT (TONNES), L-LENGTH (M), ADJUSTED FOR FLUID LOSS	942
*MEGAPTERA NOVAEANGLIAE	W = 0.0158*L**2.95	W=WET WEIGHT (TONNES),L=LENGTH (M) , ANTARCTIC, N=17	942
*MENIDIA MENIDIA	LOG(W) = 2.855*LOG(L) - 4.915	R=0.981, W=WET W. (G), L=LENGTH (CM), PATUXENT RIVER	1156
*MICHROPTERUS COOSAE	LOG(W) = -4.387 + 2.798*LOG(L)	W=WET W (G), L=LENGTH (MM), SPRING CREEK, TENNESSEE	581
*MICHROPTERUS COOSAE	LOG(W) = -4.576 + 2.870*LOG(L)	W=WET W (G), L=LENGTH (MM), ROARING R., TENNESSEE	581
*MICRASEMA SP.	W=0.013*L**3	TRICHOPTERA,W=WEIGHT,L=LENGTH	388
*MICROPTERUS SALMOIDES	LOG(W) = -4.666 + 2.9*LOG(L)	W=WET W. (G), L=LENGTH (MM), MONCOVE LAKE, W. VIRGINIA, R=0.99	1058
*MOINA MACRURA	W**(-3)=0.66*L-0.01	W=WET WT.(MG),L=LENGTH(MM)	1187
*MOLANNA SP.	W=0.02*L**3	TRICHOPTERA,W=WEIGHT,L=LENGTH	388
*MORONE AMERICANAS	LOG(W) = 3.073*LOG(L) - 4.713	R=0.967, W=WET W. (G), L=LENGTH (CM), PATUXENT RIVER	1156
*MUGIL CAPITO	LOG(W) = -4.5647 + 2.7612*LOG(L)	MALE, W=WET W (G), L=LENGTH (MM), BOROLLUS LAKE, EGYPT	403
*MUGIL CAPITO	LOG(W) = -4.9029 + 2.9392*LOG(L)	FEMALE, W=WET W (G), L=LENGTH (MM) , BOROLLUS LAKE, EGYPT	403

TABLE E14

EQUATION : BODY WEIGHT VS BODY LENGTH

ITEM:	VALUE:	CONDITIONS:	REF.:
*MUGIL CAPITO	LOG(W) = -4.4244 + 2.8071*LOG(L)	BOTH SEXES, W=WET W (G), L=LENGTH (MM), BOROLLUS LAKE, EGYPT	403
*MULLUS BARBATUS	LOG(W) = 4.6943 + 2.88*LOG(L)	BOTH SEXES, W=WET W (G), L=LENGTH (MM), ABUKIR-ROSETTA, EGYPT	403
*MULLUS SURMULETUS	LOG(W) = -5.3255 + 3.1635*LOG(L)	MALE, W=WET W (G), L=LENGTH (MM), EGYPT MEDITERRANIAN	403
*MULLUS SURMULETUS	LOG(W) = -5.3786 + 3.1941*LOG(L)	FEMALE, W=WET W (G), L=LENGTH (MM), EGYPT MEDITERRANIAN	403
*MULLUS SURMULETUS	LOG(W) = -5.3805 + 3.1924*LOG(L)	BOTH SEXES, W=WET W (G), L=LENGTH (MM), EGYPT MEDITERRANIAN	403
*NEOPERLA SP.	W=0.011*L**3	PLECOPTERA,W=WEIGHT,L=LENGTH	388
*NEPHROPS NORVEGICUS	W = 0.0002188*L**3.321	W=WET W (G), L=CARAPACE LENGTH (MM), MALE, MORAY FIRTH	1271
*NEPHROPS NORVEGICUS	W = 0.0003221*L**3.204	W=WET W (G), L=CARAPACE LENGTH (MM), MALE, MINCH	1271
*NEPHROPS NORVEGICUS	W = 0.0006840*L**2.963	W=WET W (G), L=CARAPACE LENGTH (MM), FEMALE, MORAY FIRTH VS MINCH	1271
*NEPHROPS NORVEGICUS	W = 0.0004523*L**3.117	W=WET W (G), L=CARAPACE LENGTH (MM), MALE, NE ENGLAND	1544
*NEPHROPS NORVEGICUS	W = 0.001076*L**2.849	W=WET W (G), L=CARAPACE LENGTH (MM), FEMALE, MINCH	1544
*NEPHROPS NORVEGICUS	W = 0.0002245*L**3.3345	W=WET W (G), L=CARAPACE LENGTH (MM), BOTH SEXES, IRISH SEA	445
*PEPRILUS TRICANTHUS	LOG(W) = -5.1852 + 3.2646*LOG(L)	W=WET W (G), L=LENGTH (MM), LOWER YORK R., VIRGINIA	373
*PHYSETER CATODON	W = 0.006648*L**3.18	W=WET WEIGHT (TONNES), L-LENGTH (M), ADJUSTED FOR FLUID LOSS	942
*PHYSETER CATODON	W = 0.0196*L**2.74	W=WET WEIGHT (TONNES),L=LENGTH (M), ANTARCTIC + N. PACIFIC,N=44	942
*PHYSETER CATODON	W = 0.0029*L**3.55	W=WET WEIGHT (TONNES),L=LENGTH (M), NATAL, N=3	942
*PIMEPHALES PROMELAS	LOG(W) = 5.7204 + 3.5178*LOG(L)	W=WET W. (G), L=LENGTH (MM), NORTH DAKOTA SALINE LAKES	653
*PLATYPHYLAX YOKOUCHII	W=0.026*L**3	TRICHOPTERA,W=WEIGHT,L=LENGTH	388
*POLLACHIUS VIRENS	W = 0.0699M*MM2.54	W = WET WEIGTH (G), L = LENGTH (CM), NORTH VS NORWEGIAN SEA	254
*POLYPHEMUS PEDICULUS	W**(-3)=0.44*L-0.02	W=WET WT.(MG),L=LENGTH(MM)	1424
*POTAMANTHUS KAMONIS	W=0.014*L**3	EPHEMEROPTERA,W=WEIGHT,L=LENGTH	388
*PROCOPIUM WILLIAMSONI	LOG(W) = 2.1967 + 3.2108*(LOG(X) - 2.3710)	W= WET WEIGTH (G), L=FORK LENGTH (MM), SHEEP RIVER, ALBERTA	1586
*PROTOTHERMES GRANDIS	W=0.034*L**3	NEUOPTERA,W=WEIGHT,L=LENGTH	388
*RHYACOPHILA BREVICEPHALA	W=0.02*L**3	TRICHOPTERA,W=WEIGHT,L=LENGTH	388
*RHYACOPHILA NIGROCEPHALA	W=0.0072*L**3	TRICHOPTERA,W=WEIGHT,L=LENGTH	388
*SARDINELLA AURITA	W = 4.315*10**(-6)*L**3.12	W=WET W. (G), L=LENGTH (CM), EGYPT MEDITERRANIAN	1478
*SARDINELLA MADERENSIS	W = 1.278*10**(-5)*L**2.9176	W=WET W. (G), L=LENGTH (CM), EGYPT MEDITERRANIAN	1478
*SERRANUS ALEXANDRINUS	LOG(W) = -1.79257 + 2.92925*LOG(L)	W=WET WEIGHT (G), L=LENGTH (CM)	1292
*SIEBOLDIUS JAPONICA	W=0.031*L**3	ODONATA,W=WEIGHT,L=LENGTH	388
*SIMOCEPHALUS VETULUS	W**(-3)=0.65*L-0.04	W=WET WT.(MG),L=LENGTH(MM)	1187
*SIMULIUM SP.	W=0.01*L**3	DIPTERA,W=WEIGHT,L=LENGTH	388
*STENOPSYCHE GRISEIPENNIS	W=0.013*L**3	TRICHOPTERA,W=WEIGHT,L=LENGTH	388
*STENOPSYCHE SP.	W=0.046*L**3	TRICHOPTERA,W=WEIGHT,L=LENGTH	388
*THYMALLUS THYMALLUS	LOG(W) = -2.288 + 3.219*LOG(L)	W= WET W. (G), L=LENGTH (CM), LLYN TEGID AND RIVER DEE	1750
*TILAPIA ZILLII	LOG(W) = -1.941 + 3.1849*LOG(L)	W=WET WEIGHT (G), L=LENGTH (CM), FEMALE, S=2.9 %	1372
*TILAPIA ZILLII	LOG(W) = -1.5828 + 2.8309*LOG(L)	W=WET WEIGHT (G), L=LENGTH (CM), MALE, S=2.9 %	1372
*TILAPIA ZILLII	LOG(W) = -1.825 + 3.0594*LOG(L)	W=WET WEIGHT (G), L=LENGTH (CM), BOTH SEX, S=2.9 %	1372
*TRACHURUS JAPONICUS	W = 1.318*(L**3)*10**(-5)	W=WET W (G), L=FORK LENGTH (CM), WATERS OFF CHEJU ISLAND	33

TABLE E14

EQUATION : BODY WEIGHT VS BODY LENGTH

ITEM:	VALUE:	CONDITIONS:	REF.:
*TRACHURUS TRACHURUS	W = 0.0096*L**2.95	W=WET W (G), L=LENGTH (CM), SOUTH-WEST AFRICAN WATERS	1677
*TRINECTES MACULATUS	LOG(W) = 0.0132*LOG(L) + 0.034	R=0.972, W=WET W. (G), L=LENGTH (CM), PATUXENT RIVER	1156
*TUBIFEX TUBIFEX	LOG(W) = (0.8980 +- 0.25) + (1.76 +- 0.30)*LOG(L)	W=WET W (MG),L=LENGTH (MM), 95 % CL., N=70	853
*UPENEUS MOLUCCENSIS	W = 0.0126*L**3.1526	W=WET W (G), L=FORK LENGTH (CM), CHINA SEA, FEMALE	900
*UPENEUS MOLUCCENSIS	W = 0.0146*L**3.0873	W=WET W (G), L=FORK LENGTH (CM), CHINA SEA, MALE	900

TABLE E15

EQUATION : BOWEN RATIO

ITEM:	VALUE:	CONDITIONS:	REF.:
*WATER	R = 6.19*10**(-4)*P*(T - TA)/(ES - EA)	R = BOWEN RATIO (F048)	1095

TABLE E16

EQUATION : CALORIC CONTENT AS A FUNCTION OF % ORGANIC MATTER

ITEM:	VALUE:	CONDITIONS:	REF.:
*MARINE ZOOPLANKTON	Y = -3370 + 136*X - 0.514*X**2	Y=CAL CONT (CAL/G DW),X=% ORG MATTER IN DRY PLANKTON,R=0.94	1264

TABLE E17

EQUATION : CALORIC CONTENT AS A FUNCTION OF % ORGANIC MATTER AND % ASH

ITEM:	VALUE:	CONDITIONS:	REF.:
*MARINE ZOOPLANKTON	Y = 1351 + 106*Z -21.6*A	Z=% CARBON OF DW, A=ASH CONT (% OF DW),Y=CAL CONT (CAL/G DW),R=0.98	1264

TABLE E18

EQUATION : CARBON

ITEM:	VALUE:	CONDITIONS:	REF.:
*CTENOPHORES	LOG(Y) = 0.92*LOG(X) + 0.087	Y=CARBON (UG C/ANIM.), X= EQUATORIAL DIAMETER (MM), SIZE< 4 MM	1078
*CTENOPHORES	LOG(Y) = 2.35*LOG(X)	Y=CARBON (UG C/ANIM.), X= EQUATORIAL DIAMETER (MM), SIZE> 4 MM	1078

TABLE E19

EQUATION : CD

ITEM:	VALUE:	CONDITIONS:	REF.:
*BLOODPLASMA OF DOGS	CN(T)=3100*E**(-.3308T)+830*E**(-.12166T)+20*E**(-1025T)	0.15MG CD/KG,INTRAV.INJECT.,CN(T)= CD IN BLOOD AT TIME T,T=MINUTES	487

TABLE E19

EQUATION : CD

ITEM:	VALUE:	CONDITIONS:	REF.:
*BLOODPLASMA OF DOGS	CN(T)=3350*E**(-.29267T)+510*E**(-.10613T)+15*E**(-.00383T)	0.15MG CD/KG,INTRAV.INJECT.,CN(T)= CD IN BLOOD AT TIME T,T=MINUTES	487
*BLOODPLASMA OF DOGS	CN(T)=4150*E**(-.39346T)+900*E**(-.12552T)+30*E**(-.03401T)	0.15MG CD/KG,INTRAV.INJECT.,CN(T)= CD IN BLOOD AT TIME T,T=MINUTES	487
*BLOODPLASMA OF DOGS	CN(T)=810*E**(-.46247T)+320*E**(-.12343T)+18*E**(-.00476T)	0.15MG CD/KG,INTRAV.INJECT.,CN(T)= CD IN BLOOD AT TIME T,T=MINUTES	487
*CD CONTENT OF TESTPLANT	Y=105+4.0X	R=.998,Y=CD IN PLANT,X=CD EXTRACT BY AMMONIUMACET.AT PH=7,SOIL-PH=6	31
*CD CONTENT OF TESTPLANT	Y=102+2.4X	R=.992,Y=CD IN PLANT,X=CD EXTRACT BY AMMONIUMACET.AT PH=7,SOIL-PH= 7.2	31
*HOMO SAPIENS, BODY BURDEN	B(T)=A*(1-K**T)/(1-K)	B(T)=BODY BURDEN AT TIME T,A= ABS.AMOUNT/D	487
*HOMO SAPIENS, BODY BURDEN	B(T)=A*(1-K**T)/(1-K)	K=FRACTION OF BODY BURDEN REMAINING AFTER 1DAY	487
*HOMO SAPIENS, BODY BURDEN	B(N)=A1*(1-K**365)/(1-K)*(K**365)**(N-1)+..+AN*(1-K**365)/(1-K)	B(N)=BODY BURDEN AFTER N YRS,A1= ABS.AMOUNT/D(CONST.DUR.1YR)	487
*HOMO SAPIENS, BODY BURDEN	B(N)=A1*(1-K**365)/(1-K)*(K**365)**(N-1)+..+AN*(1-K**365)/(1-K)	K=FRACTION OF BODY BURDEN REMAIN.AFTER 1DAY	487
*HOMO SAPIENSS,SMOKERS	ACCUMULATION; Y=0.188X+6.8	Y=CD IN KIDNEY,LIVER AND LUNGS(MG) ,X=CIGARETT.PACKS/D*TIME(Y) OF SMOKING	487
*MICE	Y=13.45+0.434X	R=.42,Y=CD IN FECES(NG)/24HR,X=CD IN BODY(UG), EXCR WITH FECES	487
*MICE	Y=119.0+0.274X	R=0.36,Y=CD IN FECES(NG)/24HR,X=CD IN BODY(UG), EXCR WITH FECES	487
*MICE	Y=13.45+0.434X	INJECT.OF 0.025MG CD/KG DAY(5 DAYS/W), EXCR WITH FECES	487
*MICE	Y=119.0+0.274X	INJECT.OF 0.25MG CD/KG DAY(5 DAYS/ W), EXCR WITH FECES	487
*MICE	Y=0.112X+1.785	Y=CD IN URINE(NG/24HR),X=BODY BURDEN OF CD(UG)	487
*SOIL-CD EXTRACTED IN AMMONIUMACETATE	Y=17+1.9X	Y=EXTRACTED CD,X=ADDED SEWAGE SLUDGE BETW. 0-58T/HA,SOLVENT-PH= 4.8	31
*SOIL-CD EXTRACTED IN 2 M NITRIC ACID	Y=99+2.7X	R=.995,Y=EXTRACTED CD,X=AMOUNT OF SEWAGE SLUDGE,SOLVENT-PH=4.8	31
*URINARY EXCRETION FROM MICE	Y=0.015+0.0002X	R=.98,Y=CD IN URINE(UG)/24HR,X=CD IN BODY(UG)	487
*URINARY EXCRETION FROM MICE	Y=0.011+0.000096X	R=.93,Y=CD IN URINE(UG)/24HR,X=CD IN BODY(UG)	487
*URINARY EXCRETION FROM MICE	Y=0.015+0.0002X	INJECT.OF 0.5MG CD/KG DAY(5DAYS/W)	487
*URINARY EXCRETION FROM MICE	Y=0.011+0.000096X	INJECT.OF 0.25MG CD/KG DAY(5 DAYS/ W)	487
*URINARY EXCRETION FROM MICE	Y=0.112X+1.785	APPROX. 0.25MG CD/KG DAY(5DAYS/W), SUBCUT.INJECT.	487

TABLE E20

EQUATION : COEFFICIENT OF EXPONENTIAL GROWTH AS A FUNCTION OF TEMPERATURE

ITEM:	VALUE:	CONDITIONS:	REF.:
*ARMADILLIDIUM VULGARE	LOG(K) = -2.4761 + 0.03657*T	K=COEF. OF GROWTH (1/D), T= CEL DEG., 278-304.5 K	698

TABLE E21

EQUATION : CONCENTRATION OF NITRATE NITROGEN VS OXYGEN CONCENTRATION

ITEM:	VALUE:	CONDITIONS:	REF.:
*LAKE ZURICH	Y = 41 + 0.103*X	Y=NO3-N, X=O2, UNITS = MICRO M, DEEPER WATERS	1526

TABLE E22

EQUATION : CONCENTRATION OF NITRATE NITROGEN VS PHOSPHATE PHOSPHORUS			
ITEM:	VALUE:	CONDITIONS:	REF.:
*IN WESTERN ATLANTIC	Y = 14.5*X	Y=NO3-N, X=PO4-P, UNITS=MICRO M	1526
*LAKE NORRVICKEN	Y = 145 + 16.7*X	Y=NO3-N, X=PO4-P, UNITS = MICRO M, DURING WINTER, ICE COVER	1526
*LAKE ZURICH	Y = 19 + 5.17*X	Y=NO3-N, X=PO4-P, UNITS = MICRO M, DIFFERENT DEPTHS	1526

TABLE E23

EQUATION : CONCENTRATION OF 2,4-D AS A FUNCTION OF TIME			
ITEM:	VALUE:	CONDITIONS:	REF.:
*ADENOSTOMA FASCICULATUM	LOG(Y) = 3.099 - 0.923*LOG(X+1)	Y=MG/KG WW OF PESTICIDE, X=DAYS AFTER APPL., SLOPE=5-9 %, SAN DIEGO	1265
*SOIL	LOG(Y) = 1.278 - 1.014*LOG(X+1)	Y=MG/KG WW OF PESTICIDE, X=DAYS AFTER APPL., SLOPE=5-9 %, SAN DIEGO	1265

TABLE E24

EQUATION : CONCENTRATION OF 2,4,5-T AS A FUNCTION OF TIME			
ITEM:	VALUE:	CONDITIONS:	REF.:
*ADENOSTOMA FASCICULATUM	LOG(Y) = 2.876 - 1.003*LOG(X+1)	Y=MG/KG WW OF PEST., X=TIME (D), SANDY LOAM, SLOPE=5-9 %, SAN DIEGO	1265
*SOIL	LOG(Y) = 0.979 - 0.836*LOG(X+1)	Y=MG/KG WW OF PEST., X=TIME (D), SANDY LOAM, SLOPE=5-9 %, SAN DIEGO	1265

TABLE E25

EQUATION : CONSUMATION OF FOOD AS A FUNCTION OF TIME AND FOOD CONCENTRATION			
ITEM:	VALUE:	CONDITIONS:	REF.:
*KERONOPSIS RUBRA	R = T/(A + 1/(B*P))	R= CONSUMATION, P=FOOD CONCENTRATION, A,B=CONSTANTS	1769

TABLE E26

EQUATION : CONVERSION OF OXYGEN CONSUMED TO BIOMASS OF FOOD			
ITEM:	VALUE:	CONDITIONS:	REF.:
*LIVING ORGANISMS	Y*RQ*12/32 = X	Y=MG O2 CONSUMED PER H,X=MG C UTILIZED PER H,RQ=RESPIRATORY QUOTIENT	1218
*LIVING ORGANISMS	Z*RQ*12/22.4 = X	Z=ML O2 CONS. PER H,X=MG C UTILIZED PER H,RQ=RESPIRATORY QUOTIENT	1218

TABLE E27

EQUATION : CORRELATION BETWEEN WATER AND ORGANISMS CONCENTRATION OF TCDD			
ITEM:	VALUE:	CONDITIONS:	REF.:
*DAPHNIA MAGMA	LOG(Y) = 1.377 + 0.874*LOG(X)	Y=CONC IN ORG (MG/T DW),X=CONC IN WATER (PPT),R=0.942,MODEL ECOSYSTEM	726

TABLE E27

EQUATION : CORRELATION BETWEEN WATER AND ORGANISMS CONCENTRATION OF TCDD			
ITEM:	VALUE:	CONDITIONS:	REF.:
*GAMBUSIA AFFINIS	LOG(Y) = 1.395 + 0.875*LOG(X)	Y=CONC IN ORG (MG/T DW),X=CONC IN WATER (PPT),R=0.953,MODEL ECOSYSTEM	726
*ICTALURUS PUNCTATUS	LOG(Y) = 0.893 + 0.927*LOG(X)	Y=CONC IN ORG (MG/T DW),X=CONC IN WATER (PPT),R=0.957,MODEL ECOSYSTEM	726
*LEMNA MINOR	LOG(Y) = 0.501 + 1.141*LOG(X)	Y=CONC IN ORG (MG/T DW),X=CONC IN WATER (PPT),R=0.954,MODEL ECOSYSTEM	726
*OEDOGONIUM CARDIACUM	LOG(Y) = 0.851 + 1.148*LOG(X)	Y=CONC IN ORG (MG/T DW),X=CONC IN WATER (PPT),R=0.987,MODEL ECOSYSTEM	726
*PHYSA SP.	LOG(Y) = 1.229 + 1.059*LOG(X)	Y=CONC IN ORG (MG/T DW),X=CONC IN WATER (PPT),R=0.965,MODEL ECOSYSTEM	726

TABLE E28

EQUATION : CU			
ITEM:	VALUE:	CONDITIONS:	REF.:
*DAPHINA MAGNA	Y = 8*X - 0.2	Y=RECIPROCAL SURVIVAL TIME (10**3/T),X=CUCO3 ACTIVITY (MICRO M),PH=8	32
*DAPHINA MAGNA	Y = 156*X - 0.2	Y=RECIPROCAL SURVIVAL TIME (10**3/T),X=CU ACTIVITY (MICRO M),PH=8	32

TABLE E29

EQUATION : DAILY METABOLISM AS A FUNCTION OF BODY WEIGHT			
ITEM:	VALUE:	CONDITIONS:	REF.:
*APODEMUS FLAVICOLLIS	Y = 2.364*W**(-0.50)	Y= METABOLISM (KCAL/(G*D)),W=WET WEIGHT (G)	571
*CLETHRIONOMYS GLAREOLUS	Y = 2.510*W**(-0.50)	Y= METABOLISM (KCAL/(G*D)),W=WET WEIGHT (G)	571
*CRUSTACEANS	Q = 19.24*W**0.81	Q=DAILY MET (CAL/(ANIM*D)), W=WET W (G), 293 K	1732
*MAMMALS	Q = 422*W**0.75	Q=DAILY MET (CAL/(ANIM*D)), W=WET W (G), BASAL METABOLISM	836
*POIKILOTHERMIC METAZOAN ANIMALS	Q = 16.54*W**0.75	Q=DAILY MET (CAL/(ANIM*D)), W=WET W (G), STANDARD METABOLISM	654
*PROTOZOA	Q = 2.02W**0.755	Q=DAILY MET (CAL/(ANIM*D)), W=WET W (G), 293 K	654
*TEMPERATE FISH	Q = 35.8*W**0.81	Q=DAILY MET (CAL/(ANIM*D)), W=WET W (G), RESTING, 293 K	1733

TABLE E30

EQUATION : DDT			
ITEM:	VALUE:	CONDITIONS:	REF.:
*CONCENTRATION OF DDT AS A FUNCTION OF CONCENTRATION IN SOIL	Y = 0.1*X	Y=CONC IN SOIL (PPM DW), X=CONC IN EARTHWORM (PPM WW)	389
*FISH	Y = 20 + 0.0257*X	Y= UG TOTAL DDT/KG WET WEIGHT, X= WET WEIGHT (MG)	294
*FISH	Y = -2.4 + 0.0263*X	Y= UG DDE/KG WET WEIGHT, X=WET WEIGHT (MG)	294
*IN HUMAN FAT AS A FUNCTION OF DAILY DOSE	Y = 22.5*X**0.667	Y=CONC IN FAT (PPM), X=DAILY DOSE (MG)	389

TABLE E30

EQUATION : DDT			
ITEM:	VALUE:	CONDITIONS:	REF.:
*SALMO SALAR L.	RESIDUE = 1.91/T	T=TIME(Y), AFTER 4 MONTHS TOTAL DDT=5.7 PPM, DDE=3.8 PPM WW	1467

TABLE E31

EQUATION : DEGRADATION AS A FUNCTION OF REDOX POTIENTAL AND TIME			
ITEM:	VALUE:	CONDITIONS:	REF.:
*TRIFLURALIN	Y = 96.4 - 17.6*X + 0.7*X**2	Y= % PEST. REMAINING, X=TIME (D), REDOX = -50-+50 MV, R=0.95	1722
*TRIFLURALIN	Y = 99.0 - 1.95*X	Y= % PEST. REMAINING, X=TIME (D), REDOX = +150-450 MV, R=0.86	1722

TABLE E32

EQUATION : DEGRADATION OF 1,3-DICHLOROPROPENES IN SOIL			
ITEM:	VALUE:	CONDITIONS:	REF.:
*CIS-CHLOROALLYLALCOHOL	RATE=K*CONC. (HALF LIVE 2D)	CLAY CONT. SOIL,288K	1625
*TRANS-DICHLOROALLYLALCOHOL	RATE = K* CONC. (HALF LIVE LESS THAN 1D)	CLAY CONT. SOIL,288K	1625
*1,3-DICHLOROPROPENES	RATE = K*CONC. (HALF LIVE 24D)	SANDY SOIL,288-293K,ISOMERS SAME RATE	1625
*1,3-DICHLOROPROPENES	RATE = K*CONC. (HALF LIVE 3-8D)	CLAY CONT. SOIL,288-293K	1625

TABLE E33

EQUATION : DENITRIFICATION			
ITEM:	VALUE:	CONDITIONS:	REF.:
*USED IN MODELS	MYM = MYMAX*NIT/(KNIT + NIT)	(F047)	1154
*USED IN MODELS	D(M)/DT = MYM*M	(F047)	1154
*USED IN MODELS	M = M(0) + STO(NIT(0) - NIT)	(F047)	1154
*USED IN MODELS	DENIT = DENITK*NS	(F047)	764

TABLE E34

EQUATION : DISAPEARENCE OF DDT FROM WATER (F054)			
ITEM:	VALUE:	CONDITIONS:	REF.:
*PTERONARCYS CALIFORNICA	DY/DT=K2*(Q(0)-X-N*Y)	FIRST ORDER KINETIC,Y=THE DOSE TO EACH OF THE N ORG.,K2=0.0099/DAY	527
*PTERONARCYS CALIFORNICA	DX/DT=K1*(Q(0)-X-N*Y)	NO ORG.,X=NON-BIOL.LOSS OF DDT FROM WATER,K1=0.33/DAY	527
*PTERONARCYS CALIFORNICA	DQ/DT=-(K1*Q+N*K2*Q)	Q(0)=INITIAL AMOUNT OF DDT,N= NUMBER OF ORG.,Q=AMOUNT OF DDT	527

TABLE E35

EQUATION : DISSOLVED OXYGEN OF OVERLAND FLOW

ITEM:	VALUE:	CONDITIONS:	REF.:
*LAND SURFACE	DO=14.652-0.41022T(W)+0.00791T(W) **2-0.00077774T(W)**3	SATURATION,DO=DISSOLVED OXYGEN IN PPM,	364
*LAND SURFACE	DO=14.652-0.41022T(W)+0.00791T(W) **2-0.00077774T(W)**3	T(W)=OVERLAND FLOW TEMP.IN DEGREES C	364

TABLE E36

EQUATION : DISSOLVED OXYGEN SATURATION VS BAROMETRIC AND

ITEM:	VALUE:	CONDITIONS:	REF.:
*WATER	CS' = CS*((PB - PV)/(760 - PB))	CS'=O2 SAT (MG/L),CS=CS' AT SEALEVEL (F049)	21

TABLE E37

EQUATION : DISSOLVED OXYGEN SATURATION VS SALINITY

ITEM:	VALUE:	CONDITIONS:	REF.:
*WATER	CS = CS(0)*(1 - S*10**(-5))	CS=DISSOLVED O2 SAT (MG/L),CS(0)= CS FOR S=0, S=SALINITY (MG CHLOR/ L)	440

TABLE E38

EQUATION : DISSOLVED OXYGEN SATURATION VS TEMPERATURE

ITEM:	VALUE:	CONDITIONS:	REF.:
*WATER	CS = 14.652+0.41022*T+ 0.0079910*T**2-0.000077774*T**3	CS=DISSOLVED O2 SAT (MG/L),T=WATER TEMP (CEL),STANDARD PRESSURE,S=0	415

TABLE E39

EQUATION : DISSOLVED OXYGEN SATURATION VS TEMPERATURE AND SALINITY

ITEM:	VALUE:	CONDITIONS:	REF.:
*SEAWATER	CS=14.6244-0.367134*T+ 0.0044972*T**2-0.09665+0.00205*S*T +0.00027*S**2	CS=O2 SAT (MG/L), T=TEMP (CEL), S= SALINITY (1/1000)	234

TABLE E40

EQUATION : DRIFT AS A FUNCTION OF BENTHOS DENSITY

ITEM:	VALUE:	CONDITIONS:	REF.:
*EPHEMERELLA NEEDHAMI	D=0.29X	D=DRIFT(NUMB./NIGHT),X=DENSITY (NO./0.03M**2),R=0.975,N=8,HIGH FOOD	667
*EPHEMERELLA SERRATA GP.	D=0.54X	D=DRIFT(NUMB./NIGHT),X=DENSITY (NO./0.03M**2),R=0.896,N=12,LOW FOOD	667
*EPHEMERELLA SERRATA GP.	D=0.29X	D=DRIFT(NUMB./NIGHT),X=DENSITY (NO./0.03M**2),R=0.937,N=12,HIGH FOOD	667

TABLE E40

EQUATION : DRIFT AS A FUNCTION OF BENTHOS DENSITY			
ITEM:	VALUE:	CONDITIONS:	REF.:
*EPHEMERRELLA NEEDHAMI	D=0.51X	D=DRIFT(NUMB./NIGHT),X=DENSITY (NO./0.03M**2),R=0.992,N=8,LOW FOOD	667
*HYDROPSYCHIDAE	D=0.28X	D=DRIFT(NUMB./NIGHT),X=DENSITY (NO./0.03M**2),R=0.991,N=6	667
*HYDROPSYCHIDAE	D=0.36X	D=DRIFT(NUMB./NIGHT),X=DENSITY (NO./0.03M**2),R=0.998,N=8,LOW FOOD	667
*HYDROPSYCHIDAE	D=0.22X	D=DRIFT(NUMB./NIGHT),X=DENSITY (NO./0.03M**2),R=0.980,N=7,HIGH FOOD	667
*SIMULIIDAE	D=13.45+0.056X	D=DRIFT(NUMB./NIGHT),X=DENSITY (NO./0.03M**2),R=0.950,N=8	667
*TRICORYTHODES SP.	D=0.13X	D=DRIFT(NUMB./NIGHT),X=DENSITY (NO./0.03M**2),R=0.896,N=12,LOW FOOD	667
*TRICORYTHODES SP.	D=0.13X	D=DRIFT(NUMB./NIGHT),X=DENSITY (NO./0.03M**2),R=0.896,N=12,HIGH FOOD	667

TABLE E41

EQUATION : DRY MATTER PRODUCTION AS A FUNCTION OF RAINFALL			
ITEM:	VALUE:	CONDITIONS:	REF.:
*ARID AND SEMI-ARID ZONE OF VEGETATION	Y = 9.12*X - 408	Y=PROD (KG/(HA*Y)), X=RAIN (MM), GRADIENT IN WEST AFRICA	1656

TABLE E42

EQUATION : EDDY DIFFUSION			
ITEM:	VALUE:	CONDITIONS:	REF.:
*STREAMS	LOG(K*U/V*EL/(V*H) = 6.45 - 0.762*LOG(R*V*H/P)	EL = LONG TURBULENT DIFF COEFFICIENT (F050)	78

TABLE E43

EQUATION : ENERGY BUDGET FOR SALMO TRUTTA (F054)			
ITEM:	VALUE:	CONDITIONS:	REF.:
*CHANGE IN ENERGY CONTENT	DELTA(B) = C*PP - R	B=ENERGI IN BODY (CAL), PP=ENERGY FOR GROWTH AND METABOL,R =RESP.	411
*DAILY LOSS OF ENERGY IN EXCRETORY PRODUCTS VS TEMP AND FOOD RATIO	PU = 0.0259*T**(0.580(+-0.0110)) *EXP(-0.299(+-0.0162)*C/CM))	276.8 - 293.4 K, 95 % CL.	411
*DAILY LOSS OF ENERGY IN FECES VS TEMPERATURE AND FOOD RATIO	PF = 0.212*T**(-0.222(+-0.0224)) *EXP(0.631(+-0.0350)*C/CM)	276.8 - 293.4, 95 % CL.	411
*ENERGY AVAIBLE FOR METABOLISM VS TEMPERATURE AND BODY WEIGHT	RM = 3.890*W**0.770(+-0.0254) *EXP(0.204(+-0.00394)*T)	RM=ENERGY FOR MET (CAL/D), 276.8-290.8 K, 95 % CL.	411
*ENERGY AVAIBLE FOR METABOLISM VS TEMPERATURE AND BODY WEIGHT	RM = 2.215*10**7*WW**0.757(+ -0.0208)*EXP(-0.663(+-0.0175)*T)	RM=ENERGY FOR MET (CAL/D), 290.8-292.5 K, 95 % CL.	411
*ENERGY AVAIBLE FOR METABOLISM VS TEMPERATURE AND BODY WEIGHT	RM =28.833*W**0.756(+-0.0440) *EXP(0.0325(+-0.0288)*T)	RM=ENERGY FOR MET (CAL/D), 292.5-294.7 K, 95 % CL.	411
*ENERGY AVAIBLE FOR METABOLISM VS TEMPERATURE AND BODY WEIGHT	RM = 3.890*W**0.770(+-0.0254) *EXP(0.204(+-0.00394)*T)	RM=ENERGY FOR MET (CAL/D), 276.8-290.8 K, 95 % CL.	411
*ENERGY FOR GROWTH VS ENERGY LOSS IN FECES AND EXCRETORY PRODUCTS	PP = 1 - PF - PU	SEE OTHER PART OF TABLE	411
*ENERGY FOR STANDARD METABOLISM VS TEMPERATURE AND BODY WEIGHT	RS = 4.126*W**0.734(+-0.0257) *EXP(0.192(+-0.0125)*T)	RS=ENERGY FOR STANDARD MET. (CAL/ D), 276.8-280.1 K	411

TABLE E43

EQUATION : ENERGY BUDGET FOR SALMO TRUTTA (F054)

ITEM:	VALUE:	CONDITIONS:	REF.:
*ENERGY FOR STANDARD METABOLISM VS TEMPERATURE AND BODY WEIGHT	RS = 8.277*W**0.731(+-0.0220) *EXP(0.0938(+-0.00407)*T)	RS=ENERGY FOR STANDARD MET. (CAL/D), 280.1-292.5 K	411
*ENERGY IN EXCRETORY PRODUCTS VS DAILY LOSS AND TOTAL ENERGY CONSUMED	U = C*PU	U=ENERGY IN EXCRETORY PRODUCTS (CAL/D),PU=FRACTION LOST OF CONSUMATED	411
*ENERGY IN FECES AS A FUNCTION OF DAILY LOSS AND TATAL ENERGY CONSUMED	F = C*PF	F=ENERGY IN FECES (CAL/D),PF= FRACTION LOST OF CONSUMATION=C (CAL/D)	411
*ENERGY LOSS IN WASTE PRODUCTS VS TEMPERATURE AND FOOD RATION	PT = 0.00512*T**0.812*EXP(0.523*(C/CMAX))	PT=ENERGY IN WASTE (CAL),CMAX=MAX OF C=FOOD RATIO,T=CEL DEG,277-280 K	410
*ENERGY LOSS IN WASTE PRODUCTS VS TEMPERATURE AND FOOD RATION	PT = 0.0477*T**0.801*EXP(0.223*(C/CMAX))	PT=ENERGY IN WASTE (CAL),CMAX=MAX OF C=FOOD RATIO,T=CEL DEG,280-292 K	410
*ENERGY REQIURED FOR MAINTENANCE METABOLISM VS TEMP AND BODY WEIGHT	RMAIN = 3.802*W**0.723(+-0.0998) *EXP(0.245(+-0.0811)*T)	RMAIN=ENERGY REQ. FOR MAIN. MET (CAL/D), 276.8-280.1 K, 95 % CL.	411
*ENERGY REQIURED FOR MAINTENANCE METABOLISM VS TEMP AND BODY WEIGHT	RMAIN =11.866*W**0.721(+-0.0417) *EXP(0.0915(+-0.0112)*T)	RMAIN=ENERGY REQ. FOR MAIN. MET (CAL/D), 280.1-288.0 K, 95 % CL.	411
*MAINTENANCE ENERGY INTAKE VS TEMPERATURE AND BODY WEIGHT	CMAIN = 6.169*W**0.716(+-0.107) *EXP(0.244(+-0.102)*T)	CMAIN= ENERGY INTAKE FOR MAIN. (CAL/D), 276.8-279.6 K, 95 % CL	411
*MAINTENANCE ENERGY INTAKE VS TEMPERATURE AND BODY WEIGHT	CMAIN =12.031*W**0.737(+-0.058) *EXP(0.105(+-0.016)*T)	CMAIN= ENERGY INTAKE FOR MAIN. (CAL/D), 279.6-292.5 K, 95 % CL	411
*MAX DAILY ENERGY CONTENT OF FOOD AS A FUNCTION OF BODY WEIGHT	CM = 2.902*W**(0.762(+-0.027))*EXP (0.418(+-0.035)*T)	CM=MAX CONT. (CAL/D), T=CEL DEG, W =WET W (G), 276.8-279.6 K	411
*MAX DAILY ENERGY CONTENT OF FOOD AS A FUNCTION OF BODY WEIGHT	CM =15.018*W**(0.759(+-0.023))*EXP (0.171(+-0.012)*T)	CM=MAX CONT. (CAL/D), T=CEL DEG, W =WET W (G), 279.6-286.3 K	411
*MAX DAILY ENERGY CONTENT OF FOOD AS A FUNCTION OF BODY WEIGHT	CM =26.433*W**(0.767(+-0.041))*EXP (0.126(+-0.031)*T)	CM=MAX CONT. (CAL/D), T=CEL DEG, W =WET W (G), 286.3-290.8 K	411
*MAX DAILY ENERGY CONTENT OF FOOD AS A FUNCTION OF BODY WEIGHT	CM = 3.241*10**7*(0.753(+-0.086)) *EXP(-0.662(+-0.054)*T)	CM=MAX CONT. (CAL/D), T=CEL DEG, W =WET W (G), 290.8-294.7 K	411
*OPTIMUM ENERGY INTAKE VS TEMPERATURE AND BODY WEIGHT	COPT = 2.902*W**0.762(+-0.027)*EXP (0.418(+-0.035)*T)	COPT= OPTIMUN ENERGY INTAKE (CAL/D), 276.8-279.8 K, 95 % CL.	411
*OPTIMUM ENERGY INTAKE VS TEMPERATURE AND BODY WEIGHT	COPT =15.116*W**0.767(+-0.046)*EXP (0.138(+-0.011)*T)	COPT= OPTIMUN ENERGY INTAKE (CAL/D), 279.8-288.0 K, 95 % CL.	411
*OPTIMUM ENERGY INTAKE VS TEMPERATURE AND BODY WEIGHT	COPT = 753.9*W**0.767(+-0.017)*EXP (-0.118(+-0.006)*T)	COPT= OPTIMUN ENERGY INTAKE (CAL/D), 288.0-292.5 K, 95 % CL.	411
*PROPORTION OF DIALY ENERGY INTAKE LOST BY FECES VS TEMP & FOOD RATION	PF = 0.212*T**(-0.222)*EXP (0.631*(C/CMAX))	PF=ENERGY IN FECES (CAL),CMAX=MAX OF C=FOOD RATIO,T=CEL DEG,N=76	410
*PROPORTION OF DIALY ENERGY INTAKE LOST IN EXCROTORY PRODUCTS	PU = 0.0259*T**0.580*EXP(-0.299*(C/CMAX))	PU=ENERGY IN EXC PROD. (CAL),CMAX= MAX OF C=FOOD RATIO,T=CEL DEG,N=80	410

TABLE E44

EQUATION : ENERGY USED FOR MAINTENANCE AS A FUNCTION OF ASSIMILATED ENERGY

ITEM:	VALUE:	CONDITIONS:	REF.:
*HETEROTHERMS	LOG(R) = 1.0733*LOG(P) + 0.3757	R=MAINTENANCE, P=ASS. ENERGY, UNITS=KCAL/(M2*Y)	1012
*HOMEOTHERMS	LOG(R) = 1.1740*LOG(P) + 0.1352	R=MAINTENANCE, P=ASS. ENERGY, UNITS=KCAL/(M2*Y)	1012

TABLE E45

EQUATION : ENERGY USED FOR MAINTENENCE AS A FUNCTION OF ASSIMILATED ENERGY

ITEM:	VALUE:	CONDITIONS:	REF.:
*AQUATIC HETEROTHERMS	Y = 0.811*X	Y=MAINTENANCE, X=ASSIMILATION, UNITS=KCAL/(M2*Y)	1012
*TERRESTRIAL HETEROTHERMS	Y = 0.548*X	Y=MAINTENANCE, X=ASSIMILATION, UNITS=KCAL/(M2*Y)	1012

TABLE E46

EQUATION : EXCRETION OF AMMONIUM

ITEM:	VALUE:	CONDITIONS:	REF.:
*CALANUS	Y=0.659*X	FOOD=BIDDULPHIA, Y=EXCR,X=TOT N INGESTED, UNITS=G N/(ANIM*D)	355
*CRASSOSTREA VIRGINICA	LOG(Y) = 1.22*LOG(X) + 1.33	Y=EXC. (U MOLES NH4-N/D), X=DRY W., (G), R=0.84, PH=8, 293 K, S=3 %	997
*MARINE ZOOPLANKTON	LOG(Y) = 1.237 + 0.928*LOG(X)	Y=NH4 EXC (NG-ATOM/H*ANIMAL), X= DW. (MG/ANI),N PACIFIC,298,K, MIXED	1080
*MERCENARIA MERCENARIA	LOG(Y) = 0.94*LOG(X) + 1.33	Y=EXC. (U MOLES NH4-N/D), X=DRY W., (G), R=0.67, PH=8, 293 K, S=3 %	997
*PHRONIMA SEDENTARIA	E = 0.86*W**1.08	E=EXCRETION (UG NH4-N/(MG DW*D)), W=DRY WEIGHT (MG), 286 K	997
*PHRONIMA SEDENTARIA	E = 0.57*W**1.26	E=EXCRETION (UG TOT-N/(MG DW*D)), W=DRY WEIGHT (MG), 286 K	997
*ZOOPLANKTON, AEQUOREA & ORCHISTOMA & CUNINA SP.	LOG(Y) =-1.64(SE=+-0.10) + 1.49(SE =+-0.18)*LOG(X)	Y=EXCR (UG-ATOM N/H),X=BODY PROT (MG),299 K, TROP MARINE,R=0.93,N=	124
*ZOOPLANKTON, AGALMA OKENI	LOG(Y) =-1.21(SE=+-0.07) + 0.82(SE =+-0.10)*LOG(X)	Y=EXCR (UG-ATOM N/H),X=BODY PROT (MG),299 K, TROP MARINE,R=0.89,N=	124
*ZOOPLANKTON, ATHORYBIA ROSACEA & A. SP.	LOG(Y) =-0.88(SE=+-0.08) + 0.48(SE =+-0.09)*LOG(X)	Y=EXCR (UG-ATOM N/H),X=BODY PROT (MG),299 K, TROP MARINE,R=0.88,N=	124
*ZOOPLANKTON, BATHYPHYSA SIBOGAE & RHIZOPHYSA FILIFORMIS	LOG(Y) =-1.55(SE=+-0.18) + 0.80(SE =+-0.17)*LOG(X)	Y=EXCR (UG-ATOM N/H),X=BODY PROT (MG),299 K, TROP MARINE,R=0.81,N=	124
*ZOOPLANKTON, CORELLA SPECTABILIS & GLEBA CORDATA	LOG(Y) =-1.68(SE=+-0.28) + 0.86(SE =+-0.24)*LOG(X)	Y=EXCR (UG-ATOM N/H),X=BODY PROT (MG),299 K, TROP MARINE,R=0.84,N=	124
*ZOOPLANKTON, DIPHYES DISPAR	LOG(Y) =-1.39(SE=+-0.08) + 0.68(SE =+-0.09)*LOG(X)	Y=EXCR (UG-ATOM N/H),X=BODY PROT (MG),299 K, TROP MARINE,R=0.94,N=	124
*ZOOPLANKTON, EURHAMPHEA VEXILLIGERA & CESTUM SP.	LOG(Y) =-1.22(SE=+-0.03) + 1.03(SE =+-0.06)*LOG(X)	Y=EXCR (UG-ATOM N/H),X=BODY PROT (MG),299 K, TROP MARINE,R=0.99,N=	124
*ZOOPLANKTON, FORSKALIA EDWARDSI & F. THALOIDES	LOG(Y) =-1.20(SE=+-0.04) + 1.10(SE =+-0.08)*LOG(X)	Y=EXCR (UG-ATOM N/H),X=BODY PROT (MG),299 K, TROP MARINE,R=0.96,N=	124
*ZOOPLANKTON, NANOMIA BIJUGA	LOG(Y) =-1.13(SE=+-0.07) + 0.71(SE =+-0.14)*LOG(X)	Y=EXCR (UG-ATOM N/H),X=BODY PROT (MG),299 K, TROP MARINE,R=0.91,N=	124
*ZOOPLANKTON, PEGEA CONFEDERATA (AGGREGATE)	LOG(Y) =-1.38(SE=+-0.08) + 1.22(SE =+-0.17)*LOG(X)	Y=EXCR (UG-ATOM N/H),X=BODY PROT (MG),299 K, TROP MARINE,R=0.89,N=	124
*ZOOPLANKTON, PTEROTRACHEA HIPPOCAMPUS	LOG(Y) =-0.76(SE=+-0.11) + 0.77(SE =+-0.30)*LOG(X)	Y=EXCR (UG-ATOM N/H),X=BODY PROT (MG),299 K, TROP MARINE,R=0.72,N=	124
*ZOOPLANKTON, ROSACEA CYMBIFORMIS	LOG(Y) =-1.43(SE=+-0.08) + 0.95(SE =+-0.19)*LOG(X)	Y=EXCR (UG-ATOM N/H),X=BODY PROT (MG),299 K, TROP MARINE,R=0.84,N=	124
*ZOOPLANKTON, SALPA CYLINDRICA (SOLITARY)	LOG(Y) =-0.85(SE=+-0.13) + 0.64(SE =+-0.24)*LOG(X)	Y=EXCR (UG-ATOM N/H),X=BODY PROT (MG),299 K, TROP MARINE,R=0.71,N=	124
*ZOOPLANKTON, SULCULEOLARIA QUADRIVALVIS	LOG(Y) =-0.91(SE=+-0.13) + 0.68(SE =+-0.09)*LOG(X)	Y=EXCR (UG-ATOM N/H),X=BODY PROT (MG),299 K, TROP MARINE,R=0.77,N=	124

TABLE E47

EQUATION : EXCRETION OF PHOSPHORUS

ITEM:	VALUE:	CONDITIONS:	REF.:
*DAPHNIA ROSEA	E=0.0286*E**(0.0387*T+ 0.00001*D-3.34*P)*W**(-0.383)	E=REL.RATE(UG P/MG*HR),T=TEMP.(C), D=FOOD CONC.(CELL/ML),	1013
*DAPHNIA ROSEA	E=0.0286*E**(0.0387*T+ 0.00001*D-3.34*P)*W**(-0.383)	P=FOOD-P(UG P/ML),W=DRY WT.(MG), FOOD=RHODOTORULA	1013
*EPILEMNETIC ZOOPLANKTON	R=4.5*W**(-0.69)	R=RELEASE(UG AT.P/G*HR),W=DRY WT.(UG/INDIV.),293K,EPILIMNION	130
*EPILEMNETIC ZOOPLANKTON	R=2.3*W**(-0.45)	R=RELEASE(UG AT.P/G*HR),W=DRY WT.(UG/INDIV.),278K,HYPOLIMNION	130
*HYPOLEMNETIC ZOOPLANKTON	R=22.7*W**(-0.92)	R=RELEASE(UG AT.P/G*HR),W=DRY WT.(UG/INDIV.),293K,EPILIMNION	130
*HYPOLEMNETIC ZOOPLANKTON	R=7.2*W**(-0.99)	R=RELEASE(UG AT.P/G*HR),W=DRY WT.(UG/INDIV.),278K,HYPOLIMNION	130
*MARINE ZOOPLANKTON	LOG(Y) = 0.405 + 0.915*LOG(X)	Y=PO4 EXC (NG-ATOM/H*ANIMAL), X= DW. (MG/ANIM),N PACIFIC,298 K, MIXED	1080
*ZOOPLANKTON	Y = -3.1 + 0.01481*X	Y= EXC (UG-ATOM P/(G*H)), X=RESP (UG-ATOM O2/(G*H)), MARINE	1384

TABLE E47

EQUATION : EXCRETION OF PHOSPHORUS

ITEM:	VALUE:	CONDITIONS:	REF.:
*ZOOPLANKTON	E=0.252*W**(-0.33)	E=RELEASE RATE(UG P/MG*HR),W=DRY WT.(MG)	1013
*ZOOPLANKTON	E=0.157*W**(-0.383)	E=REL.RATE(UG/MG*HR),W=DRY WT.(MG) ,293K,FOOD CONC.=10**5CELLS/ML,	1013
*ZOOPLANKTON	E=0.157*W**(-0.383)	FOOD-P=0.021UG P/ML,FOOD= RHODOTORULA(YEAST)	1013
*ZOOPLANKTON	E=0.042*W**(-0.383)	283K,FOOD CONC.=10**5CELLS/ML, FOOD-P=0.021UG P/ML	1013

TABLE E48

EQUATION : EXTRACELLULAR RELEASE AS A FUNCTION OF CHLOROPHYLL A CONCENTRATION

ITEM:	VALUE:	CONDITIONS:	REF.:
*SKELETONEMA COSTATUM	LOG(Y) = 2.83 - 0.016*X	Y=EX. RELEASE (% OF PHOTOSYNTHESIS), X=CHL A (MG/M3), R=0.95	717

TABLE E49

EQUATION : FECUNDITY

ITEM:	VALUE:	CONDITIONS:	REF.:
*CARPIODES CYPRINUS	F = 7172.5 + 5089.9*T	F= NO. OF OVA/ANIMAL, T = AGE (Y), FOUR MILE CREEK, OHIO	1745
*CARPIODES CYPRINUS	F = -15690.7 + 169.1*W - 0.1*W**2	F= NO. OF OVA/ANIMAL, L=WET W. (G), FOUR MILE CREEK, OHIO	1745
*CARPIODES CYPRINUS	F = -58755 + 2984.5*L	F= NO. OF OVA/ANIMAL, L=LENGTH (MM), FOUR MILE CREEK, OHIO	1745
*PROCOPIUM WILLIAMSONI	Y = 4347.6 + 50.7*(X - 301.5)	Y=EGGS/FISH, X=FORK LENGTH (MM), SHEEP RIVER, ALBERTA	1586
*UPENEUS MOLUCCENSIS	F = 20040*(L - 11.43)	F= NO. OF EGGS/ANIM, L=FORK L. (CM), CHINA SEA	900

TABLE E50

EQUATION : FILTRATION RATE

ITEM:	VALUE:	CONDITIONS:	REF.:
*ACARTIA SP.	Y = 0.00201*W**1.011	Y=MEAN FILT. RATE (ML/(ANIM*H)), W =DRY W. (UG), CHEASEPEAKE BAY,R= .79	14
*ACARTIA SP.	Y = 0.049*W**1.213	Y= MAX FILT. RATE (ML/(ANIM*H)), W =DRY W. (UG), CHEASEPEAKE BAY,R= .91	14
*ARCTICA ISLANDICA	F = 1.602*W**0.66	F=ML/H, W=DRY W. (G), 285 K, ALGAE, 0.2-10 G	1696
*CERIODAPHIA RETICULATA	LOG(Y) = 2.44 - 0.734*LOG(X)	Y= FILT (ML/(H*ANIM),X=NUMBER OF PART/ML (3-24MY),298 K,ANIM=0.8 MM	1148
*CRASSOSTREA GIGAS	F = 6.714*W**0.25	F=ML/H, W=DRY W. (G), 293-295 K, SESTON, 0.5-2 G	1651
*CRASSOSTREA GIGAS	F = 10.392*W**0.27	F=ML/H, W=DRY W. (G), 288 K, SESTON, 0.5-10 G G	1651
*EURYTEMORA AFFINIS	Y = -0.0043 + 0.0171*W	Y=MEAN FILT. RATE (ML/(ANIM*H)), W =DRY W. (UG), CHEASEPEAKE BAY,R= .97	14
*EURYTEMORA AFFINIS	Y = -0.0077 + 0.0373*W	Y= MAX FILT. RATE (ML/(ANIM*H)), W =DRY W. (UG), CHEASEPEAKE BAY,R= .90	14

TABLE E50

EQUATION : FILTRATION RATE			
ITEM:	VALUE:	CONDITIONS:	REF.:
*LANICE CONCHILEGA	F = 0.88*W**0.32	F=FILT (ML/H), W=DRY W. (MG), 285 K, R=0.96, FOOD=40*10**6 CELL/L	193
*LANICE CONCHILEGA	F = 1.07*W**0.52	F=FILT (ML/H), W=DRY W. (MG), 285 K, R=0.96, FOOD=40*10**6 CELL/L, MAX	193
*MODIOLUS MODIOLUS	F = 0.929*W**0.74	F=ML/H, W=DRY W. (G), 285 K, ALGAE, 0.43-4 G	1696
*MYTILUS CALIFORNIANUS	F = 2.270*W**0.29	F=ML/H, W=DRY W. (G), 293 K, GRAPHITE, 3-12 G	1296
*MYTILUS CALIFORNIANUS	F = 1.162*W**0.37	F=ML/H, W=DRY W. (G), 289 K, GRAPHITE, 2-12 G	1296
*MYTILUS CALIFORNIANUS	F = 0.599*W**0.21	F=ML/H, W=DRY W. (G), 283 K, GRAPHITE, 0.3-12 G	1296
*MYTILUS CALIFORNIANUS	F = 0.604*W**0.56	F=ML/H, W=DRY W. (G), 283 K, GRAPHITE, 0.3-2 G	1296
*MYTILUS CALIFORNIANUS	F = 1.835*W**(-0.39)	F=ML/H, W=DRY W. (G), 283 K, GRAPHITE, 3-12 G	1296
*MYTILUS CALIFORNIANUS	F = 2.069*W**0.27	F=ML/H, W=DRY W. (G), 289 K, GRAPHITE, 2-12 G	1296
*MYTILUS CALIFORNIANUS	F = 2.138*W**0.48	F=ML/H, W=DRY W. (G), 294 K, GRAPHITE, 0.5-12 G	1413
*MYTILUS CALIFORNIANUS	F = 2.156*W**0.80	F=ML/H, W=DRY W. (G), 294 K, GRAPHITE, 0.5-12 G	1413
*MYTILUS CALIFORNIANUS	F = 3.962*W**0.13	F=ML/H, W=DRY W. (G), 294 K, GRAPHITE, 3-12 G	1413
*MYTILUS CALIFORNIANUS	F = 0.527*W**0.27	F=ML/H, W=DRY W. (G), 279 K, GRAPHITE, 0.5-12 G	1413
*MYTILUS CALIFORNIANUS	F = 0.511*W**0.53	F=ML/H, W=DRY W. (G), 279 K, GRAPHITE, 0.5-2 G	1413
*MYTILUS CALIFORNIANUS	F = 0.940*W**(-0.06)	F=ML/H, W=DRY W. (G), 279 K, GRAPHITE, 3-12 G	1413
*MYTILUS EDULIS	F = 3.846*W**0.25	F=ML/H, W=DRY W. (G), 283 K, SESTON, 0.5-4 G	1651
*MYTILUS EDULIS	F = 4.788*W**0.26	F=ML/H, W=DRY W. (G), 283 K, SESTON, 0.5-2 G	1651
*MYTILUS EDULIS	F = 4.212*W**0.11	F=ML/H, W=DRY W. (G), 293-295 K, SESTON, 0.5-2 G	1651
*MYTILUS EDULIS	F = 3.046*W**0.34	F=ML/H, W=DRY W. (G), 293-295 K, SESTON, 0.2-3.2 G	1651
*MYTILUS EDULIS	F = 2.214*W**0.27	F=ML/H, W=DRY W. (G), 0.28-2.9 G	1696
*MYTILUS EDULIS	F = 1.656*W**0.49	F=ML/H, W=DRY W. (G), 288 K, GRAPHITE, 0.058-1.7 G	1572
*MYTILUS EDULIS	F = 2.410*W**0.74	F=ML/H, W=DRY W. (G), 285 K, 20*10**6 CELL/L, DUNALIELLA, 0.003-1 G	1738
*MYTILUS EDULIS	F = 1.313*W**0.73	F=ML/H, W=DRY W. (G), 285 K, 40*10**6 CEEL/L, DUNALIELLA, 0.003-1 G	1738
*MYTILUS EDULIS	F = 1.273*W**0.38	F=ML/H, W=DRY W. (G), 285-288 K, SILT, 0.7-3.1 G	1713
*OSTREA EDULIS	F = 8.796*W**0.23	F=ML/H, W=DRY W. (G), 293-295 K, SESTON, 0.5-2 G	1651
*PECTEN IRRADIANS	F = 4.742*W**0.82	F=ML/H, W=DRY W. (G), 297 K,ALGAE, 0.66-4.12 G	247
*VENERUPIS DECUSSATA	F = 4.560*W**0.28	F=ML/H, W=DRY W. (G), 293-295 K, SESTON, 0.5-1 G	1651
*VENUS MERCENARIA	F = 4.530*W**0.18	F=ML/H, W=DRY W. (G), 293-295 K, SESTON, 0.5-2 G	1651

TABLE E51

EQUATION : FILTRATION RATE AS A FUNCTION OF DISSOLVED OXYGEN			
ITEM:	VALUE:	CONDITIONS:	REF.:
*DAPHNIA GALEATA MENDOTAE	F = -0.019(+-0.077) + 0.169(+-0.016)*X	F=FILT. (ML/ANIM*H)),X=O2 (MG/L), R=0.98	651

TABLE E51

EQUATION : FILTRATION RATE AS A FUNCTION OF DISSOLVED OXYGEN			
ITEM:	VALUE:	CONDITIONS:	REF.:
*DAPHNIA MAGNA	F = -0.229 (+-0.091) + 0.554(+ -0.069)*X	F=FILT. (ML/ANIM*H)),X=O2 (MG/L), O2<3.0, SMALL ANIM=1.54 MM, R= 0.96	651
*DAPHNIA MAGNA	F = 1.015(+-0.122) + 0.032(+ -0.022)*X	F=FILT. (ML/ANIM*H)),X=O2 (MG/L), O2>3.0, SMALL ANIM=1.54 MM, R= 0.63	651
*DAPHNIA MAGNA	F = -1.036(+-0.386) + 2.259(+ -0.303)*X	F=FILT. (ML/ANIM*H)),X=O2 (MG/L), O2<3.0, LARGE ANIM=2.34 MM, R= 0.93	651
*DAPHNIA MAGNA	F = 3.482(+-0.772) + 0.101(+ -0.136)*X	F=FILT. (ML/ANIM*H)),X=O2 (MG/L), O2>3.0, LARGE ANIM=2.34 MM, R= 0.39	651

TABLE E52

EQUATION : FOOD INTAKE			
ITEM:	VALUE:	CONDITIONS:	REF.:
*ANIMALS	D(R)/DT = H*F*W**2/3	R=FOOD CON., W=BODY WEIGHT, F= FEEDING LEVEL(0-1), H=CONST.	111
*CALANUS AND THALASSIOSIRA	Y = 12.330*(1-EXP(-0.000843* (X-670)))	Y=INGESTION (10**3 CELL/(ANIM*H)), X=FOOD (10**3 CELL/ML)	494
*CALANUS AND THALASSIOSIRA	Y = 13.800*(X-767)/(785+(X-767)	Y=INGESTION (10**3 CELL/(ANIM*H)), X=FOOD (10**3 CELL/ML)	494
*CALANUS AND THALASSIOSIRA	Y = 4.07*X - 622	Y=INGESTION (10**3 CELL/(ANIM*H)), X=FOOD (10**3 CELL/ML), X<3300	494
*CALANUS AND THALASSIOSIRA	Y = -0.154*X + 13100	Y=INGESTION (10**3 CELL/(ANIM*H)), X=FOOD (10**3 CELL/ML), X>3300	494
*CALANUS HELGOLANDICUS, ADULT	I = 22.45*(1 - EXP(-0.173*P))	I=DAILY INGESTION, P=FOOD CONC., FOOD= ELINIUS MODESTUS STAGE II N	289
*CALANUS ON CENTRIC SP.	Y = 347*(1-EXP(-0.0583*(X-3.08)))	Y=INGESTION (10**3 CELL/(ANIM*H)), X=FOOD (10**3 CELL/ML)	494
*CALANUS ON CENTRIC SP.	Y = 383*(X-3.40)/((12.2 + (X-3.40)))	Y=INGESTION (10**3 CELL/(ANIM*H)), X=FOOD (10**3 CELL/ML)	494
*CALANUS ON CENTRIC SP.	Y = 9.92*X - 5.32	Y=INGESTION (10**3 CELL/(ANIM*H)), X=FOOD (10**3 CELL/ML), X<40	494
*CALANUS ON CENTRIC SP.	Y = 0.0058*X + 331	Y=INGESTION (10**3 CELL/(ANIM*H)), X=FOOD (10**3 CELL/ML), X>40	494
*CALLIOPIUS LACVIUSCULUS	I = 0.847*WC**0.748	I = INGESTION (UG CARBON/(ANIM*D)) , WC=BODY W. (UG CARBON/ANIM),281 K	319
*CALLIOPIUS LACVIUSCULUS	I = 0.525*WN**0.792	I = INGESTION (UG NITROGEN/ (ANIM*D)), WN=BODY W. (UG N/ANIM), 281 K	319
*CALLIOPIUS LACVIUSCULUS	I = 0.877*WC**0.784	I = INGESTION (UG CARBON/(ANIM*D)) , WC=BODY W. (UG CARBON/ANIM),281 K	319
*CALLIOPIUS LACVIUSCULUS	I = 0.394*WN**0.888	I = INGESTION (UG NITROGEN/ (ANIM*D)), WN=BODY W. (UG N/ANIM), 285 K	319
*CALLIOPIUS LACVIUSCULUS	I = 2.050*WC**0.695	I = INGESTION (UG CARBON/(ANIM*D)) , WC=BODY W. (UG CARBON/ANIM),281 K	319
*CALLIOPIUS LACVIUSCULUS	I = 1.031*WN**0.744	I = INGESTION (UG NITROGEN/ (ANIM*D)), WN=BODY W. (UG N/ANIM), 288 K	319
*CLUPEA HARENGUS	R = 1.465*W**0.744	R=FOOD INTAKE (G/WET WEIGHT*WEEK), W=BODY WET W.(G), 277.5 K	350
*DAPHNIA ROSEA	I=F*P/W	I=ING.RATE(UG P/MG*HR),F= FILTRAT.RATE(ML/ANIMAL*HR),	1013
*DAPHNIA ROSEA	I=F*P/W	P=P-CONC.IN FOOD(UG P/ML),W=DRY WT.(MG),FOOD=RHODOTORULA	1013
*HERBIVORES, OMNIVORES AND CARNIVORES	R = (P+T)/U	R=FOOD REQ.,1/U=EFFIC. OF ASSIM.P= MAX RATE OF PROD.,T=RATE OF EXCHAH.	1636
*TORTANUS DISCAUDATUS	I=38.5*(1-E**(-0.025*P))	FOOD=CALANUS PACIFICUS(NAU.3),I= INGEST.RATE(PREY ATTACKED),	20

TABLE E52

EQUATION : FOOD INTAKE			
ITEM:	VALUE:	CONDITIONS:	REF.:
*TORTANUS DISCAUDATUS	I=23.0*(1-E**(-0.064*P))	FOOD=CALANUS PACIFICUS(NAU.5),P= NUMBER OF PREY/L	20
*TORTANUS DISCAUDATUS	I=32.2*(1-E**(-0.046*D))	FOOD=CALANUS PACIFICUS(NAU.5),D=UG PREY (DRY WT.)/L	20
*TORTANUS DISCAUDATUS	I=23.1*(1-E**(-0.064*D))	FOOD=CALANUS PACIFICUS(NAU.3),D=UG PREY (DRY WT.)/L	20
*ZOOPLANKTON	I=IM*(1-E**(-K*P))	I=INGEST.RATE,P=PREY DENSITY,IM= MAX.INGEST.RATE,K=CONSTANT	731

TABLE E53

EQUATION : FORK LENGTH			
ITEM:	VALUE:	CONDITIONS:	REF.:
*MICROPTERUS SALMOIDES	FL = 16.615 + 0.699*S	FL=FORK LENGTH, S=ANT. SCALE RADIUS*90, APPLAGATE, CALIF	1404
*MICROPTERUS SALMOIDES	FL = 23.292 + 0.648*S	FL=FORK LENGTH, S=ANT. SCALE RADIUS*90, POTTER, CALIF	1404
*MICROPTERUS SALMOIDES	FL = 19.023 + 0.678*S	FL=FORK LENGTH, S=ANT. SCALE RADIUS*90, REUTER L., CALIF	1404
*MICROPTERUS SALMOIDES	SL = 1.7 + 0.861*FL	SL=STANDARD LENGTH, FL= FORK LENGTH	1404
*MICROPTERUS SALMOIDES	TL = 1.6 + 1.034*FL	TL=TOTAL LENGTH, FL=FORK LENGTH, CALIF	1404
*TRACHURUS JAPONICUS	L = 149*R + 0.04	L=FORK LENGTH (CM), R=ANT. SCALE RADIUS (MM), WATERS	33

TABLF E54

EQUATION : FRESHWATER AND SEDIMENTS			
ITEM:	VALUE:	CONDITIONS:	REF.:
*CHLOROPHYLL DERIVATES VS ALGAL STANDING CROP	LOG(Y) = 0.316*LOG(X) + 0.84	Y=CHL DERIV (UNITS/G ORG CARBON),X = ALGAE ST CROP (UG DW/L), R=0.83	554
*SEDIMENTS CLHOROPHYLL DERIV VS CHLOROPHYLL A	LOG(Y) = 0.59*LOG(X) + 0.36	Y=SED CHL DERIV (UNITS/G ORG C),X = EPILIMNIC CHL A (UG/L),R=0.81, UK	554
*TOTAL CAROTENOIDS VS ALGAL STANDING CROP	Y = 5.01 + 4.33*X	Y=CAROT (UNITS/G ORG CARBON),X = ALGAE ST CROP (UG DW/L), R=0.9, UK	554
*TOTAL CAROTENOIDS VS DISSOLVED CALCIUM	Y = 94.9*X - 10.0	Y=CAROT (UNITS/G ORG CARBON),X = CALCIUM (U EQ/L), R=0.93, UK	554
*TOTAL SEDIMENTARY SULFUR VS ALGAE STANDING CROP	Y = 0.088 + 0.086*X	Y=SED SUL (% OF DW),X=ALGAE BIOMASS (UG DW/L),R=0.76, UK	554
*TOTAL SEDIMENTARY SULFUR VS DISSOLVED SULFATE	Y = 3.61*X - 0.258	Y=SED SULFUR (% OF DRY WEIGHT),X= DIS SULFATE(U EQ/L),R=0.79, UK	554

TABLE E55

EQUATION : FRESHWATER SEDIMENTS			
ITEM:	VALUE:	CONDITIONS:	REF.:
*ELEMENTAL HYDROGEN VS ORGANIC CARBON	LOG(Y) = 0.528*LOG(X) - 0.333	Y=HYDROGEN,X=ORG CARBON,UNITS= % OF DW, UK	554
*EPIPHASIC VS HYPOPHASIC CAROTENOIDS	LOG(Y) = 1.04*LOG(X) - 0.24	Y=EPIPHASIC,X=HYPOPHASIC,UNITS = UNITS/G ORG C, R=0.98, UK	554
*LOSS ON IGNITION VS ORGANIC CARBON	Y = 3.47 + 1.78*X	Y=LOSS ON IGNITION,X=ORG CARBON, UNITS= % OF DRY WEIGHT,R=0.99, UK	554

TABLE E55

EQUATION : FRESHWATER SEDIMENTS			
ITEM:	VALUE:	CONDITIONS:	REF.:
*NITROGEN VS ORGANIC CARBON	Y = 0.044 + 0.0756*X	Y=NITROGEN,X=ORG CARBON,UNITS=% OF DW,R=0.97, UK	554
*TOTAL CAROTENOIDS VS CHLOROPHYLL DERIVATIVES	LOG(Y) = 1.34*LOG(X) - 0.10	Y=CAROT.,X=CHLA DERIV.,UNITS = UNITS/G ORG CARBON, R=0.93, UK	554
*TOTAL SULFUR VS CAROTENOIDS	Y = 56*X - 0.8	Y=CAROT (UNITS/G ORG CARBON),X = SULFUR (% DW), R=0.85, UK	554

TABLE E56

EQUATION : GROWTH AS A FUNCTION OF DEGREE DAYS (F051) (F054)			
ITEM:	VALUE:	CONDITIONS:	REF.:
*TREES	GR = 4*(D-DMIN)*(DMAX-D)/ (DMAX-DMIN)**2	GR=RELATIVE GROWTH,D=DG ON GROWING PLACE,DMAX=DG ON SOUTH. BOUNDARY	152
*TREES	GR = 4*(D-DMIN)*(DMAX-D)/ (DMAX-DMIN)**2	DMIN=DG ON NOTHERN BOUNDARY, NOTHERN HEMISPHERE	152

TABLE E57

EQUATION : GROWTH EQUATION (F054)			
ITEM:	VALUE:	CONDITIONS:	REF.:
*BACTERIA IN SUSPENSION	MY=LN(2)/T(D)	MY=SPECIFIC GROWTH RATE,T(D)= DOUBLING TIME	310
*BACTERIA IN SUSPENSION	MY=MY(M)*(S/(K(S)+S))	MY(M)=GROWTH RATE CONST.,S= SUBSTRATE CONC.,K(S)=SATURATION CONST.	310
*BACTERIA IN SUSPENSION	DX/DT=-Y*DS/DT	X = CONC OF ORGANISMS, T = TIME, S = SUBSTRATE CONC, Y = YIELD	310
*SULFATE REDUCING BACTERIA	R = -(K*S*X)/(KS+S)	R=VOLUMETRIC REACTION RATE(MG/L*H) ,K=MAX SPECIFIC RATE OF UTILIZATION	1034
*SULFATE REDUCING BACTERIA	R = -(K*S*X)/(KS+S)	(1/D),S=SUBSTRATE CONC (MG/L),X= BAC CONC(MG/L),KS=HALF VELOSITY COE-	1034
*SULFATE REDUCING BACTERIA	R = -(K*S*X)/(KS+S)	FFICIENT (MG/L)	1034

TABLE E58

EQUATION : GROWTH RATE			
ITEM:	VALUE:	CONDITIONS:	REF.:
*ANIMALS	D(W)/DT = H*W**(2/3) - K*W	W=BODY WEIGHT, H AND K = CONSTANTS	111
*CALLIOPIUS LACVIUSCULUS	GR = 0.112*WC*(1 - (WC/3927.3) **0.599)	GR = UG CARBON/D, WC=BODY WEIGHT (UG CARBON/ANIM), 281 K	319
*CALLIOPIUS LACVIUSCULUS	GR = 0.320*WN*(1 - (WN/962.1) **0.200)	GR = UG NITROGEN/D, WN=BODY WEIGHT (UG NITROGEN/ANIM), 281 K	319
*CALLIOPIUS LACVIUSCULUS	GR = 0.190*WC*(1 - (WC/3266.6) **0.400)	GR = UG CARBON/D, WC=BODY WEIGHT (UG CARBON/ANIM), 285 K	319
*CALLIOPIUS LACVIUSCULUS	GR = 0.270*WN*(1 - (WN/736.5) **0.200)	GR = UG NITROGEN/D, WN=BODY WEIGHT (UG NITROGEN/ANIM), 285 K	319
*CALLIOPIUS LACVIUSCULUS	GR = 0.237*WC*(1 - (WC/2562.8) **0.599)	GR = UG CARBON/D, WC=BODY WEIGHT (UG CARBON/ANIM), 288 K	319
*CALLIOPIUS LACVIUSCULUS	GR = 0.182*WN*(1 - (WN/586.5) **0.578)	GR = UG NITROGEN/D, WN=BODY WEIGHT (UG NITROGEN/ANIM), 288 K	319
*EICHHORNIA CRASSIPES	Y = 1 - 0.0037*(T - 29)**2	Y=GROWTH RATE (1/D), T=CEL DEG., IN MARSH	1049

TABLE E58

EQUATION : GROWTH RATE			
ITEM:	VALUE:	CONDITIONS:	REF.:
*HYPERA POSTICA	RT = A/(B+C*EXP(-D*T(TIME))	RT=RATE OF DEVELOPMENT,T(TIME)= TEMP AT TIME=TIME,A,B,C,D= CONSTANTS	1432

TABLE E59

EQUATION : HALF LIFE TIME OF PARATHION AS A FUNCTION OF TEMPERATURE			
ITEM:	VALUE:	CONDITIONS:	REF.:
*PARATHION	T(1/2) = 5((10**3)/K-2.8)+1	K=KELVIN DEGREES,STERILE ESTUARINE WATER, PH = 7.8, SALINITY = 2.57 %	1668

TABLE E60

EQUATION : HALF-RATE COEFFICIENT AS A FUNCTION OF TEMPERATURE			
ITEM:	VALUE:	CONDITIONS:	REF.:
*SULFATE REDUCING BACTERIA	KS = (1.65)*(10**(-39))*(10** (12082/T))	T=ABS TEMP,KS=HALF-VEL COEF.(MG/L) ,ARRHENIUS EQUATION	1034

TABLE E61

EQUATION : HATCHING TIME OF EGGS AS A FUNCTION OF TEMPERATURE			
ITEM:	VALUE:	CONDITIONS:	REF.:
*CALANUS FINMARCHICUS	D = 2232/((T+8)**1.536)	D=TIME (HOURS), T= TEMP (C), 273-293 K	1006

TABLE E62

EQUATION : HG			
ITEM:	VALUE:	CONDITIONS:	REF.:
*FERSHWATER FISH	C=K*W**0.81	FOR FISH=42CM->1.25 MG/KG; 49->1.5.MG/KG,C=HG,CONC,W=LENGTH OF FISH	1183
*HOMO SAPIENSS	Y=0.26X	Y=HG IN BLOOD(UG/L),X=HG IN URINE(UG/L)	486
*SCORPAENA PORCUS	Y = 1.39*10**(-2)*X + 1.21	Y=HG CONC (MG/KG W.W.), X= WET WEIGHT, R=0.88, MARINE, MEDITAR.	1317
*SPECIFIC METHYLATION RATE	RATE = A*B**N*(HG(TOTAL))**N (MICROGRAM/G VSS DAY)	ANAEROBIC,N=0.15,A=0.61,B=RATIO FREE HG TO TOTAL INORGANIC HG, 303K	132
*SPECIFIC METHYLATION RATE	RATE = A*B**N*(HG(TOTAL))**N (MICROGRAM/G VSS DAY)	AEROBIC,N=0.28,A=2.,B=RATIO FREE HG TO TOTAL INORGANIC HG,303K	132

TABLE E63

EQUATION : HOME RANGE SIZE AS A FUNCTION OF BODY WEIGHT			
ITEM:	VALUE:	CONDITIONS:	REF.:
*BIRDS	A = 11.63*W**1.16	A=AREA (M2),W = WET WEIGHT (G)	1617

TABLE E63

EQUATION : HOME RANGE SIZE AS A FUNCTION OF BODY WEIGHT			
ITEM:	VALUE:	CONDITIONS:	REF.:
*CARNIVORES MAMMALS	LOG(Y) = 5.4 + 0.397*X	Y=AREA (ACRES), X=WET WEIGHT (KG)	1008
*HERBIVORES MAMMALS	LOG(Y) = 0.054 + 0.397*X	Y=AREA (ACRES), X=WET WEIGHT (KG)	1008
*LIZARDS	A = 171.4*W**0.95	A=AREA (M2),W = WET WEIGHT (G)	1617
*MAMMALS	A = 352.3*W**0.63	A=AREA (M2),W = WET WEIGHT (G)	1617

TABLE E64

EQUATION : HYPOLEMNETIC OXYGEN DEFICIT VS SECCHI DEPTH			
ITEM:	VALUE:	CONDITIONS:	REF.:
*20 FRESHWATER LAKES	LOG(Y)=-1.37*LOG(X)-0.65	R=-0.85,Y=OXYGEN DEFIC.(MG/L),X= AV.SECCHI DEPTH(M)	881

TABLE E65

EQUATION : INTRACELLULAR NITROGEN CONC. AS A FUNCTION OF SPECIFIC GROWTH RATE			
ITEM:	VALUE:	CONDITIONS:	REF.:
*CHAETOCEROS GRACILIS	IN = 0.068(+-0.007) + 0.239(+ -0.013)*MY	IN=INTRA CELL CONC (MG N/MG C),MY =GROWTH (1/HOUR), NH4, 298 K	1146
*CHLORELLA PYRENOIDOSA	IN = 0.043(+-0.002) + 0.239(+ -0.016)*MY	IN=INTRA CELL CONC (MG N/MG DW),MY =GROWTH (1/HOUR), NO3, 308 K	1146
*COCCOCHLORIS STAGINA	IN = 0.065(+-0.003) + 1.78(+-0.10) *MY	IN=INTRA CELL CONC (MG N/MG C),MY =GROWTH (1/HOUR), NO3, 298 K	1146
*CYCLOTELLA NANA	IN = 0.054(+-0.012) + 1.21(+-0.23) *MY	IN=INTRA CELL CONC (MG N/MG C),MY =GROWTH (1/HOUR), NO3, 298 K	1146
*DUNALIELLA TERTIOLECTA	IN = 0.023(+-0.009) + 1.76(+-0.25) *MY	IN=INTRA CELL CONC (MG N/MG C),MY =GROWTH (1/HOUR), NH4, 298 K	1146
*MONOCHRYSIS LUTHERI	IN = 0.042(+-0.003) + 0.78(+-0.09) *MY	IN=INTRA CELL CONC (MG N/MG C),MY =GROWTH (1/HOUR), NH4, 298 K	1146
*OSCILLATORIA AGARDHII	IN = 0.040(+-0.006) + 1.45(+-0.24) *MY	IN=INTRA CELL CONC (MG N/MG DW),MY =GROWTH (1/HOUR), NO3, 293 K	1146

TABLE E66

EQUATION : LC50 OF ZINC			
ITEM:	VALUE:	CONDITIONS:	REF.:
*SALMO GAIRDNERII	LOG(ZINC) = - 0.9614 + 0.6053*LOG(TOTAL HARDNESS(AS CALCIUMCARBO.)	48H LC50 LEVEL	940

TABLE E67

EQUATION : LEAF EXTENSION AS A FUNCTION OF TEMPERATURE			
ITEM:	VALUE:	CONDITIONS:	REF.:
*FESTUCA ARUNDINACEA	Y=0.86*T-3.87	Y=LEAF EXT.(MM/DAY),T=TEMP. (DEG.CELCIUS),R=0.93,VEGETATIVE PHASE	1233
*FESTUCA ARUNDINACEA	Y=1.78*T-6.20	Y=LEAF EXT.(MM/DAY),T=TEMP. (DEG.CELCIUS),R=0.83,REPRODUCTIVE PHASE	1233

TABLE E67

EQUATION : LEAF EXTENSION AS A FUNCTION OF TEMPERATURE

ITEM:	VALUE:	CONDITIONS:	REF.:
*LOLIUM MULTIFLORUM	Y=0.64*T+1.11	Y=LEAF EXT.(MM/DAY),T=TEMP. (DEG.CELCIUS),R=0.87,VEGETATIVE PHASE	1233
*LOLIUM MULTIFLORUM	Y=2.01*T-2.84	Y=LEAF EXT.(MM/DAY),T=TEMP. (DEG.CELCIUS),R=0.59,REPRODUCTIVE PHASE	1233
*LOLIUM PERENNE	Y=0.75*T-1.13	Y=LEAF EXT.(MM/DAY),T=TEMP. (DEG.CELCIUS),R=0.90,VEGETATIVE PHASE	1233
*LOLIUM PERENNE	Y=1.75*T-8.21	Y=LEAF EXT.(MM/DAY),T=TEMP. (DEG.CELCIUS),R=0.83,REPRODUCTIVE PHASE	1233
*LOLIUM PERENNE	Y=-2.16+3.19*EXP(0.14*T)	Y=LEAF EXT.(MM/D),T=DEG.CEL.,AIR TEMP.5CM ABOVE SOIL SURFACE,R= 0.95	1232
*LOLIUM PERENNE	Y=-0.15+1.26*EXP(0.23*T)	Y=LEAF EXT.(MM/D),T=DEG.CEL.,SOIL SURFACE TEMPERATURE,R=0.96	1232
*LOLIUM PERENNE	Y=0.89+0.59*EXP(0.32*T)	Y=LEAF EXT.(MM/D),T=DEG.CEL.,SOIL TEMP.2CM BELOW SURFACE,R=0.96	1232
*LOLIUM PERENNE	Y=0.69+0.63*EXP(0.32*T)	Y=LEAF EXT.(MM/D),T=DEG.CEL.,SOIL TEMP.10CM BELOW SURFACE,R=0.93	1232
*LOLIUM PERENNE	Y=-3.8+4.6*EXP(0.13*T)	Y=LEAF EXT.(MM/D),T=DEG.CEL., HEATED GROUND,5CM AB.SOIL SURF.,R= 0.92	1231
*LOLIUM PERENNE	Y=-3.8+4.2*EXP(0.15*T)	Y=LEAF EXT.(MM/D),T=DEG.CEL., HEATED GROUND,2.5CM AB.SOIL SURF., R=0.94	1231
*LOLIUM PERENNE	Y=-3.8+3.6*EXP(0.17*T)	Y=LEAF EXT.(MM/D),T=DEG.CEL., HEATED GROUND,SOIL SURFACE,R=0.95	1231
*LOLIUM PERENNE	Y=-5.0+4.7*EXP(0.16*T)	Y=LEAF EXT.(MM/D),T=DEG.CEL., HEATED GROUND,2CM BEL.SOIL SURF.,R =0.96	1231
*LOLIUM PERENNE	Y=-5.0+4.2*EXP(0.17*T)	Y=LEAF EXT.(MM/D),T=DEG.CEL., HEATED GROUND,10CM BEL.SOIL SURF., R=0.95	1231
*PHLEUM PRATENSE	Y=0.52*T-0.76	Y=LEAF EXT.(MM/DAY),T=TEMP. (DEG.CELCIUS),R=0.88,VEGETATIVE PHASE	1233
*PHLEUM PRATENSE	Y=2.66*T-11.51	Y=LEAF EXT.(MM/DAY),T=TEMP. (DEG.CELCIUS),R=0.70,REPRODUCTIVE PHASE	1233

TABLE E68

EQUATION : LEAF WATER POTENTIAL AS A FUNCTION OF SOIL-WATER POTENTIAL

ITEM:	VALUE:	CONDITIONS:	REF.:
*LOLIUM PERENNE	Y = -6.62(+-0.29) + 1.15(+-0.14)*X	Y = LEAF POT. (PA*10**5), X=SOIL POT (PA*10**5)	895
*LYSIMACHIA NUMMULARIA	Y = -5.90(+-0.44) + 1.48(+-0.15)*X	Y = LEAF POT. (PA*10**5), X=SOIL POT (PA*10**5)	895

TABLE E69

EQUATION : LEAF WATER POTENTIAL AS A FUNCTION OF TRANSPIRATION RATE

ITEM:	VALUE:	CONDITIONS:	REF.:
*ORCHARD	Y = -1.18 - 0.334*E	Y=LEAF POT (BAR), E=TRANSPIRATION (MG H2O/(M2*SEC), R=0.97, SCOTLAND	872

TABLE E70

EQUATION : LEAF WATER POTENTIAL AS A FUNCTION OF WATER VAPOUR PRESSURE DEFICIT

ITEM:	VALUE:	CONDITIONS:	REF.:
*ORCHARD	Y = -2.50 - 0.98*DE	Y=LEAF POT (BAR), DE=VAPOUR DEF. (BAR), R=0.88, SCOTLAND	872

TABLE E71

EQUATION : LEAF-WATER POTIENTAL AS A FUNCTION OF SOIL-WATER POTENTAL			
ITEM:	VALUE:	CONDITIONS:	REF.:
*TRIFOLIUM REPENS	Y = -9.47(+-0.29) + 1.19(+-0.09)*X	Y = LEAF POT. (PA*10**5), X=SOIL POT (PA*10**5)	895

TABLE E72

EQUATION : LIGHT COMPENSATION AS A FUNCTION OF TEMPERATURE			
ITEM:	VALUE:	CONDITIONS:	REF.:
*ALNUS RUBRA	I(T) = 0 = 1/J*LN((B(T)-A(T))/B(T))	J=CONST.= 10.2, FOR A(T) AND B(T) SEE (F052)	1667

TABLE E73

EQUATION : LIGHT EXPRESSION			
ITEM:	VALUE:	CONDITIONS:	REF.:
*PHYTOPLANKTON	F(I,Z)=I(0)*EXP(-EE*Z)/IK*EXP (1-I(0)*EXP(-EE*Z)/IK)	(F047)	1503
*PHYTOPLANKTON	REF=0.08+0.02*SIN (0.12083*WEEK-1.57079)	(F047)	1225
*PHYTOPLANKTON	I(0)=IDA*CFGI/FD, IH=0.7*I(0)+ 3*10**5	UNIT CAL/(9M**2)*DAY) (F047)	1145
*PHYTOPLANKTON	F(I)=A/((AA+BB*PHYT)*V)*LN((I(0)+ IK)/(I(0)*EXP((-AA-BB*PHYT)*V/A)+ IK)	GROWTH RATE OF PHYTOPLANKTON AS F(I) (F047)	762
*PHYTOPLANKTON	F(I,Z)=2.72*FD/(EE*Z)*EXP (IDA*EXP(EE*Z)/IK)-EXP(IDA/IK)	(F047)	9
*PHYTOPLANKTON	FD=0.5+0.26*SIN(T-82), IDA=I(0)/ FD*24	LATTITUDE 60 DEGREES (F047)	9
*PHYTOPLANKTON	I(0)=350+300*SIN(T-82)-75*(1+SIN (2*(T-37))) M	UNIT=M*WH/(CM(882)*DAY) (F047)	9
*PHYTOPLANKTON	EE=0.5+14*PA(1),	(F047)	9
*PHYTOPLANKTON	I(0)(T)=IDA*K*(SIN(2*PHI/24(T-6))) **(1/2)	(F047)	1398
*PHYTOPLANKTON	IDA=INTGRL(I(0)(T)DT	K=2.6225 (F047)	1398
*PHYTOPLANKTON	F(I)=I/IK	WHEN I<IK (F047)	503
*PHYTOPLANKTON	F(I)=1	WHEN IK<I<IH (F047)	503
*PHYTOPLANKTON	F(I)=MAX(0,1-KLYS(I-IH))	WHEN IH<I (F047)	503
*PHYTOPLANKTON	I=I(0)*EXP((-AA-BB*PHYT-CC*ZOO-DD*DETRITUS)*Z)	(F047)	503

TABLE E74

EQUATION : LONGITUDINAL DISPERSION COEFFICIENT			
ITEM:	VALUE:	CONDITIONS:	REF.:
*ESTUARY	EL = CL*E**(1/3)*LE**(4/3)	EL = LONG. DISP. COEFFICIENT (LENGTH**2/TIME) (F053)	449
*ESTUARY	EL = 0.042*ABS(U)*R	SIMPLIFICATION (F053)	449
*STREAMS	DL = 3.26*U**0.607	DL=LONG DISP COEF. (FT2/MIN), U= AVERAGE VEL (FT/MIN), 0.33<U<3.30	533
*STREAMS	EL = 22.6*N*U*D**0.833	EL=LONG DISP COEFF., D=DEPTH OF FLOW, N=MANNING COEFF., U=MEAN VELOC.	449

TABLE E75

EQUATION : MAX SPECIFIC RATE OF ACETIC ACID UTILIZATION AS A FUNCTION OF TEMP.			
ITEM:	VALUE:	CONDITIONS:	REF.:
*SULFATE REDUCING BACTERIA	K = (0.695)*(10**7)*(10**(-1790/T))	T=ABS TEMP.,K=MAX SPECIFIC RATE (1/DAY),ARRHENIUS EQUATION	1034

TABLE E76

EQUATION : MEAN DRY WEIGHT PER SURVIVOR AS A FUNCTION OF FOOD CONCENTRATION			
ITEM:	VALUE:	CONDITIONS:	REF.:
*ACHIRUS LINEATUS	W = 17.51*P**0.38	W = DRY W. (UG), R=0.75, N=17, 301 K,P=FOOD CONC (NO/L)	695
*ANCHOA MITCHILLI	W = 24.22*P**0.42	W = DRY W. (UG), R=0.81, N=29, 299 K,P=FOOD CONC (NO/L)	695

TABLE E77

EQUATION : MEAN DRY WEIGHT PER SURVIVOR AS A FUNCTION OF MEAN FOOD CONCENTRATION			
ITEM:	VALUE:	CONDITIONS:	REF.:
*ACHIRUS LINEATUS	W = 17.51*P**0.38	W=DW (UG), P=FOOD (CHLORELLA/L),N= 17, R=0.75, 16 D AFTER HATCH, 299 K	695
*ANCHOA MITCHILLI	W = 24.22*P**0.42	W=DW (UG), P=FOOD (CHLORELLA/L),N= 29, R=0.81, 16 D AFTER HATCH, 299 K	695

TABLE E78

EQUATION : MEAN LENGTH OF LARVAE VS FOOD CONCENTRATION			
ITEM:	VALUE:	CONDITIONS:	REF.:
*ACHIRUS LINEATUS	L = 5.23*(1 - EXP(-0.23*LN(P)))	L=LENGTH (MM), P=FOOD CONC (NO/L), 16 D AFTER HATCH, 301 K,R=0.76	695
*ANCHOA MITCHILLI	L = 15.05*(1 - EXP(-0.18*LN(P)))	L=LENGTH (MM), P=FOOD CONC (NO/L), 16 D AFTER HATCH, 299 K,R=0.82	695

TABLE E79

EQUATION : MEAN LONGEVITY AS A FUNCTION OF TEMPERATURE			
ITEM:	VALUE:	CONDITIONS:	REF.:
*DAPHNIA MAGMA	D = 551/((T-2)**0.902)	D=LONGEVITY (DAYS), T=TEMP (C), 278-301 K	1006

TABLE E80

EQUATION : METABOLIC RATE AS A FUNCTION OF BODY WEIGHT			
ITEM:	VALUE:	CONDITIONS:	REF.:
*PASSERINE BIRDS	LOG(Y) = 1.5732 - 0.2927*LOG(X)	Y=STAND METABOLISM (CAL/(G*H)), X= WET WEIGHT (G), R=0.796	883

TABLE E81

EQUATION : METABOLISM AS A FUNCTION OF BODY WEIGHT			
ITEM:	VALUE:	CONDITIONS:	REF.:
*BIRDS	M = 0.1856*W**0.668	M= BASAL METAB. (KCAL/D), W=WET W. (G)	1617
*LIZARDS	M = 0.0944*W**0.62	M= BASAL METAB. (KCAL/D), W=WET W. (G)	1617
*MAMMALS	M = 0.1244*W**0.75	M= BASAL METAB. (KCAL/D), W=WET W. (G)	1617

TABLE E82

EQUATION : MORTALITY			
ITEM:	VALUE:	CONDITIONS:	REF.:
*PHYTOPLANKTON	MORT=MORTK*PHYT	(F047)	764
*SALMO GAIRDNERI	1/D = 0.002 + 0.0023*T	D= TIME TO DEATH (H), T=CEL DEG., HGCL=0.5 MG HG/L	962

TABLE E83

EQUATION : NET CO2 EXCHANGE AS A FUNCTION OF LIGHT INTENSITY AND TEMPERATURE (F052)			
ITEM:	VALUE:	CONDITIONS:	REF.:
*ALNUS RUBRA	EX = -A(T) + B(T)*(1-EXP(-J*I))	EX=NET EXCHANGE (MG CO2/(G DW*H)), I=LIGHT INTENSITY (CAL/(CM*2*MIN))	1667
*ALNUS RUBRA	A(T) = G*EXP(H*(T-C)-EXP(H*(T-C)))	A(T) AND B(T)= FUNCTIONS OF TEMPERATURE = T (CEL)	1667
*ALNUS RUBRA	B(T) = D - E*(T-F)**2	C,D,E,F,G,H,J = CONSTANTS	1667
*ALNUS RUBRA	A(T) = -78.4*EXP(0.096*(T-46.9) -EXP(0.096*(T-46.9)))	CALCULATED VALUE OF A(T)	1667
*ALNUS RUBRA	B(T) = 31.3 - 0.0078*(T-62.3)**2	ECALCULATED VALUE OF B(T)	1667
*ALNUS RUBRA	EX=-78.4*EXP(0.096*(T-46.9)-EXP (0.096(T-46.9)))+B(T)*(1-EXP (-10.2*I))	FOR B(T) SEE THIS TABLE	1667

TABLE E84

EQUATION : NET YIELD COEFFICIENT			
ITEM:	VALUE:	CONDITIONS:	REF.:
*SELENASTRUM CAPRICORNUTUM (F003)	Y(NET)=Y(MAX)*(1-E**(MRT/KA))	EXPON.FUNCT.,MRT=MEAN RESIDENCE TIME,Y(NET)=NET YIELD COEFFICIENT	1319
*SELENASTRUM CAPRICORNUTUM (F003)	Y(NET)=Y(MAX)*(1-E**(MRT/KA))	Y(MAX)=MAX.YIELD COEFFICIENT,KA= NUTRIENT UTILIZATION CONSTANT	1319
*SELENASTRUM CAPRICORNUTUM (F004)	Y(NET)=Y(MAX)*(MRT/(MRT+KB))	RECTANGUL.HYPERBOL.FUNCT.,Y(NET)= NET YIELD COEFF.	1319
*SELENASTRUM CAPRICORNUTUM (F004)	Y(NET)=Y(MAX)*(MRT/(MRT+KB))	Y(MAX)=MAX.YIELD COEFF.,MRT=MEAN RESID.TIME,KB=NUTRIENT UTIL.CONST.	1319

TABLE E85

EQUATION : NITRATE UPTAKE AS A FUNCTION OF AMBIENT AMMONIA CONCENTRATION			
ITEM:	VALUE:	CONDITIONS:	REF.:
*GONYAULAX POLYEDRA	Y = 0.0279 - 0.00786*X	Y=NO3-N UPTAKE (1/H), X=NH4-N CONC (UG-AT/L), OFF NV AFRICA	280

TABLE E86

EQUATION : NITRIFICATION			
ITEM:	VALUE:	CONDITIONS:	REF.:
*AMMONIA	-D(NH3)/DT = K1*(NH3)	AT 0.03MGS**(2-)/L 75% DECREASE OF K1IONHIBITION AT 0.015MGS**(2-)/L	1497
*AMMONIA	-D(NH3)/DT = K1*(NH3)	AT 4K INCREASE 5% INCREASE OF K1	1497
*NITRATE	D(NO3)/DT = K2*(NO2)	14% DECREASE OF K2 AT 0.1MGS**(2-)/L,1K INCREASE 12% INCREASE OF K2	1497
*NITRITE	-D(NO2)/DT = -K1*(NH3) + K2*(NO2)	SEE AMMONIA AND NITRATE	1497
*PHYTOPLANKTON	NITRI=NITRIK*(NS-NH4)	(F047)	766

TABLE E87

EQUATION : NITROGEN FIXATION			
ITEM:	VALUE:	CONDITIONS:	REF.:
PHYTOPLANKTON	NFIX=KNFIX(1-NS/(KN+NS)	(F047)	503
PHYTOPLANKTON	NFIX=KNFIX*PHYT(N)*(PS*5-NS) OR NFIX=KNFIX*PHYT(N)	(F047)	762
PHYTOPLANKTON	PHYT(N)/PHYT=(PS*5-NS)/(PS*5)	PS*5>NS (F047)	762

TABLE E88

EQUATION : NUMBER OF BREEDING PARS PER HECTARE AS A FUNCTION OF TIME			
ITEM:	VALUE:	CONDITIONS:	REF.:
*BIRDS	N = 2.515*LOG(T) - 0.371	N=BIRDS (NO/HA), T=AGE OF ABANDONED AGRICULTURAL FIELD (YEAR)	1011

TABLE E89

EQUATION : NUMBER OF EGGS AS FUNCTION OF LENGTH			
ITEM:	VALUE:	CONDITIONS:	REF.:
*SALMON	LOG10(N) = 2.3515*LOG10(L) - 0.582	N=NUMBER OF EGGS,L=LENGTH OF SALMON	721

TABLE E90

EQUATION : NUMBER OF LARVAE			
ITEM:	VALUE:	CONDITIONS:	REF.:
*GAMMARUS OCEANICUS ,SEGERSTR$LE	Y=7.73*X-53.7	INCUBATION TIME=20-24 D,Y=NO OF LARVAE,X=LENGTH OF FEAMALE IN MM	930

TABLE E91

EQUATION : NUMBER OF SPECIES			
ITEM:	VALUE:	CONDITIONS:	REF.:
*FRESHWATER POND	S = 34.2*LOG(T) - 38.4	S=NUMBER OF SPECIES, T=TIME (YEAR) AGE OF POND	1011

TABLE E91

EQUATION : NUMBER OF SPECIES

ITEM:	VALUE:	CONDITIONS:	REF.:
*LIZARDS	Y = 0.325*(T-28.5) + 6.64	Y=NO. OF SPECIES, T=CEL DEG, SOUTHWESTERN DESERTS OF US	1256
*LIZARDS	Y = 6.98*(X-0.40) + 7.22	Y=NO OF SPECIES,X=PLANT VOL DIVERSITY, DESERT OF SW USA	1256

TABLE E92

EQUATION : OXYGEN UPTAKE AS A FUNCTION OF BODY PROTEIN

ITEM:	VALUE:	CONDITIONS:	REF.:
*ZOOPLANKTON, AEQUOREA & ORCHISTOMA & CUNINA SP.	LOG(Y) = 0.81(SE=+-0.08) + 1.27(SE =+-0.15)*LOG(X)	Y=02 UPTAKE (ML 02/H), X=PROT.(MG) , 299 K, TROP. MARINE,R=0.93,N=14	124
*ZOOPLANKTON, AGALMA OKENI	LOG(Y) = 1.07(SE=+-0.04) + 0.87(SE =+-0.06)*LOG(X)	Y=02 UPTAKE (ML 02/H), X=PROT.(MG) , 299 K, TROP. MARINE,R=0.88,N=58	124
*ZOOPLANKTON, ATHORYBIA ROSACEA & A. SP.	LOG(Y) = 1.51(SE=+-0.05) + 0.34(SE =+-0.06)*LOG(X)	Y=02 UPTAKE (ML 02/H), X=PROT.(MG) , 299 K, TROP. MARINE,R=0.9,N=13	124
*ZOOPLANKTON, BATHYPHYSA SIBOGAE & RHIZOPHYSA FILIFORMIS	LOG(Y) = 0.80(SE=+-0.16) + 0.87(SE =+-0.15)*LOG(X)	Y=02 UPTAKE (ML 02/H), X=PROT.(MG) , 299 K, TROP. MARINE,R=0.85,N=13	124
*ZOOPLANKTON, CORELLA SPECTABILIS & GLEBA CORDATA	LOG(Y) = 1.22(SE=+-0.03) + 0.86(SE =+-0.04)*LOG(X)	Y=02 UPTAKE (ML 02/H), X=PROT.(MG) , 299 K, TROP. MARINE,R=0.98,N=17	124
*ZOOPLANKTON, DIPHYES DISPAR	LOG(Y) = 1.08(SE=+-0.05) + 0.65(SE =+-0.06)*LOG(X)	Y=02 UPTAKE (ML 02/H), X=PROT.(MG) , 299 K, TROP. MARINE,R=0.93,N=17	124
*ZOOPLANKTON, EURHAMPHEA VEXILLIGERA & CESTUM SP.	LOG(Y) = 1.28(SE=+-0.07) + 0.73(SE =+-0.13)*LOG(X)	Y=02 UPTAKE (ML 02/H), X=PROT.(MG) , 299 K, TROP. MARINE,R=0.89,N=10	124
*ZOOPLANKTON, FORSKALIA EDWARDSI & F. THALOIDES	LOG(Y) = 1.26(SE=+-0.03) + 0.89(SE =+-0.06)*LOG(X)	Y=02 UPTAKE (ML 02/H), X=PROT.(MG) , 299 K, TROP. MARINE,R=0.97,N=20	124
*ZOOPLANKTON, NANOMIA BIJUGA	LOG(Y) = 1.36(SE=+-0.02) + 0.70(SE =+-0.04)*LOG(X)	Y=02 UPTAKE (ML 02/H), X=PROT.(MG) , 299 K, TROP. MARINE,R=0.96,N=19	124
*ZOOPLANKTON, PEGEA CONFEDERATA (AGGREGATE)	LOG(Y) = 1.30(SE=+-0.04) + 1.06(SE =+-0.09)*LOG(X)	Y=02 UPTAKE (ML 02/H), X=PROT.(MG) , 299 K, TROP. MARINE,R=0.94,N=21	124
*ZOOPLANKTON, PTEROTRACHEA HIPPOCAMPUS	LOG(Y) = 1.32(SE=+-0.03) + 0.86(SE =+-0.08)*LOG(X)	Y=02 UPTAKE (ML 02/H), X=PROT.(MG) , 299 K, TROP. MARINE,R=0.97,N=8	124
*ZOOPLANKTON, ROSACEA CYMBIFORMIS	LOG(Y) = 0.75(SE=+-0.08) + 1.24(SE =+-0.19)*LOG(X)	Y=02 UPTAKE (ML 02/H), X=PROT.(MG) , 299 K, TROP. MARINE,R=0.89,N=13	124
*ZOOPLANKTON, SALPA CYLINDRICA (SOLITARY)	LOG(Y) = 1.61(SE=+-0.04) + 0.71(SE =+-0.08)*LOG(X)	Y=02 UPTAKE (ML 02/H), X=PROT.(MG) , 299 K, TROP. MARINE,R=0.94,N=13	124
*ZOOPLANKTON, SULCULEOLARIA QUADRIVALVIS	LOG(Y) = 1.61(SE=+-0.08) + 0.79(SE =+-0.15)*LOG(X)	Y=02 UPTAKE (ML 02/H), X=PROT.(MG) , 299 K, TROP. MARINE,R=0.88,N=10	124

TABLE E93

EQUATION : OXYGEN UPTAKE AS A FUNCTION OF BODY WEIGHT

ITEM:	VALUE:	CONDITIONS:	REF.:
*APIS MELLIFICA	OX = 1.78*W**0.91	OX=02 UPT. (UL/(AMIN*H)), W=WET W(MG)	377
*ARTHROPOD SPECIES	LOG(Y) = 450 - 0.31*LOG(X)	Y= 02 UPTAKE (UL 02/(G*H)), X=WET WEIGHT (MG)	795
*ASELLUS AQUATICUS	OX = 0.350*W**0.867	OX=02 UPT. (UL/(MG*H)), W=WET W (MG), 1.580-22.53 MG,FIRST 100 D, 293 K	795
*CHELA BACAILA	LOG(M) = LOG(0.4655) + 0.76*LOG(W)	M= 02 UPTAKE (MG 02/H), W = WET W. (G), FRESHWATER FISH.	943
*DAPHNIA PULEX	LOG(O) = LOG(0.0014) + 0.881*LOG (W)	O=02 UPTAKE (UL 02/(ANIM*H), W= DRY W (MG)	1328
*INSECTIVORES	Y = 29.64*W**(-0.57)	Y= 02 UPTAKE (CC 02/(G*H)), W= WET WEIGHT (G)	572
*INSECTIVORES	Y = 26.80*W**(-0.50)	Y=02 UPTAKE (CC 02/(G*H)), X= WET WEIGHT (G), EXPONENT FORSED TO .50	572
*INSECTIVORES	Y = 3.087*W**(-0.50)	Y=02 UPTAKE (KCAL/(G*D)), X= WET WEIGHT (G), EXPONENT FORSED TO .50	572

TABLE E93

EQUATION : OXYGEN UPTAKE AS A FUNCTION OF BODY WEIGHT

ITEM:	VALUE:	CONDITIONS:	REF.:
*MACROCYCLOPS ALBIDUS	OX = 1.22*W**0.84	OX= O2 UPTAKE (UL*10**(-3)/(ANIM*H)), W= WET WEIGHT (UG),ALL STAGES	838
*MACROCYCLOPS ALBIDUS	OX = 1.25*W**0.34	OX= O2 UPTAKE (UL*10**(-3)/(ANIM*H)), W= WET WEIGHT (UG), NAUPLII	838
*MACROCYCLOPS ALBIDUS	OX = 0.76*W**1.02	OX= O2 UPTAKE (UL*10**(-3)/(ANIM*H)),W=WET W (UG), ADULT & COPEPODIT	838
*MACROCYCLOPS ALBIDUS	OX = 0.372*W**0.840	OX=O2 UPT. (UL/(MG*H)), W=WET W (MG), 0.001-0.040 MG, WHOLE, 293 K	795
*PERLODES INTRICATA	OX = 1.600*W**0.659	OX=O2 UPT. (UL/(MG*H)), W=WET W (MG),90-200 MG, 2ND LIFE CYCLE, 293 K	795
*PHRONIMA SEDENTARIA	R = 2.84*W**0.57	R=O2 UPTAKE (UL O2/(MG DW*DAY), W= DRY WEIGHT (MG), 286 K	997
*POGONOMYRMEX BADIUS	OX = 8.06*W**0.91	OX=O2 UPT. (UL/(AMIN*H)), W=WET W(MG), WORKER	546
*POGONOMYRMEX BADIUS	OX = 1.63*W**0.91	OX=O2 UPT. (UL/(AMIN*H)), W=WET W(MG), SOLDIER	546
*POLYDESMUS COMPLANATUS	OX = 0.841*W**0.680	OX=O2 UPT. (UL/(MG*H)), W=WET W (MG), 5-90 MG, 2ND LIFE CYCLE, 293 K	795
*RHISOGLYPHUS ECHINOPUS	OX = 0.510*W**0.830	OX=O2 UPT. (UL/(MG*H)), W=WET W (MG), 0.001-0.050 MG, WHOLE, 293 K	795
*RODENTS	Y = 19.94*W**(-0.50)	Y=O2 UPTAKE (CC O2/(G*H)), W= WET WEIGHT (G), EXPONENT FORCED TO .50	572
*RODENTS	Y = 2.297*W**(-0.50)	Y=O2 UPTAKE (KCAL/(G*D)), W= WET WEIGHT (G), EXPONENT FORCED TO .50	572
*RODENTSVORES	Y = 17.76*W**(-0.46)	Y= O2 UPTAKE (CC O2/(G*H)), W= WET WEIGHT (G)	572
*SALMO GAIRDNERI	R = 0.192*W**0.654	R=OXYGEN CONSUMP (CC/H),W=WET WEIGHT (G), STARVATION	1499
*SALMO GAIRDNERI	R = 0.366*W**0.813	R=OXYGEN CONSUMP (CC/H),W=WET WEIGHT (G), MAXIMUM FOOD LEVEL	1499
*SALMO GAIRDNERI	R = 0.229*W**0.789	R=OXYGEN CONSUMP (CC/H),W=WET WEIGHT (G), NORMAL METABOLISM	1499
*SIMOCEPHALUS VETULUS	OX = 0.512*W**0.874	OX=O2 UPT. (UL/(MG*H)), W=WET W (MG), 0.017-0.350 MG, WHOLE, 293 K	795
*TRIBOLIUM CASTANEUM	OX = 6.653*W**0.844	OX=O2 UPT. (UL/(MG*H)), W=WET W (MG), 0.020-3.500 MG, FEEDING, 293 K	795
*TRIBOLIUM CASTANEUM	OX = 6.65*W**0.84	OX=O2 UPT. (UL/(AMIN*H)), W=WET W(MG), LARVAE 2-7, 0.022-2.26 MG	839
*TRIBOLIUM CASTANEUM	OX = 0.24*W**0.91	OX=O2 UPT. (UL/(AMIN*H)), W=WET W(MG), ANTS (PUPAE)	1734
*TRIBOLIUM CASTANEUM	R = 13.84*W**0.93	R=O2 RESP RATE (UL O2/H), W=WET W (MG), MEAN FOR DIFFERENT STAGES	839

TABLE E94

EQUATION : OXYGEN UPTAKE AS A FUNCTION OF OXYGEN CONC. IN THE WATER

ITEM:	VALUE:	CONDITIONS:	REF.:
*BENTHAL UPTAKE	DB = K*C**0.3/H*RO**(T-20)	DB=O2 UPTAKE,H=WATER DEPTH, C=O2 CONC (MG/L),RO**(T-20)=TEMP FACTOR	944
*FRESHWATER SEDIMENT (DEPTH=10M)	Y=0.13*X+1.9	Y=SEDIM.O2 UPTAKE(ML/(M**2)*H,X=O2 SAT.(%),283K,L.VIDOSTERN,R=0.99	559
*FRESHWATER SEDIMENT (DEPTH=20M)	Y=0.17*X+3.1	Y=SEDIM.O2 UPTAKE(ML/(M**2)*H,X=O2 SAT.(%),283K,L.VIDOSTERN,R=0.96	559
*FRESHWATER SEDIMENT (DEPTH=30M)	Y=0.15*X+6.0	Y=SEDIM.O2 UPTAKE(ML/(M**2)*H,X=O2 SAT.(%),283K,L.VIDOSTERN,R=0.95	559
*FRESHWATER SEDIMENT (DEPTH=38M)	Y=0.14*X+6.2	Y=SEDIM.O2 UPTAKE(ML/(M**2)*H,X=O2 SAT.(%),283K,L.VIDOSTERN,R=0.99	559

TABLE E94

EQUATION : OXYGEN UPTAKE AS A FUNCTION OF OXYGEN CONC. IN THE WATER

ITEM:	VALUE:	CONDITIONS:	REF.:
*SEDIMENTS	Y = 26.7*X**0.51/(Z**0.36)	Y=O2 UPTAKE, X=ANN. PRIMARY PROD., Z=MIXED LAYER DEPTH, SEA & LAKE	1462

TABLE E95

EQUATION : OXYGEN UPTAKE AS A FUNCTION OF TEMPERATURE

ITEM:	VALUE:	CONDITIONS:	REF.:
*BACTERIA	Y = 1.2*X**0.1	Y=O2 UPTAKE (ML O2/M2*H), X= CEL DEG, FRESHWATER SEDIMENT, SHALLOW WAT.	616
*DIAPTOMUS	U = 0.0364*T - 0.3418*LOG(W) + 0.6182	U=O2 UPTAKE (UL O2/(MG DW*H), T= TEMP (C), W=DRY WEIGHT (MG)	271
*HYALELLA AZTECA	LN(OX(MICRO L. O(2)/ (AMPHIPOD*HOUR))=-12.8+ 11.13*LN (X) + 1.98*LN(C)	OX=OXYGEN UPTAKE, X = DRY WEIGHT (MICRO G), C = TEMP IN C	617
*MYTILUS EDULIS	Y = 0.018 + 0.006084*T	Y= O2 UPTAKE (MG O2/(G WW*H)), T= TEMP (CEL)	338
*PANTERPE INSIGNIS	Y = 5 - 0.417*X	Y=O2 UPTAKE (ML O2/(G*H)), X=CEL DEG, BIRDS IN TORPOR	1741
*PANTERPE INSIGNIS	Y = 14 - 0.321*X	Y=O2 UPTAKE (ML O2/(G*H)), X=CEL DEG, RESTING BIRDS	1741

TABLE E96

EQUATION : OXYGEN UPTAKE VS % ORGANIC MATTER

ITEM:	VALUE:	CONDITIONS:	REF.:
*SEDIMENTS	LOG(Y) = -2.15 + 0.90*LOG(X)	Y=O2 UPTAKE (MG O2/(G DW*H)), X= % ORG MATTER, R=0.91, UK	618

TABLE E97

EQUATION : OXYGEN UPTAKE VS PARTICLE DIAMETER

ITEM:	VALUE:	CONDITIONS:	REF.:
*SEDIMENT DETRITUS	LOG(Y) = 1.45 - 0.52*LOG(X)	Y=O2 UPTAKE (MG O2/(GRAM DETRITUS*H)), X= PARTICLE DIA (UM), R=0.65	618
*SEDIMENT SAND	LOG(Y) = -0.20 - 0.85*LOG(X)	Y=O2 UPTAKE (MG O2/(G SAND*H)), X= PARTICLE DIA (UM), R=0.95	618
*SEDIMENTS	LOG(Y) = 2.07 - 0.69*LOG(X)	Y=O2 UPTAKE (MG O2/(G ORG MATTER*H), X= PARTICLE DIA (UM), R =0.79	618

TABLE E98

EQUATION : OXYGEN UPTAKE VS WEIGHT, DURATION OF FOOD DEPRIVATION AND TEMPERATURE (F054)

ITEM:	VALUE:	CONDITIONS:	REF.:
*FISH	Y = A + B*EXP(-C*TD)	Y = O2 UPTAKE (MG O2/H), TD = FOOD DEPRIVATION (D)	531
*FISH	Y = 0.021*T*W**0.77 + B*EXP(-C*TD)	A,B,C = CONSTANTS, T=TEMPERATURE (CELCIUS DEG.), W=WET WEIGHT (G)	531
*FISH	Y = 0.021*T*W**0.77 + (8.25*LOG(W) - 7.31)*EXP(-C*TD)	C = RATE CONSTANT (DEPENDING ON FOOD), C=-1.97 FOR 285 K	531

TABLE E98

EQUATION : OXYGEN UPTAKE VS WEIGHT, DURATION OF FOOD DEPRIVATION AND TEMPERATURE			
ITEM:	VALUE:	CONDITIONS:	REF.:
*FISH	Y = 0.021*T*W**0.77 + (8.25*LOG(W) - 7.31)*EXP(-C*TD)	C = RATE CONSTANT (DEPENDING ON FOOD), C=-1.10 FOR 293 K	531
*FISH	Y = 0.021*T*W**0.77 + (8.25*LOG(W) - 7.31)*EXP(-C*TD)	C = RATE CONSTANT (DEPENDING ON FOOD), C=-0.90 FOR 298 K	531

TABLE E99

EQUATION : PCB			
ITEM:	VALUE:	CONDITIONS:	REF.:
*OYSTERS	C(T)=CMAX(1-EXP(-K*T))	PCB CONC. IN OYSTERS AS F(T),SEE FIG.	1643

TABLE E100

EQUATION : PESTICIDES			
ITEM:	VALUE:	CONDITIONS:	REF.:
*DIRECT PHOTOLYSIS OF PESTICIDES IN DILUTE SOLUTIONN	D(P)/DT = -B*K*P	P=CONC OF PEST.,K=SP. LIGHT ABSORP. RATE,B=QUANTUM YIELD OF DISAPP.	1774

TABLE E101

EQUATION : PHOSPHORUS			
ITEM:	VALUE:	CONDITIONS:	REF.:
*PHOSPHATE REGENERATION THROUGH AUTOLYSIS	J = 0.75*Q/T	J=REG (UG/(L*H), Q=DETRITUS PHOSPHORUS, T=TIME (H), FRESHWATER	1326

TABLE E102

EQUATION : PHOTOPERIOD			
ITEM:	VALUE:	CONDITIONS:	REF.:
*ALGAE	0.5 + 0.11*SIN(0.0172(T-165)) DAY	MOSSDALE	1153
*KOOTENAY LAKE	PP=12.2+4.1*SIN(((38-W)/52)*2*PI)	PP=PHOTOPERIOD(HRS/DAY),W=WEEKS,PI =22/7	355

TABLE E103

EQUATION : PHOTOSYNTHESIS			
ITEM:	VALUE:	CONDITIONS:	REF.:
*ALGAE MATS IN THERMAL SPRING	PHOTO = 0.004*A*B*Q**(-0.29)	A=PPM CO2, B=DAYLENGTH (H), Q= ALGAE BIOMASS (G ASH FREE DW/M2)	1704
*ALGAL SUSPENSIONS	A = A(MAX)*B*I/SQRT(1+(B*I)**2)	A=PHOTOSYN, I=LIGHT ENERGY, B= CONSTANT, WITHOUT INHIBITION	1551

TABLE E103

EQUATION : PHOTOSYNTHESIS			
ITEM:	VALUE:	CONDITIONS:	REF.:
*ALGAL SUSPENSIONS	A = A(MAX)*B*I/SQRT(1+(B*I)**2) *Q(II*C)	A=PHOTOSYN, I=LIGHT ENERGY, B,C= CONST.,Q(I*C)=FUNCTION FOR INHIBITION	1551
*PHYTOPLANKTON	PHOTO=MY(T)*PHYT*MIN(I/LO,PS/PSO, NS/NSO)	(F047)	177
*PHYTOPLANKTON	PHOTO=MY(T)*(I/(I+IK)*NS/(KN+NS) *PS/(KP+PS)),MY(T)=MYMAX*F(T)	(F047)	240
*PHYTOPLANKTON	PHOTO=MY(T)*I*PS*NS*PHYT	(F047)	1215
*PHYTOPLANKTON	PHOTO=MYMAX*PHYT*(PS+NS)	(F047)	29
*PHYTOPLANKTON	PHOTO=MY(I)*F(I)*PS/(KP+PS)*NS/(KN +NS)*CS/(KC+CS),MY(T)=MYMAX*F(I)	(F047)	320
*PHYTOPLANKTON	PHOTO=MY*PHYT*PS*I*T	(F047)	741
*PHYTOPLANKTON	PHOTO=MY*PHYT*F(PS)*FD*F(I)	(F047)	885
*PHYTOPLANKTON	PHOTO=MY*PHYT*F(I)*(F(NS)+F(PS)+ F(CS)/3)	(F047)	1225
*PHYTOPLANKTON	PHOTO=MY*PHYT*MIN(F(I),F(PS),F(NS)),	(F047)	877
*PHYTOPLANKTON	UP=UPMAX*(1/(1+PK1*PA)-1/(1+ PKI*PS))	(F047)	121
*PHYTOPLANKTON	MY=MYMAX*F(I)*F(PS)*F(NS)*F(T) *FD*FAC	(F047)	503
*PHYTOPLANKTON	PHOTOSYNTHESISTO=MY(T)*I/(IOP(T)) *EXP(1-I/IOP(T))*F(PA)*F(NA)	(F047)	260
*PHYTOPLANKTON	MY(T)=0.02*EXP(0.17*T), IOP(T)= 0.06*EXP(0.22*T), F(PA)=PA/ (PA-PMIN)	(F047)	260
*PHYTOPLANKTON	F(NA)=NA/(NA-NMIN), UP=UPMAX*PS/ (KP+PS), UN=UNMAX*NS/(KN+NS)	(F047)	260
*PHYTOPLANKTON	PHOTOSYNTHESISTO=MY*F(I)*F(NA,PA), F(NA,PA)=2/(1/F(PA)+1/F(NA))	(F047)	1145
*PHYTOPLANKTON	F(PA)=(KPA+PAMAX-X)/(PAMAX-X)*(PA/ PHYT-X)/KPA+PA/PHYT-X)PA+PA/ PHYT-PA	X=PAMIN (F047)	1145
*PHYTOPLANKTON	F(NA)=(NA/PHYT-NAMIN)/ (NAMAX-NAMIN),	(F047)	1145
*PHYTOPLANKTON	DPS=PAMAX*DPHYT, DNS=NAMAX*DPHYT, UP=MY*PAMAX, UN=MY*NAMAX	EXCESS SUPPLY OF NUTRIENTS (F047)	1145
*PHYTOPLANKTON	UP AND UN EQUAL TO SUPPLY	LIMITING CONDITIONS	1145
*PHYTOPLANKTON	PHOTO=MYMAX*F(T)*F(PA)*F(NA)*F(CA) , F(PA)=(PA-PAMIN*PHYT)/PA	(F047)	762
*PHYTOPLANKTON	F(NA)=(NA-NAMIN*PHYT)/NA, F(CA)= (CA-CAMIN*PHYT/CA	(F047)	762
*PHYTOPLANKTON	UC=F(I)*UCMAX*((CAMAX*PHYT-CA)/ (CAMAX*PHYT-CAMIN*PHYT))*PHYT*CS/X	X=KC+CS (F047)	762
*PHYTOPLANKTON	UP=UPMAX*((PAMAX*PHYT-PA)/ (PAMAX*PHYT-PAMIN*PHYT))*PHYT*PS/ (KP+PS)	UN ANALOG (F047)	762
*PLANTS	PN = A*(1 - EXP(-B*LF)) + C	PN=NET PHOTOSYNTHESIS, LF=LIGHT FLUX DENSITY, A,B,C=CONST.	498
*TREES	R = A*(1-EXP(-B*(AL-C))	R=RELATIVE RATE OF PHOTOSYNTHESIS, AL=AVIABLE LIGHT,A,B,C=CONSTANTS	152

TABLE E104

EQUATION : PRODUCTION			
ITEM:	VALUE:	CONDITIONS:	REF.:
*ABOVEGROUND	P = 0.502 - 0.240*LOG(T)	P= PROD. (KCAL/(M2*Y), T=AGE OF ABANDONED AGRICULTURAL FIELD (Y)	1011
*ALGAE	Y = 0.05*X + 0.02 (MG OXYGEN/MG CHL. *H)	X IN CENTIGRADES	509
*BETULA MAXIMOWICZIANA	P = 0.145*F	P=PRODUCTION (KG/(M2 GROUND AREA*YEAR),F=LEAF AREA (M2/M2 GROUND)	1385

TABLE E104

EQUATION : PRODUCTION			
ITEM:	VALUE:	CONDITIONS:	REF.:
*BETULA MAXIMOWICZIANA	Y = 0.148*X	Y=PRODUCTION (KG/(TREE*YEAR), X= LEAF AREA (M2/TREE)	1385
*BRANCH WOOD IN TREES VS STAND DENSITY	Y = 0.18 + 0.317*X	Y=ANNUAL PRODUCTION (KG WW),X=AREA PER TREE(1/M2),PINUS SYLVESTRIS	966
*CLADOCERAN CONSUMERS AND NET PRIMARY PRODUCTIVITY	Y = -250*X - 900	Y=PROD (G C/(M2*Y), X=SHANNON DIVERSITY, IN LAKES	1697
*FRESHWATER LAKES	LOG(P) = 2.92 - 1.81*LOG(RW)	P=PRODUCTION (MG C/(100 M3*D)), RW =RENEWAL TIME (Y), CANADA, PRIM. PROD.	185
*PLANTS COMMUNITIES	LOG(P) = -1.66 + 1.66*LOG(AE)	P=PRODUC. (G DW/(M2*YEAR), AE= EVAPOTR. (MM/YEAR)	1357
*PRODUCTION AS A FUNCTION OF RESPIRATION	LOG(P) = -0.4723 + 0.6081*LOG(R)	P=PRODUCTION, R=RESPIRATION, UNITS = KCAL/(HA*Y)	572
*RODENT	LOG(P) = -2.2627 + 1.1365*LOG(R)	P=PRODUCTION, R=RESPIRATION, UNITS = KCAL/(HA*Y)	572

TABLE E105

EQUATION : PRODUCTION OF ORGANIC MATTER AND RELEASE OF DISSOLVED ORGANIC MATTER			
ITEM:	VALUE:	CONDITIONS:	REF.:
*ALGAE FROM EUPHOTIC ZONE OF WASHINGTON AND OREGON COASTS	LOG(X) = 0.68*LOG(Y) - 0.46	Y=PART. ORG. MAT.(MG C/M3*(1/2DAY)),X=DIS ORG MAT(MG C/M3*(1/2DAY))	28
*ALGAE FROM EUPHOTIC ZONE OF WASHINGTON AND OREGON COASTS	LOG(X) = 0.66*LOG(Y) + 0.03	Y=PART ORG MAT (MG C/M3*(1/2DAY)), X=DIS ORG MAT (MG C/M3*(1/2DAY))	28

TABLE E106

EQUATION : RATE OF DECLINE OF PCB VS RATE OF DECLINE OF DDE			
ITEM:	VALUE:	CONDITIONS:	REF.:
*COW MILK	Y = 0.974*X	Y= PCB RATE (1/D), X=DDE RATE (1/ D), N=31, R=0.82	490

TABLE E107

EQUATION : RATE OF DEVELOPMENT AS A FUNCTION OF TEMPERATURE			
ITEM:	VALUE:	CONDITIONS:	REF.:
*HEXARTHRA FENNICA	1/D = -0.278 + 0.0845*T	D=DURATION OF EGG DEVELOP. (D),T= CEL DEG., 287-297 K	387
*KELLICOTTIA BOSTONIENSIS	1/D = -0.083 + 0.0430*T	D=DURATION OF EGG DEVELOP. (D),T= CEL DEG., 283-293 K	387
*KERATELLA COCHLEARIS AND KELLICOTTIA LONGISPINA	1/D = +0.028 + 0.0361*T	D=DURATION OF EGG DEVELOP. (D),T= CEL DEG., 283-293 K	387
*KERATELLA QUADRATA	1/D = -0.083 + 0.0528*T	D=DURATION OF EGG DEVELOP. (D),T= CEL DEG., 283-293 K	387
*KERATELLPLOEOSOMA TRUNCATUM	1/D = -0.306 + 0.0472*T	D=DURATION OF EGG DEVELOP. (D),T= CEL DEG., 283-285 K	387
*PLOESOMA HUDSONI	1/D = -0.417 + 0.0722*T	D=DURATION OF EGG DEVELOP. (D),T= CEL DEG., 286-293 K	824
*POLYARTHRA VULGARIS	1/D = -0.083 + 0.0444*T	D=DURATION OF EGG DEVELOP. (D),T= CEL DEG., 278-293 K	824

TABLE E108

EQUATION : REAERATION COEFFICIENT AT 293 K (F055)			
ITEM:	VALUE:	CONDITIONS:	REF.:
*CHANNALS	9.9*10**(-5)*DL**(1.150*H-1.92)	DL (FEET), K2 (1/MIN), H (FEET)	858

TABLE E109

EQUATION : RELATIVE REACTION DISTANCE			
ITEM:	VALUE:	CONDITIONS:	REF.:
*COPEPODS	Y = 2*X - 8 (CM)	Y=DISTANCE,X=COPEPODS IN MILLIMETER,CL.=0.5	275
*DAPHNIA MAGNA	Y = 1.5*X + 5 (CM)	Y=DISTANCE,X=ZOOPLANKTON IN MILLIMETER,CL.=0.6	275
*DAPHNIA PULEX	Y = 1*X + 3 (CM)	Y=DISTANCE,X=ZOOPLANKTON IN MILLIMETER,CL.=0.41	275

TABLE E110

EQUATION : RELATIVE WATER CONTENT OF LEAF VS LEAF WATER POTIENTAL			
ITEM:	VALUE:	CONDITIONS:	REF.:
*ORCHARD	RW = 0.913 + 0.62*Y	Y=LEAF POT (BAR), RW=REL WATER (%) , R=0.88, SCOTLAND	872

TABLE E111

EQUATION : RELEASE OF NUTRIENTS FROM SEDIMENT			
ITEM:	VALUE:	CONDITIONS:	REF.:
*USED IN MODELS	RELEASE = 4.0*SIN + 0.08/1000* A/ V*F(T)	RELEASE OF N (F047)	764

TABLE E112

EQUATION : RELEASED DISSOLVED ORGANIC MATTER VS PARTICULATE PRODUCTIVITY			
ITEM:	VALUE:	CONDITIONS:	REF.:
*MARINE PHYTOPLANKTON	Y = 2.2*X**0.58	Y=DOM (% OF PHOTOSYNTHESIS), X= PHOTOSYNTH., R=0.94, TROPHOGENIC ZONE	105
*MARINE PHYTOPLANKTON	Y = 0.47*X**0.51	Y=DOM (% OF PHOTOSYNTHESIS), X= PHOTOSYNTH., R=0.94,ALL DATA INCL LAB	105

TABLE E113

EQUATION : REPRODUCTION RATE (F054)			
ITEM:	VALUE:	CONDITIONS:	REF.:
*YELLOW FEVER	RR=M*(A**2)*P**N/(-R*LN(P))	M=AEDES DENSITY IN REL.TO MAN,A= AV.NUMB.OF MEN BITTEN/ MOSQUITO*DAY,	278
*YELLOW FEVER	RR=M*(A**2)*P**N/(-R*LN(P))	P=PROBABILITY OF A MOSQUITO SURVIVING THROUGH ONE DAY,	278
*YELLOW FEVER	RR=M*(A**2)*P**N/(-R*LN(P))	N=TIME TAKEN FOR INCUBATION OF VIRUS IN MOSQUITO,	278
*YELLOW FEVER	RR=M*(A**2)*P**N/(-R*LN(P))	R=RECIPROCAL OF THE PERIOD OF INFECTIVITY IN MAN	278

TABLE E114

EQUATION : RESPIRATION AS A FUNCTION OF DISSOLVED OXYGEN			
ITEM:	VALUE:	CONDITIONS:	REF.:
*DAPHNIA GALEATA MENDOTAE	R = 3.998(+-1.207) + 1.411(+ -0.261)*X	R=RESP RATE (UL O2/(MG DW*H), X=O2 CONC (MG/L), R=0.83	651

TABLE E114

EQUATION : RESPIRATION AS A FUNCTION OF DISSOLVED OXYGEN			
ITEM:	VALUE:	CONDITIONS:	REF.:
*DAPHNIA MAGNA	R = 2.743(+-0.263) + 1.177(+ -0.239)*X	R=RESP.(UL O2/(MG DW*H), X=O2 (MG/ L), R=0.83	651

TABLE E115

EQUATION : RESPIRATION AS A FUNCTION OF TEMPERATURE			
ITEM:	VALUE:	CONDITIONS:	REF.:
*COREGONUS	Y = 39.8*EXP(0.085*T)	Y = RESP (UL CO2/H), T = AMBIENT TEMP (CEL)	13
*EICHHORNIA CRASSIPES	Y = 2**((T-30.7)/10)	Y=RESP RATE (1/D), T=CEL DEG., IN MARSH	1049
*EICHHORNIA CRASSIPES	Y = 2**((T-28.5)/10)	Y=RESP RATE (1/D), T=CEL DEG., IN LAKE	1049
*LEUCISCUS IDUS	Y = 16.8*EXP(0.122*T)	Y = RESP (UL CO2/H), T = AMBIENT TEMP (CEL)	13
*THYMALLUS ARCTICUS BAICALENSIS	Y = 51.6*EXP(0.085*T)	Y = RESP (UL CO2/H), T = AMBIENT TEMP (CEL)	13
*TREES	R = 0.498*T - 128	R=RESP (G CO2/(M2*D)), T=KELVIN DEG., OAK-PINE,BROOKHAVEN. 1/ 5-30/9	1748
*TREES	R = 0.127*T - 31	R=RESP (G CO2/(M2*D)), T=KELVIN DEG., OAK-PINE,BROOKHAVEN. 1/ 10-30/4	1748

TABLE E116

EQUATION : RESPIRATION VS PHOTOSYNTHETIC RATE AND STANDING DRY WEIGHT			
ITEM:	VALUE:	CONDITIONS:	REF.:
*WHITE CLOVER	R = 0.25*P + 0.015*W	R=RESP (G CH2O/(M2*D)), 293 K, P= PROD, W=DRY WEIGHT	1057

TABLE E117

EQUATION : ROOTS AS A FUNCTION OF DEPTH			
ITEM:	VALUE:	CONDITIONS:	REF.:
*VEGETABLE CROPS, CEREALS AND GRASSES	P = 100*(1-EXP(-F*X))	P=% ROOTS CONTAINED IN A DEPTH OF X CM, F =CONSTANT	517

TABLE E118

EQUATION : SALMO TRUTTA			
ITEM:	VALUE:	CONDITIONS:	REF.:
*CALORIC CONTENT AS A FUNCTION OF CONDITION FACTOR AND BODY WEIGHT	Y = 4879*EXP(+0.0953*L)*W**0.0180	Y= (CAL/G DW), L=CONDITION FACTOR, W=WET W (G), R=0.801	409
*CALORIC CONTENT AS A FUNCTION OF CONDITION FACTOR AND BODY WEIGHT	Y = 20.41*EXP(+0.0953*L)*W**0.0180	Y= (K J/G DW), L=CONDITION FACTOR, W=WET W (G), R=0.801	409
*CALORIC CONTENT AS A FUNCTION OF CONDITION FACTOR AND BODY WEIGHT	Y = 751.9*EXP(0.332*L)*W**0.0720	Y= (CAL/G WW), L=CONDITION FACTOR, W=WET W (G), R=0.854	409
*CALORIC CONTENT AS A FUNCTION OF CONDITION FACTOR AND BODY WEIGHT	Y = 3.145*EXP(+0.332*L)*W**0.0720	Y= (K J/G WW), L=CONDITION FACTOR, W=WET W (G), R=0.854	409

TABLE E118

EQUATION : SALMO TRUTTA

ITEM:	VALUE:	CONDITIONS:	REF.:
*CALORIC CONTENT AS A FUNCTION OF CONDITION FACTOR AND BODY WEIGHT	Y = 5374*L**0.0907*W**0.0178	Y= (CAL/G DW), L=CONDITION FACTOR, W=WET W (G), R=0.796	409
*CALORIC CONTENT AS A FUNCTION OF CONDITION FACTOR AND BODY WEIGHT	Y = 22.48*L**0.0907*W**0.0178	Y= (K J/G DW), L=CONDITION FACTOR, W=WET W (G), R=0.796	409
*CALORIC CONTENT AS A FUNCTION OF CONDITION FACTOR AND BODY WEIGHT	Y = 1053*L**0.317*W**0.0711	Y= (CAL/G WW), L=CONDITION FACTOR, W=WET W (G), R=0.850	409
*CALORIC CONTENT AS A FUNCTION OF CONDITION FACTOR AND BODY WEIGHT	Y = 4.406*L**0.317*W**0.0711	Y= (K J/G WW), L=CONDITION FACTOR, W=WET W (G), R=0.850	409
*CALORIC CONTENT AS A FUNCTION OF FORK LENGTH AND BODY WEIGHT	Y = 5471*EXP(+0.00323*L)*W** (-0.00195)	Y= (CAL/G DW), L=FORK LENGTH (CM), W=WET W (G), R=0.753	409
*CALORIC CONTENT AS A FUNCTION OF FORK LENGTH AND BODY WEIGHT	Y = 22.89*EXP(+0.00323*L)*W** (-0.00195)	Y= (K J/G DW), L=FORK LENGTH (CM), W=WET W (G), R=0.753	409
*CALORIC CONTENT AS A FUNCTION OF FORK LENGTH AND BODY WEIGHT	Y = 1121*EXP(+0.0113*L)*W** (0.00193)	Y= (CAL/G WW), L=FORK LENGTH (CM), W=WET W (G), R=0.815	409
*CALORIC CONTENT AS A FUNCTION OF FORK LENGTH AND BODY WEIGHT	Y = 4.690*EXP(+0.0113*L)*W** (0.00193)	Y= (K J/G WW), L=FORK LENGTH (CM), W=WET W (G), R=0.815	409
*CALORIC CONTENT AS A FUNCTION OF FORK LENGTH AND BODY WEIGHT	Y = 8148*L**(-0.271)*W**0.108	Y= (CAL/G DW), L=FORK LENGTH (CM), W=WET W (G), R=0.796	409
*CALORIC CONTENT AS A FUNCTION OF FORK LENGTH AND BODY WEIGHT	Y = 34.09*L**(-0.271) *W**0.108-0.00195)	Y= (K J/G DW), L=FORK LENGTH (CM), W=WET W (G), R=0.796	409
*CALORIC CONTENT AS A FUNCTION OF FORK LENGTH AND BODY WEIGHT	Y = 4608*L**(-0.962)*W**0.391	Y= (CAL/G WW), L=FORK LENGTH (CM), W=WET W (G), R=0.852	409
*CALORIC CONTENT AS A FUNCTION OF FORK LENGTH AND BODY WEIGHT	Y = 19.28*L**(-0.962)*W**0.391	Y= CAL (K J/G WW), L=FORK LENGTH (CM), W=WET W (G), R=0.852	409
*CONSUMATION OF FOOD PER DAY AS A FUNCTION OF BODY WEIGHT AND TEMP.	D = 0.742*W**0.786*EXP(0.366*T)	D=FOOD CON. (MG/D), W=WW (G),T=CEL ,GROWTH=90 % OF MAX,276.8-279.8 K	408
*CONSUMATION OF FOOD PER DAY AS A FUNCTION OF BODY WEIGHT AND TEMP.	D = 3.820*W**0.772*EXP(0.134*T)	D=FOOD CON. (MG/D), W=WW (G),T=CEL ,GROWTH=90 % OF MAX,279.8-288 K	408
*CONSUMATION OF FOOD PER DAY AS A FUNCTION OF BODY WEIGHT AND TEMP.	D = 5.191*W**0.794*EXP(0.107*T)	D=FOOD CON. (MG/D), W=WW (G),T=CEL ,GROWTH=90 % OF MAX,288-290.5 K	408
*CONSUMATION OF FOOD PER DAY AS A FUNCTION OF BODY WEIGHT AND TEMP.	D = 0.820*W**0.763*EXP(0.356*T)	D=FOOD CON. (MG/D), W=WW (G),T=CEL ,GROWTH=80 % OF MAX,276.8-279.8 K	408
*CONSUMATION OF FOOD PER DAY AS A FUNCTION OF BODY WEIGHT AND TEMP.	D = 3.446*W**0.780*EXP(0.130*T)	D=FOOD CON. (MG/D), W=WW (G),T=CEL ,GROWTH=80 % OF MAX,279.8-288 K	408
*CONSUMATION OF FOOD PER DAY AS A FUNCTION OF BODY WEIGHT AND TEMP.	D = 112.701*W**0.765*EXP(-0.096*T)	D=FOOD CON. (MG/D), W=WW (G),T=CEL ,GROWTH=80 % OF MAX,288-290.5 K	408
*CONSUMATION OF FOOD PER DAY AS A FUNCTION OF BODY WEIGHT AND TEMP.	D = 0.859*W**0.769*EXP(0.335*T)	D=FOOD CON. (MG/D), W=WW (G),T=CEL ,GROWTH=70 % OF MAX,276.8-279.8 K	408
*CONSUMATION OF FOOD PER DAY AS A FUNCTION OF BODY WEIGHT AND TEMP.	D = 3.184*W**0.781*EXP(0.129*T)	D=FOOD CON. (MG/D), W=WW (G),T=CEL ,GROWTH=70 % OF MAX,279.8-288 K	408
*CONSUMATION OF FOOD PER DAY AS A FUNCTION OF BODY WEIGHT AND TEMP.	D = 75.866*W**0.764*EXP(-0.076*T)	D=FOOD CON. (MG/D), W=WW (G),T=CEL ,GROWTH=70 % OF MAX,288-290.5 K	408
*CONSUMATION OF FOOD PER DAY AS A FUNCTION OF BODY WEIGHT AND TEMP.	D = 0.938*W**0.761*EXP(0.317*T)	D=FOOD CON. (MG/D), W=WW (G),T=CEL ,GROWTH=60 % OF MAX,276.8-279.8 K	408
*CONSUMATION OF FOOD PER DAY AS A FUNCTION OF BODY WEIGHT AND TEMP.	D = 3.004*W**0.775*EXP(0.129*T)	D=FOOD CON. (MG/D), W=WW (G),T=CEL ,GROWTH=60 % OF MAX,279.8-288 K	408
*CONSUMATION OF FOOD PER DAY AS A FUNCTION OF BODY WEIGHT AND TEMP.	D = 55.336*W**0.756*EXP(-0.058*T)	D=FOOD CON. (MG/D), W=WW (G),T=CEL ,GROWTH=60 % OF MAX,288-290.5 K	408
*CONSUMATION OF FOOD PER DAY AS A FUNCTION OF BODY WEIGHT AND TEMP.	D = 0.975*W**0.763*EXP(0.301*T)	D=FOOD CON. (MG/D), W=WW (G),T=CEL ,GROWTH=50 % OF MAX,276.8-279.8 K	408
*CONSUMATION OF FOOD PER DAY AS A FUNCTION OF BODY WEIGHT AND TEMP.	D = 2.948*W**0.760*EXP(0.130*T)	D=FOOD CON. (MG/D), W=WW (G),T=CEL ,GROWTH=50 % OF MAX,279.8-288 K	408
*CONSUMATION OF FOOD PER DAY AS A FUNCTION OF BODY WEIGHT AND TEMP.	D = 40.834*W**0.756*EXP(-0.043*T)	D=FOOD CON. (MG/D), W=WW (G),T=CEL ,GROWTH=50 % OF MAX,288-290.5 K	408
*CONSUMATION OF FOOD PER DAY AS A FUNCTION OF BODY WEIGHT AND TEMP.	D = 1.073*W**0.748*EXP(0.287*T)	D=FOOD CON. (MG/D), W=WW (G),T=CEL ,GROWTH=40 % OF MAX,276.8-279.8 K	408
*CONSUMATION OF FOOD PER DAY AS A FUNCTION OF BODY WEIGHT AND TEMP.	D = 2.604*W**0.771*EXP(0.130*T)	D=FOOD CON. (MG/D), W=WW (G),T=CEL ,GROWTH=40 % OF MAX,279.8-288 K	408
*CONSUMATION OF FOOD PER DAY AS A FUNCTION OF BODY WEIGHT AND TEMP.	D = 27.702*W**0.759*EXP(-0.024*T)	D=FOOD CON. (MG/D), W=WW (G),T=CEL ,GROWTH=40 % OF MAX,288-290.5 K	408
*CONSUMATION OF FOOD PER DAY AS A FUNCTION OF BODY WEIGHT AND TEMP.	D = 1.144*W**0.746*EXP(0.268*T)	D=FOOD CON. (MG/D), W=WW (G),T=CEL ,GROWTH=30 % OF MAX,276.8-279.8 K	408
*CONSUMATION OF FOOD PER DAY AS A FUNCTION OF BODY WEIGHT AND TEMP.	D = 2.367*W**0.764*EXP(0.133*T)	D=FOOD CON. (MG/D), W=WW (G),T=CEL ,GROWTH=30 % OF MAX,279.8-288 K	408
*CONSUMATION OF FOOD PER DAY AS A FUNCTION OF BODY WEIGHT AND TEMP.	D = 20.876*W**0.749*EXP(-0.007*T)	D=FOOD CON. (MG/D), W=WW (G),T=CEL ,GROWTH=30 % OF MAX,288-290.5 K	408
*CONSUMATION OF FOOD PER DAY AS A FUNCTION OF BODY WEIGHT AND TEMP.	D = 1.247*W**0.737*EXP(0.249*T)	D=FOOD CON. (MG/D), W=WW (G),T=CEL ,GROWTH=20 % OF MAX,276.8-279.8 K	408

TABLE E118

EQUATION : SALMO TRUTTA

ITEM:	VALUE:	CONDITIONS:	REF.:
*CONSUMATION OF FOOD PER DAY AS A FUNCTION OF BODY WEIGHT AND TEMP.	D = 2.294*W**0.754*EXP(0.132*T)	D=FOOD CON. (MG/D), W=WW (G),T=CEL ,GROWTH=20 % OF MAX,279.8-288 K	408
*CONSUMATION OF FOOD PER DAY AS A FUNCTION OF BODY WEIGHT AND TEMP.	D = 13.469*W**0.748*EXP(0.015*T)	D=FOOD CON. (MG/D), W=WW (G),T=CEL ,GROWTH=20 % OF MAX,288-290.5 K	408
*CONSUMATION OF FOOD PER DAY AS A FUNCTION OF BODY WEIGHT AND TEMP.	D = 1.235*W**0.740*EXP(0.238*T)	D=FOOD CON. (MG/D), W=WW (G),T=CEL ,GROWTH=10 % OF MAX,276.8-279.8 K	408
*CONSUMATION OF FOOD PER DAY AS A FUNCTION OF BODY WEIGHT AND TEMP.	D = 1.993*W**0.755*EXP(0.134*T)	D=FOOD CON. (MG/D), W=WW (G),T=CEL ,GROWTH=10 % OF MAX,279.8-288 K	408
*CONSUMATION OF FOOD PER DAY AS A FUNCTION OF BODY WEIGHT AND TEMP.	D = 8.987*W**0.749*EXP(0.036*T)	D=FOOD CON. (MG/D), W=WW (G),T=CEL ,GROWTH=10 % OF MAX,288-290.5 K	408
*FAT CONTENT AS A FUNCTION OF CONDITION FACTOR AND BODY WEIGHT	Y = 1.236*EXP(+0.781*L)*W**0.159	Y= % FAT, L=CONDITION FACTOR, W= WET W (G), R=0.862	409
*FAT CONTENT AS A FUNCTION OF CONDITION FACTOR AND BODY WEIGHT	Y = 2.732*L**0.771*W**0.157	Y= % FAT, L=CONDITION FACTOR, W= WET W (G), R=0.862	409
*FAT CONTENT AS A FUNCTION OF FORK LENGTH AND BODY WEIGHT	Y = 2.807*EXP(+0.00529*L)*W**(+ 0.124)	Y= % FAT, L=FORK LENGTH (CM), W= WET W (G), R=0.802	409
*FAT CONTENT AS A FUNCTION OF FORK LENGTH AND BODY WEIGHT	Y = 95.63*L**(-2.316)*W**0.929	Y= % FAT, L=FORK LENGTH (CM), W= WET W (G), R=0.862	409
*FAT CONTENT AS A FUNCTION OF WATER CONTENT	Y = 51.63 - 0.613(+-0.0225)*X	Y = % FAT, X = % WATER, R = 0.98	409
*PROTEIN CONTENT AS A FUNCTION OF CONDITION FACTOR AND BODY WEIGHT	Y = 12.53*EXP(+0.148*L)*W**0.0263	Y= % PROTEIN, L=CONDITION FACTOR, W=WET W (G), R=0.732	409
*PROTEIN CONTENT AS A FUNCTION OF CONDITION FACTOR AND BODY WEIGHT	Y = 14.53*L**0.148*W**0.0259	Y= % PROTEIN, L=CONDITION FACTOR, W=WET W (G), R=0.725	409
*PROTEIN CONTENT AS A FUNCTION OF FORK LENGTH AND BODY WEIGHT	Y = 15.39*EXP(+0.00987*L)*W** (-0.0344)	Y= % PROTEIN, L=FORK LENGTH (CM), W=WET W (G), R=0.729	409
*PROTEIN CONTENT AS A FUNCTION OF FORK LENGTH AND BODY WEIGHT	Y = 27.51*L**(-0.414)*W**0.164	Y= % PROTEIN, L=FORK LENGTH (CM), W=WET W (G), R=0.725	409
*PROTEIN CONTENT AS A FUNCTION OF WATER CONTENT	Y = 42.92 - 0.353(+-0.0226)*X	Y = % PROTEIN, X = % WATER, R = 0.924	409
*SPECIFIC GROWTH RATE AS A FUNCTION OF BODY WEIGHT	GR = 0.622*W**(-0.325(+-0.050))	GR=GROWTH RATE (% WW/D), W=WET W. (G), 95 % CL, 278.6+-0.5 K	407
*SPECIFIC GROWTH RATE AS A FUNCTION OF BODY WEIGHT	GR = 1.884*W**(-0.332(+-0.032))	GR=GROWTH RATE (% WW/D), W=WET W. (G), 95 % CL, 282.5+-0.7 K	407
*SPECIFIC GROWTH RATE AS A FUNCTION OF BODY WEIGHT	GR = 2.788*W**(-0.318(+-0.025))	GR=GROWTH RATE (% WW/D), W=WET W. (G), 95 % CL, 285.8+-0.7 K	407
*SPECIFIC GROWTH RATE AS A FUNCTION OF BODY WEIGHT	GR = 2.151*W**(-0.307(+-0.049))	GR=GROWTH RATE (% WW/D), W=WET W. (G), 95 % CL, 288.0+-0.6 K	407
*SPECIFIC GROWTH RATE AS A FUNCTION OF BODY WEIGHT	GR = 0.029*W**(-0.029(+-0.102))	GR=GROWTH RATE (% WW/D), W=WET W. (G), 95 % CL, 292.5+-0.8 K	407
*SPECIFIC GROWTH RATE AS A FUNCTION OF BODY WEIGHT	GR = 0.622*W**(-0.325(+-0.050))	GR=GROWTH RATE (% WW/D), W=WET W. (G), 95 % CL, 278.6+-0.5 K	407
*SPECIFIC GROWTH RATE AS A FUNCTION OF BODY WEIGHT	GR = 1.884*W**(-0.332(+-0.032))	GR=GROWTH RATE (% WW/D), W=WET W. (G), 95 % CL, 282.5+-0.7 K	407
*SPECIFIC GROWTH RATE AS A FUNCTION OF BODY WEIGHT	GR = 2.788*W**(-0.318(+-0.025))	GR=GROWTH RATE (% WW/D), W=WET W. (G), 95 % CL, 285.8+-0.7 K	407
*SPECIFIC GROWTH RATE AS A FUNCTION OF BODY WEIGHT	GR = 2.151*W**(-0.307(+-0.049))	GR=GROWTH RATE (% WW/D), W=WET W. (G), 95 % CL, 288.0+-0.6 K	407
*SPECIFIC GROWTH RATE AS A FUNCTION OF BODY WEIGHT	GR = 0.029*W**(-0.029(+-0.102))	GR=GROWTH RATE (% WW/D), W=WET W. (G), 95 % CL, 292.5+-0.8 K	407
*SPECIFIC GROWTH RATE AS A FUNCTION OF BODY WEIGHT AND TEMPERATURE	GR = 100*(-0.0100 + 0.0029(+ -0.001)*T)*W**(-0.3250(+-0.0360))	GR=GROWTH RATE (% WW/D), T=CEL DEG, W=WET W. (G),95% CL, 276.8-285.8 K	407
*SPECIFIC GROWTH RATE AS A FUNCTION OF BODY WEIGHT AND TEMPERATURE	GR = 100*(+0.0820 - 0.0042(+ -0.0002)*T)*W**(-0.2917(+-0.0760))	GR=GROWTH RATE (% WW/D), T=CEL DEG, W=WET W. (G),95% CL, 286.6-292.5 K	407
*SPECIFIC GROWTH RATE AS A FUNCTION OF BODY WEIGHT AND TEMPERATURE	GR = 100*(-0.0100 + 0.0029(+ -0.001)*T)*W**(-0.3250(+-0.0360))	GR=GROWTH RATE (% WW/D), T=CEL DEG, W=WET W. (G),95% CL, 276.8-285.8 K	407
*SPECIFIC GROWTH RATE AS A FUNCTION OF BODY WEIGHT AND TEMPERATURE	GR = 100*(+0.0820 - 0.0042(+ -0.0002)*T)*W**(-0.2917(+-0.0760))	GR=GROWTH RATE (% WW/D), T=CEL DEG, W=WET W. (G),95% CL, 286.6-292.5 K	407
*SPECIFIC GROWTH RATE AS A FUNCTION OF TEMPERATURE	GR = -0.269 + 0.081(+-0.004)*T	GR=GROWTH RATE (% WW/D), T=CEL DEG, 95 % CL, 276.8-285.8 K	407
*SPECIFIC GROWTH RATE AS A FUNCTION OF TEMPERATURE	GR = +2.568 - 0.131(+-0.006)*T	GR=GROWTH RATE (% WW/D), T=CEL DEG, 95 % CL, 286.5-292.5 K	407
*SPECIFIC GROWTH RATE AS A FUNCTION OF TEMPERATURE	GR = +4.216 - 0.216(+-0.058)*T	GR=GROWTH RATE (% WW/D), T=CEL DEG, 95 % CL, 292.5-294.7 K	407

TABLE E118

EQUATION : SALMO TRUTTA			
ITEM:	VALUE:	CONDITIONS:	REF.:
*SPECIFIC GROWTH RATE AS A FUNCTION OF TEMPERATURE	GR = -0.269 + 0.081(+-0.004)*T	GR=GROWTH RATE (% WW/D), T=CEL DEG, 95 % CL, 276.8-285.8 K	407
*SPECIFIC GROWTH RATE AS A FUNCTION OF TEMPERATURE	GR = +2.568 - 0.131(+-0.006)*T	GR=GROWTH RATE (% WW/D), T=CEL DEG, 95 % CL, 286.5-292.5 K	407
*SPECIFIC GROWTH RATE AS A FUNCTION OF TEMPERATURE	GR = +4.216 - 0.216(+-0.058)*T	GR=GROWTH RATE (% WW/D), T=CEL DEG, 95 % CL, 292.5-294.7 K	407
*WATER CONTENT AS A FUNCTION OF CONDITION FACTOR AND BODY WEIGHT	Y = 81.11*L**(-0.0705)*W**(-0.0174)	Y= % WATER, L=CONDITION FACTOR, W=WET W (G), R=0.866	409
*WATER CONTENT AS A FUNCTION OF FORK LENGTH AND BODY WEIGHT	Y = 79.83*EXP(-0.00283*L)*W**(-0.0000426)	Y= % WATER, L=FORK LENGTH (CM), W=WET W (G), R=0.838	409
*WATER CONTENT AS A FUNCTION OF FORK LENGTH AND BODY WEIGHT	Y = 57.80*L**0.223*W**(-0.0916)	Y= % WATER, L=FORK LENGTH (CM), W=WET W (G), R=0.869	409

TABLE E119

EQUATION : SPECIES DENSITY AS A FUNCTION OF TIME			
ITEM:	VALUE:	CONDITIONS:	REF.:
*BIRDS	S = 0.262*LOG(T) - 0.062	S=DENSITY (BIRDS/HA), T=AGE OF ABANDONED AGRICULTURAL FIELD	1011

TABLE E120

EQUATION : SPECIFIC DEATH RATE AS A FUNCTION OF TEMPERATURE (F054)			
ITEM:	VALUE:	CONDITIONS:	REF.:
*BACTERIA	IF T>TM THEN DR = -A*LN(K/KM)	TM=OPT. TEMP FOR GROWTH, T=AMBIENT TEMP, DR = DEATH RATE (1/D)	355
*BACTERIA	IF T>TM THEN DR = 0	K=GROWTH RATE AT T, KM=MAX. OF K, A=SCALING COEFFICIENT	355

TABLE E121

EQUATION : STOMATAL CONDUCTANCE AS A FUNCTION OF LEAF-WATER POTIENTAL			
ITEM:	VALUE:	CONDITIONS:	REF.:
*LOLIUM PERENNE	1/R = 7.55(+-0.95) - 0.55(+-0.10)*LP	R=STOMATAL COND. (CM/S), LP=LEAF POT. (PA*10**5), HUMID AIR	895
*LOLIUM PERENNE	1/R = 2.96(+-0.75) - 0.17(+-0.05)*LP	R=STOMATAL COND. (CM/S), LP=LEAF POT. (PA*10**5), DRY AIR	895
*LYSIMACHIA NUMMULARIA	1/R = 1.04(+-0.06) + 0.04(+-0.07)*LP	R=STOMATAL COND. (CM/S), LP=LEAF POT. (PA*10**5), HUMID AIR	895
*LYSIMACHIA NUMMULARIA	1/R = 0.80(+-0.07) + 0.024(+-0.06)*LP	R=STOMATAL COND. (CM/S), LP=LEAF POT. (PA*10**5), DRY AIR	895
*TRIFOLIUM REPENS	1/R = 4.20(+-0.30) - 0.16(+-0.03)*LP	R=STOMATAL COND. (CM/S), LP=LEAF POT. (PA*10**5), HUMID AIR	895
*TRIFOLIUM REPENS	1/R = 2.64(+-0.03) - 0.11(+-0.01)*LP	R=STOMATAL COND. (CM/S), LP=LEAF POT. (PA*10**5), DRY AIR	895

TABLE E122

EQUATION : STOMATAL CONDUCTANCE AS A FUNCTION OF WATER VAPOUR PRESSURE DEFICIT (F068)			
ITEM:	VALUE:	CONDITIONS:	REF.:
*PICEA SITCHENSIS	KS = 0.55 - 0.032*DE	KS=STOMATA COND. (CM/S), DE=DEFICIT (MB), 1 Y OLD-NEEDLES	1665

TABLE E123

EQUATION : STOMATAL CONDUCTANCE VS LEAF-AIR VAPOUR PRESSURE DIFFERENCE			
ITEM:	VALUE:	CONDITIONS:	REF.:
*SITKA SPRUCE	Y = -0.010*X + 0.225	Y=STOMATAL COND. (CM/S), X=PRES. DIFF. (MB), 278-293 K, JAN-MAY	1114
*SITKA SPRUCE	Y = -0.0168*X + 0.363	Y=STOMATAL COND. (CM/S), X=PRES. DIFF. (MB), 278-293 K, JULY-AUG	1114

TABLE E124

EQUATION : SULPHIDE OXIDATION			
ITEM:	VALUE:	CONDITIONS:	REF.:
*SULPHIDE	LOG((SULPHIDE)TIME/(SULPHIDE)TIME=0)=K1*T/2.303	K1=K2*(OXYGEN),K2=9.*10**(-5) (MICRO-M*MIN)**(-1),PH=8.	15
*SULPHIDE	LOG((SULPHIDE)TIME/(SULPHIDE)TIME=0)=K1*T/2.303	K1=K2*(OXYGEN),K2=6.5*10**(-5) (MICRO-M*MIN)**(-1),PH=8.5	15

TABLE E125

EQUATION : SUMMER AVERAGE CHLOROPHYLL CONC VS TOTAL PHOSPHORUS CONCENTRATION			
ITEM:	VALUE:	CONDITIONS:	REF.:
*FRESHWATER LAKES	LOG(C) = 1.583*LOG(P) - 1.134	C=CHLA A (MG/M3),P=SPRING OVERTURN P CONC (MG/M3),R=0.975, JAP LAKES	358
*FRESHWATER LAKES	LOG(C) = 1.449*LOG(P) - 1.136	C=CHLA A (MG/M3),P=SPRING OVERTURN P CONC (MG/M3),R=0.95, N=46	358

TABLE E126

EQUATION : SURVIORSHIP CURVE			
ITEM:	VALUE:	CONDITIONS:	REF.:
*APODESMUS FLAVICOLLIS	N = 104.6*EXP(-0.34*T)	N=SURVIVORS PER 100,T=TIME (MONTHS),LOW AND HIGH POPULATION DENSITY	140

TABLE E127

EQUATION : TEMPERATURE EFFECTS UPON BIOLOGICAL RATES			
ITEM:	VALUE:	CONDITIONS:	REF.:
*USED IN MODELS	K(T) = K(20)*KOT**(T-20)	(F047)	764
*USED IN MODELS	K(T) = K(OPT)*EXP(A*(T-T(OPT))*(TMAX - T/U)**(A*(U))	U = TMAX - T(OPT) (F047)	886
*USED IN MODELS	K(T) = K(OPT)*EXP(-2.3*(T-T(OPT)/(TMAX-T(OPT))**2)	FOR T>T(OPT) (F047)	764
*USED IN MODELS	K(T) = K(OPT)*EXP(-2.3*(T-T(OPT)/(TMIN-T(OPT))**2)	T<T(OPT) (F047)	764
*USED IN MODELS	K(T) = K(OPT)*EXP(-2.3*ABS((T-T(OPT)/15))	(F047)	764
*USED IN MODELS	K(T) = K(OPT)*EXP(K*T)	(F047)	764
*USED IN MODELS	K(T) = K(20)*KOT**(T-20)	(F047)	764
*USED IN MODELS	K(T) = K(OPT)*EXP(A*(T-T(OPT))*(TMAX - T/U)**(A*(U))	U = TMAX - T(OPT) (F047)	886
*USED IN MODELS	K(T) = K(OPT)*EXP(-2.3*(T-T(OPT)/(TMAX-T(OPT))**2)	FOR T>T(OPT) (F047)	764

TABLE E127

EQUATION : TEMPERATURE EFFECTS UPON BIOLOGICAL RATES

ITEM:	VALUE:	CONDITIONS:	REF.:
*USED IN MODELS	K(T) = K(OPT)*EXP(-2.3*(T-T(OPT)/ (TMIN-T(OPT))**2)	T<T(OPT) (F047)	764
*USED IN MODELS	K(T) = K(OPT)*EXP(-2.3*ABS((T-T (OPT)/15))	(F047)	764
*USED IN MODELS	K(T) = K(OPT)*EXP(K*T)	(F047)	764

TABLE E128

EQUATION : TRANSPARENCY AS A FUNCTION OF MAXIMUM PRODUCTION RATE

ITEM:	VALUE:	CONDITIONS:	REF.:
*LAKES	TP = 35/SQRT(G24)	TP=TRANSPARENCY (M), G24=MAX PRODUCTION RATE (MG C/(M3*D))	992

TABLE E129

EQUATION : UPTAKE OF CD

ITEM:	VALUE:	CONDITIONS:	REF.:
*STAPHYLLOCOCCUS AUREUS	Y = 2.0*T/(T + 2.5)	Y=CONC IN ORG. (U MOLES CD/G), T= TIME (MIN), CD-115, RESISTANT SP.	1437
*STAPHYLLOCOCCUS AUREUS	Y = 4*C/(C + 21)	Y=CONC IN ORG (U MOLES/G), C= MEDIUM (U MOLAR), 18 MIN, SENSITIVE	1437
*STAPHYLLOCOCCUS AUREUS	Y = 1.3*C/(C + 21)	Y=CONC IN ORG (U MOLES/G), C= MEDIUM (U MOLAR), 18 MIN, RESISTANT	1437

TABLE E130

EQUATION : UPTAKE OF CS-137

ITEM:	VALUE:	CONDITIONS:	REF.:
*AIX SPONSA	LN(Y) = -0.36 + 0.18*T	Y=P CI/G WW, T=TIME (D), FREE LIVING DUCKS	451

TABLE E131

EQUATION : UPTAKE OF HG AS A FUNCTION OF CONCENTRATION IN WATER AND ELIMINATION (F054)

ITEM:	VALUE:	CONDITIONS:	REF.:
*LEPOMIS MACROCHIRUS	Q(T) = (U*C/K)*(1-EXP(-K*T))	Q(T)= CONC IN FISH AT TIME T=T, U= FRACTION ABSORBED FROM WATER AT	309
*LEPOMIS MACROCHIRUS	Q(T) = ((U*C/K)+B*T)*(1-EXP(-K*T))	PASSAGE OVER GILLS,K=ELIM. RATE,C= WAT CONC,B AND K=TURNOVER RATE IN A	309
*LEPOMIS MACROCHIRUS	D(T) = U*C/K*(1-EXP(-K*T)) + U*C/ (K-G)*(EXP(-K*T)-EXP(-G*T))	COMPARTMENT D AND H,D(T)= CONC IN D AT TIME T	309
*LEPOMIS MACROCHIRUS	Q(T) = 3.84 (+-0.27 + 0.177 (+ -0.002)*T	WATER=0.2 PPB CH3HGCL, 281 K, FASTED FISH	309
*LEPOMIS MACROCHIRUS	Q(T) = 1.35 (+-0.32) + T/(3.96 (+ -0.12) + 0.00367 (+-0.000027)*T)	WATER=0.2 PPB CH3HGCL, 281 K, FASTED FISH, FISH CONC=PPB (WET)	309
*LEPOMIS MACROCHIRUS	Q(T) = C*1.37 (+-0.02)/0.00198 (+ -0.00011)*(1-EXP(0.00198*T)	WATER=0.2 PPB CH3HGCL, 281 K, FASTED FISH, FISH CONC=PPB (WET) IN A	309

TABLE E131

EQUATION : UPTAKE OF HG AS A FUNCTION OF CONCENTRATION IN WATER AND ELIMINATION (F054)			
ITEM:	VALUE:	CONDITIONS:	REF.:
*LEPOMIS MACROCHIRUS	Q(T) = (C*1.48/0.0178 + 0.142*T) *(1-EXP(-0.0178*T)	WATER=0.2 PPB CH3HGCL, 281 K, FASTED FISH, FISH CONC=PPB (WET)	309

TABLE E132

EQUATION : UPTAKE OF NITRATE AS A FUNCTION OF CONCENTRATION			
ITEM:	VALUE:	CONDITIONS:	REF.:
*PHYTOPLANKTON FROM NORTH PACIFIC	Y = -0.08 + 0.96*X	Y=SPECIFIC RATE OF ASSIMILATION (1/D),X= CONC OF NO3 (UG-ATOM/L)	429

TABLE E133

EQUATION : UPTAKE OF TOXAPHENE AS A FUNCTION OF TIME			
ITEM:	VALUE:	CONDITIONS:	REF.:
*GAMBUSIA AFFINIS	Y = 0.00068 + 0.00058*T	Y = CONC IN FISH (MG/KG WW), T= TIME (H), CONC IN MEDIUM = 2 PPM	1391

TABLE E134

EQUATION : UPTAKE OF UREA AS A FUNCTION OF CONCENTRATION			
ITEM:	VALUE:	CONDITIONS:	REF.:
*PHYTOPLANKTON FROM NORTH PACIFIC	Y = -0.003 + 0.247*X	Y=SPECIFIC RATE OF ASSIMILATION (1/D),X= CONC OF UREA (UG-ATOM/L)	429

TABLE E135

EQUATION : UPTAKE OF ZN			
ITEM:	VALUE:	CONDITIONS:	REF.:
*GAMBUSIA AFFINIS	1-UP=1.39*EXP(-0.1614*T)+16.75*EXP (-0.0316*T)+81.66*EXP(-0.003556*T)	UP=UPTAKE (%), T=TIME (D), 3 COMPARTSMENTS	301

TABLE E136

EQUATION : URINARY PROTEIN EXCRETION			
ITEM:	VALUE:	CONDITIONS:	REF.:
*HOMO SAPIENS	Y=64.4X +74	R=.76,Y=PROTEIN IN URINE(MG/G CREATIN.),X=CD IN BLOOD(NG/G)	487

TABLE E137

EQUATION : VOLUME SWEPT CLEAR AS A FUNCTION OF CELL VOLUME			
ITEM:	VALUE:	CONDITIONS:	REF.:
*CALANUS PACIFICUS	F = 2.61*LOG(V) - 4.84	F=VOL SWEPT (U**3), V=CELL VOL (U**3), N=95, R=0.79, FOOD=DIF. DIATOM	493

TABLE E138

EQUATION : WATER CONTENT AS A FUNCTION OF CONDITION FACTOR AND BODY WEIGHT			
ITEM:	VALUE:	CONDITIONS:	REF.:
*SALMO TRUTTA	Y = 87.28*EXP(-0.07280*L)*W** (-0.0175)	Y= % WATER, L=CONDITION FACTOR, W= WET W (G), R=0.867	409

TABLE E139

EQUATION : YEARLY GROWTH AS A FUNC. OF DIAMETER, HEIGHT AND LEAF AREA OR WEIGHT			
ITEM:	VALUE:	CONDITIONS:	REF.:
*TREES	DELTA(D2H) = R*LA*(1-D*H/(DM*HM)	DELTA(D2H)=CHANGE IN VOLUME OVER 1 YEAR,H=HEIGHT,D=DIAMETER,R=CONST.	152
*TREES	DELTA(D2H) = R*LA*(1-D*H/(DM*HM)	DM AND HM ARE MAX VALUES OF D AND H,LA=LEAF AREA OR WEIGHTR,R= CONST.	152

TABLE E140

EQUATION : ZN			
ITEM:	VALUE:	CONDITIONS:	REF.:
*GAMBUSIA AFFINIS	LO = 8.15*EXP(-0.2375*T)+12.96*EXP (-0.01854*T)+78.89*EXP (-0.001272*T)	LO = LOSS (%), T=TIME (D), 3 COMPARTMENTS, TOTAL LOSS	301

TABLE E141

EQUATION : ZOOPLANKTON (F054)			
ITEM:	VALUE:	CONDITIONS:	REF.:
*BODY SIZE VS FILTRATION RATE	R=0.164*L**2.52	R=FILTRATION RATE(ML/ANIMAL*HR),L= LENGTH(MM),288-298 K, DAPHNIA MAGNA	395
*BODY SIZE VS FILTRATION RATE	R=0.31*L**3.02	R=FILTRATION RATE(ML/ANIMAL*HR),L= LENGTH(MM),293 K, DAPHNIA ROSEA	395
*CARBON CONTENT VS BODY LENGTH	LOG(C) = 0.507*L + 0.190	C=CARBON CONTENT (UG), L=LENGTH (MM), DAPHNIA PULICARIA	1470
*DEVELOPMENT OF ZOOPLANKTON COPEPODIT AS A FUNCTION OF TEMPERATURE	LOG(Y) = 1.57 - 0.528*LOG(X)	Y=DEVEL OF COPEPODIT (D), X=CEL, 4 CULTURE TEMP, DIAPTOMUS PALLIOUS	513
*DEVELOPMENT OF ZOOPLANKTON EGGS AS A FUNCTION OF TEMPERATURE	LOG(Y) = 1.16 - 0.77*LOG(X)	Y=DEVEL OF EGGS (D), X=CEL, 4 CULTURE TEMP, DIAPTOMUS PALLIOUS	513
*DEVELOPMENT OF ZOOPLANKTON NAUPLIUS AS A FUNCTION OF TEMPERATURE	LOG(Y) = 1.32 - 0.455*LOG(X)	Y=DEVEL OF NAUPLIUS (D), X=CEL, 4 CULTURE TEMP, DIAPTOMUS PALLIOUS	513
*EXCRETION OF AMMONIUM-N VS PHOSPHATE-P	Y=9.76*X+0.09	Y=EXCR.(NH4-N),X=EXCR.(?068) (PO4-P),R=0.907,N=44,22HRS,288 K, MARINE	897
*EXCRETION OF AMMONIUM-N VS TOTAL-N	Y=0.75*X-0.77	Y=EXCR.(NH4-N),X=EXCR.(?068) (TOTAL-N),R=0.879,N=39,22HRS,288 K, MARINE	897
*EXCRETION OF PHOSPHATE-P VS TOTAL-P	Y=0.48*X	Y=EXCR.(PO4-P),X=EXCR.(?068) (TOTAL-P),R=0.928,N=39,22HRS,288 K, MARINE	897
*EXCRETION OF TOTAL-N VS TOTAL-P	Y=7.28*X+0.83	Y=EXCR.(TOTAL-N),X=EXCR.(?068) (TOTAL-P),R=0.922,N=41,22HRS,288 K, MARINE	897
*GRAZING RATE OF ZOOPLANKTON FEEDING ON PHYTOPLANKTON	F = V * (LN(CO) - LN(CT)/T	F=ML SWEPT OF ZOOPL.,V=VOL OF EXP WATER(ML PER DRY WEIGHT OF ZOOPLAN.	985
*GRAZING RATE OF ZOOPLANKTON FEEDING ON PHYTOPLANKTON	F = V * (LN(CO) - LN(CT)/T	CO=PHYT. CONC IN CONTROL VOL,CT= PHYT IN EXPERIMENT BOTTLE	985

TABLE E141

EQUATION : ZOOPLANKTON			
ITEM:	VALUE:	CONDITIONS:	REF.:
*GRAZING RATE, USED IN MODELS	D(PHYT)/DT= -K*PHYT*ZOO	(F047)	361
*GRAZING RATE, USED IN MODELS	GRZ=MYZ*ZOO*((PHYT-KTR)/(KZ+PHYT)) , MYZ=MYZMAX*F(T)	(F047)	1503
*GRAZING RATE, USED IN MODELS	GRZ=ZOO*(PHYT-KTR)	E.G. G(ZOO)=MYZ*ZOO (F047)	1653
*GRAZING RATE, USED IN MODELS	GRZ=ZOO*MYZ*(1-EXP(DP*(PHYT-KTR)), MYZ=MYZMAX*F(T)	(F047)	1147
*GRAZING RATE, USED IN MODELS	GRZ=MYZ=(1-ZOO/CK)*ZOO	(F047)	949
*GRAZING RATE, USED IN MODELS	GRZ=MYZ*AV*(1-ZOO/CK)*ZOO, AV=F (PHYT)*F(OX)*F(T)*F(TOX)	AV IS RELATED TO AVAIBABILITY OF FOOD (F047)	1168
*GRAZING RATE, USED IN MODELS	GRZ=MYZ*((PHYT-KTR)/(KZ+PHYT))*ZOO	PHYT>KTR (F047)	503
*GRAZING RATE, USED IN MODELS	GRZ=MYZ'*PHYT/ZOO,	PHYT<KTR (F047)	503
*MORTALITY OF FRESHWATER ZOOPLANKTON	J = 0.005*Q/T	J=ZOOPL. DEATH (UG/(L*H), Q= BIOMASS, T=TIME (H)	1326
*MORTALITY, USED IN MODELS	MORT = MORTK*ZOO	(F047)	353
*MORTALITY, USED IN MODELS	MORT = MORTK*ZOO	(F047)	353
*RESPIRATION RATE VS BODY WEIGHT	LOG(R)=-0.169*LOG(W)+0.023	R=RESP.RATE(MIC.L O2/(MG BODY WT.) *HR,MARINE, W=DRY BODY WT.,BOREAL	718
*RESPIRATION RATE VS BODY WEIGHT	LOG(R)=-0.309*LOG(W)+0.357	R=RESP.RATE(MIC.L O2/(MG BODY WT.) *HR,MARINE, W=DRY BODY WT., TEMPERATE	718
*RESPIRATION RATE VS BODY WEIGHT	LOG(R)=-0.464*LOG(W)+0.874	R=RESP.RATE(MIC.L O2/(MG BODY WT.) *HR,MARINE, W=DRY BODY WT., TROPICAL	718
*RESPIRATION VS EXCRETION OF AMMONIUM-N	Y=0.12*X+0.05	Y=RESP.(F069),X=EXCR.(F070)(NH4-N) ,R=0.911,N=44,22HRS,288 K, MARINE	897
*RESPIRATION VS EXCRETION OF PHOSPHATE-P	Y=1.18*X+0.06	Y=RESP.(F069),X=EXCR.(F070)(PO4-P) ,R=0.890,N=41,22HRS,288 K, MARINE	897
*RESPIRATION VS EXCRETION OF TOTAL NITROGEN	Y=0.08*X-0.01	Y=RESP.(F069),X=EXCR.(F070) (TOTAL-N),R=0.867,N=36,22HRS,288 K, MARINE	897
*RESPIRATION VS EXCRETION OF TOTAL PHOSPHORUS	Y=0.57*X+0.06	Y=RESP.(F069),X=EXCR.(F070) (TOTAL-P),R=0.897,N=40,22HRS,288 K, MARINE	897
*RESPIRATION, USED IN MODELS	LOG(RESP) = 0.0364*T - 0.3418*LOG(W) + 0.6182	(F047)	764
*RESPIRATION, USED IN MODELS	RESP = RESPK*F(T)*ZOO	(F047)	353
*ZOOPLANKTON VOLUME AS A FUNCTION OF DAY-TIME	LOG10(Y)=1.32+0.49*COS(T)	Y=ZOOPLANKTON VOLUME(ML/1000M**3), T=TIME OF DAY(HR)	828

TABLE E142

The surface transfer coefficient has been employed to simulate reaeration in stratified layers in lakes, oceans and estuaries. One expression is: Kl = a1 + a2*Vw. The equations used in flowing water can only be used when the oxygen concentration is constant. Another equation used in lakes is: Kl = 1E6*Dm/(200-60*Vw**(1/2)) An equation used in estuaries is Kl = 1.57 + 0.32*Vw**2, where Vw and Kl are in units of feet/day and feet/sec respectively. For shallow lakes there has been developed several equations applicable over a limited range of wind speeds. The relationships are:
Kl = 0.362*Vw**(1/2) for 0< Vw < 5.5 m/sec
and K2 = 0.0277*Vw**2 for Vw>5.5 m/sec, where K2 is in meters/day. By shallow lakes it is implied that the vertical turbulent diffusion coefficient is constant throughout the depth of the lake.
At Colorado State Univ the effect of wind on reaeration rate in flowing water has been studied. The following equation has been developed, where Kl is in feet/day.: Kl = C*U*s*h*Uc/v. The constant of proportionality C was 0.0027. The other conditions were channel slope: 0.00043 and 0.001 , air velocity 22, 30 and 38 fps for each slope, discharge 0.79,0.83,0.91 cfs at 0.001 slope and 0.58,10.63,0.75 cfs at 0.0043 slope, water depth 0.385 feet. An equation developed by O Connor for non-tidal streams and rivers has also been applied to estuaries. The formula is: K2 =(Dm*U0)**0.5/(H**1.5)
REF.: 1567

Symbol	Physical description
Kl	Surface transfer coefficient (m/sec)
Vw	Wind speed (m/day)
a1	0.005 - 0.01 m/sec
a2	1E-6 to 1E-5
Dm	Molecular diffusivity (m2/sec)
v	Kinematic viscosity of water
U	Surface shear velocity (m/sec)
s	s = 0.0185*Vw**1.5, where Vw is in m/sec
Uc	Shear velocity defined as (g*h*Sc)**0.5 (feet/sec)
h	Normal depth (with uniform flow) (feet)
Sc	pressure-adjusted channel slope, S0 * dP/dx /(r*g)
S0	slope of energy gradient (channel slope for uniform flow)
dP/dx	Air pressure gradient in the longitudinal direction
g	Acceleration of gravity
U0	mean tidal velocity over a complete tidal cycle
H	average depth over a section over the tidal cycle

TABLE E143

Description of the reaeration that occurs in small dams as the water flows over the dam. The following relationship for reaeration over dams has been developed: Da - Db = (1 - 1/(1 + 0.11*a*b*(1+0.046*T)*H))*Da This equation was developed by Gameson, A.L. 1958. in :The Effect of Temperature Aeration at Weirs. Water and water engineering. London. An alternate equation developed from data from Mohawk River and Barge Channal in the State of N.Y. is : Da - Db = 0.037*H*Da. This equation is valid for dams up 15 feet high and for temperatures in range 20-25 deg. celcius. For small dams in Kentucky under low flow another equation is: Db = Da/r, where r = exp(0.16*H) at 25 deg. celcius. The value r was corrected for temperature by: ln(r) = ln(rt)*1.022**(25-T), where T is temperature in celcius and rt is the actual r. The conditions were 2<Flow<108 cfs, 2.6<H<14.0 feet, 17.0<T<24.0 deg. celcius.

REF.: 1567

Symbol	Physical description
Da	Dissolved oxygen deficit above dam (mg/l)
Db	Dissolved oxygen deficit below dam (mg/l)
T	Temperature (deg. celcius)
a	1.25 in clear to less polluted water, 1.00 in polluted water
b	1.00 for weir with free fall, 1.3 for step weirs or cascades.
H	Height through which the water falls (ft).

TABLE E144

The following equation describes the standard light day as proposed by Vollenweider. I0 = 1/2*Imax*(1 + cos(2*pi*t/l)). Imax is calculated from the daily total solar radiation such that the integral of I0 over one day equals the total daily radiation. The day length factor, l, is given as a function of time and lattitude as follows:
l = 24/pi/(arccos(-tan(phi)*tan(rho)).
Describing the variation of light intensity with time, Tetra Tech uses the following expression in order to obtain a constant average daily intensity : LI = TSR/(0.75*DH)

REF.: 1567

Symbol	Physical description
I0	incident solar radiation
l	day length factor (hours)
phi	lattitude (radians)
rho	declination (radians = 0.409*sin(0.0172*JD-1.3762), JD=Julian dayno
LI	constant average daily light intensity
TSR	total shortwave radiation
DH	duration of daylight in hours.

TABLE E145

The convective heat exchange at a water surface is given by
Qc = 0.0001*CONVK*p/p0*u*(ta - tw)
REF.: 944

Symbol	Physical description
Qc	convective heat transfer (kcal/(m2*hour)
CONVK	coefficient of convection, range 1 - 20
p	barometric pressure at the site
p0	sea level pressure
ta	air temperature (C deg)
tw	water temperature (C deg)
u	wind velocity (m/sec)

TABLE E146

The convective heat transfer (Qc) from a water surface can be described by Qc = p/0.622*rw*Cp*Nc*U2*(t2-t1) .
REF.: 945

Symbol	Physical description
p	local barometric pressure (mb)
rw	the vapor pressure for water in millibars
Cp	specific heat of air (kcal/(kg*deg))
Nc	empirical constant
U2	wind velocity (m/sec)
t2	air temperature (C deg)
t1	water temperature (C deg)

TABLE E147

Specific rate of accumulation of heavy metals on biological sludge. The rate of accumulation is a function of both physical and chemical environment. There appears to exist an upper limit of the quantity of metal, that possibly can be associated with the biological sludge. A deterministical equation is :
r = (k1*C**a - k2*(Q/(A-Q)**b))*f(y).
For constant pH and flows that causes the therm f(y) to be maximum (no resistance by film diffusion) the equation can be simplifided to:
r = k1*C**a - k2*(Q/(A-Q)**b

REF.: 1118

Symbols	Physical description
r	Overall rate of metal accum. on biomass (mg metal/(g biomass*min))
C	Liquid phase metal concentration (mg/l)
Q	Quantity of metal associated with biomass (mg metal/g biomass)
y	a measure of physical resistance to mass transfer.
f(y)	a function of y.
k1	forward rate constant
k2	reverse rate constant
a	constant exponent
b	constant exponent
A	Max level of a heavy metal quantity on sludge biomass (mg/g)

TABLE E148

DEFINITIONS OF EFFICIENCIES THAT HAVE BEEN PROPOSED IN ECOLOGICAL LITERATURE.

Abbreviation:	Meaning:	Equivalent terminology:
$\frac{An}{In} = \frac{GPn}{In} = \frac{S5}{So}$	assimilation at n / ingestion at n	Asssimilation efficiency. Trophic level intake efficiency. Efficiency of digestion.
$\frac{S5}{S2}$	assimilation at n / ingestion at n	Efficiency of primary production
$\frac{S5}{S4}$	assililation at n / ingestion at n	Assimilation efficiency. Trophic level energy intake efficiency.
$\frac{Rn}{In} = \frac{An-NPn}{In} = \frac{S5-NPn}{So}$	respiration at n / ingestion at n	Energy degradation or respiration efficiency.
$\frac{NPn}{In} = \frac{S5-R1}{So}$	net productivity at n / ingestion at n	Energy coefficient of growth of the first order. Ecological growth efficiency. Efficiency of assimilation. Gross efficiency of growth.
$\frac{S5-R1}{S4}$	net productivity at n / ingestion at n	Efficiency of photosynthesis in relation to visible radiation avaible to photosynthetic pigments.
$\frac{Pn}{In} = \frac{P1}{So}$	energy passed to n+1 / ingestion at n	Efficiency of transfer to next level in terms of ingested energy. Ecological efficiency. Gross efficiency of yield to ingestion.
$\frac{TAn}{In} = \frac{TA1}{So}$	tissue accumulation at n / ingestion at n	Efficiency of conversion to standing crop in terms of ingested energy.
$\frac{NPn}{An}$	net productivity at n / assimilation at n	Energy coefficient of growth of the second order. Growth efficiency. Tissue growth efficiency. Net efficiency of growth.
$\frac{Rn}{NPn}$	respiration at n / net productivity at n	Respiratory coefficient.
$\frac{Pn}{NPn}$	energy passed to n+1 / net productivity at n	Efficiency of transfer to next level in terms of assimilated energy.
$\frac{TAn}{NPn}$	tissue accumulation at n / net productivity at n	Increase efficiency. Efficiency of conversion to standing crop in terms of assimilated energy.
$\frac{In}{I(n-1)} = \frac{In}{So}$	ingestion at n / ingestion at n-1	Lindemann's efficiency, ratio of intakes of trophic levels. Ecological efficiency. Efficiency of transfer to next higher level in terms of ingested energy. Gross efficiency of yield to ingestion.
$\frac{An}{A(n-1)}$	assimilation at n / assimilation at n-1	Progressive efficiency. Biological efficiency, progressive relative efficiency at a given level in terms of relative productivities. Lindemann's efficiency, ratio of intakes of trophic levels. Trophic level energy intake efficiency, trophic level assimilation efficiency.
$\frac{Rn}{An}$	respiration at n / assimilation at n	
$\frac{Pn}{An}$	energy passed to n+1 / assimilation at n	
$\frac{Rn}{R(n-1)}$	respiration at n / respiration at n-1	
$\frac{In}{NP(n-1)}$	ingestion at n / net productivity at n-1	Consumption effciency. Utilization efficiency. Ecotrophic coefficient. (approximately).
$\frac{An}{NP(n-1)}$	assimilation at n / net productivity at n-1	Efficiency of energy conversion. Trophic level production ratios. Trophic level production efficiency.

Abbrevations used:
So = incident radiation. S1 = visible radiation. S2 = light penetrating to the community (excludes albedo and absorption by water). S3 = light absorbed by photosynthetic tissues. S4 = light available to photosynthetic pigments. S5 = energy fixed in photosynthesis (= gross primary productivity, GPP, assimilation by plants). I = ingestion. A = assimilation. Df = defecation or nonassimilation (I = A + Df). R = respiration. NP = net productivity (NP = A - R). P = production; energy passed to next higher trophic level. Dc = decomposition. TA = tissue accumulation; increase in biomass. L = loss from system. The trophic level is indicated by n, n-1 and n+1; the last two relative to n.

F

EQUATIONS FOR CHEMICAL - PHYSICAL PROCESSES IN THE ENVIRONMENT

EXCHANGE RATES
HEAT CONDUCTION
DIFFUSION

TABLE F1

EQUATION : ACCUMULATION AND REMOVAL RATES OF POLLUTANT			
ITEM:	VALUE:	CONDITIONS:	REF.:
*RUNOFF FROM LAND SURFACES	COMPLEX MODEL	SEE REF FOR EQUATIONS	364

TABLE F2

EQUATION : ACCUMULATION OF DUST AND DIRT			
ITEM:	VALUE:	CONDITIONS:	REF.:
*ON LAND SURFACES	COMPLEX MODEL	SEE REF FOR EQUATIONS	364

TABLE F3

EQUATION : ACTIVITY COEFFICIENTS (F056)			
ITEM:	VALUE:	CONDITIONS:	REF.:
*INDIVIDUALS IONS, DAVIES	LOG(F) = -A*Z**2*(SQRT(I)/(1+ SQRT(I)-0.2*I)	APPLICATION WITH IONIC STRENGTH LESS THAN 0.5	1526
*INDIVIDUALS IONS, DEBYE-HUCKEL	LOG(F) = -A*Z**2*SQRT(I)	APPLICATION WITH IONIC STRENGTH LESS THAN 10**(-2.3)	1526
*INDIVIDUALS IONS, EXTENDED DEBYE-HUCKEL	LOG(F) = -A*Z**2*SQRT(I)/(1+ B*C*SQRT(I))	APPLICATION WITH IONIC STRENGTH LESS THAN 10**(-1)	1526
*INDIVIDUALS IONS, GUNTELBERG	LOG(F) = -A*Z**2*SQRT(I)/(1+SQRT (I))	APP. IN SOLUTIONS OF SEVERAL ELECTROLYTES, I < 10**(-1)	1526

TABLE F4

EQUATION : ADSORPTION			
ITEM:	VALUE:	CONDITIONS:	REF.:
*DIPROPETRYN	S = 32.5*C**0.73	COBB SAND+2 % MUCK, S=ADSORBED (UG/G),C=EQUILBR. CONC (UG/CM3)	1088
*DIPROPETRYN	S = 18.5*C**0.89	BREWER CLAY LOAM, S=ADSORBED (UG/ G),C=EQUILBR. CONC (UG/CM3)	1088
*DIPROPETRYN	S = 8.91*C**0.81	PORT SILTY CLAY, S=ADSORBED (UG/G) ,C=EQUILBR. CONC (UG/CM3)	1088
*DIPROPETRYN	S = 6.18*C**0.86	TELLER FINE SANDY LOAM, S=ADSORBED (UG/G),C=EQUILBR CONC (UG/CM3)	1088
*DIPROPETRYN	S = 1.32*C**0.86	COBB SAND, S=ADSORBED (UG/G),C= EQUILBR. CONC (UG/CM3)	1088
*DIPROPETRYN	S = 0.07*C**0.97	QUARTZ SAND, S=ADSORBED (UG/G),C= EQUILBR. CONC (UG/CM3)	1088
*MERCURIC CHLORIDE ON SYNEDRA ULNA	LOG(Y) = LOG(400.4) + (1/1.3) *LOG(X)	X = CONC. OF HG IN CULTURE,Y= CONC. OF HG IN CELLS,293K	497
*METHOXYCHLOR ON ASPERGILLUS	X/M = 0.0052*CE**0.91	X/M=PESTICIDE SORBED ON ORG. (MG/ MG DW),CE EQULIBRIUM CONC (MG/L)	1211
*METHOXYCHLOR ON BACILLUS SUBTILIS	X/M = 0.048*CE**1.2	X/M=PESTICIDE SORBED ON ORG. (MG/ MG DW),CE EQULIBRIUM CONC (MG/L)	1211
*METHOXYCHLOR ON CHLORELLA PYRENOIDOSA	X/M = 0.0084*CE**0.99	X/M=PESTICIDE SORBED ON ORG. (MG/ MG DW),CE EQULIBRIUM CONC (MG/L)	1211
*METHOXYCHLOR ON FLAVOBACTERIUM HARRISONII	X/M = 0.0012*CE**0.81	X/M=PESTICIDE SORBED ON ORG. (MG/ MG DW),CE EQULIBRIUM CONC (MG/L)	1211
*PARATHION IN SOIL	LOG(X/M) = LOG(10.899 + 3.14*X) + 1.05*LOG(C)	X/M = AMOUT ADSORBED,X=% ORG MATTER, C=EQUILIBRIUM CONCENTRATION	1645
*PHOSPHATE IN SOLUTION	Y=0.205*C**0.615	ADSORBANT=6 LAKE SEDIMENTS, SATURATION AFTER 30 H.,2-5 MY PART BEST	711

TABLE F4

EQUATION : ADSORPTION (F057)			
ITEM:	VALUE:	CONDITIONS:	REF.:
*PHOSPHATE IN SOLUTION	Y=0.205*C**0.615	Y 1 MG P/G ADSORBANT, C=P CONC IN MG/L	711
*PHOSPHORUS ON FRESHWATER SEDIMENT	SP=4.56*CE/(0.62+CE) - 0.651	ALMIND SOE,DK,EXP(F058),PH=6.7,N=19,LANGMUIR ,R=0.99	737
*PHOSPHORUS ON FRESHWATER SEDIMENT	SP=5.97*CE/(0.59+CE) - 0.651	ALMIND SOE,DK,EXP(F059),PH=7.2,N=16,LANGMUIR ,R=0.98	737
*PHOSPHORUS ON FRESHWATER SEDIMENT	SP=3.32*CE/(0.25+CE) - 0.418	BRYRUP LANGSOE,DK,EXP(F058),PH=5.5,N=19,LANGMUIR ,R=0.99	737
*PHOSPHORUS ON FRESHWATER SEDIMENT	SP=5.07*CE/(0.52+CE) - 0.418	BRYRUP LANGSOE,DK,EXP(F059),PH=6.8,N=16,LANGMUIR ,R=0.99	737
*PHOSPHORUS ON FRESHWATER SEDIMENT	SP=0.61*CE/(0.87+CE) - 0.097	KALGAARD SOE,DK,EXP(F058),PH=5.9,N=19,LANGMUIR ,R=0.98	737
*PHOSPHORUS ON FRESHWATER SEDIMENT	SP=0.91*CE/(0.81+CE) - 0.097	KALGAARD SOE,DK,EXP(F059),PH=6.3,N=16,LANGMUIR ,R=0.99	737
*PHOSPHORUS ON FRESHWATER SEDIMENT	SP=2.10*CE/(0.29+CE) - 0.152	KVIND SOE,DK,EXP(F058),PH=6.3,N=19,LANGMUIR ,R=0.99	737
*PHOSPHORUS ON FRESHWATER SEDIMENT	SP=2.71*CE/(0.29+CE) - 0.152	KVIND SOE,DK,EXP(F059),PH=7.7,N=16,LANGMUIR ,R=0.99	737
*PHOSPHORUS ON FRESHWATER SEDIMENT	SP=1.88*CE/(0.10+CE) - 0.108	ESRUM SOE,DK,EXP(F058),PH=7.9,N=19,LANGMUIR ,R=0.98	737
*PHOSPHORUS ON FRESHWATER SEDIMENT	SP=4.03*CE/(0.48+CE) - 0.108	ESRUM SOE,DK,EXP(F059),PH=8.2,N=16,LANGMUIR ,R=0.98	737
*PHOSPHORUS ON FRESHWATER SEDIMENT	SP=1.68*CE/(0.38+CE) - 0.152	FURESOE,DK,EXP(F058),PH=7.9,N=19,LANGMUIR ,R=0.99	737
*PHOSPHORUS ON FRESHWATER SEDIMENT	SP=2.62*CE/(0.55+CE) - 0.152	FURESOE,DK,EXP(F059),PH=8.2,N=16,LANGMUIR ,R=0.99	737
*PHOSPHORUS ON FRESHWATER SEDIMENT	SP=0.55*CE/(1.45+CE) - 0.140	STORE GRIBSOE,DK,EXP(F058),PH=6.5,N=19,LANGMUIR ,R=0.99	737
*PHOSPHORUS ON FRESHWATER SEDIMENT	SP=0.82*CE/(1.10+CE) - 0.140	STORE GRIBSOE,DK,EXP(F059),PH=6.2,N=16,LANGMUIR ,R=0.99	737
*PHOSPHORUS ON FRESHWATER SEDIMENT	SP=0.99*CE/(3.66+CE) - 0.166	FR.BORG SLOTSSOE,DK,EXP(F058),PH=7.9,N=19,LANGMUIR ,R=0.82	737
*PHOSPHORUS ON FRESHWATER SEDIMENT	SP=0.84*CE/(1.62+CE) - 0.166	FR.BORG SLOTSSOE,DK,EXP(F059),PH=8.1,N=16,LANGMUIR ,R=0.92	737
*PHOSPHORUS ON FRESHWATER SEDIMENT	SP=0.97*CE**0.335 - 0.651	ALMIND SOE,DK,EXP(F060),PH=6.1,N=8,FREUNFLICH ,R=0.93	737
*PHOSPHORUS ON FRESHWATER SEDIMENT	SP=0.68*CE**0.592 - 0.418	BRYRUP LANGSOE,DK,EXP(F060),PH=5.9,N=8,FREUNDLICH ,R=0.98	737
*PHOSPHORUS ON FRESHWATER SEDIMENT	SP=0.05*CE**0.759 - 0.097	KALGAARD SOE,DK,EXP(F060),PH=8.0,N=8,FREUNDLICH ,R=0.95	737
*PHOSPHORUS ON FRESHWATER SEDIMENT	SP=0.47*CE**0.655 - 0.152	KVIND SOE,DK,EXP(F060),PH=7.6,N=8,FREUNDLICH ,R=0.83	737
*PHOSPHORUS ON FRESHWATER SEDIMENT	SP=0.81*CE**0.354 - 0.108	ESRUM SOE,DK,EXP(F060),PH=8.5,N=8,FREUNDLICH ,R=0.99	737
*PHOSPHORUS ON FRESHWATER SEDIMENT	SP=0.49*CE**0.762 - 0.152	FURESOE,DK,EXP(F060),PH=8.9,N=8,FREUNDLICH ,R=0.94	737
*PHOSPHORUS ON FRESHWATER SEDIMENT	SP=0.14*CE**0.495 - 0.140	STORE GRIBSOE,DK,EXP(F060),PH=6.8,N=8,FREUNDLICH ,R=0.95	737
*PHOSPHORUS ON FRESHWATER SEDIMENT	SP=0.17*CE**0.290 - 0.166	FR.BORG SLOTSSOE,DK,EXP(F060),PH=8.4,N=8,FREUNDLICH ,R=0.89	737
*PROMETRYN	S = 28.9*C**0.79	COBB SAND+2 % MUCK, S=ADSORBED (UG/G),C=EQUILBR. CONC (UG/CM3)	1088
*PROMETRYN	S = 9.95*C**0.83	BREWER CLAY LOAM, S=ADSORBED (UG/G),C=EQUILBR. CONC (UG/CM3)	1088
*PROMETRYN	S = 4.90*C**0.76	PORT SILTY CLAY, S=ADSORBED (UG/G),C=EQUILBR. CONC (UG/CM3)	1088
*PROMETRYN	S = 3.49*C**0.80	TELLER FINE SANDY LOAM, S=ADSORBED (UG/G),C=EQUILBR. CONC (UG/CM3)	1088
*PROMETRYN	S = 0.66*C**0.77	COBB SAND, S=ADSORBED (UG/G),C=EQUILBR. CONC (UG/CM3)	1088
*PROMETRYN	S = 0.07*C**(-0.54)	QUARTZ SAND, S=ADSORBED (UG/G),C=EQUILBR. CONC (UG/CM3)	1088
*PYRIDINE ON KAOLINITE	X = 0.01*CE**1.03	X=MG PYRIDINE SORBED/(G CLAY),CE=EQUILIBRIUM CONC OF PYR. MG/L, 297 K	61
*PYRIDINE ON KAOLINITE	X = 0.03*CE**1.01	X=MG PYRIDINE SORBED/(G CLAY),CE=EQUILIBRIUM CONC OF PYR. MG/L, 297 K	61

TABLE F4

EQUATION : ADSORPTION

ITEM:	VALUE:	CONDITIONS:	REF.:
*PYRIDINE ON SODIUM MONTMORILLONITE	X = 0.06*CE**1.02	X=MG PYRIDINE SORBED/(G CLAY),CE= EQUILIBRIUM CONC OF PYR. MG/L, 297 K	61
*PYRIDINE ON SODIUM MONTMORILLONITE	X = 0.12*CE**1.04	X=MG PYRIDINE SORBED/(G CLAY),CE= EQUILIBRIUM CONC OF PYR. MG/L, 297 K	61

TABLE F5

EQUATION : ATTENUATION OF A SUBMERGED OBJECT (SECCHI DISK) (F054)

ITEM:	VALUE:	CONDITIONS:	REF.:
*WATER	CR = CO*EXP(-(A + K*COS(B))*R)	B=ANGLE TO VERTIKAL,CR=APPARENT CONTRAST AGAINST BACKGROUND,CO= INHE-	687
*WATER	CR = CO*EXP(-(A + K)*R)	RENT CONTRAST,B=90,A AND K=ATT COEF OF COLLIMATED AND DIFFUSE LIGHT	687

TABLE F6

EQUATION : BACK RADIATION (F054)

ITEM:	VALUE:	CONDITIONS:	REF.:
*WATER SURFACE	QBR = 0.97*RO*TW**4	QBR=LONGWAVE BACK RADIATION (KCAL/ (M2*SEC), TW=WATER TEMP (KELV)	944
*WATER SURFACE	QBR = 7.36*10**(-2) + 1.17*10** (-3)*(TW - 273)	273<TW<303 , RO=STEPHAN-BOLTZMAN CONST.= 1.357*10**8 CAL (M2*S*DEG**4)	1620
*WATER SURFACE WITH ICE	QBR = 0.0736 + 0.00117*TD	TD = DRY BULB AIR TEMPERATURE (CEL DEG)	1566

TABLE F7

EQUATION : CONSTANT AVERAGE DAILY LIGHT INTENSITY VS TOTAL SHORTWAVE RADIATION

ITEM:	VALUE:	CONDITIONS:	REF.:
*LIGHT	LI = TSR/(0.75*DH)	LI=DAILY LIGHT, TSR=TOTAL SHORTWAVE RAD., DH=DURATION OF DAYLIGHT (H)	1565

TABLE F8

EQUATION : CONVECTIVE HEAT TRANSFER VS BOWEN RATIO AND LOSS THROUGH VAPORATION

ITEM:	VALUE:	CONDITIONS:	REF.:
*WATER	QC = R*QE	QC=CONVECTIVE HEAT TRANSFER, R= BOWEN RATIO,QE=LOSS BY EVAPORATION	1095

TABLE F9

EQUATION : EDDY DIFFUSION COEFFICIENT

ITEM:	VALUE:	CONDITIONS:	REF.:
*LAKES	DV = 50 - 200*D(T)/DZ	DV=VERT DIFF (CM2/SEC), T=WATER TEMP (CEL), Z=DEPTH (M)	919

TABLE F10

EQUATION : EFFECTIVE MIXING COEFFECIENT VS DEPTH,VELOCITY AND SALINITY GRADIENT (F061)			
ITEM:	VALUE:	CONDITIONS:	REF.:
*ESTUARY	EL = CL*(ABS(U) + RU)*(Y + RY) + C*W	EL=EFF. MIXING COEFF. (FT2/SEC), W = SALINITY GRADIENT	1566

TABLE F11

EQUATION : EFFECTIVE VERTICAL DIFFUSION COEFFICIENT VS DENSITY GRADIENT (F054)			
ITEM:	VALUE:	CONDITIONS:	REF.:
*BETWEEN BOXES IN LAKES ($48)	IF D(R)/DZ > K THEN DV = B*E**C	K = CRITICAL DENSITY (1/M), B,C= CONSTANTS, E = DENSITY GRADIENT	1095
*BETWEEN BOXES IN LAKES ($48)	IF D(R)/DZ <= K THEN DV = 0.1*10**(-4) - 2.5*10**(-4) M**2/SEC	R = DENSITY OF WATER, Z=DEPTH, DV = EFF..VERTICAL DIFFISION COEFF.	1095

TABLE F12

EQUATION : EVAPORATION RATE VS WIND STRESS,SATURATION PRESSURE AND TEMPERATURE (F054)			
ITEM:	VALUE:	CONDITIONS:	REF.:
*WATER SURFACE	E = (A + B*V)*(ES - EA)	E=RATE OF EVAP (M/SEC), V=WIND SPEED (M/S), A,B = CONSTANTS	944
*WATER SURFACE	ES = C + D*T	ES=SATURATION VAPOUR PRESSURE AT WATER TEMPERATURE (MB)	944
*WATER SURFACE	ES = 6.05 + 0.522*T	273-278 K, EA=VAPOR PRESSURE OF THE OVERLAYING ATMOSPHERE (MB)	944
*WATER SURFACE	ES = 5.10 + 0.710*T	278-283 K, T=WATER TEMPERATURE (CEL DEG)	944
*WATER SURFACE	ES = 2.65 + 0.954*T	283-288 K, C,D=CONSTANTS	944
*WATER SURFACE	ES = -2.04 + 1.265*T	288-293 K, A IS OFTEN ZERO	944
*WATER SURFACE	ES = -9.94 + 1.¤59*T	293-298 K, B IS APP. 1*10**(-9)	944
*WATER SURFACE	ES = -22.29 + 2.151*T	298-303 K	944
*WATER SURFACE	ES = -40.63 + 2.761*T	303-308 K	944
*WATER SURFACE	ES = -66.90 + 3.511*T	308-313 K	944

TABLE F13

EQUATION : EVAPOTRANSPIRATION (F054)			
ITEM:	VALUE:	CONDITIONS:	REF.:
*CROP (J) CAPACITY	CC = -OC/WW	EQUATIONS TO HYBRID SIMULATION	314
*CROP(A), HEAT BUDGET	RN + LE + H = 0	RN=NET RADIATION PLUS HEAT INFLOW TO THE SOIL,LE=TRANSPIRATION HEAT	314
*CROP(B), SENSIBLE HEAT FLUX	H = CP(RO)*(TS-TA)/RA	FLUX,H= SENSIBLE HEAT FLUX, TS= SURFACE TEMP OF CROP,TA = AMBIENT AIR	314
*CROP(C), TRANSPIRATION HEAT FLUX	LE = L*(XT-XA)/(RC+RA)	TEMP, RA=CROP BOUNDARY LAYER RESISTANCE,CP(RO)= HEAT CAPACITY AT	314
*CROP(D), DENSITY OF WATER VAPOUR AT CROP SURFACE TEMPERATURE = XT	XT = ABS(RN+LE*RA/CP(RO)+TA)	CONSTANT PRESSURE,XA=ABSOLUTE HUMIDITY OF THE AIR AT REFERNECE HEIGTH	314
*CROP(E), GENERAL EQUATION	LE = L*(ABS(RN+LE*RA/CP(RO)+TA) -XA)/(RA+RC)	RC=STOMATAL CROP RESISTANCE,L= LATANT HEAT OF VAPORAZATION OF WATER	314
*CROP(F) BOUNDARY LAYER RESISTANCE	RA = A/U	A=CONST DEPENDING ON STRUCTURE, HEIGHT AND ROUGHNESS OF CROP,U= WIND	314

TABLE F13

EQUATION : EVAPOTRANSPIRATION (F054)

ITEM:	VALUE:	CONDITIONS:	REF.:
*CROP(G), STOMATAL RESISTANCE	RC = A + B/RN + C/(WC+WC)	SPEED,WW=PERMANENT WILTING POINT, WC=CROP WATER POTIENTAL,A,B,C= CONST.	314
*CROP(H), CAPACITY FORMULA	WC - WW = OC/CC	OC=CROP WATER CONTENT,CC=CROP CAPACITY, RST = STEM RESISTANCE	314
*CROP(I), WATERFLOW TO THE CROP	D(OC)/DT = WC - WR/RST	WR = ROOT WATER POTIENTAL	314
*CROP(K), WATER DIFFUSION FROM THE BULK SOIL TO THE ROOTS	D(O)/DT = 1/R*DELTA(R*D(O)/ DR*DELTA(O)/DR	R=ROOT DIAMETER,O=SOIL WATER CONTENT,DELTA=PARTIEL DIFFERANTATION	314

TABLE F14

EQUATION : EXCHANGE ACROSS A GAS-LIQUID PHASE

ITEM:	VALUE:	CONDITIONS:	REF.:
*CO	V = K1*A*CO(AIR) - K2*A*CO(WATER)	V= VELOCITY OF UPTAKE OF CO INTO SOLUTION (1/TIME), A=INTERPHASE AREA	1016

TABLE F15

EQUATION : EXCHANGE EPILIMNION-HYPOLIMNION

ITEM:	VALUE:	CONDITIONS:	REF.:
*USED IN MODELS	U = KU*A*(PSH - PSE)	(F407)	E(AH
*USED IN MODELS	U2 = -DELTA(V)*PSE	(F047)	E(AH
*USED IN MODELS	U = UE + SIGMA(UJ)	(F047)	1500
*USED IN MODELS	U = D*A/(DELTA(X)*VE)*SSE	(F047)	E(GA
*USED IN MODELS	U1 = DELTA(V)*PSE	(F047)	E(AH
*USED IN MODELS	MIXR(Z) = TI(Z,T) - TI(Z,0)/(TI(E, 0) - TI(Z,0))	(F047)	1500
*USED IN MODELS	VJ = VJ*(SSJ - SSE)*MIXR(Z)	(F047)	1500

TABLE F16

EQUATION : EXCHANGE OF HEAT ACROSS THE AIR-WATER INTERFACE (F062)

ITEM:	VALUE:	CONDITIONS:	REF.:
*WATER	H = SI - (60 + 6.2*TW - 5.4Z*TA) - 1634*E - 0.89*U*(TW - TA)	H = EXCHANGE ACROSS INTERFACE	945

TABLE F17

EQUATION : EXTINCTION COEFFICIENT

ITEM:	VALUE:	CONDITIONS:	REF.:
*IN LAKES	K = 1.9/D + 0.17*X	K=EXT COEF. (1/M), D=SECCHI D (M), X=ALGAE CONC (MG DW/L)	120

TABLE F17

EQUATION : EXTINCTION COEFFICIENT			
ITEM:	VALUE:	CONDITIONS:	REF.:
*IN LAKES	K = A + 0.0088*P + 0.054*P**0.66	K=EXT COEF (1/M), P=ALGAE CONC (UG/L), A=K FOR P=0, LAKE ONTARIO	944
*LAKES	K = A + SIGMA(R(I) * S(I))	K=EXT COEF (1/M), A=K OF WATER (F071)	55
*SEAWATER	K = 0.174 + 0.815*P	K=EXT COEF., P = PLANKTON (G C/M3) , AUKE BAY, ALASKA	619
*WATER	EXT = 1.9/SECCHI DEPTH	NO INDICATIONS	1577
*WATER	K=0.04+0.0088*C+0.054*C**(2/3)	EMPIRICAL,K=AV.EXTINCT.COEF.(1/M), C=CHL.A CONC.(MG/M**3)	1339

TABLE F18

EQUATION : EXTINCTION COEFFICIENTS AND CHLOROPHYLL A			
ITEM:	VALUE:	CONDITIONS:	REF.:
*UPPER 2 METER OF POND WATER	KR = 0.002*CH + 1.3	ALL DATA,N=21,R=0.56,KR=EXTINCTION COEFFICIENT(RED LIGHT)(LN UNITS/ M)	1070
*UPPER 2 METER OF POND WATER	KR = 0.001*CH + 1.4	MARCH,LARGE CONC OF INORG MATTER,R =0.25,N=23,CH=CHLOROPHYLLYA(MG/M3)	1070
*UPPER 2 METER OF POND WATER	KB = 0.001*CH + 2.5	ALL DATA,N=27,R=0.08,KB=EXT COEF (BLUE LIGHT)(LN UNITS/M)	1070
*UPPER 2 METER OF POND WATER	KB = 0.001*CH + 2.8	MARCH,LARGE AMOUNTS OF INORG MATTER,R= -0.09,N=29	1070
*UPPER 2 METER OF POND WATER	KG = 0.004*CH + 1.3	ALL DATA,R=0.63,N=26,KG=EXT COEF (GREEN LIGHT)(LN UNITS/M)	1070
*UPPER 2 METER OF POND WATER	KG = 0.004*CH + 1.3	MARCH,LARGE CONC OF INORG MATTER,R =0.64,N=28,CH=CHLOROPHYLL A(MG/M3)	1070

TABLE F19

EQUATION : FREE CARBON DIOXIDE AT EQUILIBRIUM			
ITEM:	VALUE:	CONDITIONS:	REF.:
*IN WATER	X = A*H**2/(K1*(H + 2*K2))	X = MILLIMOLES H2CO3 (INCLUDING FREE CO2)	826

TABLE F20

EQUATION : FREUNDLICH ISOTHERM K VALUE VS ORGANIC MATTER			
ITEM:	VALUE:	CONDITIONS:	REF.:
*PARATHION ADSORPTION IN SOIL	K = 10.899 + 3.14*X	K FREUND., X = % ORGANIC MATTER	1645

TABLE F21

EQUATION : FROUDE NUMBER (F076)			
ITEM:	VALUE:	CONDITIONS:	REF.:
*INFLOWING WATER TO A WATER BODY	F = (V/A)/SQRT(G*D*DELTA(RO)/RO)	F=DENSIMETRIC FROUDE NUMBER, V= FLOW RATE, RO=DENSITY OF INFLOW	866

TABLE F22

EQUATION : GAS EXCHANGE			
ITEM:	VALUE:	CONDITIONS:	REF.:
*RIVER	F=50.8*U(**0.67)*H**(-0.85)	F=EXCH.COEFF.(CM/HR)AT 293K,U= AV.VELOCITY(CM/SEC),H=MEAN DEPTH (CM)	734

TABLE F22

EQUATION : GAS EXCHANGE			
ITEM:	VALUE:	CONDITIONS:	REF.:
*USED IN MODELS	GEK = 0.96768*(T + 36)	(F047)	764
*USED IN MODELS	GEK = D/(200 - 60*WIS**0.5)*10**6	(F047)	764

TABLE F23

EQUATION : HEAT CONDUCTANCE VS TEMPERATURE			
ITEM:	VALUE:	CONDITIONS:	REF.:
*ICE	HC = 0.05*(TC - TD)/X	HC=CONDUC (KCAL/M2), TC=SURFACE ICE TEMP,TD=DRY BULB TEMP (CEL),X =ICE	1566

TABLE F24

EQUATION : HEAT CONVECTION COEFFICIENT (F064)			
ITEM:	VALUE:	CONDITIONS:	REF.:
*BLUE SPRUCE BRANCH	HC = 10.0*(DELTA(T))**(1/4)	STILL AIR	268
*BLUE SPRUCE BRANCH	HC = 20.4 + 0.2*W**0.97	MOVING AIR, ANY DIRECTION, W < 30.0 CM/SEC	268
*BLUE SPRUCE BRANCH	HC = 0.95*W**0.97	MOVING AIR, ANY DIRECTION, W > 30.0 CM/SEC	268
*CYLINDRICAL LEAVES	HC = 6.00*(DELTA(T)/W)**(1/4)	STILL AIR	268
*CYLINDRICAL LEAVES	HC = 6.17*MY**(1/3)*W**(-2/3)	MOVING AIR	268
*FLAT LEAVES	HC = 7.86*(DELTA(T)/W)**(1/4)	STIL AIR, LEAF HORIZONTAL HOTTER THAN AIR, FACING UPWARD	268
*FLAT LEAVES	HC = 3.87*(DELTA(T)/W)**(1/4)	STILL AIR, LEAF HORIZONTAL, HOTTER THAN AIR, FACING DOWNWARD	268
*FLAT LEAVES	HC = 5.86*(DELTA(T)/W)**(1/4)	STILL AIR, LEAF HORIZONTAL, AVERAGE OF BOTH SURFACES	268
*FLAT LEAVES	HC = 6.04*(MY/W)**(1/2)	STILL AIR, LEAF VERTICAL	268
*FLAT LEAVES	HC = 5.73*(MY/W)**(1/2)	MOVING AIR	268
*PONDEROSA PINE BRANCH	HC = 11 + 3.6*(DELTA(T))**(0.3)	STILL AIR	268
*PONDEROSA PINE BRANCH	HC = 18.0 + 0.71*MY	MOVING AIR	268
*WHITE FIR	HC = 10.4*(DELTA(T))**(1/4)	STILL AIR	268
*WHITE FIR	HC = 20.4 + 1.75*MY**0.75	MOVING AIR, ACROSS BRANCH	268
*WHITE FIR	HC = 20.4 + 2.79*MY**0.75	MOVING AIR, ALONG BRANCH	268

TABLE F25

EQUATION : HEAT FLUX VS TEMPERATURE AND CONDUCTIVITY OF ICE			
ITEM:	VALUE:	CONDITIONS:	REF.:
*AIR-ICE INTERFACE	QN = -5*10**(-4)*(TA - TS)/X	QN=FLUX (KCAL/(M2*SEC), X=ICE (M), TA=AIR TEMP, T=TEMP OF ICE (CEL)	55

TABLE F26

EQUATION : HEAT LOSS DUE TO EVAPORATION VS DENSITY AND LATENT HEAT (F063)			
ITEM:	VALUE:	CONDITIONS:	REF.:
*FLUIDS	QE = RO*LW*E	QE=LOSS (KCAL/(M2/SEC), RO=DENSITY (KG/M3)	944

TABLE F27

EQUATION : HORIZONTAL EDDY DIFFUSION COEFFICIENT VS LENGTH OF SCALE GRID			
ITEM:	VALUE:	CONDITIONS:	REF.:
*LAKES	DH = 0.0056*L**1.3	DH=HORIZONTAL DIFF (CM2/SEC), L= LENGTH SCALE OF GRID	869

TABLE F28

EQUATION : LATENT HEAT OF VAPORIZATION VS TEMPERATURE			
ITEM:	VALUE:	CONDITIONS:	REF.:
*WATER	LW = 597 - 0.57*T	LW=LATENT HEAT (KCAL/KG), T=WATER TEMP (CEL DEG)	944

TABLE F29

EQUATION : LIGHT			
ITEM:	VALUE:	CONDITIONS:	REF.:
*PARTIEL VS TOTAL INSOLATION	Q=Q(0)*(0.3+0.7*S(R))	Q=PARTIEL INSOL.,Q(0)=TOTAL INSOL.,S(R)=EST.OF REL.SUNSHINE, CLOUDLESS	734

TABLE F30

EQUATION : MORTALITY			
ITEM:	VALUE:	CONDITIONS:	REF.:
*FRESHWATER BACTERIA	J = 0.7*Q*2.46*S*EXP(-0.76/T)/((24*1.17+S))	J=BACT DEATH (UG/(L*H) (F072)	1326

TABLE F31

EQUATION : NET ACCUMULATION OF SEDIMENT			
ITEM:	VALUE:	CONDITIONS:	REF.:
*FROM IMPERVIOUS LAND SURFACES	COMPLEX MODEL	SEE REF FOR EQUATIONS	364

TABLE F32

EQUATION : NET LONGWAVE RADIATION VS CLOUDCOVER AND DRY BULB TEMPERATURE			
ITEM:	VALUE:	CONDITIONS:	REF.:
*AIR-WATER INTERFACE	QAT = 1.23*10**(-16)*(1+0.17*CLOUD**2)*(TA+273)**6	QAT= NET LONGWAVE RADIATON (KCAL/(M2*SEC) (F073)	944

TABLE F33

EQUATION : NET SHORTWAVE RADIATION VS REFLECTIVITY OF WATER AND SUNS ALTITUDE			
ITEM:	VALUE:	CONDITIONS:	REF.:
*FLUX IN HEAT TRANSFER OF AIR-WATER INTERFACE	QSW = (1 - R)*QS	QSW=NET SHORTWAVE RADIATION (LANGLEYS/D), R=REFLECTIVITY OF WAT.=0.03	944

TABLE F33

EQUATION : NET SHORTWAVE RADIATION VS REFLECTIVITY OF WATER AND SUNS ALTITUDE (F054)			
ITEM:	VALUE:	CONDITIONS:	REF.:
*FLUX IN HEAT TRANSFER OF AIR-WATER INTERFACE	QSW = (1 - 1.18*A**(-0.77))*QS	CLEAR SKY, R=B*A**C, B,C= CONSTANTS, A = SUNS ALTITUDE IN DEGREES	944
*FLUX IN HEAT TRANSFER OF AIR-WATER INTERFACE	QSW = (1 - 2.20*A**(-0.97))*QS	SCATTERED SKY, QS = SHORTWAVE RADIATION (LANGLEYS/D)	944
*FLUX IN HEAT TRANSFER OF AIR-WATER INTERFACE	QSW = (1 - 0.95*A**(-0.75))*QS	BROKEN SKY, QS MEASURED BY US WEATHER BUREAU	944
*FLUX IN HEAT TRANSFER OF AIR-WATER INTERFACE	QSW = (1 - 0.35*A**(-0.45))*QS	OVERCAST SKY	944

TABLE F34

EQUATION : PENETRATION OF RADIATION THROUGH CANOPY (F065)			
ITEM:	VALUE:	CONDITIONS:	REF.:
*PLANTS	PT = EXP(-LAI*SP/SIN(IS))	PT=PENETRATION OF LIGHT, LAI=LEAF AREA INDEX, IS=SOLAR ALTITUDE	378

TABLE F35

EQUATION : POLLUTANT CONTENT OF OVERLAND FLOW			
ITEM:	VALUE:	CONDITIONS:	REF.:
*RUNOFF FROM LAND SURFACES	COMPLEX MODEL	SEE REF FOR EQUATIONS	364

TABLE F36

EQUATION : POTENTIAL NEEDED TO OXIDATION OF HG(0)			
ITEM:	VALUE:	CONDITIONS:	REF.:
*IN WATER	E = 850 + 30*LOG(CONC(HG(2+)/A)	E POT. (MV), A=BINDING CAPACITY OF HG TO MATTER IN WATER	759

TABLE F37

EQUATION : PRESSURE AS A FUNCTION OF DEPTH			
ITEM:	VALUE:	CONDITIONS:	REF.:
*WATER	P(Z) = P(0) + 0.0967*Z	P(Z) AND P(0) =PRESSURE IN Z M AND 0 M (ATMOSPHERE),Z=DEPTH (M)	1689

TABLE F38

EQUATION : RADIATIVE EQUILIBRIUM (F054)			
ITEM:	VALUE:	CONDITIONS:	REF.:
*EARTH	S*PHI*R**2(1-ALPHA)=4*PHI*(R**2)*I (T)	S=SOLAR CONSTANT,R=RADIUS OF THE EARTH,ALPHA=GLOBAL ALBEDO,	191
*EARTH	S*PHI*R**2(1-ALPHA)=4*PHI*(R**2)*I (T)	I(T)=OUTWARD RADIATION INTENSITY AT TOP OF THE ATMOSPHERE	191

TABLE F39

EQUATION : REAERATION COEFFICIENT AT 293 K (F055)			
ITEM:	VALUE:	CONDITIONS:	REF.:
*DEVELOPED FROM A RECIRCULATING FLUME	K2 = 4.74*(U/H)**0.85	U (FPS), H (FEET), K2 (1/D), DEEP LEES THAN 0.5 FEET	1112
*DEVELOPED FROM LAB EXPERIMENT	K2 = 7.03*U/(H**1.5)	U (FPS), H (FEET), K2 (1/D), RECIRCULATING WATER	725
*SIX STREAMS IN ENGLAND	K2 = 21.7*U**0.67/(H**1.85)	U (FPS), H (FEET), K2 (1/D), 0.1<U<5 AND 0.4<H<11 FPS	1201
*SMALL STREAM IN KENTUCKY	K2 = 0.30 + 0.19*S**1.2	S (FEET/MI), K2 (1/D), GAS TRANSFER TECH., 0<S<42	292
*STREAMS	K2 = 7.6*U/(H**1.33)	U (FPS), H (FEET), K2 (1/D), BASED ON DATA FROM SEVERAL SOURCES	873
*STREAMS	K2 = 0.054*(DELTA(H)/T)	H (FEET), T (H), K2 (1/D), 298 K, GAS TRANSFER TECH., 25<Q<300 CFS	1615
*STREAMS	K2 = 6.86*U**0.703/(H**1.05)	U (FPS), H (FEET), K2 (1/D), REGRESSION ANALYSIS FOR 9.8<K2<28.8	533
*STREAMS	K2 = 0.000299*(1+SQRT(UX/SQRT(G*H)))*SQRT(S*G/H)	NONE	1569
*STREAMS BELOW DAMS WITH OXYGEN DEFICIENT WATER	K2 = 11.6*U**0.969/(H**1.673)	U (FPS), H (FEET), K2 (1/D), 2<H<11 AND 1.8<U<5 FPS	253
*STREAMS DISPLAYING ISOTROPIC TURBULENCE	K2 = (DM*U)**0.5/(H**1.5)	UNITS COMPATIBLE, 1<H<30 AND 0.5<U<1.6 FPS	1150
*STREAMS WITH LOW VELOCITY	K2 = SQRT(DM*U/H)	VALID UNDER ANISOTROPIC CONDITIONS	1152

TABLE F40

EQUATION : RELEASE OF NUTRIENTS FROM SEDIMENT			
ITEM:	VALUE:	CONDITIONS:	REF.:
*USED IN MODELS	RELEASE = RELK*SETTL	(F047)	764
*USED IN MODELS	RELEASE = 4.0*SIN + 0.08/1000* A/V*F(T)	RELEASE OF N (F047)	764

TABLE F41

EQUATION : RESISTANCE TO TRANSPORT IN BOUNDARY LAYER (F066)			
ITEM:	VALUE:	CONDITIONS:	REF.:
*LEAVES	RA = 1.3*SQRT(W/V)	RA=REST., W=LEAF WIDTH (CM), V=WIND SPEED (CM/S)	797
*ORCHARD	1/RA = 1/(0.0173*O**(-0.56)*SQRT(U/D))	RA=REST. (S/M),O=FOLIAGE DENSITY ($41),D=SQRT(LEAF AREA),U=WIND (M/S)	872

TABLE F42

EQUATION : SALINITY VS CHLORINITY			
ITEM:	VALUE:	CONDITIONS:	REF.:
*SEAWATER	S = 0.03 + 0.001805*C	S=SALINITY (1/1000), C=CHLORINITY (MG/L)	21

TABLE F43

EQUATION : SETTLING			
ITEM:	VALUE:	CONDITIONS:	REF.:
*PHYTOPLANKTON	SEV=SEV(P)/SEV(O)	(F047)	240

TABLE F43

EQUATION : SETTLING			
ITEM:	VALUE:	CONDITIONS:	REF.:
*PHYTOPLANKTON	SETTL=(SEPH*PHYT+SEDT*DET)*(TA)**0.5/((TR)**0.5*Z)	(F047)	762
*PHYTOPLANKTON	SETTL=SEPH*D(PHYT)/D(Z)	(F047)	351
*PHYTOPLANKTON	SEPH=MIN(0.20,0.05*Z)	UNIT=M/D (F047)	1145
*PHYTOPLANKTON	PHYT(N)=0	NS>PS*5 (F047)	762

TABLE F44

EQUATION : SETTLING VELOCITY AS A FUNCTION OF RADIUS,GRAVITY,DENSITY AND FORM (F054)			
ITEM:	VALUE:	CONDITIONS:	REF.:
*PARTICLES (PHYTOPLANKTON, ZOOPLANKTON, DETRITUS ETC.)	V = 2*G*R**2*(Q`-Q)/(9*N*O)	V=SET VELOCITY,G=ACC OF GRAVITY,R= RADIUS OF PART.,O=FORM RISISTANCE	1455
*PARTICLES (PHYTOPLANKTON, ZOOPLANKTON, DETRITUS ETC.)	V = 2*G*R**2*(Q`-Q)/(9*N*O)	Q` AND Q=DENSITIES OF PARTICLE AND LIQUID, N=VISCOSITY OF LIQUID	1455
*PARTICLES (PHYTOPLANKTON, ZOOPLANKTON, DETRITUS ETC.)	O = US/UA	US AND UA=TERMINAL VELOCITY OF SPHERE AND NON-SPH,SAME DENSITY AND VOL	1455

TABLE F45

EQUATION : SOIL EROSION			
ITEM:	VALUE:	CONDITIONS:	REF.:
*FROM LAND SURFACES	COMPLEX MODEL	SEE REF FOR EQUATIONS	364

TABLE F46

EQUATION : SOIL NITROGEN AS A FUNCTION OF YEARS CULTIVATED			
ITEM:	VALUE:	CONDITIONS:	REF.:
*PRAIRIE SOILS	LN(N) = 4.517 - 0.007*Y	N=% NITROGEN OF INITIAL, Y=YEARS OF CULTIVATION, N. AMERICA	753

TABLE F47

EQUATION : STRUCTURE OF COMMUNITIES			
ITEM:	VALUE:	CONDITIONS:	REF.:
*EVENNESS INDEX	E = H``/LOG(S)	H``=SHANNON INDEX, S = NO. OF SPECIES	1260
*INDEX OF DOMINANCE	C = SIGMA((NI/N)**2)	NI=IMPORTANCE OF I`TH SPECIES, N= SIGMA(NI)	1438
*INDEX OF SIMILARITY BETWEEN TWO SAMPLES A AND B	S = 2*C/(X + Y))**2)	X=NO. OF SPECIES IN A,Y=NO. OF SPECIES IN B, C=NO. OF SPECIES IN BOTH	1476
*SHANNON INDEX OF DIVERSITY	H`` = -SIGMA((NI/N)*LOG(NI/N))	NI=IMPORTANCE OF I`TH SPECIES, N= SIGMA(NI)	1420
*SPECIES DIVERSITY, SPECIES RICHNESS	D1 = S - 1/LOG(N)	S=NO. OF SPECIES, N= NO. OF INDIVIDUALS	980
*SPECIES DIVERSITY, SPECIES RICHNESS	D2 = S/SQRT(N)	S=NO. OF SPECIES, N= NO. OF INDIVIDUALS	1025
*SPECIES DIVERSITY, SPECIES RICHNESS	D3 = S PER 1000 INDIVIDUALS	S = NO. OF SPECIES	1169

TABLE F48

EQUATION : SULPHUR DIOXIDE			
ITEM:	VALUE:	CONDITIONS:	REF.:
*AIR CONCENTRATIONS	C = CO *(1-0.1*T)	C=SO2 CONC.(PPHM),CO=C AT 0 DEGREES,STOCKHOLM,T=TEMP C,T<0	677
*AIR CONCENTRATIONS	C = CO *(1-0.06*T)	C=SO2 CONC.(PPHM),CO=C AT 0 DEGREES, STOCKHOLM,T=TEMP IN C, T>0	677

TABLE F49

EQUATION : SURFACE TEMPERATURE (F054)			
ITEM:	VALUE:	CONDITIONS:	REF.:
*EARTH	T=((1-A)*S/4Z)**(1/4)	A=ALBEDO,S=ENERGY FLUX(=1400W/ M**2),Z=STEFAN-BOLTZMANN CONSTANT,	1441
*EARTH	T=((1-A)*S/4Z)**(1/4)	NO ATMOSPHERE AND GOOD SURFACE CONDUCTIVITY	1441
*EARTH	I(T)=I(0)-DELTA(I)=EPSILON*SIGMA*T (0)**4-DELTA(T)	I(T)=OUTWARD RADIATION INTENSITY AT TOP OF THE ATMOSPHERE,	191
*EARTH	I(T)=I(0)-DELTA(I)=EPSILON*SIGMA*T (0)**4-DELTA(T)	I(0)=EMITTED RADIATION OF EARTH-SURFA.,T(0)=MEAN EARTH-SURF.TEMP. ,	191
*EARTH	I(T)=I(0)-DELTA(I)=EPSILON*SIGMA*T (0)**4-DELTA(T)	EPSILON*SIGMA=EFFECTIVE EMISSIVITY OF SURFACE*STEFAN-BOLTZMANN CONST.	191
*EARTH	I(T)=I(0)-DELTA(I)=EPSILON*SIGMA*T (0)**4-DELTA(T)	DELTA(I)=DIFFERENCE BETW.UPWARD RAD.FROM EARTH-SURF.AND OUTWARD	191
*EARTH	I(T)=I(0)-DELTA(I)=EPSILON*SIGMA*T (0)**4-DELTA(T)	RADIATION AT TOP OF THE ATMOSPHERE	191

TABLE F50

EQUATION : SUSPENDED SOLIDS			
ITEM:	VALUE:	CONDITIONS:	REF.:
*IN RUNOFF FROM LAND SURFACES	COMPLEX MODEL	SEE REF FOR EQUATIONS	364

TABLE F51

EQUATION : TEMPERATURE OF OVERLAND FLOW			
ITEM:	VALUE:	CONDITIONS:	REF.:
*FROM LAND SURFACES	COMPLEX MODEL	SEE REF FOR EQUATIONS	364

TABLE F52

EQUATION : THICKNESS OF ICE VS HEAT LOSS TO ATMOSPHERE AND FUSION CONTENT			
ITEM:	VALUE:	CONDITIONS:	REF.:
*WATER	X = - QT/(7.97*10**(-4))	X=THICKNESS (M), QT = TOTAL HEAT LOSS TO ATMOSPHERE (KCAL/M2)	55

TABLE F53

EQUATION : TOTAL CARBONATE AS A FUNCTION OF SULFIDE SULPHUR			
ITEM:	VALUE:	CONDITIONS:	REF.:
*IN BLACK SEA	Y = 3260 + 0.36*X	Y=TOTAL CARBONATE,X=SULFIDE SULPHUR, UNITS=MICRO M	1526

TABLE F54

EQUATION : TRANSFORMATION OF DDT			
ITEM:	VALUE:	CONDITIONS:	REF.:
*ESCHERICHIA COLI	TOTAL DDT= DDT + 1.114(DDD + DDE)	NONE	808

TABLE F55

EQUATION : TRANSPORT OF SEDIMENT			
ITEM:	VALUE:	CONDITIONS:	REF.:
*FROM LAND SURFACES	COMPLEX MODEL	SEE REF FOR EQUATIONS	364

TABLE F56

EQUATION : UPTAKE OF MIREX FROM SEDIMENT			
ITEM:	VALUE:	CONDITIONS:	REF.:
*TRINECTES MACULATUS, FLOW-THROUGH EXPERIMENT	Y = 0.34*X - 0.06	Y=MUSCLE (PPB),X=WATER (PPB),PH= 7.8, 277.9 K, 4 W, SEDIMENT=543 PPB	844
*TRINECTES MACULATUS, FLOW-THROUGH EXPERIMENT	Y = 0.32*X - 0.37	Y=MUSCLE (PPB),X=WATER (PPB),PH= 7.8, 277.9 K, 4 W, SEDIMENT=1666 PPB	844
*TRINECTES MACULATUS, FLOW-THROUGH EXPERIMENT	Y = 0.82*X - 0.43	Y=MUSCLE (PPB),X=WATER (PPB),PH= 7.8, 277.9 K, 4 W, SEDIMENT=5000 PPB	844
*TRINECTES MACULATUS, FLOW-THROUGH EXPERIMENT	Y = 0.93*X - 0.20	Y=LIVER (PPB), X=WATER (PPB),PH= 7.8, 277.9 K, 4 W, SEDIMENT=1666 PPB	844
*TRINECTES MACULATUS, FLOW-THROUGH EXPERIMENT	Y = 1.42*X - 0.73	Y=LIVER (PPB), X=WATER (PPB),PH= 7.8, 277.9 K, 4 W, SEDIMENT=5000 PPB	844
*TRINECTES MACULATUS, STATIC EXPERIMENT	Y = 0.41*X - 0.34	Y=MUSCLE (PPB),X=WATER (PPB),PH= 7.8, 277.9 K, 4 W, SEDIMENT=556 PPB	844
*TRINECTES MACULATUS, STATIC EXPERIMENT	Y = 0.36*X - 0.20	Y=MUSCLE (PPB),X=WATER (PPB),PH= 7.8, 277.9 K, 4 W, SEDIMENT=1650 PPB	844
*TRINECTES MACULATUS, STATIC EXPERIMENT	Y = 0.94*X - 0.65	Y=MUSCLE (PPB),X=WATER (PPB),PH= 7.8, 277.9 K, 4 W, SEDIMENT=5000 PPB	844
*TRINECTES MACULATUS, STATIC EXPERIMENT	Y = 0.74*X - 0.65	Y=LIVER (PPB), X=WATER (PPB),PH= 7.8, 277.9 K, 4 W, SEDIMENT=556 PPB	844
*TRINECTES MACULATUS, STATIC EXPERIMENT	Y = 0.90*X - 0.34	Y=LIVER (PPB), X=WATER (PPB),PH= 7.8, 277.9 K, 4 W, SEDIMENT=1650 PPB	844
*TRINECTES MACULATUS, STATIC EXPERIMENT	Y = 2.25*X - 1.55	Y=LIVER (PPB), X=WATER (PPB),PH= 7.8, 277.9 K, 4 W, SEDIMENT=5000 PPB	844

TABLE F57

EQUATION : VERTICAL DISPERSION COEFFICIENT VS DEPTH, WIND SPEED AND THERMOCLINE (F054)			
ITEM:	VALUE:	CONDITIONS:	REF.:
*ONE-DIMENSIONAL DISPERSION	DV = A + B*V*D**(-4.6*Z/D)	DV=VERTICAL DISPERSION COEFF. (M**2/SEC), A,B=CONSTANTS (M2/S, M)	55
*ONE-DIMENSIONAL DISPERSION, AMERICAN FALLS	DV = 1*10**(-5) + 1*10**(-4)*V*D** (-4.6*Z/D)	MAX DEPTH = 18 M, WELL-MIXED, Z = DEPTH (M), V = WIND SPEED (M/S)	55
*ONE-DIMENSIONAL DISPERSION, LAKE MENDOTA	DV = 5*10**(-7) + 5*10**(-5)*V*D** (-4.6*Z/D)	MAX DEPTH = 24 M, STRATIFIED	55

TABLE F57

EQUATION : VERTICAL DISPERSION COEFFICIENT VS DEPTH, WIND SPEED AND THERMOCLINE			
ITEM:	VALUE:	CONDITIONS:	REF.:
*ONE-DIMENSIONAL DISPERSION, LAKE WASHINGTON	DV = 1*10**(-6) + 1*10**(-5)*V*D**(-4.6*Z/D)	MAX DEPTH = 65 M, STRATIFED, D = DEPTH OF THERMOCLINE	55
*ONE-DIMENSIONAL DISPERSION, LAKE WINGRA	DV = 5*10**(-5) + 2*10**(-4)*V*D**(-4.6*Z/D)	MAX DEPTH = 5 M, WELL-MIXED	55
*ONE-DIMENSIONAL DISPERSION, LONG LAKE	DV = 5*10**(-6) + 5*10**(-5)*V*D**(-4.6*Z/D)	MAX DEPTH = 54 M, LINEARLY STRATIFIED	55

TABLE F58

EQUATION : VERTICAL EDDY CONDUCTIVITY AS A FUNCTION OF WIND STRESS AND DENSITY			
ITEM:	VALUE:	CONDITIONS:	REF.:
*FRESHWATER LAKES	KV = 100*ABS(TW)/R*SQRT(1/(1 + 3.3*RI)	KV=EDDY COND,TW=WIND STRESS,RI= RICHARDSON NO.,R=DENSITY OF WATER	1565

TABLE F59

EQUATION : VERTICAL FLOW OF NUTRIENT SALTS (F067)			
ITEM:	VALUE:	CONDITIONS:	REF.:
*OCEANS	F = B*(N2-N1)/Z	F=FLOW,N2-N1= CONCENTRATION GRADIENT, Z=THICKNESS OF LAYER,B= CONST.	1636

TABLE F60

EQUATION : YEARLY CYCLE OF TEMPERATURE			
ITEM:	VALUE:	CONDITIONS:	REF.:
*KOOTENAY LAKE	TEMP=4.0+14.0*EXP(-1/2((W-34)/9)**2)	W=WEEKS,W=W+52 FOR W=<8,TEMP.IN DEG.CELCIUS	1095
*NARRAGANSETT BAY	T = 10.75 - 10.25*COS(2*3.14*(DAY-30)/365)	T=TEMP (C DEGREES), DAY = JULIAN C.D., 1/1=001	1132

G

PROCESSES IN THE ENVIRONMENT

ABSORPTION
ACTIVITY
ATM. DEPOSITION
ATM. EMISSION
DIFFUSION
DISPERSION
EQUILIBRIUM
RATE OF PROCESSES
STOICHIOMETRIC RATIO
TEMP. COEFFICIENT
TURNOVER TIME

G1

ABSORPTION AND ACTIVITY

ABSORPTION COEFFICIENT
ACTIVITY COEFFICIENT

TABLE G1

ABSORPTION COEFFICIENT : OXYGEN (VOL/VOL)			
ITEM:	VALUE:	CONDITIONS:	REF.:
*SOLUBILITY OF OXYGEN (VOL/VOL) IN WATER	48.9 ML O2/LITER	273 K, VOLUME O2 (ML) DISSOLVED IN 1 LITRE OF WATER	605
*SOLUBILITY OF OXYGEN (VOL/VOL) IN WATER	42.9 ML O2/LITER	278 K, VOLUME O2 (ML) DISSOLVED IN 1 LITRE OF WATER	605
*SOLUBILITY OF OXYGEN (VOL/VOL) IN WATER	38.0 ML O2/LITER	283 K, VOLUME O2 (ML) DISSOLVED IN 1 LITRE OF WATER	605
*SOLUBILITY OF OXYGEN (VOL/VOL) IN WATER	34.2 ML O2/LITER	288 K, VOLUME O2 (ML) DISSOLVED IN 1 LITRE OF WATER	605
*SOLUBILITY OF OXYGEN (VOL/VOL) IN WATER	31.0 ML O2/LITER	293 K, VOLUME O2 (ML) DISSOLVED IN 1 LITRE OF WATER	605
*SOLUBILITY OF OXYGEN (VOL/VOL) IN WATER	28.3 ML O2/LITER	298 K, VOLUME O2 (ML) DISSOLVED IN 1 LITRE OF WATER	605
*SOLUBILITY OF OXYGEN (VOL/VOL) IN WATER	26.1 ML O2/LITER	303 K, VOLUME O2 (ML) DISSOLVED IN 1 LITRE OF WATER	605
*SOLUBILITY OF OXYGEN (VOL/VOL) IN WATER	24.4 ML O2/LITER	308 K, VOLUME O2 (ML) DISSOLVED IN 1 LITRE OF WATER	605
*SOLUBILITY OF OXYGEN (VOL/VOL) IN WATER	23.1 ML O2/LITER	313 K, VOLUME O2 (ML) DISSOLVED IN 1 LITRE OF WATER	605

TABLE G2

ACTIVITY COEFFICIENT : IONS IN SEAWATER			
ITEM:	VALUE:	CONDITIONS:	REF.:
*ACTIVITY OF CA(2+)	0.28	CHLORINITY = 1.9 %, PH=8.15, IONIC STRENGTH = 0.7, 298 K	504
*ACTIVITY OF CL(-)	0.64	CHLORINITY = 1.9 %, PH=8.15, IONIC STRENGTH = 8.7, 298 K	504
*ACTIVITY OF CO3(2-)	0.20	CHLORINITY = 1.9 %, PH=8.15, IONIC STRENGTH = 0.7, 298 K	504
*ACTIVITY OF HCO3(-), NACO3(-), NASO4(-), KSO4(-), MGHCO3(-), CAHCO3(+)	0.68	CHLORINITY = 1.9 %, PH=8.15, IONIC STRENGTH = 0.7, 298 K	504
*ACTIVITY OF K(+)	0.64	CHLORINITY = 1.9 %, PH=8.15, IONIC STRENGTH = 0.7, 298 K	504
*ACTIVITY OF MG(2+)	0.36	CHLORINITY = 1.9 %, PH=8.15, IONIC STRENGTH = 0.7, 298 K	504
*ACTIVITY OF NA(+)	0.76	CHLORINITY = 1.9 %, PH=8.15, IONIC STRENGTH = 0.7, 298 K	504
*ACTIVITY OF NAHCO3, MGCO3, CACO3, MGSO4, CASO4	1.13	CHLORINITY = 1.9 %, PH=8.15, IONIC STRENGTH = 0.7, 298 K	504
*ACTIVITY OF SO4(2-)	0.12	CHLORINITY = 1.9 %, PH=8.15, IONIC STRENGTH = 0.7, 298 K	504

G2

THE ATMOSPHERE

DEPOSITION
AND
EMISSION

TABLE G3

ATMOSPHERE : DEPOSITION			
ITEM:	VALUE:	CONDITIONS:	REF.:
*AL, ATMOSPHERIC RAINOUT	0.7 MEGATONS/YR	GLOBAL,CONC.OF ATM.PARTIC.(HAWAII) =6.7+-7.2NG/M**3	550
*CA, ATMOSPHERIC RAINOUT	16 MEGATONS/YR	GLOBAL,CONC.OF ATM.PARTIC.(HAWAII) =160+-90NG/M**3	550
*K, ATMOSPHERIC RAINOUT	12 MEGATONS/YR	GLOBAL,CONC.OF ATM.PARTIC.(HAWAII) =120+-60NG/M**3	550
*NITROGEN, ATMOSPHERE IN ZONE 45 - 50 DEGREES N	0.35*10**12 G N/YEAR	GASEOUS,TOTAL N=4.3,NH4(+)=3.0UG N/M3,NH3=2.1PPB(V),AREA=10**12 M2	1475
*NITROGEN, ATMOSPHERE IN ZONE 45 - 50 DEGREES N	0.05*10**12 G N/YEAR	PART. N,TOTAL N=4.3,NH4(+)=3.0UG N/M3,NH3=2.1PPB(V),AREA=10**12 M2	1475
*NITROGEN, ATMOSPHERE IN ZONE 45 - 50 DEGREES N	0.3*10**12 G N/YEAR	LAND, TOTAL N (GASEOUS + PART)	1475
*NITROGEN, ATMOSPHERE IN ZONE 45 - 50 DEGREES N	0.1*10**12 G N/YEAR	SEA, TOTAL N (GASEOUS + PART)	1475
*NITROGEN, ATMOSPHERE IN ZONE 45 - 65 DEGREES N	0.86*10**12 G N/YEAR	GASEOUS DEPOSITION	1475
*NITROGEN, ATMOSPHERE IN ZONE 45 - 65 DEGREES N	0.14*10**12 G N/YEAR	PARTICULATE N DEPOSITION	1475
*NITROGEN, ATMOSPHERE IN ZONE 45 - 65 DEGREES N	0.73*10**12 G N/YEAR	DEPOSITION ON LAND	1475
*NITROGEN, ATMOSPHERE IN ZONE 45 - 65 DEGREES N	0.28*10**12 G N/YEAR	DEPOSITION ON SEA	1475
*NITROGEN, ATMOSPHERE IN ZONE 45 - 65 DEGREES N	1.00*10**12 G N/YEAR	TOTAL DEPOSITION IN NORTHWESTERN EUROPE	1475
*NITROGEN, ATMOSPHERE IN ZONE 50 - 55 DEGREES N	0.37*10**12 G N/YEAR	TOT N=4.7,NH4=2.3 UG N/M3,NH3= 3.3PPB(V),AREA=1.1*10**12 M2, GASEOUS	1475
*NITROGEN, ATMOSPHERE IN ZONE 50 - 55 DEGREES N	0.05*10**12 G N/YEAR	TOT N=4.7,NH4=2.3 UG N/M3,NH3= 3.3PPB(V),AREA=1.1*10**12 M2,PART N	1475
*NITROGEN, ATMOSPHERE IN ZONE 50 - 55 DEGREES N	0.3*10**12 G N/YEAR	TOT N=4.7,NH4=2.3 UG N/M3,NH3= 3.3PPB(V),AREA=1.1*10**12 M2,ON LAND	1475
*NITROGEN, ATMOSPHERE IN ZONE 50 - 55 DEGREES N	0.12*10**12 G N/YEAR	TOT N=4.7,NH4=2.3 UG N/M3,NH3= 3.3PPB(V),AREA=1.1*10**12 M2,ON SEA	1475
*NITROGEN, ATMOSPHERE IN ZONE 55 - 60 DEGREES N	0.08*10**12 G N/YEAR	TOT N=2.2,NH4=1.5 UG N/M3,NH3=1.1 PPB(V),AREA=10**12 M2,GASEOUS	1475
*NITROGEN, ATMOSPHERE IN ZONE 55 - 60 DEGREES N	0.02*10**12 G N/YEAR	TOT N=2.2,NH4=1.5 UG N/M3,NH3=1.1 PPB(V),AREA=10**12 M2,PARTICULATE N	1475
*NITROGEN, ATMOSPHERE IN ZONE 55 - 60 DEGREES N	0.08*10**12 G N/YEAR	TOT N=2.2,NH4=1.5 UG N/M3,NH3=1.1 PPB(V),AREA=10**12 M2,ON LAND	1475
*NITROGEN, ATMOSPHERE IN ZONE 55 - 60 DEGREES N	0.02*10**12 G N/YEAR	TOT N=2.2,NH4=1.5 UG N/M3,NH3=1.1 PPB(V),AREA=10**12 M2,ON SEA	1475
*NITROGEN, ATMOSPHERE IN ZONE 60 - 65 DEGREES N	0.06*10**12 G N/YEAR	TOT N=1.6,NH4=1.1 UG N/M3,NH3=0.8 PPB(V),AREA=0.9*10**12 M2,GASEOUS	1475
*NITROGEN, ATMOSPHERE IN ZONE 60 - 65 DEGREES N	0.02*10**12 G N/YEAR	TOT N=1.6,NH4=1.1 UG N/M3,NH3=0.8 PPB(V),AREA=0.9*10**12 M2,PART N	1475
*NITROGEN, ATMOSPHERE IN ZONE 60 - 65 DEGREES N	0.05*10**12 G N/YEAR	TOT N=1.6,NH4=1.1 UG N/M3,NH3=0.8 PPB(V),AREA=0.9*10**12 M2,ON LAND	1475
*NITROGEN, ATMOSPHERE IN ZONE 60 - 65 DEGREES N	0.04*10**12 G N/YEAR	TOT N=1.6,NH4=1.1 UG N/M3,NH3=0.8 PPB(V),AREA=0.9*10**12 M2,ON SEA	1475

TABLE G4

ATMOSPHERE : EMISSION			
ITEM:	VALUE:	CONDITIONS:	REF.:
*AEROSOLS, ANTHROPOGENIC EMISSION	20*10**7 TONS/YEAR	INDUSTRIAL EMISSION,AGRICULTURAL EMISSION	16
*AEROSOLS, NATURAL EMISSION	11*10**8 TONS/YEAR	SEA SALT,SOIL DUST,PARTICLES FROM VOLCAN.ERUPT.AND FOREST FIRES	16
*AMMONIA, ANTHROPOGENIC EMISSION	4*10**6 TONS/YEAR	WASTE TREATMENT	16

TABLE G4

ATMOSPHERE : EMISSION			
ITEM:	VALUE:	CONDITIONS:	REF.:
*AMMONIA, NATURAL EMISSION	6*10**9 TONS/YEAR	BIOLOGICAL DECAY	16
*CARBON AND FLY ASH TO ATMOSPHERE	0.25 *10**8 TONS/YR	FOSSIL FUEL COMBUSTION	550
*CARBONMONOXIDE FIRES AND BURNINGS TO ATMOSPHERE	17*10**6 TONS/YEAR	UNITED STATES, 1968	35
*CARBONMONOXIDE INDUSTRIAL PROCESSES TO ATMOSPHERE	10*10**6 TONS/YEAR	UNITED STATES, 1968	35
*CARBONMONOXIDE SOLID WASTE DISPOSAL TO ATMOSPHERE	8*10**6 TONS/YEAR	UNITED STATES, 1968	35
*CARBONMONOXIDE STATIONARY FUEL COMBUSTION TO ATMOSPHERE	2*10**6 TONS/YEAR	UNITED STATES, 1968	35
*CARBONMONOXIDE TRANSPORTATION TO ATMOSPHERE	64*10**6 TONS/YEAR	UNITED STATES, 1968	35
*CARBONMONOXIDE, ANTHROPOGENIC EMISSION	25*10**7 TONS/YEAR	AUTO EXHAUST,COMBUSTION	16
*CARBONMONOXIDE, NATURAL EMISSION	>10**9 TONS/YEAR	METHANE REACT.,FOREST FIRES, RELEASE FROM OCEANS,TERPENE REACT.	16
*CARBONMONOXIDE, TOTAL EMISSION TO ATMOSPHERE	100*10**6 TONS/Y	UNITED STATES, 1968	35
*CARBONMONOXIDE, WORLD TOTAL FROM ANTHROPOGENIC SOURCES	359 *10**6 TONS	TOTAL ANTHROPOGENIC CO-ATMOSPHERE	739
*CARBONMONOXIDE, WORLD TOTAL FROM MOBILE SOURCES	199 *10**6 TONS	SOURCE=MOTOR VEHICLES	739
*CARBONMONOXIDE, WORLD TOTAL FROM MOBILE SOURCES	197 *10**6 TONS	SOURCE=GASOLINE	739
*CARBONMONOXIDE, WORLD TOTAL FROM MOBILE SOURCES	2 *10**6 TONS	SOURCE=DIESEL	739
*CARBONMONOXIDE, WORLD TOTAL FROM MOBILE SOURCES	5 *10**6 TONS	SOURCE=AIRCRAFT(AVIATION GASOLINE, JET FUEL)	739
*CARBONMONOXIDE, WORLD TOTAL FROM MOBILE SOURCES	18 *10**6 TONS	SOURCE=WATERCRAFT	739
*CARBONMONOXIDE, WORLD TOTAL FROM MOBILE SOURCES	2 *10**6 TONS	SOURCE=RAILROADS	739
*CARBONMONOXIDE, WORLD TOTAL FROM MOBILE SOURCES	26 *10**6 TONS	SOURCE=OTHER MOTOR VEHICLES (NON-HIGHWAY)	739
*CARBONMONOXIDE, WORLD TOTAL FROM STATIONARY SOURCES	4 *10**6 TONS	SOURCE=COAL AND LIGNITE	739
*CARBONMONOXIDE, WORLD TOTAL FROM STATIONARY SOURCES	< 1 *10**6 TONS	SOURCE=RESIDUAL FUEL OIL	739
*CARBONMONOXIDE, WORLD TOTAL FROM STATIONARY SOURCES	< 1 *10**6 TONS	SOURCE=KEROSENE	739
*CARBONMONOXIDE, WORLD TOTAL FROM STATIONARY SOURCES	< 1 *10**6 TONS	SOURCE=DISTILLATE FUEL OIL	739
*CARBONMONOXIDE, WORLD TOTAL FROM STATIONARY SOURCES	< 1 *10**6 TONS	SOURCE=LIQUIFIED PETROLEUM GAS	739
*CARBONMONOXIDE, WORLD TOTAL FROM STATIONARY SOURCES	41 *10**6 TONS	SOURCE=INDUSTRIAL PROCESSES (PETROLEUM REFINERIES,STEEL MILLS, ETC.)	739
*CARBONMONOXIDE, WORLD TOTAL FROM STATIONARY SOURCES	23 *10**6 TONS	SOURCE=SOLID WASTE DISPOSAL(URBAN AND INDUSTRIAL)	739
*CARBONMONOXIDE, WORLD TOTAL FROM STATIONARY SOURCES	41 *10**6 TONS	SOURCE=MISCELLANEOUS: AGRICULT.BURNING,COAL BANK REFUSE,ETC.	739
*CO2, ANTHROPOGENIC EMISSION	4-5*10**9 TONS/YEAR AS C	COMBUSTION	16
*CO2, NATURAL EMISSION	2-3*10**10 TONS/YEAR AS C	BIOLOGICAL DECAY,RELEASE FROM OCEANS	16
*CU, ATMOSPHERIC RAINOUT	0.2 MEGATONS/YR	GLOBAL,CONC.OF ATM.PARTIC.(HAWAII) =1.9+-2.1NG/M**3	550
*HG, CEMENT MANUFACTURE TO ATMOSPHERE	1*10**8 G/YR	ANTHROPOGENIC	550
*HG, CONTINENTS TO ATMOSPHERE	8.4*10**10 G/YR	NATURAL,BASED ON PRICIPITATION WITH RAIN	550
*HG, CONTINENTS TO ATMOSPHERE	1.5*10**11 G/YR	NATURAL,BASED ON ATMOSPHERIC CONTENT	550
*HG, CONTINENTS TO ATMOSPHERE	2.5*10**10 G/YR	NATURAL,BASED ON CONTENT IN GREENLAND GLACIER	550
*HG, FOSSIL FUEL COMBUSTION TO ATMOSPHERE	1.6*10**9 G/YR	ANTHROPOGENIC	550

TABLE G4

ATMOSPHERE : EMISSION

ITEM:	VALUE:	CONDITIONS:	REF.:
*HG, LOSSES IN INDUSTRIAL AND AGRICULTURAL USAGE	4*10**9 G/YR	ANTHROPOGENIC	550
*HG, WORLD PRODUCTION	8.8*10**9 G/YR	ANTHROPOGENIC,1968	550
*HYDROCARBONS FROM AUTOMOBILES	12 *10**6 TONS/YEAR	U.S.A.,1965	1442
*HYDROCARBONS FROM ELECTRIC POWER PLANTS	1 *10**6 TONS/YEAR	U.S.A.,1965	1442
*HYDROCARBONS FROM FIRES AND BURNINGS TO ATMOSPHERE	8*10**6 TONS/YEAR	UNITED STATES, 1968	35
*HYDROCARBONS FROM INDUSTRIAL PROCESSES¤TO ATMOSPHERE	5*10**6 TONS/YEAR	UNITED STATES, 1968	35
*HYDROCARBONS FROM INDUSTRY	4 *10**6 TONS/YEAR	U.S.A.,1965	1442
*HYDROCARBONS FROM REFUSE DISPOSAL	1 *10**6 TONS/YEAR	U.S.A.,1965	1442
*HYDROCARBONS FROM SOLID WASTE DISPOSAL TO ATMOSPHERE	2*10**6 TONS/YEAR	UNITED STATES, 1968	35
*HYDROCARBONS FROM SPACE HEATING	1 *10**6 TONS/YEAR	U.S.A.,1965	1442
*HYDROCARBONS FROM STATIONARY FUEL COMBUSTION TO ATMOSPHERE	1*10**6 TONS/YEAR	UNITED STATES, 1968	35
*HYDROCARBONS FROM TRANSPORTATION TO ATMOSPHERE	17*10¤*6 TONS/YEAR	UNITED STATES, 1968	35
*HYDROCARBONS, ANTHROPOGENIC EMISSION	40*10**6 TONS/YEARR	COMBUSTION,EXHAUSTION,CHEMICAL PROCESSES	16
*HYDROCARBONS, NATURAL EMISSION	>20*10**7 TONS/YEAR	BIOLOGICAL PROCESSES	16
*HYDROCARBONS, TOTAL EMISSION TO ATMOSPHERE	32*10**6 TONS/YEAR	UNITED STATES, 1968	35
*HYDROGEN SULPHIDE, ANTHROPOGENIC EMISSION	3*10**6 TONS/YEAR AS S	CHEMICAL PROCESSES,SEWAGE TREATMENT	16
*HYDROGEN SULPHIDE, NATURAL EMISSION	10*10**7 TONS/YEAR AS S	DECAYING ORGANIC MATTER,VOLCANIC ERUPTIONS	16
*METHANE, ANTHROPOGENIC EMISSION	90*10**6 TONS/YEAR	CHEMICAL PROCESSES	16
*METHANE, NATURAL EMISSION	>10**9 TONS/YEAR	NATURAL GAS,DECAYING NATURAL MATTER	16
*MG, ATMOSPHERIC RAINOUT	40 MEGATONS/YR	GLOBAL,CONC.OF ATM.PARTIC.(HAWAII) =410+-240NG/M**3	550
*MN, ATMOSPHERIC RAINOUT	0.02 MEGATONS/YR	GLOBAL,CONC.OF ATM.PARTIC.(HAWAII) =0.19+-0.14NG/M**3	550
*NA, ATMOSPHERIC RAINOUT	350 MEGATONS/YR	GLOBAL,CONC.OF ATM.PARTIC.(HAWAII) =3500+-2000NG/M**3	550
*NITROGEN OXIDE, ANTHROPOGENIC EMISSION	30*10**6 TONS/YEAR AS N	COMBUSTION	16
*NITROGEN OXIDE, NATURAL EMISSION	15*10**7 TONS/YEAR AS N	BIOLOGICAL ACTION IN SOILS	16
*NITROGEN OXIDES FROM AUTOMOBILES	6 *10**6 TONS/YEAR	U.S.A.,1965	1442
*NITROGEN OXIDES FROM COMBUSTION OF COAL	0.5 - 8 (G NOX-N/KG)/(G/M3)	NOT ELECTRIC GENERATION, TOTAL EMISSION =0.1-1.5 G N/YEAR	1475
*NITROGEN OXIDES FROM COMBUSTION OF GAS	0.6 - 3 (G NOX-N/KG)/(G/M3)	INDUSTRIAL, TOTAL EMISSION = 0.1-0.4*10**12 G N/YEAR	1475
*NITROGEN OXIDES FROM COMBUSTION OF GAS	0.5 - 1 (G NOX-N/KG)/(G/M3)	DOMESTIC, TOTAL EMISSION = 0.02-0.03*10**12 G N/YEAR	1475
*NITROGEN OXIDES FROM COMBUSTION OF OIL	0.5 - 3 (G NOX-N/KG)/(G/M3)	TOTAL EMISSION =0.1-0.8*10**12 G N/YEAR	1475
*NITROGEN OXIDES FROM ELECTRIC POWER GENERATION	2 - 8 (G NOX-N/KG)/(G/M3)	HARD COAL, TOTAL EMISSION = 0.1-0.5*10**12 G N/YEAR	1475
*NITROGEN OXIDES FROM ELECTRIC POWER GENERATION	1.5 - 3 (G NOX-N/KG)/(G/M3)	OIL, TOTAL EMISSION =0.3-0.6 G N/ YEAR	1475
*NITROGEN OXIDES FROM ELECTRIC POWER GENERATION	10 - 15 (G NOX-N/KG)/(G/M3)	GASOLINE, TOTAL EMISSION = 0.8-1.2*10**12 G N/YEAR	1475
*NITROGEN OXIDES FROM ELECTRIC POWER GENERATION	APP 20 (G NOX-N/KG)/(G/M3)	DIESEL FUELS, TOTAL EMISSION = 1.2*10**12 G N/YEAR	1475
*NITROGEN OXIDES FROM ELECTRIC POWER GENERATION	1 - 3 (G NOX-N/KG)/(G/M3)	NATURAL GAS, TOTAL EMISSION = 0.03-0.1*10**12 G N/YEAR	1475
*NITROGEN OXIDES FROM ELECTRIC POWER GENERATION	2 - 6 (G NOX-N/KG)/(G/M3)	LIGNITE/BROWN COAL, TOTAL EMISSION =1-3*10**12 G N/YEAR	1475
*NITROGEN OXIDES FROM ELECTRIC POWER PLANTS	3 *10**6 TONS/YEAR	U.S.A.,1965	1442
*NITROGEN OXIDES FROM FIRES AND BURNINGS TO ATMOSPHERE	1*10**6 TONS/YEAR	UNITED STATES, 1968	35

TABLE G4

ATMOSPHERE : EMISSION

ITEM:	VALUE:	CONDITIONS:	REF.:
*NITROGEN OXIDES FROM INDUSTRIAL PROCESSES TO ATMOSPHERE	0.2*10**6 TONS/YEAR	UNITED STATES, 1968	35
*NITROGEN OXIDES FROM INDUSTRY	2 *10**6 TONS/YEAR	U.S.A.,1965	1442
*NITROGEN OXIDES FROM REFUSE DISPOSAL	1 *10**6 TONS/YEAR	U.S.A.,1965	1442
*NITROGEN OXIDES FROM SOLID WASTE DISPOSAL TO ATMOSPHERE	1*10**6 TONS/YEAR	UNITED STATES, 1968	35
*NITROGEN OXIDES FROM SPACE HEATING	1 *10**6 TONS/YEAR	U.S.A.,1965	1442
*NITROGEN OXIDES FROM STATIONARY FUEL COMBUSTION TO ATMOSPHERE	10*10**6 TONS/YEAR	UNITED STATES, 1968	35
*NITROGEN OXIDES INDUSTRIAL PROCESSES	0.1 (G NOX-N/KG)/(G/M3)	PETROLEUM REFINERY, TOTAL EMISSION =0.06*10**12 G N/YEAR	1475
*NITROGEN OXIDES INDUSTRIAL PROCESSES	0.1 - 0.6 (G NOX-N/KG)/(G/M3)	STEEL MANUFACTURING, TOTAL EMISSION =0.01-0.1*10**12 G N/YEAR	1475
*NITROGEN OXIDES INDUSTRIAL PROCESSES	3 KG NOX-N PER LANDING/TAKE OFF CYCLE	AVIATION, TOTAL EMMISSION LESS THAN 0.01*10**12 G N/YEAR	1475
*NITROGEN OXIDES, TOTAL EMISSION TO ATMOSPHERE	2*10**6 TONS/YEAR	UNITED STATES, 1968	35
*NITROGEN, BIOLOGICAL ACTION	234*10**6 TONS N/YR	AS NO2: 768*10**6T/YR,TOTAL GLOBAL ATMOSPHERE	1348
*NITROGEN, BIOLOGICAL ACTION	960*10**6 TONS N/YR	AS NH3: 1160*10**6T/YR,TOTAL GLOBAL ATMOSPHERE	1348
*NITROGEN, BIOLOGICAL ACTION	376*10**6 TONS N/YR	AS N2O: 592*10**6T/YR,TOTAL GLOBAL ATMOSPHERE	1348
*NITROGEN, COAL COMBUSTION	8.2*10**6 TONS N/YR	AS NO2: 26.9*10**6T/YR,SOURCE MAGNITUDE=3074*10**6T/YR,GLOBAL	1348
*NITROGEN, COMBUSTION	3.5*10**6 TONS N/YR	AS NH3: 4.2*10**6T/YR,TOTAL GLOBAL ATMOSPHERE	1348
*NITROGEN, GASOLINE COMBUSTION	2.3*10**6 TONS N/YR	AS NO2: 7.5*10**6T/YR,SOURCE MAGNITUDE=379*10**6T/YR,GLOBAL	1348
*NITROGEN, NATURAL GAS COMBUSTION	0.6*10**6 TONS N/YR	AS NO2: 2.1*10**6T/YR,SOURCE MAGNITUDE=20.56*10**12T/YR(FT**3) ,GLOBAL	1348
*NITROGEN, OTHER COMBUSTION	0.5*10**6 TONS N/YR	AS NO2: 1.6*10**6T/YR,SOURCE MAGNITUDE=1290*10**6T/YR,GLOBAL	1348
*NITROGEN, OTHER OIL COMBUSTION	4.3*10**6 TONS N/YR	AS NO2: 14.1*10**6T/YR,SOURCE MAGNITUDE=894*10**6T/YR,GLOBAL	1348
*NITROGEN, PETROLEUM REFINING	0.2*10**6 TONS N/YR	AS NO2: 0.7*10**6T/YR,SOURCE MAGNITUDE=11317*10**6T/YR(BBL), GLOBAL	1348
*NITROGEN, TOTAL EMISSION FROM ANTHROPOGENIC SOURCES	16.1*10**6 TONS N/YR	AS NO2: 52.9*10**6T/YR,GLOBAL	1348
*NITROUS OXIDE, NATURAL EMISSION	40*10**7 TONS/YEAR AS N	BIOLOGICAL ACTION IN SOILS	16
*OZONE, ANTHROPOGENIC EMISSION	SMALL	PHOTOCHEM. REACTIONS OF FACTORY EFFLUENTS	16
*OZONE, NATURAL EMISSION	2*10**9 TONS/YEAR	PHOTOCHEM.PROC.BY VOLC.ERUPT., THUNDER,FIRES,PHOTODISSOC.OF OXYGEN	16
*PARTICLE PRODUCTION FROM WIND-BLOWN DUST	250 *10**6 TONS/YR	PARTICLES<5 MY ,TOTAL(GLOBAL), NATURAL=100*10**6 TONS/YR	1043
*PARTICLE PRODUCTION, COSMIC DUST	10 *10**6 TONS/YR	ALL PARTICLE SIZES,TOTAL,GLOBAL	1043
*PARTICLE PRODUCTION, COSMIC DUST	< 1 *10**6 TONS/YR	PARTICLES<5MY,TOTAL(GLOBAL), NATURAL SOURCE	1043
*PARTICLE PRODUCTION, FOREST FIRES/ SLASH-BURN AGRICULTURE	135 *10**6 TONS/YR	ALL PARTICLE SIZES,TOTAL,GLOBAL, NATURAL=35*10**6 TONS/YR	1043
*PARTICLE PRODUCTION, FOREST FIRES/ SLASH-BURN AGRICULTURE	65 *10**6 TONS/YR	PARTICLES<5MY,TOTAL(GLOBAL), NATURAL=5*10**6 TONS/YR	1043
*PARTICLE PRODUCTION, INDUSTRIAL PROCESSES	45 *10**6 TONS/YR	ALL PARTICLE SIZES,TOTAL,GLOBAL	1043
*PARTICLE PRODUCTION, INDUSTRIAL PROCESSES	12 *10**6 TONS/YR	PARTICLES<5MY,TOTAL(GLOBAL), ANTHROPOGENIC SOURCE	1043
*PARTICLE PRODUCTION, OTHER HUMAN ACTIVITIES	30 *10**6 TONS/YR	ALL PARTICLE SIZES,TOTAL,GLOBAL	1043
*PARTICLE PRODUCTION, OTHER HUMAN ACTIVITIES	8 *10**6 TONS/YR	PARTICLES<5MY,TOTAL(GLOBAL)	1043
*PARTICLE PRODUCTION, POWER AND HEATING PLANTS	35 *10**6 TONS/YR	ALL PARTICLE SIZES,TOTAL,GLOBAL	1043

TABLE G4

ATMOSPHERE : EMISSION

ITEM:	VALUE:	CONDITIONS:	REF.:
*PARTICLE PRODUCTION, POWER AND HEATING PLANTS	10 *10**6 TONS/YR	PARTICLES<5MY,TOTAL(GLOBAL), ANTHROPOGENIC SOURCE	1043
*PARTICLE PRODUCTION, SEA SALT	1000 *10**6 TONS/YR	ALL PARTICLE SIZES,TOTAL,GLOBAL	1043
*PARTICLE PRODUCTION, SEA SALT	500 *10**6 TONS/YR	PARTICLES<5MY,TOTAL(GLOBAL), NATURAL SOURCE	1043
*PARTICLE PRODUCTION, VOLCANIC TEPHRA	250 *10**6 TONS/YR	ALL PARTICLE SIZES,TOTAL,GLOBAL	1043
*PARTICLE PRODUCTION, VOLCANIC TEPHRA	25 *10**6 TONS/YR	PARTICLES<5MY,TOTAL(GLOBAL), NATURAL SOURCE	1043
*PARTICLE PRODUCTION, WIND-BLOWN DUST	500 *10**6 TONS/YR	ALL PARTICLE SIZES,TOTAL,GLOBAL, NATURAL=200*10**6 TONS/YR	1043
*PARTICLE, TOTAL DIRECT PRODUCTION	2005 *10**6 TONS/YR	ALL PARTICLE SIZES,TOTAL(GLOBAL), NATURAL=1495*10**6 TONS/YR	1043
*PARTICLE, TOTAL DIRECT PRODUCTION	870 *10**6 TONS/YR	PARTICLES<5 MY ,TOTAL(GLOBAL), NATURAL=630*10**6 TONS/YR	1043
*PARTICLES (ATM), ACCUMULATION IN SNOWFIELDS	0.14 MM/10**3YR	GREENLAND,TRANSP.PATH=POLAR EASTERLIES	1736
*PARTICLES (ATM), ACCUMULATION IN SNOWFIELDS	0.11 MM/10**3YR	ST.ELIAS RANGE,YUKON TERRITORY, TRANSP.PATH=WESTERLIES	1736
*PARTICLES (ATM), ACCUMULATION IN SNOWFIELDS	0.21 MM/10**3YR	MT.OLYMPUS,WASHINGTON,TRANSP.PATH= WESTERLIES	1736
*PARTICLES (ATM), ACCUMULATION IN SNOWFIELDS	0.09 MM/10**3YR	MT.ORIZABA,MEXICO,TRANSP.PATH= TRADE WINDS	1736
*PARTICLES (ATM), ACCUMULATION IN SNOWFIELDS	< 0.01 MM/10**3YR	MT.POPOCATEPETL,MEXICO,TRANSP.PATH =TRADE WINDS	1736
*PARTICLES (ATM), ACCUMULATION IN SNOWFIELDS	0.80 MM/10**3YR	TASMAN GLACIER,NEW ZEALAND, TRANSP.PATH=WESTERLIES	1736
*PARTICLES (ATM), ACCUMULATION IN SNOWFIELDS	< 0.01 MM/10**3YR	NEW BYRD STATION,ANTARCTICA, TRANSP.PATH=POLAR EASTERLIES	1736
*PARTICLES (ATM), ACCUMULATION ON THE BOTTOM OF THE OCEAN	0.21-0.35 MM/10**3YR	ATLANTIC,15N-50N,TRANSP.PATH=JET STREAM,MOST PROB.VAL.	413
*PARTICLES (ATM), ACCUMULATION ON THE BOTTOM OF THE OCEAN	0.49-0.56 MM/10**3YR	ATLANTIC,15N-15S,TRANSP.PATH=TRADE WINDS,MOST PROB.VAL.	413
*PARTICLES (ATM), ACCUMULATION ON THE BOTTOM OF THE OCEAN	0.14-0.28 MM/10**3YR	ATLANTIC,15S-50S,TRANSP.PATH=JET STREAM,MOST PROB.VAL.	413
*PARTICLES (ATM), ACCUMULATION ON THE BOTTOM OF THE OCEAN	0.28-1.4 MM/10**3YR	PACIFIC,15N-50N,TRANSP.PATH=JET STREAM,MOST PROB.VAL.	413
*PARTICLES (ATM), ACCUMULATION ON THE BOTTOM OF THE OCEAN ¤	0.21-4.2 MM/10**3YR	PACIFIC,15N-15S,TRANSP.PATH=TRADE WINDS,MOST PROB.VAL.	413
*PARTICLES (ATM), ACCUMULATION ON THE BOTTOM OF THE OCEAN	0.21-0.56 MM/10**3YR	PACIFIC,15S-50S,TRANSP.PATH=JET STREAM,MOST PROB.VAL.	413
*PARTICLES (ATM), ACCUMULATION ON THE BOTTOM OF THE OCEAN	0.7 MM/10**3YR	INDIAN,15S-20N,TRANSP.PATH=TRADE WINDS,MOST PROB.VAL.	413
*PARTICLES (ATM), ACCUMULATION ON THE BOTTOM OF THE OCEAN	0.35-0.56 MM/10**3YR	INDIAN,20S-50S,TRANSP.PATH=JET STREAM,MOST PROB.VAL.	413
*PARTICLES (PRIMARY) FROM AGRICULTURAL BURNING	2.2 *10**6 TONS/YR	ANTHROPOGENIC SOURCE,U.S.A.	1102
*PARTICLES (PRIMARY) FROM AGRICULTURAL BURNING	2.18 *10**6 TONS/YR	ANTHROPOGENIC SOURCE,U.S.A.	1388
*PARTICLES (PRIMARY) FROM AGRICULTURAL BURNING	2.2 *10**6 TONS/YR	ANTHROPOGENIC SOURCE,U.S.A.	1035
*PARTICLES (PRIMARY) FROM AGRICULTURAL BURNING	21.8 *10**6 TONS/YR	ANTHROPOGENIC SOURCE,WORLDWIDE	1388
*PARTICLES (PRIMARY) FROM AGRICULTURAL BURNING	22 *10**6 TONS/YR	ANTHROPOGENIC SOURCE,WORLDWIDE	413
*PARTICLES (PRIMARY) FROM CEMENT, SAND,ETC. INDUSTRY	2.1 *10**6 TONS/YR	ANTHROPOGENIC SOURCE,U.S.A.	1102
*PARTICLES (PRIMARY) FROM CEMENT, SAND,ETC. INDUSTRY	14.9 *10**6 TONS/YR	ANTHROPOGENIC SOURCE,U.S.A.	1388
*PARTICLES (PRIMARY) FROM CEMENT, SAND,ETC. INDUSTRY	5.64 *10**6 TONS/YR	ANTHROPOGENIC SOURCE,U.S.A.	1035
*PARTICLES (PRIMARY) FROM CEMENT, SAND,ETC. INDUSTRY	60.9 *10**6 TONS/YR	ANTHROPOGENIC SOURCE,WORLDWIDE	1388
*PARTICLES (PRIMARY) FROM CEMENT, SAND,ETC. INDUSTRY	11.7 *10**6 TONS/YR	ANTHROPOGENIC SOURCE,WORLDWIDE	666
*PARTICLES (PRIMARY) FROM CEMENT, SAND,ETC. INDUSTRY	5.9 *10**6 TONS/YR	ANTHROPOGENIC SOURCE,WORLDWIDE	1347

TABLE G4

ATMOSPHERE : EMISSION			
ITEM:	VALUE:	CONDITIONS:	REF.:
*PARTICLES (PRIMARY) FROM CHEMICAL INDUSTRY	0.08 *10**6 TONS/YR	ANTHROPOGENIC SOURCE,U.S.A.	1102
*PARTICLES (PRIMARY) FROM CHEMICAL INDUSTRY	0.38 *10**6 TONS/YR	ANTHROPOGENIC SOURCE,U.S.A.	1388
*PARTICLES (PRIMARY) FROM CHEMICAL INDUSTRY	0.22 *10**6 TONS/YR	ANTHROPOGENIC SOURCE,U.S.A.	1035
*PARTICLES (PRIMARY) FROM CHEMICAL INDUSTRY	0.93 *10**6 TONS/YR	ANTHROPOGENIC SOURCE,WORLDWIDE	1347
*PARTICLES (PRIMARY) FROM FUEL PROCESSING INDUSTRY	0.26 *10**6 TONS/YR	ANTHROPOGENIC SOURCE,U.S.A.	1102
*PARTICLES (PRIMARY) FROM FUEL PROCESSING INDUSTRY	0.96 *10**6 TONS/YR	ANTHROPOGENIC SOURCE,U.S.A.	1388
*PARTICLES (PRIMARY) FROM FUEL PROCESSING INDUSTRY	0.21 *10**6 TONS/YR	ANTHROPOGENIC SOURCE,U.S.A.	1035
*PARTICLES (PRIMARY) FROM FUEL PROCESSING INDUSTRY	3.7 *10**6 TONS/YR	ANTHROPOGENIC SOURCE,WORLDWIDE	1347
*PARTICLES (PRIMARY) FROM GRAIN, FEED,ETC. INDUSTRY	1.0 *10**6 TONS/YR	ANTHROPOGENIC SOURCE,U.S.A.	1102
*PARTICLES (PRIMARY) FROM GRAIN, FEED,ETC. INDUSTRY	3.3 *10**6 TONS/YR	ANTHROPOGENIC SOURCE,U.S.A.	1388
*PARTICLES (PRIMARY) FROM GRAIN, FEED,ETC. INDUSTRY	1.65 *10**6 TONS/YR	ANTHROPOGENIC SOURCE,U.S.A.	1035
*PARTICLES (PRIMARY) FROM GRAIN, FEED,ETC. INDUSTRY	33.0 *10**6 TONS/YR	ANTHROPOGENIC SOURCE,WORLDWIDE	1388
*PARTICLES (PRIMARY) FROM GRAIN, FEED,ETC. INDUSTRY	8.4 *10**6 TONS/YR	ANTHROPOGENIC SOURCE,WORLDWIDE	1347
*PARTICLES (PRIMARY) FROM IRON AND STEEL INDUSTRY	1.7 *10**6 TONS/YR	ANTHROPOGENIC SOURCE,U.S.A.	1102
*PARTICLES (PRIMARY) FROM IRON AND STEEL INDUSTRY	13.9 *10**6 TONS/YR	ANTHROPOGENIC SOURCE,U.S.A.	1388
*PARTICLES (PRIMARY) FROM IRON AND STEEL INDUSTRY	1.3 *10**6 TONS/YR	ANTHROPOGENIC SOURCE,U.S.A.	666
*PARTICLES (PRIMARY) FROM IRON AND STEEL INDUSTRY	1.6 *10**6 TONS/YR	ANTHROPOGENIC SOURCE,U.S.A.	1035
*PARTICLES (PRIMARY) FROM IRON AND STEEL INDUSTRY	54.6 *10**6 TONS/YR	ANTHROPOGENIC SOURCE,WORLDWIDE	1388
*PARTICLES (PRIMARY) FROM IRON AND STEEL INDUSTRY	10.2 *10**6 TONS/YR	ANTHROPOGENIC SOURCE,WORLDWIDE	1347
*PARTICLES (PRIMARY) FROM MINERAL INDUSTRY	0.75 *10**6 TONS/YR	ANTHROPOGENIC SOURCE,U.S.A.	1102
*PARTICLES (PRIMARY) FROM MINERAL INDUSTRY	2.3 *10**6 TONS/YR	ANTHROPOGENIC SOURCE,U.S.A.	1388
*PARTICLES (PRIMARY) FROM MINERAL INDUSTRY	0.82 *10**6 TONS/YR	ANTHROPOGENIC SOURCE,U.S.A.	1035
*PARTICLES (PRIMARY) FROM MINERAL INDUSTRY	2.3 *10**6 TONS/YR	ANTHROPOGENIC SOURCE,WORLDWIDE	1347
*PARTICLES (PRIMARY) FROM MISCELLANEOUS SOURCES	0.45 *10**6 TONS/YR	ANTHROPOGENIC SOURCE,U.S.A.	1102
*PARTICLES (PRIMARY) FROM MISCELLANEOUS SOURCES	1.16 *10**6 TONS/YR	ANTHROPOGENIC SOURCE,U.S.A.	1035
*PARTICLES (PRIMARY) FROM MISCELLANEOUS SOURCES	28.8 *10**6 TONS/YR	ANTHROPOGENIC SOURCE,WORLDWIDE	1247
*PARTICLES (PRIMARY) FROM MISCELLANEOUS SOURCES	7.0 *10**6 TONS/YR	ANTHROPOGENIC SOURCE,WORLDWIDE	1347
*PARTICLES (PRIMARY) FROM MISCELLANEOUS SOURCES	25 *10**6 TONS/YR	ANTHROPOGENIC SOURCE,WORLDWIDE	413
*PARTICLES (PRIMARY) FROM OTHER METAL INDUSTRY	0.24 *10**6 TONS/YR	ANTHROPOGENIC SOURCE,U.S.A.	1102
*PARTICLES (PRIMARY) FROM OTHER METAL INDUSTRY	0.58 *10**6 TONS/YR	ANTHROPOGENIC SOURCE,U.S.A.	1388
*PARTICLES (PRIMARY) FROM OTHER METAL INDUSTRY	0.54 *10**6 TONS/YR	ANTHROPOGENIC SOURCE,U.S.A.	1035
*PARTICLES (PRIMARY) FROM OTHER METAL INDUSTRY	1.8 *10**6 TONS/YR	ANTHROPOGENIC SOURCE,WORLDWIDE	1347
*PARTICLES (PRIMARY) FROM PULP AND PAPER INDUSRTY	0.53 *10**6 TONS/YR	ANTHROPOGENIC SOURCE,U.S.A.	1035
*PARTICLES (PRIMARY) FROM PULP AND PAPER INDUSTRY	0.65 *10**6 TONS/YR	ANTHROPOGENIC SOURCE,U.S.A.	1102

TABLE G4

ATMOSPHERE : EMISSION			
ITEM:	VALUE:	CONDITIONS:	REF.:
*PARTICLES (PRIMARY) FROM PULP AND PAPER INDUSTRY	4.3 *10**6 TONS/YR	ANTHROPOGENIC SOURCE,U.S.A.	1388
*PARTICLES (PRIMARY) FROM PULP AND PAPER INDUSTRY	3.1 *10**6 TONS/YR	ANTHROPOGENIC SOURCE,U.S.A.	666
*PARTICLES (PRIMARY) FROM PULP AND PAPER INDUSTRY	9.5 *10**6 TONS/YR	ANTHROPOGENIC SOURCE,WORLDWIDE	1388
*PARTICLES (PRIMARY) FROM PULP AND PAPER INDUSTRY	0.4 *10**6 TONS/YR	ANTHROPOGENIC SOURCE,WORLDWIDE	1347
*PARTICLES (PRIMARY) FROM SLASH AND LITTER BURNING	15.4 *10**6 TONS/YR	ANTHROPOGENIC SOURCE,U.S.A.	1035
*PARTICLES (PRIMARY) FROM SOLID WASTES	1.0 *10**6 TONS/YR	ANTHROPOGENIC SOURCE,U.S.A.	1102
*PARTICLES (PRIMARY) FROM SOLID WASTES	1.0 *10**6 TONS/YR	ANTHROPOGENIC SOURCE,U.S.A.	1388
*PARTICLES (PRIMARY) FROM SOLID WASTES	0.84 *10**6 TONS/YR	ANTHROPOGENIC SOURCE,U.S.A.	1035
*PARTICLES (PRIMARY) FROM SOLID WASTES	2.4 *10**6 TONS/YR	ANTHROPOGENIC SOURCE,WORLDWIDE	1247
*PARTICLES (PRIMARY) FROM SOLID WASTES	3.8 *10**6 TONS/YR	ANTHROPOGENIC SOURCE,WORLDWIDE	1347
*PARTICLES (PRIMARY) FROM SOLID WASTES	2 *10**6 TONS/YR	ANTHROPOGENIC SOURCE,WORLDWIDE	413
*PARTICLES (PRIMARY) FROM STATIONARY COMBUSTION	8.1 *10**6 TONS/YR	ANTHROPOGENIC SOURCE,U.S.A.	1102
*PARTICLES (PRIMARY) FROM STATIONARY COMBUSTION	8.1 *10**6 TONS/YR	ANTHROPOGENIC SOURCE,U.S.A.	1388
*PARTICLES (PRIMARY) FROM STATIONARY COMBUSTION	5.4 *10**6 TONS/YR	ANTHROPOGENIC SOURCE,U.S.A.	1035
*PARTICLES (PRIMARY) FROM STATIONARY COMBUSTION	21 *10**6 TONS/YR	ANTHROPOGENIC SOURCE,WORLDWIDE	666
*PARTICLES (PRIMARY) FROM STATIONARY COMBUSTION	43.8 *10**6 TONS/YR	ANTHROPOGENIC SOURCE,WORLDWIDE	1247
*PARTICLES (PRIMARY) FROM STATIONARY COMBUSTION	37.8 *10**6 TONS/YR	ANTHROPOGENIC SOURCE,WORLDWIDE	1347
*PARTICLES (PRIMARY) FROM STATIONARY COMBUSTION	24 *10**6 TONS/YR	ANTHROPOGENIC SOURCE,WORLDWIDE	413
*PARTICLES (PRIMARY) FROM TOTAL INDUSTRY	6.8 *10**6 TONS/YR	ANTHROPOGENIC SOURCE,U.S.A.	1102
*PARTICLES (PRIMARY) FROM TOTAL INDUSTRY	40.6 *10**6 TONS/YR	ANTHROPOGENIC SOURCE,U.S.A.	1388
*PARTICLES (PRIMARY) FROM TOTAL INDUSTRY	16.1 *10**6 TONS/YR	ANTHROPOGENIC SOURCE,U.S.A.	666
*PARTICLES (PRIMARY) FROM TOTAL INDUSTRY	11.2 *10**6 TONS/YR	ANTHROPOGENIC SOURCE,U.S.A.	1035
*PARTICLES (PRIMARY) FROM TOTAL INDUSTRY	158 *10**6 TONS/YR	ANTHROPOGENIC SOURCE,WORLDWIDE	1388
*PARTICLES (PRIMARY) FROM TOTAL INDUSTRY	56.4 *10**6 TONS/YR	ANTHROPOGENIC SOURCE,WORLDWIDE	1247
*PARTICLES (PRIMARY) FROM TOTAL INDUSTRY	33.6 *10**6 TONS/YR	ANTHROPOGENIC SOURCE,WORLDWIDE	1347
*PARTICLES (PRIMARY) FROM TOTAL INDUSTRY	50 *10**6 TONS/YR	ANTHROPOGENIC SOURCE,WORLDWIDE	413
*PARTICLES (PRIMARY) FROM TRANSPORTATION	1.1 *10**6 TONS/YR	ANTHROPOGENIC SOURCE,U.S.A.	1102
*PARTICLES (PRIMARY) FROM TRANSPORTATION	1.1 *10**6 TONS/YR	ANTHROPOGENIC SOURCE,U.S.A.	1388
*PARTICLES (PRIMARY) FROM TRANSPORTATION	1.1 *10**6 TONS/YR	ANTHROPOGENIC SOURCE,U.S.A.	1035
*PARTICLES (PRIMARY) FROM TRANSPORTATION	2.2 *10**6 TONS/YR	ANTHROPOGENIC SOURCE,WORLDWIDE	1247
*PARTICLES (PRIMARY) FROM TRANSPORTATION	0.8 *10**6 TONS/YR	ANTHROPOGENIC SOURCE,WORLDWIDE	1347
*PARTICLES (PRIMARY) FROM TRANSPORTATION	3 *10**6 TONS/YR	ANTHROPOGENIC SOURCE,WORLDWIDE	413
*PARTICLES FORMED FROM HYDROCARBONS	90 *10**6 TONS/YR	ALL PARTICLE SIZES,TOTAL,GLOBAL, NATURAL=75*10**6 TONS/YR	1043
*PARTICLES FORMED FROM HYDROCARBONS	90 *10**6 TONS/YR	PARTICLES<5 MY ,TOTAL(GLOBAL), NATURAL=75*10**6 TONS/YR	1043

TABLE G4

ATMOSPHERE : EMISSION

ITEM:	VALUE:	CONDITIONS:	REF.:
*PARTICLES FORMED FROM NITRATES	115 *10**6 TONS/YR	ALL PARTICLE SIZES,TOTAL,GLOBAL, NATURAL=75*10**6 TONS/YR	1043
*PARTICLES FORMED FROM NITRATES	95 *10**6 TONS/YR	PARTICLES<5 MY ,TOTAL(GLOBAL), NATURAL=60*10**6 TONS/YR	1043
*PARTICLES FORMED FROM SULFATES	640 *10**6 TONS/YR	ALL PARTICLE SIZES,TOTAL,GLOBAL, NATURAL=420*10**6 TONS/YR	1043
*PARTICLES FORMED FROM SULFATES	535 *10**6 TONS/YR	PARTICLES<5 MY ,TOTAL(GLOBAL), NATURAL=335*10**6 TONS/YR	1043
*PARTICLES FROM AUTOMOBILES	1 *10**6 TONS/YEAR	U.S.A.,1965	1442
*PARTICLES FROM CONTINENTAL ROCK AND SOIL TO ATMOSPHERE	1-5 *10**8 TONS/YR	NONE	550
*PARTICLES FROM ELECTRIC POWER PLANTS	3 *10**6 TONS/YEAR	U.S.A.,1965	1442
*PARTICLES FROM FIRES AND BURNINGS TO ATMOSPHERE	10*10**6 TONS/YEAR	UNITED STATES, 1968	35
*PARTICLES FROM INDUSTRIAL PROCESSES TO ATMOSPHERE	7*10**6 TONS/YEAR	UNITED STATES, 1968	35
*PARTICLES FROM INDUSTRY	6 *10**6 TONS/YEAR	U.S.A.,1965	1442
*PARTICLES FROM REFUSE DISPOSAL	1 *10**6 TONS/YEAR	U.S.A.,1965	1442
*PARTICLES FROM SOLID WASTE DISPOSAL TO ATMOSPHERE	1*10**6 TONS/YEAR	UNITED STATES, 1968	35
*PARTICLES FROM SPACE HEATING	1 *10**6 TONS/YEAR	U.S.A.,1965	1442
*PARTICLES FROM STATIONARY FUEL COMBUSTION TO ATMOSPHERE	9*10**6 TONS/YEAR	UNITED STATES, 1968	35
*PARTICLES FROM TRANSPORTATION TO ATMOSPHERE	1*10**6 TONS/YEAR	UNITED STATES, 1968	35
*PARTICLES, TOTAL EMISSION TO ATMOSPHERE	28*10**6 TONS/YEAR	UNITED STATES, 1968	35
*PARTICLES, TOTAL PRODUCTION FORMED FROM GASES	845 *10**6 TONS/YR	ALL PARTICLE SIZES,TOTAL(GLOBAL), NATURAL=570*10**6 TONS/YR	1043
*PARTICLES, TOTAL PRODUCTION FORMED FROM GASES	720 *10**6 TONS/YR	PARTICLES<5MIC.M,TOTAL(GLOBAL), NATURAL=470*10**6 TONS/YR	1043
*PARTICULATES (INDUSTRIAL) TO ATMOSPHERE	0.54 *10**8 TONS/YR	TOTAL	550
*PARTICULATES (INDUSTRIAL) TO ATMOSPHERE	0.12 *10**8 TONS/YR	PARTICLES< 2 MY	550
*PARTICULATES (PRIMARY), TOTAL ANTHROPOGENIC EMISSION	19.6 *10**6 TONS/YR	U.S.A.	1102
*PARTICULATES (PRIMARY), TOTAL ANTHROPOGENIC EMISSION	86.8 *10**6 TONS/YR	U.S.A.	1388
*PARTICULATES (PRIMARY), TOTAL ANTHROPOGENIC EMISSION	37 *10**6 TONS/YR	U.S.A.	1035
*PARTICULATES (PRIMARY), TOTAL ANTHROPOGENIC EMISSION	179.8 *10**6 TONS/YR	WORLDWIDE	1388
*PARTICULATES (PRIMARY), TOTAL ANTHROPOGENIC EMISSION	35-100 *10**6 TONS/YR	WORLDWIDE	666
*PARTICULATES (PRIMARY), TOTAL ANTHROPOGENIC EMISSION	133.2 *10**6 TONS/YR	WORLDWIDE	1247
*PARTICULATES (PRIMARY), TOTAL ANTHROPOGENIC EMISSION	83 *10**6 TONS/YR	WORLDWIDE	1347
*PARTICULATES (PRIMARY), TOTAL ANTHROPOGENIC EMISSION	126 *10**6 TONS/YR	WORLDWIDE	413
*PARTICULATES TO ATMOSPHERE	5 %	ANTHROPOGENIC FRACTION OF TOTAL AIRBORNE PARTICULATES	1415
*PARTICULATES TO ATMOSPHERE	5 %	ANTHROPOGENIC FRACTION OF TOTAL AIRBORNE PARTICULATES	1495
*PARTICULATES TO ATMOSPHERE	20 %	ANTHROPOGENIC FRACTION OF TOTAL AIRBORNE PARTICULATES	1388
*PARTICULATES TO ATMOSPHERE	6 %	ANTHROPOGENIC FRACTION OF TOTAL AIRBORNE PARTICULATES	666
*PARTICULATES TO ATMOSPHERE	20-30 %	ANTHROPOGENIC FRACTION OF TOTAL AIRBORNE PARTICULATES	1042
*PARTICULATES TO ATMOSPHERE	5-45 %	ANTHROPOGENIC FRACTION OF TOTAL AIRBORNE PARTICULATES	1459
*PARTICULATES TO ATMOSPHERE	18 %	ANTHROPOGENIC FRACTION OF TOTAL AIRBORNE PARTICULATES	1247

TABLE G4

ATMOSPHERE : EMISSION			
ITEM:	VALUE:	CONDITIONS:	REF.:
*PARTICULATES TO ATMOSPHERE	11 %	ANTHROPOGENIC FRACTION OF TOTAL AIRBORNE PARTICULATES	1347
*PARTICULATES TO ATMOSPHERE	13.6 %	ANTHROPOGENIC FRACTION OF TOTAL AIRBORNE PARTICULATES	413
*PB, ATMOSPHERIC RAINOUT	0.3 MEGATONS/YR	GLOBAL,CONC.OF ATM.PARTIC.(HAWAII) =3.0+-2.8NG/M**3	550
*SEA SALT TO ATMOSPHERE	3 *10**8 TONS/YR	NONE	550
*SULPHUR COMPOUNDS, BACTERIAL AND PLANT PRODUCTION OF H2S	268 TONS*10**6 S/YEAR	APP. RESIDENCE TIME = 0.5 - 6 D	810
*SULPHUR COMPOUNDS, DRY FALLOUT ATMOSPHERE	30 TONS*10**6 S/YEAR	AS SO4(2-)	810
*SULPHUR COMPOUNDS, MAN-MADE PRODUCTION OF SO2 AND SO4(2-)(-2)	150 TONS*10**6 S/YEAR	FROM FOSSIL FUELS, 93 % OF TOTAL EMISSION ON N HEMISPHERE	810
*SULPHUR COMPOUNDS, PLANT UPTAKE	45 TONS*10**6 S/YEAR	AS SO2 AND SO4(2-)	810
*SULPHUR COMPOUNDS, RAIN OVER LAND	258 TONS*10**6 S/YEAR	AS SO2 AND SO4(2-)	810
*SULPHUR COMPOUNDS, RAIN OVER OCEANS	217 TONS*10**6 S/YEAR	AS SO2 AND SO4(2-)	810
*SULPHUR COMPOUNDS, VOLCANIC ACTIVITY	2 TONS*10**6 S/YEAR	EMIS. AS H2S, SO2 AND SO4(2-), RESIDENCE TIME = <1 D	810
*SULPHUR COMPOUNDS, WINDBLOWN SEA SALT	130 TONS*10**6 S/YEAR	SO4(2-) IN PRECIPATION, 10 % OF GLOBAL DEPOSITION ON LAND	810
*SULPHUR FROM COAL COMBUSTION	51*10**6 TONS S/YR	AS SO2: 102*10**6 TONS/YEARR, GLOBAL	1348
*SULPHUR FROM GLOBAL ATMOSPHERE	220*10**6 TONS S/YR	TOTAL ATMOSPHERE	1348
*SULPHUR FROM INDUSTRIAL EMISSIONS	3*10**6 TONS S/YR	AS H2S: 3*10**6 TONS/YEAR ,GLOBAL	1348
*SULPHUR FROM MARINE EMISSIONS	30*10**6 TONS S/YR	AS H2S OR ORG.S: 30*10**6 TONS/ YEARR,GLOBAL	1348
*SULPHUR FROM MARINE EMISSIONS	44*10**6 TONS S/YR	AS SO4: 130*10**6 TONS/YEAR , GLOBAL	1348
*SULPHUR FROM PETROLEUM COMBUSTION	11*10**6 TONS S/YR	AS SO2: 23*10**6 TONS/YEAR ,GLOBAL	1348
*SULPHUR FROM PETROLEUM REFINING	3*10**6 TONS S/YR	AS SO2: 6*10**6 TONS/YEAR ,GLOBAL	1348
*SULPHUR FROM SMELTING OPERATIONS	8*10**6 TONS S/YR	AS SO2: 16*10**6 TONS/YEAR ,GLOBAL	1348
*SULPHUR FROM TERRESTRIAL EMISSIONS	70*10**6 TONS S/YR	AS H2S OR ORG.S: 70*10**6 TONS/ YEAR ,GLOBAL	1348
*SULPHUR OXIDES FROM AUTOMOBILES	1 *10**6 TONS/YEAR	U.S.A.,1965	1442
*SULPHUR OXIDES FROM ELECTRIC POWER PLANTS	12 *10**6 TONS/YEAR	U.S.A.,1965	1442
*SULPHUR OXIDES FROM FIRES AND BURNINGS TO ATMOSPHERE	0.6*10**6 TONS/YEAR	UNITED STATES, 1968	35
*SULPHUR OXIDES FROM INDUSTRIAL PROCESSES TO ATMOSPHERE	7*10**6 TONS/YEAR	UNITED STATES, 1968	35
*SULPHUR OXIDES FROM INDUSTRY	9 *10**6 TONS/YEAR	U.S.A.,1965	1442
*SULPHUR OXIDES FROM REFUSE DISPOSAL	1 *10**6 TONS/YEAR	U.S.A.,1965	1442
*SULPHUR OXIDES FROM SOLID WASTE DISPOSAL TO ATMOSPHERE	0.1*10**6 TONS/YEAR	UNITED STATES, 1968	35
*SULPHUR OXIDES FROM SPACE HEATING	3 *10**6 TONS/YEAR	U.S. ,1965	1442
*SULPHUR OXIDES FROM STATIONARY FUEL COMBUSTION TO ATMOSPHERE	24*10**6 TONS/YEAR	UNITED STATES, 1968	35
*SULPHUR OXIDES FROM TRANSPORTATION TO ATMOSPHERE	1*10**6 TONS/YEAR	UNITED STATES, 1968	35
*SULPHUR OXIDES, TOTAL EMISSION TO ATMOSPHERE	33*10**6 TONS/YEAR	UNITED STATES, 1968	35
*VOLCANIC DEBRIS TO ATMOSPHERE	< 1.5 *10**8 TONS/YR	NONE	550
*VOLCANIC DEBRIS TO STRATOSPHERE	0.036 *10**8 TONS/YR	NONE	550
*WASTES DUMPED FROM SHIPS INTO OCEANS	1.4 *10**8 TONS/YR	NONE	550
*WASTES OF SOCIETY TO HYDRO-, LITHO- AND ATMOSPHERE	30 *10**8 TONS/YR	EXCLUSIVE FOSSIL FUELS	550
*WATER VAPOR, INJECTION INTO STRATOSPHERE	2.2*10**14 G/YR	FROM HADLEY CELL CIRCULATION	1443
*WATER VAPOR, INJECTION INTO STRATOSPHERE	8.0*10**14 G/YR	FROM SEVERE LOCAL STORMS	1443

TABLE G4

ATMOSPHERE : EMISSION			
ITEM:	VALUE:	CONDITIONS:	REF.:
*WATER VAPOR, INJECTION INTO STRATOSPHERE	1.1*10**14 G/YR	FROM METHANE OXIDATION	1443
*WATER VAPOR, INJECTION INTO STRATOSPHERE	5.5*10**13 G/YR	FROM METHANE OXIDATION,HUMAN RELATED	1443
*WATER VAPOR, INJECTION INTO STRATOSPHERE	7.8*10**13 G/YR	FROM SST EXHAUST,ASSUMING 500 SST'S FLYING 7HRS/DAY	1443

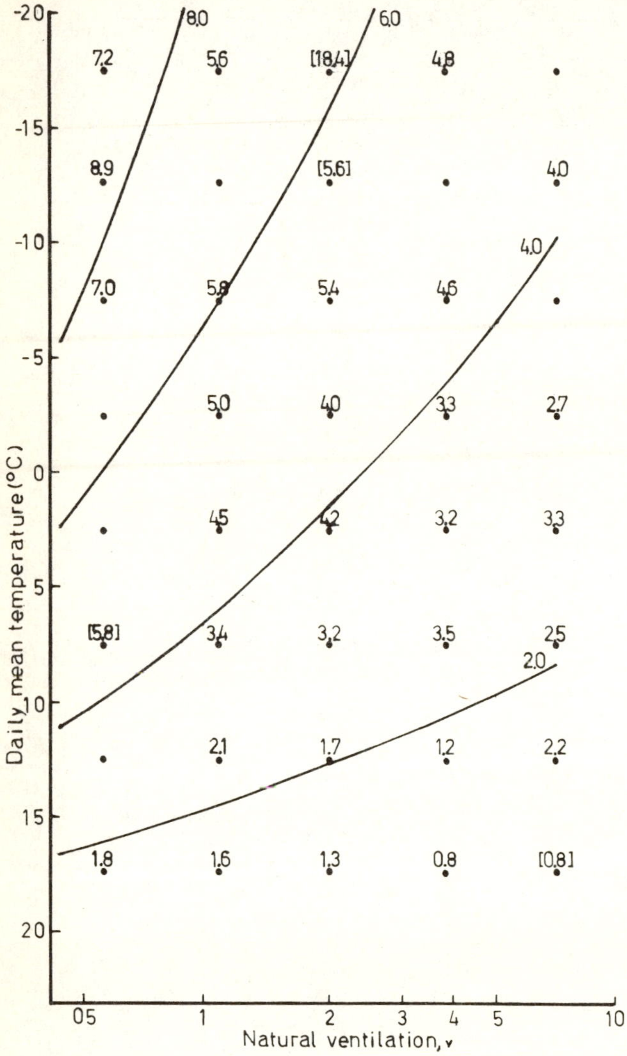

(ref: 677)

G65 SO_2-concentration in central Karlstad plotted to the dayly mean temperature and the degree of natural ventilation v. Concentration in parts per hundred million.
v is defined as U/a, where U is the 24 hours mean wind speed at standard level (10 cm above ground) and a is the mean value of the stalic stability of the atmosphere: a = 0.5 for unstable stralification. a = 1.0 for neutral and a = 2.0 for stable stralification.

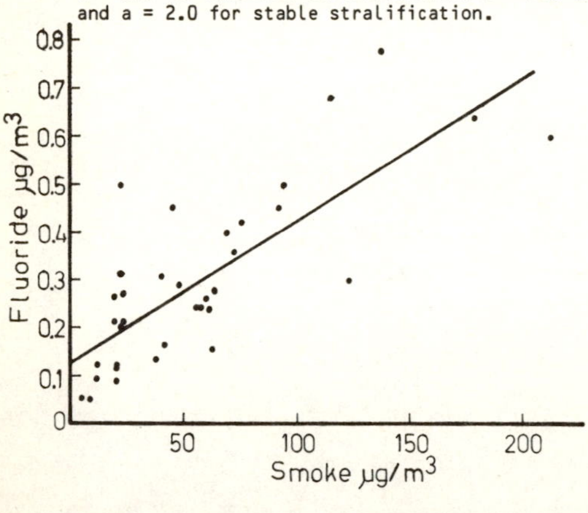

G66
Flouride aganist smoke at 35 urban sites.
µg F/m**3 versus µg/m**3.

(ref: 334)

TABLE G007

Summary of Estimates of Worldwide Emissions of Particulates other than Anthropogenic Primary Emissions. Values are in millions of metric tons/year

REF.: 413

Type	Type	Value	Method of estimation
Natural Primary	Eroded dust	1100	20*estimate for US based on fallout at 15 stations
Natural Primary	Eroded dust	500	Atm dust 5 ug/m3 blw 5 km & 40 removals/y. Oc sed rate .1 ug/cm2/y
Natural Primary	Eroded dust	500	Extrapolated from estimate for natural and agric. dust over US.
Natural Primary	Eroded dust	200	Atm conc 10 ug/m3 at land, 0.1 elsew. es. time 3d below 1km, 5-30d abv
Natural Primary	Eroded dust	70-100	Glacier sediment data. Underestimate since sampled at 3 km height.
Natural Primary	Eroded dust	380	Ocean sediment rate of 0.5 mm/1000y and density=1.5
Natural Primary	Sea salt	1000-2000	Range of values attributed by Erikson
Natural Primary	Sea salt	960	One 1 u sea salt particle/cm2/s from oceans
Natural Primary	Sea salt	300	3*Eriksons estimate of land fallout o sea salt particles.
Natural Primary	Forest Fires	200	6*estimate of 34 mio for fires in US in 1968, by US Forest service.
Natural Primary	Forest Fires	150	14*4.2E6 acres burned in US, *2E22 particle/acre, *1.3E-16 g/part.
Natural Primary	Forest Fires	35	6*estimate for US of 6.2 mio.
Natural Primary	Forest Fires	3	4*emiss. factor of 8.5 kg/tons *18 t/ac *4.5E6 ac burned in US in 1966
Natural Primary	Volcanoes	150	Montmorillonite accumulation of 0.15 kg/cm2/y in ocean sediment
Natural Primary	Volcanoes	42.5	Total of following stratospheric and tropospheric estimates.
Natural Primary	Volcanoes	6.5	1% reaches stratosphere and recidence time there is 14 months.
Natural Primary	Volcanoes	36	Average tropospheric loading divided by recidence time of 0.1 year.
Natural Primary	Volcanoes	25	Estimate of interpeak average loading based on above two estimates.
Natural Primary	Volcanoes	8	0.25*1 eruption of 11.5 mio m3/year of magma, density=2.75.
Natural Primary	Extraterrestrial	10	Upper limit of terrestrial and extraterrestrial particle flux
Natural Primary	Extraterrestrial	1.6	Average of estimates from Fe, Ni and Mo content of Greenland ice.
Natural Primary	Extraterrestrial	0-3.6	Survey of available estimates.
Natural secondary	Sulfate	33-330	Conversion of 60 % of emission of SO2 and H2S.
Natural secondary	Sulfate	244	Conversion of 2/3 of emission of SO2 and H2S to (NH4)2SO4.
Natural secondary	Sulfate	185	Conversion of biological H2S.
Natural secondary	Sulfate	170	244E6 of (NH4)2SO4 less the NH4.
Natural secondary	Sulfate	130	Assumed equivalent to anthropogenic rate.
Natural secondary	Nitrate+Ammonium	636	Conversion of 392E6 t NO3 from NOx + 44E6 t NH4 from NH3.
Natural secondary	Nitrate+Ammonium	570	Conversion of 42% of NOx to 330E6 t NH3 + 23% of NH3 to 240E6 t NH4
Natural secondary	Nitrate+Ammonium	75	Secondary sulfates divided by 5.5, measured ratio in filter.
Natural secondary	Hydrocarbon	1000	5 turnovers/day.
Natural secondary	Hydrocarbon	438	Conc = 10 ppb in 2 km, 1 turnover per day.
Natural secondary	Hydrocarbon	500	Mixing ratio of solids in precipitation.
Natural secondary	Hydrocarbon	175-200	0.1 % of dry plant production release .
Natural secondary	Hydrocarbon	200	Total conversion of hydrocarbon value
Natural secondary	Hydrocarbon	75	Assumed 50% conversion to particles.

TABLE G007

Summary of Estimates of Worldwide Emissions of Particulates other than Anthropogenic Primary Emissions. Values are in millions of metric tons/year

REF.: 413

Type	Type	Value	Method of estimation
Anthropogenic secondary	Sulfate	220	Conversion to (NH4)2SO4 of 2/3 of estimated 160 mio t/y of SO2
Anthropogenic secondary	Sulfate	200	Attributed to Peterson and Junge.
Anthropogenic secondary	Sulfate	133	Conversion of 70% of 132 mio tons of SO2
Anthropogenic secondary	Sulfate	130	Estimate from content in glacier ice.
Anthropogenic secondary	Sulfate	100	Conversion of 60% of values for SO2 and H2S.
Anthropogenic secondary	Nitrate	84	Conversion of 42 % of man produced NOX to NO3.
Anthropogenic secondary	Nitrate	35	Anthropogenic secondary sulfates divided by 5.5.
Anthropogenic secondary	Nitrate	30	Conversion of 42% of 48 mio tons of NO2.
Anthropogenic secondary	Hydrocarbon	90	Attributed by SCEP.
Anthropogenic secondary	Hydrocarbon	25	Reactive fraction of total estimate o hydrocarbon.
Anthropogenic secondary	Hydrocarbon	15	Conversion of 25 % of twice US emission of 29 mio t in 1968.

G3

DIFFUSION AND DISPERSION

TABLE G8

COEFFICIENT OF BED ACTIVITY : OXYGEN (F074)			
ITEM:	VALUE:	CONDITIONS:	REF.:
*DEOXYGENATION IN WATER	0.1	STREAM SLOPE = 2.5 FT/MI	64
*DEOXYGENATION IN WATER	0.15	STREAM SLOPE = 5.0 FT/MI	64
*DEOXYGENATION IN WATER	0.25	STREAM SLOPE = 10.0 FT/MI	64
*DEOXYGENATION IN WATER	0.4	STREAM SLOPE = 25.0 FT/MI	64
*DEOXYGENATION IN WATER	0.6	STREAM SLOPE = 50.0 FT/MI	64

TABLE G9

DIFFUSION AND DISPERSION : CH4			
ITEM:	VALUE:	CONDITIONS:	REF.:
*DIFFUSION IN LAKE KIVU	0.145 CM**2/S	35 - 40 M, IN VITRO EXPERIMENTS	740
*DIFFUSION IN LAKE KIVU	1.18 CM**2/S	40 - 45 M, IN VITRO EXPERIMENTS	740
*DIFFUSION IN LAKE KIVU	0.47 CM**2/S	45 - 50 M, IN VITRO EXPERIMENTS	740

TABLE G10

DIFFUSION AND DISPERSION : CO			
ITEM:	VALUE:	CONDITIONS:	REF.:
*DIFFUSION IN WATER	2.2*10**(-9) M**2/S	298 K	1739

TABLE G11

DIFFUSION AND DISPERSION : CO2			
ITEM:	VALUE:	CONDITIONS:	REF.:
*DIFFUSION FROM AIR TO WATER	840 MG/(M**2*DAY)	INDIRECT MEASURED	1020
*DIFFUSION FROM AIR TO WATER	0.0023 CM/S	WELL STIRRED WATER SURFACE	1020
*DIFFUSION FROM AIR TO WATER	0.01 CM/S	DURING HIG WINDS	1020
*DIFFUSION FROM AIR TO WATER	0.001 CM/S	ON CALM DAYS	1020
*DIFFUSION FROM AIR TO WATER	26 MMOL/M**2	WESTERN LAKE ERIE	1020
*DIFFUSION IN AIR	59 CM**2/S	HEIGHT = 10 CM, ARTIC TUNDRA	295
*DIFFUSION IN AIR	176 CM**2/S	HEIGHT = 20 CM, ARTIC TUNDRA	295
*DIFFUSION IN AIR	412 CM**2/S	HEIGHT = 40 CM, ARTIC TUNDRA	295
*DIFFUSION IN AIR	882 CM**2/S	HEIGHT = 80 CM, ARTIC TUNDRA	295
*DIFFUSION IN AIR	1823 CM**2/S	HEIGHT = 160 CM, ARTIC TUNDRA	295

TABLE G12

DIFFUSION AND DISPERSION : CS-137			
ITEM:	VALUE:	CONDITIONS:	REF.:
*DIFFUSION FROM SEDIMENT TO WATER	2*10**(-5) CM**2/S	LAKE ERIE, WESTERN BASIN	912
*DIFFUSION FROM SEDIMENT TO WATER	2*10**(-5) CM**2/S	LAKE ERIE, CENTRAL BASIN	912
*DIFFUSION FROM SEDIMENT TO WATER	2*10**(-5) CM**2/S	LAKE ONTARIO, WESTERN PART	912

TABLE G13

DIFFUSION AND DISPERSION : ESTUARINE WATER			
ITEM:	VALUE:	CONDITIONS:	REF.:
*DISPERSION	5 MI**2/DAY	DELAWARE RIVER, FRESHWAT INFLOW= 2500 CFS, LOW VELOC.=0.12-0.009 FPS	712
*DISPERSION	20 MI**2/DAY	HUDSON R.,FRESHWAT INFLOW=5000 CFS, LOW VELOC.=0.037 FPS	712
*DISPERSION	10 MI**2/DAY	EAST R.,FRESHWAT INFLOW=0 CFS, LOW VELOC.=0.0 FPS	712

TABLE G13

DIFFUSION AND DISPERSION : ESTUARINE WATER			
ITEM:	VALUE:	CONDITIONS:	REF.:
*DISPERSION	30 MI**2/DAY	COOPER R.,FRESHWAT INFLOW=10000 CFS, LOW VELOC.=0.25 FPS	712
*DISPERSION	10 - 20 MI**2/DAY	SAVANNAH R.,FRESHWAT INFLOW=7000 CFS, LOW VELOC.=0.7-0.17 FPS	712
*DISPERSION	5 MI**2/DAY	LOWER RARITAN R.,FRESHWAT INFLOW= 150 CFS, LOW VELOC.=0.047-0.029 FPS	712
*DISPERSION	5 MI**2/DAY	SOUTH R.,FRESHWAT INFLOW=23 CFS, LOW VELOC.=0.01 FPS	712
*DISPERSION	27 MI**2/DAY	HOUSTON SHIP CHAN.,FRESHWAT INFLOW =900 CFS, LOW VELOC.=0.05 FPS	712
*DISPERSION	2 - 10 MI**2/DAY	CAPE FEAR R.,FRESHWAT INFLOW=1000 CFS, LOW VELOC.=0.48-0.03 FPS	712
*DISPERSION	1 - 10 MI**2/DAY	POTOMAC R.,FRESHWAT INFLOW=1000 CFS, LOW VELOC.=0.48-0.03 FPS	712
*DISPERSION	1 MI**2/DAY	COMPTON CREEK,FRESHWAT INFLOW=10 CFS, LOW VELOC.=0.01-0.013 FPS	712
*DISPERSION	0.5 - 1 MI**2/DAY	WAPPINGER CREEK,FRESHWAT INFLOW=2 CFS, LOW VELOC.=0.004-0.001 FPS	712

TABLE G14

DIFFUSION AND DISPERSION : HG			
ITEM:	VALUE:	CONDITIONS:	REF.:
*DIFFUSION IN SEDIMENT	0.2*10**(-5) CM**2/SEC	FINE-GRAIN CLAY SEDIMENT IN MOBILE BAY	929

TABLE G15

DIFFUSION AND DISPERSION : IN WATER			
ITEM:	VALUE:	CONDITIONS:	REF.:
*EDDY DIFFUSION IN HORIZONTAL DIRECTION	1.4*10**6 - 5.6*10**1 CM**2/SEC	SURFACE WATER	1567
*EDDY DIFFUSION IN VERTICAL DIRECTION	2.0*10**0 - 1.0*10**(-2) CM**2/SEC	THERMOCLINE AND DEEPER REGIONS IN LAKES AND OCEANS	1567
*EDDY DIFFUSION IN VERTICAL DIRECTION	10**(-2) M**2/SEC	LAKES, UPPER MIXED LAYER	1127
*EDDY DIFFUSION IN VERTICAL DIRECTION	10**(-6) M**2/SEC	LAKES, UNDER THERMOCLINE	1127
*EDDY DIFFUSION IN VERTICAL DIRECTION	10**(-4) M**2/SEC	LAKES, DEEPS LAYERS BELOW THERMOCLINE	1127
*HEAT IN WATER	1.6*10**(-3) CM**2/SEC	NONE	1567
*MOLECULAR DIFFUSION	6.2*10**(-5) - 2.0*10**(-6) CM**2/ SEC	SALT AND GASES IN WATER	1567
*MOLECULAR DIFFUSION	1.8*10**(-6) - 1.0*10**(-7) CM**2/ SEC	PROTEINS IN WATER	1567
*MOLECULAR DIFFUSION	3.8*10**(-6) - 3.8*10**(-9) CM**2/ SEC	IONIC SOLUTES IN POROUS MEDIA (SEDIMENTS, SOILS)	1567
*THERMAL DIFFUSION	4.6*10**(-7) - 2.6*10**(-9) CM**2/ SEC	SALT IN WATER	1567

TABLE G16

DIFFUSION AND DISPERSION : NITRATE			
ITEM:	VALUE:	CONDITIONS:	REF.:
*EDDY DIFFUSION CONSTANT	5 CM**2/SEC	IN SEAWATER	71

TABLE G17

DIFFUSION AND DISPERSION : OXYGEN			
ITEM:	VALUE:	CONDITIONS:	REF.:
*DIFFUSION OF OXYGEN INTO SEAWATER	6*10**(-3) CM**2/SEC	NO OXYGEN IN WATER AT START	1648

TABLE G18

DIFFUSION AND DISPERSION : RIVER DANUBE			
ITEM:	VALUE:	CONDITIONS:	REF.:
*DISPERSION	0.034 M**2/S	FLOW=1900 M3/S, CHANGE IN MOMENT METHOD	1480
*DISPERSION	0.014 M**2/S	FLOW=1050 M3/S, CHANGE OF MONENT METHOD	1480
*DISPERSION	0.033 M**2/S	FLOW=1900 M3/S, MASS FLUX CONSTANT LINE METHOD	1480
*DISPERSION	0.0125 M**2/S	FLOW=1050 M3/S, MASS FLUX CONSTANT LINE METHOD	1480

TABLE G19

DIFFUSION AND DISPERSION : SI			
ITEM:	VALUE:	CONDITIONS:	REF.:
*DIFFUSION IN MARINE SEDIMENTS	95 CM**2/YEAR	IN PORE WATER	913

TABLE G20

DIFFUSION AND DISPERSION : SR-90			
ITEM:	VALUE:	CONDITIONS:	REF.:
*DIFFUSION FROM SEDIMENT TO WATER	4*10**(-6) CM**2/S	LAKE ERIE, WESTERN BASIN	912
*DIFFUSION FROM SEDIMENT TO WATER	4*10**(-6) CM**2/S	LAKE ERIE, CENTRAL BASIN	912
*DIFFUSION FROM SEDIMENT TO WATER	2*10**(-6) CM**2/S	LAKE ONTARIO, WESTERN PART	912

TABLE G21

EDDY DIFFUSION COEFFICIENT : PHOSPHORUS AND NITROGEN			
ITEM:	VALUE:	CONDITIONS:	REF.:
*DIFFUSION FROM META- TO EPILIMNION IN LAKE LUCERNE(HORW BAY)	1.58 CM**2/SEC.	APRIL,N-FLUX=3.06G/M**2*MONTH, P-FLUX=0.067G/M**2*MONTH	136
*DIFFUSION FROM META- TO EPILIMNION IN LAKE LUCERNE(HORW BAY)	0.48 CM**2/SEC.	MAY,N-FLUX=0.93G/M**2*MONTH,P-FLUX =0.020G/M**2*MONTH	136
*DIFFUSION FROM META- TO EPILIMNION IN LAKE LUCERNE(HORW BAY)	0.64 CM**2/SEC.	JUNE,N-FLUX=1.23G/M**2*MONTH, P-FLUX=0.027G/M**2*MONTH	136
*DIFFUSION FROM META- TO EPILIMNION IN LAKE LUCERNE(HORW BAY)	0.41 CM**2/SEC.	JULY,N-FLUX=0.80G/M**2*MONTH, P-FLUX=0.017G/M**2*MONTH	136
*DIFFUSION FROM META- TO EPILIMNION IN LAKE LUCERNE(HORW BAY)	0.14 CM**2/SEC.	AUGUST,N-FLUX=0.27G/M**2*MONTH, P-FLUX=0.006G/M**2*MONTH	136
*DIFFUSION FROM META- TO EPILIMNION IN LAKE LUCERNE(HORW BAY)	0.10 CM**2/SEC.	SEPTEMBER,N-FLUX=0.19G/M**2*MONTH, P-FLUX=0.004G/M**2*MONTH	136
*DIFFUSION FROM META- TO EPILIMNION IN ROTSEE	1.18 CM**2/SEC.	APRIL,N-FLUX=2.56G/M**2*MONTH, P-FLUX=0.61G/M**2*MONTH	136
*DIFFUSION FROM META- TO EPILIMNION IN ROTSEE	0.12 CM**2/SEC.	MAY,N-FLUX=4.85G/M**2*MONTH,P-FLUX =1.49G/M**2*MONTH	136
*DIFFUSION FROM META- TO EPILIMNION IN ROTSEE	0.13 CM**2/SEC.	JUNE,N-FLUX=9.49G/M**2*MONTH, P-FLUX=2.25G/M**2*MONTH	136
*DIFFUSION FROM META- TO EPILIMNION IN ROTSEE	0.051 CM**2/SEC.	JULY,N-FLUX=3.93G/M**2*MONTH, P-FLUX=1.06G/M**2*MONTH	136
*DIFFUSION FROM META- TO EPILIMNION IN ROTSEE	0.041 CM**2/SEC.	AUGUST,N-FLUX=4.76G/M**2*MONTH, P-FLUX=0.47G/M**2*MONTH	136
*DIFFUSION FROM META- TO EPILIMNION IN ROTSEE	0.016 CM**2/SEC.	SEPTEMBER,N-FLUX=1.68G/M**2*MONTH, P-FLUX=0.10G/M**2*MONTH	136

TABLE G22

EQUILIBRIUM CONSTANT : A HYPOTHETICAL 9 METAL , 9 LIGAND SYSTEM (F075)			
ITEM:	VALUE:	CONDITIONS:	REF.:
*FREE FREE	LOG(K) = -2.03	LOG(CONC(CA))=-2.0, 0	1063
*FREE LIGAND CITRATE	LOG(K) = -16.3	LOG(CONC(CITRATE))=-6.5	1063
*FREE LIGAND CO3	LOG(K) = -5.2	LOG(CONC(CO3))=-3.7	1063
*FREE LIGAND F	LOG(K) = -4.5	LOG(CONC(F))=-4.2	1063
*FREE LIGAND GLYCINE	LOG(K) = -9.3	LOG(CONC(GLYCINE))=-7.5	1063
*FREE LIGAND NOCARDAMINE	LOG(K) = -21.5	LOG(CONC(NOCARDAMINE))=-7.0	1063
*FREE LIGAND NTA	LOG(K) = -10.7	LOG(CONC(NTA))=-6.0	1063
*FREE LIGAND OH	LOG(K) = -6.0	NONE	1063
*FREE LIGAND PO4	LOG(K) = -10.6	LOG(CONC(PO4))=-5.7	1063
*FREE LIGAND SALICYLATE	LOG(K) = -12.5	LOG(CONC(SALICYLATE))=-7.0	1063
*FREE LIGAND SO4	LOG(K) = -1.67	LOG(CONC(SO4))=-1.6	1063
*METAL CA, FREE	LOG(K) = -2.03	LOG(CONC(CA))=-2.0	1063
*METAL CA, LIGAND CITRATE	LOG(K) = -6.9	TOTAL COMPLEX, LOG(CONC(CA))=-2.0, LOG(CONC(CITRATE))=-6.5	1063
*METAL CA, LIGAND CO3	LOG(K) = -4.7	TOTAL COMPLEX, LOG(CONC(CA))=-2.0, LOG(CONC(CO3))=-3.7	1063
*METAL CA, LIGAND F	LOG(K) = -6.0	TOTAL COMPLEX, LOG(CONC(CA))=-2.0, LOG(CONC(F))=-4.2	1063
*METAL CA, LIGAND GLYCINE	LOG(K) = -10.7	TOTAL COMPLEX, LOG(CONC(CA))=-2.0, LOG(CONC(GLYCINE))=-7.5	1063
*METAL CA, LIGAND NOCARDAMINE	LOG(K) = -20.2	TOTAL COMPLEX, LOG(CONC(CA))=-2.0, LOG(CONC(NOCARDAMINE))=-7.0	1063
*METAL CA, LIGAND NTA	LOG(K) = -6.3	TOTAL COMPLEX, LOG(CONC(CA))=-2.0, LOG(CONC(NTA))=-6.0	1063
*METAL CA, LIGAND OH	LOG(K) = -7.1	TOTAL COMPLEX, LOG(CONC(CA))=-2.0,	1063
*METAL CA, LIGAND PO4	LOG(K) = -8.0	TOTAL COMPLEX, LOG(CONC(CA))=-2.0, LOG(CONC(PO4))=-5.7, SOLID FORMAT.	1063
*METAL CA, LIGAND SO4	LOG(K) = -3.2	TOTAL COMPLEX, LOG(CONC(CA))=-2.0, LOG(CONC(SO4))=-1.6	1063
*METAL CD, FREE	LOG(K) = -9.10	LOG(CONC(CD))=-9.0, LOG(CONC(SO4))=-1.61	1063
*METAL CD, LIGAND CITRATE	LOG(K) = -13.0	TOTAL COMPLEX, LOG(CONC(CD))=-9.0, LOG(CONC(CITRATE))=-6.5	1063
*METAL CD, LIGAND CO3	LOG(K) = -12.4	TOTAL COMPLEX, LOG(CONC(CD))=-9.0, LOG(CONC(CO3))=-3.7	1063
*METAL CD, LIGAND GLYCINE	LOG(K) = -14.1	TOTAL COMPLEX, LOG(CONC(CD))=-9.0, LOG(CONC(GLYCINE))=-7.5	1063
*METAL CD, LIGAND NOCARDAMINE	LOG(K) = -21.8	TOTAL COMPLEX, LOG(CONC(CD))=-9.0, LOG(CONC(NOCARDAMINE))=-7.0	1063
METAL CD, LIGAND NTA	LOG(K) = -9.7	TOTAL COMPLEX, LOG(CONC(CD))=-9.0, LOG(CONC(NTA))=-6.0	1063
*METAL CD, LIGAND OH	LOG(K) = -10.9	TOTAL COMPLEX, LOG(CONC(CD))=-9.0,	1063
*METAL CD, LIGAND PO4	LOG(K) = -17.8	TOTAL COMPLEX, LOG(CONC(CD))=-9.0, LOG(CONC(PO4))=-5.7	1063
*ME¤AL CD, LIGAND SALICYLATE	LOG(K) = -16.1	TOTAL COMPLEX, LOG(CONC(CD))=-9.0, LOG(CONC(SALICYLATE))=-7.0	1063
*METAL CD, LIGAND SO4	LOG(K) = -9.9	TOTAL COMPLEX, LOG(CONC(CD))=-9.0, LOG(CONC(SO4))=-1.6	1063
*METAL CU(2+), FREE	LOG(K) = -9.2	LOG(CONC(CU(2+))=-7.0	1063
*METAL CU(2+), LIGAND CITRATE	LOG(K) = -7.5	TOTAL COMPLEX, LOG(CONC(CU(2+))=-7.0LOG(CONC(CITRATE))=-6.5	1063
*METAL CU(2+), LIGAND CO3	LOG(K) = -9.6	TOTAL COMPLEX, LOG(CONC(CU(2+))=-7.0LOG(CONC(CO3))=-3.7	1063
*METAL CU(2+), LIGAND F	LOG(K) = -12.9	TOTAL COMPLEX, LOG(CONC(CU(2+))=-7.0LOG(CONC(F))=-4.2	1063
*METAL CU(2+), LIGAND GLYCINE	LOG(K) = -10.3	TOTAL COMPLEX, LOG(CONC(CU(2+))=-7.0LOG(CONC(GLYCINE))=-7.5	1063
*METAL CU(2+), LIGAND NOCARDAMINE	LOG(K) = -16.1	TOTAL COMPLEX, LOG(CONC(CU(;+))=-7.0LOG(CONC(NOCARDAMINE))=-7.0	1063
*METAL CU(2+), LIGAND NTA	LOG(K) = -7.2	TOTAL COMPLEX, LOG(CONC(CU(2+))=-7.0LOG(CONC(NTA))=-6.0	1063

TABLE G22

EQUILIBRIUM CONSTANT : A HYPOTHETICAL 9 METAL, 9 LIGAND SYSTEM (F075)

ITEM:	VALUE:	CONDITIONS:	REF.:
*METAL CU(2+), LIGAND OH	LOG(K) = -9.1	TOTAL COMPLEX, LOG(CONC(CU(2+))= -7.0	1063
*METAL CU(2+), LIGAND SALICYLATE	LOG(K) = -11.1	TOTAL COMPLEX, LOG(CONC(CU(2+))= -7.0LOG(CONC(SALICYLATE))=-7.0	1063
*METAL CU(2+), LIGAND SO4	LOG(K) = -10.3	TOTAL COMPLEX, LOG(CONC(CU(2+))= -7.0LOG(CONC(SO4))=-1.6	1063
*METAL FE(3+), FREE	LOG(K) = -18.0	LOG(CONC(FE(3+))=-6.0	1063
*METAL FE(3+), LIGAND CITRATE	LOG(K) = -9.3	TOTAL COMPLEX, LOG(CONC(FE(3+))= -6.0, LOG(CONC(CITRATE))=-6.5	1063
*METAL FE(3+), LIGAND F	LOG(K) = -17.2	TOTAL COMPLEX, LOG(CONC(FE(3+))= -6.0, LOG(CONC(F))=-4.2	1063
*METAL FE(3+), LIGAND GLYCINE	LOG(K) = -17.3	TOTAL COMPLEX, LOG(CONC(FE(3+))= -6.0, LOG(CONC(GLYCINE))=-7.5	1063
*METAL FE(3+), LIGAND NOCARDAMINE	LOG(K) = -7.0	TOTAL COMPLEX, LOG(CONC(FE(3+))= -6.0, LOG(CONC(NOCARDAMINE))=-7.0	1063
*METAL FE(3+), LIGAND NTA	LOG(K) = -8.9	TOTAL COMPLEX, LOG(CONC(FE(3+))= -6.0, LOG(CONC(NTA))=-6.0	1063
*METAL FE(3+), LIGAND OH	LOG(K) = -8.0	TOTAL COMPLEX, LOG(CONC(FE(3+))= -6.0,STABLE SOLID CAN BE FORMED	1063
*METAL FE(3+), LIGAND PO4	LOG(K) = -18.6	TOTAL COMPLEX, LOG(CONC(FE(3+))= -6.0, LOG(CONC(PO4))=-5.7	1063
*METAL FE(3+), LIGAND SALICYLATE	LOG(K) = -14.0	TOTAL COMPLEX, LOG(CONC(FE(3+))= -6.0, LOG(CONC(SALICYLATE))=-7.0	1063
*METAL FE(3+), LIGAND SO4	LOG(K) = -17.4	TOTAL COMPLEX, LOG(CONC(FE(3+))= -6.0, LOG(CONC(SO4))=-1.6	1063
*METAL MG, FREE	LOG(K) = -1.33	LOG(CONC(MG))=-1.3	1063
*METAL MG, LIGAND CITRATE	LOG(K) = -6.9	TOTAL COMPLEX, LOG(CONC(MG))=-1.3, LOG(CONC(CITRATE))=-6.5	1063
*METAL MG, LIGAND CO3	LOG(K) = -4.1	TOTAL COMPLEX, LOG(CONC(MG))=-1.3, LOG(CONC(CO3))=-3.7	1063
*METAL MG, LIGAND F	LOG(K) = -4.5	TOTAL COMPLEX, LOG(CONC(MG))=-1.3, LOG(CONC(F))=-4.2	1063
*METAL MG, LIGAND GLYCINE	LOG(K) = -9.6	TOTAL COMPLEX, LOG(CONC(MG))=-1.3, LOG(CONC(GLYCINE))=-7.5	1063
*METAL MG, LIGAND NOCARDAMINE	LOG(K) = -16.8	TOTAL COMPLEX, LOG(CONC(MG))=-1.3, LOG(CONC(NOCARDAMINE))=-7.0	1063
*METAL MG, LIGAND NTA	LOG(K) = -6.6	TOTAL COMPLEX, LOG(CONC(MG))=-1.3, LOG(CONC(NTA))=-6.0	1063
*METAL MG, LIGAND OH	LOG(K) = -5.2	TOTAL COMPLEX, LOG(CONC(MG))=-1.3,	1063
*METAL MG, LIGAND PO4	LOG(K) = -7.5	TOTAL COMPLEX, LOG(CONC(MG))=-1.3, LOG(CONC(PO4))=-5.7	1063
*METAL MG, LIGAND SO4	LOG(K) = -2.5	TOTAL COMPLEX, LOG(CONC(MG))=-1.3, LOG(CONC(SO4))=-1.6	1063
*METAL MN(2+), FREE	LOG(K) = -6.73	LOG(CONC(MN(2+))=-6.7	1063
*METAL MN(2+), LIGAND CITRATE	LOG(K) = -11.7	TOTAL COMPLEX, LOG¤CONC(MN(2+))= -6.7, LOG(CONC(CITRATE))=-6.5	1063
*METAL MN(2+), LIGAND CO3	LOG(K) = -8.9	TOTAL COMPLEX, LOG(CONC(MN(2+))= -6.7, LOG(CONC(CO3))=-3.7	1063
*METAL MN(2+), LIGAND GLYCINE	LOG(K) = -14.0	TOTAL COMPLEX, LOG(CONC(MN(2+))= -6.7, LOG(CONC(GLYCINE))=-7.5	1063
*METAL MN(2+), LIGAND NOCARDAMINE	LOG(K) = -16.3	TOTAL COMPLEX, LOG(CONC(MN(2+))= -6.7, LOG(CONC(NOCARDAMINE))=-7.0	1063
*METAL MN(2+), LIGAND NTA	LOG(K) = -10.0	TOTAL COMPLEX, LOG(CONC(MN(2+))= -6.7, LOG(CONC(NTA))=-6.0	1063
*METAL MN(2+), LIGAND OH	LOG(K) = -9.9	TOTAL COMPLEX, LOG(CONC(MN(2+))= -6.7,	1063
*METAL MN(2+), LIGAND SALICYLATE	LOG(K) = -13.3	TOTAL COMPLEX, LOG(CONC(MN(2+))= -6.7, LOG(CONC(SALICYLATE))=-7.0	1063
*METAL MN(2+), LIGAND SO4	LOG(K) = -7.9	TOTAL COMPLEX, LOG(CONC(MN(2+))= -6.7, LOG(CONC(SO4))=-1.6	1063
*METAL NI, FREE	LOG(K) = -9.16	LOG(CONC(NI))=-7.0	1063
*METAL NI, LIGAND CITRATE	LOG(K) = -9.8	TOTAL COMPLEX, LOG(CONC(NI))=-7.0, LOG(CONC(CITRATE))=-6.5	1063
*METAL NI, LIGAND F	LOG(K) = -11.7	TOTAL COMPLEX, LOG(CONC(NI))=-7.0, LOG(CONC(F))=-4.2	1063

TABLE G22

EQUILIBRIUM CONSTANT : A HYPOTHETICAL 9 METAL , 9 LIGAND SYSTEM (F075)			
ITEM:	VALUE:	CONDITIONS:	REF.:
*METAL NI, LIGAND GLYCINE	LOG(K) = -11.3	TOTAL COMPLEX, LOG(CONC(NI))=-7.0, LOG(CONC(GLYCINE))=-7.5	1063
*METAL NI, LIGAND NOCARDAMINE	LOG(K) = -17.1	TOTAL COMPLEX, LOG(CONC(NI))=-7.0, LOG(CONC(NOCARDAMINE))=-7.0	1063
*METAL NI, LIGAND NTA	LOG(K) = -7.1	TOTAL COMPLEX, LOG(CONC(NI))=-7.0, LOG(CONC(NTA))=-6.0	1063
*METAL NI, LIGAND OH	LOG(K) = -8.2	TOTAL COMPLEX, LOG(CONC(NI))=-7.0,	1063
*METAL NI, LIGAND SALICYLATE	LOG(K) = -13.3	TOTAL COMPLEX, LOG(CONC(NI))=-7.0, LOG(CONC(SALICYLATE))=-7.0	1063
*METAL NI, LIGAND SO4	LOG(K) = -8.9	TOTAL COMPLEX, LOG(CONC(NI))=-7.0, LOG(CONC(SO4))=-1.6	1063
*METAL SR, FREE	LOG(K) = -3.83	LOG(CONC(SR))=-3.8	1063
*METAL SR, LIGAND CITRATE	LOG(K) = -9.4	TOTAL COMPLEX, LOG(CONC(SR))=-3.8, LOG(CONC(CITRATE))=-6.5	1063
*METAL SR, LIGAND GLYCINE	LOG(K) = -12.5	TOTAL COMPLEX, LOG(CONC(SR))=-3.8, LOG(CONC(GLYCINE))=-7.5	1063
*METAL SR, LIGAND NOCARDAMINE	LOG(K) = -21.4	TOTAL COMPLEX, LOG(CONC(SR))=-3.8, LOG(CONC(NOCARDAMINE))=-7.0	1063
*METAL SR, LIGAND NTA	LOG(K) = -9.5	TOTAL COMPLEX, LOG(CONC(SR))=-3.8, LOG(CONC(NTA))=-6.0	1063
*METAL SR, LIGAND OH	LOG(K) = -9.2	TOTAL COMPLEX, LOG(CONC(SR))=-3.8,	1063
*METAL SR, LIGAND PO4	LOG(K) = -10.4	TOTAL COMPLEX, LOG(CONC(SR))=-3.8, LOG(CONC(PO4))=-5.7	1063
*METAL SR, LIGAND SO4	LOG(K) = -5.0	TOTAL COMPLEX, LOG(CONC(SR))=-3.8, LOG(CONC(SO4))=-1.6	1063
*METAL ZN, FREE	LOG(K) = -6.8	LOG(CONC(ZN))=-6.5	1063
*METAL ZN, LIGAND CITRATE	LOG(K) = -11.7	TOTAL COMPLEX, LOG(CONC(ZN))=-6.5, LOG(CONC(CITRATE))=-6.5	1063
*METAL ZN, LIGAND F	LOG(K) = -10.5	TOTAL COMPLEX, LOG(CONC(ZN))=-6.5, LOG(CONC(F))=-4.2	1063
*METAL ZN, LIGAND GLYCINE	LOG(K) = -11.3	TOTAL COMPLEX, LOG(CONC(ZN))=-6.5, LOG(CONC(GLYCINE))=-7.5	1063
*METAL ZN, LIGAND NOCARDAMINE	LOG(K) = -16.0	TOTAL COMPLEX, LOG(CONC(ZN))=-6.5, LOG(CONC(NOCARDAMINE))=-7.0	1063
*METAL ZN, LIGAND NTA	LOG(K) = -6.9	TOTAL COMPLEX, LOG(CONC(ZN))=-6.5, LOG(CONC(NTA))=-6.0	1063
*METAL ZN, LIGAND OH	LOG(K) = -8.3	TOTAL COMPLEX, LOG(CONC(ZN))=-6.5,	1063
*METAL ZN, LIGAND SALICYLATE	LOG(K) = -12.3	TOTAL COMPLEX, LOG(CONC(ZN))=-6.5, LOG(CONC(SALICYLATE))=-7.0	1063
*METAL ZN, LIGAND SO4	LOG(K) = -7.8	TOTAL COMPLEX, LOG(CONC(ZN))=-6.5, LOG(CONC(SO4))=-1.6	1063
*SUM OF PROTONATED CONCENTRATION OF LIGAND CITRATE	LOG(K) = -8.3	LOG(CONC(CITRATE))=-6.5	1063
*SUM OF PROTONATED CONCENTRATION OF LIGAND CO3	LOG(K) = -4.3	LOG(CONC(CO3))=-3.7	1063
*SUM OF PROTONATED CONCENTRATION OF LIGAND F	LOG(K) = -9.5	LOG(CONC(F))=-4.2	1063
*SUM OF PROTONATED CONCENTRATION OF LIGAND GLYCINE	LOG(K) = -7.5	LOG(CONC(GLYCINE))=-7.5	1063
*SUM OF PROTONATED CONCENTRATION OF LIGAND NOCARDAMINE	LOG(K) = -17.6	LOG(CONC(NOCARDAMINE))=-7.0	1063
*SUM OF PROTONATED CONCENTRATION OF LIGAND NTA	LOG(K) = -8.9	LOG(CONC(NTA))=-6.0	1063
*SUM OF PROTONATED CONCENTRATION OF LIGAND PO4	LOG(K) = -7.1	LOG(CONC(PO4))=-5.7	1063
*SUM OF PROTONATED CONCENTRATION OF LIGAND SALICYLATE	LOG(K) = -7.0	LOG(CONC(SALICYLATE))=-7.0	1063
*SUM OF PROTONATED CONCENTRATION OF LIGAND SO4	LOG(K) = -7.7	LOG(CONC(SO4))=-1.6	1063

TABLE G23

EQUILIBRIUM CONSTANT : AL			
ITEM:	VALUE:	CONDITIONS:	REF.:
*AL(3+) + H2O <-> ALOH(2+) + H(+)	LOG(K) = -5.0	298 K	1435

TABLE G23

EQUILIBRIUM CONSTANT : AL			
ITEM:	VALUE:	CONDITIONS:	REF.:
*AL(3+) + 3 H2O <-> AL(OH)3(S) + 3 H(+)	LOG(K) = -9.5	298 K	1435

TABLE G24

EQUILIBRIUM CONSTANT : CA			
ITEM:	VALUE:	CONDITIONS:	REF.:
*CA(2+) + CO3(2-) <-> CACO3	LOG(K) = 3.20	FRESH WATER	1542
*CA(2+) + CO3(2-) + H(+) <-> CAHCO3 (+)	LOG(K) = 11.60	FRESH WATER	1542
*PHTHALIC ACID + CA(II)	LOG(K) = 1.07	FORMATION OF HUMIC-METAL FOR SMALL ORGANIC MOLECULES	1726
*3,4-DIHYDROXYBENZOIC ACID + CA(II)	LOG(K) = 3.71	FORMATION OF HUMIC-METAL FOR SMALL ORGANIC MOLECULES	1726

TABLE G25

EQUILIBRIUM CONSTANT : CACO3 REACTIONS			
ITEM:	VALUE:	CONDITIONS:	REF.:
*CACO3 <-> CA(2+) + CO3(2-)	LOG(K) = -8.35	IONIC STRENGTH = 0, 298 K, 1 ATM, CALCITE	1526
*CACO3 <-> CA(2+) + CO3(2-)	LOG(K) = -6.2	SEAWATER, 298 K, 1 ATM, CALCITE	1526
*CACO3 <-> CA(2+) + CO3(2-)	LOG(K) = -8.22	IONIC STRENGTH = 0, 298 K, 1 ATM, ARAGONITE	1526
*CACO3(S) <-> CA(2+) + CO3(2-)	- LOG(K) = 8.09	278 K, INFINITE DILUTION, FRESHWATER	1435
*CACO3(S) <-> CA(2+) + CO3(2-)	- LOG(K) = 8.15	283 K, INFINITE DILUTION, FRESHWATER	1435
*CACO3(S) <-> CA(2+) + CO3(2-)	- LOG(K) = 8.22	288 K, INFINITE DILUTION, FRESHWATER	1435
*CACO3(S) <-> CA(2+) + CO3(2-)	- LOG(K) = 8.28	293 K, INFINITE DILUTION, FRESHWATER	1435
*CACO3(S) <-> CA(2+) + CO3(2-)	- LOG(K) = 8.34	298 K, INFINITE DILUTION, FRESHWATER	1435
*CACO3(S) <-> CA(2+) + CO3(2-)	- LOG(K) = 8.51	313 K, INFINITE DILUTION, FRESHWATER	1435
*CACO3(S) <-> CA(2+) + CO3(2-)	- LOG(K) = 5.67	278 K, SEAWATER	1435
*CACO3(S) <-> CA(2+) + CO3(2-)	- LOG(K) = 6.19	298 K, SEAWATER	1435
*CACO3(S) + H(+) <-> HCO3(-) + CA(2 +)	- LOG(K) = -2.47	278 K, INFINITE DILUTION, FRESHWATER	1435
*CACO3(S) + H(+) <-> HCO3(-) + CA(2 +)	- LOG(K) = -2.34	283 K, INFINITE DILUTION, FRESHWATER	1435
*CACO3(S) + H(+) <-> HCO3(-) + CA(2 +)	- LOG(K) = -2.21	288 K, INFINITE DILUTION, FRESHWATER	1435
*CACO3(S) + H(+) <-> HCO3(-) + CA(2 +)	- LOG(K) = -2.10	293 K, INFINITE DILUTION, FRESHWATER	1435
*CACO3(S) + H(+) <-> HCO3(-) + CA(2 +)	- LOG(K) = -1.99	298 K, INFINITE DILUTION, FRESHWATER	1435
*CACO3(S) + H(+) <-> HCO3(-) + CA(2 +)	- LOG(K) = -1.71	313 K, INFINITE DILUTION, FRESHWATER	1435
*CACO3(S) + H(+) <-> HCO3(-) + CA(2 +)	- LOG(K) = -3.51	278 K, SEAWATER	1435
*CACO3(S) + H(+) <-> HCO3(-) + CA(2 +)	- LOG(K) = -2.77	298 K, SEAWATER	1435
*H2CO3 <-> H(+) + HCO3(-)	- LOG(K) = 5.99	298 K, SEAWATER	1435

TABLE G26

EQUILIBRIUM CONSTANT : CARBONATES AND HYDROXIDE CARBONATES			
ITEM:	VALUE:	CONDITIONS:	REF.:
*CAMG(CO3)2 <-> CA(2+) + MG(2+) + 2 CO3(2-)	LOG(K) = -16.7	IONIC STRENGTH = 0, 298 K, 1 ATM, DOLOMITE	1526

TABLE G26

EQUILIBRIUM CONSTANT : CARBONATES AND HYDROXIDE CARBONATES			
ITEM:	VALUE:	CONDITIONS:	REF.:
*CDCO3(S) + 2 H(+) <-> CD(2+) + H2O + CO2(G)	LOG(K) = 6.44	1 M NACLO4, 298 K	1526
*CU(OH)(CO3)0.5(S) + 2 H(+) <-> CU(2+) + 3/2 H2O + 1/2 CO2(G)	LOG(K) = 7.08	IONIC STRENGTH = 0, 298 K, 1 ATM	1526
*CU(OH)0.67(CO3)0.67(S) + 2 H(+) <-> CU(2+) + 4/3 H2O + 2/3 CO2(G)	LOG(K) = 7.08	IONIC STRENGTH = 0, 298 K, 1 ATM	1526
*FECO3 <-> FE(2+) + CO3(2-)	LOG(K) = 10.4	IONIC STRENGTH = 0, 298 K, 1 ATM, SIDERITE	1526
*MG(2+) + CO3(2-) <-> MGCO3	LOG(K) = 2.18	FRESH WATER	1542
*MGCO3 <-> MG(2+) + CO3(-)	LOG(K) = -5.4	IONIC STRENGTH = 0, 298 K, 1 ATM, NESQUEHONITE MGCO3	1526
*MGCO3 <-> MG(2+) + CO3(2-)	LOG(K) = -4.9	IONIC STRENGTH = 0, 298 K, 1 ATM, MAGNESITE MGCO3	1526
*MG4(CO3)3(OH)2*3H2O <-> 4 MG(2+) + 3 CO3(2-) + 2 OH(-)	LOG(K) = 29.5	IONIC STRENGTH = 0, 298 K, 1 ATM, HYDROMAGNESITE	1526
*SRCO3(S) <-> SR(2+) + CO3(2-)	LOG(K) = -9.03	IONIC STRENGTH = 0, 298 K, 1 ATM	1526
*SRCO3(S) <-> SR(2+) + CO3(2-)	LOG(K) = -6.8	SEAWATER, 298 K, 1 ATM	1526
*ZN(OH)1.2(CO3)0.4(S) + 2 H(+) <-> ZN(2+) + H2O + CO2(G)	LOG(K) = 9.8	IONIC STRENGTH = 0, 298 K, 1 ATM	1526
*ZNCO3(S) + 2 H(+) <-> ZN(2+) + H2O + CO2(G)	LOG(K) = 7.95	IONIC STRENGTH = 0, 298 K, 1 ATM	1526

TABLE G27

EQUILIBRIUM CONSTANT : CD			
ITEM:	VALUE:	CONDITIONS:	REF.:
CD((OH)(2))(SOLID) <-> CD(+2) + 2 OH(-)	LOG(K) = -13.79	COMPLEX FORMATION EQUILIBRIUM, 298 K	899
*CD(+2) + CL(-) <-> CDCL(+)	LOG(K) = 2.08	COMPLEX FORMATION EQUILIBRIUM, 298 K	899
*CD(+2) + CL(-) <-> CDCL(+)	LOG(K) = 1.32	4.5 M NACLO4, 298 K	133
*CD(+2) + CL(-) <-> CDCL(+)	K = 21.0	4.5 M NACLO4M 298 K	133
*CD(+2) + NH3 <-> CD(NH3)(+2)	LOG(K) = 2.65	2 M NH4NO3, 298 K	212
*CD(+2) + NH3 <-> CD(NH3)(+2)	K = 4.47*10**2	2 M NH4NO3, 298 K	212
*CD(+2) + OH(-) <-> CDOH(+)	LOG(K) = 4.59	COMPLEX FORMATION EQUILIBRIUM, 298 K	899
*CD(+2) + OH(-) <-> CDOH(+)	LOG(K) = 4.16	ZERO IONIC STRENGTH, 298 K	212
*CD(+2) + OH(-) <-> CDOH(+)	K = 1.45*10**4	ZERO IONIC STRENGTH, 298 K	212
*CD(+2) + OH(-) + CL(-) <-> CDOHCL(0)	LOG(K) = 5.87	COMPLEX FORMATION EQUILIBRIUM, 298 K	899
*CD(+2) + SO4(-2) <-> CDSO4(0)	LOG(K) = 2.76	COMPLEX FORMATION EQUILIBRIUM, 298 K	899
*CD(+2) + 2 CL(-) <-> CDCL2	K = 166.0	4.5 M NACLO4M 298 K	133
*CD(+2) + 2 NH3 <-> CD(NH3)2(+2)	K = 5.63*10**4	2 M NH4NO3, 298 K	212
*CD(+2) + 2 OH(-) <-> CD((OH)(2))	LOG(K) = 8.93	COMPLEX FORMATION EQUILIBRIUM, 298 K	899
*CD(+2) + 2 OH(-) <-> CD(OH)2	K = 2.46*10**8	ZERO IONIC STRENGTH, 298 K	212
*CD(+2) + 3 CL(-) <-> CDCL3(-)	K = 204.0	4.5 M NACLO4M 298 K	212
*CD(+2) + 3 NH3 <-> CD(NH3)3(+2)	K = 1.55*10**6	2 M NH4NO3, 298 K	212
*CD(+2) + 3 OH(-) <-> CD((OH)(3))(-)	LOG(K) = 9.58	COMPLEX FORMATION EQUILIBRIUM, 298 K	899
*CD(+2) + 3 OH(-) <-> HCDO2(-) + H2O	K = 1.20*10**9	ZERO IONIC STRENGTH, 298 K	133
*CD(+2) + 4 CL(-) <-> CDCL4(-2)	K = 71.5	2 M NH4NO3, 298 K	212
*CD(+2) + 4 NH3 <-> CD(NH3)4(+2)	K = 1.32*10**7	2 M NH4NO3, 298 K	212
*CD(+2) + 4 OH(-) <-> CDO2(-2) + H2O	K = 5.76*10**8	4.5 M NACLO4M 298 K	133

TABLE G27

EQUILIBRIUM CONSTANT : CD

ITEM:	VALUE:	CONDITIONS:	REF.:
*CD(+2) + 5 NH3 <-> CD(NH3)5(+2)	K = 6.31*10**6	2 M NH4NO3, 298 K	212
*CD(+2) + 6 NH3 <-> CD(NH3)6(+2)	K = 1.38*10**5	2 M NH4NO3, 298 K	212
*CD(NH3)(+2) + NH3 <-> CD(NH3)2(+2)	LOG(K) = 2.10	2 M NH4NO3, 298 K	212
*CD(NH3)2(+2) + NH3 <-> CD(NH3)3(+2)	LOG(K) = 1.44	2 M NH4NO3, 298 K	212
*CD(NH3)3(+2) + NH3 <-> CD(NH3)4(+2)	LOG(K) = 0.93	2 M NH4NO3, 298 K	212
*CD(NH3)4(+2) + NH3 <-> CD(NH3)5(+2)	LOG(K) = -0.32	2 M NH4NO3, 298 K	212
*CD(NH3)5(+2) + NH3 <-> CD(NH3)6(+2)	LOG(K) = -1.66	ZERO IONIC STRENGTH, 298 K	212
*CD(OH)2 + OH(-) <-> HCDO2(-) + H2O	LOG(K) = 0.69	ZERO IONIC STRENGTH, 298 K	212
*CD(2+) + NTA(3-) <-> CDNTA(-)	LOG(K) = 7.05	NTA=NITRILOTRIACETIC ACID, PH= 7.35,0.7 M NACLO4,CONDITIONAL EQ. CONST	1300
*CD(2+) + NTA(3-) <-> CDNTA(-)	LOG(K) = 6.45	NTA=NITRILOTRIACETIC ACID, PH= 6.45,0.59 M NACL,CONDITIONAL EQ. CONST	1300
*CD(2+) + NTA(3-) <-> CDNTA(-)	LOG(K) = 5.55	NTA=NITRILOTRIACETIC ACID, PH= 7.38, 0.01 M CACL2, CONDITIONAL EQ CON.	1300
*CD(2+) + NTA(3-) <-> CDNTA(-)	LOG(K) = 5.16	NTA=NITRILOTRIACETIC ACID, PH= 7.35,53.6 MM MGCL2, CONDITIONAL EQ CON.	1300
*CD(2+) + NTA(3-) <-> CDNTA(-)	LOG(K) = 3.96	ARTIFICIAL SEA WATER, PH=8.29, CONDITIONAL EQUILIBRIUM CONSTANT	1300
*CD(2+) + NTA(3-) <-> CDNTA(-)	LOG(K) = 3.96	ADRIATIC SEA WATER, PH=8.01, CONDITIONAL EQUILIBRIUM CONSTANT	1300
*CD(2+) + NTA(3-) <-> CDNTA(-)	LOG(K) = 9.60	NTA=NITRILOTRIACETIC ACID, PH= 7.35,0.7 M NACLO4	1300
*CD(2+) + NTA(3-) <-> CDNTA(-)	LOG(K) = 9.90	NTA=NITRILOTRIACETIC ACID, PH= 6.45,0.59 M NACL	1300
*CD(2+) + NTA(3-) <-> CDNTA(-)	LOG(K) = 9.53	NTA=NITRILOTRIACETIC ACID, PH= 7.38, 0.01 M CACL2	1300
*CD(2+) + NTA(3-) <-> CDNTA(-)	LOG(K) = 9.33	NTA=NITRILOTRIACETIC ACID, PH= 7.35,53.6 MM MGCL2	1300
*CD(2+) + NTA(3-) <-> CDNTA(-)	LOG(K) = 9.44	ARTIFICIAL SEA WATER, PH=8.29	1300
*CD(2+) + NTA(3-) <-> CDNTA(-)	LOG(K) = 9.44	ADRIATIC SEA WATER, PH=8.01	1300
*CDCL(+) + CL(-) <-> CDCL2	LOG(K) = 0.90	4.5 M NACLO4, 298 K	133
*CDCL2 + CL(-) <-> CDCL3(-)	LOG(K) = 0.09	4.5 M NACLO4, 298 K	133
*CDCL3(-) + CL(-) <-> CDCL4(-2)	LOG(K) = -0.45	2 M NH4NO3, 298 K	133
*CDCO3(SOLID) <-> CD(+2) + CO3(-2)	LOG(K) = -13.74	SOLUBILITY EQUILIBRIA, 298 K	899
*CDOH(+) + OH(-) <-> CD(OH)2	LOG(K) = 4.23	ZERO IONIC STRENGTH, 298 K	212
*HCDO2(-) + OH(-) <-> CDO2(-2) + H2O	LOG(K) = -0.32	4.5 M NACLO4, 298 K	133

TABLE G28

EQUILIBRIUM CONSTANT : CO(II)

ITEM:	VALUE:	CONDITIONS:	REF.:
*CO((OH)(2)) + H2O <-> CO((OH)(3))(-) + H(+)	LOG(K) = -12.7	WITH 4 OTHER VALUES, A SELF CONSISTENT HYDROLYSIS DATA SET, 298 K`	1435
*CO((OH)(2))(SOLID) <-> CO((OH)(2))(0)	LOG(K) = -5.7	WITH 4 OTHER VALUES, A SELF CONSISTENT HYDROLYSIS DATA SET, 298 K`	1435
*CO((OH)(2))(SOLID) <-> CO(+2) + 2 OH(-)	LOG(K) = -14.9	WITH 4 OTHER VALUES, A SELF CONSISTENT HYDROLYSIS DATA SET, 298 K`	1435
*CO(+2) + H2O <-> CO(OH)(+) + H(+)	LOG(K) = -9.6	WITH 4 OTHER VALUES, A SELF CONSISTENT HYDROLYSIS DATA SET, 298 K`	1435

TABLE G28

EQUILIBRIUM CONSTANT : CO(II)			
ITEM:	VALUE:	CONDITIONS:	REF.:
*CO(OH)(+) + H2O <-> CO((OH)(2))(0) + H(+)	LOG(K) = -9.2	WITH 4 OTHER VALUES, A SELF CONSISTENT HYDROLYSIS DATA SET, 298 K	1435

TABLE G29

EQUILIBRIUM CONSTANT : CO2-SYSTEM			
ITEM:	VALUE:	CONDITIONS:	REF.:
*CO2 (G) + AQ <-> H2CO3	-LOG(K) = 1.11	273 K, UNIT (K) = M/ATM, 1 ATM, FRESHWATER	1526
CO2 (G) + AQ <-> H2CO3	- LOG(K) = 1.19	278 K, UNIT (K) = M/ATM, 1 ATM, FRESHWATER	1526
*CO2 (G) + AQ <-> H2CO3	- LOG(K) = 1.27	283 K, UNIT (K) = M/ATM, 1 ATM, FRESHWATER	1526
*CO2 (G) + AQ <-> H2CO3	- LOG(K) = 1.32	288 K, UNIT (K) = M/ATM, 1 ATM, FRESHWATER	1526
*CO2 (G) + AQ <-> H2CO3	- LOG(K) = 1.41	293 K, UNIT (K) = M/ATM, 1 ATM, FRESHWATER	1526
*CO2 (G) + AQ <-> H2CO3	- LOG(K) = 1.47	298 K, UNIT (K) = M/ATM, 1 ATM, FRESHWATER	1526
*CO2 (G) + AQ <-> H2CO3	- LOG(K) = 1.53	303 K, UNIT (K) = M/ATM, 1 ATM, FRESHWATER	1526
*CO2 (G) + AQ <-> H2CO3	- LOG(K) = 1.64	313 K, UNIT (K) = M/ATM, 1 ATM, FRESHWATER	1526
*CO2 (G) + AQ <-> H2CO3	- LOG(K) = 1.72	323 K, UNIT (K) = M/ATM, 1 ATM, FRESHWATER	1526
*CO2 (G) + AQ <-> H2CO3	- LOG(K) = 1.51	298 K, UNIT (K) = M/ATM, 1 ATM, 1 M NACLO4	1526
*CO2 (G) + AQ <-> H2CO3	- LOG(K) = 1.59	308 K, UNIT (K) = M/ATM, 1 ATM, 1 M NACLO4	1526
*CO2 (G) + AQ <-> H2CO3	- LOG(K) = 1.19	273 K, UNIT (K) = M/ATM, 1 ATM, SEAWATER, CL=1.9 %	1526
*CO2 (G) + AQ <-> H2CO3	- LOG(K) = 1.27	278 K, UNIT (K) = M/ATM, 1 ATM, SEAWATER, CL=1.9 %	1526
*CO2 (G) + AQ <-> H2CO3	- LOG(K) = 1.34	283 K, UNIT (K) = M/ATM, 1 ATM, SEAWATER, CL=1.9 %	1526
*CO2 (G) + AQ <-> H2CO3	- LOG(K) = 1.41	288 K, UNIT (K) = M/ATM, 1 ATM, SEAWATER, CL=1.9 %	1526
*CO2 (G) + AQ <-> H2CO3	- LOG(K) = 1.47	293 K, UNIT (K) = M/ATM, 1 ATM, SEAWATER, CL=1.9 %	1526
*CO2 (G) + AQ <-> H2CO3	- LOG(K) = 1.53	298 K, UNIT (K) = M/ATM, 1 ATM, SEAWATER, CL=1.9 %	1526
*CO2 (G) + AQ <-> H2CO3	- LOG(K) = 1.58	303 K, UNIT (K) = M/ATM, 1 ATM, SEAWATER, CL=1.9 %	1526
*CO2(G) + H2O <-> H(+) + HCO3(-)	LOG(K) = -7.82	IONIC STRENGTH = 0, 298 K, 1 ATM	1526
*CO2(G) + H2O <-> H(+) + HCO3(-)	LOG(K) = -7.5	SEAWATER, 298 K, 1 ATM	1526
*CO2(G) + H2O <-> H(+) + HCO3(-)	LOG(K) = -7.3	SEAWATER, 278 K, 200 ATM	1526
*CO2(G) + H2O <-> H2CO3	- LOG(K) = 1.20	278 K, INFINITE DILUTION, FRESHWATER	1435
*CO2(G) + H2O <-> H2CO3	- LOG(K) = 1.27	283 K, INFINITE DILUTION, FRESHWATER	1435
*CO2(G) + H2O <-> H2CO3	- LOG(K) = 1.34	288 K, INFINITE DILUTION, FRESHWATER	1435
*CO2(G) + H2O <-> H2CO3	- LOG(K) = 1.41	293 K, INFINITE DILUTION, FRESHWATER	1435
*CO2(G) + H2O <-> H2CO3	- LOG(K) = 1.47	298 K, INFINITE DILUTION, FRESHWATER	1435
*CO2(G) + H2O <-> H2CO3	- LOG(K) = 1.64	313 K, INFINITE DILUTION, FRESHWATER	1435
*CO2(G) + H2O <-> H2CO3	- LOG(K) = 1.27	278 K, SEAWATER	1435
*CO2(G) + H2O <-> H2CO3	- LOG(K) = 1.53	298 K, SEAWATER	1435

TABLE G29

EQUILIBRIUM CONSTANT : CO2-SYSTEM

ITEM:	VALUE:	CONDITIONS:	REF.:
*CO2(G) + H2O <-> H2CO3	K = 3.38*10**(-2)	PK = 1.47, 298 K	412
*HCO3(-) <-> H(+) + CO3(-2)	K = 4.69*10**(-11)	PK = 10.33, 298 K	621
*HCO3(-) <-> H(+) + CO3(2-)	- LOG(K) = 10.56	278 K, INFINITE DILUTION, FRESHWATER	1435
*HCO3(-) <-> H(+) + CO3(2-)	- LOG(K) = 10.49	283 K, INFINITE DILUTION, FRESHWATER	1435
*HCO3(-) <-> H(+) + CO3(2-)	- LOG(K) = 10.43	288 K, INFINITE DILUTION, FRESHWATER	1435
*HCO3(-) <-> H(+) + CO3(2-)	- LOG(K) = 10.38	293 K, INFINITE DILUTION, FRESHWATER	1435
*HCO3(-) <-> H(+) + CO3(2-)	- LOG(K) = 10.33	298 K, INFINITE DILUTION, FRESHWATER	1435
*HCO3(-) <-> H(+) + CO3(2-)	- LOG(K) = 10.22	313 K, INFINITE DILUTION, FRESHWATER	1435
*HCO3(-) <-> H(+) + CO3(2-)	- LOG(K) = 9.18	278 K, SEAWATER	1435
*HCO3(-) <-> H(+) + CO3(2-)	- LOG(K) = 8.96	298 K, SEAWATER	1435
*HCO3(-) <-> H(+) + CO3(2-)	LOG(K) = -10.33	IONIC STRENGTH = 0, 298 K, 1 ATM	1526
*HCO3(-) <-> H(+) + CO3(2-)	LOG(K) = -9.0	SEAWATER, 1 ATM, 298 K	1526
*HCO3(-) <-> H(+) + CO3(2-)	LOG(K) = -9.1	SEAWATER, 278 K, 200 ATM	1526
*HCO3(-) <-> H(+) + CO3(2-)	- LOG(K) = 10.625	273 K, ACTIVITIES USED IN CALCULATIONS, 1 ATM, FRESHWATER	1526
*HCO3(-) <-> H(+) + CO3(2-)	- LOG(K) = 10.557	278 K, ACTIVITIES USED IN CALCULATIONS, 1 ATM, FRESHWATER	1526
*HCO3(-) <-> H(+) + CO3(2-)	- LOG(K) = 10.490	283 K, ACTIVITIES USED IN CALCULATIONS, 1 ATM, FRESHWATER	1526
*HCO3(-) <-> H(+) + CO3(2-)	- LOG(K) = 10.430	288 K, ACTIVITIES USED IN CALCULATIONS, 1 ATM, FRESHWATER	1526
*HCO3(-) <-> H(+) + CO3(2-)	- LOG(K) = 10.377	293 K, ACTIVITIES USED IN CALCULATIONS, 1 ATM, FRESHWATER	1526
*HCO3(-) <-> H(+) + CO3(2-)	- LOG(K) = 10.329	298 K, ACTIVITIES USED IN CALCULATIONS, 1 ATM, FRESHWATER	1526
*HCO3(-) <-> H(+) + CO3(2-)	- LOG(K) = 10.290	303 K, ACTIVITIES USED IN CALCULATIONS, 1 ATM, FRESHWATER	1526
*HCO3(-) <-> H(+) + CO3(2-)	- LOG(K) = 10.250	308 K, ACTIVITIES USED IN CALCULATIONS, 1 ATM, FRESHWATER	1526
*HCO3(-) <-> H(+) + CO3(2-)	- LOG(K) = 10.250	313 K, ACTIVITIES USED IN CALCULATIONS, 1 ATM, FRESHWATER	1526
*HCO3(-) <-> H(+) + CO3(2-)	- LOG(K) = 10.172	323 K, ACTIVITIES USED IN CALCULATIONS, 1 ATM, FRESHWATER	1526
*HCO3(-) <-> H(+) + CO3(2-)	- LOG(K) = 9.40	273 K, ACTIV. OF H(+) AND CONC. USED, 1 ATM, SEAWATER	1526
*HCO3(-) <-> H(+) + CO3(2-)	- LOG(K) = 9.34	278 K, ACTIV. OF H(+) AND CONC. USED, 1 ATM, SEAWATER	1526
*HCO3(-) <-> H(+) + CO3(2-)	- LOG(K) = 9.28	283 K, ACTIV. OF H(+) AND CONC. USED, 1 ATM, SEAWATER	1526
*HCO3(-) <-> H(+) + CO3(2-)	- LOG(K) = 9.23	288 K, ACTIV. OF H(+) AND CONC. USED, 1 ATM, SEAWATER	1526
*HCO3(-) <-> H(+) + CO3(2-)	- LOG(K) = 9.17	293 K, ACTIV. OF H(+) AND CONC. USED, 1 ATM, SEAWATER	1526
*HCO3(-) <-> H(+) + CO3(2-)	- LOG(K) = 9.12	295 K, ACTIV. OF H(+) AND CONC. USED, 1 ATM, SEAWATER	1526
*HCO3(-) <-> H(+) + CO3(2-)	- LOG(K) = 9.10	298 K, ACTIV. OF H(+) AND CONC. USED, 1 ATM, SEAWATER	1526
*HCO3(-) <-> H(+) + CO3(2-)	- LOG(K) = 9.02	303 K, ACTIV. OF H(+) AND CONC. USED, 1 ATM, SEAWATER	1526
*HCO3(-) <-> H(+) + CO3(2-)	- LOG(K) = 8.95	308 K, ACTIV. OF H(+) AND CONC. USED, 1 ATM, SEAWATER	1526
*HCO3(-) <-> H(+) + CO3(2-)	- LOG(K) = 9.49	295 K, ACTIV. OF H(+) AND CONC. USED, 0.75 M NACL, 1 ATM	1526
*HCO3(-) <-> H(+) + CO3(2-)	- LOG(K) = 9.57	298 K, CONCENTRATIONS USED IN CALCULATIONS, 1 M KCLO4, 1 ATM	1526
*HCO3(-) <-> OH(-) + CO2	K = 10**(-7.7)	IN WATER, 298 K	510
*H2CO3 <-> CO2	K = 10**2.8	IN WATER, 298 K	510
*H2CO3 <-> H(+) + HCO3(-)	- LOG(K) = 6.52	278 K, INFINITE DILUTION, FRESHWATER	1435

TABLE G29

EQUILIBRIUM CONSTANT : CO2-SYSTEM			
ITEM:	VALUE:	CONDITIONS:	REF.:
*H2CO3 <-> H(+) + HCO3(-)	- LOG(K) = 6.46	283 K, INFINITE DILUTION, FRESHWATER	1435
*H2CO3 <-> H(+) + HCO3(-)	- LOG(K) = 6.42	288 K, INFINITE DILUTION, FRESHWATER	1435
*H2CO3 <-> H(+) + HCO3(-)	- LOG(K) = 6.38	293 K, INFINITE DILUTION, FRESHWATER	1435
*H2CO3 <-> H(+) + HCO3(-)	- LOG(K) = 6.35	298 K, INFINITE DILUTION, FRESHWATER	1435
*H2CO3 <-> H(+) + HCO3(-)	- LOG(K) = 6.30	313 K, INFINITE DILUTION, FRESHWATER	1435
*H2CO3 <-> H(+) + HCO3(-)	- LOG(K) = 6.15	278 K, SEAWATER	1435
*H2CO3 <-> H(+) + HCO3(-)	K = 4.45*10**(-7)	PK = 6.35, 298 K	620
*H2CO3 <-> H(+) + HCO3(-)	K = 10**3.5	IN WATER, 298 K	510
*H2CO3 <-> HCO3(-) + H(+)	- LOG(K) = 6.579	273 K, ACTIVITIES USED IN CALCULATIONS, FRESHWATER, 1 ATM	1526
*H2CO3 <-> HCO3(-) + H(+)	- LOG(K) = 6.517	278 K, ACTIVITIES USED IN CALCULATIONS, FRESHWATER, 1 ATM	1526
*H2CO3 <-> HCO3(-) + H(+)	- LOG(K) = 6.464	283 K, ACTIVITIES USED IN CALCULATIONS, FRESHWATER, 1 ATM	1526
*H2CO3 <-> HCO3(-) + H(+)	- LOG(K) = 6.419	288 K, ACTIVITIES USED IN CALCULATIONS, FRESHWATER, 1 ATM	1526
*H2CO3 <-> HCO3(-) + H(+)	- LOG(K) = 6.381	293 K, ACTIVITIES USED IN CALCULATIONS, FRESHWATER, 1 ATM	1526
*H2CO3 <-> HCO3(-) + H(+)	- LOG(K) = 6.352	298 K, ACTIVITIES USED IN CALCULATIONS, FRESHWATER, 1 ATM	1526
*H2CO3 <-> HCO3(-) + H(+)	- LOG(K) = 6.327	303 K, ACTIVITIES USED IN CALCULATIONS, FRESHWATER, 1 ATM	1526
*H2CO3 <-> HCO3(-) + H(+)	- LOG(K) = 6.309	308 K, ACTIVITIES USED IN CALCULATIONS, FRESHWATER, 1 ATM	1526
*H2CO3 <-> HCO3(-) + H(+)	- LOG(K) = 6.298	313 K, ACTIVITIES USED IN CALCULATIONS, FRESHWATER, 1 ATM	1526
*H2CO3 <-> HCO3(-) + H(+)	- LOG(K) = 6.285	323 K, ACTIVITIES USED IN CALCULATIONS, FRESHWATER, 1 ATM	1526
*H2CO3 <-> HCO3(-) + H(+)	- LOG(K) = 6.15	273 K, ACTIV. OF H(+) AND CONC. USED, SEAWATER, 1 ATM,CL=1.9%	1526
*H2CO3 <-> HCO3(-) + H(+)	- LOG(K) = 6.11	278 K, ACTIV. OF H(+) AND CONC. USED, SEAWATER, 1 ATM,CL=1.9%	1526
*H2CO3 <-> HCO3(-) + H(+)	- LOG(K) = 6.08	283 K, ACTIV. OF H(+) AND CONC. USED, SEAWATER, 1 ATM,CL=1.9%	1526
*H2CO3 <-> HCO3(-) + H(+)	- LOG(K) = 6.05	288 K, ACTIV. OF H(+) AND CONC. USED, SEAWATER, 1 ATM,CL=1.9%	1526
*H2CO3 <-> HCO3(-) + H(+)	- LOG(K) = 6.02	293 K, ACTIV. OF H(+) AND CONC. USED, SEAWATER, 1 ATM,CL=1.9%	1526
*H2CO3 <-> HCO3(-) + H(+)	- LOG(K) = 6.00	295 K, ACTIV. OF H(+) AND CONC. USED, SEAWATER, 1 ATM,CL=1.9%	1526
*H2CO3 <-> HCO3(-) + H(+)	- LOG(K) = 6.00	298 K, ACTIV. OF H(+) AND CONC. USED, SEAWATER, 1 ATM,CL=1.9%	1526
*H2CO3 <-> HCO3(-) + H(+)	- LOG(K) = 5.98	303 K, ACTIV. OF H(+) AND CONC. USED, SEAWATER, 1 ATM,CL=1.9%	1526
*H2CO3 <-> HCO3(-) + H(+)	- LOG(K) = 5.97	308 K, ACTIV. OF H(+) AND CONC. USED, SEAWATER, 1 ATM,CL=1.9%	1526
*H2CO3 <-> HCO3(-) + H(+)	- LOG(K) = 6.01	278 K, CONCENTRATIONS USED IN CAL., SEAWATER, 1 ATM	1526
*H2CO3 <-> HCO3(-) + H(+)	- LOG(K) = 6.02	287 K, CONCENTRATIONS USED IN CAL., SEAWATER, 1 ATM	1526
*H2CO3 <-> HCO3(-) + H(+)	- LOG(K) = 5.89	295 K, CONCENTRATIONS USED IN CAL., SEAWATER, 1 ATM	1526
*H2CO3 <-> HCO3(-) + H(+)	- LOG(K) = 6.04	298 K, CONCENTRATIONS USED IN CALCULATIONS, 1 M NACLO4	1526

TABLE G30

EQUILIBRIUM CONSTANT : CU			
ITEM:	VALUE:	CONDITIONS:	REF.:
*CU(2+) + CL(-) <-> CUCL(+)	LOG(K) = 0.5	DILUTE WATER SOLUTION, I = 10**3	1644

TABLE G30

EQUILIBRIUM CONSTANT : CU			
ITEM:	VALUE:	CONDITIONS:	REF.:
*CU(2+) + CL(-) <-> CUCL(+)	LOG(K) = -0.11	FREŞH WATER	1542
*CU(2+) + CO3(2-) <-> CUCO3	LOG(K) = 6.77	FRESH WATER	1542
*CU(2+) + H(+) + CO3(2-) <-> CUHCO3 (2+)	LOG(K) = 12.3	FRESH WATER	1542
*CU(2+) + HCO3(-) -> CUCO3 + H(+)	2.56*10**(-4)	1*10**(-3)M HCO3(-)	1606
*CU(2+) + HCO3(-) -> CUCO3 + H(+)	2.57*10**(-4)	2*10**(-3)M HCO3(-)	1606
*CU(2+) + HCO3(-) -> CUCO3 + H(+)	2.28*10**(-4)	5*10**(-3)M HCO3(-)	1606
*CU(2+) + H2O <-> CUOH(+) + H(+)	LOG(K) = -7.53	FRESH WATER	1542
*CU(2+) + H2O <-> CUOH(+) + H(+)	LOG(K) = -7.9	DILUTE WATER SOLUTION, I = 10**3	1644
*CU(2+) + SO4(2-) <-> CUSO4	LOG(K) = 2.35	FRESH WATER	1542
*CU(2+) + 2 CL(-) <-> CUCL2	LOG(K) = 0.3	DILUTE WATER SOLUTION, I = 10**3	1644
*CU(2+) + 2 H2O <-> CU(OH)2(S) + 2 H(+)	LOG(K) = -8.4	DILUTE WATER SOLUTION, I = 10**3	1644
*CU(2+) + 2CO3(2-) <-> CU(CO3)2(2-)	LOG(K) = 10.01	FRESH WATER	1542
*CU(2+) + 2H2O <-> CU2(OH)2(2+) + 2H(+)	LOG(K) = -10.53	FRESH WATER	1542
*CU(2+) + 3 H2O <-> CU(OH)3(-) + 3 H(+)	LOG(K) = -26.8	DILUTE WATER SOLUTION, I = 10**3	1644
CU(2+) + 3CN(-) <-> CU(CN)2(-) + 1/2(CN)2	K = 10**22	NONE	1511
*CU(2+) + 4 H2O <-> CU(OH)4(2-) + 4 H(+)	LOG(K) = -39.9	DILUTE WATER SOLUTION, I = 10**3	1644
K = (CUL2)/(CU(2+)(L(-)**2))	LOG(K) = 15.4	L=LEUCINE,NEUTRAL	1511
K = (CUL2)/(CU(2+)(L(-)**2))	LOG(K) = 14.8	L=VALINE,NEUTRAL	1511
K = (CUL2)/(CU(2+)(L(-)**2))	LOG(K) = 15.5	L=ALANINE,NEUTRAL	1511
K = (CUL2)/(CU(2+)(L(-)**2))	LOG(K) = 14.6	L=SERINE,NEUTRAL	1511
K = (CUL2)/(CU(2+)(L(-)**2))	LOG B =14.8	L=GLUTAMIC ACID,ACIDIC	1511
K = (CUL2)/(CU(2+)(L(-)**2))	LOG(K) = 15.3	L=ASPARTIC ACID,ACIDIC	1511
K = (CUL2)/(CU(2+)(L(-)**2))	LOG(K) = 15.0	L=TYROSINE,ACIDIC	1511
K = (CUL2)/(CU(2+)(L(-)**2))	LOG(K) = 18.6	L=HISTIDINE,BASIC	1511
K = (CUL2)/(CU(2+)(L(-)**2))	LOG(K) = 6.9	L=GLYCINE,NEUTRAL	1511
*K = CUL(+)/(CU(2+)*L(-))	LOG K = 7.0	L=LEUCINE,NEUTRAL	1511
*K = CUL(+)/(CU(2+)*L(-))	LOG K = 7.9	L=VALINE,NEUTRAL	1511
*K = CUL(+)/(CU(2+)*L(-))	LOG K = 7.9	L=GLUTAMIC ACID,ACIDIC	1511
*K = CUL(+)/(CU(2+)*L(-))	LOG K = 8.6	L=ASPARTIC ACID,ACIDIC	1511
*K = CUL(+)/(CU(2+)*L(-))	LOG K = 10.6	L=HISTIDINE,BASIC	1511
*K = CUL(+)/(CU(2+)*L(-))	LOG K = 8.6	L=GLYCINE,NEUTRAL	1511
*K= (CUL)/(CU(+)*L(-))	LOG K = 19.5	L=CYSTEINE,S-CONTAINING	1511
*K=(A(CUCO3)*A(H+))/(A(CU(2+))*A (HCO3(-))	K = 3.0*10**(-4)	298 K,FRESH WATER, A = ACTIVITY	1511
*K=(CATECHOL*CU(2+))/ (COPPER-CATECHOL)	LOG K = 14.25	A MODEL FOR HUMIC SUBSTANCES, 1,2 DIHYDROXY GROUPING	1511
*ORGANIC MATTER FROM SMITH LAKE + CU(II)	LOG(K) = 3.68	FORMATION OF HUMIC-METAL FOR SMALL ORGANIC MOLECULES	1726
*PHTHALIC ACID + CU(II)	LOG(K) = 3.10	FORMATION OF HUMIC-METAL FOR SMALL ORGANIC MOLECULES	1726
*2 CU(2+) + 2 H2O <-> CU2(OH)2(2-) + 2 H(+)	LOG(K) = -10.3	DILUTE WATER SOLUTION, I = 10**3	1644
*2,4-DIHYDROXYBENZOIC ACID + CU(II)	LOG(K) = 1.83	FORMATION OF HUMIC-METAL FOR SMALL ORGANIC MOLECULES	1726
*3,4-DIHYDROXYBENZOIC ACID + CU(II)	LOG(K) = 14.53	FORMATION OF HUMIC-METAL FOR SMALL ORGANIC MOLECULES	1726

TABLE G31

EQUILIBRIUM CONSTANT : FE			
ITEM:	VALUE:	CONDITIONS:	REF.:
*FE(OH)2(S) <-> (FE(OH))(+) + OH(-)	K = 4*10**(-10)	10**(-3) CARBONATE SYSTEM	1526

TABLE G31

EQUILIBRIUM CONSTANT : FE			
ITEM:	VALUE:	CONDITIONS:	REF.:
*FE(OH)2(S) <-> FE(2+) + 2 OH(-)	K = 2*10**(-15)	10**(-3) CARBONATE SYSTEM	1526
*FE(OH)2(S) + OH(-) <-> (FE)OH)3)(-)	K = 8.3*10**(-6)	10**(-3) CARBONATE SYSTEM	1526
*FE(OH)2(SOLID) <-> FE(+2) + 2 OH(-)	K = 8*10**(-16)	PK = 15.1, 298 K	914
*FE(OH)2(SOLID) <-> FEOH(+) + OH(-)	K = 4*10**(-10)	PK = 9.4, 298 K	914
*FE(OH)2(SOLID) + OH(-) <-> FE(OH)3(-)	K = 8.3*10**(-6)	PK = 5.1, 298 K	511
*FE(3+) + H2O <-> FEOH(2+) + H(+)	LOG(K) = -2.2	298 K	1435
*FE(3+) + 3 H2O <-> FE(OH)3(S) + 3 H(+)	LOG(K) = -6.0	298 K	1435
*FECO3(S) <-> FE(2+) + CO3(2-)	K = 2.1*10**(-11)	10**(-3) CARBONATE SYSTEM	1526
*FECO3(S) + OH(-) <-> (FE(OH))(+) + CO3(2-)	K = 0.1*10**(-5)	10**(-3) CARBONATE SYSTEM	1526
*FECO3(SOLID) <-> FE(+2) + CO3(-2)	K = 2.1*10**(-11)	PK = 10.68, 298 K	889
*FECO3(SOLID) + H(+) <-> FE(+2) + HCO3(-)	K = 4.48*10**(-1)	PK = 0.35, 298 K	520
*FECO3(SOLID) + OH(-) <-> FEOH(+) + CO3(-2)	K = 1*10**9-5)	PK = 5, 298 K	889
*FEOH(2+) + H2O <-> FE(OH)2 + H(+)	LOG(K) = -4.7	298 K	1435

TABLE G32

EQUILIBRIUM CONSTANT : HG			
ITEM:	VALUE:	CONDITIONS:	REF.:
*HG(+2) + CL(-) <-> HGCL(+)	LOG(K) = 6.7	COMPLEX FORMATION EQUILIBRIA, 298 K	899
*HG(+2) + CYSTEINE <-> HGCYSTEINE(+2)	LOG(K) = 43.57	COMPLEX FORMATION EQUILIBRIA, 298 K	899
*HG(+2) + GLYCINE(-) <-> HGGLYCINE(+)	LOG(K) = 10.3	COMPLEX FORMATION EQUILIBRIA, 298 K	899
*HG(+2) + H2O <-> HGOH(+) + H(+)	LOG(K) = -3.4	COMPLEX FORMATION EQUILIBRIA, 298 K	899
*HG(+2) + 2 CL(-) <-> HGCL2(0)	LOG(K) = 13.2	COMPLEX FORMATION EQUILIBRIA, 298 K	899
*HG(+2) + 2 H2O <-> HG((OH)(2))(0) + 2 H(+)	LOG(K) = -6.00	COMPLEX FORMATION EQUILIBRIA, 298 K	899
*HG(+2) + 3 CL(-) <-> HGCL3(-)	LOG(K) = 14.2	COMPLEX FORMATION EQUILIBRIA, 298 K	899
*HG(+2) + 3 H2O <-> HG((OH)(3))(-) + 3 H(+)	LOG(K) = -20.7	COMPLEX FORMATION EQUILIBRIA, 298 K	899
*HG(+2) + 4 CL(-) <-> HGCL4(-2)	LOG(K) = 15.2	COMPLEX FORMATION EQUILIBRIA, 298 K	899
*HG(L) <-> HG(AQ)	LOG(K) = -6.50	SOLUBILITY EQUILIBRIA, 298 K	899
*HGCL2 + 2 H2O <-> HG((OH)(2))(0) + 2 CL(-) + 2 H(+)	LOG(K) = -19.6	COMPLEX FORMATION EQUILIBRIA, 298 K	899
*HGCL2(SOLID) <-> HG(+2) + 2CL(-)	LOG(K) = -13.8	SOLUBILITY EQUILIBRIA, 298 K	899
*HGCL2(0) + H2O <-> HGOHCL(0) + CL(-) + H(+)	LOG(K) = -9.6	COMPLEX FORMATION EQUILIBRIA, 298 K	899
*HGGLYSINE(+) + GLYCINE(-) <-> HG((GLYCINE)(2))(0)	LOG(K) = 8.9	COMPLEX FORMATION EQUILIBRIA, 298 K	899
*HGO(SOLID) + H2O <-> HG(+2) + 2 OH(-)	LOG(K) = -25.7	SOLUBILITY EQUILIBRIA, 298 K	899
*HGO(SOLID) + H2O <-> HG(OH(2))(AQ)	LOG(K) = -3.7	SOLUBILITY EQUILIBRIA, 298 K	899
*HG2CL2(SOLID) <-> HG2(+2) + 2 CL(-)	LOG(K) = -18.00	SOLUBILITY EQUILIBRIA, 298 K	899

TABLE G33

EQUILIBRIUM CONSTANT : METAL ION AND LIGANDS			
ITEM:	VALUE:	CONDITIONS:	REF.:
*METAL ION = CA(2+) , LIGAND = CO3(2-)	LOG(K) = 3.2	IONIC STRENGTH = 0, 298 K	504

TABLE G33

EQUILIBRIUM CONSTANT : METAL ION AND LIGANDS

ITEM:	VALUE:	CONDITIONS:	REF.:
*METAL ION = CA(2+) , LIGAND = CO3(2-)	LOG(K) = 3.2	IONIC STRENGTH = 0, 298 K	504
*METAL ION = CA(2+) , LIGAND = HCO3 (-)	LOG(K) = 1.26	IONIC STRENGTH = 0, 298 K	504
*METAL ION = CA(2+) , LIGAND = HCO3 (-)	LOG(K) = 1.26	IONIC STRENGTH = 0, 298 K	504
*METAL ION = CA(2+) , LIGAND = OH (-)	LOG(K) = 1.30	IONIC STRENGTH = 0, 298 K	504
*METAL ION = CA(2+) , LIGAND = OH (-)	LOG(K) = 1.30	IONIC STRENGTH = 0, 298 K	504
*METAL ION = CA(2+) , LIGAND = SO4(2-)	LOG(K) = 2.31	IONIC STRENGTH = 0, 298 K	504
*METAL ION = CA(2+) , LIGAND = SO4(2-)	LOG(K) = 2.31	IONIC STRENGTH = 0, 298 K	504
*METAL ION = H(+) , LIGAND = HCO3 (-)	LOG(K) = 6.4	IONIC STRENGTH = 0, 298 K	504
*METAL ION = H(+) , LIGAND = HCO3 (-)	LOG(K) = 6.4	IONIC STRENGTH = 0, 298 K	504
*METAL ION = H(+) , LIGAND = OH(-)	LOG(K) = 14.0	IONIC STRENGTH = 0, 298 K	504
*METAL ION = H(+) , LIGAND = OH(-)	LOG(K) = 14.0	IONIC STRENGTH = 0, 298 K	504
*METAL ION = H(+) , LIGAND = SO4 (2-)	LOG(K) = 10.33	IONIC STRENGTH = 0, 298 K	504
*METAL ION = H(+) , LIGAND = SO4 (2-)	LOG(K) = 10.33	IONIC STRENGTH = 0, 298 K	504
*METAL ION = K(+) , LIGAND = SO4 (2-)	LOG(K) = 0.96	IONIC STRENGTH = 0, 298 K	504
*METAL ION = K(+) , LIGAND = SO4 (2-)	LOG(K) = 0.96	IONIC STRENGTH = 0, 298 K	504
*METAL ION = MG(2+) , LIGAND = CO3(2-)	LOG(K) = 3.4	IONIC STRENGTH = 0, 298 K	504
*METAL ION = MG(2+) , LIGAND = CO3(2-)	LOG(K) = 3.4	IONIC STRENGTH = 0, 298 K	504
*METAL ION = MG(2+) , LIGAND = HCO3 (-)	LOG(K) = 1.16	IONIC STRENGTH = 0, 298 K	504
*METAL ION = MG(2+) , LIGAND = HCO3 (-)	LOG(K) = 1.16	IONIC STRENGTH = 0, 298 K	504
*METAL ION = MG(2+) , LIGAND = OH (-)	LOG(K) = 2.58	IONIC STRENGTH = 0, 298 K	504
*METAL ION = MG(2+) , LIGAND = OH (-)	LOG(K) = 2.58	IONIC STRENGTH = 0, 298 K	504
*METAL ION = MG(2+) , LIGAND = SO4(2-)	LOG(K) = 2.36	IONIC STRENGTH = 0, 298 K	504
*METAL ION = MG(2+) , LIGAND = SO4(2-)	LOG(K) = 2.36	IONIC STRENGTH = 0, 298 K	504
*METAL ION = NA(+) , LIGAND = CO3 (2-)	LOG(K) = 1.27	IONIC STRENGTH = 0, 298 K	504
*METAL ION = NA(+) , LIGAND = CO3 (2-)	LOG(K) = 1.27	IONIC STRENGTH = 0, 298 K	504
*METAL ION = NA(+) , LIGAND = HCO3(-)	LOG(K) = -0.25	IONIC STRENGTH = 0, 298 K	504
*METAL ION = NA(+) , LIGAND = HCO3(-)	LOG(K) = -0.25	IONIC STRENGTH = 0, 298 K	504
*METAL ION = NA(+) , LIGAND = OH(-)	LOG(K) = -0.7	IONIC STRENGTH = 0, 298 K	504
*METAL ION = NA(+) , LIGAND = OH(-)	LOG(K) = -0.7	IONIC STRENGTH = 0, 298 K	504
*METAL ION = NA(+) , LIGAND = SO4 (2-)	LOG(K) = 0.72	IONIC STRENGTH = 0, 298 K	504
*METAL ION = NA(+) , LIGAND = SO4 (2-)	LOG(K) = 0.72	IONIC STRENGTH = 0, 298 K	504

TABLE G34

EQUILIBRIUM CONSTANT : MICROBIOLOGICAL REACTION

ITEM:	VALUE:	CONDITIONS:	REF.:
*FECO3(S) + 2 H2O <-> FEOOH(S) + HCO3(-) + 2 H(+) + E	LOG(K) = -1.67	OXIDATION, 298 K	1526

TABLE G34

EQUILIBRIUM CONSTANT : MICROBIOLOGICAL REACTION			
ITEM:	VALUE:	CONDITIONS:	REF.:
*FECO3(S) + 2 H2O <-> FEOOH(S) + HCO3(-) + 2 H(+) + E	LOG(K) = -1.67	OXIDATION, 298 K	1526
*1/2 CH2O + H(+) + E <-> 1/2 CH3OH	LOG(K) = -3.01	REDUCTION, 298 K	1526
*1/2 CH2O + H(+) + E <-> 1/2 CH3OH	LOG(K) = -3.01	REDUCTION, 298 K	1526
*1/2 CH2O + 1/2 H2O <-> 1/2 HCOO(-) + 3/2 H(+) + E	LOG(K) = -7.68	OXIDATION, 298 K	1526
*1/2 CH2O + 1/2 H2O <-> 1/2 HCOO(-) + 3/2 H(+) + E	LOG(K) = -7.68	OXIDATION, 298 K	1526
*1/2 CH3OH <-> 1/2 CH2O + H(+) + E	LOG(K) = -3.01	OXIDATION, 298 K	1526
*1/2 CH3OH <-> 1/2 CH2O + H(+) + E	LOG(K) = -3.01	OXIDATION, 298 K	1526
*1/2 CH4(G) + 1/2 H2O <-> 1/2 CH3OH + H(+) + E	LOG(K) = +2.88	OXIDATION, 298 K	1526
*1/2 CH4(G) + 1/2 H2O <-> 1/2 CH3OH + H(+) + E	LOG(K) = +2.88	OXIDATION, 298 K	1526
*1/2 HCOO(-) <-> 1/2 CO2(G) + 1/2 H (+) + E	LOG(K) = -8.73	OXIDATION, 298 K	1526
*1/2 HCOO(-) <-> 1/2 CO2(G) + 1/2 H (+) + E	LOG(K) = -8.73	OXIDATION, 298 K	1526
*1/2 MNCO3(S) + 3/8 H2O <-> 1/2 MNO2(S) + 1/2 HCO3(-) + 3/2 H(+) + E	LOG(K) = -8.5	OXIDATION, 298 K	1526
*1/2 MNCO3(S) + 3/8 H2O <-> 1/2 MNO2(S) + 1/2 HCO3(-) + 3/2 H(+) + E	LOG(K) = -8.5	OXIDATION, 298 K	1526
*1/4 CH2O + 1/4 H2O <-> 1/4 CO2(G) + H(+) + E	LOG(K) = -8.20	OXIDATION, 298 K	1526
*1/4 CH2O + 1/4 H2O <-> 1/4 CO2(G) + H(+) + E	LOG(K) = -8.20	OXIDATION, 298 K	1526
*1/4 O2 + H(+) +E <-> 1/2 H2O	LOG(K) = +13.75	REDUCTION, 298 K	1526
*1/4 O2 + H(+) +E <-> 1/2 H2O	LOG(K) = +13.75	REDUCTION, 298 K	1526
*1/5 NO3(-) + 6/5 H(+) + E <-> 1/10 N2(G) + 3/5 H2O	LOG(K) = +12.65	REDUCTION, 298 K	1526
*1/5 NO3(-) + 6/5 H(+) + E <-> 1/10 N2(G) + 3/5 H2O	LOG(K) = +12.65	REDUCTION, 298 K	1526
*1/6 N2 + 4/3 H(+) + E <-> 1/3 NH4(+)	LOG(K) = -4.68	REDUCTION, 298 K	1526
*1/6 N2 + 4/3 H(+) + E <-> 1/3 NH4(+)	LOG(K) = -4.68	REDUCTION, 298 K	1526
*1/8 CO2(G) + H(+) + E <-> 1/8 CH4(G) + 1/4 H2O	LOG(K) = -4.13	REDUCTION, 298 K	1526
*1/8 CO2(G) + H(+) + E <-> 1/8 CH4(G) + 1/4 H2O	LOG(K) = -4.13	REDUCTION, 298 K	1526
*1/8 HS(-) + 1/2 H2O <-> 1/8 SO4 (2-) + 9/8 H(+) + E	LOG(K) = -3.75	OXIDATION, 298 K	1526
*1/8 HS(-) + 1/2 H2O <-> 1/8 SO4 (2-) + 9/8 H(+) + E	LOG(K) = -3.75	OXIDATION, 298 K	1526
*1/8 NH4(+) + 3/8 H2O <-> 1/8 NO3 (-) + 5/4 H(+) + E	LOG(K) = +6.16	OXIDATION, 298 K	1526
*1/8 NH4(+) + 3/8 H2O <-> 1/8 NO3 (-) + 5/4 H(+) + E	LOG(K) = +6.16	OXIDATION, 298 K	1526
*1/8 NO3(-) + 5/4 H(+) + E <-> 1/8 NH4(+) + 3/8 H2O	LOG(K) = +6.15	REDUCTION, 298 K	1526
*1/8 NO3(-) + 5/4 H(+) + E <-> 1/8 NH4(+) + 3/8 H2O	LOG(K) = +6.15	REDUCTION, 298 K	1526
*1/8 SO4(2-) + 9/8 H(+) + E <-> 1/8 HS(-) + 1/2 H2O	LOG(K) = -3.75	REDUCTION, 298 K	1526
*1/8 SO4(2-) + 9/8 H(+) + E <-> 1/8 HS(-) + 1/2 H2O	LOG(K) = -3.75	REDUCTION, 298 K	1526

TABLE G35

EQUILIBRIUM CONSTANT : MN			
ITEM:	VALUE:	CONDITIONS:	REF.:
*ORGANIC MATTER FROM SMITH LAKE + MN(II)	LOG(K) = 2.50	FORMATION OF HUMIC-METAL FOR SMALL ORGANIC MOLECULES	1726

TABLE G35

EQUILIBRIUM CONSTANT : MN			
ITEM:	VALUE:	CONDITIONS:	REF.:
*PHTHALIC ACID + MN(II)	LOG(K) = 2.74	FORMATION OF HUMIC-METAL FOR SMALL ORGANIC MOLECULES	1726
*3,4-DIHYDROXYBENZOIC ACID + MN(II)	LOG(K) = 7.22	FORMATION OF HUMIC-METAL FOR SMALL ORGANIC MOLECULES	1726

TABLE G36

EQUILIBRIUM CONSTANT : OXIDES AND HYDROXIDES			
ITEM:	VALUE:	CONDITIONS:	REF.:
*(AM) FE(OH)(S) + OH(-) <-> FE(OH)4 (-)	LOG(K) = -4.5	3 M NACLO4, 298 K	1526
*(AM) FE(OH)3(S) <-> FE(OH)2(+) + OH(-)	LOG(K) = -16.6	3 M NACLO4, 298 K	1526
*(AM) FE(OH)3(S) <-> FE(3+) + 3 OH(-)	LOG(K) = -38.7	3 M NACLO4, 298 K	1526
*(AM) FE(OH)3(S) <-> FEOH(2+) + 2 OH(-)	LOG(K) = -27.5	3 M NACLO4, 298 K	1526
*(AM) FEOOH(S) + 3 H(+) <-> FE(3+) + 2 H2O	LOG(K) = 3.55	3 M NACLO4, 298 K	1526
*(AM)AL(OH)3(S) + 3 H(+) <-> AL(3+) + 3 H2O	LOG(K) = 10.8	IONIC STRENGTH = 0, 298 K, 1 ATM	1526
*AL(3+) + 4 OH(-) <-> AL(OH)4(-)	LOG(K) = 32.5	IONIC STRENGTH = 0, 298 K, 1 ATM	1526
*ALPHA-AL(OH)3 + 3 H(+) <-> AL(3+) + 3 H2O	LOG(K) = 8.2	IONIC STRENGTH = 0, 298 K, 1 ATM, GIBBSITE	1526
*ALPHA-FEOOH(S) + 3 H(+) <-> FE(3+) + 2 H2O	LOG(K) = 1.6	3 M NACLO4, 298 K	1526
*CA(OH)2(S) <-> CA(2+) + 2 OH(-)	LOG(K) = -5.43	IONIC STRENGTH = 0, 298 K, 1 ATM	1526
*CA(OH)2(S) <-> CAOH(-) + OH(-)	LOG(K) = -4.03	IONIC STRENGTH = 0, 298 K, 1 ATM	1526
*CD(OH)2(S) + 2 H(+) <-> CD(2+) + 2 H2O	LOG(K) = 13.61	IONIC STRENGTH = 0, 298 K, 1 ATM	1526
*CD(2+) + OH(-) <-> CDOH(+)	LOG(K) = 3.8	1 M LICLO4, 298 K	1526
*CU(2+) + OH(-) <-> CUOH(+)	LOG(K) = 6.0	IONIC STRENGHT = 0, 1 ATM, 291 K	1526
*CU(2+) + 3 OH(-) <-> CU(OH)3(-)	LOG(K) = 15.2	IONIC STRENGTH = 0, 298 K, 1 ATM	1526
*CU(2+) + 4 OH(-) <-> CU(OH)4(2-)	LOG(K) = 16.1	IONIC STRENGTH = 0, 298 K, 1 ATM	1526
*CUO(S) + OH(-) <-> CU(2+) + H2O	LOG(K) = 7.65	IONIC STRENGTH = 0, 298 K, 1 ATM	1526
*FE(OH)2 <-> FE(2+) + 2 OH(-)	LOG(K) = -14.0	IONIC STRENGTH = 0, 298 K, 1 ATM, ACTIVE FE(OH)2	1526
*FE(OH)2 <-> FE(2+) + 2 OH(-)	LOG(K) = 14.5	IONIC STRENGTH = 0, 298 K, 1 ATM, INACTIVE FE(OH)2	1526
*FE(OH)2 + OH(-) <-> FE(OH)3(-)	LOG(K) = -5.5	IONIC STRENGTH = 0, 298 K, 1 ATM	1526
*GAMMA-AL(OH)3 + 3 H(+) <-> AL(3+) + 3 H2O	LOG(K) = 9.0	IONIC STRENGTH = 0, 298 K, 1 ATM, BAYERITE	1526
*MG(OH)2 <-> MG(2+) + 2 OH(-)	LOG(K) = -9.2	IONIC STRENGTH = 0, 298 K, 1 ATM, ACTIVE MG(OH)2	1526
*MG(OH)2 <-> MG(2+) + 2 OH(-)	LOG(K) = -11.6	IONIC STRENGTH = 0, 298 K, 1 ATM, BRUCITE MG(OH)2	1526
*MG(2+) + OH(-) <-> MGOH(+)	LOG(K) = 2.6	IONIC STRENGTH = 0, 298 K, 1 ATM	1526
*MN(OH)2(S) <-> MN(2+) + 2 OH(-)	LOG(K) = -12.8	IONIC STRENGTH = 0, 298 K, 1 ATM	1526
*MN(OH)2(S) + OH(-) <-> MN(OH)3(-)	LOG(K) = -5.0	IONIC STRENGTH = 0, 298 K, 1 ATM	1526
*ZN(2+) + OH(-) <-> ZNOH(-)	LOG(K) = 5.04	IONIC STRENGTH = 0, 298 K, 1 ATM	1526
*ZN(2+) + 3 OH(-) <-> ZN(OH)3(-)	LOG(K) = 13.9	IONIC STRENGTH = 0, 298 K, 1 ATM	1526
*ZN(2+) + 4 OH(-) <-> ZN(OH)4(2-)	LOG(K) = 15.1	IONIC STRENGTH = 0, 298 K, 1 ATM	1526
*ZNO(S) + 2 H(+) <-> ZN(2+) + H2O	LOG(K) = 11.81	IONIC STRENGTH = 0, 298 K, 1 ATM	1526
*2 (AM) FE(OH)3(S) <-> FE2(OH)4(4+) + 4 OH(-)	LOG(K) = -51.9	3 M NACLO4, 298 K	1526
*2 CU(2+) + 2 OH(-) <-> CU2(OH)2(2+)	LOG(K) = 17.0	IONIC STRENGHT = 0, 1 ATM, 291 K	1526

TABLE G37

EQUILIBRIUM CONSTANT : PB

ITEM:	VALUE:	CONDITIONS:	REF.:
*PB((OH)(2))(SOLID) <-> PB(+2) + 2 OH(-)	LOG(K) = -14.93	SOLUBILITY EQUILIBRIA, 298 K	899
*PB(+2) + CL(-) <-> PBCL(+)	LOG(K) = 1.62	COMPLEX FORMATION EQUILIBRIUM, 298 K	899
*PB(+2) + OH(-) <-> PBOH(+)	LOG(K) = 5.85	COMPLEX FORMATION EQUILIBRIUM, 298 K	899
*PB(+2) + 2 CL(-) <-> PBCL2(0)	LOG(K) = 1.83	COMPLEX FORMATION EQUILIBRIUM, 298 K	899
*PB(+2) + 2 OH(-) <-> PB((OH)(2)) (0)	LOG(K) = 10.80	COMPLEX FORMATION EQUILIBRIUM, 298 K	899
*PB(+2) 3 OH(-) <-> PB((OH)(3))(-)	LOG(K) = 13.92	COMPLEX FORMATION EQUILIBRIUM, 298 K	899
*PBCO3(SOLID) <-> PB(+2) + CO3(-)	LOG(K) = -13.00	SOLUBILITY EQUILIBRIA, 298 K	899
*PBSO4(SOLID) <-> PB(+2) + SO4(-2)	LOG(K) = -7.89	SOLUBILITY EQUILIBRIA, 298 K	899
*PB3((OH)(2))((CO3)(2))(SOLID) <-> 3 PB(+2) + 2 OH(-) + 2 CO3(-2)	LOG(K) = -18.80	SOLUBILITY EQUILIBRIA, 298 K	899

TABLE G38

EQUILIBRIUM CONSTANT : PHOSPHATES

ITEM:	VALUE:	CONDITIONS:	REF.:
*(CAHPO4(OH)2) <-> 2 CA(2+) + HPO4(2-) + 2 OH(-)	LOG(K) = -27	298 K, SURFACE OF CA2HPO4(OH)2	1526
*ALPO4(S) <-> AL(3+) + PO4(3-)	LOG(K) = -21	298 K	1526
*CA(2+) + HPO4(2-) <-> CAHPO4(AQ)	LOG(K) = 2.7	298 K	1435
*CA(2+) + HP2O7(3-) <-> CAHP2O7(-)	LOG(K) = 2.0	298 K	1435
*CA(2+) + P2O7(4-) <-> CAP2O7(2-)	LOG(K) = 5.6	298 K	1435
*CA(2+) + P3O10(5-) <-> FE(HP2O7) 2(3-)	LOG(K) = 8.1	298 K	1435
*CAHPO4(S) <-> CA(2+) + HPO4(2-)	LOG(K) = -7	298 K	1526
*CA2P2O7(S) <-> CA(2+) + CAP2O7	LOG(K) = -7.9	298 K	1526
*CA5OH(PO4)3(S) <-> 5 CA(2+) + 3 PO4(3-) + OH(-)	LOG(K) = -55.6	298 K	1526
*CA5OH(PO4)3(S) + 3 H2O <-> 2 (CA2HPO4(OH)2) + CA(2+) + HPO4(2-)	LOG(K) = -8.5	298 K, SURFACE OF CA2HPO4(OH)2	1526
*FE(3+) + HPO4(2-) <-> FEHPO4(+)	LOG(K) = 8.3	298 K	1435
*FE(3+) + 2 HP2O7(3-) <-> FE(HP2O7) 2(3-)	LOG(K) = 22.0	298 K	1435
*FEPO4(S) <-> FE(3+) + PO4(3-)	LOG(K) = -23	298 K	1526
*HPO4(-) <-> PO4(3-) + H(+)	LOG(K) = -12.0	298 K	1435
*HP2O7(3-) <-> P2O7(4-) + H(+)	LOG(K) = -9.3	298 K	1435
*HP3O10(4-) <-> P3O10(5-) + H(+)	LOG(K) = -9.2	298 K	1435
*HP3O9(2-) <-> P3O9(3-) + H(+)	LOG(K) = -2.1	298 K	1435
*H2PO4(-) <-> HPO4(2-) + H(+)	LOG(K) = -7.0	298 K	1435
*H2P2O7(2-) <-> HP2O7(3-) + H(+)	LOG(K) = -6.6	298 K	1435
*H2P3O10(3-) <-> HP3O10(4-) + H(+)	LOG(K) = -6.5	298 K	1435
*H3P2O7(-) <-> H2P2O7(2-) + H(+)	LOG(K) = -2.4	298 K	1435
*H3P3O10(2-) <-> H2P3O10(3-) + H(+)	LOG(K) = -2.3	298 K	1435
*MG(2+) + HPO4(2-) <-> MGHPO4(AQ)	LOG(K) = 2.5	298 K	1435
*MG(2+) + P2O7(4-) <-> MGP2O7(2-)	LOG(K) = 5.7	298 K	1435
*MG(2+) + P3O10(5-) <-> MGP3O10(3-)	LOG(K) = 8.6	298 K	1435
*NA(+) + P2O7(4-) <-> NAP2O7(3-)	LOG(K) = 2.2	298 K	1435

TABLE G39

EQUILIBRIUM CONSTANT : REDOX REACTIONS

ITEM:	VALUE:	CONDITIONS:	REF.:
*AG(+) + E <-> AG(S)	LOG(K) = +13.5	STANDARD ELECTRODE POTIENTAL = + 0.80 V, 298 K	1526

TABLE G39

EQUILIBRIUM CONSTANT : REDOX REACTIONS

ITEM:	VALUE:	CONDITIONS:	REF.:
*AGCL(S) + E <-> AG(S) + CL(-)	LOG(K) = +3.7	STANDARD ELECTRODE POTIENTAL = + 0.22 V, 298 K	1526
*CL2(G) + 2 E <-> 2 CL(-)	LOG(K) = +46	STANDARD ELECTRODE POTIENTAL = + 1.36 V, 298 K	1526
*CO(2+) + 2 E <-> CO(S)	LOG(K) = -9.5	STANDARD ELECTRODE POTIENTAL = -0.28 V, 298 K	1526
*CO(3+) + E <-> CO(2+)	LOG(K) = +31	STANDARD ELECTRODE POTIENTAL = + 1.82 V, 298 K	1526
*CU(+) + E <-> CU(S)	LOG(K) = +18.0	STANDARD ELECTRODE POTIENTAL = + 0.52 V, 298 K	1526
*CU(2+) + E <-> CU(+)	LOG(K) = +2.7	STANDARD ELECTRODE POTIENTAL = + 0.16 V, 298 K	1526
*CU(2+) + 2 E <-> CU(S)	LOG(K) = +12.0	STANDARD ELECTRODE POTIENTAL = + 0.34 V, 298 K	1526
*FE(OH)3(S) + 3 H(+) + E <-> FE(2+) + 3 H2O	LOG(K) = +18.8	STANDARD ELECTRODE POTIENTAL = + 1.06 V, 298 K	1526
*FE(2;) + 2 E <-> FE(S)	LOG(K) = -15	STANDARD ELECTRODE POTIENTAL = -0.44 V, 298 K	1526
*FE(3+) + E <-> FE(2+)	LOG(K) = +13.2	STANDARD ELECTRODE POTIENTAL = + 0.77 V, 298 K	1526
*H(+) + E <-> 1/2 H2(G)	LOG(K) = 0.0	298 K, AQUATIC PROCESSES	1526
*H(+) + E <-> 1/2 H2(G)	LOG(K) = 0.0	298 K, AQUATIC PROCESSES	1526
*H3PO4 <-> H2PO4(-) + H(+)	LOG(K) = -2.2	298 K	1435
*IO3(-) + 6 H(+) 5 E <-> 1/2 I2(S) + 3 H2O	LOG(K) = +104	STANDARD ELECTRODE POTIENTAL = + 1.23 V, 298 K	1526
*MNO2(S) + 4 H(+) + 2 E <-> MN(2+) + 2 H2O	LOG(K) = +42	STANDARD ELECTRODE POTIENTAL = + 1.23 V, 298 K	1526
*NA(-) + E <-> NA(S)	LOG(K) = -46	STANDARD ELECTRODE POTIENTAL = -2.71 V, 298 K	1526
*OXIDIZED FERREDOXIN + E <-> REDUCED FERREDOXIN	LOG(K) = -7.1	298 K, AQUATIC PROCESSES	1526
*OXIDIZED FERREDOXIN + E <-> REDUCED FERREDOXIN	LOG(K) = -7.1	298 K, AQUATIC PROCESSES	1526
*S(S) + 2 H(+) + 2 E <-> H2S	LOG(K) = +0.47	STANDARD ELECTRODE POTIENTAL = + 0.14 V, 298 K	1526
*V(3+) + E <-> V(2+)	LOG(K) = -8.8	STANDARD ELECTRODE POTIENTAL = -0.26 V, 298 K	1526
*ZN(2+) + 2 E <-> ZN(S)	LOG(K) = -26	STANDARD ELECTRODE POTIENTAL = -0.76 V, 298 K	1526
*1/2 (NADP+) + 1/2 H(+) + E <-> 1/2 (NADPH)	LOG(K) = -2.0	298 K, AQUATIC PROCESSES	1526
*1/2 (NADP+) + 1/2 H(+) + E <-> 1/2 (NADPH)	LOG(K) = -2.0	298 K, AQUATIC PROCESSES	1526
*1/2 CH2O + H(+) + E <-> 1/2 CH3OH	LOG(K) = +3.99	298 K, AQUATIC PROCESSES	1526
*1/2 CH2O + H(+) + E <-> 1/2 CH3OH	LOG(K) = +3.99	298 K, AQUATIC PROCESSES	1526
*1/2 CH3OH + H(+) + E <-> 1/2 CH4 (G) + H2O	LOG(K) = +9.88	298 K, AQUATIC PROCESSES	1526
*1/2 CH3OH + H(+) + E <-> 1/2 CH4 (G) + H2O	LOG(K) = +9.88	298 K, AQUATIC PROCESSES	1526
*1/2 CO2(G) + 1/2 H(+) + E <-> 1/2 HCOO(-)	LOG(K) = -4.83	298 K, AQUATIC PROCESSES	1526
*1/2 CO2(G) + 1/2 H(+) + E <-> 1/2 HCOO(-)	LOG(K) = -4.83	298 K, AQUATIC PROCESSES	1526
*1/2 HCOO(-) + 3/2 H(+) + E <-> 1/2 CH2O + 1/2 H2O	LOG(K) = +2.82	298 K, AQUATIC PROCESSES	1526
*1/2 HCOO(-) + 3/2 H(+) + E <-> 1/2 CH2O + 1/2 H2O	LOG(K) = +2.82	298 K, AQUATIC PROCESSES	1526
*1/2 NO3(-) + H(+) + E <-> 1/2 NO2(-) + 1/2 H2O	LOG(K) = +14.15	298 K, AQUATIC PROCESSES	1526
*1/2 NO3(-) + H(+) + E <-> 1/2 NO2(-) + 1/2 H2O	LOG(K) = +14.15	298 K, AQUATIC PROCESSES	1526
*1/2 S(S) + H(+) + E <-> 1/2 H2S(G)	LOG(K) = +2.89	298 K, AQUATIC PROCESSES	1526
*1/2 S(S) + H(+) + E <-> 1/2 H2S(G)	LOG(K) = +2.89	298 K, AQUATIC PROCESSES	1526
*1/4 CH2O + H(+) + E <-> 1/4 CH4(G) + 1/4 H2O	LOG(K) = +6.94	298 K, AQUATIC PROCESSES	1526

TABLE G39

EQUILIBRIUM CONSTANT : REDOX REACTIONS

ITEM:	VALUE:	CONDITIONS:	REF.:
*1/4 CH2O + H(+) + E <-> 1/4 CH4(G) + 1/4 H2O	LOG(K) = +6.94	298 K, AQUATIC PROCESSES	1526
*1/4 CO2(G) + H(+) + E <-> 1/24 (GLUCOSE) + 1/4 H2O	LOG(K) = -0.20	298 K, AQUATIC PROCESSES	1526
*1/4 CO2(G) + H(+) + E <-> 1/24 (GLUCOSE) + 1/4 H2O	LOG(K) = -0.20	298 K, AQUATIC PROCESSES	1526
*1/4 CO2(G) + H(+) + E <-> 1/4 CH2O + 1/4 H2O	LOG(K) = -1.20	298 K, AQUATIC PROCESSES	1526
*1/4 CO2(G) + H(+) + E <-> 1/4 CH2O + 1/4 H2O	LOG(K) = -1.20	298 K, AQUATIC PROCESSES	1526
*1/4 O2(G) + H(+) + E <-> 1/2 H2O	LOG(K) = +20.75	298 K, AQUATIC PROCESSES	1526
*1/4 O2(G) + H(+) + E <-> 1/2 H2O	LOG(K) = +20.75	298 K, AQUATIC PROCESSES	1526
*1/5 NO3(-) + 6/5 H(+) + E <-> 1/10 N2(G) + 3/5 H2O	LOG(K) = +21.05	298 K, AQUATIC PROCESSES	1526
*1/5 NO3(-) + 6/5 H(+) + E <-> 1/10 N2(G) + 3/5 H2O	LOG(K) = +21.05	298 K, AQUATIC PROCESSES	1526
*1/6 NO2(-) + 4/3 H(+) + E <-> 1/6 NH4(+) + 1/3 H2O	LOG(K) = +15.14	298 K, AQUATIC PROCESSES	1526
*1/6 NO2(-) + 4/3 H(+) + E <-> 1/6 NH4(+) + 1/3 H2O	LOG(K) = +15.14	298 K, AQUATIC PROCESSES	1526
*1/6 N2(G) + 4/3 H(+) + E <-> 1/3 NH4(+)	LOG(K) = +4.68	298 K, AQUATIC PROCESSES	1526
*1/6 N2(G) + 4/3 H(+) + E <-> 1/3 NH4(+)	LOG(K) = +4.68	298 K, AQUATIC PROCESSES	1526
*1/6 SO4(2-) + 4/3 H(+) + E <-> 1/6 S(S) + 2/3 H2O	LOG(K) = +6.03	298 K, AQUATIC PROCESSES	1526
*1/6 SO4(2-) + 4/3 H(+) + E <-> 1/6 S(S) + 2/3 H2O	LOG(K) = +6.03	298 K, AQUATIC PROCESSES	1526
*1/8 CO2(G) + H(+) + E <-> 1/8 CH4(G) + 1/4 H2O	LOG(K) = +2.87	298 K, AQUATIC PROCESSES	1526
*1/8 CO2(G) + H(+) + E <-> 1/8 CH4(G) + 1/4 H2O	LOG(K) = +2.87	298 K, AQUATIC PROCESSES	1526
*1/8 NO3(-) + 5/4 H(+) + E <-> 1/8 NH4(+) + 3/8 H2O	LOG(K) = +14.90	298 K, AQUATIC PROCESSES	1526
*1/8 NO3(-) + 5/4 H(+) + E <-> 1/8 NH4(+) + 3/8 H2O	LOG(K) = +14.90	298 K, AQUATIC PROCESSES	1526
*1/8 SO4(2-) + 5/4 H(+) + E <-> 1/ 8H2S + 1/2 H2O	LOG(K) = +5.75	298 K, AQUATIC PROCESSES	1526
*1/8 SO4(2-) + 5/4 H(+) + E <-> 1/ 8H2S + 1/2 H2O	LOG(K) = +5.75	298 K, AQUATIC PROCESSES	1526
*1/8 SO4(2-) + 9/8 H(+) + E <-> 1/8 HS(-) + 1/2 H2O	LOG(K) = +4.13	298 K, AQUATIC PROCESSES	1526
*1/8 SO4(2-) + 9/8 H(+) + E <-> 1/8 HS(-) + 1/2 H2O	LOG(K) = +4.13	298 K, AQUATIC PROCESSES	1526
*2 H(+) + 2 E <-> H2(G)	LOG(K) = 0.0	STANDARD ELECTRODE POTIENTAL = + 0.00 V, 298 K	1526

TABLE G40

EQUILIBRIUM CONSTANT : SULPHIDE AND HEAVY METALS

ITEM:	VALUE:	CONDITIONS:	REF.:
*CDS(SOLID) <-> CD(+2) + S(-2)	LOG(K) = -27.2	SOLUBILITY EQUILIBRIUM, 298 K	899
*COS(SOLID) <-> CO(+2) + S(-2)	LOG(K) = -21.3	SOLUBILITY EQUILIBRIUM, 298 K	899
*CUS(SOLID) <-> CU(+2) + S(-2)	LOG(K) = -36.2	SOLUBILITY EQUILIBRIUM, 298 K	899
*FES(SOLID) <-> FE(+2) + S(-2)	LOG(K) = -18.75	PYRITE FES, SOLUBILITY EQUILIBRIUM, 298 K	899
*FES2(SOLID) <-> FE(+2) + S2(-2)	LOG(K) = -26.1	SOLUBILITY EQUILIBRIUM, 298 K	899
*HGS(SOLID) <-> HG(+2) + S(-2)	LOG(K) = -52.2	METACINNABAR HGS, SOLUBILITY EQUILIBRIUM, 298 K	899
*HGS(SOLID) <-> HG(+2) + S(-2)	LOG(K) = -53.6	CINNABAR HGS, SOLUBILITY EQUILIBRIUM, 298 K	899

TABLE G40

EQUILIBRIUM CONSTANT : SULPHIDE AND HEAVY METALS			
ITEM:	VALUE:	CONDITIONS:	REF.:
*MNS(SOLID) <-> MN(+2) + S(-2)	LOG(K) = -12.85	CRYSYALINE MNS, SOLUBILITY EQUILIBRIUM, 298 K	899
*MNS(SOLID) <-> MN(+2) + S(-2)	LOG(K) = -15.7	PRECIPITATED MNS, SOLUBILITY EQUILIBRIUM, 298 K	899
*NIS(SOLID) <-> NI(+2) + S(-2)	LOG(K) = -20.8	ALPHA - NIS, SOLUBILITY EQUILIBRIUM, 298 K	899
*NIS(SOLID) <-> NI(+2) + S(-2)	LOG(K) = -27.75	GAMMA - NIS, SOLUBILITY EQUILIBRIUM, 298 K	899
*PBS(SOLID) <-> PB(+2) + S(-2)	LOG(K) = -28.2	SOLUBILITY EQUILIBRIUM, 298 K	899
*ZNS(SOLID) <-> ZN(+2) + S(-2)	LOG(K) = -22.05	PRECIPITATED ZNS, SOLUBILITY EQUILIBRIUM, 298 K	899
*ZNS(SOLID) <-> ZN(+2) + S(-2)	LOG(K) = -22.80	WURTZITE ZNS, SOLUBILITY EQUILIBRIUM, 298 K	899
*ZNS(SOLID) <-> ZN(+2) + S(-2)	LOG(K) = -25.15	SPHALERITE ZNS, SOLUBILITY EQUILIBRIUM, 298 K	899

TABLE G41

EQUILIBRIUM CONSTANT : ZN			
ITEM:	VALUE:	CONDITIONS:	REF.:
*ORGANIC MATTER FROM SMITH LAKE + ZN(II)	LOG(K) = 3.37	FORMATION OF HUMIC-METAL FOR SMALL ORGANIC MOLECULES	1726
*PHTHALIC ACID + ZN(II)	LOG(K) = 2.89	FORMATION OF HUMIC-METAL FOR SMALL ORGANIC MOLECULES	1726
*ZN((OH)(2))(SOLID) <-> ZN(+2) + 2 OH(-)	LOG(K) = -15.52	SOLUBILITY EQUILIBRIA, 298 K	899
*ZN(+2) + CL(-) <-> ZNCL(+)	LOG(K) = -0.56	COMPLEX FORMATION EQUILIBRIUM, 298 K	899
*ZN(+2) + OH(-) <-> ZNOH(+)	LOG(K) = 4.95	COMPLEX FORMATION EQUILIBRIUM, 298 K	899
*ZN(+2) + SO4(-2) <-> ZNSO4	LOG(K) = 2.8	COMPLEX FORMATION EQUILIBRIUM, 298 K	899
*ZN(+2) + 2 OH(-) <-> ZN((OH)(2))(0)	LOG(K) = 12.89	COMPLEX FORMATION EQUILIBRIUM, 298 K	899
*ZN(+2) + 3 OH(-) <-> ZN((OH)(3))(-)	LOG(K) = 14.22	COMPLEX FORMATION EQUILIBRIUM, 298 K	899
*ZN(+2) + 4 OH(-) <-> ZN((OH)(4))(-2)	LOG(K) = 15.48	COMPLEX FORMATION EQUILIBRIUM, 298 K	899
*ZNCO3(SOLID) <-> ZN(+2) + CO3(-2)	LOG(K) = -10.00	SOLUBILITY EQUILIBRIA, 298 K	899
*3,4-DIHYDROXYBENZOIC ACID + ZN(II)	LOG(K) = 8.13	FORMATION OF HUMIC-METAL FOR SMALL ORGANIC MOLECULES	1726

TABLE G42

EQUILIBRIUM DISTRIBUTION : THE SYSTEM C2H10NO8 AT 1 ATM AND 773 K			
ITEM:	VALUE:	CONDITIONS:	REF.:
*EQUILIBRIUM CONCENTRATION FOR XYLENE	0.17*10**(-33) MOL	PREBIOLOGICAL ATMOSPHERE	1067

TABLE G43

EQUILIBRIUM DISTRIBUTION : THE SYSTEM 2 C, 10 H, 1 N, 8 O AT 1 ATM AND 773 K			
ITEM:	VALUE:	CONDITIONS:	REF.:
*EQUILIBRIUM CONCENTRATION FOR ACETALDEHYDE	0.81*10**(-14) MOL	PREBIOLOGICAL ATMOSPHERE	1067
*EQUILIBRIUM CONCENTRATION FOR ACETAMIDE	0.11*10**(-23) MOL	PREBIOLOGICAL ATMOSPHERE	1067
*EQUILIBRIUM CONCENTRATION FOR ACETIC ACID	0.25*10**(-9) MOL	PREBIOLOGICAL ATMOSPHERE	1067
*EQUILIBRIUM CONCENTRATION FOR ACETONE	0.92*10**(-17) MOL	PREBIOLOGICAL ATMOSPHERE	1067
*EQUILIBRIUM CONCENTRATION FOR ACETYLENE	0.11*10**(-212) MOL	PREBIOLOGICAL ATMOSPHERE	1067
*EQUILIBRIUM CONCENTRATION FOR ALANINE	0.97*10**(-27) MOL	PREBIOLOGICAL ATMOSPHERE	1067
*EQUILIBRIUM CONCENTRATION FOR AMMONIA	0.15*10**(-3) MOL	PREBIOLOGICAL ATMOSPHERE	1067
*EQUILIBRIUM CONCENTRATION FOR ANTRACENE	<10**(-38) MOL	PREBIOLOGICAL ATMOSPHERE	1067
*EQUILIBRIUM CONCENTRATION FOR ASPHALT	<10**(-38) MOL	PREBIOLOGICAL ATMOSPHERE	1067
*EQUILIBRIUM CONCENTRATION FOR BENZALDEHYDE	0.12*10**(-35) MOL	PREBIOLOGICAL ATMOSPHERE	1067
*EQUILIBRIUM CONCENTRATION FOR BENZENE	0.52*10**(-25) MOL	PREBIOLOGICAL ATMOSPHERE	1067
*EQUILIBRIUM CONCENTRATION FOR BENZOIC ACID	0.65*10**(-31) MOL	PREBIOLOGICAL ATMOSPHERE	1067
*EQUILIBRIUM CONCENTRATION FOR CARBON DIOXIDE	0.88 MOL	PREBIOLOGICAL ATMOSPHERE	1067
*EQUILIBRIUM CONCENTRATION FOR CARBON MONOXIDE	0.54*10**(-4) MOL	PREBIOLOGICAL ATMOSPHERE	1067
*EQUILIBRIUM CONCENTRATION FOR CYANOGEN	0.77*10**(-31) MOL	PREBIOLOGICAL ATMOSPHERE	1067
*EQUILIBRIUM CONCENTRATION FOR ETHANE	0.34*10**(-7) MOL	PREBIOLOGICAL ATMOSPHERE	1067
*EQUILIBRIUM CONCENTRATION FOR ETHANOL	0.49*10**(-15) MOL	PREBIOLOGICAL ATMOSPHERE	1067
*EQUILIBRIUM CONCENTRATION FOR ETHYLENE	0.88*10**(-13) MOL	PREBIOLOGICAL ATMOSPHERE	1067
*EQUILIBRIUM CONCENTRATION FOR ETHYLENE GLYCOL	0.62*10**(-24) MOL	PREBIOLOGICAL ATMOSPHERE	1067
*EQUILIBRIUM CONCENTRATION FOR FORMALDEHYDE	0.13*10**(-11) MOL	PREBIOLOGICAL ATMOSPHERE	1067
*EQUILIBRIUM CONCENTRATION FOR FORMAMIDE	0.24*10**(-20) MOL	PREBIOLOGICAL ATMOSPHERE	1067
*EQUILIBRIUM CONCENTRATION FOR FORMIC ACID	0.93*10**(-9) MOL	PREBIOLOGICAL ATMOSPHERE	1067
*EQUILIBRIUM CONCENTRATION FOR FURAN	0.14*10**(-28) MOL	PREBIOLOGICAL ATMOSPHERE	1067
*EQUILIBRIUM CONCENTRATION FOR GLYCINE	0.48*10**(-21) MOL	PREBIOLOGICAL ATMOSPHERE	1067
*EQUILIBRIUM CONCENTRATION FOR HYDROGEN	0.18*10**(-1) MOL	PREBIOLOGICAL ATMOSPHERE	1067
*EQUILIBRIUM CONCENTRATION FOR HYDROGEN CYANIDE	0.73*10**(-13) MOL	PREBIOLOGICAL ATMOSPHERE	1067
*EQUILIBRIUM CONCENTRATION FOR KETENE	0.19*10**(-17) MOL	PREBIOLOGICAL ATMOSPHERE	1067
*EQUILIBRIUM CONCENTRATION FOR LACTIC ACID	0.20*10**(-23) MOL	PREBIOLOGICAL ATMOSPHERE	1067

TABLE G43

EQUILIBRIUM DISTRIBUTION : THE SYSTEM 2 C, 10 H, 1 N, 8 O AT 1 ATM AND 773 K

ITEM:	VALUE:	CONDITIONS:	REF.:
*EQUILIBRIUM CONCENTRATION FOR METHANE	0.12 MOL	PREBIOLOGICAL ATMOSPHERE	1067
*EQUILIBRIUM CONCENTRATION FOR METHANOL	0.73*10**(-11) MOL	PREBIOLOGICAL ATMOSPHERE	1067
*EQUILIBRIUM CONCENTRATION FOR METHYL ETHER	0.30*10**(-19) MOL	PREBIOLOGICAL ATMOSPHERE	1067
*EQUILIBRIUM CONCENTRATION FOR METHYLAMINE	0.64*10**(-13) MOL	PREBIOLOGICAL ATMOSPHERE	1067
*EQUILIBRIUM CONCENTRATION FOR NAPHTALENE	<10**(-38) MOL	PREBIOLOGICAL ATMOSPHERE	1067
*EQUILIBRIUM CONCENTRATION FOR NITROGEN	0.50 MOL	PREBIOLOGICAL ATMOSPHERE	1067
*EQUILIBRIUM CONCENTRATION FOR OXYGEN	<10**(-38) MOL	PREBIOLOGICAL ATMOSPHERE	1067
*EQUILIBRIUM CONCENTRATION FOR PHENOL	0.10*10**(-31) MOL	PREBIOLOGICAL ATMOSPHERE	1067
*EQUILIBRIUM CONCENTRATION FOR PYRIDINE	0.16*10**(-30) MOL	PREBIOLOGICAL ATMOSPHERE	1067
*EQUILIBRIUM CONCENTRATION FOR PYRIMIDINE	0.13*10**(-31) MOL	PREBIOLOGICAL ATMOSPHERE	1067
*EQUILIBRIUM CONCENTRATION FOR PYRROLE	0.31*10**(-30) MOL	PREBIOLOGICAL ATMOSPHERE	1067
*EQUILIBRIUM CONCENTRATION FOR PYRUVIC ACID	0.31*10**(-31) MOL	PREBIOLOGICAL ATMOSPHERE	1067
*EQUILIBRIUM CONCENTRATION FOR WATER	2.24 MOL	PREBIOLOGICAL ATMOSPHERE	1067

TABLE G44

GIBBS ENERGY : FE			
ITEM:	VALUE:	CONDITIONS:	REF.:
*ALPHA-FE2O3(S) (HAEMATETE)	-177.1 KCAL/MOLE	ENERGY OF FORMATION, 298 K	1526
*FORMATION OF (AMORPH)-FEOOH(S)	-111.1 KCAL/MOLE	ENERGY OF FORMATION, 298 K	1526
*FORMATION OF ALPHA-FEOOH(S) (GOETHITE)	-113.7 KCAL/MOLE	ENERGY OF FORMATION, 298 K	1526
*FORMATION OF ALPHA-FES(S)	-23.32 KCAL/MOLE	ENERGY OF FORMATION, 298 K	1526
*FORMATION OF FE(OH)2(S)	-115.57 KCAL/MOLE	ENERGY OF FORMATION, 298 K	1526
*FORMATION OF FE(2+) (AQ)	-20.3 KCAL/MOLE	ENERGY OF FORMATION, 298 K	1526
*FORMATION OF FE(3+) (AQ)	-2.53 KCAL/MOLE	ENERGY OF FORMATION, 298 K	1526
*FORMATION OF FECO3(S) (SIDERITE)	-160.5 KCAL/MOLE	ENERGY OF FORMATION, 298 K	1526
*FORMATION OF FEPO4*2 H2O (STRENGITE)	-279 KCAL/MOLE	ENERGY OF FORMATION, 298 K	1526
*FORMATION OF FESIO3(S)	-257 KCAL/MOLE	ENERGY OF FORMATION, 298 K	1526
*FORMATION OF FES2(S) (PYRITE)	-39.9 KCAL/MOLE	ENERGY OF FORMATION, 298 K	1526
*FORMATION OF FE2SIO4(S)	-319.8 KCAL/MOLE	ENERGY OF FORMATION, 298 K	1526
*FORMATION OF FE3O4 (MAGNETITE)	-242.4 KCAL/MOLE	ENERGY OF FORMATION, 298 K	1526
*FORMATION OF FFE(OH)3(S)	-177.1 KCAL/MOLE	ENERGY OF FORMATION, 298 K	1526

TABLE G45

GIBBS ENERGY : GLUTAMIC ACID			
ITEM:	VALUE:	CONDITIONS:	REF.:
*OXIDATION OF GLUTAMIC ACID	- 483 KCAL/MOLE	ELECTRON ACCEPTOR = O2, NH4 AS PRODUCT	259
*OXIDATION OF GLUTAMIC ACID	- 568 KCAL/MOLE	ELECTRON ACCEPTOR = O2, NO3 AS PRODUCT	259
*OXIDATION OF GLUTAMIC ACID	- 469 KCAL/MOLE	ELECTRON ACCEPTOR = NO3, NH4 AND N2 AS PRODUCT	259
*OXIDATION OF GLUTAMIC ACID	- 545 KCAL/MOLE	ELECTRON ACCEPTOR = NO3, N2 AS PRODUCT	259
*OXIDATION OF GLUTAMIC ACID	- 19 KCAL/MOLE	ELECTRON ACCEPTOR = SO4, NH4 AS PRODUCT	259
*OXIDATION OF GLUTAMIC ACID	- 19 KCAL/MOLE	ELECTRON ACCEPTOR = SO4, N2 AS PRODUCT	259

TABLE G46

GIBBS ENERGY : MN			
ITEM:	VALUE:	CONDITIONS:	REF.:
*FORMATION OF GAMMA-MNOOH(S) (MANGANITE)	-133.3 KCAL/MOLE	ENERGY OF FORMATION, 298 K	1526
*FORMATION OF GAMMA-MNO2(S) (NSITITE)	-109.1 KCAL/MOLE	ENERGY OF FORMATION, 298 K	1526
*FORMATION OF GAMMA-MN2O3(S)	-132.2 KCAL/MOLE	ENERGY OF FORMATION, 298 K	1526
*FORMATION OF MN(OH)2(S) (PYROCHROSITE)	-147.3 KCAL/MOLE	ENERGY OF FORMATION, 298 K	1526
*FORMATION OF MN(2+) (AQ)	-54.4 KCAL/MOLE	ENERGY OF FORMATION, 298 K	1526
*FORMATION OF MNCO3(S) (RHODOCHROSITE)	-194.8 KCAL/MOLE	ENERGY OF FORMATION, 298 K	1526
*FORMATION OF MNO4(-)	-111.3 KCAL/MOLE	ENERGY OF FORMATION, 298 K	1526
*FORMATION OF MNS (S)	-53 KCAL/MOLE	ENERGY OF FORMATION, 298 K	1526
*FORMATION OF MN3O4(S) (HAUSMANNITE)	-306.2 KCAL/MOLE	ENERGY OF FORMATION, 298 K	1526

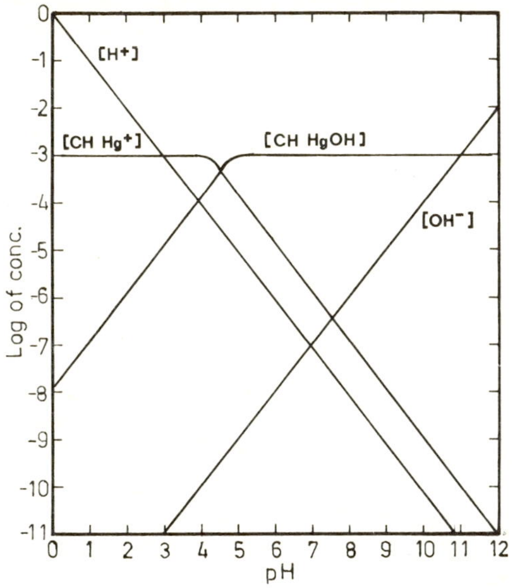

G45 Methyl mercury species versus pH.

(ref: 1313)

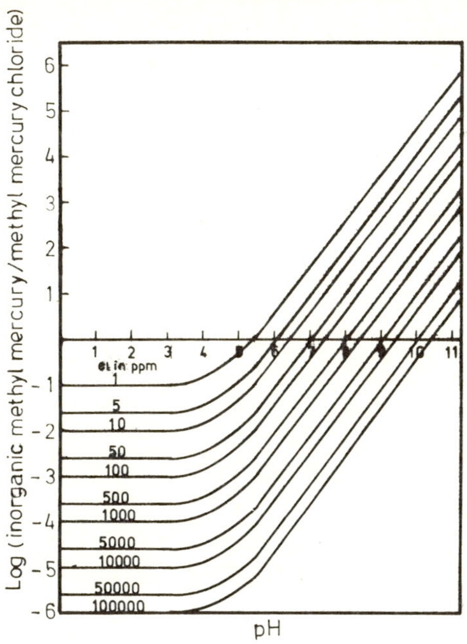

G46 Inorganic nonchloride methyl mercury/(methyl mercury chloride ratio) versus pH in water.

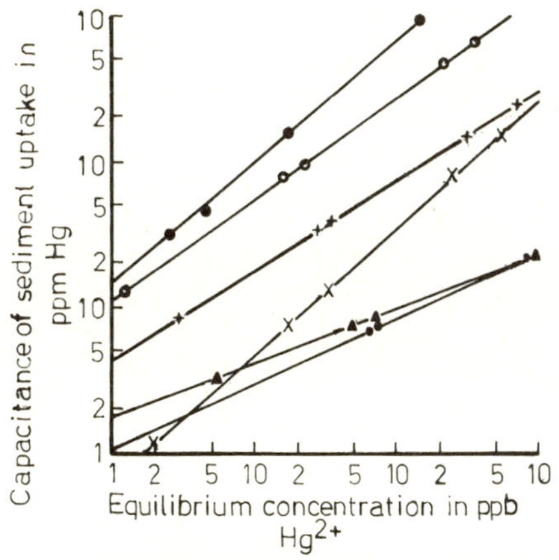

(ref: 1313)

G47 Adsorption isotherms of clays and sands for HgCl2 at pH 7.0; 0 ppm Cl(-) and 25 celcius. • Illinite, ○ Montmorillonite + Kaolinite X Fine sand ▲ Medium sand • Coarse sand.

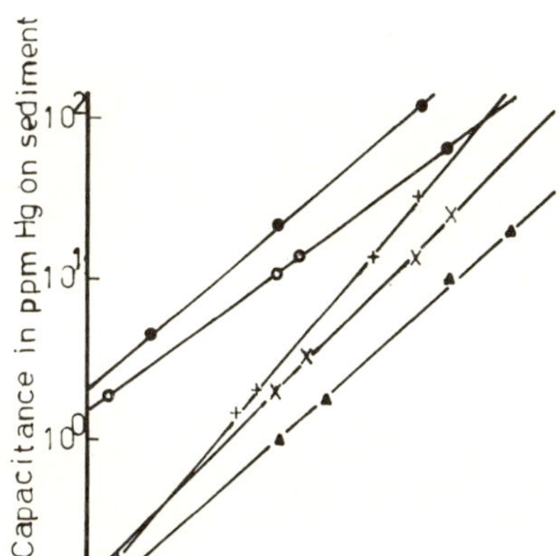

G48 Adsorptions isotherms of clays, organics and sands for HgCl2 at pH 7.0, 0 ppm Cl(-) and 25 celcius.
• Illinite ○ Montmorillonite
+ Amine X Carboxyl group. ▲ Fine sand.

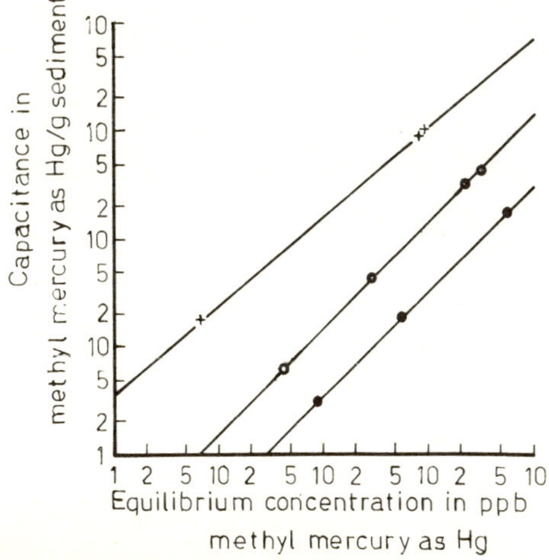

G49 Isotherms of methyl mercury adsorptions at 25 °C, pH 7 and 0 ppm Cl^-.
+ Dodecane-thiol and Sand.
○ Illite
● Montmorillonite.

(ref: 1313)

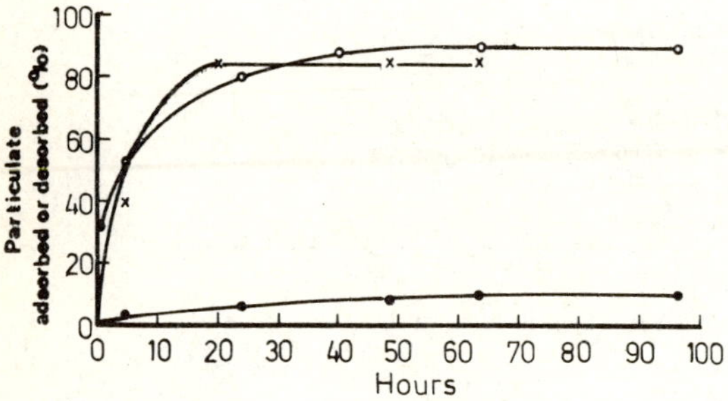

G50 Adsorption rate of Pu(IV) on sediment. (Var River water, pH = 8). Desorption in sea water medium.
∘ Adsorption. • Desorption. × Particulate formation.

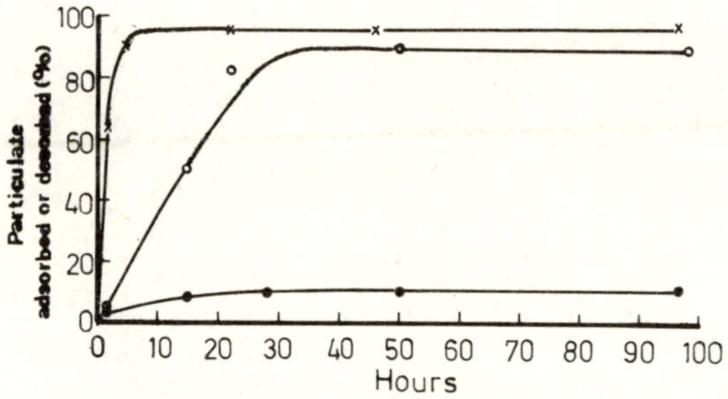

G51 Adsorption rate of Am(II) on sediment. (Var River water, pH = 8). Desorption in sea water medium.
(ref: 1087)
∘ Adsorption. • Desorption. × Particulate formation.

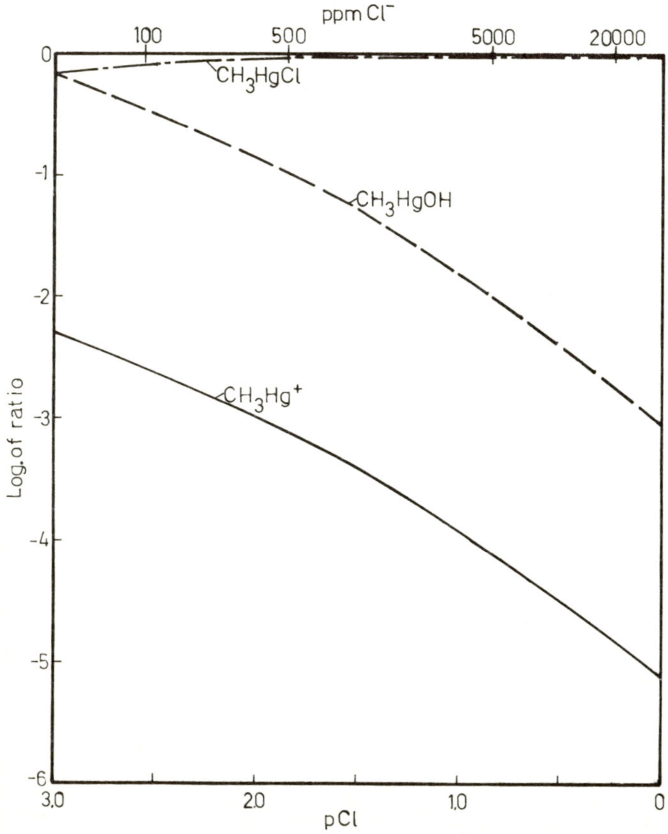

G52

Methyl mercury species: total methyl mercury ratio versus chloride. pH = 7.0.

(ref: 1431)

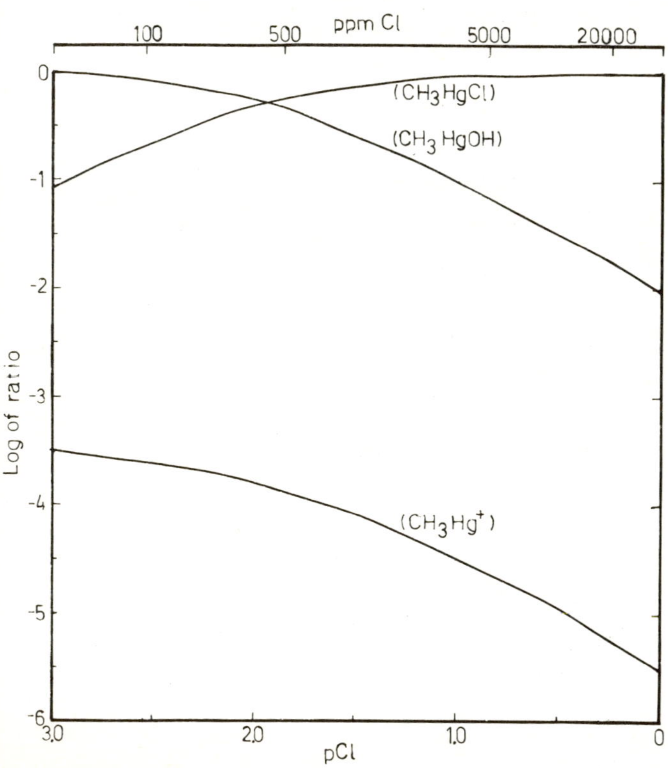

(ref: 1431)

G53

Methyl mercury species versus Cl(-) concentration. Log of ratio is plotted to ppm Cl(-) or mM Cl(-). pH = 8.

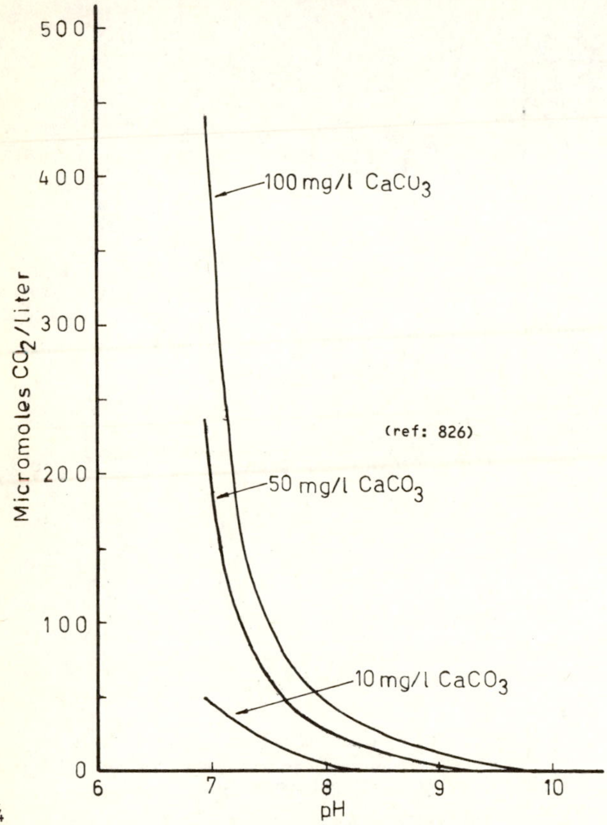

G54 Equilibrium carbon dioxide concentration under various conditions of alkalinity and pH.

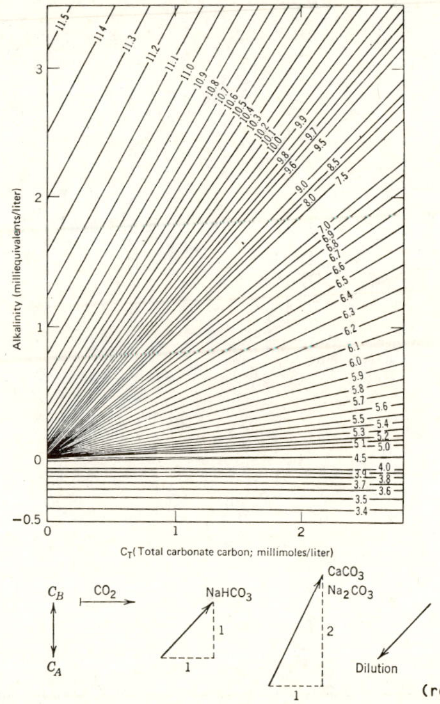

(ref: 1526)

G55 The pH contours in alkalinity versus C_t diagram. The point defining the solution composition moves as a vector in the diagram as a result of the addition (or removal) of CO_2, $NaHCO_3$, and $CaCO_3$ or C_b and C_a.

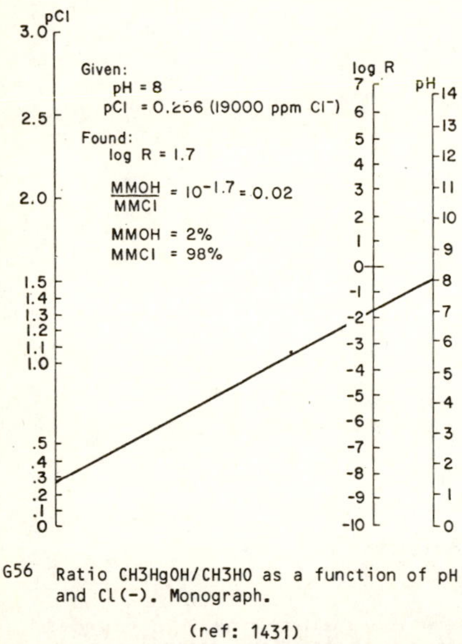

G56 Ratio CH3HgOH/CH3HO as a function of pH and Cl(-). Monograph.

(ref: 1431)

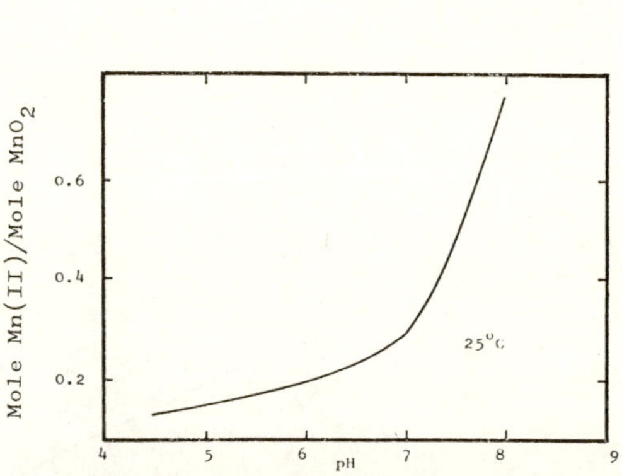

G57 Adsorption of Mn(II) on MnO2 as a function of pH at 25 celcius.

(ref: 1526)

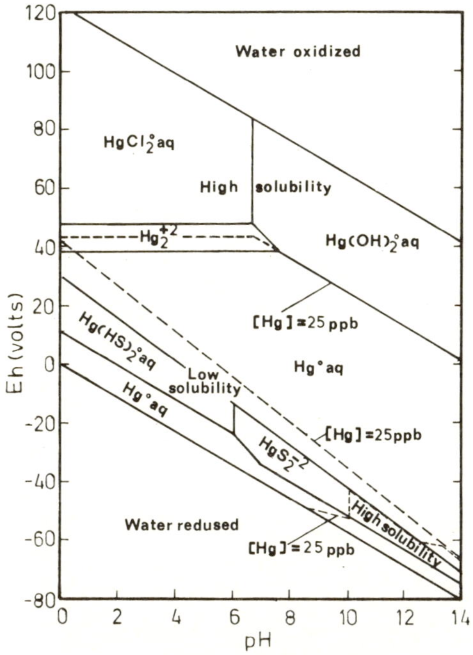

G58 Stability of mercury species at 25 celcius and 1 atm pressure. System includes water containing 36 ppm chloride and 96 ppm sulphate. Dashed line shows approximate solubility of mercury in this system.

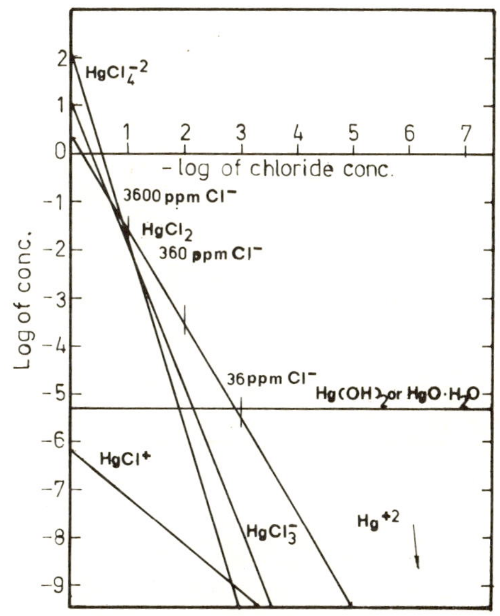

G59 (ref: 1313) Mercury complexes versus chloride concentration at 25 celcius and pH = 5.

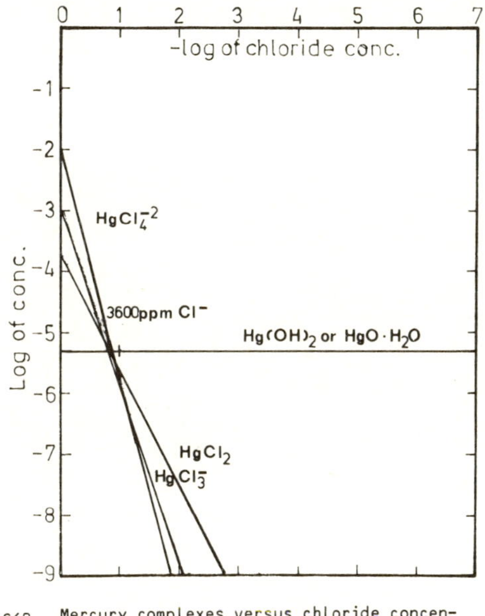

G60 Mercury complexes versus chloride concentration at 25 celcius and pH = 7.

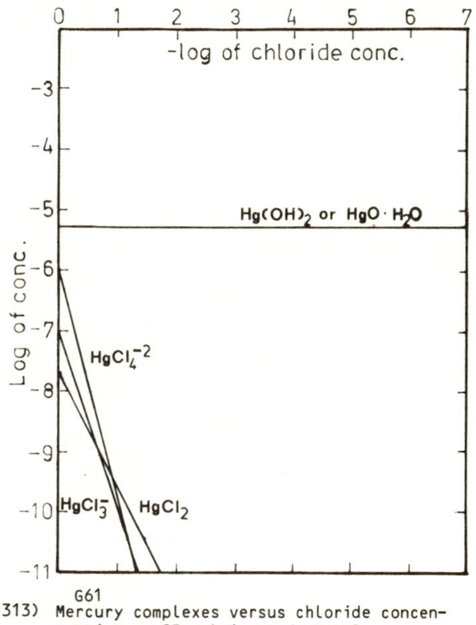

G61 (ref: 1313) Mercury complexes versus chloride concentration at 25 celcius and pH = 9.

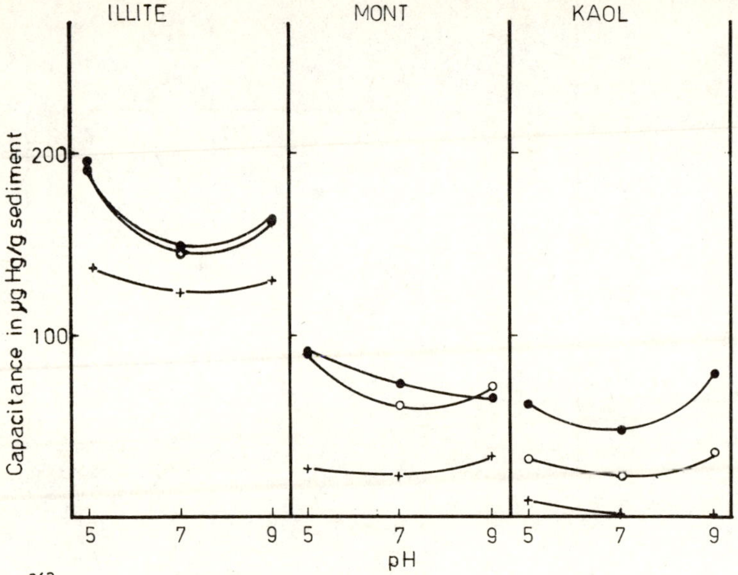

G62

Inorganic mercury adsorption capacity of clay at 25 celcius and 1 ppm Hg versus pH.

(ref: 1313)

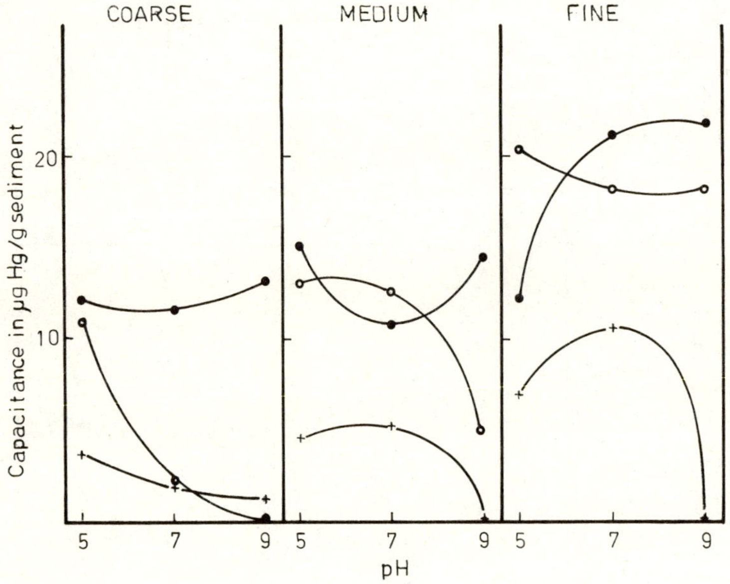

G63 Mercury inorganic adsorption capacity of sand at 25 celcius and 1 ppm Hg versus pH.

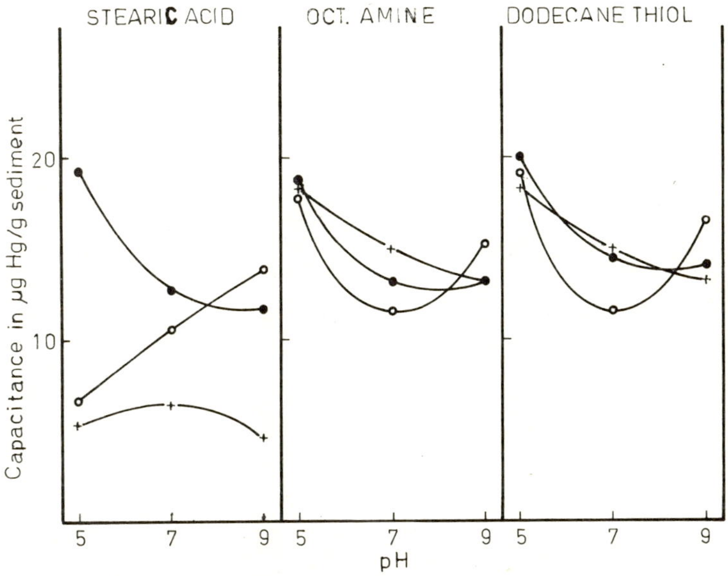

G64 Inorganic mercury adsorption capacity of organics at 25 celcius and 1 ppm Hg versus pH.

(ref: 1313)

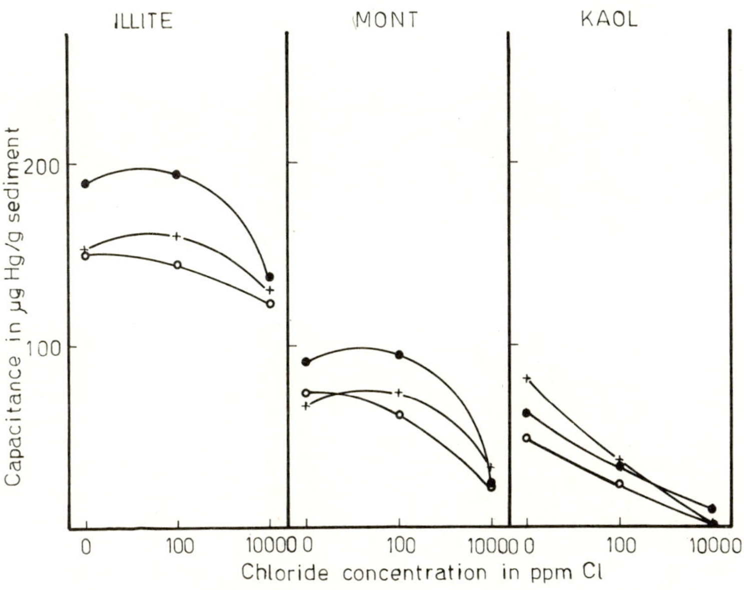

G65 Mercury inorganic adsorption capacity of clay at 25 celcius versus chloride concentration in ppm.

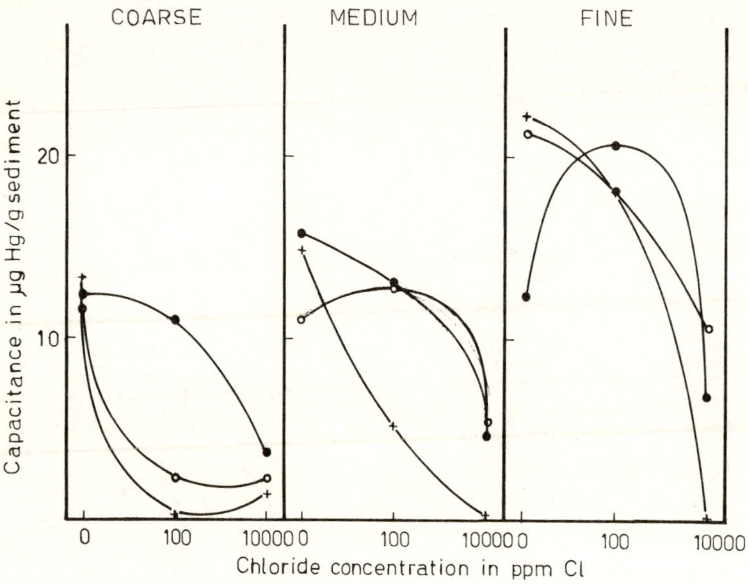

G66 Mercury inorganic adsorption capacity of sand at 25 celcius versus chloride concentration in ppm.

(ref: 1313)

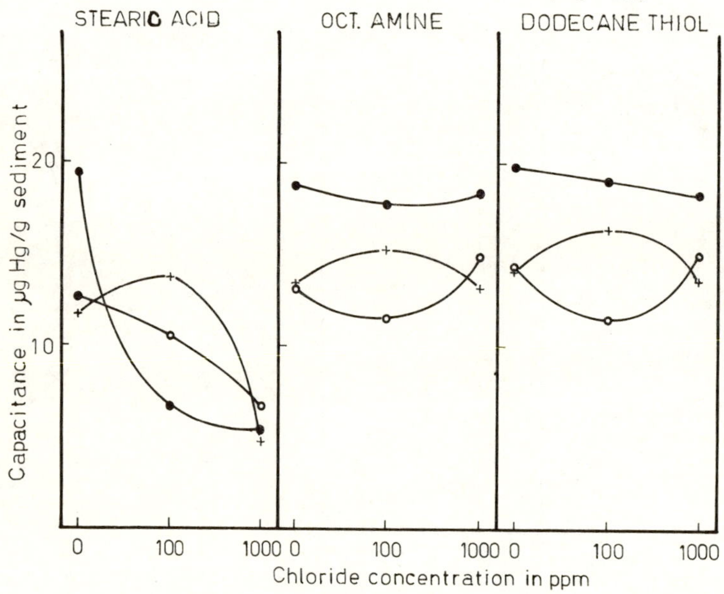

G67 Mercury inorganic adsorption capacity of organics at 25 celcius versus chloride concentration in ppm.

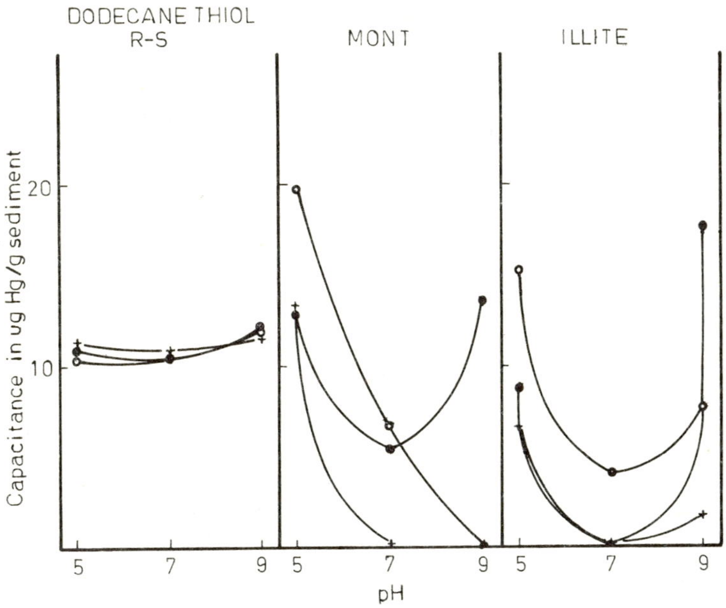

G68 Adsorption on sediment for methyl mercury at 25 celcius and 1 ppm mercury versus pH.

(ref: 1313)

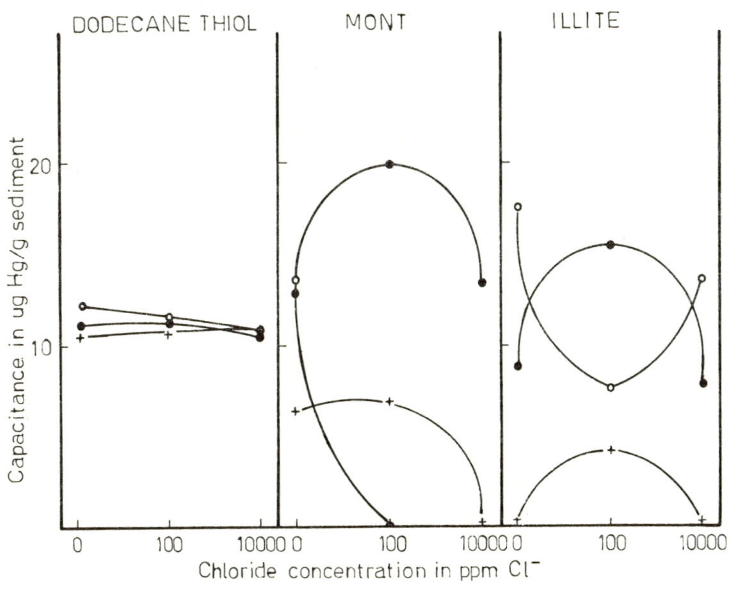

G69 Adsorption of sediment for methyl mercury at 25 celcius and 1 ppm mercury versus Cl(-) concentration.

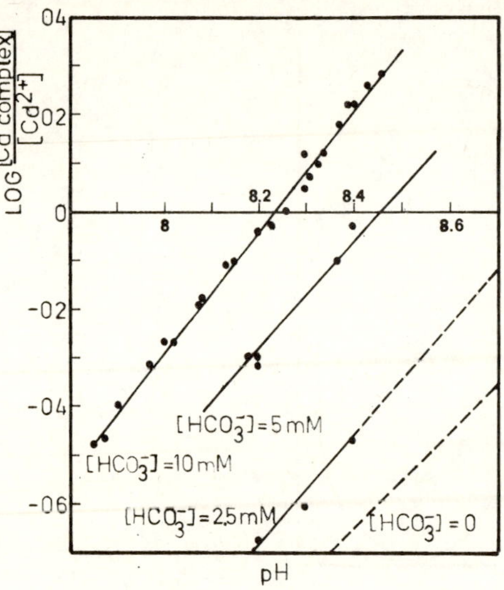

G70 Cadmium complex formation as a function of pH. Log(Cd-complex/Cd(+2)) versus pH. Hydrogen carbonate media. Total cadmium concentration 1.0*10**(-5) M. (0.9 µg/l).
(ref: 501)

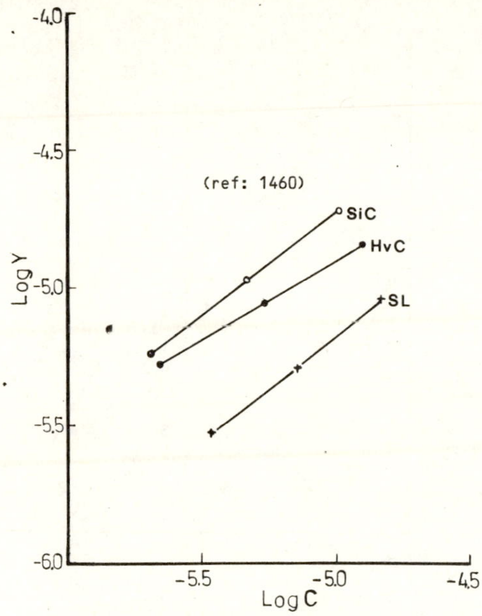

G71 Adsorption of 3,6 dichlorosalicylic acid from water to silty clay, heavy clay and sandy loam.
o silty clay, • heavy clay, + sandy loam,
Temrerature 25 +- 1 celcius.

G72
Dissolved oxygen saturation values in distilled water

Tempera- ture(^{o}C)	Dissolved oxygen (mg/l)									
	0.0	0.1	0.2	0.3	0.4	0.5	0.6	0.7	0.8	0.9
0	14.65	14.61	14.57	14.53	14.49	14.45	14.41	14.37	14.33	14.29
1	14.25	14.21	14.17	14.13	14.09	14.05	14.02	13.98	13.94	13.90
2	13.86	13.82	13.79	13.75	13.71	13.68	13.64	13.60	13.58	13.53
3	13.49	13.46	13.42	13.38	13.35	13.31	13.28	13.24	13.20	13.17
4	13.13	13.10	13.06	13.03	13.00	12.96	12.93	12.89	12.86	12.82
5	12.79	12.76	12.72	12.69	12.66	12.62	12.59	12.56	12.53	12.49
6	12.46	12.43	12.40	12.36	12.33	12.30	12.27	12.24	12.21	12.18
7	12.14	12.11	12.08	12.05	12.02	11.99	11.96	11.93	11.90	11.87
8	11.84	11.81	11.78	11.75	11.72	11.70	11.67	11.64	11.61	11.58
9	11.55	11.52	11.49	11.47	11.44	11.41	11.38	11.36	11.33	11.30
10	11.27	11.24	11.22	11.19	11.16	11.14	11.11	11.08	11.06	11.03
11	11.00	10.98	10.95	11.93	10.90	10.87	10.85	10.82	10.80	10.77
12	10.75	10.72	10.70	10.67	10.65	10.62	10.60	10.57	10.55	10.52
13	10.50	10.48	10.45	10.43	10.40	10.38	10.36	10.33	10.31	10.33
14	10.26	10.24	10.22	10.19	10.17	10.15	10.12	10.10	10.08	10.06
15	10.03	10.01	9.99	9.97	9.96	9.92	9.90	9.88	9.86	9.84
16	9.82	9.79	9.77	9.75	9.73	9.71	9.69	9.67	9.65	9.63
17	9.61	9.58	9.56	9.54	9.52	9.50	9.48	9.46	9.44	9.42
18	9.40	9.38	9.36	9.34	9.32	9.30	9.29	9.27	9.25	9.23
19	9.21	9.19	9.17	9.15	9.13	9.12	9.10	9.08	9.06	9.04
20	9.02	9.00	8.98	8.97	8.95	8.93	8.91	8.90	8.88	8.86
21	8.84	8.82	8.81	8.79	8.77	8.75	8.74	8.72	8.70	8.68
22	8.67	8.65	8.63	8.62	8.60	8.58	8.56	8.55	8.53	8.52
23	8.50	8.48	8.46	8.45	8.43	8.42	8.40	8.38	8.37	8.36
24	8.33	8.32	8.30	8.29	8.27	8.25	8.24	8.22	8.21	8.19
25	8.18	8.16	8.14	8.13	8.11	8.10	8.08	8.07	8.06	8.04
26	8.02	8.01	7.99	7.98	7.96	7.95	7.93	7.92	7.90	7.89
27	7.87	7.86	7.84	7.83	7.81	7.80	7.78	7.77	7.75	7.74
28	7.72	7.71	7.69	7.68	7.66	7.65	7.64	7.62	7.61	7.59
29	7.58	7.56	7.55	7.54	7.52	7.51	7.49	7.48	7.47	7.45
30	7.44	7.42	7.41	7.40	7.38	7.37	7.35	7.34	7.32	7.31

Oxygen concentration in water (distilled) at saturation as a function of temperature.

(ref: 765)

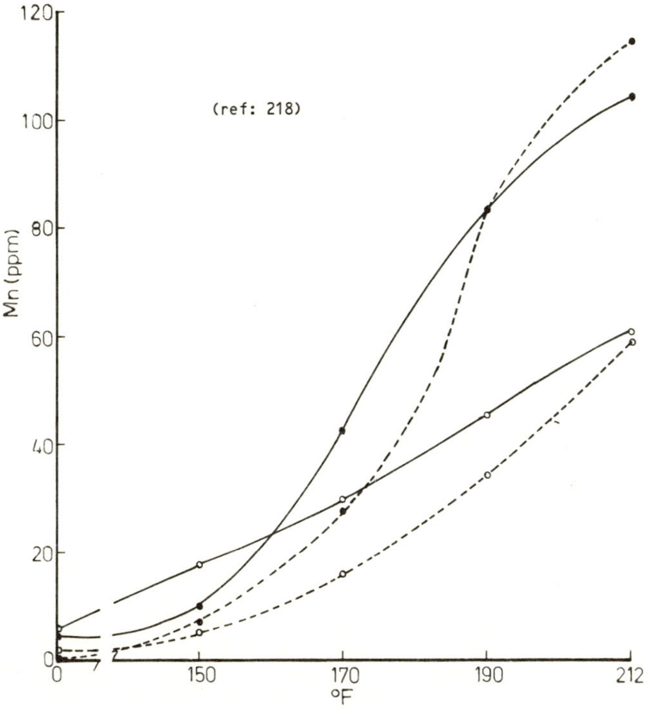

G73 Manganese (extractable) in soil at different sterilazation temperatures, ppm Mn versus F degrees.
●—— border soil (26.7.65)
●- - - border soil (9.8.65)
○—— potting mixtue
○- - - potting mixture in which plants were grown.

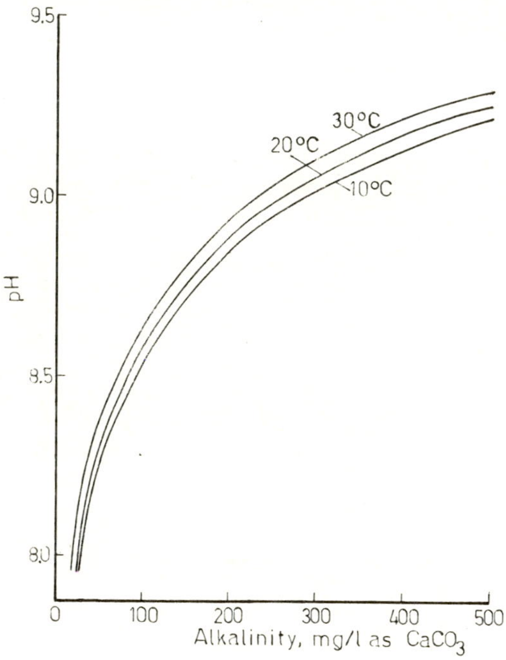

G74 Equilibrium of carbon dioxide at different alkalinity and pH.
(ref: 1606)

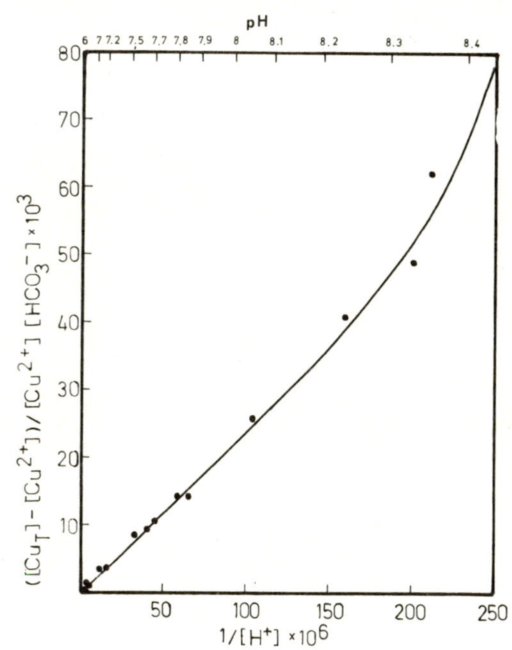

G75 Cu/HCO3(-) equilibrium in 5*10**(-3) M HCO3(-) solution.
(ref: 1606)

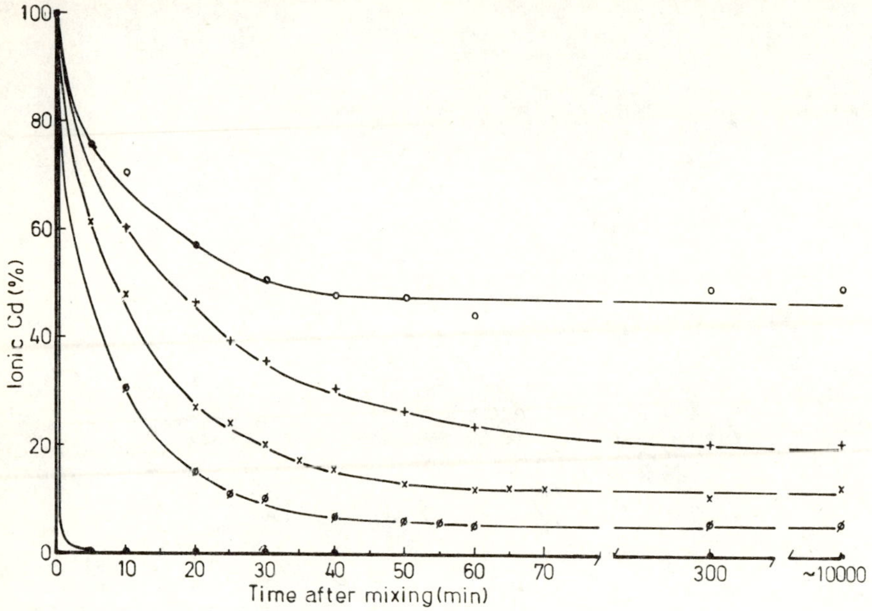

G76 Cadmium concentration (ionic) versus time after addition of different EDTA-concentrations at 25 celcius and pH - 7.8 += 0.2. All experiments in seawater.
2*10**(-6) M Cd + 2*10**(-6) M EDTA ○
2*10**(-6) M Cd + 3.5*10**(-6) M EDTA +
2*10**(-6) M Cd + 5*10**(-6) M EDTA ×
2*10**(-6) M Cd + 7*10**(-6) M EDTA ∅
2*10**(-6) M Cd + 25*10**(-6) M EDTA ●
Ionic cadmium in %. Time after mixing in minutes.

(ref: 972)

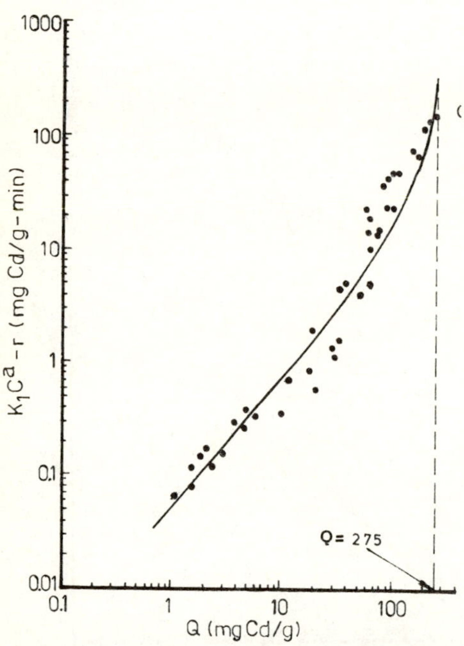

(ref: 1118)

G77 Cadmium adsorption rate versus adsorption capacity on sludge. Max capacity is indicated.

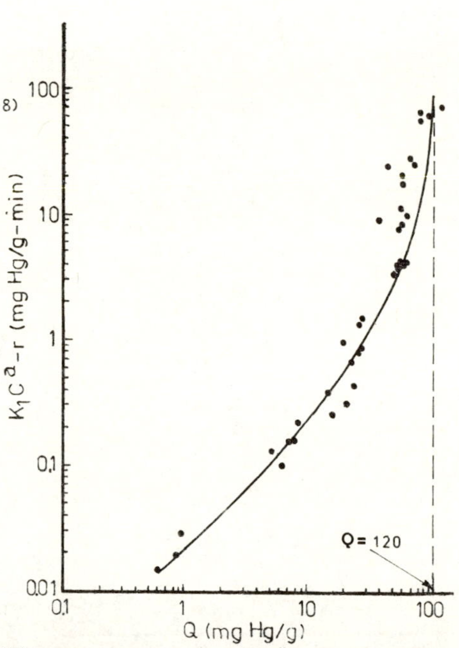

G78 Mercury adsorption rate versus adsorption capacity on sludge. Max. capacity is indicated.

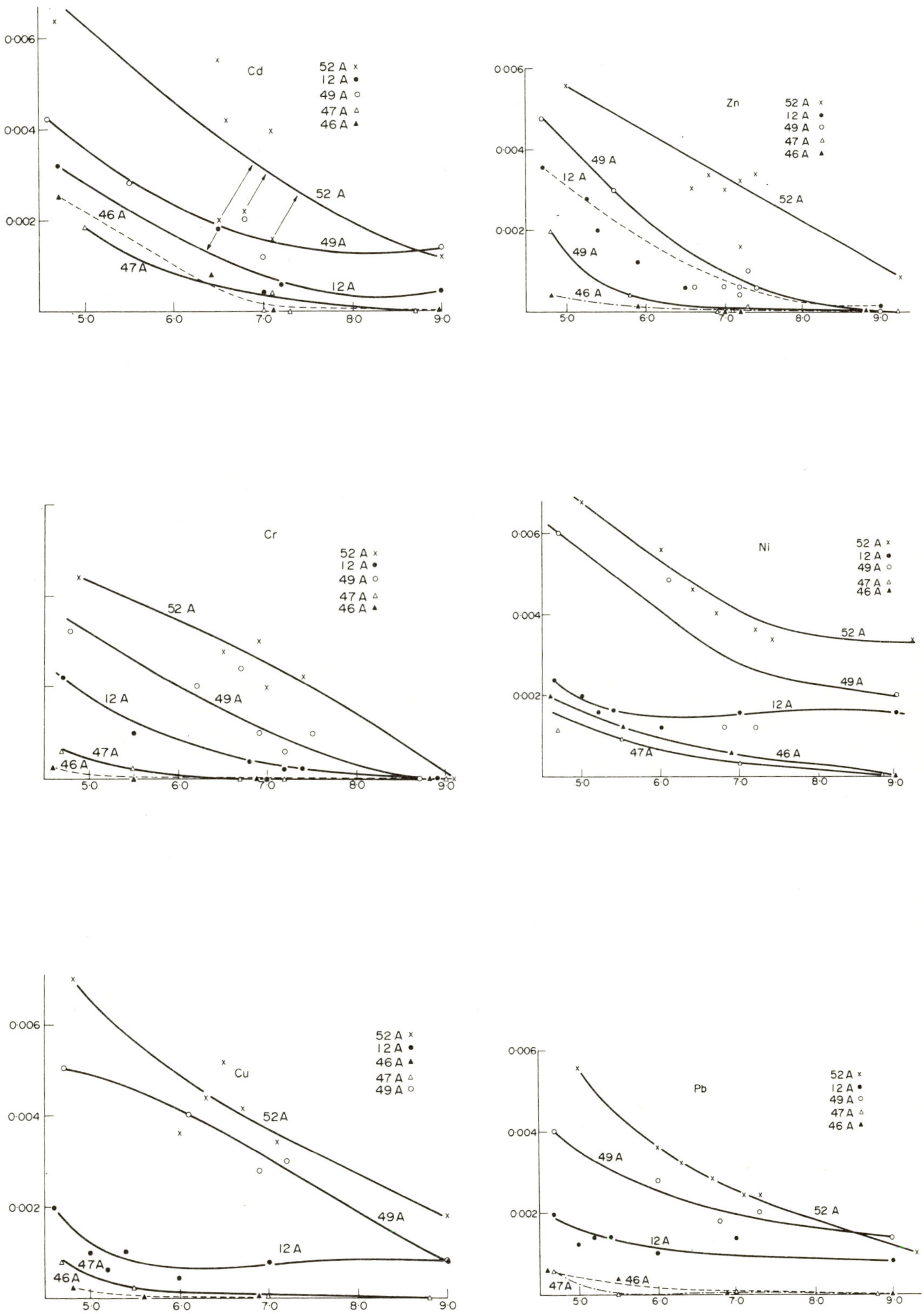
Cd
52 A ×
12 A •
49 A ○
47 A △
46 A ▲
0·006
0·004
0·002
5·0
6·0
7·0
8·0
9·0
52 A
49A
12A
46 A
47A
Zn
52 A ×
12 A •
49 A ○
47 A △
46 A ▲
49 A
12 A
52 A
46 A
Cr
52 A ×
12 A •
49 A ○
47 A △
46 A ▲
Ni
52 A ×
12 A •
49 A ○
47 A △
46 A ▲
Cu
52 A ×
12 A •
46 A ▲
47 A △
49 A ○
52A
49A
12A
47A
46A
Pb
52 A ×
12 A •
49 A ○
47 A △
46 A ▲
52 A
49 A
12A
46A
47A

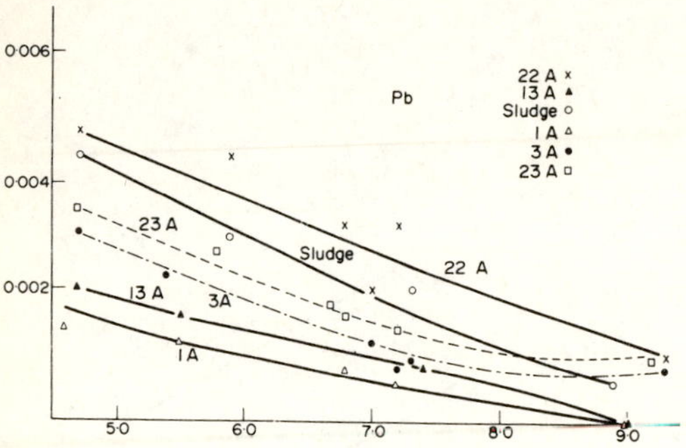
0·006
0·004
0·002
Pb
22 A x
13 A ▲
Sludge ○
1 A △
3 A ●
23 A □
23 A
Sludge
22 A
13 A
3A
1 A
5·0
6·0
7·0
8·0
9·0

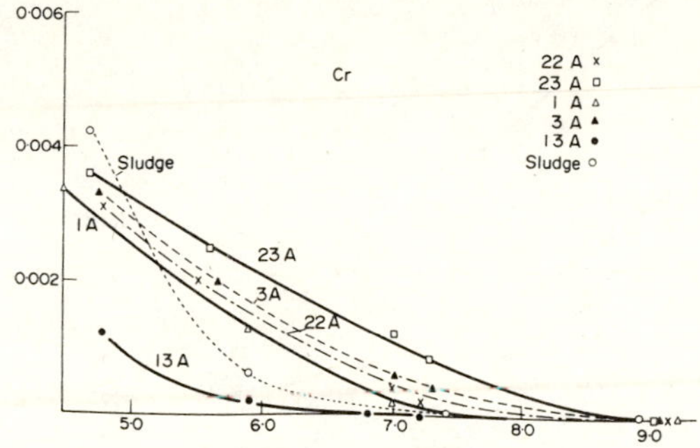
0·006
0·004
0·002
Cr
22 A x
23 A □
1 A △
3 A ▲
13 A ●
Sludge ○
Sludge
1 A
23 A
3A
22 A
13 A
5·0
6·0
7·0
8·0
9·0

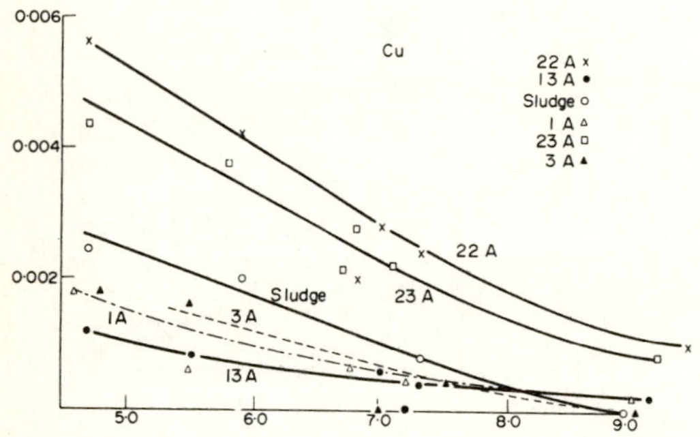
0·006
0·004
0·002
Cu
22 A x
13 A ●
Sludge ○
1 A △
23 A □
3 A ▲
22 A
Sludge
23 A
3 A
1A
13 A
5·0
6·0
7·0
8·0
9·0

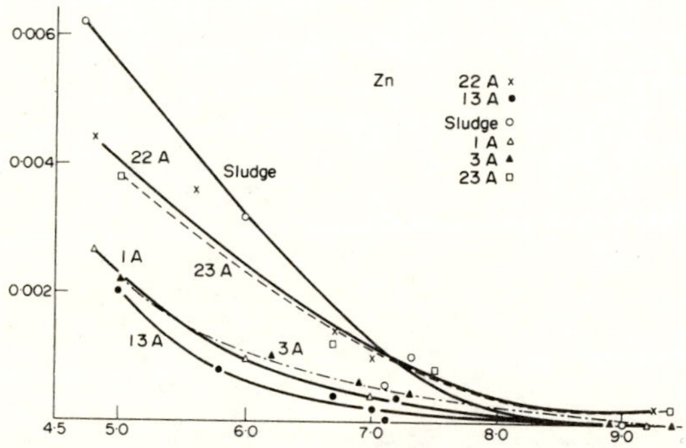
0·006
0·004
0·002
Zn
22 A x
13 A ●
Sludge ○
1 A △
3 A ▲
23 A □
22 A
Sludge
1 A
23 A
13 A
3 A
4·5
5·0
6·0
7·0
8·0
9·0

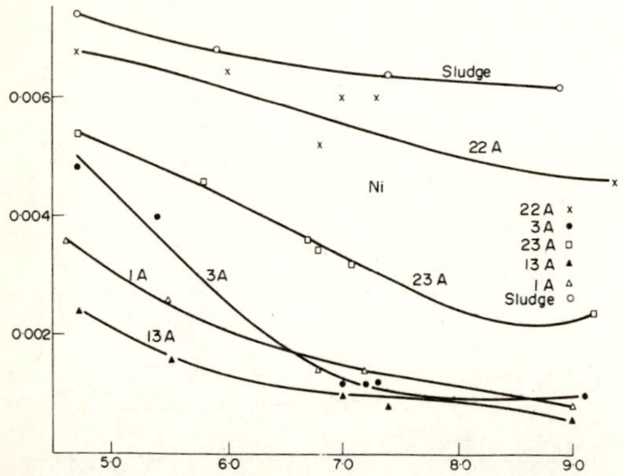
0·006
0·004
0·002
Sludge
22 A
Ni
22A x
3A ●
23 A □
13 A ▲
1 A △
Sludge ○
1A
3A
23 A
13 A
5·0
6·0
7·0
8·0
9·0

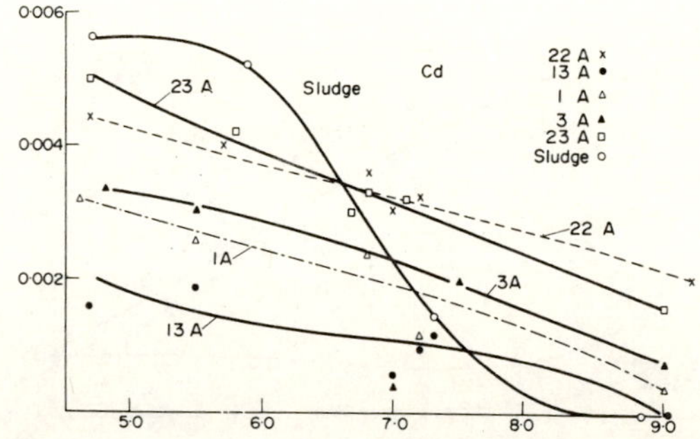
0·006
0·004
0·002
Cd
22 A x
13 A ●
1 A △
3 A ▲
23 A □
Sludge ○
23 A
Sludge
22 A
1A
3A
13 A
5·0
6·0
7·0
8·0
9·0

TABLE G79

Properties of soil and sludge used in experiment with municipal sludge adsorption
REF.: 765

Soil sample	Clay %	Humus %	CEC (meq/100 g)	Dry matter (g/l)	Ignition residue (g/kg)
47A	33.2	3.6	27.5	800	740
46A	34.4	3.9	28.0	823	757
12A	22.8	11.8	24.5	702	631
49A	11.5	3.3	13.5	908	856
52A	2.4	2.1	5.1	989	959
13a	17.8	14.3	33	700	590
1A	9.3	3.1	12.3	823	790
3A	11.3	2.3	11.7	870	830
22A	6.7	1.7	7.2	871	852
23A	7.2	2.3	9.1	829	809
Sludge				22	11
Slags				1000	1000

TABLE G080

Concentration of Ammonia (NH3 + NH4(+)) which contain an Un-ionized Ammonia Concentration of 0.025 mg NH3/l. For pH>8.9 free carbon dioxide in the water will make the values unduly low.

REF.: 435

Temperature in celcius	pH = 7.0	pH = 7.5	pH = 8.0	pH = 8.5	pH = 9.0	pH = 9.5
5	19.6	6.3	2	0.65	0.22	0.088
10	12.4	4.3	1.37	0.45	0.16	0.068
15	9.4	5.9	0.93	0.31	0.12	0.054
20	6.3	2	0.65	0.22	0.088	0.045
25	4.4	1.43	0.47	0.17	0.069	0.039
30	3.1	1	0.33	0.12	0.056	0.035

TABLE G081

Dissolved Oxygen (ppm) in Fresh, Brackish and Sea Water at different Temperatures and at different Chlorinities (%). Values are amount at Saturation.

REF.: 1493

Temperature, celcius	0 %	0.2 %	0.4 %	0.6 %	0.8 %	1 %	1.2 %	1.4 %	1.6 %	1.8 %	2.0 %
1	14.24	13.87	13.54	13.22	12.91	12.58	12.29	11.99	11.7	11.42	11.15
2	13.74	13.5	13.18	12.88	12.56	12.26	11.98	11.69	11.4	11.13	10.86
3	13.45	13.14	12.84	12.55	12.25	11.96	11.68	11.39	11.12	10.85	10.59
4	13.09	12.79	12.51	12.22	11.93	11.65	11.38	11.1	10.83	10.59	10.34
5	12.75	12.45	12.17	11.91	11.63	11.36	11.09	10.83	10.57	10.33	10.1
6	12.44	12.15	11.86	11.6	11.33	11.07	10.82	10.56	10.32	10.09	9.86
7	12.13	11.85	11.58	11.32	11.06	10.82	10.56	10.32	10.07	9.84	9.63
8	11.85	11.56	11.29	11.05	10.8	10.56	10.32	10.07	9.84	9.61	9.4
9	11.56	11.29	11.02	10.77	10.54	10.3	10.08	9.84	9.61	9.4	9.2
10	11.29	11.03	10.77	10.53	10.3	10.07	9.84	9.61	9.4	9.2	9
11	11.05	10.77	10.53	10.29	10.06	9.84	9.63	9.41	9.2	9	8.8
12	10.8	10.53	10.29	10.06	9.84	9.63	9.41	9.21	9	8.8	8.61
13	10.56	10.3	10.07	9.84	9.63	9.41	9.21	9.01	8.81	8.61	8.42
14	10.33	10.07	9.86	9.63	9.41	9.21	9.01	8.81	8.62	8.44	8.25
15	10.1	9.86	9.64	9.43	9.23	9.03	8.83	8.64	8.44	8.27	8.09
16	9.89	9.66	9.44	9.24	9.03	8.84	8.64	8.47	8.28	8.11	7.94
17	9.67	9.46	9.26	9.05	8.85	8.65	8.47	8.3	8.11	7.94	7.78
18	9.47	9.27	9.07	8.87	8.67	8.48	8.31	8.14	7.97	7.79	7.64
19	9.28	9.08	8.88	8.68	8.5	8.31	8.15	7.98	7.8	7.65	7.49
20	9.11	8.9	8.7	8.51	8.32	8.15	7.99	7.84	7.66	7.51	7.36
21	8.93	8.72	8.54	8.35	8.17	7.99	7.84	7.69	7.52	7.38	7.23
22	8.75	8.55	8.38	8.19	8.02	7.85	7.69	7.54	7.39	7.25	7.11
23	8.6	8.4	8.22	8.04	7.87	7.71	7.55	7.41	7.26	7.12	6.99
24	8.44	8.25	8.07	7.89	7.72	7.56	7.42	7.28	7.13	6.99	6.86
25	8.27	8.09	7.92	7.75	7.58	7.44	7.29	7.15	7201	6.88	6.85
26	8.12	7.94	7.78	7.62	7.45	7.31	7.16	7.03	6.89	6.86	6.63
27	7.98	7.79	7.64	7.49	7.32	7.18	7.03	6.91	6.78	6.65	6.52
28	7.84	7.65	7.51	7.36	7.19	7.06	6.92	6.79	6.66	6.53	6.4
29	7.69	7.52	7.38	7.23	7.08	6.95	6.82	6.68	6.55	6.42	6.29
30	7.56	8.39	7.25	7.12	6.96	6.83	6.7	6.58	6.45	6.32	6.19

TABLE G082

Density of Fresh, Brackish and Sea water at 15 C

REF.: 1493

Density (g/ml)	Salinity (in parts per 1000)
1.028	37.2
1.028	37.3
1.028	37.5
1.028	37.6
1.028	37.7
1.028	37.9
1.028	38
1.028	38.1
1.028	38.2
1.029	38.4
1.029	38.5
1.029	38.6
1.029	38.8
1.029	38.9
1.029	39
1.029	39.2
1.029	39.3
1.029	39.4
1.029	39.6
1.03	39.7
1.03	39.8
1.03	39.9
1.03	40.1
1.03	40.2
1.03	40.3
1.03	40.4
1.03	40.6
1.03	40.7
1.03	40.8
1.031	41
1.031	41.1
1.031	41.2
1.031	41.4
1.031	41.5
1.031	41.6
1.031	41.8
1.031	41.9
1.031	42
1.031	42.1
1.032	42.3
1.032	42.4
1.032	42.5
1.032	42.7
1.032	42.8

TABLE G082

Density of Fresh, Brackish and Sea water at 15 Celcius.

REF.: 1493

Density (g/ml)	Salinity (in parts per 1000)								
.9991	0	1.003	5.5	1.009	13.3	1.015	21.3	1.022	29.3
.9992	0.1	1.003	5.6	1.009	13.5	1.016	21.4	1.022	29.4
.9993	0.2	1.004	5.8	1.01	13.6	1.016	21.6	1.022	29.5
.9994	0.3	1.004	5.9	1.01	13.7	1.016	21.7	1.022	29.7
.9995	0.4	1.004	6	1.01	13.9	1.016	21.8	1.022	29.8
.9996	0.5	1.004	6.2	1.01	14	1.016	22	1.022	29.9
.9997	0.7	1.004	6.3	1.01	14.1	1.016	22.1	1.022	30
.9998	0.8	1.004	6.4	1.01	14.2	1.016	22.2	1.022	30.2
.9999	0.9	1.004	6.6	1.01	14.4	1.016	22.5	1.022	30.3
1	1.1	1.004	6.7	1.01	14.5	1.016	22.5	1.022	30.4
1	1.2	1.004	6.8	1.01	14.6	1.016	22.6	1.023	30.6
1	1.3	1.004	7	1.01	14.8	1.017	22.7	1.023	30.7
1	1.4	1.005	7.1	1.011	14.9	1.017	22.9	1.023	30.8
1	1.6	1.005	7.2	1.011	15	1.017	23	1.023	31
1	1.7	1.005	7.3	1.011	15.2	1.017	23.1	1.023	31.1
1.001	1.8	1.005	7.4	1.011	15.4	1.017	23.3	1.023	31.2
1.001	2	1.005	7.6	1.011	15.4	1.017	23.4	1.023	31.4
1.001	2.1	1.005	7.7	1.011	15.6	1.017	23.5	1.023	31.5
1.001	2.2	1.005	7.9	1.011	15.7	1.017	23.7	1.023	31.6
1.001	2.4	1.005	8	1.011	15.8	1.017	23.8	1.023	31.8
1.001	2.5	1.005	8.1	1.011	16	1.017	23.9	1.024	31.9
1.001	2.6	1.006	8.4	1.011	16.1	1.018	24	1.024	32
1.001	2.8	1.006	8.5	1.012	16.2	1.018	24.2	1.024	32.1
1.001	2.9	1.006	8.6	1.012	16.3	1.018	24.3	1.024	32.3
1.001	3.2	1.006	8.8	1.012	16.5	1.018	24.4	1.024	32.4
1.002	3.2	1.006	8.9	1.012	16.6	1.018	24.6	1.024	32.5
1.002	3.3	1.006	9	1.012	16.7	1.018	24.7	1.024	32.7
1.002	3.4	1.006	9	1.012	16.9	1.018	24.8	1.024	32.8
1.002	3.5	1.006	9.2	1.012	17	1.018	25	1.024	32.9
1.002	3.7	1.006	9.3	1.012	17.1	1.018	25.1	1.024	33
1.002	3.8	1.006	9.2	1.012	17.3	1.018	25.2	1.025	33.2
1.002	3.9	1.006	9.6	1.012	17.4	1.019	25.4	1.025	33.3
1.002	4.1	1.007	9.7	1.013	17.5	1.019	25.5	1.025	33.4
1.002	4.2	1.007	9.8	1.013	17.6	1.019	25.6	1.025	33.6
1.002	4.3	1.007	9.9	1.013	17.8	1.019	25.8	1.025	33.7
1.003	4.5	1.007	10.1	1.013	17.9	1.019	25.9	1.025	33.8
1.003	4.6	1.007	10.2	1.013	18	1.019	26	1.025	34
1.003	4.8	1.007	10.3	1.013	18.1	1.019	26.1	1.025	34.1
1.003	5	1.007	10.5	1.013	18.2	1.019	26.3	1.025	34.2
1.003	5.1	1.007	10.6	1.013	18.3	1.019	26.4	1.025	34.4
1.003	5.2	1.007	10.7	1.013	18.4	1.019	26.5	1.026	34.5
1.003	5.4	1.007	10.8	1.013	18.6	1.02	26.7	1.026	34.6
		1.008	11	1.013	18.7	1.02	26.8	1.026	34.7
		1.008	11.1	1.014	18.8	1.02	26.9	1.026	34.8
		1.008	11.4	1.014	19	1.02	27.1	1.026	35
		1.008	11.5	1.014	19.1	1.02	27.2	1.026	35.1
		1.008	11.6	1.014	19.2	1.02	27.3	1.026	35.3
		1.008	11.8	1.014	19.4	1.02	27.4	1.026	35.4
		1.008	11.9	1.014	19.5	1.02	27.6	1.026	35.5
		1.008	12	1.014	19.6	1.02	27.7	1.026	35.6
		1.008	12.2	1.014	19.7	1.02	27.8	1.027	35.8
		1.009	12.3	1.014	19.9	1.021	28	1.027	35.9
		1.009	12.4	1.014	20	1.021	28.1	1.027	36
		1.009	12.6	1.015	20.1	1.021	28.2	1.027	36.2
		1.009	12.7	1.015	20.3	1.021	28.4	1.027	36.3
		1.009	12.8	1.015	20.4	1.021	28.5	1.027	36.4
		1.009	12.9	1.015	20.5	1.021	28.6	1.027	36.6
		1.009	13.1	1.015	20.6	1.021	28.8	1.027	36.7
		1.009	13.2	1.015	20.8	1.021	28.9	1.027	36.8
				1.015	20.9	1.021	29	1.027	37
				1.015	21	1.021	29.1	1.028	37.1

TABLE G083

Apparent Dissociation Constant of Carbonic Acid in Seawater.
K'1 as a Function of Temperature and Salinity. Values are K'1*10**6.
REF.: 1021

Celcius.	1.9 %	2.5 %	2.7 %	2.9 %	3.1 %	3.2 %	3.3 %	3.4 %	3.5 %	3.6 %	3.7 %	3.8 %	3.9 %	4 %	4.3 %
0	0.498	0.552	0.569	0.586	0.602	0.61	0.617	0.625	0.633	0.64	0.648	0.655	0.662	0.669	0.69
1	0.512	0.567	0.585	0.601	0.618	0.626	0.634	0.642	0.65	0.657	0.665	0.672	0.68	0.687	0.708
2	0.525	0.582	0.6	0.617	0.634	0.642	0.65	0.658	0.666	0.674	0.682	0.69	0.697	0.705	0.7[illegible]7
3	0.539	0.597	0.615	0.633	0.65	0.658	0.667	0.685	0.683	0.691	0.699	0.707	0.715	0.722	0.745
4	0.552	0.612	0.63	0.648	0.666	0.675	0.683	0.692	0.7	0.708	0.716	0.724	0.732	0.74	0.763
5	0.565	0.626	0.645	0.664	0.682	0.691	0.699	0.708	0.716	0.725	0.733	0.741	0.749	0.757	0.781
6	0.579	0.641	0.66	0.679	0.698	0.707	0.716	0.724	0.733	0.741	0.75	0.758	0.766	0.775	0.798
7	0.592	0.655	0.675	0.695	0.713	0.722	0.732	0.74	0.749	0.728	0.767	0.775	0.783	0.792	0.816
8	0.615	0.67	0.69	0.71	0.729	0.738	0.747	0.757	0.766	0.774	0.783	0.795	0.8	0.819	0.833
9	0.618	0.684	0.705	0.725	0.744	0.754	0.763	0.772	0.782	0.791	0.799	0.808	0.817	0.826	0.851
10	0.631	0.698	0.819	0.74	0.729	0.769	0.779	0.788	0.797	0.807	0.816	0.825	0.833	0.842	0.868
11	0.643	0.712	0.733	0.754	0.774	0.784	0.794	0.804	0.813	0.822	0.832	0.841	0.85	0.859	0.885
12	0.656	0.726	0.747	0.769	0.789	0.799	0.809	0.819	0.828	0.838	0.847	0.857	0.866	0.875	0.901
13	0.668	0.739	0.761	0.783	0.804	0.814	0.824	0.834	0.844	0.853	0.863	0.872	0.881	0.891	0.917
14	0.68	0.752	0.775	0.797	0.818	0.828	0.838	0.848	0.858	0.868	0.878	0.887	0.897	0.906	0.933
15	0.692	0.765	0.788	0.81	0.832	0.842	0.853	0.863	0.873	0.883	0.893	0.915	0.945	0.921	0.949
16	0.704	0.778	0.804	0.824	0.846	0.856	0.867	0.877	0.887	0.897	0.907	0.917	0.927	0.93	0.964
17	0.715	0.791	0.814	0.837	0.859	0.87	0.88	0.891	0.901	0.911	0.921	0.931	0.941	0.951	0.979
18	0.726	0.803	0.827	0.85	0.872	0.883	0.894	0.904	0.915	0.925	0.935	0.945	0.955	0.965	0.994
19	0.737	0.815	0.839	0.862	0.885	0.896	0.907	0.918	0.928	0.939	0.949	0.959	0.969	0.979	1.022
20	0.748	0.826	0.851	0.874	0.897	0.908	0.919	0.93	0.941	0.952	0.962	0.972	0.982	0.992	1.022
21	0.758	0.838	0.862	0.886	0.909	0.921	0.932	0.943	0.954	0.964	0.975	0.985	0.995	1.005	1.035
22	0.768	0.848	0.873	0.898	0.921	0.932	0.944	0.955	0.966	0.976	0.987	0.998	1.008	1.018	1.048
23	0.778	0.859	0.884	0.909	0.932	0.944	0.955	0.966	0.977	0.988	0.999	1.01	1.02	1.03	1.06
24	0.787	0.869	0.895	0.919	0.943	0.955	0.966	0.977	0.989	1	1.02	1.021	1.032	1.42	1.072
25	0.796	0.879	0.905	0.929	0.953	0.965	0.977	0.988	0.999	1.011	1.021	1.032	1.043	1.053	1.084
26	0.805	0.888	0.914	0.939	0.964	0.975	0.987	0.998	1.01	1.021	1.032	1.043	1.053	1.064	1.095
27	0.813	0.897	0.923	0.949	0.973	0.985	0.997	1.008	1.02	1.031	1.042	1.053	1.064	1.074	1.105
28	0.821	0.906	0.932	0.958	0.982	0.994	1.006	1.018	1.029	1.041	1.052	1.063	1.073	1.084	1.115
29	0.829	0.914	0.941	0.966	0.991	1.003	1.015	1.027	1.038	1.05	1.061	1.072	1.083	1.093	1.125
30	0.836	0.922	0.949	0.974	0.999	1.011	1.023	1.025	1.047	1.058	1.069	1.08	1.091	1.102	1.134
31	0.843	0.929	0.956	0.982	1.007	1.019	1.031	1.043	1.055	1.066	1.077	1.089	1.1	1.11	1.142
32	0.849	0.936	0.963	0.989	1.014	1.026	1.039	1.05	1.062	1.074	1.085	1.096	1.107	1.118	1.15
33	0.855	0.943	0.97	0.996	1.021	1.033	1.045	1.057	1.069	1.081	1.092	1.103	1.121	1.125	1.157
34	0.861	0.949	0.976	1.002	1.027	1.04	1.052	1.064	1.076	1.087	1.099	1.11	1.127	1.132	1.164
35	0.866	0.954	0.981	1.008	1.033	1.045	1.058	1.07	1.082	1.093	1.105	1.116	1	1.138	1.17

TABLE G084

Apparent Dissociation Constant of Carbonic Acid in Seawater.
K'2 as a Function of Temperature and Salinity. Values are K'2*10**10.
REF.: 1021

Celcius	1.9 %	2.5 %	2.7 %	2.9 %	3.1 %	3.2 %	3.3 %	3.4 %	3.5 %	3.6 %	3.7 %	3.8 %	3.9 %	4.0 %	4.3 %
0	1.67	2.44	2.69	2.92	3.14	3.24	3.34	3.44	3.53	3.62	3.71	3.79	3.87	3.94	4.14
1	1.73	2.52	2.87	3	3.22	3.32	3.43	3.53	3.62	3.71	3.8	3.88	3.96	4.04	4.24
2	1.81	2.6	2.84	3.08	3.31	3.41	3.52	3.62	3.71	3.81	3.9	3.98	4.06	4.14	4.35
3	1.88	2.68	2.93	3.17	3.4	3.51	3.61	3.72	3.81	3.91	4	4.09	4.17	4.25	4.47
4	1.96	2.77	3.03	3.27	3.5	3.61	3.72	3.82	3.92	4.02	4.11	4.2	4.29	4.37	4.6
5	2.05	2.87	3.13	3.37	3.61	3.72	3.83	3.93	4.04	4.14	4.23	4.32	4.41	4.5	4.73
6	2.14	2.97	3.23	3.48	3.72	3.83	3.95	4.05	4.16	4.26	4.36	4.45	4.55	4.63	4.88
7	2.23	3.08	3.34	3.6	3.84	3.96	4.07	4.18	4.29	4.39	4.49	4.59	4.69	4.78	5.03
8	2.34	3.19	3.46	3.72	3.97	4.09	4.2	4.32	4.43	4.53	4.64	4.74	4.84	4.93	5.2
9	2.44	3.31	3.59	3.85	4.1	4.22	4.34	4.46	4.59	4.68	4.79	4.89	4.99	5.09	5.37
10	2.55	3.44	3.71	3.98	4.24	4.36	4.49	4.6	4.72	4.83	4.95	5.05	5.16	5.26	5.56
11	2.67	3.57	3.85	4.12	4.38	4.51	4.64	4.76	4.88	5	5.11	5.22	5.33	5.44	5.75
12	2.79	3.7	3.99	4.27	4.54	4.67	4.8	4.92	5.05	5.17	5.29	5.4	5.52	5.63	5.95
13	2.92	3.85	4.14	4.42	4.69	4.83	4.96	5.09	5.22	5.34	5.59	5.71	5.82	5.96	6.16
14	3.06	3.99	4.29	4.58	4.86	5	5.13	5.26	5.4	5.53	5.65	5.78	5.9	6.02	6.38
15	3.19	4.15	4.45	4.74	5.03	5.17	5.31	5.45	5.58	5.71	5.85	5.98	6.11	6.23	6.61
16	3.34	4.3	4.61	4.93	5.2	5.35	5.49	5.63	5.77	5.91	6.05	6.18	6.32	6.45	6.84
17	6.49	4.46	4.78	5.08	5.38	5.53	5.68	5.82	5.97	6.11	6.25	6.4	6.54	6.68	7.09
18	3.64	4.63	4.95	5.26	5.57	5.72	5.87	6.02	6.17	6.32	6.47	6.61	6.76	6.91	7.34
19	3.8	4.8	5.12	5.44	5.76	5.91	6.07	6.22	6.38	6.53	6.69	6.84	6.99	7.14	7.6
20	3.96	4.97	5.3	5.62	5.95	6.11	6.27	6.43	6.59	6.75	6.91	7.07	7.23	7.39	7.87
21	4.13	5.15	5.48	5.81	6.14	6.31	6.47	6.64	6.8	6.97	7.13	7.3	7.47	7.63	8.12
22	4.3	5.33	5.66	6	6.34	6.51	6.68	6.85	7.02	7.19	7.36	7.54	7.71	7.89	8.42
23	4.48	5.51	5.85	6.19	6.54	6.71	6.89	7.06	7.24	7.42	7.6	7.78	7.98	8.12	8.7
24	4.65	5.69	6.03	6.38	6.74	6.92	7.1	7.28	7.46	7.65	7.83	8.02	8.21	8.4	8.98
25	4.82	5.87	6.22	6.58	6.94	7.12	7.31	7.49	7.68	7.87	8.07	8.26	8.46	8.66	9.27
26	5	6.05	6.41	6.77	7.14	7.33	7.52	7.71	7.9	8.1	8.3	8.5	8.71	8.92	9.3
27	5.18	6.23	6.59	6.96	7.34	7.53	7.72	7.92	8.12	8.33	8.54	8.75	8.96	9.18	9.85
28	5.36	6.41	6.77	7.15	7.53	7.73	7.93	8.13	8.24	8.55	8.77	8.99	9.21	9.44	10.15
29	5.53	6.58	6.95	7.33	7.72	7.93	8.12	8.34	8.56	8.77	9	9.22	9.46	9.69	10.43
30	5.71	6.75	7.12	7.51	7.91	8.12	8.33	8.54	8.76	8.99	9.22	9.46	9.7	9.94	10.72
31	5.88	6.92	7.29	7.68	8.09	8.3	8.52	8.74	8.97	9.2	9.44	9.68	9.93	10.19	11
32	6.04	7.08	7.45	7.85	8.27	8.48	8.7	8.93	9.16	9.4	9.65	9.9	10.16	10.43	11.27
33	6.2	7.23	7.61	8.01	8.43	8.65	8.88	9.11	9.35	9.6	9.85	10.11	10.38	10.66	11.54
34	6.26	7.37	7.75	8.16	8.59	8.81	9.04	9.28	9.53	9.78	10.04	10.31	10.59	10.88	11.79
35	6.5	7.51	7.89	8.3	8.73	8.96	9.2	9.44	9.69	9.96	10.22	10.5	10.79	11.09	12.04

G4

RATE OF PROCESSES IN THE ENVIRONMENT

MASS BALANCES
RATE CONSTANTS
DEGRADATION IN WATER

TABLE G85

MASS BALANCE : ESTUARINE BAY ECOSYSTEM

ITEM:	VALUE:	CONDITIONS:	REF.:
*COMMERCIAL AND SPORT FISHERY HARVEST	0.006 G ORG MATTER/(M**2*DAY)	EXCHANGE RATES IN ESTUARINE BAY SYSTEM AFFECTED BY A POWER PLANT	1005
*COMMUNITY GROSS PRIMARY PRODUCTION	8.25 G ORG MATTER/(M**2*DAY)	EXCHANGE RATES IN ESTUARINE BAY SYSTEM AFFECTED BY A POWER PLANT	1005
*DEATH RATE OF PRODUCERS	3.0 G ORG MATTER/(M**2*DAY)	EXCHANGE RATES IN ESTUARINE BAY SYSTEM AFFECTED BY A POWER PLANT	1005
*MACROFAUNA CONSUMPTION OF DETRITUS	2.0 G ORG MATTER/(M**2*DAY)	EXCHANGE RATES IN ESTUARINE BAY SYSTEM AFFECTED BY A POWER PLANT	1005
*MACROFAUNA CONSUMPTION OF ZOOPLANKTON	0.025 G ORG MATTER/(M**2*DAY)	EXCHANGE RATES IN ESTUARINE BAY SYSTEM AFFECTED BY A POWER PLANT	1005
*MACROFAUNA DEATH AND FAECES PRODUCTION	1.7 G ORG MATTER/(M**2*DAY)	EXCHANGE RATES IN ESTUARINE BAY SYSTEM AFFECTED BY A POWER PLANT	1005
*MACROFAUNA GRAZING ON PHYTOPLANKTON	1.0 G ORG MATTER/(M**2*DAY)	EXCHANGE RATES IN ESTUARINE BAY SYSTEM AFFECTED BY A POWER PLANT	1005
*MACROFAUNA RESPIRATION	1.38 G ORG MATTER/(M**2*DAY)	EXCHANGE RATES IN ESTUARINE BAY SYSTEM AFFECTED BY A POWER PLANT	1005
*PHOSPHORUS RECYCLE FROM TOTAL RESPIRATION	0.08 G TOTAL P/(M**2*DAY)	EXCHANGE RATES IN ESTUARINE BAY SYSTEM AFFECTED BY A POWER PLANT	1005
*PHOSPHORUS UPTAKE BY PRODUCERS	0.08 G TOTAL P/(M**2*DAY)	EXCHANGE RATES IN ESTUARINE BAY SYSTEM AFFECTED BY A POWER PLANT	1005
*PRODUCER RESPIRATION	4.13 G ORG MATTER/(M**2*DAY)	EXCHANGE RATES IN ESTUARINE BAY SYSTEM AFFECTED BY A POWER PLANT	1005
*RESPIRATION OF DETRITUS AND BACTERIA	2.48 G ORG MATTER/(M**2*DAY)	EXCHANGE RATES IN ESTUARINE BAY SYSTEM AFFECTED BY A POWER PLANT	1005
*TOTAL COMMUNITY RESPIRATION	8.40 G ORG MATTER/(M**2*DAY)	EXCHANGE RATES IN ESTUARINE BAY SYSTEM AFFECTED BY A POWER PLANT	1005
*ZOOPLANKTON CONSUMPTION OF DETRITUS	0.08 G ORG MATTER/(M**2*DAY)	EXCHANGE RATES IN ESTUARINE BAY SYSTEM AFFECTED BY A POWER PLANT	1005
*ZOOPLANKTON DEATH AND FAECES PRODUCTION	0.025 G ORG MATTER/(M**2*DAY)	EXCHANGE RATES IN ESTUARINE BAY SYSTEM AFFECTED BY A POWER PLANT	1005
*ZOOPLANKTON GRAZING ON PHYTOPLANKTON	0.08 G ORG MATTER/(M**2*DAY)	EXCHANGE RATES IN ESTUARINE BAY SYSTEM AFFECTED BY A POWER PLANT	1005
*ZOOPLANKTON RESPIRATION	0.11 G ORG MATTER/(M**2*DAY)	EXCHANGE RATES IN ESTUARINE BAY SYSTEM AFFECTED BY A POWER PLANT	1005

TABLE G86

PROBABILITY : PROBABILITY OF OCCURING AS A SPONTANEOUS FLUCTATIONS

ITEM:	VALUE:	CONDITIONS:	REF.:
*OCCURING OF AMINO ACID	10**(-60)	TOTAL FLUCTATIONS = 10**134	1067
*OCCURING OF DIALISTER PNEUMOSINTES	10**((-10)**10)	TOTAL FLUCTATIONS = 10**134	1067
*OCCURING OF ESCHERICHIA COLI	10**((-10)**11)	TOTAL FLUCTATIONS = 10**134	1067
*OCCURING OF HEMOGLOBIN MOLECULE	10**(-4*10**6)	TOTAL FLUCTATIONS = 10**134	1067
*OCCURING OF MYCOPLASMA HOMINIS H39	10**((-5*10**9))	TOTAL FLUCTATIONS = 10**134	1067
*OCCURING OF RIBONUCLEASE MOLECULE	10**(-8000)	TOTAL FLUCTATIONS = 10**134	1067
*OCCURING OF T2 PHAGE	10**(-2*10**8)	TOTAL FLUCTATIONS = 10**134	1067

TABLE G87

RATE CONSTANT : CARBON			
ITEM:	VALUE:	CONDITIONS:	REF.:
*CARBON DIOXIDE, COMBUSTION	1.5*10**16 G/YR	GLOBAL,ATMOSPHERIC CONTENT/RATE= 150-450YR	779
*CARBON DIOXIDE, FOSSILIZATION	10**14 G/YR	GLOBAL,ATMOSPHERIC CONTENT/RATE= 2*10**4YR	779
*CARBON DIOXIDE, PHOTOSYNTHESIS	1.1*10**17 G/YR	GLOBAL,ATMOSPHERIC CONTENT/RATE= 20YR	779

TABLE G88

RATE CONSTANT : FE			
ITEM:	VALUE:	CONDITIONS:	REF.:
*2*FE(2+)+5H2O+1/2O2<-> FE(OH)3(S)+ 4H(+),K=B*(P(O2))*(OH)**2)	0.02438 1/MIN	1. ORDER, ALK = 1.12 MEQ/L,BUFFER CAPACITY =0.48 MEQ/PH	520
*2*FE(2+)+5H2O+1/2O2<-> FE(OH)3(S)+ 4H(+),K=B*(P(O2))*(OH)**2)	0.02166 1/MIN	1. ORDER, ALK = 4.4 MEQ/L,BUFFER CAPACITY =1.84 MEQ/PH	520
*2*FE(2+)+5H2O+1/2O2<-> FE(OH)3(S)+ 4H(+),K=B*(P(O2))*(OH)**2)	0.02659 1/MIN	1. ORDER, ALK = 6.0 MEQ/L,BUFFER CAPACITY =2.48 MEQ/PH	520
*2*FE(2+)+5H2O+1/2O2<-> FE(OH)3(S)+ 4H(+),K=B*(P(O2))*(OH)**2)	0.02549 1/MIN	1. ORDER, ALK = 12 MEQ/L,BUFFER CAPACITY =48 MEQ/PH	520
*2*FE(2+)+5H2O+1/2O2<-> FE(OH)3(S)+ 4H(+),K=B*(P(O2))*(OH)**2)	0.03150 1/MIN	1. ORDER, ALK = 18 MEQ/L,BUFFER CAPACITY =72 MEQ/PH	520
*2*FE(2+)+5H2O+1/2O2<-> FE(OH)3(S)+ 4H(+),K=B*(P(O2))*(OH)**2)	0.03759 1/MIN	1. ORDER, ALK = 28 MEQ/L,BUFFER CAPACITY =10.7 MEQ/PH	520
*2*FE(2+)+5H2O+1/2O2<-> FE(OH)3(S)+ 4H(+),K=B*(P(O2))*(OH)**2)	0.04200 1/MIN	1. ORDER, ALK = 40.7 MEQ/L,BUFFER CAPACITY =15.2 MEQ/PH	520

TABLE G89

RATE CONSTANT : FE-PROCESSES			
ITEM:	VALUE:	CONDITIONS:	REF.:
*2 FE(2+) + 5 H2O + 1/2 O2 <-> FE (OH)3(SOLID) + 4 H(+)	5.0+-0.4*10**16 LITER**2/ (MOL**2*ATM*MIN)	ALKALINITY=28-40.7 MEQ/L, 298 K, ACTIVATION ENERGY = 23 KCAL	520
*2 FE(2+) + 5 H2O + 1/2 O2 <-> FE (OH)3(SOLID) + 4 H(+)	3.8*10**16 LITER**3/(MOL**3*LITER)	ALKALINITY=28-40.7 MEQ/L, 298 K, ACTIVATION ENERGY = 23 KCAL	520
*2 FE(2+) + 5 H2O + 1/2 O2 <-> FE (OH)3(SOLID) + 4 H(+)	3.8*10**(-2) MOL/(MIN*LITER)	ALKALINITY=28-40.7 MEQ/L, 298 K, ACTIVATION ENERGY = 23 KCAL	520
*2 FE(2+) + 5 H2O + 1/2 O2 <-> FE (OH)3(SOLID) + 4 H(+)	14+-5 *10**13 LITER**2/ (MOL**2*ATM*MIN)	ALKALINITY=29-39 MEQ/L, 298 K, ACTIVATION ENERGY = 23 KCAL	1525
*2 FE(2+) + 5 H2O + 1/2 O2 <-> FE (OH)3(SOLID) + 4 H(+)	10.8+-4.5 *10**16 LITER**3/ (MOL**3*MIN)	ALKALINITY=29-39 MEQ/L, 298 K, ACTIVATION ENERGY = 23 KCAL	1525
*2 FE(2+) + 5 H2O + 1/2 O2 <-> FE (OH)3(SOLID) + 4 H(+)	5.4*10**(-12) MOL/(MIN*LITER)	ALKALINITY=29-39 MEQ/L, 298 K, ACTIVATION ENERGY = 23 KCAL	1525
*2 FE(2+) + 5 H2O + 1/2 O2 <-> FE (OH)3(SOLID) + 4 H(+)	2.5*10**13 LITER**2/ (MOL**2*ATM*MIN)	ALKALINITY=18-32 MEQ/L, 298 K, ACTIVATION ENERGY = 23 KCAL	1064
*2 FE(2+) + 5 H2O + 1/2 O2 <-> FE (OH)3(SOLID) + 4 H(+)	(2 - 7)*10**16 LITER**3/ (MOL**3*MIN)	ALKALINITY=18-32 MEQ/L, 298 K, ACTIVATION ENERGY = 23 KCAL	1064
*2 FE(2+)+5H2O+1/2O2<-> FE(OH)3(S)+ 4H(+)	3.07 LITER**2/MOL**2*ATM*MIN)	ALKALINITY = 1.12 MEQ/L,BUFFER CAPACITY =0.48 MEQ/PH	520
*2 FE(2+)+5H2O+1/2O2<-> FE(OH)3(S)+ 4H(+)	2.73 LITER**2/MOL**2*ATM*MIN)	ALKALINITY = 4.4 MEQ/L,BUFFER CAPACITY =1.84 MEQ/PH	520
*2 FE(2+)+5H2O+1/2O2<-> FE(OH)3(S)+ 4H(+)	3.45 LITER**2/MOL**2*ATM*MIN)	ALKALINITY = 6.0 MEQ/L,BUFFER CAPACITY =2.48 MEQ/PH	520
*2 FE(2+)+5H2O+1/2O2<-> FE(OH)3(S)+ 4H(+)	3.21 LITER**2/MOL**2*ATM*MIN)	1. ORDER, ALK = 12 MEQ/L,BUFFER CAPACITY =48 MEQ/PH	520
*2 FE(2+)+5H2O+1/2O2<-> FE(OH)3(S)+ 4H(+)	3.96 LITER**2/MOL**2*ATM*MIN)	1. ORDER, ALK = 18 MEQ/L,BUFFER CAPACITY =72 MEQ/PH	520
*2 FE(2+)+5H2O+1/2O2<-> FE(OH)3(S)+ 4H(+)	4.73 LITER**2/MOL**2*ATM*MIN)	1. ORDER, ALK = 28 MEQ/L,BUFFER CAPACITY =10.7 MEQ/PH	520

TABLE G89

RATE CONSTANT : FE-PROCESSES			
ITEM:	VALUE:	CONDITIONS:	REF.:
*2 FE(2+)+5H2O+1/2O2<-> FE(OH)3(S)+ 4H(+)	5.29 LITER**2/MOL**2*ATM*MIN)	1. ORDER, ALK = 40.7 MEQ/L,BUFFER CAPACITY =15.2 MEQ/PH	520
*2 FE(2+)+5H2O+1/2O2<-> FE(OH)3(S)+ 4H(+)	2.80 LITER**2/MOL**2*ATM*MIN)	1. ORDER, ALK=0.3 MEQ/L,BUFFER CAPACITY=1.32 EQ/PH	520
2 FE(2+)+5H2O+1/2O2<-> FE(OH)3(S)+ 4H(+)K=B(P(O2))*(OH)**2)	0.02231 1/MIN	1. ORDER, ALK=0.3 MEQ/L,BUFFER CAPACITY=1.32 EQ/PH	520

TABLE G90

RATE CONSTANT : HCO3 AND CO2			
ITEM:	VALUE:	CONDITIONS:	REF.:
*HCO3(-) <-> CO2 + OH(-)	8500 LITERS/(MOLE*SEC)	IN WATER, 298 K	510
*H2CO3(-) <-> CO2 + H2O	0.03 1/SEC	IN WATER, 298 K	510

TABLE G91

RATE CONSTANT : HYDROXYL RADICALS AND ORGANIC COMPOUNDS			
ITEM:	VALUE:	CONDITIONS:	REF.:
*OH + C2H6	1.76*10**8 L/(MOLE*SEC)	IN AIR, 300 K	18
*OH + C3H8	8.2*10**8 L/(MOLE*SEC)	IN AIR, 300 K	18
*OH + HCHO	> 4*10**9 L/(MOLE*SEC)	IN AIR, 300 K	18
*OH + HCHO	6*10**7 L/(MOLE*SEC)	IN AIR, 300 K	18
*OH + ISO-C4H10	1.28*10**9 L/(MOLE*SEC)	IN AIR, 300 K	18

TABLE G92

RATE CONSTANT : METHOXYCHLOR			
ITEM:	VALUE:	CONDITIONS:	REF.:
*SORPTION ON ASPERGILLUS SP.	1.3 (+-0.75)*10**(-2) L/(MG*HOUR)	AT EQUILIBRIUM, 1. ORDER REACTION WITH CONC OF ORGANISM AND PESTICIDE	1211
*SORPTION ON BACILLUS SUBTILIS	2.4 (+-1.8)*10**(-2) L/(MG*HOUR)	AT EQUILIBRIUM, 1. ORDER REACTION WITH CONC OF ORGANISM AND PESTICIDE	1211
*SORPTION ON CHLORELLA PYRENOIDOSA	1.2 (+-1.00)*10**(-2) L/(MG*HOUR)	AT EQUILIBRIUM, 1. ORDER REACTION WITH CONC OF ORGANISM AND PESTICIDE	1211
*SORPTION ON FLAVOBACTERIUM HARRISONII	4.2 (+-4.0)*10**(-3) L/(MG*HOUR)	AT EQUILIBRIUM, 1. ORDER REACTION WITH CONC OF ORGANISM AND PESTICIDE	1211

TABLE G93

RATE CONSTANT : OXYGEN-PROCESSES			
ITEM:	VALUE:	CONDITIONS:	REF.:
*DEGRADATION OF ORGANIC MATTER IN WATER	0.1 - 2.5 1/DAY	1. ORDER, 293 K, 1976	55
*DEGRADATION OF ORGANIC MATTER IN WATER	0.2 1/DAY	1. ORDER, 293 K, N. FORK KINGS RIVER, CALIF.	1566

TABLE G93

RATE CONSTANT : OXYGEN-PROCESSES			
ITEM:	VALUE:	CONDITIONS:	REF.:
*DEGRADATION OF ORGANIC MATTER IN WATER	0.1 - 1.5 1/DAY	1. ORDER, 293 K, 1974	1566
*DEGRADATION OF ORGANIC MATTER IN WATER	0.2 1/DAY	1. ORDER, 293 K, LAKE WASHINGTON	240
*DEGRADATION OF ORGANIC MATTER IN WATER	0.06 - 0.36 1/DAY	1. ORDER, 293 K, 1972	944
*DEGRADATION OF ORGANIC MATTER IN WATER	0.01 - 0.1 1/DAY	1. ORDER, 293 K, MERRIMACK RIVER	219
*DEGRADATION OF ORGANIC MATTER IN WATER	0.2 - 2.0 1/DAY	1. ORDER, 293 K, 1971	712
*DEGRADATION OF ORGANIC MATTER IN WATER	0.25 1/DAY	1. ORDER, 293 K, BOISE RIVER	241
*DEGRADATION OF ORGANIC MATTER IN WATER	0.2 1/DAY	1. ORDER, 293 K, SAN FRANCISCO BAY ESTUARY	242
*DEGRADATION OF ORGANIC MATTER IN WATER	0.2 1/DAY	1. ORDER, 293 K, 1974	1576
*DEGRADATION OF ORGANIC MATTER IN WATER	0.1 - 0.3 1/DAY	1. ORDER, 293 K, 1976	1566
*DEGRADATION OF ORGANIC MATTER IN WATER	0.23 1/DAY	293 K, BASE E, 1973	299
*O + CH3CHO	4.6*10**8 L/(MOLE*SEC)	ATOMIC OXYGEN AND ORGANIC COMPOUNDS, IN AIR, 300 K	18
*O + CH3CHO	2.4*10**8 L/(MOLE*SEC)	ATOMIC OXYGEN AND ORGANIC COMPOUNDS, IN AIR, 300 K	18
*O + CH4	3.3*10**6 L/(MOLE*SEC)	ATOMIC OXYGEN AND ORGANIC COMPOUNDS, IN AIR, 300 K	18
*O + CH4	0.085*10**6 L/(MOLE*SEC)	ATOMIC OXYGEN AND ORGANIC COMPOUNDS, IN AIR, 300 K	18
*O + CH4	0.7*18**4 L/(MOLE*SEC)	ATOMIC OXYGEN AND ORGANIC COMPOUNDS, IN AIR, 300 K	18
*O + C2H6	5.6*10**5 L/(MOLE*SEC)	ATOMIC OXYGEN AND ORGANIC COMPOUNDS, IN AIR, 300 K	18
*O + HCHO	2.1*10**7 L/(MOLE*SEC)	ATOMIC OXYGEN AND ORGANIC COMPOUNDS, IN AIR, 300 K	18
*O + HCHO	9*10**7 L/(MOLE*SEC)	ATOMIC OXYGEN AND ORGANIC COMPOUNDS, IN AIR, 300 K	18
*O + HCHO	2.1*10**8 L/(MOLE*SEC)	ATOMIC OXYGEN AND ORGANIC COMPOUNDS, IN AIR, 300 K	18
*O + N-C4H10	3.3*10**7 L/(MOLE*SEC)	ATOMIC OXYGEN AND ORGANIC COMPOUNDS, IN AIR, 300 K	18
*O + N-C4H10	2.5*10**7 L/(MOLE*SEC)	ATOMIC OXYGEN AND ORGANIC COMPOUNDS, IN AIR, 300 K	18
*O + N-C4H10	1.44 - 1.59*10**7 L/(MOLE*SEC)	ATOMIC OXYGEN AND ORGANIC COMPOUNDS, IN AIR, 300 K	18
*O + N-C5H12	4.02 - 4.72*10**7 L/(MOLE*SEC)	ATOMIC OXYGEN AND ORGANIC COMPOUNDS, IN AIR, 300 K	18
*O + N-C6H14	6.19 - 6.87*10**7 L/(MOLE*SEC)	ATOMIC OXYGEN AND ORGANIC COMPOUNDS, IN AIR, 300 K	18

TABLE G94

RATE CONSTANT : PHOSPHATE TRANSFORMATIONS IN SOIL			
ITEM:	VALUE:	CONDITIONS:	REF.:
*ADSORPTION OF P FROM WATER SOLUBLE FRACTION	10 1/HOUR	P APPLIED TO SOIL	979
*ADSORPTION OF P FROM WATER SOLUBLE FRACTION	10 1/HOUR	P APPLIED TO SOIL	979
*DESORPTION OF P TO WATER SOLUBLE FRACTION	1 1/HOUR	P APPLIED TO SOIL	979
*DESORPTION OF P TO WATER SOLUBLE FRACTION	1 1/HOUR	P APPLIED TO SOIL	979
*TRANSFORMATION OF ADSORBED P TO IMMOBILIZED PHOSPHATE	10**(-2) 1/HOUR	P APPLIED TO SOIL	979

TABLE G94

RATE CONSTANT : PHOSPHATE TRANSFORMATIONS IN SOIL			
ITEM:	VALUE:	CONDITIONS:	REF.:
*TRANSFORMATION OF ADSORBED P TO IMMOBILIZED PHOSPHATE	10**(-2) 1/HOUR	P APPLIED TO SOIL	979
*TRANSFORMATION OF IMMOBILIZED P TO ADSORBED PHOSPHATE	10**(-3) 1/HOUR	P APPLIED TO SOIL	979
*TRANSFORMATION OF IMMOBILIZED P TO ADSORBED PHOSPHATE	10**(-3) 1/HOUR	P APPLIED TO SOIL	979
*TRANSFORMATION OF PRECIPATED P TO WATER SOLUBLE PHOSPHATE	10**(-2) 1/HOUR	P APPLIED TO SOIL	979
*TRANSFORMATION OF PRECIPATED P TO WATER SOLUBLE PHOSPHATE	10**(-2) 1/HOUR	P APPLIED TO SOIL	979

TABLE G95

RATE CONSTANT (1. ORDER) : BOD			
ITEM:	VALUE:	CONDITIONS:	REF.:
*BOD DECAY COEFFICIENT	0.01 1/DAY-CELCIUS	POTOMAC RIVER	1153
*BOD DECAY RATE	0.1-1.5 1/DAY	NONE	1571

TABLE G96

RATE CONSTANT (1. ORDER) : CCL3F			
ITEM:	VALUE:	CONDITIONS:	REF.:
*TRANSFER BETWEEN NORTHERN AND SOUTHERN TROPOSPHERE	0.9 1/YR	SEE REF.	1108
*TRANSFER FROM AIR TO WATER	0.02 1/YR	CORRECTED FOR DEPTH	1108
*TRANSFER FROM WATER TO AIR	7.60 1/YR	CORRECTED FOR DEPTH	1108
*TRANSFER TO STRATOSPHERE	0.03 1/YR	SEE REF.	1108

TABLE G97

RATE CONSTANT (1. ORDER) : CCL4			
ITEM:	VALUE:	CONDITIONS:	REF.:
*INPUT TO NORTH TROPOSPHERE	60000 METRIC TONS/YR	INPUT-RATE CONSTANT (1. ORDER)FOR 42YRS	1108
*TRANSFER BETWEEN HEMISPHERES	0.9 1/YR	SEE REF.	1108
*TRANSFER FROM AIR TO WATER	9.28*10**(-2) 1/YR	AVERAGE MIXING DEPTH=50M	1108
*TRANSFER FROM WATER TO AIR	1.85 1/YR	AVERAGE MIXING DEPTH=50M	1108
*TRANSFER TO STRATOSPHERE	0.03 1/YR	SEE REF.	1108

TABLE G98

RATE CONSTANT (1. ORDER) : CO			
ITEM:	VALUE:	CONDITIONS:	REF.:
*CO + HO2 <-> CO2 + OH	2.5*10**(-17)	IN AIR, CONCENTRATIONS IN MOLECULES/CC	864
*CO + NO2 <-> CO2 + NO	2*10**(-11)*EXP(-16000/T)	IN AIR, CONCENTRATIONS IN MOLECULES/CC, T=KELVIN DEG.	864
*CO + NO2(2B1) <-> CO2 + NO	1*10**(-12)	IN AIR, CONCENTRATIONS IN MOLECULES/CC	864
*CO + N2O <-> CO2 + O2	2.4*10**1*EXP(-6000/T)	IN AIR, CONCENTRATIONS IN MOLECULES/CC, T=KELVIN DEG.	864
*CO + N2O <-> CO2 + N2	1*10**(-25)	IN AIR, CONCENTRATIONS IN MOLECULES/CC	864
*CO + O + M <-> CO2 + M	1*10**(-30)*EXP(-1750/T)	IN AIR, CONCENTRATIONS IN MOLECULES/CC, T=KELVIN DEG.	864
*CO + O(1D) <-> M <-> CO2 + M	LESS THAN 1*10**(-31)	IN AIR, CONCENTRATIONS IN MOLECULES/CC	864
*CO + OH <-> CO2 + H	1.1*10**(-12)*EXP(-520/T)	IN AIR, CONCENTRATIONS IN MOLECULES/CC, T=KELVIN DEG.	864
*CO + O2(DELTA(G)) <-> CO2 + O	LESS THAN 9*10**(-27)	IN AIR, CONCENTRATIONS IN MOLECULES/CC	864
*CO + O2(SIGMA(G)) <-> CO2 + O	LESS THAN 2*10**(-25)	IN AIR, CONCENTRATIONS IN MOLECULES/CC	864
*CO + O3 <-> CO2 + O2	LESS THAN 2*10**(-26)	IN AIR, CONCENTRATIONS IN MOLECULES/CC	864

TABLE G99

RATE CONSTANT (1. ORDER) : COLIFORM			
ITEM:	VALUE:	CONDITIONS:	REF.:
*COLIFORM DECAY RATE	0.5-2. 1/DAY	NONE	1571

TABLE G100

RATE CONSTANT (1. ORDER) : DDE			
ITEM:	VALUE:	CONDITIONS:	REF.:
*DECLINE RATE OF CONCENTRATION IN MILK FAT	0.0145 1/DAY	MID LACTATION, COW	490
*DECLINE RATE OF CONCENTRATION IN MILK FAT	0.0107 1/DAY	EARLY LACTATION, COW	490
*DECLINE RATE OF CONCENTRATION IN MILK FAT	0.0147 1/DAY	LATE LACTATION, COW	490
*DECLINE RATE OF CONCENTRATION IN MILK FAT	0.0134 1/DAY	MEAN LACTATION, COW	490

TABLE G101

RATE CONSTANT (1. ORDER) : DDT			
ITEM:	VALUE:	CONDITIONS:	REF.:
*LOSS OF DDT FROM ANISOPTERA	7.6*10**4 1/HOUR	FRESHWATER MARSH, LONG TERM COMPARTMENT	385
*LOSS OF DDT FROM CERATOPHYLLUM DEMERSUM	15.0*10**4 1/HOUR	FRESHWATER MARSH, LONG TERM COMPARTMENT	385
*LOSS OF DDT FROM CLADOPHORA	59.0*10**4 1/HOUR	FRESHWATER MARSH, LONG TERM COMPARTMENT	385
*LOSS OF DDT FROM CYPRINUS CARPIO	7.8*10**4 1/HOUR	FRESHWATER MARSH, LONG TERM COMPARTMENT	385
*LOSS OF DDT FROM ERPOBDELLA PUNCTATA	16.0*10**4 1/HOUR	FRESHWATER MARSH, LONG TERM COMPARTMENT	385
*LOSS OF DDT FROM LEMNA MINOR	55.0*10**4 1/HOUR	FRESHWATER MARSH, LONG TERM COMPARTMENT	385
*LOSS OF DDT FROM LEPOMIS CYANELLUS	5.9*10**4 1/HOUR	FRESHWATER MARSH, LONG TERM COMPARTMENT	385
*LOSS OF DDT FROM MYRIOPHYLLUM SP.	7.8*10**4 1/HOUR	FRESHWATER MARSH, LONG TERM COMPARTMENT	385
*LOSS OF DDT FROM NOTONECTIDAE	28.0*10**4 1/HOUR	FRESHWATER MARSH, LONG TERM COMPARTMENT	385
*LOSS OF DDT FROM ORCONECTES IMMUNIS	13.0*10**4 1/HOUR	FRESHWATER MARSH, LONG TERM COMPARTMENT	385
*LOSS OF DDT FROM PLANORBIDAE	32.0*10**4 1/HOUR	FRESHWATER MARSH, LONG TERM COMPARTMENT	385
*LOSS OF DDT FROM POTAMOGETON PECTINATUS	1.8*10**4 1/HOUR	FRESHWATER MARSH, LONG TERM COMPARTMENT	385
*LOSS OF DDT FROM POTOMOGETON CRISPUS	0.9*10**4 1/HOUR	FRESHWATER MARSH, LONG TERM COMPARTMENT	385
*LOSS OF DDT FROM POTOMOGETON FOLIOSUS	25.0*10**4 1/HOUR	FRESHWATER MARSH, LONG TERM COMPARTMENT	385
*LOSS OF DDT FROM SCIRPUS VALIDUS	39*10**4 1/HOUR	FRESHWATER MARSH, LONG TERM COMPARTMENT	385
*LOSS OF DDT FROM SPARGANIUM EURYCARPUM	5.7*10**4 1/HOUR	FRESHWATER MARSH, LONG TERM COMPARTMENT	385
*LOSS OF DDT FROM TENDIPES SP.	17.0*10**4 1/HOUR	FRESHWATER MARSH, LONG TERM COMPARTMENT	385
*LOSS OF DDT FROM UTRICULARIA VULGARIS	12.0*10**4 1/HOUR	FRESHWATER MARSH, LONG TERM COMPARTMENT	385
*UPTAKE OF DDT IN CERATOPHYLLUM DEMERSUM	13.0 1/HOUR	FRESHWATER MARSH, LONG TERM COMPARTMENT, TRANSFER COEFFICIENT	385
*UPTAKE OF DDT IN CLADOPHORA	130.0 1/HOUR	FRESHWATER MARSH, LONG TERM COMPARTMENT, TRANSFER COEFFICIENT	385

TABLE G101

RATE CONSTANT (1. ORDER) : DDT			
ITEM:	VALUE:	CONDITIONS:	REF.:
*UPTAKE OF DDT IN ERPOBDELLA PUNCTATA	77.0 1/HOUR	FRESHWATER MARSH, LONG TERM COMPARTMENT, TRANSFER COEFFICIENT	385
*UPTAKE OF DDT IN LEMNA MINOR	45.0 1/HOUR	FRESHWATER MARSH, LONG TERM COMPARTMENT, TRANSFER COEFFICIENT	385
*UPTAKE OF DDT IN MYRIOPHYLLUM SP.	8.0 1/HOUR	FRESHWATER MARSH, LONG TERM COMPARTMENT, TRANSFER COEFFICIENT	385
*UPTAKE OF DDT IN NOTONECTIDAE	30.0 1/HOUR	FRESHWATER MARSH, LONG TERM COMPARTMENT, TRANSFER COEFFICIENT	385
*UPTAKE OF DDT IN ORCONECTES IMMUNIS	27.0 1/HOUR	FRESHWATER MARSH, LONG TERM COMPARTMENT, TRANSFER COEFFICIENT	385
*UPTAKE OF DDT IN PLANORBIDAE	13.0 1/HOUR	FRESHWATER MARSH, LONG TERM COMPARTMENT, TRANSFER COEFFICIENT	385
*UPTAKE OF DDT IN POTAMOGETON PECTINATUS	9.1 1/HOUR	FRESHWATER MARSH, LONG TERM COMPARTMENT, TRANSFER COEFFICIENT	385
*UPTAKE OF DDT IN POTOMOGETON CRISPUS	7.3 1/HOUR	FRESHWATER MARSH, LONG TERM COMPARTMENT, TRANSFER COEFFICIENT	385
*UPTAKE OF DDT IN POTOMOGETON FOLIOSUS	7.5 1/HOUR	FRESHWATER MARSH, LONG TERM COMPARTMENT, TRANSFER COEFFICIENT	385
*UPTAKE OF DDT IN SCIRPUS VALIDUS	13.0 1/HOUR	FRESHWATER MARSH, LONG TERM COMPARTMENT, TRANSFER COEFFICIENT	385
*UPTAKE OF DDT IN SPARGANIUM EURYCARPUM	3.5 1/HOUR	FRESHWATER MARSH, LONG TERM COMPARTMENT, TRANSFER COEFFICIENT	385
*UPTAKE OF DDT IN TENDIPES SP.	42.0 1/HOUR	FRESHWATER MARSH, LONG TERM COMPARTMENT, TRANSFER COEFFICIENT	385
*UPTAKE OF DDT IN UTRICULARIA VULGARIS	14.0 1/HOUR	FRESHWATER MARSH, LONG TERM COMPARTMENT, TRANSFER COEFFICIENT	385

TABLE G102

RATE CONSTANT (1. ORDER) : DETRITUS			
ITEM:	VALUE:	CONDITIONS:	REF.:
*DETRITUS DECAY RATE	0.001-0.02 1/DAY	NONE	1571
*MINERALIZATION	0.03 1/DAY	SEE REF FOR DETAILS	1469
*MINERALIZATION	0.001-0.12 1/DAY	NONE	958
*MINERALIZATION	0.003-0.02 1/DAY	SEE REF FOR DETAILS	720

TABLE G103

RATE CONSTANT (1. ORDER) : DISSOLVED ORGANIC CARBON			
ITEM:	VALUE:	CONDITIONS:	REF.:
*DEGRADATION IN SEAWATER	0.25 1/DAY	PERIOD=0-2 D, DECREASE=40 %, TOKYO BAY, 298 K, DARK 0 MIN, AUG	1176
*DEGRADATION IN SEAWATER	0.098 1/DAY	PERIOD=0-3 D, DECREASE=25 %, TOKYO BAY, 298 K, DARK 20 MIN, AUG	1176
*DEGRADATION IN SEAWATER	0.051 1/DAY	PERIOD=0-5 D, DECREASE=26 %, TOKYO BAY, 298 K, DARK 0 MIN, NOV	1176
*DEGRADATION IN SEAWATER	0.038 1/DAY	PERIOD=0-5 D, DECREASE=18 %, TOKYO BAY, 298 K, DARK 16 MIN, NOV	1176
*DEGRADATION IN SEAWATER	0.046 1/DAY	PERIOD=0-5 D, DECREASE=21 %, TOKYO BAY, 298 K, DARK 0 MIN, MARCH	1176
*DEGRADATION IN SEAWATER	0.090 1/DAY	PERIOD=0-4 D, DECREASE=31 %, TOKYO BAY, 298 K, DARK 0 MIN, JUNE	1176
*DEGRADATION IN SEAWATER	0.049 1/DAY	PERIOD=0-5 D, DECREASE=21 %, TOKYO BAY, 298 K, DARK 0 MIN, AUG	1176
*DEGRADATION IN SEAWATER	0.11 1/DAY	PERIOD=0-2 D, DECREASE=20 %, TOKYO BAY, 298 K, DARK 20 MIN, AUG	1176
*DEGRADATION IN SEAWATER	0.070 1/DAY	PERIOD=0-2 D, DECREASE=14 %, TOKYO BAY, 298 K, DARK 0 MIN, AUG	1176

TABLE G103

RATE CONSTANT (1. ORDER) : DISSOLVED ORGANIC CARBON			
ITEM:	VALUE:	CONDITIONS:	REF.:
*DEGRADATION IN SEAWATER	0.033 1/DAY	PERIOD=0-7 D, DECREASE=20 %, TOKYO BAY, 298 K, DARK 0 MIN, AUG	1176
*DEGRADATION IN SEAWATER	0.0062 1/DAY	PERIOD=2-40 D, DECREASE=40 %, TOKYO BAY, 298 K, DARK 0 MIN, AUG	1176
*DEGRADATION IN SEAWATER	0.0039 1/DAY	PERIOD=3-40 D, DECREASE=25 %, TOKYO BAY, 298 K, DARK 20 MIN, AUG	1176
*DEGRADATION IN SEAWATER	0.0064 1/DAY	PERIOD=6-40 D, DECREASE=26 %, TOKYO BAY, 298 K, DARK 0 MIN, NOV	1176
*DEGRADATION IN SEAWATER	0.0045 1/DAY	PERIOD=6-40	1176
*DEGRADATION IN SEAWATER	0.0085 1/DAY	PERIOD=5-40 D, DECREASE=21 %, TOKYO BAY, 298 K, DARK 0 MIN, MARCH	1176
*DEGRADATION IN SEAWATER	0.0060 1/DAY	PERIOD=5-40 D, DECREASE=31 %, TOKYO BAY, 298 K, DARK 0 MIN, JUNE	1176
*DEGRADATION IN SEAWATER	0.0026 1/DAY	PERIOD=6-41 D, DECREASE=21 %, TOKYO BAY, 298 K, DARK 0 MIN, AUG	1176
*DEGRADATION IN SEAWATER	0.0041 1/DAY	PERIOD=3-41 D, DECREASE=20 %, TOKYO BAY, 298 K, DARK 20 MIN, AUG	1176
*DEGRADATION IN SEAWATER	0.0074 1/DAY	PERIOD=2-41 D, DECREASE=14 %, TOKYO BAY, 298 K, DARK 0 MIN, AUG	1176
*DEGRADATION IN SEAWATER	0.0017 1/DAY	PERIOD=7-40 D, DECREASE=20 %, TOKYO BAY, 298 K, DARK 0 MIN, AUG	1176
*DEGRADATION IN SEAWATER	0.15 1/DAY	PERIOD=0-2 D, DECREASE=26 %, TOKYO BAY, 298 K, DARK 0 MIN, AUG	1176
*DEGRADATION IN SEAWATER	0.095 1/DAY	PERIOD=0-1 D, DECREASE=9 %, TOKYO BAY, 298 K, DARK 20 MIN, AUG	1176
*DEGRADATION IN SEAWATER	0.070 1/DAY	PERIOD=0-2 D, DECREASE=13 %, TOKYO BAY, 298 K, DARK 40 MIN, AUG	1176
*DEGRADATION IN SEAWATER	0.0069 1/DAY	PERIOD=2-40 D, DECREASE=26 %, TOKYO BAY, 298 K, DARK 0 MIN, AUG	1176
*DEGRADATION IN SEAWATER	0.0045 1/DAY	PERIOD=1-40 D, DECREASE=9 %, TOKYO BAY, 298 K, DARK 20 MIN, AUG	1176
*DEGRADATION IN SEAWATER	0.0027 1/DAY	PERIOD=2-40 D, DECREASE=13 %, TOKYO BAY, 298 K, DARK 40 MIN, AUG	1176

TABLE G104

RATE CONSTANT (1. ORDER) : NITROGEN			
ITEM:	VALUE:	CONDITIONS:	REF.:
* ALGAE-N -> DETRITUS-N	0.12 1/DAY	POTEMAC ESTUARY,JULY-AUGUST,1968, 301K	1153
*AMMONIA -> NITRATE	0.01 1/DAY-CELCIUS	POTOMAC RIVER	1153
*AMMONIA -> NITRATE	0.002 1/DAY-CELCIUS	LAKE ONTARIO	1577
*AMMONIA DECAY RATIO	0.05-0.2 1/DAY	NONE	1571
*NH3 -> ALGAE-N	0.02 1/DAY	POTEMAC ESTUARY,JULY-AUGUST,1968, 301K	1153
* NH3 -> NO3	0.28 1/DAY	POTEMAC ESTUARY,JULY-AUGUST,1968, 301K	1153
*AMMONIUM -> NITRITE	0.01 1/DAY	280.5K,DELAWARE ESTUARY	1153
*DECAY OF ALGAE-N	0.12 1/DAY	POTEMAC ESTUARY,JULY-AUGUST,1968, 301K	1153
*NH3 -> NO3	0.3 1/DAY	POTEMAC ESTUARY,JULY-AUGUST,1968, 301K	1153
*DECAY OF NITRATE ->N	0.1 1/DAY	POTEMAC ESTUARY,JULY-AUGUST,1968, 301K	1153
*DECAY OF ORGANIC-N	0.1 1/DAY	POTOMAC RIVER	1153
*DECAY OF ORGANIC-N -> NH3	0.2 1/DAY	POTEMAC ESTUARY,JULY-AUGUST,1968, 301K	1153

TABLE G104

RATE CONSTANT (1. ORDER) : NITROGEN			
ITEM:	VALUE:	CONDITIONS:	REF.:
*DENITRIFICATION	2940 MICROGRAM-ATOM/(L*YEAR)	IN OCEANS, 149-347 M, 14 DEG N, 100 DEG W	259
*DENITRIFICATION	2755 MICROGRAM-ATOM/(L*YEAR)	IN OCEANS, 294-417 M, 19 DEG N, 105 DEG W	259
*DENITRIFICATION	60 MICROGRAM-ATOM/)L*YEAR)	DARWIN BAY, 40-120 M, 0 DEG N, 89 DEG W	259
* NO3 -> ALGAE-N	0.1 1/DAY	POTEMAC ESTUARY,JULY-AUGUST,1968, 301K	1153
*NO2 -> NO3	0.05 1/DAY	280.5K,DELAWARE ESTUARY	1153
*NO2 -> NO3	0.2-0.5 1/DAY	NONE	1571
*ORG. -> AMMONIUM	0.025 1/DAY	280.5K,DELAWARE ESTUARY	1153
*ORG.-N -> AMMONIA	0.007 1/DAY-CELCIUS	POTOMAC RIVER	1153
*ORG.-N -> AMMONIA	0.00175 1/DAY-CELCIUS	LAKE ONTARIO	1577
*ORGANICN -> NH3	0.1 1/DAY	POTEMAC ESTUARY,JULY-AUGUST,1968, 301K	1153

TABLE G105

RATE CONSTANT (1. ORDER) : ORGANIC MATTER			
ITEM:	VALUE:	CONDITIONS:	REF.:
*MINERALIZATION	0.0001-0.0008 1/DAY	UPPER SEDIMENT	958
*MINERALIZATION	0.00002 1/DAY	LOWER SEDIMENT	958
*MINERALIZATION	0.000032 1/DAY	LOWER SEDIMENT	1527

TABLE G106

RATE CONSTANT (1. ORDER) : PARTICULATE ORGANIC CARBON			
ITEM:	VALUE:	CONDITIONS:	REF.:
*ANAEROBIC DECOMPOSITION OF GREEN ALGAE	0.0088 PER DAY (SD=0.0067)	293K,0-35DAYS	1190
*ANAEROBIC DECOMPOSITION OF GREEN ALGAE	0.0051 PER DAY (SD=0.0041)	303K,10-40DAYS	1190
*DEGRADATION OF DEAD GREEN ALGAE IN WATER	0.0156 1/DAY	KILLED BY FREEZING, 273 K, FIRST 30 D	1189

TABLE G107

RATE CONSTANT (1. ORDER) : PARTICULATE ORGANIC NITROGEN			
ITEM:	VALUE:	CONDITIONS:	REF.:
*ANAEROBIC DECOMPOSITION OF GREEN ALGAE	0.0156 PER DAY (SD=0.0015)	293K,0-35DAYS	1190
*ANAEROBIC DECOMPOSITION OF GREEN ALGAE	0.0099 PER DAY (SD=0.0040)	303K,10-40DAYS	1190
*DEGRADATION OF DEAD GREEN ALGAE IN WATER	0.0296 1/DAY	KILLED BY FREEZING, 273 K, FIRST 30 D	1189

TABLE G108

RATE CONSTANT (1. ORDER) : PHOSPHORUS			
ITEM:	VALUE:	CONDITIONS:	REF.:
*DECAY OF ORG.-P	0.1 1/DAY	POTOMAC RIVER	1153

TABLE G108

RATE CONSTANT (1. ORDER) : PHOSPHORUS			
ITEM:	VALUE:	CONDITIONS:	REF.:
*ORG.-P -> INORG.-P	0.007 1/DAY-CELCIUS	POTOMAC RIVER	1153
*ORG.-P -> INORG.-P	0.007 1/DAY-CELCIUS	LAKE ONTARIO	1577

TABLE G109

RATE CONSTANT (1. ORDER) : ZN-65			
ITEM:	VALUE:	CONDITIONS:	REF.:
*ELIMINATION FROM CRASSOSTREA GIGAS	5.49*10**(-3) 1/DAY	WILLAPA BAY	312
*UPTAKE BY CRASSOSTREA GIGAS	4.93*10**(-3) 1/DAY	WILLAPA BAY	312

TABLE G110

RATE OF DEGRADATION : AUTOLYSIS			
ITEM:	VALUE:	CONDITIONS:	REF.:
*N	50 - 75 %	10-15 DAYS, CONVERSION TO AMMONIA, PART. NITROGEN	551
*P	50 %	FEW HOURS, PART. PHOSPHORUS	551

TABLE G111

RATE OF DEGRADATION : BIOLOGICAL DEGRADATION IN SOIL			
ITEM:	VALUE:	CONDITIONS:	REF.:
*BENZO(A)ANTRACENE	0 %	DEG IN 8 D, SOIL WITHOUT BACTERIA	632
*BENZO(A)ANTRACENE	53 %	DEG IN 8 D, SOIL WITH N5 BACTERIAL STRAIN	632
*BENZO(A)ANTRACENE	66 %	DEG IN 8 D, SOIL WITH N13 BACTERIAL STRAIN	632
*BENZO(A)ANTRACENE	82 %	DEG IN 8 D, SOIL WITH N13 BAC STRAIN, IN B(A)P FOR 110 D	632
*BIRCH LEAVES	46 % IN 1 YEAR	IN LITTER BAGS, PH = 4.0	656
*BIRCH LEAVES	48 % IN 1 YEAR	IN LITTER BAGS, PH = 4.3 - 5.5, NATURAL STREAM	656
*BIRCH LEAVES	52 % IN 1 YEAR	IN LITTER BAGS, PH = 6.0	656

TABLE G112

RATE OF DEGRADATION : BIOLOGICAL DEGRADATION IN WATER			
ITEM:	VALUE:	CONDITIONS:	REF.:
*ACETANILIDE	14.7 MG COD/(GRAM*HOUR)	INIT COD= 200 MG/L,NO OTHER SOURCE OF C,AEROB,293+-3 K	1263
*ALKYL BENZENE SULFONATE IN PRESENCE OF ACTIVATED SLUDGE	0.77 MG SURFACT./L*DAY	21DAYS,BATCH CULTURE,INITIAL CONC. =20MG SURFACTANT/L	332
*ALKYL BENZENE SULFONATE IN PRESENCE OF ANABAENA CYLINDRICA	0.20 MG SURFACT./L*DAY	21DAYS,BATCH CULTURE,INITIAL CONC. =20MG SURFACTANT/L	332
*ALKYL BENZENE SULFONATE IN PRESENCE OF ANABAENA VARIABILIS	0.25 MG SURFACT./L*DAY	21DAYS,BATCH CULTURE,INITIAL CONC. =20MG SURFACTANT/L	332
*ALKYL BENZENE SULFONATE IN PRESENCE OF ANACYSTIS NIDULANS	+0.15 MG SURFACT./L*DAY	21DAYS,BATCH CULTURE,INITIAL CONC. =20MG SURFACTANT/L,INCREASE	332
*ALKYL BENZENE SULFONATE IN PRESENCE OF ANKISTRODESMUS BRAUNII	0.15 MG SURFACT./L*DAY	21DAYS,BATCH CULTURE,INITIAL CONC. =20MG SURFACTANT/L	332
*ALKYL BENZENE SULFONATE IN PRESENCE OF CALOTHRIX PARIETINA	0.20 MG SURFACT./L*DAY	21DAYS,BATCH CULTURE,INITIAL CONC. =20MG SURFACTANT/L	332
*ALKYL BENZENE SULFONATE IN PRESENCE OF CHLORELLA PYRENOIDOSA	0.14 MG SURFACT./L*DAY	21DAYS,BATCH CULTURE,INITIAL CONC. =20MG SURFACTANT/L	332
*ALKYL BENZENE SULFONATE IN PRESENCE OF CHLORELLA VULGARIS	0.18 MG SURFACT./L*DAY	21DAYS,BATCH CULTURE,INITIAL CONC. =20MG SURFACTANT/L	332
*ALKYL BENZENE SULFONATE IN PRESENCE OF CYLINDROSPERMUM SP.	0.20 MG SURFACT./L*DAY	21DAYS,BATCH CULTURE,INITIAL CONC. =20MG SURFACTANT/L	332
*ALKYL BENZENE SULFONATE IN PRESENCE OF GLOEOCAPSA ALPICOLA	0.19 MG SURFACT./L*DAY	21DAYS,BATCH CULTURE,INITIAL CONC. =20MG SURFACTANT/L	332
*ALKYL BENZENE SULFONATE IN PRESENCE OF NOSTOC MUSCORUM	0.11 MG SURFACT./L*DAY	21DAYS,BATCH CULTURE,INITIAL CONC. =20MG SURFACTANT/L	332
*ALKYL BENZENE SULFONATE IN PRESENCE OF OSCILLATORIA BORNETI	0.12 MG SURFACT./L*DAY	21DAYS,BATCH CULTURE,INITIAL CONC. =20MG SURFACTANT/L	332
*ALKYL BENZENE SULFONATE IN PRESENCE OF OSCILLATORIA CHALYBIA	0.27 MG SURFACT./L*DAY	21DAYS,BATCH CULTURE,INITIAL CONC. =20MG SURFACTANT/L	332
*ALKYL BENZENE SULFONATE IN PRESENCE OF OSCILLATORIA FORMOSA	0.15 MG SURFACT./L*DAY	21DAYS,BATCH CULTURE,INITIAL CONC. =20MG SURFACTANT/L	332
*ALKYL BENZENE SULFONATE IN PRESENCE OF OSCILLATORIA TENUIS	0.19 MG SURFACT./L*DAY	21DAYS,BATCH CULTURE,INITIAL CONC. =20MG SURFACTANT/L	332

TABLE G112

RATE OF DEGRADATION : BIOLOGICAL DEGRADATION IN WATER

ITEM:	VALUE:	CONDITIONS:	REF.:
*ALKYL BENZENE SULFONATE IN PRESENCE OF PHORMIDIUM FAVEOLARUM	0.22 MG SURFACT./L*DAY	21DAYS,BATCH CULTURE,INITIAL CONC. =20MG SURFACTANT/L	332
*ALKYL BENZENE SULFONATE IN PRESENCE OF POND WATER	0.83 MG SURFACT./L*DAY	21DAYS,BATCH CULTURE,INITIAL CONC. =20MG SURFACTANT/L	332
*ALKYL BENZENE SULFONATE IN PRESENCE OF RAW WASTEWATER	0.77 MG SURFACT./L*DAY	21DAYS,BATCH CULTURE,INITIAL CONC. =20MG SURFACTANT/L	332
*ALKYL BENZENE SULFONATE IN PRESENCE OF SCENEDESMUS OBLIQUUS	0.09 MG SURFACT./L*DAY	21DAYS,BATCH CULTURE,INITIAL CONC. =20MG SURFACTANT/L	332
*ALKYL BENZENE SULFONATE IN PRESENCE OF SCENEDESMUS QUADRICAUDA	0.25 MG SURFACT./L*DAY	21DAYS,BATCH CULTURE,INITIAL CONC. =20MG SURFACTANT/L	332
*ALKYL BENZENE SULFONATE IN PRESENCE OF SOIL	0.85 MG SURFACT./L*DAY	21DAYS,BATCH CULTURE,INITIAL CONC. =20MG SURFACTANT/L	332
*ALKYL BENZENE SULFONATE IN PRESENCE OF ULOTHRIX FIMBRIATA	0.17 MG SURFACT./L*DAY	21DAYS,BATCH CULTURE,INITIAL CONC. =20MG SURFACTANT/L	332
*ALKYL POLYETHOXYLATE IN PRESENCE OF ACTIVATED SLUDGE	0.79 MG SURFACT./L*DAY	21DAYS,BATCH CULTURE,INITIAL CONC. =20MG SURFACTANT/L, AE7,0	332
*ALKYL POLYETHOXYLATE IN PRESENCE OF ACTIVATED SLUDGE	0.70 MG SURFACT./L*DAY	21DAYS,BATCH CULTURE,INITIAL CONC. =20MG SURFACTANT/L, AE7.4	332
*ALKYL POLYETHOXYLATE IN PRESENCE OF ANABAENA CYLINDRICA	0.68 MG SURFACT./L*DAY	21DAYS,BATCH CULTURE,INITIAL CONC. =20MG SURFACTANT/L, AE7,0	332
*ALKYL POLYETHOXYLATE IN PRESENCE OF ANABAENA CYLINDRICA	0.63 MG SURFACT./L*DAY	21DAYS,BATCH CULTURE,INITIAL CONC. =20MG SURFACTANT/L, AE7.4	332
*ALKYL POLYETHOXYLATE IN PRESENCE OF ANABAENA VARIABILIS	0.51 MG SURFACT./L*DAY	21DAYS,BATCH CULTURE,INITIAL CONC. =20MG SURFACTANT/L, AE7,0	332
*ALKYL POLYETHOXYLATE IN PRESENCE OF ANABAENA VARIABILIS	0.70 MG SURFACT./L*DAY	21DAYS,BATCH CULTURE,INITIAL CONC. =20MG SURFACTANT/L, AE7.4	332
*ALKYL POLYETHOXYLATE IN PRESENCE OF ANACYSTIS NIDULANS	0.75 MG SURFACT./L*DAY	21DAYS,BATCH CULTURE,INITIAL CONC. =20MG SURFACTANT/L, AE7,0	332
*ALKYL POLYETHOXYLATE IN PRESENCE OF ANACYSTIS NIDULANS	0.52 MG SURFACT./L*DAY	21DAYS,BATCH CULTURE,INITIAL CONC. =20MG SURFACTANT/L, AE7.4	332
*ALKYL POLYETHOXYLATE IN PRESENCE OF ANKISTRODESMUS BRAUNII	0.36 MG SURFACT./L*DAY	21DAYS,BATCH CULTURE,INITIAL CONC. =20MG SURFACTANT/L, AE7,0	332
*ALKYL POLYETHOXYLATE IN PRESENCE OF ANKISTRODESMUS BRAUNII	0.49 MG SURFACT./L*DAY	21DAYS,BATCH CULTURE,INITIAL CONC. =20MG SURFACTANT/L, AE7.4	332
*ALKYL POLYETHOXYLATE IN PRESENCE OF CALOTHRIX PARIETINA	0.87 MG SURFACT./L*DAY	21DAYS,BATCH CULTURE,INITIAL CONC. =20MG SURFACTANT/L, AE7,0	332
*ALKYL POLYETHOXYLATE IN PRESENCE OF CALOTHRIX PARIETINA	0.79 MG SURFACT./L*DAY	21DAYS,BATCH CULTURE,INITIAL CONC. =20MG SURFACTANT/L, AE7.4	332
*ALKYL POLYETHOXYLATE IN PRESENCE OF CHLORELLA PYRENOIDOSA	0.96 MG SURFACT./L*DAY	21DAYS,BATCH CULTURE,INITIAL CONC. =20MG SURFACTANT/L, AE7,0	332
*ALKYL POLYETHOXYLATE IN PRESENCE OF CHLORELLA PYRENOIDOSA	0.84 MG SURFACT./L*DAY	21DAYS,BATCH CULTURE,INITIAL CONC. =20MG SURFACTANT/L, AE7.4	332
*ALKYL POLYETHOXYLATE IN PRESENCE OF CHLORELLA VULGARIS	0.54 MG SURFACT./L*DAY	21DAYS,BATCH CULTURE,INITIAL CONC. =20MG SURFACTANT/L, AE7,0	332
*ALKYL POLYETHOXYLATE IN PRESENCE OF CHLORELLA VULGARIS	0.67 MG SURFACT./L*DAY	21DAYS,BATCH CULTURE,INITIAL CONC. =20MG SURFACTANT/L, AE7.4	332
*ALKYL POLYETHOXYLATE IN PRESENCE OF CYLINDROSPERMUM SP.	0.60 MG SURFACT./L*DAY	21DAYS,BATCH CULTURE,INITIAL CONC. =20MG SURFACTANT/L, AE7,0	332
*ALKYL POLYETHOXYLATE IN PRESENCE OF CYLINDROSPERMUM SP.	0.78 MG SURFACT./L*DAY	21DAYS,BATCH CULTURE,INITIAL CONC. =20MG SURFACTANT/L, AE7.4	332
*ALKYL POLYETHOXYLATE IN PRESENCE OF GLOEOCAPSA ALPICOLA	0.53 MG SURFACT./L*DAY	21DAYS,BATCH CULTURE,INITIAL CONC. =20MG SURFACTANT/L, AE7,0	332
*ALKYL POLYETHOXYLATE IN PRESENCE OF GLOEOCAPSA ALPICOLA	0.55 MG SURFACT./L*DAY	21DAYS,BATCH CULTURE,INITIAL CONC. =20MG SURFACTANT/L, AE7.4	332
*ALKYL POLYETHOXYLATE IN PRESENCE OF NOSTOC MUSCORUM	0.59 MG SURFACT./L*DAY	21DAYS,BATCH CULTURE,INITIAL CONC. =20MG SURFACTANT/L, AE7,0	332
*ALKYL POLYETHOXYLATE IN PRESENCE OF NOSTOC MUSCORUM	0.56 MG SURFACT./L*DAY	21DAYS,BATCH CULTURE,INITIAL CONC. =20MG SURFACTANT/L, AE7.4	332
*ALKYL POLYETHOXYLATE IN PRESENCE OF OSCILLATORIA BORNETI	0.31 MG SURFACT./L*DAY	21DAYS,BATCH CULTURE,INITIAL CONC. =20MG SURFACTANT/L, AE7,0	332
*ALKYL POLYETHOXYLATE IN PRESENCE OF OSCILLATORIA BORNETI	0.59 MG SURFACT./L*DAY	21 DAYS,BATCH CULTURE,INTIAL CONC. =20MG SURFACTANT/L, AE7.4	332
*ALKYL POLYETHOXYLATE IN PRESENCE OF OSCILLATORIA CHALYBIA	0.73 MG SURFACT./L*DAY	21DAYS,BATCH CULTURE,INITIAL CONC. =20MG SURFACTANT/L, AE7,0	332
*ALKYL POLYETHOXYLATE IN PRESENCE OF OSCILLATORIA CHALYBIA	0.82 MG SURFACT./L*DAY	21DAYS,BATCH CULTURE,INITIAL CONC. =20MG SURFACTANT/L, AE7.4	332

TABLE G112

RATE OF DEGRADATION : BIOLOGICAL DEGRADATION IN WATER			
ITEM:	VALUE:	CONDITIONS:	REF.:
*ALKYL POLYETHOXYLATE IN PRESENCE OF OSCILLATORIA FORMOSA	0.79 MG SURFACT./L*DAY	21DAYS,BATCH CULTURE,INITIAL CONC. =20MG SURFACTANT/L, AE7,0	332
*ALKYL POLYETHOXYLATE IN PRESENCE OF OSCILLATORIA FORMOSA	0.51 MG SURFACT./L*DAY	21DAYS,BATCH CULTURE,INITIAL CONC. =20MG SURFACTANT/L, AE7.4	332
*ALKYL POLYETHOXYLATE IN PRESENCE OF OSCILLATORIA TENUIS	0.68 MG SURFACT./L*DAY	21DAYS,BATCH CULTURE,INITIAL CONC. =20MG SURFACTANT/L, AE7,0	332
*ALKYL POLYETHOXYLATE IN PRESENCE OF OSCILLATORIA TENUIS	1.01 MG SURFACT./L*DAY	21DAYS,BATCH CULTURE,INITIAL CONC. =20MG SURFACTANT/L, AE7.4	332
*ALKYL POLYETHOXYLATE IN PRESENCE OF PHORMIDIUM FAVEOLARUM	1.02 MG SURFACT./L*DAY	21DAYS,BATCH CULTURE,INITIAL CONC. =20MG SURFACTANT/L, AE7,0	332
*ALKYL POLYETHOXYLATE IN PRESENCE OF PHORMIDIUM FAVEOLARUM	0.93 MG SURFACT./L*DAY	21DAYS,BATCH CULTURE,INITIAL CONC. =20MG SURFACTANT/L, AE7.4	332
*ALKYL POLYETHOXYLATE IN PRESENCE OF POND WATER	0.85 MG SURFACT./L*DAY	21DAYS,BATCH CULTURE,INITIAL CONC. =20MG SURFACTANT/L, AE7,0	332
*ALKYL POLYETHOXYLATE IN PRESENCE OF POND WATER	0.88 MG SURFACT./L*DAY	21DAYS,BATCH CULTURE,INITIAL CONC. =20MG SURFACTANT/L, AE7.4	332
*ALKYL POLYETHOXYLATE IN PRESENCE OF RAW WASTEWATER	0.93 MG SURFACT./L*DAY	21DAYS,BATCH CULTURE,INITIAL CONC. =20MG SURFACTANT/L, AE7,0	332
*ALKYL POLYETHOXYLATE IN PRESENCE OF RAW WASTEWATER	0.84 MG SURFACT./L*DAY	21DAYS,BATCH CULTURE,INITIAL CONC. =20MG SURFACTANT/L, AE7.4	332
*ALKYL POLYETHOXYLATE IN PRESENCE OF SCENEDESMUS OBLIQUUS	0.64 MG SURFACT./L*DAY	21DAYS,BATCH CULTURE,INITIAL CONC. =20MG SURFACTANT/L, AE7,0	332
*ALKYL POLYETHOXYLATE IN PRESENCE OF SCENEDESMUS OBLIQUUS	0.63 MG SURFACT./L*DAY	21DAYS,BATCH CULTURE,INITIAL CONC. =20MG SURFACTANT/L, AE7.4	332
*ALKYL POLYETHOXYLATE IN PRESENCE OF SCENEDESMUS QUADRICAUDA	1.03 MG SURFACT./L*DAY	21DAYS,BATCH CULTURE,INITIAL CONC. =20MG SURFACTANT/L, AE7,0	332
*ALKYL POLYETHOXYLATE IN PRESENCE OF SCENEDESMUS QUADRICAUDA	0.51 MG SURFACT./L*DAY	21DAYS,BATCH CULTURE,INITIAL CONC. =20MG SURFACTANT/L, AE7.4	332
*ALKYL POLYETHOXYLATE IN PRESENCE OF SOIL	0.87 MG SURFACT./L*DAY	21DAYS,BATCH CULTURE,INITIAL CONC. =20MG SURFACTANT/L, AE7,0	332
*ALKYL POLYETHOXYLATE IN PRESENCE OF SOIL	0.83 MG SURFACT./L*DAY	21DAYS,BATCH CULTURE,INITIAL CONC. =20MG SURFACTANT/L, AE7.4	332
*ALKYL POLYETHOXYLATE IN PRESENCE OF ULOTHRIX FIMBRIATA	0.60 MG SURFACT./L*DAY	21DAYS,BATCH CULTURE,INITIAL CONC. =20MG SURFACTANT/L, AE7,0	332
*ALKYL POLYETHOXYLATE IN PRESENCE OF ULOTHRIX FIMBRIATA	0.75 MG SURFACT./L*DAY	21DAYS,BATCH CULTURE,INITIAL CONC. =20MG SURFACTANT/L, AE7.4	332
*AMINOPHENOLSULPHONIC ACID	7.1 MG COD/(GRAM*HOUR)	INIT COD= 200 MG/L,NO OTHER SOURCE OF C,AEROB,293+-3 K	1263
*AMMONIUM OXALATE	9.3 MG COD/(GRAM*HOUR)	INIT COD= 200 MG/L,NO OTHER SOURCE OF C,AEROB,293+-3 K	1263
*ANILINE	19.0 MG COD/(GRAM*HOUR)	INIT COD= 200 MG/L,NO OTHER SOURCE OF C,AEROB,293+-3 K	1263
*BENZALDEHYDE	119.0 MG COD/(GRAM*HOUR)	INIT COD= 200 MG/L,NO OTHER SOURCE OF C,AEROB,293+-3 K	1263
*BENZENESULPHONIC ACID	10.6 MG COD/(GRAM*HOUR)	INIT COD= 200 MG/L,NO OTHER SOURCE OF C,AEROB,293+-3 K	1263
*BENZOIC ACID	88.5 MG COD/(GRAM*HOUR)	INIT COD= 200 MG/L,NO OTHER SOURCE OF C,AEROB,293+-3 K	1263
*BORNEOL	8.9 MG COD/(GRAM*HOUR)	INIT COD= 200 MG/L,NO OTHER SOURCE OF C,AEROB,293+-3 K	1263
*CAPROLACTAM	16.0 MG COD/(GRAM*HOUR)	INIT COD= 200 MG/L,NO OTHER SOURCE OF C,AEROB,293+-3 K	1263
*CARBON	1.96 1/DAY	IN STREAM, MAPLE LEAVES, 293 REFRACTTORY FRACTION OF C FROM LEAVES	355
*CYCLOHEXANOL	28.0 MG COD/(GRAM*HOUR)	INIT COD= 200 MG/L,NO OTHER SOURCE OF C,AEROB,293+-3 REFRACTTORY FRACT	1263
*CYCLOHEXANOLONE	51.5 MG COD/(GRAM*HOUR)	INIT COD= 200 MG/L,NO OTHER SOURCE OF C,AEROB,293+-3 K	1263
*CYCLOHEXANONE	30.0 MG COD/(GRAM*HOUR)	INIT COD = 200 MG/L.,NO OTHER SOURCE OF C,AEROB,293+-3 K	1263
*CYCLOPENTANOL	55.0 MG COD/(GRAM*HOUR)	INIT COD= 200 MG/L,NO OTHER SOURCE OF C,AEROB,293+-3 K	1263
*CYCLOPENTANONE	57.0 MG COD/(GRAM*HOUR)	INIT COD= 200 MG/L,NO OTHER SOURCE OF C,AEROB,293+-3 K	1263
*D-CHLORAMPHENICOL	3.3 MG COD/(GRAM*HOUR)	INIT COD= 200 MG/L,NO OTHER SOURCE OF C,AEROB,293+-3 K	1263

TABLE G112

RATE OF DEGRADATION : BIOLOGICAL DEGRADATION IN WATER

ITEM:	VALUE:	CONDITIONS:	REF.:
*DIETHANOLAMINE	19.5 MG COD/(GRAM*HOUR)	INIT COD= 200 MG/L,NO OTHER SOURCE OF C,AEROB,293+-3 K	1263
*DIETHYLENE GLYCOL	13.7 MG COD/(GRAM*HOUR)	INIT COD= 200 MG/L,NO OTHER SOURCE OF C,AEROB,293+-3 K	1263
*DIMETHYLCYCLOHEXANOL	21.6 MG COD/(GRAM*HOUR)	INIT COD= 200 MG/L,NO OTHER SOURCE OF C,AEROB,293+-3 K	1263
*ETHYLENE DIAMINE	9.8 MG COD/(GRAM*HOUR)	INIT COD= 200 MG/L,NO OTHER SOURCE OF C,AEROB,293+-3 K	1263
*ETHYLENE GLYCOL	41.7 MG COD/(GRAM*HOUR)	INIT COD= 200 MG/L,NO OTHER SOURCE OF C,AEROB,293+-3 K	1263
*FURFURYL ALCOHOL	41.1 MG COD/(GRAM*HOUR)	INIT COD= 200 MG/L,NO OTHER SOURCE OF C,AEROB,293+-3 K	1263
*FURFURYLALDEHYDE	37.0 MG COD/(GRAM*HOUR)	INIT COD= 200 MG/L,NO OTHER SOURCE OF C,AEROB,293+-3 K	1263
*GALLIC ACID	20.0 MG COD/(GRAM*HOUR)	INIT COD= 200 MG/L,NO OTHER SOURCE OF C,AEROB,293+-3 K	126;
*GENTISIC ACID	80.0 MG COD/(GRAM*HOUR)	INIT COD= 200 MG/L,NO OTHER SOURCE OF C,AEROB,293+-3 K	1263
*GLUCOSE	180.0 MG COD/(GRAM*HOUR)	INIT COD= 200 MG/L,NO OTHER SOURCE OF C,AEROB,293+-3 K	1263
*GLUCOSE	8 TO > 26 DAYS	HALF-LIFE TIME,278K,5 STATIONS AT SOUTHAMPTON	51
*GLUCOSE	3-10 DAYS	HALF-LIFE TIME,295K,5 STATIONS AT SOUTHAMPTON	51
*GLUCOSE	21 TO > 30 DAYS	HALF-LIFE TIME,300K,PORTO NOVO,2 ST.FROM RIVER VELLAR	51
*GLUCOSE	11 DAYS	HALF-LIFE TIME,300K,PORTO NOVO, KILLE BACKWATER	51
*GLUCOSE	9-10 DAYS	HALF-LIFE TIME,300K,2 STA.FROM KILLE BACKWATER AND MANGROVE SWAMP	51
*GLUCOSE	> 17 DAYS	HALF-LIFE TIME,300K,PORTO NOVO,BAY OF BENGAL	51
*GLUCOSE	45 DAYS	HALF-LIFE TIME,300K,DISTILLED WATER	51
*GLYCEROL	85.0 MG COD/(GRAM*HOUR)	INIT COD= 200 MG/L,NO OTHER SOURCE OF C,AEROB,293+-3 K	1263
*HYDROQUINONE	54.2 MG COD/(GRAM*HOUR)	INIT COD= 200 MG/L,NO OTHER SOURCE OF C,AEROB,293+-3 K	1263
*ISO-PROPANOL	52.0 MG COD/(GRAM*HOUR)	INIT COD= 200 MG/L,NO OTHER SOURCE OF C,AEROB,293+-3 K	1263
*ISOPHTHALIC ACID	76.0 MG COD/(GRAM*HOUR)	INIT COD= 200 MG/L,NO OTHER SOURCE OF C,AEROB,293+-3 K	1263
*LINEAL ALKYL BENZEN SULFONATE IN PRESENCE OF ACTIVATED SLUDGE	0.79 MG SURFACT./L*DAY	21DAYS,BATCH CULTURE,INITIAL CONC. =20MG SURFACTANT/L	332
*LINEAL ALKYL BENZEN SULFONATE IN PRESENCE OF ANABAENA CYLINDRICA	0.72 MG SURFACT./L*DAY	21DAYS,BATCH CULTURE,INITIAL CONC. =20MG SURFACTANT/L	332
*LINEAL ALKYL BENZEN SULFONATE IN PRESENCE OF ANABAENA VARIABILIS	0.92 MG SURFACT./L*DAY	21DAYS,BATCH CULTURE,INITIAL CONC. =20MG SURFACTANT/L	332
*LINEAL ALKYL BENZEN SULFONATE IN PRESENCE OF ANACYSTIS NIDULANS	0.40 MG SURFACT./L*DAY	21DAYS,BATCH CULTURE,INITIAL CONC. =20MG SURFACTANT/L	332
*LINEAL ALKYL BENZEN SULFONATE IN PRESENCE OF ANKISTRODESMUS BRAUNII	0.35 MG SURFACT./L*DAY	21DAYS,BATCH CULTURE,INITIAL CONC. =20MG SURFACTANT/L	332
*LINEAL ALKYL BENZEN SULFONATE IN PRESENCE OF CALOTHRIX PARIETINA	0.93 MG SURFACT./L*DAY	21DAYS,BATCH CULTURE,INITIAL CONC. =20MG SURFACTANT/L	332
*LINEAL ALKYL BENZEN SULFONATE IN PRESENCE OF CHLORELLA PYRENOIDOSA	0.58 MG SURFACT./L*DAY	21DAYS,BATCH CULTURE,INITIAL CONC. =20MG SURFACTANT/L	332
*LINEAL ALKYL BENZEN SULFONATE IN PRESENCE OF CHLORELLA VULGARIS	0.42 MG SURFACT./L*DAY	21DAYS,BATCH CULTURE,INITIAL CONC. =20MG SURFACTANT/L	332
*LINEAL ALKYL BENZEN SULFONATE IN PRESENCE OF CYLINDROSPERMUM SP.	0.84 MG SURFACT./L*DAY	21DAYS,BATCH CULTURE,INITIAL CONC. =20MG SURFACTANT/L	332
*LINEAL ALKYL BENZEN SULFONATE IN PRESENCE OF GLOEOCAPSA ALPICOLA	0.92 MG SURFACT./L*DAY	21DAYS,BATCH CULTURE,INITIAL CONC. =20MG SURFACTANT/L	332
*LINEAL ALKYL BENZEN SULFONATE IN PRESENCE OF NOSTOC MUSCORUM	0.46 MG SURFACT./L*DAY	21DAYS,BATCH CULTURE,INITIAL CONC. =20MG SURFACTANT/L	332
*LINEAL ALKYL BENZEN SULFONATE IN PRESENCE OF OSCILLATORIA BORNETI	0.58 MG SURFACT./L*DAY	21DAYS,BATCH CULTURE,INITIAL CONC. =20MG SURFACTANT/L	332

TABLE G112

RATE OF DEGRADATION : BIOLOGICAL DEGRADATION IN WATER			
ITEM:	VALUE:	CONDITIONS:	REF.:
*LINEAL ALKYL BENZEN SULFONATE IN PRESENCE OF OSCILLATORIA CHALYBIA	0.62 MG SURFACT./L*DAY	21DAYS,BATCH CULTURE,INITIAL CONC. =20MG SURFACTANT/L	332
*LINEAL ALKYL BENZEN SULFONATE IN PRESENCE OF OSCILLATORIA FORMOSA	0.91 MG SURFACT./L*DAY	21DAYS,BATCH CULTURE,INITIAL CONC. =20MG SURFACTANT/L	332
*LINEAL ALKYL BENZEN SULFONATE IN PRESENCE OF OSCILLATORIA TENUIS	0.28 MG SURFACT./L*DAY	21DAYS,BATCH CULTURE,INITIAL CONC. =20MG SURFACTANT/L	332
*LINEAL ALKYL BENZEN SULFONATE IN PRESENCE OF PHORMIDIUM FAVEOLARUM	0.65 MG SURFACT./L*DAY	21DAYS,BATCH CULTURE,INITIAL CONC. =20MG SURFACTANT/L	332
*LINEAL ALKYL BENZEN SULFONATE IN PRESENCE OF POND WATER	0.96 MG SURFACT./L*DAY	21DAYS,BATCH CULTURE,INITIAL CONC. =20MG SURFACTANT/L	332
*LINEAL ALKYL BENZEN SULFONATE IN PRESENCE OF RAW WASTEWATER	0.90 MG SURFACT./L*DAY	21DAYS,BATCH CULTURE,INITIAL CONC. =20MG SURFACTANT/L	332
*LINEAL ALKYL BENZEN SULFONATE IN PRESENCE OF SCENEDESMUS OBLIQUUS	0.51 MG SURFACT./L*DAY	21DAYS,BATCH CULTURE,INITIAL CONC. =20MG SURFACTANT/L	332
*LINEAL ALKYL BENZEN SULFONATE IN PRESENCE OF SCENEDESMUS QUADRICAUDA	0.30 MG SURFACT./L*DAY	21DAYS,BATCH CULTURE,INITIAL CONC. =20MG SURFACTANT/L	332
*LINEAL ALKYL BENZEN SULFONATE IN PRESENCE OF SOIL	0.82 MG SURFACT./L*DAY	21DAYS,BATCH CULTURE,INITIAL CONC. =20MG SURFACTANT/L	332
*LINEAL ALKYL BENZEN SULFONATE IN PRESENCE OF ULOTHRIX FIMBRIATA	0.70 MG SURFACT./L*DAY	21DAYS,BATCH CULTURE,INITIAL CONC. =20MG SURFACTANT/L	332
*M-AMINOPHENOL	10.6 MG COD/(GRAM*HOUR)	INIT . COD= 200 MG/L,NO OTHER SOURCE OF C,AEROB,293+-3 K	1263
*M-AMINOTOLUENE	30.0 MG COD/(GRAM*HOUR)	INIT COD= 200 MG/L,NO OTHER SOURCE OF C,AEROB,293+-3 K	1263
*M-BENZENEDISULPHONIC ACID	3.4 MG COD/(GRAM*HOUR)	INIT . COD= 200 MG/L,NO OTHER SOURCE OF C,AEROB,293+-3 K	1263
*M-CHLOROANILINE	6.2 MG COD/(GRAM*HOUR)	INIT . COD= 200 MG/L,NO OTHER SOURCE OF C,AEROB,293+-3 K	1263
*M-CRESOL	55.0 MG COD/(GRAM*HOUR)	INIT COD= 200 MG/L,NO OTHER SOURCE OF C,AEROB,293+-3 K	1263
*M-NITROANILINE	NO DEGRADATION	INIT COD= 200 MG/L,NO OTHER SOURCE OF C,AEROB,293+-3 K	1263
*M-NITROBENZALDEHYDE	10.0 MG COD/(GRAM*HOUR)	INIT COD= 200 MG/L,NO OTHER SOURCE OF C,AEROB,293+-3 K	1263
*M-NITROBENZOIC ACID	7.0 MG COD/(GRAM*HOUR)	INIT . COD= 200 MG/L,NO OTHER SOURCE OF C,AEROB,293+-3 K	1263
*M-NITROPHENOL	17.5 MG COD/(GRAM*HOUR)	INIT . COD= 200 MG/L,NO OTHER SOURCE OF C,AEROB,293+-3 K	1263
*M-NITROTOLUENE	21.0 MG COD/(GRAM*HOUR)	INIT COD= 200 MG/L,NO OTHER SOURCE OF C,AEROB,293+-3 K	1263
*M-PHENYLENDIAMINE	NO DEGRADATION	INIT . COD= 200 MG/L,NO OTHER SOURCE OF C,AEROB,293+-3 K	1263
*MENTHOL	17.7 MG COD/(GRAM*HOUR)	INIT COD= 200 MG/L,NO OTHER SOURCE OF C,AEROB,293+-3 K	1263
*METOL	0.8 MG COD/(GRAM*HOUR)	INIT COD= 200 MG/L,NO OTHER SOURCE OF C,AEROB,293+-3 K	1263
*N-BUTANOL	84.0 MG COD/(GRAM*HOUR)	INIT COD= 200 MG/L,NO OTHER SOURCE OF C,AEROB,293+-3 K	1263
*N-PHENYLANTHRANILIC ACID	NO DEGRADATION	INIT COD= 200 MG/L,NO OTHER SOURCE OF C,AEROB,293+-3 K	1263
*N-PROPANOL	71.0 MG COD/(GRAM*HOUR)	INIT COD= 200 MG/L,NO OTHER SOURCE OF C,AEROB,293+-3 K	1263
*NAPHTOIC ACID	15.5 MG COD/(GRAM*HOUR)	INIT COD= 200 MG/L,NO OTHER SOURCE OF C,AEROB,293+-3 K	1263
*NITROBENZENE	14.0 MG COD/(GRAM*HOUR)	INIT COD= 200 MG/L,NO OTHER SOURCE OF C,AEROB,293+-3 K	1263
*NITROGEN	0.14 1/DAY	POTOMAC(ESTUARY),293 K, ORGANIC N	1149
*NITROGEN	0.04 1/DAY	LAKE ERIE,293 K, ORGANIC N	1149
*NITROGEN	0.035 1/DAY	LAKE ONTARIO,293 K, ORGANIC N	1149
*NONYL PHENOL POLYETHOXYLATE IN PRESENCE OF ACTIVATED SLUDGE	1.25 MG SURFACT./L*DAY	21DAYS,BATCH CULTURE,INITIAL CONC. =20MG SURFACTANT/L	332
*NONYL PHENOL POLYETHOXYLATE IN PRESENCE OF ANABAENA CYLINDRICA	0.60 MG SURFACT./L*DAY	21DAYS,BATCH CULTURE,INITIAL CONC. =20MG SURFACTANT/L	332
*NONYL PHENOL POLYETHOXYLATE IN PRESENCE OF ANABAENA VARIABILIS	0.48 MG SURFACT./L*DAY	21DAYS,BATCH CULTURE,INITIAL CONC. =20MG SURFACTANT/L	332

TABLE G112

RATE OF DEGRADATION : BIOLOGICAL DEGRADATION IN WATER

ITEM:	VALUE:	CONDITIONS:	REF.:
*NONYL PHENOL POLYETHOXYLATE IN PRESENCE OF ANACYSTIS NIDULANS	0.72 MG SURFACT./L*DAY	21DAYS,BATCH CULTURE,INITIAL CONC. =20MG SURFACTANT/L	332
*NONYL PHENOL POLYETHOXYLATE IN PRESENCE OF ANKISTRODESMUS BRAUNII	0.60 MG SURFACT./L*DAY	21DAYS,BATCH CULTURE,INITIAL CONC. =20MG SURFACTANT/L	332
*NONYL PHENOL POLYETHOXYLATE IN PRESENCE OF CALOTHRIX PARIETINA	0.64 MG SURFACT./L*DAY	21DAYS,BATCH CULTURE,INITIAL CONC. =20MG SURFACTANT/L	332
*NONYL PHENOL POLYETHOXYLATE IN PRESENCE OF CHLORELLA PYRENOIDOSA	0.80 MG SURFACT./L*DAY	21DAYS,BATCH CULTURE,INITIAL CONC. =20MG SURFACTANT/L	332
*NONYL PHENOL POLYETHOXYLATE IN PRESENCE OF CHLORELLA VULGARIS	0.86 MG SURFACT./L*DAY	21DAYS,BATCH CULTURE,INITIAL CONC. =20MG SURFACTANT/L	332
*NONYL PHENOL POLYETHOXYLATE IN PRESENCE OF CYLINDROSPERMUM SP.	0.64 MG SURFACT./L*DAY	21DAYS,BATCH CULTURE,INITIAL CONC. =20MG SURFACTANT/L	332
*NONYL PHENOL POLYETHOXYLATE IN PRESENCE OF GLOEOCAPSA ALPICOLA	0.49 MG SURFACT./L*DAY	21DAYS,BATCH CULTURE,INITIAL CONC. =20MG SURFACTANT/L	332
*NONYL PHENOL POLYETHOXYLATE IN PRESENCE OF NOSTOC MUSCORUM	0.51 MG SURFACT./L*DAY	21DAYS,BATCH CULTURE,INITIAL CONC. =20MG SURFACTANT/L	332
*NONYL PHENOL POLYETHOXYLATE IN PRESENCE OF OSCILLATORIA BORNETI	0.43 MG SURFACT./L*DAY	21DAYS,BATCH CULTURE,INITIAL CONC. =20MG SURFACTANT/L	332
*NONYL PHENOL POLYETHOXYLATE IN PRESENCE OF OSCILLATORIA CHALYBIA	0.61 MG SURFACT./L*DAY	21DAYS,BATCH CULTURE,INITIAL CONC. =20MG SURFACTANT/L	332
*NONYL PHENOL POLYETHOXYLATE IN PRESENCE OF OSCILLATORIA FORMOSA	0.43 MG SURFACT./L*DAY	21DAYS,BATCH CULTURE,INITIAL CONC. =20MG SURFACTANT/L	332
*NONYL PHENOL POLYETHOXYLATE IN PRESENCE OF OSCILLATORIA TENUIS	0.81 MG SURFACT./L*DAY	21DAYS,BATCH CULTURE,INITIAL CONC. =20MG SURFACTANT/L	332
*NONYL PHENOL POLYETHOXYLATE IN PRESENCE OF PHORMIDIUM FAVEOLARUM	0.73 MG SURFACT./L*DAY	21DAYS,BATCH CULTURE,INITIAL CONC. =20MG SURFACTANT/L	332
*NONYL PHENOL POLYETHOXYLATE IN PRESENCE OF POND WATER	0.67 MG SURFACT./L*DAY	21DAYS,BATCH CULTURE,INITIAL CONC. =20MG SURFACTANT/L	332
*NONYL PHENOL POLYETHOXYLATE IN PRESENCE OF RAW WASTEWATER	0.81 MG SURFACT./L*DAY	21DAYS,BATCH CULTURE,INITIAL CONC. =20MG SURFACTANT/L	332
*NONYL PHENOL POLYETHOXYLATE IN PRESENCE OF SCENEDESMUS OBLIQUUS	0.69 MG SURFACT./L*DAY	21DAYS,BATCH CULTURE,INITIAL CONC. =20MG SURFACTANT/L	332
*NONYL PHENOL POLYETHOXYLATE IN PRESENCE OF SCENEDESMUS QUADRICAUDA	0.75 MG SURFACT./L*DAY	21DAYS,BATCH CULTURE,INITIAL CONC. =20MG SURFACTANT/L	332
*NONYL PHENOL POLYETHOXYLATE IN PRESENCE OF SOIL	0.84 MG SURFACT./L*DAY	21DAYS,BATCH CULTURE,INITIAL CONC. =20MG SURFACTANT/L	332
*NONYL PHENOL POLYETHOXYLATE IN PRESENCE OF ULOTHRIX FIMBRIATA	0.39 MG SURFACT./L*DAY	21DAYS,BATCH CULTURE,INITIAL CONC. =20MG SURFACTANT/L	332
*O-AMINOBENZOIC ACID	27.1 MG COD/(GRAM*HOUR)	INIT COD= 200 MG/L,NO OTHER SOURCE OF C,AEROB,293+-3 K	1263
*O-AMINOTOLUENE	15.1 MG COD/(GRAM*HOUR)	INIT COD= 200 MG/L,NO OTHER SOURCE OF C,AEROB,293+-3 K	1263
*O-CHLOROANILINE	16.7 MG COD/(GRAM*HOUR)	INIT COD= 200 MG/L,NO OTHER SOURCE OF C,AEROB,293+-3 K	1263
*O-CHLOROPHENOL	25.0 MG COD/(GRAM*HOUR)	INIT COD= 200 MG/L,NO OTHER SOURCE OF C,AEROB,293+-3 K	1263
*O-CRESOL	54.0 MG COD/(GRAM*HOUR)	INIT COD= 200 MG/L,NO OTHER SOURCE OF C,AEROB,293+-3 K	1263
*O-NITROANILINE	NO DEGRADATION	INIT COD= 200 MG/L,NO OTHER SOURCE OF C,AEROB,293+-3 K	1263
*O-NITROBENZALDEHYDE	13.8 MG COD/(GRAM*HOUR)	INIT COD= 200 MG/L,NO OTHER SOURCE OF C,AEROB,293+-3 K	1263
*O-NITROBENZOIC ACID	20.0 MG COD/(GRAM*HOUR)	INIT COD= 200 MG/L,NO OTHER SOURCE OF C,AEROB,293+-3 K	1263
*O-NITROLUENE	32.5 MG COD/(GRAM*HOUR)	INIT COD= 200 MG/L,NO OTHER SOURCE OF C,AEROB,293+-3 K	1263
*O-NITROPHENOL	14.0 MG COD/(GRAM*HOUR)	INIT COD= 200 MG/L,NO OTHER SOURCE OF C,AEROB,293+-3 K	1263
*O-PHENYLENDIAMINE	NO DEGRADATION	INIT COD= 200 MG/L,NO OTHER SOURCE OF C,AEROB,293+-3 K	1263
*P-AMINOBENZOIC ACID	12.5 MG COD/(GRAM*HOUR)	INIT COD= 200 MG/L,NO OTHER SOURCE OF C,AEROB,293+-3 K	1263
*P-AMINOPHENOL	16.7 MG COD/(GRAM*HOUR)	INIT COD= 200 MG/L,NO OTHER SOURCE OF C,AEROB,293+-3 K	1263
*P-AMINOTOLUENE	20.0 MG COD/(GRAM*HOUR)	INIT COD= 200 MG/L,NO OTHER SOURCE OF C,AEROB,293+-3 K	1263

TABLE G112

RATE OF DEGRADATION : BIOLOGICAL DEGRADATION IN WATER

ITEM:	VALUE:	CONDITIONS:	REF.:
*P-CHLOROANILINE	5.7 MG COD/(GRAM*HOUR)	INIT COD= 200 MG/L,NO OTHER SOURCE OF C,AEROB,293+-3 K	1263
*P-CHLOROPHENOL	11.0 MG COD/(GRAM*HOUR)	INIT COD= 200 MG/L,NO OTHER SOURCE OF C,AEROB,293+-3 K	1263
*P-CRESOL	55.0 MG COD/(GRAM*HOUR)	INIT COD= 200 MG/L,NO OTHER SOURCE OF C,AEROB,293+-3 K	1263
*P-HYDROXYBENZOIC ACID	100.0 MG COD/(GRAM*HOUR)	INIT COD= 200 MG/L,NO OTHER SOURCE OF C,AEROB,293+-3 K	1263
*P-NITROACETOPHENONE	5.2 MG COD/(GRAM*HOUR)	INIT COD= 200 MG/L,NO OTHER SOURCE OF C,AEROB,293+-3 K	1263
*P-NITROANILINE	NO DEGRADATION	INIT COD= 200 MG/L,NO OTHER SOURCE OF C,AEROB,293+-3 K	1263
*P-NITROBENZALDEHYDE	13.8 MG COD/(GRAM*HOUR)	INIT COD= 200 MG/L,NO OTHER SOURCE OF C,AEROB,293+-3 K	1263
*P-NITROBENZOIC ACID	19.7 MG COD/(GRAM*HOUR)	INIT COD= 200 MG/L,NO OTHER SOURCE OF C,AEROB,293+-3 K	1263
*P-NITROPHENOL	17.5 MG COD/(GRAM*HOUR)	INIT COD= 200 MG/L,NO OTHER SOURCE OF C,AEROB,293+-3 K	1263
*P-NITROTOLUENE	32.5 MG COD/(GRAM*HOUR)	INIT COD= 200 MG/L,NO OTHER SOURCE OF C,AEROB,293+-3 K	1263
*P-PHENYLENDIAMINE	MORE DEGRADABLE	INIT COD= 200 MG/L,NO OTHER SOURCE OF C,AEROB,293+-3 K	1263
*P-TOLUENESULPHONIC ACID	8.4 MG COD/(GRAM*HOUR)	INIT COD= 200 MG/L,NO OTHER SOURCE OF C,AEROB,293+-3 K	1263
*PHENOL	80.0 MG COD/(GRAM*HOUR)	INIT COD= 200 MG/L,NO OTHER SOURCE OF C,AEROB,293+-3 K	1263
*PHLOROGLUCINOL	22.1 MG COD/(GRAM*HOUR)	INIT COD= 200 MG/L,NO OTHER SOURCE OF C,AEROB,293+-3 K	1263
*PHOSPHORUS	0.14 1/DAY	POTOMAC(ESTUARY),293 K, ORGANIC P	1149
*PHOSPHORUS	0.40 1/DAY	LAKE ERIE,293 K, ORGANIC P	1149
*PHOSPHORUS	0.14 1/DAY	LAKE ONTARIO,293 K, ORGANIC P	1149
*PHTHALIC ACID	78.4 MG COD/(GRAM*HOUR)	INIT COD= 200 MG/L,NO OTHER SOURCE OF C,AEROB,293+-3 K	1263
*PHTHALIMIDE	20.8 MG COD/(GRAM*HOUR)	INIT COD= 200 MG/L,NO OTHER SOURCE OF C,AEROB,293+-3 K	1263
*PYROCATECHOL	55.5 MG COD/(GRAM*HOUR)	INIT COD= 200 MG/L,NO OTHER SOURCE OF C,AEROB,293+-3 K	1263
*PYROGALLOL	NO DEGRADATION	INIT COD= 200 MG/L,NO OTHER SOURCE OF C,AEROB,293+-3 K	1263
*RESORCINOL	57.5 MG COD/(GRAM*HOUR)	INIT COD= 200 MG/L,NO OTHER SOURCE OF C,AEROB,293+-3 K	1263
*RIBOSE	8 TO > 30 DAYS	HALF-LIFE TIME,278K,5 STATIONS AT SOUTHAMPTON	51
*RIBOSE	3-9 DAYS	HALF-LIFE TIME,295K,5 STATIONS AT SOUTHAMPTON	51
*RIBOSE	> 36 DAYS	HALF-LIFE TIME,300K,PORTO NOVO, RIVER VELLAR	51
*SALICYLIC ACID	95.0 MG COD/(GRAM*HOUR)	INIT COD= 200 MG/L,NO OTHER SOURCE OF C,AEROB,293+-3 K	1263
*SEC. BUTANOL	55.0 MG COD/(GRAM*HOUR)	INIT COD= 200 MG/L,NO OTHER SOURCE OF C,AEROB,293+-3 K	1263
*SI	0.0015 1/D	DETRITUS SI TO DISSOLVED SI	220
*SULPHANILIC ACID	4.0 MG COD/(GRAM*HOUR)	INIT COD= 200 MG/L,NO OTHER SOURCE OF C,AEROB,293+-3 K	1263
*SULPHOSALICYLIC ACID	11.3 MG COD/(GRAM*HOUR)	INIT COD= 200 MG/L,NO OTHER SOURCE OF C,AEROB,293+-3 K	1263
*TERT. BUTANOL	30.0 MG COD/(GRAM*HOUR)	INIT COD= 200 MG/L,NO OTHER SOURCE OF C,AEROB,293+-3 K	1263
*TETRAHYDROFURFURYL ALCOHOL	40.0 MG COD/(GRAM*HOUR)	INIT COD= 200 MG/L,NO OTHER SOURCE OF C,AEROB,293+-3 K	1263
*TETRAHYDROPHTHALIC ACID	NO DEGRADATION	INIT COD= 200 MG/L,NO OTHER SOURCE OF C,AEROB,293+-3 K	1263
*TETRAHYDROPHTHALIMIDE	NO DEGRADATION	INIT COD= 200 MG/L,NO OTHER SOURCE OF C,AEROB,293+-3 K	1263
*THYMOL	15.6 MG COD/(GRAM*HOUR)	INIT COD= 200 MG/L,NO OTHER SOURCE OF C,AEROB,293+-3 K	1263

TABLE G112

RATE OF DEGRADATION : BIOLOGICAL DEGRADATION IN WATER			
ITEM:	VALUE:	CONDITIONS:	REF.:
*TRIETHYLENE GLYCOL	27.5 MG COD/(GRAM*HOUR)	INIT COD= 200 MG/L,NO OTHER SOURCE OF C,AEROB,293+-3 K	1263
*1-NAPHTHALENESULFONIC ACID	18.0 MG COD/(GRAM*HOUR)	INIT COD= 200 MG/L,NO OTHER SOURCE OF C,AEROB,293+-3 K	1263
*1-NAPHTHOL	38.4 MG COD/(GRAM*HOUR)	INIT COD= 200 MG/L,NO OTHER SOURCE OF C,AEROB,293+-3 K	1263
*1-NAPHTHOL-2-SULPHONIC ACID	18.0 MG COD/(GRAM*HOUR)	INIT COD= 200 MG/L,NO OTHER SOURCE OF C,AEROB,293+-3 K	1263
*1-NAPHTHYLAMINE	NO DEGRADATION	INIT COD= 200 MG/L,NO OTHER SOURCE OF C,AEROB,293+-3 K	1263
*1-NAPHTHYLAMINE-6-SULPHONIC ACID	NO DEGRADATION	INIT COD= 200 MG/L,NO OTHER SOURCE OF C,AEROB,293+-3 K	1263
*1,2-CYCLOHEXANEDIOL	66.0 MG COD/(GRAM*HOUR)	INIT COD= 200 MG/L,NO OTHER SOURCE OF C,AEROB,293+-3 K	1263
*1,3-DINITROBENZENE	NO DEGRADATION	INIT COD= 200 MG/L,NO OTHER SOURCE OF C,AEROB,293+-3 K	1263
*1,4-BUTANEDIOL	40.0 MG COD/(GRAM*HOUR)	INIT COD= 200 MG/L,NO OTHER SOURCE OF C,AEROB,293+-3 K	1263
*1,4-DINITROBENZENE	NO DEGRADATION	INIT COD= 200 MG/L,NO OTHER SOURCE OF C,AEROB,293+-3 K	1263
*2-CHLORO-4-NITROPHENOL	5.3 MG COD/(GRAM*HOUR)	INIT COD= 200 MG/L,NO OTHER SOURCE OF C,AEROB,293+-3 K	1263
*2-NAPHTHOL	39.2 MG COD/(GRAM*HOUR)	INIT COD= 200 MG/L,NO OTHER SOURCE OF C,AEROB,293+-3 K	1263
*2,3-DIMETHYLANILINE	12.7 MG COD/(GRAM*HOUR)	INIT COD= 200 MG/L,NO OTHER SOURCE OF C,AEROB,293+-3 K	1263
*2,3-DIMETHYLPHENOL	35.0 MG COD/(GRAM*HOUR)	INIT COD= 200 MG/L,NO OTHER SOURCE OF C,AEROB,293+-3 K	1263
*2,4-DIAMINOPHENOL	12.0 MG COD/(GRAM*HOUR)	INIT COD= 200 MG/L,NO OTHER SOURCE OF C,AEROB,293+-3 K	1263
*2,4-DICHLORPHENOL	10.5 MG COD/(GRAM*HOUR)	INIT COD= 200 MG/L,NO OTHER SOURCE OF C,AEROB,293+-3 K	1263
*2,4-DIMETHYLPHENOL	28.2 MG COD/(GRAM*HOUR)	INIT COD= 200 MG/L,NO OTHER SOURCE OF C,AEROB,293+-3 K	1263
*2,4-DINITRIPHENOL	6.0 MG COD/(GRAM*HOUR)	INIT COD= 200 MG/L,NO OTHER SOURCE OF C,AEROB,293+-3 K	1263
*2,4,6-TRINITROPHENOL	NO DEGRADATION	INIT COD= 200 MG/L,NO OTHER SOURCE OF C,AEROB,293+-3 K	1263
*2,5-DIMETHYLANILINE	3.6 MG COD/(GRAM*HOUR)	INIT COD= 200 MG/L,NO OTHER SOURCE OF C,AEROB,293+-3 K	1263
*2,5-DIMETHYLPHENOL	10.6 MG COD/(GRAM*HOUR)	INIT COD= 200 MG/L,NO OTHER SOURCE OF C,AEROB,293+-3 K	1263
*2,5-DINITROPHENOL	NO DEGRADATION	INIT COD= 200 MG/L,NO OTHER SOURCE OF C,AEROB,293+-3 K	1263
*2,6-DIMETHYLPHENOL	9.0 MG COD/(GRAM*HOUR)	INIT COD= 200 MG/L,NO OTHER SOURCE OF C,AEROB,293+-3 K	1263
*2,6-DINITROPHENOL	NO DEGRADATION	INIT COD= 200 MG/L,NO OTHER SOURCE OF C,AEROB,293+-3 K	1263
*3,4-DIMETHYLANILINE	30.0 MG COD/(GRAM*HOUR)	INIT COD= 200 MG/L,NO OTHER SOURCE OF C,AEROB,293+-3 K	1263
*3,4-DIMETHYLPHENOL	13.4 MG COD/(GRAM*HOUR)	INIT COD= 200 MG/L,NO OTHER SOURCE OF C,AEROB,293+-3 K	1263
*3,5-DIMETHYLPHENOL	11.1 MG COD/(GRAM*HOUR)	INIT COD= 200 MG/L,NO OTHER SOURCE OF C,AEROB,293+-3 K	1263
*3,5-DINITROBENZOIC ACID	NO DEGRADATION	INIT COD= 200 MG/L,NO OTHER SOURCE OF C,AEROB,293+-3 K	1263
*4-METHYLCYCLOHEXANOL	40.0 MG COD/(GRAM*HOUR)	INIT COD= 200 MG/L,NO OTHER SOURCE OF C,AEROB,293+-3 K	1263
*4-METHYLCYCLOHEXANONE	62.5 MG COD/(GRAM*HOUR)	INIT COD= 200 MG/L,NO OTHER SOURCE OF C,AEROB,293+-3 K	1263

TABLE G113

RATE OF DEGRADATION : CAPTAN			
ITEM:	VALUE:	CONDITIONS:	REF.:
*HYDROLYSIS IN WATER	(1.49+-0.03)*10**(-4) 1/SEC	PH=8.39,281K,FIRST-ORDER REACTION	1743

TABLE G114

RATE OF DEGRADATION : CARBON			
ITEM:	VALUE:	CONDITIONS:	REF.:
*DEGRADATION OF LABILE FRACTION CARBON FROM LEAVES	1.96 1/DAY	IN STREAM, MAPLE LEAVES, 293 K	355

TABLE G115

RATE OF DEGRADATION : DEHYDROCHLORINATION			
ITEM:	VALUE:	CONDITIONS:	REF.:
*DDT	2.48*10**(-2) 1/M*SEC	293.11K	1743
*DDT	6.96*10**(-2) 1/M*SEC	303.37K	1743
*METHOXYCHLOR	9.18*10**(-5) 1/M*SEC	293.11K	1743
*METHOXYCHLOR	3.06*10**(-4) 1/M*SEC	303.37K	1743

TABLE G116

RATE OF DEGRADATION : HYDROLYSIS			
ITEM:	VALUE:	CONDITIONS:	REF.:
*BAYGON	30.4 1/M*MIN.	293K,PH=4-10,UNBUFFERED SECOND-ORDER REACTION	1743
*DIMETILAN	3.4*10**(-3) 1/M*MIN.	293K,PH=4-10,UNBUFFERED SECOND-ORDER REACTION	1743
*PYROLAN	0.7 1/M*MIN.	293K,PH=4-10,UNBUFFERED SECOND-ORDER REACTION	1743
*2,4-D METHYL ESTER	(1.73+-0.23)*10 1/M*SEC	301 K, BASE HYDROLYSIS	1743
*2,4-D METHYL ESTER	17.3 1/M*SEC	301K	1743
*2,4-D N-BUTOXYETHYL ESTER	(6.59+-0.13)*10**(-4) 1/M*SEC	DELTA(H)=17.6+-0.9 KCAL/MOLE, ACID HYDROLYSIS,340K	1743
*2,4-D N-BUTOXYETHYL ESTER	(2.96+-0.18)*10**(-3) 1/M*SEC	DELTA(H)=17.6+-0.9 KCAL/MOLE, ACID HYDROLYSIS,360K	1743
*2,4-D N-BUTOXYETHYL ESTER	(2.0+-0.5)*10**(-5) 1/M*SEC	DELTA(H)=17.6+-0.9 KCAL/MOLE, ACID HYDROLYSIS,301K,EXTRAPOLATED VAL.	1743
*2,4-D N-BUTOXYETHYL ESTER	(3.02+-0.23)*10 1/M*SEC	DELTA(H)=20.1+-1.4 KCAL/MOLE, BASE HYDROLYSIS,301K	1743
*2,4-D N-BUTOXYETHYL ESTER	(2.35+-0.13)*10**2 1/M*SEC	DELTA(H)=20.1+-1.4 KCAL/MOLE, BASE HYDROLYSIS,320K	1743
*2,4-D 1-BUTYL ESTER	3.7 1/M*SEC	301K	1743
*2,4-D 1-OCTYL ESTER	3.7 1/M*SEC	301K	1743
*2,4-D 2-BUTOXY-METHYLETHYL ESTER	4.3 1/M*SEC	301K	1743
*2,4-D 2-BUTOXYETHYL ESTER	30.2 1/M*SEC	301K	1743
*2,4-D 2-OCTYL ESTER	0.52 1/M*SEC	301K	1743
*2,4-D 2-PROPYL ESTER	1.1 1/M*SEC	301K	1743

TABLE G117

RATE OF DEGRADATION : HYDROLYSIS IN WATER			
ITEM:	VALUE:	CONDITIONS:	REF.:
*CAPTAN	(1.79+-0.06)*10**(-5) 1/SEC	PH=1.97,301K,FIRST-ORDER REACTION	1743
*CAPTAN	(1.67+-0.08)*10**(-5) 1/SEC	PH=2.48,301K,FIRST-ORDER REACTION	1743
*CAPTAN	(1.87+-0.08)*10**(-5) 1/SEC	PH=3.39,301K,FIRST-ORDER REACTION	1743
*CAPTAN	(1.97+-0.08)*10**(-5) 1/SEC	PH=4.83,301K,FIRST-ORDER REACTION	1743

TABLE G117

RATE OF DEGRADATION : HYDROLYSIS IN WATER			
ITEM:	VALUE:	CONDITIONS:	REF.:
*CAPTAN	(1.87+-0.09)*10**(-5) 1/SEC	PH=5.16,301K,FIRST-ORDER REACTION	1743
*CAPTAN	(4.6+-0.3)*10**(-5) 1/SEC	PH=6.10,301K,FIRST-ORDER REACTION	1743
*CAPTAN	(6.5+-0.5)*10**(-5) 1/SEC	PH=7.07,301K,FIRST-ORDER REACTION	1743
*CAPTAN	(1.5+-0.1)*10**(-4) 1/SEC	PH=7.45,301K,FIRST-ORDER REACTION	1743
*CAPTAN	(2.8+-0.2)*10**(-4) 1/SEC	PH=7.66,301K,FIRST-ORDER REACTION	1743
*CAPTAN	(3.9+-0.1)*10**(-4) 1/SEC	PH=7.82,301K,FIRST-ORDER REACTION	1743
*CAPTAN	(4.8+-0.2)*10**(-4) 1/SEC	PH=7.93,301K,FIRST-ORDER REACTION	1743
*CAPTAN	(6.5+-0.1)*10**(-4) 1/SEC	PH=8.08,301K,FIRST-ORDER REACTION	1743
*CAPTAN	(1.1+-0.1)*10**(-3) 1/SEC	PH=8.25,381K,FIRST-ORDER REACTION	1743
*CAPTAN	(3.57+-0.09)*10**(-4) 1/SEC	PH=3.3,321K,FIRST-ORDER REACTION	1743
*CAPTAN	(1.49+-0.03)*10**(-4) 1/SEC	281K,PH=8.39,H2BO3/NAOH, PSEUDO-FIRST-ORDER RATE CONST.	1743
*CAPTAN	(1.79+-0.06)*10**(-5) 1/SEC	301K,PH=1.97,HCL, PSEUDO-FIRST-ORDER RATE CONST.	1743
*CAPTAN	(1.67+-0.08)*10**(-5) 1/SEC	301K,PH=2.48,HCL, PSEUDO-FIRST-ORDER RATE CONST.	1743
*CAPTAN	(1.87+-0.08)*10**(-5) 1/SEC	301K,PH=3.39,HCL, PSEUDO-FIRST-ORDER RATE CONST.	1743
*CAPTAN	(1.97+-0.08)*10**(-5) 1/SEC	301K,PH=4.83,NAOAC/HOAC, PSEUDO-FIRST-ORDER RATE CONST.	1743
*CAPTAN	(1.87+-0.09)*10**(-5) 1/SEC	301K,PH=5.16,NAOAC/HOAC, PSEUDO-FIRST-ORDER RATE CONST.	1743
*CAPTAN	(4.6+-0.3)*10**(-5) 1/SEC	301K,PH=6.10,NA2HPO4/NAH2PO4, PSEUDO-FIRST-ORDER RATE CONST.	1743
*CAPTAN	(6.5+-0.5)*10**(-5) 1/SEC	301K,PH=7.07,NA2HPO4/NAH2PO4, PSEUDO-FIRST-ORDER RATE CONST.	1743
*CAPTAN	(1.5+-0.1)*10**(-4) 1/SEC	301K,PH=7.45,NA2HPO4/NAH2PO4, PSEUDO-FIRST-ORDER RATE CONST.	1743
*CAPTAN	(2.81+-0.2)*10**(-4) 1/SEC	301K,PH=7.66,NA2HPO4/NAH2PO4, PSEUDO-FIRST-ORDER RATE CONST.	1743
*CAPTAN	(3.9+-0.1)*10**(-4) 1/SEC	301K,PH=7.82,H2BO3/NAOH, PSEUDO-FIRST-ORDER RATE CONST.	1743
*CAPTAN	(4.8+-0.2)*10**(-4) 1/SEC	301K,PH=7.93,H2BO3/NAOH, PSEUDO-FIRST-ORDER RATE CONST.	1743
*CAPTAN	(6.5+-0.1)*10**(-4) 1/SEC	301K,PH=8.08,H2BO3/NAOH, PSEUDO-FIRST-ORDER RATE CONST.	1743
*CAPTAN	(1.1+-0.1)*10**(-3) 1/SEC	301K,PH=8.25,H2BO3/NAOH, PSEUDO-FIRST-ORDER RATE CONST.	1743
*CAPTAN	(3.57+-0.09)*10**(-4) 1/SEC	321K,PH=3.3,HCL,PSEUDO-FIRST-ORDER RATE CONST.,AV.OF TWO VAL.	1743
*CARBARYL	2.04*10**2 1/M*MIN.	293K,PH=4-10,UNBUFFERED SECOND-ORDER REACTION	1743
*METHOXYCHLOR	3.3*10**(-6) 1/SEC	PSEUDO-FIRST-ORDER RATE CONST.,PH= 5,338K	1743
*METHOXYCHLOR	4.0*10**(-6) 1/SEC	PSEUDO-FIRST-ORDER RATE CONST.,PH= 7,338K	1743
*METHOXYCHLOR	2.6*10**(-6) 1/SEC	PSEUDO-FIRST-ORDER RATE CONST.,PH= 9,338K	1743
*METHOXYCHLOR	2.1*10**(-5) 1/SEC	PSEUDO-FIRST-ORDER RATE CONST.,PH= 3,358K	1743
*METHOXYCHLOR	2.6*10**(-5) 1/SEC	PSEUDO-FIRST-ORDER RATE CONST.,PH= 5,358K	1743
*METHOXYCHLOR	2.4*10**(-5) 1/SEC	PSEUDO-FIRST-ORDER RATE CONST.,PH= 7,358K	1743
*METHOXYCHLOR	2.6*10**(-5) 1/SEC	PSEUDO-FIRST-ORDER RATE CONST.,PH= 9,358K	1743
*METHOXYCHLOR	4.5*10**(-5) 1/SEC	PSEUDO-FIRST-ORDER RATE CONST.,PH= 11,358K	1743

TABLE G118

RATE OF DEGRADATION : ISOPROPYL CARBANILATE			
ITEM:	VALUE:	CONDITIONS:	REF.:
*MICROBIEL DEGRADATION IN WATER	3.40 1/HOUR	PSEUDOMONAS STRICTA, 293 K, GROWTH =0.18 DIV/H, INITIAL=50 PPM	1054

TABLE G118

RATE OF DEGRADATION : ISOPROPYL CARBANILATE			
ITEM:	VALUE:	CONDITIONS:	REF.:
*MICROBIEL DEGRADATION IN WATER	1.91 1/HOUR	PSEUDOMONAS STRICTA, 293 K, GROWTH =0.18 DIV/H, INITIAL=100 PPM	1054
*MICROBIEL DEGRADATION IN WATER	1.83 1/HOUR	PSEUDOMONAS STRICTA, 293 K, GROWTH =0.17 DIV/H, INITIAL=200 PPM	1054
*MICROBIEL DEGRADATION IN WATER	4.12 1/HOUR	PSEUDOMONAS STRICTA, 303 K, GROWTH =0.17 DIV/H, INITIAL=50 PPM	1054
*MICROBIEL DEGRADATION IN WATER	2.46 1/HOUR	PSEUDOMONAS STRICTA, 303 K, GROWTH =0.18 DIV/H, INITIAL=100 PPM	1054
*MICROBIEL DEGRADATION IN WATER	2.33 1/HOUR	PSEUDOMONAS STRICTA, 303 K, GROWTH =0.15 DIV/H, INITIAL=200 PPM	1054

TABLE G119

RATE OF DEGRADATION : ISOPROPYL M-CHLOROCARBANILATE			
ITEM:	VALUE:	CONDITIONS:	REF.:
*MICRODIEL DEGRADATION IN WATER	4.81 1/HOUR	PSEUDOMONAS STRICTA, 293 K, GROWTH =0.12 DIV/H, INITIAL=50 PPM	1054
*MICROBIEL DEGRADATION IN WATER	2.00 1/HOUR	PSEUDOMONAS STRICTA, 293 K, GROWTH =0.10 DIV/H, INITIAL=100 PPM	1054
*MICROBIEL DEGRADATION IN WATER	1.79 1/HOUR	PSEUDOMONAS STRICTA, 293 K, GROWTH =0.13 DIV/H, INITIAL=200 PPM	1054
*MICROBIEL DEGRADATION IN WATER	6.33 1/HOUR	PSEUDOMONAS STRICTA, 303 K, GROWTH =0.11 DIV/H, INITIAL=50 PPM	1054
*MICROBIEL DEGRADATION IN WATER	2.26 1/HOUR	PSEUDOMONAS STRICTA, 303 K, GROWTH =0.09 DIV/H, INITIAL=100 PPM	1054
*MICROBIEL DEGRADATION IN WATER	2.03 1/HOUR	PSEUDOMONAS STRICTA, 303 K, GROWTH =0.10 DIV/H, INITIAL=200 PPM	1054

TABLE G120

RATE OF DEGRADATION : MALATHION			
ITEM:	VALUE:	CONDITIONS:	REF.:
*DEGRADATION OF MALATHION IN ACID WATER	(2.88+-0.01)*10**(-2) 1/M*SEC	SECOND-ORDER REACTION,PH=2.5,DELTA (H)=22.3KCAL/MOLE,360K	1743
*DEGRADATION OF MALATHION IN ACID WATER	(4.36+-0.04)*10**(-2) 1/M*SEC	SECOND-ORDER REACTION,PH=2.5,DELTA (H)=22.3KCAL/MOLE,340K	1743
*DEGRADATION OF MALATHION IN ACID WATER	(4.8+-0.20)*10**(-5) 1/M*SEC	SECOND-ORDER REACTION,PH=2.5,DELTA (H)=22.3KCAL/MOLE,300K,EXTRAPOL.	1743
*DISAPPEARENCE OF MALATHION IN ALKALINE WATER	0.067+-0.003 1/M*SEC	SECOND-ORDER REACTION,PH=8,273K,	1743
*DISAPPEARENCE OF MALATHION IN ALKALINE WATER	5.5+-0.3 1/M*SEC	SECOND-ORDER REACTION,PH=8,300K,	1743
*DISAPPEARENCE OF MALATHION IN ALKALINE WATER	47+-1 1/M*SEC	SECOND-ORDER REACTION,PH=8,320K,	1743
*DISAPPEARENCE OF MALATHION IN ALKALINE WATER	730+-20 1/M*SEC	SECOND-ORDER REACTION,PH=8,340K	1743
*DISAPPEARENCE OF MALATHION MONOACID IN WATER	(3.1+-0.2)*10**(-1) 1/M*SEC	PH=8,300K	1743
*ELIMINATION OF MALATHION	4.0*10**(-2) 1/M*SEC	DELTA(H)=27.8KCAL/MOLE,273K	1743
*ELIMINATION OF MALATHION	0.78 1/M*SEC	DELTA(H)=27.8KCAL/MOLE,300K	1743
*ELIMINATION OF MALATHION	41 1/M*SEC	DELTA(H)=27.8KCAL/MOLE,320K	1743
*ELIMINATION OF MALATHION	6.6*10**2 1/M*SEC	DELTA(H)=27.8KCAL/MOLE,340K	1743
*ELIMINATION OF MALATHION DIACID IN WATER	(1.8+-0.2)*10**(-2) 1/M*SEC	PH=8,300K	1743
*ELIMINATION OF MALATHION MONOACID IN WATER	(2.0+-0.1)*10**(-1) 1/M*SEC	PH=8,300K	1743
*HYDROLYSIS OF MALATHION CARBOXYL ESTER	2.5*10**(-2) 1/M*SEC	DELTA(H)=20.7KCAL/MOLE,273K	1743

TABLE G120

RATE OF DEGRADATION : MALATHION

ITEM:	VALUE:	CONDITIONS:	REF.:
*HYDROLYSIS OF MALATHION CARBOXYL ESTER	0.14 1/M*SEC	DELTA(H)=20.7KCAL/MOLE,300K	1743
*HYDROLYSIS OF MALATHION CARBOXYL ESTER	6.7 1/M*SEC	DELTA(H)=20.7KCAL/MOLE,320K	1743
*HYDROLYSIS OF MALATHION CARBOXYL ESTER	68 1/M*SEC	DELTA(H)=20.7KCAL/MOLE,340K	1743
*HYDROLYSIS OF MALATHION MONOACID CARBOXYL ESTER IN WATER	(1.3+-0.1)*10**(-1) 1/M*SEC	PH=8,300K	1743

TABLE G121

RATE OF DEGRADATION : O-AMINOPHENOL

ITEM:	VALUE:	CONDITIONS:	REF.:
*BIOLOGICAL DEGRADATION IN WATER	21.1 MG COD/(GRAM*HOUR)	INIT COD= 200 MG/L,NO OTHER SOURCE OF C,AEROB,293+-3 K	1263

TABLE G122

RATE OF DEGRADATION : ORGANIC MATTER

ITEM:	VALUE:	CONDITIONS:	REF.:
*LOSS OF DRY ORGANIC MATTER FROM MACROPHYTE	5 %	EMERGENT, 7-14 D, JUNCUS EFFUSUS	161
*LOSS OF DRY ORGANIC MATTER FROM MACROPHYTE	30 %	EMERGENT, 30 W, JUNCUS EFFUSUS	161
*LOSS OF DRY ORGANIC MATTER FROM MACROPHYTE	60 %	EMERGENT, 15 - 16 W, JUNCUS EFFESUS	161
*LOSS OF DRY ORGANIC MATTER FROM MACROPHYTE	65 %	EMERGENT, 24 W, JUNCUS EFFESUS	161
*LOSS OF DRY ORGANIC MATTER FROM MACROPHYTE	5 %	EMERGENT, 7 - 14 D, TYPHA LATIFOLIA	159
*LOSS OF DRY ORGANIC MATTER FROM MACROPHYTE	25 %	EMERGENT, 12 W, TYPHA LATIFOLIA	159
*LOSS OF DRY ORGANIC MATTER FROM MACROPHYTE	30 %	EMERGENT, 15 - 16 W, TYPHA LATIFOLIA	159
*LOSS OF DRY ORGANIC MATTER FROM MACROPHYTE	45 %	EMERGENT, 24 W, TYPHA LATIFOLIA	159
*LOSS OF DRY ORGANIC MATTER FROM MACROPHYTE	2 %	EMERGENT, 7 - 14 D, SPARTINA ALTERNIFLORA	200
*LOSS OF DRY ORGANIC MATTER FROM MACROPHYTE	10 %	EMERGENT, 12 W, SPARTINA ALTERNIFLORA	200
*LOSS OF DRY ORGANIC MATTER FROM MACROPHYTE	30 %	EMERGENT, 15 - 16 W, SPARTINA ALTERNIFLORA	200
*LOSS OF DRY ORGANIC MATTER FROM MACROPHYTE	50 %	EMERGENT, 24 W, SPARTINA ALTERNIFLORA	200
*LOSS OF DRY ORGANIC MATTER FROM MACROPHYTE	85 %	EMERGENT, 43 W, SPARTINA ALTERNIFLORA	200
*LOSS OF DRY ORGANIC MATTER FROM MACROPHYTE	42 %	EMERGENT, 4 W, CAREX GRACILIS	850
*LOSS OF DRY ORGANIC MATTER FROM MACROPHYTE	42 %	EMERGENT, 6 W, CAREX GRACILIS	850
*LOSS OF DRY ORGANIC MATTER FROM MACROPHYTE	55 %	EMERGENT, 9 W, CAREX GRACILIS	850
*LOSS OF DRY ORGANIC MATTER FROM MACROPHYTE	22 %	EMERGENT, 7 - 14 D, SPARGANIUM SIMPLEX, OVERWINTERED	851
*LOSS OF DRY ORGANIC MATTER FROM MACROPHYTE	48 %	EMERGENT, 6 W, SPARGANIUM SIMPLEX, OVERWINTERED	851
*LOSS OF DRY ORGANIC MATTER FROM MACROPHYTE	58 %	EMERGENT, 12 W, SPARGANIUM SIMPLEX, OVERWINTERED	851

TABLE G122

RATE OF DEGRADATION : ORGANIC MATTER

ITEM:	VALUE:	CONDITIONS:	REF.:
*LOSS OF DRY ORGANIC MATTER FROM MACROPHYTE	100 %	EMERGENT, 15 - 16 W, SPARGANIUM SIMPLEX, OVERWINTERED	851
*LOSS OF DRY ORGANIC MATTER FROM MACROPHYTE	17 - 56 %	EMERGENT, 7 - 14 D, SALIX LEAVES	1259
*LOSS OF DRY ORGANIC MATTER FROM MACROPHYTE	40 %	EMERGENT, 7 - 14 D, PHRAGMITES COMMUNIS, OVERWINTERED PLANTS	851
*LOSS OF DRY ORGANIC MATTER FROM MACROPHYTE	42 %	EMERGENT, 6 W, PHRAGMITES COMMUNIS, OVERWINTERED PLANTS	851
*LOSS OF DRY ORGANIC MATTER FROM MACROPHYTE	50 %	EMERGENT, 12 W, PHRAGMITES COMMUNIS, OVERWINTERED PLANTS	851
*LOSS OF DRY ORGANIC MATTER FROM MACROPHYTE	53 %	EMERGENT, 15 - 16 W, PHRAGMITES COMMUNIS, OVERWINTERED PLANTS	851
*LOSS OF DRY ORGANIC MATTER FROM MACROPHYTE	43 %	EMERGENT, 4 W, PHRAGMITES COMMUNIS, FRESH PLANTS	850
*LOSS OF DRY ORGANIC MATTER FROM MACROPHYTE	48 %	EMERGENT, 6 W, PHRAGMITES COMMUNIS, FRESH PLANTS	850
*LOSS OF DRY ORGANIC MATTER FROM MACROPHYTE	63 %	EMERGENT, 9 W, PHRAGMITES COMMUNIS, FRESH PLANTS	850
*LOSS OF DRY ORGANIC MATTER FROM MACROPHYTE	20 %	FLOATING-LEAVED, 7 - 14 D, LEMNACEAE	890
*LOSS OF DRY ORGANIC MATTER FROM MACROPHYTE	90 %	FLOATING-LEAVED, 12 W, LEMNACEAE	890
*LOSS OF DRY ORGANIC MATTER FROM MACROPHYTE	2 %	FLOATING-LEAVED, 7 - 14 D, BRASENIA SCHREIBERI, LEAVES IN SURFACE WAT	852
*LOSS OF DRY ORGANIC MATTER FROM MACROPHYTE	45 %	FLOATING-LEAVED, 12 W, BRASENIA SCHREIBERI, LEAVES IN SURFACE WATER	852
*LOSS OF DRY ORGANIC MATTER FROM MACROPHYTE	50 %	FLOATING-LEAVED, 24 W, BRASENIA SCHREIBERI, LEAVES IN SURFACE WATER	852
*LOSS OF DRY ORGANIC MATTER FROM MACROPHYTE	95 %	FLOATING-LEAVED, 43 W, BRASENIA SCHREIBERI, LEAVES IN SURFACE WATER	852
*LOSS OF DRY ORGANIC MATTER FROM MACROPHYTE	20 %	FLOATING-LEAVED, 7 - 14 D, BRASENIA SCHREIBERI, LEAVES IN SEDIMENT	852
*LOSS OF DRY ORGANIC MATTER FROM MACROPHYTE	20 %	FLOATING-LEAVED, 12 W, BRASENIA SCHREIBERI, LEAVES IN SEDIMENT	852
*LOSS OF DRY ORGANIC MATTER FROM MACROPHYTE	45 %	FLOATING-LEAVED, 24 W, BRASENIA SCHREIBERI, LEAVES IN SEDIMENT	852
*LOSS OF DRY ORGANIC MATTER FROM MACROPHYTE	95 %	FLOATING-LEAVED, 43 W, BRASENIA SCHREIBERI, LEAVES IN SEDIMENT	852
*LOSS OF DRY ORGANIC MATTER FROM MACROPHYTE	15 %	FLOATING-LEAVED, 7 - 14 D, NYMPHAEA ODORATA, LEAVES IN SURFACE WATER	852
*LOSS OF DRY ORGANIC MATTER FROM MACROPHYTE	45 %	FLOATING-LEAVED, 12 W, NYMPHAEA ODORATA, LEAVES IN SURFACE WATER	852
*LOSS OF DRY ORGANIC MATTER FROM MACROPHYTE	70 %	FLOATING-LEAVED, 24 W, NYMPHAEA ODORATA, LEAVES IN SURFACE WATER	852
*LOSS OF DRY ORGANIC MATTER FROM MACROPHYTE	98 %	FLOATING-LEAVED, 43 W, NYMPHAEA ODORATA, LEAVES IN SURFACE WATER	852
*LOSS OF DRY ORGANIC MATTER FROM MACROPHYTE	20 %	FLOATING-LEAVED, 7 - 14 D, NYMPHAEA ODORATA, LEAVES IN SEDIMENT	850
*LOSS OF DRY ORGANIC MATTER FROM MACROPHYTE	58 %	FLOATING-LEAVED, 12 W, MYMPHAEA ODORATA, LEAVES IN SEDIMENT	852
*LOSS OF DRY ORGANIC MATTER FROM MACROPHYTE	75 %	FLOATING-LEAVED, 24 W, MYMPHAEA ODORATA, LEAVES IN SEDIMENT	852
*LOSS OF DRY ORGANIC MATTER FROM MACROPHYTE	95 %	FLOATING-LEAVED, 43 W, MYMPHAEA ODORATA, LEAVES IN SEDIMENT	852
*LOSS OF DRY ORGANIC MATTER FROM MACROPHYTE	25 %	FLOATING-LEAVED, 7 - 14 D, POLYGONUM AMPHIBIUM, OVERWINTERED PLANTS	850
*LOSS OF DRY ORGANIC MATTER FROM MACROPHYTE	35 %	FLOATING-LEAVED, 6 W, POLYGONUM AMPHIBIUM, OVERWINTERED PLANTS	850
*LOSS OF DRY ORGANIC MATTER FROM MACROPHYTE	35 %	FLOATING-LEAVED, 12 W, POLYGONUM AMPHIBIUM, OVERWINTERED PLANTS	850
*LOSS OF DRY ORGANIC MATTER FROM MACROPHYTE	37 %	FLOATING-LEAVED, 15 - 16 W, POLYGONUM AMPHIBIUM, OVERWINTERED PLANTS	850

TABLE G122

RATE OF DEGRADATION : ORGANIC MATTER			
ITEM:	VALUE:	CONDITIONS:	REF.:
*LOSS OF DRY ORGANIC MATTER FROM MACROPHYTE	42 %	FLOATING-LEAVED, 4 W, POLYGONUM AMPHIBIUM, FRESH PLANTS	850
*LOSS OF DRY ORGANIC MATTER FROM MACROPHYTE	46 %	FLOATING-LEAVED, 6 W, POLYGONUM AMPHIBIUM, FRESH PLANTS	850
*LOSS OF DRY ORGANIC MATTER FROM MACROPHYTE	55 %	FLOATING-LEAVED, 9 W, POLYGONUM AMPHIBIUM, FRESH PLANTS	850
*LOSS OF DRY ORGANIC MATTER FROM MACROPHYTE	6 - 92 %	SUBMERSED, 7 - 14 D, POTAMOGETON LUCENS, SUBMERSED	1259
*LOSS OF DRY ORGANIC MATTER FROM MACROPHYTE	6 - 95 %	SUBMERSED, 7 - 14 D, POTAMOGETON PERFOLIATUS, SUBMERSED	1259

TABLE G123

RATE OF DEGRADATION : P-AMINOACETANILIDE			
ITEM:	VALUE:	CONDITIONS:	REF.:
*BIOLOGICAL DEGRADATION IN WATER	11.3 MG COD/(GRAM*HOUR)	INIT COD= 200 MG/L,NO OTHER SOURCE OF C,AEROB,293+-3 K	1263

TABLE G124

RATE OF DEGRADATION : P-AMINOBENZOIC ACID			
ITEM:	VALUE:	CONDITIONS:	REF.:
*BIOLOGICAL DEGRADATION IN WATER	7.0 MG COD/(GRAM*HOUR)	INIT COD= 200 MG/L,NO OTHER SOURCE OF C,AEROB,293+-3 K	1263

TABLE G125

RATE OF DEGRADATION : PHOTOCHEMICAL DEGRADATION			
ITEM:	VALUE:	CONDITIONS:	REF.:
*2,2',3,3',4,4',5,5'-OCTACHLOROBIPHENYL	< 1% OF INITIAL AMOUNT	% REMAINING AFTER 24HRS IN HEXANE, 310NM	709
*2,2',4,4',5,5'-HEXACHLOROBIPHENYL	3.8% OF INITIAL AMOUNT	% REMAINING AFTER 24HRS IN HEXANE, 310NM	709
*2,2',5,5'-TETRACHLOROBIPHENYL	33% OF INITIAL AMOUNT	% REMAINING AFTER 24HRS IN HEXANE, 310NM	709
*2,2',6,6'-TETRACHLOROBIPHENYL	29% OF INITIAL AMOUNT	% REMAINING AFTER 24HRS IN HEXANE, 310NM	709
*3,3',4,4'-TETRACHLOROBIPHENYL	32% OF INITIAL AMOUNT	% REMAINING AFTER 24HRS IN HEXANE, 310NM	709

TABLE G126

RATE OF DEGRADATION : 3,5-DINITROSALICYLIC ACID			
ITEM:	VALUE:	CONDITIONS:	REF.:
*NO DEGRADATION	BIOLOGICAL DEGRADATION IN WATER	INIT COD= 200 MG/L,NO OTHER SOURCE OF C,AEROB,293+-3 K	1263

TABLE G127

REACTION PERCENTAGE : PARATHION TO P-NITROPHENOL			
ITEM:	VALUE:	CONDITIONS:	REF.:
*HYDROLYSIS OF PARATHION TO P-NITROPHENOL	45 %	296 K,ESTUARINE WATER, SALINITY= 2.57%, PH=7.8	1668
*HYDROLYSIS OF PARATHION TO P-NITROPHENOL	24 %	323 K,ESTUARINE WATER, SALINITY= 2.57%, PH=7.8	1668
*HYDROLYSIS OF PARATHION TO P-NITROPHENOL	21 %	343 K,ESTUARINE WATER, SALINITY= 2.57%, PH=7.8	1668
*HYDROLYSIS OF PARATHION TO P-NITROPHENOL	17 %	363 K,ESTUARINE WATER, SALINITY= 2.57%, PH=7.8	1668
*HYDROLYSIS OF PARATHION TO P-NITROPHENOL	14 %	343 K,SEAWATER, SALINITY = 3.25 %, PH =8.0	1668
*HYDROLYSIS OF PARATHION TO P-NITROPHENOL	41 %	343 K, SEAWATER, SALINITY =3.25 %, PH = 1.0	1668
*HYDROLYSIS OF PARATHION TO P-NITROPHENOL	9 %	343 K, DISTILLED WATER, PH = 5.9	1668
*HYDROLYSIS OF PARATHION TO P-NITROPHENOL	3 %	343 K, DISTILLED WATER,30 G NACL/ L, PH = 5.5	1668
*HYDROLYSIS OF PARATHION TO P-NITROPHENOL	21 %	343 K, DISTILLED WATER + DIL NAOH, PH = 8.6	1668
*HYDROLYSIS OF PARATHION TO P-NITROPHENOL	95 %	343 K, DISTILLED WATER + DIL NAOH, PH = 12	1668
*HYDROLYSIS OF PARATHION TO P-NITROPHENOL	24 %	343 K, BUFFER, SALT 30 G/L, PH = 7.9	1668
*HYDROLYSIS OF PARATHION TO P-NITROPHENOL	<10 %	343 K, BUFFER, SALT 0.55 G/L, PH = 7.5	1668

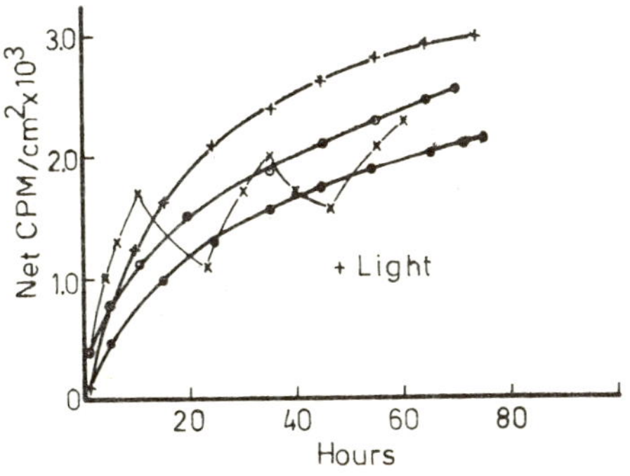

G128 Uptake of Zn-65 by periphyton.
Continous light + .Continous dark o .Killed community • . 12 h photoperiod x
pH 7.6-8.1.

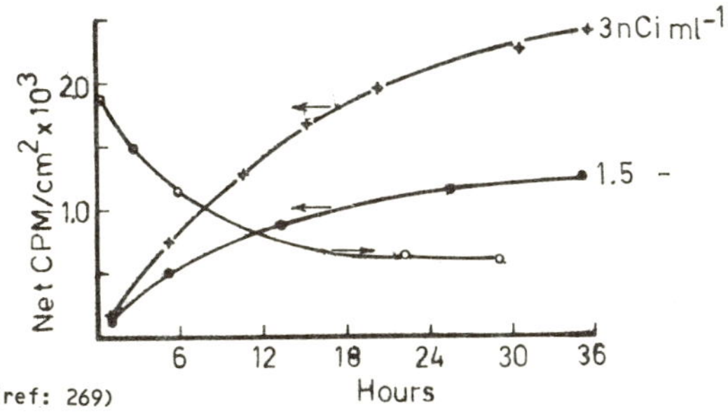

(ref: 269)

G129 Uptake of Zn-65 by periphyton.
Under varying initial ambient concentrations and decline of ambient concentration during experiment (dash curve, right ordinate).

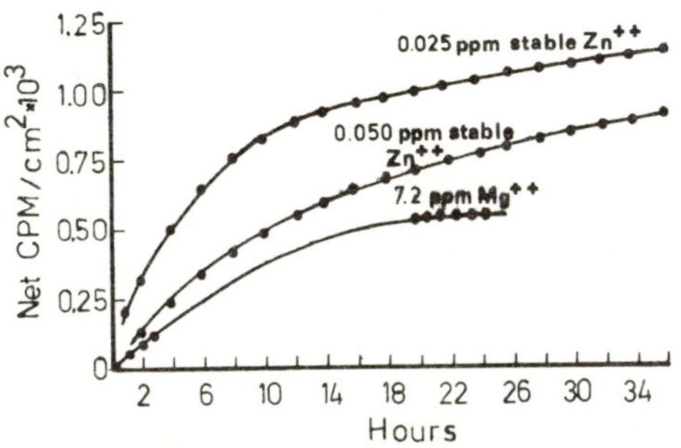

G130 Uptake of Zn-65 by periphytes.

(ref: 269)

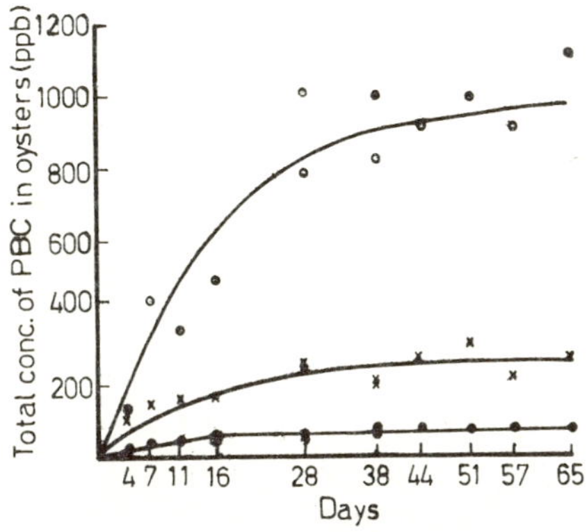

(ref: 1643)

G131 Accumulation of PCB in oysters.
Seawater concentration • 5.5 part x 17 part o 60 part in 10**12.

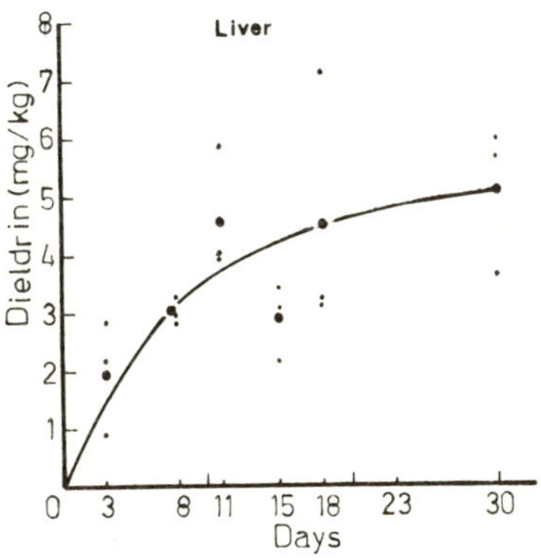

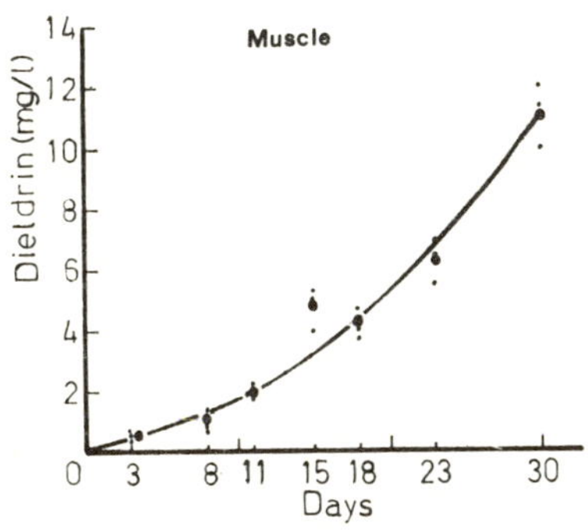

(ref: 1634)

G132 Uptake of dieldrin by plaice (Pleuronectes platessa). Concentration in liver versus time during 30 days exposure to 0.4 µg/l seawater.
(ref: 1634)

G133 Uptake of dieldrin by plaice (pleuronectes platessa). Concentration in muscle tissue versus time during 30 days exposure to 0.4 µg/l seawater.

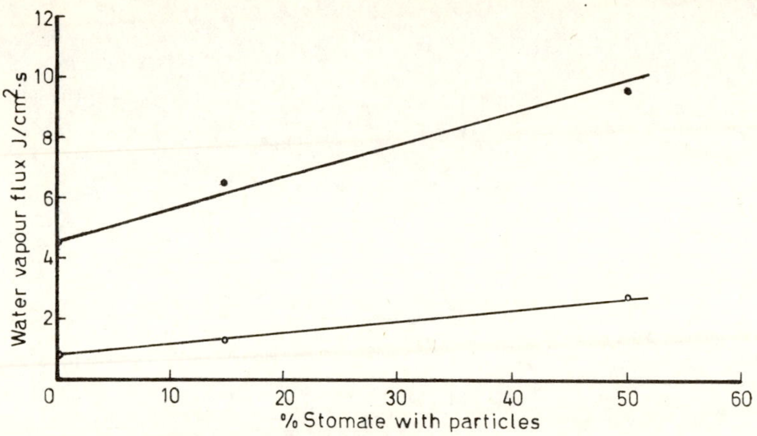

G134 Water vapour flux versus percentage stomata with particles. (ref: 1332)

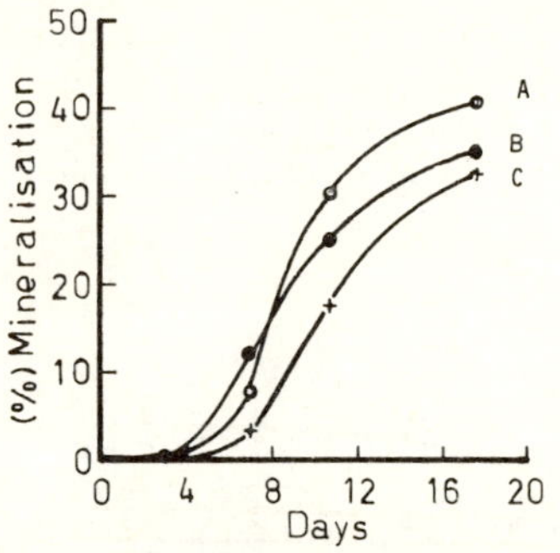

G135 Mineralisation of crude oil by microbial populations of seawater.
Swedish crude oil. The seawater was supplemented with N and P. A,B and D different stations. Mineralisation measured by CO2-evolution.

(ref: 49)

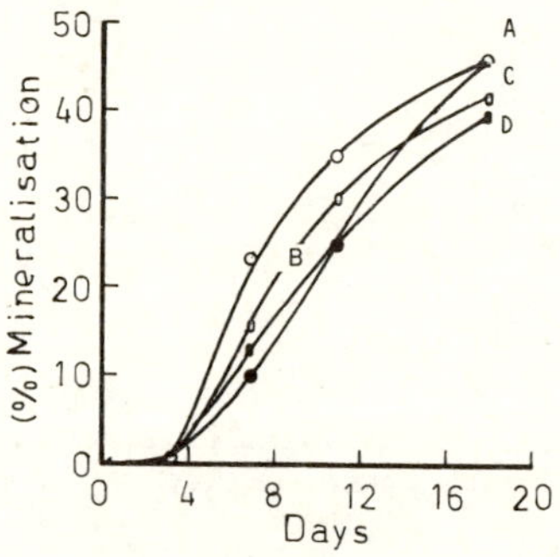

G136 Mineralisation of crude oil by microbial populations of seawater.
South Louisianna crude oil. The seawater was supplemented with N and P. A,B, C and D different stations. Mineralisation measured by CO2-evolution.

G137

Influence of pH on half life of ozone in water

pH	Half-life(min)
7.6	41
8.5	11
8.9	7
9.2	4
9.7	2
10.4	0.5

Ozone, half life in water influenced by pH. (ref: 765)

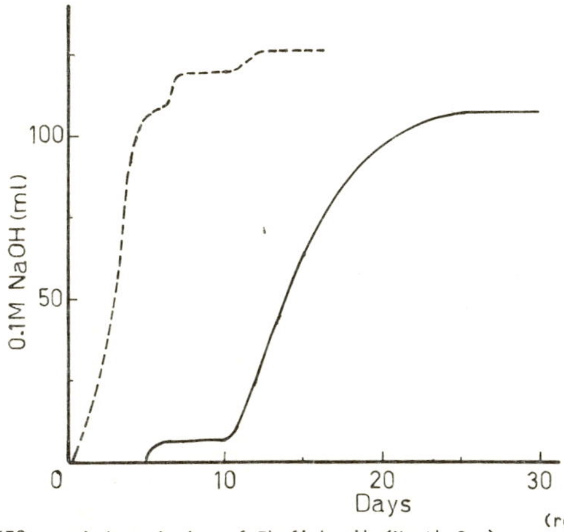

(ref: 1513)

G138 Biodegradation of Ekofish oil (North Sea) at pH = 8.1 by Saccharomycopsis lipolytica. ml 0.1 M NaOH is plotted to time to maintaine desired pH-value for 5 ml oil in 1 l seawater. 10 mg K2PHO4 and 1.0 g (NH4)2SO4 was added the flasksample was shaked at 120 strokes per minute. Unbroken line at 8 celcius, broken line 20 celcius.

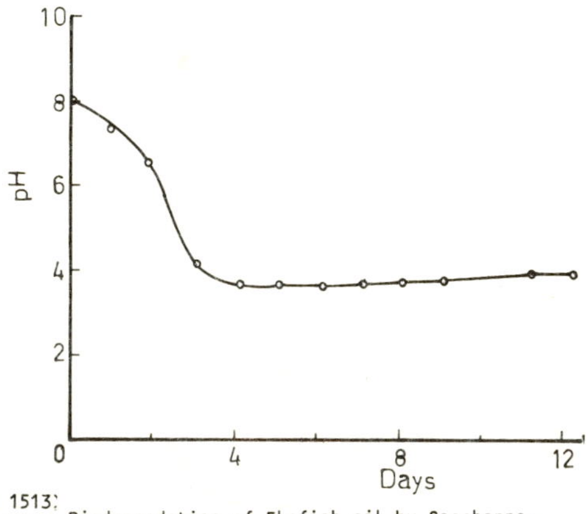

G139 Biodegradation of Ekofish oil by Saccharromycopsis lipolytica indicated as pH-variation. Grown as indicated above.

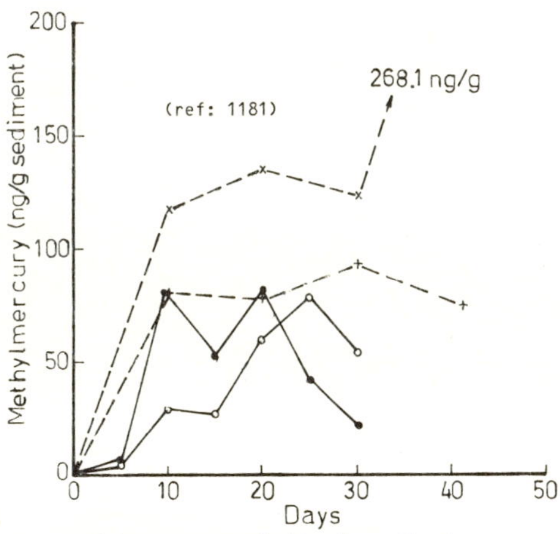

G140 Methyl mercury production by sediment. Sediment type A (see table with properties) from San Francisco Bay.
× 100 µg/g and + 10 µg/g Hg as HgCl2 under anaerobic conditions.
• 100 µg/g and ○ 10 µg/g Hg as HgCl under aerobic conditions.

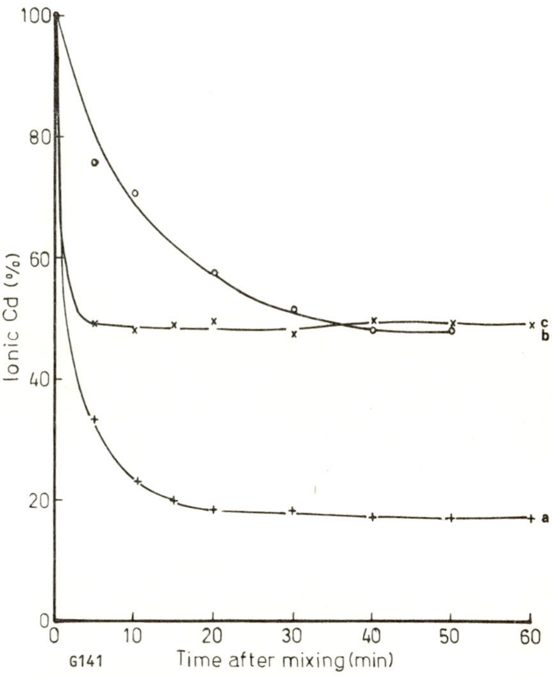

G141

Cadmium concentration (ionic) versus time after addition of different EDTA-concentrations at 25 celcius and pH = 7.8 +- 0.2. (ref: 972)
2*10**(-5) M Cd + seawater + 2*10**(-5) EDTA +
2*10**(-6) M Cd + seawater + 2*10**(-6) EDTA ○
2*10**(-6) M Cd + 0.59 M NaCl + 10**(-2) M Ca/ + 2*10**(-6) EDTA. ×
Ionic Cd in %. Time after mixing in minutes.

TABLE G140

Physical and chemical properties of San Francisco Bay sediments.
REF.: 1181

Properties	Sediment type A	Sediment type B	Sediment type C
Volatile solids (%)	4.9	2.9	1.4
Chemical Oxygen Demand (mg/g)	45.3	39.4	12.5
Chlorides (mg/g)	3.8	5.5	4.8
Ammonia-Nitrogen (ug/g)	30.4	54.3	38.2
Total-Nitrogen (ug/g)	833	750	480
Total Mercury (ug/g)	0.2	1.3	0.1
Total Methylmercury (ng/g)	1.9	0.4	1
Sand (%)	55	63	84
Silt (%)	35	28	12
Clay (%)	10	9	4

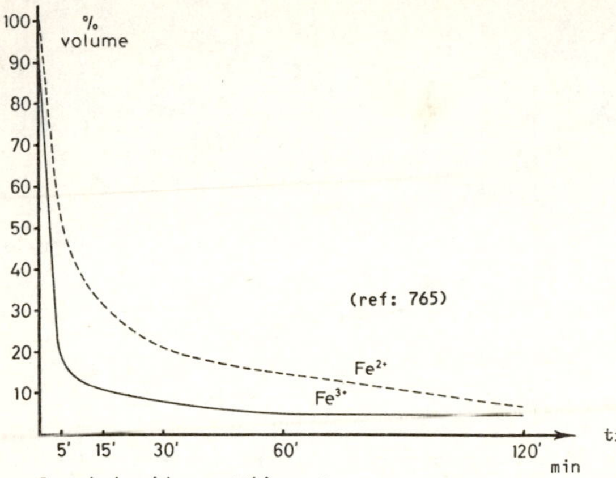

G142 Iron hydroxides, settling rate.

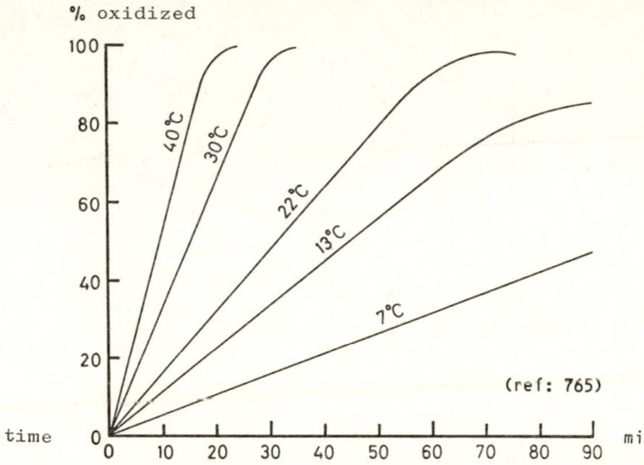

G143 Iron (II) oxidation versus time for different temperatures. Complete aeration.

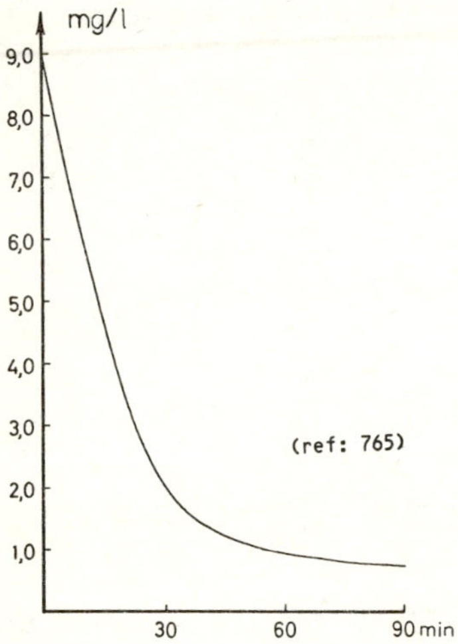

G144 Phenol, oxidation plotted to aeration time.

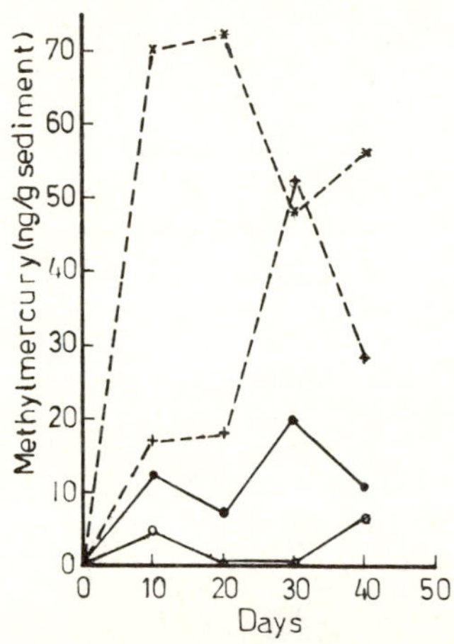

(ref: 1181)

G145 Methyl mercury production by sediment. Sediment type B (see table with properties) from San Francisco Bay.
× 100 μg/g and+10 μg/g Hg as HgCl2 under anaerobic conditions.
• 100 μg/g and ∘ 10 μg/g Hg as HgCl under aerobic conditions.

see table G140

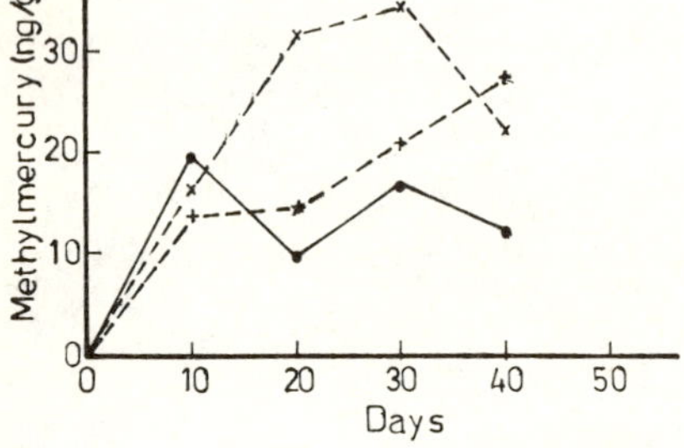

G146

Methyl mercury production by sediment. Sediment type C (see table with properties) from San Francisco Bay.
× 100 μg/g and+10 μg/g Hg as HgCl2 under anaerobic conditions.
• 100 μg/g $HgCl_2$ under aerobic conditions.

(ref: 1181) see table G140

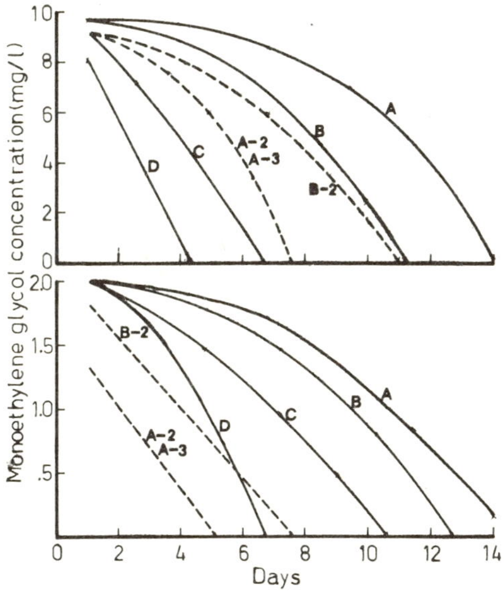

G147 Biodegradation of monoethylene glycol. A: a major water course. B: a major tributary. C: fast running, upland streams. D: slow running river through agriculturel land. A2 and A3: A with added bottom mud with resp. without aeration. B2 and B3: B abstracted for flood conditions with resp. without aeration. At 8 celcius.

(ref: 436)

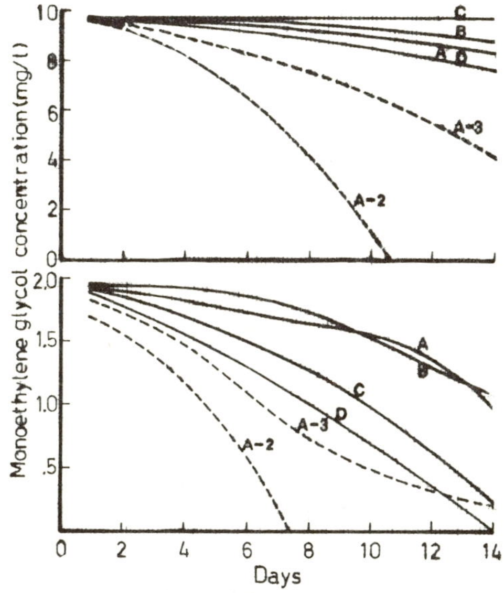

G148 Biodegradation of monoethylene glycol. A: a major water course. B: a major tributary. C: fast running, upland streams. D: slow running river through agriculturel land. A2: and A3: A with added bottom mud with resp. without aeration. B2 and B3: B abstracted for flood conditions with resp. without aeration. At 4 celcius.

(ref: 436)

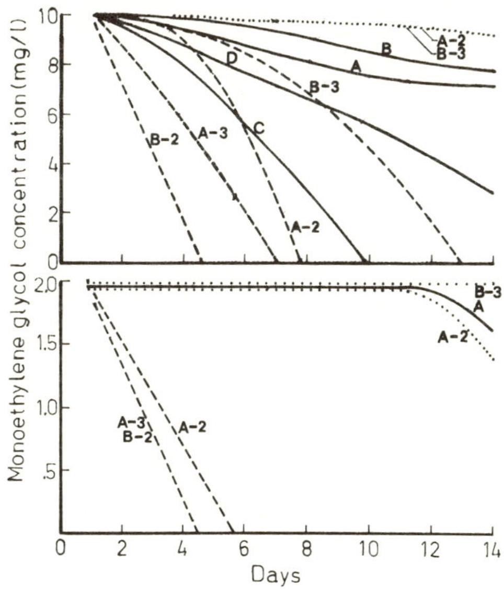

G149 Biodegradation of diethylene glycol at 20 celcius (solid and dashed line) , at 8 celcius (pointed line). Symbols see figure above.

(ref: 436)

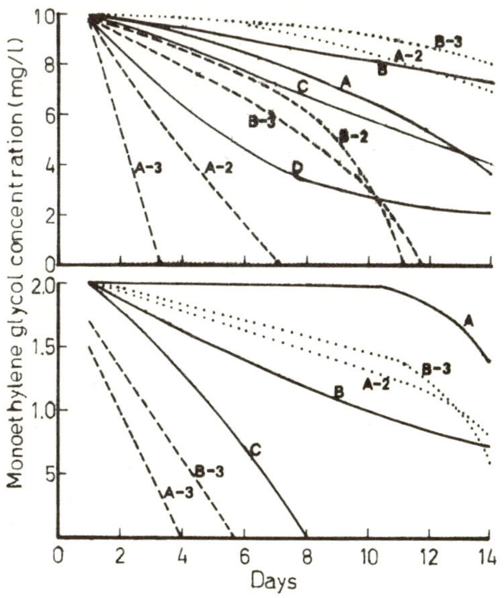

G150 Biodegradation of triethylene glycol at 20 celcius (solid and dashed lines) at 8 celcius (pointed line). Sympols see preceeding figure.

(ref: 436)

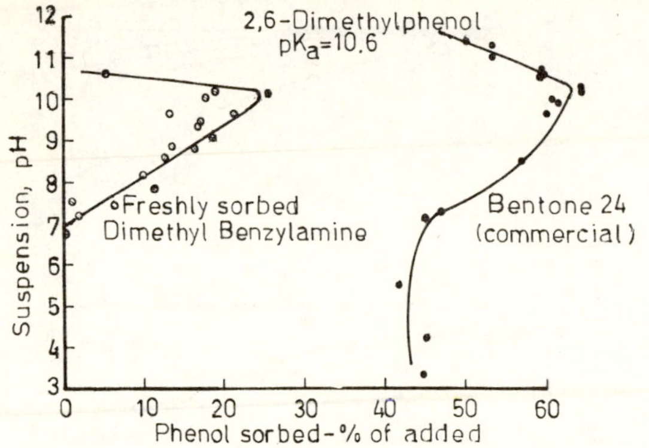

(ref: 1037)

G151 Adsorption of phenols on clay pH effect is shown. Bentone 24 (bentonite) • and freshly sorbed clay ○ (without organics) are used. % phenol adsorbed versus pH for 2,6 dimethylphenol (pK = 10.6). 5 g clay and 0.5 mmoles of phenols per liter is used at 20 celcius.

(ref: 1037)

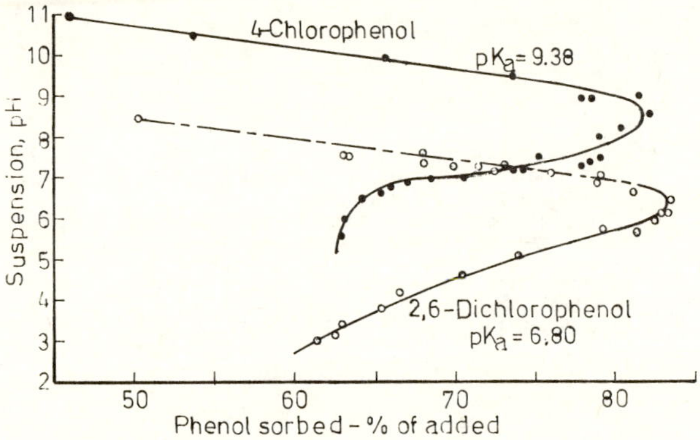

G152 Adsorption of phenols on clay. pH effect is shown. % phenol adsorbed versus pH for 4-chlorophenols • (pK = 9.38) and 2,6-dichlorophenol ○ (pK = 6.80). 20 celcius, 5 g clay and 0.5 mmoles of phenols per liter.

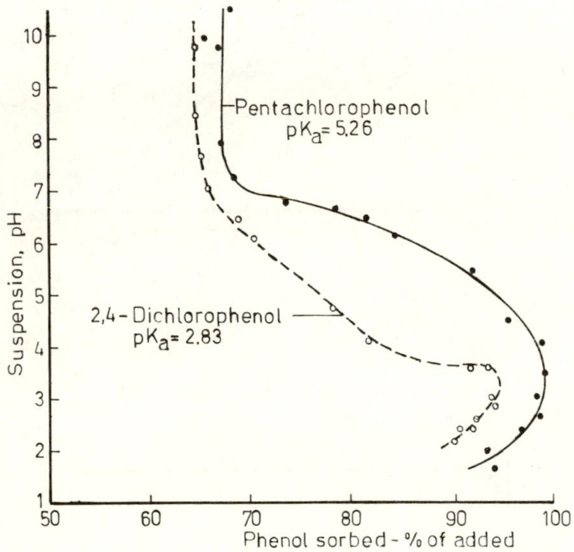

(ref: 1037

G153 Adsorption of phenols on clay. pH effect is shown. % phenol versus pH for penthachlorophenol•(pK = 5.26) and 2,4 dichlorophenol○ (pK = 2.83). Bentone 24 (bentonite) 5 g, is used at 20 celcius, 0.5 mmoles of phenoles per liter.

TABLE G154

Lake Erie food chain Model. Reactions rate for a 7 Compartment Model.
REF.: 1578

Type of Flux	Range of reaction rate (1/days) in compartments
Algae uptake from water	0.117 - 0.253
Algae	.304 - 0.111
Excretion from algae	0.175 - 0.082
Zooplankton grazing on algae	0.128 - 0.028
Zooplankton	0.100
Excretion from zooplankton	0.020
Fish grazing on zooplankton	0.060
Fish	0.020
Excretion from fish	0.002
Lake birds uptake of fish	0.000018
Birds	0.000002
Excretion from birds	0.00000002

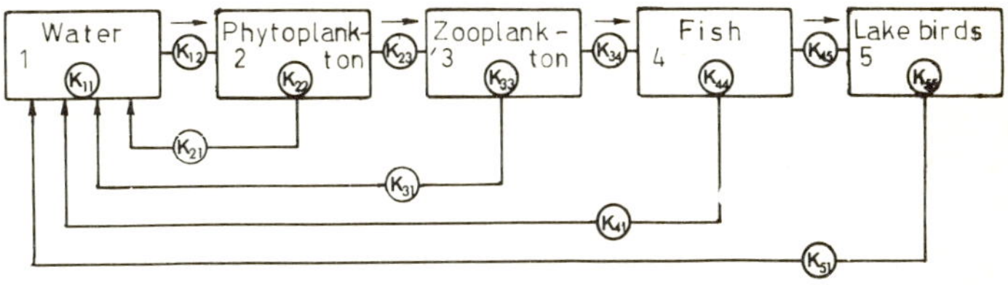

TABLE G155

Values of a Pine-Mor Model. C = degrees celcius, M = mor or Mesostigmata S = Symphypleona, P = Poduromorpha, E = Entomobryomorpha, A = Araneae, T = Trombidiformes, O = Oribatei, F = Formicidae, L = Lithobiomorpha.

REF.: 1703

Symbol	Physical description	Units	Value
k1	Fraction of precipitation appearing as throughfall	number	0.8
k2	Fraction of precipitation appearing as stemflow	number	0.04
k3	Scale factor for potential evaporation	g H2O/(m2*d)	101.6
k4	Temperature sensitivity of potential evaporation	1/C	0.065
k5	Maximum mor moisture fraction	number	2.3
k6	Minimum mor moisture fraction	number	0.4
k7	Temperature sensitivity of k8	1/c	0.069
k8	Standard mor respiration rate	g M/(g m*day)	.0012
k9	Temperature at which k8 determined	C	20
k10	Mor moisture fraction at which k8 determined	number	1
k11	Standard Oribatei ingestion rate of mor	g M/(g O*day)	0.25
k13	Temperature at which k11 determined	C	20
k12	Temperature sensitivity of Oribatei	1/C	0.139
k14	Standard Symphypleona ingestion of mor	g M/(g S*day)	0.14
k15	Temperature sensitivity of Symphypleona	1/C	0.139
k16	Temperature at which k14 determined	C	15
k17	Standard Poduromorpha ingestion rate of mor	g M/(g P*day)	0.14
k18	Temperature sensitivity of Poduromorpha	1/C	0.139
k19	Temperature at which k17 determined	C	15
k20	Standard Entomobrymorpha ingestion of mor	g M/(g E*day)	0.14
k21	Temperature sensitivity of k20	1/C	0.139
k22	Temperature at which k20 determined	C	15
k23	Temperature sensitivity of Araneae	1/C	0.139
k24	Temperature sensitivity of Mesostimata	1/C	0.139
k25	Temperature sensitivity of Trombidiformes	1/C	0.139
k26	Standard Araneae respiration rate	g A/(g A*day)	.0047
k27	Temperature at which k26 determined	C	16
k28	Standard Mesostigmata respiration rate	g M/(g M*day)	0.05
k29	Temperature at which k28 determined	C	16
k30	Standard Trombidiformes respiration rate	g T/(g T*day)	0.05
k31	Temperature at which k30 determined	C	16
k32	Standard Oribatei respiration rate	g O/(g O*day)	.0018
k33	Temperature at which k32 determined	C	16

TABLE G157

Global Stratospheric effects of Supersonic Transports.
Reactions used in Johnson model stratosphere, together with photochemical (j) and kinetic (k) rate constants. Parentheses used with rate constants indicates concentrations.
REF.: 355

Reaction	Rate constant
0 + 02 + M -> 03 + M	k = (0)*(02)*(M)
03 + hv (190 to 350 nm, 450 to 650 nm) -> 0 + 03	j = (03)
03 + hv (above 313 nm) -> 02 + 0(3P)	j = (03)
03 + hv (below 313 nm) -> 02 + 0 (1D)	j = (03)
02 + hv (below 242 nm) -> 0 + 0	j = (02)
0 + 0 + M -> 02 + M	k = (0)**2*(M)
0 + 03 -> 2 02	k = (0)*(03)
NO + 03 -> NO2 + 02	k = (NO)*(03)
NO2 + 0 -> NO + 02	k = (NO2)*(0)
NO2 + hv (260 to 400 nm) -> NO + 0	j = (NO2)
2 NO + 02 -> 2 NO2	k = (NO)**2*(02)
NO + 0 + M -> NO2 + M	k = (NO)*(0)*(M)
NO2 + 03 -> NO3 + 02	k = (NO2)*(02)
NO3 + hv(visible) -> NO + 02	j = (NO3)
NO2 + NO3 + M -> N205 + M	k = (NO2)*(NO3)*(M)
N205 + M -> NO2 + NO3 + M	k = (M)*(N205)
N205 + hv -> 2 NO2 + 0	j = (N205)
N205 + 0 -> 2 NO3 + 02	k = (N205)*(0)
N205 + H20 -> 2 HNO3	k = (N205)*(H20)
HO + NO2 M -> HNO3 + M	k = (HO)*(NO2)*(M)
HNO3 + hv(below 300 nm) -> HO + NO2	j = (HNO3)
HO + HNO3 -> H20 + NO3	k = (HO)*(HNO3)
0(1D) + M -> 0(3P) + M	k = (M)*(0(1D))
0(1D) + H20 -> HOO + 02	k = (H20)*(0(1D))
HO + 03 -> HOO + 02	k = (HO)*(03)
0 + HOO -> HO + 02	k = (0)*(HOO)
0 + HO -> H + 02	k = (0)*(HO)
H + 02 + M -> HOO + M	k = (H)*(02)*(M)
HOO + HO -> H20 + 02	k = (HOO)*(HO)
HOO + NO -> HNO3	k = (HOO)*(NO)
HOO + NO -> HO + NO3	k = (HOO)*(NO)

TABLE G156

Chemical reactions in air. Confirmed SO2/NH3 reaction system.

REF.: 433

Phase	Reaction
Gas phase	NH3 + SO2 <-> NH3:SO2 <-> SONH + H20
Gas phase	x SONH <-> (SONH)x
Gas phase	NH3 + SO2 + H20 <-> NH4HSO4
Gas phase	(SONH)6 + 2 H20 <-> S4H4 + 2 NH4HSO4
Gas phase	2 SO2 + 02 <-> 2 SO3 (slow)
Gas phase	2 SO3 + NH3 <-> NH(NH4SO3)2
Gas phase	NH(NH4SO3)2 <-> NH2SO3H + NH2SO3NH4
Gas phase	NH2HSO3 + NH3 <-> NHSO3NH4
Liquid phase	NH3 + H20 <-> NH4(+) + OH(-)
Liquid phase	SO3 + y H20<->SO2*xH20 + (y-x) H20
Liquid phase	SO2 + 2 NH4(+) 1/2 02 <-> 2 NH4(+) + SO4(-2)
Liquid phase	SO2 + H20 <-> HSO3(-) + H(+)
Liquid phase	SO3 + H20 <-> HSO4(-) + H(+)
Liquid phase	H2SO4 + 2 NH3 <-> (NH4)2SO4
Liquid phase	NH2SO3NH4 + H20 -> (NH4)2SO4
Liquid phase	2 HSO3(-) <-> S205(-2) + H20
Liquid phase	2 S204 + H20 <-> S206(-2) + SO4(-2)
Liquid phase	S4N4 + H20 -> NH3 + elemental S + various oxides

TABLE G155

Values of a Pine-Mor Model. C = degrees celcius, M = mor or Mesostigmata S = Symphypleona, P = Poduromorpha, E = Entomobryomorpha, A = Araneae, T = Trombidiformes, O = Oribatei, F = Formicidae, L = Lithobiomorpha.

REF.: 1703

Symbol	Physical description	Units	Value
k46	Poduromorpha assimilation fraction of mor	number	0.55
k47	Entomobryomorpha assimilation fraction of mor	number	0.55
k57	Araneae ingestion of prey	g P ingested/(g A*g P*day)	0.359
k56	Temperature at which k47 determined	C	15
k59	Araneae assimilation fraction of prey	number	0.93
k81	Mesostigmata ingestion rate of prey	g P ingested/(g M*gP*d) 0.364	.
k83	Mesostigmata assimilation fraction of prey	number	0.47
k105	Trombidiformes ingestion of prey	g P ingested/(g T*g P*d) 0.368	.
k107	Trombidiformes assimilation fraction of prey	number	0.47
k120	Standard Formicidae ingestion of mor	g M/(g F*day)	0
k121	Temperature at which k120 determined	C	15
k122	Formicidae assimilation of mor	number	0.55
k132	Formicidae ingestion of prey	g P ingested/(g F*g P*d)	0.417
k130	Temperature at which k132 determined	C	15
k134	Formicidae assimilation fraction of prey	number	0.93
k147	Temperature sensitivity of Formicidae	1/C	0.139
k157	Lithobiomorpha ingestion of prey	g P ingested/(g L*g P*d)	0.417
k155	Temperature at which k157 determined	C	15
k159	Lithobiomorpha assimilation fraction of prey	number	1
k172	Temperature sensitivity of Lithobiomorpha	1/C	0.13

G5

STOICHIOMETRIC RATIO
TEMPERATURE COEFFICIENTS
TRANSFER RATES
WATER EMISSIONS

TABLE G158

STOICHIOMETRIC RATIO : CARBON			
ITEM:	VALUE:	CONDITIONS:	REF.:
*CARBONDIOXIDE/BOD DECAY	0.2	NONE	1571
*OXYGEN/ALGAL GROWTH	1.6	NONE	1571
*OXYGEN/BIOMASS RESPIRATION	2.	NONE	1571
*OXYGEN/DETRITUS DECAY	2.	NONE	1571

TABLE G159

STOICHIOMETRIC RATIO : NITROGEN			
ITEM:	VALUE:	CONDITIONS:	REF.:
*OXYGEN/AMMONIA DECAY	3.5	NONE	1571
*OXYGEN/NITRITE DECAY	1.2	NONE	1571

TABLE G160

TEMPERATURE COEFFICIENT : FILTRATION			
ITEM:	VALUE:	CONDITIONS:	REF.:
*ARCTICA ISLANDICA	2.05	Q10,277-287K	1737
*ARCTICA ISLANDICA	1.23	Q10,283-293K	1737
*HIATELLA ARCTICA	6.2	Q10,278-288K	1403
*HIATELLA ARCTICA	0.75	Q10,283-293K	1403
*MYTILUS EDULIS	2.15	Q10,277-287K	1572
*MYTILUS EDULIS	4.96	Q10,278-288K	1403
*MYTILUS EDULIS	1.13	Q10,283-293K	1572
*MYTILUS EDULIS	1.22	Q10,283-293K	1403
*MYTILUS EDULIS	1.001	Q10,288-298K	1403
*MYTILUS EDULIS	0.059	Q10,293-303K	1403

TABLE G161

TEMPERATURE COEFFICIENT : NITROGEN			
ITEM:	VALUE:	CONDITIONS:	REF.:
*AMMONIA OXIDATION	1.0876	ACCORDING TO X = X(20)*1.0876** (T-20), T=CEL DEG	1519
*AMMONIA OXIDATION	1.0997	ACCORDING TO X = X(20)*1.0997** (T-20), T=CEL DEG	842
*AMMONIA OXIDATION	1.0953	ACCORDING TO X = X(20)*1.0953** (T-20), T=CEL DEG	1567
*AMMONIA OXIDATION	1.0757	ACCORDING TO X = X(20)*1.0757** (T-20), T=CEL DEG	1095
*AMMONIA OXIDATION	1.0548	ACCORDING TO X = X(20)*1.0548** (T-20), T=CEL DEG	1707
*AMMONIUM -> NITRITE	1.08	TEMP. FACTOR IS 1 AT 293K,DELAWARE ESTUARY	1153
*NITRIFICATION	1.7-1.86	Q10,293-303K,NITROSOMONAS SP., CULTURE	209
*NITRIFICATION	CA. 2	Q10,293-303K,NITROSOMONAS SP. IN SLUDGE	1040
*NITRITE -> NITRATE	1.06	TEMP. FACTOR IS 1 AT 293K,DELAWARE ESTUARY	1153
*ORG. N -> AMMONIUM	1.08	TEMP. FACTOR IS 1 AT 293K,DELAWARE ESTUARY	1153

TABLE G162

TEMPERATURE COEFFICIENT : OXYGEN			
ITEM:	VALUE:	CONDITIONS:	REF.:
*BENTHIC OXYGEN UPTAKE	1.130	278 - 288 K EXPONENT VAL	1567
*BENTHIC OXYGEN UPTAKE	1.080	283 - 293 K EXPONENT VAL	1567
*BENTHIC OXYGEN UPTAKE	1.040	288 - 298 K EXPONENT VAL	1567
*BENTHIC OXYGEN UPTAKE	1.077	283 - 293 K EXPONENT VAL	1567
*BENTHIC OXYGEN UPTAKE	1.090	275 - 295 K EXPONENT VAL	1567
*BENTHIC OXYGEN UPTAKE	1.067	278 - 298 K EXPONENT VAL	1567
*BENTHIC OXYGEN UPTAKE	1.088	278 - 283 K EXPONENT VAL	1567
*BENTHIC OXYGEN UPTAKE	1.041	278 - 288 K EXPONENT VAL	1567
*BENTHIC OXYGEN UPTAKE	1.065	283 - 303 K EXPONENT VAL	1567
*DEGRADATION OF ORGANIC MATTER IN WATER	1.02 - 1.09	1976 EXPONENT VAL	55

TABLE G162

TEMPERATURE COEFFICIENT : OXYGEN

ITEM:	VALUE:	CONDITIONS:	REF.:
*DEGRADATION OF ORGANIC MATTER IN WATER	1.075	Q10, 1972	944
*DEGRADATION OF ORGANIC MATTER IN WATER	1.047	Q10, BOISE RIVER	241
*DEGRADATION OF ORGANIC MATTER IN WATER	1.047	Q10, SAN FRANCISCO BAY ESTUARY	242
*DEGRADATION OF ORGANIC MATTER IN WATER	1.05	Q10, 1973	299
*EXCRETION OF NH4 NITROGEN FROM ZOOPLANKTON	4.34	278.0-283.5 K, PHRONIMA SEDENTARIA = 36.3 MG, Q10	997
*EXCRETION OF NH4 NITROGEN FROM ZOOPLANKTON	1.96	283.5-288.1 K, PHRONIMA SEDENTARIA = 36.3 MG, Q10	997
*EXCRETION OF NH4 NITROGEN FROM ZOOPLANKTON	1.76	288.1-292.5 K, PHRONIMA SEDENTARIA = 36.3 MG, Q10	997
*EXCRETION OF TOTAL NITROGEN FROM ZOOPLANKTON	2.57	278.0-283.5 K, PHRONIMA SEDENTARIA = 36.3 MG, Q10	997
*EXCRETION OF TOTAL NITROGEN FROM ZOOPLANKTON	1.49	283.5-288.1 K, PHRONIMA SEDENTARIA = 36.3 MG, Q10	997
*EXCRETION OF TOTAL NITROGEN FROM ZOOPLANKTON	1.20	288.1-292.5 K, PHRONIMA SEDENTARIA = 36.3 MG, Q10	997
*MYTILUS EDULIS	2.36	Q10, 275 - 295 K, OXYGEN COMSUMPTION	338
*OXYGEN UPTAKE IN SEDIMENTSENTS	3.2	Q10, NATURAL RANGE, LIMFJORDEN	768
*REAREATION OF STREAMS	1.047	298 K, EXPONENT VAL., SURFACE TRANSFER COEF.=0.01 CM/MIN, CHANNAL	1029
*REAREATION OF STREAMS	1.0241	298 K, EXPONENT VAL., SURFACE TRANSFER COEF.=0.02 CM/MIN, STIRRED	1029
*REAREATION OF STREAMS	1.0226	AERATION SYSTEM STIRRED	1029
*REAREATION OF STREAMS	1.020	298 K, EXPONENT VAL., SURFACE TRANSFER COEF.=0.02 CM/MIN, STIRRED	1029
*REAREATION OF STREAMS	1.024	298 K, EXPONENT VAL., SURFACE TRANSFER COEF.=0.03 CM/MIN, STIRRED	1029
*REAREATION OF STREAMS	1.016	298 K, EXPONENT VAL., SURFACE TRANSFER COEF.=0.06 CM/MIN, STIRRED	1029
*REAREATION OF STREAMS	1.016	298 K, EXPONENT VAL., SURFACE TRANSFER COEF.=0.15 CM/MIN, STREAM	1029
*REAREATION OF STREAMS	1.018	298 K, EXPONENT VAL., SURFACE TRANSFER COEF.=0.028 CM/MIN, CHANNAL	1029
*REAREATION OF STREAMS	1.015	298 K, EXPONENT VAL., SURFACE TRANSFER COEF.=0.042 CM/MIN, CHANNAL	1029
*REAREATION OF STREAMS	1.008	298 K, EXPONENT VAL., SURFACE TRANSFER COEF.=0.37 CM/MIN, CHANNAL	1029
*RESPIRATION OF ZOOPLANKTON	3.28	278.0-283.5 K, PHRONIMA SEDENTARIA = 36.3 MG, Q10	997
*RESPIRATION OF ZOOPLANKTON	1.05	283.5-288.1 K, PHRONIMA SEDENTARIA = 36.3 MG, Q10	997
*RESPIRATION OF ZOOPLANKTON	1.77	288.1-292.5 K, PHRONIMA SEDENTARIA = 36.3 MG, Q10	997

TABLE G163

TEMPERATURE COEFFICIENT : OXYGEN UPTAKE

ITEM:	VALUE:	CONDITIONS:	REF.:
*RESPIRATION OF COMMUNITY	1.7 - 1.8	Q10, ARTIC TUNDRA, COMBINED RESPIRATION	295
*RESPIRATION OF COMMUNITY	1.6 - 2.3	Q10, MISSOURI TALLGRASS PRAIRIE SOIL, 283-303 K	859

TABLE G163

TEMPERATURE COEFFICIENT : OXYGEN UPTAKE			
ITEM:	VALUE:	CONDITIONS:	REF.:
*RESPIRATION OF COMMUNITY	2.5	Q10, MIXED HARDWOOD FOREST, COMBINED RESPIRATION, IN LITTER BAGS	1740
*RESPIRATION OF COMMUNITY	3	Q10, MIXED HARDWOOD FOREST, COMBINED RESPIRATION, IN LITTER LAYER	1740
*RESPIRATION OF COMMUNITY	1.7 - 2.0	Q10, SOFT AND HARDWOOD FOREST, SOIL RESP., 293-313 K, COMBINED RESP.	1701
*RESPIRATION OF COMMUNITY	3.5	Q10, CROP LAND	1622
*RESPIRATION OF COMMUNITY	1.8 - 2.1	Q10, CROP LAND, DARK RESP., ARIEAL PART OF BARROW, COMB. RESP.	295

TABLE G164

TEMPERATURE COEFFICIENT : PHOTOSYNTHESIS			
ITEM:	VALUE:	CONDITIONS:	REF.:
*VALUE FOR LINEAR SATURATION LEVEL OF PHOTOSYNTHESIS	2.1	Q10	1553

TABLE G165

TEMPERATURE COEFFICIENT : RESPIRATION			
ITEM:	VALUE:	CONDITIONS:	REF.:
*RESPIRATION IN FRESHWATER FISH	3.28	LEUCISCUS IDUS, EURYTHERMIC Q10	13
*RESPIRATION IN FRESHWATER FISH	2.34	THYMALLUS ARCTICUS BAICALENSIS, Q10 STENOTHERMIC	13
*RESPIRATION IN FRESHWATER FISH	2.18	COREGONUS AUTUMNALIS MIGRATORIUS, STENOTHERMIC Q10	13

TABLE G166

TEMPERATURE COEFFICIENT : SULFATE			
ITEM:	VALUE:	CONDITIONS:	REF.:
*SULFATA REDUCTION IN SEDIMENTS	3.4	Q10, NATURAL RANGE, LIMFJORDEN	768
*SULFATA REDUCTION IN SEDIMENTS	3.9	Q10, NATURAL RANGE, WADDEN SEA	1642

TABLE G167

TEMPERATURE CONSTANT : BACTERIES

ITEM:	VALUE:	CONDITIONS:	REF.:
*NITRIFICATION	0.06 1/DEG.CELCIUS	284-294K,SUSPENDED CULTURE	699
*NITRIFICATION	0.05 1/DEG.CELCIUS	283-293K,SUSPENDED CULTURE	1075
*NITRIFICATION	0.03 1/DEG.CELCIUS	291-302K,ATTACHED CULTURE	1018
*NITRIFICATION	0.05 1/DEG.CELCIUS	283-293K,SUSPENDED CULTURE	1507
*NITRIFICATION	0.07 1/DEG.CELCIUS	283-293K,SUSPENDED CULTURE	1075
*NITRIFICATION	0.05 1/DEG.CELCIUS	278-300K,SUSPENDED CULTURE	335
*NITRIFICATION	0.05 1/DEG.CELCIUS	283-313K,SUSPENDED CULTURE	716
*NITRIFICATION	0.03 1/DEG.CELCIUS	279-298K,SUSPENDED CULTURE	1083
*NITRIFICATION	0.03 1/DEG.CELCIUS	279-298K,ATTACHED CULTURE	1083
*NITRIFICATION	0.04 1/DEG.CELCIUS	278-298K,SUSPENDED CULTURE	1534
*NITRIFICATION	0.06 1/DEG.CELCIUS	280-295K,SUSPENDED CULTURE	251
*NITRIFICATION	0.07 1/DEG.CELCIUS	288-297K,SUSPENDED CULTURE	1214
*NITRIFICATION	0.02 1/DEG.CELCIUS	279-292K,ATTACHED CULTURE	623
*NITRIFICATION	0.04-0.08 1/DEG.CELCIUS	291-297K,SUSPENDED CULTURE,RAW SERV.CARB.ENDOGEN.	70

TABLE G168

TRANSFER RATE : CARBONMONOXIDE

ITEM:	VALUE:	CONDITIONS:	REF.:
*EXCHANGE FROM GAS TO LIQUID	21*10**7 M/SEC	STIRRING = 225 RPM, CO (AIR) = 1.8 MG/CM3, DIST WATER, 298 K	1016
*EXCHANGE FROM GAS TO LIQUID	3.9*10**7 M/SEC	STIRRING = 190 RPM, CO (AIR) = 1.3 MG/CM3, DIST WATER, 298 K	1016
*EXCHANGE FROM GAS TO LIQUID	19*10**7 M/SEC	STIRRING = 160 RPM, CO (AIR) = 5.5 MG/CM3, DIST WATER, 298 K	1016
*EXCHANGE FROM GAS TO LIQUID	13*10**7 M/SEC	STIRRING = 100 RPM, CO (AIR) = 3.5 MG/CM3, DIST WATER, 298 K	1016
*EXCHANGE FROM GAS TO LIQUID	7.2*10**7 M/SEC	STIRRING = 90 RPM, CO (AIR) = 1.6 MG/CM3, DIST WATER, 298 K	1016
*EXCHANGE FROM GAS TO LIQUID	5.1*10**7 M/SEC	STIRRING = 75 RPM, CO (AIR) = 0.6 MG/CM3, DIST WATER, 298 K	1016
*EXCHANGE FROM GAS TO LIQUID	11*10**7 M/SEC	STIRRING = 55 RPM, CO (AIR) = 1.8 MG/CM3, DIST WATER, 298 K	1016
*EXCHANGE FROM GAS TO LIQUID	2.5*10**7 M/SEC	STIRRING = 0 RPM, CO (AIR) = 1.8 MG/CM3, DIST WATER, 298 K	1016
*EXCHANGE FROM GAS TO LIQUID	10*10**7 M/SEC	STIRRING = 230 RPM, CO (AIR) = 3.7 MG/CM3, SEAWATER, 298 K	1016
*EXCHANGE FROM GAS TO LIQUID	5.2*10**7 M/SEC	STIRRING = 150 RPM, CO (AIR) = 3.6 MG/CM3, SEAWATER, 298 K	1016
*EXCHANGE FROM GAS TO LIQUID	4.1*10**7 M/SEC	STIRRING = 110 RPM, CO (AIR) = 2.0 MG/CM3, SEAWATER, 298 K	1016
*EXCHANGE FROM GAS TO LIQUID	7.5*10**7 M/SEC	STIRRING = 75 RPM, CO (AIR) = 1.8 MG/CM3, SEAWATER, 298 K	1016
*EXCHANGE FROM GAS TO LIQUID	1.5*10**7 M/SEC	STIRRING = 0 RPM, CO (AIR) = 1.8 MG/CM3, SEAWATER, 298 K	1016
*EXCHANGE FROM LIQUID TO GAS	10.6*10**6 M/SEC	STIRRING = 225 RPM, CO (AIR) = 1.8 MG/CM3, DIST WATER, 298 K	1016
*EXCHANGE FROM LIQUID TO GAS	2.0*10**6 M/SEC	STIRRING = 190 RPM, CO (AIR) = 1.3 MG/CM3, DIST WATER, 298 K	1016
*EXCHANGE FROM LIQUID TO GAS	9.8*10**6 M/SEC	STIRRING = 160 RPM, CO (AIR) = 5.5 MG/CM3, DIST WATER, 298 K	1016
*EXCHANGE FROM LIQUID TO GAS	6.6*10**6 M/SEC	STIRRING = 100 RPM, CO (AIR) = 3.5 MG/CM3, DIST WATER, 298 K	1016
*EXCHANGE FROM LIQUID TO GAS	3.7*10**6 M/SEC	STIRRING = 90 RPM, CO (AIR) = 1.6 MG/CM3, DIST WATER, 298 K	1016
*EXCHANGE FROM LIQUID TO GAS	2.6*10**6 M/SEC	STIRRING = 75 RPM, CO (AIR) = 0.6 MG/CM3, DIST WATER, 298 K	1016
*EXCHANGE FROM LIQUID TO GAS	5.7*10**6 M/SEC	STIRRING = 55 RPM, CO (AIR) = 1.8 MG/CM3, DIST WATER, 298 K	1016
*EXCHANGE FROM LIQUID TO GAS	1.3*10**6 M/SEC	STIRRING = 0 RPM, CO (AIR) = 1.8 MG/CM3, DIST WATER, 298 K	1016
*EXCHANGE FROM LIQUID TO GAS	6.1*10**6 M/SEC	STIRRING = 230 RPM, CO (AIR) = 3.7 MG/CM3, SEAWATER, 298 K	1016
*EXCHANGE FROM LIQUID TO GAS	3.2*10**6 M/SEC	STIRRING = 150 RPM, CO (AIR) = 3.6 MG/CM3, SEAWATER, 298 K	1016
*EXCHANGE FROM LIQUID TO GAS	2.5*10**6 M/SEC	STIRRING = 110 RPM, CO (AIR) = 2.0 MG/CM3, SEAWATER, 298 K	1016
*EXCHANGE FROM LIQUID TO GAS	4.6*10**6 M/SEC	STIRRING = 75 RPM, CO (AIR) = 1.8 MG/CM3, SEAWATER, 298 K	1016
*EXCHANGE FROM LIQUID TO GAS	0.9*10**6 M/SEC	STIRRING = 0 RPM, CO (AIR) = 1.8 MG/CM3, SEAWATER, 298 K	1016

TABLE G169

TRANSFER RATE : GAS

ITEM:	VALUE:	CONDITIONS:	REF.:
*MASS TRANSFER FROM ATMOSPHERE TO OCEANS AND LAKES	5.5 METER/DAY	WHOLE OCEAN, BEST ESTIMATE, BASED ON C-14	1081
*MASS TRANSFER FROM ATMOSPHERE TO OCEANS AND LAKES	8.8 METER/DAY	HIGH LAT., BEST ESTIMATE, BASED ON C-14	1081
*MASS TRANSFER FROM ATMOSPHERE TO OCEANS AND LAKES	1.9 METER/DAY	LOW LAT., BEST ESTIMATE, BASED ON C-14	1081

TABLE G169

TRANSFER RATE : GAS

ITEM:	VALUE:	CONDITIONS:	REF.:
*MASS TRANSFER FROM ATMOSPHERE TO OCEANS AND LAKES	1.9 METER/DAY	TROPICAL N. ATL., BEST ESTIMATE, BASED ON RN-222	1081
*MASS TRANSFER FROM ATMOSPHERE TO OCEANS AND LAKES	1.8 METER/DAY	N. ATL., BEST ESTIMATE, BASED ON RN-222	1237
*MASS TRANSFER FROM ATMOSPHERE TO OCEANS AND LAKES	3.9 METER/DAY	N. PAC.,BEST ESTI.,MAX=6.8, MIN= 1.8 METER/DAY, BASED ON RN-222	1237
*MASS TRANSFER FROM ATMOSPHERE TO OCEANS AND LAKES	1.9 METER/DAY	S. PAC., BEST ESTIMATE, BASED ON RN-222	1237
*MASS TRANSFER FROM ATMOSPHERE TO OCEANS AND LAKES	5.0 METER/DAY	ANTARCTIC,BEST EST., MAX=18, MIN= 2.3 METER/DAY, BASED ON RN-222	1237
*MASS TRANSFER FROM ATMOSPHERE TO OCEANS AND LAKES	1.6 METER/DAY	GREAT SALT L.,BEST EST., MAX=2.2, MIN=0.4 METER/DAY, BASED ON C-14	1592
*MASS TRANSFER FROM ATMOSPHERE TO OCEANS AND LAKES	0.8 METER/DAY	PYRAMID L.,BEST EST., MAX=2.7, MIN =0.8 METER/DAY, BASED ON C-14	1592
*MASS TRANSFER FROM ATMOSPHERE TO OCEANS AND LAKES	1.9 METER/DAY	WALKER L., BEST EST., MAX=2.7, MIN =1.1 METER/DAY, BASED ON C-14	1592
*MASS TRANSFER FROM ATMOSPHERE TO OCEANS AND LAKES	<0.6 METER/DAY	MONO L., BEST EST., MAX=3.1, MIN= <0.6 METER/DAY, BASED ON C-14	1592
*MASS TRANSFER FROM ATMOSPHERE TO OCEANS AND LAKES	0.4 METER/DAY	LAKE 227, BEST EST., MAX=0.6, MIN= 0.2 METER/DAY, BASED ON RN-222	417
*MASS TRANSFER FROM ATMOSPHERE TO OCEANS AND LAKES	0.16 METER/DAY	LAKE 261, BEST ESTIMATE, BASED ON RN-222	416
*MASS TRANSFER FROM ATMOSPHERE TO OCEANS AND LAKES	0.16 METER/DAY	LAKE 227, BEST ESTIMATE, BASED ON RN-222	416
*MASS TRANSFER FROM ATMOSPHERE TO OCEANS AND LAKES	0.20 METER/DAY	LAKE 304, BEST ESTIMATE, BASED ON RN-222	416

TABLE G170

WATER : EMISSION

ITEM:	VALUE:	CONDITIONS:	REF.:
*DUMPING IN ATLANTIC OCEAN	15808 *10**3 TONS	DREDGE SPOILS,1968	814
*DUMPING IN ATLANTIC OCEAN	3013.2 *10**3 TONS	INDUSTRIAL WASTES,1968	814
*DUMPING IN ATLANTIC OCEAN	4477 *10**3 TONS	SEWAGE SLUDGE,1968	814
*DUMPING IN ATLANTIC OCEAN	574 *10**3 TONS	CONSTRUCTION AND DEMOLITION DEBRIS,1968	814
*DUMPING IN ATLANTIC OCEAN	0 *10**3 TONS	SOLID WASTES,1968	814
*DUMPING IN ATLANTIC OCEAN	15.2 *10**3 TONS	EXPLOSIVES,1968	814
*DUMPING IN GULF OCEAN	15300 *10**3 TONS	DREDGE SPOILS,1968	814
*DUMPING IN GULF OCEAN	696 *10**3 TONS	INDUSTRIAL WASTES,1968	814
*DUMPING IN GULF OCEAN	0 *10**3 TONS	SEWAGE SLUDGE,1968	814
*DUMPING IN GULF OCEAN	0 *10**3 TONS	CONSTRUCTION AND DEMOLITION DEBRIS,1968	814
*DUMPING IN GULF OCEAN	0 *10**3 TONS	SOLID WASTES,1968	814
*DUMPING IN GULF OCEAN	0 *10**3 TONS	EXPLOSIVES,1968	814
*DUMPING IN PACIFIC OCEAN	7320 *10**3 TONS	DREDGE SPOILS,1968	814
*DUMPING IN PACIFIC OCEAN	981.3 *10**3 TONS	INDUSTRIAL WASTES,1968	814
*DUMPING IN PACIFIC OCEAN	0 *10**3 TONS	SEWAGE SLUDGE,1968	814
*DUMPING IN PACIFIC OCEAN	0 *10**3 TONS	CONSTRUCTION AND DEMOLITION DEBRIS,1968	814
*DUMPING IN PACIFIC OCEAN	26 *10**3 TONS	SOLID WASTE,1968	814
*DUMPING IN PACIFIC OCEAN	0 *10**3 TONS	EXPLOSIVES,1968	814
*HG FROM CONTINENTS TO OCEANS	< 3.8*10**9 G/YR	NATURAL,RIVER TRANSPORT TO OCEANS	550
*OIL TO MARINE ENVIRONMENT	0.5*10**12 G/YEAR	FROM TANKER OPERATIONS	1388
*OIL TO MARINE ENVIRONMENT	0.5*10**12 G/YEAR	FROM OTHER SHIP OPERATIONS	1388
*OIL TO MARINE ENVIRONMENT	0.1*10**12 G/YEAR	FROM OFFSHORE OIL PRODUCTION	1388
*OIL TO MARINE ENVIRONMENT	0.2*10**12 G/YEAR	FROM ACCIDENTAL SPILLS	1388
*OIL TO MARINE ENVIRONMENT	0.3*10**12 G/YEAR	FROM REFINERY OPERATIONS	1388
*OIL TO MARINE ENVIRONMENT	0.45*10**12 G/YEAR	FROM INDUSTRIAL AND AUTOMOTIVE WASTES	1388
*OIL TO MARINE ENVIRONMENT	2.1*10**12 G/YEAR	TOTAL, TOTAL PRODUCTION (1969) = 1.82*10**15 G/YEAR	1388
*OIL TO MARINE ENVIRONMENT	0.118*10**12 G	FROM TORREY CANYON DISCHARGE	1388
*OIL TO MARINE ENVIRONMENT	0.003 - 0.011*10**12 G	FROM SANTA BARBARA BLOWOUT	1388
*OIL TO MARINE ENVIRONMENT	LESS THAN 0.1*10**12 G/YEAR	FROM NATURAL SEEPAGE INTO OCEAN, TRANSPORT BY SHIP= 1.18*10**15 G/ YEAR	1388
*OIL TO MARINE ENVIRONMENT	90*10**12 G/YEAR	FROM ATMOSPHERE, OIL VAPORATED FROM CONTINENTS	1388
*SEDIMENT DISCHARGE FROM CONTINENT TO OCEAN	245 TONS/MI**2	NORTH AMERICA, TOTAL DISC= 1.96*10**9 T, DRAINAGE AREA= 8.0*10**6 MI2	682
*SEDIMENT DISCHARGE FROM CONTINENT TO OCEAN	160 TONS/MI**2	SOUTH AMERICA, TOTAL DISC= 1.20*10**9 T, DRAINAGE AREA= 7.5*10**6 MI2	682
*SEDIMENT DISCHARGE FROM CONTINENT TO OCEAN	70 TONS/MI**2	AFRICA, TOTAL DISC=0.54*10**9 T, DRAINAGE AREA=7.7*10**6 MI2	682
*SEDIMENT DISCHARGE FROM CONTINENT TO OCEAN	115 TONS/MI**2	AUSTRALIA , TOTAL DISC=0.23*10**9 T, DRAINAGE AREA=2.0*10**6 MI2	682
*SEDIMENT DISCHARGE FROM CONTINENT TO OCEAN	0.32 TONS/MI**2	EUROPE, TOTAL DISC=0.32*10**9 T, DRAINAGE AREA=3.6*10**6 MI2	682
*SEDIMENT DISCHARGE FROM CONTINENT TO OCEAN	1530 TONS/MI**2	ASIA, TOTAL DISC=15.91*10**9 T, DRAINAGE AREA=10.4*10**6 MI2	682
*SEDIMENT DISCHARGE FROM CONTINENT TO OCEAN	514 TONS/MI**2	TOTAL, TOTAL DISC=20.16*10**9 T, DRAINAGE AREA=39.2*10**6 MI2	682
*SOLIDS, DISSOLVED RIVER SOLIDS INTO OCEANS	39 *10**8 TONS/YR	NONE	550
*SULPHATE FROM RIVERS TO OCEANS	17 *10**6 TONS/YR	EUROPE,WATER FLOW=2.5*10**12TONS/ YR,NATURAL	550
*SULPHATE FROM RIVERS TO OCEANS	37 *10**6 TONS/YR	N.AMERICA,WATER FLOW= 4.6*10**12TONS/YR,NATURAL	550

TABLE G170

WATER : EMISSION			
ITEM:	VALUE:	CONDITIONS:	REF.:
*SULPHATE FROM RIVERS TO OCEANS	39 *10**6 TONS/YR	S.AMERICA,WATER FLOW= 8.2*10**12TONS/YR,NATURAL	550
*SULPHATE FROM RIVERS TO OCEANS	81 *10**6 TONS/YR	AFRICA,WATER FLOW=6.0*10**12TONS/ YR,NATURAL	550
*SULPHATE FROM RIVERS TO OCEANS	94 *10**6 TONS/YR	ASIA,WATER FLOW=11.2*10**12TONS/ YR,NATURAL	550
*SULPHATE FROM RIVERS TO OCEANS	268 *10**6 TONS/YR	GLOBAL,WATER FLOW=32.5*10**12TONS/ YR,NATURAL	550
*SULPHATE FROM RIVERS TO OCEANS	45 *10**6 TONS/YR	EUROPE,WATER FLOW=2.5*10**12TONS/ YR,POLLUTIONAL PART	550
*SULPHATE FROM RIVERS TO OCEANS	55 *10**6 TONS/YR	N.AMERICA,WATER FLOW= 4.6*10**12TONS/YR,POLLUTIONAL PART	550
*SULPHATE FROM RIVERS TO OCEANS	0 *10**6 TONS/YR	S.AMERICA,WATER FLOW= 8.2*10**12TONS/YR,POLLUTIONAL PART	550
*SULPHATE FROM RIVERS TO OCEANS	0 *10**6 TONS/YR	AFRICA,WATER FLOW=6.0*10**12TONS/ YR,POLLUTIONAL PART	550
*SULPHATE FROM RIVERS TO OCEANS	0 *10**6 TONS/YR	ASIA,WATER FLOW=11.2*10**12TONS/ YR,POLLUTIONAL PART	550
*SULPHATE FROM RIVERS TO OCEANS	100 *10**6 TONS/YR	GLOBAL,WATER FLOW=32.5*10**12TONS/ YR,POLLUTIONAL PART	550
*SUSPENDED RIVER SOLIDS INTO OCEANS	180 *10**8 TONS/YR	NONE	550

Footnotes

F001 No growth of algae without addition of nitrogen, and no growth whit a mixture of amino acids in dark.

F002 The uptake rate of nutrients is a modifided equation, so than the uptake rate becomes zero for a posetive value of nutrient concentration. This concentration of nutrient is given by SI.

F003 See TABLE A81 and TABLE E84 together

F004 See TABLE A81 and TABLE E84 together

F005 Fecal pellets from: Meganyctipphanus norvegia, Euphasia krohnii and nematoscelis magalops.

F006 Fooo = juvenile Artemia sp. or Phaoedactylum tricornatum.

F007 pT50 = -log(potential toxicity)50

F008 Numbers indicate comparision with growth in control culteres:
4 = full potential similar to control.
3 = slight inhibition, turbidity and chlorophyll decrease
2 = moderate inhibition, turbidity and chlorophyll decrese with yellowing of cells.
1 = marked inhibition, turbidity decrese, white to yellow coloration, with few living cells.

F009 Bacterial cells: Bacillus sp., Sarcina sp., Achromobacter sp., Flavobacterium sp., Brevibacterium sp., Chromobacterium sp., Pseudomonas sp., Azobacter sp., and Streptomyces sp.

F010 TWEEN 80 = polyoxyethylene (80) monooleate

F011 Bacterial colonies: Clear lake : Pseooomonas sp., Flavobacterium sp., Brevibacterium sp., Enterobacter sp. Achromobacter sp., Escherichia sp., Sarcina sp., Caulobacter sp., Proteus sp., Micrococcus sp., Streptomyces sp.
Lake Houston: Pseudomonas sp., Brevibacterium sp., Bacillus sp., Flavobacterium sp., Sarcina sp., Achromobacterium sp., Micrococcus sp., Enterobacter sp.

F012 Weed, unpolluted soil. Eclipta prostrata, Gnaphalium affine, Lapsana apogonoides, Dopatrium junceum, Vandellia procumbens, Mazus japonicus, Rotala indica, Vicia angustifolia, Stellaria alsine, Chenopodium ambrosiodes, Cyperus difformis, Fimbristylis miliacea, Echinochloa crusgalli, Setaria viridis, Digitaria adscendens.

F013 The first value in the collumn for values gives the mortality in %; the second value gives the concentration in the animal. If there is only one value this value gives the mortality in %.

F014 Pseudomonas sp., Xanthomonas sp., Acromobacter sp., and Flavobacterium sp.

F015 Biowellii sp., Gloeosporium sp. etc.

F016 Number of isolates per genus expressed as % of total numbers of colonies isolated.

F017 ZY CL = earthtype

F018 TWEEN 20 = polyoxyethylene (20) sorbitan monolaurate

F019 TWEEN 40 = polyoxyethylene (40) monopalmitate

F020 TWEEN 60 = polyoxyethylene (60) monostearate

F021 The range given in conditions shows the temperature for zero mortality for the first 20 observations. The lower and upper limit is given. Outside this range there is a mortality from the temperature stress.

F022 This table shows the effect of PCB on E. coli. The effect is based on laboratory experiments. The incubation time is 24 hours and a volume of 500 ml is used. The mean harvest weight is determinated from this volume.

F023 Concentration factor is based on :
CF(element) = ((mol % conc of element in coral)/(mol % conc of Ca in coral))/((mol % conc of element in seawater)/(mol % conc of Ca in seawater)).

F024 Halowax 1000 = 60 % mono- and 40 % dichloronaphthalene, 26 % chlorine (weight).

F025 Halowax 1013 = 10 % tri-, 50 % tetra- plus 40 % pentachloronaphthalene, 56 % chlorine (weight)

F026 Halowax 1014 = 20 % tetra-, 40 % penta- plus 40 % hexachloronaphthalene, 62 % chlorine (weight)

F027 Conc factor = specific activity of larvae (cpm/g)/specific activity of the water (cpm/g). Initial 203-Hg(+) 6.89*10**5 cpm/g.

F028 Conc factor = specific activity of bacterial cells (cpm/g)/specific activity of media (cpm/g), initial 203-Hg(+) acitivity = 1.63*10**5 cpm/g.

F029 Conc factor = conc in org (p Ci/kg wet wt)/conc in water (p Ci/l).

F030 Podzol - Horizons: F = 2.5 - 5 cm, H = 10 - 20 cm, A1A2 = 25.5 - 33 cm, B1(pan) = 33 cm, B2 = 33 - 43 cm, B3 = 61 - 71 cm

F031 Background sources of Hg: water = 0.028 ppb, fish = 0.102 ppm, food suppl. = 0.4 ug/day, for Pagophilus groenlandicus.

F032 Food = Asperagus, broccoli, brussels sprouts, cabbage, cucumber, eggplant, horseradish, onions, parsnips, peppers, pimentoes, radish, Raddish tops.

F033 Food = Potatoes, sweet potatoes, rutabages, turnips, sugar beet, radish.

F034 Asparagus, broccoli, Brussels sprouts, cabbage, celery, cauliflower, mustard greens, spinach.

F035 Beans, peas, eggplant, tomatoes, collards (=coleworts).

F036 Wheat, rye, oats, rice (polished), maize, popcorn, Sorghum.

F037 Cantalopes, cucumbers, pumpkins, squash, watermelon.

F038 Almonds, banaes, figs, filberts, guavas, mangos, olives, passionfruit, papayas, pecans, pomegranates, pineapples, strawberries, walnuts.

F039 Citrus, pome and stone fruits.

F040 Crude soybean and linseed oils, Crude cottenseed oil, ediable cottenseed oil, edible soybean oil.

F041 Apples, pears, peaches, apricots, small fruits (except strawberries), vegetables (except root vegetables), meat and poultry (on fat basis).

F042 Nuts (shelled), strawberries, root vegetables.

F043 Cherries, plums, citrus and tropical fruit.

F044 Edible cottenseed and maize oil.

F045 Apples, wheat, barley, sorghum, rice (husked and/or polished).

F046 % = ((conc in whole plant) - (conc in processed plant))/(conc in whole plant)*100

F047 This generel footnote indicates as follows:
a=a constant; A=area; ASSE=assimilability of food; AV=availability of food; CA=conc. of phytoplankton carbon, g/m**3; CFGI=conversion factor from global irradiance to light, active in photosynthesis; CK=carrying capacity, g/m**3; CS=conc. of inorganic carbon, g/m**3; D=diffusion constant; Dp=constant, DENIT=rate of denitrification, g/(m**3*d); DENITK=rate coefficient of denitrification, 1/d; DERV=detritus mineralization rate, g/(m**3:d); DET=detritus concentration, g/m**3; DETR=detritus conc., g/m**3; FD=relative day length; FMAX=maximum fractional reduction in daily specific growth rate over euphoric depth; GEK=gas exchange coefficient, m/s; GRZ=grazing rate, g/(m**3*d); I=irradiance; I(0)=irradiance, surface intensity; IDA=irradiance, daily average; IOP(T)=optimum irradiance, IH=light saturation parameter, IK=light saturation parameter; K=constant, unspecific; K20=constant or coefficient at 20 celcius; K(T)=constant or coefficient at T celcius; KOPT=constant at optimum temperature; KC=Half saturation constant of CS-uptake, g/m**3; KDET=Saturation constant for detritus mineralization, g/m**3; KLYS=constant for light inhibition; KN=half saturation constant for NS-uptake, g/m**3; KNFIX=nitrogen fixation coefficient, 1/d; KNIT=nitrogen coefficient for denitrification, g/m**3; KOT=temperature coefficient; KP=half saturation constant for PS-uptake, g/m**3; KPA=saturation constant for intracellular phosphorus; KTP=threshold concentration for grazing, g/m**3; KZ=half saturation constant for grazing, g/m**3; LO=constant; M=concentration of denitrification microorganisms, g/m**3; M(0)=conc. of denitrification microorganisms at time=0, g/m**3; MIXR(Z)=mixing ratios at depth; MORT=mprtality; MORTK=mortality coefficient, 1/d; MY=growth rate of phytoplankton, 1/d; MYDET=detritus mineralization rate coefficient, 1/d; MYM=growth rate coefficient of denitrification microorganisms, 1/d; MYZ=growth rate of zooplankton; NA=concentration of nitrogen in phytoplamkton, g/m**3; NH4-NS=ammonia-nitrogen concentration, g/m**3; NIT=concentration of nitrate-nitrogen, g/m**3; NIT(0)=concentration of nitrate-nitrogen at time=0, g/m**3; NITRIK=nitrification rate, g/(m**3*d); NITRIK=nitrification rate coefficient, 1/d; NS=concentration of soluble inorganic nitrogen, g/m**3; NSO=constant; OX=oxygen concentration, g/m**3; PA=Concentration of phosphorus in phytoplankton, g/m**3; PHOTO=photosynthetic rate, g/(m**3*d); PHYT=phytoplankton concentration, g/m**3; PHYT(N)=concentration of phytoplankton with nitrogen fixation, g/m**3; PKI=equilibrium constant for reaction between phosphorus and carrier, litre/mol; PS=concentration of soluble inorganic phosphorus, g/m**3; PSE=solu

ble inorganic phosphorus in epilimnion, g/m**3; PSH=soluble inorganic phosphorus in hypolimnion, g/m**3; PSO=constant; REF=reflected light; R(T)= total daily radiation; RELK=release coefficient; RESPV=respiration coefficient, g/(m**3*d); RESP= respiration, 1/d, SEDT=settling rate of detritus, m/d; SEPH=settling rate of phytoplankton, m/d; SETTL=removed by settling, 1/d; SEV= settling rate, m/d; SEV(O)=surface overflow rate, m/d; SEV(p)=settling rate of particles, m/d; SN=nitrogen in upper layer of sediment, g/l; SSE=average concentration of solute S in epilimnion, g/m**3; SSJ=average concentration of s in depth interval J, g/m**3; STO=stochiometric constant mg bacterial mass per mg substrate; t=time, days; T=temperature, celcius; TA=temperature, kelvin; TMAX,TMIN=temperature, maximum and minimum. rate coefficient=0 at TMAX and TMIN, celcius; TOP=temperature, optimum, celcius; TR=temperature, reference; TOX=concentration of toxic material, g/m**3; U=mass flow, m/d; UJ=mass flow from the J depth interval; UC=uptake rate of inorganic carbon, g/(m**3*d); UCMAX=max. uptake rate of inorganic carbon, g/(m**3*d); UE=Mass transport due to epilimnion expansion, kg/d; UN=nitrogen uptake rate, g/(m**3*d); UNMAX=maximum nitrogen uptake rate, g/(m**3*d); UP=phosphorus uptake rate, g/(m**3*d); UPMAX=maximum phosphorus uptake rate, g/(m**3*d); V=volume, m**3; DELTA(V)=change of volume, m**3; VE=volume of epilimnion, m**3; W=weight, kg; WIS=wind speed, m/sec; DELTA(X)= distance between centres of two boxes, m,Z=depth, m; ZOO=concentration of zooplankton, g/m**3; AA= attenuation of water, 1/m; BB=specific attenuation coefficient of phytoplankton, g/m**2; CC=specific attenuation coefficient of zooplankton, g/m**2; CC=specific attenuation coefficient of detratis, g/m**2; EE=attenuation coefficient, 1/m.

F048 The Bowen ratio is used when calculation the heat loss from a water surface; and it is defined by Qc = R*Qe, where R is the Bowen ratio, Qc = convective heat transfer and Qe = heat loss due to evaporation (kcal/(m**2* sec)). The symbols used are T = water surface temperature (Celcius), TA = dry bulb air temperature (Celcius), ES = saturation vapour pressure at the water surface temperature (mb), EA = vapor pressure of the overlaying atmosphere (mb).

F049 CS = saturation value at sea level, at the water temperature (mg/l), CS' = corrected value at the altitude of the river ((mg/l), PB = barometric pressure at river altitude (mm Hg), PV = saturation vapor pressure of water at the river temperature (mm Hg)

F050 K = regional dispersion factor (equal to 1 for Big Blue River), U = average velocity, V = x/tp, where x = reach length and tp = time to peak arrival of constituent concentration at sampling station.
h = A/B, where A = cross-sectional area and B = top width of flow.
R = density, P = coefficient of viscosity.

F051 DG = degrees days calculated as the number of days of the year for which the mean temperature excreeds 40 F.

F052 From the table can expression for the light compensation curve be derived.
I(T) = 0 = 1/C*ln((B(T) = A(T))/B(T))
See TABLE E72 and TABLE E83 together

F053 EL = longtidual dispersion coefficient, (length**2/time), E = rate of energy dissipation per unit mass, LE = mean size of eddies participating in the mixing process, CL = function of relative channal roughness.

F054 See all or more than one of the lines in the condition, to get a explanation of the parameters.

F055 All expressions given to base e and for 20 celcius unless otherwise noted. Defination of symbols:
DM: Molecular diffusion coefficient (8.1*10**(-5) hour (at 20 celcius). U : stream velocity. H : Average stream depth. DELTA(H) : change in stream elevation between two points. T = travel time between the two points for which DELTA(H) is measured. S = Slope (change in water surface elevation divided by distance). Q : Specific discharge (cfs per square mile of drainage area). DL = Longitudinal dispersion cogfficient. UX = Bed shear velocity.

F056 I = Ionic strength = 1/2*SIGMA(Ci*Z**2), A = 2*10**6*(e*T)**(2/3), e = dielectric constant, A = 0.5 for water at 25 celcius, Z = charge of ion, B = 0.33 in water at 25 celcius, c = adjustable parameter (Jngstroem units) corresponding to the size of the ion.

F057 Langmuir isotherm: SP = PCS*CE/(K+CE) - NAP.
SP: sorbed amount of sorbant (mg/g dry wt). PSC: sorbant sorption capacity (mg/g dry wt). CE: equilibrium concentration of sorbant (mg/l). K: half saturation constant (mg/l). NAP: native amount of adsorbed sorbant (mg/g dry wt).
Freundlich isotherm: SP = KC*CE**p - NAP.
SP: sorbed amount of sorbant (mg/g dry weight). CE: equilibrium concentration (mg/l), KC and p are constants. NAP: native amount of adsorbed sorbant (mg/g dry wt).

F058 Experiment 1: oxidized mud and demineralized water

F059 Experinent 2: oxidized mud and lake water.

F060 Experiment 3: reduced mud and demineralized water.

F061 El = effective mixing coefficent (ft**2/sec), CL = tidally-induced mixing coefficient (no dimension), y = tidally averaged depth, U = tidally averaged absolute value of the velocity, RU = standard deviation of the velocity, RY = standard deviation of the depth, C = density-induced mixing coefficient (ft**3/sec/mg/l-salinity).

F062 SI = incident solar radiation, TA = air temperature and TW = water temperature (both in F degrees), E = evaporation (inches/hour) = 0.000073*U(rw - ra)
RW = the vapor pressure for water in millibars
RA = the vapor pressure for air in millibars.

F063 LW = latent heat of vaporization (kcal/(m**2*sec), E = evaporation rate (m/sec). E is approximated by E = (A+B*V)*(ES-EA), where A and B are empirical coefficients, V = wind speed over water surface (m/sec), ES = saturation vapor pressure at the surface water temperature (mb), Ea = vapor pressure of the overlying atmosphere. A is ofen zero and B about 10**(-9)

F064 Wind velocity U is given in cm/sec, the width of diameter of organosms (W) in cm, T in cel degrees and HC is in 10**(-3) cal/(cm**2*min*(cel deg)).

F065 PT = penetration of radiation, LAI = leaf area index, SP = shadow projection of leaf in axis of radiation, IS = solar altitude.

F066 Ra = resistance (s/m), O = foliage density = ratio of leaf plan to leaf frontal area, here = 4.0, D = sqrt(leaf area), U = wind speed (m/s).

F067 F = Flow, N2 - N1 = Concentration gradient, z = thickness of layer, B = 200 cm**2/sec in the upper layer from the surface to the upper boundary of the thermocline; in the thermocline B = 20 cm**2/sec, and in the waters below the thermocline B = 150 cm**2/sec.

F068 KS = stomata conductance (cm/s), DE = vapour pressure deficit, 1 year old needles, Temp > 288 K, light > 500 uE/(m**2*sec).

F069 Respiration measured in: ml O2/(mg dry wt*day)

F070 Excretion measured in: ug atom N or P/(mg dry wt*day).

F071 KE = extinction coefficient (1/m), AO = light extinction coefficient of water (1/m), R(I) = self-shading coefficient of constituent I, S(I) = concentration of suspended matter , I, mg/l (detritus, sediment, algae), Sigma denotes the summation over the different I.

F072 The constant M is species specific. M has the values 207, 115 and 55 kg/(ha*day) for sorghun, wheat and alfalfa respectively.

F073 QAT = net long wave atmospheric radiation (kcal/(m**2*s) CLOUD = fraction of cloud cover, TA = dry bulb air temperature (celcius).

F074 The coefficient of bed activity in a river is defined according to KD = K1 + n*(V/D), where n is this coefficient. KD is deoxygenation constant in the river, K1 is the BOD (Biological Oxygen Demand) measured in the laboratory, V is the stream velocity (ft/sec) and D is the stream depth. The formula and the table enables one ot transfer laboratory values the the field.

F075 The system has some, but not all the properties of seawater. These properties are TOTAL METAL CONCENTRATION: the value is given in conditions and TOTAL LIGAND CONCENTRATION: the value is given in condition.

F076 Used in models with densimetric flow in vertical direction. D = half the thickness of the withdrawal of the inflow flow field (meters), Q = inflow or outflow rate (m**3/sec), W = the width of the resevoir at inflow or withdrawal depth (meters), B = the density gradient at the point of inflow or withdrawal (kg.m**4).

REFERENCES

A

1 Aakesson, B. 1975. In; J.H. Koeman et al.(eds), Sublethal Effects of Toxic Chemicals on Aquatic Animals, Elsevier Scientific Publishing Company, Amsterdam, Oxford, N.Y., pp 121-135.

2 Abernathy, A.R. and P.M. Cumbie. 1977. Bull. Environ. Contam. Toxicol. 17(5), pp 595-602.

3 Adams, F.S., H. Cole and L.B. Massie. 1974. Environ. Pollut., 5, pp 117-147.

4 Adams, J.A. and J.H. Steele. 1966. Some contemporary studies in Marine Science. Shipboard Experiment on the feeding of Calanus Finmarchicus, H. Barnes(ed), G. Allen and Unwin LTD., London, pp 19-85.

5 Admiraal, W. 1977. Mar. Biol., 42(1), pp 1-8.

6 Aguayo, H.M. 1974. Invest. Pesq. Inst. Fom. Pesq. Chile, (No 19), 17 p.

7 Aharonson, N. and U. Kafkafi. 1975. J. agric. Fd. Chem., 23(3), pp 434-437.

8 Aharonson, N. and V. Kafkafi. 1975. J. Agric. Food. Chem. 23(4), pp 720-724.

9 Ahlgren, I. 1973. Scripta Limn. Upsaliensa, No 334.

10 Albrecht, M.-L. 1964. Deutsche Fischerei-Zeitung 11, pp 371-376.

11 Albright, L.J. and E.L. Wilson. 1974. Water Res., 8(2), pp 101-105.

12 Alexander, M. and P.W. Wilson. 1954. Appl. Microbiol., 2, pp 135-140.

13 Alikin, Y.S. 1976. Zh. Obshch. Biol. 37(1), pp 127-134.

14 Allan, J.D., S. Richman, D.R. Heinle and R. Huff. 1977. Mar. Biol., 43(4), pp 317-331.

15 Almgren, T. and I. Hagstroem. 1974. Water Res. 8(7), pp 395-400.

16 Almquist, E. 1974. Ambio, 3(5), pp 161-167.

17 Altman, P.L. and D.S. Dittmar. 1964. Biology Data Book. Wash. D.C. Federation of Amer. Soc. Exp. Biol.

18 Altshuller, A.P. and J.J. Bufalini. 1971. Environ. Sci. Technol., 5(1), pp 39-64.

19 Alverson, D.L. and M.J. Carney. 1975. J. Cons. int. Explor. Mer., 36(2), pp 133-143.

20 Ambler, J.W. and B.W. Frost. 1974. Limnol. Oceanogr. 19(3), pp 446-451.

21 American Public Health Association. Standard Methods for the Examination of Water and Wastewater. 1973. 13tn Edition. Wash, D.C.

22 Amiard-Triquet, C. 1974. Mar. Biol. 26, pp 161-165.

23 Amiard, J.-C. 1976. Health Phys., 31(4), pp 371-373.

24 Ancellin, J. and A. Vilquin. 1966. in: Disposal of Radioactive Wastes into Seas, Ocean and Surface Water. pp 583-604. IAEA. Vienna.

25 Ancellin, J. and A. Vilquin. 1968. Radioprot. 3(3), pp 185-213.

26 Andersen, J.M. 1974. Arch. Hydrobiol., 74, pp 528-550.

27 Anderson, A. and K.O. Nilsson. 1972. Ambio, 1(5), pp 176-179.

28 Anderson, G.E. and R. P. Zeutschel. 1970. Limnol. Oceanogr., 15(3), pp 402-407.

29 Anderson, J.M. 1973. In; D. Meadows (ed); Towards Global Equlibrium. M.I.T.-Press.

30 Anderson, L.W.J. and B.M. Sweeney. 1977. Limnol. Oceanogr., 22(3), pp 539-551.

31 Andersson, A. and K.O. Nilsson. 1974. Ambio, 3(5), pp 198-199.

32 Andrew, R.W., K.E. Biesinger and G.E. Glass. 1977. Water Res., 11(3), pp 309-315.

33 Ann, H.B. 1973. Bull. Fish. Res. Dev. Agency, Pusan 10, pp 73-85.

34 Annett, C.S., M.P. Fadow, F.M. D'Itri and M.E.Stephenson. 1972. Mich. Acad., 4(3), pp 325-337.

35 Annual Report of the Council on Environmental Quality. 1971. U.S. Government Printing Office.

36 Anraku, M. and M. OMori. 1963. Limnol. Oceanogr., 8(1), pp 116-126.

37 Anraku, M. 1964. J. Oceanogr. Soc. Japan, 20(5), pp 19-29.

38 Anthonisen, A.C. 1974. The Effects of Free Ammonia and Free Nitrous Acid on the Nitrification Process. Ph. D. Thesis. Cornell University. Ithaca. New York.

39 Antia, N.J., C.D. McAllister, T.R. Parsons, K. Stephens and J.H.D. Strickland. 1963. Limnol. Oceanogr., 8, pp 166-183.

40 Archer, F.C. 1971. In: Tech. Bull. 21, pp 150-158. Trace Elements in Soils and Crops.

41 Armstrong, D.E., G. Chester and R.F. Harris. 1967. Soil Sci. Soc. Amer. Proc., 31, pp 61.

42 Arnold, D.E. 1971. Limnol. Oceanogr., 16(6), pp 906-920.

43 Aruga, Y. 1965. Bot. Mag., Tokyo, 78, pp 280-288.

44 Aruga, Y. 1965. Bot. Mag., Tokyo, 78, pp 360-365.

45 Arvik, J.H. and R.L. Zimdahl. 1974. J. Environ. Qual., 3(4), pp 374-376.

46 Aston, A.S. and I. Thornton. 1977. Sci. Total Environ. 7(3), pp 247-260.

47 Aston, R.J. 1974. Environ. Pollut., 5, pp 1-10.

48 Aston, S.R., I. Thornton and J.S. Webb. 1975. Sci. Total Environ., 4(4), pp 345-358.

49 Atlas, R.M. and R.Bartha. 1974. Environ. Pollut., 4, pp 291-300.

50 Austin, E.R. and F.G. Lee. 1973. J. Water Poll. Cont. Fed. 45, pp 870-879.

51 Austin, J. and J.D. Redmond. 1972. Mar. Poll. Bull. 3(3), pp 45-47.

52 Autian, J. 1973. Environ. Helth Perspect., 4, pp 3-26.

53 Avargues, M., L. Foulquier and A. Vilquin. 1973. Radio- protection, 8, pp 19.

54 Azim, M.A. and E.R. Roberts. 1956. Biochim. Biophys. Acta, 21, pp 308-312.

B

55 Baca and Arnett. 1976. A Limnological Model For Eutrophic lakes and Impoundments. Battle Pacific NW Lab. EPA.

56 Bachmann, R.W. 1963. in: Schutz, V. and A.W. Klement, Jr. (eds.); Radioecology. pp 485-496. Reinhold Publ. Corp. New York.

57 Bailey, G.W., A.D. Barnett, W.R. Payne, Jr and C.N. Smith. 1974. EPA-6602-74-017, U.S. Environ. Protection Agency. Washington D.C..

58 Bailey, S. P.J. Bunyan ,P.M. Jennings, J.D. Norris. P.I. Stanley and J.H. Williams. 1974. Agro-Ecosyst., 1(4), pp 323-328.

59 Bailey, T.F. 1970. J. San. Eng. Div, ASCE. 96, SA2, pp 279.

60 Baity, H.G. 1938. Sewage Works J., 10, pp 539-568.

61 Baker, R.A. and M.-D. Luh. 1971. Water Res., 5(10), pp 839-848.

62 Balch, N. 1972. Limnol. Oceanogr., 17(6), pp 906-908.

63 Baldini, L.E. and F. Cugurra. 1974. Water Res., 8(5), pp 323-324.

64 Bansal, M.K. 1975. Wat. Res. Bull. 11(3), pp 491-504.

65 Baptist, J.P. and T.J. Price. 1962. Fishery Bull. Fish Wildl Serv. U.S., 206, pp 177-187.

66 Baptist, J.P., D.E. Hoss and C.W. Lewis. 1970. Health Phys., 18(2), pp 141-148.

67 Barber, R.T. 1968. Nature, 220, pp 274-275.

68 Bargmann, R.F. and E. Halfon. 1977. Ecol. Modell., 3(3), pp 211-226.

69 Baric, J. 1974. Acta Adriat., Inst. Oceanogr. Ribar. 15(10). pp 1-12.

70 Barnard. 1974. Water and Wastes Engng., 11(7-8), pp 33-44.

71 Barnes, C.A., A.C. Duxbury and B.A. Morse. 1972. in: Prutter, A.T. and D.L. Alverson. (eds); The Columbia River Estuary and adjacent Ocean Waters: Bio-Enviromental Studies. pp 41-80. Univ. of Washington Press. Seattle.

72 Barnes, R. 1972. Environ. Pollut., 3, pp 133-138.

73 Barnett, A.P., E.W. Hauser, A.W. White and J.H. Holliday. 1967. Weeds, 15, pp 133-137.

74 Barrett, G.W. 1968. Ecology, 49, pp 1019-1035.

75 Bartholomew, G.A. 1968. in: Gordon, M.S. (ed.); Animal Functions: Principles and Adaptations. MacMillan. New York.

76 Bates, S.S. 1976. Limnol. Oceanogr., 21(2), pp 212-218.

77 Batey, T. 1971. In: Tech. Bull. 21, pp 137-148. Trace Elements in Soils and Crops.

78 Battelle. 1974. Development of a Mathematical Water Quality Model for Gray's Harbor and the Chehalis River, Wash. Documentation Report. Pacific NW Lab. Richland, Wash.

79 Baude, F.J., H.L. Pease and R.F. Holt. 1974. J. agric. Fd. Chem. 22(3), pp 413-418.

80 Baudin, M.F. and O.Ravera. 1972. Limnol. Oceanogr., 17(4), pp 645-649. Ambio, 4, p 166.

81 Baumeister, W. 1958. In; Ruhland, W. (ed); Encyclopedia of Plant Physiology, 4, p 5.

82 Bayogbona, E.O. 1969. in: Proc. Symp. Oceanogr. Fish. resources in trop Atlantic. pp 349-359. Paris. Unesco.

83 Bays, L.R. 1971. Environ. Pollut., 1(3), pp 205-234.

84 Beasley, T.M. and S.W. Fowler. 1976. Mar. Biol., 38(2), pp 95-100.

85 Beaumariage, D.S. 1973. Fla. Mar. Res. Publ., 1, pp 45.

86 Bedrosian, P.H. 1962. Dissertation Abstracts 22, pp 3146.

87 Beers, J.R. 1964. J. Cons. perm. inst. Explor. Mer., 29, pp 123-129.

88 Beers, J.R. 1966. Limnol. Oceanogr., 11, pp 520-528.

89 Beeton, A.M. 1960. J. Fish. Res. Bd Can., 17, pp 517-539.

90 Bell, H.L. 1971. Water Res., 5(6), pp 313-319.

91 Bell, R.K. and F.J. Ward. 1970. Limnol. Oceanogr., 15(5), pp 715-726.

92 Benedek, P. and F. Horvath. 1967. Water Res., 1, p 663.

93 Bengtson, B.E. and C.H. Carlin, A.A. Larson and O. Svanberg.

94 Bengtson, B.-E., C.H. Carlin, A.A. Larsson and O. Svanberg. 1975. Ambio, 4(4), pp 166-168.

95 Bengtsson, B.E. 1974. Bull. Environ. Contam. Toxicol., 12, p 654.

96 Benijts-Claus, C. and F. Benijts. 1975. In; Koeman, J.H. et al (eds); Sublethal effect of Toxic Chemicals on Aquatic Animals. Elsevier . Amsterdam. -Oxford, N.Y., pp 43-52.

97 Bennett, J. 1960. Heredity Lond.,15, pp 65-77.

98 Benoit, D.A. 1976. Water Res. 10(6), pp 497-500.

99 Benoit, R.J. 1957. Limnol. Oceanogr., 2, pp 233-240.

100 Berg, A.R. and G.M. Weiss. 1975. Impacts of Nuclear releases into the Aquatic Environment. IAEA-SM-198/47, pp 121-132. Vienna.

101 Bergamini, P.G., G. Palmas, F. Piantelli and M. Rigato. 1970. Health Phys., 19(4), pp 521-528.

102 Berger, W.H. and D.J.W. Piper. 1972. Limnol. Oceanogr., 17(2), pp 275-287.

103 Berglund, H. 1969. Vatten., 25(2), pp 157-165.

104 Berland, B.R., P.J. Bonin, P.L. Laborde and S.Y. Maestrini. 1972. Mar. biol., 13, pp 338-345.

105 Berman, T. and O. Holm-Hansen. 1974. Mar. biol., 28(4), pp 305-310.

106 Berman, T. and U. Pollingher. 1974. Limnol. Oceanogr., 19(1), pp 31-54.

107 Bernard, J.M. 1973. Pol. Arch. Hydrobiol., 20, pp 207-214.

108 Berner, R.A. 1964. Geochim. cosmochim. Acta, 28, pp 1497-1503.

109 Berthet, P. 1963. In: Doeksen, J. and J. van der Drift. (eds); Soil Organism. North-Holland Publ. Co. Amsterdam.

110 Betzer, S.B. and M.E.Q. Pilson. 1974. Biol. Bull. 146(2), pp 165-175.

111 Beyer, J.E. and H. Lassen. 1976. Multispecies Models of an exploited Sea. IMSOR. Tech. Univ. Denmark.

112 Beyer, J.E. 1976. Fish II. Survival of Fish Larvae - a Single server Queue approch. IMSOR. Tech. Univ. Denmark.

113 Beyer, J.H. 1976. Fish I. Survival of Fish Larvae - a single server Queue approch. IMSOR. Tech Univ. Denmark.

114 Beyer, J.E. 1976. Ecosystems, an Operational Research Approach. IMSOR. Tech. Univ. Denmark.

115 Bhatt, Y.M. 1966. Curr. Sci., 35, p 512.

116 Bhatt, Y.M., V.N. Sastry, S.M. Shah and T.M. Krishnamoorthy. 1968. Proc. Nat. Inst. Sci., India, pt B, Biol. Sec. 34(B,6), pp 283-287.

117 Bhovichitra, M. and E. Swift, 1977. Limnol. Oceanogr., 22(1), pp 73-83.

118 Bienfang, P. and K. Gundersen. 1977. Mar Biol., 43(3), pp 187-199.

119 Bienfang, P.K. 1975. Limnol. Oceanogr., 20(3), pp 402-411.

120 Bierman, V.J. Jr. 1976. In: Canale, R.P. (ed); Mathematical Modeling of Biochemical processes in aquatic Ecosystems. Ann Arbor Science Press.

121 Bierman, V.J, F.H. Verhoff, T.C. Poulson and M.W. Tenney. 1974. In; Middlebrooks, E.J., D.H. Falkenborg and T.E. Moloney (eds). Modelling the eutrophication Process. Ann Arbor Publishers, Ann Arbor, Mich. pp 15-33.

122 Biesinger, K.E. and G.M. Christensen. 1972. J. Fish. Res. bd. Can., 29(12), 1691-1700.

123 Biesinger, K.E., A.E. Lemka , W.E. Smith and R.M. Tyo. 1976. Water Poll. Control, 48(1), pp 183-187.

124 Biggs, D.C. 1977. Limnol. Oceanogr., 22(1), pp 108-117.

125 Billen, G., C. Joiris and R. Wolast. 1974. Water Res., 8(4), pp 219-225.

126 Billings, W.D. and R.J. Morris. 1951. Am. J. Bot., 33, pp 327-331.

127 Bingham, F.T., A.L. Page, R.J. Mahler and T.J. Ganje. 1976. J. Environ. Qual., 5(4), pp 501-513.

128 Birch, L.C. 1953. Ecology 34, pp 698-711.

129 Birch, W.R. 1975. J. appl. Ecol., 12(1), pp 201-212.

130 Bishop, J.W. and J.P. Barlow. 1975. Limnol. Oceanogr., 20(1), pp 148-149.

131 Bishop, J.W. 1968. Limnol. Oceanogr., 13(1), pp 58-62.

132 Bisogni, J.J. and A.W. Lawrence. 1975. Water Pollut. Control, 47, pp 135-152.

133 Bjerrum, J., G. Schwarzenbach and L.G. Sillen. 1957. Stability Constants, London, Chem. Soc..

134 Bjoerkman, O. and D. Holmgren. 1963. Physiol. Plant., 16, pp 889-914.

135 Black, C.C. 1971. Adv. Ecol. Res., 7, pp 87-114.

136 Bloesch, J., P. Stadelmann and H. Boehrer. 1977. Limnol. Oceanogr., 22(3), pp 511-526.

137 Bloodworth, M.E., C.A. Burleson and W.R. Cowey. 1958. Agron. J., 50, pp 317-320.

138 Blum, J.J. 1966. J. Gen. Physiol., 49, pp 1125-1137.

139 Boaze, J.L. and R.T. Lackey. 1974. Prog. Fish Cult. 36(6), pp 163-164.

140 Bobek, B. 1969. Acta Theriol. 14, pp 191-210.

141 Boehm, H.H. and H. Muller. 1976. Naturwisenschaftens 63(3), pp 296.

142 Bogdan, K.G. and D.C. McNaught. 1975. Verh. Int. Ver. Limnol., 19, pp 29-35.

143 Bolin, B. 1970. Scientific American., 223(3), p 130.

144 Bolin, B. and R.J. Charlson. 1976. Ambio, 5(2), pp 47-54.

145 Bolin, B. and E. Arrhenius. Ambio, 6(2-3), pp 96-105.

146 Bonner, J.T. 1965. Size and Cycle: an essay on the Structure of Biology. 219 p. Princetown Univ. Press.

147 Boon, B. and H. Laudelout. 1962. Biochem. J. 85, pp 440-447.

148 Booth, B.O. 1975. Limnol. Oceanogr., 20(5), pp 865-868.

149 Bornebusch, C.H. 1930. Forstl. Forsoegsvaesen Copenhagen. 11, pp 1-224.

150 Boroughs, H., S.J. Townsley and R.W. Hiatt. 1956. Biol. Bull. 111, pp 336-351.

151 Boroughs, H., S.J. Townsley and R.W. Hiatt. 1956. Biol. Bull. 111, pp 352-357.

152 Botkin, D.B. 1977. in: Hall, C.A.S. and J.W. Day, Jr. (eds.); Ecosystem Modeling in theory and practice. pp 214-233. John Wiley and Sons. New York.

153 Bottenheim, J.W., S.E. Braslavsky and O.P. Strausz. 1977. Environ. Sci. Technol., 11(8), pp 801-808.

154 Boucher, F.R. and G.F. Lee. 1972. Environ. Sci. Technol., 6(6), pp 538-543.

155 Bougis, P. 1965. C.r. hebd. Seanc. Acad. Sci. Paris 260, pp 2929-2931.

156 Bovey,R.W., E. Burnett, C. Richardson, M.G. Merkle, J.R. Baur and W.G. Knisel. 1974. J. Environ. Quality, 3, pp 61-64.

157 Bowen, H.J.M. 1966. Trace Elements in Biochemistry, Ac. Press.

158 Boyd, C.E. and C.P. Goodyear. 1971. Arch. Hydrobiol., 69(2), pp 256-270.

159 Boyd, C.E. 1970. Arch. Hydrobiol., 66, pp 511-517.

160 Boyd, C.E. 1970. Arch. Hydrobiol., 67(1), pp 78-85.

161 Boyd, C.E. 1971. Am. Midland Nat., 86, pp 28-45.

162 Bradley, J.R.Jr., T.J. Sheets and M.D. Jackson. 1972. J. Environ. Quality, 1, pp 102-105.

163 Braham, H.W. 1974. Environ. Pollut., 5, pp 253-258.

164 Brandt, C.J. and R.W. Rhodes. 1973. Environ. Pollut., 4, pp 207-214.

165 Bray. J.R., D.B. Lawrence and L.C. Pearson. 1959. Oikos, 10, pp 38-49.

166 Bremner, J.M. and K. Shaw. 1958. J. agric. Sci. 51, pp 22-39.

167 Brereton, A., H. Lord, I. Thornton and J.S. Webb. 1973. Mar. biol. 19(2), pp 96-101.

168 Brett, J.R. 1952. Bull. Fish Res. Board Can., 9, pp 265-323.

169 Brewer, P.G. and J.C. Goldman. 1976. Limnol. Oceanogr. 21(1), pp 108-117.

170 Brezonik, P.L. and G.F. Lee. 1968. Environ. Sci. Tech. 2, pp 120-125.

171 Brezonik, P.L. 1975. Denitrification in Natural Waters. in: Conf. on Nitrogen as a Water Poll. Vol 2, session B3. Copenhagen.

172 Bricker, O.P. III and B.N. Troup. 1975. In: Cronin, L.E. (ed.); Estuarine Research. Vol 1 pp 3-27. Academic Press. New York.

173 Brinkhurst, R.O., K.E. Chua and N.K. Kaushik. 1972. Limnol. Oceanogr., 17(1), pp 122-133.

174 Brisbin, I.L. 1968. Ecology 49, pp 792-794.

175 Brock, T.D. 1966. Principles of Microbial Ecology. Prentice-Hall. New Jersey.

176 Bromley, P.J. 1971. J. Cons. Int. Explor. Mer. 34(1), pp 131-134.

177 Broquist, S. 1971. Matematisk modell foer Ekosystemet i en Sjoe. Matematiske Institution, Tekniska Hoegskolan, Stockholm.

178 Brown, B.E. 1976. Water Res., 10(6), pp 555-559.

179 Brown, C.M., P.S. MacDonald-Brown and S.O Stanley. 1975. Mar. Biol., 31(1), pp 7-13.

180 Brown, E.R. 1961. Wash. State Game Dept. Biol. Bull, No 13, pp 124.

181 Brown, R.A., C.M. Nazario, R.S. de Tirado, J. Castrillon and E.T. Agard. 1977. Environ. Res., 13(1), pp 56-61.

182 Brown, V.M., T.L. Shaw and D.G. Shurben. 1974. Water Res., 8 (10), pp 797-802

183 Browning, E. 1969. Toxicity of Industrial Metals. Butterwoths. London.

184 Brungs, W.A., J.R. Geckler and M. Gast. 1976. Water Res. 10(1), pp 37-44.

185 Brunskill, G.J. and D.W. Schindler. 1971. J. Fish. Res. bd. Can., 28, pp 134-155.

186 Bryan, G.W., A. Preston and W.L. Templeton. 1966. In: Disposal of Radioactive Wastes into Seas, Oceans and Surface Waters, pp 623-667.

187 Bryan, G.W. and E. Ward. 1965. J. mar. biol. Ass. U. K. 45, pp 65-95.

188 Bryan, G.W. and L.G. Hummerstone. 1973. J. mar. biol. Ass. U.K. 53, pp 839-857.

189 Bryan, G.W. 1961. J. Mar. Biol. Ass. U.K., 41, p 551.

190 Bryan, G.W. 1963. J. Mar. Biol. Ass. U.K., 43, pp 519.

191 Bryson, R.A. and W.M. Wendland, 1975. In: S.F. Singer (ed.); The Changing Global Environment, pp 139-147. D. Reidel Publ. Comp.

192 Budyko, M.I. 1971. Climate and Life. Hydrological Publ. Co. Leningrad.

193 Buhr, K.-J. 1976. Mar. Biol., 38(4), pp 373-383.

194 Buikema, A.L. 1973. Hydrobiol., 41(4), pp 515-527

195 Bullen, W.A. and D.S. Frear. 1957. Arch. Biochem. Biophys., 66, pp 502-503.

196 Bump, G., R.W. Darrow, F.C. Edminster and W.F. Crissley. 1947. The Ruffed Grouse: Life History - propagation - management. Albany. New York Conservation Dept.

197 Bunt, J.S. 1964. Antarctic Res. Ser., 1, pp 27-31.

198 Burgis, M.J. 1970: Limnol. Oceanogr., 15(5), pp 742-747.

199 Burkholder, P.R. and E.F. Mandelli. 1965. Science, 149, pp 872-874.

200 Burkholder, P.R. and G.H. Bornside. 1957. Bull. Torrey. bot. Club, 84, pp 366-383.

201 Burkholder, P.R., L.M. Burkholder and J.A. Rivero. 1969. Bull. Torreya bot. Club, 86, pp 88-93.

202 Burns, C.W. 1967. Limnol. Oceanogr., 14(5), pp 693-700.

203 Burns, G.P. 1923. Ecology 4.

204 Burns, C.W. and F.H. Rigler. 1967. Limnol. Oceanogr. 12. pp 492-502.

205 Burris, J.E. 1977. Mar. Biol., 39(4), pp 371-379.

206 Burrows, W.D. and P.A. Krenkel. 1973. Environ. Sci. Technol., 7(13), pp 1127-1130.

207 Burton, J.D. and K.S. Massie. 1971. J. mar. biol. Ass. U.K. 51, pp 675-683.

208 Busby, W.F. and J. Lewin. 1967. J. Phydrocol. 3, pp 127-131.

209 Buswell, A.M., T. Shiota, N. Lawrence and I.V. Meter. 1954. Appl. Microbiol., 2, pp 21-25.

210 Butler, E.I., E.D.S. Corner and S.M. Marshall. 1970. J. mar. biol. Ass. U.K. 50, pp 525-560.

211 Butler, E.I, E.D.S. Corner and S.M. Marshall. 1969. J. mar. biol. Ass. U.K. 49, pp 977-1001.

212 Butler, J.N. 1964. In; Wesley, A. (ed); Ionic Equilibrium - A Mathematical Approach. Reading, Mass.

213 Butterworth. J. et al. 1972. Mar. Poll. Bull., 3(5), pp 72-74.

214 Buttery, B.R. and J.M. Lambert. 1973. J. Ecol., 53, pp 163-181.

C

215 Cairns, J. Jr. 1971. Water Pollut. Control., 43(1), pp 55-66.

216 Calabrise, A., M.A. Dawson, F.P. Thurberg and D.R. Wenzloff. 1975. In; Koeman, J.H. et al. (eds); Sublethal Effects of Toxic Chemicals on Aquatic Animals. Elsevier Amsterdam. N.Y. pp 15-21.

217 Calapaj, G.G. 1974. Ig. Mod. 67(2), pp 136-145.

218 Caldwell, T.H. 1971. In: Tech. Bull. 21, pp 62-87. Trace Elements in Soils and Crops.

219 Camp, T.R. 1965. J. San. Eng. Div, ASCE. 91, SA5, pp 1.

220 Canale, R.P., L.M. De Palma and A.H. Vogel. 1976. in: Canale, R.P. (ed.); Mathematical modelling of Biochemical processes in Aquatic Ecosystems. Ann Arbor Science Press.

221 Canton, J.H. 1976. Bull. Environ. Contam. Toxicol. 16(2), pp 214-218.

222 Canton, J.H. and G.J. Vanesch. 1976. Bull. Environ. Contam. Toxicol. 15(6), pp 720-279.

223 Canton, J.H. and W.Sloff. 1977. Water Res., 11(1), pp 117-121.

224 Canton, J.H., G.J. van Fisch and P.A. Greve, 1977. In: S.H. Jenkins, Vol. 2(1), pp 111-115.

225 Caperon, J. and J. Meyer. 1972. Deep Sea Res., 19, pp 601-632.

226 Capuzzo, J.M. 1977. Water Res. 11(12), pp 1021-1024.

227 Capuzzo, J.M., S.A. Lawrence and J.A. Davidson. 1976. Water Res., 10(12), pp 1093-1099.

228 Carnahan, J.E., L.E. Mortensen and J.E. Castle. 1960. J. Bact., 80, pp 311-315.

229 Caro, J.H. and A.W. Taylor. 1971. J. agric. Fd. Chem., 19, pp 379-384.

230 Carpenter, E.J. and R.R.L. Guillard. 1971. Ecology, 52, pp 183-185.

231 Carpenter, E.J. and J.J. McCarthy. 1975. Limnol. Oceanogr., 20(3), pp 389-401.

232 Carpenter, E.J. and J.L. Cox. 1974. Limnol. Oceanogr., 19(3), pp 429-436.

233 Carringer, R.D., J.B. Weber and T.J. Monaco. 1975. J. agric. Fd Chem., 23(3), pp 568-572.

234 Carritt, D.E. and E.J. Green. 1967. Mar. Res. 25(2)

235 Caspers, N. 1975. Arch. Hydrobiol., 75(4), pp 484-489.

236 Cavari, B. 1976. Limnol. Oceanogr. 21(2), pp 231-236.

237 Cearley, J.E. and R.L. Coleman. 1973. Bull. Environ. Contam. Toxicol., 9, pp 100-101.

238 Chamberlin, C. and R. Mitchell. 1978. in: Mitchell, R. (ed.); Water Pollution Microbiology, Vol 2.

239 Chappelle, E.W. and G.V. Levin. 1968. Biochem. Med. 2, pp 41-52.

240 Chen, C.W. and G.T. Orlob. 1975. in: Patten, B.C. (ed.); Systems Analysis and Simulation in Ecology, vol 3. pp 476-537. Acad. Press. N.Y.

241 Chen, C.W. and Wells. 1975. Boise River Water Quality-Ecological Model for Urban Planning Study. US Army Eng. Walla Walla, Idaho Water Res. Board.

242 Chen, C.W. 1975. in: Russell, C.S.; Ecological Modeling, in a Resource Management Framework. NOAA.

243 Chen, R.L., D.R. Keeney and J.G. Konrad. 1972. J. Environ. Qual., 1(2), pp 151-154.

244 Cherry, D.S., R.K. Guthrie and R.S. Harvey. 1974. Water Res., 8(3), pp 149-155.

245 Chi, K.-S. and R.-T. Yang. Acta Oceanogr. Taiwan. 3, pp 199-222.

246 Childress, J.J. 1971. Limnol. Oceanogr., 16(1), pp 104-106.

247 Chipman, W.A. and J.G. Hopkins. 1954. Biol. Bull. mar. biol.Lab., Woods Hole, 107, pp 80-91.

248 Chipman, W.A. 1967. Proc. Int. Symp. Radioecol. Concent. Progress, 1966. Stockholm, pp 931-941.

249 Chisholm, S.W., R.G. Struss and P.A. Nobbs. 1975. J. Phycol., 11, pp 307-373.

250 Chittleborough, R.G. 1976. Aust. J. Mar. Freshwater Res. 27(2), pp 279-295.

251 Christensen, M.H., P. Harremoes and O. R.-Jensen. 1975. Combined Sludge Denitrification of Sewage utilizing internal Carbon sources. in: Conf. on Nitrogen as a Water Poll. Vol 3. Session A3. Copenhagen.

252 Chughtai, A., R. Marshall and G.H. Nancollas. 1968. J. Phys. Chem., 72, p 208.

253 Churchill, M., H. Elmore and R. Buckingham. 1962. The Prediction of Stream Reareation Rates. J. San. Eng. Div.

254 Cieglewicz, W. and B. Draganik. 1969. Pr. Morsk. Inst. Ryb. Gdynia, A, 15, pp 133-152.

255 Claesson, A. 1975. Vattan, 31(4), pp 333-338.

256 Clasen, J. and H. Bernhardt. 1974. Water Res. 8(1), pp 31-44.

257 Clemenson, H. 1963. Schweiz. Z. Hydrol., 25, pp 157-165.

258 Clendenning, K. A., T.E. Brown and H.C. Eyster. 1956. Can. J. Bot., 34, pp 943-966.

259 Cline, J.D. and F.A. Richards. 1972. Limnol. Oceanogr., 17(6), pp 885-900.

260 Cloern, J.E. 1978. Ecol. Model. 4. (in press).

261 Clubb, R.W., A.R. Gaufin and J.L. Lords. 1975. Environ. Res., 9(3), pp 332-341.

262 Clutter, R.I. and G.H. Thielacker. 1971. Fish. Bull., U.S. 69, pp 93-115.

263 Clymo, R.S. and E.J.F. Reddaway. 1974. J. Ecol., 62, pp 191-196.

264 Cocks, J.A. 1974. Environ. Pollut., 5, pp 149-151.

265 Codispoti, L.A. and F.A. Richards. 1976. Limnol. Oceanogr., 21(3), pp 379-388.

266 Coffin, C.C., F.R. Hayes, L.A. Jodrey and S.G. Whiteway. 1949. Can. J. Res., 37, pp 207-222.

267 Coffmann, W.P., K.W. Cummins and J.C. Wuycheck. 1970. Arch. Hydrobiol., 68(2), pp 232-276.

268 Collier, B.D., G.W. Cox, A.W. Johnson and P.C. Miller. 1974. Dynamic Ecology. Prentice-Hall. New Jersey.

269 Collos, Y. and J. Lewin. 1976. Limnol. Oceanogr., 21(2), pp 219-225.

270 Comita, G.W. 1964. Verh. Int. Ver. Limnol., 15, pp 646-653.

271 Comita, G.W. 1968. Limnol. Oceanogr., 13, pp 51-57.

272 Comita, G.W. 1972. Arch. Hydrobiol. 70, pp 14-66.

273 Committee on Biologic Effects of Atmospheric Pollunants, Chromium, Medical and Biologic Effects of Environmental Pollutants, National Academy of Sci. Wash. D.C. 1974.

274 Commoner, B. 1977. Ambio, 6(2-3), pp 157-161.

275 Confer, J.L. and P.I. Blades. 1975. Limnol. Oceanogr., 20, pp 571-579.

276 Conover, R.J. and E.D.S. Corner. 1968. J. mar. biol. Ass. U.K., 48, pp 49-75.

277 Conover, R.J. 1961. J. mar. biol. Ass. U.K., 41, pp 484-488.

278 Conway, G.R. and G. Murdie. 1971. in: Jeffers, J.N.R. (ed.); Mathematical Models in Ecology. pp 209. Symp. Blackwell Scientific Publ.

279 Conway, H.L. and P.J. Harrison. 1977. Mar. Biol., 43(1), pp 33-43.

280 Conway, H.L. 1977. Mar. Biol. 39(3), pp 221-232.

281 Cooke, A.S. 1971. Environ. Pollut., 1(1), pp 57-71.

282 Cooke, A.S. 1972. Environ. Pollut., 3,

pp 51-68.

283 Cooke, A.S. 1973. Environ. Pollut., 4, pp 85-148.

284 Cooke, A.S. 1974. Environ. Pollut., 5, pp 259-273.

285 Corbet, S.A., J. Green and E. Betney. 1973. Environ. Pollut., 4, pp 193-206.

286 Corner, E.D.S. and B.S. Newell. 1967. J. mar. biol. Ass. U.K., 47, pp 113-120.

287 Corner, E.D.S., C.B. Cowey and S.M. Marshall. 1965. J. mar. biol. Ass. U.K., 45, pp 429-442.

288 Corner, E.D.S., R.N. Head and C.C. Kilvington. 1972. J. mar. biol. Ass. U.K., 52, pp 847-861.

289 Corner, E.D.S., R.N. Head, C.C. Kilvington and L. Pennycuick. J. Mar. Biol. Assoc. U.K., 56(2), pp 345-358.

290 Cornwell, D.A., J. Zoltek Jr., C.D. Patrinely, T. deS. Furman and J.I. Kim. 1977. Water Pollut. Control, 49(1), pp 57-65.

291 Cottrell, T.L. 1958. The Strength of Chemical Bonds. Butterworth. London and Wash. D.C.

292 Covar, A.P. 1976. Selecting the Proper Reaeration Coefficient for Use in Water Quality Models. Presented at the US. Epa Conference on Environmental Simulation and Modeling, April 19-22.

293 Cox, J.L. 1970. Bull Environ. Contam. Toxicol., 5, pp 218-221.

294 Cox, J.L. 1970. Nature, 227, pp 192-193.

295 Coyne, P.I. and J.J. Kelly. 1975. J. appl. Ecol. 12, pp 587-611.

296 Craig, R.G. and W.F. Baksi. 1977. Water Res., 11(8), pp 621-626.

297 Craker, L.E. 1971. Environ. Pollut., 1(4), pp 299-304.

298 Cranmore, G. and F.L. Harrison. 1975. Health Phys., 28, pp 319-333.

299 Crim, R.L. and N.L. Lovelace. 1973. Auto-Qual Modeling Systems, EPA Report 440/9-73-003.

300 Crosby, D.G. and R.K. Tucker. 1971. Environ. Sci. Technol., 5(8), pp 714-716.

301 Cross, F.A. J.N. Willis, L.H. Hardy, N.Y. Jones and J.M. Lewis 1975. in: Cronin, L.E. (ed.); Estuarine Research. Vol 1 pp 45-63. Academic Press. New York.

302 Crowder, L.A. and E.F. Dindal. 1974. Bull. Environ. Contam. Toxicol., 12(3), pp 320-327.

303 Crutzen, P.J. 1972. Ambio, 1(2), pp 41-51.

304 Cummins, K.W. 1973. Ann. Rev. Entomol, 18, pp 183-206.

305 Cummins, K.W., R.R. Costa, R.E. Rowe, G.A. Moshiri, R.M. Scanlon and R.K. Zajdel. 1969. Oikos, 20, pp 189-223.

306 Cunningham, P.A. and M.R. Tripp. 1975. Mar. Biol., 31(4), pp 311-319.

307 Cunningham, P.A. and M.R.Tripp. 1973. Mar. biol. 20, pp 14-19.

308 Curl, H. 1962. Rapp. Proc.-Verb. Cons. Int. Explor. Mer., 153, pp 183-189.

309 Curtis, E.H., J.J. Beauchamp and B.C. Blaylock. 1977. Ecol. Modell., 3(4), pp 273-284.

310 Curtis, E.J.C., K. Durrant and M.M.I. Harman. 1975. Water Res., 9(3), pp 255-268.

311 Cushing, C.E. and L.A. Rancitelli. 1972. Northwest Science, 46(2), pp 115-121.

312 Cutshall, N.1974. Health Phys., 26, pp 327-331.

D

313 D'Elia, C.F. 1977. Limnol. Oceanogr., 22(2), pp 301-315.

314 D'Hollander, E. and I. Impens. 1975. in: Vansteenkiste, G.C. (ed.); Computer Simulation of Water Resources systems. American Elsevier. New York.

315 Daan, N. 1974. Neth. J. Sea Res. 8(1), pp 27-48.

316 Daborn, G.R. 1974. Can. J. Zool., 52(III), pp 1303-1310

317 Dabrowski, K.R.. 1976. Water Res., 10(9), pp 793-796.

318 Dagg, M. 1977. Limnol. Oceanogr., 22(1), pp 99-107.

319 Dagg, M.J. 1976. Int. Rev. Gesamt. Hydrobiol. 61(3), pp 297-357.

320 Dahl-Madsen, K.I. and S. Nielsen. 1974. Vand, 5, pp 24-31. Cph.

321 Dales, R.P. 1957. J. mar. biol. Ass. U.K., 36, pp 309-316.

322 Dales, R.P. 1961. Biol. Bull. mar. Biol. Lab., Woods Hole, 121, pp 82-91.

323 Daniels, F. and R.A. Alberty. 1967. Physical Chemistry. 3rd Edition. John Wiley and Sons. New York.

324 Daubenmire, R. 1972. J. appl. Ecol., 9, pp 11-23.

325 Dave, G. A. Buman, M-L. Johansson-Sjoebeck and U. Lidman. 1975. In; Koeman, J.H. et al. (eds); Sublethal Effects of Toxic Chemicals on Aquatic Animals, pp 197-205. Elsevier. Amsterdam.

326 Davies, A.G. 1970. J. mar. biol. Ass. U.K., 50, pp 65-86.

327 Davies, B.E. and J. Lewin. 1974. Environ. Pollut., 6, pp 49-57.

328 Davis, B.N.K. and M.E. French. 1969. Soil Biol. Biochem. 1, p 45.

329 Davis, B.N.K. 1968. Am. Appl. Biol. 61, pp 45.

330 Davis, C.O. 1973. Effects of changes in light intensity and photoperiod on the silicate-limited continuous culture of the marine diatom Skeletomena costatum (Grev.) Cleve. 122 p. Ph. D. Dissertation. Univ. of Wash. Seattle.

331 Davis, D.E. and F.B. Golley. 1963. Principles in Mammalogy. Reinh. Publ. Corp. London.

332 Davis, E.M. and E.F. Gloyna. 1969. Water Pollut. Control. 41 (8), pp 1494-1504.

333 Davis, J.B. and R.L. Barnes. 1974. Environ. Pollut., 5, pp 35-44.

334 Davison, A.W., A.W. Rand and W.E. Betts. 1974. Environ. Pollut., 5, pp 23-33.

335 Dawson and Murphy. 1972. Water res., 6, pp 71-83.

336 Dayhoff, M.O., E.R. Lippincott and R.V. Eck. 1964. Science 146, p 1461.

337 Dawson, F.H. 1973. Ann. Limnol. Stn. Biol. Lac Oredon, 9(3), pp 231-240.

338 De Vooys, C.G.N. 1976. Mar. Biol., 36, pp 25-30.

339 Dedolph,R. et al. 1970. Environ. Sci. Technol., 4(3), pp 217-223.

340 Deer, S.K. 1974. Bull. Environ. Contam. Toxicol., 11(6), pp 500-502.

341 Deevey, E.S. 1970. Scientific American. 223(3), p 150.

342 Delwiche, C.C. 1970. Scientific American, 223(3), pp 137-140.

343 Delwiche, C.C. 1977. Ambio, 6(2-3), pp 106-111.

344 Demayo, A. 1972. Bull. Environ. Contam. Toxicol., 8(4), pp 234-237.

345 Demerjian, K.L., J.A. Kerr and J.G. Calvert. 1974. Adv. Environ. Sci. Technol. 4, pp 1.

346 Denman, F.B. 1973. J. Phys. Oceanogr., 3(2), pp 173-184.

347 Department of Scientific and Industrial Research. 1964. Water Pollution research. 1963. H.M.S.O. London.

348 Deustachio, A.J. and G.V. Levin. 1967. Bact. Proc., 20, pp 121.

349 Dewitt, J.B., C.M. Menzie, V.A. Adomaitis and W.L. Reickel. 1960. Trans. 25th N. Am. Wildl. Conf., pp 277.

350 De Silva, S.S. and F. Balbontin. 1974. J. Fish. Biol., 6, pp 645-658.

351 Di Giano, F.A. 1978. Ecol Modell. 4. (in press)

352 Di Salvo. L.H., H.E. Guard and L. Hunter. 1975. Environ. Sci. Technol., 9, pp 247.

353 Di Toro, D.M, D.J. O'Connor and J.L. Mancini. 1975. In: Patten, B.C. (ed.); Systems Analysis and Simulation in Ecology. Ac. Prщss. N.Y.

354 Di Toro, D.M., D.J. O'Connor and R.V. Thomann. 1971. A Dynamic Model of the Phytoplankton Population in the Sacramento-San Joaquin Delta. Advances in Chemistry Series. Vol 106.

355 Different sources and Internal Reports.

356 Dilling, W.L., C.J. Bredeweg and N.B. Tefertiller. 1976. Environ. Sci. Technol., 10(4), pp 351-356.

357 Dilling, W.L., N.B. Tefertiller and G.J. Kallos. 1975. Environ. Sci. Technol. 9(9), pp 833-837.

358 Dillon, P.J. and F.H. Riegler. 1974. Limnol. Oceanogr., 19(5), pp 767-773.

359 Do Chi, T. and G. Lasserre. 1974. Bull. Ecol. 5(2), pp 91-104.

360 Doblen, van W.H. 1962. Neth. J. agric. Sci. 10, pp 377-389.

361 Dodson, S.I. 1975. Limnol. Oceanogr., 20(3), pp 426-433.

362 Doig, M.T. III. and D.F. Martin. 1974. Water Res., 8(9), pp 601-606.

363 Dolan, J.M and A.C. Hendricks. 1976. Water Pollut. Control, 48(1), pp 2570-2577.

364 Donigian, A.S. and N.H. Crawford. 1976. Modeling nonpoint Pollution from the Land Surface, EPA-600/3-76-083. July.

365 Dorland, R.E. and F.W. Went. 1947. Am. J. Bot., 34, pp 393-401.

366 Doudoroff, P. and M. Katz. 1953. Swedish. ind. Waters, 25, p 802.

367 Dowd, G. and J.L. Monkman. 1976. Sci. Total Environ., 6(1), pp 1-13.

368 Downing, A.L., H.A. Painter and G. Knowles. 1964. J. Proc. Inst. Sew. Purif., pp 130-153.

369 Droop, M.R. 1970. Helgol Wiss. Meeresunters., 20, pp 629-636.

370 Droop, M.R. 1973. J. Phycol., 9, pp 264-272.

371 Drozd, J.W. Jr., R.S. Tubb and J.R. Postgate. 1972. J. gen. Microbiol., 73, pp 221-232.

372 Drozdz, A. 1967. in: Petrusewicz, K. (ed). Secondary Productivity of Terrestrial Ecosystems, Vol I, Naukowe Warszawa.

373 Du Paul, W.D. and J.D. McEachran. 1973. Cheasepeake Sci. 14(3), pp 205-207.

374 Duffer, W.R. and T.C. Dorris. 1966. Limnol. Oceanogr., 11, pp 143-151.

375 Duggan, M.J. and S. Willams. 1977. Sci. Total Environ., 7(11), pp 91-97.

376 Duggan, R.E. and J.R.. Weatherway. 1967. Science, 157, pp 1006-1010.

377 Duncan, A. and R.Z. Klekowski. 1975. in: Grodzinsky, W., R.Z. Klekowski and A. Duncan (eds.); Methods for Ecological

Bioenergeticts. IBP Handbook No. 24, pp 97-147. Blackwell. Oxford.

378 Duncan, W.B., R.S. Loomis, W.A. Williams and R. Hanau. 1967. Hilgardia 38, pp 181-205.

379 Dunn, B.P. and H.F. Stich. 1976. Bull. Environ. Contam. Toxicol., 15(4), pp 398-401.

380 Dunn, I,G. 1970. Limnol. Oceanogr. 15(3), pp 373-379.

381 Dunson, W.A. 1969. Proc. 2nd Nat. Symp. Radioecology, USAEC. Conf. 670503, pp 191-197.

382 Durbin, P.W., K.G. Scott and J.G. Hamilton. 1957. The Distribution of Radioisotopes of some heavy Metals in Rats. Univ. of Calif. Publ. in Pharmacology.

383 Duval, W.S. and G.H. Geen. 1976. Limnol. Oceanogr., 21, pp 823-829.

384 Dykyjova, D. and D. Hradecka. 1973. Pol. Arch. Hydrobiol., 20, pp 111-119.

E

385 Eberhardt, L.L., R.L. Meeks and T.J. Peterle. 1971. Nature, 230, pp 60-62.

386 Edgington, D.N., S.A. Gordon, M.M. Thommes and L.R. Almodovar. 1970 Limnol. Oceanogr.,, 15(6), pp 945-955.

387 Edmonson, W.T. 1965. Ecol. Monogr., 35, pp 61-111.

388 Edmonson, W.T. 1971. In: W.T. Edmondson and G.G. Winberg (eds.); A Manual on Methods for the Assessment of Secondary Productivity in Fresh Waters. IBP Handbook No 17, p 145.

389 Edwards, C.A. 1975. Persistent Pesticides in the Environment, 2nd edition, CRC-Press.

390 Edwards, C.A., D.E. Reichle and D.A. Crossley, Jr. 1970. in: Reichle, D.E. (ed); Analysis of Temperate Forest Ecosystems. pp 147-172. Springer-Verlag. Berlin.

391 Edwards, R.W. and H.L.J. Rolley. 1965. J. Ecol., 53, pp 1-19.

392 Edwards, R.W. and M. Owens. 1962. J. Ecol., 50, pp 207-220.

393 Edwards, R.W. and M. Owens. 1960. J. Ecol., 48, pp 151-180.

394 Edwards, W.M. and B.L. Glass. 1971. Bull. Environ. Contam. Toxicol., 6, pp 81-84.

395 Egloff. D.A. and D.S. Palmer. 1971. Limnol. Oceanogr., 16(6), pp 900-905.

396 Eichhorn, M. 1969. Flora, Jena (Abt.A.) 159, pp 494-506.

397 Eifac. 1970. Water Quality Criteria for European Freshwater Fish. Report on Ammonia and Inland fisheries Prepared by Eifac, working Party on Water Quality Criteria for European Freshwater Fish, FAO, Roma, no 11.

398 Eisler, R. G.E. Zaaroogian and R.J. Hennekey. 1972. J. Fish. Res. bd. Canada 29, pp 1367-1369.

399 El Sharkawy, M.A., R.S. Loomis and W.A. Williams. 1967. Physiol. Plant., 20, pp 171-186.

400 El Zorgani, G.A. 1976. Bull. Environ. Contam. Toxicol., 15(3), pp 378-382.

401 El-Dib, M.A. and O.A. Acy. 1976. Water Res., 10(12), pp 1051-1053.

402 El-Dib, M.A. and O.A. Acy. 1976. Water Res., 10(12), pp 1047-1050.

403 El-Maghraby, A.M., M.T. Hashem and H.M. El-sedfy. 1973. Bull. Inst. Oceanogr. Fish. Cairo. 3, pp 55-82.

404 Elder, J.F. and M. Whitfield. 1975. Limnol. Oceanogr., 20(1), pp 103-107.

405 Elderfield, H. et al. 1971. Mar. Poll. Bull., 2(3), pp 44-47.

406 Elgmork, K., A. Hagen and A. Langeland. 1973. Environ. Pollut., 4, pp 41-52.

407 Elliott, J.M. 1975. J. Anim. Ecol. 44, pp 805-821.

408 Elliott, J.M. 1975, J. Anim. Ecol. 44, pp 823-842.

409 Elliott, J.M. 1976. J. Anim. Ecol. 45, pp 273-289.

410 Elliott, J.M. 1976. J. Anim. Ecol. 45, pp 561-580.

411 Elliott, J.M. 1976. J. Anim. Ecol. 45, pp 923-948.

412 Ellis, A.J. 1959. Am. J. Sci. 257, pp 219.

413 Ellsaesser, H.W. 1975. In: S.F. Singer (ed.): The Changing Global Environment, pp 235-269. D. Reidel Publ. Comp.

414 Elmi, H. 1975. A study of Assimilative Capacities for Receiving Streams. Research Report. Florida Tech. Univ.

415 Elmore, H.L. and T.W. Hayes. 1960. J. San. Eng. Div. ASCE. 86, SA4, pp 41.

416 Emerson, S. 1975. Limnol. Oceanogr., 20(5), pp 754-761.

417 Emerson, S., W.S. Broecker and D.W. Schindler. 1973. J. Fish. Bd Can., 30 pp 1475-1484.

418 EPA. 1976. Fate and Biological Effects of Mercury introduced into artificial Streams. EPA-600/3-76-060.

419 EPA. Water Quality Crit. Data Book. 2rd. Inorganic Pollution of Freshwater. 18010. DPU 07/71.

420 EPA. 1976. Areawide Assessment Procedures Manual. Cincinnati. Ohio. Report EPA-600/9-76-014.

421 EPA. 1976. Fate and Biological Effects

of Mercury introduced into Artificial streams. EPA-600/3-76-060.

422 Epifano, C.E. 1973. Mar. Biol., 19, pp 320-322.

423 Eppley, R.W. and E.H. Renger. 1974. J. Phycol., 10, pp 15-23.

424 Eppley, R.W. and J.L. Coatsworth. 1968. J. Phycol., 4. pp 151-156.

425 Eppley, R.W. and J.N. Rogers and J.J. McCarthy. 1969. Limnol. Oceanogr. 14. pp 912-920.

426 Eppley, R.W. and W.H. Thomas. 1969. J. Phycol., 5. pp 375-379.

427 Eppley, R.W. 1972. Fish Bull., 70, pp 1063-1085.

428 Eppley, R.W., A.F. Carlucci, O.H.-Hansen, D. Kiefer, J.J. McCarthy, E. Venrick and P.M. Williams. 1971. Limnol. Oceanogr., 16(5), pp 741-757.

429 Eppley, R.W., E.H. Renger, E.L. Venrick and M.M. Mullin. 1973. Limnol. Oceanogr., 18(4), pp 534-551.

430 Eppley, R.W., F.M.H. Reid and J.D.H. Strickland. 1970. Bull. Scripps Inst. Oceanogr. 17, pp 33-42.

431 Eppley, R.W., R.W. Holmes and J.D.H. Strickland. 1967. J. Explor. Mar. Biol. Ecol., 1, pp 191-208.

432 Epstein, E. and W.J. Grant. 1968. Soil Sci Soc. Am. Proc., 32 pp 423-426.

433 Esmen, N. A. and R.B. Fergus. 1977. Sci. Total Environ. 8(1), pp 69-77.

434 Ettenhuber, E. and W. R|hnsch. 1975. Impacts of Nuclear Releases into the Aquatic Environment. IAEA-SM-198/15. pp 473-481. Vienna.

435 European Inland Fisheries Advisory Commission Working Party on Water Quality Criteria for European Fershwater Fish. 1973. Water Res., 7(7), pp 1011-1022.

436 Evans, W.H. and E.J David. 1974. Water Res., 8(2), pp 97-100.

437 Eyman, L.D. and T.R. Trabalka. 1977. Health Phys., 32(6), pp 475-478.

F

438 Fabacher, D.L. and H. Chambers. 1976. Bull. Environ. Contam. Toxicol., 16(2), pp 203-207.

439 Faber, R.B., R.W. Risenbough and H.M. Pratt. 1972. Environ. Pollut., 3, pp 111-122.

440 Fair, G.M., J.C. Geyer and D.A. Okun. 1968. Water and Wastewater Engineering, Vol 2. Water Purification and Wastewater Treatment and Disposal. John Wiley and Sons. New York

441 Falco, J.W., P.C. Kerr, M.B. Barron and D.L. Brockway. 1975. Ecol. Modell., 1(2), pp 117-131.

442 Falk, D.L. and W.A. Dunson. 1977. Water Res., 11(1), pp 13-15.

443 Fang, S.C. 1974. Environ. Res., 8(1), pp 112-118.

444 FAO. United Nations. 1966. Protection Yearbook, 21.

445 Farmer, A.S. 1974. J. Nat. Hist., 8(6), pp 605-620.

446 Faust, S.D. and H.M. Gomaa. 1972. Environ. Letters, 3, pp 171.

447 Feder, W.A. 1971. Environ. Pollut., 1(1), pp 73-79.

448 Fehrenbacher, J.B. and J.D. Alexander. 1955. Agron. J., 47, pp 468-472.

449 Feigner, K.D. and H.S. Harris. 1970. Documentation Report FWQA Dynamic Estuary Model. Federal Water Quality Administration. Wash. D.C. 248 p.

450 Fenchel, T. 1970. Limnol. Oceanogr. 15(1), pp 14-20.

451 Fendley, T.T., M.N. Manlove and I.L. Brisbin, Jr. 1977. Health Phys., 32(5), pp 415-422.

452 Fenerich, N.A., M.Y. Narahara and H.M. Godinho. 1975. Bol. Inst. Pesca, Sao Paulo, 4(1), pp 15-28.

453 Ferguson, R.L. and M.B. Murdoch. 1975. in: Cronin, L.E. (ed.); Estuarine Research. Vol 1 pp 229-250. Academic Press. New York.

454 Ferrante, J.G. 1976. Hydrobiologia, 49(3), pp 203-204.

455 Fiala-Mediont, A. 1973. Mar Biol., 23, pp 137-145,

456 Finenko, Z.Z. and D.K. Krupatkina-Aki-Nina. 1974. Mar. Biol., 26, pp 193-201.

457 Fischer, Z. 1972. Pol. Arch. Hydrobiol., 14, pp 59-71.

458 Fischer, Z. 1972. Pol. Arch. Hydrobiol., 19, pp 65-82.

459 Fisher, N.S. and C.F. Wurster. 1974. Environ. Pollut., 5, pp 205-212.

460 Floehn, H. 1973. Naturwissenschaften 60, pp 340-348.

461 Florkin, M. 1949. Biochemical Evolution. Academic Press. New York.

462 Florkin and Stutz (ed). 1972. Comprehensive Biochemistry, Elsevier Publ. Comp., 21(2).

463 Foehrenbach, J. 1972. Water Pollut. Control, 44(4), pp 619-624.

464 Foerstner, V. 1977. Arch. Hydrobiol., 80(2), pp 172-191.

465 Fogg, G.E. 1958. Advmt. Sci., 14, pp 395.

466 Fogg, G.E. 1966. Algal Cultures and Phytoplankton Ecology. Univ. of Wisconsin

Press. Wis.

467 Fogg, G.E. 1968. Photosynthesis. English Univ. Press, London.

468 Fogg, G.E. 1973. Water Res., 7(1/2), pp 77-91.

469 Foreman, E.E. and W.L. Templeton. 1958. U.S. Atomic Energy Authority, R and DB(W) TN-187, pp 16.

470 Fortescue, J.A.C. and G.G. Marten. 1970. in: Richle, D.E. (ed.); Analysis of Temperate Forest Ecosystems. pp 173-198. Springer-Verlag. Berlin.

471 Fossato, V.U. 1975. Mar. Poll. Bull., 6, p 7.

472 Fossato, V.U. and W.J. Canzonier. 1976. Mar Biol., 36, pp 243-250.

473 Foster, M., M. Neushul and R. Zingmark. 1971. Environ. Pollut., 2, pp 115-134.

474 Foster, R.F. and J.F. Honstead. 1967. Health Physics 13, pp 39-43.

475 Foth, H.D. 1962. Agron. J., 54, pp 49-52.

476 Fowler, S.W. and G. Benayoun. 1974. in: Comparative Studies of Food and Environmental Contamination. IAEA, pp 159-178. Vienna.

477 Fowler, S.W. and G. Benayoun. 1976. Bull. Environ. Contam. Toxicol., 16(3), pp 339-346.

478 Fowler, S.W. and G. Benayoun. 1976. Mar. Biol., 37, pp 59-68.

479 Fowler, S.W. and L.F. Small. 1972. Limnol. Oceanogr., 17, pp 293-296.

480 Fowler, S.W., L.F. Small and J.M. Dean. 1971. Mar. Biol., 8, pp 224-231.

481 Fowler, S.W., M. Heyraud, L.F. Small and G. Benayoun. 1973. Mar. Biol., 21, pp 317-325.

482 Fraizer, A. and J.C. Ancellin. 1975. Impacts of Nuclear releases into the Aquatic Environment. IAEA-sm-198/8, pp 49-61. Vienna.

483 Fraizier, A. and A. Vilquin. 1971. Mar. Biol., 10, pp 154-156.

484 Frank, R., K. Ronald and H.E. Braun. 1973. Bull. Fish Res. Board Can., 30, pp 1053.

485 French, C.E., S.A. Liscinsky and D.R. Miller. 1957. J. Wildl. Mgmt. 21, pp 348.

486 Friberg, L. and J. Vostal. 1974. Mercury in the Environment. CRC-Press.

487 Friberg, L., M. Piscater, G.F. Nordberg and T. Kjellstroem. Cadmium in the Environment 2nd edition, CRC-Press.

488 Fries, G.F. and G.S. Marrow. 1975. J. agric. Fd. Chem. 23(2), pp 265-269.

489 Fries, G.F., G.S. Marrow, Jr. and C.H. Cordon. 1973. J. Agric. Food Chem. 21, pp 117.

490 Fries, G.F., G.S. Marrow, Jr. and C.H. Gordon. 1972 Bull. Environ. Contam. Toxicol., 7(4), pp 57-62.

491 Fromm, P.O. and R.H. Schiffman. 1959. Stream Pollut., 31 pp 205.

492 Fromm, P.O. and R.H.J. Schiffman. 1958. Wild. Managment 22, pp 40.

493 Frost, B.W. 1972. Limnol. Oceanogr., 17(6), pp 805-815.

494 Frost, B.W. 1975. Limnol. Oceanogr., 29(2), pp 263-266.

495 Fuhs, G.W. 1969. J. Phycol., 5, pp 312-321.

496 Fuhs, G.W., S.P. Demmerle, E. Canelli and M. Chen. 1972. in; G.E. Likens (ed); Nutrients and eutrophication. Amer. Soc. Limnol. Oceangr. Spec. Symp. 1, pp 113-132.

497 Fujita, M. and K. Hashizume. 1974. Water Res., 9(10), pp 889-894.

498 Fukai, S., S. Koh and A. Kumura. 1976. J. appl. Ecol. 13(3), pp 877-887.

499 Furr, A.K., W.C. Kelly, C.A. Bache. W.H. Gutenmann and D.J. Lisk. 1976. Bull. Environ. Contam. Toxicol., 16(6), pp 756-763.

500 FAO/WHO

G

501 Gardiner, J. 1974. Water Res., 8(1), pp 23-30.

502 Gardiner, J. 1974. Water Res., 8(3), pp 157-164.

503 Gargas, E. 1976. A Three box eutrophication Model of a Mesotrophic Danish Lake. Water Quality Institute.

504 Garrels, R.M. and M.E. Thompson. 1962. Am. J. Sci. 260., p 57.

505 Garrels, R.M. and M.E. Thompson. 1962. J. Phys. Chem., 72, p 208.

506 Garrigan, G.A. 1977. Water Pollut. Control 49(12), pp 2380-2389.

507 Gates, D.M. 1962. Energy Exchange in the Biosphere. Harper and Row Publ. New York.

508 Gates, D.M., H.J. Keegan and V.R. Weidner. 1966. Scientific Experiments for Manned Orbital Flight. Science and Technology Series 4, pp 71-86.

509 Gauf, G.G. 1975. Ecologia (Berl.), 15, pp 17-32.

510 Gavis, J. and J.F. Ferguson. 1975. Limnol. Oceanogr., 20(2), pp 211-221.

511 Gayer, K.H. and L. Woontner. 1956. J. Phys. Chem., 60, pp 1509.

512 Geiger, R. 1965 The Climate near the

Ground. Harvard Univ. Press. Cambridge. Mass.

513 Geiling, W.T. and R.S. Campbell. 1972. Limnol. Oceanogr., 17(2), pp 304-306.

514 George, D.G. and R.W. Edwards. 1974. Freshwater Biol., 4, pp 445-465.

515 Gerloff, S. 1957 Ecology 38, pp 556-561.

516 Gerwits, A. 1972. Rep. Natn. Veg. Res. St. 1971. pp 36.

517 Gerwitz, A. and E.R. Page. 1974. J. Appl. Ecol., 11, pp 773-781.

518 Gessner, F. 1959. Hydrobotanik. II. Stoffhauhalt. VEB.

519 Getzin, L. W. 1977. Bull. Environ. Contam. Toxicol. 18(4), pp 499-502.

520 Ghosh, M.M. 1974. in: Rubin, A.J. (ed.): Aqueous-Environmental Chemistry of Metals. pp 193-217. Ann Arbor sci. Michigan.

521 Gibb, J. 1957. Bird study, 4, pp 207-215.

522 Giblin, F.J. and E.J. Massaro. 1973. Toxicol. Appl. Pharmacol. 24(1), pp 81-91.

523 Gibson, C.E. 1975. A Field and Laboratory study of Oxygen uptake by Planktonic Blue-green Algae.

524 Giese, A.C. 1962. Cell Physiology. 2nd edition. Saunders. Philadelphia.

525 Gilat, Y., Laichter and N.H. Shafrir. 1975. Impacts of Nuclear releases into the Aquatic Environment. IAEA-SM-198/18, pp 63-75. Vienna.

526 Gilbert, O.L. 1965. In; Goodman, G.T. et al. (eds); Ecology and Industrial Society. Blackwell. Oxford., pp 35-47.

527 Gillespie, D.M., J.D. Eldredge and C.K. Thompson. 1975. Water Res., 9(9), pp 817-819.

528 Gish, C.D. 1970. Pestic. Monit. J. 3, pp 241.

529 Gissel-Nielsen, G. 1973. Ambio, 2, pp 114-117.

530 Gjessing, Y. and E. Gjessing. 1973. Vatten, 3, pp 233-237.

531 Glass, N.R. 1971. in: Patten, B.C. (ed); System Analysis and Simulation in Ecology. 1, pp 325-363. Academic Press. New York.

532 Gleuckauf, E. 1951. Compendium of Meteorology. Am. Meteorological Soc. Boston.

533 Gloyna, E.A. 1971. Transport of Organic and Inorganic Materials in Small-scale Ecosystems. Advances in Chem. Series. Vol. 106.

534 Gloyna, E.F., Y.A. Yosef and T.J. Padden. 1971. U.S. Atom. Ener. Comm. Contract AT(11-1)-490, Tech. Rept. 21, CRWR-75, EHE-71-1, ORO-21, pp 1-66.

535 Goering, J.J. and P.L. Parker. 1972. Limnol. Oceanogr., 17(2), pp 320-323.

536 Goering, J.J. and V.A. Dugdale. 1966. Limnol. Oceanogr., 3, pp 291-298.

537 Goerke, H. and W. Ernst. 1974. in: Barnes, H. (ed.); Ninth European Marina Biology Symposium. Oban. UK. 2 Oct 1974.

538 Goldberg, E.D. 1962. Limnol Oceanogr., 7, pp 72.

539 Goldberg, E.D., 1963. The Sea, Interscience London, 2, pp 3.

540 Goldman, C.R. and R.G. Wetzel. 1963. Ecology, 44, pp 283-294.

541 Goldman, J.C. and E.J. Carpenter. 1974. Limnol. Oceanogr., 19(5), pp 756-766.

542 Goldman, J.C., D.B. Porcella, E.J. Middlebrooks and D.F. Toerien. 1972. Water Res., 6(6), pp 637-680.

543 Goldsmith, C.D. Jr., P.F. Scancon and W.R. Pieie. 1976. Bull. Environ. Contam. Toxicol. 16(1), pp 66-70.

544 Golley, F.B. 1961. Ecology 42, pp 581-584.

545 Golley, F.B. 1969. Ecology 50, pp 517-519.

546 Golley, F.B. and J.B. Gentry. 1964. Ecology 45, pp 217-225.

547 Golley, F.B. 1959. Mimeogr. Univ. Georgia, pp 7.

548 Golley, F.B. 1967. in: Petrusewicz, K. (ed.); Secondary Productivity of Terrestrial Ecosystems. pp 99-124. Warszawa. Krakow.

549 Golley, F.B. 1971. in: Wienes, J.A. (ed.); Ecosystem Structure and Function. pp 69-88. Oregon Sta. Univ. Press. Oregon.

550 Golterman, H.L. 1975. Physiological Limnol., Elsevier Amsterdam.

551 Golterman, H.L. 1960. Acta bot. neerl. 9, pp 1-58.

552 Gorecki, A. 1969. Acta theriol. 18, pp 333-352.

553 Gorecki, A. 1975. in: Grodzinsky, W., R.Z. Klekowski and A. Duncan (eds.); Methods for Ecological Bioenergetics. IBP Handbook No 24, pp 275-292. Blackwell. Oxford.

554 Gorham, E.J., J.W.G. Lund, J.E. Sanger and W.E. Dean, Jr. 1974. Limnol. Oceanogr., 19(4), pp 601-617.

555 Gosztonyi, A.E. 1974. Physis 33(86), pp 1-8.

556 Goudrian, J. 1977. Crop Micrometeorology: A Simulation Study. Chap 4. Pudoc. Wageningen.

557 Graetz, D.A., D.R. Keeney and R.B. Aspiras. 1973. Limnol. Oceanogr., 18, pp 908-917.

558 Gramms, L.C. and W.C. Boyle. 1972. Water Res., 6(12), pp 1433-1444.

559 Granek, W. 1977. Vatten 3, pp 251-265.

560 Granmo, Aa. and S. Kollberg. 1976. Water Res., 10(3), pp 189-194.

561 Greenberg, D.H. and W.W. Campbell. 1940. Proc. National Acad. Sciences Wash., 24.

562 Greenwald, I. et. al. 1940. J. Biol. Sci., 135, pp 65.

563 Greenwald, I. 1941. J. Biol. Chem., 141, pp 789.

564 Greichus, Y.A., A. Greichus and R.J. Emerick. 1973. Bull. Environ. Contam. Toxicol. 9, pp 321-327.

565 Greig, R.A., A. Adams and D.R. Wenzloff. 1977. Bull. Environ. Contam. and Toxic., 18(1), pp 3-8.

566 Grenney, W.J., D.V. Bella and H.C. Curl,jr. 1974. Water Pollut. Control. 46(7), pp 1751-1760.

567 Greze, V.N. 1973. Okeanologiya, 13(1), pp 107-113.

568 Gribble, G.W. 1974. TCDD. Chemistry, 47, pp 15-18.

569 Grim, J. 1939. Int. Rev. Hydrobiol. 39. pp 139-315.

570 Grodzinsky, W. 1971. Acta Theriol. 16, pp 231-275.

571 Grodzinsky, W., B. Bobek, A. Drozdz and A. Gorecki. 1970. in: Petrusewitz, K. and L. Ryszkowsk (eds). Energy Flow through Small Mammal Populations, pp 291-298. PWN. Warszawa.

572 Grodzwsky, W. 1975. in: Grodzinsky, W., R.Z. Klekowsky and A. Duncan (eds). Methods for Ecological Bioenergetics IBP Handbook No. 24, pp 65-94. Blackwell, Oxford.

573 Groth, P. 1971. Arch. Hydrobiol. 68(3), pp 305-375.

574 Grylls, F.S.M. 1961. in: Long, C. (ed.): Biochemists Handbook. Van Nostrand Princeton. New Jersey.

575 Guary, J.C. and A. Fraizier. 1977. Health Phys., 32(1), pp 21-28.

576 Guary, J.C. and A. Fraizier. 1977. Mar. Biol. 41(3), pp 263-267.

577 Guha, M.M. and R.L. Mitchell. 1966. Plant and Soil 24, pp 90-113.

578 Guillard, R.R.L. and V. Cassie. 1963. Limnol. Oceanogr., 8, pp 161-165.

579 Guthrie, R.K., F.L. Singleton and D.S. Ckerry. 1977. Water Res., 11(8), pp 643-646.

580 Gutierrez, M.J. and V.I. Otsu. 1975. Rev. Biol. Mar. Dep. Oceanogr. Univ. Chile 15(2), pp 179-199.

581 Gwinner, H.R., H.J. Cathey and F.J. Bulow. 1975. J. Tenn. Acad. Sci. 50(3), pp 102-105.

H

582 Haakanson, L. 1977. Vatten 3, pp 266-289.

583 Haan, C.T.. Am. Soc. Agr. Eng. Trans., 14, pp 445-449

584 Hadley, E.B. and L.C. Bliss. 1964. Ecol. Monogr., 34, pp 331-357.

585 Haedrich, R.L. 1975. Water Res., 9(11), pp 945-952.

586 Haegele, M.A. and R.H. Hudson. 1973. Environ. Pollut., 4, pp 53-57.

587 Haffty, J. 1960. Residu Method for common minor Elements, U.S. Geological Survey Water Supply Paper 150-A, U.S. Goverment Printing Office, Wash. D.C., pp 9.

588 Hagedorn, H. 1971. Arch. Hydrobiol., 68, pp 382-399.

589 Haegerhaell, B. 1973. Bot. Mar., 16, p 53.

590 Haghiri, F. 1974. J. Environ. Qual., 3(2), pp 180-183.

591 Hakonson, T.E. and F.W. Whicker. 1975. Health Physics., 28, pp 699-706.

592 Hakonson, T.E., A.F. Gallegos and F.W. Whieker. 1975. Health Phys., 29, pp 301-306.

593 Halcrow, W. et al. 1972. Mar. Poll. Bull., 3(1), pp 7-10.

594 Hall, D.J. 1964. Ecology, 45, pp 94-112.

595 Hall, D.J., W.E. Cooper and E.E. Werner. 1970. Limnol. Oceanogr., 15(3), pp 838-928.

596 Hall, I. 1974. Water Pollut Control, 3, pp 538-547.

597 Hall, J.K. 1972. J. Environ. Quality, 1, pp 172-176.

598 Hall, J.K. 1974. J. Environ. Qual., 3(2), pp 174-180.

599 Hall, J.K. 1974. J. Environ. Quality, 3, pp 174-180.

600 Halmann, M. and M. Stiller. 1974. Limnol. Oceanogr., 19(5), pp 774-783.

601 Hamilton, E.I. 1976. Sci. Total Environ. 5(1), pp 1-62.

602 Hamnes, J.K. and J.F. Bartz. 1963. Agron. J. 55, pp 329-333.

603 Hampson, M.A. 1967. J. Exp. Bot. 18, pp 17-53.

604 Hampson, R.F. Jr. and D. Garvin. 1975. Chemical Kinetic and Photochemical Data for Modeling Atmospheric Chemistry. NBS Teeh. Note 866. U.S.

605 Handbook of Chemistry and Physics., 1967. The Chemical Rubber Company.

606 Haney, A. and R.L Lipsey. 1973. Environ. Pollut., 5, pp 305-316.

607 Haney, J.F. 1973. Arch. Hydrobiol., 72, pp 87-132.

608 Hannan, H.H. and T.C. Dorris. 1970. Limnol. Oceanogr., 15(3) pp 442-453.

609 Hannerz, L. 1966. Proc. 1st Int. Congr. Radiat. Prot., pp 417.

610 Hannerz, L. 1968. Rep. Inst. Freshwat. Res. Drottningholm 48, pp 120-176.

611 Hansen, J.C. 1970. Vatten. 26(4), pp 365-369.

612 Hanson, D.L. and T.F. Waters. 1974. Trans. Am. Fish. Soc. 103(3), pp 431-439.

613 Haq, S.M. 1967. Limnol. Oceanogr., 12, pp 40-51.

614 Hardy, R.W.F. and V.D. Havelka. 1975. Science, 188, pp 633.

615 Hargis, J.R. 1977. Limnol. Oceanogr., 22(5), pp 942-944.

616 Hargrave, B.T. 1969. J. Fish. Res. bd. Can. 26, pp 2003-2026.

617 Hargrave, B.T. 1970. Limnol. Oceanogr., 15(1), pp 21-30.

618 Hargrave, B.T. 1972. Limnol. Oceanogr., 17(4), pp 583-596.

619 Hargrave, B.T. 1973. J. Fish. Res., 30(9), pp 1317-1326.

620 Harned, H.S. and R. Davies.Jr. 1943. J. Amer. Chem. Soc., 65, p 2030.

621 Harned, H.S. and S.R. Scholes. 1941. J Am. Chem. Soc., 63, p 1706.

622 Harp, M.J. and F.I. Scoular. 1952. J. of Natr., 47, pp 67-72.

623 Harremoes, P. and M. Riemer. 1975. Pilot-scale expiriments on downflow filter denetrification. in: Conf. on Nitrogen as a Water Poll. Vol 3, session A3. Copenhagen.

624 Harris, E. 1957. Can. J. Res., 35, pp 769-782.

625 Harris, E. 1959. Bull. Bingham oceanogr. Coll., 17, pp 31-65.

626 Harris, R.P. 1973. J. mar. biol. Ass. U.K., 53, pp 785-800.

627 Harrison, F.L. 1973. In; Radioactive contamination of the Marine Environment. Vienna, IAEA.

628 Harrison, F.L. 1969. Proc. 2nd Nat. Symp. on Radioecology. USAEC Conf. 670503, pp 198-220.

629 Harrison, F.L. 1973. Health Physics, 24, pp 331-343.

630 Harrison, P.J., H.L. Conway and R.C. Dugdale. 1976. Mar. biol. 35, pp 177-186.

631 Harrison, P.J., H.L. Conway, R.W. Holmes and C.O. Davis. 1977. Mar. Biol., 43(1), pp 19-31.

632 Harrison, R.M., R. Perry and R.A. Wellings. 1975. Water Res., 9(4), pp 331-346.

633 Harriss, R.C. et al. 1972. Mar. Poll. Bull. 3(12), pp 182-183.

634 Harriss, R.C., D.B. White and R.B. MacFarlane. 1970. Science 170(3959), pp 736-737.

635 Harrod, M.F. 1971. In: Tech. Bull. 21, pp 176-189. Trace Elements in Soils and Crops.

636 Hart, J.S. 1947. Lethal Temparature relations of certain Fish of the Toronto Region. Trans. R. Soc. Can., 41, pp 57-64.

637 Hart, J.S. 1952. Geographic variations of some Physiological and Morphological Characters in certain Freshwater Fish. Publ. Int. Fish Res. Lab., 72, pp 79.

638 Hartmann, M. and H. Nielsen. 1969. Geol. Rdsch. 58, pp 621-655.

639 Harvey, R.S. 1969. Health Physics, 17, pp 149-154.

640 Hasanen, E., S. Kolehmainen and J.K. Miettinen. 1966. Proc. 1st Int. Congr. Radiat. Prot., pp 401-406.

641 Hashem, M.T. 1973. Bull. Inst. Oceanogr. Fish. Cairo. 3, pp 95-115.

642 Hashem, M.T. 1973. Bull. Inst. Oceanogr. Fish. Cairo. 3, pp 163-182.

643 Hattori, A. 1962. Plant Cell Physiol. 3, pp 355-369.

644 Haurwitz, B. and J.M. Austin 1944. Climatology. McGraw-Hill New York.

645 Hayes, F.R. and J.E. Phillips. 1958. Limnol. Oceanogr., 3, pp 459-475.

646 Hayes, F.R., J.A. McCarter, M.L. Cameron and D.A. Livingstone. 1952. J. Ecol. 40, pp 203-216.

647 Hayward, R.S. and D.N. Gallup. 1976. Arch. Hydrobiol., 77(2), pp 139-163.

648 Healey, I.N. 1967. in: Petrusewicz, K. (ed.); Secondary Productivity of Temperate Terrestrial Ecosystems. pp 695-708. Panstwowe Wydawnictwo Naukowe. Warsaw.

649 Hecky, R.E. and P. Kilkam. 1974. Limnol. Oceanogr., 19(2), pp 361-365.

650 Heip, C. 1972. Mar. Biol., 12, pp 219-221.

651 Heisey, D. and K.G. Porter. 1977. Limnol. Oceanogr., 22(5), pp 839-845.

652 Heisinger, J.F. and W. Green. 1975. Bull. Environ. Contam. Toxicol. 14(6), pp 665-673.

653 Held, J.W. and J.J. Paterka. 1974. Trans. Am. Fish. Soc., 103(4), pp 743-756.

654 Hemmingsen, A.M. 1960. Rep. Steno meml. Hosp. 9, part 2, pp 1-110.

655 Henderson, C. and Q.H. Pickering. 1966. Air Water Pollut. 10, pp 453.

656 Hendrey, G.R., K. Baalsrud, T.S. Traaen, M. Laake and G. Raddum. 1976. Ambio. 5(5-6), pp 224-227.

657 Herbert, D.W.M. and D.S. Shurben. 1964. Water Waste Treat. J. 10, pp 141-142.

658 Herrmann, H., M. Ruf, K. Hubel and W. Lunsmann. 1975. Impacts of Nuclear Releases into the Aquatic Environment. IAEA-SM-198/14. pp 461-470. Vienna.

659 Hervey, R.J. 1949. Bot. Gaz., 111, pp 1.

660 Hesketh, J.D. and D. Baker. 1967. Crop. Sci. 7, pp 285-293.

661 Hethering, J.A., D.F. Jefferies and M.B. Lovett. 1975. Impacts of Nuclear Releases into the Aquatic Environment. IAEA-SM-198/29, pp 193-210. vienna.

662 Hewett, C.J. and D.F. Jefferies. 1976. J. Fish. Biol. 9, pp 479-489.

663 Hibiya, T. and M. Oguri. 1961. Bull. Jap. Soc. Sci. Fish. 27, pp 996-1000.

664 Hichs, R.M. 1972. Chem. Biol. Interact. 5, pp 361-390.

665 Hickey, J.J., J.A. Keith and F.B. Coon. 1966. J. Appl. Ecol. 3 (Suppl), pp 141-154.

666 Hidy, G.M. and J.R. Brock, 1971. In: H.M. Englund and W.T. Beery (eds.), Proc. 2nd Int. Clean Air Congr. p 1088. Ac. Press, N.Y., Lond.

667 Hildebrand, S.G. 1974. Limnol. Oceanogr., 19(6), pp 951-957.

668 Hill, A.C. 1971. J.Air Pollut. Control, 21, pp 341-346.

669 Hinesly, T.D. R.L. Jones , J.J. Tyler and E.L. Ziegler. 1976. Water Pollut. Control., 48(9), pp 2137-2152.

670 Hirst, S.M. 1975. Savanna Ecosystem project. Progress report 1974/75.

671 Hiyama, Y. and M. Shimizu. 1964. Rec. Oceanogr. Works. Jap. 7(2), pp 43-77.

672 Hobbie, J.E. and G.E. Likens. 1973. Limnol. Oceanogr., 734-742.

673 Hobson, L.A. and R.J. Pariser. 1971. J. exp. mar. biol. Ecol. 6, pp 71-78.

674 Hobson, L.A. 1971. Invest. pesq. 35, pp 195-208.

675 Hobson, L.A., D.W. Menzel and R.T. Barber. 1973. Mar. Biol. 19, pp 298-306.

676 Hodge, V.F., F.L. Hoffman and T.R. Folson. 1974. Health Phys., 27, pp 29-35.

677 Hoegstroem, V. 1975. Ambio. 4(3), pp 120-125.

678 Hoesslin, R.v. 1955. Festschr. 25 jaehr Bestehen Gaern. Hochschulstud., Deutchland, 1945/5, pp 47-62.

679 Hogetsu, K., M. Sakamoto and H. Samikawa. 1959. Bot. Mag., Tokyo, 72, pp 421-422.

680 Hohendorf, K. 1966. Kieler Meeresforsch., 22, pp 70-90.

681 Holden, A.V. 1973. J. Food. Tech. 8, pp 1-25.

682 Holeman, J.N. 1968. Water Res., 4, pp 737-747.

683 Holland, G.A. Lasater, J.E. Neumann and W.E. Eldridge. 1960. State of Wash. Dep. of Fish. Res. Bull. 5.

684 Holleman, D.F. and J.R. Luick. 1976. Health Phys., 30(2), pp 243-245.

685 Hollister, T.A., G.E. Walsh and J. Forester. 1975. Bull. Environ. Contam. Toxicol. 14(6), pp 753-759.

686 Holm-Hansen, O., W.H. Sutcliffe, Jr. and J. Sharp. 1968. Limnol. Oceanogr., 13, pp 507-514.

687 Holmes, R.W. 1970. Limnol. Oceanogr., 15(5), pp 689-694.

688 Hoogenhant, H. and J. Amesz. 1965. Arch. Microbiol. 50, pp 10-24.

689 Hook, R.I. van and A.J. Yates. 1975. Environ. Res. 9(2), pp 76-83.

690 Hook, R.I. van Jr., B.G. Blaylock, E.A. Bondietti, C.W. Francis, J.W. Huckabee, D.E. Reichle, F.H. Sweeton and J.P. Witherspoon. 1976. Environ. Qual. Saf. 5, pp 167-182.

691 Hook, R.I. van Jr., W.F. Haris and G.S. Henderson. 1977. Ambio, 6(5), pp 281-286.

692 Horner, R. and V. Alexander. 1972. Limnol. Oceanogr., 17(3), pp 454-457.

693 Horne, A.J. and G.E. Fogg. 1970. Proc. R. ent Soc., London, Ser. B. 175, pp 351-366.

694 Hoskin, C.M. 1959. Publ. Inst. Mar. Sci. Univ Tex. 6, pp 186-192.

695 Houde, E.D. 1977. Mar. Biol., 43(4), pp 333-341.

696 Howard-Williams, C. and W.J. Junk. 1977. Arch. Hydrobiol., 79(4), pp 446-464.

697 Howard, D.L., J.I. Frea, R.M. Pfister and P.R. Dugan. 1970. Science 169, pp 61-62.

698 Hubbel, S.P. 1971. in: Patten, B.C. (ed.); Systems Analysis and Simulation in Ecology. 1, pp 269-324. Academic Press. New York.

699 Huenerberg and Sarfert. 1967. Ges u. Wasserfach, Wasser-Abwasser, 108, pp 966-969 and 1197-1205.

700 Hubbert, K. 1971. Scientific American , 224(3), pp 60-87.

701 Huckabee, J.W. and B.G. Blaylock. 1972. in: Dhar, S.K. (ed.); Metal Ions in Biological Systems. pp 125-160. Plenum Publ. Corp. New York.

702 Huggins, M.L. 1953. J. Am. Chem. Soc. 75, pp 4123.

703 Hughes, J.C. and J.W.G. Lund. 1962. Arch. Microbiol. 42. pp 117-129.

704 Hunnego, J.N. and D.L. Harrison. 1971. New Zealand. J. Agr. Res. 14, pp 406.

705 Hunter, J.V., M.A. Hartnett and A.P.

Cryan. 1973. A study of the Factors Determination the Oxygen Uptake of Benthal Stream Deposits. Dept. of Env. Sci. Rutgers Univ.

706 Hutchinson, G.E. 1957. A Treatise on Limnology. I. Geography, Physics and Chemistry. John Wiley and Sons. New York.

707 Hutchinson, G.E. 1957. A Treatise on Limnology, vol II. Introduction of Lake Biology and the Limnoplankton, Wiley, N.Y., pp 306-354.

708 Hutchinson, J.C., R.J. Millington and D.B. Peters. 1972. Science, N.Y., 175, pp 771-772.

709 Hutzinger, O., S. Safe and V. Zitko. 1974. The Chemistry of PCBs, CRC-Press.

710 Huun, J.B. 1963. Uptake, Distribution and excreation Of the radionuclide I-131 by rainbow trout. Thesis, Michigan Statee Univ. East Lansing, Mich. 126 p.

711 Hwang, C.P., T.H. Lackie and P.M. Huang. 1976. Water Pollut. Control. 48(12), pp 2754-2760.

712 Hydroscience, Inc. 1971. Simplified Mathematical Modeling of Water Quality. EPA Water Programs.

I

713 Ichimura, S. and Y. Aruga. 1964. in; Y. Miyake and T. Koyoma (eds.); Recent Researches in the Fields of Hydrosphere, Atmosphere and Nuclear Geochemistry, Tokyo, pp 13-37.

714 Ichimura, S., S. Nagasawa and T. Tanaka. 1968. Bot. Mag., Tokyo, 81, pp 1-10.

715 Ichimura, S, Y. Saijo and Y. Aruga. 1962. Bot. Mag., Tokyo, 75, pp 212-220.

716 Ide, Tohya, Suzuki, Matsuo and Osanai. 1972. Removal of Nitrogen and Phosphorus in Waste Water Treatment. Pachec 72. Session 16-3.

717 Ignatiades, L. and G.E. Fogg. 1973. J. mar. biol. Ass. U.K., 53, pp 937-956.

718 Ikeda, T. 1970. Bull. Fac. Fish. Hokkaido Univ., 21, pp 91-112.

719 Ikeda, T. 1974. Mem. Fac. Fish. Hokkaido Univ. 22, pp 1-97.

720 Imboden, D.M. 1974. Limnol. Oceanogr., 19, pp 297-304.

721 Information from the Swedish Salmon Research Institute. 1975.

722 Ingestad, T. 1977. Ambio, 6(2-3), pp 146-151.

723 Inman, R.E., R.B. Ingersoll and F.A. Levy. 1971. Science, N.Y., 172, pp 1220-1231.

724 Ireland, M.P. 1973. Environ. Pollut., 4, pp 27-35.

725 Isaacs, W.P. and A.F. Gaudy. 1968. J. San. Eng. Div. ASCE, 94, SA2, pp 319.

726 Isensee, A.R. and G.E. Jones. 1975. Environ. Sci. Technol. 9(7), pp 668-672.

727 Isensee, A.R. 1977. Bioaccumulation of TCDD. Draft Subject to Revision, Conf. paper. Stockholm.

728 Ivanov, M.V. 1968. Microbiological processes in the Formation of Sulfur Deposits. 298 pp. Israel Program for Scientific Translations. Jerusalem.

729 Iversen, T.M. and J.Jessen. 1977. Fresh. Biol. 7, pp 287-298.

730 Ivie, G.W., H.W. Dorough and H.E. Bryant. 1974. Bull. Environ. Contam., 11, pp 129-135.

731 Ivlev, V.S. 1961.

732 Ivo Cerrea, C.T. 1973. Arq. Cienc. Mar., 13(1), pp 19-23.

733 Iwata, I., M.E. Dusch, W.E. Westlake and F.A. Gunther. 1975. Bull. Environ. Contam. Toxicol. 14(1), pp 49-56.

734 IBP Handbook no 12. 1974. in: R.A. Vollenweider (ed.): A Manual on Methods for Measuring Primary Production in Aquatic Environments, 2 edition.

J

735 Jackim, E., G. Morrison and R. Steele. 1977. Mar. Biol. 40(3), pp 303-308

736 Jackson, G.A. 1976. Bull. Environ. Contam. Toxicol. 16(5), pp 505-507.

737 Jacobsen, O.S. 1977. Vatten 3, pp 290-297.

738 Jacques, G., H.J. Minas, M. Minas and P. Nival. 1973. Mar. Biol. 23, pp 251-265.

739 Jaffe, L.S. 1975. In: S.F. Singer (ed.): The Changing Global Environment, pp 83-110. D. Reidel Publ. Comp.

740 Jannasch, H.W. 1975. Limnol. Oceanogr. 20(5), pp 860-868.

741 Jansson, B.O. 1971. Ecosystem Approach to Baltic Problem. Bulletins from the Ecological Research Committes, no 16, Swedish Natural Science Research Council.

742 Jarvenpaa, T., M. Tillander and J.K. Miettinen. 1970. Suom. Kemistilehti. 43, pp 439-442.

743 Jawed, M. 1973. Mar biol. 23, pp 115-120.

744 Jeanjean, R. 1969. Bull. Soc. Fr. Physiol. reg.15, pp 150-171.

745 Jefferies, D.F. and C.J. Hewett. 1971. J. mar. biol. Ass. U.K. 51, pp 411-422.

746 Jefferies, D.J. and M.C. French. 1971. Environ. Pollut., 1(3), pp 235-242.

747 Jefferies, D.J. and M.C. French. 1972. Environ. Pollut., 3, pp 147-156.

748 Jefferies, P.F. and C.J. Hewett. 1971. J Mar. Biol. Ass. U.K., 51, pp 411.

749 Jeffrey Brandt, C. and R.W. Rhodes. 1972. Environ. Pollut., 3, pp 217-225.

750 Jeffrey, S.W. and M.B. Allen. 1964. J. Gen. Microbiol., 36, pp 277-288.

751 Jellinek, H.H.G. and S.P. Sangal. 1972. Water Res. 6(3), pp 305-314.

752 Jennett, J.C. and S.M. Linnemann. 1977. Water Pollut. Control, 49(8), pp 1842-1856.

753 Jenny, H.1933. Mo. Agr. Exp. Sta. Bull. Res. 324.

754 Jensen, S. and O. Petterson. 1971. Environ. Pollut., 2, pp 145-155.

755 Jensen, A.A. 1977. Phthalater i Miljoeet. Publ. from the danish environ. agency, Miljoestyrelsen. Cph.

756 Jerneloev, A. 1970. Limnol. Oceanogr. 15(6), pp 958-960.

757 Jerneloev, A., R. Rosenberg and S. Jensen. 1972. Water Res. 6(10), pp 1181-1191.

758 Jerneloev, A., R. Rosenberg and S. Jensen. 1972. Water Res. 6(10), pp 1181-1191.

759 Jerneloev, A. 1969. in: Miller, M.W. (ed); Chemical Fallout. pp 68-74. Springfield. Ill.

760 Joergensen, E.G. 1964. Physiol. Plant.,17, pp 136-145.

761 Joergensen, S.E. 1975. Water Res. 9(2), pp 163-170.

762 Joergensen, S.E. 1976. Ecol. Model. 2, pp 147-165.

763 Joergensen, S.E. 1979. in: Orlob, G. (ed.): State of the Art of Water Quality

764 Joergensen, S.E. 1979. Modelling the distribution and effect of heavy metals. Ecol Modell., 6(3), (In press).

765 Joergensen, S.E. 1979. Industrial Waste Water Managment. Elsevier. Amsterdam.

766 Joergensen, S.E., L. Bengtson and M. Friis. 1978. Ecol. Model. 4 (In press).

767 Joergensen, B.B. and T. Fenchel. 1974. Mar biol. 24, pp 189-201.

768 Joergensen, B.B. 1977. Limnol. Oceanogr. 22(5), pp 814-832.

769 Joergensen, C.B. 1949. J. Mar. Biol. Ass. U.K., 28, pp 333-344.

770 Joergensen, C.B. 1960. J. Cons. Int. Explor.Mer., 26, pp 94-116.

771 Johannes, R.E. 1964. Limnol. Oceanogr. 9, pp 235-242.

772 Johanson, N., J.E. Kihlstroem and A. Wahlberg. 1973. Ambio. II(1-2), pp 42-43.

773 Johansson-Sjoebeck, M-L., G. Dave and U. Lidman. 1975. In: J.H. Koeman et al.(eds.), Sublethal Effects of Toxic Chemicals on Aquatic Animals, Elsevier Scientific Publishing Company, Amsterdam, Oxford, N.Y., pp 189-195.

774 John, M.K. 1972. Bull. Environ. Contam. Toxicol. 8(2), pp 77-80.

775 John, M.K. 1973. Environ. Pollut., 4, pp 7-15.

776 Johnels, A.G., T. Westermark, W. Berg., P.I. Persson and B. Sjoestrand. 1967. Oikos 18, pp 323-333.

777 Johnson, F.S. 1954. J. Meteorol., 11, p 431.

778 Johnson, F.S. 1970. Biol. Cons. 2, pp 83-89.

779 Johnson, F.S. 1975. In: S.F. Singer (ed.), The Changing Global Environment, pp 49-56. D. Reichel Publ. Copm.

780 Johnson, L.E., L.R. Albee, R.O. Smith and A.L. Moxon. 1951. South Dakota Agr. Expt. St. Bull. no. 412.

781 Johnson, N.M., G.E. Likens, F.H. Bormann and R.S. Pierce. 1968. Geochem. et Cosmochim. Acta 32, pp 531-545.

782 Jonasson, P.M. and H. Mathiesen. 1959. Oikos 10, pp 137-167.

783 Jonasson, P.M. 1972. Oikos Suppl. 14, pp 1-148.

784 Jones, A.S. and C.S. Hodges. 1974. J. Agric. Food. Chem. 22(3), pp 435-441.

785 Jones, D., K. Ronald, D.H. Lavinge, R. Frank, M. Holdrinet and J.F. Uthe. 1976. Sci. Total Environ., 5(2), pp 181-196.

786 Jones, G.E., L.G. Royle and L. Murray.1976. Applied Science Publ., 5(12), pp 889-895.

787 Jones, J.R. and R.W. Bachmann. 1976. Water Pollut. Control, 48(9), pp 2176-2182.

788 Jones,R. 1954. Mar. Res. 2.

789 Journal of Agric. Food Chem. 1977. 25(3), pp 545.

790 Junge, C.E. 1963. Air Chemistry and Radioactivity, Ac. Press.

K

791 Kale, H.W. 1965. Publ. Nutt. Ornithol. Club. Cambridge. Massachusettes.

792 Kalff, J. and H.E. Welch. 1974. J. Fish. Res. bd. Can. 31, pp 621-636.

793 Kalinin, G.P. and V.P. Bykov. 1969. Impact 19, pp 135-150.

794 Kameda, K., M. Shimizu and Y. Hiyama.

1968. J. Rad. Res. 9, pp 50-62.

795 Kamler, E. 1970. Pol. Arch. Hydrobiol. 17, pp 201-216.

796 Kanazawa, J., A.R. Isensee and P.C. Kearney. 1975. J. Agric. Food. Chem. 23(4), pp 760-763.

797 Kanemasu, E.T., G.W. Thurtell and C.B. Tanner. 1969. Pl. Physiol. 44, pp 881-885.

798 Kaplan, I.R., K.O. Emery and S.C. Rittenberg. 1963. Geochim. cosmochim. Acta 27, pp 297-331.

799 Karbe, L. 1972. Verh. Int. Ver. Limnol. 18, pp 911-918.

800 Karien, J.F., J.G. Lamberton, N.E. Stewart and L.C. Terriere. 1967. J. Agric. Food. Chem. 15, pp 148.

801 Kashiwada, K., A. Kanazawa and S. Tachibanazono. 1963. Mem. Fac. Fish., Kagoshima Univ. 12, pp 153-157.

802 Katz, S.E., R.F. Struze and C.A. Fassbender. 1970. Bull. Environ. Contam. Toxicol., 5, pp 50-55.

803 Kawamura, A. 1974. Sci. Rep. Whales Res. Inst. Tokyo. 26, pp 25-144.

804 Kearney, P.C., J.E. Oliver, C.S. Helling, A.R. Isensee and A. Kontson. 1977. J. Agric. Food. Chem. 25(5), pp 1177-1181.

805 Keeney, D.R., R.L. Chen and D.A. Graetz. 1971. Nature 233, pp 66-67.

806 Keeney, W.L., W.G. Breck , G.W. Vanloon and J.A. Page. 1976. Water Res., 10(11)., pp 981-984.

807 Kehoe, R.A. 1961. J. Royal. Inst. Publ. Helth. Hyg. 24, pp 101-143.

808 Keil, J.E., S.H. Sandifer, C.D. Graber and L.E. Priester. 1972 Water Res. 6(7), pp 837-842.

809 Keirn, M.A. and P.A. Brezonik. 1971. Limnol. Oceanogr. 16, pp 720-731.

810 Kellog, W.W., R.D. Cadle, E.R. Allen, A.L. Lazrus and E. A. Martell. 1972. Science 175, pp 587-596.

811 Kersting, K. 1972. Limnol. Oceanogr. 17(4), pp 643-644.

812 Kersting, K. 1975. Effects of Diuron on the energy Budget of a Daphnia Magna Population. In J.H. Koeman et al.(ed), Sublethal Effects of Toxic Chemicals on Aquatic Animals, Elsevier Scientific Publishing Company, Amsterdam, Oxford, N.Y., pp 159-166.

813 Ketchum, B.H. 1939. Am. J. Bot. 26. pp 399-407.

814 Ketchum, B.H. 1975. in: Singer, S.F. (ed.); The Changing Global Environment. pp 311-328. D. Reidel Publ. Comp. New York.

815 Keulen, H. van. 1975. Simulation of water use and herbage growth in arid regions. Center for Agricultural Publishing and Documentation. Wageningen.

816 Khan, R.A. and A.Q. Siddiqui. 1973. Proc. Indian Natl. Sci. Acad., Part B. 39(5), pp 582-597.

817 Khan, S.U. and P.B. Marriage. 1977. J. Agric. Food. Chem. 25(6), pp 1408-1413.

818 Kibby. H.V. 1971. Ecol. Monogr. 41, pp 311-327.

819 Kibby, H.V. and F.H. Rigler. 1973. Verh Int. Ver. Limnol. 18, pp 1457-1461.

820 Kihlstroem, K.I., J. Oerberg, C. Lundberg and P.B. Danielsson. 1974. Ambio. 3(6), pp 231-233.

821 Kiigemagi, U. and L.C. Terrierre. 1972. Bull. Environ. Contam. Toxicol. 7(6), pp 348-352.

822 Kimerle. R.A. and N.H. Anderson. 1971. Limnol. Oceanogr. 16(4), pp 646-659.

823 Kimura, S., H. Kumada and Y. Matida. 1973. Bull. Freshwater Fish. Res. Lab. Tokyo 23(2), pp 115-123.

824 King, C.E. 1967. Ecology 48, pp 111-128.

825 King, D.L. and J.T. Novak. 1974. Water Pollut. Control. 46(7) pp 1812-1815.

826 King, D.L. 1970. Water Pollut. Control. 42(12), pp 2035-2051.

827 King, F.D. and T.T. Packard. 1975. Limnol. Oceanogr. 20(5), pp 849-853.

828 King, J.E. and T.S. Hilda. 1954. U.S. Fish. and Wildl. Serv. Spec. Sci. Rep. Fish. 18, pp 66.

829 King, R.J. and W. Schramm. 1976. Mar. Biol. 37, pp 209-213.

830 Kingma Boltjes, T.Y. 1936. Arch. Mikrobiol. 6, p 79.

831 Kinsey,D.W. 1973. Environ. Pollut., 4, pp 167-182.

832 Kirby, E.J.M. and O. Rackham. 1971. J. appl. Ecol. 8, pp 919-924.

833 Klaassen, H.E. and A.H. Townsend. 1973. Trans. Kans. Acad. Sci. 76(3), pp 248-253.

834 Klaassen, H.E. and F.W. Cook, Jr. 1973. Trans. Kans. Acad. Sci. 76(3), pp 244-247.

835 Klebahn, H. 1922. Jh. wiss. Bot. 61, p 535.

836 Kleiber, M. 1961. The Fire Of Life. An Introduction to Animal Energeticts. J. Wiley. New York.

837 Klekowski, R.Z. 1975. in: Vinogradov, M.E. (ed.); Pelagic Ecosystem of the Pacific Ocean, pp 372-378.

838 Klekowski, R.Z. and E.A. Shushkina. 1966. Verh. int. Ver. Limnol. 16, pp 399-418.

839 Klekowsky, R.Z., T. Prus and H. Zyromska-Rudzka. 1967. in: Petrusewitz, K. (ed). Secondary Productivity of Terrestrial Ecosystems, pp 859-879, PWN, Warszawa.

840 Kloeckner, K. 1976. Oekologie von Pomatoceros triqueter (L.) (Serpulidae, Polycheata), Dissertation, Universitaet Tubingen.

841 Knaver, G.A. and J.H. Martin. 1972. Limnol. Oceanogr. 17(6), pp 868-876.

842 Knowles, C., A.L. Downing and M.J. Barrett. 1965. J. Gen. Microbiol. 38, pp 263-278.

843 Knudsen, G. 1965. Biochim. Biophys. Acta. 103, pp 495-502.

844 Kobylinski, G.J. and R.J. Livingston. 11975. Bull. Environ. Contam. Toxicol. 14(6), pp 692-698.

845 Koeppe, D.E. 1977. Sci. Total Environ. 7(3), pp 197-206.

846 Koike, I., E. Wade, T. Tsuji and A. Hattori. 1972. Arch. Hydrobiol. 69, pp 508-520.

847 Kolenbrander, G.J. 1975. Nitrogen in Organic Matter and Fertilizer as a source of Pollution. in: Conf. on Nitrogen as a Water Poll. Vol 1, session C1. Copenhagen.

848 Kolisko, A. 1938. Arch. Hydrobiol. 33, pp 165-207.

849 Kontogiannis, J.E. and G.J. Barnett 1973. Environ. Pollut., 4, pp 69-79.

850 Koreliakova, I.L. 1959. Bull. Inst. Biol. Vodochranilishch 3, pp 13-16.

851 Koreliakova, I.L. 1968. Bull. Biol. Vodochranilishch 1, pp 22-25.

852 Kormondy, E.J. 1968. Limnol. Oceanogr. 13, pp 522-526.

853 Kosiorek, D. 1974. Pol. Arch. Hydrobiol. 21(3-4), pp 411-422.

854 Kovda, V.A. 1977. Agro-Ecosyst. 3(3), pp 205-224.

855 Kowal, N.E. 1971. in: Patten, B.C. (ed.); Systems Analysis and Simulation in Ecology. 1, pp 123-196. Academic Press. New York.

856 Kozlovsky, D.G. 1968. Ecology 49, pp 48-60.

857 Kreitzer, J.F. and J.W. Spann. 1973. Bull. Environ. Contam., 9, pp 250-256.

858 Krenkel, P.A. and G.T. Orlob. 1963. Amer. Soc. Civ. Eng. Transactions. 128, pp 293.

859 Kucera, C.L. and D.R. Kirkham. 1971. Ecology 52, pp 912-915.

860 Kudo, A. D.R. Miller and D.R. Townsed. 1977. Prog. Water Tech. 9, pp 925-935.

861 Kudo, A. 1976. J. Environ. Qual. 5(4), pp 427-430.

862 Kuenzler, E.J. 1961. Limnol. Oceanogr. 6, pp 400-415.

863 Kuenzler, E.J. and B.H. Ketchum. 1962. Biol. Bull. 125, pp 134-145.

864 Kummler, R.H., M.H. Bortner and L.S. Jaffe. 1971. Environ. Sci. Technol. 5(11), pp 1140-1141.

865 Kuniyuki, K., M. Yano and M. Kaanishi. 1975. Bull. Nansei Reg. Fish. Res. Lab., pp 81-100.

866 Kuznetsov, S.I. 1970. Microflora ozeriee geokhimicheskaya deyatel'nost'. Izdatel'stvo Nauka. Leningrad.

L

867 Lafcur, K.S. 1976. J. Environ. Qual. 5(1), pp 91-92.

868 Lagerwerff, J.V. 1972. in: Morvedt, Giordano and Lindsay (eds.) ; Micronutrients in Agriculture. Madison.

869 Lam, D.C.L. and J.M. Jacquet. 1976. J. Fish. Res. bd. Canada. 33(3), pp 550-563.

870 Landry, M.R. 1975. Limnol. Oceanog. 20(5), pp 854-857.

871 Landsberg, H.E. 1970. Meteorological Monographs 11, pp 91.

872 Landsberg, J.J. et al. 1975. J. appl. Ecol. 12, pp 659-684.

873 Langbien, W.B. and W.H. Durum. 1967. The Reaeration Capacity of Streams. U.S. Geological Survey Circular S42.

874 Lange, O.L., E.-D. Schulze, M. Evenari, L. Kappen and U. Buschbom. 1974. Oecologia 17, pp 97-110.

875 Langley, D.C. 1973. Water Pollut. Control, 48(9), pp 44-51.

876 Langmuir, D. 1968. Geochim. Cosmochim. Acta. 32, pp 835.

877 Larsen, D.P., H.T. Mercier and K.W. Malveg. 1974. in: E.J. Middlebrooks, D.H. Falkenborg and T.E. Moloney (eds). Modeling the eutrophication Process. Ann Arbor Publishers, Ann Arbor, Mich. pp. 15-33.

878 Larsen, V. 1974. Nitrogen transformations in some Danish Lakes with special reference to denitrification. Det Danske Hedeselskab. Forsoegsvirksomheden. Beretning no 14.

879 Larson, Aa. The National Swedish Environment Protection Board, Brackish Water. Some Biochemical effects of Cadmium on Fish. In J.H. Koeman et al.(ed), Sublethal Effects of Toxic Chemicals on Aquatic Animals, Elsevier Scientific Publishing Company, Amsterdam, Oxford, N.Y., pp 15-21.

880 Lasenby, D.C. and R.R. Langford. 1972. J. Fish. Res. bd. Can. 29, pp 1701-1708.

881 Lasenby, D.C. 1975. Limnol. Oceanogr. 20(6), pp 993-999.

882 Lasker, R. 1966. J. Fish. Res. bd. Can. 23, pp 1291-1317.

883 Lasiewski. R.C. and W.R. Dawson. 1967. Condor 69, pp 13-23.

884 Lasker, R., J.B.J. Wells and A.D. McIntyre. 1970. J. mar. biol. Ass. U.K. 50, pp 147-160.

885 Lassen, H. and P.B. Nielsen. 1972. Simple Mathematical Model for the primary production as a Function of the phosphate Concentration and the incoming solar Energy applied to the North Sea. Danmarks Hav- og Fiskeriundersoegelser. International Council for the Exploration of the Sea.

886 Lassiter, R.R. 1975. Modeling Dynamics of Biological and Chemical Components of Aquatic Ecosystems. EPA-660/3-75-012.

887 Lassiter, R.R. and D.K. Kearns. 1974. In: Middlebrooks, E.J., D.H. Falkenborg and T.E. Maloney (eds.); Modeling the Eutrophication Process. Ann Arbor. Mich.

888 Lassiter, R.R. Principles and Constrains for Predicting Exposure to Environmental Pollutants. EPA=report. (in Press)

889 Latimer, W.E. 1952. in E. Cliffs(ed), Oxidation Potentials, 2 2nd ed, pp 222.

890 Laube, H.R. and J.R. Wohler. 1973. Bull. Torr. Bot. Club. 100, pp 238-240.

891 Laughlin, R. 1965. J. Anim. Ecol. 34, pp 77-91.

892 Laumond, F., M. Neuburger, B. Donnier, A. Fourcy, R. Bittel and M. Aubert. 1973. Rev. Int. Oceanogr. Med. CERBOM., 31-32, pp 47-53.

893 Laveskog, A., A. Lindskog and V. Stenberg. 1976. Om Metaller, Statens Naturvaardsverk, Stockholm.

894 Laville, H. 1975. Ann. Limnol., 11(1), pp 67-77.

895 Lawlor, D.W. and J.V. Lake. 1976. J. appl. Ecol. 13(2), pp 639-638.

896 Lawrence, A.W. and P.L. McCarty. 1969. Water Pollut. Control, 41, R1.(12)

897 Le Borgne, R.P. 1973. Mar. Biol. 19, pp 249-257.

898 Leatherland, T.M., J.D. Burton, F. Culkin, M.J. McCartney and R.J. Morris. 1973. Deep-Sea Res. 20, pp 679-685.

899 Leckie, J.O. and R.O. James. 1974. in: Rubin, A.J. (ed.): Aqueous-Environmental Chemistry of Metals. pp 1-76. Ann Arbor sci. Michigan.

900 Lee, C.K.C. 1974. in: The Kuroshio 3. pp 389. National Research Council of Thailand. Bangkok.

901 Lee, J.J. and W.A. Muller. 1973. Am. Zool. 13(1), pp 215-223.

902 Lee, R.F., R. Sauerheber and A.A. Benson. 1972. Science 177, pp 344.

903 Leeuwangh, P., H. Bult and L. Schneiders. 1975. in: Koeman, J.H. et al.(ed): Sublethal Effects of Toxic Chemicals on Aquatic Animals, Elsevier Scientific Publishing Company, Amsterdam, Oxford, N.Y., pp 646-658.

904 Lehman, J.T., J. Sarvala, I. Hakala and M.-L. Viljanen. 1976. Limnol. Oceanogr. 21(5), pp 724-730.

905 Leidy, G.R. and R.M. Jenkins. 1977. The development of Fishery Compartments and Population Rate Coefficients for Use in Resevoir Ecosystem Modeling. Contract Report Y-77-1, USDI Fish and Wildlife Service, National Reservoir Res. Program, Fayetteville, Arkansas. Prepared for Chief of Engineers, US Army. Wash, D.C.

906 Leonard, R.A., G.W. Baily and R.R. Swank(JR). 1976. Land Application of Water Materials, pp 48-78.

907 Leoni, V. and S.U. D'Arca. 1976. Sci. Total Environ. 5(3), pp 253-272.

908 Leopold, A.C. 1964. Plant Growth and Development. McGraw-Hill, New York.

909 Leopold, G. E. Furakawa, W. Forth and W. Rummel. 1969. Naunym Schmiedeberg Arch. Pharm., 263, pp 275-276.

910 Leopold, L.B. and T. Maddock. 1953. U.S. Geol. Survey. Professional Paper 252.

911 Lerman, A. and G.J. Brunkill. 1971. Limnol. Oceanogr., 16(6), pp 880-890.

912 Lerman, A. and T.A. Lietzke. 1975. Limnol. Oceanogr. 20(4), pp 497-510.

913 Lerman, A. 1975. Am. J. Sci. 275, pp 609-635.

914 Leussing, D.L. and I.M. Kolthoff. 1953. J. Amer. Chem. Soc., 75, pp 2476.

915 Lewin, J. and C. Chen. 1968. J. Phycol. 4, pp 161-166.

916 Lewis, J.T., L.W. Grossman, J.G.Kereiakes and E.L.Saenger 1976. Health Physics. 30(3), pp 318-320.

917 Lewis, W.M. Jr. 1974. Ecol. Monogr. 44, pp 337-409.

918 LeGall, P. and J. Ancellin. 1971. Available from Nat. Tech. Inf. Ser., Springfield, Va. as CEA-N-1488, pp 1-17.

919 Lick, W., J. Paul and Y.P. Sheng. 1976. in: Canale, R. P. (ed.); Mathematical Modeling of Biochemical Processes in Aquatic Ecosystems. Ann Arbor Science Press

920 Licktenstein, E.P., C.H. Muller, G.R. Myrdal and K.R. Schulz. 1962. J. Econ. Entomol., 55, pp 215.

921 Licktenstein, E.P., G.R. Myrdal and K.R. Schulz. 1964. J. Econ. Entomol., 57(1), pp 133.

922 Lidman, U., G. Dave and M-L. Johansson-Sjoeberg. 1975. In: Koeman et al. (eds.): Sublethal Effects of Toxic Chemicals on Aquatic Animals, Elsevier Scientific Publishing Company, Amsterdam, Oxford, N.Y., pp 207-220.

923 Liere, L. van, W. Zevenboom and L.R. Mur. 1975. Nitrogen as a Limiting factor for the Growth of the Blue-green alga Oscillatoria agardhii. in: Conf. on nitrogen as a water Pollutant, Vol 2, session B2. Copenhagen.

924 Lieth, H. 1963. J. Geophys. Res. 68(13), pp 3887-3898.

925 Likens, G.E. 1975. in: Lieth, H. and R.H. Whittaker (eds.); The Primary Productivity of the Biosphere. Springer-Verlag. New York.

926 Lillelund, K. and R. Lasker. 1971. Fish. Bull., 69, pp 655-667.

927 Lind, D.T., L.L. Smith(JR) and S.J. Broderius. 1977. Water Pollut. Control, 49(2), pp 262-268.

928 Lindberg, S.E. and R.C. Harris. 1977. Water Pollut. Control 49(12), pp 2479-2487.

929 Lindberg, S.E., A.W. Andren and R.C. Harriss. 1975. in: Cronin, L.E. (ed.); Estuarine Research. Vol 1 pp 64-107. Academic Press. New York.

930 Linden, O. 1976. Ambio. V(1), pp 36-37.

931 Linger, J.L. and D.B. Peakall. 1973. Environ. Pollut., 4, pp 59-68.

932 Lipton, W.V. and A.S. Goldin. 1976. Health Physics. 31(5), pp 425-430.

933 Lisemu. 1975. in: Conf. on Nitrogen as a water pollutant, Vol 2, session B2. Copenhagen.

934 List, R.J. 1963. Smithsonian Meteorological Tables. Smithsonian Inst. Wash. D.C.

935 Little, P.W. and M.H. Martin. 1972. Environ. Pollut., 3, pp 241-254.

936 Little, P. 1974. Environ. Pollut., 5, pp 159-172.

937 Liu, S.T. and G.H. Nancollas. 1970. J. Chrystal Growth, 6, pp 281.

938 Liu, S.T. and G.H. Nancollas. 1973. Desalination 12, pp 75.

939 Livingston, H.D. and G. Thompson. 1971. Limnol. Oceanogr. 16(5), pp 786-¤96.

940 Lloyd, R. and D.W.M. Herbert. 1967. Water Res., 6(8)., pp 132-145.

941 Lock, R.A.C. 1975. in: Koeman et al (eds.): Sublethal Effects of Toxic Chemicals on Aquatic Animals, Elsevier Scientific Publishing Company, Amsterdam, Oxford, N.Y., pp 61-79.

942 Lockyer, C. 1975. J. Cons. Int. Explor. Mer., 36(3), pp 259-273.

943 Lohgaonker, A.L. 1974. Curr. Sci., 43(18), pp 584-585.

944 Lombardo, P.S. 1972. Hydrocomp, Inc. Mathematical Model of Water Quality in Rivers and Impoundsments.

945 Lombardo, P.S. 1973. Hydrocomp, Inc. Critical Review of Currently Available Water Quality Models. Contract No. 14-31-0001-3751.

946 Long, J.N. and J. Turner. 1975. J. appl. Ecol., 12, pp 179-188.

947 Lopez-Zavala, A. A.S. de Aluja,B. Eliasl L.B. Manjarrez, L.Buchmann, I.A. Mercado and S. Caltenco. 1975. Progress Water Technol. 7(2), pp 73-82.

948 Lord, R.D., Jr. 1963. The Cottentail rabbit in Illinois. Carbondale. State Ill. Univ. Press.

949 Lotka, A.J. 1956. Elements of Mathematical Biology. Reprinted. Dover. New York.

950 Lounamaa, J. 1967. in: Kvauheim, A. (ed.); Geochemical Prospecting in Fennoscandia. pp 287-317. Interscience Publ. New York.

951 Loveless, J.E. and H.A. Painter. 1968. J. Gen. Microbiol. 52, pp 1-14.

952 Lowman, F.G. 1963. in: Schultz, V. and A.W. Klement, Jr. (eds.); Radioecology. pp 561-567. Reinhold Publ. Corp. New York.

953 Luckey, T.D. and B. Venugopal. 1977. Metal Toxicity in Mammals, 1. Plenum Press.

954 Lui, N.S.T. and O.A. Roels. 1972. J. Phycol. 8, pp 259-263.

955 Lund, J.W.G. 1964. Verh. Int. Ver. Limnol. 15, pp 37-56.

956 Lund, J.W.G. 1965. Biol Rev. 40, pp 231-293.

957 Lund, J.W.G. 1970. Water Treat. Exam. 19, pp 332-358.

958 Lung, W.S., R.P. Canele and P.L. Freedman. 1976. Water Res., 10(12), pp 1101-1114.

959 Luoma, S.N. 1977. Mar. Biol. 41(3), pp 269-273.

M

960 Mackay, D. and P.J. Leinonen, 1975. Environ. Sci. Tech. 9(13), pp 1178-1180.

961 Mackkereth, F.J. 1953. J. Expt. Bot. 4. pp 296-313.

962 Macleod, J.C. and E. Pessah. 1973. J. Fish. Res. bd. Can. 30(4), pp 485-492.

963 MacFadyen, A. 1963. Animal Ecology, Aims and Methods. 2nd edition. Pitman. London.

964 MacIsaac, J.J. and R.C. Dagdale. 1972. Deep Sea Res. 16. pp 415-422.

965 Madsen, B.L., P.N. Markmann and G.Rasmussen. 1977. Vand, 8(1), pp 4-8.

966 Madwick, H.A.I. 1970. in: Reichle, D.E. (ed.); Analysis of Temperate Forest Ecosystem. pp 47-54. Springer-Verlag. Berlin.

967 Mague, T.H., F.C. Mague and O. Holm-Hansen. 1977. Mar. Biol. 41(3), pp 213-227.

968 Maguire, L.A. and J.J. Porter. 1977.

Ecol. Modell. 3(4), pp 249-271.

969 Mahoney, J.P. and W.J. Small. 1968. J. Clinical Investigation, 47, pp 643.

970 Majernik, D. and T.A. Mansfield. 1972. Environ. Pollut., 3, pp 1-7.

971 Majori, L. and F. Petronio. 1973. Rev. Int. Oceanogr. Med. CERBOM., 31-32, pp 55-90.

972 Maljkovic, D. and M. Branica, 1971. Limnol. Oceanogr. 16(5), pp 779-785.

973 Malone, T.C. 1971. Mar. Biol. 10(4), pp 285-289.

974 Malovitskaia, L.M. and J. Sorekin. 1961. Trudy Inst. Biol. Vodokhranilishch 4, pp 262-272.

975 Manguld, K. and S. von Boletzky. 1973. Mar. biol. 19, pp 7-12.

976 Mann, K.H. 1971. in: Edmondson, W.T. (ed.); A Manual on Methods for the Assessment of Secondary Productivity of Fresh Waters. pp 204. IBP Handbook No 17.

977 Mann, K.H. 1972. Mar. Biol. 14, pp 199-209.

978 Manning, W.J. 1971. Environ. Pollut., 2, pp 69-76.

979 Mansell, R.S. et al. 1977. Fertilizer and Pesticide Movements from Citrus groves in Florida flatwood Soils. EPA-600/2-77-177.

980 Margalef, R. 1958. Gen. Syst. 3, pp 36-71.

981 Marshall, C.E. 1964. The Physical Chemistry and Mineralogy of Soils, Wiley and Sons, N.Y., vol 1.

982 Marshall, R. W. and G.H. Nancollas. 1970. J. Phys. Chem. 73, pp 3838.

983 Marshall, S.M. and A.P. Orr. 1961. J. Mar. Biol. Ass. U.K. 41, pp 463-488.

984 Martin, J.-L.M. 1974. Mar. biol., 28(4), pp 245-251.

985 Martin, J.H 1970. Limnol. Oceanogr. 15(5), pp 756-761.

986 Martin, J.H. and G.A. Knauer. 1973. Geochim. Cosmochim. Acta., 37, p 1639.

987 Martin, J.H. 1968. Limnol. Oceanogr. 13, pp 63-71.

988 Martin, J.H. 1970. Limnol. Oceanogr. 15(3), pp 413-418.

989 Mason, B. 1966. Principles of Geochemistry. 3rd ed. John Wiley and Sons. New York.

990 Mason, C.F. and R.J. Bryant. 1975. Freshwater biol. 5(3), pp 271-277.

991 Mathews, C.P. 1975. J. Fish. Biol. 7(1), pp 83-94.

992 Mathiesen, H. 1970. Vatten 26, pp 146-173.

993 Mathis, B.J. and T.F. Cummins. 1973. Water Pollut. Control 45 pp 1573.

994 Mauchline, J. and W.L Templeton. 1964. Oceanogr. Mar. Biol. Ass. Rev., 2, pp 229.

995 Mauchline, J. 1963. The Biological and Geographical distribution in the Irish Sea of Radioactive Effluent from Windscale Works, 1959-1960. UKAEA AHSB(RP) R.27,

996 Mayer, 1976. J. Fish. Res. Bd Can. 33, pp 2610-2613.

997 Mayzaud, P. and S. Pallot. 1973. Mar biol. 19, pp 307-314.

998 Mayzaud, P. 1973. Mar. Biol. 21, pp 19-28.

999 Mazur, T. 1967. Ekol. pol. 15, pp 607-613.

1000 McCarthy, J.J. and R.W. Eppley. 1972. Limnol. Oceanogr. 17(3), pp 371-382.

1001 McConnell, G., D.M. Ferguson and G.R. Pearson. 1975. Endeavour 13.

1002 McCorle, F.M., J.E. Chambers and J.D. Yarbrough. 1977. Bull. Environ. Contam. Toxicol. 18(3), pp 267-270.

1003 McCully, K.A., D.C. Villeneuve, W.P. McKinley, W.E.J. Phillips and M. Hidiroglou. 1966. J. Assoc. Official Agr. Chem. 49, pp 966.

1004 McKee, J.R. and H.W. Wolf. 1963. Water Quality Criteria, Publication no. 3A, Calif. State Water Res. Control Board 2nd Edition.

1005 McKellar, H.N. Jr. 1977. Ecol. Modell. 3(2), pp 85-118.

1006 McLaren, I.A. 1963. J. Fish. Res. bd. Can., 20, pp 685 -727.

1007 McMahon, J.W. and F.H. Rigler. 1965. Limnol. Oceanogr. 10(1), pp 105-113.

1008 McNab, B.K. 1963. Amer. Nat. 97, pp 133-140.

1009 McNaught, D.C. and A.D. Hasler. 1964. J. Fish. Res. Bd Can. 21, pp 291-318.

1010 McNaught, D.C. and A.D. Hasler. 1966. Verh. Int. Ver. Limnol., 16, pp 194-203.

1011 McNaughton, S.J. and L.L. Wolf. 1975. General Ecology. Holt, Rinehart and Winston, Int., New York.

1012 McNeill, S. and J.H. Lawton. 1970. Nature 225, pp 472-474

1013 McRoy, C.P. and R.J. Barsdate. 1970. Limnol. Oceanogr. 15(1) pp 6-13.

1014 McRoy, C.P., R.J. Barsdate and M. Nebert. 1972. Limnol. Oceanogr., 17(1), pp 58-67.

1015 McRoy, C.P. 1970. J. Fish. Res. bd. Can. 27, pp 1811-1829.

1016 Meadows, R.W. and D.J. Spedding. 1974. TELLUS., 26(1-2), pp 143-150.

1017 Meats, A. 1974. Oecologia 16, pp

119-138.

1018 Mechalas, A. and Matyskiela. 1970. A Study on Nitrification. Water Poll. Control Res. Ser. (17010 ORD07) 70. U.S. Dep. Interior. FWQA.

1019 Megard, R.O. 1972. Limnol. Oceanogr. 17, 68-87.

1020 Megard, R.O. 1975. Limnol. Oceanogr. 20(6), p 1053.

1021 Mehrbach, C., C.H. Culberson, J.E. Hawley and R.M. Pytkowicz. 1973. Limnol. Oceanogr. 18, pp 172-179.

1022 Meiklejohn, J. and G. Wimmer. 1952. Z. Hyg., 48, p 135.

1023 Meiklejohn, J. 1952. Nature 170, p 1132.

1024 Meiklejohn, J. 1954. Some Aspects of the Physiology of the Nitrifying Bacteria. In B.A. Fry and J.L. Peel(ed), The fourth Symposium of the Society for General Microbiology

1025 Menhinick, E.F. 1964. Ecology 45, pp 859-861.

1026 Mercier, M.J. 1977. 2,3,7,8-Tetrachlorodibenzo-p-dioxin, An Overview, Draft paper, Belgium.

1027 Merlini, M. 1967. Proc. Int. Symp. Radioecol. Concent. Process., 1966, Stockholm, pp 977-982.

1028 Merlini, M., C. Rigliocca, A. Berg and G. Pozzi. 1970. IAEA/ SM, 142/27.

1029 Metzger, I. 1968. J. San. Eng. Div. ASCE. 94, pp 1153.

1030 Meyer, B.S., D.B. Anderson and R.H. BOhning. 1960. Introduction to Plant Physiology. Van Nostrand. Princeton New York.

1031 Meyerhof, O. and P. Oesper. 1949. J. Biol. Chem. 179, pp 1371.

1032 Middlebrock, E.J. 1977. J. Wat. Pollut. Cont. Fed., 49, pp 1659-1660.

1033 Middlebrooks, E.J., D.B. Porcella, E.A. Pearson, P.H. McGauhey and G.A. Rohlich. 1971. Water Pollut. Control. 43(3), pp 454-473.

1034 Middleton, A.C. and A.W. Lawrence. 1977. J. Water Pollut. Control, 49(7), pp 1659-1670.

1035 Midwest Research Inst. (MRI), 1971. Partic. Pollut. System Study. Vol. 1: Mass Emissions. Durham NC.

1036 Miller, H.G. and J.D. Miller. 1976. J. appl. Ecol. 13(1), pp 249-256.

1037 Miller, R.W. and S.D. Faust. 1973. Environ. Lett. 4(3), pp 211-223.

1038 Miller, S.L. and H.C. Urey. 1959. Science 130, p 245.

1039 Milner, H.W. 1953. in: Burlew, J.S. (ed.): Algal Culture from Laboratory to Pilot Plant. Publ. 600. Carnegie Inst of Wash. Wash. D.C.

1040 Ministry of Technology. 1968. Water Pollution Research, 1967. H.M.S.O., London.

1041 Mitchell, A.L. (ed.). 1974. pp 427-434. EPA, Gulf Breeze environ. Res. Lab.

1042 Mitchell, J.H. Jr. 1971. Air Pollution and Climate Changes. 64th annual meeting. Am. Inst. of Chem. eng., San Francisco.

1043 Mitchell J.M. 1975. In: S.F. Singer (ed.), The Changing Global Environment, pp 149-173. D. Reidel Publ. Comp. New York.

1044 Mitchell, N.T. 1968. Radioactivity in Surface and Coastal Waters of the British Isles. 1968. M.A.F.F. FRL 2, Lowestoft, Suffolk.

1045 Mitchell, N.T. 1969. Radioactivity in Surface and Coastal Waters of the British Isles. 168. M.A.F.F. FRL 5, Lowestoft, Suffolk.

1046 Mitchell, N.T. 1971. Radioactivity in Surface and Coastal Water of the British Isles, 1970 M.A.F.F. FRL 8, Lowertoft, Suffolk.

1047 Mitchell, N.T. 1971. Radioctivity in Surface and Coastal Waters of the British Isles . 1969. M.A.F.F. FRL 7, Lo westoft, Suffolk.

1048 Mitchell, R. and C. Chamberlin. 1978. in: Berg. G. (ed.); Indicators of Enteric Contamination in Natural Waters.

1049 Mitsch, W.J. 1976. Ecol. Modell. 2(1), pp 69-89.

1050 Miyamoto, J. 1975. Env. Qual. Saf. 4, pp 134-135.

1051 Miyazaki, S., H.C. Sikka and R.S. Lynch. 1975. J. Agric. Food. Chem. 23(3), pp 365-368.

1052 Mizotani, T., K. Hidaka, T. Ohe and M. Matsumoto. 1977. Bull. Environ. Contam. Toxicol. 18(4), pp 452-461.

1053 Mochnacka-Lawacz, H. 1974. Pol. Arch. Hydrobiol. 21(314), pp 381-386.

1054 Moe, P.G. 1970. Environ. Sci. Technol. 4(5), pp 429-431.

1055 Moghissi, A.A., M.W. Carter and R. Lieberman. 1971. Health Physics. 21(1), pp 57-60.

1056 Mommaerts-Billet, F. 1973. Environ. Pollut., 4, pp 261-282.

1057 Monteith, J.L. 1972. J. appl. Ecol. 9, pp 747-766.

1058 Montgomery, W.V. 1975. Proc. W. Va. Acad. Sci. 46(3-4), pp 265-273.

1059 Mooney, H.A. and F. Shropdhire. 1967. Oecol. Plant. 2, pp 1-13.

1060 Moore, M.E. 1977. Sci. Total Environ., 7(2), pp 109-115.

1061 Moore, S.F. and R.L. Dwyer. 1974. Water Res. 8(10), pp 819-827

1062 Mootz, C.A. and C.E. Epifanio. 1974. Biol. Bull. mar. Biol. Lab. Woods Hole 146, pp 44-55.

1063 Morel, F. and J.J. Morgan. 1968. A Numerical Method for Solution of Chemical Equilibria in Aqueous Systems. Calif. Inst. Tech., Calif.

1064 Morgan, J.J. and F.B. Birkner. 1966. J. San. Eng. Div., ASCE 92, p 137.

1065 Morishima, H. and H.I Oka. 1977. Agro-Ecosyst., 3(2), pp 131-145.

1066 Morrison, C.R. 1971. in: Edmondson, W.T. and G.G. Winberg. (eds.): A Manual on Methods for the Assesment of Secondary Productivity in Fresh Water. p 204, IBP Handbook no 17.

1067 Morowitz, H.J. 1968. Energy Flow in Biology. Academic Press. New York.

1068 Mortimer, C.H. 1961. Verh. Int. Ver. Limnol. 14, pp 79-83.

1069 Morton, H.L., E.D. Robison and R.E. Meyer. 1967. Weeds 15, pp 268-271.

1070 Moss, B. 1970. Limnol. Oceanogr. 15(4), pp 504-513.

1071 Motto, H.L. et al. 1970. Environ. Sci. Technol. 4(3), pp 231-237.

1072 Moulder, B.C., D.E. Reichle and S.I. Auerbach. 1970. Oak Ridge National Lab. USA, ORNL 4452.

1073 Mowrer, J., J. Calambokidis, N. Musgrove, B. Drager, M.W. Beug and S.G. Herman. 1977. Bull. Environ. Contam. Toxicol. 18(5), pp 588-594.

1074 Mukherjee, S. and S. Bhattacharya. 1977. Water Res., 11(1), pp 71-74.

1075 Mulbarger. 1971. J. Wat. Poll. Cont. Fed. 43, pp 2059-2070.

1076 Muller, D. 1960. in: Rukland, W. (ed.): Encyclopedia of Plant Physiology, 12(2), pp 934.

1077 Muller, H. von. 1972. Arch. Hydrobiol. Suppl. 38(4). pp 399-484.

1078 Mullin, M.M. and E.R. Brooks. 1970 Limnol. Oceanogr. 15(5), pp 748-755.

1079 Mullin, M.M. 1963. Limnol. Oceanogr., 8(2), pp 239-250.

1080 Mullin, M.M., H.J. Perry, E.H. Renger and P.M. Evans. 1975. Mar. Sci. Commun., 1(1), pp 1-13.

1081 Munnick, K.O. and W. Roether, 1967. In: Radioactive dating and low level counting, pp 93-104. IAEA Vienna.

1082 Munro, I.G. 1974. Oecologia 16, pp 355-367.

1083 Murphy and Sutton. 1974. Pilot Scale Studies on Biological Denitrification. Presented at 7th IAWPR Conf. Paris.

1084 Murphy, G.F. 1971. Limnol. Oceanogr., 16(6), pp 981-983.

1085 Murphy, P.G. 1971. Bull. Environ. Contam. Toxicol. 6(1), pp 20-23.

1086 Murphy, P.C. 1970. Bull. Environ. Contam. Toxicol., 5, pp 404-409.

1087 Murray, C.N. and R. Fucai. 1975. Impacts of Nuclear releases into the Aquatic Environment. IAEA-SM-198/25, pp 179-191. Vienna.

1088 Murray, D.S., P.W. Santelmann and J. M. Davidson. 1975. J. Agric. Food. Chem. 23(3), pp 578-582.

1089 Myers, J. 1964. in: Burlew, J.S. (ed.): Dynamic water quality, Forecasting and Manegement, Carnegie inst. of Wash. D.C., Publ. 600, pp 37-54.

1090 Myrcha, A. and W. Walkowa. 1968. Acta theriol 13, pp 391-400.

1091 Myrcha, A. 1969. Acta theriol. 14, pp 211-227.

1092 Myrcha, A. 1967. Small Mammals Newslett. 3, pp 10.

1093 Myttenaere, C., M. Merlini, R.Bittel, P. Dabin, J.M. Mousny and G. Pozzi. 1975. Impacts of Nuclear releases into the Aquatic Environment. IAEA-SM-198/16, pp 29-47. Vienna. +

N

1094 Nadeau, P.J. and R.A. Davis. 1976. Bull. Environ. Contam. Toxicol. 16(4), pp 436-444.

1095 Nagaya, Y. and K. Nakamura. 1975. Impacts of Nuclear releases into the Aquatic Environment. IAEA-SM-198/21, pp 313-328. Vienna.

1096 Nakaniski, M. and M. Monji. 1965. J. Fac. Sci., Univ. Tokyo, Sec III Botany, 9, pp 19-42. >PARA 10 'I(NAKA) ' Nakahara, M., T. Koganayi and M. Saiki. 1972. Impacts of Nuclear releases into the Aquatic Environment. IAEA-SM-198/20, pp 301-311. Vienna.

1097 Nakayama, F.S. and C.H.M.v. Bavel. 1963. Agron. J. 55, pp 271-274.

1098 Nancollas, G.H. Interactions in Electrolyte Solutions, Amsterdam, Elsvier, 1966.

1099 Nash, R.G. and C.A. Woolson. 1967. Science 157, pp 924-927.

1100 Nash, R.G., M.L. Ball, Jr. and W.L. Harris. 1977. J. Agric. Food. Chem. 25(2), pp 336-341.

1101 Nasim, A.L., M.M.H. Paig and K.A. Lord. 1971. Environ. Pollut., 2, pp 7-12.

1102 National Air Pollution Control Adm. (NAPCA). Nationwide Inventory of Air Pollut. Emiss. 1968. Publ. no AP-73, NAPCA, Raleigh NC. 1970.

1103 Naturvaardsverkets Brackvattan Toxikologiska Lab.. Raport Jan. 1974.

1104 Nauwerck, A. 1959. Arch. Hydrobiol. Suppl. 25, pp 83-101.

1105 Nauwerck, A. 1963. Symb. Bot. Ups. 17(5), pp 1-163.

1106 Nebeker, A.V. 1971. Water Res. 5(9), pp 779-783.

1107 Nedwell, D.B. 1975. Water Res. 9, pp 221-232.

1108 Neely, W.B. 1977. Sci. Total Environ. 8(3), pp 267-274.

1109 Neely, W.B. 1977. Sci. Total Environ., 8(3), pp 267-274.

1110 Neff, J.M. and J.W. Anderson. 1976. Proceedings of the 1975 Conference in Prevention and Control of Oil Pollution. 469 pp, American Petrolium Institute. Wash. D.C.

1111 Neff, J.M., B.A. Lox, D. Dixit and J.W. Anderson. 1976. Mar. Biol. 38(3), pp 279-289.

1112 Negulescu, M. and V. Rojanski. 1969. Wat. Res. 3(3), pp 189.

1113 Nehring, R.B. 1976. Bull. Environ. Contam. Toxicol. 15(2), pp 147-153.

1114 Neilson, R.E. and P.G. Jarvis. 1975. J. appl. Ecol. 12, pp 879-891.

1115 Nelson, D.J. 1967. Proc. Int. Symp. Radioecol. Concent. Process., 1966, Stockholm, pp 843-851.

1116 Netzer, A., P. Wilkinson and S. Beszedits. 1974. Water Res. 8(10), pp 813-817.

1117 Neudorf, S. and M.A.Q. Khan. 1975. Bull. Environ. Contam. Toxicol. 31, pp 443-450.

1118 Neufeld, R.D.,J. Gutierrez and R.A. Novak. 1977. Water Pollut. Control. 49(3), pp 480-498.

1119 Neveu, A. 1973. Ann. Hydrobiol. Inst. Natl. Rech. Agron. Paris, 4(2), pp 183-199.

1120 Newbould, P., F.B. Ellis, B.T. Barnes, K.R. Howse and F.G.H. Lupton. 1970. Rep. agric. Res. Coun. Letcombe Lab., 1969, pp 38-40.

1121 Newbould, P., F.B. Ellis, B.T. Barnes, K.R. Howse, T.G. Elliott and N. Squires. 1970. Rep. agric. Res. Coun. Letcombe Lab., 1969, pp 40-43.

1122 Newbould, P., F.B. Ellis, B.T. Barnes, K.R. Howse, T.E. Williams and E.A. Garwood. 1970. Rep. agric. Res. Coun. Letcombe lab., 1969, pp 45-48.

1123 Newsome, W.H. 1974. Bull. Environ. Contam.,11, pp 174-176.

1124 Nicholson, H.P. 1969. Proc. Wash. Acad. Sci., 59(4-5), pp 77.

1125 Nielsen, C.O. 1961. Oikos 12, pp 17-35.

1126 Nielsen, E.S. 1958. Rapp. p.-v. Reun. Cons. Perm. Int. Explor. Mer., 144, pp 92.

1127 Nihoul, J.C.J. 1975. Modelling of Marine Systems. pp 69-95. Elsevier. Amsterdam.

1128 Nilles, G.P. and M.J. Zabik. 1974. J. Agric. Food. Chem. 22(4), pp 684-688.

1129 Nilsson, I. 1972. Oikos, 22, pp 366.

1130 Nimmo, D.R., J. Forester, P.T. Heitmuller and G.H. Cook. 1974. Bull. Environ. Contam., 11, pp 303-308.

1131 Nival, P., G. Malara, R. Charra, I. Palazzoli and S. Nival. 1974. J. Exp. Mar. Biol. Ecol. 15, pp 231-260

1132 Nixon, S.W. and J.N. Kremer. 1977. in: Hall, C.A.S. and J.W. Day, Jr. (eds.); Ecosystem Modeling in theory and practice. pp 622-673. John Wiley and Sons. New York.

1133 Noddack, W. 1937. Angew. Chem., 50, pp 505.

1134 Norbac, D.H. and J.R. Allen. 1973. Environ. health Persp., 9, 233-240.

1135 Nordisk Hygienisk Tidsskrift, 1970. Suppl. 3, Stockholm.

1136 Norris, L.A. and R.A. Miller. 1974. Bull. Environ. Contam. Toxicol. 12(1), pp 76-80.

1137 North, B.B. 1975. Limnol. Oceanogr. 20(1), pp 20-27.

1138 Noshkin, V.E. 1972. Health Physics. 21(6), pp 537-549.

1139 Noyes, R. 1969. Dis. Nerv. Syst., 30, pp 318-321.

1140 Nuorteva, P. and E. Hasanen. 1972. Ann. Zool. Fenn. 9, pp 23-27.

1141 Nutman, P S.(ed). Symbiotic Nitrogen fixation in Plants, International Biological Program. Cambridge University Press, Cambridge 1975, vol. 5, pp 211.

1142 Nuttall, P.M. and G.H. Bielby. 1974. Environ. Pollut., 5, pp 77-86.

1143 Nygaard, G. 1955. Verh. Int. Ver. Limnol. 12, pp 123-133.

1144 Nyholm, N. 1977. Biotech. Bioeng. 19, pp 467-492.

1145 Nyholm, N. 1978. Ecol. Model. 4 (in press).

1146 Nyholm,N. 1977. Prog. Wat. Tech. 8(4/5), pp 347-358.

O

1147 O'Brien, J.J. and J.S. Wroblewski. 1972. An Ecological Model of the Lower Marine Trophic Levels on the Continental Shelf of West Florida. Technical Report, Geophysical Fluid Dynamics Institute. Florida State Univ. Tallahassee. Fla. 170 p.

1148 O'Brien, W.J. 1974. Am. Midl. Nat. 91(2), pp 509-512.

1149 O'Connor, D.J., D.M. DiToro and R.V. Thomann. 1975. in: Russel, C.S. (ed.); Ecological Modeling in a Resource Management Framework. pp 149-209.

1150 O'Connor, D.J. and D.M. Di Toro. 1970. J. San. Eng. Div, ASCE. 96, SA2.

1151 O'Connor, D.J. and Hydroscience, Inc. 1968. Water Quality Analysis of The Mohawk River-Barge Canal. New York state Dep. of Health.

1152 O'Connor, D.J. and W.E. Dobbins. 1958. Amer. Soc. Civ. Eng. Transactions. 123, pp 641.

1153 O'Connor, D.J., R.V. Thomann and D.M. DiToro. 1973. Dynamic Water Quality Forecasting and Magnament. Office of Research and Development U.S. Environmental Protection Agency, Wash. D.C. 20460.

1154 O'Connor, D.J., R.V. Thomann and D.M. DiToro. 1976. In: A.V. Biswas (ed.), Systems Approach to Water Management, pp 299-333. McGraw Hill N.Y.

1155 O'Connor, F.B. 1963. in: Doeksen, J. and J. van der Drift (eds.); Soil Organisms. pp 32-48. North-Holland publ. Co. Amsterdam.

1156 O'Connor, S.G. and A.J. McErlean. 1975. in: Cronin, L.E. (ed.); Eustarine Research. pp 494-518. Academic Press. New York.

1157 O'Donnell, J.W., R.K. Severs and S.M. Pier. 1976. Sci. Total Environ. 5(3), pp 231-242.

1158 Odell, B.L., B.J. Cambell. 1972. Trace Elements: Metabolism and Metabolic Function Ch. 2. In Comprehensive Chemistry., vol.21, Elsevier.

1159 Odum, E.P. 1968. Amer. Zool. 8, pp 11-18.

1160 Odum, E.P. 1969. Science 164, pp 262-270.

1161 Odum, E.P. 1971. Fundamentals of Ecology. Third edition. W.B. Saunders , Philadelphia.

1162 Odum, E.P., G.S. Marshall and T.G. Marples. 1965. Ecology 46, pp 901-904.

1163 Odum, H.T. and r.f Pigeon. 1970. A tropical rain forest. Nat Tech. Info. Srevice. Springfield.

1164 Odum, H.T. 1957. Ecol. Monogr., 27, pp 55-112.

1165 Odum, H.T. 1956. Limnol. Oceanogr. 1, pp 102-117.

1166 Odum, H.T. 1957. Ecol. Monogr. 27, pp 55-112.

1167 Odum, H.T. 1971. Environment, Power and Society. Wiley-Interscience. New York.

1168 Odum, H.T. 1972. in: Patten, B.C. (ed.); Systems Analysis and Simulation in Ecology. 1, pp 139-211. Academic Press. New York.

1169 Odum, H.T., J.E. Cantlon and L.S. Korniller. 1960. Ecology 41, pp 395-399.

1170 Oehme, F.W. 1972. Clinical Toxicology, 5(2), pp 151-167.

1171 Oerberg, J., N. Johansson, J.E. Kihlstroem and C. Lundberg. 1972. Ambio. 1(4), pp 148-151.

1172 Ogura, N. 1970. Deep-Sea Res., 17, pp 221-231.

1173 Ogura, N. 1970. Limnol. Oceanogr., 15, pp 476-479.

1174 Ogura, N. 1970. Nature. 227, pp 1335-1336.

1175 Ogura, N. 1972. In K. Sugawara(ed), the Kuroshio II, Tokyo Saikon., pp 201-205.

1176 Ogura, N. 1975. Mar. Biol., 31(2), pp 101-111.

1177 Ogura, N., N. Nakamoto, M. Funakoshi, A. Kamatani and S. Iwata. 1975. J. Oceanogr. Soc. Japan, 31, pp 43-47.

1178 Ohwada, K. and N. Taga. 1972. Limnol. Oceanogr. 17, pp 315-320.

1179 Ohwada, K., M. Otsuhata and N. Taka. 1972. Bull. Japan Soc. Sci. Fish. 38, pp 817-823.

1180 Old, S.M. Ecol Monogr. 39, pp 355-383.

1181 Olson, B.H. and R.C. Cooper. 1976. Water Res. 10(2), pp 113-116.

1182 Olson, L.E., J.L. Allen. and W.L. Mauck. 1975. J. Agric. Food. Chem. 23(3), pp 437-439.

1183 Olsson, M. 1976. Ambio.2, pp 73-76.

1184 Omori, M. 1969. Mar. Biol., 3, pp 4-10.

1185 Ophel, I.L. 1963. in: Schultz V. and W. Klement(eds). Radioecology, pp 213-216. Reinhold Publ. Comp., New York.

1186 Orhon, D. 1975. Water Res. 9(5/6), pp 467-472.

1187 Osmera, S. 1966. Zool. Listy 15, pp 79-83.

1188 Osterberg, C., A.G. Carey and H. Curl, Jr. 1963. Nature. 200, pp 1276-1277.

1189 Otsuki, A and T. Hanya. 1972. Limnol. Oceanogr. 17(2), pp 248-257.

1190 Otsuki, A. and T. Hanya. 1972. Limnol. Oceanogr. 17(2), pp 258-264.

1191 Otterman, J., Y. Waisel and E. Rosenberg. 1975. Agro-Ecosyst. 2(1), pp 47-59.

1192 Otto, C. 1975. Oikos 26(2), pp 159-169.

1193 Overbeck, J. 1965. Naturwissenschaften 51, pp 145.

1194 Ovington, J.D. 1961 Ann. Bot. 25, pp 12-20.

1195 Ovington, J.D. and D. Heitkamp, 1960. J. Ecol. 48, pp 639-646.

1196 Ovington, J.D. and D.M. Lawrence. 1967. Ecology 48, pp 515-524.

1197 Ovington, J.D. 1956. J. Ecol. 44, pp 597-604.

1198 Ovington, J.D. 1962. Adv. Ecol. Res. 1, pp 103-192.

1199 Owens, M. and R.W. Edwards. 1961. J. Ecol. 49, pp 119-126.

1200 Owens, M. and R.W. Edwards. 1962. J. Ecol. 50, pp 157-162.

1201 Owens, M., R.W. Edwards and J.W. Gibbs. 1964. Int. J. Air Water Poll. 8, pp 469.

P

1202 Paasche, E. 1973. Mar. Biol., 19, pp 117-126.

1203 Paasche, F. 1973. Mar. Biol. 19. pp 262-269.

1204 Paffenhoefer, G.-A. 1976. Limnol. Oceanogr. 21(1), pp 39-50.

1205 Page, E.R. and A. Gerwitz. 1969. J. Sci. Fd. Agric. 20, pp 85-90.

1206 Painter, H.A. and K. Jones. 1963. J. Appl. Bact. 26, pp 471-483.

1207 Painter, H.A. 1975. Microbial Transformations of inorganic Nitrogen. in: Conf. on Nitrogen as a Water Poll. Vol 1, Session A1. Copenhagen.

1208 Pamatmat, M.M. 1971. Limnol. Oceanogr. 16(4), pp 536-550.

1209 Pandian, T.J. and S. Mathavan. 1974. Odonatologica, 3(4), pp 241-248.

1210 Paris, D.F. and D.L. Lewis. 1974. Environ. Qual. Saf., Supp III, pp 288-291.

1211 Paris, D.F. and D.L. Lewis. 1976. Bull. Environ. Contam. Toxicol. 15(1), pp 24-32.

1212 Paris, D.F., D.L. Lewis and J.T. Barnett. 1977. Bull. Environ. Contam. Toxicol. 17(5), pp 564-572.

1213 Parker, B.C. and M.A. Wachtel. 1971. in: Cairns, J. (ed.); The Structure and Function of Freshwater Microbial Communities. Res. Div. Monogr. Virginia Polytech. Inst. 3, pp 195-207.

1214 Parker, D.S., R.C. Aberley and D.H. Caldwell, 1975. In: Conf. on Nitrogen as a Water Pollutant. Vol. 3, Session A3. Cph.

1215 Parker, R.A. 1976. Int. J. System Sci. 7. 957-962.

1216 Parslow, J.L.E. 1974. Environ. Pollut., 5, pp 295-304.

1217 Parslow, J.L.F. and D.J. Jefferies. 1974. Environ. Pollut., 5, pp 87-101.

1218 Parsons, T. and M. Takahashi. 1973. Biological Oceanographic Processesp. p 186 Pergamon Press.

1219 Parsons, T.R. and M. Takahashi. 1973. Limnol. Oceanogr., 18(4), pp 511-515.

1220 Parsons, T.R., K. Stephens and J.D.H. Strickland. 1961. J. Fish. Res. Bd. Can., 18, pp 1001-1016.

1221 Paschal, D.C. and M.E. Neville. 1976. J. Environ. Qual. 5(4), pp 441-443.

1222 Patin, S.A. and N.I. Popov. 1966. Proc. Symp. Int. Atom. Energ. Agency, pp 443-449.

1223 Patrick, F.M. and M. Loutit. 1976. Water Res. 10(4), pp 333-335.

1224 Patrick, F.M. and M.W. Loutit. 1977. Water Res., 11(8), pp 699-703.

1225 Patten, B.C., P.A. Egloff and T.H. Richardson. 1975. Total Ecosystem Model for a Cove in Lake Texoma. in: B.C. Patten (ed), Systems analysis in Ecology. Vol III. Academic Press. N.Y. pp. 206-423.

1226 Patterson, J.B.E. 1971. Tech. Bull., 21, pp 193-207.

1227 Pattison, F.L.M. 1959. Toxic Aliphatic Fluorine Compounds

1228 Patty, F.A.(ed). 1963. Industrial Hygiene and Toxicology. II, Interscience Publ., N.Y..

1229 Pauling, L. 1942. The Nature Of Chemical Bond. Cornell Univ. Press. Ithaca. New York.

1230 Payne, A.I.I. 1976. Collect. Sci. Pap. ICESEAF/Recl. Doc. Sci. CIPASE/Coleco. Doc. Cient. Cipaso, v 3, pp 159-163.

1231 Peacock, J.M. 1975. J. appl Ecol., 12, pp 115-123.

1232 Peacock, J.M. 1975. J. appl. Ecol. 12, p 99-114.

1233 Peacock, J.M. 1976. J. appl. Ecol. 13(1), pp 225-232.

1234 Pechen, G.A. 1965. Gidrobiol. Zh. 1, pp 19-26.

1235 Penfound, W.T. and T.T. Earle. 1948. Ecol. Monogr. 18, pp 447-472.

1236 Penfound, W.T. 1956. Limnol. Oceanogr. 1(1), pp 92-101.

1237 Peng, T.H. 1973. Determination of gas exchange rates across sea-air interface by the Radon method. Ph.D. Thesis, Columbia Univ.

1238 Pennak, R.W. 1944. Ecology 25, pp 387-403.

1239 Penning, F.W.T. de Vries. 1972. J. appl. Ecol. 9, pp 57-77.

1240 Pentreath, R.J. and D.F. Jefferies. 1971. J. mar. biol. Ass. U.K. 51, pp 963-976.

1241 Pentreath, R.L. 1973. J. Mar. Biol. Ass. U.K., 53, pp 127.

1242 Penumalli, B.R. R.H. Flake and E.G. Fruh. 1976 Ecol. Modell. 2(2), pp 101-115.

1243 Perry, M.J. 1976. Limnol. Oceanogr. 21(1), pp 88-107.

1244 Petal, J. 1964. in: Petrusewicz, K. (ed.); Secondary Productivity of Terrestrial Ecosystems. pp 841-855. Krakow. Warszawa.

1245 Peters, D.S. and M.A. Kjelson. 1975. in: Cronin, L.E. (ed.); Estuarine Research. Vol 1 pp 448-472. Academic Press. New York.

1246 Peters, R.H. and F.H. Rigler. 1973. Limnol. Oceanogr., 18(6), pp 821-339.

1247 Peterson, J.T. and C.E. Junge, 1971. In: W.H. Matthews, W.W. Kellogg and G.D. Robinson (eds.), Man's Impact on the Climate, pp 310-320. MIT Press, Cambr. Mass., Lond.

1248 Peterson, P.J. 1962. N. Z. J. Sci. Technol. 5, pp 141-161.

1249 Petipa, T.S., E.V. Pavlova and G.N. Mironov. 1970. in ; Steele, J.H. (ed.): Marine Food Chains. pp 143-167 Oliver and Boyd. Edinburgh.

1250 Petrides, G.A. and W.G. Swank. 1966. Proc. Internat. Grassland Cong., Sao Paulo, Brazil. 9, pp 831-942.

1251 Petrocelli, S.K., J.W. Anderson and A.R. Hanks. 1975. Mar. Biol., 31, pp 215-218.

1252 Petrocelli, S.R. and J.W. Anderson. 1975. Bull. Environ. Contam. Toxicol. 31, pp 108-116.

1253 Petrusewitz, K. and MacFadyen. 1970. Productivity of Terrestrial Animals. IBP Handbook No. 13, Blackwell, Oxford.

1254 Philipson, J. 1964. Oikos 15, pp 130-139.

1255 Phillips, J.E. 1964. in: Heukelekain,H. and N.C. Dondero. (eds.); Principles and Applications in Aquatic Microbiology. Wiley. New York.

1256 Pianka, E.R. 1967. Ecology. 48, pp 333-351.

1257 Picer,M., N. Picer and P. Strohal. 1977. Sci. Total Environ., 8(2), pp 159-164.

1258 Pickering, Q.H. 1974. Water Pollut. Control, 46(4), pp 760-765.

1259 Pieczynska, E. 1972. Wiadomosci Ekologiczne 18, pp 131-140

1260 Pielou, E.C. 1966. J. Theoret. Biol. 13, pp 131-144.

1261 Pillai, K.C., N.N. Dey, E. Mathew and B.U. Kothari. 1975. Impacts of Nuclear releases into the Aquatic Environment. IAEA-SM-198/16, pp 277-299. Vienna.

1262 Pitcairn, C.E.R. and H.A. Hawkes. 1973. Water Res. 7(1/2), pp 159-171.

1263 Pitter, P. 1976. Water Res., 10(3), pp 231-236.

1264 Platt, T., V.M. Brawn and B. Irwin. 1969. J. Fish. Res. bd. Can., 26, pp 2345-2349.

1265 Plumb, T.R., L.A. Norris and M.L. Montgomery. 1977. Bull. Environ. Contam. Toxicol. 17(1), pp 1-8.

1266 Poels, C.L.M. and J.J.T.W.A. Strik. 1975. In: J.H. Koeman et al.(eds.): Sublethal Effects of Toxic Chemicals on Aquatic Animals, Elsevier Scientific Publishing Company. Amsterdam, Oxford, N.Y., pp 81-91.

1267 Polikarpov, G.G. 1966. Radioecolegy of Aquatic Organisms. N.Y. Reinhold.

1268 Pomeroy, L. 1963. in: Schultz, V. and A.W. Klement, Jr. (eds.); Radioecology. pp 163-166. Reinhold Co. New York.

1269 Pomeroy, L., R. Henry, M. Mathews and H. Shik Min. 1963. Limnol. Oceanogr. 8, pp 50-55.

1270 Poon, C.P.C. and H. Campbell. 1967. Wat. Sewage Works. 114, pp 461.

1271 Pope, J.A. and H.J. Thomas. 1967. J. Cons. perm. Int. Explor. Mer., 31, pp 265-291.

1272 Portmann, J.E. and K.W. Wilson. 1971. Shellfish Information Leaflet no 22.

1273 Potzger, J.E. and W.A. Engel. 1942. Trans. Wis. Acad. Sci. Arts lett. 34, pp 149-166.

1274 Powell, D. 1975. Florida Mar. Res. Publ. (no5), 21 p.

1275 Prabhu, N.V. and M.K. Hamdy. 1977. Bull. Environ. Contam. Toxicol. 18(4), pp 409-417.

1276 Pree, D.J. and D.K.R. Stewart. 1975. Bull. Environ. Contam. Toxicol. 14(1), pp 117-121.

1277 Preston, A. and D.F. Jeffries. 1969. In Environmental Contamination by Radioactive Materials, Vienna, FAO/IAEA, pp 183-214.

1278 Preston, A., D.R. Jefferies , J.W.R. Harvey and A.K. Steele. 1972. Environ. Pollut., 3, pp 183-214.

1279 Prestt, I., P.J. Jefferies and N.W. Moore. 1971. Environ. Pollut., 1(1), pp 3-26.

1280 Pringle, B.H., D.E. Hissong, E.L. Katz and S.T. Mulawka. 1968. J. San. Eng. Div. Am. Soc. Civ. Eng., 94(5970), pp 455-475.

1281 Provasoli, L. and A.F. Carlucci. 1974. in: Stewart, W.D.P. (ed.); Algal Physiology and Biochemistry. pp 741-787. Univ. Calif. Press. Berkeley.

1282 Prus, T. 1972. Pol. Arch. Hydrobiol 19, pp 97-112.

1283 Purves, D. 1972. Environ. Pollut., 3, pp 17-24.

1284 Purvis, M.J., D.C. Collier and D. Wallo. 1966. Laboratory Techniques in Botany. Butterworths. London.

1285 Putnam, H.D. and T.A. Olson. 1961. Studies on the Productivity and Plankton of Lake Superior. Rep. Scholl Pub. Health. Univ. Minesota.

Q

86 Qasim, S.Z. and D.C.V. Easterson. 1974. Indian J. Mar. Sci. 3(2), pp 131-134.

1287 Qasim, S.Z., P.M.A. Bhattathiri and V.P. Devassy. 1972. Mar. Biol. 16, pp 22-27.

1288 Quasim, S.Z., P.M.A. Bhattathiri and V.P. Devassy. 1973. Mar. Biol. 21, pp 299-304.

1289 Quistad, G.B., L.E. Staiger and D.A. Schooley. 1975. J. Agric. Food. Chem. 23(2), pp 299-303.

R

1290 Rachlin, J.W. and M. Farran. 1974. Water Res., 8(8), pp 575-577.

1291 Radford, P.J. 1971. in: Jeffecs , J.N.R. (ed.): Mathematical Modeld in Ecology. 289 p. Balckwell sci. Publ. Symposium.

1292 Rafail, S.Z. 1971. Mar. Biol., 10, pp 13-21.

1293 Rakusa-Suszcewski, S., M.A. McWhite and M.O. Cahoon. 1976. Limnol. Oceanogr. 21(5), pp 763-765.

1294 Ramamoorthy, S., S. Springthorpe and D.J. Kushner. 1977. Bull. Environ. Contam. Toxicol. pp 505-511.

1295 Randers and Meadows. 1971. System to Simulation to test environmental Policy: a sample study of DDT movements in environment. In; System Dymanics Group. MIT. Cambrigde.

1296 Rao, K.P. 1953. Biol. Bull. Mar. Biol. Lab., Woods Hole, 104, pp 171-181.

1297 Rao, S.S and B.J. Dutka. 1974. Water Res. 8(8), pp 525-528.

1298 Rappaport, J. 1938. Meded. Landb-Hoogesch. Opzoek. Stns. Gent. 6, pp 121-204.

1299 Raspopov, I.M. 1971. Hidrobiologia 12, pp 241-247.

1300 Raspor,B., P. Valenta, H.W. Nurnberg and M. Branica. 1977. Sci. Total Envr. 9, pp 87-109.

1301 Ray, G.L. and M.R. Tripp. 1976. J. Environ. Qual. 5(2), pp 193-197.

1302 Raymand, L.N. 1964. U.S. Geological survey.

1303 Reakall, D.B. 1974. J. Environ. Qual. 52, pp 1-39.

1304 Reay, P.F. 1972. J. appl. Ecol. 9, pp 557-565.

1305 Reed, J.R. and D.J. Nelson. 1969. Proc. 2nd Nat. Symp. on Radioecology. USAEC. Conf. 670503, pp 226-233.

1306 Reeve, M.R. 1964. Nature, 201, pp 211-213.

1307 Reeve, M.R., J.E.G. Raymont and J.K.B. Raymont. 1970. Mar. Biol. 6, pp 357-364.

1308 Rehwoldt, R. and D. Karimian-Teherani. 1976. Bull. Environ. Contam. Toxicol. 15(4), pp 442-446.

1309 Reichle, D.E. 1967. Ecology 48, pp 351-366.

1310 Reichle, D.E. 1968. Ecology 49, pp 538-542.

1311 Reichle, D.E., M.H. Shanks and D.A. Crossey, Jr. 1969. Ann. Entomol. Soc. Amer. 62, pp 57-62.

1312 Reiff, B. 1975. in: Koeman, J.H. et al.(eds.): Sublethal Effects of Toxic Chemicals on Aquatic Animals, Elsevier Scientific Publishing Company, Amsterdam, Oxford, N.Y., pp 53-60.

1313 Reimers, R. S. and P.A. Krenkel. 1974. Water Pollut. Control 46(20) pp 352-365.

1314 Renfro, W.C. and C. Osterberg. 1969. Proc. 2nd Nat. Symp. on Radioecology. USAEC. Conf. 670503, pp 372-379.

1315 Renfro, W.C. 1973. Mar. biol. 21, pp 305-316.

1316 Renfro, W.C., S.W. Fowler, M. Heyraud and J. La Rosa. 1975. J. Fish. Res. bd. Canada. 32(8), pp 1339-1345.

1317 Renzoni, A., E. Bacci and L. Falciat. 1973. Rev. Int. Oceanogr. Med. CERBOM., 31-32, pp 17-45.

1318 Revelle, R. and H.E. Suess. 1957. Tellus 9, pp 18.

1319 Reynolds, J.H., E.J. Middlebrooks, D.B. Porcella and W.J. Grenney, 1975. J. Wat. Poll. Control Fed. 47(10), pp 2420-2436.

1320 Reynolds, P., I.L. Lindahl, H.C. Cecil and J. Bitman. 1976. Bull. Environ. Contam. Toxicol. 16(2), pp 240-247.

1321 Rhee, G.-Y. 1973. J. Phycol. 9. pp 495-506.

1322 Rhodes, R.C. and J.D. Long. 1974. Bull. Environ. Contam. Toxicol. 12(4), pp 385-393.

1323 Rice, C.P. and H.C. Sikka. 1973. Bull. Environ. Contam. Toxicol. 9, pp 116-123.

1324 Rice, T.R. and V.M. Willis. 1959. Limnol. Oceanogr. 4, pp 227-290.

1325 Rich, P.H., R.G. Wetzel and N.V. Thury. 1971. Freshwater Biol. 1, pp 3-21.

1326 Richey, J.E. 1977. in: Hall, C.A.S. and J.W. Day, Jr. (eds.); Ecosystem Modeli g in theory and practice. John Wiley and Sons. New York.

1327 Richman, S. 1958. Ecol. Monogr. 28, pp 273-291.

1328 Richman, S. 1958. Ecol. Monogr. 28, pp 273-291.

1329 Richman, S. 1966. Verh. Int. Ver. Limnol. 16, pp 392-398.

1330 Richman, S. 1966. Verh. Int. Ver. Limnol. 16, pp 392-398.

1331 Rickett, H.W. 1921. Trans. Wis. Acad. Sci. Arts Lett. 20, pp 501-527.

1332 Ricks, G.R. and R.J.W. Willams. 1974. Environ. Pollut., 6, pp 87-110.

1333 Rigers, H.E. and D.E. Davis. 1972. Weed Sci. 20(5), pp 423-428.

1334 Rigler, F.H. 1956. Ecology 37, pp 550-562.

1335 Rigler, F.H. 1961. Limnol. Oceanogr. 6, pp 165-174.

1336 Rigler, F.H. 1964. Limnol. Oceanogr. 9, pp 511-518.

1337 Riley, G.A. and R. von Arx. 1949. J. marine res., 8(11), pp 60-72.

1338 Riley, G.A. 1956. Bull. Bingham Oceanogr. Coll. 15, pp 324-344.

1339 Riley, G.A. 1956. Bull. Bingham Oceanogr. Coll. 15, pp 15-46.

1340 Riley, G.A. 1963. in: Hill. M.N. (ed.): The Sea Theory of Foodchains relations in the Ocean. pp 438-463. Interscience. N.Y.

1341 Riley, G.A. 1965. Limnol. Oceanogr., 10(suppl.), pp 202-215.

1342 Riley, J.P. and G. Skirrow. 1965. Chemical Oceanography. Vol 2. Academic Press. N.Y.

1343 Risebrough, R.W., B.W. De Lappe and T.T. Schmidt. 1976. Mar. Pollut. Bull. 7(12), pp 225-228.

1344 Ritter, W.F., H.P. Johnson, W.G. Lovely and M. Molnad 1974. Environ. Sci. Technol., 8 pp 38-42.

1345 Roberts, D. 1975. Bull. Environ. Contam. Toxicol. 31, pp 170-176.

1346 Robinette, W.L. and G.F.T. Child. 1964. Puku 2, pp 84-117.

1347 Robinson, E. and R.C. Robbins, 1971. Emissions, Concentration and Fate of Particulate Atmospheric Pollutants. Report to Am. Petr. Inst., Menlo Park, Calif.

1348 Robinson, E. and R.C. Robbins, 1975. In: S.F. Singer (ed.), The Changing Global Environment, pp 111-123. D. Reidel Publ. Comp.

1349 Robinson, K., S.R. Draper and A.L. Gelman. 1971. Environ. Pollut., 2, pp 49-56.

1350 Rodhe, W. 1958. Verh. Int. Ver. Limnol. 13, pp 121-141.

1351 Rodhe, W. 1969. Nat. Acad. Sci. (U.S.) Symp. Eutrophication, pp 50-64.

1352 Roff, J.C. and J.C.H. Carter. 1972. Limnol. Oceanogr., 17(3), pp 363-370.

1353 Rohatgi, N. and K.Y. Chen. 1975. Water Pollut. Control, 47(9), pp 2298-2316.

1354 Roman, M.R. 1977. Mar. Biol. 42(2), pp 149-155.

1355 Ronald, K., S.V. Tessaro, J.F. Uthe, H.C. Freeman and R. Frank. 1977. Sci. Total Environ. 8, pp 1-11.

1356 Rosenberg, R. 1972. Environ. Pollut., 3, pp 313-318.

1357 Rosenzweig, M.L. 1968. Am. Nat. 102, pp 67-74.

1358 Ross, P.J., E.F. Henzell and D.R. Ross. 1972. J. appl. Ecol., 9, pp 535-556.

1359 Rowland, D.F.S. and M.J. Molina. 1975. Rev. Geophys. Spach. Phys. 13, pp 1-35.

1360 Rubey, W.W. 1951. Geol. Soc. Amer. Bull. 62, pp 1120-1121.

1361 Ryther, J.H. 1954. Ecology, 35, pp 522-33.

1362 Ryther, J.H. 1956. Limnol. Oceanogr. 1, pp 61-70.

1363 Ryther, J.H. 1959. Science, N.Y., 130, pp 602.

1364 Ryther, J.H. 1963. in: Hill, M. N. (ed). The Sea, Vol 2, Wiley Intèrscience, New York.

1365 Ryther, J.H. 1969. Science 166, pp 72-76.

S

1366 Sabbioni, E. and F. Girardi. 1977. Sci Total Environ., 7(2), pp 145-179.

1367 Saenko, G.N., M.D. Koryakov, V.F. Makienko and I.G. Dobrosmy. 1976. Mar. biol., 34(2), pp 169-176.

1368 Saha, J.G. and W.W.A. Stewart. 1967. Can. J. Plant. Sci., 47, pp 79.

1369 Sakamoto, M. 1966. Arch. Hydrobiol. 62, pp 1-25.

1370 Sakamoto, M. 1971. J. Fish. Res. bd. Can. 28, pp 203-213.

1371 Sakamoto, T. and T. Suzuki. 1974. Bull. Jap. Soc. Sci. Fish. 40(6), pp 551-560.

1372 Saleh, H.H. 1972. Mar. biol. 12, pp 255-260.

1373 Sallivan, G.W., A.L. Buikema and J. Cairns, Jr. 1973. Va. J. Sci. 24(3), pp 120.

1374 Salo, E.O. and W.L. Leet. 1969. Proc. 2nd Nat. Symp. on Radioecology. USAEC. Conf. 670503, pp 363-371.

1375 Salonen, K., J. Sarvala, I. Hakala and M.-L. Viljanen. 1976. Limnol. Oceanogr., 21(5), pp 724-730.

1376 Salonen, L. and H.A. Vaajakorpi. 1976. Environ. Qual. Saf. 5, pp 130-140.

1377 Saltzman, S., U. Mingelgrin and B. Yaron. 1976. J. Agric. Food. Chem. 24(4), pp 739-743.

1378 Samuels, S., N.M. Shah and G.E. Fogg. 1971. J. mar. biol. Ass. U.K. 51(4), pp

793-798.

1379 Sanborn, J.R. and C.-C. Yu. 1973. Bull. Environ. Contam. Toxicol. 10(6), pp 340-346.

1380 Sanborn, J.R., W.F. Childers and L.G. Hansen. 1977. J. Agric. Food. Chem. 25(3), pp 551-553.

1381 Sanders, H.O. and J.H. Chandler. 1972. Bull. Environ. Contam. Toxicol. 7(5), pp 257-263.

1382 Sandstroem, O. and S. Soedergren. 1974. Vatten, 30, pp 229-34

1383 Satchell, J.E. 1967. in: Burges, A. and F. Raw. (eds.); Soil Biology. pp 259-322. Academic Press. London.

1384 Satomi, M. and L.R. Pomeroy. 1965. Ecology 46, pp 877-881.

1385 Satoo, T. 1970. in: Reichle, D.E. (ed.); Analysis of Temperate Forest Ecosystem. pp 55-72. Springer-Verlag. Berlin.

1386 Sauchez, I. and G.F. Lee. 1973. Water Res. 7(4), pp 587-593.

1387 Scavia, D. and B.J. Eadie. 1976. Ecol. Modell., 2, pp 315-319.

1388 Scep: Report on the Study on Critical Environmental Problems. 1970. in: Mans inpact on the Global Environment. MIT-press. Cambridge,. Mass.

1389 Schaefer, E.D. and W.O. Pipes. 1973. Water Res., 7(12), pp 1781-1790.

1390 Schaefer, C.H. and E.F. Dupras, Jr. 1976. J. Agric. Food. Chem. 24(4), pp 733-739.

1391 Schaper, R.A. and L.A. Crowder. 1976. Bull. Environ. Contam. Toxicol. 15(5), pp 581-587.

1392 Schell, W.R. and R.S. Barnes. 1974. in: Rubin, A.J. (ed.): Aqueous-Environmental Chemistry of Metals. pp 155-158.

1393 Schlesinger, W.H., W.A. Reiners and D.S. Knopman. 1973. Environ. Pollut., 6, pp 39-47.

1394 Schmidt-Nielsen, K. 1954. in: Cloudsley-Thompson, J.B. (ed.); Biology of Deserts. pp 182-187. Inst. of Biol. London.

1395 Schoene, H.K. 1972. Mar. biol. 13, pp 284-291.

1396 Schoener, A. 1972. Limnol. Oceanogr. 17(2), pp 193-199.

1397 Schofield, C.L. 1976. Ambio. 5(5-6), pp 228-230.

1398 Schofields, W.K. and R.G. Krutchoff. 1974. Bulletin 60 Stochastic Model for a Dynamic Ecosystem. VPI-WRRC-Bull, 60 Virginia Water Resources Research Center.

1399 Scholander, P.F. 1955. Evolution 9, pp 15-26.

1400 Schooley, D.A., Bergot, L.L. Dunham and J.B. Siddall. 1975. J. Agric. Food. Chem. 23(2), pp 293-298.

1401 Schroeder, H. 1919. Naturwissenshaften, 7, pp 8.

1402 Schultz, R.R., G.A. Tompkin, L. Leventhal and K.L. Babcock. 1976. J. Environ. Qual. 5(4), pp 406-410.

1403 Schultze, E.H. 1975. Mar. biol., 30(4), pp 331-341.

1404 Schultze, R.F. and C.D. Vanicek. 1974. Calif. Fish. Game, 60(2), pp 94-96.

1405 Schulz-Baldes, M. 1973.. Mar. Biol., 21, pp 98-102.

1406 Schuth, C.K., A.R. Isensee, E.A. Woolson and P.C. Kearney. 1974. J. Agric. Food. Chem. 22(6), pp 999-1003.

1407 Schuurman, J.J. and M.A.J. Goedewaagen. 1971. Methods for Examination of Roots Systems and Roots. 2ed. Wageningen.

1408 Schewets, B.A. et al. 1973. Environ. health Persp., 9. pp 233-240.

1409 Scott, L.M. and C.M. West. 1975. Health Physics, 28, pp 563-565.

1410 Sebetich, M.J. 1975. Ecology 56(6), pp 1262-1280.

1411 Seeger, W. 1970. Arch. Hydrobiol. 68(4), pp 528-574.

1412 Seelinger, U. and P. Edwards. 1977. Mar. Pollut. Bull. 8(1), pp 16-19.

1413 Segal, E., K.P. Rao and T.W. James. 1953. Nature, 172, pp 1108-1109.

1414 Sekavee, G.B. 1974. Trans. Am. Fish. Soc., 103(3), pp 472-476.

1415 Selezneva, E.S. 1976. Atmosfernge aerozoli (Atmospheric Aerosols). Gidrometeorologisckeskoi Izdatelstro. Leningrad, translated and publ by: Indian Natioanl Scientific Documentation Center. New Dehli. 1970.

1416 Sellers, C.M., A.G. Heath and M.L. Bass. 1975. Water Res. 9(4), pp 401-408.

1417 Sellers, D.D. 1965. Physical Climatology. Univ. of Chicago Press. Chicago.

1418 Serchuk, F.M. and C.F. Cole. 1974. Cheasepeake Sci. 15(4), pp 205-213.

1419 Seymour, A. 1966. Disposal of Radioactive Wastes into Seas, Oceans and Surface Waters, IAEA, Vienna, pp 605-619.

1420 Shannon, C.E. and W. Weaver. 1963. The Mathematical Theory of Communication. Univ. of Illinois Press. Urbana.

1421 Shannon, L.R. 1977. Bull. Environ. Contam. Toxicol. 17(6), pp 637-644.

1422 Sharp, J.H. 1977. Limnol. Oceanogr. 22(3), pp 381-399.

1423 Shaw, T.L. and V.M. Brown. 1974. Water Res., 8(6), pp 377-382.

1424 Shcherbakov, A.P. 1952. Dokl. Akad. Nauk SSSR, 84, pp 153-157.

1425 Shelef, G., W.J. Oswald and C.C. Golueke. 1972. Adv. Wat. Poll. Res., Proc. of the 5th Int. Conf. on Water oll., Paper III, 25, pp 1-9.

1426 Shelef, G., W.J. Oswald and C.C. Golveke. 1970. In S.H. Jenkins(ed) Proc. Int. Conf. Water Pollut. Res. 5th, pp 13-25. Pergamon.

1427 Sheler, G. 1968. Kinetics in Algal Systems in Waste Treatment. Light Intensity and Nitrogen Concentration as Growth limiting Factors. Ph. D. Thesis. Univ. Calif. Berkeley.

1428 Shimizu, M., T. Kajihara and Y. Hiyama. 1970. Rec. Oceanogr. Works. Jap. 10(2), pp 137-145.

1429 Shimizu, M., T. Kajihara, I. Suyama and Y. Hiyama. 1971. J. Rad. Res. 12(1), pp 17-28.

1430 Shimwell, D.W. and A.E. Laurie. 1972. Environ. Pollut., 3, pp 291-301.

1431 Shin, E.-B. and P.A. Krenkel. 1976. Water Pollut. Control. 48 (3), pp 473-501.

1432 Shoemaker, C.A. 1977. in: Hall, C.A.S. and J.W. Day, Jr. (eds.); Ecosystem Modeling in theory and Practice. pp 546-574. John Wiley and Sons. New York.

1433 Short, Z.F., P.R. Olson, R.F. Palumbo, J.R. Donaldson and F.G. Lowman. 1971. Radionuclides in Ecosystems, Proc, Third Nat. Symp. Radioecology 1, pp 474-485.

1434 Sidorovski, M.N. 1974. Izd. Zavod. Ribar. SR Marked. 5(2), pp 15.

1435 Sillen, L.G. and A.E. Martell. 1964. Stability Constants of Metal-ion Complexes, Special publ No 17. The Chemical Society. London.

1436 Silva Bezerra, J.W. 1975. Notes Doc. Peche Piscic. (Nouv. Ser.), 11, pp 65-74.

1437 Silver, S., J. Schottel and A. Weiss. 1976. In J.M. Sharpley and A.M. Kaplan (eds.): Procedings of the Third International Biodegradation Symposium, 5(12), pp 899-915.

1438 Simpson, E.H. 1949. Nature 163, pp 688.

1439 Sims, J.L. and F.K. Pfaender. 1975. Bull. Environ. Contam. Toxicol. 14(2), pp 214-220.

1440 Singer, A. and R.M. Brown. 1956. Trans Amer. Geophysical Union 37, pp 746.

1441 Singer, S.F. 1975. In: S.F. Singer (ed.), The Changing Global Environment, pp 8-12. D. Reidel Publ. Comp.

1442 Singer, S.F. 1975. In: S.F. Singer (ed.), The Changing Global Environment, pp 25-43. D. Reidel Publ. Comp.

1443 Singer, S.F. 1975. In: S.F. Singer (ed.), The Changing Global Environment, pp 125-134. D. Reidel Publ Comp.

1444 Singleton, F.L. and R.K. Guthrie. 1977. Water Res., 11(8), pp 639-642.

1445 Sisler, F.D. 1975. In: S.F. Singer (ed.), The Changing Global Environment, pp 57-69. D. Reidel Publ. Comp.

1446 Skinner, M. and N. Walker. 1961. Arch. Microbiol. 38, pp 339-349.

1447 Slobodkin, L.B. and S. Richman. 1961. Nature 191, pp 299.

1448 Slobodkin, L.B. 1962. Advan. Ecol. Res. 1, pp 69-101.

1449 Slonim, A.R. 1973. Water Pollut. Control, 45(10), pp 2110-2122.

1450 Slonim, A.R. 1977. Water Res. 11(10), pp 889-896.

1451 Smayda, T.J. and B.J. Boleyn. 1965. Limnol. Oceanogr. 10, pp 499-509.

1452 Smayda, T.J. and B.J. Boleyn. 1966. Limnol. Oceanogr. 11, pp 18-34.

1453 Smayda, T.J. and B.J. Boleyn. 1966. Limnol. Oceanogr. 11, pp 35-43.

1454 Smayda, T.J. 1969. Limnol. Oceanogr. 14, pp 621-625.

1455 Smayda, T.J. 1970. Oceanogr. Mar. Biol. Ann. Rev., 8, pp 353-414.

1456 Smayda, T.J. 1971. Mar. Geol. 11, pp 105-122.

1457 Smayda, T.J. 1974. Limnol. Oceanogr. 19(4), pp 628-635.

1458 Smelt, J.H., L.G.M. th. Maenhout, Tuinstra and W. Luten 1975. Agro-Ecosyst., 2(2), pp 133-143.

1459 Smic. 1971. Report of the Study of Mans impact on Climate. in: Climate Modification. MIT-Press. Cambridge. England.

1460 Smith, A.E. 1974. J. Agric. Food. Chem. 22(4), pp 601-605.

1461 Smith, G.A., R.G. Thomas and J.K. Scott. 1960. Health Phys., 4, pp 101-108.

1462 Smith, K.L. Jr. 1974. Limnol. Oceanogr. 19(6), pp 939-944.

1463 Smith, T., C.J. Edmonds and C.F. Barnaby. 1972 Health Phys., 22, pp 359-367.

1464 Smith, W.H. 1973. Metal contamination of Urban Woody Plants, Science

1465 Smock, L.A. 1976. Water Res., 10(6), pp 537-543.

1466 Smyly, W.J.P. 1973. Oecologia 11, pp 163-186.

1467 Snedecor, G.W. and W.G. Corhran. 1967. Environ. Pollut., 1, pp 203.

1468 Snodgrass, W.J. and C.R. O'Melia. 1974. Limnol. Oceanogr. 304.

1469 Snodgrass, W.J. and C.R. O'Melia. 1975. Environ. Sci. Technol., 9, pp 937-944.

1470 Snow, N.B. 1972. Limnol. Oceanogr. 17(6), pp 909-913.

1471 Snyder, G.R. and T.H. Blohm. 1971. Water Pollut. Control. 43 (5), pp 890-899.

1472 Soedergren, A. 1973. Vatten 2, pp 90-108.

1473 Soedergren, A., Bj. Svensson and S. Ulfstrand. 1972. Environ. Pollut., 3, pp 25-36.

1474 Soeder, C.J., H. Muller, H.D. Payer and H. Schulle. 1971. Mitt. Int. Ver. Theor. Angew. Limnol., 19, pp 39-58.

1475 Soederlund, R. 1977. Ambio, 6(2-3), pp 118-122.

1476 Soerensen, T. 1948. K. Danske Vidensk. Selsk. 5, pp 1-34.

1477 Solbe, J.F., G. Del and V.A. Cooper. 1976. Water Res. 10(6), pp 523-527.

1478 Soliman, I.A.M., G.A. Botros and A.M.El-Maghraby. 1970. Bull. Inst. Oceanogr. Fish., Cairo 1, pp 27-45.

1479 Solbe, J.F. de L.G. 1974. Water Res., 8(6), pp 389-391.

1480 Somlyody, L. 1977. Water Res., 11(4),pp 411-417.

1481 Sorokin, C. and R.W. Krauss. 1958. The effects of light intensity on the growth rates of Green Algae, Plant Physiol, 33, pp 109-113.

1482 Sorokin, C. and R.W. Krauss. 1962. Effects of temperature and Illuminance on Chorella growth uncoupled from cell division, Plant Physiol, 37, pp 109-113.

1483 Sorokin, Y.I. 1962. Mikrobiologiya 31, pp 402-410.

1484 Sorokin, Y.I. 1963. Hydrobiologia 22, pp 306-316.

1485 Southworth, G.R. and C.W. Gehrs. 1976. Water Res., 10(11), pp 967-971.

1486 Sparrow, A.H., L.A. Schairer and R.C. Sparrow. 1963. Science 141, pp 163-166.

1487 Special Rept. Ser. Med. Res. Com. London No. 297. 1960. The Composition of Foods.

1488 Spector, W.S. 1956. Handbook of Biological Data, Saunders.

1489 Spence, D.H.N., R.M. Campbell and J. Chrystal. 1971. Hydrobiologia 12, pp 169-176.

1490 Spencer, C.P. 1954. Studies of the culture of a Marine Diatom, J. Marine Biol. Assoc. U.K., Quoted by Harvey H.W., 1966. The Chemistry and Fertility of Sea Water, Cambridge Univ. Press, pp 94.

1491 Spencer, W.F. and M.M. Cliath. 1974. J. Agric. Food. Chem. 22(6), pp 987-991.

1492 Spies, R.B. 1975. Health Physics, 29, pp 695-699.

1493 Spotte. 1970. Fish and Invertebrate Culture. Wiley Interscience. New York.

1494 Sprague, J.B. 1970. Water Res., 4(1), pp 3-32.

1495 Squires, P. 1966. J. Rech. Atmos. 2, pp 297-308.

1496 Sridharen, N. and G.F. Lee. 1977. Water Res. 11(10), pp 849-858.

1497 Srna, R.F. and A. Baggaley. 1975. Water Pollut. Control, 47(3), pp 472-486.

1498 Stanley, P.M. and J.T. Stanley. 1977. Limnol. Oceanogr. 22(1), pp 26-37.

1499 Staples, D.J. and M. Nomura. 1976. J. Fish. Biol. 9, pp 29-43.

1500 Stauffer, R.E. and G.F. Lee, 1974. The Role of Thermocline Migration in regulating Algal Blooms, pp 73-82.

1501 Stearn, A.E. 1949. Advan. Enzymol. 9, pp 25.

1502 Stedman, D.H. and H. Niki. 1973. Environ. Sci. Technol. 7, pp 735.

1503 Steele, J.H. 1974. The Structure of Marine Ecosystems. Blackwell Scientific Publications Ltd. Oxford. pp. 74-91.

1504 Steemann Nielsen, E. and V.K. Hansen. 1975. Plant Physiol., 12, pp 353-370.

1505 Steemann Nielsen, E. 1968. Plant Physiol., 21, pp 401-413.

1506 Steemann Nielsen, E. 1961. Plant Physiol., 14, pp 868-876.

1507 Stensel. 1971. Biological Kinetics of suspended Growth denitrification Process. Ph.D. thesis. Cornell Univ.

1508 Stensel, H.D. et. al. 1973. Water Pollut. Control, 45, pp 243.

1509 Stewart, J. and M. Schulz-Baldes. 1976. Mar. biol. 36, pp 19-24.

1510 Stiff, M.J. 1971. Water Res. 5(8), pp 585-600.

1511 Stiff, M.J. 1971. Water Res., 5(8), pp 585-600.

1512 Stirling, I. and E.H. McEwan. 1975. Can. J. Zool. 53(8), pp 1021-1027.

1513 Stoermer, F.C. and A. Vinsjansen. 1976. Ambio. 5(3), pp 141-142.

1514 Stoller, E.W., L.M. Wax, L.C. Haderlie and F.W. Slife. 1975. J. Agric. Food. Chem. 23(4), pp 682-684.

1515 Strabskraba, M. 1968. Mitt. Int. Ver. Limnol. 14, pp 212-230.

1516 Straheer, A.N. and A.H. Strahler. 1974. Introduction to Environmental Science, John Wiley and Sons, California.

1517 Strahler, A.N. 1971. The Earth Sciences. 2nd ed. Harper and Row. New York.

1518 Strathmann, R.R. 1967. Limnol. Oceanogr., 12, pp 411- 418.

1519 Stratton, f. 1966. Nitrification Effects on Oxygen Resources in Streams. Ph.D. Thesis. Stanford Univ.

1520 Strickland, J.D.H. 1960. Fish. Res. Bd. Can. Bull., 122, pp 172.

1521 Strickland, J.D.H. 1965. Chemical Oceanog., 1, pp 503.

1522 Strickland, J.D.H., O. Holm-Hansen, R.W. Eppley and R. J. Linn. 1969. Limnol. Oceanogr., 14, pp 23-24.

1523 Strik, J.J.T.W.A., H.H. Iongh, J.W.A. van Rijn van Alkemade and T.P. Wuite.

1975. in: Koeman, J.H. et al. (eds.): Sublethal Effects of Toxic Chemicals on Aquatic Animals, Elsevier Scientific Publishing Company, Amsterdam, Oxford, N.Y., pp 31-41

1524 Strohal, P. and T. Pinter. 1973. Limnol. Oceanogr. 18(2), pp 250-253.

1525 Stumm, W. and G.F. Lee. 1961. Ind. Eng. Chem., 53, p 143.

1526 Stumm, W. and J.J. Morgan, 1970. Aquatic Chemistry. Wiley N.Y.

1527 Stumm, W. and J.O. Leckie. 1971. Water Res., 2(3), pp 16.

1528 Styron, C.E., T.M. Hagan, D.R. Campbell, J. Harvin, N.K. Whittenburg, G.A. Baughman, M.E. Bransford, W.H. Sanders, D.C. Willians, C. Woodle, N.K. Dixon and C.R. McNeil. 1976. j. mar. biol. Ass. UK., 56, pp 13-20.

1529 Sturevant, J.M. 1962. in: Skinner, H.A. (ed.); Experimental Thermochemistry. Vol II. Chap. 19. Wiley. New York.

1530 Suess, M.J. 1976. Sci. Total Environ. 6(3), pp 239-250.

1531 Summers-smith, D. 1963. The House Sparrow. Collins. London.

1532 Sunada, J.S. 1974. Calif. Fish. Game, 60(2), pp 64-73.

1533 Sundstrand, G., O. Hutzinger, S. Safe, L. Ruzo and D. Jones. 1975 Methods for the Study of Metabolism of Toxic and Persistent Chemicals in Aquatic Organisms, Exemplied by Chloronaphthalenes. In J.H. Koeman et al.(ed), Sublethal Effects of Toxic Chemicals on Aquatic Animals, Elsevier Scientific Publishing Company, Amsterdam, Oxford, N.Y., pp 177-187.

1534 Sutton, P.M., K.L. Murphy and B.E. Jank. 1975. Nitrogen Control: A Basis for Design with Activated Sludge Systems. Presented at: Conf. on Nitrogen as a Water Poll. Vol 3. Session A2. Copenhagen.

1535 Suzuki, H., T. Koganage and M. Saki. 1975. Impacts of Nuclear Releases into the Aquatic Environment. IAEA-SM-198/19, pp 77-89. Vienna.

1536 Suzucki, I., U. Dular and S.C. Kwok. 1974. J. Bact. 120, pp 556-558.

1537 Swanson, Hollaender and Kaufman. 1948. Genetics 33, pp 431.

1538 Sweers, H.E. 1970. Limnol. Oceanogr. 15(2), pp 273-280.

1539 Swift, D.G. 1967. Growth of Vitamin B(12) limited Cultures: Cyclotella Nana, Monochrysis Lutheri and Isochrysis Galbana. MS. Thesis, John Hopkins Univ., Baltimore.

1540 Swift, E. and W.R. Taylor. 1960. Biol. Bull. 119, pp 342.

1541 Swift, M.C. and A.Y. Fedorenko, 1975. Limnol. Oceanogr. 20(3), pp 418-425.

1542 Sylva, R.N. 1976. Water Res. 10(9), pp 789-792.

1543 Sylvester, J.R. 1972. Environ. Pollut., 3, pp 205-215.

1544 Symonds, D.J. 1972. Fishery Invest. Ser. II, Lond., 27(3), pp 1-35.

1545 Szczepanska, W. 1973. Pol. Arch. Hydrobiol. 15, pp 51-77.

1546 Szwykowska, M.M. 1969. Ekol. pol. 17, pp 795-809.

T

1547 Taguchi, S. and H. Ishii. 1972. in: Takenouti, A.Y.(ed). Biological Oceanography of the northern North Pacific Ocean. pp 419-431. Idemitsu Shoten. Tokyo.

1548 Takahashi, M. and S. Ichimura. 1970. Limnol. Oceanogr. 15(6) pp 929-944.

1549 Takahaski, M., K. Satake and N. Nakamoto. 1972. J. Oceanogr. Soc. Japan, 28, pp 27-36.

1550 Talekar, N.S., L.-t. Sun, E.-M. Lee and J.-S. Chen. 1977. J. Agric. Food. Chem. 25(2), pp 348-352.

1551 Talling, J.F. 1957. New Phytol 56, pp 29-50.

1552 Talling, J.F. 1965. Int. Rev. ges. Hydrobiol. 50, pp 1-32.

1553 Talling, J.F. 1966. J. Ecol. 54, pp 99-127.

1554 Tamiya, H., E. Hase, K. Shibata et. al. 1964. Algal Culture from Laboratory to Pilot Plant, Kinetics Growth of Chlorella with special reference to its dependence on Quality of available Light and on Temperature. J.S. Burlew(ed), Publ. 600, Carnegie Inst. of Washington D.C., pp 204-234.

1555 Taylor, A.W., D.E. Glotfelty, B.C. Turner, R.E. Silver, H.P. Freeman and A. Weiss. 1977. J. Agric. Food. Chem. 25(3), pp 542-548.

1556 Taylor, A.W., D.E. Glottelty, B.L. Glass, H.P. Freeman and W.M. Edwards. 1974. J. Agric. Food. Chem. 24(3), pp 625-631.

1557 Taylor, D. 1976. Estuarine Coastal Mar. Sci. 2, pp 417.

1558 Taylor, D. 1976. Sci. Toatal Environ. 6(3), pp 259-264.

1559 Taylor, W.R. and E.P. Odum. 1960. Biol. Bull. 119, p 343.

1560 Taylor, W.R. 1964. Helgd. Wiss. Meeresun¤ers. 10, pp 29-37.

1561 Templeton, W.L. 1962. In Proc. Seminar on Aspects of Radioactive Contamination in normal and emergency situations, Rome, FAO, pp 42-64.

1562 Terborgh, J. and J.S. Weske. 1969. Ecology 50, pp 765-782.

1563 Terhaar, C.J., W.S. Ewell, S.P. Dziuba, W.W. White and P.J. Murphy. 1977. Water Res. 11(1), pp 101-111.

1564 Terre, J.J. 1976. Collect. Sci. Pap. IC-SEAF/Recl. Doc. Sci. CIPASE/Coleco. Doc. Cient. CIPASO, v. 3, pp 39-44.

1565 Tetra Tech, Inc. 1975. A Comprehensive Water Quality Ecological Model for Lake Ontario. NOAA. Ann Arbor. Mich.

1566 Tetra Tech, Inc. 1976. Estuary Water Quality Models, Lond Island, New York - User's Guide. For Nassau- Suffolk Regional Planning Board. Hauppauge, New York. 11787.

1567 Tetra Tech, Inc. 1977. Draft Report. (In press).

1568 Tett, P., J.C. Cottrelll, D.O. Trew and B.J.B. Wood. 1975. Limnol. Oceanogr. 20(4), pp 587-603.

1569 Thackston, E.L. and P.A. Krenkel. 1969. J. San. Eng. Div. ASCE. 95, SA1, pp 65.

1570 Thakur, N.K. and N.K. das. 1975. J. Inland Fish. Soc. India. 6(1974), pp 95-96.

1571 The Hydrological Engineering Center Corps of Engineers U.S. Army. 1975. Water Quality for river-reservoir systems, Davis, Cal. 9 5 6 1 6 (916) 499-2329.

1572 Theede, H. 1963. Kieler Meeresforsch., 19, pp 20-41.

1573 Therkelsen, O. 1977. Vatten 3, pp 299-317.

1574 This table is made from severel different references.

1575 Thomann, R.V. 1972. System Analysis and Water Quality Measurement. Environ. Research and Applications. Inc. New York.

1576 Thomann, R.V., D.M. Di Toro and D.J. O'Connor. 1974. Preliminary Model of Potomac Estuary Phytoplankton. J. Environ. Eng. Div. ASCE. 100, EE3.

1577 Thomann, R.V., D.M. Di Toro, R.P. Winfield and D.J. O'Connor. 1975. Mathematical Modeling Of Phytoplankton in Lake Otario: 1. Model Development and Verification. National Environmental Reseach Center. Office of Research and Development. U.S. Environmental Protection Agency Corvallis, Oregon 9 73 30.

1578 Thomann, R.V., D.S. Szumski, D.M. DiToro and D.J. O'Conner. 1974. Water Res. 8(10), pp 841-849..

1579 Thomann, R.V. 1978. EPA-report 600-3-78-036.

1580 Thomas, M.D. and G.R. Hill. 1949. in: Franck, J and W E. Loomis(eds). Photosynthesis in Plants, Iowa State College Press, pp 19-52.

1581 Thomas, W.H. and A.N. Dodson. 1968. Biol. Bull. Mar. Biol. Lab. , Woods Hole 134, pp 199-208.

1582 Thomas, W.H. and A.N. Dodson. 1972. Limnol. Oceanogr. 17, pp 515-523.

1583 Thomas, W.H. and A.N. Dodson. 1974. Mar biol. 24, pp 213-217.

1584 Thomas, W.H. 1970. Limnol. Oceanogr. 15(3), pp 380-385.

1585 Thomas, W.H. 1970. Limnol. Oceanogr. 15(3), pp 386-394.

1586 Thompson, G.E. and R.W. Davies. 1976. Trans. Am. Fish. Soc. 105(2), pp 208-219.

1587 Thompson, J.A. 1971. Brit. J. Industr. Med. 27, pp 189-194.

1588 Thompson, J.A.J. and J.C. Davis. 1976. Water Res., 10(10), pp 869-875.

1589 Thompson, M.A. 1975. Impacts of Nuclear Releases into the Aquatic Environment. IAEA-SM-198/38, pp 213-225. Vienna.

1590 Thorne, L. and G.P. Hanson. 1972. Environ Pollut., 3, pp 303-312.

1591 Thorpe, J.E. 1974. Proc. R. Soc. Edinb. B. 1974, pp 299-313.

1592 Thurber, D.L. and W.S. Broecker, 1970. In: Olson, I.V. (ed.); Nobel Symposium 12, pp 379-400. Wiley.

1593 Thurow, F. 1973. Ber. dt. wiss. Kommn. Meeresforsch., 22, pp 445-451.

1594 Tietjen, J.H., J.J. Lee, J. Rullman, A. Greengart and J. Trompeter. 1970. Limnol. Oceanogr. 15(4), pp 535-543.

1595 Tillander, M., J.K. Miettinen and I. Koivisto. 1972. in: Ruivo, M. (ed.); Marine Pollution and Sea Life. pp 303-305. Fishing Trading news (books) Ltd. London.

1596 Tilzer, M.M., H.W. Paerl and C.R. Goldman. 1977. Limnol. Oceanogr. 22(1), pp 84-91.

1597 Timofeeva-Resovskaya, E.A., T-R. N.V. and E.A. Gileva. 1961. Dokl. Akad. Nauk. SSSR, 140, pp 1437.

1598 Timoshuck, V.I. 1970. Available from Nat. Tech. Inf. Ser. Springfield, Va. as Part of Aec-tr-7299, pp 185-196.

1599 Tipton, I.H., P.L. Stewart and P.G. Martin. 1966. Health Phys., 12, pp 1683-1689¤

1600 Titman, D. and P. Kilham. 1976. Limnol. Oceanogr. 21(3), pp 409-417.

1601 Todd, R.L., W.T. Swank, J.E. Douglass, P.C. Kerr, D.L. Brockway and C.D. Monk. 1975. Agro-Ecosyst., 2(2), pp 127-132.

1602 Toerien, D.F. and C.H. Huang. Water Res. 7(11), pp 1673-1681.

1603 Toerien, D.F., C.H. Huang, J. Radimsky, E.A. Pearson and J. Scherfig. 1971. Final Report: Provisional Algal assay Procedures Sanit. Eng. Res. Lab., SERL rep., Univ. Calif. Berkeley, pp 71-76.

1604 Toetz, D., L. Varga and M. Pierce. 1977. Water Res. 11(3), pp 253-258.

1605 Toet¤, D.W. 1971. Limnol. Oceanogr. 16(5), pp 819-821.

1606 Tomkiewicz, S.M. JR. and W.A. Dunson.

1977. Water Res., 11(4), pp 397-402.

1607 Tomlinson, T.G., A.G. Boon and C.N.A. Trotman. 1966. J. appl. Bacteriol. 29, pp 266-291.

1608 Torma, A.E. 1976. App. Sci. Publ., pp 937-945.

1609 Torvik, E., E. Pfitzer, J.G.Kereiakes and R. Blanchard. 1974. Health Physics, 26, pp 81-87.

1610 Tovell, P.W.A., C. Newsome and D. Howes. 1974. Water Res., 8(5), pp 291-296.

1611 Toxic Substances List 1973, U.S. Department of Health, Education and Welfare, Public Health Service, National Institute for Occupational Safety and Health, Rockville Maryland, USA, June 1973.

1612 Trichell, D.W., H.L. Morton and M.G. Merkle. 1968. Weed Sci., 16, pp 447-449.

1613 Tsai, S.-C., M. Buush and F. Matsumura. 1975. Bull. Environ. Contam. Toxicol. 31, pp 188-193.

1614 Tsivoglou, E.C. and L.A. Neal. 1976. Wat. Poll. Cont. Fed. 48(12), pp 2669.

1615 Tsivoglou, E.C. et al. 1968. Wat. Poll. Cont. Fed. 40(2), pp 285.

1616 Turekian, K.K. and K.H. Wedepohl. 1961. Amer. Geol. Soc. Bull. 72, pp 175-192.

1617 Turner, F.B., R.I. Jennrich and J.D. Weintraub. 1969. Ecology 50, pp 1076-1081.

1618 Turner, J.T. 1977. Mar. Biol. 40(3), pp 249-259.

1619 Tyler, A.V. 1973. Mar biol. 19, pp 258-261.

U

1620 U.S. Army Corps of Engineers. 1974. Water Quality for River-Resevoirs Systems.

1621 U.S. Army Engineer District, Walla Walla Corps of Engineers. 1973. Water Quality Report: Lower Granite Lock and Dam Snake River, Washington-Idaho, vol 3.

1622 Uchijima, Z. 1970. Prediction and Measurement of Photosynthetic Productivity. pp 178-198. Wageningen.

1623 United States Department of Agriculture. 1963. Agriculture Handbook, 8.

1624 Urey, J.C., J.C. Kricher and J.M. Boylan. 1976. Bull. Environ. Contam. Toxicol. 16(1), pp 81-85.

V

1625 Van Dijk, H. 1974. Agro-Ecosyst., 1(3), pp 193-204.

1626 Van Droogenbroeck, R. and H. Laudelout. 1967. Antonie van Leeuwenhoek, 33, p 287.

1627 Van Emden, H.M., C.C.M. Kroon, E.N. Schoeman and H.A. van Seventer. 1974. Environ. Pollut. 6, pp 297-308.

1628 Vance, B.D. and A.W. Maki. 1976. Bull. Environ. Contam. Toxicol. 15(5), pp 601-607.

1629 Varty, I.W. and W.N. Yule. 1976. Bull. Environ. Contam. Toxicol. 15(3), pp 257-264.

1630 Verduin, J. 1975. Limnol. Oceanogr. 20(6), pp 1052.

1631 Verheij, E.W.H. 1970. Neth. J. Agric. Sci. 18, pp 89-104.

1632 Vernadsky, W. 1933. Z. Krist. Mineral. Petrog. Abt. B Mineral. Petrog. Mitt., 44, p 191.

1633 Viner, A.B. and Horne, A.J. 1971. Nature 232, pp 417-418.

1634 Vink, G.J. 1975. pp 145-157. in: Koeman, J.H. and J.S.T. Strik. (eds.): Sublethal Effects of Toxic Chemicals on Aquatic Animals, Elsevier Scientific Publishing Company, Amsterdam.

1635 Vinogradov, A.P. 1953. The Elementary Chemical Composition of Marine Organisms, Sears Foundation New Haven.

1636 Vinogradov, M.E., V.V. Menshutkin and E.A. Shushkina. 1972. Mar. biol. 16, pp 261-268.

1637 Vivekane, E., M.M. Haniffa, T.J. Pandian and R. Raghuram. 1974. Freshwat. biol. 4(3), pp 275-279.

1638 Vlymen, W.J. 1970. Limnol. Oceanogr. 15(3), pp 348-356.

1639 Vollenweider, R.A. 1960. Mem. Ist. Ital. Idrobiol., 12 pp 201-244.

1640 Vollenweider, R.A. 1968. Scientific Fundamentals of the Eutrophication of Lakes and Flowing Waters, With Particular Reference to Nitrogen and Phosphorus as Factors in Eutrophication. OECD. DAS/CSI/68.27. Paris.

1641 Vollenweider, R.A., M. Munawar and P. Stadelmann. 1974. J. Fish. Res. bd. Canada 31, pp 739-762.

1642 Vosjan, J.H. 1975. Ecological and Physiological Aspects of Bacterial Sulphate reduction in the Wadden Sea. Ph. D. Thesis Groningen, pp 68.

1643 Vreeland, V. 1974. Environ. Pollut., 6, pp 135-140.

1644 Vuceta, J. and J.J. Morgan. 1977. Limnol. Oceanogr. 22(4), pp 742-745.

W

1645 Wahid, P.A. and N. Sethunathan. 1978. J. Agric. Food. Chem. 26(1)., pp 101-105.

1646 Waiwood, K.G. and P.H. Johansen. 1974. Water Res., 8(7), pp 401-406.

1647 Walch, G.E., T.A. Hollister and J. Forester. 1974. Bull. Environ. Contam. Toxicol. 12(2), pp 129-135.

1648 Waldichuk, M. 1975. in: Ocean 75. IEEE-75-Cho-995-1-Oec, pp 907-912. New york.

1649 Waldron, H.A. and D. Stoeten. 1974. Subclinical Lead Poisoning.

1650 Walker, A. 1974. J. Environ. Qual. 3(4), pp 396-401.

1651 Walne, P.R. 1972. J. Mar. Biol. Ass. U.K., 52, pp 345-374.

1652 Walsh, G.E., K.A. Ainsworth and L. Paas. 1977. Bull. Environ. Contam. Toxicol. 18(3), pp 297-302.

1653 Walsh,J.J. and R.C. Dugdale. 1971. Invest. Pesq.35, pp 309-330

1654 Walter, C.M. F.C. June and H.G. Brown. 1973, Water Pollut. Control., 45(10), pp 2203-2210.

1655 Walter, C.M., H.G. Brown and C.P. Hensley. 1974. Water Res., 8(7), pp 413-418.

1656 Walter, H. 1954. Les Divisions Ecologiques du Monde, pp 27-39.

1657 Wang, W.C., G.F. Lee and D. Spyvidakis. 1972. Water Res. 6(10), pp 1219-1228.

1658 Ware, G.W., B. Estesen, W.C. Kronland and W.P. Cahill. 1977. Bull. Environ. Contam. Toxicol., 17(3), pp 317-322.

1659 Ware, G.W., W.P. Cahill and B.J. Esresen. 1975. Bull. Environ. Contam. Toxicol., 14(1), pp 88-97.

1660 Water Quality Criteria. 1968. Federal Water Pollution Control Administration Washington D.C., pp 234.

1661 Waters, T.F. 1977. Adv. Ecol. Res. 10, pp 91-164.

1662 Watt, B.K. and A.L. Merrill. 1963. Composition of Foods. U.S. Dept. Agriculture. Handbook 8. Wash. D.C.

1663 Watt, W.D. and F.R. Hayes. 1963. Limnol. Oceanogr. 8, pp 276-285.

1664 Watters, R.L. and W.R. Hansen. 1970. Health Physics. 18(4), pp. 409-413.

1665 Watts, W.R., R.E. Neilson and P.G. Jarvis. 1976. J. appl. Ecol. 13(2), pp 623-638.

1666 Webb, J.E. and C.M. Earle. 1972. Environ. Pollut., 3 pp 157-169.

1667 Webb, W.L., M. Newton and D. Starr. 1974. Oecologia 17, pp 281-291.

1668 Webber, K. 1976. Water Res. 10(3), pp 239-243.

1669 Weber, W.J. and W. Stumm. 1963. J. Chemo. engng. Data. 8, pp 464-468.

1670 Wedepoki, K.H.(ed). 1969. Handbook of Geochemistry, Springer-Verlag Berlin.

1671 Weers, A.W. van and N.H. Petten. 1975. Impacts of Nuclear Releases into the Aquatic Environment. IAEA-SM-198/44. pp 349-359. Vienna.

1672 Weers, A.W. van. 1973. In Radioactive Contamination of the Marine Environment. Vienna, IAEA.

1673 Weinig, E. and B. Boerner. 1961. Arch. Toxic. 19, pp 34-48.

1674 Weiss, R.L. 1973. Limnol. Oceanogr. 18(6), pp 877-883.

1675 Welch, H. 1968. Ecology 49, pp 755-759.

1676 Wells, L. 1960. Fish. Bull. U.S. Fish. Wildl. Serv. 60(172), pp 343-369.

1677 Wengrzyn, J. 1976. Collect. Sci. Pap. ICSEAF/Recl. Doc. Sci. CIPASE/Coleco. Doc. Cient. CIPASO, v 3., pp 51-57.

1678 Westlake, D.F. 1961. Int. Ver. Theor. Angew. Limnol. Verh. 15, pp 499-504.

1679 Westlake, D.F. 1963. Biol. Rev. 38, pp 385-425.

1680 Westlake, D.F. 1968. Methods of Productivity Studies in Roots Systems and Rhizosphere Organisms. Publ. House Nauka. Leningrad.

1681 Westman, I., M-L. Johansson-Sjoebeck and R. Faenge. 1975. in: Koeman, J.H. and J.J.W.A. Strik. (eds.): Sublethal Effects of Toxic Chemicals on Aquatic Animals, Elsevier Scientific Publishing Company, Amsterdam, Oxford, N.Y., pp 111-119.

1682 Wettern, M., D.W. Lorch and A. Weber. 1976. Arch. Hydrobiol. 77(3), pp 267-276.

1683 Wetzel. R.G. 1964. Int. Rev. ges. Hydrobiol. 49, pp 1-64.

1684 Wetzel. R.G. 1966. Invest. Indiana Lakes Streams 7, pp 1-32.

1685 Wetzel. R.G. 1966. Verh. Int. Ver. Limnol., 16, pp 321-332.

1686 Wetzel. R.G. 1973. Hydrobiol. Stud. 3, pp 91-143.

1687 Wetzel, R.G. and M. Owens. 1960. J. Ecol. 48, pp 151-160.

1688 Wetzel, R.G. Int. Rev. ges. Hydrobiol 49, pp 1-64.

1689 Wetzel, R.G. 1975. Limnology. Saunders. Philadelphia.

1690 Wheatley, G.A., J.A. Hardman and A.H. Strickland. 1962. Plant Pathol., 11, pp 81.

1691 Wheelen, W.B. 1970. Bull. Environ. Contam. Tixicol. 5, pp 463-467.

1692 Wheeler, P.A., B.B. North and G.C. Stephens. 1974. Limnol. Oceanogr. 19(2), pp 249-259.

1693 White, A.W., A.P. Barnett, B.C. Wright and J.H. Holliday. 1967. Environ. Sci

Technol., 1, pp 740-744.

1694 White, A.W., L.E. Asmussen, E.W. Hauser and J.W. Turnbull. 1976. J. Environ. Qual. 5(4), pp 487-490.

1695 White, J.D. and J.A. Dracup. 1977. Water Pollut. Control 49(10), pp 2179-2189.

1696 White, W.R. 1968. Lab. Note. cent. Elect. Res. Lab. RD/L/N 116/68, pp 1-7.

1697 Whiteside, M.C. 1970. Ecol. Monogr. 40, pp 79-118.

1698 Whitleg, L.S. and R.A. Sikora. 1970. Water Pollut. Control. 42(2), pp r57-r66.

1699 Whittaker, R.A. and G.M. Woodwell. 1969. J. Ecol. 57, pp 103-174.

1700 Whittaker, R.H. 1970. Communities and Ecosystems. Macmillan. New York.

1701 Wiant, H.V. 1967. J. For. 65, pp 489-490.

1702 Wickstroem, G. 1972. Work-Environmental-Health, 9, pp 2-8.

1703 Wiegert, R.G. 1975. In: B.C. Patten (ed.), Systems Analysis and Simulation in Ecology, Vol. 3, pp 157-181.

1704 Wiegert, R.G. 1977. in: Hall, C.A.S. and J.W. Day, Jr. (eds.); Ecosystem Modeling in theory and practice. pp 290-315. John Wiley and Sons. New York.

1705 Wiegert, R.G. 1964. Ecol. Monogr. 34, pp 225-241.

1706 Wiegert, R.G. 1965. BioScience 15, pp 543-545.

1707 Wild, H.E., C.N. Sawyer and T.C. McMahon. 1971. J. Wat. Poll. Cont. Fed.

1708 Wildish, D.J. 1972. Water Res., 6, pp 759-762.

1709 Wildish, P.J. 1974. Water Res., 8(7), pp 433-437.

1710 Wildung, R.E. and T.R. Garland. 1974. J. Agric. Food. Chem. 22(5), pp 836-838.

1711 Wildung, R.E., T.R. Garland and D.A. Cataldo. 1977. Health Physics. 32(4), pp 314-316.

1712 Wilkniss, P.E., J.W. Swinnerton, R.A. Lamontague and D.J. Bressan. 1975. Science 187, pp 832.

1713 Willemsen, J. 1952. Archs. neerl. Zool., 10, pp 152- 160.

1715 Williams, D.T. and B.J. Blanchfield. 1974. Bull. Environ. Contam. Toxicol. 12(1), pp 109-112.
Williams, D.T., H.M. Cunningham and B.J. Blanchfield. 1972. Bull. Environ. Contam. Toxicol. 7(1), pp 57-62.

1716 Williams, F.M. 1965. Population Growth and regulation in continuously Culture Algae. Ph. D. Thesis. Univ. Yale.

1717 Williams, I.H., M.J. Brown and P. Whitehead. 1976. Bull. Environ. Contam. Toxicol., 15(2), pp 242-243.

1718 Williams, J.H. 1971. In: Tech. Bull. 21, pp 119-135. Trace Elements in Soils and Crops.

1719 Williams, P.M. 1971. In S.J. Faust and J.V. Hunter. (eds.): Organic Compounds in Aquatic Environments, N.Y. Marcel Dekker Inc., pp 145-163.

1720 Williams, R.J.H., M.M. Lloyd and G.R. Ricks. 1971, Environ. Pollut., 2, pp 57-68.

1721 Willis, G.H. and R.A. Hamilton. 1973. J. Environ.Quality 2, pp 463-466.

1722 Willis, G.H., R.C. Wander and L.M. Southwick. 1974. J. Environ. Qual. 3(3), pp 262-265.

1723 Willis, G.H., R.L. Rogers and L.M. Southwick. 1975. J. Environ. Qual. 4(3), pp 399-402.

1724 Willis, G.H., R.L. Rogers and L.M. Southwick. 1976. J. Environ. Quality, pp **

1725 Willis, J.N. and N.Y. Jones. 1977. Health Physics. 32(5), pp 381-387.

1726 Wilson, D.E. and P. Kinney. 1977. Limnol. Oceanogr. 22(2), pp 281-289.

1727 Wilson, L.R. 1937. Bull. Torr. Bot. Club 64, pp 199-208.

1728 Wilson, L.R. 1941. Trans. Wis. Acad. Sci. Arts Lett. 33, pp 135-146.

1729 Wimmer,G. 1904. Z. Hyg. Infekt. (Ger.), 48, pp 135.

1730 Winberg, G.G. et al. 1970. in: Kajak, Z. and A. Hillbricht-Ilkowska (eds.); Productivity Problems of Freshwater. pp 383-404. PWN Polish Scientific Publishers. Warsaw.

1731 Winberg, G.G. et al. 1973. Hydrobiol. J. 9, pp 77-84.

1732 Winberg, G.G. 1950. Z. Obshch. Biol. 11, pp 367-380.

1733 Winberg, G.G. 1956. Fish. Res. Bd. Can. Transl. Ser., No. 194.

1734 Winberg, G.G. 1956. Uc. Zap. Beloruss. gos. Univ., Ser. Biol. 26, pp 243-254.

1735 Winberg, G.G. 1971. Methods for the Estimation of Production of Aquatic Animals. Academic Press.

1736 Windom, H.L. 1969. Geol. Soc. Amer. Bull. 80, pp 761-782.

1737 Winter, J.E. 1969. Mar. Biol., 4, pp 87-135.

1738 Winter, J.E. 1973. Mar. Biol., 22, pp 317-328.

1739 Wise, D.L. and G. Houghton. 1968. Chem. Eng. Sci., 23, pp 1211.

1740 Witkamp. M. 1969. Ecology 50, pp 922-924.

1741 Wolf, L.L. and F.R. Hainsworth. 1972. Comp. Biochem. Physiol., 41, pp 167-173.

1742 Wolfe, 1975. In: Cronin, L.E. (ed.); Estuarine Research, Vol. 1. Ac. Press N.Y.

1743 Wolfe, N.L., R.G. Zepp, G.L. Banghman, R.C. Fincher and J.A. Gordon. 1976. Chemical and Photochemical Transformations of selected Pesticides in Aquatic Systems, EPA-600/3-76-067.

1744 Wong, P.T.S., Y.K. Chau and P.L. Luxon. 1975. Nature 253(5489), pp 263-264.

1745 Woodward, R.L. and T.E. Wissing. 1976. Trans. Am. Fish. Soc. 105(3), pp 411-415.

1746 Woodwell, G.M. and R. Whittaker. 1968. Amer. Zool. 8, pp 19-30.

1747 Woodwell. G.M., P.P. Graig and H.A. Johnson. 1975. in: Singer, S.F. (ed.); The Changing Global Environment. pp 295-309. D. Reidel Publ. Comp. New York.

1748 Woodwell, G.M. and P.F. Bourdeau. 1966. Science 154, pp 1031-1034.

1749 Woodwell, G.M., C.F. Wurster and P.A. Isaacson. 1967. Science 156, pp 821-824.

1750 Woolland, J.V. and J.W. Jones. 1975. J. Fish. Biol., 7(6), pp 749-773.

1751 Worthington, E.B. 1931. Int. Rev. ges. Hydrobiol. 25, pp 394-436.

1752 Wright, J.C. 1958. Limnol. Oceanogr., 3(2), pp 150-159.

1753 Wright, J.C. 1965. Limnol. Oceanogr. 10, pp 583-590.

1754 Wright, R.T. 1975. Limnol. Oceanogr. 20(4), pp 626-633.

1755 Wright, S.J.L. 1972. Chemosphere 1(1), pp 11-14.

1756 Wu, C.H. and H. Niki. 1975. Environ. Sci. Technol. 9(1), pp 46-52.

1757 Wunderlich, W.O. 1971. in: Hall, G.E.(ed.); Reservoir Fisheries and Limnology. pp 219-231. Amer. Fish. Soc., Spec. Publ. 8. Wash. D.C.

1758 Wurster, C.F. 1968. Science 159, pp 1474-1475.

Y

1759 Yaron, B., H. Bielorai and L. Klinger. 1974. J. Environ. Qual. 3(4), pp 413-417.

1760 Yentsch, C.S. and R.W. Lee. 1966. J. Mar. Res., 24, pp 319-337.

1761 Young, D.R. and T.R. Folson. 1967. Biol. Bull. 133, pp 438-447.

1762 Young, D.R., D.J McDermott and T. Heesen. 1976. Bull. Environ. Contam. Toxicol. 16(5), pp 604-611.

1763 Young, H.E. and V.P. Guinn. 1966. Tappi 49, pp 190-197.

1764 Young, M.L. 1975. J. Mar. Biol. Ass. U.K., 55, pp 583-610.

1765 Young, R.G. and D.J. Lisk. 1972. Water Pollut. Control, 44(8), pp 1643-1647.

1766 Young, W.D., W. H. Gutenmann and D.J. Lisk. 1972. Environ. Sci. Technol. 6(5), pp 451-452.

1767 Yull Rhee, G. 1973. J. Phycol. 9, pp 495-506.

Z

1768 Zabat, M. 1970. Kinetics of Phosphorus removal by Algae Ph. D. Thesis. Univ. Calif. Berkeley.

1769 Zaika, V.Y., T.V. Pavlovskaya and N.A. Ostrovskaya. 1975. Biol. Morya. 33, pp 18-22.

1770 Zaret, T.M. and J.S. Suffern. 1976. Limnol. Oceanogr. 21(6), pp 804-813.

1771 Zaroogian, G.E. and S. Cheer. 1976. Nature 261, pp 408-410.

1772 Zeitschel, B. 1970. Mar. Biol. 7, pp 305-318.

1773 Zepp, R.G., N.L. Wolfe, G.L. Baughman and J.A. Gordon. 1974. Environ. Qual. Saf. Suppl III, pp 313-319.

1774 Zepp, R.G., N.L. Wolfe, J.A. Gordon and R.C. Fincher. 1976. Agric. Fd Chem., 24(4), pp 727-732.

1775 Zirino, A. and M.C. Healy. 1970 Limnol. Oceanogr. 15(6), pp 956-958.

1776 Zitko, V. and O. Hutzinger. 1976. Bull. Environ. Contam. Toxicol. 16(6), pp 665-673.

1777 Zitko, V. and W.V. Carson. 1975. Bull. Environ. Contam. Toxicol. 14(5), pp 530-533.

1778 Zitko, V. 1977. Bull. Environ. Contam. Toxicol. 17(3), pp 285-292.

1779 Zook, E.G., J.J. Powell, B.M. Hackley, J.A. Emerson, J.R. Brooker and G.M. Knobl, Jr. 1976. J. Agric. Food. Chem. 24(1), pp 47-53.

Latin - English List
English - Latin List

The following systematical organism-index presents a list of the organisms included in the tables of the handbook. The purpose is to give the user a systematical index of the organisms so that it is more easy to get answers to questions as: 'What kind of organism is it?' and 'How many different species of a certain group or genus are representet?'.

The organisms are seperated into two main groups: 1) animals and 2) plants. Within each group the organisms are subdivided according to their respective phyla or convenient division. The index is separated on two parts:

1) The first list gives the english name to a given latin name. If it is impossible to give an english name to a latin name, the latin name only is listed. Latin names are presented in alphabetical order.

2) The second list gives the latin name to a given english name (when possible).

To find tables, where a certain organism is located, it is necessary to use the subject index in the final part of the handbook. Please notice that although most of the names are given in latin, there are incidents, where the name is in english only or in both english and latin. To be sure to find all the tables where a certain organism is represented, it may be necessary to search the subject index for both the latin and the english name. Corresponding english and latin names

It will be noticed that some of the names are marked with an asterix , which indicates that this name is mentioned in a footnote only, and thus cannot be found in the subjec index.

****

A N I M A L S

****

ANNELIDA

ALLOLOBOPHORA ROSEA	LUMBRICID
APHRODITA HASTATA	POLYCHAETE
COGNETTIA COGNETTII	ENCHYTRAE
EISENIA ARBOREA	OLIGOCHAETE
EISENIA BICYCLIS	OLIGOCHAETE
ENCHYTRACIDAE	OLIGOCHAETES
ERPOBDELLA OCTOCULATA	LEECH
EUPOMATUS SP.	POLYCHAETE
GLOSSIPHONIA COMPLANATA	LEECH
GLOSSIPHONIA HETEROCLITA	LEECH
HELOBDELLA STAGNALIS	LEECH
HIRUDINEA	LEECHES
HYDROIDES NORVEGICA	POLYCHAETE
ILYODRILUS HAMMONIENSIS	OLIGOCHAETE
LANICE CONCHILEGA	POLYCHAETE
LIMNODRILUS HOFFMEISTERI	OLIGOCHAETE
LIMNODRILUS HOFFMEISTERI CLARAPAREDE	OLIGOCHAETE
LIMNODRILUS SP.	OLIGOCHAETE
LUMBRICIDAE	OLIGOCHAETES
LUMBRICUS SP.	OLIGOCHAETE
LUMBRICUS TERRESTRIS	OLIGOCHAETE EARTHWORM
LUMBRINERIS FRAGILIS	POLYCHAETE
MYXICOLA INFUNDIBULUM	POLYCHAETE
NEANTHES SUCCINEA	POLYCHAETE
NEANTHES VIRENS	POLYCHAETE
NEPHTYS INCISA	POLYCHAETE
NEPHTYS SP.	POLYCHAETE
NEREIS DIVERSICOLOR	POLYCHAETE
NEREIS SP.	POLYCHAETE
OLIGOCHAETA	OLIGOCHAETE WORMS
PELOSCOLEX MULTISETOSUS	OLIGOCHAETE
POLYCHAETA	POLYCHAETE WORMS
POMATOCEROS TRIQUETER	POLYCHAETE
POTAMOTHRIX HAMMONIENSIS	OLIGOCHAETE
PRAXILLELLA PRAETERMISSA	POLYCHAETE
SABELLA PAVONINA	POLYCHAETE
SALMACINA DYSTERI	POLYCHAETE
SCHIZOBRANCHIA INSIGNIS	POLYCHAETE
SERPULA SP.	POLYCHAETE
SPIRORBIS BOREALIS	POLYCHAETE
SPIRORBIS SP.	POLYCHAETE
TOMOPTERIS SEPTENTRIONALIS	POLYCHAETE
TUBIFEX SP.	OLIGOCHAETE
TUBIFEX TUBIFEX	OLIGOCHAETE
TUBIFICIDAE	OLIGOCHAETES

ARTHROPODA

ABLABESMYIA SP.	MIDGE
ACANTHOCYCLOPS VERNALIS	COPEPOD
ACARTIA CLAUSII	COPEPOD
ACARTIA SP.	COPEPOD
ACARTIA TONSA	COPEPOD
ACHETA DOMESTICUS	HOUSE CRICKET
ACHYNOPTERYX CURVATA	STONEFLY
ACRONEURIA LYCORIAS	STONEFLY
ACRONEURIA SP.	STONEFLY
ACROPERUS HARPAE	WATER FLEA
AEDES AEGYPTI	MOSQUITO (YELLOW FEVER)
AESCHNA CYANEA	DRAGONFLY
AGRYPNIA OBSOLETA	CADDISFLY
ALLOPERLA MEDIANA	STONEFLY
ALONA COSTATA	WATER FLEA
ALONA GUTTATA	WATER FLEA
AMMOPHILA BREVILIGULATA	DIGGER-WASP
AMPHIPODA	AMPHIPODS
ANATOPYNIA DYARI	MIDGE
ANTOCHA SP.	CRANEFLY
APHELOCHEIRUS VITTATUS	HONEY BEE
ARACHNIDA	SPIDERS SCORPIONS AND MITES
ARANEAE	SPIDERS
ARCHTODIAPTOMUS SP.	COPEPOD
ARCTIIDAE	BUTTERFLIES
ARMADILLIDIUM VULGARE	ISOPOD
ARTEMIA SP.	SHRIMP
ASELLOPSIS INTERMEDIA	CRUSTACEAN
ASELLUS AQUATICUS	HOG LOUSE
ASELLUS MERIDIANUS	HOG LOUSE
ASELLUS MILIARIS	HOG LOUSE
ASTACUS SP.	CRAYFISH
ATHERIX SP.	FLY
ATHRIPSOIDES ANCYLUS	CADDISFLY
BAETIS RHODANI	MAYFLY
BAETIS SP.	MAYFLY
BAETIS VAGANS	MAYFLY
BALANUS BALANOIDES	BARNACLE
BALANUS SP.	BARNACLE
BOSMINA COREGONI	WATER FLEA
BOSMINA LONGIROSTRIS	WATER FLEA
BOSMINA SP.	WATER FLEA
BOSMINOPSIS DEITERSI	WATER FLEA
BOYERIA VINOSA	DRAGONFLY
BRACHINECTA GIGAS	CRUSTACEAN
BRACHINECTA MACKINI	CRUSTACEAN
BRACHORIA INITIALIS	MILLIPEDE
BRACHYCENTRUS AMERICANUS	CADDISFLY
BRACHYCENTRUS SP.	CADDISFLY
BRACHYURA	CRUSTACEAN
CAENIS SIMULANS	MAYFLY
CALANOIDES CARINATUS	COPEPOD
CALANUS CRISTATUS	COPEPOD
CALANUS DISCAUDATUS	COPEPOD
CALANUS FINMARCHICUS	COPEPOD
CALANUS HELGOLANDICUS	COPEPOD
CALANUS HYPERBOREUS	COPEPOD
CALANUS PACIFICUS	COPEPOD
CALANUS PLUMCHRUS	COPEPOD
CALANUS SP.	COPEPOD
CALLIBAETIS FERRUGINEUS	MAYFLY
CALLINECTES SAPIDUS	BLUE CRAB
CALLIOPIUS LAEVIUSCULUS	AMPHIPOD
CALOPSECTRA DIVES	MIDGE
CAMBARINAE	CRAYFISH
CAMPONOTUS AMERICANUS	ANT
CAMPONOTUS PENNSYLVANICUS	ANT
CANCER IRRORATUS	ROCK CRAB
CANCER PAGURUS	CRAB
CANCER PRODUCTUS	CRAB
CANUELLA PERPLEXA	COPEPOD
CARCINUS MAENAS	GREEN CRAB
CARCINUS MEDITERRANEUS	CRAB
CARCINUS SP.	CRAB
CATACLYSTA MIDAS	BUTTERFLY
CENTROPAGES HAMATUS	COPEPOD
CENTROPAGES TYPICUS	COPEPOD
CENTROPTILUM LUTEOLUM	MAYFLY
CERATOPOGONIDAE	BITING MIDGES
CERIODAPHNIA CORNUTA	WATER FLEA
CERIODAPHNIA PULCHELLA	WATER FLEA
CERIODAPHNIA QUADRANGULA	WATER FLEA
CERIODAPHNIA RETICULATA	WATER FLEA
CERIODAPHNIA SP.	WATER FLEA
CEUTHOPHILUS GRACILIPES	CAVE CRICKET
CHAETOPSIS SP.	DIPTERAN
CHAOBORUS FLAVICANS	PHANTOM MIDGE
CHAOBORUS PUNCTIPENNIS	PHANTOM MIDGE
CHAOBORUS SP.	PHANTOM MIDGE
CHAOBORUS TRIVITTATUS	PHANTOM MIDGE
CHARYBDIS LONGICOLLIS	CRAB
CHEUMATOPSYCHE SP.	CADDISFLY
CHIRONOMIDAE	NON-BITING MIDGES
CHIRONOMUS ANTHRACINUS	MIDGE
CHIRONOMUS COMMUTATUS	MIDGE
CHIRONOMUS PLUMOSUS	MIDGE
CHIRONOMUS SP.	MIDGE
CHIRONOMUS TENTANS	MIDGE
CHOROTERPES MEXICANUS	MAYFLY
CHORTHIPPUS PARALLELLUS	GRASSHOPPER
CHTHAMALUS FISSUS	BARNACLE
CHTHAMALUS SP.	BARNACLE
CHYDORUS EURYNOTUS	WATER FLEA
CHYDORUS SPHAERICUS	WATER FLEA
CLADOCERA	WATER FLEAS
CLADOTANYTARSUS MANCUS	MIDGE
CLIBANARIUS VIRESCENS	HERMIT CRAB
COLEOPTERA	BEETLES
COLLEMBOLA	SPRINGTAILS
COMPTOCERUS RECTIROSTRIS	CRUSTACEAN
COPEPODA	COPEPODS
CORDULIA AENEA	DRAGONFLY

CORIXIDAE	WATER BOATMEN
CORYDALIS CORNUTA	LACEWING OR DOBSONFLY
CORYDALUS CORNUTUS	DOBSONFLY
CORYNONEURA SP.	MIDGE
CRANGON ALLMANI	SHRIMP
CRANGON CRANGON	SAND SHRIMP
CRANGON SEPTEMSPINOSA	SHRIMP
CRANGONYX PSEUDOGRACILIS	AMPHIPOD
CRANGONYX RICHMONDENSIS	AMPHIPOD
CREMATOGASTER CLARA	ANT
CRUSTACEA	CRUSTACEANS: CRABS SHRIMPS LOBSTERS ETC
CULEX TARSALIS	MOSQUITO
CYCLOPS BICUSPIDATUS	COPEPOD
CYCLOPS SP.	COPEPOD
CYCLOPS STRENUUS	COPEPOD
CYCLOPS VERICANS RUBELLUS	COPEPOD
CYCLOPS VERNALIS	COPEPOD
CYCLOPS VICINUS	COPEPOD
CYLISTICUS CONVEXUS	ISOPOD
CYRNUS TRIMACULATUS	CADDISFLY
DAPHNIA CRISTATA	WATER FLEA
DAPHNIA GALEATA	WATER FLEA
DAPHNIA GALEATA MENDOTAE	WATER FLEA
DAPHNIA HYALINA	WATER FLEA
DAPHNIA LONGISPINA	WATER FLEA
DAPHNIA MAGNA	WATER FLEA
DAPHNIA MIDDENDORFFIANA	WATER FLEA
DAPHNIA PARVULA	WATER FLEA
DAPHNIA PULEX	WATER FLEA
DAPHNIA PULICARIA	WATER FLEA
DAPHNIA RETROCURVA	WATER FLEA
DAPHNIA ROSEA	WATER FLEA
DAPHNIA SCHODLERI	WATER FLEA
DAPHNIA SP.	WATER FLEA
DELEATIDIUM SP.	MAYFLY
DIAPHANOSOMA BRACHYURUM	WATER FLEA
DIAPHANOSOMA LEUCHTENBERG	WATER FLEA
DIAPHANOSOMA SP.	WATER FLEA
DIAPTOMUS ALASKAENSIS	COPEPOD
DIAPTOMUS ARCTICUS	COPEPOD
DIAPTOMUS CLACIPES	COPEPOD
DIAPTOMUS GATUNENSIS	COPEPOD
DIAPTOMUS GRACILIS	COPEPOD
DIAPTOMUS GRACILOIDES	COPEPOD
DIAPTOMUS LEPTOPUS	COPEPOD
DIAPTOMUS NEVADENSIS	COPEPOD
DIAPTOMUS OREGONENSIS	COPEPOD
DIAPTOMUS PALLIDUS	COPEPOD
DIAPTOMUS SHOSHONE	COPEPOD
DIAPTOMUS SICILOIDES	COPEPOD
DIAPTOMUS SP.	COPEPOD
DINARTHRODES SP.	CADDISFLY
DIPLACODES TRIVIALIS	DRAGONFLY
DIPTERA	TRUE FLIES
DRASSYLLUS VIRGINIANA	SPIDER
DYTISCIDAE	WATER-BEETLES
ECDYONURUS SP.	MAYFLY
ECDYONURUS YOSHIDAE	MAYFLY
ELMIS SP.	RIFFLE BEETLE
EPEORUS HIEMALIS	MAYFLY
EPHEMERA LINEATA	MAYFLY
EPHEMERA STRIGATA	MAYFLY
EPHEMERELLA GRANDIS	MAYFLY
EPHEMERELLA NEEDHAMI	MAYFLY
EPHEMERELLA SERRATA	MAYFLY
EPHEMERELLA SUBVARIA	MAYFLY
EPHEMERELLA TRISPINA	MAYFLY
EPILABIDOCERA AMPHITRITES	COPEPOD
EPIOPHLEBIA SP.	DRAGONFLY
ERIOCERA SP.	CRANEFLY
EUBOSMINA TUBICENS	WATER FLEA
EUBRANCHIPUS BUNDYI	BRANCHIOPOD
EUBRIANAX SP.	BEETLE
EUCYCLOPS AGILIS	COPEPOD
EUDIAPTOMUS GRACILIS	COPEPOD
EUDIAPTOMUS GRACILOIDES	COPEPOD
EUPHAUSIA KROHNII (*)	EUPHAUSIID
EUPHAUSIA PACIFICA	EUPHAUSIID
EUPHAUSIACEA	EUPHAUSIIDS
EUROSTUS HILLERI	BEETLE
EURYCERCUS LAMELLATUS	WATER FLEA
EURYTEMORA AFFINIS	COPEPOD
EUSIMULIUM SP.	BLACK FLY
EUTERPINA ACUTIFRONS	COPEPOD
EVARTHRUS SODALIS	CARABID BEETLE
GAMMARACANTHUS LACUSTRIS	AMPHIPOD
GAMMARUS FOSSARUM	SCUD
GAMMARUS LACUSTRIS	SCUD
GAMMARUS OCEANICUS	SCUD
GAMMARUS PSEUDOLIMNAEUS	SCUD
GAMMARUS PULEX	SCUD
GAMMARUS TIGRINUS	SCUD
GEOTRUPES SP.	DUNG BEETLE
GIBBIUM PSYLLOIDES	BEETLE
GLYPTOTENDIPES BARBIPES	MIDGE
GLYPTOTENDIPES PARIPES	MIDGE
GLYPTOTENDIPES SP.	MIDGE
GOERA JAPONICA	CADDISFLY
GOMPHUS SP.	DRAGONFLY
GOTTIDIA PYRAMIDATA	BRANCHIOPOD
GRYLLUS ASSIMILIS	CRICKET
GYRINIDAE	WHIRLIGIGS
HALICYCLOPS MAGNICEPS	COPEPOD
HALIPLIDAE	BEETLES
HEPTAGENIA FUSCOGRISEA	MAYFLY
HERMIONE HYSTRIX	SOLDIER FLY
HETEROCOPE SEPTENTRIONALIS	COPEPOD
HEXAGENIA BILINEATA	MAYFLY
HEXAGENIA LIMBATA	MAYFLY
HOLOPEDIUM GIBBERUM	WATER FLEA
HOLOPEDIUM SP.	WATER FLEA
HOMARUS AMERICANUS	NORTHERN LOBSTER
HOMARUS VULGARIS	LOBSTER
HYALELLA AZTECA	AMPHIPOD
HYAS ARANEUS	CRAB
HYDROPHILIDAE	DIVING BEETLES
HYDROPHILUS TRIANGULARIS	DIVING BEETLE
HYDROPSYCHE BETTENI	CADDISFLY
HYDROPSYCHE SP.	CADDISFLY
HYDROPSYCHIDAE	CADDISFLIES
HYPERA POSTICA	BEETLE
INSECTA	INSECTS
ISOGENUS FRONTALIS	STONEFLY
ISONYCHIA JAPONICA	MAYFLY
ISOPERLA GORTZI	STONEFLY
ISOPODA	ISOPODS
ISOPTERA	TERMITES
JAPONARIA LAMINATA ARMIGERA	DIPLOPOD
LABIDOCERA JOLLAE	COPEPOD
LASIODERMA SERRICORNE	CIGARETTE BEETLE
LEANDER ADSPERSUS	SHRIMP
LEANDER PACIFICUS	SHRIMP
LEIOBUNUM FLAVUM	SPIDER
LEPTOCHEIRUS PINGUIS	CRUSTACEAN
LEPTODIUS FLORIDANUS	CRAB
LEPTODORA KINDTII	WATER FLEA
LEPTOPHLEBIA MARGINATA	MAYFLY
LEPTOPHLEBIA SP.	MAYFLY
LEPTOTERNA DOLABRATA	PLANT BUG
LESTES SPONSA	DAMSELFLY
LEYDIGIA ACANTHOCEROIDES	WATER FLEA
LEYDIGIA QUADRANGULARIS	WATER FLEA
LIGIDIUM BLUERIDGENSIS	ISOPOD
LIGIDIUM JAPONICA	ISOPOD
LIMNEPHILIDAE	CADDISFLIES
LIMNOCALANUS MACRURUS	COPEPOD
LIMNOCHIRONOMUS PULSUS	MIDGE
LIMNOCHIRONOMUS SP.	MIDGE
LITHOBIUS CRASSIPES	CENTIPEDE
LITHOBIUS FORTIFICATUS	CENTIPEDE
LUCIOLA CRUCIATA	FIREFLY
LYCOSA SP.	SPIDER
LYSMATA SETICAUDATA	SHRIMP
MACROCYCLOPS ALBIDUS	COPEPOD
MACRONEMA SP.	CADDISFLY
MACROTHRIX LATICORNIS	WATER FLEA
MACROTHRIX ROSEA	WATER FLEA
MEGANYCTIPHANES NORVEGICA	CRUSTACEAN
MENIPPE MERCENARIA	CRUSTACEAN
MESOCYCLOPS ALBIDUS	COPEPOD
MESOCYCLOPS EDAX	COPEPOD
MESOCYCLOPS LEUCKARTI	COPEPOD
METAMYSIDOPSIS ELONGATA	OPOSSUM SHRIMP
METAPENAEUS MONOCEROS	SHRIMP
METRIDIA LUCENS	COPEPOD
MEZIUM AFFINE	BEETLE
MICRASEMA SP.	CADDISFLY
MICROTENDIPES CHLORIS	MIDGE
MICROTENDIPES SP.	MIDGE
MITOPUS MORIO	SPIDER
MIXODIAPTOMUS LACINIATUS	COPEPOD
MOINA MACRURA	WATER FLEA

MOINA MINUTUS	WATER FLEA
MOLANNA SP.	CADDISFLY
MYRMICA SP.	ANT
MYSIDAE	MYSIDS
MYSIS RELICTA	MYSID
MYSIS STENOLEPIS	MYSID
NEMATOSCELIS ATLANTICA	EUPHAUSIID
NEMATOSCELIS MEGALOPS	EUPHAUSIID
NEMOURA SP.	STONEFLY
NEODIAPTOMUS SCHMACKERI	COPEPOD
NEOMYSIS INTEGER	MYSID
NEOPERLA SP.	STONEFLY
NEPHROPS NORVEGICUS	NORWAY LOBSTER
NEZARA VIRIDULA	STINK BUG
NIPTUS HOLOLECUS	BEETLE
NOTONECTA GLAUCA	BACKSVIMMER
NOTONECTA SP.	BACKSVIMMER
NOTONECTIDAE	BACKSVIMMERS
OITHONA SIMILIS	CRUSTACEAN
ONYCHIURUS PROCAMPATUS	SPRINGTAIL
OPHIOGOMPHUS RUPINSULENSIS	DRAGONFLY
OPILIONES	HARVESTMEN OR HARVEST SPIDERS
OPLOPHORUS SP.	DECAPOD
ORCHELIMUM FIDICINIUM	MEADOW KATYDID
ORCHELIMUM SP.	KATYDID
ORCONECTES NAIS	CRAYFISH
ORCONECTES VIRILIS	CRAYFISH
ORTHOPTERA	GRASSHOPPERS
OTOCRYPTOPS SEXSPINOSUS	CENTIPEDE
PACHYLOMERIDES AUDOUINI	TRAPDOOR SPIDER
PALAEMON DEBILIS	SHRIMP
PALAEMONETES KADIAKENSIS	GRASS SHRIMP
PALAEMONETES PUGIO	GRASS SHRIMP
PALAEMONETES VULGARIS	GRASS SHRIMP
PALLASEA QUADRISPINOSA	AMPHIPOD
PANDALUS MONTAGUI	CRUSTACEAN
PANULIRUS LAEVICAUDA	ROCK LOBSTER
PANULIRUS LONGIPES CYGNUS	ROCK LOBSTER
PARACHIRONOMUS TENER	MIDGE
PARACYCLOPS FIMBRIATUS POPPEI	COPEPOD
PARAPENAEUS LONGIROSTRIS	SHRIMP
PARATHEMISTO PACIFICA	AMPHIPOD
PARCOBLATTA SP.	COCKROACH
PARHYALLA WHELPLEYI	AMPHIPOD
PARONYCHOCAMPTUS NANUS	COPEPOD
PEDICULUS HUMANUS	HEAD AND BODY LOUSE
PENAEUS AZTECUS	BROWN SHRIMP
PENAEUS DUORARUM	SHRIMP
PENAEUS JAPONICUS	SHRIMP
PERLODES INTRICATA	STONEFLY
PHILAENUS LEUCOPTHALMUS	CICADA
PHILAENUS SPUMARIUS	CICADA
PHRONIMA SEDENTARIA	AMPHIPOD
PHRYGANEA SP.	CADDISFLY
PHYLLOPERTHA HORTICOLA	BEETLE
PIMPLA INSTIGATOR	ICHNEUMON FLY
PLATYNOTHRUS PELTIFER	MITE
PLATYPHYLAX YOKOUCHII	CADDISFLY
PLECTROCNEMIA CONSPERSA	CADDISFLY
PLEURONCODES PLANIPES	CRUSTACEAN
PLEUROXUS DENTICULATUS	WATER FLEA
PLEUROXUS PROCURVIS	WATER FLEA
POGONOMYRMEX BADIUS	MITE
POLLICIPES POLYMERUS	GOOSENECK BARNACLE
POLYARTEMIELLA HAZENI	CRUSTACEAN
POLYCENTROPUS SP.	CADDISFLY
POLYDESMUS COMPLANATUS	MILLIPEDE
POLYPHEMUS PEDICULUS	COPEPOD
PONTOPOREIA AFFINIS	AMPHIPOD
POPILIUS DISJUNCTUS	BEETLE
PORTUNAS PUBER	CRUSTACEAN
POTAMANTHUS KAMONIS	MAYFLY
POTAMOPHYLAX CINGULATUS	CADDISFLY
PROCAMBARUS CLARKI	CRAYFISH
PROCAMBARUS SP.	CRAYFISH
PROCLADIUS CHOREUS	MIDGE
PROCLADIUS PECTINATUS	MIDGE
PROCLADIUS RUFOVITTATUS	MIDGE
PROCLADIUS SP.	MIDGE
PROTOTHERMES GRANDIS	NEUROPTERAN
PSEUDOCALANUS MINUTUS	COPEPOD
PTEROMALUS PUPARUM	ICHNEUMON FLY
PTERONARCELLA BADIA	STONEFLY
PTERONARCYS CALIFORNICA	STONEFLY
PTERONARCYS DORSATA	STONEFLY
PTERONEMOBIUS FASCIATUS	CRICKET

PTINUS FUR	SPIDER-BEETLE
PTINUS SEXPUNCTATUS	SPIDER-BEETLE
PTINUS TECTUS	SPIDER-BEETLE
PTYOIULUS IMPRESSUS	MILLIPEDE
PYCNOPSYCHE GUTTIFER	CADDISFLY
PYCNOPSYCHE LEPIDO	CADDISFLY
PYRRHOSOMA NYMPHULA	DRAGONFLY
RHINCALANUS GIGAS	COPEPOD
RHINCALANUS NASUTUS	COPEPOD
RHITHROGENA SEMICOLORATA	MAYFLY
RHIZOGLYPHUS ECHINOPUS	MITE
RHYACOPHILA BREVICEPHALA	CADDISFLY
RHYACOPHILA FASCIATA	CADDISFLY
RHYACOPHILA NIGROCEPHALA	CADDISFLY
SARCOPHAGA BULLATA	FLESH FLY
SCHISTOCERCA SP.	LOCUST
SCHIZOCOSA SP.	SPIDER
SCIOMYZIDAE	DIPTERANS
SCYTONOTUS GRANULATUS	MILLIPEDE
SIALIS MITSUHASHII	ALDERFLY
SIALIS SP.	ALDERFLY
SIEBOLDIUS JAPONICA	DRAGONFLY
SIGARA SP.	WATER BOATMAN
SIMOCEPHALUS SERRULATUS	WATER FLEA
SIMOCEPHALUS VETULUS	WATER FLEA
SIMULIIDAE	BLACKFLIES
SIMULIUM BREVICAULE	BLACKFLY
SIMULIUM MONTICOLA	BLACKFLY
SIMULIUM ORNATUM	BLACKFLY
SIMULIUM SP.	BLACKFLY
SIMULIUM VITTATUM	BLACKFLY
SPHAERODERUS STENOSTOMUS	CARABID BEETLE
SPIRONTOCARIS SP.	SHRIMP
STENOPERLA PRASINA	STONEFLY
STENOPSYCHE GRISEIPENNIS	CADDISFLY
STENOPSYCHE SP.	CADDISFLY
STETHOMEZIUM SQUAMOSUM	BEETLE
STICTOCHIRONOMUS SP.	MIDGE
SYSTELLASPIS DEBILIS	DECAPOD
TAENIOPTERYX MAURA	STONEFLY
TANYTARSINI	MIDGES
TANYTARSUS HOLOCHLORUS	MIDGE
TANYTARSUS INOPERTUS	MIDGE
TANYTARSUS LUGENS	MIDGE
TEMORA LONGICORNIS	COPEPOD
TEMORA STYLIFERA	COPEPOD
TENDIPEDIDAE	DIPTERANS
TENEBRIO MOLITOR	YELLOW MEALWORM BEETLE
TETRACLITA SP.	BARNACLE
THERMOCYCLOPS HYALINUS	COPEPOD
THERMOCYCLOPS NEGLECTUS	COPEPOD
TIGRIOPUS BREVICORNIS	CRUSTACEAN
TIPULA SP.	DADDY LONG LEG OR CRANEFLY
TOMOCERUS SP.	SPRINGTAIL
TORTANUS DISCAUDATUS	COPEPOD
TRIBOLIUM CASTANEUM	RED FLOUR BEETLE
TRICHOPTERA	CADDISFLIES
TRICORYTHODES SP.	MAYFLY
TRIGONOGENIUS GLOBULUS	BEETLE
TRIGONOTYLUS SP.	BUG
TROPOCYCLOPS PRASINUS	COPEPOD
TYROGLYPHUS INTERNA	MITE
TYROGLYPHUS LINTNERI	MITE
UCA ININAX	FIDDLER CRAB
UCA SP.	FIDDLER CRAB
UNCIOLA LEUCOPIS	CRUSTACEAN
ZYGOPTERA	DAMSELFLIES

BRACHIOPODA

TEREBRATALIA SP.
TEREBRATULINA SP.

BRYOZOA

ANTROPORA SP.
DIAPEROCCIA CALIFORNICA
MEMBRANIPORA SP.
PHIDOLOPORA SP.
SCRUPOCELLARIA SP.

CHAETOGNATHA

SAGITTA ELEGANS	
SAGITTA HISPIDA	
SAGITTA SETOSA	

COELENTERATA

ACONTIUM VELATUM	
AEQUOREA SP.	HYDROZOAN
AGALMI OKENI	HYDROZOAN
ASTRAEA UNDOSA	CORAL
ATHORYBIA ROSACEA	HYDROZOAN
ATHORYBIA SP.	HYDROZOAN
BALANOPHYLLIA SP.	
BATHYPHYSA SIBOGAE	HYDROZOAN
CHLOROHYDRA VIRIDISSIMA	
CUNINA SP.	HYDROZOAN
DIPHYES DISPAR	HYDROZOAN
FORSKALIA EDWARDSI	HYDROZOAN
FORSKALIA THOLOIDES	HYDROZOAN
FUNGIA SCUTARIA	CORAL
HYDRA LITTORALIS	HYDROZOAN
HYDROMEDUSAE	
LEUCKARTIARA OCTONA	HYDROZOAN
MURICEA SP.	CORAL
NANOMIA BIJUGA	HYDROZOAN
ORCHISTOMA SP.	HYDROZOAN
PAVONA CLIVOSA	CORAL
PAVONA GIGANTEA	CORAL
PENNATULA RUBRA	SEA PEN
PHIALIDIUM GREGARIUM	HYDROZOAN
POCILLOPORA DAMICORNIS	CORAL
POCILLOPORA DANAE	CORAL
POCILLOPORA ROBUSTA	CORAL
PORITES PANAMENSIS	STONY CORAL
PSAMMOCORA STELLATA	CORAL
RHIZOPHYSA FILIFORMIS	HYDROZOAN
ROSACEA CYMBIFORMIS	HYDROZOAN
SIPHONOPHORA	HYDROZOANS
STOMOTOCA ARTRA	
SULCULEOLARIA QUADRIVALVIS	HYDROZOAN

CTENOPHORA

CESTUM SP.	
EURHAMPHEA VEXILLIGERA	
PLEUROBRACHIA BACHEII	

ECHINODERMATA

AMPHIPHOLIS SQUAMATA	SERPENT STAR
AMPHIURA ANGULARIS	OPHIUROID
AMPHIURA ANNULIFERA	OPHIUROID
AMPHIURA BELGICAE	OPHIUROID
AMPHIURA FILIFORMIS	OPHIUROID
AMPHIURA MICROPLAX	OPHIUROID
AMPHIURA MONORIMA	OPHIUROID
ASTERINA GIBBOSA	SEA STAR
ASTROCHLAMYS BRUNEUS	OPHIUROID
ASTROMETIS SERTULIFERA	SEA STAR
ASTROPECTEN ARANCIACUS	SEA STAR
ASTROPECTEN ARMATUS	SEA STAR
BRISSOPSIS LYRIFERA	SEA URCHIN
DERMASTERIAS IMBRICATA	SEA STAR
HENRICIA LEVIUSCULA	SEA STAR
HOLOTHURIA FORSKALI	SEA CUCUMBER
LINCKIA COLUMBIAE	SEA STAR
LOVENIA CORDIFORMIS	HEART URCHIN
LOXECHINUS ALBUS	SEA URCHIN
LYTECHINUS ANAMESUS	SEA URCHIN
OPHIACANTHA DENSISPINA	OPHIUROID
OPHIACANTHA VIVIPARA	OPHIUROID
OPHIOCERES INCIPIENS	OPHIUROID
OPHIOCHONDRUS STELLIGER	OPHIUROID
OPHIOCOMA ECHINATA	BRITTLE STAR
OPHIOCOMA ERINACEUS	BRITTLE STAR
OPHIOLEBELLA BISCUTIFERA	OPHIUROID
OPHIOMITRELLA CLAVIGERA	OPHIUROID
OPHIOMITRELLA FALKLANDICA	OPHIUROID
OPHIOMITRELLA INGRATA	OPHIUROID
OPHIOMYXA BREVIRIMA	OPHIUROID
OPHIONOTUS HEXACTIS	SERPENT STAR
OPHIOPHOLIS ACULEATA	OPHIUROID
OPHIOPLOCUS ESMARKI	OPHIUROID
OPHIOSCOLEX NUTRIX	OPHIUROID
OPHIOSTEIRA ANTARCTICA	OPHIUROID
OPHIOZONELLA FALKLANDI	OPHIUROID
OPHIURA MERIDIONALIS	OPHIUROID
OPHIURA ROUCHI	OPHIUROID
OPHIUROIDEA	BRITTLE SERPENT AND BASKET STARS
OPHIUROLEPIS MARTENSI	OPHIUROID
PARACENTROTUS LIVIDUS	SEA URCHIN
PATIRIA MINIATA	SEA STAR
PECTINURA CYLINDRICA	OPHIUROID
PECTINURA GRACILIS	OPHIUROID
PISASTER SP.	SEA STAR
PORANIA PULVILLUS	SEA STAR
STEGOPHIURA NODOSA	OPHIUROID
STEGOPHIURA SCULPTA	OPHIUROID
STRONGYLOCENTROTUS FRANCISCANUS	SEA URCHIN
STRONGYLOCENTROTUS PURPURATUS	SEA URCHIN

MOLLUSCA

ACANTHINA SPIRATA	MARINE SNAIL
ACMAEA SP.	MARINE LIMPET
AEQUIPECTEN IRRADIANS	SCALLOP
AGRIOLIMAX RETICULATUS	LAND SLUG
ALETES SQUAMIGERUS	MARINE SNAIL
ANCYLUS LACUSTRIS	FRESHWATER LIMPET
ANODONTA ANATINA	UNIONID FRESHWATER CLAM
ANODONTA NUTTALLIANA	UNIONID FRESHWATER CLAM
ANODONTA PISCINALIS	UNIONID FRESHWATER CLAM
APLYSIA SP.	SEA HARE
ARCTICA ISLANDICA	MARINE BIVALVE
ARION SP.	LAND SLUG
ASTARTE UNDATA	BIVALVE
BELLAMYA UNICOLOR	BIVALVE
BIOMPHALARIA GLABRATA	PULMONATE SNAIL
BITHYNIA TENTACULATA	PROSOBRANCH SNAIL
BRACHYDONTES DEMISSUS PLICATULUS	
BRACHYDONTES RECURVUS	
BUCCINUM UNDATUM	CONCH
BURSA CALIFORNICUS	MARINE GASTROPOD
BUSYCON CANALICULATUS	WHELK
CAMPELOMA SP.	FRESHWATER SNAIL
CARDIUM EDULE	MARINE MUSSEL
CHAMA PELLUCIDA	BIVALVE
CHIONE UNDATELLA	MARINE BIVALVE
CHLAMYS OPERCULARIS	BIVALVE
CLEOPATRA BULIMOIDES	SNAIL
COLUS STIMPSONI	SNAIL
CONUS CALIFORNICUS	GASTROPOD
CORBICULA AFRICANA	CLAM
CORBICULA MANILENSIS	CLAM
CRASSOSTREA ANGULATA	OYSTER
CRASSOSTREA GIGAS	PACIFIC OYSTER
CRASSOSTREA GRYPHOIDES	OYSTER
CRASSOSTREA VIRGINICA	EASTERN OYSTER
CREPIDULA FORNICATA	SLIPPER SHELL
CREPIDULA SP.	SLIPPER SHELL
CYPRAEA SPADICEA	SNAIL
DENTALIUM ENTALE	TOOTH SHELL
DEROCERAS SP.	LAND SLUG
ENSIS DIRECTUS	RAZOR CLAM
ENSIS ENSIS	RAZOR CLAM
EPITONIUM SP.	MARINE GASTROPOD
FISSURELLA VOLCANO	KEYHOLE LIMPET
GLEBA CORDATA	OPISTHOBRANCH GASTROPOD
GONIDEA SP.	UNIONID MUSSEL
HALIOTIS DISCUS	ABALONE
HALIOTIS RUFESCENS	ABALONE
HELISOMA SP.	PULMONATE SNAIL
HIATELLA ARCTICA	CLAM
KELLETTIA KELLETTII	GASTROPOD
LAMPSILIS RADIATA	FRESHWATER CLAM
LIGUMIA SP.	BIVALVE
LIMAX SP.	LAND SLUG
LITHOPHAGA PLUMULA	MARINE MUSSEL
LITTORINA IRRORATA	PERIWINKLE

LITTORINA LITTORALIS	PERIWINKLE
LITTORINA LITTOREA	COMMON PERIWINKLE
LITTORINA OBTUSATA	NORTHERN YELLOW PERIWINKLE
LITTORINA SAXATILIS	ROUGH PERIWINKLE
LITTORINA SP.	PERIWINKLE
LOTTIA GIGANTEA	OWL LIMPET
LYMNAEA PEREGER	POND SNAIL
LYMNAEA STAGNALIS	POND SNAIL
MACOMA BALTICA	MARINE MUSSEL
MARGARITIFERA SP.	PEARL MUSSEL
MEGATHURA CRENULATA	KEYHOLE LIMPET
MELAMPUS BIDENATUS	MARSH SNAIL
MELANIA TUBERCULATA	FRESHWATER SNAIL
MERCENARIA MERCENARIA	QUAHOG
MERETRIX LUSORIA	BIVALVE
MESODON THYROIDES	LAND SNAIL
MESOMPHIX VULGATUS	GASTROPOD
MODIOLUS DEMISSUS	MUSSEL
MODIOLUS MODIOLUS	MUSSEL
MULINIA LATERALIS	BIVALVE
MUSSEL	
MYTILUS CALIFORNIANUS	MARINE MUSSEL
MYTILUS EDULIS	EDIBLE MUSSEL
MYTILUS GALLOPROVINCIALIS	MARINE MUSSEL
MYTILUS SP.	MARINE MUSSEL
NASSARIUS OBSOLETUS	MUD SNAIL
NUCELLA LAPILLUS	DOG WHELK
NUCULA PROXIMA	NUT SHELL
NUCULA SP.	NUT SHELL
NUCULA SULCATA	NUT SHELL
NUCULANA SP.	BIVALVE
NUTTALLINA CALIFORNICA	CHITON
OCTOPUS VULGARIS	SEA SQUID
OSTREA EDULIS	EDIBLE OYSTER
OSTREA EQUESTRIS	OYSTER
OSTREA LURIDA	OLYMPIA OYSTER
PATELLA VULGATA	LIMPET
PECTEN IRRADIANS	SCALLOP
PENITELLA CONRADI	BIVALVE
PHYSA GYRINA	PULMONATE SNAIL
PHYSA PSEUDOLIMNAEUS	PULMONATE SNAIL
PHYSA SP.	PULMONATE SNAIL
PISIDIUM CASERTANUM	SEED CLAM
PISIDIUM VIRGINICUM	SEED CLAM
PLANORBIDAE	RAMSHORN SNAILS
PLANORBIS CONTORTUS	RAMSHORN SNAIL
PLANORBIS VORTEX	RAMSHORN SNAIL
PODODESMUS PERNOIDES	BIVALVE
PROTOTHACA STAMINEA	LITTLE-NECKED CLAM
PTEROPOD	SEA BUTTERFLY
PTEROTRACHEA HIPPOCAMPUS	GASTROPOD
RADIX PEREGRA OVATA	SNAIL
RANGIA CUNEATA	MARSH CLAM
SAXIDONUS GIGANTEUS	BIVALVE
SCROBICULARIA PLANA	MUSSEL
SEA SQUID	
SEMELE SP.	MARINE BIVALVE
SEPIA OFFICINALIS	CUTTLEFISH
SPHAERIUM CORNEUM	BIVALVE
SUCCINEA OVALIS	SNAIL
TACHIDIUS DISCIPEA	BIVALVE
TEGULA FUNEBRALIS	BLACK TURBAN SNAIL
UNIO MANCUS	UNIONID FRESHWATER CLAM
UNIO PICTORUM	UNIONID FRESHWATER CLAM
UNIO TUMIDUS	UNIONID FRESHWATER CLAM
VENERICARDIA BOREALIS	BIVALVE
VENERUPIS DECUSSATA	BIVALVE
VENUS MERCENARIA	MARINE MUSSEL
VIVIPARUS VIVIPARUS	PROSOBRANCH SNAIL
YOLDIA SAPOTILLA	MARINE BIVALVE

NEMATODA

RHABDITIS MARINA

PLATYHELMINTHES

BIPALIUM KEWENSE	
CRENOBIA ALPINA	PLANARIA
DUGESIA GONOCEPHALA	PLANARIA
DUGESIA TIGRINA	PLANARIA

PROTOZOA

ALLOGROMIA LATICOLLARIS	
FORAMINIFERA	
KERONOPSIS RUBRA	CILIATE
PARAMECIUM SP.	CILIATE
ROSALINA LEEI	FORAMINIFERAN
SPIROLOCULINA HYALINA	FORAMINIFERAN
TETRAHYMENA PYRIFORMIS	

ROTIFERA

ASCOMORPHA SP.
ASPLANCHNA GIRODI
ASPLANCHNA PRIODONTA
ASPLANCHNA SP.
BRACHIONUS CALYCIFLORUS
BRACHIONUS PLICATILIS
BRACHIONUS SP.
EUCHLANIS DILATATA
FILINIA LONGISETA
GASTROPUS SP.
HEXARTHRA FENNICA
KELLICOTTIA BOSTONIENSIS
KELLICOTTIA LONGISPINA
KERATELLA COCHLEARIS
KERATELLA HIEMALIS
KERATELLA QUADRATA
KERATELLA VALGA
PHILODINA ROSEOLA
PHILODINA SP.
PLOEOSOMA HUDSONI
PLOEOSOMA TRUNCATUM
POLYARTHRA SP.
POLYARTHRA VULGARIS
SYNCHAETA PECTINATA
SYNCHAETA SP.

UROCHORDATA

ASCIDIACEA	SEA SQUIRTS
ASCIDIELLA ASPERSA	SEA SQUIRT
CIONA INTESTINALIS	SEA SQUIRT
CORELLA SPECTABILIS	SEA SQUIRT
PEGEA CONFEDERATA	
PHALLUSIA MAMMILLATA	SEA SQUIRT
SALPA CYLINDRICA	THALIACEAN
SALPA FUSIFORMIS	THALIACEAN
STYELA PLICATA	SEA SQUIRT
THALIA DEMOCRATICA	THALIACEAN

VERTEBRATA

ABRAMIS BRAMA	BREAM
ACHIRUS LINEATUS	LINED SOLE
ADENOTA KOB THOMASI	UGANDA KOB
AIX SPONSA	CAROLINA WOOD DUCK
ALBURNUS ALBURNUS	BLEAK
ALOPEX LAGOPUS	ARCTIC FOX
ALOSA PSEUDOHARENGUS	ALEWIFE
AMBLOPLITES RUPESTRIS	ROCK BASS
AMEIURUS NEBULOSUS	BROWN BULLHEAD
AMIA SP.	BOWFIN
ANAS PLATYRHYNCHOS	MALLARD
ANAS RUBRIPES	BLACK DUCK
ANCHOA MITCHILLI	BAY ANCHOVY
ANGUILLA ANGUILLA	EUROPEAN EEL
ANGUILLA JAPONICA	EEL
ANGUILLA SP.	EEL
APLODINOTUS GRUNNIENS	DRUM
APODEMUS AGRARIUS	STRIPED FIELD-MOUSE
APODEMUS FLAVICOLLIS	YELLOW NECKED FIELD MOUSE
ARCHOSARGUS RHOMBOIDALIS	SEA BREAM
ASPIUS ASPIUS	ASPEN
ASPIUS RAPAX	FISH
BALAENOPTERA ACUTOROSTRATA	MINKE WHALE
BALAENOPTERA BOREALIS	SEI WHALE
BALAENOPTERA BRYDEI	BRYDE WHALE

BALAENOPTERA MUSCULUS	BLUE WHALE
BALAENOPTERA MUSCULUS BREVICAUDA	PIGMY BLUE WHALE
BALAENOPTERA PHYSALUS	FIN WHALE
BLENNIUS PHOLIS	SHANNY
BONASA UMBELLUS	RUFFED GROUSE
BOOPS BOOPS	BOQUE
BRACHYDANIO RERIO	ZEBRA FISH
BREVOORTIA SP.	MENHADEN
BREVOORTIA TYRANNUS	ATLANTIC MENHADEN
BUFO BUFO	COMMON TOAD
CALANDRA ORYZAE	LARK
CALIDRIS ALPINA	DUNLIN
CALIDRIS CANUTUS	KNOT
CANIS FAMILIARIS	DOG
CANIS LATRANS	COYOTE
CANIS LUPUS	WOLF
CAPREOLUS CAPREOLUS	ROE DEER
CARASSIUS AURATUS	GOLDFISH
CARASSIUS CARASSIUS	CRUCIAN CARP
CARCHARHINUS OBSURUS	DUSKY SHARK
CARPIODES CYPRINUS	QUILLBACK CARPSUCKER
CARPIODES SP.	CARPSUCKER
CARPIODES VELIFER	CARPSUCKER
CASMERODIUS ALBUS	COMMON EGRET
CATOSTOMUS COMMERSONI	WHITE SUCKER
CAVIA PORCELLUS	GUINEA PIG
CHELA BACAILA	FISH
CHONDROSTOMA NASUS PRESPENSE	HOTU
CHRYSOPHRYS MAJOR	RED SEA BREAM
CIRCUS HUDSONIUS	MARSH HAWK
CLETHRIONOMYS GLAREOLUS	BANK VOLE
CLETHRIONOMYS RUTILUS	RUDDY VOLE
CLETHRIONOMYS RUTILUS DAWSONI	RED-BACKED VOLE
CLUPEA HARENGUS	HERRING
CLUPEA SP.	HERRING
COLINUS VIRGINIANUS	BOBWHITE QUAIL
COLOLABIS SAIRA	SAURY
COLOMBA LIVIA	ROCK-DOVE
COREGONUS ALBULA TYPICA	WHITE FISH
COREGONUS SP.	WHITE FISH
COTTUS CAROLINAE	BULLHEAD
COTTUS COGNATUS	BULLHEAD
COTURNIX COTURNIX	COTURNIX QUAIL
COTURNIX COTURNIX JAPONICA	JAPANESE QUAIL
CTENOPHARYNGODON IDELLA	GRASS CARP
CTENOPHARYNGODON SP.	GRASS CARP
CYANOCITTA CRISTATA	BLUE JAY
CYANOCITTA STELLERI	STELLER'S JAY
CYPRINODON VARIEGATUS	SHEEPSHEAD MINNOW
CYPRINUS CARPIO	CARP
DIEMICTYLUS SP.	NEWT
DIPODOMYS MERRIAMI	KANGUROO RAT
DOROSOMA CEPEDIANUM	GIZZARD SHAD
ELEGINOPS MACLOVINUS	PATAGONIAN BLENNY
ELOPS SAURUS	TENPOUNDER
ENGRAULIS JAPONICUS	ANCHOVY
ENGRAULIS RINGENS	ANCHOVY
ENTEROLOBIUM SAMAN	BOVINE
EOPSETTA JORDANI	FLOUNDER
EQUUS CABALLUS	HORSE
ESCHRICHTIUS ROBUSTUS	GREY WHALE
ESOX LUCIUS	PIKE
ESOX MASQUINONGY	MUSKELLUNG
ETHEOSTOMA SPECTABILE	ORANGE-THROAT DARTER
EUBALAENA GLACIALIS SIEBOLDII	PACIFIC RIGHT WHALE
FALCO MEXICANUS	PRAIRIE FALCON
FALCO SPARVERIUS	AMERICAN KESTREL OR SPARROW HAWK
FELIS SILVESTRIS	CAT
FUNDULUS HETEROCLITUS	MUMMICHOG (KILLIFISH)
FUNDULUS MAJALIS	STRIPED KILLIFISH
GADUS MACROCEPHALUS	PACIFIC COD
GADUS MORHUA	ATLANTIC COD
GADUS SP.	COD
GALLUS DOMESTICUS	CHICKEN
GAMBUSIA AFFINIS	MOSQUITO FISH
GENYPTERUS CAPENSIS	KING KLIP
GILLICHTHYS MIRABILIS	GOLDFISH
GLAUCOMYS SABRINUS YUKONENSIS	FLYING SQUIRREL
GOBIO GOBIO	GUDGEON
GOBIUS SP.	GOBY
HALIAETUS LEUCOCEPHALUS	BALD EAGLE OR WHITE HEADED EAGLE

HALICHOERUS GRYPUS	GREY SEAL
HETEROPNEUSTES FOSSILIS	CATFISH
HIPPOGLOSSUS SP.	HALIBUT
HOMO SAPIENS	HUMAN BEING
HYLOCICHLA MUSTELLINA	THRUSH
HYPENTELIUM SP.	HOG SUCKER
ICTALURIDAE	CATFISHES
ICTALURUS LACUSTRIS	CATFISH
ICTALURUS PUNCTATUS	CHANNEL CATFISH
ISURUS OXYRINCHUS	MAKO SHARK
JORDANELLA FLORIDAE	FLAGFISH
KATSUWONUS PELAMIS	SKIPJACK TUNA
KOBUS LECHE	WATERBUCK
LABEO ROHITA	ROHU
LARUS ARGENTATUS	HERRING GULL
LEBISTES RETICULATUS	GUPPY
LEIOSTOMUS SP.	SPOT
LEIOSTOMUS XANTHURUS	SPOT
LEPIDOPSETTA BILINEATA	ROCK SOLE
LEPOMIS CYANELLUS	GREEN SUNFISH
LEPOMIS MACROCHIRUS	BLUEGILL SUNFISH
LEPUS CALIFORNICUS	BLACK-TAILED JACKRABBIT
LEPUS EUROPAEUS	EUROPEAN HARE
LEUCISCUS CEPHALUS	CHUB
LEUCISCUS IDUS	IDE OR SILVER ORFE
LEUCISCUS LEUCISCUS	DACE
LEUCISCUS RUTILUS	
LIMANDA LIMANDA	DAB
LONCHURA STRIATA	BENGALESE FINCH
LOXODONTA AFRICANA	AFRICAN ELEPHANT
LUCIOPERCA LUCIOPERCA	ZANDER
LUTJANIDAE	SNAPPERS
MEGAPTERA NOVAEANGLIAE	HUMBACK WHALE
MELANOGRAMMUS AEGLEFINUS	HADDOCK
MELEAGRIS GALLOPAVO	COMMON TURKEY
MENIDIA MENIDIA	ATLANTIC SILVERSIDE
MENIDIA SP.	SILVERSIDE
MERLUCCIUS ANGUSTIMANUS	HAKE
MERLUCCIUS PRODUCTUS	HAKE
MERLUCCIUS SP.	HAKE
MICROPOGON UNDULATUS	ATLANTIC CROAKER
MICROPTERUS COOSAE	BASS
MICROPTERUS DOLOMIEUI	SMALLMOUTH BASS
MICROPTERUS SALMOIDES	LARGEMOUTH BASS
MICROTUS AGRESTIS	SHORT TAILED VOLE OR FIELD VOLE
MICROTUS ARVALIS	COMMON VOLE
MICROTUS OECONOMUS	TUNDRA VOLE
MICROTUS OECONOMUS MACFARLANI	VOLE
MICROTUS ORCADENSIS	VOLE
MICROTUS PENNSYLVANICUS	MEADOW VOLE
MICROTUS SP.	VOLE
MOLA MOLA	OCEAN SUNFISH
MOLOTHRUS SP.	COWBIRD
MORONE AMERICANAS	BASS
MOXOSTOMA SP.	REDHORSE (FISH)
MUGIL CAPITO	THIN-LIPPED GREY MULLET
MUGIL CEPHALUS	GREY MULLET
MUGIL CHELO	THICK LIPPED GREY MULLET
MULLUS BARBATUS	MULLET
MULLUS SURMULETUS	RED MULLET
MUS MUSCULUS	HOUSE MOUSE
MUSTELA RIXOSA ALLEGHNIENSI	LEAST WEASEL
NOEMACHEILUS BARBATULUS	STONE LOACH
NOTROPIS CORNUTUS	COMMON SHINER
OCHOTONA SP.	PIKA OR PIPING HARE
ODOCOILEUS HEMIONUS COLUMBIANUS	BLACK-TAILED DEER
ODOCOILEUS SP.	DEER
ODOCOILEUS VIRGINIANUS	WHITE-TAILED DEER
ONCORHYNCHUS GORBUSCHA	PINK SALMON
ONCORHYNCHUS KETA	CHUM SALMON
ONCORHYNCHUS KISUTCH	COHO SALMON
ONCORHYNCHUS NERKA	SOCKEYE SALMON
ONCORHYNCHUS TSCHAWYTSCHA	CHINOOK SALMON
OPSANUS TAU	OYSTER TOADFISH
ORYZIAS LATIPES	JAPANESE MEDAKA
ORYZOMYS PALUSTRIS	MARSH RICE RAT
PAGOPHILUS GROENLANDICUS	HARP SEAL
PANDION HALIAETUS	FISH HAWK OR OSPREY
PANTERPE INSIGNIS	HUMMINGBIRD
PARALICHTHYS DENTATUS	SUMMER FLOUNDER
PARALICHTHYS OLIVACEUS	PLAICE
PARULA AMERICANA	NORTHERN PARULA
PASSER DOMESTICUS	HOUSE SPARROW
PASSERCULUS SANDWICHENSIS	SAVANNA SPARROW
PASSERIFORMES	PASSERINE BIRDS

PEPRILUS TRICANTHUS	FISH
PERCA FLAVESCENS	YELLOW PERCH
PERCA FLUVIATILIS	PERCH
PERCIDAE	PERCHES
PERDIX PERDIX	PARTRIDGE
PEROMYSCUS POLIONOTUS	DEER MOUSE
PHALACROCORAX CARBO	CORMORANT
PHOCA HISPIDA = PUSA HISPIDA	RINGED SEAL
PHOXINUS PHOXINUS	MINNOW
PHYSETER CATODON	SPERM WHALE
PIMELODUS MACULATUS	FISH
PIMEPHALES PROMELAS	FATHEAD MINNOW
PIRANGA OLIVACEA	SCARLET TANAGER
PITYMYS SUBTERRANEUS	PINE VOLE
PLATICHTHYS STELLATUS	STARRY FLOUNDER
PLEURONECTES FLESUS	FLOUNDER
PLEURONECTES PLATESSA	PLAICE
PLEURONECTIDAE	FLATFISHES
POECILIA RETICULATUS	GUPPY
POLLACHIUS POLLACHIUS	POLLOCK
POLLACHIUS VIRENS	COALFISH
POMATOMUS SALTATRIX	BLUEFISH
POMOXIS ANNULARIS	CRAPPIE
POMOXIS SP.	CRAPPIE
PREONACE GLAUCA	BLUE SHARK
PROSOPIUM WILLIAMSONI	MOUNTAIN WHITEFISH
PSEUDEMYS SP.	TURTLE
PSEUDOPLEURONECTES AMERICANUS	WINTHER FLOUNDER
PSEUDOTOLITHUS SENEGALENSIS	SCIAENIDAE (FISH)
PSEUDOTOLITHUS TYPUS	SCIAENIDAE (FISH)
PTYCHOCHEILUS OREGONENSIS	NORTHERN SQUAWFISH
PUSA HISPIDA = PHOCA HISPIDA	RINGED SEAL
RAJA CLAVATA	THORNBACK SKATE
RALLUS LONGIROSTRIS	RAIL
RANA ARVALIS	FROG
RANA PIPIENS	LEOPARD FROG
RANA TEMPORARIA	COMMON FROG
RANGIFER TARANDUS	CARIBOU OR REINDEER
RATTUS NORVEGICUS	NORWAY RAT
RICHMODENA CARDINALIS	COMMON CARDINAL
ROCCUS AMERICANUS	WHITE PERCH OR SILVER BASS
RODENTIA	RODENTS
RUTILUS RUTILUS	ROACH
RUTILUS SP.	ROACH
RYNCHOPS NIGRA NIGRA	BLACK SKIMMER
SAIMIRI SCIUREUS	SQUIRREL MONKEY
SALMO CLARKI	CUT-THROAT TROUT
SALMO GAIRDNERII	RAINBOW TROUT OR STEELHEAD TROUT
SALMO SALAR	ATLANTIC SALMON
SALMO TRUTTA	BROWN TROUT
SALMONIDAE	SALMONOIDS
SALVELINUS FONTINALIS	BROOK TROUT
SALVELINUS MALMA	BULLHEAD TROUT
SALVELINUS NAMAYCUSH	LAKE TROUT
SARDINA PILCHARDUS	PILCHARD
SARDINELLA AURITA	FISH
SARDINELLA MADERENSIS	FISH
SCOMBER JAPONICUS	MACKEREL
SCOMBER SCOMBRUS	ATLANTIC MACKEREL
SCOMBER SP.	MACKEREL
SCOMBEROMORUS CAVALLA	MACKEREL
SCOMBEROMORUS MACULATUS	SPANISH MACKEREL
SCORPAENA PORCUS	SMALL SCALED SCORPION FISH
SEBASTODES ALUTUS	PACIFIC OCEAN PERCH
SERIOLA QUINQUERADIATA	YELLOW-TAIL
SERRANUS ALEXANDRINUS	PERCH
SIGANUS LURIDUS	SPINEFOOT (RABBITFISH)
SIGMODON HISPIDUS	COTTON RAT
SILLAGINIDAE	WHITINGS
SOLEA SOLEA	SOLE
SOLEA SP.	SOLE
SOREX ARANEUS	COMMON SHREW
SOREX MINUTUS	PIGMY SHREW
SPERMOPHILUS UNDULATUS	ARCTIC GROUND SQUIRREL
SPHYRAENA SP.	BARRACUDA
SPIZELLA PASSERINA	SPARROW
STERNA HIRUNDO	COMMON TERN
STERNA SP.	TERN
STIZOSTEDION LUCIPERCA	WALLEYE
STIZOSTEDION VITREUM VITREUM	WALLEYE
STREPTOPELIA RISORIA	BENGALESE FINCH
SYLVILAGUS FLORIDANUS	COTTONTAIL RABBIT
TAMIASCIURUS HUDSONI PREBLEI	RED SQUIRREL
TAMIASCIURUS SP.	SQUIRREL
TAUTOGA SP.	TAUTOG
TAUTOGOLABRUS ADSPERSUS	CUNNER
TELMATODYTES PALUSTRIS	LONG-BILLED MARSH WREN
THALARCTOS MARITIMUS	POLAR BEAR
THALEICHTHYS PACIFICUS	CANDLEFISH
THUNNUS ALBACORES	YELLOW-FIN TUNA
THUNNUS SP.	TUNA
THUNNUS THYNNUS	BLUE-FIN TUNA
THYMALLUS ARCTICUS BAICALENSIS	GRAYLING
THYMALLUS THYMALLUS	GRAYLING
TILAPIA MELANOPLEURA	WHITE TILAPIA
TILAPIA MOSSAMBICA	FISH
TILAPIA ZILLII	FISH
TINCA TINCA	GOLDEN TENCH
TINCA VULGARIS	TENCH
TRACHURUS JAPONICUS	HORSE MACKEREL
TRACHURUS TRACHURUS	SILVERY SIDED HORSE MACKEREL
TRINECTES MACULATUS	FISH
TRINGA TOTANUS	REDSHANK
UPENEUS MOLUCCENSIS	GOATFISH
VULPES VULPES	RED FOX
ZALOPHUS CALIFORNIANUS	SEA LION
ZENAIDURA MACROURA	MOURNING DOVE

****

P L A N T S

****

KORMOPHYTA

ABIES BALSAMEA	BALSAM FIR
ABIES GRANDIS	GRAND FIR
ABIES MARIESII (*)	FIR
ABIES SACHALINENSIS	FIR
ABIES VEITCHII (*)	FIR
ACER CAMPESTRIS	HEDGE MAPLE
ACER PENNSYLVANICUM	PENNSYLVANIAN MAPLE
ACER RUBRUM	RED MAPLE
ACER SACCHARUM	SUGAR MAPLE
ACER SP.	MAPLE
ADENOSTOMA FASCICULATUM	CHAMISE
AESCULUS OCTANDRA	HORSE CHESTNUT
AGATHIS AUSTRALIS	KAURI PINE
ALISMA PLANTAGO	
ALISMA PLANTAGO AQUATICA	WATER PLANTAIN
ALLIUM CEPA	ONION, GARDEN
ALLIUM PORRUM	LEEK
ALNUS RUBRA	RED ALDER
AMARANTHUS EDULIS	GIANT PIGWEED
AMBLYSTEGIUM RIPARIUM	MOSS
ANANAS COMOSUS	PINEAPPLE
ANGIOSPERMAE	ANGIOSPERMS
APIUM GRAVEOLENS	CELERY
ARACHIS HYPOGAEA	PEANUT
ARCTOSTAPHYLOS UVA-URSI	BEARBERRY
ARISTIDA ARGENTEA	GRASS
ARTEMISIA HERBA-ALBA	
AVENA SATIVA	OAT
AVENA STERILIS	
AVICENNIA TOMENTOSA	MANGROVE
AZOLLA SP.	AQUATIC FERN
BETA SACCHARIFERA	SUGAR BEET
BETA VULGARIS	COMMON BEET
BETULA ALLEGHENIENSIS	YELLOW BIRCH
BETULA ERMANII	ERMAN'S BIRCH
BETULA ERNAII (*)	BIRCH
BETULA MAXIMOWICZIANA	BIRCH
BETULA PAPYRIFERA	PAPER BIRCH OR CANOE BIRCH
BETULA PLATYPHYLLA (*)	BIRCH
BETULA SP.	BIRCH
BRASENIA SCHREBERI	WATERSHIELD
BRASSICA NAPUS VAR NAPO BRASSICA	RUTABAGA
BRASSICA OLERACEA	CABBAGE
BRASSICA OLERACEA VAR BOTRYTIS	CAULIFLOWER
BRASSICA OLERACEA VAR CAPITATA	CABBAGE
BRASSICA OLERACEA VAR GEMMIFERA	BRUSSEL'S SPROUT
BRASSICA OLERACEA VAR GONGYLODES	KOHL-RABI
BRASSICA OLERACEA VAR ITALICA	BROCCOLI
BRASSICA RAPA	TURNIP
BRYUM PSEUDOTRIQUETRUM	MOSS
BURKEA AFRICANA	LEGUME
CABOMBA CAROLINIANA	
CABOMBA SP.	
CAJANUS CAJAN	PIGEON PEA
CALOPOGONIUM MUCUNOIDES	LEGUME
CAMELLIA JAPONICA	COMMON CAMELLIA
CAMPTOTHECIUM LUTESCENS	MOSS
CANAVALIA ENSIFORMIS	JACK BEAN OR SWORD BEAN
CARYA SP.	
CASSIA MIMOSOIDES	SENNA
CASTANEA DENTATA	AMERICAN CHESTNUT
CASTANOPSIS CUSPIDATA	
CEANOTHUS CRASSIFOLIUS	
CENTROSEMA PUBESCENS	LEGUME
CERATOPHYLLUM DEMERSUM	RIGID HORNWORT
CERATOPHYLLUM SP.	HORNWORT
CERCIS CANADENSIS	JUDAS TREE
CHAMAECYPARIS OBTUSA	
CHAMAEDAPHNE CALYCULATA	
CHENOPODIUM AMBROSIOIDES	WORMWOOD
CICER ARIETINUM	GRAM CHICK-PEA
CICHORIUM ENDIVIA	ENDIVE
CLADIUM JAMAICENSIS	SAW GRASS
COCCOLOBA UVIFERA	SEA GRAPE
CONOCARPUS ERECTUS	BUTTON-WOOD
CORNUS FLORIDA	FLOWERING DOGWOOD
CRATAEGUS MONOGYNA	HAWTHORN
CUCUMIS SATIVUS	CUCUMBER
CYAMOPSIS TETRAGONOLOBUS	GUAR
CYMODOCEA SP.	TROPICAL SEA-GRASS
CYPERUS DIFFORMIS (*)	
CYPERUS FUSCUS	
DACTYLIS GLOMERATA	COCK'S FOOT
DAUCUS CAROTA	CARROT
DAUCUS CAROTA VAR SATIVA	CARROT
DESMODIUM CANUM	TICK CLOVER
DESMODIUM INTORTUM	TICK CLOVER
DICHAPETALUM CYMOSUM	
DICRANELLA VARIA	MOSS
DIGITARIA ADSENDENS (*)	GRASS
DIGITARIA ERIANTHA	GRASS
DIOSCOREA SP.	YAM
DOPATRIUM JUNCEUM (*)	
ECHINOCHLOA CRUSGALLI	BARNYARD GRASS
ECHINOCHLOA CRUSGALLI VAR FRUMENTACEA	JAPANESE MILLET
ECLIPTA PROSTRATA (*)	
EICHHORNIA CRASSIPES	
ELEOCHARIS ACICULARIS	NEEDLE SPIKE-RUSH
ELEOCHARIS CELLULOSA	SPIKE-RUSH
ELEOCHARIS QUADRANGULATA	SPIKE-RUSH
ELEPHANTORRHIZA SP.	LEGUME
ELODEA CANADENSIS	CANADIAN WATERWEED
ELODEA DENSA	WATERWEED
ELODEA SP.	WATErWEED
ENCALYPTA VULGARIS	MOSS
ENHALUS SP.	TROPICAL SEA-GRASS
EQUISETUM FLUVIATILE	WATER HORSETAIL
ERAGROSTIS PALLENS	GRASS
FAGUS CRENATA	BEECH
FAGUS GRANDIFOLIA	AMERICAN BEECH
FAGUS SP.	BEECH
FESTUCA ARUNDINACEA	FESCUE
FILICES	FERNS
FIMBRISTYLIS MILIACEA	
FONTINALIS SQUAMOSA	MOSS
FRAXINUS AMERICANA	AMERICAN ASH
FRAXINUS EXCELSIOR	EUROPEAN ASH
GLYCERIA MAXIMA	REED SWEET-GRASS
GLYCINE MAX	SOYBEAN
GNAPHALIUM AFFINE (*)	
GOSSYPIUM BARBADENSE	SEA ISLAND COTTON
GOSSYPIUM HIRSUTUM	UPLAND COTTON
GOSYPIUM SP.	COTTON
GRAMINEAE	GRASSES
GYMNOSPERMAE	GYMNOSPERMS
HALODULE SP.	TROPICAL SEA-GRASS
HALOPHILA DECIPIENS	TROPICAL SEA-GRASS
HALOPHILA MINOR	TROPICAL SEA-GRASS
HALOPHILA OVALIS	TROPICAL SEA-GRASS
HALOPHILA SPINULOSA	TROPICAL SEA-GRASS
HAMAMELIS VIRGINIANA	WITCH HAZEL
HAMMADA SCOPARIA	
HELIANTHUS ANNUUS	COMMON SUNFLOWER
HETERANTHERA DUBIA	
HIBISCUS PALUSTRIS	KETMIA
HORDEUM MURINUM	BARLEY
HORDEUM SATIVUM	BARLEY
HORDEUM SP.	BARLEY
HYDROCOTYLE SP.	
HYDROTRIDA CAROLINENSIS	
HYPARRHENIA RUFA	TUFTED PERENNIAL GRASS
HYPERTHELIA DISSOLUTA	
INDIGOFERA OXYTROPIS	LEGUME
INDIGOFERA SORDIDA	LEGUME
IPOMOEA BATATAS	SWEET POTATO
JUNCUS EFFUSUS	SOFT RUSH
JUNIPERUS VIRGINIANA	EASTERN RED CEDAR
LACTUCA SATIVA	LETTUCE
LACTUCA SATIVA VAR CRISPA	LEAF LETTUCE
LAGAROSIPHON MAJOR	
LAGUNCULARIA RACEMOSA	WHITE MANGROVE
LAPSANA APOGONOIDES (*)	
LARIX LEPTOLEPIS	JAPANESE LARCH
LARIX SP.	LARCH

LAURUS NOBILIS	GRECIAN LAUREL
LEMNA MINOR	COMMON DUCKWEED
LEMNA SP.	DUCKWEED
LEMNA TRISULCA	STAR DUCKWEED
LENS CULINARIS	COMMON LENTIL
LESPEDEZA SP.	
LEUCAENA GLAUCA	JUMPY BEAN
LINDERNIA PROCUMBENS	
LIRIODENDRON TULIPIFERA	TULIP TREE
LITTORELLA UNIFLORA	SHORE-WEED
LOLIUM MULTIFLORUM	ITALIAN RYEGRASS
LOLIUM PERENNE	PERENNIAL RYEGRASS
LOTONOSIS BAINESII	
LOTUS CORNICULATUS	BIRDSFOOT TREFOIL
LUDWIGIA PALUSTRIS	
LUPINUS SP.	LUPINE
LYCOPERSICON ESCULENTUM	TOMATO
LYSIMACHIA NUMMULARIA	MONEYWORT
MAGNOLIA ACUMINATA	MOUNTAIN MAGNOLIA
MALVA SYLVESTRIS	
MAZUS JAPONICUS (*)	
MEDICAGO HISPIDA	BUR CLOVER
MEDICAGO SATIVA	ALFALFA
MEDICAGO SP.	CLOVER
MELILOTUS SP.	SWEET CLOVER
MIKANIA CORDATA	
MITELLA POLYMERUS	
MYRIOPHYLLUM BRASILIENSE	WATER MILFOIL
MYRIOPHYLLUM HETEROPHYLLUM	WATER MILFOIL
MYRIOPHYLLUM SP.	WATER MILFOIL
NAJAS GUADALUPENSIS	
NAJAS SP.	
NELUMBO LUTEA	
NELUMBO SP.	
NUPHAR ADVENA	YELLOW POND LILY
NUPHAR LUTEUM	YELLOW WATER-LILY
NUPHAR POLYSEPALUM	WATER-LILY
NUPHAR PUMILUM	LEAST WATER-LILY
NUPHAR SP.	WATER-LILY
NYMPHAEA MARLIACEA	
NYMPHAEA MEXICANA	
NYMPHAEA ODORATA	
NYSSA SYLVATICA	BLACK TUPELO
OCHNA PULCHRA	
OCHROMA LAGOPUS	BALSA
ORONTIUM AQUATICUM	GOLDEN CLUB
ORYZA SATIVA	RICE
OSTRYA VIRGINIANA	AMERICAN HOP HORNBEAM
OXYDENDRUM ARBOREUM	SORREL TREE OR SOUR WOOD
PANICUM HEMITONIUM	
PARINARI CAPENSE	TREE
PETROSELINUM CRISPUM	PARSLEY
PHALARIS MINOR	GRASS
PHASEOLUS ATRAPURPUREA	BEAN
PHASEOLUS AUREUS	MUNG BEAN
PHASEOLUS VULGARIS	KIDNEY BEAN
PHILONOTIS FONTANA	MOSS
PHLEUM PRATENSE	TIMOTHY-GRASS
PHRAGMITES COMMUNIS	COMMON REED
PHYLLOSPADIX IWATENSIS	SURF-GRASS
PHYLLOSPADIX TORREYI	SURF-GRASS
PICEA ABIES	NORWAY SPRUCE
PICEA ENGELMANNI	ENGELMANN'S SPRUCE
PICEA RUBRA	RED PINE
PICEA SITCHENSIS	SITKA SPRUCE
PICEA SP.	SPRUCE
PINUS DENSIFLORA	PINE
PINUS ELLIOTTII	SLASH PINE
PINUS NIGRA	AUSTRIAN PINE
PINUS PONDEROSA	PONDEROSA PINE
PINUS RESINOSA	RED PINE
PINUS SEROTINA	POND PINE
PINUS STROBUS	EASTERN WHITE PINE
PINUS TAEDA	LOBLOLLY PINE
PINUS THUNBERGII	PINE
PINUS VIRGINIANA	VIRGINIA PINE
PISUM SATIVUM	GARDEN PEA
PISUM SP.	PEA
PLANTAGO AQUATICA	
PONTEDERIA CORDATA	PICKEREL WEED
PONTEDERIA LANCEOLATA	
POPULUS DAVIDIANA	POPLAR
POPULUS DELTOIDES	EASTERN POPLAR
POPULUS GRANDIDENTATA	LARGE-TOOTHED ASPEN
POPULUS TREMULOIDES	QUAKING ASPEN OR TREMBLING ASPEN
POTAMOGETON CRISPUS	CURLED PONDWEED
POTAMOGETON DIVERSIFOLIUS	PONDWEED

POTAMOGETON FOLIOSUS	NARROW-LEAVED PONDWEED
POTAMOGETON ILLINOEASIS	PONDWEED
POTAMOGETON NATANS	BROAD-LEAVED PONDWEED
POTAMOGETON PECTINATUS	FENNEL PONDWEED OR SAGO PONDWEED
POTAMOGETON PRAELONGUS	LONG-STALKED PONDWEED
POTAMOGETON SP.	PONDWEED
PRUNUS ARMENIACA	APRICOT
PSEUDOTSUGA MENZIESSII	DOUGLAS FIR
PTERIDIUM AQUILINUM	EAGLE FERN OR BRACKEN
PUERARIA PHASEOLOIDES	LEGUME
PYGMAEOTHAMNUS ZEYHERI	
QUERCUS ALBA	WHITE OAK
QUERCUS CERRIS	EUROPEAN TURKEY OAK
QUERCUS COCCINEA	SCARLET OAK
QUERCUS DUMOSA	OAK
QUERCUS ILEX	HOLY OAK
QUERCUS MUEHLENBERGII	OAK
QUERCUS PETRAEA	DURMAST OAK
QUERCUS PRINUS	CHESTNUT OAK
QUERCUS ROBUR	ENGLISH OAK
QUERCUS RUBRA	RED OAK
QUERCUS SP.	OAK
QUERCUS VELUTINA	BLACK OAK
RAPHANUS SATIVUS	RADISH, GARDEN
REAUMURIA NEGEVENSIS	SHRUB
REBOUDIA PINNATA	
RHIZOPHORA MANGLE	RED MANGROVE
RHODODENDRON MAXIMUM	
RHYNCHOSIA MONOPHYLLA	LEGUME
RICCIA FLUITANS	AQUATIC LIWERWORT
ROBINIA PSEUDOACACIA	BLACK LOCUST
ROSA CANINA	DOG ROSE OR HIP ROSE
ROTALA INDICA (*)	
RYE-GRASS	
SACCHARUM OFFICINALE	SUGAR CANE
SAGITTARIA LATIFOLIA	ARROWLEAF
SAGITTARIA PLATYPHYLLA	ARROWLEAF
SALIX ALBA	WHITE WILLOW
SALIX RUBRUM	WILLOW
SALSOLA INERMIS	
SALVINIA AURICULATA	AQUATIC FERN
SASSAFRAS ALBIDUM	SASSAFRAS OR AUGUE TREE
SATIVUM LEGUMINOSAE	
SCHOENOPLECTUS LACUSTRIS	COMMON CLUBRUSH OR GREATER RUSH
SCIRPUS AMERICANUS	
SCIRPUS SP.	BULRUSH OR CLUBRUSH
SCIRPUS VALIDUS	SOFT STEM BULRUSH OR GREAT BULRUSH
SECALE CEREALE	RYE
SERRATIA MARCESCENS	
SESBANIA CANNABINA	DHAINCHA
SETARIA VIRIDIS (*)	
SOLANUM TUBEROSUM	POTATO
SOLIDAGO VIRGAUREA	EUROPEAN GOLDEN ROD
SORGHUM VULGARE	SORGHUM
SPARGANIUM EURYCAZPUM	GIANT BUR-REED
SPARGANIUM SP.	BUR-REED
SPARTINA SP.	CORDGRASS
SPINACIA OLERACEA	SPINACH
STELLARIA ALSINE (*)	CHICKWEED
STIPA CAPENSIS	GRASS
STYLOSANTHES SP.	LEGUME
SWIETENIA MAHOGANI	MAHOGANY
SYRINGA VULGARIS	LILAC
SYRINGODIUM SP.	TROPICAL SEA-GRASS
TEPHROSIA LUPINIIFOLIA	LEGUME
TERMINALIA SERICEA	ASSEGAI WOOD MANGWE OR YELLOW WOOD
THALASSIA SP.	SEA GRASS
THALASSIA TESTUDIUNUM	SEA GRASS
THUJA OCCIDENTALIS	EASTERN ARBORVITAE OR WHITE CEDAR
TILIA AMERICANA	AMERICAN LINDEN
TRIFOLIUM PERENNE	CLOVER
TRIFOLIUM PRATENSE	RED CLOVER
TRIFOLIUM REPENS	WHITE CLOVER
TRIGONELLA ARABICA	LEGUME
TRIGONELLA FOENUMGRAECUM	FENUGREEK
TRITICUM AESTIVUM = T SATIVUM	COMMON WHEAT
TRITICUM DURUM	DURUM WHEAT
TRITICUM SATIVUM = T AESTIVUM	COMMON WHEAT
TSUGA CANADENSIS	EASTERN HEMLOCK
TYPHA ANGUSTIFOLIA	LESSER CAT'S TAIL OR LESSER REEDMACE
TYPHA LATIFOLIA	GREATER REEDMACE OR CAT'S TAIL

TYPHA ORIENTALIS	
TYPHA SP.	REEDMACE OR CAT'S TAIL
ULMUS RUBRA	RED ELM
UTRICULARIA INFLATA	BLADDERWORT
UTRICULARIA SP.	BLADDERWORT
UTRICULARIA SUBULATA	BLADDERWORT
UTRICULARIA VULGARIS	GREATER BLADDERWORT
VALLISNERIA AMERICANA	EEL-GRASS OR TAPE-GRASS
VALLISNERIA SP.	EEL-GRASS OR TAPE-GRASS
VANDELLIA ANGUSTIFOLIA	
VIBURNUM ACERIFOLIUM	ARROW WOOD OR DOCKMACKIE
VIBURNUM PRUNIFOLIUM	BLACK HAW OR STAG BUSH
VICIA ANGUSTIFOLIA (*)	NARROW-LEAVED VETCH
VICIA FABA	HORSE BEAN
VIGNA SINENSIS	COMMON COW-PEA
VIOLA TRICOLOR	PANSY
VITIS SP.	
VITIS VULPINA	WILD GRAPE
XEROPHYTA RETINERVIS	
ZANNICHELLIA PALUSTRIS	HORNED PONDWEED
ZEA MAYS	MAIZE
ZIZANIA AQUATICA	WILD RICE OR INDIAN RICE
ZOSTERA ASIATICA	GRASS WEED OR GRASS WRACK
ZOSTERA MARINA	GRASS WEED OR GRASS WRACK
ZOSTERA SP.	GRASS WEED OR GRASS WRACK
ZYGOPHYLLUM DUMOSUM	

THALLOPHYTA: ALGAE

ACROSIPHONIA CENTRALIS	GREEN ALGA
AGARUM CRIBROSUM	BROWN ALGA
AGMENELLUM QUADRUPLICATUM	BLUE-GREEN ALGA
AHNFELTIA PARADOXA	RED ALGA
AMPHIDINIUM CARTERI	DINOFLAGELLATE
AMPHIDINIUM SP.	DINOFLAGELLATE
AMPHIPLEURA RUTILANS	DIATOM
AMPHIPRORA ORNATA	DIATOM
ANABAENA CYLINDRICA	BLUE-GREEN ALGA
ANABAENA DOMINANCE	BLUE-GREEN ALGA
ANABAENA FLOS-AQUAE	BLUE-GREEN ALGA
ANABAENA SCHEREMETIEVI	BLUE-GREEN ALGA
ANABAENA SP.	BLUE-GREEN ALGA
ANABAENA VARIABILIS	BLUE-GREEN ALGA
ANABAENOPSIS SP.	BLUE-GREEN ALGA
ANACYSTIS NIDULANS	BLUE-GREEN ALGA
ANKISTRODESMUS AMALLOIDES	GREEN ALGA
ANKISTRODESMUS BRAUNII	GREEN ALGA
ANKISTRODESMUS FALCATUS	GREEN ALGA
APHANIZOMENON FLOS AQUAE F GRACILE	BLUE-GREEN ALGA
APHANIZOMENON SP.	BLUE-GREEN ALGA
APHANOCAPSA DELICATISSIMA	BLUE-GREEN ALGA
APHANOTHECE CLATHRATA	BLUE-GREEN ALGA
APHANOTHECE NIDULANS	BLUE-GREEN ALGA
ASTERIONELLA FORMOSA	DIATOM
ASTERIONELLA JAPONICA	DIATOM
ASTERIONELLA SP.	DIATOM
BACILLARIOPHYCEAE	DIATOMS
BACTERIASTRUM HYALINUM	DIATOM
BELLOCHIA SP.	
BIDDULPHIA AURITA	DIATOM
BIDDULPHIA SINENSIS	DIATOM
BLIDINGIA MINIMA	GREEN ALGA
BOSSIELLA SP.	RED ALGA
BOTRYOCOCCUS BRAUNII	GREEN ALGA
CACHONINA SP.	DINOFLAGELLATE
CALLIARTHRON SP.	
CALOTHRIX PARIETINA	BLUE-GREEN ALGA
CERAMIUM STRICTUM	RED ALGA
CERATAULINA BERGONII	DIATOM
CERATIUMOFURCA	DINOFLAGELLATE
CERATIUM HIRUNDINELLA	DINOFLAGELLATE
CERATIUM TRIPOS	DINOFLAGELLATE
CHAETOCEROS CURVISETUS	DIATOM
CHAETOCEROS DEBILIS	DIATOM
CHAETOCEROS GRACILIS	DIATOM
CHAETOCEROS LAUDERI	DIATOM
CHAETOCEROS LORENZIANUS	DIATOM
CHAETOCEROS SOCIALIS	DIATOM
CHAETOCEROS SP.	DIATOM
CHAETOMORPHA SP.	GREEN ALGA
CHARA SP.	GREEN ALGA
CHLAMYDOMONAS MAGNUSII	GREEN ALGA
CHLAMYDOMONAS REINHARDTII	GREEN ALGA
CHLAMYDOMONAS SP.	GREEN ALGA
CHLAMYDOMONAS SUBCOMPLETA	GREEN ALGA
CHLORELLA ELLIPSOIDEA	GREEN ALGA
CHLORELLA PYRENOIDOSA	GREEN ALGA
CHLORELLA SP.	GREEN ALGA
CHLORELLA VULGARIS	GREEN ALGA
CHLOROCOCCUM SP.	GREEN ALGA
CHLOROPHYCEAE	GREEN ALGAE
CHLOROPHYTA	
CHODATELLA SP.	GREEN ALGA
CHONDRUS OCELLATUS	RED ALGA
CHONDRUS YENDOI	RED ALGA
CHORDA FILUM	BROWN ALGA
CHROMULINA PYRIFORMIS	GOLDEN-BROWN ALGA
CHROOCOCCUS LIMNETICUS	BLUE-GREEN ALGA
CHROOCOCCUS SP.	BLUE-GREEN ALGA
CHROOCOCCUS TURGIDUS	BLUE-GREEN ALGA
CHROOMONAS NORDSTEDTII	CRYPTOPHYCEAE
CHRYSOPHYCEAE	GOLDEN-BROWN ALGAE
CLADOPHORA GLOMERATA	GREEN ALGA
CLADOPHORA GRACILIS	GREEN ALGA
CLADOPHORA SP.	GREEN ALGA
CLOSTERIUM ACICÜLARE	GREEN ALGA
CLOSTERIUM ACICULARE VAR SUBPRONUM	GREEN ALGA
COCCOCHLORIS SP.	BLUE-GREEN ALGA
COCCOCHLORIS STAGNINA	BLUE-GREEN ALGA
COCCOLITHUS HUXLEYI	HAPTOPHYCEAE
COCCOLITHUS SP.	HAPTOPHYCEAE
CODIUM YESSOENSIS	GREEN ALGA
COELOSPHAERIUM NAEGELIANUM	BLUE-GREEN ALGA
COENOCOCCUS PLANKTONICUS	GREEN ALGA
COMPSOPOGON HOOKERI	RED ALGA
CORALLINA OFFICINALIS	RED ALGA
CORALLINA SP.	RED ALGA
COSCINODISCUS ANGSTII	DIATOM
COSCINODISCUS ANGSTII VAR GRANULOMARGINATUS	DIATOM
COSCINODISCUS EXCENTRICUS	DIATOM
COSCINODISCUS GIGAS	DIATOM
COSCINODISCUS GRANII	DIATOM
COSCINODISCUS LINEATUS	DIATOM
COSCINODISCUS RADIATUS	DIATOM
COSCINODISCUS SP.	DIATOM
COSCINODISCUS WAILESII	DIATOM
COSMARIUM PHASEOLUS	GREEN ALGA
COSMARIUM RENIFORME	GREEN ALGA
COSMARIUM SP.	GREEN ALGA
COSTARIA COSTATA	BROWN ALGA
CRICOSPHAERA GALBANA	HAPTOPHYCEAE
CRICOSPHAERA SP.	HAPTOPHYCEAE
CRYPTOMONAS EROSA	CRYPTOPHYCEAE
CRYPTOMONAS OVATA	CRYPTOPHYCEAE
CRYPTOMONAS PSEUDOBALTICA	CRYPTOPHYCEAE
CRYPTOMONAS SP.	CRYPTOPHYCEAE
CYANOPHYCEAE	BLUE-GREEN ALGAE
CYCLOTELLA BODANICA	DIATOM
CYCLOTELLA COMENSIS	DIATOM
CYCLOTELLA COMTA	DIATOM
CYCLOTELLA GLOMERATA	DIATOM
CYCLOTELLA MENEGHINIANA	DIATOM
CYCLOTELLA NANA	DIATOM
CYCLOTELLA SOCIALIS	DIATOM
CYCLOTELLA SP.	DIATOM
CYCLOTELLA STELLIGERA	DIATOM
CYLINDROSPERMUM SP.	BLUE-GREEN ALGA
CYLINDROTHECA CLOSTERIUM	DIATOM
CYMATOPLEURA SOLEA	DIATOM
CYSTOSEIRA OSMUNDACEA	BROWN ALGA
DACTYLOCOCCOPSIS SMITHII	BLUE-GREEN ALGA
DELESSERIA SANGUINEA	RED ALGA
DIATOMA ELONGATUM	DIATOM
DIATOMA VULGARE	DIATOM
DICRATERIA SP.	
DICTYOCHA SP.	GOLDEN-BROWN ALGA
DICTYOSPHAERIUM PULCHELLUM	GREEN ALGA
DINOBRYON BORGEI	GOLDEN-BROWN ALGA
DINOBRYON CYLINDRICUM	GOLDEN-BROWN ALGA
DINOBRYON DIVERGENS	GOLDEN-BROWN ALGA
DINOBRYON SOCIALE	GOLDEN-BROWN ALGA
DINOBRYON SOCIALE VAR AMERICANUM	GOLDEN-BROWN ALGA
DINOBRYON SP.	GOLDEN-BROWN ALGA
DINOPHYCEAE	DINOFLAGELLATES
DINOPHYSIS MILES	DINOFLAGELLATE

DIPLOPSALIS ACUTA	DINOFLAGELLATE
DISSODINIUM SP.	DINOFLAGELLATE
DITYLUM BRIGHTWELLII	DIATOM
DUMONTIA INCRASSATA	RED ALGA
DUNALIELLA MARINA (*)	GREEN ALGA
DUNALIELLA PEIRCEI	GREEN ALGA
DUNALIELLA SALINA	GREEN ALGA
DUNALIELLA SP.	GREEN ALGA
DUNALIELLA TERTIOLECTA	GREEN ALGA
ECTOCARPUS CONFERVOIDES	BROWN ALGA
EGREGIA MENZIESII	BROWN ALGA
ELAKATOTHRIX GELATINOSA	GREEN ALGA
ENTEROMORPHA COMPRESSA	GREEN ALGA
ENTEROMORPHA INTESTINALIS	GREEN ALGA
ENTEROMORPHA LINZA	GREEN ALGA
ENTEROMORPHA NANA	GREEN ALGA
ENTEROMORPHA PROLIFERA	GREEN ALGA
ENTEROMORPHA SP.	GREEN ALGA
ERKENIA SUBAEQUICILIATA	DIATOM
EUGLENA GRACILIS	EUGLENOPHYCEAE
EUGLENA SP.	EUGLENOPHYCEAE
EXUVIAELLA SP.	DINOFLAGELLATE
FRAGILARIA CROTONENSIS	DIATOM
FRAGILARIA PINNATA	DIATOM
FRAGILARIA SP.	DIATOM
FUCUS SERRATUS	BROWN ALGA
FUCUS VESICULOSUS	BROWN ALGA
FURCELLARIA FASTIGIATA	RED ALGA
GIGARTINA AGARDHII	RED ALGA
GIGARTINA LEPTORYNCHOS	RED ALGA
GLENODINIUM SP.	DINOFLAGELLATE
GLOEOCAPSA ALPICOLA	BLUE-GREEN ALGA
GLOEOCAPSA RUPESTRIS	BLUE-GREEN ALGA
GLOEOCOCOCCUS SCHROETERI	GREEN ALGA
GOMPHOSPHAERIA LACUSTRIS	BLUE-GREEN ALGA
GONYAULAX POLYEDRA	DINOFLAGELLATE
GYMNODINIUM FUSCUM	DINOFLAGELLATE
GYMNODINIUM HELVETICUM	DINOFLAGELLATE
GYMNODINIUM ORDINATUM	DINOFLAGELLATE
GYMNODINIUM SP.	DINOFLAGELLATE
GYMNODINIUM SPLENDENS	DINOFLAGELLATE
HAEMATOCOCCUS PLUVIALIS	GREEN ALGA
HEMIAULUS SINENSIS	DIATOM
HETEROCHORDARIA ABIETINA	BROWN ALGA
HETEROTHRIX SP.	YELLOW-GREEN ALGA
HIZIKIA FUSIFORME	
HYDRODICTYON SP.	GREEN ALGA
HYMENOMONAS SP.	HAPTOPHYCEAE
ISOCHRYSIS GALBANA	GOLDEN-BROWN ALGA
ISOCHRYSIS SP.	GOLDEN-BROWN ALGA
KEPHYRION SP.	GOLDEN-BROWN ALGA
LAMINARIA CHICHORIOIDES	BROWN ALGA
LAMINARIA DIGITATA	BROWN ALGA
LAMINARIA JAPONICA	BROWN ALGA
LAMINARIA LONGICRURIS	BROWN ALGA
LAMINARIA SACCHARINA	BROWN ALGA
LAUDERIA BOREALIS	DIATOM
LAUDERIA SP.	DIATOM
LEPTOCYLINDRUS DANICUS	DIATOM
LICMOPHORA SP.	DIATOM
LITHOTHAMNIOID ALGA	RED ALGA
LYNGBYA SP.	BLUE-GREEN ALGA
MACROCYSTIS PYRIFERA	BROWN ALGA
MALLOMONAS CAUDATA	GOLDEN-BROWN ALGA
MALLOMONAS URNAFORMIS	GOLDEN-BROWN ALGA
MELOBESIA SP.	RED ALGA
MELOSIRA AGASSIZII	DIATOM
MELOSIRA GRANULATA	DIATOM
MELOSIRA ISLANDICA	DIATOM
MELOSIRA ISLANDICA SUBSP HELVETICA	DIATOM
MELOSIRA ITALICA	DIATOM
MELOSIRA ITALICA SUBSP SUBARCTICA	DIATOM
MELOSIRA SP.	DIATOM
MERISMOPEDIA TENUISSIMA	BLUE-GREEN ALGA
MICROCYSTIS AERUGINOSA	BLUE-GREEN ALGA
MICROCYSTIS FLOS-AQUAE	BLUE-GREEN ALGA
MICROCYSTIS SP.	BLUE-GRFEN ALGA
MONOCHRYSIS LUTHERI	GOLDEN-BROWN ALGA
MONODUS SUBTERRANEUS	YELLOW-GREEN ALGA
MOUGEOTIA SP.	GREEN ALGA
MYXOMYPHYCEAE CYANOPHYCEAE	BLUE-GREEN ALGAE
NANNOCHLORIS ATOMUS	GREEN ALGA
NAVICULA ARENARIA	DIATOM
NAVICULA INCERTA	DIATOM
NAVICULA MINIMA	DIATOM
NAVICULA PELLICULOSA	DIATOM
NAVICULA SALINARUM	DIATOM
NAVICULA SPP.	DIATOM
NAVICULA TORQUATUM	DIATOM
NITELLA FLEXILIS	GREEN ALGA
NITELLA OPACA	GREEN ALGA
NITZSCHIA ACTINASTROIDES	DIATOM
NITZSCHIA ALBA	DIATOM
NITZSCHIA CLOSTERIUM	DIATOM
NITZSCHIA FRUSTULUM	DIATOM
NITZSCHIA GRACILIS	DIATOM
NITZSCHIA OVALIS	DIATOM
NITZSCHIA SERIATA	DIATOM
NITZSCHIA SP.	DIATOM
NOSTOC MUSCORUM	BLUE-GREEN ALGA
OEDOGONIUM CARDIACUM	GREEN ALGA
OEDOGONIUM SP.	GREEN ALGA
OIKOMONOS TERMO	GOLDEN-BROWN ALGA
OLISTHODISCUS LUTEUS	FLAGELLATE
OOCYSTIS SOLITARIA	GREEN ALGA
OOCYSTIS SP.	GREEN ALGA
OSCILLATORIA AGARDHII	BLUE-GREEN ALGA
OSCILLATORIA AGARDHII VAR ISOTHRIX	BLUE-GREEN ALGA
OSCILLATORIA BORNETI	BLUE-GREEN ALGA
OSCILLATORIA BOURRELLYI	BLUE-GREEN ALGA
OSCILLATORIA CHALYBIA	BLUE-GREEN ALGA
OSCILLATORIA FORMOSA	BLUE-GREEN ALGA
OSCILLATORIA LIMNETICA	BLUE-GREEN ALGA
OSCILLATORIA REDEKEI	BLUE-GREEN ALGA
OSCILLATORIA RUBESCENS	BLUE-GREEN ALGA
OSCILLATORIA SP.	BLUE-GREEN ALGA
OSCILLATORIA TENUIS	BLUE-GREEN ALGA
OSCILLATORIA THIEBAUTII	BLUE-GREEN ALGA
PANDORINA MORUM	GREEN ALGA
PAULSCHULZIA TENERA	GREEN ALGA
PEDIASTRUM DUPLEX	GREEN ALGA
PEDIASTRUM SPP.	GREEN ALGA
PEDIASTRUM TETRAS	GREEN ALGA
PELVETIA WRIGHTII	BROWN ALGA
PERIDINIUM CINCTUM	DINOFLAGELLATE
PERIDINIUM CINCTUM FA WESTII	DINOFLAGELLATE
PERIDINIUM SPP.	DINOFLAGELLATE
PERIDINIUM WILLEI	DINOFLAGELLATE
PETALONIA FASCIA	BROWN ALGA
PHAEODACTYLUM TRICORNUTUM	DIATOM
PHAEOPHYCEAE	BROWN ALGAE
PHAEOPHYTA	BROWN ALGAE
PHORMIDIUM FAVEOLARUM	BLUE-GREEN ALGA
PHORMIDIUM LURIDUM	BLUE-GREEN ALGA
PHYCODRYS RUBENS	RED ALGA
PHYLLOPHORA BRODIAEI	RED ALGA
PITHOPHORA SP.	GREEN ALGA
PLANKTONIELLA SOL	DIATOM
PLATYMONAS SP.	PRASINOPHYCEAE
PLATYMONAS SUBCORDIFORMIS	PRASINOPHYCEAE
PLECTONEMA NOSTOCORUM	BLUE-GREEN ALGA
POLYSIPHONIA JAPONICA	RED ALGA
POLYSIPHONIA NIGRESCENS	RED ALGA
PORPHYRA LEUCOSTICTA	RED ALGA
PORPHYRA SP.	RED ALGA
PORPHYRA UMBILICALIS	RED ALGA
PORPHYRIDIUM AERUGINEUM	RED ALGA
PORPHYRIDIUM SP.	RED ALGA
PROROCENTRUM MICANS	DINOFLAGELLATE
PTERYGOPHORA CALIFORNICA	BROWN ALGA
PTILOTA FILICINA	RED ALGA
PYROCYSTIS FUSIFORMIS	DINOFLAGELLATE
PYROCYSTIS NOCTILUCA	DINOFLAGELLATE
RADIOCYSTIS GEMINATA	BLUE-GREEN ALGA
RHIZOCHRYSIS LIMNETICA	GOLDEN-BROWN ALGA
RHIZOCLONIUM HIEROGLYPHICUM	GREEN ALGA
RHIZOCLONIUM SP.	GREEN ALGA
RHIZOSOLENIA ROBUSTA	DIATOM
RHIZOSOLENIA SETIGERA	DIATOM
RHIZOSOLENIA STOLTERFOTHII	DIATOM
RHIZOSOLENIA STYLIFORMIS	DIATOM
RHODOMELA LARIX	RED ALGA
RHODOMONAS LACUSTRIS	CRYPTOPHYCEAE
RHODOMONAS MINUTA	CRYPTOPHYCEAE
RHODOPHYTA	RED ALGAE
SARGASSUM FUL^ELLUM	BROWN ALGA
SARGASSUM PALLIDUM	BROWN ALGA
SARGASSUM RINGGOLDIANUM	BROWN ALGA
SARGASSUM SP.	BROWN ALGA
SARGASSUM THUNBERGII	BROWN ALGA
SCENEDESMUS ACUTUS	GREEN ALGA
SCENEDESMUS BASILENSIS	GREEN ALGA
SCENEDESMUS GRACILE	GREEN ALGA

SCENEDESMUS OBLIQUUS	GREEN ALGA
SCENEDESMUS QUADRICAUDA	GREEN ALGA
SCENEDESMUS SP.	GREEN ALGA
SCHIZOTHRIX CALCICOLA	BLUE-GREEN ALGA
SCHIZYMENIA EPIPHYTICA	RED ALGA
SCYTOSIPHON LOMENTARIA	BROWN ALGA
SELENASTRUM CAPRICORNUTUM	GREEN ALGA
SELENASTRUM SP.	GREEN ALGA
SKELETONEMA COSTATUM	DIATOM
SKELETONEMA SP.	DIATOM
SPIROGYRA SP.	GREEN ALGA
STAURASTRUM PARADOXUM	GREEN ALGA
STEPHANODISCUS ASTRAEA	DIATOM
STEPHANODISCUS HANTZSCHII	DIATOM
STEPHANODISCUS HANTZSCHII VAR PUSILLUS	DIATOM
STEPHANODISCUS NIAGAREA	DIATOM
STEPHANOPYXIS SP.	DIATOM
STICHOCOCCUS BACILLARIS	GREEN ALGA
STIGEOCLONIUM PACHYDERUM	GREEN ALGA
SYNECHOCOCCUS AERUGINOSUS	BLUE-GREEN ALGA
SYNEDRA ACUS	DIATOM
SYNEDRA ACUS VAR ANGUSTISSIMA	DIATOM
SYNEDRA ACUS VAR RADIANS	DIATOM
SYNEDRA CAPITATA	DIATOM
SYNEDRA DELICATISSIMA	DIATOM
SYNEDRA SP.	DIATOM
SYNEDRA ULNA	DIATOM
SYRACOSPHAERA CARTERI	HAPTOPHYCEAE
TABELLARIA FENESTRATA	DIATOM
TABELLARIA FLOCCULOSA	DIATOM
TABELLARIA FLOCCULOSA VAR ASTERIONELLOIDES	DIATOM
TABELLARIA FLOCCULOSA VAR FLOCCULOSA	DIATOM
TABELLARIA SP.	DIATOM
TETRAEDRON MINIMUM	GREEN ALGAE
TETRAEDRON SP.	GREEN ALGAE
TETRASELMIS CARTERIFORMIS	PRASINOPHYCEAE
TETRASELMIS CHUII	PRASINOPHYCEAE
TETRASELMIS GRACILIS	PRASINOPHYCEAE
TETRASELMIS MACULATA	PRASINOPHYCEAE
TETRASELMIS STRICIA	PRASINOPHYCEAE
THALASSIOSIRA DECIPIENS	DIATOM
THALASSIOSIRA FLUVIATILIS	DIATOM
THALASSIOSIRA GRAVIDA	DIATOM
THALASSIOSIRA NANA	DIATOM
THALASSIOSIRA PSEUDONANA	DIATOM
THALASSIOSIRA ROTULA	DIATOM
THALASSIOSIRA SP.	DIATOM
TRACHELOMONAS HISPIDA	EUGLENOPHYCEAE
TRACHELOMONAS VOLVOCINA	EUGLENOPHYCEAE
TRICERATIUM FAVUS	DIATOM
TRICHODESMIUM SP.	BLUE-GREEN ALGAE
ULOTHRIX FIMBRIATA	GREEN ALGAE
ULOTHRIX SPECIOSA	GREEN ALGAE
ULOTHRIX ZONATA	GREEN ALGAE
ULVA ANGUSTA	GREEN ALGAE
ULVA FENESTRATA	GREEN ALGAE
ULVA LACTUCA	GREEN ALGAE
ULVA PERTUSA	GREEN ALGAE
ULVA SP.	GREEN ALGAE
ULVARIA SPENDENS	GREEN ALGAE
ULVOPSIS GREVILLEI	GREEN ALGAE
UNDARIA PINNATIFIDA	BROWN ALGAE
UROGLENA AMERICANA	GOLDEN-BROWN ALGAE
VALONIA VENTRICULOSA	GREEN ALGAE
ZOOXANTHELLAE	DINOFLAGELLATE

THALLOPHYTA: FUNGI AND LICHENES

ALTERNARIA SP.	FUNGUS
ASPERGILLUS SP.	FUNGUS
BASISPORIUM GALLARUM	FUNGUS
CANDIDA SP.	FUNGUS
CLADOSPORIUM SP.	FUNGUS
COLLETOTRICHUM SP.	FUNGUS
CORNICULARIA MURICATA	LICHENE
CRYPTOCOCCUS SP.	FUNGUS
CURVULARIA SP.	FUNGUS
MONILINIA FRUCTICOLA	FUNGUS
NEUROSPORA CRASSA	FUNGUS
PELTIGERA CANINA	LICHENE
PERICONIELLA SP.	FUNGUS
PESTALOTIA SP.	FUNGUS
PINICILLIUM SP.	FUNGUS
PIRICAUDA SP.	FUNGUS
RHODOTORULA SP.	FUNGUS
SACCHAROMYCES CEREVISEAE	YEAST
SACCHAROMYCES SP.	YEAST
TORULOPSIS UTILIS	YEAST

THALLOPHYTA: SCHIZOMYCETES

ACHROMOBACTER SP. (*)
AEROBACTER AEROGENES
ANCALOMICROBIUM SP.
ARTHROBACTER MARINUS
AZOTOBACTER CHROOCOCCUM
AZOTOBACTER SP.
BACILLUS CEREUS
BACILLUS COAGULANS
BACILLUS FLUORESCENS NON LIQUEFACIENS
BACILLUS GLOBIGII
BACILLUS LICHENIFORMIS
BACILLUS MESENTERICUS VULGATUS
BACILLUS SP. (*)
BACILLUS SUBTILIS
BREVIBACTERIUM HELVOLUM
BREVIBACTERIUM SP. (*)
CAULOBACTER SP. (*)
CHROMOBACTER SP. (*)
CLOSTRIDIUM PASTEURIANUM
ENTEROBACTER SP. (*)
ERWINIA CAROTOVORA
ESCHERICHIA COLI
ESCHERICHIA SP. (*)
FLAVOBACTERIUM ARBORESCENS
FLAVOBACTERIUM HARRISONII
FLAVOBACTERIUM SP. (*)
GAFFKYA TETRAGENA
KLEBSIELLA PNEUMONIAE
MICROCOCCUS LYSODEIKTICUS
MICROCOCCUS SP. (*)
MYCOBACTERIUM PHLEI
MYCOBACTERIUM SMEGMATIS
MYCOPLASMA GALLISEPTICUM
NITROBACTER SP.
NITROSOMONAS EUROPAEA
NITROSOMONAS SP.
PNEUMOCOCCUS BACTERIUM
PROTEUS SP. (*)
PROTEUS VULGARIS
PSEUDOMONAD
PSEUDOMONAS AEROGINOSA
PSEUDOMONAS CUPRODURANS
PSEUDOMONAS FLUORESCENS
PSEUDOMONAS SP. (*)
SALMONELLA TYPHIMURIUM
SARCINA LUTEA
SARCINA SP. (*)
SPHAEROTILUS SP.
STAPHYLOCOCCUS AUREUS
STREPTOMYCES SP. (*)
SULFOLOBUS SP.
THIOBACILLUS FERROOXIDANS
THIOBACILLUS THIOOXIDANS
XANTHOMONAS SP. (*)

THALLOPHYTA: VIRA

T-EVEN BACTERIOPHAGE
TOBACCO MOSAIC VIRUS
T1-BACTERIOPHAGE

****

A N I M A L S

****

ANNELIDA

EARTHWORMS	
ENCHYTRAE	COGNETTIA COGNETTII
LEECH	ERPOBDELLA OCTOCULATA
	GLOSSIPHONIA COMPLANATA
	GLOSSIPHONIA HETEROCLITA
	HELOBDELLA STAGNALIS
LEECHES	HIRUDINEA
LUMBRICID	ALLOLOBOPHORA ROSEA
OLIGOCHAETE	EISENIA ARBOREA
	EISENIA BICYCLIS
	ILYODRILUS HAMMONIENSIS
	LIMNODRILUS HOFFMEISTERI
	LIMNODRILUS HOFFMEISTERI CLARAPAREDE
	LIMNODRILUS SP.
	LUMBRICUS SP.
	PELOSCOLEX MULTISETOSUS
	POTAMOTHRIX HAMMONIENSIS
	TUBIFEX SP.
	TUBIFEX TUBIFEX
OLIGOCHAETE EARTHWORM	LUMBRICUS TERRESTRIS
OLIGOCHAETE WORMS	OLIGOCHAETA
OLIGOCHAETES	ENCHYTRACIDAE
	LUMBRICIDAE
	TUBIFICIDAE
POLYCHAETE	APHRODITA HASTATA
	EUPOMATUS SP.
	HYDROIDES NORVEGICA
	LANICE CONCHILEGA
	LUMBRINERIS FRAGILIS
	MYXICOLA INFUNDIBULUM
	NEANTHES SUCCINEA
	NEANTHES VIRENS
	NEPHTYS INCISA
	NEPHTYS SP.
	NEREIS DIVERSICOLOR
	NEREIS SP.
	POMATOCEROS TRIQUETER
	PRAXILLELLA PRAETERMISSA
	SABELLA PAVONINA
	SALMACINA DYSTERI
	SCHIZOBRANCHIA INSIGNIS
	SERPULA SP.
	SPIRORBIS BOREALIS
	SPIRORBIS SP.
	TOMOPTERIS SEPTENTRIONALIS
POLYCHAETE WORMS	POLYCHAETA

ARTHROPODA

ALDERFLY	SIALIS MITSUHASHII
	SIALIS SP.
AMPHIPOD	CALLIOPIUS LAEVIUSCULUS
	CRANGONYX PSEUDOGRACILIS
	CRANGONYX RICHMONDENSIS
	GAMMARACANTHUS LACUSTRIS
	HYALELLA AZTECA
	PALLASEA QUADRISPINOSA
	PARATHEMISTO PACIFICA
	PARHYALLA WHELPLEYI
	PHRONIMA SEDENTARIA
	PONTOPOREIA AFFINIS
AMPHIPODS	AMPHIPODA
ANT	CAMPONOTUS AMERICANUS
	CAMPONOTUS PENNSYLVANICUS
	CREMATOGASTER CLARA
	MYRMICA SP.
BACKSVIMMER	NOTONECTA GLAUCA
	NOTONECTA SP.
BACKSVIMMERS	NOTONECTIDAE
BARNACLE	
	BALANUS BALANOIDES
	BALANUS SP.
	CHTHAMALUS FISSUS
	CHTHAMALUS SP.
	TETRACLITA SP.
BEETLE	EUBRIANAX SP.
	EUROSTUS HILLERI
	GIBBIUM PSYLLOIDES
	HYPERA POSTICA
	MEZIUM AFFINE
	NIPTUS HOLOLECUS
	PHYLLOPERTHA HORTICOLA
	POPILIUS DISJUNCTUS
	STETHOMEZIUM SQUAMOSUM
	TRIGONOGENIUS GLOBULUS
BEETLES	COLEOPTERA
	HALIPLIDAE
BITING MIDGES	CERATOPOGONIDAE
BLACK FLY	EUSIMULIUM SP.
BLACKFLIES	SIMULIIDAE
BLACKFLY	SIMULIUM BREVICAULE
	SIMULIUM MONTICOLA
	SIMULIUM ORNATUM
	SIMULIUM SP.
	SIMULIUM VITTATUM
BLOODWORM	
BLUE CRAB	CALLINECTES SAPIDUS
BRANCHIOPOD	EUBRANCHIPUS BUNDYI
	GOTTIDIA PYRAMIDATA
BROWN SHRIMP	PENAEUS AZTECUS
BUG	TRIGONOTYLUS SP.
BUTTERFLIES	ARCTIIDAE
BUTTERFLY	CATACLYSTA MIDAS
CADDISFLIES	HYDROPSYCHIDAE
	LIMNEPHILIDAE
	TRICHOPTERA
CADDISFLY	AGRYPNIA OBSOLETA
	ATHRIPSOIDES ANCYLUS
	BRACHYCENTRUS AMERICANUS
	BRACHYCENTRUS SP.
	CHEUMATOPSYCHE SP.
	CYRNUS TRIMACULATUS
	DINARTHRODES SP.
	GOERA JAPONICA
	HYDROPSYCHE BETTENI
	HYDROPSYCHE SP.
	MACRONEMA SP.
	MICRASEMA SP.
	MOLANNA SP.
	PHRYGANEA SP.
	PLATYPHYLAX YOKOUCHII
	PLECTROCNEMIA CONSPERSA
	POLYCENTROPUS SP.
	POTAMOPHYLAX CINGULATUS
	PYCNOPSYCHE GUTTIFER
	PYCNOPSYCHE LEPIDO
	RHYACOPHILA BREVICEPHALA
	RHYACOPHILA FASCIATA
	RHYACOPHILA NIGROCEPHALA

	STENOPSYCHE GRISEIPENNIS
	STENOPSYCHE SP.
CALANOID COPEPODID	
CARABID BEETLE	EVARTHRUS SODALIS
	SPHAERODERUS STENOSTOMUS
CAVE CRICKET	CEUTHOPHILUS GRACILIPES
CENTIPEDE	LITHOBIUS CRASSIPES
	LITHOBIUS FORTIFICATUS
	OTOCRYPTOPS SEXSPINOSUS
CICADA	PHILAENUS LEUCOPTHALMUS
	PHILAENUS SPUMARIUS
CIGARETTE BEETLE	LASIODERMA SERRICORNE
COCKROACH	PARCOBLATTA SP.
COPEPOD	ACANTHOCYCLOPS VERNALIS
	ACARTIA CLAUSII
	ACARTIA SP.
	ACARTIA TONSA
	ARCHTODIAPTOMUS SP.
	CALANOIDES CARINATUS
	CALANUS CRISTATUS
	CALANUS DISCAUDATUS
	CALANUS FINMARCHICUS
	CALANUS HELGOLANDICUS
	CALANUS HYPERBOREUS
	CALANUS PACIFICUS
	CALANUS PLUMCHRUS
	CALANUS SP.
	CANUELLA PERPLEXA
	CENTROPAGES HAMATUS
	CENTROPAGES TYPICUS
	CYCLOPS BICUSPIDATUS
	CYCLOPS SP.
	CYCLOPS STRENUUS
	CYCLOPS VERICANS RUBELLUS
	CYCLOPS VERNALIS
	CYCLOPS VICINUS
	DIAPTOMUS ALASKAENSIS
	DIAPTOMUS ARCTICUS
	DIAPTOMUS CLACIPES
	DIAPTOMUS GATUNENSIS
	DIAPTOMUS GRACILIS
	DIAPTOMUS GRACILOIDES
	DIAPTOMUS LEPTOPUS
	DIAPTOMUS NEVADENSIS
	DIAPTOMUS OREGONENSIS
	DIAPTOMUS PALLIDUS
	DIAPTOMUS SHOSHONE
	DIAPTOMUS SICILOIDES
	DIAPTOMUS SP.
	EPILABIDOCERA AMPHITRITES
	EUCYCLOPS AGILIS
	EUDIAPTOMUS GRACILIS
	EUDIAPTOMUS GRACILOIDES
	EURYTEMORA AFFINIS
	EUTERPINA ACUTIFRONS
	HALICYCLOPS MAGNICEPS
	HETEROCOPE SEPTENTRIONALIS
	LABIDOCERA JOLLAE
	LIMNOCALANUS MACRURUS
	MACROCYCLOPS ALBIDUS
	MESOCYCLOPS ALBIDUS
	MESOCYCLOPS EDAX
	MESOCYCLOPS LEUCKARTI
	METRIDIA LUCENS
	MIXODIAPTOMUS LACINIATUS
	NEODIAPTOMUS SCHMACKERI
	PARACYCLOPS FIMBRIATUS POPPEI
	PARONYCHOCAMPTUS NANUS
	POLYPHEMUS PEDICULUS
	PSEUDOCALANUS MINUTUS
	RHINCALANUS GIGAS
	RHINCALANUS NASUTUS
	TEMORA LONGICORNIS
	TEMORA STYLIFERA
	THERMOCYCLOPS HYALINUS
	THERMOCYCLOPS NEGLECTUS
	TORTANUS DISCAUDATUS
	TROPOCYCLOPS PRASINUS
COPEPODS	COPEPODA
CRAB	
	CANCER PAGURUS
	CANCER PRODUCTUS
	CARCINUS MEDITERRANEUS
	CARCINUS SP.
	CHARYBDIS LONGICOLLIS
	HYAS ARANEUS
	LEPTODIUS FLORIDANUS
CRANEFLY	ANTOCHA SP.
	ERIOCERA SP.
CRAYFISH	
	ASTACUS SP.
	CAMBARINAE
	ORCONECTES NAIS
	ORCONECTES VIRILIS
	PROCAMBARUS CLARKI
	PROCAMBARUS SP.
CRICKET	GRYLLUS ASSIMILIS
	PTERONEMOBIUS FASCIATUS
CRUSTACEAN	ASELLOPSIS INTERMEDIA
	BRACHINECTA GIGAS
	BRACHINECTA MACKINI
	BRACHYURA
	COMPTOCERUS RECTIROSTRIS
	LEPTOCHEIRUS PINGUIS
	MEGANYCTIPHANES NORVEGICA
	MENIPPE MERCENARIA
	OITHONA SIMILIS
	PANDALUS MONTAGUI
	PLEURONCODES PLANIPES
	POLYARTEMIELLA HAZENI
	PORTUNAS PUBER
	TIGRIOPUS BREVICORNIS
	UNCIOLA LEUCOPIS
CRUSTACEANS: CRABS SHRIMPS LOBSTERS ETC	CRUSTACEA
CUTWORM	
CYCLOPOID COPEPODID	
DADDY LONG LEG OR CRANEFLY	TIPULA SP.
DAMSELFLIES	ZYGOPTERA
DAMSELFLY	LESTES SPONSA
DECAPOD	OPLOPHORUS SP.
	SYSTELLASPIS DEBILIS
DIGGER-WASP	AMMOPHILA BREVILIGULATA
DIPLOPOD	JAPONARIA LAMINATA ARMIGERA
DIPTERAN	CHAETOPSIS SP.
DIPTERANS	SCIOMYZIDAE
	TENDIPEDIDAE
DIVING BEETLE	HYDROPHILUS TRIANGULARIS
DIVING BEETLES	HYDROPHILIDAE
DOBSONFLY	CORYDALUS CORNUTUS
DRAGONFLY	AESCHNA CYANEA
	BOYERIA VINOSA
	CORDULIA AENEA
	DIPLACODES TRIVIALIS
	EPIOPHLEBIA SP.
	GOMPHUS SP.
	OPHIOGOMPHUS RUPINSULENSIS
	PYRRHOSOMA NYMPHULA
	SIEBOLDIUS JAPONICA
DUNG BEETLE	GEOTRUPES SP.
EUPHAUSIID	EUPHAUSIA KROHNII (*)
	EUPHAUSIA PACIFICA
	NEMATOSCELIS ATLANTICA

	NEMATOSCELIS MEGALOPS
EUPHAUSIIDS	EUPHAUSIACEA
FIDDLER CRAB	UCA ININAX
	UCA SP.
FIREFLY	LUCIOLA CRUCIATA
FLESH FLY	SARCOPHAGA BULLATA
FLY	
	ATHERIX SP.
GOOSENECK BARNACLE	POLLICIPES POLYMERUS
GRASS SHRIMP	PALAEMONETES KADIAKENSIS
	PALAEMONETES PUGIO
	PALAEMONETES VULGARIS
GRASSHOPPER	CHORTHIPPUS PARALLELLUS
GRASSHOPPERS	ORTHOPTERA
GREEN CRAB	CARCINUS MAENAS
HARVESTMEN OR HARVEST SPIDERS	OPILIONES
HEAD AND BODY LOUSE	PEDICULUS HUMANUS
HERMIT CRAB	CLIBANARIUS VIRESCENS
HOG LOUSE	ASELLUS AQUATICUS
	ASELLUS MERIDIANUS
	ASELLUS MILIARIS
HONEY BEE	APHELOCHEIRUS VITTATUS
HOUSE CRICKET	ACHETA DOMESTICUS
ICHNEUMON FLY	PIMPLA INSTIGATOR
	PTEROMALUS PUPARUM
INSECTS	INSECTA
ISOPOD	ARMADILLIDIUM VULGARE
	CYLISTICUS CONVEXUS
	LIGIDIUM BLUERIDGENSIS
	LIGIDIUM JAPONICA
ISOPODS	ISOPODA
KATYDID	ORCHELIMUM SP.
LACEWING OR DOBSONFLY	CORYDALIS CORNUTA
LOBSTER	HOMARUS VULGARIS
LOCUST	SCHISTOCERCA SP.
MAYFLY	BAETIS RHODANI
	BAETIS SP.
	BAETIS VAGANS
	CAENIS SIMULANS
	CALLIBAETIS FERRUGINEUS
	CENTROPTILUM LUTEOLUM
	CHOROTERPES MEXICANUS
	DELEATIDIUM SP.
	ECDYONURUS SP.
	ECDYONURUS YOSHIDAE
	EPEORUS HIEMALIS
	EPHEMERA LINEATA
	EPHEMERA STRIGATA
	EPHEMERELLA GRANDIS
	EPHEMERELLA NEEDHAMI
	EPHEMERELLA SERRATA
	EPHEMERELLA SUBVARIA
	EPHEMERELLA TRISPINA
	HEPTAGENIA FUSCOGRISEA
	HEXAGENIA BILINEATA
	HEXAGENIA LIMBATA
	ISONYCHIA JAPONICA
	LEPTOPHLEBIA MARGINATA
	LEPTOPHLEBIA SP.
	POTAMANTHUS KAMONIS
	RHITHROGENA SEMICOLORATA
	TRICORYTHODES SP.
MEADOW KATYDID	ORCHELIMUM FIDICINIUM

MIDGE	ABLABESMYIA SP.
	ANATOPYNIA DYARI
	CALOPSECTRA DIVES
	CHIRONOMUS ANTHRACINUS
	CHIRONOMUS COMMUTATUS
	CHIRONOMUS PLUMOSUS
	CHIRONOMUS SP.
	CHIRONOMUS TENTANS
	CLADOTANYTARSUS MANCUS
	CORYNONEURA SP.
	GLYPTOTENDIPES BARBIPES
	GLYPTOTENDIPES PARIPES
	GLYPTOTENDIPES SP.
	LIMNOCHIRONOMUS PULSUS
	LIMNOCHIRONOMUS SP.
	MICROTENDIPES CHLORIS
	MICROTENDIPES SP.
	PARACHIRONOMUS TENER
	PROCLADIUS CHOREUS
	PROCLADIUS PECTINATUS
	PROCLADIUS RUFOVITTATUS
	PROCLADIUS SP.
	STICTOCHIRONOMUS SP.
	TANYTARSUS HOLOCHLORUS
	TANYTARSUS INOPERTUS
	TANYTARSUS LUGENS
MIDGES	TANYTARSINI
MILLIPEDE	BRACHORIA INITIALIS
	POLYDESMUS COMPLANATUS
	PTYOIULUS IMPRESSUS
	SCYTONOTUS GRANULATUS
MITE	PLATYNOTHRUS PELTIFER
	POGONOMYRMEX BADIUS
	RHIZOGLYPHUS ECHINOPUS
	TYROGLYPHUS INTERNA
	TYROGLYPHUS LINTNERI
MOSQUITO	CULEX TARSALIS
MOSQUITO (YELLOW FEVER)	AEDES AEGYPTI
MYSID	MYSIS RELICTA
	MYSIS STENOLEPIS
	NEOMYSIS INTEGER
MYSIDS	MYSIDAE
NEUROPTERAN	PROTOTHERMES GRANDIS
NON-BITING MIDGES	CHIRONOMIDAE
NORTHERN LOBSTER	HOMARUS AMERICANUS
NORWAY LOBSTER	NEPHROPS NORVEGICUS
OPOSSUM SHRIMP	METAMYSIDOPSIS ELONGATA
ORIBATID MITES	
PHANTOM MIDGE	CHAOBORUS FLAVICANS
	CHAOBORUS PUNCTIPENNIS
	CHAOBORUS SP.
	CHAOBORUS TRIVITTATUS
PLANT BUG	LEPTOTERNA DOLABRATA
RED FLOUR BEETLE	TRIBOLIUM CASTANEUM
RIFFLE BEETLE	ELMIS SP.
ROCK CRAB	CANCER IRRORATUS
ROCK LOBSTER	PANULIRUS LAEVICAUDA
	PANULIRUS LONGIPES CYGNUS
SAND SHRIMP	CRANGON CRANGON
SARCOSAPROPHAGOUS FLIES	
SCUD	GAMMARUS FOSSARUM
	GAMMARUS LACUSTRIS
	GAMMARUS OCEANICUS
	GAMMARUS PSEUDOLIMNAEUS
	GAMMARUS PULEX
	GAMMARUS TIGRINUS

SHRIMP	
	ARTEMIA SP.
	CRANGON ALLMANI
	CRANGON SEPTEMSPINOSA
	LEANDER ADSPERSUS
	LEANDER PACIFICUS
	LYSMATA SETICAUDATA
	METAPENAEUS MONOCEROS
	PALAEMON DEBILIS
	PARAPENAEUS LONGIROSTRIS
	PENAEUS DUORARUM
	PENAEUS JAPONICUS
	SPIRONTOCARIS SP.
SILKWORM	
SOLDIER FLY	HERMIONE HYSTRIX
SPIDER	DRASSYLLUS VIRGINIANA
	LEIOBUNUM FLAVUM
	LYCOSA SP.
	MITOPUS MORIO
	SCHIZOCOSA SP.
SPIDER-BEETLE	PTINUS FUR
	PTINUS SEXPUNCTATUS
	PTINUS TECTUS
SPIDERS	ARANEAE
SPIDERS SCORPIONS AND MITES	ARACHNIDA
SPRINGTAIL	ONYCHIURUS PROCAMPATUS
	TOMOCERUS SP.
SPRINGTAILS	COLLEMBOLA
STINK BUG	NEZARA VIRIDULA
STONEFLY	ACHYNOPTERYX CURVATA
	ACRONEURIA LYCORIAS
	ACRONEURIA SP.
	ALLOPERLA MEDIANA
	ISOGENUS FRONTALIS
	ISOPERLA GORTZI
	NEMOURA SP.
	NEOPERLA SP.
	PERLODES INTRICATA
	PTERONARCELLA BADIA
	PTERONARCYS CALIFORNICA
	PTERONARCYS DORSATA
	STENOPERLA PRASINA
	TAENIOPTERYX MAURA
TERMITES	ISOPTERA
TRAPDOOR SPIDER	PACHYLOMERIDES AUDOUINI
TRUE FLIES	DIPTERA
WATER BOATMAN	SIGARA SP.
WATER BOATMEN	CORIXIDAE
WATER FLEA	ACROPERUS HARPAE
	ALONA COSTATA
	ALONA GUTTATA
	BOSMINA COREGONI
	BOSMINA LONGIROSTRIS
	BOSMINA SP.
	BOSMINOPSIS DEITERSI
	CERIODAPHNIA CORNUTA
	CERIODAPHNIA PULCHELLA
	CERIODAPHNIA QUADRANGULA
	CERIODAPHNIA RETICULATA
	CERIODAPHNIA SP.
	CHYDORUS EURYNOTUS
	CHYDORUS SPHAERICUS
	DAPHNIA CRISTATA
	DAPHNIA GALEATA
	DAPHNIA GALEATA MENDOTAE
	DAPHNIA HYALINA
	DAPHNIA LONGISPINA
	DAPHNIA MAGNA
	DAPHNIA MIDDENDORFFIANA
	DAPHNIA PARVULA
	DAPHNIA PULEX
	DAPHNIA PULICARIA
	DAPHNIA RETROCURVA
	DAPHNIA ROSEA
	DAPHNIA SCHODLERI
	DAPHNIA SP.
	DIAPHANOSOMA BRACHYURUM
	DIAPHANOSOMA LEUCHTENBERG
	DIAPHANOSOMA SP.
	EUBOSMINA TUBICENS
	EURYCERCUS LAMELLATUS
	HOLOPEDIUM GIBBERUM
	HOLOPEDIUM SP.
	LEPTODORA KINDTII
	LEYDIGIA ACANTHOCEROIDES
	LEYDIGIA QUADRANGULARIS
	MACROTHRIX LATICORNIS
	MACROTHRIX ROSEA
	MOINA MACRURA
	MOINA MINUTUS
	PLEUROXUS DENTICULATUS
	PLEUROXUS PROCURVIS
	SIMOCEPHALUS SERRULATUS
	SIMOCEPHALUS VETULUS
WATER FLEAS	CLADOCERA
WATER-BEETLES	DYTISCIDAE
WHIRLIGIGS	GYRINIDAE
YELLOW MEALWORM BEETLE	TENEBRIO MOLITOR

COELENTERATA

CORAL	
	ASTRAEA UNDOSA
	FUNGIA SCUTARIA
	MURICEA SP.
	PAVONA CLIVOSA
	PAVONA GIGANTEA
	POCILLOPORA DAMICORNIS
	POCILLOPORA DANAE
	POCILLOPORA ROBUSTA
	PSAMMOCORA STELLATA
HYDROZOAN	AEQUOREA SP.
	AGALMI OKENI
	ATHORYBIA ROSACEA
	ATHORYBIA SP.
	BATHYPHYSA SIBOGAE
	CUNINA SP.
	DIPHYES DISPAR
	FORSKALIA EDWARDSI
	FORSKALIA THOLOIDES
	HYDRA LITTORALIS
	LEUCKARTIARA OCTONA
	NANOMIA BIJUGA
	ORCHISTOMA SP.
	PHIALIDIUM GREGARIUM
	RHIZOPHYSA FILIFORMIS
	ROSACEA CYMBIFORMIS
	SULCULEOLARIA QUADRIVALVIS
HYDROZOANS	SIPHONOPHORA
SEA PEN	PENNATULA RUBRA
STONY CORAL	PORITES PANAMENSIS

ECHINODERMATA

BRITTLE STAR	OPHIOCOMA ECHINATA
	OPHIOCOMA ERINACEUS
BRITTLE SERPENT AND BASKET STARS	OPHIUROIDEA
HEART URCHIN	LOVENIA CORDIFORMIS
KIRK'S OPHIUROID	

OPHIUROID	AMPHIURA ANGULARIS
	AMPHIURA ANNULIFERA
	AMPHIURA BELGICAE
	AMPHIURA FILIFORMIS
	AMPHIURA MICROPLAX
	AMPHIURA MONORIMA
	ASTROCHLAMYS BRUNEUS
	OPHIACANTHA DENSISPINA
	OPHIACANTHA VIVIPARA
	OPHIOCERES INCIPIENS
	OPHIOCHONDRUS STELLIGER
	OPHIOLEBELLA BISCUTIFERA
	OPHIOMITRELLA CLAVIGERA
	OPHIOMITRELLA FALKLANDICA
	OPHIOMITRELLA INGRATA
	OPHIOMYXA BREVIRIMA
	OPHIOPHOLIS ACULEATA
	OPHIOPLOCUS ESMARKI
	OPHIOSCOLEX NUTRIX
	OPHIOSTEIRA ANTARCTICA
	OPHIOZONELLA FALKLANDI
	OPHIURA MERIDIONALIS
	OPHIURA ROUCHI
	OPHIUROLEPIS MARTENSI
	PECTINURA CYLINDRICA
	PECTINURA GRACILIS
	STEGOPHIURA NODOSA
	STEGOPHIURA SCULPTA
SEA CUCUMBER	HOLOTHURIA FORSKALI
SEA STAR	ASTERINA GIBBOSA
	ASTROMETIS SERTULIFERA
	ASTROPECTEN ARANCIACUS
	ASTROPECTEN ARMATUS
	DERMASTERIAS IMBRICATA
	HENRICIA LEVIUSCULA
	LINCKIA COLUMBIAE
	PATIRIA MINIATA
	PISASTER SP.
	PORANIA PULVILLUS
SEA URCHIN	
	BRISSOPSIS LYRIFERA
	LOXECHINUS ALBUS
	LYTECHINUS ANAMESUS
	PARACENTROTUS LIVIDUS
	STRONGYLOCENTROTUS FRANCISCANUS
	STRONGYLOCENTROTUS PURPURATUS
SERPENT STAR	AMPHIPHOLIS SQUAMATA
	OPHIONOTUS HEXACTIS
STARFISH	

MOLLUSCA

ABALONE	HALIOTIS DISCUS
	HALIOTIS RUFESCENS
BIVALVE	ASTARTE UNDATA
	BELLAMYA UNICOLOR
	CHAMA PELLUCIDA
	CHLAMYS OPERCULARIS
	LIGUMIA SP.
	MERETRIX LUSORIA
	MULINIA LATERALIS
	NUCULANA SP.
	PENITELLA CONRADI
	PODODESMUS PERNOIDES
	SAXIDONUS GIGANTEUS
	SPHAERIUM CORNEUM
	TACHIDIUS DISCIPEA
	VENERICARDIA BOREALIS
	VENERUPIS DECUSSATA
BLACK TURBAN SNAIL	TEGULA FUNEBRALIS
CHITON	
	NUTTALLINA CALIFORNICA
CLAM	
	CORBICULA AFRICANA
	CORBICULA MANILENSIS
	HIATELLA ARCTICA
COMMON PERIWINKLE	LITTORINA LITTOREA
CONCH	BUCCINUM UNDATUM
CUTTLEFISH	SEPIA OFFICINALIS
DOG WHELK	NUCELLA LAPILLUS
EASTERN OYSTER	CRASSOSTREA VIRGINICA
EDIBLE MUSSEL	MYTILUS EDULIS
EDIBLE OYSTER	OSTREA EDULIS
FRESHWATER CLAM	LAMPSILIS RADIATA
FRESHWATER LIMPET	ANCYLUS LACUSTRIS
FRESHWATER SNAIL	CAMPELOMA SP.
	MELANIA TUBERCULATA
GASTROPOD	CONUS CALIFORNICUS
	KELLETTIA KELLETTII
	MESOMPHIX VULGATUS
	PTEROTRACHEA HIPPOCAMPUS
KEYHOLE LIMPET	FISSURELLA VOLCANO
	MEGATHURA CRENULATA
LAND SLUG	AGRIOLIMAX RETICULATUS
	ARION SP.
	DEROCERAS SP.
	LIMAX SP.
LAND SNAIL	MESODON THYROIDES
LIMPET	PATELLA VULGATA
LITTLE-NECKED CLAM	PROTOTHACA STAMINEA
MARINE BIVALVE	ARCTICA ISLANDICA
	CHIONE UNDATELLA
	SEMELE SP.
	YOLDIA SAPOTILLA
MARINE GASTROPOD	BURSA CALIFORNICUS
	EPITONIUM SP.
MARINE LIMPET	ACMAEA SP.
MARINE MUSSEL	CARDIUM EDULE
	LITHOPHAGA PLUMULA
	MACOMA BALTICA
	MYTILUS CALIFORNIANUS
	MYTILUS GALLOPROVINCIALIS
	MYTILUS SP.
	VENUS MERCENARIA
MARINE SNAIL	ACANTHINA SPIRATA
	ALETES SQUAMIGERUS
MARSH CLAM	RANGIA CUNEATA
MARSH SNAIL	
	MELAMPUS BIDENATUS
MUD SNAIL	NASSARIUS OBSOLETUS
MUSSEL	MODIOLUS DEMISSUS
	MODIOLUS MODIOLUS
	SCROBICULARIA PLANA
NORTHERN YELLOW PERIWINKLE	LITTORINA OBTUSATA
NUT SHELL	NUCULA PROXIMA
	NUCULA SP.
	NUCULA SULCATA
OLYMPIA OYSTER	OSTREA LURIDA
OPISTHOBRANCH GASTROPOD	GLEBA CORDATA
OWL LIMPET	LOTTIA GIGANTEA
OYSTER	

	CRASSOSTREA ANGULATA CRASSOSTREA GRYPHOIDES OSTREA EQUESTRIS
PACIFIC OYSTER	CRASSOSTREA GIGAS
PEARL MUSSEL	MARGARITIFERA SP.
PERIWINKLE	LITTORINA IRRORATA LITTORINA LITTORALIS LITTORINA SP.
POND SNAIL	LYMNAEA PEREGER LYMNAEA STAGNALIS
PROSOBRANCH SNAIL	BITHYNIA TENTACULATA VIVIPARUS VIVIPARUS
PULMONATE SNAIL	BIOMPHALARIA GLABRATA HELISOMA SP. PHYSA GYRINA PHYSA PSEUDOLIMNAEUS PHYSA SP.
QUAHOG	MERCENARIA MERCENARIA
RAMSHORN SNAIL	PLANORBIS CONTORTUS PLANORBIS VORTEX
RAMSHORN SNAILS	PLANORBIDAE
RAZOR CLAM	ENSIS DIRECTUS ENSIS ENSIS
ROUGH PERIWINKLE	LITTORINA SAXATILIS
SCALLOP	
	AEQUIPECTEN IRRADIANS PECTEN IRRADIANS
SEA BUTTERFLY	PTEROPOD
SEA HARE	APLYSIA SP.
SEA SQUID	OCTOPUS VULGARIS
SEED CLAM	PISIDIUM CASERTANUM PISIDIUM VIRGINICUM
SLIPPER SHELL	CREPIDULA FORNICATA CREPIDULA SP.
SLUG	
SNAIL	
	CLEOPATRA BULIMOIDES COLUS STIMPSONI CYPRAEA SPADICEA RADIX PEREGRA OVATA SUCCINEA OVALIS
SOFT-SHELLED CLAM	
TOOTH SHELL	DENTALIUM ENTALE
UNIONID FRESHWATER CLAM	ANODONTA ANATINA ANODONTA NUTTALLIANA ANODONTA PISCINALIS UNIO MANCUS UNIO PICTORUM UNIO TUMIDUS
UNIONID MUSSEL	GONIDEA SP.
WHELK	BUSYCON CANALICULATUS

PLATYHELMINTHES

PLANARIA	CRENOBIA ALPINA DUGESIA GONOCEPHALA DUGESIA TIGRINA

PROTOZOA

CILIATE	KERONOPSIS RUBRA PARAMECIUM SP.
FORAMINIFERAN	ROSALINA LEEI SPIROLOCULINA HYALINA

UROCHORDATA

SEA SQUIRT	ASCIDIELLA ASPERSA CIONA INTESTINALIS CORELLA SPECTABILIS PHALLUSIA MAMMILLATA STYELA PLICATA
SEA SQUIRTS	ASCIDIACEA
THALIACEAN	SALPA CYLINDRICA SALPA FUSIFORMIS THALIA DEMOCRATICA

VERTEBRATA

AFRICAN ELEPHANT	LOXODONTA AFRICANA
ALEWIFE	ALOSA PSEUDOHARENGUS
AMERICAN KESTREL OR SPARROW HAWK	FALCO SPARVERIUS
ANCHOVY	
	ENGRAULIS JAPONICUS ENGRAULIS RINGENS
ARCTIC FOX	ALOPEX LAGOPUS
ARCTIC GROUND SQUIRREL	SPERMOPHILUS UNDULATUS
ASPEN	ASPIUS ASPIUS
ATLANTIC COD	GADUS MORHUA
ATLANTIC CROAKER	MICROPOGON UNDULATUS
ATLANTIC MACKEREL	SCOMBER SCOMBRUS
ATLANTIC MENHADEN	BREVOORTIA TYRANNUS
ATLANTIC SALMON	SALMO SALAR
ATLANTIC SILVERSIDE	MENIDIA MENIDIA
BALD EAGLE OR WHITE HEADED EAGLE	HALIAETUS LEUCOCEPHALUS
BANK VOLE	CLETHRIONOMYS GLAREOLUS
BARRACUDA	SPHYRAENA SP.
BASS	
	MICROPTERUS COOSAE MORONE AMERICANAS
BAY ANCHOVY	ANCHOA MITCHILLI
BENGALESE FINCH	LONCHURA STRIATA STREPTOPELIA RISORIA
BLACK BASS	
BLACK BULLHEAD	
BLACK CRAPPIE	
BLACK DUCK	ANAS RUBRIPES
BLACK SKIMMER	RYNCHOPS NIGRA NIGRA
BLACK-TAILED DEER	ODOCOILEUS HEMIONUS COLUMBIANUS

BLACK-TAILED JACKRABBIT	LEPUS CALIFORNICUS
BLEAK	ALBURNUS ALBURNUS
BLUE FISH	
BLUE JAY	CYANOCITTA CRISTATA
BLUE SHARK	PREONACE GLAUCA
BLUE WHALE	BALAENOPTERA MUSCULUS
BLUE-FIN TUNA	THUNNUS THYNNUS
BLUEFISH	POMATOMUS SALTATRIX
BLUEGILL SUNFISH	LEPOMIS MACROCHIRUS
BOBWHITE QUAIL	COLINUS VIRGINIANUS
BONITO	
BOQUE	BOOPS BOOPS
BOVINE	ENTEROLOBIUM SAMAN
BOWFIN	AMIA SP.
BREAM	ABRAMIS BRAMA
BROOK TROUT	SALVELINUS FONTINALIS
BROWN BULLHEAD	AMEIURUS NEBULOSUS
BROWN TROUT	SALMO TRUTTA
BRYDE WHALE	BALAENOPTERA BRYDEI
BUFFALO	
BUFFALOFISH	
BULLHEAD	
	COTTUS CAROLINAE
	COTTUS COGNATUS
BULLHEAD MINNOW	
BULLHEAD TROUT	SALVELINUS MALMA
CALF	
CANDLEFISH	THALEICHTHYS PACIFICUS
CARIBOU OR REINDEER	RANGIFER TARANDUS
CAROLINA WOOD DUCK	AIX SPONSA
CARP	
	CYPRINUS CARPIO
CARPSUCKER	CARPIODES SP.
	CARPIODES VELIFER
CAT	FELIS SILVESTRIS
CATFISH	
	HETEROPNEUSTES FOSSILIS
	ICTALURUS LACUSTRIS
CATFISHES	ICTALURIDAE
CHANNEL CATFISH	ICTALURUS PUNCTATUS
CHICKEN	
	GALLUS DOMESTICUS
CHINOOK SALMON	ONCORHYNCHUS TSCHAWYTSCHA
CHUB	LEUCISCUS CEPHALUS
CHUM SALMON	ONCORHYNCHUS KETA
CLAWED TOAD	
COALFISH	POLLACHIUS VIRENS
COD	GADUS SP.
COHO SALMON	ONCORHYNCHUS KISUTCH

COMMON CARDINAL	RICHMODENA CARDINALIS
COMMON EGRET	CASMERODIUS ALBUS
COMMON FROG	RANA TEMPORARIA
COMMON SHINER	NOTROPIS CORNUTUS
COMMON SHREW	SOREX ARANEUS
COMMON TERN	STERNA HIRUNDO
COMMON TOAD	BUFO BUFO
COMMON TURKEY	MELEAGRIS GALLOPAVO
COMMON VOLE	MICROTUS ARVALIS
CORMORANT	PHALACROCORAX CARBO
COTTON RAT	SIGMODON HISPIDUS
COTTONTAIL RABBIT	SYLVILAGUS FLORIDANUS
COTURNIX QUAIL	COTURNIX COTURNIX
COW	
COWBIRD	MOLOTHRUS SP.
COYOTE	CANIS LATRANS
CRAPPIE	POMOXIS ANNULARIS
	POMOXIS SP.
CROAKERS	
CRUCIAN CARP	CARASSIUS CARASSIUS
CUNNER	TAUTOGOLABRUS ADSPERSUS
CUT-THROAT TROUT	SALMO CLARKI
DAB	LIMANDA LIMANDA
DACE	LEUCISCUS LEUCISCUS
DALL SHEEP	
DEER	ODOCOILEUS SP.
DEER MOUSE	PEROMYSCUS POLIONOTUS
DOG	CANIS FAMILIARIS
DRUM	APLODINOTUS GRUNNIENS
DUNLIN	CALIDRIS ALPINA
DUSKY SHARK	CARCHARHINUS OBSURUS
EEL	ANGUILLA JAPONICA
	ANGUILLA SP.
EGRET	
EUROPEAN EEL	ANGUILLA ANGUILLA
EUROPEAN HARE	LEPUS EUROPAEUS
FALL BIRDS	
FATHEAD MINNOW	PIMEPHALES PROMELAS
FERAL PIGEON	
FIN WHALE	BALAENOPTERA PHYSALUS
FISH	ASPIUS RAPAX
	CHELA BACAILA
	PEPRILUS TRICANTHUS
	PIMELODUS MACULATUS
	SARDINELLA AURITA
	SARDINELLA MADERENSIS
	TILAPIA MOSSAMBICA
	TILAPIA ZILLII
	TRINECTES MACULATUS
FISH HAWK OR OSPREY	PANDION HALIAETUS

FLAGFISH	JORDANELLA FLORIDAE
FLATFISHES	PLEURONECTIDAE
FLOUNDER	
	EOPSETTA JORDANI
	PLEURONECTES FLESUS
FLYING SQUIRREL	GLAUCOMYS SABRINUS YUKONENSIS
FRESHWATER DRUM	
FROG	
	RANA ARVALIS
GAR	
GIZZARD SHAD	DOROSOMA CEPEDIANUM
GOAT	
GOATFISH	UPENEUS MOLUCCENSIS
GOBY	GOBIUS SP.
GOLDEN TENCH	TINCA TINCA
GOLDFISH	
	CARASSIUS AURATUS
	GILLICHTHYS MIRABILIS
GRASS CARP	CTENOPHARYNGODON IDELLA
	CTENOPHARYNGODON SP.
GRAYLING	THYMALLUS ARCTICUS BAICALENSIS
	THYMALLUS THYMALLUS
GREEN SUNFISH	LEPOMIS CYANELLUS
GREY MULLET	MUGIL CEPHALUS
GREY SEAL	HALICHOERUS GRYPUS
GREY WHALE	ESCHRICHTIUS ROBUSTUS
GUDGEON	GOBIO GOBIO
GUINEA PIG	CAVIA PORCELLUS
GUPPY	LEBISTES RETICULATUS
	POECILIA RETICULATUS
HADDOCK	MELANOGRAMMUS AEGLEFINUS
HAKE	MERLUCCIUS ANGUSTIMANUS
	MERLUCCIUS PRODUCTUS
	MERLUCCIUS SP.
HALIBUT	HIPPOGLOSSUS SP.
HARP SEAL	PAGOPHILUS GROENLANDICUS
HERON	
HERRING	CLUPEA HARENGUS
	CLUPEA SP.
HERRING GULL	LARUS ARGENTATUS
HOG SUCKER	HYPENTELIUM SP.
HORSE	EQUUS CABALLUS
HORSE MACKEREL	TRACHURUS JAPONICUS
HOTU	CHONDROSTOMA NASUS PRESPENSE
HOUSE MOUSE	MUS MUSCULUS
HOUSE SPARROW	PASSER DOMESTICUS
HUMAN BEING	HOMO SAPIENS
HUMBACK WHALE	MEGAPTERA NOVAEANGLIAE
HUMMINGBIRD	PANTERPE INSIGNIS

HYBRID SUNFISH	
IDE OR SILVER ORFE	LEUCISCUS IDUS
JAPANESE MEDAKA	ORYZIAS LATIPES
JAPANESE QUAIL	COTURNIX COTURNIX JAPONICA
KANGUROO RAT	DIPODOMYS MERRIAMI
KESTREL	
KING KLIP	GENYPTERUS CAPENSIS
KNOT	CALIDRIS CANUTUS
LAKE CHUBSUCKER	
LAKE TROUT	SALVELINUS NAMAYCUSH
LAMB	
LARGEMOUTH BASS	MICROPTERUS SALMOIDES
LARGESCALE SUCKER	
LARK	CALANDRA ORYZAE
LEAST WEASEL	MUSTELA RIXOSA ALLEGHNIENSIS
LEOPARD FROG	RANA PIPIENS
LINED SOLE	ACHIRUS LINEATUS
LIZARD	
LONG-BILLED MARSH WREN	TELMATODYTES PALUSTRIS
LONGFIN SMELT	
MACKEREL	SCOMBER JAPONICUS
	SCOMBER SP.
	SCOMBEROMORUS CAVALLA
MADTOM	
MAKO SHARK	ISURUS OXYRINCHUS
MALLARD	ANAS PLATYRHYNCHOS
MARSH HAWK	CIRCUS HUDSONIUS
MARSH RICE RAT	ORYZOMYS PALUSTRIS
MEADOW VOLE	MICROTUS PENNSYLVANICUS
MENHADEN	BREVOORTIA SP.
MERGANSER	
MINKE WHALE	BALAENOPTERA ACUTOROSTRATA
MINNOW	
	PHOXINUS PHOXINUS
MONKEY	
MOSQUITO FISH	
	GAMBUSIA AFFINIS
MOUNTAIN WHITEFISH	PROSOPIUM WILLIAMSONI
MOURNING DOVE	ZENAIDURA MACROURA
MOUSE	
MULLET	MULLUS BARBATUS
MUMMICHOG (KILLIFISH)	FUNDULUS HETEROCLITUS
MUSKELLUNG	ESOX MASQUINONGY
NEEDLEFISH	
NEWT	
	DIEMICTYLUS SP.
NORTHERN PARULA	PARULA AMERICANA

NORTHERN SQUAWFISH	PTYCHOCHEILUS OREGONENSIS
NORWAY RAT	RATTUS NORVEGICUS
OCEAN SUNFISH	MOLA MOLA
ORANGE-THROAT DARTER	ETHEOSTOMA SPECTABILE
OYSTER TOADFISH	OPSANUS TAU
PACIFIC COD	GADUS MACROCEPHALUS
PACIFIC OCEAN PERCH	SEBASTODES ALUTUS
PACIFIC RIGHT WHALE	EUBALAENA GLACIALIS SIEBOLDII
PARTRIDGE	PERDIX PERDIX
PASSERINE BIRDS	PASSERIFORMES
PATAGONIAN BLENNY	ELEGINOPS MACLOVINUS
PERCH	PERCA FLUVIATILIS SERRANUS ALEXANDRINUS
PERCHES	PERCIDAE
PHEASANT	
PICKEREL	
PIG	
PIGMY BLUE WHALE	BALAENOPTERA MUSCULUS BREVICAUDA 0
PIGMY SHREW	SOREX MINUTUS
PIKA OR PIPING HARE	OCHOTONA SP.
PIKE	ESOX LUCIUS
PILCHARD	SARDINA PILCHARDUS
PINE VOLE	PITYMYS SUBTERRANEUS
PINFISH	
PINK SALMON	ONCORHYNCHUS GORBUSCHA
PLAICE	PARALICHTHYS OLIVACEUS PLEURONECTES PLATESSA
POLAR BEAR	THALARCTOS MARITIMUS
POLLOCK	POLLACHIUS POLLACHIUS
PRAIRIE FALCON	FALCO MEXICANUS
QUAIL	
QUILLBACK CARPSUCKER	CARPIODES CYPRINUS
RABBIT	
RAIL	RALLUS LONGIROSTRIS
RAINBOW TROUT OR STEELHEAD TROUT	SALMO GAIRDNERII
RAT	
RED FOX	VULPES VULPES
RED MULLET	MULLUS SURMULETUS
RED SEA BREAM	CHRYSOPHRYS MAJOR
RED SQUIRREL	TAMIASCIURUS HUDSONI PREBLEI
RED-BACKED VOLE	CLETHRIONOMYS RUTILUS DAWSONI
REDHORSE (FISH)	MOXOSTOMA SP.
REDSHANK	TRINGA TOTANUS
RETICULATE SCULPIN	
RINGED SEAL	PHOCA HISPIDA = PUSA HISPIDA PUSA HISPIDA = PHOCA HISPIDA
ROACH	RUTILUS RUTILUS RUTILUS SP.
ROCK BASS	AMBLOPLITES RUPESTRIS
ROCK SOLE	LEPIDOPSETTA BILINEATA
ROCK-DOVE	COLOMBA LIVIA
RODENTS	RODENTIA
ROE DEER	CAPREOLUS CAPREOLUS
ROHU	LABEO ROHITA
RUDDY VOLE	CLETHRIONOMYS RUTILUS
RUFFED GROUSE	BONASA UMBELLUS
SACRAMENTO BLACKFISH	
SALMON	
SALMONOIDS	SALMONIDAE
SAURY	COLOLABIS SAIRA
SAVANNA SPARROW	PASSERCULUS SANDWICHENSIS
SCARLET TANAGER	PIRANGA OLIVACEA
SCIAENIDAE (FISH)	PSEUDOTOLITHUS SENEGALENSIS PSEUDOTOLITHUS TYPUS
SEA BREAM	ARCHOSARGUS RHOMBOIDALIS
SEA LION	ZALOPHUS CALIFORNIANUS
SEI WHALE	BALAENOPTERA BOREALIS
SHANNY	BLENNIUS PHOLIS
SHEEP	
SHEEPSHEAD MINNOW	CYPRINODON VARIEGATUS
SHORT TAILED VOLE OR FIELD VOLE	MICROTUS AGRESTIS
SHREW	
SILVER FOX MUTANT OF VULPES VULPES	
SILVERSIDE	MENIDIA SP.
SILVERSIDE MINNOW	
SILVERY SIDED HORSE MACKEREL	TRACHURUS TRACHURUS
SKIPJACK TUNA	KATSUWONUS PELAMIS
SMALL SCALED SCORPION FISH	SCORPAENA PORCUS
SMALLMOUTH BASS	MICROPTERUS DOLOMIEUI
SNAPPERS	LUTJANIDAE
SOCKEYE SALMON	ONCORHYNCHUS NERKA
SOLE	SOLEA SOLEA SOLEA SP.
SPANISH MACKEREL	SCOMBEROMORUS MACULATUS
SPARROW	SPIZELLA PASSERINA
SPERM WHALE	PHYSETER CATODON
SPINEFOOT (RABBITFISH)	SIGANUS LURIDUS

SPOT	LEIOSTOMUS SP. LEIOSTOMUS XANTHURUS
SPRING BIRDS	
SQUIRREL	TAMIASCIURUS SP.
SQUIRREL MONKEY	SAIMIRI SCIUREUS
STARRY FLOUNDER	PLATICHTHYS STELLATUS
STEER	
STELLER'S JAY	CYANOCITTA STELLERI
STONE LOACH	NOEMACHEILUS BARBATULUS
STRIPED FIELD-MOUSE	APODEMUS AGRARIUS
STRIPED KILLIFISH	FUNDULUS MAJALIS
SUCKER	
SUMMER FLOUNDER	PARALICHTHYS DENTATUS
SUNFISH	
TAUTOG	TAUTOGA SP.
TENCH	TINCA VULGARIS
TENPOUNDER	ELOPS SAURUS
TERN	STERNA SP.
THICK LIPPED GREY MULLET	MUGIL CHELO
THIN-LIPPED GREY MULLET	MUGIL CAPITO
THORNBACK SKATE	RAJA CLAVATA
THREADFIN SHAD	
THRUSH	HYLOCICHLA MUSTELLINA
TROUT	
TUNA	THUNNUS SP.
TUNDRA VOLE	MICROTUS OECONOMUS
TURBOT	
TURTLE	PSEUDEMYS SP.
UGANDA KOB	ADENOTA KOB THOMASI
VOLE	MICROTUS OECONOMUS MACFARLANI MICROTUS ORCADENSIS MICROTUS SP.
WALLEYE	STIZOSTEDION LUCIPERCA STIZOSTEDION VITREUM VITREUM
WATERBUCK	KOBUS LECHE
WEASEL	
WHITE CATFISH	
WHITE FISH	COREGONUS ALBULA TYPICA COREGONUS SP.
WHITE FOX MUTANT OF VULPES VULPES	
WHITE PERCH OR SILVER BASS	ROCCUS AMERICANUS
WHITE SUCKER	CATOSTOMUS COMMERSONI
WHITE TILAPIA	TILAPIA MELANOPLEURA
WHITE-TAILED DEER	ODOCOILEUS VIRGINIANUS
WHITING	
WHITINGS	SILLAGINIDAE
WINTHER FLOUNDER	PSEUDOPLEURONECTES AMERICANUS
WOLF	CANIS LUPUS
WOODCOCK	
YELLOW PERCH	PERCA FLAVESCENS
YELLOW-FIN TUNA	THUNNUS ALBACORES
YELLOW NECKED FIELD MOUSE	APODEMUS FLAVICOLLIS
YELLOW-TAIL	SERIOLA QUINQUERADIATA
ZANDER	LUCIOPERCA LUCIOPERCA
ZEBRA FISH	BRACHYDANIO RERIO

****

P L A N T S

****

KORMOPHYTA

ALFALFA	MEDICAGO SATIVA
AMERICAN ASH	FRAXINUS AMERICANA
AMERICAN BEECH	FAGUS GRANDIFOLIA
AMERICAN CHESTNUT	CASTANEA DENTATA
AMERICAN HOP HORNBEAM	OSTRYA VIRGINIANA
AMERICAN LINDEN	TILIA AMERICANA
ANGIOSPERMS	ANGIOSPERMAE
APRICOT	PRUNUS ARMENIACA
AQUATIC FERN	AZOLLA SP. SALVINIA AURICULATA
AQUATIC LIWERWORT	RICCIA FLUITANS
ARROW WOOD OR DOCKMACKIE	VIBURNUM ACERIFOLIUM
ARROWLEAF	SAGITTARIA LATIFOLIA SAGITTARIA PLATYPHYLLA
ASSEGAI WOOD MANGWE OR YELLOW WOOD	TERMINALIA SERICEA
AUSTRIAN PINE	PINUS NIGRA
BALSA	OCHROMA LAGOPUS
BALSAM FIR	ABIES BALSAMEA
BARLEY	HORDEUM MURINUM HORDEUM SATIVUM HORDEUM SP.
BARNYARD GRASS	ECHINOCHLOA CRUSGALLI
BEAN	PHASEOLUS ATRAPURPUREA
BEARBERRY	ARCTOSTAPHYLOS UVA-URSI
BEECH	FAGUS CRENATA FAGUS SP.
BIRCH	BETULA ERNAII (*) BETULA MAXIMOWICZIANA BETULA PLATYPHYLLA (*) BETULA SP.
BIRDSFOOT TREFOIL	LOTUS CORNICULATUS
BLACK HAW OR STAG BUSH	VIBURNUM PRUNIFOLIUM
BLACK LOCUST	ROBINIA PSEUDOACACIA
BLACK OAK	QUERCUS VELUTINA
BLACK TUPELO	NYSSA SYLVATICA
BLADDERWORT	UTRICULARIA INFLATA UTRICULARIA SP. UTRICULARIA SUBULATA
BLUE SPRUCE	
BROAD-LEAVED PONDWEED	POTAMOGETON NATANS
BROCCOLI	BRASSICA OLERACEA VAR ITALICA
BRUSSEL'S SPROUT	BRASSICA OLERACEA VAR GEMMIFERA
BULRUSH OR CLUBRUSH	SCIRPUS SP.
BUR CLOVER	MEDICAGO HISPIDA
BUR-REED	SPARGANIUM SP.
BUTTON-WOOD	CONOCARPUS ERECTUS
CABBAGE	BRASSICA OLERACEA BRASSICA OLERACEA VAR CAPITATA
CANADIAN WATERWEED	ELODEA CANADENSIS
CARROT	DAUCUS CAROTA DAUCUS CAROTA VAR SATIVA
CAULIFLOWER	BRASSICA OLERACEA VAR BOTRYTIS
CELERY	APIUM GRAVEOLENS
CEREAL	
CHAMISE	ADENOSTOMA FASCICULATUM
CHESTNUT OAK	QUERCUS PRINUS
CHICKWEED	STELLARIA ALSINE (*)
CLOVER	MEDICAGO SP. TRIFOLIUM PERENNE
COCK'S FOOT	DACTYLIS GLOMERATA
COMMON BEET	BETA VULGARIS
COMMON CAMELLIA	CAMELLIA JAPONICA
COMMON CLUBRUSH OR GREATER RUSH	SCHOENOPLECTUS LACUSTRIS
COMMON COW-PEA	VIGNA SINENSIS
COMMON DUCKWEED	LEMNA MINOR
COMMON LENTIL	LENS CULINARIS
COMMON REED	PHRAGMITES COMMUNIS
COMMON SUNFLOWER	HELIANTHUS ANNUUS
COMMON WHEAT	TRITICUM AESTIVUM = T SATIVUM TRITICUM SATIVUM = T AESTIVUM
CONIFEROUS TREES	
CONIFERS	
CORDGRASS	SPARTINA SP.
CORN	
COTTON	GOSYPIUM SP.
CUCUMBER	CUCUMIS SATIVUS
CURLED PONDWEED	POTAMOGETON CRISPUS
CYPRESS	
DHAINCHA	SESBANIA CANNABINA
DOG ROSE OR HIP ROSE	ROSA CANINA
DOUGLAS FIR	PSEUDOTSUGA MENZIESSII
DUCKWEED	LEMNA SP.
DURMAST OAK	QUERCUS PETRAEA
DURUM WHEAT	TRITICUM DURUM
EAGLE FERN OR BRACKEN	PTERIDIUM AQUILINUM
EASTERN ARBORVITAE OR WHITE CEDAR	THUJA OCCIDENTALIS

EASTERN HEMLOCK	TSUGA CANADENSIS
EASTERN POPLAR	POPULUS DELTOIDES
EASTERN RED CEDAR	JUNIPERUS VIRGINIANA
EASTERN WHITE PINE	PINUS STROBUS
EEL-GRASS OR TAPE-GRASS	VALLISNERIA AMERICANA VALLISNERIA SP.
ENDIVE	CICHORIUM ENDIVIA
ENGELMANN'S SPRUCE	PICEA ENGELMANNI
ENGLISH OAK	QUERCUS ROBUR
ERMAN'S BIRCH	BETULA ERMANII
EUROPEAN ASH	FRAXINUS EXCELSIOR
EUROPEAN GOLDEN ROD	SOLIDAGO VIRGAUREA
EUROPEAN TURKEY OAK	QUERCUS CERRIS
FENNEL PONDWEED OR SAGO PONDWEED	POTAMOGETON PECTINATUS
FENUGREEK	TRIGONELLA FOENUMGRAECUM
FERNS	FILICES
FESCUE	FESTUCA ARUNDINACEA
FIR	ABIES MARIESII (*) ABIES SACHALINENSIS ABIES VEITCHII (*)
FLAX	
FLOWERING DOGWOOD	CORNUS FLORIDA
FODDER RAPE	
GARDEN PEA	PISUM SATIVUM
GIANT BUR-REED	SPARGANIUM EURYCARPUM
GIANT PIGWEED	AMARANTHUS EDULIS
GOLDEN CLUB	ORONTIUM AQUATICUM
GRAM CHICK-PEA	CICER ARIETINUM
GRAND FIR	ABIES GRANDIS
GRASS	ARISTIDA ARGENTEA DIGITARIA ADSENDENS (*) DIGITARIA ERIANTHA ERAGROSTIS PALLENS PHALARIS MINOR STIPA CAPENSIS
GRASS WEED OR GRASS WRACK	ZOSTERA ASIATICA ZOSTERA MARINA ZOSTERA SP.
GRASSES	GRAMINEAE
GREATER BLADDERWORT	UTRICULARIA VULGARIS
GREATER REEDMACE OR CAT'S TAIL	TYPHA LATIFOLIA
GRECIAN LAUREL	LAURUS NOBILIS
GUAR	CYAMOPSIS TETRAGONOLOBUS
GYMNOSPERMS	GYMNOSPERMAE
HAWTHORN	CRATAEGUS MONOGYNA
HEDGE MAPLE	ACER CAMPESTRIS
HEMLOCK	
HOLY OAK	QUERCUS ILEX
HORNED PONDWEED	ZANNICHELLIA PALUSTRIS
HORNWORT	CERATOPHYLLUM SP.
HORSE BEAN	VICIA FABA
HORSE CHESTNUT	AESCULUS OCTANDRA
ITALIAN RYEGRASS	LOLIUM MULTIFLORUM
JACK BEAN OR SWORD BEAN	CANAVALIA ENSIFORMIS
JAPANESE LARCH	LARIX LEPTOLEPIS
JAPANESE MILLET	ECHINOCHLOA CRUSGALLI VAR FRUMENTACEA
JUDAS TREE	CERCIS CANADENSIS
JUMPY BEAN	LEUCAENA GLAUCA
KAURI PINE	AGATHIS AUSTRALIS
KETMIA	HIBISCUS PALUSTRIS
KIDNEY BEAN	PHASEOLUS VULGARIS
KOHL-RABI	BRASSICA OLERACEA VAR GONGYLODES
LARCH	LARIX SP.
LARGE-TOOTHED ASPEN	POPULUS GRANDIDENTATA
LEAF LETTUCE	LACTUCA SATIVA VAR CRISPA
LEAST WATER-LILY	NUPHAR PUMILUM
LEEK	ALLIUM PORRUM
LEGUME	BURKEA AFRICANA CALOPOGONIUM MUCUNOIDES CENTROSEMA PUBESCENS ELEPHANTORRHIZA SP. INDIGOFERA OXYTROPIS INDIGOFERA SORDIDA PUERARIA PHASEOLOIDES RHYNCHOSIA MONOPHYLLA STYLOSANTHES SP. TEPHROSIA LUPINIIFOLIA TRIGONELLA ARABICA
LESSER CAT'S TAIL OR LESSER REEDMACE	TYPHA ANGUSTIFOLIA
LETTUCE	LACTUCA SATIVA
LILAC	SYRINGA VULGARIS
LIVERWORT	
LOBLOLLY PINE	PINUS TAEDA
LONG-STALKED PONDWEED	POTAMOGETON PRAELONGUS
LUPINE	LUPINUS SP.
MAHOGANY	SWIETENIA MAHOGANI
MAIZE	ZEA MAYS
MANGROVE	AVICENNIA TOMENTOSA
MAPLE	ACER SP.
MONEYWORT	LYSIMACHIA NUMMULARIA
MOSS	AMBLYSTEGIUM RIPARIUM BRYUM PSEUDOTRIQUETRUM CAMPTOTHECIUM LUTESCENS DICRANELLA VARIA ENCALYPTA VULGARIS FONTINALIS SQUAMOSA PHILONOTIS FONTANA
MOUNTAIN MAGNOLIA	MAGNOLIA ACUMINATA
MUNG BEAN	PHASEOLUS AUREUS

MUSTARD	
NARROW-LEAVED PONDWEED	POTAMOGETON FOLIOSUS
NARROW-LEAVED VETCH	VICIA ANGUSTIFOLIA (*)
NEEDLE SPIKE-RUSH	ELEOCHARIS ACICULARIS
NORTHERN WHITE CEDAR	
NORWAY SPRUCE	PICEA ABIES
OAK	QUERCUS DUMOSA QUERCUS MUEHLENBERGII QUERCUS SP.
OAT	AVENA SATIVA
ONION, GARDEN	ALLIUM CEPA
PANSY	VIOLA TRICOLOR
PAPER BIRCH OR CANOE BIRCH	BETULA PAPYRIFERA
PARSLEY	PETROSELINUM CRISPUM
PEA	PISUM SP.
PEA VINE	
PEANUT	ARACHIS HYPOGAEA
PENNSYLVANIAN MAPLE	ACER PENNSYLVANICUM
PEPPER	
PERENNIAL RYEGRASS	LOLIUM PERENNE
PICKEREL WEED	PONTEDERIA CORDATA
PIGEON PEA	CAJANUS CAJAN
PINE	PINUS DENSIFLORA PINUS THUNBERGII
PINEAPPLE	ANANAS COMOSUS
POND PINE	PINUS SEROTINA
PONDEROSA PINE	PINUS PONDEROSA
PONDWEED	POTAMOGETON DIVERSIFOLIUS POTAMOGETON ILLINOEASIS POTAMOGETON SP.
POPLAR	POPULUS DAVIDIANA
POTATO	SOLANUM TUBEROSUM
QUAKING ASPEN OR TREMBLING ASPEN	POPULUS TREMULOIDES
RADISH, GARDEN	RAPHANUS SATIVUS
RED ALDER	ALNUS RUBRA
RED CLOVER	TRIFOLIUM PRATENSE
RED ELM	ULMUS RUBRA
RED MANGROVE	RHIZOPHORA MANGLE
RED MAPLE	ACER RUBRUM
RED OAK	QUERCUS RUBRA
RED PINE	PICEA RUBRA PINUS RESINOSA
REED SWEET-GRASS	GLYCERIA MAXIMA
REEDMACE OR CAT'S TAIL	TYPHA SP.
RICE	ORYZA SATIVA
RIGID HORNWORT	CERATOPHYLLUM DEMERSUM
RUTABAGA	BRASSICA NAPUS VAR NAPO BRASSICA

RYE	SECALE CEREALE
SASSAFRAS OR AUGUE TREE	SASSAFRAS ALBIDUM
SAW GRASS	CLADIUM JAMAICENSIS
SCARLET OAK	QUERCUS COCCINEA
SEA GRAPE	COCCOLOBA UVIFERA
SEA GRASS	THALASSIA SP. THALASSIA TESTUDIUNUM
SEA ISLAND COTTON	GOSSYPIUM BARBADENSE
SENNA	CASSIA MIMOSOIDES
SHORE-WEED	LITTORELLA UNIFLORA
SHRUB	REAUMURIA NEGEVENSIS
SITKA SPRUCE	PICEA SITCHENSIS
SLASH PINE	PINUS ELLIOTTII
SOFT RUSH	JUNCUS EFFUSUS
SOFT STEM BULRUSH OR GREAT BULRUSH	SCIRPUS VALIDUS
SORGHUM	SORGHUM VULGARE
SORREL TREE OR SOUR WOOD	OXYDENDRUM ARBOREUM
SOYBEAN	GLYCINE MAX
SPIKE-RUSH	ELEOCHARIS CELLULOSA ELEOCHARIS QUADRANGULATA
SPINACH	SPINACIA OLERACEA
SPRING BARLEY	
SPRING WHEAT	
SPRUCE	PICEA SP.
STAR DUCKWEED	LEMNA TRISULCA
SUGAR BEET	BETA SACCHARIFERA
SUGAR CANE	SACCHARUM OFFICINALE
SUGAR MAPLE	ACER SACCHARUM
SURF-GRASS	PHYLLOSPADIX IWATENSIS PHYLLOSPADIX TORREYI
SWEET CLOVER	MELILOTUS SP.
SWEET POTATO	IPOMOEA BATATAS
TICK CLOVER	DESMODIUM CANUM DESMODIUM INTORTUM
TIMOTHY-GRASS	PHLEUM PRATENSE
TOMATO	LYCOPERSICON ESCULENTUM
TREE	PARINARI CAPENSE
TROPICAL SEA-GRASS	CYMODOCEA SP. ENHALUS SP. HALODULE SP. HALOPHILA DECIPIENS HALOPHILA MINOR HALOPHILA OVALIS HALOPHILA SPINULOSA SYRINGODIUM SP.
TUFTED PERENNIAL GRASS	HYPARRHENIA RUFA
TULIP TREE	LIRIODENDRON TULIPIFERA
TURNIP	BRASSICA RAPA
UPLAND COTTON	GOSSYPIUM HIRSUTUM

VETCH	
VIRGINIA PINE	PINUS VIRGINIANA
WATER HORSETAIL	EQUISETUM FLUVIATILE
WATER MILFOIL	MYRIOPHYLLUM BRASILIENSE
	MYRIOPHYLLUM HETEROPHYLLUM
	MYRIOPHYLLUM SP.
WATER PLANTAIN	ALISMA PLANTAGO AQUATICA
WATER-LILY	NUPHAR POLYSEPALUM
	NUPHAR SP.
WATERSHIELD	BRASENIA SCHREBERI
WATERWEED	ELODEA DENSA
	ELODEA SP.
WHITE BIRCH	
WHITE CLOVER	TRIFOLIUM REPENS
WHITE FIR	
WHITE MANGROVE	LAGUNCULARIA RACEMOSA
WHITE OAK	QUERCUS ALBA
WHITE PINE	
WHITE POTATO	
WHITE WILLOW	SALIX ALBA
WILD GRAPE	VITIS VULPINA
WILD RICE OR INDIAN RICE	ZIZANIA AQUATICA
WILLOW	SALIX RUBRUM
WINTER WHEAT	
WITCH HAZEL	HAMAMELIS VIRGINIANA
WORMWOOD	CHENOPODIUM AMBROSIOIDES
YAM	DIOSCOREA SP.
YELLOW BIRCH	BETULA ALLEGHENIENSIS
YELLOW POND LILY	NUPHAR ADVENA
YELLOW WATER-LILY	NUPHAR LUTEUM

THALLOPHYTA: ALGAE

BLUE-GREEN ALGA	AGMENELLUM QUADRUPLICATUM
	ANABAENA CYLINDRICA
	ANABAENA DOMINANCE
	ANABAENA FLOS-AQUAE
	ANABAENA SCHEREMETIEVI
	ANABAENA SP.
	ANABAENA VARIABILIS
	ANABAENOPSIS SP.
	ANACYSTIS NIDULANS
	APHANIZOMENON FLOS AQUAE F GRACILE
	APHANIZOMENON SP.
	APHANOCAPSA DELICATISSIMA
	APHANOTHECE CLATHRATA
	APHANOTHECE NIDULANS
	CALOTHRIX PARIETINA
	CHROOCOCCUS LIMNETICUS
	CHROOCOCCUS SP.
	CHROOCOCCUS TURGIDUS
	COCCOCHLORIS SP.
	COCCOCHLORIS STAGNINA
	COELOSPHAERIUM NAEGELIANUM
	CYLINDROSPERMUM SP.
	DACTYLOCOCCOPSIS SMITHII
	GLOEOCAPSA ALPICOLA
	GLOEOCAPSA RUPESTRIS
	GOMPHOSPHAERIA LACUSTRIS
	LYNGBYA SP.
	MERISMOPEDIA TENUISSIMA
	MICROCYSTIS AERUGINOSA
	MICROCYSTIS FLOS-AQUAE
	MICROCYSTIS SP.
	NOSTOC MUSCORUM
	OSCILLATORIA AGARDHII
	OSCILLATORIA AGARDHII VAR ISOTHRIX
	OSCILLATORIA BORNETI
	OSCILLATORIA BOURRELLYI
	OSCILLATORIA CHALYBIA
	OSCILLATORIA FORMOSA
	OSCILLATORIA LIMNETICA
	OSCILLATORIA REDEKEI
	OSCILLATORIA RUBESCENS
	OSCILLATORIA SP.
	OSCILLATORIA TENUIS
	OSCILLATORIA THIEBAUTII
	PHORMIDIUM FAVEOLARUM
	PHORMIDIUM LURIDUM
	PLECTONEMA NOSTOCORUM
	RADIOCYSTIS GEMINATA
	SCHIZOTHRIX CALCICOLA
	SYNECHOCOCCUS AERUGINOSUS
BLUE-GREEN ALGAE	CYANOPHYCEAE
	MYXOMYPHYCEAE CYANOPHYCEAE
	TRICHODESMIUM SP.
BROWN ALGA	AGARUM CRIBROSUM
	CHORDA FILUM
	COSTARIA COSTATA
	CYSTOSEIRA OSMUNDACEA
	ECTOCARPUS CONFERVOIDES
	EGREGIA MENZIESII
	FUCUS SERRATUS
	FUCUS VESICULOSUS
	HETEROCHORDARIA ABIETINA
	LAMINARIA CHICHORIOIDES
	LAMINARIA DIGITATA
	LAMINARIA JAPONICA
	LAMINARIA LONGICRURIS
	LAMINARIA SACCHARINA
	MACROCYSTIS PYRIFERA
	PELVETIA WRIGHTII
	PETALONIA FASCIA
	PTERYGOPHORA CALIFORNICA
	SARGASSUM FULVELLUM
	SARGASSUM PALLIDUM
	SARGASSUM RINGGOLDIANUM
	SARGASSUM SP.
	SARGASSUM THUNBERGII
	SCYTOSIPHON LOMENTARIA
BROWN ALGAE	PHAEOPHYCEAE
	PHAEOPHYTA
	UNDARIA PINNATIFIDA
CRYPTOPHYCEAE	CHROOMONAS NORDSTEDTII
	CRYPTOMONAS EROSA
	CRYPTOMONAS OVATA
	CRYPTOMONAS PSEUDOBALTICA
	CRYPTOMONAS SP.
	RHODOMONAS LACUSTRIS
	RHODOMONAS MINUTA
DIATOM	AMPHIPLEURA RUTILANS
	AMPHIPRORA ORNATA
	ASTERIONELLA FORMOSA
	ASTERIONELLA JAPONICA
	ASTERIONELLA SP.
	BACTERIASTRUM HYALINUM
	BIDDULPHIA AURITA
	BIDDULPHIA SINENSIS
	CERATAULINA BERGONII
	CHAETOCEROS CURVISETUS
	CHAETOCEROS DEBILIS
	CHAETOCEROS GRACILIS
	CHAETOCEROS LAUDERI
	CHAETOCEROS LORENZIANUS
	CHAETOCEROS SOCIALIS
	CHAETOCEROS SP.
	COSCINODISCUS ANGSTII
	COSCINODISCUS ANGSTII VAR GRANULOMARGINATUS

	COSCINODISCUS EXCENTRICUS
	COSCINODISCUS GIGAS
	COSCINODISCUS GRANII
	COSCINODISCUS LINEATUS
	COSCINODISCUS RADIATUS
	COSCINODISCUS SP.
	COSCINODISCUS WAILESII
	CYCLOTELLA BODANICA
	CYCLOTELLA COMENSIS
	CYCLOTELLA COMTA
	CYCLOTELLA GLOMERATA
	CYCLOTELLA MENEGHINIANA
	CYCLOTELLA NANA
	CYCLOTELLA SOCIALIS
	CYCLOTELLA SP.
	CYCLOTELLA STELLIGERA
	CYLINDROTHECA CLOSTERIUM
	CYMATOPLEURA SOLEA
	DIATOMA ELONGATUM
	DIATOMA VULGARE
	DITYLUM BRIGHTWELLII
	ERKENIA SUBAEQUICILIATA
	FRAGILARIA CROTONENSIS
	FRAGILARIA PINNATA
	FRAGILARIA SP.
	HEMIAULUS SINENSIS
	LAUDERIA BOREALIS
	LAUDERIA SP.
	LEPTOCYLINDRUS DANICUS
	LICMOPHORA SP.
	MELOSIRA AGASSIZII
	MELOSIRA GRANULATA
	MELOSIRA ISLANDICA
	MELOSIRA ISLANDICA SUBSP HELVETICA
	MELOSIRA ITALICA
	MELOSIRA ITALICA SUBSP SUBARCTICA
	MELOSIRA SP.
	NAVICULA ARENARIA
	NAVICULA INCERTA
	NAVICULA MINIMA
	NAVICULA PELLICULOSA
	NAVICULA SALINARUM
	NAVICULA SPP.
	NAVICULA TORQUATUM
	NITZSCHIA ACTINASTROIDES
	NITZSCHIA ALBA
	NITZSCHIA CLOSTERIUM
	NITZSCHIA FRUSTULUM
	NITZSCHIA GRACILIS
	NITZSCHIA OVALIS
	NITZSCHIA SERIATA
	NITZSCHIA SP.
	PHAEODACTYLUM TRICORNUTUM
	PLANKTONIELLA SOL
	RHIZOSOLENIA ROBUSTA
	RHIZOSOLENIA SETIGERA
	RHIZOSOLENIA STOLTERFOTHII
	RHIZOSOLENIA STYLIFORMIS
	SKELETONEMA COSTATUM
	SKELETONEMA SP.
	STEPHANODISCUS ASTRAEA
	STEPHANODISCUS HANTZSCHII
	STEPHANODISCUS HANTZSCHII VAR PUSILLUS
	STEPHANODISCUS NIAGAREA
	STEPHANOPYXIS SP.
	SYNEDRA ACUS
	SYNEDRA ACUS VAR ANGUSTISSIMA
	SYNEDRA ACUS VAR RADIANS
	SYNEDRA CAPITATA
	SYNEDRA DELICATISSIMA
	SYNEDRA SP.
	SYNEDRA ULNA
	TABELLARIA FENESTRATA
	TABELLARIA FLOCCULOSA
	TABELLARIA FLOCCULOSA VAR ASTERIONELLOIDES
	TABELLARIA FLOCCULOSA VAR FLOCCULOSA
	TABELLARIA SP.
	THALASSIOSIRA DECIPIENS
	THALASSIOSIRA FLUVIATILIS
	THALASSIOSIRA GRAVIDA
	THALASSIOSIRA NANA
	THALASSIOSIRA PSEUDONANA
	THALASSIOSIRA ROTULA
	THALASSIOSIRA SP.
	TRICERATIUM FAVUS
DIATOMS	BACILLARIOPHYCEAE
DINOFLAGELLATE	AMPHIDINIUM CARTERI
	AMPHIDINIUM SP.
	CACHONINA SP.
	CERATIUM FURCA
	CERATIUM HIRUNDINELLA
	CERATIUM TRIPOS
	DINOPHYSIS MILES
	DIPLOPSALIS ACUTA
	DISSODINIUM SP.
	EXUVIAELLA SP.
	GLENODINIUM SP.
	GONYAULAX POLYEDRA
	GYMNODINIUM FUSCUM
	GYMNODINIUM HELVETICUM
	GYMNODINIUM ORDINATUM
	GYMNODINIUM SP.
	GYMNODINIUM SPLENDENS
	PERIDINIUM CINCTUM
	PERIDINIUM CINCTUM FA WESTII
	PERIDINIUM SPP.
	PERIDINIUM WILLEI
	PROROCENTRUM MICANS
	PYROCYSTIS FUSIFORMIS
	PYROCYSTIS NOCTILUCA
	ZOOXANTHELLAE
DINOFLAGELLATES	DINOPHYCEAE
EUGLENOPHYCEAE	EUGLENA GRACILIS
	EUGLENA SP.
	TRACHELOMONAS HISPIDA
	TRACHELOMONAS VOLVOCINA
FLAGELLATE	OLISTHODISCUS LUTEUS
GOLDEN-BROWN ALGA	CHROMULINA PYRIFORMIS
	DICTYOCHA SP.
	DINOBRYON BORGEI
	DINOBRYON CYLINDRICUM
	DINOBRYON DIVERGENS
	DINOBRYON SOCIALE
	DINOBRYON SOCIALE VAR AMERICANUM
	DINOBRYON SP.
	ISOCHRYSIS GALBANA
	ISOCHRYSIS SP.
	KEPHYRION SP.
	MALLOMONAS CAUDATA
	MALLOMONAS URNAFORMIS
	MONOCHRYSIS LUTHERI
	OIKOMONOS TERMO
	RHIZOCHRYSIS LIMNETICA
GOLDEN-BROWN ALGAE	CHRYSOPHYCEAE
	UROGLENA AMERICANA
GREEN ALGA	ACROSIPHONIA CENTRALIS
	ANKISTRODESMUS AMALLOIDES
	ANKISTRODESMUS BRAUNII
	ANKISTRODESMUS FALCATUS
	BLIDINGIA MINIMA
	BOTRYOCOCCUS BRAUNII
	CHAETOMORPHA SP.
	CHARA SP.
	CHLAMYDOMONAS MAGNUSII
	CHLAMYDOMONAS REINHARDTII
	CHLAMYDOMONAS SP.
	CHLAMYDOMONAS SUBCOMPLETA
	CHLORELLA ELLIPSOIDEA
	CHLORELLA PYRENOIDOSA
	CHLORELLA SP.
	CHLORELLA VULGARIS
	CHLOROCOCCUM SP.
	CHODATELLA SP.
	CLADOPHORA GLOMERATA
	CLADOPHORA GRACILIS
	CLADOPHORA SP.
	CLOSTERIUM ACICULARE
	CLOSTERIUM ACICULARE VAR SUBPRONUM
	CODIUM YESSOENSIS
	COENOCOCCUS PLANKTONICUS

	COSMARIUM PHASEOLUS
	COSMARIUM RENIFORME
	COSMARIUM SP.
	DICTYOSPHAERIUM PULCHELLUM
	DUNALIELLA MARINA (*)
	DUNALIELLA PEIRCEI
	DUNALIELLA SALINA
	DUNALIELLA SP.
	DUNALIELLA TERTIOLECTA
	ELAKATOTHRIX GELATINOSA
	ENTEROMORPHA COMPRESSA
	ENTEROMORPHA INTESTINALIS
	ENTEROMORPHA LINZA
	ENTEROMORPHA NANA
	ENTEROMORPHA PROLIFERA
	ENTEROMORPHA SP.
	GLOEOCOCOCCUS SCHROETERI
	HAEMATOCOCCUS PLUVIALIS
	HYDRODICTYON SP.
	MOUGEOTIA SP.
	NANNOCHLORIS ATOMUS
	NITELLA FLEXILIS
	NITELLA OPACA
	OEDOGONIUM CARDIACUM
	OEDOGONIUM SP.
	OOCYSTIS SOLITARIA
	OOCYSTIS SP.
	PANDORINA MORUM
	PAULSCHULZIA TENERA
	PEDIASTRUM DUPLEX
	PEDIASTRUM SPP.
	PEDIASTRUM TETRAS
	PITHOPHORA SP.
	RHIZOCLONIUM HIEROGLYPHICUM
	RHIZOCLONIUM SP.
	SCENEDESMUS ACUTUS
	SCENEDESMUS BASILENSIS
	SCENEDESMUS GRACILE
	SCENEDESMUS OBLIQUUS
	SCENEDESMUS QUADRICAUDA
	SCENEDESMUS SP.
	SELENASTRUM CAPRICORNUTUM
	SELENASTRUM SP.
	SPIROGYRA SP.
	STAURASTRUM PARADOXUM
	STICHOCOCCUS BACILLARIS
	STIGEOCLONIUM PACHYDERUM
GREEN ALGAE	CHLOROPHYCEAE
	TETRAEDRON MINIMUM
	TETRAEDRON SP.
	ULOTHRIX FIMBRIATA
	ULOTHRIX SPECIOSA
	ULOTHRIX ZONATA
	ULVA ANGUSTA
	ULVA FENESTRATA
	ULVA LACTUCA
	ULVA PERTUSA
	ULVA SP.
	ULVARIA SPENDENS
	ULVOPSIS GREVILLEI
	VALONIA VENTRICULOSA
HAPTOPHYCEAE	COCCOLITHUS HUXLEYI
	COCCOLITHUS SP.
	CRICOSPHAERA GALBANA
	CRICOSPHAERA SP.
	HYMENOMONAS SP.
	SYRACOSPHAERA CARTERI
PRASINOPHYCEAE	PLATYMONAS SP.
	PLATYMONAS SUBCORDIFORMIS
	TETRASELMIS CARTERIFORMIS
	TETRASELMIS CHUII
	TETRASELMIS GRACILIS
	TETRASELMIS MACULATA
	TETRASELMIS STRICIA
RED ALGA	AHNFELTIA PARADOXA
	BOSSIELLA SP.
	CERAMIUM STRICTUM
	CHONDRUS OCELLATUS
	CHONDRUS YENDOI
	COMPSOPOGON HOOKERI
	CORALLINA OFFICINALIS
	CORALLINA SP.
	DELESSERIA SANGUINEA
	DUMONTIA INCRASSATA
	FURCELLARIA FASTIGIATA
	GIGARTINA AGARDHII
	GIGARTINA LEPTORYNCHOS
	LITHOTHAMNIOID ALGA
	MELOBESIA SP.
	PHYCODRYS RUBENS
	PHYLLOPHORA BRODIAEI
	POLYSIPHONIA JAPONICA
	POLYSIPHONIA NIGRESCENS
	PORPHYRA LEUCOSTICTA
	PORPHYRA SP.
	PORPHYRA UMBILICALIS
	PORPHYRIDIUM AERUGINEUM
	PORPHYRIDIUM SP.
	PTILOTA FILICINA
	RHODOMELA LARIX
	SCHIZYMENIA EPIPHYTICA
RED ALGAE	RHODOPHYTA
YELLOW-GREEN ALGA	HETEROTHRIX SP.
	MONODUS SUBTERRANEUS

THALLOPHYTA: FUNGI AND LICHENES

BLACK MYCELIUM	
BROWN MYCELIUM	
FUNGUS	ALTERNARIA SP.
	ASPERGILLUS SP.
	BASISPORIUM GALLARUM
	CANDIDA SP.
	CLADOSPORIUM SP.
	COLLETOTRICHUM SP.
	CRYPTOCOCCUS SP.
	CURVULARIA SP.
	MONILINIA FRUCTICOLA
	NEUROSPORA CRASSA
	PERICONIELLA SP.
	PESTALOTIA SP.
	PINICILLIUM SP.
	PIRICAUDA SP.
	RHODOTORULA SP.
GREY MYCELIUM	
LICHENE	CORNICULARIA MURICATA
	PELTIGERA CANINA
MUSHROOM	
WHITE MYCELIUM	
YEAST	SACCHAROMYCES CEREVISEAE
	SACCHAROMYCES SP.
	TORULOPSIS UTILIS

THALLOPHYTA: SCHIZOMYCETES

AMMONIA OXIDATING BACTERIA	
COLIFORM BACTERIA	
DENITRIFYING BACTERIA	
FECAL STREPTOCOCCI	
GREEN SULPHUR BACTERIA	
INFLUENZA BACILLUS	
IRON BACTERIA	
METHANE FERMENTING BACTERIA	
NITRIFYING BACTERIA	
PURPLE SULPHUR BACTERIA	

SULPHATE REDUCING
BACTERIA

TYPHOID BACILLUS

SUBJECT INDEX